S0-AWR-545

ENCLOSED FREE WITH EVERY NEW BOOK!

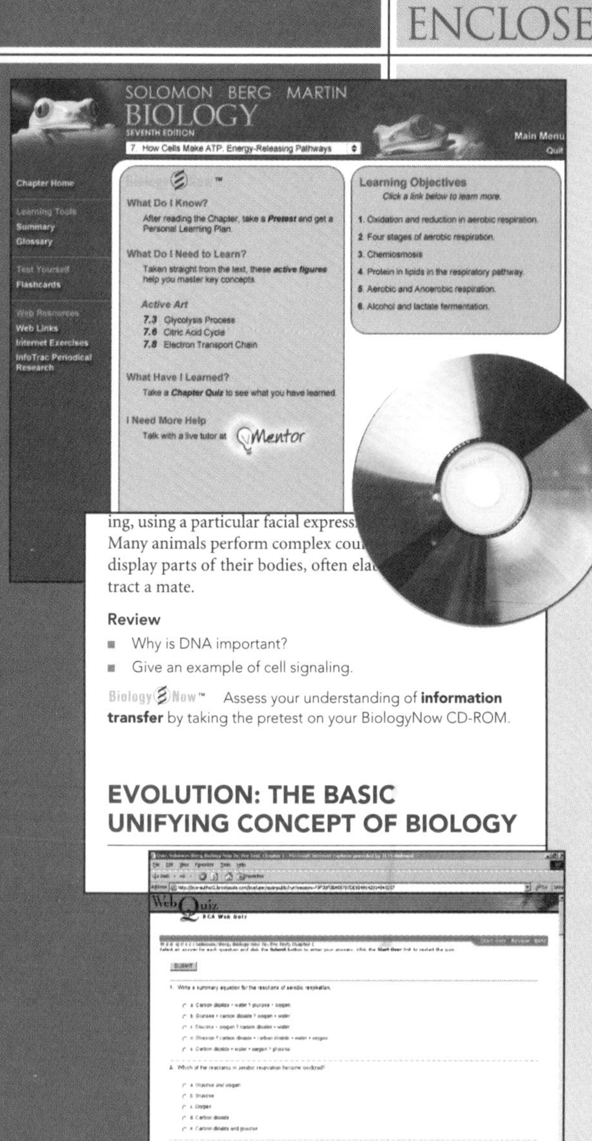

NEW! *BiologyNow™ CD-ROM*
Your personal chapter-by-chapter biology tutor!

BiologyNow™ The *BiologyNow CD-ROM* partners with this textbook, helping you to reinforce your understanding of every chapter's key concepts. Utilizing interactive diagnostic tests, animations, and much more, *BiologyNow* takes you on an exciting multimedia journey, helping you to study smarter, retain what you read, and excel in this course!

It's an easy step from this book to the CD-ROM!

BiologyNow™ This prompt, found at the beginning and throughout every chapter, tells you when to visit the CD-ROM for topic-specific reinforcement.

Diagnostic "Pretests" that give you a personalized learning plan

At the end of each major chapter section, you'll find a *BiologyNow* prompt, guiding you to the CD-ROM where you'll complete a **"Pretest"** (diagnostic questions on the chapter section you've just read). Based on your answers, *BiologyNow* will provide you with a customized learning plan that links you to the text, CD-ROM animations, and other resources for focused study that maximizes your learning. Each **"Pretest"** reinforces the chapter's **"Learning Objectives"** to help you review and retain. In addition, the CD-ROM's interactive **"Post-test"** for each chapter is correlated with the chapter-end **"Post-test"** in the book—giving you another opportunity to review.

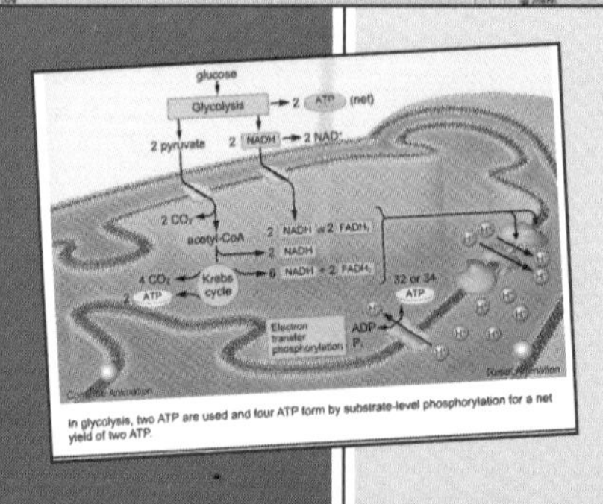

Colorful, instructive animations and "Active Figures"

Even biology's most complex concepts become understandable with the help of the animations you'll find on the *BiologyNow CD-ROM*. The CD-ROM also includes "Active Figures" that utilize the power of multimedia to bring biological processes to life.

vMentor . . . one-on-one tutoring by experienced biology teachers!

FREE online tutoring via the CD-ROM's live link! With *vMentor,* you'll interact with experienced tutors right at your computer. *vMentor* offers a virtual classroom environment with two-way voice, a shared whiteboard, chat, and more! All *vMentor* tutors have advanced degrees in biology and extensive teaching experience. When you access *vMentor,* you can ask as many questions as you want—and you don't need to set up an appointment in advance!

www.brookscole.com

brookscole.com is the World Wide Web site for Brooks/Cole and is your direct source to dozens of online resources.

At *brookscole.com* you can find out about supplements, demonstration software, and student resources. You can also send e-mail to many of our authors and preview new publications and exciting new technologies.

brookscole.com
Changing the way the world learns®

Biology

Seventh Edition

Eldra P. Solomon
University of South Florida

Linda R. Berg
St.Petersburg College

Diana W. Martin
Rutgers University

Australia • Canada • Mexico • Singapore • Spain
United Kingdom • United States

Executive Editor: Nedah Rose
Editor-in-Chief: Michelle Julet
Development Editors: Shelley Parlante, Betsy Dilernia
Assistant Editors: Christopher Delgado, Kari Hopperstead
Editorial Assistants: Jennifer Keever, Sarah Lowe
Technology Project Manager: Travis Metz
Marketing Manager: Ann Caven
Marketing Assistant: Leyla Jowza
Advertising Project Manager: Linda Yip
Project Manager, Editorial Production: Teri Hyde
Print/Media Buyer: Kris Waller
Permissions Editor: Joohee Lee
Production Service: Thomas E. Dorsaneo, Publishing Consultant
Text Designer: John Walker
Photo Researcher: Meyers Photo Art
Copy Editor: Linda Purrington
Illustrator: Elizabeth Morales
Cover Designer: Larry Didona
Cover Image: © Chase Swift/CORBIS
Cover Printer: Quebecor World Versailles
Compositor: Thompson Type
Printer: Quebecor World Versailles

COPYRIGHT © 2005 Brooks/Cole, a division of Thomson Learning, Inc. Thomson Learning™ is a trademark used herein under license.

ALL RIGHTS RESERVED. No part of this work covered by the copyright hereon may be reproduced or used in any form or by any means—graphic, electronic, or mechanical, including but not limited to photocopying, recording, taping, Web distribution, information networks, or information storage and retrieval systems—without the written permission of the publisher.

Printed in the United States of America
1 2 3 4 5 6 7 07 06 05 04 03

For more information about our products, contact us at:
Thomson Learning Academic Resource Center
1-800-423-0563

For permission to use material from this text, contact us by:
Phone: 1-800-730-2214
Fax: 1-800-730-2215
Web: http://www.thomsonrights.com

ExamView® and *ExamView Pro*® are registered trademarks of FSCreations, Inc. Windows is a registered trademark of the Microsoft Corporation used herein under license. Macintosh and Power Macintosh are registered trademarks of Apple Computer, Inc. Used herein under license.

COPYRIGHT 2005 Thomson Learning, Inc. All Rights Reserved. Thomson Learning WebTutorTM is a trademark of Thomson Learning, Inc.

About the Cover

Two red-eyed tree frogs (*Agalychnis callidryas*) peer over a leaf. Native to Central and South America, these brightly colored frogs inhabit lowland tropical rain forests near water. They are nocturnal animals and sleep attached to leaves during the day. They blend into the foliage quite well, because they cover the colorful parts of their bodies when sleeping. The bulging red eyes, red feet, and blue and yellow stripes along the sides of their bodies are thought to startle potential predators who come upon them during the day when they are sleeping. The little frogs open their eyes and leap, exposing all their colors and for a brief moment confusing the would-be predator. When red-eyed tree frogs mate, the female deposits her eggs on a leaf that overhangs the water. When the eggs hatch, the tadpoles drop into the water, where they live and continue to develop.

Brooks/Cole-Thomson Learning
10 Davis Drive
Belmont, CA 94002
USA

Asia
Thomson Learning
5 Shenton Way #01-01
UIC Building
Singapore 068808

Australia/New Zealand
Thomson Learning
102 Dodds Street
Southbank, Victoria 3006
Australia

Canada
Nelson
1120 Birchmount Road
Toronto, Ontario M1K 5G4
Canada

Europe/Middle East/Africa
Thomson Learning
High Holborn House
50/51 Bedford Row
London WC1R 4LR
United Kingdom

Latin America
Thomson Learning
Seneca, 53
Colonia Polanco
11560 Mexico D.F., Mexico

Spain/Portugal
Paraninfo
Calle/Magallanes, 25
28015 Madrid, Spain

Library of Congress Control Number: **2003107210**

Student Edition: ISBN 0-534-49276-2
Instructor's Edition: ISBN 0-534-49547-8

DEDICATION

To our families, friends, and colleagues who gave freely of their love, support, knowledge, and time as we prepared this seventh edition of *Biology* . . .

Especially to . . .

Rabbi Theodore and **Freda Brod**
Alan, **Jennifer**, and **Corey**
Chuck and **Margaret**

In Memoriam

Claude A. Villee, Andelot Professor Emeritus of Biological Chemistry, Harvard Medical School; *Biology* Co-Author, editions 1–4
Yuichiro Hiraizumi, Emeritus Professor of Zoology, University of Texas at Austin

ABOUT THE AUTHORS

ELDRA P. SOLOMON has written several leading college-level textbooks in biology and in human anatomy and physiology. Her books have been translated into more than 10 languages. Dr. Solomon earned an M.S. from the University of Florida and an M.A. and Ph.D. from the University of South Florida. She is an adjunct professor and member of the Graduate Faculty at the University of South Florida. Dr. Solomon taught biology and nursing students for more than 20 years.

Dr. Solomon is a biopsychologist as well as a biologist with a special interest in the neurophysiology of traumatic experience. Her research has focused on the relationships among stress, emotions, and health. In her clinical work, she specializes in health psychology and Post-traumatic Stress Disorder. Dr. Solomon has served as Clinical Director of the Center for Mental Health Education, Assessment, and Therapy since 1992.

Dr. Solomon has been recognized nationally and internationally. She was an Invited Scientist to the XVth Congress of Scientific Investigation, sponsored by Interamerican University of Puerto Rico, where she presented the Plenary Session, "Mind-Body Connections—An Introduction to Psychoneuroimmunology." She has been profiled more than 20 times in leading publications, including *Who's Who in America, Who's Who in Science and Engineering, Who's Who in Medicine and Healthcare, Who's Who in American Education, Who's Who of American Women,* and *Who's Who in the World.*

LINDA R. BERG is an award-winning teacher and textbook author. She received a B.S. in science education, an M.S. in botany, and a Ph.D. in plant physiology from the University of Maryland. Her research focused on the evolutionary implications of steroid biosynthetic pathways in various organisms.

Dr. Berg taught at the University of Maryland at College Park for 17 years and is presently an Adjunct Professor at St. Petersburg College in Florida. During her career, she has taught introductory courses in biology, botany, and environmental science to thousands of students. At the University of Maryland, she received numerous teaching and service awards. Dr. Berg is also the recipient of many national and regional awards, including the National Science Teachers Association Award for Innovations in College Science Teaching, the Nation's Capital Area Disabled Student Services Award, and the Washington Academy of Sciences Award in University Science Teaching.

During her career as a professional science writer, Dr. Berg has authored or co-authored several leading college science textbooks. Her writing reflects her teaching style and love of science.

DIANA W. MARTIN is the Director of General Biology, Division of Life Sciences, at Rutgers University, New Brunswick Campus. She received an M.S. at Florida State University, where she studied the chromosomes of related plant species to understand their evolutionary relationships. She earned a Ph.D. at the University of Texas at Austin, where she studied the genetics of the fruit fly, *Drosophila melanogaster,* and then conducted postdoctoral research at Princeton University. She has taught general biology and other courses at Rutgers for more than 20 years and has been involved in writing textbooks since 1988. She is immensely grateful that her decision to study biology in college has led to a career that allows her many ways to share her excitement about all aspects of biology.

Preface

Biology is an exciting and dynamic science that affects every aspect of our lives from our health and behavior to the challenging environmental issues that confront us. Recent discoveries in the biological sciences have increased our understanding of both the unity and diversity of life's processes and adaptations. With this understanding, we have become more aware of our interdependence with the vast diversity of organisms with which we share planet Earth.

BIOLOGY IS A BOOK FOR STUDENTS

One of our principal goals in developing *Biology* has been to share with beginning biology students our sense of excitement about biology. We seek to help students better appreciate Earth's diverse organisms, their remarkable adaptations to the environment, and their evolutionary and ecological relationships. We want students to understand the dynamic way that science works and to appreciate the contributions of scientists whose discoveries not only expand our knowledge of biology but also help shape and protect the future of our planet.

Since the earliest edition of *Biology,* we have focused on presenting the principles of biology in a way that is accurate, interesting and accessible to the student. In this seventh edition of *Biology,* we continue this tradition. We have worked hard to write in a student-friendly style. Throughout the text, we spark interest by relating concepts to experience within the student's frame of reference. By helping students make such connections, we facilitate the integration of concepts. In addition, we use Focus On boxes to explore issues of special relevance to students (such as the effects of smoking or alcohol abuse). These boxes also provide a forum for discussing certain topics of current interest in more detail, such as Alzheimer's disease and seed banks. We include numerous tables, many illustrated, to help the student organize and summarize material presented in the text. We hope the combined effect of an engaging writing style and interesting features will fascinate students and encourage them to continue their study of biology.

INTRODUCING OUR LEARNING SYSTEM

We have developed a text that is enjoyable to read, with a well-developed art program, and in the seventh edition we introduce our new Learning System, which focuses on learning outcomes. Learning the principles of biology is challenging. To help students master this complex subject, we provide Learning Objectives both for the course and for each major section of every chapter. At the end of each section, we provide Review questions based on the learning objectives so students can assess their mastery of the material presented in the section. Throughout the book, students are directed to BiologyNow, a powerful diagnostic tool on the free CD-ROM, which helps students assess their study needs and master the chapter objectives. After taking a pretest on BiologyNow, students receive feedback based on their answers, and links to animations and other resources keyed to their specific learning needs. Select illustrations in the text are also keyed to Active Figures on the BiologyNow CD-ROM.

Course Learning Objectives

The student can demonstrate mastery of the principles of biology by responding accurately to the following Course Learning Objectives:

- Design an experiment to test a given hypothesis, using the procedure and terminology of the scientific method.
- Cite the cell theory, and relate structure to function in both prokaryotic and eukaryotic cells.
- Describe the theory of evolution, explain why it is the principal unifying concept in biology, and discuss natural selection as the primary agent of evolutionary change.
- Explain the role of genetic information in all species, and discuss applications of genetics that affect society.
- Describe several mechanisms by which cells and organisms transfer information, including the use of nucleic acids, chemical signals (such as hormones and pheromones), electrical signals (for example, neural transmission), signal transduction, sounds, and visual displays.

- Argue for or against the classification of organisms in three domains and six kingdoms, characterizing each of these clades; based on your knowledge of genetics and evolution, give specific examples of the unity and diversity of these organisms.
- Compare the structural adaptations, life processes, and life cycles of a prokaryote, protist, fungus, plant, and animal.
- Define *homeostasis,* and give examples of regulatory mechanisms, including feedback systems.
- Trace the flow of matter and energy through a photosynthetic cell and a nonphotosynthetic cell, and through the biosphere, comparing the roles of producers, consumers, and decomposers.
- Describe the study of ecology at the levels of an individual organism, a population, a community, and an ecosystem.

Learning System Strategies

We use numerous learning strategies to increase the student's success:

- ***Learning Objectives*** at the beginning of each major section in the chapter indicate, in behavioral terms, what the student must do to demonstrate mastery of the material in that section.
- Each major section of the chapter is followed by a series of ***Review* questions** that assess comprehension by asking the student to describe, explain, compare, contrast, or illustrate important concepts. The review questions are based on the section Learning Objectives.
- A list of the major section headings at the beginning of each chapter provides a ***Chapter Overview***. Chapter outlines, including heads and subheads, are posted on our Web site at **http://biology.brookscole.com/solomon7**.
- ***Concept Statement Subheads*** introduce sections, previewing and summarizing the key idea or ideas to be discussed in that section.
- ***Sequence Summaries*** within the text simplify and summarize information presented in paragraph form. For example, paragraphs describing blood circulation through the body or the steps by which cells take in certain materials are followed by a Sequence Summary listing the structures or steps.
- A ***Summary with Key Terms*** at the end of each chapter is organized around the chapter Learning Objectives. This summary provides a review of the material, and because selected key terms are boldfaced in the summary, students are provided with the opportunity to study vocabulary words within the context of related concepts.
- **End-of-chapter questions** provide students with the opportunity to evaluate their understanding of the material in the chapter. The ***Post-Test*** consists of multiple-choice questions, some of which are based on the recall of important terms, whereas others challenge students to integrate their knowledge. Answers to the Post-Test questions are provided at the end of the chapter. A series of ***Critical Thinking* questions** encourages the student to make connections among important concepts. To answer these questions, the student must apply the concepts just learned to new situations or relate concepts learned in previous chapters to concepts in the current chapter.
- The ***Glossary*** at the end of the book, the most comprehensive glossary found in any biology text, provides precise definitions of terms. The Glossary is especially useful because it is extensively cross-referenced and includes pronunciations. The vertical purple bar along the margin facilitates rapid access to the Glossary. The companion Web Site also includes glossary flash cards with audio pronunciations.
- An ***art program that is fully integrated with the text*** brings to life, reinforces, and expands concepts discussed in the text. New in this edition are figures featured to illustrate key concepts. Many figures have numbered, multiple parts that show sequences of events in important processes or life cycles. Numerous photographs, both alone and combined with line art, help students understand concepts by connecting the "real" to the "ideal." The line art uses devices such as ***orientation icons*** to help the student put the detailed figures into the broad context. We use symbols and colors consistently throughout the book, to help students connect concepts. For example, the same four colors and shapes are used throughout the book to identify guanine, cytosine, adenine, and thymine.

Emphasis on the Process of Science

Understanding how scientific knowledge is derived is crucial for scientists and nonscientists alike. *Biology* provides insight into what science is, how scientists work, the roles of the many scientists who have contributed to our current understanding of biology, and how scientific knowledge affects daily life. Two features clearly reflect this emphasis:

- A ***"Process of Science"* icon**, embedded at relevant points in the text, highlights discussions about the work that scientists do and helps students to understand that the process of scientific discovery is ongoing and indefinite. In addition to the discussion on the process and method of science in Chapter 1, we give many examples of this dynamic process within the context of each particular subject discussed.
- ***On the Cutting Edge* boxes** present exciting research areas, such as how some plant-eating insects "eavesdrop" on plant defensive signals and respond by activating their own defenses (see Chapter 36). The abstract for the Cutting Edge box is formatted like a much-abbreviated science report, which further reinforces the student's understanding of the scientific process being reported.

AN OVERVIEW OF *BIOLOGY*, SEVENTH EDITION

Three themes provide the structural foundation for *Biology:* the evolution of life, the transmission of biological information, and the flow of energy through living systems. As we introduce the concepts of modern biology, we explain how these themes are connected and how life depends on them.

Educators present the major topics of an introductory biology course in a variety of orders. For this reason, we carefully designed the eight parts of this book so that they do not depend heavily on preceding chapters and parts. The instructor can present the eight parts and their 55 chapters in any number of sequences with pedagogical success. Chapter 1, which introduces the student to the major principles of biology, provides a good springboard for future discussions, whether the professor prefers to start with the "big picture" and work down, or vice versa.

In this edition, as in previous editions, we examined every line of every chapter for accuracy and currency, and we made a serious attempt to update every topic and verify all new material. The following brief survey provides a general overview of the eight parts of *Biology* and some changes made to the seventh edition.

Part 1: The organization of life

The five chapters that make up Part 1 give the student basic background knowledge. We begin Chapter 1 by discussing the Human Genome Project and then introduce the main themes of the book—evolution, energy transfer, and information transfer. Chapter 1 examines several fundamental concepts: the characteristics and similarities of living things; the organization of life on individual and ecological levels; information transfer; evolution as the main unifying concept in biology; the diversity of life and how biologists classify organisms; energy transfer; and how science works. Chapters 2 and 3, which focus on the molecular level of organization, establish the foundations in chemistry necessary for understanding biological processes. Chapters 4 and 5 focus on the cell level of organization. Research on receptors and signal transduction is shedding light on many life processes on a cell level. We introduce these concepts in Chapter 5. However, we are convinced that this information is best delivered within a context, so we integrate many research findings on receptors and signal transduction in relevant chapters throughout the book.

Part 2: Energy transfer through living systems

Because all living cells need energy for life processes, the flow of energy through living systems—that is, capturing energy and converting it to usable forms—is a basic theme of *Biology.* Chapter 6 examines how cells capture, transfer, store, and use energy. Chapters 7 and 8, which can be taught in either order, discuss the metabolic adaptations by which organisms obtain and use energy through cellular respiration and photosynthesis.

Part 3: The continuity of life: genetics

We have completely revised and updated the eight chapters of Part 3 for the seventh edition. There are many new illustrations, seven new ones in Chapter 10 alone. In these chapters we explore the science of genetics, giving students the tools they need to grasp the important new findings reported almost daily. We begin this unit by discussing mitosis and meiosis (Chapter 9), thus providing a foundation for considering Mendelian genetics and related patterns of inheritance in Chapter 10. We then turn our attention to the flow of information in cells, beginning with the structure and replication of DNA (in Chapter 11), followed by a discussion of RNA and protein synthesis (Chapter 12). Gene regulation is discussed in Chapter 13, and in Chapter 14, we focus on genetic engineering. These chapters build the necessary foundation for exploring the human genome in Chapter 15. In Chapter 16, we introduce the role of genes in development, emphasizing studies on specific model organisms that have led to spectacular advances in this field. Changes in these chapters include new material on how genes interact with the environment to determine phenotype, latest findings on telomeres and telomerase, new material on mammalian cloning, and a new Cutting Edge box on studying aging in mice. New tables show a timeline of selected historical DNA discoveries, present color-coded data that support Chargaff's rules, and summarize the enzymes involved in DNA replication.

Part 4: The continuity of life: evolution

Although we explore evolution as the cornerstone of biology throughout the book, Part 4 delves into the subject in depth. We provide the history behind the discovery of the theory of evolution, the mechanism by which it occurs, and the methods by which it is studied and tested. Chapter 17 introduces the Darwinian concept of evolution and presents several kinds of evidence that support the theory of evolution. In Chapter 18, we examine evolution at the population level. Chapter 19 describes the evolution of new species and discusses aspects of macroevolution. Chapter 20 summarizes the evolutionary history of life on Earth. In Chapter 21 we recount the evolution of the primates, including humans. Many topics and examples have been added to the seventh edition, including new material on molecular clocks, the evolutionary species concept, sexual selection, recent fossil discoveries of human ancestors, and a new Cutting Edge box on the origin of flight in birds.

Part 5: The diversity of life

In this edition of *Biology,* we emphasize the cladistic approach. Based on recent developments and on reviewer input, we have replaced the phylogenetic trees of past editions with cladograms. We use an evolutionary framework to discuss each group of organisms, presenting current hypotheses of how groups of organisms are related. By focusing on evolutionary relationships and the structural and functional adaptations of each group of organisms, we avoid the traditional parade through the phyla that is characteristic of many biology textbooks.

In Chapter 22, we discuss *why* organisms are classified and provide insight into the scientific process of deciding *how* they are classified. Chapter 23, focuses on the viruses and prokarotes, compares the three domains, and discusses viroids, prions, and emerging diseases. We have revised Chapter 24 to reflect the developing consensus on protist diversity. We summarize the eight major eukaryote groups, and include a new table and new figure showing evolutionary relationships among the groups. Chapter 25 describes the fungi. Chapters 26 and 27 present the members of the plant kingdom. In Chapters 28 through 30, which cover the diversity of animals, we present the most recent classification of animal phyla, including division into Lophotrochozoa, Ecdysozoa, and Deuterostomia clades.

Part 6: Structure and life processes in plants

Part 6 introduces students to the fascinating plant world. It stresses relationships between structure and function in plant cells, tissues, organs, and individual organisms. In Chapter 31, we introduce plant structure, growth, and differentiation. Chapters 32 through 34 discuss the structural and physiological adaptations of leaves, stems, and roots. Chapter 35 describes reproduction in flowering plants, including asexual reproduction, flowers, fruits, and seeds. Chapter 36 focuses on growth responses and regulation of growth. In the seventh edition, we present the latest findings generated by the continuing explosion of knowledge in plant biology, particularly at the molecular level. Some of the new topics include an updated section on stomatal opening and closing, new information on floral meristem identity genes, updated research on self-incompatibility in plant reproduction, and new data on nastic movements and tropisms.

Part 7: Structure and life processes in animals

In Part 7, we emphasize the structural, functional, and behavioral adaptations that help animals meet environmental challenges. We use a comparative approach to examine how various animal groups have solved similar and diverse problems. In Chapter 37, we discuss the basic tissues and organ systems of the animal body, homeostasis, and how animals regulate their body temperature. Chapter 38 focuses on body coverings, skeletons, and muscles. In Chapters 39 through 41, we discuss neural signaling, neural regulation, and sensory reception. In Chapters 42 through 49, we compare how different animal groups carry on specific life processes, such as internal transport, internal defense, gas exchange, digestion, reproduction, and development. Each chapter in this part considers the human adaptations for the life processes being discussed. The unit ends with a discussion of behavioral adaptations in Chapter 50, which includes a reorganized and updated sections on sexual selection and on helping behavior.

Reflecting recent research findings, we have added material on glial cells, new findings on a neurotrophin that opens certain sodium channels, and a Cutting Edge box on the neurophysiology of traumatic experience. In the immunology chapter, we have added a discussion of the danger hypothesis and have included new findings on Toll signaling receptors and pathogen-associated molecular patterns. We introduce new findings on mechanisms of hormone action and have added a brief section on enzyme-linked receptors.

Part 8: The interactions of life: ecology

Part 8 focuses on the dynamics of populations, communities, and ecosystems and on the application of ecological principles to disciplines such as conservation biology. Chapters 51 through 54 give the student an understanding of the ecology of populations, communities, ecosystems, and the biosphere, whereas the final chapter (55) focuses on environmental problems humans have caused. We continue to interweave ecological theory and the scientific process by giving clear, concrete examples of ecological studies to illustrate conceptual points. Among the many changes in this unit, the authors have updated Vitousek's work on human appropriation of global NPP, expanded the definition of biological diversity, and added a new section on dominant species.

A COMPREHENSIVE PACKAGE FOR LEARNING AND TEACHING

To further facilitate learning, a carefully designed supplement package is available. In addition to the usual print resources, we are pleased to present student multimedia tools that have been developed in conjunction with the text.

Resources for Students

Study Guide to Accompany *Biology,* Seventh Edition, by Ronald S. Daniel of California State Polytechnic University, Pomona; Sharon C. Daniel of Orange Coast College; and Ronald L. Taylor. Extensively updated for this edition, the study guide provides the student with many opportunities to review chapter concepts. Multiple-choice study questions, coloring book exercises, vocabulary-building exercises, and many other types of active-learning tools are provided to suit different cognitive learning styles.

A Problem-Based Guide to Basic Genetics by Donald Cronkite of Hope College. This brief guide provides students with a systematic approach to solving genetics problems, along with numerous solved problems and practice problems.

Web Site. The content-rich companion Web site that accompanies *Biology,* seventh edition, gives students access to a wealth of high-quality resources, including focused quizzing, a Glossary complete with pronunciations, *InfoTrac College Edition* (with questions), Internet Activities (with questions), Chapter Summaries, Learning Objectives, further readings for each chapter, and annotated Web links. The site also includes an all-new genetics resource, including specialized genetics problems. Finally, Career Visions interviews on the Web site help students become aware of the many doors that a biology degree can open by sharing the experiences of young people who have found fulfilling careers in which they use their knowledge of biology.

Additional Resources for Instructors

The instructors' Examination Copy for this edition lists a comprehensive package of print and multimedia supplements, including online resources, that are available to qualified adopters. Please ask your local sales representative for details.

ACKNOWLEDGMENTS

The development and production of the seventh edition of *Biology* required extensive interaction and cooperation among the authors and many individuals in our home and professional environments. We thank our editors, colleagues, students, family, and friends for their help and support.

Preparing a book of this complexity is challenging and requires a cohesive, talented, and hardworking professional team whose members believe in the project. We were fortunate to have just such a team, and appreciate the contributions of everyone on the editorial and production staff at Brooks/Cole/Thomson Learning who worked on this seventh edition of *Biology*. We thank Michelle Julet, Vice President and Editor-in-Chief, and Executive Editor, Nedah Rose, for their commitment to this book and for their support in making the seventh edition happen. We appreciate Ann Caven, our Marketing Manager, whose expertise ensured that you would know about our new edition.

We appreciate the hard work of our dedicated Developmental Editor, Shelley Parlante, who provided us with valuable input as she guided the seventh edition through its many phases. Developmental Editor Betsy Dilernia carefully reviewed selected chapters and gave us many helpful suggestions for improving the manuscript. We appreciate the help of Teri Hyde, Senior Production Project Manager, and Project Editor Tom Dorsaneo, who expertly shepherded the project. We thank Editorial Assistant Jennifer Keever for quickly providing us with resources whenever we needed them. We also appreciate the help of editorial assistant Sarah Lowe.

We thank Elizabeth Morales for sharing her artistic talent and for her great ideas for visual presentations. We appreciate the efforts of photo editors Don Murie and Joan Murie. We thank Art Director Rob Hugel, Text Designer John Walker, and Cover Designer Larry Didona. We also thank Joy Westberg for developing the Instructor's Preface.

We are grateful to Travis Metz, Technology Project Manager, who coordinated the many high-tech components of the computerized aspects of our Learning System. We thank Assistant Editor Kari Hopperstead for coordinating the print supplements.

These dedicated professionals and many others at Brooks/Cole provided the skill, attention, and good humor needed to produce *Biology*, Seventh Edition. We thank them for their help and support throughout this project.

We greatly appreciate the expert assistance of Mary Kay Hartung of Florida Gulf Coast University, who came to our rescue whenever we had difficulty finding information. Whenever asked, she quickly helped us find needed research studies from the Internet. We thank doctoral student Lois Ball of the University of South Florida, Department of Biology, who reviewed several chapters and offered helpful suggestions.

We thank our families and friends for their understanding, support, and encouragement as we struggled through many revisions and deadlines. We especially thank Dr. Amy Solomon, Dr. Kathleen M. Heide, Mical Solomon, Alan Berg and Jennifer Brookhouzen, and Dr. Charles Martin and Margaret Martin for their support and input.

Our colleagues and students who have used our book have provided valuable input by sharing their responses to past editions of *Biology*. We thank them and ask again for their comments and suggestions as they use this new edition. We can be reached through the Internet at our Web site **http://biology.brookscole.com/solomon7** or through our editors at Brooks/Cole, a division of Thomson Learning.

We express our thanks to the many biologists who have read the manuscript during various stages of its development and provided us with valuable suggestions for improving it. Seventh-edition reviewers include the following:

Hema Bandaranayake, Xavier University of Louisiana; Gayle Birchfield, University of Missouri; Judy Bluemer, Morton College; Paul Bottino, University of Maryland; Nancy Boury, Iowa State University; Robert Boyd, Auburn University; Jeff Carmichael, University of North Dakota; Linda Collins, University of Tennessee Chattanooga; Elizabeth Cowles, Eastern Connecticut State University; Andrew Crain, Maryville College; Karen Dalton, Community College of Baltimore County; Robert Evans, Rutgers University; Daniel Fairbanks, Brigham Young University; Christopher Harendza, Montgomery County Community College; Harriette Howard-Lee Block, Prairie View A&M University; Joan Hudson, Sam Houston State University; John B. Jenkins, Swarthmore College; Craig Martin, University of Kansas; Allan Nelson, Tarleton State University; Nancy L. Pencoe, State University of West Georgia; Chris E. Petersen, College of DuPage; Sylvia Christie Saunders, Borough of Manhattan Community College and City University of New York; Gerald Shields, Carroll College; Rob Snetsinger, Queen's University; Louisa Stark, University of Utah Genetic Science; Mary White, Southeastern Louisiana University; Heather Weber, Los Angeles City College.

Reviewers of Previous Editions

Faculty

Dawn Adams *(Baylor University)*, James Adams *(Dalton College)*, John H. Adler *(Michigan Technical University)*, Surinder Aggarwal *(Michigan State University)*, Julie Aires *(Florida Community College, Jacksonville)*, David E. Alexander *(University of Kansas)*, Venita Allison *(Southern Methodist University)*, Sylvester Allred *(Northern Arizona University)*, Jane Aloi *(Saddleback College)*, Marvin Alvarez *(University of South Florida)*, David Asch *(Youngstown State University)*, Edward Ashworth *(Purdue University)*, Sonya Baird *(University of Georgia)*, Susan Bandoni Muench *(State University of New York, Geneseo)*, James Barron *(Ohio University, Lancaster)*, Lisa G. Bates *(Florida Community College at Jacksonville)*, Penny Bauer *(Colorado State University)*, Lester Bazinet *(Community College of Philadelphia)*, Chris Beard *(Carnegie Museum of Natural History)*, J.T. Beatty *(University of British Columbia)*, Ed Bedecarrax *(City College of San Francisco)*, David Begun *(University of Texas at Austin)*, Vincent Bellis *(East Carolina University)*, David Benner *(Eastern Tennessee State University)*, Todd Bennethum *(Purdue University)*, Gerald Bergtrom *(University of Wisconsin, Milwaukee)*, Dorothy Berner *(Temple University)*, Charles Biggers *(University of Memphis)*, William L. Bischoff *(University of Toledo)*, Del Blackburn *(Clark College)*, Gary Booth *(Brigham Young University)*, Nicole Bournias *(California State University at San Bernardino)*, Nancy Boury *(Iowa State University)*, George Bowes *(University of Florida)*, Barry Bowman *(University of California at Santa Cruz)*, George Boyajian *(University of Pennsylvania)*, Robert Boyd *(Auburn University)*, Dean Bratis *(Delaware Community College)*, W.H. Breazeale, Jr. *(Francis Marion College)*, Anne Britt *(University of California, Davis)*, William Brooks *(Florida Atlantic University)*, George Brown *(Iowa State University)*, Gary Brusca *(Humboldt State University)*, Arthur Buikema, Jr. *(Virginia Tech)*, Ruth Buskirk *(University of Texas at Austin)*, Warren R. Buss *(University of Northern Colorado)*, Vicki Cameron *(Ithaca College)*, David Carlberg *(California State University, Long Beach)*, David Carr *(University at Maryland at College Park)*, Barry Chess *(Pasadena City College)*, W. Dennis Clark *(Arizona State University)*, Keith Clay *(Indiana University)*, Robert E. Cleland *(University of Washington)*, William Cohen *(University of Kentucky)*, Jim Colbert *(Iowa State University)*, Gary Cole *(University of Texas at Austin)*, Linda T. Collins *(University of Tennessee at Chattanooga)*, Bruce Condon *(Seattle Pacific University)*, Mark Condon *(Dutchess Community College)*, Amy Cook *(East Carolina University)*, Rebecca A. Cook *(Lambuth University*, Joyce Corban *(Wright State University)*, Jeffrey Corden *(The Johns Hopkins University)*, Robert Cordero *(St. Joseph's University)*, William Cordes *(Loyola University)*, Harry O. Corwin *(University of Pittsburgh)*, David Cotter *(Georgia College)*, Elizabeth A. Cowles *(Eastern Connecticut State University)*, James T. Cronin *(University of North Dakota)*, Kenneth Curry *(University of Southern Mississippi)*, Anne Cusic *(University of Alabama, Birmingham)*, Stan Dalton *(Jones County Community College)*, Henry Daniell *(Auburn University)*, Peter J. Davies *(Cornell University)*, Thomas Davis *(University of New Hampshire)*, Jonathan Day *(Pennsylvania State University)*, John V. Dean *(DePaul University)*, Patricia DeLeon *(University of Delaware)*, Daniel V. DerVartanian *(University of Georgia)*, Jean DeSaix *(University of North Carolina at Chapel Hill)*, Laura DiCaprio *(Ohio University)*, David Dilcher *(University of Florida)*, Stephen J. Dina *(St. Louis University)*, Linda Dion *(University of Delaware)*, Peter Dixon *(University of California, Irvine)*, Penny Dobbins *(Syracuse University)*, Andrew Dobson *(University of Rochester)*, Warren Dolphin *(Iowa State University)*, Rob Dorit *(Harvard University)*, Lee C. Drickamer *(Williams College* and *Southern Illinois University at Carbondale)*, Ernest F. DuBrul *(University of Toledo)*, Peter Ducey *(State University of New York, Cortland)*, Janice Edgerly-Rooks *(Santa Clara University)*, Inge Eley *(Hudson Valley Community College)*, David H. Evans*(University of Florida)*, John Evans *(Memorial University of Newfoundland)*, Robert C. Evans *(Rutgers University, Camden)*, Sharon Eversman *(Montana State University)*, Guy Farish *(Adams State College)*, Dale Fast *(St. Xavier University)*, Craig S. Feibel *(Rutgers University)*, Millicent Ficken *(University of Wisconsin, Milwaukee)*, Milton Fingerman *(Tulane University)*, David Firmage *(Colby College)*, Steven K. Fisher *(University of California, Santa Barbara)*, Malinda Fitzgerald *(Christian Brothers University)*, Jim Florini *(Syracuse University)*, Kathy Foltz *(University of California, Santa Barbara)*, Bruce Fowles *(Colby College)*, James French *(Rutgers University, New Brunswick)*, David Fromson *(California State University, Fullerton)*, Bernard Frye *(University of Texas at Arlington)*, Douglas Gaffin *(University of Oklahoma)*, Michael Gaines *(University of Kansas)*, Gary Galbreath *(Northwestern University)*, Darrell Galloway *(Ohio State University)*, Daniel L. Gebo *(Northern Illinois University)*, Patricia Gensel *(University of North Carolina, Chapel Hill)*, Robert P. George *(University of Wyoming)*, Michael Ghedotti *(University of Kansas)*, George W. Gilchrist *(Clarkson University)*, Florence Gleason *(University of Minnesota)*, Gene Godbold *(University of Alabama, Huntsville)*, Elizabeth A. Godrick *(Boston University)*, David Goldstein *(Wright State University)*, Paul Goldstein *(University of Texas at El Paso)*, Judith Goodenough *(University of Massachusetts, Amherst)*, Wayne Goodey *(University of British Columbia)*, H. Thomas Goodwin *(Andrews University)*, John S. Graham *(Bowling Green State University)*, Nels Granholm *(South Dakota State University)*, Edward J. Greding, Jr. *(Del Mar College)*, Katharine B. Gregg *(West Virginia Wesleyan College)*, Peter Gregory *(Cornell University)*, Floyd Grimm *(Harford Community College)*, Mark Gromko *(Bowling Green University)*, Thaddeus Grudzien *(Oakland University)*, David Hale *(Texas A&M University)*, Thomas Hanson *(Temple University)*, Alexander Harcourt *(University of California at Davis)*, Jeff Hardin *(University of Wisconsin, Madison)*, Michael B. Harvey *(East Tennessee State University)*, Paul K. Hayes *(University of Bristol)*, James Hayward *(Andrews University)*, Steven Heidemann *(Michigan State University, East Lansing)*, Jean Heitz *(University of Wisconsin at Madison)*, Louis Held *(Texas Tech University)*, Jean Helgeson *(Collin County Community College)*, Wiley Henderson *(Alabama A&M Univeristy)*, Charles Henry *(University of Connecticut)*, Fritz Hertel *(University of California, Los Angeles)*, Martinez Hewlett *(University of Arizona)*, Linden Higgins *(University of Texas at Austin)*, Andrew Hill *(Yale University)*, Betsy Hirsch *(University of Minnesota)*, Helmut Hirsch *(State University of New York, Albany)*, Ricky Hirschorn *(Hood College)*, Carl Hoagstrom *(Ohio Northern University)*, Donna Hoefner *(Delaware County Community College)*, Dan Hoffman *(Bucknell University)*, Luke Holbrook *(Rowan University)*, Rebecca Holburton *(University of Mississippi)*, Robert Holmberg *(Athabasca University)*, Jill VanWort Hood *(University of Texas at Arlington)*, Linda Hsu *(Seton Hall University)*, Joan Hudson *(Sam Houston State University)*, Stephen Hudson *(Furman University)*, Pat Humphrey *(Ohio University)*, Robert Hurst *(Purdue University)*, Gerard Iwantsch *(Fordham University)*, Alice C. Jacklet *(State University of New York, Albany)*, Mark Jacobs *(Swarthmore College)*, Charles Janson *(State University of New York, Stony Brook)*, Paul Jarrell, *Pasadena City College)*, Dan Johnson *(Eastern Tennessee State University)*, Randal Johnston *(University of Calgary)*, Claudia Jolls *(East Carolina University)*, Thomas C. Kane *(University of Cincinnati)*, Kenneth Kardong *(Washington State University)*, Richard Karp *(University of Cincinnati)*, Glenn Kasparian *(Brookhaven Community College)*, Alan J. Katz *(Illinois State University)*, Donald Keefer *(Loyola College, Maryland)*, Phil Keeting *(West Virginia University)*, Tasneem Khaleel *(Montana State University)*, M.A.Q. Khan *(University of Illinois, Chicago)*, Joanne M. Kilpatrick *(Auburn University at Montgomery)*, William Kimbel *(Institute of Human Origins)*, Robert Kitchin *(University of Wyoming)*, Loren W. Knapp *(University of South Carolina)*, Ross Koning *(Eastern Connecticut State University)*, Robert W. Korn *(Bellarmine College)*, Dan E. Krane *(Wright State University)*, James B. Kring *(Roane State Community College)*, William Kroen *(Wesley College)*, Paul Kugrens *(Colorado State University)*, Paul Lago *(University of Mississippi)*, Zhi-Chun Lai *(Pennsylvania State University)*, Vaughn A. Langman *(Louisiana*

State University), Ralph Larson *(San Francisco State University)*, Virginia Latta *(Jefferson State College)*, Brenda Leicht *(University of Iowa)*, Joe Leverich *(St. Louis University)*, Harvey Lillywhite *(University of Florida)*, Graeme Lindbeck *(Valencia Community College)*, Roger M. Lloyd *(Florida Community College, Jacksonville)*, Marion B. Lobstein *(Northern Virginia Community College)*, Heather Lorimer *(Youngstown State University)*, Victor Lotrich *(University of Delaware)*, Jennifer Lundmark *(California State University, Sacramento)*, Karl Maddox *(Miami University of Ohio)*, Sharook Madon *(Pace University, Westchester)*, Charles Mallery *(University of Miami)*, Arthur Mange *(University of Massachusetts, Amherst)*, Ronald H. Matson *(Kennesaw State University)*, Dennis Matthews *(University of New Hampshire)*, James Mauseth *(University of Texas at Austin)*, Jeffrey May *(Marshall University)*, Tim McDowell *(East Tennessee State University)*, Henry M. McHenry *(University of California, Davis)*, Roger McMacken *(The Johns Hopkins University)*, Michael Meighan *(University of California, Berkeley)*, Lee Meserve *(Bowling Green State University)*, Joseph Michalewicz *(Holy Family College)*, Ann Mickle *(LaSalle University)*, James E. Mickle *(North Carolina State University)*, Roger Milkman *(University of Iowa)*, Lillian Miller *(Florida Community College)*, Charles Mims *(University of Georgia)*, Manuel Molles *(University of New Mexico)*, Marion Monahan *(Immaculata College)*, John Moner *(University of Massachusetts, Amherst)*, Russell Monson *(University of Colorado, Boulder)*, Darrell Moore *(East Tennessee State University)*, James Moore *(University of California, San Diego)*, Robert E. Moore *(Montana State University)*, Edward Morgan *(Temple Junior College, Texas)*, Michael Morgan *(University of Wisconsin, Green Bay)*, Jim Morrone *(Louisiana State University)*, Anthony G. Moss *(Auburn University)*, Alison Mostrom *(Philadelphia College of Pharmacy and Science)*, Alan Muchlinski *(California State University, Los Angeles)*, Debbie Mueler *(Cardinal Stritch College)*, Darrel L. Murray *(University of Illinois at Chicago)*, James Murray *(University of Virginia)*, John Murray *(University of Pennsylvania)*, Patrick M. Muzzall *(Michigan State University)*, Richard Myers *(Southwest Missouri State University)*, Thomas L. Naples *(Delaware Community College)*, William H. Nelson *(Morgan State University)*, Anne Penney Newton *(Temple Junior College, Texas)*, Dan Nickrent *(Southern Illinois University)*, Frank G. Nordlie *(University of Florida)*, David O. Norris *(University of Colorado, Boulder)*, Stephen F. Norton *(East Carolina University)*, Gary Ogden *(Moorpark College)*, Carolyn Ogren *(Parkland College)*, John Olsen *(Rhodes College)*, Beulah Parker *(North Carolina State University)*, Glenn R. Parsons *(University of Mississippi)*, Robert Patterson *(San Francisco State University)*, Greg Paulson *(Shippensburg University)*, Daniel M. Pavuk *(Bowling Green State University)*, David Pennock *(Miami University of Ohio)*, Jerome Perry *(North Carolina State University)*, Chris E. Petersen *(College of DuPage)*, Greg Phillips *(Blinn Bryan College)*, Richard E. Phillips *(University of Minnesota)*, Ronald Phillips *(Seattle Pacific University)*, Ruth Pitkin *(Shippensburg University)*, Thomas Pitzer *(Florida International University)*, Jeanne S. Poindexter *(The Public Health Research Institute of the City of New York, Inc.)*, Dave Polcyn *(California State University at San Bernardino)*, Shirley Porteus-Gafford *(Fresno City College)*, Trevor Price *(University of California, San Diego)*, Susan Pross *(University of South Florida)*, Jerry Purcell *(San Antonio College)*, James Pushnik *(Chico State University)*, Richard Racusen *(University of Maryland)*, Peggy Redshaw *(Austin College)*, Arthur Repak *(Quinnipiac College)*, Eric Ribbens *(Western Illinois University)*, Florence Ricciuti *(Albertus Magnus College)*, Robert Roberson *(Arizona State University)*, Laurel Roberts *(University of Pittsburgh)*, Martin Roeder *(Florida State University)*, Rodney A. Rogers *(Drake University)*, John Romeo *(University of South Florida)*, Marvin J. Rosenberg *(California State University, Fullerton)*, Wayne Rowley *(Iowa State University)*, Lori S. Rynd *(Pacific University)*, Jean Saillant *(University of Michigan, Dearborn)*, Ted Sargent *(University of Massachusetts, Amherst)*, Mimi A. Sayed *(Michigan State University)*, Louis A. Scala *(Monmouth University)*, Carl Schlicting *(Pennsylvania State University, University Park)*, John A. Schmidt *(Ohio State University)*, Edward Schneider *(University of California, Santa Barbara)*, Janet L. Schottel *(University of Minnesota)*, Karen Schumaker *(University of Arizona)*, Brian Schwartz *(Columbus State University)*, Kathleen Scott *(Rutgers University, New Brunswick)*, William A. Searcy *(University of Miami)*, Duane W. Sears *(University of California, Santa Barbara)*, David Seigler *(University of Illinois)*, Mary Colavito Shepanski *(Santa Monica College)*, Mark Sheridan *(North Dakota State University)*, Lisa Shimeld *(Crafton Hills College)*, David Shomay *(University of Illinois, Chicago)*, Jane Shoup *(Purdue University at Calumet)*, J. Kenneth Shull, Jr. *(Appalachian State University)*, James Siedow *(Duke University)*, Paul Small *(Eureka College)*, Bruce Smith *(Brigham Young University)*, Deborah Smith *(University of Kansas)*, Dennis M. Smith *(Wellesley College)*, Phillip Snider *(University of Houston)*, Richard C. Snyder *(University of Washington)*, Nancy G. Solomon *(Miami University of Ohio)*, Bruce Stallsmith *(University of Massachusetts, Boston)*, Martha L. Stauderman *(University of San Diego)*, Karen Steudel *(University of Wisconsin, Madison)*, Charles L. Stevens *(University of Pittsburgh)*, Robert Stockhouse *(Pacific University)*, Gerald Summers *(University of Missouri)*, Marshall Sundberg *(Louisiana State University)*, Daryl Sweeney *(University of Illinois at Urbana, Champaign)*, Chris Tarp *(Contra Costa College)*, Walter Taylor *(University of Central Florida)*, Robert M. Timm *(University of Kansas)*, Ian Tizard *(Texas A&M University)*, Kenneth Thomulka *(Philadelphia College of Pharmacy)*, Sylvia D. Torti *(University of Utah)*, Nathan Tublitz *(University of Oregon)*, John Tudor *(St. Joseph's University)*, Mary S. Tyler *(University of Maine)*, Gordon Uno *(University of Oklahoma)*, Frederick Utech *(Carnegie Museum of Natural History)*, John Utley *(University of New Orleans)*, Joseph W. Vanable *(Purdue University)*, Steve Vessey *(Bowling Green State University)*, Darrel Vodopich *(Baylor University)*, Thomas C. Vogelmann *(University of Wyoming)*, Jack Waber *(West Chester State University)*, Charles Walcott *(Cornell University)*, Elizabeth Waldorf *(Mississippi Gulf Community College)*, Eileen Walsh *(Westchester Community College)*, Fred Wasserman *(Boston University)*, Judy Watts *(Cleveland State Community College)*, Jacqueline F. Webb *(Villanova University)*, Michael Weber *(Carleton University)*, David Whetstone *(Jacksonville State University, Alabama)*, Mary E. White *(Southeastern Louisiana University)*, Matthew White *(Ohio University)*, John Whitmarsh *(University of Illinois, Urbana)*, Donald Whitmore *(University of Texas, Arlington)*, Varley Wiedeman *(University of Louisville)*, Leslie Williams *(Mountain View College)*, David Wilson *(University of Miami)*, Lawrence Winship *(Hampshire College)*, Dwayne Wise *(Mississippi State University)*, Steven Woeste *(Scholl College)*, Clarence Wolfe *(North Virginia Community College)*, Drew H. Wolfe *(Hillsborough Community College)*, David Woodruff *(University of California, San Diego)*, Stephen Yazulla *(State University of New York, Stony Brook)*, Robert Yost *(Indiana University/Purdue University at Indianapolis)*, Roger Young *(Texas A&M University)*, John L. Zimmerman *(Kansas State University)*, William Zimmerman *(Amherst College)*

Students

Felisha Avery *(Mountain View College, Texas Women's University)*, Leslie Baker *(Eastfield College, Texas Tech University)*, Sreddevi Chittineni *(University of Delaware)*, Chris Churchman *(Furman University, North Lake Community College)*, Beverly Cimino *(Montgomery County Community College)*, Alan Cohen *(University of Michigan, Ann Arbor)*, Christopher David Colson *(Ohio University)*, Jana A. Damphousse *(Oakland University)*, Karen Davis *(Tarrant County Junior College)*, Anjali Cherise D'Souza *(North Lake Community College, University of Texas at Austin)*, Estelle S. D'Souza *(University of Toledo)*, Kimberly Dunham *(University of Delaware)*, Katharine Edmund *(University of Michigan, Ann Arbor)*, Cory Fajardo *(East Carolina University)*, Michael A. Fox *(North Lake Community College, Parker Chiropractic College)*, Michele France *(University of Delaware)*, Hannah A. M. Gilkenson *(University of Michigan, Ann Arbor)*, Shuaib A. Gill *(Wayne State University)*, Beth Glaze *(Orange Coast College)*, Lindsay Goodman *(Eastfield College, University of Texas at Austin)*, Kelly B. Hall *(North Lake Community College, Texas A&M University)*, Cory Hinchman *(Orange Coast College)*, Stacy Hirth *(Ohio University)*, Shelli Hornberger *(Mountain View College, Stephen F. Austin University)*, William DeVaughn Hunt *(East Carolina University)*, Rebekah L. Hunter *(Eastern Michigan University)*, James Patrick Jarvis *(East Carolina University)*, Jennifer Ray Jones *(North Lake Community College, Texas Tech University)*, Jeffrey L. Kacsandi *(Ohio University)*, Michael Kane *(Delaware County Community College)*, Jenny Kerekles *(University of Michigan, Ann Arbor)*, Evelyn Knox *(East Carolina University)*, Elizabeth Kucera *(Ohio University)*, Mike Tien Minh Le *(Orange Coast College)*, Nancy Lee *(Orange Coast College)*, Charles W. Luce *(East Carolina University)*, Nasser Mahaud *(Delaware County Community College)*, Emedio Marchozzi *(Montgomery County Community College)*, Joe Matthews *(Delaware County Community College)*, Glenda McCourt *(Delaware County Community College)*, Marianna J. McSweeney *(University of Delaware)*, Beth L. Measamer *(Brookhaven College)*, Jami Miller *(Ohio University)*, Kimberly S. Miller *(Eastfield Community College, Abilene Christian University)*, Michael Jason Miller *(East Carolina University)*, Ngoc Quang Nguyen *(University of North Texas, Texas Women's University)*, Mark Nolan *(University of Texas at Arlington)*, Anthony Orlando *(University of Michigan, Ann Arbor)*, Stacy Peebles *(Tarrant County Junior College)*, Rick Poce *(Delaware County Community College)*, Mary A. Radlick *(Oakland University)*, Cynthia Rainey *(Texas A&M University)*, Tanga M. Ray *(University of Delaware)*, Gus Reese *(University of Texas at Arlington)*, Renee Sandora *(Eastern Michigan University)*, Jennifer Schklair *(Mountain View Community College)*, Heather Slater *(Montgomery County Community College)*, Katherine Strafford *(Ohio University)*, Theresa Tidd *(Delaware County Community College)*, Nicholas John Urbanczyk *(University of Michigan, Dearborn)*, Travis Vaughn *(University of Delaware)*, Jill Wauldron *(Oakland University)*, Irene Wedderien *(Orange Coast College)*, Alex M. Zadeh *(Orange Coast College)*

We would also like to thank the Introductory Biology Students at Ohio University and Montgomery County Community College.

To the Student

Biology is a challenging subject. The thousands of students we have taught have differed in their life goals and learning styles. Some have had excellent backgrounds in science, others poor ones. Regardless of their backgrounds, it is common for students taking their first college biology course to find that they must work harder than they expected. You can make the task easier by using approaches to learning that have been successful for a broad range of our students over the years. Be sure to use the Learning System we use in this book. It is described in the Preface.

Make a Study Schedule

Many college professors suggest that students study three hours for every hour spent in class. This major investment in study time is one of the main differences between high school and college. To succeed academically, college students must learn to manage their time effectively. The actual number of hours you spend studying biology will vary depending on how quickly you learn the material, as well as on your course load and personal responsibilities, such as work schedules and family commitments.

The most successful students are often those who are best organized. At the beginning of the semester, make a detailed daily calendar. Mark off the hours you are in each class, along with travel time to and from class if you are a commuter. After you get your course syllabi, add to your calendar the dates of all exams, quizzes, papers, and reports. As a reminder, it also helps to add an entry for each major exam or assignment one week before the test or due date. Now add your work schedule and other personal commitments to your calendar. Using a calendar helps you find convenient study times.

Many of our successful biology students set aside 2 hours a day to study biology, rather than depending on a weekly marathon session for 8 or 10 hours during the weekend (when that kind of session rarely happens). Put your study hours into your daily calendar, and stick to your schedule.

Determine Whether the Professor Emphasizes Text Material or Lecture Notes

Some professors test almost exclusively on material covered in lecture. Others rely on their students' learning most, or even all, of the content in assigned chapters. Find out what your professor's requirements are, because the way you study will vary accordingly.

How to study when professors test lecture material

If lectures are the main source of examination questions, make your lecture notes as complete and organized as possible. Before going to class, skim over the chapter, identifying key terms and examining the main figures, so that you can take effective lecture notes. Spend no more than one hour on this.

Within 24 hours after class, rewrite (or type) your notes. Before rewriting, however, read the notes and make marginal notes about anything that is not clear. Then read the corresponding material in your text. Highlight or underline any sections that clarify questions you had in your notes. Read the *entire* chapter, including parts that are not covered in lecture. This extra information will give you breadth of understanding and will help you grasp key concepts.

After reading the text, you are ready to rewrite your notes, incorporating relevant material from the text. It also helps to use the Glossary to help define unfamiliar terms. Many students develop a set of flash cards of key terms and concepts as a way to study. Flash cards are a useful tool to help you learn scientific terminology. They are portable and can be used at times when other studying is not possible, for example, when riding a bus.

Flash cards are not effective when the student tries to second-guess the professor. ("She won't ask this, so I won't make a flash card of it.") Flash cards are also a hindrance when students rely on them exclusively. Studying flash cards instead of reading the text is a bit like reading the first page of each chapter in a mystery novel: It's hard to fill in the missing parts, because you are learning the facts in a disconnected way.

How to study when professors test material in the book

If the assigned readings in the text are going to be tested, you must use your text intensively. After reading the chapter introduction, read the list of Learning Objectives for the first section. These objectives are written in behavioral terms; that is, they ask you to "do" something in order to demonstrate mastery. The objectives give you a concrete set of goals for each section of the chapter. At the end of each section, you will find Review questions keyed to the Learning Objectives. Test yourself, going back over the material to check your responses.

Read each chapter section *actively.* Many students read and study passively. An active learner always has questions in mind and is constantly making connections. For example, there are many processes that must be understood in biology. Don't try to blindly memorize these; instead, think about causes and effects, so that every process becomes a story. Eventually you'll see that many processes are connected by common elements.

You will probably have to read each chapter two or three times before mastering the material. The second and third times through will be much easier than the first, because you'll be reinforcing concepts that you have already partially learned.

After reading the chapter, write a four- to six-page chapter outline by using the subheads as the body of the outline (first-level heads are boldface, in color, and all caps; second-level heads are in color and not all caps). Flesh out your outline by adding important concepts and boldface terms with definitions. Use this outline when preparing for the exam.

Now it is time to test yourself. Answer the Post-Test questions, and check your answers. Write answers to each of the Critical Thinking questions. Finally, review the Learning Objectives in the Chapter Summary and try to answer them before reading the summary provided. If your professor has told you that some or all of the exam will be short-answer or essay format, write out the answer for each Learning Objective. Remember, this is a self-test. If you do not know an answer to a question, find it in the text. If you can't find the answer, use the Index.

Learn the Vocabulary

One stumbling block for many students is learning the many terms that make up the language of biology. In fact, it would be much more difficult to learn and communicate if we did not have this terminology, because words are really tools for thinking. Learning terminology generally becomes easier if you realize that most biological terms are modular. They consist of mostly Latin and Greek roots; once you learn many of these, you will have a good idea of the meaning of a new word even before it is defined. For this reason, we have included an Appendix on Understanding Biological Terms. To be sure you understand the precise definition of a term, use the Index and Glossary. The more you use biological terms in speech and writing, the more comfortable you will be.

Form a Study Group

Active learning is facilitated if you do some of your studying in a small group. In a study group, the roles of teacher and learner can be interchanged: A good way to learn material is to teach. A study group lets you meet challenges in a nonthreatening environment and can provide some emotional support. Study groups are effective learning tools when combined with individual study of text and lecture notes. If, however, you and other members of your study group have not prepared for your meetings by studying individually in advance, the study session can be a waste of time.

Prepare for the Exam

Your calendar tells you it is now one week before your first biology exam. If you have been following these suggestions, you are well prepared and will need only some last-minute reviewing. No all-nighters will be required.

During the week prior to the exam, spend two hours each day actively studying your lecture notes or chapter outlines. It helps many students to read these notes out loud (most people listen to what they say!). Begin with the first lecture/chapter covered on the exam, and continue in the order on the lecture syllabus. Stop when you have reached the end of your two-hour study period. The following day, begin where you stopped the previous day. When you reach the end of your notes, start at the beginning and study them a second time. The material should be very familiar to you by the second or third time around. At this stage, use your textbook only to answer questions or clarify important points.

The night before the exam, do a little light studying, eat a nutritious dinner, and get a full night's sleep. That way, you'll arrive in class on exam day with a well-rested body (and brain) and the self-confidence that goes with being well prepared.

Eldra P. Solomon
Linda R. Berg
Diana W. Martin

Brief Contents

Contents

3 THE CHEMISTRY OF LIFE: ORGANIC COMPOUNDS 41

4 ORGANIZATION OF THE CELL 66

5 BIOLOGICAL MEMBRANES 95

Part 6 STRUCTURE AND LIFE PROCESSES IN PLANTS

39 NEURAL SIGNALING 744

40 NEURAL REGULATION 762

41 SENSORY RECEPTION 786

42 INTERNAL TRANSPORT 807

43 INTERNAL DEFENSE 831

44 GAS EXCHANGE 857

45 PROCESSING FOOD AND NUTRITION 875

46 OSMOREGULATION AND DISPOSAL OF METABOLIC WASTES 896

47 ENDOCRINE REGULATION 913

Part 8 THE INTERACTIONS OF LIFE: ECOLOGY

A View of Life

Jim Olive/Peter Arnold Inc.

Human genome research. A gel containing DNA is loaded into a sequencing machine. On the screen is the information from the gel which is ordered into the DNA base sequences. This image was taken at the Baylor College of Medicine at the Texas Medical Center in Houston.

CHAPTER OUTLINE

- **Characteristics of Life**
- **Biological Organization**
- **Information Transfer**
- **Evolution: The Basic Unifying Concept of Biology**
- **The Energy of Life**
- **The Process and Method of Science**

This is an exciting time to begin studying **biology,** the science of life. Almost daily, biologists are making remarkable new discoveries about the human species and about the millions of other organisms with which we share this planet. One of the most rapidly expanding areas of biological research is genetics, the biologic science that focuses on the mechanisms of heredity.

For 13 years, an international group of scientists worked in 20 sequencing centers in six countries to map the chain of 3 billion letters that make up the **human genome,** the complete set of genes that make up human genetic material. Genes, which are made up of segments of DNA, control specific characteristics, such as eye color and height. In 2003, the International Consortium for the Sequencing of the Human Genome announced the completion of the **Human Genome Project.** One stunning finding of the project has been that the DNA sequences that make up the estimated 30,000 genes of the human genome are 99.9% identical in all humans. Scientists have hailed the completion of the human genome project as a brilliant achievement, a big step toward deciphering the "book of life."

Locating the genes is just the first step, however. Scientists need to refine what has been done to determine which genes do what and how they function. The next level of research will focus on the proteins for which the genes code. Certain proteins make up the structural framework of an organism. Others are enzymes, catalysts that regulate the biochemical reactions essential to life. Geneticists are also carrying out detailed analyses of the genomes of bacteria and other organisms, including primates.

Dr. Francis Collins, director of the genome center at the National Institutes of Health, has said that the completion of the Human Genome Project "marks the start of an exciting new era—the era of the genome in medicine and health." Genome research

BiologyNow™ Seeing BiologyNow throughout the text indicates an opportunity for you to test yourself on key concepts, and to explore animations and interactions on your BiologyNow CD-ROM.

is already contributing to the new science of gene therapy and is opening new avenues for preventing, diagnosing, and treating many human disorders. The science of **genomics,** the analysis of the complete DNA sequence of an organism, will have a worldwide effect on health by increasing knowledge of genetic susceptibility and the body's responses to infectious diseases.

In 1990, geneticists had identified fewer than 100 genes associated with human disease. By 2003, they had identified more than 1400. Using genomics, researchers are identifying hereditary factors in diseases such as cardiovascular disease, cancer, and diabetes. Knowing the locations of genes involved in disease is an important step toward understanding their molecular mechanisms. In turn, this understanding will lead to improved methods for diagnosis and new therapeutic approaches. Researchers are working toward "individualized medicine," in which treatment is tailored to each person's genetic profile.

In addition to its promise in health and medicine, genomics has important implications for agriculture, environmental science, and many other arenas. For example, as they gain knowledge about the genetics of plants, researchers can develop tools to increase crop production. The U.S. Department of Energy's Genomes to Life Program focuses on understanding the molecular biology of thousands of microbe species. Scientists in this program plan to use the new findings to solve major environmental problems such as removing excess carbon dioxide from the atmosphere, cleaning up environments contaminated with toxic wastes, and developing clean fuel sources.

The era of the genome brings with it ethical concerns and responsibilities. How do people safeguard the privacy of genetic information? How can we be certain knowledge of our individual genetic codes would not be used against us when we seek employment or health insurance? Scientists must be ethically responsible and must help educate people about their work, including its benefits relative to its risks. Interestingly, at the very beginning of the Human Genome Project, part of its budget was allocated for research on the ethical, legal, and social implications of its findings. Appropriate legislation may help reduce society's fears about misuse of genetic information.

Genetics is only one of many exciting areas of biology that have an impact on our lives. Whatever your college major or career goals, knowledge of biological concepts is a vital tool for understanding this world and for meeting many of the personal, societal, and global challenges that confront us. Among these challenges are decreasing biological diversity, diminishing natural resources, the expanding human population, and prevention and cure of diseases, such as cancer, Alzheimer's disease, malaria, and acquired immunodeficiency syndrome (AIDS). Meeting these challenges will require the combined efforts of biologists and other scientists, politicians, and biologically informed citizens.

This book is a starting point for an exploration of biology. It provides you with the basic knowledge and the tools to become a part of this fascinating science and a more informed member of society. In this first chapter we introduce three basic themes of biology: (1) the evolution of life, (2) transmission of information, and (3) the flow of energy through living systems. Scientists have accumulated a wealth of evidence showing that the diverse life forms on this planet are related and that organisms have evolved through time from earlier forms of life. The process of evolution is the framework for the science of biology and is a major theme of this book.

Evolution, as well as the survival and function of every organism, depends on the orderly transmission of information. In turn, transmitting information, and all other life processes, including the thousands of chemical transactions that maintain life's organization, require a continuous input of energy. Evolution, information transmission, and energy flow are the forces that give life its unique characteristics. We begin this study of biology by developing a more precise understanding of the fundamental characteristics of living systems. ■

CHARACTERISTICS OF LIFE

Learning Objective

1. Distinguish between living and nonliving things by describing the features that characterize living organisms.

We easily recognize that a pine tree, a butterfly, and a horse are living things, whereas a rock is not. Despite their diversity, the organisms that inhabit our planet share a common set of characteristics that distinguish them from nonliving things. These features include a precise kind of organization, growth and development, self-regulated metabolism, the ability to respond to stimuli, reproduction, and adaptation to environmental change.

Organisms are composed of cells

Although they vary greatly in size and appearance, all organisms consist of basic units called **cells.** New cells are formed only by the division of previously existing cells. These concepts are expressed in the **cell theory** (discussed in Chapter 4), a fundamental unifying concept of biology. Some of the simplest life forms, such as protozoa, are *unicellular organisms,* meaning that each consists of a single cell (Fig. 1-1). In contrast, the body of a cat or a maple tree is made of billions of cells. In such complex *multicellular organisms,* life processes depend on the coordinated functions of component cells that may be organized to form tissues, organs, and organ systems.

Every cell is enveloped by a protective **plasma membrane** that separates it from the surrounding external environment. The plasma membrane regulates passage of materials between cell and environment. Cells have specialized molecules that

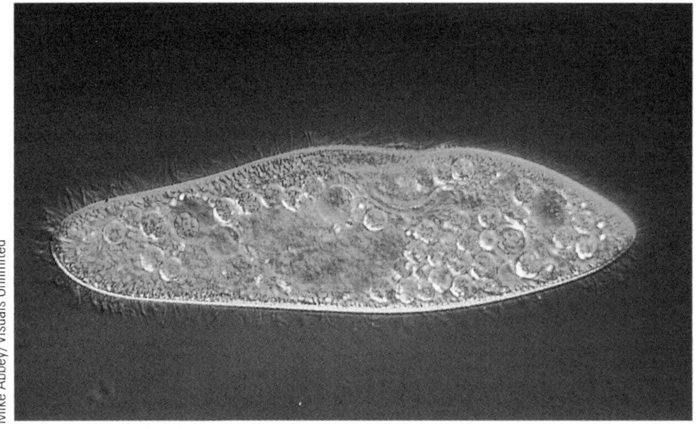

Mike Abbey/Visuals Unlimited

(a)

McMurray Photography

(b)

FIGURE **1-1** **Unicellular and multicellular life forms.**
(a) Unicellular organisms are generally smaller than multicellular organisms and consist of one intricate cell that performs all the functions essential to life. Ciliates, such as this *Paramecium,* move about by beating their hairlike cilia. **(b)** Multicellular organisms, such as this African buffalo *(Syncerus caffer),* and the plants on which it grazes, may consist of billions of cells specialized to perform specific functions.

contain genetic instructions. In most cells, the genetic instructions are encoded in deoxyribonucleic acid, more simply known as **DNA.** Cells typically have internal structures called **organelles** that are specialized to perform specific functions.

There are two fundamentally different types of cells: prokaryotic and eukaryotic. **Prokaryotic cells** are exclusive to bacteria and to microscopic organisms called *archaea.* All other organisms are characterized by their **eukaryotic cells.** These cells typically contain a variety of organelles enclosed by membranes, including a **nucleus,** which houses DNA. Prokaryotic cells are structurally simpler: They do not have a nucleus or other membrane-enclosed organelles.

Organisms grow and develop

Biological growth involves an increase in the size of individual cells of an organism, in the number of cells, or in both. Growth may be uniform in the various parts of an organism, or it may be greater in some parts than in others, causing the body proportions to change as growth occurs. Some organisms—most trees, for example—continue to grow throughout their lives. Many animals have a defined growth period that terminates when a characteristic adult size is reached. An intriguing aspect of the growth process is that each part of the organism typically continues to function as it grows.

Living organisms develop as well as grow. **Development** includes all the changes that take place during an organism's life. Just like many other organisms, every human begins life as a fertilized egg that then grows and develops. The structures and body form that develop are exquisitely adapted to the functions the organism must perform.

Organisms regulate their metabolic processes

Within all organisms, chemical reactions and energy transformations occur that are essential to nutrition, the growth and repair of cells, and the conversion of energy into usable forms. The sum of all the chemical activities of the organism is its **metabolism.** Metabolic processes occur continuously in every living organism, and they must be carefully regulated to maintain **homeostasis,** an appropriate, balanced internal environment. When enough of a cell product has been made, its manufacture must be decreased or turned off. When a particular substance is needed, cell processes that produce it must be turned on. These *homeostatic mechanisms* are self-regulating control systems that are remarkably sensitive and efficient.

The regulation of glucose (a simple sugar) concentration in the blood of complex animals is a good example of a homeostatic mechanism. Your cells require a constant supply of glucose, which they break down to obtain energy. The circulatory system delivers glucose and other nutrients to all the cells. When the concentration of glucose in the blood rises above normal limits, glucose is stored in the liver and in muscle cells. When the concentration begins to fall (between meals), stored nutrients are converted to glucose so that the concentration in the blood returns to normal levels. When glucose becomes depleted, you also feel hungry and restore nutrients by eating.

Organisms respond to stimuli

All forms of life respond to **stimuli,** physical or chemical changes in their internal or external environment. Stimuli that evoke a response in most organisms are changes in the color, intensity, or direction of light; changes in temperature, pressure, or sound; and changes in the chemical composition of the surrounding soil, air, or water. Responding to stimuli involves movement, though not always locomotion (moving from one place to another).

In simple organisms, the entire individual may be sensitive to stimuli. Certain unicellular organisms, for example, respond to bright light by retreating. In some organisms, locomotion is achieved by the slow oozing of the cell, the process of *amoeboid movement.* Other organisms move by beating tiny, hairlike extensions of the cell called **cilia** or longer structures known as **flagella** (Fig. 1-2). Some bacteria move by means of rotating flagella.

Most animals move very obviously. They wiggle, crawl, swim, run, or fly by contracting muscles. Sponges, corals, and oysters have free-swimming larval stages but do not move from place to place as adults. Even though these adults are sessile, meaning they remain firmly attached to a surface, they may have cilia or flagella. These structures beat rhythmically, moving the surrounding water, which contains needed food and oxygen. In complex animals such as polar bears and humans, certain highly specialized cells of the body respond to specific types of stimuli. For example, cells in the retina of the eye respond to light.

Although their responses may not be as obvious as those of animals, plants do respond to light, gravity, water, touch, and other stimuli. For example, plants orient their leaves to the sun and grow toward light. Many plant responses involve different growth rates of various parts of the plant body.

A few plants, such as the Venus flytrap of the Carolina swamps, are very sensitive to touch and catch insects (Fig. 1-3). Their leaves are hinged along the midrib, and they have a scent that attracts insects. Trigger hairs on the leaf surface detect the arrival of an insect and stimulate the leaf to fold. When the edges come together, the hairs interlock, preventing the insect's escape. The leaf then secretes enzymes that kill and digest the insect. The Venus flytrap usually grows in soil deficient in nitrogen. The plant obtains part of the nitrogen required for its growth from the insect it "eats."

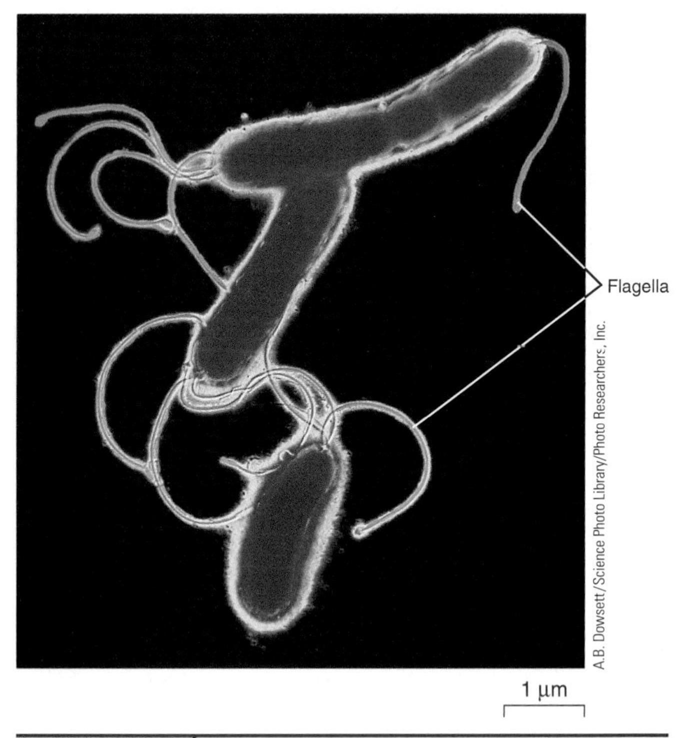

FIGURE **1-2** **Biological movement.**

These bacteria *(Helicobacter pylori),* equipped with flagella for locomotion, have been linked to stomach ulcers. The photograph is a color-enhanced scanning electron micrograph.

Organisms reproduce

At one time, people thought worms arose spontaneously from horsehair in a water trough, maggots from decaying meat, and frogs from the mud of the Nile. Thanks to the work of several

(a)

(b)

FIGURE **1-3** **Plants respond to stimuli.**

(a) Hairs on the leaf surface of the Venus flytrap *(Dionaea muscipula)* detect the touch of an insect, and the leaf responds by folding. **(b)** The edges of the leaf come together and interlock, preventing the fly's escape. The leaf then secretes enzymes that kill and digest the insect.

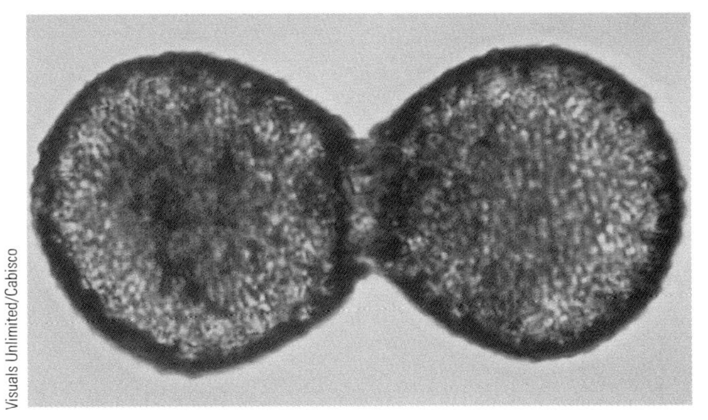

(a) Asexual reproduction

(b) Sexual reproduction

FIGURE **1-4** **Asexual and sexual reproduction.**
(a) Asexual reproduction in *Difflugia*, a unicellular amoeba. One individual gives rise to two or more offspring that are similar to the parent. **(b)** A pair of tropical flies mating. In sexual reproduction, two parents each contribute a gamete (sperm or egg). Gametes fuse to produce the offspring, which has a combination of the traits of both parents.

scientists, including the Italian physician Francesco Redi in the 17th century and French chemist Louis Pasteur in the 19th century, we now know that an organism can come only from previously existing organisms.

Simple organisms, such as amoebas, perpetuate themselves by **asexual reproduction,** without the fusion of egg and sperm to form a fertilized egg (Fig. 1-4a). When an amoeba has grown to a certain size, it reproduces by splitting in half to form two new amoebas. Before an amoeba divides, its hereditary material (set of genes) duplicates, and one complete set is distributed to each new cell. Except for size, each new amoeba is similar to the parent cell. The only way that variation occurs among asexually reproducing organisms is by genetic *mutation,* a permanent change in the genes.

In most plants and animals, **sexual reproduction** is carried out by the fusion of egg and sperm cells to form a fertilized egg (Fig. 1-4b). The new organism develops from the fertilized egg. Offspring produced by sexual reproduction are the product of the interaction of various genes contributed by the mother and the father. This genetic variation is important in the vital processes of evolution and adaptation.

FIGURE **1-5** **Adaptations.**
These Burchell's zebras *(Equus burchelli)*, photographed at Ngorongoro Crater in Tanzania, are behaviorally adapted to position themselves to watch for lions and other predators. Stripes are thought to be an adaptation for visual protection against predators. They serve as camouflage or to break up form when spotted from a distance. The zebra stomach is adapted for feeding on coarse grass passed over by other grazers, an adaptation that helps the animal survive when food is scarce.

Populations evolve and become adapted to the environment

The ability of a population to evolve (change over time) and adapt to its environment equips it to survive in a changing world. **Adaptations** are characteristics that enhance an organism's ability to survive in a particular environment. The long, flexible tongue of the frog is an adaptation for catching insects, the feathers and lightweight bones of birds are adaptations for flying, and the thick fur coat of the polar bear is an adaptation for surviving frigid temperatures. Adaptations may be structural, physiological, behavioral, or a combination of all three (Fig. 1-5). Every biologically successful organism is a complex collection of coordinated adaptations produced through evolutionary processes.

Review

- What characteristics distinguish a living organism from a nonliving object?
- What would be the consequences to an organism if its homeostatic mechanisms failed? Explain your answer.

BiologyNow™ Assess your understanding of **characteristics of life** by taking the pretest on your BiologyNow CD-ROM.

BIOLOGICAL ORGANIZATION

Learning Objective

2 Construct a hierarchy of biological organization, including levels of an individual organism and ecological levels.

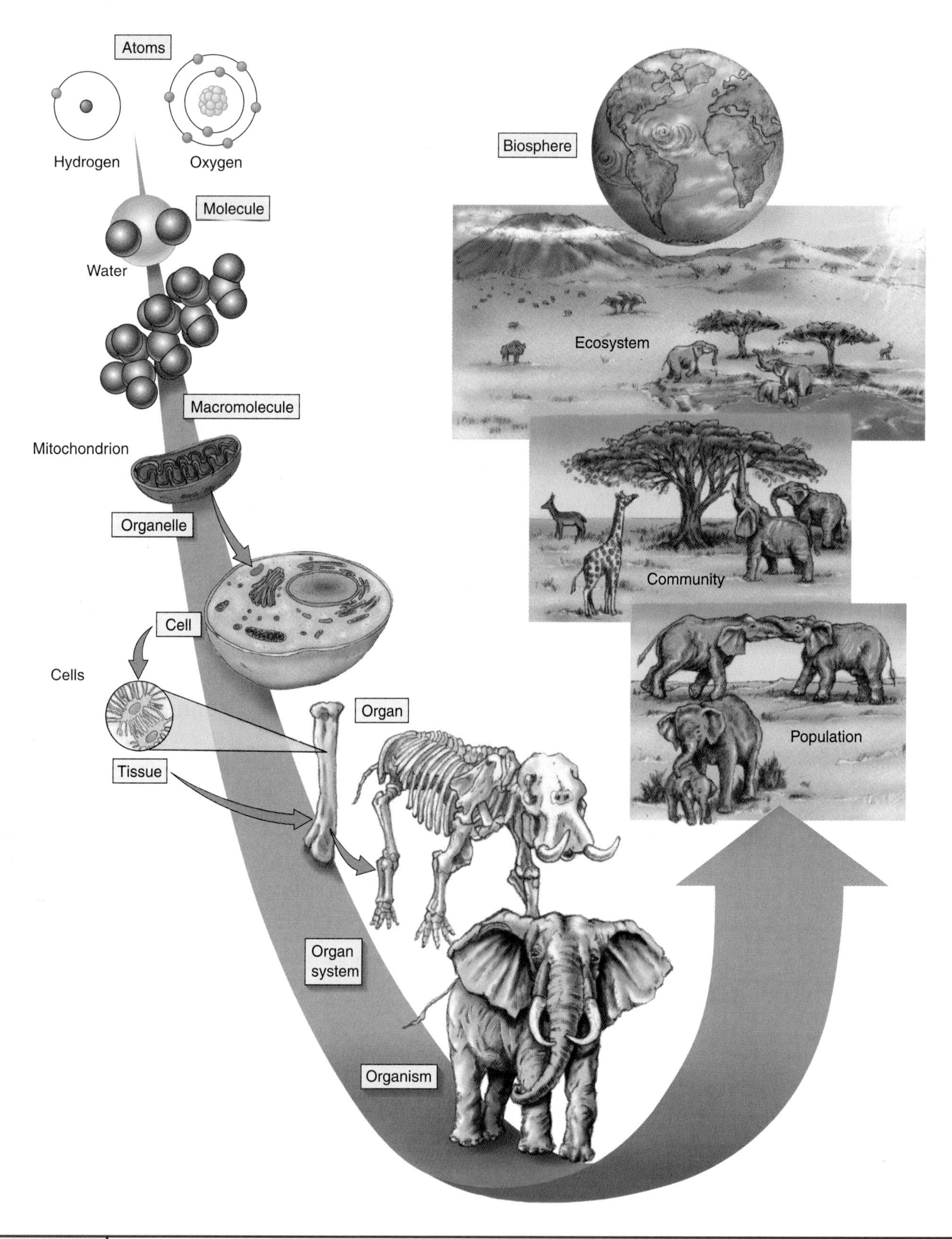

FIGURE **1-6** **The hierarchy of biological organization.**

Atoms join to form molecules of varying size, including very large macromolecules such as proteins and DNA. Atoms and molecules form organelles, such as the cell's nucleus or mitochondria (the site of energy transformations). Many organelles work together to perform the various functions of the cell. Cells associate to form tissues, such as bone tissue. Tissues form organs, such as bones, that in turn comprise organ systems. The skeletal system and other organ systems work together to make up the functioning organism. A population consists of organisms of the same species. The populations of different species that inhabit a particular area make up a community, which together with the nonliving environment form an ecosystem. Earth and all its communities constitute the biosphere.

Whether we study a single complex organism or the world of life as a whole, we can identify a hierarchy of biological organization (Fig. 1-6). At every level, structure and function are precisely coordinated. One way to study a particular level is by looking at its components. For example, biologists can learn about cells by studying atoms and molecules. Learning about a structure by studying its parts is called **reductionism.** However, the whole is more than the sum of its parts. Each level has **emergent properties,** characteristics not found at lower levels. For example, populations have emergent properties such as population density, age structure, and birth and death rates. The individuals that make up a population lack these characteristics.

Organisms have several levels of organization

The chemical level, the most basic level of organization, includes atoms and molecules. An **atom** is the smallest unit of a chemical element that retains the characteristic properties of that element. For example, an atom of iron is the smallest possible amount of iron. Atoms combine chemically to form **molecules.** Two atoms of hydrogen combine with one atom of oxygen to form a single molecule of water. Although composed of two types of atoms that are gases, water is a liquid with very different properties, an example of emergent properties.

At the cell level many different types of atoms and molecules associate with one another to form cells. However, a cell is much more than a heap of atoms and molecules. Its emergent properties make it the basic structural and functional unit of life, the simplest component of living matter that can carry on all the activities necessary for life.

During the evolution of multicellular organisms, cells associated to form **tissues.** For example, most animals have muscle tissue and nervous tissue, and plants have epidermis, a tissue that serves as a protective covering. In most complex organisms, tissues organize into functional structures called **organs,** such as the heart and stomach in animals and roots and leaves in plants. In animals, each major group of biological functions is performed by a coordinated group of tissues and organs called an organ system. The circulatory and digestive systems are examples of organ systems. Functioning together with great precision, organ systems make up a complex, multicellular **organism.** Again, emergent properties are evident. An organism is much more than its component organ systems.

Several levels of ecological organization can be identified

Organisms interact to form still more complex levels of biological organization. All the members of one species that live in the same geographic area at the same time make up a **population.** The populations of organisms that inhabit a particular area and interact with one another form a **community.** A community can consist of hundreds of different types of organisms. As populations within a community evolve, the community changes.

A community together with its nonliving environment is referred to as an **ecosystem.** An ecosystem can be as small as a pond (or even a puddle) or as vast as the Great Plains of North America or the Arctic tundra. All of Earth's ecosystems together are known as the **biosphere.** The biosphere includes all of Earth that is inhabited by living organisms—the atmosphere, the hydrosphere (water in any form), and the lithosphere (Earth's crust). The study of how organisms relate to one another and to their physical environment is called **ecology** (derived from the Greek *oikos,* meaning "house").

Review

- What are the levels of organization within an organism?
- What are the levels of ecological organization?

BiologyNow™ Assess your understanding of hierarchical **biological organization** by taking the pretest on your BiologyNow CD-ROM.

INFORMATION TRANSFER

Learning Objective

3 Summarize the importance of information transfer to living systems, giving specific examples.

For an organism to grow, develop, carry on self-regulated metabolism, respond to stimuli, and reproduce, it must have precise instructions and its cells must be able to communicate. The information an organism needs to carry on these life processes is coded and delivered in the form of chemical substances and electrical impulses. Organisms must also communicate information to each other.

DNA transmits information from one generation to the next

Humans give birth only to human babies, not to giraffes or rose bushes. In organisms that reproduce sexually, each offspring is a combination of the traits of its parents. In 1953, James Watson and Francis Crick worked out the structure of **DNA,** the large molecule that makes up the **genes,** the units of hereditary material (Fig. 1-7). Watson and Crick's work led to the understanding of the genetic code that transmits genetic information from generation to generation. This code works somewhat like an alphabet; it can "spell" an amazing variety of instructions for making organisms as diverse as bacteria, frogs, and redwood trees. The genetic code is a dramatic example of the unity of life because it is used to specify instructions for making every living organism.

Information is transmitted by chemical and electrical signals

Genes control the development and functioning of every organism. DNA contains the "recipes" for making all the proteins needed by the organism. **Proteins** are large molecules important in determining the structure and function of cells and tissues. Brain cells differ from muscle cells in large part because they have different types of proteins. Some proteins are important in communication within and among cells. Certain proteins on

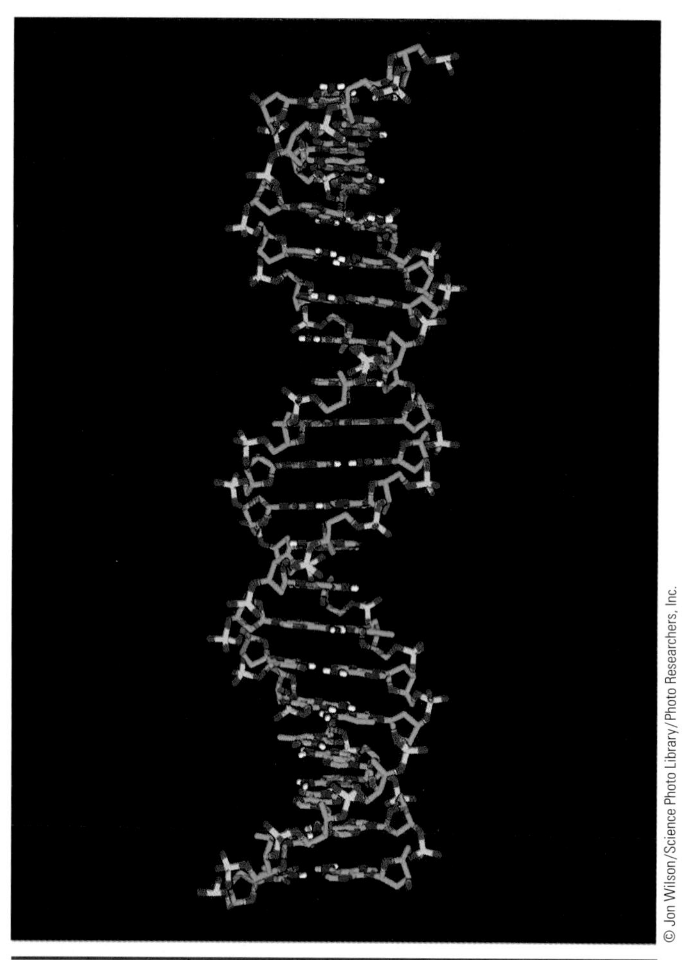

© Jon Wilson/Science Photo Library/Photo Researchers, Inc.

FIGURE **1-7** **DNA.**

Organisms transmit information from one generation to the next by way of its DNA, the hereditary material. As shown in this model, DNA consists of two chains of atoms twisted into a helix. Each chain consists of subunits called nucleotides. The sequence of nucleotides makes up the genetic code.

the surface of a cell serve as markers so that other cells "recognize" them. Some cell surface proteins serve as receptors that combine with chemical messengers.

Cells use proteins and many other types of molecules to communicate with one another. In a multicellular organism, chemical compounds secreted by cells help regulate growth, development, and metabolic processes in other cells. The mechanisms involved in **cell signaling** are complex, often involving multistep biochemical sequences, and cell signaling is currently an area of intense research. A major focus has been the transfer of information among cells of the immune system. A better understanding of how cells communicate promises new insights into how the body protects itself against disease organisms. Learning to manipulate cell signaling may lead to new methods of delivering drugs into cells and new treatments for cancer and other diseases. Throughout this book we discuss examples of cell signaling.

Hormones are molecules that function as chemical messengers that transmit information from one part of an organism to another. A hormone can signal cells to produce or secrete a certain protein or other substance.

Many organisms use electrical signals to transmit information. Most animals have nervous systems that transmit information by way of both electrical impulses and chemical compounds known as **neurotransmitters.** Information transmitted from one part of the body to another is important in regulating life processes. In complex animals, the nervous system transmits signals from sensory receptors such as the eyes and ears to the brain, giving the animal information about its outside environment.

Information must also be transmitted from one organism to another. Mechanisms for this type of communication include the release of chemicals, visual displays, and sounds. Typically, organisms use a combination of several types of communication signals. For example, a dog may signal aggression by growling, using a particular facial expression, and laying its ears back. Many animals perform complex courtship rituals in which they display parts of their bodies, often elaborately decorated, to attract a mate.

Review

- Why is DNA important?
- Give an example of cell signaling.

Biology Now™ Assess your understanding of **information transfer** by taking the pretest on your BiologyNow CD-ROM.

EVOLUTION: THE BASIC UNIFYING CONCEPT OF BIOLOGY

Learning Objectives

4 Demonstrate the binomial system of nomenclature using several specific examples, and classify an organism (such as a human) in its domain, kingdom, phylum, class, order, family, genus, and species.

5 Identify the six kingdoms of living organisms, and give examples of organisms assigned to each group.

6 Give a brief overview of the theory of evolution, and explain why it is the principal unifying concept in biology.

7 Apply the theory of natural selection to any given adaptation, suggesting a logical explanation of how the adaptation may have evolved.

The theory of **evolution,** which explains how populations of organisms have changed over time, has become the most important unifying concept of biology. Some element of an evolutionary perspective is present in every specialized field within biology. Biologists try to understand the structure, function, and behavior of organisms and their interactions with one another by considering them in light of the long, continuing process of evolution. Although we discuss evolution in depth in Chapters 17 through 21, we present a brief overview here in Chapter 1 to give you the background necessary to understand other aspects of biology. First we examine how biologists organize the millions of organisms that have evolved, and then we summarize the mechanisms that drive evolution.

Biologists use a binomial system for naming organisms

About 1.7 million species of extant (currently living) organisms have been scientifically identified, and biologists estimate that several million more remain to be discovered. To study life, we need a system for organizing, naming, and classifying its myriad forms. **Systematics** is the field of biology that studies the diversity of organisms and their evolutionary relationships. **Taxonomy,** a subspecialty of systematics, is the science of naming and classifying organisms.

In the 18th century Carolus Linnaeus, a Swedish botanist, developed a hierarchical system of naming and classifying organisms that, with some modification, is still used today. The lowest category of classification is the **species,** a group of organisms with similar structure, function, and behavior; in nature, they breed only with each other. Members of a species have a common gene pool and share a common ancestry. Closely related species are grouped together in the next higher category of classification, the **genus** (pl. *genera*).

The Linnaean system of naming species is known as the **binomial system of nomenclature** because each species is assigned a two-part name. The first part of the name is the genus, and the second part, the **species epithet,** designates a particular species belonging to that genus. The species epithet is often a descriptive word expressing some quality of the organism. It is always used together with the full or abbreviated generic name preceding it. The generic name is always capitalized; the species epithet is generally not capitalized. Both names are always italicized or underlined. For example, the domestic dog, *Canis familiaris* (abbreviated *C. familiaris*), and the timber wolf, *Canis lupus (C. lupus),* belong to the same genus. The domestic cat, *Felis catus,* belongs to a different genus. The scientific name of the American white oak is *Quercus alba,* whereas the name of the European white oak is *Quercus robur.* Another tree, the white willow, *Salix alba,* belongs to a different genus. The scientific name for our own species is *Homo sapiens* ("wise man").

Taxonomic classification is hierarchical

Just as closely related species may be grouped together in a common genus, related genera can be grouped in a more inclusive group, a **family.** Families are grouped into **orders,** orders into **classes,** and classes into **phyla** (sing., *phylum*). Biologists group phyla into **kingdoms,** and kingdoms are assigned to **domains.** Each formal grouping at any given level is a **taxon** (pl., *taxa*). Note that each taxon is more inclusive than the taxon below it. Together they form a hierarchy ranging from species to domain (Table 1-1; Fig. 1-8).

Consider a specific example. The family Canidae, which includes all doglike carnivores (animals that eat mainly meat), consists of 12 genera and about 34 living species. Family Canidae, along with family Ursidae (bears), family Felidae (catlike animals), and several other families that eat mainly meat, are all placed in order Carnivora. Order Carnivora, order Primates (to

TABLE 1-1 Taxonomic Classification

Category	Cat	Human	White Oak
Domain	Eukarya	Eukarya	Eukarya
Kingdom	Animalia	Animalia	Plantae
Phylum	Chordata	Chordata	Anthophyta
Subphylum	Vertebrata	Vertebrata	None
Class	Mammalia	Mammalia	Dicotyledones
Order	Carnivora	Primates	Fagales
Family	Felidae	Hominidae	Fagaceae
Genus and Species	*Felis catus*	*Homo sapiens*	*Quercus alba*

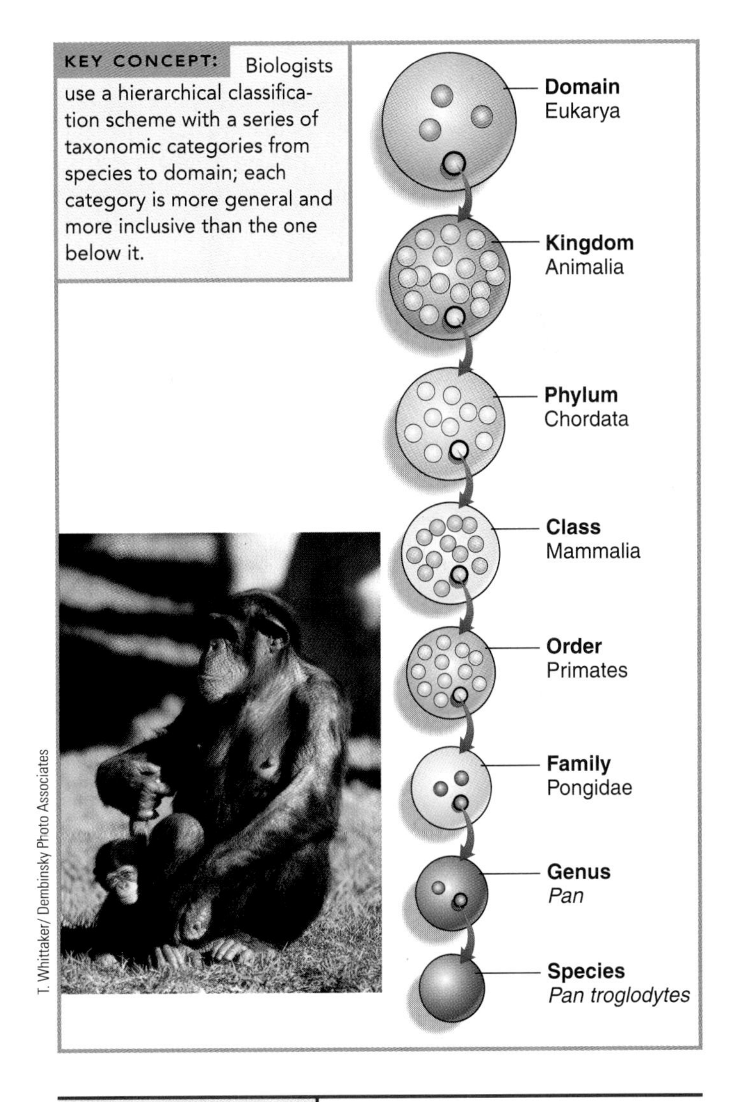

ACTIVE FIGURE 1-8 Classification of the chimpanzee *(Pan troglodytes).*

As illustrated by this example, the classification scheme used by biologists is hierarchical. The smaller circles within each large circle represent the categories below it. For example, the four smaller circles in domain Eukarya represent the four kingdoms in this domain.

BiologyNow™ Learn more about **biological classification** by clicking on this figure on your BiologyNow CD-ROM.

which chimpanzees and humans belong), and several other orders belong to class Mammalia (mammals). Class Mammalia is grouped with several other classes that include fishes, amphibians, reptiles, and birds in subphylum Vertebrata. The vertebrates belong to phylum Chordata, which is part of kingdom Animalia. Animals are assigned to domain Eukarya.

Organisms can be assigned to three domains and six kingdoms

Systematics has itself evolved as scientists have developed new molecular techniques. As researchers report new data, the classification of organisms changes. Although not all biologists agree on how organisms are related or on how to classify them, many biologists now assign organisms to three domains and six kingdoms.

Bacteria and archaebacteria are unicellular prokaryotic cells; they differ from all other organisms in that they are **prokaryotes.** Two distinct groups have been recognized among the prokaryotes, and biologists assign them to two domains: **Eubacteria** and **Archaea.** The **eukaryotes,** organisms with eukaryotic cells, are classified in domain **Eukarya.**

In the classification system used in this book, every organism is also assigned to one of six **kingdoms** (Fig. 1-9). Two kingdoms correspond to the prokaryotic domains: Kingdom **Archaebacteria** corresponds to domain Archaea, and kingdom Eubacteria corresponds to domain Eubacteria. The remaining four kingdoms are assigned to domain Eukarya. Kingdom **Protista** consists of protozoa, algae, water molds, and slime molds. These are unicellular or simple multicellular organisms. Some protists are adapted to carry out photosynthesis, the process in which light energy is converted to the chemical energy of food molecules. Kingdom **Fungi** is composed of the yeasts, mildews, molds, and mushrooms. These organisms do not photosynthesize. They obtain their nutrients by secreting digestive enzymes into food and then absorbing the predigested food.

Members of kingdom **Plantae** are complex multicellular organisms adapted to carry out photosynthesis. Among characteristic plant features are the *cuticle* (a waxy covering over aerial parts that reduces water loss), *stomata* (tiny openings in stems and leaves for gas exchange), and multicellular *gametangia* (organs that protect developing reproductive cells). Kingdom Plantae includes both nonvascular plants (mosses) and vascular plants (ferns, conifers, and flowering plants), those that have tissues specialized for transporting materials throughout the plant body.

Kingdom **Animalia** is made up of multicellular organisms that eat other organisms for nutrition. Complex animals exhibit considerable tissue specialization and body organization. These characters have evolved along with complex sense organs, nervous systems, and muscular systems.

We discuss the diversity of life in more detail in Chapters 22 through 30, and we summarize classification in Appendix B. We refer to these groups repeatedly throughout this book, as we consider the many kinds of challenges living organisms face and the various adaptations that have evolved in response to them.

Species adapt in response to changes in their environment

Every organism is the product of complex interactions between environmental conditions and the genes of its ancestors. If all individuals of a species were exactly alike, any change in the environment might be disastrous to all, and the species would become extinct. Adaptations to changes in the environment occur as a result of evolutionary processes that take place over time and involve many generations.

Natural selection is an important mechanism by which evolution proceeds

Although philosophers and naturalists discussed the concept of evolution for centuries, Charles Darwin and Alfred Wallace first brought a theory of evolution to general attention and suggested a plausible mechanism, **natural selection,** to explain it. In his book *The Origin of Species by Natural Selection,* published in 1859, Darwin synthesized many new findings in geology and biology. He presented a wealth of evidence that the present forms of life descended, with modifications, from previously existing forms. Darwin's book raised a storm of controversy in both religion and science, some of which still lingers.

Darwin's theory of evolution has helped shape the biological sciences to the present day. His work generated a great wave of scientific observation and research that has provided much additional evidence that evolution governs the great diversity of organisms on our planet. Even today, the details of the process of evolution are a major focus of investigation and discussion.

Darwin based his theory of natural selection on the following four observations: (1) Individual members of a species show some variation from one another. (2) Organisms produce many more offspring than will survive to reproduce (Fig. 1-10). (3) Organisms compete for necessary resources such as food, sunlight, and space. Individuals with characteristics that enable them to obtain and use resources are more likely to survive to reproductive maturity and thus produce offspring. (4) The survivors that reproduce pass their adaptations for survival on to their offspring. Thus the best adapted individuals of a population leave, on average, more offspring than do other individuals. Because of this differential reproduction, a greater proportion of the population becomes adapted to the prevailing environmental conditions. The environment *selects* the best adapted organisms for survival. Note that adaptation involves changes in populations rather than in individual organisms.

Darwin did not know about DNA or understand the mechanisms of inheritance. Scientists now understand that most variations among individuals are a result of different varieties of genes that code for each characteristic. The ultimate source of these variations is random **mutations,** chemical or physical changes in DNA that persist and can be inherited. Mutations modify genes; by this process they provide the raw material for evolution.

Domains:

Eubacteria | Archaea | Eukarya

Kingdoms:

Animalia

Fungi

Archaebacteria

Plantae

Protista

Eubacteria

Common ancestor

(a)

R. Robinson/Visuals Unlimited

(b) 5 μm

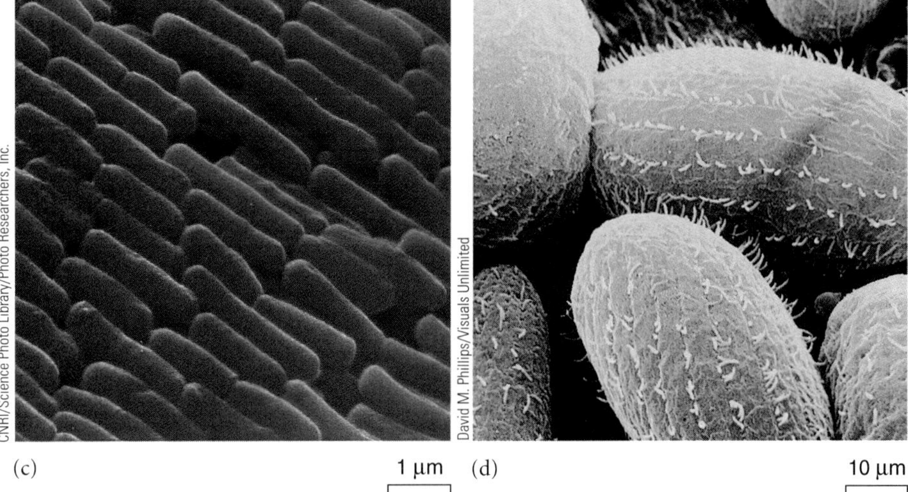

CNRI/Science Photo Library/Photo Researchers, Inc.

(c) 1 μm

David M. Phillips/Visuals Unlimited

(d) 10 μm

Ulf Sjostedt/FPG International

(e)

John Arnaldi

(f)

McMurray Photography

(g)

FIGURE **1-9** **A survey of the kingdoms of life.**

(a) In this book organisms are assigned to three domains and six kingdoms. **(b)** These archaebacteria *(Methanosarcina mazei)*, members of kingdom Archaebacteria, produce methane. **(c)** The large, rod-shaped bacterium *Bacillus anthracis*, a member of kingdom Eubacteria, causes anthrax, a cattle and sheep disease that can infect humans. **(d)** Unicellular protozoa *(Tetrahymena)* are classified in kingdom Protista. **(e)** Mushrooms, such as these fly agaric mushrooms *(Amanita muscaria)*, belong to kingdom Fungi. The fly agaric is poisonous and causes delirium, raving, and profuse sweating when ingested. **(f)** The plant kingdom claims many beautiful and diverse forms, such as the lady's slipper *(Phragmipedium caricinum)*. **(g)** Among the fiercest members of the animal kingdom, lions *(Panthera leo)* are also among the most sociable. The largest of the big cats, lions live in prides (groups).

J. Serrao/Photo Researchers, Inc.

FIGURE **1-10** **Egg masses of the wood frog *(Rana sylvatica).***

Many more eggs are produced than can possibly develop into adult frogs. Random events are largely responsible for determining which of these developing frogs will hatch, reach adulthood, and reproduce. However, certain traits of each organism also contribute to its probability for success in its environment. Not all organisms are as prolific as the frog, but the generalization that more organisms are produced than survive is true throughout the living world.

Populations evolve as a result of selective pressures from changes in the environment

All the genes present in a population make up its **gene pool.** By virtue of its gene pool, a population is a reservoir of variation. Natural selection acts on individuals within a population. Selection favors individuals with genes that specify traits that enable them to respond effectively to pressures exerted by the environment. These organisms are most likely to survive and produce offspring. As these successful organisms pass on their genetic recipe for survival, their traits become more widely distributed in the population. Over time, as organisms continue to change (and as the environment itself changes, bringing different selective pressures), the members of the population become better adapted to their environment and less like their ancestors.

As members of a population adapt to environmental pressures and exploit new opportunities for finding food, maintaining safety, and avoiding predators, the population diversifies and new species may evolve. The Hawaiian honeycreepers, a group of related birds, are a good example. When honeycreeper ancestors first reached Hawaii, few other birds were present, so there was little competition. Honeycreepers moved into a variety of food zones, and evolved various types of bills (Fig. 1-11; see also Chapter 19 and Fig. 19-15). Some honeycreepers now have long, curved bills, adapted for feeding on nectar from tubular flowers. Others have short, thick bills for foraging for insects, and still others have adapted for eating seeds.

Review

- What is the binomial system of nomenclature?
- How might you explain the sharp claws and teeth of tigers in terms of natural selection?

BiologyNow™ Assess your understanding of **evolution as the basic unifying concept of biology** by taking the pretest on your BiologyNow CD-ROM.

THE ENERGY OF LIFE

Learning Objective

8 Summarize the flow of energy through ecosystems, contrasting the roles of producers, consumers, and decomposers.

Life depends on a continuous input of energy from the sun, because every activity of a living cell or organism requires energy. Whenever energy is used to perform biological work, some is converted to heat and dispersed into the environment.

Jack Jeffrey, Inc.

(a)

Jack Jeffrey, Inc.

(b)

Jack Jeffrey, Inc.

(c)

FIGURE **1-11** **Adaptation and diversification in Hawaiian honeycreepers.**

(a) The bill of this 'Akiapola'au male *(Hemignathus munroi)* is adapted for extracting insect larvae from bark. The lower mandible (jaw) is used to peck at and pull off bark, whereas the upper mandible and tongue remove the prey. **(b)** 'I'iwi *(Vestiaria coccinea)* in 'ohi'a blossoms. The bill is adapted for feeding on nectar in tubular flowers. **(c)** Palila *(Loxiodes bailleui)* in mamane tree. This finch-billed honeycreeper feeds on immature seeds in pods of the mamane tree. It also eats insects, berries, and young leaves. All three species shown here are endangered, mainly because their habitats have been destroyed by humans.

Energy flows through cells and organisms

Recall that all the energy transformations and chemical processes that occur within an organism are referred to as its *metabolism*. Energy is necessary to carry on the metabolic activities essential for growth, repair, and maintenance. Each cell of an organism requires nutrients that contain energy. Certain nutrients are used as fuel for **cellular respiration,** a process during which some of the energy stored in the nutrient molecules is released for use by the cells (Fig. 1-12). This energy can be used for cell work or for the synthesis of needed materials, such as new cell components. Virtually all cells carry on cellular respiration.

Energy flows through ecosystems

Like individual organisms, ecosystems depend on a continuous input of energy. A self-sufficient ecosystem contains three types of organisms—producers, consumers, and decomposers—and has a physical environment appropriate for their survival. These organisms depend on each other and on the environment for nutrients, energy, oxygen, and carbon dioxide. However, there is a one-way flow of energy through ecosystems. Organisms can neither create energy nor use it with complete efficiency. During every energy transaction, some energy disperses into the environment as heat and is no longer available to the organism (Fig. 1-13).

Producers, or **autotrophs,** are plants, algae, and certain bacteria that produce their own food from simple raw materials. Most of these organisms use sunlight as an energy source and carry out **photosynthesis,** the process in which producers synthesize complex molecules from carbon dioxide and water. The light energy is transformed into chemical energy, which is stored within the chemical bonds of the food molecules produced. Oxygen, which is required not only by plant cells but also by the cells of most other organisms, is produced as a by-product of photosynthesis:

Carbon dioxide + water + light energy
⟶ sugars (food) + oxygen

Animals are **consumers,** or **heterotrophs**—that is, organisms that depend on producers for food, energy, and oxygen.

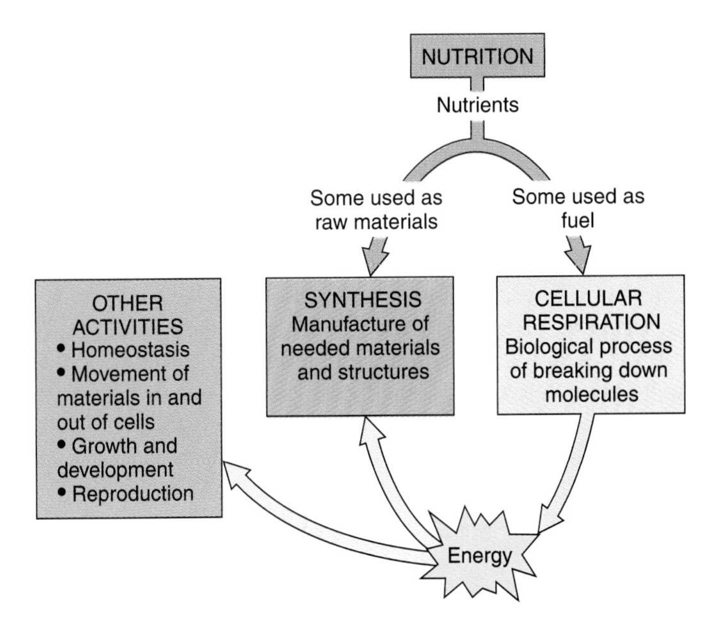

FIGURE **1-12** **Relationships among metabolic processes.**
These processes occur continuously in the cells of living organisms. Cells use some of the nutrients in food to synthesize needed materials and cell parts. Cells use other nutrients as fuel for cellular respiration, a process that releases energy stored in food. This energy is needed for synthesis and for other forms of cell work.

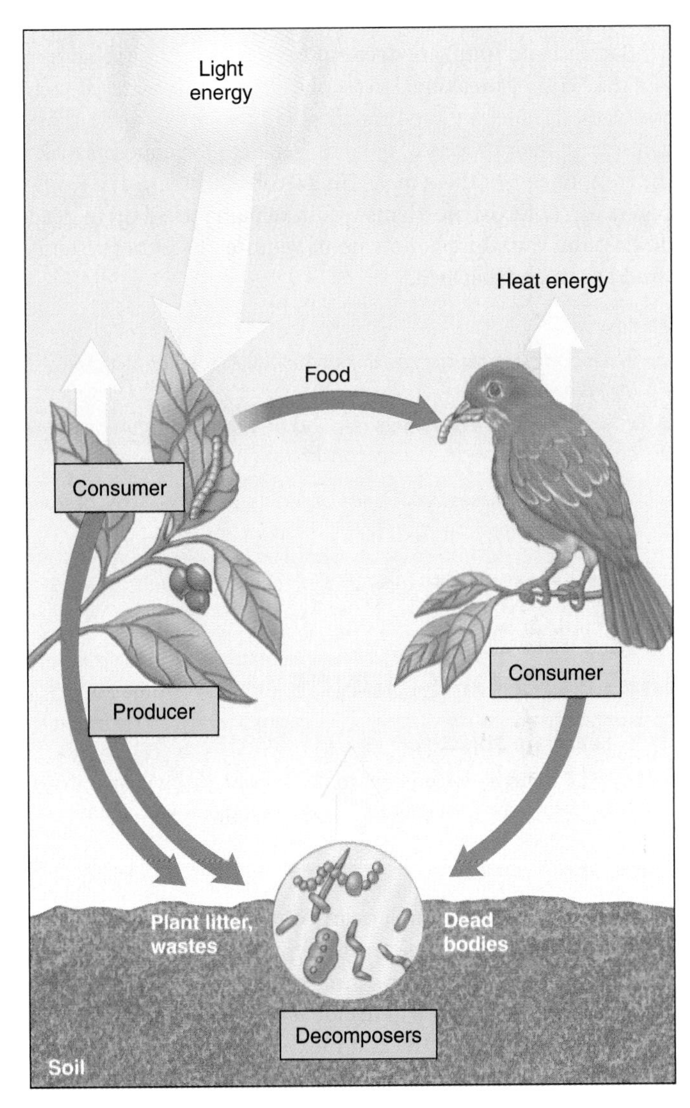

FIGURE **1-13** **Energy flow.**
Continuous energy input from the sun operates the biosphere. During photosynthesis, producers use the energy from sunlight to make complex molecules from carbon dioxide and water. Consumers, such as the caterpillar and robin shown here, obtain energy, carbon, and other needed materials when they eat producers or consumers that have eaten producers. Wastes and dead organic material supply decomposers with energy and carbon. During every energy transaction, some energy is lost to biological systems, dispersing into the environment as heat.

Consumers obtain energy by breaking down sugars and other food molecules originally produced during photosynthesis. When chemical bonds are broken during this process of cellular respiration, their stored energy is made available for life processes:

Sugars (and other food molecules) + oxygen
⟶ carbon dioxide + water + energy

Consumers contribute to the balance of the ecosystem. For example, consumers produce carbon dioxide needed by producers. (Note that producers also carry on cellular respiration.) The metabolism of consumers and producers helps maintain the life-sustaining mixture of gases in the atmosphere.

Bacteria and fungi are **decomposers,** heterotrophs that obtain nutrients by breaking down nonliving organic material such as wastes, dead leaves and branches, and the bodies of dead organisms. In their process of obtaining energy, decomposers make the components of these materials available for reuse. If decomposers did not exist, nutrients would remain locked up in dead bodies, and the supply of elements required by living systems would soon be exhausted.

Review

- What components do you think a balanced forest ecosystem might have?
- In what ways do consumers depend on producers? On decomposers? Include energy considerations in your answer.

BiologyNow™ Assess your understanding of **the energy of life** by taking the pretest on your BiologyNow CD-ROM.

THE PROCESS AND METHOD OF SCIENCE

Learning Objective

9 Design an experiment to test a given hypothesis, using the procedure and terminology of the scientific method.

PROCESS OF SCIENCE

Biologists work in laboratories and out in the field (Fig. 1-14). Their investigations range from the study of molecular biology and viruses to the interactions of the communities of our biosphere. Perhaps you will decide to become a research biologist and help unravel the complexities of the human brain, discover new hormones that cause plants to flower, identify new species of animals or bacteria, or develop new stem cell strategies to treat cancer, AIDS, or heart disease. Applications of basic biological research have provided the technology to transplant kidneys, livers, and hearts, manipulate genes, treat many diseases, and increase world food production. Biology has been a powerful force in providing the quality of life that most of us enjoy. You may choose to enter an applied field of biology, such as environmental science, dentistry, medicine, pharmacology, or veterinary medicine. Several interesting careers in the biological sciences are discussed in the Career Visions on our Web site.

Mark Moffett/Minden Pictures

FIGURE 1-14 | **Biologist at work.**
This biologist studying the rainforest canopy in Costa Rica is part of an international effort to study and preserve tropical rain forests. Researchers study the interactions of organisms and the effects of human activities on the rain forests.

Biology is a science. The word *science* comes from a Latin word meaning "to know." Science is a way of thinking and a method of investigating the world around us in a systematic manner. Science enables us to uncover ever more about the world we live in and leads us to an expanded appreciation of our universe.

The **process of science** is investigative, dynamic, and often controversial. Because it is influenced by cultural, social, and historical contexts, as well as by the personalities of scientists themselves, the process changes over time. The observations made, the range of questions posed, and the design of experiments depend on the creativity of the individual scientist. In contrast, the **scientific method** involves a series of ordered steps and is a framework that most scientists use.

Using the scientific method, scientists make careful observations, ask critical questions, and develop **hypotheses,** which are testable statements. Using their hypotheses, scientists make predictions that can be tested, and test their predictions by making further observations or by performing experiments (Fig. 1-15; also see On the Cutting Edge: New Possibilities for Environmentally Friendly Pest-Control Strategies). They interpret the results of their experiments and draw conclusions from them. Even results that do not support the hypothesis may be valuable and may lead to new hypotheses. If the results do support a hypothesis, a scientist may use them to generate related hypotheses.

Science is systematic. Scientists organize, and often quantify, knowledge, making it readily accessible to all who wish to build on its foundation. In this way, science is both a personal and a social endeavor. Science is not mysterious. Anyone who understands its rules and procedures can take on its challenges. What distinguishes science is its insistence on rigorous methods to examine a problem. Science seeks to give precise knowledge about those aspects of the world that are accessible to its methods of inquiry. It is not a replacement for philosophy, religion, or art. Being a scientist does not prevent one from participating in other fields of human endeavor, just as being an artist does not prevent one from practicing science.

Science requires systematic thought processes

Two types of systematic thought processes scientists use are deduction and induction. With **deductive reasoning,** we begin with supplied information, called *premises,* and draw conclusions on the basis of that information. Deduction proceeds from general principles to specific conclusions. For example, if you accept the premise that all birds have wings and the second premise that sparrows are birds, you can conclude deductively that sparrows have wings. Deduction helps us discover relationships among known facts.

Inductive reasoning is the opposite of deduction. We begin with specific observations and draw a conclusion or discover a general principle. For example, if you know sparrows have wings and are birds, and you know robins, eagles, pigeons, and hawks have wings and are birds, you might induce that all birds have wings. In this way, you can use the inductive method to organize raw data into manageable categories by answering the question, What do all these facts have in common?

A weakness of inductive reasoning is that conclusions generalize the facts to all possible examples. When we formulate the general principle, we go from many observed examples to all possible examples. This is known as an *inductive leap.* Without it, we could not arrive at generalizations. However, we must be sensitive to exceptions and to the possibility that the conclusion is not valid. For example, the kiwi bird of New Zealand does *not* have functional wings! The generalizations in inductive conclusions come from the creative insight of the human mind, and creativity, however admirable, is not infallible.

Scientists make careful observations and ask critical questions

Chance and luck are often involved in recognizing a phenomenon or problem, but significant discoveries are usually made by those who are in the habit of looking critically at nature. Necessary technology for investigating the problem must also be available. In 1928, British bacteriologist Alexander Fleming observed that a blue mold had invaded one of his bacterial cultures. He almost discarded it, but then he noticed that the area contaminated by the mold was surrounded by a zone where bacterial colonies did not grow well.

The bacteria were disease organisms of the genus *Staphylococcus,* which can cause boils and skin infections. Anything that could kill them was interesting! Fleming saved the mold, a variety of *Penicillium* (blue bread mold). Later, scientists discovered that the mold produced a substance that slowed reproduction of the bacterial population but was usually harmless to laboratory animals and humans. The substance was penicillin, the first antibiotic.

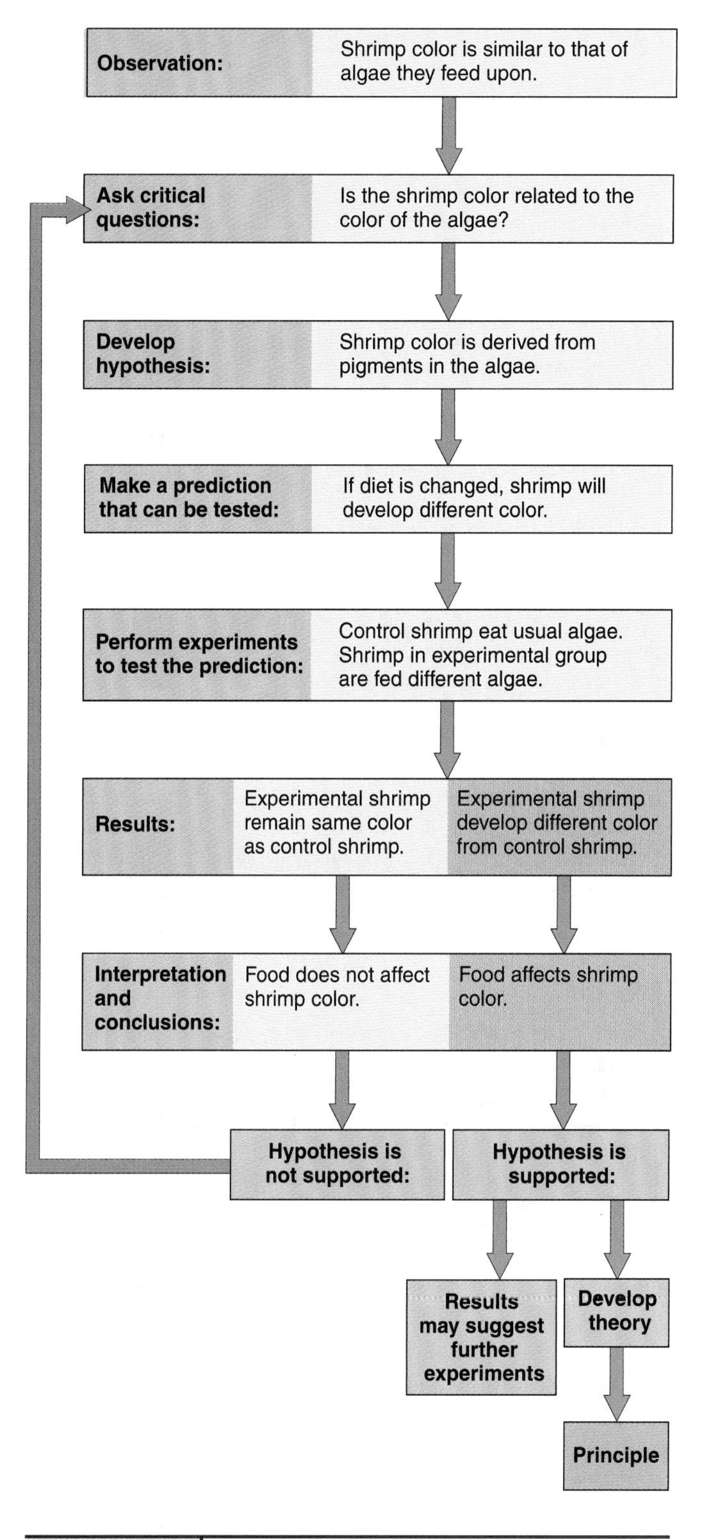

FIGURE **1-15** **The scientific method.**
Scientists use the scientific method as a framework for their research.

You may wonder how many times the same type of mold grew on the cultures of other bacteriologists who failed to make the connection and simply threw away their contaminated cultures. Fleming benefited from chance, but his mind was prepared

ON THE CUTTING EDGE New Possibilities for Environmentally Friendly Pest-Control Strategies

Hypothesis: When attacked by insects that eat them, plants emit airborne chemicals that reduce the number of herbivores (animals that eat plants) feeding on them.

Method: Researchers conducted a field study in which they examined volatile chemicals released by wild tobacco plants during attack by insect herbivores. The researchers also mimicked the release of each of five commonly released volatile chemicals.

Results: The plants emitted volatile chemicals that reduced the number of eggs laid by some herbivores and increased predation of the herbivore eggs by carnivorous insects.

Conclusion: By releasing certain volatile chemicals, plants significantly reduce the number of herbivores feeding on them.

In 1983, while still an undergraduate at Dartmouth College in the United States, Ian Baldwin and his mentor, biologist Jack Schultz, published a controversial hypothesis stating that airborne chemical signals from damaged maple and poplar trees appear to increase the chemical defenses of undamaged trees nearby. Other biologists were not receptive to the idea that plants could communicate with one another, and for some time Baldwin and Schultz had difficulty obtaining funding for their research. Eventually these researchers produced experimental results that changed their colleagues' attitudes. During the past few years, several research teams have studied plant signaling in laboratory or agricultural settings.

In 2001, Andre Kessler, a graduate student at the Max Planck Institute for Chemical Ecology in Jena, Germany, and Ian Baldwin, now director of molecular ecology research at the Max Planck Institute, reported in *Science** that they had studied volatile chemicals released by wild tobacco plants (*Nicotiana attenuate*) in a natural environment—the desert of southwestern Utah. These researchers quantified volatile chemicals released by the tobacco plants during attack by three species of herbivorous insects. They also studied five of these volatile chemicals individually by mimicking their release by plants.

Kessler and Baldwin first established that wild tobacco releases volatile chemicals in response to attack by herbivorous insects. Then they studied predation of the herbivore eggs by leaf bugs. They glued the eggs of herbivorous insects onto tobacco leaves that they treated with a single synthetic volatile chemical similar to the compounds released by the plants. In some experiments, they treated the leaves with *jasmonate,* a plant hormone that stimulates the release of volatile chemicals.

Kessler and Baldwin reported that some volatile chemicals discouraged herbivorous insects from laying eggs. Furthermore, in response to certain volatile chemicals, predators ate more of the herbivore eggs. The researchers found that discouraging herbivores and attracting their predators, reduced the number of herbivore eggs laid on the tobacco leaves by about 90%. Thus the results of Kessler and Baldwin's experiments strongly support the hypothesis that when attacked by herbivores, plants release volatile chemicals that reduce the number of herbivores feeding on them. Natural selection has resulted in plant signals that herbivorous insects detect and avoid and that carnivorous insects detect and approach. This system of information transfer protects plants from attacking herbivorous insects.

In 2003, Jorg Degenhardt, of the Max Planck Institute, and his colleagues reported that humans now have the technology to genetically engineer crop plants to release volatile chemicals for attracting enemies of herbivores.† These researchers suggest that plant breeders can modify the types and amounts of volatile chemical signals released, increasing the ability of attacked plants to defend themselves. These findings are exciting because they may lead to environmentally friendly pest-control strategies. At present, farmers depend mainly on pesticides, which also kill beneficial insects, birds, and other animal species. Pesticides have also been associated with long-term environmental contamination and with human disease. Pesticides poison about 67,000 people annually in the United States alone.

In addition to its potential importance to agriculture, plant signaling has gained the attention of the military. Jack Schultz and Ramesh Raina of Pennsylvania State University have received a $3.5-million grant from the U.S. Department of Defense to study how plants respond to various environmental stressors. Their goal is to genetically engineer plants to detect the use of biological or chemical weapons, and to alert people by emitting volatile signals.

*A. Kessler and I.T. Baldwin, "Defensive Function of Herbivore-Induced Plant Volatile Emissions in Nature," *Science,* Vol. 291, 16 March 2001.

†J. Degenhardt, J. Gershenzon, I.T. Baldwin, and A. Kessler, "Attracting Friends to Feast on Foes: Engineering Terpene Emission to Make Crop Plants More Attractive to Herbivore Enemies," *Current Opinion in Biotechnology, 14,* (2003), 169–176.

to make observations and formulate critical questions, and his pen was prepared to publish them. However, even though Fleming recognized the potential practical benefit of penicillin, he did not develop the chemical techniques needed to purify it, and more than 10 years passed before the drug was put to significant use.

In 1939, Sir Howard Florey and Ernst Boris Chain developed chemical procedures to extract and produce the active agent penicillin from the mold. Florey took the process to laboratories in the United States, and penicillin was first produced to treat wounded soldiers in World War II. In 1945, Fleming, Florey, and Chain shared the Nobel Prize in Medicine.

A hypothesis is a testable statement

In the early stages of an investigation, a scientist typically thinks of many possible hypotheses. Hypotheses have many potential sources, including preliminary direct observations or even computer simulations. Increasingly in biology, hypotheses may be derived from *models* that scientists have developed to provide a comprehensive explanation for a large number of previous observations. Examples of such testable models include the model of the structure of DNA and the model of the structure of the plasma membrane (see Chapter 5). After generating hypotheses, the scientist decides which, if any, could and should be subjected to experimental test. Why not test them all? Time and money are important considerations in conducting research. Scientists must establish priority among the hypotheses to decide which to test first. Fortunately, some guidelines exist. A good hypothesis exhibits the following: (1) It is reasonably consistent with well-established facts. (2) It is capable of being tested; that is, it should generate definite predictions, whether the results are positive or negative. Test results should also be repeatable by independent observers; (3) it is falsifiable, which means it can be proven false.

A hypothesis cannot really be proven true, but in theory (though not necessarily in practice) a well-stated hypothesis can be proved false. Belief in an unfalsifiable hypothesis (such as the existence of invisible and undetectable angels) must be rationalized on grounds other than scientific ones.

Consider the following hypothesis: All female mammals (animals that have hair and produce milk for their young) bear live young. The hypothesis is based on the observations that dogs, cats, cows, lions, and humans all are mammals and all bear live young. Consider further that a new species, species X, is identified as a mammal. Biologists predict that females of species X will bear live young. When a female of the new species gives birth to offspring, this supports the hypothesis. Yet it does not really *prove* the hypothesis.

Before the Southern Hemisphere was explored, most people would probably have accepted the hypothesis without question, because all known furry, milk-producing animals did, in fact, bear live young. But biologists discovered that two Australian animals (the duck-billed platypus and the spiny anteater) had fur and produced milk for their young but laid eggs (Fig. 1-16). The hypothesis, as stated, was false no matter how many times it had previously been supported. As a result, biologists either had to consider the platypus and the spiny anteater as nonmammals or had to broaden their definition of mammals to include them. (They chose to do the latter.)

A hypothesis is not true just because some of its predictions (the ones people happen to have thought of or have thus far been able to test) have been shown to be true. After all, they could be true by coincidence. Failure to observe a predicted outcome does not make a hypothesis false, but neither does it show the hypothesis is true.

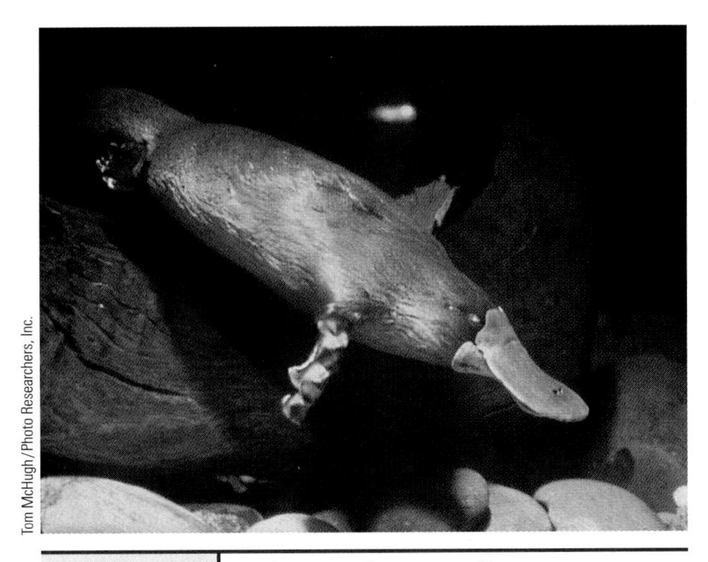

Tom McHugh/Photo Researchers, Inc.

FIGURE **1-16** **Is this animal a mammal?**

The duck-billed platypus *(Ornithorhynchus anatinus)* is classified as a mammal because it has fur and produces milk for its young. However, unlike most mammals, it lays eggs.

Predictions can be tested by experiment

A hypothesis is an abstract idea, so there is no way to test it directly. But hypotheses suggest certain logical consequences, that is, observable things that cannot be false if the hypothesis is true. In contrast, if the hypothesis is in fact false, other definite predictions should disclose that. As used here, then, a **prediction** is a deductive, logical consequence of a hypothesis. It does not have to be a future event.

A prediction can be tested by controlled experiments. Early biologists observed that the nucleus was the most prominent part of the cell, and they hypothesized that it might be essential for the well-being of the cell. They predicted that if the nucleus were removed from the cell, the cell would die. Biologists then experimented, surgically removing the nucleus of a unicellular amoeba. The amoeba continued to live and move, but it did not grow, and after a few days it died. These results suggested that the nucleus is necessary for the metabolic processes that provide for growth and cell reproduction.

But, the investigators asked, what if the operation itself, not the loss of the nucleus, caused the amoeba to die? They performed a controlled experiment, subjecting two groups of amoebas to the same operative trauma (Fig. 1-17). However, in the experimental group the nucleus was removed; in the control group, it was not. An **experimental group** differs from a control group only with respect to the variable being studied. In the **control group,** the researcher inserted a microloop into each amoeba and pushed it around inside the cell to simulate removal of the nucleus; then the needle was withdrawn, leaving the nucleus inside. Amoebas treated with such a sham operation recovered and subsequently grew and divided, but the amoebas without nuclei died. This experiment showed that the removal of the nucleus, not simply the operation, caused the death of the amoebas. The data supported the hypothesis that the nucleus is essential for the well-being of the cell.

In scientific studies, researchers must avoid bias. For example, to prevent bias most medical experiments today are carried out in a double-blind fashion. When a drug is tested, one group of patients receives the new medication, whereas a second similar group of patients (the control group) receives a placebo (a harmless starch pill similar in size, shape, color, and taste to the pill being tested). This is a *double-blind study,* because neither the patient nor the physician knows who is getting the experimental drug and who is getting the placebo. The pills or treatments are coded in some way, and the code is broken only after the experiment is over and the results are recorded. Not all experiments can be so neatly designed; for one thing, it is often difficult to establish appropriate controls.

Scientists interpret the results of experiments and make conclusions

Scientists gather data in an experiment, interpret their results, and then formulate conclusions. For example, in the amoeba experiment described earlier, investigators concluded the nucleus was essential for the cell's well-being.

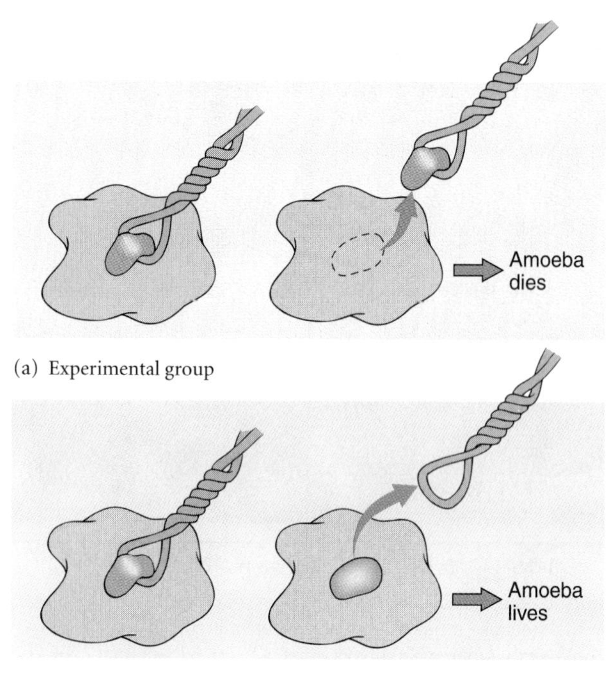

(a) Experimental group

(b) Control group

FIGURE **1-17** | **Testing a prediction.**

An early controlled experiment tested the prediction that if the nucleus is removed from a cell, the cell would die. The data gathered from this and similar experiments supported the hypothesis that the nucleus is essential for the cell's well-being. **(a)** When its nucleus is surgically removed with a microloop, the amoeba dies. **(b)** Control amoebas subjected to similar surgical procedures (including insertion of a microloop), but without actual removal of the nucleus, do not die.

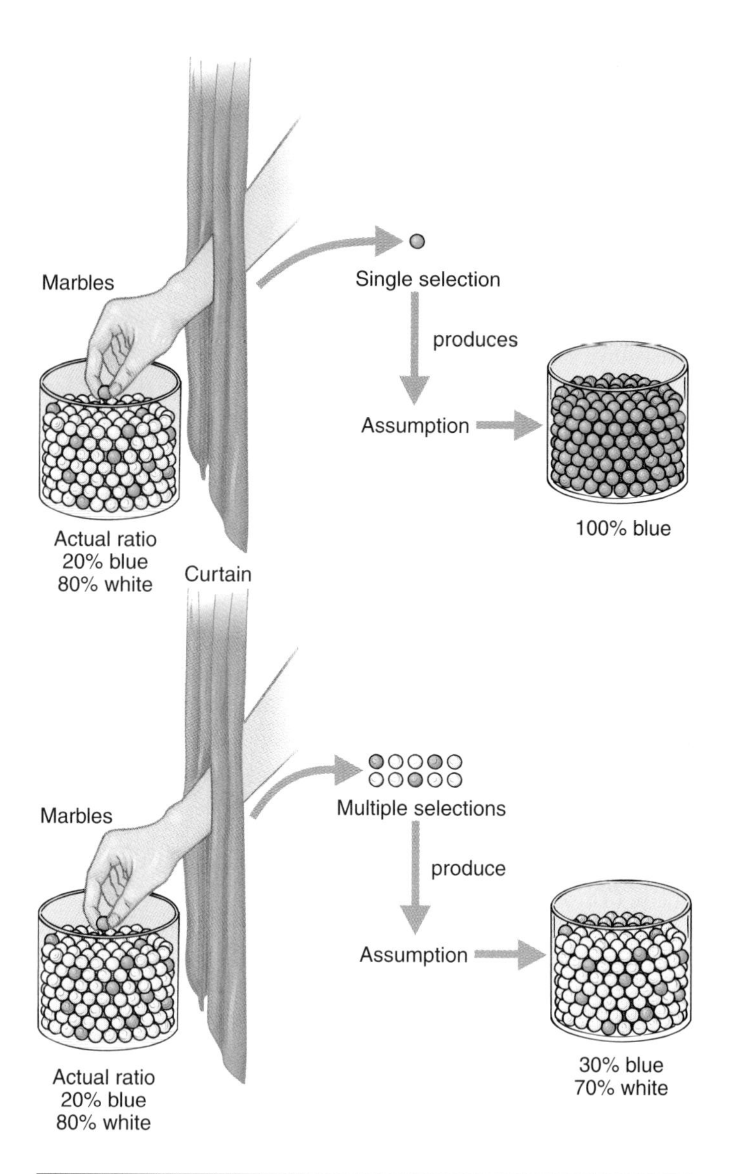

ACTIVE FIGURE **1-18** | **Statistical probability.**

Taking a single selection can result in sampling error. If the only marble selected is blue, we might assume all the marbles are blue. The greater the number of selections we take of an unknown, the more likely we can make valid assumptions about it.

BiologyNow™ Do your own random sampling by clicking on this figure on your BiologyNow CD-ROM.

One reason for inaccurate conclusions is *sampling error.* Because not *all* cases of what is being studied can be observed or tested (scientists cannot study every amoeba), scientists must be content with a sample. Yet how can you know whether that sample is truly representative of whatever you are studying? In the first place, if the sample is too small it may be different owing to random factors. A study with only two, or even nine, amoebas may not yield reliable data that can be generalized to other amoebas. If you test a large number of subjects, you are more likely to draw accurate scientific conclusions (Fig. 1-18). The scientist seeks to state with some level of confidence that any specific conclusion has a certain statistical probability of being correct.

Experiments must also be repeatable. When researchers publish their findings in a scientific journal, they typically describe their methods and procedures so other scientists can repeat the experiments. When the findings are replicated, the conclusions are, of course, strengthened.

A well-supported hypothesis may lead to a theory

Nonscientists often use the word *theory* incorrectly to refer to a hypothesis. A **theory** is actually an integrated explanation of a number of hypotheses, each supported by consistent results from many observations or experiments. A theory relates data that previously appeared unrelated. A good theory grows, building on additional facts as they become known. It predicts new facts and suggests new relationships among phenomena. It may even suggest practical applications.

A good theory, by showing the relationships among classes of facts, simplifies and clarifies our understanding of natural phenomena. As Einstein wrote, "In the whole history of science from Greek philosophy to modern physics, there have been constant attempts to reduce the apparent complexity of natural phenomena to simple, fundamental ideas and relations."

Science has ethical dimensions

Scientific investigation depends on a commitment to practical ideals, such as truthfulness and the obligation to communicate results. Honesty is particularly important in science. Consider the great (though temporary) damage done whenever an unprincipled or even desperate researcher, whose career may depend on the publication of a research study, knowingly disseminates false data. Until the deception is uncovered, researchers may devote thousands of dollars and hours of precious professional labor to futile lines of research inspired by erroneous reports. Deception can also be dangerous, especially in medical research. Fortunately, science tends to correct itself through consistent use of the scientific process. Sooner or later, someone's experimental results are sure to cast doubt on false data.

In addition to being ethical about their own work, scientists face many broad ethical issues surrounding areas such as genetic research, stem cell research, cloning, and human and animal experimentation. For example, some stem cells that show the greatest potential for treating human disease come from early embryos. The cells can be taken from 5- or 6-day-old human embryos and then cultured in laboratory glassware. Such cells could be engineered to treat failing hearts or brains harmed by stroke, injury, Parkinson's disease, or Alzheimer's disease. They could save the lives of burn victims and perhaps be engineered to treat specific cancers. Scientists, and the larger society, will need to determine whether the potential benefits of any type of research outweigh its ethical risks.

Review

- What is meant by a "controlled" experiment?
- What are the characteristics of a good hypothesis?

BiologyNow™ Assess your understanding of **the process and method of science** by taking the pretest on your BiologyNow CD-ROM.

SUMMARY WITH KEY TERMS

1 Distinguish between living and nonliving things by describing the features that characterize living organisms.

- A living organism can grow and develop, carry on self-regulated metabolism, respond to stimuli, and reproduce. Species evolve and adapt to their environment.
- All living organisms are composed of one or more **cells.**
- Organisms grow by increasing the size and/or number of their cells.
- **Metabolism** includes all the chemical activities that take place in the organism, including the chemical reactions essential to nutrition, growth and repair, and conversion of energy to usable forms. **Homeostasis** is the tendency of organisms to maintain an appropriate, balanced internal environment.
- Organisms respond to **stimuli,** physical or chemical changes in their external or internal environment. Responses typically involve movement. Some organisms use tiny extensions of the cell, called **cilia,** or longer **flagella** to move from place to place. Some organisms are sessile and remain rooted to some surface.
- In **asexual reproduction,** offspring are typically identical to the single parent; in **sexual reproduction,** offspring are the product of the fusion of gametes, and genes are typically contributed by two parents.
- Populations evolve and become adapted to their environment. **Adaptations** are traits that increase an organism's ability to survive in its environment.

2 Construct a hierarchy of biological organization, including levels of an individual organism and ecological levels.

- Biologic organization is hierarchical. A complex organism is organized at the chemical, cell, **tissue, organ,** and **organ system** levels.
- The basic unit of ecological organization is the **population.** Various populations form **communities;** a community and its physical environment are an **ecosystem;** all of Earth's ecosystems together make up the **biosphere.**

3 Summarize the importance of information transfer to living systems, giving specific examples.

- Organisms transmit information chemically, electrically, and behaviorally.
- DNA, which makes up the **genes,** contains the instructions for the development of an organism and for carrying out life processes. DNA codes for **proteins,** which are important in determining the structure and function of cells and tissues.
- Information encoded in **DNA** is transmitted from one generation to the next.
- **Hormones,** chemical messengers that transmit messages from one part of an organism to another, are an important type of **cell signaling.**
- Many organisms use electrical signals to transmit information; most animals have nervous systems that transmit electrical impulses and release **neurotransmitters.**

4 Demonstrate the binomial system of nomenclature using several specific examples, and classify an organism (such as a human) in its domain, kingdom, phylum, class, order, family, genus, and species.

- Millions of species have evolved. A **species** is a group of organisms with similar structure, function, and behavior that, in nature, breed only with each other. Members of a species have a common **gene pool** and share a common ancestry.
- Biologists use a **binomial system of nomenclature** in which the name of each species includes a **genus** name and a **specific epithet.**
- Taxonomic classification is hierarchical; it includes species, genus, **family, order, class, phylum, kingdom,** and **domain.** Each grouping is referred to as a **taxon.**

5 Identify the six kingdoms of living organisms, and give examples of organisms assigned to each group.

- Bacteria and archaebacteria have **prokaryotic cells;** all other organisms have **eukaryotic cells.**
- Organisms can be classified into three domains: **Archaea, Eubacteria,** and **Eukarya,** and six kingdoms: **Archaebacteria, Eubacteria, Protista** (protozoa, algae, water molds, and slime molds), **Fungi** (molds and yeasts), **Plantae,** and **Animalia.**

6 Give a brief overview of the theory of evolution, and explain why it is the principal unifying concept in biology.

- **Evolution** is the process by which populations change over time in response to changes in the environment. The theory of evolution explains how millions of species came to be and helps us understand the structure, function, behavior, and interactions of organisms.
- **Natural selection,** the mechanism by which evolution proceeds, favors individuals with traits that enable them to cope with environmental changes. These individuals are most likely to survive and to produce offspring.
- Charles Darwin based his theory of natural selection on his observations that individuals of a species vary; organisms produce more offspring than survive to reproduce; individuals that are best adapted to their environment are more likely to survive and reproduce; as successful organisms pass on their hereditary information, their traits become more widely distributed in the population.
- The source of variation in a population is random **mutation.**

7 Apply the theory of natural selection to any given adaptation, suggesting a logical explanation of how the adaptation may have evolved.

- When the ancestors of Hawaiian honeycreepers first reached Hawaii, few other birds were present, so there was little competition for food. Through many generations, honeycreepers with longer, more curved bills became adapted for feeding on nectar from tubular flowers. Perhaps those with the longest, most curved bills were best able to survive in this food zone and lived to transmit their genes to their offspring. Those with shorter, thicker bills were more successful foraging for insects and passed their genes to new generations of offspring. Eventually different species evolved, adapted to specific food zones.

8 Summarize the flow of energy through ecosystems, contrasting the roles of producers, consumers, and decomposers.

- Activities of living cells require energy; life depends on continuous energy input from the sun.
- During **photosynthesis** plants, algae, and certain bacteria use the energy of sunlight to synthesize complex molecules from carbon dioxide and water.
- Virtually all cells carry on **cellular respiration,** a biochemical process in which they capture the energy stored in nutrients by producers. Some of that energy is then used to synthesize needed materials or to carry out other cell activities.
- A self-sufficient ecosystem includes **producers,** or **autotrophs,** which make their own food; **consumers,** which eat producers or organisms that have eaten producers; and **decomposers,** which obtain energy by breaking down wastes and dead organisms. Consumers and decomposers are **heterotrophs,** organisms that depend on producers as an energy source and for food and oxygen.

9 Design an experiment to test a given hypothesis, using the procedure and terminology of the scientific method.

- The **process of science** is a dynamic approach to investigation. The **scientific method** is a framework that scientists use in their work; it includes observing, recognizing a problem or stating a critical question, developing a hypothesis, making a prediction that can be tested, performing experiments, interpreting results, and drawing conclusions that support or falsify the hypothesis.
- Deductive reasoning and inductive reasoning are two categories of systematic thought used in the scientific method. **Deductive reasoning** proceeds from general principles to specific conclusions and helps people discover relationships among known facts. **Inductive reasoning** begins with specific observations and draws conclusions from them. Inductive reasoning helps people discover general principles.
- A **hypothesis** is a testable statement about the nature of an observation or relationship.
- A properly designed scientific experiment includes both a **control group** and an **experimental group,** and must be as free as possible from bias. The experimental group differs from a control group only with respect to the variable being studied.
- When a number of related hypotheses have been supported by conclusions from many experiments, scientists may develop a **theory** based on them.
- Science has important ethical dimensions.

POST-TEST

1. Metabolism (a) is the sum of all the chemical activities of an organism (b) results from an increase in the number of cells (c) is characteristic of plant and animal kingdoms only (d) refers to chemical changes in an organism's environment (e) does not take place in producers
2. Homeostasis (a) is the tendency of organisms to maintain an appropriate, balanced internal environment (b) generally depends on the action of cilia (c) is the long-term response of organisms to changes in their environment (d) occurs at the ecosystem level, not in cells or organisms (e) may be sexual or asexual
3. Structures used by some organisms for locomotion are (a) cilia and nuclei (b) flagella and DNA (c) nuclei and membranes (d) cilia and sessiles (e) cilia and flagella
4. The splitting of an amoeba into two is best described as an example of (a) locomotion (b) neurotransmission (c) asexual reproduction (d) sexual reproduction (e) metabolism
5. Cells (a) are the building blocks of living organisms (b) always have nuclei (c) are not found among the bacteria (d) answers a, b, and c are correct (e) only answers a and b are correct
6. An increase in the size or number of cells best describes (a) homeostasis (b) biological growth (c) chemical level of organization (d) asexual reproduction (e) adaptation
7. DNA (a) makes up the genes (b) transmits information from one species to another (c) cannot be changed (d) is a neurotransmitter (e) is produced during cellular respiration
8. Cellular respiration (a) is a process whereby sunlight is used to synthesize cell components with the release of energy (b) occurs in heterotrophs only (c) is carried on by both autotrophs and heterotrophs (d) causes chemical changes in DNA (e) occurs in response to environmental changes
9. Which of the following is a correct sequence of levels of biological organization? (a) cell, organ, tissue, organ system (b) chemical,

POST-TEST *(continued)*

cell, organ, tissue (c) chemical, cell, tissue, organ (d) tissue, organ, cell, organ system (e) chemical, cell, population, species

10. Which of the following is a correct sequence of levels of biological organization? (a) organism, population, ecosystem, community (b) organism, population, community, ecosystem (c) population, biosphere, ecosystem, community (d) species, population, ecosystem, community (e) ecosystem, population, community, biosphere
11. Protozoa are assigned to kingdom (a) Protista (b) Fungi (c) Archaebacteria (d) Animalia (e) Plantae
12. Yeasts and molds are assigned to kingdom (a) Protista (b) Fungi (c) Archaebacteria (d) Animalia (e) Plantae
13. In the binomial system of nomenclature, the first part of an organism's name designates the (a) species epithet (b) genus (c) class (d) kingdom (e) phylum
14. Which of the following is a correct sequence of levels of classification (a) genus, species, family, order, class, phylum, kingdom (b) genus, species, order, phylum, class, kingdom (c) genus, species, order, family, class, phylum, kingdom (d) species, genus, family, order, class, phylum, kingdom (e) species, genus, order, family, class, kingdom, phylum
15. Darwin suggested that evolution takes place by (a) mutation (b) changes in the individuals of a species (c) natural selection (d) interaction of hormones (e) homeostatic responses to each change in the environment
16. A testable statement is a(an) (a) theory (b) hypothesis (c) principle (d) inductive leap (e) critical question
17. Ideally, an experimental group differs from a control group (a) only with respect to the hypothesis being tested (b) only with respect to the variable being studied (c) by being less subject to bias (d) in that it is less vulnerable to sampling error (e) in that its subjects are more reliable

CRITICAL THINKING

1. How might a firm understanding of evolutionary processes help a biologist doing research in (a) animal behavior, (b) ocean ecology, or (c) the development of a vaccine against human immunodeficiency virus, which causes AIDS?
2. Make a prediction and devise a suitably controlled experiment to test each of the following hypotheses: (a) A type of mold found in your garden does not produce an effective antibiotic. (b) The growth rate of a bean seedling is affected by temperature. (c) Estrogen alleviates symptoms of Alzheimer's disease in elderly women.

- Visit our Web site at **http://biology.brookscole.com/solomon7** for links to chapter-related resources on the World Wide Web. Additional online materials relating to this chapter can also be found on our Web site.

BIOLOGY NOW RESOURCES

BiologyNow™ The BiologyNow CD-ROM packaged free with your text, uses a learning system that allows you to review your general understanding of a concept. First you answer a series of diagnostic review questions. Based on your answers, BiologyNow will provide you with a customized learning plan that links you to the text, study guide, animations, Genetics Problem-Solving Guide, and CNN Video for focused study that maximizes your learning. You can also connect to V-Mentor for one-on-one tutoring help from experienced biology teachers.

Web Site

The Web site for this book contains a wealth of helpful study aids, as well as many ideas for further reading and research. Log on to: **http://biology.brookscole.com/solomon7**

- For study and review, **Chapter Outline** gives you an outline of the chapter, **Chapter Summary** allows you to review the chapter's main ideas, and **Glossary** lists concepts and terms for the chapter along with their definitions.
- To test your mastery of important terminology for this chapter, you can use the electronic **Flash Cards,** which may be sorted by definition or by term.
- For testing your knowledge and preparing for in-class examinations, our **Quizzes** pose multiple choice and/or true-false questions based on each chapter.
- **Hypercontents** takes you to an extensive list of current links to Internet sites with news, research, and images related to individual subjects in the chapter.
- **Internet Exercises** are critical thinking questions that involve research on the Internet with starter URLs provided.
- **InfoTrac Exercises** leads you to Critical Thinking Projects that use InfoTrac College Edition® as a research tool. For more readings, go to *InfoTrac College Edition,* your online research library, at: **http://infotrac.thomsonlearning.com**

Active Figures

1-8: Biological classification

1-18: Random sampling

Preparing for an exam? Take a diagnostic test on your BiologyNow CD-ROM.

Post-Test Answers

1. a	2. a	3. e	4. c
5. a	6. b	7. a	8. c
9. c	10. b	11. a	12. b
13. b	14. d	15. c	16. b
17. b			

2

Atoms and Molecules: The Chemical Basis of Life

Frans Lanting/Minden Pictures

A jaguar (*Panthera onca*), the largest cat in the Western Hemisphere, pauses to drink water from a rainforest stream. Water is a basic requirement for all life.

CHAPTER OUTLINE

A knowledge of chemistry is essential for understanding organisms and how they function. This jaguar and the plants of the tropical rain forest, as well as abundant unseen insects and microorganisms, share fundamental similarities in their chemical composition and basic metabolic processes. These chemical similarities provide strong evidence for the evolution of all organisms from a common ancestor and explain why much of what biologists learn from studying bacteria or rats in laboratories can be applied to other organisms, including humans. Furthermore, the basic chemical and physical principles governing organisms are not unique to living things, for they apply to nonliving systems as well.

The success of the Human Genome Project (introduced in Chapter 1) relied heavily on biochemistry and **molecular biology,** the chemistry and physics of the molecules that constitute living things. A biochemist may investigate the precise interactions among a cell's atoms and molecules that maintain the energy flow essential to life, and a molecular biologist may study how proteins interact with deoxyribonucleic acid (DNA) in ways that control the expression of certain genes. However, an understanding of chemistry is essential to *all* biologists. An evolutionary biologist may study evolutionary relationships by comparing the DNA of different types of organisms. An ecologist may study how energy is transferred among the organisms living in an estuary or monitor the biological effects of changes in the salinity of the water. A botanist may study unique compounds produced by plants and may even be a "chemical prospector," seeking new sources of medicinal agents.

In this chapter we lay a foundation for understanding how the structure of atoms determines the way they form chemical bonds to produce complex compounds. Most of our discussion focuses on small, simple substances known as **inorganic compounds.** Among the biologically important groups of inorganic compounds are water, many simple acids and bases, and simple salts. We pay particular attention to water, the most abundant substance in organisms and on Earth's surface, and we examine how its unique properties affect living things as well as their nonliving environment. In Chapter 3 we extend our discussion to **organic compounds,** carbon-containing compounds that are generally large and complex. In all but the simplest organic compounds, two or more carbon atoms are bonded to each other to form the backbone, or skeleton, of the molecule. ■

ELEMENTS AND ATOMS

Learning Objectives

1. Name the principal chemical elements in living things, and give an important function of each.
2. Compare the physical properties (mass and charge) and locations of electrons, protons, and neutrons. Distinguish between the atomic number and the mass number of an element.
3. Define the terms *orbital* and *electron shell.* Relate electron shells to principal energy levels.

Elements are substances that cannot be broken down into simpler substances by ordinary chemical reactions. Each element has a **chemical symbol:** usually the first letter or first and second letters of the English or Latin name of the element. For example, O is the symbol for oxygen, C for carbon, H for hydrogen, N for nitrogen, and Na for sodium (Latin *natrium*).

Just four elements—oxygen, carbon, hydrogen, and nitrogen—are responsible for more than 96% of the mass of most organisms. Others, such as calcium, phosphorus, potassium, and magnesium, are also consistently present but in smaller quantities. Some elements, such as iodine and copper, are known as *trace elements,* because they are required only in minute amounts. Table 2-1 lists the elements that make up organisms, and briefly explains the importance of each in typical plants and animals.

TABLE 2-1 Functions of Elements in Organisms

Element (chemical symbol)	Functions
Oxygen (O)	Required for cellular respiration; present in most organic compounds; component of water
Carbon (C)	Forms backbone of organic molecules; each carbon atom can form four bonds with other atoms
Hydrogen (H)	Present in most organic compounds; component of water; hydrogen ion (H^+) is involved in some energy transfers
Nitrogen (N)	Component of proteins and nucleic acids; component of chlorophyll in plants
Calcium (Ca)	Structural component of bones and teeth; calcium ion (Ca^{2+}) is important in muscle contraction, conduction of nerve impulses, and blood clotting; associated with plant cell wall
Phosphorus (P)	Component of nucleic acids and of phospholipids in membranes; important in energy transfer reactions; structural component of bone
Potassium (K)	Potassium ion (K^+) is a principal positive ion (cation) in interstitial (tissue) fluid of animals; important in nerve function; affects muscle contraction; controls opening of stomata in plants
Sulfur (S)	Component of most proteins
Sodium (Na)	Sodium ion (Na^+) is a principal positive ion (cation) in interstitial (tissue) fluid of animals; important in fluid balance; essential for conduction of nerve impulses; important in photosynthesis in plants
Magnesium (Mg)	Needed in blood and other tissues of animals; activates many enzymes; component of chlorophyll in plants
Chlorine (Cl)	Chloride ion (Cl^-) is principal negative ion (anion) in interstitial (tissue) fluid of animals; important in water balance; essential for photosynthesis
Iron (Fe)	Component of hemoglobin in animals; activates certain enzymes

*Other elements found in very small (trace) amounts in animals, plants, or both include iodine (I), manganese (Mn), copper (Cu), zinc (Zn), cobalt (Co), fluorine (F), molybdenum (Mo), selenium (Se), boron (B), silicon (Si), and a few others.

An **atom** is defined as the smallest portion of an element that retains its chemical properties. Atoms are much smaller than the tiniest particle visible under a light microscope. By scanning tunneling microscopy, magnified as high as 5 million times, researchers have been able to photograph the positions of some large atoms in molecules.

Physicists have discovered a number of subatomic particles, but for our purposes we need consider only three: electrons, protons, and neutrons. An **electron** is a particle that carries a unit of negative electrical charge; a **proton** carries a unit of positive charge; and a **neutron** is an uncharged particle. In an electrically neutral atom, the number of electrons is equal to the number of protons.

Clustered together, protons and neutrons compose the **atomic nucleus.** Electrons, however, have no fixed locations and move rapidly through the mostly empty space surrounding the atomic nucleus.

An atom is uniquely identified by its number of protons

Every element has a fixed number of protons in the atomic nucleus, known as the **atomic number.** It is written as a subscript to the left of the chemical symbol. Thus, $_1H$ indicates that the hydrogen nucleus contains 1 proton, and $_8O$ means that the oxygen nucleus contains 8 protons. The atomic number determines an atom's identity and defines the element.

The **periodic table** is a chart of the elements arranged in order by atomic number (Fig. 2-1 and Appendix A). The periodic table is useful because it lets us simultaneously correlate many of the relationships among the various elements.

Figure 2-1 includes representations of the **electron configurations** of several elements important in organisms. These *Bohr models,* which show the electrons arranged in a series of concentric circles around the nucleus, are convenient to use but inaccurate. The space outside the nucleus is actually extremely large compared to the nucleus, and, as you will see, electrons do not actually circle the nucleus in fixed concentric pathways.

Protons plus neutrons determine atomic mass

The mass of a subatomic particle is exceedingly small, much too small to be conveniently expressed in grams or even micro-

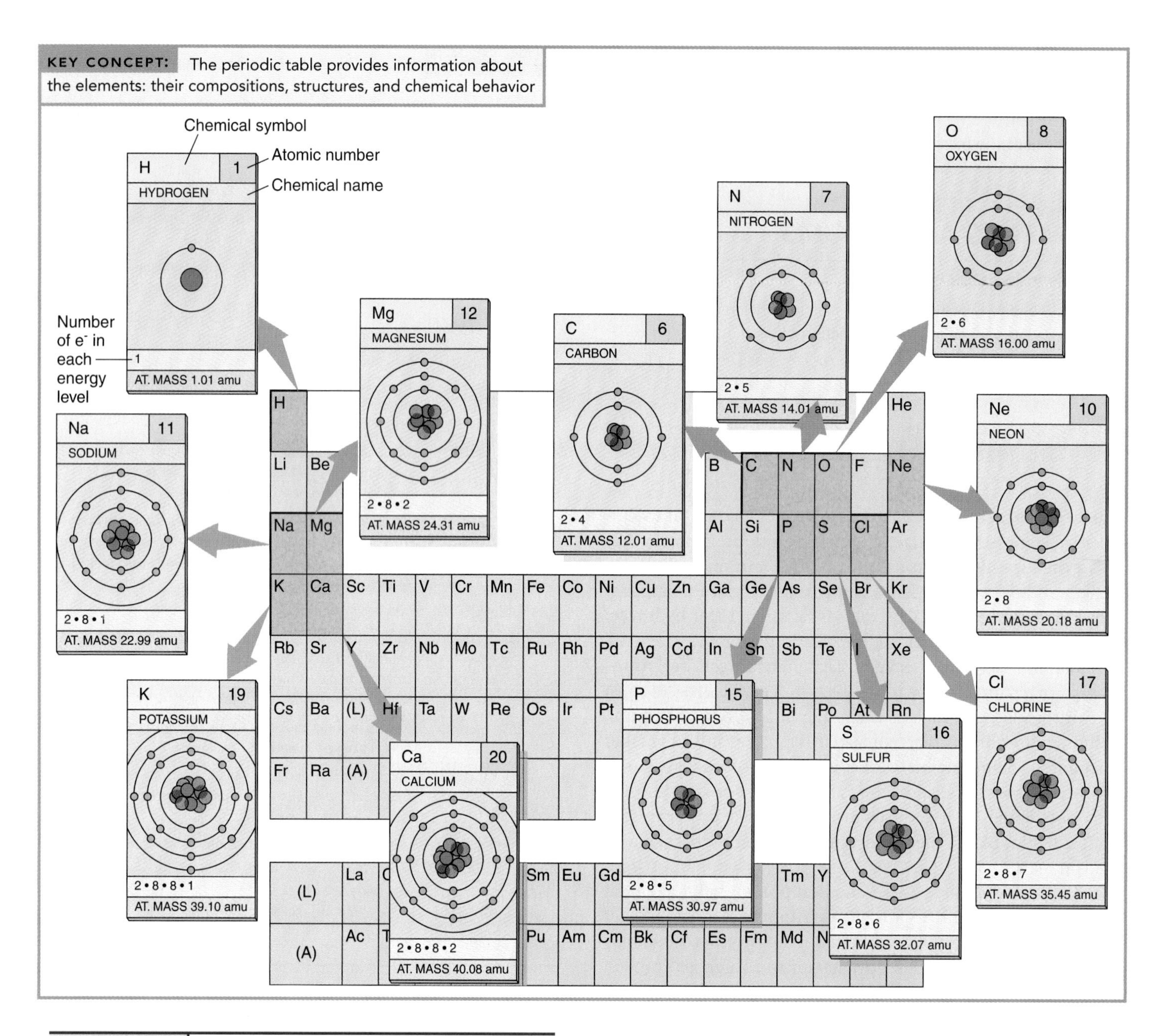

FIGURE **2-1** **The periodic table.**

Note the Bohr models depicting the electron configuration of atoms of some biologically important elements. Although the Bohr model does not depict electron configurations accurately, it is commonly used because of its simplicity and convenience. A complete periodic table is given in Appendix B.

grams.[1] Such masses are expressed in terms of the **atomic mass unit (amu),** also called the **dalton** in honor of John Dalton, who formulated an atomic theory in the early 1800s. One amu is equal to the approximate mass of a single proton or a single neutron. Protons and neutrons make up almost all the mass of an atom. The mass of a single electron is only about 1/1800 the mass of a proton or neutron.

The **atomic mass** of an atom is a number that indicates approximately how much matter it contains compared with another atom. This value is determined by adding the number of protons to the number of neutrons and expressing the result in atomic mass units or daltons.[2] The mass of the electrons is ignored because it is so small. The atomic mass number is indicated by a superscript to the left of the chemical symbol. The common form of the oxygen atom, with 8 protons and 8 neutrons in its nucleus, has an atomic number of 8 and a mass of 16 atomic mass units. It is indicated by the symbol $^{16}_{8}O$.

[1] Tables of commonly used units of scientific measurement are printed inside the back cover of this text.

[2] Unlike weight, mass is independent of the force of gravity. For convenience, however, we consider mass and weight equivalent. Atomic weight has the same numerical value as atomic mass, but it has no units.

The characteristics of protons, electrons, and neutrons are summarized in the following table:

Particle	Charge	Approximate Mass	Location
Proton	Positive	1 amu	Nucleus
Neutron	Neutral	1 amu	Nucleus
Electron	Negative	Approx. 1/1800 amu	Outside nucleus

Isotopes of an element differ in number of neutrons

Most elements consist of a mixture of atoms with different numbers of neutrons and thus different masses. Such atoms are called **isotopes.** Isotopes of the same element have the same number of protons and electrons; only the number of neutrons varies. The three isotopes of hydrogen, $^{1}_{1}H$ (ordinary hydrogen), $^{2}_{1}H$ (deuterium), and $^{3}_{1}H$ (tritium), contain 0, 1, and 2 neutrons, respectively. Figure 2-2 shows Bohr models of two isotopes of carbon, $^{12}_{6}C$ and $^{14}_{6}C$. The mass of an element is expressed as an average of the masses of its isotopes (weighted by their relative abundance in nature). For example, the atomic mass of hydrogen is not 1.0 amu, but 1.0079 amu, reflecting the natural occurrence of small amounts of deuterium and tritium in addition to the more abundant ordinary hydrogen.

Because they have the same number of electrons, all isotopes of a given element have essentially the same chemical characteristics. However, some isotopes are unstable and tend to break down, or decay, to a more stable isotope (usually becoming a different element); such **radioisotopes** emit radiation when they decay. For example, the radioactive decay of $^{14}_{6}C$ occurs as a neutron decomposes to form a proton and a fast-moving electron, which is emitted from the atom as a form of radiation known as a beta (β) particle. The resulting stable atom is the common form of nitrogen, $^{14}_{7}N$. Using sophisticated instruments, scientists can detect and measure β particles and other types of radiation. Radioactive decay can also be detected by a method known as **autoradiography,** in which radiation causes the appearance of dark silver grains in photographic film (Fig. 2-3).

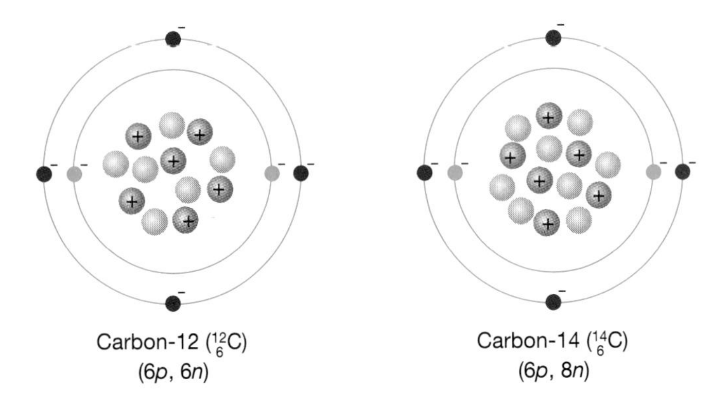

FIGURE **2-2** **Isotopes.**
Carbon-12 ($^{12}_{6}C$) is the most common isotope of carbon. Its nucleus contains 6 protons and 6 neutrons, so its atomic mass is 12. Carbon-14 ($^{14}_{6}C$) is a rare radioactive carbon isotope. It contains 8 neutrons, so its atomic mass is 14.

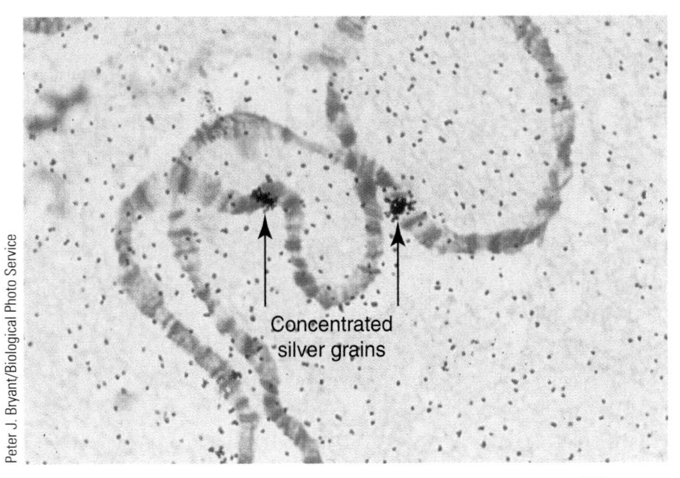

FIGURE **2-3** **Autoradiography.**
The chromosomes of the fruit fly, *Drosophila melanogaster,* shown in this light micrograph, have been covered with photographic film in which silver grains *(dark spots)* are produced when tritium (^{3}H) that has been incorporated into DNA undergoes radioactive decay. The concentrations of silver grains *(arrows)* mark the locations of specific DNA molecules.

Because the different isotopes of a given element have the same chemical characteristics, they are essentially interchangeable in molecules. Molecules containing radioisotopes are usually metabolized and/or localized in the organism in a similar way to their nonradioactive counterparts, and they can be substituted. For this reason, radioisotopes such as ^{3}H (tritium), ^{14}C, and ^{32}P are extremely valuable research tools used in areas such as dating fossils (see Fig. 17-9), tracing biochemical pathways, determining the sequence of genetic information in DNA (see Fig. 14-10), and understanding sugar transport in plants.

In medicine, radioisotopes are used for both diagnosis and treatment. The location and/or metabolism of a substance such as a hormone or drug can be followed in the body by labeling the substance with a radioisotope such as carbon-14 or tritium. Radioisotopes are used to test thyroid gland function, to provide images of blood flow in the arteries supplying the heart muscle, and to study many other aspects of body function and chemistry. Because radiation can interfere with cell division, radioisotopes have been used therapeutically in treating cancer, a disease often characterized by rapidly dividing cells.

Electrons move in orbitals corresponding to energy levels

Electrons move through characteristic regions of 3-D space, or **orbitals.** Each orbital contains a maximum of 2 electrons. Because it is impossible to know an electron's position at any given time, orbitals are most accurately depicted as "electron clouds," shaded areas whose density is proportional to the probability that an electron is present there at any given instant. The energy of an electron depends on the orbital it occupies. Electrons in orbitals with similar energies, said to be at the same **principal energy level,** make up an **electron shell** (Fig. 2-4).

In general, electrons in a shell distant from the nucleus have greater energy than those in a shell close to the nucleus. This is because to move a negatively charged electron farther away from the positively charged nucleus, energy is required. The most energetic electrons, known as **valence electrons,** are said to occupy the **valence shell.** The valence shell is represented as the outermost concentric ring in a Bohr model.

An electron can move to an orbital farther from the nucleus by receiving more energy, or it can give up energy and sink to a lower energy level in an orbital nearer the nucleus. Changes in electron energy levels are important in energy conversions in organisms. For example, during photosynthesis, light energy absorbed by chlorophyll molecules causes electrons to move to a higher energy level (see Fig. 8-3).

Review

- Do all atoms of an element have the same atomic number? The same atomic mass?
- What is a radioisotope? What are some ways radioisotopes are used in biological research?
- How do electrons in different orbitals of the same electron shell compare with respect to their energy?

BiologyNow™ Assess your understanding of **elements and atoms** by taking the pretest on your BiologyNow CD-ROM.

CHEMICAL REACTIONS

Learning Objectives

4 Explain how the number of valence electrons of an atom is related to its chemical properties.

5 Distinguish among simplest, molecular, and structural chemical formulas.

6 Explain why the mole concept is so useful to chemists.

The chemical behavior of an atom is determined primarily by the number and arrangement of its valence electrons. The valence shell of hydrogen or helium is full (stable) when it contains 2 electrons. The valence shell of any other atom is full when it contains 8 electrons. When the valence shell is not full, the atom tends to lose, gain, or share electrons to achieve a full outer shell. The valence shells of all isotopes of an element are identical; this is why they have similar chemical properties and can substitute for each other in chemical reactions (for example, tritium can substitute for ordinary hydrogen).

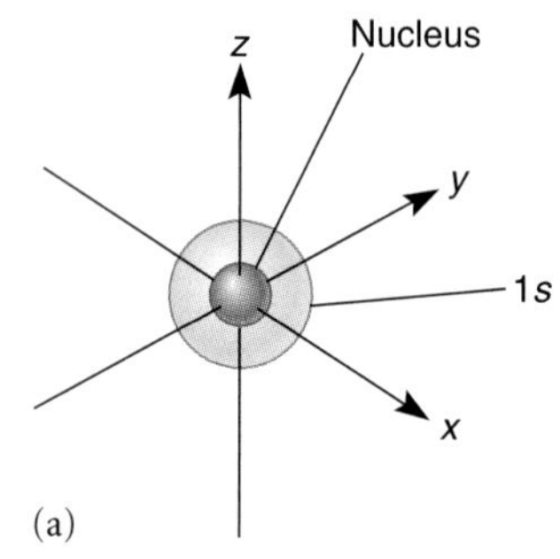

FIGURE 2-4 **Atomic Orbitals**

Each orbital is represented as an "electron cloud." The arrows labeled *x, y,* and *z* establish the imaginary axes of the atom. **(a)** The first principal energy level contains a maximum of 2 electrons, occupying a single spherical orbital (designated 1*s*). The electrons depicted in the diagram could be present anywhere in the blue area. **(b)** The second principal energy level includes four orbitals, each with a maximum of 2 electrons: one spherical (2*s*) and three dumbbell-shaped (2*p*) orbitals at right angles to each other. **(c)** Orbitals of the first and second principal energy levels are shown superimposed. Compare this more realistic view of the atomic orbitals with the **(d)** Bohr model of a neon atom. Note that the single 2*s* orbital plus three 2*p* orbitals make up neon's full valence shell of 8 electrons.

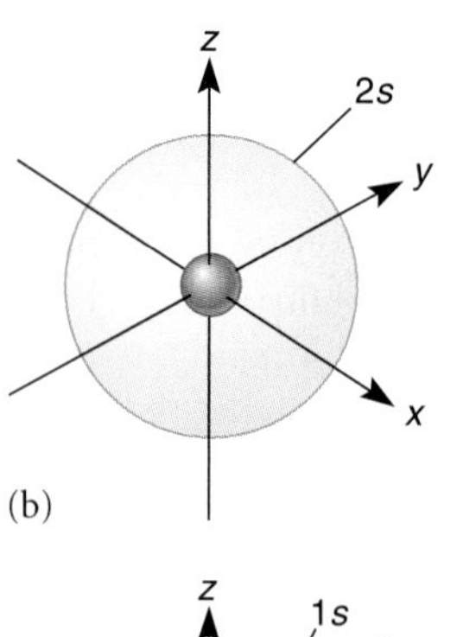

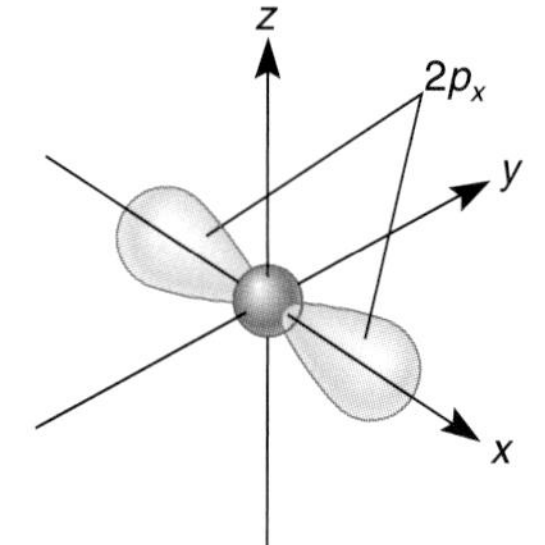

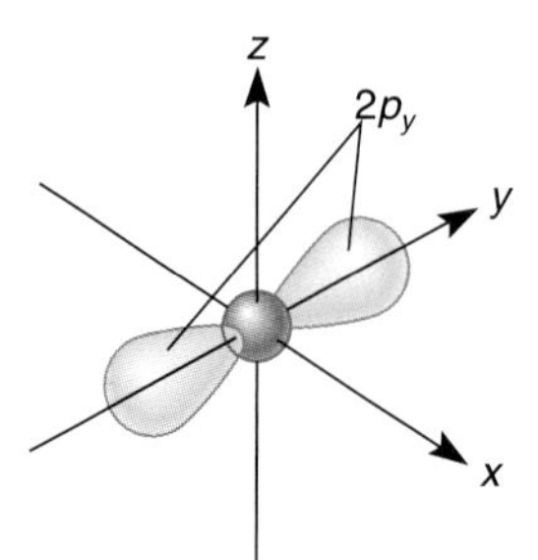

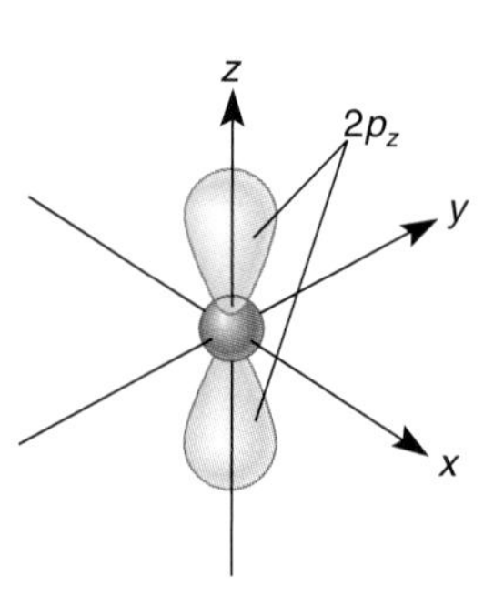

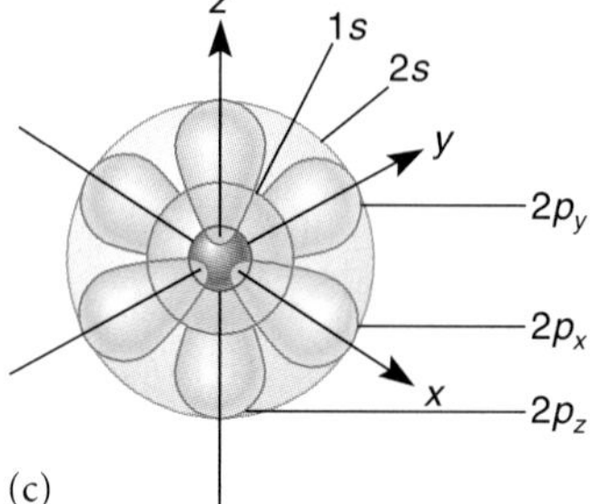

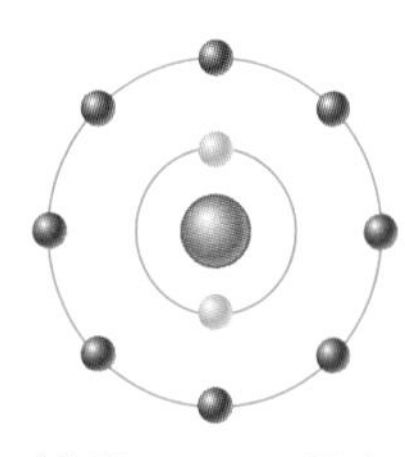

(d) Neon atom (Bohr model)

Elements in the same vertical column (belonging to the same *group*) of the periodic table have similar chemical properties because their valence shells have similar tendencies to lose, gain, or share electrons. For example, chlorine and bromine, included in a group commonly known as the *halogens,* are highly reactive. Because their valence shells have 7 electrons, they tend to gain an electron in chemical reactions. By contrast, hydrogen, sodium, and potassium each have a single valence electron, which they tend to give up or share with another atom. Helium (He) and neon (Ne) belong to a group referred to as the "noble gases." They are quite unreactive, because their valence shells are full. Notice the incomplete valence shells of some of the elements important in organisms, including carbon, hydrogen, oxygen, and nitrogen, in Figure 2-1, and compare them with the full valence shell of neon in Figure 2-4d.

Atoms form compounds and molecules

Two or more atoms may combine chemically. When atoms of *different* elements combine, the result is a chemical compound. A **chemical compound** consists of atoms of two or more different elements combined in a fixed ratio. For example, water is a chemical compound composed of hydrogen and oxygen in a ratio of 2:1. Common table salt, sodium chloride, is a chemical compound made up of sodium and chlorine in a 1:1 ratio.

Two or more atoms may become joined very strongly to form a stable particle called a **molecule.** For example, when two atoms of oxygen combine chemically, a molecule of oxygen is formed. Water is a molecular compound, with each molecule consisting of two atoms of hydrogen and one of oxygen. However, as you will see, not all compounds are made up of molecules. Sodium chloride is an example of a compound that is not molecular.

Simplest, molecular, and structural chemical formulas give different information

A **chemical formula** is a shorthand expression that describes the chemical composition of a substance. Chemical symbols indicate the types of atoms present, and subscript numbers indicate the ratios among the atoms. There are several types of chemical formulas, each providing specific kinds of information.

In a **simplest formula** (also known as an *empirical formula*), the subscripts give the smallest whole-number ratios for the atoms present in a compound. For example, the simplest formula for hydrazine is NH_2, indicating a 1:2 ratio of nitrogen to hydrogen. (Note that when a single atom of a type is present, the subscript number 1 is never written.)

In a **molecular formula,** the subscripts indicate the actual numbers of each type of atom per molecule. The molecular formula for hydrazine is N_2H_4, which indicates that each molecule of hydrazine consists of two atoms of nitrogen and four atoms of hydrogen. The molecular formula for water, H_2O, indicates that each molecule consists of two atoms of hydrogen and one atom of oxygen.

A **structural formula** shows not only the types and numbers of atoms in a molecule but also their arrangement. For example, the structural formula for water is H—O—H. As you will learn in Chapter 3, it is common for complex organic molecules with different structural formulas to share the same molecular formula.

One mole of any substance contains the same number of units

The **molecular mass** of a compound is the sum of the atomic masses of the component atoms of a single molecule; thus, the molecular mass of water, H_2O, is (hydrogen: 2×1 amu) + (oxygen: 1×16 amu), or 18 amu. (Because of the presence of isotopes, atomic mass values are not whole numbers, but for easy calculation each atomic mass value has been rounded off to a whole number.) Similarly, the molecular mass of glucose ($C_6H_{12}O_6$), a simple sugar that is a key compound in cell metabolism, is (carbon: 6×12 amu) + (hydrogen: 12×1 amu) + (oxygen: 6×16 amu), or 180 amu.

The amount of an element or compound whose mass in grams is equivalent to its atomic or molecular mass is 1 **mole (mol).** Thus 1 mol of water is 18 grams (g), and 1 mol of glucose has a mass of 180 g. The mole is an extremely useful concept, because it lets us make meaningful comparisons between atoms and molecules of very different mass. This is because *1 mol of any substance always has exactly the same number of units,* whether they are small atoms or large molecules. The very large number of units in a mole, 6.02×10^{23}, is known as **Avogadro's number,** named for the Italian physicist Amadeo Avogadro, who first calculated it. Thus 1 mol (180 g) of glucose contains 6.02×10^{23} molecules, as does 1 mol (2 g) of molecular hydrogen (H_2). Although it is impossible to count atoms and molecules individually, a scientist can calculate them simply by weighing a sample. Molecular biologists usually deal with smaller values, either millimoles (mmol, one thousandth of a mole) or micromoles (μmol, one millionth of a mole).

The mole concept also lets us make useful comparisons among solutions. A 1-molar solution, represented by 1 *M,* contains 1 mol of that substance dissolved in a total volume of 1 liter (L). For example, we can compare 1 L of a 1-*M* solution of glucose with 1 L of a 1-*M* solution of sucrose (table sugar, a larger molecule). They differ in the mass of the dissolved sugar (180 g and 340 g, respectively), but they each contain 6.02×10^{23} sugar molecules.

Chemical equations describe chemical reactions

During any moment in the life of an organism—a bacterial cell, a mushroom, or a butterfly—many complex chemical reactions are taking place. Chemical reactions, such as the reaction between glucose and oxygen, can be described by means of chemical equations:

$$\underset{\text{Glucose}}{C_6H_{12}O_6} + \underset{\text{Oxygen}}{6\,O_2} \longrightarrow \underset{\text{Carbon dioxide}}{6\,CO_2} + \underset{\text{Water}}{6\,H_2O} + \text{energy}$$

In a chemical equation, the **reactants,** the substances that participate in the reaction, are generally written on the left side, and the **products,** the substances formed by the reaction, are written on the right side. The arrow means "yields" and indicates the direction in which the reaction proceeds.

Chemical compounds react with each other in quantitatively precise ways. The numbers preceding the chemical symbols or formulas (known as *coefficients*) indicate the relative number of atoms or molecules reacting. For example, 1 mol of glucose burned in a fire or metabolized in a cell reacts with 6 mol of oxygen to form 6 mol of carbon dioxide and 6 mol of water.

Many reactions can proceed simultaneously in the reverse direction (to the left) as well as in the forward direction (to the right). At **dynamic equilibrium,** the rates of the forward and reverse reactions are equal (see Chapter 6). Reversible reactions are indicated by double arrows:

$$\underset{\text{Carbon dioxide}}{CO_2} + \underset{\text{Water}}{H_2O} \rightleftharpoons \underset{\text{Carbonic acid}}{H_2CO_3}$$

In this example, the arrows are drawn in different lengths to indicate that when the reaction reaches equilibrium, there will be more reactants (CO_2 and H_2O) than product (H_2CO_3).

Review

- Why is a radioisotope able to substitute for an ordinary (nonradioactive) atom of the same element in a molecule?
- Which kind of chemical formula provides the most information?
- How many atoms would be included in 1 gram of hydrogen atoms? in 2 grams of hydrogen molecules?

BiologyNow™ Assess your understanding of **chemical reactions** by taking the pretest on your BiologyNow CD-ROM.

CHEMICAL BONDS

Learning Objective

7 Distinguish among covalent bonds, ionic bonds, and hydrogen bonds. Compare them in terms of the mechanisms by which they form and their relative bond strengths.

The atoms of a compound are held together by forces of attraction called **chemical bonds.** Each bond represents a certain amount of chemical energy. **Bond energy** is the energy necessary to break a chemical bond. The valence electrons dictate how many bonds an atom can participate in. The two principal types of strong chemical bonds are covalent bonds and ionic bonds.

In covalent bonds electrons are shared

Covalent bonds involve the sharing of electrons between atoms in a way that results in each atom having a filled valence shell. A compound consisting mainly of covalent bonds is called a **covalent compound.** A simple example of a covalent bond is the joining of two hydrogen atoms in a molecule of hydrogen gas, H_2. Each atom of hydrogen has 1 electron, but 2 electrons are required to complete its valence shell. The hydrogen atoms have equal capacities to attract electrons, so neither donates an electron to the other. Instead, the two hydrogen atoms share their single electrons so that the two electrons are attracted simultaneously to the 2 protons in the two hydrogen nuclei. The 2 electrons thus whirl around both atomic nuclei, joining the two atoms.

A simple way of representing the electrons in the valence shell of an atom is to use dots placed around the chemical symbol of the element. Such a representation is called the *Lewis structure* of the atom, named for G. N. Lewis, an American chemist who developed this type of notation. In a water molecule, two hydrogen atoms are covalently bonded to an oxygen atom:

$$H\cdot + H\cdot + \cdot\ddot{\underset{..}{O}}\cdot \longrightarrow H:\ddot{\underset{..}{O}}:H$$

Oxygen has 6 valence electrons; by sharing electrons with two hydrogen atoms, it completes its valence shell of 8. At the same time each hydrogen atom obtains a complete valence shell of 2. (Note that in the structural formula H—O—H, each pair of shared electrons constitutes a covalent bond, represented by a solid line. Unshared electrons are usually omitted in a structural formula.)

The carbon atom has 4 electrons in its valence shell, all of which are available for covalent bonding:

When one carbon and four hydrogen atoms share electrons, a molecule of methane, CH_4, is formed:

```
   H                  H
   ..                 |
H : C : H    or    H—C—H
   ..                 |
   H                  H
```

Lewis structure   Structural formula

The nitrogen atom has 5 electrons in its valence shell. Recall that each orbital can hold a maximum of 2 electrons. Usually 2 electrons occupy one orbital, leaving 3 available for sharing with other atoms:

$$\cdot\ddot{\underset{\cdot}{N}}\cdot$$

When a nitrogen atom shares electrons with three hydrogen atoms, a molecule of ammonia, NH_3, is formed:

```
   ..
H : N : H    or    H—N—H
   ..                 |
   H                  H
```

Lewis structure   Structural formula

When one pair of electrons is shared between two atoms, the covalent bond is called a **single covalent bond** (Fig. 2-5a). Two oxygen atoms may achieve stability by forming covalent bonds with one another. Each oxygen atom has 6 electrons in its outer shell. To become stable, the two atoms share two pairs of electrons, forming molecular oxygen (Fig. 2-5b). When two pairs of electrons are shared in this way, the covalent bond is called a **double covalent bond,** which is represented by two parallel solid lines. Similarly, a **triple covalent bond** is formed when three pairs of electrons are shared between two atoms (represented by three parallel solid lines).

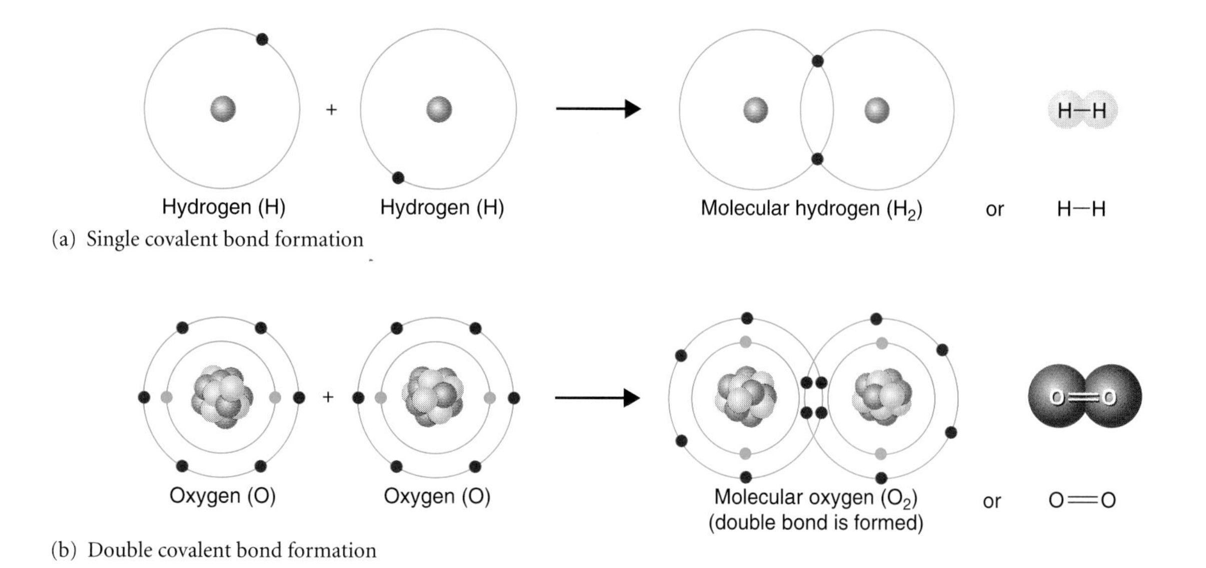

FIGURE 2-5 **Electron sharing in covalent compounds.**
(a) Two hydrogen atoms achieve stability by sharing a pair of electrons, thereby forming a molecule of hydrogen. In the structural formula on the right, the straight line between the hydrogen atoms represents a single covalent bond. **(b)** In molecular oxygen, two oxygen atoms share two pairs of electrons, forming a double covalent bond.

The number of covalent bonds usually formed by the atoms in biologically important molecules is summarized as follows:

Atom	Symbol	Covalent Bonds
Hydrogen	H	1
Oxygen	O	2
Carbon	C	4
Nitrogen	N	3
Phosphorus	P	5
Sulfur	S	2

The function of a molecule is related to its shape

In addition to being composed of atoms with certain properties, each kind of molecule has a characteristic size and a general overall shape. Although the shape of a molecule may change (within certain limits), the functions of molecules in living cells are dictated largely by their geometric shapes. A molecule that consists of two atoms is linear. Molecules composed of more than two atoms may have more complicated shapes. The geometric shape of a molecule provides the optimal distance between the atoms to counteract the repulsion of electron pairs.

When an atom forms covalent bonds with other atoms, the orbitals in the valence shell may become rearranged in a process known as **orbital hybridization,** thereby affecting the shape of the resulting molecule. For example, when four hydrogen atoms combine with a carbon atom to form a molecule of methane (CH_4), the hybridized valence shell orbitals of the carbon form a geometric structure known as a *tetrahedron,* with one hydrogen atom present at each of its four corners (Fig. 2-6; see Fig. 3-2b).

Covalent bonds can be nonpolar or polar

Atoms of different elements vary in their affinity for electrons. **Electronegativity** is a measure of an atom's attraction for shared electrons in chemical bonds. Very electronegative atoms such as oxygen, nitrogen, fluorine, and chlorine are sometimes called "electron greedy." When covalently bonded atoms have similar electronegativities, the electrons are shared equally, and the covalent bond is described as **nonpolar.** The covalent bond of the hydrogen molecule is nonpolar, as are the covalent bonds of molecular oxygen and methane.

In a covalent bond between two different elements, such as oxygen and hydrogen, the electronegativities of the atoms may be different. If so, electrons are pulled closer to the atomic nucleus of the element with the greater electron affinity (in this case, oxygen). A covalent bond between atoms that differ in electronegativity is called a **polar covalent bond.** Such a bond has two dissimilar ends (or poles), one with a partial positive charge

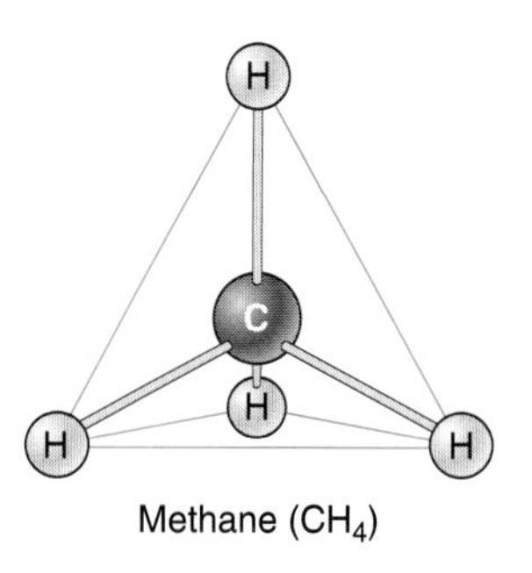

FIGURE 2-6 **Orbital hybridization in methane.**
The four hydrogens are located at the corners of a tetrahedron owing to hybridization of the valence shell orbitals of carbon.

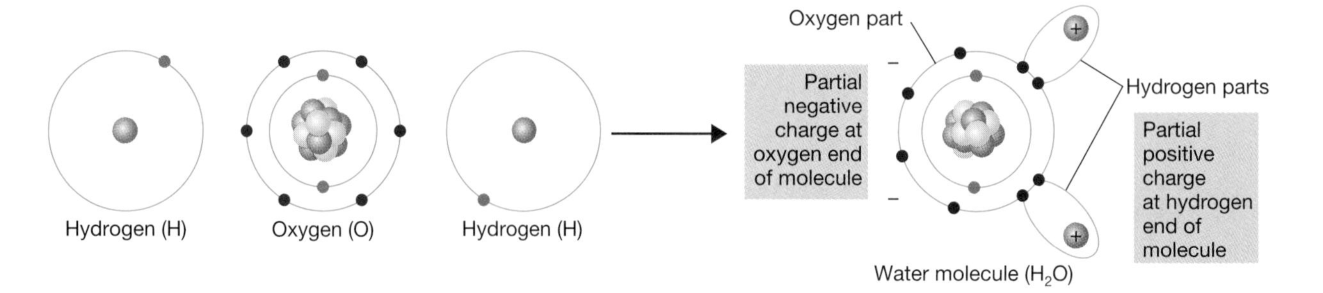

FIGURE **2-7** **Water, a polar molecule.**

Note that the electrons tend to stay closer to the nucleus of the oxygen atom than to the hydrogen nuclei. This results in a partial negative charge on the oxygen portion of the molecule and a partial positive charge at the hydrogen end. Although the water molecule as a whole is electrically neutral, it is a polar covalent compound.

and the other with a partial negative charge. Each of the two covalent bonds in water is polar, because there is a partial positive charge at the hydrogen end of the bond and a partial negative charge at the oxygen end, where the "shared" electrons are more likely to be.

Covalent bonds differ in their degree of polarity, ranging from those in which the electrons are equally shared (as in the nonpolar hydrogen molecule) to those in which the electrons are much closer to one atom than to the other (as in water). Oxygen is quite electronegative and forms polar covalent bonds with carbon, hydrogen, and many other atoms. Nitrogen is also strongly electronegative, although less so than oxygen.

A molecule with one or more polar covalent bonds can be polar even though it is electrically neutral as a whole. This is because a **polar molecule** has one end with a partial positive charge and another end with a partial negative charge. One example is water (Fig. 2-7). The polar bonds between the hydrogens and the oxygen are arranged in a V shape, rather than linearly. The oxygen end constitutes the negative pole of the molecule, and the end with the two hydrogens is the positive pole.

Ionic bonds form between cations and anions

Some atoms or groups of atoms are not electrically neutral. A particle with 1 or more units of electrical charge is called an **ion.** An atom becomes an ion if it gains or loses 1 or more electrons. An atom with 1, 2, or 3 electrons in its valence shell tends to lose electrons to other atoms. Such an atom then becomes positively charged, because its nucleus contains more protons than the number of electrons orbiting around the nucleus. These positively charged ions are termed **cations.** Atoms with 5, 6, or 7 valence electrons tend to gain electrons from other atoms and become negatively charged **anions.**

The properties of ions are quite different from those of the electrically neutral atoms from which they were derived. For example, although chlorine gas is a poison, chloride ions (Cl^-) are essential to life (see Table 2-1). Because their electrical charges provide a basis for many interactions, cations and anions are involved in energy transformations within the cell, the transmission of nerve impulses, muscle contraction, and many other biological processes (Fig. 2-8).

A group of covalently bonded atoms can also become an ion *(polyatomic ion).* Unlike a single atom, a group of atoms can lose or gain protons (derived from hydrogen atoms) as well as electrons. Therefore, a group of atoms can become a cation if it loses 1 or more electrons or gains 1 or more protons. A group of atoms becomes an anion if it gains 1 or more electrons or loses 1 or more protons.

An **ionic bond** forms as a consequence of the attraction between the positive charge of a cation and the negative charge of an anion. An **ionic compound** is a substance consisting of anions and cations bonded together by their opposite charges.

A good example of how ionic bonds are formed is the attraction between sodium ions and chloride ions. A sodium atom has 1 electron in its valence shell. It cannot fill its valence shell by obtaining 7 electrons from other atoms, because it would then have a large unbalanced negative charge. Instead, it gives up its single valence electron to a very electronegative atom, such as chlorine, which acts as an electron acceptor (Fig. 2-9). Chlorine cannot give up the seven electrons in its valence shell,

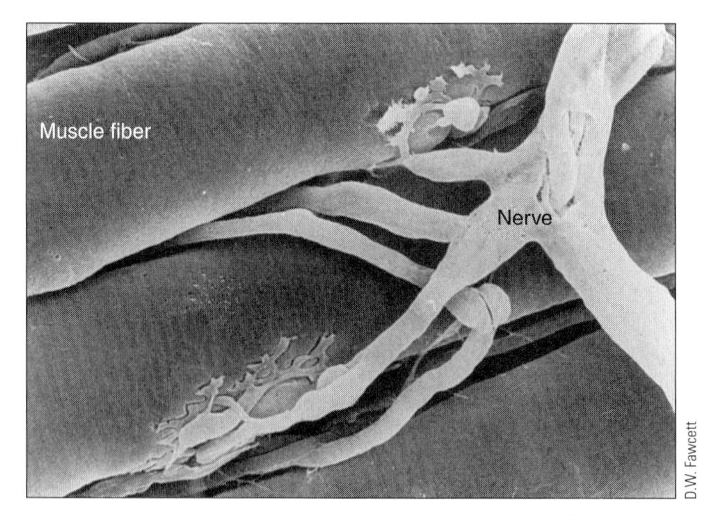

FIGURE **2-8** **Ions and biological processes.**

Sodium, potassium, and chloride ions are essential for this nerve cell to stimulate these muscle fibers, initiating a muscle contraction. Calcium ions in the muscle cell are required for muscle contraction.

because it would then have a large positive charge. Instead, it strips an electron from an electron donor (sodium, in this example) to complete its valence shell.

When sodium reacts with chlorine, sodium's valence electron is transferred completely to chlorine. Sodium becomes a cation, with 1 unit of positive charge (Na^+). Chlorine becomes an anion, a chloride ion with 1 unit of negative charge (Cl^-). These ions attract each other as a result of their opposite charges. This electrical attraction in ionic bonds holds them together to form NaCl, sodium chloride, or common table salt.

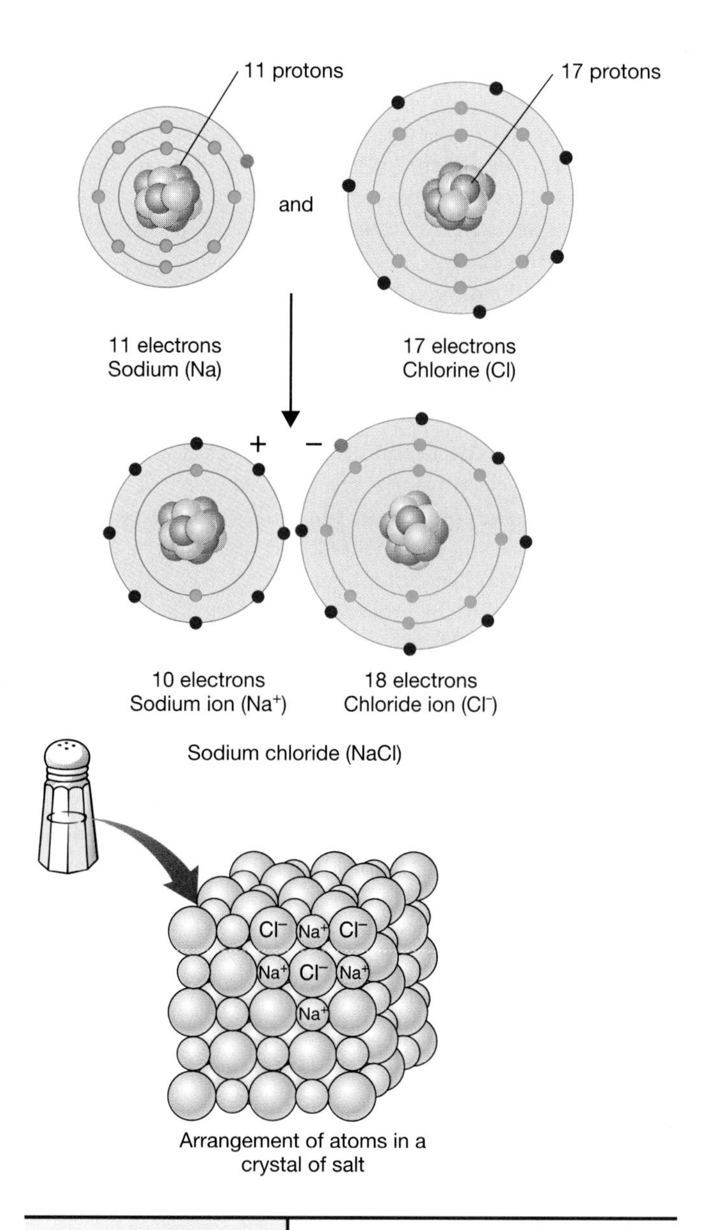

ACTIVE FIGURE **2-9** **Ionic bonding.**

Sodium becomes a positively charged ion when it donates its single valence electron to chlorine, which has 7 valence electrons. With this additional electron, chlorine completes its valence shell and becomes a negatively charged chloride ion. These sodium and chloride ions are attracted to one another by their unlike electrical charges, forming the ionic compound sodium chloride.

Biology Now™ Learn more about **ionic bonding** by clicking on this figure on your Biology Now CD-ROM.

The term *molecule* does not adequately explain the properties of ionic compounds such as NaCl. When NaCl is in its solid crystal state, each ion is actually surrounded by six ions of opposite charge. The simplest formula, NaCl, indicates that sodium ions and chloride ions are present in a 1:1 ratio, but the actual crystal has no discrete molecules composed of one Na^+ and one Cl^- ion.

Compounds joined by ionic bonds, such as sodium chloride, have a tendency to *dissociate* (separate) into their individual ions when placed in water:

$$\underset{\text{Sodium chloride}}{NaCl} \xrightarrow{\text{in } H_2O} \underset{\text{Sodium ion}}{Na^+} + \underset{\text{Chloride ion}}{Cl^-}$$

In the solid form of an ionic compound (that is, in the absence of water), the ionic bonds are very strong. Water, however, is an excellent **solvent;** as a liquid it is capable of dissolving many substances, particularly those that are polar or ionic. This is because of the polarity of water molecules. The localized partial positive charge (on the hydrogen atoms) and partial negative charge (on the oxygen atom) on each water molecule attract and surround the anions and cations, respectively, on the surface of an ionic solid. As a result, the solid dissolves. A dissolved substance is referred to as a **solute.** In solution, each cation and anion of the ionic compound is surrounded by oppositely charged ends of the water molecules. This process is known as **hydration** (Fig. 2-10). Hydrated ions still interact with each other to some extent, but the transient ionic bonds formed are much weaker than those in a solid crystal.

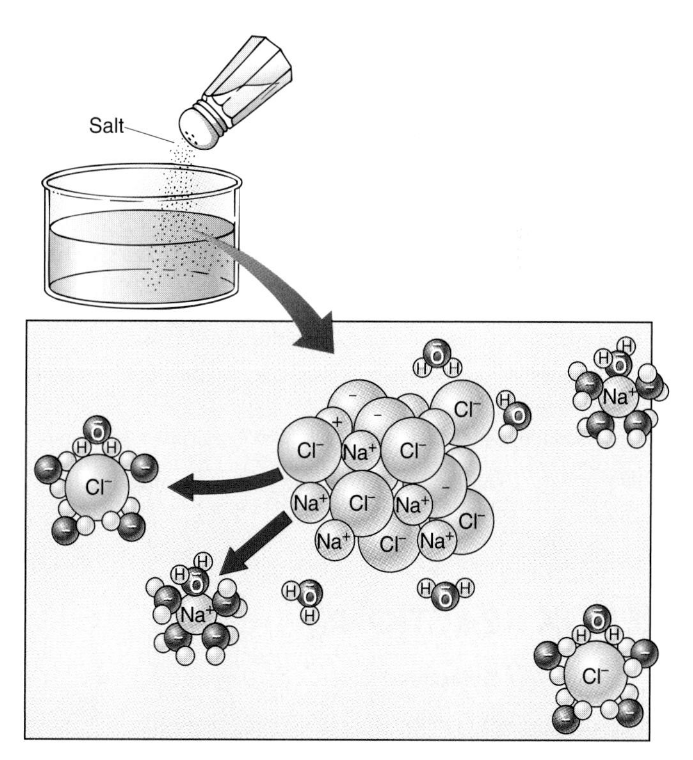

FIGURE **2-10** **Hydration of an ionic compound.**

When the crystal of NaCl is added to water, the sodium and chloride ions are pulled apart. When the NaCl is dissolved, each Na^+ and Cl^- is surrounded by water molecules electrically attracted to it.

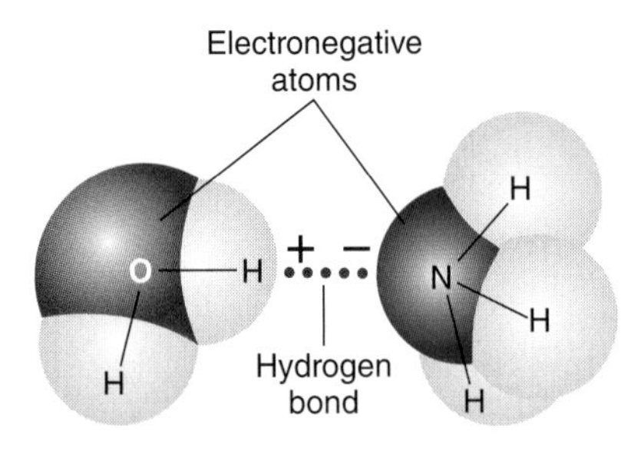

ACTIVE FIGURE **2-11** | **Hydrogen bonding.**

A hydrogen bond (indicated by a dotted line) can form between two molecules with regions of unlike partial charge. Here, the nitrogen atom of an ammonia molecule is joined by a hydrogen bond to a hydrogen atom of a water molecule.

BiologyNow™ Learn more about **hydrogen bonding** by clicking on this figure on your BiologyNow CD-ROM.

Hydrogen bonds are weak attractions

Another type of bond important in organisms is the **hydrogen bond.** When hydrogen combines with oxygen (or with another relatively electronegative atom such as nitrogen), it acquires a partial positive charge because its electron spends more time closer to the electronegative atom. Hydrogen bonds tend to form between an atom with a partial negative charge and a hydrogen atom that is covalently bonded to oxygen or nitrogen (Fig. 2-11). The atoms involved may be in two parts of the same large molecule or in two different molecules. Water molecules interact with each other extensively through hydrogen bond formation.

Hydrogen bonds are readily formed and broken. Although individually relatively weak, hydrogen bonds are collectively strong when present in large numbers. Furthermore, they have a specific length and orientation. As you will see in Chapter 3, these features are very important in determining the 3-D structure of large molecules such as DNA and proteins.

Review

- Are all compounds composed of molecules? Explain.
- What are the ways an atom or molecule can become an anion or a cation?
- How do ionic and covalent bonds differ?

BiologyNow™ Assess your understanding of **chemical bonds** by taking the pretest on your BiologyNow CD-ROM.

REDOX REACTIONS

Learning Objective

8 Distinguish between the terms *oxidation* and *reduction*, and relate these processes to the transfer of energy.

Many energy conversions that go on in a cell involve reactions in which an electron transfers from one substance to another. This is because the transfer of an electron also involves the transfer of the energy of that electron. Such an electron transfer is known as an oxidation-reduction, or **redox reaction.** Oxidation and reduction always occur together. **Oxidation** is a chemical process in which an atom, ion, or molecule *loses* electrons. **Reduction** is a chemical process in which an atom, ion, or molecule *gains* electrons. (The term refers to the fact that the gain of an electron results in the reduction of any positive charge that might be present.)

Rusting—the combining of iron (symbol Fe) with oxygen—is a simple illustration of oxidation and reduction:

$$4\ Fe + 3\ O_2 \longrightarrow \underset{\text{Iron (III) oxide}}{2\ Fe_2O_3}$$

In rusting, each iron atom becomes oxidized as it loses 3 electrons.

$$4\ Fe \rightarrow 4\ Fe^{3+} + 12e^-$$

The e^- represents an electron; the + superscript in Fe^{3+} represents an electron deficit. (When an atom loses an electron, it acquires 1 unit of positive charge from the excess of 1 proton. In our example, each iron atom loses 3 electrons and acquires 3 units of positive charge.)

Recall that the oxygen atom is very electronegative, able to remove electrons from other atoms. In this reaction, oxygen becomes reduced when it accepts electrons from the iron.

$$3\ O_2 + 12e^- \rightarrow 6\ O^{2-}$$

Redox reactions occur simultaneously because one substance must accept the electrons that are removed from the other. In a redox reaction, one component, the *oxidizing agent,* accepts one or more electrons and becomes reduced. Oxidizing agents other than oxygen are known, but oxygen is such a common one that its name was given to the process. Another reaction component, the *reducing agent,* gives up one or more electrons and becomes oxidized.

In our example, there was a complete transfer of electrons from iron (the reducing agent) to oxygen (the oxidizing agent). Similarly, in Figure 2-9 an electron was transferred from sodium (the reducing agent) to chlorine (the oxidizing agent).

Electrons are not easily removed from covalent compounds unless an entire atom is removed. In cells, oxidation often involves the removal of a hydrogen atom (an electron plus a proton that "goes along for the ride") from a covalent compound; reduction often involves the addition of the equivalent of a hydrogen atom (see Chapter 6).

Review

- Why must oxidation and reduction occur simultaneously?
- Why are redox reactions important in some energy transfers?

BiologyNow™ Assess your understanding of **redox reactions** by taking the pretest on your BiologyNow CD-ROM.

WATER

Learning Objective

9 Explain how hydrogen bonds between adjacent water molecules govern many of the properties of water.

A large part of the mass of most organisms is water. In human tissues the percentage of water ranges from 20% in bones to

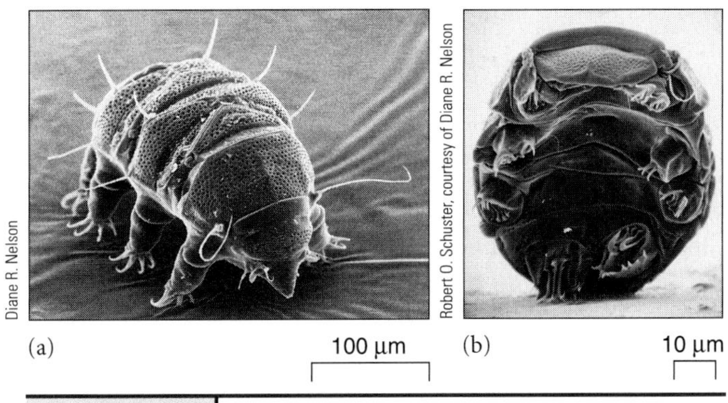

FIGURE **2-12** **Water's essential role.**

(a) Commonly known as "water bears," tardigrades, such as these members of the genus *Echiniscus,* are small animals (less than 1.2 mm long) that normally live in moist habitats, such as thin films of water on mosses. **(b)** When subjected to desiccation (dried out), tardigrades assume a barrel-shaped form known as a *tun,* remaining in this state, motionless but alive, for as long as 100 years. When rehydrated, they assume their normal appearance and activities.

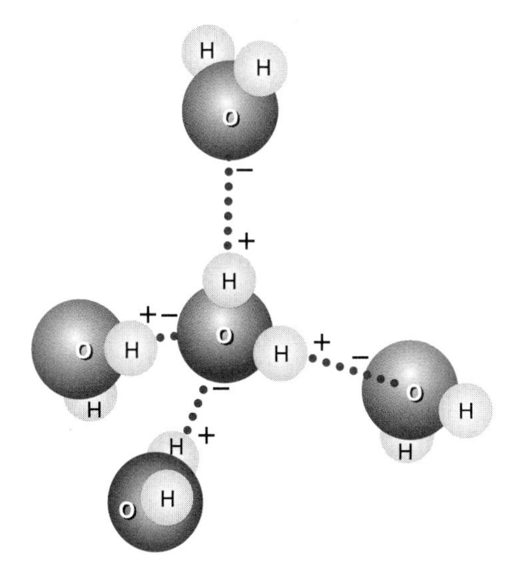

FIGURE **2-13** **Hydrogen bonding of water molecules.**

Each water molecule can form hydrogen bonds *(dotted lines)* with as many as four neighboring water molecules.

85% in brain cells; about 70% of our total body weight is water. As much as 95% of a jellyfish and certain plants is water. Water is the source, through photosynthesis, of the oxygen in the air we breathe, and its hydrogen atoms become incorporated into many organic compounds. Water is also the solvent for most biological reactions and a reactant or product in many chemical reactions.

Water is important not only as an internal constituent of organisms but also as one of the principal environmental factors affecting them (Fig. 2-12). Many organisms live in the ocean or in freshwater rivers, lakes, or puddles. Water's unique combination of physical and chemical properties is considered to have been essential to the origin of life, as well as to the continued survival and evolution of life on Earth.

As discussed, water molecules are polar; that is, one end of each molecule bears a partial positive charge and the other a partial negative charge (see Fig. 2-7). The water molecules in liquid water and in ice associate by hydrogen bonds. The hydrogen atom of one water molecule, with its partial positive charge, is attracted to the oxygen atom of a neighboring water molecule, with its partial negative charge, forming a hydrogen bond. An oxygen atom in a water molecule has two regions of partial negative charge, and each of the two hydrogen atoms has a partial positive charge. Each water molecule can therefore form hydrogen bonds with a maximum of four neighboring water molecules (Fig. 2-13).

Because its molecules are polar, water is an excellent solvent, a liquid capable of dissolving many different kinds of substances, especially polar and ionic compounds. Earlier we discussed how polar water molecules pull the ions of ionic compounds apart so that they dissociate (see Fig. 2-10). Because of its solvent properties and the tendency of the atoms in certain compounds to form ions in solution, water plays an important role in facilitating chemical reactions. Substances that interact readily with water are **hydrophilic** ("water-loving"). Examples include table sugar (sucrose, a polar compound) and table salt (NaCl, an ionic compound), which dissolve readily in water. Not all substances in organisms are hydrophilic, however. Many **hydrophobic** ("water-fearing") substances found in living things are especially important, because of their ability to form associations or structures that are not disrupted or dissolved by water. Examples are fats and other nonpolar substances (see Chapter 3).

Water molecules have a strong tendency to stick to each other, a property known as **cohesion.** This is due to the hydrogen bonds among the molecules. Because of the cohesive nature of water molecules, any force exerted on part of a column of water is transmitted to the column as a whole. The major mechanism of water movement in plants (see Chapter 33) depends on the cohesive nature of water. Water molecules also display **adhesion,** the ability to stick to many other kinds of substances, most notably those with charged groups of atoms or molecules on their surfaces. These adhesive forces explain how water makes things wet.

A combination of adhesive and cohesive forces accounts for **capillary action,** which is the tendency of water to move in narrow tubes, even against the force of gravity (Fig. 2-14). For example, water moves through the microscopic spaces between soil particles to the roots of plants by capillary action.

Water has a high degree of **surface tension** because of the cohesion of its molecules, which have a much greater attraction for each other than for molecules in the air. Thus water molecules at the surface crowd together, producing a strong layer as they are pulled downward by the attraction of other water molecules beneath them (Fig. 2-15).

Water helps maintain a stable temperature

Hydrogen bonding explains the way water responds to changes in temperature. Water exists in three forms, which differ in their

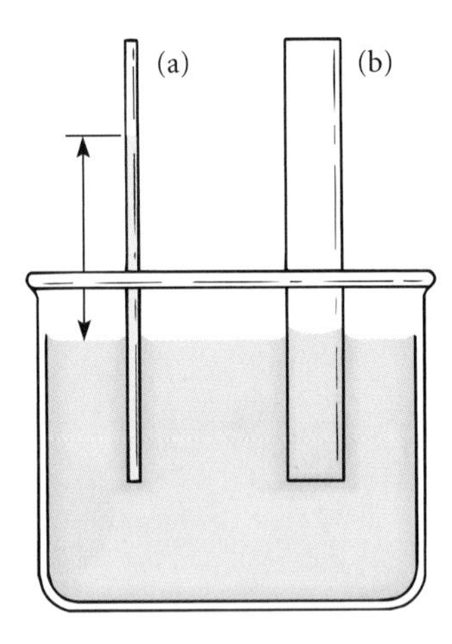

FIGURE **2-14** **Capillary action.**

(a) In a narrow tube, there is adhesion between the water molecules and the glass wall of the tube. Other water molecules inside the tube are then "pulled along" because of cohesion, which is due to hydrogen bonds between the water molecules. **(b)** In the wider tube, a smaller percentage of the water molecules line the glass wall. As a result, the adhesion is not strong enough to overcome the cohesion of the water molecules beneath the surface level of the container, and water in the tube rises only slightly.

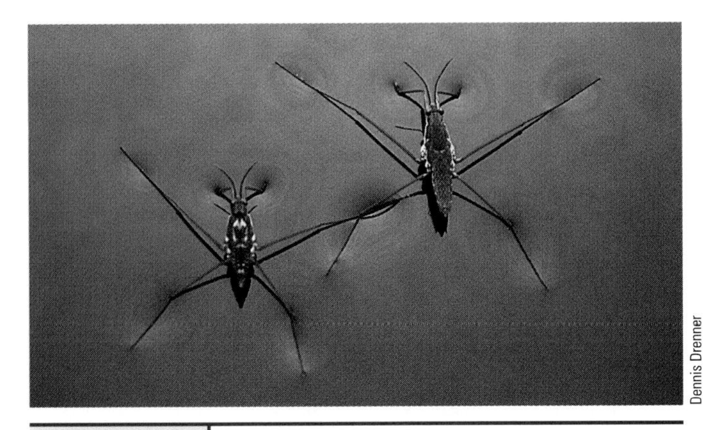

Dennis Drenner

FIGURE **2-15** **Surface tension of water.**

Hydrogen bonding between water molecules is responsible for the surface tension of water, which is strong enough to support these water striders (*Gerris*) and causes the dimpled appearance of the surface. Although water striders are denser than water, these insects can walk on the surface of a pond, because fine hairs at the ends of their legs spread their weight over a large area.

degree of hydrogen bonding: gas (vapor), liquid, and ice, a crystalline solid (Fig. 2-16). Hydrogen bonds are formed or broken as water changes from one state to another.

Raising the temperature of a substance involves adding heat energy to make its molecules move faster, that is, to increase the energy of motion—**kinetic energy**—of the molecules (see Chapter 6). The term **heat** refers to the *total* amount of kinetic energy in a sample of a substance; **temperature** is a measure of the *average* kinetic energy of the particles. For the molecules to move more freely, some of the hydrogen bonds of water must be broken. Much of the energy added to the system is used up in breaking the hydrogen bonds, and only a portion of the heat energy is available to speed the movement of the water molecules, thereby increasing the temperature of the water. Conversely, when liquid water changes to ice, additional hydrogen bonds must be formed, making the molecules less free to move, and liberating a great deal of heat into the environment.

Heat of vaporization, the amount of heat energy required to change 1 g of a substance from the liquid phase to the vapor phase, is expressed in units called *calories.* A **calorie (cal)** is the amount of heat energy (equivalent to 4.184 joules [J]) required to raise the temperature of 1 g of water 1 degree Celsius (C). Water has a high heat of vaporization—540 cal—because its molecules are held together by hydrogen bonds. The heat of vaporization of most other common liquid substances is much less. As a sample of water is heated, some molecules are moving much faster than others (they have more heat energy). These faster-moving molecules are more likely to escape the liquid phase and enter the vapor phase (see Fig. 2-16a). When they do, they take their heat energy with them, lowering the temperature of the sample, a process called **evaporative cooling.** For this reason, the human body can dissipate excess heat as sweat evaporates from the skin, and a leaf can keep cool in the bright sunlight as water evaporates from its surface.

Hydrogen bonding is also responsible for water's high **specific heat;** that is, the amount of energy required to raise the temperature of water is quite large. The specific heat of water is 1 cal/g of water per degree Celsius. Most other common substances such as metals, glass, and ethyl alcohol have much lower specific heat values. The specific heat of ethyl alcohol, for example, is 0.59 cal/g/1°C (2.46 J/g/1°C).

Because so much heat input is required to raise the temperature of water (and so much heat is lost when the temperature is lowered), the oceans and other large bodies of water have relatively constant temperatures. Thus, many organisms living in the ocean are provided with a relatively constant environmental temperature. The properties of water are crucial in stabilizing temperatures on Earth's surface. Although surface water is only a thin film relative to Earth's volume, the quantity is enormous compared to the exposed land mass. This relatively large mass of water resists both the warming effect of heat and the cooling effect of low temperatures.

Hydrogen bonding causes ice to have unique properties with important environmental consequences. Liquid water expands as it freezes because the hydrogen bonds joining the water molecules in the crystalline lattice keep the molecules far enough apart to give ice a density about 10% less than the density of liquid water (see Fig. 2-16c). When ice has been heated enough to raise its temperature above 0°C (32°F), the hydrogen bonds are broken, freeing the molecules to slip closer together. The density of water is greatest at 4°C, above which water begins to expand again as the speed of its molecules increases. As a result, ice floats on the denser cold water.

This unusual property of water has been important to the evolution of life. If ice had a greater density than water, it would sink; eventually all ponds, lakes, and even the ocean would freeze

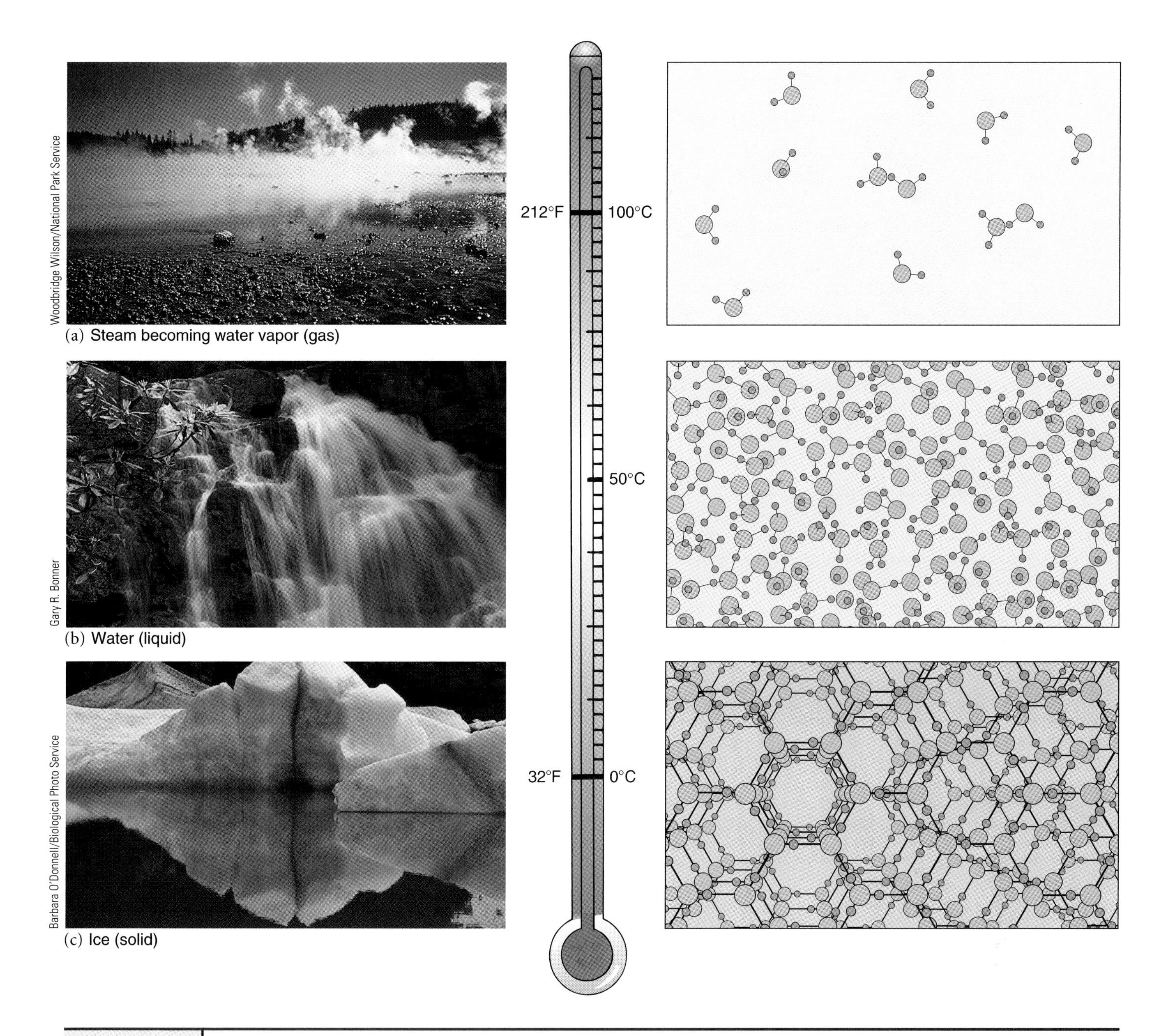

(a) Steam becoming water vapor (gas)

(b) Water (liquid)

(c) Ice (solid)

FIGURE **2-16** | **Three forms of water.**

(a) When water boils, as in this hot spring at Yellowstone National Park, many hydrogen bonds are broken, causing steam, consisting of minuscule water droplets, to form. If most of the remaining hydrogen bonds break, the molecules move more freely as water vapor (a gas). **(b)** Water molecules in a liquid state continually form, break, and re-form hydrogen bonds with each other. **(c)** In ice, each water molecule participates in four hydrogen bonds with adjacent molecules, resulting in a regular, evenly distanced crystalline lattice structure.

solid from the bottom to the surface, making life impossible. When a deep body of water cools, it becomes covered with floating ice. The ice insulates the liquid water below it, retarding freezing and permitting organisms to survive below the icy surface.

The high water content of organisms helps them maintain relatively constant internal temperatures. Such minimizing of temperature fluctuations is important because biological reactions can take place only within a relatively narrow temperature range.

Review

- Why does water form hydrogen bonds?
- What are some properties of water that result from hydrogen bonding? How do these properties contribute to the role of water as an essential component of organisms?
- How can weak forces, such as hydrogen bonds, have significant effects in organisms?

BiologyNow™ Assess your understanding of **water** by taking the pretest on your BiologyNow CD-ROM.

ACIDS, BASES, AND SALTS

Learning Objectives

10 Contrast acids and bases, and discuss their properties.

11 Convert the hydrogen ion concentration (moles per liter) of a solution to a pH value, and describe how buffers help minimize changes in pH.

12 Describe the composition of a salt, and explain why salts are important in organisms.

Water molecules have a slight tendency to **ionize,** that is, to dissociate into hydrogen ions (H^+) and hydroxide ions (OH^-). The H^+ immediately combines with a negatively charged region of a water molecule, forming a hydronium ion (H_3O^+). However, by convention, H^+, rather than the more accurate H_3O^+, is used. In pure water, a small number of water molecules ionize. This slight tendency of water to dissociate is reversible as hydrogen ions and hydroxide ions reunite to form water:

$$HOH \rightleftharpoons H^+ + OH^-$$

Because each water molecule splits into one hydrogen ion and one hydroxide ion, the concentrations of hydrogen ions and hydroxide ions in pure water are exactly equal (0.0000001 or 10^{-7} mol/L for each ion). Such a solution is said to be neutral, that is, neither acidic nor basic (alkaline).

An **acid** is a substance that dissociates in solution to yield hydrogen ions (H^+) and an anion.

$$\text{Acid} \rightarrow H^+ + \text{Anion}$$

An acid is a proton *donor.* (Recall that a hydrogen ion, or H^+, is nothing more than a proton.) Hydrochloric acid (HCl) is a common organic acid.

A **base** is defined as a proton *acceptor.* Most bases are substances that dissociate to yield a hydroxide ion (OH^-) and a cation when dissolved in water. A hydroxide ion can act as a base by accepting a proton (H^+) to form water. Sodium hydroxide (NaOH) is a common inorganic base.

$$NaOH \rightarrow Na^+ + OH^-$$

$$OH^- + H^+ \rightarrow H_2O$$

Some bases do not dissociate to yield hydroxide ions directly. For example, ammonia (NH_3) acts as a base by accepting a proton from water, producing an ammonium ion (NH_4^+) and releasing a hydroxide ion.

$$NH_3 + H_2O \rightarrow NH_4^+ + OH^-$$

pH is a convenient measure of acidity

The degree of a solution's acidity is generally expressed in terms of **pH,** defined as the negative logarithm (base 10) of the hydrogen ion concentration (expressed in moles per liter):

$$pH = -\log_{10}[H^+]$$

The brackets refer to concentration; therefore, the term $[H^+]$ means "the concentration of hydrogen ions," which is expressed in moles per liter because we are interested in the *number* of hydrogen ions per liter. Because the range of possible pH values is broad, a logarithmic scale (with a 10-fold difference between successive units) is more convenient than a linear scale.

TABLE 2-2 Calculating pH Values and Hydroxide Ion Concentrations from Hydrogen Ion Concentrations

Substance	$[H^+]$*	log $[H^+]$	pH	$[OH^-]$†
Gastric juice	$0.01, 10^{-2}$	−2	2	10^{-12}
Pure water, neutral solution	$0.0000001, 10^{-7}$	−7	7	10^{-7}
Household ammonia	$0.00000000001, 10^{-11}$	−11	11	10^{-3}

*$[H^+]$ = hydrogen ion concentration (mol/L)

Hydrogen ion concentrations are nearly always less than 1 mol/L. One gram of hydrogen ions dissolved in 1 L of water (a 1-*M* solution) may not sound impressive, but such a solution would be extremely acidic. The logarithm of a number less than 1 is a negative number; thus the *negative* logarithm corresponds to a *positive* pH value. (Solutions with pH values less than zero can be produced but do not occur under biological conditions.)

Whole-number pH values are easy to calculate. For instance, consider our example of pure water, which has a hydrogen ion concentration of 0.0000001 (10^{-7}) mol/L. The logarithm is −7. The negative logarithm is 7; therefore, the pH is 7. Table 2-2 shows how to calculate pH values from hydrogen ion concentrations, and the reverse. For comparison, the table also includes the hydroxide ion concentrations, which can be calculated because the product of the hydrogen concentration and the hydroxide ion concentration is 1×10^{-14}:

$$[H^+][OH^-] = 1 \times 10^{-14}$$

Pure water is an example of a **neutral solution;** with a pH of 7, it has equal concentrations of hydrogen ions and hydroxide ions (10^{-7} moles per liter). An **acidic solution** has a hydrogen ion concentration that is higher than its hydroxide ion concentration and has a pH value of less than 7. For example, the hydrogen ion concentration of a solution with pH 1 is 10 times that of a solution with pH 2. A **basic solution** has a hydrogen ion concentration that is lower than its hydroxide ion and has a pH greater than 7.

The pH values of some common substances are shown in Figure 2-17. Although some very acidic compartments exist within cells (see Chapter 4), most of the interior of an animal or plant cell is neither strongly acidic nor strongly basic but an essentially neutral mixture of acidic and basic substances. Although certain bacteria are adapted to life in extremely acidic environments (see Chapter 23), a substantial change in pH is incompatible with life for most cells. The pH of most types of plant and animal cells (and their environment) ordinarily ranges from around 7.2 to 7.4.

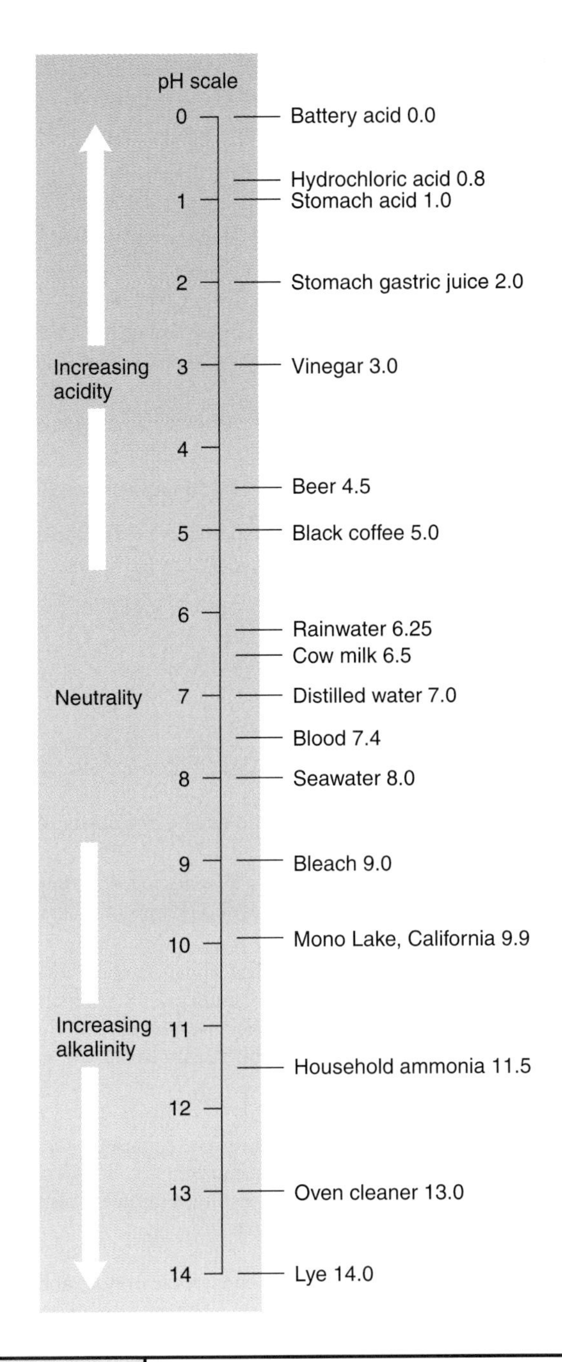

FIGURE **2-17** **pH values of some common solutions.**

A neutral solution (pH 7) has equal concentrations of H^+ and OH^-. Acidic solutions, which have a higher concentration of H^+ than OH^-, have pH values lower than 7; pH values higher than 7 characterize basic solutions, which have an excess of OH^-.

Buffers minimize pH change

Many homeostatic mechanisms operate to maintain appropriate pH values. For example, the pH of human blood is about 7.4 and must be maintained within very narrow limits. If the blood becomes too acidic (for example, as a result of respiratory disease), coma and death may result. Excessive alkalinity can result in overexcitability of the nervous system and even convulsions. Organisms contain many natural buffers. A **buffer** is a substance or combination of substances that resists changes in pH when an acid or base is added. A buffering system includes a weak acid or a weak base. A weak acid or weak base does not ionize completely. At any given instant, only a fraction of the molecules are ionized; most are not dissociated.

One of the most common buffering systems functions in the blood of vertebrates (see Chapter 44). Carbon dioxide, produced as a waste product of cell metabolism, enters the blood, the main constituent of which is water. The carbon dioxide reacts with the water to form carbonic acid, a weak acid that dissociates to yield a hydrogen ion and a bicarbonate ion. The following expression describes the buffering system:

$$\underset{\text{Carbon dioxide}}{CO_2} + \underset{\text{Water}}{H_2O} \rightleftharpoons \underset{\text{Carbonic acid}}{H_2CO_3} \rightleftharpoons H^+ + \underset{\text{Bicarbonate ion}}{HCO_3^-}$$

As the double arrows indicate, all the reactions are reversible. Because carbonic acid is a weak acid, undissociated molecules are always present, as are all the other components of the system. The expression describes the system when it is at *dynamic equilibrium*, that is, when the rates of the forward and reverse reactions are equal and the relative concentrations of the components are not changing. A system at dynamic equilibrium tends to stay at equilibrium unless a stress is placed on it, which causes it to shift to reduce the stress until it attains a new dynamic equilibrium. A change in the concentration of any component is one such stress. Therefore, the system can be "shifted to the right" by adding reactants or removing products. Conversely, it can be "shifted to the left" by adding products or removing reactants.

Hydrogen ions are the important products to consider in this system. The addition of excess hydrogen ions temporarily shifts the system to the left, as they combine with the bicarbonate ions to form carbonic acid. Eventually a new dynamic equilibrium is established. At this point the hydrogen ion concentration is similar to the original concentration, and the product of the hydrogen ion and hydroxide ion concentrations is restored to the equilibrium value of 1×10^{-14}.

If hydroxide ions are added, they combine with the hydrogen ions to form water, effectively removing a product and thus shifting the system to the right. More carbonic acid then ionizes, replacing the hydrogen ions that were removed.

Organisms contain many weak acids and weak bases, which allows them to maintain an essential reserve of buffering capacity and helps them avoid pH extremes.

An acid and a base react to form a salt

When an acid and a base are mixed together in water, the H^+ of the acid unites with the OH^- of the base to form a molecule of water. The remainder of the acid (an anion) combines with the remainder of the base (a cation) to form a salt. For example, hydrochloric acid reacts with sodium hydroxide to form water and sodium chloride:

$$HCl + NaOH \rightarrow H_2O + NaCl$$

A **salt** is a compound in which the hydrogen ion of an acid is replaced by some other cation. Sodium chloride, NaCl, is a salt in which the hydrogen ion of HCl has been replaced by the cation Na^+.

When a salt, an acid, or a base is dissolved in water, its dissociated ions can conduct an electrical current; these substances are called **electrolytes.** Sugars, alcohols, and many other substances do not form ions when dissolved in water; they do not conduct an electrical current and are referred to as **nonelectrolytes.**

Cells and extracellular fluids (such as blood) of animals and plants contain a variety of dissolved salts that are the source of the many important mineral ions essential for fluid balance and acid–base balance. The concentrations and relative amounts of the various cations and anions are kept remarkably constant. Any marked change results in impaired cell functions and may lead to death. Nitrate and ammonium ions from the soil are the important nitrogen sources for plants. In animals, nerve and muscle function, blood clotting, bone formation, and many other aspects of body function depend on ions. Sodium, potassium, calcium, and magnesium are the chief cations present; chloride, bicarbonate, phosphate, and sulfate are important anions.

Review

- A solution has a hydrogen ion concentration of 0.01 mol/L. What is its pH? What is its hydroxide ion concentration? Is it acidic, basic, or neutral? How does this solution differ from one with a pH of 1?
- What would be the consequences of adding or removing a reactant or a product from a reversible reaction that is at dynamic equilibrium?
- Why are buffers important in organisms? Why can't strong acids or bases work as buffers?
- Why are acids, bases, and salts referred to as *electrolytes?*

BiologyNow™ Assess your understanding of **acids, bases, and salts** by taking the pretest on your BiologyNow CD-ROM.

SUMMARY WITH KEY TERMS

1 Name the principal chemical elements in living things, and give an important function of each.

- An **element** is a substance that cannot be decomposed into simpler substances by normal chemical reactions. About 96% of an organism's mass consists of carbon, the backbone of organic molecules; hydrogen and oxygen, the components of water; and nitrogen, a component of proteins and nucleic acids.

2 Compare the physical properties (mass and charge) and locations of electrons, protons, and neutrons; distinguish between the atomic number and the mass number of an element.

- Each **atom** is composed of a **nucleus** containing positively charged **protons** and uncharged **neutrons.** Negatively charged **electrons** encircle the nucleus.
- An atom is identified by its number of protons **(atomic number).** The **atomic mass** of an atom is equal to the sum of its protons and neutrons.
- A single proton or a single neutron each has a mass equivalent to one **atomic mass unit.** The mass of a single electron is only about 1/1800 amu.

3 Define the terms *orbital* and *electron shell,* and relate electron shells to principal energy levels.

- In the space outside the nucleus, electrons move rapidly in electron **orbitals.** An **electron shell** consists of electrons in orbitals at the same **principal energy level.** Electrons in a shell distant from the nucleus have greater energy than those in a shell closer to the nucleus.

4 Explain how the number of valence electrons of an atom is related to its chemical properties.

- The chemical properties of an atom are determined chiefly by the number and arrangement of its most energetic electrons, known as **valence electrons.** The valence shell of most atoms is full when it contains 8 electrons; that of hydrogen or helium is full when it contains 2. An atom tends to lose, gain, or share electrons to fill its valence shell.

5 Distinguish among simplest, molecular, and structural chemical formulas.

- Different atoms are joined by chemical bonds to form **compounds.** A **chemical formula** gives the types and relative numbers of atoms in a substance.
- A **simplest formula** gives the smallest whole-number ratio of the component atoms. A **molecular formula** gives the actual numbers of each type of atom in a molecule. A **structural formula** shows the arrangement of the atoms in a molecule.

6 Explain why the mole concept is so useful to chemists.

- One **mole** (the atomic or molecular mass in grams) of any substance contains 6.02×10^{23} atoms, molecules, or ions, enabling scientists to "count" particles by weighing a sample. This number is known as **Avogadro's number.**

7 Distinguish among covalent bonds, ionic bonds, and hydrogen bonds. Compare them in terms of the mechanisms by which they form and their relative bond strengths.

- **Covalent bonds** are strong, stable bonds formed when atoms share **valence electrons,** forming molecules. When covalent bonds are formed, the orbitals of the valence electrons may become rearranged in a process known as **orbital hybridization.** Covalent bonds are **nonpolar** if the electrons are shared equally between the two atoms. Covalent bonds are **polar** if one atom is more **electronegative** (has a greater affinity for electrons) than the other.
- An **ionic bond** is formed between a positively charged **cation** and a negatively charged **anion.** Ionic bonds are strong in the absence of water but relatively weak in aqueous solution.
- **Hydrogen bonds** are relatively weak bonds formed when a hydrogen atom with a partial positive charge is attracted to an atom (usually oxygen or nitrogen) with a partial negative charge already bonded to another molecule or in another part of the same molecule.

8 Distinguish between the terms *oxidation* and *reduction,* and relate these processes to the transfer of energy.

SUMMARY WITH KEY TERMS *(continued)*

- **Oxidation** and **reduction** reactions (**redox reactions**) are chemical processes in which electrons (and their energy) are transferred from a reducing agent to an oxidizing agent. In oxidation, an atom, ion, or molecule loses electrons (and their energy). In reduction, an atom, ion, or molecule gains electrons (and their energy).

9 Explain how hydrogen bonds between adjacent water molecules govern many of the properties of water.

- Water is a **polar molecule** because one end has a partial positive charge and the other has a partial negative charge. Because its molecules are polar, water is an excellent **solvent** for ionic or polar **solutes.**
- Water molecules exhibit the property of **cohesion** because they form hydrogen bonds with each other; they also exhibit **adhesion** through hydrogen bonding to substances with ionic or polar regions.
- Because hydrogen bonds must be broken to raise its temperature, water has a high **specific heat,** which helps organisms maintain a relatively constant internal temperature; this property also helps keep the ocean and other large bodies of water at a constant temperature.
- Water has a high **heat of vaporization.** Hydrogen bonds must be broken for molecules to enter the vapor phase. These molecules carry a great deal of heat, which accounts for **evaporative cooling.**
- The hydrogen bonds between water molecules in ice cause it to be less dense than liquid water. The fact that ice floats makes the aquatic environment less extreme than it would be if ice sank to the bottom.

10 Contrast acids and bases, and discuss their properties.

- **Acids** are proton (hydrogen ion, H^+) donors; **bases** are proton acceptors. An acid dissociates in solution to yield H^+ and an anion. Many bases dissociate in solution to yield hydroxide ions (OH^-), which then accept protons to form water.

11 Convert the hydrogen ion concentration (moles per liter) of a solution to a pH value, and describe how buffers help minimize changes in pH.

- The **pH scale** is the negative log of the hydrogen ion concentration of a solution (expressed in moles per liter). A **neutral solution** with equal concentrations of H^+ and OH^- (10^{-7} moles per liter) has a pH of 7, an **acidic solution** has a pH less than 7, and a **basic solution** has a pH greater than 7.
- A buffering system is based on a weak acid or a weak base. A **buffer** resists changes in the pH of a solution when acids or bases are added.

12 Describe the composition of a salt, and explain why salts are important in organisms.

- A **salt** is a compound in which the hydrogen atom of an acid is replaced by some other cation. Salts provide the many mineral ions essential for life functions.

POST-TEST

1. Which of the following elements is *mismatched* with its properties or function? (a) carbon—forms the backbone of organic compounds (b) nitrogen—component of proteins (c) hydrogen—very electronegative (d) oxygen—can participate in hydrogen bonding (e) all of the above are correctly matched
2. Which of the following applies to a neutron? (a) positive charge and located in an orbital (b) negligible mass and located in the nucleus (c) positive charge and located in the nucleus (d) uncharged and located in the nucleus (e) uncharged and located in an orbital
3. $^{32}_{15}P$, a radioactive form of phosphorus, has (a) an atomic number of 32 (b) an atomic mass of 15 (c) an atomic mass of 47 (d) 32 electrons (e) 17 neutrons
4. Which of the following facts allows you to determine that atom A and atom B are isotopes of the same element? (a) they each have 6 protons (b) they each have 4 neutrons (c) in each, the sum of their electrons and neutrons is 14 (d) they each have 4 valence electrons (e) they each have an atomic mass of 14
5. $^{1}_{1}H$ and $^{3}_{1}H$ have (a) different chemical properties, because they have different atomic numbers (b) the same chemical properties, because they have the same number of valence electrons (c) different chemical properties, because they differ in their number of protons and electrons (d) the same chemical properties, because they have the same atomic mass (e) the same chemical properties, because they have the same number of protons, electrons, and neutrons.
6. Sodium and potassium atoms behave similarly in chemical reactions. This is because (a) they have the same number of neutrons (b) each has a single valence electron (c) they have the same atomic mass (d) they have the same number of electrons (e) they have the same number of protons
7. The orbitals comprising an atom's valence electron shell (a) are arranged as concentric spheres (b) contain the atom's least energetic electrons (c) may change shape when covalent bonds are formed (d) never contain more than 1 electron each (e) more than one of the preceding is correct
8. Which of the following bonds and properties are correctly matched? (a) ionic bonds—strong only if the participating ions are hydrated (b) hydrogen bonds—responsible for bonding oxygen and hydrogen to form a single water molecule (c) polar covalent bonds—can occur between two atoms of the same element (d) covalent bonds—may be single, double, or triple (e) hydrogen bonds—stronger than covalent bonds
9. In a redox reaction (a) energy is transferred from a reducing agent to an oxidizing agent (b) a reducing agent becomes oxidized as it accepts an electron (c) an oxidizing agent accepts a proton (d) a reducing agent donates a proton (e) the electrons in an atom move from its valence shell to a shell closer to its nucleus
10. Water has the property of adhesion because (a) hydrogen bonds form between adjacent water molecules (b) hydrogen bonds form between water molecules and hydrophilic substances (c) it has a high specific heat (d) covalent bonds hold an individual water molecule together (e) it has a great deal of kinetic energy
11. The high heat of vaporization of water accounts for (a) evaporative cooling (b) the fact that ice floats (c) the fact that heat is liberated when ice forms (d) the cohesive properties of water (e) capillary action

POST-TEST *(continued)*

12. Water has a high specific heat because (a) hydrogen bonds must be broken to raise its temperature (b) hydrogen bonds must be formed to raise its temperature (c) it is a poor insulator (d) it has low density considering the size of the molecule (e) it can ionize
13. A solution at pH 7 is considered neutral because (a) its hydrogen ion concentration is 0 mol/L (b) its hydroxide ion concentration is 0 mol/L (c) the product of its hydrogen ion concentration and its hydroxide ion concentration is 0 mol/L (d) its hydrogen ion concentration is equal to its hydroxide ion concentration (e) it is nonpolar
14. A solution with a pH of 2 has a hydrogen ion concentration that is _______________ the hydrogen ion concentration of a solution with a pH of 4. (a) 1/2 (b) 1/100 (c) 2 times (d) 10 times (e) 100 times
15. Which of the following cannot function as a buffer? (a) phosphoric acid, a weak acid (b) sodium hydroxide, a strong base (c) sodium chloride, a salt that ionizes completely (d) a and c (e) b and c
16. NaOH and HCl react to form Na^+, Cl^-, and water. Which of the following statements is true? (a) Na^+ is an anion, and Cl^- is a cation (b) Na^+ and Cl^- are both anions (c) a hydrogen bond can form between Na^+ and Cl^- (d) Na^+ and Cl^- are electrolytes (e) Na^+ is an acid, and Cl^- is a base
17. Which of the following statements is true? (a) the number of individual particles (atoms, ions, or molecules) contained in one mole varies depending on the substance (b) Avogadro's number is the number of particles contained in one mole of a substance (c) Avogadro's number is 10^{23} particles (d) one mole of ^{12}C has a mass of 12 g (e) both b and d are true

CRITICAL THINKING

1. Element A has 2 electrons in its valence shell (which is complete when it contains 8 electrons). Would you expect element A to share, donate, or accept electrons? What would you expect of element B, which has 4 valence electrons, and element C, which has 7?
2. A hydrogen bond formed between two water molecules is only about 1/20 as strong as a covalent bond between hydrogen and oxygen. In what ways would the physical properties of water be different if these hydrogen bonds were stronger (for example, 1/10 the strength of covalent bonds)?
3. Consider the following reaction (in water).

   $$\text{Hcl} \longrightarrow H^+ + Cl^-$$

   Name the reactant(s) and product(s). Does the expression indicate the reaction is reversible? Could HCl be used as a buffer?

- Visit our Web site at **http:biology.brookscole.com/solomon7** for links to chapter-related resources on the World Wide Web. Additional on-line materials relating to this chapter can be found on our Web site.

BIOLOGY NOW RESOURCES

Active Figures

2-9: Ionic bonding

2-11: Hydrogen bonding

Preparing for an exam? Take a diagnostic test on your BiologyNow CD-ROM.

Post-Test Answers

1. c	2. d	3. e	4. a
5. b	6. b	7. c	8. d
9. a	10. b	11. a	12. a
13. d	14. e	15. e	16. d
17. e			

The Chemistry of Life: Organic Compounds

© Momatiuk Eastcott/The Image Works

This young girl is using a leaf to feed her baby brother.

CHAPTER OUTLINE

Both inorganic and organic forms of carbon occur widely in nature. Many types of organic compounds will become incorporated into the body of the baby in the photograph as he grows. **Organic compounds** are those in which carbon atoms are covalently bonded to each other to form the backbone of the molecule. Some very simple carbon compounds are considered inorganic if the carbon is not bonded to another carbon or to hydrogen. The carbon dioxide we exhale as a waste product from the breakdown of organic molecules to obtain energy is an example of inorganic carbon. Organic compounds are so named because at one time it was thought that they could be produced only by living (organic) organisms. In 1928 the German chemist Friedrich Wühler synthesized urea, a metabolic waste product. Since that time, scientists have learned to synthesize many organic molecules and have discovered organic compounds not found in any organism.

Organic compounds are extraordinarily diverse; in fact, more than 5 million have been identified. There are many reasons for this diversity. Organic compounds can be produced in a wide variety of three-dimensional (3-D) shapes. Furthermore, the carbon atom can form bonds with a greater number of different elements than any other type of atom. The addition of chemical groups containing atoms of other elements—especially oxygen, nitrogen, phosphorus, and sulfur—can profoundly change the properties of an organic molecule.

Diversity also results from the fact that a great many organic compounds found in organisms are extremely large *macromolecules,* which cells construct from simpler modular subunits. For example, protein molecules are built from smaller compounds called *amino acids.*

As you study this chapter, you will develop an understanding of the major groups of organic compounds found in organisms, including carbohydrates, lipids, proteins, and nucleic acids (DNA and RNA). Why are these compounds of central importance to all living things? The most obvious answer is that they constitute the structures of cells and tissues. However, they are also responsible for a wide range of other equally important roles as they participate in and regulate metabolic reactions, transmit information, and provide energy for life processes. ■

CARBON ATOMS AND MOLECULES

Learning Objectives

1. Describe the properties of carbon that make it the central component of organic compounds.
2. Define the term *isomer,* and distinguish among the three principal isomer types.
3. Identify the major functional groups present in organic compounds, and describe their properties.
4. Explain the relationship between polymers and macromolecules.

Carbon has unique properties that allow the formation of the carbon backbones of the large, complex molecules essential to life (Fig. 3-1). Because a carbon atom has 4 valence electrons, it can complete its valence shell by forming a total of four covalent bonds (see Fig. 2-2). Each bond can link it to another carbon atom or to an atom of a different element. Carbon is particularly well suited to serve as the backbone of a large molecule because carbon-to-carbon bonds are strong and not easily broken. However, they are not so strong that it would be impossible for cells to break them. Carbon-to-carbon bonds are not limited to single bonds (based on sharing one electron pair). Two carbon atoms can share two electron pairs with each other, forming double bonds:

$$>C=C<$$

In some compounds, triple carbon-to-carbon bonds are formed:

$$-C\equiv C-$$

As shown in Figure 3-1, **hydrocarbons,** organic compounds consisting only of carbon and hydrogen, can exist as unbranched or branched chains, or as rings. Rings and chains are joined in some compounds.

The molecules in the cell are analogous to the components of a machine. Each component has a shape that allows it to fill certain roles and to interact with other components (often with a complementary shape). Similarly, the shape of a molecule is important in determining its biological properties and function. Carbon atoms are able to link to each other and to other atoms, to produce a wide variety of 3-D molecular shapes. This is because the four covalent bonds of carbon do not form in a single plane. Instead, as discussed in Chapter 2, the valence electron orbitals become elongated and project from the carbon atom toward the corners of a tetrahedron (Fig. 3-2). The structure is highly symmetrical, with an angle of about 109.5 degrees between any two of these bonds. Keep in mind that, for simplicity, many of the figures in this book are drawn as two-dimensional (2-D) graphic representations of 3-D molecules. Even the simplest hydrocarbon chains, such as those in Figure 3-1, are not actually straight but have a 3-D zigzag structure.

Generally, there is freedom of rotation around each carbon-to-carbon single bond. This property permits organic molecules to be flexible and to assume a variety of shapes, depending on the extent to which each single bond is rotated. Double and triple

FIGURE **3-1** **Organic molecules.**

Note that each carbon atom forms four covalent bonds, producing a wide variety of shapes. **(a)** Chains. **(b)** Double bonds. **(c)** Branched chains. **(d)** Rings. **(e)** Joined rings and chains.

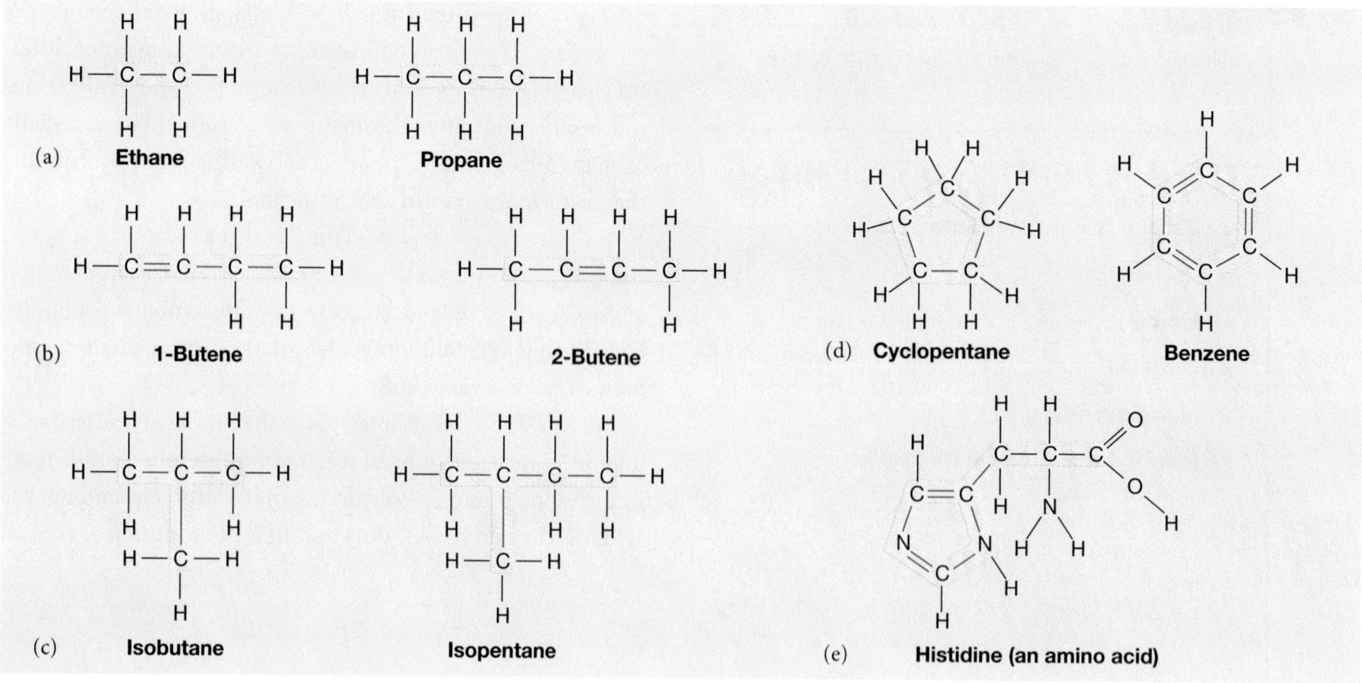

(a) **Ethane** **Propane**

(b) **1-Butene** **2-Butene**

(c) **Isobutane** **Isopentane**

(d) **Cyclopentane** **Benzene**

(e) **Histidine (an amino acid)**

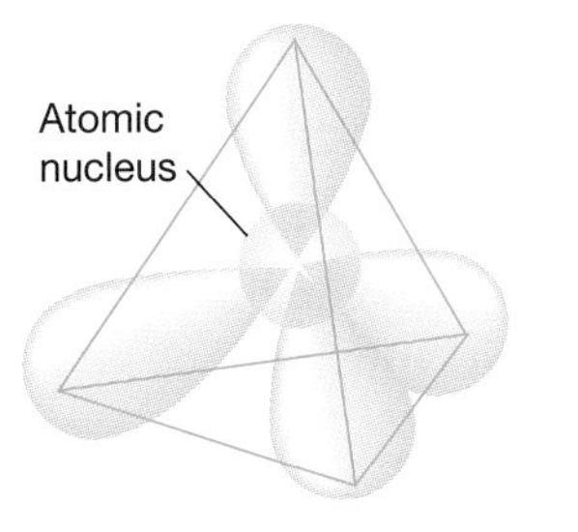

(a) Carbon (C)

(b) Methane (CH_4)

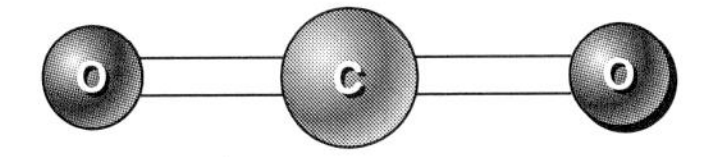

(c) Carbon dioxide (CO_2)

FIGURE **3-2** **Carbon bonding.**

(a) The 3-D arrangement of the bonds of a carbon atom is responsible for **(b)** the tetrahedral architecture of methane. **(c)** In carbon dioxide, oxygen atoms are joined linearly to a central carbon by polar double bonds.

bonds do not allow rotation, so regions of a molecule with such bonds tend to be inflexible.

Isomers have the same molecular formula, but different structures

One reason for the great number of possible carbon-containing compounds is the fact that the same components usually can link together in more than one pattern, generating an even wider variety of molecular shapes. Compounds with the same molecular formulas, but different structures and thus different properties, are called **isomers.** Isomers do not have identical physical or chemical properties and may have different common names. Cells can distinguish between isomers. Usually, one isomer is biologically active and the other is not. Three types of isomers are structural isomers, geometric isomers, and enantiomers.

Structural isomers are compounds that differ in the covalent arrangements of their atoms. For example, Figure 3-3a illustrates two structural isomers with the molecular formula C_2H_6O. Similarly, there are two structural isomers of the four-carbon hydrocarbon butane (C_4H_{10}), one with a straight chain and the other with a branched chain (isobutane). Large compounds have more possible structural isomers. There are only two structural isomers of butane, but there can be up to 366,319 isomers of $C_{20}H_{42}$.

Geometric isomers are compounds that are identical in the arrangement of their covalent bonds but different in the spatial arrangement of atoms or groups of atoms. Geometric isomers are present in some compounds with carbon-to-carbon double bonds. Because double bonds are not flexible like single bonds, atoms joined to the carbons of a double bond cannot rotate freely about the axis of the bonds. These *cis–trans* isomers may be represented as shown in Figure 3-3b. The designation *cis* (Latin, "on this side") indicates that the two larger components are on the same side of the double bond. If they are on opposite sides of the double bond, the compound is designated a *trans* (Latin, "across") isomer.

Enantiomers are molecules that are mirror images of one another (Fig. 3-3c). Recall that the four groups bonded to a single carbon atom are arranged at the vertices of a tetrahedron. If the four bonded groups are all different, the central carbon is described as asymmetrical. Figure 3-3c illustrates that the four

FIGURE **3-3** **Isomers.**

Isomers have the same molecular formula, but their atoms are arranged differently. **(a)** Structural isomers differ in the covalent arrangement of their atoms. **(b)** Geometric, or *cis–trans,* isomers have identical covalent bonds but differ in the order in which groups of atoms are arranged in space. **(c)** Enantiomers are isomers that are mirror images of one another. The central carbon is asymmetrical because it is bonded to four different groups. Because of their 3-D structure, the two figures cannot be superimposed no matter how they are rotated.

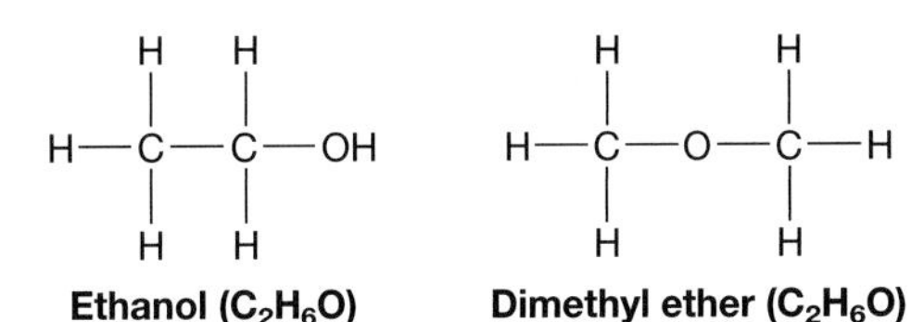

(a) Structural isomers

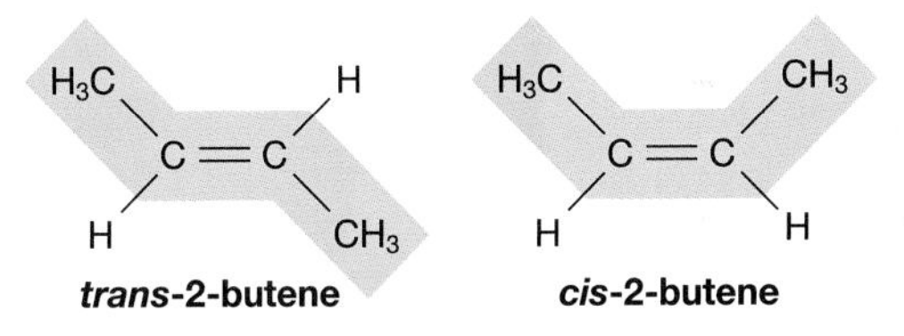

(b) Geometric isomers

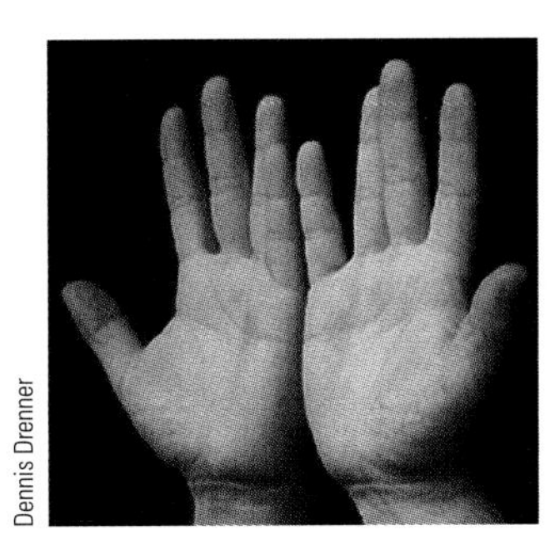

Dennis Drenner

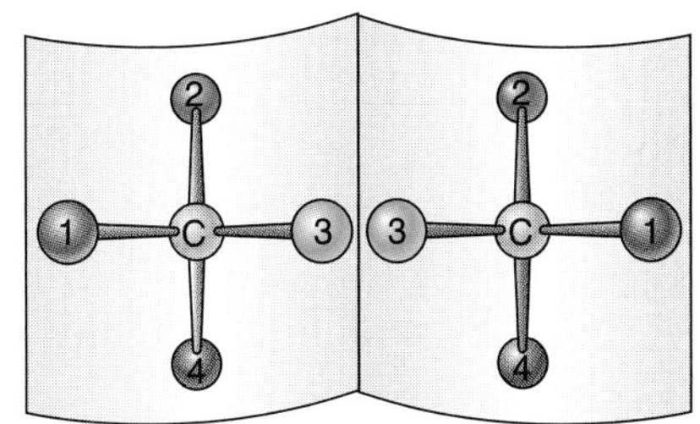

(c) Enantiomers

groups can be arranged about the asymmetrical carbon in two different ways that are mirror images of each other. The two molecules are enantiomers if they cannot be superimposed on one another no matter how they are rotated in space. Although enantiomers have similar chemical properties and most of their physical properties are identical, cells recognize the difference in shape, and usually only one form is found in organisms.

Functional groups change the properties of organic molecules

The existence of isomers is not the only source of variety among organic molecules. The addition of various combinations of atoms generates a vast array of molecules with differing properties.

TABLE 3-1 Some Biologically Important Functional Groups

Functional Group and Description	Structural Formula		Class of Compound Characterized by Group
Hydroxyl Polar because electronegative oxygen attracts covalent electrons	$R{-}OH$		Alcohols $H{-}C(H)(H){-}C(H)(H){-}OH$ Example, ethanol
Carbonyl Aldehydes: Carbonyl group carbon is bonded to at least one H atom; polar because electronegative oxygen attracts covalent electrons Ketones: Carbonyl group carbon is bonded to two other carbons; polar because electronegative oxygen attracts covalent electrons	$R{-}C(=O){-}H$ $R{-}C(=O){-}R$		Aldehydes $H{-}C(=O){-}H$ Example, formaldehyde Ketones $H{-}C(H)(H){-}C(=O){-}C(H)(H){-}H$ Example, acetone
Carboxyl Weakly acidic; can release an H^+ ion	$R{-}C(=O){-}OH$ Non-ionized	$R{-}C(=O){-}O^- + H^+$ Ionized	Carboxylic acids (organic acids) $R{-}C(=O){-}R$ Example, amino acid
Amino Weakly basic; can accept an H^+ ion	$R{-}N(H){-}H$ Non-ionized	$R{-}N^+(H)(H){-}H$ Ionized	Amines $R{-}C(NH_2)(H){-}C(=O){-}OH$ Example, amino acid
Phosphate Weakly acidic; one or two H^+ ions can be released	$R{-}O{-}P(=O)(OH){-}OH$ Non-ionized	$R{-}O{-}P(=O)(O^-){-}O^-$ Ionized	Organic Phosphates $HO{-}P(=O)(OH){-}O{-}R$ Example, phosphate ester (as found in ATP)
Sulfhydryl Helps stabilize internal structure of proteins	$R{-}SH$		Thiols $H{-}C(H)(SH){-}C(H)(NH_2){-}C(=O){-}OH$ Example, cysteine

Because covalent bonds between hydrogen and carbon are nonpolar, hydrocarbons lack distinct charged regions. For this reason, hydrocarbons are insoluble in water and tend to cluster together, through **hydrophobic** interactions. "Water fearing," the literal meaning of the term *hydrophobic,* is somewhat misleading. Hydrocarbons interact with water, but much more weakly than the water molecules cohere to each other through hydrogen bonding. Hydrocarbons interact weakly with each other, but the main reason for hydrophobic interactions is that they are driven together in a sense, having been excluded by the hydrogen-bonded water molecules. However, the characteristics of an organic molecule can be changed dramatically by replacing one of the hydrogens with one or more **functional groups,** groups of atoms that determine the types of chemical reactions and associations in which the compound participates. Most functional groups readily form associations, such as ionic and hydrogen bonds, with other molecules. Polar and ionic functional groups are **hydrophilic** because they associate strongly with polar water molecules.

The properties of the major classes of biologically important organic compounds—carbohydrates, lipids, proteins, and nucleic acids—are largely a consequence of the types and arrangement of functional groups they contain. When we know what kinds of functional groups are present in an organic compound, we can predict its chemical behavior. Note that the symbol *R* is used to represent the *remainder* of the molecule of which each functional group is a part. For example, the **methyl group,** a common nonpolar hydrocarbon group, is abbreviated $R—CH_3$. As you read the rest of this section, refer to Table 3-1 for the complete structural formulas of other important functional groups, as well as additional information.

The **hydroxyl group** (abbreviated *R*—OH) is polar because of the presence of a strongly electronegative oxygen atom. (Do not confuse it with the hydroxide ion, OH^-, discussed in Chapter 2.) If a hydroxyl group replaces one hydrogen of a hydrocarbon, the resulting molecule can have significantly altered properties. For example, ethane (see Fig. 3-1a) is a hydrocarbon that is a gas at room temperature. If a hydroxyl group replaces a hydrogen atom, the resulting molecule is ethyl alcohol, or ethanol, which is found in alcoholic beverages (Fig. 3-3a). Ethanol is somewhat cohesive, because the polar hydroxyl groups of adjacent molecules interact; it is therefore liquid at room temperature. Unlike ethane, ethyl alcohol dissolves in water because the polar hydroxyl groups interact with the polar water molecules.

The **carbonyl group** consists of a carbon atom that has a double covalent bond with an oxygen atom. This double bond is polar because of the electronegativity of the oxygen; thus the carbonyl group is hydrophilic. The position of the carbonyl group in the molecule determines the class to which the molecule belongs. An **aldehyde** has a carbonyl group positioned at the end of the carbon skeleton (abbreviated *R*—CHO); a **ketone** has an internal carbonyl group (abbreviated *R*—CO—*R*).

The **carboxyl group** (abbreviated *R*—COOH) in its non-ionized form consists of a carbon atom joined by a double covalent bond to an oxygen atom, and by a single covalent bond to another oxygen, which is in turn bonded to a hydrogen atom. Two electronegative oxygen atoms in such close proximity establish an extremely polarized condition, which can cause the hydrogen atom to be stripped of its electron and released as a hydrogen ion (H^+). The resulting ionized carboxyl group has 1 unit of negative charge ($R—COO^-$):

$$R—C(=O)—O—H \longrightarrow R—C(=O)—O^- + H^+$$

Carboxyl groups are weakly acidic; only a fraction of the molecules ionize in this way. This group therefore exists in one of two hydrophilic states: ionic or polar. Carboxyl groups are essential constituents of amino acids.

An **amino group** (abbreviated $R—NH_2$) in its non-ionized form includes a nitrogen atom covalently bonded to two hydrogen atoms. Amino groups are weakly basic because they are able to accept a hydrogen ion (proton). The resulting ionized amino group has one unit of positive charge ($R—NH_3^+$). Amino groups are components of amino acids and of nucleic acids.

A **phosphate group** (abbreviated $R—PO_4H_2$) is weakly acidic. The attraction of electrons by the oxygen atoms can result in the release of one or two hydrogen ions, producing ionized forms with one or two units of negative charge. Phosphates are constituents of nucleic acids and certain lipids.

The **sulfhydryl group** (abbreviated *R*—SH), consisting of an atom of sulfur covalently bonded to a hydrogen atom, is found in molecules called *thiols.* As you will see, amino acids that contain a sulfhydryl group can make important contributions to the structure of proteins.

Many biological molecules are polymers

Many biological molecules such as proteins and nucleic acids are very large, consisting of thousands of atoms. Such giant molecules are known as **macromolecules.** Most macromolecules are **polymers,** produced by linking small organic compounds called **monomers** (Fig. 3-4). Just as all the words in this

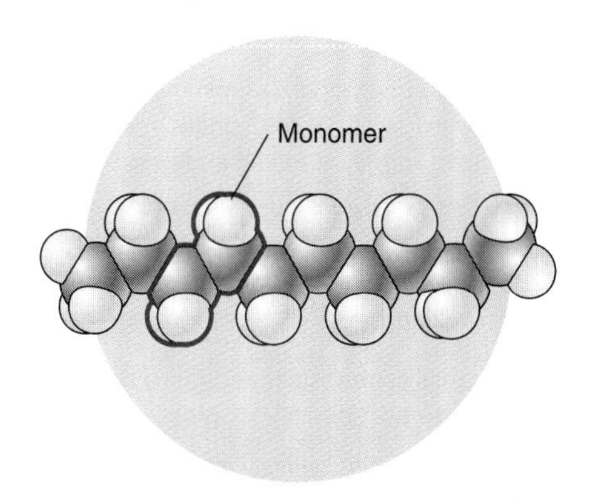

FIGURE **3-4** | **A simple polymer.**
This small polymer of polyethylene is formed by linking two-carbon ethylene (C_2H_4) monomers. One such monomer is outlined in red. The structure is represented by a space-filling model, which accurately depicts the actual 3-D shape of the molecule.

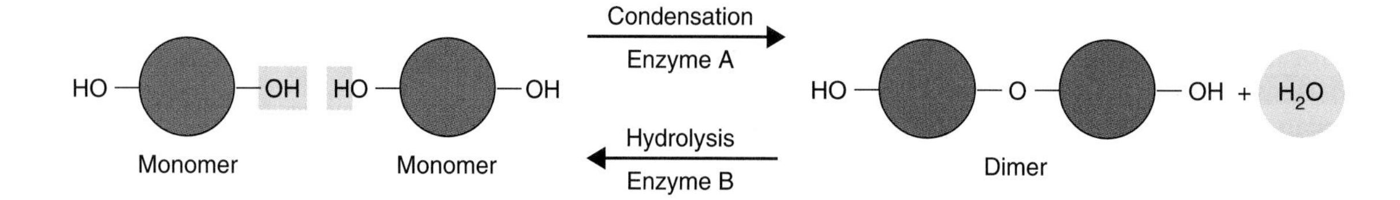

FIGURE **3-5** **Condensation and hydrolysis reactions.**

Joining two monomers yields a dimer; incorporating additional monomers produces a polymer. Note that condensation and hydrolysis reactions are catalyzed by different enzymes.

book have been written by arranging the 26 letters of the alphabet in various combinations, monomers can be grouped together to form an almost infinite variety of larger molecules. The thousands of different complex organic compounds present in organisms are constructed from about 40 small, simple monomers. For example, the 20 monomers called *amino acids* can be linked end-to-end in countless ways to form the polymers known as *proteins.*

Polymers can be degraded to their component monomers by **hydrolysis reactions** ("to break with water"). In a reaction regulated by a specific enzyme, (biological catalyst), a hydrogen from a water molecule attaches to one monomer, and a hydroxyl from water attaches to the adjacent monomer (Fig. 3-5).

Monomers are covalently linked by **condensation reactions.** Because the *equivalent* of a molecule of water is removed during the reactions that combine monomers, the term *dehydration synthesis* is sometimes used to describe condensation (see Fig. 3-5). However, in biological systems the synthesis of a polymer is not simply the reverse of hydrolysis, even though the net effect is the opposite of hydrolysis. Synthetic processes such as condensation require energy and are regulated by different enzymes.

In the following sections we examine carbohydrates, lipids, proteins, and nucleic acids. Our discussion begins with the smaller, simpler forms of these compounds and extends to the linking of these monomers to form macromolecules.

Review

- What are some of the ways that the features of carbon-to-carbon bonds influence the stability and 3-D structure of organic molecules?
- Draw pairs of simple sketches comparing two (a) structural isomers, (b) geometric isomers, and (c) enantiomers. Why are these differences biologically important?
- Sketch the following functional groups: methyl, amino, carbonyl, hydroxyl, carboxyl, and phosphate. Include both nonionized and ionized forms for acidic and basic groups.
- How is the fact that a group is nonpolar, polar, acidic, or basic related to its hydrophilic or hydrophobic properties?
- Why is the equivalent of a water molecule important to both condensation reactions and hydrolysis reactions?

BiologyNow™ Assess your understanding of **carbon atoms and molecules** by taking the pretest on your BiologyNow CD-ROM.

CARBOHYDRATES

Learning Objective

5 Distinguish among monosaccharides, disaccharides, and polysaccharides; compare storage polysaccharides with structural polysaccharides.

Sugars, starches, and cellulose are **carbohydrates.** Sugars and starches serve as energy sources for cells; cellulose is the main structural component of the walls that surround plant cells. Carbohydrates contain carbon, hydrogen, and oxygen atoms in a ratio of approximately one carbon to two hydrogens to one oxygen $(CH_2O)_n$. The term *carbohydrate,* meaning "hydrate (water) of carbon," reflects the 2:1 ratio of hydrogen to oxygen, the same ratio found in water (H_2O). Carbohydrates contain one sugar unit (monosaccharides), two sugar units (disaccharides), or many sugar units (polysaccharides).

Monosaccharides are simple sugars

Monosaccharides typically contain from three to seven carbon atoms. In a monosaccharide, a hydroxyl group is bonded to each carbon except one; that carbon is double-bonded to an oxygen atom, forming a carbonyl group. If the carbonyl group is at the end of the chain, the monosaccharide is an aldehyde; if the carbonyl group is at any other position, the monosaccharide is a ketone. (By convention, the numbering of the carbon skeleton of a sugar begins with the carbon at or nearest the carbonyl end of the open chain.) The large number of polar hydroxyl groups, plus the carbonyl group, gives a monosaccharide hydrophilic properties.

Figure 3-6 shows simplified, 2-D representations of some common monosaccharides. The simplest carbohydrates are the three-carbon sugars (trioses): glyceraldehyde and dihydroxyacetone. Ribose and deoxyribose are common pentoses, sugars that contain five carbons; they are components of nucleic acids (DNA, RNA, and related compounds). Glucose, fructose, galactose, and other six-carbon sugars are called **hexoses.** (Note that the names of carbohydrates typically end in *-ose.*)

Glucose ($C_6H_{12}O_6$), the most abundant monosaccharide, is used as an energy source in most organisms. During cellular respiration (see Chapter 7), cells oxidize glucose molecules, converting the stored energy to a form that can be readily used for cell work. Glucose is also used as a component in the synthesis of other types of compounds such as amino acids and fatty acids. Glucose is so important in metabolism that mechanisms have evolved to maintain its concentration at relatively constant levels in the blood of humans and other complex animals (see Chapter 47).

Glucose and fructose are structural isomers: They have identical molecular formulas, but their atoms are arranged differently. In fructose (a ketone) the double-bonded oxygen is linked

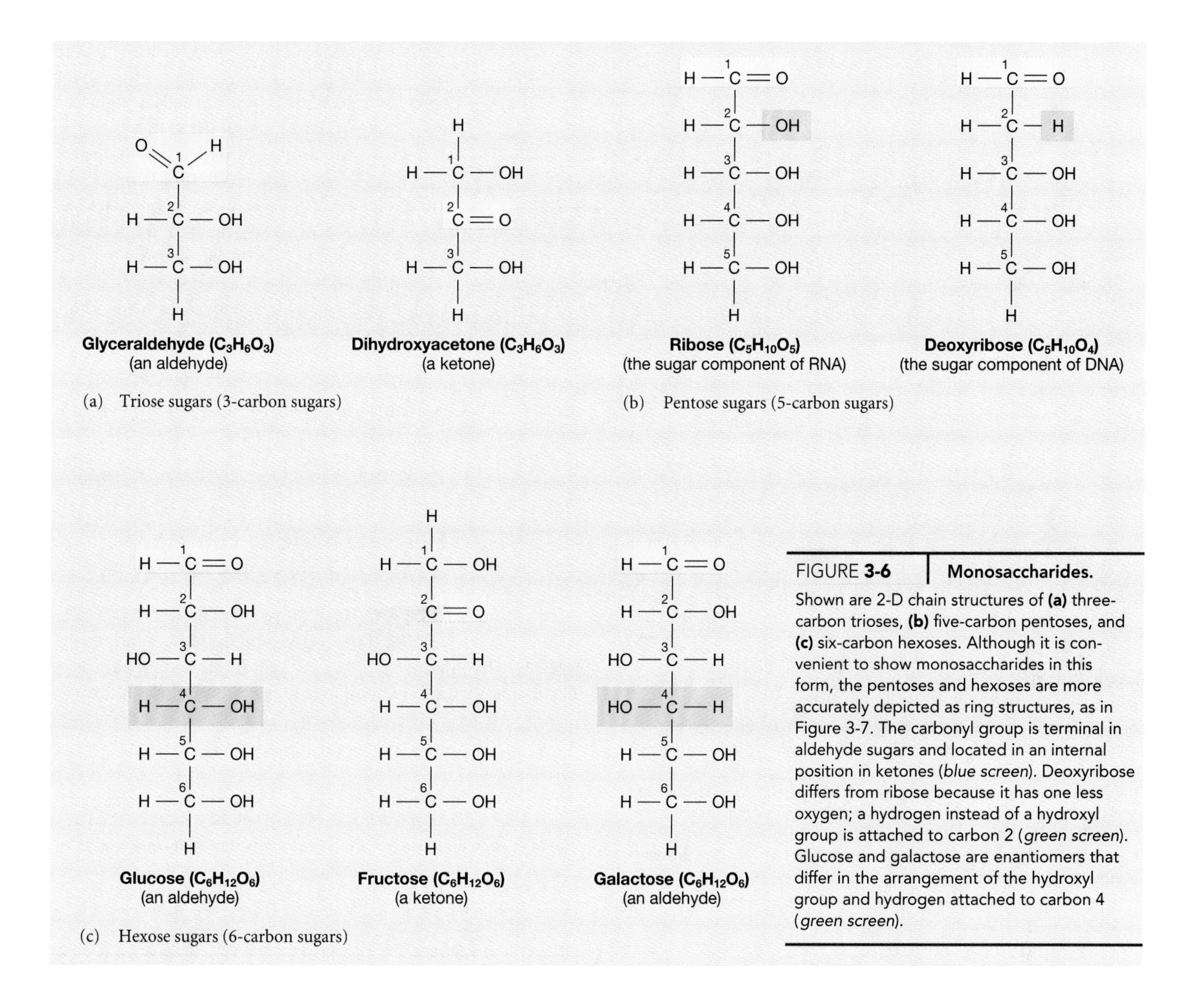

FIGURE **3-6** | **Monosaccharides.** Shown are 2-D chain structures of **(a)** three-carbon trioses, **(b)** five-carbon pentoses, and **(c)** six-carbon hexoses. Although it is convenient to show monosaccharides in this form, the pentoses and hexoses are more accurately depicted as ring structures, as in Figure 3-7. The carbonyl group is terminal in aldehyde sugars and located in an internal position in ketones (*blue screen*). Deoxyribose differs from ribose because it has one less oxygen; a hydrogen instead of a hydroxyl group is attached to carbon 2 (*green screen*). Glucose and galactose are enantiomers that differ in the arrangement of the hydroxyl group and hydrogen attached to carbon 4 (*green screen*).

to a carbon within the chain, rather than to a terminal carbon as in glucose (an aldehyde). Because of these differences, the two sugars have different properties. For example, fructose, found in honey and some fruits, tastes sweeter than glucose.

Glucose and galactose are both hexoses and aldehydes. However, they are mirror images (enantiomers) because they differ in the arrangement of the atoms attached to asymmetrical carbon atom 4.

The linear formulas in Figure 3-6 give a clear but somewhat unrealistic picture of the structures of some common monosaccharides. As we have mentioned, molecules are not 2-D; in fact, the properties of each compound depend largely on its 3-D structure. Thus, 3-D formulas are helpful in understanding the relationship between molecular structure and biological function. Molecules of glucose and other pentoses and hexoses in solution are actually rings, rather than extended straight carbon chains.

Glucose in solution (as in the cell) typically exists as a ring of five carbons and one oxygen. It assumes this configuration when its atoms undergo a rearrangement, permitting a covalent bond to connect carbon 1 to the oxygen attached to carbon 5 (Fig. 3-7). When glucose forms a ring, two isomeric forms are possible, differing only in orientation of the hydroxyl (—OH) group attached to carbon 1. When this hydroxyl group is on the same side of the plane of the ring as the —CH_2OH side group, the glucose is designated beta glucose (β-glucose). When it is on the side (with respect to the plane of the ring) opposite the —CH_2OH side group, the compound is designated alpha glucose (α-glucose). Although the differences between these isomers may seem small, they have important consequences when the rings join to form polymers.

Disaccharides consist of two monosaccharide units

A **disaccharide** (two sugars) contains two monosaccharide rings joined by a **glycosidic linkage,** consisting of a central oxygen covalently bonded to two carbons, one in each ring (Fig. 3-8). The glycosidic linkage of a disaccharide generally forms between

(a) Forms of glucose

(b) Simplified ring structure

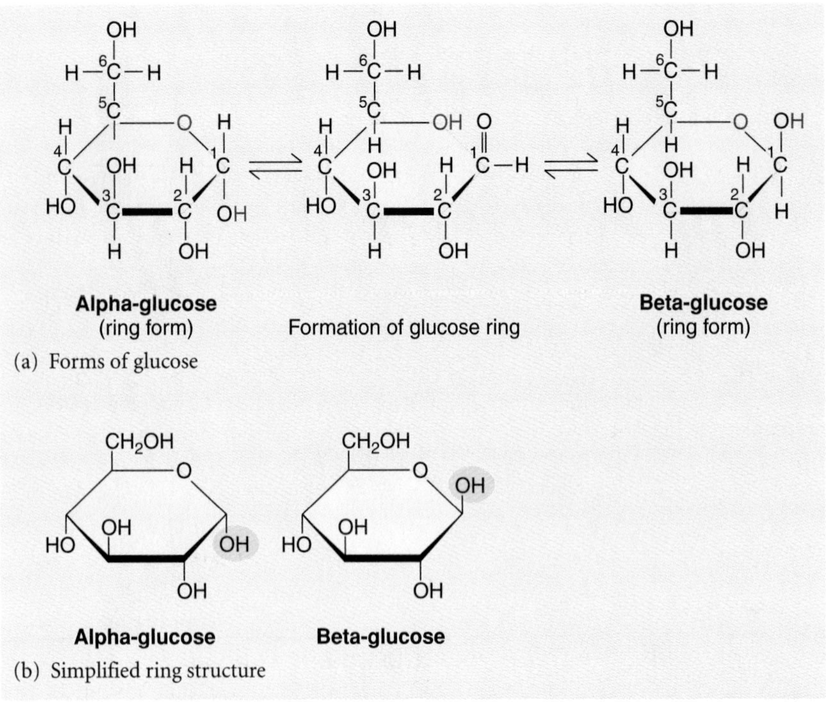

FIGURE 3-7 | α and β forms of glucose.

(a) When dissolved in water, glucose undergoes a rearrangement of its atoms, forming one of two possible ring structures: α-glucose or β-glucose. Although the drawing does not show the complete 3-D structure, the thick, tapered bonds in the lower portion of each ring represent the part of the molecule that would project out of the page toward you. **(b)** The essential differences between α-glucose and β-glucose are more readily apparent in these simplified structures. By convention, a carbon atom is assumed to be present at each angle in the ring unless another atom is shown. Most hydrogen atoms have been omitted.

carbon 1 of one molecule and carbon 4 of the other molecule. The disaccharide maltose (malt sugar) consists of two covalently linked α-glucose units. Sucrose, common table sugar, consists of a glucose unit combined with a fructose unit. Lactose (the sugar present in milk) consists of one molecule of glucose and one of galactose.

As shown in Figure 3-8, a disaccharide can be hydrolyzed, that is, split by the addition of water, into two monosaccharide units. During digestion, maltose is hydrolyzed to form two molecules of glucose:

$$\text{Maltose} + \text{water} \longrightarrow \text{glucose} + \text{glucose}$$

Similarly, sucrose is hydrolyzed to form glucose and fructose:

$$\text{Sucrose} + \text{water} \longrightarrow \text{glucose} + \text{fructose}$$

Polysaccharides can store energy or provide structure

A **polysaccharide** is a macromolecule consisting of repeating units of simple sugars, usually glucose. The polysaccharides are the most abundant carbohydrates and include starches, glycogen, and cellulose. Although

FIGURE 3-8 | Hydrolysis of disaccharides.

(a) Maltose may be broken down (as during digestion) to form two molecules of glucose. The glycosidic linkage is broken in a hydrolysis reaction, which requires the addition of water. **(b)** Sucrose can be hydrolyzed to yield a molecule of glucose and a molecule of fructose. Note that an enzyme is needed to promote these reactions.

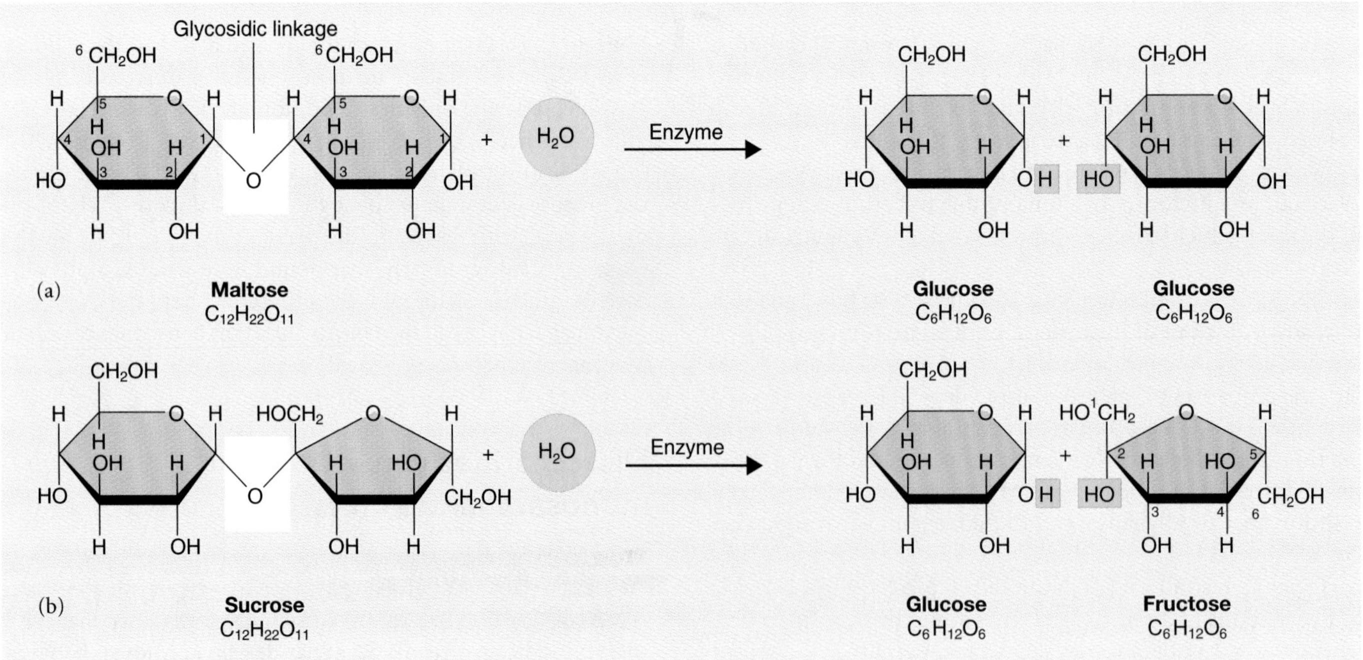

the precise number of sugar units varies, thousands of units are typically present in a single molecule. A polysaccharide may be a single long chain or a branched chain. Because they are composed of different isomers and because the units may be arranged differently, polysaccharides vary in their properties. Those that can be easily broken down to their subunits are well suited for energy storage, whereas the macromolecular 3-D architecture of others makes them particularly well suited to form stable structures.

Starch, the typical form of carbohydrate used for energy storage in plants, is a polymer consisting of α-glucose subunits. These monomers are joined by α 1—4 linkages, which means that carbon 1 of one glucose is linked to carbon 4 of the next glucose in the chain (Fig. 3-9). Starch occurs in two forms: amylose and amylopectin. Amylose, the simpler form, is unbranched. Amylopectin, the more common form, usually consists of about 1000 glucose units in a branched chain.

Plant cells store starch mainly as granules within specialized organelles called **amyloplasts** (Fig. 3-9a); some cells, such as those of potatoes, are very rich in amyloplasts. Virtually all organisms have enzymes that can break α 1—4 linkages. When energy is needed for cell work, the plant hydrolyzes the starch, releasing the glucose subunits. Humans and other animals that eat plant foods also have enzymes to hydrolyze starch.

Glycogen (sometimes referred to as *animal starch*) is the form in which glucose subunits, joined by α 1—4 linkages, are stored as an energy source in animal tissues. Glycogen is similar in structure to plant starch but more extensively branched and more water soluble. Glycogen is stored mainly in liver and muscle cells.

Carbohydrates are the most abundant group of organic compounds on Earth, and **cellulose** is the most abundant carbohydrate; it accounts for 50% or more of all the carbon in plants (Fig. 3-10). Cellulose is a structural carbohydrate. Wood is about half cellulose, and cotton is at least 90% cellulose. Plant cells are surrounded by strong supporting cell walls consisting mainly of cellulose.

Cellulose is an insoluble polysaccharide composed of many glucose molecules joined together. The bonds joining these sugar units are different from those in starch. Recall that starch is composed of α-glucose subunits, joined by α 1—4 glycosidic linkages. Cellulose contains β-glucose monomers joined by β 1—4 linkages. These bonds cannot be split by the enzymes that hydrolyze the α linkages in starch. Because humans, like other animals, lack enzymes that digest cellulose, we cannot use it as a nutrient. The cellulose found in whole grains and vegetables remains fibrous and provides bulk that helps keep our digestive tract functioning properly.

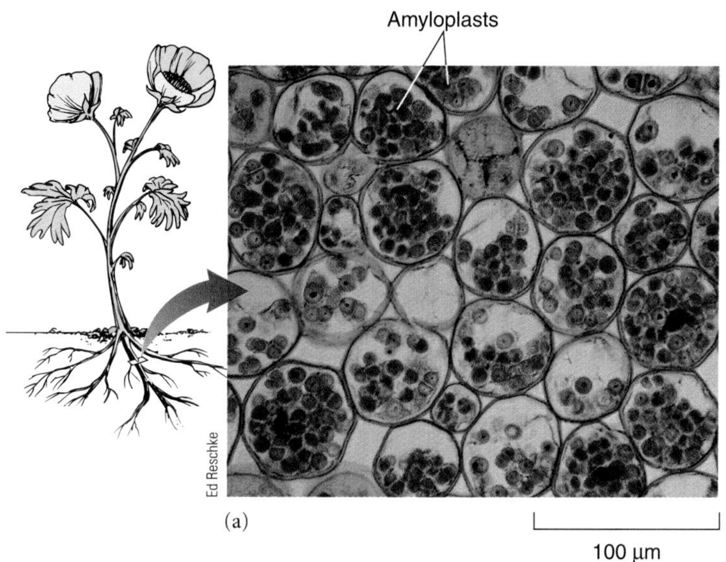

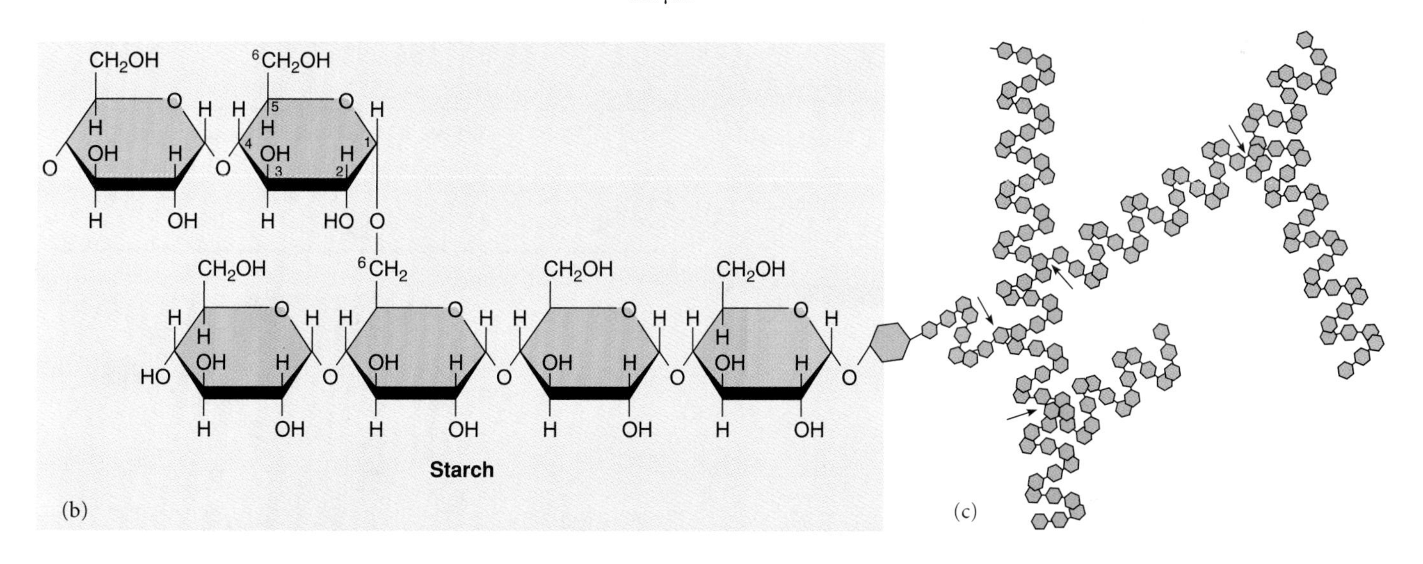

FIGURE **3-9** **Starch, a storage polysaccharide.**

(a) Starch (*stained purple*) is stored in specialized organelles, called *amyloplasts,* in these cells of a buttercup root. **(b)** Starch is composed of α-glucose molecules joined by glycosidic bonds. At the branch points are bonds between carbon 6 of the glucose in the straight chain and carbon 1 of the glucose in the branching chain. **(c)** Starch consists of highly branched chains; the arrows indicate the branch points. Each chain is actually a coil or helix, stabilized by hydrogen bonds between the hydroxyl groups of the glucose subunits.

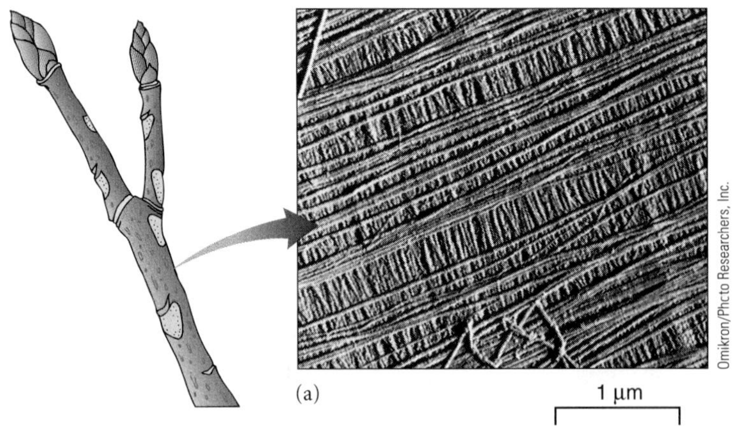

(b) Cellulose

FIGURE 3-10 **Cellulose, a structural polysaccharide.**
(a) An electron micrograph of cellulose fibers from a cell wall. The fibers consist of bundles of cellulose molecules, interacting through hydrogen bonds. **(b)** The cellulose molecule is an unbranched polysaccharide consisting of about 10,000 β-glucose units joined by glycosidic bonds.

Some microorganisms digest cellulose to glucose. In fact, cellulose-digesting bacteria live in the digestive systems of cows and sheep, enabling these grass-eating animals to obtain nourishment from cellulose. Similarly, the digestive systems of termites contain microorganisms that digest cellulose (see Fig. 24-5b).

Cellulose molecules are well suited for a structural role. The β-glucose subunits are joined in a way that allows extensive hydrogen bonding among different cellulose molecules, and they aggregate in long bundles of fibers (Fig. 3-10a).

Some modified and complex carbohydrates have special roles

Many derivatives of monosaccharides are important biological molecules. Some form important structural components. The amino sugars galactosamine and glucosamine are compounds in which a hydroxyl group (—OH) is replaced by an amino group ($—NH_2$). Galactosamine is present in cartilage, a constituent of the skeletal system of vertebrates. *N*-acetyl glucosamine (NAG) subunits, joined by glycosidic bonds, compose **chitin,** a main component of the cell walls of fungi and of the external skeletons of insects, crayfish, and other arthropods (Fig. 3-11). Chitin forms very tough structures because, as in cellulose, its molecules interact through multiple hydrogen bonds. Some chitinous structures, such as the shell of a lobster, are further hardened by the addition of calcium carbonate ($CaCO_3$, an inorganic form of carbon).

Carbohydrates may also combine with proteins to form **glycoproteins,** compounds present on the outer surface of cells other than bacteria. Some of these carbohydrate chains allow

FIGURE **3-11** **Chitin, a structural polysaccharide.**
(a) Chitin is a polymer composed of *N*-acetyl glucosamine subunits. **(b)** Chitin is an important component of the exoskeleton (outer covering) this dragonfly is shedding.

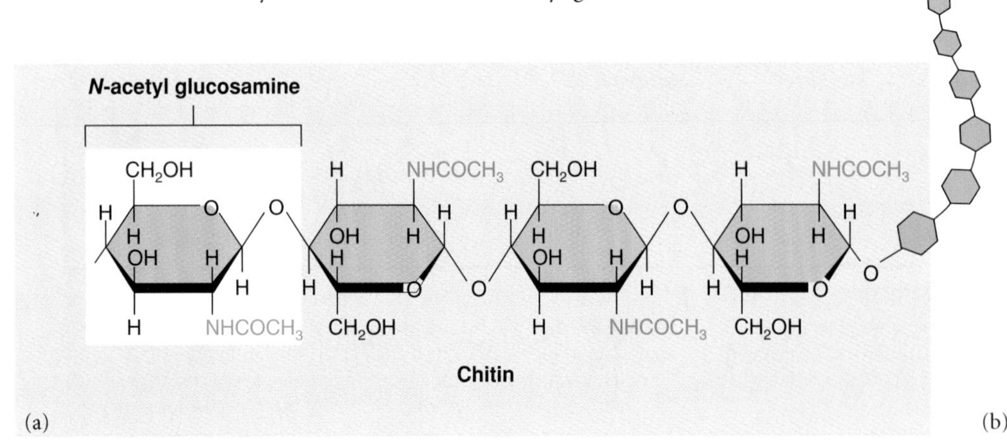

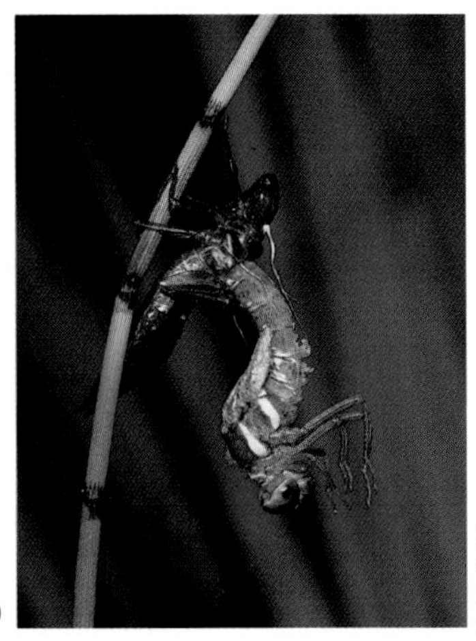

cells to adhere to one another, whereas others provide protection. Most proteins secreted by cells are glycoproteins. These include the major components of mucus, a complex protective material secreted by the mucous membranes of the respiratory and digestive systems. Carbohydrates combine with lipids to form **glycolipids,** compounds on the surfaces of animal cells that allow cells to recognize and interact with one another.

Review

- What features related to hydrogen bonding give storage polysaccharides, such as starch and glycogen, very different properties from structural polysaccharides, such as cellulose and chitin?
- Why can't humans digest cellulose?

BiologyNow™ Assess your understanding of **carbohydrates** by taking the pretest on your BiologyNow CD-ROM.

LIPIDS

Learning Objective

6 Distinguish among fats, phospholipids, and steroids, and describe the composition, characteristics, and biological functions of each.

Unlike carbohydrates, which are defined by their structure, **lipids** are a heterogeneous group of compounds that are categorized by the fact that they are soluble in nonpolar solvents (such as ether and chloroform) and are relatively insoluble in water. Lipid molecules have these properties because they consist mainly of carbon and hydrogen, with few oxygen-containing functional groups. Hydrophilic functional groups typically contain oxygen atoms; therefore lipids, with little oxygen, tend to be hydrophobic. Among the biologically important groups of lipids are fats, phospholipids, carotenoids (orange and yellow plant pigments), steroids, and waxes. Some lipids are used for energy storage, others serve as structural components of cellular membranes, and some are important hormones.

Triacylglycerol is formed from glycerol and three fatty acids

The most abundant lipids in living organisms are triacylglycerols. These compounds, commonly known as *fats,* are an economical form of reserve fuel storage because, when metabolized, they yield more than twice as much energy per gram as do carbohydrates. Carbohydrates and proteins can be transformed by enzymes into fats and stored within the cells of adipose (fat) tissue of animals and in some seeds and fruits of plants.

A **triacylglycerol** molecule (also known as a *triglyceride*) consists of glycerol joined to three fatty acids (Fig. 3-12). **Glycerol** is a three-carbon alcohol that contains three hydroxyl (—OH) groups, and a **fatty acid** is a long, unbranched hydrocarbon chain with a carboxyl group (—COOH) at one end. A triacylglycerol molecule is formed by a series of three condensation reactions. In each reaction, the equivalent of a water molecule is removed as one of the glycerol's hydroxyl groups reacts with the carboxyl group of a fatty acid, resulting in the formation of a covalent linkage known as an **ester linkage** (Fig. 3-12b). The first reaction yields a **monoacylglycerol** (*monoglyceride*); the second, a diacylglycerol (*diglyceride*); and the third, a triacylglycerol. During digestion triacylglycerols are hydrolyzed to produce fatty acids and glycerol (see Chapter 45). Diacylglycerol is an important molecule for sending signals within the cell (see Chapter 47).

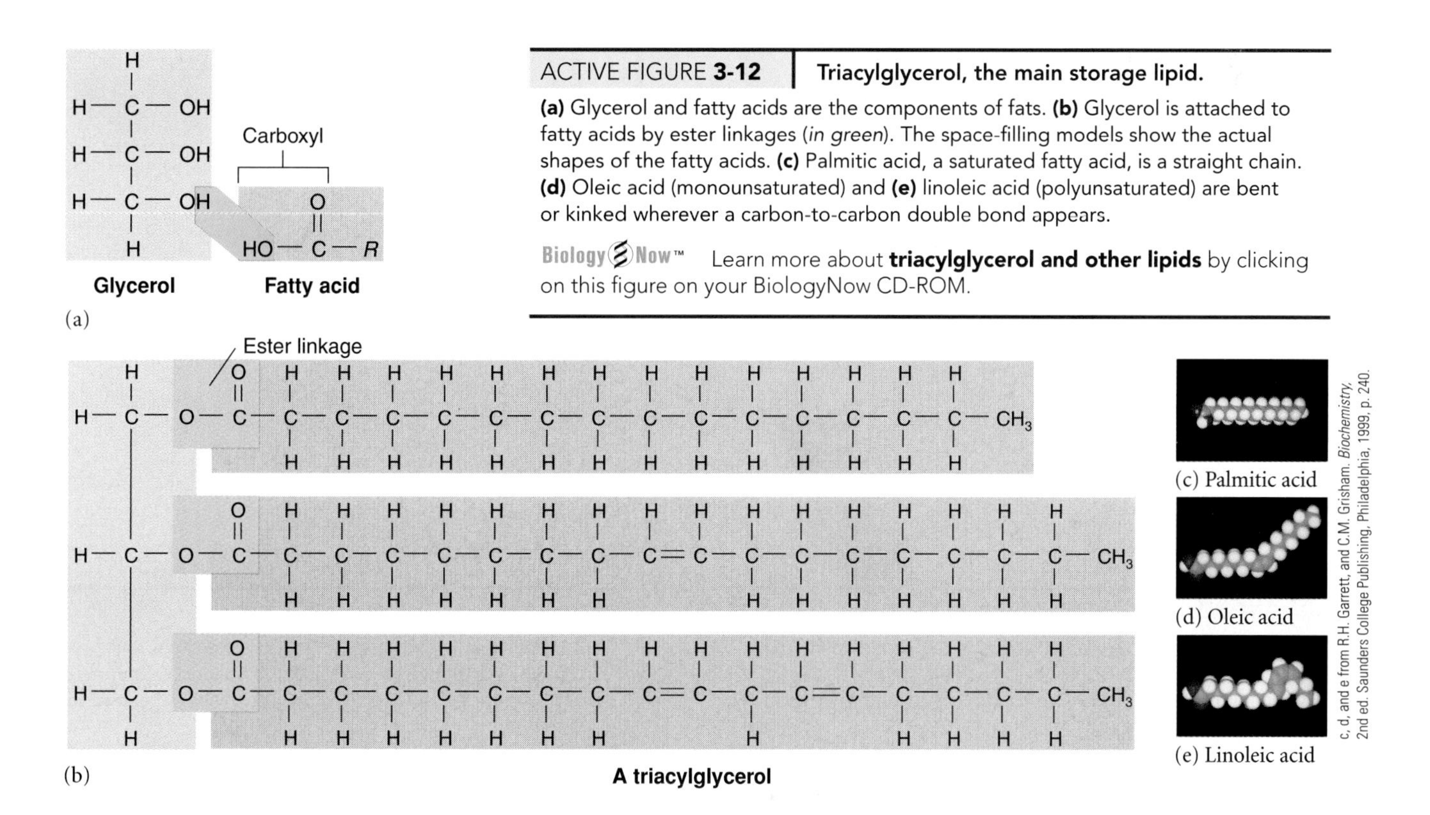

ACTIVE FIGURE 3-12 **Triacylglycerol, the main storage lipid.**

(a) Glycerol and fatty acids are the components of fats. **(b)** Glycerol is attached to fatty acids by ester linkages (*in green*). The space-filling models show the actual shapes of the fatty acids. **(c)** Palmitic acid, a saturated fatty acid, is a straight chain. **(d)** Oleic acid (monounsaturated) and **(e)** linoleic acid (polyunsaturated) are bent or kinked wherever a carbon-to-carbon double bond appears.

BiologyNow™ Learn more about **triacylglycerol and other lipids** by clicking on this figure on your BiologyNow CD-ROM.

c, d, and e from R.H. Garrett, and C.M. Grisham. *Biochemistry,* 2nd ed. Saunders College Publishing, Philadelphia, 1999, p. 240.

Saturated and unsaturated fatty acids differ in physical properties

About 30 different fatty acids are commonly found in lipids, and they typically have an even number of carbon atoms. For example, butyric acid, present in rancid butter, has four carbon atoms. Oleic acid, with 18 carbons, is the most widely distributed fatty acid in nature and is found in most animal and plant fats.

Saturated fatty acids contain the maximum possible number of hydrogen atoms. Palmitic acid, a 16-carbon fatty acid, is a common saturated fatty acid (Fig. 3-12c). Fats high in saturated fatty acids, such as animal fat and solid vegetable shortening, tend to be solid at room temperature. This is because even electrically neutral, nonpolar molecules can develop transient regions of weak positive charge and weak negative charge. This occurs as the constant motion of their electrons causes some regions to have a temporary excess of electrons, whereas others have a temporary electron deficit. These slight opposite charges result in attractions, known as **van der Waals interactions,** between adjacent molecules. Although van der Waals interactions are individually weak, they can be strong when many occur among long hydrocarbon chains.

Unsaturated fatty acids include one or more adjacent pairs of carbon atoms joined by a double bond. Therefore they are not fully saturated with hydrogen. Fatty acids with one double bond are **monounsaturated fatty acids,** whereas those with more than one double bond are **polyunsaturated fatty acids.** Oleic acid is a monounsaturated fatty acid, and linoleic acid is a common polyunsaturated fatty acid (Fig. 3-12d, e). Fats containing a high proportion of monounsaturated or polyunsaturated fatty acids tend to be liquid at room temperature. This is because each double bond produces a bend in the hydrocarbon chain that prevents it from aligning closely with an adjacent chain, thereby limiting van der Waals interactions.

Food manufacturers commonly hydrogenate or partially hydrogenate cooking oils to make margarine and other foodstuffs, converting unsaturated fatty acids to saturated fatty acids and making the fat more solid at room temperature. This process makes the fat less healthful because saturated fatty acids in the diet are known to increase the risk of cardiovascular disease (see Chapter 42). The hydrogenation process has yet another effect. Note that in the naturally occurring unsaturated fatty acids oleic acid and linoleic acid shown in Figure 3-12, the two hydrogens flanking each double bond are on the same side of

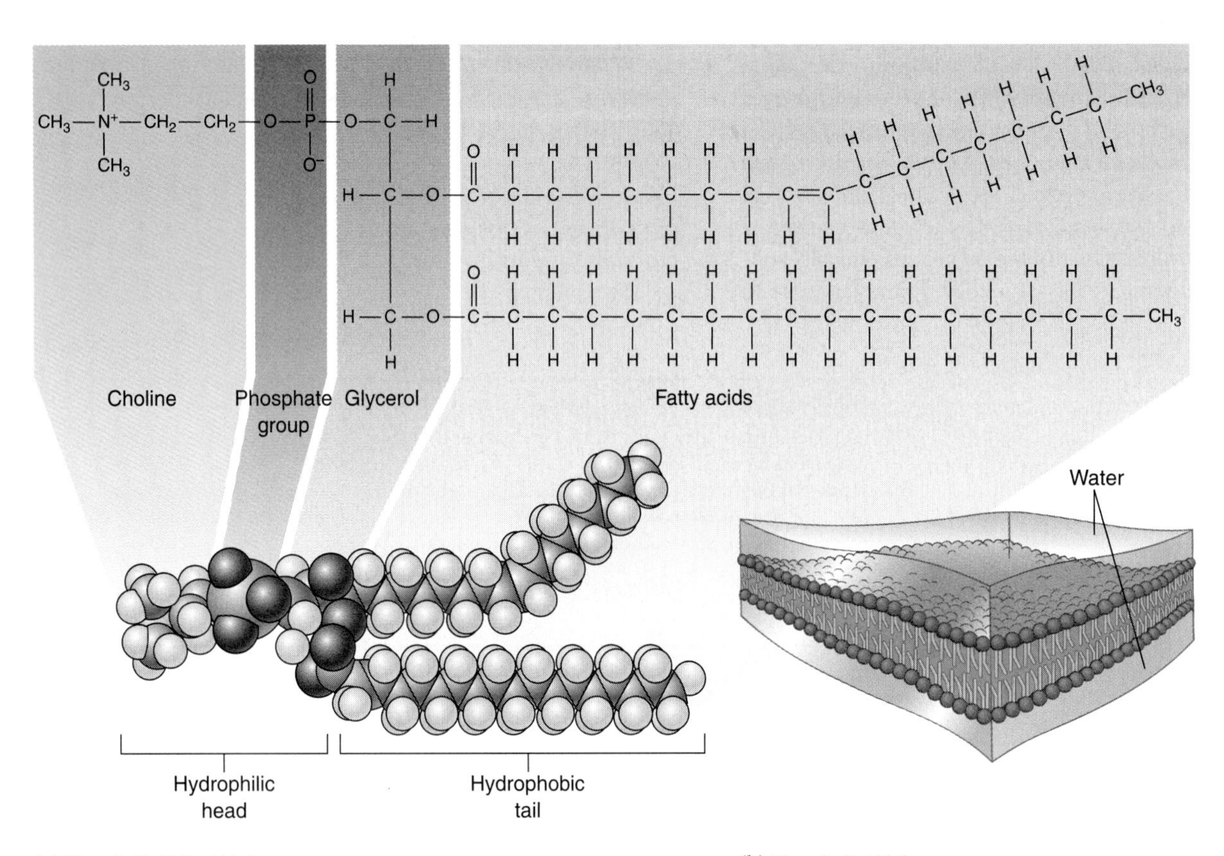

(a) Phospholipid (lecithin)

(b) Phospholipid bilayer

FIGURE **3-13** **A phospholipid and a phospholipid bilayer.**

(a) A phospholipid consists of a hydrophobic tail, made up of two fatty acids, and a hydrophilic head, which includes a glycerol bonded to a phosphate group, which is in turn bonded to an organic group that can vary. Choline is the organic group in lecithin (or phosphatidylcholine), the molecule shown. The fatty acid at the top of the figure is monounsaturated; it contains one double bond that produces a characteristic bend in the chain. **(b)** Phospholipids form lipid bilayers in which the hydrophilic heads interact with water and the hydrophobic tails are in the bilayer interior.

the hydrocarbon chain (the *cis* configuration). When fatty acids are artificially hydrogenated, the double bonds can become rearranged, resulting in a *trans* configuration, analogous to the arrangement shown in Fig. 3-3b. *Trans* fatty acids are technically unsaturated, but they mimic many of the properties of saturated fatty acids. Because the *trans* configuration does not produce a bend at the site of the double bond, *trans* fatty acids are more solid at room temperature and, like saturated fatty acids, they increase the risk of cardiovascular disease.

At least two unsaturated fatty acids (linoleic acid and arachidonic acid) are essential nutrients that must be obtained from food because the human body cannot synthesize them. However, the amounts required are small, and deficiencies are rarely seen. There is no dietary requirement for saturated fatty acids.

Phospholipids are components of cell membranes

Phospholipids belong to a group of lipids, called **amphipathic lipids,** in which one end of each molecule is hydrophilic and the other end is hydrophobic (Fig. 3-13). The two ends of a phospholipid differ both physically and chemically. A **phospholipid** consists of a glycerol molecule attached at one end to two fatty acids, and at the other end to a phosphate group linked to an organic compound such as choline. The organic compound usually contains nitrogen. (Note that phosphorus and nitrogen are absent in triacylglycerols as shown in Fig. 3-12b.) The fatty acid portion of the molecule (containing the two hydrocarbon "tails") is hydrophobic and not soluble in water. However, the portion composed of glycerol, phosphate, and the organic base (the "head" of the molecule) is ionized and readily water soluble. The amphipathic properties of phospholipids cause them to form lipid bilayers in aqueous (watery) solution. Thus they are uniquely suited to function as the fundamental components of cell membranes (see Chapters 4 and 5).

Carotenoids and many other pigments are derived from isoprene units

The orange and yellow plant pigments called **carotenoids** are classified with the lipids because they are insoluble in water and have an oily consistency. These pigments, found in the cells of all plants, play a role in photosynthesis. Carotenoid molecules, such as β-carotene, and many other important pigments, consist of five-carbon hydrocarbon monomers known as **isoprene units** (Fig. 3-14).

Most animals convert carotenoids to vitamin A, which can then be converted to the visual pigment **retinal.** Three different groups of animals—the mollusks, insects, and vertebrates—have eyes and use retinal in the process of light reception.

Notice that carotenoids, vitamin A, and retinal all have a pattern of double bonds alternating with single bonds. The electrons that make up these bonds can move about relatively easily when light strikes the molecule. Such molecules are pigments; they tend to be highly colored because the mobile electrons cause them to strongly absorb light of certain wavelengths and reflect light of other wavelengths.

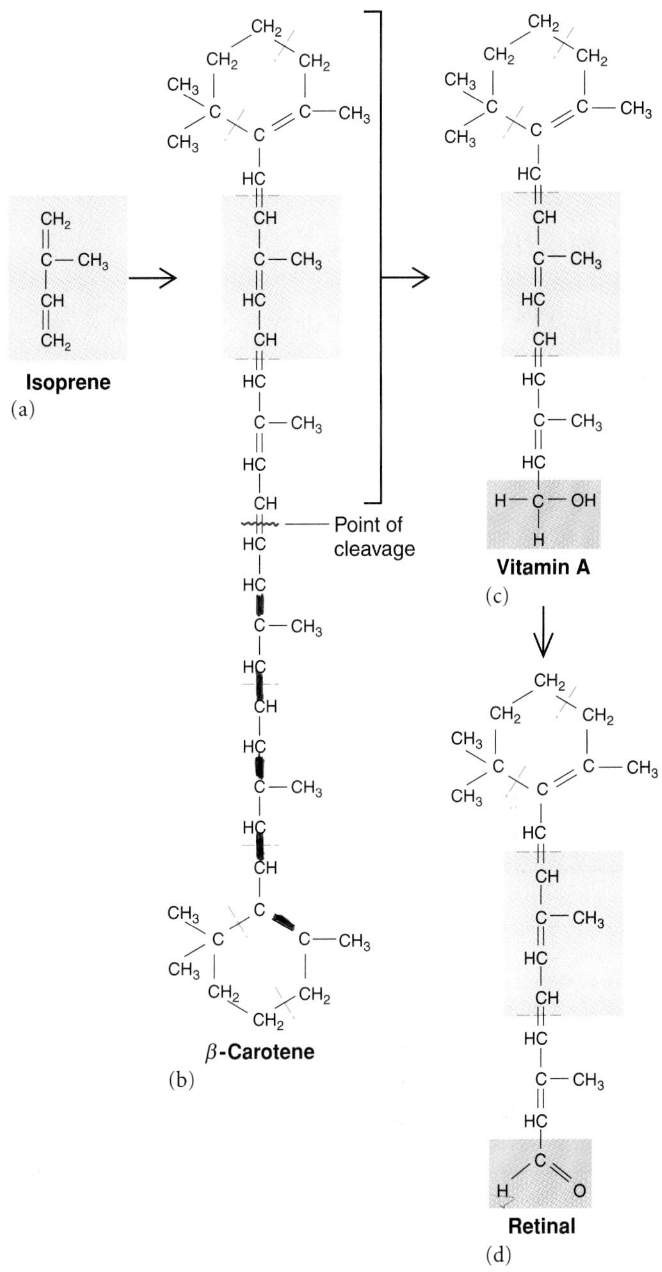

FIGURE **3-14** **Isoprene-derived compounds.**
(a) An isoprene subunit. **(b)** β-carotene, with dashed lines indicating the boundaries of the individual isoprene units within. The wavy line is the point at which most animals cleave the molecule to yield two molecules of **(c)** vitamin A. Vitamin A is converted to the visual pigment **(d)** retinal.

Steroids contain four rings of carbon atoms

A **steroid** consists of carbon atoms arranged in four attached rings; three of the rings contain six carbon atoms, and the fourth contains five (Fig. 3-15). The length and structure of the side chains that extend from these rings distinguish one steroid from another. Like carotenoids, steroids are synthesized from isoprene units.

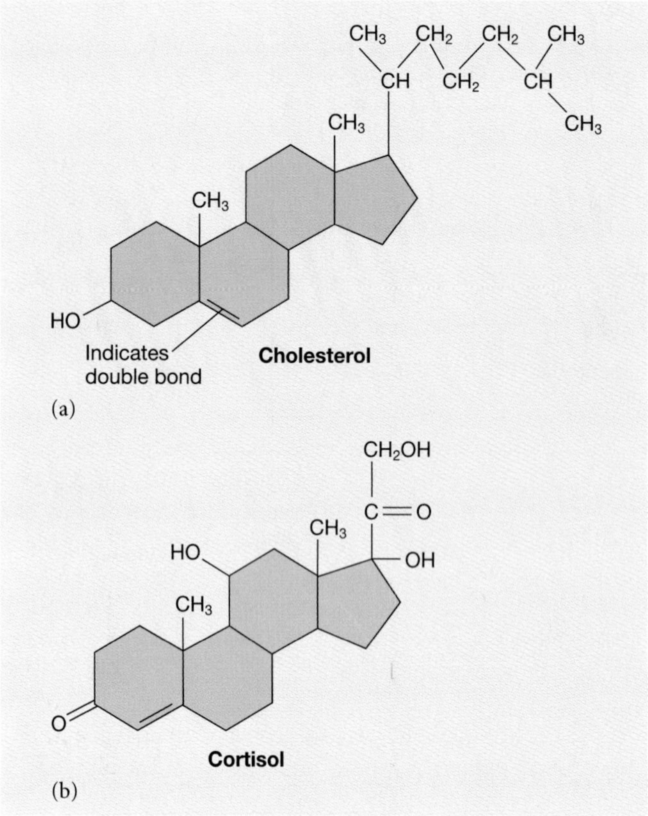

FIGURE 3-15 | **Steroids.**

Four attached rings—three six-carbon rings and one with five carbons—make up the fundamental structure of a steroid *(shown in green).* Note that some carbons are shared by two rings. In these simplified structures, a carbon atom is present at each angle of a ring; the hydrogen atoms attached directly to the carbon atoms have not been drawn. **(a)** Cholesterol is an essential component of animal cell membranes. **(b)** Cortisol is a steroid hormone secreted by the adrenal glands. Cortisol differs from cholesterol in its attached functional groups.

Among the steroids of biological importance are cholesterol, bile salts, reproductive hormones, and cortisol and other hormones secreted by the adrenal cortex. Cholesterol is an essential structural component of animal cell membranes, but when excess cholesterol in blood forms plaques on artery walls, it leads to an increased risk of cardiovascular disease (see Chapter 42). Plant cell membranes contain molecules similar to cholesterol. Interestingly, some of these plant steroids are able to block the intestine's absorption of cholesterol. Bile salts emulsify fats in the intestine so they can be enzymatically hydrolyzed. Steroid hormones regulate certain aspects of metabolism in a variety of animals, including vertebrates, insects, and crabs.

Some chemical mediators are lipids

Animal cells secrete chemicals to communicate with each other or to regulate their own activities. Some chemical mediators are produced by the modification of fatty acids that have been removed from membrane phospholipids. These include *prostaglandins,* which have varied roles, including promoting inflammation and smooth muscle contraction. Certain hormones, such as the juvenile hormone of insects, are also fatty acid derivatives (see Chapter 47).

Review

- Why do saturated, unsaturated, and *trans* fatty acids differ in their properties?
- Why do phospholipids form lipid bilayers in aqueous conditions?

BiologyNow™ Assess your understanding of **lipids** by taking the pretest on your BiologyNow CD-ROM.

PROTEINS

Learning Objectives

7 Give an overall description of the structure and functions of proteins.

8 Describe the features that are shared by all amino acids, and explain how amino acids are grouped into classes based on the characteristics of their side chains.

9 Distinguish among the four levels of organization of protein molecules.

Proteins, macromolecules composed of amino acids, are the most versatile cell components. As discussed in Chapter 15, scientists have succeeded in sequencing virtually all the genetic information in a human cell, and the genetic information of many other organisms is being studied. Some people might think that the sequencing of genes is the end of the story, but it is actually only the beginning. Most genetic information is used to specify the structure of proteins, and it has been predicted that most of the 21st century will be devoted to understanding this extraordinarily multifaceted group of macromolecules that are of central importance in the chemistry of life. In a real sense, proteins are involved in virtually all aspects of metabolism because most **enzymes** (molecules that accelerate the thousands of different chemical reactions that take place in an organism) are proteins. Proteins are assembled into a variety of shapes, allowing them to serve as major structural components of cells and tissues. For this reason, growth and repair, as well as maintenance of the organism, depend on proteins. As shown in Table 3-2, proteins perform many other specialized functions.

The protein constituents of a cell are the clues to its lifestyle. Each cell type contains characteristic forms, distributions, and amounts of protein that largely determine what the cell looks like and how it functions. A muscle cell contains large amounts of the proteins myosin and actin, which are responsible for its appearance as well as its ability to contract. The protein hemoglobin, found in red blood cells, is responsible for the specialized function of oxygen transport.

Amino acids are the subunits of proteins

Amino acids, the constituents of proteins, have an amino group ($—NH_2$) and a carboxyl group ($—COOH$) bonded to the same asymmetrical carbon atom, known as the **alpha carbon.** Twenty amino acids are commonly found in proteins, each uniquely

TABLE 3-2	Major Classes of Proteins and Their Functions
Protein Class	**Functions and Examples**
Enzymes	Catalyze specific chemical reactions
Structural proteins	Strengthen and protect cells and tissues (e.g., collagen strengthens animal tissues)
Storage proteins	Store nutrients; particularly abundant in eggs (e.g., ovalbumin in egg white) and seeds (e.g., zein in corn kernels)
Transport proteins	Transport specific stubstances between cells (e.g., hemoglobin transports oxygen in red blood cells; move specific substances (e.g., ions, glucose, amino acids) across cell membranes
Regulatory proteins	Some are protein hormones (e.g., insulin); some control the expression of specific genes
Motile proteins	Participate in cellular movements (e.g., actin and myosin are essential for muscle contraction)
Protective proteins	Defend against foreign invaders (e.g., antibodies play a role in the immune system)

identified by the variable side chain (*R* group) bonded to the α carbon (Fig. 3-16). Glycine, the simplest amino acid, has a hydrogen atom as its R group; alanine has a methyl ($—CH_3$) group.

Amino acids in solution at neutral pH are mainly dipolar ions. This is generally how amino acids exist at cell pH. Each carboxyl group (—COOH) donates a proton and becomes ionized ($—COO^-$), whereas each amino group ($—NH_2$) accepts a proton and becomes $—NH_3^+$ (Fig. 3-17). Because of the ability of their amino and carboxyl groups to accept and release protons, amino acids in solution resist changes in acidity and alkalinity and therefore are important biological buffers.

The amino acids are grouped in Figure 3-16 by the properties of their side chains. These broad groupings actually include amino acids with a fairly wide range of properties. Amino acids classified as having *nonpolar* side chains tend to have hydrophobic properties, whereas those classified as *polar* are more hydrophilic. An acidic amino acid has a side chain that contains a carboxyl group. At cell pH the carboxyl group is dissociated, giving the R group a negative charge. A basic amino acid becomes positively charged when the amino group in its side chain accepts a hydrogen ion. Acidic and basic side chains are ionic at cell pH and therefore hydrophilic.

In addition to the 20 common amino acids, some proteins have unusual ones. These rare amino acids are produced by the modification of common ones after they have become part of a protein. For example, after they have been incorporated into collagen, lysine and proline may be converted to hydroxylysine and hydroxyproline. These amino acids can form cross links between the peptide chains that make up collagen. Such cross links produce the firmness and great strength of the collagen molecule, which is a major component of cartilage, bone, and other connective tissues.

With some exceptions, bacteria and plants synthesize all their needed amino acids from simpler substances. If the proper raw materials are available, the cells of animals can manufacture some, but not all, of the biologically significant amino acids. **Essential amino acids** are those an animal cannot synthesize in amounts sufficient to meet its needs and must obtain from the diet. Animals differ in their biosynthetic capacities; what is an essential amino acid for one species may not be for another. The essential amino acids for humans are isoleucine, leucine, lysine, methionine, phenylalanine, threonine, tryptophan, valine, and histidine. For children arginine is added to the list because they do not synthesize enough to support growth.

Peptide bonds join amino acids

Amino acids combine chemically with one another by a condensation reaction that bonds the carboxyl carbon of one molecule to the amino nitrogen of another (Fig. 3-18). The covalent carbon-to-nitrogen bond linking two amino acids together is a **peptide bond.** When two amino acids combine, a **dipeptide** is formed; a longer chain of amino acids is a **polypeptide.** A protein consists of one or more polypeptide chains. Each polypeptide has a free amino group at one end and a free carboxyl group (belonging to the last amino acid added to the chain) at the opposite end. The other amino and carboxyl groups of the amino acid monomers (except those in side chains) are part of the peptide bonds. The complex process by which polypeptides are synthesized is discussed in Chapter 12.

A polypeptide may contain hundreds of amino acids joined in a specific linear order. The backbone of the polypeptide chain includes the repeating sequence

$$N—C—C—N—C—C—N—C—C$$

plus all other atoms *except those in the* R *groups.* The *R* groups of the amino acids extend from this backbone.

An almost infinite variety of protein molecules is possible, differing from one another in the number, types, and sequences of amino acids they contain. The 20 types of amino acids found in proteins may be thought of as letters of a protein alphabet; each protein is a very long sentence made up of amino acid letters.

Proteins have four levels of organization

The polypeptide chains making up a protein are twisted or folded to form a macromolecule with a specific *conformation,* or 3-D shape. Some polypeptide chains form long fibers. **Globular** proteins are tightly folded into compact, roughly spherical shapes. There is a close relationship between a protein's conformation and its function. For example, a typical enzyme is a globular protein with a unique shape that allows it to catalyze a specific chemical reaction. Similarly, the shape of a protein hormone enables it to combine with receptors on its target cell (the cell the hormone acts on). Scientists recognize four main levels of protein organization: primary, secondary, tertiary, and quaternary.

Primary structure is the amino acid sequence

The sequence of amino acids, joined by peptide bonds, is the **primary structure** of a polypeptide chain. As discussed in Chap-

FIGURE 3-16

The 20 common amino acids.

Polar amino acids have relatively hydrophilic side chains, whereas nonpolar amino acids have side chains that are relatively hydrophobic. Carboxyl groups and amino groups are electrically charged at cell pH; therefore, acidic and basic amino acids are hydrophilic. The three-letter abbreviations appear below the amino acid names.

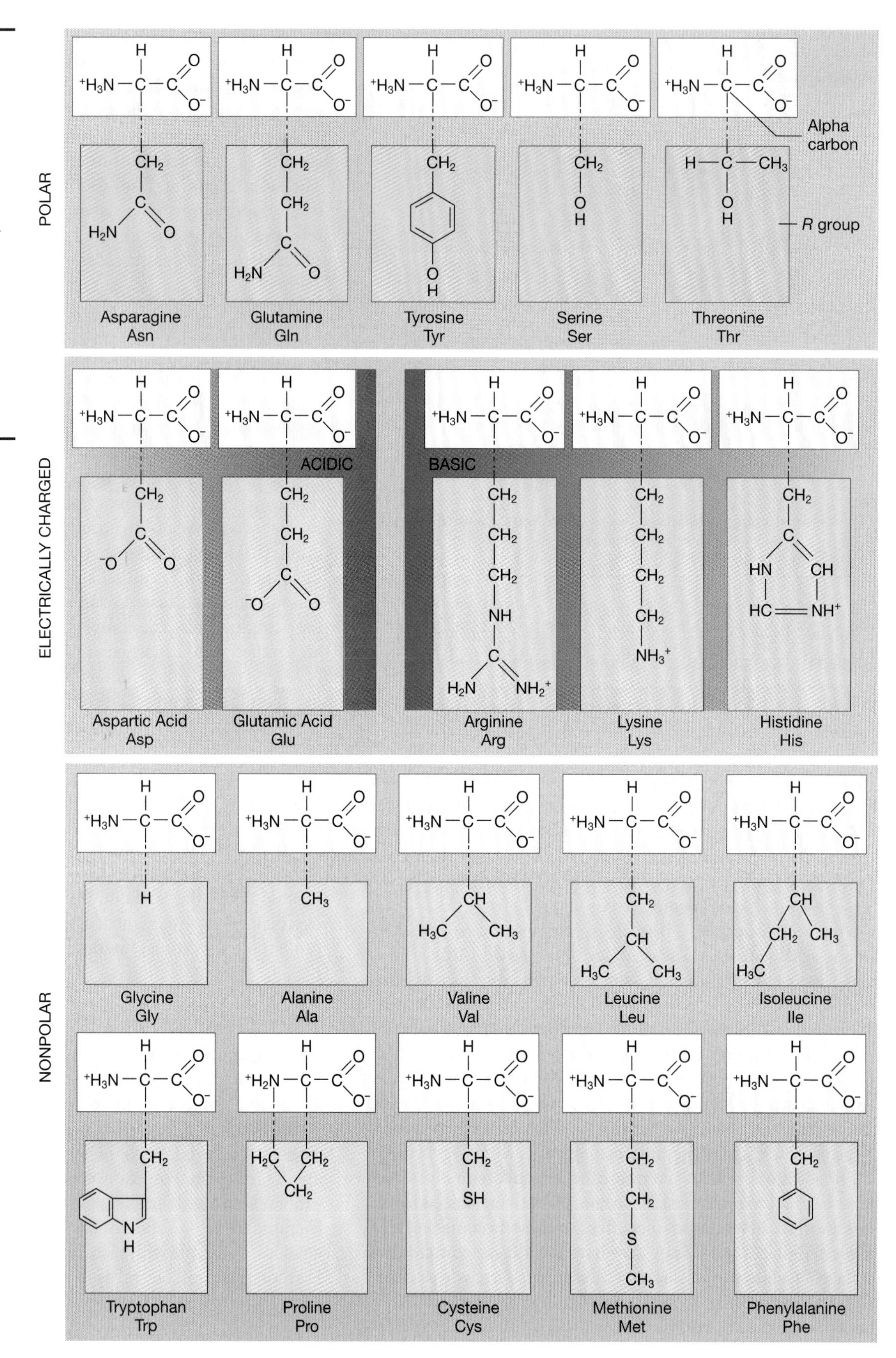

ter 12, this sequence is specified by the instructions in a gene. Using analytical methods investigators can determine the exact sequence of amino acids in a protein molecule. The primary structures of thousands of proteins are known. For example, glucagon, a hormone secreted by the pancreas, is a small polypeptide, consisting of only 29 amino acid units (Fig. 3-19).

FIGURE **3-17** | **An amino acid at pH 7.**

In living cells, amino acids exist mainly in their ionized form, as dipolar ions.

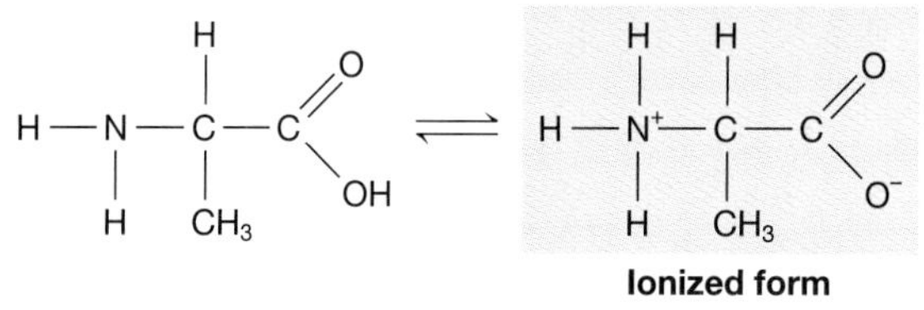
Ionized form

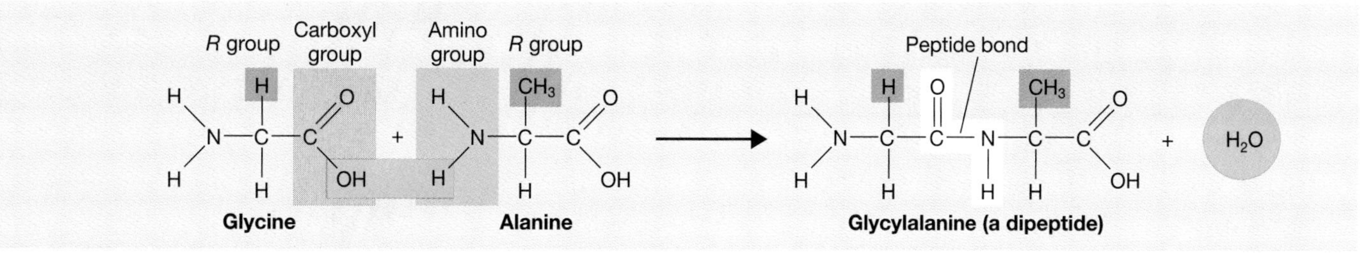
R group
Carboxyl group
Amino group
R group
Glycine
Alanine
Peptide bond
Glycylalanine (a dipeptide)

FIGURE **3-18** | **Peptide bonds.**

(a) A dipeptide is formed by a condensation reaction, that is, by the removal of the equivalent of a water molecule from the carboxyl group of one amino acid and the amino group of another amino acid. The resulting peptide bond is a covalent, carbon-to-nitrogen bond. Note that the carbon is also part of a carbonyl group, and that the nitrogen is also covalently bonded to a hydrogen. Additional amino acids can be added to form a long polypeptide chain with a free amino group at one end and a free carboxyl group at the other.

Primary structure is always represented in a simple, linear, "beads-on-a-string" form. However, the overall conformation of a protein is far more complex, involving interactions among the various amino acids that comprise the primary structure of the molecule. Therefore, the higher orders of structure—secondary, tertiary, and quaternary—ultimately derive from the specific amino acid sequence (the primary structure).

Secondary structure results from hydrogen bonding involving the backbone

Some regions of a polypeptide exhibit **secondary structure,** which is highly regular. The two most common types of secondary structure are the α-helix and the β-pleated sheet; the designations α and β refer simply to the order in which these two types of secondary structure were discovered. An **α-helix** is a region where a polypeptide chain forms a uniform helical coil (Fig. 3-20a). The helical structure is determined and maintained by the formation of hydrogen bonds between the backbones of the amino acids in successive turns of the spiral coil. Each hydrogen bond forms between an oxygen with a partial negative charge and a hydrogen with a partial positive charge. The oxygen is part of the remnant of the carboxyl group of one amino acid; the hydrogen is part of the remnant of the amino group of the fourth amino acid down the chain. Thus 3.6 amino acids are included in each complete turn of the helix. Every amino acid in an α-helix is hydrogen bonded in this way.

The α-helix is the basic structural unit of some fibrous proteins that make up wool, hair, skin, and nails. The elasticity of these fibers is due to a combination of physical factors (the helical shape) and chemical factors (hydrogen bonding). Although hydrogen bonds maintain the helical structure, these bonds can be broken, allowing the fibers to stretch under tension (like a telephone cord). When the tension is released, the fibers recoil and hydrogen bonds reform. This is why you can stretch the hairs on your head to some extent and they will snap back to their original length.

The hydrogen bonding in a **β-pleated sheet** takes place between different polypeptide chains, or different regions of a polypeptide chain that has turned back on itself (Fig. 3-20b). Each chain is fully extended, but because each has a zigzag structure the resulting "sheet" has an overall pleated conformation (much like a sheet of paper that has been folded to make a fan). Although the pleated sheet is strong and flexible, it is not elastic. This is because the distance between the pleats is fixed, determined by the strong covalent bonds of the polypeptide backbones. Fibroin, the protein of silk, is characterized by a β-pleated sheet structure, as are the cores of many globular proteins.

It is not uncommon for a single polypeptide chain to include both α-helical regions and regions with β-pleated sheet conformations. The properties of some complex biological materials result from such combinations. A spider's web is composed of a material that is extremely strong, flexible, and elastic. Once again we see function and structure working together, as these properties derive from the fact that spider silk is a com-

FIGURE **3-19** | **Primary structure of a polypeptide.**

Glucagon is a very small polypeptide made up of 29 amino acids. The linear sequence of amino acids is indicated by ovals containing their abbreviated names (see Fig. 3-16).

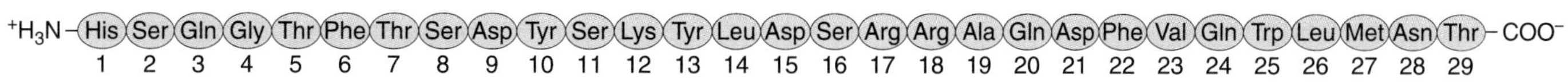
+H3N–His–Ser–Gln–Gly–Thr–Phe–Thr–Ser–Asp–Tyr–Ser–Lys–Tyr–Leu–Asp–Ser–Arg–Arg–Ala–Gln–Asp–Phe–Val–Gln–Trp–Leu–Met–Asn–Thr–COO–
1 2 3 4 5 6 7 8 9 10 11 12 13 14 15 16 17 18 19 20 21 22 23 24 25 26 27 28 29

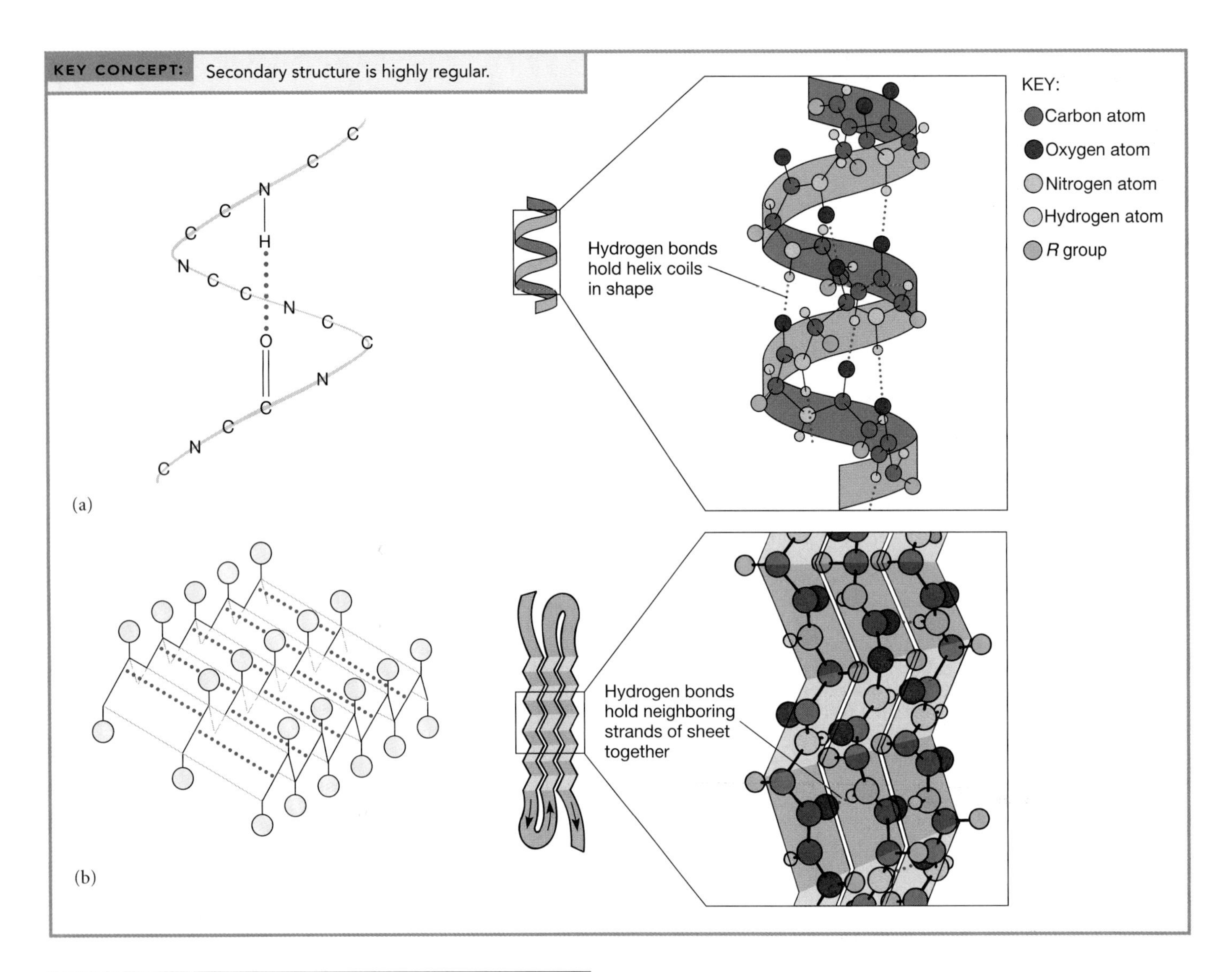

FIGURE **3-20** **Secondary structure of a protein.**

(a) In an α-helix the *R* groups project out from the sides. (The *R* groups have been omitted in the simplified diagram at left.)
(b) A β-pleated sheet forms when a polypeptide chain folds back on itself *(arrows)*; half the *R* groups project above the sheet, and the other half project below it.

posite of proteins with α-helical conformations (providing elasticity) and others with β-pleated sheet conformations (providing strength).

Tertiary structure depends on interactions among side chains

The **tertiary structure** of a protein molecule is the overall shape assumed by each individual polypeptide chain (Fig. 3-21). This 3-D structure is determined by four main factors that involve interactions among *R* groups (side chains) belonging to the same polypeptide chain. These include both weak interactions (hydrogen bonds, ionic bonds, and hydrophobic interactions) and strong covalent bonds.

1. Hydrogen bonds form between *R* groups of certain amino acid subunits.
2. An ionic bond can occur between an *R* group with a unit of positive charge and one with a unit of negative charge.
3. Hydrophobic interactions result from the tendency of nonpolar *R* groups to be excluded by the surrounding water and therefore to associate in the interior of the globular structure.
4. Covalent bonds known as *disulfide bonds* or *disulfide bridges* (—S—S—) may link the sulfur atoms of two cysteine subunits belonging to the same chain. A disulfide bridge forms when the sulfhydryl groups of two cysteines react; the two hydrogens are removed, and the two sulfur atoms that remain become covalently linked.

Quaternary structure results from interactions among polypeptides

Many functional proteins are composed of two or more polypeptide chains, interacting in specific ways to form the biologically active molecule. **Quaternary structure** is the resulting 3-D

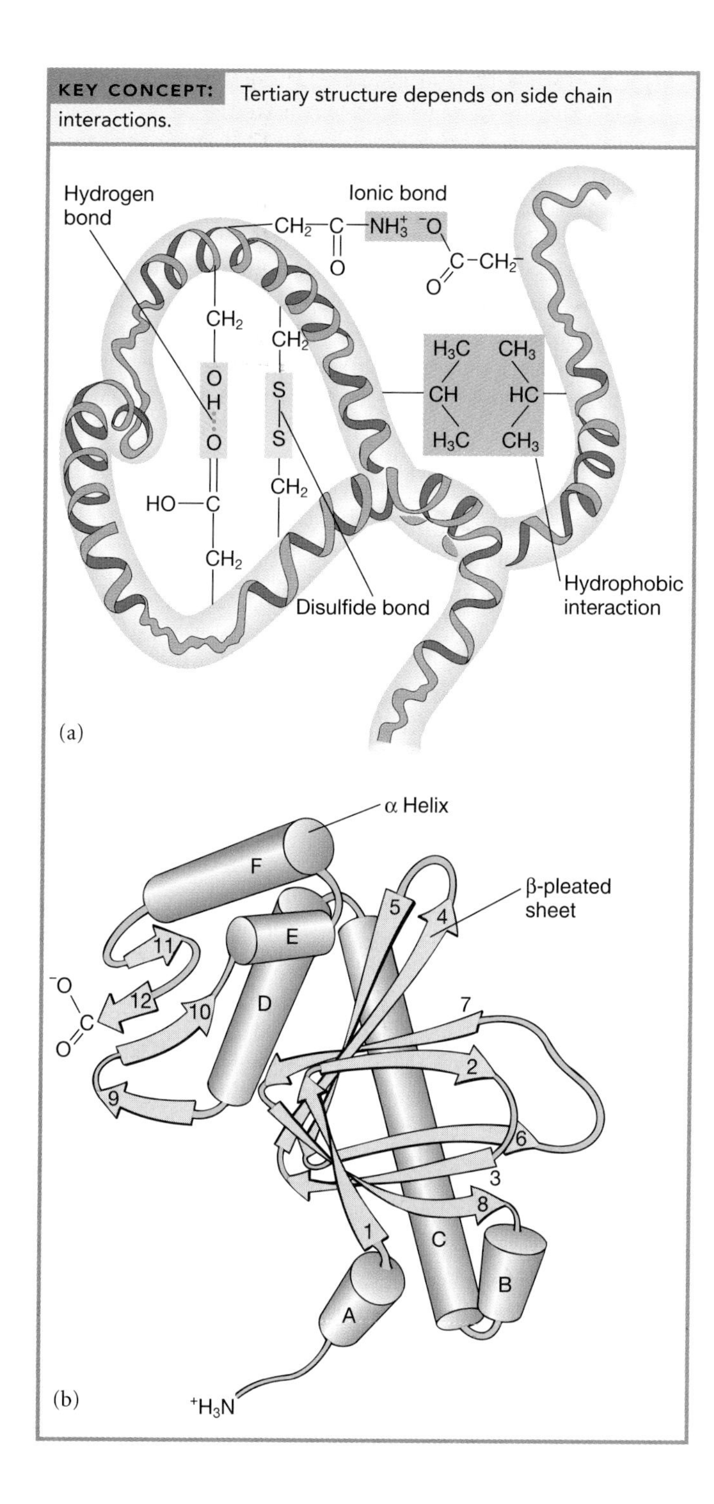

ACTIVE FIGURE **3-21** **Tertiary structure of a protein.**

(a) Hydrogen bonds, ionic bonds, hydrophobic interactions, and disulfide bridges between *R* groups hold the parts of the molecule in the designated shape. **(b)** In this drawing, α-helical regions are represented as blue tubes lettered A through F; β-pleated sheets are the gray arrows numbered 1 through 12. Green lines represent connecting regions. Although the molecule seems very complicated, it is a single polypeptide chain, starting at the amino end *(bottom left)* and terminating at the carboxyl end *(upper left)*. Most of the bends and foldbacks that give the molecule its overall conformation (tertiary structure) are stabilized by *R*-group interactions. This polypeptide is a subunit of a DNA-binding protein (known as CAP) from the bacterium *Escherichia coli*.

Biology Now™ Learn more about the **structure of a protein** by clicking on this figure on your BiologyNow CD-ROM.

Hemoglobin, the protein in red blood cells responsible for oxygen transport, is an example of a globular protein with a quaternary structure (Fig. 3-22a). Hemoglobin consists of 574 amino acids arranged in four polypeptide chains: two identical chains called *alpha chains* and two identical chains called *beta chains*.

Collagen, mentioned previously, has a fibrous type of quaternary structure that allows it to function as the major strengthener of animal tissues. It consists of three polypeptide chains, architecture of these polypeptide chains, each with its own primary, secondary, and tertiary structure. The same types of interactions that produce secondary and tertiary structure also contribute to quaternary structure; these include hydrogen bonding, ionic bonding, hydrophobic interactions, and disulfide bridges.

A functional antibody molecule, for example, consists of four polypeptide chains joined by disulfide bridges (see Chapter 43). Disulfide bridges are a common feature of proteins secreted from cells, such as antibodies. These strong bonds stabilize the molecules in the extracellular environment.

FIGURE **3-22** **Quaternary structure of a protein.**

(a) Hemoglobin, a globular protein, consists of four polypeptide chains, each joined to an iron-containing molecule, a heme. **(b)** Collagen, a fibrous protein, is a triple helix consisting of three long polypeptide chains.

KEY CONCEPT: Proteins with two or more polypeptide chains have quaternary structure.

Heme
Beta chain (β-globin)
Alpha chain (α-globin)
Alpha chain (α-globin)
Beta chain (β-globin)
(a) Hemoglobin
(b) Collagen

wound about each other and bound by cross links between their amino acids (Fig. 3-22b).

The amino acid sequence of a protein determines its conformation

PROCESS OF SCIENCE

In 1996, researchers at the University of Illinois at Champaign-Urbana devised a test of the hypothesis that the conformation of a protein is dictated by its amino acid sequence. They conducted an experiment in which they completely unfolded myoglobin, a polypeptide that stores oxygen in muscle cells, and then used sophisticated technology to track the refolding process. They found that within a few fractions of a microsecond the molecule had coiled up to form α-helices, and within 4 microseconds formation of the tertiary structure was completed. Thus these researchers demonstrated that, at least under defined experimental conditions in vitro (outside a living cell), a polypeptide can spontaneously undergo folding processes that yield its normal, functional conformation.

This and other types of evidence support the widely held conclusion that amino acid sequence is the ultimate determinant of protein conformation. However, because conditions in vivo (in the cell) are quite different from defined laboratory conditions, proteins do not necessarily always fold spontaneously. On the contrary, scientists have learned that proteins known as **molecular chaperones** mediate the folding of other protein molecules. Molecular chaperones are thought to make the folding process more orderly and efficient and to prevent partially folded proteins from becoming inappropriately aggregated. However, there is no evidence that molecular chaperones actually dictate the folding pattern. For this reason, the existence of chaperones is not an argument against the idea that amino acid sequence determines conformation.

Protein conformation determines function

The overall structure of a protein helps determine its biological activity. A single protein may have more than one distinct structural region, each with its own function. Many proteins are modular, consisting of two or more globular regions, called *domains,* connected by less compact regions of the polypeptide chain. Each domain may have a different function. For example, a protein may have one domain that attaches it to a membrane and another that allows it to act as an enzyme.

The biological activity of a protein can be disrupted by a change in amino acid sequence that results in a change in conformation. For example, the genetic disease known as *sickle cell anemia* is due to a mutation that causes the substitution of the amino acid valine for glutamic acid at position 6 (the sixth amino acid from the amino end) in the beta chain of hemoglobin. The substitution of valine (which has a nonpolar side chain) for glutamic acid (which has a charged side chain) makes the hemoglobin less soluble and more likely to form crystal-like structures. This alteration of the hemoglobin affects the red blood cells, changing them to the crescent or sickle shapes that characterize this disease (see Fig. 15-8).

The biological activity of a protein may be affected by changes in its 3-D structure. When a protein is heated, subjected to significant pH changes, or treated with any of a number of chemicals, its structure becomes disordered and the coiled peptide chains unfold, yielding a more random conformation. This unfolding, which is mainly due to the disruption of hydrogen bonds and ionic bonds, is typically accompanied by a loss of normal function. Such changes in shape and the accompanying loss of biological activity are termed **denaturation** of the protein. For example, a denatured enzyme would lose its ability to catalyze a chemical reaction. An everyday example of denaturation occurs when we fry an egg. The consistency of the egg white protein, known as *albumin,* changes to a solid. Denaturation generally cannot be reversed (you can't "unfry" an egg). However, under certain conditions, some proteins have been denatured and have returned to their original shape and biological activity when normal environmental conditions have been restored.

Protein conformation is studied through a variety of methods

PROCESS OF SCIENCE

The architecture of a protein can be ascertained directly through sophisticated types of analysis, such as the x-ray diffraction studies discussed in Chapter 11. Because these studies are tedious and costly, researchers are developing alternative approaches, which rely heavily on the enormous databases generated by the Human Genome Project and related initiatives. Today a protein's primary structure can be determined rapidly through the application of genetic engineering techniques (see Chapter 14), or by the use of sophisticated technology such as mass spectrometry. Researchers use a variety of techniques to effectively use these amino acid sequence data to predict a protein's higher levels of structure. As you have seen, side chains interact in relatively predictable ways, such as through ionic and hydrogen bonds. In addition, regions with certain types of side chains appear more likely to form α-helices or β-pleated sheets. Complex computer programs make such predictions, but these are imprecise because of the many possible combinations of folding patterns.

Computers are an essential part of yet another strategy. Once the amino acid sequence of a polypeptide has been determined, researchers use computers to search databases to find polypeptides with similar sequences. If the conformations of any of those polypeptides or portions have already been determined directly by x-ray diffraction or other techniques, this information can be extrapolated to make similar correlations between amino acid sequence and 3-D structure for the protein under investigation. These predictions are increasingly reliable, as more information is added to the databases every day.

Review

- Draw the structural formula of a simple amino acid. What is the importance of the carboxyl group, amino group, and *R* group?
- How does the primary structure of a polypeptide influence its secondary and tertiary structures?

■ How can the conformation of a protein be disrupted?

BiologyNow™ Assess your understanding of **proteins** by taking the pretest on your BiologyNow CD-ROM.

NUCLEIC ACIDS

Learning Objective

10 Describe the components of a nucleotide. Name some nucleic acids, and discuss the importance of these compounds in living organisms.

Nucleic acids transmit hereditary information and determine what proteins a cell manufactures. Two classes of nucleic acids are found in cells: ribonucleic acid and deoxyribonucleic acid. **Deoxyribonucleic acid (DNA)** comprises the genes, the hereditary material of the cell, and contains instructions for making all the proteins, as well as all the RNA the organism needs. **Ribonucleic acid (RNA)** participates in the complex process in which amino acids are linked to form polypeptides. Some types of RNA, known as **ribozymes,** can even act as specific biological catalysts. Like proteins, nucleic acids are large, complex molecules. The name *nucleic acid* reflects the fact that they are acidic and were first identified, by Friedrich Miescher in 1870, in the nuclei of pus cells.

Nucleic acids are polymers of **nucleotides,** molecular units that consist of (1) a five-carbon sugar, either **deoxyribose** (in DNA) or **ribose** (in RNA); (2) one or more phosphate groups, which make the molecule acidic; and (3) a nitrogenous base, a ring compound that contains nitrogen. The nitrogenous base may be either a double-ring **purine** or a single-ring **pyrimidine** (Fig. 3-23).

DNA commonly contains the purines adenine (A) and guanine (G), the pyrimidines cytosine (C) and thymine (T), the sugar deoxyribose, and phosphate. RNA contains the purines adenine and guanine, and the pyrimidines cytosine and uracil (U), together with the sugar ribose, and phosphate.

The molecules of nucleic acids are made of linear chains of nucleotides, which are joined by **phosphodiester linkages,** each consisting of a phosphate group and the covalent bonds that attach it to the sugars of adjacent nucleotides (Fig. 3-24). Note that each nucleotide is defined by its particular base and that nucleotides can be joined in any sequence. A nucleic acid molecule

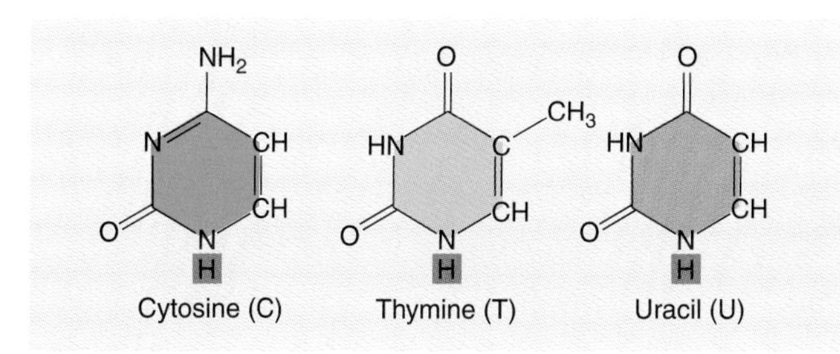

(a) Pyrimidines

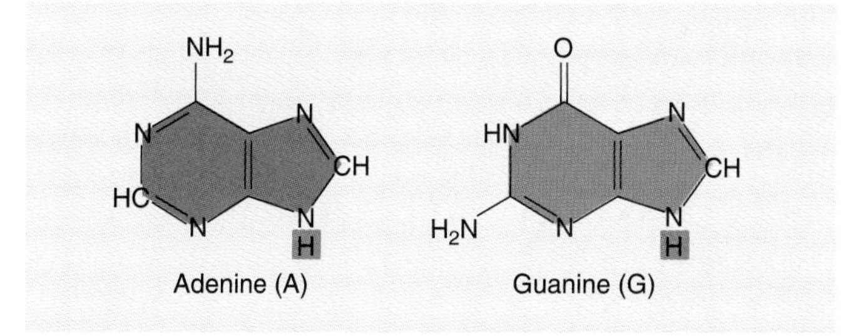

(b) Purines

FIGURE **3-23** **Components of nucleotides.**

(a) The three major pyrimidine bases found in nucleotides are cytosine, thymine (in DNA only), and uracil (in RNA only). **(b)** The two major purine bases found in nucleotides are adenine and guanine. The hydrogens indicated by the boxes are removed when the base is attached to a sugar.

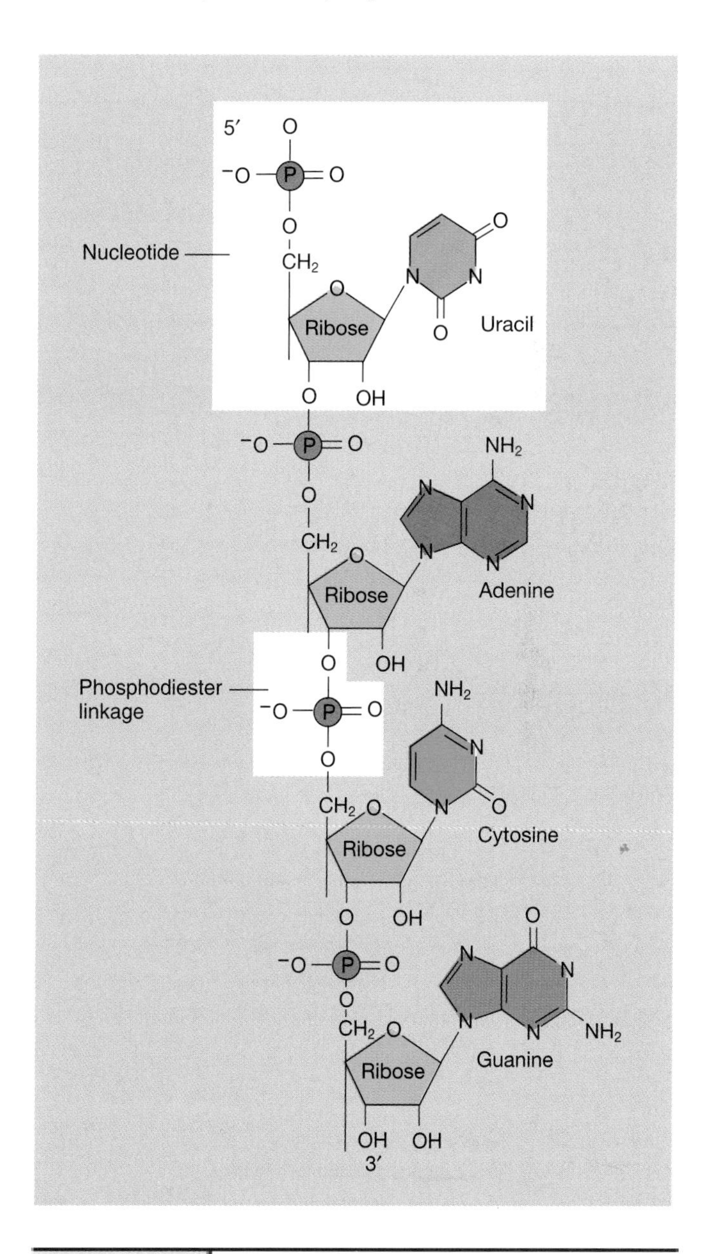

FIGURE **3-24** **RNA, a nucleic acid.**

Nucleotides, each with a specific base, are joined by phosphodiester linkages.

TABLE 3-3 Classes of Biologically Important Organic Compounds

Class and Component Elements	Description	How to Recognize	Principal Function in Living Systems
Carbohydrates C, H, O	Contain approximately 1 C:2 H:1 O (but make allowance for loss of oxygen when sugar units as nucleic acids and glycoproteins are linked)	Count the carbons, hydrogens, and oxygens.	Cell fuel; energy storage; structural component of plant cell walls; component of other compounds such as nucleic acids and glycoproteins
	1. Monosaccharides (simple sugars). Mainly five-carbon (pentose) molecules such as ribose or six-carbon (hexose) molecules such as glucose and fructose	Look for the ring shapes:	Cell fuel; components of other compounds
	2. Disaccharides. Two sugar units linked by a glycosidic bond, e.g., maltose, sucrose	Count sugar units	Components of other compounds; form of sugar transported in plants
	3. Polysaccharides. Many sugar units linked by glycosidic bonds, e.g., glycogen, cellulose	Count sugar units	Energy storage; structural components of plant cell walls
Lipids C, H, O (sometimes N, P)	Contain much less oxygen relative to carbon and hydrogen than do carbohydrates		Energy storage; cellular fuel, components of cells; thermal insulation
	1. *Fats.* Combination of glycerol with one to three fatty acids. Monoacylglycerol contains one fatty acid; diacylglycerol contains two fatty acids; triacylglycerol contains three fatty acids. If fatty acids contain double carbon-to-carbon linkages (C = C), they are unsaturated; otherwise they are saturated	Look for glycerol at one end of molecule: H, H—C—O—, H—C—O—, H—C—O—, H	Cell fuel; energy storage
	2. *Phospholipids.* Composed of glycerol attached to one or two fatty acids and to an organic base containing phosphorus	Look for glycerol and side chain containing phosphorus and nitrogen.	Components of cell membranes
	3. *Steroids.* Complex molecules containing carbon atoms arranged in four attached rings (Three rings contain six carbon atoms each, and the fourth ring contains five.)	Look for four attached rings:	Some are hormones, others include cholesterol, bile salts, vitamin D, components of cell membranes
	4. *Carotenoids.* Orange and yellow pigments; consist of isoprene units	Look for isoprene units. H CH_3 $H_2C{=}C{-}C{=}CH_2$	Retinal (important in photoreception) and vitamin A are formed from carotenoids
Proteins C, H, O, N (usually S)	One or more polypeptides (chains of amino acids) coiled or folded in characteristic shapes	Look for amino acid units joined by C—N bonds.	Serve as enzymes; structural components; muscle proteins; hemoglobin.
Nucleic acids C, H, O, N, P	Backbone composed of alternating pentose and phosphate groups, from which nitrogenous bases project. DNA contains the sugar deoxyribose and the bases guanine, cytosine, adenine, and thymine. RNA contains the sugar ribose and the bases guanine, cytosine, adenine, and uracil. Each molecular subunit, called a nucleotide, consists of a pentose, a phosphate, and a nitrogenous base.	Look for a pentose-phosphate backbone. DNA forms a double helix.	Storage, transmission, and expression of genetic information

is uniquely defined by its specific sequence of nucleotides, which constitutes a kind of code (see Chapter 12). Whereas RNA is usually composed of one nucleotide chain, DNA consists of two nucleotide chains held together by hydrogen bonds and entwined around each other in a double helix (see Fig. 1-7).

Some nucleotides are important in energy transfers and other cell functions

In addition to their importance as subunits of DNA and RNA, nucleotides perform other vital functions in living cells. **Adeno-**

sine triphosphate (ATP)**, composed of adenine, ribose, and three phosphates (see Fig. 6-5), is of major importance as the primary energy currency of all cells (see Chapter 6). The two terminal phosphate groups are joined to the nucleotide by covalent bonds. These are traditionally indicated by wavy lines, which indicate that ATP can transfer a phosphate group to another molecule, making that molecule more reactive. In this way ATP is able to donate some of its chemical energy. Most of the readily available chemical energy of the cell is associated with the phosphate groups of ATP. Like ATP, **guanosine triphosphate (GTP),** a nucleotide that contains the base guanine, can transfer energy by transferring a phosphate group and also has a role in cell signaling (see Chapter 5).

A nucleotide may be converted to an alternative form with specific cellular functions. ATP, for example, is converted to **cyclic adenosine monophosphate (cyclic cAMP)** by the enzyme adenylyl cyclase (Fig. 3-25). Cyclic AMP regulates certain cell functions and is important in the mechanism by which some hormones act (see Chapters 13, 39, and 47). A related molecule, **cyclic guanosine monophosphate (cGMP),** also plays a role in certain cell signaling processes.

Cells contain several dinucleotides, which are of great importance in metabolic processes. For example, as discussed in Chapter 6, **nicotinamide adenine dinucleotide** has a primary role in biological oxidation and reduction reactions in cells. It can exist in an oxidized form (**NAD^+**) that is converted to a reduced form (**NADH**) when it accepts electrons (in association with hydrogen; see Fig. 6-7). These electrons, along with their energy, are transferred to other molecules.

Review

- Compare the functions of proteins and nucleic acids. How are their structures related to these functions?

BiologyNow™ Assess your understanding of **nucleic acids** by taking the pretest on your BiologyNow CD-ROM.

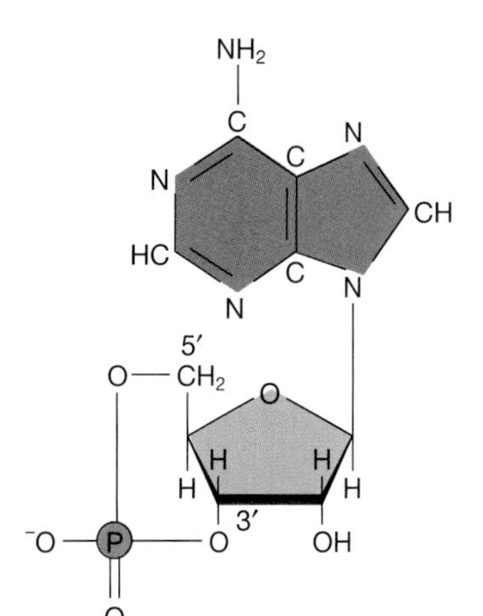

FIGURE **3-25**

Cyclic adenosine monophosphate (cAMP).

The single phosphate is part of a ring connecting two regions of the ribose.

IDENTIFYING BIOLOGICAL MOLECULES

Learning Objective

11 Compare the functions and chemical compositions of the major groups of organic compounds: carbohydrates, lipids, proteins, and nucleic acids.

Although the fundamental classes of biological molecules may seem overwhelming at first, you will learn to distinguish them readily by understanding their chief attributes. These are summarized in Table 3-3.

Review

- How can you distinguish a pentose sugar from a hexose sugar? A disaccharide from a sterol? An amino acid from a monosaccharide? A phospholipid from a triacylglycerol? A protein from a polysaccharide? A nucleic acid from a protein?

BiologyNow™ Assess your understanding of **biological molecules** by taking the pretest on your BiologyNow CD-ROM.

SUMMARY WITH KEY TERMS

1 Describe the properties of carbon that make it the central component of organic compounds.

- Each carbon atom forms four covalent bonds with up to four other atoms; these bonds are single, double, or triple bonds.
- Carbon atoms form straight or branched chains or join into rings.
- Carbon forms covalent bonds with a greater number of different elements than does any other type of atom.

2 Define the term *isomer,* and distinguish among the three principal isomer types.

- **Isomers** are compounds with the same molecular formula but different structures.
- **Structural isomers** differ in the covalent arrangements of their atoms. **Geometric isomers,** or *cis–trans* isomers, differ in the spatial arrangements of their atoms. **Enantiomers** are isomers that are mirror images of each other. Cells can distinguish between these configurations.

3 Identify the major functional groups present in organic compounds, and describe their properties.

- Hydrocarbons, organic compounds consisting of only carbon and hydrogen, are nonpolar and hydrophobic. The **methyl group** is a hydrocarbon group.
- Polar and ionic functional groups interact with each other and are hydrophilic.
- Partial charges on atoms at opposite ends of a bond are responsible for the polar property of a functional group. **Hydroxyl** and **carbonyl groups** are polar.
- **Carboxyl** and **phosphate groups** are acidic, becoming negatively charged when they release hydrogen ions. The **amino group** is basic, becoming positively charged when it accepts a hydrogen ion.

4 Explain the relationship between polymers and macromolecules.

- Long chains of **monomers** (similar organic compounds) linked together through **condensation reactions** are called **polymers.**
- Large polymers such as polysaccharides, proteins, and DNA are referred to as **macromolecules.** They can be broken down by **hydrolysis reactions.**

SUMMARY WITH KEY TERMS *(continued)*

5 Distinguish among monosaccharides, disaccharides, and polysaccharides; compare storage polysaccharides with structural polysaccharides.

- **Carbohydrates** contain carbon, hydrogen, and oxygen in a ratio of approximately one carbon to two hydrogens to one oxygen.
- **Monosaccharides** are simple sugars such as glucose, fructose, and ribose.
- Two monosaccharides join by a **glycosidic linkage** to form a **disaccharide** such as maltose or sucrose.
- Most carbohydrates are **polysaccharides,** long chains of repeating units of a simple sugar. Carbohydrates are typically stored in plants as the polysaccharide **starch** and in animals as the polysaccharide **glycogen.** The cell walls of plants are composed mainly of the structural polysaccharide **cellulose.**

6 Distinguish among fats, phospholipids, and steroids, and describe the composition, characteristics, and biological functions of each.

- **Lipids** are composed mainly of hydrocarbon-containing regions, with few oxygen-containing (polar or ionic) functional groups. Lipids have a greasy or oily consistency and are relatively insoluble in water.
- **Triacylglycerol,** the main storage form of fat in organisms, consists of a molecule of **glycerol** combined with three **fatty acids. Monoacylglycerols** and **diacylglycerols** contain one and two fatty acids, respectively. A fatty acid can be either **saturated** with hydrogen, or **unsaturated.**
- **Phospholipids** are structural components of cell membranes. A phospholipid consists of a glycerol molecule attached at one end to two fatty acids and at the other end to a phosphate group linked to an organic compound such as choline.
- **Steroid** molecules contain carbon atoms arranged in four attached rings. Cholesterol, bile salts, and certain hormones are important steroids.

7 Give an overall description of the structure and functions of proteins.

- **Proteins** are large, complex molecules made of simpler subunits, called **amino acids,** joined by **peptide bonds.**
- Proteins are the most versatile class of biological molecules, serving a variety of functions, such as **enzymes,** structural components, and cell regulators.
- Proteins are composed of various linear sequences of 20 different amino acids. Two amino acids combine to form a **dipeptide.** A longer chain of amino acids is a **polypeptide.**

8 Describe the features that are shared by all amino acids, and explain how amino acids are grouped into classes based on the characteristics of their side chains.

- All amino acids contain an amino group and a carboxyl group.
- Amino acids vary in their side chains, which dictate their chemical properties—nonpolar, polar, acidic or basic.
- Amino acids generally exist as dipolar ions at cell pH and serve as important biological buffers.

9 Distinguish among the four levels of organization of protein molecules.

- **Primary structure** is the linear sequence of amino acids in the polypeptide chain.
- **Secondary structure** is a regular conformation, such as an **α-helix** or a **β-pleated sheet;** it is due to hydrogen bonding between elements of the backbones of the amino acids.
- **Tertiary structure** is the overall shape of the polypeptide chains, as dictated by chemical properties and interactions of the side chains of specific amino acids. Hydrogen bonds, ionic bonds, hydrophobic interactions, and disulfide bridges contribute to tertiary structure.
- **Quaternary structure** is determined by the association of two or more polypeptide chains.

10 Describe the components of a nucleotide. Name some nucleic acids and nucleotides, and discuss the importance of these compounds in living organisms.

- **Nucleotides** are composed of a two-ring **purine** or one-ring **pyrimidine** nitrogenous base, a five-carbon sugar (**ribose** or **deoxyribose**), and one or more phosphate groups.
- The **nucleic acids DNA** and **RNA,** composed of long chains of nucleotide subunits, store and transfer information that governs the sequence of amino acids in proteins and ultimately the structure and function of the organism.
- **ATP** (**adenosine triphosphate**) is a nucleotide of special significance in energy metabolism. **NAD$^+$** is also involved in energy metabolism through its role as an electron (hydrogen) acceptor in biological oxidation and reduction reactions.

11 Compare the functions and chemical compositions of the major groups of organic compounds: carbohydrates, lipids, proteins, and nucleic acids.

- Review Table 3-3.

POST-TEST

1. Which of the following is generally considered an inorganic form of carbon? (a) CO_2 (b) C_2H_4 (c) CH_3COOH (d) b and c (e) all of the preceding are inorganic
2. Carbon is particularly well suited to be the backbone of organic molecules because (a) it can form both covalent bonds and ionic bonds (b) its covalent bonds are very irregularly arranged in three-dimensional space (c) its covalent bonds are the strongest chemical bonds known (d) it can bond to atoms of a large number of other elements (e) all the bonds it forms are polar
3. The structures depicted are

```
  CH3 CH3          H   CH3
   |   |           |   |
H—C———C—H       H—C———C—H
   |   |           |   |
   H   H          CH3  H
```

(a) enantiomers (b) different views of the same molecule (c) geometric (*cis–trans*) isomers (d) both geometric isomers and enantiomers (e) structural isomers

POST-TEST *(continued)*

4. Which of the following are generally hydrophobic? (a) polar molecules and hydrocarbons (b) ions and hydrocarbons (c) nonpolar molecules and ions (d) polar molecules and ions (e) none of the above
5. Which of the following is a nonpolar molecule? (a) water, H_2O (b) ammonia, NH_3 (c) methane, CH_4 (d) ethane, C_2H_6 (e) more than one of the preceding
6. Which of the following functional groups normally acts as an acid? (a) hydroxyl (b) carbonyl (c) sulfhydryl (d) phosphate (e) amino
7. The synthetic process by which monomers are covalently linked is (a) hydrolysis (b) isomerization (c) condensation (d) glycosidic linkage (e) ester linkage
8. A monosaccharide designated as an aldehyde sugar contains (a) a terminal carboxyl group (b) an internal carboxyl group (c) a terminal carbonyl group (d) an internal carbonyl group (e) a terminal carboxyl group and an internal carbonyl group
9. Structural polysaccharides typically (a) have extensive hydrogen bonding between adjacent molecules (b) are much more hydrophilic than storage polysaccharides (c) have much stronger covalent bonds than do storage polysaccharides (d) consist of alternating α-glucose and β-glucose subunits (e) form helical structures in the cell
10. A carboxyl group is always found in (a) organic acids and sugars (b) sugars and fatty acids (c) fatty acids and amino acids (d) alcohols (e) glycerol
11. Fatty acids are components of (a) phospholipids and carotenoids (b) carotenoids and triacylglycerol (c) steroids and triacylglycerol (d) phospholipids and triacylglycerol (e) carotenoids and steroids
12. Saturated fatty acids are so named because they are saturated with (a) hydrogen (b) water (c) hydroxyl groups (d) glycerol (e) double bonds
13. Fatty acids in phospholipids and triacylglycerols interact with each other by (a) disulfide bridges (b) van der Waals interactions (c) covalent bonds (d) hydrogen bonds (e) actually, fatty acids do not interact with each other
14. Which pair of amino acid side groups would be most likely to associate with each other by an ionic bond?
    1. $—CH_3$
    2. $—CH_2—COO^-$
    3. $—CH_2—CH_2—NH_3^+$
    4. $—CH_2—CH_2—COO^-$
    5. $—CH_2—OH$

    (a) 1 and 2 (b) 2 and 4 (c) 1 and 5 (d) 2 and 5 (e) 3 and 4
15. Which of the following levels of protein structure may be affected by hydrogen bonding? (a) primary and secondary (b) primary and tertiary (c) secondary, tertiary, and quaternary (d) primary, secondary, and tertiary (e) primary, secondary, tertiary, and quaternary
16. Which of the following associations between *R* groups are the strongest? (a) hydrophobic interactions (b) hydrogen bonds (c) ionic bonds (d) peptide bonds (e) disulfide bridges
17. Each phosphodiester linkage in DNA or RNA includes a phosphate joined by covalent bonds to (a) two bases (b) two sugars (c) two additional phosphates (d) a sugar, a base, and a phosphate (e) a sugar and a base

CRITICAL THINKING

1. Like oxygen, sulfur forms two covalent bonds. However, sulfur is far less electronegative. In fact, it is approximately as electronegative as carbon. How would the properties of the various classes of biological molecules be altered if you were to replace all the oxygen atoms with sulfur atoms?
2. In what ways are all species alike biochemically? How do species differ from one another biochemically?
3. Hydrogen bonds and van der Waals interactions are much weaker than covalent bonds, yet they are vital to organisms. Why?

- Visit our Web site at **http://biology.brookscole.com/solomon7** for links to chapter-related resources on the World Wide Web. Additional online materials relating to this chapter can also be found on our Web site.

BIOLOGY NOW RESOURCES

Active Figures

3-12: Triacylglycerol and other lipids

3-21: Structure of a protein

Preparing for an exam? Take a diagnostic test on your BiologyNow CD-ROM.

Post-Test Answers:

1. a	2. d	3. b	4. e
5. e	6. d	7. c	8. c
9. a	10. c	11. d	12. a
13. b	14. e	15. c	16. e
17. b			

4

Organization of the Cell

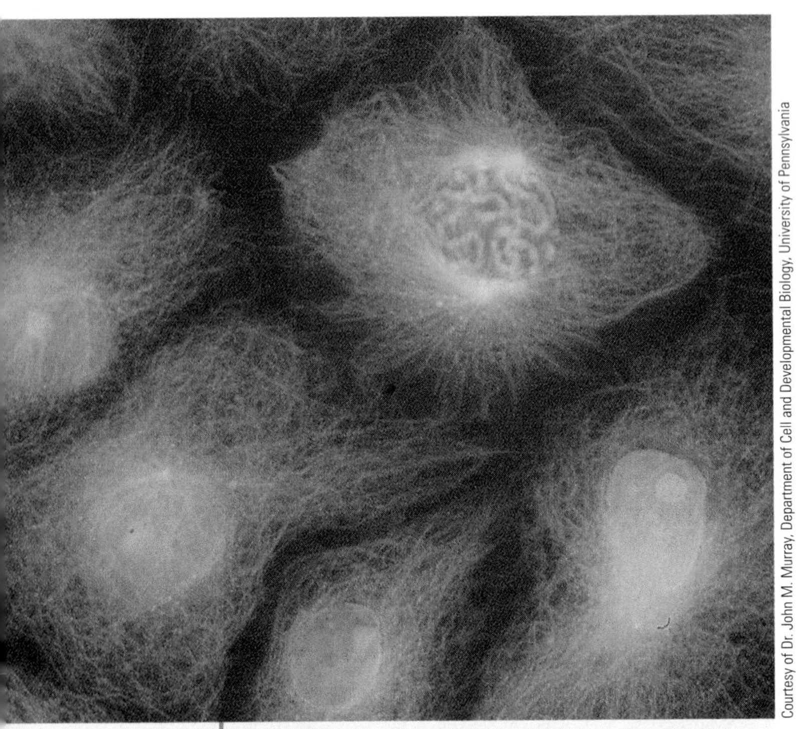

Courtesy of Dr. John M. Murray, Department of Cell and Developmental Biology, University of Pennsylvania

Microtubules, key components of the cytoskeleton. The cells shown here were stained with fluorescent antibodies (specific proteins) that bind to proteins associated with DNA *(orange)* and to a protein (tubulin) in microtubules *(green)*. This type of microscopy, known as *confocal fluorescence microscopy,* shows the extensive distribution of microtubules in these cells.

CHAPTER OUTLINE

- Cell Organization and Size
- Methods for Studying Cells
- Prokaryotic and Eukaryotic Cells
- Cell Membranes
- The Cell Nucleus
- Organelles in the Cytoplasm
- The Cytoskeleton
- Cell Coverings

Cells are dramatic examples of the underlying unity of all living things. This idea was first expressed by two German scientists, botanist Matthias Schleiden in 1838 and zoologist Theodor Schwann in 1839. Using their own observations and those of other scientists, these early investigators used inductive reasoning to conclude that all plants and animals consist of cells. Later, Rudolf Virchow, another German scientist, observed cells dividing and giving rise to daughter cells. In 1855, Virchow proposed that new cells form only by the division of previously existing cells.

The work of Schleiden, Schwann, and Virchow contributed greatly to the development of the **cell theory,** the unifying concept that (1) cells are the basic living units of organization and function in all organisms and (2) that all cells come from other cells. About 1880 another German biologist, August Weismann, added an important corollary to Virchow's concept by pointing out that the ancestry of all the cells alive today can be traced back to ancient times. Evidence that all living cells have a common origin is provided by the basic similarities in their structures and in the molecules of which they are made. When we examine a variety of diverse organisms, ranging from simple bacteria to the most complex plants and animals, we find striking similarities at the cell level. Careful studies of shared cell characteristics help us trace the evolutionary history of various groups of organisms and furnish powerful evidence that all organisms alive today had a common origin.

Each cell is a microcosm of life. It is the smallest unit that can carry out all activities we associate with life. When provided with essential nutrients and an appropriate environment, some cells can be kept alive and growing in the laboratory for many years. By contrast, no isolated part of a cell is capable of sustained survival. Composed of a vast array of inorganic and organic ions and molecules, including water, salts, carbohydrates, lipids, proteins, and nucleic acids, most cells have all the physical and chemical components needed for their own maintenance, growth, and division. Genetic information is stored in DNA molecules and is faithfully replicated and passed to each new generation of cells during cell division. Information in DNA codes for specific proteins that in turn determine cell structure and function. In this chapter and those that follow, we discuss how cells use many of the chemical materials we introduced in Chapters 2 and 3.

Cells exchange materials and energy with the environment. All living cells need one or more sources of energy, but a cell rarely obtains energy in a form that is immediately usable. Cells convert energy from one form to another, and that energy is used to carry out various activities, ranging from mechanical work to chemical synthesis. Cells convert energy to a convenient form, usually chemical energy stored in adenosine triphosphate, or ATP (see Chapter 3). Although the specifics vary, the basic strategies cells use for energy conversion are very similar. The chemical reactions that convert energy from one form to another are essentially the same in all cells, from bacteria to those of complex plants and animals.

Cells are the building blocks of complex multicellular organisms. Although they are basically similar, cells are also extraordinarily diverse and versatile. They can be modified in a variety of ways to carry out specialized functions.

Thanks to advances in technology, cell biologists use increasingly sophisticated tools in their search to better understand the structure and function of cells. For example, investigation of the cytoskeleton (cell skeleton), currently an active and exciting area of research, has been greatly enhanced by advances in microscopy. In the photograph, we see the extensive distribution of microtubules in cells. Microtubules are key components of the cytoskeleton. They help maintain cell shape, function in cell movement, and facilitate transport of materials within the cell. Proteins associated with DNA are also stained in the photomicrograph, and chromosomes are visible in the upper cell. As biologists continue to unlock the secrets of DNA, many new doors are opening to development of medical treatments as well as to better understanding of the organisms that share our planet. ■

CELL ORGANIZATION AND SIZE

Learning Objectives

1. Summarize the relationship between cell organization and homeostasis.
2. Explain the relationship between cell size and maintaining homeostasis.

The organization of cells and their small size allow them to maintain **homeostasis,** an appropriate internal environment. Cells experience constant changes in their environments, such as deviations in salt concentration, pH, and temperature. They must work continuously to restore and maintain the internal conditions that enable their biochemical mechanisms to function.

The organization of all cells is basically similar

To maintain homeostasis, the contents of the cell must be separated from the external environment. The **plasma membrane** is a structurally distinctive surface membrane that surrounds all cells. By making the interior of the cell an enclosed compartment, the plasma membrane allows the chemical composition of the cell to be quite different from that outside the cell. The plasma membrane serves as an extremely selective barrier between the cell contents and the outer environment. Cells exchange materials with the environment and can accumulate needed substances and energy stores.

Typically, cells have internal structures, called **organelles,** that are specialized to carry out metabolic activities such as converting energy to usable forms, synthesizing needed compounds, and manufacturing structures necessary for functioning and reproduction. Each cell has genetic instructions coded in its DNA, which is concentrated in a limited region of the cell.

Cell size is limited

Although their sizes vary over a wide range (Fig. 4-1), most cells are microscopic, and must be measured by very small units. The basic unit of linear measurement in the metric system (see inside back cover) is the meter (m), which is just a little longer than a yard. A millimeter (mm) is 1/1000 of a meter and is about as long as the bar enclosed in parentheses (-). The micrometer (μm) is the most convenient unit for measuring cells. A bar 1 μm long is 1/1,000,000 (one millionth) of a meter, or 1/1000 of a millimeter—far too short to be seen with the unaided eye. Most of us have difficulty thinking about units that are too small to see, but it is helpful to remember that a micrometer has the same relationship to a millimeter that a millimeter has to a meter (1/1000).

As small as it is, the micrometer is actually too large to measure most cell components. For this purpose biologists use the nanometer (nm), which is 1/1,000,000,000 (one billionth) of a meter, or 1/1000 of a micrometer. To mentally move down to the world of the nanometer, recall that a millimeter is 1/1000 of a meter, a micrometer is 1/1000 of a millimeter, and a nanometer is 1/1000 of a micrometer.

A few specialized algae and animal cells are large enough to be seen with the naked eye. A human egg cell, for example, is about 130 μm in diameter, or approximately the size of the period at the end of this sentence. The largest cells are birds' eggs, but they are atypical because both the yolk and the egg white consist of food reserves. The functioning part of the cell is a small mass on the surface of the yolk. Most cells are small and can only be seen with a microscope.

Why are most cells so small? If you consider what a cell must do to maintain homeostasis and to grow, it may be easier to understand the reasons for its small size. A cell must take in food and other materials and must rid itself of waste products generated by metabolic reactions. Everything that enters or leaves a cell must pass through its plasma membrane. The plasma membrane contains specialized "pumps" and channels with "gates"

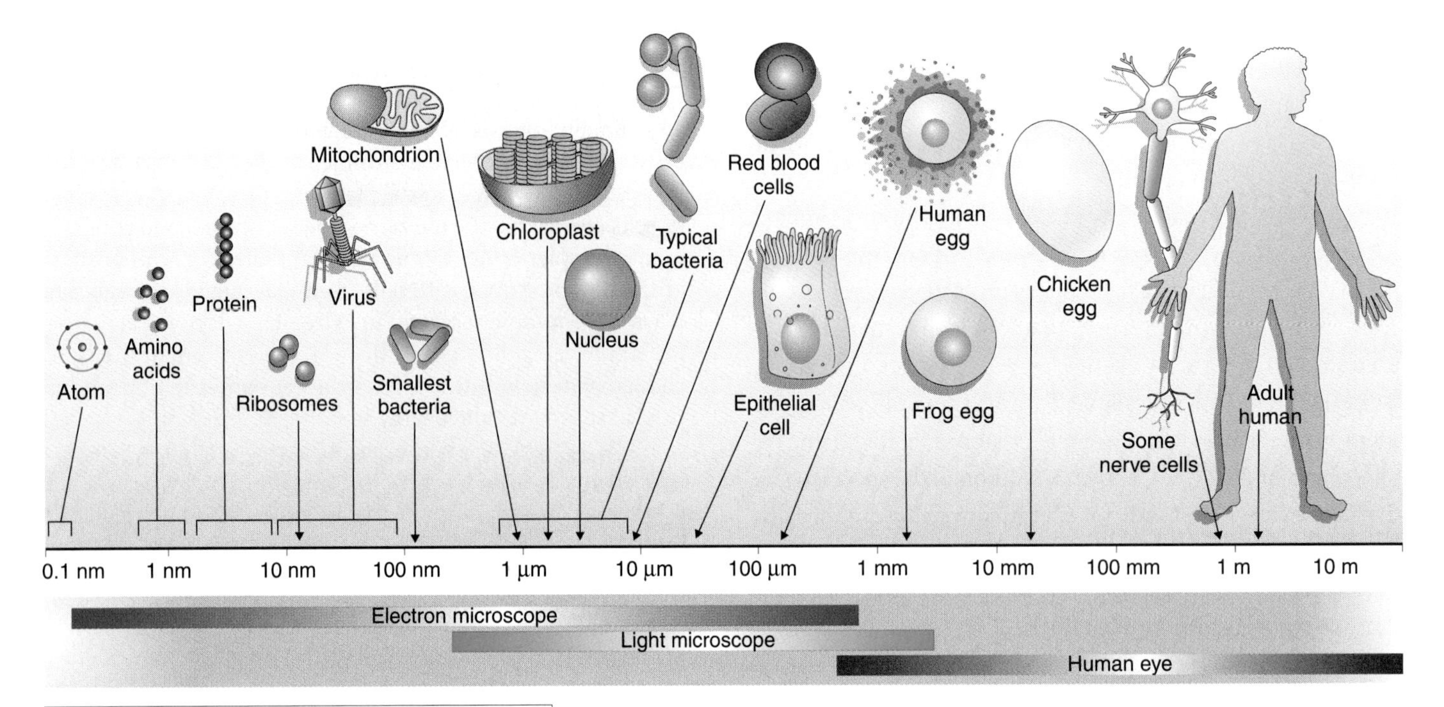

Measurements		
1 meter	=	1000 millimeters (mm)
1 millimeter	=	1000 micrometers (µm)
1 micrometer	=	1000 nanometers (nm)

FIGURE **4-1** **Biological size and cell diversity.**

We can compare relative size from the chemical level to the organismic level, using a logarithmic scale (multiples of 10). The prokaryotic cells of bacteria typically range in size from less than 1 to 10 µm long; their small size enables them to grow and divide rapidly. Eukaryotic cells are typically 10 to 100 µm in diameter; most are between 10 and 30 µm. The nuclei of animal and plant cells range from about 3 to 10 µm in diameter. Mitochondria are about the size of small bacteria, whereas chloroplasts are usually larger, about 5 µm long. Ova (egg cells) are among the largest cells. Although microscopic, some nerve cells are very long. The cells shown here are not drawn to scale.

that selectively regulate the passage of materials into and out of the cell. The plasma membrane must be large enough relative to the cell volume to keep up with the demands of regulating the passage of materials. Thus a critical factor in determining cell size is the ratio of its surface area (the plasma membrane) to its volume (Fig. 4-2).

As a cell becomes larger, its volume increases at a greater rate than its surface area (its plasma membrane), which effectively places an upper limit on cell size. Above some critical size, the number of molecules required by the cell could not be transported into the cell fast enough to sustain its needs. In addition, the cell would not be able to regulate its concentration of various ions or efficiently export its wastes.

Of course, not all cells are spherical or cuboid. Because of their shapes, some very large cells have relatively favorable ratios of surface area to volume. In fact, some variations in cell shape represent a strategy for increasing the ratio of surface area to volume. For example, many large plant cells are long and thin, which increases their surface-to-volume ratio. Some cells, such as epithelial cells lining the small intestine, have fingerlike projections of the plasma membrane, called **microvilli,** that significantly increase the surface area for absorbing nutrients and other materials (see Fig. 45-10c).

Another reason for the small size of cells is that, once inside, molecules must be trans-

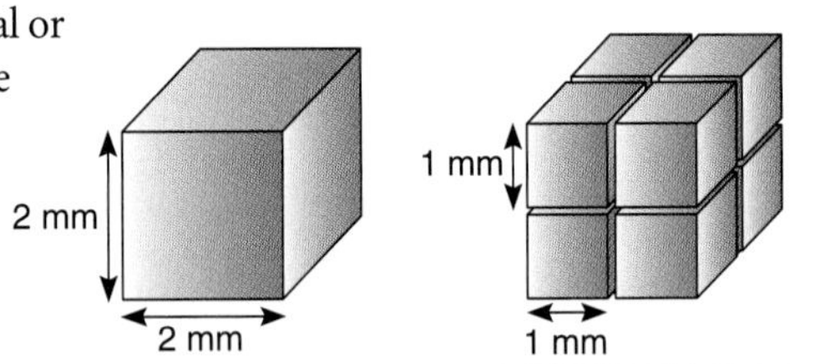

Surface Area (mm)	Surface area = height × width × number of sides × number of cubes	24 (2 × 2 × 6 × 1)	48 (1 × 1 × 6 × 8)
Volume (mm)	Volume = height × width × length × number of cubes	8 (2 × 2 × 2 × 1)	8 (1 × 1 × 1 × 8)
Surface Area/ Volume Ratio	Surface area/ volume	3 (24:8)	6 (48:8)

FIGURE **4-2** **Surface area-to-volume ratio.**

The surface area of a cell must be large enough relative to its volume to allow adequate exchange of materials with the environment. Although their volumes are the same, eight small cells have a much greater surface area (plasma membrane) in relation to their total volume than one large cell. In the example shown, the ratio of the total surface area to total volume of eight 1-mm cubes is double the surface-to-volume ratio of the single large cube.

ported to the locations where they are converted into other forms. Because cells are small, the distances molecules travel within them are relatively short, which speeds up many cell activities.

Cell size and shape are related to function

The sizes and shapes of cells are related to the functions they perform. Some cells, such as the amoeba and the white blood cell, change their shape as they move about. Sperm cells have long, whiplike tails, called *flagella,* for locomotion. Nerve cells have long, thin extensions that enable them to transmit messages over great distances. The extensions of some nerve cells in the human body may be as long as 1 m. Other cells, such as certain epithelial cells, are almost rectangular and are stacked much like building blocks to form sheetlike structures.

Review

- How does the plasma membrane help maintain homeostasis?
- Why is the relationship between surface area and volume of a cell important in determining cell size limits?

BiologyNow™ Assess your understanding of **cell organization and size** by taking the pretest on your BiologyNow CD-ROM.

METHODS FOR STUDYING CELLS

Learning Objective

3 Describe methods that biologists use to study cells, including microscopy and cell fractionation.

PROCESS OF SCIENCE

One of the most important tools biologists use for studying cell structures is the microscope. Cells were first described in 1665 by the English scientist Robert Hooke in his book *Micrographica.* Using a microscope he had made, Hooke examined a piece of cork, and drew and described what he saw. Hooke chose the term *cell* because the tissue reminded him of the small rooms monks lived in. Interestingly, what Hooke saw were not actually living cells, but the walls of dead cork cells (Fig. 4-3a). Much later, scientists recognized that the interior enclosed by the walls is the important part of living cells.

A few years later, inspired by Hooke's work, the Dutch naturalist Anton van Leeuwenhoek viewed living cells with small lenses that he made. Leeuwenhoek was highly skilled at grinding lenses and was able to magnify images more than 200 times. Among his important discoveries were bacteria, protists, blood cells, and sperm cells. Leeuwenhoek was among the first scientists to report cells in animals. Leeuwenhoek was a merchant, and not formally trained as a scientist. However, his skill, curiosity, and diligence in sharing his discoveries with scientists at the Royal Society of London brought an awareness of microscopic life to the scientific world. Unfortunately, Leeuwenhoek did not share his techniques, and not until more than 100 years later, in the late 19th century, were microscopes sufficiently developed for biologists to seriously focus their attention on the study of cells.

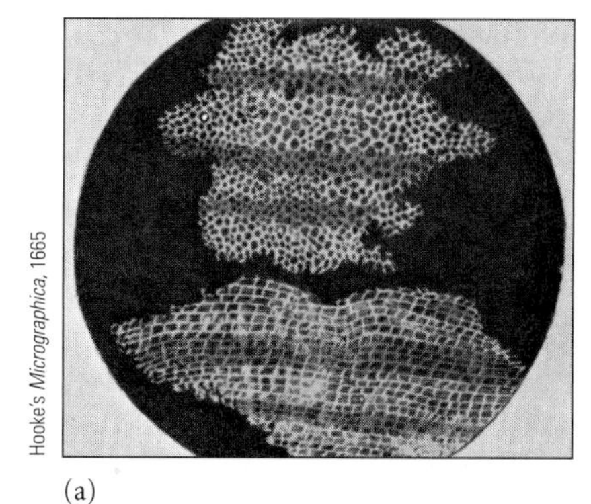

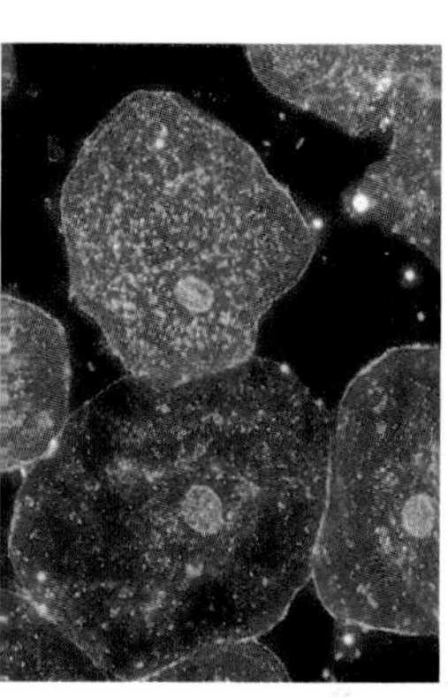

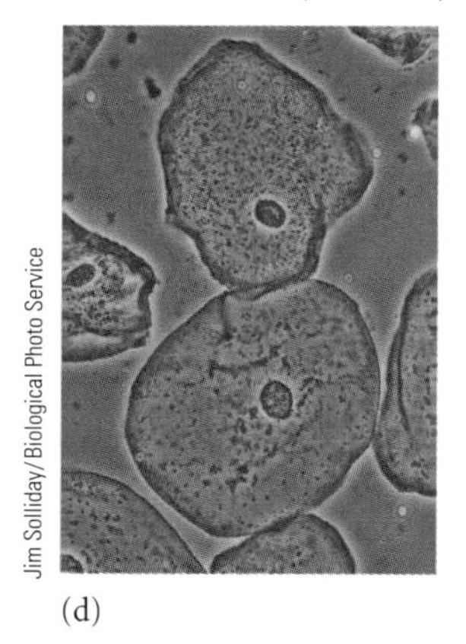

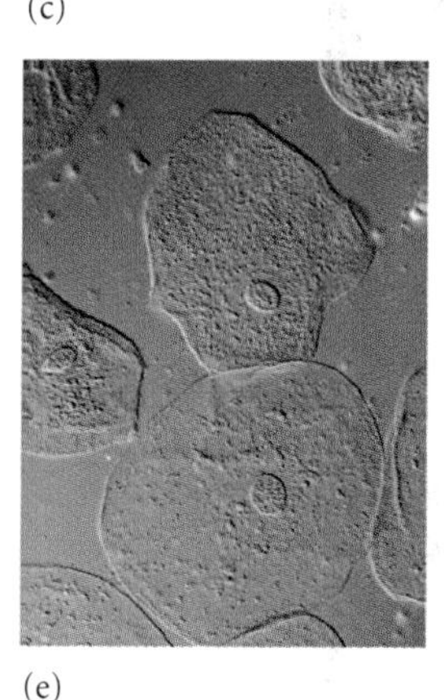

FIGURE **4-3** **Viewing cells with various types of microscopes.**

(a) Using a crude microscope that he constructed, Robert Hooke looked at a thin slice of cork and drew what he saw. More sophisticated microscopes and techniques enable biologists to view cells in more detail. Unstained epithelial cells from the skin of a human cheek are compared using **(b)** bright-field (transmitted light), **(c)** dark-field, **(d)** phase-contrast, and **(e)** Nomarski differential interference microscopy. Bright-field can be enhanced by staining. The phase-contrast and differential interference microscopes enhance detail by increasing the differences in optical density in different regions of the cells.

Light microscopes are used to study stained or living cells

The **light microscope (LM),** the type used by most students, consists of a tube with glass lenses at each end. Because it contains several lenses, the modern light microscope is referred to as a *compound microscope.* Visible light passes through the specimen being observed and through the lenses. Light is refracted (bent) by the lenses, magnifying the image.

Two features of a microscope determine how clearly a small object can be viewed: magnification and resolving power. **Magnification** is the ratio of the size of the image seen with the microscope to the actual size of the object. The best light microscopes usually magnify an object no more than 1000 times. **Resolution,** or **resolving power,** is the capacity to distinguish fine detail in an image; it is defined as the minimum distance between two points at which they can both be seen separately rather than as a single, blurred point. Resolving power depends on the quality of the lenses and the wavelength of the illuminating light. As the wavelength decreases, the resolution increases. The visible light used by light microscopes has wavelengths ranging from about 400 nm (violet) to 700 nm (red); this limits the resolution of the light microscope to details no smaller than the diameter of a small bacterial cell (about 1 μm).

By the early 20th century, refined versions of the light microscope, as well as certain organic compounds that specifically stain different cell structures, became available. Using these tools, biologists discovered that cells contain many different internal structures, the organelles. The contribution of organic chemists in developing biological stains was essential to this understanding, because the interior of many cells is transparent. Most methods used to prepare and stain cells for observation, however, also kill them in the process.

Living cells can now be studied using light microscopes with special optical systems. In *bright-field microscopy,* an image is formed by transmitting light through a cell in culture (Fig. 4-3b). Because there is little contrast, the details of cell structure are not visible. In *dark-field microscopy,* rays of light are directed from the side and only scattered light enters the lenses. The cell is visible as a bright object against a dark background (Fig. 4-3c). *Phase contrast microscopy* and *differential-interference-contrast microscopy (Nomarski)* take advantage of variations in density within the cell (Fig. 4-3d and e). (These variations in density cause differences in the way various regions of the cytoplasm refract (bend light). Using these microscopes, scientists can observe living cells in action, with numerous internal structures that are constantly changing shape and location.

Cell biologists use a *fluorescence microscope* to detect the locations of specific molecules in cells. Fluorescent stains (like paints that glow under black light) are molecules that absorb light energy of one wavelength and then release some of that energy as light of a longer wavelength. One such stain binds specifically to DNA molecules and emits green light after absorbing ultraviolet light. Cells can be stained, and the location of the DNA can be determined, by observing the source of the green fluorescent light within the cell.

Some fluorescent stains are chemically bonded to *antibodies,* protein molecules important in internal defense. The antibody then binds to a highly specific region of a molecule in the cell. A single type of antibody molecule binds to only one type of structure, such as a part of a specific protein or some of the sugars in a specific polysaccharide. Purified fluorescent antibodies known to bind to a specific protein isolated from a cell are used to determine where that protein is located. Powerful computer imaging methods have allowed the development of the *confocal fluorescence microscope,* which greatly improves the resolution of structures labeled by fluorescent dyes (see the micrograph in the chapter introduction).

Cell biologists are developing new techniques for viewing cells using computers, lasers, and photodetectors. Computer-based image processing combines multiple images to produce 3-D views.

Electron microscopes provide a high-resolution image that can be greatly magnified

Even with improved microscopes and techniques for staining cells, ordinary light microscopes can distinguish only the gross details of many cell parts (Fig. 4-4a). In most cases, you can clearly see only the outline of an organelle and its ability to be stained by some dyes and not by others. With the development of the **electron microscope (EM),** which came into wide use in the 1950s, researchers began to study the fine details, or **ultrastructure,** of cells.

Whereas the best light microscopes have about 500 times more resolution than the human eye, the electron microscope multiplies the resolving power by more than 10,000. This is because electrons have very short wavelengths, on the order of about 0.1 to 0.2 nm. Although such resolution is difficult to achieve with biological material, researchers can approach that resolution when examining isolated molecules such as proteins and DNA. This high degree of resolution permits very high magnifications of 250,000 times or more as compared to typical magnifications of no more than 1000 times in light microscopy.

The image formed by the electron microscope is not directly visible. The electron beam itself consists of energized electrons, which, because of their negative charge, can be focused by electromagnets just as images are focused by glass lenses in a light microscope (see Fig. 4-4b). For **transmission electron microscopy (TEM),** the specimen is embedded in plastic and then cut into extraordinarily thin sections (50 to 100 nm thick) with a glass or diamond knife. A section is then placed on a small metal grid. The electron beam passes through the specimen and then falls onto a photographic plate or a fluorescent screen.

When you look at TEMs in this chapter (and elsewhere), keep in mind that each represents only a thin cross section of a cell. To reconstruct a 3-D view of the cell interior, the cell biologist studies many consecutive sectional views (called *serial sections*) through the object. (To understand the enormity of such a task, imagine trying to reconstruct an image of the contents of your home from a set of hundreds of consecutive 5-cm sections.)

Researchers detect certain specific molecules in electron microscope images by using antibody molecules to which very tiny gold particles are bound. The dense gold particles block the electron beam and identify the location of the proteins recognized by the antibodies as precise black spots on the electron micrograph.

In another type of electron microscope, the **scanning electron microscope (SEM),** the electron beam does not pass through

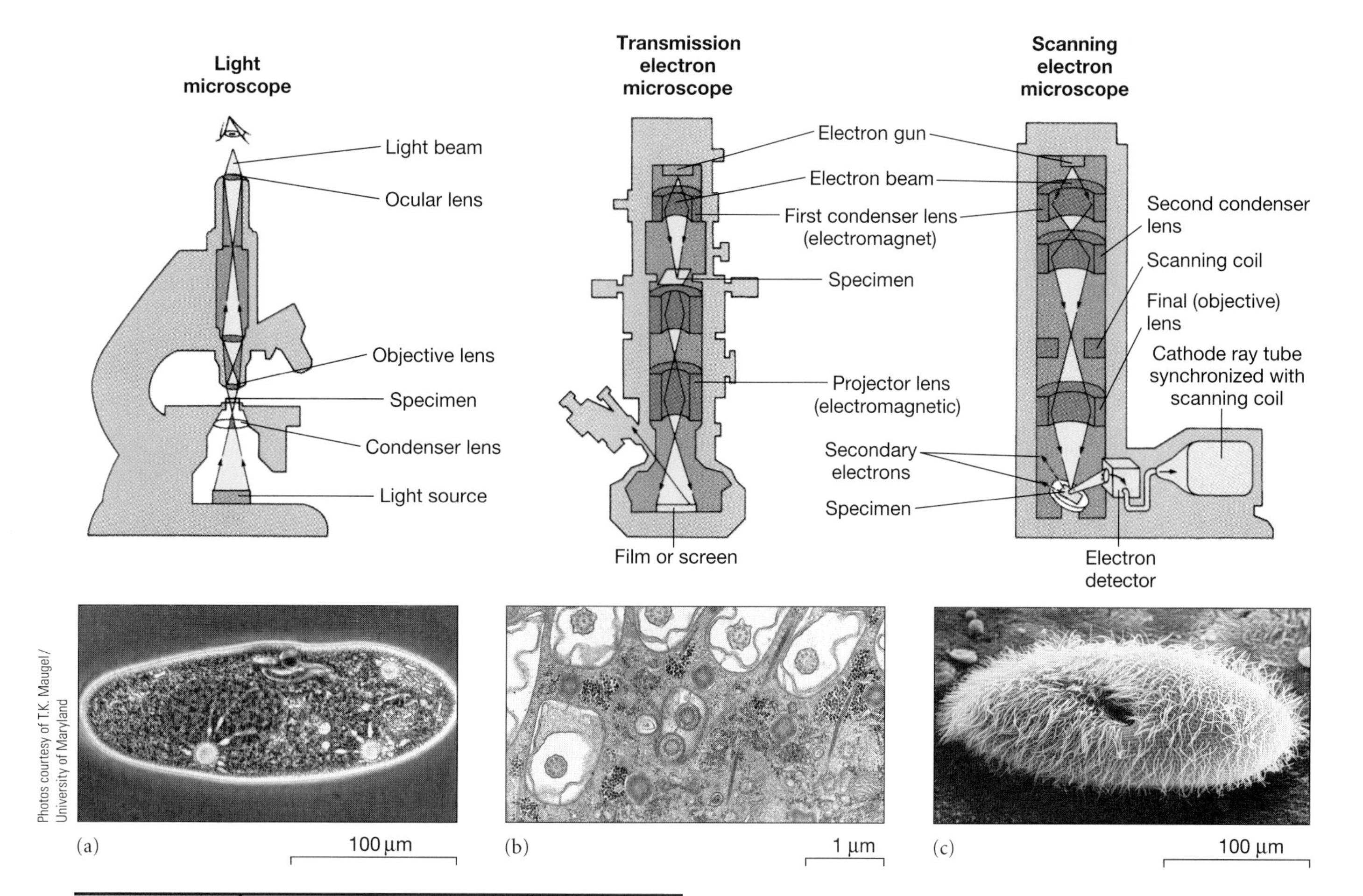

FIGURE **4-4** **Comparing light and electron microscopy.**

Distinctive images of cells, such as the protist *Paramecium* shown in these photomicrographs, are provided by three types of microscopes. **(a)** A phase contrast light microscope can be used to view stained or living cells, but at relatively low resolution. **(b)** The transmission electron microscope (TEM) produces a high-resolution image that can be greatly magnified. A small part of a thin slice through the *Paramecium* is shown. **(c)** The scanning electron microscope (SEM) is used to provide a clear view of surface features.

the specimen. Instead, the specimen is coated with a thin film of gold or some other metal. When the electron beam strikes various points on the surface of the specimen, secondary electrons are emitted whose intensity varies with the contour of the surface. The recorded emission patterns of the secondary electrons give a 3-D picture of the surface (see Fig. 4-4c). The SEM provides information about the shape and external features of the specimen that cannot be obtained with the TEM.

Note that the LM, TEM, and SEM are focused by similar principles. A beam of light or an electron beam is directed by the condenser lens onto the specimen and is magnified by the objective lens and the eyepiece in the light microscope or by the objective lens and the projector lens in the TEM. The TEM image is focused onto a fluorescent screen, and the SEM image is viewed on a type of television screen. Lenses in electron microscopes are actually magnets that bend the beam of electrons.

Cell fractionation enables the study of cell components

PROCESS OF SCIENCE

The EM is a powerful tool for studying cell structure, but it has limitations. The methods used to prepare cells for electron microscopy kill them and may alter their structure. Furthermore, electron microscopy provides few clues about the functions of organelles and other cell components. To determine what organelles actually do, researchers purify different parts of cells so that they can be studied by physical and chemical methods.

Cell fractionation is a technique for purifying organelles. Generally, cells are broken apart as gently as possible, and the mixture, referred to as the *cell extract,* is subjected to centrifugal force by spinning in a device called a **centrifuge** (Fig. 4-5 top). An ultracentrifuge, a very powerful centrifuge, can spin at speeds exceeding 100,000 revolutions per minute (rpm), generating a centrifugal force of 500,000 × G (a G is equal to the force of gravity). Centrifugal force separates the extract into two fractions: a pellet and a supernatant. The *pellet* that forms at the bottom of the tube contains heavier materials, such as nuclei, packed together. The *supernatant,* the liquid above the pellet, contains lighter particles, dissolved molecules, and ions.

The supernatant can be centrifuged again at a higher speed to obtain a pellet that contains the next heaviest cell components, for example, mitochondria and chloroplasts. In **differential centrifugation,** the supernatant is spun at successively higher

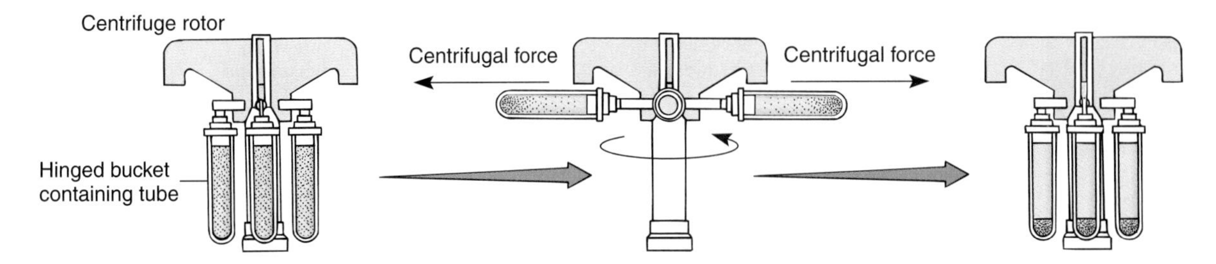

(a) Centrifugation

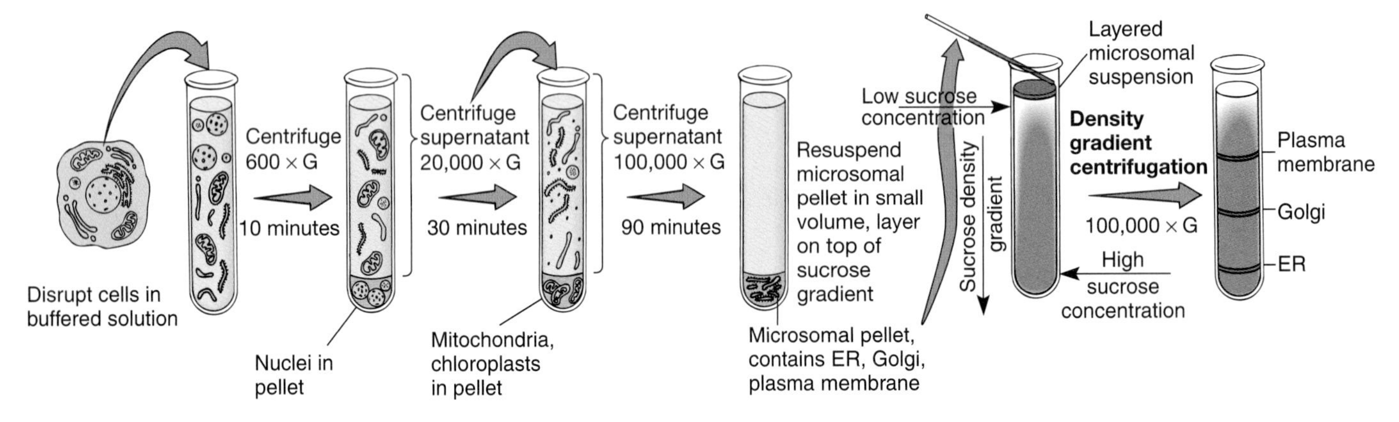

(b) Differential Centrifugation

FIGURE **4-5** **Cell fractionation.**

(a) (Top) In centrifugation, large or very dense particles move toward the bottom of a tube. **(b)** (Bottom) Differential centrifugation enables cell biologists to separate cell structures into various fractions by spinning the suspension at increasing revolutions per minute. Membranes and organelles from the resuspended pellets can then be further purified by density gradient centrifugation, shown as the last step in the figure. G is the force of gravity. ER is the endoplasmic reticulum.

speeds, permitting various cell components to be separated on the basis of their different sizes and densities (Fig. 4-5 bottom).

Cell components in the resuspended pellets are further purified by **density gradient centrifugation.** In this procedure, the resuspended pellet is placed in a layer on top of a density gradient, usually made up of a solution of sucrose (table sugar) and water. The concentration of sucrose is highest at the bottom of the tube and decreases gradually so that it is lowest at the top. Because the densities of organelles differ, each will migrate during centrifugation and form a band at the position in the gradient where its own density equals that of the sucrose solution.

Purified organelles are examined to determine what kinds of proteins and other molecules they might contain, as well as the nature of the chemical reactions that take place within them. Cell biologists typically use a combination of experimental approaches to study the functions of cell structures.

Review

- What is the main advantage of the electron microscope? Explain.
- What is cell fractionation? Describe the process.

BiologyNow™ Assess your understanding of **methods for studying cells** by taking the pretest on your BiologyNow CD-ROM.

PROKARYOTIC AND EUKARYOTIC CELLS

Learning Objective

4 Compare and contrast the general characteristics of prokaryotic and eukaryotic cells, and contrast plant and animal cells.

Recall from Chapter 1 that two basic types of cells are known: prokaryotic and eukaryotic. Bacteria and archaea are prokaryotic cells. All other known organisms consist of eukaryotic cells. Prokaryotic cells are typically smaller than eukaryotic cells. In fact, the average prokaryotic cell is only about one 10th the diameter of the average eukaryotic cell. In prokaryotic cells, the DNA is not enclosed in a nucleus. Instead, the DNA is located in a limited region of the cell called a **nuclear area,** or **nucleoid,** which is not enclosed by a membrane (Fig. 4-6). The term *prokaryotic,* meaning "before the nucleus" refers to this major difference between prokaryotic and eukaryotic cells. Other types of internal membrane–enclosed organelles are also absent in prokaryotic cells.

Like eukaryotic cells, prokaryotic cells have a plasma membrane that confines the contents of the cell to an internal compartment. In some prokaryotic cells the plasma membrane may be folded inward to form a complex of membranes along which many of the cell's metabolic reactions take place. Most prokaryotic cells have **cell walls,** which are extracellular structures that enclose the entire cell, including the plasma membrane. Many prokaryotes have **flagella** (sing., *flagellum*), long fibers that project from the surface of the cell. Prokaryotic flagella, which operate like propellers, are important in locomotion.

The dense internal material of the bacterial cell contains **ribosomes,** small complexes of ribonucleic acid (RNA) and protein that synthesize polypeptides. The ribosomes of prokaryotic

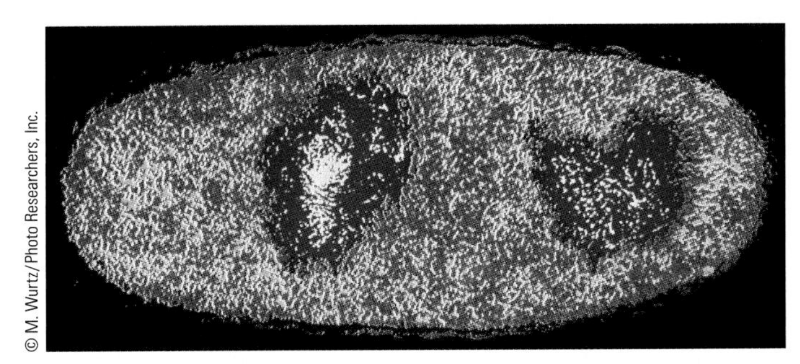

© M. Wurtz/Photo Researchers, Inc.

FIGURE **4-6** **TEM of a prokaryotic cell.**

This bacterium *(E. coli)*, has two nuclear areas (blue areas) because it is preparing to divide. The nucelar material (DNA) appears as pale fibrils within the blue patches.

cells are smaller than those found in eukaryotic cells. Prokaryotic cells also contain storage granules that hold glycogen, lipid, or phosphate compounds. This chapter focuses primarily on eukaryotic cells. Prokaryotes are discussed in more detail in Chapter 23.

Eukaryotic cells are characterized by highly organized membrane-enclosed organelles, including a prominent *nucleus,* which contains the hereditary material, DNA. The term *eukaryotic* means "true nucleus." Early biologists thought cells consisted of a homogeneous jelly, which they called *protoplasm.* With the electron microscope and other modern research tools, perception of the environment within the cell has been greatly expanded. We now know the cell is highly organized and complex (Figs. 4-7 and 4-8). The eukaryotic cell has its own control center, internal transportation system, power plants, factories for making needed materials, packaging plants, and even a "self-destruct" system. Biologists refer to the part of the cell outside the nucleus as **cytoplasm** and the part of the cell within the nucleus as **nucleoplasm.** Various organelles are suspended within the fluid component of the cytoplasm, which is called the **cytosol.** Therefore, the term *cytoplasm* includes both the cytosol and all the organelles other than the nucleus.

The many specialized organelles of eukaryotic cells solve some of the problems associated with large size, so eukaryotic cells can be much larger than prokaryotic cells. Eukaryotic cells also differ from prokaryotic cells in having a supporting framework, or cytoskeleton, important in maintaining shape and transporting materials within the cell.

Some organelles are only in specific cells. For example, *chloroplasts,* structures that trap sunlight for energy conversion, are only in cells that carry on photosynthesis, such as certain plant or algal cells. Most bacteria, fungi, and plant cells are surrounded by a *cell wall* external to the plasma membrane. Plant cells also contain a large, membrane-enclosed *vacuole.* We discuss these and other differences among major types of cells throughout this chapter. Plant and animal cells are compared in Figures 4-7 and 4-8 and also in Figures 4-9 and 4-10.

Review

- What are two important differences between prokaryotic and eukaryotic cells?
- How might we explain the larger size of eukaryotic cells compared to prokaryotic cells?

BiologyNow™ Assess your understanding of **prokaryotic and eukaryotic cells** by taking the pretest on your BiologyNow CD-ROM.

CELL MEMBRANES

Learning Objective

5 Describe three functions of cell membranes.

Membranes divide the eukaryotic cell into compartments, and their unique properties enable membranous organelles to carry out a wide variety of functions. For example, cell membranes never have free ends; therefore, a membranous organelle always contains at least one enclosed internal space or compartment. These membrane-enclosed compartments allow certain cell activities to be localized within specific regions of the cell. Reactants located in only a small part of the total cell volume are far more likely to come in contact, dramatically increasing the rate of the reaction. Membrane-enclosed compartments keep certain reactive compounds away from other parts of the cell that they might adversely affect. Compartmentalizing also allows many different activities to go on simultaneously.

Membranes allow cells to store energy. The membrane serves as a barrier that is analogous to a dam on a river. A difference in the concentration of some substance on the two sides of a membrane is a form of stored energy or *potential energy* (see Chapter 6). As particles of the substance move across the membrane from the side of higher concentration to the side of lower concentration, the cell converts some of this potential energy to the chemical energy of ATP molecules. This process of energy conversion (discussed in Chapters 7 and 8) is a basic mechanism that cells use to capture and convert the energy necessary to sustain life.

Membranes also serve as important work surfaces. For example, many chemical reactions in cells are carried out by enzymes that are bound to membranes. Because the enzymes that carry out successive steps of a series of reactions are organized close together on a membrane surface, certain series of chemical reactions occur more rapidly.

In a eukaryotic cell, several types of membranes are generally considered part of the internal membrane system, or **endomembrane system.** In Figures 4-9 and 4-10 on page 76 (also see Figs. 4-7 and 4-8), notice how membranes divide the cell into many compartments: the nucleus, endoplasmic reticulum (ER), Golgi complex, lysosomes, vesicles, and vacuoles. Although it is not internal, the plasma membrane is also included because it participates in the activities of the endomembrane system. (Mitochondria and chloroplasts are also separate compartments but are not generally considered part of the endomembrane system, because they function somewhat independently of other membranous organelles.)

Some organelles have direct connections between their membranes and compartments. Others transport materials in **vesicles,** small, membrane-enclosed sacs formed by "budding" from the membrane of another organelle. Vesicles also carry materials from one organelle to another. Through a complex

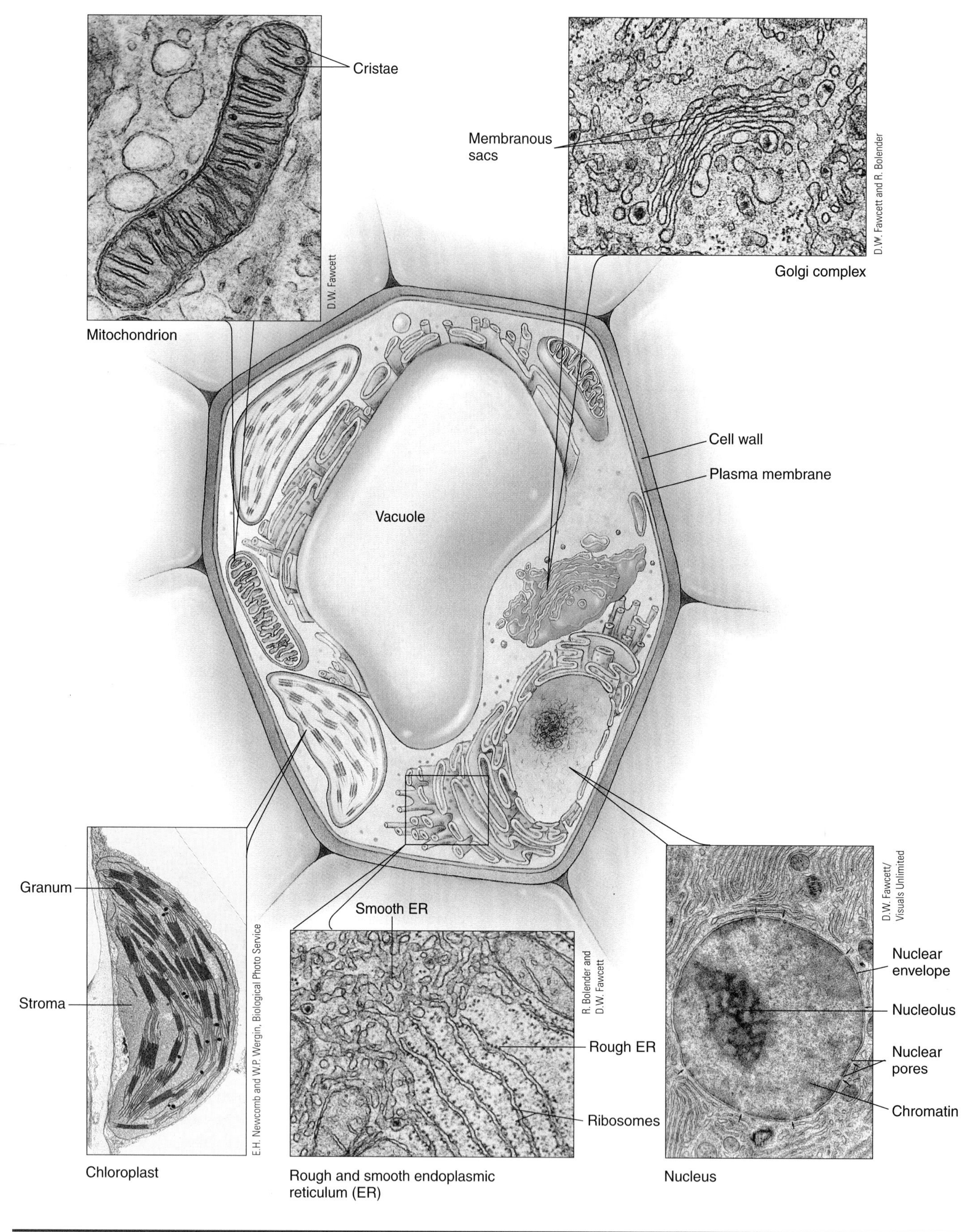

ACTIVE FIGURE **4-7** | **Composite diagram of a plant cell.**

Chloroplasts, a cell wall, and prominent vacuoles are characteristic of plant cells. The TEMs show certain structures or areas of the cell. Some plant cells do not have all the organelles shown here. For example, leaf and stem cells that carry on photosynthesis contain chloroplasts, whereas root cells do not. Many of the organelles, such as the nucleus, mitochondria, and endoplasmic reticulum (ER), are characteristic of all eukaryotic cells.

BiologyNow™ Test yourself on the **structure of the eukaryotic cell** by clicking on this figure on your BiologyNow CD-ROM.

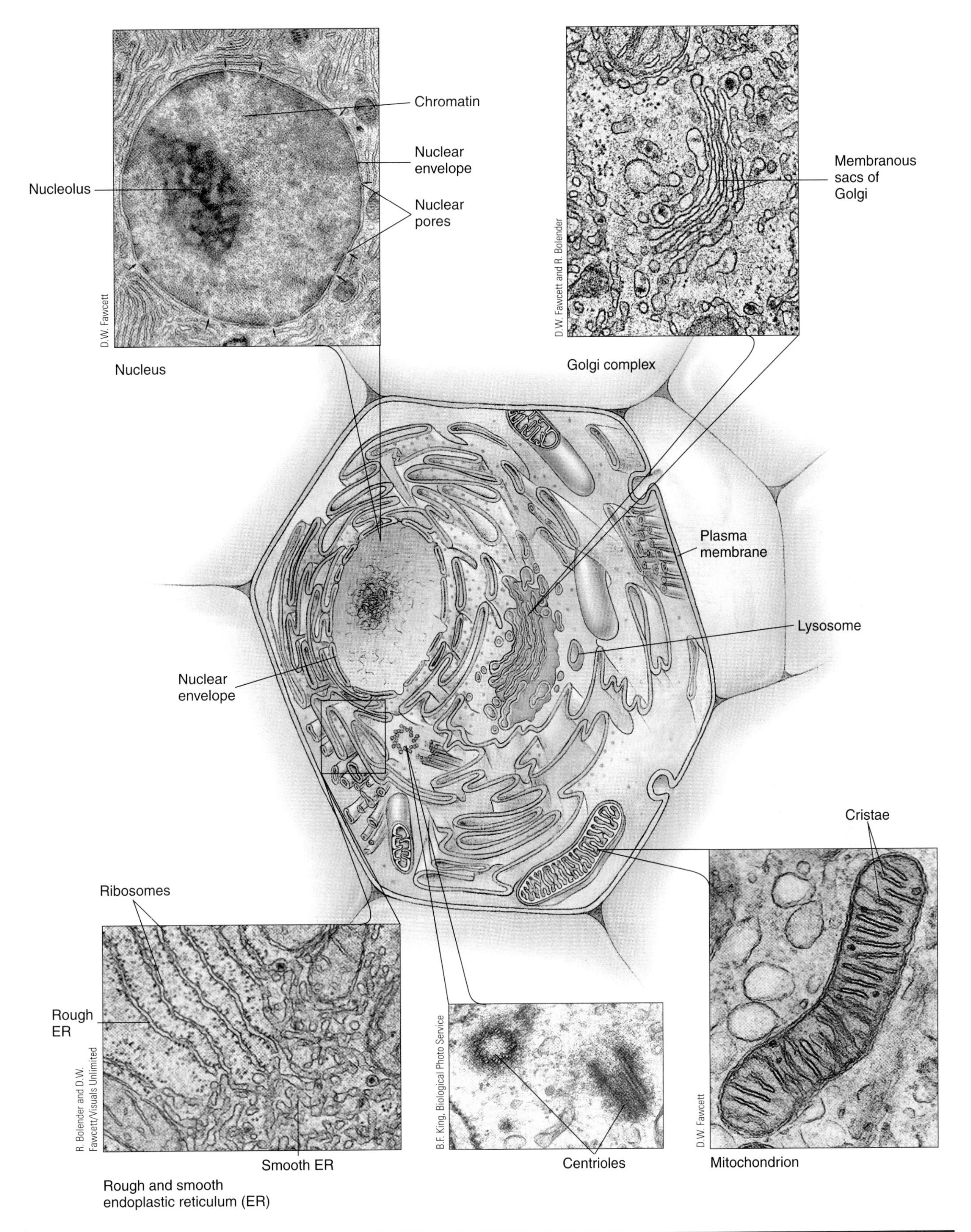

FIGURE **4-8** | **Composite diagram of an animal cell.**

This generalized animal cell is shown in a realistic context surrounded by adjacent cells, which cause it to be slightly compressed. The TEMs show the structure of various organelles. Depending on the cell type, certain organelles may be more or less prominent.

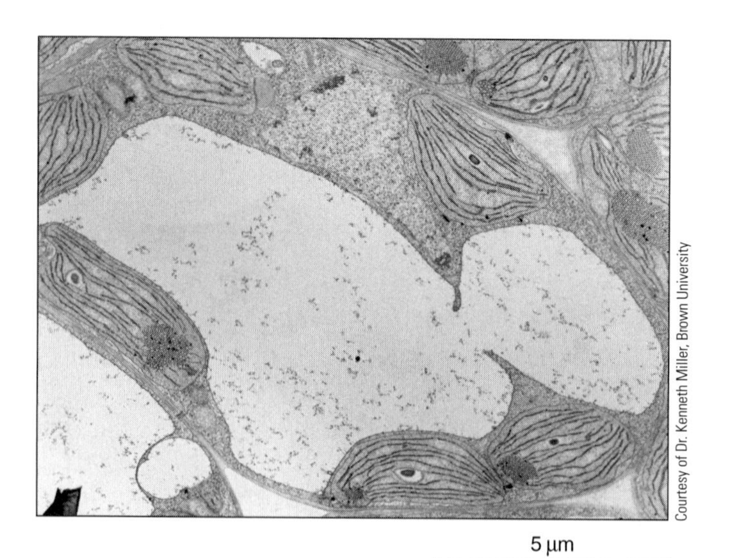

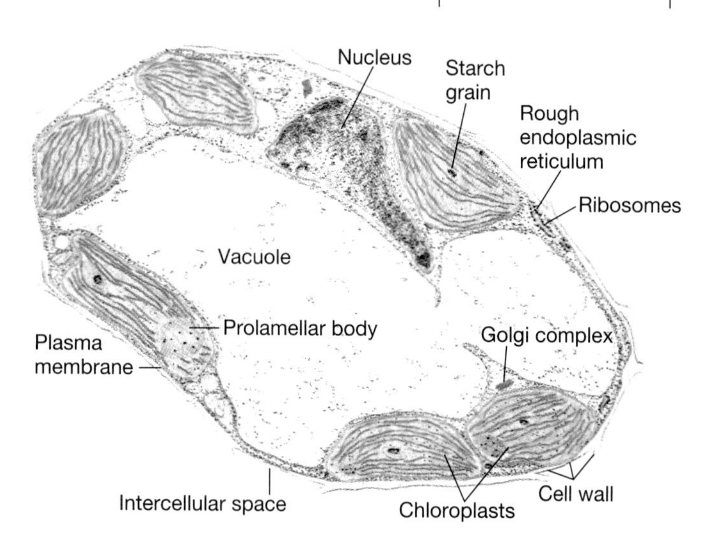

FIGURE **4-9** **TEM of a plant cell and an interpretive drawing.**

Most of this cross section of a cell from the leaf of a young bean plant (*Phaseolus vulgaris*) is dominated by a vacuole. Prolamellar bodies are membranous regions typically seen in developing chloroplasts.

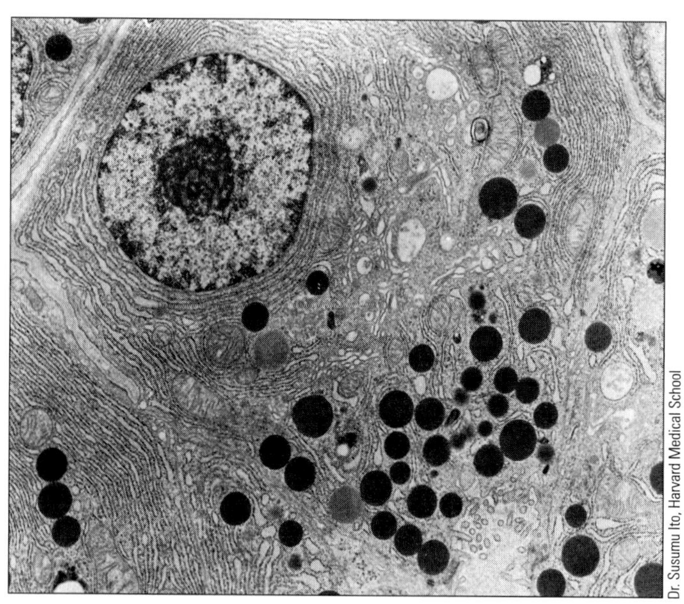

FIGURE **4-10** **TEM of a human pancreas cell and an interpretive drawing.**

Most of the structures of a typical animal cell are present. However, like most cells, this one has certain structures associated with its specialized functions. Pancreas cells such as the one shown here secrete large amounts of digestive enzymes. The large, dark, circular bodies in the TEM and the corresponding structures in the drawing are zymogen granules containing inactive enzymes. When released from the cell, the enzymes catalyze chemical reactions such as the breakdown of peptide bonds of ingested proteins in the intestine. Most of the membranes visible in this section are part of the rough endoplasmic reticulum, an organelle specialized to manufacture protein.

series of steps, a vesicle can form as a "bud" from one membrane and then be transported to another membrane to which it fuses, thus delivering its contents into another compartment.

Review

- How do membrane-enclosed organelles facilitate cell metabolism?
- What organelles belong to the endomembrane system?

BiologyNow™ Assess your understanding of **cell membranes** by taking the pretest on your BiologyNow CD-ROM.

THE CELL NUCLEUS

Learning Objective

6 Describe the structure and functions of the nucleus.

Typically, the **nucleus** is the most prominent organelle in the cell. It is usually spherical or oval in shape and averages 5 μm in diameter. Because of its size and the fact that it often occupies a relatively fixed position near the center of the cell, some early investigators guessed long before experimental evidence was available that the nucleus served as the control center of the cell (see the *Focus On:* Acetabularia and the Control of Cell Activities). Most cells have one nucleus, although there are exceptions.

The **nuclear envelope** consists of two concentric membranes that separate the nuclear contents from the surrounding cytoplasm (Fig. 4-11). These membranes are separated by about 20 to 40 nm. At intervals the membranes come together to form **nuclear pores,** which consist of protein complexes. Nuclear pores regulate the passage of materials between nucleoplasm and cytoplasm. How materials are transported through nuclear pores and how the process is regulated are areas of active research.

The cell stores information in the form of DNA, and most of the cell's DNA is located inside the nucleus. When a cell divides, the information stored in DNA must be reproduced and passed intact to the two daughter cells. DNA has the unique ability to make an exact duplicate of itself through a process called *replication.* Recall from Chapter 3 that DNA molecules consist of sequences of nucleotides called **genes,** which contain the chemically coded instructions for producing the proteins needed by the cell. The nucleus controls protein synthesis by transcribing its information in *messenger RNA* molecules. Messenger RNA moves into the cytoplasm where proteins are manufactured.

DNA is associated with proteins, forming a complex known as **chromatin,** which appears as a network of granules and strands in cells that are not dividing. Although chromatin appears disorganized, it is not. Because DNA molecules are extremely long and thin, they must be packed inside the nucleus in a very regular fashion as part of structures called **chromosomes.** In dividing cells, the chromosomes become visible as distinct threadlike structures. If the DNA in the 46 chromosomes of one human cell could be stretched end to end, it would extend for 2 m!

Most nuclei have one or more compact structures called **nucleoli** (sing., *nucleolus*). A nucleolus, which is *not* enclosed by a membrane, usually stains differently from the surrounding chromatin. Each nucleolus contains a nucleolar organizer, made up of chromosomal regions containing instructions for making the type of RNA in ribosomes. This ribosomal RNA is synthesized in the nucleolus. The proteins needed to make ribosomes

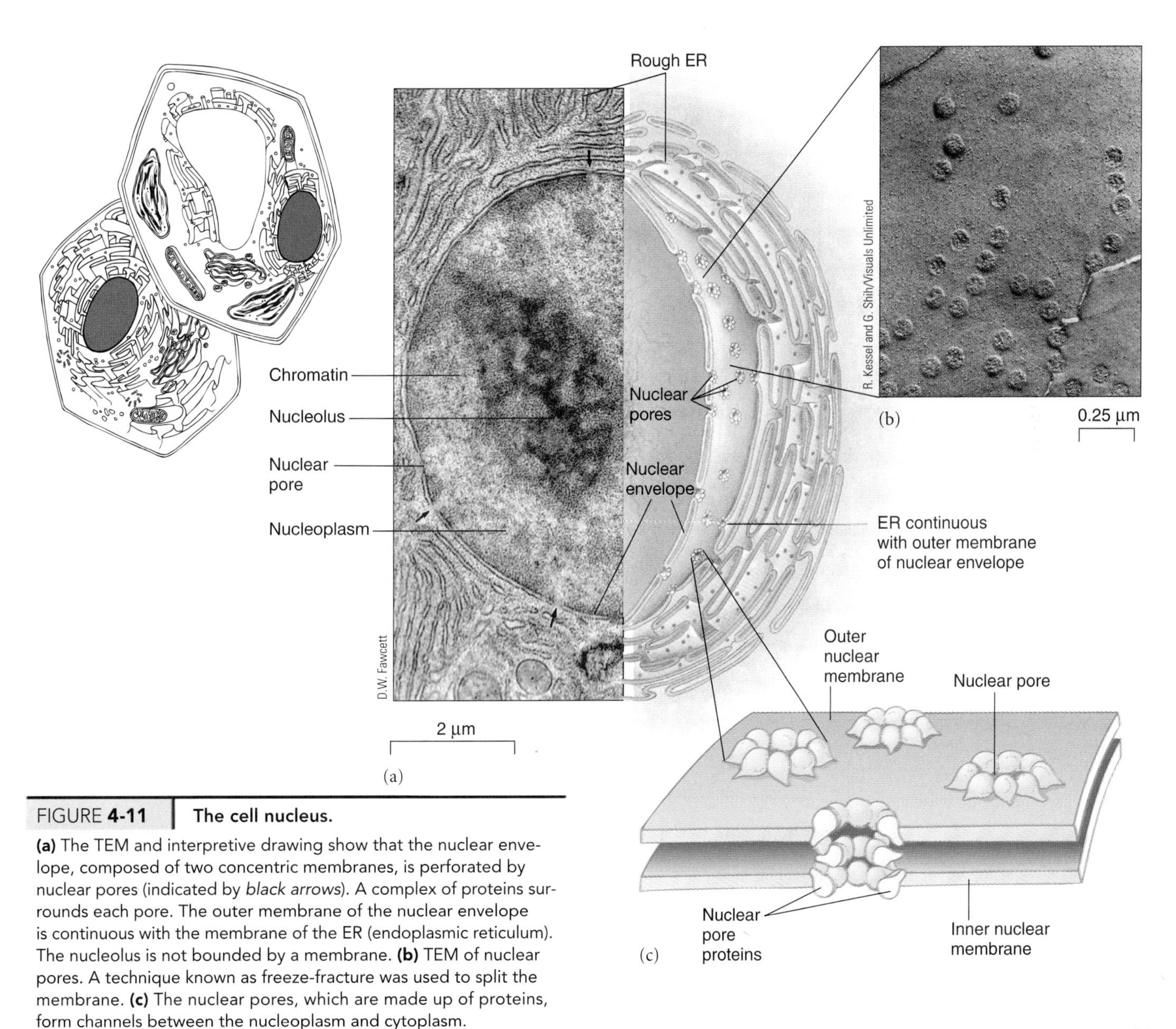

FIGURE **4-11** **The cell nucleus.**

(a) The TEM and interpretive drawing show that the nuclear envelope, composed of two concentric membranes, is perforated by nuclear pores (indicated by *black arrows*). A complex of proteins surrounds each pore. The outer membrane of the nuclear envelope is continuous with the membrane of the ER (endoplasmic reticulum). The nucleolus is not bounded by a membrane. **(b)** TEM of nuclear pores. A technique known as freeze-fracture was used to split the membrane. **(c)** The nuclear pores, which are made up of proteins, form channels between the nucleoplasm and cytoplasm.

PROCESS OF SCIENCE

To the romantically inclined, the little seaweed *Acetabularia* resembles a mermaid's wineglass, although the literal translation of its name, "vinegar cup," is somewhat less elegant (Fig. A). In the 19th century, biologists discovered that this marine eukaryotic alga consists of a single cell. At up to 5 cm (2 in) in length, *Acetabularia* is small for a seaweed but gigantic for a cell. It consists of a rootlike **holdfast;** a long, cylindrical **stalk;** and a cuplike **cap.** The nucleus is in the holdfast, about as far away from the cap as it can be. Because it is a single giant cell, *Acetabularia* is easy for researchers to manipulate.

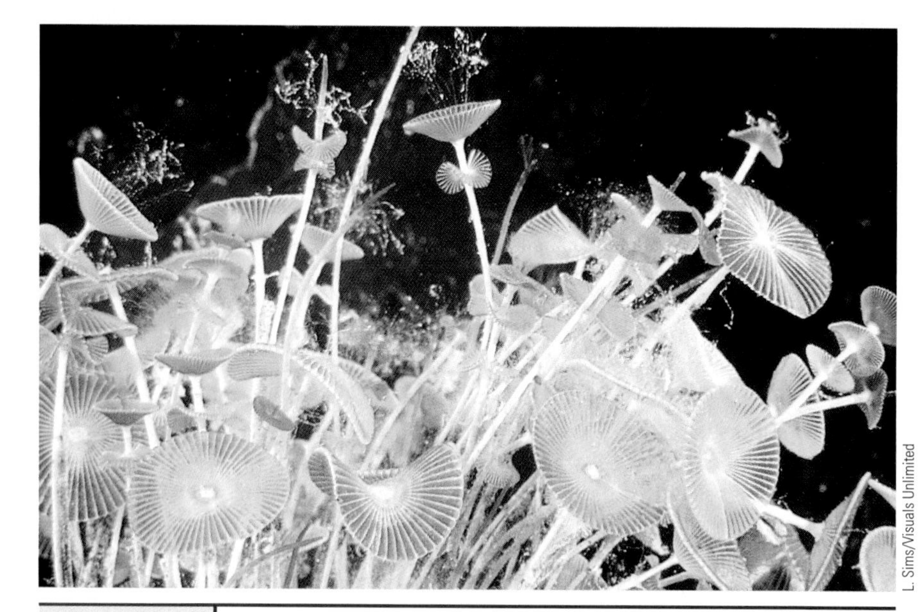

FIGURE **A** **Light micrograph of *Acetabularia*.**

What controls the cap shape?

If the cap of *Acetabularia* is removed experimentally, another one grows after a few weeks. Such a response, common among simple organisms, is called **regeneration.** This fact attracted the attention of investigators, especially Danish biologist J. Hämmerling and Belgian biologist J. Brachet, who became interested in whether a relationship exists between the nucleus and the physical characteristics of the alga. Because of its great size, *Acetabularia* could be subjected to surgery that would be impossible with smaller cells. During the 1930s and 1940s, these researchers performed brilliant experiments that in many ways laid the foundation for much of our modern knowledge of the nucleus. Two species were used for most experiments: *A. mediterranea,* which has a smooth cap, and *A. crenulata,* which has a cap divided into a series of finger-like projections.

The kind of cap that is regenerated depends on the species of *Acetabularia* used in the experiment. As you might expect, *A. crenulata* regenerates a "cren" cap, and *A. mediterranea* regenerates a "med" cap. But it is possible to graft together two capless algae of different species. Through this union, they regenerate a common cap that has characteristics intermediate between those of the two species involved (Fig. B).

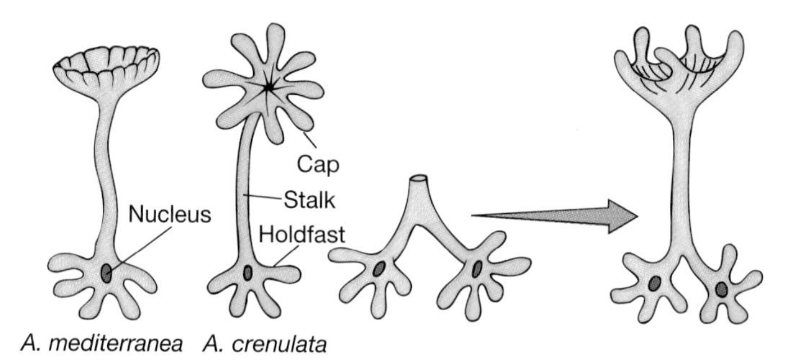

FIGURE **B**

Thus, it is clear that something in the stalk or holdfast controls cap shape.

Stalk exchange experiments.

By telescoping the cell walls of the two into one another, it is possible to attach a section of *Acetabularia* to a holdfast that is not its own. In this way the stalks and holdfasts of different species may be intermixed.

First, we take *A. mediterranea* and *A. crenulata* and remove their caps. Then we sever the stalks from the holdfasts. Finally, we exchange the parts (Fig. C). What happens? Not, perhaps, what you would expect! The caps that regenerate are characteristic not of the species donating the holdfasts but of those donating the stalks!

However, if the caps are removed once again, this time the caps that regenerate are characteristic of the species that donated the holdfasts. This continues to be the case no matter how many more times the regenerated caps are removed.

From all these results Hämmerling and Brachet deduced that the ultimate control of the *Acetabularia* cell is associated with the holdfast. Because there is

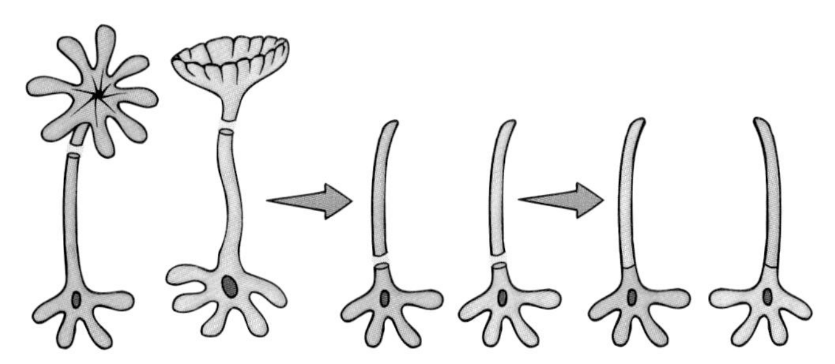
Stalks and holdfasts exchanged

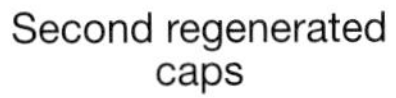

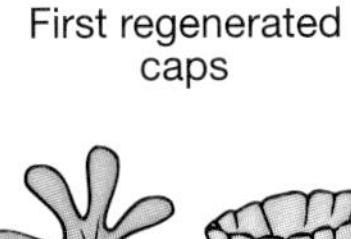

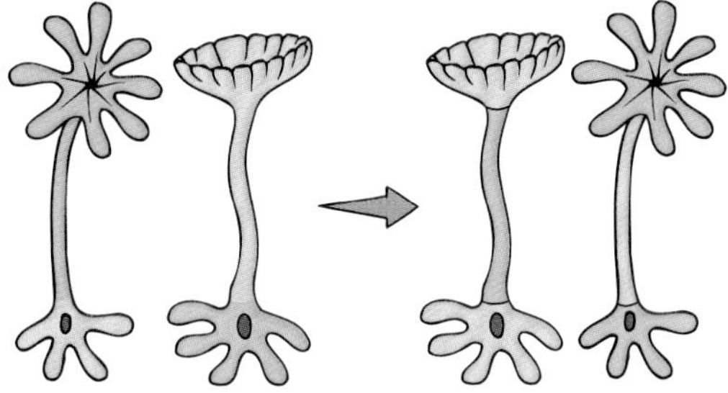
First regenerated caps
Second regenerated caps

FIGURE **C**

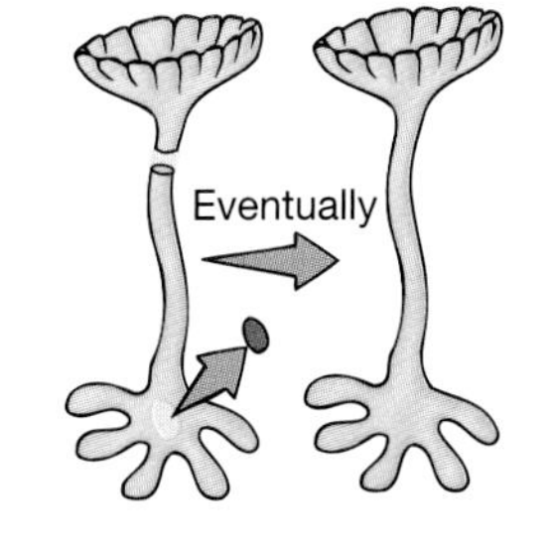
Eventually

FIGURE **D**

Eventually

FIGURE **E**

a time lag before the holdfast appears to take over, they hypothesized it produces some temporary cytoplasmic messenger substance whereby it exerts its control. They further hypothesized that the grafted stalks initially contain enough of the substance from their former holdfasts to regenerate a cap of the former shape. But this still leaves the question of how the holdfast exerts its apparent control. An obvious suspect is the nucleus.

Nuclear exchange experiments.

If the nucleus is removed and the cap cut off, a new cap regenerates (Fig. D). *Acetabularia,* however, can usually regenerate only once without a nucleus. If the nucleus of another species is now inserted and the cap is cut off once again, a new cap regenerates that is characteristic of the species of the nucleus (Fig. E)! If two kinds of nuclei are inserted, the regenerated cap is intermediate in shape between those of the species that donated the nuclei.

As a result of these and other experiments, biologists began to develop some basic ideas about the control of cell activities. The holdfast controls the cell because the nucleus is located there. Further, the nucleus is the apparent source of some "messenger substance" that temporarily exerts control but is limited in quantity and cannot be produced without the nucleus (Fig. F). This information helped provide a starting point for research on the role of nucleic acids in the control of all cells.

Cell biologists extended these early findings as they developed our modern view of information flow and control in the cell. We now know that the nucleus of eukaryotes controls the cell's activities because it contains DNA, the ultimate source of biological information. DNA passes its information to successive generations because it is able to precisely replicate itself. The information in DNA specifies the sequence of amino acids in all the proteins of the cell. To carry out its mission, DNA uses a type of ribonucleic acid (RNA) as a cytoplasmic messenger substance.

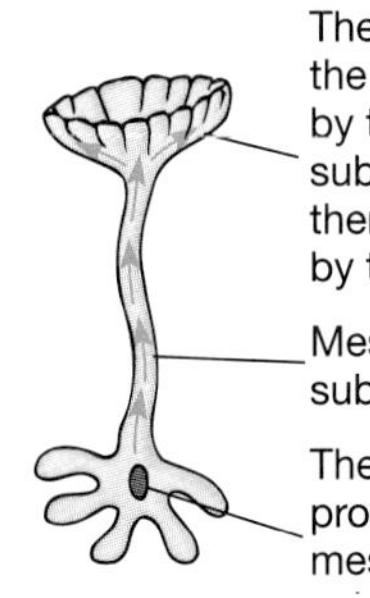
The characteristics of the cell are governed by the messenger substance, and therefore ultimately by the nucleus.
Messenger substance
The nucleus produces the messenger

FIGURE **F**

are synthesized in the cytoplasm and imported into the nucleolus. Ribosomal RNA and proteins are then assembled into ribosomal subunits that leave the nucleus through the nuclear pores.

Review

- How does the nucleus store information?
- What is the function of the nuclear envelope?

Biology Now™ Assess your understanding of the **cell nucleus** by taking the pretest on your BiologyNow CD-ROM.

ORGANELLES IN THE CYTOPLASM

Learning Objectives:

7 Distinguish between smooth and rough endoplasmic reticulum in terms of both structure and function.

8 Trace the path of proteins synthesized in the rough endoplasmic reticulum as they are subsequently processed, modified, and sorted by the Golgi complex and then transported to specific destinations.

9 Describe the functions of lysosomes and peroxisomes.

10 Compare the functions of mitochondria and chloroplasts, and discuss ATP synthesis by each of these organelles.

Cell biologists have identified many types of organelles in the cytoplasm of eukaryotic cells. Among them are the endoplasmic reticulum, ribosomes, Golgi complex, lysosomes, peroxisomes, vacuoles, mitochondria, and chloroplasts. Eukaryotic cell structures and functions are summarized in Table 4-1.

The endoplasmic reticulum and ribosomes manufacture proteins

One of the most prominent features in the electron micrographs in Figures 4-7 and 4-8 is a maze of parallel internal membranes that encircle the nucleus and extend into many regions of the cytoplasm. This complex of membranes, the **endoplasmic reticulum (ER),** forms a network that makes up a significant part of the total volume of the cytoplasm in many cells. A higher-magnification TEM of the ER is shown in Figure 4-12. Remember that a TEM represents only a thin cross section of the cell, so there is a tendency to interpret the ER as a series of tubes. In fact, many ER membranes consist of a series of tightly packed and flattened, saclike structures that form interconnected compartments within the cytoplasm.

The internal space the membranes enclose is called the **ER lumen.** In most cells the ER lumen forms a single internal compartment that is continuous with the compartment formed between the outer and inner membranes of the nuclear envelope (see Fig. 4-11). The membranes of other organelles are not directly connected to the ER and appear to form distinct and separate compartments within the cytoplasm.

The ER membranes and lumen contain enzymes that catalyze many different types of chemical reactions. In some cases the membranes serve as a framework for systems of enzymes that carry out sequential biochemical reactions. The two surfaces of the membrane contain different sets of enzymes and represent regions of the cell with different synthetic capabilities, just as different regions of a factory are used to make different parts of a particular product. Still other enzymes are located within the ER lumen.

Two distinct regions of the ER can be distinguished in TEMs: rough ER and smooth ER. Although these regions have different functions, their membranes are connected and their internal spaces are continuous. **Smooth ER** has a tubular appearance and its outer membrane surfaces appear smooth. The smooth ER is the primary site of phospholipid, steroid, and fatty acid metabolism. Whereas the smooth ER may be a minor membrane component in some cells, extensive amounts of smooth ER are present in others. For example, extensive smooth ER is present in human liver cells, where it synthesizes and processes cholesterol and other lipids and serves as a major detoxification site. Enzymes located along the smooth ER of liver cells break down toxic chemicals such as carcinogens (cancer-causing agents). The cell then converts these compounds to water-soluble products that it excretes.

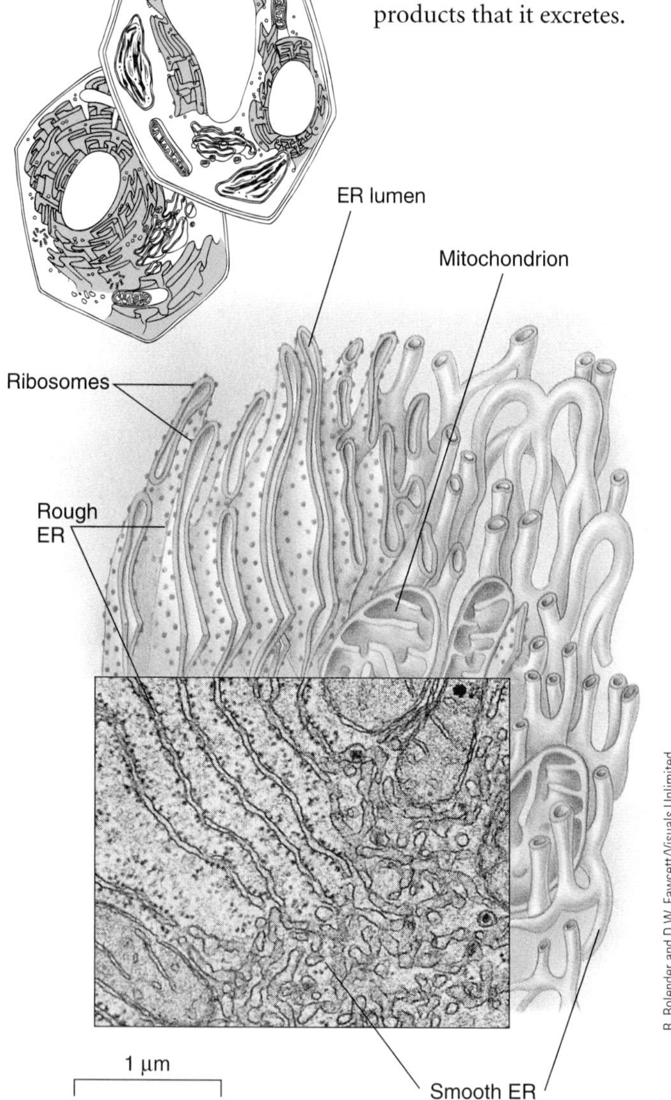

FIGURE **4-12** **Endoplasmic reticulum (ER).**
The TEM shows both rough and smooth ER in a liver cell.

TABLE 4-1 Eukaryotic Cell Structures and Their Functions

Structure	Description	Function
Cell Nucleus		
Nucleus	Large structure surrounded by double membrane; contains nucleous and chromosomes	Information in DNA is transcribed in RNA synthesis; specifies cell proteins
Nucleous	Granular body within nucleus; consists of RNA and protein	Site of ribosomal RNA synthesis; ribosome subunit assembly
Chromosomes	Composed of a complex DNA and protein known as *chromatin;* condense during cell division, becoming visible as rodlike structures	Contain genes (units of hereditary information) that govern structure and activity of cell
Cytoplasmic Organelles		
Plasma membrane	Membrane boundary of cell	Encloses cell contents; regulates movement of materials in and out of cell; helps maintain cell shape; communicates with other cells (also present in prokaryotes)
Endoplasmic reticulum (ER)	Network of internal membranes extending through cytoplasm	Synthesizes lipids and modifies many proteins; origin of intracellular transport vesicles that carry proteins
Smooth	Lacks ribosomes on outer surface	Lipid biosynthesis; drug detoxification
Rough	Ribosomes stud outer surface	Manufacture of many proteins destined for secretion or for incorporation into membranes
Ribosomes	Granules composed of RNA and protein; some attached to ER, some free in cytosol	Synthesize polypeptides in both prokaryotes and eukaryotes
Golgi complex	Stacks of flattened membrane sacs	Modifies proteins; packages secreted proteins; sorts other proteins to vacuoles and other organelles
Lysosomes	Membranous sacs (in animals)	Contain enzymes to break down ingested materials, secretions, wastes
Vacuoles	Membranous sacs (mostly in plants, fungi, algae)	Store materials, wastes, water; maintain hydrostatic pressure
Peroxisomes	Membranous sacs containing a variety of enzymes	Site of many diverse metabolic reactions
Mitochondria	Sacs consisting of two membranes; inner membrane is folded to form cristae and encloses matrix	Site of most reactions of cellular respiration; transformation of energy originating from glucose or lipids into ATP energy
Plastids (e.g., chloroplasts)	Double-membrane structure enclosing internal thylakoid membranes; chloroplasts contain chlorophyll in thylakoid membranes	Chloroplasts are site of photosynthesis; chlorophyll captures light energy; ATP and other energy-rich compounds are formed and then used to convert CO_2 to carbohydrate
Cytoskeleton		
Microtubules	Hollow tubes made of subunits of tubulin protein	Provide structural support; have role in cell and organelle movement and cell division; components of cilia, flagella, centrioles, basal bodies
Microfilaments	Solid, rodlike structures consisting of actin protein	Provide structural support; play role in cell and organelle movement and cell division
Intermediate filaments	Tough fibers made of protein	Help strengthen cytoskeleton; stabilize cell shape
Centrioles	Pair of hollow cylinders located near nucleus; each centriole consists of nine microtubule triplets (9×3 structure)	Mitotic spindle forms between centrioles during animal cell division; may anchor and organize microtubule formation in animal cells; absent in most plants
Cilia	Relatively short projections extending from surface of cell; covered by plasma membrane; made of two central and nine pairs of peripheral microtubules ($9 + 2$ structure)	Movement of some unicellular organisms; used to move materials on surface of some tissues
Flagella	Long projections made of two central and nine pairs of peripheral microtubules ($9 + 2$ structure); extend from surface of cell; covered by plasma membrane	Cell locomotion by sperm cells and some unicellular eukaryotes

The outer surface of **rough ER** is studded with **ribosomes** that appear as dark granules. Notice in Figure 4-12 the lumen side of the rough ER appears bare, whereas the outer surface (the cytosolic side) looks rough. Ribosomes contain the enzyme necessary to form peptide bonds (see Chapter 3), and they function as manufacturing plants that assemble proteins. The ribosomes attached to the rough ER are known as *bound ribosomes; free ribosomes* are suspended in the cytosol.

Ribosomes consist of RNA and protein. Each eukaryotic ribosome is actually a knot of three *ribosoma RNA* strands in association with about 75 different proteins. Each ribosome has two main components: a large subunit and a small subunit.

The rough ER plays a central role in the synthesis and assembly of proteins. Many proteins that are exported from the cell (such as digestive enzymes), and those destined for other organelles, are synthesized on ribosomal bound to the ER membrane. The ribosome forms a tight seal with the ER membrane. A tunnel within the ribosome connects to an ER pore, or *translocon.* Proteins are transported through the tunnel and the pore in the ER membrane into the ER lumen. In the ER lumen, proteins may be modified by enzymes that add complex carbohydrates or lipids to them. Other enzymes, called **molecular chaperones,** in the ER lumen catalyze the efficient folding of proteins into proper conformations. The proteins are then transferred to other compartments within the cell by small **transport vesicles,** which bud off the ER membrane and then fuse with the membrane of some target organelle.

The Golgi complex processes, sorts, and modifies proteins

The **Golgi complex** (also known as the *Golgi body* or *Golgi apparatus*) was first described in 1898 by the Italian microscopist Camillo Golgi, who found a way to specifically stain this organelle. However, many investigators thought the Golgi was an artifact, and its legitimacy as a cell organelle was not confirmed until cells were studied with the electron microscope in the 1950s.

In many cells, the Golgi complex consists of stacks of flattened membranous sacs called **cisternae** (sing., *cisterna*). In certain regions, cisternae may be distended because they are filled with cell products (Fig. 4-13). Each of the flattened sacs has an internal space, or lumen. However, unlike the ER, most of these internal spaces of the Golgi complex and the membranes that form them are not continuous. The Golgi complex contains a number of separate compartments, as well as some that are interconnected.

Each Golgi stack has three areas referred to as the *cis face,* the *trans face,* and a *medial region* between. Typically, the *cis* face is located nearest the nucleus and receives materials from transport vesicles from the ER. The *trans* face, closest to the plasma membrane, packages molecules in vesicles and transports them out of the Golgi.

In a cross-sectional view like that in the TEM in Figure 4-13, many ends of the sheetlike layers of Golgi membranes are distended, an arrangement characteristic of well-developed Golgi complexes in many cells. In some animal cells, the Golgi complex lies at one side of the nucleus; other animal cells and plant cells have many Golgi complexes, usually consisting of separate stacks of membranes dispersed throughout the cell. Cells that secrete large amounts of *glycoproteins* have large numbers of Golgi stacks. (Recall from Chapter 3 that a glycoprotein is a protein with a covalently attached carbohydrate.) Golgi complexes of plant cells produce extracellular polysaccharides that are used as components of the cell wall.

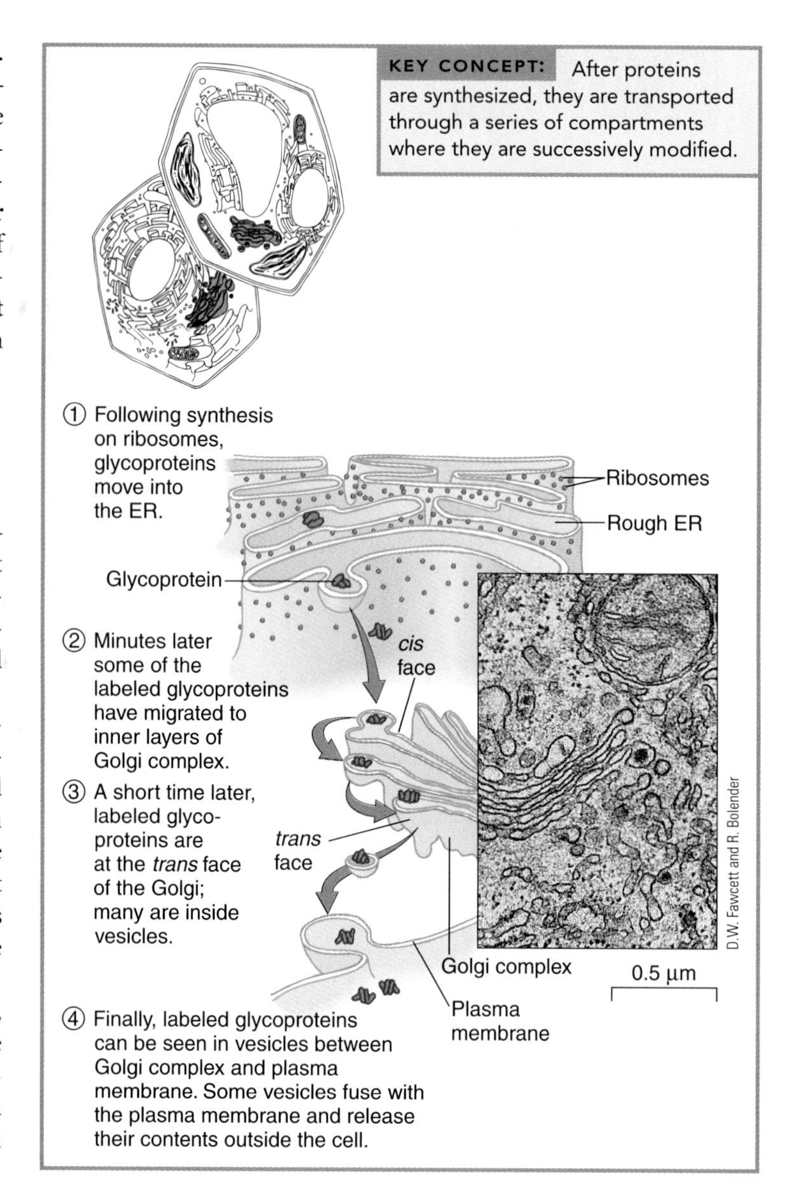

KEY CONCEPT: After proteins are synthesized, they are transported through a series of compartments where they are successively modified.
① Following synthesis on ribosomes, glycoproteins move into the ER.
Ribosomes
Rough ER
Glycoprotein
② Minutes later some of the labeled glycoproteins have migrated to inner layers of Golgi complex.
cis face
③ A short time later, labeled glycoproteins are at the trans face of the Golgi; many are inside vesicles.
trans face
Golgi complex
0.5 µm
Plasma membrane
④ Finally, labeled glycoproteins can be seen in vesicles between Golgi complex and plasma membrane. Some vesicles fuse with the plasma membrane and release their contents outside the cell.
D.W. Fawcett and R. Bolender

FIGURE **4-13** **TEM and an interpretive drawing of the Golgi complex.**

Glycoproteins are transported from the rough ER to the Golgi, where they are modified. This diagram shows the passage of glycoproteins through the Golgi complex during the secretory cycle of a mucus-secreting goblet cell that lines the intestine. Mucus is a complex mixture of covalently linked proteins and carbohydrates.

PROCESS OF SCIENCE

Cell biologists have demonstrated that the Golgi complex processes, sorts, and modifies proteins. Researchers have studied the function of the Golgi complex by radioactively labeling newly manufactured amino acids or carbohydrates and observing their movement. Glycoproteins are synthesized and are first located in the rough ER (see Figure 4-13). The proteins are transported from the rough ER to the *cis* face of the Golgi complex in small transport vesicles formed from the ER membrane. Until recently, researchers thought glycoprotein molecules released into the Golgi complex became enclosed in new vesicles that shuttle them from one compartment to another within the Golgi. A competing hypothesis, now the focus of research, holds that the cisternae themselves may move from *cis* to *trans* positions. The vesicles may move backward to recycle materials.

Regardless of how proteins are moved through the Golgi complex, while there they are modified in different ways, resulting in the formation of complex biological molecules. For example, the carbohydrate part of a glycoprotein (first added to proteins in the rough ER) may be modified. In some cases the carbohydrate component may be a "sorting signal," a kind of zip code that routes the protein to a specific organelle.

Glycoproteins are packaged in secretory vesicles in the *trans* face. These vesicles pinch off from the Golgi membrane and transport their contents to a specific destination. Vesicles transporting products for export from the cell fuse with the plasma membrane. The vesicle membrane becomes part of the plasma membrane, and the glycoproteins are secreted from the cell. Other vesicles may store glycoproteins for secretion at a later time, and still others are routed to various organelles of the endomembrane system. In animal cells, the Golgi complex also manufactures lysosomes.

In summary, here is a typical sequence followed by a protein destined for secretion from the cell:

protein synthesized on ribosomes ⟶ carbohydrate component added in lumen of ER ⟶ transport vesicles move glycoprotein to Golgi (*cis* face) ⟶ protein further modified in Golgi ⟶ vesicle transports glycoprotein from Golgi (*trans* face) to plasma membrane ⟶ contents released from cell

Lysosomes are compartments for digestion

Lysosomes are small sacs of digestive enzymes dispersed in the cytoplasm of most eukaryotic cells (Fig. 4-14). Researchers have identified about 40 different digestive enzymes in lysosomes. Most lysosomal enzymes are active under rather acidic conditions (about pH 5) and the lysosome maintains a pH of about 5 in its interior. Lysosomal enzymes break down complex molecules in bacteria and debris that scavenger cells ingest. The powerful enzymes and low pH that the lysosome maintains provide an excellent example of the importance of separating functions within the cell into different compartments. Under most normal conditions, the lysosome membrane confines its enzymes and their actions. However, some forms of tissue damage have been related to "leaky" lysosomes.

Primary lysosomes are formed by budding from the Golgi complex. Their hydrolytic enzymes are synthesized in the rough ER. As these enzymes pass through the lumen of the ER, sugars attach to each molecule, identifying it as bound for a lysosome. This signal permits the Golgi complex to appropriately sort the enzyme to the lysosomes rather than to export it from the cell.

When scavenger cells ingest bacteria (or debris), they are enclosed in a vesicle formed from part of the plasma membrane. One or more primary lysosomes fuse with the vesicle containing the ingested material, forming a larger vesicle called a *secondary lysosome.* In the secondary lysosome the powerful enzymes come in contact with the ingested molecules and degrade them into their components. Under some conditions lysosomes break down organelles so their components can be recycled or used as an energy source.

In certain genetic diseases of humans, known as *lysosomal storage diseases,* one of the normally present digestive enzymes is absent. Its substrate (a substance the enzyme would normally break down) accumulates in the lysosomes, ultimately interfering with cell activities. An example is Tay-Sachs disease (see Chapter 15), in which a normal lipid cannot be broken down in brain cells. The lipid accumulates in the cells, resulting in mental retardation and death.

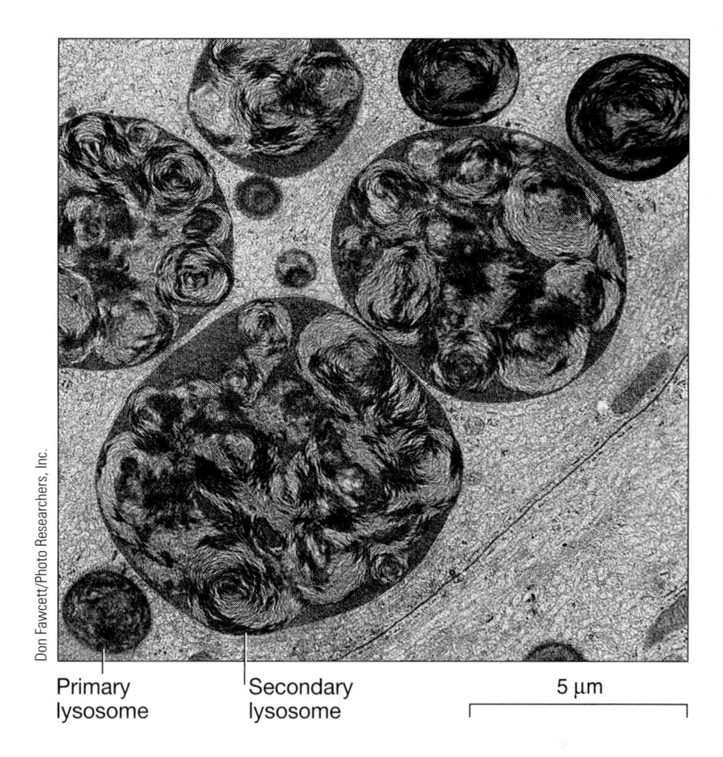

FIGURE **4-14** **Lysosomes.**
The dark vesicles in this TEM are lysosomes, compartments that separate powerful digestive enzymes from the rest of the cell. Primary lysosomes bud off from the Golgi complex. After a lysosome encounters and takes in material to be digested, it is known as a *secondary lysosome*. The large vesicles shown here are secondary lysosomes containing various materials being digested.

Peroxisomes metabolize small organic compounds

Peroxisomes are membrane-enclosed organelles containing enzymes that catalyze an assortment of metabolic reactions in which hydrogen is transferred from various compounds to oxygen (Fig. 4-15). During these reactions, they produce hydrogen peroxide (H_2O_2), which they use to detoxify certain compounds. Too much hydrogen peroxide is toxic to the cell; peroxisomes contain the enzyme catalase that splits excess hydrogen peroxide, rendering it harmless.

Peroxisomes are found in large numbers in cells that synthesize, store, or degrade lipids. For example, they synthesize certain phospholipids that are components of the insulating covering of nerve cells. In fact, certain neurological disorders occur when peroxisomes do not perform this function.

When yeast cells are grown in an alcohol-rich medium, they manufacture large peroxisomes, containing an enzyme that degrades the alcohol. Peroxisomes in human liver and kidney cells detoxify certain toxic compounds, including ethanol, the alcohol in alcoholic beverages.

In plant seeds, specialized peroxisomes, called *glyoxysomes,* contain enzymes that convert stored fats to sugars. The sugars are used by the young plant as an energy source and as a component for synthesizing other compounds. Animal cells lack glyoxysomes and cannot convert fatty acids into sugars.

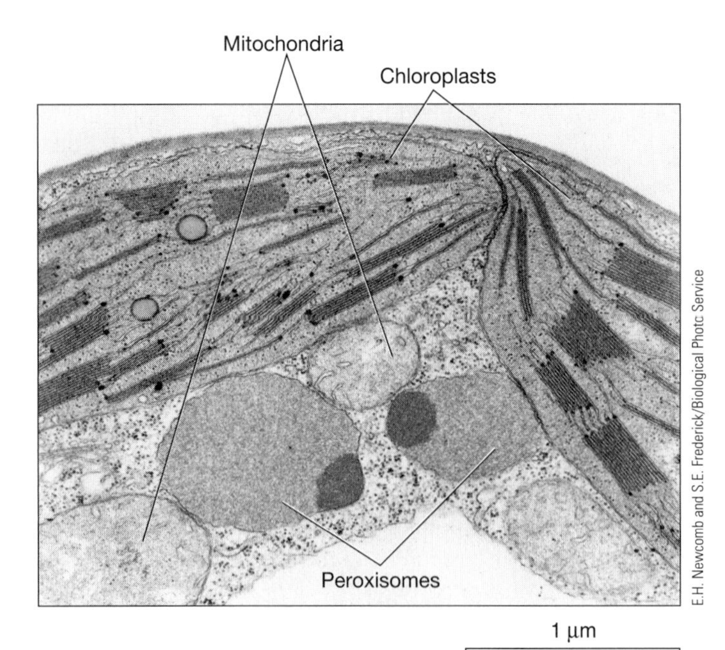

FIGURE **4-15** | **Peroxisomes.**

In this TEM of a tobacco (*Nicotiana tabacum*) leaf cell, peroxisomes are seen in close association with chloroplasts and mitochondria. These organelles may cooperate in carrying out some metabolic processes.

Vacuoles are large, fluid-filled sacs with a variety of functions

Although lysosomes have been identified in almost all kinds of animal cells, their occurrence in plant and fungal cells is open to debate. Many of the functions carried out in animal cells by lysosomes are performed in plant cells by a large, single, membrane-enclosed sac referred to as a **vacuole.** The vacuolar membrane, part of the endomembrane system, is called a **tonoplast.** The term *vacuole,* which means "empty," refers to the fact that these organelles have no internal structure. Although some biologists use the terms *vacuole* and *vesicle* interchangeably, vacuoles are usually larger structures, sometimes produced by the merging of many vesicles. Some biologists define a *vesicle* as a small, membrane-enclosed structure that holds cargo.

Vacuoles play a significant role in plant growth and development. Immature plant cells are generally small and contain numerous small vacuoles. As water accumulates in these vacuoles, they tend to coalesce, forming a large central vacuole. A plant cell increases in size mainly by adding water to this central vacuole.

As much as 90% of the volume of a plant cell may be occupied by a large central vacuole containing water, as well as stored food, salts, pigments, and metabolic wastes (see Figs. 4-7 and 4-9). The vacuole may serve as a storage compartment for inorganic compounds and for molecules such as proteins in seeds. Plants lack organ systems for disposing of toxic metabolic waste products. Wastes may be recycled in the vacuole, or they may aggregate and form small crystals inside the vacuole. Compounds that are noxious to herbivores (animals that eat plants) may also be stored in some plant vacuoles as a means of defense. Plant vacuoles are like lysosomes in their ability to break down unneeded organelles and other cell components. The vacuole is also important in maintaining hydrostatic (turgor) pressure in the plant cell.

Vacuoles have numerous other functions and are also present in many types of animal cells and in unicellular protists. Most protozoa have **food vacuoles,** which fuse with lysosomes so that the food they contain can be digested (Fig. 4-16). Some types of protozoa also have **contractile vacuoles,** which remove excess water from the cell (see Chapter 24).

Mitochondria and chloroplasts are energy-converting organelles

When a cell obtains energy from its environment, it is usually in the form of chemical energy in food molecules (such as glucose) or in the form of light energy. These types of energy must be converted to forms that cells can use more conveniently. Some energy conversions occur in the cytosol, but other types take place in mitochondria and chloroplasts, organelles specialized to facilitate the conversion of energy from one form to another. Chemical energy is most commonly stored in ATP. Recall from Chapter 3 that the chemical energy of ATP can be used to drive a variety of chemical reactions in the cell. Figure 4-17 summarizes the main activities that take place in mitochondria, found in almost all eukaryotic cells (including algae and plants), and in chloroplasts, found only in algae and certain plant cells.

Mitochondria and chloroplasts grow and reproduce themselves. They contain small amounts of DNA that code for a small number of the proteins found in these organelles. These proteins are synthesized by mitochondrial or chloroplast ribosomes, which are similar to the ribosomes of prokaryotes. The existence

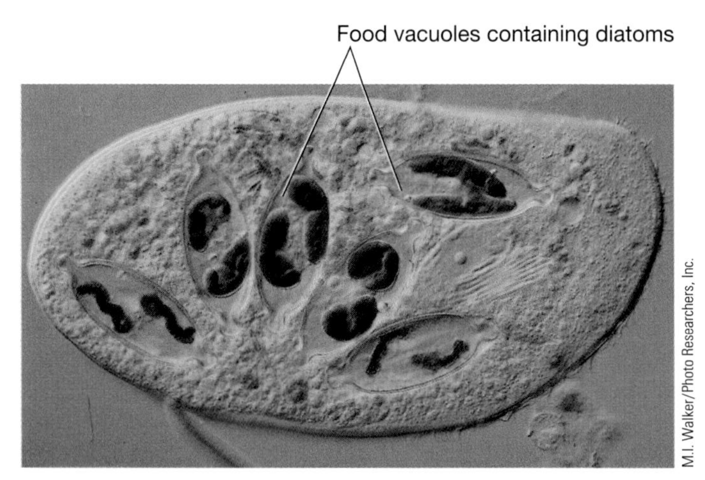

FIGURE **4-16** | **LM of food vacuoles.**

This protist, *Chilodonella*, has ingested many small, photosynthetic protists called *diatoms (dark areas)* that have been enclosed in food vacuoles. From the number of diatoms scattered about its cell, one might judge that *Chilodonella* has a rather voracious appetite.

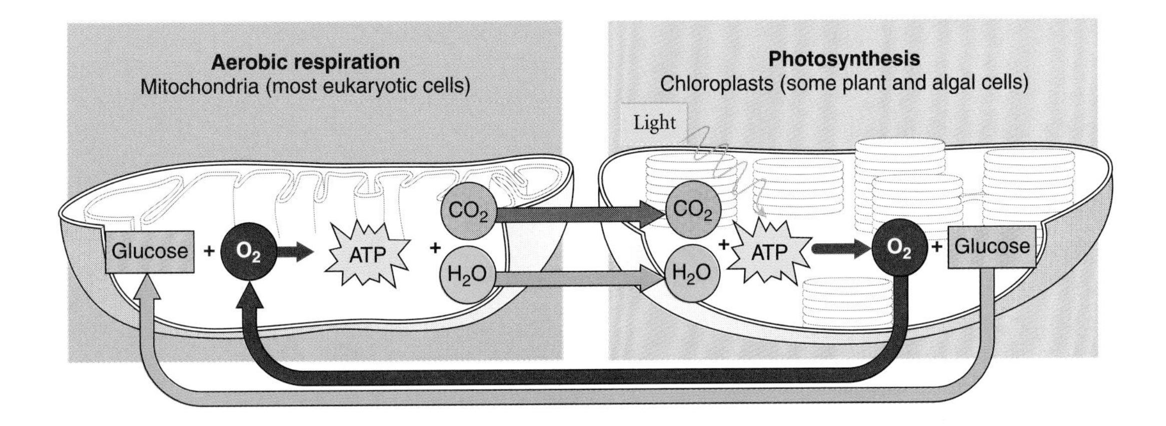

FIGURE **4-17** **Cellular respiration and photosynthesis.**

Cellular respiration takes place in the mitochondria of virtually all eukaryotic cells. In this process, some of the chemical energy in glucose is transferred to ATP. Photosynthesis, which is carried out in chloroplasts in some plant and algal cells, converts light energy to ATP and to other forms of chemical energy. This energy is used to synthesize glucose from carbon dioxide and water.

of a separate set of ribosomes and DNA molecules in mitochondria and chloroplasts and their similarity in size to many bacteria provide support for the **endosymbiont theory** (discussed in Chapters 20 and 24; see Figs. 20-7 and 24-2). According to this theory, mitochondria and chloroplasts evolved from prokaryotic organisms that took up residence inside larger cells and eventually lost the ability to function as autonomous organisms.

Mitochondria make ATP through cellular respiration

Virtually all eukaryotic cells (plant, animal, fungal, and protist) contain complex organelles called **mitochondria** (sing., *mitochondrion*). These organelles are the site of **aerobic respiration,** an oxygen-requiring process that includes most of the reactions that convert the chemical energy present in certain foods to ATP (see Chapter 7). During aerobic respiration, carbon and oxygen atoms are removed from food molecules, such as glucose, and converted to carbon dioxide and water.

Mitochondria are most numerous in cells that are very active and therefore have high energy requirements. More than 1000 mitochondria have been counted in a single liver cell! These organelles vary in size, ranging from 2 to 8 μm in length, and change size and shape rapidly. Mitochondria usually give rise to other mitochondria by growth and subsequent division.

Each mitochondrion is enclosed by a double membrane, which forms two *different* compartments within the organelle: the intermembrane space and the matrix (Fig. 4-18; see Chapter 7 for more detailed descriptions of mitochondrial structure). The **intermembrane space** is the compartment formed between the outer and inner mitochondrial membranes. The **matrix,** the compartment enclosed by the inner mitochondrial membrane, contains enzymes that break down food molecules and convert their energy to other forms of chemical energy.

The **outer mitochondrial membrane** is smooth and allows many small molecules to pass through it. By contrast, the **inner mitochondrial membrane** has numerous folds and strictly regulates the types of molecules that can move across it. The folds, called **cristae** (sing., *crista*), extend into the matrix. Cristae greatly increase the surface area of the inner mitochondrial membrane, providing a surface for the chemical reactions that transform the chemical energy in food molecules into the energy of ATP. The membrane contains the complex series of enzymes and other proteins needed for these reactions.

In a mammalian cell, each mitochondrion has 5 to 10 identical, circular molecules of DNA, accounting for up to 1% of the total DNA in the cell. Mutations in mitochondrial DNA have been associated with certain genetic diseases, including a form of young adult blindness, and certain types of progressive muscle degeneration. Mitochondrial DNA mutates far more frequently than nuclear DNA, and an accumulation of mutations may interfere

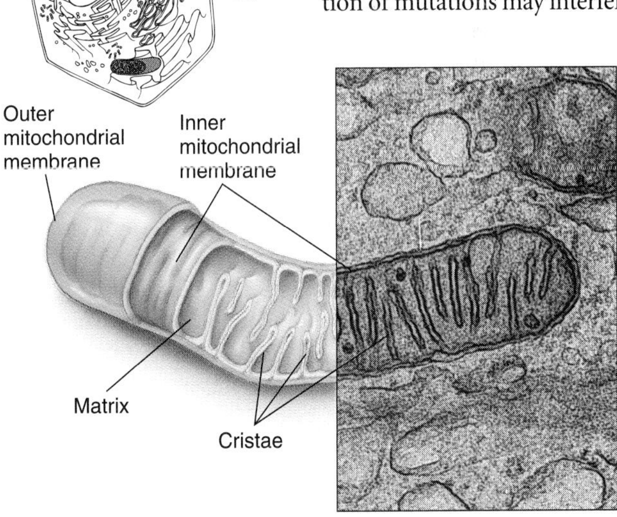

FIGURE **4-18** **Mitochondria.**

Aerobic respiration takes place within mitochondria. Cristae are evident in the TEM as well as in the drawing. The drawing shows the relationship between the inner and outer mitochondrial membranes.

with mitochondrial function. A diminished capacity to generate energy may contribute to the aging process.

Mitochondria also affect health and aging by leaking electrons. These electrons form **free radicals,** which are toxic, highly reactive compounds with unpaired electrons. These electrons bond with other compounds in the cell, interfering with normal function.

Mitochondria play an important role in programmed cell death, or **apoptosis.** Unlike **necrosis,** which is uncontrolled cell death that causes inflammation and damages other cells, apoptosis is a normal part of development and maintenance. For example, during the metamorphosis of a tadpole to a frog the cells of the tadpole tail must die. The hand of a human embryo is webbed until apoptosis destroys the tissue between the fingers. Cell death also occurs in the adult. For example, cells in the upper layer of human skin and in the intestinal wall are continuously destroyed, and replaced by new cells.

Mitochondria initiate cell death in several different ways. For example, they can interfere with energy metabolism or activate enzymes that mediate cell destruction. When a mitochondrion is injured, large pores open in its membrane, and cytochrome *c*, a protein important in energy production, is released into the cytoplasm. Cytochrome *c* triggers apoptosis by activating a group of enzymes known as **caspases,** which cut up vital compounds in the cell. Inappropriate initiation or inhibition of apoptosis may contribute to a variety of diseases, including cancer, acquired immunodeficiency syndrome (AIDS), and Alzheimer's disease. Pharmaceutical companies are developing drugs that block apoptosis. However, cell dynamics are extremely complex, and blocking apoptosis could lead to a worse fate, including necrosis.

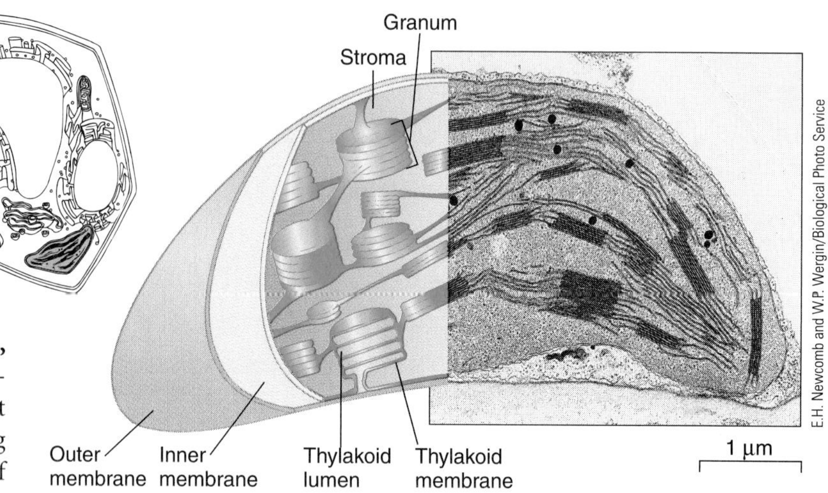

FIGURE **4-19** **A chloroplast, the organelle of photosynthesis.**

The TEM shows part of a chloroplast from a corn leaf cell. Chlorophyll and other photosynthetic pigments are found in the thylakoid membranes. One granum has been cut open to show the thylakoid lumen. The inner chloroplast membrane may or may not be continuous with the thylakoid membrane (as shown).

Chloroplasts convert light energy to chemical energy through photosynthesis

Certain plant and algal cells carry out **photosynthesis,** a complex set of reactions during which light energy is transformed into the chemical energy of glucose and other carbohydrates. Carbon dioxide and water are used as raw materials (see Chapters 1 and 8). **Chloroplasts** are organelles that contain **chlorophyll,** a green pigment that traps light energy for photosynthesis. Chloroplasts also contain a variety of light-absorbing yellow and orange pigments known as **carotenoids** (see Chapter 3). A unicellular alga may have only a single large chloroplast, whereas a leaf cell may have 20 to 100. Chloroplasts tend to be somewhat larger than mitochondria, with lengths typically ranging from about 5 to 10 μm or longer.

Chloroplasts are typically disc-shaped structures and, like mitochondria, have a complex system of folded membranes (Fig. 4-19; see Chapter 8 for more detailed descriptions of chloroplast structure). Two membranes, separated by a small space, separate the chloroplast from the cytosol. The inner membrane encloses a fluid-filled space called the **stroma,** which contains enzymes responsible for producing carbohydrates from carbon dioxide and water, using energy trapped from sunlight. A system of internal membranes, suspended in the stroma, consists of an interconnected set of flat, disclike sacs called **thylakoids.** The thylakoids are arranged in stacks called **grana** (sing., *granum*).

The thylakoid membranes enclose a third, innermost compartment within the chloroplast, called the **thylakoid lumen.** Chlorophyll is present in the thylakoid membranes, which are similar to the inner mitochondrial membranes in that they are involved in the formation of ATP. Energy absorbed from sunlight by the chlorophyll molecules excites electrons; the energy in these excited electrons is then used to produce ATP and other molecules that transfer chemical energy.

Chloroplasts belong to a group of organelles, known as **plastids,** that produce and store food materials in cells of plants and algae. All plastids develop from **proplastids,** precursor organelles found in less specialized plant cells, particularly in growing, undeveloped tissues. Depending on the special functions a cell will eventually have, its proplastids can mature into a variety of specialized mature plastids. These are extremely versatile organelles; in fact, under certain conditions even mature plastids can convert from one form to another.

Chloroplasts are produced when proplastids are stimulated by exposure to light. **Chromoplasts** contain pigments that give certain flowers and fruits their characteristic colors; these attract animals that serve as pollinators or as seed dispersers. **Leukoplasts** are unpigmented plastids; they include **amyloplasts** (see Fig. 3-9), which store starch in the cells of many seeds, roots, and tubers (such as white potatoes).

Review

- How do the structure and function of rough ER differ from the structure of smooth ER?
- What are the functions of the Golgi complex?
- What sequence of events must take place for a protein to be manufactured and then secreted from the cell?

- How are chloroplasts like mitochondria? How are they different? Draw a chloroplast and a mitochondrion.

BiologyNow™ Assess your understanding of **organelles in the cytoplasm** by taking the pretest on your BiologyNow CD-ROM.

THE CYTOSKELETON

Learning Objectives

11 Describe the structure and functions of the cytoskeleton.

12 Compare cilia and flagella, and describe their functions.

Scientists watching cells growing in the laboratory see that they frequently change shape and that many types of cells move about. The **cytoskeleton,** a dense network of protein fibers, gives cells mechanical strength, shape, and their ability to move (Fig. 4-20). The cytoskeleton also functions in cell division and in the transport of materials within the cell.

The cytoskeleton is highly dynamic and constantly changing. Its framework is made of three types of protein filaments: microtubules, microfilaments, and intermediate filaments. Both microfilaments and microtubules are formed from beadlike, globular protein subunits, which can be rapidly assembled and disassembled. Intermediate filaments are made from fibrous protein subunits and are more stable than microtubules and microfilaments.

Microtubules are hollow cylinders

Microtubules, the thickest filaments of the cytoskeleton, are about 25 nm in outside diameter and up to several micrometers in length. In addition to playing a structural role in the formation of the cytoskeleton, these extremely adaptable structures are involved in the movement of chromosomes during cell division. They serve as tracks for several other kinds of intracellular movement and are the major structural components of cilia and flagella—specialized structures used in some cell movements.

Microtubules consist of two very similar proteins: **α-tubulin** and **β-tubulin** that combine to form a dimer. (Recall from Chapter 3 that a dimer forms from the association of two similar, simpler units, referred to as monomers.) A microtubule elongates by the addition of tubulin dimers (Fig. 4-21). Microtubules are disassembled by the removal of dimers, which are recycled to form microtubules in other parts of the cell. Each

FIGURE **4-20** **The cytoskeleton.**
Eukaryotic cells have a cytoskeleton consisting of networks of several types of fibers, including microtubules, microfilaments, and intermediate filaments. The cytoskeleton contributes to the shape of the cell, anchors organelles, and sometimes rapidly changes shape during cell locomotion.

Plasma membrane

Microtubule

Intermediate filament

Microfilament

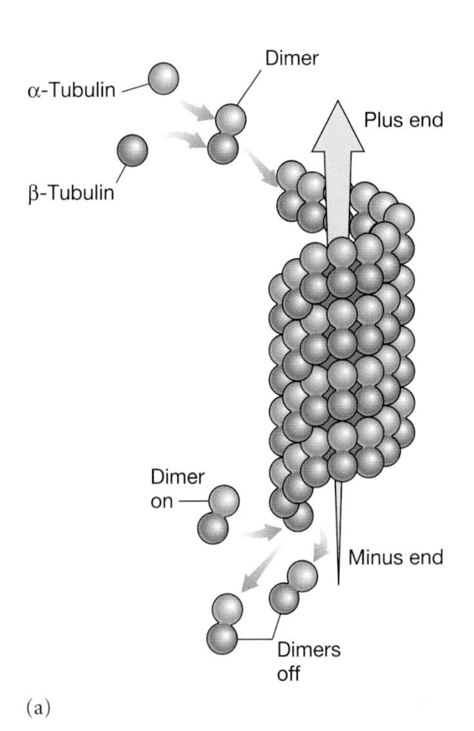

(a)

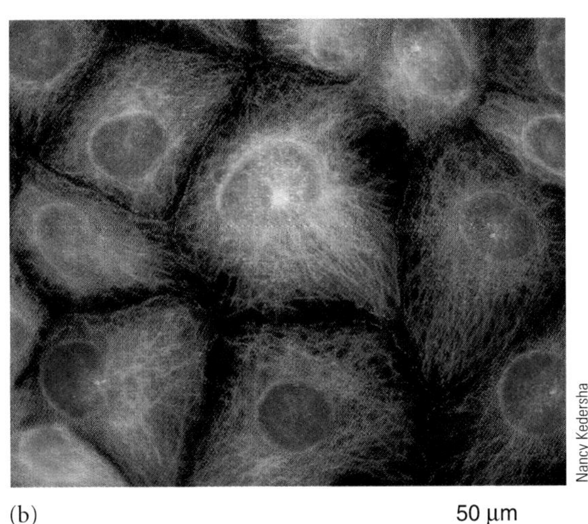

(b)

FIGURE **4-21** **Organization of microtubules.**
(a) Microtubules are manufactured in the cell by adding dimers of α-tubulin and β-tubulin to an end of the hollow cylinder. Notice that the cylinder has polarity. The end shown at the top of the figure is the fast-growing, or plus, end; the opposite end is the minus end. Each turn of the spiral requires 13 dimers. **(b)** Confocal fluorescence LM showing microtubules in green. A microtubule-organizing center *(pink dot)* is visible beside or over most of the cell nuclei *(blue)*.

microtubule has polarity, and its two ends are referred to as *plus* and *minus*. The plus end elongates more rapidly.

For microtubules to act as a structural framework or participate in cell movement, they must be anchored to other parts of the cell. In nondividing cells, the minus ends of microtubules appear to be anchored in regions called **microtubule-organizing centers (MTOCs).** In animal cells, the main MTOC is the cell center or **centrosome,** a structure that is important in cell division. In many cells, including almost all animal cells, the centrosome contains two structures called **centrioles** (Fig. 4-22). These structures, which are oriented within the centrosome at right angles to each other, are known as *9 × 3 structures;* they consist of nine sets of three attached microtubules arranged to form a hollow cylinder. The centrioles are duplicated before cell division and may play a role in some types of microtubule assembly. Most plant cells and fungal cells have an MTOC but lack centrioles. This suggests either that centrioles are not essential to most microtubule assembly processes or that alternative assembly mechanisms are present.

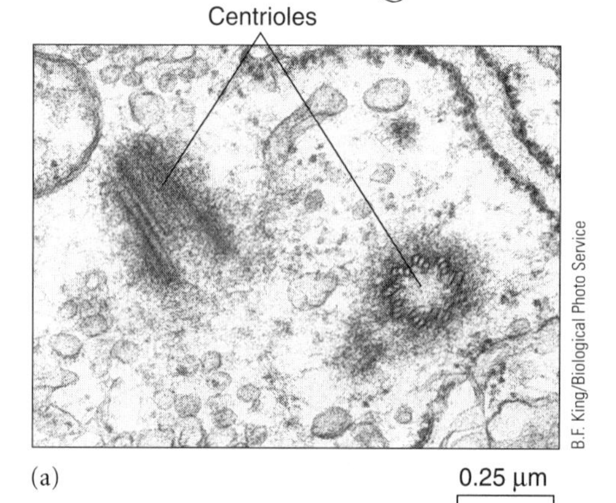

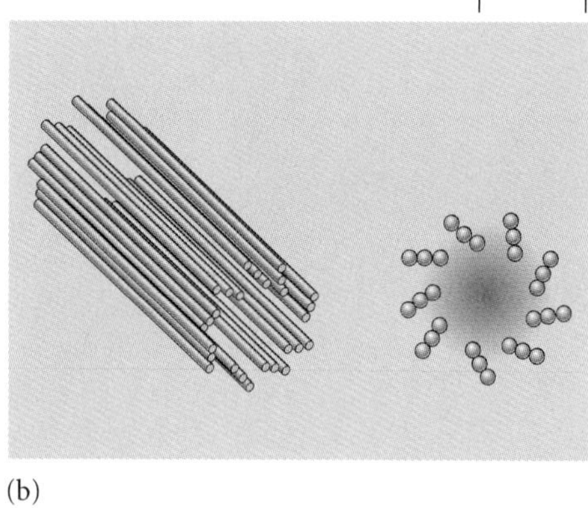

FIGURE **4-22** **Centrioles.**

(a) In the TEM, the centrioles are positioned at right angles to each other, near the nucleus of a nondividing animal cell. **(b)** Note the 9 × 3 arrangement of microtubules. The centriole on the right has been cut transversely.

The ability of microtubules to assemble and disassemble rapidly is seen during cell division, when much of the cytoskeleton appears to break down (see Chapter 9). Many of the tubulin subunits organize into a structure called the **spindle,** which serves as a framework for the orderly distribution of chromosomes during cell division.

Microtubule-associated proteins (MAPs) are classified into two groups: structural MAPs and motor MAPs. *Structural MAPs* may help regulate microtubule assembly, and they cross-link microtubules to other cytoskeletal polymers. *Motor MAPs* use ATP energy to produce movement.

Investigators are studying the mechanisms by which organelles and other materials move within the cell. Nerve cells typically have long extensions called *axons* that transmit signals to other nerve cells, muscle cells, or cells that produce hormones. Because of its length and accessibility and because other cells use similar transport mechanisms, researchers have used the axon as a model for studying the transport of organelles within the cell. They have found that mitochondria, transport and secretory vesicles, and other organelles may attach to microtubules, which then serve as tracks along which organelles move to different cell locations.

One motor protein, *kinesin,* moves organelles toward the plus end of a microtubule (Fig. 4-23). *Dynein,* another motor protein, transports organelles in the opposite direction, toward the minus end. This dynein movement is referred to as *retrograde transport.* A protein complex called *dynactin* is also required for retrograde transport. Dynactin binds to both microtubules and dynein and may function in transport, linking the organelle, microtubule, and dynein.

Cilia and flagella are composed of microtubules

Thin, movable structures, important in cell movement, project from surfaces of many cells. If a cell has one, or only a few, of these appendages and if they are long (typically about 200 μm) relative to the size of the cell, they are called **flagella** (sing., *flagellum*). If the cell has many short (typically 2–10 μm long) appendages, they are called **cilia** (sing., *cilium*). Cells use both cilia and flagella to move through a watery environment, and some cells use cilia to move liquids and particles across the cell surface. Cilia and flagella are commonly found on unicellular and small multicellular organisms. In animals and certain plants, flagella serve as the tails of sperm cells. In animals, cilia commonly occur on the surfaces of cells that line internal ducts of the body (such as respiratory passageways).

Eukaryotic cilia and flagella are structurally alike (but different from bacterial flagella). Each consists of a slender, cylindrical stalk covered by an extension of the plasma membrane. The core of the stalk contains a group of microtubules arranged so there are nine attached pairs of microtubules around the circumference and two unpaired microtubules in the center (Fig. 4-24). This *9 + 2 arrangement* of microtubules is characteristic of virtually all eukaryotic cilia and flagella.

The microtubules in cilia and flagella move by sliding in pairs past each other. The sliding force is generated by dynein proteins, which are attached to the microtubules like small

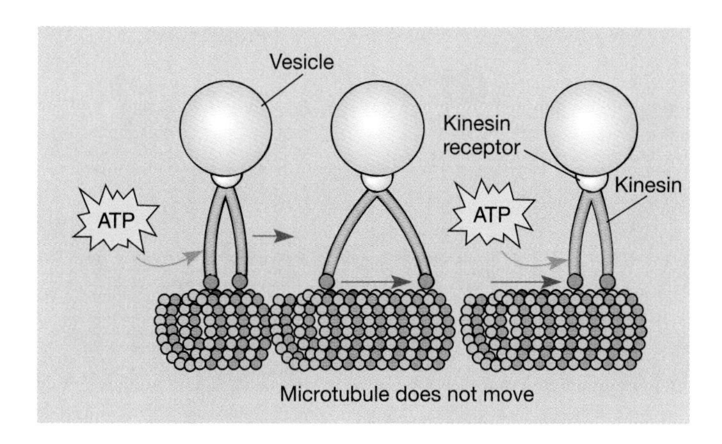

ACTIVE FIGURE **4-23** **A hypothetical model of a kinesin motor.**

A kinesin molecule attaches to a specific receptor on the vesicle. Energy from ATP allows the kinesin molecule to change its conformation and "walk" along the microtubule, carrying the vesicle along.

BiologyNow™ Learn more about **kinesin motors** by clicking on this figure on your BiologyNow CD-ROM.

arms. These proteins use the energy from ATP to power the cilia or flagella. The dynein proteins (arms) on one pair of tubules change their shape and "walk" along the adjacent microtubule pair. Thus, the microtubules on one side of a cilium or a flagellum extend farther toward the tip than those on the other side. This sliding of microtubules translates into a bending motion (Fig. 4-24b). Cilia typically move like oars, alternating power and recovery strokes and exerting a force that is parallel to the cell surface. A flagellum moves like a whip, exerting a force perpendicular to the cell surface.

Each cilium or flagellum is anchored in the cell by a **basal body,** which has nine sets of three attached microtubules in a cylindrical array (*9 × 3 structure*). The basal body appears to be the organizing structure for the cilium or flagellum when it first begins to form. However, experiments have shown that as growth proceeds, the tubulin subunits are added much faster to the tips of the microtubules than to the base. Basal bodies and centrioles may be functionally related as well as structurally similar. In fact, centrioles are typically found in the cells of organisms that produce flagellated or ciliated cells; these include animals, certain protists, a few fungi, and a few plants. Both basal bodies and centrioles replicate themselves.

Microfilaments consist of intertwined strings of actin

Microfilaments, also called *actin filaments,* are flexible, solid fibers about 7 nm in diameter. Each microfilament consists of two intertwined polymer chains of beadlike **actin** molecules (Fig. 4-25). Actin filaments are cross-linked with one another and with other proteins by linker proteins. They form bundles of fibers that provide mechanical support for various cell structures. In many cells, a network of microfilaments is visible in the cytosol just inside the plasma membrane.

In muscle cells, actin is associated with another protein, **myosin,** to form fibers that generate the forces that contract muscles (see Chapter 38). In nonmuscle cells, actin can also associate with myosin, forming contractile structures involved in various cell movements. Actin filaments themselves cannot contract, but they can generate movement by rapidly assembling and disassembling. Actin filaments associated with myosin are involved in certain transient functions. For example, in animal cell division, contraction of a ring of actin associated with myosin constricts the cell, forming two daughter cells (see Chapter 9). Certain organelles in the giant axons of the squid move along microfilaments. A type of myosin appears to be the motor for this transport.

As mentioned earlier in the chapter, some types of cells have microvilli, projections of the plasma membrane that increase

FIGURE **4-24** **Structure of cilia.**

(a) This 3-D representation shows nine attached microtubule pairs (doublets) arranged in a cylinder, with two unattached microtubules in the center. The dynein "arms," shown widely spaced for clarity, are actually much closer together along the longitudinal axis. **(b)** The dynein arms move the microtubules by forming and breaking cross bridges on the adjacent microtubules, so that one microtubule "walks" along its neighbor. **(c)** TEM of cross sections through cilia showing the 9 + 2 arrangement of microtubules. **(d)** TEM of a longitudinal section of three cilia of the protist *Tetrahymena*, an organism often used in genetic research. Some of the interior microtubules are visible.

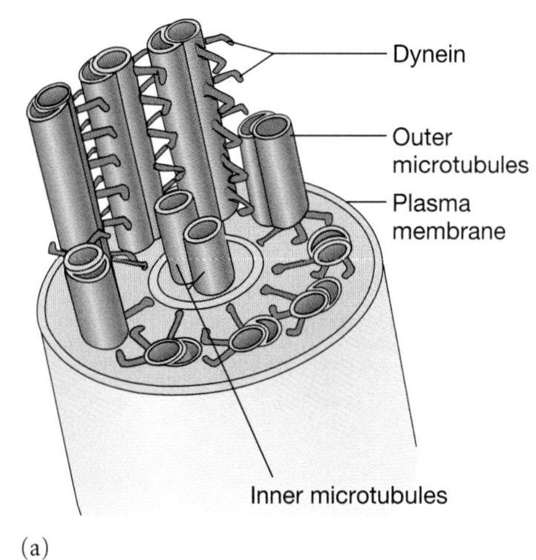

(a)

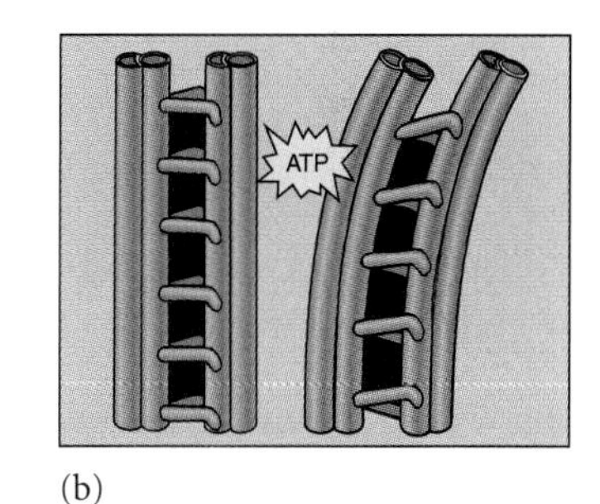

(b)

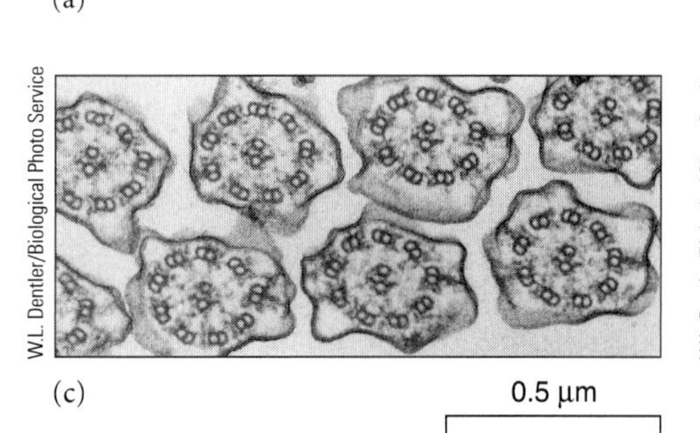

(c)

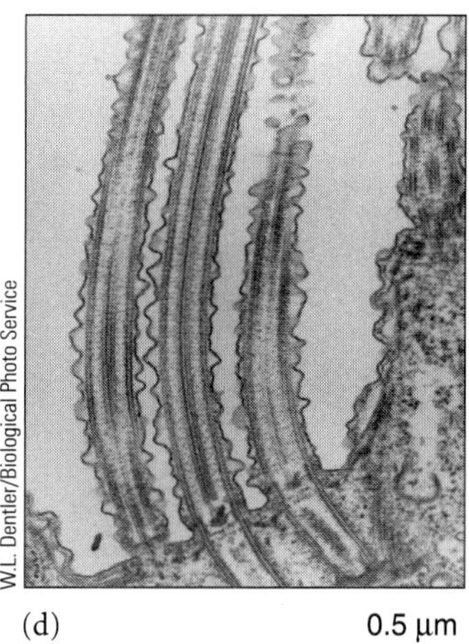

(d)

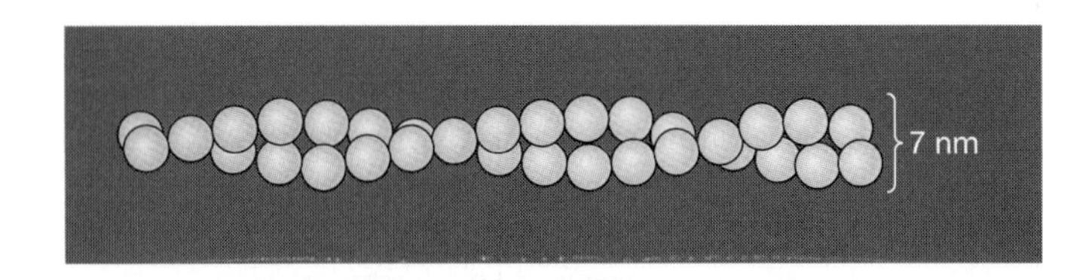

(a)

(b)

100 µm

Nancy Kedersha/ImmunoGen, Inc.

FIGURE 4-25 | **Microfilaments**

(a) An individual microfilament consists of two intertwined strings of beadlike actin molecules. **(b)** Many bundles of aggregated microfilaments *(green)* are evident in this confocal fluorescence LM of fibroblasts, cells found in connective tissue.

the surface area of the cell for transporting materials across the plasma membrane. Composed of bundles of microfilaments, microvilli extend and retract as the microfilaments assemble and disassemble.

Intermediate filaments help stabilize cell shape

Intermediate filaments are tough, flexible fibers, about 10 nm in diameter (Fig. 4-26). They provide mechanical strength and help stabilize cell shape. These filaments are abundant in regions of a cell that may be subject to mechanical stress applied from outside the cell. Certain proteins cross-link intermediate filaments with other types of filaments and mediate interactions between them.

All eukaryotic cells have microtubules and microfilaments, but only some animal groups, including vertebrates, are known to have intermediate filaments. Even when present, intermediate filaments vary widely in protein composition and size among different cell types and different organisms. Examples of intermediate filaments are the keratins found in the epithelial cells of the vertebrate skin and neurofilaments found in vertebrate nerve cells.

Certain mutations in genes coding for intermediate filaments weaken the cell and have been associated with several diseases. For example, in the neurodegenerative disease amyotrophic lateral sclerosis (ALS, or Lou Gehrig's disease), abnormal neurofilaments have been identified in nerve cells that control muscles. This condition interferes with normal transport of materials in the nerve cells and degeneration of the nerve cells. The resulting loss of muscle function is typically fatal.

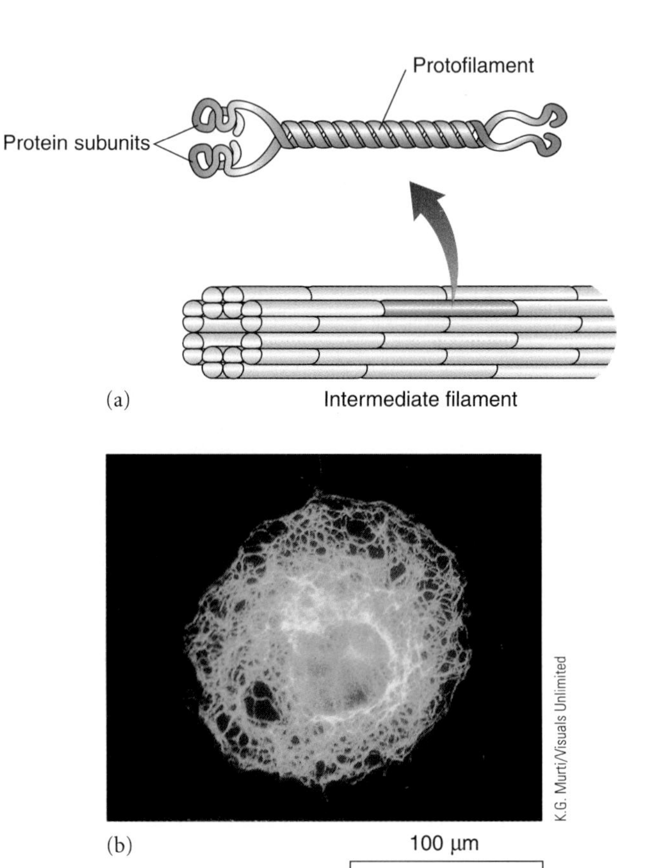

FIGURE 4-26 | **Intermediate filaments.**

(a) Intermediate filaments are flexible rods about 10 nm in diameter. Each intermediate filament consists of components, called *protofilaments,* that are made up of coiled protein subunits. **(b)** Intermediate filaments are stained green in this human cell isolated from a tissue culture.

Review

- What are the main functions of the cytoskeleton?
- How are microfilaments and microtubules similar? How are they different?
- How are cilia and flagella similar? How are they different?

BiologyNow™ Assess your understanding of the **cytoskeleton** by taking the pretest on your BiologyNow CD-ROM.

CELL COVERINGS

Learning Objective

13 Describe the glycocalyx, extracellular matrix, and cell wall.

Most eukaryotic cells are surrounded by a **glycocalyx,** or **cell coat,** formed by polysaccharide side chains of proteins and lipids that are part of the plasma membrane. The glycocalyx protects the cell and may help keep other cells at a distance. Certain molecules of the glycocalyx enable cells to recognize one an-

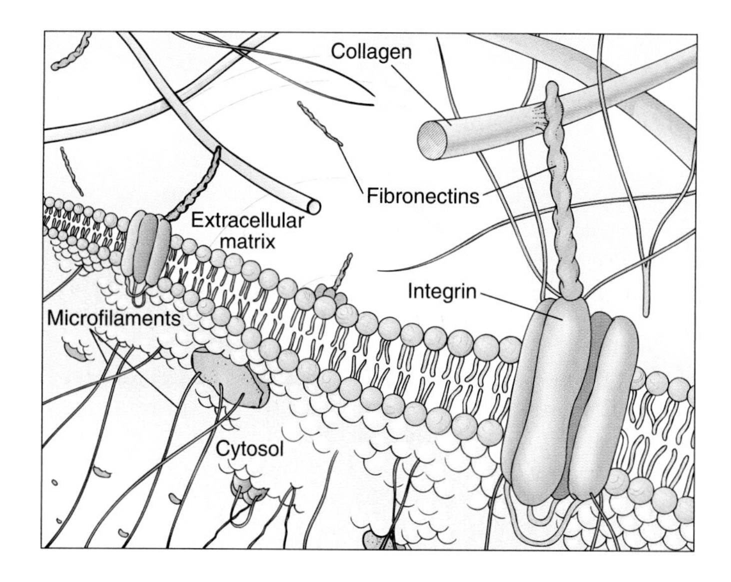

FIGURE **4-27** | **The extracellular matrix (ECM).**
Fibronectins, glycoproteins of the ECM, bind to integrins and other receptors in the plasma membrane.

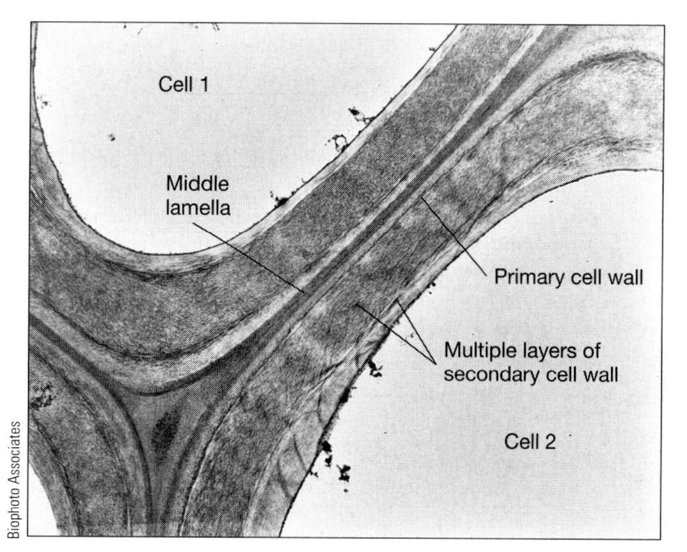

FIGURE **4-28** | **Plant cell walls.**
The cell walls of two adjacent plant cells are labeled in this TEM. The cells are cemented together by the middle lamella, a layer of gluelike polysaccharides called pectins. A growing plant cell first secretes a thin primary wall that is flexible and can stretch as the cell grows. The thicker layers of the secondary wall are secreted inside the primary wall after the cell stops elongating.

other, to make contact, and in some cases to form adhesive or communicating associations. Other molecules of the cell coat contribute to the mechanical strength of multicellular tissues.

Many animal cells are also surrounded by an **extracellular matrix (ECM),** which they secrete. It consists of a gel of carbohydrates and fibrous proteins (Fig. 4-27). The main structural protein in the ECM is **collagen,** which forms very tough fibers. Certain glycoproteins of the ECM, called **fibronectins,** help organize the matrix and help cells attach to it. Fibronectins bind to protein receptors that extend from the plasma membrane.

Integrins are proteins that serve as membrane receptors for the ECM. These proteins activate many **cell signaling** pathways that communicate information to the cell from the ECM. Integrins appear to be important in cell movement and in organizing the cytoskeleton so that cells assume a definite shape. In many types of cells, integrins anchor the external ECM to the microfilaments of the internal cytoskeleton. When these cells are not appropriately anchored, apoptosis results. Cancer cells apparently lose this requirement to be anchored to the ECM.

Most bacteria, fungi, and plant cells are surrounded by a **cell wall** and proteins. Plant cells have thick cell walls that contain multiple layers of the polysaccharide **cellulose** (see Fig. 3-10). Other polysaccharides in the plant cell wall form cross links between the bundles of cellulose fibers. Each cellulose fiber layer runs in a different direction from the adjacent layer, giving the cell wall great mechanical strength.

A growing plant cell secretes a thin, flexible *primary cell wall,* which stretches and expands as the cell increases its size (Fig. 4-28). After the cell stops growing, either new wall material is secreted that thickens and solidifies the primary wall or multiple layers of a *secondary cell wall* with a different chemical composition are formed between the primary wall and the plasma membrane. Wood is made mainly of secondary cell walls. Between the primary cell walls of adjacent cells lies the **middle lamella,** a layer of gluelike polysaccharides called *pectins.* The middle lamella causes the cells to adhere tightly to one another. (For more information on plant cell walls, see Chapter 31's discussion of the ground tissue system.)

Review

- What are the functions of the glycocalyx?
- How do the functions of fibronectins and integrins differ?
- What is the main component of plant cell walls?
- How are cell walls formed?

BiologyNow™ Assess your understanding of **cell coverings** by taking the pretest on your BiologyNow CD-ROM.

SUMMARY WITH KEY TERMS

1 Summarize the relationship between cell organization and homeostasis.

- The **cell** must maintain **homeostasis,** an appropriate internal environment.
- Every cell is surrounded by a **plasma membrane** that forms a cytoplasmic compartment. The plasma membrane helps maintain homeostasis by allowing the cell to exchange materials with its external environment and to maintain internal conditions that may be very different from those of the outer environment.
- Cells have **organelles,** internal structures that carry out specific functions that help maintain homeostasis.

2 Explain the relationship between cell size and maintaining homeostasis.

- Most cells are microscopic. Most prokaryotic cells are smaller than eukaryotic cells.
- A critical factor in determining cell size is the ratio of the plasma membrane (surface area) to the cell's volume; the plasma membrane must be large enough to regulate the passage of materials into and out of the cell.
- Cell size and shape are related to function and are limited by the need to maintain homeostasis.

3 Describe methods that biologists use to study cells, including microscopy and cell fractionation.

- Biologists have learned about cell structure by studying cells with **light** and **electron microscopes** and by using a variety of chemical methods.
- The electron microscope has superior **resolving power,** enabling investigators to see details of cell structures not observable with conventional microscopes.
- Cell biologists use **cell fractionation** methods for purifying organelles, to gain information about the function of cell structures.

4 Compare and contrast the general characteristics of prokaryotic and eukaryotic cells, and contrast plant and animal cells.

- **Prokaryotic cells** are bounded by a plasma membrane but have little or no internal membrane organization. They have a **nuclear area** rather than a membrane-enclosed nucleus. Prokaryotes typically have a **cell wall** and **ribosomes** and may have propeller-like **flagella.**
- **Eukaryotic cells** have a membrane-enclosed **nucleus** and **cytoplasm,** which contains a variety of organelles; the fluid component of the cytoplasm is the **cytosol.**
- Plant cells differ from animal cells in that they have rigid cell walls, **plastids,** and large **vacuoles;** cells of most plants lack centrioles. Vacuoles are important in plant growth and development.

5 Describe three functions of cell membranes.

- Membranes divide the cell into compartments, allowing it to conduct specialized activities within small areas of the cytoplasm, concentrate molecules, and organize metabolic reactions. Membranes are also important in energy storage and conversion.
- A system of interacting membranes forms the **endomembrane system.**
- Small membrane-bounded sacs, called **vesicles,** transport materials between compartments.

6 Describe the structure and functions of the nucleus.

- The nucleus, the control center of the cell, contains genetic information coded in DNA.
- The nucleus is bounded by a **nuclear envelope** consisting of a double membrane perforated with **nuclear pores** that communicate with the cytoplasm.
- DNA in the nucleus associates with protein to form **chromatin** which is organized into **chromosomes.** During cell division, the chromosomes condense and become visible as thread-like structures.
- The **nucleolus** is a region in the nucleus that is the site of ribosomal RNA synthesis and ribosome assembly.

7 Distinguish between smooth and rough endoplasmic reticulum in terms of both structure and function.

- The **endoplasmic reticulum (ER)** is a network of folded internal membranes in the cytosol. **Smooth ER** is the site of lipid synthesis and detoxifying enzymes.
- **Rough ER** is studded along its outer surface with ribosomes that manufacture proteins. Proteins synthesized on rough ER may be moved into the **ER lumen,** where they are modified by the addition of a carbohydrate or lipid.

8 Trace the path of proteins synthesized in the rough endoplasmic reticulum as they are subsequently processed, modified, and sorted by the Golgi complex and then transported to specific destinations.

- The **Golgi complex** consists of stacks of flattened membranous sacs called **cisternae** that process, sort, and modify proteins synthesized on the ER. The Golgi complex also manufactures lysosomes.
- Glycoproteins are transported from the ER to the *cis* face of the Golgi complex by **transport vesicles,** formed by membrane budding.
- The Golgi modifies carbohydrates and lipids that were added to proteins by the ER, and packages them in vesicles. Glycoproteins exit the Golgi at its *trans* face. The Golgi routes some proteins to the plasma membrane for export from the cell. Others are transported to lysosomes or other organelles within the cytoplasm.

9 Describe the functions of lysosomes and peroxisomes.

- **Lysosomes** contain enzymes that break down worn-out cell structures, bacteria, and other substances taken into cells.
- **Peroxisomes** contain enzymes that produce and degrade hydrogen peroxide. They are involved in lipid metabolism and detoxify harmful compounds.

10 Compare the functions of mitochondria and chloroplasts, and discuss ATP synthesis by each of these organelles.

- **Mitochondria,** the sites of aerobic respiration, are organelles enclosed by a double-membrane. The inner membrane is folded, forming **cristae** that increase its surface area. Mitochondria contain DNA that codes for some of its proteins.
- Mitochondria play an important role in **apoptosis,** or programmed cell death.

- The cristae and the compartment enclosed by the inner membrane, the **matrix,** contain enzymes for the reactions of aerobic respiration. During aerobic respiration, nutrients are broken down in the presence of oxygen. Energy captured from nutrients is packaged in ATP, and carbon dioxide and water are produced as by-products.
- **Chloroplasts** are plastids that carry out **photosynthesis.**
- The inner membrane of the chloroplast encloses a fluid-filled space, the **stroma. Grana,** stacks of disclike membranous sacs called **thylakoids,** are suspended in the stroma.
- During photosynthesis, **chlorophyll,** the green pigment found in the thylakoid membranes, traps light energy. This energy is converted to chemical energy in ATP and used to synthesize carbohydrates from carbon dioxide and water.

11 Describe the structure and functions of the cytoskeleton.

- The **cytoskeleton** is a dynamic internal framework made of microtubules, microfilaments, and intermediate filaments. The cytoskeleton provides structural support and functions in various types of cell movement, including transport of materials in the cell.
- **Microtubules** are hollow cylinders assembled from subunits of the protein **tubulin.** In cells that are not dividing, the minus ends of microtubules appear to be anchored in **microtubule-organizing centers (MTOCs).**
- The main MTOC of animal cells is the **centrosome,** which usually contains two **centrioles.** Each centriole has a 9×3 arrangement of microtubules.
- **Microtubule-associated proteins (MAPs)** include structural MAPs and motor MAPs. Two motor MAPs are kinesin and dynein.
- **Microfilaments,** or actin filaments, formed from subunits of the protein **actin,** are important in cell movement.
- **Intermediate filaments** strengthen the cytoskeleton and stabilize cell shape.

12 Compare cilia and flagella, and describe their functions.

- **Cilia** and **flagella** are thin, movable structures that project from the cell surface and function in movement. Each consists of a $9 + 2$ arrangement of microtubules, and each is anchored in the cell by a **basal body** that has a 9×3 organization of microtubules. Cilia are short and flagella are long.

13 Describe the glycocalyx, extracellular matrix, and cell wall.

- Most cells are surrounded by a **glycocalyx,** or **cell coat,** formed by polysaccharides extending from the plasma membrane.
- Many animal cells are also surrounded by an **extracellular matrix (ECM)** consisting of carbohydrates and protein. **Fibronectins** are glycoproteins of the ECM that bind to **integrins,** receptor proteins in the plasma membrane.
- Most bacteria, fungi, and plant cells are surrounded by a cell wall made of carbohydrates. Plant cells secrete **cellulose** and other polysaccharides that form rigid cell walls.

POST-TEST

1. The ability of a microscope to reveal fine detail is known as (a) magnification (b) resolving power (c) cell fractionation (d) scanning electron microscopy (e) phase contrast
2. A plasma membrane is characteristic of (a) all cells (b) prokaryotic cells only (c) eukaryotic cells only (d) animal cells only (e) eukaryotic cells except for plant cells
3. Detailed information about the shape and external features of a specimen can best be obtained by using a (a) differential centrifuge (b) fluorescence microscope (c) transmission electron microscope (d) scanning electron microscope (e) light microscope
4. In eukaryotic cells, DNA is found in (a) chromosomes (b) chromatin (c) mitochondria (d) answers a, b, and c are correct (e) only answers a and b are correct
5. Which of the following structures would *not* be found in prokaryotic cells? (a) cell wall (b) ribosomes (c) nuclear area (d) nucleus (e) propeller-like flagellum
6. Which of the following is/are most closely associated with protein synthesis? (a) ribosomes (b) smooth ER (c) mitochondria (d) microfilaments (e) lysosomes
7. Which of the following is/are most closely associated with the breakdown of ingested material? (a) ribosomes (b) smooth ER (c) mitochondria (d) microfilaments (e) lysosomes
8. Which of the following are most closely associated with photosynthesis? (a) basal bodies (b) smooth ER (c) cristae (d) thylakoids (e) MTOCs
9. A $9 + 2$ arrangement of microtubules best describes (a) cilia (b) centrosomes (c) basal bodies (d) microfilaments (e) microvilli
10. Which sequence most accurately describes information flow in the eukaryotic cell? (a) DNA in nucleus ⟶ messenger RNA ⟶ ribosomes ⟶ protein synthesis (b) DNA in nucleus ⟶ ribosomal RNA ⟶ mitochondria ⟶ protein synthesis (c) RNA in nucleus ⟶ messenger DNA ⟶ ribosomes ⟶ protein synthesis (d) DNA in nucleus ⟶ messenger RNA ⟶ Golgi complex ⟶ protein synthesis (e) DNA in nucleus ⟶ messenger RNA ⟶ smooth ER ⟶ protein synthesis
11. Which sequence most accurately describes glycoprotein processing in the eukaryotic cell? (a) smooth ER ⟶ transport vesicle ⟶ *cis* region of Golgi ⟶ *trans* region of Golgi ⟶ plasma membrane or other organelle (b) rough ER ⟶ transport vesicle ⟶ *cis* region of Golgi ⟶ *trans* region of Golgi ⟶ plasma membrane or other organelle (c) rough ER ⟶ transport vesicle ⟶ *trans* region of Golgi ⟶ *cis* region of Golgi ⟶ plasma membrane or other organelle (d) rough ER ⟶ nucleus ⟶ *cis* region of Golgi ⟶ *trans* region of Golgi ⟶ plasma membrane or other organelle (e) smooth ER ⟶ transport vesicle ⟶ *cis* region of Golgi ⟶ chloroplast
12. Which of the following is/are part of the cytoskeleton? (a) microfilaments (b) lysosomes (c) peroxisomes (d) ribosomes (e) endoplasmic reticulum
13. Which of the following function(s) in cell movement? (a) microtubules (b) cristae (c) grana (d) smooth ER (e) rough ER

14. Which of the following is/are *not* associated with mitochondria? (a) cristae (b) aerobic respiration (c) apoptosis (d) free radicals (e) thylakoids
15. The extracellular matrix (a) consists mainly of myosin and RNA (b) projects to form microvilli (c) houses the centrioles (d) contains fibronectins that bind to integrins (e) has an elaborate system of cristae
16. Label the diagrams of the animal and plant cells. How is the structure of each organelle related to its function? Use Figures 4-7 and 4-9 to check your answers.

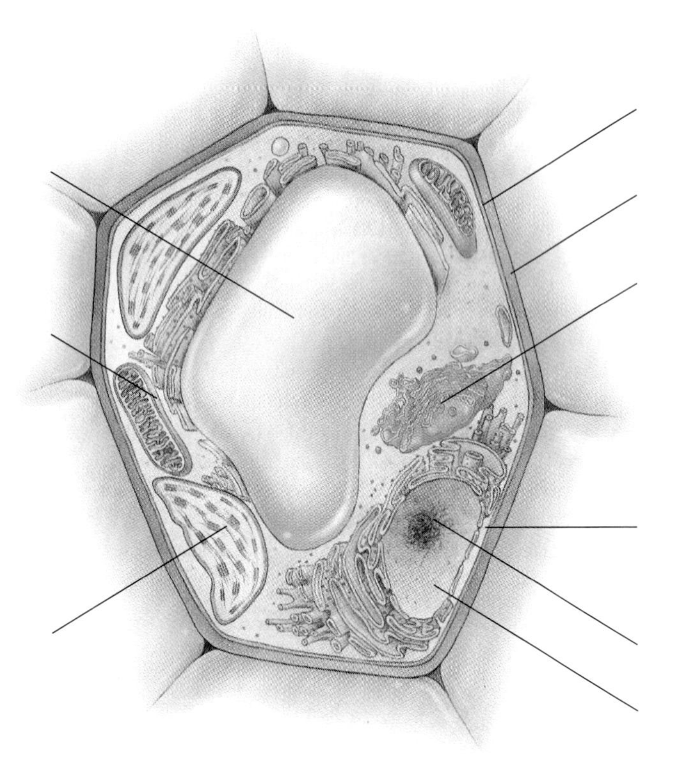

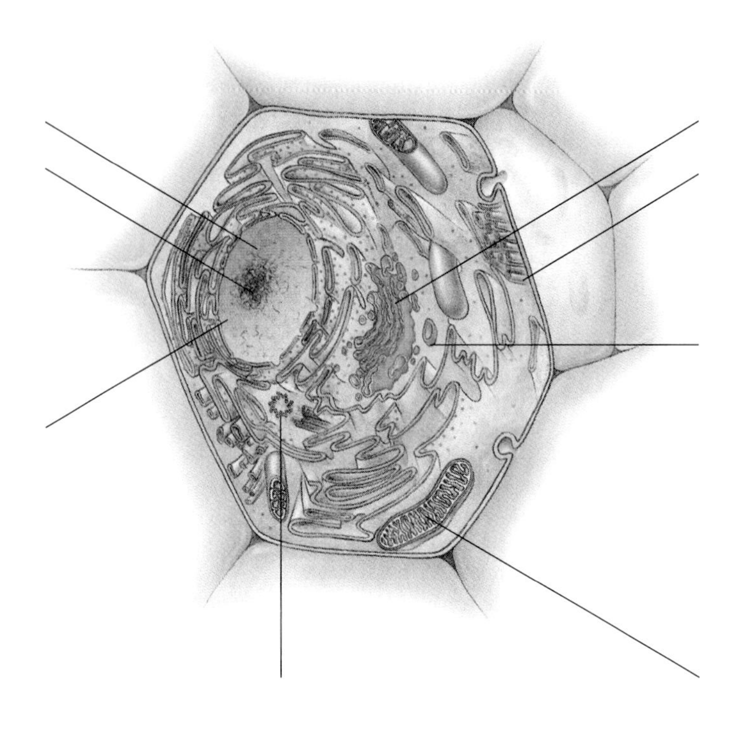

CRITICAL THINKING

1. Explain why the cell is considered the basic unit of life, and discuss some of the implications of the cell theory.
2. Why does a eukaryotic cell need both membranous organelles and fibrous cytoskeletal components?
3. Describe a specific example of the correlation between cell structure and function. (*Hint:* Think of mitochondrial structure.)
4. The *Acetabularia* experiments described in this chapter suggest that DNA is much more stable in the cell than is messenger RNA. Is this advantageous or disadvantageous to the cell? Why? How can *Acetabularia* continue to live for a few days after its nucleus is removed?

- Visit our Web site at **http:biology.brookscole.com/solomon7** for links to chapter-related resources on the World Wide Web. Additional online materials relating to this chapter can be found on our Web site.

BIOLOGY NOW RESOURCES

Active Figures

4-7: Structure of the eukaryotic cell

Preparing for an exam? Take a diagnostic test on your BiologyNow CD-ROM.

4-23: Kinesin motors

Post-Test Answers

1. b	2. a	3. d	4. d
5. d	6. a	7. e	8. d
9. a	10. a	11. b	12. a
13. a	14. e	15. d	

Biological Membranes

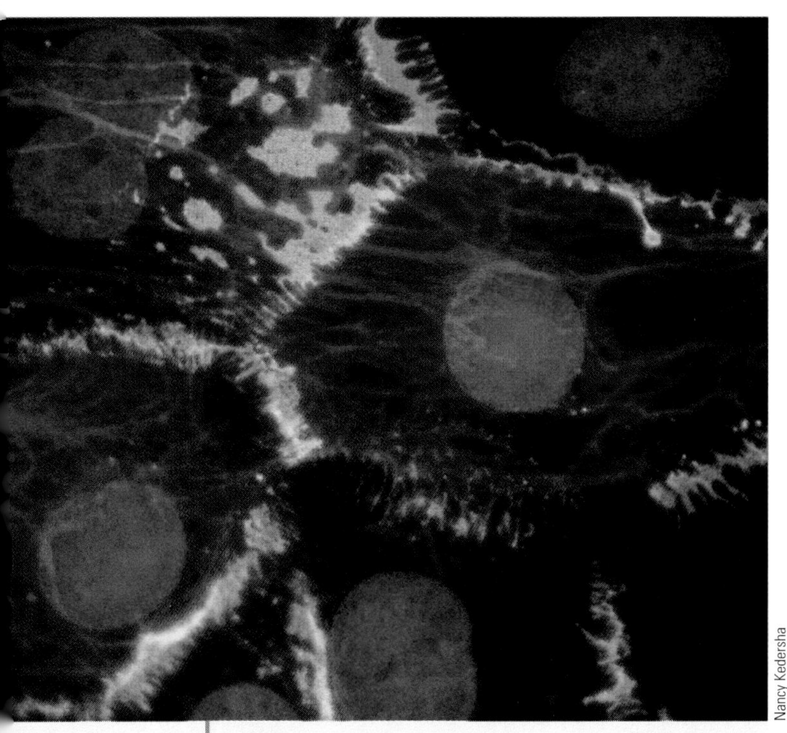
Nancy Kedersha

Cadherins. The human skin cells shown in this LM were grown in culture and stained with fluorescent antibodies. Cadherins, a group of membrane proteins, are seen as green belts around each cell in this sheet of cells. The nuclei appear as blue spheres; myosin in the cells appears red.

CHAPTER OUTLINE

The evolution of biological membranes was an essential step in the origin of life. Later, membranes made the evolution of complex cells possible, because the extensive internal membranes of eukaryotes form multiple compartments with unique environments for highly specialized activities. In Chapter 4, we discussed the importance of cell membranes in maintaining homeostasis. Biological membranes are not inanimate barriers; they are complex, dynamic structures made of lipid and protein molecules that are in constant motion. The unusual properties of membranes allow them to perform many functions in addition to defining the cell as a compartment and regulating the passage of materials. These functions include participating in many chemical reactions, transmitting signals and information between the environment and the interior of the cell, and acting as an essential part of energy transfer and storage systems (see Chapters 7 and 8).

The plasma membrane that surrounds every cell physically separates the cell from the outside world and defines the cell as a distinct entity. The plasma membrane helps maintain a life-supporting internal environment by regulating passage of materials into and out of the cell. To carry out the many chemical reactions necessary to sustain life, the cell must maintain an appropriate internal environment.

One exciting area of cell membrane research focuses on membrane proteins. Many proteins associated with the plasma membrane are enzymes. Others transport materials or transfer information. Still others, known as *cell adhesion molecules,* are important in connecting cells to one another to form tissues. Researchers are studying how membrane proteins function in health and disease.

The principal cell adhesion molecules in vertebrates and in many invertebrates are **cadherins.** These molecules are responsible for calcium-dependent adhesion between cells that form multicellular sheets. For example, cadherins form cell junctions important in maintaining the structure of the epithelium that makes up human skin (see photograph). An absence of these membrane proteins is associated with the invasiveness of some malignant tumors. Certain cadherins mediate the way cells adhere in the early embryo, and thus they are important in development.

In this chapter, we first consider what is known about the composition and structure of biological membranes. We survey how various materials, ranging from ions to complex molecules and even bacteria, move across membranes. We then consider how information crosses the plasma membrane through a signal relay system. Finally, we examine specialized structures that enable membranes of different cells to interact. Although much of our discussion centers on the structure and functions of plasma membranes, many of the concepts apply to other cell membranes. ■

THE STRUCTURE OF BIOLOGICAL MEMBRANES

Learning Objectives

1. Evaluate the importance of membranes to the homeostasis of the cell, emphasizing their various functions.
2. Describe the fluid mosaic model of cell membrane structure.
3. Explain how the properties of the lipid bilayer govern many properties of the cell membrane and of the cell.
4. Describe how membrane proteins associate with the lipid bilayer, and discuss the functions of membrane proteins.

In Chapter 4 and in the introduction to this chapter, we discussed the importance of membranes to the cell in maintaining homeostasis. How do the properties of cell membranes enable the cell to carry on such varied functions as regulating passage of materials, compartmentalizing the cell, serving as a surface for chemical reactions, adhering to and communicating with other cells, and receiving information from the environment?

Long before the development of the electron microscope, scientists knew that membranes consist of both lipids and proteins. Work by researchers in the 1920s and 1930s had provided clues that the core of cell membranes consist of lipids, mostly phospholipids (see Chapter 3).

Phospholipids form bilayers in water

Phospholipids are primarily responsible for the physical properties of biological membranes. This is because certain phospholipids have unique attributes, including features that allow them to form bilayered structures. A phospholipid contains two fatty acid chains linked to two of the three carbons of a glycerol molecule (see Fig. 3-13). The fatty acid chains make up the nonpolar, *hydrophobic* ("water-fearing") portion of the phospholipid. Bonded to the third carbon of the glycerol is a negatively charged, *hydrophilic* ("water-loving") phosphate group, which in turn is linked to a polar, hydrophilic organic group. Molecules of this type, which have distinct hydrophobic and hydrophilic regions, are called **amphipathic molecules.** All lipids that make up the core of biological membranes have amphipathic characteristics.

Because one end of each phospholipid associates freely with water and the opposite end does not, the most stable orientation for them to assume in water results in the formation of a bilayer structure (Fig. 5-1a). This arrangement allows the hydrophilic heads of the phospholipids to be in contact with the aqueous medium while their oily tails, the hydrophobic fatty acid chains, are buried in the interior of the structure away from the water molecules.

Amphipathic properties alone do not predict the ability of lipids to associate as a bilayer. Shape is also important. Phospholipids tend to have uniform widths; their roughly cylindrical shapes, together with their amphipathic properties, are responsible for bilayer formation. In summary, phospholipids form bilayers because the molecules have (1) two distinct regions, one strongly hydrophobic and the other strongly hydrophilic (making them strongly amphipathic); and (2) cylindrical shapes that allow them to associate with water most easily as a bilayer structure.

Do you know why detergents remove grease from your hands or from dirty dishes? Many common detergents are amphipathic molecules, each containing a single hydrocarbon chain (like a fatty acid) at one end and a hydrophilic region at the other. These molecules are roughly cone shaped, with the hydrophilic end forming the broad base and the hydrocarbon tail leading to the point. Because of their shapes, these molecules do not associate as bilayers but instead tend to form spherical structures in water (see Fig. 5-1b). Detergents can "solubilize" oil because the oil molecules associate with the hydrophobic interiors of the spheres.

Current data support a fluid mosaic model of membrane structure

PROCESS OF SCIENCE

By examining the plasma membrane of the mammalian red blood cell and comparing its surface area with the total number of lipid molecules per cell, early investigators calculated that the membrane is no more than two phospholipid molecules thick. These findings, together with other data, led Hugh Davson and James Danielli, working at London's University College, in

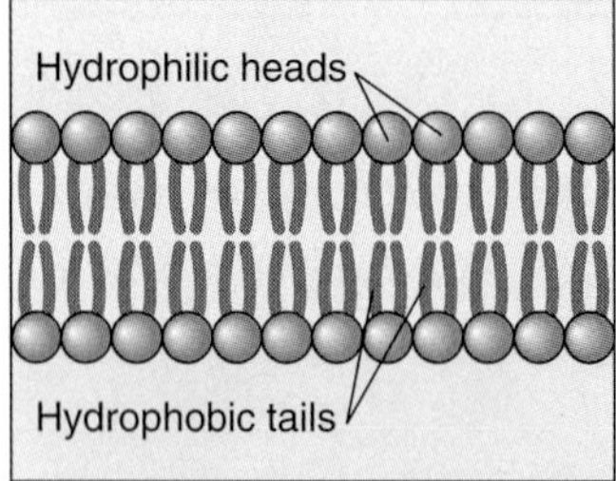

(a) Phospholipids in water

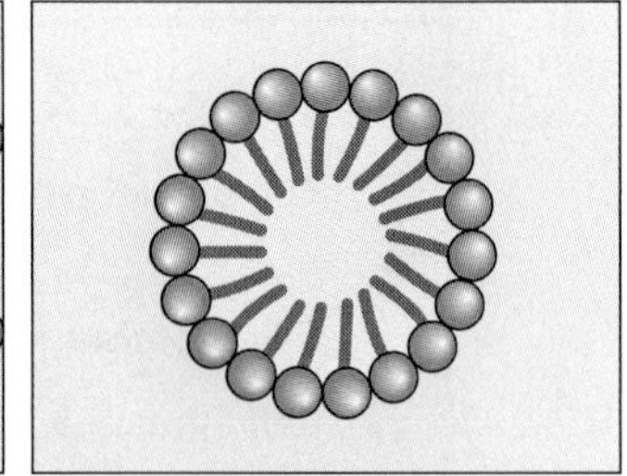

(b) Detergent in water

FIGURE **5-1** **Lipid membranes.**

(a) Phospholipids associate as bilayers in water because they are roughly cylindrical amphipathic molecules. The hydrophobic fatty acid chains associate with each other and are not exposed to water. The hydrophilic phospholipid heads are in contact with water. **(b)** Detergent molecules are roughly cone-shaped amphipathic molecules that associate in water as spherical structures.

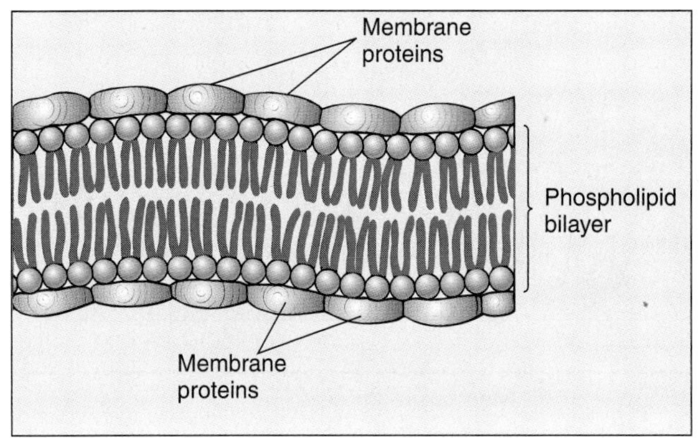

(a) The Davson-Danielli "sandwich" model

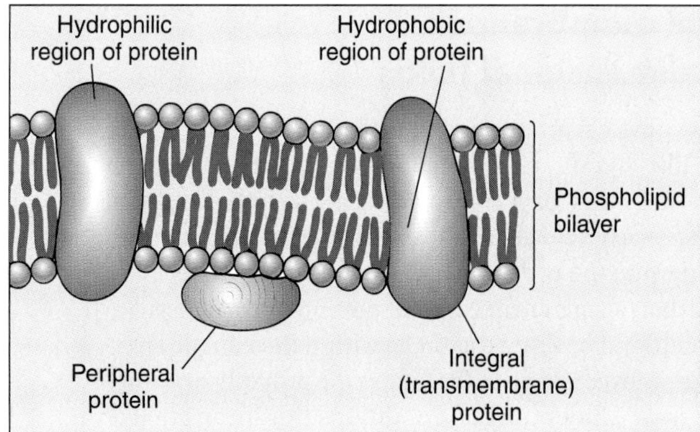

(b) Fluid mosaic model

ACTIVE FIGURE **5-2** **Two models of membrane structure.**

(a) According to the Davson-Danielli model, the membrane is a sandwich of phospholipids spread between two layers of protein. Although accepted for more than 20 years, this model was shown to be incorrect. **(b)** According to the fluid mosaic model, a cell membrane is a fluid lipid bilayer with a constantly changing "mosaic pattern" of associated proteins.

BiologyNow™ Watch protein movement in the **fluid mosaic model** by clicking on this figure on your BiologyNow CD-ROM.

1935 to propose a model in which they envisioned a membrane as a kind of "sandwich" consisting of a *lipid bilayer* (a double layer of lipid) between two protein layers (Fig. 5-2a). This useful model had a great influence on the direction of membrane research for more than 20 years. Models are important in the scientific process; good ones not only explain the available data but are testable. Scientists use the model to help them develop hypotheses that can be tested experimentally (see Chapter 1).

With the development of the electron microscope in the 1950s, cell biologists were able to see the plasma membrane for the first time. One of their most striking observations was how uniform and thin the membranes are. The plasma membrane is no more than 10 nm thick. The electron microscope revealed a three-layered structure, something like a railroad track, with two dark layers separated by a lighter layer (Fig. 5-3). Their findings seemed to support the protein-lipid-protein sandwich model.

During the 1960s, a paradox emerged regarding the arrangement of the proteins. Biologists assumed membrane proteins were uniform and had shapes that would allow them to lie like thin sheets on the membrane surface. But when purified by cell fractionation, the proteins were far from uniform; in fact, they varied widely in composition and size. Some proteins are quite large. How could they fit within a surface layer of a membrane less than 10 nm thick? At first, some researchers tried to answer this question by modifying the model with the hypothesis that the proteins on the membrane surfaces were a flattened, extended form, perhaps a β-pleated sheet (see Figure 3-20b).

Other cell biologists found that instead of having sheetlike structures, many membrane proteins are rounded, or globular. Studies of a number of membrane proteins showed that one region (or domain) of the molecule could always be found on one side of the bilayer, whereas another part of the protein might be located on the opposite side. Rather than forming a thin surface layer, many membrane proteins extended completely through the lipid bilayer. Thus the evidence suggested that membranes contain many different types of proteins of different shapes and sizes that are associated with the bilayer in a mosaic pattern.

In 1972, S. Jonathan Singer and Garth Nicolson of the University of California, at San Diego, proposed a model of membrane structure that represented a synthesis of the known properties of biological membranes. According to their **fluid mosaic model,** a cell membrane consists of a fluid bilayer of phospholipid molecules in which the proteins are embedded or otherwise associated, much like the tiles in a mosaic picture. This mosaic pattern is not static, however, because the positions of the proteins are constantly changing as they move about like icebergs in a fluid sea of phospholipids. This model has provided great impetus to research; it has been repeatedly tested and has been shown to accurately predict the properties of many kinds of cell membranes. Figure 5-2b depicts the plasma membrane of a eukaryotic cell according to the fluid mosaic model; prokaryotic plasma membranes are discussed in Chapter 23.

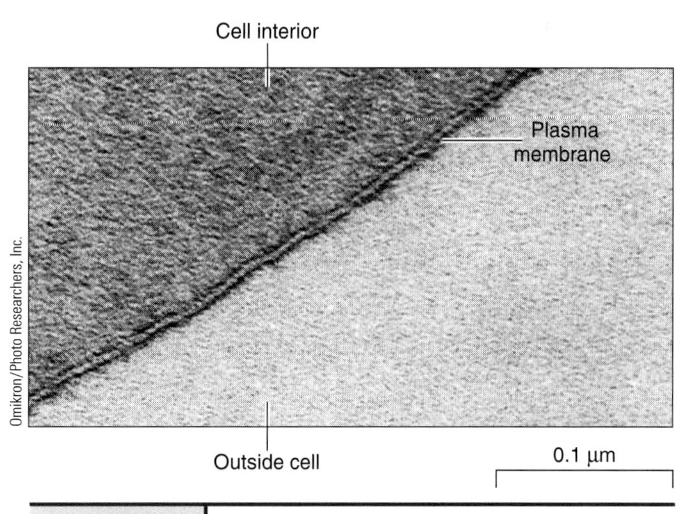

FIGURE **5-3** **TEM of the plasma membrane of a mammalian red blood cell.**

The plasma membrane separates the cytoplasm *(darker region)* from the external environment *(lighter region).* The hydrophilic heads of the phospholipids are seen as the parallel dark lines, whereas the hydrophobic tails are visible as the light zone between them.

Biological membranes are two-dimensional fluids

An important physical property of phospholipid bilayers is that they behave like *liquid crystals.* The bilayers are crystal-like in that the lipid molecules form an ordered array with the heads on the outside and fatty acid chains on the inside; they are liquid-like in that, despite the orderly arrangement of the molecules, their hydrocarbon chains are in constant motion. Thus molecules are free to rotate and can move laterally within their single layer (Fig. 5-4). Such movement gives the bilayer the property of a *two-dimensional fluid.* Under normal conditions this means that a single phospholipid molecule can travel laterally across the surface of a eukaryotic cell in seconds.

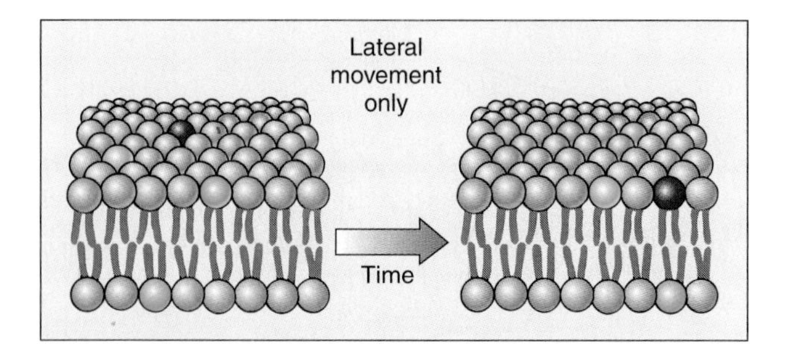

FIGURE **5-4** **Membrane fluidity.**
The ordered arrangement of phospholipid molecules makes the cell membrane a liquid crystal. The hydrocarbon chains are in constant motion, allowing each molecule to move laterally on the same side of the bilayer.

PROCESS OF SCIENCE

The fluid qualities of lipid bilayers also allow molecules embedded in them to move along the plane of the membrane (as long as they are not anchored in some way). This was elegantly demonstrated by David Frye and Michael Edidin in 1970. They conducted experiments in which they followed the movement of membrane proteins on the surface of two cells that had been joined (Fig. 5-5). When the plasma membranes of a mouse cell and a human cell are fused, within minutes, at least some of the membrane proteins from each cell migrate and become randomly distributed over the single continuous plasma membrane that surrounds the joined cells. Frye and Edidin showed that the fluidity of the lipids in the membrane allows many of the proteins to move, producing an ever-changing configuration.

For a membrane to function properly, its lipids must be in a state of optimal fluidity. The membrane's structure is weakened if its lipids are too fluid. However, many membrane functions, such as the transport of certain substances, are inhibited or cease if the lipid bilayer is too rigid. At normal temperatures, cell membranes are fluid, but at low temperatures the motion of the fatty acid chains is slowed. If the temperature decreases to a critical point, the membrane is converted to a more solid gel state.

Certain properties of membrane lipids have significant effects on the fluidity of the bilayer. Recall from Chapter 3 that molecules are free to rotate around single carbon-to-carbon covalent bonds. Because most of the bonds in hydrocarbon chains are single bonds, the chains themselves twist more and more rapidly as the temperature rises.

The fluid state of the membrane depends on its component lipids. You have probably noticed that when melted butter is left at room temperature, it solidifies. Vegetable oils, however, remain liquid at room temperature. Recall from our discussion of fats in Chapter 3 that animal fats such as butter are high in saturated fatty acids that lack double bonds. In contrast, a vegetable oil may be polyunsaturated, with most of its fatty acid chains having two or more double bonds. At each double bond there is a bend in the molecules that prevents the hydrocarbon chains from coming close together and interacting through van der Waals interactions. In this way, unsaturated fats lower the temperature at which oil or membrane lipids solidify.

Many organisms have regulatory mechanisms for maintaining cell membranes in an optimally fluid state. For example, some organisms compensate for temperature changes by altering the fatty acid content of their membrane lipids. When the outside temperature is cold, the membrane lipids contain relatively high proportions of unsaturated fatty acids.

Some membrane lipids stabilize membrane fluidity within certain limits. One such "fluidity buffer" is cholesterol, a steroid found in animal cell membranes. A cholesterol molecule is largely hydrophobic but is slightly amphipathic owing to the presence of a single hydroxyl group (see Fig. 3-15a). This hydroxyl group associates with the hydrophilic heads of the phospholipids; the hydrophobic remainder of the cholesterol molecule fits between the fatty acid hydrocarbon chains (Fig. 5-6).

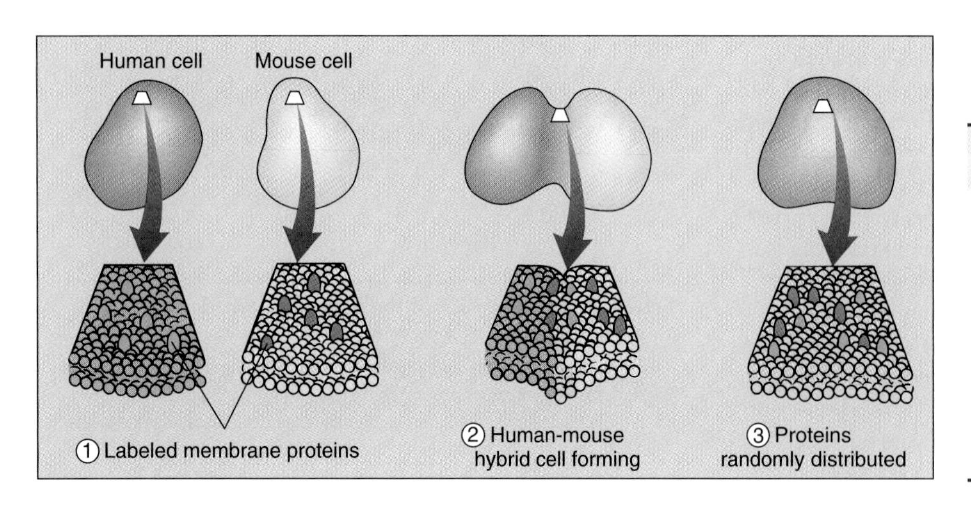

FIGURE **5-5** **Frye and Edidin's experiment.**
① Membrane proteins of mouse cells and human cells were labeled with fluorescent dye markers in two different colors. ② When the plasma membranes of a mouse cell and a human cell were fused, mouse proteins migrated to the human side and human proteins to the mouse side. ③ After a short time, mouse and human proteins became randomly distributed on the cell surface.

At low temperatures cholesterol molecules act as "spacers" between the hydrocarbon chains, restricting van der Waals interactions that would promote solidifying. Cholesterol also helps prevent the membrane from becoming weakened or unstable at higher temperatures. This is because the cholesterol molecules interact strongly with the portions of the hydrocarbon chains closest to the phospholipid head. This interaction restricts motion in these regions. Plant cells have steroids other than cholesterol that carry out similar functions.

Biological membranes fuse and form closed vesicles

Lipid bilayers, particularly those in the liquid-crystalline state, have additional important physical properties. Bilayers tend to resist forming free ends; as a result, they are self-sealing and under most conditions spontaneously round up to form closed vesicles. Lipid bilayers are also flexible, allowing cell membranes to change shape without breaking. Under appropriate conditions lipid bilayers fuse with other bilayers.

Membrane fusion is an important cell process. When a vesicle fuses with another membrane, both membrane bilayers and their compartments become continuous. Various transport and secretory vesicles form from and also merge with membranes of the ER and Golgi complex, facilitating the transfer of materials from one compartment to another. A secretory vesicle fuses with the plasma membrane when a product is secreted from the cell.

Membrane proteins include integral and peripheral proteins

The two major classes of membrane proteins, integral proteins and peripheral proteins, are defined by how tightly they are associated with the lipid bilayer (see Fig. 5-6). **Integral membrane**

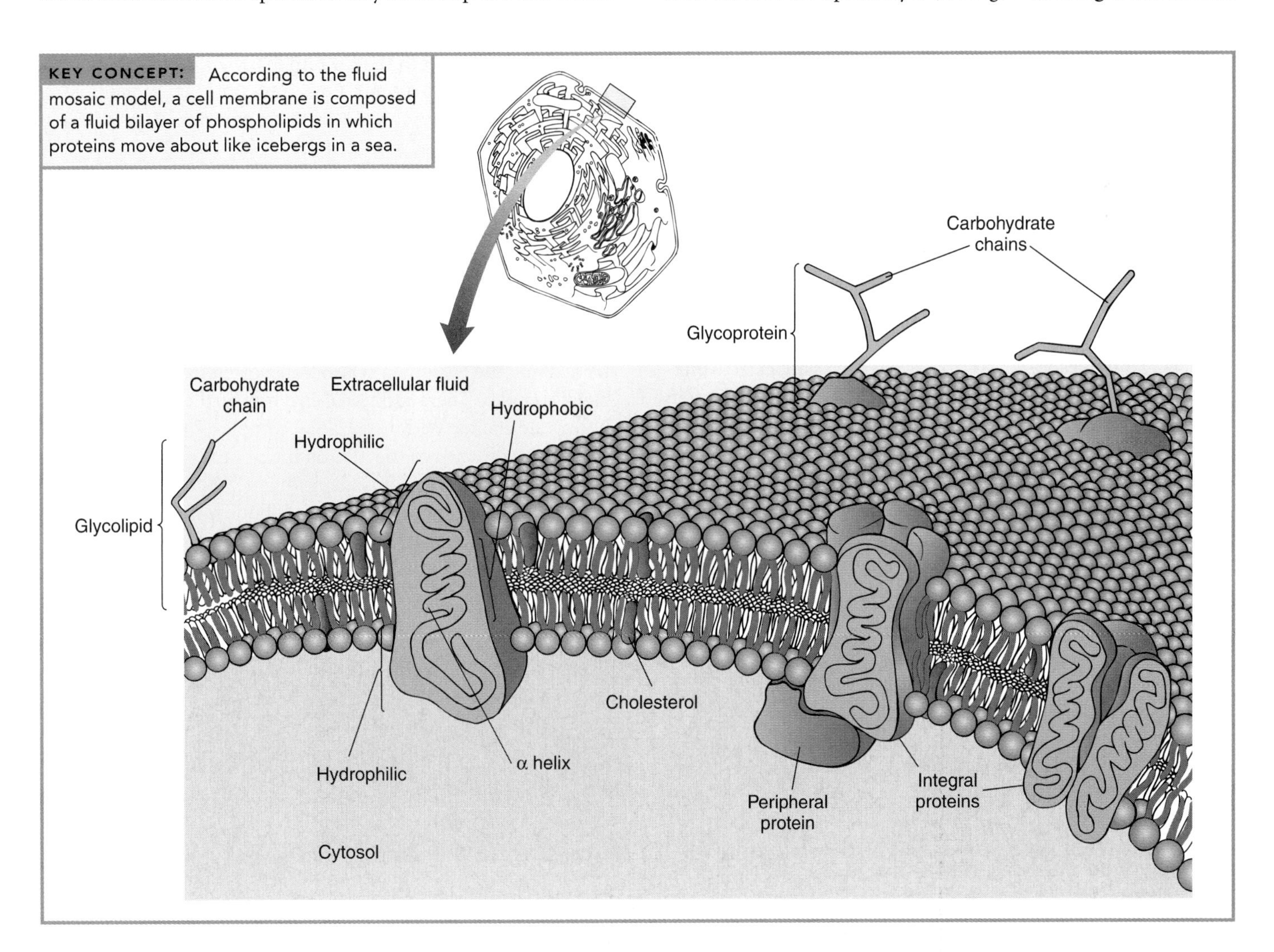

FIGURE **5-6** **Detailed structure of the plasma membrane.**
Although the lipid bilayer consists mainly of phospholipids, other lipids, such as cholesterol and glycolipids, are present. Peripheral proteins are loosely associated with the bilayer, whereas integral proteins are tightly bound. The integral proteins shown here are transmembrane proteins that extend through the bilayer. They have hydrophilic regions on both sides of the bilayer, connected by a membrane-spanning α-helix. Glycolipids (carbohydrates attached to lipids) and glycoproteins (carbohydrates attached to proteins) are exposed on the extracellular surface; they play roles in cell recognition and adhesion.

proteins are firmly bound to the membrane. Cell biologists usually can release them only by disrupting the bilayer with detergents. These proteins are amphipathic. Their hydrophilic regions extend out of the cell or into the cytoplasm, while their hydrophobic regions interact with the fatty acid tails of the membrane phospholipids.

Some integral proteins do not extend all the way through the membrane. Many others, called **transmembrane proteins,** extend completely through the membrane. Some span the membrane only once, whereas others wind back and forth as many as 24 times. The most common kind of transmembrane protein is an α-helix (see Chapter 3) with hydrophobic amino acid side chains projecting out from the helix into the hydrophobic region of the lipid bilayer.

Peripheral membrane proteins are not embedded in the lipid bilayer. They are located on the inner or outer surface of the plasma membrane, usually bound to exposed regions of integral proteins by noncovalent interactions. Peripheral proteins can be easily removed from the membrane without disrupting the structure of the bilayer.

Proteins are oriented asymmetrically across the bilayer

PROCESS OF SCIENCE

One of the most remarkable demonstrations that proteins are actually embedded in the lipid bilayer comes from *freeze-fracture* electron microscopy, a technique that enables a researcher to literally see the membrane from "inside out." When cell biologists examine membranes in this way, they observe numerous particles on the fracture faces. These particles are clearly integral membrane proteins, because researchers never see them in freeze-fractured artificial lipid bilayers. These findings profoundly influenced Singer and Nicolson in developing the fluid mosaic model.

When we compare the two sides of a membrane, large numbers of particles are found on one side and very few on the other (Fig. 5-7). This does not necessarily mean more proteins are on one side of the membrane than on the other but rather that most are more firmly attached to a given side. Thus the protein molecules are *asymmetrically oriented.* Each side of a membrane has different characteristics because each type of protein is oriented in the bilayer in only one way. Proteins are not randomly placed into membranes; asymmetry is produced by the highly specific way in which each protein is inserted into the bilayer.

Membrane proteins that will become part of the inner surface of the plasma membrane are manufactured by free ribosomes and move to the membrane through the cytoplasm. Membrane proteins that will be associated with the cell's outer surface are manufactured like proteins destined to be exported from the cell. As discussed in Chapter 4, these proteins are initially formed by ribosomes on the rough ER. They pass through the ER membrane into the ER lumen, where sugars are added, making them **glycoproteins.** Only a part of each protein passes through the ER membrane, so each completed protein has some regions that are located in the ER lumen and other regions that remain in the cytosol. Enzymes that attach the sugars to certain amino acids on the protein are found only in the lumen of the ER. Thus carbohydrates can be added only to the parts of proteins that are located in that compartment.

In Figure 5-8, follow from top to bottom the vesicle budding and membrane fusion events that are part of the transport process. You can see that the same region of the protein that protruded into the ER lumen is also transported to the lumen of the Golgi complex. There additional enzymes further modify the carbohydrate chains. Within the Golgi complex, the glycoprotein is sorted and directed to the plasma membrane. The modified region of the protein remains inside the membrane compartment of a transport (secretory) vesicle as it buds from the Golgi com-

Extracellular fluid
E-face
P-face
Cytosol

(a)

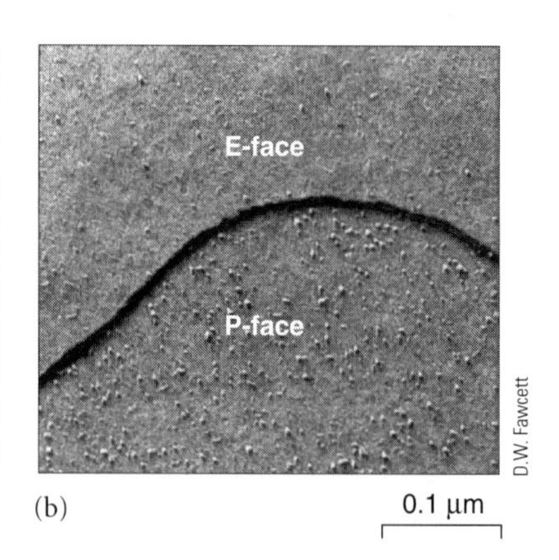

(b)

FIGURE **5-7** **Asymmetry of the plasma membrane.**

(a) In the freeze-fracture method, the path of membrane cleavage is along the hydrophobic interior of the lipid bilayer. Two complementary fracture faces result. The inner half-membrane presents the P-face (or protoplasmic face), from which project most of the membrane proteins. A relatively smooth, outer half-membrane presents the E-face (or external face), which shows fewer protein particles. In a good fracture, particles are visible on both of the inside faces of the fractured membrane, as shown here. These particles are transmembrane proteins inserted into the lipid bilayer. Freeze-fractured bilayers of lipids alone do not have particles on the fracture planes. **(b)** A freeze-fracture TEM. Notice the greater number of proteins on the P-face of the membrane.

plex. When the transport vesicle fuses with the plasma membrane, the carbohydrate chain becomes the part of the membrane protein that extends to the exterior of the cell surface.

In summary, this is the sequence:

ER lumen ⟶ transport vesicle ⟶ vesicles in Golgi (transport to successive compartments) ⟶ transport (secretory) vesicle ⟶ plasma membrane

Membrane proteins function in transport, in information transfer, and as enzymes

Why does the plasma membrane require so many different proteins? This diversity reflects the multitude of activities that take place in or on the membrane. Generally, plasma membrane proteins fall into several broad functional categories as shown in Figure 5-9. Some membrane proteins anchor the cell to its substrate. For example, integrins, described in Chapter 4, attach to the extracellular matrix while simultaneously binding to microfilaments inside the cell (Fig. 5-9a). They also serve as receptors, or docking sites, for proteins of the extracellular matrix.

Many membrane proteins are involved in the transport of molecules across the membrane. Some form channels that selectively allow the passage of specific ions or molecules (Fig. 5-9b). Other proteins form pumps that use ATP to actively transport solutes across the membrane (Fig. 5-9c).

Certain membrane proteins are enzymes that catalyze reactions near the cell surface (Fig. 5-9d). For example, in mitochondrial or chloroplast membranes, enzymes may be organized in a sequence to regulate a series of reactions, as in cellular respiration or photosynthesis.

Some membrane proteins are receptors that receive information from other cells. For example, cells receive hormonal signals from endocrine cells. This information may be transmitted to the cell interior by *signal transduction*, discussed later in this chapter (Fig. 5-9e).

Some membrane proteins serve as identification tags that other cells recognize. Cells that recognize one another may connect to form a tissue. Human cells have distinctive receptors that identify them as part of a particular individual. Certain cells recognize the surface proteins, or antigens, of bacterial cells as foreign. Antigens stimulate immune defenses that destroy the bacteria (Fig. 5-9f). Some membrane proteins form junctions between adjacent cells (Fig. 5-9g). These proteins may also serve as anchoring points for networks of cytoskeletal elements.

Review

- What molecules are responsible for the physical properties of a cell membrane?
- How might a transmembrane protein be positioned in a lipid bilayer? How do the hydrophilic and hydrophobic regions of the protein affect its orientation?
- What is the pathway used by cells to place carbohydrates on plasma membrane proteins? How does this pathway result in the carbohydrate groups being exposed on only one side of the lipid bilayer?
- What are some functions of the plasma membrane?

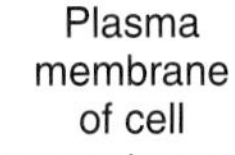

Assess your understanding of the **structure of biological membranes** by taking the pretest on your BiologyNow CD-ROM.

FIGURE **5-8** **The formation of a membrane protein.**

The orientation of a protein in the plasma membrane results from the pathway of its synthesis and transport in the cell. The surface of the rough ER membrane that faces the lumen of the rough ER also faces the lumen of the Golgi complex and vesicles. When a vesicle fuses with the plasma membrane, its inner surface becomes the extracellular surface of the plasma membrane. Carbohydrates added to proteins in the ER and then modified in the Golgi complex are associated with the extracellular surface of the plasma membrane.

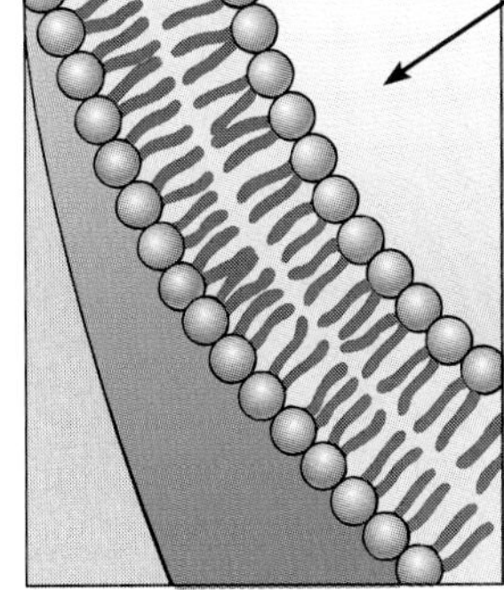

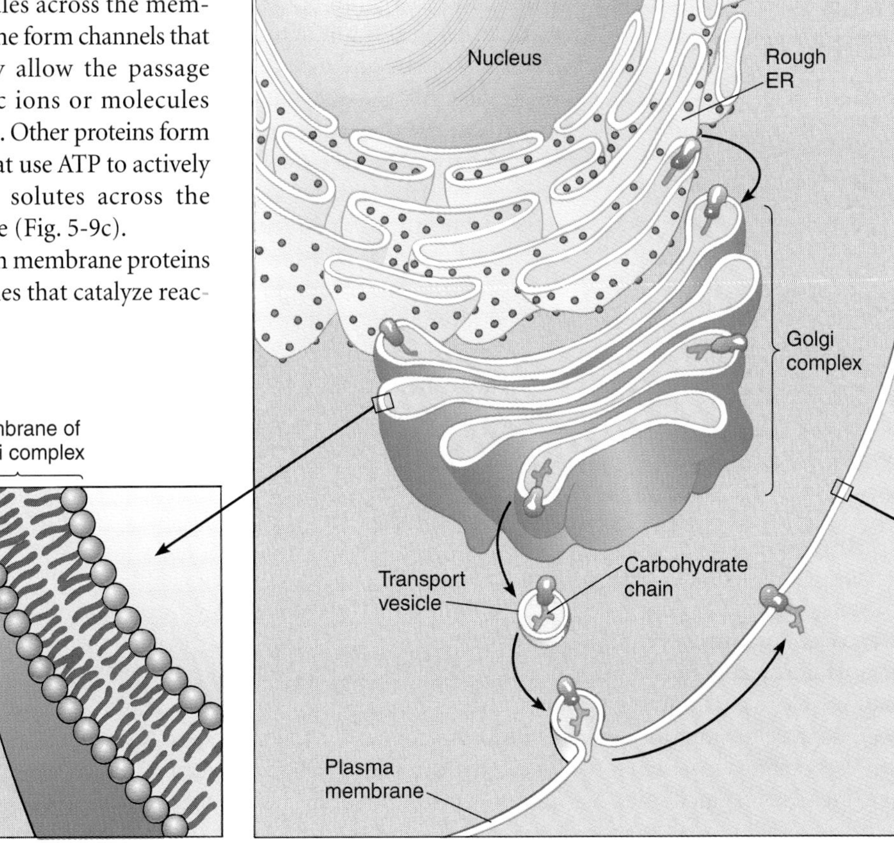

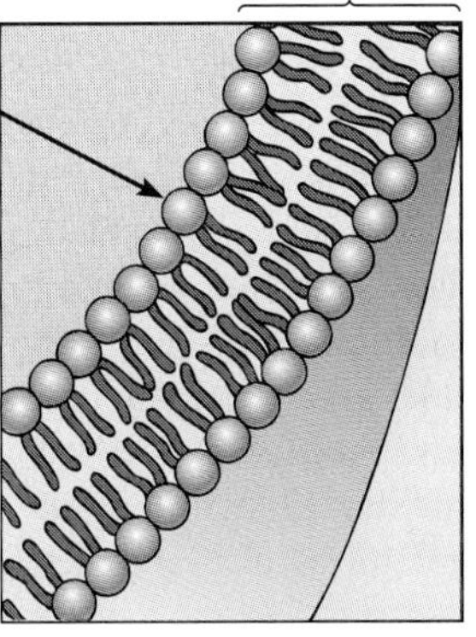

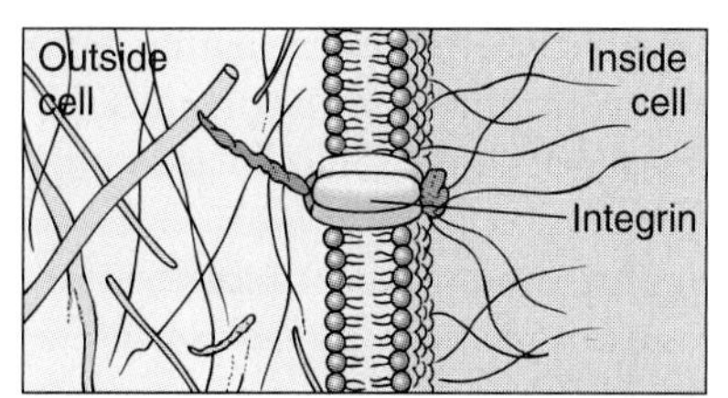

(a) Anchoring. Some membrane proteins, such as integrins, anchor the cell to the extracellular matrix; they also connect to microfilaments within the cell.

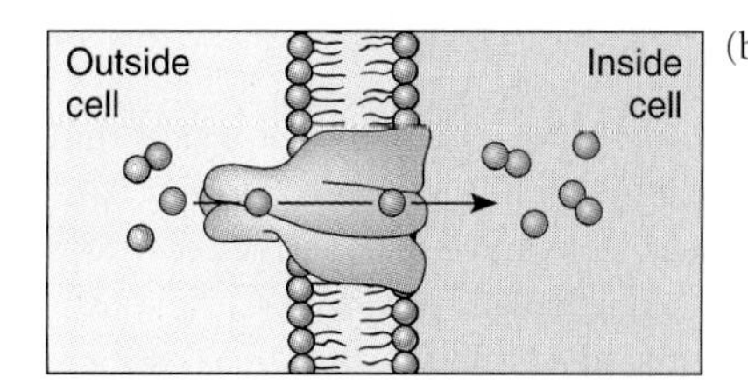

(b) Passive transport. Certain proteins form channels for selective passage of ions or molecules.

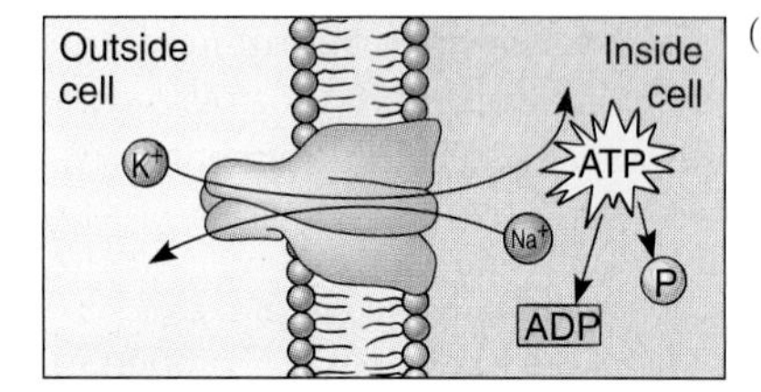

(c) Active transport. Some transport proteins pump solutes across the membrane, which requires a direct input of energy.

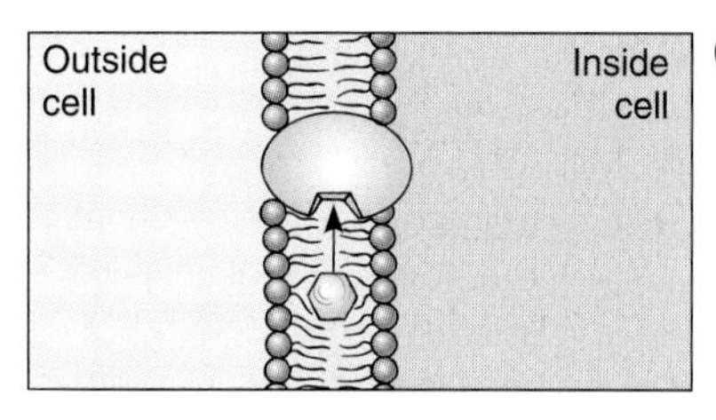

(d) Enzymatic activity. Many membrane-bound enzymes catalyze reactions that take place within or along the membrane surface.

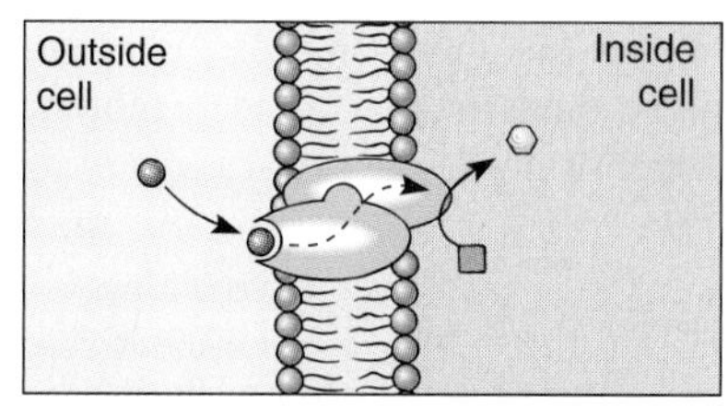

(e) Signal transduction. Some receptors bind with signal molecules such as hormones and transmit information into the cell.

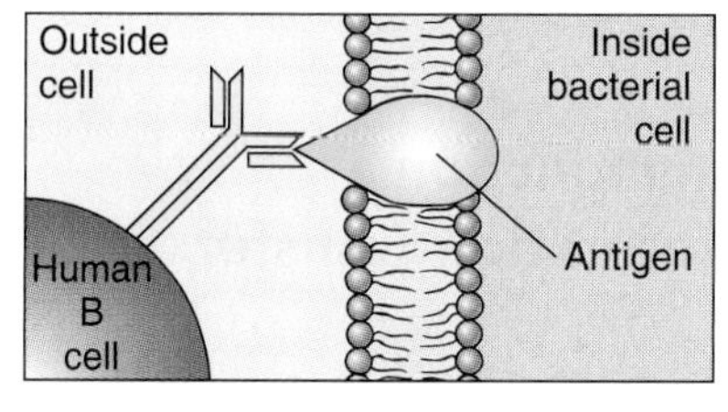

(f) Cell recognition. Some receptor proteins function as identification tags. For example, bacterial cells have surface proteins, or antigens, that human cells recognize as foreign.

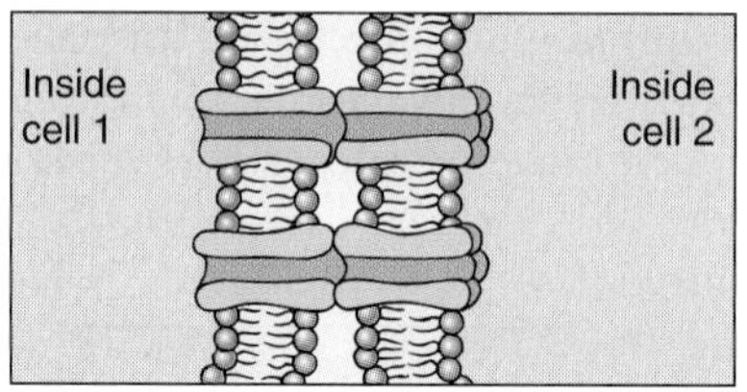

(g) Intercellular junction. Cell adhesion proteins attach membranes of adjacent cells.

FIGURE **5-9** **Functions of membrane proteins.**

PASSAGE OF MATERIALS THROUGH CELL MEMBRANES

Learning Objectives

5 Contrast the physical processes of simple diffusion and osmosis with the carrier-mediated physiological processes by which materials are transported across cell membranes.

6 Solve simple problems involving osmosis; for example, predict whether cells will swell or shrink under various osmotic conditions.

7 Differentiate between the processes of facilitated diffusion and active transport, and identify energy sources for each process.

8 Compare endocytotic and exocytotic transport mechanisms.

A membrane is *permeable* to a given substance if it allows that substance to pass through, and impermeable if it does not. Biological membranes are **selectively permeable membranes**—they allow some but not all substances to pass through them. In general, biological membranes are most permeable to small molecules and to lipid-soluble substances able to pass through the hydrophobic interior of the bilayer. Because the interior of its lipid bilayer is hydrophobic, biological membranes present a barrier to most polar (and therefore not lipid-soluble) molecules.

In response to varying environmental conditions or cell needs, a membrane may be a barrier to a particular substance at one time and actively promote its passage at another time. By regulating chemical traffic across its plasma membrane, a cell controls its volume and its internal ionic and molecular composition, which can be quite different from the outside.

Although they are polar, water molecules rapidly cross a lipid bilayer. They are small enough to pass through gaps that occur as a fatty acid chain momentarily moves out of the way. The following also cross the lipid bilayer rapidly: gases such as oxygen, carbon dioxide, and nitrogen; small polar molecules like glycerol; and larger, nonpolar (hydrophobic) substances such as hydrocarbons. Slightly larger polar molecules, such as glucose, and charged ions of any size pass through the bilayer slowly.

The bilayer is relatively impermeable to ions, but cells must move ions, as well as amino acids, sugars, and other water-soluble molecules, across membranes. The permeability of membranes to those substances is due primarily to the activities of specialized membrane **transport proteins.** In the following sections, we will discuss the roles of two main types of membrane transport proteins: carrier proteins and channel proteins.

Some ions and molecules move through membranes by passive processes such as diffusion. Passive transport does not require the cell to expend metabolic energy. Many materials must be actively transported across the membrane. Active transport mechanisms include active transport, exocytosis, and endocytosis (discussed later in this chapter). These processes require a direct expenditure of metabolic energy by the cell.

Random motion of particles leads to diffusion

Some substances pass into or out of cells and move about within cells by simple **diffusion,** a physical process based on random motion. All atoms and molecules possess kinetic energy, or energy of motion, at temperatures above absolute zero (0°Kelvin, −273°Celsius, or −459.4°F). Matter may exist as a solid, liquid, or gas, depending on the freedom of movement of its constituent particles. The particles of a solid are closely packed, and the forces of attraction between them let them vibrate, but not move around. In a liquid the particles are farther apart; the intermolecular attractions are weaker, and the particles move about with considerable freedom. In a gas the particles are so far apart that intermolecular forces are negligible; molecular movement is restricted only by the walls of the container that encloses the gas. Atoms and molecules in liquids and gases move in a kind of "random walk," changing directions as they collide.

Although the movement of individual particles is undirected and unpredictable, we can nevertheless make predictions about the behavior of groups of particles. If the particles (atoms, ions, or molecules) are not evenly distributed, then at least two regions exist: one with a higher concentration of particles and the other with a lower concentration. Such a difference in the concentration of a substance from one place to another establishes a **concentration gradient.**

In diffusion, the random motion of particles results in their net movement "down" their own concentration gradient, from the region of higher concentration to the one of lower concentration. This does not mean individual particles are prohibited from moving "against" the gradient. However, because there are initially more particles in the region of high concentration, it logically follows that more particles move randomly from there into the low-concentration region than the reverse (Fig. 5-10).

Diffusion occurs rapidly over very short distances. The rate of diffusion is determined by the movement of the particles, which in turn is a function of their size and shape, their electrical charges, and the temperature. As the temperature rises, particles move faster and the rate of diffusion increases.

Particles of different substances in a mixture diffuse independently of each other. If particles are not added to or removed from the system, a state of **equilibrium,** a condition of no net change in the system, is ultimately reached. At equilibrium the particles are uniformly distributed.

In organisms, equilibrium is rarely attained. For example, carbon dioxide continually forms within a human cell as sugars and other molecules are metabolized during aerobic respiration. Carbon dioxide readily diffuses across the plasma membrane but then is rapidly removed by the blood. This limits the opportunity for the molecules to re-enter the cell, so a sharp concentration gradient of carbon dioxide molecules always exists across the plasma membrane.

Osmosis is diffusion of water across a selectively permeable membrane

Osmosis is a special kind of diffusion that involves the net movement of water (the principal *solvent* in biological systems) through a selectively permeable membrane from a region of higher concentration to a region of lower concentration. Water molecules pass freely in both directions, but, as in all types of diffusion, *net* movement is from the region where the water molecules are more concentrated to the region where they are less concentrated. Most *solute molecules* (for example, sugar and salt) cannot diffuse freely through the selectively permeable membranes of the cell.

The principles involved in osmosis can be illustrated using an apparatus called a U-tube (Fig. 5-11). The U-tube is divided into two sections by a selectively permeable membrane that allows solvent (water) molecules to pass freely but excludes solute molecules). A water/solute solution is placed on one side, and pure water is placed on the other. The side containing the solute has a lower effective concentration of water than the pure water side. This is because the solute particles, which are charged (ionic) or polar, interact with the partial electrical charges on the polar water molecules. Many of the water molecules are thus "bound up" and no longer free to diffuse across the membrane.

Because of the difference in effective water concentration, there is net movement of water molecules from the pure water side (with a high effective concentration of water) to the water/solute side (with a lower effective concentration of water). As a result, the fluid level drops on the pure water side and rises on the water/solute side. Because the solute molecules do not diffuse across the membrane, equilibrium is never attained. Net movement of water continues, and the fluid level rises on the side containing the solute. The weight of the rising column of fluid eventually exerts enough pressure to stop further changes in fluid levels, although water molecules continue to pass through the selectively permeable membrane in both directions.

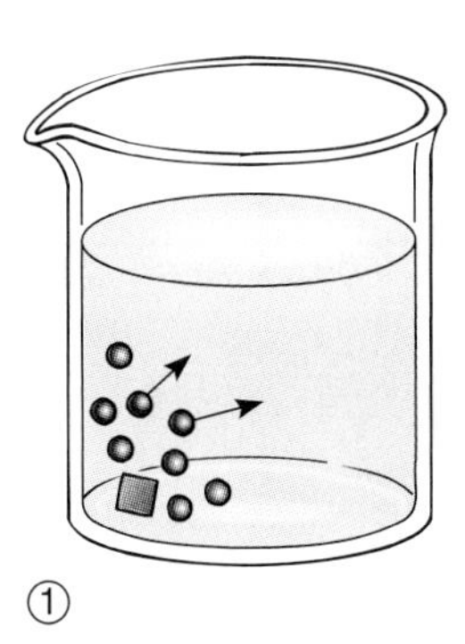

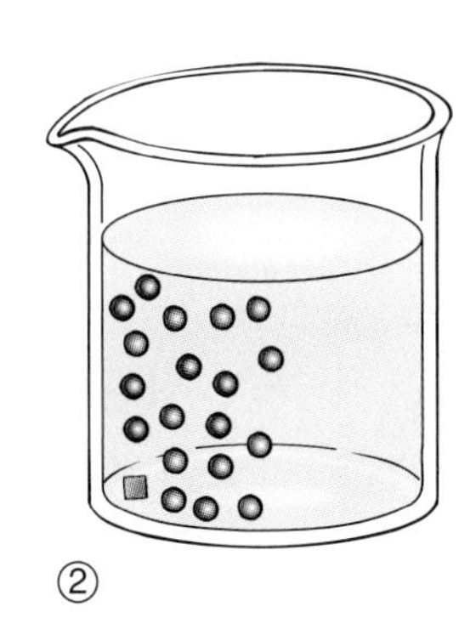

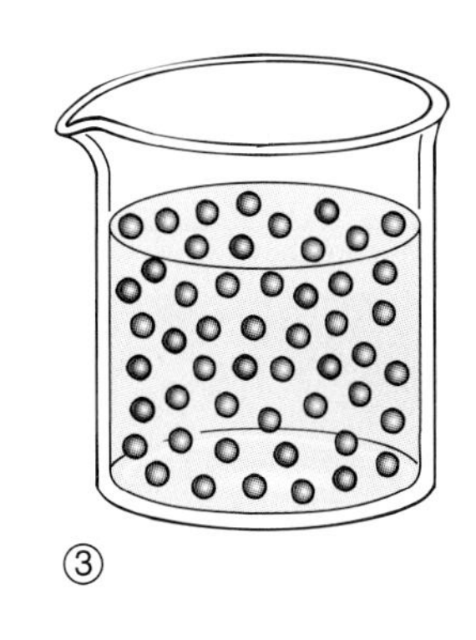

FIGURE **5-10** | **Diffusion.**

① When a lump of sugar is dropped into a beaker of pure water, ② its molecules dissolve and begin to diffuse through the water. ③ Eventually the sugar molecules become distributed equally throughout the water.

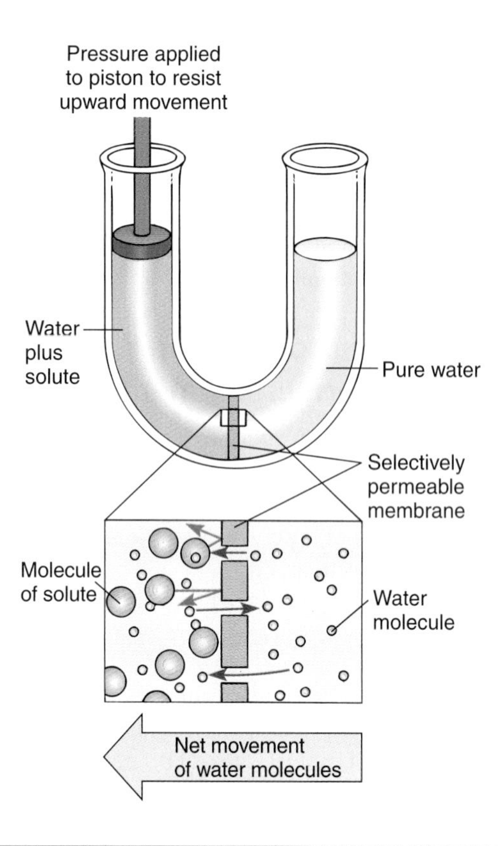

FIGURE **5-11** | **Osmosis.**

The U-tube contains pure water on the right and water plus a solute on the left, separated by a selectively permeable membrane. Water molecules cross the membrane in both directions *(red arrows)*. Solute molecules cannot cross *(green arrows)*. The fluid level rises on the left and falls on the right because net movement of water *(blue arrow)* is to the left. The force that must be exerted by the piston to prevent the rise in fluid level is equal to the osmotic pressure of the solution.

We define the **osmotic pressure** of a solution as the pressure that must be exerted on the side of a selectively permeable membrane containing the higher concentration of solute, to prevent the diffusion of water (by osmosis) from the side containing the lower solute concentration. In the U-tube example, you could measure the osmotic pressure by inserting a piston on the water/solute side of the tube and measuring how much pressure must be exerted by the piston to prevent the rise of fluid on that side of the tube. A solution with a high solute concentration has a low effective water concentration and a high osmotic pressure; conversely, a solution with a low solute concentration has a high effective water concentration and a low osmotic pressure.

Two solutions may be isotonic, or one may be hypertonic and the other hypotonic

Salts, sugars, and other substances are dissolved in the fluid compartment of every cell. These solutes give the cytosol a specific osmotic pressure. Table 5-1 summarizes the movement of water into and out of a solution (or cell) depending on relative solute concentrations. When a cell is placed in a fluid with exactly the same osmotic pressure, no net movement of water molecules occurs, either into or out of the cell. The cell neither swells nor shrinks. Such a fluid is **isotonic,** of equal solute concentration, to the fluid within the cell. Normally, your blood plasma (the fluid component of blood) and all your other body fluids are isotonic to your cells; they contain a concentration of water equal to that in the cells. A solution of 0.9% sodium chloride (sometimes called *physiological saline*) is isotonic to the cells of humans and other mammals. Human red blood cells placed in 0.9% sodium chloride neither shrink nor swell (Fig. 5-12a).

If the surrounding fluid has a concentration of dissolved substances greater than the concentration within the cell, it has a higher osmotic pressure than the cell and is said to be **hypertonic** to the cell. Because a hypertonic solution has a lower effective water concentration, a cell placed in such a solution shrinks as it loses water by osmosis. Human red blood cells placed in a solution of 1.3% sodium chloride shrivel and die (Fig. 5-12b). If the surrounding fluid contains a lower concentration of dissolved materials than does the cell, it has a lower osmotic pressure and is said to be **hypotonic** to the cell; water then enters the cell and causes it to swell. Red blood cells placed in a solution of 0.6% sodium chloride gain water, swell (Fig. 5-12c), and may eventually burst. Many cells that normally live in hypotonic environments have adaptations to prevent excessive water accumulation. For example, certain protists such as *Paramecium* have contractile vacuoles that expel excess water (see Fig. 24-7).

Turgor pressure is the internal hydrostatic pressure usually present in walled cells

The relatively rigid cell walls of plant cells, algae, bacteria, and fungi enable these cells to withstand, without bursting, an external medium that is very dilute, containing only a very low concentration of solutes. Because of the substances dissolved in the cytoplasm, the cells are hypertonic to the outside medium (conversely, the outside medium is hypotonic to the cytoplasm). Water moves into the cells by osmosis, filling their central vac-

TABLE **5-1** **Osmotic Terminology**

Solute Concentration In Solution A	Solute Concentration in Solution B	Tonicity	Direction of Net Movement of Water
Greater	Less	A hypertonic to B; B hypotonic to A	B to A
Less	Greater	B hypertonic to A; A hypotonic to B	A to B
Equal	Equal	A and B are isotonic to each other	No net movement

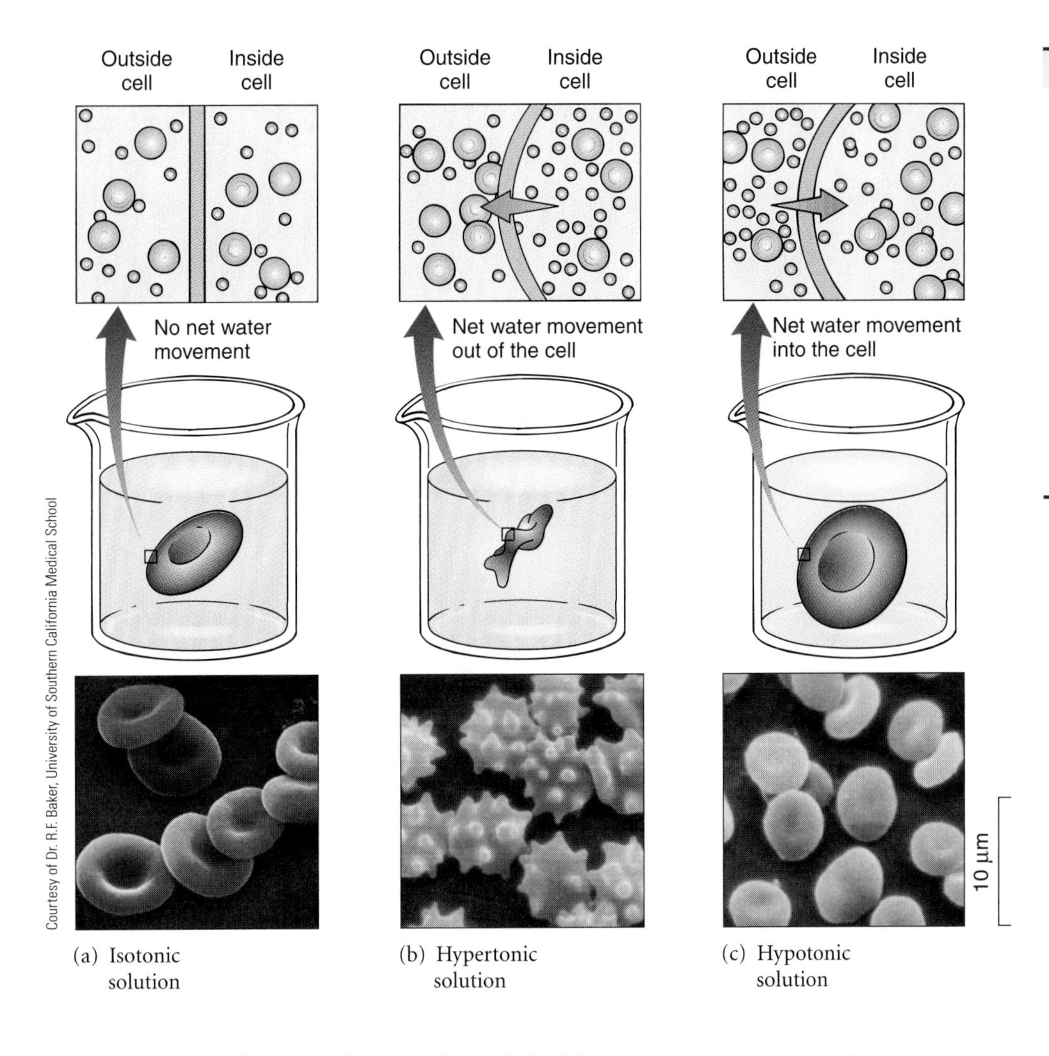

(a) Isotonic solution (b) Hypertonic solution (c) Hypotonic solution

FIGURE **5-12**

The responses of animal cells to osmotic pressure differences.

(a) When a cell is placed in an isotonic solution, water molecules pass in and out of the cell, but the net movement is zero. **(b)** When a cell is placed in a hypertonic solution, there is a net movement of water out of the cell *(arrow)*, and the cell becomes dehydrated and shrunken, and may die. **(c)** When a cell is placed in a hypotonic solution, the net movement of water molecules into the cell *(arrow)* causes the cell to swell or even burst.

uoles and distending the cells. The cells swell, building up **turgor pressure** against the rigid cell walls (Fig. 5-13a). The cell walls stretch only slightly, and a steady state is reached when their resistance to stretching prevents any further increase in cell size and thereby halts the *net movement* of water molecules into the cells. (Of course, molecules continue to move back and forth across the plasma membrane.) Turgor pressure in the cells is an important factor in supporting the body of nonwoody plants.

If a cell that has a cell wall is placed in a hypertonic medium, it loses water to its surroundings. Its contents shrink, and the plasma membrane separates from the cell wall, a process known as **plasmolysis** (Fig. 5-13b and c). Plasmolysis occurs in plants when the soil or water around them contains high concentrations of salts or fertilizers. It also explains why lettuce becomes limp in a salty salad dressing, and a picked flower wilts from lack of water.

Channel proteins and carrier proteins affect membrane permeability

Cell biologists have identified transmembrane proteins called **aquaporins** that function as gated water channels. These **channel proteins** facilitate the rapid transport of water through the plasma membrane in response to osmotic gradients. Aquaporins have been identified in a wide range of cells from bacteria to humans. In some cells, such as those lining the kidney tubules of mammals, aquaporins respond to specific signals from hormones.

Ions play important roles in cell signaling and many other physiological processes. By controlling the influx and efflux of ions, the cell directly or indirectly regulates many metabolic activities. For example, changes in the cytoplasmic concentration of calcium ions trigger changes in a number of cell processes, including muscle contraction (see Chapter 38). Because of their electric charges, ions cannot cross a lipid bilayer by simple diffusion. Certain integral membrane proteins form channels through which specific ions pass. Some ions are transported through the membrane by integral membrane proteins known as **carrier proteins.**

Cells must continually acquire essential polar nutrient molecules, such as glucose and amino acids. The lipid bilayer of the plasma membrane is relatively impermeable to most large polar molecules. This is advantageous to cells for a number of reasons. Most of the compounds required in metabolism are polar, and the impermeability of the plasma membrane prevents their loss by diffusion.

Systems of carrier proteins that transport ions and nutrients through membranes apparently evolved early in the origin of cells. Transfer of solutes by proteins located within the membrane is called **carrier-mediated transport.** The two forms of carrier-mediated transport—facilitated diffusion and carrier-mediated active transport—differ in their capabilities and energy sources.

Facilitated diffusion occurs down a concentration gradient

In all processes in which substances move across membranes by passive diffusion, the net transfer of those molecules from one side to the other occurs as a result of a concentration gradient (see Fig. 6-4). If the membrane is permeable to a substance, there is net movement from the side of the membrane where it is more highly concentrated to the side where it is less concentrated. Such a gradient across the membrane is actually

FIGURE **5-13** **Turgor pressure and plasmolysis.**

(a) In hypotonic surroundings, the vacuole of a plant cell fills, but the rigid cell walls prevent the cell from expanding. The cells of this healthy begonia plant are turgid. **(b and c)** When the begonia plant is exposed to a hypertonic solution, its cells become plasmolyzed as they lose water. The plant wilts and eventually dies.

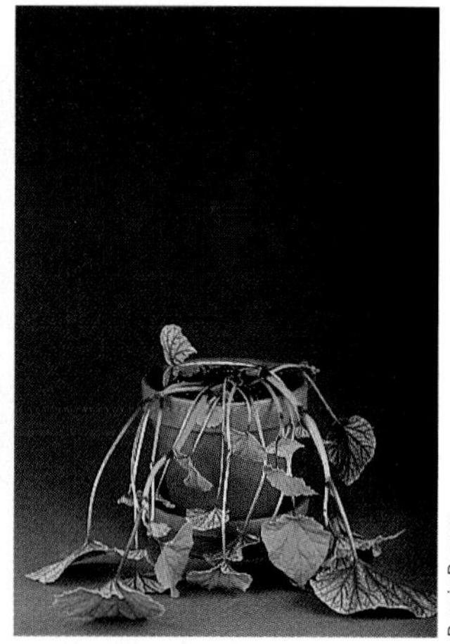

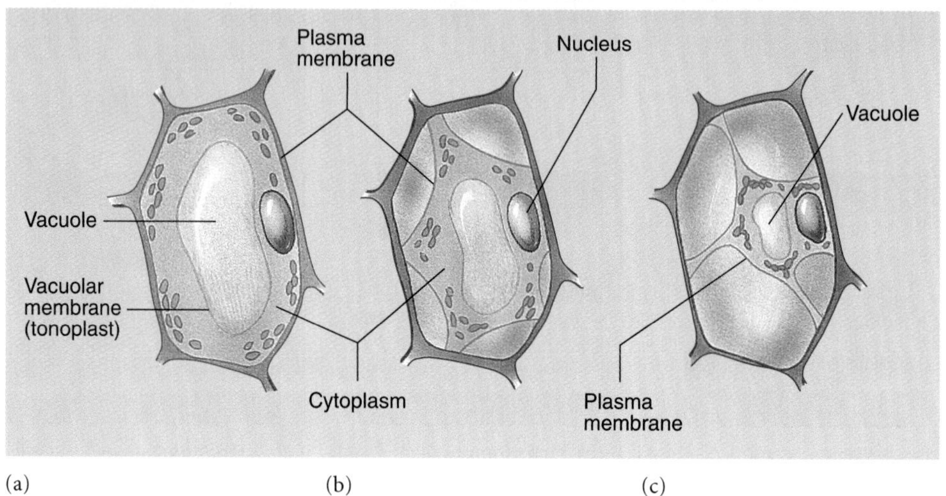

a form of stored energy. A concentration gradient occurs as a result of many different processes that take place in cells. The stored energy of the concentration gradient is released when molecules move from a region of high concentration to one of low concentration; movement down a concentration gradient is therefore spontaneous. (These types of energy and spontaneous processes are discussed in greater detail in Chapter 6.)

In the type of transport known as **facilitated diffusion,** the membrane may be made permeable to a solute, such as an ion or a polar molecule, by a specific carrier protein (Fig. 5-14). For example, *glucose permease* is a transmembrane carrier protein that transports glucose into red blood cells. These cells keep the internal concentration of glucose low by immediately adding a phosphate group to entering glucose molecules, converting them to highly charged glucose phosphates that cannot pass back through the membrane. Because glucose phosphate is a different molecule, it does not contribute to the glucose concentration gradient. Thus a steep concentration gradient for glucose is continually maintained, and glucose rapidly diffuses into the cell, only to be immediately changed to the phosphorylated form.

Researchers have studied facilitated diffusion for glucose using **liposomes,** artificial vesicles surrounded by phospholipid bilayers. The phospholipid membrane of a liposome does not allow the passage of glucose unless glucose permease is present in the liposome membrane. Glucose permease and similar carrier proteins temporarily bind to the molecules they transport. This mechanism appears to be similar to the way an enzyme binds with its substrate, the molecule on which it acts (see Chapter 6). In addition, as in enzyme action, binding apparently changes the shape of the carrier protein. This change allows the glucose molecule to be released on the inside of the cell. According to this model, when the glucose is released into the cytoplasm, the carrier protein reverts to its original shape and is available to bind another glucose molecule on the outside of the cell.

ACTIVE FIGURE **5-14** **Facilitated diffusion.**

A carrier protein in the membrane binds a solute particle. The protein shape changes, opening a channel through the membrane. A specific solute can be transported from the inside of the cell to the outside or from the outside to the inside, but net movement is always from a region of higher solute concentration to a region of lower concentration. Facilitated diffusion requires the potential energy of a concentration gradient.

BiologyNow™ Learn more about **membrane transport** by clicking on this figure on your BiologyNow CD-ROM.

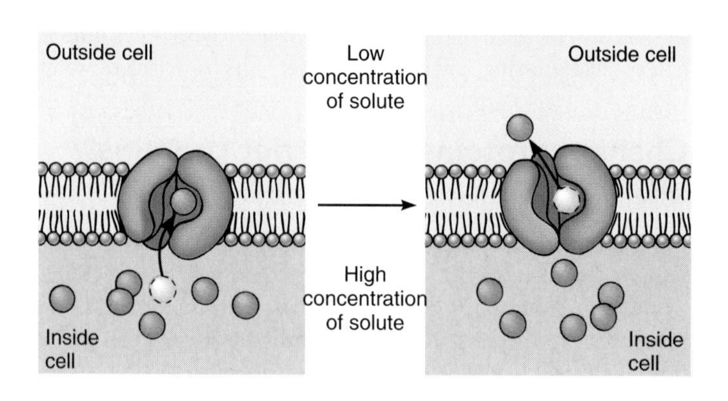

Another similarity to enzyme action is that carrier proteins become saturated when the transported molecule is at high concentration. This may be because a finite number of carrier proteins are available and they operate at a defined maximum rate. When the concentration of solute molecules to be transported reaches a certain level, all the carrier proteins are working at their maximum rate.

Some carrier-mediated active transport systems "pump" substances against their concentration gradients

Although adequate amounts of some substances move across cell membranes by diffusion, a cell often needs to transport solutes against a concentration gradient. The cell requires many substances in concentrations higher than those outside the cell. **Carrier-mediated active transport** mechanisms move these molecules across cell membranes. Because this active transport requires that particles be "pumped" from a region of low concentration to a region of high concentration—*against a concentration gradient*—transport must be coupled to an energy source such as ATP.

One of the most striking examples of an active transport mechanism is the **sodium-potassium pump** found in virtually all animal cells (Fig. 5-15). The pump is a group of specific carrier proteins in the plasma membrane that uses energy in the form of ATP to exchange sodium ions on the inside of the cell for potassium ions on the outside of the cell. The exchange is unequal, so that usually only two potassium ions are imported for every three sodium ions exported. Because these particular concentration gradients involve ions, an electrical potential (separation of electrical charges) is generated across the membrane, and we say that the membrane is *polarized.*

Both sodium and potassium ions are positively charged, but because there are fewer potassium ions inside relative to the sodium ions outside, the inside of the cell is negatively charged relative to the outside. The unequal distribution of ions establishes an **electrical gradient** that drives ions across the plasma membrane. Sodium-potassium pumps help maintain a charge separation across the plasma membrane. The separation of charges across a plasma membrane is called a **membrane potential.** Because there is both an electrical charge difference and a concentration difference on the two sides of the membrane, the gradient is called an **electrochemical gradient.** Such gradients store energy (like water stored behind a dam) that is used to drive other transport systems. So important is the electrochemical gradient produced by these pumps that some cells (such as nerve cells) expend 70% of their total energy just to power this one transport system.

Sodium-potassium pumps (as well as all other ATP-driven pumps) are transmembrane proteins that extend entirely through the membrane. By undergoing a series of conformational changes (changes in shape), the pumps exchange sodium for potassium across the plasma membrane. Unlike facilitated diffusion, at least one of the conformational changes in the pump cycle requires energy, which is provided by ATP. The shape of the pump protein changes as a phosphate group from ATP first binds to it and is subsequently removed later in the pump cycle.

The use of electrochemical potentials for energy storage is not confined to the plasma membranes of animal cells. Plant and fungal cells use ATP-driven plasma membrane pumps to transfer protons from the cytoplasm of their cells to the outside. Removal of positively charged protons from the cytoplasm of these cells results in a large difference in the concentration of protons, such that the outside of the cells is relatively positively charged and the inside of the plasma membrane is relatively negatively charged. The energy stored in these electrochemical gradients can be used to do many kinds of cell work.

Other proton pumps are used in "reverse" to synthesize ATP. Bacteria, mitochondria, and chloroplasts use energy from food or sunlight to establish proton concentration gradients (see Chapters 7 and 8). When the protons diffuse through the proton carriers from a region of high proton concentration to one of low concentration, ATP is synthesized. These electrochemical gradients form the basis for the major energy conversion systems in virtually all cells.

Ion pumps have other important roles. For example, they are instrumental in the ability of an animal cell to equalize the osmotic pressures of its cytoplasm and its external environment. If an animal cell does not control its internal osmotic pressure, its contents become hypertonic relative to the exterior. Water will enter the cell by osmosis, causing it to swell and possibly burst (see Fig. 5-12c). By controlling the ion distribution across the membrane, the cell indirectly controls the movement of water, because when ions are pumped out of the cell, water leaves by osmosis.

Linked cotransport systems indirectly provide energy for active transport

The electrochemical concentration gradients generated by the sodium-potassium pump (and other pumps) provide sufficient energy to power the active transport of other essential substances. In these systems, a transport protein **cotransports** the required molecules *against* their concentration gradient, while sodium, potassium, or hydrogen ions move *down* their gradient. Energy from ATP may be used indirectly in this process. ATP produces the ion gradient; the energy of this gradient then drives the active transport of a required substance against its gradient.

In some cells, more than one system may work to transport a given substance. For example, the transport of glucose from the intestine to the blood occurs through a thin sheet of epithelial cells that line the intestine. The surface that is exposed to the intestine has many **microvilli** (sing., *microvillus*), finger-like extensions that effectively increase the surface area of the membrane available for absorption (see Fig. 45-10c). The glucose transport protein on that region of the cell surface is part of an active transport system for glucose that is "driven" by the cotransport of sodium. The sodium concentration inside the cell is kept low by an ATP-requiring sodium-potassium pump that transports sodium out of the cell and into the blood. Because of its high concentration inside the cell (relative to the blood), glucose is transported to the blood by facilitated diffusion.

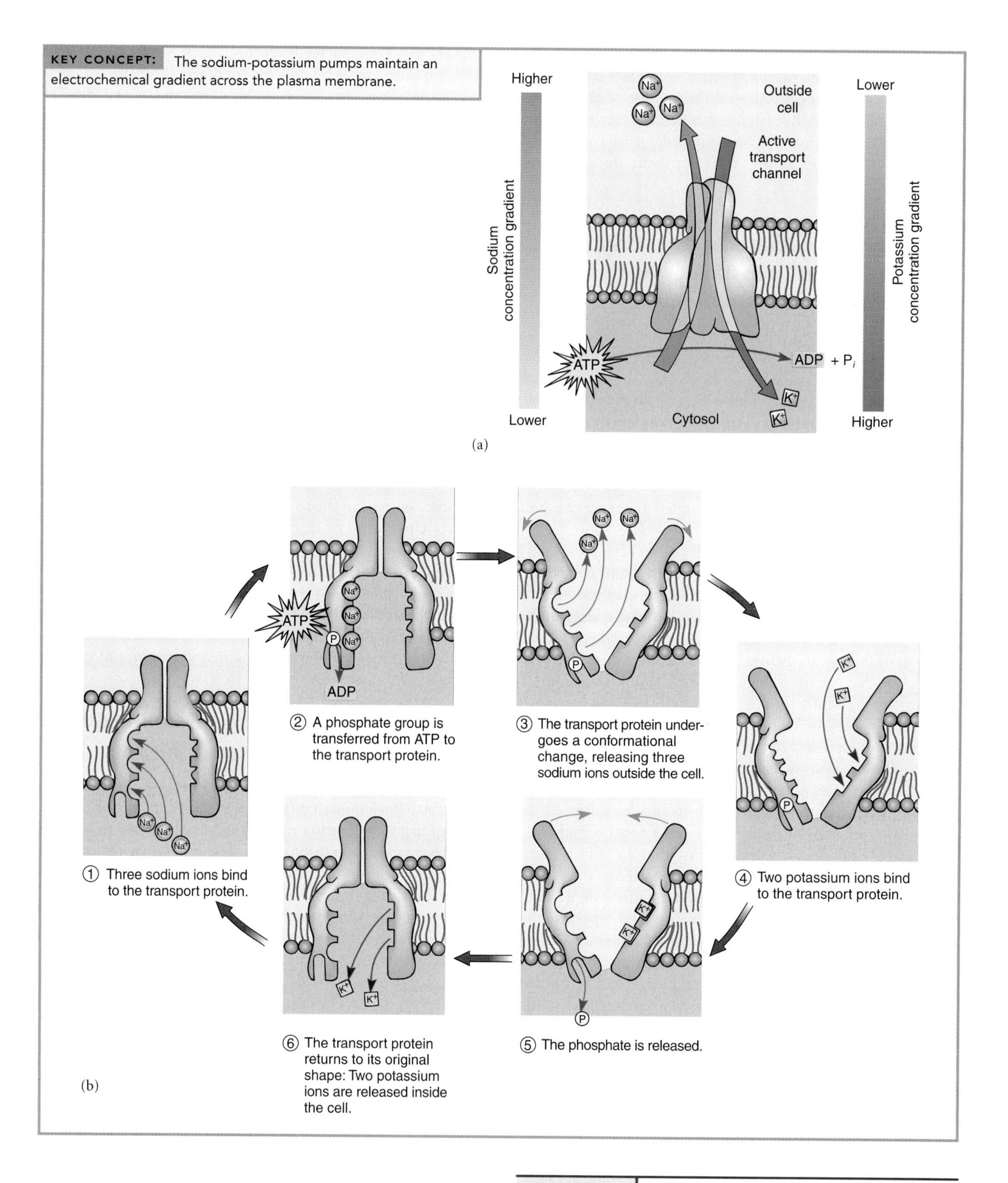
KEY CONCEPT: The sodium-potassium pumps maintain an electrochemical gradient across the plasma membrane.
Higher
Sodium concentration gradient
Lower
Outside cell
Active transport channel
ATP
ADP + P_i
Cytosol
Lower
Potassium concentration gradient
Higher
(a)
ADP
① Three sodium ions bind to the transport protein.
② A phosphate group is transferred from ATP to the transport protein.
③ The transport protein undergoes a conformational change, releasing three sodium ions outside the cell.
④ Two potassium ions bind to the transport protein.
⑤ The phosphate is released.
⑥ The transport protein returns to its original shape: Two potassium ions are released inside the cell.
(b)

What are the signals that target each transport protein to its appropriate region in the plasma membrane? Some cell biologists are focusing their research on understanding mechanisms such as those that enable the cell to place different transport proteins in separate regions of the same plasma membrane.

FIGURE **5-15** **The sodium-potassium pump.**

(a) The sodium-potassium pump is an active transport system that requires energy from ATP. Each complete pumping cycle uses one molecule of ATP; three sodium ions are exported, and two potassium ions are imported. **(b)** How the sodium-potassium pump works.

Facilitated diffusion is powered by a concentration gradient; active transport requires another energy source

It is a common misconception that diffusion, whether simple or facilitated, is somehow "free of cost" and that only active transport mechanisms require energy. Because diffusion always involves the net movement of a substance down its concentration gradient, we say that the concentration gradient "powers" the process. However, energy is required to do the work of establishing and maintaining the gradient. Think back to the example of facilitated diffusion of glucose. The cell maintains a steep concentration gradient (high outside, low inside) by phosphorylating the glucose molecules once they enter the cell. One ATP molecule is spent for every glucose molecule phosphorylated, not to mention such additional costs as the energy required to make the enzymes that carry out the reaction.

An active transport system works against a concentration gradient, pumping materials from a region of low concentration to a region of high concentration. The energy stored in the concentration gradient is not only unavailable to the system but actually works against it. For this reason, the cell needs some other source of energy. As we have seen, in many cases cells use ATP energy directly. In a cotransport system, a concentration gradient provides energy for some other substance (such as an ion), but the cell may indirectly require ATP to power the pump that produces the ion gradient.

To summarize, both diffusion and active transport require energy. The energy for diffusion is provided by a concentration gradient for the substance being transported. Active transport requires some other, usually more direct, expenditure of metabolic energy.

The patch clamp technique has revolutionized the study of ion channels

PROCESS OF SCIENCE

Because ions cannot cross a lipid bilayer by simple diffusion, every membrane of every cell contains numerous *ion channels.* The movement of ions across a membrane can result in a charge difference, or electrical gradient. If the cell is large enough, this charge difference (usually expressed in millivolts, mV) can be measured by using two microelectrodes connected to an extremely sensitive oscilloscope or voltmeter (Fig. 5-16a). One of the microelectrodes is inserted into the cell, and the other is placed just outside the plasma membrane. Although valuable, these techniques have serious limitations, because they cannot be used on smaller cells and do not provide information on the function of individual ion channels.

In the mid-1970s, Erwin Neher and Bert Sakmann, both of the Max Planck Institute in Germany, developed a method, known as the **patch clamp technique,** that enables researchers to study single ion channels of very small cells. In this technique, the tip of a micropipette is tightly sealed to a patch of membrane so small that it generally contains only a single ion channel (Fig. 5-16b). The flow of ions through the channel is measured using an extremely sensitive recording device.

Using this patch clamp technique, cell biologists study the action of a single ion channel over time. They have found that the current flow is intermittent and corresponds to the opening and closing of the ion channel. The permeability of the channel affects the magnitude of the current. The patch clamp technique has been modified in many ways and has been

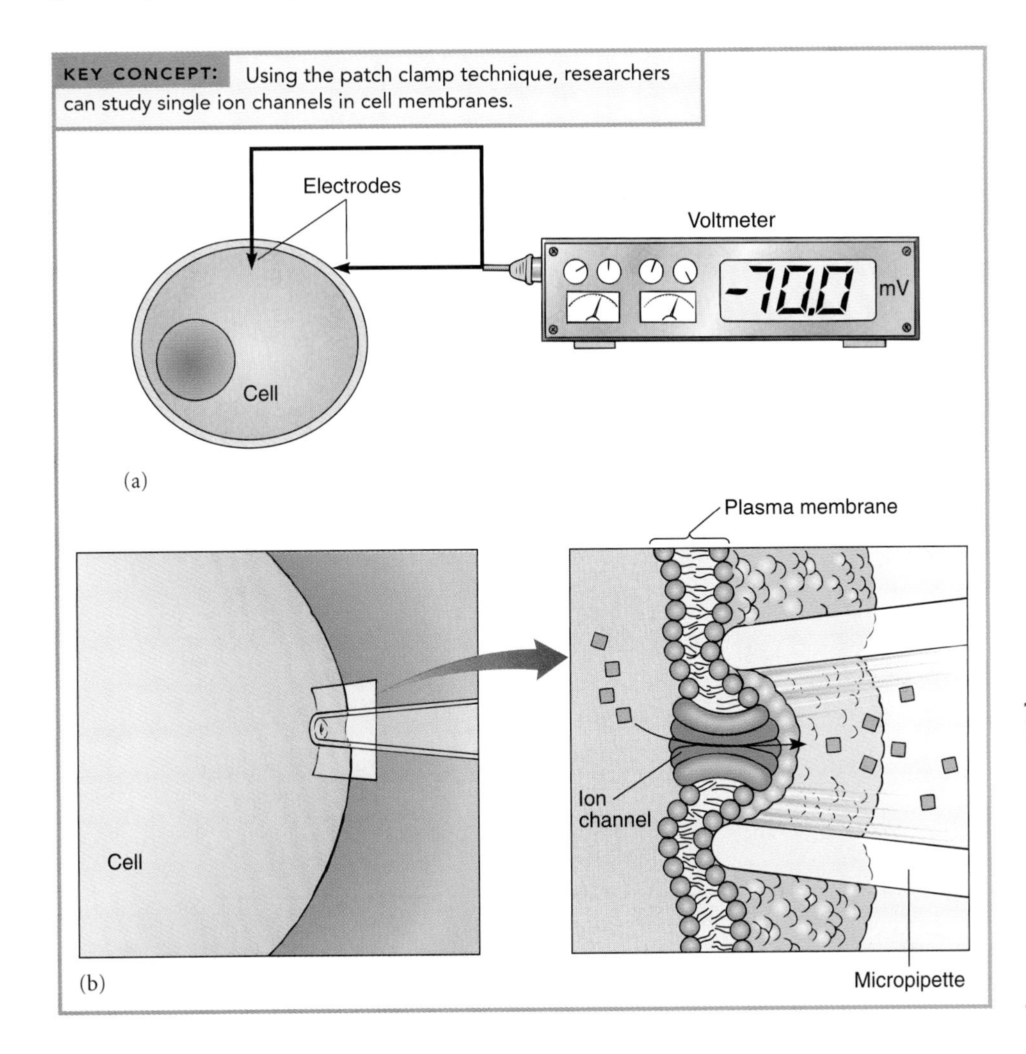

FIGURE **5-16** **The patch clamp technique.**

(a) Microelectrodes and a voltmeter measure the difference in electrical charge across a membrane. **(b)** A micropipette forms a tight seal with a patch of plasma membrane. The membrane is pulled away from the rest of the cell, enabling researchers to study the flow of ions through a single ion channel.

applied to studies of the roles of ion channels in a wide range of cell processes in both plants and animals. For example, studies of single ion channels enabled researchers to demonstrate that the genetic disease cystic fibrosis (see Chapter 15) is caused by a defect in a specific type of chloride ion channel. Because of the far-reaching implications of their work, Neher and Sakmann were awarded the 1991 Nobel Prize in Physiology or medicine.

In exocytosis and endocytosis, vesicles or vacuoles transport large particles

In both simple and facilitated diffusion, and in carrier-mediated active transport, individual molecules and ions pass through the plasma membrane. Some larger materials, such as large molecules, particles of food, and even small cells, are also moved into or out of cells. Such work requires cells to expend energy directly, making it a form of active transport.

In **exocytosis,** a cell ejects waste products, or specific products of secretion such as hormones, by the fusion of a vesicle with the plasma membrane (Fig. 5-17). Exocytosis results in the incorporation of the membrane of the secretory vesicle into the plasma membrane, as the contents of the vesicle are released from the cell. This is also the primary mechanism by which plasma membranes grow larger.

In **endocytosis,** materials are taken into the cell. Several types of endocytotic mechanisms operate in biological systems, including phagocytosis, pinocytosis, and receptor-mediated endocytosis. In **phagocytosis** (literally, "cell eating"), the cell ingests large solid particles such as bacteria and food (Fig. 5-18). Phagocytosis is used by certain protists and by several types of vertebrate white blood cells to ingest particles, some of which are as large as an entire bacterium. During ingestion, folds of the plasma membrane enclose the particle, which has bound to the surface of the cell, forming a large membranous sac, or vacuole. When the membrane has encircled the particle, it fuses at the point of contact. The vacuole then fuses with lysosomes, and the ingested material is degraded.

In the form of endocytosis known as **pinocytosis** ("cell drinking"), the cell takes in dissolved materials (Fig. 5-19). Tiny droplets of fluid are trapped by folds in the plasma membrane, which pinch off into the cytosol as tiny vesicles. The liquid contents of these vesicles are then slowly transferred into the cytosol; the vesicles become progressively smaller.

In a third type of endocytosis, **receptor-mediated endocytosis,** specific molecules combine with receptor proteins embedded in the plasma membrane. Cells take up cholesterol from the blood by this process. Cholesterol is transported in the blood as part of particles called *low-density lipoproteins (LDLs).* Cells use cholesterol as a component of cell membranes and as a precursor of steroid hormones. Much of the receptor-mediated endocytosis pathway was detailed through studies by Michael Brown and Joseph Goldstein on the LDL receptor. In 1985, these researchers, both of the University of Texas Health Sci-

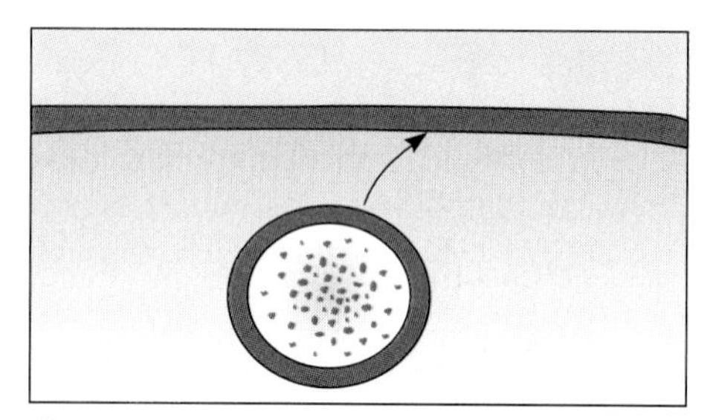

① A vesicle approaches the plasma membrane,

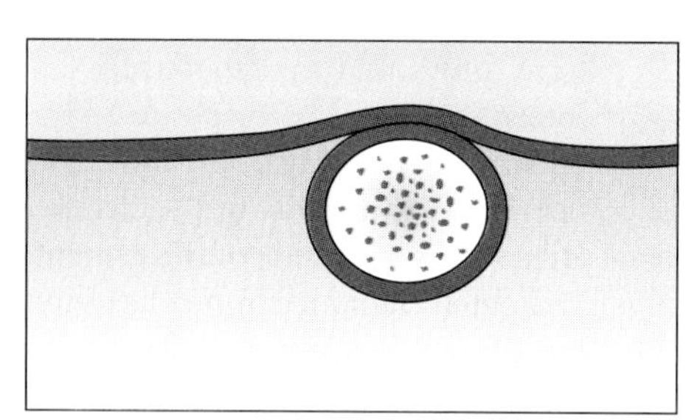

② fuses with it, and

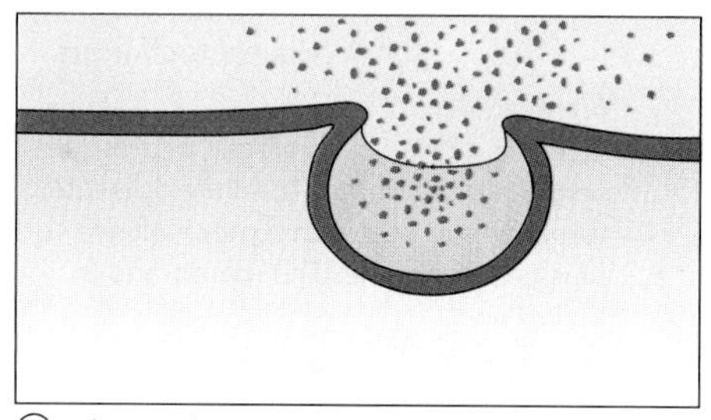

③ releases its contents outside the cell.

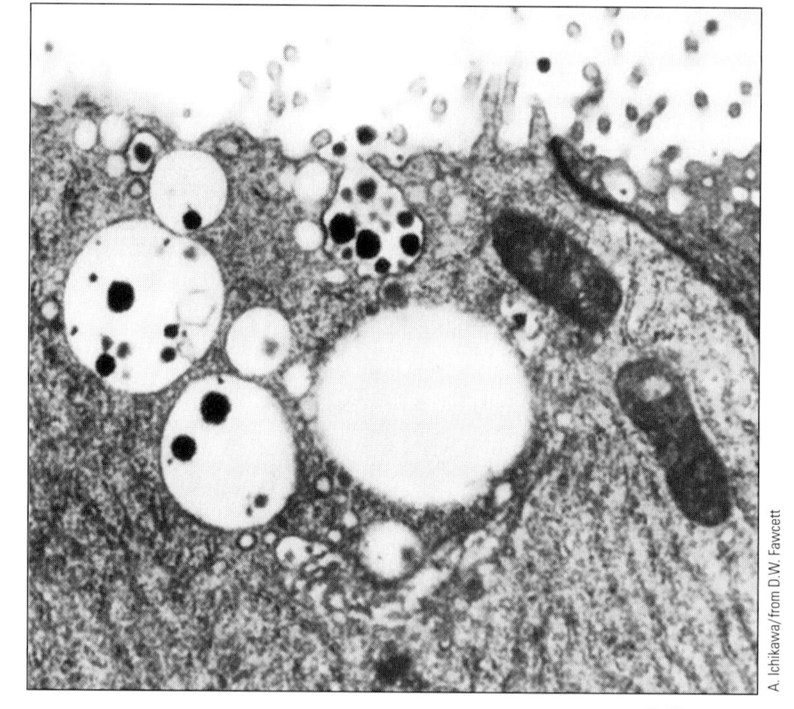

FIGURE **5-17** **Exocytosis.**

The TEM shows exocytosis of the protein components of milk by a mammary gland cell.

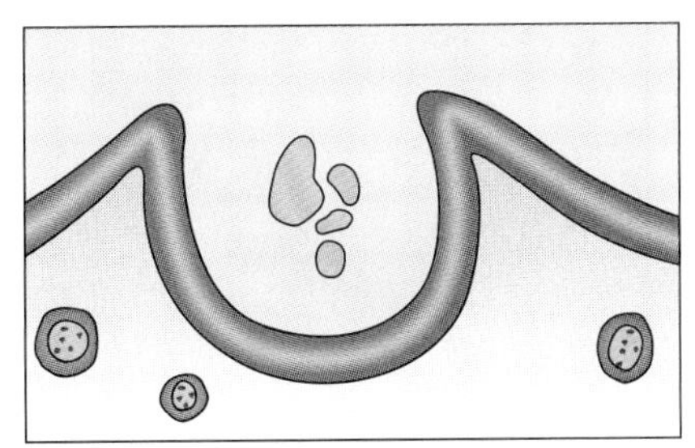
1 Folds of the plasma membrane surround the particle to be ingested, forming a small vacuole around it.

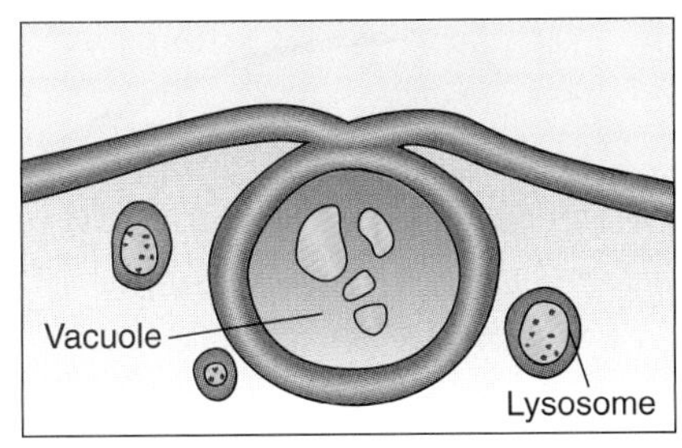

2 The vacuole then pinches off inside the cell.

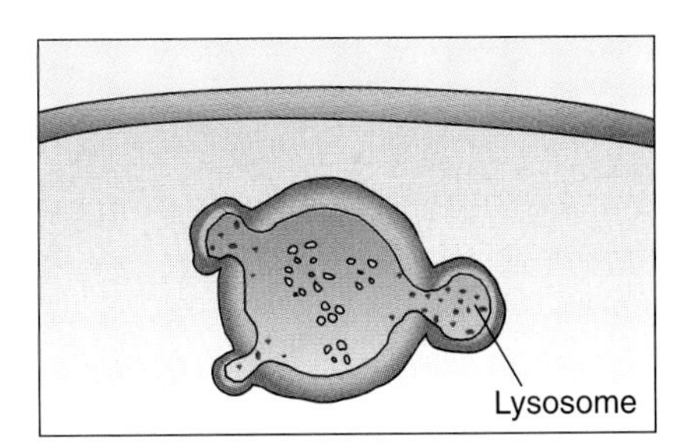

3 Lysosomes may fuse with the vacuole and pour their potent enzymes onto the ingested material.

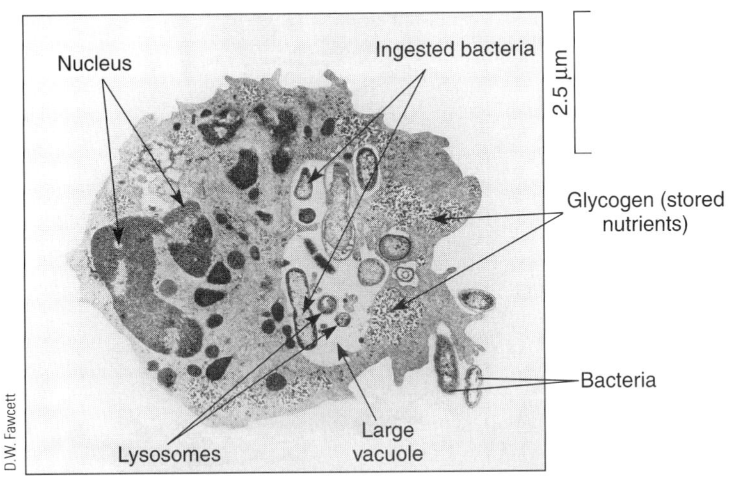

FIGURE **5-18** **Phagocytosis.**

In this type of endocytosis, a cell ingests relatively large solid particles. The white blood cell (a neutrophil) shown in the TEM is phagocytizing bacteria. The vacuoles contain bacteria that have already been ingested, whereas other bacteria are still outside the cell. Lysosomes in the cytosol contain digestive enzymes.

ence Center, were awarded the Nobel Prize in Physiology or medicine for their pioneering work. Their findings have important medical implications, because cholesterol that remains in the blood instead of entering the cells can become deposited in the artery walls, increasing the risk of cardiovascular disease.

When a cell needs cholesterol, it makes LDL receptors. The receptors are concentrated in *coated pits,* depressed regions on the cytoplasmic surface of the plasma membrane. Each pit is coated by a layer of a protein, called *clathrin,* found just below the plasma membrane. A molecule that binds specifically to a receptor is called a **ligand.** In this case, LDL is the ligand. After the LDL binds with a receptor, the coated pit forms a *coated vesicle* by endocytosis.

Figure 5-20 shows the uptake of an LDL particle. Seconds after the vesicle moves into the cytoplasm, the coating dissociates from it, leaving an uncoated vesicle. The vesicles deliver their contents to compartments called *endosomes.* LDL separates from its receptor and is transferred to a lysosome. There, the LDL is broken down and cholesterol is released into the cytosol for use by the cell. The LDL receptors are transported to the plasma membrane, where they are recycled. A simplified summary of receptor-mediated endocytosis follows:

Ligand binds to receptors in coated pits of plasma membrane ⟶ coated vesicle forms by endocytosis ⟶ coating detaches from vesicle ⟶ contents transferred to endosome ⟶ ligand separates from its receptor:

- ⟶ receptors are transported to plasma membrane and recycled
- ⟶ endosome fuses with lysosome ⟶ contents are digested and released into the cytosol

FIGURE **5-19** Pinocytosis or "cell-drinking."

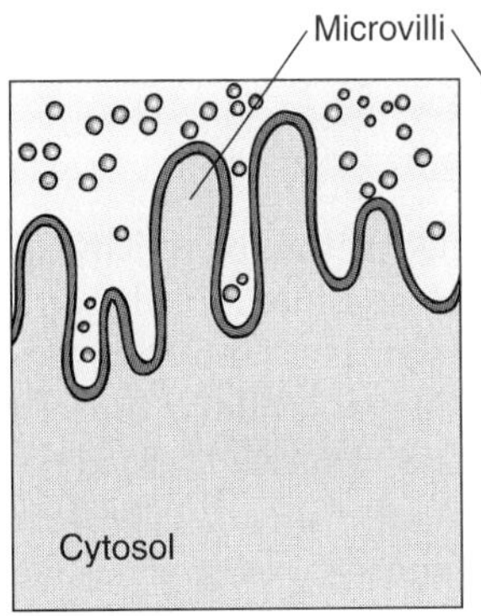

1 Tiny droplets of fluid are trapped by folds of the plasma membrane.

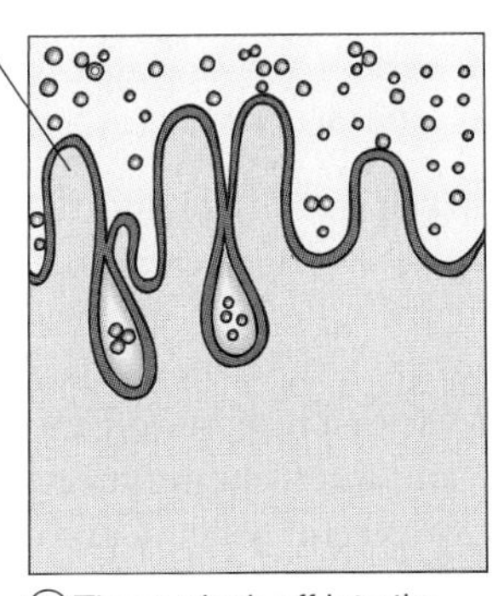
2 These pinch off into the cytosol as small fluid-filled vesicles.

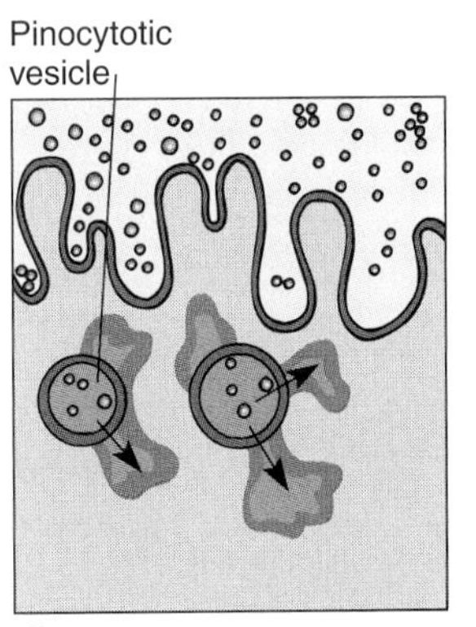

3 The contents of these vesicles are then slowly transferred to the cytosol.

The recycling of LDL receptors to the plasma membrane through vesicles illustrates a problem common to all cells that use endocytotic and exocytotic mechanisms. A type of phagocytic cell known as a *macrophage,* for example, ingests the equivalent of its entire plasma membrane in about 30 minutes, requiring an equivalent amount of recycling or new membrane synthesis for the cell to maintain its surface area. In cells that are constantly involved in secretion, an equivalent amount of membrane must be returned to the interior of the cell for each vesicle that fuses with the plasma membrane; if it is not,

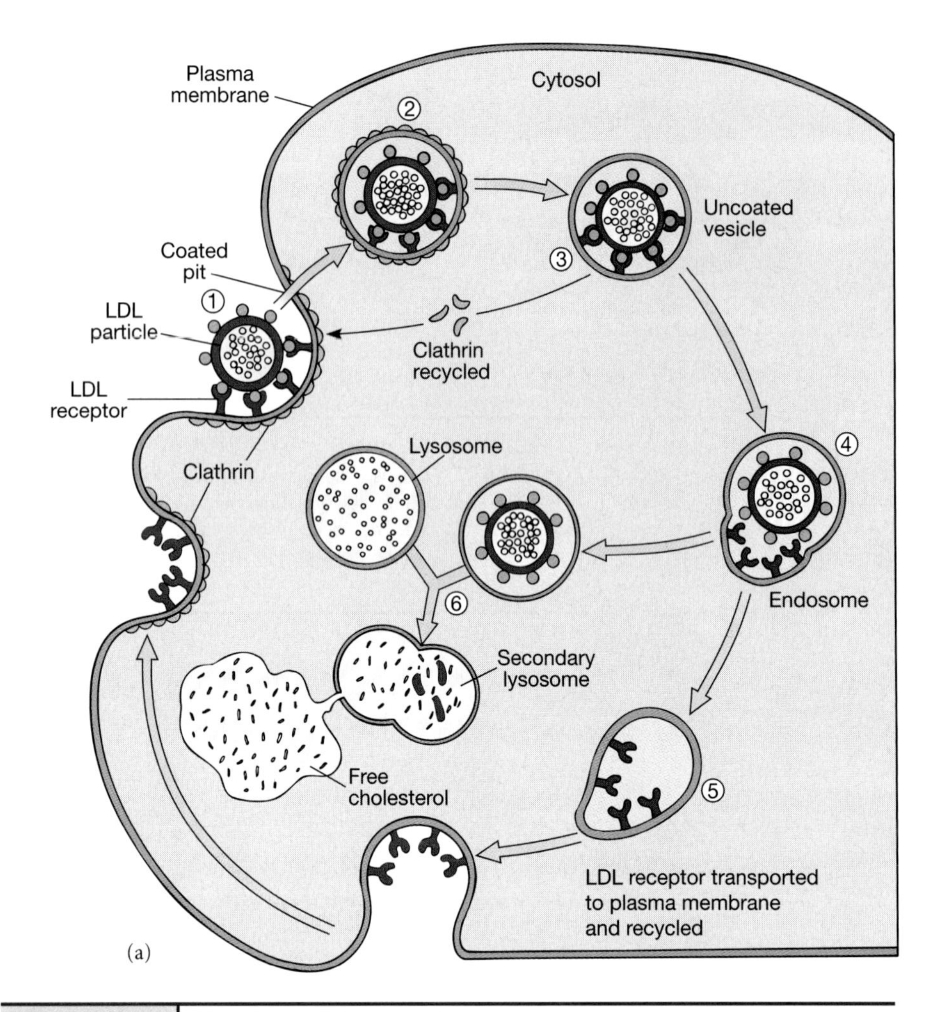

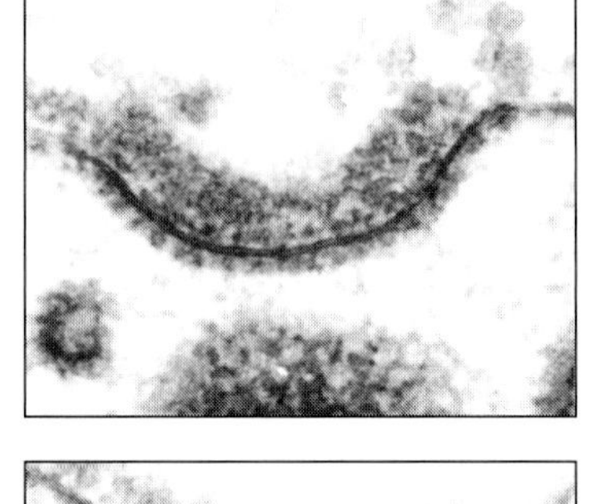

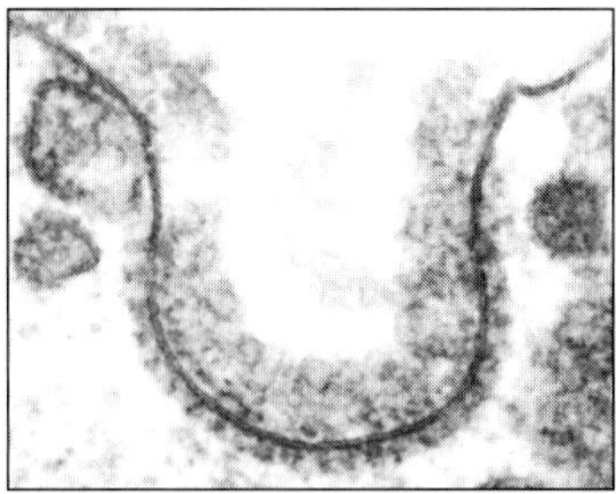

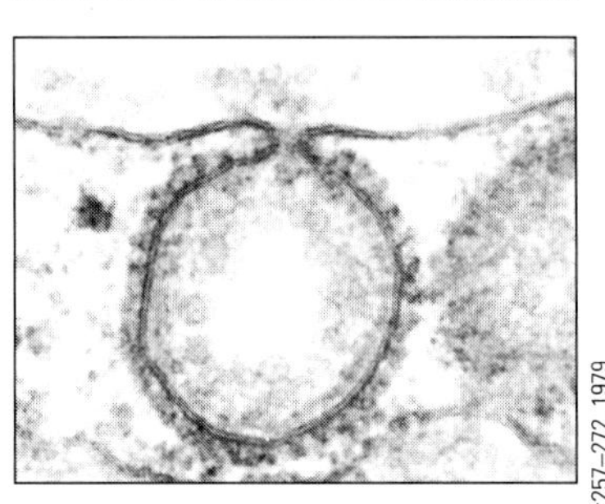

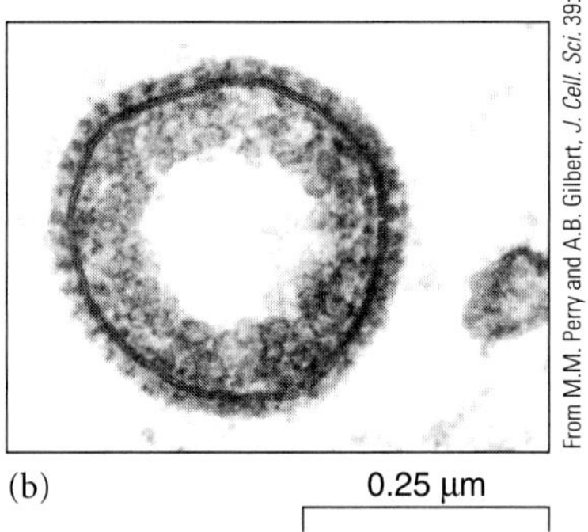

From M.M. Perry and A.B. Gilbert, *J. Cell. Sci.* 39:257–272, 1979

FIGURE 5-20 Receptor-mediated endocytosis.

(a) Uptake of low-density lipoprotein (LDL) particles,which transport cholesterol in the blood: ① LDL attaches to specific receptors in coated pits on the plasma membrane. ② Endocytosis results in the formation of a coated vesicle in the cytosol. ③ Seconds later the coat is removed. ④ The vesicle transfers its contents to an endosome. ⑤ The receptors are returned to the plasma membrane and recycled. ⑥ The vesicle containing LDL particles fuses with a lysosome forming a secondary lysosome. Hydrolytic enzymes then digest the cholesterol from the LDL particles for use by the cell. **(b)** This series of TEMs shows the formation of a coated vesicle from a coated pit.

the cell surface will keep expanding even though the growth of the cell itself may be arrested.

Review

- In what direction do particles move along their concentration gradient? Would your answers be different for facilitated diffusion compared with simple diffusion?
- What is the immediate source of energy for simple diffusion? For facilitated diffusion? For active transport?
- What would happen if a plant cell were placed in a relatively (a) isotonic, (b) hypertonic, or (c) hypotonic environment? How would you modify your predictions for an animal cell?
- How are exocytosis and endocytosis similar?
- How are the processes of phagocytosis and pinocytosis different?

BiologyNow™ Assess your understanding of the **passage of materials through cell membranes** by taking the pretest on your BiologyNow CD-ROM.

CELL SIGNALING

Learning Objective

9 Describe the generalized process of cell signaling.
10 Explain how an extra cellular signal is converted to an intracellular signal in signal transduction.

The term **cell signaling** refers to the mechanisms by which cells communicate with one another. Most commonly, cells communicate with chemical signals. Cells signal one another by secreting certain molecules, or a signaling molecule on one cell combines with a receptor on another cell. Unicellular bacteria, protists, and fungi communicate with other members of their species by secreting chemical compounds. For example, when food is scarce, the amoeba-like cellular slime mold *Dictyostelium* secretes **cyclic adenosine monophosphate (cAMP)** (see Fig. 3-25). This chemical compound diffuses through the cell's environment and induces nearby slime molds to come together and form a multicellular slug-shaped colony (see Fig. 24-24). Yeast cells identify cells of compatible mating types by chemical communication. In 2003, Marc Spehr of Ruhr University Bochum in Germany and his colleagues reported that human sperm have receptors that respond to chemical signals that guide the sperm to the egg.

About a billion years ago, when cells began to associate to form multicellular organisms, elaborate systems of cell signaling evolved. The development and functioning of complex organisms require precise internal communication, as well as effective responses to the outside environment. In plants and animals, hormones serve as important chemical signals between various cells and organs. Animals have evolved nervous systems in which neurons transmit information electrically and chemically.

The process of cell signaling includes

- Synthesis and release of the signaling molecules
- Transport to target cells
- Reception of the information by target cells
- Signal transduction
- Response by the cell
- Termination of signaling

Signaling molecules may be neurotransmitters (produced by nerve cells), hormones, or other regulatory molecules. They may be synthesized by neighboring cells or by specialized tissues some distance away from the target cells. These molecules reach target cells by diffusion or via the circulatory system. In some cases, neurons transport signaling molecules from one location to another. Reception typically depends on receptor proteins in the plasma membrane of target cells. The signaling molecules are ligands that bind with the receptors.

Many regulatory molecules transmit information to the cell interior without physically crossing the plasma membrane. These signal molecules rely on systems of interacting integral membrane proteins to transmit the information. **Signal transduction** is the process in which cells convert and amplify an extracellular signal into an intracellular signal. Each component of a signal transduction system acts as a relay "switch," which can be in an activated ("on") state or an inactive ("off") state.

The first component in a signal transduction system is typically the receptor, which may be a transmembrane protein with a domain (a structural and functional component of a protein) exposed on the extracellular surface. A receptor generally has at least three domains. The external domain is a docking site for a signaling molecule. A second domain extends through the plasma membrane, and a third domain is a "tail" that extends into the cytoplasm.

In a typical signaling pathway, when the ligand binds with the receptor, it activates it by changing the shape of the receptor tail that extends into the cytoplasm. The activated receptor changes the conformation of a second protein, which then becomes activated. The signal may be relayed through a sequence of proteins (Fig. 5-21). Ultimately these interactions result in the activation of a specific enzyme bound to the membrane. That enzyme may itself catalyze the production of large numbers of intracellular signaling molecules, or it may activate intracellular enzymes. In this way the original signal received by the receptor protein is amplified many times, and the metabolism of the cell may be profoundly altered.

The ligand that acts as a signaling molecule is sometimes referred to as the **first messenger.** Some ligand-receptor complexes bind to and activate specific integral membrane proteins, referred to as **G proteins.** In 1994, Alfred G. Gilman, of the University of Texas, and Martin Rodbell, of the National Institute of Environmental Health Sciences, were awarded the Nobel Prize for Physiology or medicine for their research on G proteins. These proteins are so named because the active form is bound to **guanosine triphosphate (GTP),** a molecule similar to ATP but containing the base guanine instead of adenine. G proteins catalyze the hydrolysis of GTP to guanosine diphosphate (GDP), a process that releases energy.

In a complex sequence of events, a G protein relays the message from the receptor to an enzyme that catalyzes the production of a **second messenger,** which is an intracellular signal. Often the second messenger is cyclic AMP. The enzyme adenylyl cyclase, which is bound to the plasma membrane, catalyzes the formation of cyclic AMP from ATP. Typically, the second messenger activates **protein kinases,** enzymes that activate specific proteins by transferring phosphate groups to them from ATP. This sequence of reactions, beginning with the binding of the signaling molecule to the receptor, leads to a change in some cell function.

G proteins are involved in a number of important signal transductions, including the action of many hormones (see Chapter 47). Some G proteins regulate channels that allow ions to cross the plasma membrane, and others play important roles in the senses of sight, smell, and taste (see Chapter 41).

Ras proteins, a group of GTP-binding proteins that function somewhat like G proteins, are thought to be important in signal transduction necessary for many cell activities. Fibroblasts (a type of connective tissue cell) require the presence of two *growth factors,* epidermal growth factor and platelet-derived growth factor, for DNA synthesis. Investigators conducted an experiment in which they injected fibroblasts with antibodies that bind to Ras proteins thereby inactivating them. These fibroblasts with inactivated Ras proteins no longer synthesized DNA in response to growth factors. Data from this and similar experiments led to the conclusion that Ras proteins are important in signal transduction involving growth factors.

Cell biologists have demonstrated that when certain ligands bind to integrins (transmembrane proteins that connect the cell to the extracellular matrix) in the plasma membrane, specific signal transduction pathways are activated. Growth factors also turn on signaling pathways. Interestingly, growth factors and certain molecules of the extracellular matrix may modulate each other's messages. Integrins also respond to information received from inside the cell. This inside-out signaling affects how selective integrins are with respect to the molecules to which they bind and how strongly they bind to them.

Cell biologists are only beginning to identify the many ways that proteins interact in cell signaling pathways. The relay sequences we have described here are oversimplified. Some proteins in signaling pathways probably come together briefly, producing large molecular complexes that may be shared among

FIGURE 5-21 Signal transduction.

① A signal molecule binds with a receptor in the plasma membrane. ② The signal molecule-receptor complex activates a G protein. ③ The G protein activates an enzyme that catalyzes the production of a second messenger such as cyclic AMP (cAMP). ④ cAMP then activates one or more enzymes such as protein kinases. The enzymes may phosphorylate proteins, which then alter the activity of the cell in some way.

KEY CONCEPT: In signal transduction, cells convert an extracellular signal into an intracellular signal.

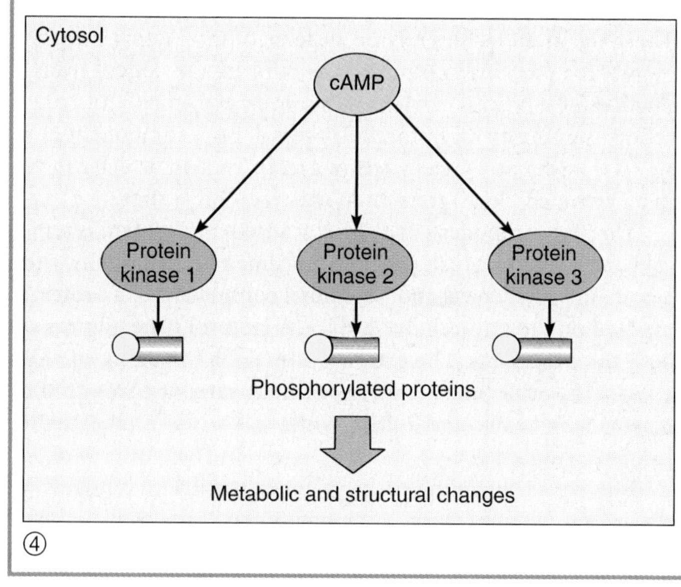

several pathways. We have much to learn about how cells "talk" to one another.

Review

- What generalized sequence of events takes place in cell signaling?
- What are G proteins?
- What is signal transduction? Explain the process.

BiologyNow™ Assess your understanding of **cell signaling** by taking the pretest on your BiologyNow CD-ROM.

CELL JUNCTIONS

Learning Objective

11 Compare the structures and functions of anchoring junctions, tight junctions, gap junctions, and plasmodesmata.

Cells in close contact with each other typically develop specialized intercellular junctions. These structures may allow neighboring cells to form strong connections with each other, prevent the passage of materials, or establish rapid communication between adjacent cells. Several types of junctions connect animal cells, including anchoring junctions, tight junctions, and gap junctions. Plant cells are connected by plasmodesmata.

Anchoring junctions connect cells of an epithelial sheet

Adjacent epithelial cells, such as those found in the outer layer of the skin, are so tightly bound to each other by **anchoring junctions** that strong mechanical forces are required to separate them. Cadherins, transmembrane proteins shown in the chapter opening photograph, are important components of anchoring junctions. These junctions do not affect the passage of materials between adjacent cells. Two common types of anchoring junctions are desmosomes and adhering junctions.

Desmosomes are points of attachment between cells (Fig. 5-22). They hold cells together at one point like a rivet or a spot weld. As a result, cells form strong sheets, and substances still pass freely through the spaces between the plasma membranes. Each desmosome is made up of regions of dense material associated with the cytosolic sides of the two plasma membranes, plus protein filaments that cross the narrow intercellular space between them. Desmosomes are anchored to systems of intermediate filaments inside the cells. Thus the intermediate filament networks of adjacent cells are connected so that mechanical stresses are distributed throughout the tissue.

Adhering junctions cement cells together. Cadherins form a continuous adhesion belt around each cell, binding the cell to neighboring cells. These junctions connect to microfilaments of the cytoskeleton. The cadherins of adhering junctions are a potential path for signals from the outside environment to be transmitted to the cytoplasm.

Tight junctions seal off intercellular spaces between some animal cells

Tight junctions are literally areas of tight connections between the membranes of adjacent cells. These connections are so tight that no space remains between the cells and substances cannot leak between them. TEMs of tight junctions show that in the region of the junction the plasma membranes of the two cells are in actual contact with each other, held together by proteins linking the two cells. However, as shown in Figure 5-23, tight junctions are located intermittently. The plasma membranes of the two cells are not fused over their entire surface.

Cells connected by tight junctions seal off body cavities. For example, tight junctions between cells lining the intestine prevent substances in the intestine from entering the body or the blood by passing around the cells. The sheet of cells thus acts as a selective barrier. Food substances must be transported across the plasma membranes and through the intestinal cells before they enter the blood. This arrangement helps prevent toxins and other unwanted materials from entering the blood and also prevents nutrients from leaking out of the intestine. Tight junctions are also present between the cells that line capillaries in the brain. They help form the *blood–brain barrier,* which prevents many substances in the blood from passing into the brain.

Gap junctions allow the transfer of small molecules and ions

A **gap junction** is like a desmosome in that it bridges the space between cells; however, the space it spans is somewhat narrower (Fig. 5-24). Gap junctions also differ in that they are communicating junctions. They not only connect the membranes but also contain channels connecting the cytoplasm of adjacent cells. Gap junctions are composed of *connexin,* an integral membrane protein. Groups of six connexin molecules cluster to form a cylinder that spans the plasma membrane. The connexin cylinders on adjacent cells become tightly joined. The two cylinders form a channel, about 1.5 nm in diameter. Small inorganic molecules (such as ions) and some regulatory molecules (such as cyclic AMP) pass through the channels, but larger molecules are excluded. When appropriate marker substances are injected into one of a group of cells connected by gap junctions, the marker passes rapidly into the adjacent cells but does not enter the space between the cells.

Gap junctions provide for rapid chemical and electrical communication between cells. Cells control the passage of materials through gap junctions by opening and closing the channels (Fig. 5-24d). Cells in the pancreas, for example, are linked together by gap junctions in such a way that if one of a group of cells is stimulated to secrete insulin, the signal is passed through the junctions to the other cells in the cluster, ensuring a coordinated response to the initial signal. Gap junctions allow some

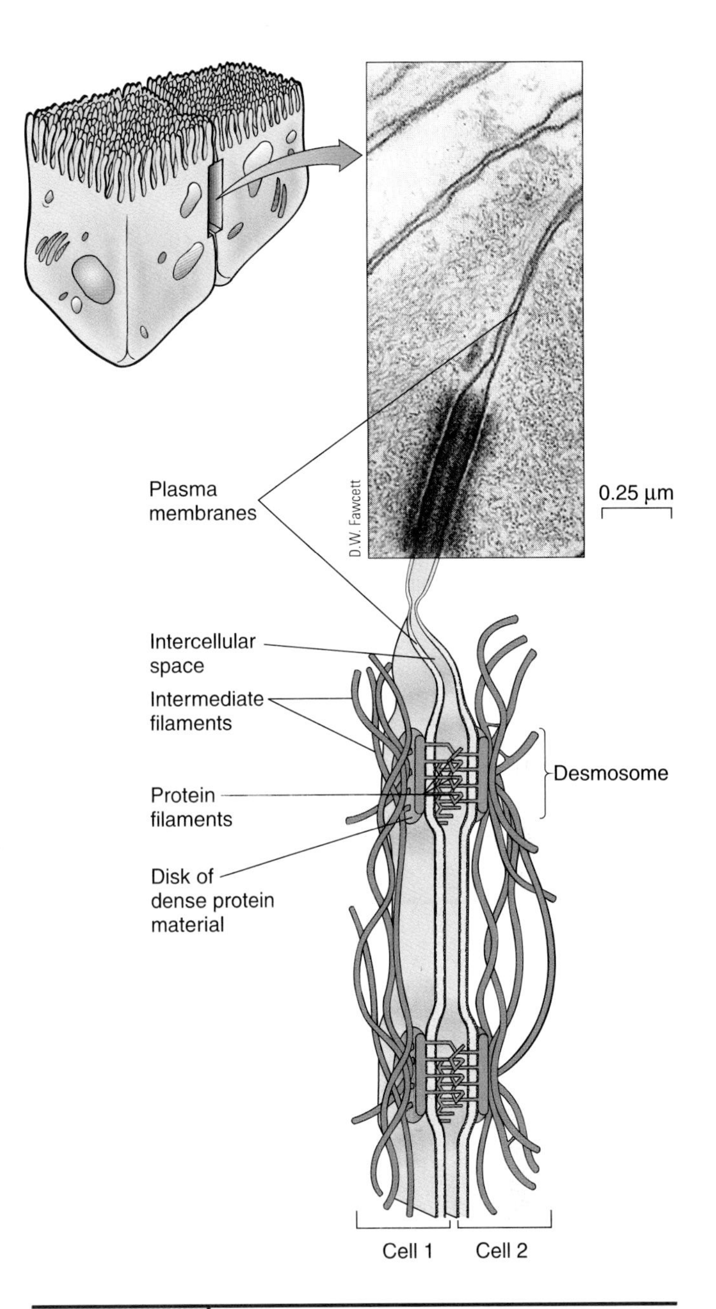

FIGURE **5-22** **Desmosomes.**

The dense structure in the TEM is a desmosome. Each desmosome consists of a pair of button-like discs associated with the plasma membranes of adjacent cells, plus the intercellular protein filaments that connect them. Intermediate filaments in the cells are attached to the discs and are connected to other desmosomes.

nerve cells to be electrically coupled. Heart muscle cells are linked by gap junctions that permit the flow of ions necessary to synchronize contractions.

Plasmodesmata allow certain molecules and ions to move between plant cells

Plant cells do not need desmosomes for strength because they have cell walls. However, these same walls would isolate the cells,

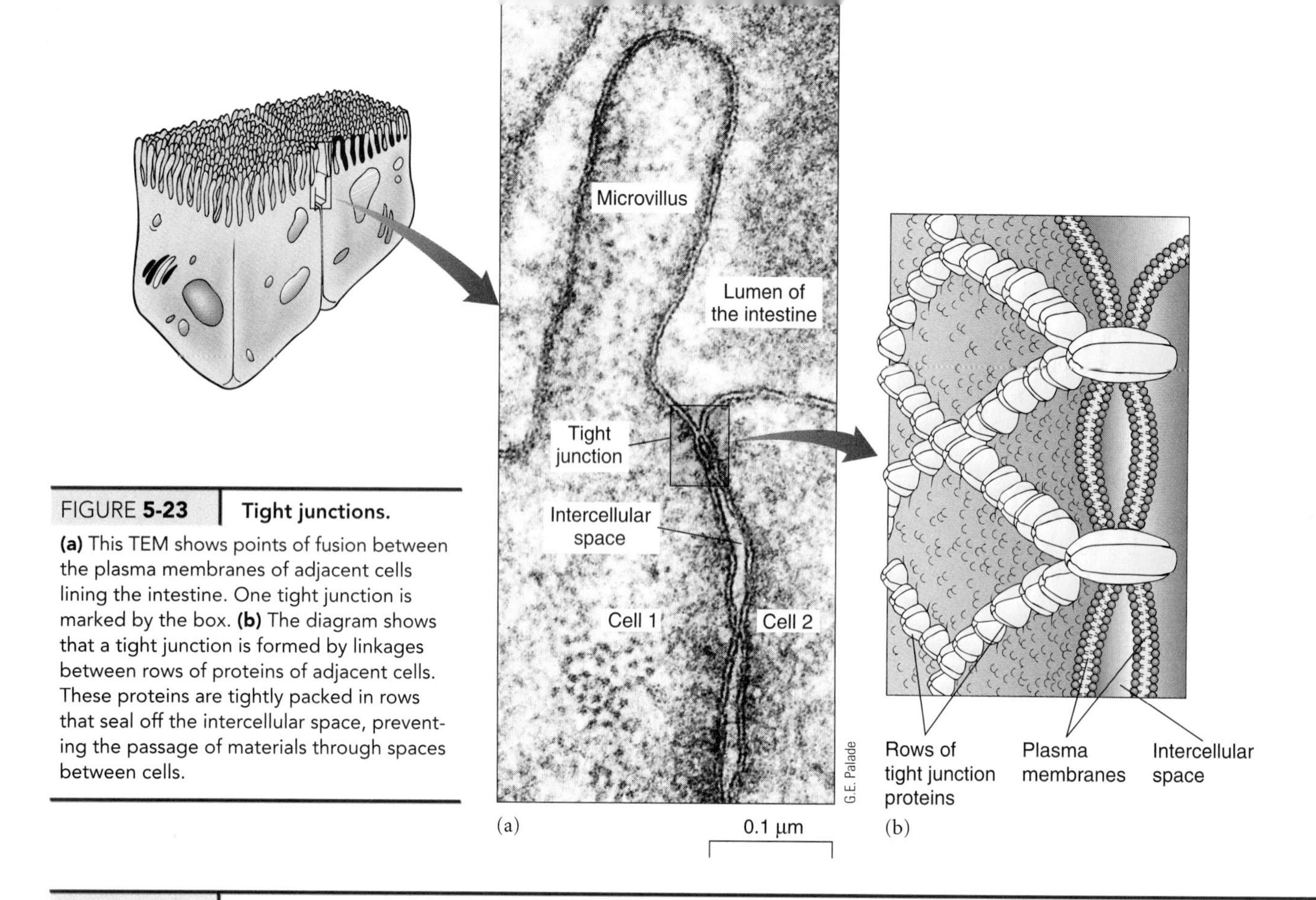

FIGURE 5-23 | **Tight junctions.**

(a) This TEM shows points of fusion between the plasma membranes of adjacent cells lining the intestine. One tight junction is marked by the box. **(b)** The diagram shows that a tight junction is formed by linkages between rows of proteins of adjacent cells. These proteins are tightly packed in rows that seal off the intercellular space, preventing the passage of materials through spaces between cells.

FIGURE 5-24 | **Gap junctions.**

These connections allow the transfer of small molecules and ions between adjacent cells. **(a)** A TEM of a gap junction *(between the arrows)*. **(b)** This model of a gap junction is based on electron microscopic and x-ray diffraction data. The two membranes contain cylinders composed of six connexin molecules. Two cylinders from opposite membranes join to form a channel connecting the cytoplasmic compartments of the two cells. **(c)** A freeze-fracture replica of the P-face of a gap junction between two ovarian cells of a mouse. Each particle corresponds to a connexin cylinder. **(d)** This model shows how a gap junction pore might open and close.

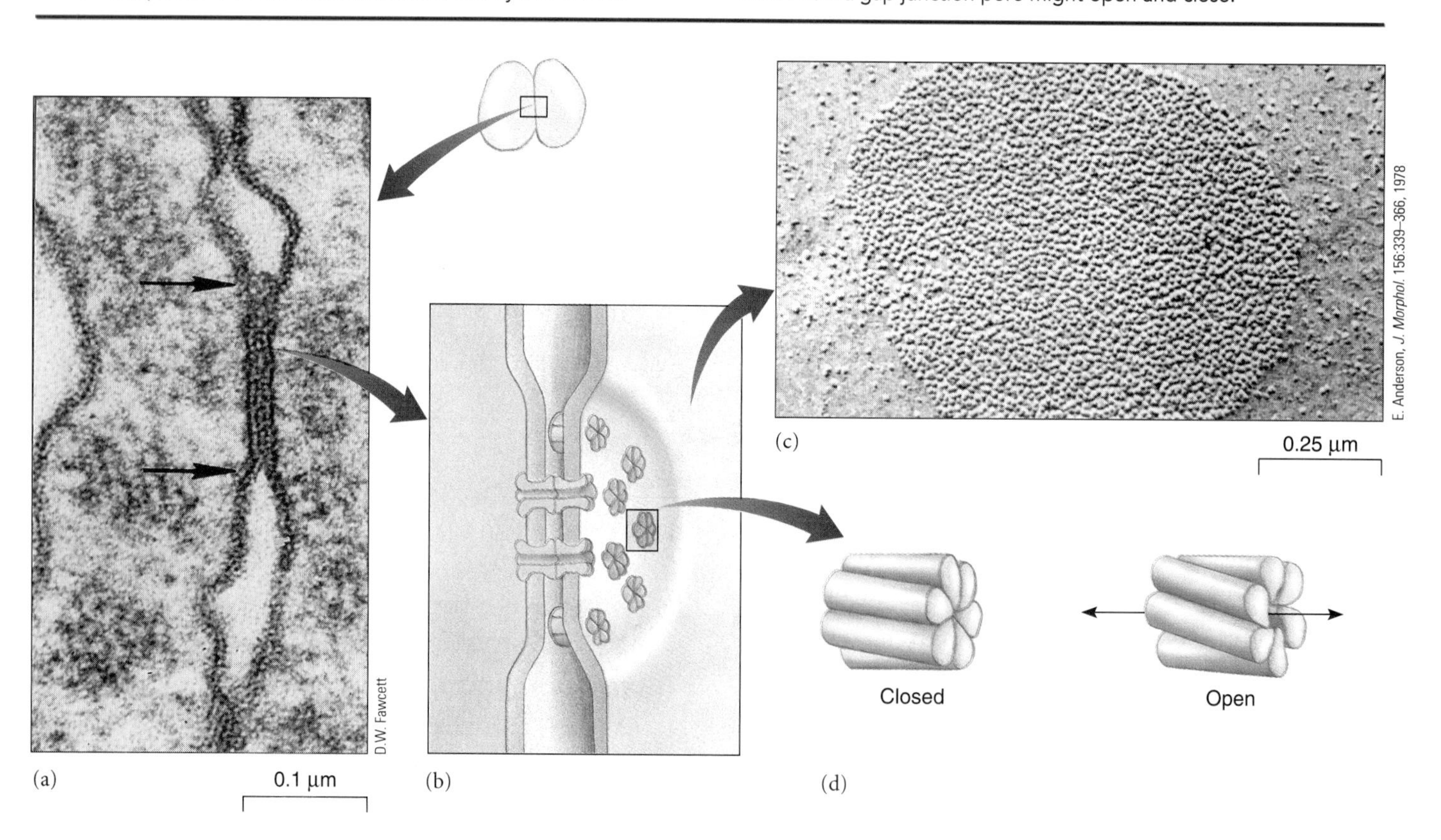

FIGURE **5-25** | **Plasmodesmata.**

TEM and line art of cytoplasmic channels through the cell walls of adjacent plant cells *(wide arrows)* that allow passage of water, ions, and small molecules. The channels are lined with the fused plasma membranes of the two adjacent cells. Desmotubules are not shown.

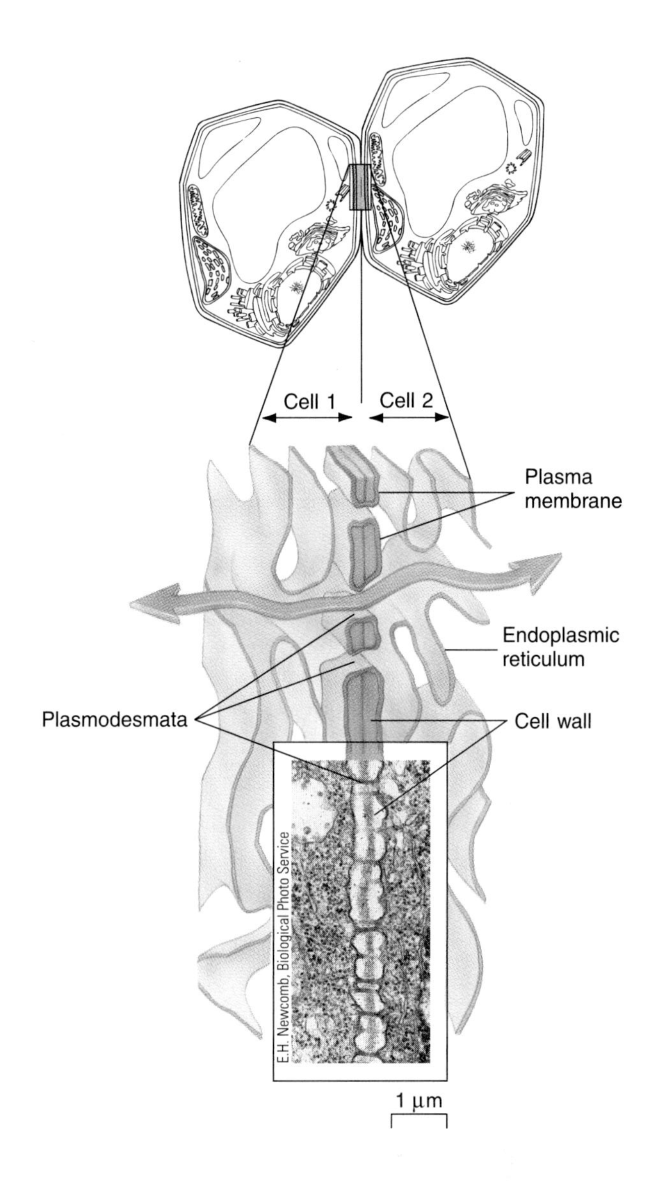

preventing them from communicating. For this reason, plant cells require connections that are functionally equivalent to the gap junctions of some animal cells. **Plasmodesmata** (sing., *plasmodesma*) are channels, 20- to 40-nm wide, through adjacent cell walls, connecting the cytoplasm of neighboring cells (Fig. 5-25). The plasma membranes of adjacent cells are continuous with each other through the plasmodesmata. Most plasmodesmata contain a cylindrical membranous structure, called the *desmotubule,* which also runs through the opening and connects the ER of the two adjacent cells.

Plasmodesmata generally allow molecules and ions, but not organelles, to pass through the openings from cell to cell. The movement of ions through the plasmodesmata allows for a very slow type of electrical signaling in plants. Whereas the channels of gap junctions have a fixed diameter, plants cells can dilate the plasmodesmata channels.

Review

- How are desmosomes and tight junctions functionally similar? How do they differ?
- What is the justification for considering gap junctions and plasmodesmata to be functionally similar? How do they differ structurally?

BiologyNow™ Assess your understanding of **cell junctions** by taking the pretest on your BiologyNow CD-ROM.

SUMMARY WITH KEY TERMS

1 Evaluate the importance of membranes to the homeostasis of the cell, emphasizing their various functions.

- Biological membranes (1) physically separate the interior of the cell from the extracellular environment and (2) form compartments within eukaryotic cells, allowing a variety of separate functions.
- The **plasma membrane** regulates the passage of materials into and out of the cell, participates in and serve as surfaces for biochemical reactions, receives information about changes in the environment, and communicates with other cells.

2 Describe the fluid mosaic model of cell membrane structure.

- According to the **fluid mosaic model,** membranes consist of a fluid phospholipid bilayer in which a variety of proteins are embedded.
- The phospholipid molecules are **amphipathic:** They have hydrophobic and hydrophilic regions. The hydrophilic heads of the phospholipids are at the two surfaces of the bilayer, and their hydrophobic fatty acid chains are in the interior.

3 Explain how the properties of the lipid bilayer govern many of the properties of a cell membrane and of the cell.

- In almost all biological membranes, the lipids of the bilayer are in a fluid or liquid-crystalline state, which allows the molecules to move rapidly in the plane of the membrane. Proteins move within the membrane.
- Lipid bilayers are flexible and self-sealing, and can fuse with other membranes. These properties are the basis for transport of materials from one part of the cell to another in vesicles that bud from various cell membranes and then fuse with some other membrane.

4 Describe how membrane proteins associate with the lipid bilayer, and discuss the functions of membrane proteins.

- **Integral membrane proteins** are embedded in the bilayer with their hydrophilic surfaces exposed to the aqueous environment and their hydrophobic surfaces in contact with the hydrophobic interior of the bilayer. **Transmembrane proteins** are integral proteins that extend completely through the membrane.

- **Peripheral membrane proteins** are associated with the surface of the bilayer, usually bound to exposed regions of integral proteins, and are easily removed without disrupting the structure of the membrane.
- Membrane proteins, lipids, and carbohydrates are asymmetrically positioned with respect to the bilayer so that one side of the membrane has a different composition and structure from the other.
- Membrane proteins have many functions, including transport of materials, acting as enzymes or receptors, cell recognition, and structurally linking cells together.

5 Contrast the physical processes of simple diffusion and osmosis with the carrier-mediated physiological processes by which materials are transported across cell membranes.

- Biological membranes are selectively permeable membranes; that is, they allow the passage of some substances but not others. **Diffusion** is the net movement of a substance down its **concentration gradient** from a region of greater concentration to one of lower concentration.
- Osmosis is a kind of diffusion in which molecules of water pass through a selectively permeable membrane from a region where water has a higher effective concentration to a region where its effective concentration is lower.
- Diffusion and osmosis are physical processes that do not require the cell to directly expend metabolic energy.
- Membrane **transport proteins** facilitate the passage of certain ions and molecules through biological membranes. **Channel proteins** are transport proteins that form passageways through which water and certain ions travel through the membrane. **Carrier proteins** are transport proteins that undergo a series of conformational changes as they bind and transport a specific solute.

6 Solve simple problems involving osmosis; for example, predict whether cells will swell or shrink under various osmotic conditions.

- The concentration of dissolved substances (solutes) in a solution determines its **osmotic pressure.** Cells regulate their internal osmotic pressures to prevent shrinking or bursting.
- An **isotonic** solution has an equal solute concentration compared to another fluid, for example, the fluid within the cell.
- When placed in a **hypertonic** solution, one that has a greater solute concentration than the cell, cells lose water to the surroundings; plant cells undergo **plasmolysis,** a process in which the plasma membrane separates from the cell wall.
- When cells are placed in a **hypotonic** solution, one with a lower concentration of dissolved materials relative to the cell, water enters the cells and causes them to swell.
- Plant cells withstand high internal hydrostatic pressure because their cell walls prevent them from expanding and bursting. When water moves into cells by osmosis, it fills the central vacuoles. The cells swell, building up **turgor pressure** against the rigid cell walls.

7 Differentiate between the processes of facilitated diffusion and active transport, and identify energy sources for each process.

- In **carrier-mediated transport,** specific carrier proteins move ions or molecules across a membrane. **Facilitated diffusion** is a form of carrier-mediated transport that uses the energy of a concentration gradient to transport compounds across a membrane. Facilitated diffusion cannot work against a gradient.
- In **carrier-mediated active transport,** the cell expends metabolic energy to move ions or molecules across a membrane against a concentration gradient. For example, the **sodium-potassium pump** uses ATP to pump sodium ions out of the cell and potassium ions into the cell.
- In **cotransport,** an ATP-powered pump such as the sodium-potassium pump transports ions or some other solute and indirectly powers the transport of other solutes by maintaining a concentration gradient.

8 Compare endocytotic and exocytotic transport mechanisms.

- The cell expends metabolic energy to carry on physiological processes, such as carrier-mediated active transport, exocytosis, and endocytosis.
- In **exocytosis,** the cell ejects waste products or secretes substances such as hormones or mucus by fusion of vesicles with the plasma membrane. In this process, the surface area of the plasma membrane increases.
- In **endocytosis** materials such as food may be moved into the cell; a portion of the plasma membrane envelops the material, enclosing it in a vesicle or vacuole that is then released inside the cell. In this process, the surface area of the plasma membrane decreases. Three types of endocytosis are phagocytosis, pinocytosis, and receptor-mediated endocytosis.
- In **phagocytosis,** the plasma membrane encloses a particle such as a bacterium or protist, forms a vacuole around it, and moves it into the cell.
- In **pinocytosis,** the cell takes in dissolved materials by forming tiny vesicles around droplets of fluid trapped by folds of the plasma membrane.
- In **receptor-mediated endocytosis,** specific receptors in coated pits along the plasma membrane bind **ligands.** These pits, coated by the protein clathrin, form coated vesicles by endocytosis.

9 Describe the generalized process of cell signaling.

- Cells communicate by **cell signaling.** Signaling molecules include neurotransmitters, hormones, and other regulatory molecules. Cell signaling involves synthesis and release of the signaling molecule, transport to target cells, reception of information by target cells, signal transduction, response by the cell, and termination of the signal.

10 Explain how an extracellular signal is converted to an intracellular signal in signal transduction.

- In **signal transduction,** a receptor converts an extracellular signal into an intracellular signal that causes some change in the cell. Signal transduction typically involves a series of molecules that relay information from one to another.
- Signal transduction often involves activation of **G proteins** by binding of a ligand to a receptor; a second messenger such as **cyclic AMP;** and **protein kinases,** enzymes that activate specific proteins by phosphorylating them. The phosphorylated protein then alters some cell functions.

11 Compare the structures and functions of anchoring junctions, tight junctions, gap junctions, and plasmodesmata.

- Cells in close contact with one another may develop intercellular junctions. **Anchoring junctions** include desmosomes and adhering junctions; they are found between cells that form a sheet of tissue. **Desmosomes** spot-weld adjacent animal cells together.

Adhering junctions are formed by cadherins that cement cells together.

- **Tight junctions** seal membranes of adjacent animal cells together, preventing substances from moving through the spaces between the cells.
- **Gap junctions,** composed of the protein connexin, form channels, allowing communication between the cytoplasm of adjacent animal cells.
- **Plasmodesmata** are channels connecting adjacent plant cells. Openings in the cell walls allow the plasma membranes and cytoplasm to be continuous; certain molecules and ions pass from cell to cell.

POST-TEST

1. Which of the following statements is *not* true? Biological membranes (a) are composed partly of amphipathic lipids (b) have hydrophobic and hydrophilic regions (c) are typically in a fluid state (d) are made mainly of lipids and of proteins that lie like thin sheets on the membrane surface (e) function in signal transduction
2. According to the fluid mosaic model, membranes consist of (a) a lipid-protein sandwich (b) mainly phospholipids with scattered nucleic acids (c) a fluid phospholipid bilayer in which proteins are embedded (d) a fluid phospholipid bilayer in which carbohydrates are embedded (e) a protein bilayer that behaves as a liquid crystal
3. Transmembrane proteins (a) are peripheral proteins (b) are receptor proteins (c) extend completely through the membrane (d) extend along the surface of the membrane (e) are secreted from the cell
4. Which of the following is *not* a function of the plasma membrane? (a) transports materials (b) helps to structurally link cells together (c) manufactures proteins (d) anchors the cell to the extracellular matrix (e) has receptors that relay signals
5. Which of the following processes requires the cell to expend metabolic energy directly (for example, from ATP)? (a) active transport (b) facilitated diffusion (c) all forms of carrier-mediated transport (d) osmosis (e) simple diffusion
6. Which of the following is an example of carrier-mediated transport? (a) simple diffusion (b) facilitated diffusion (c) movement of water through aquaporins (d) osmosis (e) osmosis when a cell is in a hypertonic solution
7. The action of sodium-potassium pumps is an example of (a) carrier-mediated active transport (b) pinocytosis (c) aquaporin transport (d) exocytosis (e) facilitated diffusion
8. The patch clamp technique (a) cannot be applied to plant cells (b) is mainly used to study exocytosis (c) allows researchers to study single ion channels (d) helped researchers understand signal transduction involving G proteins (e) was developed by Singer and Nicolson
9. A cell takes in dissolved materials by forming tiny vesicles around fluid droplets trapped by folds of the plasma membrane. This process is (a) carrier-mediated active transport (b) pinocytosis (c) receptor-mediated endocytosis (d) exocytosis (e) facilitated diffusion
10. When plant cells are in a hypotonic medium, they (a) undergo plasmolysis (b) build up turgor pressure (c) wilt (d) decrease pinocytosis (e) lose water to the environment
11. After a ligand binds to receptors in coated pit (a) the ligand binds to receptors in coated vesicle (b) a coated vesicle forms by endocytosis (c) a vesicle enters the cytosol by facilitated diffusion (d) lysosomes destroy protein coating of the pit (e) G proteins signal phagocytosis
12. In signal transduction (a) an extracellular signal is converted to an intracellular signal (b) a signal is relayed through a series of molecules in the membrane (c) signal molecules are destroyed before target cells can respond to the signal (d) answers a, b, and c are correct (e) only answers a and b are correct
13. When a ligand binds with a receptor (a) tight junctions develop (b) a third messenger is activated (c) cell signaling is stopped (d) it activates the receptor (e) a G protein is destroyed
14. G proteins (a) relay a message from the activated receptor to an enzyme that activates a second messenger (b) are GTP molecules (c) stop cell signaling (d) directly activate protein kinases (e) are hormones that function as first messengers
15. Anchoring junctions that hold cells together at one point like a spot weld are (a) tight junctions (b) adhering junctions (c) desmosomes (d) gap junctions (e) plasmodesmata
16. Junctions that permit the transfer of water, ions, and molecules between adjacent plant cells are (a) tight junctions (b) adhering junctions (c) desmosomes (d) gap junctions (e) plasmodesmata
17. Junctions that help form the blood–brain barrier are (a) tight junctions (b) adhering junctions (c) desmosomes (d) gap junctions (e) plasmodesmata

CRITICAL THINKING

1. Why can't larger polar molecules and ions diffuse through the plasma membrane? Would it be advantageous to the cell if they could? Explain.
2. Most adjacent plant cells are connected by plasmodesmata, whereas only certain adjacent animal cells are associated through gap junctions. Why?
3. Evaluate the importance of membranes to the cell, discussing their various functions.

- Visit our Web site at **http://biology.brookscole.com/solomon7** for links to chapter-related resources on the World Wide Web. Additional online materials relating to this chapter can also be found on our Web site.

BIOLOGY NOW RESOURCES

Active Figures

5-2: Protein movement in the fluid mosaic model
5-14: Membrane transport

Preparing for an exam? Take a diagnostic test on your BiologyNow CD-ROM.

Post-Test Answers

1. d	2. c	3. c	4. c
5. a	6. b	7. a	8. c
9. b	10. b	11. b	12. e
13. d	14. a	15. c	16. e
17. a			

6 Energy and Metabolism

Barbara Gerlach/Visuals Unlimited

Black-tailed prairie dog (*Cynomys ludovicianus*). The chemical energy produced by photosynthesis and stored in seeds and leaves transfers to the black-tailed prairie dog as the animal eats.

CHAPTER OUTLINE

All living things require energy to carry out life processes. It may seem obvious that cells need energy to grow and reproduce, but even nongrowing cells need energy simply to maintain themselves. The sun is the ultimate source of almost all the energy that powers life; this *radiant energy* flows from the sun as electromagnetic waves. Plants and other photosynthetic organisms capture about 0.02% of the sun's energy that reaches Earth. In the process of photosynthesis, plants convert radiant energy to *chemical energy* in the bonds of organic molecules. The chemical energy captured by photosynthesis and stored in seeds and leaves is transferred to animals, such as the black-tailed prairie dog in the photograph, when they eat. Plants, animals, and other organisms need the energy stored in these organic molecules, and the process of cellular respiration breaks them apart and converts their energy to more immediately usable forms.

Cells obtain energy in many forms, but that energy can seldom be used directly to power cell processes. For this reason, cells have mechanisms that convert energy from one form to another. Because most components of these energy conversion systems evolved very early in the history of life, many aspects of energy metabolism tend to be similar in a wide range of organisms.

This chapter focuses on some of the basic principles that govern how cells capture, transfer, store, and use energy. We discuss the functions of adenosine triphosphate (ATP) and other molecules used in energy conversions, including those that transfer electrons in oxidation-reduction (redox) reactions. We also pay particular attention to the essential role of enzymes in cell energy dynamics. In Chapter 7 we explore some of the main metabolic pathways used in cellular respiration, and in Chapter 8 we discuss the energy transformations of photosynthesis. The flow of energy in ecosystems is discussed in Chapter 53.

BIOLOGICAL WORK

Learning Objectives

1. Define *energy*, emphasizing how it is related to work and to heat.
2. Use examples to contrast potential energy and kinetic energy.

Energy, one of the most important concepts in biology, can be understood in the context of **matter,** which is anything that has mass and takes up space. **Energy** is defined as the capacity to do work, which is any change in the state or motion of matter. Technically, mass is a form of energy, which is the basis behind the energy generated by the sun and other stars. More than 4 billion kg of matter per second are converted into energy in the sun.

Biologists generally express energy in units of work—**kilojoules (kJ).** It can also be expressed in units of *heat energy*—**kilocalories, kcal**—thermal energy that flows from an object with a higher temperature to an object with a lower temperature. One kcal is equal to 4.184 kJ. Heat energy cannot do cell work, because a cell is too small to have regions that differ in temperature. For that reason, the unit most biologists prefer today is the kilojoule. However, we will use both units because references to the kilocalorie are common in the scientific literature.

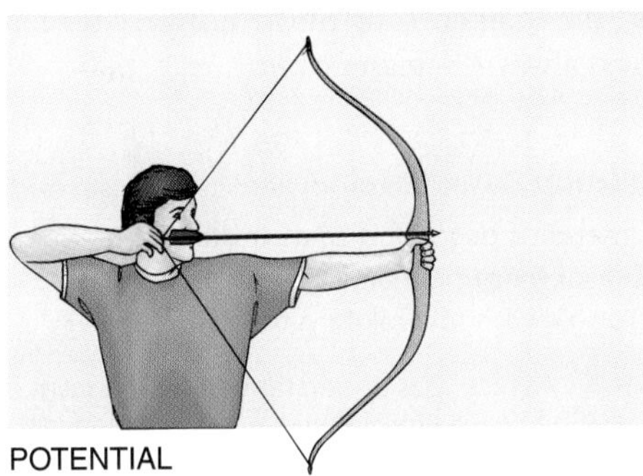

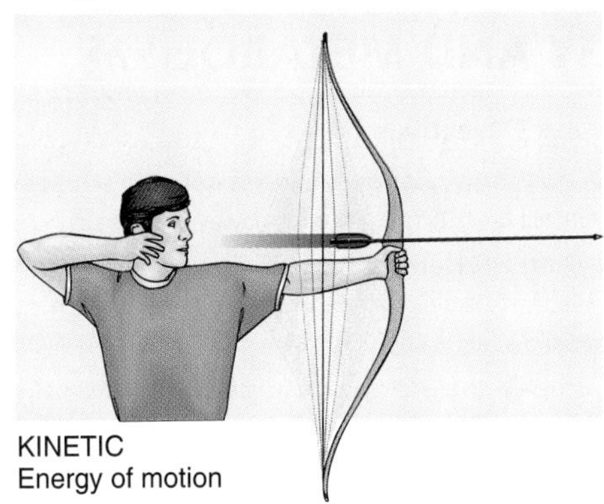

FIGURE **6-1** **Potential versus kinetic energy.**

The potential chemical energy released by cellular respiration is converted to kinetic energy in the muscles, which do the work of drawing the bow. The potential energy stored in the drawn bow is transformed into kinetic energy as the bowstring pushes the arrow toward its target.

Organisms carry out conversions between potential energy and kinetic energy

When an archer draws a bow, **kinetic energy,** the energy of motion, is used and work is performed (Fig. 6-1). The resulting tension in the bow and string represents stored, or potential, energy. **Potential energy** is the capacity to do work owing to position or state. When the string is released, this potential energy is converted to kinetic energy in the motion of the bow, which propels the arrow.

Most actions of an organism involve a complex series of energy transformations that occur as kinetic energy is converted to potential energy or as potential energy is converted to kinetic energy. **Chemical energy,** potential energy stored in chemical bonds, is of particular importance to organisms. In our example, the chemical energy of food molecules is converted to kinetic energy in the muscle cells of the archer. The contraction of the archer's muscles, like many of the activities performed by an organism, is an example of *mechanical energy,* which performs work by moving matter.

Review

- You exert tension on a spring and then release it. How do these actions relate to work, potential energy, and kinetic energy?

BiologyNow™ Assess your understanding of **biological work** by taking the pretest on your BiologyNow CD-ROM.

THE LAWS OF THERMODYNAMICS

Learning Objective

3. State the first and second laws of thermodynamics, and discuss the implications of these laws as they relate to organisms.

Thermodynamics, the study of energy and its transformations, governs all activities of the universe, from the life and death of cells to the life and death of stars. When considering thermodynamics, scientists use the term *system* to refer to an object that they are studying, whether a cell, an organism, or planet Earth. The rest of the universe other than the system being studied constitutes the *surroundings.* A **closed system** does not exchange energy with its surroundings, whereas an **open system** can exchange energy with its surroundings (Fig. 6-2). Biological systems are open systems.

Two laws about energy apply to all things in the universe: the first and second laws of thermodynamics.

The total energy in the universe does not change

According to the **first law of thermodynamics,** energy cannot be created or destroyed, although it can be transferred or converted from one form to another, including conversions between matter and energy. As far as we know, the total mass-energy present in the universe when it formed, almost 14 billion years ago, equals the amount of energy present in the uni-

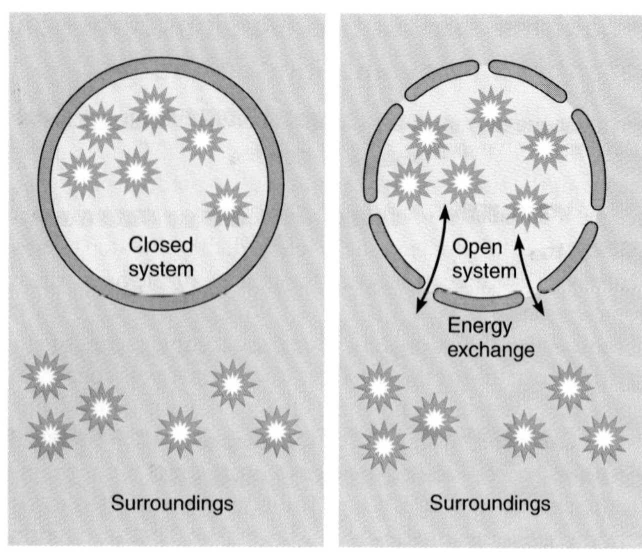

(a) Closed system (b) Open system

FIGURE **6-2** **Closed and open systems.**

(a) Energy is not exchanged between a closed system and its surroundings. **(b)** Energy is exchanged between an open system and its surroundings.

verse today. This is all the energy that can ever be present in the universe. Similarly, the energy of any system plus its surroundings is constant. A system may absorb energy from its surroundings, or it may give up some energy to its surroundings, but the total energy content of that system plus its surroundings is always the same.

As specified by the first law of thermodynamics, then, organisms cannot create the energy they require in order to live. Instead, they must capture energy from the environment and transform it to a form that can be used for biological work.

The entropy of the universe is increasing

The **second law of thermodynamics** is as follows: When energy is converted from one form to another, some usable energy—that is, energy available to do work—is converted into heat that disperses into the surroundings (see Fig. 53-1). As you learned in Chapter 2, **heat** is the kinetic energy of randomly moving particles. Unlike *heat energy,* which flows from an object with a higher temperature to one with a lower temperature, this random motion cannot perform work. As a result, the amount of usable energy available to do work in the universe decreases over time.

It is important to understand that the second law of thermodynamics is consistent with the first law; that is, the total amount of energy in the universe is *not* decreasing with time. However, the total amount of energy in the universe that is available to do work is decreasing over time.

Less usable energy is more diffuse, or disorganized. **Entropy (*S*)** is a measure of this disorder, or randomness; organized, usable energy has a low entropy, whereas disorganized energy, such as heat, has a high entropy.

Entropy is continuously increasing in the universe in all natural processes. Maybe at some time, billions of years from now, all energy will exist as heat uniformly distributed throughout the universe. If that happens, the universe will cease to operate, because no work will be possible. Everything will be at the same temperature, so there will be no way to convert the thermal energy of the universe into usable mechanical energy.

As a consequence of the second law of thermodynamics, no process requiring an energy conversion is ever 100% efficient, because much of the energy is dispersed as heat, increasing entropy. For example, an automobile engine, which converts the chemical energy of gasoline to mechanical energy, is between 20% and 30% efficient. Thus only 20% to 30% of the original energy stored in the chemical bonds of the gasoline molecules is actually transformed into mechanical energy; the other 70% to 80% dissipates as waste heat. Energy use in your cells is about 40% efficient, with the remaining energy given to the surroundings as heat.

Organisms have a high degree of organization, and at first glance they appear to refute the second law of thermodynamics. As organisms grow and develop, they maintain a high level of order and do not appear to become more disorganized. However, organisms maintain their degree of order over time only with the constant input of energy from their surroundings. That is why plants must photosynthesize and animals must eat. Although the order within organisms may tend to increase temporarily, the total entropy of the universe (organisms plus surroundings) always increases over time.

Review

- What is the first law of thermodynamics? The second law?
- Life is sometimes described as a constant struggle against the second law of thermodynamics. How do organisms succeed in this struggle without violating the second law?

BiologyNow™ Assess your understanding of the **laws of thermodynamics** by taking the pretest on your BiologyNow CD-ROM.

ENERGY AND METABOLISM

Learning Objectives

4 Discuss how changes in free energy in a reaction are related to changes in entropy and enthalpy.

5 Distinguish between exergonic and endergonic reactions, and give examples of how they may be coupled.

6 Compare the energy dynamics of a reaction at equilibrium with the dynamics of a reaction not at equilibrium.

The chemical reactions that enable an organism to carry on its activities—to grow, move, maintain and repair itself, reproduce, and respond to stimuli—together make up its metabolism. Recall from Chapter 1 that **metabolism** is the sum of all the chemical activities taking place in an organism. An organism's metabolism consists of many intersecting series of chemical reactions, or pathways, which are of two main types: anabolism and catabolism. **Anabolism** includes the various pathways in which complex molecules are synthesized from simpler substances, such as in the linking of amino acids to form proteins. **Catabolism** includes the pathways in which larger molecules are broken down into smaller ones, such as in the degradation of starch to form monosaccharides.

As you will see, these changes involve not only alterations in the arrangement of atoms but also various energy transformations. Catabolism and anabolism are complementary processes; catabolic pathways involve an overall release of energy, some of which powers anabolic pathways, which have an overall energy requirement. In the following sections we discuss how to predict whether a particular chemical reaction requires energy or releases it.

Enthalpy is the total potential energy of a system

In the course of any chemical reaction, including the metabolic reactions of a cell, chemical bonds break, and new and different bonds may form. Every specific type of chemical bond has a certain amount of *bond energy,* defined as the energy required to break that bond. The total bond energy is essentially equivalent to the total potential energy of the system, a quantity known as **enthalpy** (***H***).

Free energy is available to do cell work

Entropy and enthalpy are related by a third type of energy, termed **free energy (*G*),** which is the amount of energy available to do work under the conditions of a biochemical reaction. (*G*, also known as "Gibbs free energy," is named for J.W. Gibbs, a Yale professor who was one of the founders of the science of thermodynamics.) Free energy, the only kind of energy that can do cell work, is the aspect of thermodynamics of greatest interest to a biologist.

Enthalpy, free energy, and entropy are related by the following equation:

$$H = G + TS$$

in which H is enthalpy, G is free energy, S is entropy, and T is the absolute temperature of the system, expressed in degrees Kelvin. Disregarding temperature (T) for the moment, enthalpy (the total energy of a system) is equal to free energy (the usable energy) plus entropy (the unusable energy).

A rearrangement of the equation shows that as entropy increases, the amount of free energy decreases:

$$G = H - TS$$

If we assume that entropy is zero, the free energy is simply equal to the total potential energy (enthalpy); an increase in entropy reduces the amount of free energy.

What is the significance of the temperature (T)? Remember that as the temperature increases, the increase in random molecular motion contributes to disorder and multiplies the effect of the entropy term.

Chemical reactions involve changes in free energy

Biologists analyze the role of energy in the many biochemical reactions of metabolism. Although the total free energy of a system (G) cannot be effectively measured, the equation $G = H - TS$ can be extended to predict whether a particular chemical reaction will release energy or require an input of energy. This is because *changes* in free energy can be measured. Scientists use the Greek capital letter delta (Δ) to denote any change that occurs in the system between its initial state before the reaction and its final state after the reaction. To express what happens with respect to energy in a chemical reaction, the equation becomes

$$\Delta G = \Delta H - T\Delta S$$

Notice that the temperature does not change; it is held constant during the reaction. Thus the change in free energy (ΔG) during the reaction is equal to the change in enthalpy (ΔH) minus the product of the absolute temperature (T) multiplied by the change in entropy (ΔS). Scientists express ΔG and ΔH in kilojoules or kilocalories per mole; they express ΔS in kilojoules or kilocalories per degree.

Free energy decreases during an exergonic reaction

An **exergonic reaction** releases energy and is said to be a spontaneous or a "downhill" reaction, from higher to lower free energy (Fig. 6-3a). Because the total free energy in its final state is less than the total free energy in its initial state, ΔG is a negative number for exergonic reactions.

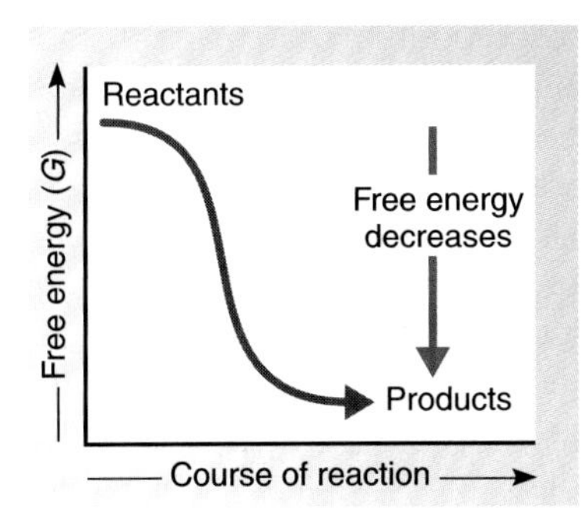

(a) Exergonic reaction (spontaneous; energy-releasing)

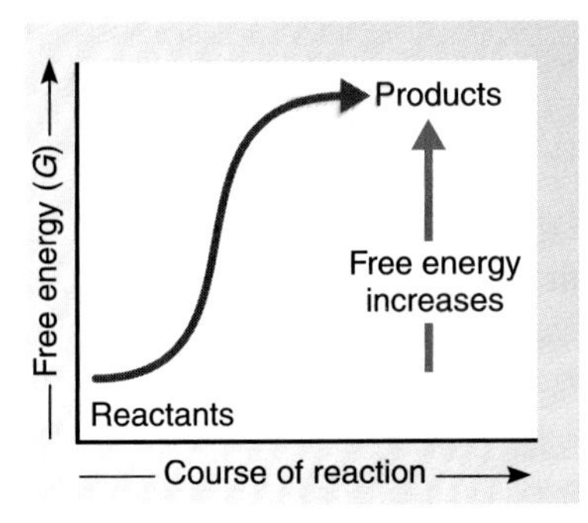

(b) Endergonic reaction (not spontaneous; energy-requiring)

FIGURE **6-3**

(a) In an exergonic reaction, there is a net loss of free energy. The products have less free energy than was present in the reactants, and the reaction proceeds spontaneously. **(b)** In an endergonic reaction, there is a net gain have more free energy than was present in the reactants. An endergonic reaction occurs only if energy is supplied by an exergonic reaction.

The term *spontaneous* may give the false impression that such reactions are always instantaneous. In fact, spontaneous reactions do not necessarily occur readily; some are extremely slow. This is because energy, known as *activation energy*, is required to initiate every reaction, even a spontaneous one. We discuss activation energy later in the chapter.

Free energy increases during an endergonic reaction

An **endergonic reaction** is a reaction in which there is a gain of free energy (Fig. 6-3b). Because the free energy of the products is greater than the free energy of the reactants, ΔG has a positive value. Such a reaction cannot take place in isolation. Instead, it must occur in such a way that energy can be supplied from the surroundings. Of course, many energy-requiring reactions take place in cells, and as you will see, metabolic mechanisms have evolved that supply the energy needed to "drive" these nonspontaneous cell reactions in a particular direction.

Diffusion is an exergonic process

In Chapter 5, you saw that randomly moving particles diffuse down their own concentration gradient (Fig. 6-4). Although the movements of the individual particles are random, net movement of the group of particles seems to be directional. What provides energy for this apparently directed process? A **concentration gradient,** with a region of higher concentration and another region of lower concentration, is an orderly state. A cell must expend energy to produce a concentration gradient. Because work is done to produce this order, a concentration gradient is a form of potential energy. As the particles move about randomly, the gradient becomes degraded. Thus free energy decreases as entropy increases.

In cellular respiration and photosynthesis, the potential energy stored in a concentration gradient of hydrogen ions (H^+) is transformed into chemical energy in adenosine triphosphate (ATP) as the hydrogen ions pass through a membrane down their concentration gradient. This important concept, known as *chemiosmosis,* is discussed in detail in Chapters 7 and 8.

Free energy changes depend on the concentrations of reactants and products

According to the second law of thermodynamics, any process that increases entropy can do work. As we have discussed, differences in the concentration of a substance, such as between two different parts of a cell, represent a more orderly state than when the substance is diffused homogeneously throughout the cell. Free energy changes in any chemical reaction depend mainly on the difference in bond energies (enthalpy, H) between reactants and products. Free energy also depends on *concentrations* of both reactants and products. The change in molecules from a more concentrated to a less concentrated state increases entropy because it is movement from a more orderly to a less orderly state.

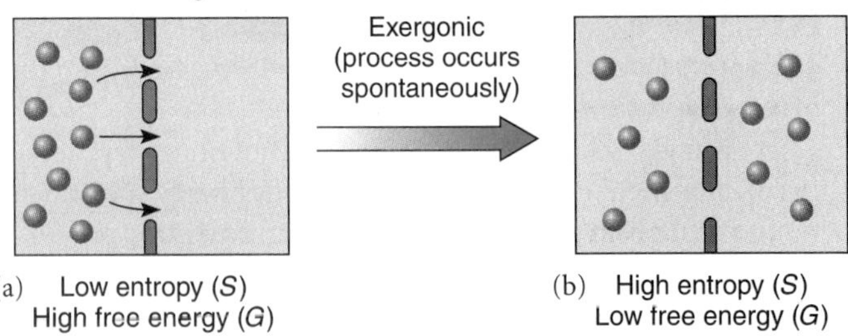

FIGURE **6-4** **Entropy and diffusion.**
The tendency of entropy to increase can be used to produce work, in this case, diffusion. **(a)** A concentration gradient is a form of potential energy. **(b)** When molecules are evenly distributed, they have high entropy.

In most biochemical reactions there is little intrinsic free energy difference between reactants and products. Such reactions are reversible, indicated by drawing double arrows ($\rightleftharpoons$).

$$A \rightleftharpoons B$$

At the beginning of a reaction, only the reactant molecules (A) may be present. As the reaction proceeds, the concentration of the reactant molecules decreases, and the concentration of the product molecules (B) increases. As the concentration of the product molecules increases, they may have enough free energy to initiate the reverse reaction. The reaction thus proceeds in both directions simultaneously; if undisturbed, it eventually reaches a state of **dynamic equilibrium,** in which the rate of the reverse reaction equals the rate of the forward reaction. At equilibrium there is no net change in the system; a reverse reaction balances every forward reaction.

At a given temperature and pressure, each reaction has its own characteristic equilibrium. For any given reaction, chemists can perform experiments and calculations to determine the relative concentrations of reactants and products present at equilibrium. If the reactants have much greater intrinsic free energy than the products, the reaction goes almost to completion; that is, it reaches equilibrium at a point at which most of the reactants have been converted to products. Reactions in which the reactants have much less intrinsic free energy than the products reach equilibrium at a point where very few of the reactant molecules have been converted to products.

If you increase the initial concentration of A, then the reaction will "shift to the right," and more A will be converted to B. A similar effect can be obtained if B is removed from the reaction mixture. The reaction always shifts in the direction that reestablishes equilibrium, so that the proportions of reactants and products characteristic of that reaction at equilibrium are restored. The opposite effect occurs if the concentration of B increases or if A is removed; here the system "shifts to the left." The actual free energy change that occurs during a reaction is defined mathematically to include these effects, which stem from the relative initial concentrations of reactants and products.

Cells manipulate the relative concentrations of reactants and products of almost every reaction. Cell reactions are virtually never at equilibrium. By displacing their reactions far from equilibrium, cells supply energy to endergonic reactions and direct their metabolism according to their needs.

Cells drive endergonic reactions by coupling them to exergonic reactions

Many metabolic reactions, such as protein synthesis, are anabolic and endergonic. Because an endergonic reaction cannot take place without an input of energy, endergonic reactions are coupled to exergonic reactions. In **coupled reactions,** the thermodynamically favorable exergonic reaction provides the energy required to drive the thermodynamically unfavorable endergonic reaction. The endergonic reaction proceeds only if it absorbs free energy released by the exergonic reaction to which it is coupled.

Consider the free energy change, ΔG, in the following reaction:

$$(1)\ A \longrightarrow B \qquad \Delta G = +20.9 \text{ kJ/mol } (+5 \text{ kcal/mol})$$

Because ΔG has a positive value, you know that the product of this reaction has more free energy than the reactant. This is an endergonic reaction. It is not spontaneous and does not take place without an energy source.

By contrast, consider the following reaction:

$$(2)\ C \longrightarrow D \qquad \Delta G = -33.5 \text{ kJ/mol } (-8 \text{ kcal/mol})$$

The negative value of ΔG tells you that the free energy of the reactant is greater than the free energy of the product. This exergonic reaction proceeds spontaneously.

You can sum up reactions 1 and 2 as follows:

(1) $A \longrightarrow B$	$\Delta G = +20.9$ kJ/mol (+5 kcal/mol)
(2) $C \longrightarrow D$	$\Delta G = -33.5$ kJ/mol (−8 kcal/mol)
Overall	$\Delta G = -12.6$ kJ/mol (−3 kcal/mol)

Because thermodynamics considers the overall changes in these two reactions, which show a net negative value of ΔG, the two reactions taken together are exergonic.

The fact that scientists can write reactions this way is a useful bookkeeping device, but it does not mean that an exergonic reaction mysteriously transfers energy to an endergonic "bystander" reaction. However, these reactions are coupled if their pathways are altered so a common intermediate links them. Reactions 1 and 2 might be coupled by an intermediate (I) in the following way:

(3) $A + C \longrightarrow I$	$\Delta G = -8.4$ kJ/mol (−2 kcal/mol)
(4) $I \longrightarrow B + D$	$\Delta G = -4.2$ kJ/mol (−1 kcal/mol)
Overall	$\Delta G = -12.6$ kJ/mol (−3 kcal/mol)

Note that reactions 3 and 4 are sequential. Thus the reaction pathways have changed, but overall the reactants (A and C) and products (B and D) are the same, and the free energy change is the same.

Generally, for each endergonic reaction occurring in a living cell there is a coupled exergonic reaction to drive it. Often the exergonic reaction involves the breakdown of ATP. Now let's examine specific examples of the role of ATP in energy coupling.

Review

- Consider the free energy change in a reaction in which enthalpy decreases and entropy increases. Is ΔG zero, or does it have a positive value or a negative value? Is the reaction endergonic or exergonic?
- Why can't a reaction at equilibrium do work?

BiologyNow™ Assess your understanding of **energy and metabolism** by taking the pretest on your BiologyNow CD-ROM.

ATP, THE ENERGY CURRENCY OF THE CELL

Learning Objective

7 Explain how the chemical structure of ATP allows it to transfer a phosphate group. Discuss the central role of ATP in the overall energy metabolism of the cell.

In all living cells, energy is temporarily packaged within a remarkable chemical compound called **adenosine triphosphate (ATP),** which holds readily available energy for very short periods. We may think of ATP as the energy currency of the cell. When you work to earn money, you might say your energy is symbolically stored in the money you earn. The energy the cell requires for immediate use is temporarily stored in ATP, which is like cash. When you earn extra money, you may deposit some in the bank; similarly, a cell may deposit energy in the chemical bonds of lipids, starch, or glycogen. Moreover, just as you dare not make less money than you spend, the cell must avoid energy bankruptcy, which would mean its death. Finally, just as you probably don't keep money you make very long, the cell continuously spends its ATP, which must be replaced immediately.

ATP is a nucleotide consisting of three main parts: adenine, a nitrogen-containing organic base; ribose, a five-carbon sugar; and three phosphate groups, identifiable as phosphorus atoms surrounded by oxygen atoms (Fig. 6-5). Notice that the phosphate groups are bonded to the end of the molecule in a series, rather like three cars behind a locomotive, and, like the cars of a train, they can be attached and detached.

ATP donates energy through the transfer of a phosphate group

When the terminal phosphate is removed from ATP, the remaining molecule is **adenosine diphosphate (ADP)** (see Fig. 6-5). If the phosphate group is not transferred to another molecule, it is released as inorganic phosphate (P_i). This is an exergonic reaction. ATP is sometimes called a "high-energy" compound because the hydrolysis reaction that releases a phosphate has a relatively large negative value of ΔG. (Calculations of the free energy of ATP hydrolysis vary somewhat, but range between about −28 and −37 kJ/mol, or −6.8 to −8.7 kcal/mol.)

$$(5)\ ATP + H_2O \longrightarrow ADP + P_i \qquad \Delta G = -32 \text{ kJ/mol (or } -7.6 \text{ kcal/mol)}$$

Reaction 5 can be coupled to endergonic reactions in cells. Consider the following endergonic reaction, in which two monosaccharides, glucose and fructose, form the disaccharide sucrose.

$$(6)\ \text{Glucose} + \text{fructose} \longrightarrow \text{sucrose} + H_2O \qquad \Delta G = +27\ \text{kJ/mol (or } +6.5\ \text{kcal/mol)}$$

With a free energy change of -32 kJ/mol (-7.6 kcal/mol), the hydrolysis of ATP in reaction 5 can drive reaction 6, but only if the reactions are coupled through a common intermediate.

The following series of reactions is a simplified version of an alternative pathway that some bacteria use:

$$(7)\ \text{Glucose} + \text{ATP} \longrightarrow \text{glucose-P} + \text{ADP}$$

$$(8)\ \text{Glucose-P} + \text{fructose} \longrightarrow \text{sucrose} + P_i$$

Reaction 7 is a **phosphorylation reaction,** one in which a phosphate group is transferred to some other compound. Glucose is phosphorylated to form glucose phosphate (glucose-P), the intermediate that links the two reactions. Glucose-P, which corresponds to I in reactions 3 and 4, reacts exergonically with fructose to form sucrose. For energy coupling to work in this way, reactions 7 and 8 must occur in sequence.

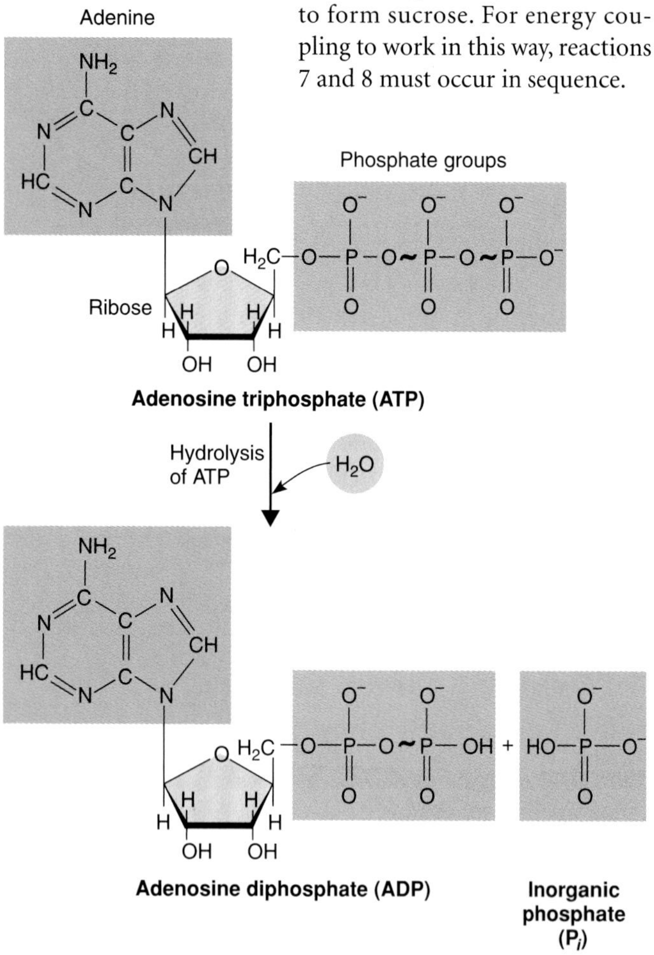

Adenine

Phosphate groups

Ribose

Adenosine triphosphate (ATP)

Hydrolysis of ATP

H_2O

Adenosine diphosphate (ADP)

Inorganic phosphate (P_i)

FIGURE **6-5** **ATP and ADP.**

The energy currency of all living things, ATP consists of adenine, ribose, and three phosphate groups. The hydrolysis of ATP, an exergonic reaction, yields ADP and inorganic phosphate. (The black wavy lines indicate unstable bonds. These bonds allow the phosphates to be transferred to other molecules, making them more reactive.)

It's convenient to summarize the reactions thus:

$$(9)\ \text{Glucose} + \text{fructose} + \text{ATP} \longrightarrow \text{sucrose} + \text{ADP} + P_i \qquad \Delta G = -5\ \text{kJ/mol} \ (-1.2\ \text{kcal/mol})$$

When you encounter an equation written in this way, remember that it is actually a summary of a series of reactions and that transitory intermediate products (in this case, glucose-P) are sometimes not shown.

ATP links exergonic and endergonic reactions

We have just discussed how the transfer of a phosphate group from ATP to some other compound is coupled to endergonic reactions in the cell. Conversely, adding a phosphate group to adenosine monophosphate, or AMP (forming ADP) or to ADP (forming ATP) requires coupling to exergonic reactions in the cell.

$$\text{AMP} + P_i + \text{energy} \longrightarrow \text{ADP}$$

$$\text{ADP} + P_i + \text{energy} \longrightarrow \text{ATP}$$

Thus ATP occupies an intermediate position in the metabolism of the cell and is an important link between exergonic reactions, which are generally components of *catabolic pathways,* and endergonic reactions, which are generally part of *anabolic pathways* (Fig. 6-6).

The cell maintains a very high ratio of ATP to ADP

The cell maintains a ratio of ATP to ADP far from the equilibrium point. ATP constantly forms from ADP and inorganic

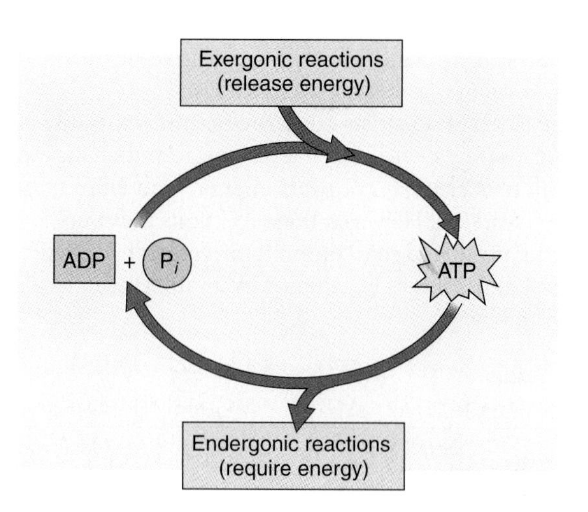

FIGURE **6-6** **ATP links exergonic and endergonic reactions.**

Exergonic reactions in catabolic pathways (*top*) supply energy to drive the endergonic formation of ATP from ADP. Conversely, the exergonic hydrolysis of ATP supplies energy to endergonic reactions in anabolic pathways (*bottom*).

phosphate as nutrients break down in cellular respiration or as photosynthesis traps the radiant energy of sunlight. At any time, a typical cell contains more than 10 ATP molecules for every ADP molecule. The fact that the cell maintains the ATP concentration at such a high level (relative to the concentration of ADP) makes its hydrolysis reaction even more strongly exergonic and more able to drive the endergonic reactions to which it is coupled.

Although the cell maintains a high ratio of ATP to ADP, the cell cannot store large quantities of ATP. The concentration of ATP is always very low, less than 1 mmol/L. In fact, studies suggest a bacterial cell has no more than a 1-second supply of ATP. Thus it uses ATP molecules almost as quickly as they are produced. A healthy adult human at rest uses about 45 kg (100 lb) of ATP each day, but the amount present in the body at any given moment is less than 1 g (0.035 oz). Every second in every cell, an estimated 10 million molecules of ATP are made from ADP and phosphate, and an equal number of ATPs transfer their phosphate groups, along with their energy, to whatever chemical reactions need them.

Review

- Why do coupled reactions typically have common intermediates? Give a generalized example involving ATP, distinguishing between the exergonic and endergonic reactions.
- Why is the ATP concentration in a cell about 10 times the concentration of ADP?

BiologyNow™ Assess your understanding of **ATP** by taking the pretest on your BiologyNow CD-ROM.

ENERGY TRANSFER IN REDOX REACTIONS

Learning Objective

8 Relate the transfer of electrons (or hydrogen atoms) to the transfer of energy.

You have seen that cells transfer energy through the transfer of a phosphate group from ATP. Energy is also transferred through the transfer of electrons. As discussed in Chapter 2, **oxidation** is the chemical process in which a substance loses electrons, whereas **reduction** is the complementary process in which a substance gains electrons. Because electrons released during an oxidation reaction cannot exist in the free state in living cells, every oxidation reaction must be accompanied by a reduction reaction, in which the electrons are accepted by another atom, ion, or molecule. Oxidation and reduction reactions are often called **redox reactions,** because they occur simultaneously. The substance that becomes oxidized gives up energy as it releases electrons, and the substance that becomes reduced receives energy as it gains electrons.

Redox reactions often occur in a series as electrons are transferred from one molecule to another. These electron transfers, which are equivalent to energy transfers, are an essential part of cellular respiration, photosynthesis, and many other chemical reactions. Redox reactions, for example, release the energy stored in food molecules so that ATP can be synthesized using that energy.

Most electron carriers transfer hydrogen atoms

Generally it is not easy to remove one or more electrons from a covalent compound; it is much easier to remove a whole atom. For this reason, redox reactions in cells usually involve the transfer of a hydrogen atom rather than just an electron. A hydrogen atom contains an electron, plus a proton that does not participate in the oxidation-reduction reaction.

When an electron, either singly or as part of a hydrogen atom, is removed from an organic compound, it takes with it some of the energy stored in the chemical bond of which it was a part. That electron, along with its energy, is transferred to an acceptor molecule. An electron progressively loses free energy as it is transferred from one acceptor to another.

One of the most frequently encountered acceptor molecules is **nicotinamide adenine dinucleotide (NAD^+).** When NAD^+ becomes reduced, it temporarily stores large amounts of free energy. Here is a generalized equation showing the transfer of hydrogen from a compound we call X, to NAD^+:

$$\underset{\text{Oxidized}}{XH_2 + NAD^+} \longrightarrow \underset{\text{Reduced}}{X + NADH + H^+}$$

Notice that the NAD^+ becomes reduced when it combines with hydrogen. NAD^+ is an ion with a net charge of + 1. When 2 electrons and 1 proton are added, the charge is neutralized and the reduced form of the compound, **NADH,** is produced (Fig. 6-7). (Although the correct way to write the reduced form of NAD^+ is $NADH + H^+$, for simplicity we present the reduced form as NADH in this book.) Some energy stored in the bonds holding the hydrogen atoms to molecule X has been transferred by this redox reaction and is temporarily held by NADH. When NADH transfers the electrons to some other molecule, some of their energy is transferred. This energy is usually then transferred through a series of reactions that ultimately result in the formation of ATP (see Chapter 7).

Nicotinamide adenine dinucleotide phosphate ($NADP^+$) is a hydrogen acceptor that is chemically similar to NAD^+ but has an extra phosphate group. Unlike NADH, the reduced form of $NADP^+$, abbreviated **NADPH,** is not involved in ATP synthesis. Instead, the electrons of NADPH are used more directly to provide energy for certain reactions, including certain essential reactions of photosynthesis (see Chapter 8).

Other important hydrogen acceptors or electron acceptors are FAD and the cytochromes. **Flavin adenine dinucleotide (FAD)** is a nucleotide that accepts hydrogen atoms and their electrons; its reduced form is **$FADH_2$.** The **cytochromes** are proteins that contain iron; the iron component accepts electrons from hydrogen atoms and then transfers these electrons to some other compound. Like NAD^+ and $NADP^+$, FAD and the cytochromes are electron transfer agents. Each exists in a *reduced state,* in which it has more free energy, or in an *oxidized state,* in which it has less. Each is an essential component of many redox reaction sequences in cells.

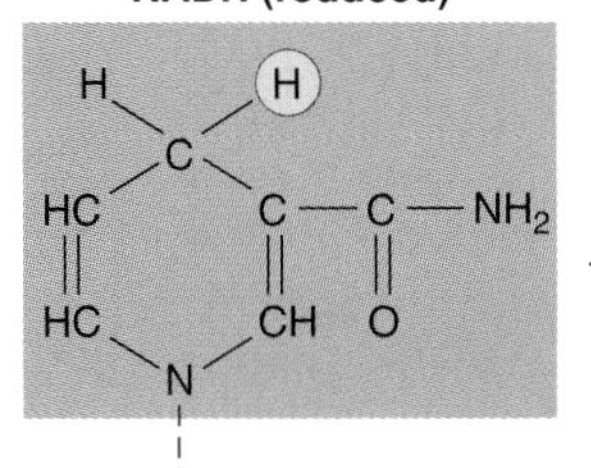

FIGURE **6-7** NAD$^+$.

NAD$^+$ consists of two nucleotides, one with adenine and one with nicotinamide, that are joined at their phosphate groups. The oxidized form (NAD$^+$, *purple screen at top*) becomes reduced (NADH, *pink screen*) by the transfer of 2 electrons and 1 proton from another organic compound (XH_2), which becomes oxidized (to X) in the process.

Review

- Which has the most energy, the oxidized form of a substance, or its reduced form? Why?

BiologyNow™ Assess your understanding of **energy transfer in redox reactions** by taking the pretest on your BiologyNow CD-ROM.

ENZYMES

Learning Objectives

9 Explain how an enzyme lowers the required energy of activation for a reaction.

10 Describe specific ways enzymes are regulated.

The principles of thermodynamics help us predict whether a reaction can occur, but they tell us nothing about the speed of the reaction. The breakdown of glucose, for example, is an exergonic reaction, yet a glucose solution stays unchanged virtually indefinitely in a bottle if it is kept free of bacteria and molds and not subjected to high temperatures or strong acids or bases. Cells cannot wait for centuries for glucose to break down, nor can they use extreme conditions to cleave glucose molecules. Cells regulate the rates of chemical reactions with **enzymes,** which are biological **catalysts** that increase the speed of a chemical reaction without being consumed by the reaction. Although most enzymes are proteins, scientists have learned that some types of RNA molecules have catalytic activity as well (see Chapter 12).

Cells require a steady release of energy, and they must regulate that release to meet metabolic energy requirements. Metabolism generally proceeds by a series of steps such that a molecule may go through as many as 20 or 30 chemical transformations before it reaches some final state. Even then, the seemingly completed molecule may enter yet another chemical pathway and become totally transformed or consumed to release energy. The changing needs of the cell require a system of flexible metabolic control. The key directors of this control system are enzymes.

The catalytic ability of some enzymes is truly impressive. For example, hydrogen peroxide (H_2O_2) breaks down extremely slowly if the reaction is uncatalyzed, but a single molecule of the enzyme **catalase** brings about the decomposition of 40 million molecules of hydrogen peroxide per second! Catalase has the highest catalytic rate known for any enzyme. It protects cells by destroying hydrogen peroxide, a poisonous substance produced as a byproduct of some cell reactions. The bombardier beetle uses the enzyme catalase as a defense mechanism (Fig. 6-8).

Thomas Eisner and Daniel Aneshansley/Cornell University

FIGURE **6-8** **Catalase as a defense mechanism.**

When threatened, a bombardier beetle (*Stenaptinus insignis*) uses the enzyme catalase to decompose hydrogen peroxide. The oxygen gas formed in the decomposition ejects water and other chemicals with explosive force. Because the reaction releases a great deal of heat, the water comes out as steam. (A wire attached by a drop of adhesive to the beetle's back immobilizes it. The researcher prodded its leg with the dissecting needle on the left to trigger the ejection.)

All reactions have a required energy of activation

All reactions, whether exergonic or endergonic, have an energy barrier known as the **energy of activation (E_A)**, or **activation energy,** which is the energy required to break the existing bonds and begin the reaction. In a population of molecules of any kind, some have a relatively high kinetic energy, whereas others have a lower energy content. Only molecules with a relatively high kinetic energy are likely to react to form the product.

Even a strongly exergonic reaction, one that releases a substantial quantity of energy as it proceeds, may be prevented from proceeding by the activation energy required to begin the reaction. For example, molecular hydrogen and molecular oxygen can react explosively to form water:

$$2\ H_2 + O_2 \longrightarrow 2\ H_2O$$

This reaction is spontaneous, yet hydrogen and oxygen can be safely mixed as long as all sparks are kept away. This is because the required activation energy for this particular reaction is relatively high. A tiny spark provides the activation energy that allows a few molecules to react. Their reaction liberates so much heat that the rest react, producing an explosion. Such an explosion occurred on the space shuttle *Challenger* on January 28, 1986 (Fig. 6-9). The failure of a rubber O-ring to properly seal caused the liquid hydrogen in the tank attached to the shuttle to leak and start burning. When the hydrogen tank ruptured a few seconds later, the resulting force burst the nearby oxygen tank as well, mixing hydrogen and oxygen and igniting a huge explosion.

FIGURE **6-9** | **The space shuttle *Challenger* explosion.**
This disaster resulted from an explosive exergonic reaction between hydrogen and oxygen. All seven crew members died in the accident on January 28, 1986.

An enzyme lowers a reaction's activation energy

Like all catalysts, enzymes affect the rate of a reaction by lowering the activation energy (E_A) necessary to initiate a chemical reaction (Fig. 6-10). If molecules need less energy to react because the activation barrier is lowered, a larger fraction of the reactant molecules reacts at any one time. As a result, the reaction proceeds more quickly.

Although an enzyme lowers the activation energy for a reaction, it has no effect on the overall free energy change; that is, an enzyme can only promote a chemical reaction that could proceed without it. If the reaction goes to equilibrium, no catalyst can cause a reaction to proceed in a thermodynamically unfavorable direction, or can influence the final concentrations of reactants and products. Enzymes simply speed up reaction rates.

An enzyme works by forming an enzyme-substrate complex

An uncatalyzed reaction depends on random collisions among reactants. Because of its ordered structure, an enzyme reduces this reliance on random events and thereby controls the reaction. The enzyme accomplishes this by forming an unstable intermediate complex with the **substrate,** the substance on which it acts. When the **enzyme-substrate complex,** or **ES complex,**

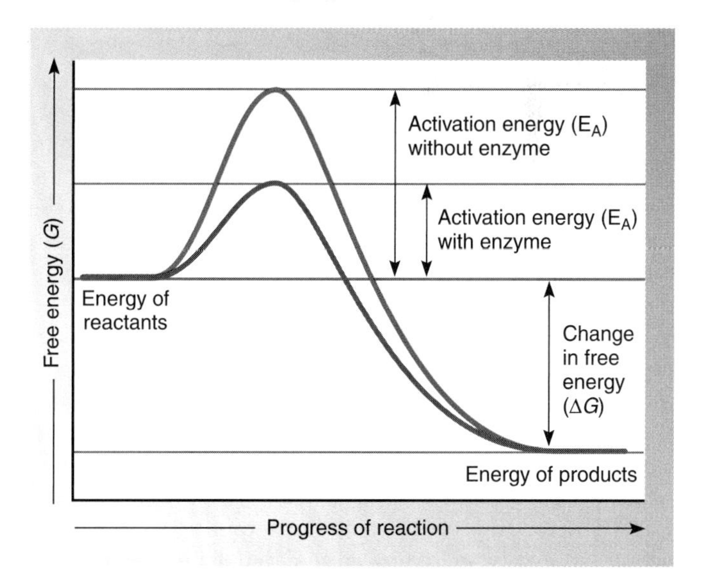

ACTIVE FIGURE **6-10** | **Activation energy and enzymes.**
An enzyme speeds up a reaction by lowering its activation energy (E_a). In the presence of an enzyme, reacting molecules require less kinetic energy to complete a reaction.

BiologyNow™ Learn more about **activation energy** by clicking on this figure on your BiologyNow CD-ROM.

FIGURE 6-11

An enzyme-substrate complex.

This computer graphic model shows the enzyme hexokinase (*blue*) and its substrate, glucose (*red*). **(a)** Prior to forming an ES complex, the enzyme's active site is the furrow where glucose will bind. **(b)** The binding of glucose to the active site induces a changes in the enzyme's active site.

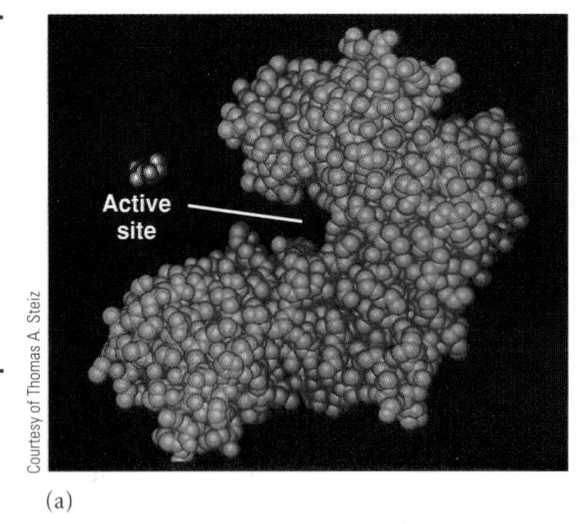

(a)

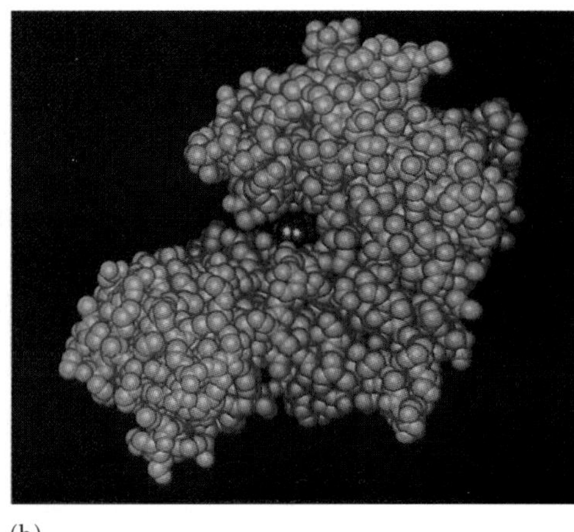

(b)

breaks up, the product is released; the original enzyme molecule is regenerated and is free to form a new ES complex:

$$\text{Enzyme} + \text{substrate(s)} \longrightarrow \text{ES complex}$$
$$\text{ES complex} \longrightarrow \text{enzyme} + \text{product(s)}$$

The enzyme itself is not permanently altered or consumed by the reaction and can be reused.

As shown in Figure 6-11a, every enzyme contains one or more **active sites,** regions to which the substrate binds, forming the ES complex. The active sites of some enzymes are grooves or cavities in the enzyme molecule, formed by amino acid side chains. The active sites of most enzymes are located close to the surface. During the course of a reaction, substrate molecules occupying these sites are brought close together and react with one another.

The shape of the enzyme does not seem exactly complementary to that of the substrate. When the substrate binds to the enzyme molecule, it causes a change, known as **induced fit,** in the shape of the enzyme (Fig. 6-11b). Usually the shape of the substrate also changes slightly, in a way that may distort its chemical bonds. The proximity and orientation of the reactants, together with strains in their chemical bonds, facilitate the breakage of old bonds and the formation of new ones. Thus the substrate is changed into a product, which moves away from the enzyme. The enzyme is then free to catalyze the reaction of more substrate molecules to form more product molecules.

Scientists usually name enzymes by adding the suffix *-ase* to the name of the substrate. The enzyme sucrase, for example, splits sucrose into glucose and fructose. A few enzymes retain traditional names that do not end in *-ase;* some of these end in *-zyme.* For example, lysozyme (from the Greek *lysis,* "a loosening") is an enzyme found in tears and saliva; it breaks down bacterial cell walls. Other examples of enzymes with traditional names are pepsin and trypsin, which break peptide bonds in proteins.

Enzymes are specific

Enzymes catalyze virtually every chemical reaction that takes place in an organism. Because the shape of the active site is closely related to the shape of the substrate, most enzymes are highly specific. Most catalyze only a few closely related chemical reactions or, in many cases, only one particular reaction. For example, the enzyme urease, which decomposes urea to ammonia and carbon dioxide, attacks no other substrate. The enzyme sucrase splits only sucrose; it does not act on other disaccharides, such as maltose or lactose. A few enzymes are specific only to the extent that they require the substrate to have a certain kind of chemical bond. For example, lipase, secreted by the pancreas, splits the ester linkages connecting the glycerol and fatty acids of a wide variety of fats.

Scientists classify into groups enzymes that catalyze similar reactions, although each particular enzyme in the group may catalyze only one specific reaction. Table 6-1 describes the six classes of enzymes that biologists recognize. Each class is divided into many subclasses. For example, sucrase, mentioned earlier, is called a glycosidase, because it cleaves a glycosidic linkage (see Chapter 3). Glycosidases are a subclass of the hydrolases.

Many enzymes require cofactors

Some enzymes consist only of protein. For example, the enzyme pepsin, which is secreted by the animal stomach and digests dietary protein by breaking certain peptide bonds, is exclusively a protein molecule. Other enzymes have two components: a protein called the *apoenzyme* and an additional chemical component called a **cofactor.** Neither the apoenzyme nor the cofactor alone has catalytic activity; only when the two are combined

TABLE 6-1 **Important Classes of Enzymes**

Enzyme Class	Function
Oxidoreductases	Catalyze oxidation-reduction reactions
Transferases	Catalyze the transfer of a functional group from a donor molecule to an acceptor molecule
Hydrolases	Catalyze hydrolysis reactions
Isomerases	Catalyze conversion of a molecule from one isomeric form to another
Ligases	Catalyze certain reactions in which two molecules join in a process coupled to the hydrolysis of ATP
Lyases	Catalyze certain reactions in which double bonds form or break

does the enzyme function. A cofactor may be inorganic, or it may be an organic molecule.

Some enzymes require a specific metal ion as a cofactor. Two very common inorganic cofactors are magnesium ions and calcium ions. Most of the trace elements, such as iron, copper, zinc, and manganese, all of which organisms require in very small amounts, function as cofactors.

An organic, nonpolypeptide compound that binds to the apoenzyme and serves as a cofactor is called a **coenzyme.** Most coenzymes are carrier molecules that transfer electrons or part of a substrate from one molecule to another. We have already introduced some examples of coenzymes in this chapter. NADH, NADPH, and $FADH_2$ are coenzymes; they transfer electrons. ATP functions as a coenzyme; it is responsible for transferring phosphate groups. Yet another coenzyme, **coenzyme A,** is involved in the transfer of groups derived from organic acids. Most vitamins, which are organic compounds that an organism requires in small amounts but cannot synthesize itself, are coenzymes or components of coenzymes (see Table 45-3).

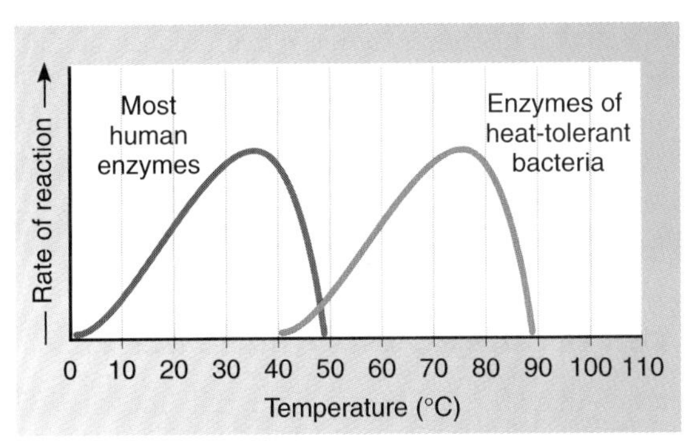

(a)

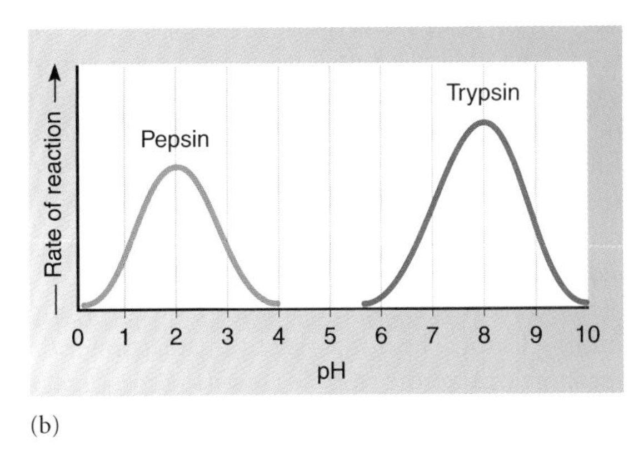

(b)

FIGURE **6-12** **The effect of temperature and pH on enzyme activity.**

Substrate and enzyme concentrations are held constant in the reactions illustrated. **(a)** Generalized curves for the effect of temperature on enzyme activity. As temperature increases, enzyme activity increases until it reaches an optimal temperature. Enzyme activity abruptly falls after it exceeds the optimal temperature because the enzyme, being a protein, denatures. **(b)** Enzyme activity is very sensitive to pH. Pepsin is a protein-digesting enzyme in the very acidic stomach juice. Trypsin, secreted by the pancreas into the slightly basic small intestine, digests polypeptides.

Enzymes are most effective at optimal conditions

Enzymes generally work best under certain narrowly defined conditions, such as appropriate temperature, pH, and ion concentration. Any departure from optimal conditions adversely affects enzyme activity.

Each enzyme has an optimal temperature

Most enzymes have an optimal temperature, at which the rate of reaction is fastest. For human enzymes, the temperature optima are near the human body temperature (35° to 40°C). Enzymatic reactions occur slowly or not at all at low temperatures. As the temperature increases, molecular motion increases, resulting in more molecular collisions. The rates of most enzyme-controlled reactions therefore increase as the temperature increases, within limits (Fig. 6-12a). High temperatures rapidly denature most enzymes. The molecular conformation (3-D shape) of the protein becomes altered as the hydrogen bonds responsible for its secondary, tertiary, and quaternary structures are broken. Because this inactivation is usually not reversible, activity is not regained when the enzyme is cooled.

Most organisms are killed by even a short exposure to high temperature; their enzymes are denatured, and they are unable to continue metabolism. There are a few stunning exceptions to this rule. Certain species of archaea (see Chapter 1) can survive in the waters of hot springs, such as those in Yellowstone Park, where the temperature is almost 100°C; these organisms are responsible for the brilliant colors in the terraces of the hot springs (Fig. 6-13). Still other archaea live at temperatures much above that of boiling water, near deep-sea vents, where the

From R.B. Smith, and L.J. Siegel: *Windows into the Earth: The Geologic Story of Yellowstone and Grand Teton Parks.* Oxford University Press, Oxford, 2000.

FIGURE **6-13** **Yellowstone National Park's Grand Prismatic Spring.**

The world's third largest spring, about 61 m (200 ft) in diameter, the Grand Prismatic Spring teems with heat-tolerant bacteria. The rings around the perimeter, where the water is slightly cooler, get their distinctive colors from the various kinds of bacteria living there.

extreme pressure keeps water in its liquid state (see Chapter 23; see also Chapter 53, *Focus On: Life Without the Sun*).

Each enzyme has an optimal pH

Most enzymes are active only over a narrow pH range and have an optimal pH, at which the rate of reaction is fastest. The optimal pH for most human enzymes is between 6 and 8. (Recall from Chapter 2 that buffers minimize pH changes in cells so that the pH is maintained within a narrow limit.) Pepsin, a protein-digesting enzyme secreted by cells lining the stomach, is an exception; it works only in a very acidic medium, optimally at pH 2 (Fig. 6-12b). In contrast, trypsin, a protein-splitting enzyme secreted by the pancreas, functions best under the slightly basic conditions found in the small intestine.

The activity of an enzyme may be markedly changed by any alteration in pH, which in turn alters electrical charges on the enzyme. Changes in charge affect the ionic bonds that contribute to tertiary and quaternary structure, thereby changing the protein's conformation and activity. Many enzymes become inactive, and usually irreversibly denatured, when the medium is made very acidic or very basic.

Enzymes are organized into teams in metabolic pathways

Enzymes play an essential role in reaction coupling because they usually work in sequence, with the product of one enzyme-controlled reaction serving as the substrate for the next. You can picture the inside of a cell as a factory with many different assembly (and disassembly) lines operating simultaneously. An assembly line consists of a number of enzymes. Each enzyme carries out one step, such as changing molecule A into molecule B. Then molecule B is passed along to the next enzyme, which converts it into molecule C, and so on. Such a series of reactions is called a **metabolic pathway.**

$$A \xrightarrow{\text{Enzyme 1}} B \xrightarrow{\text{Enzyme 2}} C$$

Theoretically, each of these reactions is reversible, and that an enzyme catalyzes it does not change that fact. An enzyme does not itself determine the direction of the reaction it catalyzes. However, the overall reaction sequence is portrayed as proceeding from left to right. Recall that if there is little intrinsic free energy difference between the reactants and products for a particular reaction, its direction is determined mainly by the relative concentrations of reactants and products.

In biological pathways, both intermediate and final products are often removed and converted to other chemical compounds. Such removal drives the sequence of reactions in a particular direction. Let's assume that reactant A is continually supplied and that its concentration remains constant. Enzyme 1 converts reactant A to product B. The concentration of B is always lower than the concentration of A, because B is removed as it is converted to C in the reaction catalyzed by enzyme 2. If C is removed as quickly as it is formed (perhaps by leaving the cell), the entire reaction pathway is "pulled" toward C.

The cell regulates enzymatic activity

Enzymes regulate the chemistry of the cell, but what controls the enzymes? One regulatory mechanism depends simply on controlling the amount of enzyme produced. A specific gene directs the synthesis of each type of enzyme. The gene, in turn, may be switched on by a signal from a hormone or by some other type of cell product. When the gene is switched on, the enzyme is synthesized. The total amount of enzyme present then influences the overall cell reaction rate.

If the pH and temperature are kept constant (as they are in most cells), the rate of the reaction can be affected by the substrate concentration or by the enzyme concentration. If an excess of substrate is present, the enzyme concentration is the rate-limiting factor. The initial rate of the reaction is then directly proportional to the enzyme concentration (Fig. 6-14a).

If the enzyme concentration is kept constant, the rate of an enzymatic reaction is proportional to the concentration of substrate present. Substrate concentration is the rate-limiting factor at lower concentrations; the rate of the reaction is therefore directly proportional to the substrate concentration. However, at higher substrate concentrations, the enzyme molecules become saturated with substrate; that is, substrate molecules are bound to all available active sites of enzyme molecules. In this situation, increasing the substrate concentration does not increase the net reaction rate (Fig. 6-14b).

The product of one enzymatic reaction may control the activity of another enzyme, especially in a com-

FIGURE **6-14**

The effect of enzyme concentration and substrate concentration on the rate of a reaction.

(a) In this example, the rate of reaction is measured at different enzyme concentrations, with an excess of substrate present. (Temperature and pH are constant.) The rate of the reaction is directly proportional to the enzyme concentration. **(b)** In this example, the rate of the reaction is measured at different substrate concentrations, and enzyme concentration, temperature, and pH are constant. If the substrate concentration is relatively low, the reaction rate is directly proportional to substrate concentration. However, higher substrate concentrations do not increase the reaction rate, because the enzymes become saturated with substrate.

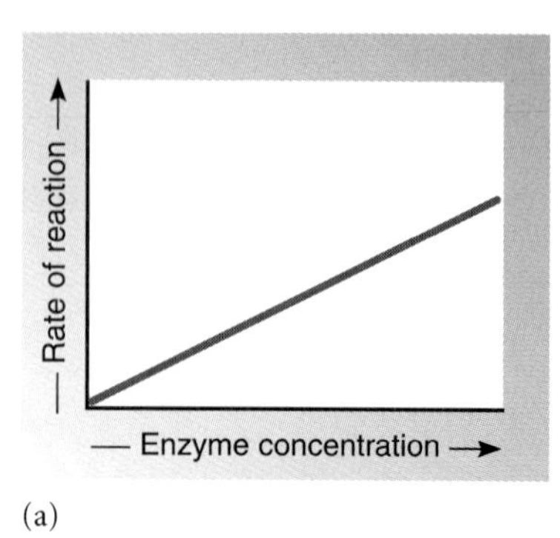

(a)

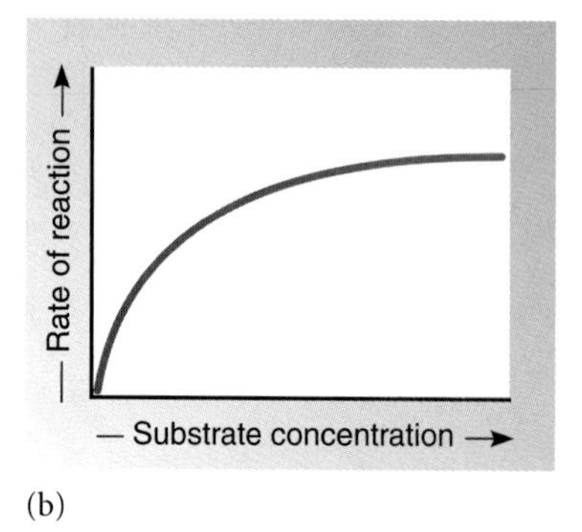

(b)

plex sequence of enzymatic reactions. For example, consider the following metabolic pathway:

$$A \xrightarrow{\text{Enzyme 1}} B \xrightarrow{\text{Enzyme 2}} C \xrightarrow{\text{Enzyme 3}} D \xrightarrow{\text{Enzyme 4}} E$$

A different enzyme catalyzes each step, and the final product E may inhibit the activity of enzyme 1. When the concentration of E is low, the sequence of reactions proceeds rapidly. However, an increasing concentration of E serves as a signal for enzyme 1 to slow down and eventually to stop functioning. Inhibition of enzyme 1 stops the entire reaction sequence. This type of enzyme regulation, in which the formation of a product inhibits an earlier reaction in the sequence, is called **feedback inhibition** (Fig. 6-15).

Another important method of enzymatic control focuses on the activation of enzyme molecules. In their inactive form, the active sites of the enzyme are inappropriately shaped, so that the substrates do not fit. Among the factors that influence the shape of the enzyme are pH, the concentration of certain ions, and the addition of phosphate groups to certain amino acids in the enzyme.

Some enzymes have a receptor site, called an **allosteric site,** on some region of the enzyme molecule other than the active site. (The word *allosteric* means "another space.") When a substance binds to an enzyme's allosteric site, the conformation of the enzyme's active site changes, thereby modifying the enzyme's activity. Substances that affect enzyme activity by binding to allosteric sites are called **allosteric regulators.** Some allosteric regulators are allosteric inhibitors that keep the enzyme in its inactive shape. Conversely, the activities of allosteric activators result in an enzyme with a functional active site.

The enzyme *cyclic AMP-dependent protein kinase* is an allosteric enzyme regulated by a protein that binds reversibly to the allosteric site and inactivates the enzyme. Protein kinase is in this inactive form most of the time (Fig. 6-16). When protein kinase activity is needed, the compound cyclic AMP (cAMP; see Fig. 3-25) contacts the enzyme-inhibitor complex and removes the inhibitory protein, thereby activating the protein kinase. Activation of protein kinases by cAMP is an important aspect of the mechanism of action of certain hormones (see Chapters 5 and 47).

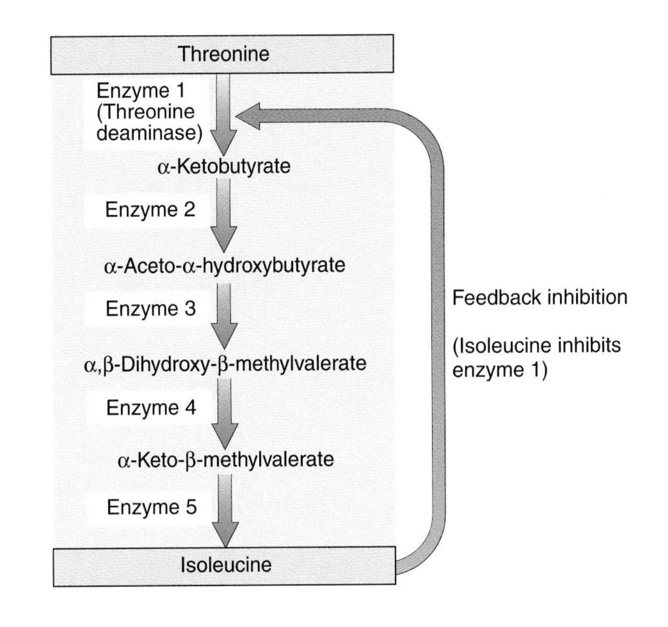

ACTIVE FIGURE **6-15** **Feedback inhibition.**

Bacteria synthesize the amino acid isoleucine from the amino acid threonine. The isoleucine pathway involves five steps, each catalyzed by a different enzyme. When enough isoleucine accumulates in the cell, the isoleucine inhibits threonine deaminase, the enzyme that catalyzes the first step in this pathway.

BiologyNow™ Learn more about **feedback inhibition** by clicking on this figure on your BiologyNow CD-ROM.

Enzymes are inhibited by certain chemical agents

Most enzymes are inhibited or even destroyed by certain chemical agents. Enzyme inhibition may be reversible or irreversible. **Reversible inhibition** occurs when an inhibitor forms weak chemical bonds with the enzyme. Reversible inhibition can be competitive or noncompetitive.

In **competitive inhibition,** the inhibitor competes with the normal substrate for binding to the active site of the enzyme (Fig. 6-17a). Usually a competitive inhibitor is structurally similar to the normal substrate and fits into the active site and combines with the enzyme. However, it is not similar enough to substitute fully for the normal substrate in the chemical reaction, and the enzyme cannot con-

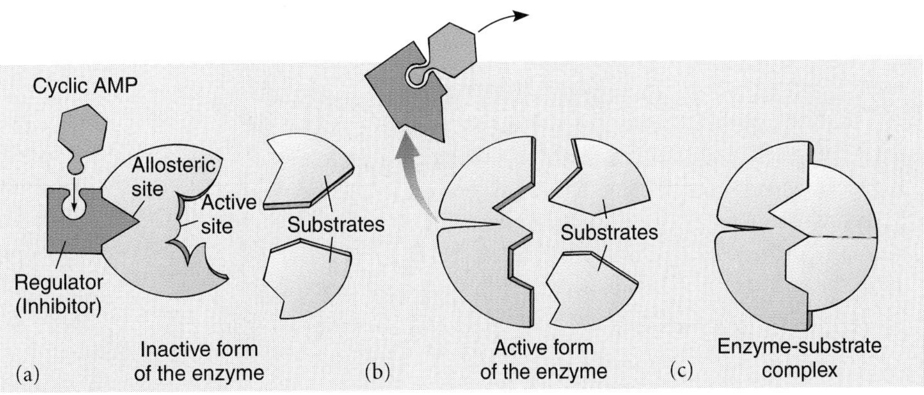

ACTIVE FIGURE **6-16** **An allosteric enzyme**

(a) The enzyme protein kinase is inhibited by a regulatory protein that binds reversibly to its allosteric site. When the enzyme is in this inactive form, the shape of the active site is modified so that the substrate cannot combine with it. **(b)** Cyclic AMP removes the allosteric inhibitor and activates the enzyme. **(c)** The substrate can then combine with the active site.

BiologyNow™ Learn more about **allosteric enzymes** by clicking on this figure on your BiologyNow CD-ROM.

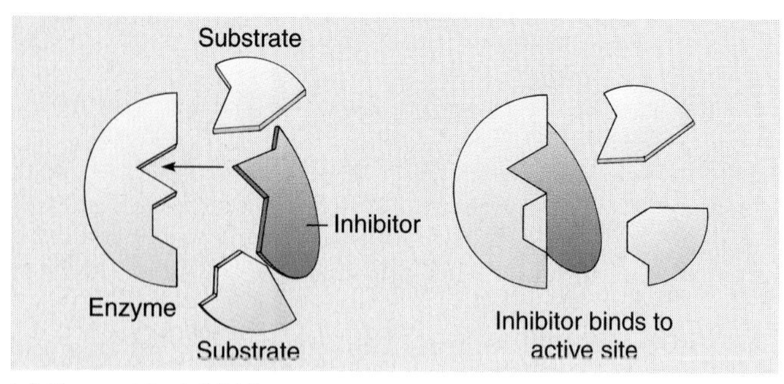

(a) Competitive inhibition

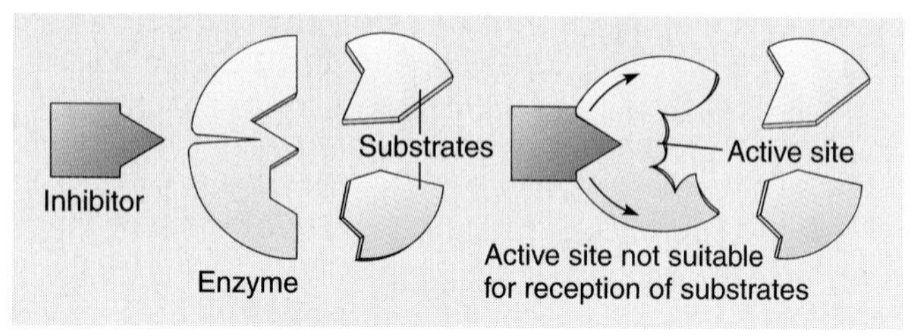

(b) Noncompetitive inhibition

FIGURE **6-17** **Competitive and noncompetitive inhibition.**

(a) In competitive inhibition, the inhibitor competes with the normal substrate for the active site of the enzyme. A competitive inhibitor occupies the active site only temporarily. **(b)** In noncompetitive inhibition, the inhibitor binds with the enzyme at a site other than the active site, altering the shape of the enzyme and thereby inactivating it.

vert it to product molecules. A competitive inhibitor occupies the active site only temporarily and does not permanently damage the enzyme. In competitive inhibition, an active site is occupied by the inhibitor part of the time and by the normal substrate part of the time. If the concentration of the substrate is increased relative to the concentration of the inhibitor, the active site is usually occupied by the substrate. Scientists demonstrate competitive inhibition experimentally by showing that increasing the substrate concentration reverses competitive inhibition.

In **noncompetitive inhibition,** the inhibitor binds with the enzyme at a site other than the active site (Fig. 6-17b). Such an inhibitor inactivates the enzyme by altering its shape so that the active site cannot bind with the substrate. Many important noncompetitive inhibitors are metabolic substances that regulate enzyme activity by combining reversibly with the enzyme. Noncompetitive inhibition has some features in common with allosteric inhibition.

In **irreversible inhibition,** an inhibitor permanently inactivates or destroys an enzyme when it combines with one of its functional groups, either at the active site or elsewhere. Many poisons are irreversible enzyme inhibitors. For example, heavy metals such as mercury and lead bind irreversibly to and denature many proteins, including enzymes. Certain nerve gases poison the enzyme acetylcholinesterase, which is important for the functioning of nerves and muscles. Cytochrome oxidase, one of the enzymes that transports electrons in cellular respiration, is especially sensitive to cyanide. Death results from cyanide

H_2N–C$_6$H$_4$–C(=O)OH

Para-aminobenzoic acid (PABA)

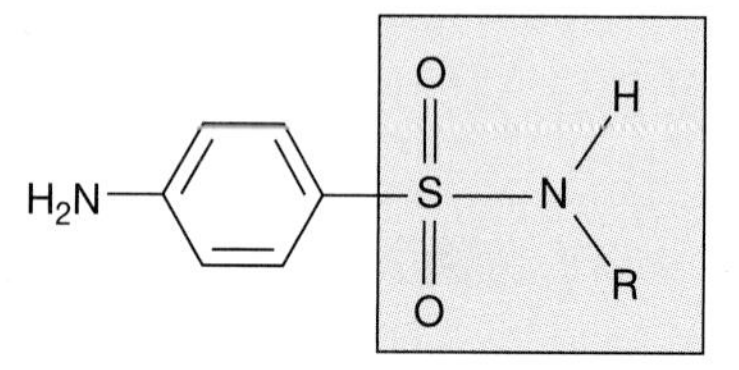

Generic sulfonamide (Sulfa drug)

FIGURE **6-18**

Para-aminobenzoic acid and sulfonamides.

Sulfa drugs inhibit an enzyme in bacteria necessary for the synthesis of folic acid, an important vitamin required for growth. (Note the unusual structure of the sulfonamide molecule, in which sulfur, which commonly forms two covalent bonds, forms six instead.)

poisoning because cytochrome oxidase is irreversibly inhibited and no longer transfers electrons from its substrate to oxygen.

Some drugs are enzyme inhibitors

Physicians treat many bacterial infections with drugs that directly or indirectly inhibit bacterial enzyme activity. For example, sulfa drugs have a chemical structure similar to that of the nutrient *para-aminobenzoic acid (PABA)* (Fig. 6-18). When PABA is available, microorganisms can synthesize the vitamin *folic acid,* which is necessary for growth. Humans do not synthesize folic acid from PABA, and that's why sulfa drugs selectively affect bacteria. When a sulfa drug is present, the drug competes with PABA for the active site of the bacterial enzyme. When bacteria use the sulfa drug instead of PABA, they synthesize a compound that cannot be used to make folic acid. Therefore, the bacterial cells are unable to grow.

Penicillin and related antibiotics irreversibly inhibit a bacterial enzyme called *transpeptidase.* This enzyme establishes some of the chemical linkages in the bacterial cell wall. Susceptible bacteria cannot produce properly constructed cell walls and are prevented from multiplying effectively. Human cells do not have cell walls and therefore do not use this enzyme. Thus, except for individuals allergic to it, penicillin is harmless to humans. Unfortunately, during the years since it was introduced, many bacterial strains have evolved resistance to penicillin. The resistant bacteria fight back with an enzyme of their own, penicillinase, which breaks down the penicillin and renders it ineffective. Because bacteria evolve at such a rapid rate, drug resistance is a growing problem in medical practice.

Review

- What effect does an enzyme have on the required activation energy of a reaction?
- How does the function of the active site of an enzyme differ from that of an allosteric site?
- How are temperature and pH optima of an enzyme related to its structure and function?
- Is allosteric inhibition competitive or noncompetitive?

BiologyNow™ Assess your understanding of **enzymes** by taking the pretest on your BiologyNow CD-ROM.

SUMMARY WITH KEY TERMS

1 Define *energy,* emphasizing how it is related to work and to heat.

- **Energy** is the capacity to do work (expressed in **kilojoules, kJ**).
- Energy can be conveniently measured as **heat energy,** thermal energy that flows from an object with a higher temperature to an object with a lower temperature; the unit of heat energy is the **kilocalorie (kcal),** which is equal to 4.184 kilojoules. Heat energy cannot do cell work.

2 Use examples to contrast potential energy and kinetic energy.

- **Potential energy** is stored energy; **kinetic energy** is energy of motion.
- All life depends on a continuous input of energy. All forms of energy are interconvertible. For example, photosynthetic organisms capture radiant energy and convert some of it to **chemical energy,** a form of potential energy that powers many life processes.

3 State the first and second laws of thermodynamics, and discuss the implications of these laws as they relate to organisms.

- A **closed system** does not exchange energy with its surroundings. Organisms are **open systems.**
- The **first law of thermodynamics** states that energy cannot be created or destroyed but can be transferred and changed in form. The first law explains why organisms cannot produce energy, but as open systems they continuously capture it from the surroundings.
- The **second law of thermodynamics** states that disorder (entropy) in the universe, a closed system, is continuously increasing. No energy transfer is 100% efficient; some energy is dissipated as **heat,** random motion that contributes to **entropy** or disorder. Organisms maintain their ordered states at the expense of their surroundings.

4 Discuss how changes in free energy in a reaction are related to changes in entropy and enthalpy.

- As entropy increases, the amount of **free energy** decreases, as shown in the equation $G = H - TS$, in which G is the free energy, H is the **enthalpy** (total potential energy of the system), T is the absolute temperature (expressed in degrees Kelvin), and S is entropy.
- The equation $\Delta G = \Delta H - TS$ indicates that the change in free energy (ΔG) during a chemical reaction is equal to the change in enthalpy (ΔH) minus the product of the absolute temperature (T) multiplied by the change in entropy (ΔS).

5 Distinguish between exergonic and endergonic reactions, and give examples of how they may be coupled.

- An **exergonic reaction** has a negative value of ΔG; that is, free energy decreases. Such a reaction is spontaneous; it released free energy that can perform work.
- Free energy increases in an **endergonic reaction.** Such a reaction has a positive value of ΔG, and is nonspontaneous. In a **coupled reaction,** the input of free energy required to drive an endergonic reaction is supplied by an exergonic reaction.

6 Compare the energy dynamics of a reaction at equilibrium with the dynamics of a reaction not at equilibrium.

- When a chemical reaction is in a state of **dynamic equilibrium,** the rate of change in one direction is exactly the same as the rate of change in the opposite direction; the system can do no work because the free energy difference between the reactants and products is zero.
- When the concentration of reactant molecules is increased, the reaction shifts to the right, and more product molecules are formed until equilibrium is re-established.

7 Explain how the chemical structure of ATP allows it to transfer a phosphate group. Discuss the central role of ATP in the overall energy metabolism of the cell.

- **Adenosine triphosphate, ATP,** is the immediate energy currency of the cell. It donates energy by means of its terminal phosphate group, which is easily transferred to an acceptor molecule. ATP is formed by the **phosphorylation** of **adenosine diphosphate, ADP,** an endergonic process that requires an input of energy.
- ATP is the common link between exergonic and endergonic reactions and between **catabolism** (degradation of large complex molecules into smaller, simpler molecules) and **anabolism** (synthesis of complex molecules from simpler molecules)

8 Relate the transfer of electrons (or hydrogen atoms) to the transfer of energy.

- Energy is transferred in **oxidation-reduction (redox) reactions.** A substance becomes oxidized as it gives up one or more electrons to a substance that has become reduced. Electrons are typically transferred as part of hydrogen atoms.
- NAD^+ and $NADP^+$ accept electrons as part of hydrogen atoms and become reduced to form **NADH** and **NADPH,** respectively. These electrons (along with some of their energy) can be transferred to other acceptors.

9 Explain how an enzyme lowers the required energy of activation for a reaction.

- An **enzyme** is a biological **catalyst;** it greatly increases the speed of a chemical reaction without being consumed.
- An enzyme works by lowering the **activation energy,** the kinetic energy necessary to get a reaction going. The **active site** of an enzyme is a 3-D region where **substrates** come into close contact and thereby react more readily. When a substrate binds to an active site, an **enzyme-substrate complex** forms in which the shapes of the enzyme and substrate change slightly. This **induced fit** facilitates the breaking of bonds and formation of new ones.

10 Describe specific ways enzymes are regulated.

- Enzymes work best at specific temperature and pH conditions.
- A cell can regulate enzymatic activity by controlling the amount of enzyme produced and by regulating metabolic conditions that influence the shape of the enzyme.
- Some enzymes have **allosteric sites,** noncatalytic sites to which an **allosteric regulator** binds, changing the enzyme's activity. Some allosteric enzymes are subject to **feedback inhibition,** in which the formation of an end product inhibits an earlier reaction in the **metabolic pathway.**
- **Reversible inhibition** occurs when an inhibitor forms weak chemical bonds with the enzyme. Reversible inhibition may be **competitive,** in which the inhibitor competes with the substrate for the active site, or **noncompetitive,** in which the inhibitor binds with the enzyme at a site other than the active site. **Irreversible** inhibition occurs when an inhibitor combines with an enzyme and permanently inactivates it.

POST-TEST

1. According to the first law of thermodynamics (a) energy may be changed from one form to another but is neither created nor destroyed (b) much of the work an organism does is mechanical work (c) the disorder of the universe is increasing (d) free energy is available to do cell work (e) a cell is in a state of dynamic equilibrium
2. According to the second law of thermodynamics (a) energy may be changed from one form to another but is neither created nor destroyed (b) much of the work an organism does is mechanical work (c) the disorder of the universe is increasing (d) free energy is available to do cell work (e) a cell is in a state of dynamic equilibrium
3. In thermodynamics, ______________ is a measure of the amount of disorder in the system. (a) bond energy (b) catabolism (c) entropy (d) enthalpy (e) work
4. The ______________ energy of a system is that part of the total energy available to do cell work. (a) activation (b) bond (c) kinetic (d) free (e) heat
5. A reaction that requires a net input of free energy is described as (a) exergonic (b) endergonic (c) spontaneous (d) both a and c (e) both b and c
6. A reaction that releases energy is described as (a) exergonic (b) endergonic (c) spontaneous (d) both a and c (e) both b and c
7. A spontaneous reaction is one in which the change in free energy (ΔG) has a ______________ value. (a) positive (b) negative (c) positive or negative (d) none of these (ΔG has no measurable value)
8. To drive a reaction that requires an input of energy (a) an enzyme-substrate complex must form (b) the concentration of ATP must be decreased (c) the activation energy must be increased (d) some reaction that yields energy must be coupled to it (e) some reaction that requires energy must be coupled to it
9. Which of the following reactions could be coupled to an endergonic reaction with $\Delta G = +3.56$ kJ/mol? (a) $A \longrightarrow B$, $\Delta G = +6.08$ kJ/mol (b) $C \longrightarrow D$, $\Delta G = +3.56$ kJ/mol (c) $E \longrightarrow F$, $\Delta G = 0$ kJ/mol (d) $G \longrightarrow H$, $\Delta G = -1.22$ kJ/mol (e) $I \longrightarrow J$, $\Delta G = -5.91$ kJ/mol
10. Consider this reaction: Glucose $+ 6\ O_2 \longrightarrow 6\ CO_2 + 6\ H_2O$ ($\Delta G = -2880$ kJ/mol). Which of the following statements about this reaction is *not* true? (a) the reaction is spontaneous in a thermodynamic sense (b) a small amount of energy (activation energy) must be supplied to start the reaction, which then proceeds with a release of energy (c) the reaction is exergonic (d) the reaction can be coupled to an endergonic reaction (e) the reaction must be coupled to an exergonic reaction
11. The kinetic energy required to initiate a reaction is called (a) activation energy (b) bond energy (c) potential energy (d) free energy (e) heat energy
12. A biological catalyst that affects the rate of a chemical reaction without being consumed by the reaction is a(an) (a) product (b) cofactor (c) coenzyme (d) substrate (e) enzyme
13. The region of an enzyme molecule that combines with the substrate is the (a) allosteric site (b) reactant (c) active site (d) coenzyme (e) product
14. Which inhibitor binds to the active site of an enzyme? (a) noncompetitive inhibitor (b) competitive inhibitor (c) irreversible inhibitor (d) allosteric regulator (e) PABA
15. In the following reaction series, which enzyme(s) is/are most likely to have an allosteric site to which the end product E binds?

$$A \xrightarrow{\text{Enzyme 1}} B \xrightarrow{\text{Enzyme 2}} C \xrightarrow{\text{Enzyme 3}} D \xrightarrow{\text{Enzyme 4}} E$$

(a) enzyme 1 (b) enzyme 2 (c) enzyme 3 (d) enzyme 4 (e) enzymes 3 and 4

CRITICAL THINKING

1. Reactions 1 and 2 happen to have the same free energy change: $\Delta G = -41.8$ kJ/mol (-10 kcal/mol). Reaction 1 is at equilibrium, but reaction 2 is far from equilibrium. Is either reaction capable of performing work? If so, which one?
2. Let's say you are performing an experiment in which you are measuring the rate at which succinate is converted to fumarate by the enzyme succinic dehydrogenase. You decide to add a little malonate to make things interesting. You observe that the reaction rate slows markedly and hypothesize that malonate is inhibiting the reaction. Design an experiment that will help you decide whether malonate is acting as a competitive inhibitor or a noncompetitive inhibitor.
3. Given what you have learned in this chapter, explain why an extremely high fever (body temperature above 105°F or 40°C) is often fatal.

- Visit our Web site **http://biology.brookscole.com/solomon7** for links to chapter-related resources on the World Wide Web. Additional online materials relating to this chapter can also be found on our Web site.

BIOLOGY NOW RESOURCES

Active Figures

6-10: Enzyme activation energy

6-17: Allosteric enzymes

Preparing for an exam? Take a diagnostic test on your BiologyNow CD-ROM.

Post Test Answers

1. a	2. c	3. c	4. d
5. b	6. d	7. b	8. d
9. e	10. e	11. a	12. e
13. c	14. b	15. a	

How Cells Make ATP: Energy-Releasing Pathways

Renee Lynn/Photo Researchers, Inc.

Female gerenuks. Gerenuks (*Litocranius walleri*) live in the dry brush country of East Africa, where they browse on leaves, fruits, and flowers of thorny trees and shrubs.

CHAPTER OUTLINE

Cells are tiny factories that process materials on the molecular level, through thousands of metabolic reactions. Cells exist in a dynamic state and are continuously building up and breaking down the many different cell constituents. As you learned in Chapter 6, metabolism has two complementary components: **catabolism,** which releases energy by splitting complex molecules into smaller components, and **anabolism,** the synthesis of complex molecules from simpler building blocks. Anabolic reactions produce proteins, nucleic acids, lipids, polysaccharides, and other complex molecules that help maintain the cell or the organism. Most anabolic reactions are endergonic and require ATP or some other energy source to drive them.

Every organism must extract energy from the organic food molecules that it either manufactures by photosynthesis or captures from the environment. The gerenuks in the photograph, for example, obtain organic molecules when they eat the leaves of thorny shrubs and trees. How do they obtain energy from these organic molecules? First the complex food molecules are broken down by digestion into simpler components that are absorbed into the blood and transported to all the cells. The catabolic processes that convert the energy in the chemical bonds of nutrients to chemical energy stored in ATP then occur inside cells, usually through a process known as **cellular respiration.** (The term *cellular respiration* is used to distinguish it from *organismic respiration,* the exchange of oxygen and carbon dioxide with the environment by animals that have special organs, such as lungs or gills, for gas exchange.)

Cellular respiration may be either aerobic or anaerobic. **Aerobic** respiration requires molecular oxygen (O_2), whereas **anaerobic** pathways, which include anaerobic respiration and fermentation, do not require oxygen. A steady supply of oxygen enables your cells to capture energy through aerobic respiration, which is by far the most common pathway and the main subject of this chapter. All three pathways—aerobic respiration, anaerobic respiration, and fermentation—are exergonic and release free energy. ■

REDOX REACTIONS

Learning Objective

1 Write a summary reaction for aerobic respiration, showing which reactant becomes oxidized and which becomes reduced.

Most eukaryotes and prokaryotes carry out **aerobic respiration,** a form of cellular respiration requiring oxygen. During aerobic respiration, nutrients are catabolized to carbon dioxide and water. Most cells use aerobic respiration to obtain energy from glucose, which enters the cell though a specific transport protein in the plasma membrane (see discussion of facilitated diffusion in Chapter 5). The overall reaction pathway for the aerobic respiration of glucose is summarized as follows:

$$C_6H_{12}O_6 + 6\ O_2 + 6\ H_2O \longrightarrow 6\ CO_2 + 12\ H_2O + \text{energy (in the chemical bonds of ATP)}$$

Note that water is shown on both sides of the equation; this is because it is a reactant in some reactions and a product in others.

For purposes of discussion, the equation for aerobic respiration can be simplified to indicate that there is a net yield of water:

$$C_6H_{12}O_6 + 6\ O_2 \rightarrow 6\ CO_2 + 6\ H_2O + \text{energy (in the chemical bonds of ATP)}$$

(Oxidation: $C_6H_{12}O_6 \rightarrow CO_2$; Reduction: $O_2 \rightarrow H_2O$)

If we analyze this summary reaction, it appears CO_2 is produced by the removal of hydrogen atoms from glucose. Conversely, water seems to be formed as oxygen accepts the hydrogen atoms. Because the transfer of hydrogen atoms is equivalent to the transfer of electrons, this is a **redox reaction** in which glucose becomes **oxidized** and oxygen becomes **reduced** (see Chapters 2 and 6).

The products of the reaction would be the same if the glucose were simply placed in a test tube and burned in the presence of oxygen. However, if a cell were to burn glucose its energy would be released all at once as heat, which not only would be unavailable to the cell but also would actually destroy it. For this reason, cells do not transfer hydrogen atoms directly from glucose to oxygen. Aerobic respiration includes a series of redox reactions in which electrons associated with the hydrogen atoms in glucose are transferred to oxygen in a series of steps (Fig. 7-1). During this process, the free energy of the electrons is coupled to ATP synthesis.

Review

- What is the specific role of oxygen in most cells?

BiologyNow™ Assess your understanding of **redox reactions** by taking the pretest on your BiologyNow CD-ROM.

THE FOUR STAGES OF AEROBIC RESPIRATION

Learning Objectives

2 List and give a brief overview of the four stages of aerobic respiration.

3 Indicate where each stage of aerobic respiration takes place in a eukaryotic cell.

4 Add up the energy captured (as ATP, NADH, and $FADH_2$) in each stage of aerobic respiration.

5 Define *chemiosmosis,* and explain how a gradient of protons is established across the inner mitochondrial membrane.

6 Describe the process by which the proton gradient drives ATP synthesis in chemiosmosis.

The chemical reactions of the aerobic respiration of glucose are grouped into four stages (Fig. 7-2, Table 7-1; see also the summary equations at the end of the chapter). In eukaryotes, the first stage (glycolysis) takes place in the cytosol, and the remaining stages take place inside mitochondria. Most bacteria and archaea also carry out these processes, but because prokaryotic cells lack mitochondria the reactions of aerobic respiration occur in the cytosol and in association with the plasma membrane.

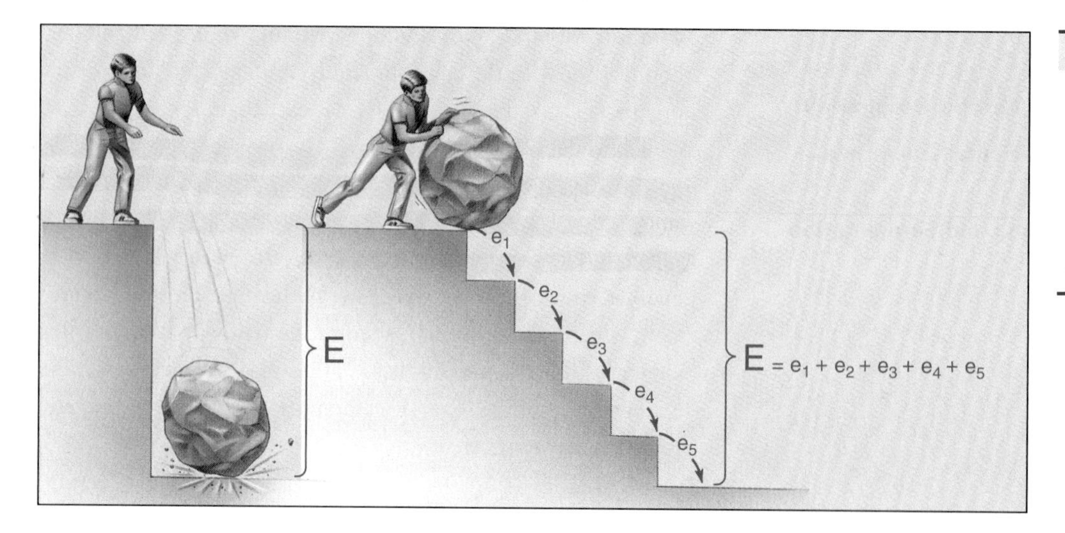

FIGURE **7-1**

Changes in free energy.

The release of energy from a glucose molecule is analogous to the liberation of energy by a falling object. The total energy released (*E*) is the same whether it occurs all at once or in a series of steps.

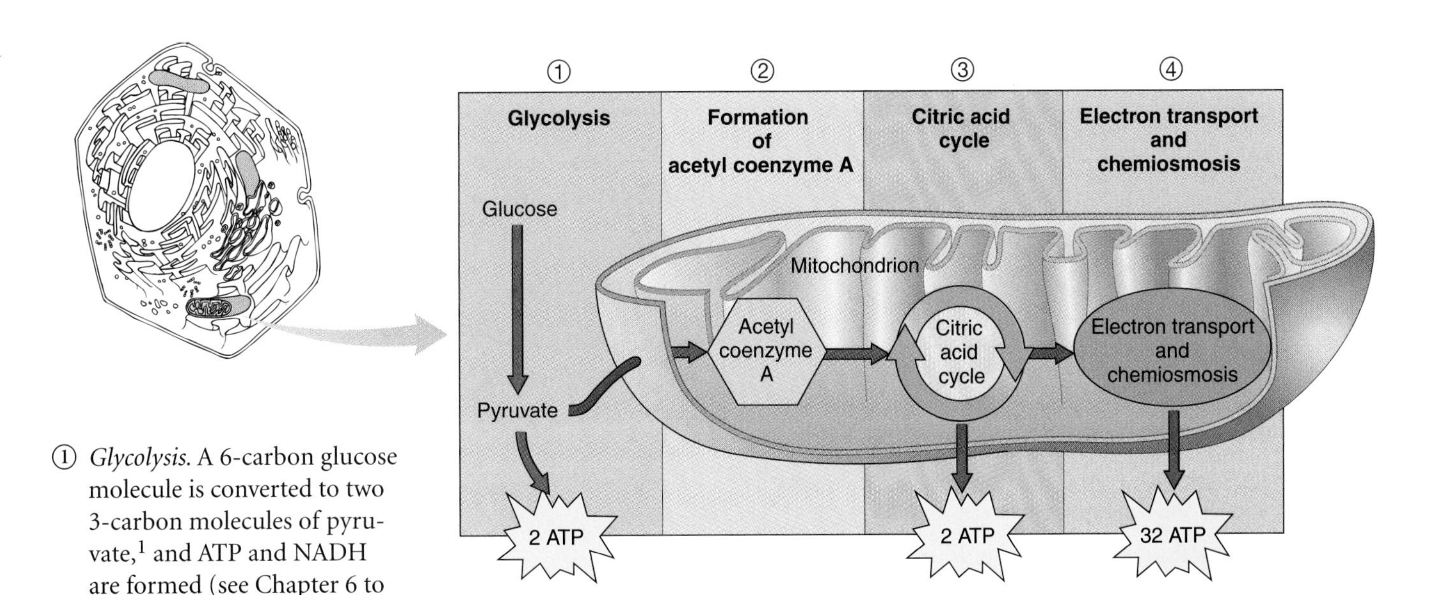

① *Glycolysis.* A 6-carbon glucose molecule is converted to two 3-carbon molecules of pyruvate,[1] and ATP and NADH are formed (see Chapter 6 to review ATP and NADH).[2]

② *Formation of acetyl coenzyme A.* Each pyruvate enters a mitochondrion and is oxidized to a 2-carbon group (acetate) that combines with coenzyme A, forming acetyl coenzyme A. NADH is produced, and carbon dioxide is released as a waste product.

③ *The citric acid cycle.* The acetate group of acetyl coenzyme A combines with a four-carbon molecule (oxaloacetate) to form a 6-carbon molecule (citrate). In the course of the cycle, citrate is recycled to oxaloacetate, and carbon dioxide is released as a waste product. Energy is captured as ATP and the reduced, high-energy compounds NADH and $FADH_2$ (see Chapter 6 to review $FADH_2$).

[1] Pyruvate and many other compounds in cellular respiration exist as anions at the pH found in the cell. They sometimes associate with H^+ to form acids. For example, pyruvate forms pyruvic acid. In some textbooks these compounds are presented in the acid form.

[2] Although the correct way to write the reduced form of NAD^+ is $NADH + H^+$, for simplicity we present the reduced form as NADH throughout the book.

FIGURE **7-2** **The four stages of aerobic respiration.**
① Glycolysis, the first stage of aerobic respiration, occurs in the cytosol. ② Pyruvate, the product of glycolysis, enters a mitochondrion, where cellular respiration continues with the formation of acetyl CoA, ③ the citric acid cycle, and ④ electron transport/chemiosmosis. Most ATP is synthesized by chemiosmosis.

④ *The electron transport chain and chemiosmosis.* The electrons removed from glucose during the preceding stages are transferred from NADH and $FADH_2$ to a chain of electron acceptor compounds. As the electrons are passed from one electron acceptor to another, some of their energy is used to transport hydrogen ions (protons) across the inner mitochondrial membrane, forming a proton gradient. In a process known as *chemiosmosis* (described later), the energy of this proton gradient is used to produce ATP.

Most reactions involved in aerobic respiration are one of three types: dehydrogenations, decarboxylations, and those we infor-

TABLE **7-1** **Summary of Aerobic Respiration**

Stage	Summary	Some Starting Materials	Some End Products
1. Glycolysis (in cytosol)	Series of reactions in which glucose is degraded to pyruvate; net profit of 2 ATPs; hydrogen atoms are transferred to carriers; can proceed anaerobically	Glucose, ATP, NAD^+, ADP, P_i	Pyruvate, ATP, NADH
2. Formation of acetyl CoA (in mitochondria)	Pyruvate is degraded and combined with coenzyme A to form acetyl CoA; hydrogen atoms are transferred to carriers; CO_2 is released	Pyruvate, coenzyme A, NAD^+	Acetyl CoA, CO_2, NADH
3. Citric acid cycle (in mitochondria)	Series of reactions in which the acetyl portion of acetyl CoA is degraded to CO_2; hydrogen atoms are transferred to carriers; ATP is synthesized	Acetyl CoA, H_2O, NAD^+, FAD, ADP, P_i	CO_2 NADH, $FADH_2$, ATP
4. Electron transport and chemiosmosis (in mitochondria)	Chain of several electron transport molecules; electrons are passed along chain; released energy is used to form a proton gradient; ATP is synthesized as protons diffuse down the gradient; oxygen is final electron acceptor	NADH, $FADH_2$, O_2, ADP, P_i	ATP, H_2O, NAD^+, FAD

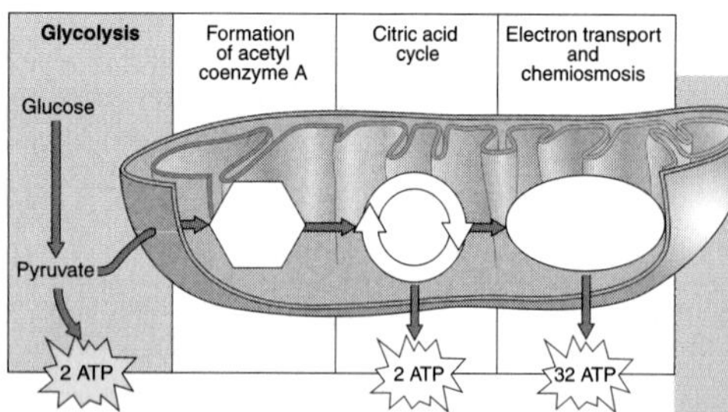

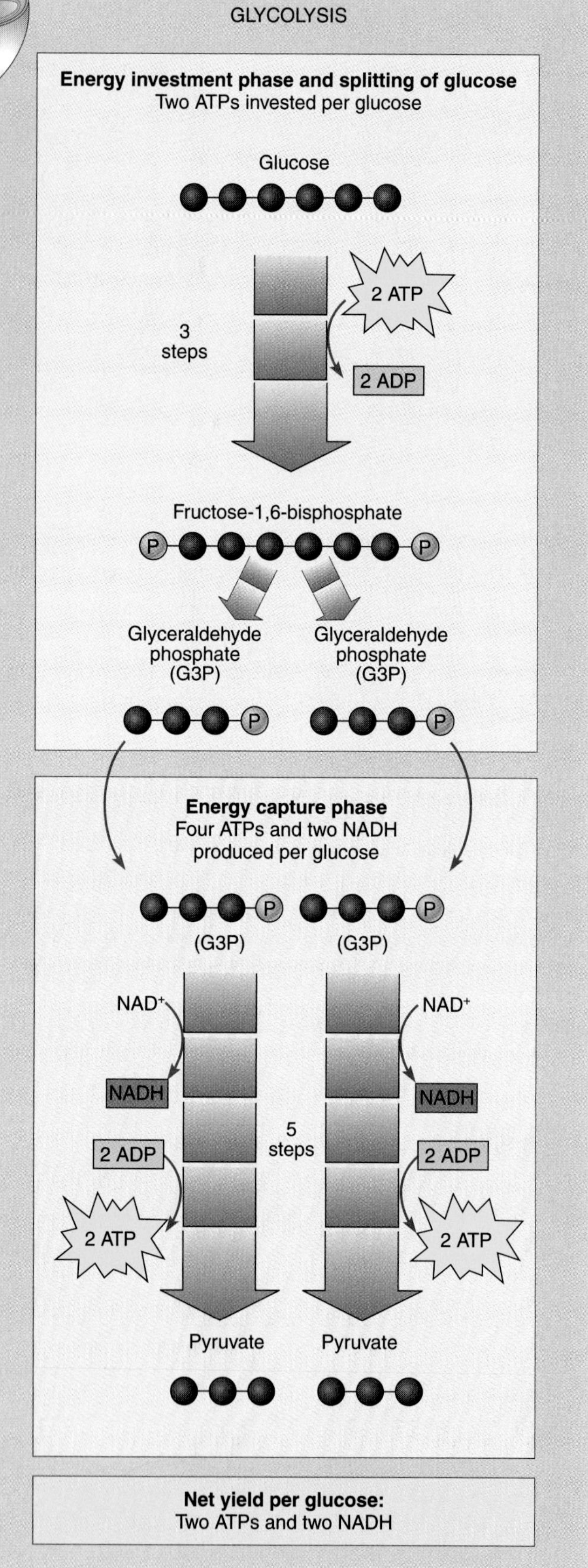

ACTIVE FIGURE **7-3** **An overview of glycolysis.**

The black spheres represent carbon atoms. The energy investment phase of glycolysis leads to the splitting of sugar; ATP and NADH are produced during the energy capture phase. During glycolysis, each glucose molecule is converted to two pyruvates, with a net yield of two ATP molecules and two NADH molecules.

BiologyNow™ See the process of **glycolysis** unfold by clicking on this figure on your BiologyNow CD-ROM.

mally categorize as preparation reactions. **Dehydrogenations** are reactions in which two hydrogen atoms (actually, 2 electrons plus 1 or 2 protons) are removed from the substrate and transferred to NAD^+ or FAD. **Decarboxylations** are reactions in which part of a carboxyl group (—COOH) is removed from the substrate as a molecule of CO_2. The carbon dioxide you exhale with each breath is derived from decarboxylations that occur in your cells. The rest of the reactions are preparation reactions in which molecules undergo rearrangements and other changes so that they can undergo further dehydrogenations or decarboxylations. As you examine the individual reactions of aerobic respiration, you will encounter many examples of these three basic types.

In following the reactions of aerobic respiration, it helps to do some bookkeeping as you go along. Because glucose is the starting material, it is useful to express changes on a per glucose basis. We will pay particular attention to changes in the number of carbon atoms per molecule and to steps in which some type of energy transfer takes place.

In glycolysis, glucose yields two pyruvates

The word **glycolysis** comes from Greek words meaning "sugar splitting," which refers to the fact that the sugar glucose is metabolized. Glycolysis does not require oxygen and proceeds under aerobic or anaerobic conditions. Figure 7-3 shows a simplified overview of glycolysis, in which a glucose molecule consisting of six carbons is converted to two molecules of **pyruvate,** a three-carbon molecule. Some of the energy in the glucose is captured; there is a net yield of two ATP molecules and two NADH molecules. The reactions of glycolysis take place in the cytosol, where the necessary reactants, such as ADP, NAD^+, and inorganic phosphates, float freely and are used as needed.

The glycolysis pathway consists of a series of reactions, each of which is catalyzed by a specific enzyme (Fig. 7-4). Glycolysis is divided into two major phases: The first includes endergonic reactions that require ATP, and the second includes exergonic reactions that yield ATP and NADH.

The first phase of glycolysis requires an investment of ATP

The first phase of glycolysis is sometimes called the *energy investment phase* (Fig. 7-4, steps ① – ⑤). Glucose is a relatively stable molecule and is not easily broken down. In two separate **phosphorylation reactions,** a phosphate group is transferred from ATP to the sugar. The resulting phosphorylated sugar (fructose-1,6-bisphosphate) is less stable and is broken enzymatically into two 3-carbon molecules, dihydroxyacetone phosphate and glyceraldehyde-3-phosphate (G3P). The dihydroxyacetone phosphate is enzymatically converted to G3P, so the products at this point in glycolysis are two molecules of G3P. We can summarize this portion of glycolysis as follows:

$$\underset{\text{Six-carbon compound}}{\text{Glucose}} + 2\text{ ATP} \longrightarrow \underset{\text{Three-carbon compound}}{2\text{ G3P}} + 2\text{ ADP}$$

The second phase of glycolysis yields NADH and ATP

The second phase of glycolysis is sometimes called the *energy capture phase* (Fig. 7-4, steps ⑥ – ⑩). Each G3P is converted to pyruvate. In the first step of this process, each G3P is oxidized by the removal of 2 electrons (as part of two hydrogen atoms). These immediately combine with the hydrogen carrier molecule, NAD^+:

$$\underset{\text{Oxidized}}{NAD^+} + \underset{\text{(From G3P)}}{2\text{ H}} \longrightarrow \underset{\text{Reduced}}{NADH} + H^+$$

Because there are two G3P molecules for every glucose, two NADH are formed. The energy of the electrons carried by NADH is used to form ATP later. This process is discussed in conjunction with the electron transport chain.

In two of the reactions leading to the formation of pyruvate, ATP forms when a phosphate group is transferred to ADP from a phosphorylated intermediate (see Fig. 7-4). This process is called **substrate-level phosphorylation.** Note that in the energy investment phase of glycolysis two molecules of ATP are consumed, but in the energy capture phase four molecules of ATP are produced. Thus glycolysis yields a net energy profit of *two* ATPs per glucose.

We can summarize the energy capture phase of glycolysis as follows:

$$2\text{ G3P} + 2\text{ NAD}^+ + 4\text{ ADP} \longrightarrow 2\text{ pyruvate} + 2\text{ NADH} + 4\text{ ATP}$$

Pyruvate is converted to acetyl CoA

In eukaryotes, the pyruvate molecules formed in glycolysis enter the mitochondria, where they are converted to **acetyl coenzyme A (acetyl CoA).** These reactions occur in the cytosol of aerobic prokaryotes. In this series of reactions, pyruvate undergoes a process known as **oxidative decarboxylation.** First, a carboxyl group is removed as carbon dioxide, which diffuses out of the cell (Fig. 7-5). Then the remaining two-carbon fragment is oxidized, and NAD^+ accepts the electrons removed during the oxidation. Finally, the oxidized two-carbon fragment, an acetyl group, becomes attached to **coenzyme A,** yielding acetyl CoA. *Pyruvate dehydrogenase,* the enzyme that catalyzes these reactions, is an enormous multienzyme complex consisting of 72 polypeptide chains!

Recall from Chapter 6 that coenzyme A transfers groups derived from organic acids. In this case, coenzyme A transfers an acetyl group, which is related to acetic acid. Coenzyme A is manufactured in the cell from one of the B vitamins, pantothenic acid.

The overall reaction for the formation of acetyl coenzyme A is

$$2\text{ Pyruvate} + 2\text{ NAD}^+ + 2\text{ CoA} \longrightarrow 2\text{ Acetyl CoA} + 2\text{ NADH} + 2\text{ CO}_2$$

Note that the original glucose molecule has now been partially oxidized, yielding two acetyl groups and two CO_2 molecules. The electrons removed have reduced NAD^+ to NADH. At this point in aerobic respiration, four NADH molecules have been formed as a result of the catabolism of a single glucose molecule: two during glycolysis and two during the formation of acetyl CoA from pyruvate. Keep in mind that these NADH molecules will be used later (during electron transport) to form additional ATP molecules.

The citric acid cycle oxidizes acetyl CoA

The **citric acid cycle** is also known as the **tricarboxylic acid (TCA) cycle** and the **Krebs cycle,** after Hans Krebs, the German biochemist who assembled the accumulated contributions of many scientists and worked out the details of the cycle in the 1930s. He received the Nobel Prize for Medicine in 1953 for this contribution. A simplified overview of the citric acid cycle, which takes place in the matrix of the mitochondria, is given in Figure 7-6. The eight steps of the citric acid cycle are shown in Figure 7-7. A specific enzyme catalyzes each reaction.

The first reaction of the cycle occurs when acetyl CoA transfers its two-carbon acetyl group to the four-carbon acceptor compound **oxaloacetate,** forming **citrate,** a six-carbon compound (step ①):

$$\underset{\text{Four-carbon compound}}{\text{Oxaloacetate}} + \underset{\text{Two-carbon compound}}{\text{acetyl CoA}} \longrightarrow \underset{\text{Six-carbon compound}}{\text{citrate}} + \text{CoA}$$

The citrate then goes through a series of chemical transformations, losing first one and then a second carboxyl group as CO_2 (steps ②, ③, and ④). Most of the energy made available by the oxidative steps of the cycle is transferred as energy-rich electrons to NAD^+, forming NADH. For each acetyl group that enters the citric acid cycle, three molecules of NADH are produced (steps ③, ④, and ⑧). Electrons are also transferred to the electron acceptor FAD, forming $FADH_2$ (step ⑥).

In the course of the citric acid cycle, two molecules of CO_2 and the equivalent of eight hydrogen atoms (8 protons and 8 electrons) are removed, forming three NADH and one $FADH_2$. You may wonder why more hydrogen is generated by these reactions than entered the cycle with the acetyl CoA molecule. These hydrogen atoms come from water molecules that are added during the reactions of the cycle. The CO_2 produced

Energy investment phase and splitting of glucose
Two ATPs invested per glucose

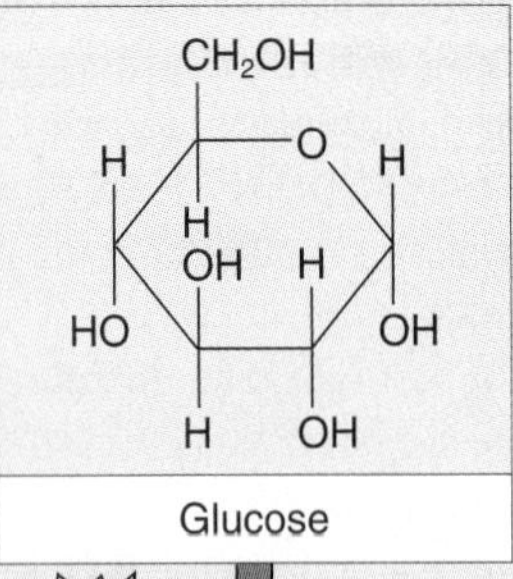

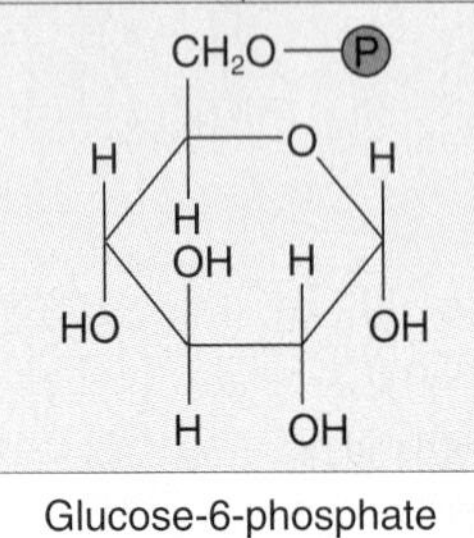

① Glycolysis begins with a preparation reaction in which glucose receives a phosphate group from an ATP molecule. The ATP serves as a source of both phosphate and the energy needed to attach the phosphate to the glucose molecule. (Once the ATP is spent, it becomes ADP and joins the ADP pool of the cell until turned into ATP again.) The phosphorylated glucose is known as glucose-6-phosphate. (Note the phosphate attached to its carbon atom 6.) Phosphorylation of the glucose makes it more chemically reactive.

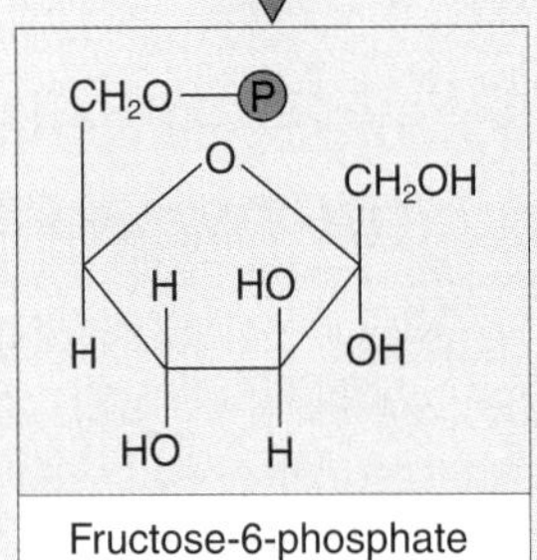

② Glucose-6-phosphate undergoes another preparation reaction, the rearrangement of its hydrogen and oxygen atoms. In this reaction glucose-6-phosphate is converted to its isomer, fructose-6-phosphate.

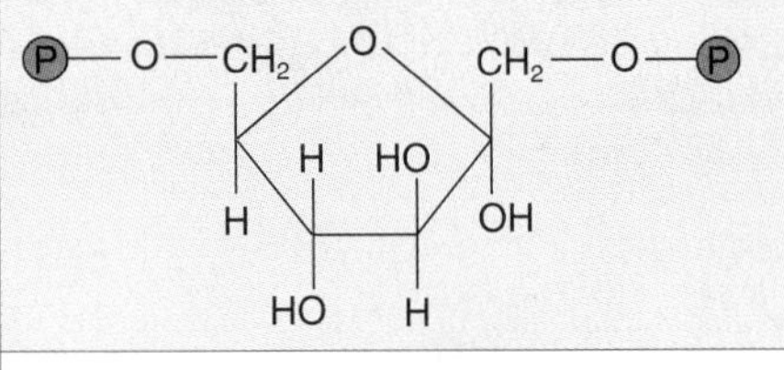

③ Next, another ATP donates a phosphate to the molecule, forming fructose-1,6-bisphosphate. So far, two ATP molecules have been invested in the process without any being produced. Phosphate groups are now bound at carbons 1 and 6, and the molecule is ready to be split.

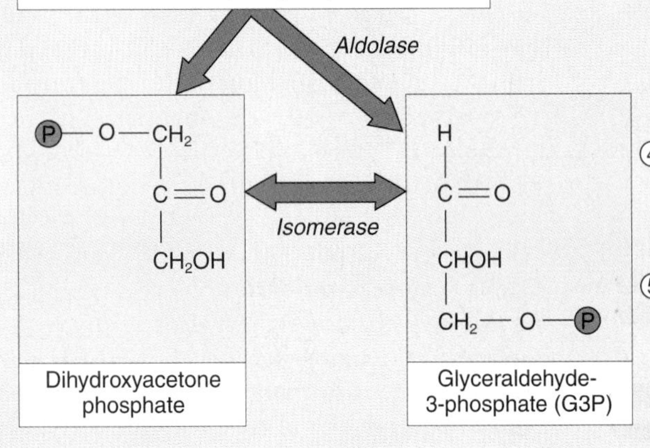

④ Fructose-1,6-bisphosphate is then split into two 3-carbon sugars, glyceraldehyde-3-phosphate (G3P) and dihydroxyacetone phosphate.

⑤ Dihydroxyacetone phosphate is enzymatically converted to its isomer, glyceraldehyde-3-phosphate, for further metabolism in glycolysis.

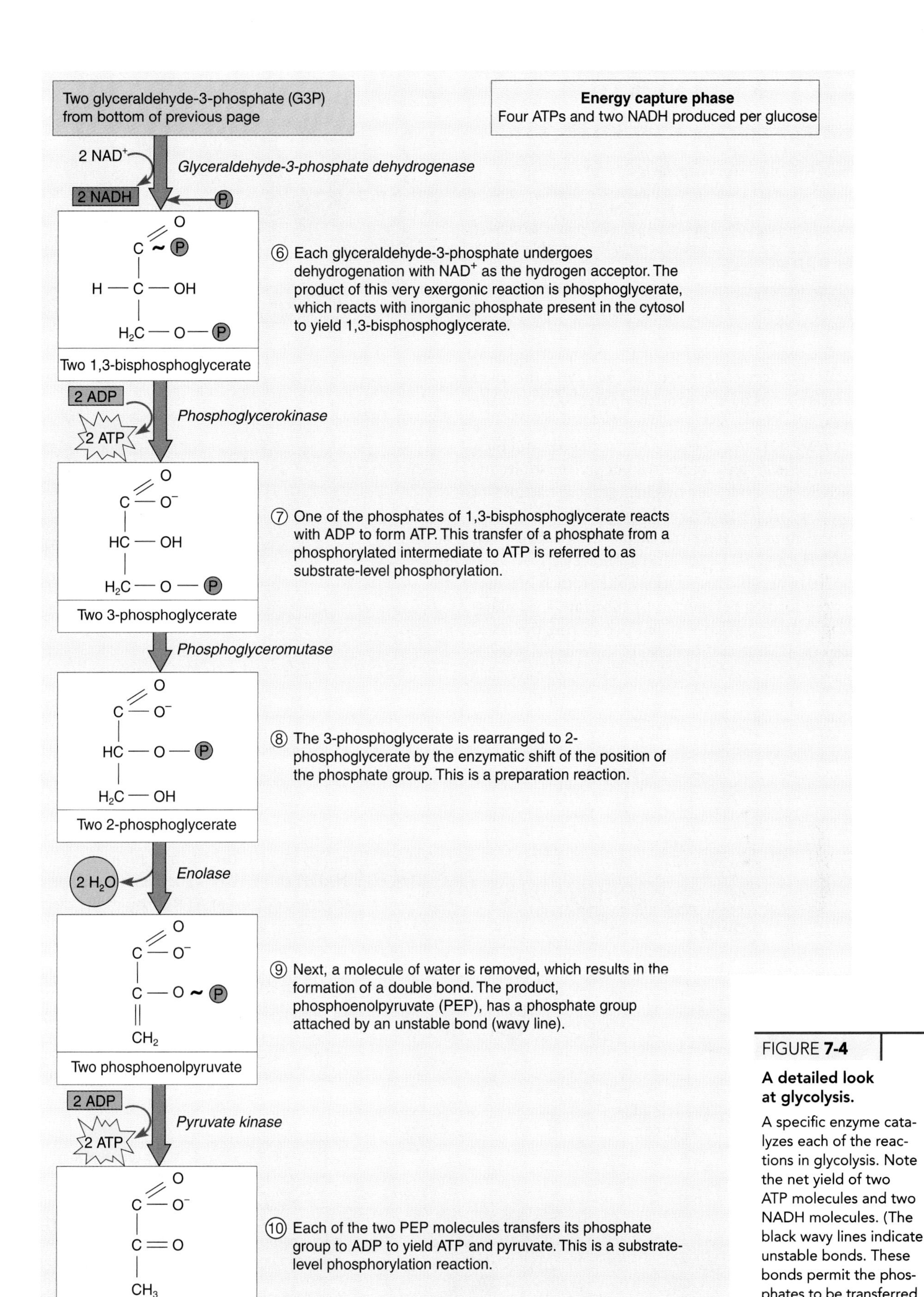

Two glyceraldehyde-3-phosphate (G3P) from bottom of previous page
Energy capture phase
Four ATPs and two NADH produced per glucose
2 NAD⁺
2 NADH
Glyceraldehyde-3-phosphate dehydrogenase
Two 1,3-bisphosphoglycerate
⑥ Each glyceraldehyde-3-phosphate undergoes dehydrogenation with NAD⁺ as the hydrogen acceptor. The product of this very exergonic reaction is phosphoglycerate, which reacts with inorganic phosphate present in the cytosol to yield 1,3-bisphosphoglycerate.
2 ADP
2 ATP
Phosphoglycerokinase
Two 3-phosphoglycerate
⑦ One of the phosphates of 1,3-bisphosphoglycerate reacts with ADP to form ATP. This transfer of a phosphate from a phosphorylated intermediate to ATP is referred to as substrate-level phosphorylation.
Phosphoglyceromutase
Two 2-phosphoglycerate
⑧ The 3-phosphoglycerate is rearranged to 2-phosphoglycerate by the enzymatic shift of the position of the phosphate group. This is a preparation reaction.
2 H₂O
Enolase
Two phosphoenolpyruvate
⑨ Next, a molecule of water is removed, which results in the formation of a double bond. The product, phosphoenolpyruvate (PEP), has a phosphate group attached by an unstable bond (wavy line).
2 ADP
2 ATP
Pyruvate kinase
Two pyruvate
⑩ Each of the two PEP molecules transfers its phosphate group to ADP to yield ATP and pyruvate. This is a substrate-level phosphorylation reaction.

FIGURE **7-4**

A detailed look at glycolysis.

A specific enzyme catalyzes each of the reactions in glycolysis. Note the net yield of two ATP molecules and two NADH molecules. (The black wavy lines indicate unstable bonds. These bonds permit the phosphates to be transferred to other molecules, in this case ADP.)

FIGURE **7-5** **The formation of acetyl CoA.**

This series of reactions is catalyzed by the enzyme pyruvate dehydrogenase. Pyruvate, a three-carbon molecule that is the end product of glycolysis, enters the mitochondrion and undergoes oxidative decarboxylation. First, the carboxyl group is split off as carbon dioxide. Then the remaining two-carbon fragment is oxidized, and its electrons are transferred to NAD^+. Finally, the oxidized two-carbon group, an acetyl group, is attached to coenzyme A. CoA has a sulfur atom that forms a very unstable bond, shown as a black wavy line, with the acetyl group.

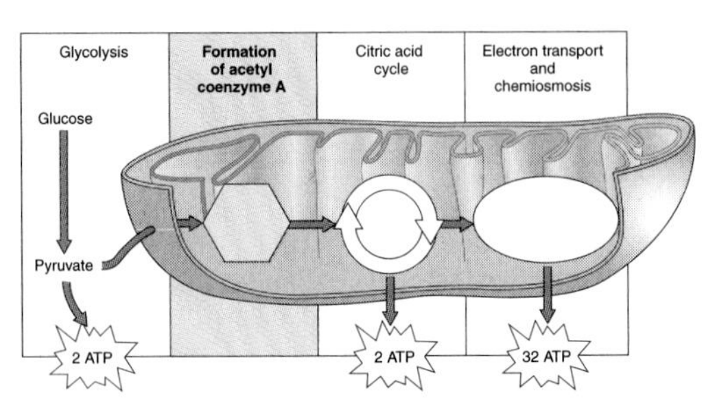

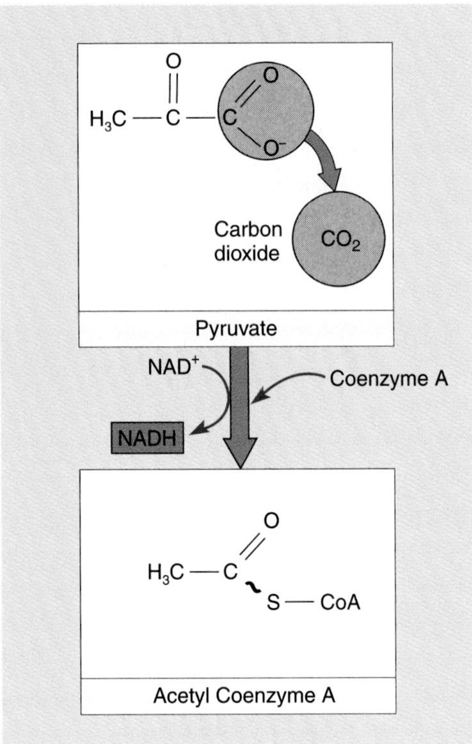

accounts for the two carbon atoms of the acetyl group that entered the citric acid cycle. At the end of each cycle, the four-carbon oxaloacetate has been regenerated (step ⑧), and the cycle continues.

Because two acetyl CoA molecules are produced from each glucose molecule, two cycles are required per glucose molecule. After two turns of the cycle, the original glucose has lost all its carbons and may be regarded as having been completely consumed. To summarize, the citric acid cycle yields 4 CO_2, 6 NADH, 2 $FADH_2$, and 2 ATP per glucose molecule.

At this point in aerobic respiration, only four molecules of ATP have been formed per glucose by substrate-level phosphorylation: two during glycolysis and two during the citric acid cycle (step ⑤). Most of the energy of the original glucose molecule is in the form of high-energy electrons in NADH and $FADH_2$. Their energy will be used to synthesize additional ATP through the electron transport chain and chemiosmosis.

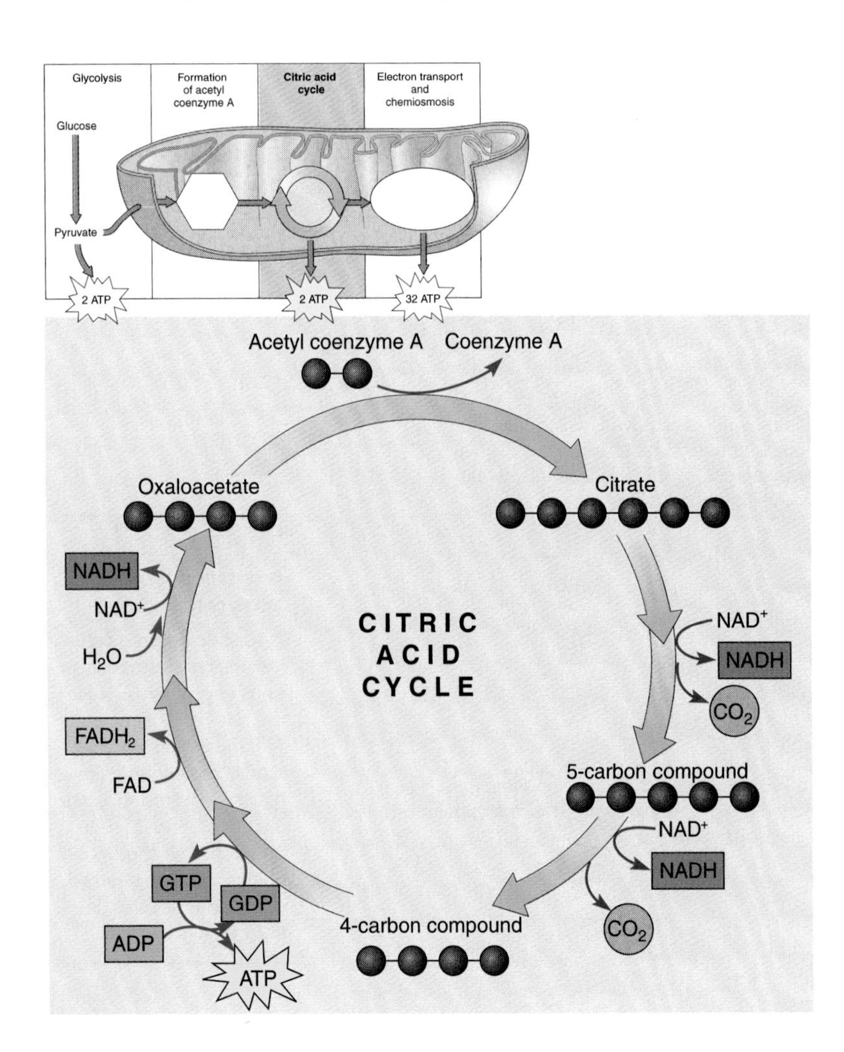

The electron transport chain is coupled to ATP synthesis

Let's consider the fate of all the electrons removed from a molecule of glucose during glycolysis, acetyl

ACTIVE FIGURE **7-6** **Overview of the citric acid cycle.**

For every glucose, two acetyl groups enter the citric acid cycle *(top)*. Each two-carbon acetyl group combines with a four-carbon compound, oxaloacetate, to form the six-carbon compound citrate. Two CO_2 molecules are removed, and energy is captured as one ATP, three NADH, and one $FADH_2$ per acetyl group (or two ATPs, six NADH, and two $FADH_2$ per glucose molecule).

BiologyNow™ Interact with the **citric acid cycle** by clicking on this figure on your BiologyNow CD-ROM.

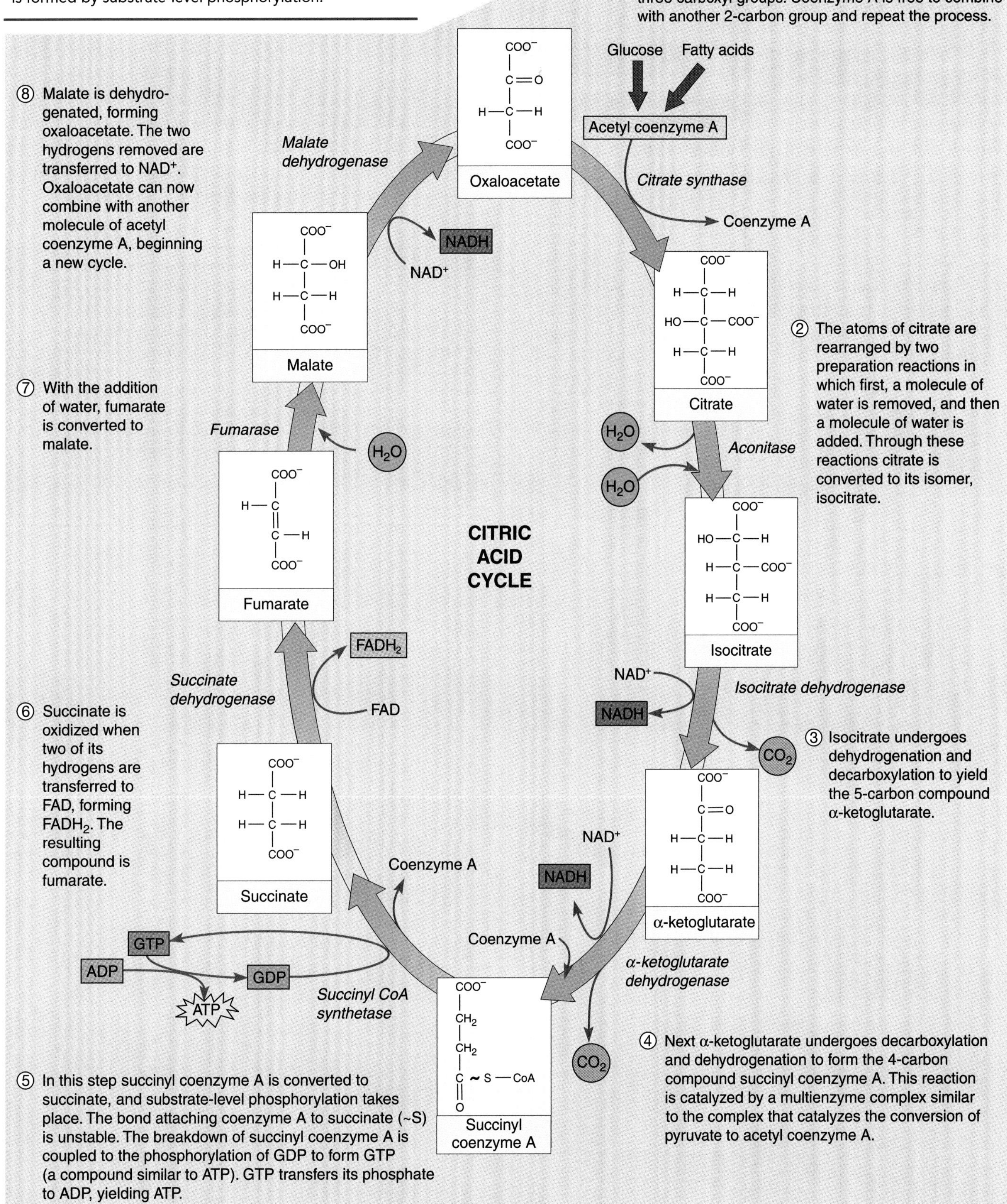

① The unstable bond attaching the acetyl group to coenzyme A breaks. The 2-carbon acetyl group becomes attached to a 4-carbon oxaloacetate molecule, forming citrate, a 6-carbon molecule with three carboxyl groups. Coenzyme A is free to combine with another 2-carbon group and repeat the process.
Glucose
Fatty acids
Acetyl coenzyme A
Citrate synthase
Coenzyme A
Oxaloacetate
Citrate
② The atoms of citrate are rearranged by two preparation reactions in which first, a molecule of water is removed, and then a molecule of water is added. Through these reactions citrate is converted to its isomer, isocitrate.
H_2O
Aconitase
H_2O
Isocitrate
NAD^+
Isocitrate dehydrogenase
NADH
CO_2
③ Isocitrate undergoes dehydrogenation and decarboxylation to yield the 5-carbon compound α-ketoglutarate.
α-ketoglutarate
NAD^+
NADH
Coenzyme A
α-ketoglutarate dehydrogenase
CO_2
④ Next α-ketoglutarate undergoes decarboxylation and dehydrogenation to form the 4-carbon compound succinyl coenzyme A. This reaction is catalyzed by a multienzyme complex similar to the complex that catalyzes the conversion of pyruvate to acetyl coenzyme A.
Succinyl coenzyme A
Coenzyme A
Succinyl CoA synthetase
GTP
ADP
GDP
ATP
⑤ In this step succinyl coenzyme A is converted to succinate, and substrate-level phosphorylation takes place. The bond attaching coenzyme A to succinate (~S) is unstable. The breakdown of succinyl coenzyme A is coupled to the phosphorylation of GDP to form GTP (a compound similar to ATP). GTP transfers its phosphate to ADP, yielding ATP.
Succinate
⑥ Succinate is oxidized when two of its hydrogens are transferred to FAD, forming $FADH_2$. The resulting compound is fumarate.
Succinate dehydrogenase
$FADH_2$
FAD
Fumarate
⑦ With the addition of water, fumarate is converted to malate.
Fumarase
H_2O
Malate
⑧ Malate is dehydrogenated, forming oxaloacetate. The two hydrogens removed are transferred to NAD^+. Oxaloacetate can now combine with another molecule of acetyl coenzyme A, beginning a new cycle.
Malate dehydrogenase
NADH
NAD^+
CITRIC ACID CYCLE

FIGURE 7-7 | **A detailed look at the citric acid cycle.**

Begin with step (1), in the upper right corner, where acetyl coenzyme A attaches to oxaloacetate. During the citric acid cycle, the entry of a two-carbon acetyl group is balanced by the release of two molecules of CO_2. Electrons are transferred to NAD^+ or FAD, yielding NADH and $FADH_2$, respectively, and ATP is formed by substrate-level phosphorylation.

CoA formation, and the citric acid cycle. Recall that these electrons were transferred as part of hydrogen atoms to the acceptors NAD^+ and FAD, forming NADH and $FADH_2$. These reduced compounds now enter the **electron transport chain,** where the high-energy electrons of their hydrogen atoms are shuttled from one acceptor to another. As the electrons are passed along in a series of exergonic redox reactions, some of their energy is used to drive the synthesis of ATP, which is an endergonic process. Because ATP synthesis (by phosphorylation of ADP) is coupled to the redox reactions in the electron transport chain, the entire process is known as **oxidative phosphorylation.**

The electron transport chain transfers electrons from NADH and $FADH_2$ to oxygen

The electron transport chain is a series of electron carriers embedded in the inner mitochondrial membrane of eukaryotes and in the plasma membrane of aerobic prokaryotes. Like NADH and $FADH_2$, each carrier exists in an oxidized form or a reduced form. Electrons pass down the electron transport chain in a series of redox reactions that works much like a bucket brigade, the old-time chain of people who passed buckets of water from a stream to each other, to a building that was on fire. In the electron transport chain, each acceptor molecule becomes alternately reduced as it accepts electrons and oxidized as it gives them up. The electrons entering the electron transport chain have a relatively high energy content. They lose some of their energy at each step as they pass along the chain of electron carriers (just as some of the water spills out of the bucket as it is passed from one person to another).

Members of the electron transport chain include the flavoprotein *flavin mononucleotide (FMN),* the lipid *ubiquinone* (also called *coenzyme Q* or *CoQ*), several *iron-sulfur proteins,* and a group of closely related iron-containing proteins called *cytochromes* (Fig. 7-8). Each electron carrier has a different mechanism for accepting and passing electrons. As cytochromes accept and donate electrons, for example, the charge on the iron atom, which is the electron carrier portion of the cytochromes, alternates between Fe^{2+} (reduced) and Fe^{3+} (oxidized).

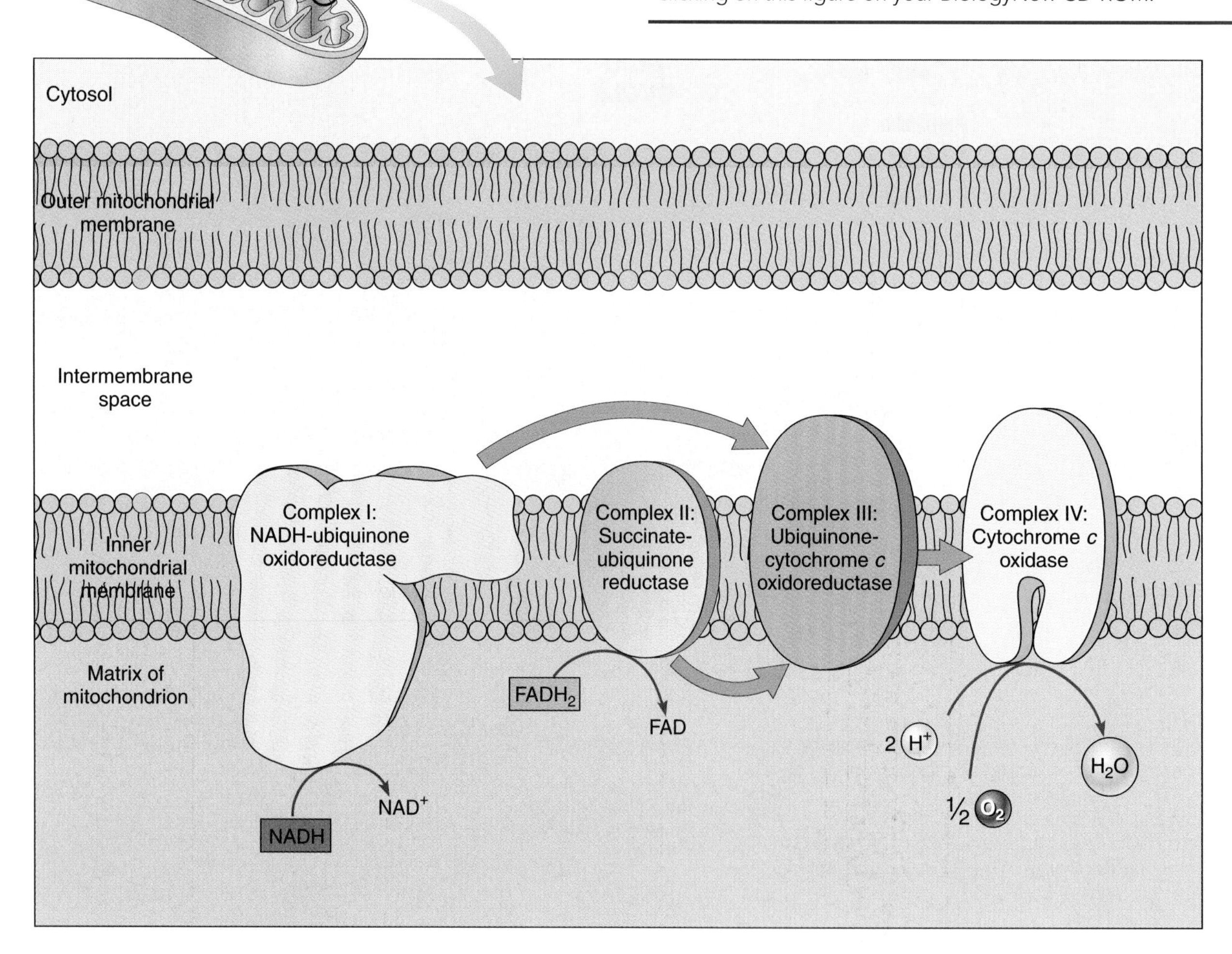

ACTIVE FIGURE **7-8** **An overview of the electron transport chain.**

Electrons fall to successively lower energy levels as they are passed along the four complexes of the electron transport chain located in the inner mitochondrial membrane. (The orange arrows indicate the pathway of electrons.) The carriers within each complex become alternately reduced and oxidized as they accept and donate electrons. The terminal acceptor is oxygen; one of the two atoms of an oxygen molecule (written as $\frac{1}{2}\ O_2$) accepts 2 electrons, which are added to 2 protons from the surrounding medium to produce water.

BiologyNow™ See the **electron transport chain** in action by clicking on this figure on your BiologyNow CD-ROM.

What is the source of our body heat? Essentially, it is a byproduct of various exergonic reactions, especially those involving the electron transport chains in our mitochondria. Some cold-adapted animals, hibernating animals, and newborn animals produce unusually large amounts of heat by uncoupling electron transport from ATP production. These animals have **adipose tissue** (tissue in which fat is stored) that is brown. The brown color comes from the large number of mitochondria found in the brown adipose tissue cells. The inner mitochondrial membranes of these mitochondria contain an *uncoupling protein* that produces a passive proton channel through which protons flow into the mitochondrial matrix. As a consequence, most of the energy of glucose is converted to heat rather than to chemical energy in ATP.

Certain plants, which are not generally considered "warm" organisms, also have the ability to produce large amounts of heat. Skunk cabbage (*Symplocarpus foetidus*), for example, lives in North American swamps and wet woodlands and generally flowers during February and March when the ground is still covered with snow (see figure). Its uncoupled mitochondria generate large amounts of heat, enabling the plant to melt the snow and attract insect pollinators by vaporizing certain odiferous molecules into the surrounding air. The flower temperature of skunk cabbage is 15° to 22°C (59° to 72°F) when the air surrounding it is −15° to 10°C (5° to 50°F). Skunk cabbage flowers maintain this temperature for two weeks or more. Other plants, such as splitleaf philodendron (*Philodendron selloum*) and sacred lotus (*Nelumbo nucifera*), also generate heat when they bloom and maintain their temperatures within precise limits.

Leonard Lee Rue III/Earth Scenes

Skunk cabbage (*Symplocarpus foetidus*).
This plant not only produces a significant amount of heat when it flowers but also regulates its temperature within a specific range.

Some plants generate as much or more heat per gram of tissue than animals in flight, which have long been considered the greatest heat producers in the living world. The European plant lords-and-ladies (*Arum maculatum*), for example, produces 0.4 joules (0.1 cal) of heat per second per gram of tissue, whereas a hummingbird in flight produces 0.24 J (0.06 cal) per second per gram of tissue.

The electron transport chain has been isolated and purified from the inner mitochondrial membrane as four large, distinct protein complexes, or groups, of acceptors. *Complex I (NADH-ubiquinone oxidoreductase)* accepts electrons from NADH molecules that were produced during glycolysis, the formation of acetyl CoA, and the citric acid cycle. *Complex II (succinate-ubiquinone reductase)* accepts electrons from $FADH_2$ molecules that were produced during the citric acid cycle. Complexes I and II both produce the same product, reduced ubiquinone, which is the substrate of *complex III (ubiquinone-cytochrome* c *oxidoreductase)*. That is, complex III accepts electrons from reduced ubiquinone and passes them on to cytochrome *c*. *Complex IV (cytochrome* c *oxidase)* accepts electrons from cytochrome *c* and uses these electrons to reduce molecular oxygen, forming water in the process. The electrons simultaneously unite with protons from the surrounding medium to form hydrogen, and the chemical reaction between hydrogen and oxygen produces water.

Because oxygen is the final electron acceptor in the electron transport chain, organisms that respire aerobically require oxygen. What happens when cells that are strict aerobes are deprived of oxygen? When no oxygen is available to accept them, the last cytochrome in the chain is stuck with its electrons. When that occurs, each acceptor molecule in the chain remains stuck with electrons (each is reduced), and the entire chain is blocked all the way back to NADH. Because oxidative phosphorylation is coupled to electron transport, no further ATPs are produced by way of the electron transport chain. Most cells of multicellular organisms cannot live long without oxygen, because the small amount of ATP they produce by glycolysis alone is insufficient to sustain life processes.

Lack of oxygen is not the only factor that interferes with the electron transport chain. Some poisons, including cyanide, inhibit the normal activity of the cytochromes. Cyanide binds tightly to the iron in the last cytochrome in the electron transport chain (cytochrome a_3), making it unable to transport electrons to oxygen. This blocks the further passage of electrons through the chain, halting ATP production.

Although the flow of electrons in electron transport is usually tightly coupled to the production of ATP, some organisms uncouple the two processes to produce heat (see *Focus On: Electron Transport and Heat*).

The chemiosmotic model explains the coupling of ATP synthesis to electron transport in aerobic respiration

PROCESS OF SCIENCE

For decades, scientists were aware that oxidative phosphorylation occurs in mitochondria, and many experiments had shown that the transfer of 2 electrons from each NADH to oxygen (via the electron transport chain) usually results in the production of up to three ATP molecules. However, for a long time, the connection between ATP synthesis and electron transport remained a mystery.

In 1961 Peter Mitchell, a British biochemist, proposed the *chemiosmotic model,* based on his experiments with bacteria. Because the respiratory electron transport chain is located in the plasma membrane of an aerobic bacterial cell, the bacterial plasma membrane can be considered comparable to the inner mitochondrial membrane. Mitchell demonstrated that if bacterial cells are placed in an acidic environment (that is, an environment with a high hydrogen ion, or proton, concentration), the cells synthesized ATP even if electron transport was not taking place. On the basis of these and other experiments, Mitchell proposed that electron transport and ATP synthesis are coupled by means of a proton gradient across the inner mitochondrial membrane in eukaryotes (or across the plasma membrane in bacteria). His model was so radical it was not immediately accepted. By 1978 so much evidence had accumulated in support of **chemiosmosis** that Peter Mitchell was awarded a Nobel Prize for Chemistry.

The electron transport chain establishes the proton gradient; some of the energy released as electrons pass down the electron transport chain is used to move protons (H^+) across a membrane. In eukaryotes the protons are moved across the inner mitochondrial membrane into the intermembrane space (Fig. 7-9). Hence the inner mitochondrial membrane separates a space with a higher concentration of protons (the intermembrane space) from a space with a lower concentration of protons (the mitochondrial matrix).

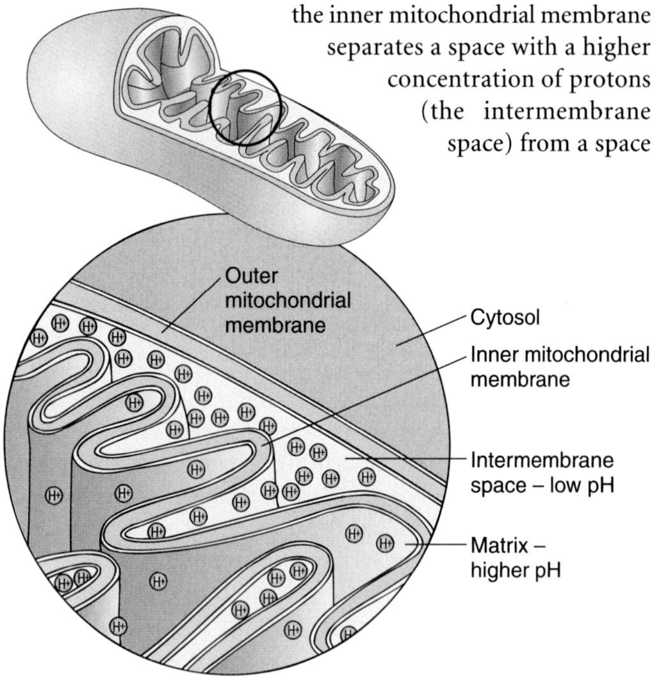

Protons are moved across the inner mitochondrial membrane by three of the four electron transport complexes (complexes I, III, and IV) (Fig. 7-10a). Like water behind a dam, the resulting proton gradient is a form of potential energy that can be harnessed to provide the energy for ATP synthesis.

Diffusion of protons from the intermembrane space, where they are highly concentrated, through the inner mitochondrial membrane to the matrix of the mitochondrion is limited to specific channels formed by a fifth enzyme complex, **ATP synthase,** a transmembrane protein. Portions of these complexes project from the inner surface of the membrane (the surface that faces the matrix) and are visible by electron microscopy (Fig. 7-10b). Diffusion of the protons down their gradient, through the ATP synthase complex, is exergonic because the entropy of the system increases. This exergonic process provides the energy for ATP production, although the exact mechanism by which ATP synthase catalyzes the phosphorylation of ADP is still not completely understood. In 1997, Paul Boyer of the University of California at Los Angeles and John Walker, of the Medical Research Council Laboratory of Molecular Biology, Cambridge, United Kingdom, shared the Nobel Prize for chemistry for the discovery that ATP synthase functions in an unusual way. Experimental evidence strongly suggests ATP synthase acts like a highly efficient molecular motor: During the production of ATP from ADP and inorganic phosphate, a central structure of ATP synthase rotates, possibly in response to the force of protons moving through the enzyme complex. The rotation apparently alters the conformation of the catalytic subunits in a way that allows ATP synthesis.

Chemiosmosis is a fundamental mechanism of energy coupling in cells; it allows exergonic redox reactions to drive the endergonic reaction in which ATP is produced by phosphorylating ADP. In photosynthesis (see Chapter 8), ATP is produced by a comparable process.

FIGURE **7-9** **The accumulation of protons (H^+) within the intermembrane space.**

As electrons move down the electron transport chain, the electron transport complexes move protons (H^+) from the matrix to the intermembrane space, creating a proton gradient. The high concentration of H^+ in the intermembrane space lowers the pH.

Aerobic respiration of one glucose yields a maximum of 36–38 ATPs

Let's now review where biologically useful energy is captured in aerobic respiration and calculate the total energy yield from the complete oxidation of glucose. Figure 7-11 summarizes the arithmetic involved.

1. In glycolysis, glucose is activated by the addition of phosphates from 2 ATP molecules and converted ultimately to 2 pyruvates + 2 NADH + 4 ATPs, yielding a net profit of 2 ATPs.
2. The 2 pyruvates are metabolized to 2 acetyl CoA + 2 CO_2 + 2 NADH.
3. In the citric acid cycle the 2 acetyl CoA molecules are metabolized to 4 CO_2 + 6 NADH + 2 $FADH_2$ + 2 ATPs.

Because the oxidation of NADH in the electron transport chain yields up to 3 ATPs per molecule, the total of 10

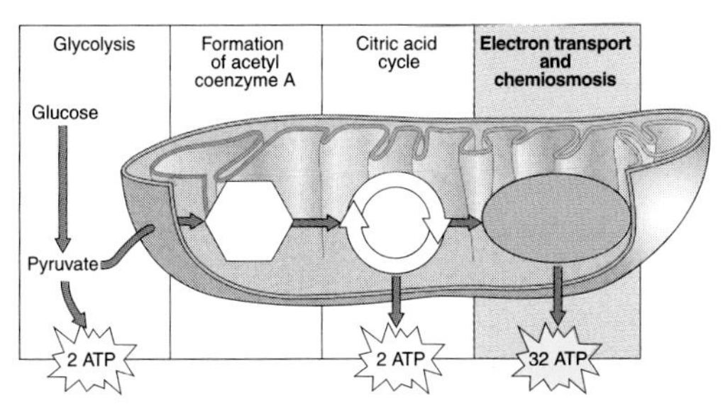

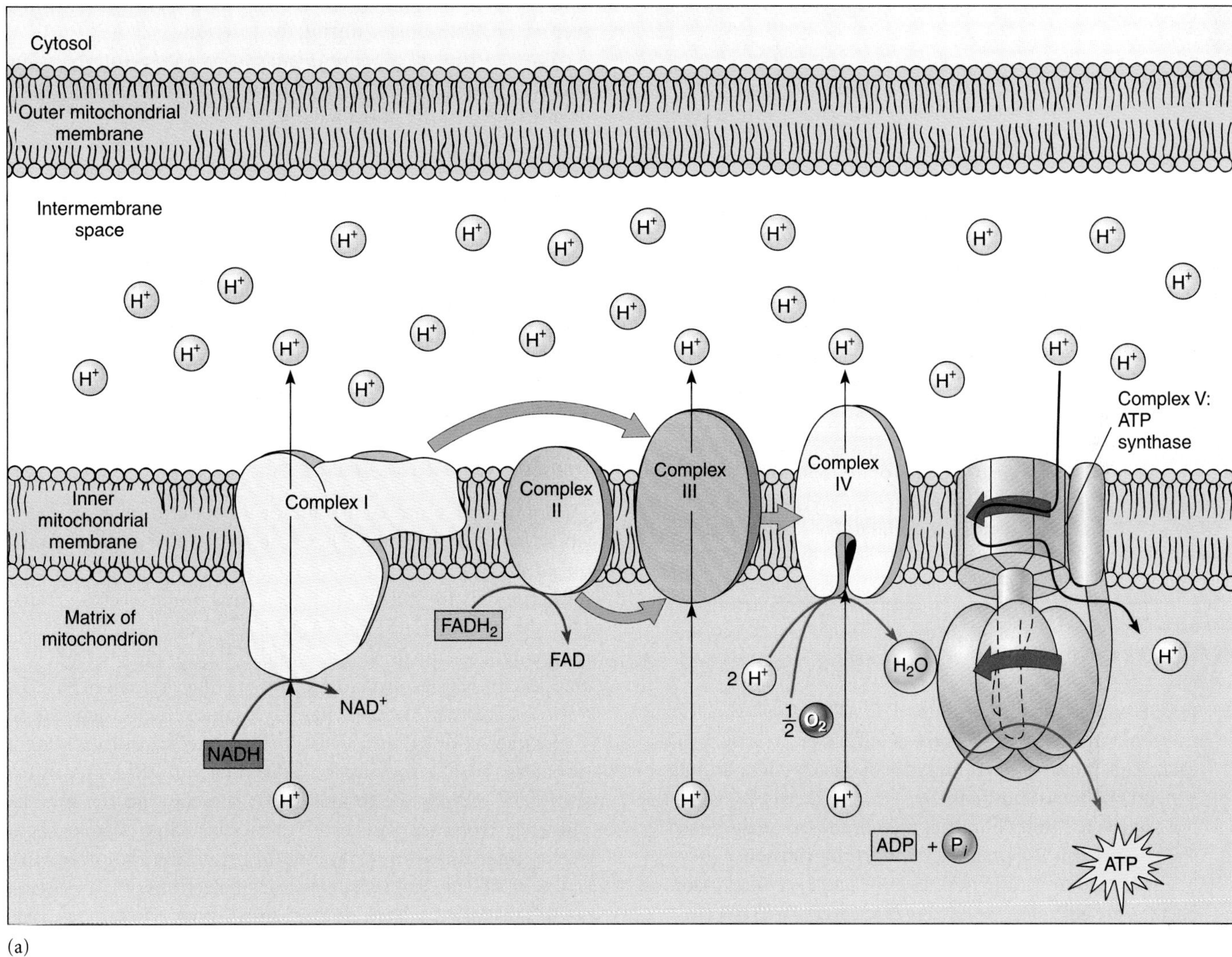

(a)

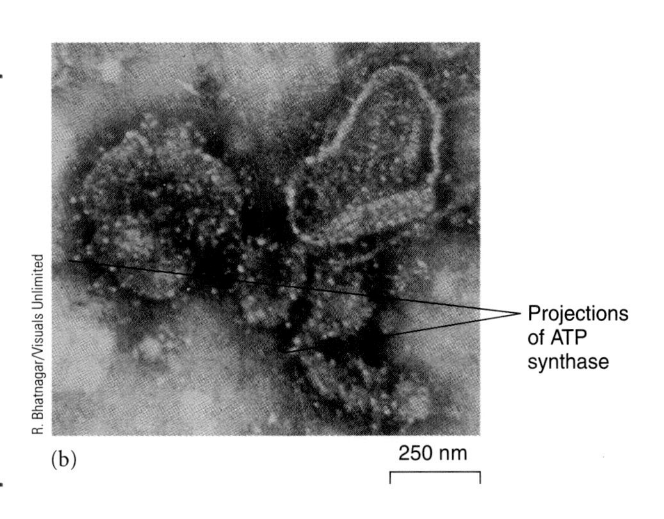

R. Bhatnagar/Visuals Unlimited

(b)

FIGURE **7-10** **A detailed look at electron transport and chemiosmosis.**

(a) The electron transport chain in the inner mitochondrial membrane includes three proton pumps that are located in three of the four electron transport complexes. (The orange arrows indicate the pathway of electrons, and the black arrows the pathway of protons.) The energy released during electron transport is used to transport protons (H^+) from the mitochondrial matrix to the intermembrane space, where a high concentration of protons accumulates. The protons cannot diffuse back into the matrix except through special channels in ATP synthase in the inner membrane. The flow of the protons through ATP synthase provides the energy for generating ATP from ADP and P_i. In the process, the inner part of ATP synthase rotates *(thick red arrows)* like a motor. **(b)** This TEM shows hundreds of projections of ATP synthase complexes along the surface of the inner mitochondrial membrane.

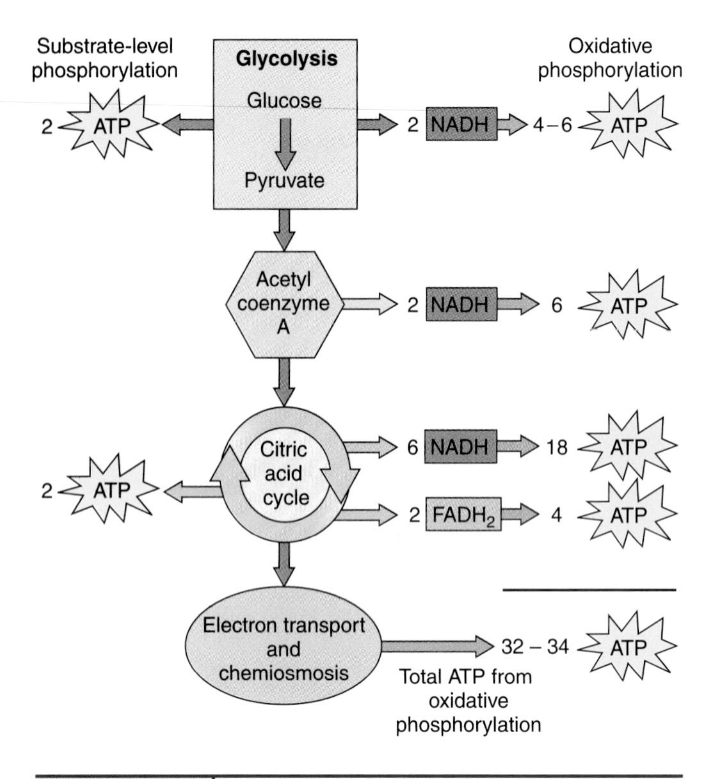

FIGURE **7-11** **Energy yield from the complete oxidation of glucose by aerobic respiration.**

A maximum of 36 to 38 ATPs are produced per glucose molecule. Of these ATPs, four are produced by substrate-level phosphorylation, and the remainder by oxidative phosphorylation (that is, electron transport and chemiosmosis).

NADH molecules can yield up to 30 ATPs. The 2 NADH molecules from glycolysis, however, yield either 2 or 3 ATPs each. This is because certain types of eukaryotic cells must expend energy to shuttle the NADH produced by glycolysis across the mitochondrial membrane (to be discussed shortly). Prokaryotic cells lack mitochondria; hence they have no need to shuttle NADH molecules. For this reason, bacteria are able to generate 3 ATPs for every NADH, even those produced during glycolysis. Thus, the maximum number of ATPs formed using the energy from NADH is 28 to 30.

The oxidation of $FADH_2$ yields 2 ATPs per molecule (recall that $FADH_2$ enters the electron transport chain at a different location from NADH), so the 2 $FADH_2$ molecules produced in the citric acid cycle yield 4 ATPs.

4. Summing all the ATPs (2 from glycolysis, 2 from the citric acid cycle, and 32 to 34 from electron transport and chemiosmosis), you can see that the complete aerobic metabolism of one molecule of glucose yields a maximum of 36 to 38 ATPs. Note that most of the ATPs are generated by oxidative phosphorylation, which involves the electron transport chain and chemiosmosis. Only 4 ATPs are formed by substrate-level phosphorylation in glycolysis and the citric acid cycle.

We can analyze the efficiency of the overall process of aerobic respiration by comparing the free energy captured as ATP to the total free energy in a glucose molecule. Recall from Chapter 6 that, although heat energy cannot power biological reactions, it is convenient to measure energy as heat. This is done through the use of a calorimeter, an instrument that measures the heat of a reaction. A sample is placed in a compartment surrounded by a chamber of water. As the sample burns (becomes oxidized), the temperature of the water rises, providing a measure of the heat released during the reaction.

When 1 mol of glucose is burned in a calorimeter, some 686 kcal (2870 kJ) are released as heat. The free energy temporarily held in the phosphate bonds of ATP is about 7.6 kcal (31.8 kJ) per mole. When 36 to 38 ATPs are generated during the aerobic respiration of glucose, the free energy trapped in ATP amounts to 7.6 kcal/mol × 36, or about 274 kcal (1146 kJ) per mole. Thus the efficiency of aerobic respiration is 274/686, or about 40%. (By comparison, a steam power plant has an efficiency of 35% to 36% in converting its fuel energy into electricity.) The remainder of the energy in the glucose is released as heat.

Mitochondrial shuttle systems harvest the electrons of NADH produced in the cytosol

The inner mitochondrial membrane is not permeable to NADH, which is a large molecule. Therefore, the NADH molecules produced in the cytosol during glycolysis cannot diffuse into the mitochondria to transfer their electrons to the electron transport chain. Unlike ATP and ADP, NADH does not have a carrier protein to transport it across the membrane. Instead, several systems have evolved to transfer just the *electrons* of NADH, not the NADH molecules themselves, into the mitochondria.

In liver, kidney, and heart cells, a special shuttle system transfers the electrons from NADH through the inner mitochondrial membrane to an NAD^+ molecule in the matrix. These electrons are transferred to the electron transport chain in the inner mitochondrial membrane, and up to three molecules of ATP are produced per pair of electrons.

In skeletal muscle, brain, and some other types of cells, another type of shuttle operates. Because this shuttle requires more energy than the shuttle in liver, kidney, and heart cells, the electrons are at a lower energy level when they enter the electron transport chain. They are accepted by ubiquinone rather than by NAD^+ and so generate a maximum of 2 ATP molecules per pair of electrons. This is why the number of ATPs produced by aerobic respiration of 1 molecule of glucose in skeletal muscle cells is 36 rather than 38.

Review

- How much ATP is made available to the cell from a single glucose molecule by the operation of (a) glycolysis, (b) the formation of acetyl CoA, (c) the citric acid cycle, and (d) the electron transport chain and chemiosmosis?
- Why is each of the following essential to chemiosmotic ATP synthesis? (a) electron transport chain (b) proton gradient (c) ATP synthase complex

- What are the roles of NAD^+ and FAD, and oxygen in aerobic respiration?

Biology Now™ Assess your understanding of the **four stages of aerobic respiration** by taking the pretest on your BiologyNow CD-ROM.

ENERGY YIELD OF NUTRIENTS OTHER THAN GLUCOSE

Learning Objective

7 Summarize how the products of protein and lipid catabolism enter the same metabolic pathway that oxidizes glucose.

Many organisms, including humans, depend on nutrients other than glucose as a source of energy. In fact, you usually obtain more of your energy by oxidizing fatty acids than by oxidizing glucose. Amino acids derived from protein digestion are also used as fuel molecules. Such nutrients are transformed into one of the metabolic intermediates that are fed into glycolysis or the citric acid cycle (Fig. 7-12).

Amino acids are metabolized by reactions in which the amino group ($—NH_2$) is first removed, a process called **deamination.** In mammals and some other animals, the amino group is converted to urea (see Fig. 46-1) and excreted, but the carbon chain is metabolized and eventually is used as a reactant in one of the steps of aerobic respiration. The amino acid alanine, for example, undergoes deamination to become pyruvate, the amino acid glutamate is converted to α-ketoglutarate, and the amino acid aspartate yields oxaloacetate. Pyruvate enters aerobic respiration as the end product of glycolysis, and α-ketoglutarate and oxaloacetate both enter aerobic respiration as intermediates in the citric acid cycle. Ultimately, the carbon chains of all the amino acids are metabolized in this way.

Each gram of lipid in the diet contains 9 kcal (38 kJ), more than twice as much energy as 1 g of glucose or amino acids, which have about 4 kcal (17 kJ) per gram. Lipids are rich in energy because they are highly reduced; that is, they have many hydrogen atoms and few oxygen atoms. When completely oxidized in aerobic respiration, a molecule of a six-carbon fatty acid generates up to 44 ATPs (compared with 36 to 38 ATPs for a molecule of glucose, which also has 6 carbons).

Both the glycerol and fatty acid components of a triacylglycerol (see Chapter 3) are used as fuel; phosphate is added to glycerol, converting it to G3P or another compound that enters glycolysis. Fatty acids are oxidized and split enzymatically into two-carbon acetyl groups that are bound to coenzyme A; that is, fatty acids are converted to acetyl CoA. This process, which occurs in the mitochondrial matrix, is called **beta-oxidation,** or **β-oxidation.** Acetyl CoA molecules formed by β-oxidation enter the citric acid cycle.

Review

- How can a person obtain energy from a low-carbohydrate diet?

Biology Now™ Assess your understanding of the **energy yield of nutrients other than glucose** by taking the pretest on your BiologyNow CD-ROM.

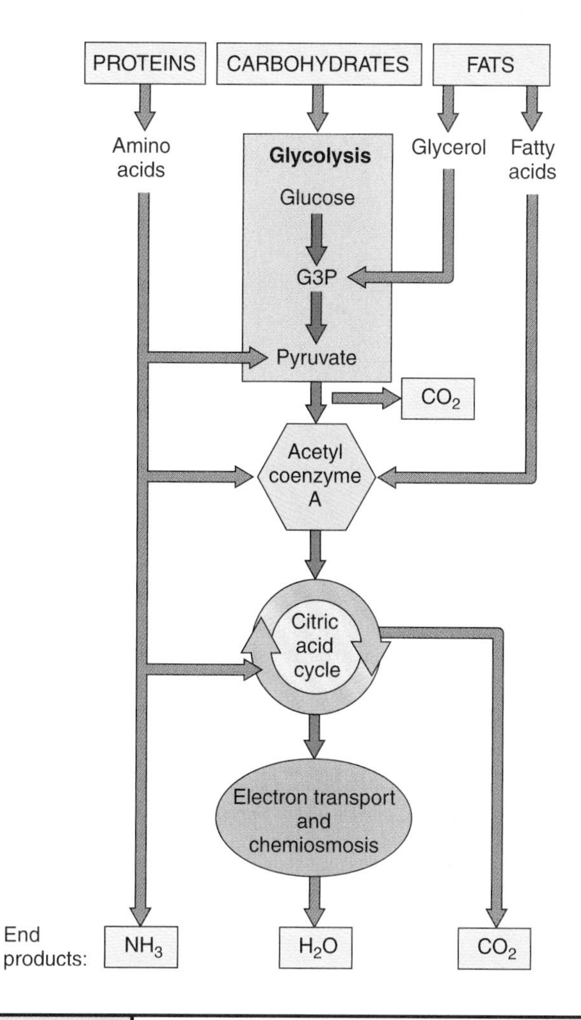

FIGURE **7-12** **Energy from carbohydrates, proteins, and fats.**

Products of the catabolism of carbohydrates, proteins, and fats enter glycolysis or the citric acid cycle at various points. This diagram is greatly simplified and illustrates only a few of the principal catabolic pathways.

ANAEROBIC RESPIRATION AND FERMENTATION

Learning Objective

8 Compare and contrast anaerobic respiration and fermentation; include the mechanism of ATP formation, the final electron acceptor, and the end products.

Anaerobic respiration, which does not use oxygen as the final electron acceptor, is performed by some prokaryotes that live in anaerobic environments such as waterlogged soil, stagnant ponds, or animal intestines. As in aerobic respiration, electrons are transferred in anaerobic respiration from glucose to NADH; they then pass down an electron transport chain that is coupled to ATP synthesis by chemiosmosis. However, an inorganic substance such as nitrate (NO_3^-) or sulfate (SO_4^{2-}) replaces molecular oxygen as the terminal electron acceptor. The end products of this type of anaerobic respiration are carbon dioxide, one or more reduced inorganic substances, and ATP. One representative type of anaerobic respiration, which is part of the biogeochemical cycle known as the **nitrogen cycle** (see Chapter 53) is summarized below.

$$C_6H_{12}O_6 + 12\ KNO_3 \longrightarrow$$

Potassium nitrate

$$6\ CO_2 + 6\ H_2O + 12\ KNO_2 + \text{energy}$$

Potassium nitrite — (in the chemical bonds of ATP)

Certain other bacteria, as well as some fungi, regularly use **fermentation,** an anaerobic pathway that does not involve an electron transport chain. During fermentation only two ATPs are formed per glucose (by substrate-level phosphorylation during glycolysis). One might expect that a cell that obtains energy from glycolysis would produce pyruvate, the end product of glycolysis. However, this cannot happen because every cell has a limited supply of NAD^+, and NAD^+ is required for glycolysis to continue. If virtually all NAD^+ becomes reduced to NADH during glycolysis, then glycolysis stops, and no more ATP is produced.

In fermentation, NADH molecules transfer their hydrogen atoms to organic molecules, thus regenerating the NAD^+ needed to keep glycolysis going. The resulting relatively reduced organic molecules (commonly alcohol or lactate) tend to be toxic to the cells and are essentially waste products.

Table 7-2 compares the three processes: aerobic respiration, anaerobic respiration, and fermentation.

Alcohol fermentation and lactate fermentation are inefficient

Yeasts are **facultative anaerobes** that carry out aerobic respiration when oxygen is available but switch to **alcohol fermentation** when deprived of oxygen (Fig. 7-13a). These eukaryotic, unicellular fungi have enzymes that decarboxylate pyruvate, releasing carbon dioxide and forming a two-carbon compound called *acetaldehyde.* NADH produced during glycolysis transfers hydrogen atoms to acetaldehyde, reducing it to *ethyl alcohol* (Fig. 7-13b). Alcohol fermentation is the basis for the production of beer, wine, and other alcoholic beverages. Yeast cells are also used in baking to produce the carbon dioxide that causes dough to rise; the alcohol evaporates during baking.

Certain fungi and bacteria perform **lactate (lactic acid) fermentation.** In this alternative pathway, NADH produced during glycolysis transfers hydrogen atoms to pyruvate, reducing it to *lactate* (Fig. 7-13c). The ability of some bacteria to produce lactate is exploited by humans, who use these bacteria to make yogurt and to ferment cabbage for sauerkraut.

Lactate is also produced by muscle cells. Exercise can cause fatigue and muscle cramps possibly due to insufficient oxygen, the depletion of fuel molecules, and the accumulation of lactate during strenuous activity. This buildup of lactate occurs because muscle cells shift briefly to lactate fermentation if the amount of oxygen delivered to muscle cells is insufficient to support aerobic respiration. The shift is only temporary, however, and oxygen is required for sustained work. About 80% of the lactate is eventually exported to the liver, where it is used to regenerate more glucose for the muscle cells. The remaining 20% of the lactate is metabolized in muscle cells in the presence of oxygen. This explains why you continue to breathe heavily after you have stopped exercising: The additional oxygen is needed to oxidize lactate, thereby restoring the muscle cells to their normal state.

Although humans use lactate fermentation to produce ATP for only a few minutes, a few animals can live without oxygen for much longer periods. The red-eared slider, a freshwater turtle, remains under water for as long as two weeks. During this time, it is relatively inactive and therefore does not expend a great deal of energy. It relies on lactate fermentation for ATP production.

Both alcohol fermentation and lactate fermentation are highly inefficient, because the fuel is only partially oxidized. Alcohol, the end product of fermentation by yeast cells, can be burned and is even used as automobile fuel; obviously, it contains a great deal of energy that the yeast cells cannot extract using anaerobic methods. Lactate, a three-carbon compound, contains even more energy than the two-carbon alcohol. In contrast, all available energy is removed during aerobic respiration, because the fuel molecules become completely oxidized to CO_2. A net profit of only 2 ATPs is produced by the fermentation of one molecule of glucose, compared with up to 36 to 38 ATPs when oxygen is available.

The inefficiency of fermentation necessitates a large supply of fuel. To perform the same amount of work, a cell engaged in fermentation must consume up to 20 times more glucose or other carbohydrate per second than a cell using aerobic respi-

TABLE 7-2 A Comparison of Aerobic Respiration, Anaerobic Respiration, and Fermentation

	Aerobic Respiration	Anaerobic Respiration	Fermentation
Immediate Fate of Electrons in NADH	Transferred to electron transport chain	Transferred to electron transport chain	Transferred to organic molecule
Terminal Electron Acceptor of Electron Transport Chain	O_2	Inorganic substances such as NO_3^- or SO_4^{2-}	No electron transport chain
Reduced Product(s) Formed	Water	Relatively reduced inorganic substances	Relatively reduced organic compounds (commonly, alcohol or lactate)
Mechanism of ATP Synthesis	Oxidative phosphorylation/ chemiosmosis; also substrate-level phosphorylation	Oxidative phosphorylation/chemiosmosis; also substrate-level phosphorylation	Substrate-level phosphorylation only (during glycolysis)

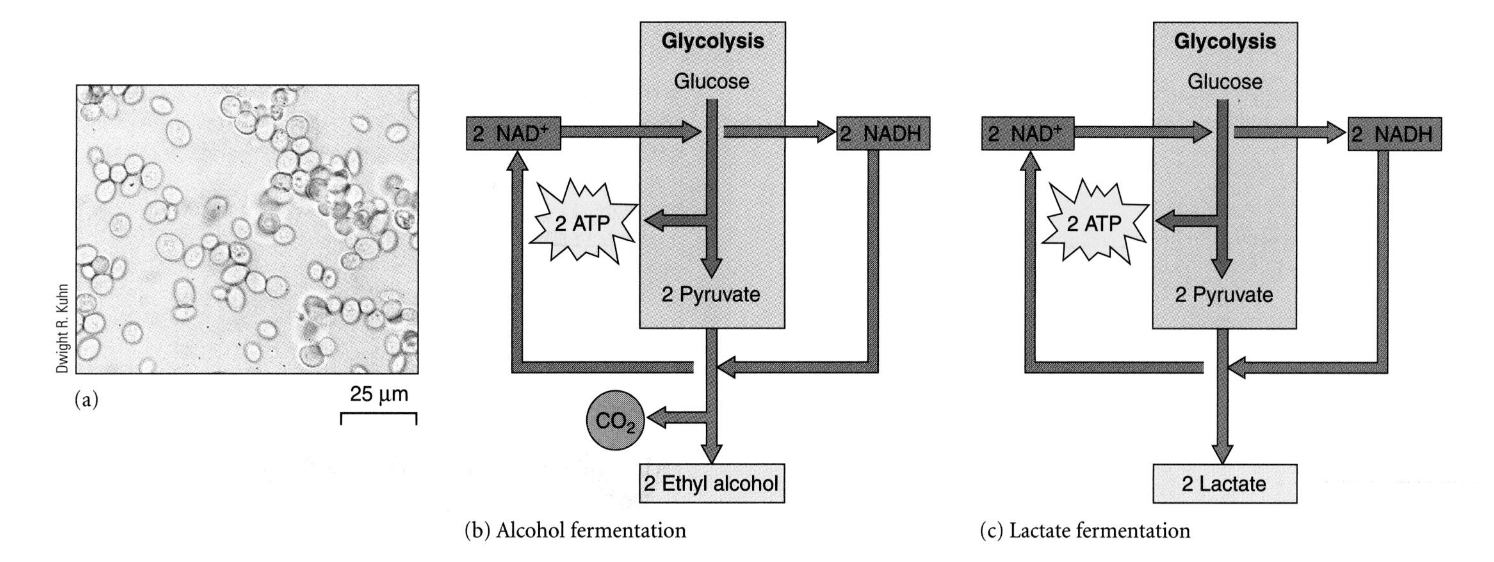

FIGURE 7-13 Fermentation.

(a) Light micrograph of live brewer's yeast (*Saccharomyces cerevisiae*). Yeast cells have mitochondria and carry on aerobic respiration when O_2 is present. In the absence of O_2, yeasts carry on alcohol fermentation. **(b, c)** Glycolysis is the first part of fermentation pathways. **(b)** In alcohol fermentation, CO_2 is split off, and the two-carbon compound ethyl alcohol is the end product. **(c)** In lactate fermentation, the final product is the three-carbon compound lactate. In both alcohol and lactate fermentation, there is a net gain of only two ATPs per molecule of glucose. Note that the NAD^+ used during glycolysis is regenerated during both alcohol fermentation and lactate fermentation.

ration. For this reason, your skeletal muscle cells store large quantities of glucose in the form of glycogen, enabling them to metabolize anaerobically for short periods.

Review

- What is the fate of hydrogen atoms removed from glucose during glycolysis when oxygen is present in muscle cells? How does this compare to the fate of hydrogen atoms removed from glucose when the amount of available oxygen is insufficient to support aerobic respiration?
- Why is the ATP yield of fermentation only a tiny fraction of the yield from aerobic respiration?

BiologyNow™ Assess your understanding of **anaerobic respiration and fermentation** by taking the pretest on your BiologyNow CD-ROM.

SUMMARY WITH KEY TERMS

1 Write a summary reaction for aerobic respiration, showing which reactant becomes oxidized and which becomes reduced.

- **Aerobic respiration** is a catabolic process in which a fuel molecule such as glucose is broken down to form carbon dioxide and water. It includes **redox reactions** that result in the transfer of electrons from glucose (which becomes **oxidized**) to oxygen (which becomes **reduced**)
- $$C_6H_{12}O_6 + 6\ O_2 \longrightarrow 6\ CO_2 + 6\ H_2O + \text{energy}$$ (oxidation: $C_6H_{12}O_6 \rightarrow CO_2$; reduction: $O_2 \rightarrow H_2O$)
- Energy released during aerobic respiration is used to produce up to 36 to 38 ATPs per molecule of glucose.

2 List and give a brief overview of the four stages of aerobic respiration.

- The chemical reactions of aerobic respiration occur in four stages: glycolysis, formation of acetyl CoA, the citric acid cycle, and the electron transport chain/chemiosmosis.
- During **glycolysis,** a molecule of glucose is degraded to two molecules of **pyruvate.** Two ATP molecules (net) are produced by **substrate-level phosphorylation** during glycolysis. Four hydrogen atoms are removed and used to produce two NADH.
- During the formation of **acetyl CoA,** the two pyruvate molecules each lose a molecule of carbon dioxide, and the remaining acetyl groups each combine with **coenzyme A,** producing two molecules of acetyl CoA; one NADH is produced per pyruvate.
- Each acetyl CoA enters the **citric acid cycle** by combining with a four-carbon compound, **oxaloacetate,** to form **citrate,** a six-carbon compound. Two acetyl CoA molecules enter the cycle for every glucose molecule. For every two carbons that enter the cycle as part of an acetyl CoA molecule, two leave as carbon dioxide. For every acetyl CoA, hydrogen atoms are transferred to three NAD^+ and one FAD; only one ATP is produced by substrate-level phosphorylation.
- Hydrogen atoms (or their electrons) removed from fuel molecules are transferred from one electron acceptor to another down an **electron transport chain** located in the mitochondrial inner membrane; ultimately these electrons reduce molecular oxygen, forming water. In **oxidative phosphorylation,** the redox reac-

tions in the electron transport chain are coupled to synthesis of ATP through the mechanism of **chemiosmosis.**

3 Indicate where each stage of aerobic respiration takes place in a eukaryotic cell.

- Glycolysis occurs in the cytosol, and the remaining stages of aerobic respiration take place in the mitochondrion.

4 Add up the energy captured (as ATP, NADH, and $FADH_2$) in each stage of aerobic respiration.

- In glycolysis, each glucose molecule produces 2 NADH and 2 ATP (net). The conversion of 2 pyruvates to acetyl CoA results in the formation of 2 NADH. In the citric acid cycle, the 2 acetyl CoA molecules are metabolized to form 6 NADH, 2 $FADH_2$, and 2 ATP. Adding up, we have 4 ATP, 10 NADH, and 2 $FADH_2$.
- When the 10 NADH and 2 $FADH_2$ pass through the electron transport chain, 32 to 34 ATP are produced by chemiosmosis. Therefore, each glucose molecule produces a total of up to 36 to 38 ATP.

5 Define *chemiosmosis,* and explain how a gradient of protons is established across the inner mitochondrial membrane.

- In chemiosmosis, some of the energy of the electrons in the electron transport chain is used to pump protons across the inner mitochondrial membrane into the intermembrane space. This pumping establishes a proton gradient across the inner mitochondrial membrane. Protons (H^+) accumulate within the intermembrane space, lowering the pH.

6 Describe the process by which the proton gradient drives ATP synthesis in chemiosmosis.

- The diffusion of protons through channels formed by the enzyme **ATP synthase,** through the inner mitochondrial membrane from the intermembrane space to the mitochondrial matrix, provides the energy to synthesize ATP.

7 Summarize how the products of protein and lipid catabolism enter the same metabolic pathway that oxidizes glucose.

- Amino acids are **deaminated,** and their carbon skeletons are converted to metabolic intermediates of aerobic respiration.
- Both the glycerol and fatty acid components of lipids are oxidized as fuel. Fatty acids are converted to acetyl CoA molecules by the process of **β-oxidation.**

8 Compare and contrast anaerobic respiration and fermentation; include the mechanism of ATP formation, the final electron acceptor, and the end products.

- In **anaerobic respiration,** electrons are transferred from fuel molecules to an electron transport chain; the final electron acceptor is an inorganic substance such as nitrate or sulfate, not molecular oxygen.
- **Fermentation** is an anaerobic process that does not use an electron transport chain. There is a net gain of only two ATPs per glucose; these are produced during glycolysis. To maintain the supply of NAD^+ essential for glycolysis, hydrogen atoms are transferred from NADH to an organic compound derived from the initial nutrient.
- Yeast cells carry out **alcohol fermentation,** in which ethyl alcohol and carbon dioxide are the final waste products.
- Certain fungi, prokaryotes, and animal cells carry out **lactate (lactic acid) fermentation,** in which hydrogen atoms are added to pyruvate to form lactate, a waste product.

Summary Reactions for Aerobic Respiration

Summary reaction for the complete oxidation of glucose:

$$C_6H_{12}O_6 + 6\ O_2 + 6\ H_2O \longrightarrow 6\ CO_2 + 12\ H_2O + \text{Energy (36 to 38 ATP)}$$

Summary reaction for glycolysis:

$$C_6H_{12}O_6 + 2\ ATP + 2\ ADP + 2\ P_i + 2\ NAD^+ \longrightarrow 2\ \text{Pyruvate} + 4\ ATP + 2\ NADH + H_2O$$

Summary reaction for the conversion of pyruvate to acetyl CoA:

$$2\ \text{Pyruvate} + 2\ \text{Coenzyme A} + 2\ NAD^+ \longrightarrow 2\ \text{Acetyl CoA} + 2\ CO_2 + 2\ NADH$$

Summary reaction for the citric acid cycle:

$$2\ \text{Acetyl CoA} + 6\ NAD^+ + 2\ FAD + 2\ ADP + 2\ P_i + 2\ H_2O \longrightarrow 4\ CO_2 + 6\ NADH + 2\ FADH_2 + 2\ ATP + 2\ CoA$$

Summary reactions for the processing of the hydrogen atoms of NADH and $FADH_2$ in the electron transport chain:

$$NADH + 3\ ADP + 3\ P_i + \tfrac{1}{2}O_2 \longrightarrow NAD^+ + 3\ ATP + H_2O$$

$$FADH_2 + 2\ ADP + 2\ P_i + \tfrac{1}{2}O_2 \longrightarrow FAD + 2\ ATP + H_2O$$

Summary Reactions for Fermentation

Summary reaction for lactate fermentation:

$$C_6H_{12}O_6 \longrightarrow 2\ \text{Lactate} + \text{Energy (2 ATP)}$$

Summary reactions for alcohol fermentation:

$$C_6H_{12}O_6 \longrightarrow 2\ CO_2 + 2\ \text{Ethyl alcohol} + \text{Energy (2 ATP)}$$

POST-TEST

1. The process of splitting larger molecules into smaller ones is an aspect of metabolism called (a) anabolism (b) fermentation (c) catabolism (d) oxidative phosphorylation (e) chemiosmosis
2. The synthetic aspect of metabolism is called (a) anabolism (b) fermentation (c) catabolism (d) oxidative phosphorylation (e) chemiosmosis
3. A chemical process during which a substance gains electrons is called (a) oxidation (b) oxidative phosphorylation (c) deamination (d) reduction (e) dehydrogenation
4. The pathway through which glucose is degraded to pyruvate is called (a) aerobic respiration (b) the citric acid cycle (c) the oxidation of pyruvate (d) alcohol fermentation (e) glycolysis
5. The reactions of ________ take place within the cytosol of eukaryotic cells. (a) glycolysis (b) oxidation of pyruvate (c) the citric acid cycle (d) chemiosmosis (e) the electron transport chain
6. Before pyruvate enters the citric acid cycle, it is decarboxylated, oxidized, and combined with coenzyme A, forming acetyl CoA, carbon dioxide, and one molecule of (a) NADH (b) $FADH_2$ (c) ATP (d) ADP (e) $C_6H_{12}O_6$

POST-TEST *(continued)*

7. In the first step of the citric acid cycle, acetyl CoA reacts with oxaloacetate to form (a) pyruvate (b) citrate (c) NADH (d) ATP (e) CO_2
8. Dehydrogenase enzymes remove hydrogen atoms from fuel molecules and transfer them to acceptors such as (a) O_2 and H_2O (b) ATP and FAD (c) NAD^+ and FAD (d) CO_2 and H_2O (e) CoA and pyruvate
9. Which of the following is a major source of electrons for the electron transport chain? (a) H_2O (b) ATP (c) NADH (d) ATP synthase (e) coenzyme A
10. In the process of ___________, electron transport and ATP synthesis are coupled by a proton gradient across the inner mitochondrial membrane. (a) chemiosmosis (b) deamination (c) anaerobic respiration (d) glycolysis (e) decarboxylation
11. Which of the following is a common energy flow sequence in aerobic respiration, starting with the energy stored in glucose? (a) glucose ⟶ NADH ⟶ pyruvate ⟶ ATP (b) glucose ⟶ ATP ⟶ NADH ⟶ electron transport chain (c) glucose ⟶ NADH ⟶ electron transport chain ⟶ ATP (d) glucose ⟶ oxygen ⟶ NADH ⟶ water (e) glucose ⟶ $FADH_2$ ⟶ NADH ⟶ coenzyme A
12. Which multiprotein complex in the electron transport chain is responsible for reducing molecular oxygen? (a) complex I (NADH-ubiquinone oxidoreductase) (b) complex II (succinate-ubiquinone reductase) (c) complex III (ubiquinone-cytochrome *c* oxidoreductase) (d) complex IV (cytochrome *c* oxidase) (e) complex V (ATP synthase)
13. A net profit of only 2 ATPs can be produced anaerobically from the ________ of one molecule of glucose, compared with a maximum of 38 ATPs produced in ________. (a) fermentation; anaerobic respiration (b) aerobic respiration; fermentation (c) aerobic respiration; anaerobic respiration (d) dehydrogenation; decarboxylation (e) fermentation; aerobic respiration
14. When deprived of oxygen, yeast cells obtain energy by fermentation, producing carbon dioxide, ATP, and (a) acetyl CoA (b) ethyl alcohol (c) lactate (d) pyruvate (e) citrate
15. During strenuous muscle activity, the pyruvate in muscle cells may accept hydrogen from NADH to become ________. (a) acetyl CoA (b) ethyl alcohol (c) lactate (d) pyruvate (e) citrate
16. Label the ten blank lines in the figure. Use Figure 7-2 to check your answers.

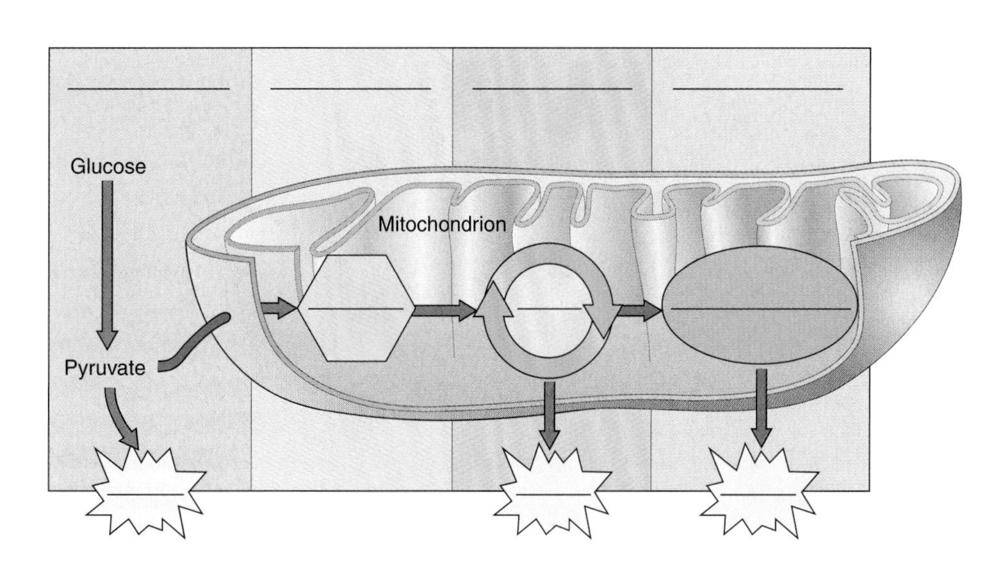

CRITICAL THINKING

1. The reactions of glycolysis are identical in *all* organisms—prokaryotes, protists, fungi, plants, and animals—that obtain energy from glucose catabolism. What does this universality suggest about the evolution of glycolysis?
2. How are the endergonic reactions of the first phase of glycolysis coupled to the hydrolysis of ATP, which is exergonic? How are the exergonic reactions of the second phase of glycolysis coupled to the endergonic synthesis of ATP and NADH?
3. In what ways is the inner mitochondrial membrane essential to the coupling of electron transport and ATP synthesis? Could the membrane carry out its function if its lipid bilayer were readily permeable to hydrogen ions (protons)?
4. Based on what you have learned in this chapter, explain why a schoolchild can run 17 miles per hour in a 100-yard dash, but a trained athlete can run only about 11.5 miles per hour in a 26-mile marathon.

- Visit our Web site **http://biology.brookscole.com/solomon7** for links to chapter-related resources on the World Wide Web. Additional online materials relating to this chapter can also be found on our Web site.

BIOLOGY NOW RESOURCES

Active Figures

7-3: Process of glycolysis
7-6: Citric acid cycle
7-8: Electron transport chain

Preparing for an exam? Take a diagnostic test on your BiologyNow CD-ROM.

Post-Test Answers

1. c	2. a	3. d	4. e
5. a	6. a	7. b	8. c
9. c	10. a	11. c	12. d
13. e	14. b	15. c	

8

Photosynthesis: Capturing Energy

Skip Moody/Dembinsky Photo Associates

Photoautotrophs. These blue lupines (*Lupinus hirsutus*), and the trees behind them, use CO_2 as a carbon source and light as an energy source. This photograph was taken in southern Michigan.

CHAPTER OUTLINE

- **Light**
- **Chloroplasts**
- **Overview of Photosynthesis**
- **The Light-Dependent Reactions**
- **The Carbon Fixation Reactions**

Photosynthesis, the process by which light energy is converted into the stored chemical energy of organic molecules, is the first step in the flow of energy through most of the living world. All organisms—from microscopic prokaryotes to dolphins to palm trees—can be classified according to nutritional factors: their carbon and energy requirements. Carbon atoms are required for the carbon skeletons of an organism's organic molecules, and as you learned in Chapters 6 and 7, energy powers all life processes, from growth, to movement, to repair of worn or injured tissues.

Organisms obtain carbon in one of two ways. **Autotrophs** (from the Greek *auto,* "self," and *trophos,* "nourishing") are able to carry out carbon fixation; they convert carbon that is in gaseous form in carbon dioxide (CO_2) into carbon that has a fixed position in a carbon skeleton. **Heterotrophs** (from the Greek *heter,* "other," and *trophos,* "nourishing") cannot fix carbon; they use organic molecules produced by other organisms as the building blocks from which they synthesize the carbon compounds they need.

Organisms obtain energy in one of two ways. **Phototrophs** are photosynthetic organisms that use light as their energy source. In contrast, **chemotrophs** use organic compounds, such as glucose, or inorganic substances, such as iron, nitrate, ammonia, or sulfur, as sources of energy. Chemotrophs typically obtain energy from these materials by redox reactions (Chapters 6 and 7).

All organisms fall into one of four groups based on carbon and energy requirements. Land plants (see photograph), algae, and certain prokaryotes are **photoautotrophs** (that is, both phototrophs and autotrophs). Photoautotrophs use light energy to make ATP and other molecules that temporarily hold chemical energy but are unstable and cannot be stockpiled in the cell. Their energy drives the the anabolic pathway by which a photosynthetic cell synthesizes stable organic molecules from the simple inorganic compounds CO_2 and water. These organic compounds are used not only as starting materials to synthesize all the other organic compounds the photosynthetic organism needs (such as complex carbohydrates, amino acids, and lipids) but also for energy storage. Glucose and other carbohydrates produced during photosynthesis are relatively reduced compounds that can be subsequently oxidized by aerobic respiration or by some other catabolic pathway (see Chapter 7).

A few bacteria, known as nonsulfur purple bacteria, are **photoheterotrophs** (both phototrophs and heterotrophs). Photoheterotrophs are able to use light energy but unable to carry out carbon fixation. Photoheterotrophs must obtain carbon from organic compounds (as "food").

Some bacteria are **chemoautotrophs**—both chemotrophs and autotrophs. These prokaryotes obtain their energy from the oxidation of reduced inorganic molecules such as hydrogen sulfide (H_2S), nitrite (NO_2^-), or ammonia (NH_3). Some of this energy is then used to carry out carbon fixation.

All animals, fungi, and most bacteria are **chemoheterotrophs;** that is, they are both chemotrophs and heterotrophs. Chemoheterotrophs use preformed organic molecules as a source of both energy and carbon. Plants and other photosynthetic organisms produce almost all the preformed organic molecules used by chemoheterotrophs.

Photosynthesis is the process that captures the vast majority of the energy that living organisms use. Photosynthesis not only sustains plants and other photoautotrophs but also indirectly supports almost all animals and other chemoheterotrophs in the biosphere. Each year plants and other photosynthetic organisms convert CO_2 into billions of tons of organic molecules. The chemical energy stored in these molecules fuels the metabolic reactions that sustain almost all life. ■

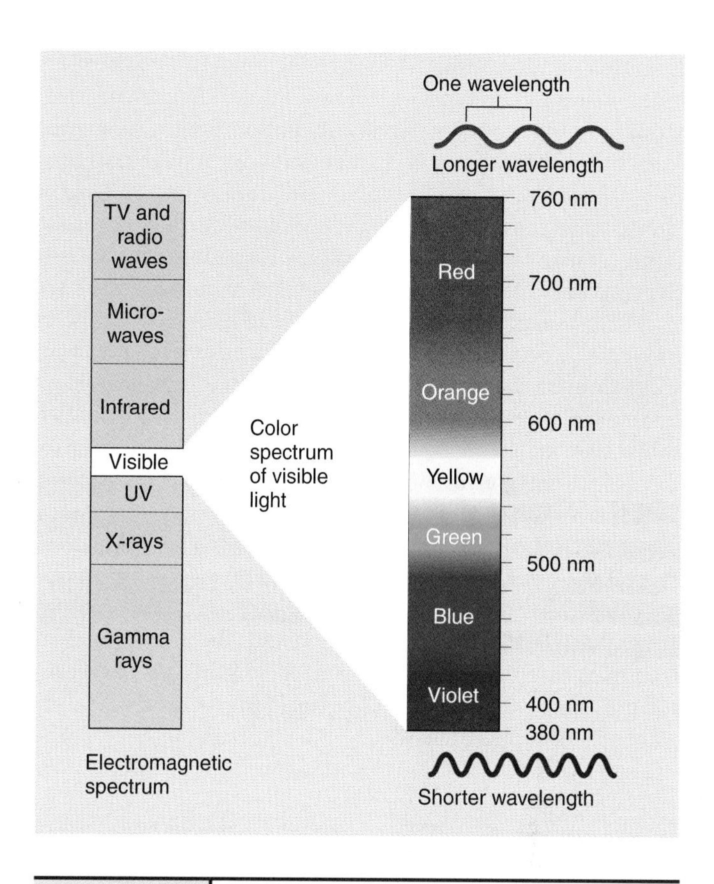

FIGURE **8-1** **The electromagnetic spectrum.**

Waves in the electromagnetic spectrum have similar properties but different wavelengths. Radio waves are the longest (and least energetic) waves, with wavelengths as long as 20 km. Gamma rays are the shortest (and most energetic) waves. Visible light represents a small fraction of the electromagnetic spectrum and consists of a mixture of wavelengths ranging from about 380 to 760 nm. The energy from visible light is used in photosynthesis.

LIGHT

Learning Objective

1 Describe the physical properties of light, and explain the relationship between a wavelength of light and its energy.

Because most life on this planet depends on light, either directly or indirectly, it is important to understand the nature of light and its essential role in photosynthesis. Visible light represents a very small portion of a vast, continuous range of radiation called the *electromagnetic spectrum* (Fig. 8-1). All radiation in this spectrum travels as waves. A **wavelength** is the distance from one wave peak to the next. At one end of the electromagnetic spectrum are gamma rays, which have very short wavelengths measured in fractions of nanometers, or nm (1 nanometer equals 10^{-9} m, one billionth of a meter). At the other end of the spectrum are radio waves, with wavelengths so long they can be measured in kilometers. The portion of the electromagnetic spectrum from 380 to 760 nm is called the *visible spectrum,* because we humans can see it. The visible spectrum includes all the colors of the rainbow (Fig. 8-2); violet has the shortest wavelength, and red has the longest.

Light is composed of small particles, or packets, of energy called **photons.** The energy of a photon is inversely proportional to its wavelength: Shorter-wavelength light has more energy per photon than longer-wavelength light.

Why does photosynthesis depend on light detectable by the human eye (visible light) rather than on some other wavelength of radiation? We can only speculate on the answer. Perhaps it is

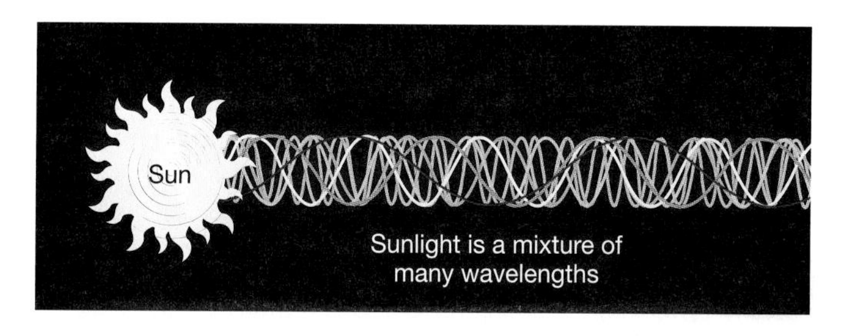

FIGURE **8-2** **Visible radiation emitted from the sun.**

Electromagnetic radiation from the sun includes ultraviolet radiation and visible light of varying colors and wavelengths.

because radiation within the visible light portion of the spectrum excites certain types of biological molecules, moving electrons into higher energy levels. Radiation with wavelengths longer than those of visible light doesn't have enough energy to excite these biological molecules. Radiation with wavelengths shorter than those of visible light is so energetic it disrupts the bonds of many biological molecules. Thus visible light has just the right amount of energy to be useful in photosynthesis.

When a molecule absorbs a photon of light energy, one of its electrons becomes energized, which means that the electron shifts from a lower-energy atomic orbital to a high-energy orbital that is more distant from the atomic nucleus. One of two things then happens, depending on the atom and its surroundings (Fig. 8-3). The atom may return to its **ground state,** which is the condition in which all its electrons are in their normal, lowest-energy levels. When an electron returns to its ground state, its energy dissipates as heat or as an emission of light of a longer wavelength than the absorbed light; this emission of light is called **fluorescence.** Alternatively, the energized electron may leave the atom and be accepted by an electron acceptor molecule, which becomes reduced in the process; this is what occurs in photosynthesis.

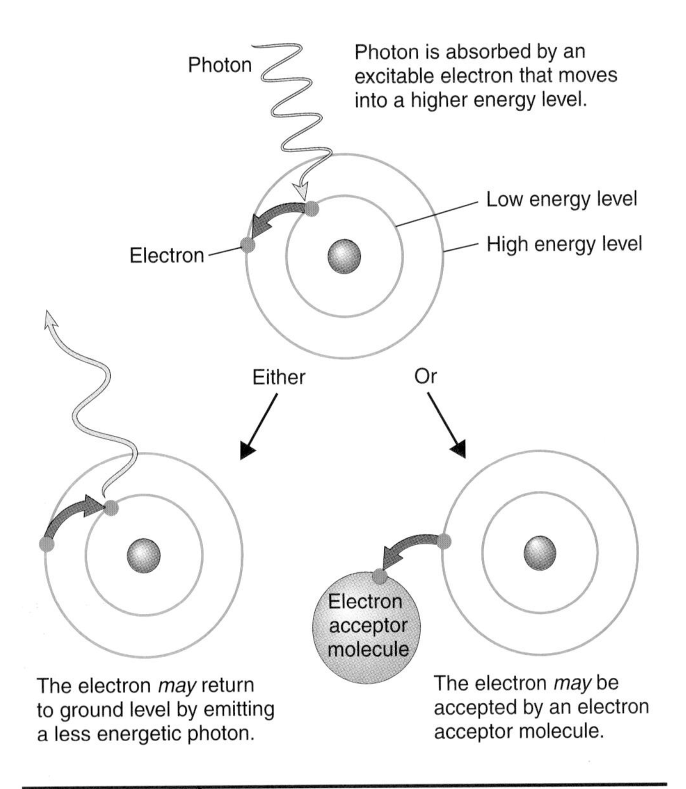

Photon
Photon is absorbed by an excitable electron that moves into a higher energy level.
Low energy level
High energy level
Electron
Either
Or
Electron acceptor molecule
The electron *may* return to ground level by emitting a less energetic photon.
The electron *may* be accepted by an electron acceptor molecule.

FIGURE **8-3** **Interactions between light and atoms or molecules.**

(Top) When a photon of light energy strikes an atom, or a molecule of which the atom is a part, the energy of the photon may push an electron to an orbital farther from the nucleus (that is, into a higher energy level). *(Lower left)* If the electron returns to the lower, more stable energy level, the energy may be released as a less energetic, longer-wavelength photon, or fluorescence *(shown)*, or as heat. *(Lower right)* If the appropriate electron acceptors are available, the electron may leave the atom. During photosynthesis, an electron acceptor captures the energetic electron and passes it to a chain of acceptors.

Now that you understand some of the properties of light, let's consider the organelles that use light for photosynthesis.

Review

- Why does photosynthesis require visible light?

BiologyNow™ Assess your understanding of **light** by taking the pretest on your BiologyNow CD-ROM.

CHLOROPLASTS

Learning Objectives

2 Diagram the internal structure of a chloroplast, and explain how its components interact and facilitate the process of photosynthesis.

3 Describe what happens to an electron in a biological molecule such as chlorophyll when a photon of light energy is absorbed.

If you examine a section of leaf tissue in a microscope, you see that the green pigment, **chlorophyll,** is not uniformly distributed in the cell but is confined to organelles called **chloroplasts** (Fig. 8-4). In plants, chloroplasts lie mainly inside the leaf, in the cells of the **mesophyll,** a layer with many air spaces and a very high concentration of water vapor (Fig. 8-4a). The interior of the leaf exchanges gases with the outside through microscopic pores, called **stomata** (sing., *stoma*). Each mesophyll cell has 20 to 100 chloroplasts (Fig. 8-4b).

The chloroplast, like the mitochondrion, is enclosed by outer and inner membranes (Fig. 8-4c). The inner membrane encloses a fluid-filled region called the **stroma,** which contains most of the enzymes required to produce carbohydrate molecules. Suspended in the stroma is a third system of membranes that forms an interconnected set of flat, disclike sacs called **thylakoids.** The thylakoid membrane encloses a fluid-filled interior space, the **thylakoid lumen.** In some regions of the chloroplast, thylakoid sacs are arranged in stacks called **grana** (sing., *granum*). Each granum looks something like a stack of coins, with each "coin" being a thylakoid. Some thylakoid membranes extend from one granum to another. These membranes, like the inner mitochondrial membrane (see Chapter 7), are involved in ATP synthesis. (Photosynthetic prokaryotes have no chloroplasts, but thylakoid membranes are often arranged around the periphery of the cell as infoldings of the plasma membrane.)

Chlorophyll is found in the thylakoid membrane

Thylakoid membranes contain several kinds of *pigments,* which are substances that absorb visible light. Different pigments absorb light of different wavelengths. **Chlorophyll,** the main pigment of photosynthesis, absorbs light primarily in the blue and red regions of the visible spectrum. Green light is not appreciably absorbed by chlorophyll. Plants usually appear green because some of the green light that strikes them is scattered or reflected.

A chlorophyll molecule has two main parts, a complex ring and a long side chain (Fig. 8-5). The ring structure, called a *porphyrin ring,* is made up of joined smaller rings composed of

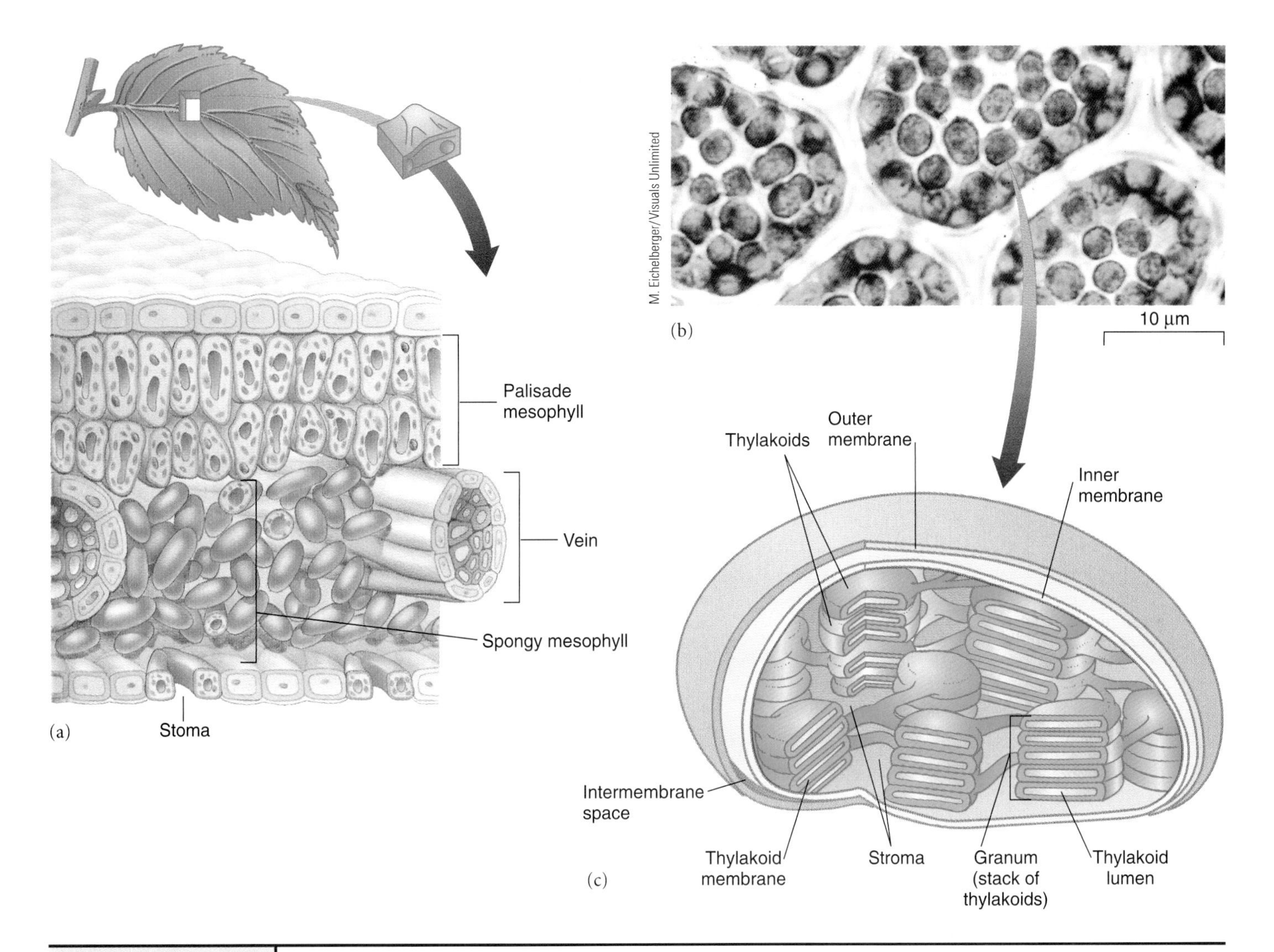

ACTIVE FIGURE **8-4** **The site of photosynthesis.**

(a) This leaf cross section reveals that the mesophyll is the photosynthetic tissue. CO_2 enters the leaf through tiny pores or stomata, and H_2O is carried to the mesophyll in veins. **(b)** Notice the numerous chloroplasts in this LM of plant cells. **(c)** In the chloroplast, pigments necessary for the light-capturing reactions of photosynthesis are part of thylakoid membranes, whereas the enzymes for the synthesis of carbohydrate molecules are in the stroma.

BiologyNow™ Learn more about **photosynthesis in plants** by clicking on this figure on your BiologyNow CD-ROM.

carbon and nitrogen atoms; the porphyrin ring absorbs light energy. The porphyrin ring of chlorophyll is strikingly similar to the heme portion of the red pigment hemoglobin in red blood cells. However, unlike heme, which contains an atom of iron in the center of the ring, chlorophyll contains an atom of magnesium in that position. The chlorophyll molecule also contains a long, hydrocarbon side chain that makes the molecule extremely nonpolar.

All chlorophyll molecules in the thylakoid membrane are associated with specific *chlorophyll-binding proteins;* biologists have identified about 15 different kinds. Each thylakoid membrane is filled with precisely oriented chlorophyll molecules and chlorophyll-binding proteins to facilitate the transfer of energy from one molecule to another.

There are several kinds of chlorophyll. The most important is **chlorophyll *a*,** the pigment that initiates the light-dependent reactions of photosynthesis. **Chlorophyll *b*** is an accessory pigment that also participates in photosynthesis. It differs from chlorophyll *a* only in a functional group on the porphyrin ring: The methyl group (—CH_3) in chlorophyll *a* is replaced in chlorophyll *b* by a terminal carbonyl group (—CHO). This difference shifts the wavelengths of light absorbed and reflected by chlorophyll *b,* making it appear yellow-green, whereas chlorophyll *a* appears bright green.

Chloroplasts have other accessory photosynthetic pigments, such as **carotenoids,** which are yellow and orange (see Fig. 3-14). Carotenoids absorb different wavelengths of light from chlorophyll, thereby expanding the spectrum of light that provides energy for photosynthesis. Chlorophyll may be excited by light directly, by energy passed to it from the light source, or indirectly, by energy passed to it from accessory pigments that have become excited by light. When a carotenoid molecule is excited, its energy can be transferred to chlorophyll *a.* Carotenoids also protect chlorophyll and other parts of the thylakoid membrane

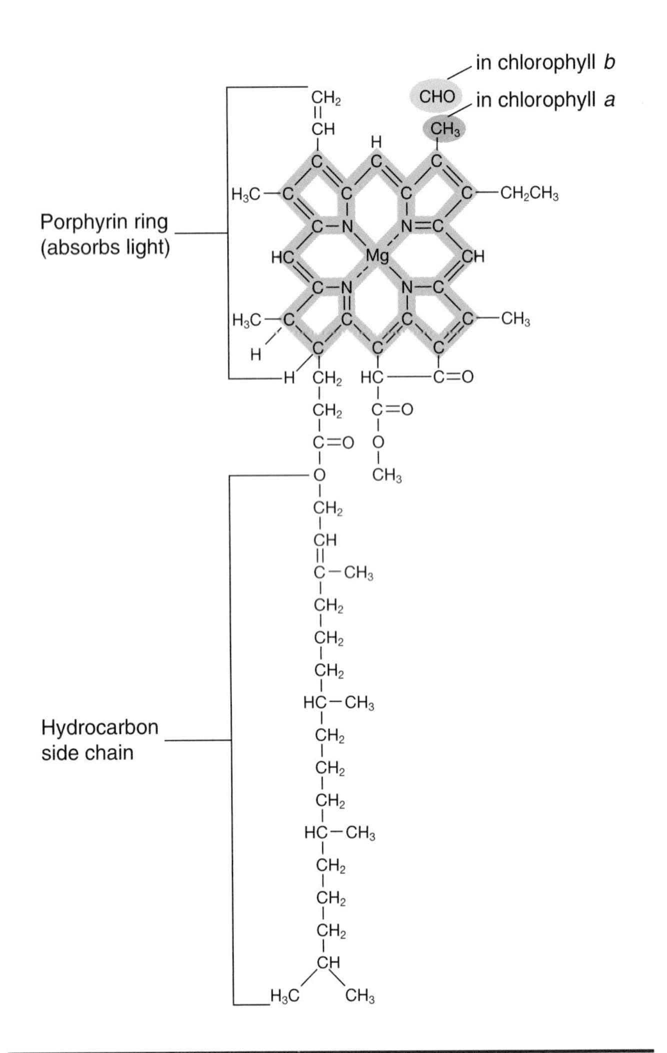

FIGURE **8-5** **The structure of chlorophyll.**

Chlorophyll consists of a porphyrin ring and a hydrocarbon side chain. The porphyrin ring, with a magnesium atom in its center, contains a system of alternating double and single bonds; these are commonly found in molecules that strongly absorb visible light. Notice that at the top right corner of the diagram, the methyl group (—CH_3) distinguishes chlorophyll *a* from chlorophyll *b*, which has a carbonyl group (—CHO) in this position.

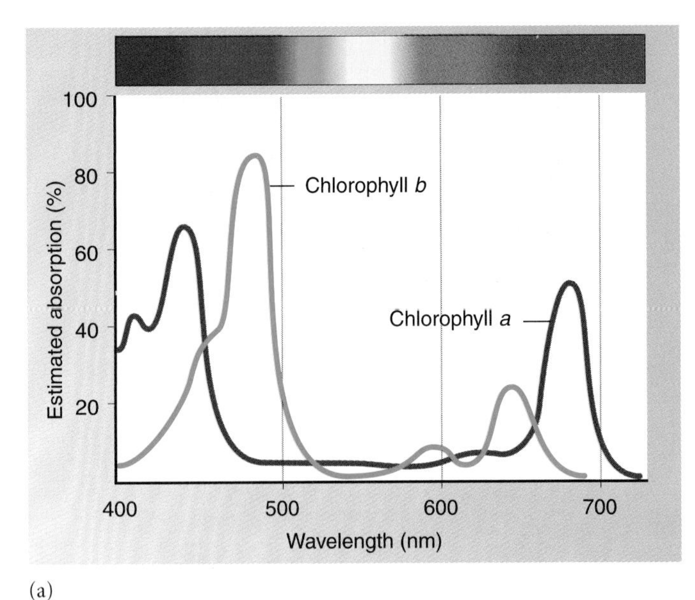

(a)

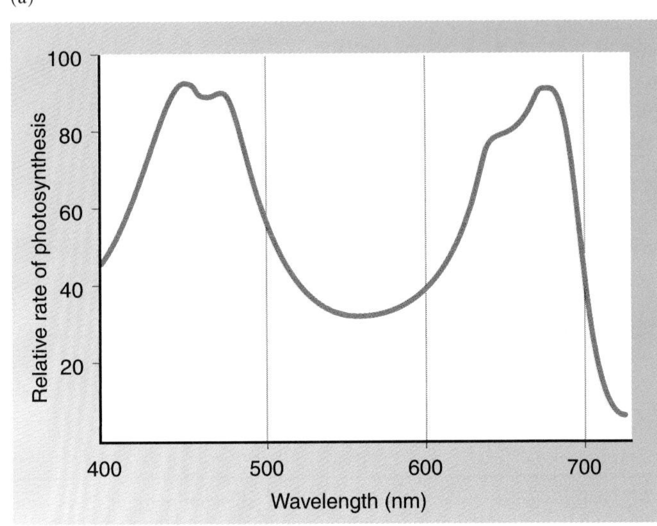

(b)

FIGURE **8-6**

The absorption spectra for chlorophylls *a* and *b* and the action spectrum for photosynthesis.

(a) Chlorophylls *a* and *b* absorb light mainly in the blue (422 to 492 nm) and red (647 to 760 nm) regions. **(b)** The action spectrum of photosynthesis indicates the effectiveness of various wavelengths of light in powering photosynthesis. Many plant species have action spectra for photosynthesis that resemble the generalized action spectrum shown here.

from excess light energy that could easily damage the photosynthetic components. (High light intensities often occur in nature.)

Chlorophyll is the main photosynthetic pigment

As you have seen, the thylakoid membrane contains more than one kind of pigment. An instrument called a *spectrophotometer* measures the relative abilities of different pigments to absorb different wavelengths of light. The **absorption spectrum** of a pigment is a plot of its absorption of light of different wavelengths. Figure 8-6a shows the absorption spectra for chlorophylls *a* and *b*.

An **action spectrum** of photosynthesis is a graph of the relative effectiveness of different wavelengths of light. To obtain an action spectrum, scientists measure the rate of photosynthesis at each wavelength for leaf cells or tissues exposed to monochromatic light (light of one wavelength) (Fig. 8-6b).

PROCESS OF SCIENCE

In a classic biology experiment, the German biologist T.W. Engelmann obtained the first action spectrum in 1883. Engelmann's experiment took advantage of the shape of the chloroplast in a species of the green alga *Spirogyra* (Fig. 8-7a). Its long, filamentous strands are found in freshwater habitats, especially slow-moving or still waters. *Spirogyra* cells each contain a long,

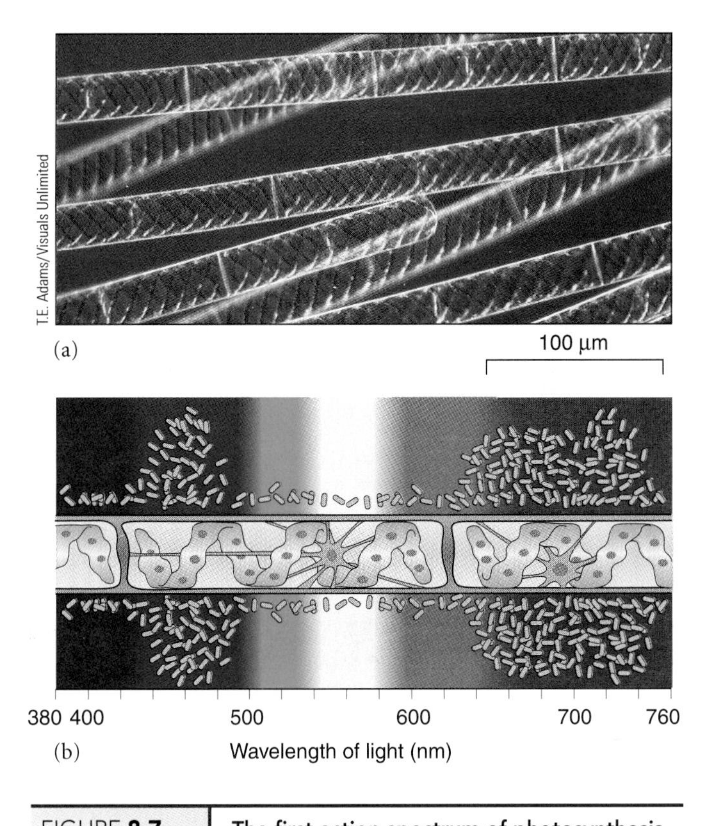

FIGURE **8-7** **The first action spectrum of photosynthesis.**
(a) Light micrograph of filaments of *Spirogyra*, the green alga Engelmann used in his classic experiment. **(b)** The density of bacteria in the blue and red regions of the spectrum indicates the effectiveness of blue and red light for photosynthesis.

spiral-shaped, emerald-green chloroplast embedded in the cytoplasm. Engelmann exposed these cells to a color spectrum produced by passing light through a prism. He hypothesized that if chlorophyll were indeed responsible for photosynthesis, that process would take place most rapidly in the areas where the chloroplast was illuminated by the colors most strongly absorbed by chlorophyll.

Yet how could photosynthesis be measured in those techno logically unsophisticated days? Engelmann knew that photosynthesis produces oxygen and that certain motile bacteria are attracted to areas of high oxygen concentration (Fig. 8-7b). He determined the action spectrum of photosynthesis by observing that the bacteria swam toward the parts of the *Spirogyra* filaments in the blue and red regions of the spectrum. How did Engelmann know bacteria were not simply attracted to blue or red light? As a control, Engelmann exposed bacteria to the spectrum of visible light in the absence of *Spirogyra*. The bacteria showed no preference for any particular wavelength of light. Because the action spectrum of photosynthesis closely matched the absorption spectrum of chlorophyll, Engelmann concluded that chlorophyll in the chloroplasts (and not another compound in another organelle) is responsible for photosynthesis. Numerous studies using sophisticated instruments have since confirmed Engelmann's conclusions.

The action spectrum of photosynthesis does not parallel the absorption spectrum of chlorophyll exactly (see Fig. 8-6). This difference occurs because accessory pigments, such as carotenoids, transfer some of the energy of excitation produced by green light to chlorophyll molecules. The presence of these accessory photosynthetic pigments can be demonstrated by chemical analysis of almost any leaf, although it is obvious in temperate climates when leaves change color in the fall. Toward the end of the growing season, chlorophyll breaks down (and its magnesium is stored in the permanent tissues of the tree), leaving orange and yellow accessory pigments in the leaves.

Review

- What chloroplast membrane is most important in photosynthesis? What two spaces does it separate?
- What is the significance of the fact that the combined absorption spectra of chlorophyll *a* and *b* roughly match the action spectrum of photosynthesis? Why do they not coincide exactly?

Biology Now™ Assess your understanding of **chloroplasts** by taking the pretest on your BiologyNow CD-ROM.

OVERVIEW OF PHOTOSYNTHESIS

Learning Objectives

4 Describe photosynthesis as a redox process.
5 Distinguish between the light-dependent reactions and carbon fixation reactions of photosynthesis.

During photosynthesis, a cell uses light energy captured by chlorophyll to power the synthesis of carbohydrates. The overall reaction of photosynthesis can be summarized as follows:

$$\underset{\text{Carbon dioxide}}{6\ CO_2} + \underset{\text{Water}}{12\ H_2O} \xrightarrow[\text{Chlorophyll}]{\text{Light energy}} \underset{\text{Glucose}}{C_6H_{12}O_6} + \underset{\text{Oxygen}}{6\ O_2} + \underset{\text{Water}}{6\ H_2O}$$

The equation is typically written in the form just given, with H_2O on both sides, because water is a reactant in some reactions and a product in others. Furthermore, all the oxygen produced comes from water, so 12 molecules of water are required to produce 12 oxygen atoms. However, because there is no net yield of H_2O, we can simplify the summary equation of photosynthesis for purposes of discussion:

$$6\ CO_2 + 6\ H_2O \xrightarrow[\text{Chlorophyll}]{\text{Light}} C_6H_{12}O_6 + 6\ O_2$$

When you analyze this process, it appears that hydrogen atoms are transferred from H_2O to CO_2 to form carbohydrate, so you can recognize it as a redox reaction. Recall from Chapter 6 that in a **redox reaction** one or more electrons, usually as part of one or more hydrogen atoms, are transferred from an electron donor (a reducing agent) to an electron acceptor (an oxidizing agent).

$$6\ CO_2 + 6\ H_2O \xrightarrow[\text{Chlorophyll}]{\text{Light}} C_6H_{12}O_6 + 6\ O_2$$

Reduction ($CO_2 \rightarrow C_6H_{12}O_6$); Oxidation ($H_2O \rightarrow O_2$)

When the electrons are transferred, some of their energy is transferred as well. However, the summary equation of photo-

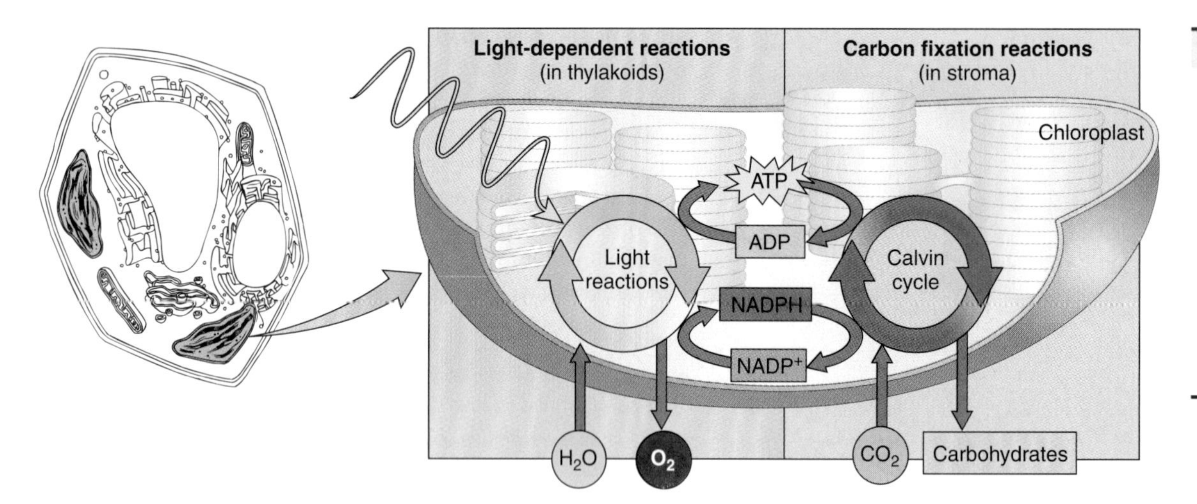

FIGURE **8-8**

An overview of photosynthesis.

Photosynthesis consists of light-dependent reactions, which occur in association with the thylakoids, and carbon fixation reactions, which occur in the stroma.

synthesis is somewhat misleading, because no direct transfer of hydrogen atoms actually occurs. The summary equation describes *what* happens but not *how* it happens. The "how" is more complex and involves multiple steps, many of which are redox reactions.

The reactions of photosynthesis are divided into two phases: the light-dependent reactions (the *photo* part of photosynthesis) and the carbon fixation reactions (the *synthesis* part of photosynthesis). Each set of reactions occurs in a different part of the chloroplast: the light-dependent reactions in association with the thylakoids, and the carbon fixation reactions in the stroma (Fig. 8-8).

ATP and NADPH are the products of the light-dependent reactions: An overview

Light energy is converted to chemical energy in the **light-dependent reactions,** which are associated with the thylakoids. The light-dependent reactions begin as chlorophyll captures light energy, which causes one of its electrons to move to a higher energy state. The energized electron is transferred to an acceptor molecule and is replaced by an electron from H_2O. When this happens, H_2O is split and molecular oxygen is released (Fig. 8-9). Some energy of the energized electrons is used to phosphorylate **adenosine diphosphate (ADP),** forming **adenosine triphosphate (ATP).** In addition, the coenzyme **nicotinamide adenine dinucleotide phosphate ($NADP^+$)** becomes reduced, forming **NADPH.**[1] The products of the light-dependent reactions, ATP and NADPH, are both needed in the endergonic carbon fixation reactions.

Carbohydrates are produced during the carbon fixation reactions: An overview

The ATP and NADPH molecules produced during the light-dependent phase are suited for transferring chemical energy but not for long-term energy storage. For this reason, some of their energy is transferred to chemical bonds in carbohydrates, which can be produced in large quantities and stored for future use. Known as **carbon fixation,** these reactions "fix" carbon atoms from CO_2 to existing skeletons of organic molecules. Because the carbon fixation reactions have no direct requirement for light, they were previously referred to as the "dark" reactions. However, they do not require darkness; in fact, many of the enzymes involved in carbon fixation are much more active in the light than in the dark. Furthermore, carbon fixation reactions depend on the products of the light-dependent reactions. Carbon fixation reactions take place in the stroma of the chloroplast.

Now that we have presented an overview of photosynthesis, let's examine the entire process more closely.

Review

- Which is more oxidized, oxygen that is part of a water molecule, or molecular oxygen?
- In what ways do the carbon fixation reactions depend on the light-dependent reactions?

BiologyNow™ Assess your understanding of **photosynthesis** by taking the pretest on your BiologyNow CD-ROM.

Bernd Wittich/Visuals Unlimited

FIGURE **8-9** **Oxygen produced by photosynthesis.**

On sunny days, the oxygen released by aquatic plants is sometimes visible as bubbles in the water. This plant (*Elodea*) is actively carrying on photosynthesis.

[1] Although the correct way to write the reduced form of $NADP^+$ is $NADPH + H^+$, for simplicity's sake we present the reduced form as NADPH throughout the book.

THE LIGHT-DEPENDENT REACTIONS

Learning Objectives

6 Describe the flow of electrons through photosystems I and II in the noncyclic electron transport pathway; contrast this with cyclic electron transport.

7 Explain how a proton (H^+) gradient is established across the thylakoid membrane and how this gradient functions in ATP synthesis.

In the light-dependent reactions, the radiant energy from sunlight phosphorylates ADP, producing ATP, and reduces $NADP^+$, forming NADPH. The light energy that chlorophyll captures is temporarily stored in these two compounds. The light-dependent reactions are summarized as follows:

$$12\ H_2O + 12\ NADP^+ + 18\ ADP + 18\ P_i \xrightarrow[\text{Chlorophyll}]{\text{Light}} 6\ O_2 + 12\ NADPH + 18\ ATP$$

Photosystems I and II each consist of a reaction center and multiple antenna complexes

The light-dependent reactions of photosynthesis begin when chlorophyll *a* and/or accessory pigments absorb light. According to the currently accepted model, chlorophylls *a* and *b* and accessory pigment molecules are organized with pigment-binding proteins in the thylakoid membrane into units called **antenna complexes.** The pigments and associated proteins are arranged as highly ordered groups of about 250 chlorophyll molecules associated with specific enzymes and other proteins. Each antenna complex absorbs light energy and transfers it to the **reaction center,** which consists of chlorophyll molecules and proteins, including electron transfer components, that participate directly in photosynthesis (Fig. 8-10). Light energy is converted to chemical energy in the reaction centers by a series of electron transfer reactions.

Two types of photosynthetic units, designated photosystem I and photosystem II, are involved in photosynthesis. Their reaction centers are distinguishable because they are associated with proteins in a way that causes a slight shift in their absorption spectra. Ordinary chlorophyll *a* has a strong absorption peak at about 660 nm. In contrast, the chlorophyll *a* molecule that makes up the reaction center associated with **photosystem I** has an absorption peak at 700 nm and is referred to as **P700.** The reaction center of **photosystem II** is made up of a chlorophyll *a* molecule with an absorption peak of about 680 nm and is referred to as **P680.**

When a pigment molecule absorbs light energy, that energy is passed from one pigment molecule to another until it reaches the reaction center. When the energy reaches a molecule of P700 (in a photosystem I reaction center) or P680 (in a photosystem II reaction center), an electron is then raised to a higher energy level. As we explain in the next section, this energized electron can be donated to an electron acceptor that becomes reduced in the process.

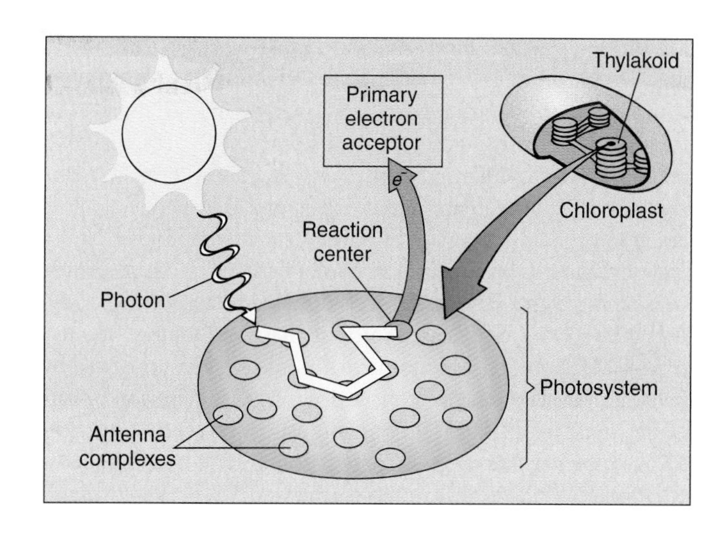

FIGURE **8-10** **A photosystem.**
Chlorophyll molecules and accessory pigments are arranged in light-harvesting arrays, or antenna complexes. When a molecule in an antenna complex absorbs a photon, the photon's energy is funneled into the reaction center. When this energy reaches the P700 (or P680) chlorophyll molecule in the reaction center, an electron becomes energized and is accepted by a primary electron acceptor.

Noncyclic electron transport produces ATP and NADPH

Let's begin our discussion of **noncyclic electron transport** with the events associated with photosystem I (Fig. 8-11). A pigment molecule in an antenna complex associated with photosystem I absorbs a photon of light. The absorbed energy is transferred to the reaction center, where it excites an electron in a molecule of P700. This energized electron is transferred to a primary electron acceptor, which is the first of several electron acceptors in a series. (Uncertainty exists regarding the exact chemical nature of the primary electron acceptor for photosystem I.) The energized electron is passed along an **electron transport chain** from one electron acceptor to another, until it is passed to *ferredoxin,* an iron-containing protein. Ferredoxin transfers the electron to $NADP^+$ in the presence of the enzyme *ferredoxin–$NADP^+$ reductase.* When $NADP^+$ accepts 2 electrons, they unite with a proton (H^+); thus the reduced form of $NADP^+$ is NADPH, which is released into the stroma. P700 becomes positively charged when it gives up an electron to the primary electron acceptor; the missing electron is replaced by one donated by photosystem II.

Like photosystem I, photosystem II is activated when a pigment molecule in an antenna complex absorbs a photon of light energy. The energy is transferred to the reaction center, where it causes an electron in a molecule of P680 to move to a higher energy level. This energized electron is accepted by a primary electron acceptor (a highly modified chlorophyll molecule known as *pheophytin*) and then passes along an electron transport chain until it is donated to P700 in photosystem I.

How is the electron that has been donated to the electron transport chain replaced? This occurs through **photolysis** (light splitting) of water, a process that not only yields electrons, but is also the source of almost all the oxygen in Earth's atmosphere. A molecule of P680 that has given up an energized electron to the primary electron acceptor is positively charged. This P680 molecule is an oxidizing agent so strong that it pulls electrons away from an oxygen atom that is part of a H_2O molecule. In a reaction probably catalyzed by a unique, manganese-containing enzyme, water is broken into its components: 2 electrons, 2 protons, and oxygen. Each electron is donated to a P680 molecule, and the protons are released into the thylakoid lumen. Because oxygen does not exist in atomic form, the oxygen produced by splitting one H_2O molecule is written $\frac{1}{2}$ O_2. Two water molecules must be split to yield one molecule of oxygen. The photolysis of water is a remarkable reaction, but its name is somewhat misleading because it implies that water is broken by light. Actually, light splits water indirectly, by causing P680 to become oxidized.

Noncyclic electron transport is a continuous linear process

In the presence of light, there is a continuous, one-way flow of electrons from the ultimate electron source, H_2O, to the terminal electron acceptor, $NADP^+$. Water undergoes enzymatically catalyzed photolysis to replace energized electrons donated to the electron transport chain by molecules of P680 in photosystem II. These electrons travel down the electron transport chain that connects photosystem II with photosystem I. Thus they provide a continuous supply of replacements for energized electrons that have been given up by P700.

As electrons are transferred along the electron transport chain that connects photosystem II with photosystem I, they lose energy. Some of the energy released is used to pump protons across the thylakoid membrane, from the stroma to the thylakoid lumen, producing a proton gradient. The energy of this proton gradient is harnessed to produce ATP from ADP by chemiosmosis, which we will discuss shortly. ATP and NADPH, the products of the light-dependent reactions, are released into the stroma, where both are required in the carbon fixation reactions.

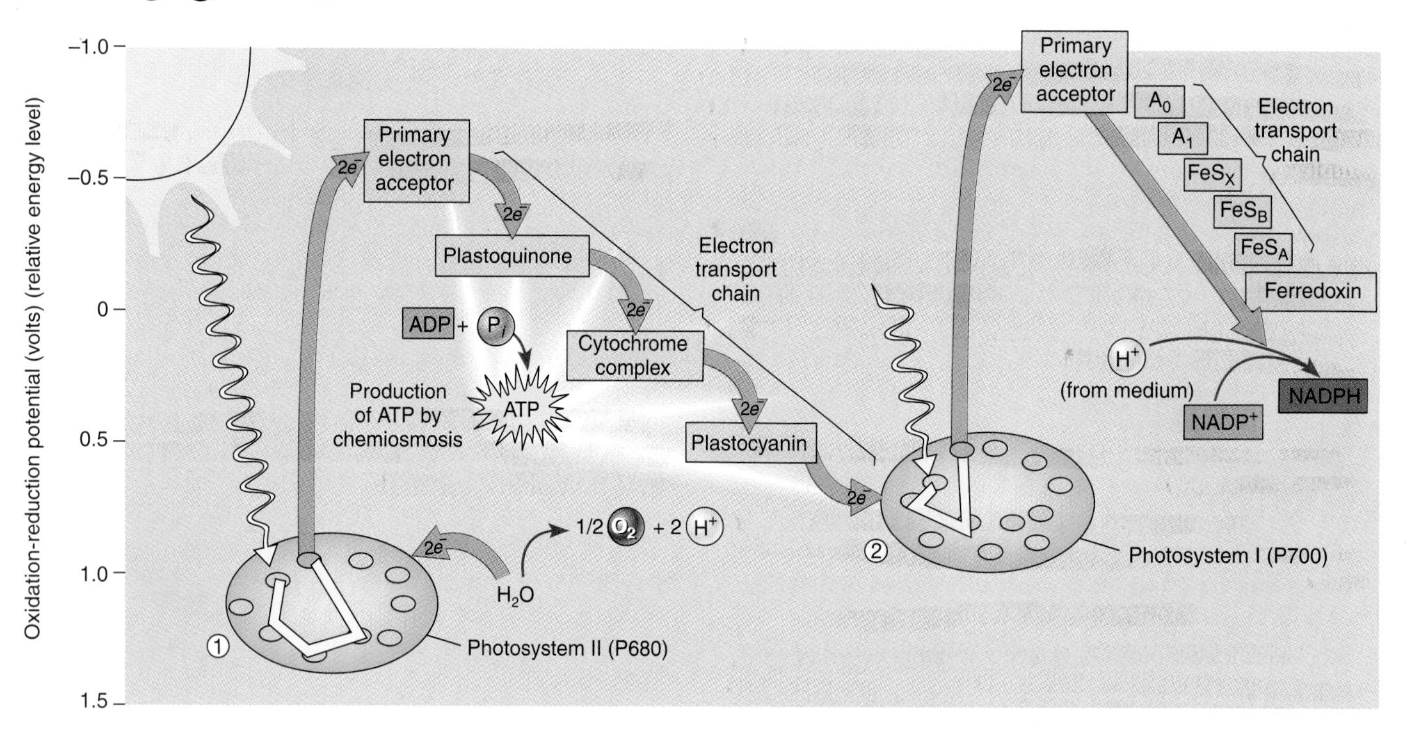

ACTIVE FIGURE **8-11** | **Noncyclic electron transport.**

In noncyclic electron transport, the formation of ATP is coupled to the one-way flow of energized electrons *(orange arrows)* from H_2O *(lower left)* to $NADP^+$ *(middle right)*. Single electrons actually pass down the electron transport chain; 2 are shown in this figure because 2 electrons are required to form one molecule of NADPH. ① Electrons are supplied to the system from the splitting of H_2O by photosystem II, with the release of O_2 as a byproduct. When photosystem II is activated by absorbing photons, electrons are passed along an electron transport chain and are eventually donated to photosystem I. ② Electrons in photosystem I are "re-energized" by the absorption of additional light energy and are passed to $NADP^+$, forming NADPH.

BiologyNow™ Experience the process of **noncyclic electron transport** by clicking on this figure on your BiologyNow CD-ROM.

Cyclic electron transport produces ATP but no NADPH

Only photosystem I is involved in **cyclic electron transport**, the simplest light-dependent reaction. The pathway is cyclic because energized electrons that originate from P700 at the reaction center eventually return to P700. In the presence of light, electrons flow continuously through an electron transport chain within the thylakoid membrane. As they pass from one acceptor to another, the electrons lose energy, some of which is used to pump protons across the thylakoid membrane. An enzyme (ATP synthase, discussed shortly) in the thylakoid membrane uses the energy of the proton gradient to manufacture ATP. NADPH is not produced, H_2O is not split, and oxygen is not generated. By itself, cyclic electron transport could not serve as the basis of photosynthesis because, as we explain later in the chapter, NADPH is required to reduce CO_2 to carbohydrate.

The significance of cyclic electron transport to photosynthesis in plants is unclear. Cyclic electron transport may occur in plant cells when there is too little $NADP^+$ to accept electrons from ferredoxin. Biologists generally agree that ancient bacteria used this process to produce ATP from light energy. A reaction pathway analogous to cyclic electron transport in plants is present in some modern photosynthetic prokaryotes. Noncyclic and cyclic electron transport are compared in Table 8-1.

ATP synthesis occurs by chemiosmosis

Each member of the electron transport chain that links photosystem II to photosystem I can exist in an oxidized (lower energy) form and a reduced (higher energy) form. The electron accepted from P680 by the primary electron acceptor is highly energized; it is passed from one carrier to the next in a series of exergonic redox reactions, losing some of its energy at each step. Some of the energy given up by the electron is not lost by the system, however; it is used to drive the synthesis of ATP, an endergonic reaction. Because the synthesis of ATP (that is, the phosphorylation of ADP) is coupled to the transport of electrons that have been energized by photons of light, the process is called **photophosphorylation.**

TABLE 8-1 A Comparison of Noncyclic and Cyclic Electron Transport

	Noncyclic Electron Transport	Cyclic Electron Transport
Electron source	H_2O	None—electrons cycle through the system
Oxygen released?	Yes (from H_2O)	No
Terminal electron acceptor	$NADP^+$	None—electrons cycle through the system
Form in which energy is temporarily captured	ATP (by chemiosmosis); NADPH	ATP (by chemiosmosis)
Photosystem(s) required	PS I (P700) and PS II (P680)	PS I (P700) only

The chemiosmotic model explains the coupling of ATP synthesis and electron transport

As discussed earlier, the pigments and electron acceptors of the light-dependent reactions are embedded in the thylakoid membrane. Energy released from electrons traveling through the chain of acceptors is used to pump protons from the stroma, across the thylakoid membrane, and into the thylakoid lumen (Fig. 8-12). Thus the pumping of protons results in the formation of a proton gradient across the thylakoid membrane. Protons also accumulate in the thylakoid lumen as water is split during noncyclic electron transport. Because protons are actually hydrogen ions (H^+), the accumulation of protons causes the pH of the thylakoid interior to fall to a pH of about 5 in the thylakoid lumen, compared to a pH of about 8 in the stroma. This difference of about 3 pH units across the thylakoid membrane means there is an approximately 1000-fold difference in hydrogen ion concentration.

The proton gradient has a great deal of free energy because of its state of low entropy. How does the chloroplast convert that energy to a more useful form? According to the general principles of diffusion, the concentrated protons inside the thylakoid might be expected to diffuse out readily. However, they

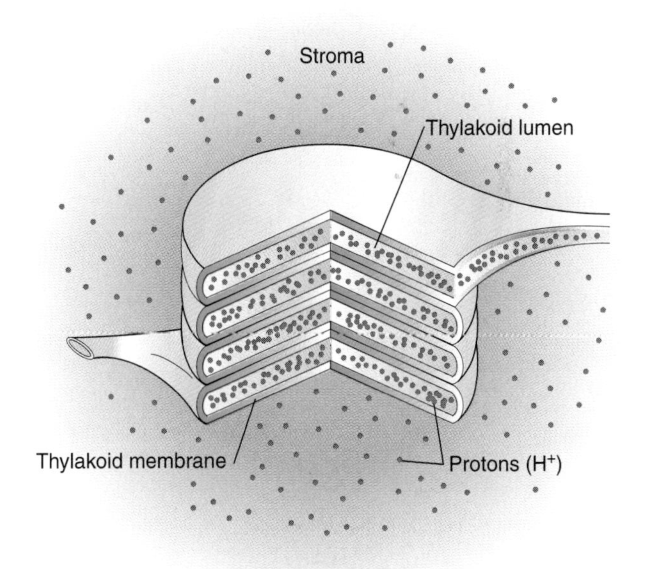

FIGURE 8-12 The accumulation of protons in the thylakoid lumen.

As electrons move down the electron transport chain, protons (H^+) move from the stroma to the thylakoid lumen, creating a proton gradient. The greater concentration of H^+ in the thylakoid lumen lowers the pH.

are prevented from doing so because the thylakoid membrane is impermeable to H^+ except through certain channels formed by the enzyme **ATP synthase.** This enzyme, a transmembrane protein, forms complexes so large they can be seen in electron micrographs; these complexes project into the stroma. As the protons diffuse through an ATP synthase complex, free energy decreases as a consequence of an increase in entropy. Each ATP synthase complex couples this exergonic process of diffusion down a concentration gradient to the endergonic process of phosphorylation of ADP to form ATP, which is released into the stroma (Fig. 8-13). The movement of protons through ATP synthase is thought to induce changes in the conformation of the enzyme that are necessary for the synthesis of ATP. It is estimated that for every 4 protons that move through ATP synthase, one ATP molecule is synthesized.

The mechanism by which the phosphorylation of ADP is coupled to diffusion down a proton gradient is called **chemiosmosis.** As the essential connection between the electron transport chain and the phosphorylation of ADP, chemiosmosis is a basic mechanism of energy coupling in cells. You may recall from Chapter 7 that chemiosmosis also occurs in aerobic respiration (see Table 8-2).

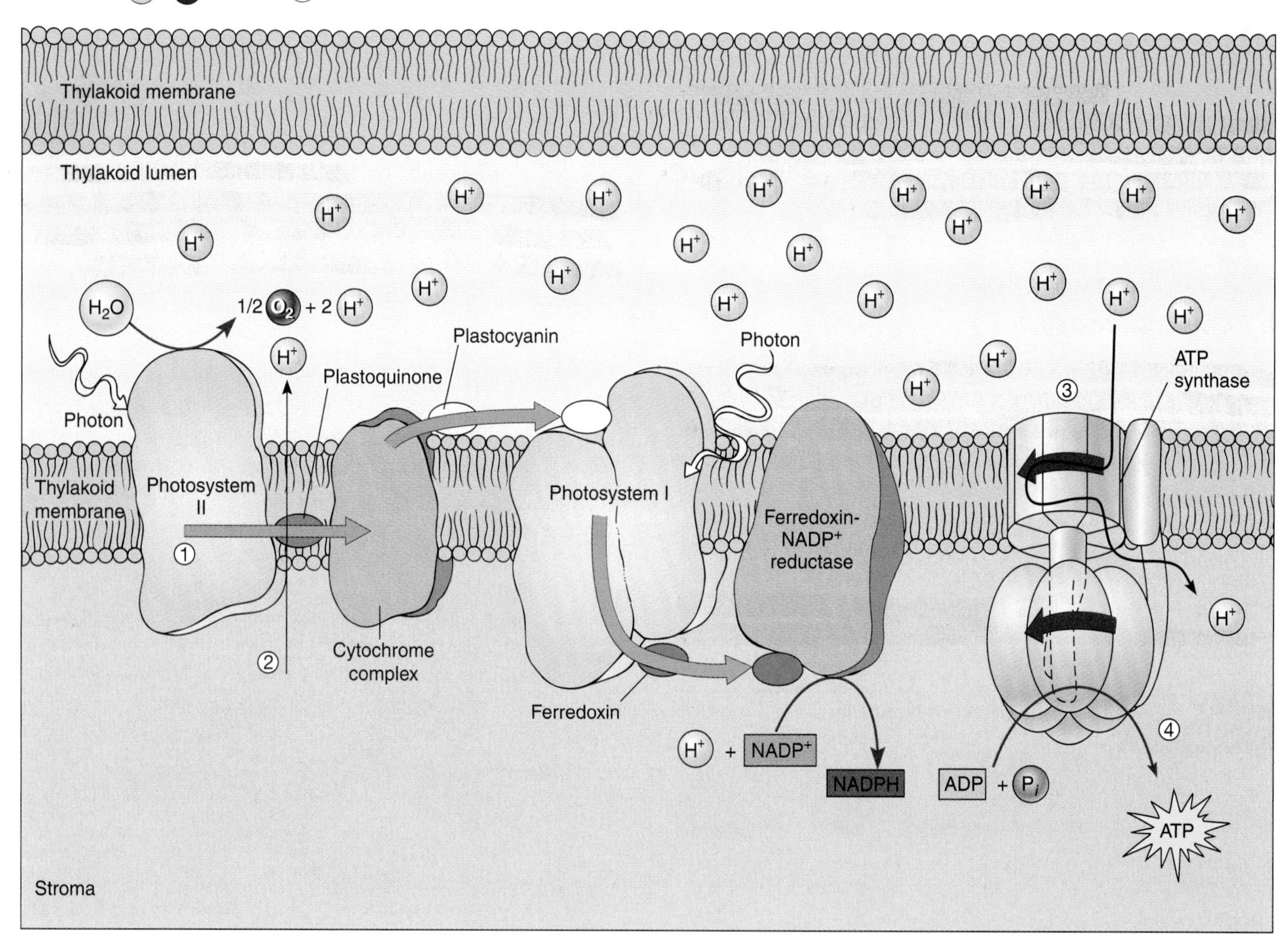

FIGURE **8-13** | **A detailed look at electron transport and chemiosmosis.**

① The orange arrows indicate the pathway of electrons along the electron transport chain in the thylakoid membrane. The electron carriers within the membrane become alternately reduced and oxidized as they accept and donate electrons. ② The energy released during electron transport is used to transport H^+ from the stroma to the thylakoid lumen, where a high concentration of H^+ accumulates. ③ The H^+ are prevented from diffusing back into the stroma except through special channels in ATP synthase in the thylakoid membrane. ④ The flow of the H^+ through ATP synthase generates ATP.

TABLE 8-2 A Comparison of Photosynthesis and Aerobic Respiration

	Photosynthesis	Aerobic Respiration
Type of metabolic reaction	Anabolism	Catabolism
Raw materials	CO_2, H_2O	$C_6H_{12}O_6$, O_2
End products	$C_6H_{12}O_6$, O_2	CO_2, H_2O
Which cells have these processes?	Cells that contain chlorophyll (certain cells of plants, algae, and some bacteria)	Every actively metabolizing cell has aerobic respiration or some other energy-releasing pathway
Sites involved (in eukaryotic cells)	Chloroplasts	Cytosol (glycolysis); mitochondria
ATP production	By photophosphorylation (a chemiosmotic process)	By substrate-level phosphorylation and by oxidative phosphorylation (a chemiosmotic process)
Principal electron transfer compound	$NADP^+$ is reduced to form NADPH*	NAD^+ is reduced to form NADH*
Location of electron transport chain	Thylakoid membrane	Mitochondrial inner membrane (cristae)
Source of electrons for electron transport chain	In noncyclic electron transport: H_2O (undergoes photolysis to yield electrons, protons, and oxygen)	Immediate source: NADH, $FADH_2$ Ultimate source: glucose or other carbohydrate
Terminal electron acceptor for electron transport chain	In concyclic electron transport: $NADP^+$ (becomes reduced to form NADPH)	O_2 (becomes reduced to form H_2O)

*NADPH and NADH are very similar hydrogen (i.e., electron) carriers, differing only in a single phosphate group. However, NADPH generally works with enzymes in anabolic pathways, such as photosynthesis. NADH is associated with catabolic pathways, such as cellular respiration.

Review

- Why is molecular oxygen a necessary byproduct of photosynthesis?
- What process is the actual mechanism of photophosphorylation?
- Why are both photosystems I and II required for photosynthesis? Can cyclic phosphorylation alone support photosynthesis?

BiologyNow™ Assess your understanding of **light-dependent reactions** by taking the pretest on your BiologyNow CD-ROM.

THE CARBON FIXATION REACTIONS

Learning Objectives

8 Summarize the three phases of the Calvin cycle, and indicate the roles of ATP and NADPH in the process.

9 Discuss how photorespiration reduces photosynthetic efficiency.

10 Compare the C_4 and CAM pathways.

In carbon fixation, the energy of ATP and NADPH is used in the formation of organic molecules from CO_2. The carbon fixation reactions may be summarized as follows:

$$12\ NADPH + 18\ ATP + 6\ CO_2 \longrightarrow C_6H_{12}O_6 + 12\ NADP^+ + 18\ ADP + 18\ P_i + 6\ H_2O$$

Most plants use the Calvin cycle to fix carbon

Carbon fixation occurs in the stroma through a sequence of 13 reactions known as the **Calvin cycle**. During the 1950s, University of California researchers Melvin Calvin, Andrew Benson, and others elucidated the details of this cycle. Calvin was awarded a Nobel Prize for chemistry in 1961.

The 13 reactions of the Calvin cycle are divided into three phases: CO_2 uptake, carbon reduction, and RuBP regeneration (Fig. 8-14). All 13 enzymes that catalyze steps in the Calvin cycle are located in the stroma of the chloroplast. Ten of the enzymes also participate in glycolysis (see Chapter 7). These enzymes catalyze reversible reactions, degrading carbohydrate molecules in cellular respiration and synthesizing carbohydrate molecules in photosynthesis.

1. *CO_2 uptake.* The first phase of the Calvin cycle consists of a single reaction in which a molecule of CO_2 reacts with a phosphorylated five-carbon compound, *ribulose bisphosphate (RuBP).* This reaction is catalyzed by the enzyme *ribulose bisphosphate carboxylase/oxygenase,* also known as **rubisco.** More rubisco enzyme is present in the chloroplast than any other protein, and it may be one of the most abundant proteins in the biosphere. The product of this reaction is an unstable, six-carbon intermediate, which immediately breaks down into two molecules of **phosphoglycerate (PGA)** with three carbons each. The carbon that was originally part of a CO_2 molecule is now part of a carbon skeleton; the carbon has been "fixed." The Calvin cycle is also known as the **C_3 pathway** because the product of the initial carbon fixation reaction is a three-carbon compound. Plants that initially fix carbon in this way are called **C_3 plants.**

2. *Carbon reduction.* The second phase of the Calvin cycle consists of two steps in which the energy and reducing power from ATP and NADPH (both produced in the light-dependent reactions) are used to convert the PGA molecules to **glyceraldehyde-3-phosphate (G3P).** As shown in Figure 8-14, for every six carbons that enter the cycle as CO_2, six carbons can leave the system as two molecules of G3P, to be used in carbohydrate synthesis. Each of these three carbon molecules of G3P is essentially half a hexose (six-carbon sugar) molecule. (In fact, you may recall that G3P is a key intermediate in the splitting of sugar in glycolysis; see Figs. 7-3 and 7-4.)

The reaction of two molecules of G3P is exergonic and leads to the formation of glucose or fructose. In some plants, glucose and fructose are then joined to produce sucrose (common table sugar). Sucrose is harvested from sugar cane, sugar beets, and maple sap. The plant cell also uses glucose to produce starch or cellulose.

3. *RuBP regeneration.* Notice that although 2 G3P molecules are removed from the cycle, 10 G3P molecules remain; this represents 30 carbon atoms in all. Through a series of 10 reactions that make up the third phase of the Calvin cycle, these 30 carbons and their associated atoms become rearranged into six molecules of ribulose phosphate, each of which becomes phosphorylated to produce RuBP, the 5-carbon compound with which the cycle started. These RuBP molecules begin the process of CO_2 fixation and eventual G3P production once again.

In summary, the inputs required for the carbon fixation reactions are six molecules of CO_2, phosphates transferred from ATP, and electrons (as hydrogen) from NADPH. In the end, the six carbons from the CO_2 are accounted for by the harvest of a hexose molecule. The remaining G3P molecules are used to

ACTIVE FIGURE **8-14** | **A detailed look at the Calvin cycle.**

① This diagram, in which carbon atoms are black balls, shows that six molecules of CO_2 must be "fixed" (incorporated into pre-existing carbon skeletons) in the CO_2 uptake phase to produce one molecule of a six-carbon sugar such as glucose. ② Glyceraldehyde-3-phosphate (G3P) is formed in the carbon reduction phase. Two G3P molecules "leave" the cycle for every glucose formed. ③ Ribulose bisphosphate (RuBP) is regenerated and a new cycle can begin. Although these reactions do not require light directly, the energy that drives the Calvin cycle comes from ATP and NADPH, which are the products of the light-dependent reactions.

Biology Now™ See the **Calvin cycle in action** by clicking on this figure on your BiologyNow CD-ROM.

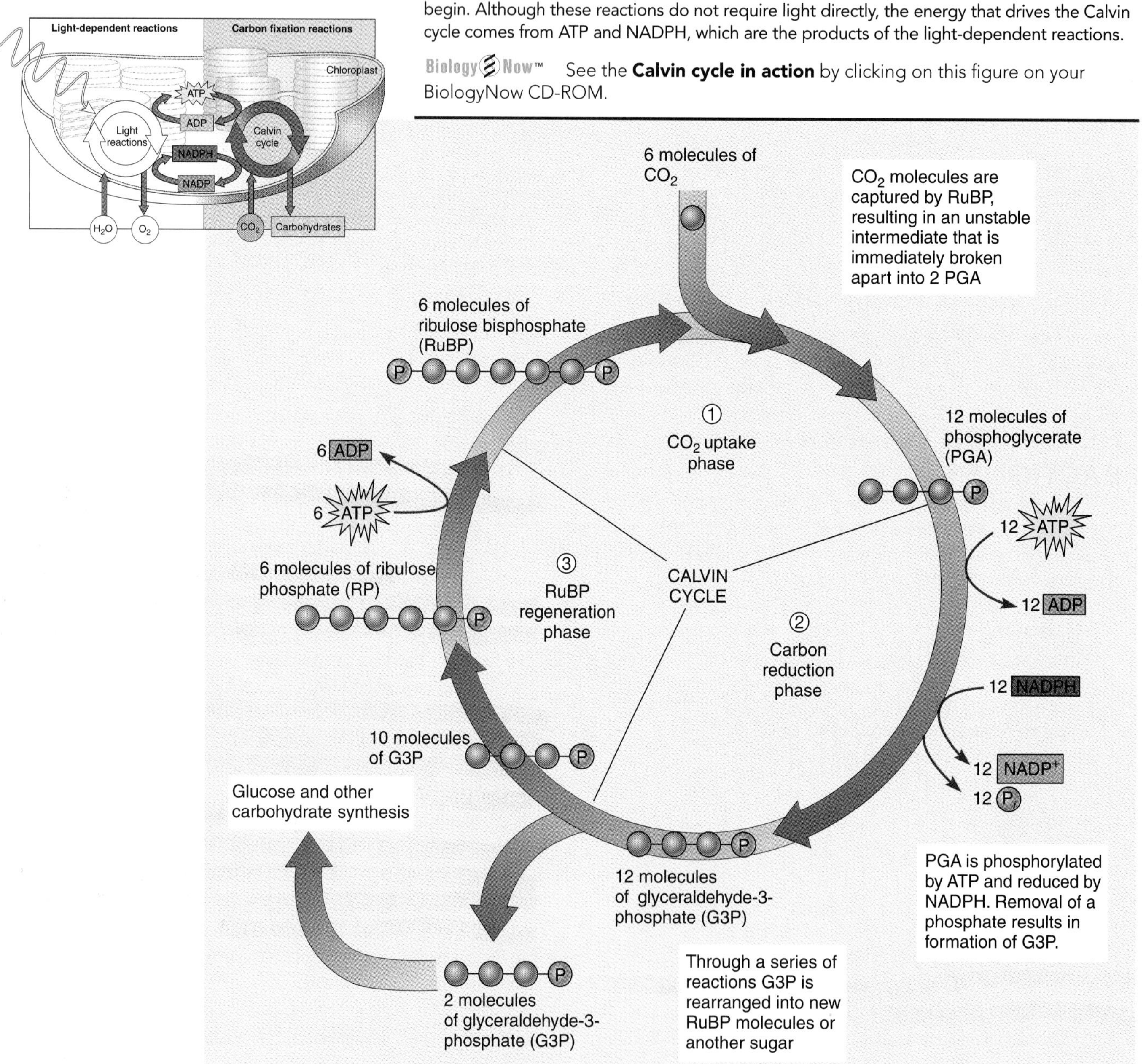

TABLE 8-3 Summary of Photosynthesis

Reaction Series	Summary of Process	Needed Materials	End Products
Light-dependent reactions (take place in thylakoid membranes)	Energy from sunlight used to split water, manufacture ATP, and reduce $NADP^+$		
Photochemical reactions	Chlorophyll-activated; reaction center gives up photoexcited electron to electron acceptor	Light energy; pigments (chlorophyll)	Electrons
Electron transport	Electrons transported along chain of electron acceptors in thylakoid membranes; electrons reduce $NADP^+$; splitting of water provides some H^+ that accumulates inside thylakoid space	Electrons, $NADP^+$, H_2O, electron acceptors	NADPH, O_2
Chemiosmosis	H^+ permitted to diffuse across the thylakoid membrane down their gradient; they cross the membrane through special channels in ATP synthase complex; energy released is used to produce ATP	Proton gradient, ADP + P_i	ATP
Carbon fixation reactions (take place in stroma)	Carbon fixation: Carbon dioxide used to make carbohydrate	Ribulose bisphosphate, CO_2, ATP, NADPH, necessary enzymes	Carbohydrates, ADP + P_i, $NADP^+$

synthesize the RuBP molecules with which more CO_2 molecules may combine. Table 8-3 provides a summary of photosynthesis.

Photorespiration reduces photosynthetic efficiency

Many C_3 plants, including certain agriculturally important crops such as soybeans, wheat, and potatoes, do not yield as much carbohydrate from photosynthesis as might be expected, especially during very hot spells in summer. This phenomenon is a consequence of tradeoffs between the plant's need for CO_2 and its need to prevent water loss. Recall that most photosynthesis occurs in mesophyll cells inside the leaf and that the entry and exit of gases from the interior of the leaf is regulated by stomata, tiny pores concentrated on the underside of the leaf (see Fig. 8-4a). On hot, dry days, plants close their stomata to conserve water. Once the stomata close, photosynthesis rapidly uses up the CO_2 remaining in the leaf and produces O_2, which accumulates in the chloroplasts.

Recall that the enzyme RuBP carboxylase/oxygenase (rubisco) catalyzes CO_2 fixation in the Calvin cycle by attaching CO_2 to RuBP. As its full name implies, rubisco acts not only as a carboxylase, but also as an oxygenase because high levels of O_2 compete with CO_2 for the active site of rubisco. Some of the intermediates involved in the Calvin cycle are degraded to CO_2 and H_2O in a process that is called **photorespiration,** because (1) it occurs in the presence of light; (2) it requires oxygen, like aerobic respiration; and (3) it produces CO_2 and H_2O, like aerobic respiration. However, photorespiration does not produce ATP, and it reduces photosynthetic efficiency because it removes some of the intermediates used in the Calvin cycle.

The reasons for photorespiration are incompletely understood, although scientists hypothesize that it reflects the origin of rubisco at an ancient time when CO_2 levels were high and molecular oxygen levels were low. Genetic engineering to produce plants with Rubisco that has a much lower affinity for oxygen is a promising area of research to improve yields of certain valuable crop plants.

The initial carbon fixation step differs in C_4 plants and in CAM plants

Photorespiration is not the only problem faced by plants engaged in photosynthesis. Because CO_2 is not a very abundant gas (composing only about 0.03% of the atmosphere), it is not easy for plants to obtain the CO_2 they need. As you have learned, when conditions are hot and dry, the stomata close to reduce the loss of water vapor, greatly diminishing the supply of CO_2. Ironically, CO_2 is potentially less available at the very times when maximum sunlight is available to power the light-dependent reactions.

Many plant species living in hot, dry environments have adaptations that facilitate carbon fixation. **C_4 plants** first fix CO_2 into a four-carbon compound, oxaloacetate. **CAM plants** initially fix carbon at night through the formation of oxaloacetate. These special pathways found in C_4 and CAM plants precede the Calvin cycle (C_3 pathway); they do not replace it.

The C_4 pathway efficiently fixes CO_2 at low concentrations

The **C_4 pathway,** in which CO_2 is fixed through the formation of oxaloacetate, occurs not only before the C_3 pathway but also in different cells. Leaf anatomy is usually distinctive in C_4 plants. The photosynthetic mesophyll cells are closely associated with prominent, chloroplast-containing **bundle sheath cells,** which tightly encircle the veins of the leaf (Fig. 8-15). The C_4 pathway occurs in the mesophyll cells, whereas the Calvin cycle takes place within the bundle sheath cells.

The key component of the C_4 pathway is a remarkable enzyme that has an extremely high affinity for CO_2, binding it effectively even at unusually low concentrations. This enzyme, **PEP carboxylase,** catalyzes the reaction by which CO_2 reacts

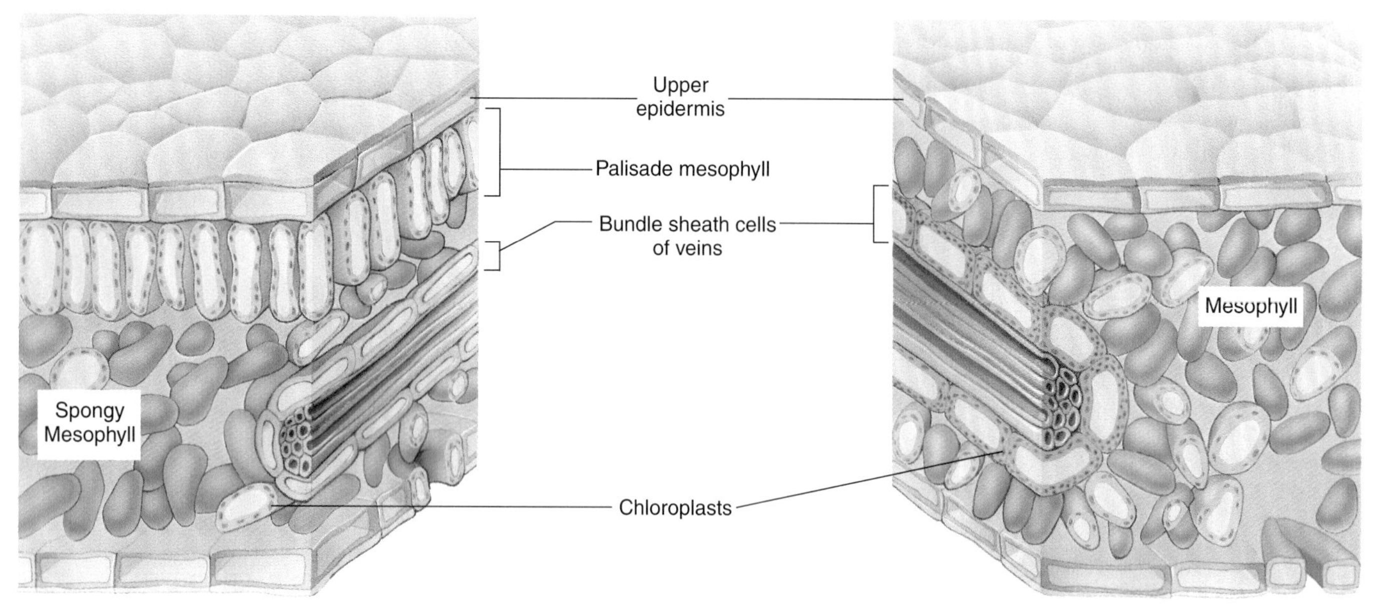

(a) Arrangement of cells in a C_3 leaf

(b) Arrangement of cells in a C_4 leaf

FIGURE **8-15** | **C_3 and C_4 plant structure compared.**
(a) In C_3 plants, the Calvin cycle takes place in the mesophyll cells, and the bundle sheath cells are nonphotosynthetic. **(b)** In C_4 plants, reactions that fix CO_2 into four-carbon compounds take place in the mesophyll cells. The four-carbon compounds are transferred from the mesophyll cells to the photosynthetic bundle sheath cells, where the Calvin cycle takes place.

with the three-carbon compound **phosphoenolpyruvate (PEP),** forming **oxaloacetate** (Fig. 8-16).

In a step that requires NADPH, oxaloacetate is converted to some other four-carbon compound, usually malate. The malate then passes to chloroplasts within bundle sheath cells, where a different enzyme catalyzes the decarboxylation of malate to yield pyruvate (which has three carbons) and CO_2. NADPH is formed, replacing the one used earlier.

$$\text{Malate} + \text{NADP}^+ \longrightarrow \text{pyruvate} + CO_2 + \text{NADPH}$$

The CO_2 released in the bundle sheath cell combines with ribulose bisphosphate in a reaction catalyzed by rubisco, and goes through the Calvin cycle in the usual manner. The pyruvate formed in the decarboxylation reaction returns to the mesophyll cell, where it reacts with ATP to regenerate phosphoenolpyruvate.

Because the C_4 pathway captures CO_2 and provides it to the bundle sheath cells so efficiently, CO_2 concentration within the bundle sheath cells is about 10 to 60 times greater than its concentration in the mesophyll cells of plants having only the C_3 pathway. Photorespiration is negligible in C_4 plants, because the concentration of CO_2 in bundle sheath cells (where rubisco is present) is always high.

The combined C_3–C_4 pathway involves the expenditure of 30 ATPs per hexose, rather than the 18 ATPs used by the C_3

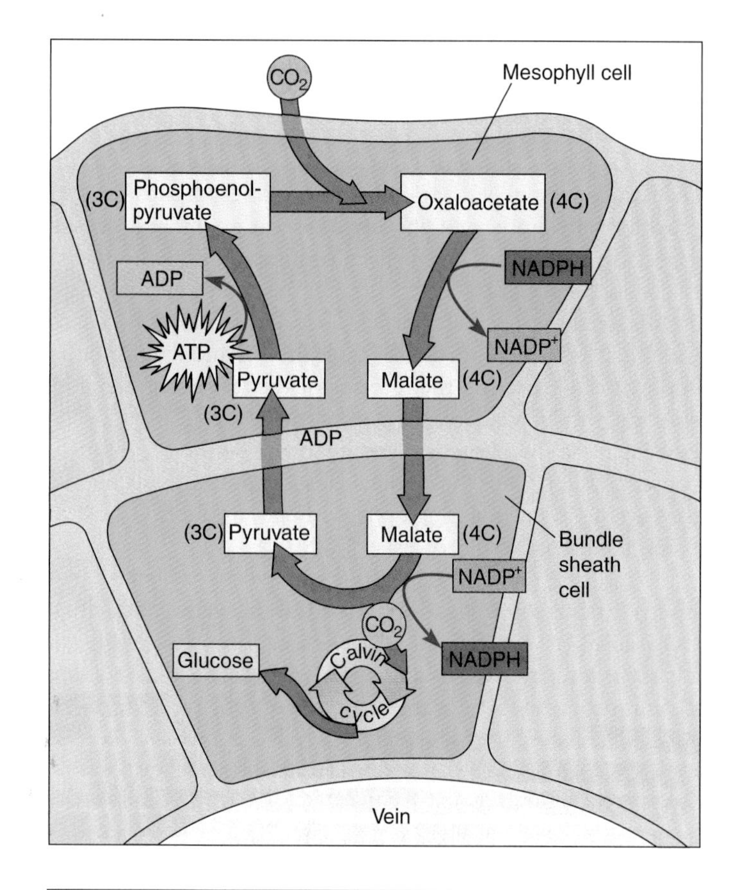

FIGURE **8-16** | **Summary of the C_4 pathway.**
CO_2 combines with phosphoenolpyruvate (PEP) in the chloroplasts of mesophyll cells, forming a four-carbon compound that is converted to malate. Malate goes to the chloroplasts of bundle sheath cells, where it is decarboxylated. The CO_2 released in the bundle sheath cell is used to make carbohydrate by way of the Calvin cycle.

pathway alone. The extra energy expense required to regenerate PEP from pyruvate is worthwhile at high light intensities because it ensures a high concentration of CO_2 in the bundle sheath cells and permits them to carry on photosynthesis at a rapid rate. At lower light intensities and temperatures, C_3 plants are favored. For example, winter rye, a C_3 plant, grows lavishly in cool weather, when crabgrass cannot because it requires more energy to fix CO_2.

CAM plants fix CO_2 at night

Plants living in very dry, or *xeric,* conditions have a number of structural adaptations that enable them to survive. Many xeric plants have physiological adaptations as well, including a special carbon fixation pathway, the **CAM pathway,** or **crassulacean acid metabolism.** The name comes from the stonecrop plant family (the Crassulaceae), which possesses the CAM pathway, although it has evolved independently in some members of more than 25 other plant families, including the cactus family (Cactaceae), the lily family (Liliaceae), and the orchid family (Orchidaceae) (Fig. 8-17).

Unlike most plants, CAM plants open their stomata at night, admitting CO_2 while minimizing water loss. They use the enzyme PEP carboxylase to fix CO_2, forming oxaloacetate, which is converted to malate and stored in cell vacuoles. During the day, when stomata are closed and gas exchange cannot occur between the plant and the atmosphere, CO_2 is removed from malate by a decarboxylation reaction. Now the CO_2 is available within the leaf tissue to be fixed into sugar by the Calvin cycle (C_3 pathway).

The CAM pathway is very similar to the C_4 pathway but with important differences. C_4 plants initially fix CO_2 into four-carbon organic acids in mesophyll cells. The acids are later decarboxylated to produce CO_2, which is fixed by the C_3 pathway in the bundle sheath cells. In other words, the C_4 and C_3 pathways occur in *different locations* within the leaf of a C_4 plant. In CAM plants, the initial fixation of CO_2 occurs at night. Decarboxylation of malate and subsequent production of sugar from CO_2 by the normal C_3 photosynthetic pathway occur during the day. In other words, the CAM and C_3 pathways occur at *different times* within the same cell of a CAM plant.

Robert W. Domm/Visuals Unlimited

FIGURE **8-17** **A typical CAM plant.**
Prickly pear cactus (*Opuntia*) is a CAM plant. The more than 200 species of *Opuntia* living today originated in various xeric habitats in North and South America.

Although it does not promote rapid growth the way that the C_4 pathway does, the CAM pathway is a very successful adaptation to xeric conditions. CAM plants can exchange gases for photosynthesis and to reduce water loss significantly. Plants with CAM photosynthesis survive in deserts where neither C_3 nor C_4 plants can.

Review

- Which phase of the Calvin cycle requires both ATP and NADPH?
- In what ways does photorespiration differ from aerobic respiration?
- Do C_3, C_4, and CAM plants all have rubisco? PEP carboxylase?

BiologyNow™ Assess your understanding of the **carbon fixation reactions** by taking the pretest on your BiologyNow CD-ROM.

SUMMARY WITH KEY TERMS

1 Describe the physical properties of light and explain the relationship between a wavelength of light and its energy.

- Light consists of particles called **photons** that move as waves. Photons with shorter **wavelengths** have more energy than those with longer wavelengths.

2 Diagram the internal structure of a chloroplast, and explain how its components interact and facilitate the process of photosynthesis.

- In plants, **photosynthesis** occurs in chloroplasts, which are located mainly within **mesophyll** cells inside the leaf.
- **Chloroplasts** are organelles enclosed by a double membrane; the inner membrane encloses the **stroma** in which membranous, saclike **thylakoids** are suspended. Each thylakoid encloses a **thylakoid lumen.** Thylakoids arranged in stacks are called **grana.**
- **Chlorophyll *a*, chlorophyll *b*, carotenoids,** and other photosynthetic pigments are components of the thylakoid membranes of chloroplasts.

3 Describe what happens to an electron in a biological molecule such as chlorophyll when a photon of light energy is absorbed.

- Photons excite biological molecules such as chlorophyll and other photosynthetic pigments, causing one or more electrons to become energized. These energized electrons may be accepted by electron acceptor compounds.
- The combined **absorption spectra** of chlorophylls *a* and *b* are similar to the **action spectrum** for photosynthesis.

4 Describe photosynthesis as a redox process.

- During photosynthesis, light energy is captured and converted to the chemical energy of carbohydrates; hydrogens from water are used to reduce carbon, and oxygen derived from water becomes oxidized, forming molecular oxygen.

5 Distinguish between the light-dependent reactions and carbon fixation reactions of photosynthesis

- In the **light-dependent reactions,** electrons energized by light are used to generate **ATP** and **NADPH;** these compounds provide energy for the formation of carbohydrate during the **carbon fixation reactions.**

6 Describe the flow of electrons through photosystems I and II in the noncyclic electron transport pathway; contrast this with cyclic electron transport.

- **Photosystems I** and **II** are the two types of photosynthetic units involved in photosynthesis. Each photosystem includes chlorophyll molecules and accessory pigments organized with pigment-binding proteins into **antenna complexes.**
- Only a special chlorophyll *a* in the **reaction center** of an antenna complex gives up its energized electrons to a nearby electron acceptor. **P700** is the reaction center for photosystem I; **P680** is the photosystem II reaction center.
- During the noncyclic light-dependent reactions, known as **noncyclic electron transport, ATP** and **NADPH** are formed. Electrons in photosystem I are energized by the absorption of light and passed through an **electron transport chain** to $NADP^+$, forming NADPH. Electrons given up by P700 in photosystem I are replaced by electrons from P680 in photosystem II. A series of redox reactions takes place as energized electrons are passed along the electron transport chain from photosystem II to photosystem I. Electrons given up by P680 in photosystem II are replaced by electrons made available by the **photolysis** of H_2O; oxygen is released in the process.
- During **cyclic electron transport,** electrons from photosystem I are eventually returned to photosystem I. ATP is produced by chemiosmosis, but no NADPH or oxygen is generated.

7 Explain how a proton (H^+) gradient is established across the thylakoid membrane and how this gradient functions in ATP synthesis.

- **Photophosphorylation** is the synthesis of ATP coupled to the transport of electrons energized by photons of light. Some of the energy of the electrons is used to pump protons across the thylakoid membrane, providing the energy to generate ATP by **chemiosmosis.**
- As protons diffuse through **ATP synthase,** an enzyme complex in the thylakoid membrane, ADP is phosphorylated to form ATP.

8 Summarize the three phases of the Calvin cycle, and indicate the roles of ATP and NADPH in the process.

- The **carbon fixation reactions** proceed by way of the **Calvin cycle,** also known as the **C_3 pathway.**
- In the CO_2 uptake phase of the Calvin cycle, CO_2 is combined with ribulose bisphosphate (RuBP), a five-carbon sugar, by the enzyme ribulose bisphosphate carboxylase/oxygenase, commonly known as **rubisco,** forming the three-carbon molecule **phosphoglycerate (PGA).**
- In the carbon reduction phase of the Calvin cycle, the energy and reducing power of ATP and NADPH are used to convert PGA molecules to **glyceraldehyde-3-phosphate (G3P).** For every 6 CO_2 molecules fixed, 12 molecules of G3P are produced, and 2 molecules of G3P leave the cycle to produce the equivalent of 1 molecule of glucose.
- In the RuBP regeneration phase of the Calvin cycle, the remaining G3P molecules are modified to regenerate RuBP.

9 Discuss how photorespiration reduces photosynthetic efficiency.

- In **photorespiration,** C_3 plants consume oxygen and generate CO_2 by degrading Calvin cycle intermediates but do not produce ATP. Photorespiration is significant on bright, hot, dry days when plants close their stomata, conserving water but preventing the passage of CO_2 into the leaf.

10 Compare the C_4 and CAM pathways.

- In the **C_4 pathway,** the enzyme **PEP carboxylase** binds CO_2 effectively, even when CO_2 is at a low concentration. C_4 reactions take place within mesophyll cells. The CO_2 is fixed in **oxaloacetate,** which is then converted to malate. The malate moves into a **bundle sheath cell,** and CO_2 is removed from it. The released CO_2 then enters the Calvin cycle.
- The **crassulacean acid metabolism (CAM)** pathway is similar to the C_4 pathway. PEP carboxylase fixes carbon at night in the mesophyll cells, and the Calvin cycle occurs during the day in the same cells.

Summary Reactions for Photosynthesis

The light-dependent reactions (noncyclic electron transport):

$$12\,H_2O + 12\,NADP^+ + 18\,ADP + 18\,P_i \xrightarrow[\text{Chlorophyll}]{\text{Light}} 6\,O_2 + 12\,NADPH + 18\,ATP$$

The carbon fixation reactions (Calvin cycle):

$$12\,NADPH + 18\,ATP + 6\,CO_2 \longrightarrow C_6H_{12}O_6 + 12\,NADP^+ + 18\,ADP + 18\,P_i + 6\,H_2O$$

By canceling out the common items on opposite sides of the arrows in these two coupled equations, we obtain the simplified overall equation for photosynthesis:

$$\underset{\text{Carbon dioxide}}{6\,CO_2} + \underset{\text{Water}}{12\,H_2O} \xrightarrow[\text{Chlorophyll}]{\text{Light energy}} \underset{\text{Glucose}}{C_6H_{12}O_6} + \underset{\text{Oxygen}}{6\,O_2} + \underset{\text{Water}}{6\,H_2O}$$

POST-TEST

1. Where is chlorophyll located in the chloroplast? (a) thylakoid membranes (b) stroma (c) matrix (d) thylakoid lumen (e) between the inner and outer membranes

2. In photolysis, some of the energy captured by chlorophyll is used to split (a) CO_2 (b) ATP (c) NADPH (d) H_2O (e) both b and c

POST-TEST (continued)

3. Light is composed of particles of energy called (a) carotenoids (b) reaction centers (c) photons (d) antenna complexes (e) photosystems
4. The relative effectiveness of different wavelengths of light in photosynthesis is demonstrated by (a) an action spectrum (b) photolysis (c) carbon fixation reactions (d) photoheterotrophs (e) an absorption spectrum
5. In plants, the final electron acceptor in the light-dependent reactions is (a) $NADP^+$ (b) CO_2 (c) H_2O (d) O_2 (e) G3P
6. In addition to chlorophyll, most plants contain accessory photosynthetic pigments such as (a) PEP (b) G3P (c) carotenoids (d) PGA (e) $NADP^+$
7. The part of a photosystem that absorbs light energy is its (a) antenna complexes (b) reaction center (c) terminal quinone electron acceptor (d) pigment-binding protein (e) thylakoid lumen
8. In ____________, electrons that have been energized by light contribute their energy to add phosphate to ADP, producing ATP. (a) crassulacean acid metabolism (b) the Calvin cycle (c) photorespiration (d) C_4 pathways (e) photophosphorylation
9. In ____________, there is a one-way flow of electrons to $NADP^+$, forming NADPH. (a) crassulacean acid metabolism (b) the Calvin cycle (c) photorespiration (d) cyclic electron transport (e) noncyclic electron transport
10. The mechanism by which electron transport is coupled to ATP production by means of a proton gradient is called (a) chemiosmosis (b) crassulacean acid metabolism (c) fluorescence (d) the C_3 pathway (e) the C_4 pathway
11. In photosynthesis in eukaryotes, the transfer of electrons through a sequence of electron acceptors provides energy to pump protons across the (a) chloroplast outer membrane (b) chloroplast inner membrane (c) thylakoid membrane (d) inner mitochondrial membrane (e) plasma membrane
12. The inputs for ____________ are CO_2, NADPH, and ATP. (a) cyclic electron transport (b) the carbon fixation reactions (c) noncyclic electron transport (d) photosystems I and II (e) chemiosmosis
13. The Calvin cycle begins when CO_2 reacts with (a) phosphoenolpyruvate (b) glyceraldehyde-3-phosphate (c) ribulose bisphosphate (d) oxaloacetate (e) phosphoglycerate
14. The enzyme directly responsible for almost all carbon fixation on Earth is (a) Rubisco (b) PEP carboxylase (c) ATP synthase (d) phosphofructokinase (e) ligase
15. In C_4 plants, C_4 and C_3 pathways occur at different ____________, whereas in CAM plants CAM and C_3 pathways occur at different ____________. (a) times of day; locations within the leaf (b) seasons; locations (c) locations; times of day (d) locations; seasons (e) times of day; seasons
16. Label the figure. Use Figure 8-8 to check your answers.

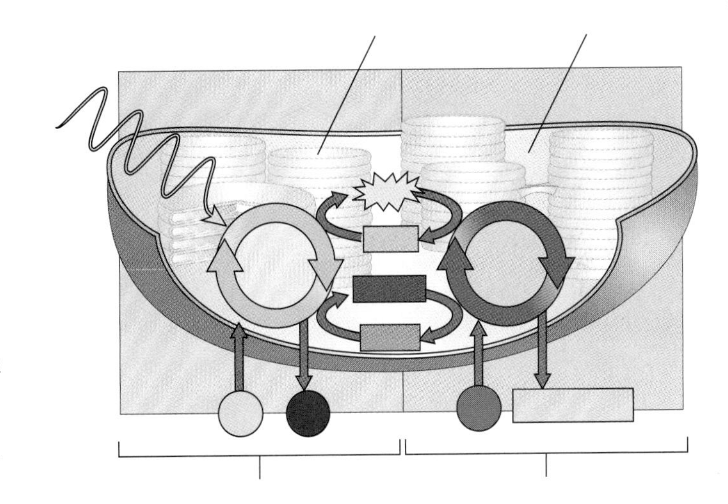

CRITICAL THINKING

1. Must all autotrophs use light energy? Explain.
2. Only some plant cells have chloroplasts, but all actively metabolizing plant cells have mitochondria. Why?
3. Explain why the proton gradient formed during chemiosmosis represents a state of low entropy. (You may wish to refer to the discussion of entropy in Chapter 6.)
4. The electrons in glucose have relatively high free energies. How did they become so energetic?
5. What strategies may be employed in the future to increase world food supply? Base your answer on your knowledge of photosynthesis and related processes.

- Visit our Website at **http://biology.brookscole.com/solomon7** for links to chapter-related resources on the World Wide Web. Additional online materials relating to this chapter can also be found on our Web site.

BIOLOGY NOW RESOURCES

Active Figures

8-4: Photosynthesis in plants
8-11: Noncyclic electron transport
8-14: Calvin cycle

Preparing for an exam? Take a diagnostic test on your BiologyNow CD-ROM.

Post-Test Answers

1. a	2. d	3. c	4. a
5. a	6. c	7. a	8. e
9. e	10. a	11. c	12. b
13. c	14. a	15. c	

9

Chromosomes, Mitosis, and Meiosis

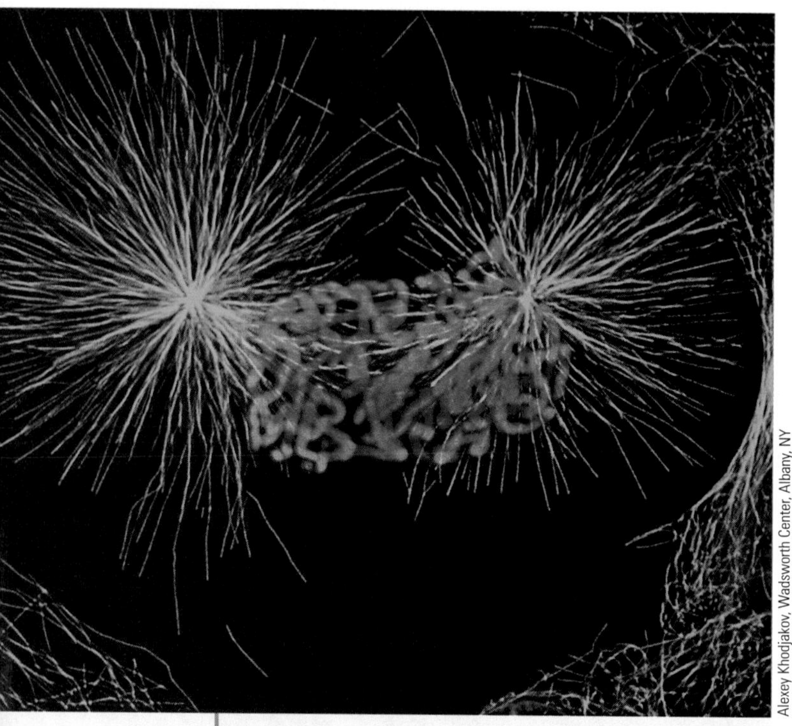
Alexey Khodjakov, Wadsworth Center, Albany, NY

Fluorescence LM of a cultured newt lung cell in early mitosis. The nuclear envelope has broken down, and the microtubules of the mitotic spindle *(green)* now interact with the chromosomes *(blue)*.

CHAPTER OUTLINE

Pre-existing cells divide to form new cells. This remarkable process enables an organism to grow, repair damaged parts, and reproduce. Cells serve as the basic units of life and the essential link between generations. Even the simplest cell contains a massive amount of precisely coded genetic information in the form of deoxyribonucleic acid (DNA), collectively called the organism's **genome.** Genomes are organized into informational units called *genes,* which control the activities of the cell and are passed on to its descendants.

When a cell divides, the information contained in the DNA must be faithfully duplicated and the copies then transmitted to each daughter cell through a precisely choreographed series of steps (see photograph). DNA is a very long, thin molecule that could easily become tangled and broken, and a eukaryotic cell's nucleus contains a huge amount of DNA. In this chapter, we consider how eukaryotes accommodate the genetic material by packaging each DNA molecule with proteins to form a structure called a *chromosome,* each of which contains hundreds or thousands of genes.

We then consider *mitosis,* the process that ensures a parent cell transmits one copy of every chromosome to each of its two daughter cells. In this way, the chromosome number is preserved through successive mitotic divisions. Most body cells of eukaryotes divide by mitosis.

Finally we discuss *meiosis,* a process that reduces the chromosome number by half. Sexual life cycles in eukaryotes require meiosis. Sexual reproduction involves the fusion of two sex cells, or *gametes,* to form a single cell called a *zygote.* Meiosis makes it possible for each gamete to contain only half the number of parent chromosomes, preventing the zygotes from having twice as many chromosomes as the parents.

Bacterial reproduction is described in Chapter 23. Prokaryotic cells contain much less DNA than do most eukaryotic cells. Their DNA is usually circular and is packaged with associated proteins. Although the distribution of genetic material in dividing prokaryotic cells is a simpler process than mitosis, it nevertheless is very precise, to ensure that the daughter cells are genetically identical to the parent cell. ■

EUKARYOTIC CHROMOSOMES

Learning Objectives

1. Discuss the significance of chromosomes in terms of their information content.
2. Compare the organization of DNA in prokaryotic and eukaryotic cells.

The major carriers of genetic information in eukaryotes are the **chromosomes,** which lie within the cell nucleus. Although *chromosome* means "colored body," chromosomes are virtually colorless; the term refers to their ability to be stained by certain dyes. In the 1880s, light microscopes had been improved to the point that biologists such as the German biologist Walther Fleming began to observe chromosomes during cell division. In 1903, American biologist Walter Sutton and German biologist Theodor Boveri noted independently that chromosomes were the physical carriers of genes, the genetic factors Gregor Mendel discovered in the 19th century (see Chapter 10).

Chromosomes are made of **chromatin,** a material consisting of DNA and associated proteins. When a cell is not dividing, the chromosomes are present but in an extended, partially unraveled form. Chromatin consists of long, thin threads that are somewhat aggregated, giving them a granular appearance when viewed with the electron microscope (see Fig. 4-11). During cell division, the chromatin fibers condense and the chromosomes become visible as distinct structures (Fig. 9-1).

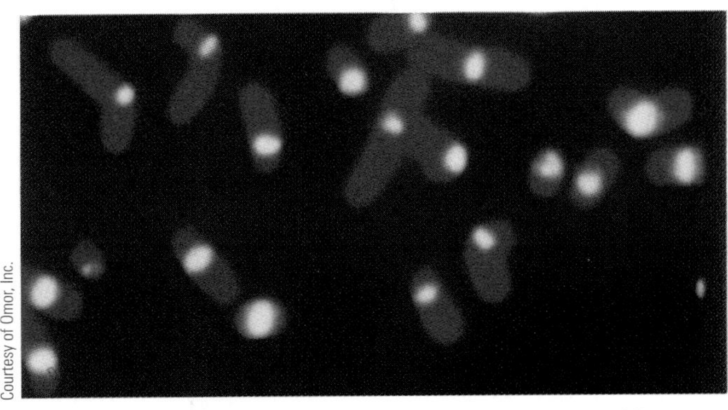

FIGURE **9-1** **Chromosomes.**
The replicated human chromosomes shown in this LM have been stained with a fluorescent antibody that binds to the centromere regions *(yellow dots).*

DNA is organized into informational units called genes

An organism's genome may contain hundreds or even thousands of genes. For example, the **Human Genome Project** estimates that humans have less than 30,000 genes that code for proteins (see Chapter 15). As you will see in later chapters, the concept of the gene has changed considerably since the science of genetics began, but our definitions have always centered on the gene as an informational unit. By providing information needed to carry out one or more specific cell functions, a **gene** ultimately affects some characteristic of the organism. For example, genes govern eye color in humans, wing length in flies, and seed color in peas.

DNA is packaged in a highly organized way in chromosomes

Prokaryotic and eukaryotic cells differ markedly in their DNA content as well as in the organization of DNA molecules. The bacterium *E. coli* normally contains about 4×10^6 base pairs (almost 1.35 mm) of DNA in its single circular DNA molecule. In fact, the total length of its DNA is about 1000 times longer than the length of the cell itself. Therefore, the DNA molecule is, with the help of proteins, twisted and folded compactly to fit inside the bacterial cell.

A typical eukaryotic cell contains much more DNA than a bacterium does, and it is organized in the nucleus as multiple chromosomes; these vary widely in size and number among different species. Although a human cell nucleus is about the size of a large bacterial cell, it contains more than 1000 times the amount of DNA found in *E. coli.* The DNA content of a human sperm cell is about 3×10^9 base pairs; stretched end to end, it would measure almost 1 m long.

How does a eukaryotic cell pack its DNA into the chromosomes? Chromosome packaging is facilitated by certain proteins known as **histones.**[1] Histones have a positive charge because they have a high proportion of amino acids with basic side chains (see Chapter 3). The positively charged histones associate with DNA, which has a negative charge because of its phosphate groups, to form structures called **nucleosomes.** The fundamental unit of each nucleosome consists of a beadlike structure with 146 base pairs of DNA wrapped around a disc-shaped core of eight histone molecules (two each of four different histone types) (Fig. 9-2). Although the nucleosome was originally defined as a bead plus a DNA segment that links it to an adjacent bead, today the term more commonly refers only to the bead itself (that is, the eight histones and the DNA wrapped around them).

Nucleosomes function like tiny spools, preventing DNA strands from becoming tangled. You can see the importance of this role in Figure 9-3, which illustrates the enormous number of DNA fibers that unravel from a mouse chromosome after researchers have removed the histones. **Scaffolding proteins** are nonhistone proteins that help maintain chromosome structure. But the role of histones is more than structural, because their arrangement also affects the activity of the DNA with which they are associated.

[1] A few types of eukaryotic cells lack histones. Conversely, histones occur in one group of prokaryotes, the archaea (see Chapter 23).

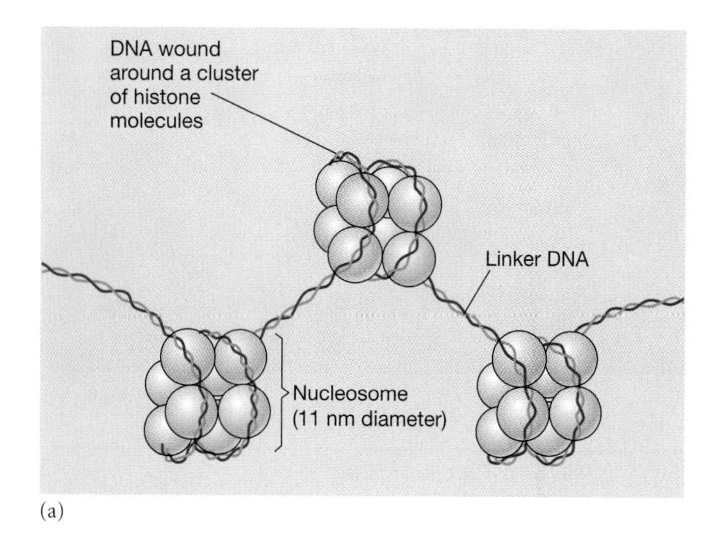

(a)

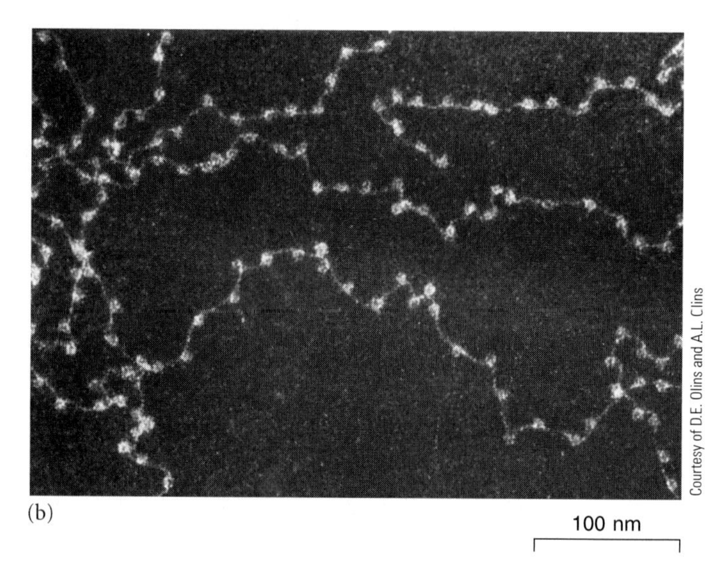

(b)

FIGURE **9-2** **Nucleosomes.**

(a) A model for the structure of a nucleosome. Each nucleosome bead contains a set of eight histone molecules, forming a protein core around which the double-stranded DNA winds. The DNA surrounding the histones consists of 146 nucleotide pairs; another segment of DNA, about 60 nucleotide pairs long, links nucleosome beads. **(b)** TEM of nucleosomes from the nucleus of a chicken cell. Normally nucleosomes are packed more closely together, but the preparation procedure has spread them apart, revealing the DNA linkers.

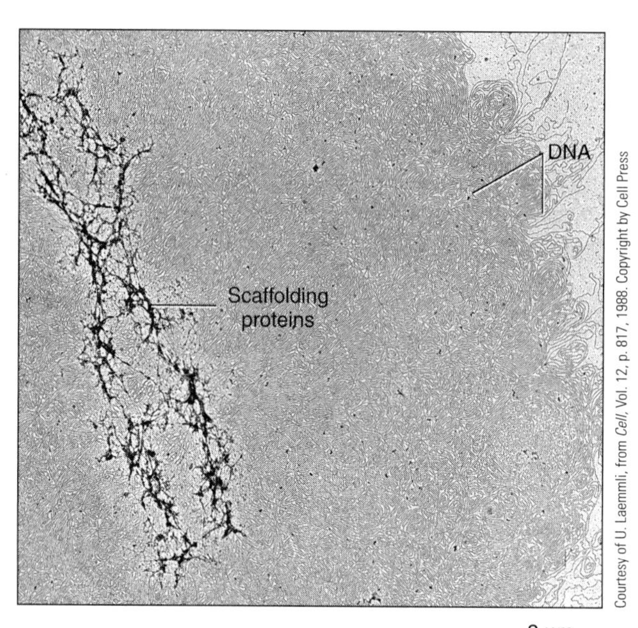

FIGURE **9-3** **TEM of a mouse chromosome depleted of histones.**

Notice how densely packed the DNA strands are, even though they have been released from the histone proteins that organize them into tightly coiled structures.

Nucleosomes are part of the chromatin. Figure 9-4 shows the higher-order structures of chromatin leading to the formation of a condensed chromosome. The nucleosomes themselves are 11 nm in diameter. The packed nucleosome state occurs when a fifth type of histone, known as *histone H1,* associates with the linker DNA, packing adjacent nucleosomes together to form a 30-nm-diameter fiber. In extended chromatin, these fibers form large, coiled loops held together by scaffolding proteins. The loops then interact to form the condensed chromatin found in a metaphase chromosome.

Chromosome number and informational content differ among species

Every individual of a given species has a characteristic number of chromosomes in most nuclei of its body cells. However, it is not the number of chromosomes that makes each species unique, but the information the genes specify.

Most human body cells have exactly 46 chromosomes, but humans are not humans merely because they have 46 chromosomes. In fact, some individuals have an abnormal chromosome constitution, or **karyotype,** with more or fewer than 46 (see Fig. 15-5). Humans are not unique in having 46 chromosomes; some other species—the olive tree, for example—also have 46.

Other species have different chromosome numbers. A certain species of roundworm has only 2 chromosomes in each cell, whereas some crabs have as many as 200, and some ferns have more than 1000. Most animal and plant species have between 8 and 50 chromosomes per body cell. Numbers above and below these are uncommon.

Review

- What are the informational units on chromosomes called? Of what do these informational units consist?
- How are bacterial DNA molecules and eukaryotic chromosomes similar? How do they differ?

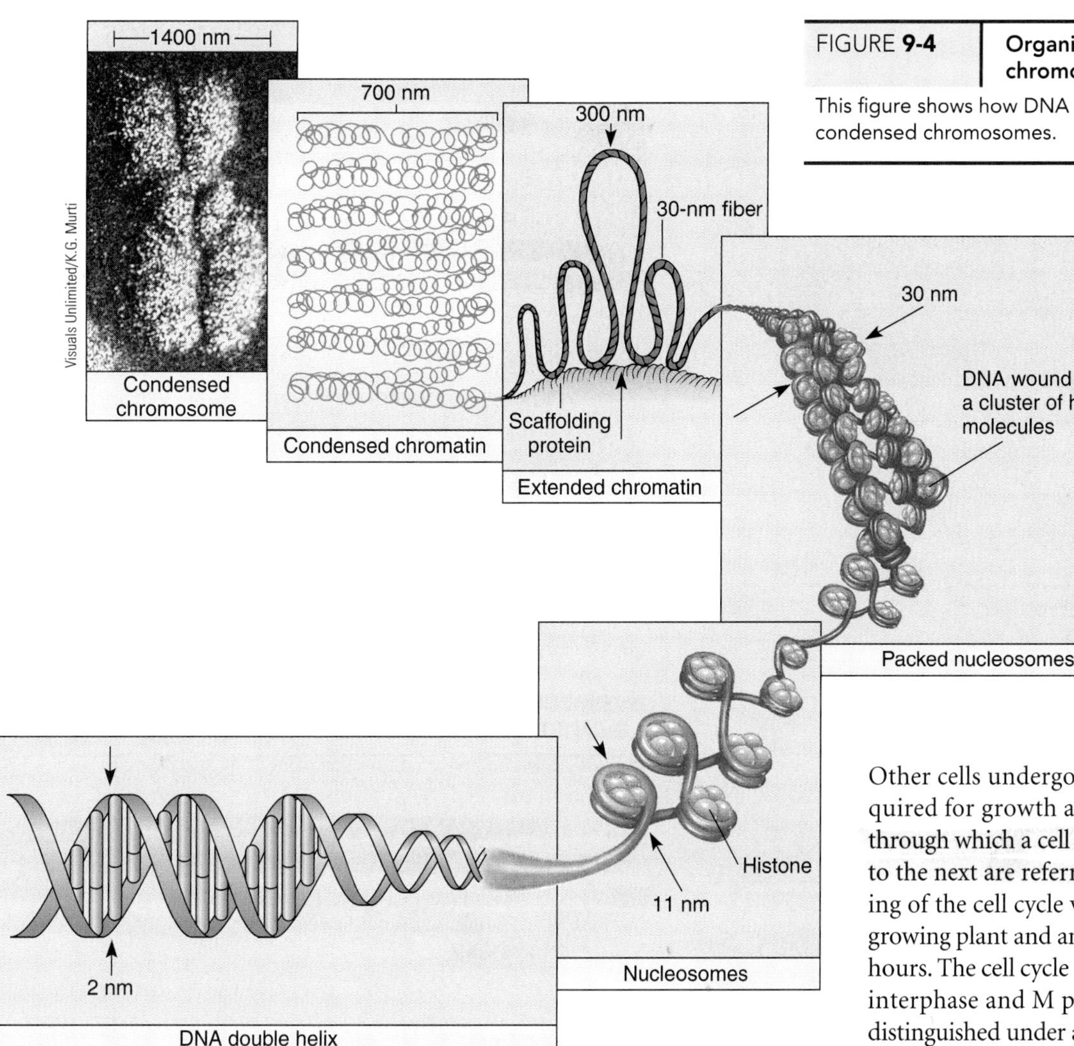

FIGURE **9-4** **Organization of a eukaryotic chromosome.**

This figure shows how DNA is packaged into highly condensed chromosomes.

- How is the large discrepancy between DNA length and nucleus size addressed in eukaryotic cells?
- How can two species have the same chromosome number, yet have very different attributes?

BiologyNow™ Assess your understanding of **eukaryotic chromosomes** by taking the pretest on your BiologyNow CD-ROM.

THE CELL CYCLE AND MITOSIS

Learning Objectives

3 Identify the stages in the eukaryotic cell cycle, describe their principal events, and point out some ways in which the cycle is controlled.

4 Describe the structure of a duplicated chromosome, including the sister chromatids, centromeres, and kinetochores.

5 Explain the significance of mitosis, and describe the process.

When cells reach a certain size, they usually either stop growing or divide. Not all cells divide; some, such as nerve, skeletal muscle, and red blood cells, do not normally divide once they are mature. Other cells undergo a sequence of activities required for growth and cell division. The stages through which a cell passes from one cell division to the next are referred to as the **cell cycle.** Timing of the cell cycle varies widely, but in actively growing plant and animal cells, it is about 8 to 20 hours. The cell cycle consists of two main phases, interphase and M phase, both of which can be distinguished under a light microscope (Fig. 9-5).

M phase involves two main processes, mitosis and cytokinesis. *Mitosis,* a process involving the nucleus, ensures that each new nucleus receives the same number and types of chromosomes as were present in the original nucleus. *Cytokinesis,* which generally begins before mitosis is complete, is the division of the cell cytoplasm to form two cells. Multinucleated cells form if mitosis is not followed by cytokinesis; this is a normal condition for certain cell types.

Chromosomes duplicate during interphase

Most of a cell's life is spent in **interphase,** the time when no cell division is occurring. Although the appearance of the nucleus is generally unremarkable (see Fig. 4-11), a cell that is capable of dividing is typically very active during this time, synthesizing needed materials and growing. The cell synthesizes most proteins, lipids, and other biologically important materials throughout interphase. Here is the sequence of interphase and M phase in the eukaryotic cell cycle:

G_1 phase → S phase → G_2 phase → mitosis and cytokinesis

(Interphase: G_1 phase, S phase, G_2 phase; M phase: mitosis and cytokinesis)

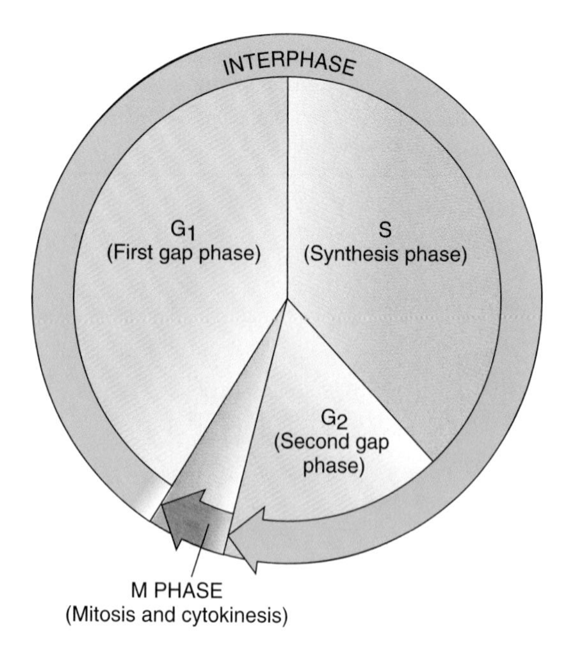

FIGURE **9-5** **The eukaryotic cell cycle.**

The cell cycle includes interphase (G_1, S, and G_2) and M phase (mitosis and cytokinesis). Proportionate amounts of time spent at each stage vary among species, cell types, and growth conditions. If the cell cycle were a period of 12 hours, G_1 would be about 5 hours, S would be 4.5 hours, G_2 would be 2 hours, and M phase would be 30 minutes.

The time between the end of mitosis and the beginning of the S phase is termed the **G_1 phase** (*G* stands for *gap*, an interval during which no DNA synthesis occurs). Growth and normal metabolism take place during the G_1 phase, which is typically the longest phase. Cells that are not dividing usually become arrested in this part of the cell cycle and are said to be in a state called G_0. Toward the end of G_1, the enzymes required for DNA synthesis become more active. Synthesis of these enzymes, along with proteins needed to initiate cell division (discussed later in the chapter), enable the cell to enter the S phase.

PROCESS OF SCIENCE

During the **synthesis phase,** or **S phase,** DNA replicates and histone is synthesized, as the cell makes duplicate copies of its chromosomes. How did researchers identify the S phase of the cell cycle? In the early 1950s, researchers demonstrated that cells preparing to divide duplicate their chromosomes at a relatively restricted interval during interphase and not during early mitosis, as previously hypothesized. These investigators used isotopes, such as ^3H, to synthesize radioactive thymidine, a nucleotide that is incorporated specifically into DNA as it is synthesized. After radioactive thymidine was supplied for a brief period (such as 30 minutes) to actively growing cells, autoradiography (see Chapter 2) on exposed film showed that a fraction of the cells had silver grains over their chromosomes. The nuclei of these cells were radioactive, because during the experiment they had replicated DNA. DNA replication was not occurring in the cells that did not have radioactively labeled chromosomes. Researchers therefore inferred that the proportion of labeled cells out of the total number of cells provides a rough estimate of the length of the S phase relative to the rest of the cell cycle.

After it completes the S phase, the cell enters a second gap phase, the **G_2 phase.** At this time, increased protein synthesis occurs, as the final steps in the cell's preparation for division take place. For many cells, the G_2 phase is short relative to the G_1 and S phases.

Mitosis, the nuclear division that produces two nuclei identical to the parental nucleus, begins at the end of the G_2 phase. Mitosis is a continuous process, but for descriptive purposes, it is divided into four stages:

Prophase ⟶ metaphase ⟶ anaphase ⟶ telophase

Look at Figure 9-6 while you read the following descriptions of these stages as they occur in a typical plant or animal cell.

During prophase, duplicated chromosomes become visible with the microscope

The first stage of mitosis, **prophase,** begins with **chromosome compaction,** when the long chromatin fibers begin a coiling process that makes them shorter and thicker (see Fig. 9-4). The chromatin can then be distributed to the daughter cells without tangling. Cell biologists have identified a group of proteins, collectively called **condensin,** required for chromosome compaction. Using the energy of ATP hydrolysis, condensin binds to DNA and wraps it into coiled loops that are compacted into a mitotic chromosome.

When stained with certain dyes and viewed through the light microscope, chromosomes become visible as darkly staining bodies as prophase progresses. Each chromosome was duplicated during the preceding S phase and consists of a pair of **sister chromatids,** which contain identical, double-stranded DNA sequences. Each chromatid includes a constricted region called the **centromere.** Sister chromatids are tightly associated in the vicinity of their centromeres (Fig. 9-7). Precise DNA sequences and proteins that bind to those DNA sequences are the chemical basis for this close association at the centromeres. Attached to each centromere is a **kinetochore,** a structure formed from proteins to which **microtubules** can bind. These microtubules function in chromosome distribution during mitosis.

A dividing cell is usually described as a globe, with an equator that determines the midplane (equatorial plane) and two opposite poles. This terminology is used for all cells regardless of

▶ ACTIVE FIGURE **9-6** **Interphase and the stages of mitosis.**

The drawings depict generalized animal cells with a diploid chromosome number of 4; the sizes of the nuclei and chromosomes are exaggerated to show the structures more clearly. The left column of LMs shows cells of the whitefish (*Coregonus*). The right column of LMs shows sectioned cells of the onion (*Allium cepa*).

BiologyNow™ Walk step-by-step through the **stages of mitosis** by clicking on this figure on your BiologyNow CD-ROM.

Animal | **Plant**

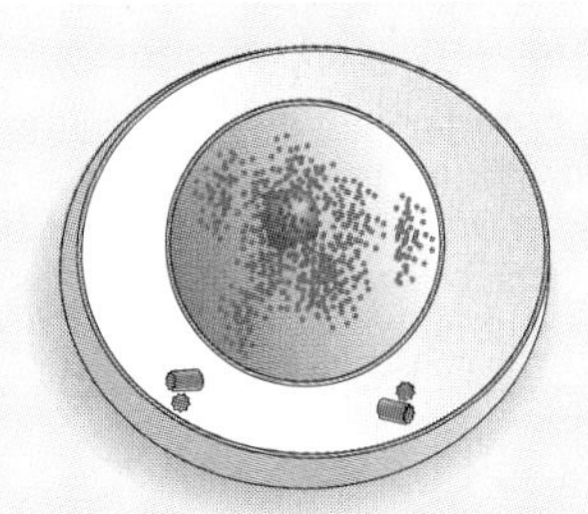

INTERPHASE

Cell is carrying out its normal life activities. Chromosomes become duplicated.

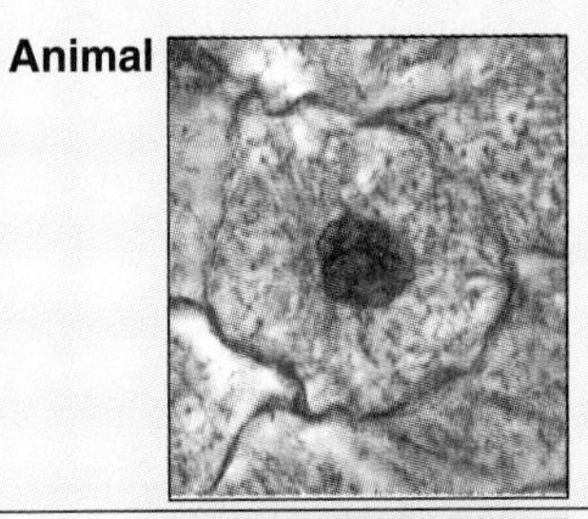

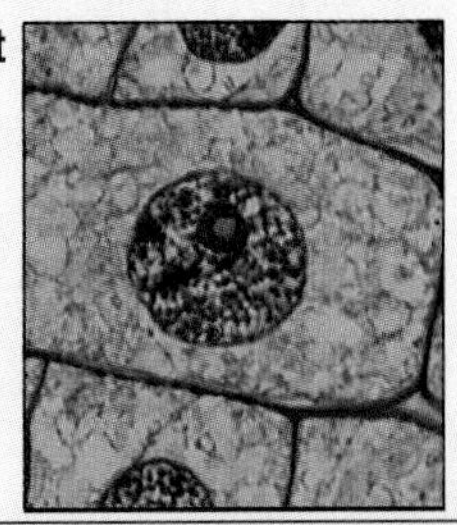

EARLY PROPHASE

Nuclear envelope begins to disappear. Nucleolus disappears. Long fibers of chromatin become evident and begin to condense as visible chromosomes.

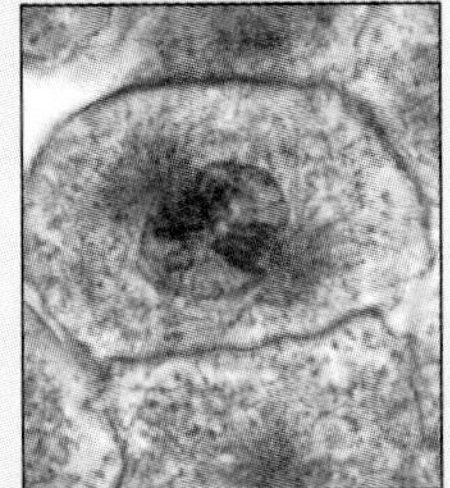

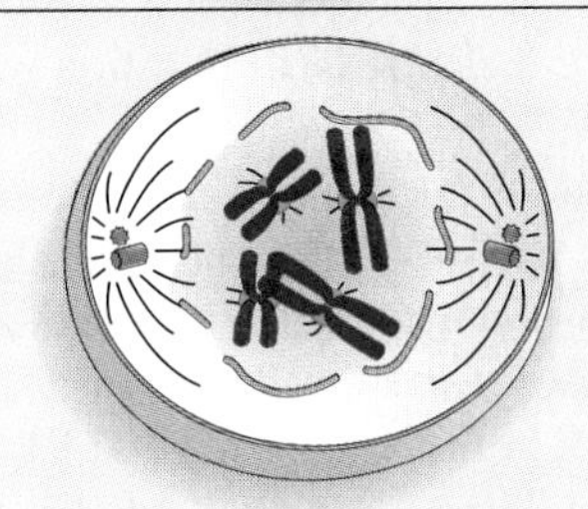

LATE PROPHASE

Chromosomes continue to shorten and thicken. Spindle forms between centrioles, which have moved to the poles of the cell. Kinetochores begin attaching to microtubules.

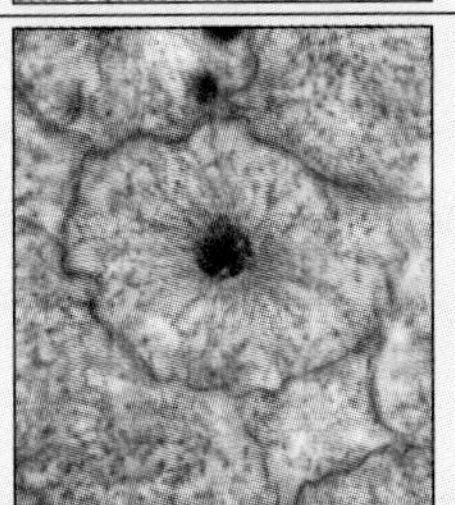

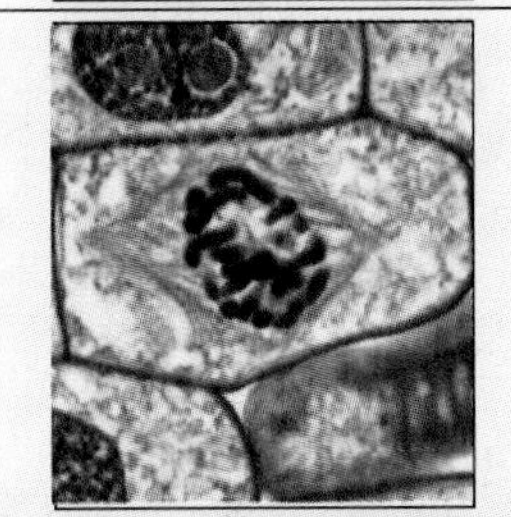

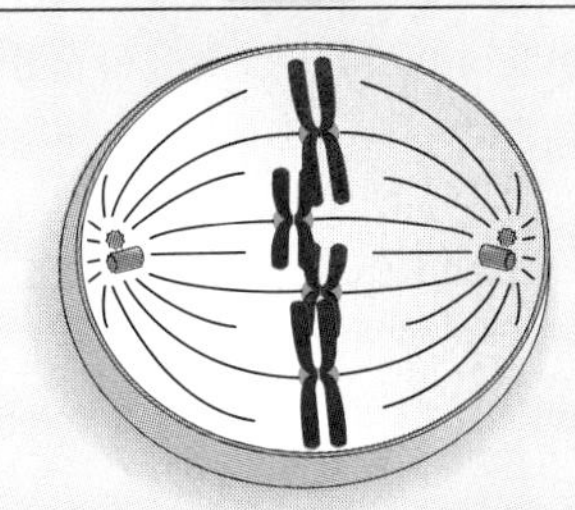

METAPHASE

Spindle fibers attach to the kinetochores of the chromosomes, which line up along the cell's midplane.

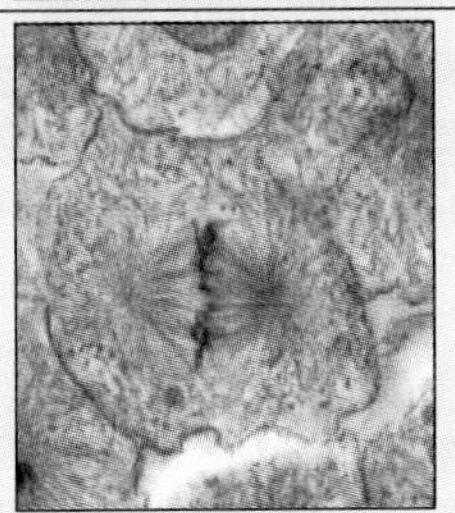

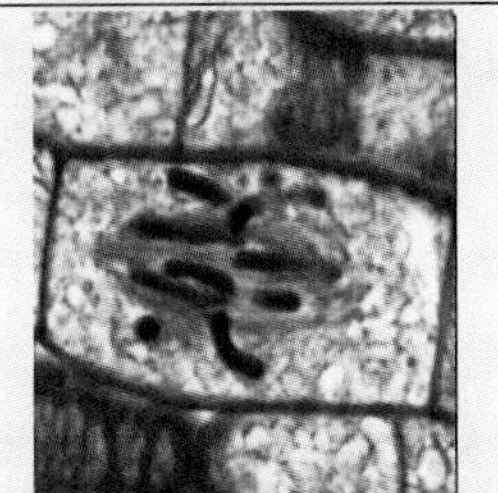

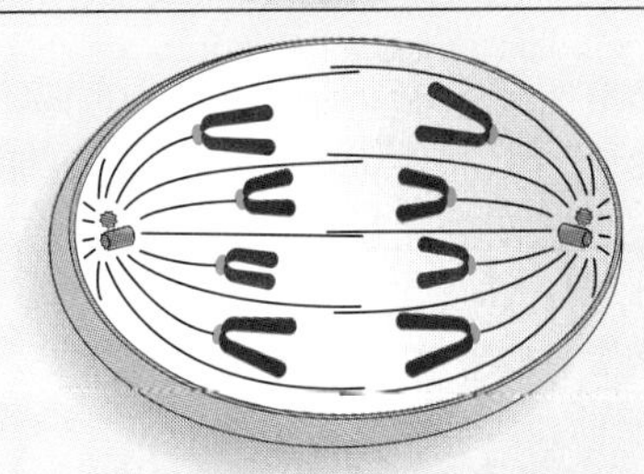

ANAPHASE

Chromatids separate at centromeres, and one group of chromosomes moves toward each pole.

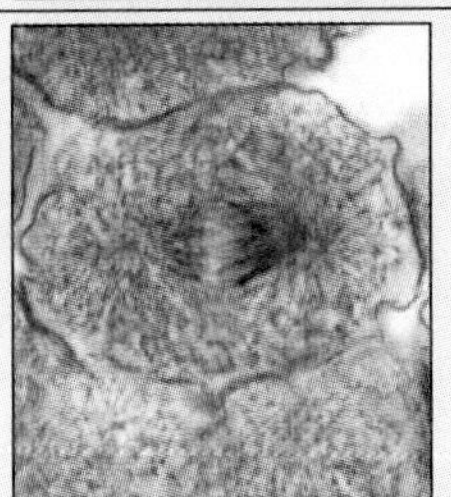

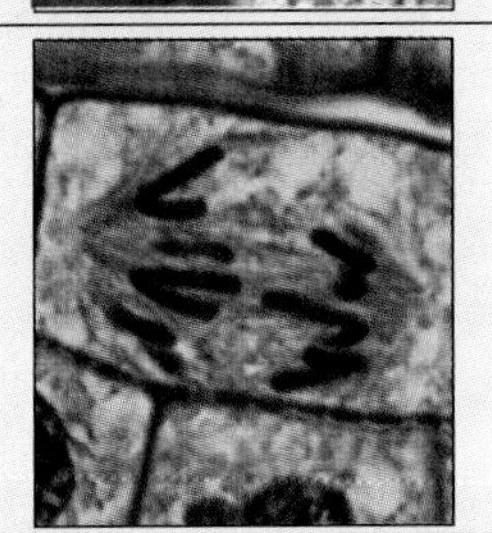

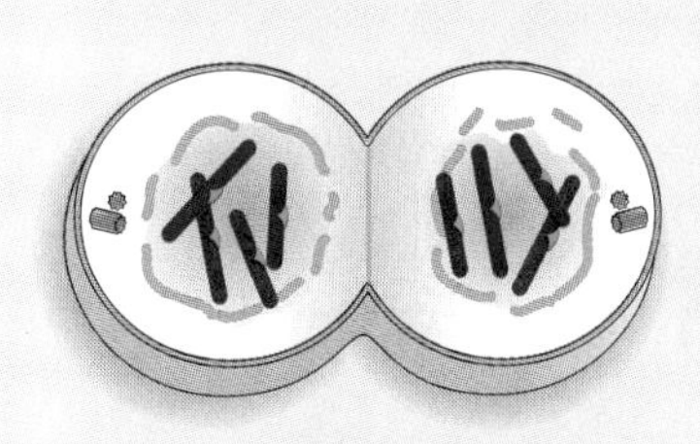

TELOPHASE

Chromosomes have arrived at the poles, and the nuclear envelopes begin to form. Cytokinesis produces two daughter cells.

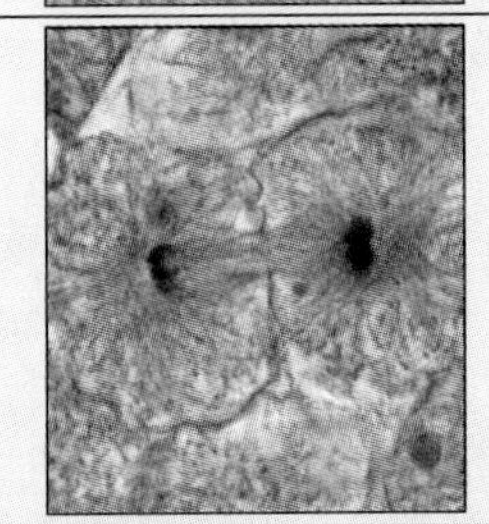

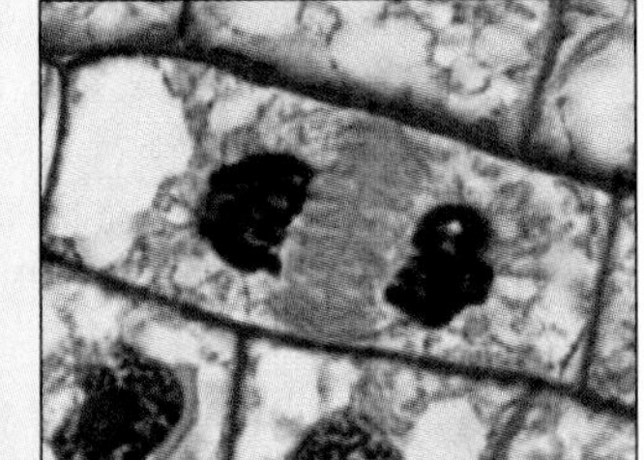

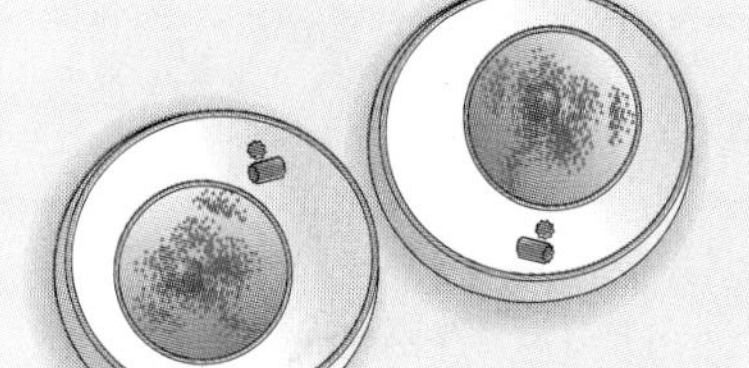

INTERPHASE

Daughter cells formed are genetically identical to the parent cell.

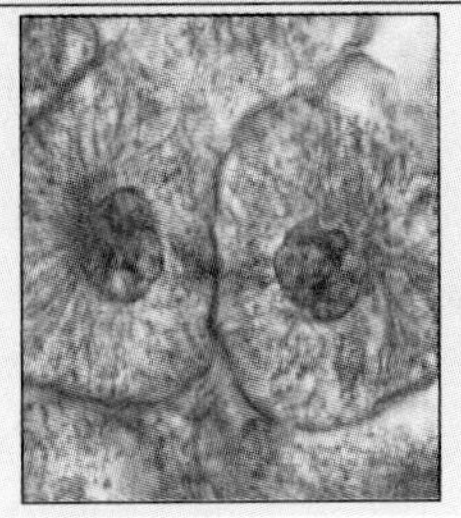

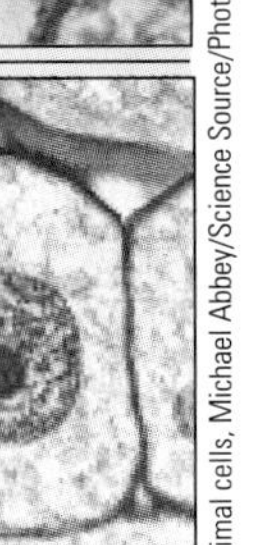

25 µm

Animal cells, Michael Abbey/Science Source/Photo Researchers, Inc.; Plant cells, first interphase through telophase, Ed Reschke; second interphase, Carolina Biological Supply Company/Phototake

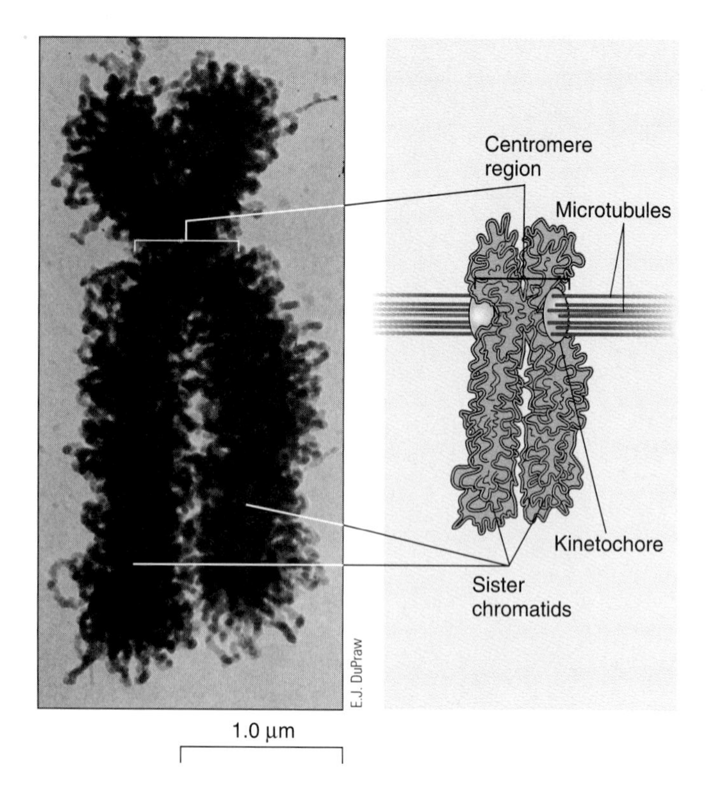

FIGURE **9-7** **Sister chromatids and centromeres.**
The sister chromatids, each consisting of tightly coiled chromatin fibers, are tightly associated at their centromere regions, indicated by the brackets. Associated with each centromere is a kinetochore, which serves as a microtubule attachment site. Kinetochores and microtubules are not visible in this SEM of a metaphase chromosome.

their actual shape. Microtubules radiate from each pole, and some of these protein fibers elongate toward the chromosomes, forming the **mitotic spindle,** which separates the chromosomes during anaphase (Fig. 9-8).

Animal cells differ from plant cells in the details of mitotic spindle formation. In both types of dividing cells, each pole contains a region, the **microtubule-organizing center,** from which extend the microtubules that form the mitotic spindle. The electron microscope shows that microtubule-organizing centers in plant cells consist of fibrils with little or no discernible structure.

In contrast, animal cells have a pair of **centrioles** in the middle of each microtubule-organizing center (see Fig. 4-22). The centrioles are surrounded by fibrils that make up the **pericentriolar material.** The spindle microtubules terminate in the pericentriolar material, but they do not actually touch the centrioles. Although cell biologists once thought spindle formation in animal cells required centrioles, their involvement is probably coincidental. Current evidence suggests centrioles organize the pericentriolar material and ensure its duplication when the centrioles duplicate.

Each of the two centrioles is duplicated during interphase, yielding two centriole pairs. Late in prophase, microtubules radiate from the pericentriolar material surrounding the centrioles; these clusters of microtubules are called **asters.** The two asters move to opposite sides of the nucleus, establishing the two poles of the mitotic spindle.

During prophase, the nucleolus shrinks and usually disappears. Toward the end of prophase, the nuclear envelope breaks

FIGURE **9-8** **The mitotic spindle.**
(a) One end of each microtubule of this animal cell is associated with one of the poles. Astral microtubules *(green)* radiate in all directions, forming the aster. Kinetochore microtubules *(red)* connect the kinetochores to the poles, and polar microtubules *(blue)* overlap at the midplane. **(b)** This fluorescence LM of an animal cell at metaphase shows a well-defined spindle and asters (chromosomes, *orange;* microtubules, *green*).

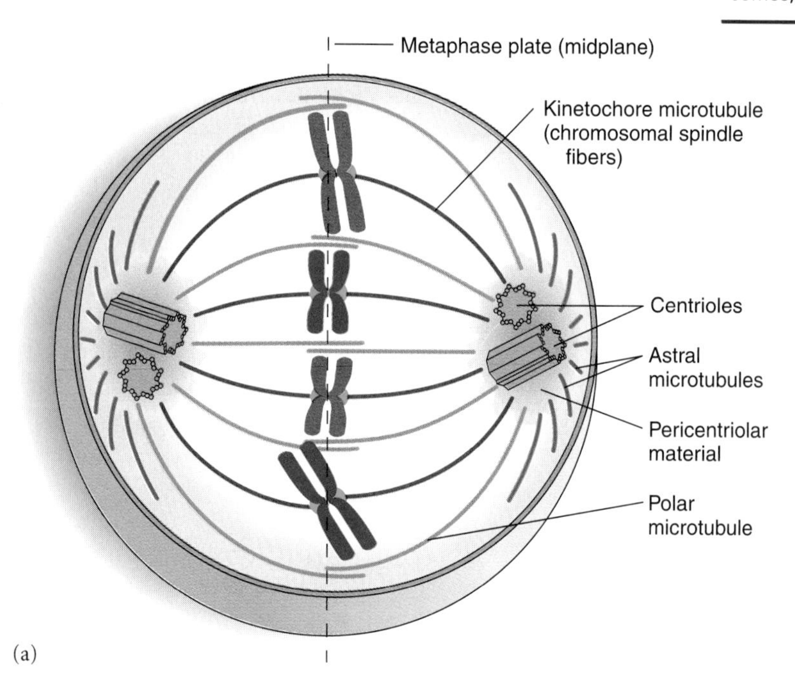

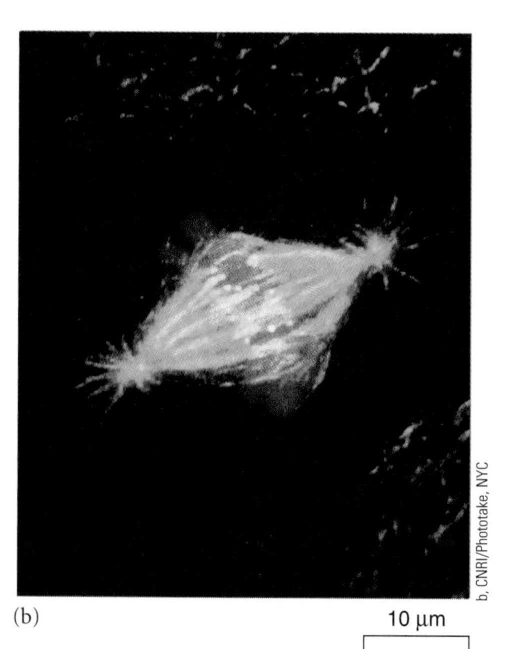

down, and each sister chromatid attaches to some of the spindle microtubules at its kinetochore. Each chromosome consists of a pair of sister chromatids associated along their entire length and joined tightly at their centromeres.

Duplicated chromosomes line up on the midplane during metaphase

During **metaphase,** all the cell's chromosomes align at the cell's midplane, or **metaphase plate.** One chromatid of each chromosome attaches by its kinetochore to microtubules from one pole, and its sister chromatid attaches by its kinetochore to microtubules from the opposite pole. The mitotic spindle has two types of microtubules: polar microtubules and kinetochore microtubules (see Fig. 9-8). **Polar microtubules** extend from each pole to the equatorial region, where they generally overlap. **Kinetochore microtubules,** also called *chromosomal spindle fibers,* extend from each pole and attach to the kinetochores.

During metaphase each chromatid is completely condensed and appears quite thick and distinct. Because individual chromosomes can be seen more distinctly at metaphase than at any other time, they are usually photographed for a karyotype at this stage when chromosomal abnormalities are suspected, (see Chapter 15).

During anaphase, chromosomes move toward the poles

Anaphase begins as the sister chromatids separate. Once the chromatids are no longer attached to their duplicates, each chromatid is called a *chromosome.* The now separate chromosomes move to opposite poles, using the spindle microtubules as tracks. The kinetochores, still attached to kinetochore microtubules, lead the way, with the chromosome arms trailing behind. Anaphase ends when all the chromosomes have reached the poles.

PROCESS OF SCIENCE

The overall mechanism of chromosome movement in anaphase is still poorly understood, although researchers are making significant progress in this area. Microtubules lack elastic or contractile properties. Then how do the chromosomes move apart? Are they pushed or pulled, or do other forces operate?

Chromosome movements are studied in several ways. Researchers determine the number of microtubules at a particular stage or after certain treatments, by carefully analyzing electron micrographs. Researchers also physically perturb living cells that are dividing, using laser beams or mechanical devices known as *micromanipulators.* Skilled researchers can move chromosomes, break their connections to microtubules, and even remove them from the cell entirely.

Microtubules are dynamic structures, with *tubulin* subunits being constantly removed from their ends and others being added. Evidence indicates that kinetochore microtubules shorten during anaphase. Therefore, current hypotheses to explain anaphase movement include the idea that chromosomes move poleward because they remain anchored to the kinetochore microtubules even as tubulin subunits are being removed at the kinetochore. Multiple types of motor proteins, including forms of *kinesin* and *dynein,* probably play a role in this movement.

A second phenomenon also plays a role in chromosome separation. During anaphase the spindle as a whole elongates, at least partly because polar microtubules originating at opposite poles are associated with motors that let them slide past one another at the midplane. The sliding decreases the degree of overlap, thereby "pushing" the poles apart. This mechanism indirectly causes the chromosomes to move apart because they are attached to the poles by kinetochore microtubules.

During telophase, two separate nuclei form

The final stage of mitosis, **telophase,** is characterized by the arrival of the chromosomes at the poles and, in its final stage, by a return to interphase-like conditions. The chromosomes decondense by partially uncoiling. A new nuclear envelope forms around each set of chromosomes, made at least in part from small vesicles and other components derived from the old nuclear envelope. The spindle microtubules disappear, and the nucleoli reorganize.

Cytokinesis forms two separate daughter cells

Cytokinesis, the division of the cytoplasm to yield two daughter cells, usually overlaps mitosis, generally beginning during telophase. Cytokinesis of an animal cell begins as a ring of actin microfilaments associated with the plasma membrane encircles the cell in the equatorial region, at right angles to the spindle (Fig. 9-9a). The ring contracts, producing a **cleavage furrow** that gradually deepens and separates the cytoplasm into two daughter cells, each with a complete nucleus.

In plant cells, cytokinesis occurs by forming a **cell plate** (Fig. 9-9b), a partition constructed in the equatorial region of the spindle and growing laterally toward the cell wall. The cell plate forms as a line of vesicles originating in the *Golgi complex.* The vesicles contain materials to construct both a primary cell wall for each daughter cell and a middle lamella that cements the primary cell walls together. The vesicle membranes fuse to become the plasma membrane of each daughter cell.

Mitosis produces two cells genetically identical to the parent cell

The remarkable regularity of the process of cell division ensures that each daughter nucleus receives exactly the same number and kinds of chromosomes that the parent cell had. Thus, with a few exceptions, every cell of a multicellular organism has the same genetic makeup. If a cell receives more or fewer than the characteristic number of chromosomes through some malfunction of the cell division process, the resulting cell may show marked abnormalities and often cannot survive.

Mitosis provides for the orderly distribution of chromosomes (and of centrioles, if present), but what about the various cytoplasmic organelles? For example, all eukaryotic cells,

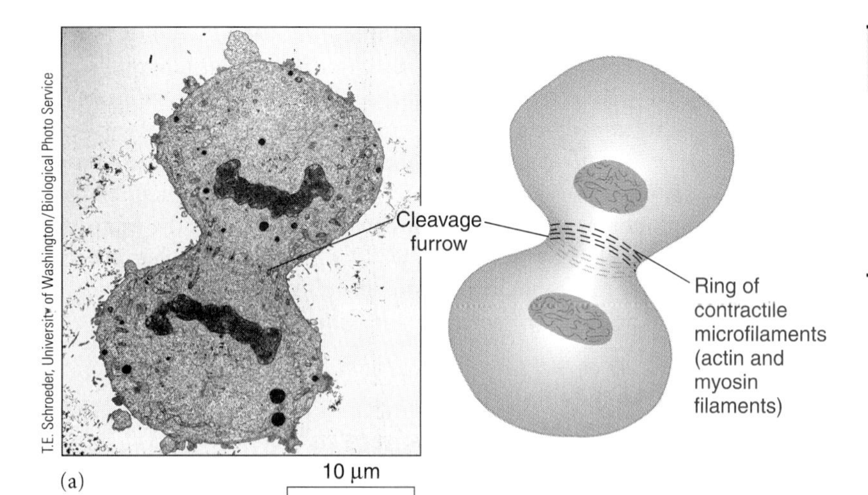

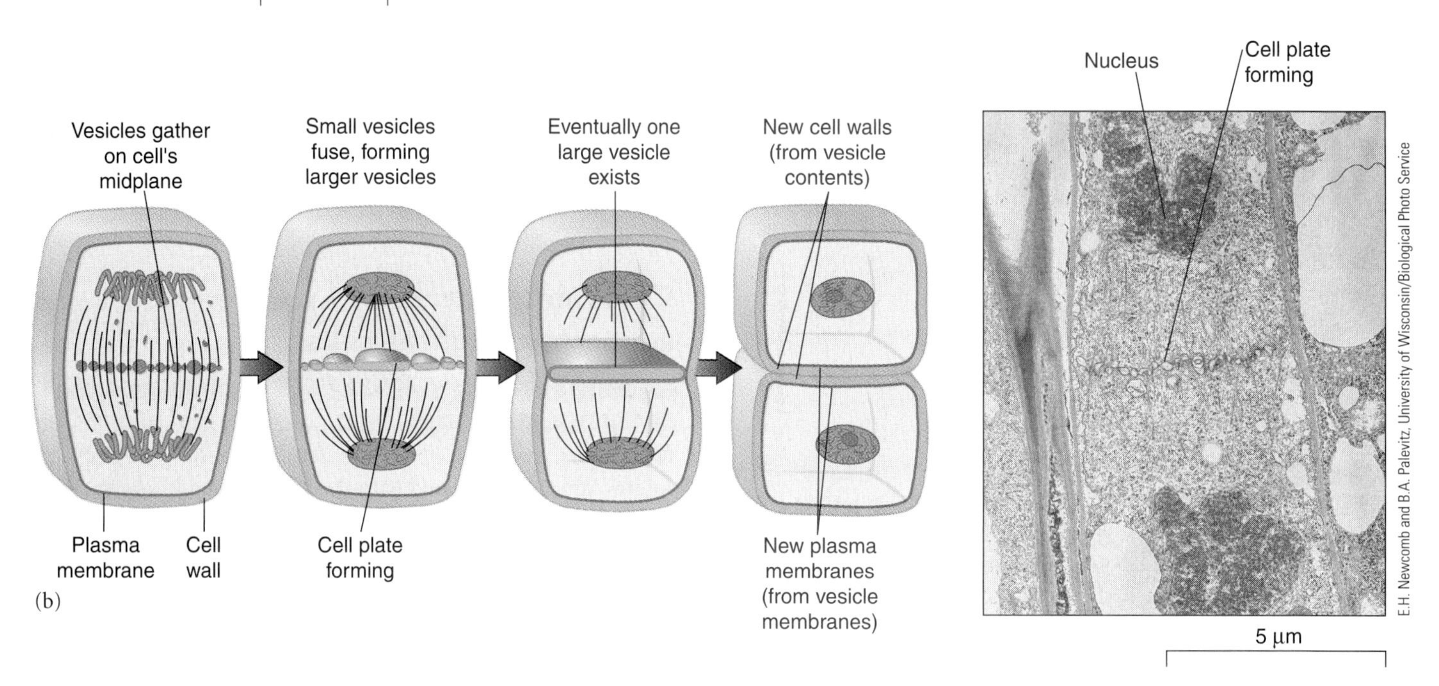

FIGURE 9-9 Cytokinesis in plant and animal cells.

The nuclei in both TEMs are in telophase. The drawings show 3-D relationships. **(a)** TEM of the equatorial region of a cultured animal cell undergoing cytokinesis shows a cleavage furrow. **(b)** TEM of a maple leaf cell (*Acer saccharinum*) undergoing cytokinesis shows cell plate formation.

including plant cells, require mitochondria. Likewise, photosynthetic plant cells cannot carry out photosynthesis without chloroplasts. These organelles contain their own DNA and appear to form by the division of previously existing mitochondria or plastids or their precursors. This nonmitotic division process is similar to prokaryotic cell division (see Chapter 23) and generally occurs during interphase, not when the cell divides. Because many copies of each organelle are present in each cell, organelles are apportioned with the cytoplasm that each daughter cell receives during cytokinesis.

An internal genetic program interacting with external signals regulates the cell cycle

When conditions are optimal, some prokaryotic cells can divide every 20 minutes. The generation times of eukaryotic cells are generally much longer, although the frequency of cell division varies widely among different species and among different tissues of the same species. Some cells in the central nervous system usually stop dividing after the first few months of life, whereas blood-forming cells, digestive tract cells, and skin cells divide frequently throughout the life of the organism. Under optimal conditions of nutrition, temperature, and pH, the length of the eukaryotic cell cycle is constant for any given cell type. Under less favorable conditions, however, the generation time may be longer.

Certain regulatory molecules that control the cell cycle are common to all eukaryotes. Genetically programmed in the cell's nucleus, these regulatory molecules are found in organisms as diverse as yeast (a unicellular fungus), clams, frogs, humans, and plants. Molecular regulators trigger a specific sequence of events during the cell cycle. Because a failure to carefully control cell division can have disastrous consequences, signals in the genetic program, called **cell cycle checkpoints,** ensure that all the events of a particular stage have been completed before the next stage begins. For example, if a cell produces damaged or unreplicated DNA, the cell cycle halts and the cell will not undergo mitosis.

Figure 9-10 shows the key molecules involved in regulating the cell cycle. Among them are **protein kinases,** enzymes that

FIGURE 9-10 **Molecular control of the cell cycle.**

Different cyclins associate with Cdks (cyclin-dependent kinases), triggering the onset of the different stages of the cell cycle. This diagram is a simplified view of the control system that triggers the cell to move from G_2 to M phase. ① Cyclin is synthesized and accumulates. ② Cdk associates with cyclin, forming a cyclin-Cdk complex, M-Cdk. ③ M-Cdk phosphorylates proteins, activating those that facilitate mitosis and inactivating those that inhibit mitosis. ④ An activated enzyme complex recognizes a specific amino acid sequence in cyclin and targets it for destruction. When cyclin is degraded, M-Cdk activity is terminated, and the cells formed by mitosis enter G_1. ⑤ Cdk is not degraded but is recycled and reused.

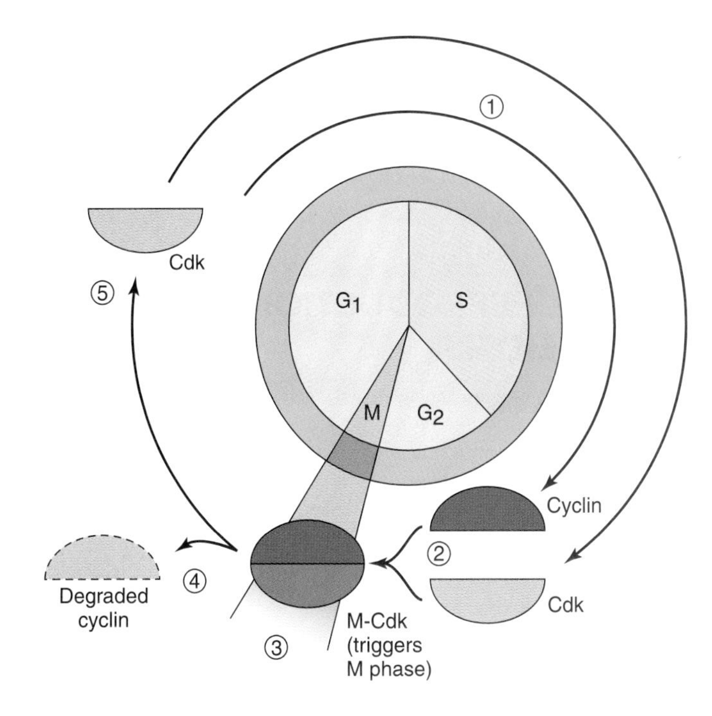

activate or inactivate other proteins by *phosphorylating* (adding phosphate groups to) them. The protein kinases involved in controlling the cell cycle are **cyclin-dependent kinases (Cdks).** The activity of various Cdks increases and then decreases as the cell moves through the cell cycle. Cdks are active only when they bind tightly to regulatory proteins called **cyclins.** The cyclins are so named because their levels fluctuate predictably during the cell cycle (that is, they "cycle," or are alternately synthesized and degraded as part of the cell cycle).

Three scientists who began their research in the 1980s on the roles of protein kinases and cyclins in the cell cycle (American Leland Hartwell, Briton Paul Nurse, and Briton Tim Hunt) won the Nobel Prize in Physiology or medicine in 2001. Their discoveries were cited as important not only in working out the details of the fundamental cell process of mitosis but also in understanding why cancer cells divide when they should not. For example, cyclin levels are often higher than normal in human cancer cells.

When a specific Cdk associates with a specific cyclin, it forms a **cyclin-Cdk complex.** Cyclin-Cdk complexes phosphorylate enzymes and other proteins. Some of these proteins become activated when they are phosphorylated, and others become inactivated. For example, phosphorylation of the protein **p27,** known to be a major inhibitor of cell division, is thought to initiate degradation of the protein. As various enzymes are activated or inactivated by phosphorylation, the activities of the cell change. Thus, a decrease in a cell's level of p27 causes a nondividing cell to resume division.

Eukaryotic cells form four major cyclin-Cdk complexes (G_1-Cdk, G_1/S-Cdk, S-Cdk, and M-Cdk), and each cyclin-Cdk complex phosphorylates a different group of proteins. G_1-Cdk prepares the cell to pass from the G_1 phase to the S phase, and then G_1/S-Cdk commits the cell to undergo DNA replication. S-Cdk initiates DNA replication. M-Cdk promotes the events of mitosis, including chromosome condensation, nuclear envelope breakdown, and mitotic spindle formation.

M-Cdk also activates another enzyme complex, the **anaphase-promoting complex (APC),** toward the end of metaphase. APC initiates anaphase by allowing degradation of the proteins that hold the sister chromatids together during metaphase. As a result, the sister chromatids separate into two daughter chromosomes. At this point, cyclin is degraded to negligible levels and M-Cdk activity drops, allowing the mitotic spindle to disassemble and the cell to exit mitosis.

Certain drugs can stop the cell cycle at a specific checkpoint. Some of these prevent DNA synthesis, whereas others inhibit the synthesis of proteins that control the cycle or inhibit the synthesis of structural proteins that contribute to the mitotic spindle. Because one of the distinguishing features of most cancer cells is their high rate of cell division relative to most normal body cells, they can be most affected by these drugs. Many side effects of certain anticancer drugs (such as nausea and hair loss) are due to the drugs' effects on rapidly dividing normal cells in the digestive system and hair follicles.

In plant cells, certain hormones stimulate mitosis. These include the **cytokinins,** a group of plant hormones that promote mitosis both in normal growth and in wound healing (see Chapter 36).

Similarly, animal hormones, such as certain steroids, stimulate growth (see Chapter 47). Protein **growth factors,** which are active at extremely low concentrations, stimulate mitosis in some animal cells by forming G_1-Cdk. Of the approximately 50 protein growth factors known, some act only on specific types of cells, whereas others act on a broad range of cell types. For example, the effects of the growth factor erythropoietin are limited to cells that will develop into red blood cells, but epidermal growth factor stimulates many cell types to divide. Many types of cancer cells divide in the absence of growth factors.

Review

- What are the stages of the cell cycle? During which stage does DNA replicate?
- What are sister chromatids?
- How does the DNA content of the cell change from the beginning of interphase to the end of interphase? Does the number of chromatids change? Does the number of chromosomes change?
- Assume an animal has a chromosome number of 10. (a) How many chromosomes would it have in a typical body cell, such as a skin cell during G_1? (b) How many chromosomes would

be present in each daughter cell produced by mitosis? Assuming the daughter cells are in G_1, are these duplicated chromosomes?

BiologyNow™ Assess your understanding of the **cell cycle and mitosis** by taking the pretest on your BiologyNow CD-ROM.

SEXUAL REPRODUCTION AND MEIOSIS

Learning Objectives

6 Differentiate between asexual and sexual reproduction.
7 Distinguish between haploid and diploid cells, and define *homologous chromosomes*.
8 Explain the significance of meiosis, and diagram the process.
9 Contrast mitosis and meiosis, emphasizing the different outcomes.
10 Compare the roles of mitosis and meiosis in various generalized life cycles.

Although the details of the reproductive process vary greatly among different kinds of eukaryotes, biologists distinguish two basic types of reproduction: asexual and sexual. In **asexual reproduction** a single parent splits, buds, or fragments to produce two or more individuals. In most forms of eukaryotic asexual reproduction, all the cells are the result of mitotic divisions, so their genes and inherited traits are like those of the parent. Such a group of genetically identical organisms is termed a **clone.** In asexual reproduction, organisms well adapted to their environment produce new generations of similarly adapted organisms. Asexual reproduction occurs rapidly and efficiently, partly because the organism does not need to expend time and energy finding a mate.

In contrast, **sexual reproduction** involves the union of two sex cells, or **gametes,** to form a single cell called a **zygote.** Usually two different parents contribute the gametes, but in some cases a single parent furnishes both gametes. In the case of animals and plants, the egg and sperm cells are the gametes, and the fertilized egg is the zygote.

Sexual reproduction results in genetic variation among the offspring. (How this genetic variation arises is discussed later in this chapter and in Chapter 10.) Because the offspring produced by sexual reproduction are not genetically identical to their parents or to each other, some offspring may be able to survive environmental changes better than either parent does. However, one disadvantage of sexual reproduction is that some offspring with a different combination of traits may be less likely to survive than their parents.

There is a potential problem in eukaryotic sexual reproduction: If each gamete had the same number of chromosomes as the parent cell that produced it, then the zygote would have twice as many chromosomes. This doubling would occur generation after generation. How do organisms avoid producing zygotes with ever-increasing chromosome numbers? To answer this question, we need more information about the types of chromosomes found in cells.

Each chromosome in a somatic (body) cell of a plant or animal normally has a partner chromosome. The two partners, known as **homologous chromosomes,** are similar in size, shape, and the position of their centromeres. Futhermore, special chromosome-staining procedures make a characteristic pattern of bands evident in the members of each chromosome pair. In most species, chromosomes vary enough in their structure that cytologists can distinguish the different chromosomes and match up the homologous pairs. The 46 chromosomes in human cells constitute 23 homologous pairs. The most important feature of homologous chromosomes is that they carry very similar, but not necessarily identical, genetic information. For example, each member of a homologous pair may carry a gene that specifies hemoglobin structure. But one member may have the information for the normal hemoglobin β chain (see Fig. 3-22a), whereas the other may specify the abnormal form of hemoglobin associated with sickle cell anemia (see Chapter 15). Homologous chromosomes can therefore be contrasted with the two members of a pair of sister chromatids, which are precisely identical to each other.

A *set* of chromosomes has one of each kind of chromosome; in other words, it contains one member of each homologous pair. If a cell or nucleus contains two sets of chromosomes, it is said to have a **diploid** chromosome number. If it has only a single set of chromosomes, it has the **haploid** number. In humans, the diploid chromosome number is 46 and the haploid number is 23. When a sperm and egg fuse at fertilization, each gamete is haploid, contributing one set of chromosomes; the diploid number is thereby restored in the fertilized egg (zygote). When the zygote divides by mitosis to form the first two cells of the embryo, each daughter cell receives the diploid number of chromosomes, and subsequent mitotic divisions repeat this. Thus, most human body cells are diploid.

If an individual's cells have three or more sets of chromosomes, we say that it is **polyploid.** Polyploidy is relatively rare among animals but quite common among plants (see Chapter 19). In fact, polyploidy has been important in plant evolution. As many as 80% of all flowering plants are polyploid. Polyploid plants are often larger and hardier than diploid members of the same group. Many commercially important plants, such as wheat and cotton, are polyploid.

The chromosome number found in the gametes of a particular species is represented as n, and the zygotic chromosome number is represented as $2n$. If the organism is not polyploid, the haploid chromosome number is equal to n, and the diploid number is equal to $2n$; thus in humans, $n = 23$ and $2n = 46$. For simplicity, in the rest of this chapter we assume the organisms used as examples are not polyploid. We use diploid and $2n$ interchangeably, and haploid and n interchangeably, although the terms are not strictly synonymous.

Meiosis produces haploid cells with unique gene combinations

We have examined the process of mitosis, which ensures that each daughter cell receives exactly the same number and kinds of chromosomes as the parent cell. A diploid cell that undergoes mitosis produces two diploid cells. Similarly, a haploid cell that undergoes mitosis produces two haploid cells. (Some eukaryotic organisms—certain yeasts, for example—are haploid.)

A division that reduces chromosome number is called **meiosis.** The term means "to make smaller," and the chromosome number is reduced by one half. In meiosis a diploid cell undergoes two cell divisions, potentially yielding four haploid cells.

The events of meiosis are similar to the events of mitosis, with four important differences:

1. Meiosis involves two successive nuclear and cytoplasmic divisions, producing up to four cells.
2. Despite two successive nuclear divisions, the DNA and other chromosomal components duplicate only once—during the interphase preceding the first meiotic division.
3. Each of the four cells produced by meiosis contains the haploid chromosome number, that is, only one chromosome set containing only one representative of each homologous pair.
4. During meiosis, the genetic information from both parents is shuffled, so each resulting haploid cell has a virtually unique combination of genes.

Meiosis typically consists of two nuclear and cytoplasmic divisions, designated the *first* and *second meiotic divisions*, or simply **meiosis I** and **meiosis II.** Each includes prophase, metaphase, anaphase, and telophase stages. During meiosis I, the members of each homologous chromosome pair first join together and then separate and move into different nuclei. In meiosis II, the sister chromatids that make up each chromosome separate and are distributed to two different nuclei. The following discussion describes meiosis in an organism with a diploid chromosome number of 4. Refer to Figures 9-11 and 9-12 as you read.

Prophase I includes synapsis and crossing-over

As in mitosis, the chromosomes duplicate during the S phase of interphase, before meiosis actually begins. Each duplicated chromosome consists of two chromatids. During **prophase I,** while the chromatids are still elongated and thin, the homologous chromosomes come to lie lengthwise side by side. This process is called **synapsis,** which means "fastening together." In our example, because the diploid number is 4, synapsis results in two homologous pairs.

One member of each homologous pair is called the **maternal homologue,** because it was originally inherited from the female parent during the formation of the zygote; the other member of a homologous pair is the **paternal homologue,** because it was inherited from the male parent. Because each chromosome duplicated during interphase and now consists of two chromatids, synapsis results in the association of four chromatids. The resulting association is a **tetrad.** The number of tetrads per prophase I cell is equal to the haploid chromosome number. In our example of an animal cell with a diploid number of 4, there are 2 tetrads (and a total of 8 chromatids); in a human cell at prophase I, there are 23 tetrads (and a total of 92 chromatids).

Homologous chromosomes become closely associated during synapsis. Electron microscopic observations reveal that a characteristic structure, the **synaptonemal complex,** forms be-

FIGURE **9-11** **Meiosis in the trumpet lily (*Lilium longiflorum*).**

The chromosomes shown in these LMs have been stained and the cells flattened on microscope slides. **(a)** Mid-prophase I. **(b)** Late prophase I. **(c)** Metaphase I. **(d)** Anaphase I. **(e)** Prophase II. **(f)** Metaphase II. **(g)** Anaphase II. **(h)** Four daughter cells.

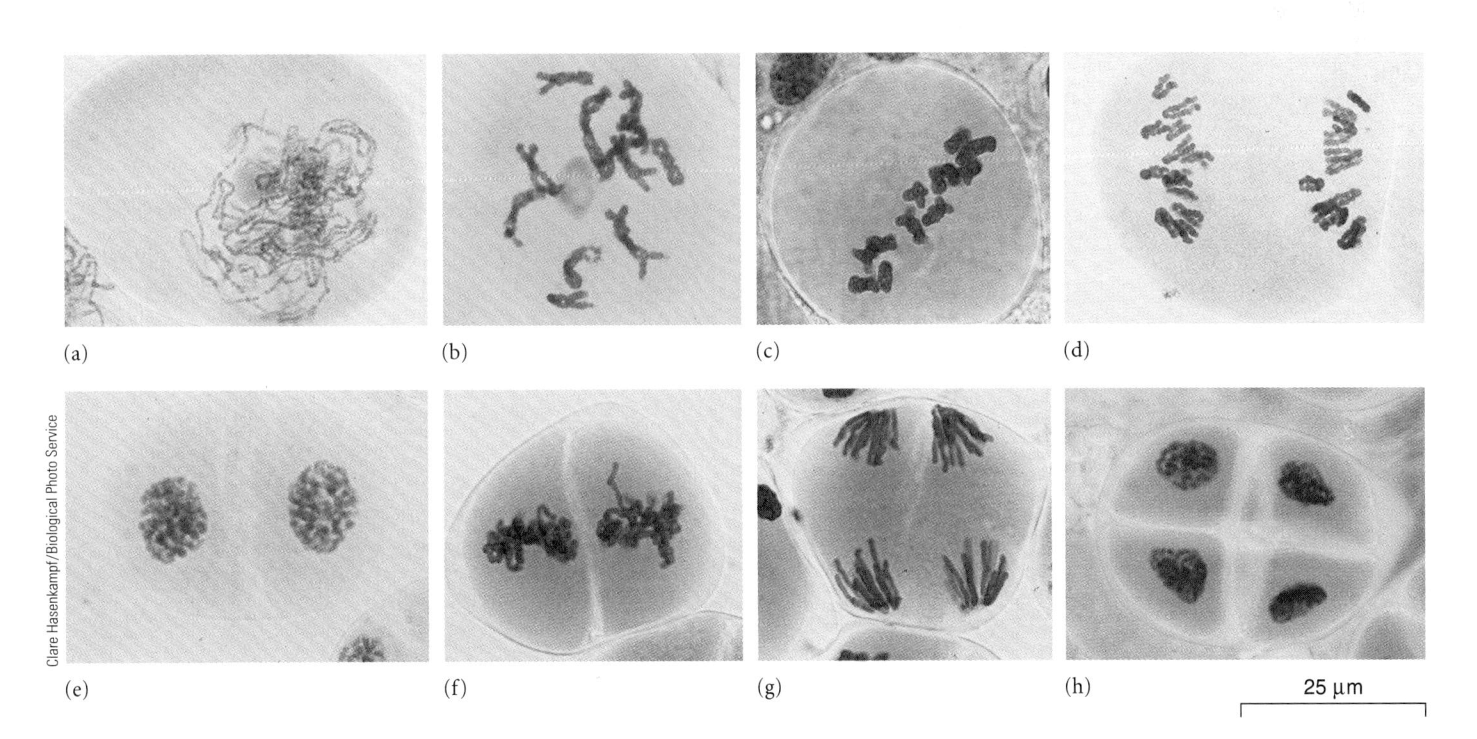

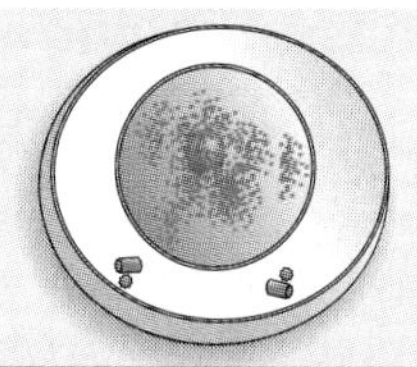

INTERPHASE
Interphase preceding meiosis; DNA replicates.

MEIOSIS I

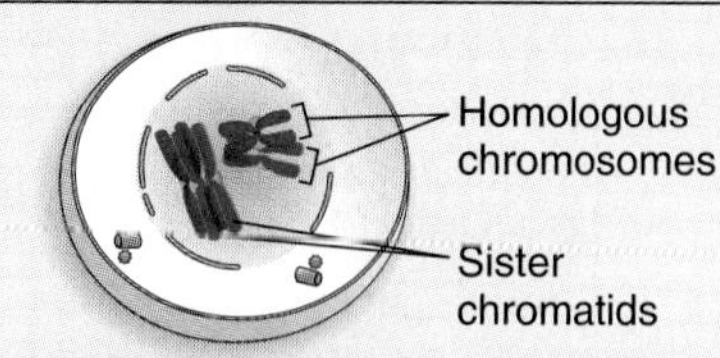

PROPHASE I
Homologous chromosomes synapse, forming tetrads; nuclear envelope breaks down.

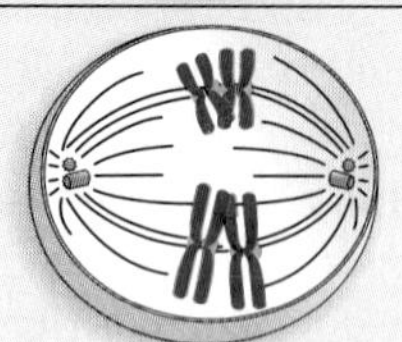

METAPHASE I
Tetrads line up on cell's midplane. Tetrads held together at chiasmata (sites of prior crossing-over).

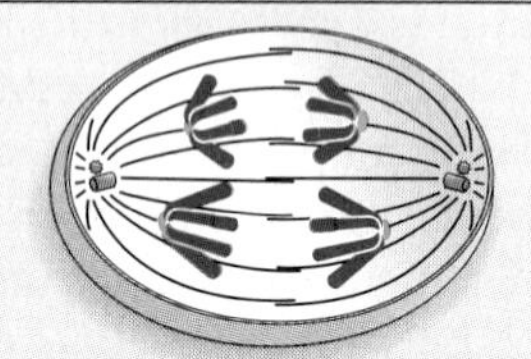

ANAPHASE I
Homologous chromosomes separate and move to opposite poles. Note that sister chromatids remain attached at their centromeres.

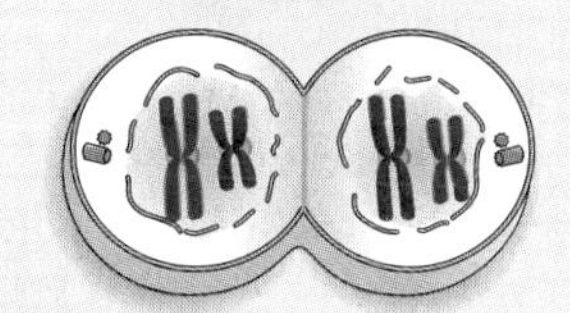

TELOPHASE I
One of each pair of homologous chromosomes is at each pole. Cytokinesis occurs.

MEIOSIS II

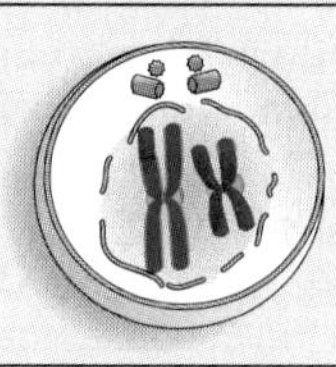

PROPHASE II
Chromosomes condense again following a brief period of interkinesis. DNA does *not* replicate again.

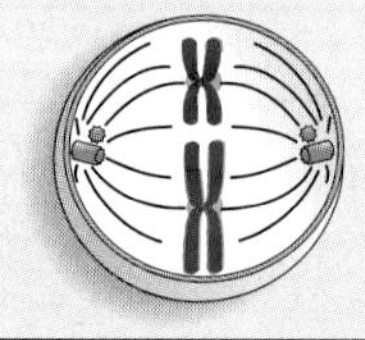
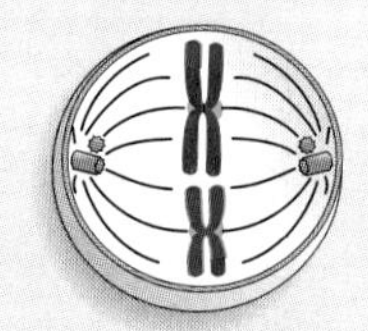

METAPHASE II
Chromosomes line up along cell's midplane.

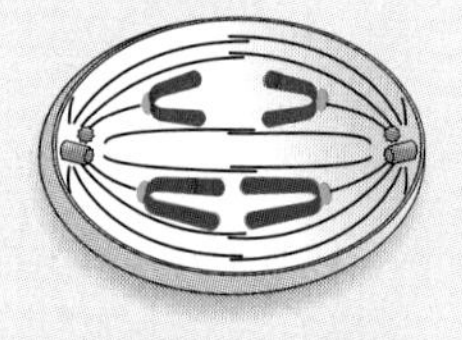
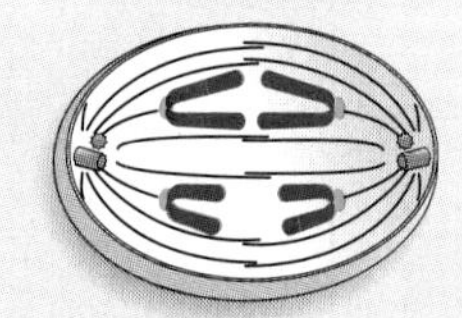

ANAPHASE II
Sister chromatids separate, and chromosomes move to opposite poles.

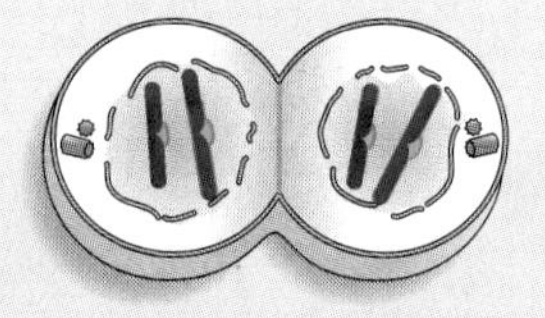
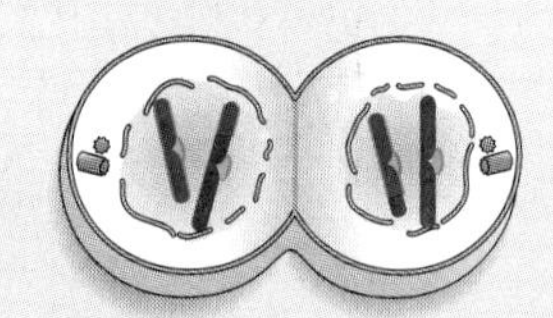

TELOPHASE II
Nuclei formed at opposite poles of each cell. Cytokinesis occurs.

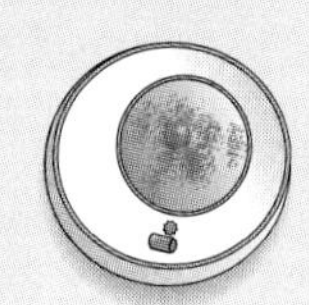

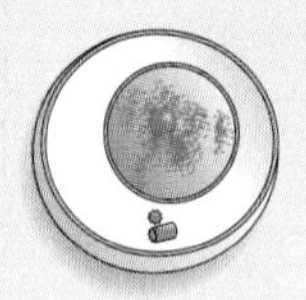

HAPLOID CELLS
Four gametes (animal) or four spores (plant) are produced.

◀ FIGURE **9-12** **Meiosis.**

Meiosis consists of two nuclear divisions, meiosis I and meiosis II. Shown here is an animal cell with a diploid chromosome number of 4. The maternal chromosomes are shown in red; the paternal chromosomes are blue. Meiosis ends with the formation of four haploid cells with two chromosomes each.

tween the synapsed homologues (Fig. 9-13). This structure holds the synapsed homologues together and is thought to play a role in chromosomal **crossing-over,** a process in which paired homologous chromosomes exchange genetic material (DNA). Crossing-over produces new combinations of genes. The **genetic recombination** from crossing-over greatly enhances the genetic variation—that is, new combinations of traits—among sexually produced offspring.

In addition to the unique processes of synapsis and crossing-over, events similar to those in mitotic prophase also occur during prophase I. A spindle forms consisting of microtubules and other components. In animal cells, one pair of centrioles moves to each pole, and astral microtubules form. The nuclear envelope disappears in late prophase I, and in cells with large and distinct chromosomes, the structure of the tetrads can be seen clearly with the microscope. The sister chromatids remain closely aligned along their lengths. However, the centromeres (and kinetochores) of the homologous chromosomes become separated from one another. In late prophase I, the homologous chromosomes are held together only at specific regions, termed **chiasmata.** Each chiasma originates at a crossing-over site, that is, a site at which homologous chromatids exchanged genetic material, and rejoined, producing an X-shaped configuration (Fig. 9-14). The consequences of crossing-over and genetic recombination are discussed in Chapter 10.

During meiosis I, homologous chromosomes separate

Prophase I ends when the tetrads align on the midplane. The cell is now said to be at **metaphase I.** Both sister kinetochores of one duplicated chromosome are attached by spindle fibers to the same pole, and both sister kinetochores of the duplicated homologous chromosome are attached to the opposite pole. (By contrast, in mitosis sister kinetochores are attached to opposite poles.)

During **anaphase I,** the paired homologous chromosomes separate, or disjoin, and move toward opposite poles. Each pole receives a random mixture of maternal and paternal chromosomes, but only one member of each homologous pair is present at each pole. The sister chromatids are united at their centromere regions. Again, this differs from mitotic anaphase, in which the sister chromatids separate and move to opposite poles.

During **telophase I,** the chromatids generally decondense somewhat, the nuclear envelope may reorganize, and cytokinesis may take place. Each telophase I nucleus contains the haploid number of chromosomes, but each chromosome is a duplicated chromosome (it consists of a pair of chromatids). In our example, two duplicated chromosomes lie at each pole, for a total of four chromatids; humans have 23 duplicated chromosomes (46 chromatids) at each pole.

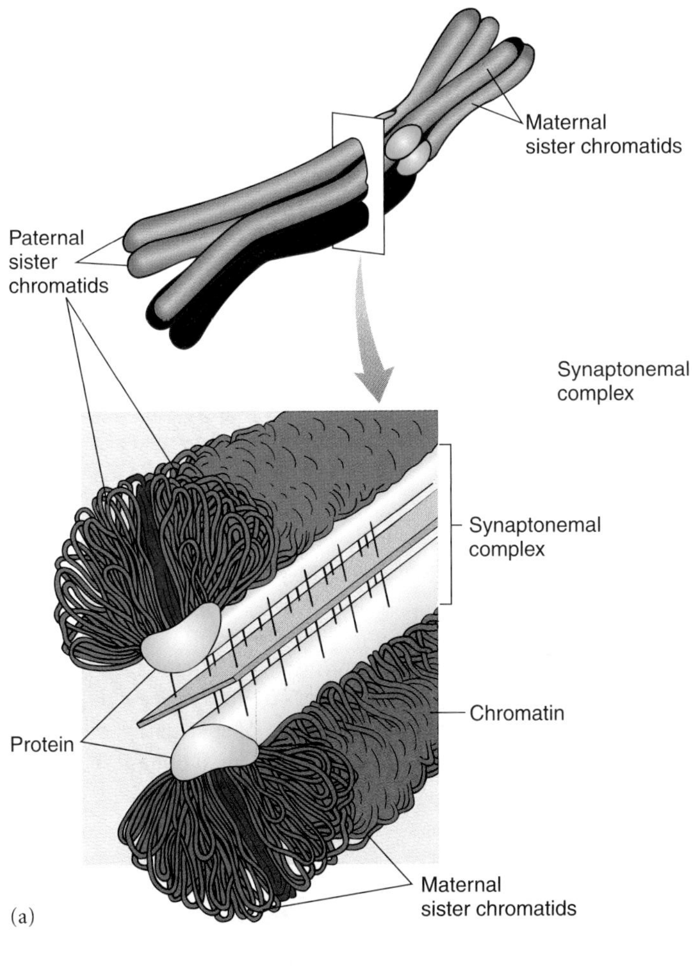

Maternal sister chromatids
Paternal sister chromatids
Synaptonemal complex
Chromatin
Protein
Maternal sister chromatids
(a)

Synaptonemal complex

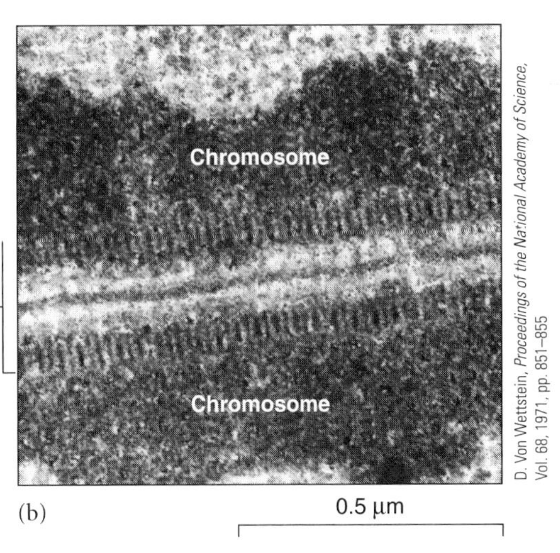

(b) 0.5 μm

FIGURE **9-13** **A synaptonemal complex.**

Synapsing homologous chromosomes in meiotic prophase I are held together by a synaptonemal complex, composed mainly of protein. **(a)** A 3-D model of a tetrad with a complete synaptonemal complex. **(b)** TEM of a synaptonemal complex.

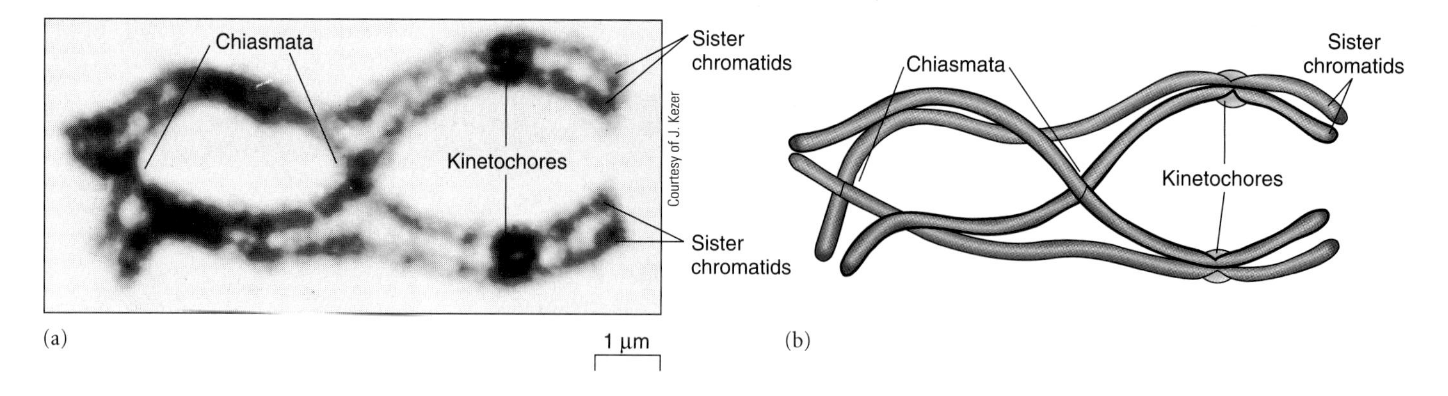

FIGURE **9-14** **A meiotic tetrad with two chiasmata.**

The two chiasmata are the result of separate crossing-over events. **(a)** This LM is of a tetrad during late prophase I of a male meiotic cell (spermatocyte) from a salamander. **(b)** This drawing shows the structure of the tetrad. The paternal chromatids are purple, and the maternal chromatids are pink.

An interphase-like stage usually follows. Because it is not a true interphase—there is no S phase and therefore no intervening DNA replication—it is called **interkinesis.** Interkinesis is very brief in most organisms and absent in some.

Chromatids separate in meiosis II

Because the chromosomes usually remain partially condensed between divisions, the prophase of the second meiotic division is brief. **Prophase II** is similar to mitotic prophase in many respects. There is no pairing of homologous chromosomes (indeed, only one member of each pair is present in each nucleus) and no crossing-over.

During **metaphase II,** the chromosomes line up on the midplanes of their cells. You can easily distinguish the first and second metaphases in diagrams; at metaphase I the chromatids are arranged in bundles of four (tetrads), and at metaphase II they are in groups of two (as in mitotic metaphase). This is not always so obvious in living cells.

During **anaphase II** the chromatids, attached to spindle fibers at their kinetochores, separate and move to opposite poles, just as they would at mitotic anaphase. As in mitosis, each former chromatid is now referred to as a *chromosome.* Thus, at **telophase II** there is one representative for each homologous pair at each pole. Each is an unduplicated (single) chromosome. Nuclear envelopes then re-form, the chromosomes gradually elongate to form chromatin fibers, and cytokinesis occurs.

The two successive divisions of meiosis yield four haploid nuclei, each containing *one* of each kind of chromosome. Each resulting haploid cell has a different combination of genes. This genetic variation has two sources: (1) During meiosis, the maternal and paternal chromosomes of homologous pairs separate independently. The chromosomes are "shuffled" so that each member of a pair becomes randomly distributed to one of the poles at anaphase I. (2) DNA segments are exchanged between maternal and paternal homologues during crossing-over.

Mitosis and meiosis lead to contrasting outcomes

Although mitosis and meiosis share many similar features, specific distinctions between these processes result in the formation of different types of cells (Fig. 9-15). Mitosis is a single nuclear division in which sister chromatids separate from each other. If cytokinesis occurs, they are distributed to the two daughter cells, which are genetically identical to each other and to the original cell. Homologous chromosomes do not associate physically at any time in mitosis.

In meiosis, a diploid cell undergoes two successive nuclear divisions, meiosis I and meiosis II. In prophase I of meiosis, the homologous chromosomes undergo synapsis to form tetrads. Homologous chromosomes separate during meiosis I, and sister chromatids separate during meiosis II. Meiosis ends with the formation of four genetically different, haploid daughter cells. The fates of these cells depend on the type of life cycle; in animals they differentiate as gametes, whereas in plants they become spores.

The timing of meiosis in the life cycle varies among species

Because sexual reproduction is characterized by the fusion of two haploid sex cells to form a diploid zygote, it follows that in a sexual life cycle, meiosis must occur before gametes can form.

In animals and a few other organisms, meiosis leads directly to gamete production (Fig. 9-16a). An organism's **somatic cells** (body cells) multiply by mitosis and are diploid; the only haploid cells produced are the gametes. Gametes develop when **germ line cells,** which give rise to the next generation, undergo meiosis.

The formation of gametes is known as **gametogenesis.** Male gametogenesis, termed **spermatogenesis,** forms four haploid sperm cells for each cell that enters meiosis. (See Chapter 48 and Fig. 48-5 for a detailed description of spermatogenesis.)

In contrast, female gametogenesis, termed **oogenesis,** forms a single egg cell, or *ovum,* for every cell that enters meiosis. In

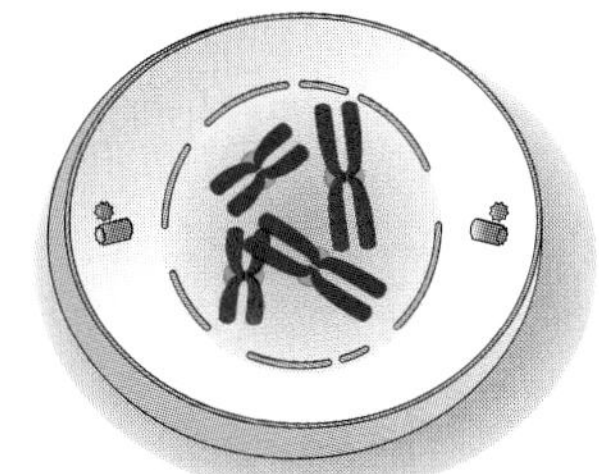

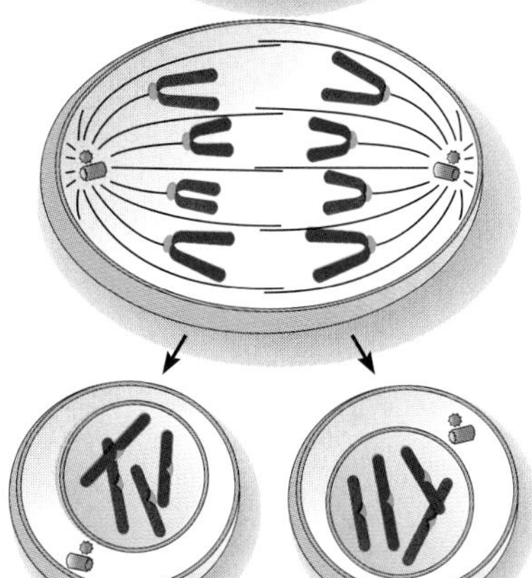

ACTIVE FIGURE **9-15** | **Mitosis and meiosis.**

This drawing compares the events and outcomes of mitosis and meiosis, in each case beginning with a diploid cell with four chromosomes (two pairs of homologous chromosomes). Because the chromosomes duplicated in the previous interphase, each chromosome consists of two sister chromatids. The chromosomes derived from one parent are shown in blue, and those from the other parent are red. Homologous pairs are similar in size and shape. Chiasmata are not shown, and some of the stages have been omitted for simplicity.

BiologyNow™ Watch a movie that features **living cells undergoing mitosis and meiosis** by clicking on this figure on your BiologyNow CD-ROM.

this process, all the cytoplasm goes to only one of the two cells produced during each meiotic division. At the end of the first meiotic division, one nucleus is retained and the other, called the first *polar body,* degenerates. Similarly, at the end of the second division one nucleus becomes the second polar body and the other nucleus survives. In this way, one haploid nucleus receives most of the accumulated cytoplasm and nutrients from the original meiotic cell. (See Chapter 48 and Fig. 48-11 for a detailed description of oogenesis.)

Although meiosis occurs at some point in a sexual life cycle, it does not always *immediately* precede gamete formation. Many simple eukaryotes, including some fungi and algae, remain haploid (their cells dividing mitotically) throughout most of their life cycles, with individuals being unicellular or multicellular. Two haploid gametes (produced by mitosis) fuse to form a diploid zygote that undergoes meiosis to restore the haploid state (Fig. 9-16b). Examples of these types of life cycles are found in Figures 24-19 and 25-7.

Plants, some algae, and some fungi have some of the most complicated life cycles (Fig. 9-16c): These life cycles, characterized by an **alternation of generations,** consist of a multicellular diploid stage, the **sporophyte generation,** and a multicellular haploid stage, the **gametophyte generation.** Diploid sporophyte cells undergo meiosis to form haploid spores, each of which then divides mitotically to produce a multicellular haploid gametophyte. Gametophytes produce gametes by mitosis. The female and male gametes (egg and sperm cells) then fuse to form a diploid zygote that divides mitotically to form a multicellular, diploid sporophyte.

In ferns, conifers, and flowering plants, the diploid sporophyte—which includes the roots, stems, and leaves of the plant body—is the dominant form. The gametophytes are small and in-

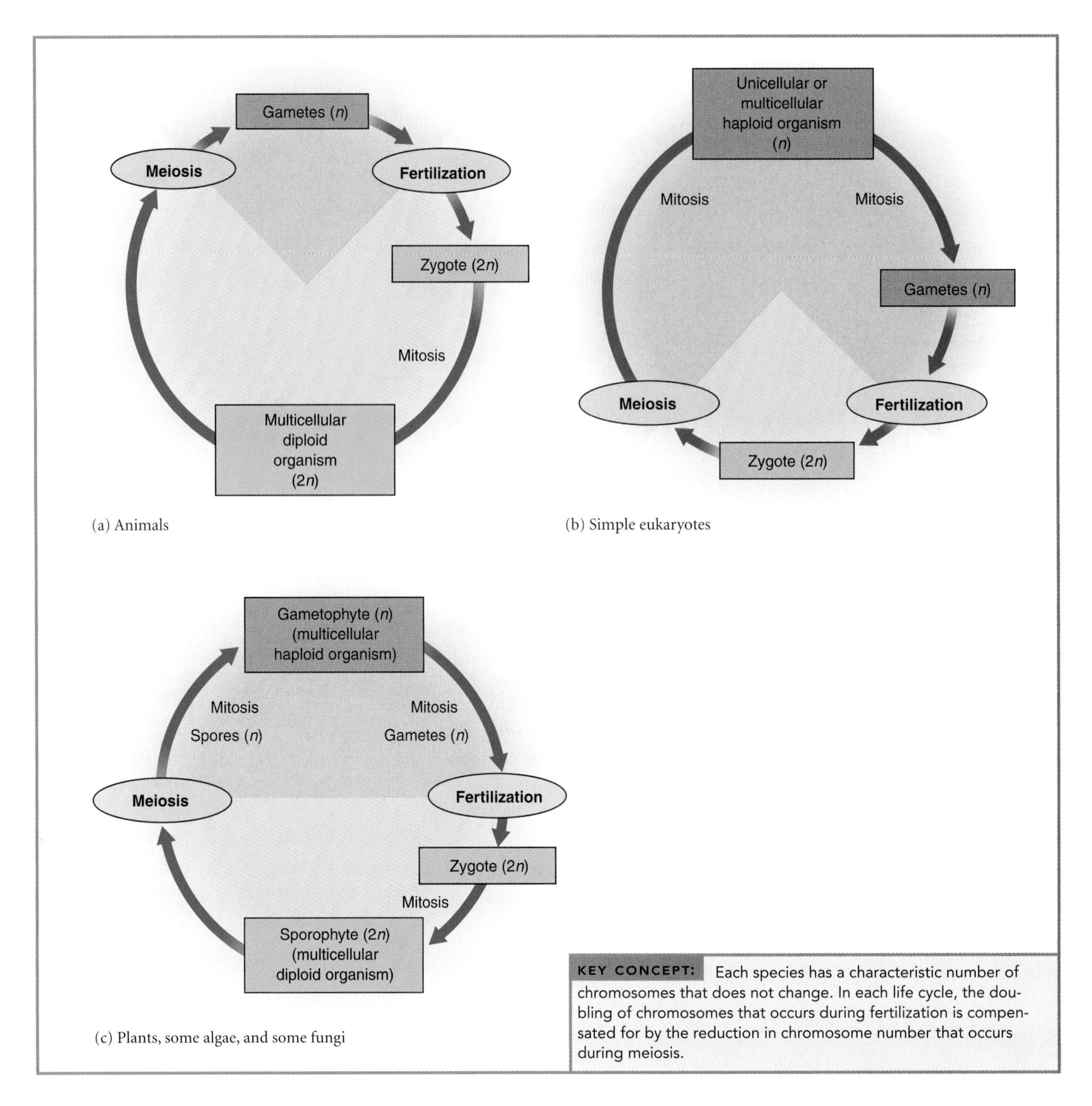

KEY CONCEPT: Each species has a characteristic number of chromosomes that does not change. In each life cycle, the doubling of chromosomes that occurs during fertilization is compensated for by the reduction in chromosome number that occurs during meiosis.

FIGURE **9-16** **Representative life cycles.**

The color code and design here is used throughout the rest of the book. For example, in all life cycles the haploid (*n*) generation is shown in purple, and the diploid (2*n*) generation is gold. The processes of meiosis and fertilization always link the haploid and diploid generations.

conspicuous. For example, in flowering plants, a microscopic pollen grain contains a haploid male gametophyte that forms haploid sperm cells by mitosis. You can find more detailed descriptions of alternation of generations in plants in Chapters 26 and 27.

Review

- Are homologous chromosomes present in a diploid cell? Are they present in a haploid cell?
- Assume an animal cell has a diploid chromosome number of 10. (a) How many tetrads would form in prophase I of meiosis? (b) How many chromosomes would be present in each gamete? Are these duplicated chromosomes?
- How does the outcome of meiosis differ from the outcome of mitosis?
- Can haploid cells divide by mitosis? By meiosis?

BiologyNow™ Assess your understanding of the **sexual reproduction and mitosis** by taking the pretest on your BiologyNow CD-ROM.

SUMMARY WITH KEY TERMS

1 Discuss the significance of chromosomes in terms of their information content.

- **Genes,** the cell's informational units, are made of DNA. In eukaryotes, DNA associates with protein to form the **chromatin** fibers that make up **chromosomes.** The organization of eukaryotic DNA into chromosomes allows the DNA, which is much longer than a cell's nucleus, to be accurately replicated and sorted into daughter cells without tangling.

2 Compare the organization of DNA in prokaryotic and eukaryotic cells.

- Prokaryotic cells usually have circular DNA molecules.
- Eukaryotic chromosomes have several levels of organization. The DNA is associated with **histones** (basic proteins) to form **nucleosomes,** each of which consists of a histone bead with DNA wrapped around it. The nucleosomes are organized into large, coiled loops held together by nonhistone **scaffolding proteins.**

3 Identify the stages in the eukaryotic cell cycle, describe their principal events, and point out some ways in which the cycle is controlled.

- The eukaryotic **cell cycle** is the period from the beginning of one division to the beginning of the next. The cell cycle consists of interphase and M phase.
- **Interphase** consists of the first gap phase (G_1), the synthesis phase (S), and the second gap phase (G_2). During the **G_1 phase,** the cell grows and prepares for the S phase. During the **S phase,** DNA and the chromosomal proteins are synthesized, and chromosome duplication occurs. During the **G_2 phase,** protein synthesis increases in preparation for cell division.
- **Cyclin-dependent kinases (Cdks)** are **protein kinases** involved in controlling the cell cycle. Cdks are active only when they bind tightly to regulatory proteins called **cyclins.** Cyclin levels fluctuate predictably during the cell cycle.
- **M phase** consists of **mitosis,** the nuclear division that produces two nuclei identical to the parental nucleus, and **cytokinesis,** the division of the cytoplasm to yield two daughter cells.

4 Describe the structure of a duplicated chromosome, including the sister chromatids, centromeres, and kinetochores.

- A duplicated chromosome consists of a pair of **sister chromatids,** which contain identical DNA sequences. Each chromatid includes a constricted region called the **centromere.** Sister chromatids are tightly associated in the region of their centromeres. Attached to each centromere is a **kinetochore,** a structure formed from proteins to which microtubules can bind.

5 Explain the significance of mitosis, and describe the process.

- In mitosis, identical chromosomes are distributed to each pole of the cell, and a nuclear envelope forms around each set.
- During **prophase,** duplicated chromosomes, each composed of a pair of sister chromatids, become visible with the microscope. The nucleolus disappears, the nuclear envelope breaks down, and the **mitotic spindle** begins to form.
- During **metaphase,** the chromosomes are aligned on the **metaphase plate** of the cell; the mitotic spindle is complete and the kinetochores of the sister chromatids are attached by microtubules to opposite poles of the cell.
- During **anaphase,** the sister chromatids separate and move to opposite poles. Each former chromatid is now a chromosome.
- During **telophase,** a nuclear envelope re-forms around each set of chromosomes, nucleoli become apparent, the chromosomes uncoil, and the spindle disappears. Cytokinesis generally begins in telophase.

6 Differentiate between asexual and sexual reproduction.

- Offspring produced by **asexual reproduction** usually have hereditary traits identical to those of the single parent. Mitosis is the basis for asexual reproduction in eukaryotic organisms.
- In **sexual reproduction,** two haploid sex cells, or **gametes,** fuse to form a single diploid **zygote.** In a sexual life cycle, **meiosis** must occur before gametes can be produced.

7 Distinguish between haploid and diploid cells, and define *homologous chromosomes.*

- A **diploid** cell has a characteristic number of chromosome pairs per cell. The members of each pair, called **homologous chromosomes,** are similar in length, shape, and other features and carry genes affecting the same kinds of attributes of the organism.
- A **haploid** cell contains only one member of each homologous chromosome pair.

8 Explain the significance of meiosis, and diagram the process.

- A diploid cell undergoing **meiosis** completes two successive cell divisions, yielding four haploid cells.
- **Meiosis I** begins with **prophase I,** in which the members of a homologous pair of chromosomes physically join by the process of **synapsis. Crossing-over** is a process of **genetic recombination** during which homologous (nonsister) chromatids exchange segments of DNA strands.
- At **metaphase I, tetrads**—each consisting of a pair of homologous chromosomes held together by one or more **chiasmata**—line up on the **metaphase plate.** The members of each pair of homologous chromosomes separate during meiotic **anaphase I** and are distributed to different nuclei. Each nucleus contains the haploid number of chromosomes; each chromosome consists of two chromatids.
- During **meiosis II,** the two chromatids of each chromosome separate, and one is distributed to each daughter cell. Each former chromatid is now a chromosome.

9 Contrast mitosis and meiosis, emphasizing the different outcomes.

- Mitosis involves a single nuclear division in which the two daughter cells formed are genetically identical to each other and to the original cell. Synapsis of homologous chromosomes does not occur during mitosis.
- Meiosis involves two successive nuclear divisions and forms four haploid cells. Synapsis of homologous chromosomes occurs during prophase I of meiosis.

10 Compare the roles of mitosis and meiosis in various generalized life cycles.

- The **somatic cells** of animals are diploid and are produced by mitosis. The only haploid cells are the gametes, produced by **gametogenesis,** which in animals occurs by meiosis.

SUMMARY WITH KEY TERMS *(continued)*

- Simple eukaryotes may be haploid. The only diploid stage is the zygote, which undergoes meiosis to restore the haploid state.
- The life cycle of plants and some algae includes an **alternation of generations.** The multicellular diploid **sporophyte generation** forms haploid spores by meiosis. Each spore divides mitotically to form a multicellular haploid **gametophyte generation** which produces gametes by mitosis. Two haploid gametes then fuse to form a diploid zygote, which divides mitotically to produce a new diploid sporophyte generation.

POST-TEST

1. Chromatin fibers include (a) DNA and structural polysaccharides (b) RNA and phospholipids (c) protein and carbohydrate (d) DNA and protein (e) triacylglycerol and steroids
2. A nucleosome consists of (a) DNA and scaffolding proteins (b) scaffolding proteins and histones (c) DNA and histones (d) DNA, histones, and scaffolding proteins (e) histones only
3. The term *S phase* refers to (a) DNA synthesis during interphase (b) synthesis of chromosomal proteins during prophase (c) gametogenesis in animal cells (d) synapsis of homologous chromosomes (e) fusion of gametes in sexual reproduction
4. At which of the following stages do human skin cell nuclei have the same DNA content? (a) early mitotic prophase; late mitotic telophase (b) G_1; G_2 (c) G_1; early mitotic prophase (d) G_1; late mitotic telophase (e) G_2; late mitotic telophase
5. In a cell at ____________, each chromosome consists of a pair of attached chromatids. (a) mitotic prophase (b) meiotic prophase II (c) meiotic prophase I (d) meiotic anaphase I (e) all of the preceding
6. In an animal cell at mitotic metaphase, you would expect to find (a) two pairs of centrioles located on the metaphase plate (b) a pair of centrioles inside the nucleus (c) a pair of centrioles within each microtubule-organizing center (d) a centriole within each centromere (e) no centrioles
7. Cell plate formation usually begins during (a) telophase in a plant cell (b) telophase in an animal cell (c) G_2 in a plant cell (d) G_2 in an animal cell (e) a and b are correct
8. A particular plant species has a diploid chromosome number of 20. A haploid cell of that species at mitotic prophase contains a total of ____________ chromosomes and ____________ chromatids. (a) 20; 20 (b) 20; 40 (c) 10; 10 (d) 10; 20 (e) none of the preceding, because haploid cells cannot undergo mitosis
9. A diploid nucleus at early mitotic prophase has ____________ set(s) of chromosomes; a diploid nucleus at mitotic telophase has ____________ set(s) of chromosomes. (a) 1; 1 (b) 1; 2 (c) 2; 2 (d) 2; 1 (e) not enough information has been given
10. The life cycle of a sexually reproducing organism includes (a) mitosis (b) meiosis (c) fusion of sex cells (d) b and c (e) a, b, and c
11. Which of the following are genetically identical? (a) two cells resulting from meiosis I (b) two cells resulting from meiosis II (c) four cells resulting from meiosis I followed by meiosis II (d) two cells resulting from a mitotic division (e) all of the preceding
12. You would expect to find a synaptonemal complex in a cell at (a) mitotic prophase (b) meiotic prophase I (c) meiotic prophase II (d) meiotic anaphase I (e) meiotic anaphase II
13. A chiasma links a pair of (a) homologous chromosomes at prophase II (b) homologous chromosomes at late prophase I (c) sister chromatids at metaphase II (d) sister chromatids at mitotic metaphase (e) sister chromatids at metaphase I

CRITICAL THINKING

Decide whether each of the following is an example of sexual or asexual reproduction, and state why.

1. A diploid queen honeybee produces haploid eggs by meiosis. Some of these eggs are never fertilized and develop into haploid male honeybees (drones).
2. Seeds develop after a flower has been pollinated with pollen from the same plant.
3. After it has been placed in water, a cutting from a plant develops roots. After it is transplanted to soil, the plant survives and grows.

- Visit our Web site at **http://biology.brookscole.com/solomon7** for links to chapter-related resources on the World Wide Web. Additional online materials relating to this chapter can also be found on our Web site.

BIOLOGY NOW RESOURCES

Active Figures

9-6: The stages of mitosis

9-15: Living cells undergoing mitosis

Preparing for an exam? Take a diagnostic test on your BiologyNow CD-ROM.

Post-Test Answers

1. d	2. c	3. a	4. d
5. e	6. c	7. a	8. d
9. c	10. e	11. d	12. b
13. b			

10

The Basic Principles of Heredity

Gregor Mendel. This painting shows Mendel with his pea plants in the monastery garden at Brünn, Austria (now Brüno, Czech Republic).

CHAPTER OUTLINES

- **Mendel's Principles of Inheritance**
- **Mendelian Inheritance and Chromosomes**
- **Extensions of Mendelian Genetics**

Do you have your father's height and your mother's eye color and freckles? You have inherited these and a multitude of other characteristics, passed on from one generation to another. **Heredity,** the transmission of genetic information from parent to offspring, generally follows predictable patterns in organisms as diverse as humans, penguins, baker's yeast, and sunflowers. **Genetics,** the science of heredity, studies both genetic similarities and **genetic variation,** the differences, between parents and offspring or among individuals of a population.

The study of inheritance as a modern branch of science began in the mid-19th century with the work of Gregor Mendel (1822–1884), a monk who bred pea plants. Mendel was the first scientist to effectively apply quantitative methods to the study of inheritance. He didn't merely describe his observations; he planned his experiments carefully, recorded the data, and analyzed the results mathematically. Although unappreciated during his lifetime, his work was rediscovered in 1900. The science of genetics is based on his major findings, including those now known as *Mendel's principles of segregation and independent assortment.*

During the decades following the rediscovery of Mendel's findings, geneticists initially extended Mendel's principles by correlating the transmission of genetic information from generation to generation with the behavior of chromosomes during meiosis. They also refined his methods and, by studying a variety of organisms, both verified Mendel's findings and added to a growing list of so-called exceptions to his principles. These exceptions include such phenomena as linkage, X linkage, and pleiotropy.

Some geneticists were very active in developing the science of statistical analysis, which was emerging during Mendel's time. Using statistics, scientists could analyze and interpret experimental data in increasingly sophisticated ways. Statistical analysis was also essential for studying the genetic makeup of natural populations of organisms. Scientists eventually combined the genetics of populations with Charles Darwin's theory of evolution by natural selection, to develop a unified modern theory of evolution, firmly based on genetic principles (see Chapters 17 and 18).

Geneticists study not only the transmission of genes but also the expression of genetic information. As you will see in this chapter and those that follow, understanding of the relationship between an organism's genes and its characteristics has become increasingly sophisticated as people have learned more about the flow of information in cells. ■

MENDEL'S PRINCIPLES OF INHERITANCE

Learning Objectives

1. Define the terms *phenotype, genotype, locus, allele, dominant allele, recessive allele, homozygous,* and *heterozygous.*
2. Describe Mendel's principles of segregation and independent assortment.
3. Solve genetics problems involving monohybrid, dihybrid, and test crosses.
4. Apply the product rule and sum rule appropriately when predicting the outcomes of genetic crosses.

PROCESS OF SCIENCE

Gregor Mendel was not the first plant breeder. At the time he began his work, breeders had long recognized the existence of **hybrid** plants and animals, the offspring of two genetically dissimilar parents. When Mendel began his breeding experiments in 1856, two main concepts about inheritance were widely accepted: (1) All hybrid plants that are the offspring of genetically pure, or **true-breeding,** parents are similar in appearance. (2) When these hybrids mate with each other, they do not breed true; their offspring show a mixture of traits. Some look like their parents, and some have features like their grandparents.

Mendel's genius lay in his ability to recognize a pattern in the way the parental traits reappear in the offspring of hybrids. Before Mendel, no one had categorized and counted the offspring and analyzed these regular patterns over several generations to the extent he did. Just as do geneticists today, Mendel chose the organism for his experiments very carefully. The garden pea, *Pisum sativum,* had several advantages. Pea plants are easy to grow, and many varieties were commercially available. Another advantage of pea plants is that controlled pollinations are relatively easy to conduct. Pea flowers have both male and female parts and naturally self-pollinate (Fig. 10-1). However, the anthers (the male parts of the flower that produce pollen) can be removed to prevent self-fertilization. Pollen from a different source can then be applied to the stigma (the receptive surface of the female part). Pea flowers are easily protected from other sources of pollen because the petals completely enclose the reproductive structures.

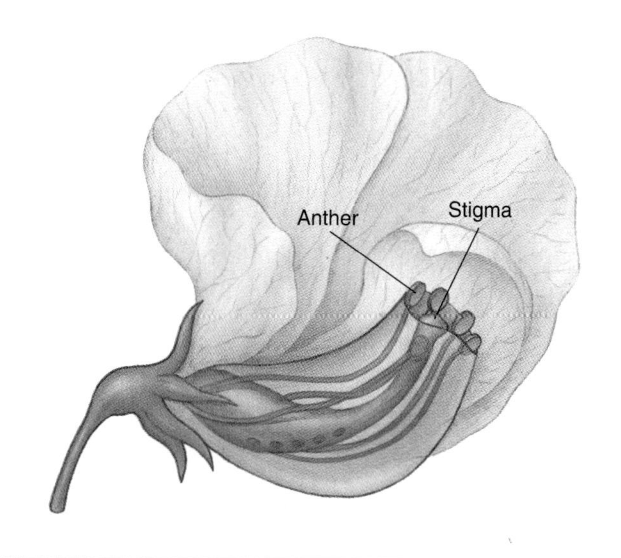

FIGURE **10-1** **Reproductive structures of a pea flower.**
This cutaway view shows the pollen-producing anthers and the stigma, the portion of the female part of the flower that receives the pollen.

Mendel obtained his original pea seeds from commercial sources and did some important preliminary work before starting his actual experiments. For two years he verified that the varieties were true-breeding lines for various inherited features. Today scientists use the term **phenotype** to refer to the physical appearance of an organism. A true-breeding line produces only offspring expressing the same phenotype (for example, round seeds or tall plants), generation after generation. During this time Mendel apparently chose those traits of his pea strains that he could study most easily. He probably made the initial observations that later formed the basis of his theories.

Mendel eventually chose strains representing seven **characters,** the attributes (such as seed color) for which heritable differences, or **traits,** are known (such as yellow seeds and green seeds). The characters Mendel selected had clearly contrasting phenotypes (Fig. 10-2). Mendel's results were easy to analyze because he chose easily distinguishable phenotypes and limited the genetic variation studied in each experiment.

Mendel began his experiments by crossing plants from two different true-breeding lines with contrasting phenotypes; these genetically pure individuals constituted the **parental generation,** or **P generation.** In every case, the members of the first generation of offspring all looked alike and resembled one of the two parents. For example, when he crossed tall plants with short plants, all the offspring were tall (Fig. 10-3). These offspring were the first filial generation, or the **F_1 generation** (*filial* is from the Latin for "sons and daughters"). The second filial generation, or **F_2 generation,** resulted from a cross between F_1 individuals, or by self-pollination of F_1 individuals. Mendel's F_2 generation in this experiment included 787 tall plants and 277 short plants.

Most breeders in Mendel's time thought fluids "blended" together to control inheritance in hybrids. One implication of this idea is that a hybrid should be intermediate between the two parents, and in fact plant breeders had obtained such hybrids. Although Mendel observed some intermediate types of hybrids, he chose for further study those F_1 hybrids in which "hereditary factors" (as he called them) from one of the parents apparently masked the expression of those factors from the other parent. Other breeders had also observed these types of hybrids, but they had not explained them. Using modern terms, the factor expressed in the F_1 generation (tallness, in our example) is

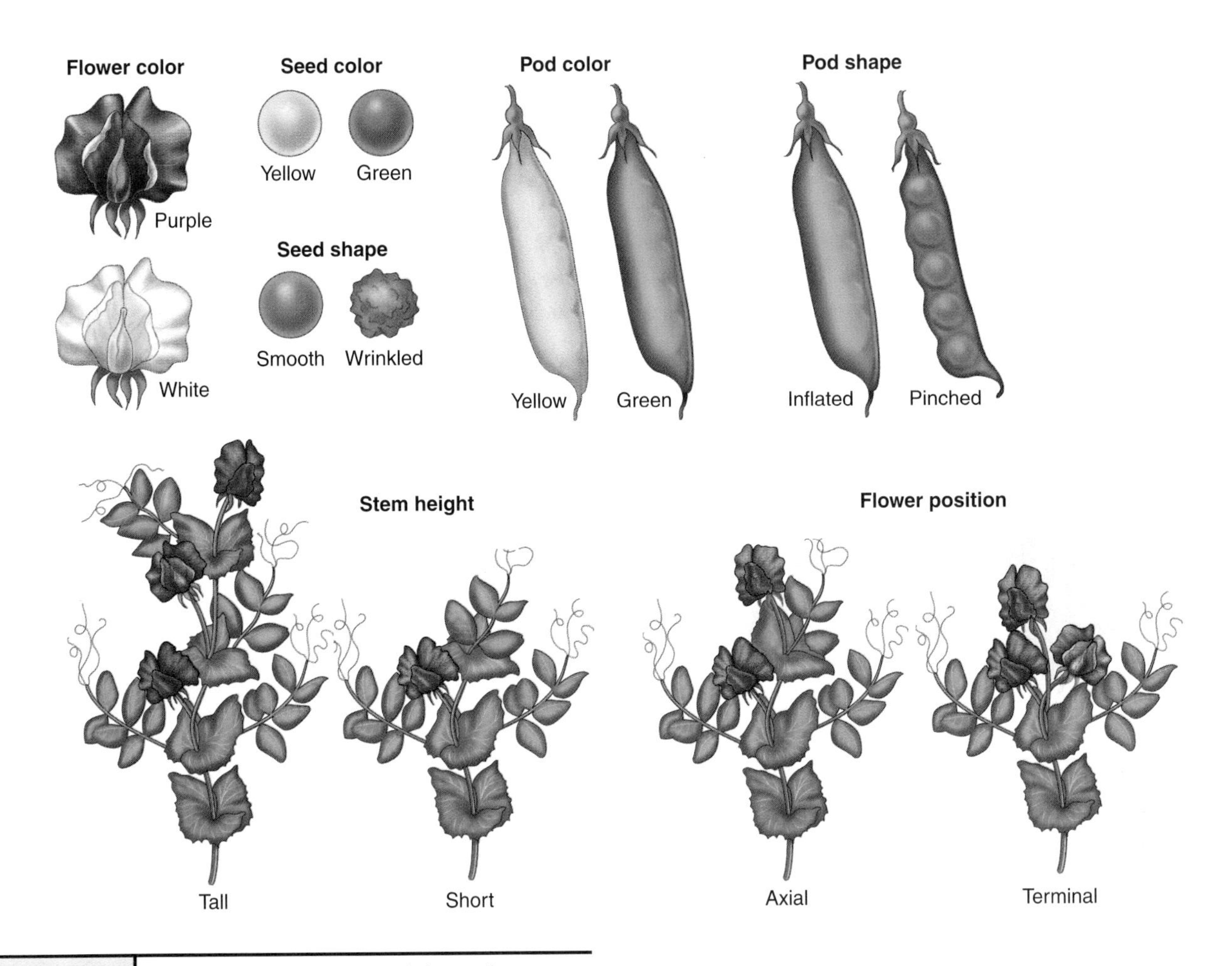

FIGURE **10-2** **Seven characters in Mendel's study of pea plants.**

Each character had two clearly distinguishable phenotypes.

said to be **dominant;** the one hidden in the F_1 (shortness) is **recessive.** Dominant traits mask recessive ones when both are present in the same individual. Although scientists know today that dominance is not always observed (we'll explore exceptions later in this chapter), the fact that dominance can occur was not entirely consistent with the notion of blending inheritance.

Mendel's results also argued against blending inheritance in a more compelling way. Once two fluids have blended, it is very difficult to imagine how they can separate. However, in the preceding example, in the F_1 generation the hereditary factor(s) that controlled shortness clearly were not lost or blended inseparably with the hereditary factor(s) that controlled tallness, because shortness reappeared in the F_2 generation. Mendel was very comfortable with the theoretical side of biology, because he was also a student of physics and mathematics. He therefore proposed that each kind of inherited feature of an organism is controlled by two factors that behave like discrete particles and are present in every individual. To Mendel these hereditary factors were abstractions—he knew nothing about chromosomes and DNA. These factors are essentially what scientists today call **genes.**

Mendel's experiments led to his discovery and explanation of the major principles of heredity, which we now know as the principles of segregation and independent assortment. We discuss the first principle next and the second later in the chapter.

Alleles separate before gametes are formed

The term **alleles** refers to the alternative forms of a gene. In the example in Figure 10-3, each F_1 generation tall plant had two different alleles that control plant height: a **dominant allele** for tallness (which we designate *T*) and a **recessive allele** for shortness (designated *t*), but because the tall allele was dominant these plants were tall. To explain his experimental results, Mendel proposed an idea now known as the principle of segregation. Using modern terminology, the **principle of segregation** states that before sexual reproduction occurs, the two alleles carried by an individual parent must become separated (segregated). As a result, each sex cell (egg or sperm) formed contains only one allele of each pair. An essential feature of the process is that the alleles remain intact (one does not mix with or eliminate the other); thus recessive alleles are not lost and can reappear in the F_2 generation.

In our example, before the F_1 plants formed gametes, the allele for tallness separated (segregated) from the allele for shortness, so that half the gametes contained a *T* allele and the other half a *t* allele. The random process of fertilization led to three possible combinations of alleles in the F_2 offspring: one fourth with two tallness alleles *(TT)*, one fourth with two shortness alleles *(tt)*, and one half with one allele for tallness and one for shortness (Tt). Because both TT and Tt plants are tall, on average Mendel expected approximately three fourths (787 of the 1064 plants he obtained) to express the phenotype of the dominant allele (tall) and about one fourth (277/1064) the phenotype of the recessive allele (short). (We will explain the mathematical reasoning behind these predictions shortly.)

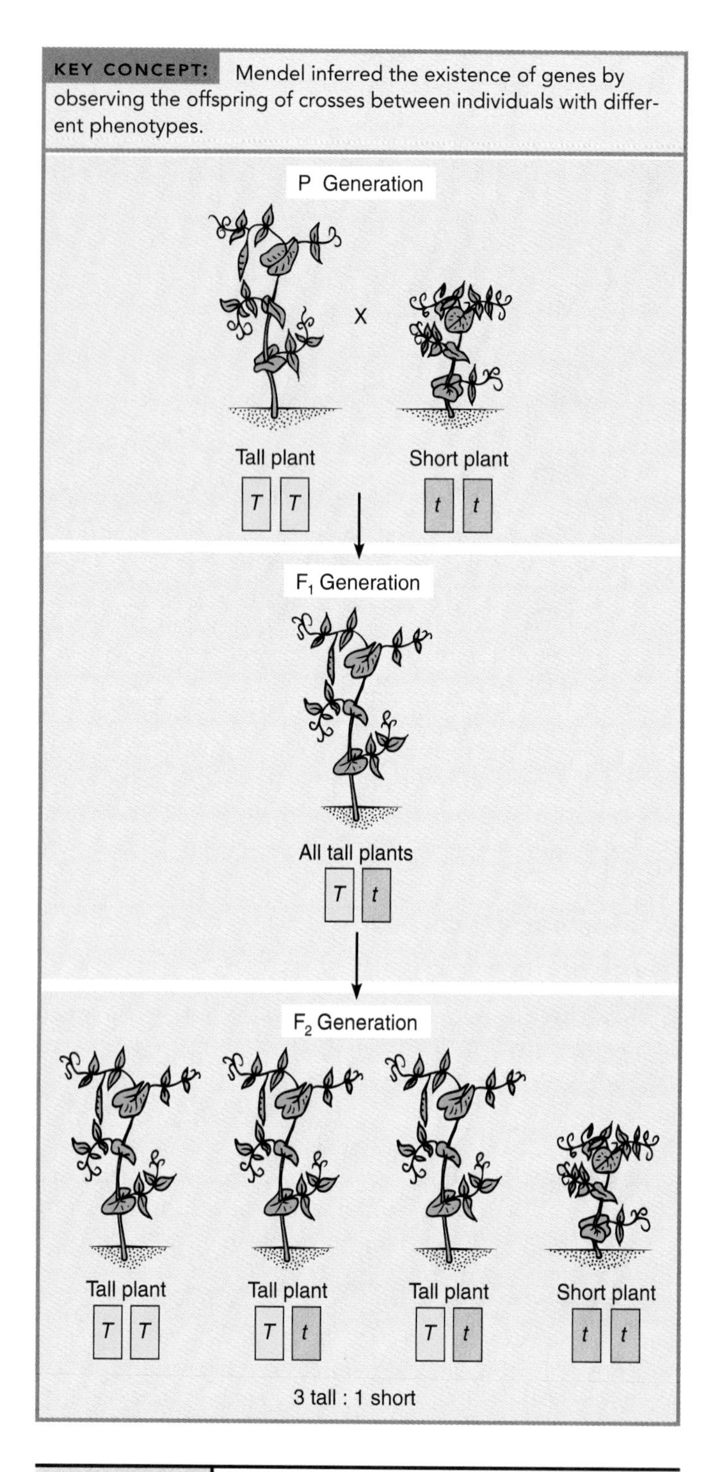

FIGURE **10-3** **One of Mendel's pea crosses.**

Crossing a true-breeding tall pea plant with a true-breeding short pea plant yielded only tall offspring in the F_1 generation. But when these F_1 individuals self-pollinated, or when two F_1 individuals were crossed, the resulting F_2 generation included tall and short plants in a ratio of about 3:1.

Mendel reported these and other findings at a meeting of the Brünn Society for the Study of Natural Science; he published his results in the society's report in 1866. At that time biology was largely a descriptive science, and biologists had little interest in applying quantitative and experimental methods such as Mendel had used. Other biologists of the time did not appreciate the importance of his results and his interpretations of those results. For 34 years his findings were largely neglected.

In 1900, Hugo DeVries in Holland, Carl Correns in Germany, and Erich von Tschermak in Austria recognized Mendel's principles in their own experiments; they later discovered Mendel's paper and found it explained their own research observations. Correns gave credit to Mendel by naming the basic laws of inheritance after him. By this time biologists had a much greater appreciation of the value of quantitative experimental methods. The details of mitosis, meiosis, and fertilization had been described, and in 1902, German biologist Theodor Boveri and American biologist Walter Sutton independently pointed out the connection between Mendel's segregation of alleles and the separation of homologous chromosomes during meiosis. The time was right for wider acceptance and extension of these ideas and their implications.

Alleles occupy corresponding loci on homologous chromosomes

Today scientists know that each unduplicated chromosome consists of one long linear DNA molecule and that each gene is actually a segment of that DNA molecule. We also know homologous chromosomes are similar not only in size and shape, but they also usually have the same genes (often with different alleles) located in corresponding positions. The term **locus** (pl., *loci*) originally designated the location of a particular gene on the chromosome (Fig. 10-4). We are actually referring to a segment of the DNA that has the information for controlling some aspect of the structure or function of the organism. One locus may govern seed color, another seed shape, still another the shape of the pods, and so on. Traditional genetic methods can infer the existence of a particular locus only if at least two allelic variants of that locus, producing contrasting phenotypes (for example, yellow peas versus green peas), are available for study. In the simplest cases an individual can express one (yellow) or the other (green) but not both.

Alleles are, therefore, genes that govern variations of the same character (yellow versus green seed color) and occupy corresponding loci on homologous chromosomes. Geneticists assign each allele of a locus a single letter or group of letters as its symbol. Although they often use more complicated forms of notation, it is customary when working simple genetics problems to indicate a dominant allele with a capital letter and a recessive allele with the same letter in lowercase.

Remember that the term *locus* designates not only a position on a chromosome but also a type of gene controlling a particular character; thus, Y (yellow) and y (green) represent a specific pair of alleles of a locus involved in determining seed color in peas. Although you may initially be uncomfortable with the fact that geneticists sometimes use the term *gene* to specify a locus and at other times to specify one of the alleles of that locus, the meaning is usually clear from the context.

A monohybrid cross involves individuals with different alleles of a given locus

The basic principles of genetics and the use of genetics terms are best illustrated by examples. In the simplest case, a **monohybrid cross,** the inheritance of two different alleles of a single locus is studied. Figure 10-5 illustrates a monohybrid cross featuring a locus that governs coat color in guinea pigs. The female comes from a true-breeding line of black guinea pigs. We say she is **homozygous** for black because the two alleles she carries for this locus are identical. The brown male is also from a true-breeding line and is homozygous for brown. What color would you expect the F_1 offspring to be? Dark brown? Spotted? It is impossible to make such a prediction without more information.

In this particular case, the F_1 offspring are black, but they are **heterozygous,** meaning they carry two different alleles for this locus. The brown allele influences coat color only in a homozygous brown individual; it is a recessive allele. The black allele influences coat color in both homozygous black and heterozygous individuals; it is a dominant allele. On the basis of this information, we can use standard notation to designate the dominant black allele *B* and the recessive brown allele *b.*

During meiosis in the female parent *(BB),* the two *B* alleles separate, according to Mendel's principle of segregation, so each egg has only one *B* allele. In the male *(bb)* the two *b* alleles separate, so each sperm has only one *b* allele. The fertilization of each *B* egg by a *b* sperm results in heterozygous F_1 offspring, each with the alleles *Bb;* that is, each individual has one allele for brown coat and one for black coat. Because this is the only possible combination of alleles present in the eggs and sperm, all the F_1 offspring are *Bb.*

A Punnett square predicts the ratios of the various offspring of a cross

During meiosis in heterozygous black guinea pigs *(Bb),* the chromosome containing the *B* allele becomes separated from its homologue (the chromosome containing the *b* allele), so each normal sperm or egg contains *B* or *b* but never both. Heterozygous *Bb* individuals form gametes containing *B* alleles and gametes containing *b* alleles in equal numbers. Because no special attraction or repulsion occurs between an egg and a sperm containing the same allele, fertilization is a random process.

As you can see in Figure 10-5, the possible combinations of eggs and sperm at fertilization can be represented in the form of a grid known as a **Punnett square,** devised by the early English geneticist Sir Reginald Punnett. The types of gametes (and their expected frequencies) from one parent are listed across the top, and those from the other parent are listed along the left side. The squares are then filled in with the resulting F_2 zygote

A gamete has one set of chromosomes, the *n* number. It carries *one* chromosome of *each* homologous pair. A given gamete can only have *one* gene of any particular pair of alleles.

When the gametes fuse, the resulting zygote is diploid (2*n*) and has homologous pairs of chromosomes. For purposes of illustration, these are shown physically paired.

(a)

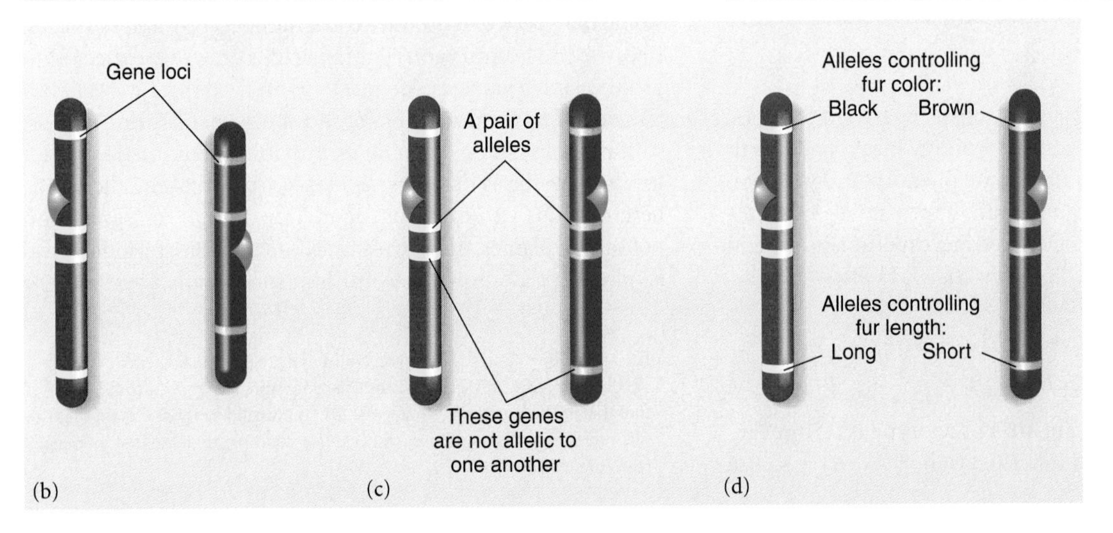

FIGURE 10-4 | Gene loci and their alleles.
(a) One member of each pair of homologous chromosomes is of maternal origin (red), and the other is paternal (blue). **(b)** These chromosomes are nonhomologous. Each chromosome is made up of thousands of genes. A locus is the specific place on a chromosome where a gene is located. **(c)** These chromosomes are homologous. Alleles are members of a gene pair that occupy corresponding loci on homologous chromosomes. **(d)** Alleles govern the same character but do not necessarily contain the same information.

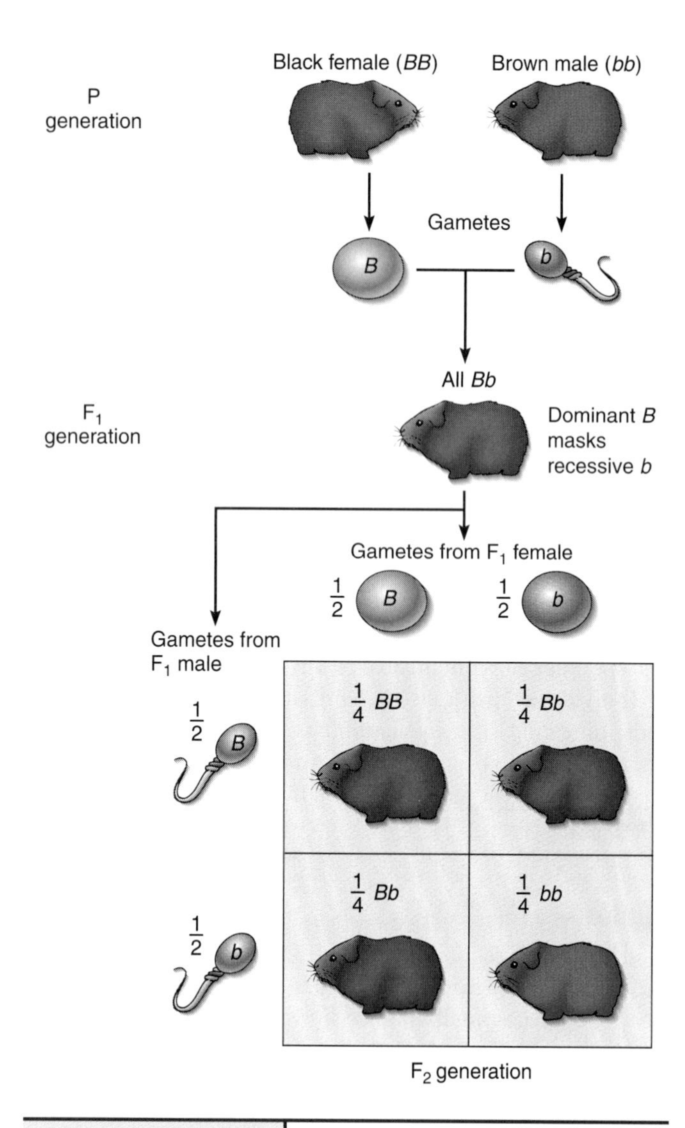

ACTIVE FIGURE **10-5** | **A monohybrid cross in guinea pigs.**

In this example, a homozygous black guinea pig is mated with a homozygous brown guinea pig. The F_1 generation includes only black individuals. However, the mating of two of these F_1 offspring yields F_2 generation offspring in the expected ratio of 3 black to 1 brown, indicating that the F_1 individuals are heterozygous.

Biology Now™ Learn more from an interactive tutorial on **monohybrid crosses** by clicking on this figure on your BiologyNow CD-ROM.

combinations. Three fourths of all F_2 offspring have the genetic constitution *BB* or *Bb* and are phenotypically black; one fourth have the genetic constitution *bb* and are phenotypically brown. The genetic mechanism that governs the approximate 3:1 F_2 ratios obtained by Mendel in his pea-breeding experiments is again evident. These ratios are called *monohybrid F_2 phenotypic ratios.*

The phenotype of an individual does not always reveal its genotype

As mentioned earlier, an organism's phenotype is its appearance with respect to a certain inherited trait. However, because some alleles may be dominant and others recessive, we can't always determine, simply by examining its phenotype, which alleles are carried by an organism. The *genetic constitution* of that organism, most often expressed in symbols, is its **genotype.** In the cross we have been considering, the genotype of the female parent is homozygous dominant, *BB*, and her phenotype is black. The genotype of the male parent is homozygous recessive, *bb*, and his phenotype is brown. The genotype of all the F_1 offspring is heterozygous, *Bb*, and their phenotype is black. To prevent confusion we always indicate the genotype of a heterozygous individual by writing the symbol for the dominant allele first and the recessive allele second (always *Bb*, never *bB*).

The phenomenon of dominance partly explains why an individual may resemble one parent more than the other, even if the two parents contribute equally to their offspring's genetic constitution. Dominance is not predictable and can be determined only by experiment. In one species of animal, black coat may be dominant to brown; in another species, brown may be dominant to black. In a population, the dominant phenotype is not necessarily more common than the recessive phenotype.

A test cross can detect heterozygosity

Guinea pigs with the genotypes *BB* and *Bb* are alike phenotypically; they both have black coats. How, then, can we know the genotype of a black guinea pig? Geneticists can accomplish this by performing a **test cross,** in which an individual of unknown genotype is crossed with a homozygous recessive individual (Fig. 10-6). In a test cross, the alleles carried by the gametes from the parent of unknown genotype are never "hidden" in the offspring by dominant alleles contributed by the other parent. Therefore, you can deduce the genotypes of all offspring directly from their phenotypes. If all the offspring were black, what inference would you make about the genotype of the black parent? If any of the offspring were brown, what conclusion would you draw regarding the genotype of the black parent? Would you be more certain about one of these inferences than about the other?[1]

PROCESS OF SCIENCE

Mendel conducted several test crosses; for example, he bred F_1 (tall) pea plants with homozygous recessive *(tt)* short ones. He reasoned that the F_1 individuals were heterozygous *(Tt)* and would be expected to produce equal numbers of *T* and *t* gametes. Because the homozygous short parents *(tt)* were expected to produce only *t* gametes, Mendel hypothesized that he would obtain equal numbers of tall *(Tt)* and short *(tt)* offspring. His results agreed with his hypothesis, providing additional evidence for the hypothesis that there is 1:1 segregation of the alleles of a heterozygous parent. Thus Mendel's principle of segregation not only explained the known facts, such as the 3:1 monohybrid F_2 phenotypic ratio, but also let him successfully anticipate the

[1] If all the offspring were black, you could infer that the black parent is probably homozygous, *BB*. If any of the offspring were brown, you could infer that the black parent is heterozygous, *Bb*. You would be more certain that the second inference (about the *Bb* individual) is correct than the first inference (the *BB* individual).

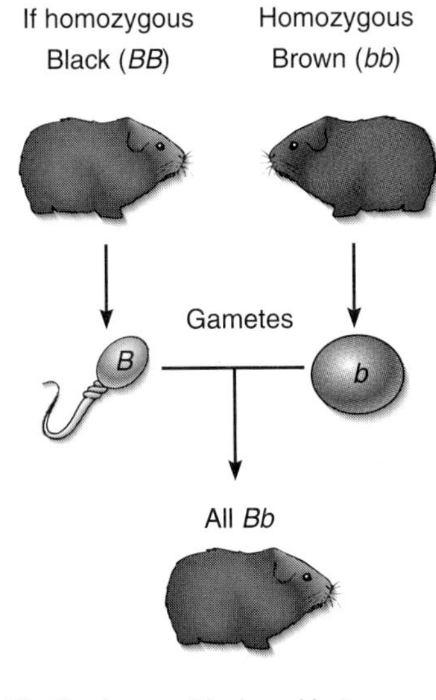

(a)

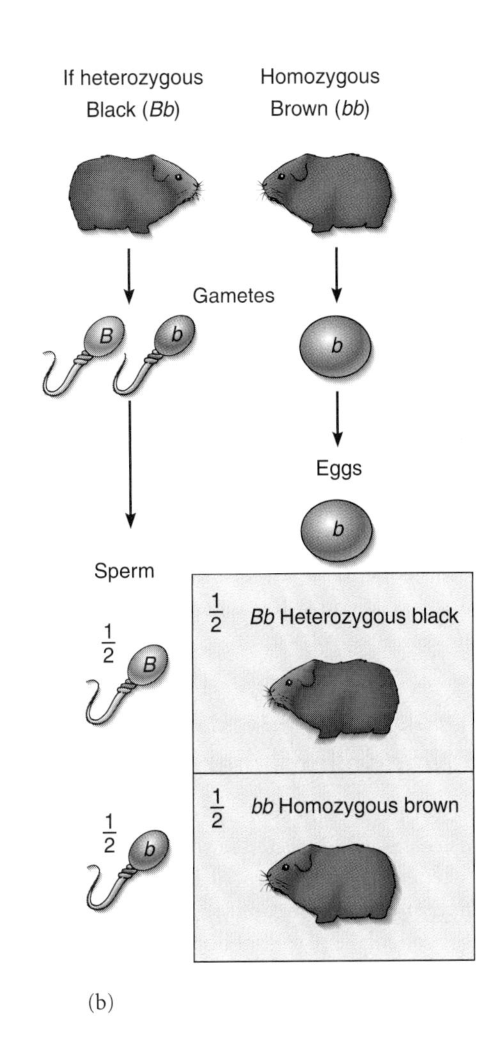

(b)

FIGURE **10-6** | **A test cross in guinea pigs.**

In this illustration, a test cross is used to determine the genotype of a black guinea pig. **(a)** If a black guinea pig is mated with a brown guinea pig and all the offspring are black, the black parent probably has a homozygous genotype. **(b)** If any of the offspring is brown, the black guinea pig must be heterozygous. The expected phenotypic ratio is 1 black to 1 brown.

results of other experiments—in this case, the 1:1 test cross phenotypic ratio.

A dihybrid cross involves individuals that have different alleles at two loci

Simple monohybrid crosses involve a pair of alleles of a single locus. Mendel also analyzed crosses involving alleles of two or more loci. A mating between individuals with different alleles at two loci is called a **dihybrid cross.** Consider the case when two pairs of alleles lie in nonhomologous chromosomes (that is, one pair of alleles is in one pair of homologous chromosomes, and the other pair of alleles is in a different pair of homologous chromosomes). Each pair of alleles is inherited independently; that is, each pair segregates during meiosis independently of the other.

An example of a dihybrid cross carried through the F_2 generation is shown in Figure 10-7. In this example, black is dominant to brown, and short hair is dominant to long hair. When a homozygous, black, short-haired guinea pig *(BBSS)* and a homozygous, brown, long-haired guinea pig *(bbss)* are mated, the *BBSS* animal produces gametes that are all *BS,* and the *bbss* individual produces gametes that are all *bs.* Each gamete contains one allele for each of the two loci. The union of the *BS* and *bs* gametes yields only individuals with the genotype *BbSs.* All these F_1 offspring are heterozygous for hair color and for hair length, and all are phenotypically black and short-haired.

Alleles on nonhomologous chromosomes are randomly distributed into gametes

Each F_1 guinea pig produces four kinds of gametes with equal probability: *BS, Bs, bS,* and *bs.* Hence, the Punnett square has 16 (that is, 4^2) squares representing the zygotes, some of which are genotypically or phenotypically alike. There are 9 chances in 16 of obtaining a black, short-haired individual; 3 chances in 16 of obtaining a black, long-haired individual; 3 chances in 16 of obtaining a brown, short-haired individual; and 1 chance in 16 of obtaining a brown, long-haired individual. This 9:3:3:1 phenotypic ratio is expected in a dihybrid F_2 if the hair color and hair length loci are on nonhomologous chromosomes.

On the basis of similar results, Mendel formulated the principle of inheritance, now known as Mendel's **principle of independent assortment,** which states that members of any gene pair segregate from one another independently of the members of the other gene pairs. This mechanism occurs in a regular way to ensure that each gamete contains one allele for each locus, but the alleles of different loci are assorted at random with respect to each other in the gametes.

FIGURE 10-7 | **A dihybrid cross in guinea pigs.**

When a black, short-haired guinea pig is crossed with a brown, long-haired one, all the offspring are black and have short hair. However, when two members of the F_1 generation are crossed, the ratio of phenotypes is 9:3:3:1.

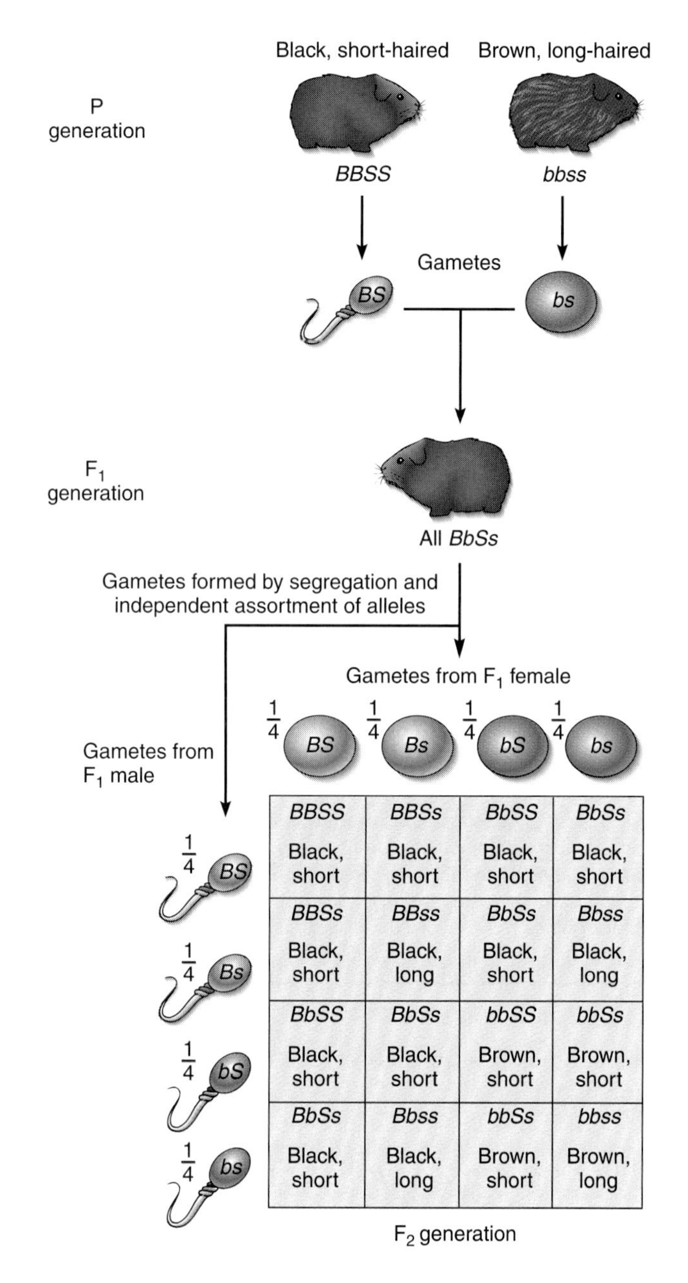

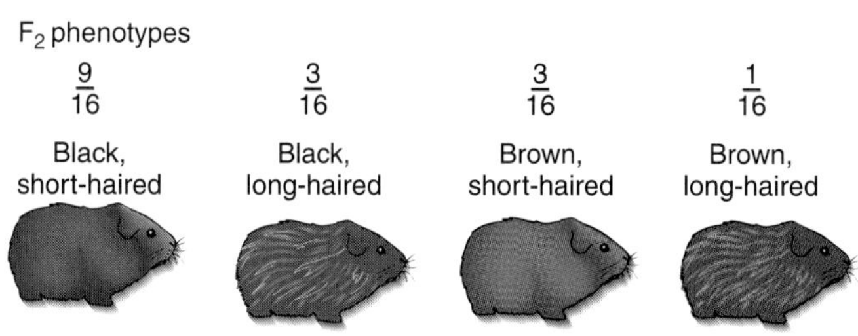

Today we recognize independent assortment is related to the events of meiosis. It occurs because two pairs of homologous chromosomes can be arranged in two different ways at metaphase I of meiosis. These arrangements occur randomly, with approximately half the meiotic cells oriented one way and the other half oriented the opposite way. The orientation of the homologous chromosomes on the metaphase plate then determines the way they subsequently separate and disperse into the haploid cells (Fig. 10-8). (As you will soon see, however, independent assortment does not always occur.)

The rules of probability are useful in predicting Mendelian inheritance

All genetic ratios are properly expressed in terms of probabilities. In monohybrid crosses, the expected ratio of the dominant and recessive phenotypes is 3:1. The probability of an event is its expected frequency. Therefore, we can say there are 3 chances in 4 (or $\frac{3}{4}$) that any particular individual offspring of two heterozygous individuals will express the dominant phenotype and 1 chance in 4 (or $\frac{1}{4}$) that it will express the recessive phenotype. Although we sometimes speak in terms of percentages, probabilities are calculated as fractions (such as $\frac{3}{4}$) or decimal fractions (such as 0.75). If an event is certain to occur, its probability is 1; if it is certain *not* to occur, its probability is 0. A probability can be 0, 1, or some number between 0 and 1.

The Punnett square lets you combine two or more probabilities. When you use a Punnett square, you are following two important statistical principles known as the product rule and the sum rule. The **product rule** predicts the combined probabilities of independent events. Events are *independent* if the occurrence of one does not affect the probability that the other will occur. For example, the probability of obtaining heads on the first toss of a coin is $\frac{1}{2}$; the probability of obtaining heads on the second toss (an independent event) is also $\frac{1}{2}$. If two or more events are independent of each other, the probability of both occurring is the product of their individual probabilities. If this seems strange, keep in mind that when we multiply two numbers that are less than 1, the product is a smaller number. Therefore, the probability of obtaining heads two times in a row is $\frac{1}{2} \times \frac{1}{2} = \frac{1}{4}$, or 1 chance in 4 (Fig. 10-9).

Similarly, we can apply the product rule to genetic events. If both parents are *Bb*, what is the probability they will produce a child who is *bb*? For the child to be *bb*, he or she must receive a *b* gamete from each parent. The probability of a *b* egg is $\frac{1}{2}$, and the probability of a *b* sperm is also $\frac{1}{2}$. Like the outcomes of the coin tosses, these probabilities are independent, so we combine them by the product rule ($\frac{1}{2} \times \frac{1}{2} = \frac{1}{4}$). You might like to check this result using a Punnett square.

The **sum rule** predicts the combined probabilities of *mutually exclusive* events. In some cases, there is more than one way to obtain a specific outcome. These different ways are mutually exclusive; if one occurs, the other(s) cannot. For example, if both parents are *Bb*, what is the probability that their first child will also have the *Bb* genotype? There are two different ways these parents can have a *Bb* child: Either a *B* egg combines with a *b* sperm (probability $\frac{1}{4}$), or a *b* egg combines with a *B* sperm (probability $\frac{1}{4}$).

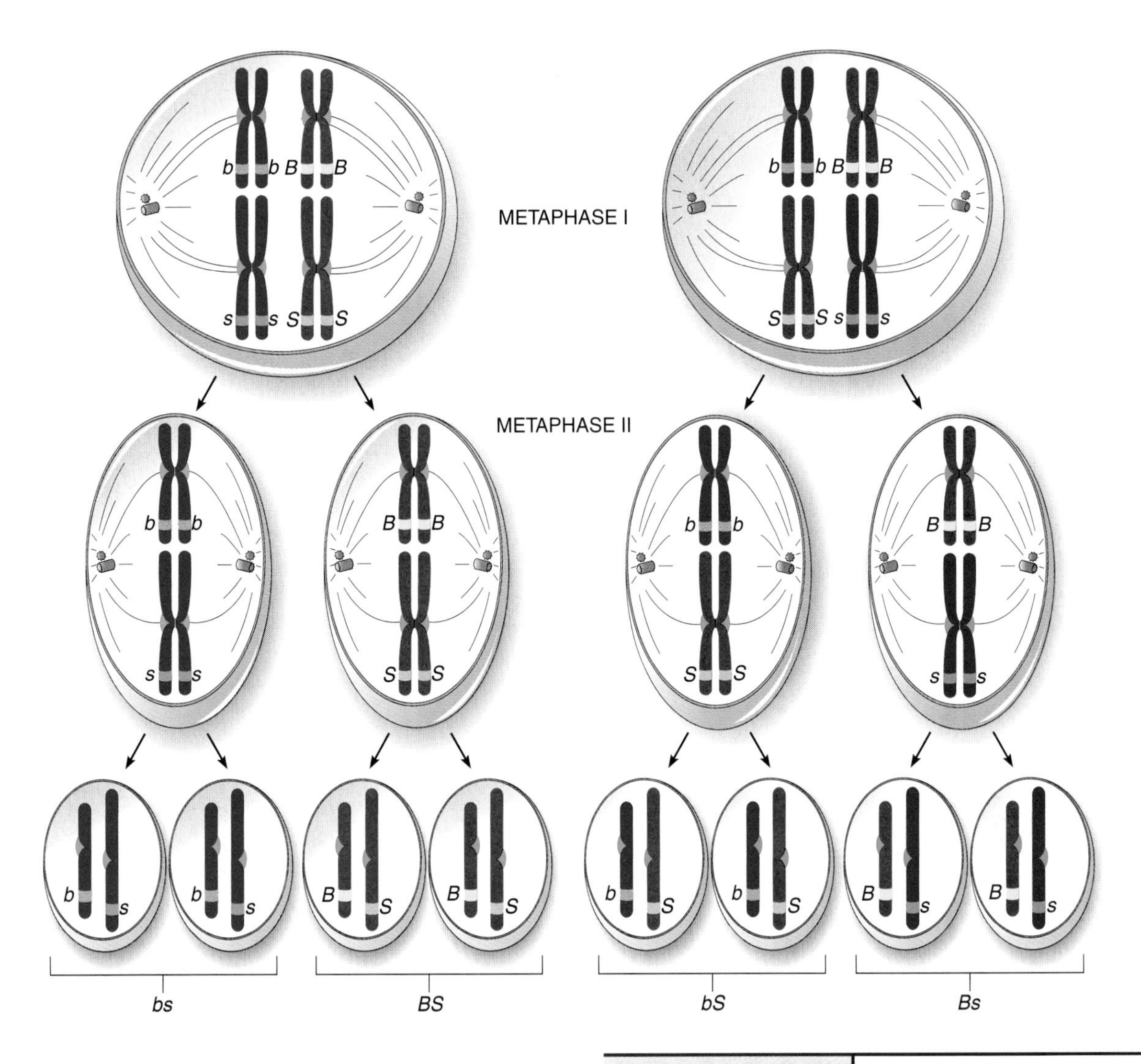

ACTIVE FIGURE 10-8 **Meiosis and independent assortment.**

Two different pairs of homologous chromosomes can line up two different ways at metaphase I and be subsequently distributed. A cell with the orientation shown at the left produces half *BS* and half *bs* gametes. Conversely, the cell at the right produces half *Bs* and half *bS* gametes. Because approximately half of the meiotic cells at metaphase I are of each type, the ratio of the four possible types of gametes is 1:1:1:1.

BiologyNow™ Learn more about **independent assortment** by clicking on this figure on your BiologyNow CD-ROM.

Naturally, if there is more than one way to get a result, the chances of its being obtained improve; we therefore combine the probabilities of mutually exclusive events by summing (adding) their individual probabilities. The probability of obtaining a *Bb* child in our example is therefore $1/4 + 1/4 = 1/2$. (Because there is only one way these heterozygous parents can produce a homozygous recessive child, *bb*, that probability is only $1/4$. The probability of a homozygous dominant child, *BB*, is likewise $1/4$.)

The rules of probability can be applied to a variety of calculations

The rules of probability have wide applications. For example, what are the probabilities that a family with two (and only two) children will have two girls, two boys, or one girl and one boy? For purposes of discussion, let's assume male and female births are equally probable. The probability of having a girl first is $1/2$, and the probability of having a girl second is also $1/2$. These are independent events, so we combine their probabilities by multiplying: $1/2 \times 1/2 = 1/4$. Similarly, the probability of having two boys is $1/4$.

In families with both a girl and a boy, the girl can be born first or the boy can be born first. The probability that a girl will be born first is $1/2$, and the probability that a boy will be born second is also $1/2$. We use the product rule to combine the probabilities of these two independent events: $1/2 \times 1/2 = 1/4$. Similarly, the probability that a boy will be born first and a girl second is also $1/4$. These two kinds of families represent mutually exclusive outcomes, that is, two different ways of obtaining a family with one boy and one girl. Having two different ways of obtaining the desired result improves our chances, so we use the sum rule to combine the probabilities: $1/4 + 1/4 = 1/2$.

In working with probabilities, keep in mind a point that many gamblers forget: Chance has no memory. If events are truly independent, past events have no influence on the probability of the occurrence of future events. When working probability problems,

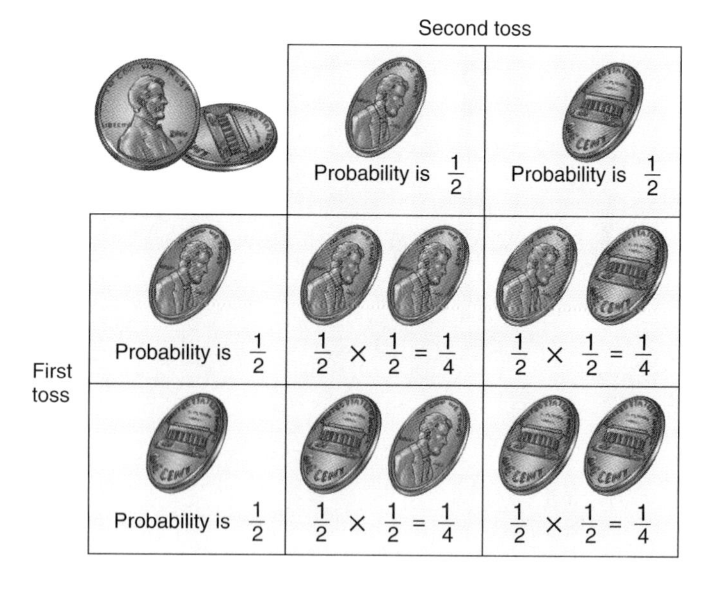

FIGURE **10-9** **The rules of probability.**

For each coin toss, the probability of heads is ½ and the probability of tails is also ½. Because the outcome of the first toss is independent of the outcome of the second, the combined probabilities of the outcomes of successive tosses are calculated by multiplying their individual probabilities (according to the product rule: ½ × ½ = ¼). These same rules of probability predict genetic events.

common sense is more important than blindly memorizing rules. Examine your results to see whether they appear reasonable; if they don't, re-evaluate your assumptions. (See *Focus On: Solving Genetics Problems* on page 206 for step-by-step procedures to solve genetics problems, including when to use the rules of probability.)

Review

- What are the relationships among loci, genes, and alleles?
- What is Mendel's principle of segregation?
- What is Mendel's principle of independent assortment?
- How is probability used to predict the outcome of genetic crosses?

BiologyNow™ Assess your understanding of **Mendel's principles of inheritance** by taking the pretest on your BiologyNow CD-ROM.

MENDELIAN INHERITANCE AND CHROMOSOMES

Learning Objectives

5. Explain Mendel's principles of segregation and independent assortment, given what scientists know about genes and chromosomes.
6. Define *linkage*, and relate it to specific events in meiosis.
7. Show how data from a test cross involving alleles of two loci can be used to distinguish between independent assortment and linkage.
8. Discuss the genetic determination of sex and the inheritance of X-linked genes in mammals.

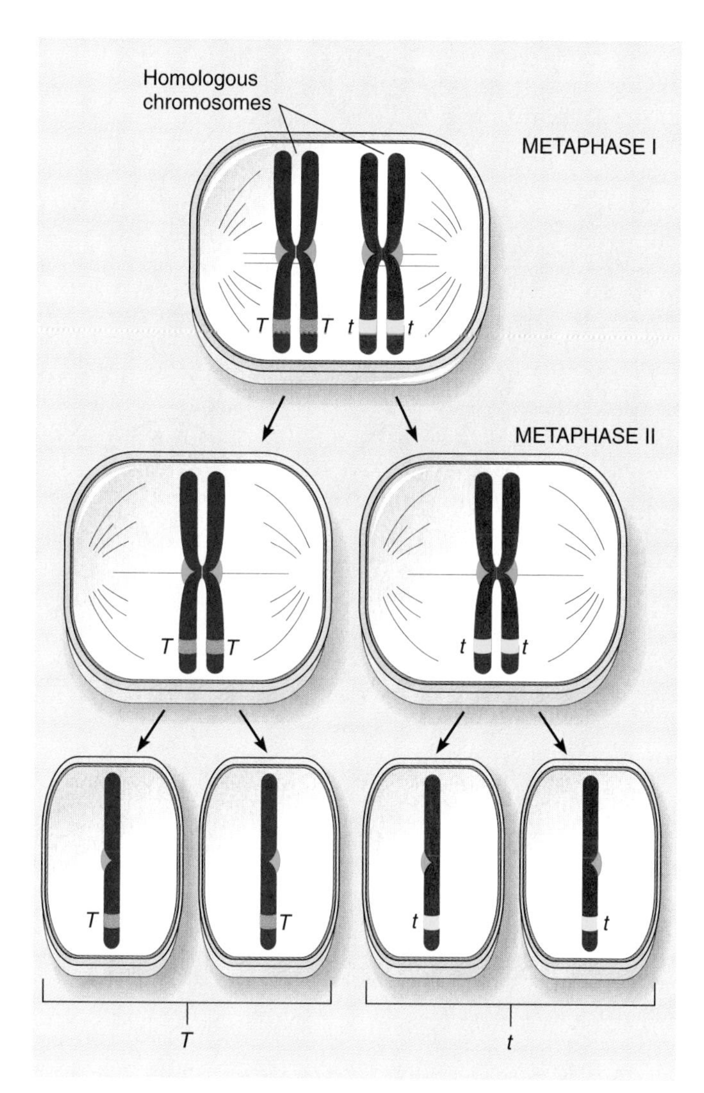

FIGURE **10-10** **The chromosomal basis for segregation.**

The separation of homologous chromosomes during meiosis results in the segregation of alleles in a heterozygote. Note that half of the gametes will carry *T* and half will carry *t*.

It is a measure of Mendel's genius that he worked out the principles of segregation and independent assortment without knowing anything about the chromosomal basis of inheritance. Today we know the segregation of alleles is a direct result of homologous chromosomes separating during meiosis (Fig. 10-10; also see Fig. 9-12). (Recall from Chapter 9 that in all sexual life cycles, meiosis must occur at some point before gamete formation.) Later, at the time of fertilization, each haploid gamete contributes one chromosome from each homologous pair and therefore one allele for each gene pair. Although gametes and fertilization were known at the time Mendel carried out his research, mitosis and meiosis had not yet been discovered. It is truly remarkable that Mendel formulated his ideas mainly on the basis of mathematical abstractions. Today his principles are much easier to understand, because we relate the transmission of genes to the behavior of chromosomes.

The chromosomal basis of inheritance also helps explain certain apparent exceptions to Mendelian inheritance. One of these so-called exceptions involves linked genes.

Linked genes do not assort independently

Beginning around 1910, the research of American geneticist Thomas Hunt Morgan and his graduate students extended the concept of the chromosomal basis of inheritance. Morgan's research organism was the fruit fly (*Drosophila melanogaster*). By carefully analyzing the results of crosses involving fruit flies, Morgan and his students demonstrated that genes are arranged in a linear order on each chromosome.

Morgan also showed that independent assortment does not apply if the two loci lie close together in the same pair of homologous chromosomes. In fruit flies there is a locus controlling wing shape (the dominant allele *V* for normal wings and the recessive allele *v* for abnormally short, or vestigial, wings) and another locus controlling body color (the dominant allele *B* for gray body and the recessive allele *b* for black body). If a homozygous *BBVV* fly is crossed with a homozygous *bbvv* fly, the F_1 flies all have gray bodies and normal wings, and their genotype is *BbVv*.

Because these loci happen to lie close to one another in the *same pair* of homologous chromosomes, their alleles do not assort independently; instead, they are **linked genes** that tend to be inherited together. **Linkage** is the tendency for a group of genes on the same chromosome to be inherited together in successive generations. You can readily observe linkage in the results of a test cross in which heterozygous F_1 flies *(BbVv)* are mated with homozygous recessive *(bbvv)* flies (Fig. 10-11). Because heterozygous individuals are mated to homozygous recessive individuals, this test cross is similar to the test cross described earlier. However, it is called a **two-point test cross** because alleles of two loci are involved.

If the loci governing these traits were unlinked—that is, on different chromosomes—the heterozygous parent in a test cross would produce four kinds of gametes (*BV, Bv, bV,* and *bv*) in equal numbers. This independent assortment would produce offspring with new gene combinations not present in the parental generation. Any process that leads to new gene combinations is called **recombination.** In our example, gametes *Bv* and *bV* are **recombinant types.** The other two kinds of gametes, *BV* and *bv,* are **parental types** because they are identical to the gametes produced by the P generation. Of course, the homozygous recessive parent produces only one kind of gamete, *bv.* Thus if independent assortment were to occur in the F_1 flies, approximately 25% of the test-cross offspring would be gray-bodied and normal-winged *(BbVv),* 25% black-bodied and normal-winged *(bbVv),* 25% gray-bodied and vestigial-winged *(Bbvv),* and 25% black-bodied and vestigial-winged *(bbvv).* Notice that the two-point test cross

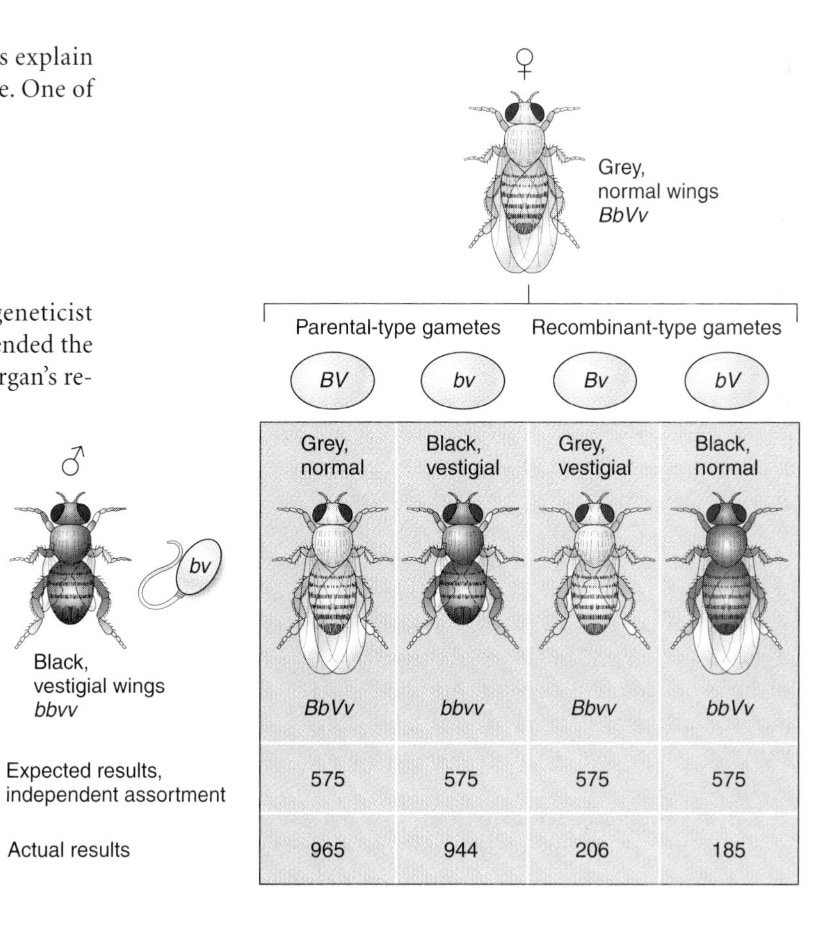

FIGURE **10-11** **A two-point test cross to detect linkage in fruit flies.**

Linkage can be recognized when an excess of parental-type offspring and a deficiency of recombinant-type offspring are produced in a two-point test cross. In this example using data from an actual cross, loci for wing length and body color are linked; they are located on a homologous chromosome pair. This is evident in the 2300 offspring (bottom row). About 920 of the offspring (or 40%) belong to each of the two parental classes (80% total), and 230 offspring (or 10%) belong to each of the two recombinant classes (20% total). The row above the actual results lets you contrast the data with the expected numbers for independent assortment.

lets us determine the genotypes of the offspring directly from their phenotypes.

By contrast, the alleles of the loci in our example do not undergo independent assortment, because they are linked. Alleles at different loci but close to one another on a given chromosome tend to be inherited together; because chromosomes pair and separate during meiosis as units, they therefore tend to be inherited as units. If linkage were complete, only parental-type flies would be produced, with approximately 50% having gray bodies and normal wings *(BbVv),* and 50% having black bodies and vestigial wings *(bbvv).* However, in our example, the offspring also include some gray-bodied, vestigial-winged flies and some black-bodied, normal-winged flies. These are recombinant flies, having received a recombinant-type gamete from the heterozygous F_1 parent. Each recombinant-type gamete arose by **crossing-over** between these loci in a meiotic cell of a heterozygous female fly. (Fruit flies are unusual in that crossing-over occurs only in females and not in males; it is far more common for crossing-over to occur in both sexes of a species.) Recall from Chapter 9 that when chromosomes pair and undergo synapsis, crossing-over occurs as homologous (nonsister) chromatids exchange segments of chromosomal material by a process of breakage and rejoining (Fig. 10-12; also see Fig. 9-14).

Calculating the frequency of crossing-over reveals the linear order of linked genes on a chromosome

In our example (see Fig. 10-11), 391 of the offspring are recombinant types: gray flies with vestigial wings, *Bbvv* (206 of the total); and black flies with normal wings, *bbVv* (185 of the total). The remaining 1909 offspring are parental types. These data can be used to calculate the percentage of crossing-over between the loci. You can do this by adding the number of individuals in the two recombinant classes of offspring (206 +185), dividing by the *total number of offspring* (965 + 944 + 206 + 185), and multiplying by 100: $391 \div 2300 = 0.17; 0.17 \times 100 = 17\%$. Thus the *V* locus and the *B* locus have 17% recombination between them.

During a single meiotic division, crossing-over may occur at several different points along the length of each homologous chromosome pair. In general, crossing-over is more likely to occur between two loci if they lie far apart on the chromosome and less likely to occur if they lie close together. Because of this rough correlation between the frequency of recombination of two loci and the linear distance between them, a genetic map of the chromosome can be generated by converting the percentage of recombination to **map units.** By convention, 1% recombination between two loci equals a distance of 1 map unit, so the loci in our example are 17 map units apart.

Scientists have determined the frequencies of recombination between specific linked loci in many species. All the experimental results are consistent with the hypothesis that genes are present in a linear order in the chromosomes. Figure 10-13 illustrates the traditional method for determining the linear order of genes in a chromosome.

More than one crossover between two loci in a single tetrad can occur in a given cell undergoing meiosis. (Recall from Chapter 9 that a *tetrad* is a group of four chromatids that make up a pair of synapsed homologous chromosomes.) We can observe only the frequency of offspring receiving recombinant-type gametes from the heterozygous parent, not the actual number of crossovers. In fact, the actual frequency of crossing-over is slightly more than the observed frequency of recombinant-type gametes. This is because the simultaneous occurrence of two crossovers involving the same two homologous chromatids reconstitutes the original combination of genes (Fig. 10-14). When two loci are relatively close together, the effect of double crossing-over is minimized.

By putting together the results of many crosses, scientists developed detailed linkage maps for many eukaryotes, including the fruit fly, the mouse, yeast, *Neurospora* (a fungus), and many plants, especially those that are important crops. In addition, researchers have used genetic methods to develop a detailed map for *Escherichia coli*, a bacterium with a single, circular DNA molecule, and many other prokaryotes and viruses. They have made much more sophisticated maps of chromosomes by means of recombinant DNA technology (see Chapter 14). Using these techniques, the *Human Genome Project* has produced maps of human chromosomes (see Chapter 15).

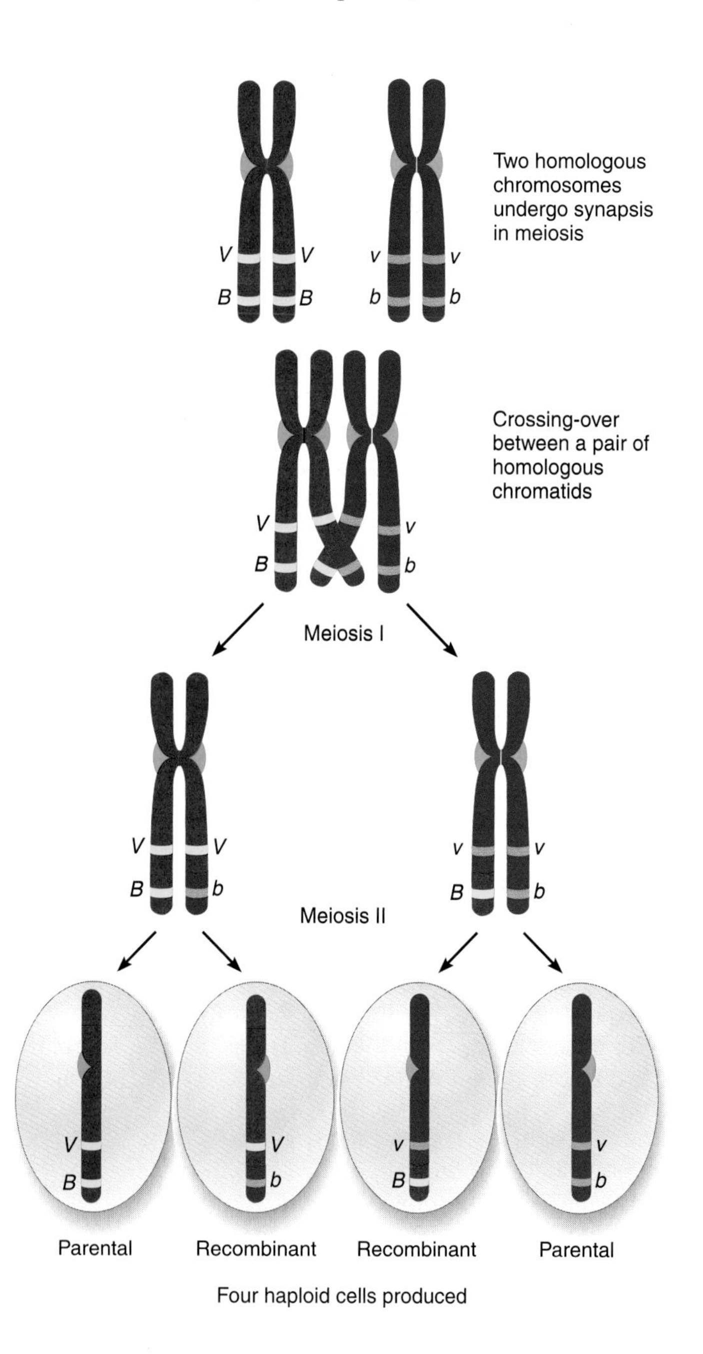

FIGURE **10-12** | **Crossing-over.**

The exchange of segments between chromatids of homologous chromosomes facilitates the recombination of linked genes. Genes located far apart on a chromosome have a greater probability of being separated by crossing-over than do genes that are closer together.

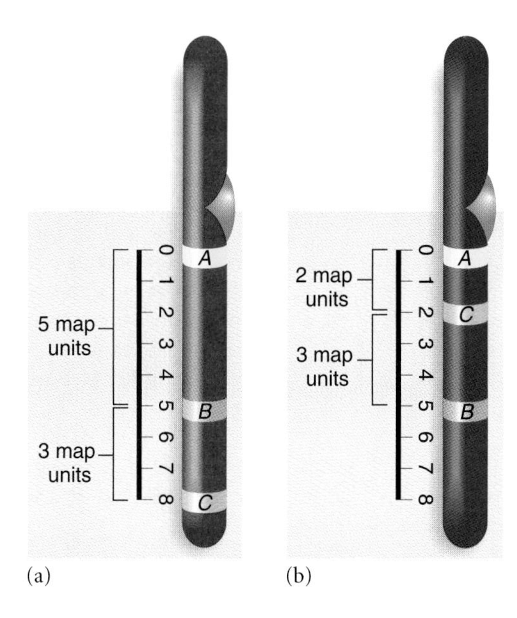

FIGURE **10-13** | **Gene mapping.**

Gene order (that is, which locus lies between the other two) is determined by the percentage of recombination between each of the possible pairs. In this hypothetical example, the percentage of recombination between locus *A* and locus *B* is 5% (corresponding to five map units) and that between *B* and *C* is 3% (three map units). There are two alternatives for the linear order of these alleles. **(a)** If the recombination between *A* and *C* is 8% (8 map units), *B* must be in the middle. **(b)** If the recombination between *A* and *C* is 2%, then *C* must be in the middle.

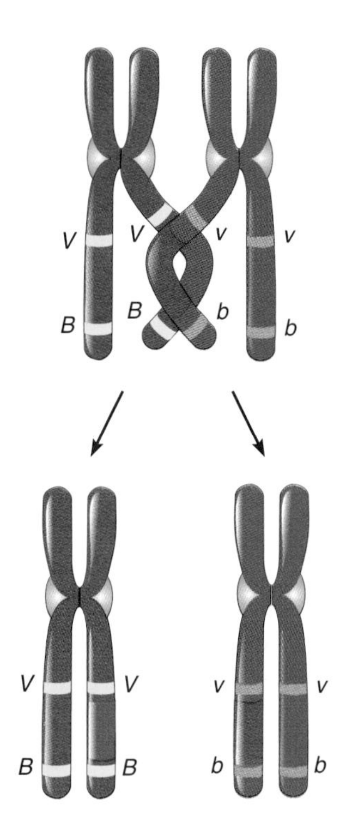

FIGURE **10-14** | **Double crossing-over.**

If the same homologous chromatids undergo double crossing-over between the genes of interest, the gametes formed are not recombinant for these genes.

Sex is generally determined by sex chromosomes

In some species, environmental factors exert a large control over an individual's sex. However, genes are the most important sex determinants in most organisms. The major sex-determining genes of mammals, birds, and many insects are carried by **sex chromosomes.** Typically, members of one sex have a pair of similar sex chromosomes and produce gametes that are all identical in sex chromosome constitution. The members of the other sex have two different sex chromosomes and produce two kinds of gametes, each bearing a single kind of sex chromosome.

The cells of females of many animal species, including humans, contain two **X chromosomes.** In contrast, the males have a single X chromosome and a smaller **Y chromosome.** For example, human males have 22 pairs of **autosomes,** which are chromosomes other than the sex chromosomes, plus one X chromosome and one Y chromosome; females have 22 pairs of autosomes plus a pair of X chromosomes. Domestic cats have 19 pairs of autosomes, to which are added a pair of X chromosomes in females, or an X plus a Y in males.

The Y chromosome determines male sex in most species of mammals

Do male humans have a male phenotype because they have only one X chromosome, or because they have a Y chromosome? Much of the traditional evidence bearing on this question comes from studies of people with abnormal sex chromosome constitutions (see Chapter 15). A person with an XXY constitution is a nearly normal male in external appearance, although his testes are underdeveloped (Klinefelter syndrome). A person with one X but no Y chromosome has the overall appearance of a female but has defects such as short stature and undeveloped ovaries (Turner syndrome). An embryo with a Y but no X does not survive. Thus all individuals require at least one X, and the Y is the male-determining chromosome.

Geneticists have identified several genes on the Y chromosomes that are involved in male determination. The *sex reversal on Y (SRY) gene,* the major male-determining gene on the Y chromosome, acts as a "genetic switch" that causes testes to develop in the fetus. The developing testes then secrete the hormone **testosterone,** which causes other male characteristics to develop. A few other genes on the Y chromosome also play a role in sex determination, as do many genes on the X chromosome, which explains why an XXY individual does not have a completely normal male phenotype. Some genes on the autosomes also affect sex development.

The X and Y chromosomes are thought to have originated as a homologous pair. However, they are not truly homologous in their present forms, because they are not similar in size, shape, or genetic constitution. Nevertheless, they have retained a short homologous "pairing region" that lets them synapse and separate from one another during meiosis. Half the sperm contain an X

FOCUS ON: SOLVING GENETICS PROBLEMS

Simple Mendelian genetics problems are like puzzles. They can be fun and easy to work if you follow certain conventions and are methodical in your approach.

1. Always use standard designations for the generations. The generation in which a particular genetic experiment is begun is called the P, or parental, generation. Offspring of this generation are called the F_1, or first filial, generation. The offspring resulting when two F_1 individuals are bred constitute the F_2, or second filial, generation.
2. Write down a key for the symbols you are using for the allelic variants of each locus. Use an uppercase letter to designate a dominant allele and a lowercase letter to designate a recessive allele. Use the same letter of the alphabet to designate both alleles of a particular locus. If you are not told which allele is dominant and which is recessive, the phenotype of the F_1 generation is a good clue.
3. Determine the genotypes of the parents of each cross by using the following types of evidence:
   - Are they from true-breeding lines? If so, they should be homozygous.
   - Can their genotypes be reliably deduced from their phenotypes? This is usually true if they express the recessive phenotype.
   - Do the phenotypes of their offspring provide any information? Exactly how this is done is discussed shortly.
4. Indicate the possible kinds of gametes formed by each of the parents. It is helpful to draw a circle around the symbols for each kind of gamete.
   - If it is a monohybrid cross, we apply the principle of segregation; that is, a heterozygote *Aa* forms two kinds of gametes, *A* and *a*. A homozygote, such as *aa*, forms only one kind of gamete, *a*.
   - If it is a dihybrid cross, we apply the principles of segregation and independent assortment. For example, an individual heterozygous for two loci would have the genotype *AaBb*. Allele *A* segregates from *a*, and *B* segregates from *b*. The assortment of *A* and *a* into gametes is independent of the assortment of *B* and *b*. *A* is equally likely to end up in a gamete with *B* or *b*. The same is true for *a*. Thus an individual with the genotype *AaBb* produces four kinds of gametes in equal amounts: *AB*, *Ab*, *aB*, and *ab*.
5. Set up a Punnett square, placing the possible types of gametes from one parent down the left side and the possible types from the other parent across the top.
6. Fill in the Punnett square. Avoid confusion by consistently placing the dominant allele first and the recessive allele second in heterozygotes (*Aa*, never *aA*). If it is a dihybrid cross, always write the two alleles of one

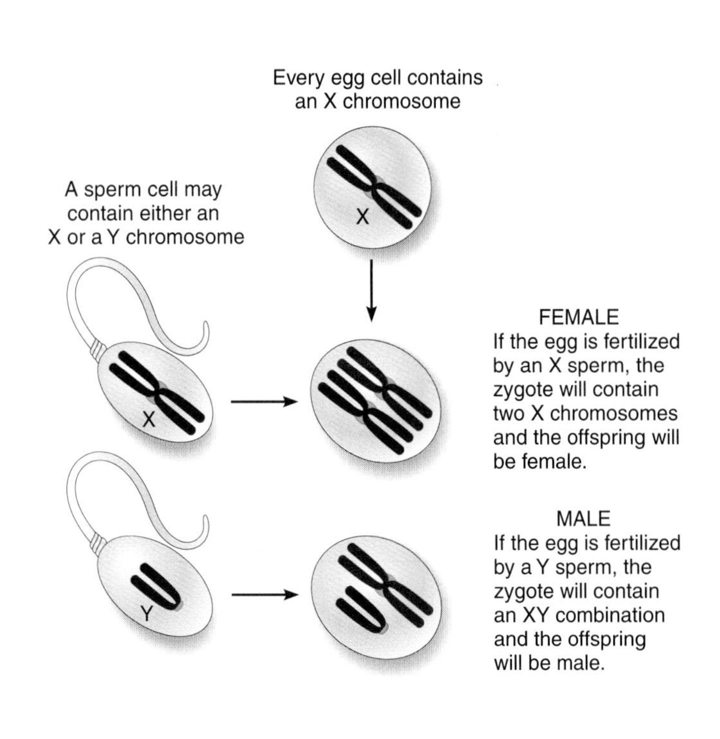
Every egg cell contains an X chromosome
A sperm cell may contain either an X or a Y chromosome
X
X
Y
FEMALE
If the egg is fertilized by an X sperm, the zygote will contain two X chromosomes and the offspring will be female.
MALE
If the egg is fertilized by a Y sperm, the zygote will contain an XY combination and the offspring will be male.

FIGURE 10-15 | Sex determination in mammals.

The sperm determines sex in mammals. An X-bearing sperm produces a female, and a Y-bearing sperm produces a male.

chromosome and half contain a Y chromosome (Fig. 10-15). All normal eggs bear a single X chromosome. Fertilization of an X-bearing egg by an X-bearing sperm results in an XX (female) zygote; fertilization by a Y-bearing sperm results in an XY (male) zygote.

You might expect to have equal numbers of X- and Y-bearing sperm and a 1:1 ratio of females to males. However, in humans more males are conceived than females, and more males die before birth. Even at birth the ratio is not 1:1; about 106 boys are born for every 100 girls. Although we do not know why this occurs, Y-bearing sperm seem to have some selective advantage.

An XX/XY sex chromosome mechanism, similar to that of humans, operates in many species of animals. However, it is not universal, and many of the details may vary. For example, the fruit fly, *Drosophila*, has XX females and XY males, but the Y does not determine maleness; a fruit fly with only one X chromosome and no Y chromosome has a male phenotype. In birds and butterflies the mechanism is reversed, with males having the equivalent of XX and females the equivalent of XY.

X-linked genes have unusual inheritance patterns

The human X chromosome contains many loci that are required in both sexes. Genes located in the X chromosome, such

locus first and the two alleles of the other locus second. It does not matter which locus you choose to write first, but once you have decided on the order, it is crucial to maintain it consistently. This means that if the individual is heterozygous for both loci, you will always use the form *AaBb*. Writing this particular genotype as *aBbA*, or even as *BbAa*, would cause confusion.

7. If you do not need to know the frequencies of all the expected genotypes and phenotypes, you can use the rules of probability as a shortcut. For example, if both parents are *AaBb*, what is the probability of an *AABB* offspring? To be *AA*, the offspring must receive an *A* gamete from each parent. The probability that a given gamete is *A* is $\frac{1}{2}$ and each gamete represents an independent event, so combine their probabilities by multiplying ($\frac{1}{2} \times \frac{1}{2} = \frac{1}{4}$). The probability of *BB* is calculated similarly and is also $\frac{1}{4}$. The probability of *AA* is independent of the probability of *BB*, so again, use the product rule to obtain their combined probabilities ($\frac{1}{4} \times \frac{1}{4} = \frac{1}{16}$).

Very often the genotypes of the parents can be deduced from the phenotypes of their offspring. In peas, for example, the allele for yellow seeds (*Y*) is dominant to the allele for green seeds (*y*). Suppose plant breeders cross two plants with yellow seeds, but you don't know whether the yellow-seeded plants are homozygous or heterozygous. You can designate the cross as: *Y_* × *Y_*. The offspring of the cross are allowed to germinate and grow, and their seeds are examined: Of the offspring, 74 produce yellow seeds, and 26 produce green seeds. Knowing that green-seeded plants are recessive, *yy*, the answer is obvious: Each parent contributed a *y* allele to the green-seeded offspring, so the parents' genotypes must have been *Yy* × *Yy*.

Sometimes it is impossible to determine from the data given whether an individual is homozygous dominant or heterozygous. For example, suppose a cross between a yellow-seeded plant and a green-seeded plant yields offspring that all produce yellow seeds. From the information given about the parents, you can designate the cross as: *Y_* × *yy*. Because all the offspring have yellow seeds, they have the genotype *Yy*. You can deduce this because each parent contributes one allele. (The offspring have yellow seeds so they must have at least one *Y* allele; however, the green seed parent can only contribute a *y* allele to the offspring.) Therefore, the cross is most likely (*YY* × *yy*), but you cannot be absolutely certain about the genotype of the yellow-seeded parent. Further crosses would be necessary to determine this.

as those governing color perception and blood clotting, are sometimes called **sex-linked genes.** It is more appropriate, however, to refer to them as **X-linked genes** because they follow the transmission pattern of the X chromosome and, strictly speaking, are not linked to the sex of the organism per se.

A female receives one X from her mother and one X from her father. A male receives his Y chromosome, which makes him male, from his father. From his mother he inherits a single X chromosome and therefore all his X-linked genes. In the male, every allele present on the X chromosome is expressed, whether that allele was dominant or recessive in the female parent. A male is neither homozygous nor heterozygous for his X-linked alleles; instead, he is always **hemizygous**—that is, he has only one copy of each X-linked gene.

We will use a simple system of notation for problems involving X linkage, indicating the X chromosome and incorporating specific alleles as superscripts. For example, the symbol X^c signifies a recessive X-linked allele for colorblindness and X^C the dominant X-linked allele for normal color vision. The Y chromosome is written without superscripts because it does not carry the locus of interest. Two recessive X-linked alleles must be present in a female for the abnormal phenotype to be expressed (X^cX^c), whereas in the hemizygous male a single recessive allele (X^cY) causes the abnormal phenotype. As a consequence, these abnormal alleles are much more frequently expressed in male offspring. A heterozygous female may be a *carrier*, an individual who possesses one copy of a mutant recessive allele but does not express it in the phenotype (X^CX^c).

To be expressed in a female, a recessive X-linked allele must be inherited from both parents. A colorblind female, for example, must have a colorblind father and a mother who is homozygous or heterozygous for a recessive colorblindness allele (Fig. 10-16). The homozygous combination is unusual, because the frequency of alleles for colorblindness is relatively low. In contrast, a colorblind male need only have a mother who is heterozygous for colorblindness; his father can have normal vision. Therefore, X-linked recessive traits are generally much more common in males than in females, a fact that may partially explain why human male embryos are more likely to die.

Dosage compensation equalizes the expression of X-linked genes in males and females

The X chromosome contains numerous genes required by both sexes, yet a normal female has two copies ("doses") for each locus, whereas a normal male has only one. **Dosage compensation** is a mechanism that makes equivalent the two doses in the female and the single dose in the male. Male fruit flies accomplish this by making their single X chromosome more active. In most tissues, the metabolic activity of a single male X chromosome is equal to the combined metabolic activity of the two X chromosomes present in the female.

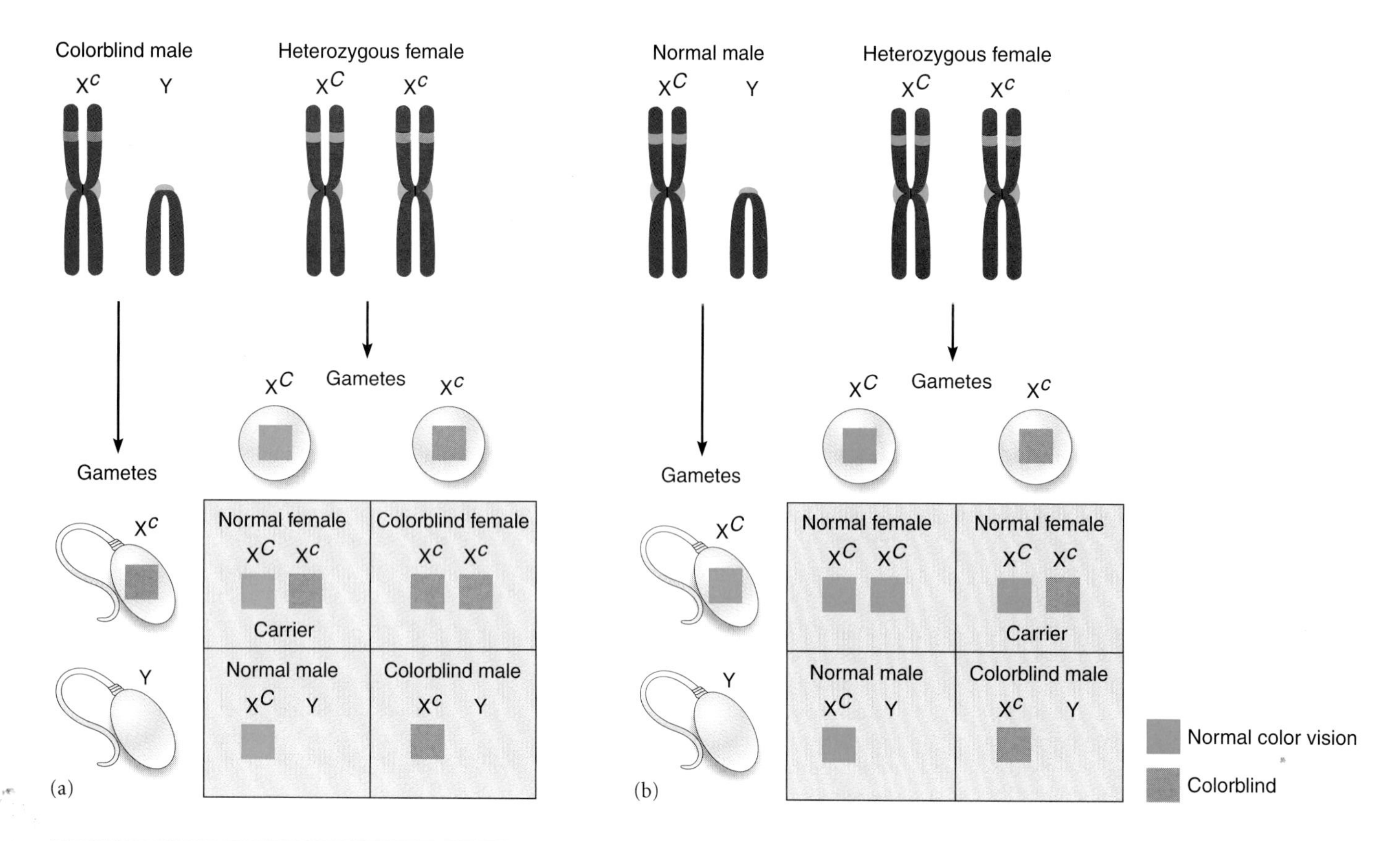

FIGURE 10-16 | X-linked red–green colorblindness.

Note that the Y chromosome does not carry a gene for color vision. **(a)** To be colorblind, a female must inherit alleles for colorblindness from both parents. **(b)** If a normal male mates with a carrier (heterozygous) female, half of their sons would be expected to be colorblind and half of their daughters would be expected to be carriers.

Dosage compensation in mammals generally involves the inactivation of one of the two X chromosomes in the female (Fig. 10-17a). During interphase, a dark spot of chromatin is visible at the edge of the nucleus of each female mammalian cell when stained and observed with a microscope. This dark spot, known as a **Barr body,** is a dense, metabolically inactive X chromosome. The other X chromosome resembles the metabolically active autosomes; during interphase, it is a greatly extended thread that is not evident in light microscopy. From this and other evidence, the British geneticist Mary Lyon hypothesized in 1961 that in most cells of a female mammal, only one of the two X chromosomes is active; the other is inactive and is condensed as a Barr body. (Actually, X chromosome inactivation is never complete; a small fraction of the genes are active.)

X chromosome inactivation is a random event in each somatic (body) cell of the female embryo. A female mammal that is heterozygous at an X-linked locus expresses one of the alleles in about half her cells and the other allele in the other half. This is sometimes evident in the phenotype. Mice and cats have several X-linked genes for certain coat colors. Females that are heterozygous for such genes may show patches of one coat color in the middle of areas of the other coat color. This phenomenon, termed **variegation,** is evident in tortoiseshell and calico cats (Fig. 10-17b). Early in development, when relatively few cells are present, X chromosome inactivation occurs randomly in each cell. When any one of these cells divides by mitosis, the cells of the resulting *clone* (a group of genetically identical cells) all have the same active X chromosome, and, therefore, a patch develops of cells that all express the same color.

Why, you might ask, is variegation not always apparent in females heterozygous at X-linked loci? The answer is that, although variegation usually occurs, we may need to use special techniques to observe it. For example, colorblindness is caused by a defect involving the pigments in the cone cells in the retina of the eye (see Chapter 41). In at least one type of red-green colorblindness, the retina of a heterozygous female actually contains patches of abnormal cones, but the patches of normal cones are enough to provide normal color vision. Variegation can be very hard to observe in cases where cell products become mixed in bodily fluids. For instance, in females heterozygous for the allele that causes hemophilia, only half the cells responsible for producing a specific blood-clotting factor do so, but they produce enough to ensure that the blood clots normally.

Review

- What are linked genes?
- How do geneticists use a two-point test cross to detect linkage?
- Which chromosome determines the male sex in humans and most species of mammals?
- What is dosage compensation, and what does it generally involve in mammals?

BiologyNow™ Assess your understanding of **Mendelian inheritance and chromosomes** by taking the pretest on your BiologyNow CD-ROM.

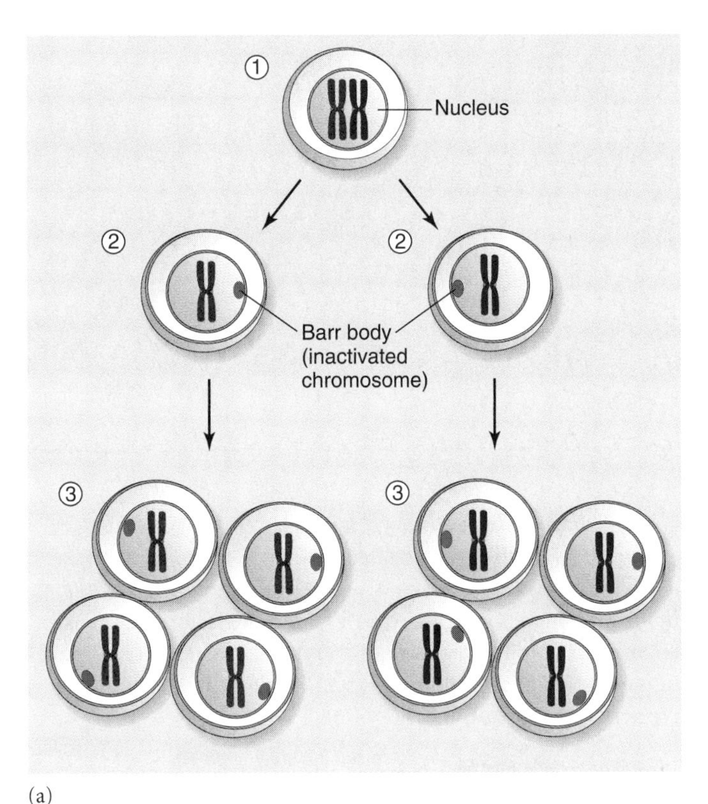

(a)

(b)

FIGURE **10-17** **Dosage compensation in female mammals.**

(a) Inactivation of one X chromosome in female cells. ① The zygote and early embryonic cells have two X chromosomes, one from each parent. ② The random inactivation of one X chromosome occurs early in development. Roughly half the cells inactivate one X chromosome, like the left cell, and the other half inactivate the second X chromosome, as in the right cell. The inactive X chromosome is visible as a Barr body near the nuclear envelope. ③ Chromosomal inactivation persists through subsequent mitotic divisions, resulting in patches of cells in the adult body. **(b)** Persian calico cat. A calico cat has X-linked genes for both black and yellow (or orange) pigmentation of the fur, but because of random X chromosome inactivation, black is expressed in some groups of cells and orange is expressed in others. (The patches of white fur are due to the presence of other genes that affect fur color.)

EXTENSIONS OF MENDELIAN GENETICS

Learning Objectives

9 Explain some of the ways genes may interact to affect the phenotype; discuss how a single gene can affect many features of the organism simultaneously.

10 Solve genetics problems involving incomplete dominance, codominance, multiple alleles, epistasis, and polygenes.

11 Describe *norm of reaction,* and give an example.

The relationship between a given locus and the trait it controls may or may not be simple. A single pair of alleles of a locus may regulate the appearance of a single trait (such as tall versus short in garden peas). Alternatively, a pair of alleles may participate in the control of several traits, or alleles of many loci may interact to affect the phenotypic expression of a single character. Not surprisingly, these more complex relationships are quite common.

You can assess the phenotype on one or many levels. It may be a morphological trait, such as shape, size, or color. It may be a physiological trait or even a biochemical trait, such as the presence or absence of a specific enzyme required for the metabolism of some specific molecule. In addition, changes in the environmental conditions under which the organism develops may alter the phenotypic expression of genes.

Dominance is not always complete

Studies of the inheritance of many traits in a wide variety of organisms have shown that one member of a pair of alleles may not be completely dominant over the other. In such instances, it is inaccurate to use the terms *dominant* and *recessive.*

PROCESS OF SCIENCE

The plants commonly known as four o'clocks (*Mirabilis jalapa*) may have red or white flowers. Each color breeds true when these plants are self-pollinated. What flower color might you expect in the offspring of a cross between a red-flowering plant and a white-flowering one? Without knowing which is dominant, you might predict that all would have red flowers or all would have white flowers. German botanist Carl Correns, one of the rediscoverers of Mendel's work, first performed this cross and found that all F_1 offspring have pink flowers! Does this result in any way indicate that Mendel's assumptions about inheritance are wrong? Did the parental traits blend inseparably in the offspring? Quite the contrary, for when two of these pink-flowered plants are crossed, red-flowered, pink-flowered, and white-flowered offspring appear in a ratio of 1:2:1 (Fig. 10-18). In this instance, as in all other aspects of the scientific process, results that differ from those hypothesized prompt scientists to re-examine and modify their hypotheses to account for the exceptional results. The pink-flowered plants are clearly the heterozygous individuals, and neither the red allele nor the white allele is completely dominant. When the heterozygote has a phenotype intermediate between those of its two parents, the genes are said to show **incomplete dominance.** In these crosses, the genotypic and phenotypic ratios are identical.

Incomplete dominance is not unique to four o'clocks, and additional examples of incomplete dominance are known in both plants and animals. For example, true-breeding white chickens and true-breeding black chickens produce bluish-gray offspring, known as Andalusian blues, when crossed.

In both cattle and horses, reddish coat color is not completely dominant to white coat color. Heterozygous individuals have a mixture of reddish hairs and white hairs, which is called *roan.* If you saw a white mare nursing a roan foal, what would you guess was the coat color of the foal's father? Because the reddish and white colors are expressed independently (hair by hair) in the roan heterozygote, scientists sometimes refer to this as a case of **codominance.** Strictly speaking, the term *incomplete dominance* refers to instances in which the heterozygote is intermediate in phenotype, and *codominance* refers to instances in which the heterozygote simultaneously expresses the phenotypes of both types of homozygotes.

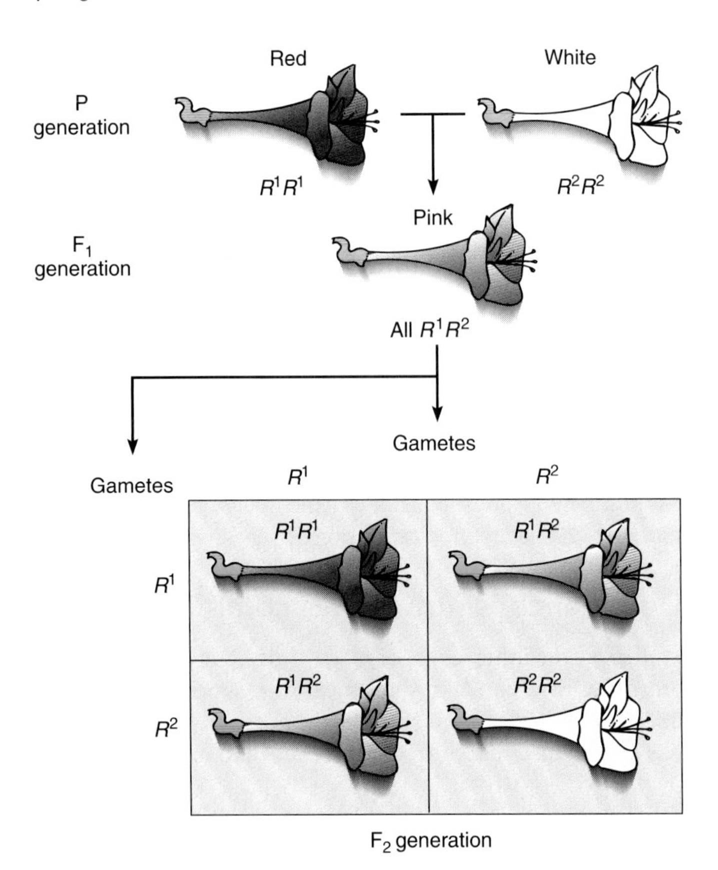

FIGURE **10-18** | **Incomplete dominance in four o'clocks.**

If a pair of alleles is incompletely dominant to each other, a heterozygote has a phenotype intermediate between its parents. Two incompletely dominant alleles, R^1 and R^2, are responsible for red, white, and pink flower colors. Red-flowered plants are R^1R^1, white-flowered plants are R^2R^2, and heterozygotes (R^1R^2) are pink. Note that uppercase letters are used for both alleles, because neither is recessive to the other.

TABLE **10-1** **ABO Blood Types***

Phenotype (blood type)	Genotypes	Antigen on RBC	Antibodies to A or B Antigens in Plasma
A	I^AI^A, I^Ai	A	Anti-B
B	I^BI^B, I^Bi	B	Anti-A
AB	I^AI^B	A, B	None
0	ii		Anti-A, anti-B

*This table and the discussion of the ABO system have been simplified somewhat. Note that the body produces antibodies against the antigens *lacking* on its own red blood cells (RBCs). Because of their specificity for the corresponding antigens, these antibodies are used in standard tests to determine blood types.

Humans have four blood types (A, B, AB, and O), collectively called the **ABO blood group.** The human ABO blood group is an excellent example of codominant alleles. Blood types A, B, AB, and O are controlled by three alleles representing a single locus (Table 10-1). Allele I^A codes for the synthesis of a specific glycoprotein, antigen A, which is expressed on the surface of red blood cells. (Immune responses are discussed in Chapter 43; for now, we define antigens simply as substances capable of stimulating an immune response.) Allele I^B leads to the production of antigen *B*, a different (but related) glycoprotein. Allele *i* is allelic to I^A and I^B, but it does not code for an antigen. Individuals with the genotype I^AI^A or I^Ai have blood type A. People with genotype I^BI^B or I^Bi have blood type B. Those with genotype I^AI^B have blood type AB, whereas those with genotype *ii* have blood type O. These results show that neither allele I^A nor allele I^B is dominant to the other. Both alleles are expressed phenotypically in the heterozygote and are therefore codominant to each other, although each is dominant to allele *i.*

At one time, determining blood types was used to settle cases of disputed parentage. Although blood type tests can *exclude* someone as a possible parent of a particular child, they can never prove a certain person is the parent; they only determine he or she *could* be. Could a man with blood type AB be the father of a child with blood type O? Could a woman with blood type O be the mother of a child with blood type AB? Could a type B child with a type A mother have a type A father or a type O father?[2]

Multiple alleles for a locus may exist in a population

Most of our examples so far have dealt with situations in which each locus was represented by a maximum of two allelic variants. It is true that a single diploid individual has a maximum of two

[2] The answer to all these questions is no.

FIGURE **10-19** | **Multiple alleles in rabbits.**

In rabbits the locus for coat color has four alleles, designated *C*, c^{ch}, c^h, and *c*. An individual rabbit, being diploid, may have any two alleles of this series. The phenotypes and their associated genotypes are shown.

different alleles for a particular locus and that a haploid gamete has only one allele for each locus. However, if you survey a population you may find more than two alleles for a particular locus, as you saw with the ABO blood group. If three or more alleles for a given locus exist within the population, we say that locus has **multiple alleles.** Research has shown that many loci have multiple alleles. Some alleles can be identified by the activity of a certain enzyme or by some other biochemical feature but do not produce an obvious phenotype. Others produce a readily recognizable phenotype, and certain patterns of dominance can be discerned when the alleles are combined in various ways.

In rabbits, four alleles occur at the locus for coat color (Fig. 10-19). A *C* allele causes a fully colored dark gray coat. The homozygous recessive genotype, *cc,* causes albino (white) coat color. There are two additional allelic variants of the same locus, c^{ch} and c^h. An individual with the genotype $c^{ch}c^{ch}$ has the chinchilla pattern, in which the entire body has a light, silvery gray color. The genotype c^hc^h causes the Himalayan pattern, in which the body is white but the tips of the ears, nose, tail, and legs are colored, like the color pattern of a Siamese cat. On the basis of the results of genetic crosses, these alleles can be arranged in the following series:

$$C > c^{ch} > c^h > c$$

Each allele is dominant to those following it and recessive to those preceding it. For example, a $c^{ch}c^{ch}$, $c^{ch}c^h$, or $c^{ch}c$ rabbit has the chinchilla pattern, whereas a c^hc^h or c^hc rabbit has the Himalayan pattern. (We revisit the Himalayan coat pattern later in the chapter when we discuss the influence of environment on gene expression.)

In other series of multiple alleles, certain alleles may be codominant and others incompletely dominant. In such cases, the heterozygotes commonly have phenotypes intermediate between those of their parents.

A single gene may affect multiple aspects of the phenotype

In our examples, the relationship between a gene and its phenotype has been direct, precise, and exact, and the loci have controlled the appearance of single traits. However, the relationship of gene to trait may not have such a simple genetic basis. Most genes have several effects on different characters. The ability of a single gene to have multiple effects is known as **pleiotropy.** Most cases of pleiotropy can be traced to a single fundamental cause. For example, a defective enzyme may affect the functioning of many types of cells. Pleiotropy is evident in many genetic diseases in which a single pair of alleles causes multiple symptoms. For example, people who are homozygous for the recessive allele that causes cystic fibrosis produce abnormally thick mucus in many parts of the body, including the respiratory, digestive, and reproductive systems (see Chapter 15).

Alleles of different loci may interact to produce a phenotype

Several pairs of alleles may interact to affect a single phenotype, or one pair may inhibit or reverse the effect of another pair. One example of gene interaction is illustrated by the inheritance of combs in chickens, where two genes may interact to produce a novel phenotype (Fig. 10-20). The allele for a rose comb, *R,* is dominant to that for a single comb, *r.* A second, unlinked pair of alleles governs the inheritance of a pea comb, *P,* versus a single comb, *p.* A single-comb chicken is homozygous for the recessive allele at both loci *(pprr).* A rose comb chicken is either *ppRR* or *ppRr,* and a pea comb chicken is either *PPrr* or *Pprr.* When an *R* and *P* occur in the same individual, the phenotype is neither a pea nor a rose comb but a completely different type, a walnut comb. The walnut comb phenotype is produced whenever a chicken has one or two *R* alleles, plus one or two *P* alleles (that is, *PPRR, PpRR, PPRr,* or *PpRr*). What would you hypothesize about the types of combs among the offspring of two heterozygous walnut comb chickens, *PpRr?* How does this form of gene interaction affect the ratio of phenotypes in the F_2 generation? Is it the typical Mendelian 9:3:3:1 ratio?[3]

[3] The offspring of two heterozygous walnut comb chickens will have four genotypes in what appears to be a Mendelian 9:3:3:1 ratio: 9 walnut comb, 3 pea comb, 3 rose comb, and 1 single comb. This example is not a typical Mendelian 9:3:3:1 ratio because it involves a single character (comb shape) coded by alleles at two different loci. In Mendelian inheritance, the 9:3:3:1 ratio involves *two* characters (such as seed color and seed shape) coded by alleles at two loci.

FIGURE **10-20** | **Gene interaction in chickens.**

Two gene pairs govern four chicken comb phenotypes. Chickens with walnut combs have the genotype *P_R_*. Chickens with pea combs have the genotype *P_rr*, and those with rose combs have the genotype *ppR_*. Chickens that are recessive for both genes, *pprr*, have a single comb. (The blanks can be either dominant or recessive alleles.)

Epistasis is a common type of gene interaction in which the presence of certain alleles of one locus can prevent or mask the expression of alleles of a different locus and express its own phenotype instead. (The term epistasis means "standing on.") Unlike the chicken example, no novel phenotypes are produced in epistasis.

Coat color in Labrador retrievers is an example of epistasis that involves a gene for pigment and a gene for depositing color in the coat (Fig. 10-21). The two alleles for the pigment gene are *B* for black coat and its recessive counterpart *b* for brown coat. The gene for depositing color in the coat has two alleles, *E* for the expression of black and brown coats and *e*, which is epistatic and blocks the expression of the *B/b* gene. The epistatic allele is recessive and therefore is only expressed as a yellow coat in the homozygous condition (*ee*), regardless of the combination of *B* and *b* alleles in the genotype.

Polygenes act additively to produce a phenotype

Many human characters—such as height, body form, and skin pigmentation—are not inherited through alleles at a single locus. The same holds true for many commercially important characters in domestic plants and animals, such as milk and egg production. Alleles at several, perhaps many, loci affect each character. The term **polygenic inheritance** is applied when multiple independent pairs of genes have similar and additive effects on the same character.

Polygenes—as many as 60 loci—account for the inheritance of skin pigmentation in humans. To keep things simple in this example, we illustrate the principle of polygenic inheritance in human skin pigmentation with pairs of alleles at three unlinked loci (Fig. 10-22). These can be designated *A* and *a*, *B* and *b*, and *C* and *c*. The capital letters represent incompletely dominant alleles producing dark skin. The more capital letters, the darker the skin, because the alleles affect skin pigmentation in an additive fashion. A person with the darkest skin possible would have the genotype *AABBCC*, and a person with the lightest skin possible would have the genotype *aabbcc*. The F_1 offspring of an *aabbcc* person and an *AABBCC* person are all *AaBbCc* and have an intermediate skin color. The F_2 offspring of two such triple heterozygotes (*AaBbCc* × *AaBbCc*) would have skin pigmentation ranging from very dark to very light.

Polygenic inheritance is characterized by an F_1 generation that is intermediate between the two completely homozygous parents and by an F_2 generation that shows wide variation between the two parental types. When the number of individuals in a population is plotted against the amount of skin pigmentation and the points are connected, the result is a bell-shaped curve, a **normal distribution curve.** Most of the F_2 generation individuals have one of the intermediate phenotypes; only a few show the extreme phenotypes of the original P generation (that is, the grandparents). On average, only 1 of 64 is as dark as the very dark grandparent, and only 1 of 64 is as light as the very light grandparent. The alleles *A*, *B*, and *C* each produce about the same amount of darkening of the skin; hence the genotypes *AaBbCc*, *AABbcc*, *AAbbCc*, *AaBBcc*, *aaBBCc*, *AabbCC*, and *aaBbCC* all produce similar intermediate phenotypes.

Genes interact with the environment to shape phenotype

Imagine taking two cuttings from a coleus plant. You place one cutting in a growth chamber where it receives a lot of light and the other in a growth chamber programmed to deliver dim light. Although the cuttings are genetically identical (they came from the same plant), the appearance of the two plants would be dramatically different in a very short period. The plant with adequate light would thrive and grow vigorously, whereas the plant with inadequate light would be spindly and weak. During an organism's development, both genotype and environmental factors affect the expression of phenotype, and genetically identical individuals may develop differently in different environments.

Consider another example of the effect of environment on gene expression, this one involving the Himalayan rabbits discussed earlier. Recall that the phenotype of these rabbits is white fur except for dark patches on the ears, nose, and paws. The local surface temperature of a rabbit's ears, nose, and paws is colder in the rabbit's natural environment, and this temperature difference causes the production of dark fur. If you raise rabbits

FIGURE **10-21** **Epistasis in Labrador retrievers.**

Two gene pairs interact to govern coat color in Labrador retrievers. Black Labs have the genotype *B_E_*; yellow Labs have the genotype *B_ee* or *bbee*; and chocolate Labs have the genotype *bbE_*. The blanks represent either dominant or recessive alleles.

with the Himalayan genotype at a warm temperature (30°C), the rabbits are completely white, with no dark patches on the ears, nose, or paws. If you raise Himalayan rabbits at a cooler temperature (25°C), however, they develop the characteristic dark patches of fur.

Now let's examine a human example—height—in the context of genes and the environment. The inheritance of height in humans is polygenic and involves alleles representing 10 or more loci. Because many genes are involved and because height

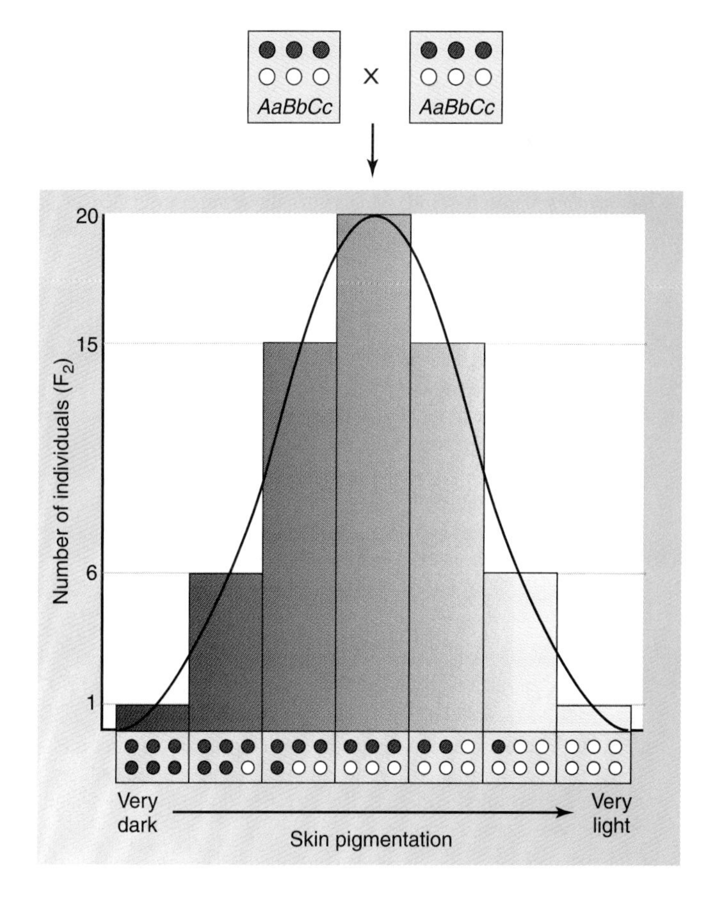

FIGURE **10-22** **Polygenic inheritance in human skin pigmentation.**

This simplified example assumes that skin pigmentation in humans is governed by alleles of three unlinked loci. The alleles producing dark skin (*A*, *B*, and *C*) are represented by capital letters, but they are not dominant. Instead their effects are additive. The number of dark dots, each signifying an allele producing dark skin, is counted to determine the phenotype. A wide range of phenotypes is possible when individuals of intermediate phenotype mate and have offspring (*AaBbCc* × *AaBbCc*). The expected distribution of phenotypes is consistent with the superimposed normal distribution curve.

is modified by a variety of environmental conditions, such as diet and general health, the height of most adults ranges from 4 ft 2 in (1.25 m) to 7 ft 2 in (2.15 m). The genes that affect height set limits for the phenotype—no human is 12 ft tall, for example—but the environment shapes the phenotype within its genetic limits. The range of phenotypic possibilities that can develop from a single genotype under different environmental conditions is known as the **norm of reaction.** In certain genotypes, such as the human ABO blood group, the norm of reaction is quite limited. In other genotypes, such as those involved in human height, the norm of reaction is quite broad.

Although the phenotypes of many characters are influenced by interactions between genes and the environment, it is impossible to determine the exact contributions of genes and the environment to a given phenotype. This subject, known as the *nature–nurture question,* is particularly controversial, because it relates to human characters ranging from intelligence to the tendency to abuse alcohol. For example, intelligence has a genetic component, but environment is also an important influence. To test for the relative effects of genes and environment on intelligence, we would have to separate and raise genetically identical children in different environments. Because it is ethically unacceptable to manipulate humans in this way, such tests cannot be performed. All we can say with certainty is that an individual's genes provide the potential for developing a particular phenotype but that environmental influences also shape the phenotype.

Review

- What is the difference between incomplete dominance and codominance?
- What is the difference between multiple alleles and polygenes?
- What is the difference between pleiotropy and epistasis?
- What is a norm of reaction?

BiologyNow™ Assess your understanding of **extensions of Mendelian genetics** by taking the pretest on your BiologyNow CD-ROM.

SUMMARY WITH KEY TERMS

1 Define the terms *phenotype, genotype, locus, allele, dominant allele, recessive allele, homozygous,* and *heterozygous.*

- **Genes** are in chromosomes; the site a gene occupies in the chromosome is its **locus.** Different forms of a particular gene are **alleles;** they occupy corresponding loci on homologous chromosomes. Therefore, genes exist as pairs of alleles in diploid individuals.
- An individual that carries two identical alleles is said to be **homozygous** for that locus. If the two alleles are different, that individual is said to be **heterozygous** for that locus.
- One allele, the **dominant allele,** may mask the expression of the other allele, the **recessive allele,** in a heterozygous individual. For this reason two individuals with the same appearance, or **phenotype,** may differ from each other in genetic constitution, or **genotype.**

2 Describe Mendel's principles of segregation and independent assortment.

- According to Mendel's **principle of segregation,** during meiosis the alleles for each locus separate, or segregate, from each other. When haploid gametes are formed, each contains only one allele for each locus.
- According to Mendel's **principle of independent assortment,** alleles of different loci are distributed randomly into the gametes. This can result in **recombination,** the production of new gene combinations that were not present in the **parental (P) generation.**

3 Solve genetics problems involving monohybrid, dihybrid, and test crosses.

- A cross between homozygous parents (P generation) that differ from each other with respect to their alleles at one locus is called a **monohybrid cross;** if they differ at two loci, it is called a **dihybrid cross.**
- The first generation of offspring is heterozygous and is called the first filial, or **F_1, generation;** the generation produced by a cross of two F_1 individuals is the second filial, or **F_2, generation.**
- A **test cross** is between an individual of unknown genotype and a homozygous recessive individual.

4 Apply the product rule and sum rule appropriately when predicting the outcomes of genetic crosses.

- Genetic ratios can be expressed in terms of probabilities. Any probability is expressed as a fraction or decimal fraction, calculated as the number of favorable events divided by the total number of events. This can range from 0 (an impossible event) to 1 (a certain event).
- According to the **product rule,** the probability of two independent events occurring together can be calculated by multiplying the probabilities of each event occurring separately.
- According to the **sum rule,** the probability of an outcome that can be obtained in more than one way can be calculated by adding the separate probabilities.

5 Explain Mendel's principles of segregation and independent assortment, given what scientists know about genes and chromosomes.

- Segregation of alleles is a direct result of homologous chromosomes separating during meiosis.
- Independent assortment occurs because there are two ways in which two pairs of homologous chromosomes can be arranged at metaphase I of meiosis. The orientation of homologous chromosomes on the metaphase plate determines the way chromosomes are distributed into haploid cells.

6 Define *linkage,* and relate it to specific events in meiosis.

- Each chromosome behaves genetically as if it consisted of genes arranged in a linear order. **Linkage** is the tendency for a group of genes on the same chromosome to be inherited together. Groups of genes on the same chromosome are **linked genes.** Independent assortment does not apply if two loci are linked close together on the same pair of homologous chromosomes.

- Recombination of linked genes can result from **crossing-over** (breaking and rejoining of homologous chromatids) in meiotic prophase I.
- By measuring the frequency of recombination between linked genes, it is possible to construct a linkage map of a chromosome.

7 Show how data from a test cross involving alleles of two loci can be used to distinguish between independent assortment and linkage.

- To distinguish between independent assortment and linked genes, perform a **two-point test cross** between an individual that is heterozygous for the linked genes and an individual that is homozygous recessive for both characters.
- Linkage is recognized when an excess of parental type offspring and a deficiency of recombinant type offspring are produced in a two-point test cross.

8 Discuss the genetic determination of sex and the inheritance of X-linked genes in mammals.

- The sex of humans and many other animals is determined by the **X** and **Y chromosomes** or their equivalents. Normal female mammals have two X chromosomes; normal males have one X and one Y. The fertilization of an X-bearing egg by an X-bearing sperm results in a female (XX) zygote. The fertilization of an X-bearing egg by a Y-bearing sperm results in a male (XY) zygote.
- The Y chromosome determines male sex in mammals. The X chromosome contains many important genes unrelated to sex determination that are required by both males and females. A male receives all his **X-linked genes** from his mother. A female receives X-linked genes from both parents.
- A female mammal shows **dosage compensation** of X-linked genes. Only one of the two X chromosomes is expressed in each cell; the other is inactive and is seen as a dark-staining **Barr body** at the edge of the interphase nucleus.

9 Explain some of the ways genes may interact to affect the phenotype; discuss how a single gene can affect many features of the organism simultaneously.

- The relationship between genotype and phenotype may be quite complex. **Pleiotropy** is the ability of one gene to have several effects on different characters. Most cases of pleiotropy can be traced to a single cause, such as a defective enzyme. Alternatively, alleles of many loci may interact to affect the phenotypic expression of a single character.

10 Solve genetics problems involving incomplete dominance, codominance, multiple alleles, epistasis, and polygenes.

- Dominance does not always apply, and some alleles demonstrate **incomplete dominance,** in which the heterozygote is intermediate in phenotype, or **codominance,** in which the heterozygote simultaneously expresses the phenotypes of both homozygotes.
- **Multiple alleles,** three or more alleles that can potentially occupy a particular locus, may exist in a population. A diploid individual has any two of the alleles; a haploid individual or gamete has only one.
- In **epistasis,** an allele of one locus can mask the expression of alleles of a different locus.
- In **polygenic inheritance,** multiple independent pairs of genes may have similar and additive effects on the phenotype.

11 Describe *norm of reaction,* and give an example.

- The range of phenotypic possibilities that can develop from a single genotype under different environmental conditions is known as the **norm of reaction.**
- Many genes are involved in the inheritance of height in humans. Also, height is modified by a variety of environmental conditions, such as diet and general health. The genes that affect height set the norm of reaction—that is, the limits—for the phenotype, and the environment molds the phenotype within its norm of reaction.

POST-TEST

1. One reason Mendel discovered the basic principles of inheritance is that he (a) understood the behavior of chromosomes in mitosis and meiosis (b) studied a wide variety of experimental organisms (c) began by establishing true-breeding lines (d) studied various types of linkage (e) studied hybrids between parents that differed in many, often not clearly defined, ways
2. One of the autosomal loci controlling eye color in fruit flies has two alleles, one for brown eyes and the other for red eyes. Fruit flies from a true-breeding line with brown eyes were crossed with flies from a true-breeding line with red eyes. The F_1 flies had red eyes. What conclusion can be drawn from this experiment? (a) these alleles underwent independent assortment (b) these alleles underwent segregation (c) these genes are X-linked (d) the allele for red eyes is dominant to the allele for brown eyes (e) all of the above are true
3. The F_1 flies described in question 2 were mated with brown-eyed flies from a true-breeding line. What phenotypes would you expect the offspring to have? (a) all red eyes (b) all brown eyes (c) half red eyes and half brown eyes (d) red-eyed females and brown-eyed males (e) brown-eyed females and red-eyed males
4. The type of cross described in question 3 is a(an) (a) F_2 cross (b) dihybrid cross (c) test cross (d) two-point test cross (e) none of the preceding

Use the following information to answer questions 5 through 8:

In peas, the allele for round seeds *(R)* is dominant to that for wrinkled seeds *(r);* the allele for yellow seeds *(Y)* is dominant to that for green seeds *(y).* These loci are unlinked. Plants from a true-breeding line with round, green seeds are crossed with plants from a true-breeding line with wrinkled, yellow seeds. These parents constitute the P generation.

5. The genotypes of the P generation are (a) *RRrr* and *Yyyy* (b) *RrYy* (c) *RRYY* and *rryy* (d) *RRyy* and *rrYY* (e) *RR* and *YY*
6. What are the expected genotypes of the F_1 hybrids produced by the described cross? (a) *RRrr* and *YYyy* (b) all *RrYy* (c) *RRYY* and *rryy* (d) *RRyy* and *rrYY* (e) *RR* and *YY*
7. What kinds of gametes can the F_1 individuals produce? (a) *RR* and *YY* (b) *Rr* and *Yy* (c) *RR, rr, YY,* and *yy* (d) *R, r, Y,* and *y* (e) *RY, Ry, rY,* and *ry*
8. What is the expected proportion of F_2 wrinkled, yellow seeds? (a) $^9/_{16}$ (b) $^1/_{16}$ (c) $^3/_{16}$ (d) $^1/_4$ (e) zero

9. Individuals of genotype *AaBb* were crossed with *aabb* individuals. Approximately equal numbers of the following classes of offspring were produced: *AaBb, Aabb, aaBb,* and *aabb.* These results illustrate Mendel's principle(s) of (a) linkage (b) independent assortment (c) segregation (d) a and c (e) b and c
10. Assume that the ratio of females to males is 1:1. A couple already has two daughters and no sons. If they plan to have a total of six children, what is the probability they will have four more girls? (a) 1⁄4 (b) 1⁄8 (c) 1⁄16 (d) 1⁄32 (e) 1⁄64
11. Red-green colorblindness is an X-linked recessive disorder in humans. Your friend is the daughter of a colorblind father. Her mother had normal color vision, but her maternal grandfather was colorblind. What is the probability your friend is colorblind? (a) 1 (b) 1⁄2 (c) 1⁄4 (d) 3⁄4 (e) zero
12. When homozygous, a particular allele of a locus in rats causes abnormalities of the cartilage throughout the body, an enlarged heart, slow development, and death. This is an example of (a) pleiotropy (b) polygenic inheritance (c) epistasis (d) codominance (e) dosage compensation
13. In peas, yellow seed color is dominant to green. Determine the phenotypes (and their proportions) of the offspring of the following crosses: (a) homozygous yellow X green (b) heterozygous yellow X green (c) heterozygous yellow X homozygous yellow (d) heterozygous yellow X heterozygous yellow
14. If two animals heterozygous for a single pair of alleles are mated and have 200 offspring, about how many would be expected to have the phenotype of the dominant allele (that is, to look like the parents)?
15. When two long-winged flies were mated, the offspring included 77 with long wings and 24 with short wings. Is the short-winged condition dominant or recessive? What are the genotypes of the parents?
16. Outline a breeding procedure whereby a true-breeding strain of red cattle could be established from a roan bull and a white cow.
17. What is the probability of rolling a seven with a pair of dice? Which is a more likely outcome, rolling a six with a pair of dice or rolling an eight?
18. In rabbits, spotted coat *(S)* is dominant to solid color *(s)*, and black *(B)* is dominant to brown *(b)*. These loci are not linked. A brown, spotted rabbit from a pure line is mated to a solid black one, also from a pure line. What are the genotypes of the parents? What would be the genotype and phenotype of an F_1 rabbit? What would be the expected genotypes and phenotypes of the F_2 generation?
19. The long hair of Persian cats is recessive to the short hair of Siamese cats, but the black coat color of Persians is dominant to the brown-and-tan coat color of Siamese. Make up appropriate symbols for the alleles of these two unlinked loci. If a pure black, long-haired Persian is mated to a pure brown-and-tan, short-haired Siamese, what will be the appearance of the F_1 offspring? If two of these F_1 cats are mated, what is the chance that a long-haired, brown-and-tan cat will be produced in the F_2 generation? (Use the shortcut probability method to obtain your answer; then check it with a Punnett square.)
20. Mr. and Mrs. Smith are concerned because their own blood types are A and B respectively, but their new son, Richard, is blood type O. Could Richard be the child of these parents?
21. The expression of an allele called *frizzle* in fowl causes abnormalities of the feathers. As a consequence, the animal's body temperature is lowered, adversely affecting the functions of many internal organs. When one gene affects many traits of the organism in this way, we say that gene is ____________.
22. A walnut comb rooster is mated to three hens. Hen A, which is walnut comb, has offspring in the ratio of 3 walnut to 1 rose. Hen B, which has a pea comb, has offspring in the ratio of 3 walnut to 3 pea to 1 rose to 1 single. Hen C, which has a walnut comb, has only walnut comb offspring. What are the genotypes of the rooster and the three hens?
23. What kinds of matings result in the following phenotypic ratios? (a) 3:1 (b) 1:1 (c) 9:3:3:1 (d) 1:1:1:1
24. The weight of the fruit in a certain variety of squash is determined by allelic pairs for two loci: *AABB* produces fruits that average 2 kg each, and *aabb* produces fruits that weigh 1 kg each. Each allele represented by a capital letter adds 0.25 kg. When a plant that produces 2-kg fruits is crossed with a plant that produces 1-kg fruits, all the offspring produce fruits that weigh 1.5 kg each. If two of these F_1 plants were crossed, how much would the fruits produced by the F_2 plants weigh?
25. The X-linked *barred* locus in chickens controls the pattern of the feathers, with the alleles *B* for barred pattern and *b* for no bars. If a barred female (X^BY) is mated to a nonbarred male (X^bX^b), what will be the appearance of the male and female progeny? (Recall that in birds males are XX and females are XY.) Do you see any commercial usefulness for this result? (*Hint:* It is notoriously difficult to determine the sex of newly hatched chicks.)
26. Individuals of genotype *AaBb* were mated to individuals of genotype *aabb.* One thousand offspring were counted, with the following results: 474 *Aabb,* 480 *aaBb,* 20 *AaBb,* and 26 *aabb.* What type of cross is this known as? Are these loci linked? What are the two parental classes and the two recombinant classes of offspring? What is the percentage of recombination between these two loci? How many map units apart are they?
27. Genes *A* and *B* are 6 map units apart, and *A* and *C* are 4 map units apart. Which gene is in the middle if *B* and *C* are 10 map units apart? Which is in the middle if *B* and *C* are 2 map units apart?

CRITICAL THINKING

1. Would the science of genetics in the 20th century have developed any differently if Gregor Mendel had never lived?
2. Sketch a series of diagrams showing each of the following, making sure to end each series with haploid gametes:
   a. How a pair of alleles for a single locus segregates in meiosis
   b. How the alleles of two unlinked loci assort independently in meiosis
   c. How the alleles of two linked loci undergo genetic recombination
3. Can you always ascertain an organism's genotype for a particular locus if you know its phenotype? Conversely, if you are given an organism's genotype for a locus, can you always reliably predict its phenotype? Explain.

- Visit our Web site at **http://biology.brookscole.com/solomon7** for links to chapter-related resources on the World Wide Web. Additional online materials relating to this chapter can also be found on our Web site.

BIOLOGY NOW RESOURCES

Active Figures

10-5: Monohybrid crosses

10-8: Independent assortment

Preparing for an exam? Take a diagnostic test on your BiologyNow CD-ROM.

Post-Test Answers

1. c	2. d	3 c	4. c
5. d	6. b	7. e	8. c
9. e	10. c	11. b	12. a

13. a. all yellow b. ½ yellow: ½ green
    c. all yellow d. ¾ yellow: ¼ green
14. 150
15. The short-winged condition is recessive. Both parents are heterozygous.
16. Repeated matings of the roan bull and the white cow will yield an approximate 1:1 ratio of roan-to-white offspring. Repeated matings among roan offspring will yield red, roan, and white offspring in an approximate 1:2:1 ratio. The mating of two red individuals will yield only red offspring.
17. There are 36 possible outcomes when a pair of dice is rolled. There are six ways of obtaining a seven: 1,6; 6,1; 2,5; 5,2; 3,4; and 4,3. There are five ways of rolling a six: 1,5; 5,1; 2,4; 4,2; and 3,3. There are also five ways of rolling an eight: 2,6; 6,2; 3,5; 5,3; and 4,4.
18. The genotype of the brown spotted rabbit is *bbSS*. The genotype of the black, solid rabbit is *BBss*. An F_1 rabbit would be black, spotted *(BbSs)*. The F_2 is expected to be 9/16 black, spotted *(B_S_)*, 3/16 black, solid *(B_ss)*, 3/16 brown, spotted *(bbS_)*, and 1/16 brown, solid *(bbss)*.
19. The F_1 offspring are expected to be black, short-haired. There is a 1/16 chance of brown and tan, long-haired offspring in the F_2 generation.
20. Yes
21. Pleiotropic
22. The rooster is *PpRr;* hen A is *PpRR;* hen B is *Pprr;* and hen C is *PPRR.*
23. a. Both parents are heterozygous for a single locus.
    b. One parent is heterozygous and the other parent homozygous recessive (for a single locus).
    c. Both parents are heterozygous (for two loci).
    d. One parent is heterozygous and the other is homozygous recessive (for two loci).
24. The expected types of F_2 plants are 1/16 of the plants produce 2-kg fruits; ¼ of the plants produce 1.75-kg fruits, 3/8 of the plants produce 1.5-kg fruits, ¼ of the plants produce 1.25-kg fruits, and 1/16 of the plants produce 1-kg fruits.
25. All male offspring of this cross are barred, and all females are nonbarred, thus allowing the sex of the chicks to be determined by their phenotypes.
26. This is a two-point test cross involving linked loci. The parental class of offspring are *Aabb* and *aaBb;* the recombinant classes are *AaBb* and *aabb.* There is 4.6% recombination, which corresponds to 4.6 map units between the loci.
27. If genes B and C are 10 map units apart, gene A is in the middle. If genes B and C are 2 map units apart, gene C is in the middle.

11

DNA: The Carrier of Genetic Information

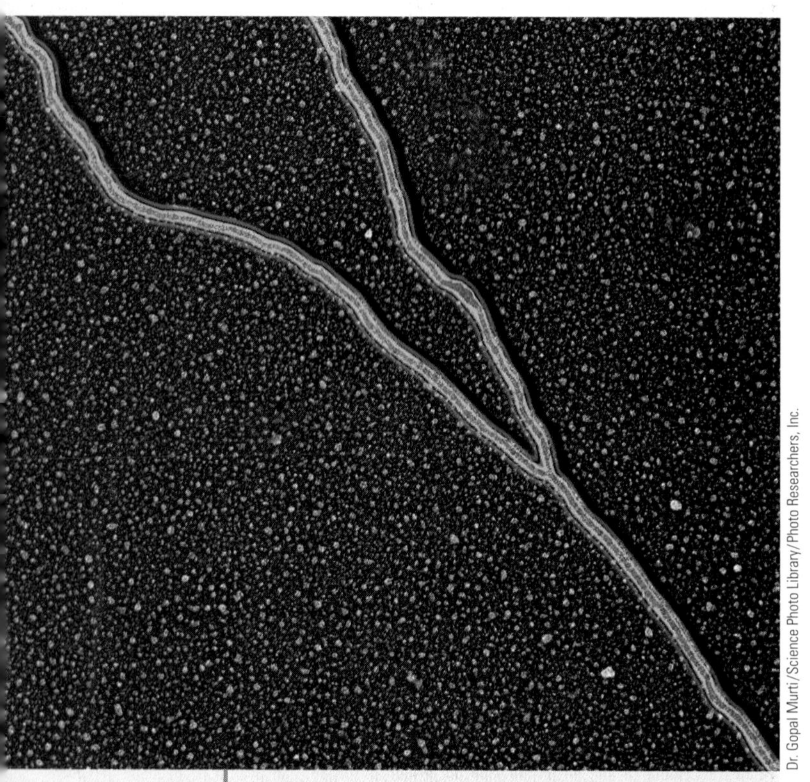

Dr. Gopal Murti/Science Photo Library/Photo Researchers, Inc.

Electron micrograph of DNA replication. During replication, two DNA molecules are synthesized from the original parent molecule. Replication is occurring at the Y-shaped structure, which is called a *replication fork.*

CHAPTER OUTLINE

After the rediscovery of Mendel's principles in 1900, geneticists conducted a variety of elegant experiments to learn how genes are arranged in chromosomes and how they are transmitted from generation to generation. However, the basic questions remained unanswered through most of the first half of the 20th century: What are genes made of? How do genes work? The studies of inheritance patterns described in Chapter 10 did not answer these questions. However, they provided a foundation of knowledge that enabled scientists to make predictions about the molecular (chemical) nature of genes and how genes function.

Scientists generally agreed that the function of genes must be to provide information; therefore, the molecules of which genes are made would have to store information in a form that the cell could retrieve and use. But genes had other properties that also had to be accounted for. For example, countless experiments on a variety of organisms had demonstrated that genes are usually stable, being passed unchanged from generation to generation. However, occasionally a gene converted to a different form; such genetic changes, called *mutations,* were then transmitted unchanged to future generations.

As the science of genetics was developing, the science of biochemistry was flourishing as well, and scientists were making a growing effort to correlate the known properties of genes with the nature of various biological molecules. What kind of molecule could store information? How could that information be retrieved and used to direct cell functions? What kind of molecule could be relatively stable but have the capacity to change, resulting in a mutation, under certain conditions?

Some scientists thought they could never answer these questions. They thought the information a cell requires is so complex that no one type of molecule could function as the genetic material. As they learned more about the central role of proteins in virtually every aspect of cell structure and metabolism, other scientists considered proteins the prime candidates for the genetic material. However, protein did not turn out to be the molecule that governs inheritance. In this chapter we discuss how researchers discovered that **deoxyribonucleic acid (DNA),** a nucleic acid once thought unremarkable, is the molecular basis of inheritance. We explore the unique features of DNA, including its replication (see photograph), that enable it to carry out this role. ■

EVIDENCE OF DNA AS THE HEREDITARY MATERIAL

Learning Objective

1 Summarize the evidence that accumulated during the 1940s and early 1950s demonstrating that DNA is the genetic material.

PROCESS OF SCIENCE

During the 1930s and early 1940s, most geneticists paid little attention to DNA, convinced that the genetic material must be protein. Given the accumulating evidence that genes control production of proteins (discussed in Chapter 12), it certainly seemed likely that genes themselves must also be proteins. Scientists knew proteins consisted of more than 20 different kinds of amino acids in many different combinations, which conferred unique properties on each type of protein. Given their complexity and diversity compared with other molecules, proteins seemed to be the "stuff" of which genes are made.

In contrast, scientists had established that DNA and other nucleic acids were made of only four nucleotides, and what was known about their arrangement made them relatively uninteresting to most researchers. For this reason, several early clues to the role of DNA were not widely noticed.

DNA is the transforming principle in bacteria

One of these clues had its origin in 1928, when Frederick Griffith, a British medical officer, made a curious observation concerning two strains of pneumococcus bacteria (Fig. 11-1). A smooth (S) strain, named for its formation of smooth colonies on a solid growth medium, was known to exhibit **virulence,** the ability to cause disease, and often death, in its host. When living cells of this strain were injected into mice, the animals contracted pneumonia and died. Not surprisingly, the injected animals survived if the cells were first killed with heat. A related rough (R) strain of bacteria, which forms colonies with a rough surface, was known to exhibit **avirulence,** or inability to produce pathogenic effects; mice injected with either living or heat-killed cells of this strain survived. However, when Griffith injected mice with a mixture of *heat-killed* virulent S cells and live avirulent R cells, a high proportion of the mice died. Griffith then isolated living S cells from the dead mice.

Because neither the heat-killed S strain nor the living R strain could be converted to the living virulent form when injected by itself in the control experiments, it seemed that something in the heat-killed cells converted the avirulent cells to the lethal form. This type of permanent genetic change in which the properties of one strain of dead cells are conferred on a different strain of living cells became known as **transformation.** Scientists hypothesized that some chemical substance (the "transforming principle") was transferred from the dead bacteria to the living cells and caused transformation.

In 1944, American biologist Oswald T. Avery and his colleagues Colin M. MacLeod and Maclyn McCarty chemically identified Griffith's transforming principle as DNA. They did this through a series of careful experiments in which they lysed (split open) S cells and separated the cell contents into several fractions: lipids, proteins, polysaccharides, and nucleic acids (DNA and RNA). They tested each fraction to see if it could transform living R cells into S cells. The experiments using lipids, polysaccharides, and proteins did not cause transformation. However,

FIGURE **11-1** | **Griffith's transformation experiments.**

Although neither the rough (R) strain nor the heat-killed smooth (S) strain could kill a mouse, a combination of the two did. Autopsy of the dead mouse showed the presence of living S-strain pneumococci. These results indicated that some substance in the heat-killed S cells had transformed the living R cells into a virulent form. Avery and his colleagues later showed that purified DNA isolated from the S cells confers virulence on the R cells, establishing that DNA carries the necessary information for bacterial transformation.

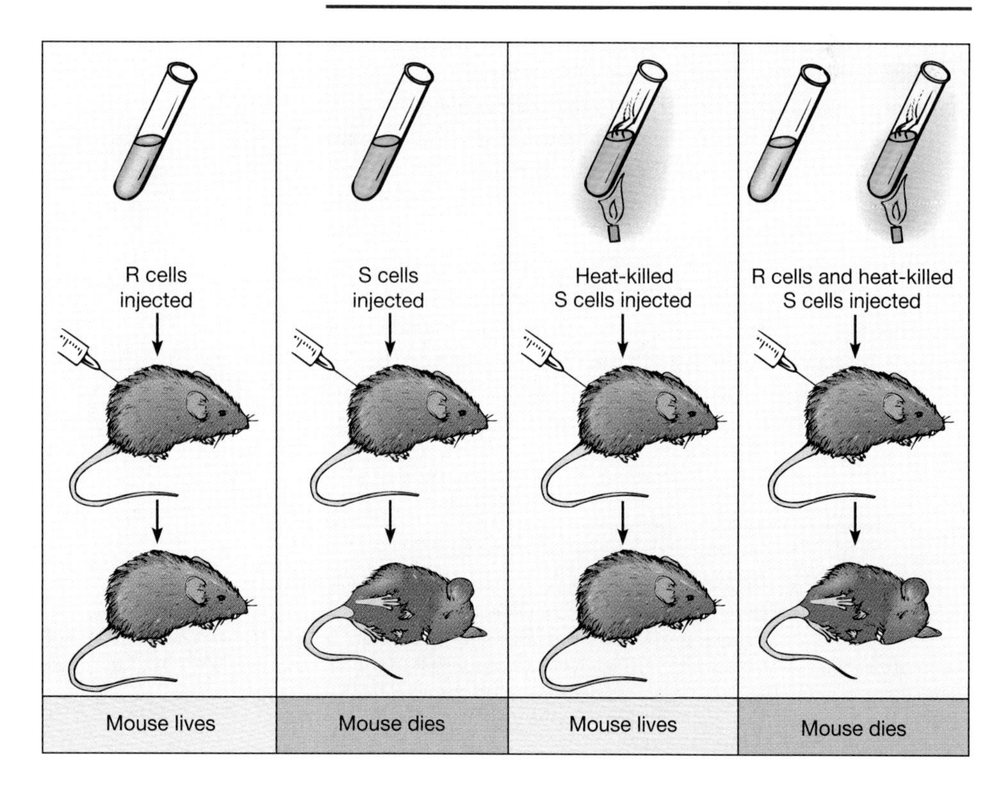

when Avery treated living R cells with nucleic acids extracted from S cells, the R cells were transformed into S cells.

Although today scientists consider their results to be the first definitive demonstration that DNA is the genetic material, not all scientists of the time were convinced. Many thought the findings might apply only to bacteria and might not have any relevance for the genetics of eukaryotes. During the next few years, new evidence accumulated that the haploid nuclei of pollen grains and gametes such as sperm contain only half the amount of DNA found in diploid somatic cells of the same species. (Somatic cells are body cells and never become gametes.) Because scientists generally accepted that genes are on chromosomes, these findings correlating DNA content with chromosome number provided strong circumstantial evidence for DNA's importance in eukaryotic inheritance.

DNA is the genetic material in certain viruses

In 1952, American geneticists Alfred Hershey and Martha Chase performed a series of elegant experiments (Fig. 11-2) on the reproduction of viruses that infect bacteria, known as **bacteriophages** (see Chapter 23). When they planned their experiments, they knew bacteriophages reproduce inside a bacterial cell, eventually causing the cell to break open, releasing large numbers of new viruses. Because electron microscopic studies had shown that only part of an infecting bacteriophage actually enters the cell, they reasoned the genetic material should be included in that portion.

They labeled the viral protein of one sample of bacteriophages with ^{35}S, a radioactive isotope of sulfur, and the viral DNA of a second sample with ^{32}P, a radioactive isotope of phosphorus. (Recall from Chapter 3 that proteins contain sulfur as part of the amino acids cysteine and methionine and that nucleic acids contain phosphate groups.) The bacteriophages in each sample attached to bacteria, and the researchers then shook them off by agitating them in a blender. Then they centrifuged the cells.

In the sample in which they had labeled the proteins with ^{35}S, they subsequently found the label in the supernatant, indicating that the protein had not entered the cells. In the sample in which they had labeled the DNA with ^{32}P, they found the label associated with the bacterial cells (in the pellet): DNA had actually entered the cells. Hershey and Chase concluded that bacteriophages inject their DNA into bacterial cells, leaving most of their protein on the outside. This finding emphasized the significance of DNA in viral reproduction, and many scientists saw it as an important demonstration of DNA's role as the hereditary material.

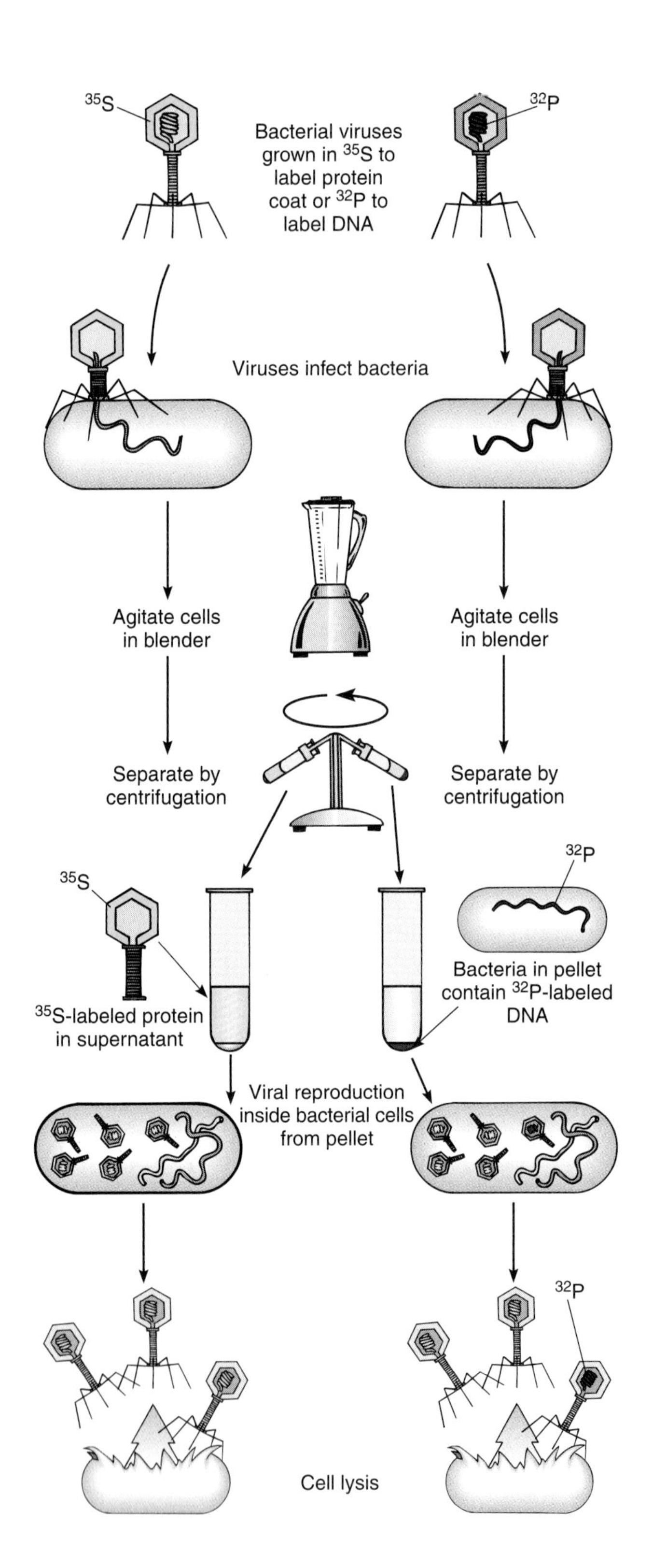

FIGURE **11-2** **The Hershey-Chase experiments.**

The researchers could separate bacteriophage protein coats labeled with the radioactive isotope ^{35}S *(left side)* from infected bacterial cells without affecting viral reproduction. However, they could not separate viral DNA labeled with the radioactive isotope ^{32}P *(right side)*, from infected bacterial cells. This demonstrated that viral DNA enters the bacterial cells and is required for the synthesis of new viral particles.

Review

- How did the experiments of Avery and his colleagues point to DNA as the essential genetic material?
- Did the Hershey-Chase experiment establish that DNA is the genetic material in all organisms? Explain your answer.

BiologyNow™ Assess your understanding of the **evidence of DNA as the hereditary material** by taking the pretest on your BiologyNow CD-ROM.

THE STRUCTURE OF DNA

Learning Objectives

2 Sketch how nucleotide subunits link together to form a single DNA strand.

3 Illustrate how the two strands of DNA are oriented with respect to each other.

4 State the base-pairing rules for DNA, as determined by Chargaff and coworkers, and describe how complementary bases bind to each other.

Scientists did not generally accept DNA as the genetic material until 1953, when American scientist James Watson and British scientist Francis Crick, both working in England, proposed a model for its structure that had extraordinary explanatory power. The story of how the structure of DNA was figured out is one of the most remarkable chapters in the history of modern biology (Table 11-1). As you will see in the following discussion, scientists already knew a great deal about DNA's physical and chemical properties when Watson and Crick became interested in the problem; in fact, they did not conduct any experiments or gather any new data. Their all-important contribution was to integrate all the available information into a model that demonstrated how the molecule can both carry information for making proteins and serve as its own **template** (pattern or guide) for its duplication.

Nucleotides can be covalently linked in any order to form long polymers

As discussed in Chapter 3, each DNA building block is a **nucleotide** consisting of the pentose sugar **deoxyribose,** a phosphate, and one of four nitrogenous bases (Fig. 11-3). It is conventional to number the atoms in a molecule using a system devised by organic chemists. Accordingly, in nucleic acid chemistry the individual carbons in each sugar and each base are numbered. The carbons in a base are designated by numerals, but the carbons in a sugar are distinguished from those in the base by prime symbols, such as 2′. The nitrogenous base is attached to the 1′ carbon of the sugar, and the phosphate is attached to the 5′ carbon. The bases include two **purines, adenine (A)** and **guanine (G),** and two **pyrimidines, thymine (T)** and **cytosine (C).**

The nucleotides are linked by covalent bonds to form an alternating sugar–phosphate backbone. The 3′ carbon of one sugar is bonded to the 5′ phosphate of the adjacent sugar to form a 3′,5′ **phosphodiester linkage.** A polymer of indefinite length forms, with the nucleotides linked in any order. Scientists now know that

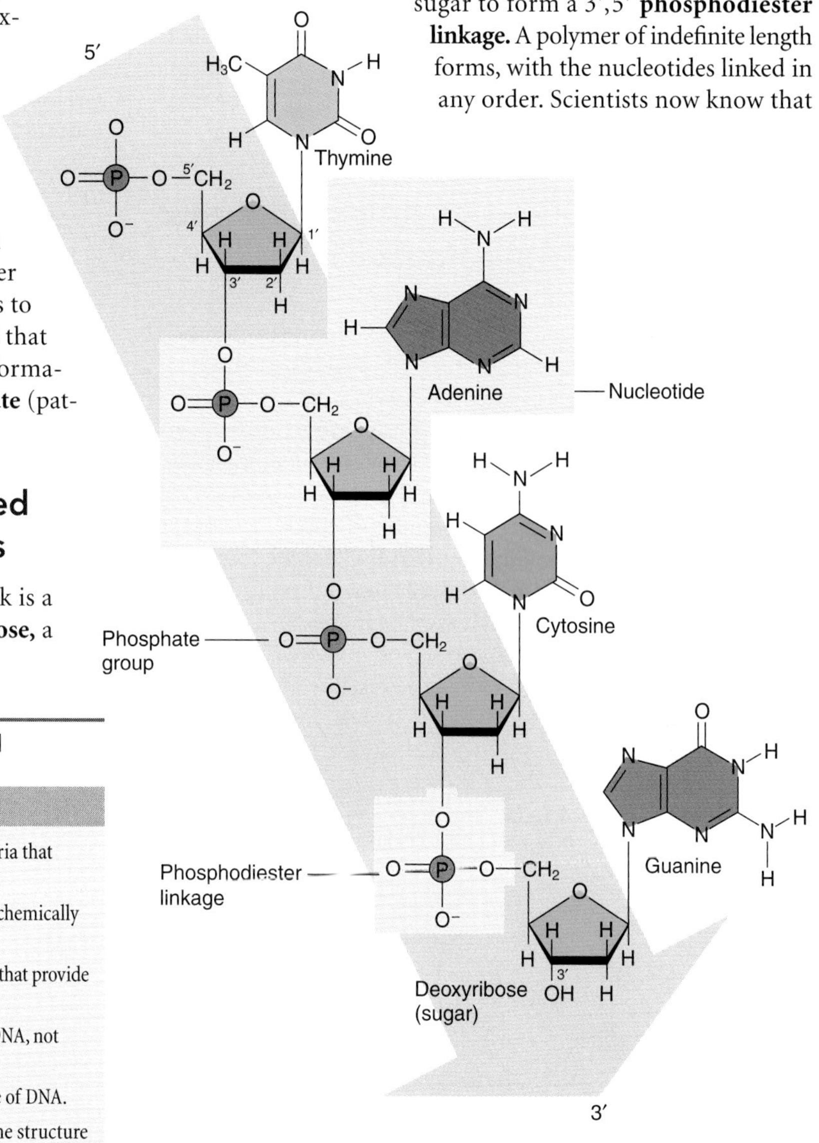

PROCESS OF SCIENCE

TABLE 11-1 A Timeline of Selected Historical DNA Discoveries

Date	Discovery
1928	**Federick Griffith** finds a substance in heat-killed bacteria that "transforms" living bacteria.
1944	**Oswald Avery, Colin MacLeod,** and **Maclyn McCarty** chemically identify Griffith's transforming principle as DNA.
1949	**Erwin Chargaff** reports relationships among DNA bases that provide a clue to the structure of DNA.
1952	**Alfred Hershey** and **Martha Chase** demonstrate that DNA, not protein, is involved in viral reproduction.
1952	**Rosalind Franklin** produces an x-ray diffraction image of DNA.
1953	**James Watson** and **Francis Crick** propose a model of the structure of DNA.
1958	**Matthew Meselson** and **Franklin Stahl** demonstrate that DNA replication is semiconservative.
1962	**James Watson, Francis Crick,** and **Maurice Wilkins** are awarded the Nobel Prize in Medicine for discoveries about the molecular structure of nucleic acids.
1969	**Alfred Hershey** is awarded the Nobel Prize in Medicine for discovering the replication mechanism and genetic structure of viruses.

FIGURE 11-3 The nucleotide subunits of DNA.

A single strand of DNA consists of a backbone made of phosphate groups alternating with the sugar deoxyribose *(green).* Phosphodiester linkages *(pink)* join sugars of adjacent nucleotides. Linked to the 1′ carbon of each sugar is one of four nitrogenous bases: adenine, guanine, thymine, and cytosine. Note the polarity of the polynucleotide chain, with the 5′ end at the top of the figure and the 3′ end at the bottom.

most DNA molecules found in cells are millions of bases long. Figure 11-3 also shows that a single polynucleotide chain is directional. No matter how long the chain may be, one end, the **5′ end,** has a 5′ carbon attached to a phosphate and the other, the **3′ end,** has a 3′ carbon attached to a hydroxyl group.

In 1949, Erwin Chargaff and his colleagues at Columbia University had determined the base composition of DNA from several organisms and tissues. They found a simple relationship among the bases that turned out to be an important clue to the structure of DNA. Regardless of the source of the DNA, in Chargaff's words the "ratios of purines to pyrimidines and also of adenine to thymine and of guanine to cytosine were not far from 1" (Table 11-2). In other words, in double-stranded DNA molecules, the number of purines equals the number of pyrimidines, the number of adenines equals the number of thymines (A equals T), and the number of guanines equals the number of cytosines (G equals C). These equalities became known as **Chargaff's rules.**

TABLE **11-2** Base Compositions in DNA from Selected Organisms

	Percentage of DNA Bases				Ratios	
Source of DNA	**A**	**T**	**G**	**C**	**A/T**	**G/C**
E. coli	26.1	23.9	24.9	25.1	1.09	0.99
Yeast	31.3	32.9	18.7	17.1	0.95	1.09
Sea urchin sperm	32.5	31.8	17.5	18.2	1.02	0.96
Herring sperm	27.8	27.5	22.2	22.6	1.01	0.98
Human liver	30.3	30.3	19.5	19.9	1.00	0.98
Corn (*zea mays*)	25.6	25.3	24.5	24.6	1.01	1.00

DNA is made of two polynucleotide chains intertwined to form a double helix

PROCESS OF SCIENCE

Key information about the structure of DNA came from x-ray diffraction studies on crystals of purified DNA, carried out by British scientist Rosalind Franklin from 1951 to 1953 in the laboratory of Maurice Wilkins in England. **X-ray diffraction,** a powerful method for elucidating the 3-D structure of a molecule, can determine the distances between the atoms of molecules arranged in a regular, repeating crystalline structure (Fig. 11-4a). X-rays have such short wavelengths that they can be scattered by the electrons surrounding the atoms in a molecule. Atoms with dense electron clouds (such as phosphorus and oxygen) tend to deflect electrons more strongly than atoms with lower atomic numbers. Exposing a crystal to an intense beam of x-rays causes the regular arrangement of its atoms to diffract, or scatter, the x-rays in specific ways. The pattern of diffracted x-rays appears on photographic film as dark spots. Mathematical analysis of the pattern and distances between the spots yields the precise distances between atoms and their orientation within the molecules.

Franklin had already produced x-ray crystallographic films of DNA patterns when Watson and Crick began to pursue the problem of DNA structure. Her pictures clearly showed that DNA has a type of helical structure, and three major types of regular, repeating patterns in the molecule (with the dimensions 0.34 nm, 3.4 nm, and 2.0 nm) were evident (Fig. 11-4b). Franklin and Wilkins had inferred from these patterns that the nucleotide bases (which are flat molecules) are stacked like the rungs of a ladder. Using this information, Watson and Crick began to build scale models of the DNA components and then fit them together to correlate with the experimental data.

After several trials, they worked out a model that fit the existing data (Fig. 11-5). The nucleotide chains conformed to the dimensions of the x-ray data only if each DNA molecule consisted of *two* polynucleotide chains arranged in a coiled **double helix.** In their model, the sugar–phosphate backbones of the two chains form the out-

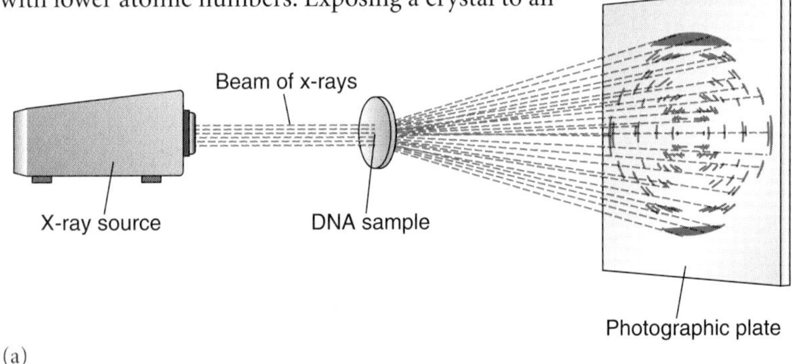

(a)

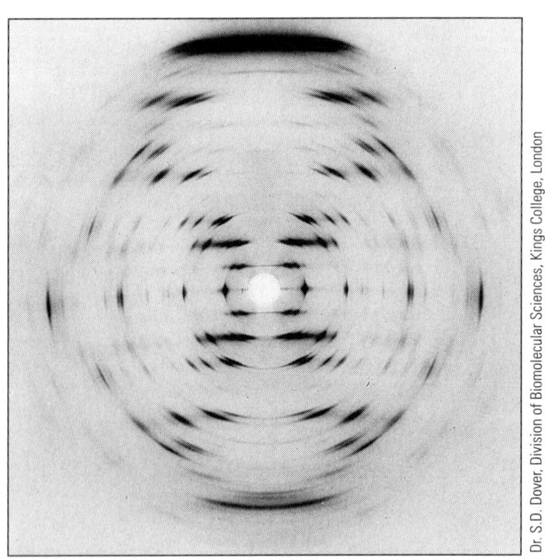

(b)

FIGURE **11-4** X-ray diffraction of DNA.

(a) The basic technique. Researchers direct a narrow beam of x-rays at a single crystal of DNA. Important clues about DNA structure are provided by detailed mathematical analysis of measurements of the spots, which are formed by x-rays hitting the photographic plate. **(b)** X-ray diffraction image of DNA, which Watson and Crick used. The diagonal pattern of spots stretching from 11 to 5 and from 1 to 7 (as on a clock face) provides evidence for the helical structure of DNA. The elongated horizontal patterns at the top and bottom indicate that the purine and pyrimidine bases are stacked 0.34 nm apart and are perpendicular to the axis of the DNA molecule.

side of the helix. The bases belonging to the two chains associate as pairs along the central axis of the helix. The reasons for the repeating patterns of 0.34-nm and 3.4-nm measurements are readily apparent from the model: Each pair of bases is exactly 0.34 nm from the adjacent pairs above and below. Because exactly 10 base pairs are present in each full turn of the helix, each turn constitutes 3.4 nm of length. To fit the data, the two chains must run in opposite directions; therefore, each end of the double helix must have an exposed 5′ phosphate on one strand and an exposed 3′ hydroxyl group (—OH) on the other. Because the two strands run in opposite directions, they are said to be **antiparallel** to each other (Fig. 11-6a).

In double-stranded DNA, hydrogen bonds form between A and T and between G and C

Other features of the Watson and Crick model integrated critical information about the chemical composition of DNA with the x-ray diffraction data. The x-ray diffraction studies indicated that the double helix has a precise and constant width, as shown by the 2.0-nm measurements. This finding is actually consistent with Chargaff's rules. As Figure 11-3 shows, each pyrimidine (cytosine or thymine) contains only one ring of atoms, whereas each purine (guanine or adenine) contains two rings. Study of the models made it clear to Watson and Crick that if each rung of the ladder contained one purine and one pyrimidine, the width of the helix at each base pair would be exactly 2.0 nm. By contrast, the combination of two purines (each of which is 1.2 nm wide) would be wider than 2.0 nm and that of two pyrimidines would be narrower, so the diameter would not be constant. Further examination of the model showed that adenine can pair with thymine (and guanine with cytosine) in such a way that hydrogen bonds form between them; the opposite combinations, cytosine with adenine and guanine with thymine, do not lead to favorable hydrogen bonding.

The nature of the hydrogen bonding between adenine and thymine and between guanine and cytosine is shown in Figure 11-6b. Two hydrogen bonds form between adenine and thymine, and three between guanine and cytosine. This concept of specific base pairing neatly explains Chargaff's rules. The amount of cytosine must equal the amount of guanine, because every cytosine in one chain must have a paired guanine in the other chain. Similarly, every adenine in the first chain must have a thymine in the second chain. The sequences of bases in the two chains are **complementary** to each other—that is, the sequence of nucleotides in one chain dictates the complementary sequence of nucleotides in the other. For example, if one strand has this sequence,

3′—AGCTAC—5′

then the other strand has the complementary sequence:

5′—TCGATG—3′

The double-helix model strongly suggested that the sequence of bases in DNA provides for the storage of genetic information and that this sequence ultimately relates to the sequences of amino acids in proteins. Although restrictions limit how the bases pair with each other, the number of possible sequences of bases in a strand is virtually unlimited. Because a DNA molecule

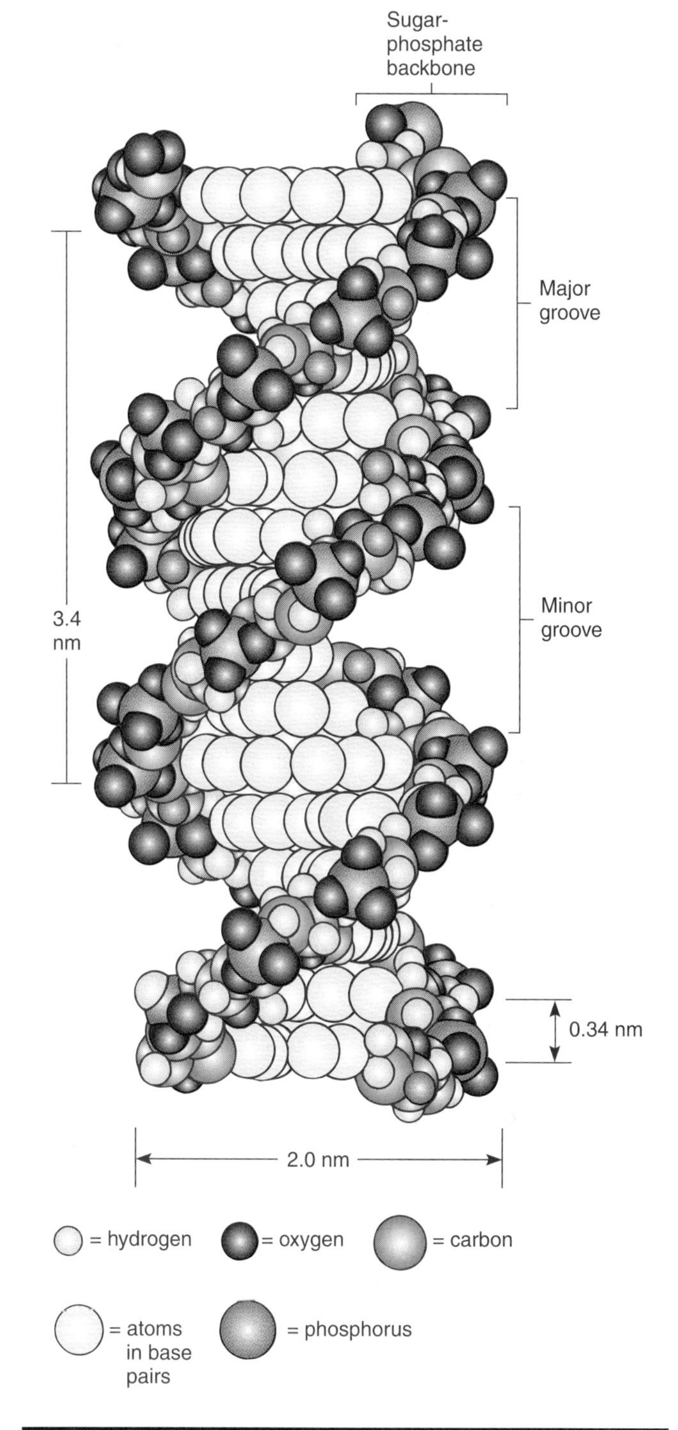

ACTIVE FIGURE **11-5** **A three-dimensional model of the DNA double helix.**

The measurements match those derived from x-ray diffraction images.

BiologyNow™ Interact with the structure of the **DNA double helix** by clicking on this figure on your BiologyNow CD-ROM.

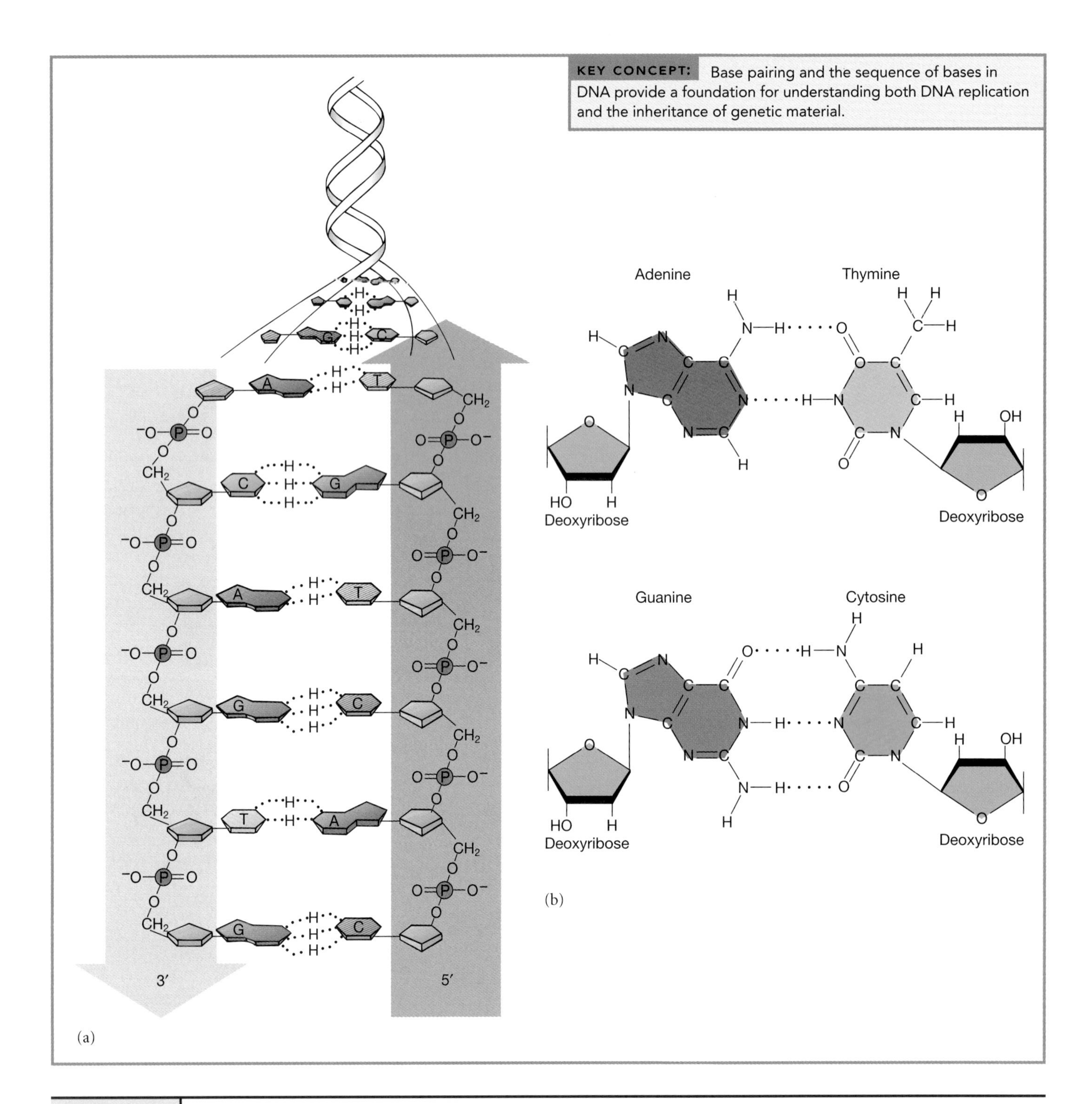

FIGURE **11-6** **Base pairing and hydrogen bonding.**

The two strands of a DNA double helix are hydrogen-bonded between the bases. **(a)** The two sugar–phosphate chains run in opposite directions. This orientation permits the complementary bases to pair. **(b)** Hydrogen bonding between base pairs adenine (A) and thymine (T) *(top)* and guanine (G) and cytosine (C) *(bottom)*. The AT pair has two hydrogen bonds; the GC pair has three.

in a cell can be millions of nucleotides long, it can store enormous amounts of information, usually consisting of hundreds of genes.

Review

- What types of subunits make up a single strand of DNA? How are the subunits linked?
- What is the structure of double-stranded DNA, as determined by Watson and Crick?
- Does a single strand of DNA follow Chargaff's rules? How do Chargaff's rules relate to the structure of DNA?

BiologyNow™ Assess your understanding of the **structure of DNA** by taking the pretest on your BiologyNow CD-ROM.

DNA REPLICATION

Learning Objectives

5 Cite experimental evidence that enabled scientists to differentiate between semiconservative replication of DNA and alternative models (conservative and dispersive replication).

6 Summarize how DNA replicates, and identify some unique features of the process.

7 Explain the complexities of DNA replication that make it (a) bidirectional and (b) continuous in one strand and discontinuous in the other.

8 Define *telomere;* describe the possible connections between telomerase and cell aging and between telomerase and cancer.

Two immediately apparent and distinctive features of the Watson-Crick model made it seem plausible that DNA is the genetic material. We have already mentioned that the sequence of bases in DNA can carry coded information. The model also suggested a way in which the sequence of nucleotides in DNA could be precisely copied, a process known as **DNA replication.** The connection between DNA replication and the behavior of chromosomes in mitosis was obvious to Watson and Crick. A chromosome becomes duplicated so that it consists of two identical sister chromatids that later separate at anaphase; the genetic material must be precisely duplicated and distributed to the daughter cells. They noted, in a classic and now famous understatement at the end of their first brief paper, "It has not escaped our notice that the specific pairing we have postulated immediately suggests a possible copying mechanism for the genetic material."

The model suggested that, because the nucleotides pair with each other in complementary fashion, each strand of the DNA molecule could serve as a template for synthesizing the opposite strand. It would simply be necessary for the hydrogen bonds between the two strands to break (recall that hydrogen bonds are relatively weak) and the two chains to separate. Each strand of the double helix could then pair with new complementary nucleotides to replace its missing partner. The result would be two DNA double helices, each identical to the original one and consisting of one original strand from the parent molecule and one newly synthesized complementary strand. This type of information copying is known as **semiconservative replication** (Fig. 11-7a).

Meselson and Stahl verified the mechanism of semiconservative replication

PROCESS OF SCIENCE

Although the semiconservative replication mechanism suggested by Watson and Crick was (and is) a simple and compelling model, experimental proof was needed to establish that DNA in fact duplicates in that manner. Researchers first needed to rule out other possibilities. With *conservative replication,* both parent (or old) strands would remain together, and the two newly synthesized strands would form a second double helix (Fig. 11-7b). As a third hypothesis, the parental and newly synthesized strands might become randomly mixed during the replication process; this possibility was known as *dispersive replication* (Fig. 11-7c). To discriminate among semiconservative replication and the other models, investigators had to distinguish between old and newly synthesized strands of DNA.

One technique is to use a heavy isotope of nitrogen, ^{15}N (ordinary nitrogen is ^{14}N), to label the bases of the DNA strands, making them more dense. Using **density gradient centrifugation,** scientists can separate large molecules such as DNA on the basis of differences in their density. When DNA is mixed with a solution containing cesium chloride (CsCl) and centrifuged at high speed, the solution forms a density gradient in the centrifuge tube, ranging from a region of lowest density at the top to one of highest density at the bottom. During centrifugation, the DNA molecules migrate to the region of the gradient identical to their own density.

In 1958, Matthew Meselson and Franklin Stahl at the California Institute of Technology grew the bacterium *Escherichia coli* on a medium that contained ^{15}N in the form of ammonium chloride (NH_4Cl). The cells used the ^{15}N to synthesize bases, which then became incorporated into DNA (Fig. 11-7d). The resulting DNA molecules, which contained heavy nitrogen, were extracted from some of the cells. When the researchers subjected them to density gradient centrifugation, they accumulated in the high-density region of the gradient. The team transferred the rest of the bacteria (which also contained ^{15}N-labeled DNA) to a different growth medium in which the NH_4Cl contained the naturally abundant, lighter ^{14}N isotope and then allowed them to undergo additional cell divisions.

Meselson and Stahl expected the newly synthesized DNA strands to be less dense, because they incorporated bases containing the lighter ^{14}N isotope. Indeed, double-stranded DNA from cells isolated after one generation had an intermediate density, indicating they contained half as many ^{15}N atoms as the "parent" DNA. This finding supported the semiconservative replication model, which predicted that each double helix would contain a previously synthesized strand (heavy in this case) and a newly synthesized strand (light in this case). It was also consistent with the dispersive model, which would also yield one class of molecules, all with intermediate density. It was inconsistent with the conservative model, which predicted two classes of double-stranded molecules, those with two heavy strands and those with two light strands.

After another cycle of cell division in the medium with the lighter ^{14}N isotope, two types of DNA appeared in the density gradient, exactly as predicted by the semiconservative replication model. One consisted of DNA with a density intermediate between ^{15}N-labeled DNA and ^{14}N-labeled DNA, whereas the other contained only DNA with a density of ^{14}N-labeled DNA. This finding refuted the dispersive model, which predicted that all strands would have an intermediate density.

Semiconservative replication explains the perpetuation of mutations

The recognition that DNA could be copied by a semiconservative mechanism suggested how DNA could explain a third

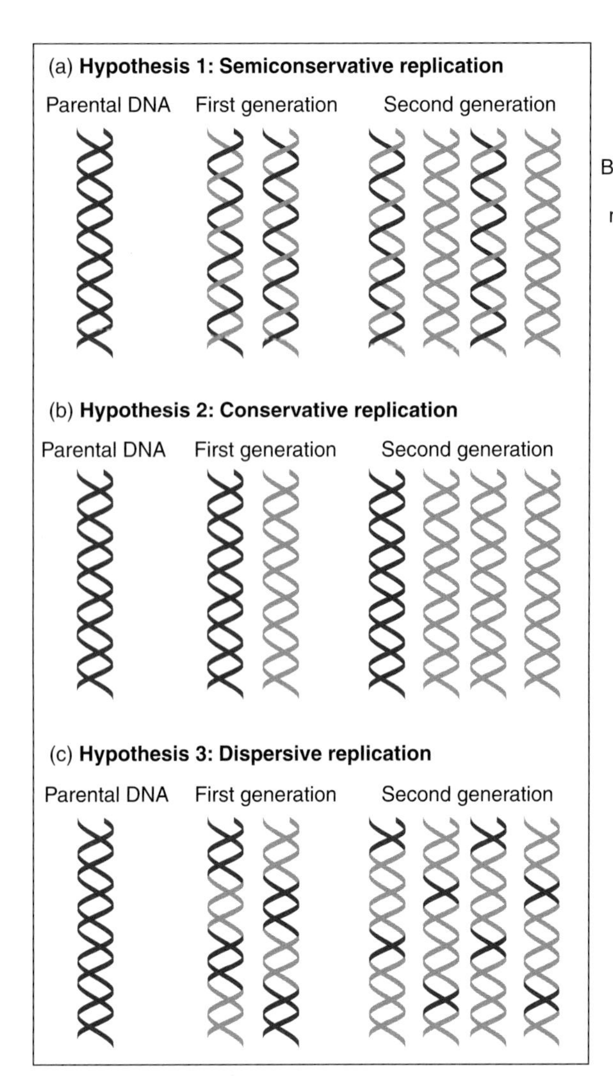
(a) Hypothesis 1: Semiconservative replication
Parental DNA
First generation
Second generation
(b) Hypothesis 2: Conservative replication
Parental DNA
First generation
Second generation
(c) Hypothesis 3: Dispersive replication
Parental DNA
First generation
Second generation

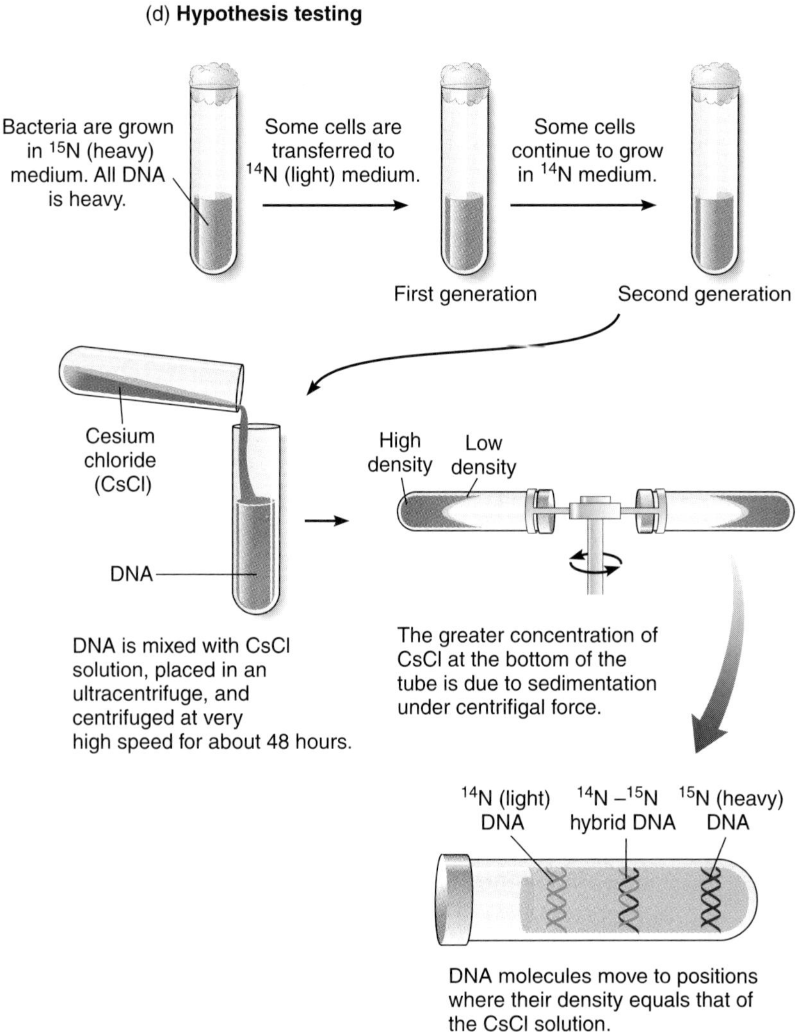
(d) Hypothesis testing
Bacteria are grown in ^{15}N (heavy) medium. All DNA is heavy.
Some cells are transferred to ^{14}N (light) medium.
Some cells continue to grow in ^{14}N medium.
First generation
Second generation
Cesium chloride (CsCl)
DNA
High density
Low density
DNA is mixed with CsCl solution, placed in an ultracentrifuge, and centrifuged at very high speed for about 48 hours.
The greater concentration of CsCl at the bottom of the tube is due to sedimentation under centrifigal force.
^{14}N (light) DNA
$^{14}N-^{15}N$ hybrid DNA
^{15}N (heavy) DNA
DNA molecules move to positions where their density equals that of the CsCl solution.

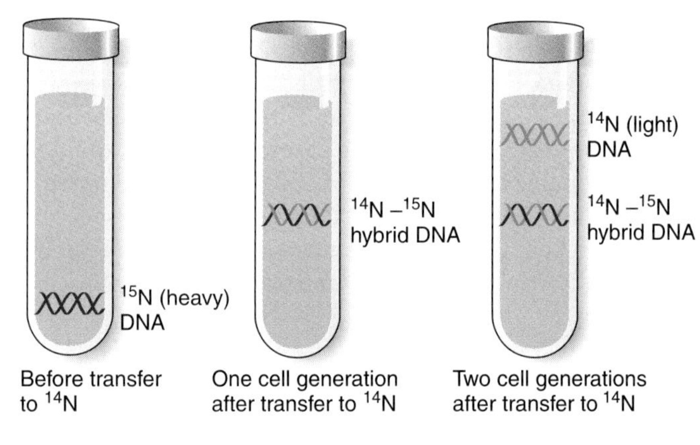
(e) Results
^{15}N (heavy) DNA
$^{14}N-^{15}N$ hybrid DNA
^{14}N (light) DNA
$^{14}N-^{15}N$ hybrid DNA
Before transfer to ^{14}N
One cell generation after transfer to ^{14}N
Two cell generations after transfer to ^{14}N
The location of DNA molecules within the centrifuge tube can be determined by UV optics. DNA solutions absorb strongly at 260 nm.

FIGURE **11-7** **Replication models and the Meselson-Stahl experiment.**

The hypothesized arrangement of old *(dark blue)* and newly synthesized *(light blue)* DNA strands after one and two generations, according to **(a)** the semiconservative model, **(b)** the conservative model, and **(c)** the dispersive model. **(d)** Meselson and Stahl grew bacteria (*E. coli*) in heavy nitrogen (^{15}N) growth medium for many generations. They then transferred some of the cells to light nitrogen (^{14}N) medium. They isolated DNA from bacterial cells one and two generations after transfer to the ^{14}N medium. **(e)** The density of the DNA molecules in each generation matched the pattern expected for semiconservative replication.

essential characteristic of genetic material—the ability to mutate. It was long known that **mutations,** or genetic changes, could arise in genes and then be transmitted faithfully to succeeding generations. For example, a mutation in the fruit fly (*Drosophila melanogaster*) produces vestigial wings (see Fig. 10-11).

When the double-helix model was proposed, it seemed plausible that mutations could represent a change in the sequence of bases in the DNA. You could predict that if DNA is copied by a mechanism involving complementary base pairing, any change in the sequence of bases on one strand would produce a new sequence of complementary bases during the next replication cycle. The new base sequence would then transfer to daughter molecules by the same mechanism used to copy the original genetic material, as if no change had occurred.

For the example in Figure 11-8, an adenine base in one of the DNA strands has been changed to guanine. This could occur by a rare error in DNA replication or by one of several other known mechanisms. Certain enzymes correct errors when they occur, but not all errors are corrected properly. By one estimate, the rate of uncorrected errors that occurs during DNA replication is equal to about one nucleotide in a billion. When the DNA molecule containing an error replicates (left side of Fig. 11-8), one of the strands gives rise to a molecule exactly like its parent strand; the other (mutated) strand gives rise to a molecule with a new combination of bases that is transmitted generation after generation.

DNA replication is a complex process that requires protein "machinery"

Although semiconservative replication by base pairing appears simple and straightforward, the actual process is highly regulated and requires a "replication machine" containing many protein and enzyme molecules. Many essential features of DNA replication are common to all organisms, although prokaryotes and eukaryotes differ somewhat because their DNA is organized differently. In most bacterial cells, such as *E. coli,* most or all of the DNA is in the form of a single, *circular,* double-stranded DNA molecule. In contrast, each unreplicated eukaryotic chromosome contains a single, *linear,* double-stranded molecule associated with at least as much protein (by mass) as DNA.

We now present DNA replication, based on the current understanding of the process. Figure 11-9 shows a simplified overview, and Figures 11-10 and 11-11 outline the steps. Although scientists know much about DNA replication, many aspects of the process remain unclear. For example, in the unicellular yeast *Saccharomyces cerevisiae,* which is considered a relatively "simple" eukaryote, 88 genes are involved in DNA replication! Determining the roles and interactions of all those genes in DNA replication will require the efforts of many scientists over a long time.

DNA strands must be unwound during replication

Watson and Crick recognized that in their double-helix model, the two DNA strands wrap around each other like the strands of a rope. If you tried to pull the strands apart, the rope must either rotate or twist into tighter coils. You could expect similar results when complementary DNA strands are separated for replication. **DNA helicases** are enzymes that separate the two strands of DNA by traveling along the helix, opening the double helix like a zipper as they move (Table 11-3). Once the strands are separated, **single-strand binding proteins (SSBs),** also called *helix-destabilizing proteins,* bind to single DNA strands and stabilize them; this prevents the double helix from reforming until the strands are copied. As the DNA strands unwind to open for replication, excessive twisting occurs in another part of the DNA molecule. Enzymes called **topoisomerases** produce breaks in the DNA molecules and then rejoin the strands, relieving strain and effectively preventing excessive coiling and knot formation during replication.

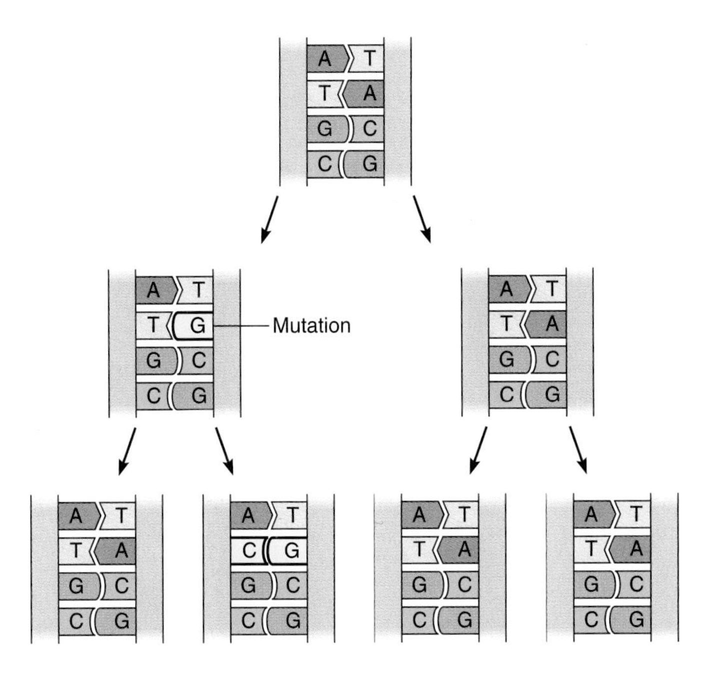

FIGURE **11-8** | **The perpetuation of a mutation.**
The process of DNA replication can stabilize a mutation *(bright yellow)* so that it is transmitted to future generations.

DNA synthesis always proceeds in a 5′ ⟶ 3′ direction

The enzymes that catalyze the linking together of the nucleotide subunits are called **DNA polymerases.** They add nucleotides only to the 3′ end of a growing polynucleotide strand, and this strand must be paired with the strand being copied (see Fig. 11-9). Nucleotides with three phosphate groups, as in ATP and GTP, are substrates for the polymerization reaction. As the nucleotides link together, two of the phosphates are removed. These reactions are strongly exergonic and do not need additional energy. Because the polynucleotide chain is elongated by the linkage of the 5′ phosphate group of the next nucleotide subunit to the 3′ hydroxyl group of the sugar at the end of the pre-existing strand, the new strand of DNA always grows in the 5′ ⟶ 3′ direction.

TABLE **11-3** **Enzymes Involved in DNA Replication**

Enzyme	Function
DNA helicases	Open the double helix by disrupting the hydrogen bonds that hold the two strands together.
Topoisomerases	Break one or both DNA strands, preventing excessive coiling during replication, and rejoin them.
DNA polymerases	Link nucleotide subunits together.
DNA primase	Synthesizes short RNA primers on the lagging strand. Begins replication of the leading strand.
DNA ligase	Links Okazaki fragments by joining the 3′ end of the new DNA fragment to the 5′ end of the adjoining DNA.
Telomerase	Lengthens telomeric DNA.

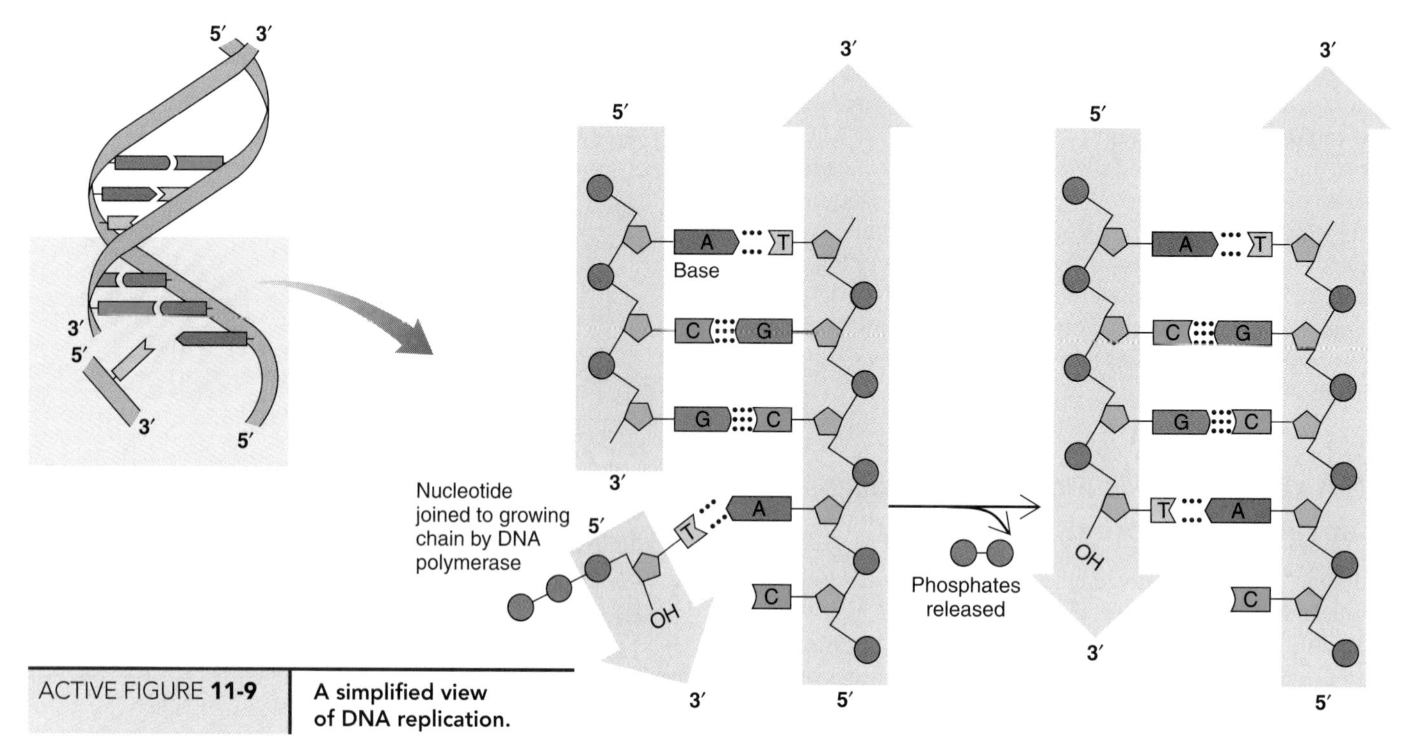

ACTIVE FIGURE **11-9** **A simplified view of DNA replication.**

DNA polymerase adds one nucleotide at a time to the 3′ end of a growing chain.

Biology Now™ Watch the **process of replication** by clicking on this figure on your BiologyNow CD-ROM.

FIGURE **11-10** **An overview of DNA replication.**

This process requires several steps involving several enzymes, proteins, and RNA primers.

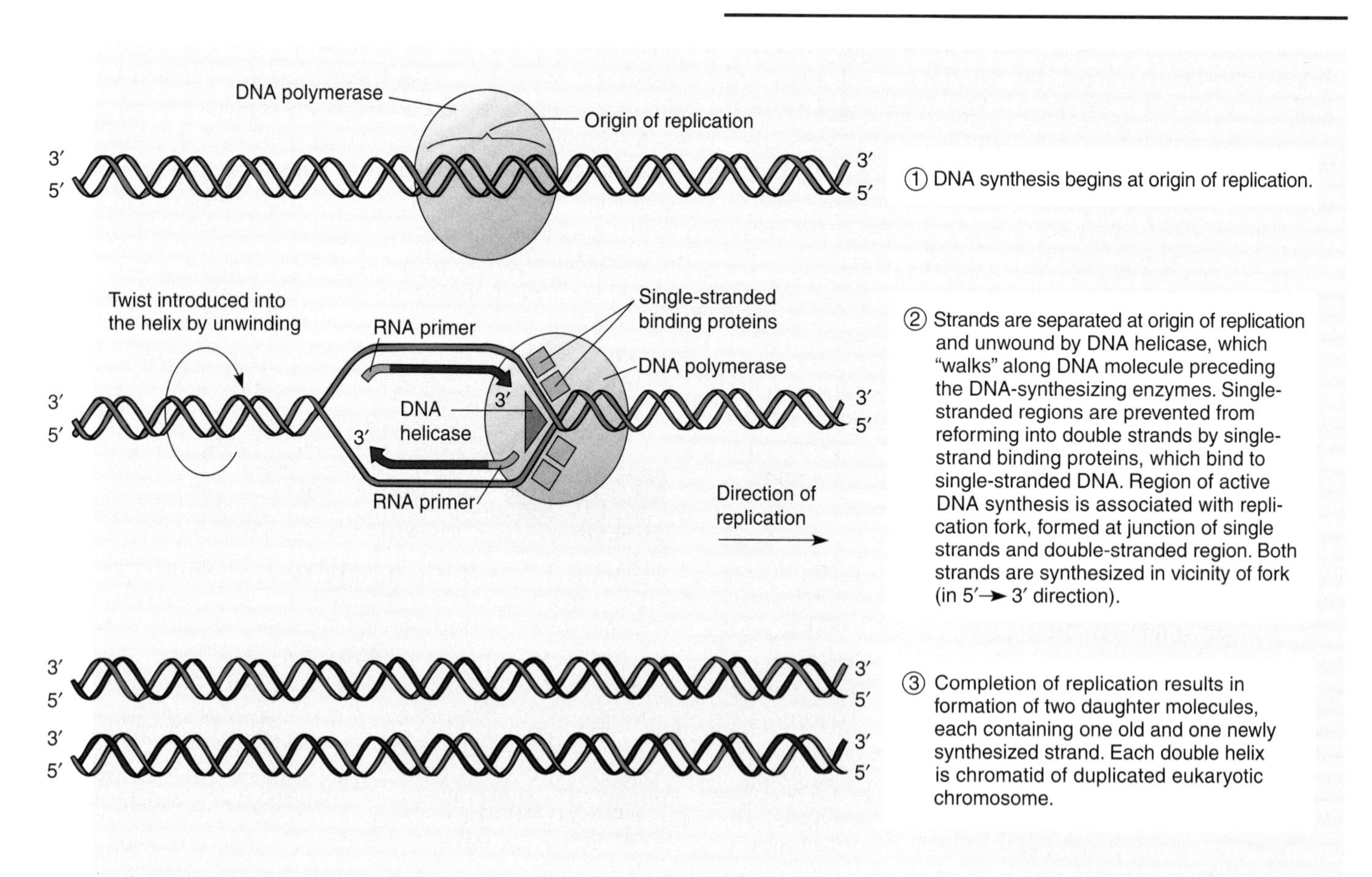

DNA polymerase
Origin of replication
3′
5′
① DNA synthesis begins at origin of replication.
Twist introduced into the helix by unwinding
RNA primer
Single-stranded binding proteins
DNA polymerase
DNA helicase
RNA primer
Direction of replication
② Strands are separated at origin of replication and unwound by DNA helicase, which "walks" along DNA molecule preceding the DNA-synthesizing enzymes. Single-stranded regions are prevented from reforming into double strands by single-strand binding proteins, which bind to single-stranded DNA. Region of active DNA synthesis is associated with replication fork, formed at junction of single strands and double-stranded region. Both strands are synthesized in vicinity of fork (in 5′→ 3′ direction).
③ Completion of replication results in formation of two daughter molecules, each containing one old and one newly synthesized strand. Each double helix is chromatid of duplicated eukaryotic chromosome.

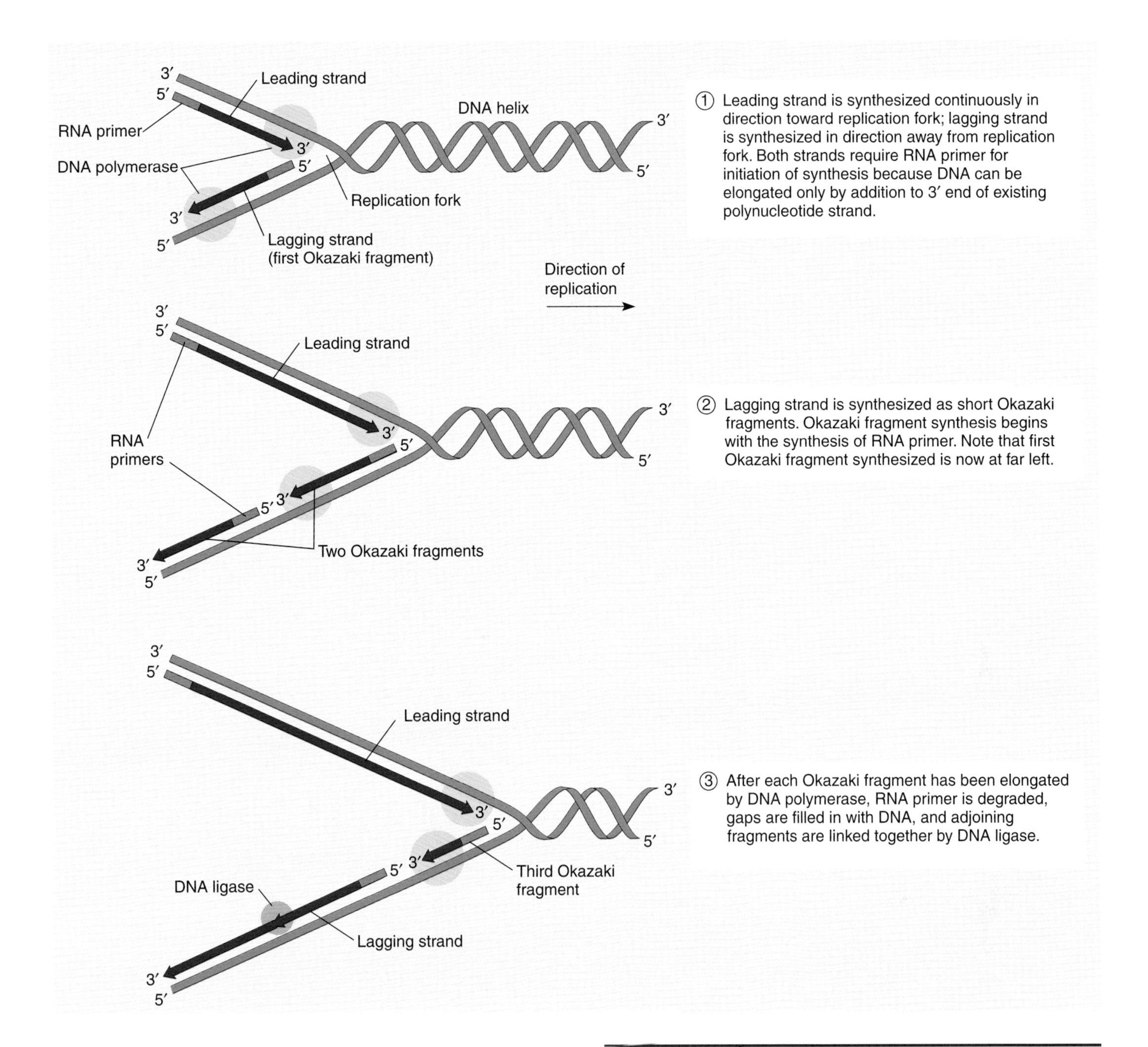
3′
5′
Leading strand
RNA primer
DNA polymerase
DNA helix
3′
5′
3′
5′
Replication fork
3′
5′
Lagging strand
(first Okazaki fragment)
Direction of
replication
① Leading strand is synthesized continuously in direction toward replication fork; lagging strand is synthesized in direction away from replication fork. Both strands require RNA primer for initiation of synthesis because DNA can be elongated only by addition to 3′ end of existing polynucleotide strand.
3′
5′
Leading strand
RNA
primers
3′
5′
5′ 3′
Two Okazaki fragments
3′
5′
3′
5′
② Lagging strand is synthesized as short Okazaki fragments. Okazaki fragment synthesis begins with the synthesis of RNA primer. Note that first Okazaki fragment synthesized is now at far left.
3′
5′
Leading strand
3′
5′
5′ 3′
Third Okazaki
fragment
DNA ligase
Lagging strand
3′
5′
3′
5′
③ After each Okazaki fragment has been elongated by DNA polymerase, RNA primer is degraded, gaps are filled in with DNA, and adjoining fragments are linked together by DNA ligase.

FIGURE **11-11** **Leading and lagging DNA strands.**

Because elongation can proceed only in the $5' \longrightarrow 3'$ direction, the two strands at the replication fork are copied in different ways, each by a separate DNA polymerase molecule.

DNA synthesis requires an RNA primer

As mentioned, DNA polymerases add nucleotides only to the 3′ end of an *existing* polynucleotide strand. Then how can DNA synthesis be initiated once the two strands are separated? The answer is that first a short piece of RNA (5 to 14 nucleotides) called an **RNA primer** is synthesized at the point where replication begins (Fig. 11-10).

RNA, or **ribonucleic acid** (see Chapters 3 and 12), is a nucleic acid polymer consisting of nucleotide subunits that can associate by complementary base pairing with the single-stranded DNA template. The RNA primer is synthesized by **DNA primase,** an enzyme that starts a new strand of RNA opposite a DNA strand. After a few nucleotides have been added, DNA polymerase displaces the primase and subsequently adds subunits to the 3′ end of the short RNA primer. Specific enzymes later degrade the primer, and the space fills in with DNA.

DNA replication is discontinuous in one strand and continuous in the other

We mentioned earlier that the complementary DNA strands are antiparallel. DNA synthesis proceeds only in the $5' \longrightarrow 3'$ direction, which means that the strand being copied is being read in the $3' \longrightarrow 5'$ direction. Thus it may seem necessary to copy one of the strands starting at one end of the double helix and the other strand starting at the opposite end. Some viruses replicate their DNA in this way, but this replication method is not workable in the extremely long DNA molecules in eukaryotic chromosomes.

Instead, DNA replication begins at specific sites on the DNA molecule, called **origins of replication,** and both strands replicate at the same time at a Y-shaped structure called the **replication fork** (see Fig. 11-11 and chapter introduction photograph). The position of the replication fork is constantly moving as replication proceeds. Two identical DNA polymerase molecules catalyze replication. One of these adds nucleotides to the 3′ end of the new strand that is always growing *toward* the replication fork. Because this strand can be formed smoothly and continuously, it is called the **leading strand.**

A separate (but identical) DNA polymerase molecule adds nucleotides to the 3′ end of the other new strand. Called the **lagging strand,** it is always growing *away* from the replication fork. Only short pieces can be synthesized, because if the DNA polymerase were to add continuously to the 3′ end of that strand, it would need to move far away from the replication fork. These 100- to 2000-nucleotide pieces are called **Okazaki fragments** after their discoverer, Japanese molecular biologist Reiji Okazaki.

A separate RNA primer initiates each Okazaki fragment, which DNA polymerase then extends toward the 5′ end of the previously synthesized fragment. When the RNA primer of the previously synthesized fragment is reached, DNA polymerase degrades and replaces the primer with DNA. The fragments are then joined together by **DNA ligase,** an enzyme that links the 3′ hydroxyl of one Okazaki fragment to the 5′ phosphate of the DNA immediately next to it, forming a phosphodiester linkage.

DNA synthesis is bidirectional

When double-stranded DNA separates, two replication forks form, and the molecule replicates in both directions from the origin of replication. In bacteria, each circular DNA molecule usually has only one origin of replication, so the two replication forks proceed around the circle and eventually meet at the other side to complete the formation of two new DNA molecules (Fig. 11-12a).

A eukaryotic chromosome consists of one long, linear DNA molecule, so having multiple origins of replication expedites the replication process (Fig. 11-12b,c). Synthesis continues at each replication fork until it meets a newly synthesized strand coming from the opposite direction. This results in a chromosome containing two DNA double helices, each corresponding to a chromatid.

Telomeres cap eukaryotic chromosome ends

Unlike bacterial DNA, which is circular, eukaryotic chromosomes have free ends. Because DNA replication is discontinu-

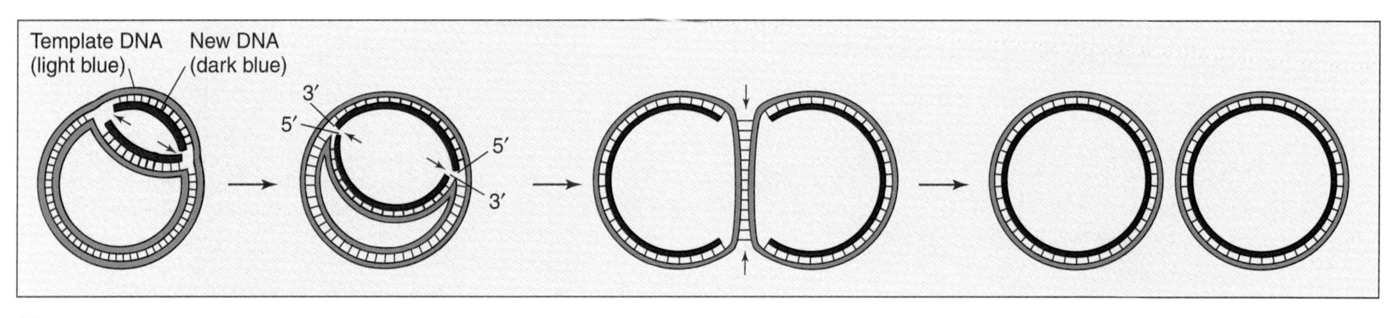

(a)

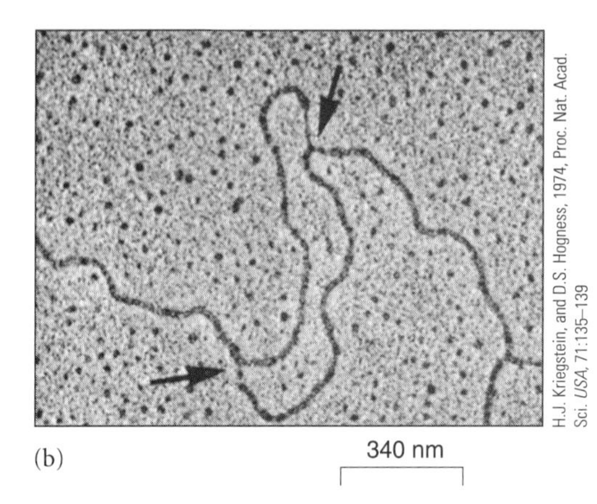

(b)

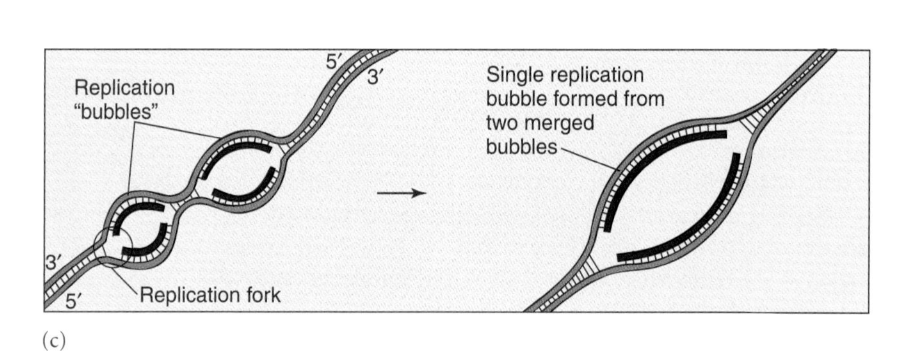

(c)

FIGURE **11-12** **Bidirectional DNA replication in bacteria and eukaryotes.**

The illustrations do not show leading strands and lagging strands. **(a)** The circular DNA in the prokaryote *E. coli* has only one origin of replication. Because DNA synthesis proceeds from that point in both directions, two replication forks form *(small black arrows),* travel around the circle, and eventually meet. **(b)** This TEM shows two replication forks *(arrows)* in a segment of a eukaryotic chromosome that has partly replicated. **(c)** Eukaryotic chromosomal DNA contains multiple origins of replication. DNA synthesis proceeds in both directions from each origin until adjacent replication "bubbles" eventually merge.

ous in the lagging strand, DNA polymerases do not complete replication of the strand neatly. At the end of the DNA, a small portion is left unreplicated, and a small, single-stranded segment of the DNA is lost with each cell cycle (Fig. 11-13a). The important genetic information is retained because chromosomes have end caps known as **telomeres** that do not contain protein-coding genes. Telomeres consist of short, simple, noncoding DNA sequences that repeat many times (Fig. 11-13b). Therefore, although a small amount of telomeric DNA fails to replicate each time the cell divides, a cell can divide many times before it starts losing essential genetic information.

Telomerase, a special DNA replication enzyme, can lengthen telomeric DNA. This enzyme, which researchers discovered in 1984, is typically present in cells that divide an unlimited number of times, including protozoa and other unicellular eukaryotes, and most types of **cancer cells,** which proliferate in a rapid and uncontrolled manner. In humans and other mammals, active telomerase is usually present in germ line cells (the cells that give rise to eggs and sperm) but not in normal somatic cells.

Research evidence suggests that the shortening of telomeres may contribute to **cell aging** and *apoptosis,* which is programmed cell death. Scientists have analyzed cell aging since the 1960s, following the pioneering studies of American biologist Leonard Hayflick, who showed that when normal somatic cells of the human body are grown in culture, they lose their ability to divide after a limited number of cell divisions. Furthermore, the number of cell divisions is determined by the age of the individual from whom the cells were taken. Cells from a 70-year-old divide only 20 to 30 times, compared with those from an infant, which divide 80 to 90 times.

PROCESS OF SCIENCE

Many investigators have observed correlations between telomerase activity and the ability of cells to undergo unlimited divisions without showing signs of the aging process. However, they couldn't find evidence of a causal relationship until the early 2000s, when scientists at the Geron Corporation and the University of Texas Southwestern Medical Center conducted a direct test. Using recombinant DNA technology (see Chapter 14), they infected cultured normal human cells with a virus that carried the genetic information coding for the catalytic subunit of telomerase. The infected cells not only produced active telomerase, which elongated the telomeres significantly, but also continued to divide long past the point at which cell divisions would normally cease.

Telomeres and telomerase are an active focus of current research, for both theoretical and practical reasons. The ability to give somatic cells telomerase so they can divide many times more than they ordinarily would, has many potential therapeutic applications, especially if lost or injured cells must be replaced. However, giving such cells the property of unlimited cell division also has potentially serious consequences. For example, if transplanted into the body, cells with active telomerase might behave like cancer cells.

Cancer cells have the ability to divide hundreds of times in culture; in fact, they are virtually immortal. Most cancer cells, including human cancers of the breast, lung, colon, prostate gland, and pancreas, have telomerase to maintain telomere length and possibly to resist apoptosis. Research is underway to develop anticancer drugs that inhibit telomerase activity. An alternative treatment for cancer is also under study—directing the body's immune system to recognize and attack cancer cells that contain telomerase.

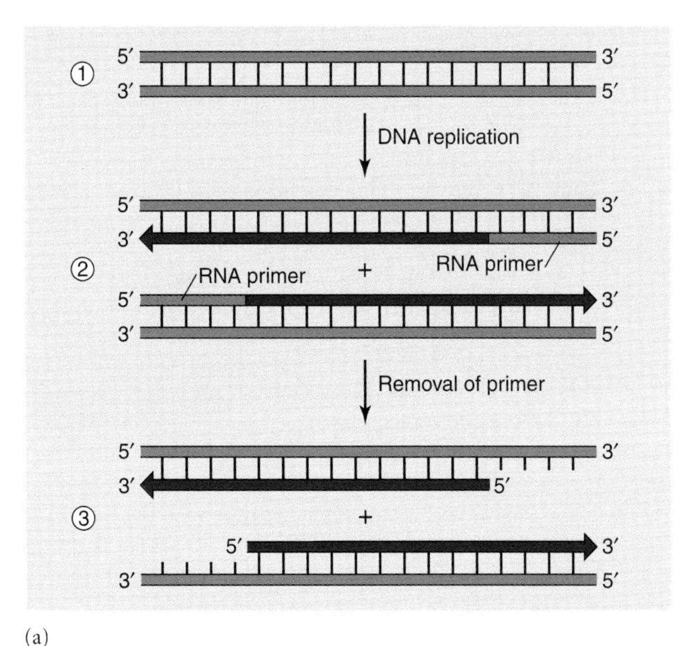

(a)

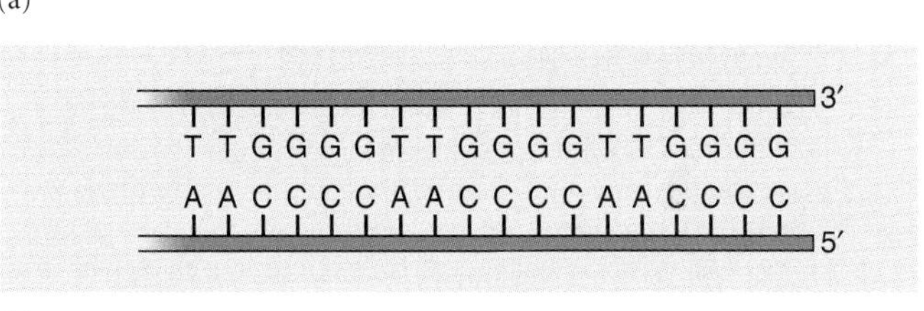

(b)

FIGURE 11-13 **Replication at chromosome ends.**

(a) Why one strand of DNA is shorter than that of the preceding generation. ① This is the unwound DNA of a eukaryotic chromosome before replication. ② When DNA replication occurs, the 5′ end of each newly synthesized strand contains a short segment of RNA that functioned as a primer on the lagging strand. ③ Removal of the 5′ RNA primer results in DNA molecules with shorter 5′ ends than in the original chromosome because there is no way to prime the last section at the 5′ end. **(b)** The telomeres of chromosomes have short noncoding DNA sequences that repeat many times. Shown is TTGGGG, which is the segment repeated some 20 to 70 times in telomeres of the protozoon *Tetrahymena,* the unicellular organism in which telomerase was first discovered.

Review

- How did Meselson and Stahl verify that DNA replication is semiconservative?
- What are some of the complexities of DNA replication?
- Why is DNA replication continuous for one strand but discontinuous for the other?

BiologyNow™ Assess your understanding of **DNA replication** by taking the pretest on your BiologyNow CD-ROM.

SUMMARY WITH KEY TERMS

1 Summarize the evidence that accumulated during the 1940s and early 1950s demonstrating that DNA is the genetic material.

- Many early geneticists thought genes were made of proteins. They knew proteins are complex and variable, whereas they thought of nucleic acids as simple molecules with a limited ability to store information.
- Several lines of evidence supported the idea that **DNA (deoxyribonucleic acid)** is the genetic material. In **transformation** experiments, the DNA of one strain of bacteria can endow related bacteria with new genetic characteristics.
- When a bacterial cell becomes infected with a **bacteriophage** (virus), only the DNA from the virus enters the cell; this DNA is sufficient for the virus to reproduce and form new viral particles.
- Watson and Crick's model of the structure of DNA demonstrated how information can be stored in the molecule's structure and how DNA molecules can serve as **templates** for their own replication.

2 Sketch how nucleotide subunits link together to form a single DNA strand.

- DNA is a very regular polymer of **nucleotides.** Each nucleotide subunit contains a nitrogenous base, which may be one of the **purines (adenine** or **guanine)** or one of the **pyrimidines** (**thymine** or **cytosine**). Each base covalently links to a pentose sugar, **deoxyribose,** which covalently bonds to a phosphate group.
- The backbone of each single DNA chain is formed by alternating sugar and phosphate groups, joined by covalent **phosphodiester linkages.** Each phosphate group attaches to the 5′ carbon of one deoxyribose and to the 3′ carbon of the neighboring deoxyribose.

3 Illustrate how the two strands of DNA are oriented with respect to each other.

- Each DNA molecule consists of two polynucleotide chains that associate as a **double helix.** The two chains are **antiparallel** (running in opposite directions); at each end of the DNA molecule one chain has a phosphate attached to a 5′ deoxyribose carbon, the **5′ end,** and the other has a hydroxyl group attached to a 3′ deoxyribose carbon, the **3′ end.**

4 State the base-pairing rules for DNA, as determined by Chargaff and coworkers, and describe how complementary bases bind to each other.

- Hydrogen bonding between specific base pairs holds together the two chains of the helix. Adenine (A) forms two hydrogen bonds with thymine (T); guanine (G) forms three hydrogen bonds with cytosine (C).
- **Complementary** base pairing between A and T and between G and C is the basis of Chargaff's rules, which state that A equals T and G equals C.
- Because complementary base pairing holds together the two strands of DNA, it is possible to predict the base sequence of one strand if you know the base sequence of the other strand.

5 Cite experimental evidence that enabled scientists to differentiate between semiconservative replication of DNA and alternative models (conservative and dispersive replication).

- When *E. coli* cells are grown for many generations in a medium containing heavy nitrogen (^{15}N), they incorporate the ^{15}N into their DNA. When researchers transfer cells from a ^{15}N medium to a ^{14}N medium and isolate them after either one or two generations, the density of the DNA in each group is what would be expected if DNA replication were semiconservative. In **semiconservative replication,** each daughter double helix consists of an original strand from the parent molecule and a newly synthesized complementary strand.

6 Summarize how DNA replicates, and identify some unique features of the process.

- During **DNA replication,** the two strands of the double helix unwind. Each strand serves as a template for forming a new, complementary strand. The enzyme that adds new nucleotide subunits to a growing DNA strand is **DNA polymerase.**
- Additional enzymes and other proteins are required to unwind and stabilize the separated DNA helix. **DNA helicases** open the double helix and **topoisomerases** prevent tangling and knotting. **DNA primase** synthesizes short **RNA primers** on the lagging strand, and **DNA ligase** links together **Okazaki fragments** of newly synthesized DNA.

7 Explain the complexities of DNA replication that cause it to be (a) bidirectional and (b) discontinuous in one strand and continuous in the other.

- DNA replication is bidirectional, starting at the **origin of replication** and proceeding in both directions from that point. A eukaryotic chromosome may have multiple origins of replication and may be replicating at many points along its length at any one time.
- DNA synthesis always proceeds in a 5′ ⟶ 3′ direction. This requires that one DNA strand, the **lagging strand,** be synthesized discontinuously, as short Okazaki fragments. The opposite strand, the **leading strand,** is synthesized continuously.

8 Define *telomere;* describe the possible connections between telomerase and cell aging and between telomerase and cancer.

- Eukaryotic chromosome ends, known as **telomeres,** are short, noncoding, repetitive DNA sequences. Telomeres shorten slightly with each cell cycle but can be extended by the enzyme **telomerase.**
- The absence of telomerase activity in certain cells may be a cause of **cell aging,** in which cells lose their ability to divide after a limited number of cell divisions.
- Most **cancer cells,** including human cancers of the breast, lung, colon, prostate gland, and pancreas, have telomerase to maintain telomere length and possibly to resist apoptosis.

POST-TEST

1. When Griffith injected mice with a combination of live rough-strain and heat-killed smooth-strain pneumococci, he discovered that (a) the mice were unharmed (b) the dead mice contained living rough-strain bacteria (c) the dead mice contained living smooth-strain bacteria (d) DNA had been transferred from the smooth-strain bacteria to the mice (e) DNA had

POST-TEST *(continued)*

been transferred from the rough-strain bacteria to the smooth-strain bacteria

2. Which of the following inspired Avery and his colleagues to perform the experiments demonstrating that the transforming principle in bacteria is DNA? (a) the fact that A is equal to T, and G is equal to C (b) Watson and Crick's model of DNA structure (c) Meselson and Stahl's studies on DNA replication in *E. coli* (d) Griffith's experiments on smooth and rough strains of pneumococci (e) Hershey and Chase's experiments on the reproduction of bacteriophages
3. In the Hershey-Chase experiment with bacteriophages (a) harmless bacterial cells permanently transformed into virulent cells (b) DNA was shown to be the transforming principle of earlier bacterial transformation experiments (c) the replication of DNA was conclusively shown to be semiconservative (d) viral DNA was shown to enter bacterial cells and cause the production of new viruses within the bacteria (e) viruses injected their proteins, not their DNA, into bacterial cells
4. The two complementary strands of the DNA double helix are held to one another by (a) ionic bonds between deoxyribose molecules (b) ionic bonds between phosphate groups (c) covalent bonds between nucleotide bases (d) covalent bonds between deoxyribose molecules (e) hydrogen bonds between nucleotide bases
5. If a segment of DNA is 5′—CATTAC—3′, the complementary DNA strand is
   (a) 3′—CATTAC—5′ (b) 3′—GTAATG—5′
   (c) 5′—CATTAC—3′ (d) 5′—GTAATG—3′
   (e) 5′—CATTAC—5′
6. Each DNA strand has a backbone that consists of alternating (a) purines and pyrimidines (b) nucleotide bases (c) hydrogen bonds and phosphodiester linkages (d) deoxyribose and phosphate (e) phosphate and phosphodiester linkages
7. The experiments in which Meselson and Stahl grew bacteria in heavy nitrogen conclusively demonstrated that DNA (a) is a double helix (b) replicates semiconservatively (c) consists of repeating nucleotide subunits (d) has complementary base pairing (e) is always synthesized in the 5′ to 3′ direction
8. The statement "DNA replicates by a semiconservative mechanism" means that (a) only one DNA strand is copied (b) first one DNA strand is copied, and then the other strand is copied (c) the two strands of a double helix have identical base sequences (d) some portions of a single DNA strand are old, and other portions are newly synthesized (e) each double helix consists of one old and one newly synthesized strand
9. What technique did Franklin use to determine many of the physical characteristics of DNA? (a) x-ray diffraction (b) transformation (c) radioisotope labeling (d) density gradient centrifugation (e) transmission electron microscopy
10. Multiple origins of replication (a) speed up replication of eukaryotic chromosomes (b) enable the lagging strands and leading strands to be synthesized at different replication forks (c) help to relieve strain as the double helix unwinds (d) prevent mutations (e) are necessary for the replication of a circular DNA molecule in bacteria
11. Topoisomerases (a) synthesize DNA (b) synthesize RNA primers (c) join Okazaki fragments (d) break and rejoin DNA to resolve knots that have formed (e) prevent single DNA strands from joining to form a double helix
12. A phosphate in DNA (a) hydrogen-bonds to a base (b) covalently links to two bases (c) covalently links to two deoxyriboses (d) hydrogen-bonds to two additional phosphates (e) covalently links to a base, a deoxyribose, and another phosphate
13. Which of the following depicts the relative arrangement of the complementary strands of a DNA double helix?
    (a) 5′—5′ / 3′—3′ (b) 3′—5′ / 3′—5′ (c) 3′—3′ / 3′—3′ (d) 5′—5′ / 5′—5′ (e) 3′—5′ / 5′—3′
14. A lagging strand forms by (a) joining primers (b) joining Okazaki fragments (c) joining leading strands (d) breaking up a leading strand (e) joining primers, Okazaki fragments, and leading strands
15. The immediate source of energy for DNA replication is (a) the hydrolysis of the nucleotides, with the release of two phosphates (b) the oxidation of NADPH (c) the hydrolysis of ATP (d) electron transport (e) the breaking of hydrogen bonds
16. Which of the following statements about eukaryotic chromosomes is *false?* (a) Eukaryotic chromosomes have free ends (b) Telomeres contain protein-coding genes (c) Telomerase lengthens telomeric DNA (d) Telomere shortening may contribute to cell aging (e) Cells with active telomerase may undergo many cell divisions

CRITICAL THINKING

1. What characteristics must a molecule have to function as genetic material?
2. What important features of the structure of DNA are consistent with its role as the chemical basis of heredity?

- Visit our Web site at **http://biology.brookscole.com/solomon7** for links to chapter-related resources on the World Wide Web. Additional online materials relating to this chapter can also be found on our Web site.

BIOLOGY NOW RESOURCES

Active Figures

11-5: The structure of the DNA double helix

11-9: Replication

Preparing for an exam? Take a diagnostic test on your BiologyNow CD-ROM.

Post-Test Answers

1. c	2. d	3. d	4. e
5. b	6. d	7. b	8. e
9. a	10. a	11. d	12. c
13. e	14. b	15. a	16. b

12

Gene Expression

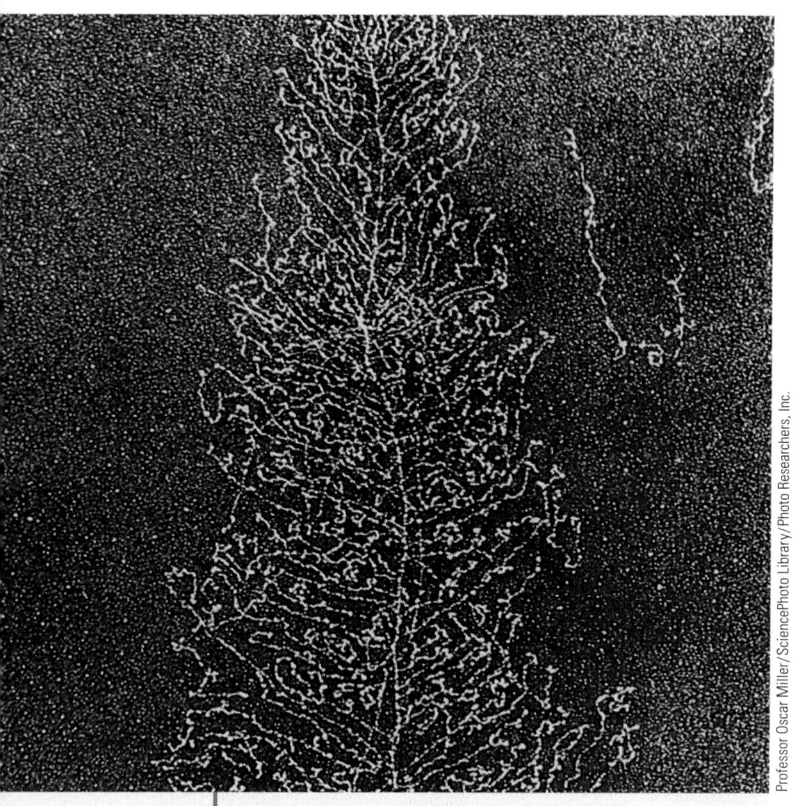

Professor Oscar Miller/SciencePhoto Library/Photo Researchers, Inc.

Visualizing transcription. In this TEM, RNA molecules *(lateral strands)* are synthesized as complementary copies of a DNA template *(central axis).*

CHAPTER OUTLINE

In Chapter 11, we discussed how the cell accurately replicates the nucleotide sequence in DNA to pass the genetic material unaltered to the next generation. Watson and Crick originally described the basic features of the DNA double helix, which scientists now know to be the same in all cells studied to date—from the simplest bacteria to highly complex human cells.

By the mid-1950s, researchers had determined that the sequence of bases in DNA contains the information that specifies all the proteins the cell needs. However, more than a decade of intense investigation by many scientists preceded a fundamental understanding of how cells convert DNA information into the amino acid sequences of proteins. Much of that understanding came from studying the functions of bacterial genes. After Watson and Crick discovered the structure of DNA, researchers chose bacterial cells as the organisms of choice for these studies because bacteria grew quickly and easily, and because they contained the minimal amount of DNA needed for growth and reproduction. As researchers learned that all organisms have fundamental genetic similarities, the validity, as well as the utility, of this approach was repeatedly confirmed.

In this chapter we examine the evidence that accumulated in the first half of the 20th century indicating that most genes specify the structure of proteins. We then consider how DNA affects the phenotype of the organism at the molecular level through the process of gene expression. **Gene expression** involves a series of steps in which the information in the sequence of bases in DNA specifies the makeup of the cell's proteins. The proteins affect the phenotype in some way; these effects range from readily observable physical traits to subtle changes detectable only at the biochemical level. The first major step of gene expression is **transcription,** the synthesis of RNA molecules complementary to the DNA (see photograph). The second major step is **translation,** in which RNA becomes a coded template to direct protein synthesis. In Chapter 13 we consider some of the ways gene expression is regulated. ■

DISCOVERY OF THE GENE–PROTEIN RELATIONSHIP

Learning Objective

1 Summarize the early evidence indicating that most genes specify the structure of proteins.

PROCESS OF SCIENCE

The idea that genes and proteins are connected originated early in the 20th century, shortly after scientists rediscovered Mendel's principles. In the first edition of his book, *Inborn Errors of Metabolism* (1908), Archibald Garrod, an English physician and biochemist, discussed a rare genetic disease called **alkaptonuria,** which scientists hypothesized had a simple recessive inheritance pattern. The condition involves the metabolic pathway that breaks down the amino acid tyrosine, ultimately converting it to carbon dioxide and water. An intermediate in this pathway, homogentisic acid, accumulates in the urine of affected people, turning it black when exposed to air (Fig. 12-1). Other symptoms of alkaptonuria include the later development of arthritis and, in men, stones in the prostate gland.

In Garrod's time, scientists knew about enzymes but didn't recognize they were proteins. Garrod hypothesized that people with alkaptonuria lack the enzyme that normally oxidizes homogentisic acid. Before the second edition of his book had been published, in 1923, researchers found that affected people do indeed lack the enzyme that oxidizes homogentisic acid. Garrod's hypothesis was correct: A mutation in this specific gene is associated with the absence of a specific enzyme. Shortly thereafter, in 1926, U.S. biochemist James Sumner purified a different enzyme, urease, and showed it to be a protein. This was the first clear identification of an enzyme as a protein. In 1946 Sumner was awarded the Nobel Prize in Chemistry for being the first to crystallize an enzyme.

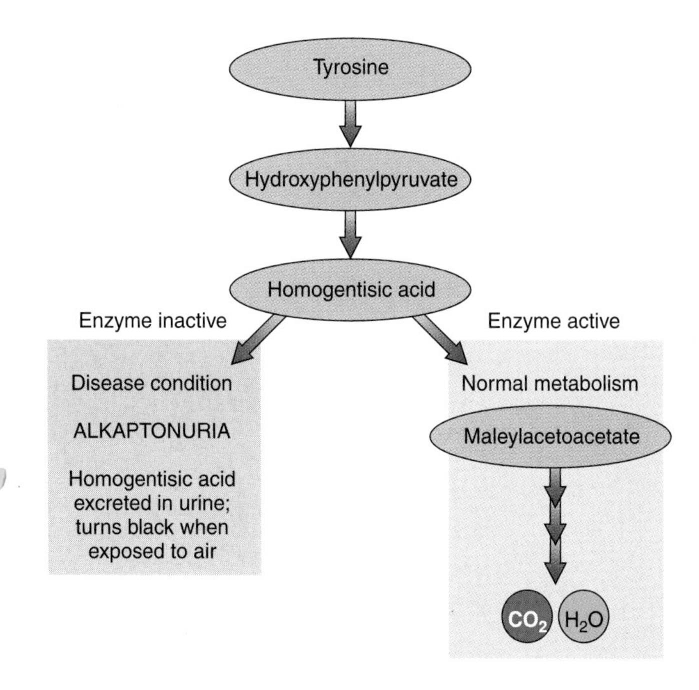

FIGURE **12-1** **An "inborn error of metabolism."**

Garrod proposed that the alkaptonuria allele causes the inactivation of the enzyme homogentisic acid oxidase, which normally converts homogentisic acid to maleylacetoacetate. Thus the acid accumulates in the blood and is excreted in the urine, which turns black on contact with the air.

Beadle and Tatum proposed the one-gene, one-enzyme hypothesis

A major advance in understanding the relationship between genes and enzymes came in the early 1940s, when U.S. geneticists George Beadle and Edward Tatum and their associates developed a new approach to the problem. Most efforts until that time had focused on studying known loci and seeking to determine what biochemical reactions they affected. Researchers examined previously identified loci, such as those controlling eye color in *Drosophila* or pigments in plants. They found that a series of biosynthetic reactions control specific phenotypes, but it was not clear whether the genes themselves were acting as enzymes or if they determined the workings of the enzymes in more complex ways.

Beadle and Tatum decided to take the opposite approach. Rather than try to identify the enzymes affected by single genes, they decided to look for mutations interfering with known metabolic reactions that produce essential molecules, such as amino acids and vitamins. They chose a fungus, *Neurospora,* a common bread mold, as an experimental organism, for several important reasons. First, wild-type *Neurospora* is easy to grow in culture. The adjective **wild-type** refers to an individual with the normal phenotype. *Neurospora* manufactures all its essential biological molecules when grown on a simple growth medium (minimal medium) containing only sugar, salts, and the vitamin biotin. However, a mutant *Neurospora* strain that cannot make a substance such as an amino acid, can still grow if that substance is simply added to the growth medium.

Second, the life cycle of *Neurospora* includes both sexual and asexual reproduction, which facilitates certain types of manipulations and genetic analysis. Third, *Neurospora* grows primarily as a haploid organism, allowing a recessive mutant allele to be immediately identified because there is no homologous chromosome that could carry a dominant allele that would mask its expression. (For an illustration of the generalized life cycles of simple organisms such as fungi, see Fig. 9-16b; a more detailed life cycle of organisms similar to *Neurospora* is given in Fig. 25-9.)

Beadle and Tatum began by exposing thousands of haploid wild-type *Neurospora* asexual spores to x-rays or ultraviolet radiation to induce mutant strains. They first grew each irradiated strain on a complete growth medium, which contained all the amino acids and vitamins normally made by *Neurospora.* Then they tested each strain on the minimal medium described previously. About 1% to 2% of the strains that grew on the complete medium failed to grow after transfer to the minimal medium. Beadle and Tatum reasoned that such a strain carried a mutation preventing the fungus from producing one of the chemicals essential for growth. Further testing of the mutant strain on media containing different combinations of amino acids, purines, and vitamins enabled the investigators to determine the exact compound required (Fig. 12-2).

The work on *Neurospora* revealed that each mutant strain had a mutation in only one gene locus and that each gene locus affected only one enzyme. Beadle and Tatum stated this one-to-one correspondence between genes and enzymes as the *one-gene, one-enzyme hypothesis.* In 1958, they received the Nobel Prize in Medicine for discovering that genes regulate specific chemical events. The work of Beadle, Tatum, and others led to a more precise understanding of what a gene is and to additional predictions about the chemical nature of genes. The idea that a gene encodes the information to produce a single enzyme held for almost a decade, until additional findings required a modification of this definition.

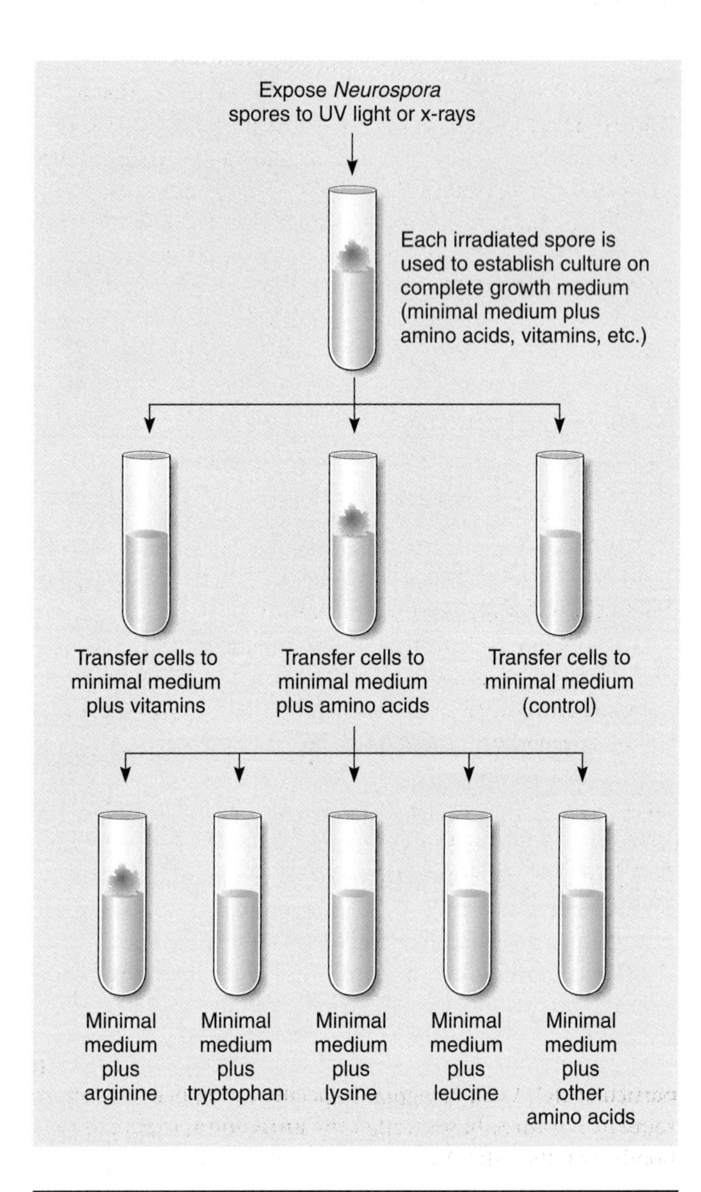

FIGURE **12-2** **Mutations affecting biochemical pathways.**
Beadle and Tatum irradiated *Neurospora* spores to induce mutations. Cultures derived from these spores were grown on a complete growth medium that contained all the amino acids, vitamins, and other nutrients that *Neurospora* normally makes for itself. They hypothesized that any strain that failed to grow when transferred to a minimal medium carried a mutation that blocked a step in a biochemical pathway. They identified the specific nutritional requirement in the mutant strain by testing for growth on minimal media supplemented with individual vitamins or amino acids. In this example, only the medium containing the amino acid arginine supports growth, indicating the mutation affects some part of the arginine biosynthetic pathway.

In the late 1940s, researchers began to understand that genes control not only enzymes but other proteins as well. The U.S. chemist Linus Pauling and his colleagues demonstrated in 1949 that a mutation of a single locus alters the structure of hemoglobin. This particular mutant form of hemoglobin is associated with the genetic disease sickle cell anemia (see Chapter 15). In 1954, Pauling received the Nobel Prize in Chemistry for his work on the genetic aspects of the molecular structure of hemoglobin. In addition, studies by other scientists showed that many proteins are constructed from two or more polypeptide chains, each of which is under the control of a different locus. For example, hemoglobin contains two types of polypeptide chains, the α and β subunits (see Fig. 3-22a). Sickle cell anemia results from a mutation affecting the β subunits.

Scientists therefore extended the definition of a gene to say that one gene is responsible for one polypeptide chain. Even this definition has proved only partially correct, although as you will see later in this chapter, scientists still define a gene in terms of its product.

Although the elegant work of Beadle and Tatum and others demonstrated that genes are expressed in the form of proteins, the mechanism was completely unknown. After Watson and Crick's discovery of the structure of DNA, many scientists worked to understand exactly how gene expression takes place. We begin with an overview of gene expression and then consider the various steps in more detail.

Review

- What is the one-gene, one-enzyme hypothesis?
- What were the contributions of each of the following scientists—A. Garrod, G. Beadle and E. Tatum, and L. Pauling—to our understanding of the relationship between genes and proteins?

BiologyNow™ Assess your understanding of **the discovery of the gene–protein relationship** by taking the pretest on your BiologyNow CD-ROM.

INFORMATION FLOW FROM DNA TO PROTEIN: AN OVERVIEW

Learning Objectives

2 Outline the flow of genetic information in cells, from DNA to protein.

3 Compare the structures of DNA and RNA.

4 Explain why the genetic code is said to be redundant and virtually universal, and discuss how these features may reflect its evolutionary history.

Although the sequence of bases in DNA determines the sequence of amino acids in polypeptides, cells do not use the information in DNA directly. Instead, a related nucleic acid, **ribonucleic acid (RNA),** is the link between DNA and protein. When a protein-coding gene is expressed, first an RNA copy is made of the information in the DNA. It is this RNA copy that provides the information that directs protein synthesis.

Like DNA, RNA is a polymer of nucleotides, but it has some important differences (Fig. 12-3). RNA is usually single-stranded, although internal regions of some RNA molecules may have complementary sequences that allow the strand to fold back and pair to form short, double-stranded segments. As shown in Figure 12-3, the sugar in RNA is **ribose,** which is similar to deoxyribose of DNA but has a hydroxyl group at the 2′ position. (Compare ribose with the deoxyribose of DNA, shown in Fig. 11-3, which has only a hydrogen at the 2′ position.) The base **uracil** substitutes for thymine and, like thymine, is a pyrimidine that can form two hydrogen bonds with adenine. Hence uracil and adenine are a complementary pair.

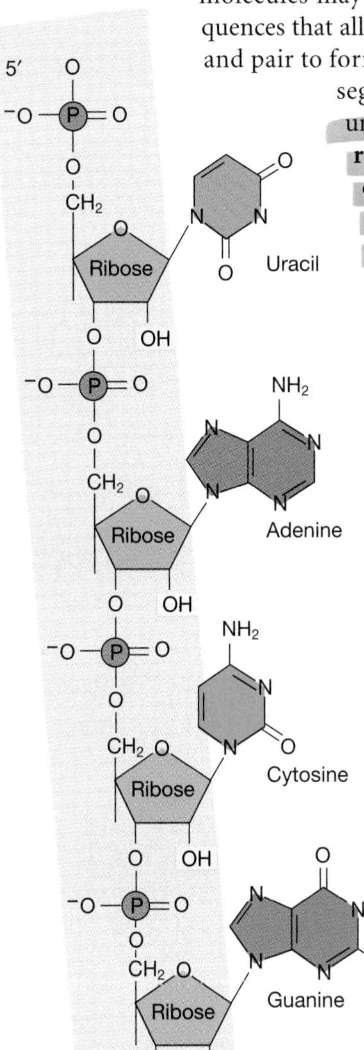

FIGURE **12-3** **The nucleotide structure of RNA.**
The nucleotide subunits of RNA are joined by 5′ ⟶ 3′ phosphodiester linkages, like those found in DNA. Adenine, guanine, and cytosine are present, as in DNA, but the base uracil replaces thymine. All four nucleotide types contain the five-carbon sugar ribose, which has a hydroxyl group *(yellow)* on its 2′ carbon atom.

DNA is transcribed to form RNA

The process by which RNA is synthesized resembles DNA replication in that the sequence of bases in the RNA strand is determined by complementary base-pairing with one of the DNA strands, the **template strand** (Fig. 12-4). Because RNA synthesis takes the information in one kind of nucleic acid (DNA) and copies it as another nucleic acid (RNA), this process is called **transcription** ("copying").

Three specific kinds of RNA molecules are transcribed: messenger RNA, transfer RNA, and ribosomal RNA. **Messenger RNA (mRNA)** is a single, uncoiled strand of RNA that carries the specific information for making a protein. Each of the 20 or so **transfer RNAs (tRNAs)** is a single strand of RNA that folds back on itself to form a specific shape. Each tRNA bonds with only one specific amino acid and carries it to the ribosome. **Ribosomal RNA (rRNA),** which is in a globular form, is an important part of the structure of ribosomes and has catalytic functions needed during protein synthesis.

RNA is translated to form a polypeptide

Figure 12-4 also shows the second step of gene expression, in which the transcribed information in the mRNA is used to specify the amino acid sequence of a polypeptide. This process is called **translation,** because it involves conversion of the "nucleic acid language" in the mRNA molecule into the "amino acid language" of protein.

A sequence of three consecutive bases in mRNA, called a **codon,** specifies one amino acid. For example, one codon that corresponds to the amino acid threonine is 5′—ACG—3′. Because each codon consists of three nucleotides, the code is described as a **triplet code.** The assignments of codons for amino acids and for start and stop signals are collectively named the **genetic code** (Fig. 12-5).

Transfer RNAs are crucial parts of the decoding machinery, because they act as "adapters" that connect amino acids and nucleic acids. This mechanism is possible because each tRNA can (1) link with a specific amino acid and (2) recognize the appropriate mRNA codon for that particular amino acid. A particular tRNA can recognize a specific codon because it has a sequence of three bases, called the **anticodon,** that hydrogen bonds with the mRNA codon by complementary base-pairing. The exact anticodon that is complementary to the codon for threonine in our example is 3′—UGC—5′.

Translation requires that (1) each tRNA anticodon is hydrogen-bonded to the complementary mRNA codon, and (2) the amino acids carried by the tRNAs are linked together in the order specified by the sequence of codons in the mRNA. **Ribosomes,** the site of translation, are organelles composed of two different subunits, each containing protein and rRNA. Ribosomes attach to one end of the mRNA and travel along it, allowing the tRNAs to attach sequentially to the codons of mRNA. In this way the amino acids carried by the tRNAs take up the proper position to be joined by *peptide bonds* in the correct sequence to form a polypeptide.

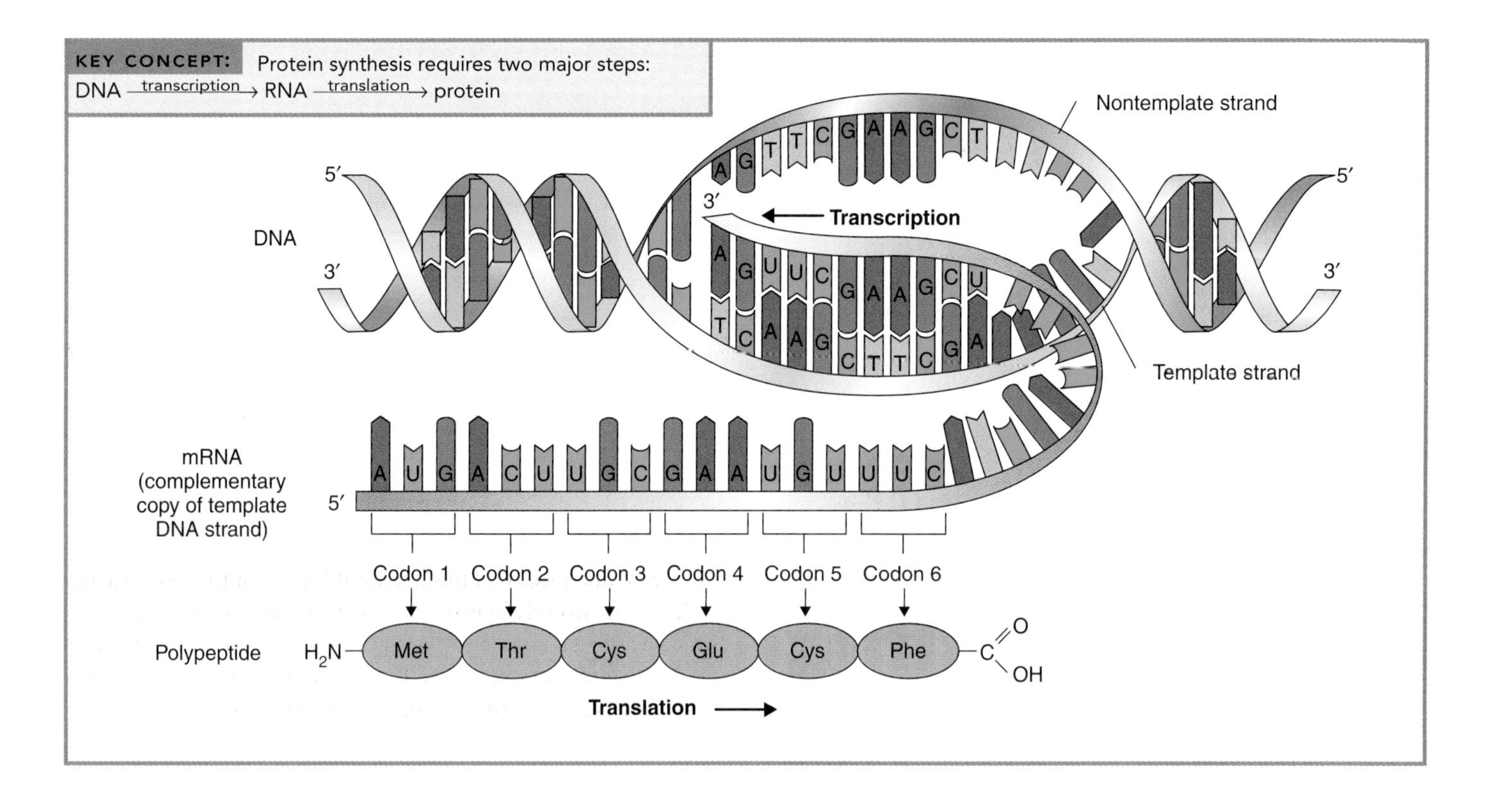

FIGURE **12-4** **An overview of transcription and translation.**

In transcription, messenger RNA is synthesized as a complementary copy of one of the DNA strands, the template strand. Messenger RNA carries genetic information in the form of sets of three bases, or codons, each of which specifies one amino acid. Codons are translated consecutively, thus specifying the linear sequence of amino acids in the polypeptide chain. Translation requires tRNA and ribosomes *(not shown)*. The figure depicts transcription and translation in bacteria. In eukaryotes, transcription takes place in the nucleus and translation occurs in the cytosol.

Biologists cracked the genetic code in the 1960s

PROCESS OF SCIENCE

Before the genetic code was deciphered, scientists had become interested in how a genetic code might work. The Watson and Crick model of DNA showed it to be a linear sequence of four nucleotides. If each nucleotide coded for a single amino acid, the genetic code could specify only 4 amino acids, not the 20 found in the vast variety of proteins in the cell. Scientists saw that the DNA bases could serve as a four-letter "alphabet" and hypothesized that three-letter combinations of the four bases would make it possible to form a total of 64 "words," more than enough to specify all the naturally occurring amino acids. In 1961, Crick and British scientist Sydney Brenner concluded from experimental evidence that the code used nonoverlapping triplets of bases. They predicted the code is read, one triplet at a time, from a fixed starting point that establishes the **reading frame.** Because no "commas" separate the triplets, an alteration in the reading frame results in the incorporation of incorrect amino acids.

Marshall Nirenberg, a U.S. biochemist, and his postdoctoral researcher, Heinrich Matthaei, first obtained experimental evidence indicating the assignment of specific triplets to specific amino acids. By constructing artificial mRNA molecules with known base sequences, they determined which amino acids would be incorporated into protein in purified in vitro protein synthetic systems derived from *E. coli.* For example, when they added the synthetic mRNA polyuridylic acid (UUUUUUUUU . . .) to a mixture of purified ribosomes, aminoacyl-tRNAs (amino acids linked to their respective tRNAs), and essential cofactors needed to synthesize protein, only phenylalanine was incorporated into the resulting polypeptide chain. The inference was inescapable that the UUU triplet codes for phenylalanine. Similar experiments showed that polyadenylic acid (AAAAAAAAA . . .) codes for a polypeptide of lysine and polycytidylic acid (CCCCCCCCC . . .) codes for a polypeptide of proline. In 1968, Nirenberg received the Nobel Prize in Medicine for his work in deciphering the genetic code.

By using mixed nucleotide polymers (such as a random polymer of A and C) as artificial messengers, researchers assigned the other codons to specific amino acids. However, three of the codons—UAA, UGA, and UAG—did not specify any amino acids. These codons are now known to be the signals that terminate the coding sequence for a polypeptide chain. By 1967 the genetic code was completely "cracked," and scientists had identified the coding assignments of all 64 possible codons shown in Figure 12-5.

Remember that the genetic code we define and use is an mRNA code. The tRNA anticodon sequences, as well as the

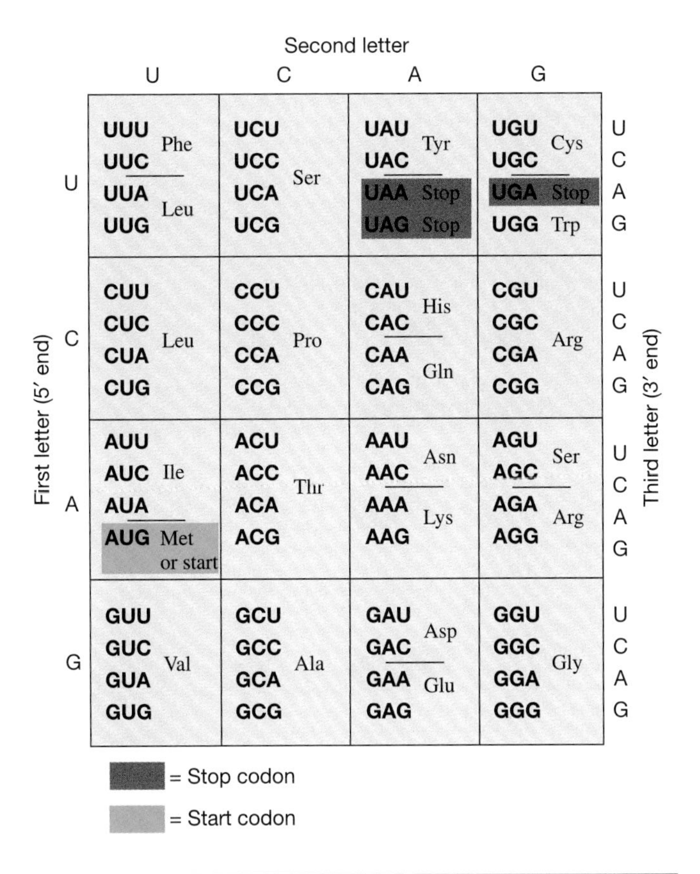

Second letter
U
C
A
G
UUU Phe
UUC
UUA Leu
UUG
UCU
UCC Ser
UCA
UCG
UAU Tyr
UAC
UAA Stop
UAG Stop
UGU Cys
UGC
UGA Stop
UGG Trp
CUU
CUC Leu
CUA
CUG
CCU
CCC Pro
CCA
CCG
CAU His
CAC
CAA Gln
CAG
CGU
CGC Arg
CGA
CGG
AUU
AUC Ile
AUA
AUG Met or start
ACU
ACC Thr
ACA
ACG
AAU Asn
AAC
AAA Lys
AAG
AGU Ser
AGC
AGA Arg
AGG
GUU
GUC Val
GUA
GUG
GCU
GCC Ala
GCA
GCG
GAU Asp
GAC
GAA Glu
GAG
GGU
GGC Gly
GGA
GGG
First letter (5′ end)
Third letter (3′ end)
= Stop codon
= Start codon

FIGURE 12-5 The genetic code.

The genetic code specifies all possible combinations of the three bases that comprise codons in mRNA. Of the 64 possible codons, 61 specify amino acids (see Fig. 3-16 for an explanation of abbreviations). The codon AUG specifies the amino acid methionine and also signals the ribosome to initiate translation ("start"). Three codons—UAA, UGA, and UAG—do not specify amino acids; they terminate protein synthesis ("stop").

DNA sequence from which the message is transcribed, are complementary to the sequences in Figure 12-5. For example, the mRNA codon for the amino acid methionine is 5′—AUG—3′. It is transcribed from the DNA base sequence 3′—TAC—5′, and the corresponding tRNA anticodon is 3′—UAC—5′.

The genetic code is virtually universal

Perhaps the single most remarkable feature of the code is that it is essentially universal. Over the years, biologists have examined the genetic code in a diverse array of species and found it the same in organisms as different as the bacterium *E. coli,* redwood trees, jellyfish, and humans. These findings strongly suggest the code is an ancient legacy that evolved very early in the history of life (see Chapter 20).

Scientists have discovered some very minor exceptions to the universality of the genetic code. In several unicellular protozoa, UAA and UGA code for the amino acid glutamine instead of functioning as stop signals. Other exceptions are found in mitochondria, which contain their own DNA and protein synthesis machinery for some genes (see Chapters 4 and 20). These slight coding differences vary with the organism, but keep in mind that in each case, all the other coding assignments are identical to the standard genetic code.

The genetic code is redundant

Given 64 possible codons and only 20 common amino acids, it is not surprising that more than 1 codon specify certain amino acids. This redundancy in the genetic code has certain characteristic patterns. The codons CCU, CCC, CCA, and CCG are "synonymous" in that they all code for the amino acid proline. The only difference among the 4 codons involves the nucleotide at the 3′ end of the triplet. Although the code may be read three nucleotides at a time, only the first two nucleotides seem to contain specific information for proline. A similar pattern can be seen for several other amino acids. Only methionine and tryptophan are specified by single codons. Each of the other amino acids is specified by 2 to 6 different codons.

Crick first proposed this apparent breach of the base-pairing rules as the **wobble hypothesis.** He reasoned that the third nucleotide of a tRNA anticodon (which is the 5′ base of the anticodon sequence) may sometimes form hydrogen bonds with more than one kind of third nucleotide (the 3′ base) of a codon. Investigators later established this experimentally by determining the anticodon sequences of tRNA molecules and testing their specificities in artificial systems. Some tRNA molecules can recognize as many as three separate codons differing in their third nucleotide but specifying the same amino acid.

Review

- Sketch a simple flow diagram that shows the relationships among the following: RNA, translation, DNA, transcription, and protein.
- How was the genetic code deciphered?

BiologyNow™ Assess your understanding of **information flow from DNA to protein** by taking the pretest on your BiologyNow CD-ROM.

TRANSCRIPTION

Learning Objective

5 Compare the processes of transcription and DNA replication, identifying both similarities and differences.

Now that we have presented an overview of information flow from DNA to RNA to protein, let's examine the entire process more closely. In transcription, most RNA is synthesized by **DNA-dependent RNA polymerases,** enzymes present in all cells. These enzymes require DNA as a template and have many similarities to the DNA polymerases discussed in Chapter 11. Like DNA polymerases, they carry out synthesis in the 5′ ⟶ 3′ direction; that is, they begin at the 5′ end of the RNA molecule being synthesized and then continue to add nucleotides at the 3′ end until the molecule is complete (Fig. 12-6). DNA-dependent RNA polymerases use nucleotides with three phosphate groups as substrates, removing two of the phosphates as the nucleotides are covalently linked to the 3′ end of the RNA. Like DNA replication and the hydrolysis of ATP, these reactions are strongly exergonic (see Chapter 6).

Whenever nucleic acid molecules associate by complementary base pairing, the two strands are **antiparallel.** Just as the two paired strands of DNA are antiparallel, the template strand of the DNA and the complementary RNA strand are also antiparallel. Therefore, as shown in Fig. 12-6, when transcription takes place, as RNA is synthesized in its 5′ ⟶ 3′ direction, the DNA template is read in its 3′ ⟶ 5′ direction.

Scientists conventionally refer to a sequence of bases in a gene or the mRNA sequence transcribed from it as upstream or downstream of some reference point. **Upstream** is toward the 5′ end of the mRNA sequence or the 3′ end of the template DNA strand. **Downstream** is toward the 3′ end of the RNA or the 5′ end of the template DNA strand.

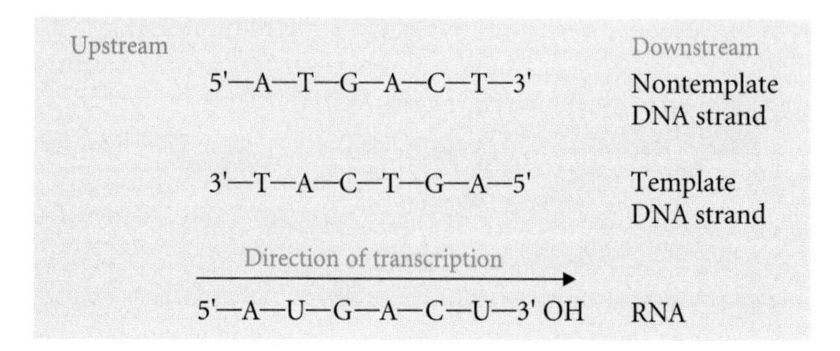

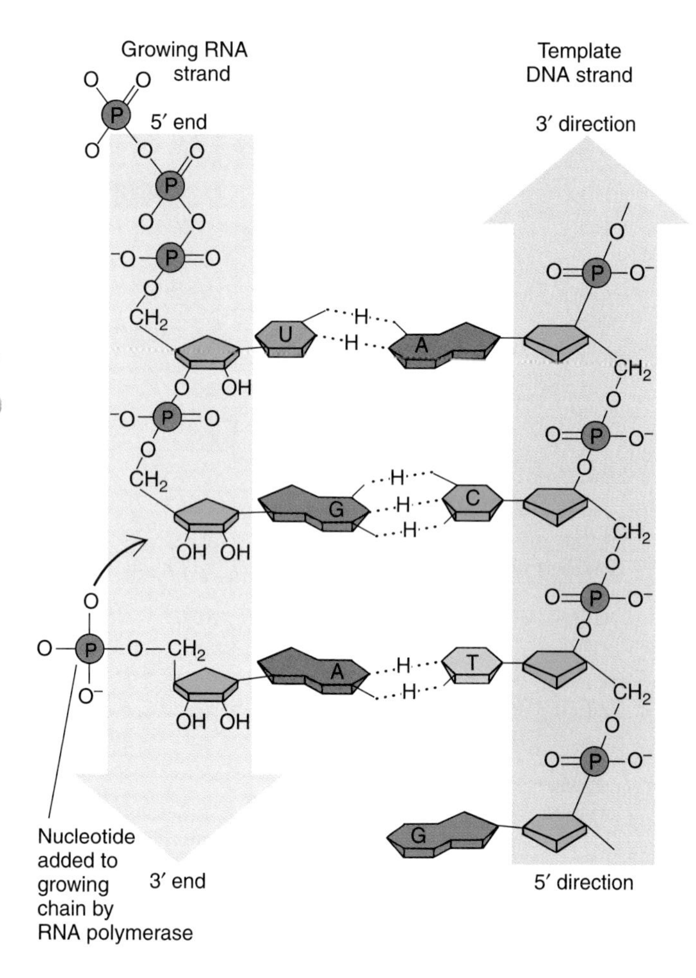

FIGURE **12-6** **Transcription.**

Incoming nucleotides with three phosphates pair with complementary bases on the template DNA strand *(right).* RNA polymerase cleaves two phosphates *(not shown)* from each nucleotide and covalently links the remaining phosphate to the 3′ end of the growing RNA chain. Thus, RNA, like DNA, is synthesized in the 5′ ⟶ 3′ direction.

The synthesis of mRNA includes initiation, elongation, and termination

In both bacteria and eukaryotes, the nucleotide sequence in DNA to which RNA polymerase initially binds is called the **promoter.** Because the promoter is not transcribed, RNA polymerase moves past the promoter to begin transcription of the protein-coding sequence of DNA. Different genes may have slightly different promoter sequences, so the cell can direct which genes are transcribed at any one time. Bacterial promoters are usually about 40 bases long and lie in the DNA just upstream of the point at which transcription will begin. Once RNA polymerase has recognized the correct promoter, it unwinds the DNA double helix and *initiates* transcription. Unlike DNA synthesis, RNA synthesis does not require a primer. However, transcription requires several proteins in addition to RNA polymerase; we discuss these in Chapter 13.

As Figure 12-6 illustrates, the first nucleotide at the 5′ end of a new mRNA chain initially retains its triphosphate group. However, during the *elongation* stage of transcription, as each additional nucleotide is incorporated at the 3′ end of the growing RNA molecule, two of its phosphates are removed in an exergonic reaction that leaves the remaining phosphate to become part of the sugar-phosphate backbone (as in DNA replication). The last nucleotide to be incorporated has an exposed 3′-hydroxyl group.

Elongation of RNA continues until RNA polymerase recognizes a stop signal termination sequence consisting of a set of specific base sequences on the DNA template. This signal leads to a separation of the enzyme from both the template DNA and the newly synthesized RNA. Different mechanisms in bacteria and eukaryotes terminate transcription. In bacteria, transcription stops at the end of the stop signal. When RNA polymerase comes to the stop signal, it releases both the DNA template and the new RNA strand. In eukaryotic cells, RNA polymerase adds nucleotides to the mRNA molecule after it passes the stop signal. The 3′ end of the mRNA becomes separated from RNA polymerase about 10 to 35 nucleotides past the stop signal.

Usually only one strand in a protein-coding region of DNA is used as a template (Fig. 12-7a). For example, consider a segment of DNA that contains the following DNA base sequence in the template strand:

3′—TAACGGTCT—5′

If the complementary DNA strand

5′—ATTGCCAGA—3′

were used as a template, a message would be produced specifying an entirely different (and generally nonfunctional) protein. However, that only one strand is transcribed does not mean the same strand is always the template for all genes throughout the length of a chromosome-sized DNA molecule. Instead, a particular strand may serve as the template strand for some genes and the nontemplate strand for others (Fig. 12-7b).

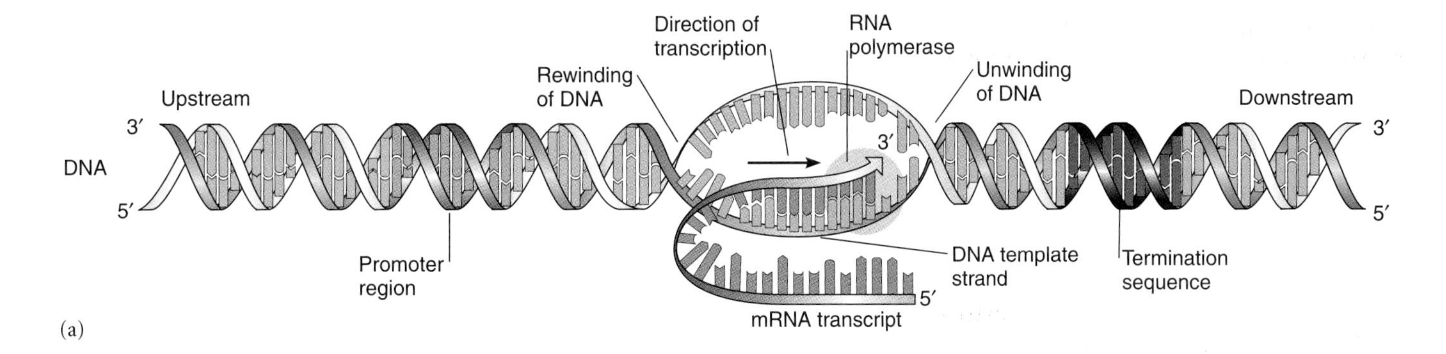

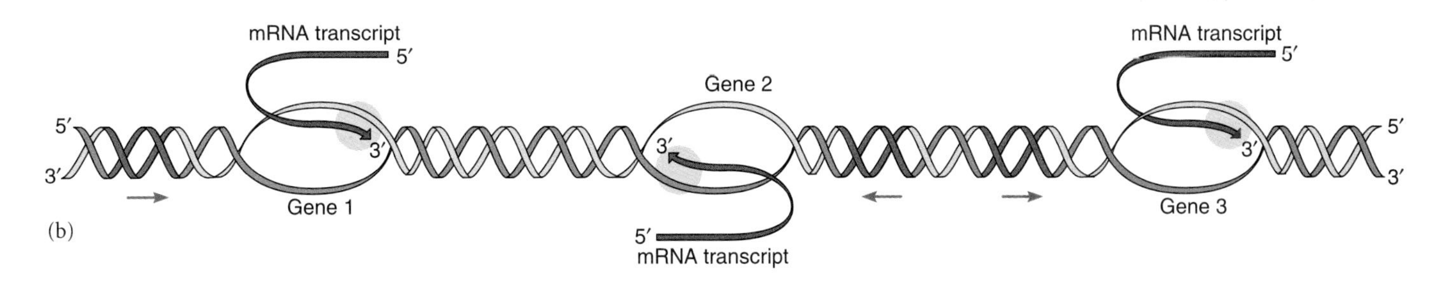

ACTIVE FIGURE **12-7** | **The synthesis of mRNA.**

(a) The mRNA is synthesized in the 5′ ⟶ 3′ direction from the template strand of the DNA molecule. Transcription starts downstream from a DNA promoter sequence, to which the RNA polymerase initially binds. Termination signals downstream from the protein-coding sequences, signal the RNA polymerase to stop transcription and be released from the DNA. **(b)** Usually only one of the two strands is transcribed for a given gene, but the opposite strand may be transcribed for a neighboring gene. Each transcript starts at its own promoter *(orange region).*

BiologyNow™ Watch **transcription** unfold by clicking on this figure on your BiologyNow CD ROM.

Messenger RNA contains base sequences that do not directly code for protein

A completed mRNA contains more than the nucleotide sequence that codes for a protein. Figure 12-8 shows a typical bacterial mRNA. (The unique features of eukaryotic mRNA are discussed later in the chapter.) In both bacteria and eukaryotes, RNA polymerase starts transcription of a gene upstream of the protein-coding DNA sequence. As a result, the mRNA has a noncoding **leader sequence** at its 5′ end. The leader has recognition signals for ribosome binding, which properly position the ribosomes to translate the message. The **start codon** follows the leader sequence and signals the beginning of the **coding sequence** that contains the actual message for the polypeptide. Unlike eukaryotic cells, it is common for one or more polypeptides to be encoded by a single mRNA molecule in bacterial cells (see Chapter 13). At the end of each coding sequence, a **stop codon** signals the end of the protein. The stop codons—UAA, UGA, and UAG— end both bacterial and eukaryotic messages. They are followed by noncoding 3′ **trailing sequences,** which vary in length.

Review

- In what ways are DNA polymerase and RNA polymerase similar? How do they differ?
- A certain template DNA strand has the following nucleotide sequence:

  3′—TACTGCATAATGATT— 5′

  What would be the sequence of codons in the mRNA transcribed from this strand? What would be the nucleotide sequence of the complementary nontemplate DNA strand?

BiologyNow™ Assess your understanding of **transcription** by taking the pretest on your BiologyNow CD-ROM.

TRANSLATION

Learning Objectives

6 Identify the features of tRNA that are important in decoding genetic information and converting it into "protein language."

7 Explain how ribosomes function in protein synthesis.

8 Diagram the processes of initiation, elongation, and termination in protein synthesis.

In eukaryotes, mRNA must move from the nucleus (the site of transcription) to the cytosol (the site of translation or polypeptide synthesis). No such movement occurs in bacteria, which lack a nucleus. Translation adds another level of complexity to the process of information transfer because it involves conversion of the triplet nucleic acid code to the 20–amino-acid alphabet of proteins. The structural differences between a polynucleotide chain and a polypeptide chain prohibit amino acids from directly interacting with an mRNA molecule to make a protein. Translation requires the coordinated functioning of more than 100 kinds of macromolecules, including the protein and RNA components of the ribosomes, mRNA, and amino acids linked to tRNAs.

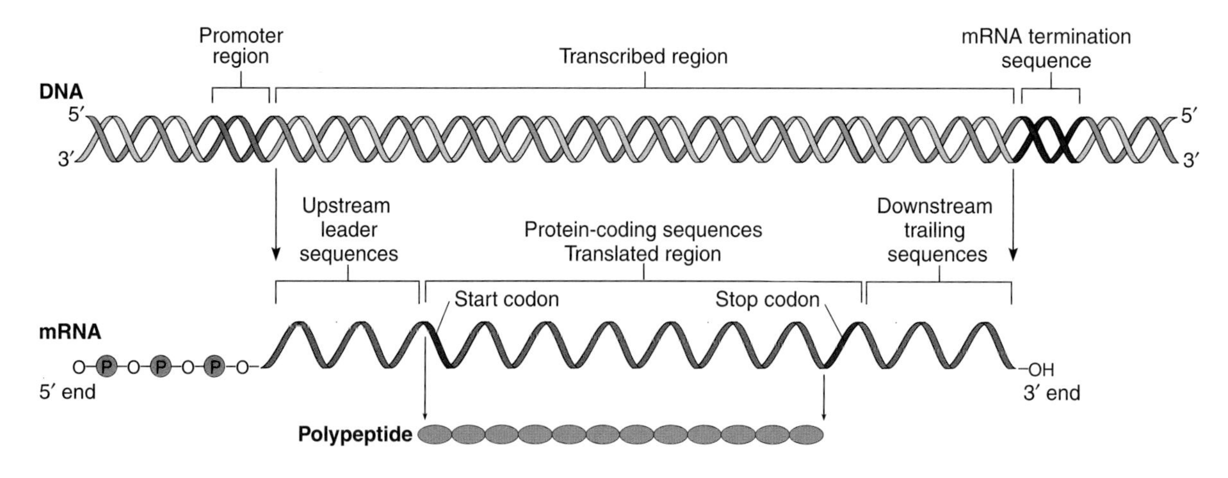

FIGURE **12-8** **Bacterial mRNA.**

This figure compares a bacterial mRNA with the region of DNA from which it was transcribed. RNA polymerase recognizes, but does not transcribe, promoter sequences in the DNA. Initiation of RNA synthesis occurs five to eight bases downstream from the promoter. Ribosome recognition sites are located in the 5′ mRNA upstream leader sequences. Protein-coding sequences begin at a start codon, which follows the leader sequences, and end at a downstream stop codon near the 3′ end of the molecule. Downstream trailing sequences, which vary in length, follow the protein-coding sequences.

Amino acids are joined by *peptide bonds* to form proteins (see Chapter 3). This joining links the amino and carboxyl groups of adjacent amino acids. The translation process ensures both that peptide bonds form and that the amino acids link in the correct sequence specified by the codons in the mRNA.

An amino acid is attached to tRNA before incorporation into a polypeptide

How do the amino acids align in the proper sequence so they can link together? Crick proposed that a molecule was needed to serve as an "adapter" in protein synthesis and bridge the gap between mRNA and proteins. Crick's adapters turned out to be tRNA molecules, as mentioned earlier. DNA contains tRNA genes that are transcribed to form the tRNAs. Each kind of tRNA molecule binds to a specific amino acid. Amino acids are covalently linked to their respective tRNA molecules by specific enzymes, **aminoacyl-tRNA synthetases,** which use ATP as an energy source. The resulting complexes, called **aminoacyl-tRNAs,** bind to the mRNA coding sequence to align the amino acids in the correct order to form the polypeptide chain.

The tRNAs are polynucleotide chains 70 to 80 nucleotides long, each with several unique base sequences, as well as some sequences that are common to all (Fig. 12-9). Although considerably smaller than mRNA or rRNA molecules, tRNA molecules have a complex structure. A tRNA molecule has several properties: (1) It has an anticodon, a specific complementary binding sequence for the correct mRNA codon; (2) it is recognized by a specific aminoacyl-tRNA synthetase that adds the correct amino acid; (3) it has an attachment site for the specific amino acid specified by the anticodon (the site lies about 180 degrees from the anticodon) and (4) it is recognized by ribosomes.

Complementary base-pairing within each tRNA molecule causes it to be doubled back and folded. Three or more loops of unpaired nucleotides are formed, one of which contains the anticodon. The amino acid–binding site is at the 3′ end of the molecule. The carboxyl group of the amino acid is bound to the exposed 3′ hydroxyl group of the terminal nucleotide, leaving the amino group of the amino acid free to participate in peptide bond formation. The pattern of folding in tRNAs keeps a constant distance between the anticodon and amino acid, allowing precise positioning of the amino acids during translation.

The components of the translational machinery come together at the ribosomes

The importance of ribosomes and of protein synthesis in cell metabolism is exemplified by a rapidly growing *E. coli* cell, which contains some 15,000 ribosomes—nearly one third of the total mass of the cell. Although bacterial and eukaryotic ribosomes are not identical, ribosomes from all organisms share many fundamental features and basically work in the same way. Ribosomes consist of two subunits, both made up of protein (about 40% by weight) and rRNA (about 60% by weight). Unlike mRNA and tRNA, rRNA does not transfer specific information but instead has catalytic functions. The ribosomal proteins do not appear to be catalytic but instead contribute to the overall structure of the ribosome.

Researchers have isolated each ribosomal subunit intact in the laboratory, then separated each into its RNA and protein constituents. In bacteria, the smaller of these subunits contains 21 proteins and one rRNA molecule, and the larger subunit contains 34 proteins and two rRNA molecules. Under certain conditions researchers can reassemble each subunit into a functional form by adding each component in its correct order. Using this approach, together with sophisticated electron microscopic studies, researchers have determined the 3-D structure of the ribo-

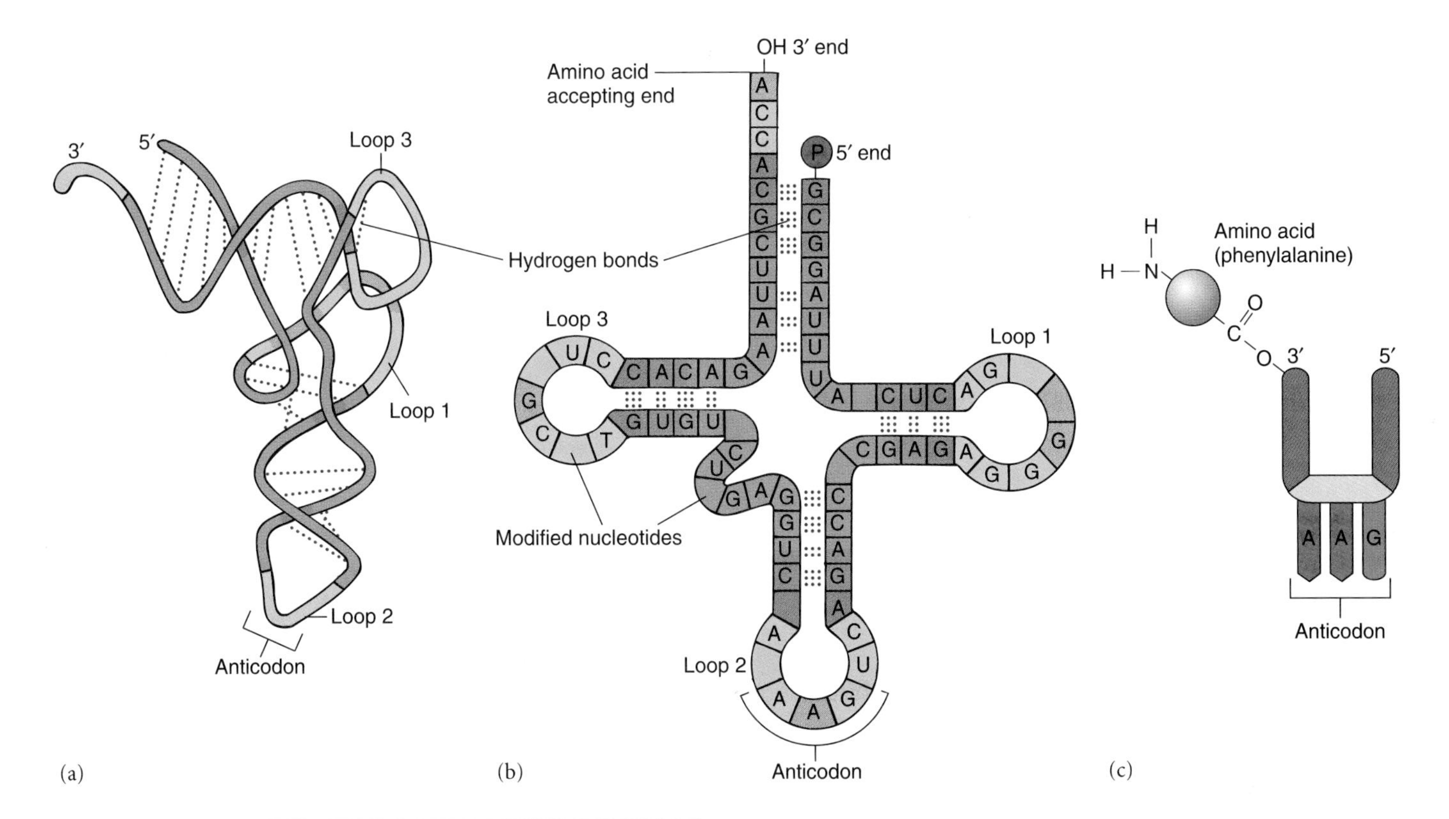

FIGURE **12-9** **Three representations of a tRNA molecule.**
(a) The 3-D shape of a tRNA molecule is determined by hydrogen bonds formed between complementary bases. **(b)** One loop contains the anticodon; these unpaired bases pair with a complementary mRNA codon. The amino acid attaches to the terminal nucleotide at the hydroxyl (OH) 3′ end. **(c)** This diagram of an aminoacyl-tRNA shows that the amino acid attaches to tRNA by its carboxyl group, leaving its amino group exposed for peptide bond formation.

some (Fig. 12-10a), as well as how it is assembled in the living cell. The large subunit contains a depression on one surface into which the small subunit fits. During translation, the mRNA fits in a groove between the contact surfaces of the two subunits.

The structure of the ribosome has four binding sites, one for mRNA and three for tRNAs. Thus the ribosome holds not only the mRNA template but also the aminoacyl-tRNA molecules and the growing polypeptide chain in the correct orientation so the genetic code can be read and the next peptide bond formed. Transfer RNA molecules attach to three depressions on the ribosome, the A, P, and E binding sites (Fig. 12-10b). The **P site,** or peptidyl site, is so named because the tRNA holding the growing polypeptide chain occupies the P site. The **A site** is named the aminoacyl site because the aminoacyl-tRNA delivering the next amino acid in the sequence binds at this location. The **E site** (for *exit*) is where tRNAs that have delivered their amino acids to the growing polypeptide chain exit the ribosome.

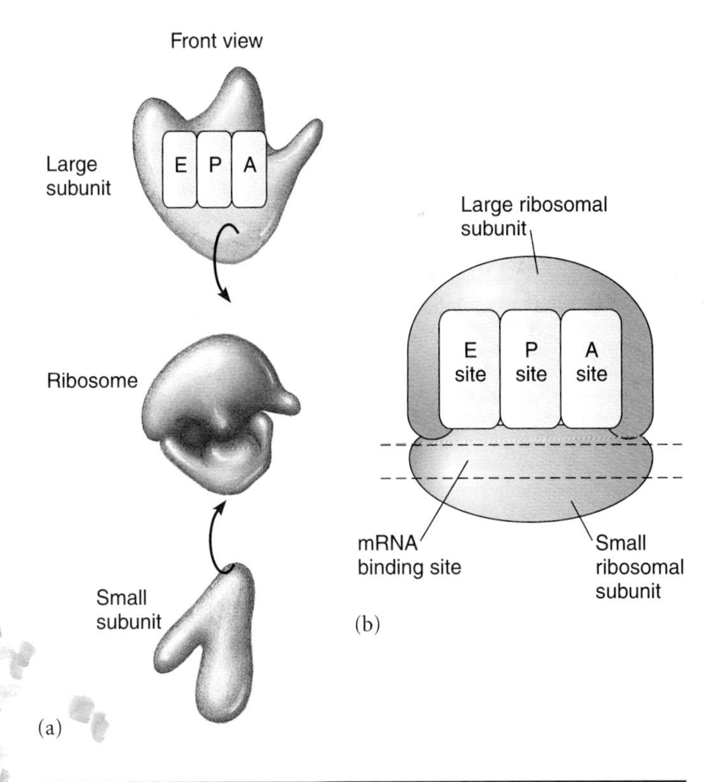

FIGURE **12-10** **Ribosome structure.**
(a) This model of a ribosome is based on a 3-D reconstruction of electron microscopic images. A ribosome consists of two subunits, one large and one small. **(b)** The mRNA passes through a groove between the two ribosomal subunits. Each ribosome contains three binding sites—the A, P, and E sites—that play an important role in protein synthesis.

Translation begins with the formation of an initiation complex

The process of protein synthesis has three distinct stages: initiation, repeating cycles of elongation, and termination. Initiation of protein synthesis is essentially the same in all organisms. Here we describe initiation in bacteria and then briefly discuss some differences between bacterial and eukaryotic initiation.

The **initiation** of translation uses proteins called **initiation factors,** which become attached to the small ribosomal subunit. In bacteria, three different initiation factors attach to the small subunit, which then binds to the mRNA molecule in the region of the AUG start codon (Fig. 12-11, step ①). A leader sequence upstream (toward the 5′ end) of the AUG helps the ribosome identify the AUG sequence that signals the beginning of the mRNA code for a protein.

The tRNA that bears the first amino acid of the polypeptide is the **initiator tRNA.** The amino acid methionine is bound to the initiator tRNA, and as a result, the first amino acid on new polypeptides is methionine. (Quite often, the methionine is removed later). In bacteria, the initiator methionine is modified by the addition of a one-carbon group derived from formic acid, and is designated *fMet* for *N*-formylmethionine. The fMet-initiator tRNA, which has the anticodon 3′—UAC—5′, binds to the AUG start codon, releasing one of the initiation factors in the process (Fig. 12-11, step ②). The **initiation complex** is complete when the large ribosomal subunit binds to the small subunit, and the remaining initiation factors are released (Fig. 12-11, step ③).

In eukaryotes, initiation of translation differs in three respects. One, the methionine of the initiator tRNA is unmodified. Two, instead of a leader sequence to help identify the start codon on mRNA, the start codon is *embedded* within a short sequence indicating the site of translation initiation. Three, the initiation complex in eukaryotes, which contains perhaps 10 protein factors, is more complex than in bacteria.

During elongation amino acids are added to the growing polypeptide chain

Figure 12-12 outlines the four steps in a cycle of **elongation,** the stage of translation in which other amino acids are added to the growing polypeptide. Elongation is essentially the same in bacteria and eukaryotes. In step ①, the appropriate aminoacyl-tRNA recognizes the codon in the A site and ② binds there by specific base-pairing of its anticodon with the complementary mRNA codon. This binding step requires several proteins called **elongation factors.** It also requires energy from **guanosine triphosphate (GTP),** an energy transfer molecule similar to ATP.

The amino group of the amino acid at the A site is now aligned with the carboxyl group of the preceding amino acid at the P site. In step ③ of elongation, a peptide bond forms between the amino group of the new amino acid and the carboxyl group of the preceding amino acid. In this way, the amino acid attached at the P site is released from its tRNA and attaches to the aminoacyl-tRNA at the A site. This reaction is spontaneous (it does not require additional energy), because ATP transferred energy during formation of the aminoacyl-tRNA. Peptide bond formation does, however, require the enzyme **peptidyl transferase.** Remarkably, this enzyme is not a protein but an rRNA component of the large ribosomal subunit. Such an RNA catalyst is known as a **ribozyme.**

Recall from Chapter 3 that polypeptide chains have direction, or polarity. The amino acid on one end of the polypeptide chain has a free amino group (the amino end), and the amino acid at the other end has a free carboxyl group (the carboxyl end). Protein synthesis always proceeds from the amino end to the carboxyl end of the growing polypeptide chain:

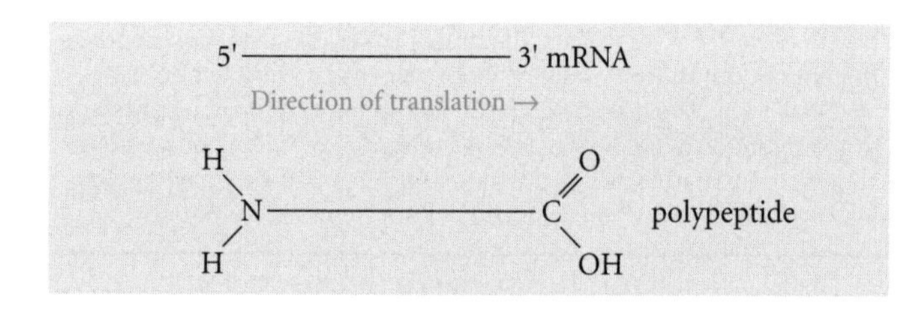

In step ④ of elongation, known as **translocation,** the ribosome moves down the mRNA by one codon. As a result, the

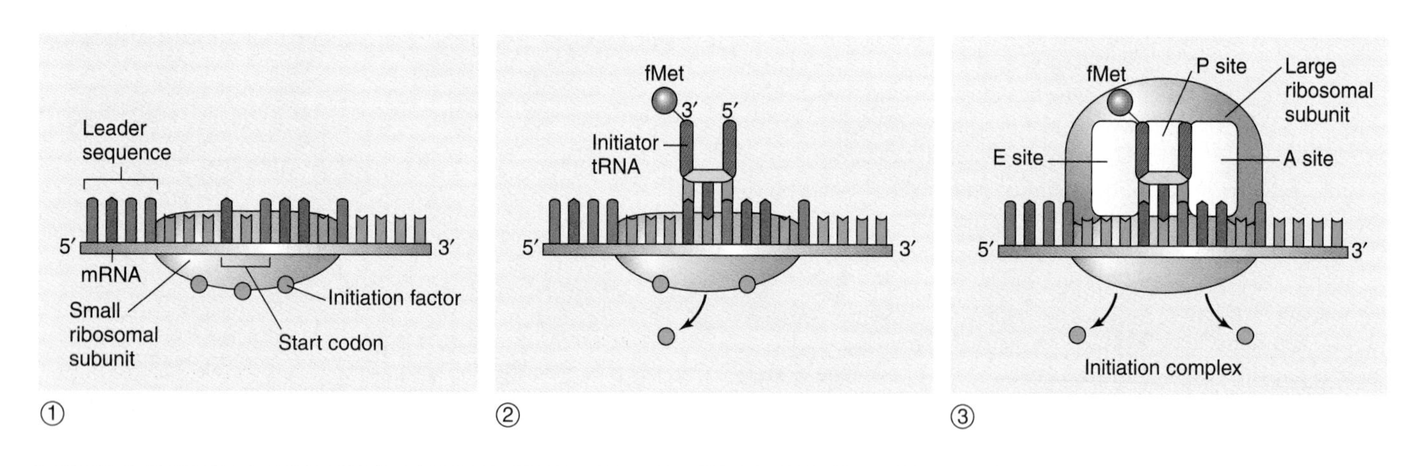

FIGURE **12-11** **The initiation of translation in bacteria.**
Initiation involves a ribosome, an mRNA, an initiator tRNA to which formylated methionine (fMet) is bound, and three different protein initiation factors. ① The small ribosomal subunit binds to the mRNA at the AUG start codon. The leader sequence upstream of the AUG sequence helps the ribosome identify the AUG sequence. ② The initiator tRNA binds to the start codon, and one of the initiation factors is released. ③ When the large ribosomal subunit binds to the small subunit and the remaining initiation factors are released, the initiation complex is complete.

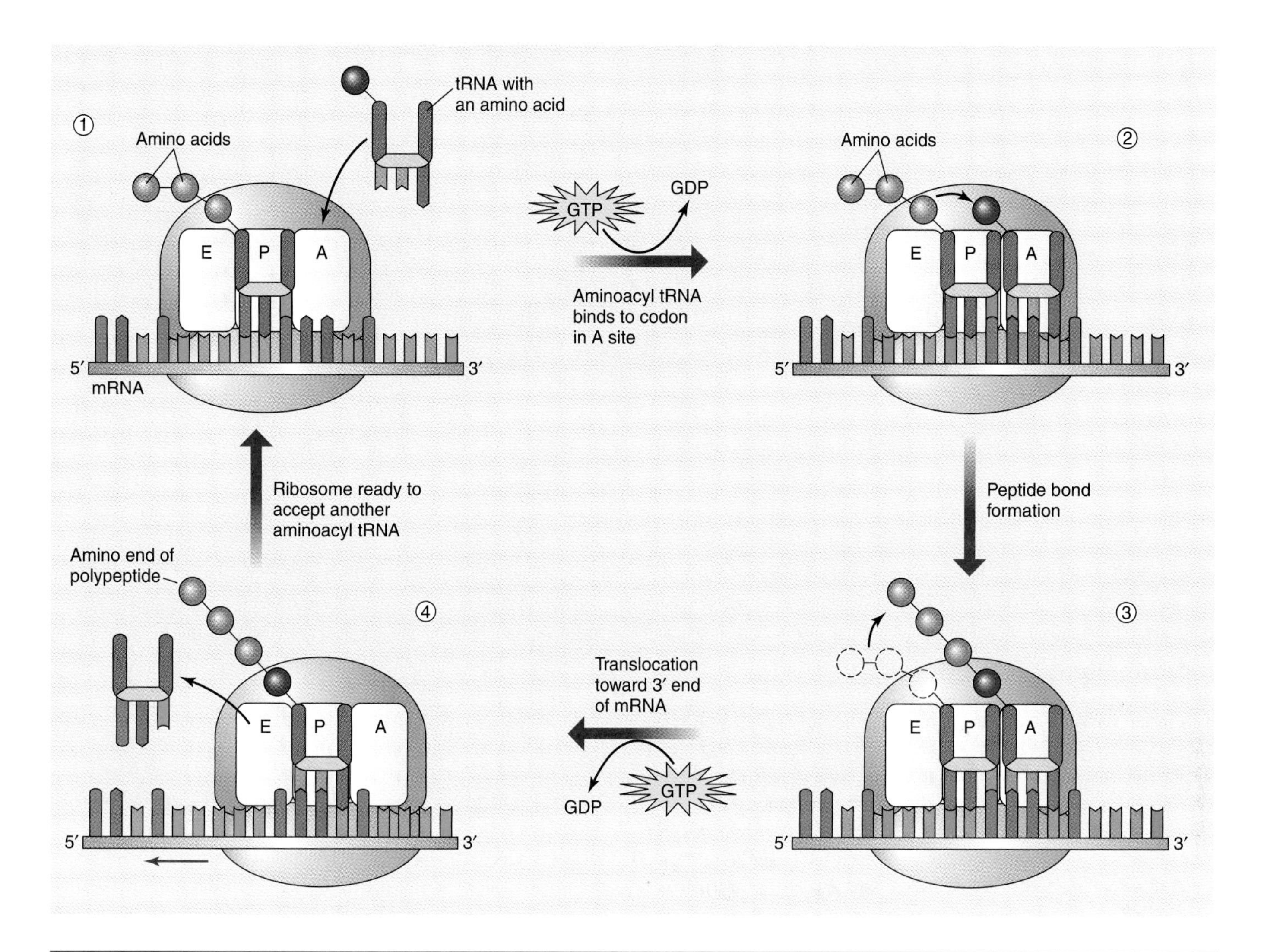

ACTIVE FIGURE **12-12** | **An elongation cycle in translation.**

Each repetition of the elongation process adds one amino acid to the growing polypeptide chain. This illustration begins after a short chain of amino acids has formed. ① The polypeptide chain is covalently bonded to the tRNA that carries the amino acid most recently added to the chain. This tRNA is in the P site of the ribosome. ② An aminoacyl-tRNA binds to the A site by complementary base-pairing between the tRNA's anticodon and the mRNA's codon. ③ The growing polypeptide chain detaches from the tRNA molecule in the P site and becomes attached by a peptide bond to the amino acid linked to the tRNA at the A site. ④ In the translocation step, the ribosome moves one codon toward the 3′ end of mRNA. As a result, the growing polypeptide chain is transferred to the P site. The uncharged tRNA in the E site exits the ribosome.

BiologyNow™ Learn more about **protein synthesis** by clicking on this figure on your BiologyNow CD ROM.

mRNA codon specifying the next amino acid in the polypeptide chain becomes positioned in the unoccupied A site. The translocation process requires energy, which again GTP supplies. The uncharged tRNA—that is, the tRNA without an attached amino acid—is moved from the P site to the E site. From there it exits the ribosome and joins the cytosolic pool of tRNAs.

Because translocation involves movement of the ribosome in the 3′ direction along the mRNA molecule, translation always proceeds in the 5′ to 3′ direction. Each peptide bond forms in a fraction of a second; by repeating the elongation cycle, an average-sized protein of about 360 amino acids is assembled by a bacterium in about 18 seconds, and by a eukaryotic cell in a little over 1 minute.

One of three stop codons signals the termination of translation

In **termination,** the final stage of translation, the synthesis of the polypeptide chain is terminated by a **release factor,** a protein that recognizes the stop codon at the end of the coding sequence. (No tRNA molecule binds to a stop codon, so the codon is available for binding by the release factor.) When the A site binds to the release factor, the bond between the tRNA in the P site and the last amino acid of the polypeptide chain breaks (Fig. 12-13). This hydrolysis reaction frees the newly synthesized polypeptide and also separates the translation complex into its parts: the mRNA molecule, the release factor, the tRNA molecule in the P site, and the large and small ribosomal

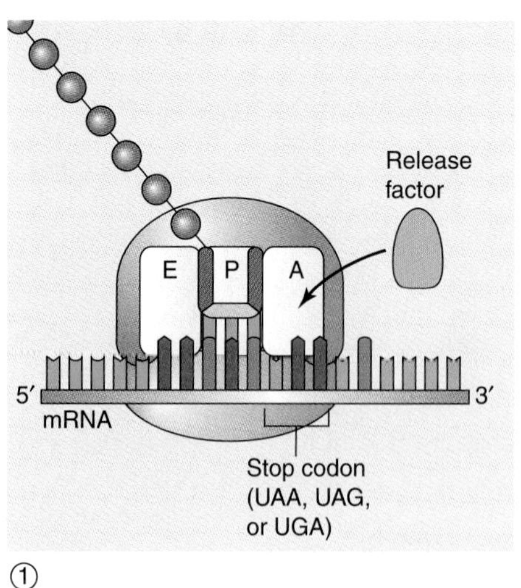

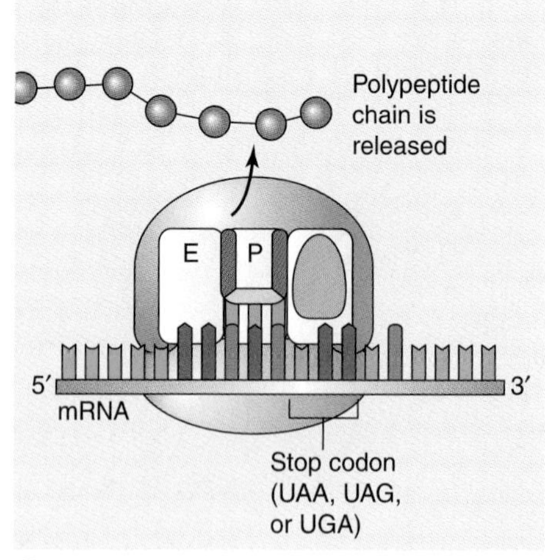

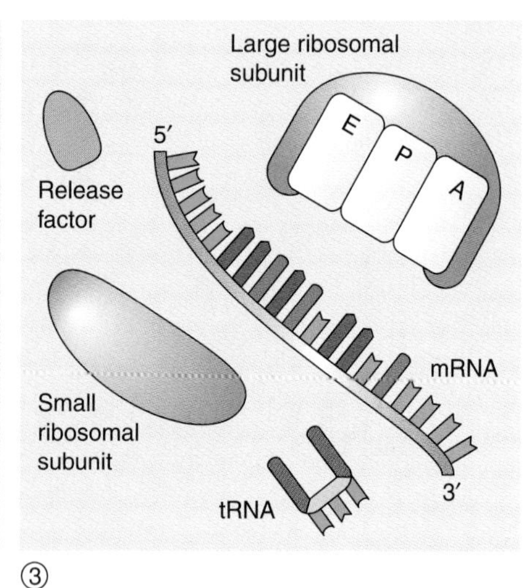

FIGURE 12-13 | The termination of translation.

No tRNA molecules exist with an anticodon complementary to a stop codon. ① When the ribosome reaches a stop codon, the A site binds to a protein release factor. ② The release factor hydrolyzes the bond between the polypeptide chain and the tRNA, causing the release of the polypeptide chain from the tRNA molecule in the P site. ③ The remaining parts of the translation complex dissociate.

subunits. The two ribosomal subunits can be used to form a new initiation complex with another mRNA molecule.

A polyribosome is a complex of one mRNA and many ribosomes

In *E. coli* and other bacteria, transcription and translation are *coupled* (Fig. 12-14). The end of the mRNA molecule synthesized first during transcription is also the first translated to form a polypeptide. Ribosomes bind to the 5′ end of the growing mRNA and initiate translation long before the message is fully synthesized. As many as 15 ribosomes may bind to a single mRNA. Molecules of mRNA bound to clusters of ribosomes are called **polyribosomes,** or *polysomes.* Polyribosomes also occur in eukaryotic cells.

Although many polypeptide chains can be actively synthesized on a single mRNA at any one time, the half-life of mRNA molecules in bacterial cells is only about 2 minutes. (Half-life is the time it takes for half the molecules to be degraded.) Usually, degradation of the 5′ end of the mRNA begins even before the first polypeptide is complete. Once the ribosome recognition sequences at the 5′ end of the mRNA have been degraded, no more ribosomes attach and initiate protein synthesis.

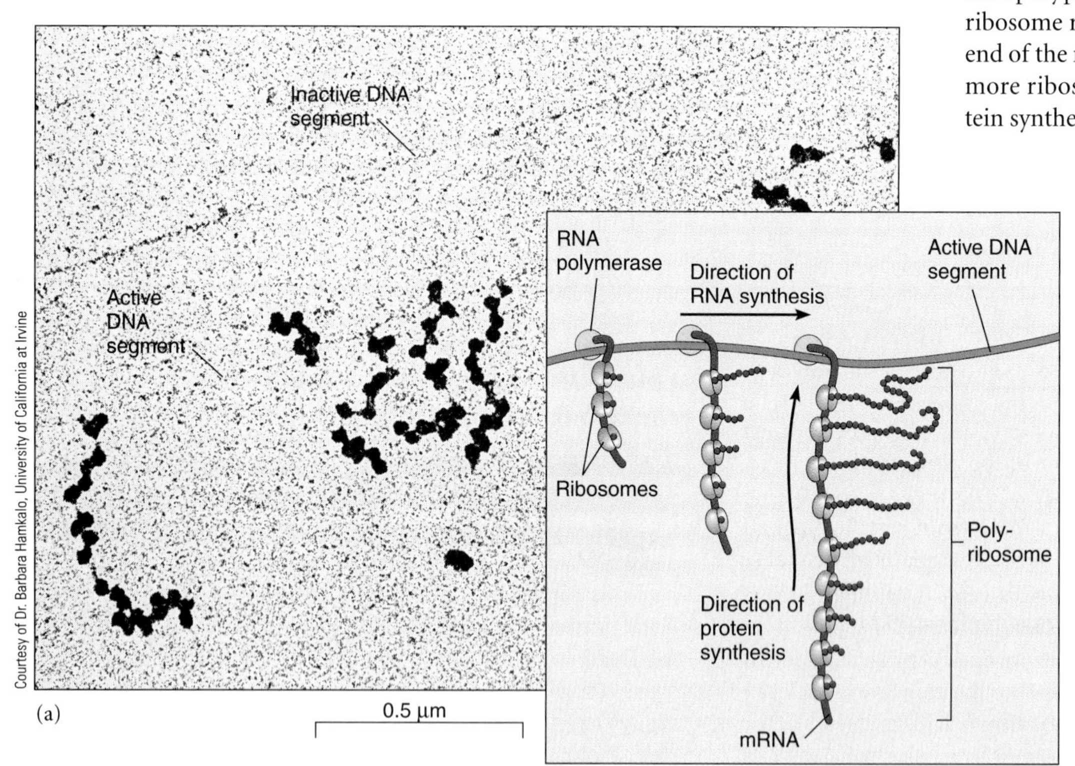

FIGURE 12-14 |

Coupled transcription and translation in bacteria.

(a) This TEM shows two strands of *E. coli* DNA, one inactive and the other actively producing mRNA. Protein synthesis begins while the mRNA is being completed, as multiple ribosomes attach to the mRNA to form a polyribosome. **(b)** A sequence of coupled transcription and translation.

Review

- What are ribosomes made of? Do ribosomes carry information to specify the amino acid sequence of proteins?
- How would you describe the initiation, elongation, and termination stages of protein synthesis?
- A certain mRNA strand has the following nucleotide sequence:

  5′—ATG—ACG—UAU—UAC—UTT—3′

  What is the anticodon for each codon? What is the amino acid sequence of the polypeptide? (Use Fig. 12-5.)

BiologyNow™ Assess your understanding of **translation** by taking the pretest on your BiologyNow CD-ROM.

VARIATIONS IN PROTEIN SYNTHESIS IN DIFFERENT ORGANISMS

Learning Objectives

9 Compare bacterial and eukaryotic mRNAs, and explain the functional significance of their structural differences.

10 Describe the differences in translation in bacterial and eukaryotic cells.

11 Describe retroviruses and the enzyme reverse transcriptase.

Although the basic mechanisms of transcription and translation are quite similar in all organisms, some significant differences separate bacteria and eukaryotes. Some result from differences in cell structure. Bacterial mRNA is translated as it is transcribed from DNA. This cannot occur in eukaryotes, because eukaryotic chromosomes are confined to the cell nucleus, and protein synthesis takes place in the cytosol. Before it is translated, eukaryotic mRNA is transported through the pores in the nuclear envelope and into the cytosol.

Many other differences separate bacterial and eukaryotic mRNA. Bacterial mRNAs are used immediately after transcription, without further processing. In eukaryotes, the original transcript, known as **precursor mRNA,** or **pre-mRNA,** is modified in several ways while it is still in the nucleus. These *posttranscriptional modification and processing* activities prepare mature mRNA for transport and translation (Fig. 12-15).

Modification of the eukaryotic message begins when the growing RNA transcript is about 20 to 30 nucleotides long. At that point, enzymes add a **5′ cap** to the 5′ end of the mRNA chain. The cap is in the form of an unusual nucleotide, 7-methylguanylate, which is guanosine monophosphate (GMP) with a methyl group added to one of the nitrogens in the base. Eukaryotic ribosomes cannot bind to an uncapped message.

Capping may also protect the RNA from certain types of degradation and may therefore partially explain why eukaryotic mRNAs are much more stable than bacterial mRNAs. Eukaryotic mRNAs have half-lives ranging from 30 minutes to as long as 24 hours. The average half-life of an mRNA molecule in a mammalian cell is about 10 hours, compared with 2 minutes in a bacterial cell.

A second modification of eukaryotic mRNA, known as **polyadenylation,** may occur at the 3′ end of the molecule. Near the 3′ end of a completed message usually lies a sequence of bases that serves as a signal for adding many adenine-containing nucleotides, known as a **poly-A tail** (for polyadenylated). Enzymes in the nucleus recognize the poly-A tail and cut the mRNA molecule at that site. This is followed by the enzymatic addition of a string of 100 to 250 adenine nucleotides to the 3′ end. Scientists do not completely understand the function of polyadenylation, although evidence indicates it helps export the mRNA from the nucleus, stabilizes some mRNAs against degradation in the cytosol, and facilitates initiation of translation.

Both noncoding and coding sequences are transcribed from eukaryotic genes

One of the greatest surprises in the history of molecular biology was the finding that most eukaryotic genes have **interrupted coding sequences;** that is, long sequences of bases within the protein-coding sequences of the gene do not code for amino acids in the final protein product. The noncoding regions within the gene are called **introns** (intervening sequences), as opposed to **exons** (expressed sequences), which are parts of the protein-coding sequence.

A typical eukaryotic gene may have multiple exons and introns, although the number varies. For example, the β-globin gene, which produces one component of hemoglobin, contains 2 introns; the ovalbumin gene of egg white contains 7; and the gene specifying another egg white protein, conalbumin, contains 16. In many cases, the combined lengths of the intron sequences are much greater than the combined lengths of the exons. For instance, the ovalbumin gene contains about 7700 base pairs, whereas the total of all the exon sequences is only 1859 base pairs.

When a gene that contains introns is transcribed, the entire gene is copied as a large RNA transcript, the pre-mRNA (see Fig. 12-15). A pre-mRNA molecule contains both exon and intron sequences. (Note that *intron* and *exon* refer to corresponding nucleotide sequences in both DNA and RNA.) For the pre-mRNA to be made into a functional message, it must be capped and have a poly-A tail added, and the introns must be removed and the exons spliced together to form a continuous protein-coding message.

Splicing itself occurs by several different mechanisms, depending on which type of RNA is involved. In many instances, splicing involves the association of **small nuclear ribonucleoprotein complexes (snRNPs),** which bind to the introns and catalyze the excision and splicing reactions. In some cases, the RNA within the intron acts as a ribozyme, splicing itself without the use of protein enzymes.

Posttranscriptional modification in eukaryotes is summarized as follows:

Pre-mRNA containing introns and exons ⟶ 5′ end of pre-mRNA capped with modified nucleotide ⟶ poly-A tail added to 3′ end ⟶ introns removed and exons spliced together ⟶ mature mRNA transported into cytosol ⟶ translation at ribosome

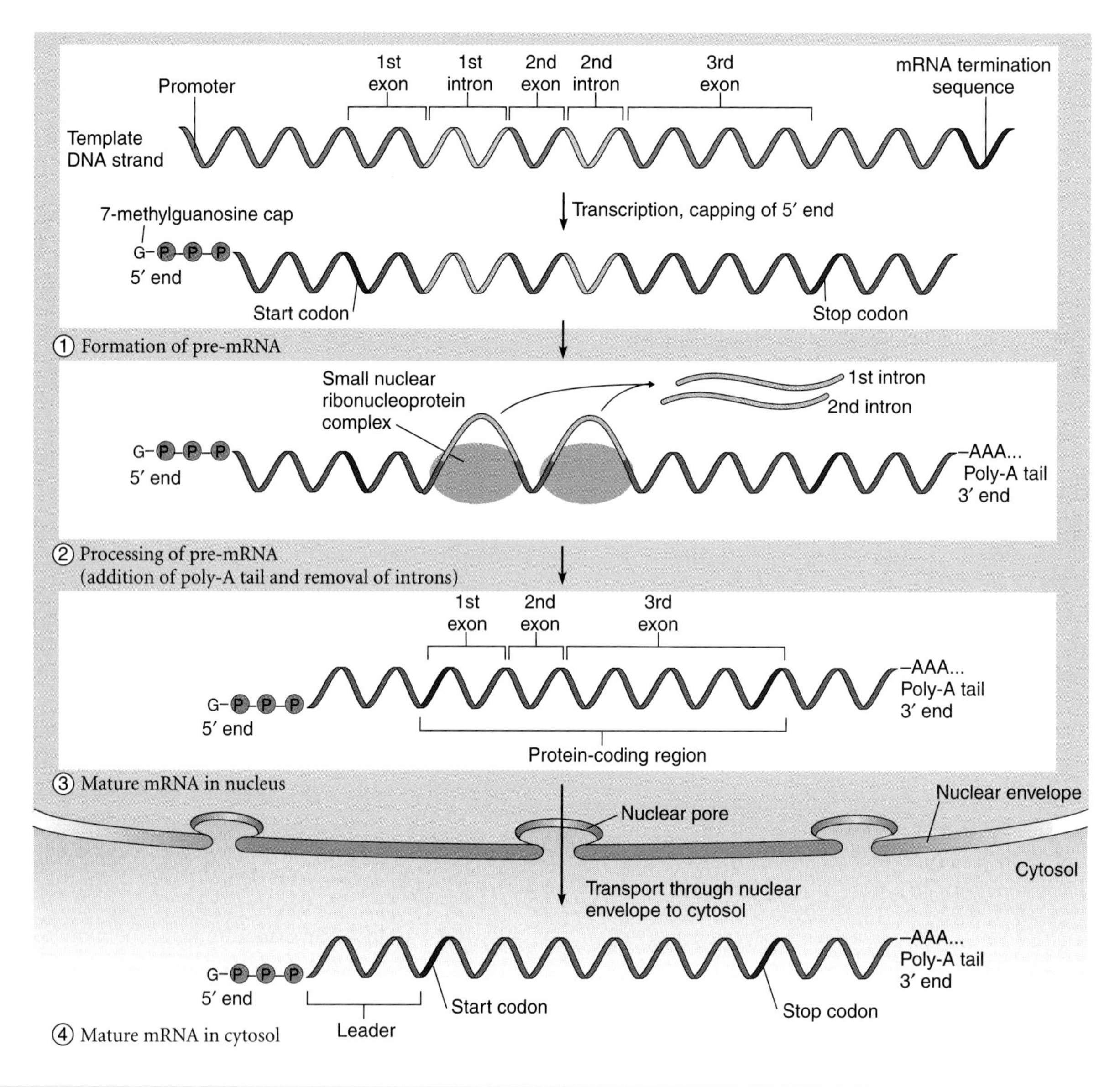

FIGURE **12-15** **Posttranscriptional modification of eukaryotic RNA.**

① A DNA sequence containing both exons and introns is transcribed by RNA polymerase to make the primary transcript, or pre-mRNA. As it is synthesized, the pre-mRNA is capped by the addition of a modified base attached "backward" (by a 5′—5′ linkage) to its 5′ end. ② A poly-A tail (100 to 250 nucleotides long) is added to the 3′ end; introns are removed, and the exons are spliced together. ③ The mature mRNA is transported through the nuclear envelope and ④ into the cytosol to be translated by a ribosome.

The evolution of eukaryotic gene structure is not completely understood

The reason for the complex structure of eukaryotic genes is a matter of ongoing debate among molecular biologists. Why do introns occur in most eukaryotic nuclear genes but not in the genes of most bacteria (or of mitochondria and chloroplasts)? How did this remarkable genetic system with interrupted coding sequences ("split genes") evolve, and why has it survived? It seems incredible that as much as 75% of the original transcript of a eukaryotic nuclear gene must be *removed* to make a working message.

In the early 1980s, Walter Gilbert of Harvard University proposed that exons are nucleotide sequences that code for **protein domains,** regions of protein tertiary structure that may have specific functions. For example, the active site of an enzyme may comprise one domain. A different domain may enable that enzyme to bind to a particular cell structure, and yet another may be a site involved in allosteric regulation (see Chapter 6). Analyses of the DNA and amino acid sequences of many eukaryotic genes have shown that most exons are too small to code for an entire protein domain, although a block of several exons can code for a domain.

Gilbert further postulated that new proteins with new functions emerge rapidly when genetic recombination produces novel combinations of exons that code for different proteins. This hypothesis has become known as *evolution by exon shuf-*

fling. It has been supported by examples such as the low-density lipoprotein (LDL) receptor protein, a protein found on the surface of human cells that binds to cholesterol transport molecules (see Chapter 5). The LDL receptor protein has domains that are related to parts of other proteins with totally different functions. However, many other genes and their corresponding proteins show no evidence of exon shuffling.

Some scientists hypothesize that introns first evolved in the nucleus of an early eukaryote and were propagated as movable genetic elements, known as *transposons* (to be discussed shortly). Regardless of how split genes originated, intron excision provides one of many ways in which present-day eukaryotes regulate expression of their genes (see Chapter 13). This opportunity for control, together with the fact that eukaryotic RNAs are far more stable than those of bacteria, may balance the energy cost of maintaining a large load of noncoding DNA.

The usual direction of information flow has exceptions

PROCESS OF SCIENCE

For several decades, a central premise of molecular biology was that genetic information always flows from DNA to RNA to protein. Through his studies of viruses, U.S. biologist Howard Temin discovered an important exception to this rule in 1964. Although viruses are not cellular organisms, they contain a single type of nucleic acid and reproduce in a host cell. Temin was studying unusual, cancer-causing tumor viruses that have RNA, rather than DNA, as their genetic material. He found that infection of a host cell by one of these viruses is blocked by inhibitors of DNA synthesis and also by inhibitors of transcription. These findings suggested that DNA synthesis and transcription are required for RNA tumor viruses to multiply and that there must be a way for information to flow in the "reverse" direction—from RNA to DNA.

Temin proposed that a **DNA provirus** forms as an intermediary in the replication of RNA tumor viruses. This hypothesis required a new kind of enzyme that would synthesize DNA using RNA as a template. In 1970, Temin and U.S. biologist David Baltimore discovered just such an enzyme, and they shared the Nobel Prize in Medicine in 1975 for their discovery. This RNA-directed DNA polymerase, also known as **reverse transcriptase,** is found in all RNA tumor viruses. (Some RNA viruses that do not produce tumors, however, replicate directly without using a DNA intermediate.) Figure 12-16 shows the steps of RNA tumor virus reproduction. Because they reverse the usual direction of information flow, viruses that require reverse transcriptase are called **retroviruses.** HIV-1, the virus that causes AIDS, is the most widely known retrovirus. As you will see in Chapter 14, the reverse transcriptase enzyme has become an extremely important research tool for molecular biologists.

Review

- What features do mature eukaryotic mRNA molecules have that bacterial mRNAs lack?
- How do retroviruses use the enzyme reverse transcriptase?

BiologyNow™ Assess your understanding of **variations in protein synthesis in different organisms** by taking the pretest on your BiologyNow CD-ROM.

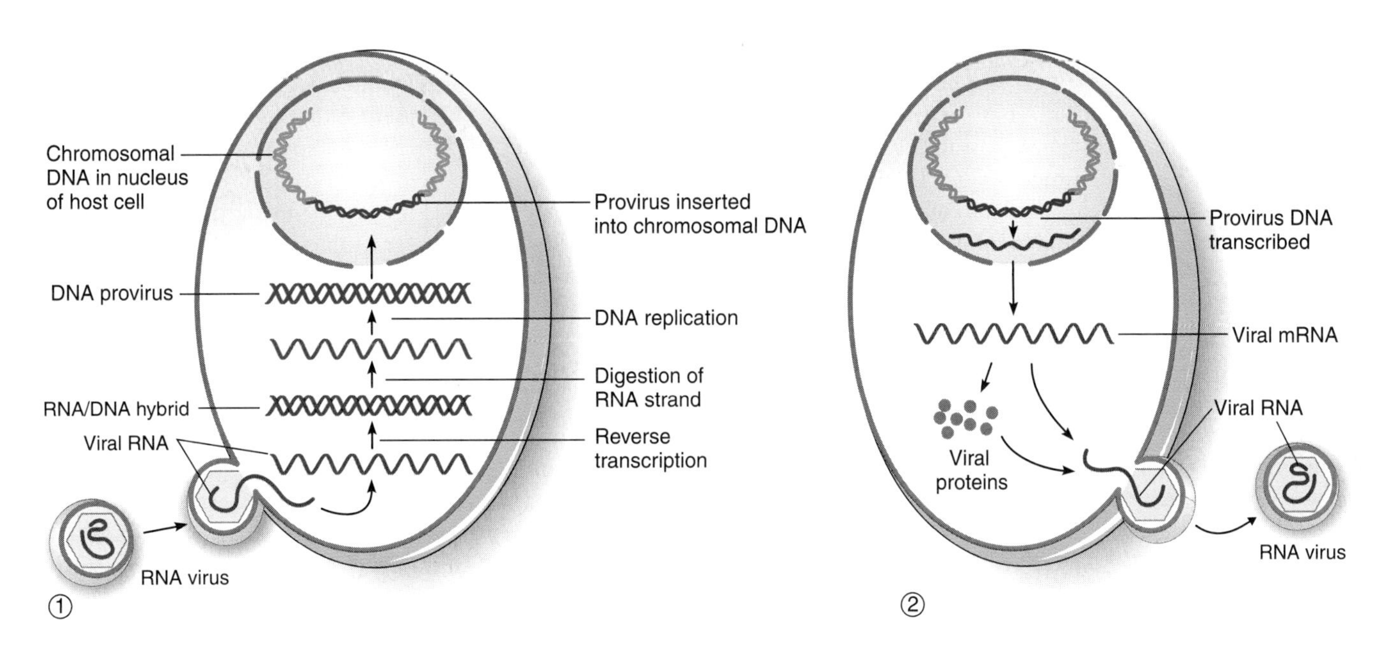

FIGURE **12-16** **The infection cycle of an RNA tumor virus.**
① After an RNA tumor virus enters the host cell, the viral enzyme reverse transcriptase synthesizes a DNA strand that is complementary to the viral RNA. Next, the RNA strand is degraded and a complementary DNA strand is synthesized, thus completing the double-stranded DNA provirus, which is then integrated into the host cell's DNA. ② The provirus DNA is transcribed, and the resulting viral mRNA is translated to form viral proteins. Additional viral RNA molecules are produced and then incorporated into mature viral particles enclosed by protein coats.

MUTATIONS AND GENES

Learning Objective

12 Give examples of the different classes of mutations that affect the base sequence of DNA, and explain the effects that each has on the protein produced.

One of the first major discoveries about genes was that they undergo **mutations,** changes in the nucleotide sequence of the DNA. However, the overall observed mutation rate is much lower than the frequency of damage to DNA, because all organisms have systems of enzymes that repair certain kinds of DNA damage.

As explained in Chapter 11, once the DNA sequence has been changed, DNA replication copies the altered sequence just as it would copy a normal sequence, making the mutation stable over an indefinite number of generations. In most cases, the mutant allele has no greater tendency than the original allele to mutate again. Mutations provide the diversity of genetic material that enables researchers to study inheritance and the molecular nature of genes. As you will see in later chapters, mutations also provide the variation necessary for evolution within a given species.

Mutation alters genes in several ways. Scientists now determine where a specific mutation occurs in a gene by using recombinant DNA methods to isolate the gene and determine its sequence of bases (see Chapter 14).

Base substitution mutations result from the exchange of one base pair for another

The simplest type of mutation, called a **base substitution,** involves a change in only one pair of nucleotides (Fig. 12-17a). Often these mutations result from errors in base pairing during the replication process. For example, a GC, CG, or TA pair might replace an AT base pair. Such a mutation may cause the altered DNA to be transcribed as an altered mRNA. The altered mRNA may then be translated into a polypeptide chain with only one amino acid different from the normal sequence.

Base substitutions that result in replacement of one amino acid by another are sometimes called **missense mutations.** Missense mutations have a wide range of effects. If the amino acid substitution occurs at or near the active site of an enzyme, the activity of the altered protein may decrease or even be destroyed. Some missense mutations involve a change in an amino acid that is not part of the active site. Others may result in the substitution of a closely related amino acid (one with very similar chemical characteristics). Such silent mutations may be undetectable if one simply examines their effect on the whole organism. Because silent mutations occur relatively frequently, the true number of mutations in an organism or a species is much greater than is actually observed.

Nonsense mutations are base substitutions that convert an amino acid–specifying codon to a stop codon. A nonsense mutation usually destroys the function of the gene product; in the case of a protein-specifying gene, the part of the polypeptide chain that follows the mutant stop codon is missing.

Frameshift mutations result from the insertion or deletion of base pairs

In **frameshift mutations,** one or two nucleotide pairs are inserted into or deleted from the molecule, altering the reading frame (Fig. 12-17b). As a result, codons downstream of the insertion or deletion site specify an entirely new sequence of amino acids. Depending on where the insertion or deletion occurs in the gene, different effects are generated. In addition to producing an entirely new polypeptide sequence immediately after the change, frameshift mutations may produce a stop codon within a short distance of the mutation. This codon terminates the already altered polypeptide chain. A frame shift in a gene specifying an enzyme usually results in a complete loss of enzyme activity.

Some mutations involve larger DNA segments

Some types of mutations are due to a change in chromosome structure. These changes usually have a wide range of effects because they involve many genes.

PROCESS OF SCIENCE

One type of mutation is caused by DNA sequences that "jump" into the middle of a gene. These movable sequences of DNA are known as **transposons,** or **transposable genetic elements.** They not only disrupt the functions of some genes but also inactivate some previously active genes. Transposons were discovered in maize (corn) by the U.S. geneticist Barbara McClintock in the 1950s. She observed that certain genes appeared to be "turned off" and "turned on" spontaneously. She deduced that the mechanism involved a segment of DNA that moved from one region of a chromosome to another, where it would inactivate a gene. However, biologists did not understand this phenomenon until the development of recombinant DNA methods and the discovery of transposons in a wide variety of organisms. We now know that transposons are segments of DNA that range from a few hundred to several thousand bases. For incorporation into a new location within the chromosome, transposons require the enzyme **transposase.** In recognition of her insightful findings, McClintock was awarded the Nobel Prize in Physiology or Medicine in 1983.

Most transposons are **retrotransposons,** which replicate by forming an RNA intermediate. Reverse transcriptase converts them to their original DNA sequence before jumping into a gene. Many retrotransposons enlist the host cell's gene expression machinery to produce reverse transcriptase. Because retrotransposons use reverse transcriptase, some biologists hypothesize that certain retrotransposons evolved from retroviruses, and vice versa. Scientists estimate that human cells have fewer than 100 retrotransposons actively jumping around. Nevertheless, retrotransposons are extremely important agents of mutation.

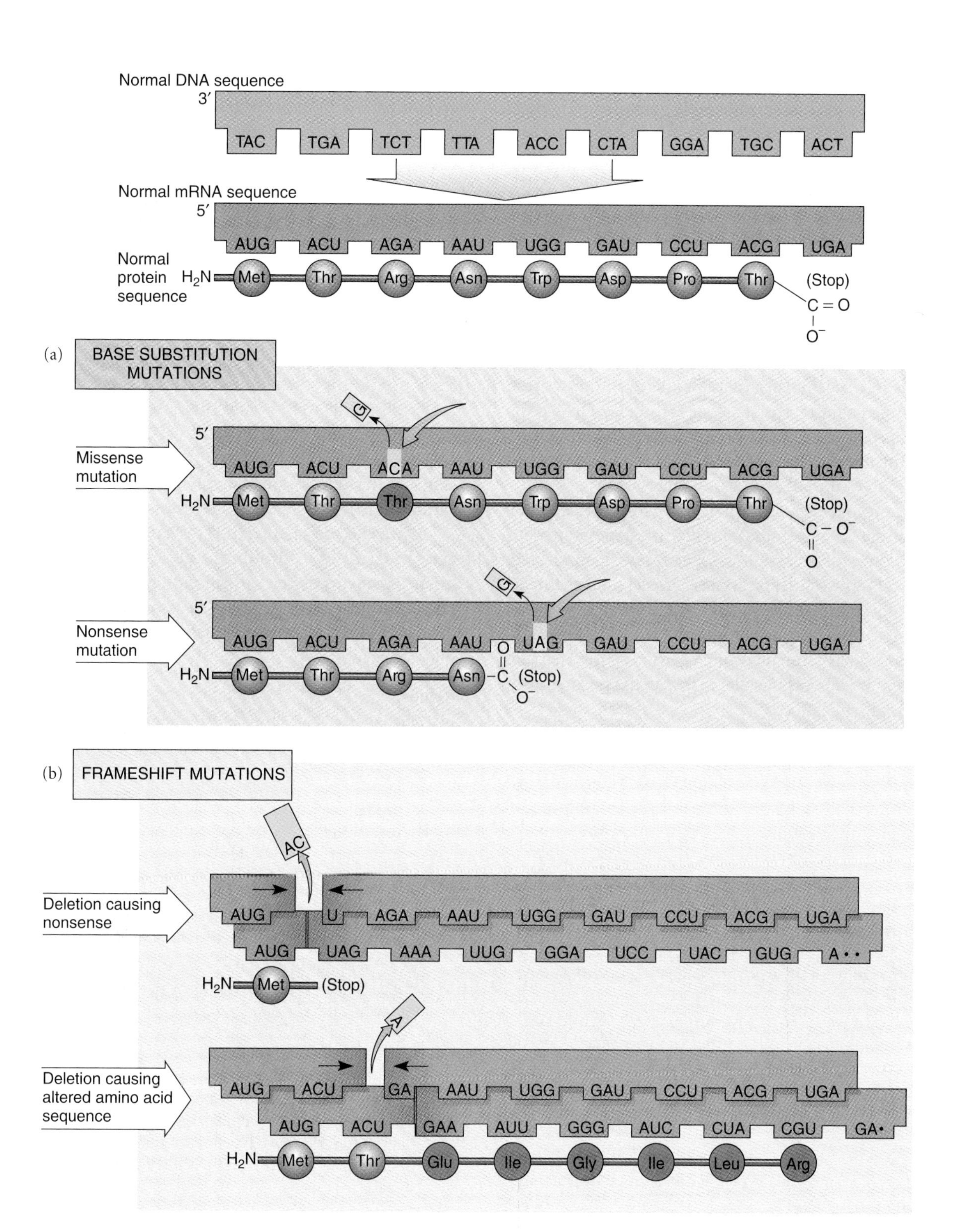

FIGURE 12-17 | Mutations.

(a) Missense and nonsense mutations are types of base substitution mutations. A missense mutation results in a protein of normal length, but with an amino acid substitution. A nonsense mutation, caused by conversion of an amino acid–specifying codon to a stop codon, results in the production of a truncated (shortened) protein, which is usually not functional. **(b)** A frameshift mutation results from the deletion or insertion (not shown) of one or two bases, causing the base sequence following the mutation to shift to a new reading frame. A frame shift may produce a stop codon downstream of the mutation (which would have the same effect as a nonsense mutation caused by base substitution), or it may produce an entirely new amino acid sequence.

By one estimate, retrotransposons cause about 10% of naturally occurring mutations that cause noticeable phenotypic changes.

Mutations have various causes

Most types of mutations occur infrequently but spontaneously from mistakes in DNA replication or defects in the mitotic or meiotic separation of chromosomes. Some regions of DNA, known as mutational *hot spots,* are much more likely than others to undergo mutation. An example is a short stretch of repeated nucleotides, which can cause DNA polymerase to "slip" while reading the template during replication.

Mutations in certain genes increase the overall mutation rate. For example, a mutation in a gene coding for DNA polymerase may make DNA replication less precise, or a mutation in a gene coding for a repair enzyme may allow more mutations to arise as a result of unrepaired DNA damage.

Not all mutations occur spontaneously. **Mutagens,** which cause many of the types of mutations discussed previously, include various types of radiation, such as x-rays, gamma rays, cosmic rays, and ultraviolet rays. Some chemical mutagens react with and modify specific bases in the DNA, leading to mistakes in complementary base pairing when the DNA molecule is replicated. Other mutagens cause nucleotide pairs to be inserted into or deleted from the DNA molecule, changing the normal reading frame during replication.

Despite the presence of enzymes that repair damage to DNA, some new mutations do arise. In fact, each of us has some mutant alleles that either of our parents didn't have. Although some of these mutations alter the phenotype, most are not noticeable, because they are recessive.

Mutations that occur in the cells of the body (somatic cells) are not passed on to offspring. However, these mutations are of concern because somatic mutations and cancer are closely related. Many mutagens are also **carcinogens,** agents that cause cancer.

A gene is a functional unit

At the beginning of this chapter, we traced the development of ideas regarding the nature of the gene. For a time scientists found it useful to define a gene as a sequence of nucleotides that codes for one polypeptide chain. As biologists have learned more about how genes work, they have revised the definition. They now know some genes are transcribed to produce RNA molecules such as rRNA and tRNA, whereas others specify the RNA component of the snRNPs used to modify mRNA molecules. Studies have also shown that in eukaryotic cells, a single gene may produce more than one polypeptide chain by modifications in the way the mRNA is processed.

As we end this chapter, it is perhaps most useful to define a gene in terms of its product. A **gene** is a DNA nucleotide sequence that carries the information needed to produce a specific RNA or protein product. As you will see in Chapter 13, a gene includes noncoding regulatory sequences, as well as coding sequences.

Review

- What are the main types of mutations?
- What effects does each type of mutation have on the protein product produced?

BiologyNow™ Assess your understanding of **mutations and genes** by taking the pretest on your BiologyNow CD-ROM.

SUMMARY WITH KEY TERMS

1 Summarize the early evidence indicating that most genes specify the structure of proteins.

- Garrod's early work on inborn errors of metabolism and that of Beadle and Tatum in the 1940s with *Neurospora* mutants provided strong evidence that genes specify proteins.

2 Outline the flow of genetic information in cells, from DNA to protein.

- The process by which information encoded in DNA specifies the sequences of amino acids in proteins involves two steps: transcription and translation.
- During **transcription,** an RNA molecule complementary to the template DNA strand is synthesized. **Messenger RNA (mRNA)** molecules contain information that specifies the amino acid sequences of polypeptide chains.
- During **translation,** a polypeptide chain specified by the mRNA is synthesized. Each sequence of three nucleotide bases in the mRNA constitutes a **codon,** which specifies one amino acid in the polypeptide chain, or a start or stop signal. Translation requires tRNAs and cell machinery, including ribosomes.

3 Compare the structures of DNA and RNA.

- RNA is formed from nucleotide subunits, each of which contains the sugar ribose, a base (uracil, adenine, guanine, or cytosine), and three phosphates. Like DNA, RNA subunits are covalently joined by a 5′—3′ linkage to form an alternating sugar-phosphate backbone.

4 Explain why the genetic code is said to be redundant and virtually universal, and discuss how these features may reflect its evolutionary history.

- The **genetic code** is read from mRNA as a series of nonoverlapping codons that specify a single sequence of amino acids. Of the 64 codons, 61 code for amino acids, and 3 codons serve as stop signals.
- The genetic code is redundant because some amino acids are specified by more than one codon. The genetic code is virtually universal, strongly suggesting that all organisms are descended from a common ancestor. Only a few minor variations are exceptions to the standard code.

5 Compare the processes of transcription and DNA replication, identifying both similarities and differences.

- **DNA-dependent RNA polymerases,** involved in RNA synthesis, have many similarities to DNA polymerases involved in DNA replication. Both enzymes carry out synthesis in the 5′ ⟶ 3′ direction. Both use nucleotides with three phosphate groups as substrates, removing two of the phosphates as the nucleotides are covalently linked to the 3′ end of the newly synthesized strand.

- Just as the paired strands of DNA are **antiparallel,** the template strand of the DNA and its complementary RNA strand are also antiparallel. As a result, the DNA template strand is read in its 3′ ⟶ 5′ direction as RNA is synthesized in its 5′ ⟶ 3′ direction.
- The same base-pairing rules are followed as in DNA replication, except that uracil is substituted for thymine.

6 Identify the features of tRNA that are important in decoding genetic information and converting it into "protein language."

- **Transfer RNAs (tRNAs)** are the "decoding" molecules in the translation process.
- Each tRNA molecule is specific for only one amino acid. One part of the molecule contains an **anticodon,** which is complementary to a codon of mRNA. Attached to one end of the tRNA molecule is the amino acid specified by the complementary mRNA codon.
- Amino acids are covalently bound to tRNA by **aminoacyl-tRNA synthetase** enzymes.

7 Explain how ribosomes function in protein synthesis.

- **Ribosomes** bring together all the mechanical machinery necessary for translation. They couple the tRNAs to their proper codons on the mRNA, catalyze the formation of peptide bonds between amino acids, and translocate the mRNA so the next codon can be read.
- Each ribosome is made of a large and a small subunit; each subunit contains **ribosomal RNA (rRNA)** and many proteins.

8 Diagram the processes of initiation, elongation, and termination in protein synthesis.

- **Initiation** is the first stage of translation. **Initiation factors** bind to the small ribosomal subunit, which then binds to mRNA in the region of AUG, the **start codon.** The **initiator tRNA** binds to the start codon, followed by binding of the large ribosomal subunit.
- **Elongation** is a cyclic process in which amino acids are added one by one to the growing polypeptide chain. Elongation proceeds in the 5′ ⟶ 3′ direction along the mRNA. The polypeptide chain grows from its amino end to its carboxyl end.
- **Termination,** the final stage of translation, occurs when the ribosome reaches one of three **stop codons.** The A site binds to a **release factor,** which triggers the release of the completed polypeptide chain and dissociation of the translation complex.

9 Compare bacterial and eukaryotic mRNAs, and explain the functional significance of their structural differences.

- Eukaryotic genes and their mRNA molecules are more complex than those of bacteria.
- After transcription, a **5′ cap** (a modified guanosine triphosphate) is added to the 5′ end of a eukaryotic mRNA molecule. Many also have a **poly-A tail** of adenine-containing nucleotides added at the 3′ end. These modifications seem to protect eukaryotic mRNA molecules from degradation, giving them a longer life span than bacterial mRNA.
- In many eukaryotic genes the coding regions, called **exons,** are interrupted by noncoding regions, called **introns.** Both introns and exons are transcribed, but the introns are later removed from the original transcript, or **pre-mRNA,** and the exons are spliced together to produce a continuous protein-coding sequence.

10 Describe the differences in translation in bacterial and eukaryotic cells.

- Unlike eukaryotic cells, in bacterial cells transcription and translation are coupled. Translation of the bacterial mRNA molecule usually begins before the 3′ end of the transcript is completed.

11 Describe retroviruses and the enzyme reverse transcriptase.

- The flow of genetic information is reversed by **reverse transcriptase,** an enzyme associated with **retroviruses,** which synthesize DNA from an RNA template. HIV-1, the virus that causes AIDS, is a retrovirus.

12 Give examples of the different classes of mutations that affect the base sequence of DNA, and explain the effects that each has on the protein produced.

- Types of **mutations** range from disruption of a chromosome's structure to a change in only a single pair of nucleotide bases. A **base substitution** destroys the function of a protein if a codon changes so that it specifies a different amino acid (a **missense mutation,**) or becomes a stop codon (a **nonsense mutation**). A base substitution has minimal effects if the amino acid is not altered or if the codon is changed to specify a chemically similar amino acid.
- The insertion or deletion of one or two base pairs in a gene invariably destroys the function of that protein because it results in a **frameshift mutation,** which changes the codon sequences downstream from the mutation.
- One type of mutation is caused by movable DNA sequences, known as **transposons,** that "jump" into the middle of a gene. Most transposons are **retrotransposons,** which replicate by forming an RNA intermediate; reverse transcriptase converts them to their original sequence before they jump into a gene.

POST-TEST

1. Beadle and Tatum (a) predicted that tRNA molecules would have anticodons (b) discovered the genetic disease alkaptonuria (c) showed that the genetic disease sickle cell anemia is caused by a change in a single amino acid in a hemoglobin polypeptide chain (d) worked out the genetic code (e) studied the relationship between genes and enzymes in *Neurospora*
2. What is the correct order of information flow in bacterial and eukaryotic cells? (a) DNA ⟶ mRNA ⟶ protein (b) protein ⟶ mRNA ⟶ DNA (c) DNA ⟶ protein ⟶ mRNA (d) protein ⟶ DNA ⟶ mRNA (e) mRNA ⟶ protein ⟶ DNA
3. During transcription, how many RNA nucleotide bases would usually be encoded by a sequence of 99 DNA nucleotide bases? (a) 297 (b) 99 (c) 33 (d) 11 (e) answer is impossible to determine with the information given
4. The genetic code is defined as a series of _______________ in _______________. (a) anticodons; tRNA (b) codons; DNA (c) anticodons; mRNA (d) codons; mRNA (e) codons and anticodons; rRNA

POST-TEST *(continued)*

5. Transcription is the process by which _______________ is/are synthesized. (a) mRNA only (b) mRNA and tRNA (c) mRNA, tRNA, and rRNA (d) protein (e) mRNA, tRNA, rRNA, and protein
6. RNA differs from DNA in that the base _______________ is substituted for _______________. (a) adenine; uracil (b) uracil; thymine (c) guanine; uracil (d) cytosine; guanine (e) guanine; adenine
7. RNA grows in the _______________ direction, as DNA-dependent RNA polymerase moves along the template DNA strand in the _______________ direction. (a) $5' \longrightarrow 3'$; $3' \longrightarrow 5'$ (b) $3' \longrightarrow 5'$; $3' \longrightarrow 5'$ (c) $5' \longrightarrow 3'$; $5' \longrightarrow 3'$ (d) $3' \longrightarrow 3'$; $5' \longrightarrow 5'$ (e) $5' \longrightarrow 5'$; $3' \longrightarrow 3'$
8. Which of the following is/are *not* found in a bacterial mRNA molecule? (a) stop codon (b) upstream leader sequences (c) downstream trailing sequences (d) start codon (e) promoter sequences
9. Which of the following is/are typically removed from pre-mRNA during nuclear processing in eukaryotes? (a) upstream leader sequences (b) poly-A tail (c) introns (d) exons (e) all the preceding are removed
10. Which of the following is a spontaneous process, with no direct requirement for ATP or GTP? (a) formation of a peptide bond (b) translocation of the ribosome (c) formation of aminoacyl-tRNA (d) a and b (e) all the preceding
11. The role of tRNA is to transport (a) amino acids to the ribosome (b) amino acids to the nucleus (c) initiation factors to the ribosome (d) mRNA to the ribosome (e) release factors to the ribosome
12. Select the steps of an elongation cycle during protein synthesis from the following list and place them in the proper sequence:
    1. Peptide bond formation
    2. Binding of the small ribosomal subunit to the 5′ end of the mRNA
    3. Binding of aminoacyl-tRNA to the A site
    4. Translocation of the ribosome
    (a) $1 \longrightarrow 3 \longrightarrow 2 \longrightarrow 4$ (b) $3 \longrightarrow 1 \longrightarrow 4$ (c) $3 \longrightarrow 1 \longrightarrow 3 \longrightarrow 2$ (d) $1 \longrightarrow 3 \longrightarrow 4$ (e) $4 \longrightarrow 2 \longrightarrow 1 \longrightarrow 3$
13. Suppose you mix the following components of protein synthesis in a test tube: amino acids from a rabbit; ribosomes from a dog; tRNAs from a mouse; mRNA from a chimpanzee; and necessary enzymes plus an energy source from a giraffe. If protein synthesis occurs, which animal's protein will be made? (a) rabbit (b) dog (c) mouse (d) chimpanzee (e) giraffe
14. During elongation in translation, the growing polypeptide chain attaches to a tRNA in the _____ site, and an incoming aminoacyl-tRNA attaches to the _____ site. (a) E; P (b) P; E (c) E; A (d) A; E (e) P; A
15. What is the minimum number of tRNA molecules needed to produce a polypeptide chain 50 amino acids long but that has only 16 different kinds of amino acids? (a) 50 (b) 25 (c) 20 (d) 16 (e) 8
16. The statement "the genetic code is redundant" means that (a) some codons specify punctuation (stop and start signals) rather than amino acids (b) some codons specify more than one amino acid (c) certain amino acids are specified by more than one codon (d) the genetic code is read one triplet at a time (e) all organisms have essentially the same genetic code
17. A nonsense mutation (a) causes one amino acid to be substituted for another in a polypeptide chain (b) results from the deletion of one or two bases, leading to a shift in the reading frame (c) results from the insertion of one or two bases, leading to a shift in the reading frame (d) results from the insertion of a transposon (e) usually results in the formation of an abnormally short polypeptide chain

CRITICAL THINKING

1. Compare and contrast the formation of mRNA in bacterial and eukaryotic cells. How do the differences affect the way in which each type of mRNA is translated? Does one system have any obvious advantage in terms of energy cost? Which system offers greater opportunities for control of gene expression?
2. Biologists hypothesize that transposons eventually lose the ability to replicate themselves and therefore remain embedded in DNA without moving around. Based on what you have learned in this chapter, suggest a possible reason for this loss.

- Visit our Web site at **http://biology.brookscole.com/solomon7** for links to chapter-related resources on the World Wide Web. Additional online materials relating to this chapter can also be found on our Web site.

BIOLOGY NOW RESOURCES

Active Figures

12-7: Transcription

12-12: Protein synthesis

Preparing for an exam? Take a diagnostic test on your BiologyNow CD-ROM.

Post-Test Answers

1. e	2. a	3. b	4. d
5. c	6. b	7. a	8. e
9. c	10. a	11. a	12. b
13. d	14. e	15. d	16. c
17. e			

13

Gene Regulation

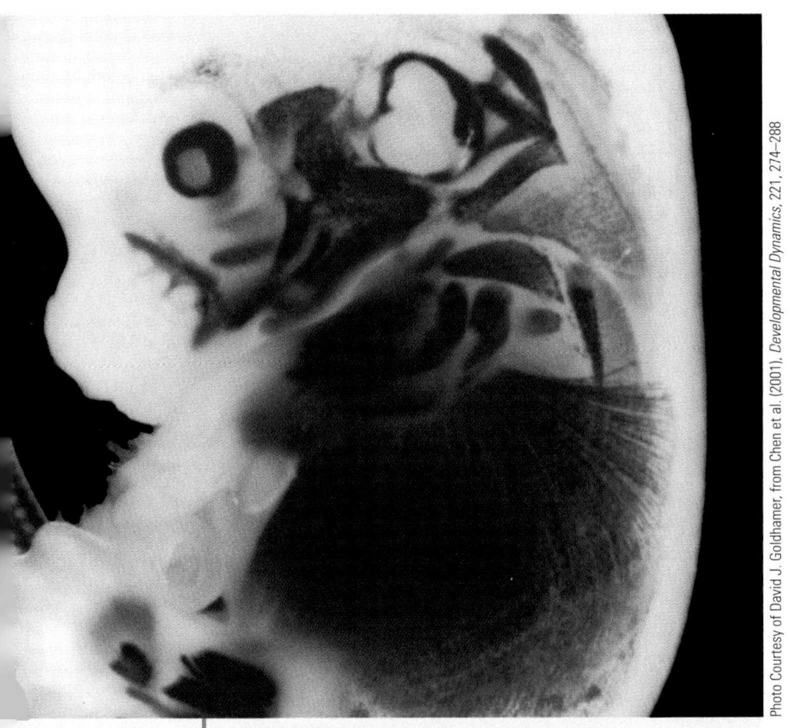

Photo Courtesy of David J. Goldhamer, from Chen et al. (2001), *Developmental Dynamics*, 221, 274–288

Specific gene expression in a mouse embryo. This 13.5-day-old mouse embryo has been genetically engineered so that cells that transcribe and express the myogenin gene stain blue. The myogenin gene is expressed only in those cells of the embryo that will give rise to muscle tissue.

CHAPTER OUTLINE

Each type of cell in a multicellular organism has a characteristic shape, carries out very specific activities, and makes a distinct set of proteins. Yet with few exceptions, all an organism's cells contain the same genetic information. Why, then, aren't they identical in structure and molecular composition? Cells differ because gene expression is regulated, and only certain subsets of the total genetic information are expressed in any given cell (see photograph).

What mechanisms control gene expression? Let's consider a gene coding for a protein that is an enzyme. Expressing that gene involves three steps: transcribing the gene to form messenger RNA (mRNA), translating the mRNA into protein, and activating the protein so it carries out its role to catalyze a specific reaction in the cell. Thus gene expression results from a series of processes, each of which is regulated in many different ways. The control mechanisms use information in the form of various signals, some originating within the cell and others coming from other cells or from the environment, that interact with DNA, RNA, or protein.

Some of the main mechanisms of regulating gene expression are controls on the amount of mRNA transcribed from a gene, on the rate of translation of the mRNA, and on the activity of the protein product. These controls are accomplished in a variety of ways. For example, the rate of transcription and the rate of mRNA degradation both control the amount of available mRNA.

Although bacteria are not multicellular, regulation of gene expression is essential for their survival. Energy efficiency and the economical use of resources, for example, are crucial to bacterial functions. Therefore, gene regulation in bacteria often involves controlling the transcription of genes whose products are involved in resource use. In eukaryotes, by contrast, fine-tuning the control systems occurs at *all* levels of gene regulation. Involving all levels of gene regulation is consistent with the greater complexity of eukaryotic cells and the need for developmental controls in multicellular organisms (see Chapter 16). ■

GENE REGULATION IN BACTERIA AND EUKARYOTES: AN OVERVIEW

Learning Objective

1. Explain why bacterial and eukaryotic cells have different mechanisms of gene regulation.

Bacterial and eukaryotic cells use distinctly different mechanisms of gene regulation, based on the specific requirements of the organism. Bacterial cells exist independently, and each cell performs all its own essential functions. Because they grow rapidly and have relatively short times between cell division events, bacteria carry fewer chemical components.

The primary requirement of bacterial gene regulation is *economy,* and **transcriptional-level control** is the most efficient mechanism. The organization of related genes into groups that are rapidly turned on and off as units allows the synthesis of only the gene products needed at any particular time. This type of regulation requires rapid turnover of mRNA molecules, to prevent messages from accumulating and continuing to be translated when they are not needed. Bacteria rarely regulate enzyme levels by degrading proteins. Once the synthesis of a protein ends, the previously synthesized protein molecules are diluted so rapidly in subsequent cell divisions that breaking them down is usually not necessary. Only when cells are starved or deprived of essential amino acids do protein-digesting enzymes recycle amino acids by breaking down proteins no longer needed for survival.

Eukaryotic cells have different regulatory requirements. Although transcriptional-level control predominates, control at other levels of gene expression is also very important, especially in multicellular organisms, in which groups of cells cooperate with each other in a division of labor. Because a single gene is regulated in different ways in different types of cells, eukaryotic gene regulation is complex.

Eukaryotic cells usually have a long life span during which they may need to respond repeatedly to many different stimuli. New enzymes are not synthesized each time the cells respond to a stimulus; instead, preformed enzymes and other proteins are rapidly converted from an inactive to an active state. Some cells have a large store of inactive messenger RNA; for example, the mRNA of an egg cell becomes activated when it is fertilized.

Much gene regulation in multicellular organisms is focused on *specificity* in the form and function of the cells in each tissue. Each type of cell has certain genes that are active and other genes that may never be used (see Chapter 16). For example, developing red blood cells produce the oxygen transport protein hemoglobin, whereas muscle cells never produce hemoglobin but instead produce myoglobin, a related protein that stores oxygen in muscle tissues. Apparently the selective advantages of cell cooperation in eukaryotes far outweigh the detrimental effects of carrying a load of inactive genes through many cell divisions.

Review

- What is the overall mechanism of gene regulation in bacteria? In eukaryotes?

BiologyNow™ Assess your understanding of **gene regulation in bacteria and eukaryotes** by taking the pretest on your BiologyNow CD-ROM.

GENE REGULATION IN BACTERIA

Learning Objectives

2. Define *operon;* sketch the main elements of the *lac* operon, and explain the functions of the operator and promoter regions.
3. Distinguish among inducible, repressible, and constitutive genes.
4. Differentiate between positive and negative control, and show how both types of control operate in regulating the *lac* operon.
5. Describe the types of posttranscriptional control in bacteria.

The *E. coli* bacterium is common in the intestines of humans and other mammals. It has 4288 genes that code for proteins, approximately 60% of which have known functions. Some of these genes encode proteins that are always needed (such as enzymes involved in glycolysis). These genes, which are constantly transcribed, are **constitutive genes;** biologists say they are *constitutively expressed.* Other proteins are needed only when the bacterium is growing under certain conditions.

For instance, bacteria living in the colon of an adult cow are not normally exposed to the milk sugar lactose, a disaccharide. If those cells ended up in the colon of a calf, however, they would have lactose available as a source of energy. This situation presents a dilemma. Should a bacterial cell invest energy and materials to produce lactose-metabolizing enzymes just in case it ends up in the digestive system of a calf? Given that the average life span of an actively growing *E. coli* cell is about 30 minutes, such an evolutionary strategy appears wasteful. Yet if *E. coli* cells couldn't produce those enzymes, they might starve in the middle of an abundant food supply. Thus *E. coli* functions by regulating many enzymes to efficiently use available organic molecules.

Cell metabolic activity is controlled in two ways, by regulating the *activity* of certain enzymes (how effectively an enzyme molecule works), and/or by regulating the *number* of enzyme molecules present in each cell. Some enzymes are regulated in both ways in the same type of cell.

An *E. coli* cell growing on glucose needs about 800 different enzymes. Some are present in large numbers, whereas only a few molecules of others are required. For the cell to function properly, the quantity of each enzyme must be efficiently controlled.

Bacteria respond to changing environmental conditions. If lactose is added to a culture of *E. coli* cells, they rapidly synthesize the three enzymes needed to metabolize lactose. Bacteria respond so efficiently because functionally related genes—such as the three genes involved in lactose metabolism—are regulated together in gene complexes called *operons.*

Operons in bacteria facilitate the coordinated control of functionally related genes

The French researchers François Jacob and Jacques Monod are credited with the first demonstration, in 1961, of gene regulation at the biochemical level, through their studies on the genes that code for the enzymes that metabolize lactose. In 1965, they received the Nobel Prize in Medicine for their discoveries relating to genetic control of enzymes.

To use lactose as an energy source, *E coli* cells first cleave the sugar into the monosaccharides glucose and galactose, using the enzyme β-galactosidase. Another enzyme converts galactose to glucose, and enzymes in the glycolysis pathway further break down the resulting two glucose molecules (see Chapter 7).

E. coli cells growing on glucose produce very little β-galactosidase enzyme. However, each cell grown on lactose as the sole carbon source has several thousand β-galactosidase molecules, accounting for about 3% of the cell's total protein. Amounts of two other enzymes, galactose permease and galactoside transacetylase, also increase when the cells are grown on lactose. The cell needs permease to transport lactose efficiently across the bacterial plasma membrane; without it, only small amounts of lactose enter the cell. The transacetylase may function in a minor aspect of lactose metabolism, although its role is not clear.

Jacob and Monod identified mutant strains of *E. coli* in which a single genetic defect wiped out all three enzymes. This finding, along with other information, led the researchers to conclude that the DNA coding sequences for all three enzymes are linked together as a unit on the bacterial DNA and are controlled by a common mechanism. Each protein-coding sequence is a **structural gene.** Jacob and Monod coined the term **operon** for a gene complex consisting of a group of structural genes with related functions, plus the closely linked DNA sequences responsible for controlling them. The structural genes of the *lac* operon (lactose operon)—*lac Z, lac Y,* and *lac A*—code for β-galactosidase, galactose permease, and galactoside transacetylase, respectively.

Transcription of the *lac* operon begins as RNA polymerase binds to a single **promoter** region upstream from the coding sequences. It then proceeds to transcribe the DNA, forming a single mRNA molecule that contains the coding information for all three enzymes. Each enzyme-coding sequence on this mRNA contains its own start and stop codons; thus the mRNA is translated to form three separate protein molecules. Because all three enzymes are translated from the same mRNA molecule, their synthesis is coordinated by turning a single molecular "switch" on or off.

The switch that controls mRNA synthesis is the **operator,** which is a sequence of bases upstream from the first structural gene in the operon. In the absence of lactose, a **repressor protein** called the **lactose repressor** binds tightly to the operator. RNA polymerase binds to the promoter but is blocked from transcribing the protein-coding genes of the *lac* operon (Fig. 13-1a).

The lactose repressor protein is encoded by a **repressor gene,** which in this case is an adjacent structural gene located upstream from the promoter site. Unlike the *lac* operon genes, the repressor gene is constitutively expressed and is therefore constantly transcribed; the cell continually produces small amounts of the repressor protein.

The repressor protein binds specifically to the *lac* operator sequence. When cells grow in the absence of lactose, a repressor molecule nearly always occupies the operator site. When the operator site is briefly free of the repressor, the cell synthesizes a small amount of mRNA. However, the cell synthesizes very few enzyme molecules, because *E. coli* mRNA is degraded rapidly (it has a half-life of about 2 to 4 minutes).

Lactose "turns on," or *induces,* the transcription of the *lac* operon because the lactose repressor protein contains a second functional region separate from its DNA binding site. This second region is an allosteric binding site for allolactose, a structural isomer made from lactose (Fig. 13-1b). Recall from Chapter 6 that an **allosteric regulator,** such as allolactose, binds to a region in a protein other than its active site, changing its shape and thereby altering its function. If lactose is in the growth medium, a few molecules enter the cell and are converted to allolactose by the few β-galactosidase molecules present. When a molecule of allolactose binds to the repressor at the allosteric site, it alters the shape of the protein so that its DNA-binding site no longer recognizes the operator. When all the repressor molecules have allolactose bound to them and are therefore inactivated, RNA polymerase actively transcribes the structural genes of the operon.

The *E. coli* cell continues to produce β-galactosidase and the other *lac* operon enzymes until it uses up virtually all the lactose. When intracellular levels of lactose drop, allolactose dissociates from the repressor protein, which then takes a shape that lets it bind to the operator region and shut down transcription of the operon.

Jacob and Monod isolated genetic mutants to study the lac *operon*

PROCESS OF SCIENCE

How did Jacob and Monod elucidate the functioning of the *lac* operon? Their approach involved the use of mutant strains, which even today play an essential role in enabling researchers to unravel the components of a regulatory system. Mutant strains allow investigators to carry out genetic crosses to determine the map positions (linear order) of the genes on the DNA and to infer normal gene functions by studying what happens when they are missing or altered. Researchers usually combine this information with the results of direct biochemical studies.

To understand Jacob and Monod's reasoning, follow the branching steps in Figure 13-2 as you read. They divided their mutant strains into two groups, based on whether a particular mutation affected only one enzyme or all three. In one group, only one enzyme of the three—β-galactosidase, galactose permease, or galactoside transacetylase—was affected. Subsequent gene mapping studies showed these were mutations in structural genes located next to each other in a linear sequence (*left side,* Fig. 13-2).

Jacob and Monod also studied strains they classified as regulatory mutants because a single mutation affected the expression of all three enzymes (*right side,* Fig. 13-2). Some of these regulatory mutants were constitutive; in these the structural genes of the *lac* operon were always transcribed at a significant rate, even in the absence of lactose, causing the cell to waste energy in producing unneeded enzymes. One group of constitutive gene mutations had map positions just outside the *lac* operon itself. Using certain genetic strains, it was possible to show that these particular mutations always caused constitutive expression, regardless of their location in the genome. On the basis of these findings, Jacob and Monod hypothesized the existence of a repressor gene that codes for a repressor molecule (later found to be a protein). Although the specific defect may vary, the members of this group of constitutive mutants do not produce active repressor proteins; hence, no binding to the *lac* operator and promoter takes place, and the *lac* operon is constitutively expressed.

The genes responsible for the behavior of a second group of constitutive mutants had map positions within the *lac* operon but did not directly involve any of the three structural genes. Jacob and Monad hypothesized that the members of this group produced normal repressor molecules but had abnormal operator sequences incapable of binding to the repressor.

In contrast to the constitutive mutants, other mutants failed to transcribe the *lac* operon even when lactose was present. Some of these abnormal genes had the same map position as the hypothesized regulatory gene. Researchers eventually found these abnormal genes had an altered binding site on the repressor protein that prevented allolactose from binding, although the repressor could still bind to the operator. Once bound to the operator, such a mutant repressor remains bound, keeping the operon "turned off."

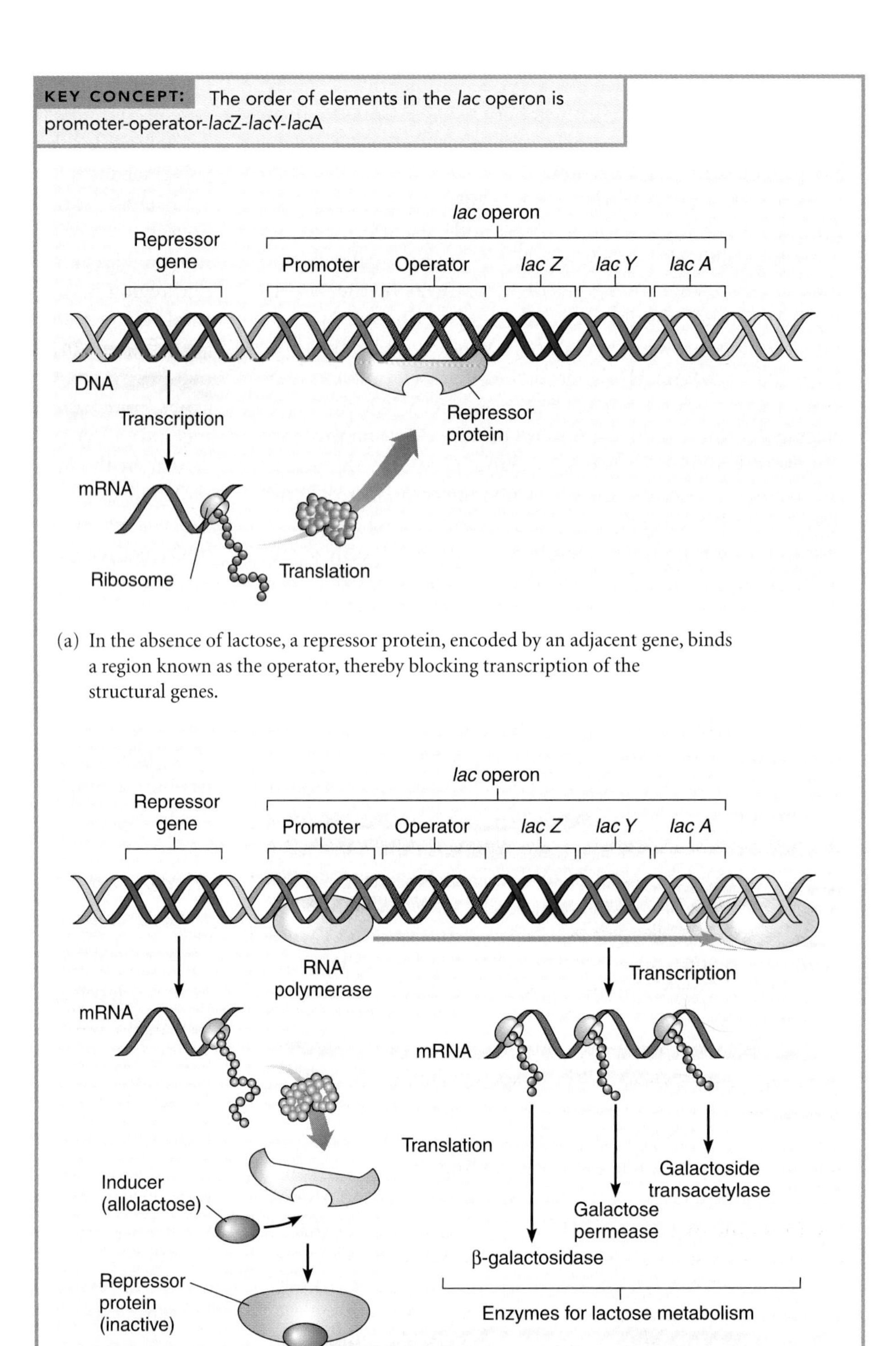

(a) In the absence of lactose, a repressor protein, encoded by an adjacent gene, binds a region known as the operator, thereby blocking transcription of the structural genes.

(b) When lactose is present, it is converted to allolactose, which binds to the repressor at an allosteric site, altering the structure of the protein so it no longer binds to the operator. This allows RNA polymerase to transcribe the structural genes.

ACTIVE FIGURE **13-1** | **The *lac* operon.**

The structural genes coding for the three enzymes used by *E. coli* to metabolize the disaccharide lactose are transcribed as part of a single mRNA molecule.

BiologyNow™ Watch the **lactose operon** in action by clicking on this figure on your BiologyNow CD-ROM.

An inducible gene is not transcribed unless a specific inducer inactivates its repressor

Geneticists call the *lac* operon an **inducible operon.** A repressor usually controls an inducible gene or operon by keeping it turned "off." The presence of an **inducer** (in this case, allolactose) inactivates the repressor, permitting the gene or operon to be transcribed. Inducible genes or operons usually code for enzymes that are part of catabolic pathways, which break down molecules to provide both energy and components for anabolic reactions. This type of regulatory system lets the cell save the energy cost of making enzymes when no substrates are available on which they can act.

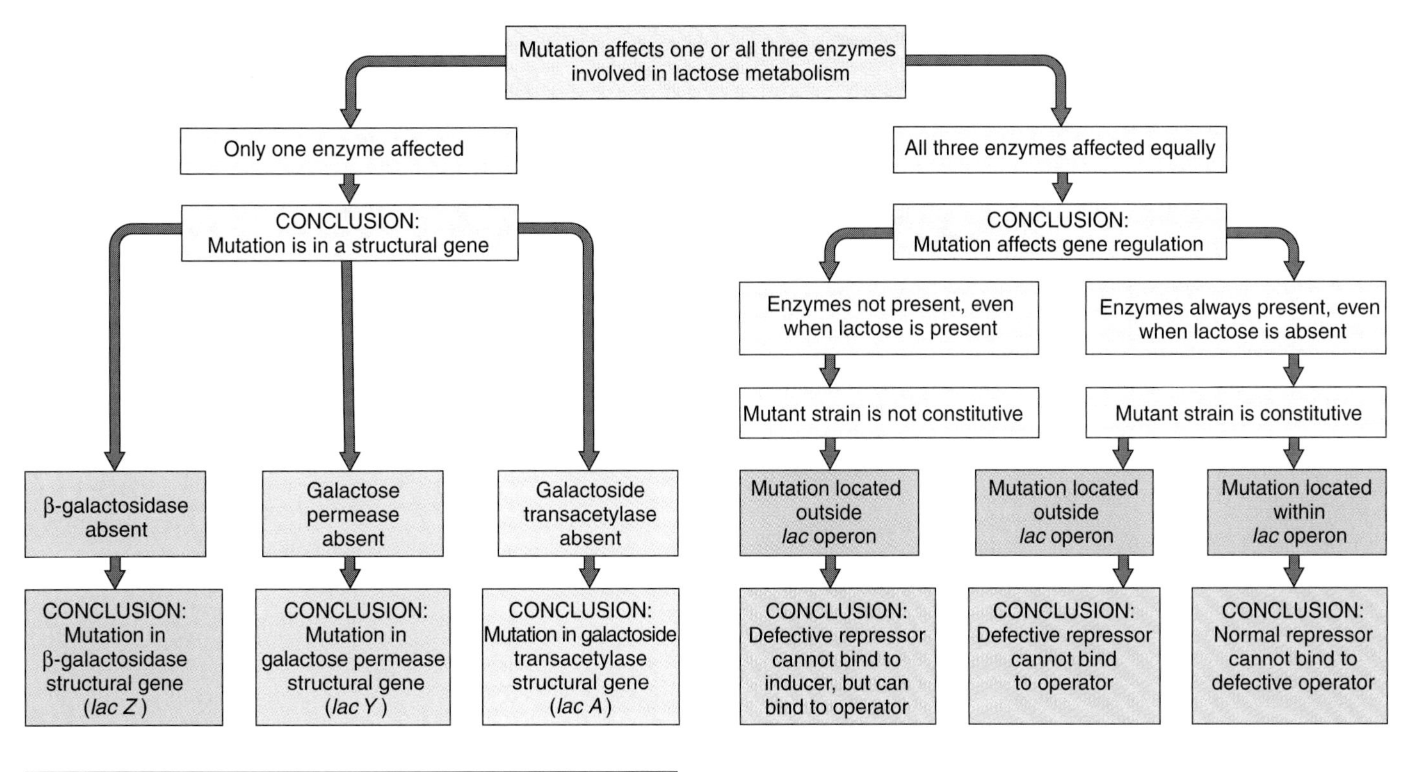

FIGURE **13-2** **Genetic and biochemical characterization of the *lac* operon.**

Jacob and Monod analyzed the properties of various mutant strains of *E. coli* to deduce the structure and function of the *lac* operon.

A repressible gene is transcribed unless a specific repressor-corepressor complex is bound to the DNA

Another type of gene regulation system in bacteria is associated mainly with anabolic pathways such as those in which cells synthesize amino acids, nucleotides, and other essential biological molecules from simpler materials. Enzymes coded by repressible genes normally regulate these pathways.

Repressible operons and genes are usually turned "on"; they are turned off only under certain conditions. In most cases the molecular signal for regulating these genes is the end product of the anabolic pathway. When the supply of the end product (such as an amino acid) becomes low, all enzymes in the pathway are actively synthesized. When intracellular levels of the end product rise, enzyme synthesis is repressed. Because the growing cell continuously needs compounds such as amino acids, the most effective mechanism for the cell is to keep the genes that control their production turned on, except when a large supply of the amino acid is available. The ability to turn the genes off prevents cells from overproducing amino acids and other molecules that are essential but expensive to make, in terms of energy.

The *trp* operon (tryptophan operon) is an example of a repressible system. In both *E. coli* and a related bacterium, *Salmonella,* the *trp* operon consists of five structural genes that code for the enzymes the cell needs to synthesize the amino acid tryptophan; these are clustered together as a transcriptional unit with a single promoter and a single operator (Fig. 13-3a). A distant repressor gene codes for a diffusible repressor protein, which differs from the lactose repressor in that the cell synthesizes it in an inactive form that cannot bind to the operator region of the *trp* operon.

The DNA-binding site of the repressor becomes effective only when tryptophan, its **corepressor,** binds to an allosteric site on the repressor (Fig. 13-3b). When intracellular tryptophan levels are low, the repressor protein is inactive and cannot bind to the operator region of the DNA. The enzymes required for tryptophan synthesis are produced, and the intracellular concentration of tryptophan increases. Some of it binds to the allosteric site of the repressor, altering the repressor's shape so it binds tightly to the operator. This switches the operon off, thereby blocking transcription.

Negative regulators inhibit transcription; positive regulators stimulate transcription

The features of the *lac* and *trp* operons we have described so far are examples of **negative control,** a regulatory mechanism in which the DNA binding regulatory protein is a repressor that turns off transcription of the gene. **Positive control** is regulation by **activator proteins** that bind to DNA and thereby stimulate the transcription of a gene. The *lac* operon is controlled by both a negative regulator (the lactose repressor) and a positively acting activator protein.

Positive control of the *lac* operon requires the cell to recognize the absence of the sugar glucose, which is the initial substrate in the glycolysis pathway. Lactose, like glucose, undergoes stepwise breakdown to yield energy. However, because glucose is a product of the catabolic hydrolysis of lactose, it is most effi-

FIGURE **13-3**

The *trp* operon.

Genes coding for enzymes that synthesize the amino acid tryptophan are organized in a repressible operon.

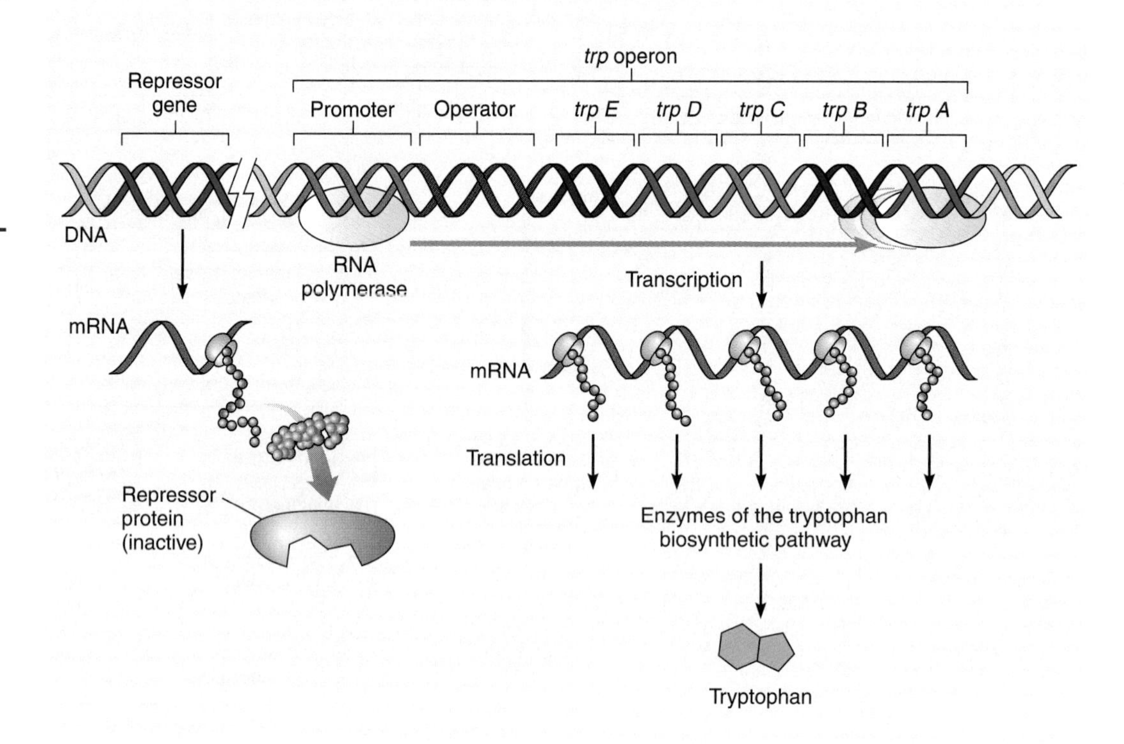

(a) Intracellular tryptophan levels low. Repressor protein is unable to prevent transcription because it cannot bind to the operator.

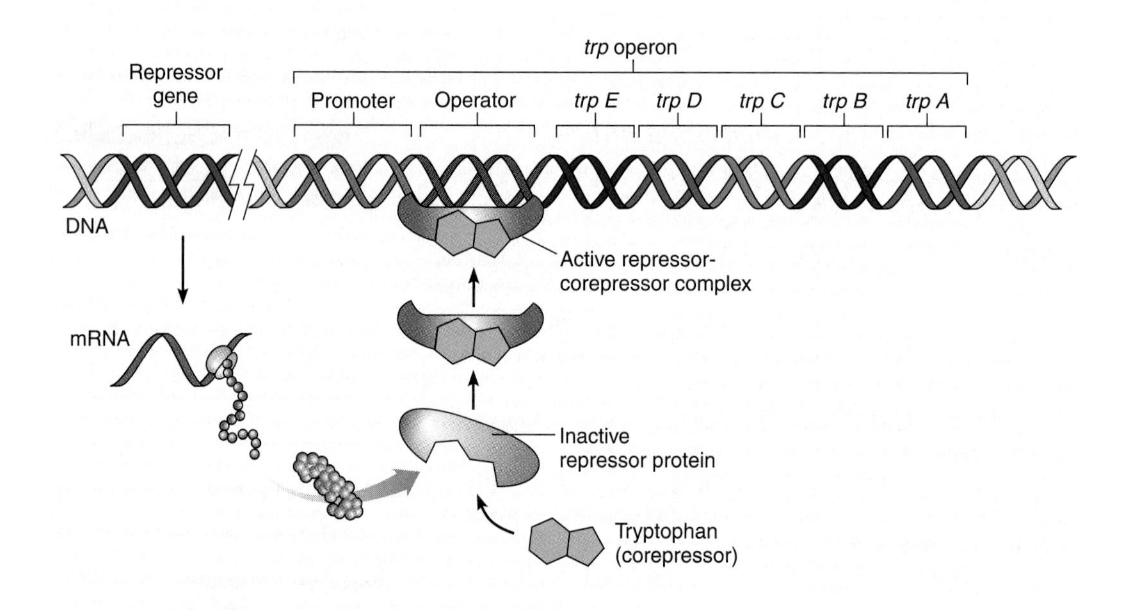

(b) Intracellular tryptophan levels high. The amino acid tryptophan binds to an allosteric site on the repressor protein, changing its conformation. The resulting active form of the repressor binds to the operator region, blocking transcription of the operon until tryptophan is again required by the cell.

cient for *E. coli* cells to use the available supply of glucose first, sparing the cell the considerable energy cost of making additional enzymes such as β-galactosidase (Fig. 13-4a).

The *lac* operon has a very inefficient promoter element—that is, it has a low affinity for RNA polymerase, even when the repressor protein is inactivated. However, a DNA sequence adjacent to the promoter site is a binding site for another regulatory protein, the **catabolite activator protein (CAP).**

In its active form, CAP has **cyclic AMP,** or **cAMP,** an alternative form of adenosine monophosphate (see Fig. 3-25), bound

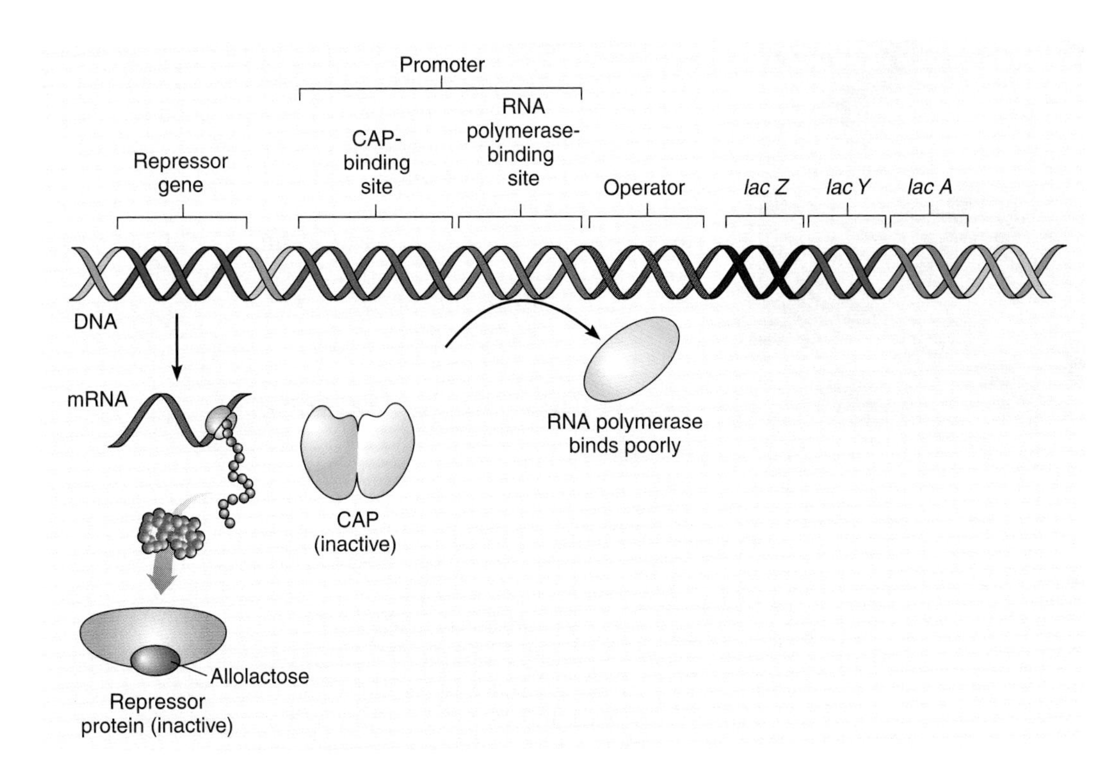

(a) Lactose high, glucose high, cAMP low. When glucose levels are high, cAMP is low. CAP is in an inactive form and cannot stimulate transcription. Transcription occurs at a low level or not at all.

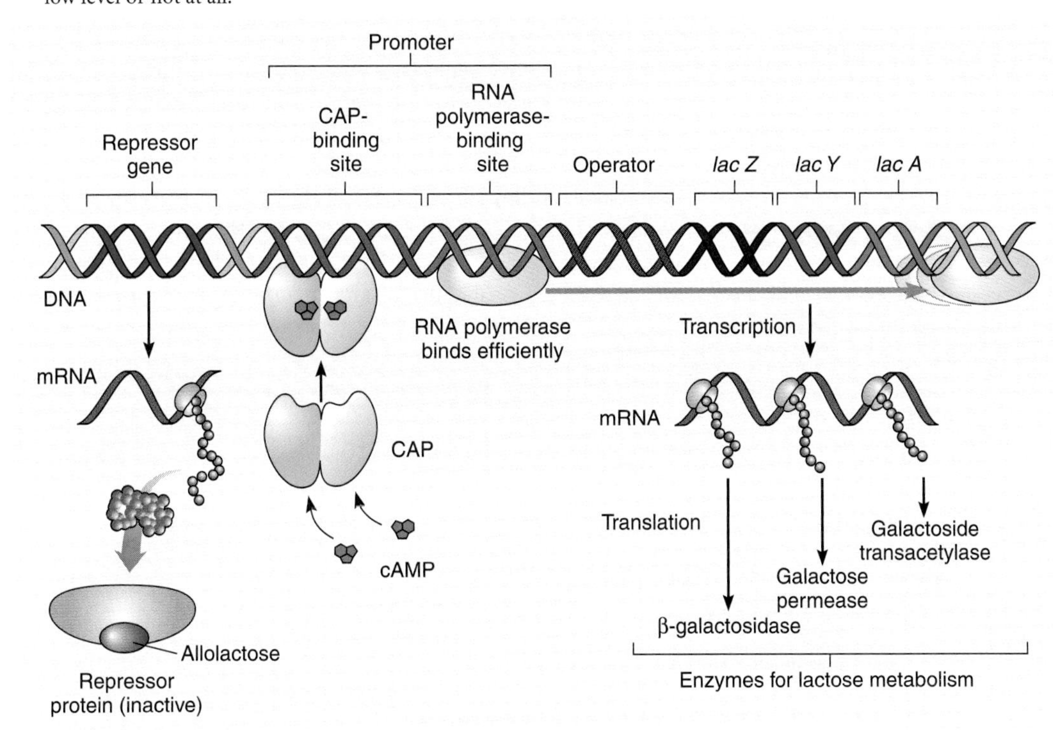

(b) Lactose high, glucose low, cAMP high. When glucose concentrations are low, each CAP polypeptide has cAMP bound to its allosteric site. The active form of CAP binds to the DNA sequence, and transcription becomes activated.

FIGURE **13-4**

Positive control of the *lac* operon.

The lactose promoter by itself is weak and binds RNA polymerase inefficiently even when the lactose repressor is inactive. CAP is an allosteric regulator whose active form binds to a sequence of bases in the promoter, allowing RNA polymerase to bind efficiently, thereby stimulating transcription of the operon. The CAP molecule contains two polypeptides, both of which bind to cAMP at allosteric sites. The cell's cAMP concentration is inversely proportional to the glucose concentration.

to an allosteric site. As the cells become depleted of glucose, cAMP levels increase (Fig. 13-4b). The cAMP molecules bind to CAP, and the resulting active complex then binds to the CAP-binding site near the *lac* operon promoter. This binding of active CAP bends DNA's double helix (Fig. 13-5), strengthening the affinity of the promoter region for RNA polymerase so that the rate of transcriptional initiation accelerates in the presence of lactose. Thus the *lac* operon is fully active only if lactose is

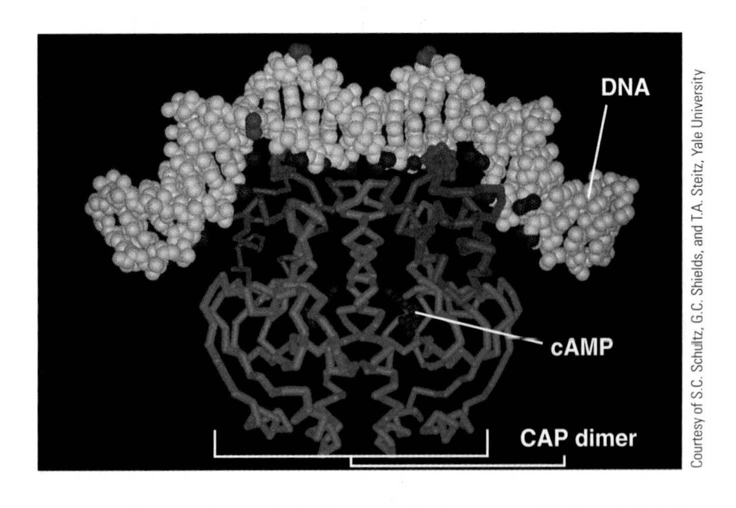

FIGURE **13-5** **The binding of CAP to DNA.**
This computer-generated image shows the bend formed in the DNA double helix when it binds to CAP. CAP is a dimer consisting of two identical polypeptide chains, each of which binds one molecule of cAMP.

available and intracellular glucose levels are low. Table 13-1 summarizes negative and positive controls in bacteria.

A regulon is a group of functionally related operons controlled by a common regulator

CAP differs from the lactose and tryptophan repressors in that it controls the transcription of operons involved in the metabolism of several catabolites, such as the sugars galactose, arabinose, and maltose, as well as of lactose. A group of operons controlled by a single regulator is called a **regulon** (Fig. 13-6).

Other multigene systems in bacteria are also controlled this way. For example, genes involved in nitrogen and phosphate metabolism are organized into regulons that consist of multiple sets of operons controlled by one or more combinations of regulatory genes. Other multigene systems respond to changes in environmental conditions, such as rapid shifts in temperature, exposure to radiation, changes in osmotic pressure, and changes in oxygen levels. Specific mutants often provide clues to the existence of a regulon system. A single mutation that destroys CAP activity, for example, prevents the cell from efficiently metabolizing not only lactose but also many other sugars that CAP regulates.

Constitutive genes are transcribed at different rates

Many gene products encoded by *E. coli* DNA are needed only under certain environmental or nutritional conditions. As you have seen, these genes are generally regulated at the level of transcription. They are turned on and off as metabolic and environmental conditions change. By contrast, constitutive genes are continuously transcribed, but they are not necessarily transcribed (or their mRNAs translated) at the same rate. Some enzymes work more effectively or are more stable than others and thus are present in smaller amounts. Constitutive genes that encode proteins required in large amounts generally have greater expression—that is, are transcribed more rapidly—than genes coding for proteins required in smaller amounts. The promoter elements of these genes control their transcriptional rate. Constitutive genes with efficient ("strong") promoters bind RNA polymerase more frequently and consequently transcribe more mRNA molecules than those with inefficient ("weak") promoters.

Genes coding for repressor or activator proteins that regulate metabolic enzymes are usually constitutive and produce their protein products constantly. Because each cell usually needs relatively few molecules of any specific repressor or activator protein, promoters for those genes tend to be relatively weak.

TABLE **13-1** Transcriptional Control in Bacteria*

Negative Control	Result
Inducible genes	
Repressor protein alone	**Active repressor turns off regulated gene(s)**
Lactose repressor alone	*lac* operon not transcribed
Repressor protein + inducer	**Inactive repressor-inducer complex fails to turn off regulated gene(s)**
Lactose repressor + allolactose	*lac* operon transcribed
Repressible genes	
Repressor protein alone	**Inactive repressor fails to turn off regulated gene(s)**
Tryptophan repressor alone	*trp* operon transcribed
Repressor protein + corepressor	**Active repressor-corepressor complex turns off regulated gene(s)**
Tryptophan repressor + tryptophan	*trp* operon not transcribed
Positive Control	**Result**
Activator protein alone	**Activator alone cannot stimulate transcription of regulated gene(s)**
CAP alone	Transcription of *lac* operon not stimulated
Activator protein + coactivator	**Functional activator-coactivator complex stimulates transcription of regulated gene(s)**
CAP + cAMP	Transcription of *lac* operon stimulated

*A general description of each type is followed by a specific example. A negative regulator is a repressor that turns off transcription of the regulated gene(s). Conversely, a positive regulator is an activator that stimulates transcription.

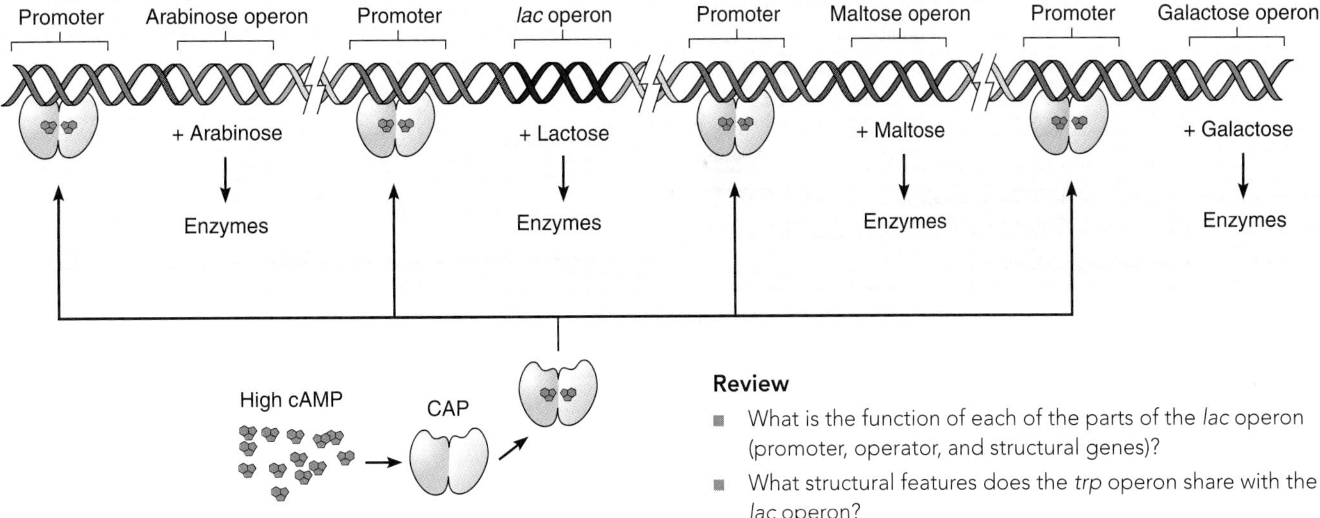

FIGURE **13-6** | **A regulon.**

Operons that convert several different sugars to glucose in *E. coli* make up a regulon under positive control by CAP. When glucose levels are low, cAMP levels rise, activating CAPs, which bind to their recognition sites in the promoter regions of all operons in the regulon. If the inducer for an operon is available, its repressor is inactivated, and transcription of the message occurs at a rapid rate.

Some posttranscriptional regulation occurs in bacteria

As you have seen, much of the variability in protein levels in *E. coli* is determined by transcriptional-level control. However, regulatory mechanisms after transcription, known as **posttranscriptional controls,** also occur at various levels of gene expression.

Translational controls are posttranscriptional controls that regulate the rate at which an mRNA molecule is translated. Because the life span of an mRNA molecule in a bacterial cell is very short, one that is translated rapidly produces more proteins than one that is translated slowly. Some mRNA molecules in *E. coli* are translated as much as 1000 times faster than other mRNAs. Most of the differences appear to be due to the rate at which ribosomes attach to the mRNA and begin translation.

Posttranslational controls act as switches that activate or inactivate one or more existing enzymes, thereby letting the cell respond to changes in the intracellular concentrations of essential molecules, such as amino acids. A common posttranslational control adjusts the rate of synthesis in a metabolic pathway through **feedback inhibition** (see Fig. 6-15). The end product of a metabolic pathway binds to the first enzyme in the pathway at an allosteric site, temporarily inactivating the enzyme. When the first enzyme in the pathway does not function, all the succeeding enzymes are deprived of substrates. Notice that feedback inhibition differs from the repression caused by tryptophan discussed previously. In that case, the end product of the pathway (tryptophan) prevented the formation of *new* enzymes. Feedback inhibition acts as a fine-tuning mechanism that regulates the activity of the *existing* enzymes in a metabolic pathway.

Review

- What is the function of each of the parts of the *lac* operon (promoter, operator, and structural genes)?
- What structural features does the *trp* operon share with the *lac* operon?
- Why do scientists define the *trp* operon as repressible and the *lac* operon as inducible?
- How is glucose involved in positive control of the *lac* operon?

BiologyNow™ Assess your understanding of **gene regulation in prokaryotic cells** by taking the pretest on your BiologyNow CD-ROM.

GENE REGULATION IN EUKARYOTIC CELLS

Learning Objectives

6 Discuss the structure of a typical eukaryotic gene and the DNA sequences involved in regulating that gene.

7 Give examples of some of the ways eukaryotic DNA-binding proteins bind to DNA.

8 Illustrate how a change in chromosome structure may affect the activity of a gene.

9 Explain how a gene in a multicellular organism may produce different products in different types of cells.

10 Identify some of the types of regulatory controls that operate in eukaryotes after mature mRNA is formed.

Like bacteria, eukaryotic cells respond to changes in their environment. In addition, multicellular eukaryotes require regulation modes that let individual cells commit to specialized roles and groups of cells organize into tissues and organs. This is mainly done by regulating transcription, but posttranscriptional (translational and posttranslational) controls are also important. In Chapters 11 and 12, we observed that in eukaryotes all aspects of information transfer—including replication, transcription, and translation—are far more complicated. Not surprisingly, this complexity offers more ways to control gene expression.

Unlike many bacterial genes, eukaryotic genes do not normally form operon-like clusters. (A notable exception is the nematodes, or roundworms, in which some genes are organized into operons.) However, each eukaryotic gene has specific regulatory sequences that are essential in controlling transcription.

Many "housekeeping" enzymes (those needed by virtually all cells) are encoded by constitutive genes and are expressed at all times. Some inducible genes also respond to environmental threats or stimuli, such as heavy metal ingestion, viral infection, and heat shock. For example, when a cell is exposed to high temperature, many proteins fail to fold properly. These unfolded proteins elicit a survival response in which *heat-shock genes* are transcribed, and heat-shock proteins are formed. Although the functions of most heat-shock proteins are unknown, some are **molecular chaperones,** which help newly synthesized proteins fold into their proper shape (see Chapter 3).

Some genes are inducible only during certain periods in the life of the organism; they are controlled by **temporal regulation** mechanisms. Finally, some genes are under **tissue-specific regulation.** For example, a gene involved in production of a particular enzyme may be regulated by one stimulus (such as a hormone) in muscle tissue, by an entirely different stimulus in pancreatic cells, and by a third stimulus in liver cells. We explore these types of regulation in more detail in Chapters 16 and 47.

Table 13-2 summarizes the regulation of gene expression in eukaryotes. You may want to refer to it as you read the following discussion.

Eukaryotic transcription is controlled at many sites and by many different regulatory molecules

Most genes of multicellular eukaryotes are controlled at the transcriptional level. As you will see, various base sequences in the DNA are important in transcriptional control. In addition, regulatory proteins and the way the DNA is organized in the chromosome affect the rate of transcription.

TABLE 13-2 Gene Regulation In Eukaryotes

Level of Regulation	Description
Transcriptional control (main level of gene regulation in eukaryotes)	Selective transcription: promoter and enhancer elements in DNA interact with protein transcription factors to activate or repress transcription
	Chromatin structure regulates transcription; heterochromatin cannot be transcribed
	DNA methylation regulates transcription; methylated DNA is inaccessible to transcription machinery
Posttranscriptional control: mRNA processing and transport	Control mechanisms, such as rate of intron/exon splicing, regulate mRNA processing
	Differential mRNA processing (alternate splicing of exons) produces different proteins from same mRNA
	Controlling access to, or efficiency of, transport through nuclear pores regulates mRNA transport from nucleus to cytosol
Translational control	Translational controls determine how often and how long specific mRNA is translated
	Translational controls determine degree to which mRNA is protected from destruction; proteins that bind to mRNA in cytosol affect stability
Posttranslational control of protein product	Chemical modifications, such as phosphorylation, affect activity of protein after it is produced
	Selective degradation targets specific proteins for destruction

Eukaryotic promoters vary in efficiency, depending on their upstream promoter elements

In eukaryotic as well as bacterial cells, the transcription of any gene requires a base pair where transcription begins, known as **transcription initiation site,** plus a sequence of bases, the promoter, to which RNA polymerase binds. In multicellular eukaryotes, RNA polymerase binds to a promoter called a **TATA box,** located about 25 to 35 base pairs upstream from the transcription initiation site (Fig. 13-7a). The TATA box is required for transcription to begin.

Other eukaryotic promoter elements have a regulatory function and control the expression of the gene. Many eukaryotic promoters contain one or more sequences of 8 to 12 bases within a short distance upstream of the RNA polymerase-binding site. These promoters are called **upstream promoter elements (UPEs).** Specific regulatory proteins bind to the UPEs to regulate expression of the gene. The efficient initiation of transcription seems related to the number and type of UPEs. A constitutive gene containing only one UPE is generally weakly expressed, whereas one with five or six UPEs is transcribed much more actively (Fig. 13-7b and c). Thus a given gene may have one, a few, or many promoter elements.

Enhancers are DNA sequences that increase the transcription rate

Regulated eukaryotic genes commonly need not only UPEs but also DNA sequences called **enhancers.** Whereas the promoter elements are required for accurate and efficient initiation of mRNA synthesis, enhancers increase the *rate* of RNA synthesis after initiation, often by several orders of magnitude (Fig. 13-7d). Specific regulatory proteins bind to enhancer elements and activate transcription by interacting with the proteins bound to the promoters.

Enhancer elements are remarkable in many ways. Although present in all cells, a particular enhancer is functional only in certain types of cells. An enhancer regulates a gene on the same DNA molecule from very long distances—up to thousands of base pairs away from the promoter—and is either upstream or downstream of the promoter it controls. Furthermore, if an enhancer element is experimentally cut out of the DNA and inverted, it still regulates the gene it normally controls. As you will see, evidence suggests that at least some enhancers work by interacting with proteins that regulate transcription.

Transcription factors are regulatory proteins with several functional domains

We previously discussed some DNA-binding proteins that regulate transcription in bacteria. These include the lactose repressor, the tryptophan repressor, and the catabolite activator protein (CAP). Researchers have identified many more DNA-binding

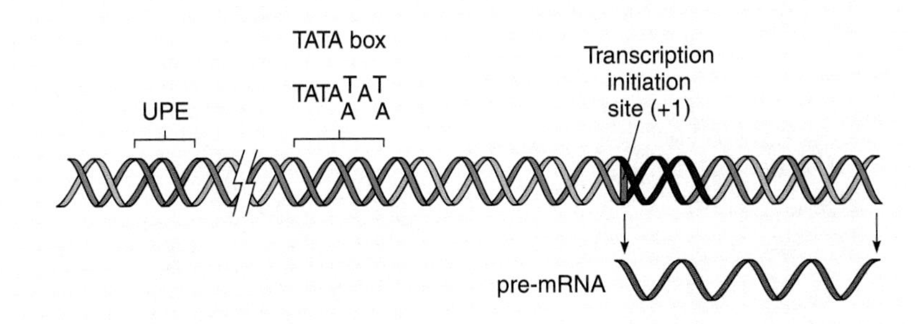

(a) Eukaryotic promoter elements. A eukaryotic promoter usually contains a TATA box located 25 to 35 base pairs upstream from the transcription initiation site. The most commonly found base sequence is shown (either T or A can be present at the positions where they appear together). One or more upstream promoter elements (UPEs) are usually present.

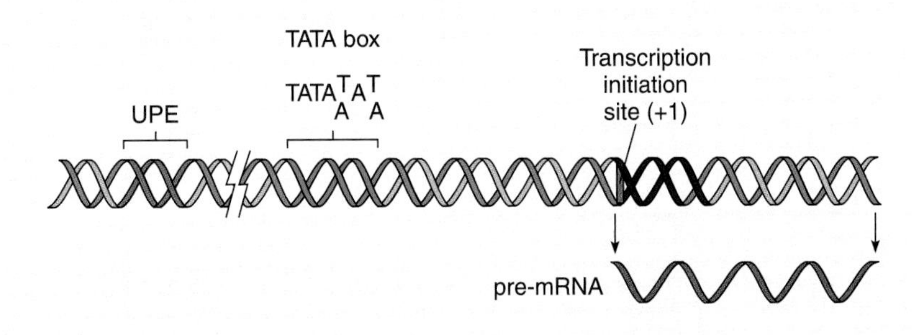

(b) A weak eukaryotic promoter. A weakly expressed gene contains only one UPE.

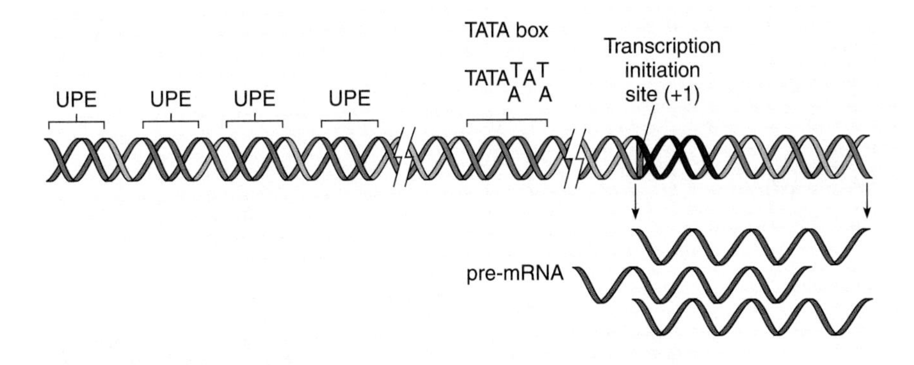

(c) A strong eukaryotic promoter. A strongly expressed gene is likely to contain several UPEs.

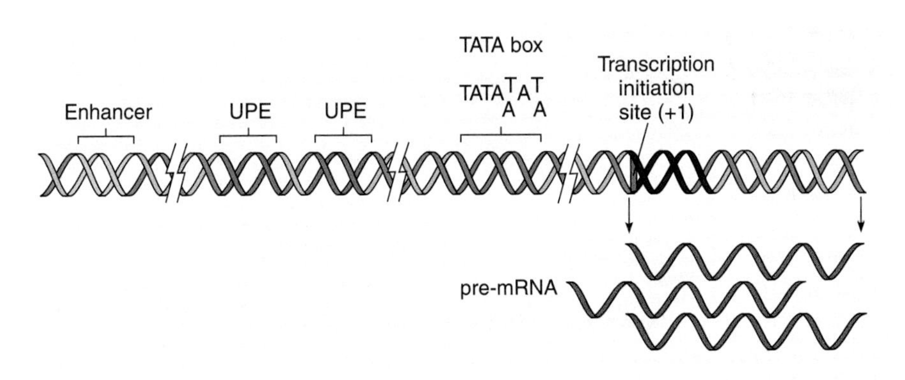

(d) A strong eukaryotic promoter plus an enhancer. Transcription of this eukaryotic gene is stimulated by an enhancer, located several thousand bases from the promoter.

FIGURE **13-7**

The regulation of transcription in eukaryotes.

The DNA double helix and other elements are not drawn to scale.

proteins that regulate transcription in eukaryotes than in bacteria; these eukaryotic proteins are known as **transcription factors.** In humans, for example, researchers have identified more than 2000 transcription factors

It is useful to compare regulatory proteins in bacteria and eukaryotes. Many regulatory proteins are modular molecules; that is, they have more than one **domain,** a region with its own tertiary structure, and each domain has a different function. Each eukaryotic transcription factor, like the regulatory proteins of bacteria, has a DNA-binding domain, plus at least one other domain that is either an activator or a repressor of transcription for a given gene.

Many transcription factors in eukaryotes (and regulatory proteins in bacteria) have similar DNA-binding domains. One example is the *helix-turn-helix* arrangement, consisting of two α-helical segments. One of these, the recognition helix, inserts into a groove of the DNA without unwinding the double helix. The other helps hold the first one in place (Fig. 13-8a). The "turn" is a sequence of amino acids that forms a sharp bend in the molecule.

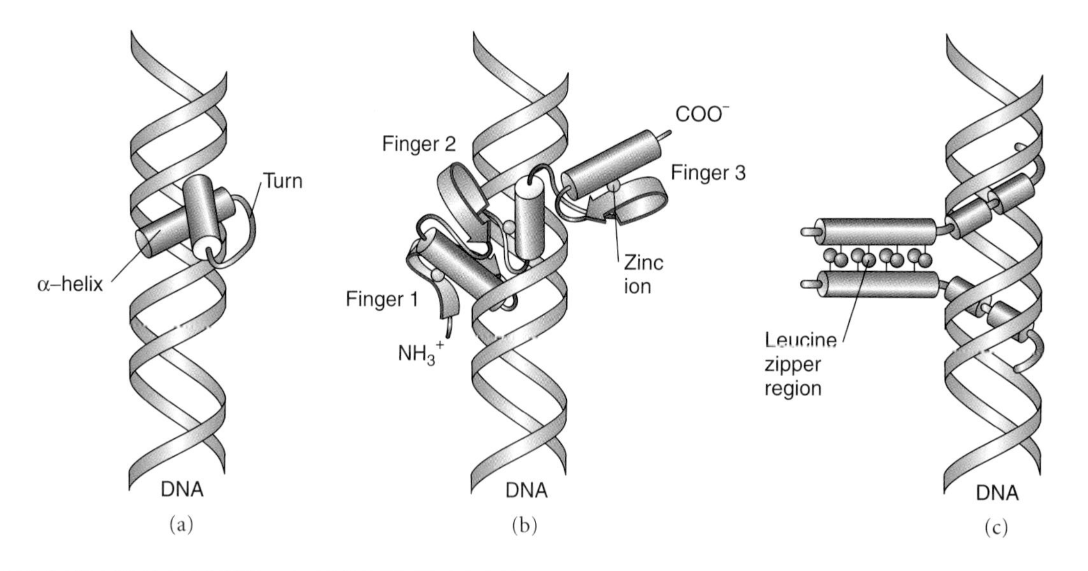

FIGURE **13-8** **Regulatory proteins.**

In these illustrations, cylinders represent α-helical regions of regulatory proteins and ribbons represent β-pleated sheets. **(a)** This portion of a regulatory protein *(purple)* contains the helix-turn-helix arrangement. The recognition helix is inserted into the groove of the DNA and is connected to a second helix that helps hold it in place by a sequence of amino acids that form a sharp bend. **(b)** Regions of certain transcription factors form projections known as "zinc fingers," which insert into the grooves of the DNA and bind to specific base sequences. **(c)** This leucine zipper protein is a dimer, held together by hydrophobic interactions involving side chains of leucine and other amino acids.

Other regulatory proteins have DNA binding domains with multiple "zinc fingers," loops of amino acids held together by zinc ions. Each loop includes an α-helix that fits into a groove of the DNA (Fig. 13-8b). Certain amino acid functional groups exposed in each finger have been shown to recognize specific DNA sequences.

Many regulatory proteins have DNA-binding domains that function only as pairs, or *dimers.* Many of these transcription factors are known as **leucine zipper proteins** because they are held together by the side chains of leucine and other hydrophobic amino acids (Fig. 13-8c). In some cases the two polypeptides that make up the dimer are identical and form a *homodimer.* In other instances they are different, and the resulting *heterodimer* may have very different regulatory properties from a homodimer. For a simple and speculative example, let's assume that three regulatory proteins—A, B, and C—are involved in controlling a particular set of genes. These three proteins might associate as dimers in six different ways: three kinds of homodimers (AA, BB, and CC) and three kinds of heterodimers (AB, AC, and BC). Such multiple combinations of regulatory proteins greatly increase the number of possible ways that transcription is controlled.

Transcription in eukaryotes requires multiple regulatory proteins that are bound to different parts of the promoter. The general transcriptional machinery is a protein complex that binds to the TATA box of the promoter near the transcription initiation site. That complex is required for RNA polymerase to bind, thereby initiating transcription.

Both enhancers and UPEs apparently become functional when specific transcription factors bind to them. Figure 13-9

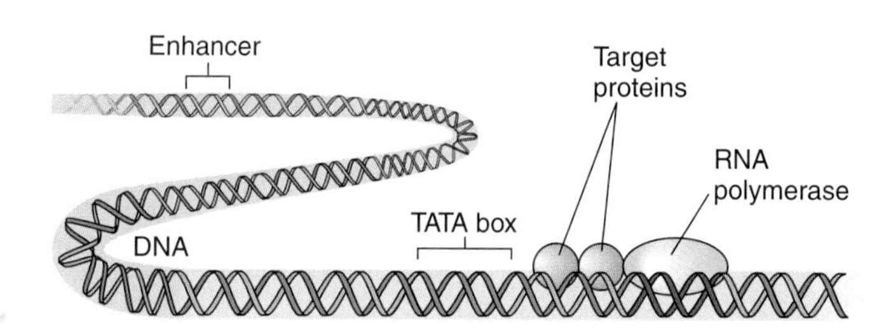

(a) Little or no transcription

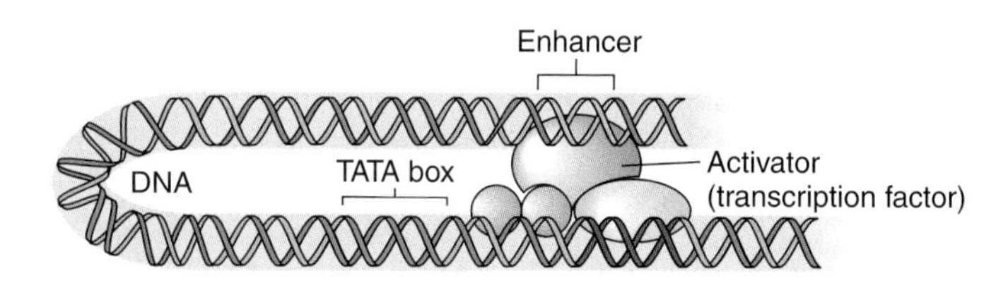

(b) High rate of transcription

FIGURE **13-9** **The stimulation of transcription by an enhancer.**

(a) This gene is transcribed at a very low rate or not at all, even though the general transcriptional machinery, including RNA polymerase, is bound to the promoter. **(b)** A transcription factor that functions as an activator binds to an enhancer. The intervening DNA forms a loop, allowing the transcription factor to contact one or more target proteins in the general transcriptional machinery, thereby increasing the rate of transcription. This diagram is highly simplified, and many more target proteins than the two shown are involved.

shows interactions involving an enhancer and a transcription factor that acts as an activator. Each activator has at least two functional domains: a DNA recognition site that usually binds to an enhancer or UPE, and a gene activation site that contacts the target in the general transcriptional machinery. The DNA between the enhancer and promoter elements forms a loop that lets an activator bound to an enhancer come in contact with the target proteins associated with the general transcriptional machinery. When this occurs, the rate of transcription increases.

Chromosome organization may affect the expression of some genes

A chromosome is not only a bearer of genes. Various arrangements of a chromosome's ordered components (see Fig. 9-4) increase or decrease expression of the genes it contains. In multicellular eukaryotes, only a subset of the genes in a cell are active at any one time. The inactivated genes differ among cell types and in many cases are irreversibly dormant. Recent evidence suggests that certain small RNA molecules, 21 to 28 nucleotides long, affect chromatin structure and can permanently shut down sections of chromosomes. This inhibition of gene expression is known as *RNA interference.*

Some of the inactive genes lie in highly compacted chromatin, visible microscopically as densely staining regions of chromosomes during cell division. These regions of chromatin remain tightly coiled and bound to chromosome proteins throughout the cell cycle; even during interphase, they are visible as darkly staining fibers called **heterochromatin** (Fig. 13-10a). Evidence suggests that most of the heterochromatin DNA is not transcribed. When researchers inactivate one of the two X chromosomes in female mammals, most of the inactive X chromosome becomes heterochromatic and is seen as a *Barr body* (see Fig. 10-17).

Active genes are associated with a more loosely packed chromatin structure called **euchromatin** (Fig. 13-10b). The exposure of the DNA in euchromatin lets it interact with transcription factors and other regulatory proteins.

Gene inactivation by DNA methylation Inactive genes of vertebrates and some other organisms typically show a pattern of **DNA methylation** in which the DNA has been chemically altered by enzymes that add methyl groups to certain cytosine nucleotides in DNA. (The resulting 5-methylcytosine still pairs with guanine in the usual way.) Evidence indicates that certain regulatory proteins selectively bind to methylated DNA and make it inaccessible to the general transcriptional machinery.

DNA methylation probably reinforces gene inactivation rather than serving as the initial mechanism to silence genes. Apparently once a gene has been turned off by some other means, DNA methylation ensures it will remain inactive. For example, the DNA of the inactive X chromosome of a female mammal becomes methylated after the chromosome has become a condensed Barr body. Each time the DNA replicates, methylation enzymes perpetuate the pre-existing methylation pattern; hence, the DNA continues to be transcriptionally inactive in both daughter cells.

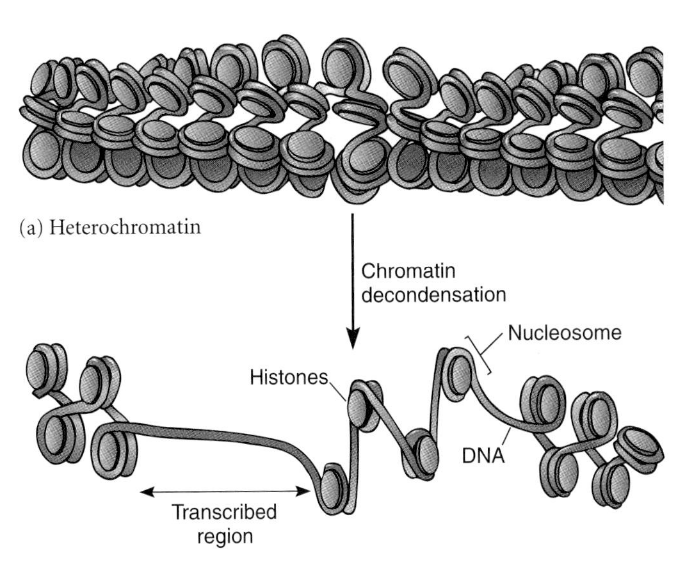

FIGURE **13-10** **The effect of chromatin structure on transcription.**

(a) An inactive region of DNA, heterochromatin is organized into tightly associated nucleosomes. **(b)** Active genes are found in decondensed chromatin called euchromatin. Chromatin decondensation is often a response to specific inducing signals. Loosely packed chromatin increases the accessibility to RNA polymerase required for transcription of the region. The histones are physically removed from the DNA in the region where transcription occurs.

Multiple copies of genes A single gene cannot always provide enough copies of its mRNA to meet the cell's needs. The requirement for high levels of certain products is met if multiple copies of the genes that encode them are present in the chromosome. Genes of this type, whose products are essential for all cells, may occur as multiple copies arranged one after another along the chromosome, in *tandemly repeated gene sequences.* Other genes, which are required only by a small group of cells, are selectively replicated in those cells in a process called **gene amplification** (see Fig. 16-5).

Within an array of repeated genes, all the copies are almost identical. Histone genes, which code for the proteins that associate with DNA to form nucleosomes (see Chapter 9), are usually multiple copies of 50 to 500 genes in cells of multicellular organisms. Similarly, multiple copies (150 to 450) of genes for rRNA and tRNA are present in cells.

The mRNAs of eukaryotes have many types of posttranscriptional control

The half-life of bacterial mRNA is usually minutes long; eukaryotic mRNA, even when it turns over rapidly, is far more stable. Bacterial mRNA is transcribed in a form that is translated immediately. In contrast, eukaryotic mRNA molecules undergo further modification and processing before use in protein synthesis. The message is capped, polyadenylated, spliced, and then transported from the nucleus to the cytoplasm to initiate translation (see Chapter 12). These events represent potential control points for translation of the message and the production of its encoded protein.

The addition of a **poly-A tail** to eukaryotic mRNA, for example, appears necessary to initiate translation. Researchers have shown that mRNAs with long poly-A tails are efficiently translated, whereas mRNAs with short poly-A tails are essentially dormant. *Polyadenylation,* which commonly occurs in the nucleus, can also take place in the cytosol. When the short poly-A tail of an mRNA is elongated, the mRNA becomes activated and is then translated.

Some pre-mRNAs are processed in more than one way

Investigators have discovered several forms of regulation involving mRNA processing. In some instances, the same gene produces one type of protein in one tissue and a related but somewhat different type of protein in another tissue. This is possible because some genes produce pre-mRNA molecules that have alternative splicing patterns; that is, they are spliced in different ways depending on the tissue. Typically, such a gene includes at least one segment that can be either an intron or an exon. As an intron, the sequence is removed, but as an exon, it is retained. Through **differential mRNA processing,** the cells in each tissue produce their own version of mRNA corresponding to the particular gene (Fig. 13-11). For example, this mechanism produces different forms of troponin, a protein that regulates muscle contraction, in different muscle tissues.

The stability of mRNA molecules varies

Controlling the life span of a particular kind of mRNA molecule permits control over the number of protein molecules translated from it. In some cases, mRNA stability is under hormonal control. This is true for mRNA that codes for vitellogenin, a protein synthesized in the liver of certain female animals, such as frogs and chickens. Vitellogenin is transported to the oviduct, where it is used in forming egg yolk proteins.

The hormone estradiol regulates vitellogenin synthesis. When estradiol levels are high, the half-life of vitellogenin mRNA in frog liver is about 500 hours. When researchers deprive cells of estradiol, the half-life of the mRNA drops rapidly, to less than 165 hours. This rapidly lowers cell vitellogenin mRNA levels and decreases synthesis of the vitellogenin protein. In addition to affecting mRNA stability, the hormone seems to control the rate at which the mRNA is synthesized.

Posttranslational chemical modifications may alter the activity of eukaryotic proteins

Another way to control the ultimate phenotypic expression of a gene is by regulating the activity of the gene product. As in bacteria, many metabolic pathways in eukaryotes contain allosteric enzymes regulated through feedback inhibition. In addition, after they are synthesized many eukaryotic proteins are extensively modified.

In *proteolytic processing,* proteins are synthesized as inactive precursors, which are converted to an active form by the removal of a portion of the polypeptide chain. For example, proinsulin contains 86 amino acids. Removing 35 amino acids yields the hormone insulin, which consists of two polypeptide chains containing 30 and 21 amino acids, respectively, linked by disulfide bridges. Other proteins are regulated in part by *selective degradation,* which keeps their numbers constant within the cell.

Chemical modification, by adding or removing functional groups, reversibly alters the activity of an enzyme. One common way to modify the activity of an enzyme or other protein is to add or remove phosphate groups. Enzymes that add phosphate groups are called **kinases;** those that remove them are **phosphatases.** For example, the cyclin-dependent kinases discussed in Chapter 9 help control the cell cycle by adding phosphate groups to certain key proteins, causing them to become activated or inactivated. Chemical modifications such as protein phosphorylation also let the cell respond rapidly to certain hormones (see Chapter 47), or to fast-changing environmental or nutritional conditions.

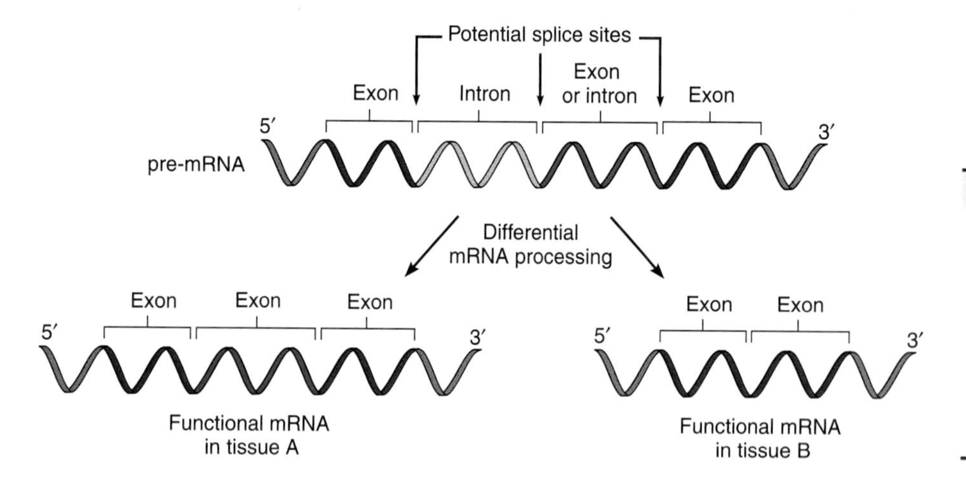

FIGURE **13-11**

Differential mRNA processing.

In some cases, a pre-mRNA molecule is processed in more than one way to yield two or more mature mRNAs, each of which encodes a related, but different, protein. In this generalized example, the gene contains a segment that can be an exon in tissue A *(left),* but an intron in tissue B *(right).*

Review

- How does regulation of eukaryotic genes differ from regulation of bacterial genes?
- Why must certain genes in eukaryotic cells be present in multiple copies?
- How does chromosome structure affect the activity of some eukaryotic genes?
- How does differential mRNA processing give rise to different proteins?

BiologyNow™ Assess your understanding of **gene regulation in eukaryotic cells** by taking the pretest on your BiologyNow CD-ROM.

SUMMARY WITH KEY TERMS

1 Explain why bacterial and eukaryotic cells have different mechanisms of gene regulation.

- Bacterial cells grow rapidly and have a relatively short life span. Controlling transcription is the most economical way for the cell to regulate gene expression.
- Eukaryotic cells have different regulatory requirements, in part because eukaryotes usually have a long life span, during which they respond to many different stimuli. Also, a single gene is regulated in different ways in different types of cells. Although transcriptional-level control predominates, control at other levels of gene expression is also important.

2 Define *operon;* sketch the main elements of the *lac* operon, and explain the functions of the operator and promoter regions.

- Most regulated genes in bacteria are organized into operons. Each **operon** is a gene complex consisting of a group of structural genes with related functions, plus the closely linked DNA sequences responsible for controlling them.
- Each operon has a single **promoter** region upstream from the protein-coding regions.
- The **operator** serves as the regulatory switch for **transcriptional-level control** of the operon. When a **repressor protein** binds to the operator sequence, it prevents transcription; although RNA polymerase binds to the promoter, it is blocked from transcribing the structural genes. When the repressor is not bound to the operator, transcription proceeds.

3 Distinguish among inducible, repressible, and constitutive genes.

- An **inducible operon,** such as the *lac* operon, is normally turned off. The repressor protein is synthesized in an active form that binds to the operator. If lactose is present, it is converted to allolactose, the **inducer,** which binds an allosteric site on the repressor protein, changing its shape. The altered repressor cannot bind to the operator, and the operon is transcribed.
- A **repressible operon,** such as the *trp* operon, is normally turned on. The repressor protein is synthesized in an inactive form that cannot bind to the operator. A metabolite (usually the end product of a metabolic pathway) acts as a **corepressor.** When intracellular corepressor levels are high, a corepressor molecule binds to an allosteric site on the repressor, changing its shape so that it binds to the operator and thereby turns off transcription of the operon.
- **Constitutive genes** are neither inducible nor repressible; they are active at all times. Regulatory proteins such as **catabolite activator protein (CAP)** and the repressor proteins are produced constitutively. These proteins work by recognizing and binding to specific base sequences in the DNA. The activity of constitutive genes is controlled by how efficiently RNA polymerase binds to their promoter regions.

4 Differentiate between positive and negative control, and show how both types of control operate in regulating the *lac* operon.

- Repressible and inducible operons are under **negative control.** When the repressor protein binds to the operator, transcription of the operon is turned off.
- Some inducible operons are also under **positive control.** A separate protein binds to the DNA and stimulates transcription of the gene. CAP activates the lactose operon; CAP binds to the promoter region, stimulating transcription by binding RNA polymerase tightly. To bind to the *lac* operon, CAP requires **cyclic AMP (cAMP).** Levels of cAMP increase as levels of glucose decrease.

5 Describe the types of posttranscriptional control in bacteria.

- Some **posttranscriptional controls** operate in bacteria. A **translational control** is a posttranscriptional control that regulates the rate of translation of a particular mRNA. **Posttranslational controls** include **feedback inhibition** of key enzymes in some metabolic pathways.

6 Discuss the structure of a typical eukaryotic gene and the DNA sequences involved in regulating that gene.

- Eukaryotic genes are not normally organized into operons. Regulation of eukaryotic genes occurs at the levels of transcription, mRNA processing, translation, and the protein product.
- The promoter of a regulated eukaryotic gene consists of an RNA polymerase-binding site and short DNA sequences known as **upstream promoter elements (UPEs).** The number and types of UPEs within the promoter region determine the efficiency of the promoter.
- Inducible eukaryotic genes are controlled by **enhancers,** which are located thousands of bases away from the promoter. Proteins that bind to enhancers seem to facilitate the binding of RNA polymerase to the promoter.

7 Give examples of some of the ways eukaryotic DNA-binding proteins bind to DNA.

- Eukaryotic genes are controlled by DNA-binding protein regulators called **transcription factors.** Many are transcriptional activators; others are transcriptional repressors.
- Each transcription factor has a DNA-binding **domain.** Some transcription factors have a helix-turn-helix arrangement and insert one of the helices into the DNA. Other transcription

factors have loops of amino acids held together by zinc ions; each loop includes an α-helix that fits into the DNA. Some transcription factors are **leucine zipper proteins** that have dimers that insert into the DNA.

8 Illustrate how a change in chromosome structure may affect the activity of a gene.

- Genes are inactivated by changes in chromosome structure. Densely packed regions of chromosomes called **heterochromatin** contain inactive genes. Active genes are associated with a loosely packed chromatin structure called **euchromatin.**
- **DNA methylation** is a mechanism that perpetuates gene inactivation.
- Some genes whose products are required in large amounts exist as multiple copies in the chromosome. In the process of **gene amplification,** some cells selectively amplify genes by DNA replication.

9 Explain how a gene in a multicellular organism may produce different products in different types of cells.

- As a result of **differential mRNA processing,** a single gene produces different forms of a protein in different tissues, depending on how the pre-mRNA is spliced. Typically, such a gene contains a segment that can be either an intron or an exon. As an intron, the sequence is removed, and as an exon, the sequence is retained.

10 Identify some of the types of regulatory controls that operate in eukaryotes after mature mRNA is formed.

- Certain regulatory mechanisms increase the stability of mRNA, allowing more protein molecules to be synthesized before mRNA degradation. In certain cases mRNA stability is under hormonal control.
- Posttranslational control of eukaryotic genes occurs by feedback inhibition or by modification of the protein structure. The function of a protein is changed by **kinases** adding phosphate groups or by **phosphatases** removing phosphates.

POST-TEST

1. The regulation of most bacterial genes occurs at the level of (a) transcription (b) translation (c) replication (d) posttranslation (e) postreplication
2. The operator of an operon (a) encodes information for the repressor protein (b) is the binding site for the inducer (c) is the binding site for the repressor protein (d) is the binding site for RNA polymerase (e) encodes the information for the CAP
3. A mutation that inactivates the repressor gene of the *lac* operon results in (a) the continuous transcription of the structural genes (b) no transcription of the structural genes (c) the binding of the repressor to the operator (d) no production of RNA polymerase (e) no difference in the rate of transcription
4. At a time when the *lac* operon is actively transcribed, (a) the operator is bound to the inducer (b) the lactose repressor is bound to the promoter (c) the operator is not bound to the promoter (d) the gene coding for the repressor is not expressed constitutively (e) the lactose repressor is bound to the inducer
5. A repressible operon codes for the enzymes of the following pathway. Which component of the pathway is most likely to be the corepressor for that operon?

   $$A \xrightarrow{\text{Enzyme 1}} B \xrightarrow{\text{Enzyme 2}} C \xrightarrow{\text{Enzyme 3}} D$$

   (a) substance A (b) substance B or C (c) substance D (d) enzyme 1 (e) enzyme 3
6. An mRNA molecule transcribed from the *lac* operon contains nucleotide sequences complementary to (a) structural genes coding for enzymes (b) the operator region (c) the promoter region (d) the repressor gene (e) introns
7. Feedback inhibition is an example of control at the level of ____________. (a) transcription (b) translation (c) posttranslation (d) replication (e) all of the preceding
8. Which of the following control mechanisms is generally the most economical in terms of conserving energy and resources? (a) control by means of operons and regulons (b) feedback inhibition (c) selective degradation of mRNA (d) selective degradation of enzymes (e) gene amplification
9. A repressible operon, such as the *trp* operon, is "off" when (a) the gene that codes for the repressor is expressed constitutively (b) the repressor-corepressor complex binds to the operator (c) the repressor binds to the structural genes (d) the corepressor binds to RNA polymerase (e) CAP binds to the promoter
10. Which of the following is an example of positive control? (a) transcription occurs when a repressor binds to an inducer (b) transcription cannot occur when a repressor binds to a corepressor (c) transcription is stimulated when an activator protein binds to DNA (d) a and b (e) a and c
11. Which of the following are typically absent in bacteria? (a) enhancers (b) proteins that regulate transcription (c) repressors (d) promoters (e) operators
12. The "zipper" of a leucine zipper protein attaches (a) specific amino acids to specific DNA base pairs (b) two polypeptide chains to each other (c) one DNA region to another DNA region (d) amino acids to zinc atoms (e) RNA polymerase to the operator
13. Inactive genes tend to be found in (a) highly condensed chromatin, known as euchromatin (b) decondensed chromatin, known as euchromatin (c) highly condensed chromatin, known as heterochromatin (d) decondensed chromatin, known as heterochromatin (e) chromatin that is not organized as nucleosomes
14. Which of the following is characteristic of genes and gene regulation in bacteria, but *not* eukaryotes? (a) presence of enhancers (b) capping of mRNAs (c) many chromosomes per cell (d) extensive binding of DNA to many kinds of regulatory proteins (e) exon splicing not required
15. Which of the following is characteristic of genes and gene regulation in *both* bacteria and eukaryotes? (a) promoter (b) noncoding

POST-TEST *(continued)*

DNA within coding sequences (c) enhancer (d) operons (e) DNA located in a nucleus

16. Which of the following is characteristic of genes and gene regulation in eukaryotes, but *not* bacteria? (a) enhancers (b) transcription factors (c) promoters (d) a and b only (e) a, b, and c are found in eukaryotes but not bacteria

17. Through differential mRNA processing, eukaryotes (a) reinforce gene inactivation (b) prevent transcription of heterochromatin (c) produce related but different proteins in different tissues (d) amplify genes to meet the requirement of high levels of a gene product (e) bind transcription factors to enhancers to activate transcription

CRITICAL THINKING

1. Develop a simple hypothesis that would explain the behavior of each of the following types of mutants in *E. coli:*
   Mutant a: The map position of this mutation is in the *trp* operon. The mutant cells are constitutive; that is, they produce all the enzymes coded for by the *trp* operon, even if large amounts of tryptophan are present in the growth medium.
   Mutant b: The map position of this mutation is in the *trp* operon. The mutant cells do not produce any enzymes coded for by the *trp* operon under any conditions.
   Mutant c: The map position of this mutation is some distance from the *trp* operon. The mutant cells are constitutive; that is, they produce all the enzymes coded for by the *trp* operon, even if the growth medium contains large amounts of tryptophan.
   Mutant d: The map position of this mutation is some distance from the *trp* operon. The mutant cells do not produce any enzymes coded for by the *trp* operon under any conditions.

2. Compare the types of bacterial genes associated with inducible operons, those associated with repressible operons, and those that are constitutive. Predict the category into which each of the following would most likely fit: (a) a gene that codes for RNA polymerase, (b) a gene that codes for an enzyme required to break down maltose, (c) a gene that codes for an enzyme used in the synthesis of adenine.

3. The regulatory gene that codes for the tryptophan repressor is not tightly linked to the *trp* operon. Would it be advantageous if it were? Explain your answer.

- Visit our Web site at **http://biology.brookscole.com/solomon7** for links to chapter-related resources on the World Wide Web. Additional online materials relating to this chapter can also be found on our Web site.

BIOLOGY NOW RESOURCES

Active Figure

13-1: Lactose operon

Preparing for an exam? Take a diagnostic test on your BiologyNow CD-ROM.

Post-Test Answers

1. a	2. c	3. a	4. e
5. c	6. a	7. c	8. a
9. b	10. c	11. a	12. b
13. c	14. e	15. a	16. d
17. c			

14

DNA Technologies

Keith V. Wood/Visuals Unlimited

Genetically engineered tobacco plant. This plant glows as it expresses the luciferase gene from a firefly. Luciferase is an enzyme that catalyzes a reaction that produces a flash of light.

CHAPTER OUTLINE

During the mid-1970s, as the development of new ways of studying DNA led to radically new research approaches, a revolution in the field of biology began. These techniques have had a major impact, and not just in genetic studies. DNA technologies affect a wide range of areas from cell biology and evolution to ethical and societal issues.

This chapter begins with a consideration of **recombinant DNA technology,** in which researchers splice together DNA from different organisms in the laboratory. The primary goal of this technology is to enable scientists to obtain many copies of any specific DNA segment for the purpose of studying it biochemically. Using recombinant DNA technology, scientists introduce foreign DNA into the cells of microorganisms. Under the right conditions, when a cell divides this DNA is replicated and transmitted to the daughter cells. In this way, a particular DNA sequence is amplified, or **cloned,** to provide millions of identical copies that are isolated in pure form. Today these methods have been supplemented by extremely valuable techniques for the cloning of DNA in vitro, outside a living organism. Because new recombinant DNA methods are continually emerging, we do not attempt to explore them all here. Instead, we discuss some of the major approaches that have provided a foundation for the technology.

We also consider the ways in which studies of **cloned DNA sequences** have been of immense value in helping scientists understand the organization of genes and the relationship between genes and their products. In fact, most current knowledge of the structure and control of eukaryotic genes, and of the roles of genes in development, comes from applying these methods.

This chapter also explores some of the many practical applications of DNA technologies. One of the rapidly advancing areas of study is **genetic engineering**—modifying the DNA of an organism to produce new genes with new traits. Genetic engineering can take many forms, ranging from basic research (see photograph) to the production of strains of bacteria that manufacture useful protein products, to the development of plants and animals that express foreign genes.

This wide range of applications of ongoing discoveries in molecular genetics is transforming people's view of **biotechnology,** the use of organisms to benefit humanity. Traditional forms of biotechnology include such familiar examples as the selective breeding of plants and the use of yeast to make alcoholic beverages or cause bread to rise. However, the examples of biotechnology most frequently cited today are applications of genetic engineering in such diverse areas as medicine and the pharmaceutical industry, foods and agriculture, and forensic science. Biotechnology does not stand alone; its advances are greatly facilitated by other kinds of technology, including powerful computer programs and automated systems. ■

RECOMBINANT DNA METHODS

Learning Objectives

1. Explain how a typical restriction enzyme cuts DNA molecules, and give examples of the ways in which these enzymes are used in recombinant DNA technology.
2. Distinguish among a genomic library, a chromosome library, and a complementary DNA (cDNA) library; explain why one would clone the same eukaryotic gene from both a genomic library and a cDNA library.
3. Identify some uses of DNA hybridization probes.
4. Describe how the polymerase chain reaction amplifies DNA in vitro.
5. Describe the chain termination method of DNA sequencing.

Recombinant DNA technology was not developed quickly. It actually had its roots in the 1940s with genetic studies of bacteria and **bacteriophages** ("bacteria eaters"), the viruses that infect them (see Chapters 11 and 23). After decades of basic research and the accumulation of extensive knowledge, the technology became feasible and available to the many scientists who now use these methods.

In recombinant DNA technology, enzymes from bacteria, known as **restriction enzymes,** are used to cut DNA molecules only in specific places. Restriction enzymes enable researchers to cut DNA into manageable segments. Each fragment is then incorporated into a suitable **vector** molecule, a carrier capable of transporting the DNA fragment into a cell. Bacteriophages and DNA molecules called plasmids are two examples of vectors. Bacterial DNA is circular; a **plasmid** is a separate, much smaller, circular DNA molecule that may be present and replicate inside a bacterial cell, such as *E. coli.* Researchers introduce plasmids into bacterial cells by a method called **transformation,** the uptake of foreign DNA by cells (see Chapter 11). For transformation to be efficient, the researcher alters the bacterial cell walls to make them permeable to the plasmid DNA molecules. Once a plasmid enters a cell, it is replicated and distributed to its daughter cells during cell division. When a *recombinant plasmid* (one that has foreign DNA spliced into it) replicates in this way, many copies of the foreign DNA are made—that is, the foreign DNA is cloned.

Restriction enzymes are "molecular scissors"

Discovering restriction enzymes was a major breakthrough in developing recombinant DNA technology. Today large numbers of different types of restriction enzymes, each with its own characteristics, are readily available to researchers. For example, a restriction enzyme known as *Hin*dIII recognizes and cuts a DNA molecule at the restriction site 5′—AAGCTT—3′, whereas the sequence 5′—GAATTC—3′ is cut by another, known as *Eco*RI. (The names of restriction enzymes are generally derived from the names of the bacteria from which they were originally isolated. Thus, *Hin*dIII and *Eco*RI are derived from *Hemophilus influenzae* and *E. coli,* respectively.)

Why do bacteria produce such enzymes? During infection, a bacteriophage injects its DNA into a bacterial cell. The bacterium can defend itself if it has restriction enzymes that can attack the bacteriophage DNA. The cell protects its own DNA from breakdown by modifying it after replication. An enzyme adds a methyl group to one or more bases in each restriction site so that the restriction enzyme does not recognize and cut the bacterial DNA.

Restriction enzymes enable scientists to cut DNA from chromosomes into shorter fragments in a controlled way. Many of the restriction enzymes used for recombinant DNA studies cut **palindromic sequences,** which means the base sequence of one strand reads the same as its complement when both are read in the 5′ to 3′ direction. Thus, in our *Hin*dIII example, both strands read 5′—AAGCTT—3′, which as a double-stranded molecule is diagrammed as

5′—AAGCTT—3′
3′—TTCGAA—5′

By cutting both strands of the DNA, but in a staggered fashion, these enzymes produce fragments with identical, complementary, single-stranded ends:

5′—A AGCTT—3′
3′—TTCGA A—5′

These ends are called *sticky ends* because they pair (by hydrogen bonding) with the complementary, single-stranded ends of other DNA molecules that have been cut with the same enzyme (Fig. 14-1). Once the sticky ends of two molecules have been joined together in this way, they are treated with **DNA ligase,** an enzyme that covalently links the two DNA fragments to form a stable recombinant DNA molecule.

Recombinant DNA forms when DNA is spliced into a vector

In recombinant DNA technology, geneticists cut both foreign DNA and plasmid DNA with the same restriction enzyme. The two types of DNA are then mixed together under conditions that facilitate hydrogen bonding between the complementary

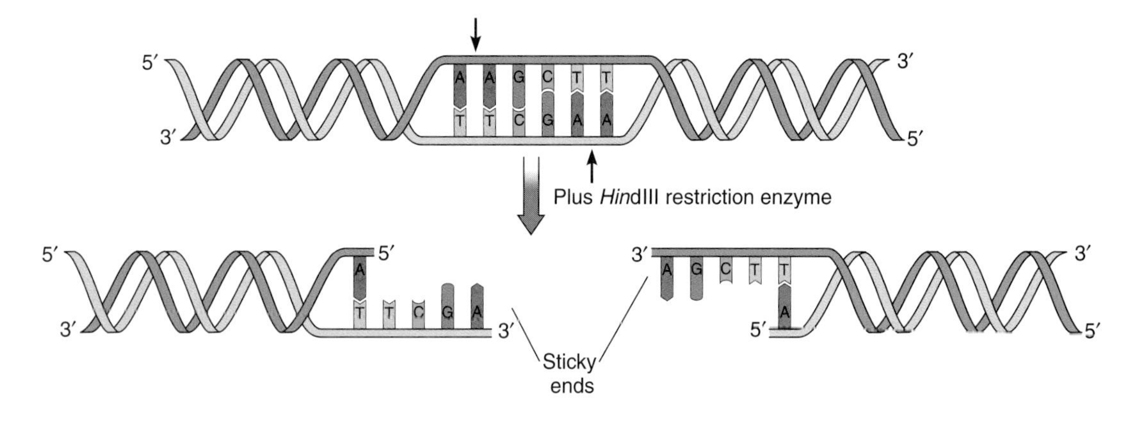

FIGURE **14-1** **Cutting DNA with a restriction enzyme.**
Many restriction enzymes, like *Hin*dIII, cut DNA at sequences that are palindromic, producing complementary sticky ends. The small black arrows designate the enzyme's cleavage sites.

bases of the sticky ends, and the nicks in the resulting recombinant DNA are sealed by DNA ligase (Fig. 14-2).

The plasmids now used in recombinant DNA work have been extensively manipulated in the laboratory to include features helpful in isolating and analyzing cloned DNA (Fig. 14-3). Among these are an origin of replication (see Chapter 11), one or more restriction sites, and genes that let researchers select cells transformed by recombinant plasmids. These genes cause transformed cells to grow under specified conditions that do not allow untransformed cells to grow. In this way, the researchers use features that are also common in naturally occurring plasmids. Typically, plasmids do not contain genes essential to the bacterial cells under normal conditions. However, they often carry genes useful under specific environmental conditions, such as genes that confer resistance to particular antibiotics or let the cells use a specific nutrient. For example, cells transformed with a plasmid that includes a gene for resistance to the antibiotic tetracycline can grow in a medium that contains tetracycline, whereas untransformed cells cannot.

A limiting property of any vector, however, is the size of the DNA fragment it can effectively carry. The size of a DNA segment is often given in kilobases, with 1 kilobase (kb) being

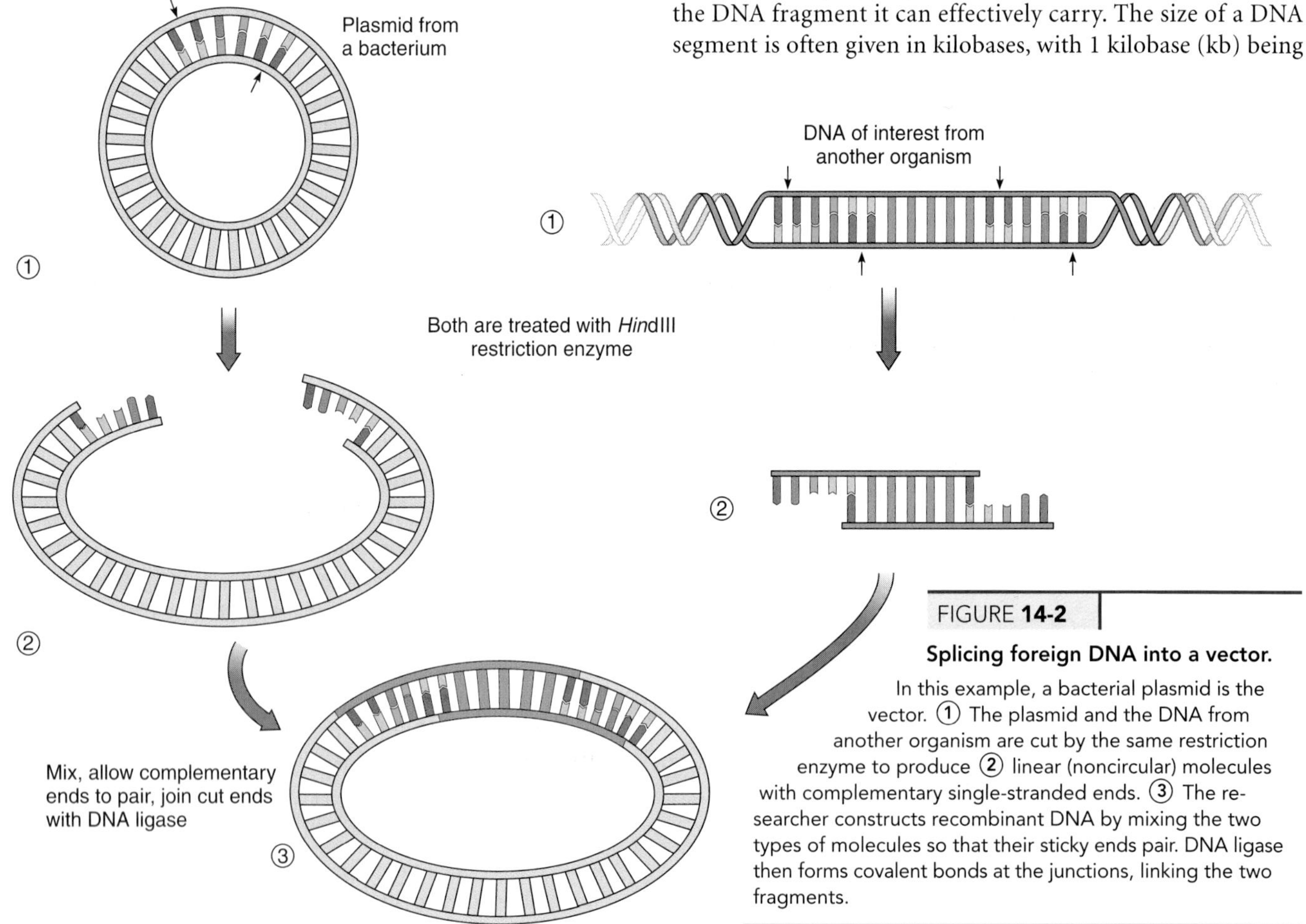

FIGURE **14-2**
Splicing foreign DNA into a vector.
In this example, a bacterial plasmid is the vector. ① The plasmid and the DNA from another organism are cut by the same restriction enzyme to produce ② linear (noncircular) molecules with complementary single-stranded ends. ③ The researcher constructs recombinant DNA by mixing the two types of molecules so that their sticky ends pair. DNA ligase then forms covalent bonds at the junctions, linking the two fragments.

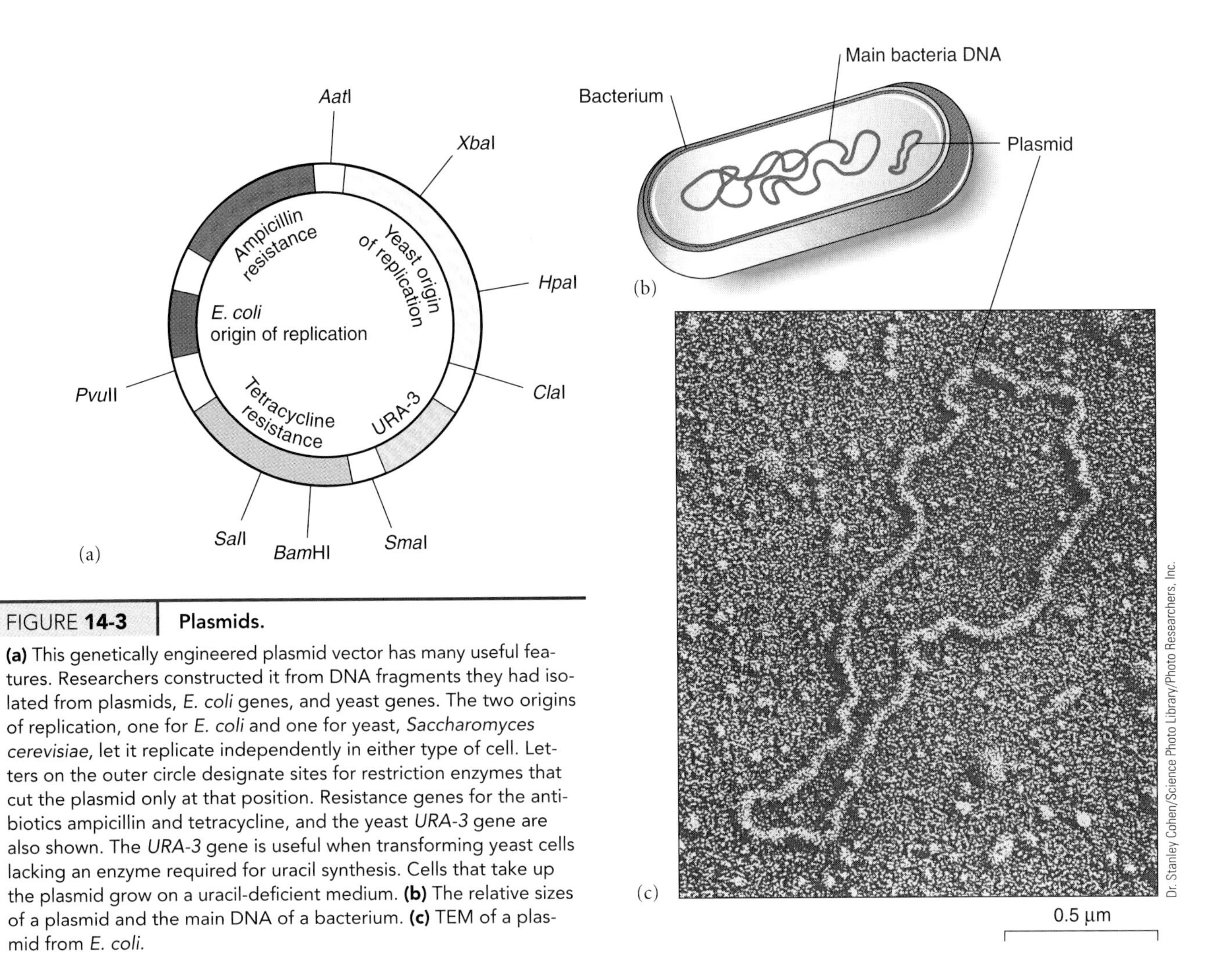

FIGURE **14-3** **Plasmids.**

(a) This genetically engineered plasmid vector has many useful features. Researchers constructed it from DNA fragments they had isolated from plasmids, *E. coli* genes, and yeast genes. The two origins of replication, one for *E. coli* and one for yeast, *Saccharomyces cerevisiae*, let it replicate independently in either type of cell. Letters on the outer circle designate sites for restriction enzymes that cut the plasmid only at that position. Resistance genes for the antibiotics ampicillin and tetracycline, and the yeast *URA-3* gene are also shown. The *URA-3* gene is useful when transforming yeast cells lacking an enzyme required for uracil synthesis. Cells that take up the plasmid grow on a uracil-deficient medium. **(b)** The relative sizes of a plasmid and the main DNA of a bacterium. **(c)** TEM of a plasmid from *E. coli*.

equal to 1000 base pairs. Fragments smaller than 10 kb can usually be inserted into plasmids for use in *E. coli*. However, larger fragments require the use of bacteriophage vectors, which can handle up to 23 kb of DNA. Other vectors are **cosmid cloning vectors,** which are combination vectors with features from both bacteriophages and plasmids, and **bacterial artificial chromosomes (BACs),** which accommodate much larger fragments of DNA. For example, a BAC can include up to about 200 kb of extra DNA, a feature that made BACs especially useful in the Human Genome Project (see Chapter 15).

Recombinant DNA can also be introduced into cells of eukaryotic organisms. For example, geneticists use engineered viruses as vectors in mammal cells. These viruses are disabled so they do not kill the cells they infect. Instead, the viral DNA, as well as any foreign DNA they carry, becomes incorporated into the cell's chromosomes after infection. As discussed later, other methods do not require a biological vector.

DNA can be cloned inside cells

Because a single gene is only a small part of the total DNA in an organism, isolating the piece of DNA containing that particular gene is like finding a needle in a haystack: A powerful detector is needed. Today many methods enable biologists to isolate a specific nucleotide sequence from an organism. We start by discussing methods in which DNA is cloned inside bacterial cells. We use the cloning of human DNA as an example, although the procedure is applicable for any organism.

A genomic library contains fragments of all DNA in the genome

The total DNA in a cell is called the **genome.** For example, if DNA is extracted from human cells, we refer to it as human genomic DNA. A **genomic library** is a collection of thousands of DNA fragments that represent all the DNA in the genome. Each fragment is inserted into a plasmid, which is usually incorporated into a bacterial cell. Thus the human genomic library is stored in a collection of recombinant bacteria, each with a different fragment of human DNA. Genomic libraries are used to isolate and study specific genes.

Individual chromosomes can also be isolated to make a **chromosome library** containing all the DNA fragments in that specific chromosome. If a gene of interest is known to be associated with a particular chromosome, it is easier to isolate that gene from a chromosome library than from a genomic library.

The first step in producing a genomic or chromosome library is to cut the DNA with a restriction enzyme, generating

FIGURE **14-4**

Producing a genomic or chromosome library.

Only a small part of one chromosome is shown. A great many more DNA fragments would be produced from an entire chromosome or genome. The numbered steps in the figure are explained in the text.

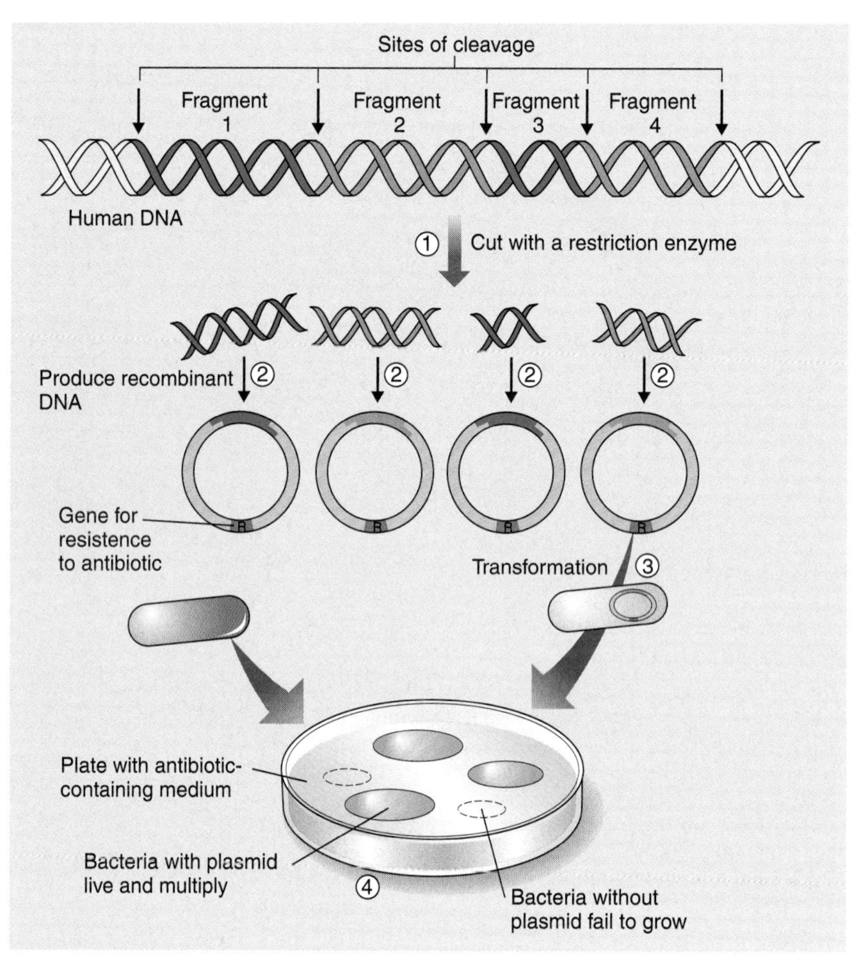

a population of DNA fragments (Fig. 14-4, step ①). These fragments vary in size and in the genetic information they carry, but they all have identical sticky ends. Geneticists treat plasmid DNA that will be used as a vector with the same restriction enzyme, which converts the circular plasmids into linear molecules with sticky ends complementary to those of the human DNA fragments. Recombinant plasmids are produced by mixing the two kinds of DNA (human and plasmid) together under conditions that promote hydrogen bonding of complementary bases. Then DNA ligase is used to covalently bond the paired ends of the plasmid and human DNA, forming recombinant DNA (Fig. 14-4, step ②). Unavoidably, nonrecombinant plasmids also form, because some plasmids revert to their original circular shape without incorporating foreign DNA.

The geneticists insert plasmids into antibiotic-sensitive bacterial cells by transformation (Fig. 14-4, step ③). Because the ratio of plasmids to cells is kept very low, it is rare for a cell to receive more than one plasmid molecule, and not all cells receive a plasmid. The researchers incubate normally antibiotic-sensitive cells on a nutrient medium that includes antibiotics, so only those cells grow that have incorporated a plasmid (which contains a gene for antibiotic resistance) (Fig. 14-4, step ④). In addition, the plasmid has usually been engineered in ways that enable researchers to identify those cells containing recombinant plasmids.

Genomic and chromosome libraries contain redundancies; that is, certain human DNA sequences have been inserted into plasmids more than once, purely by chance. However, each individual recombinant plasmid (analogous to a book in the library) contains only a single fragment of the total human genome.

To identify a plasmid containing a sequence of interest, the researcher clones each plasmid until there are millions of copies to work with. This process occurs as the bacterial cells grow and divide. A dilute sample of the bacterial culture is spread on solid growth medium, so the cells are widely separated. Each cell divides many times, yielding a visible **colony,** which is a clone of genetically identical cells originating from a single cell. All the cells of a particular colony contain the same recombinant plasmid, so during this process a specific sequence of human DNA is also cloned. The major task is to determine which colony (out of thousands) contains a cloned fragment of interest. Specific DNA sequences are identified in various ways.

A complementary genetic probe detects a specific DNA sequence

Suppose a researcher wishes to screen the thousands of recombinant DNA molecules in bacterial cells to find the one that contains a gene coding for a specific protein. A common approach to detecting the DNA of interest involves using a **genetic probe,** which is usually a radioactively labeled segment of single-stranded DNA that can **hybridize**—become attached by base pairing—to complementary base sequences in the target gene. If at least part of the amino acid sequence of that protein is known, it is possible to synthesize a radioactive, single-stranded DNA fragment to code for that sequence. However, because of the existence of synonymous codons, a specific amino acid sequence could potentially be coded for by many different base sequences (see Fig. 12-5). One approach to solving this problem has been to synthesize a complete mixture of all possible probes that could code for the desired amino acid sequence.

Biologists use genetic probes in a variety of ways. For example, they transfer cells from bacterial colonies containing recombinant plasmids to a nitrocellulose or nylon membrane, which then becomes a *replica* of the colonies (Fig. 14-5). They treat the cells on the membrane chemically to lyse them, making the DNA single-stranded. Then they incubate the membrane with the radioactive probe mixture to let the probes hybridize with any complementary strands of DNA that may be

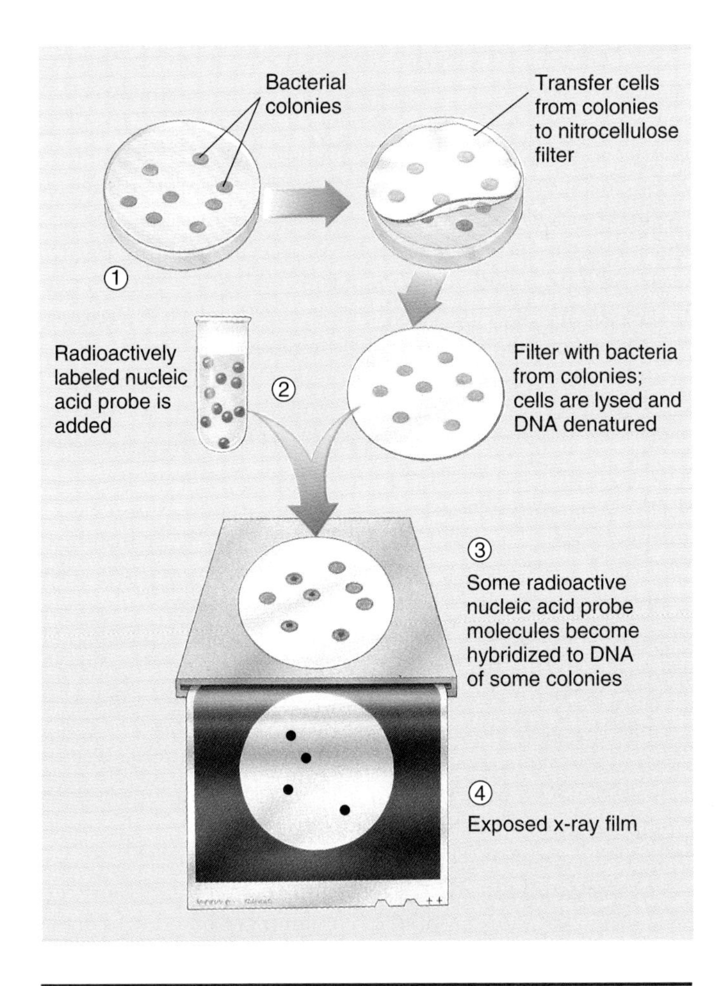

FIGURE **14-5** **Using a genetic probe to find bacterial cells with a specific recombinant DNA molecule.**
A radioactive nucleic acid probe, which is usually single-stranded DNA, reveals the presence of complementary sequences of DNA.

present. Each spot on the membrane containing DNA complementary to that particular probe becomes radioactive and is detected by autoradiography (see Chapter 2), using x-ray film. Each spot on the film therefore identifies a colony containing a plasmid that includes the DNA of interest. Cells from the de sired colonies are then cultured to produce millions of new bacteria, all containing the recombinant DNA of interest.

A cDNA library is complementary to mRNA and does not contain introns

Researchers frequently wish to clone intact genes while avoiding **introns,** which are regions of eukaryotic genes that do not code for proteins. Scientists also may wish to clone only genes that are expressed in a particular cell type. In such cases, they construct libraries of DNA copies of mature mRNA from which introns have been removed. The copies, known as **complementary DNA (cDNA)** because they are complementary to mRNA, also lack introns. The researchers use the enzyme **reverse transcriptase** (see Chapter 12) to synthesize single-stranded cDNA, which they separate from the mRNA and make double-stranded with DNA polymerase (Fig. 14-6). A **cDNA library** is formed using mRNA from a single cell type as the starting material. The double-stranded cDNA molecules are inserted into plasmid or virus vectors, which then multiply in bacterial cells.

Analyzing cDNA clones lets investigators determine certain characteristics of the protein encoded by the gene, including its amino acid sequence. They can also study the structure of the mature mRNA. Because the cDNA copy of the mRNA does

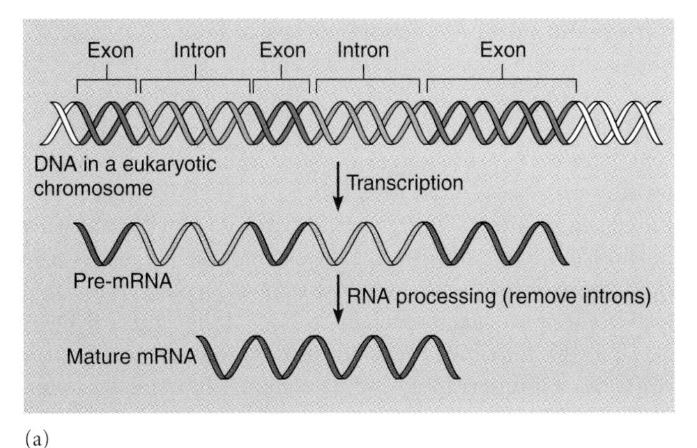

(a)

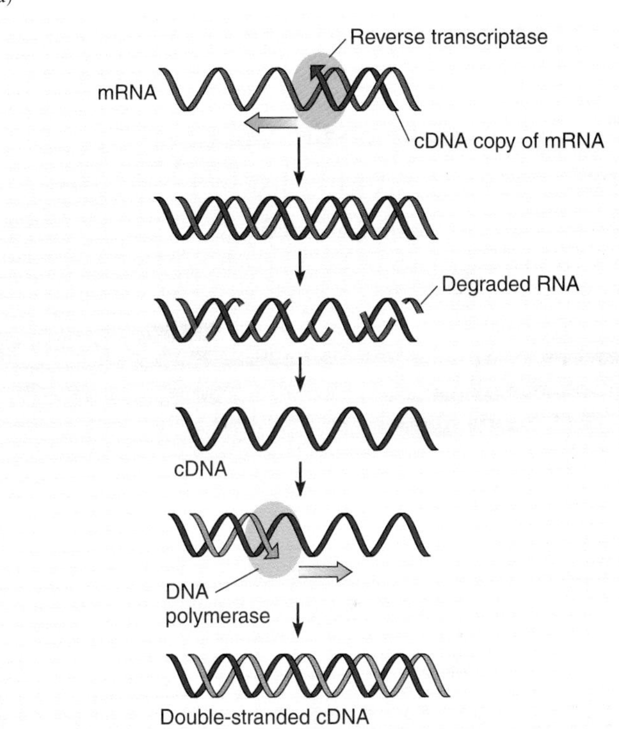

(b)

FIGURE **14-6** **The formation of cDNA.**
(a) The formation of cDNA relies on RNA processing that occurs in the nucleus to yield mature mRNA. **(b)** The mature mRNA is extracted and purified. To make a DNA copy of the mature mRNA, researchers use reverse transcriptase to produce a single-stranded cDNA complementary to the mRNA. Specific enzymes then degrade the mRNA, and DNA polymerase is used to synthesize a second strand of DNA. The end result is a double-stranded cDNA molecule.

not contain intron sequences, comparing the DNA base sequences in cDNA and the DNA in genomic or chromosome libraries reveals the locations of intron and exon coding sequences in the gene.

Cloned cDNA sequences are also useful when geneticists want to produce a eukaryotic protein in bacteria. When they introduce an intron-containing human gene, such as the gene for human growth hormone, into a bacterium, it can't remove the introns from the transcribed RNA to make a functional mRNA for producing its protein product. If they insert a cDNA clone of the gene into the bacterium, however, its transcript contains an uninterrupted coding region. A functional protein is synthesized if the geneticist inserts the gene downstream of an appropriate bacterial promoter and if he or she places the appropriate bacterial translation initiation sequences in it.

The polymerase chain reaction is a technique for amplifying DNA in vitro

The methods for amplifying a specific DNA sequence just described all involve cloning DNA in cells, usually those of bacteria. These processes are time consuming and require an adequate DNA sample as starting material. The **polymerase chain reaction (PCR)** technique, which U.S. biochemist Kary Mullis developed in 1985, lets researchers amplify a tiny sample of DNA millions of times in a few hours. In 1993, Mullis received the Nobel Prize in Chemistry for his work.

In PCR, DNA polymerase uses nucleotides and primers to replicate a DNA sequence in vitro, thereby producing two DNA molecules (Fig. 14-7). Then the researchers denature (separate by heating) the two strands of each molecule and replicate them again, yielding four double-stranded molecules. After the next cycle of heating and replication, there are eight molecules, and so on, with the number of DNA molecules doubling in each cycle. After only 20 heating and cooling cycles (done using automated equipment), this exponential process yields 2^{20}, or more than 1 million, copies of the target sequence!

Because the reaction can be carried out efficiently only if the DNA polymerase can remain stable through many heating cycles, researchers use a heat-resistant DNA polymerase, known as *Taq* polymerase. The name of this enzyme reflects its source, *Thermus aquaticus,* a bacterium that lives in hot springs in Yellowstone National Park. Because the water in this environment is close to the boiling point, all enzymes in *T. aquaticus* have evolved to be stable at high temperatures. Bacteria living in deep-sea thermal vents have similar enzymes (see *Focus On: Life without the Sun,* Chapter 53).

The PCR technique has virtually limitless applications. It enables researchers to amplify and analyze tiny DNA samples from a variety of sources, ranging from crime scenes to archaeological remains. For example, in 1997 investigators reported the first analysis of mitochondrial DNA obtained from the bones of Neandertals (see Chapter 21).

A limitation of the PCR technique is that it is almost too sensitive. Even a tiny amount of contaminant DNA in a sample

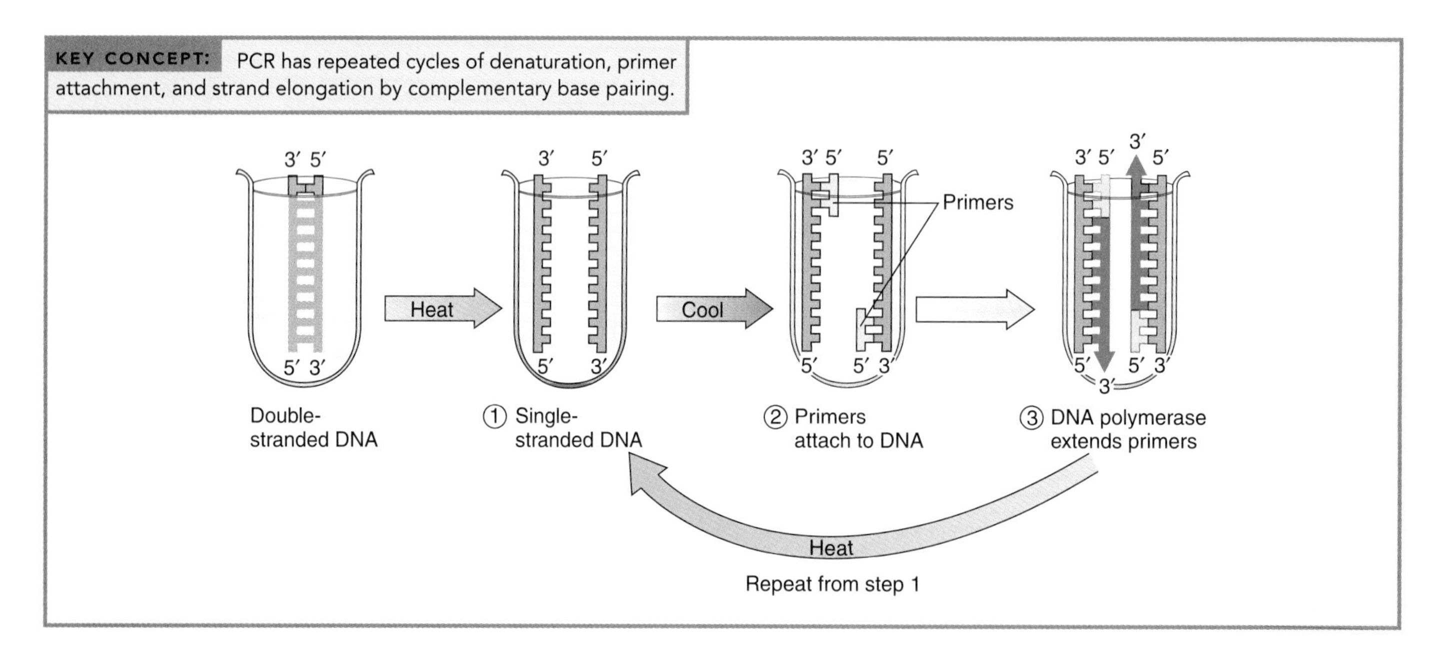

FIGURE **14-7** **Amplification of DNA by PCR.**

The initial reaction mixture includes a very small amount of double-stranded DNA, DNA precursors (deoxyribonucleotides), specific nucleic acid primers, and heat-resistant *Taq* DNA polymerase.
① The DNA is denatured (separated into single strands) by heat.
② Primers attach to the primer-binding site on each DNA strand.
③ Each DNA strand acts as a template for DNA synthesis catalyzed by the *Taq* DNA polymerase. The number of double-stranded DNA molecules doubles each time the cycle of heating and cooling is repeated.

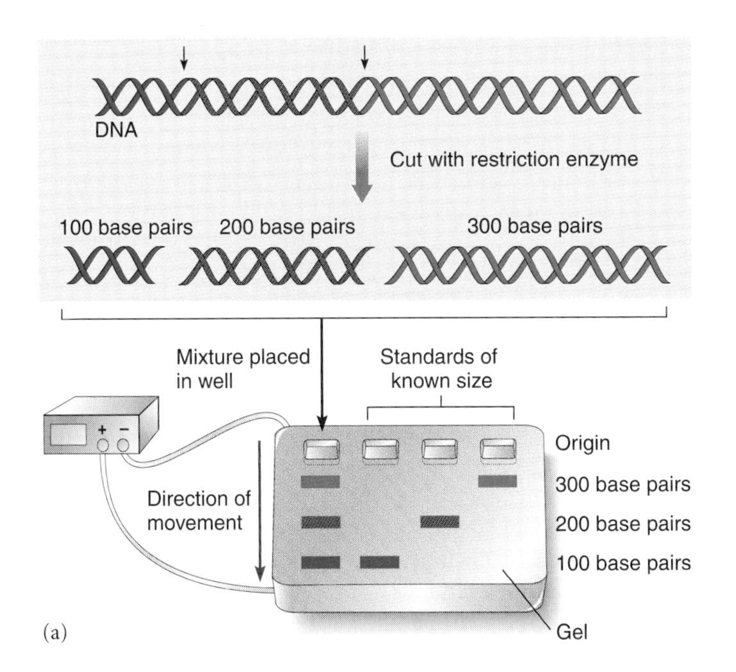

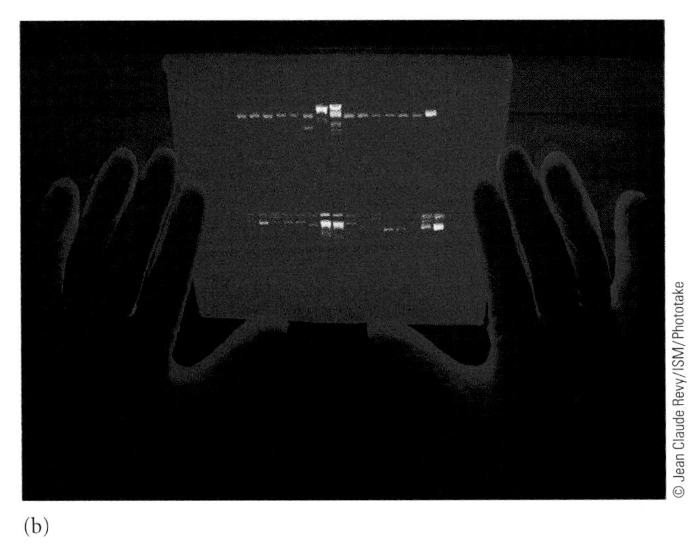

© Jean Claude Revy/ISM/Phototake

FIGURE **14-8** **Gel electrophoresis.**

(a) A researcher sets up an electrical field in a gel material, consisting of agarose or polyacrylamide, which is poured as a thin slab on a glass or Plexiglas holder. After the gel has solidified, the researcher loads samples containing a mixture of macromolecules of different sizes in wells formed at one end of the gel, then applies an electrical current. The smallest DNA fragments *(green)* travel the longest distance. **(b)** A gel containing separated DNA fragments. The gel is stained with ethidium bromide, a dye that binds to DNA and is fluorescent under UV light.

may become amplified if it includes a DNA sequence complementary to the primers, potentially leading to an erroneous conclusion. Researchers take appropriate precautions to avoid this and other technical pitfalls.

Gel electrophoresis is used for separating macromolecules

Mixtures of certain macromolecules—proteins, polypeptides, DNA fragments, or RNA—are separated by **gel electrophoresis,** a method that exploits the fact that these molecules carry charged groups that cause them to migrate in an electrical field. Figure 14-8 illustrates gel electrophoresis of DNA molecules. Both DNA and RNA migrate through the gel toward the positive pole of the electrical field because they are negatively charged, due to their phosphate groups. Because the gel slows down the large molecules more than the small molecules, they travel at a rate inversely proportional to their length (that is, molecular weight). Including DNA fragments of known size as standards ensures accurate measurements of the molecular weights of the unknown fragments.

Geneticists can identify specific DNA fragments that hybridize with a complementary genetic probe. However, it is impossible to hybridize a probe to DNA fragments contained in a gel. For this reason the DNA is usually denatured and then transferred to a nitrocellulose or nylon membrane, which picks up the DNA like a blotter picks up ink. The resulting blot, which is essentially a replica of the gel, is incubated with a DNA probe, which hybridizes with any complementary DNA fragments. If the probe is radioactive, the blot is subjected to autoradiography. The resulting spots on the x-ray film correspond to the locations of the fragments in the gel that are complementary to the probe. Today many radioactive probes are detected by chemical luminescence, which is analyzed by computer scanners, eliminating the need for autoradiography.

This method of detecting DNA fragments—separating them by gel electrophoresis and then transferring them to a nitrocellulose or nylon membrane—is called a **Southern blot,** named for its inventor, Edward M. Southern of Edinburgh University in Scotland. The procedure has widespread applications. It is often used to diagnose certain types of genetic disorders. For example, in some cases the DNA of a mutant allele is detected because it migrates differently in the gel from the DNA of its normal counterpart.

Similar blotting techniques are used to study RNA and proteins. When RNA molecules separated by electrophoresis are transferred to a membrane, the result is called, rather in jest, a **Northern blot.** In the same spirit, the term **Western blot** is applied to a blot consisting of proteins or polypeptides previously separated by gel electrophoresis. (So far, no one has invented a type of blot that could be called an Eastern blot.) In the case of Western blotting, scientists recognize the polypeptides of interest by labeled antibody molecules that bind to them specifically. For example, Western blotting is used diagnostically to detect the presence of proteins specific to HIV-1, the AIDS virus.

One way to characterize DNA is to determine its sequence of nucleotides

Investigators use a cloned piece of DNA as a research tool for many different applications. For example, they may clone a gene to obtain the encoded protein for some industrial or pharmaceutical process. Regardless of the particular application, before engineering a gene researchers must know a great deal about the gene and how it functions. The usual first step is determining the sequence of nucleotides.

DNA sequencing is based on methods that Sanger and Gilbert developed

In the 1990s, the advent of automated **DNA-sequencing** machines connected to powerful computers let scientists sequence huge amounts of DNA quickly and reliably. Automated DNA sequencing relies on the *chain termination method* that British biochemist Fred Sanger and U.S. biochemist Walter Gilbert developed in 1974. In 1980, Sanger and Gilbert shared the Nobel Prize in Chemistry for this contribution. (This was Sanger's second Nobel Prize. His first was for his work on the structure of the protein insulin.) Although DNA sequencing is now fully automated, we briefly consider the essential steps of the chain termination method.

This method of DNA sequencing is based on the fact that a replicating DNA strand that has incorporated a modified synthetic nucleotide, known as a *dideoxynucleotide,* cannot elongate beyond that point. Unlike a "normal" deoxynucleotide, which lacks a hydroxyl group on its 2′ carbon, a dideoxynucleotide also lacks a hydroxyl group on its 3′ carbon (Fig. 14-9). Recall from Chapter 11 that a 3′ hydroxyl group reacts each time a phosphodiester linkage is formed. Thus, dideoxynucleotides terminate elongation during DNA replication.

FIGURE **14-9** Dideoxynucleotide.

Dideoxynucleotides are modified nucleotides that lack a 3′ hydroxyl group and thus block further elongation of a new DNA chain.

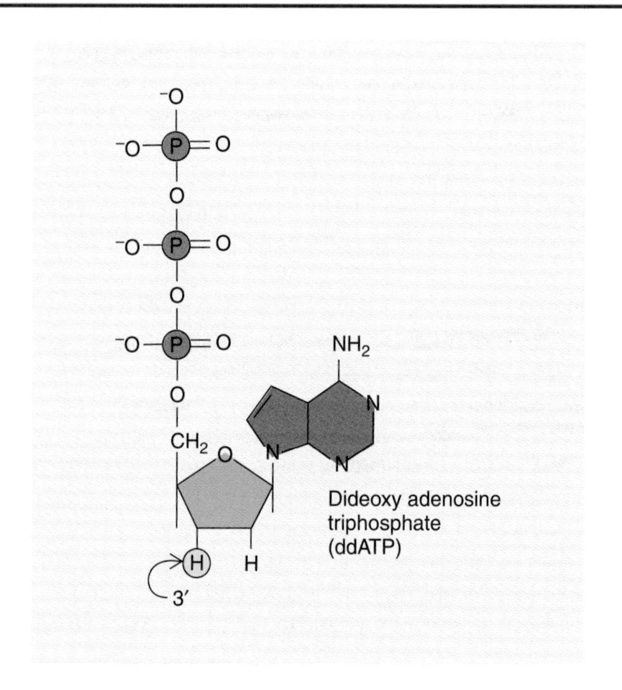

The researcher prepares four different reaction mixtures. Each contains multiple single-stranded copies of the DNA to be sequenced, DNA polymerase, appropriate radioactively labeled primers, and all four deoxynucleotides needed to synthesize DNA: dATP, dCTP, dGTP, and dTTP. Each mixture also includes a small amount of only one of the four dideoxynucleotides: ddATP, ddCTP, ddGTP, or ddTTP (Fig. 14-10a). (The prefix "dd" is for dideoxynucleotides, to distinguish them from deoxynucleotides, which are designated "d.")

Here's how the reaction proceeds in the mixture that includes ddATP (Fig. 14-10b). At each site where adenine is specified, occasionally a growing strand incorporates a ddATP and does not elongate farther. So a mixture of DNA fragments of varying lengths forms in the reaction mixture. Each fragment that contains a ddATP marks a specific location where adenine would normally be found in the newly synthesized strand. Similarly, in the reaction mixture that includes ddCTP, each fragment that contains ddCTP marks the position of a cytosine in the newly synthesized strand, and so on.

The radioactive fragments from each reaction are denatured and then separated by gel electrophoresis, with each reaction mixture (corresponding to A, T, G, or C) occupying its own lane in the gel. The positions of the newly synthesized fragments in the gel can then be determined by autoradiography (Fig. 14-10c, d). Because the high resolution of the gel makes it possible to distinguish between fragments that differ in length by only a single nucleotide, the researcher can read off the sequence in the newly synthesized DNA one base at a time. For example, consider a film of a DNA-sequencing gel that shows the following:

Starting at the bottom, the sequence would be read as: 5′—C—G—T—A—3′.

The chain termination method is still used in automated DNA-sequencing machines. However, the method of detection has changed in that the nucleotide sequence is no longer visualized by radioactivity. Instead, researchers label each of the four ddNTPs with a differently colored fluorescent dye. The computer uses a laser to read the fluorescence of the dye markers as the bases emerge from the end of a lane of electrophoresis medium (Fig. 14-11).

FIGURE **14-11** Automated DNA-sequencing results. ▶

The computer produces the series of peaks shown as it reads the fluorescing bands on the gel. The nucleotide base adenine is shown in green, guanine in black, cytosine in blue, and thymine in red. The DNA sequence assigned by the computer appears at the top of the printout.

FIGURE **14-10**

The chain termination method of DNA sequencing.

Single-stranded DNA fragment to be sequenced

5′ 3′

A T G C T A T G C T C C

+ddATP +ddCTP +ddGTP +ddTTP

(a) Four different reaction mixtures are used to sequence a DNA fragment; each contains a small amount of a single dideoxynucleotide, such as ddATP. Larger amounts of the four normal deoxynucleotides (dATP, dCTP, dGTP, and dTTP) plus DNA polymerase and radioactively labeled primers are also included.

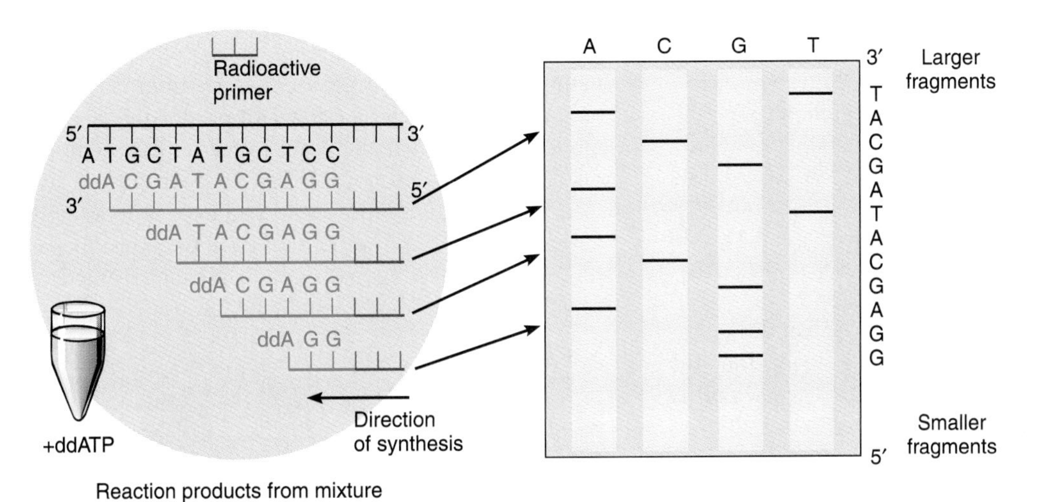

(b) The random incorporation of dideoxyATP (ddATP) into the growing chain generates a series of smaller DNA fragments ending at all the possible positions where adenine is found in the newly synthesized fragments. These correspond to positions where thymine occurs in the original template strand.

(c) The radioactive products of each reaction mixture are separated by gel electrophoresis and located by exposing the gel to x-ray film. The nucleotide sequence of the newly synthesized DNA is read directly from the film (5′ → 3′). The sequence in the original template strand is its complement (3′ → 5′).

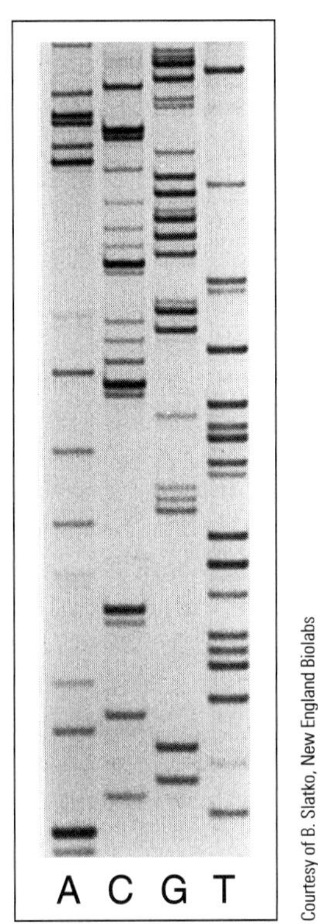

Courtesy of B. Slatko, New England Biolabs

(d) An exposed x-ray film of a DNA sequencing gel. The four lanes represent A, C, G, and T dideoxy reaction mixtures, respectively.

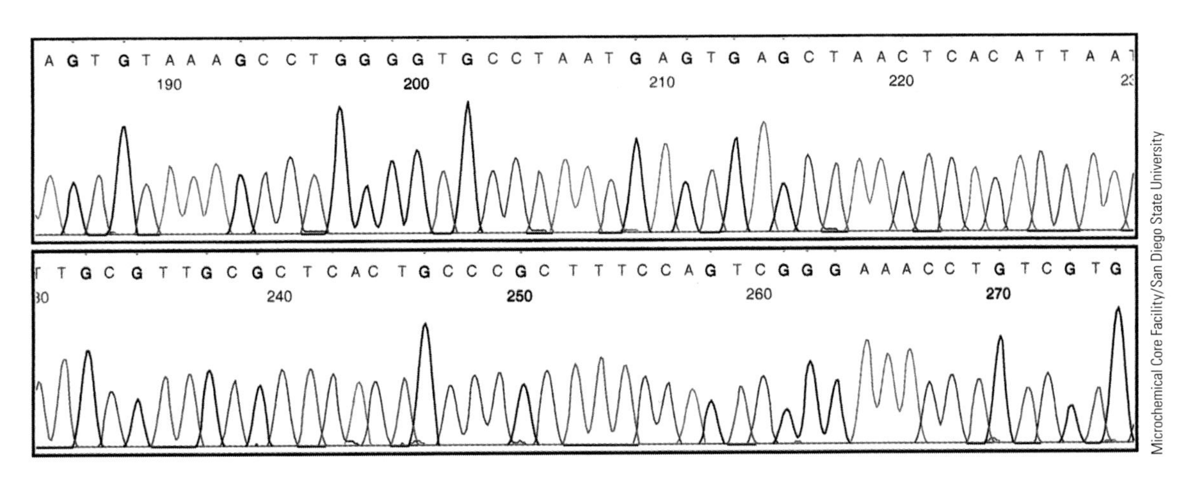

Microchemical Core Facility/San Diego State University

Entire genomes have been sequenced using automated DNA sequencing

Today sequencing machines are very powerful and can decode about 1.5 million bases in a 24-hour period. These advances in sequencing technology have made it possible for researchers to study the nucleotide sequences of entire genomes in a wide variety of organisms, both prokaryotic and eukaryotic. Much of this research received its initial impetus by the Human Genome Project, which began in 1990. The sequencing of the 3 billion base pairs of the human genome was essentially completed in 2001. By 2003, the genomes of more than 100 different organisms had been sequenced, and several hundred more were in various stages of planning or completion. We are in the middle of an extraordinary explosion of gene sequence data, largely due to automated sequencing methods.

DNA sequence information is now kept in large computer databases, many of which are accessed through the Internet. Examples are databases maintained by the National Center for Biotechnology Information (a service of the U.S. National Library of Medicine and the National Institutes of Health) and by the Human Genome Organization (HUGO). Geneticists use these databases to compare newly discovered sequences with those already known, to identify genes, and to access many other kinds of information. By searching for DNA (and amino acid) sequences in a database, researchers can gain a great deal of insight into the function and structure of the gene product, the evolutionary relationships among genes, and the variability among gene sequences within a population.

Review

- What are restriction enzymes? How are they used in recombinant DNA research?
- What are the relative merits of genomic libraries, chromosome libraries, and cDNA libraries?
- How are genetic probes used?
- Why is the PCR technique valuable?
- What technique does automated DNA sequencing use?

BiologyNow™ Assess your understanding of **recombinant DNA methods** by taking the pretest on your BiologyNow CD-ROM.

APPLICATIONS OF DNA TECHNOLOGIES

Learning Objective

6 Describe at least one important application of recombinant DNA technology in each of the following fields: medicine and pharmacology, DNA typing, and transgenic organisms.

Recombinant DNA technology has provided not only a new and unique set of tools for examining fundamental questions about cells, but also new approaches to applied problems in many other fields. In some areas, the production of genetically engineered proteins and organisms has begun to have considerable impact on our lives. The most striking of these has been in the fields of pharmacology and medicine.

DNA technology has revolutionized medicine and pharmacology

In increasing numbers of cases, doctors perform **genetic tests** to determine whether an individual has a particular genetic mutation associated with such disorders as Huntington's disease, hemophilia, cystic fibrosis, Tay-Sachs disease, and sickle cell anemia. **Gene therapy,** the use of specific DNA to treat a genetic disease by correcting the genetic problem, is another application of DNA technology that is currently in its infancy. Because genetic testing and gene therapy focus almost exclusively on humans, these applications of DNA technology are discussed in Chapter 15. We focus our discussion here on the use of DNA technology to produce pharmaceutical products (Fig. 14-12).

Human insulin produced by *E. coli* became one of the first genetically engineered proteins approved for use by humans. Before the use of recombinant DNA techniques to generate genetically altered bacteria capable of producing the human hormone, insulin was derived exclusively from other animals. Many diabetic patients become allergic to the insulin from animal sources, because its amino acid sequence differs slightly from human insulin. The ability to produce the human hormone by recombinant DNA methods has resulted in significant medical benefits to insulin-dependent diabetics.

Rosenfeld Images Ltd/Science Photo Library/Photo Researchers, Inc.

FIGURE **14-12** **A bioreactor.**

This automated bioreactor produces an optimal environment for genetically engineered bacteria that make proteins for use in pharmaceutical products.

Genetically engineered human growth hormone (GH) is available to children who need it to overcome growth deficiencies, specifically pituitary dwarfism (see Chapter 47). In the past, human GH was obtainable only from cadavers. Only small amounts were available, and evidence suggested that some of the preparations from human cadavers were contaminated with infectious agents similar to those that cause bovine spongiform encephalopathy (mad cow disease; see Chapter 23).

The list of products produced by genetic engineering is continually growing. For example, *tissue plasminogen activator (TPA),* a protein that prevents or dissolves blood clots, is used to treat cardiovascular disease. If administered shortly after a heart attack, TPA reduces the risk of a subsequent attack. *Tissue growth factor–beta (TGF-β)* promotes the growth of blood vessels and skin and is used in wound and burn healing. Researchers also use TGF-β in **tissue engineering,** a developing technology to meet the pressing need for human tissues and, eventually, organs for transplantation by growing them from cell cultures. The U.S. Food and Drug Administration (FDA) has approved tissue-engineered skin grafts, and tissue-engineered cartilage is in clinical trials. Hemophilia A is treated with *human blood clotting factor VIII.* Before the development of recombinant DNA techniques, factor VIII was available only from human- or animal-derived blood, which posed a risk of transmitting of infectious agents, such as HIV. And *Dornase Alpha (DNase)* improves respiratory function and general well-being in people with cystic fibrosis.

Recombinant DNA technology is increasingly used to produce vaccines that provide safe and effective immunity against infectious diseases in humans and animals. One way to develop a recombinant vaccine is to clone a gene for a surface protein produced by the pathogen (the disease-causing agent); the researcher then introduces the gene into a nonpathogenic vector. When the vaccine is delivered into the human or animal host, it stimulates an immune response to the surface-exposed protein; as a result, if the pathogen carrying that specific surface protein is encountered, the immune system targets it for destruction. Human examples of antiviral recombinant vaccines are vaccines for influenza A, hepatitis B, and polio. Recombinant vaccines are also being developed against certain bacterial diseases and human cancers.

DNA typing has applications ranging from forensics to analyzing ancient DNA

The analysis of DNA extracted from an individual, which is unique to that individual, is known as **DNA typing,** also called *DNA fingerprinting* or *DNA profiling.* DNA typing has many applications, such as investigating crime, identifying mass disaster victims, identifying human cancer cell lines, tracking tainted foods, studying the genetic ancestry of human populations, and clarifying disputed parentage (Fig. 14-13).

DNA typing relies on PCR amplification and gel electrophoresis to detect molecular markers. The most useful molecular markers are highly polymorphic within the human population. In molecular biology, a **polymorphism** is the presence of detectable variation in the genomes of different individuals in a population; a marker that is highly polymorphic has a great deal of variation. **Short tandem repeats (STRs)** are molecular markers that are short sequences of repetitive DNA—up to 200 nucleotide bases with a simple repeat pattern such as GTGTGTGTGT or CAGCAGCAGCAG. STRs are highly polymorphic because they vary in length from one individual to another, and this characteristic makes them useful in identifying individuals with a high degree of certainty.

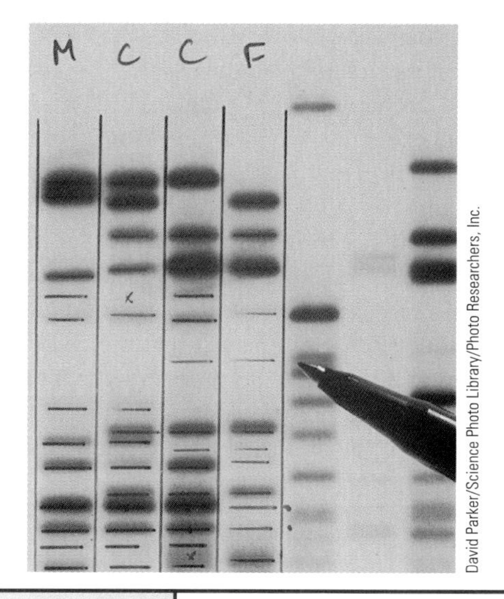

David Parker/Science Photo Library/Photo Researchers, Inc.

ACTIVE FIGURE **14-13** | **DNA typing.**

These gels show DNA profiles of a mother (lane labeled M), father (F), and their two children (C). Note that every band present in one of the children is also found in at least one of the parents.

BiologyNow™ Solve a murder case with **DNA fingerprinting** by clicking on this figure on your BiologyNow CD ROM.

If enough markers are compared, the odds that two people taken at random from the general population would have identical DNA profiles may be as low as one in several billion. The FBI uses a set of STRs from 13 different markers to establish a unique DNA profile for an individual. Such a profile distinguishes that person from every other individual in the United States, except for an identical twin.

Recall that DNA is also found in the cell's mitochondria. Whereas nuclear DNA occurs in two copies per cell (one on each of the homologous chromosomes), mitochondrial DNA has as many as 100,000 copies per cell. Therefore, mitochondrial DNA is the molecule of choice for DNA typing in which biological samples have been damaged—identifying exhumed human remains, for example.

DNA typing has revolutionized law enforcement. The FBI established the Combined DNA Index System (CODIS) in 1990, consisting of DNA databases from all 50 states. Today a DNA profile of an unknown suspect can be compared to millions of DNA profiles of convicted offenders in the database, often resulting in identification of the suspect. (The DNA from the unknown suspect may come from blood, semen, bones, teeth, hair, saliva, urine, or feces left at the crime scene. Tiny amounts of human DNA have even been extracted from cigarette butts, licked envelopes or postage stamps, dandruff, fingerprints, razor blades, chewing gum, wrist watches, ear wax, debris from under fingernails, and toothbrushes.)

If applied properly, DNA typing has the power to identify the guilty and exonerate the innocent. Hundreds of convicted individuals have won new trials and have been subsequently released from incarceration, based on correlating DNA profiles with physical evidence from the crime scene. Such evidence has been ruled admissible in many court cases, including certain trials that have received a great deal of attention in recent years. One limitation arises from the fact that the DNA samples are usually small and may have been degraded. Obviously, great care must be taken to prevent contamination of the samples. This is especially crucial if the PCR technique is used to amplify DNA.

Transgenic organisms have incorporated foreign DNA into their cells

Plants and animals in which foreign genes have been incorporated are referred to as **transgenic organisms.** Researchers use varied approaches to insert foreign genes into plant or animal cells. They often use viruses as vectors, although other methods, such as the direct injection of DNA into cells, have also been applied.

FIGURE **14-14** **A transgenic mouse.**
The mouse on the right is normal, whereas the mouse on the left is a transgenic animal that expresses rat growth hormone.

Transgenic animals are valuable in research

PROCESS OF SCIENCE

Transgenic animals are usually produced by injecting the DNA of a particular gene into the nucleus of a fertilized egg cell or of **embryonic stem cells (ES cells)** (see Chapter 1). The researcher then implants the eggs into the uterus of a female and lets them develop. Alternatively, the researcher injects genetically modified ES cells into isolated *blastocysts,* a stage in embryonic development, and then implants them into a foster mother.

Injecting DNA into cells is not the only way to produce transgenic animals. Researchers may use viruses as recombinant DNA vectors. RNA viruses called **retroviruses** make DNA copies of themselves by reverse transcription. Sometimes the DNA copies become integrated into the host chromosomes, where they are replicated along with host DNA.

Transgenic animals provide valuable applications over a wide range of research areas, such as regulation of gene expression, immune system function, genetic diseases, viral diseases, and genes involved in the development of cancer. The laboratory mouse has become a particularly important model organism for these studies.

In a classic study of the control of gene expression, University of Pennsylvania geneticist Ralph L. Brinster in 1983 produced transgenic mice carrying a gene for rat growth hormone (see Fig. 16-17). Brinster and his coworkers wanted to understand the controls that let certain genes be expressed in some tissues and not in others. The pituitary gland of a mouse normally produces small amounts of GH, and these researchers reasoned that other tissues might also be capable of producing the hormone. First they isolated the GH gene from a library of genomic rat DNA. They combined it with the promoter region of a mouse gene that normally produces metallothionein, a protein that is active in the liver and whose synthesis is stimulated by the presence of toxic amounts of heavy metals such as zinc. The researchers used metallothionein regulatory sequences as a switch to turn the production of rat GH on and off at will. After injecting the engineered gene into mouse embryo cells, they implanted the embryos into the uterus of a mouse and let them develop. Because of the difficulty in manipulating the embryos without damaging them, the gene transplant succeeded in only a small percentage of the animals. When exposed to small amounts of zinc, these transgenic mice produced large amounts of growth hormone, because the liver is a much larger organ than the pituitary gland. The mice grew rapidly, and one mouse, which developed from an embryo that had received two copies of the GH gene, grew to more than double the normal size (Fig. 14-14). As might be expected, such mice often transmit their increased growth capability to their offspring.

Gene targeting reveals gene function

PROCESS OF SCIENCE

One extremely powerful research tool is **gene targeting,** a procedure in which the researcher chooses and "knocks out" (inactivates) a single gene in an organism. *Knockout mice* are common models for studying the roles of genes of unknown function in mammals, including humans. The roles of the inactivated gene are determined by observing the phenotype of the mice bearing the knockout gene. If the gene codes for a protein, for example, studying individuals who lack that protein helps the researcher identify the function of the protein. Because at least 99% of the loci of mice have human counterparts (although the specific alleles are usually different), information about knockout genes in mice also provides details about human genes.

Gene targeting, pioneered by Mario R. Capecchi, a molecular geneticist at the University of Utah School of Medicine, is a lengthy procedure; it takes about a year to develop a new strain of knockout mice. A nonfunctional (knockout) gene is introduced into mouse ES cells. ES cells are particularly easy to handle because, like cancer cells, they grow in culture indefinitely. Most importantly, if placed in a mouse embryo, ES cells divide and produce all the cell types normally found in the mouse. In a tiny fraction of these ES cells, the introduced knockout gene becomes physically associated with the corresponding gene in a chromosome. If this occurs, the chromosomal gene and the knockout gene tend to exchange DNA segments in a process called *homologous recombination.* In this way, the knockout allele replaces the normal allele in the mouse chromosome.

Researchers inject into early mouse embryos ES cells they hope are carrying a knockout gene and let the mice develop to maturity. If the gene is not lethal when inactivated, the researchers generally study animals that carry the knockout gene in every cell. However, because many genes are essential to life, researchers have modified the knockout technique to develop strains in which a specific gene is selectively inactivated in only one cell type. Research laboratories around the world have developed about 3000 different strains of knockout mice, each displaying its own characteristic phenotype, and the number continues to grow.

Gene targeting in mice is providing answers to basic biological questions relating to the development of embryos, the development of the nervous system, and the normal functioning of the immune system. This technique has great potential for revealing more about various human diseases, especially as scientists have learned that many thousands of diseases have a genetic component. Geneticists are using gene targeting to study cancer, heart disease, sickle cell anemia, respiratory diseases such as cystic fibrosis, and other disorders.

Mutagenesis screening reveals the genes involved in a particular phenotype

Many large-scale mutagenesis screening projects in mice are currently underway. In **mutagenesis screening,** researchers treat male mice with chemical *mutagens* that cause mutations in DNA. They then breed the males and screen their offspring for unusual phenotypes. Unlike gene targeting, mutagenesis does not disable a gene completely; instead it causes small, random mutations that change the properties of the proteins that the DNA encodes. The scale of mutagenesis screening is enormous; one screening project in Germany has screened some 28,000 different mouse mutants for changes in phenotypes.

Transgenic animals can produce genetically engineered proteins

Certain transgenic animal strains produce foreign proteins of therapeutic or commercial importance that are secreted into milk. For example, researchers have introduced into sheep the gene for the human protein TPA. Producing transgenic livestock, such as pigs, sheep, cows, and goats, that secrete foreign proteins in their milk is known informally as "pharming," a combination of "pharmaceuticals" and "farming" (Fig. 14-15).

Pentti Vänskä, Helsingin Sanomat International Edition

FIGURE **14-15** **Transgenic "pharm" cows.**
These cows contain a human gene that codes for lactoferrin, a protein found in human mothers' milk and in secretions such as tears, saliva, bile, and pancreatic fluids. Lactoferrin is one of the immune system's lines of defense against disease-causing organisms. The cows secrete human lactoferrin in their milk.

In pharming, recombinant genes are fused to the regulatory sequences of the milk protein genes, and such genes are therefore activated only in mammary tissues involved in milk production. The advantages of obtaining the protein from milk are that potentially it can be produced in large quantities and that it can be harvested simply by milking the animal. The protein is then purified from the milk. The introduction of the gene does not harm the animals, and because the offspring of the transgenic animal usually produce the recombinant protein, transgenic strains are established.

Transgenic plants are increasingly important in agriculture

People have selectively bred plants for thousands of years. The success of such efforts depends on desirable traits in the variety of plant selected, or in closely related wild or domesticated plants whose traits are transferred by cross-breeding. Local varieties or closely related species of cultivated plants often have traits, such as disease resistance, that agriculturalists could advantageously introduce into varieties more suited to modern human needs. If genes are introduced into plants from strains or species with which they do not ordinarily interbreed, the possibilities for improvement increase greatly.

The most widely used vector system for introducing recombinant genes into many types of plant cells is the crown gall bacterium, *Agrobacterium tumefaciens.* This bacterium normally produces plant tumors by introducing a plasmid, called the *Ti plasmid,* into the cells of its host (*Ti* stands for "tumor-inducing"). The Ti plasmid induces abnormal growth by forcing the plant cells to produce elevated levels of a plant growth hormone called *cytokinin* (see Chapter 36).

Geneticists "disarm" the Ti plasmid so that it does not induce tumor formation and then use it as a vector to insert genes into plant cells. The cells into which the altered plasmid is introduced are essentially normal except for the inserted genes. Genes placed in the plant genome in this fashion may be transmitted sexually, via seeds, to the next generation, but they can also be propagated asexually (for example, by taking cuttings).

Unfortunately, not all plants take up DNA readily, particularly the cereal grains that are a major food source for humans. One useful approach has been the development of a genetic "shotgun." Researchers coat microscopic gold or tungsten fragments with DNA and then shoot them into plant cells, penetrating the cell walls. Some of the cells retain the DNA, which transforms them. Those cells can then be cultured and used to regenerate an entire plant (see Chapter 16). For example, geneticists have successfully used such an approach to transfer a gene for resistance to a bacterial disease into cultivated rice from one of its wild relatives.

An additional complication of plant genetic engineering is that about 120 plant genes lie in the DNA of the chloroplasts. (The other 3000 or so genes that plastids require to function are in the nucleus.) Chloroplasts are essential in photosynthesis, which is the basis for plant productivity. Because great agricultural potential hinges on improving the productivity of photosynthesis, developing methods for changing the part of the plant's DNA within the chloroplast is desirable. Dozens of labs are currently studying methods of chloroplast engineering, although progress has been slow. In 2001 Australian plant physiologists reported that they had altered **Rubisco,** the key carbon-fixing enzyme of photosynthesis (see Chapter 8) by changing one of the genes in the chloroplast genome.

Selected applications of transgenic plants Agricultural geneticists are developing transgenic plants, known as **genetically modified (GM) crops,** that are resistant to insect pests, viral diseases, drought, heat, cold, herbicides, and salty or acidic soil. For example, consider the European corn borer, which is the most damaging insect pest in corn in the United States and Canada (Fig. 14-16a). Efforts to control the European corn borer cost farmers more than $1 billion each year. Corn has been genetically modified to contain the *Bt* gene, a bacterial gene that codes for a protein with insecticidal properties (*Bt* stands for the scientific name, *Bacillus thuringiensis*). *Bt* corn, introduced in the United States in 1996, doesn't need periodic sprays of chemical insecticides to control the European corn borer.

DNA technology also has the potential to develop crops that are more nutritious. For example, in the 1990s geneticists engineered rice to produce high quantities of β-carotene, which the human body uses to make vitamin A (Fig. 14-16b; see also Fig. 3-14). In developing countries, vitamin A deficiency is a leading cause of blindness in children. According to the World Health Organization, 275 million children are vitamin A deficient, and 250,000 to 500,000 become irreversibly blind each year. Vitamin A deficiency also makes children more susceptible to measles and other infectious diseases. Because rice is the staple diet in many countries with vitamin A deficiency, the widespread use of GM rice with β-carotene has the potential to prevent vitamin A deficiency in many of the world's children.

Like some transgenic animals, certain transgenic plants can potentially be "pharmed" to produce large quantities of medically important proteins, such as antibodies against the herpes virus. Methods for developing transgenic plants are well established, but it has been difficult to get plants to produce the de-

U.S. Department of Agriculture

Courtesy of Peter Beyer

ACTIVE FIGURE **14-16** **Uses of transgenic plants.**

(a) The European corn borer, shown in the larval form, is the most destructive pest on corn in North America. Genetic engineers have designed *Bt* corn to control the European corn borer without heavy use of chemical insecticides. **(b)** "Golden rice," shown here intermixed with regular white rice, contains high concentrations of β-carotene. To make golden rice, scientists took the gene for β-carotene in daffodil flowers and inserted it into the endosperm cells of rice.

BiologyNow™ Learn more about the development of **transgenic rice** by clicking on this figure on your BiologyNow CD ROM.

sired protein in large enough quantities. To date, most transgenic plants developed for pharming have produced foreign proteins equal to only about 1% of the plant's total protein output. The developers of this technology are working to increase the production of foreign proteins, to demonstrate the feasibility of these production methods.

Some people are concerned about the health effects of consuming foods derived from GM crops and think that such foods should be restricted. For example, critics say some consumers may develop food allergies. Scientists also recognize this concern and routinely screen new GM crops for allergenicity. There is also an ongoing controversy as to whether GM foods should be labeled. The FDA and most scientists think such labeling would be counterproductive, because it would increase public anxiety over a technology that is as safe as traditional breeding methods. In 1996, the U.S. Court of Appeals upheld the FDA position that labeling should not be required.

Review

- Why has the production of human insulin by recombinant DNA methods had significant medical advantages for diabetics?
- What are short tandem repeats, and why are they so useful in DNA typing?
- Why do gene targeting and mutagenesis screening in mice have potential benefit for humans?

BiologyNow™ Assess your understanding of **applications of DNA technologies** by taking the pretest on your BiologyNow CD-ROM.

SAFETY GUIDELINES FOR DNA TECHNOLOGY

Learning Objective

7 Describe at least two safety issues associated with recombinant DNA technology, and explain how these issues are being addressed.

When recombinant DNA technology was introduced in the early 1970s, many scientists considered the potential misuses at least as significant as the possible benefits. The possibility that an organism with undesirable environmental effects might be accidentally produced was of great concern, because scientists feared that totally new strains of bacteria or other organisms, with which the world has no previous experience, might be difficult to control. The scientists who developed the recombinant DNA methods insisted on stringent guidelines for making the new technology safe.

Experiments in thousands of university and industrial laboratories over more than 30 years have shown that recombinant DNA manipulations can be carried out safely. Geneticists have designed laboratory strains of *E. coli* to die in the outside world. Researchers carry out experiments that might present unusual risks in facilities designed to hold pathogenic organisms; this precaution ensures that researchers can work with them safely. So far no evidence suggests that researchers have accidentally cloned hazardous genes or have released dangerous organisms into the environment. However, malicious *intentional* manipulations of dangerous genes certainly remain a possibility.

As the safety of the experiments has been established, scientists have relaxed many of the restrictive guidelines for using recombinant DNA. Stringent restrictions still exist, however, in certain areas of recombinant DNA research where there are known dangers, or where questions about possible effects on the environment are still unanswered.

These restrictions are most evident in research that proposes to introduce transgenic organisms into the wild, such as agricultural strains of plants whose seeds or pollen might spread in an uncontrolled manner. A great deal of research now focuses on determining the effects of introducing transgenic organisms into a natural environment. Carefully conducted tests have shown that transgenic organisms are not dangerous to the environment simply because they are transgenic.

However, it is important to assess the risks of each new recombinant organism. Scientists determine whether the organism has characteristics that might cause it to be environmentally hazardous under certain conditions. For example, if geneticists have engineered a transgenic crop plant to resist an herbicide, could that gene be transferred, via pollen or some other route, to that plant's weedy relatives, generating herbicide-resistant "superweeds"? In 2003, ecologists at the University of Tennessee, Knoxville, announced the crossing of transgenic oilseed rape plants that contained the *Bt* gene with its wild relative (a weed). They crossed the resulting hybrids with the wild relative again and then tested its ability to compete with other weeds in a field of wheat. The transgenic weed was a poor competitor and had less effect on wheat production than its wild relatives in a control field. These results, although encouraging, must be interpreted with care. Scientists must evaluate each transgenic crop plant individually to see if there is gene flow to wild relatives, and, if so, the resulting effect.

Other concerns relate to plants engineered to produce pesticides, such as the *Bt* toxin. The future of the *Bt* toxin in transgenic crops is not secure, because low levels of the insecticide could potentially provide ideal conditions for selection for resistant individuals in the insect population. It appears certain insects may evolve genetic resistance to the *Bt* toxin in transgenic plants in the same way they evolve genetic resistance to chemical insecticides.

Another concern is that non-pest species could be harmed. For example, people paid a great deal of attention to the finding that monarch butterfly larvae raised in the laboratory are harmed if they are fed pollen from *Bt* corn plants. Although more recent studies suggest that monarch larvae living in a natural environment do not consume enough pollen to cause damage, such concerns persist and will have to be addressed individually.

Environmental concerns about transgenic animals also exist. Several countries are in the process of developing fast-growing transgenic fish, usually by inserting a gene that codes for a growth hormone. Transgenic Atlantic salmon, for example, grow up to six times faster than nontransgenic salmon grown for human consumption. (The transgenic fish do not grow larger than other fish, just faster.) The benefits of such genetically enhanced fish include reduced pressure on wild fisheries and less pollution from fish farms. However, if the transgenic fish escaped from the fish farm, what effect would they have on wild relatives? (To address this concern, the transgenic salmon being developed in the United States are all nonreproducing females.)

To summarize, DNA technology in agriculture offers many potential benefits, including higher yields (by providing disease resistance), more nutritious foods, and the reduced use of chemical pesticides. However, like other kinds of technology, genetic engineering poses some risks, such as the risk that genetically modified plants and animals could pass their foreign genes to wild relatives, causing unknown environmental problems. The science of risk assessment, which uses statistical methods to quantify risks so they can be compared and contrasted, will help society decide whether to ignore, reduce, or eliminate specific risks of genetically engineered organisms.

Review

- Why are some people concerned about plants that have been genetically engineered to produce insecticides?
- What is the potential problem with transgenic salmon? How are scientists addressing this concern?

BiologyNow™ Assess your understanding of **safety guidelines for DNA technology** by taking the pretest on your BiologyNow CD-ROM.

SUMMARY WITH KEY TERMS

1 Explain how a typical restriction enzyme cuts DNA molecules, and give examples of the ways in which these enzymes are used in recombinant DNA technology.

- **Recombinant DNA technology** isolates and amplifies specific sequences of DNA by incorporating them into **vector** DNA molecules. Researchers then propagate and amplify the resulting recombinant DNA in organisms such as *E. coli.*
- Researchers use **restriction enzymes** to cut DNA into specific fragments. Each type of restriction enzyme recognizes and cuts DNA at a highly specific base sequence. Many restriction enzymes cleave DNA sequences to produce complementary, single-stranded sticky ends.
- Geneticists construct the most common recombinant DNA vectors from naturally occurring circular bacteria DNA molecules called **plasmids,** or from bacterial viruses called **bacteriophages.**
- Geneticists often construct recombinant DNA molecules by allowing the ends of a DNA fragment and a plasmid (both cut with the same restriction enzyme) to associate by complementary base pairing. Then **DNA ligase** covalently links the DNA strands to form a stable recombinant molecule.

2 Distinguish among a genomic library, a chromosome library, and a complementary DNA (cDNA) library; explain why one would clone the same eukaryotic gene from both a genomic library and a cDNA library.

- A **genomic library** contains thousands of DNA fragments that represent the total DNA of an organism, and a **chromosome library** contains all the DNA fragments from a specific chromosome. Each DNA fragment of a genomic or chromosome library is stored in a specific bacterial strain. Analyzing DNA fragments in genomic and chromosome libraries yields useful information about genes and their encoded proteins.
- A **cDNA library** is produced using **reverse transcriptase** to make DNA copies of mature mRNA isolated from eukaryotic cells. These copies, known as **complementary DNA (cDNA),** are then incorporated into recombinant DNA vectors.
- Genes present in genomic and chromosome libraries from eukaryotes contain **introns,** regions that do not code for protein. Those genes are amplified in bacteria, but the protein is not properly expressed. Because the introns have been removed from mRNA molecules, eukaryotic genes in cDNA libraries can sometimes be expressed in bacteria, which produce functional protein products.

3 Identify some uses of DNA hybridization probes.

- Researchers use a radioactive DNA or RNA sequence as a **genetic probe** to identify complementary nucleic acid sequences. Each spot on the x-ray film identifies a colony containing a plasmid that includes the DNA of interest.
- In the **Southern blot technique,** researchers separate DNA fragments by **gel electrophoresis,** denature them, and then blot them onto a nitrocellulose or nylon membrane. A radioactive probe is then **hybridized** by complementary base pairing to the DNA bound to the membrane, and the radioactive band or bands of DNA are identified by autoradiography or chemical luminescence.

4 Describe how the polymerase chain reaction amplifies DNA in vitro.

- The **polymerase chain reaction (PCR)** is a widely used, automated, in vitro technique in which researchers target a particular DNA sequence by specific primers and then **clone** it by a heat-resistant DNA polymerase.
- Using PCR, scientists amplify and analyze tiny DNA samples taken from various sites, from crime scenes to archaeological remains.

5 Describe the chain termination method of DNA sequencing.

- **DNA sequencing** yields information about the structure of a gene and the probable amino acid sequence of its encoded proteins. Geneticists compare DNA sequences with other sequences stored in massive databases.
- Automated DNA sequencing is based on the chain termination method, which uses dideoxynucleotides, each tagged with a differently colored fluorescent dye, to terminate elongation during DNA replication. Gel electrophoresis separates the resulting fragments, and a laser identifies the nucleotide sequence.

6 Describe at least one important application of recombinant DNA technology in each of the following fields: medicine and pharmacology, DNA typing, and transgenic organisms.

- Genetically altered bacteria produce many important human protein products, including insulin, growth hormone, tissue plasminogen activator (TPA), tissue growth factor-beta (TGF-β, blood clotting factor VIII, and Dornas Alpha (DNase).
- **DNA typing** is the analysis of an individual's DNA. It is based on a variety of **short tandem repeats (STRs),** molecular markers that are highly **polymorphic** within the human population. DNA typing has applications in law enforcement, issues of disputed parentage, and tracking tainted foods, to name a few.
- **Transgenic organisms** have foreign DNA incorporated into their genetic material. **Gene targeting** and **mutagenesis screening** in mice help identify the function of a gene and its protein product. Transgenic livestock produce foreign proteins in their milk. Transgenic plants have great potential in agriculture.

SUMMARY WITH KEY TERMS *(continued)*

7 Describe at least two safety issues associated with recombinant DNA technology, and explain how these issues are being addressed.

- Some consumers are concerned about the safety of genetically engineered organisms. To address these concerns, scientists carry out recombinant DNA technology under specific safety guidelines.
- The introduction of transgenic plants and animals into the natural environment, where they may spread in an uncontrolled manner, is an ongoing concern that must be evaluated on a case-by-case basis.

POST-TEST

1. A plasmid (a) is used as a DNA vector (b) is a type of bacteriophage (c) is a type of cDNA (d) is a retrovirus (e) b and c
2. DNA molecules with complementary sticky ends associate by (a) covalent bonds (b) hydrogen bonds (c) ionic bonds (d) disulfide bonds (e) phosphodiester linkages
3. Human DNA and a particular plasmid both have sites that are cut by the restriction enzymes *Hind*III and *Eco*RI. To make recombinant DNA, the scientist should (a) cut the plasmid with *Eco*RI and the human DNA with *Hind*III (b) use *Eco*RI to cut both the plasmid and the human DNA (c) use *Hind*III to cut both the plasmid and the human DNA (d) a or b (e) b or c
4. Which of the following sequences is *not* palindromic?
   (a) 5′—AAGCTT—3′ / 3′—TTCGAA—5′
   (b) 5′—GATC—3′ / 3′—CTAG—5′
   (c) 5′—GAATTC—3′ / 3′—CTTAAG—5′
   (d) 5′—CTAA—3′ / 3′—GATT—5′
   (e) b and d
5. The PCR technique uses (a) heat-resistant DNA polymerase (b) reverse transcriptase (c) DNA ligase (d) restriction enzymes (e) b and c
6. A cDNA clone contains (a) introns (b) exons (c) anticodons (d) a and b (e) b and c
7. The dideoxynucleotides ddATP, ddTTP, ddGTP, and ddCTP are important in DNA sequencing because they (a) cause premature termination of a growing DNA strand (b) are used as primers (c) cause the DNA fragments that contain them to migrate more slowly through a sequencing gel (d) are not affected by high temperatures (e) have more energy than deoxynucleotides
8. In the Southern blot technique, ______________ is/are transferred from a gel to a nitrocellulose or nylon membrane. (a) protein (b) RNA (c) DNA (d) bacterial colonies (e) reverse transcriptase
9. Gel electrophoresis separates nucleic acids on the basis of differences in (a) length (molecular weight) (b) charge (c) nucleotide sequence (d) relative proportions of adenine and guanine (e) relative proportions of thymine and cytosine
10. The Ti plasmid, carried by *Agrobacterium tumefaciens,* is especially useful for introducing genes into (a) bacteria (b) plants (c) animals (d) yeast (e) all eukaryotes
11. A genomic library (a) represents all the DNA in a specific chromosome (b) is made using reverse transcriptase (c) is stored in a collection of recombinant bacteria (d) is a DNA copy of mature mRNAs (e) allows researchers to amplify a tiny sample of DNA
12. Tissue growth factor-beta (a) is a genetic probe for recombinant plasmids (b) is a product of DNA technology used in tissue engineering (c) is necessary to make a cDNA library (d) cannot be synthesized without a heat-resistant DNA polymerase (e) is isolated by the Southern blot technique
13. These highly polymorphic molecular markers are useful in DNA typing: (a) short tandem repeats (b) cloned DNA sequences (c) palindromic DNA sequences (d) cosmid cloning vectors (e) complementary DNAs

CRITICAL THINKING

1. What are some of the problems that might arise if you were trying to produce a eukaryotic protein in a bacterium? How might using transgenic plants or animals help solve some of these problems?
2. Would genetic engineering be possible if we did not know a great deal about the genetics of bacteria? Explain.
3. What are some of the environmental concerns regarding transgenic organisms? What kinds of information does society need to determine if these concerns are valid?

- Visit our Web site at **http://biology.brookscole.com/solomon7** for links to chapter-related resources on the World Wide Web. Additional online materials relating to this chapter can also be found on our Web site.

BIOLOGY NOW RESOURCES

Active Figures

14-13: DNA fingerprinting

14-16: Development of transgenic rice

Preparing for an exam? Take a diagnostic test on your BiologyNow CD-ROM.

Post-Test Answers

1. a	2. b	3. e	4. d
5. a	6. b	7. a	8. c
9. a	10. b	11. c	12. b
13. a			

15 The Human Genome

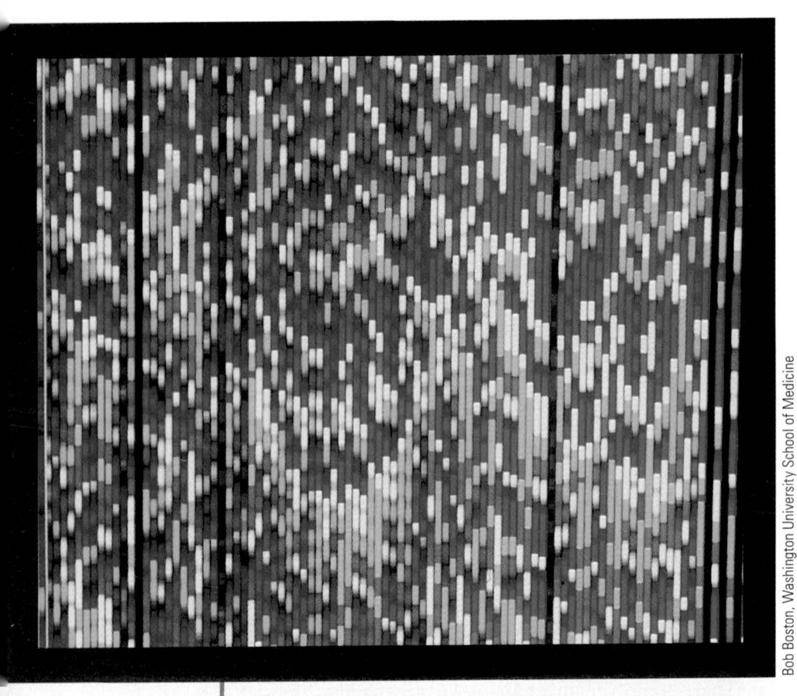
Bob Boston, Washington University School of Medicine

DNA sequencing. This is a display on a computer monitor attached to a DNA-sequencing machine. The machine determines the base sequence of a strand of DNA using the chain termination method and tags each of the four bases with a different colored fluorescent dye.

CHAPTER OUTLINE

The principles of genetics apply to all organisms, including humans. However, some important differences separate genetic research on humans and genetic research on other organisms. To study aspects of inheritance in other species, geneticists ideally have standard stocks of genetically similar individuals—**true-breeding strains** whose traits remain unchanged for many generations because they are homozygous at virtually all loci. Geneticists conduct *controlled matings* between members of different true-breeding strains and raise the offspring under carefully controlled conditions.

Of course, experimental matings under controlled conditions are not feasible in the human population. In addition, most human families are small, and 20 to 30 years or more elapse between generations. It is therefore virtually impossible to conduct genetic research on humans in the same way as with other sexually reproducing species.

Despite the inherent difficulties of studying inheritance in humans, understanding of **human genetics,** the science of inherited variation in humans, is progressing very rapidly. Researchers traditionally examined human genetics using such approaches as population studies of large extended families. More recently, the field of human genetics has been greatly facilitated by the medical attention given to human genetic diseases. The extensive medical records of diseases serve as a useful database for testing hypotheses. Genetic studies of other organisms also have provided invaluable insights. Indeed, many genetic traits in humans that were initially puzzling have been explained using model organisms, such as bacteria, yeasts, worms, fruit flies, and mice to address analogous inheritance questions.

The **human genome,** which represents the totality of genetic information in human cells, has been mapped and sequenced. In **DNA sequencing,** researchers have identified the order of nucleotides in DNA to help us understand the genetic basis of human similarities and differences (see figure). The human genome includes the DNA content of both the nucleus and the mitochondria. However, nuclear DNA accounts for almost all the genetic information in the human genome. Like genomes of other eukaryotic organisms, some of the human genome specifies the synthesis of polypeptides or RNA molecules. However, much of the genome consists of noncoding DNA, repetitive sequences (multiple copies) of DNA, and gene segments whose functions remain unknown. Relating specific genes to the proteins they code for, and determining the role these proteins play in the body, are some of the avenues of human genetic research that scientists will pursue during the 21st century.

In this chapter, we first examine how the human genome is studied, including advances in new fields that are developing as a result of the Human Genome Project. Then we discuss a variety of human genetic disorders. We explore the use of gene therapy for some of these disorders, as well as the application of genetic testing, screening, and counseling for families at risk. The chapter concludes with a discussion of ethical issues that relate to the expanding knowledge of the human genome. ■

STUDYING HUMAN GENETICS

Learning Objectives

1. Distinguish between karyotyping and pedigree analysis.
2. Discuss the implications of the Human Genome Project, including the emerging fields of bioinformatics, pharmacogenetics, and proteomics.
3. Discuss the mouse model for studying cystic fibrosis.

Human geneticists use a variety of methods that enable them to make inferences about a trait's mode of inheritance. We consider three of these methods—the identification of chromosomes by karyotyping, the analysis of family inheritance patterns using pedigrees, and DNA sequencing and mapping of genes by genome projects. Investigators often study human inheritance most effectively by combining these and other approaches.

Human chromosomes are studied by karyotyping

Cytogenetics is the study of chromosomes and their role in inheritance. Researchers working with simpler organisms discovered many of the basic principles of genetics. In such organisms, it is often possible to relate genetic data to the number and structure of specific chromosomes. Some organisms used in genetics, such as the fruit fly *Drosophila melanogaster*, have very few chromosomes; the fruit fly has only four pairs. In *Drosophila* larval salivary glands, the chromosomes are large enough that their structural details are readily evident. This organism, therefore, has provided unique opportunities for correlating certain inherited phenotypic changes with alterations in chromosome structure.

PROCESS OF SCIENCE

The normal number of chromosomes for the human species is 46: 44 **autosomes** (22 pairs) and 2 sex chromosomes (one pair). A **karyotype** (from the Greek, for nucleus) is an individual's chromosome composition. Until the mid-1950s, when biologists adopted modern methods of karyotyping, the accepted number of chromosomes for the human species was 48, based on a study of human chromosomes published in 1923. The reason researchers counted 48 human chromosomes was the difficulty in separating the chromosomes so they could be accurately counted. In 1952, University of Texas cell biologist T.C. Hsu treated cells with a hypotonic salt solution by mistake; this caused the cells to swell and the chromosomes to spread apart, thereby making them easier to count. Other techniques were also developed, and in 1956 researchers Jo Hin Tjio and Albert Levan, working in Sweden, reported that humans have 46 chromosomes, not 48. Other researchers subsequently verified this report. The story of the human chromosome number is a valuable example of the "self-correcting" nature of science (although such corrections may take time). The re-evaluation of established facts and ideas, often by using improved techniques or new methods, is an essential part of the scientific process.

Human chromosomes are visible only in dividing cells (see Chapter 9), and it is difficult to obtain dividing cells directly from the human body. Researchers typically use blood, because white blood cells can be induced to divide in a culture medium by treating them with chemicals. Other sources of dividing cells include skin and, for prenatal chromosome studies, chorionic villi or fetal cells shed into the amniotic fluid (discussed later in the chapter).

In karyotyping, biologists culture dividing human cells and then treat them with the drug *colchicine*, which arrests the cells at mitotic metaphase or late prophase, when the chromosomes are most highly condensed. Next the researchers put the cells into a hypotonic solution; the cells swell, and the chromosomes spread out so they are easily observed. The cells are then flattened on microscope slides, and the chromosomes are stained to reveal the patterns of bands unique for each homologous pair. After the microscopic image has been scanned into a computer, the homologous pairs are electronically matched and placed together (Fig. 15-1a).

By convention, geneticists identify chromosomes by length; position of the centromere; banding patterns, which are produced by staining chromosomes with dyes that produce dark and light cross-bands of varying widths; and other features such as *satellites*, tiny knobs of chromosome material at the tips of certain chromosomes. With the exception of the sex chromosomes, the chromosomes are numbered and lined up in order of size, except that chromosome 21 is smaller than chromosome 22. The largest human chromosome (chromosome 1) is about five times as long as the smallest one (chromosome 21), but there are only slight size differences among some of the intermediate-sized chromosomes. The X and Y chromosomes of a normal male are homologous only at their tips; normal females have two X chromosomes and no Y chromosome. Differences from the normal karyotype—that is, deviations in chromosome number or structure—are associated with certain disorders such as Down syndrome (discussed later in the chapter).

Another way to distinguish chromosomes in a karyotype is by **fluorescent in situ hybridization (FISH).** A geneticist tags a

© SIU/Peter Arnold, Inc.

(a)

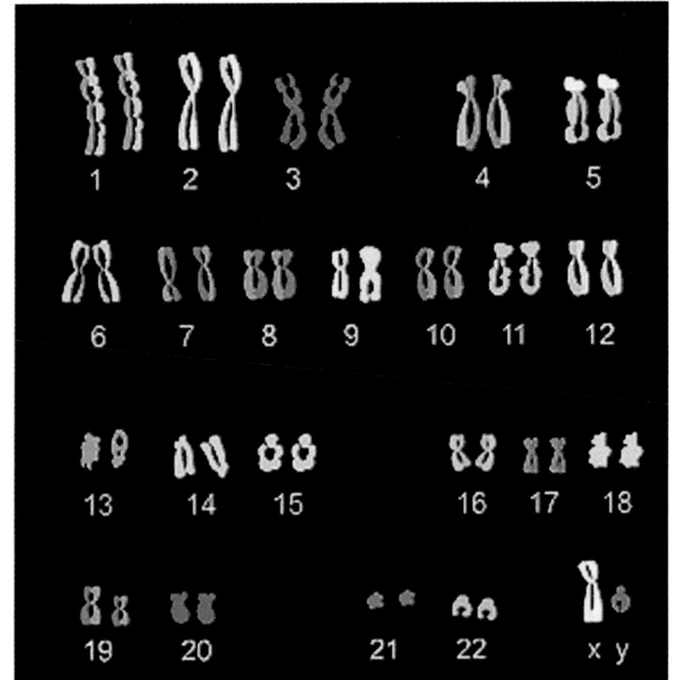

© Oklahoma State University

(b)

FIGURE 15-1 Karyotyping.

(a) Using an image analysis computer to prepare a karyotype, the biologist matches homologous chromosomes and organizes them by size. Before computers, researchers prepared karyotypes by cutting and pasting chromosomes from photographs. **(b)** A normal human male karyotype. These chromosomes have been "painted." DNA molecules, to which different colored fluorescent dyes are attached, hybridize to specific pairs of homologous chromosomes. Painting helps the researcher identify chromosomes under the light microscope.

DNA strand complementary to DNA in a specific chromosome with a fluorescent dye. The DNA in the chromosome is denatured—that is, the two strands are separated—so the tagged strand can bind to it. A different dye is used for each chromosome, "painting" each with a different color (Fig. 15-1b). A chromosome that is multicolored (not shown) indicates breakage and fusion of chromosomes, an abnormality associated with certain genetic diseases and many types of cancer.

Family pedigrees help identify certain inherited conditions

Early studies of human genetics usually dealt with readily identified pairs of contrasting traits and their distribution among members of a family. A "family tree" that shows inheritance patterns, the transmission of genetic traits within a family over several generations, is known as a **pedigree.** Pedigree analysis remains widely used, even in today's world of powerful molecular genetic techniques, because it helps molecular geneticists determine the exact interrelationships of the DNA molecules they analyze from related individuals. Pedigree analysis is also an important tool of genetic counselors and clinicians. However, because human families tend to be small and information about certain family members, particularly deceased relatives, may not be available, pedigree analysis has limitations.

Pedigrees are produced using standardized symbols. Examine Figure 15-2, which shows a pedigree for **albinism,** a lack of the pigment melanin in the skin, hair, and eyes. Each horizontal row represents a separate generation, with the earliest generation (roman numeral I) at the top and the most recent generation at the bottom. Within a given generation, the individuals are usually numbered consecutively, from left to right, using arabic numbers. A horizontal line connects two parents, and a vertical line drops from the parents to their children. For example, individuals II-3 and II-4 are parents of four offspring (III-1, III-2, III-3, and III-4). Note that individuals in a given generation can be genetically unrelated. For example, II-1, II-2, and II-3 are unrelated to II-4 and II-5. Within a group of siblings, the oldest is on the left, and the youngest is on the right.

Studying pedigrees enables human geneticists to predict how phenotypic traits that are governed by the genotype at a single locus are inherited. About 10,000 traits have been described in humans. Pedigree analysis most often identifies three modes of single-locus inheritance: autosomal dominant, autosomal recessive, and X-linked recessive. We define and discuss examples of these inheritance modes later in the chapter.

Certain traits that do not show a simple Mendelian inheritance pattern can also be characterized using pedigree analysis. Some of these traits are the result of **genomic imprinting,** in which the expression of a gene in a given tissue or developmental stage is based on its parental origin—that is, whether the individual inherits the gene from the male or female parent. For some imprinted genes, the paternally inherited allele is always repressed (not expressed); for other imprinted genes, the maternally inherited allele is always repressed.

Two rare genetic disorders provide a fascinating demonstration of genomic imprinting. In *Prader-Willi syndrome (PWS),* individuals become compulsive overeaters and obese; they are also short in stature and mildly to moderately retarded. In *Angelman syndrome (AS),* affected individuals are hyperactive, mentally retarded, unable to speak, and suffer from seizures. Both PWS and AS are caused by a small *deletion* of several loci from

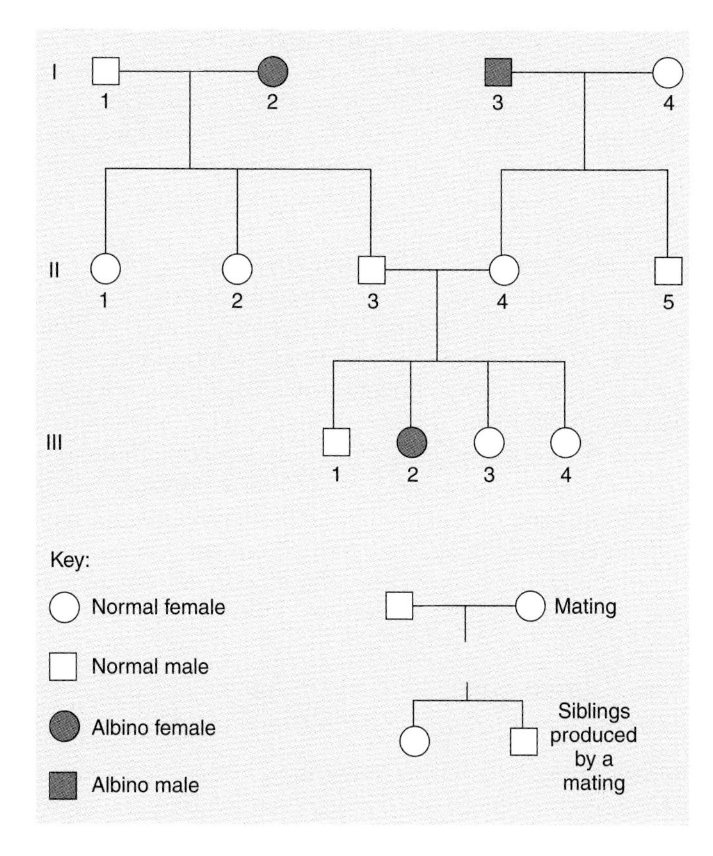

FIGURE **15-2** **A pedigree for albinism.**

By studying family histories, a researcher can determine the genetic mechanism of an inherited trait. In this example, III-2 represents an albino girl with two phenotypically normal parents, II-3 and II-4. The allele for albinism cannot be dominant because if it were, at least one of III-2's parents would have to be an albino. Also, albinism cannot be an X-linked recessive allele because if it were, her father would have to be an albino (and her mother would have to be a heterozygous carrier). This pedigree is explained if albinism is inherited as an autosomal recessive allele (not carried on a sex chromosome). In such cases, two phenotypically normal parents could produce an albino offspring because they are heterozygotes and could each transmit a recessive allele.

the same region of chromosome 15. One of these deleted loci is responsible for PWS, another for AS. (Chromosome deletions are discussed later in the chapter.)

Pedigree analysis has shown that when the person inherits the deletion from the father, PWS occurs, whereas when the person inherits the deletion from the mother, AS occurs. This inheritance pattern suggests that the normal PWS gene is expressed only in the paternal chromosome, and the normal AS gene is expressed only in the maternal chromosome. PWS occurs because the repressed (imprinted) PWS gene from the mother cannot make up for the absent PWS gene in the paternal chromosome. Similarly, AS occurs because the repressed (imprinted) AS gene from the father cannot make up for the absent AS gene in the maternal chromosome.

The Human Genome Project sequenced the DNA on all human chromosomes

Workers in the **Human Genome Project** have sequenced the DNA in the entire nuclear human genome—about 2.9 billion base pairs. (The human mitochondrial genome was sequenced in 1981.) This international undertaking, based on the DNA from 6 to 10 anonymous individuals, was essentially completed in 2000 and published in 2001 by hundreds of researchers comprising two independent teams, the government-funded International Human Genome Sequencing Consortium and Celera Genomics, a privately funded company. The final completion of the Human Genome Project in 2003 was a significant milestone in genetics (Table 15-1).

Scientists hope to eventually identify where all the genes are located in the sequenced DNA. We do not yet know how many genes are in the human genome. In 2001, the National Human Genome Research Institute (NHGRI) estimated there were 35,000 to 45,000 genes in the human genome, but by 2003 the NHGRI estimate was lowered to under 30,000.

Gene identification represents a formidable challenge. How do you identify a gene in a DNA sequence if you know nothing about it? The human genome is extremely complex. Only 1% to 2% of human DNA codes for protein or RNA. The rest either is noncoding regulatory elements or has some function researchers have yet to discover. In human chromosome 22, for example, the 545 genes identified so far were determined by computers that scanned DNA sequences for identifying markers such as promoter elements typically associated with genes, conserved intron sequences, and coding regions (exons). Many of these identified genes are "potential" genes in that their messenger RNAs (mRNAs) and protein products have not yet been isolated.

PROCESS OF SCIENCE

TABLE **15-1** **Major Milestones in Genetics**

Year	Scientific Advance
1866	Mendel proposed existence of hereditary factors now known as genes
1869	Nucleic acids discovered
1953	Structure of DNA determined
1960s	Genetic code explained (how proteins are made from DNA)
1977	DNA sequencing began
1986	DNA sequencing automated
1995	Sequencing of first genome (bacterium *Haemophilus influenzae*) completed
1996	Sequencing of first eukaryotic genome (yeast *Saccharomyces cerevisiae*) completed
1998	Sequencing of first multicellular eukaryotic genome (nematode worm *Caenorhabditis elegans*) completed
2001	Draft sequence of entire human genome published
2003	Final completion of DNA sequencing of human genome

Other genes were identified because they code for proteins similar to those previously identified in humans or other organisms. Despite the difficulties, progress is being made, and mapping studies will help scientists understand the physical and functional relationships among genes and groups of genes as revealed by their order on the chromosomes.

Now that the human genome has been sequenced, researchers will be busy for many decades analyzing the growing body of human molecular data. In addition to identifying genes, scientists want to understand each gene's role, how each gene interacts with other genes, and how its expression is regulated in different tissues. Eventually, all the thousands of proteins produced in human cells will be identified, their 3-D structures determined, and their properties and functions evaluated.

Researchers also want to study sequence variations within the human genome, to elucidate differences that might be related to illness or disease. The potential medical applications of the Human Genome Project are extremely promising. Genes on human chromosome 22, for example, are associated with at least 27 diseases known to have a genetic component. The causative genes of many of these disorders have not yet been identified. For example, a gene involved in schizophrenia is strongly linked to human chromosome 22, but scientists do not know its exact location or function.

The Human Genome Project stimulated the genome sequencing of other species

To aid in analyzing the human genome, investigators carried out sequencing and mapping studies simultaneously on the mouse and rat genomes. Mice and rats were obvious choices for DNA sequencing because they have been studied for almost a century and much is known about their biology. Mice and rats are sufficiently different from humans that any conserved sequences found in both rodents and humans are probably functionally important.

In addition, mice are similar enough to humans that both share many physiological traits, including some of the same diseases. To demonstrate the shared similarity between the mouse and human genomes, a consortium of scientists published a study in 2002 that compared the mouse chromosome 16 with the human genome. Only 14 of the 731 known genes on the mouse chromosome had no human counterpart; the remaining 717 genes appeared in one form or another in the human genome.

Comparison of the DNA sequences and chromosome organization of related genes from different species is a powerful tool for identifying the elements essential for their functions. If a human gene has an unknown function, researchers can often deduce clues about its role by studying the equivalent gene in another species, such as a mouse or rat.

Investigators have also sequenced the genomes of other model vertebrate organisms, such as the pufferfish (which has the smallest known genome of all vertebrates) and zebrafish. By comparing the human genome to that of the pufferfish, researchers identified shared sequences that have not changed in several hundred million years. Scientists hypothesize that these sequences may be genes or regulatory elements that are essential in all vertebrates. Further study of the pufferfish genome may help scientists identify the genes that direct vertebrate development. Biologists have used the zebrafish for many years as a model of vertebrate development because it is easy to maintain in the lab; its embryonic stages are transparent and therefore convenient to examine.

Genome analysis of commercially important organisms, such as salmon, chickens, pigs, and rice, is also a priority. Scientists published the rice genome in 2002. Researchers published the genome of the human parasite *Plasmodium,* which causes malaria, in 2002, and are currently sequencing the genome of the parasite *Trypanosoma,* which causes African sleeping sickness. One of the most significant genome-sequencing efforts now underway focuses on one or more nonhuman primates, such as the chimpanzee. Comparing the DNA sequences of humans and our closest living relatives will help biologists understand the genetic changes that occurred during human evolution, including what genes govern mental and linguistic capabilities.

The Human Genome Project has enormous implications for the future

Knowledge of the human genome may revolutionize human health care, and some experts predict that in highly developed countries, the average human life expectancy at birth will be 90 to 95 years by 2050. (As a comparison, the current average life expectancy in the United States is 77 years.) Nearly every week investigators announce new health-related information about the human genome, such as identifying genes associated with hypertension or with specific cancers.

Several scientific fields have emerged in the wake of the Human Genome Project. These include bioinformatics, pharmacogenetics, and proteomics.

Bioinformatics. The discipline known as **bioinformatics,** or *biological computing,* includes the storage, retrieval, and comparison of DNA sequences within a given species and among different species. Bioinformatics uses powerful computers and sophisticated software to manage and analyze large amounts of data generated by sequencing and other technologies. For example, as researchers determine new DNA sequences, automated computer programs scan the sequences for patterns typically found in genes. By comparing the databases of DNA sequences from different organisms, bioinformatics has already led to insights about gene identification, gene function, and evolutionary relationships.

Pharmacogenetics. The new science of gene-based medicine known as **pharmacogenetics** customizes drugs to match a patient's genetic makeup. Currently, a physician does not know in advance whether a particular medication will benefit a patient or cause severe side effects. An individual's genes, especially those that code for drug-metabolizing enzymes, largely determine that person's response to a specific drug. Pharmacogenetics takes into account the subtle genetic differences among individuals. In as few as 5 to 10 years, patients may take routine genetic screening tests before a physician prescribes a drug.

Many of these diagnostic tests may involve **DNA microarrays** (also known as *DNA chips* or *gene chips*), in which thousands of different DNA molecules are placed on a glass slide or chip. Figure 15-3 shows how a DNA microarray might be used. In step ①, a mechanical robot prepares the microarray. It spots each location on the grid with thousands to millions of copies of a specific *complementary DNA (cDNA)* strand. The single-stranded cDNA molecules for each spot, known as a *microdot,* are made using *reverse transcriptase* from a specific mRNA and then amplified using the *polymerase chain reaction* (see Chapter 14).

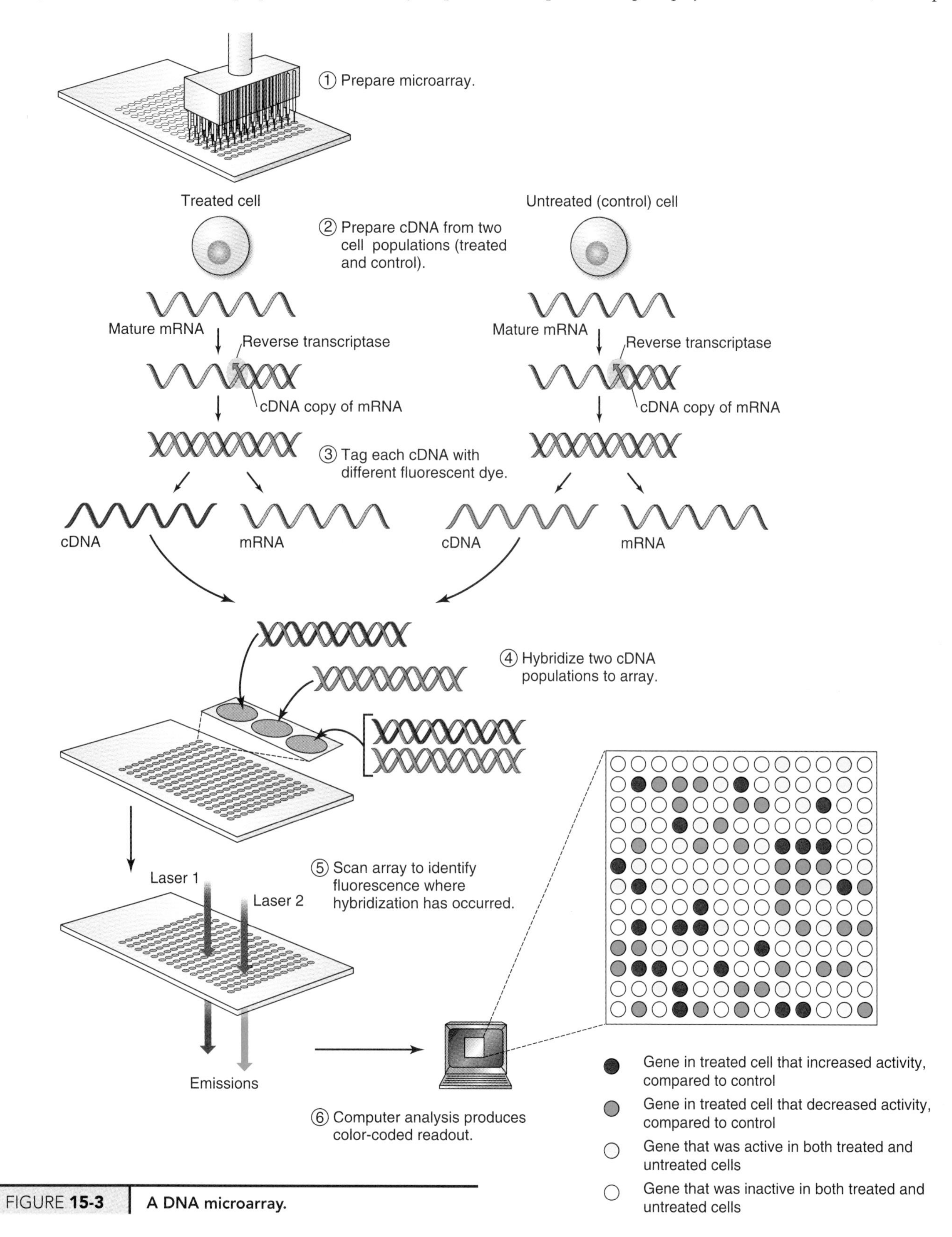

① Prepare microarray.
Treated cell
Untreated (control) cell
② Prepare cDNA from two cell populations (treated and control).
Mature mRNA
Reverse transcriptase
cDNA copy of mRNA
Mature mRNA
Reverse transcriptase
cDNA copy of mRNA
③ Tag each cDNA with different fluorescent dye.
cDNA
mRNA
cDNA
mRNA
④ Hybridize two cDNA populations to array.
Laser 1
Laser 2
⑤ Scan array to identify fluorescence where hybridization has occurred.
Emissions
⑥ Computer analysis produces color-coded readout.
Gene in treated cell that increased activity, compared to control
Gene in treated cell that decreased activity, compared to control
Gene that was active in both treated and untreated cells
Gene that was inactive in both treated and untreated cells

FIGURE **15-3** A DNA microarray.

In step ②, researchers isolate mature mRNA molecules from two cell populations—for example, liver cells treated with a newly developed drug and control liver cells not treated with the drug. (The liver produces many enzymes that metabolize foreign molecules, including drug molecules. Drugs the liver cannot metabolize, or metabolizes weakly, may be too toxic to be used.) Researchers use the isolated mRNA molecules to prepare cDNA molecules for each cell population, which ③ are then tagged with different colored fluorescent dyes. For example, the treated cells' cDNA molecules might be labeled with red dye and the untreated cells' cDNA molecules with green dye. In step ④, researchers add the two cDNA populations to the array, and some of the cDNA subsequently hybridizes (forms base pairs with) the cDNA on the array.

After washing it to remove any cDNA that has not hybridized to the array, ⑤ investigators scan the array with lasers to identify red and green fluorescence where hybridization has occurred. In step ⑥, computer analysis of the ratio of red to green fluorescence at each spot in the array produces a color-coded readout that researchers can further analyze using statistics and bioinformatics. For example, a medical researcher might compare the experimental drug's overall pattern of gene activity with that of known drugs and toxins. If the gene activity for the treated cells matches that of a toxin that damages the liver, the drug will probably not go into clinical trials.

DNA microarrays enable researchers to compare the activities of thousands of genes in normal and diseased cells from tissue samples. Because cancer and other diseases exhibit altered patterns of gene expression, DNA microarrays have the potential to identify genes or the proteins they code for, which can then be targeted by therapeutic drugs.

Here's an example of the application. In 2002 medical researchers identified 17 genes that are active in different kinds of diffuse large B-cell lymphoma. Using DNA microarrays, researchers determined which combination of those 17 genes is active in each subtype of this form of cancer. Armed with the knowledge of which subtype of cancer a patient has, physicians can choose the treatment that will probably be most effective for that patient. Examining patterns of gene activity with DNA microarrays also helps medical scientists identify which patients will probably remain free of cancer after treatment and which will probably relapse.

Pharmacogenetics, like other fields in human genetics, presents difficult ethical questions. The genetic testing that will be an everyday part of pharmacogenetics raises issues of privacy, genetic bias, and potential discrimination. We consider these issues later in the chapter.

Together, the Human Genome Project and pharmacogenetics will advance knowledge of the role of genes in human health and disease. However, individualized genetic testing may cause people to worry unnecessarily about a genetic disease for which they may never develop symptoms. Most common diseases result from a complex interplay between multiple genes and nongenetic, or environmental, factors. Recall from Chapter 10 that the environment is an important factor influencing gene expression. In humans, healthy environmental factors include proper diet, adequate exercise, and not smoking (see Chapter 37, *Focus On: Unwelcome Tissues: Cancers*).

Proteomics. The study of all the proteins encoded by the human genome and produced in a person's cells and tissues is **proteomics** or *functional genomics.* Scientists want to identify all the proteins made by a given kind of cell, but the process is much more complicated than sequencing the human genome. For one thing, some genes encode several different proteins (see Fig. 13-11, differential mRNA processing). Also, every somatic cell in the human body has essentially the same genome, but cells in different tissues vary greatly in the kinds of proteins they produce. Protein expression patterns vary not only in different tissues but also at different stages in the development of a single cell.

Scientists want to understand the role of each protein in a cell, how the various proteins interact, and the 3-D structure for each protein. While advancing biological knowledge, these goals also promise advances in medicine. If they know the shape of a protein associated with a type of cancer or other disease, pharmacologists may be able to develop drugs that bind to active sites on that protein, turning its activity off. Today, the pharmaceutical industry has drugs that target about 500 proteins in the cell. However, they estimate proteomics may yield 10,000 to 20,000 additional protein targets.

Researchers use mouse models to study human genetic diseases

Many questions relating to human genetic diseases are difficult to answer because of the ethical issues involved in using humans as test subjects. However, research on any disease is greatly facilitated if an animal model is used for experimentation. A good example is *cystic fibrosis,* a genetic disease caused by a single gene mutation inherited as a recessive allele (discussed later in the chapter). In 1994, researchers used **gene targeting** to produce strains of mice that were either homozygous or heterozygous for cystic fibrosis (see Chapter 14).

The allele that causes cystic fibrosis is a mutant form of a locus involved in controlling the body's water and electrolyte balance. Geneticists have cloned this gene and found it codes for a protein, the *CFTR protein,* that serves as a chloride ion channel in the plasma membrane. (*CFTR* stands for *cystic fibrosis transmembrane conductance regulator.*) This ion channel transports chloride ions out of the cells lining the digestive tract and the respiratory system. When the chloride ions leave the cells, water follows by osmosis. Thus the normal secretions of these cells are relatively watery. Because the cells of individuals with cystic fibrosis lack normal chloride ion channels, their secretions have a very low water content and their sweat is very salty. Cells of heterozygous individuals have only half the usual number of functional CFTR ion channels, but these are enough to maintain the normal fluidity of their secretions.

Some researchers now focus their efforts on understanding the way in which the CFTR channel is activated or inactivated in mice. They hope to use this information to design drugs that enhance chloride transport through the CFTR channel. Such drugs have the potential to treat cystic fibrosis in humans by activating the mutant channel.

Review

- What kinds of information can a human karyotype provide?
- What is pedigree analysis?
- What are two possible benefits scientists hope to obtain by further study of the human genome?
- How does using a mouse model for a genetic disease overcome some of the difficulties in studying human inheritance?

BiologyNow™ Assess your understanding of **studying human genetics** by taking the pretest on your BiologyNow CD-ROM.

ABNORMALITIES IN CHROMOSOME NUMBER AND STRUCTURE

Learning Objectives

4 Explain how nondisjunction in meiosis is responsible for chromosome abnormalities such as Down syndrome, Klinefelter syndrome, and Turner syndrome.

5 Distinguish among the following structural abnormalities in chromosomes: translocations, deletions, and fragile sites.

Polyploidy, the presence of multiple sets of chromosomes, is common in plants but rare in animals. It may arise from the failure of chromosomes to separate during meiosis or from the fertilization of an egg by more than one sperm. When it occurs in all the cells of the body, polyploidy is lethal in humans and many other animals. For example, *triploidy* ($3n$) is sometimes found in human embryos that have been spontaneously aborted in early pregnancy.

Abnormalities caused by the presence of a single extra chromosome or the absence of a chromosome—**aneuploidies**—are more common than polyploidy. **Disomy** is the normal state: two of each kind of chromosome. In **trisomy,** a person has an extra chromosome, that is, three of one kind. In **monosomy,** an individual lacks one member of a pair of chromosomes. Table 15-2 summarizes some disorders that aneuploidies produce.

Aneuploidies generally arise as a result of an abnormal meiotic (or, rarely, mitotic) division in which chromosomes fail to separate at anaphase. This phenomenon, called **nondisjunction,** can occur with the autosomes or with the sex chromosomes. In meiosis, chromosome nondisjunction may occur during the first or second meiotic division (or both). For example, two X chromosomes that fail to separate at either the first or the second meiotic division may both enter the egg nucleus. Alternatively, the two joined X chromosomes may go into a *polar body,* leaving the egg with no X chromosome. (Recall from Chapter 9 that a polar body is a nonfunctional haploid cell produced during oogenesis; also see Fig. 48-11.)

Nondisjunction of the XY pair during the first meiotic division in the male may lead to the formation of a sperm with both X and Y chromosomes or a sperm with neither an X nor a Y chromosome (Fig. 15-4). Similarly, nondisjunction at the second meiotic division can produce sperm with two Xs or two Ys. When an abnormal gamete unites with a normal one, the resulting zygote has a chromosome abnormality that will be present in every cell of the body.

Meiotic nondisjunction results in an abnormal chromosome number at the zygote stage of development, so that all cells in the individual have an abnormal chromosome number. In contrast, nondisjunction during a mitotic division occurs sometime later in development and leads to the establishment of a clone of abnormal cells in an otherwise normal individual. Such a mixture of cells with different chromosome numbers may or may not affect somatic (body) or germ-line (reproductive) tissues.

Down syndrome is usually caused by trisomy 21

Down syndrome is one of the most common chromosome abnormalities in humans. (The term *syndrome* refers to a set of symptoms that usually occur together in a particular disorder.) It was named after J. Langdon Down, the British physician who

TABLE 15-2 Chromosome Abnormalities: Disorders Produced by Aneuploidies

Karyotype	Common Name	Clinical Description
Trisomy 13	Patau syndrome	Multiple defects, with death typically by age 3 months.
Trisomy 18	Edwards syndrome	Ear deformities, heart defects, spasticity, and other damage; death typically by age 1 year, but some survive much longer.
Trisomy 21	Down syndrome	Overall frequency is about 1 in 800 live births. True trisomy is most often found among children of older (age 35+) mothers, but translocation resulting in the equivalent of trisomy is not age-related. Trisomy 21 is characterized by a fold of skin above the eye, varying degrees of mental retardation, short stature, protruding furrowed tongue, transverse palmar crease, cardiac deformities, and increased risk of leukemia and Alzheimer's disease.
XO	Turner syndrome	Short stature, webbed neck, sometimes slight mental retardation; ovaries degenerate in late embryonic life, leading to rudimentary sexual characteristics; gender is female; no Barr bodies.
XXY	Klinefelter syndrome	Male with slowly degenerating testes, enlarged breasts; one Barr body per cell.
XYY	XYY karotype	Many males have no unusual symptoms; others are unusually tall, with heavy acne, and some tendency to mild mental retardation.
XXX	Triplo-X	Despite three X chromosomes, usually fertile females with normal intelligence; two Barr bodies per cell.

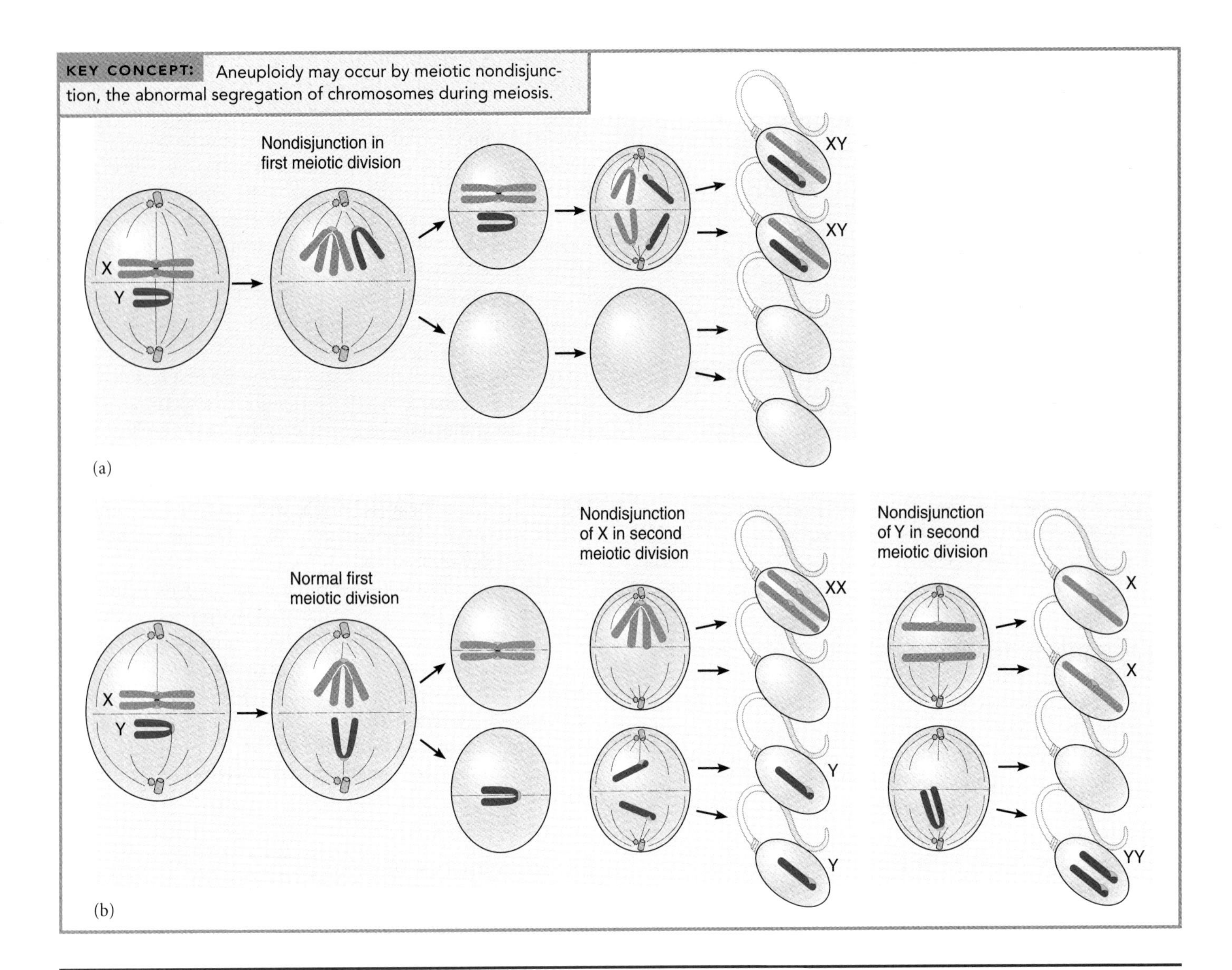

FIGURE **15-4** | **Meiotic nondisjunction.**

In these examples of nondisjunction of the sex chromosomes in the human male, only the X *(purple)* and Y *(blue)* chromosomes are shown. **(a)** Nondisjunction in the first meiotic division results in two XY sperm and two sperm with neither an X nor a Y. **(b)** Second-division nondisjunction of the X chromosome results in one sperm with two X chromosomes, two with one Y each, and one with no sex chromosomes. Nondisjunction of the Y chromosome results in one sperm with two Y chromosomes, two with one X each, and one with no sex chromosome *(box on right).*

in 1866 first described the condition. Affected individuals have abnormalities of the face, eyelids, tongue, hands, and other parts of the body and are often mentally and physically retarded (Fig. 15-5a). They are also unusually susceptible to certain diseases, such as leukemia and Alzheimer's disease.

Cytogenetic studies have revealed that most people with Down syndrome have 47 chromosomes because of *autosomal trisomy:* they are trisomic for chromosome 21 (Fig. 15-5b). This condition is known as **trisomy 21.** Nondisjunction during meiosis is responsible for the presence of the extra chromosome. Although no genetic information is missing in these individuals, the extra copies of chromosome 21 genes bring about some type of genetic imbalance that causes abnormal physical and mental development. Down syndrome is quite variable in expression, with some individuals far more severely affected than others. Researchers are using DNA technologies to attempt to pinpoint genes on chromosome 21 that affect mental development, as well as possible *oncogenes* (cancer-causing genes) and genes that may be involved in Alzheimer's disease.

Down syndrome occurs in all ethnic groups in about 1 out of 800 live births. Its incidence increases markedly with increasing maternal age. The occurrence of Down syndrome is not affected by the father's age (although other disorders are, including schizophrenia and achondroplasia, the most common form of dwarfism). Down syndrome is 68 times more likely in the offspring of mothers of age 45 than in the offspring of mothers age 20. However, most babies with Down syndrome in the United States are born to mothers younger than 35, in part because

Richard Hutchings/Photo Researchers, Inc.

(a)

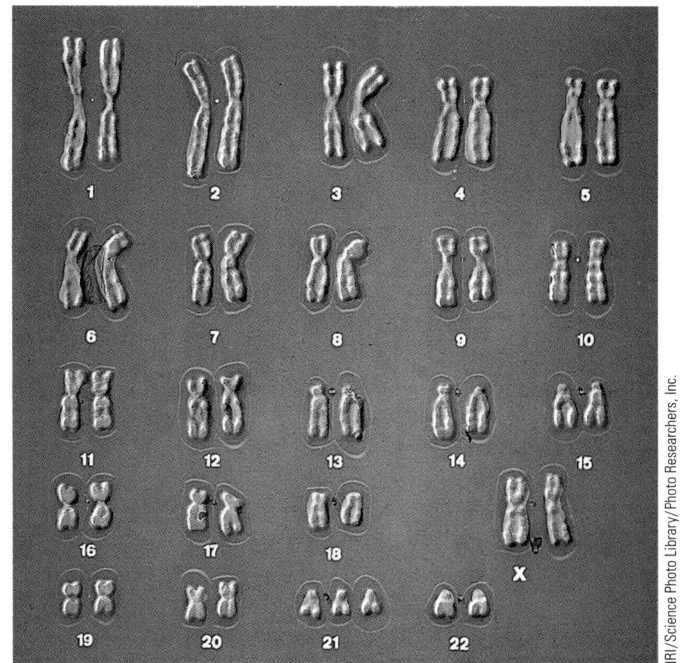

CNRI/Science Photo Library/Photo Researchers, Inc.

(b)

ACTIVE FIGURE **15-5** | **Down syndrome.**

(a) This boy with Down syndrome is working on a science experiment in his kindergarten class. Some individuals with Down syndrome learn to read and write. **(b)** Note the presence of an extra chromosome 21 in this colorized karyotype of a female with Down syndrome.

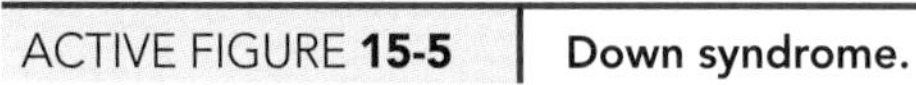 Learn more about **normal and abnormal karyotypes** by clicking on this figure on your BiologyNow CD-ROM.

these greatly outnumber older mothers, and in part because about 90% of older women who undergo prenatal testing terminate the pregnancy if Down syndrome is diagnosed.

The relationship between increased incidence in Down syndrome and maternal age has been studied for decades, but there is no explanation. Several hypotheses have been proposed to explain the maternal age effect, but none are supported unequivocally. One explanation is that older women have held eggs in suspended meiosis too long, leading to a deterioration of the meiotic spindle apparatus. Another possibility is that an aging womb is less likely to reject an abnormal fetus.

Most sex chromosome aneuploidies are less severe than autosomal aneuploidies

Sex chromosome aneuploidies are tolerated relatively well (see Table 15-2) This is true, at least in part, because of the mechanism of **dosage compensation:** mammalian cells compensate for extra X chromosome material by rendering all but one X chromosome inactive. The inactive X is seen as a **Barr body,** a region of darkly staining, condensed chromatin next to the nuclear envelope of an interphase nucleus (see Fig. 10-17). Investigators have used the presence of the Barr body in the cells of normal females (but not normal males) as an initial screen to determine whether an individual is genetically female or male. However, as you'll see shortly in the context of sex chromosome aneuploidies, the Barr body test has limitations.

Individuals with **Klinefelter syndrome** are males with 47 chromosomes, including two Xs and one Y. They have small testes, produce few or no sperm, and are therefore sterile. The hypothesis that the Y chromosome is the major determinant of the male phenotype has been substantiated by the fact that at least one gene on the Y chromosome acts as a genetic switch, directing male development. Males with Klinefelter syndrome tend to be unusually tall and have female-like breast development. About half show some mental retardation, but many live relatively normal lives. However, each of their cells have one Barr body. On the basis of such a test, they would be erroneously classified as females. About 1 in 600 to 1000 live-born males has Klinefelter syndrome.

The sex chromosome composition for **Turner syndrome,** in which an individual has only one sex chromosome, an X chromosome, is designated X0. The 0 refers to the absence of a second sex chromosome. Because they lack the male-determining effect of the Y chromosome, individuals with Turner syndrome develop as females. However, both their internal and their external genital structures are underdeveloped, and they are sterile. Apparently a second X chromosome is necessary for normal development of ovaries in a female embryo. Examination of their cells reveals no Barr bodies, because there is no extra X chromosome to be inactivated. Using the standards of the Barr body test, such an individual would be classified erroneously as a male. About 1 in 2500 live-born females has Turner syndrome.

People with an X chromosome plus two Y chromosomes are phenotypically males, and they are fertile. Other characteristics of these individuals (tall, often with severe acne) hardly qualify as a syndrome; hence the designation **XYY karotype.** Some years ago, several widely publicized studies suggested that people with this condition are more likely to display criminal tendencies and thus to be imprisoned. However, these studies were flawed because they were based on small numbers of XYY males, without adequate or well-matched control studies of XY males. The prevailing opinion in medical genetics today is that many undiagnosed XYY males in the general population do not have overly aggressive or criminal behaviors and are not incarcerated.

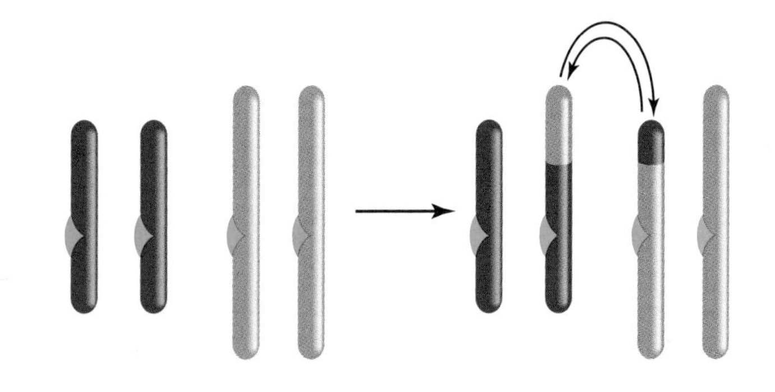

FIGURE **15-6** **Reciprocal translocation, the most common type of translocation.**

Two nonhomologous chromosomes exchange segments. About half of the gametes produced following translocation and meiosis are abnormal and have duplications or deletions. (Unduplicated chromosomes are shown for simplicity.)

Aneuploidies usually result in prenatal death

Recognizable chromosome abnormalities are seen in less than 1% of all live births, but substantial evidence suggests the rate at conception is much higher. At least 17% of pregnancies recognized at 8 weeks will end in spontaneous abortion (miscarriage). Approximately half of these spontaneously aborted embryos have major chromosome abnormalities, including autosomal trisomies (such as trisomy 21), triploidy, tetraploidy, and Turner syndrome (X0). Autosomal monosomies are exceedingly rare, possibly because they induce a spontaneous abortion very early in the pregnancy, before a woman is even aware she is pregnant. Some investigators give surprisingly high estimates (50% or more) for the loss rate of very early embryos. Chromosome abnormalities probably induce many of these spontaneous abortions.

Abnormalities in chromosome structure cause certain disorders

Chromosome abnormalities are caused not only by changes in chromosome number but also by distinct changes in the structure of one or more chromosomes. Here we consider three simple examples of structural abnormalities: translocations, deletions, and fragile sites.

Translocation is the attachment of part of one chromosome to another

In some cases of **translocation,** a chromosome fragment breaks off and attaches to a nonhomologous chromosome. In a **reciprocal translocation,** two nonhomologous chromosomes exchange parts (Fig. 15-6). The consequences of translocations vary considerably. They include deletions, in which some genes are missing, and **duplications,** which are extra copies of certain genes.

In about 4% of individuals with Down syndrome, only 46 chromosomes are present, but one is abnormal. The large arm of chromosome 21 has been translocated to the large arm of another chromosome, usually chromosome 14. Individuals with **translocation Down syndrome** have one chromosome 14, one combined 14/21 chromosome, and two normal copies of chromosome 21. All or part of the genetic material from chromosome 21 is thus present in triplicate. When geneticists study the karyotypes of the parents in such cases, they usually find either the mother or the father has only 45 chromosomes, although she or he is generally phenotypically normal. The parent with 45 chromosomes has one chromosome 14, one combined 14/21 chromosome, and one chromosome 21; although the karyotype is abnormal, there is no extra genetic material. In contrast to trisomy 21, translocation Down syndrome can run in families, and its incidence is not related to maternal age.

A deletion is the loss of part of a chromosome

In the chromosome abnormality known as a **deletion,** part of a chromosome is missing. Sometimes chromosomes break and fail to rejoin. Such breaks result in deletions of as little as a few base pairs to as much as an entire chromosome arm. As you might expect, large deletions are generally lethal, whereas small deletions may have no effect or may cause recognizable human disorders.

One deletion disorder (1 in 50,000 live births) is **cri du chat syndrome,** in which part of the short arm of chromosome 5 is deleted. As in most deletions, the exact point of breakage in chromosome 5 varies from one individual to another; some cases of cri du chat involve a small loss, whereas others involve a more substantial deletion of base pairs. Infants born with cri du chat syndrome typically have a small head with altered features described as a "moon face" and a distinctive cry that sounds like a kitten mewing. (The name literally means "cry of the cat" in French.) Affected individuals usually survive beyond childhood but exhibit severe mental retardation.

Fragile sites are weak points at specific locations in chromatids

A **fragile site** is a place where part of a chromatid appears to be attached to the rest of the chromosome by a thin thread of DNA. Fragile sites may occur at a specific location on both chromatids of a chromosome. They have been identified on the X chromosome as well as on certain autosomes. The location of a fragile site is exactly the same in all of an individual's cells, as well as in cells of other family members. Scientists report growing evi-

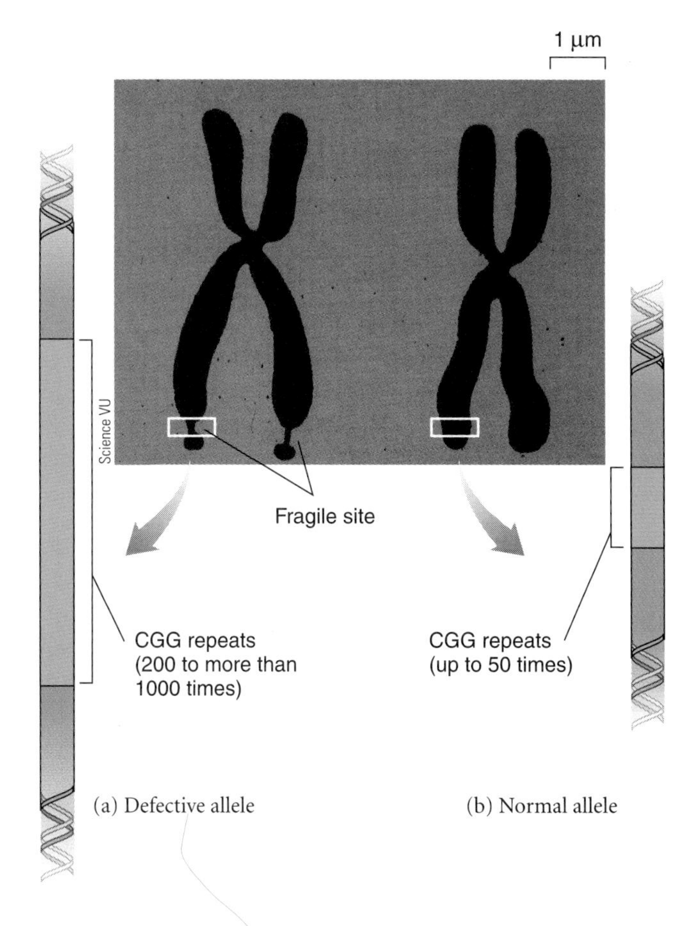

FIGURE **15-7** **Fragile X syndrome.**

This colorized SEM shows an X chromosome with a fragile site and a normal X chromosome. **(a)** The defective allele at the tips of both chromatids of the X chromosome repeats CGG 200 to 1000 times. **(b)** The normal allele repeats CGG up to 50 times.

dence that cancer cells may have breaks at these fragile sites. Whether cancer destabilizes the fragile sites, leading to breakage, or the fragile sites themselves contain genes that contribute to cancer is unknown at this time.

In **fragile X syndrome** a fragile site occurs near the tip of the X chromosome, where the fragile X gene contains a nucleotide triplet CGG that repeats 200 to more than 1000 times (Fig. 15-7a). In a normal chromosome, CGG repeats up to 50 times (Fig. 15-7b).

Fragile X syndrome is the most common cause of inherited mental retardation. The effects of fragile X syndrome, which are more pronounced in males than in females, range from mild learning and attention deficit to severe mental retardation and hyperactivity. According to the National Fragile X Foundation, about 80% of boys and 35% of girls with fragile X syndrome are at least mildly mentally retarded. Females with fragile X syndrome are usually heterozygous (because their other X chromosome is normal) and are therefore more likely to have normal intelligence.

The discovery of the fragile X gene in 1991, as well as the development of the first fragile X mouse model in 1994, have provided researchers with ways to develop and test potential treatments, including gene therapy. At the microscopic level, the nerve cells of individuals with fragile X syndrome have malformed dendrites (the part of the nerve cell that receives nerve impulses from other nerve cells). At the molecular level, the triplet repeats associated with fragile X syndrome disrupt the functioning of a gene that codes for a certain protein, designated *fragile X mental retardation protein (FMRP)*. In normal cells, FMRP binds to dozens of different mRNA molecules (exactly why it binds is not yet understood), but in cells of individuals with fragile X syndrome the mutated allele does not produce functional FMRP.

Review

- What are the specific chromosome abnormalities in Down syndrome, Kleinfelter syndrome, and Turner syndrome?
- What is the chromosome abnormality in cri du chat syndrome?
- What is the chromosome abnormality in fragile X syndrome?

BiologyNow™ Assess your understanding of **abnormalities in chromosome number and structure** by taking the pretest on your BiologyNow CD-ROM.

GENETIC DISEASES CAUSED BY SINGLE-GENE MUTATIONS

Learning Objective

6 State whether each of the following genetic defects is inherited as an autosomal recessive, autosomal dominant, or X-linked recessive: phenylketonuria (PKU), sickle cell anemia, cystic fibrosis, Tay-Sachs disease, Huntington's disease, and hemophilia A.

You have seen that several human disorders involve chromosome abnormalities. Hundreds of human disorders, however, involve enzyme defects caused by mutations of single genes. *Phenylketonuria (PKU)* and *alkaptonuria* (see Chapter 12) are examples of these disorders, which are sometimes referred to as an **inborn error of metabolism,** a metabolic disorder caused by the mutation of a gene that codes for an enzyme needed in a biochemical pathway. Both PKU and alkaptonuria involve blocks in the metabolism of the amino acid phenylalanine.

Most genetic diseases are inherited as autosomal recessive traits

Many human genetic diseases have a simple autosomal recessive inheritance pattern and therefore appear only in the homozygous state. Why are these traits recessive? Most recessive mutations result in a mutant allele that encodes a product that no longer works (either there is not enough gene product, or it is a defective gene product). In the heterozygous state, there is one functional copy of the gene and one mutated, nonfunctional copy. The normal copy of the gene generally produces enough protein to meet the cell's needs. In homozygous recessive individuals, *both* alleles of the gene are nonfunctional, and the cell's needs are not met. As a result, the person shows symptoms of disease.

Phenylketonuria results from an enzyme deficiency

Phenylketonuria (PKU), which is most common in individuals of western European descent, is an autosomal recessive disease caused by a defect of amino acid metabolism. It affects about 1 in 10,000 live births in North America. Homozygous recessive individuals lack an enzyme that converts the amino acid phenylalanine to another amino acid, tyrosine. They accumulate high levels of phenylalanine, phenylpyruvic acid, and similar compounds. The accumulating phenylalanine is converted to phenylketones, which damage the central nervous system, including the brain, in children. The ultimate result in untreated cases is severe mental retardation. An infant with PKU is usually healthy at birth because its mother, who is heterozygous, breaks down excess phenylalanine for both herself and her fetus. However, during infancy and early childhood the accumulation of toxic products eventually causes irreversible damage to the central nervous system.

In the 1950s, infants with PKU were identified early and placed on a low-phenylalanine diet, dramatically alleviating their symptoms. The diet is difficult to adhere to because it contains no meat, fish, dairy products, breads, or nuts. Also, individuals with PKU should not consume the sugar substitute aspartame, found in many diet drinks and foods, because it contains phenylalanine. Biochemical tests for PKU have been developed, and screening of newborns through a simple blood test is required in the United States. Because of these screening programs and the availability of effective treatment, thousands of PKU-diagnosed children have not developed severe mental retardation. Most must continue the diet through at least adolescence. Doctors now recommend that patients stay on the diet throughout life, because some adults who have discontinued the low-phenylalanine diet experience certain mental problems, such as concentration difficulty and short-term memory loss.

Ironically, the success of PKU treatment in childhood presents a new challenge today. If a homozygous female who has discontinued the special diet becomes pregnant, the high phenylalanine levels in her blood can damage the brain of the fetus she is carrying, even though that fetus is heterozygous. Therefore, she must resume the diet, preferably before becoming pregnant. This procedure is usually (although not always) successful in preventing the effects of **maternal PKU.** It is especially important for women with PKU to be aware of maternal PKU and to obtain appropriate counseling and medical treatment during pregnancy.

Sickle cell anemia results from a hemoglobin defect

Sickle cell anemia is inherited as an autosomal recessive trait. The disease is most common in people of African descent (approximately 1 in 500 African Americans), and about 1 in 12 African Americans is heterozygous. Under low oxygen conditions, the red blood cells of an individual with sickle cell anemia are shaped like sickles, or half-moons, whereas normal red blood cells are biconcave discs.

The mutation that causes sickle cell anemia was first identified more than 50 years ago. The sickled cells contain abnormal hemoglobin molecules, which have the amino acid valine instead of glutamic acid at position 6 (the sixth amino acid from the amino terminal end) in the β-globin chain (see Chapter 3). The substitution of valine for glutamic acid makes the hemoglobin molecules stick to one another to form fiber-like structures that change the shape of the red blood cells. This sickling occurs in the veins after the oxygen has been released from the hemoglobin. The blood cells' abnormal sickled shape slows blood flow and blocks small blood vessels (Fig. 15-8), resulting in tissue damage from lack of oxygen and essential nutrients and episodes of pain. Sickled red blood cells also have a shorter life span than normal red blood cells, leading to severe anemia in many affected individuals.

Treatments for sickle cell anemia include pain relief measures, transfusions, and, more recently, medicines such as hydroxyurea, which activates the gene for the production of normal fetal hemoglobin (this gene is generally not expressed after birth). The presence of normal fetal hemoglobin in the red blood cells dilutes the sickle cell hemoglobin, thereby minimizing the painful episodes and reducing the need for blood transfusions. The long-term effects of hydroxyurea are not known at this time, but there are concerns that it may induce tumor formation.

Ongoing research is directed toward providing gene therapy for sickle cell anemia. The development of a mouse model for studying sickle cell anemia has enabled researchers to test gene therapy. The first gene therapy treatments in mice used a mouse

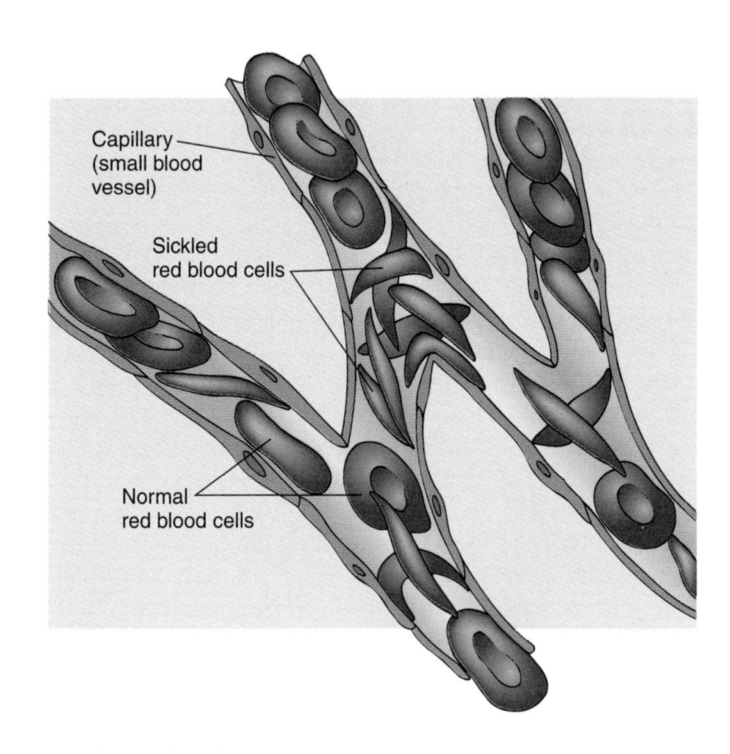

FIGURE **15-8** | **Sickle cell anemia.**

Sickled red blood cells do not pass through small blood vessels as easily as normal red blood cells. The sickled cells can cause blockages that prevent oxygen from being delivered to tissues.

retrovirus as a **vector,** a carrier that transfers the genetic information. However, the retrovirus did not effectively transport the normal gene for hemoglobin into the bone marrow, where stem cells produce new blood cells. In 2001, researchers cured sickle cell anemia in mice using a modified HIV as a vector. Before this treatment can be tested in humans, however, researchers must demonstrate that the HIV vector is safe. Bone marrow transplants are also a promising future treatment for seriously ill individuals.

The reason the sickle cell allele occurs at a higher frequency in parts of Africa is well established. Individuals who are heterozygous (Hb^AHb^S) and carry alleles for both normal hemoglobin (Hb^A) and sickle cell hemoglobin (Hb^S) are more resistant to the malarial parasite, *Plasmodium falciparum,* which causes a severe, often fatal, form of malaria. The malarial parasite, which spends part of its life cycle inside red blood cells, does not thrive when sickle cell hemoglobin is present. (An individual heterozygous for sickle cell anemia produces both normal and sickle cell hemoglobin.) Areas in Africa where falciparum malaria occurs correlate well with areas in which the frequency of the sickle cell allele is more common in the human population. Thus, Hb^AHb^S individuals, who possess one copy of the mutant sickle cell allele, have a selective advantage over homozygous individuals, both Hb^AHb^A (who may die of malaria) and Hb^SHb^S (who may die of sickle cell anemia). This phenomenon, known as **heterozygote advantage,** is discussed further in Chapter 18 (see Fig. 18-6).

FIGURE **15-9** | **Treating cystic fibrosis.**
One traditional treatment for cystic fibrosis has been chest percussion, or gentle pounding on the chest, to clear mucus from clogged airways in the lungs.

Abraham Menashe

Cystic fibrosis results from defective ion transport

Cystic fibrosis is the most common autosomal recessive disorder in children of European descent (1 in 2500 births). About 1 in 25 individuals in the United States is a heterozygous carrier of the mutant cystic fibrosis allele. Abnormal secretions characterize this disorder. Its most severe effect is on the respiratory system, where abnormally viscous mucus clogs the airways. The cilia that line the bronchi (see Chapter 44) cannot easily remove the mucus, and it thus becomes a growth medium for dangerous bacteria. These bacteria or their toxins attack the surrounding tissues, leading to recurring pneumonia and other complications. The heavy mucus also occurs elsewhere in the body, causing digestive difficulties and other effects.

As discussed earlier, the gene responsible for cystic fibrosis codes for CFTR, the protein that regulates the transport of chloride ions across cell membranes. The defective protein, found in plasma membranes of epithelial cells lining the passageways of the lungs, intestines, pancreas, liver, sweat glands, and reproductive organs, results in the production of an unusually thick mucus that eventually leads to tissue damage. Although many forms of cystic fibrosis exist that vary somewhat in the severity of symptoms, the disease is almost always very serious.

Antibiotics are used to control bacterial infections, and daily physical therapy is required to clear mucus from the respiratory system (Fig. 15-9). Treatment with *Dornase Alpha (DNase)*, an enzyme produced by recombinant DNA technology, helps break down the mucus. Without treatment, death would occur in infancy. With treatment, the average life expectancy for individuals with cystic fibrosis is now about 30 years. Because of the serious limitations of available treatments, gene therapy for cystic fibrosis is under development.

The most severe mutant allele for cystic fibrosis predominates in northern Europe, and another, somewhat less serious, mutant allele is more prevalent in southern Europe. Presumably these mutant alleles are independent mutations that have been maintained by natural selection. Some experimental evidence supports the hypothesis that heterozygous individuals are less likely to die from infectious diseases that cause severe diarrhea, another possible example of heterozygote advantage.

Tay-Sachs disease results from abnormal lipid metabolism in the brain

Tay-Sachs disease is an autosomal recessive disease that affects the central nervous system and results in blindness and severe mental retardation. The symptoms begin within the first year of life and result in death before the age of 5 years. Because of the absence of an enzyme, a normal membrane lipid in brain cells fails to break down properly and accumulates in intracellular organelles called *lysosomes.* The lysosomes swell and cause the nerve cells to malfunction. Although research is ongoing, no effective treatment for Tay-Sachs disease is available at this time. However, an effective treatment strategy in a mouse model was reported in 1997: Oral administration of an inhibitor reduced the synthesis of the lipid that accumulates in the lysosomes. This treatment offers the hope of future breakthroughs to deal more effectively with Tay-Sachs disease in humans.

The abnormal allele is especially common in the United States among Jews whose ancestors came from eastern and central Europe (Ashkenazi Jews). The disease occurs in about 1 in 4000 live births in the North American Jewish population. In contrast, Jews whose ancestors came from the Mediterranean region (Sephardic Jews) have a very low frequency of the allele.

Some genetic diseases are inherited as autosomal dominant traits

Huntington's disease (HD), named after George Huntington, the U.S. physician who first described it in 1872, is caused by a rare autosomal dominant allele that affects the central nervous system. The disease causes severe mental and physical deterioration, uncontrollable muscle spasms, and personality changes; death ultimately results. No effective treatment has been found. Every child of an affected individual has a 50% chance of also being affected (and, if affected, of passing the abnormal allele to his or her offspring). Ordinarily we would expect a dominant allele with such devastating effects to occur only as a new mutation and not to be transmitted to future generations. Because HD symptoms do not appear until relatively late in life (most people do not develop the disease until they are in their thirties or forties), a person may have children before the disease develops (Fig. 15-10). In North America, HD occurs in 1 in 20,000 live births.

The gene responsible for HD was identified in 1993. It is found at one end of chromosome 4. The mutation is a nucleotide triplet (CAG) that is repeated many times; the normal allele repeats CAG from 6 to 35 times, whereas the mutant allele repeats CAG from 40 to more than 150 times. Because CAG codes for the amino acid glutamine, the resulting protein, called *huntingtin,* has a long strand of glutamines. The number of nucleotide triplet repeats seems to be important in determining the age of onset and the severity of the disease; larger numbers of repeats correlate with an earlier age of onset and greater severity.

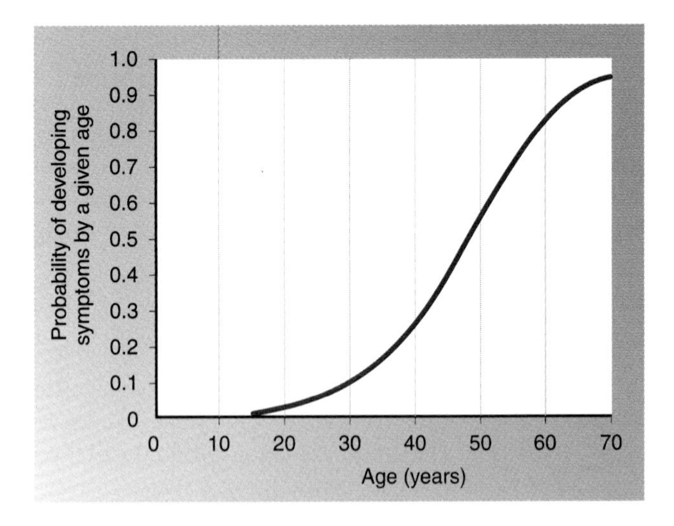

FIGURE **15-10** **The age of onset of Huntington's disease.** The graph shows the cumulative probability that an individual affected with a Huntington's disease allele will have developed symptoms at a given age. *(Adapted from P. S. Harper,* Genetic Counseling, *5th edition, Butterworth-Heinemann, Oxford, 1998)*

Much research now focuses on how the mutation is linked to neurodegeneration in the brain. A mouse model of HD is providing valuable clues about the development of the disease. Using this model, researchers have demonstrated that the defective version of huntingtin binds to enzymes called *acetyltransferases* in brain cells, blocking their action. Acetyltransferases are involved in turning genes on for expression, so in the brain cells of HD individuals, much of normal transcription cannot occur. Once neurologists better understand HD's mechanism of action on nerve cells, it may be possible to develop effective treatments to slow the progression of the disease.

Cloning of the *HD* allele became the basis for tests that allow those at risk to learn presymptomatically if they carry the allele. The decision to be tested for any genetic disease is understandably a highly personal one. The information is, of course, invaluable for those who must decide whether or not to have children. However, someone who tests positive for the *HD* allele must then live with the virtual certainty of eventually developing this devastating and incurable disease. Researchers hope that information from affected individuals who choose to be identified before the onset of symptoms may ultimately contribute to the development of effective treatments.

Some genetic diseases are inherited as X-linked recessive traits

Hemophilia A was once referred to as a disease of royalty because of its high incidence among male descendants of Queen Victoria, but it is also found in many nonroyal pedigrees. Caused by the absence of blood-clotting factor VIII, **hemophilia A** is characterized by severe internal bleeding in the head, joints, and other areas from even a slight wound. The mode of inheritance is X-linked recessive. Thus, affected individuals are almost exclusively male, having inherited the abnormal allele on the X chromosome from their heterozygous carrier mothers. (For a female to be affected by an X-linked trait, she would have to inherit the defective allele from both parents, whereas an affected male need only inherit one defective allele from his mother.)

Treatments for hemophilia A consist of blood transfusions and the administration of clotting factor VIII (the missing gene product) by injection. Unfortunately, these treatments are costly. During the 1980s, many clotting factor VIII preparations made from human plasma were contaminated with HIV, and many men with hemophilia subsequently died from AIDS. Since 1992, virus-free clotting factor VIII has been available from both human plasma and recombinant DNA technology.

Review

- Which of the following genetic diseases is/are inherited as an autosomal recessive: phenylketonuria, Huntington's disease, Tay-Sachs disease?

- Which of the following genetic diseases is/are inherited as an autosomal dominant: sickle cell anemia, hemophilia A, Huntington's disease?
- Which of the following genetic diseases is/are inherited as an X-linked recessive: hemophilia A, cystic fibrosis, Tay-Sachs disease?

BiologyNow™ Assess your understanding of **genetic diseases caused by single-gene mutations** by taking the pretest on your BiologyNow CD-ROM.

GENE THERAPY

Learning Objective

7 Briefly discuss the process of gene therapy, including some of its technical challenges.

Because serious genetic diseases are difficult to treat, scientists have dreamed of developing actual cures. Today, advances in genetics may bring these dreams closer to reality. One strategy is **gene therapy,** which aims to replace a mutant allele in certain body cells with a normal allele. The rationale is that, although a particular allele may be present in all cells, it is expressed only in some. Expression of the normal allele in only the cells that require it may be sufficient to yield a normal phenotype.

This approach presents several technical problems. The solutions to these problems must be tailored to the nature of the gene itself, as well as to its product and the types of cells in which it is expressed. First the gene is cloned and the DNA introduced into the appropriate cells. One of the most successful techniques is packaging the normal allele in a viral vector, a virus that moves the normal allele into target cells that currently have a mutant allele. Ideally the virus should infect a high percentage of the cells. Most importantly, the virus should do no harm, especially over the long term.

Although many obstacles must be overcome, gene therapies for several genetic diseases are under development or are being tested on individuals in clinical trials. Scientists are currently addressing some of the unique problems presented by each disease.

Gene therapy programs are carefully scrutinized

Until recently, major technical advances caused the number of clinical studies involving gene therapy to grow dramatically. However, the death of a young man in a gene therapy trial in 1999, and two more recent cases of cancer (leukemia) in children, led to a shutdown of many trials, pending the outcome of investigations about health risks. The main safety concern in these inquiries is the potential toxicity of viral vectors. The vector used in the young man who died was an adenovirus (see Fig. 23-1b), a virus required in large doses to transfer enough copies of the normal alleles for effective therapy. Unfortunately, the high viral doses triggered a fatally strong immune response in the patient's body. Both children who developed leukemia were being treated for severe combined immunodeficiency disease (SCID). The vector in these cases was a retrovirus. The virus inserted itself into the children's DNA near or in a gene that can cause childhood leukemia.

Performing clinical trials on humans always has inherent risks. Researchers carefully select patients and thoroughly explain the potential benefits and risks, as far as they are known, so the patient—or, in the case of children, the parents—can give informed consent for the procedure. However, the problems in gene therapy trials in recent years have researchers busy developing safer alternatives to viral vectors.

Review

- How are viruses used in human gene therapy?
- Why are viral vectors of potential concern in human gene therapy?

BiologyNow™ Assess your understanding of **gene therapy** by taking the pretest on your BiologyNow CD-ROM.

GENETIC TESTING AND COUNSELING

Learning Objectives

8 State the relative advantages and disadvantages of amniocentesis, chorionic villus sampling, and preimplantation genetic diagnosis in the prenatal diagnosis of human genetic abnormalities.

9 Distinguish between genetic screening programs for newborns and adults, and discuss the scope and implications of genetic counseling.

Geneticists have made many advances in detecting genetic disorders in individuals in recent years, including in prenatal diagnosis and genetic screening. With these advances comes increased information for couples at risk of having children with genetic diseases. Helping couples understand and deal with the genetic information now available is part of the rapidly expanding field of genetic counseling.

Prenatal diagnosis detects chromosome abnormalities and gene defects

Health care professionals are increasingly successful at diagnosing genetic diseases prenatally. In the diagnostic technique called **amniocentesis,** a sample of the *amniotic fluid* surrounding the fetus is obtained. A technician inserts a needle through the pregnant woman's abdomen, into the uterus, and then into the amniotic sac surrounding the fetus. Some of the amniotic fluid is withdrawn from the amniotic cavity into a syringe (Fig. 15-11). The fetus is normally safe from needle injuries because **ultrasound imaging** helps determine the positions of the fetus, placenta, and the needle (see Fig. 49-18). However, there is a 0.5%, or 1 in 200, chance that amniocentesis will induce a miscarriage.

Amniotic fluid contains living cells sloughed off the body of the fetus and hence genetically identical to the cells of the fetus. After cells grow for about 2 weeks in culture in the lab, technicians karyotype dividing cells to detect chromosome abnormalities. Other DNA tests have also been developed to identify

FIGURE **15-11** **Amniocentesis.**

Certain genetic diseases and other abnormal conditions are diagnosed prenatally by amniocentesis.

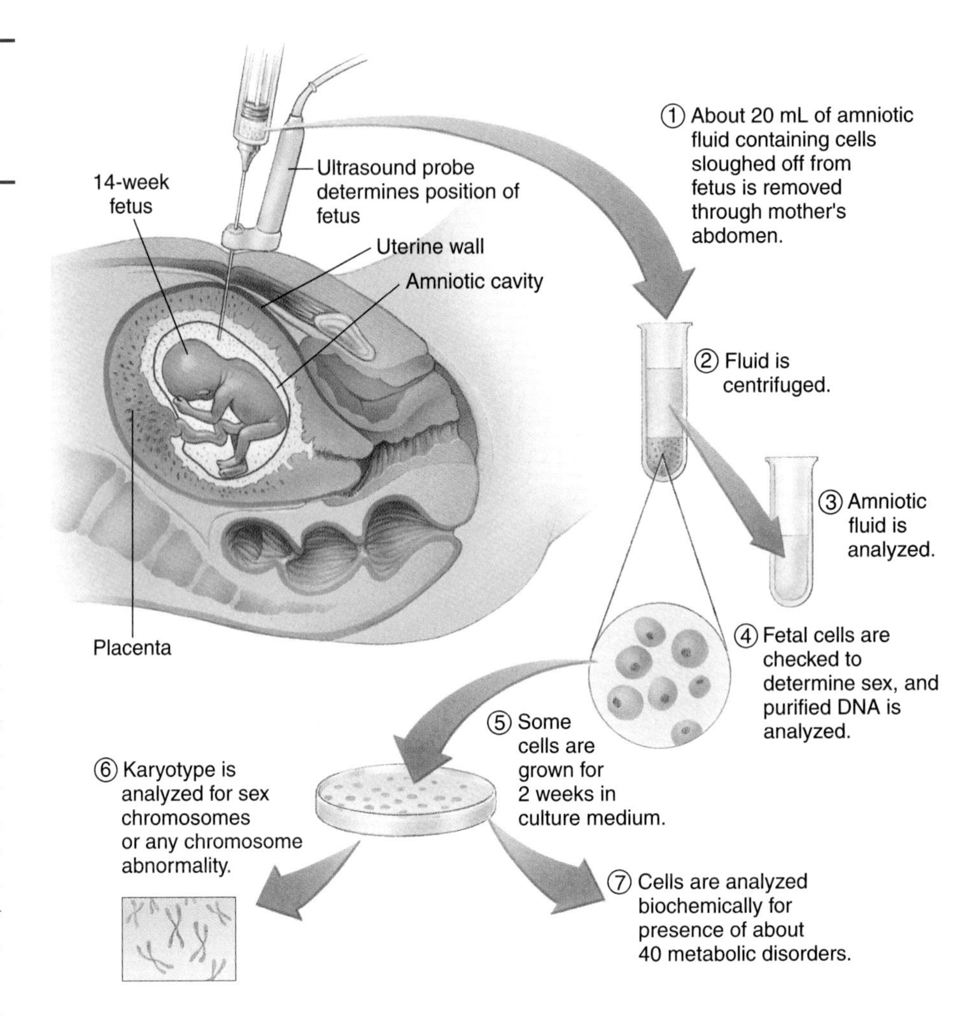

14-week fetus
Ultrasound probe determines position of fetus
Uterine wall
Amniotic cavity
Placenta
① About 20 mL of amniotic fluid containing cells sloughed off from fetus is removed through mother's abdomen.
② Fluid is centrifuged.
③ Amniotic fluid is analyzed.
④ Fetal cells are checked to determine sex, and purified DNA is analyzed.
⑤ Some cells are grown for 2 weeks in culture medium.
⑥ Karyotype is analyzed for sex chromosomes or any chromosome abnormality.
⑦ Cells are analyzed biochemically for presence of about 40 metabolic disorders.

most chromosome abnormalities. Amniocentesis, which has been performed since the 1960s, is routinely offered for pregnant women older than age 35 because their fetuses have a higher-than-normal risk of Down syndrome.

Other prenatal tests have been developed to detect many genetic disorders with a simple inheritance pattern, but these disorders are rare enough that physicians usually order the tests performed only if they suspect a particular problem. Enzyme deficiencies can often be detected by incubating cells recovered from amniotic fluid with the appropriate substrate and measuring the product; this technique has been useful in prenatal diagnosis of disorders such as Tay-Sachs disease. The tests for several other diseases, including sickle cell anemia, Huntington's disease, and cystic fibrosis, involve directly testing the individual's DNA for the mutant allele.

Amniocentesis is also useful in detecting a condition known as *spina bifida,* in which the spinal cord does not close properly during development. A relatively common malformation (about 1 in 300 births), this birth defect is associated with abnormally high levels of a normally occurring protein, *α-fetoprotein,* in the amniotic fluid. Some of this protein crosses the placenta into the mother's blood, which is tested for *maternal serum α-fetoprotein (MSAFP)* as a screen for spinal cord defects. If an elevated level of MSAFP is detected, diagnostic tests, such as ultrasound imaging and amniocentesis, are performed. (Interestingly, abnormally *low* levels of MSAFP are associated with Down syndrome and other trisomies.)

One problem with amniocentesis is that most of the conditions it detects are unpreventable and incurable, and the results are generally not obtained until well into the second trimester, when terminating the pregnancy is both psychologically and medically more difficult than earlier. Therefore, tests that yield results earlier in the pregnancy have been developed. **Chorionic villus sampling (CVS)** involves removing and studying cells that will form the fetal contribution to the placenta (Fig. 15-12). CVS, which has been performed in the United States since about 1983, is associated with a slightly greater risk of infection or miscarriage than amniocentesis, but its advantage is that results are obtained earlier than in amniocentesis, usually within the first trimester.

A relatively new embryo-screening process, known as **preimplantation genetic diagnosis (PGD),** is available for couples who carry alleles for Tay-Sachs disease, hemophilia, sickle cell anemia, and dozens of other inherited genetic conditions. Conception is by in vitro fertilization (see *Focus On: Novel Origins,* in Chapter 48). The embryos are then screened for one or more genetic diseases before a physician places a healthy embryo into the woman's uterus. PGD differs from amniocentesis and CVS in that the test is performed *before* a woman is pregnant, so it eliminates the decision of whether or not to terminate the pregnancy if an embryo has a genetic abnormality. However, PGD is not as accurate as amniocentesis or CVS, and it is more expensive. Moreover, PGD is sometimes controversial, because some couples use it to choose the gender of their offspring, not to screen for genetic diseases.

Although using amniocentesis, CVS, and PGD can help physicians diagnose certain genetic disorders with a high degree of accuracy, they are not foolproof, and many disorders cannot be diagnosed at all. Therefore, the lack of an abnormal finding is no guarantee of a normal pregnancy.

Genetic screening searches for genotypes or karyotypes

Genetic screening is a systematic search through a population for individuals with a genotype or karyotype that might cause a serious genetic disease in themselves or their offspring. There are

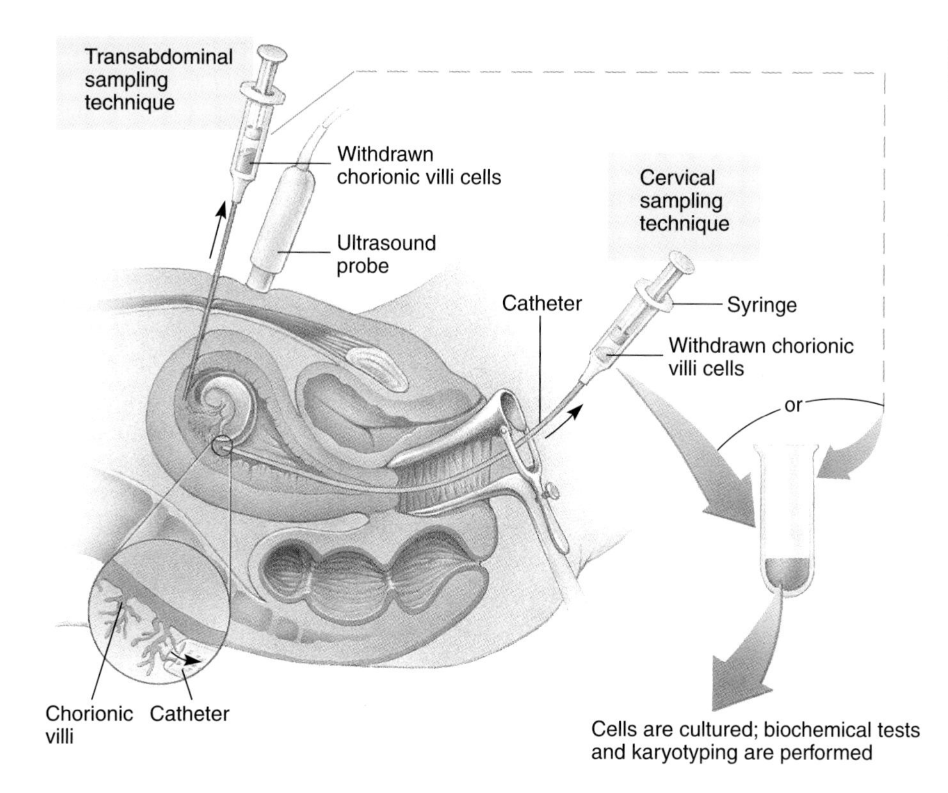

FIGURE **15-12** **Chorionic villus sampling (CVS).**

This test allows the early diagnosis of some genetic abnormalities. Samples may be obtained by inserting a needle through the uterine wall or the cervical opening.

two main types of genetic screening, for newborns and for adults, and each serves a different purpose. Newborns are screened primarily as the first step in preventive medicine, and adults are screened to help them make informed reproductive decisions.

Newborns are screened to detect and treat certain genetic diseases before the onset of serious symptoms. The routine screening of infants for PKU began in 1962 in Massachusetts. Laws in all 50 states of the United States, as well as in many other countries, currently require PKU screening. Sickle cell anemia is also more effectively treated with early diagnosis. Screening newborns for sickle cell anemia reduces infant mortality by about 15% because doctors can administer daily doses of antibiotics, thereby preventing bacterial infections common to newborns with the disease. The number of genetic disorders that can be screened in newborns is rapidly increasing, and many countries are trying to decide how best to implement the additional procedures.

Genetic screening of adults identifies carriers (heterozygotes) of recessive genetic disorders. If both prospective parents are heterozygous, the carriers are counseled about the risks involved in having children. Since the 1970s, about 1 million young Jewish adults in the United States, Israel, and other countries have been screened voluntarily for Tay-Sachs disease, and about 1 in 30 has been identified as a carrier. Tay-Sachs screening programs have reduced the incidence of Tay-Sachs disease by more than 90%.

Genetic counselors educate people about genetic diseases

Couples who are concerned about the risk of abnormality in their children, because they have either an abnormal child or a relative affected by a hereditary disease, may seek **genetic counseling** for medical and genetic information as well as support and guidance. Genetic clinics, available in most major metropolitan centers, are usually affiliated with medical schools.

Genetic counselors, who have received training in counseling, medicine, and human genetics, provide people with the information they need to make reproductive decisions. They offer advice, tempered with respect and sensitivity, in terms of risk estimates—that is, the *probability* that any given offspring will inherit a particular condition. The counselor uses the complete family histories of both the man and the woman, and a clinical geneticist (a doctor who specializes in genetics) may screen for the detection of heterozygous carriers of certain conditions.

When a disease involves only a single gene locus, probabilities can usually be easily calculated. For example, if one prospective parent is affected with a trait that is inherited as an autosomal dominant disorder, such as Huntington's disease, the probability that any given child will have the disease is 0.5, or 50%. The birth to phenotypically normal parents of a child affected with an autosomal recessive trait, such as albinism or PKU, establishes that both parents are heterozygous carriers, and the probability that any subsequent child will be affected is therefore 0.25, or 25%. For a disease inherited through a recessive allele on the X chromosome, such as hemophilia A, a normal woman and an affected man will have daughters who are carriers and sons who are normal. The probability that the son of a carrier mother and a normal father will be affected is 0.5, or 50%; the probability that their daughter will be a carrier is also 0.5.

It is important for identified carriers to receive appropriate genetic counseling. A genetic counselor is trained not only to provide information pertaining to reproductive decisions but also to help individuals understand their situation and avoid feeling stigmatized.

Review

- What are the relative advantages and disadvantages of amniocentesis, chorionic villus sampling, and preimplantation genetic diagnosis?
- What is the purpose of genetic screening for newborns?
- What is the purpose of genetic screening for adults?

BiologyNow™ Assess your understanding of **genetic testing and counseling** by taking the pretest on your BiologyNow CD-ROM.

HUMAN GENETICS, SOCIETY, AND ETHICS

Learning Objective

10 Discuss the controversies of genetic discrimination.

Many misconceptions exist about genetic diseases and their effects on society. Some people erroneously think of certain individuals or populations as genetically unfit and thus responsible for many of society's ills. They argue, for example, that medical treatment of people affected with genetic diseases, especially those who reproduce, increases the frequency of abnormal alleles in the population. However, such notions are incorrect. Genetic disorders are so rare that modern medical treatments will have only a negligible effect on their incidence.

Recessive mutant alleles are present in *all* individuals and *all* ethnic groups; no one is exempt. According to one estimate, each of us is heterozygous for several (3 to 15) very harmful recessive alleles, any of which could cause debilitating illness or death in the homozygous state. Why, then, are genetic diseases relatively uncommon? Each of us has many thousands of essential genes, any of which can be mutated. It is very unlikely that the abnormal alleles that one individual carries are also carried by that individual's mate. Of course, this possibility is more likely if the harmful allele is a relatively common one, such as the one responsible for cystic fibrosis.

Relatives are more likely than nonrelatives to carry the same harmful alleles, having inherited them from a common ancestor. In fact, a greater-than-normal frequency of a particular genetic disease among offspring of **consanguineous matings,** matings between genetically related individuals, is often the first clue that the mode of inheritance is autosomal recessive. The offspring of consanguineous matings have a small but significantly increased risk of genetic disease. In fact, they account for a disproportionately high percentage of those individuals in the population with autosomal recessive disorders. Because of this perceived social cost, marriages of close relatives, including first-cousins, are prohibited by about half the states in the United States. However, consanguineous marriages are still relatively common in many other countries.

Genetic discrimination provokes heated debate

One of the fastest growing areas of medical diagnostics is genetic screening and testing, and the number of new genetic tests that screen for diseases such as cystic fibrosis, sickle cell anemia, HD, colon cancer, and breast cancer increases each year. However, genetic testing raises many social, ethical, and legal issues that we as a society must address.

One of the most difficult issues is whether genetic information should be available to health insurance and life insurance companies. Many people think genetic information should not be given to insurance companies, but others, including employers, insurers, and many organizations representing people affected by genetic disorders, say such a view is unrealistic. If people use genetic tests to help them decide when to buy insurance and how much, then insurers insist they should also have access to this information. Insurers say they need access to genetic data to help calculate equitable premiums (insurance companies average risks over a large population). However, some are concerned that insurers might use the results of genetic tests to discriminate against people with genetic diseases or to deny them coverage.

Doctors argue that people at risk for a particular genetic disease might delay being tested because they fear genetic discrimination from insurers and employers. **Genetic discrimination** is discrimination against an individual or family members because of differences from the "normal" genome in that individual. The perception of genetic discrimination already exists in society. A 1996 study found that 25% of 332 people with family histories of one or more genetic disorders thought they had been refused life insurance, 22% thought they had been refused health insurance, and 13% thought they had been denied employment because of genetic discrimination. In a 1998 survey by the National Center for Genetic Resources, 63% of respondents said they probably or definitely would not take a genetic test if the results could be disclosed to either their employers or insurers.

Complicating the issue even more, genetic tests are sometimes difficult to interpret, in part because of the many complex interactions between genes and the environment. If a woman tests positive for an allele that has been linked to breast cancer, for example, she is at significant risk, but testing positive does not necessarily mean she will develop breast cancer. These uncertainties also make it hard to decide what form of medical intervention, from frequent mammograms to surgical removal of healthy breasts, is appropriate.

The Ethical, Legal, and Social Implications (ELSI) Research Program of the National Human Genome Research Institute has developed principles designed to protect people against genetic discrimination. The Health Insurance Portability and Accountability Act of 1996 provides some safeguards against genetic discrimination, and the Americans with Disabilities Act may also apply to genetic discrimination. As this book goes to press, bills that extend significant protection against workplace discrimination, health discrimination, and invasion of privacy based on genetic information are up for consideration by both federal and state legislatures. These issues will be debated for years to come.

Many ethical issues related to human genetics must be addressed

Genetic discrimination is only one example of ethical issues arising from our expanding knowledge of human genetics. Consider the following questions, all of which deal with the broad ethical issue of individual rights: What is the youngest age at which genetic testing should be permitted for adult-onset diseases, such as Huntington's disease? What are the emotional and psychological effects on individuals who are told they have tested positive for an incurable genetic disease? Should testing be performed when some family members want testing and others do not? Should parents be able to test their minor children? Should access to genetic test data be permitted in cases of paternity or kinship testing? Should states be able to collect genetic data on their residents? Should school administrators or law enforcement agencies have access to genetic data? These questions are only a sample of the many issues that ethicists must consider, both now and in the future. As human genetics assumes an increasingly important role in society, issues of genetic privacy and the confidentiality of genetic information must be addressed.

Review

- Why is it incorrect to assume that certain individuals or populations carry most abnormal alleles found in humans?
- To be expressed, an autosomal recessive genetic disease must be homozygous. What relationship does this fact have to consanguineous matings?
- Why do health and life insurance companies want genetic information about their clients?

BiologyNow™ Assess your understanding of **human genetics, society, and ethics** by taking the pretest on your BiologyNow CD-ROM.

SUMMARY WITH KEY TERMS

1 Distinguish between karyotyping and pedigree analysis.

- Studies of an individual's **karyotype,** the number and kinds of chromosomes present in the nucleus, enable researchers to identify various chromosome abnormalities.
- A **pedigree** is a "family tree" that shows the transmission of genetic traits within a family over several generations. Pedigree analysis is useful in detecting autosomal dominant mutations, autosomal recessive mutations, X-linked recessive mutations, and defects due to **genomic imprinting,** expressions of a gene based on its parental origin, within a family.

2 Discuss the implications of the Human Genome Project, including the emerging fields of bioinformatics, pharmacogenetics, and proteomics.

- **Human genetics** is the science of inherited variation in humans. The **human genome** is the total genetic information in human cells. The **Human Genome Project** sequenced all the DNA in the nuclear human genome.
- **Bioinformatics** includes the storage, retrieval, and comparison of DNA sequences within a given species and among different species.
- The Human Genome Project has given rise to the emerging field of **pharmacogenetics,** in which drugs are customized to match a patient's genetic makeup.
- **Proteomics** is the study of all the proteins encoded by the human genome and produced in a person's cells and tissues.

3 Discuss the mouse model for studying cystic fibrosis.

- The use of animal models greatly helps researchers investigate human disease. Researchers used **gene targeting** to produce strains of mice that are either homozygous or heterozygous for **cystic fibrosis.** Results from these studies may yield more effective drugs for treating the disease.

4 Explain how nondisjunction in meiosis is responsible for chromosome abnormalities such as Down syndrome, Klinefelter syndrome, and Turner syndrome.

- In **aneuploidy,** there are either missing or extra copies of certain chromosomes. Aneuploidies include **trisomy,** in which an individual possesses an extra chromosome, and **monosomy,** in which one member of a pair of chromosomes is missing.
- **Trisomy 21,** the most common form of **Down syndrome,** and **Klinefelter syndrome** (XXY) are examples of trisomy. **Turner syndrome** (X0) is an example of monosomy.
- Trisomy and monosomy are caused by meiotic **nondisjunction,** in which sister chromatids or homologous chromosomes fail to move apart properly during meiosis.

5 Distinguish among the following structural abnormalities in chromosomes: translocations, deletions, and fragile sites.

- In a **translocation,** part of one chromosome becomes attached to another. About 4% of individuals with Down syndrome have a translocation in which the long arm of chromosome 21 is attached to the long arm of one of the larger chromosomes, such as chromosome 14.
- A **deletion** can result in chromosome breaks that fail to rejoin. The deletion may range in size from a few base pairs to an entire chromosome arm. One deletion disorder in humans is **cri du chat syndrome,** in which part of the short arm of chromosome 5 is deleted.
- **Fragile sites** may occur at specific locations on both chromatids of a chromosome. In **fragile X syndrome,** a fragile site occurs near the tip on the X chromosome, where the nucleotide triplet CGG is repeated many more times than is normal. Fragile X syndrome is the most common cause of inherited mental retardation.

6 State whether each of the following genetic defects is inherited as an autosomal recessive, autosomal dominant, or X-linked recessive: phenylketonuria (PKU), sickle cell anemia, cystic fibrosis, Tay-Sachs disease, Huntington's disease, and hemophilia A.

- Most human genetic diseases that show a simple inheritance pattern are transmitted as autosomal recessive traits. An **inborn error of metabolism** is a metabolic disorder caused by the mutation of a gene that codes for an enzyme needed for a biochemical pathway.
- **Phenylketonuria (PKU)** is an autosomal recessive disorder in which toxic phenylketones damage the developing nervous system. **Sickle cell anemia** is an autosomal recessive disorder in which abnormal hemoglobin (the protein that transports oxygen in the blood) is produced.
- **Cystic fibrosis** is an autosomal recessive disorder in which abnormal secretions are produced primarily in organs of the respiratory and digestive systems. **Tay-Sachs disease** is an autosomal recessive disorder caused by abnormal lipid metabolism in the brain.
- **Huntington's disease** has an autosomal dominant inheritance pattern that results in mental and physical deterioration, usually beginning in adulthood.

SUMMARY WITH KEY TERMS *(continued)*

- **Hemophilia A** is an X-linked recessive disorder that results in a defect in a blood component required for clotting.

7 Briefly discuss the process of gene therapy, including some of its technical challenges.

- In **gene therapy,** the normal allele is cloned and the DNA introduced into certain body cells, where its expression may be sufficient to yield a normal phenotype.
- One technical challenge in gene therapy is finding a safe, effective **vector,** usually a virus, to deliver the gene of interest into the cells. Ideally, the virus should infect a high percentage of the cells and do no harm, especially over the long term.

8 State the relative advantages and disadvantages of amniocentesis, chorionic villus sampling, and preimplantation genetic diagnosis in the prenatal diagnosis of human genetic abnormalities.

- In **amniocentesis,** the amniotic fluid surrounding the fetus is sampled and the fetal cells suspended in the fluid are cultured and screened for genetic defects. Amniocentesis provides results in the second trimester of pregnancy.
- In **chorionic villus sampling (CVS),** some fetal cells are removed and studied. CVS provides results in the first trimester of pregnancy but is associated with a slightly greater risk of infection and miscarriage than amniocentesis.
- In **preimplantation genetic diagnosis (PGD),** couples conceive by in vitro fertilization. The embryos are screened for one or more genetic diseases before placing a healthy embryo into the woman's uterus. PGD is not as accurate as amniocentesis or CVS, and it is more expensive.

9 Distinguish between genetic screening programs for newborns and adults, and discuss the scope and implications of genetic counseling.

- **Genetic screening** identifies individuals who might carry a serious genetic disease. Screening of newborns is the first step in preventive medicine, and screening of adults helps them make informed reproductive decisions.
- Couples who are concerned about the risk of abnormality in their children may seek **genetic counseling.** Genetic counselors provide medical and genetic information.

10 Discuss the controversies of genetic discrimination.

- **Genetic discrimination** is discrimination against an individual or family members because of differences from the "normal" genome in that individual.
- One of the most difficult issues in avoiding genetic discrimination is whether genetic information should be available to employers and to health and life insurance companies. Physicians are concerned that people at risk for a particular genetic disease might delay being tested because they fear genetic discrimination from insurers and employers.
- As human genetics assumes an increasingly important role in human society, issues of genetic privacy and the confidentiality of genetic information must be addressed.

POST-TEST

1. The most important tool in bioinformatics is (a) controlled matings (b) karyotyping (c) pedigree analysis (d) a computer (e) chorionic villus sampling
2. A diagram of a pedigree shows (a) controlled matings between members of different true-breeding strains (b) the total genetic information in human cells (c) a comparison of DNA sequences among genomes of humans and other species (d) the subtle genetic differences among unrelated people (e) the expression of genetic traits in the members of two or more generations of a family
3. The Human Genome Project (a) sequenced all the DNA in the nuclear human genome (b) was exclusively concerned with the comparisons of DNA sequences between human DNA and DNA of other species (c) customized drugs to match an individual's genetic makeup (d) searched for individuals with a genotype that might cause a serious genetic disease in them or their offspring (e) provided risk estimates on human genetic diseases
4. An abnormality in which there is one more or one fewer than the normal number of chromosomes is called a(an) (a) karyotype (b) fragile site (c) aneuploidy (d) trisomy (e) translocation
5. An individual with one extra chromosome (three of one kind) is said to be (a) monosomic (b) triploid (c) trisomic (d) consanguineous (e) true-breeding
6. An individual who is missing one chromosome, having only one member of a given pair, is said to be (a) monosomic (b) haploid (c) trisomic (d) consanguineous (e) true-breeding
7. The failure of chromosomes to separate normally during cell division is called (a) a fragile site (b) an inborn error of metabolism (c) a satellite knob (d) a translocation (e) nondisjunction
8. The transfer of a part of one chromosome to a nonhomologous chromosome is called a(an) (a) karyotype (b) inborn error of metabolism (c) pedigree (d) translocation (e) nondisjunction
9. A photomicrograph of the stained metaphase chromosomes present in a given cell is called a (a) karyotype (b) nucleotide triplet repeat (c) pedigree (d) DNA microarray (e) translocation
10. Individuals with trisomy 21, or ______________, are mentally and physically retarded and have abnormalities of the face, tongue, and eyelids. (a) Down syndrome (b) Klinefelter syndrome (c) Turner syndrome (d) Huntington's disease (e) Tay-Sachs disease
11. An inherited disorder caused by a defective or absent enzyme is called a(an) (a) karyotype (b) trisomy (c) reciprocal translocation (d) inborn error of metabolism (e) aneuploidy
12. In ______________, a genetic mutation codes for an abnormal hemoglobin molecule that is less soluble than usual and more likely than normal to deform the shape of the red blood cell. (a) Down syndrome (b) Tay-Sachs disease (c) sickle cell anemia (d) PKU (e) hemophilia A
13. In an individual with ______________, the mucus is abnormally viscous and tends to plug the ducts of the pancreas and liver and to accumulate in the lungs. (a) Down syndrome (b) Tay-Sachs disease (c) sickle cell anemia (d) PKU (e) cystic fibrosis

14. During this procedure, a sample of the fluid that surrounds the fetus is obtained by inserting a needle through the walls of the abdomen and uterus. (a) DNA marker (b) chorionic villus sampling (c) ultrasound imaging (d) preimplantation genetic diagnosis (e) amniocentesis
15. For which of the following situations would a genetic counselor *not* recommend prenatal diagnosis involving amniocentesis or chorionic villus sampling? (a) an increased risk of a chromosomal abnormality (b) an increased risk of a single-locus (Mendelian) disease (c) an increased risk of a spinal cord defect (d) a desire to know the sex of the fetus (e) a pregnant woman is older than 35
16. A DNA microarray (a) represents the totality of genetic information in human cells (b) can compare the activities of thousands of genes in normal and diseased cells (c) is the study of the role of chromosomes in inheritance (d) is the chromosome composition of an individual (e) charts the transmission of genetic traits within a family
17. Examine the following pedigrees, and decide whether each disorder is most likely inherited by an autosomal recessive, an autosomal dominant, or an X-linked recessive allele. Determine the probable genotypes for all individuals shown.

(a) 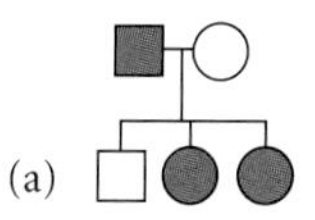

(b) 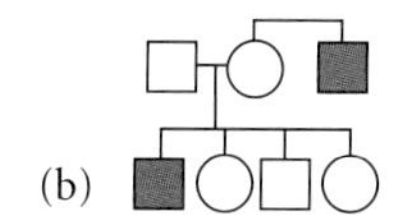

(c) 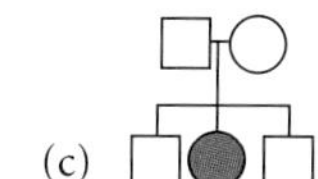

18. Complete the table by checking the correct box for each genetic disorder.

Disease	Chromosome Abnormality	Autosomal Recessive	Autosomal Dominant	X-Linked Recessive
Down syndrome				
Tay-Sachs disease				
Phenylketonuria				
Hemophilia A				
Sickle cell anemia				
Turner syndrome				
Huntington's disease				
Klinefelter syndrome				
Cri du chat syndrome				
Fragile X syndrome				

CRITICAL THINKING

1. Imagine you're a genetic counselor. What advice or suggestions might you give in the following situations?
   a. A couple has come for advice because the woman had a sister who died of Tay-Sachs disease.
   b. A young man and woman who are not related are engaged to be married. However, they have learned that the man's parents are first cousins. They are worried about the possibility of increased risk of genetic defects in their own children.
   c. A young woman's paternal uncle (her father's brother) has hemophilia A. Her father is free of the disease, and there has never been a case of hemophilia A in her mother's family. Should she be concerned about the possibility of hemophilia A in her own children?
   d. A 20-year-old man is seeking counseling because his father was recently diagnosed with Huntington's disease.
2. A common belief about human genetics is that an individual's genes alone determine his or her destiny. Explain why this is a misconception.

- Visit our Web site at **http://biology.brookscole.com/solomon7** for links to chapter-related resources on the World Wide Web. Additional online materials relating to this chapter can also be found on our Web site.

BIOLOGY NOW RESOURCES

Active Figure

15-5: Normal and abnormal karyotypes

Preparing for an exam? Take a diagnostic test on your BiologyNow CD-ROM.

Post-Test Answers

1. d	2. e	3. a	4. c
5. c	6. a	7. e	8. d
9. a	10. a	11. d	12. c
13. e	14. e	15. d	16. b

17. (a) autosomal recessive, autosomal dominant, or X-linked recessive (cannot be more precise with the information given); (b) autosomal recessive or X-linked recessive; (c) autosomal recessive

16

Genes and Development

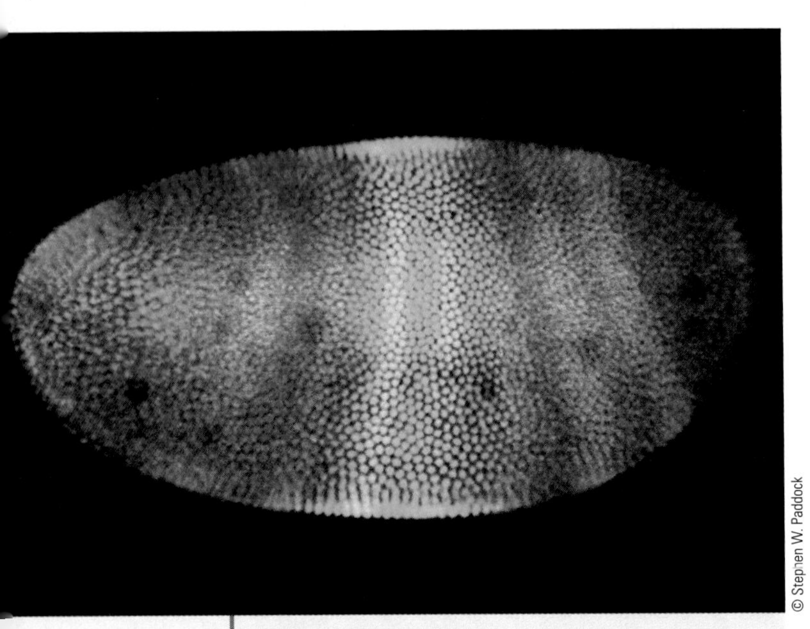

© Stephen W. Paddock

Differential gene expression in the model organism *Drosophila*. The bands of color show the production of several different proteins in different locations in this developing embryo.

CHAPTER OUTLINE

- **Cell Differentiation and Nuclear Equivalence**
- **The Genetic Control of Development**
- **Cancer and Cell Development**

The study of **development,** which is broadly defined as all the changes that occur in the life of an individual, encompasses some of the most fascinating and difficult problems in biology today. Of particular interest is the process by which cells specialize and organize into a complex organism. During the many cell divisions required for a single cell to develop into a multicellular organism, groups of cells become gradually committed to specific patterns of gene activity through the process of **cell determination.** The final step leading to cell specialization is **cell differentiation.** A differentiated cell has a characteristic appearance and characteristic activities.

An even more intriguing part of the developmental puzzle is the building of the body. In **morphogenesis,** the development of form, cells in specific locations differentiate and become spatially organized into recognizable structures. Morphogenesis proceeds through the multistep process of **pattern formation,** which includes signaling between cells, changes in cell shape, and cell migrations.

Until the late 1970s, biologists knew little about how genes interact with signals from within the organism and from the environment to control development. Although certain genes affecting developmental pathways had been identified, their specific functions in the organism were not well understood. Unraveling genetic interactions that take place during development was an intractable problem using traditional methods. However, rapid progress in recombinant DNA technology led scientists to search for developmental mutants and to apply the most sophisticated techniques to study them.

Today scientists interested in development study a variety of genetic mutants of model organisms with altered developmental patterns. They use the tools of genetic engineering, combined with more conventional descriptive and experimental approaches, to derive fresh insights about the role of genetic information in the control of development. The organism in the photograph is a developing embryo of the fruit fly *Drosophila melanogaster.* Geneticists used the technique of **immunofluorescence,** in which a fluorescent dye is joined to an anti-

body that binds to a specific protein, enabling the protein to be localized. In this case researchers bound three different antibodies—one red, one blue, and one yellow—to three specific proteins. The patterns of colored bands indicate that different cells of the embryo have differential gene expression—that is, different genes active at the same time in development.

Drosophila has many attributes that make it unusually attractive for developmental studies. Work with *Drosophila* has profound implications for understanding both normal human development and the kinds of malfunctions that can lead to birth defects and even "normal" aging. Although fruit flies may seem to have little in common with humans, scientists are learning that many genes important in development are quite similar in a wide range of organisms. These similarities have led to new ways of unraveling evolutionary relationships, through the study of developmentally important genetic mechanisms that appear deeply rooted in the evolutionary history of multicellular organisms.

New model organisms, each with characteristics that uniquely suit it for developmental studies, are being added to the list of well-characterized experimental systems. Biologists are now learning how genes are activated, inactivated, and modified to control development. These activities involve the interaction of batteries of master **regulatory genes,** which turn the transcription of other genes on or off. Eventually scientists expect to understand not only how differentiation and morphogenesis are controlled but also how the basic control systems have evolved. The identification of certain features common to many organisms, such as homeobox genes (discussed in the chapter), makes the task easier, but the work has just begun. Many interactions remain to be explored, and many stunning surprises await us. ■

CELL DIFFERENTIATION AND NUCLEAR EQUIVALENCE

Learning Objectives

1. Distinguish between cell determination and cell differentiation, and between nuclear equivalence and totipotency.
2. Define *stem cells,* and describe some of the promising areas of research involving stem cells.
3. Point out some of the known exceptions to the general principle of nuclear equivalence.

The human body, like that of other vertebrates, contains more than 200 recognizably different types of cells (Fig. 16-1). Combinations of these specialized cells, known as **differentiated cells,** are organized into diverse and complex structures—such as the eye, hand, and brain—each capable of carrying out many sophisticated activities. Most remarkable of all is the fact that all the structures of the body and the different cells within them descend from a single **zygote,** a fertilized egg.

All multicellular organisms undergo complex patterns of development. The root cells of plants, for example, have structures and functions very different from those of the various types of cells located in leaves. Diversity is also found at the molecular level; most strikingly, each type of plant or animal cell makes a highly specific set of proteins. In some cases, such as the protein hemoglobin in red blood cells, one cell-specific protein may make up more than 90% of the cell's total mass of protein. Other cells may have a complement of cell-specific proteins, each of which is present in small amounts but still plays an essential role. However, because certain proteins are required in every type of cell (all cells, for example, require the same enzymes for glycolysis), cell-specific proteins usually make up only a fraction of the total number of different kinds of proteins.

When researchers first discovered that each type of differentiated cell makes a unique set of proteins, some scientists hypothesized that each group of cells loses the genes it does not need and retains only those required. However, this does not generally seem true. According to the principle of **nuclear equivalence,** the nuclei of essentially all differentiated adult cells of an individual are genetically (though not necessarily metabolically) identical to each other and to the nucleus of the zygote from which they descended. This means that virtually all somatic cells in an adult have the same genes. However, different cells express different subsets of these genes. **Somatic cells** are all the cells of the body other than **germ line cells,** which ultimately give rise to a new generation. In animals, germ line cells—whose descendants ultimately undergo meiosis and differentiate into gametes—are generally set aside early in development. In plants, the difference between somatic cells and germ line cells is not as distinct, and the determination that certain cells undergo meiosis is made much later in development.

The evidence for nuclear equivalence comes from cases in which differentiated cells or their nuclei have been found to retain the potential of directing the development of the entire organism. Such cells or nuclei are said to exhibit **totipotency.**

A totipotent nucleus contains all the instructions for development

PROCESS OF SCIENCE

In plants, at least some differentiated cells can be induced to become the equivalent of embryonic cells. Biologists use *tissue culture techniques* to isolate individual cells from certain plants and to allow them to grow in a nutrient medium.

In the 1950s F.C. Steward and his coworkers at Cornell University conducted some of the first experiments investigating cell totipotency in plants (Fig. 16-2). They induced root cells from a carrot to divide in a liquid nutrient medium, forming groups of cells called *embryoid* (embryo-like) *bodies.* These clumps of dividing cells were then transferred to an agar medium, which provided nutrients and a solid supporting structure for the developing plant cells. Some of the cells of the embryoid bodies gave rise to roots, stems, and leaves. The resulting small plants, called *plantlets* to distinguish them from true seedlings,

FIGURE 16-1 **Vertebrate cell lineages.**

Repeated divisions of the zygote *(bottom of figure)* result in the establishment of tissues containing groups of specialized cells. Germ line cells (cells that produce the gametes) are set aside early in development. Somatic cells progress along various developmental pathways, undergoing a series of commitments that progressively determine their fates.

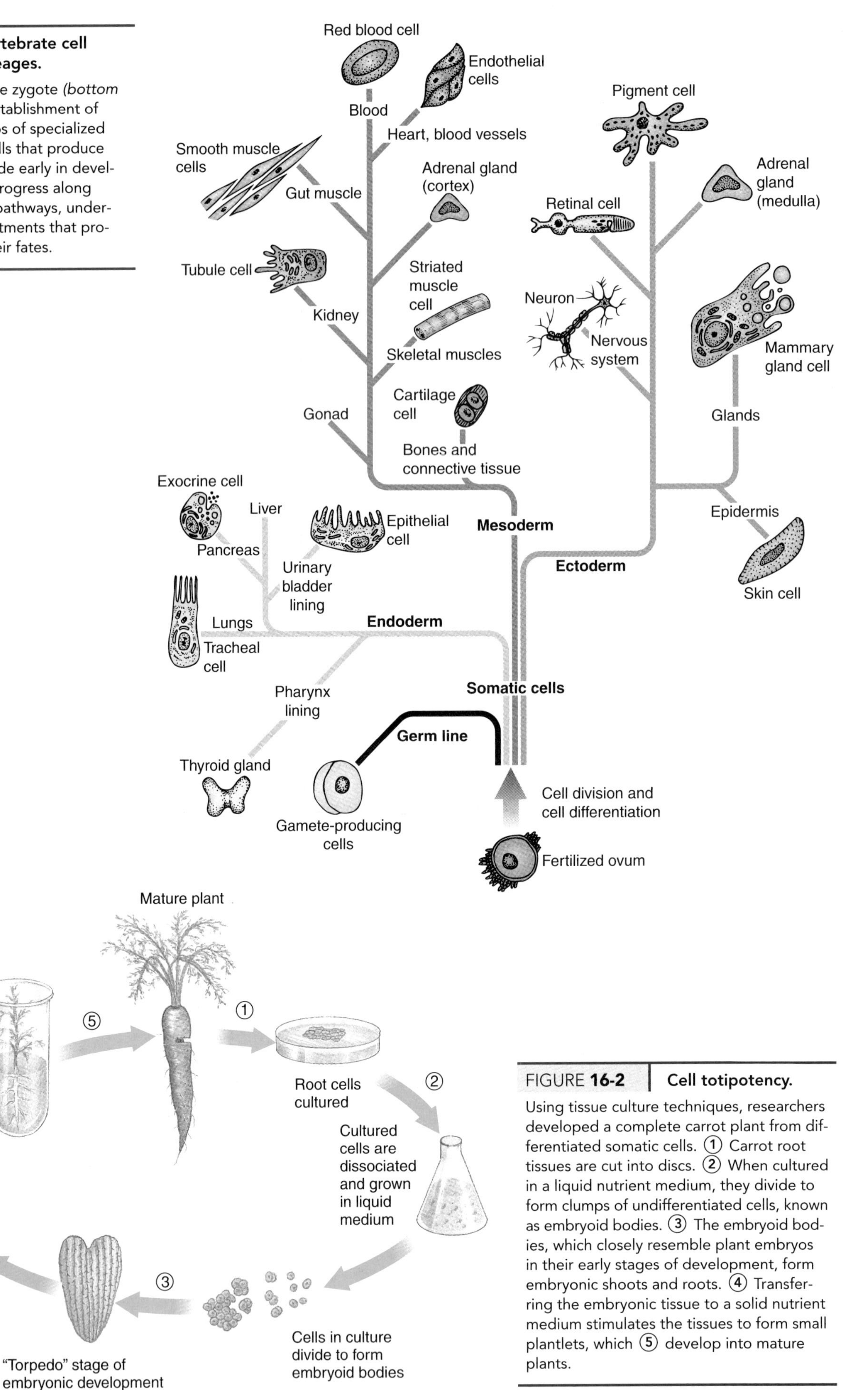

FIGURE 16-2 **Cell totipotency.**

Using tissue culture techniques, researchers developed a complete carrot plant from differentiated somatic cells. ① Carrot root tissues are cut into discs. ② When cultured in a liquid nutrient medium, they divide to form clumps of undifferentiated cells, known as embryoid bodies. ③ The embryoid bodies, which closely resemble plant embryos in their early stages of development, form embryonic shoots and roots. ④ Transferring the embryonic tissue to a solid nutrient medium stimulates the tissues to form small plantlets, which ⑤ develop into mature plants.

were then transplanted to soil, where they ultimately developed into adult plants capable of producing flowers and viable seeds. If these plants are all derived from the same parent plant, they are essentially genetically alike and therefore constitute a **clone.** The methods of plant tissue culture are now extensively used to produce genetically engineered plants, because they enable researchers to regenerate whole plants from individual cells that have incorporated recombinant DNA molecules (see Chapter 36, *Focus On: Cell and Tissue Culture*).

Similar experiments have been attempted with animal cells, but thus far researchers have not been able to induce a fully differentiated somatic cell to behave like a zygote. Instead, they have tested whether steps in the process of determination are reversible by transplanting the *nucleus* of a cell in a relatively late stage of development into an egg cell that has been *enucleated* (that is, its own nucleus has been destroyed).

In the 1950s Robert Briggs and Thomas J. King of the Institute for Cancer Research in Pennsylvania pioneered *nuclear transplantation experiments.* They transplanted nuclei from frog cells at different stages of development into egg cells whose nuclei had been removed. Some of the transplants proceeded normally through several developmental stages, and a few even developed into normal tadpoles. As a rule, the nuclei transplanted from cells at earlier stages were most likely to support development to the tadpole stage. As the fate of the cells became more and more determined, the probability quickly declined that a transplanted nucleus could control normal development.

British biologist John Gurdon carried out experiments on nuclear transplantation in frogs during the 1960s. In a few cases he demonstrated that nuclei isolated from the intestinal epithelial cells of a tadpole directed development up to the tadpole stage (Fig. 16-3). This occurred infrequently (about 1.5% of the time); however, in these kinds of experiments success counts more than failure. Therefore, he could safely conclude that at least some nuclei of differentiated animal cells are in fact totipotent.

For many years these successes with frogs could not be repeated with mammalian embryos, leading many developmental biologists to conclude that some fundamental feature of mammalian reproductive biology might be an impenetrable barrier to mammalian cloning. This perception changed markedly in 1996 and 1997 with the first reports of the birth of cloned mammals.

The first cloned mammal was a sheep

PROCESS OF SCIENCE

In 1996, Ian Wilmut and his coworkers at the Roslin Institute in Edinburgh, Scotland, reported that they had succeeded in cloning sheep, using nuclei from early sheep embryos (the blastocyst stage; see Chapter 49). These scientists received worldwide attention in early 1997 when they announced the birth of a lamb, named Dolly (after the singer Dolly Parton). Dolly's genetic material was derived from a cultured mammary gland

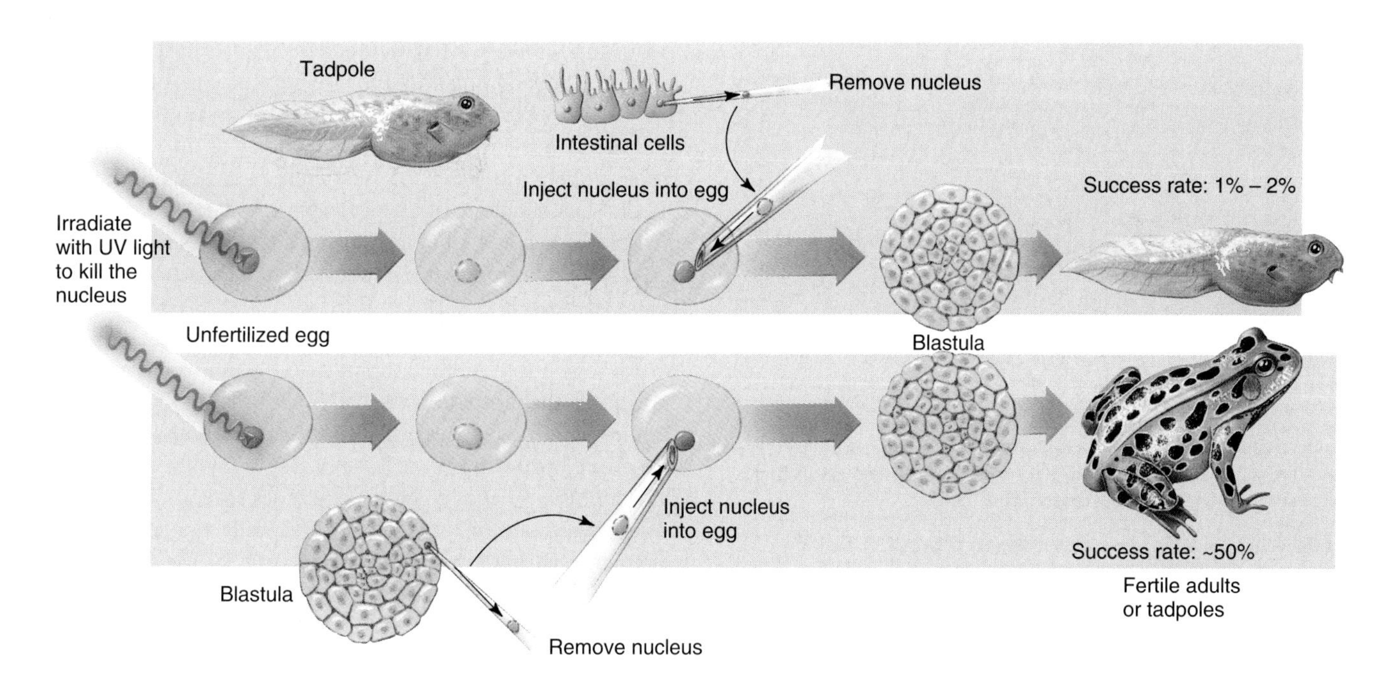

FIGURE **16-3** **Nuclear totipotency.**

In nuclear transplantation experiments conducted during the 1960s on frogs, biologists injected the nuclei of differentiated cells at different stages of development into eggs whose own nuclei were destroyed by ultraviolet radiation. *Upper panel:* Using nuclei from tadpole intestinal cells (a relatively late developmental stage), normal development proceeded to the tadpole stage in a small number (1% to 2%) of trials. This result indicated that the genes for programming development up to that point were still present and appropriately activated. *Lower panel:* When researchers used nuclei from earlier developmental stages, the success rate improved dramatically. With a nucleus from a blastula (a ball of about 1000 cells), in about half the cases the transplanted nucleus could successfully program normal development to form a tadpole or fertile adult.

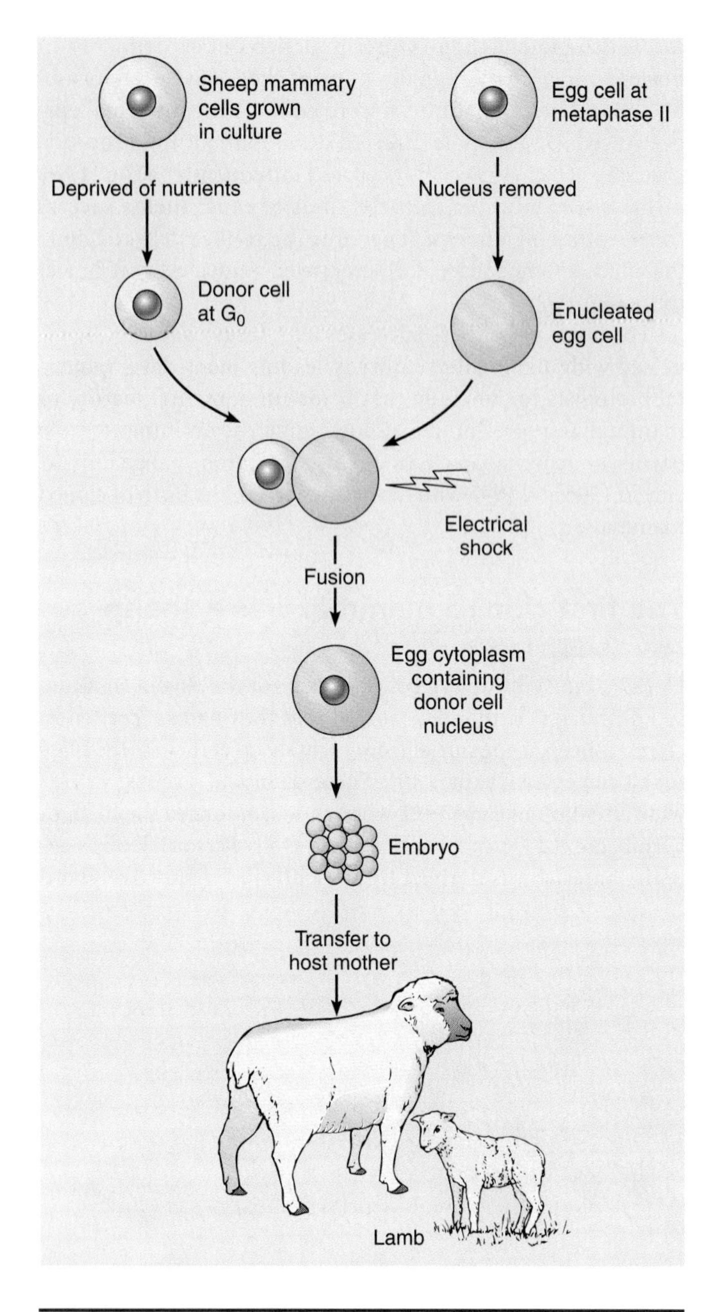

ACTIVE FIGURE **16-4** | **Mammalian cloning.**

An embryo produced by fusing a cultured adult sheep mammary cell with an enucleated sheep's egg is implanted into the uterus of a host mother. The embryo develops into a lamb.

BiologyNow™ Learn more about **sheep cloning** by clicking on this figure on your BiologyNow CD-ROM.

cell, from an adult sheep, that was fused with an enucleated sheep's egg. The resulting cell divided and developed into an embryo that was then cultured in vitro until it reached a stage at which it could be transferred to a host mother (Fig. 16-4). Not surprisingly, the overall success rate was low: Out of 277 fused cells, only 29 developed into embryos that could be transferred, and Dolly was the only live lamb produced. These researchers have also produced cloned lambs derived from fetal cells.

Why did Wilmut's team succeed when so many other researchers had failed? Applying the basic principles of cell biology, they recognized that the **cell cycles** of the egg cytoplasm and the donor nucleus were not synchronous. The egg cell is arrested at metaphase II of meiosis, whereas the actively growing donor somatic cell is usually in the DNA synthesis phase (S), or in G_2. By withholding certain nutrients from the mammary gland cells used as donors, they cause these cells to enter a nondividing state referred to as G_0 (see Chapter 9). This had the effect of synchronizing the cell cycles of the donor nucleus and the egg. They then used an electrical shock to fuse the donor cell with the egg and initiate embryo development.

Although an extremely high level of technical expertise is required, these and other researchers have modified and extended these techniques to produce cloned calves, goats, pigs, and mice. The list of mammals successfully cloned will likely continue to grow. However, the success rate for each set of trials continues to be low, around 1% to 2%, and the incidence of genetic defects is high. Dolly was euthanized at age 6 because she was suffering from a virus-induced lung cancer that infected several sheep where she was housed. However, she developed arthritis at 5½ years, which is relatively young for a sheep to have this degenerative disease. Some biologists speculate that using adult genetic material to produce a clone might produce an animal with prematurely old cells (see discussion of telomeres and cell aging in Chapter 11). Further research may provide some answers to this potential problem.

The production of **transgenic organisms,** in which foreign genes have been incorporated, continues to be the main focus of cloning research (see Chapter 14). Researchers are actively pursuing new techniques to improve the efficiency of the cloning process. Only then will it be possible to produce large numbers of cloned transgenic animals for a variety of uses, such as increasing the populations of endangered species. For example, the first healthy clone of an endangered species, a wild relative of cattle known as a *banteng,* was born in 2003. The nucleus for this clone came from a frozen skin cell of a banteng that died in 1980 at the San Diego Zoo.

Cloning research continues to fuel an ongoing debate regarding the potential for human cloning and its ethical implications. In the United States, the National Bioethics Advisory Commission has been established to study this and other questions. In considering these issues, it is important to recognize that *cloning* is a broad term that includes several different processes involved in producing biological cells, tissues, organs, or organisms. **Human reproductive cloning** has the goal of making a newborn human that is genetically identical to another, usually adult, human. It would involve placing a human embryo produced by a process other than fertilization into a woman's body. Many countries are opposed to human reproductive cloning.

In contrast, **human therapeutic cloning** involves duplication of human cells for scientific study or medical purposes; no newborn human is formed. In human therapeutic cloning, scientists would take the nucleus from—for example, a skin cell—and place it in an enucleated egg cell, which would then be treated to develop into an embryo. *Stem cells* could be extracted from the embryo to provide a supply of replacement tissues for revolutionary medical procedures.

Stem cells divide and give rise to differentiated cells

Stem cells are undifferentiated cells that can divide to produce differentiated descendants, yet retain the ability to divide to reproduce themselves, thereby maintaining the stem cell population. The most versatile stem cells are totipotent and have the potential to give rise to all tissues of the body. Other stem cells, known as **pluripotent stem cells,** appear more specialized; they can give rise to many, but not all, of the types of cells in an organism. For example, neural stem cells are pluripotent and differentiate to form all types of brain cells, and stem cells in the bone marrow form various types of blood cells. However, recent studies have shown that even specialized stem cells may be more versatile than once thought. For example, neural stem cells form blood cells when transplanted into bone marrow, and bone marrow stem cells can differentiate into muscle cells.

Stem cells are potential sources for cell transplantation into patients to treat serious degenerative conditions. For example, Parkinson's disease results from a progressive loss of cells that produce the neurotransmitter dopamine in a specific region of the brain. Transplantation of stem cells that have been induced to differentiate as dopamine-producing cells holds great promise as an effective long-term treatment. Similarly, stem cells may become a source of insulin-producing cells for transplantation into the pancreas of individuals with diabetes mellitus. Stem cells might also provide replacement nerve cells in people with spinal cord injury or other types of neurological damage.

Researchers ultimately hope to establish lines of human pluripotent stem cells that can grow indefinitely in culture, be induced to differentiate under controlled conditions and stably maintain their differentiated state, and be manipulated genetically. They particularly want to develop embryonic stem cell lines from patients with cancer, diabetes, cardiovascular disease, and neurogenerative disorders (such as Parkinson's disease); such cell lines would be invaluable in research on these disorders. Although work on stem cells in mice and other mammals has been conducted for many years, similar studies in humans have progressed slowly, despite the great promise of stem cells as a therapeutic tool.

Private companies currently fund these studies, largely because of government restrictions on public funding due to ethical considerations related to the origins of stem cells. Thus far, the only known source of totipotent stem cells is early human embryos left over from **in vitro fertilization** (see Chapter 48). Fetal tissues are also a source of some types of specialized stem cells. For example, pluripotent blood stem cells can be obtained from a newborn's umbilical cord. More recent findings that certain types of specialized stem cells are also present in tissues of adult mice, as well as human adults and children, may alleviate some ethical concerns. Unfortunately, these cells are rare and lack many of the advantages of embryonic stem cells.

Most cell differences are due to differential gene expression

Because genes do not seem to be lost regularly during development (and thus nuclear equivalence is present in different cell types), differences in the molecular composition of cells must be regulated by the activities of different genes. The process of developmental gene regulation is often referred to as **differential gene expression.** As discussed in Chapter 13, the expression of eukaryotic genes is regulated in many different ways and at many levels. For example, a particular enzyme may be produced in an inactive form and then be activated later. However, much of the regulation that is important in development occurs at the transcriptional level. The transcription of certain sets of genes is repressed, whereas that of other sets is activated. Even the expression of genes that are *constitutive*—that is, constantly transcribed—is regulated during development so that the *quantity* of each product varies from one tissue type to another.

We can think of differentiation as a series of pathways leading from a single cell to cells in each of the different specialized tissues, arranged in an appropriate pattern. At times a cell makes genetic commitments to the developmental path its descendants will follow. These commitments gradually restrict the development of the descendants to a limited set of final tissue types. Determination, then, is a progressive fixation of the fate of a cell's descendants.

As the development of a cell becomes determined along a differentiation pathway, its physical appearance may not change significantly. Nevertheless, when a stage of determination is complete, the changes in the cell usually become self-perpetuating and are not easily reversed. Cell differentiation is usually the last stage in the developmental process. At this stage, a precursor cell becomes structurally and functionally recognizable as a bone cell, for example, and its pattern of gene activity differs from that of a nerve cell, or any other cell type.

DNA microarrays track gene expression

PROCESS OF SCIENCE

The science of determining the roles of genes in cells, or **functional genomics,** includes the analysis of patterns of gene expression in different cell types. One approach is the use of **DNA microarrays,** a powerful tool for research as well as for diagnosing and treating human diseases (see Fig. 15-3). Thousands of tiny spots of DNA, known as *microdots,* are spotted on a chip, which is usually a glass microscope slide. Each microdot contains many single-stranded copies of a fragment of DNA from a particular tissue, and collectively all the microdots on a chip are a microarray that contains representative segments of all or most of the DNA in that organism's genome.

The researchers then extract RNA from cells they wish to study, and they synthesize DNA complementary to it (cDNA) using *reverse transcriptase* (see Chapter 14). The single-stranded cDNA molecules are tagged with a fluorescent dye and incubated with the DNA on the chip under conditions that promote complementary base pairing. If a particular segment of DNA on the microarray corresponds to a gene that was actively transcribed in the cells being studied, the corresponding cDNA molecules bind to that microdot, causing it to fluoresce very brightly. Conversely, a microdot that fluoresces dimly or not at all is one that contains DNA that was not actively transcribed in those cells. Instruments that detect the fluorescence patterns then scan the microarrays, and computers analyze the data. These methods allow researchers to compare patterns of gene expression, as measured by RNA synthesis, in various cell types, or in the same cell type under different conditions.

Some exceptions to the principle of nuclear equivalence have been found

Although the principle of nuclear equivalence applies to most cells in multicellular organisms, certain types of developmental regulation can involve physical changes in DNA. Such changes in the structure of the genome are not common. Exceptions to nuclear equivalence include genomic rearrangements and gene amplification.

The activity of some genes may be modified during development by different types of **genomic rearrangements** that lead to actual physical changes in the structure of the gene. In some cases, parts of genes are rearranged to make new coding sequences. Genomic rearrangement is an important mechanism for the development of the immune system. Cells of the immune system rearrange their genetic elements into functional genes that encode a diversity of antibodies (see Chapter 43). Genomic rearrangement of the several hundred known genetic elements has the potential to produce more than 200 million different antibody molecules!

Some gene products are required in such large quantities during development that a single copy of a gene cannot be transcribed, nor can its mRNA be translated, rapidly enough to meet the needs of the developing cells. In some cases, the number of gene copies may be increased, through a process known as **gene amplification,** to meet the demand. For example, the *Drosophila* chorion (eggshell) gene product is a protein made specifically in cells of the female insect's reproductive tract. These cells make massive amounts of the protein that envelops and protects the zygote. Amplifying the chorion protein gene by DNA replication meets the demand for chorion mRNA. In other words, the DNA in that small region of the chromosome is copied many times (Fig. 16-5). In other cells of the insect body, however, the gene appears to exist as a single copy in the chromosome.

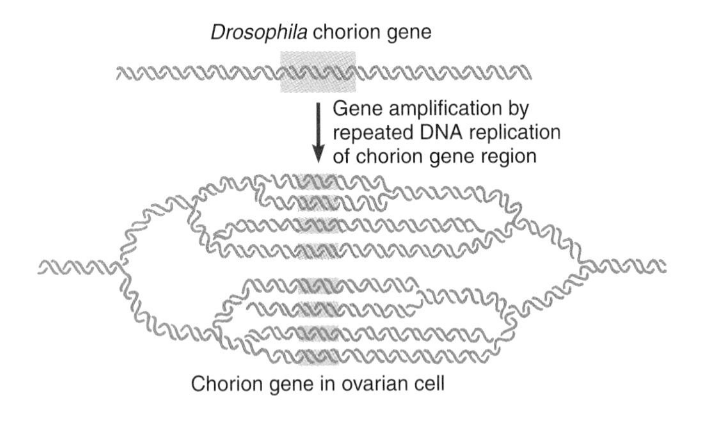

FIGURE **16-5** **Gene amplification.**

In *Drosophila,* multiple replications of a small region of the chromosome amplify the chorion (eggshell) protein genes. Replication is initiated at a discrete chromosome origin of replication *(pink box)* for each copy of the gene that is produced. Replication is randomly terminated, resulting in a series of forked structures in the chromosome.

Review

- What lines of evidence support the principle of nuclear equivalence?
- Why was Wilmut's team successful in mammalian cloning where others had failed?
- What are stem cells?
- Under what conditions do cells use gene amplification?

BiologyNow™ Assess your understanding of **cell differentiation and nuclear equivalence** by taking the pretest on your BiologyNow CD-ROM.

THE GENETIC CONTROL OF DEVELOPMENT

Learning Objectives

4 Indicate the features of *Drosophila melanogaster, Caenorhabditis elegans, Mus musculus,* and *Arabidopsis thaliana* that have made these organisms valuable models in developmental genetics.

5 Distinguish among maternal effect genes, segmentation genes, and homeotic genes in *Drosophila.*

6 Explain the relationship between transcription factors and genes that control development.

7 Define *induction* and *apoptosis,* and give examples of the roles they play in development.

Development has been an important area of research for many years, and researchers have spent considerable time studying the development of invertebrate and vertebrate animals. By investigating patterns of morphogenesis in different species, researchers have identified similarities, as well as differences, in the basic plan of development from a zygote to an adult in organisms ranging from the sea urchin to mammals (see Chapter 49).

In addition to descriptive studies, many classic experiments have demonstrated how groups of cells differentiate and undergo pattern formation. Researchers have developed elaborate screening programs to detect mutations that let them identify many developmental genes in both plants and animals. They then exploit molecular genetic techniques and other sophisticated methodologies to determine how those genes work and how they interact to coordinate developmental processes.

In studies of the genetic control of development, the choice of organism to use as an experimental model is important. One of the most powerful approaches involves isolating mutants with arrested or abnormal development at a particular stage. Not all organisms have useful characteristics that allow researchers to isolate and maintain developmental mutants for future study. Geneticists so thoroughly understand the genetics of the fruit fly, *Drosophila melanogaster,* that this organism has become one of the most important systems for such studies. Other organisms—the nematode worm, *Caenorhabditis elegans;* the laboratory mouse, *Mus musculus;* certain plants, including *Arabidopsis thaliana,* a tiny weed with many convenient features; and some simple eukaryotes, such as the yeast *Saccharomyces cerevisiae*—have also become important models in developmental genetics. Each of these organisms has attributes that make it particularly useful for examining certain aspects of development.

The maternal genome controls early development in *Drosophila melanogaster*

Undoubtedly the most extensive and spectacular examples of genes that control development have been identified in the fruit fly *Drosophila.* The *Drosophila* genome sequence, which was completed in late 1999, includes about 13,600 protein-coding genes. One of the traditional advantages of *Drosophila* as a research organism is the abundance of mutant alleles, including those of developmental genes, available for study and the relative ease with which a new mutation is mapped on the chromosomes.

Genetic analysis in *Drosophila* is greatly facilitated by **polytene chromosomes,** found in certain fly larval tissues with large, metabolically active cells, including the salivary glands. Polytene ("many-stranded") chromosomes are formed when the DNA replicates many times but without mitosis and cytokinesis. A typical polytene chromosome may consist of more than 1000 DNA double helices (along with associated histones and other proteins) aligned side by side. Polytene chromosomes are therefore quite large and show a pattern of bands that is very useful in assigning a particular gene to a specific locus on the chromosome. When a gene is active, the chromosome band in which it resides uncoils and forms a "puff," a site of intense RNA synthesis.

Studies of *Drosophila* are also facilitated by the fact that foreign DNA injected into eggs becomes incorporated into the fly's DNA. This process is called **transformation,** analogous to transformation in prokaryotes.

The Drosophila *life cycle includes egg, larval, pupal, and adult stages*

Development in *Drosophila* consists of several distinct stages (Fig. 16-6). After the egg is fertilized, a period of embryogenesis occurs during which the zygote develops into a sexually immature form known as a **larva** (pl., *larvae*). After hatching from the egg, each larva undergoes several molts (shedding of the external covering or cuticle). Each molt results in a size increase until the larva is ready to become a **pupa.** Pupation involves a molt and the hardening of the new external cuticle, so that the pupa is completely encased. The insect then undergoes **metamorphosis,** a complete change in form. During that time, most of the larval tissues degenerate and other tissues differentiate to form the body parts of the sexually mature adult fly.

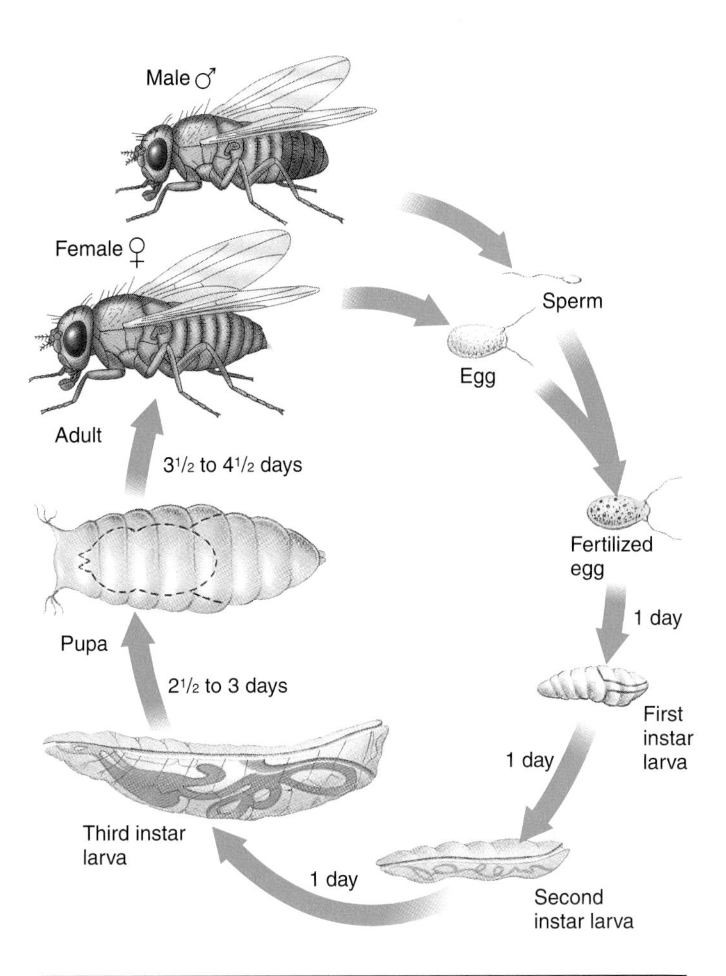

FIGURE **16-6** **The life cycle of *Drosophila.***
As it develops from a fertilized egg to a sexually mature adult fly, a fruit fly passes through several stages. It takes about 12 days, at 25°C, to complete the life cycle. A juvenile form (larva) hatches from the egg and undergoes a series of molts as it grows. The periods between molts are called instars. The dotted lines within the pupa represent the animal undergoing metamorphosis.

The larvae are wormlike in appearance and look nothing like the adult flies. However, very early in embryogenesis of the developing larvae, precursor cells of many of the adult structures are organized as relatively undifferentiated, paired structures called **imaginal discs.** The name comes from *imago,* the adult form of the insect. Each imaginal disc occupies a definite position in the larva and will form a specific structure, such as a wing or a leg, in the adult body (Fig. 16-7). The discs are formed by the time embryogenesis is complete and the larva is ready to begin feeding. In some respects the larva is a developmental stage that feeds and nurtures the precursor cells that give rise to the adult fly, which is the only form that reproduces.

The organization of the precursors of the adult structures, including the imaginal discs, is under genetic control. Thus far, more than 50 genes have been identified that specify the formation of the imaginal discs, their positions within the larva, and their ultimate functions within the adult fly. Those genes were identified through mutations that either prevent certain discs from forming or alter their structure or ultimate fate.

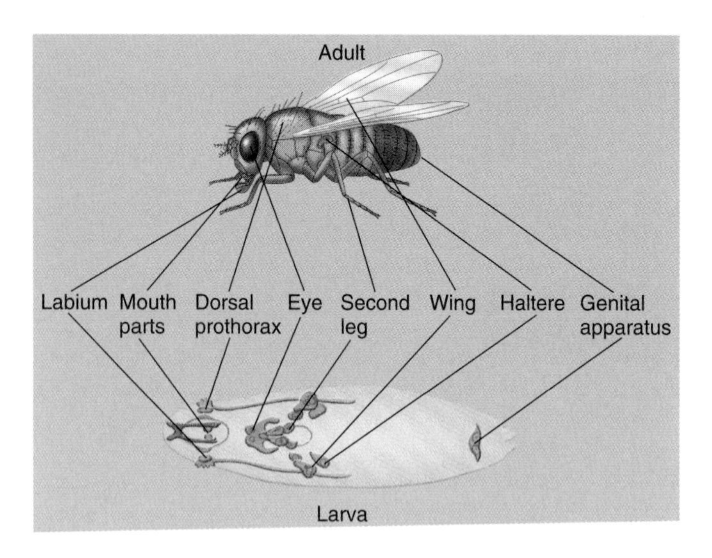

FIGURE **16-7** | **The location of imaginal discs.**
Each pair of discs in a *Drosophila* larva *(bottom)* develops into a specific pair of structures in the adult fly.

Drosophila *developmental mutations affect the body plan*

Many developmental mutations have been identified in *Drosophila.* Researchers have examined their effects on development in various combinations and have studied them extensively at the molecular level. In our discussion we pay particular attention to mutations that affect the segmented body plan of the organism, in both the larva and the adult.

Early *Drosophila* development occurs in the following way. The structure of the egg becomes organized as it develops in the ovary of the female. Stores of mRNA, along with yolk proteins and other cytoplasmic molecules, pass from the surrounding maternal cells into the egg. Immediately after fertilization, the zygote nucleus divides, beginning a series of 13 mitotic divisions.

Each division takes only 5 or 10 minutes, which means that the DNA in the nuclei replicates constantly and at a very rapid rate. During that time, the nuclei do not synthesize RNA. Cytokinesis does not take place, and the nuclei produced by the first seven divisions remain at the center of the embryo until the eighth division occurs. At that time, most of the nuclei migrate out from the center and become localized at the periphery of the embryo. This is known as the *syncytial blastoderm* stage, because the nuclei are not surrounded by individual plasma membranes. (A *syncytium* is a structure containing many nuclei residing in a common cytoplasm.) Subsequently, plasma membranes form, and the embryo becomes known as a *cellular blastoderm.*

Maternal effect genes The genes that organize the structure of the egg cell are called **maternal effect genes.** These genes in the surrounding maternal tissues are transcribed to produce mRNA molecules that are transported into the developing egg. Analysis of mutant flies with defective maternal effect genes has revealed that many of the genes are involved in establishing the polarity of the embryo—such as what part of the embryo will become the head and what part will become the tail. Polarity dictates those parts of the egg that are dorsal or ventral and those that are anterior or posterior (see Chapter 28); thus, these genes are known as *egg polarity genes.*

Figure 16-8a illustrates concentration gradients for two specific maternal mRNA molecules in the very early embryo. These mRNA transcripts of some of the maternal effect genes are identified by their ability to hybridize with radioactive DNA probes derived from cloned genes. Alternatively, researchers use fluorescently tagged antibodies (as in the chapter opening photograph) to bind to specific protein products of the maternal effect genes. These protein gradients organize the early pattern of development in the embryo by determining anterior and posterior regions.

A combination of protein gradients may provide positional information that specifies the fate—that is, the developmental path—of each nucleus within the embryo. For example, mutations in certain maternal effect genes cause the absence of specific signals, resulting in an embryo with two heads or two posterior ends.

In many cases, injecting normal maternal mRNA into the mutant embryo reverses the phenotype associated with a mutation in a given maternal effect gene. When this is done, the fly

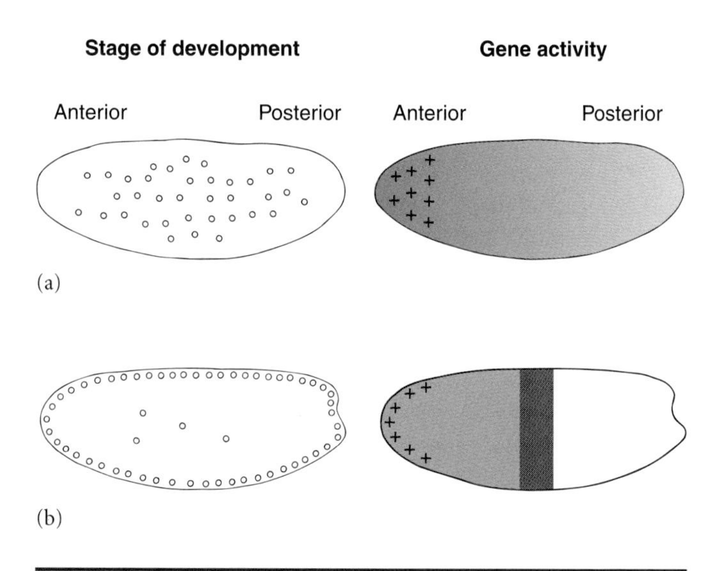

FIGURE **16-8** | **Early development in *Drosophila.***
Longitudinal sections on the left show two early stages of development of the *Drosophila* embryo. These are matched with the simplified patterns of gene activity at each stage on the right. **(a)** At 1.25 hours after fertilization, the embryo consists of a common cytoplasm with about 128 nuclei *(small circles).* Maternal effect genes divide the embryo into anterior and posterior sections. The crosses in the anterior region represent mRNA molecules transcribed from one maternal effect gene. The pink shading represents a different maternal mRNA molecule that is more concentrated in the anterior region. **(b)** At 2 hours after fertilization, about 1500 nuclei have migrated into the periphery of the embryo and started to make their own mRNA molecules. The gap genes divide the embryo into anterior, middle, and posterior sections. The maternal RNA shown in pink is now transcribed from segmentation genes *(crosses)* in nuclei in the anterior region. The mRNA from another gap gene is transcribed in the middle of the embryo *(black region). (Adapted from M.E. Akam, "The Molecular Basis for Metameric Pattern in the* Drosophila *embryo,"* Development, *Vol. 101, 1987)*

TABLE 16-1 Classes of Genes Involved in Pattern Formation of Embryonic Segments in *Drosophila*

Class of Gene	Site of Gene Activity	Effects of Mutant Alleles and Proposed Functions of Genes
Maternal effect genes	Maternal tissues surrounding egg	Many maternal effect mutations alter polarity of embryo; initiate pattern formation by activating regulatory genes in nuclei in certain locations in embryo
Segmentation genes		
Gap genes	Embryo	Mutant alleles cause one or more segments to be missing; some may influence activity of pair-rule genes, segment polarity genes, and homeotic genes
Pair-rule genes	Embryo	When mutated, cause alternate segments to be missing; some may influence activity of segment polarity genes and homeotic genes
Segment polarity genes	Embryo	Mutant alleles delete part of every segment and replace it with mirror image of remaining structure; may influence activity of homeotic genes
Homeotic genes	Embryo	Homeotic mutations cause parts of fly to form structures normally formed in other segments; control identities of segments

develops normally, indicating that the gene product is needed only for a short time in the earliest stages of development.

Segmentation genes As the nuclei start to migrate to the periphery of the embryo, **segmentation genes** in those nuclei begin to produce embryonic mRNA. Thus far, geneticists have identified at least 24 segmentation genes that are responsible for generating a repeating pattern of body segments within the embryo and adult fly. Based on a study of mutant phenotypes, the segmentation genes fall into three classes—gap genes, pair-rule genes, and segment polarity genes.

Gap genes are the first set of segmentation genes to act. These genes interpret the maternal anterior–posterior information in the egg and begin organization of the body into anterior, middle, and posterior regions (Fig. 16-8b). A mutation in one of the gap genes usually causes the absence of one or more body segments in an embryo (Fig.16-9a).

The other two classes of segmentation genes do not act on small groups of body segments; instead, they affect all segments. Mutations in **pair-rule genes** delete every other segment, producing a larva with half the normal number of segments (Fig. 16-9b). Mutations in **segment polarity genes** produce segments in which one part is missing and the remaining part is duplicated as a mirror image (Fig. 16-9c). Table 16-1 summarizes the effects of the different classes of mutants.

Each segmentation gene has distinctive times and places in the embryo in which it is most active (Fig. 16-10). The observed pattern of expression of maternal effect genes and segmentation genes indicates that a progressive series of developmental events determines cells destined to form adult structures. First,

Mutant
Wild-type
(a) Gap genes

Mutant
Wild-type
(b) Pair-rule genes

Mutant
Wild-type
(c) Segment polarity genes

FIGURE 16-9 A comparison of mutations in *Drosophila* segmentation genes.

(a) Gap genes, **(b)** pair-rule genes, and **(c)** segment polarity genes control the pattern of body segments in a *Drosophila* embryo. The blue bands mark the regions in which the protein products of these genes are normally expressed in wild-type embryos. These same regions are absent in embryos in which the gene is mutated, and the resulting phenotype is characteristic of the class to which the gene belongs. *(Adapted from C. Nüsslein-Volhard and E. Wieschaus, "Mutations Affecting Segment Number and Polarity in* Drosophila,*"* Nature, *Vol. 287, 1980)*

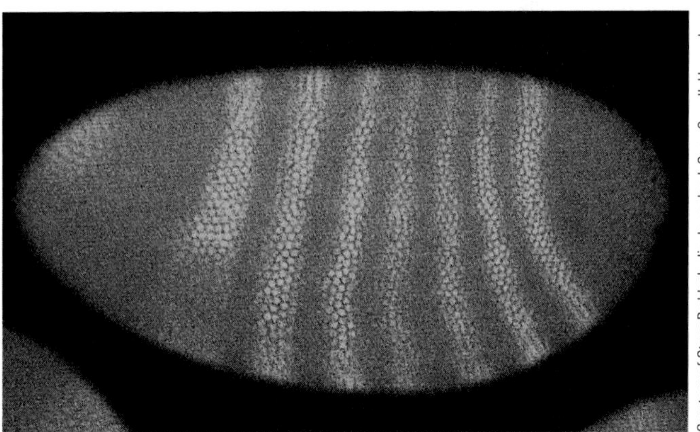

Courtesy of Steve Paddock, Jim Langeland, Sean Carroll, Howard Hughes Medical Institute, University of Wisconsin

FIGURE 16-10 Segmentation gene activity.

The bright bands in this fluorescence LM reveal the presence of mRNA transcribed from one of the pair-rule genes known as *fushi tarazu* (Japanese for "not enough segments"). When this locus is mutated, the segments of the larva that are normally derived from these bands are absent.

maternal effect genes that form gradients of morphogens in the egg determine the anterior–posterior (head-to-tail) axis and the dorsal and ventral regions of the embryo. A **morphogen** is a chemical agent that affects the differentiation of cells and the development of form. Morphogen gradients are signals that help cells determine their location within the embryo and their eventual differentiation into specialized tissues and organs.

Next, segmentation genes respond to the morphogens at each location to regulate the production of segments from the head to the posterior region. Within each segment, other genes are then activated to specify which body part that segment will become. Every cell's position is further specified with a specific "address" designated by combinations of the activities of the regulatory genes.

The segmentation genes act in sequence, with the gap genes acting first, then the pair-rule genes, and finally the segment polarity genes. In addition, the genes of the three groups interact. Each time a new group of genes acts, cells for that group become more determined in their development. As the embryo develops, each region is progressively subdivided into smaller regions.

Most segmentation genes code for **transcription factors,** DNA-binding proteins that regulate gene transcription in eukaryotic cells (see Chapter 13). For example, some segmentation genes code for a "zinc-finger" type of DNA-binding regulatory protein (see Fig. 13-8b). *Homeotic genes,* discussed next, also code for transcription factors. The fact that many of the genes involved in controlling development code for transcription factors indicates those proteins act as genetic switches regulating the expression of other genes.

Once researchers have identified proteins that function as transcription factors, they can use the purified proteins to determine the DNA target sequences to which the proteins bind. This approach has been increasingly useful in identifying additional parts of the regulatory pathway involved in different stages of development. Transcription factors also play a role in cancer (discussed later in the chapter).

Homeotic Genes After segmentation genes have established the basic pattern of segments in the fly body, **homeotic genes** specify the developmental plan for each segment. Mutations in homeotic genes cause one body part to be substituted for another and therefore produce some peculiar changes in the adult. A striking example is the *Antennapedia* mutant fly, which has legs that grow from the head where the antennae would normally be (Fig. 16-11).

Homeotic genes in *Drosophila* were originally identified by the altered phenotypes produced by mutant alleles. When geneticists analyzed the DNA sequences of several homeotic genes, they discovered a short DNA sequence of approximately 180 base pairs, which characterizes many homeotic genes as well as some other genes that play a role in development. This DNA sequence is called the **homeobox.** Each homeobox codes for a protein functional region called a **homeodomain,** consisting of 60 amino acids that form four α-helices. One of these serves as a recognition helix that binds to specific DNA sequences and thereby affects transcription. Thus the products of homeotic genes, like those of the earlier acting segmentation genes, are transcription factors. In fact, some segmentation genes also contain homeoboxes.

FIGURE 16-11 The *Antennapedia* locus.

Antennapedia mutations cause homeotic transformations in *Drosophila* in which legs or parts of legs replace the antennae. **(a)** The head of a normal fly and a fly with an *Antennapedia* mutation. **(b)** SEM of the head of a fly with an *Antennapedia* mutation.

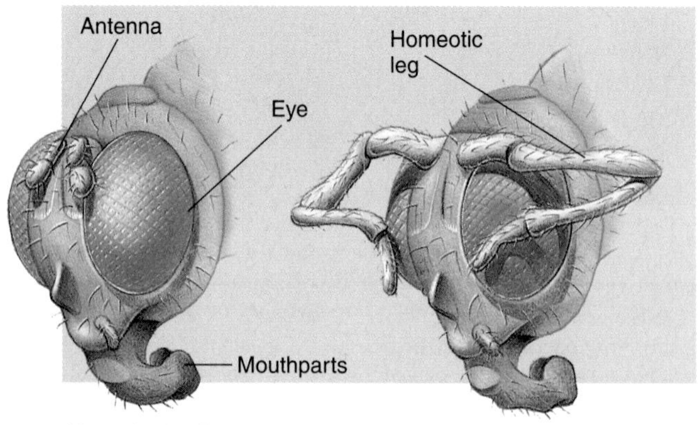

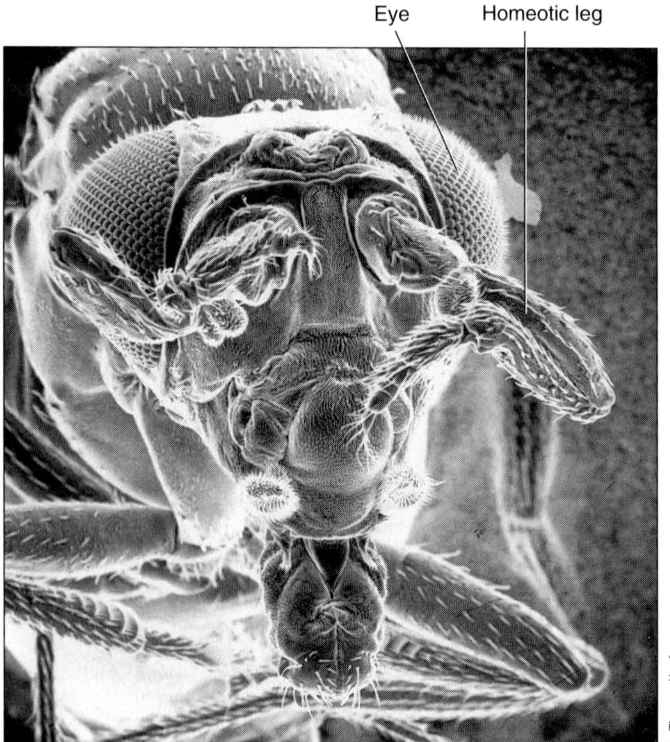

Studies of ***Hox* genes,** clusters of homeobox-containing genes that specify the anterior–posterior axis during development, provide insights about evolutionary relationships. *Hox* genes were initially discovered in *Drosophila,* where they are arranged in two adjacent groups on the chromosome: the *Antennapedia complex* and the *bithorax complex.* As *Hox* genes have been identified in other animals, including other arthropods, annelids (segmented worms), roundworms, and vertebrates, researchers have found that these genes are also clustered and that their organization is remarkably similar to that in *Drosophila.*

Figure 16-12 compares the organization of the *Hox* gene clusters of *Drosophila,* the laboratory mouse, and the roundworm *Caenorhabditis elegans.* The *Drosophila* and mouse *Hox* genes are located in the same order along the chromosome, although the correlation is less exact for *C. elegans.* Moreover, the linear order of the genes on the chromosome reflects the order of the corresponding regions they control (from anterior to posterior) in the animal. This organization apparently reflects the need for these genes to be transcribed in a specific temporal sequence.

KEY CONCEPT: *Hox* genes are arranged in the same order on the chromosome as they are expressed along the anterior-posterior axis of the embryo.

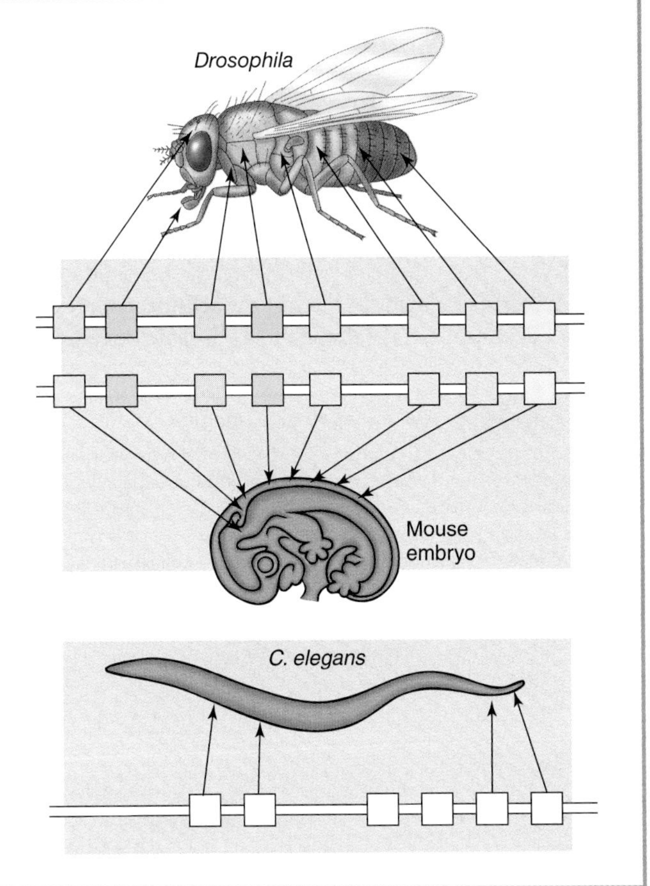

FIGURE **16-12** | ***Hox* gene clusters.**

Clusters of *Hox* are found in all animal groups except for sponges and cnidarians (jellyfish and their relatives). Note that in each organism, the order of these developmental genes *(squares)* on the chromosome *(white line)* reflects their spatial order of expression in the embryo. (Only one of the four mouse chromosomes with *Hox* gene clusters is shown.) *C. elegans* also has similar *Hox* gene clusters, although the gene order is not identical to that in *Drosophila* and the mouse. *(Adapted from C. Kenyon and B. Wang,* Science, *Vol. 253, 2 Aug, 1991; and K. Van Auken, et al.,* Proc. Natl. Acad. Sci., *Vol. 97, 25 Apr. 2000)*

Drosophila has only one *Antennapedia-bithorax* complex. However, humans and other vertebrates have four similar *Hox* gene clusters, each located in a different chromosome. These complexes probably arose through gene duplication. The fact that extra copies of these genes are present helps explain why mutations causing homeotic-like transformations are seldom seen in vertebrate animals. However, one particular type of *Hox* mutation that has been described in both mice and humans causes abnormalities in the limbs and genitalia. The involvement of the genitalia provides a further explanation for the rarity of these mutant alleles, because affected individuals are unlikely to reproduce.

The fact that very similar developmental controls are seen in organisms as diverse as insects, roundworms, and vertebrates (including humans) indicates that the basic mechanism evolved early and has been highly conserved in all animals that have an anterior–posterior axis, even those that are not segmented. Clearly, once a successful way of regulating groups of genes and integrating their activities evolved, it was retained, although it has apparently been modified to provide for alterations of the body plan.

The finding of homeobox-like genes in plants suggests these genes originated early during eukaryotic evolution. With further investigations, researchers hope to develop an overall model of how the rudiments of morphogenesis are controlled in all multicellular eukaryotes. These systems of master genes that regulate development are a rich source of "molecular fossils" that may illuminate evolutionary history in new and exciting ways.

Developmental studies of *C. elegans* elucidated apoptosis

The nematode worm *Caenorhabditis elegans* is an ideal model organism because its system for the genetic control of development is relatively easy to study. Sydney Brenner, a British molecular geneticist, began studying molecular development in this animal at Cambridge University in the 1960s. He selected *C. elegans* because it is small, has a short life cycle, and is genetically simple. Consisting of about 19,700 protein-coding genes, it was the first animal genome sequenced. Today, *C. elegans* is an important tool for answering basic questions about the development of individual cells within a multicellular organism.

As an adult, *C. elegans* is only 1.5 mm long and contains only 959 somatic cells. Individuals are either males or **hermaphrodites,** organisms with both sexes in the same individual. Hermaphroditic individuals are capable of self-fertilization, which makes it easy to obtain offspring that are homozygous for newly induced recessive mutations. The availability of males that can mate with the hermaphrodites makes it possible to perform genetic crosses as well.

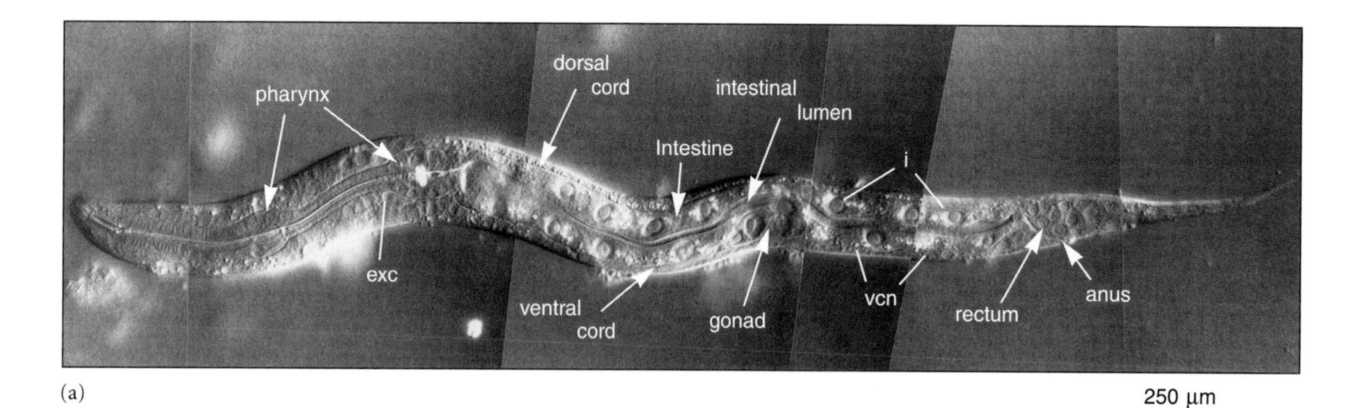

(a)

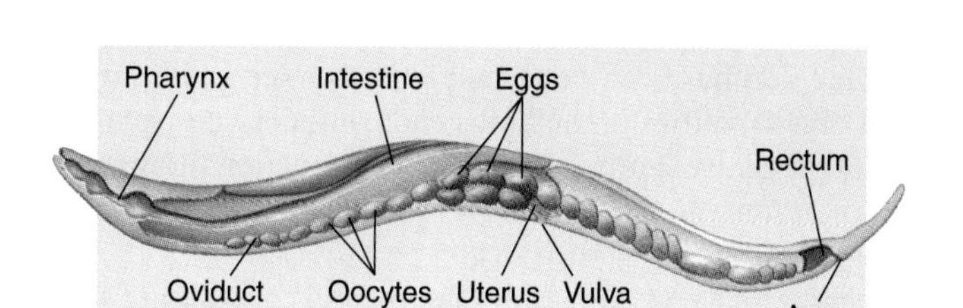

(b)

FIGURE **16-13** ***Caenorhabditis elegans.***

This transparent organism has a fixed number of somatic cells. **(a)** Nomarski interference LM of the adult hermaphrodite nematode. (The labels "exc," "i," and "vcn" refer to cells of the excretory, digestive and nervous system, respectively.) **(b)** Structures in the adult hermaphrodite. The sperm-producing structures are not shown. *(Adapted from Figure 22-6a in V. Walbot and N. Holder,* Developmental Biology, *1987. Used with permission from McGraw-Hill Companies.)*

Because the worm's body is transparent, researchers can follow the development of literally every one of its somatic cells using a Nomarski differential interference microscope, which provides contrast in transparent specimens (Fig. 16-13). As a result of efforts by several research teams, the lineage of each somatic cell in the adult has now been determined. Those studies have shown that the nematode has a very rigid, or fixed, developmental pattern. After fertilization, the egg undergoes repeated divisions, producing about 550 cells that make up the small, sexually immature larva. After the larva hatches from the egg case, further cell divisions give rise to the adult worm.

The lineage of each somatic cell in the adult can be traced to a single cell in a small group of **founder cells,** which are formed early in development (Fig. 16-14a). If a particular founder cell is destroyed or removed, the adult structures that would normally develop from that cell are missing. Such a rigid developmental pattern, in which the fates of the cells become restricted early in development, is referred to as **mosaic development.**

FIGURE **16-14** **Cell lineages of *C. elegans.***

(a) All somatic cells of *C. elegans* are derived from five somatic founder cells (*shown in blue*), produced during the early cell divisions of the embryo. The cell shown in white gives rise to germ-line cells. **(b)** This lineage map traces the development of the cells that form the intestine. The dashed lines represent many cell divisions of a particular lineage.

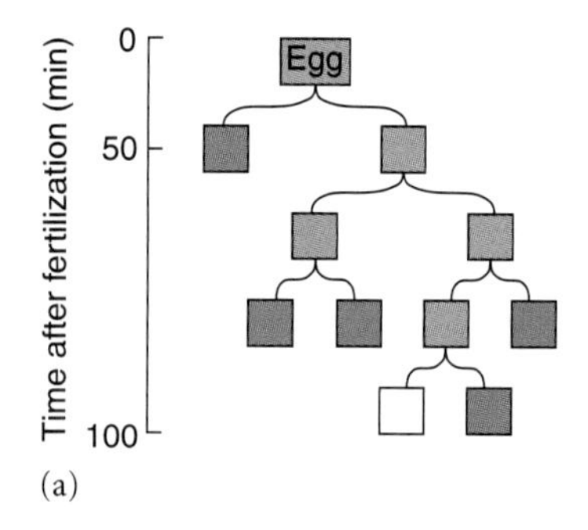

(a)

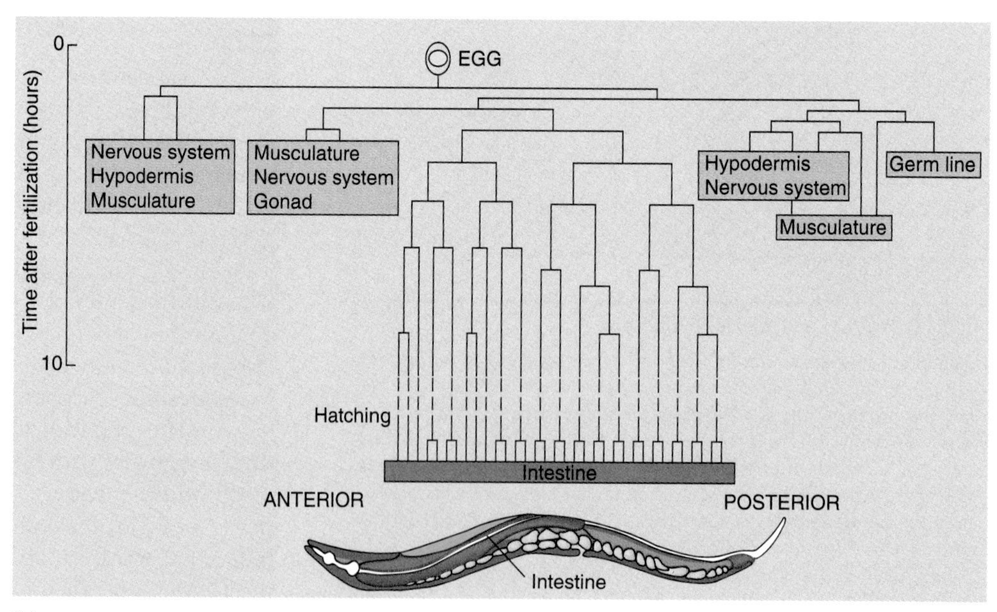

(b)

Each cell has a specific fate in the embryo, just as each tile in a mosaic design forms part of the pattern.

Scientists initially hypothesized that every organ system in *C. elegans* might be derived from only one founder cell. Detailed analysis of cell lineages, however, reveals that many of the structures found in the adult, such as the nervous system and the musculature, are in fact derived from more than one founder cell (Fig. 16-14b). A few lineages have been identified in which a nerve cell and a muscle cell are derived from the division of a single cell. Mutations affecting cell lineages have been isolated, and many of them appear to have properties that would be expected of genes involved in regulating developmental events.

A researcher who uses microscopic laser beams small enough to destroy individual cells can determine the influence one cell has on the development of a neighboring cell. Consistent with the rigid pattern of cell lineages, the destruction of an individual cell in *C. elegans* results, in most cases, in the absence of all of the structures derived from that cell, but with the normal differentiation of all the neighboring somatic cells. This observation suggests that development in each cell is regulated through its own internal program.

However, the developmental pattern of *C. elegans* is not entirely mosaic. In some cases, cell differentiation is influenced by interactions with particular neighboring cells, a phenomenon known as **induction.** One example is the formation of the vulva (pl., *vulvae*), the reproductive structure through which the eggs are laid. A single nondividing cell, called the *anchor cell,* is a part of the ovary, the structure in which the germ line cells undergo meiosis to produce the eggs. The anchor cell attaches to the ovary and to a point on the outer surface of the animal, triggering the formation of a passage through which the eggs pass to the outside. When the anchor cell is present, it induces cells on the surface to form the vulva and its opening. If the anchor cell is destroyed by a laser beam, however, the vulva does not form, and the cells that would normally form the vulva remain as surface cells (Fig. 16-15).

The analysis of mutations has contributed to our understanding of inductive interactions. For example, several types of mutations cause more than one vulva to form. In such mutant animals, multiple vulvae form even if the anchor cell is destroyed. Thus the mutant cells do not require an inductive signal from an anchor cell to form a vulva. Evidently in these mutants the gene or genes responsible for vulva formation are constitutive. Conversely, mutants lacking a vulva are also known. In some of these, the cells that would normally form the vulva apparently do not respond to the inducing signal from the anchor cell.

Chronogenes are genes involved in developmental timing. Researchers have identified recessive alleles that cause certain cells to adopt fates that would ordinarily occur later in development. Dominant alleles of the same locus cause certain cells to adopt fates that would usually occur earlier. Chronogenes are good candidates for master switches that control developmental timing.

During development in *C. elegans,* there are instances in which cells die by apoptosis shortly after they are produced. **Apoptosis,** or programmed cell death, has been observed in a wide variety of organisms, both plant and animal (see Chapter 4). For example, the human hand is formed as a webbed structure, but the fingers become individualized when the cells between them undergo apoptosis. In *C. elegans,* as in other organisms, apoptosis is under genetic control. In 1986, U.S. molecular geneticist Robert Horvitz discovered mutant worms that do not lose cells to apoptosis. Homologous genes were subsequently identified in other organisms, including humans. The loci code for a family of proteins in mammals known as **caspases,** proteolytic enzymes that are active in the early stages of apoptosis. The molecular events that trigger apoptosis are an area of intense research interest, the results of which will shed considerable light on the general processes of cell aging and apoptosis.

Robert Horvitz and British scientists Sydney Brenner and John Sulston shared the Nobel Prize in 2002 for their work on the genetic regulation of organ development and apoptosis in *C. elegans.* Sulston was one of Brenner's students. He traced the lineage of how a single fertilized egg gives rise to the 959 cells in the nematode adult and also observed that some cells die during normal development. Continuing Sulston's work on apoptosis, Horvitz, as mentioned earlier, was the first to discover genes involved in apoptosis.

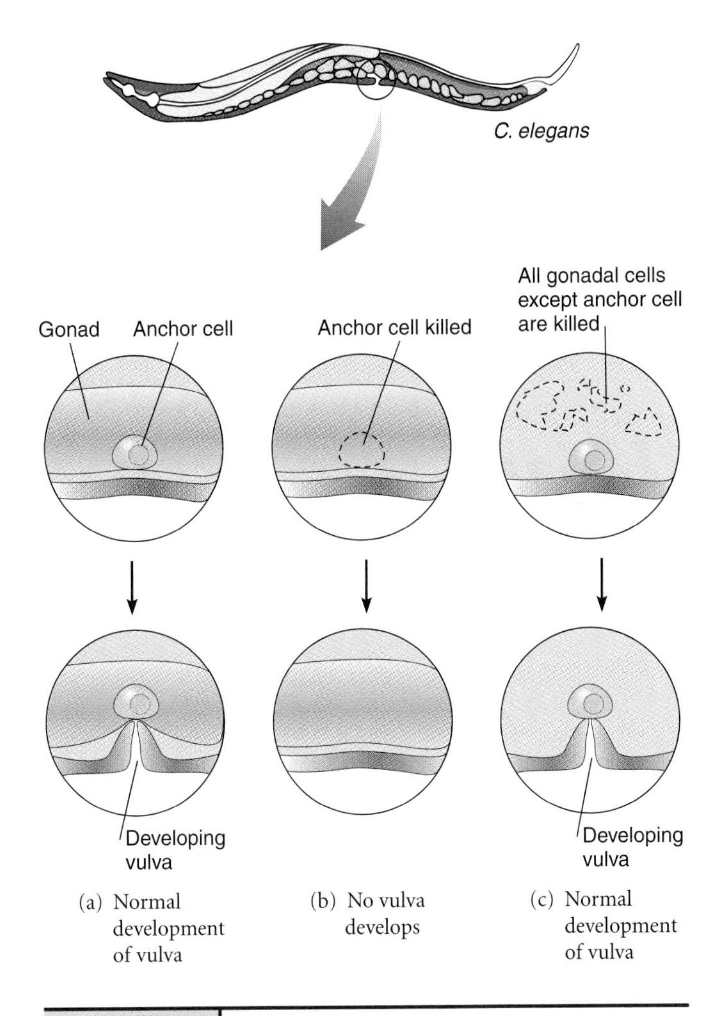

FIGURE **16-15** **Induction.**

A single anchor cell induces neighboring cells to form the vulva in *C. elegans.* This diagram shows how laser destruction of single cells or group of cells demonstrates the influence of a cell on its neighbors.

The mouse is a model for mammalian development

Mammalian embryos develop in markedly different ways from the embryos of *Drosophila* and *C. elegans.* The laboratory mouse, *Mus musculus,* is the best-studied example of mammalian development. Researchers have identified numerous genes affecting development in the mouse. The mouse genome sequence, which was published in 2002, contains 27,000 to 30,000 genes, similar to the number of protein-coding genes in the human genome. Indeed, 99% of the genes in the mouse have counterparts in humans.

The early development of the mouse and other mammals is similar in many ways to human development (see Chapter 49). During the early developmental period, the embryo lives free in the reproductive tract of the female. It then implants itself in the wall of the uterus, after which the mother meets its nutritional and respiratory needs. Consequently, mammalian eggs are very small and contain little in the way of food reserves. Almost all research on mouse development has concentrated on the stages leading to implantation, because in those stages the embryo is free-living and can be experimentally manipulated. During that period, developmental commitments that have a significant effect on the future organization of the embryo take place.

After fertilization, a series of cell divisions gives rise to a loosely packed group of cells. Research shows that in the very early mouse embryo all cells are equivalent. For example, at the two-cell stage of mouse embryogenesis, if researchers destroy one cell and implant the remaining cell into the uterus of a surrogate mother, a normal mouse usually develops.

Conversely, if two embryos at the eight-cell stage of development are fused together and implanted into a surrogate mother, a normal-sized mouse develops (Fig. 16-16). By using two embryos with different genetic characters, such as coat color, researchers demonstrated that the resulting mouse has four genetic parents. These chimeric mice have fur with patches of different colors derived from clusters of genetically different cells. A **chimera** is an organism containing two or more kinds of genetically dissimilar cells arising from different zygotes. (The term is derived from the name of a mythical beast that was said to have the head of a lion, the body of a goat, and a snakelike tail.) Chimeras let researchers use genetically marked cells to trace the fates of certain cells during development.

The responses of mouse embryos to such manipulations contrast with the mosaic or predetermined nature of early *C. elegans* development, in which destroying one of the founder cells results in the loss of a significant portion of the embryo. For this reason, biologists say that the mouse has highly **regulative development**—the early embryo acts as a self-regulating whole that accommodates missing or extra parts. However, thus far researchers have not demonstrated the totipotency of cells from slightly later stages of mouse development. Nevertheless, experiments in which nuclei have been transplanted into mouse eggs, leading to live births, demonstrate that at least some nuclei of differentiated mouse cells are totipotent.

Transgenic mice are used in studies of developmental regulation

In transformation experiments similar to those done with *Drosophila,* foreign DNA injected into fertilized mouse eggs is incorporated into the chromosomes and expressed (Fig. 16-17). The resulting transgenic mice provide insights into how genes are activated during development. In addition, mouse genes can be inactivated (knocked out) by the technique of **gene targeting** (see Chapter 14). For example, when a mouse gene encoding insulin-like growth receptors is knocked out, the resulting mouse dies at birth and exhibits many developmental abnormalities (see *On the Cutting Edge: Studying Aging in Mice* on page 328).

Scientists can identify a *transgene* (foreign gene) that has been introduced into a mouse and determine whether it is active by marking the gene in several ways. Sometimes a similar

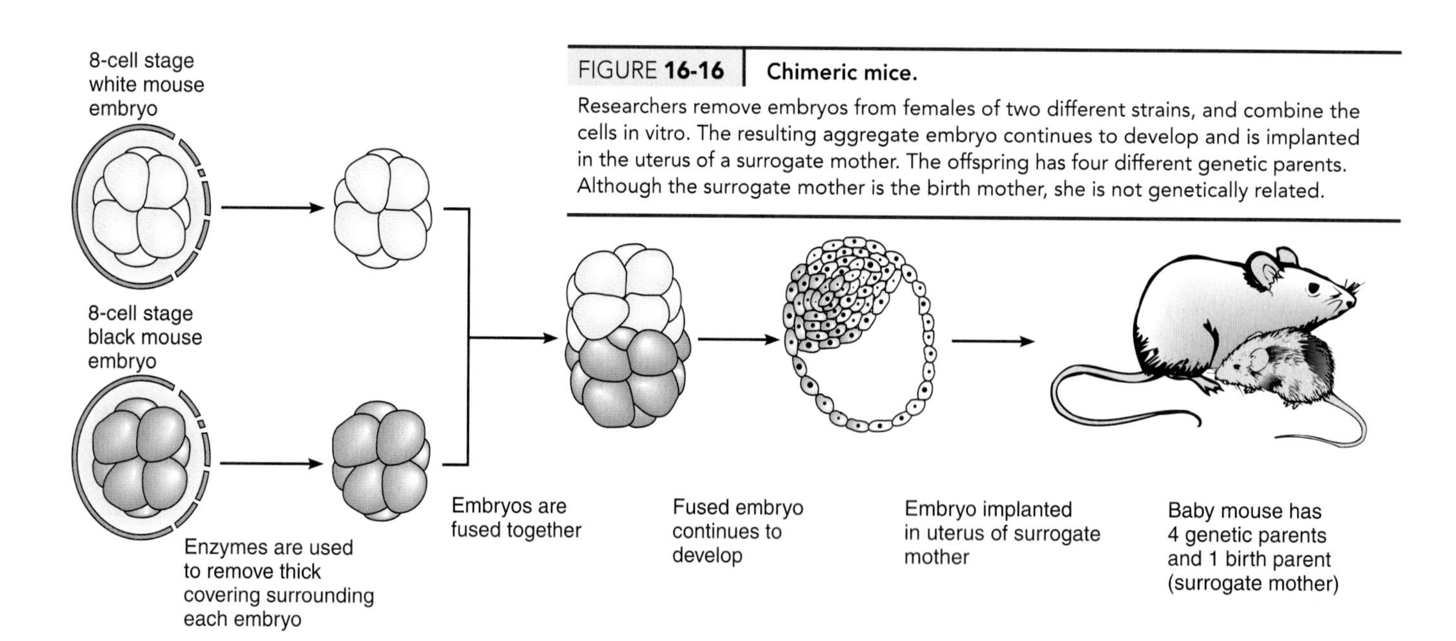
8-cell stage white mouse embryo
8-cell stage black mouse embryo
Enzymes are used to remove thick covering surrounding each embryo
Embryos are fused together
Fused embryo continues to develop
Embryo implanted in uterus of surrogate mother
Baby mouse has 4 genetic parents and 1 birth parent (surrogate mother)

FIGURE **16-16** **Chimeric mice.**

Researchers remove embryos from females of two different strains, and combine the cells in vitro. The resulting aggregate embryo continues to develop and is implanted in the uterus of a surrogate mother. The offspring has four different genetic parents. Although the surrogate mother is the birth mother, she is not genetically related.

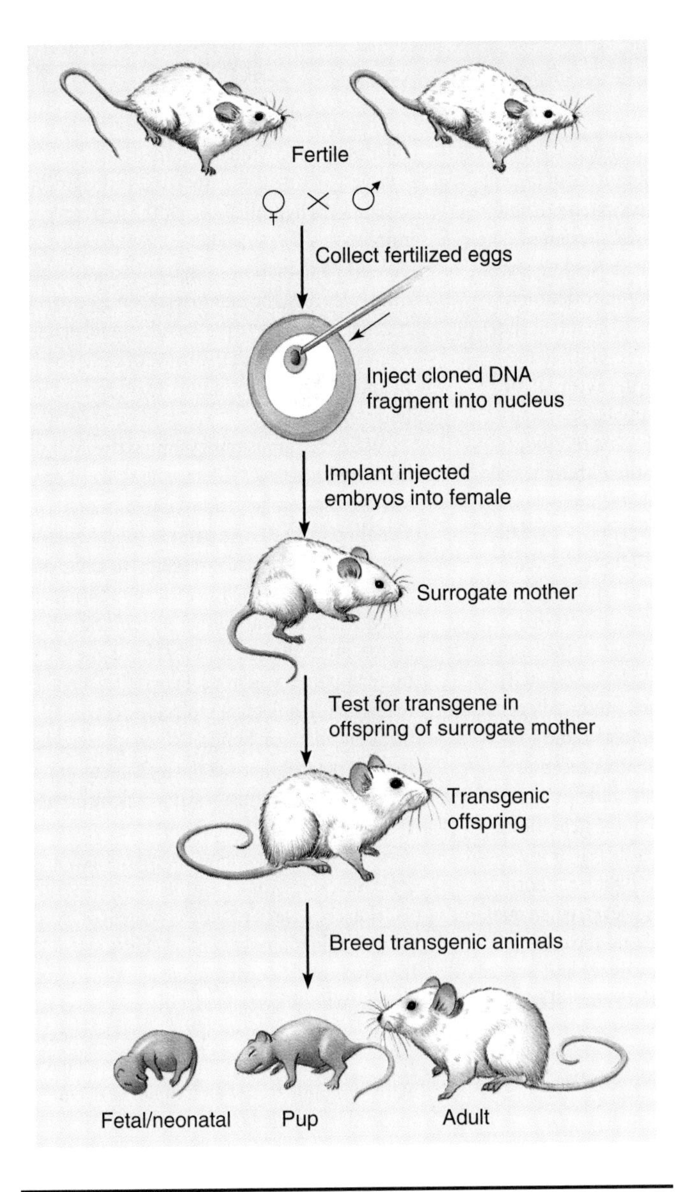

FIGURE **16-17** **Producing a transgenic mouse.**

Researchers inject cloned DNA fragments into the nucleus of a fertilized mouse egg. They then implant the injected eggs into a female, which becomes the surrogate mother. The researchers verify the presence of the foreign gene in the transgenic animal or breed the animal to establish a transgenic line of mice.

gene from a different species is used; its protein is distinguished from the mouse protein by specific antibodies. It is also possible to construct a hybrid gene consisting of the regulatory elements of a mouse gene and part of another, nonmouse gene that codes for a "reporter" protein. For example, the reporter protein could be an enzyme not normally found in mice. Such studies have been important in showing which DNA sequences of a mouse homeobox gene determine where the gene is expressed in the embryo.

Many developmentally controlled genes introduced into mice have yielded important information about gene regulation. When researchers introduce developmentally controlled genes from other species, such as humans or rats, into mice, the genes are regulated the same way they normally are in the donor animal. When introduced into the mouse, for example, human genes encoding insulin, globin, and crystallin—which are normally expressed in cells of the pancreas, blood, and eye lens, respectively—are expressed only in those same tissues in the mouse. That these genes are correctly expressed in their appropriate tissues indicates the signals for tissue-specific gene expression are highly conserved through evolution. Information on the regulation of genes controlling development in one organism can have valuable applications to other organisms, including humans.

Arabidopsis is a model for studying plant development, including transcription factors

Botanists use certain well-characterized plants in studying the genetic control of development. Many of these are economically important crop plants such as corn, *Zea mays.* Genes with developmental effects are known in corn, including some that are analogous (but not homologus) to the homeotic genes of *Drosophila.*

Arabidopsis thaliana, the organism most widely used to study genetics and development in plants, is a member of the mustard family. Although *Arabidopsis* itself is a weed of no economic importance, it has several advantages for research. The plant completes its life cycle in just a few weeks and is small enough to be grown in a petri dish, yielding thousands of individuals in limited space. Botanists use chemical mutagens to produce mutant strains and have isolated many developmental mutants. When they insert cloned foreign genes into *Arabidopsis* cells, the genes integrate into the chromosomes and are expressed. The researchers then induce these transformed cells to differentiate into transgenic plants.

In 2000, the *Arabidopsis* genome became the first plant genome sequenced. Although its genome is relatively small (the rice genome is about four times larger), it includes about 26,000 protein-coding genes. In comparison, *Drosophila* has about 13,600 genes and *C. elegans* about 19,700. Many genes in *Arabidopsis* are functionally equivalent to genes in *Drosophila, C. elegans,* and other animal species.

Of particular importance to development are the more than 1500 genes that code for transcription factors in *Arabidopsis* (compared to 635 at last count for *Drosophila*). Not surprisingly, many genes known to specify identities of parts of the *Arabidopsis* flower, code for transcription factors. During the development of *Arabidopsis* flowers, four distinct flower parts differentiate: sepals, petals, stamens, and carpels (Fig. 16-18a). Sepals cover and protect the flower when it is a bud, petals help attract animal pollinators to the flower, stamens produce pollen grains, and the pistil produces ovules, which develop into seeds following fertilization (see Chapter 27).

The **ABC model** is a working hypothesis that may explain the molecular biology behind how these four organs develop. The *A* gene is needed to specify sepals, both *A* and *B* genes are needed to specify petals, both *B* and *C* genes are needed to spec-

PROCESS OF SCIENCE

Hypothesis: Insulin-like growth factor (IGF) and its membrane receptor influence aging in mammals.

Method: The effects of the IGF receptor on aging were studied in mice using gene targeting to knock out one or both alleles of the gene *igf1r*.

Results: Mice that had one allele of the gene *igf1r* knocked out had an extended life span, whereas mice that had both alleles knocked out died at birth.

Conclusion: Reduced signaling from IGF, caused by the presence of fewer IGF receptors on the cell surface, increases longevity in mice.

Aging, which is defined as a progressive decline in the performance of various parts of the body, is an important field of developmental biology. The study of aging has great practical potential because age-related diseases represent one of the biggest challenges of biomedical research today. Scientists have demonstrated that many environmental factors influence aging. For example, severe calorie restriction in rodents and other mammals delays aging. However, how the mammalian genome interacts with the cell environment during the aging process is not well understood.

As you have learned during your study of genetics, organisms as diverse as worms, fruit flies, and mice provide valuable insights about the biological functioning of other organisms, including humans. Researchers working with model organisms, such as *Drosophila* and *C. elegans,* have found that hundreds of protein-coding genes, when mutated, extend the life span of the organism to variable degrees. One of the most potent life span–extending genes, present in the worm *C. elegans,* codes for a protein that is very similar to the membrane receptor that allows cells to respond to the peptide, **insulin-like growth factor (IGF),** in humans and other mammals (see Chapter 47). The binding of IGF to its receptor triggers **signal transduction** within the cell.

When the allele for this receptor protein mutates in *C. elegans,* the worm has a greatly extended life span. Researchers wondered whether the IGF receptor influences aging in mammals, including humans. To test this hypothesis, Martin Holzenberger at the French National Institute for Biomedical Research (INSERM) and his coworkers produced knockout mice in which they inactivated the gene for the insulin-like growth factor receptor (IGF-1R).[1] Because mice are diploid organisms, they have two copies of the gene, designated *igf1r.* When both alleles of *igf1r* are inactivated, the mice develop many abnormalities and die at birth. However, when one allele of *igf1r* is inactivated and the other left in its normal state, the mice thrive and live an average of 26% longer than control mice.

[1] M. Holzenberger, J. Dupont, B. Ducos, P. Leneuve, A. Géloën, P. Even, P. Cervera, and Y. LeBouc, "IGF-1 Receptor Regulates Lifespan and Resistance to Oxidative Stress in Mice," *Nature,* Vol. 421, 9 Jan. 2003.

Because both genetically manipulated mice and control mice eventually died of a variety of diseases associated with aging, researchers hypothesize that the heterozygous mice have a *decreased rate* of aging, as opposed to a reduction in aging diseases. In all other respects, the heterozygous mice are normal. They are virtually indistinguishable from control mice in their rate of development, metabolic rate, ability to reproduce, and body size.

Mice that are heterozygous at the *igf1r* locus produce fewer IGF receptors, and therefore the cells are not exposed to as much signaling from IGF. As a result, the heterozygous cells may be more resistant to environmental stressors, such as oxygen radicals. (Oxygen radicals and other *free radicals* are atoms, molecules, or ions that have one or more unpaired electrons. Free radicals react readily with biological molecules, often breaking chemical bonds.) When mouse cells that are heterozygous at the *igf1r* locus are grown in culture, they resist attack by oxygen radicals. The same is true at the organism level, for mice that are heterozygous at the *igf1r* locus.

Holzenberger's research on the IGF receptor and aging in mice implies that some of the hundreds of other mutant genes in *C. elegans* that have been associated with aging may also be involved in mammalian aging. Thus the subject of the genetic control of mammalian aging has enough research topics to occupy molecular geneticists for many years.

(a)

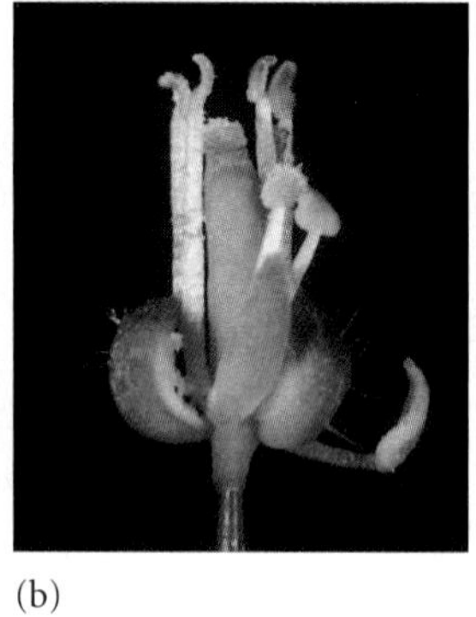
(b)

(c)

(d)

Photos courtesy of Jose Luis Riechmann and Elliot Meyerowitz

FIGURE **16-18** **Homeotic mutants in *Arabidopsis* flowers.**
(a) A normal *Arabidopsis* flower has four outer leafy green sepals, four white petals, six stamens (the male reproductive structures), and a central pistil (the female reproductive structure). **(b)** This homeotic mutant is missing petals. **(c)** This homeotic mutant has only sepals and carpels. **(d)** This homeotic mutant has sepals and petals but no other floral structures. The homeotic genes in all these plants code for transcription factors.

ify stamens, and the *C* gene is needed to specify the pistil. Mutations in the *A*, *B*, or *C* organ-identity genes, all of which are homeotic and code for transcription factors, cause one flower part to be substituted for another. For example, class *C* homeotic mutants (which have an inactive *C* gene) have petals in place of stamens and sepals in place of the pistil. Therefore, the entire flower consists of only sepals and petals. Figure 16-18b, c, and d shows three homeotic mutants of *Arabidopsis*.

The ABC model is not the entire explanation for floral development. Another class of genes, designated *SEPALLATA*, interacts with the *B* and *C* genes to specify the development of petals, stamens, and carpels. Remarkably, when *SEPALLATA* genes are turned on permanently in *Arabidopsis*, the resulting plants have white petals growing where the leaves should be.

These findings in *Arabidopsis* vastly increase the number of molecular probes available from plants. Researchers use these probes to identify other genes that control development in various plant species, and to compare them with genes from a wide range of organisms. The success of the *Arabidopsis* sequencing project led to an international initiative by plant biologists, the 2010 Project, whose goal is to understand the functions of all genes in *Arabidopsis* by the year 2010. This functional genomic information will lead to a far deeper understanding of plant development and evolutionary history.

Review

- What are the relative merits of *Drosophila*, *C. elegans*, *Mus musculus*, and *Arabidopsis* as model organisms for the study of development?
- What are transcription factors, and how do they influence development?
- What role does induction play in development?
- What is apoptosis?

BiologyNow™ Assess your understanding of **the genetic control of development** by taking the pretest on your BiologyNow CD-ROM.

CANCER AND CELL DEVELOPMENT

Learning Objective

8 Discuss the relationship between cancer and mutations that affect cell developmental processes.

Cancer cells lack normal biological inhibitions. Normal cells are tightly regulated by control mechanisms that prompt them to divide when necessary and prevent them from growing and dividing inappropriately. Cells of many tissues in the adult are normally prevented from dividing; they reproduce only to replace a neighboring cell that has died or become damaged. Cancer cells have escaped such controls and can divide continuously.

As a consequence of their abnormal growth pattern, some cancer cells eventually form a mass of tissue called a **tumor.** If the tumor remains at the spot where it originated, it can usually be removed by surgery. One of the major problems with certain forms of cancer is that the cells can escape from the controls that maintain them in one location. **Metastasis** is the spreading of cancer cells to different parts of the body. Cancer cells invade other tissues and form multiple tumors. Lung cancer, for example, is particularly deadly because its cells are highly metastatic; they enter the bloodstream, spread, and form tumors in other parts of the lungs and in other organs, such as the liver and the brain. Tumors with cells that can metastasize are referred to as **malignant tumors.**

Biologists now know that cancer is caused by the altered expression of specific genes critical for cell division. Using recombinant DNA methods, researchers have identified many of the genes that, when they function abnormally, transform normal cells into cancer cells. The traits of each kind of cancer cell come from at least one, and probably several, **oncogenes,** or cancer-causing genes. Oncogenes arise from changes in the expression of certain genes called **proto-oncogenes,** which are *normal* genes found in all cells and are involved in the control of growth and development.

Investigators first discovered oncogenes in viruses that infect mammalian cells and transform them into cancer cells *(malignant transformation)*. Such viruses have oncogenes as part of their genomes. When these viruses infect a cell, the viral oncogenes are expressed, causing the cell to divide. The viral oncogenes resemble proto-oncogenes normally present in the cell.

A proto-oncogene in a cell that has not been infected by a virus can also mutate and become an oncogene. One of the first oncogenes researchers identified was isolated from a tumor in a human urinary bladder. In the cell that gave rise to the tumor, a proto-oncogene had undergone a single base-pair mutation; the result was that the amino acid valine replaced the amino acid glycine in the protein product of the gene. This subtle change was apparently a critical factor in converting the normal cell into a cancer cell.

Some of the control mechanisms of cell growth are illustrated in greatly simplified form in Figure 16-19. One or more external signal molecules trigger the growth and division of cells. Some of these substances are **growth factors** that bind to specific **growth factor receptors** associated with the cell surface, initiating a cascade of events inside the cell. Often the growth factor receptor complex acts as a **protein kinase,** an enzyme that phosphorylates proteins, which then phosphorylates specific amino acids of several cytoplasmic proteins. This post-translational modification usually results in the activation of previously inactive enzymes. These activated enzymes then catalyze the activation of certain nuclear proteins, many of which are transcription factors. Activated transcription factors bind to their DNA targets and stimulate the transcription of specific sets of genes that initiate growth and cell division (see Chapter 9).

Even in the simplified scenario presented in the figure, it is evident that multiple steps are required to control cell proliferation. Researchers have identified the proto-oncogenes that encode the products responsible for many of these steps. The current list of known proto-oncogenes includes genes that code for various growth factors or growth factor receptors and genes that respond to stimulation by growth factors, including many transcription factors. When a proto-oncogene mutates or is expressed inappropriately (when it becomes an oncogene), the cell may misinterpret the signal and respond by growing and di-

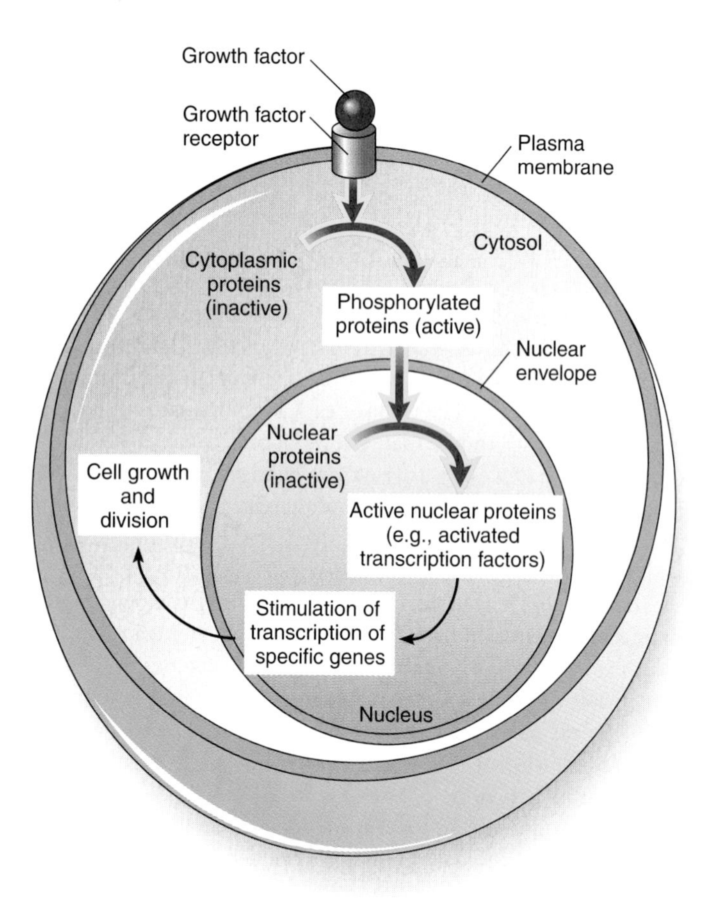

FIGURE **16-19** | **A cell growth control cascade.**

In this example, a growth factor stimulates cell growth. The growth factor receptor and some of the other components of the system are coded for by proto-oncogenes. When a proto-oncogene mutates, becoming an oncogene, the cell grows and divides even in the absence of the growth factor.

viding. For example, in some cases a proto-oncogene encoding a growth factor receptor mutates in a way that compromises regulation of the receptor. It is always switched "on," even in the absence of the growth factor that normally controls it.

Not all genes that cause cancer when mutated are proto-oncogenes. About half of all cancers are caused by a mutation in a **tumor suppressor gene.** These genes, also known as *anti-oncogenes,* normally interact with growth-inhibiting factors to block cell division. When mutated they lose their ability to "put on the brakes," and uncontrolled growth ensues.

Currently, more than 100 oncogenes and 15 tumor suppressor genes have been identified. A change in a single proto-oncogene is usually insufficient to cause a cell to become malignant. The development of cancer is usually a multistep process involving both mutations that activate oncogenes and mutations that inactivate tumor suppressor genes. Additional factors, such as the inappropriate activation of the enzyme responsible for the maintenance of telomeres (see Chapter 11) or chromosomal translocation events (see Chapter 15), may also play a role. The Cancer Genome Project, a long-term international initiative based in Great Britain, is currently examining every human gene for cancer-related mutations. As more of these genes are discovered and their interactions unraveled, we will gain a fuller understanding of the control of growth and development. This understanding will improve the diagnosis and treatments of various types of cancer.

Review

- What are oncogenes? Tumor suppressor genes?
- How are oncogenes and tumor suppressor genes related to genes involved in the control of normal growth and development?
- How is a growth factor involved in a growth control cascade?

BiologyNow™ Assess your understanding of **cancer and cell development** by taking the pretest on your BiologyNow CD-ROM.

SUMMARY WITH KEY TERMS

1 Distinguish between cell determination and cell differentiation, and between nuclear equivalence and totipotency.

- **Development,** all the changes that occur in the life of an individual, includes the processes by which the descendants of a single cell specialize and organize into a complex organism.
- An organism contains many types of cells that are specialized both structurally and metabolically to carry out specific functions. These cells are the product of a process of gradual commitment, called **cell determination,** which ultimately leads to the final step in cell specialization, called **cell differentiation.** Differences among various cell types are apparently due to **differential gene expression.**
- There is no evidence that genes are normally lost during most developmental processes. At least some nuclei from differentiated plant and animal cells contain all the genetic material that would be present in the nucleus of a zygote. **Nuclear equivalence** is the concept that, with a few exceptions, all the nuclei of the differentiated **somatic cells** of an organism are identical to each other and to the nucleus of the single cell from which they descended. **Totipotency** is the capability of cells to direct the development of an entire organism.

2 Define *stem cells,* and describe some of the promising areas of research involving stem cells.

- **Stem cells** can divide to produce differentiated descendants, yet retain the ability to divide to maintain the stem cell population. Totipotent stem cells give rise to all cell types, whereas **pluripotent stem cells** give rise to many, but not all, types of cells in an organism.
- Stem cells show promise in treating diseases, such as Parkinson's disease and diabetes mellitus.

3 Point out some of the known exceptions to the principle of nuclear equivalence.

- **Genomic rearrangement** results in physical changes in the structure of a gene.
- **Gene amplification** provides more copies of certain genes for transcription.

SUMMARY WITH KEY TERMS

4 Indicate the features of *Drosophila melanogaster, Caenorhabditis elegans, Mus musculus,* and *Arabidopsis thaliana* that have made these organisms valuable models in developmental genetics.

- Developmental mutations have been identified in the fruit fly, *Drosophila melanogaster,* many of which affect the organism's segmented body plan.
- *Caenorhabditis elegans* is a roundworm with **mosaic development,** a rigid developmental pattern in which the fates of cells are restricted early in development. The lineage of every somatic cell in the adult is known, and each can be traced to a single **founder cell** in the early embryo.
- The laboratory mouse, *Mus musculus,* is extensively used in studies of mammalian development. In contrast to *C. elegans,* the mouse shows **regulative development;** the very early embryo is a self-regulating whole and can develop normally even if it has extra or missing cells. **Transgenic** mice, in which foreign genes have been incorporated, have helped researchers determine how genes are activated and regulated during development.
- Genes affecting development have also been identified in certain plants, including *Arabidopsis* and *Zea mays* (corn). The **ABC model** of interactions among three kinds of genes hypothesizes how floral organs develop in *Arabidopsis.* Mutations in these genes cause one flower part to be substituted for another.

5 Distinguish among maternal effect genes, segmentation genes, and homeotic genes in *Drosophila.*

- The earliest developmental events to operate in the egg are established by **maternal effect genes** in the surrounding maternal tissues, which are active prior to fertilization. Some produce gradients of **morphogens,** chemical agents that affect the differentiation and development of form. Maternal effect genes establish polarity in the embryo.
- **Segmentation genes** generate a repeating pattern of body segments within the embryo. **Gap genes** are segmentation genes that begin organizing the body into anterior, middle, and posterior regions. **Pair-rule genes** and **segment polarity genes** affect all segments instead of small groups of body segments.
- The later-acting **homeotic genes** are responsible for specifying the identity of each segment.

6 Explain the relationship between transcription factors and genes that control development.

- **Transcription factors** are DNA-binding proteins that regulate transcription in eukaryotes. Some genes that code for transcription factors contain a DNA sequence called a **homeobox,** which codes for a protein with a DNA-binding region called a **homeodomain.**
- Some homeobox genes are organized into complexes that appear to be systems of master genes specifying an organism's body plan. Parallels exist between the homeobox complexes of *Drosophila* and those of other animals.

7 Define *induction* and *apoptosis,* and give examples of the roles they play in development.

- **Induction** is developmental interactions with neighboring cells. During development in *C. elegans,* the anchor cell induces cells of the surface to organize to form the vulva, the structure through which the eggs are laid.
- **Apoptosis** is programmed cell death. During human development, the human hand forms as a webbed structure, but the fingers become individualized when the cells between them die.

8 Discuss the relationship between cancer and mutations that affect cell developmental processes.

- The traits of cancer cells are due to cancer-causing **oncogenes.** Oncogenes arise from changes in the expression of normal genes called **proto-oncogenes,** which exist in all cells and are involved in the control of growth and development.
- Known proto-oncogenes include genes that code for various **growth factors** or **growth factor receptors** and genes that respond to stimulation by growth factors, including many transcription factors. When a proto-oncogene is expressed inappropriately (becomes an oncogene), the cell may misinterpret the signal and respond by growing and dividing.
- **Tumor suppressor genes** normally interact with growth-inhibiting factors to block cell division. A mutation in a tumor suppressor gene may inactivate it, leading to cancer.

POST-TEST

1. Morphogenesis occurs through the multistep process of (a) differentiation (b) determination (c) pattern formation (d) totipotency (e) selection
2. The cloning experiments carried out on frogs demonstrated that (a) all differentiated frog cells are totipotent (b) some differentiated frog cells are totipotent (c) all nuclei from differentiated frog cells are totipotent (d) some nuclei from differentiated frog cells are totipotent (e) the mechanism of cell differentiation always requires the loss of certain genes
3. *Drosophila* is a particularly good model for developmental studies because (a) a large number of developmental mutants are available (b) it has a fixed number of somatic cells in the adult (c) its embryos are transparent (d) it is a vertebrate (e) all of the preceding
4. The anterior–posterior axis of a *Drosophila* embryo is first established by certain (a) homeotic genes (b) maternal effect genes (c) segmentation genes (d) chronogenes (e) pair-rule genes
5. You discover a new *Drosophila* mutant in which mouthparts appear where the antennae are normally found. You predict that the mutated gene is most likely a (a) homeotic gene (b) gap gene (c) pair-rule gene (d) maternal effect gene (e) segment polarity gene
6. Most segmentation genes code for (a) transfer RNAs (b) enzymes (c) transcription factors (d) histones (e) transport proteins
7. The developmental pattern of *C. elegans* is said to be mosaic because (a) development is controlled by gradients of morphogens (b) part of the embryo fails to develop if a founder cell is destroyed (c) some individuals are self-fertilizing hermaphrodites (d) all development is controlled by maternal effect genes (e) apoptosis never occurs
8. The formation of the vulva, the structure through which eggs are laid, in *C. elegans* involves (a) maternal effect genes that organize the egg cytoplasm (b) gradients of morphogens in the eggs (c) groups of *Hox* genes that form the *Antennapedia* complex and

bithorax complex (d) induction of surface cells by the anchor cell (e) mutations in chronogenes, which control developmental timing

9. Which of the following illustrates the regulative nature of early mouse development? (a) the mouse embryo is free-living prior to implantation in the uterus (b) it is possible to produce a transgenic mouse (c) it is possible to produce a mouse in which a specific gene has been knocked out (d) genes related to *Drosophila* homeotic genes have been identified in mice (e) a chimeric mouse can be produced by fusing two mouse embryos
10. When the human gene that codes for insulin is introduced into fertilized mouse eggs that are subsequently allowed to develop, the insulin gene is correctly expressed in the mouse's pancreatic cells. This indicates that (a) the gene that codes for insulin is analogous to the homeotic genes of *Drosophila* (b) the signals for tissue-specific gene expression are highly conserved through evolution (c) like humans, the mouse has polytene chromosomes (d) unlike the rigid developmental pattern of *C. elegans,* the development of mice and humans is highly regulative (e) genomic rearrangements have occurred in the mouse embryo
11. *Arabidopsis* is useful as a model organism for the study of plant development because (a) it is of great economic importance (b) it has large polytene chromosomes (c) many developmental mutants have been isolated (d) it contains a large amount of DNA per cell (e) it has a rigid developmental pattern
12. According to the ABC model of floral organ development in *Arabidopsis,* the A gene is needed to specify sepals, the A and B genes to specify petals, the B and C genes to specify stamens, and the C gene to specify the pistil. If a mutation occurs in one of the B genes, rendering it inactive, the resulting flowers will consist of (a) sepals, petals, stamens, and pistils, (b) sepals, stamens, and pistils (c) petals, stamens, and pistils (d) sepals and pistils (e) petals and stamens
13. Pluripotent stem cells (a) lose genetic material during development (b) give rise to many, but not all, types of cells in an organism (c) organize into recognizable structures through pattern formation (d) cannot grow in tissue culture (e) have been used to clone a sheep and several other mammals
14. The genetic material for Dolly, the first cloned sheep, was a nucleus from (a) an early sheep embryo (b) cultured cancer cells (c) intestinal epithelial cells (d) a mouse–sheep chimera (e) a cultured mammary gland cell
15. Genomic rearrangement is an exception to the principle of (a) nuclear equivalence (b) pattern formation (c) morphogenesis (d) differential gene expression (e) mosaic development
16. Which of the following statements about cancer is *false?* (a) Oncogenes arise from mutations in proto-oncogenes. (b) Tumor suppressor genes normally interact with growth-inhibiting factors to block cell division (c) More than 100 oncogenes and 15 tumor suppressor genes have been identified (d) Oncogenes were first discovered in mouse models for cancer (e) The development of cancer is usually a multistep process involving both oncogenes and mutated tumor suppressor genes
17. Proto-oncogenes code for (a) morphogens (b) antibodies for immune responses (c) growth factor receptors and other components of the growth control cascade (d) enzymes such as reverse transcriptase (e) proteins such as chorion protein

CRITICAL THINKING

1. Why is an understanding of gene regulation in eukaryotes crucial to an understanding of developmental processes?
2. Why is it necessary for scientists to study development in more than one type of organism?
3. Could a gene be involved in the growth of both stem cells and some kinds of cancer? Explain your answer.

- Visit our Web site at **http://biology.brookscole.com/solomon7** for links to chapter-related resources on the World Wide Web. Additional online materials relating to this chapter can also be found on our Web site.

BIOLOGY NOW RESOURCES

Active Figure

16-4: Sheep cloning

Preparing for an exam? Take a diagnostic test on your BiologyNow CD-ROM.

Post-Test Answers

1. c	2. d	3. a	4. b
5. a	6. c	7. b	8. d
9. e	10. b	11. c	12. d
13. b	14. e	15. a	16. d
17. c			

17

Introduction to Darwinian Evolution

The Granger Collection, New York

Charles Darwin. This portrait was made shortly after Darwin returned to England from his voyage around the world.

CHAPTER OUTLINE

- **Pre-Darwinian Ideas about Evolution**
- **Darwin and Evolution**
- **Evidence for Evolution**

The biological diversity represented by the millions of species currently living on our planet may have evolved from a single ancestor during Earth's long history. Thus organisms that are radically different from each other, such as slime molds and crocodiles, are in fact distantly related, linked through numerous intermediate ancestors to a single, common ancestor. The British naturalist Charles Darwin (1809–1882), shown here at age 31, developed a remarkably simple, scientifically testable mechanism to explain this. He argued persuasively that all the species that exist today, as well as the countless extinct species that existed in the past, arose from earlier ones by a process of gradual *divergence* (splitting into separate evolutionary pathways), or evolution.

Evolution is defined as the accumulation of inherited changes within populations over time. A **population** is a group of individuals of one species that live in the same geographic area at the same time. The term *evolution* does not refer to changes that occur in an individual within its lifetime. Instead, it refers to changes in the characteristics of populations over the course of generations. These changes may be so small they are difficult to detect or so great that the population differs markedly from its ancestral population. Eventually, two populations may diverge to such a degree that we refer to them as different species. (The concept of species is developed extensively in Chapter 19. For now, a simple working definition is that a **species** is a group of organisms, with similar structure, function, and behavior, that are capable of interbreeding with one another.) Thus evolution has two main perspectives—the minor evolutionary changes of populations usually viewed over a few generations (*microevolution,* see Chapter 18), and the major evolutionary events usually viewed over a long period, such as formation of different species from common ancestors (*macroevolution*—see Chapter 19).

The concept of evolution is the cornerstone of biology, because it links all fields of the life sciences into a unified body of knowledge. As geneticist Theodosius Dobzhansky put it, "Nothing in biology makes sense except in the light of evolution."[1] Biologists seek to understand both the remarkable variety as well as the fundamental similarities of organisms within the context of evolution. The science of evolution allows biologists to compare common threads among organisms as seemingly different as bacteria, whales, lilies, and tapeworms. Behavioral evolution, evolutionary developmental biology, evolutionary genetics, evolutionary ecology, evolutionary systematics, and molecular evolution are examples of some of the biological disciplines that focus on evolution.

Evolution also has important practical applications. Agriculture must deal with the evolution of pesticide resistance in insects and other pests, and agricultural scientists strive to develop techniques that slow the evolutionary process in pest organisms. Likewise, medicine must respond to the rapid evolutionary potential of disease-causing organisms such as bacteria. (Significant evolutionary change occurs in a very short time period in insects, bacteria, and other organisms with short life spans.) Medical researchers use evolutionary principles to predict which flu strains are evolving more quickly, information the scientists need to make next year's flu vaccine. The conservation management of rare and endangered species makes use of the evolutionary principles of population genetics (see Chapter 18). The rapid evolution of bacteria and fungi in polluted soils is used in the field of **bioremediation,** in which microorganisms are employed to clean up hazardous waste sites. Evolution even has applications beyond biology. For example, certain computer applications make use of algorithms that mimic natural selection in biological systems.

This chapter discusses Darwin and the scientific development of the theory of evolution by natural selection. It also presents several kinds of evidence that support evolution, including fossils, comparative anatomy, biogeography, developmental biology, molecular biology, and experimental studies of ongoing evolutionary change, both in the laboratory and in nature. ■

PRE-DARWINIAN IDEAS ABOUT EVOLUTION

Learning Objective

1 Discuss the historical development of the theory of evolution.

Although Darwin is universally associated with evolution, ideas of evolution predate Darwin by centuries. Aristotle (384–322 B.C.E.) saw much evidence of natural affinities among organisms. This led him to arrange all the organisms he knew in a "scale of nature" that extended from the exceedingly simple to the most complex. Aristotle visualized organisms as being imperfect but "moving toward a more perfect state." Some scientific historians have interpreted this idea as the forerunner of evolutionary theory, but Aristotle was vague on the nature of this "movement toward perfection" and certainly did not propose that natural processes drove the process of evolution. Furthermore, modern evolutionary theory now recognizes that evolution does not move toward more "perfect" states, nor even necessarily toward greater complexity.

Long before Darwin, fossils had been discovered embedded in rocks. Some of these corresponded to parts of familiar species, but others were strangely unlike any known species. Fossils were often found in unexpected contexts. Marine invertebrates (sea animals without backbones) were sometimes discovered in rocks high on mountains. Leonardo da Vinci (1452–1519) was among the first to correctly interpret these unusual finds as the remains of animals that had existed in previous ages but had become extinct.

The French naturalist Jean Baptiste de Lamarck (1744–1829) was the first scientist to propose that organisms undergo change over time as a result of some natural phenomenon rather than divine intervention. According to Lamarck, a changing environment caused an organism to alter its behavior, thereby using some organs or body parts more and others less. Over several generations, a given organ or body part would increase in size if it was used a lot, or shrink and possibly disappear if it was used less. His hypothesis required that organisms pass traits they acquired during their lifetimes to their offspring. For example, Lamarck suggested that the long neck of the giraffe developed when a short-necked ancestor stretched its neck to browse on the leaves of trees. Its offspring inherited the longer neck, which stretched still further as they ate. This process, repeated over many generations, resulted in the long necks of modern giraffes. In his *Philosophie Zoologique,* published in 1809, Lamarck presented this explanation for how organisms evolved. Lamarck also thought all organisms were endowed with a vital force that drove them to change toward greater complexity and "perfection" over time.

Lamarck's proposed mechanism of evolution is quite different from the mechanism later proposed by Darwin. However, Lamarck's hypothesis remained a reasonable explanation for evolution until Mendel's basis of heredity was rediscovered at the beginning of the 20th century. At that time, Lamarck's ideas were largely discredited.

Review

- Why is Aristotle linked to early evolutionary thought?
- What were Jean Baptiste de Lamarck's ideas concerning evolution?

BiologyNow™ Assess your understanding of **pre-Darwinian ideas about evolution** by taking the pretest on your BiologyNow CD-ROM.

[1] *American Biology Teacher,* Vol. 35, No. 125 (1973).

DARWIN AND EVOLUTION

Learning Objectives

2 Define *evolution,* and explain the four premises of evolution by natural selection as proposed by Charles Darwin.

3 Compare the modern synthesis with Darwin's original theory of evolution.

PROCESS OF SCIENCE

Darwin, the son of a prominent physician, was sent at the age of 15 to study medicine at the University of Edinburgh. Finding himself unsuited for medicine, he transferred to Cambridge University to study theology. During that time, he became the protégé of the Rev. John Henslow, who was a professor of botany. Henslow encouraged Darwin's interest in the natural world. Shortly after receiving his degree, Darwin embarked on the H.M.S. *Beagle,* which was taking a five-year exploratory cruise around the world to prepare navigation charts for the British Navy.

The *Beagle* left Plymouth, England, in 1831 and cruised along the east and west coasts of South America (Fig. 17-1). While other members of the crew mapped the coasts and harbors, Darwin spent many weeks ashore studying the animals, plants, fossils, and geological formations of both coastal and inland regions, areas that had not been extensively explored. He collected and cataloged thousands of plant and animal specimens and kept notes of his observations, information that became essential in the development of his theory.

The *Beagle* spent almost two months at the Galapagos Islands, 965 km (600 mi) west of Ecuador, where Darwin continued his observations and collections. He compared the animals and plants of the Galapagos with those of the South American mainland. He was particularly impressed by their similarities and wondered why the organisms of the Galapagos should resemble those from South America more than those from other islands in different parts of the world. Moreover, although there were similarities between Galapagos and South American species, there were also distinct differences. There were even recognizable differences in the reptiles and birds from one island to the next. After he returned home, Darwin pondered these observations and attempted to develop a satisfactory explanation for the distribution of species among the islands.

Darwin drew on several lines of evidence when considering how species might have originated. Despite the work of Lamarck, the general notion in the mid-1800s was that Earth was too young for organisms to have changed significantly since they had first appeared. During the early 19th century, however, geologists advanced the idea that mountains, valleys, and other physical features of Earth's surface did not originate in their present forms. Instead, these features developed slowly over long periods by the geological processes of volcanic activity, uplift, erosion, and glaciation. Darwin took *Principles of Geology,* published by English geologist Charles Lyell in 1830, with him on his voyage and studied it carefully. Lyell provided an important concept for Darwin—that the slow pace of geological processes, which still occur today, indicated that Earth was extremely old.

Other important evidence that influenced Darwin was the fact that breeders and farmers could develop many varieties of domesticated animals in just a few generations (Fig. 17-2). They did so by choosing certain traits and breeding only individuals that exhibited the desired traits, a procedure known as **artificial selection.** Breeders, for example, have produced numerous dog varieties—bloodhounds, Dalmatians, Airedales, border collies, and Pekinese, to name a few—by artificial selection.

Many plant varieties were also produced by artificial selection. For example, cabbage, broccoli, brussels sprouts, cauliflower, collard greens, kale, and kohlrabi are distinct vegetable crops that are all members of the same species, *Brassica oleracea* (Fig. 17-3). Selective breeding of the colewort, or wild cabbage, a leafy plant native to Europe and Asia, produced all seven vegetables. Beginning more than 4000 years ago, some farmers artificially selected wild cabbage plants that formed overlapping leaves. Over time, these leaves became so prominent that the plants, which resembled modern cabbages, became recognized as separate and distinct from their wild cabbage ancestor. Other farmers selected different features of the wild cabbage, giving rise to the other modifications. For example, kohlrabi was pro-

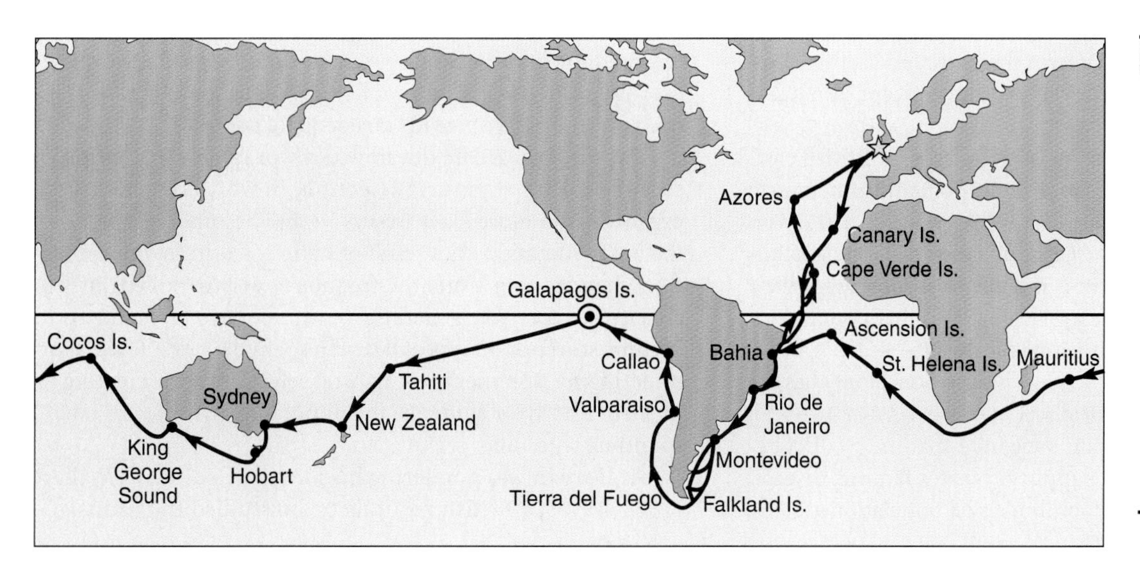

FIGURE **17-1** **The voyage of H.M.S. *Beagle.*** The five-year voyage began in Plymouth, England *(star),* in 1831. Observations made in the Galapagos Islands *(bull's-eye)* off the western coast of South America helped Darwin discover a satisfactory mechanism to explain how a population of organisms could change over time.

Eric Sander

FIGURE **17-2** | **Artificial selection in chickens.**

Shown is a chicken that was deliberately bred to resemble Big Bird on *Sesame Street.* Show breeds of chickens exhibit a great deal of variation. Domestic chickens are not a recognizable breed but are hybrids bred for meat or egg production.

John Amaldi

FIGURE **17-3** | **Artificial selection in *Brassica oleracea.***

An enlarged terminal bud (the "head") was selected in cabbage *(lower left),* flower clusters in broccoli *(upper left)* and cauliflower *(middle right),* axillary buds in brussels sprouts *(bottom middle),* leaves in collards *(upper right)* and kale *(lower right),* and stems in kohlrabi *(middle).*

duced by selection for an enlarged storage stem, and brussels sprouts by selection for enlarged axillary buds. Thus humans are responsible for the evolution of *B. oleracea* into seven distinct vegetable crops. Darwin was impressed by the changes induced by artificial selection and hypothesized that a similar selective process occurred in nature. Darwin therefore used artificial selection as a model when he developed the concept of natural selection.

The ideas of Thomas Malthus (1766–1834), a British clergyman and economist, were another important influence on Darwin. In *An Essay on the Principle of Population as It Affects the Future Improvement of Society,* published in 1798, Malthus noted that population growth is not always desirable—a view contrary to the beliefs of his day. He observed that populations have the capacity to increase geometrically ($1 \longrightarrow 2 \longrightarrow 4 \longrightarrow 8 \longrightarrow 16$) and thus outstrip the food supply, which only has the capacity to increase arithmetically ($1 \longrightarrow 2 \longrightarrow 3 \longrightarrow 4 \longrightarrow 5$). In the case of humans, Malthus suggested that the conflict between population and food supply generates famine, disease, and war, which serve as inevitable brakes on population growth.

Darwin proposed that evolution occurs by natural selection

Malthus's idea that there is a strong and constant check on human population growth strongly influenced Darwin's explanation of evolution. Darwin's years of observing the habits of animals and plants had introduced him to the struggle for existence described by Malthus. It occurred to Darwin that in this struggle inherited variations favorable to survival would tend to be preserved, whereas unfavorable ones would be eliminated. The result would be **adaptation,** an evolutionary modification that improves the chances of survival and reproductive success in a given environment. Eventually, the accumulation of modifications might result in a new species. Time was the only thing required for new species to originate, and the geologists of the era, including Lyell, had supplied evidence that Earth was indeed old enough to provide an adequate period.

Darwin had at last developed a workable explanation of evolution, that of **natural selection,** in which better adapted organisms are more likely to survive and become the parents of the next generation. As a result of natural selection, the population changes over time; the frequency of favorable traits increases in successive generations, whereas less favorable traits become scarce or disappear. Darwin spent the next 20 years formulating his arguments for natural selection, accumulating an immense body of evidence to support his theory, and corresponding with other scientists.

As Darwin was pondering his ideas, Alfred Russel Wallace (1823–1913), a British naturalist who studied the plants and animals of the Malay Archipelago for eight years, was similarly struck by the diversity of species and the peculiarities of their distribution. He wrote a brief essay on this subject and sent it to Darwin, by then a world-renowned biologist, asking his opinion. Darwin recognized his own theory and realized Wallace had independently arrived at the same conclusion—that evolution occurs by natural selection. Darwin's colleagues persuaded him to present Wallace's manuscript along with an abstract of his own work, which he had prepared and circulated to a few friends several years earlier. Both papers were presented in July 1858 at a London meeting of the Linnaean Society. Darwin's monumental book, *The Origin of Species by Natural Selection; or, The Preservation of Favored Races in the Struggle for Life,* was published in 1859. Wallace's book, *Contributions to the Theory of Natural Selection,* was published in 1870, eight years after he returned from the Malay Archipelago.

Darwin's mechanism of evolution by natural selection consists of observations on four aspects of the natural world: variation, overproduction, limits on population growth, and differential reproductive success.

1. *Variation.* The individuals in a population exhibit variation. Each individual has a unique combination of traits, such as size, color, and ability to tolerate harsh environmental conditions. Some traits improve an individual's chances of survival and reproductive success, whereas others do not. Remember, the variation necessary for evolution by natural selection must be inherited (Fig. 17-4). Although Darwin recognized the importance to evolution of inherited variation, he did not know the mechanism of inheritance.
2. *Overproduction.* The reproductive ability of each species has the potential to cause its population to geometrically increase over time. A female frog lays about 10,000 eggs, and a female cod produces perhaps 40 million eggs! In each case, however, only about two offspring survive to reproduce. Thus in every generation each species has the capacity to produce more offspring than can survive.
3. *Limits on population growth,* or a struggle for existence. There is only so much food, water, light, growing space, and other resources available to a population, so organisms compete with one another for these limited resources. Because there are more individuals than the environment can support, not all survive to reproduce. Other limits on population growth include predators, disease organisms, and unfavorable weather conditions.
4. *Differential reproductive success.* Those individuals that have the most favorable combination of characteristics (those that make individuals better adapted to their environment) are more likely to survive and reproduce. Offspring tend to resemble their parents, because the next generation inherits the parents' genetically based traits. Successful reproduction is the key to natural selection: The best adapted individuals produce the most offspring, whereas individuals that are less well adapted die prematurely or produce fewer or inferior offspring.

Over time, enough changes may accumulate in geographically separated populations (often with slightly different environments) to produce new species. Darwin noted that the Galapagos finches appeared to have evolved in this way. The 13 species are all descended from a single species that found its way from the South American mainland. (The blue-black grassquit, a bird that lives in western South America, may be a close relative of the Galapagos finches.) The different islands of the Galapagos kept the finches isolated from one another, thereby allowing them to diverge into 13 separate species. Of the 13 species, 6 are seed eaters, 6 are insect eaters, and 1 is a woodpecker-type insect eater. The evolution of new species is considered in greater detail in Chapter 19.

BIOS/Peter Arnold, Inc.

FIGURE **17-4** **Genetic variation in emerald tree boas.** These snakes, all of the same species (*Corallus caninus*), were caught in a small section of forest in French Guiana. Many snake species exhibit considerable variation in their coloration and patterns.

The modern synthesis combines Darwin's theory with genetics

PROCESS OF SCIENCE

One of the premises on which Darwin based his theory of evolution by natural selection is that individuals transmit traits to the next generation. However, Darwin was unable to explain *how* this occurs or *why* individuals vary within a population. Although he was a contemporary of Gregor Mendel (see Chapter 10), who elucidated the basic patterns of inheritance, Darwin was apparently not acquainted with Mendel's work, which was not recognized by the scientific community until the early part of the 20th century.

Beginning in the 1930s and 1940s, biologists experienced a conceptual breakthough when they combined the principles of Mendelian inheritance with Darwin's theory of natural selection. The result was a unified explanation of evolution known as the **modern synthesis,** or the **synthetic theory of evolution.** In this context, *synthesis* refers to combining parts of several previous theories to form a unified whole. Some of the founders of the modern synthesis were Russian geneticist Theodosius Dobzhansky (1900–1975), British geneticist and statistician Ronald Fisher (1890–1962), British geneticist J.B.S. Haldane (1892–1964), British biologist Julian Huxley (1887–1975), German taxonomist Ernst Mayr (1904–), U.S. paleontologist George Gaylord Simpson (1902–1984), U.S. botanist G. Ledyard Stebbins (1906–2000) and U.S. geneticist Sewell Wright (1889–1988).

Today, the modern synthesis incorporates our expanding knowledge in genetics, systematics, paleontology, developmental biology, behavior, and ecology. It explains Darwin's observation of variation among offspring in terms of **mutation,** or changes in DNA, such as nucleotide substitutions. Mutations provide the genetic variability on which natural selection acts during evolution. The modern synthesis, which emphasizes the genetics of populations as the central focus of evolution, has held up well since it was developed. It has dominated the thinking and research of biologists working in many areas and has resulted in an enormous accumulation of new discoveries that validate evolution by natural selection.

Most biologists not only accept the basic principles of the modern synthesis but also try to better understand the causal processes of evolution. For example, what is the role of chance in evolution? How rapidly do new species evolve? These and other questions have arisen in part from a re-evaluation of the fossil record and in part from new discoveries in molecular aspects of inheritance. Such critical analyses are an integral part of the scientific process because they stimulate additional observation and experimentation, along with re-examination of previous evidence. Science is an ongoing process, and information obtained in the future may require modifications to certain parts of the modern synthesis.

We now consider one of the many evolutionary questions currently being addressed by biologists: the relative effects of chance and natural selection on evolution.

Biologists study the effect of chance on evolution

Biologists have wondered, if we were able to repeat evolution by starting with similar organisms exposed to similar environmental conditions, would we get the same results? That is, would the same kinds of changes evolve, as a result of natural selection? Or would the organisms be quite different, as a result of random events? Several recently reported examples of evolution in action suggest that chance may not be as important as natural selection, at least at the population level.

A fruit fly species (*Drosophila subobscura*) native to Europe inhabits areas from Denmark to Spain. Biologists had noted that the northern flies have larger wings than southern flies (Fig. 17-5). The same fly species was accidentally introduced to North and South America in the late 1970s, in two separate introductions. Ten years after its introduction to the Americas, biologists determined that no statistically significant changes in wing size had occurred in the different regions of North America. However, 20 years after its introduction, the fruit flies in North America exhibited the same type of north–south wing changes as in Europe. (It is not known why larger wings evolve in northern areas and smaller wings in southern climates.)

George Gilchrist

FIGURE **17-5** **Wing size in female fruit flies.**
In Europe, female fruit flies (*Drosophila subobscura*) in northern countries have larger wings than flies in southern countries. Shown are two flies: one from Denmark *(right)* and the other from Spain *(left).* The same evolutionary pattern emerged in North America after the accidental introduction of *D. subobscura* to the Americas.

A study of the evolution of fish known as *sticklebacks* in three coastal lakes of west Canada yielded intriguingly similar results to the fruit fly study. Molecular evidence indicates that when the lakes first formed several thousand years ago, they were populated with the same ancestral species. (Analysis of the mitochondrial DNA of sticklebacks in the three lakes supports the hypothesis of a common ancestor.) In each lake, the same two species have evolved from the common ancestral fish: One species is large and consumes invertebrates along the bottom of the lake, whereas the other species is smaller and consumes plankton at the lake's surface. Members of the two species within a single lake do not interbreed with one another, but individuals of the larger species from one lake interbreed in captivity with individuals of the larger species from the other lakes. Similarly, smaller individuals from one lake interbreed with smaller individuals from the other lakes.

In these examples, natural selection appears to be a more important agent of evolutionary change than chance. If chance were the most important factor influencing the direction of evolution, then fruit fly evolution would not have proceeded the same way on two different continents, and stickleback evolution would not have proceeded the same way in three different lakes. However, just because we have many examples of the importance of natural selection in evolution, it does not necessarily follow that random events should be discounted as a factor in directing evolutionary change. Proponents of the role of chance think chance is more important in the evolution of major taxonomic groups (macroevolution) than in the evolution of populations (microevolution). It also may be that random events take place but their effects on evolution are simply harder to demonstrate than natural selection.

Review

- What is natural selection?
- Why are only inherited variations important in the evolutionary process?
- What part of Darwin's theory was he unable to explain? How does the modern synthesis fill this gap?

BiologyNow™ Assess your understanding of **Darwin and evolution** by taking the pretest on your BiologyNow CD-ROM.

EVIDENCE FOR EVOLUTION

Learning Objectives

4 Summarize the evidence for evolution obtained from the fossil record.

5 Summarize the evidence for evolution derived from comparative anatomy.

6 Define *biogeography,* and summarize how the distribution of organisms supports evolution.

7 Briefly explain how developmental biology provides insights into the evolutionary process.

8 Give an example of how evolutionary hypotheses are tested experimentally.

A vast body of scientific evidence supports evolution, including observations from the fossil record, comparative anatomy, biogeography, developmental biology, and molecular biology. In addition, evolutionary hypotheses are increasingly being tested experimentally.

Tom Till

FIGURE **17-6** **Exposed layers of sedimentary rock.** Shown are weathered rock layers near the Paria River in the Grand Staircase–Escalante National Monument, Utah. The younger layers overlie the older layers. Many layers date to the Mesozoic era, from 248 to 65 mya. Characteristic fossils are associated with each layer.

The fossil record provides strong evidence for evolution

Perhaps the most direct evidence for evolution comes from the discovery, identification, and interpretation of **fossils,** which are the remains or traces typically left in sedimentary rock by previously existing organisms. (The term *fossil* comes from the Latin word *fossilis,* meaning "something dug up.") Sedimentary rock forms by the accumulation and solidification of particles (pebbles, sand, silt, or clay) produced by the weathering of older rocks, such as volcanic rocks. The sediment particles, which are usually deposited on a riverbed, lake bottom, or the ocean floor, accumulate over time and exhibit distinct layers (Fig. 17-6). In an undisturbed rock sequence, the oldest layer is at the bottom, and upper layers are successively younger. The study of sedimentary rock layers, including their composition, arrangement, and correlation (similarity) from one location to another, enables geologists to place events recorded in rocks in their correct sequence.

The fossil record shows a progression from the earliest unicellular organisms to the many unicellular and multicellular organisms living today. The fossil record therefore demonstrates that life has evolved through time. To date, paleontologists (scientists who study extinct species) have described and named about 300,000 fossil species, and more are being discovered all the time.

Although most fossils are preserved in sedimentary rock, some more recent remains have been exceptionally well preserved in bogs, tar, amber (ancient tree resin), or ice (Fig. 17-7). For example, the remains of a woolly mammoth deep-frozen in Siberian ice for more than 25,000 years were so well preserved that part of its DNA could be analyzed.

The formation and preservation of a fossil require that an organism be buried under conditions that slow or prevent the decay process. This is most likely to occur if an organism's remains are covered quickly by a sediment of fine soil particles suspended in water. In this way remains of aquatic organisms may be trapped in bogs, mud flats, sandbars, or deltas. Remains of terrestrial organisms that lived on a flood plain may also be covered by water-borne sediments or, if the organism lived in an arid region, by wind-blown sand. Over time, the sediments harden to form sedimentary rock, and minerals usually replace the organism's remains so that many details of its structure, even cellular details, remain.

The fossil record is not a random sample of past life but instead is biased toward aquatic organisms and those living in the few terrestrial habitats conducive to fossil formation. Relatively few fossils of tropical rainforest organisms have been found, for example, because their remains decay extremely rapidly on the forest floor, before fossils can develop. Another reason for bias in the fossil record is that organisms with hard body parts such

FIGURE **17-7** **Fossils develop in different ways.**

(a) Although some fossils contain traces of organic matter, all that remains in this fossil of a seed fern leaf is an impression, or imprint, in the rock. **(b)** Petrified wood from the Petrified Forest National Park in Arizona consists of trees that were buried and infiltrated with minerals. The logs were exposed as the mudstone layers in which they were buried weathered. **(c)** A 2-million-year-old insect fossil (a midge) was embedded in amber. **(d)** A cast fossil of ancient echinoderms called *crinoids* formed when the crinoids decomposed, leaving a mold that later filled with dissolved minerals that hardened. **(e)** Dinosaur footprints, each 75 to 90 cm (2.5 to 3 ft) in length, provide clues about the posture, gait, and behavior of these extinct animals.

as bones and shells are more likely to form fossils than are those with soft body parts.

PROCESS OF SCIENCE

Because of the nature of the scientific process, each fossil discovery represents a separate "test" of the theory of evolution. If any of the tests fail, the theory would have to be modified to fit the existing evidence. The verifiable discovery, for example, of fossil remains of modern humans (*Homo sapiens*) in Precambrian rocks, which are more than 570 million years old, would falsify the theory of evolution as currently proposed. However, Precambrian rocks examined to date contain only fossils of simple organisms, such as algae and small, soft-bodied animals, that evolved early in the history of life. The earliest fossils of *H. sapiens* with anatomically modern features do not appear in the fossil record until approximately 100,000 years ago (see Chapter 21).

Fossils provide a record of ancient organisms and some understanding of where and when they lived. Using fossils of organisms from different geological ages, the lines of descent (evolutionary relationships) that gave rise to modern-day organisms can sometimes be inferred. In many instances, fossils provide direct evidence of the origin of new species from preexisting species, including many transitional forms.

Transitional fossils document whale evolution

Over the past century biologists have found evidence suggesting that whales and other cetaceans (an order of marine mammals) evolved from land-dwelling mammals. During the 1980s and 1990s, paleontologists discovered several fossil intermediates in whale evolution that document the whales' transition from land to water. Fossil evidence suggests that one candidate for the ancestor of whales is a now-extinct group of four-legged, land-dwelling mammals called *mesonychians* (Fig. 17-8a). These animals had unusually large heads and teeth that were remarkably similar to those of the earliest whales. About 50 million to 60 million years ago (mya), some descendants of mesonychians had adapted to swimming in shallow seas.

Fossils of *Ambulocetus natans*, a 50-million-year-old whale discovered in Pakistan, have many features of modern whales but also possess hind limbs and feet (Fig. 17-8b). (Modern whales do not have hindlimbs, although *vestigial* pelvic and hindlimb bones persist. Vestigial structures are discussed later in the chapter.) The vertebrae of *Ambulocetus'* lower back were very flexible, allowing the back to move dorsoventrally (up and down) during swimming and diving, as with modern whales. In addition to swimming, this ancient whale moved about on land, perhaps as sea lions do today.

Rodhocetus is a fossil whale found in slightly younger rocks in Pakistan (Fig. 17-8c). The vertebrae of *Rodhocetus* were even more flexible than those found in *Ambulocetus*. The flexible vertebrae allowed *Rodhocetus* a more powerful dorsoventral movement during swimming. *Rodhocetus* may have been totally aquatic.

By 40 mya, the whale transition from land to ocean was almost complete. Egyptian fossils of *Basilosaurus* show a whale with a streamlined body and front flippers for steering, like modern-day whales (Fig. 17-8d). *Basilosaurus* retained vestiges

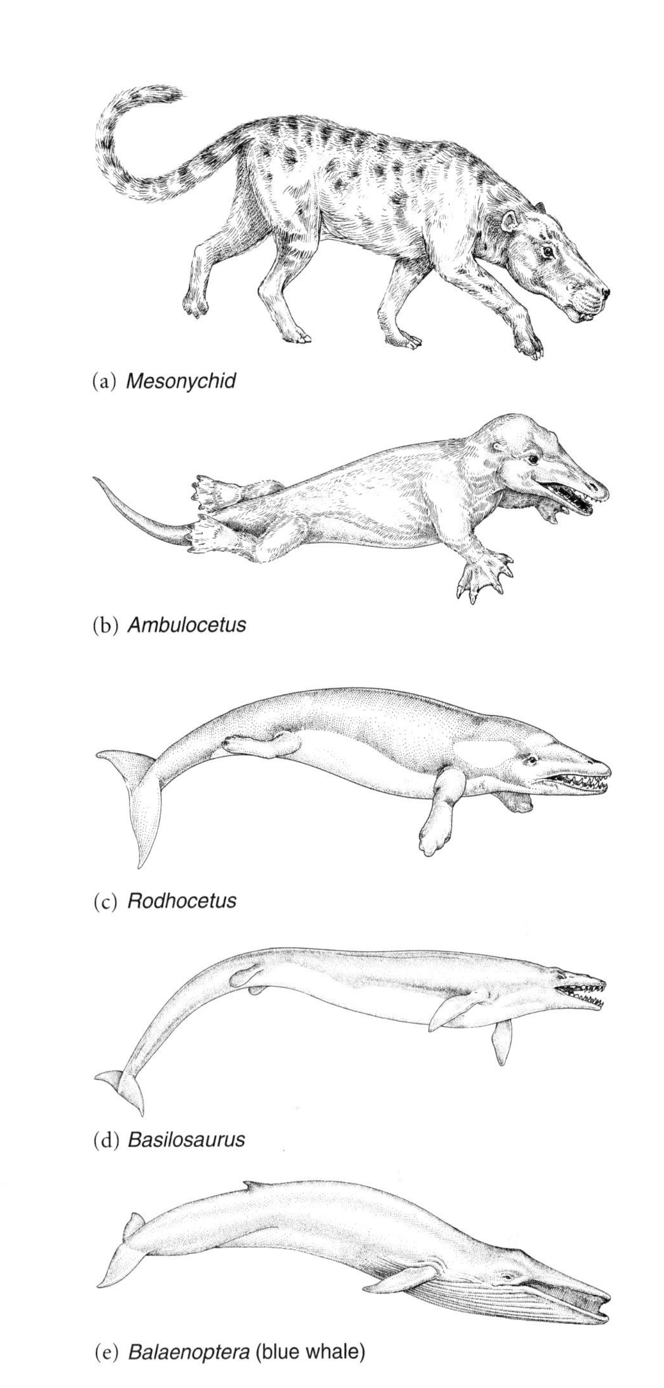

ACTIVE FIGURE **17-8** **Fossil intermediates in whale evolution.**

Figures are not drawn to scale. **(a)** *Mesonychid,* an extinct terrestrial mammal, may have been the ancestor of whales. **(b)** *Ambulocetus natans,* a transitional form between modern whale descendants and their terrestrial ancestors, possessed several recognizable whale features yet retained the hind limbs of its four-legged ancestors. **(c)** The more recent *Rodhocetus* had flexible vertebrae that permitted a powerful dorsoventral movement during swimming. **(d)** *Basilosaurus* was more streamlined and possessed tiny nonfunctional hindlimbs. **(e)** *Balaenoptera,* the modern blue whale, contains vestiges of pelvis and leg bones embedded in its body. *(a–d: Adapted with permission from Fig. 2 on pages 260–261 in Futuyma, D.J.* Science on Trial: The Case for Evolution, *Sinauer Associates, Inc., Sunderland, MA, 1995)*

BiologyNow™ Learn about **fossil intermediates in the evolution of a different species** by clicking on this figure on your BiologyNow CD-ROM.

of its land-dwelling ancestors—a pair of reduced hindlimbs that were disjointed from the backbone and probably not used in locomotion. Reduction in the hindlimbs continued to the present. The modern blue whale has vestigial pelvis and femur bones embedded in its body (Fig. 17-8e).

Various methods determine the age of fossils

Because layers of sedimentary rock occur naturally in the sequence of their deposition, with the more recent layers on top of the older, earlier ones, most fossils are dated by their relative position in sedimentary rock. However, geological events occurring after the rocks were initially formed have occasionally changed the relationships of some rock layers. Geologists identify specific sedimentary rocks not only by their positions in layers but also by features such as mineral content and by the fossilized remains of certain organisms, known as **index fossils,** that characterize a specific layer over large geographical areas. Index fossils are fossils of organisms that existed for a relatively short geological time but were preserved as fossils in large numbers. With this information, geologists can arrange rock layers and the fossils they contain in chronological order and identify comparable layers in widely separated locations.

Radioactive isotopes, also called **radioisotopes,** present in a rock provide a means to accurately measure its age (see Chapter 2). Radioisotopes emit invisible radiations. As a radioisotope emits radiation, its nucleus changes into the nucleus of a different element in a process known as **radioactive decay.** The radioactive nucleus of uranium-235, for example, decays into lead-207.

Each radioisotope has its own characteristic rate of decay. The time required for one half of the atoms of a radioisotope to change into a different atom is known as its **half-life** (Fig. 17-9). Radioisotopes differ significantly in their half-lives. For example, the half-life of iodine-132 is only 2.4 hours, whereas the half-life of uranium-235 is 704 million years. The half-life of a particular radioisotope is constant and does not vary with temperature, pressure, or any other environmental factor.

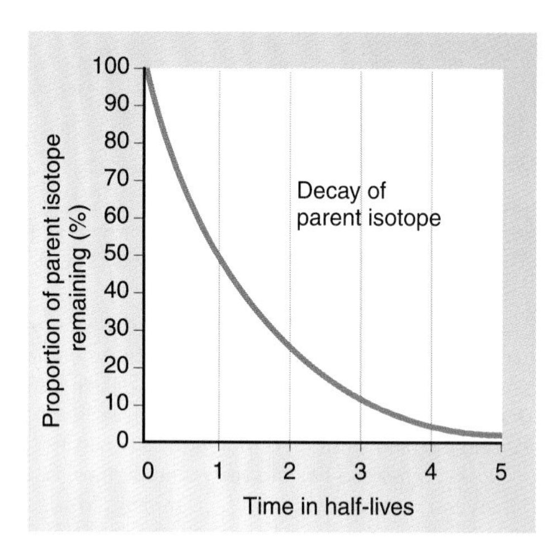

FIGURE **17-9** **Radioisotope decay.**

At time zero, the sample is composed entirely of the radioisotope, and the radioactive clock begins ticking. After one half-life, only 50% of the original radioisotope remains. During each succeeding half-life, half of the remaining radioisotope is converted to decay product(s).

The age of a fossil in sedimentary rock is usually estimated by measuring the relative proportions of the original radioisotope and its decay product in volcanic rock intrusions that penetrate the sediments. For example, the half-life of potassium-40 is 1.3 billion years, meaning that in 1.3 billion years half of the radioactive potassium will have decayed into its decay product, argon-40. The radioactive clock begins ticking when the magma solidifies into volcanic rock. The rock initially contains some potassium but no argon. Because argon is a gas, it escapes from hot rock as soon as it forms, but when potassium decays in rock that has cooled and solidified, the argon accumulates in the crystalline structure of the rock. If the ratio of potassium-40 to argon-40 in the rock being tested is 1:1, the rock is 1.3 billion years old.

Several radioisotopes are commonly used to date fossils. These include potassium-40 (half-life 1.3 billion years), uranium-235 (half-life 704 million years), and carbon-14 (half-life 5730 years). Potassium-40, with its long half-life, is used to date fossils that are many hundreds of millions of years old. Radioisotopes other than carbon-14 are used to date the *rock* in which fossils are found, whereas carbon-14 is used to date the *carbon remains* of anything that was once living, such as wood, bones, and shells. Whenever possible, the age of a fossil is independently verified using two or more different radioisotopes.

Carbon-14, which is continuously produced in the atmosphere from nitrogen-14 (by cosmic radiation), subsequently decays back to nitrogen-14. Because the formation and the decay of carbon-14 occur at constant rates, the ratio of carbon-14 to carbon-12 (a more abundant, stable isotope of carbon) is constant in the atmosphere. Organisms absorb carbon from the atmosphere either directly (by photosynthesis) or indirectly (by consuming photosynthetic organisms). Since each organism absorbs carbon from the atmosphere, its ratio of carbon-14 to carbon-12 is the same as the atmosphere. When an organism dies, however, it no longer absorbs carbon, and the proportion of carbon-14 in its remains declines as carbon-14 decays to nitrogen-14. Because of its relatively short half-life, carbon-14 is useful for dating fossils that are 50,000 years old or less. It is particularly useful for dating archaeological sites.

Comparative anatomy of related species demonstrates similarities in their structures

Comparing the structural details of features found in different but related organisms reveals a basic similarity. Such features that are derived from the same structure in a common ancestor are termed **homologous features;** the condition is known as **homology.** For example, consider the limb bones of mammals. A human arm, a cat forelimb, a whale front flipper, and a bat wing, although quite different in appearance, have strikingly similar arrangements of bones, muscles, and nerves. Figure 17-10 shows a comparison of their skeletal structures. Each has a single

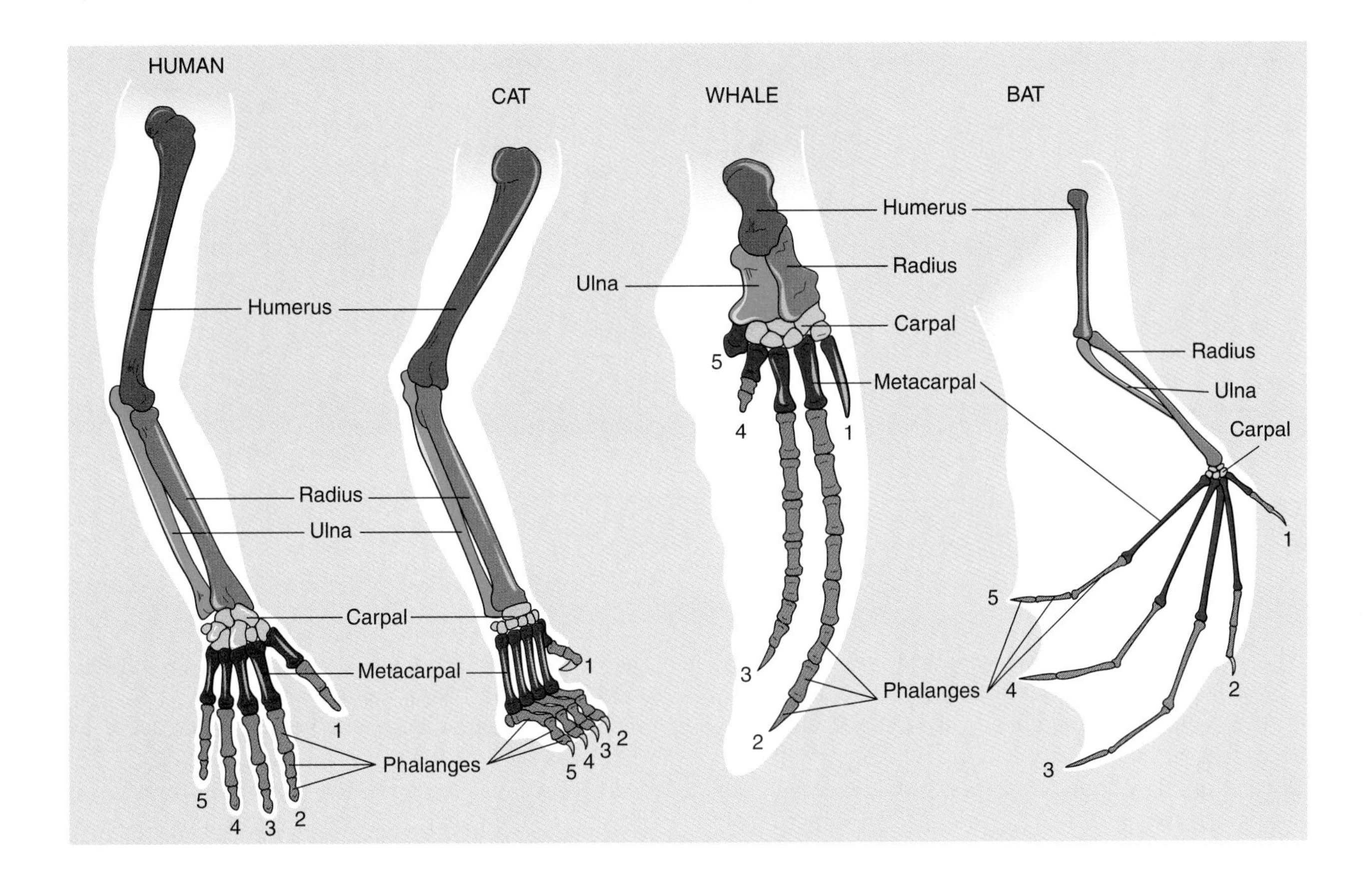

FIGURE **17-10** | **Homology in animals.**

The human arm, cat forelimb, whale flipper, and bat wing have a basic underlying similarity of structure because they are derived from a common ancestor. The five digits are numbered in each drawing.

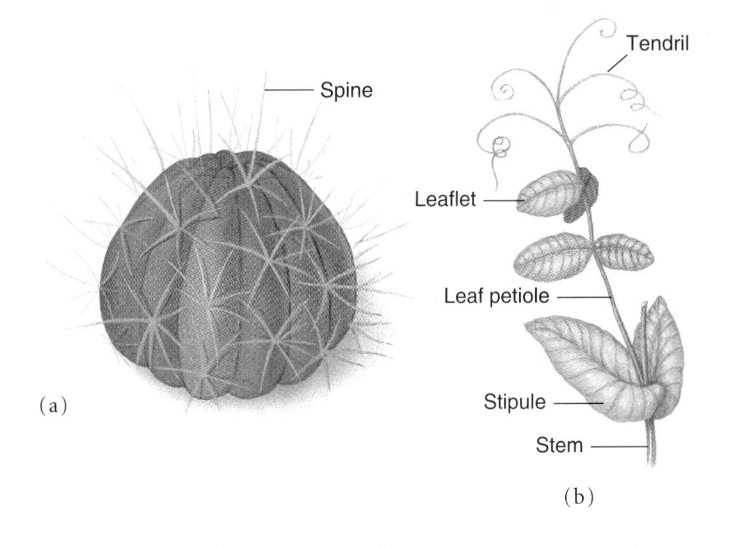

FIGURE **17-11** | **Homology in plants.**

(a) The spines of the fishhook cactus (*Ferocactus wislizenii*) are modified leaves, as are **(b)** the tendrils of the garden pea (*Pisum sativum*). Leaves of the garden pea are compound, and the terminal leaflets are modified into tendrils that are frequently branched.

bone (the humerus) in the part of the limb nearest the trunk of the body, followed by the two bones (radius and ulna) of the forearm, a group of bones (carpals) in the wrist, and a variable number of digits (metacarpals and phalanges). This similarity is particularly striking because arms, forelimbs, flippers, and wings are used for different types of locomotion, and there is no overriding mechanical reason for them to be so similar structurally. Similar arrangements of parts of the forelimb are evident in ancestral reptiles and amphibians and even in the first fishes that came out of water onto land hundreds of millions of years ago.

Leaves are an example of homology in plants. In many plant species, leaves have been modified for functions other than photosynthesis. A cactus spine and a pea tendril, although quite different in appearance, are homologous because both are modified leaves (Fig. 17-11). The spine protects the succulent stem tissue of the cactus, whereas the tendril, which winds around a small object once it makes contact, helps support the climbing stem of the pea plant. Such modifications in organs used in different ways are the expected outcome of a common evolutionary origin. The basic structure present in a common ancestor was modified in different ways for different functions as various descendants subsequently evolved.

Not all species with "similar" features have descended from a recent common ancestor, however. Structurally similar features that are not homologous but simply have similar functions that evolved independently in distantly related organisms are said to be **homoplastic features.** Such a condition is called **homoplasy.**[2] For example, the wings of various distantly re-

[2] An older, less precise term that some biologists still use for nonhomologous features with similar functions is *analogy.*

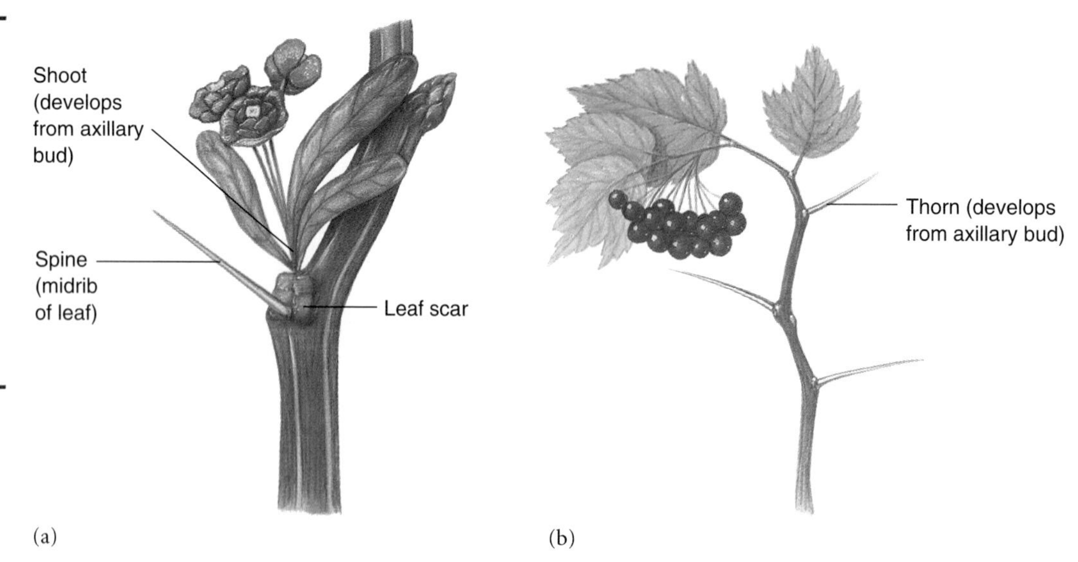

FIGURE **17-12**

Homoplasy.

(a) A spine of Japanese barberry (*Berberis thunbergii*) is a modified leaf. (In this example, the spine is actually the midrib of the original leaf, which has been shed.) **(b)** Thorns of downy hawthorn (*Crataegus mollis*) are modified stems that develop from axillary buds.

lated flying animals, such as insects and birds, resemble each other superficially; they are homoplastic features that evolved over time to meet the common function of flight, although they differ in more fundamental aspects. Bird wings are modified forelimbs supported by bones, whereas insect wings may have evolved from gill-like appendages present in the aquatic ancestors of insects. Spines, which are modified leaves, and thorns, which are modified stems, are an example of homoplasy in plants. Spines and thorns resemble one another superficially but are homoplastic features that evolved independently to solve the common need for protection from herbivores (Fig. 17-12).

Like homology, homoplasy offers crucial evidence of evolution. Homoplastic features are of evolutionary interest because they demonstrate that organisms with separate ancestries may adapt in similar ways to similar environmental demands. Such independent evolution of similar structures in distantly related organisms is known as **convergent evolution.** Aardvarks, anteaters, and pangolins are an excellent example of convergent evolution (Fig. 17-13). They resemble one another in lifestyle and certain structural features. All have strong, sharp claws to dig open ant and termite mounds and elongated snouts with long, sticky tongues to catch these insects. Yet aardvarks, anteaters, and pangolins evolved from three distantly related orders of mammals. (See Chapter 22 for further discussion of homology and homoplasy. Also, see Figure 30-25, which shows several examples of convergent evolution in placental and marsupial mammals.)

Comparative anatomy reveals the existence of **vestigial structures.** Many organisms contain organs or parts of organs that are seemingly nonfunctional and degenerate, often undersized or lacking some essential part. Vestigial structures are remnants of more developed structures that were present and functional in ancestral organisms. In the human body, more than 100 structures are considered vestigial, including the coccyx (fused tailbones), third molars (wisdom teeth), and the muscles that move our ears. Whales and pythons have vestigial hindlimb bones (Fig. 17-14); pigs have vestigial toes that do not touch the ground; wingless birds such as the kiwi have vestigial wing bones; and many blind, burrowing or cave-dwelling animals have nonfunctioning, vestigial eyes.

The occasional presence of a vestigial structure is to be expected as a species adapts to a changing mode of life. Some structures become much less important for survival and may end up as vestiges. When a structure no longer confers a selective advantage, it may become smaller and lose much or all of its function with the passage of time. Because the presence of the vestigial structure is usually not harmful to the organism, however, selective pressure for completely eliminating it is weak, and the vestigial structure is found in many subsequent generations.

The distribution of plants and animals supports evolution

The study of the past and present geographic distribution of organisms is called **biogeography.** The geographic distribution of organisms affects their evolution. Darwin was interested in biogeography, and he considered why the species found on ocean islands tend to resemble species of the nearest mainland, even if the environment is different. He also observed that species on ocean islands do not tend to resemble species on islands with similar environments in other parts of the world. Darwin studied the plants and animals of two sets of arid islands—the Cape Verde Islands, nearly 640 km (400 mi) off western Africa, and the Galapagos Islands, about 960 km (600 mi) west of Ecuador, South America. On each group of islands, the plants and terrestrial animals were indigenous (native), but those of the Cape Verde Islands resembled African species and those of the Galapagos resembled South American species. The similarities of Galapagos species to South American species were particularly striking considering that the Galapagos Islands are dry and rocky and the nearest part of South America is humid and has a lush tropical rain forest. Darwin concluded that species from the neighboring continent migrated or were carried to the islands, where they subsequently adapted to the new environment and, in the process, evolved into new species.

FIGURE **17-13** **Convergent evolution.**

Three distantly related mammals adapted independently to eat ants and termites in similar grassland/forest environments in different parts of the world. **(a)** The aardvark (*Orycteropus afer*) is native to central, southern, and eastern Africa. **(b)** A giant anteater (*Myrmecophaga tridactyla*) at a termite mound. The anteater is native to Latin America, from southern Mexico to northern Argentina. **(c)** The pangolin (*Manis crassicaudata*) is native to Africa and to southern and southeastern Asia.

If evolution were not a factor in the distribution of species, we would expect to find a given species everywhere that it could survive. However, the geographic distribution of organisms that actually exists makes sense in the context of evolution. For example, Australia, which has been a separate land mass for millions of years, has distinctive organisms. Australia has populations of egg-laying mammals (monotremes) and pouched mammals (marsupials) not found anywhere else. Two hundred million years ago, Australia and the other continents were joined together in a major land mass. Over the course of millions of years, the Australian continent gradually separated from the others. The monotremes and marsupials in Australia continued to thrive and diversify. The isolation of Australia also prevented placental mammals, which arose elsewhere at a later time, from competing with its monotremes and marsupials. In other areas of the world where placental mammals occurred, most monotremes and marsupials became extinct.

We now consider how Earth's dynamic geology has affected biogeography and evolution.

FIGURE **17-14** **Vestigial structures.**

(a) An African rock python (*Python sebae*) **(b)** Closeup of part of a python skeleton showing the hindlimb bones. All pythons have remnants of hindlimb bones embedded in their bodies.

Earth's geological history is related to biogeography and evolution

In 1915 the German scientist Alfred Wegener, who had noted a correspondence between the geographical shapes of South America and Africa, proposed that all the landmasses had at one time been joined into one huge supercontinent, which he called Pangaea (Fig. 17-15a). He further suggested that Pangaea had subsequently broken apart and that the various landmasses had separated in a process known as **continental drift.** Wegener did not know of any mechanism that could have caused continental drift, so his idea, although debated initially, was largely ignored.

In the 1960s, scientific evidence accumulated that provided the explanation for continental drift. Earth's crust is composed of seven large plates (plus a few smaller ones) that float on the mantle, which is the mostly solid layer of Earth lying beneath

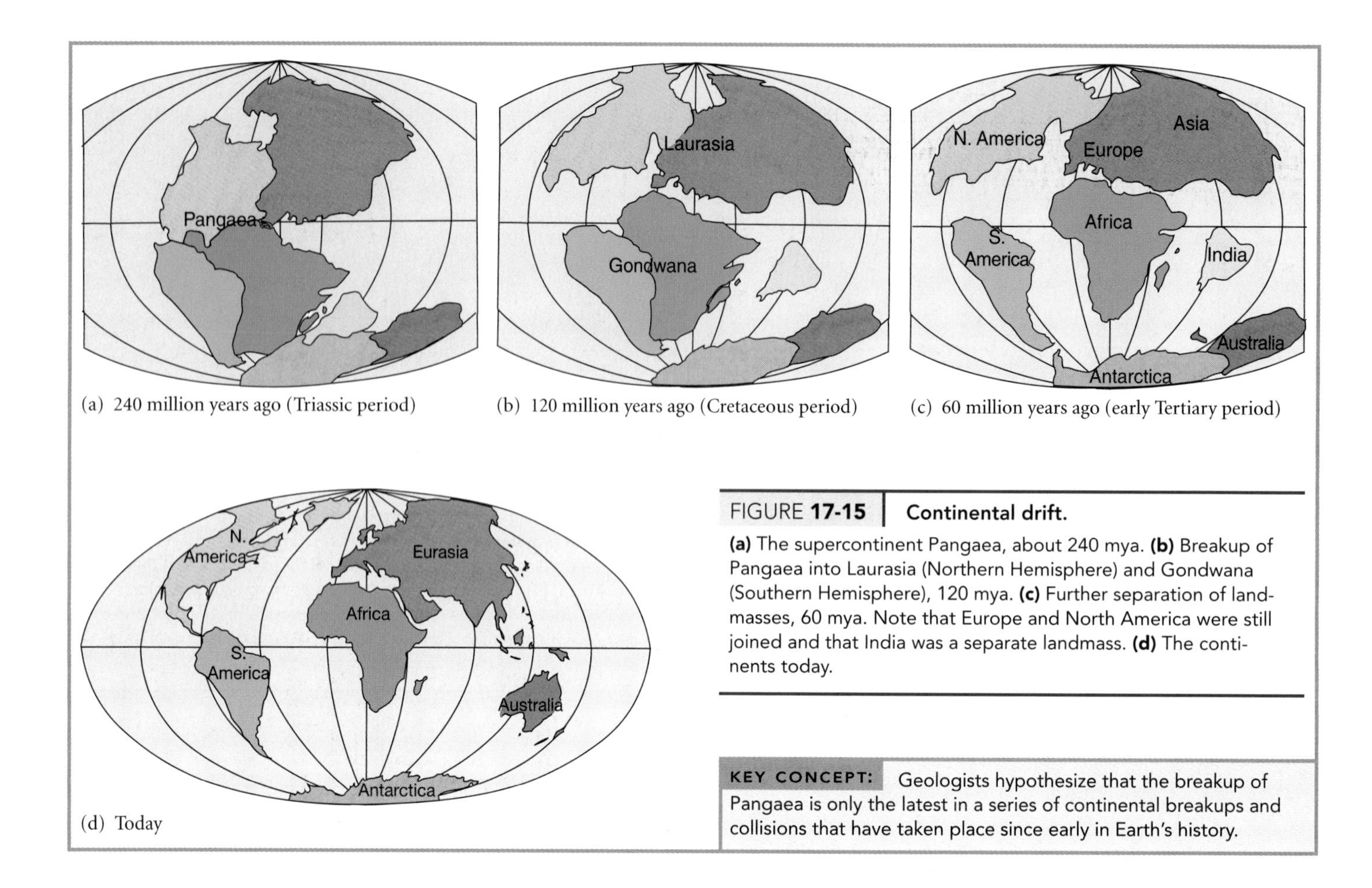

(a) 240 million years ago (Triassic period) (b) 120 million years ago (Cretaceous period) (c) 60 million years ago (early Tertiary period) (d) Today

FIGURE **17-15** **Continental drift.**
(a) The supercontinent Pangaea, about 240 mya. **(b)** Breakup of Pangaea into Laurasia (Northern Hemisphere) and Gondwana (Southern Hemisphere), 120 mya. **(c)** Further separation of landmasses, 60 mya. Note that Europe and North America were still joined and that India was a separate landmass. **(d)** The continents today.

KEY CONCEPT: Geologists hypothesize that the breakup of Pangaea is only the latest in a series of continental breakups and collisions that have taken place since early in Earth's history.

the crust and above the core.[3] The landmasses are situated on some of these plates. As the plates move, the continents change their relative positions (Fig. 17-15b, c, and d). The movement of the crustal plates is termed **plate tectonics.**

Any area where two plates meet is a site of intense geological activity. Earthquakes and volcanoes are common in such a region. Both San Francisco, noted for its earthquakes, and the Mount Saint Helens volcano are situated where two plates meet. If landmasses lie on the edges of two adjacent plates, mountains may form. The Himalayas formed when the plate carrying India rammed into the plate carrying Asia. When two plates grind together, one of them is sometimes buried under the other in a process known as *subduction.* When two plates move apart, a ridge of lava forms between them. The Atlantic Ocean is getting larger because of the expanding zone of lava along the Mid-Atlantic Ridge, where two plates are separating.

Knowledge that the continents were at one time connected and have since drifted apart is useful in explaining certain aspects of biogeography (Fig. 17-16). Likewise, continental drift has played a major role in the evolution of different organisms. When Pangaea originally formed during the late Permian period, it brought together terrestrial species that had evolved separately from one another, leading to competition and some extinctions. Marine life was adversely affected, in part because, with the continents joined as one large mass, less coastline existed. (Because coastal areas are shallow, they contain high concentrations of marine species.)

Pangaea separated into several land masses approximately 180 mya. As the continents began to drift apart, populations became geographically isolated in different environmental conditions and began to diverge along separate evolutionary pathways. As a result, the plants, animals, and other organisms of previously connected continents—South America and Africa, for example—differ. Continental drift also caused gradual changes in ocean and atmospheric currents that have profoundly influenced the biogeography and evolution of organisms. (Biogeography is discussed further in Chapter 54.)

Developmental biology helps unravel evolutionary patterns

How snakes became elongated and lost their limbs has long intrigued evolutionary biologists. Comparative anatomy indicates, for example, that pythons have vestigial hindlimb bones embedded in their bodies (see Fig. 17-14). Other snakes have lost their hindlimbs entirely.

Increasingly, developmental biology, particularly at the molecular level, is providing answers to such questions. In many cases, evolutionary changes such as limblessness in snakes, occur as a result of changes in genes that affect the orderly sequence of events that occur during development. In pythons, for exam-

[3] Most of the rock in the upper portion of the mantle is solid, although 1% or 2% is melted. Because of its higher temperature, the solid rock of the mantle is more plastic than the solid rock of the Earth's crust above it.

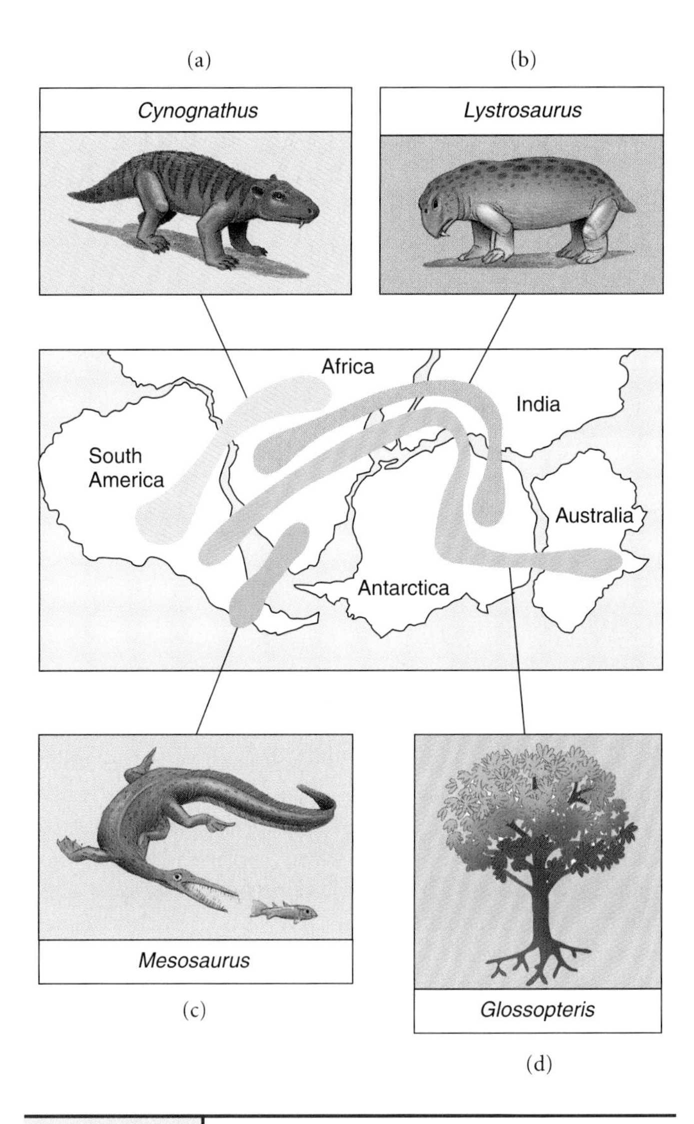

FIGURE **17-16**

Distribution of fossils on continents that were joined during the Permian and Triassic periods (286–213 mya).

(a) *Cynognathus* was a carnivorous reptile found in Triassic rocks in South America and Africa. **(b)** *Lystrosaurus* was a large, herbivorous reptile with beaklike jaws that lived during the Triassic period. Fossils of *Lystrosaurus* have been found in Africa, India, and Antarctica. **(c)** *Mesosaurus* was a small freshwater reptile found in Permian rocks in South America and Africa. **(d)** *Glossopteris* was a seed-bearing tree dating from the Permian period. *Glossopteris* fossils have been found in South America, Africa, India, Antarctica, and Australia. *(Adapted from Colbert, E.H.* Wandering Lands and Animals, *Hutchinson, London, 1973)*

ple, the loss of forelimbs and elongation of the body are linked to mutations in several *Hox* genes that affect the expression of body patterns and limb formation in a wide variety of animals (see Chapter 16 discussion of *Hox* gene clusters). Apparently the hindlimbs do not develop because python embryonic tissue does not respond to internal signals that trigger leg elongation.

Scientific evidence overwhelmingly demonstrates that development in different animals is controlled by the same kinds of genes; these genetic similarities in a wide variety of organisms reflect a shared evolutionary history. For example, all vertebrates have similar patterns of embryological development that indicate they share a common ancestor. All vertebrate embryos have segmented muscles, pharyngeal (gill) pouches, a tubular heart without left and right sides, a system of arteries known as *aortic arches* in the gill region, and many other shared features. All these structures are necessary and functional in the developing fish. The small, segmented muscles of the fish embryo give rise to the segmented muscles used by the adult fish in swimming. The gill pouches break through to the surface as gill slits. The adult fish heart remains undivided and pumps blood forward to the gills that develop in association with the aortic arches.

Because none of these embryonic features persists in the adults of reptiles, birds, or mammals, why are these fishlike structures present in their embryos? Evolution is a conservative process, and natural selection builds on what has come before rather than starting from scratch. The evolution of new features often does not require the evolution of new developmental genes but instead depends on a modification in developmental genes that already exist (see Chapter 19 discussion of preadaptations). Terrestrial vertebrates are thought to have evolved from fishlike ancestors; therefore, they share some of the early stages of development still found in fish today. The accumulation of genetic changes over time in these vertebrates has modified the basic body plan laid out in fish development.

Molecular comparisons among organisms provide evidence for evolution

Similarities and differences in the biochemistry and molecular biology of various organisms provide evidence for evolutionary relationships. Lines of descent based solely on biochemical and molecular characters often resemble lines of descent based on structural and fossil evidence. Molecular evidence for evolution includes the universal genetic code and the conserved sequences of amino acids in proteins and of nucleotides in DNA.

The genetic code is virtually universal

Organisms owe their characteristics to the types of proteins they possess, which in turn are determined by the sequence of nucleotides in their messenger ribonucleic acid (mRNA), as specified by the order of nucleotides in their DNA. Evidence that all life is related comes from the fact that all organisms use a genetic code that is virtually identical.[4] Recall from Chapter 12 that the genetic code specifies a triplet (a sequence of three nucleotides in DNA) that codes for a particular codon (a sequence of three nucleotides in mRNA). The codon then codes for a particular amino acid in a polypeptide chain. For example, "AAA" in DNA codes for "UUU" in mRNA, which codes for the amino acid phenylalanine in organisms as diverse as shrimp, humans, bacteria, and tulips. In fact, "AAA" codes for phenylalanine in all organisms examined to date.

The universality of the genetic code—no other code has been found in any organism—is compelling evidence that all organ-

[4] There is some minor variation in the genetic code. For example, mitochondria have some deviations from the standard code.

isms arose from a common ancestor. The genetic code has been maintained and transmitted through all branches of the evolutionary tree since its origin in some extremely early (and successful) organism.

Proteins and DNA contain a record of evolutionary change

Thousands of comparisons of protein and DNA sequences from various species have been done during the past 25 years or so. In many cases, sequence-based relationships agree with earlier studies, which based evolutionary relationships on similarities in structure among living organisms and on fossil data of extinct organisms.

Investigations of the sequence of amino acids in proteins that play the same roles in many species have revealed both great similarities and certain specific differences. Even organisms that are only remotely related, such as humans, fruit flies, sunflowers, and yeasts, share some proteins, such as cytochrome *c*, which is part of the electron transport chain in aerobic respiration. To survive, all aerobic organisms need a respiratory protein with the same basic structure and function as the cytochrome *c* of their common ancestor. Consequently, not all amino acids that confer the structural and functional features of cytochrome *c* are free to change. Any mutations that changed the amino acid sequence at structurally important sites of the cytochrome *c* molecule would have been harmful, and natural selection would have prevented such mutations from being passed to future generations. However, in the course of the long, independent evolution of different organisms, mutations have resulted in the substitution of many amino acids at less important locations in the cytochrome *c* molecule. The greater the differences in the amino acid sequences of their cytochrome *c* molecules, the longer it has been since two species diverged.

Because a protein's amino acid sequences are coded in DNA, the differences in amino acid sequences indirectly reflect the nature and number of underlying DNA base-pair changes that must have occurred during evolution. Of course, not all DNA codes for proteins (witness introns and transfer RNA genes). **DNA sequencing**—that is, determining the order of nucleotide bases in DNA—of both protein-coding DNA and non–protein-coding DNA is useful in determining evolutionary relationships.

Generally, the more closely species are considered related on the basis of other scientific evidence, the greater the percentage of nucleotide sequences that their DNA molecules have in common. By using the DNA sequence data in Table 17-1, for example, you can conclude that the closest living relative of humans is the chimpanzee (because its DNA has the lowest percentage differences in the sequence examined). Which of the primates in Table 17-1 is the most distantly related to humans? (Primate evolution is discussed in Chapter 21.)

PROCESS OF SCIENCE

In some cases, molecular evidence challenges traditional evolutionary ideas that were based on structural comparisons among living species and/or on studies of fossil skeletons. Consider artiodactyls, an order of even-toed hoofed mammals such as pigs, camels, deer, antelope, cattle, and hippopotamuses. Traditionally, whales, which do not have toes, are not classified as artiodactyls (although early fossil whales had an even number of toes on their appendages).

TABLE 17-1 Differences in Nucleotide Sequences in DNA as Evidence of Phylogenetic Relationships

Primate Species Pairs	Percent Divergence in a Selected DNA Sequence
Human–chimpanzee	1.7
Human–gorilla	1.8
Human–orangutan	3.3
Human–gibbon	4.3
Human–rhesus monkey (Old World monkey)	7.0
Human–spider monkey (New World monkey)	10.8
Human–tarsier	24.6

Source: From M., Goodman, et al., "Primate Evolution at the DNA Level and a Classification of Hominoids," *Journal of Molecular Evolution*, Vol. 30, 1990.

Note: Percent divergence refers to how different the base sequences are for the same gene in different organisms. In this example, humans and chimpanzees have a 1.7% difference in their DNA base sequences, which means that 98.3% of the DNA examined is identical. The data shown are for the noncoding sequence of the β-globin gene.

Figure 17-17 depicts a hypothetical phylogenetic tree for whales and selected artiodactyls based on molecular data. Such **phylogenetic trees**—diagrams showing lines of descent—can be derived from differences in a given DNA nucleotide sequence. This diagram suggests whales should be classified as artiodactyls and shows that hippopotamuses are more closely related to whales than any other artiodactyl. The branches representing whales and hippopotamuses probably diverged relatively recently because of the close similarity of DNA sequences in these species. In contrast, camels, which have DNA sequences that are less similar to whales, diverged much earlier. The molecular evidence indicates that whales and hippopotamuses share a recent common ancestor, a hippo-like artiodactyl that split from the rest of the artiodactyl line some 55 mya.

However, available fossil evidence does not currently provide support for the molecular hypothesis; a fossil ancestor common to both whales and hippos has not yet been discovered. (Recall from earlier in the chapter that most paleontologists currently suggest that the mesonychians, which are not ancient artiodactyls, may have been the ancestor of whales.) Scientists hope that future fossil discoveries will help clarify this discrepancy between molecular and fossil data.

DNA sequencing is used to estimate the time of divergence between two closely related species or taxonomic groups

Imagine you know the distance from Miami, Florida, to New York City, and you also know how long it takes to drive that distance. Now imagine you do *not* know the distance from Miami to Chicago, but you do know how long it takes to drive that distance. Based on the knowledge you have, you can infer the distance from Miami to Chicago. Similar reasoning is used to estimate the divergence between closely related taxonomic groups.

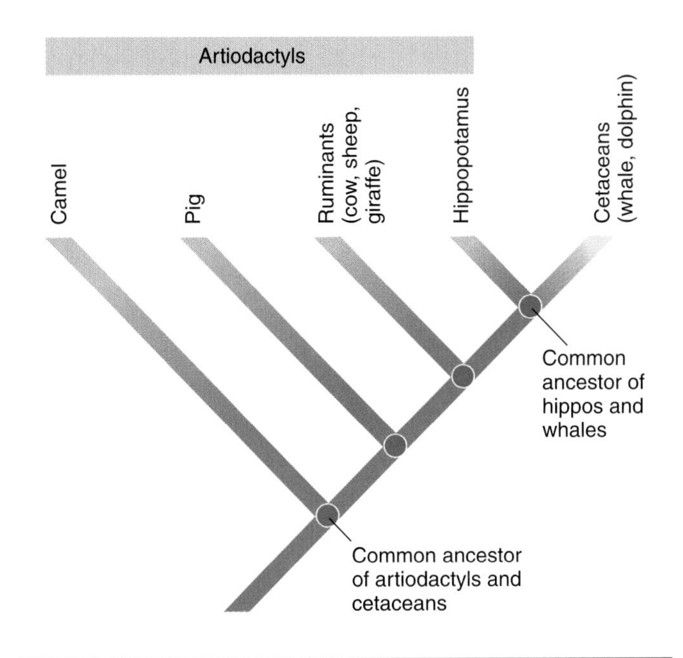

FIGURE 17-17 Phylogenetic tree of whales and their closest living relatives.

This branching diagram, called a *cladogram*, shows hypothetical evolutionary relationships. Based on DNA sequence differences among selected mammals, it suggests that artiodactyls are the close relatives of whales, that the hippopotamus is the closest living artiodactyl relative of whales, and that artiodactyls and whales share a common ancestor in the distant past. The nodes (circles) represent branch points where a species splits into two or more lineages. (Ruminants are artiodactyls that have a multichambered stomach and chew regurgitated plant material to make it more digestible.) *(Adapted from M., Nikaido, et al., "Phylogenetic Relationships among Cetartiodactyls Based on Insertions of Short and Long Interspersed Elements: Hippopotamuses Are the Closest Extant Relatives of Whales,"* Proceedings of the National Academy of Sciences, *Vol. 96, 31 Aug. 1999)*

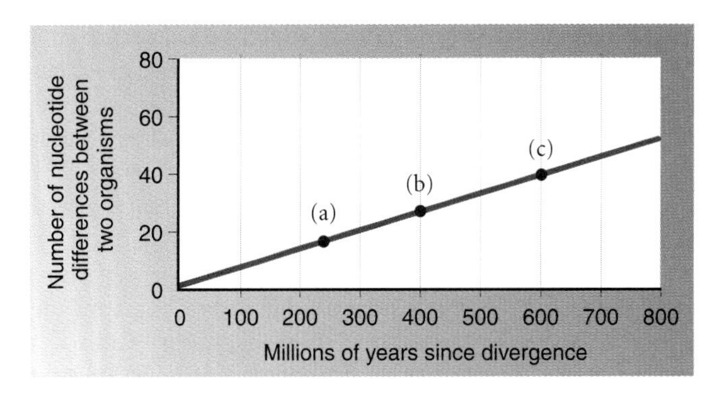

FIGURE 17-18 Calibration and use of a molecular clock.

In this hypothetical example, the DNA of a specific gene is sequenced for birds, reptiles, fish, and insects. The number of nucleotide differences between birds and reptiles (a) is placed on a graph at the time at which birds and reptiles are thought to have diverged based on fossil evidence. Likewise, the number of nucleotide differences between reptiles and fish (b) is placed on the graph at the time of branching indicated by the fossil record. A line is drawn through points (a) and (b) and extended, enabling scientists to estimate a much earlier time (c) at which the insect line diverged from the vertebrate line.

Within a given taxonomic group, mutations are assumed to have occurred at a fairly steady rate over millions of years. Thus, if more differences occur in homologous sequences of DNA of one species compared with another, more time elapsed since the two species diverged from a common ancestor.

From the number of alterations in homologous DNA sequences taken from different species, we can develop a **molecular clock** to estimate the time of divergence between two closely related species or higher taxonomic groups. A molecular clock makes use of the average rate at which a particular gene evolves. The clock is calibrated by comparing the number of nucleotide differences between two organisms with the dates of evolutionary branch points that are known from the fossil record. Once a molecular clock is calibrated, past evolutionary events whose timing is not known with certainty are estimated (Fig. 17-18).

Molecular clocks can be used to complement geological estimates of the divergence of species or to assign tentative dates to evolutionary events that lack fossil evidence. Where there is no fossil record of an evolutionary event, molecular clocks are the only way to estimate the timing of that event. Molecular clocks are also used, along with fossil evidence and structural data, to help reconstruct **phylogeny,** which is the evolutionary history of a group of related species (see Chapter 22). By assigning tentative dates to the divergence of species, molecular clocks show the relative order of branch points in phylogeny.

Molecular clocks must be developed and interpreted with care. Mutation rates vary among different genes and among distantly related taxonomic groups, causing molecular clocks to tick at different rates. Some genes, such as the gene for the respiratory protein cytochrome *c*, code for proteins that lose their function if the amino acid sequence changes slightly; these genes evolve slowly. Other genes, such as genes for blood-clotting proteins, code for proteins that are less constrained by changes in amino acid sequence; these genes evolve rapidly. Scientists consider molecular clocks that are based on several genes to be more accurate than clocks based on a single gene.

Although many dates estimated by molecular clocks agree with fossil evidence, some discrepancies between molecular clocks and fossils exist. In most of these cases, the molecular clock's estimates of divergence times of particular organisms are much older than the dates at which the groups are first observed in the fossil record. Resolving these differences will require additional research in both molecular biology and paleontology.

Evolutionary hypotheses are tested experimentally

PROCESS OF SCIENCE

Increasingly, biologists are designing imaginative experiments, often in natural settings, to test evolutionary hypotheses. David Reznick from the University of California at Santa Barbara and John Endler from James Cook University in Australia have studied evolution in guppy populations in Venezuela and in Trinidad, a small island in the southern Caribbean.

Reznick and Endler observed that different streams have different kinds and numbers of fishes that prey on guppies.

Predatory fish that prey on larger guppies are present in all streams at lower elevations; these areas of intense predation pressure are known as *high-predation habitats.* Predators are often excluded from tributaries or upstream areas by rapids and waterfalls. The areas above such barriers are known as *low-predation habitats* because they contain only one species of small predatory fish that occasionally eats smaller guppies.

Differences in predation are correlated with many differences in the guppies, such as male coloration, behavior, and attributes known as *life history traits.* These traits include age and size at sexual maturity, the number of offspring per reproductive event, the size of the offspring, and the frequency of reproduction. For example, guppy adults are larger in streams found at higher elevations and smaller in streams found at lower elevations.

Do predators actually cause these differences to evolve? Reznick and colleagues tested this evolutionary hypothesis by conducting field experiments in Trinidad. Taking advantage of waterfalls that prevent upstream movement of guppies, guppy predators, or both, they moved either guppies or guppy predators over such barriers. For example, guppies from a high-predation habitat were introduced into a low-predation habitat by moving them over a barrier waterfall into a section of stream that was free of guppies and large predators. The only fish species that lived in this section of stream before the introduction was the predator that occasionally preyed on small guppies.

Eleven years later, the researchers captured adult females from the introduction site (low-predation habitat) and the control site below the barrier waterfall (high-predation habitat). They bred these females in their laboratory and compared the life history traits of succeeding generations. The descendants of guppies introduced into the low-predation habitat matured at an older age and larger size than did the descendants of guppies from the control site below the waterfall (Fig. 17-19). They also produced fewer, but larger, offspring. The life histories of the introduced fish had therefore evolved to be similar to those of fish typically found in such low-predation habitats. Similar studies have demonstrated that predators have played an active role in the evolution of other traits, such as average number of offspring produced during the lifetime of an individual female (fecundity), male coloration, and behavior.

These and other experiments demonstrate not only that evolution is real but also that it is occurring now, driven by selective environmental forces, such as predation, that can be experimentally manipulated. Darwin incorrectly assumed evolution is so gradual that humans cannot observe it. As Jonathan Weiner, author of *The Beak of the Finch: A Story of Evolution in Our Time,* puts it, "Darwin did not know the strength of his own theory. He vastly underestimated the power of natural selection. Its action is neither rare nor slow. It leads to evolution daily and hourly, all around us, and we can watch."

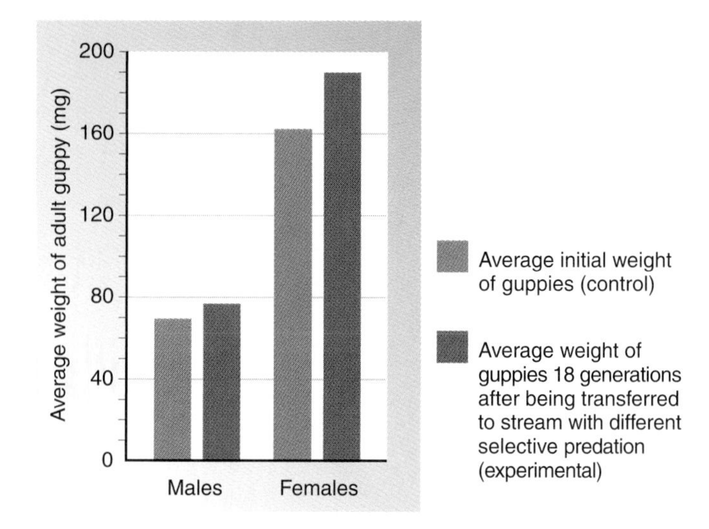

FIGURE **17-19** **Experimental evidence of natural selection in guppies.**

Male and female guppies from a stream in which the predators preferred large adult guppies as prey *(brown bars)* were transferred to a stream in which the predators preferred juveniles and small adults. After 11 years, the descendants of these guppies *(pink bars)* were measurably larger in size, compared with their ancestors. *(Data used with permission from D.N. Reznick, et al., "Evaluation of the Rate of Evolution in Natural Populations of Guppies* (Poecilia reticulata)*,"* Science, *Vol. 275, 28 Mar. 1997)*

Review

- How do scientists date fossils?
- How do homologous and homoplastic features provide evidence of evolution?
- Why are fossils of *Mesosaurus,* an extinct reptile that could not swim across open water, found in the southern parts of both Africa and South America?
- How does developmental biology provide evidence of a common ancestry for vertebrates as diverse as reptiles, birds, pigs, and humans?
- How do predator preferences drive the evolution of size in guppies?

BiologyNow™ Assess your understanding of **evidence for evolution** by taking the pretest on your BiologyNow CD-ROM.

SUMMARY WITH KEY TERMS

1 Discuss the historical development of the theory of evolution.

- Jean Baptiste de Lamarck was the first scientist to propose that organisms undergo change over time as a result of some natural phenomenon rather than divine intervention. Lamarck thought organisms were endowed with a vital force that drove them to change toward greater complexity over time. He thought organisms could pass traits acquired during their lifetimes to their offspring.
- Charles Darwin's observations while voyaging on the H.M.S. *Beagle* was the basis for his theory of evolution. Darwin tried to explain his observations of the similarities between animals and plants of the arid Galapagos Islands and the humid South American mainland.

SUMMARY WITH KEY TERMS *(continued)*

- Darwin was influenced by **artificial selection,** in which breeders develop many varieties of domesticated plants and animals in just a few generations. Darwin applied Thomas Malthus's ideas on the natural increase in human populations to natural populations. Darwin was influenced by the idea that Earth was extremely old, an idea promoted by Charles Lyell and other geologists.

2 Define *evolution,* and explain the four premises of evolution by natural selection as proposed by Charles Darwin.

- **Evolution,** the accumulation of inherited changes within a **population** over time, is the unifying concept of biology. Evolution is the cornerstone of biology because it links all fields of the life sciences into a unified body of knowledge.
- Charles Darwin and Alfred Wallace independently proposed the theory of evolution by **natural selection,** which is based on four observations. First, genetic variation exists among the individuals in a population. Second, the reproductive ability of each species causes its populations to geometrically increase in number over time. Third, organisms compete with one another for the resources needed for life, such as food, living space, water, and light. Fourth, offspring with the most favorable combination of characteristics are most likely to survive and reproduce, passing those genetic characteristics to the next generation.
- Natural selection results in **adaptations,** evolutionary modifications that improve the chances of survival and reproductive success in a particular environment. Over time, enough changes may accumulate in geographically separated populations to produce new species.

3 Compare the modern synthesis with Darwin's original theory of evolution.

- The **modern synthesis,** or **synthetic theory of evolution** combines Darwin's theory of evolution by natural selection with modern genetics to explain how species adapt to their environment.
- **Mutation** provides the genetic variability that natural selection acts on during evolution.

4 Summarize the evidence for evolution obtained from the fossil record.

- Direct evidence of evolution comes from **fossils,** the remains or traces of ancient organisms.
- Layers of sedimentary rock normally occur in their sequence of deposition, with the more recent layers on top of the older, earlier ones. **Index fossils** characterize a specific layer over large geographical areas. **Radioisotopes** present in a rock provide a way to accurately measure the rock's age.

5 Summarize the evidence for evolution derived from comparative anatomy.

- **Homologous features** have basic structural similarities, even though the structures may be used in different ways, because homologous features derive from the same structure in a common ancestor. Homologous features indicate evolutionary affinities among the organisms possessing them.
- **Homoplastic features** have similar functions in quite different, distantly related organisms. Homoplastic features demonstrate **convergent evolution,** in which organisms with separate ancestries adapt in similar ways to comparable environmental demands.
- **Vestigial structures** are nonfunctional or degenerate remnants of structures that were present and functional in ancestral organisms. Structures occasionally become vestigial as species adapt to different modes of life.

6 Define *biogeography,* and summarize how the distribution of organisms supports evolution.

- **Biogeography,** the geographical distribution of organisms, affects their evolution. Areas that have been separated from the rest of the world for a long time have organisms unique to those areas.
- At one time the continents were joined to form a supercontinent. **Continental drift,** which caused the various landmasses to break apart and separate, has played a major role in evolution.

7 Briefly explain how developmental biology provides insights into the evolutionary process.

- Evolutionary changes are often the result of mutations in genes that affect the orderly sequence of events during development. Development in different animals is controlled by the same kinds of genes, which indicates these animals have a shared evolutionary history.
- The accumulation of genetic changes since organisms diverged, or took separate evolutionary pathways, has modified the pattern of development in more complex vertebrate embryos.

8 Give an example of how evolutionary hypotheses are tested experimentally.

- Reznick and Endler have studied the effects of predation intensity on the evolution of guppy populations in the laboratory and in nature. Such experiments are a powerful way for investigators to test the underlying processes of natural selection.

POST-TEST

1. Evolution is based on which of the following concepts? (a) organisms share a common origin (b) over time, organisms have diverged from a common ancestor (c) an animal's body parts can change over its lifetime, and these acquired changes are passed to the next generation (d) a and b are correct (e) a, b, and c are correct
2. Evolution is the accumulation of genetic changes within ___________ over time. (a) individuals (b) populations (c) communities (d) a and b (e) a and c
3. Charles Darwin proposed that evolution could be explained by the differential reproductive success of organisms that resulted from their naturally occurring variation. Darwin called this process (a) coevolution (b) convergent evolution (c) natural selection (d) artificial selection (e) homoplasy
4. Which of the following statements is *false?* (a) Darwin was the first to supply convincing evidence for biological evolution (b) Darwin was the first to propose that organisms change over time (c) Wallace independently developed the same theory as Darwin (d) Darwin's theory is based on four observations about the natural world (e) Darwin's studies in the Galapagos strongly influenced his ideas about evolution
5. Which of the following is *not* part of Darwin's mechanism of evolution? (a) differential reproductive success (b) variation in a

population (c) inheritance of acquired (nongenetic) traits (d) overproduction of offspring (e) struggle for existence

6. The modern synthesis (a) is based on the sequence of fossils in rock layers (b) uses genetics to explain the source of hereditary variation that is essential to natural selection (c) was first proposed by ancient Greek scholars (d) considers the influence of the geographic distribution of organisms on their evolution (e) is reinforced by homologies that are explained by common descent
7. Jewish and Muslim men have been circumcised for many generations, yet this practice has had no effect on the penile foreskin of their offspring. This observation is inconsistent with evolution as envisioned by (a) Lamarck (b) Darwin (c) Wallace (d) Lyell (e) Malthus
8. Which of the following is *least* likely to have occurred after a small population of finches reached the Galapagos Islands from the South American mainland? (a) after many generations, the finches became increasingly different from the original population (b) over time, the finches adapted to their new environment (c) after many generations, the finches were unchanged and unmodified in any way (d) the finches were unable to survive in their new home and died out (e) the finches survived by breeding with other species
9. The fossil record (a) usually occurs in sedimentary rock (b) sometimes appears fragmentary (c) is relatively complete for tropical rainforest organisms but incomplete for aquatic organisms (d) a and b are correct (e) a, b, and c are correct
10. The molecular record found inside cells suggests that evolutionary changes are caused by an accumulation of (a) traits acquired through need (b) alterations in the order of nucleotides in DNA (c) characters acquired during an individual's lifetime (d) hormones (e) environmental changes
11. In __________, the selecting agent is the environment, whereas in __________, the selecting agent is humans. (a) natural selection; convergent evolution (b) mutation; artificial selection (c) homoplasy; homology (d) artificial selection; natural selection (e) natural selection; artificial selection
12. Features that are similar in underlying form in different species because of a common evolutionary origin, are called (a) homoplastic (b) homologous (c) vestigial (d) convergent (e) synthetic
13. Aardvarks, anteaters, and pangolins are only distantly related but are similar in structure and form as a result of (a) homology (b) convergent evolution (c) biogeography (d) vestigial structures (e) artificial selection
14. The species of the Galapagos Islands (a) are similar to those on other islands at the same latitude (b) are similar to those on the South American mainland (c) are identical to those on other islands at the same latitude (d) are identical to those on the South American mainland (e) are similar to those on both the African and South American mainlands

CRITICAL THINKING

1. The use of model organisms such as the mouse for research and biomedical testing of human diseases is based on the assumption that all organisms share a common ancestor. On what evidence is this assumption based?
2. What adaptations must an animal possess to swim in the ocean? Why are such genetically different organisms as porpoises, which are mammals, and sharks, which are fish, so similar in form?
3. The human fetus grows a coat of fine hair (the lanugo) that is shed before or shortly after birth. Fetuses of chimpanzee and other primates also grow coats of hair, but they are not shed. Explain these observations based on what you have learned in this chapter.
4. Charles Darwin once said, "It is not the strongest of the species that survive, nor the most intelligent, but the one most responsive to change." Explain what he meant.
5. Write short paragraphs explaining each of the following statements:
   a. Natural selection chooses from among the individuals in a population those most suited to *current* environmental conditions. It does not guarantee survival under future conditions.
   b. Individuals do not evolve, but populations do.
   c. The organisms that exist today do so because their ancestors had traits that allowed them and their offspring to thrive.
   d. At the molecular level, evolution can take place by the replacement of one nucleotide by another.
   e. Evolution is said to have occurred within a population when measurable genetic changes are detected.

- Visit our Web site at **http://biology.brookscole.com/solomon7** for links to chapter-related resources on the World Wide Web. Additional online materials relating to this chapter can also be found on our Web site.

BIOLOGY NOW RESOURCES

Active Figure

17-8: Fossil intermediates

Preparing for an exam? Take a diagnostic test on your BiologyNow CD-ROM.

Post-Test Answers

1. d	2. b	3. c	4. b
5. c	6. b	7. a	8. c
9. d	10. b	11. e	12. b
13. b	14. b		

18

Evolutionary Change in Populations

G.I. Bernard/Animals Animals

Genetic variation in snail shells. Shown are the shell patterns and colors in a single snail species (*Cepaea nemoralis*), native to Scotland. Variation in shell color may have adaptive value in these snails, because some colors predominate in cooler environments, whereas other colors are more common in warmer habitats.

CHAPTER OUTLINE

As you learned in Chapter 17, evolution occurs in populations, not individuals. Although natural selection results from differential survival and reproduction of individuals, individuals do not evolve during their lifetimes. Evolutionary change, which includes modifications in structure, physiology, ecology, and behavior, is inherited from one generation to the next. Although Darwin recognized that evolution occurs in populations, he did not understand how the attributes of organisms are passed to successive generations. One of the most significant advances in biology since Darwin's time has been the demonstration of the genetic basis of evolution. As you will see in this chapter, Gregor Mendel's principles of inheritance (see Chapter 10) underlie Darwinian evolution.

Recall from Chapter 17 that a **population** consists of all individuals of the same species that live in a particular place at the same time. Individuals within a population vary in many recognizable characters. A population of snails, for example, may vary in shell size, weight, or color (see photograph). Some of this variation is due to the environment, and some is due to heredity.

Biologists study variation in a particular character by taking measurements of that character in a population. By comparing the character in parents and offspring, it is possible to estimate the amount of observed variation that is genetic, as represented by the number, frequency, and kinds of alleles in a population. (Recall from Chapter 10 that an **allele** is one of two or more alternate forms of a gene. Alleles occupy corresponding positions, or **loci,** on homologous chromosomes.)

This chapter will help you develop an understanding of the importance of genetic variation as the raw material for evolution and of the basic concepts of **population genetics,** the study of genetic variability within a population and of the forces that act on it. Population genetics represents an extension of Mendelian inheritance. You will learn how to distinguish genetic equilibrium from evolutionary change and to assess the roles of the five factors responsible for evolutionary change: nonrandom mating, mutation, genetic drift, gene flow, and natural selection. ■

GENOTYPE, PHENOTYPE, AND ALLELE FREQUENCIES

Learning Objectives

1 Define what is meant by a population's gene pool.

2 Distinguish among genotype, phenotype, and allele frequencies.

Each population possesses a **gene pool,** which includes all the alleles for all the loci present in the population. Because diploid organisms have a maximum of two different alleles at each genetic locus, a single individual typically has only a small fraction of the alleles present in a population's gene pool. The genetic variation that is evident among individuals in a given population indicates that each individual has a different subset of the alleles in the gene pool.

The evolution of populations is best understood in terms of genotype, phenotype, and allele frequencies. Suppose, for example, that all 1000 individuals of a hypothetical population have their genotypes tested, with the following results:

Genotype	Number	Genotype Frequency
AA	490	0.49
Aa	420	0.42
aa	90	0.09
Total	1000	1.00

Each **genotype frequency** is the proportion of a particular genotype in the population. Genotype frequency is usually expressed as a decimal fraction, and the sum of all genotype frequencies is 1.0 (somewhat like probabilities, which were discussed in Chapter 10). For example, the genotype frequency for the *Aa* genotype is $420 \div 1000 = 0.42$.

A **phenotype frequency** is the proportion of a particular phenotype in the population. If each genotype corresponds to a specific phenotype, then the phenotype and genotype frequencies are the same. If allele *A* is dominant over allele *a,* however, the phenotype frequencies in our hypothetical population would be the following:

Phenotype	Number	Phenotype Frequency
Dominant	910	0.91
Recessive	90	0.09
Total	1000	1.00

(In this example, the dominant phenotype is the sum of two genotypes, *AA* and *Aa,* and so the number 910 is obtained by adding $490 + 420$.)

An **allele frequency** is the proportion of a specific allele (that is, of *A* or *a*) in a particular population. As mentioned earlier, each individual, being diploid, has two alleles at each genetic locus. Because we started with a population of 1000 individuals, we must account for a total of 2000 alleles. The 490 *AA* individuals have 980 *A* alleles, whereas the 420 *Aa* individuals have 420 *A* alleles, making a total of 1400 *A* alleles in the population. The total number of *a* alleles in the population is $420 + 90 + 90 = 600$. Now it is easy to calculate allele frequencies:

Allele	Number	Allele Frequency
A	1400	0.7
a	600	0.3
Total	2000	1.0

Review

- Does the term *gene pool* apply to individuals, populations, or both?
- Can the frequencies of all genotypes in a population be determined directly with respect to a locus that has only two alleles, one dominant and the other recessive?
- In a human population of 1000, 840 are tongue rollers (*TT* or *Tt*), and 160 are not tongue rollers (*tt*). What is the frequency of the dominant allele (*T*) in the population?

BiologyNow™ Assess your understanding of **genotype, phenotype, and allele frequencies** by taking the pretest on your BiologyNow CD-ROM.

THE HARDY-WEINBERG PRINCIPLE

Learning Objectives

3 Discuss the significance of the Hardy-Weinberg principle as it relates to evolution, and list the five conditions required for genetic equilibrium.

4 Use the Hardy-Weinberg principle to solve problems involving populations.

In the example just discussed, we observe that only 90 of the 1000 individuals in the population exhibit the recessive phenotype characteristic of the genotype *aa.* The remaining 910 individuals exhibit the dominant phenotype and are either *AA* or *Aa.* You might assume that, after many generations, genetic recombination during sexual reproduction would cause the dominant allele to become more common in the population. You might also assume that the recessive allele would eventually disappear altogether. These were common assumptions of many biologists early in the 20th century. However, these assumptions were incorrect, because the frequencies of alleles and genotypes do not change from generation to generation unless influenced by outside factors (discussed later).

A population whose allele and genotype frequencies do not change from generation to generation is said to be at **genetic equilibrium.** Such a population, with no net change in allele or genotype frequencies over time, is not undergoing evolutionary change. A population that is at genetic equilibrium is not evolving with respect to the locus being studied. However, if allele frequencies change over successive generations, evolution is occurring.

The explanation for the stability of successive generations in populations at genetic equilibrium was provided independently by Godfrey Hardy, an English mathematician, and Wilhelm Weinberg, a German physician, in 1908. They pointed out that the expected frequencies of various genotypes in a population can be described mathematically. The resulting **Hardy-Weinberg principle** shows that if the population is large, the process of inheritance does not by itself cause changes in allele frequencies.

Genotypes	*AA*	*Aa*	*aa*
Frequency of genotypes in population	0.49	0.42 (0.21 + 0.21)	0.09
Frequency of alleles in gametes	*A* = 0.49 + 0.21 = 0.7		*a* = 0.21 + 0.09 = 0.3

(a) Genotype and allele frequencies

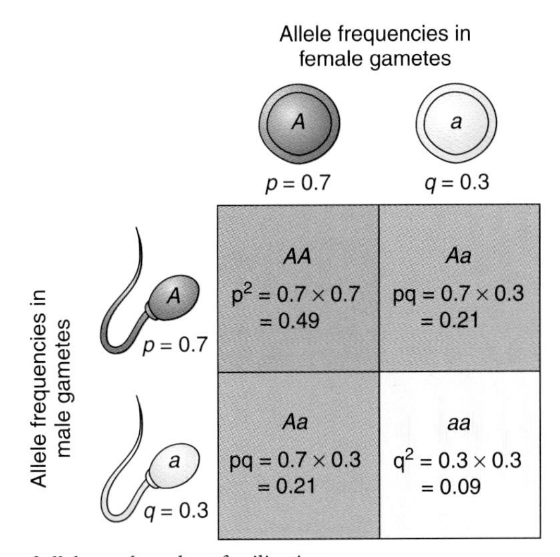

(b) Segregation of alleles and random fertilization

ACTIVE FIGURE **18-1** | **Hardy-Weinberg principle**

(a) How to calculate frequencies of the alleles *A* and *a* in the gametes. **(b)** When eggs and sperm containing *A* or *a* alleles unite randomly, the frequency of each of the possible genotypes (*AA, Aa, aa*) among the offspring is calculated by multiplying the frequencies of the alleles *A* and *a* in eggs and sperm.

BiologyNow™ Interact with the **Hardy-Weinberg principle** by clicking on this figure on your BiologyNow CD-ROM.

It also explains why dominant alleles are not necessarily more common than recessive ones. The Hardy-Weinberg principle represents an ideal situation that seldom occurs in the natural world. However, it is useful because it provides a model to help us understand the real world. Knowledge of the Hardy-Weinberg principle is essential to understanding the mechanisms of evolutionary change in sexually reproducing populations.

We now expand our original example to illustrate the Hardy-Weinberg principle. Keep in mind as we go through these calculations that in most cases we only know the phenotype frequencies. When alleles are dominant and recessive, it is usually impossible to visually distinguish heterozygous individuals from homozygous dominant individuals. The Hardy-Weinberg principle lets us use phenotype frequencies to calculate the expected genotype frequencies and allele frequencies, assuming we have a clear understanding of the genetic basis for the character under study.

As mentioned earlier, the frequency of either allele, *A* or *a*, is represented by a number that ranges from 0 to 1. An allele that is totally absent from the population has a frequency of zero. If all the alleles of a given locus are the same in the population, then the frequency of that allele is 1.

Because only two alleles, *A* and *a*, exist at the locus in our example, the sum of their frequencies must equal 1. If we let p represent the frequency of the dominant (*A*) allele in the population, and q the frequency of the recessive (*a*) allele, then we can summarize their relationship with a simple binomial equation, $p + q = 1$. When we know the value of either p or q, we can calculate the value of the other: $p = 1 - q$ and $q = 1 - p$.

Squaring both sides of $p + q = 1$ results in $(p + q)^2 = 1$. This equation can be expanded to describe the relationship of the allele frequencies to the genotypes in the population. When it is expanded, we obtain the frequency of the offspring genotypes:

$$\underset{\text{Frequency of } AA}{p^2} + \underset{\text{Frequency of } Aa}{2pq} + \underset{\text{Frequency of } aa}{q^2} = \underset{\text{All individuals in the population}}{1}$$

We always begin Hardy-Weinberg calculations by determining the frequency of the homozygous recessive genotype. From the fact that we had 90 homozygous recessive individuals in our population of 1000, we infer that the frequency of the *aa* genotype, q^2, is 90/1000, or 0.09. Because q^2 equals 0.09, q (the frequency of the recessive *a* allele) is equal to the square root of 0.09, or 0.3. From the relationship between p and q, we conclude that the frequency of the dominant *A* allele, p, equals $1 - q = 1 - 0.3 = 0.7$.

Given this information, we can calculate the frequency of homozygous dominant (*AA*) individuals: $p^2 = 0.7 \times 0.7 = 0.49$ (Fig. 18-1). The expected frequency of heterozygous individuals (*Aa*) would be $2pq = 2 \times 0.7 \times 0.3 = 0.42$. Thus, approximately 490 individuals are expected to be homozygous dominant, and 420 are expected to be heterozygous. Note that the sum of homozygous dominant and heterozygous individuals equals 910, the number of individuals showing the dominant phenotype we started with.

Any population in which the distribution of genotypes conforms to the relation $p^2 + 2pq + q^2 = 1$, whatever the absolute values for p and q may be, is at genetic equilibrium. The Hardy-Weinberg principle allows biologists to calculate allele frequencies in a given population if we know the genotype frequencies, and vice versa. These values provide a basis of comparison with a population's allele or genotype frequencies in succeeding generations. During that time, if the allele or genotype frequencies deviate from the values predicted by the Hardy-Weinberg principle, then the population is evolving.

Genetic equilibrium occurs if certain conditions are met

The Hardy-Weinberg principle of genetic equilibrium tells us what to expect when a sexually reproducing population is not evolving. The relative proportions of alleles and genotypes in successive generations will always be the same, provided the following five conditions are met:

1. Random mating. In unrestricted random mating, each individual in a population has an equal chance of mating with

any individual of the opposite sex. In our example, the individuals represented by the genotypes *AA*, *Aa*, and *aa* must mate with one another at random and must not select their mates on the basis of genotype or any other factors that result in nonrandom mating.

2. No net mutations. There must be no mutations that convert *A* into *a* or vice versa. That is, the frequencies of *A* and *a* in the population must not change because of mutations.
3. Large population size. Allele frequencies in a small population are more likely to change by random fluctuations (that is, by genetic drift, which is discussed later) than are allele frequencies in a large population.
4. No migration. There can be no exchange of alleles with other populations that might have different allele frequencies. In other words, there can be no migration[1] of individuals into or out of a population.
5. No natural selection. If natural selection is occurring, certain phenotypes (and their corresponding genotypes) are favored over others. Consequently, the allele frequencies will change, and the population will evolve.

Human MN blood groups are a valuable illustration of the Hardy-Weinberg principle

Humans have dozens of antigens on the surfaces of their blood cells. (An *antigen* is a molecule, usually a protein or carbohydrate, that is recognized as foreign by cells of another organism's immune system.) One group of antigens, designated the MN blood group, stimulates the production of antibodies when injected into rabbits or guinea pigs. However, humans do not produce antibodies for M and N, so the MN blood group is not medically important, for example, when giving blood transfusions. (Recall the discussion of the medically important ABO alleles in Chapter 10.) The MN blood group is of interest to population geneticists because the alleles for the MN blood group, usually designated *M* and *N*, are codominant (genotype *MM* produces antigen M only, genotype *NN* produces antigen N only, and the heterozygous genotype *MN* produces both antigens.) This allows all three possible genotype frequencies to be observed directly and compared with calculated frequencies. The following data are typical of the MN blood group in people in the United States:

Genotype	Observed
MM	320
MN	480
NN	200
Total	1000

Because 1000 diploid individuals are in the sample, there are a total of 2000 alleles. The frequency of *M* alleles in the population $= p = (2 \times 320 + 480) \div 2000 = 0.56$. The frequency of *N* alleles in the population $= q = (2 \times 200 + 480) \div 2000 = 0.44$. As a quick check, the sum of the frequencies should equal 1. Does it?

If this population is in genetic equilibrium, then the expected *MM* genotype frequency $= p^2 = (0.56)^2 = 0.31$. The expected *MN* genotype frequency $= 2pq = 2 \times 0.56 \times 0.44 = 0.49$. The expected *NN* genotype frequency $= q^2 = (0.44)^2 = 0.19$. As a quick check, the sum of the three genotype frequencies should equal 1. Does it? You can use the calculated genotype frequencies to determine how many individuals in a population of 1000 should have the expected genotype frequencies. By comparing the expected numbers with the actual results observed, you see how closely the population is to genetic equilibrium. Simply multiply each genotype frequency by 1000:

Genotype	Observed	Expected
MM	320	313.6
MN	480	492.8
NN	200	193.6
Total	1000	1000.0

The expected numbers closely match the observed numbers, indicating that the *MN* blood groups in the human population are almost at genetic equilibrium. This is not surprising, because the lack of medical significance suggests that the *MN* characteristic is not subject to natural selection and that it does not produce a visible phenotype that might affect random mating.

Review

- In a population at genetic equilibrium, the frequency of the homozygous recessive genotype (*tt*) is 0.16. What are the allele frequencies of *T* and *t*, and what are the expected frequencies of the *TT* and *Tt* genotypes?
- In a population at genetic equilibrium, the frequency of the dominant phenotype is 0.96. What are the frequencies of the dominant (*A*) and recessive (*a*) alleles, and what are the expected frequencies of the *AA*, *Aa*, and *aa* genotypes?
- The genotype frequencies of a population are determined to be 0.6 *BB*, 0.0 *Bb*, and 0.4 *bb*. Is it likely that this population meets all the conditions required for genetic equilibrium?

BiologyNow™ Assess your understanding of the **Hardy-Weinberg principle** by taking the pretest on your BiologyNow CD-ROM.

MICROEVOLUTION

Learning Objectives

5 Define *microevolution.*

6 Discuss how each of the following microevolutionary forces alters allele frequencies in populations: nonrandom mating, mutation, genetic drift, gene flow, and natural selection.

7 Distinguish among stabilizing selection, directional selection, and disruptive selection, and give an example of each.

Evolution represents a departure from the Hardy-Weinberg principle of genetic equilibrium. The degree of departure between the observed allele or genotype frequencies and those expected by

[1] Note that evolutionary biologists use the term *migration,* not in its ordinary sense of periodic or seasonal movement of individuals from one location to another, but instead to refer to a movement of individuals that results in a transfer of alleles from one population to another.

the Hardy-Weinberg principle indicates the amount of evolutionary change. This type of evolution—generation-to-generation changes in allele or genotype frequencies *within* a population—is sometimes referred to as **microevolution,** because it often involves relatively small or minor changes, usually over a few generations. Changes in the allele frequencies of a population result from five microevolutionary processes: nonrandom mating, mutation, genetic drift, gene flow, and natural selection. These microevolutionary processes are the opposite of the conditions that must be met if a population is in genetic equilibrium. When one or more of these processes acts on a population, allele or genotype frequencies change from one generation to the next.

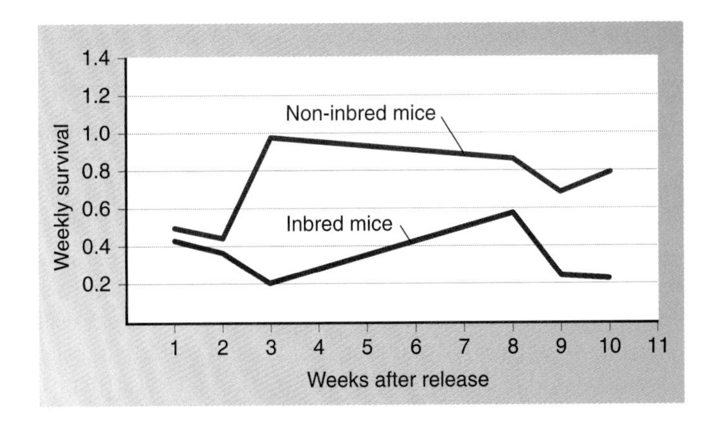

FIGURE **18-2** **Survival of inbred and non-inbred mice.**

The mouse population was sampled six times (each for a three-day span) during a 10-week period. Non-inbred mice *(red)* had a higher survival rate than inbred mice *(blue)*. Values on the Y-axis are the estimated proportion of mice that survived from one week to the next. Hence, a value of 0.6 means that 60% of the mice alive at the beginning of the week survived through that week. *(Adapted with permission from "An Experimental Study of Inbreeding Depression in a Natural Habitat," by J. A. Jiménez, et al., in* Science, *Vol. 266, Oct. 14, 1994. Copyright © 1994 American Association for the Advancement of Science.)*

Nonrandom mating changes genotype frequencies

When individuals select mates on the basis of phenotype (thereby selecting the corresponding genotype), they bring about evolutionary change in the population. Two examples of nonrandom mating are inbreeding and assortative mating.

In many populations, individuals mate more often with close neighbors than with more distant members of the population. As a result, neighbors tend to be more closely related—that is, genetically similar—to one another. The mating of genetically similar individuals that are more closely related than if they had been chosen at random from the entire population is known as **inbreeding.** Although inbreeding does not change the overall allele frequency, the frequency of homozygous genotypes increases with each successive generation of inbreeding. The most extreme example of inbreeding is self-fertilization, which is particularly common in certain plants.

Inbreeding does not appear to be detrimental in some populations, but in others it causes **inbreeding depression,** in which inbred individuals have lower fitness than those not inbred. **Fitness** is the relative ability of a given genotype to make a genetic contribution to subsequent generations; fitness is usually measured as the average number of surviving offspring of one genotype compared to the average number of surviving offspring of competing genotypes. Inbreeding depression, as evidenced by fertility declines and high juvenile mortality, is thought to be caused by the expression of harmful recessive alleles as homozygosity increases with inbreeding.

Several studies in the 1990s provided direct evidence of the deleterious consequences of inbreeding in nature. For example, white-footed mice (*Peromyscus leucopus*) were taken from a field and used to develop both inbred and non-inbred populations in the laboratory. When these laboratory-bred populations were returned to nature, their survivorship was estimated from release–recapture data. The non-inbred mice had a statistically significant higher rate of survival (Fig. 18-2). It is not known why the inbred mice had a lower survival rate. Some possibilities include higher disease susceptibility, poorer ability to evade predators, less ability to find food, and less ability to win fights with other white-footed mice.

Assortative mating, in which individuals select mates by their phenotypes, is another example of nonrandom mating. For example, biologists selected two phenotypes—high bristle number and low bristle number—in a fruit fly (*Drosophila melanogaster*) population. Although they made no effort to control mating, they observed that the flies preferentially mated with those of similar phenotypes. Females with high bristle number tended to mate with males with high bristle number, and females with low bristle number tended to mate with males with low bristle number. Such selection of mates with the same phenotype is known as *positive assortative mating* (as opposed to the less common phenomenon, *negative assortative mating,* in which mates with opposite phenotypes are selected).

Positive assortative mating is practiced in many human societies, in which men and women tend to marry individuals like themselves in such characteristics as height or intelligence. Like inbreeding, assortative mating usually increases homozygosity at the expense of heterozygosity in the population and does not change the overall allele frequencies in the population. However, assortative mating changes genotype frequencies only at the loci involved in mate choice, whereas inbreeding affects genotype frequencies in the entire genome.

Mutation increases variation within a population

Variation is introduced into a population through **mutation,** which is an unpredictable change in deoxyribonucleic acid (DNA). As discussed in Chapter 12, mutations, which are the source of all new alleles, result from (1) a change in the nucleotide base pairs of a gene, (2) a rearrangement of genes within chromosomes so that their interactions produce different effects, or (3) a change in chromosome structure. Mutations occur unpredictably and spontaneously. The rate of mutation appears relatively constant for a particular locus, but may vary by several orders of magnitude among genes within a single species and among different species.

Not all mutations pass from one generation to the next. Those occurring in somatic (body) cells are not inherited. When an individual with such a mutation dies, the mutation is lost. Some mutations, however, occur in reproductive cells. These mutations may or may not overtly affect the offspring, because most of the DNA in a cell is "silent" and does not code for specific polypeptides or proteins that are responsible for physical characteristics. Even if a mutation occurs in the DNA that codes for a polypeptide, it may still have little effect on the structure or function of that polypeptide (we discuss such *neutral variation* later in the chapter). However, when a polypeptide is sufficiently altered to change its function, the mutation is usually harmful. By acting against seriously abnormal phenotypes, natural selection eliminates or reduces to low frequencies the most harmful mutations. Mutations with small phenotypic effects, even if slightly harmful, have a better chance of being incorporated into the population, where at some later time, under different environmental conditions, they may produce phenotypes that are useful or adaptive.

Mutations do not determine the *direction* of evolutionary change. Consider a population living in an increasingly dry environment. A mutation producing a new allele that helps an individual adapt to dry conditions is no more likely to occur than one for adapting to wet conditions or one with no relationship to the changing environment. The production of new mutations simply increases the genetic variability that is acted on by natural selection and, therefore, increases the potential for new adaptations.

Mutation by itself causes small deviations in allele frequencies from those predicted by the Hardy-Weinberg principle. Although allele frequencies may be changed by mutation, these changes are typically several orders of magnitude smaller than changes caused by other evolutionary forces, such as genetic drift. As an evolutionary force, mutation is usually negligible, but it is important as the ultimate source of variation for evolution.

In genetic drift, random events change allele frequencies

The size of a population has important effects on allele frequencies because random events, or chance, tend to cause changes of relatively greater magnitude in a small population. If a population consists of only a few individuals, an allele present at a low frequency in the population could be completely lost by chance. Such an event would be unlikely in a large population. For example, consider two populations, one with 10,000 individuals and one with 10 individuals. If an uncommon allele occurs at a frequency of 10%, or 0.1, in both populations, then 1900 individuals in the large population have the allele.[2] That same frequency, 0.1, in the smaller population means that only about two individuals have the allele.[3] From this exercise, it is easy to see that there is a greater likelihood of losing the rare allele from the smaller population than from the larger one. Predators, for example, might happen to kill one or two individuals possessing the uncommon allele in the smaller population purely by chance, so these individuals would leave no offspring.

The production of random evolutionary changes in small breeding populations is known as **genetic drift.** Genetic drift results in changes in allele frequencies in a population from one generation to another. One allele may be eliminated from the population purely by chance, regardless of whether that allele is beneficial, harmful, or of no particular advantage or disadvantage. Thus, genetic drift decreases genetic variation *within* a population, although it tends to increase genetic differences *among* different populations.

When bottlenecks occur, genetic drift becomes a major evolutionary force

Because of fluctuations in the environment, such as depletion in food supply or an outbreak of disease, a population may rapidly and markedly decrease from time to time. The population is said to go through a **bottleneck** during which genetic drift can occur in the small population of survivors. As the population again increases in size, many allele frequencies may be quite different from those in the population preceding the decline.

Scientists hypothesize that genetic variation in the cheetah (see Figs. 50-9 and 50-20d) was considerably reduced by a bottleneck at the end of the last Ice Age, some 10,000 years ago. Cheetahs nearly became extinct, perhaps from overhunting by humans. The few surviving cheetahs had greatly reduced genetic variability, and as a result, the cheetah population today is so genetically uniform that unrelated cheetahs can accept skin grafts from one another. (Normally, only identical twins accept skin grafts so readily.)

The founder effect occurs when a few "founders" establish a new colony

When one or a few individuals from a large population establish, or found, a colony (as when a few birds separate from the rest of the flock and fly to a new area), they bring with them only a small fraction of the genetic variation present in the original population. As a result, the only alleles among their descendants will be those of the colonizers. Typically, the allele frequencies in the newly founded population are quite different from those of the parent population. The genetic drift that results when a small number of individuals from a large population found a new colony is called the **founder effect.**

The Finnish people may illustrate the founder effect (Fig. 18-3). Geneticists who sampled DNA from Finns and from the European population at large found that Finns exhibit considerably less genetic variation than other Europeans. This evidence supports the hypothesis that Finns are descended from a small group of people who settled about 4000 years ago in that area that is now Finland and, because of the geography, remained separate from other European societies for centuries.

The founder effect is sometimes of medical importance. For example, by chance one of the approximately 200 founders of

[2] $2pq + q^2 = 2(0.9)(0.1) + (0.1)^2 = 0.18 + 0.01 = 0.19; 0.19 \times 10{,}000 = 1900.$

[3] $0.19 \times 10 = 1.9.$

FIGURE 18-3 **Finns and the founder effect.**
The Finnish people are thought to have descended from a small founding population that remained separate from the rest of Europe for centuries.

the Amish population of Pennsylvania carried a recessive allele that is responsible for a form of dwarfism, Ellis-van Creveld syndrome, when homozygous. Although this allele is rare in the general population (frequency about 0.001), today it is relatively common in the Amish population (frequency about 0.07).

Gene flow generally increases variation within a population

Individuals of a species tend to be distributed in local populations that are genetically isolated to some degree from other populations. For example, the bullfrogs of one pond form a population separated from those in an adjacent pond. Some exchanges occur by migration between ponds, but the frogs in one pond are much more likely to mate with those in the same pond. Because each population is isolated to some extent from other populations, they have distinct genetic traits and gene pools.

The migration of breeding individuals between populations causes a corresponding movement of alleles, or **gene flow,** that has significant evolutionary consequences. As alleles flow from one population to another, they usually increase the amount of genetic variability within the recipient population. If sufficient gene flow occurs between two populations, they become more similar genetically. Because gene flow reduces the amount of variation between two populations, it tends to counteract the effects of natural selection and genetic drift, both of which often cause populations to become increasingly distinct.

If migration by members of a population is considerable, and if populations differ in their allele frequencies, then significant genetic changes occur in local populations. For example, by 10,000 years ago modern humans occupied almost all of Earth's major land areas except a few islands. Because the population density was low in most locations, the small, isolated human populations underwent random genetic drift and natural selection. More recently (during the past 300 years or so), major migrations have increased gene flow, significantly altering allele frequencies within previously isolated human populations.

Natural selection changes allele frequencies in a way that increases adaptation

Natural selection is the mechanism of evolution first proposed by Darwin in which members of a population that are more successfully adapted to the environment are more likely to survive and reproduce (see Chapter 17). Over successive generations, the proportion of favorable alleles increases in the population. In contrast with other microevolutionary processes (nonrandom mating, mutation, genetic drift, and gene flow), natural selection leads to adaptive evolutionary change. Natural selection not only explains why organisms are well adapted to the environments in which they live but also helps account for the remarkable diversity of life. Natural selection enables populations to change, thereby adapting to different environments and different ways of life.

Natural selection is the differential reproduction of individuals with different traits, or phenotypes (and therefore different genotypes), in response to the environment. Natural selection preserves individuals with favorable phenotypes and eliminates those with unfavorable phenotypes. Individuals that survive and produce fertile offspring have a selective advantage.

The mechanism of natural selection does not develop a "perfect" organism. Rather, it weeds out those individuals whose phenotypes are less adapted to environmental challenges, while allowing better adapted individuals to survive and pass their alleles to their offspring. By reducing the frequency of alleles that result in the expression of less favorable traits, the probability is increased that favorable alleles responsible for an adaptation will come together in the offspring.

Natural selection operates on an organism's phenotype

Natural selection does not act directly on an organism's genotype. Instead, it acts on the phenotype, which is, at least in part, an expression of the genotype. The phenotype represents an interaction between the environment and all the alleles in the organism's genotype. It is rare that alleles of a single locus determine the phenotype, as Mendel originally observed in garden peas. Much more common is the interaction of alleles of several loci for the expression of a single phenotype (see Chapter 10). Many plant and animal characteristics are under this type of polygenic control.

When characters (characteristics) are under polygenic control (as is human height), a range of phenotypes occurs, with most of the population located in the median range and fewer at either extreme. This is a normal distribution or standard bell curve (Fig. 18-4a; see also Fig. 10-22). Three kinds of selection cause changes in the normal distribution of phenotypes in a population: stabilizing, directional, and disruptive selection. Although we consider each process separately, in nature their influences generally overlap.

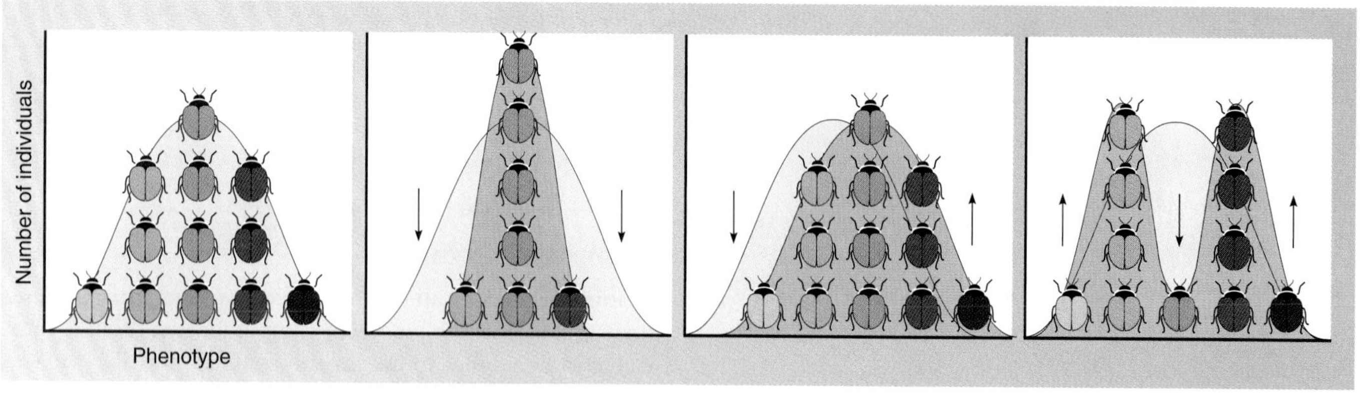

(a) No selection (b) Stabilizing selection (c) Directional selection (d) Disruptive selection

ACTIVE FIGURE **18-4** | **Modes of selection.**

The blue screen represents the distribution of individuals by phenotype (in this example, color variation) in the original population. The purple screen represents the distribution by phenotype in the evolved population. The arrows represent the pressure of natural selection on the phenotypes. **(a)** A character that is under polygenic control (in this example, wing colors in a hypothetical population of beetles) exhibits a normal distribution of phenotypes in the absence of selection. **(b)** As a result of stabilizing selection, which trims off extreme phenotypes, variation about the mean is reduced. **(c)** Directional selection shifts the curve in one direction, changing the average value of the character. **(d)** Disruptive selection, which trims off intermediate phenotypes, results in two or more peaks.

BiologyNow™ Watch **natural selection** in action by clicking on this figure on your BiologyNow CD-ROM.

Stabilizing selection The process of natural selection associated with a population well adapted to its environment is known as **stabilizing selection.** Most populations are probably influenced by stabilizing forces most of the time. Stabilizing selection selects against phenotypic extremes. In other words, individuals with an average, or intermediate, phenotype are favored.

Because stabilizing selection tends to decrease variation by favoring individuals near the mean of the normal distribution at the expense of those at either extreme, the bell curve narrows (Fig. 18-4b). Although stabilizing selection decreases the amount of variation in a population, variation is rarely eliminated by this process, because other microevolutionary processes act against a decrease in variation. For example, mutation is slowly but continually adding to the genetic variation within a population.

One of the most widely studied cases of stabilizing selection involves human birth weight, which is under polygenic control and is also influenced by environmental factors. Extensive data from hospitals have shown that infants born with intermediate weights are most likely to survive (Fig. 18-5). Infants at either extreme (too small or too large) have higher rates of mortality. When newborn infants are too small, their body systems are immature, and when they are too large, they have difficult deliveries because they cannot pass as easily through the cervix and vagina. Stabilizing selection operates to reduce variability in birth weight so it is close to the weight with the minimum mortality rate.

Directional selection If an environment changes over time, **directional selection** may favor phenotypes at one of the extremes of the normal distribution (Fig. 18-4c). Over successive generations, one phenotype gradually replaces another. So, for example, if greater size is advantageous in a new environment, larger individuals will become increasingly common in the population. Directional selection only occurs, however, if alleles favored under the new circumstances are already present in the population.

Darwin's Galapagos finches provide an excellent example of directional selection. Since 1973, Peter Grant and Rosemary Grant of Princeton University have studied the Galapagos

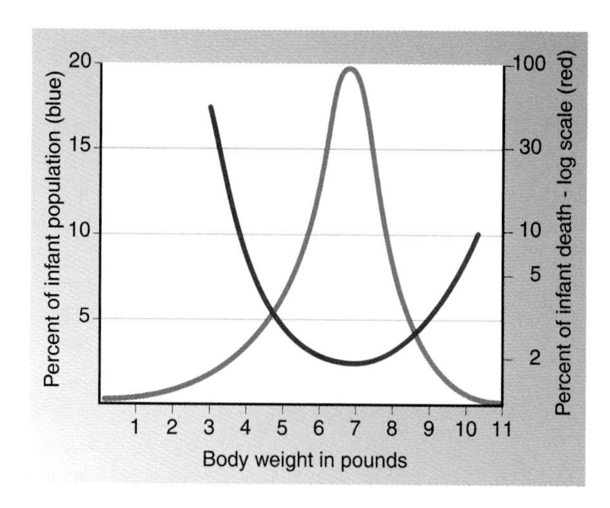

FIGURE **18-5** | **Stabilizing selection.**

The blue curve shows the distribution of birth weights in a sample of 13,730 infants. The red curve shows mortality (death) at each birth weight. Infants with very low or very high birth weights have higher death rates than infants of average weight. The optimum birth weight, that is, the one with the lowest mortality, is close to the average birth weight (about 7 pounds). *(Adapted from, L.L. Cavalli-Sforza and W.F. Bodmer,* The Genetics of Human Populations, *W.H. Freeman and Company, San Francisco, 1971)*

TABLE 18-1 Population Changes in *Geospiza fortis* Before and After the 1976–1977 Drought

Character	Average Before Drought (634)*	Average After Drought (135)*	Difference
Weight (g)	16.06	17.13	+1.07
Wing length (mm)	67.88	68.87	+0.99
Tarsus (leg, just above the foot) length (mm)	19.08	19.29	+0.21
Bill length (mm)	10.63	10.95	+0.32
Bill depth (mm)	9.21	9.70	+0.49
Bill width (mm)	8.58	8.83	+0.25

*Number of birds in sample.

From P.R. Grant and B.R. Grant, "Predicting Microevolutionary Responses to Directional Selection on Heritable Variation," *Evolution*, Vol. 49, 1995.

finches. The Grants did a meticulous analysis of finch eating habits and beak sizes on Isla Daphne Major during three extended droughts (1977–1978, 1980, and 1982), one of which was followed by an extremely wet El Niño event (1983). During the droughts, the number of insects and small seeds declined, and large, heavy seeds became the finches' primary food source. Many finches died during this time, and most of the survivors were larger birds whose beaks were larger and deeper. In a few generations, these larger birds became more common in the population (Table 18-1). After the wet season, however, smaller seeds became the primary food source, and smaller finches with average-sized beaks were favored. In this example, natural selection is directional: During the drought, natural selection operated in favor of the larger phenotype, whereas after the wet period selection occurred in the opposite direction, favoring the smaller phenotype. The guppy populations studied in Venezuela and Trinidad (see Chapter 17) are another example of directional selection.

Disruptive selection Sometimes extreme changes in the environment may favor two or more different phenotypes at the expense of the mean. That is, more than one phenotype may be favored in the new environment. **Disruptive selection** is a special type of directional selection in which there is a trend in several directions rather than just one (Fig. 18-4d). It results in a divergence, or splitting apart, of distinct groups of individuals within a population. Disruptive selection, which is relatively rare, selects against the average, or intermediate, phenotype.

Limited food supply during a severe drought caused a population of finches on another island in the Galapagos to experience disruptive selection. The finch population initially exhibited a variety of beak sizes and shapes. Because the only foods available on this island during the drought were wood-boring insects and seeds from cactus fruits, natural selection favored birds with beaks suitable for obtaining these types of food. Finches with longer beaks survived because they could open cactus fruits, and those with wider beaks survived because they could strip off tree bark to expose insects. However, finches with intermediate beaks could not use either food source efficiently, and had a lower survival rate.

Natural selection induces change in the types and frequencies of alleles in populations only if there is pre-existing inherited variation. Genetic variation is the raw material for evolutionary change, because it provides the diversity on which natural selection acts. Without genetic variation, evolution cannot occur. In the next section we explore the genetic basis for variation that is acted on by natural selection.

Review

- Which microevolutionary force leads to adaptive changes in allele frequencies?
- Why is mutation important to evolution if it is the microevolutionary force that generally has the smallest effect on allele frequencies?
- Which microevolutionary forces are most associated with an increase in variation within a population? Among populations?
- Which microevolutionary force typically changes genotype frequencies without changing allele frequencies? Explain.

BiologyNow™ Assess your understanding of **microevolution** by taking the pretest on your BiologyNow CD-ROM.

GENETIC VARIATION IN POPULATIONS

Learning Objective

8 Describe the nature and extent of genetic variation, including genetic polymorphism, balanced polymorphism, neutral variation, and geographic variation.

Populations contain abundant genetic variation that was originally introduced by mutation. Sexual reproduction, with its associated crossing-over, independent assortment of chromosomes during meiosis, and random union of gametes, also contributes to genetic variation. The sexual process allows the variability introduced by mutation to be combined in new ways, which may be expressed as new phenotypes.

Genetic polymorphism exists among alleles and the proteins for which they code

One way of evaluating genetic variation in a population is to examine **genetic polymorphism,** which is the presence in a population of two or more alleles for a given locus. Genetic polymorphism is extensive in populations, although many of the alleles are present at low frequencies. Much of genetic polymorphism is not evident, because it does not produce distinct phenotypes.

One way biologists estimate the total amount of genetic polymorphism in populations is by comparing the different forms of a particular protein. Each form consists of a slightly different amino acid sequence that is coded for by a different allele. For example, tissue extracts containing a particular enzyme may be analyzed by gel electrophoresis for different individuals. In gel electrophoresis, the enzymes are placed in slots on an agarose gel, and an electric current is applied, causing each enzyme to migrate across the gel (see Fig. 14-8). Slight variations in amino acid sequences in the different forms of a particular enzyme cause each to migrate at a different rate, which can be detected using special stains or radioactive labels. Table 18-2 shows the degree of polymorphism in selected plant and animal groups based on gel electrophoresis of several enzymes. Note that genetic polymorphism tends to be greater in plants than in animals.

Determining the sequence of nucleotides in DNA from individuals in a population provides a *direct* estimate of genetic polymorphism. DNA sequencing is shown in Figures 14-9 and 14-10. DNA sequencing of specific alleles in an increasing number of organisms, including humans, indicates that genetic polymorphism is extensive in most populations.

TABLE 18-2 Genetic Polymorphism of Selected Enzymes Within Plant and Animal Species

Organism	Number of Species Examined	Percentage of Enzymes Studied That Are Polymorphic
Plants		
Gymnosperms	55	70.9
Flowering plants (monocots)	111	59.2
Flowering plants (dicots)	329	44.8
Invertebrates		
Marine snails	5	17.5
Land snails	5	43.7
Insects	23	32.9
Vertebrates		
Fishes	51	15.2
Amphibians	13	26.9
Reptiles	17	21.9
Birds	7	15.0
Mammals	46	14.7

Plant data adapted from J.L. Hamrick and M.J. Godt. "Allozyme Diversity in Plant Species," In A.H.D. Brown, M.T. Clegg, A.L. Kahler, and B.J. Weir (eds.), *Plant Population Genetics, Breeding, and Genetic Resources,* Sunderland, MA, Sinauer Associates, 1990. Animal data adapted from D. Hartl, *Principles of Population Genetics,* Sunderland, MA, Sinauer Associates, 1980, and P.W. Hedrick, *Genetics of Populations,* Boston, Science Books International, 1983.

Balanced polymorphism exists for long periods

Balanced polymorphism is a special type of genetic polymorphism in which two or more alleles persist in a population over many generations as a result of natural selection. Heterozygote advantage and frequency-dependent selection are mechanisms that preserve balanced polymorphism.

Genetic variation may be maintained by heterozygote advantage

We have seen that natural selection often eliminates unfavorable alleles from a population, whereas favorable alleles are retained. However, natural selection sometimes helps maintain genetic diversity in a population, including alleles that are unfavorable in the homozygous state. This happens, for example, when the heterozygote, *Aa,* has a higher degree of fitness than either homozygote, *AA* or *aa.* This phenomenon, known as **heterozygote advantage,** is demonstrated in humans by the selective advantage of heterozygous carriers of the sickle cell allele.

The mutant allele (Hb^S) for sickle cell anemia produces an altered hemoglobin that deforms or sickles the red blood cells, making them more likely to form dangerous blockages in capillaries and to be destroyed in the liver, spleen, or bone marrow (see Chapter 15). People who are homozygous for the sickle cell allele (Hb^SHb^S) usually die at an early age if medical treatment is not available.

Heterozygous individuals carry alleles for both normal (Hb^A) and sickle cell hemoglobin. The heterozygous condition (Hb^AHb^S) makes a person more resistant to a type of severe malaria (caused by the parasite *Plasmodium falciparum*) than people who are homozygous for the normal hemoglobin allele (Hb^AHb^A). In a heterozygous individual, each allele produces its own specific kind of hemoglobin, and the red blood cells contain the two kinds in roughly equivalent amounts. Not only do such cells sickle much less readily than cells containing only the Hb^S allele, but also they are more resistant to infection by the malaria-causing parasite, which lives in red blood cells, than are the red blood cells containing only normal hemoglobin.

Where malaria is a problem, each of the two types of homozygous individuals is at a disadvantage. Those homozygous for the sickle cell allele are likely to die of sickle cell anemia, whereas those homozygous for the normal allele may die of malaria. The heterozygote is therefore more fit than either homozygote. In parts of Africa, the Middle East, and southern Asia where falciparum malaria is prevalent, heterozygous individuals survive in greater numbers than either homozygote (Fig. 18-6). The Hb^S allele is maintained at a high frequency in the population, even though the homozygous recessive condition is almost always lethal.

What happens to the frequency of Hb^S alleles in Africans and others who possess it when they migrate to the United States and other nonmalarial countries? As might be expected, the frequency of the Hb^S allele gradually declines in such populations, possibly because it confers a selective disadvantage by causing sickle cell anemia in homozygous individuals but no longer confers a selective advantage by preventing malaria in heterozygous individuals. The Hb^S allele never disappears from the pop-

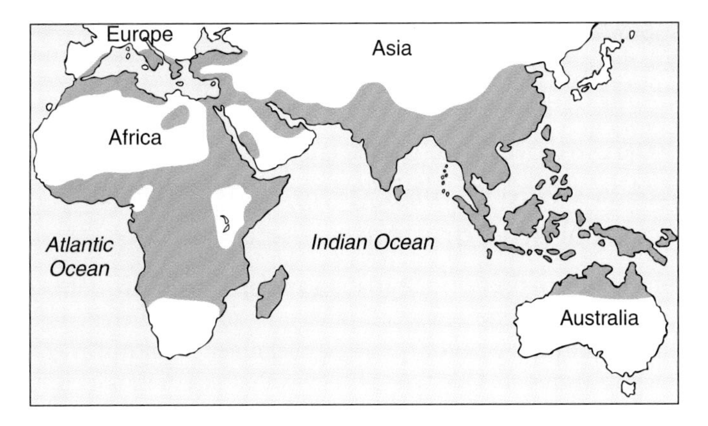

(a) *P. falciparum* malaria

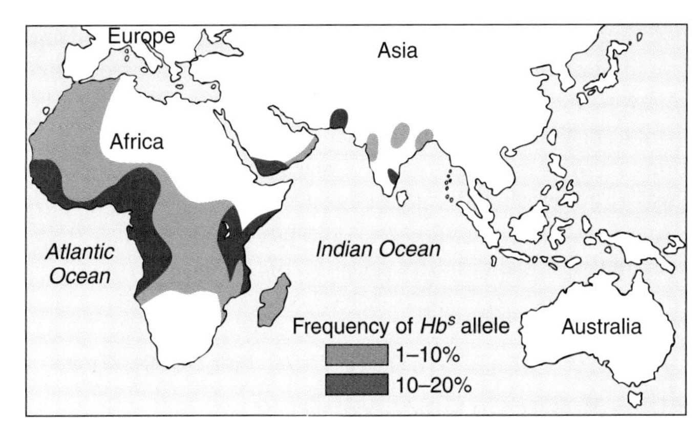

(b) Sickle cell anemia

FIGURE **18-6** **Heterozygote advantage.**

The geographic distribution of **(a)** falciparum malaria *(green)* is compared with that of **(b)** sickle cell anemia *(red and orange)*. The greater fitness of heterozygous individuals in malarial regions supports the hypothesis of heterozygote advantage, which can be seen by the large area of codistribution. *(Adapted from A.C. Allison, "Protection Afforded by Sickle-Cell Traits against Subtertian Malarial Infection."* British Medical Journal, *Vol. 1, 1954)*

ulation, however, because it is "hidden" from selection in heterozygous individuals and because it is reintroduced into the population by gene flow from the African population.

Genetic variation may be maintained by frequency-dependent selection

Thus far in our discussion of natural selection, we have assumed the fitness of particular phenotypes (and their corresponding genotypes) is independent of their frequency in the population. However, in cases of **frequency-dependent selection** the fitness of a particular phenotype depends on how frequently it appears in the population. Often a phenotype has a greater selective value when rare than when common in the population. Such phenotypes lose their selective advantage as they become more common.

Frequency-dependent selection often acts to maintain genetic variation in populations of prey species. In this case, the predator catches and consumes the more common phenotype but may ignore the rarer phenotypes. Consequently, the less common phenotype produces more offspring and therefore makes a greater relative contribution to the next generation.

Frequency-dependent selection is demonstrated in scale-eating fish (cichlids of the species *Perissodus microlepsis*) from Lake Tanganyika in Africa. The scale-eating fish, which obtain food by biting scales off other fish, have either left-pointing or right-pointing mouths. A single locus with two alleles determines this characteristic; the allele for right-pointing mouth is dominant over the allele for left-pointing mouth. These fish attack their prey from behind, and from a single direction, depending on mouth morphology. Those with left-pointing mouths always attack the right flanks of their prey, whereas those with right-pointing mouths always attack the left flanks (Fig. 18-7a).

The prey species are more successful at evading attacks from the more common form of scale-eating fish. For example, if the cichlids with right-pointing mouths are more common than those with left-pointing mouths, the prey are attacked more often on their left flanks. They therefore become more wary against such attacks, conferring a selective advantage to the less common cichlids with left-pointing mouths. The cichlids with left-

FIGURE **18-7** **Frequency-dependent selection in scale-eating cichlids.**

(a) Scale-eating cichlids have two forms, right-pointing mouths and left-pointing mouths. **(b)** The frequency of fish with left-pointing mouths over a 10-year period. Frequency-dependent selection maintains the frequencies of left-pointing and right-pointing fish in approximately equal numbers, that is, at about 0.5. *(b, Adapted with permission from "Frequency-Dependent Natural Selection in the Handedness of Scale-Eating Cichlid Fish," by M. Hori in* Science, *Vol. 260, April 9, 1993. Copyright © 1993 American Association for the Advancement of Science.)*

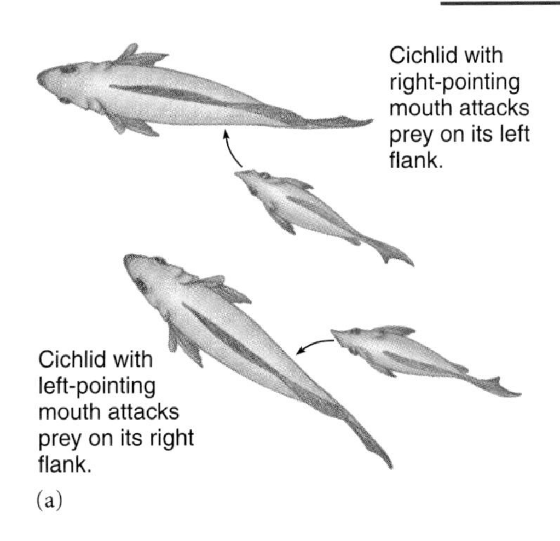

(a)

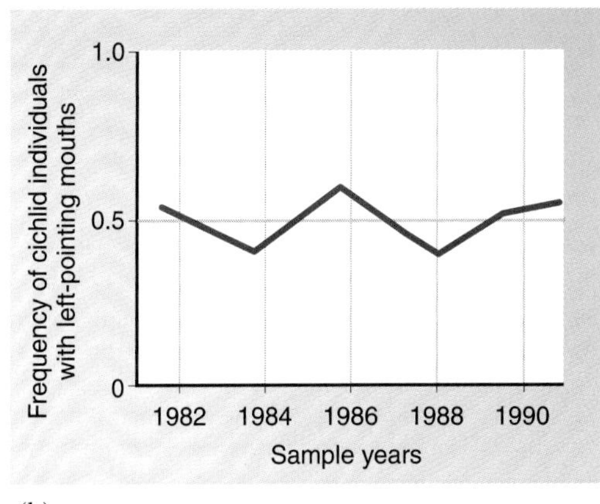

(b)

pointing mouths would be more successful at obtaining food and would therefore have more offspring. Over time, the frequency of fish with left-pointing mouths would increase in the population, until their abundance causes frequency-dependent selection to work against them and confer an advantage on the now less-common fish with right-pointing mouths. Thus frequency-dependent selection maintains both populations of fish at approximately equal numbers (Fig. 18-7b).

Neutral variation may give no selective advantage or disadvantage

Some of the genetic variation observed in a population may confer no apparent selective advantage or disadvantage in a particular environment. For example, random changes in DNA that do not alter protein structure usually do not affect the phenotype. Variation that does not alter the ability of an individual to survive and reproduce and is, therefore, not adaptive is called **neutral variation.**

The extent of neutral variation in organisms is difficult to determine. It is relatively easy to demonstrate that an allele is beneficial or harmful if its effect is observable. But the variation in alleles that involves only slight differences in the proteins they code for may or may not be neutral. These alleles may be influencing the organism in subtle ways that are difficult to measure or assess. Also, an allele that is neutral in one environment may be beneficial or harmful in another.

Populations in different geographic areas often exhibit genetic adaptations to local environments

In addition to the genetic variation among individuals within a population, genetic differences often exist among different populations within the same species, a phenomenon known as *geographic variation.* One type of geographic variation is a **cline,** which is a gradual change in a species' phenotype and genotype frequencies through a series of geographically separate populations as a result of an environmental gradient. A cline exhibits variation in the expression of such attributes as color, size, shape, physiology, or behavior. Clines are common among species with continuous ranges over large geographic areas. For example, the body sizes of many widely distributed birds and mammals increase gradually as the latitude increases, presumably because larger animals are better able to withstand the colder temperatures of winter.

The common yarrow (*Achillea millefolium*), a wildflower that grows in a variety of North American habitats from lowlands to mountain highlands, exhibits clinal variation in height in response to different climates at different elevations. Although

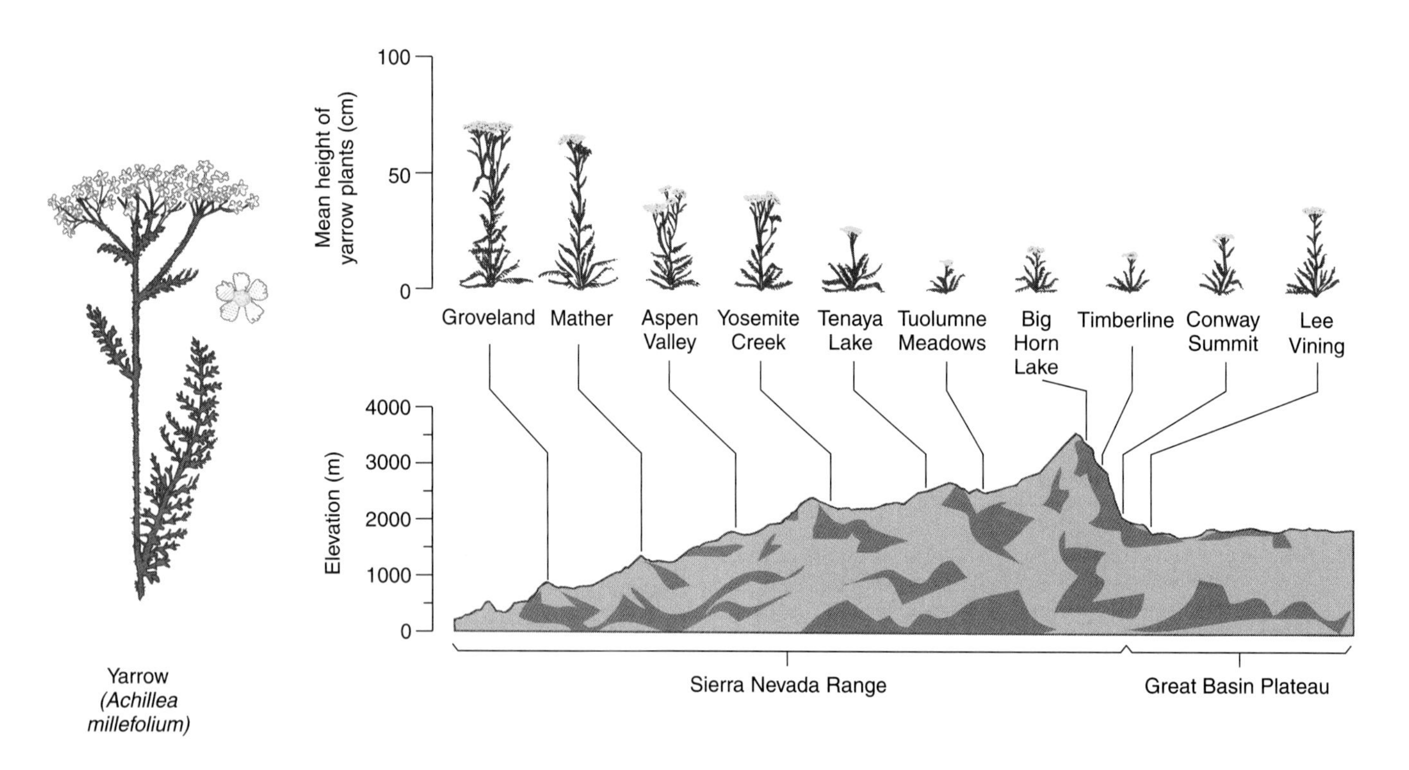

FIGURE **18-8** | **Clinal variation in yarrow (*Achillea millefolium*).**

Seeds from widely dispersed populations in the Sierra Nevada of California and Nevada were collected and grown for several generations under identical conditions in the same test garden at Stanford, California. The plants retained their distinctive heights, revealing genetic differences related to the elevation where the seeds were collected. *(After J. Clausen, D.D. Keck, and W.M. Hiesey, "Experimental Studies on the Nature of Species: III. Environmental Responses of Climatic Races of* Achillea," Carnegie Institute Washington Publication, *Vol. 58, 1948)* substantial variation exists among individuals within each population, individuals in populations at higher elevations are, on average, shorter than those at lower elevations. The genetic basis of these clinal differences was experimentally demonstrated in a set of classical experiments in which series of populations from different geographic areas were grown in the same environment (Fig. 18-8). Despite being exposed to identical environmental conditions, each experimental population exhibited the traits characteristic of the elevation from which it was collected.

Review

- How does the sickle cell allele illustrate heterozygote advantage?
- How does frequency-dependent selection affect genetic variation within a population over time?
- How can researchers test the hypothesis that clinal variation in a particular population has a genetic basis?

BiologyNow™ Assess your understanding of **genetic variation in populations** by taking the pretest on your BiologyNow CD-ROM.

SUMMARY WITH KEY TERMS

1 Define what is meant by a population's gene pool.

- All the individuals that live in a particular place at the same time make up a **population.** Each population has a **gene pool,** which includes all the **alleles** for all the **loci** present in the population. **Population genetics** is the study of genetic variability within a population and of the forces that act on it.

2 Distinguish among genotype, phenotype, and allele frequencies.

- A **genotype frequency** is the proportion of a particular genotype in the population.
- A **phenotype frequency** is the proportion of a particular phenotype in the population.
- An **allele frequency** is the proportion of a specific allele of a given genetic locus in the population.

3 Discuss the significance of the Hardy-Weinberg principle as it relates to evolution, and list the five conditions required for genetic equilibrium.

- The **Hardy-Weinberg principle** states that allele and genotype frequencies do not change from generation to generation (no evolution is occurring) in a population at **genetic equilibrium.**
- The Hardy-Weinberg principle applies only if mating is random in the population, there are no net mutations that change the allele frequencies, the population is large, individuals do not migrate between populations, and natural selection does not occur.

4 Use the Hardy-Weinberg principle to solve problems involving populations.

- In the Hardy-Weinberg equation, p = the frequency of the dominant allele, q = the frequency of the recessive allele: $p + q = 1$.
- The genotype frequencies of a population are described by the relationship $p^2 + 2pq + q^2 = 1$, where p^2 is the frequency of the homozygous dominant genotype, $2pq$ the frequency of the heterozygous genotype, and q^2 the frequency of the homozygous recessive genotype.

5 Define *microevolution.*

- **Microevolution** is a change in allele or genotype frequencies within a population over successive generations.

6 Discuss how each of the following microevolutionary forces alters allele frequencies in populations: nonrandom mating, mutation, genetic drift, gene flow, and natural selection.

- In nonrandom mating, individuals select mates on the basis of phenotype, indirectly selecting the corresponding genotype(s). **Inbreeding** is the mating of genetically similar individuals that are more closely related than if they had been chosen at random from the entire population. Inbreeding in some populations causes **inbreeding depression,** in which inbred individuals have lower **fitness** than non-inbred individuals. In **assortative mating** individuals select mates by their phenotypes. Both inbreeding and assortative mating increase the frequency of homozygous genotypes.
- **Mutations,** unpredictable changes in DNA, are the source of new alleles. Mutations increase the genetic variability acted on by natural selection.
- **Genetic drift** is a random change in the allele frequencies of a small population. Genetic drift decreases genetic variation within a population, and the changes caused by genetic drift are usually not adaptive. A sudden decrease in population size caused by adverse environmental factors is known as a **bottleneck.** The **founder effect** is genetic drift that occurs when a small population colonizes a new area.
- **Gene flow,** a movement of alleles caused by the migration of individuals between populations, causes changes in allele frequencies.
- **Natural selection** causes changes in allele frequencies that lead to adaptation. Natural selection operates on an organism's phenotype, but it changes the genetic composition of a population in a favorable direction for a particular environment.

7 Distinguish among stabilizing selection, directional selection, and disruptive selection, and give an example of each.

- **Stabilizing selection** favors the mean at the expense of phenotypic extremes.
- **Directional selection** favors one phenotypic extreme over another, causing a shift in the phenotypic mean.
- **Disruptive selection** favors two or more phenotypic extremes.

8 Describe the nature and extent of genetic variation, including genetic polymorphism, balanced polymorphism, neutral variation, and geographic variation.

- **Genetic polymorphism** is the presence in a population of two or more alleles for a given locus.
- **Balanced polymorphism** is a special type of genetic polymorphism in which two or more alleles persist in a population over many generations as a result of natural selection. **Heterozygote advantage** occurs when the heterozygote exhibits greater fitness than either homozygote. In **frequency-dependent selection,** a genotype's selective value varies with its frequency of occurrence.
- **Neutral variation** is genetic variation that confers no detectable selective advantage.
- Geographic variation is genetic variation that exists among different populations within the same species. A **cline** is a gradual change in a species' phenotype and genotype frequencies through a series of geographically separate populations.

POST-TEST

1. The genetic description of an individual is its genotype, whereas the genetic description of a population is its (a) phenotype (b) gene pool (c) genetic drift (d) founder effect (e) changes in allele frequencies
2. In a diploid species, each individual possesses (a) one allele for each locus (b) two alleles for each locus (c) three or more alleles for each locus (d) all the alleles found in the gene pool (e) half of the alleles found in the gene pool
3. The MN blood group is of interest to population geneticists because (a) people with genotype *MN* cannot receive blood transfusions from either *MM* or *NN* people (b) the *MM, MN,* and *NN* genotype frequencies can be observed directly and compared with calculated expected frequencies (c) the *M* allele is dominant to the *N* allele (d) people with the *MN* genotype exhibit frequency-dependent selection (e) people with the *MN* genotype exhibit heterozygote advantage
4. If all copies of a given locus have the same allele throughout the population, then the allele frequency is (a) 0 (b) 0.1 (c) 0.5 (d) 1.0 (e) 10.0
5. If a population's allele and genotype frequencies remain constant from generation to generation (a) the population is undergoing evolutionary change (b) the population is said to be at genetic equilibrium (c) microevolution has taken place (d) directional selection is occurring, but only for a few generations (e) genetic drift is a significant evolutionary force
6. Comparing the different forms of a particular protein in a population provides biologists with an estimate of (a) genetic drift (b) genetic polymorphism (c) gene flow (d) heterozygote advantage (e) frequency-dependent selection
7. The continued presence of the allele that causes sickle cell anemia in areas where falciparum malaria is prevalent demonstrates which of the following phenomena? (a) inbreeding depression (b) frequency-dependent selection (c) heterozygote advantage (d) genetic drift (e) a genetic bottleneck
8. Frequency-dependent selection maintains ______________ in a population. (a) assortative mating (b) genetic drift (c) gene flow (d) genetic variation (e) stabilizing selection
9. According to the Hardy-Weinberg principle, (a) allele frequencies are not dependent on dominance or recessiveness but remain essentially unchanged from generation to generation (b) the sum of allele frequencies for a given locus is always greater than 1 (c) if a locus has a single allele, its frequency must be zero (d) allele frequencies change from generation to generation (e) the process of inheritance, by itself, causes changes in allele frequencies
10. What is the correct equation for the Hardy-Weinberg principle? (a) $p^2 + pq + 2q^2 = 1$ (b) $p^2 + 2pq + 2q^2 = 1$ (c) $2p^2 + 2pq + 2q^2 + 1$ (d) $p^2 + pq + q^2 = 1$ (e) $p^2 + 2pq + q^2 = 1$
11. The Hardy-Weinberg principle is applicable if (a) population size is small (b) migration only occurs at the beginning of the breeding season (c) mutations occur at a constant rate (d) matings occur exclusively between individuals of the same genotype (e) natural selection does not occur
12. Which of the following is *not* an evolutionary agent that causes change in allele frequencies? (a) mutation (b) natural selection (c) genetic drift (d) random mating (e) gene flow from migration
13. Mutation (a) leads to adaptive evolutionary change (b) adds to the genetic variation of a population (c) is the result of genetic drift (d) almost always benefits the organism (e) a and b are correct
14. Which of the following is *not* true of natural selection? (a) Natural selection acts to preserve favorable traits and eliminate unfavorable traits. (b) The offspring of individuals that are better adapted to the environment will make up a larger proportion of the next generation. (c) Natural selection directs the course of evolution by preserving the traits acquired during an individual's lifetime. (d) Natural selection acts on a population's genetic variability, which arises through mutation. (e) Natural selection may result in changes in allele frequencies in a population.
15. In ______________, individuals with a phenotype near the phenotypic mean of the population are favored over those with phenotypic extremes. (a) microevolution (b) stabilizing selection (c) directional selection (d) disruptive selection (e) genetic equilibrium

CRITICAL THINKING

1. Why are mutations almost always neutral or harmful?
2. Explain this apparent paradox: People discuss evolution in terms of *genotype* fitness (the selective advantage that a particular genotype confers on an individual), yet natural selection acts on an organism's *phenotype.*
3. Recall that the allele for right-pointing mouth in cichlids *(A)* is dominant over that for left-pointing mouth *(a)*. Use the Hardy-Weinberg principle to calculate the allele frequencies that would correspond to equal frequencies of the two phenotypes. Which allele has the higher frequency? Why? How are these scale-eating fish an example of balanced polymorphism?

- Visit our Web site at **http://biology.brookscole.com/solomon7** for links to chapter-related resources on the World Wide Web. Additional online materials relating to this chapter can also be found on our Web site.

BIOLOGY NOW RESOURCES

Active Figures

18-1: Hardy-Weinberg principle

18-4: Natural selection

Preparing for an exam? Take a diagnostic test on your BiologyNow CD-ROM.

Post-Test Answers

1. b	2. b	3. b	4. d
5. b	6. b	7. c	8. d
9. a	10. e	11. e	12. d
13. b	14. c	15. b	

19

Speciation and Macroevolution

Gary Retherford/Photo Researchers, Inc.

Interbreeding between different species. Shown is a "zebrass," a sterile hybrid formed by a cross between a zebra and a donkey that retains features of both parental species. Although such matings may occur under artificial conditions, such as the wildlife ranch in Texas where this cross took place, zebras and donkeys do not interbreed in the wild.

CHAPTER OUTLINE

A Brazilian rain forest contains thousands of different species of insects, amphibians, reptiles, birds, and mammals. The Great Barrier Reef off the coast of Australia has thousands of species of sponges, corals, mollusks, crustaceans, sea stars, and fishes. We don't know exactly how many species exist today, but biologists estimate there may be something on the order of 10 million to 100 million![1] About 1.8 million of these species have been scientifically named and described. These include 250,000 plant species, 42,000 vertebrate animals, and some 750,000 insects. How did all these species evolve?

In Chapters 17 and 18, we examined natural selection and how populations evolve. We now focus on how species and higher taxonomic categories evolve. The taxonomic groups above the level of species are artificial constructs that humans use to indicate degrees of relatedness among organisms. As described in Chapter 1, scientists group closely related species into the same genus (pl., *genera*), similar genera into the same family, similar families into the same order, similar orders into the same class, and similar classes into the same phylum (pl., *phyla*).

The concept of distinct kinds of organisms, known as **species** (from Latin, meaning "kind") is not new. However, every definition of species has some sort of limitation. Linnaeus, the 18th-century biologist—who founded modern **taxonomy,** the science of naming, describing, and classifying organisms—classified plants and other organisms into separate species based on their structural differences (see Chapter 22). This method, known as the *morphological species concept,* is still used to help characterize species, but structure alone is not adequate to explain what constitutes a species.

Population genetics did much to clarify the concept of species. According to the **biological species concept,** first expressed by evolutionary biologist Ernst Mayr in 1942, a species consists of one or more populations whose members interbreed in nature to produce fertile offspring and do not interbreed with—that is, are reproductively isolated from—members of different

[1] M. L. Reaka-Kudla, D. E. Wilson, and E. O. Wilson, eds., *Biodiversity II* (Washington, D.C.: Joseph Henry Press, 1997).

species. In other words, each species has a *gene pool* separate from that of other species, and reproductive barriers restrict each species from interbreeding with other species.

One problem with the biological species concept is that it applies only to sexually reproducing organisms. Organisms that reproduce asexually do not interbreed, so they cannot be considered in terms of reproductive isolation. These organisms and extinct organisms are classified on the basis of structural and biochemical characteristics. Another potential problem with the biological species concept is that although individuals assigned to different species do not normally interbreed, they may *sometimes* successfully interbreed (see photograph).

Some biologists prefer the evolutionary species concept instead of the biological species concept. In the **evolutionary species concept,** also called the *phylogenetic species concept,* to be declared a separate species a population must have undergone evolution long enough for statistically significant differences in diagnostic traits to emerge. This approach has the advantage of being testable. However, many biologists do not want to abandon the biological species concept, in part because the evolutionary species concept cannot be applied if the evolutionary history of a taxonomic group has not been carefully studied, and most groups have not been rigorously analyzed. Also, if the evolutionary species concept were universally embraced, that would probably double the number of named species. This increase would occur because many closely related populations that are classified as subspecies or varieties under the biological species concept would fit the requirements of separate species under the evolutionary species concept. Thus the exact definition of a species remains fuzzy. Unless we clearly state otherwise, when we mention *species* in this text we mean the biological species concept.

This chapter discusses reproductive barriers that isolate species from one another, the possible evolutionary mechanisms that explain how the millions of species that live today or lived in the past originated from ancestral species, and the rates of evolutionary change. We then examine macroevolution. ■

REPRODUCTIVE ISOLATION

Learning Objective

1 Explain the significance of reproductive isolating mechanisms, and distinguish among the different prezygotic and postzygotic barriers.

Various **reproductive isolating mechanisms** prevent interbreeding between two different species whose *ranges* (areas where each lives) overlap. These mechanisms preserve the genetic integrity of each species, because gene flow between the two species is prevented. Most species have two or more mechanisms that block a chance occurrence of individuals from two closely related species overcoming a single reproductive isolating mechanism. Most occur before mating or fertilization occurs *(prezygotic),* whereas others work after fertilization has taken place *(postzygotic).*

Prezygotic barriers interfere with fertilization

Prezygotic barriers are reproductive isolating mechanisms that prevent fertilization from taking place. Because male and female gametes never come into contact, an *interspecific zygote*—that is, a fertilized egg formed by the union of an egg from one species and a sperm from another species—is never produced. Prezygotic barriers include temporal isolation, habitat isolation, behavioral isolation, mechanical isolation, and gametic isolation.

Sometimes genetic exchange between two groups is prevented because they reproduce at different times of the day, season, or year. Such examples demonstrate **temporal isolation.** For example, two very similar species of fruit flies, *Drosophila pseudoobscura* and *Drosophila persimilis,* have ranges that overlap to a great extent, but they do not interbreed. *D. pseudoobscura* is sexually active only in the afternoon, and *D. persimilis* only in the morning. Similarly, two frog species have overlapping ranges in eastern Canada and the United States. The wood frog (*Rana sylvatica*) usually mates in late March or early April, when the water temperature is about 7.2°C (45°F), whereas the northern leopard frog (*Rana pipiens*) usually mates in mid-April, when the water temperature is 12.8°C (55°F) (Fig. 19-1).

Although two closely related species may be found in the same geographic area, they usually live and breed in different habitats in that area. This provides **habitat isolation** between the two species. For example, the five species of small birds known as flycatchers are nearly identical in appearance and have overlapping ranges in the eastern part of North America. They exhibit habitat isolation because, during the breeding season, each species stays in a particular habitat within its range, so potential mates from different species do not meet. The least flycatcher (*Empidonax minimus*) frequents open woods, farms, and orchards, whereas the acadian flycatcher (*E. virescens*) is found in deciduous forests, particularly in beech trees, and swampy woods. The alder flycatcher (*E. alnorum*) prefers wet thickets of alders, the yellow-bellied flycatcher (*E. flaviventris*) nests in conifer woods, and the willow flycatcher (*E. traillii*) frequents brushy pastures and willow thickets.

Many animal species exchange a distinctive series of signals before mating. Such courtship behaviors illustrate **behavioral isolation** (also known as **sexual isolation**). Bowerbirds, for example, exhibit species-specific courtship patterns. The male satin bowerbird of Australia constructs an elaborate bower of twigs, adding decorative blue parrot feathers and white flowers at the entrance (Fig. 19-2). When a female approaches the bower, the male dances about her, holding a particularly eye-catching decoration in his beak. While dancing, he puffs his feathers, extends his wings, and sings a courtship song that consists of a variety of sounds, including loud buzzes and laughlike hoots. These specific courtship behaviors keep similar bird species reproductively isolated from the satin bowerbird. If a male and female of two different species begin courtship, it stops when one member does

L. & D. Klein/Photo Researchers, Inc.

(a)

Rod Planck/Photo Researchers, Inc.

(b)

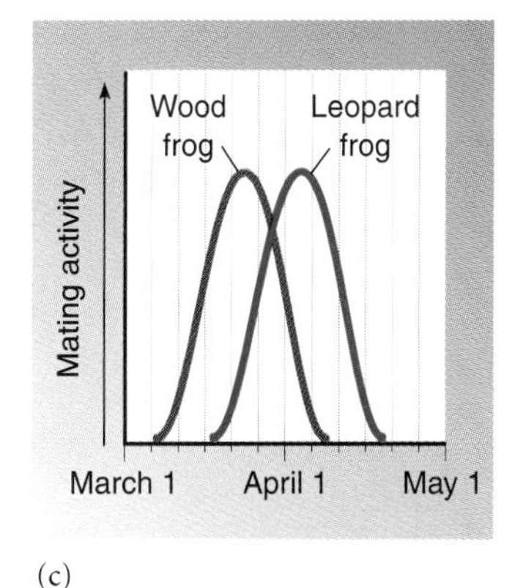

(c)

ACTIVE FIGURE 19-1 Temporal isolation in wood and leopard frogs.

(a) The wood frog (*Rana sylvatica*) mates in early spring, often before the ice has completely melted in the ponds. **(b)** The leopard frog (*R. pipiens*) typically mates a few weeks later. **(c)** Graph of peak mating activity in wood and leopard frogs. In nature, wood and leopard frogs do not interbreed, although they have done so in the laboratory.

BiologyNow™ Explore **reproductive isolation in another species** by clicking on this figure on your BiologyNow CD ROM.

not recognize or respond to the signals of the other. Another example of behavioral isolation involves the wood frogs and northern leopard frogs (just discussed as an example of temporal isolation). Males of these two species have very specific vocalizations to attract females of their species for breeding. These vocalizations reinforce each species' reproductive isolation.

Sometimes members of different species court and even attempt copulation, but the incompatible structures of their genital organs prevent successful mating. Structural differences that inhibit mating between species produce **mechanical isolation.** For example, many flowering plant species have physical differences in their flower parts that help them maintain their reproductive isolation from one another. In such plants, the flower parts are adapted for specific insect pollinators. Two species of sage, for example, have overlapping ranges in southern California. Black sage (*Salvia mellifera*), which is pollinated by small bees, has a floral structure different from that of white sage (*Salvia apiana*), which is pollinated by large carpenter bees (Fig. 19-3). Interestingly, black sage and white sage are also prevented from mating by a temporal barrier: black sage flowers in early spring, and white sage flowers in late spring and early summer. Presumably, mechanical isolation prevents insects from cross-pollinating the two species should they happen to flower at the same time.

R. Brown/VIREO

FIGURE 19-2 Behavioral isolation in bowerbirds.

Each bowerbird species has highly specialized courtship patterns that prevent its mating with another species. The male satin bowerbird (*Ptilonorhynchus violacens*) constructs an enclosed place, or bower, of twigs to attract a female. (The bower is the dark "tunnel" on the left.) Note the white flowers and blue decorations, including human-made objects such as bottle caps, that he has arranged at the entrance to his bower.

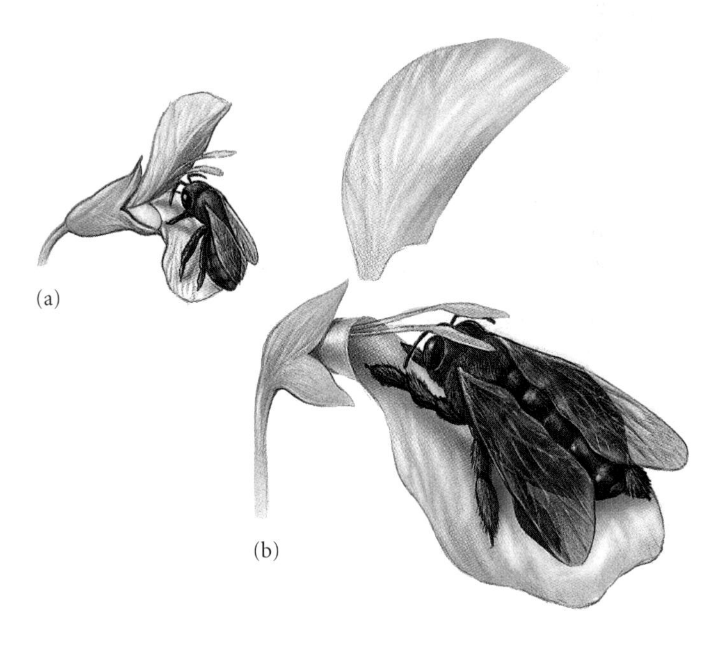

FIGURE 19-3 Mechanical isolation in black sage and white sage.

Their differences in floral structures allow black sage and white sage to be pollinated by different insects. Because the two species exploit different pollinators, they cannot interbreed. **(a)** The petal of the black sage functions as a landing platform for small bees. Larger bees cannot fit on this platform. **(b)** The larger landing platform and longer stamens of white sage allow pollination by larger California carpenter bees (a different species). If smaller bees land on white sage, their bodies do not brush against the stamens. (In the figure, the upper part of the white sage flower has been removed.)

If mating takes place between two species, their gametes may still not combine. Molecular and chemical differences between species cause **gametic isolation,** in which the egg and sperm of different species are incompatible. In aquatic animals that release their eggs and sperm into the surrounding water simultaneously, interspecific fertilization is extremely rare. The surface of the egg contains specific proteins that bind only to complementary molecules on the surface of sperm cells of the same species (see Chapter 49). A similar type of molecular recognition often occurs between pollen grains and the stigma (receptive surface of the female part of the flower) so that pollen does not germinate on the stigma of a different plant species.

Postzygotic barriers prevent gene flow when fertilization occurs

Fertilization sometimes occurs between gametes of two closely related species despite the existence of prezygotic barriers. When this happens, **postzygotic barriers** that increase the likelihood of reproductive failure come into play. Generally, the embryo of an interspecific hybrid spontaneously aborts. Embryonic development is a complex process requiring the precise interaction and coordination of many genes. Apparently the genes from parents belonging to different species do not interact properly in regulating the mechanisms for normal development. In this case, reproductive isolation occurs by **hybrid inviability.** For example, nearly all the hybrids die in the embryonic stage when the eggs of a bullfrog are fertilized artificially with sperm from a leopard frog. Similarly, in crosses between different species of irises the embryos die before reaching maturity.

If an interspecific hybrid does live, it may not reproduce. There are several reasons for this. Hybrid animals may exhibit courtship behaviors incompatible with those of either parental species, and as a result they will not mate. More often, **hybrid sterility** occurs when problems during meiosis cause the gametes of an interspecific hybrid to be abnormal. Hybrid sterility is particularly common if the two parental species have different chromosome numbers. For example, a mule is the offspring of a female horse ($2n = 64$) and a male donkey ($2n = 62$) (Fig. 19-4). This type of union almost always results in sterile offspring ($2n = 63$), because synapsis (the pairing of homologous chromosomes during meiosis) and segregation cannot occur properly. The zebrass shown in the chapter introduction also exhibits hybrid sterility because of different chromosome numbers. Donkeys have 62 chromosomes, and zebras, depending on the species, have 32, 44, or 46 chromosomes.

John Eastcott/Yva Momatiuk/Animals Animals

FIGURE **19-4** **Hybrid sterility in mules.**

Mules are interspecific hybrids formed by mating a female horse with a male donkey. Although the mule *(center)* exhibits valuable characteristics of each of its parents, it is sterile.

Occasionally, a mating between two F_1 hybrids produces a second hybrid generation (F_2). The F_2 hybrid may exhibit **hybrid breakdown,** the inability of a hybrid to reproduce because of some defect. For example, hybrid breakdown in the F_2 generation of a cross between two sunflower species was 80%. In other words, 80% of the F_2 generation were defective in some way and could not reproduce successfully. Hybrid breakdown also occurs in the F_3 and later generations.

Table 19-1 summarizes the various prezygotic and postzygotic barriers that prevent interbreeding between two species.

Biologists are discovering the genetic basis of isolating mechanisms

Progress has been made in identifying some of the genes involved in reproductive isolation. For example, scientists have determined the genetic basis for prezygotic isolation in species of abalone, large mollusks found along the Pacific coast of North America. In abalone, the fertilization of eggs by sperm requires lysin, a sperm protein that attaches to a lysin receptor protein located on the egg envelope. After attachment, the lysin produces a hole in the egg envelope that permits the sperm to penetrate the egg. Scientists cloned the lysin receptor gene and demonstrated that this gene varies among abalone species. Differences in the lysin receptor protein in various abalone species determine sperm compatibility with the egg. Sperm of one abalone species do not attach to a lysin receptor protein of an egg of a different abalone species.

TABLE **19-1** **Reproductive Isolating Mechanisms**

Mechanism	How It Works
Prezygotic Barriers	Prevent fertilization
Temporal isolation	Similar species reproduce at different times
Habitat isolation	Similar species reproduce in different habitats
Behavioral isolation	Similar species have distinctive courtship behaviors
Mechanical isolation	Similar species have structural differences in their reproductive organs
Gametic isolation	Gametes of similar species are chemically incompatible
Postzygotic Barriers	Reduce viability or fertility of hybrid
Hybrid inviability	Interspecific hybrid dies at early stage of embryonic development
Hybrid sterility	Interspecific hybrid survives to adulthood but is unable to reproduce successfully
Hybrid breakdown	Offspring of interspecific hybrid are unable to reproduce successfully

Review

- What barriers prevent wood frogs and leopard frogs from interbreeding in nature?
- How is temporal isolation different from behavioral isolation?
- How is mechanical isolation different from gametic isolation?
- Of which type of postzygotic barrier is the mule an example?

BiologyNow™ Assess your understanding of **reproductive isolation** by taking the pretest on your BiologyNow CD-ROM.

SPECIATION

Learning Objectives

2 Explain the mechanism of allopatric speciation, and give an example.

3 Explain the mechanisms of sympatric speciation, and give both plant and animal examples.

We are now ready to consider how entirely new species may arise from previously existing ones. The evolution of a new species is called **speciation.** The formation of two species from a single species occurs when a population becomes reproductively isolated from other populations of the species, and the gene pools of the two separated populations begin to diverge in genetic composition. When a population is sufficiently different from its ancestral species that no genetic exchange occurs between them, we say speciation has occurred. Speciation occurs in two ways: allopatric speciation and sympatric speciation.

Long physical isolation and different selective pressures result in allopatric speciation

Speciation that occurs when one population becomes geographically separated from the rest of the species and subsequently evolves by natural selection and/or genetic drift is known as **allopatric speciation** (from the Greek *allo,* "different," and *patri,* "native land"). Recall from Chapter 17 that *natural selection* occurs as individuals that possess favorable adaptations to the environment survive and become parents of the next generation. *Genetic drift* is a random change in allele frequency resulting from the effects of chance on the survival and reproductive success of individuals in a small breeding population (see Chapter 18). Both natural selection and genetic drift result in changes in allele frequencies in a population, but only in natural selection is the change in allele frequency adaptive.

Allopatric speciation is the most common method of speciation, and the evolution of new animal species has been almost exclusively by allopatric speciation. The geographic isolation required for allopatric speciation may occur in several ways. Earth's surface is in a constant state of change. Such change includes rivers shifting their courses; glaciers migrating; mountain ranges forming; land bridges developing that separate previously united aquatic populations; and large lakes diminishing into several smaller, geographically separated pools.

What might be an imposing geographic barrier to one species may be of no consequence to another. Birds and cattails, for example, do not become isolated when a lake subsides into smaller pools; birds easily fly from one pool to another, and cattails disperse their pollen and fruits by air currents. Fish, in contrast, usually cannot cross the land barriers between the pools and so become reproductively isolated.

In the Death Valley region of California and Nevada, large interconnected lakes formed during wetter climates of the last Ice Age. These lakes were populated by one or several species of pupfish. Over time, the climate became drier, and the large lakes dried up, leaving isolated pools. Presumably, each pool contained a small population of pupfish that gradually diverged from the common ancestral species by genetic drift and natural selection in response to the high temperatures, high salt concentrations, and low oxygen levels characteristic of desert springs. Today, there are 20 or so distinct species, subspecies, and populations in the area of Death Valley, California, and Ash Meadows, Nevada. Many, such as the Devil's Hole pupfish (*Cyprinodon diabolis*) and the Owens pupfish (*Cyprinodon radiosus*), are restricted to one or two isolated springs (Fig. 19-5).

Allopatric speciation also occurs when a small population migrates or is dispersed (such as by a chance storm) and colonizes a new area away from the range of the original species. This colony is geographically isolated from its parental species, and the small *microevolutionary* changes that accumulate in the isolated gene pool over many generations may eventually be sufficient to form a new species. Because islands provide the geographic isolation required for allopatric speciation, they offer excellent opportunities to study this evolutionary mechanism. A few individuals of a few species probably colonized the Galapagos Islands and the Hawaiian Islands, for example. The hundreds of unique species presently found on each island presumably descended from these original colonizers (Fig. 19-6).

Speciation is more likely to occur if the original isolated population is small. Recall that genetic drift, including the founder

Steinhart Aquarium, Tom McHugh/Photo Researchers, Inc.

FIGURE **19-5** **Allopatric speciation of pupfish (*Cyprinodon*).**

Shown is one of the more than 20 pupfish species that apparently evolved when larger lakes in southern Nevada dried up about 10,000 years ago, leaving behind small, isolated desert pools fed by springs. The pupfish's short, stubby body is characteristic of fish that live in springs; fish that live in larger bodies of water are more streamlined.

Victoria McCormick/Animals Animals

FIGURE **19-6** **Allopatric speciation of the Hawaiian goose (the nene).**

Nene (pronounced "nay–nay"; *Branta sandvicensis*) are geese originally found only on volcanic mountains on the geographically isolated islands of Hawaii and Maui, which are some 4200 km (2600 mi) from the nearest continent. Compared with other geese, the feet of Hawaiian geese are not completely webbed, their toenails are longer and stronger, and their foot pads are thicker; these adaptations enable Hawaiian geese to walk easily on lava flows. Nene are thought to have evolved from a small population of geese that originated in North America. Photographed in Hawaii Volcanoes National Park, Hawaii.

effect, is more consequential in small populations (see Chapter 18). Genetic drift tends to result in rapid changes in allele frequencies in the small, isolated population. The different selective pressures of the new environment to which the population is exposed further accentuate the divergence caused by genetic drift.

The Kaibab squirrel may be an example of allopatric speciation in progress

About 10,000 years ago, when the American Southwest was less arid, the forests in the area supported a tree squirrel with conspicuous tufts of hair sprouting from its ears. A small tree squirrel population living on the Kaibab Plateau of the Grand Canyon became geographically isolated when the climate changed, causing areas to the north, west, and east to become desert. Just a few miles to the south lived the rest of the squirrels, known as Abert squirrels, but the two groups were separated by the Grand Canyon. With changes over time in both its appearance and its ecology, the Kaibab squirrel is on its way to becoming a new species.

During its many years of geographic isolation, the small population of Kaibab squirrels has diverged from the widely distributed Abert squirrels in several ways. Perhaps most evident are changes in fur color. The Kaibab squirrel now has a white tail and a gray belly, in contrast to the gray tail and white belly of the Abert squirrel (Fig. 19-7). Biologists think these striking changes arose in Kaibab squirrels as a result of genetic drift.

Some scientists consider the Kaibab squirrel and the Abert squirrel as distinct populations of the same species (*Sciurus aberti*). Because the Kaibab and Abert squirrels are reproductively isolated from each other, however, some scientists have classified the Kaibab squirrel as a different species (*Sciurus kaibabensis*).

Porto Santo rabbits may be an example of extremely rapid allopatric speciation

Allopatric speciation has the potential to occur quite rapidly. Early in the 15th century, a small population of rabbits was released on Porto Santo, a small island off the coast of Portugal. Because there were no other rabbits or competitors and no predators on the island, the rabbits thrived. By the 19th century, these rabbits were markedly different from their European ancestors. They were only half as large (weighing slightly more than 500 g, or 1.1 lb), with a different color pattern and a more nocturnal lifestyle. Most significantly, attempts to mate Porto Santo rabbits with mainland European rabbits failed. Many biologists concluded that, within 400 years, an extremely brief period in evolutionary history, a new species of rabbit had evolved.

Not all biologists agree that the Porto Santo rabbit is a new species. The objection stems from a more recent breeding experiment and is based on biologists' lack of a consensus about the definition of a species. In the experiment, foster mothers of the wild Mediterranean rabbit raised newborn Porto Santo rabbits. When they reached adulthood, these Porto Santo rabbits

(a) Tom and Pat Leeson

(b) Kent and Donna Dannen

ACTIVE FIGURE **19-7**

Allopatric speciation in progress.

(a) The Kaibab squirrel, with its white tail and gray belly, is found north of the Grand Canyon. **(b)** The Abert squirrel, with its gray tail and white belly, is found south of the Grand Canyon.

BiologyNow™ Explore **allopatric speciation in another species** by clicking on this figure on your BiologyNow CD ROM.

mated successfully with Mediterranean rabbits to produce healthy, fertile offspring. To some biologists, this experiment clearly demonstrated that Porto Santo rabbits are not a separate species but, instead, are an example of speciation in progress, much like the Kaibab squirrels just discussed. Other biologists think the Porto Santo rabbit is a separate species, because it does not interbreed with other rabbits under natural conditions. They point out that the breeding experiment was successful only after the baby Porto Santo rabbits were raised under artificial conditions that probably modified their natural behavior.

Two populations diverge in the same physical location by sympatric speciation

Although geographic isolation is an important factor in many cases of evolution, it is not an absolute requirement. In **sympatric speciation** (from the Greek *sym*, "together," and *patri*, "native land"), a new species evolves within the same geographic region as the parental species. The divergence of two populations in the same geographic range occurs when reproductive isolating mechanisms evolve at the *start* of the speciation process. Sympatric speciation is especially common in plants. The role of sympatric speciation in animal evolution is probably much less important than allopatric speciation; until recently, sympatric speciation in animals has been difficult to demonstrate in nature.

Sympatric speciation occurs in at least two ways: a change in **ploidy** (the number of chromosome sets making up an organism's genome) and a change in ecology. We now examine each of these mechanisms.

Allopolyploidy is an important mechanism of sympatric speciation in plants

As a result of reproductive isolating mechanisms discussed earlier, the union of two gametes from different species rarely forms viable offspring; if offspring are produced, they are usually sterile. Before gametes form, meiosis reduces the chromosome number (see Chapter 9). For the chromosomes to be parceled correctly into the gametes, homologous chromosome pairs must come together (a process called *synapsis*) during prophase I. This cannot usually occur in interspecific hybrid offspring, because not all the chromosomes are homologous. However, if the chromosome number doubles *before* meiosis, then homologous chromosomes undergo synapsis. Although not common, this spontaneous doubling of chromosomes has been documented in a variety of plants and a few animals. It produces nuclei with multiple sets of chromosomes.

Polyploidy, the possession of more than two sets of chromosomes, is a major factor in plant evolution. Reproductive isolation occurs in a single generation when a polyploid species with multiple sets of chromosomes arises from diploid parents. There are two kinds of polyploidy: autopolyploidy and allopolyploidy. An **autopolyploid** contains multiple sets of chromosomes from a single species, and an **allopolyploid** contains multiple sets of chromosomes from two or more species. We discuss only allopolyploidy, because it is much more common in nature.

Allopolyploidy occurs in conjunction with **hybridization,** which is sexual reproduction between individuals from closely related species. Allopolyploidy produces a fertile interspecific hybrid because the polyploid condition provides the homologous chromosome pairs necessary for synapsis during meiosis. As a result, gametes may be viable (Fig. 19-8). An allopolyploid, that is, an interspecific hybrid produced by allopolyploidy, reproduces with itself (self-fertilization) or with a similar individual. However, allopolyploids are reproductively isolated from both parents, because their gametes have a different number of chromosomes from those of either parent.

If a population of allopolyploids (that is, a new species) becomes established, selective pressures cause one of three outcomes. First, the new species may not compete successfully against species that are already established, so it becomes extinct. Second, the allopolyploid individuals may assume a new role in the environment and so coexist with both parental species. Third, the new species may successfully compete with either or both of its parental species. If it has a combination of traits that confers greater fitness than one or both parental species for all or part of the original range of the parent(s), the hybrid species may replace the parent(s).

Although allopolyploidy is extremely rare in animals, it is significant in the evolution of flowering plant species. As many as 80% of all flowering plant species are polyploids, and most of these are allopolyploids. Moreover, allopolyploidy provides a mechanism for extremely rapid speciation. A single generation is all that is needed to form a new, reproductively isolated species. Allopolyploidy may explain the rapid appearance of many flowering plant species in the fossil record and their remarkable diversity (about 235,000 species) today.

The kew primrose (*Primula kewensis*) is an example of sympatric speciation that was documented at the Royal Botanic Gardens at Kew, England, in 1898 (Fig. 19-9). The interspecific hybrid of two primrose species, *Primula floribunda* ($2n = 18$) and *Primula verticillata* ($2n = 18$), *P. kewensis* had a chromosome number of 18 but was sterile. Then, at three different times, it was reported to have spontaneously formed a fertile branch, which was an allopolyploid ($2n = 36$) that produced viable seeds of *P. kewensis.*

The mechanism of sympatric speciation has been experimentally verified for many plant species. One example is a group of species, collectively called *hemp nettles,* that occurs in temperate parts of Europe and Asia. One hemp nettle, *Galeopsis tetrahit* ($2n = 32$), is a naturally occurring allopolyploid hypothesized to have formed by the hybridization of two species, *Galeopsis pubescens* ($2n = 16$) and *Galeopsis speciosa* ($2n = 16$). This process occurred in nature but was experimentally reproduced. *Galeopsis pubescens* and *G. speciosa* were crossed to produce F_1 hybrids, most of which were sterile. Nevertheless, both F_2 and F_3 generations were produced. The F_3 generation included a polyploid plant with $2n = 32$ that self-fertilized to yield fertile F_4 offspring that could not mate with either of the parental species. These allopolyploid plants had the same appearance and chromosome number as the naturally occurring *G. tetrahit.* When the experimentally produced plants were crossed with the naturally occurring *G. tetrahit,* a fertile F_1 generation was formed. Thus, the experiment duplicated the speciation process that occurred in nature.

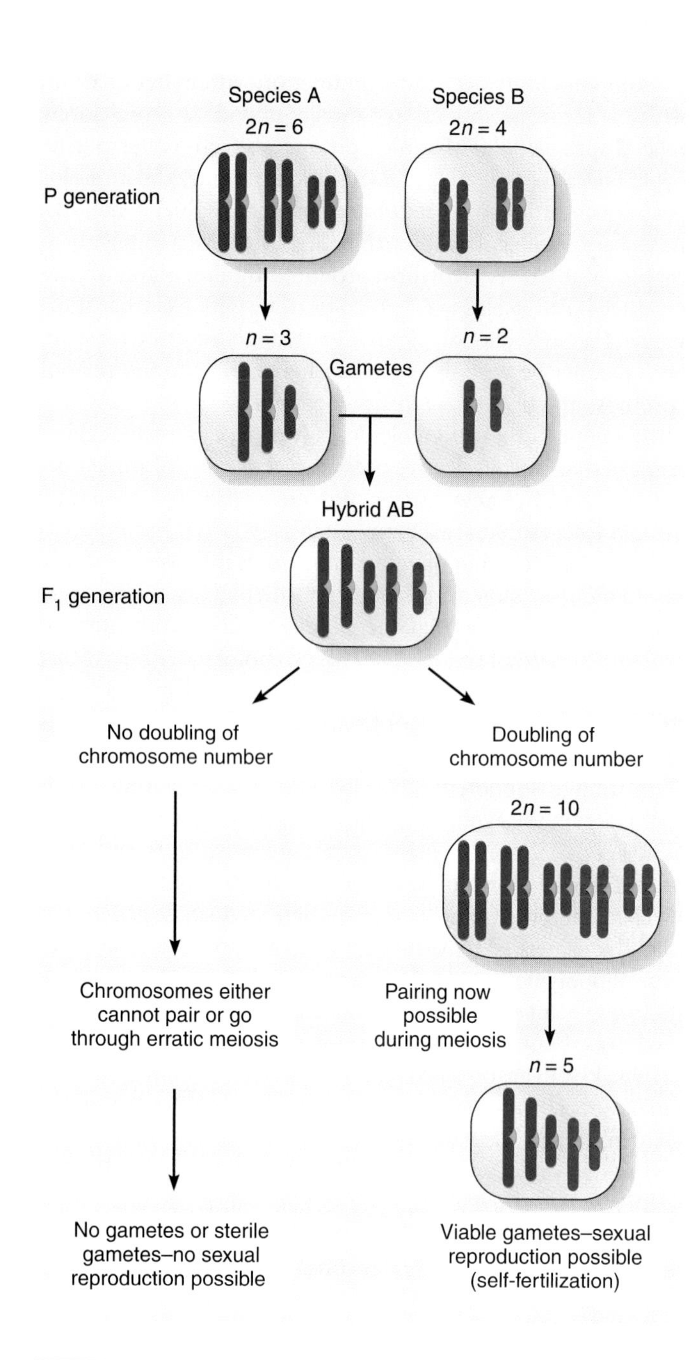

FIGURE **19-8** **Sympatric speciation by allopolyploidy in plants.**

When two species (designated the P generation) successfully interbreed, the interspecific hybrid offspring (the F_1 generation) are almost always sterile *(bottom left).* If the chromosomes double, proper synapsis and segregation of the chromosomes occur, and viable gametes are produced *(bottom right).* (Unduplicated chromosomes are shown for clarity.)

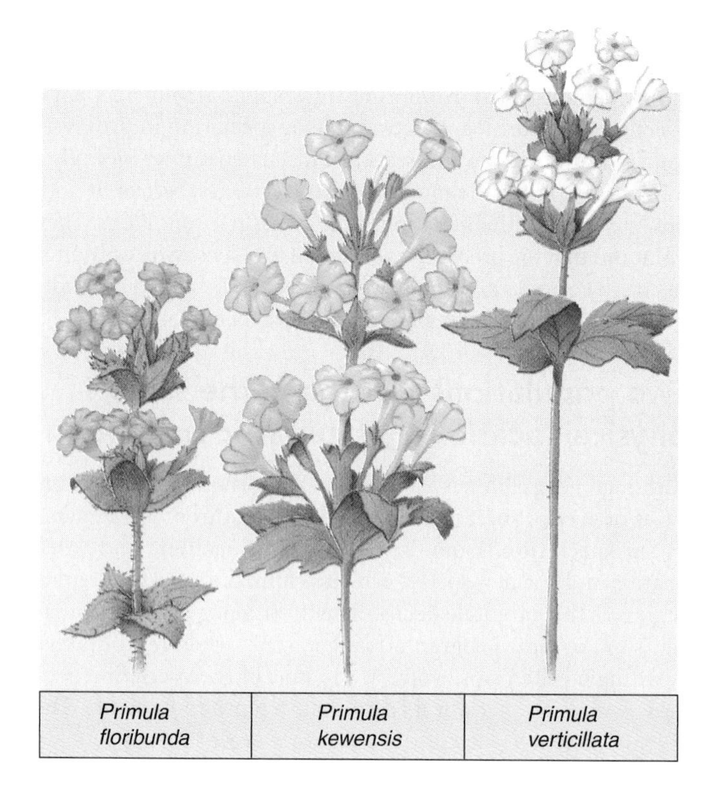

FIGURE **19-9** **Sympatric speciation of a primrose.**

An allopolyploid primrose, *Primula kewensis,* arose in 1898 as an allopolyploid derived from the interspecific hybridization of *P. floribunda* and *P. verticillata.* Today *P. kewensis* is a popular houseplant.

Changing ecology causes sympatric speciation in animals

Biologists have observed the occurrence of sympatric speciation in animals, but its significance—how often it occurs and under what conditions—is still debated. Many examples of sympatric speciation in animals involve parasitic insects and rely on genetic mechanisms other than polyploidy. For example, in the 1860s in the Hudson River Valley of New York, a population of fruit maggot flies (*Rhagoletis pomonella*) parasitic on the small red fruits of native hawthorn trees was documented to have switched to a new host, domestic apples, which had been introduced from Europe. Although the sister populations (hawthorn maggot flies and apple maggot flies) continue to occupy the same geographic area, no gene flow occurs between them because they eat, mate, and lay their eggs on different hosts (Fig. 19-10). In other words, because the hawthorn and apple maggot flies have diverged and are reproductively isolated from each other, they have effectively become separate species. Most entomologists, however, still recognize hawthorn and apple maggot flies as a single species, because their appearance is virtually identical.

In situations like the fruit maggot flies, a mutation arises in an individual and spreads through a small group of insects by sexual reproduction. The particular mutation leads to *disruptive selection* (see Chapter 18), in which both the old and new phenotypes are favored. This occurs because the original population is still favored when it parasitizes the original host, and the mutants are favored by a new ecological opportunity—in this case, to parasitize a different host species. Additional mutations may occur that cause the sister populations to diverge even further.

Biologists have studied the speciation of colorful fishes known as *cichlids* (pronounced "sick-lids") in several East African lakes. The different species of cichlids in a given lake have remarkably

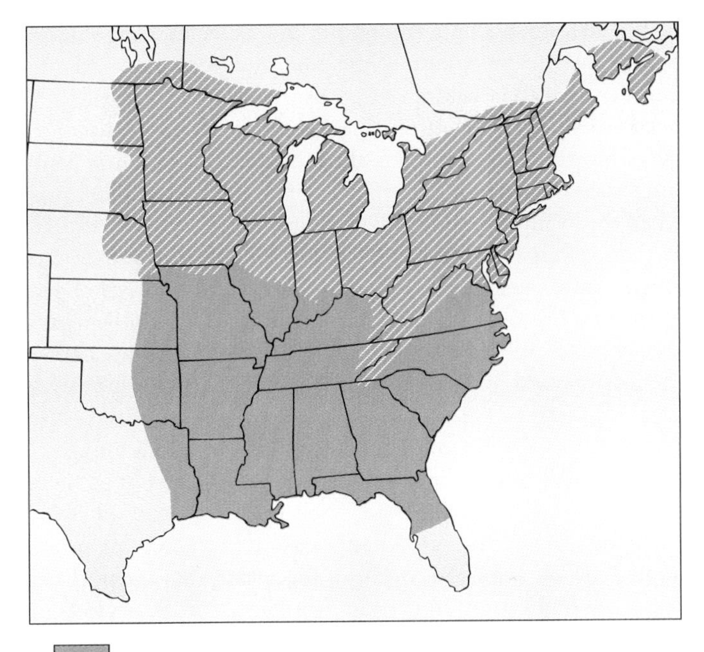

Hawthorn maggot fly range

Apple maggot fly and hawthorn maggot fly range

FIGURE **19-10** **Ranges of apple and hawthorn maggot flies.**

Apple and hawthorn maggot flies are sympatric throughout the northern half of the hawthorn maggot fly's range. *(Adapted from G. L. Bush, "Sympatric Host Race Formation and Speciation in Frugivorous Flies of the Genus* Rhagoletis *[Diptera, Tephritidae]," Evolution, Vol. 23, No. 2, Jun. 1969)*

different eating habits, which are reflected in the shapes of their jaws and teeth. Some graze on algae, some consume dead organic material at the bottom of the lake, and others are predatory and eat plankton (microscopic aquatic organisms), insect larvae, the scales off fish, or even other cichlid species. In some cichlids food preferences are related to size (smaller cichlids consume plankton), which in turn is related to mating preference (small, plankton-eating cichlids mate only with other small, plankton-eating cichlids).

In other cichlids related species do not differ in size but only in color. Each species has its own distinct male coloration, which matches the preferences of females of that species in choosing mates (Fig. 19-11). Charles Darwin called choosing a mate based on its color or some other characteristic **sexual selection.** He recognized that females' preferences for certain males, if it was based on *inherited* variation, could result in certain male variants becoming more common over time. Darwin suggested that sexual selection, like natural selection, could eventually result in the evolution of new species. (Sexual selection is discussed in greater detail in Chapter 50.)

Recent DNA sequence data indicate the cichlid species within a single lake are more closely related to one another than to fishes in nearby lakes or rivers. These molecular data suggest that cichlid species evolved sympatrically, or at least within the confines of a lake, rather than by repeated colonizations by fish populations in nearby rivers.

How rapidly did sympatric speciation occur in cichlids? In 1996 scientists published seismic and drill core data suggesting that the more than 500 endemic (that is, found nowhere else) cichlid species in Lake Victoria evolved in a remarkably short time—less than 12,400 years. This inference is based on evidence that Lake Victoria dried up completely during the late Pleistocene (from about 17,000 to 12,400 years ago), when much of north and equatorial Africa was arid. It appears the cichlids evolved after the climate became wetter and Lake Victoria refilled. If future data substantiate this conclusion, it means the evolution of Lake Victoria's cichlids is the fastest known for such a large number of vertebrate species.

Reproductive isolation breaks down in hybrid zones

When two populations have significantly diverged as a result of geographic separation, there is no easy way to determine if the

Courtesy of Ole Seehausen/Leiden University, The Netherlands, and Hull University, United Kingdom

(a)

(b)

(c)

FIGURE **19-11** **Color variation in Lake Victoria cichlids.**

(a) *Pundamilia pundamilia* males have bluish-silver bodies. **(b)** *Pundamilia nyererei* males have red backs. **(c)** *Pundamilia* "red head" males have a red "chest." (The "red head" species has not yet been scientifically named.) Some evidence suggests that changes in male coloration may be the first step in speciation of Lake Victoria cichlids. Later, other traits, including ecological characteristics, diverge. Female cichlids generally have cryptic coloration; their drab colors help them blend into their surroundings.

speciation process is complete (recall the disagreement about whether Porto Santo rabbits are a separate species or a *subspecies,* which is a taxonomic subdivision of a species). If such populations, subspecies, or species come into contact, they may hybridize where they meet, forming a **hybrid zone,** or area of overlap in which interbreeding occurs. Hybrid zones are typically narrow, presumably because the hybrids are not well adapted for either parental environment, and the hybrid population is typically very small compared with the parental populations.

On the Great Plains of North America, red-shafted and yellow-shafted flickers (types of woodpeckers) meet and interbreed. The red-shafted flicker, named for the male's red underwings and tail, is found in the western part of North America, from the Great Plains to the Pacific Ocean. Yellow-shafted flicker males, which have yellow underwings and tails, range east of the Rockies. Hybrid flickers, which form a stable hybrid zone from Texas to southern Alaska, are varied in appearance, although many have orange underwings and tails.

PROCESS OF SCIENCE

Biologists do not agree about whether the red-shafted and yellow-shafted flickers are separate species or geographic subspecies. According to the biological species concept, if red-shafted and yellow-shafted flickers are two species, they should maintain their reproductive isolation. However, the flicker hybrid zone has not expanded, that is, the two types of flickers have maintained their distinctiveness and have not rejoined into a single, freely interbreeding population.

The study of hybrid zones has made important contributions to what is known about speciation. As in other fields of science, disagreements and differences of opinion are an important part of the scientific process because they stimulate new ideas, hypotheses, and experimental tests that expand our base of scientific knowledge.

Review

- What are five geographic barriers that might lead to allopatric speciation?
- How do hybridization and polyploidy cause a new plant species to form in as little as one generation?
- What is the likely mechanism of speciation for pupfish? For cichlids?

BiologyNow™ Assess your understanding of **speciation** by taking the pretest on your BiologyNow CD-ROM.

THE RATE OF EVOLUTIONARY CHANGE

Learning Objective

4 Take either side in a debate on the pace of evolution, by representing the opposing views of gradualism and punctuated equilibrium.

We have seen that speciation is hard for us to directly observe as it occurs. Does the fossil record provide clues about how rapidly new species arise? Biologists have long recognized that the fossil record lacks many transitional forms; the starting points (ancestral species) and the end points (new species) are present, but the intermediate stages in the evolution from one species to another are often lacking. This observation has traditionally been explained by the incompleteness of the fossil record. Biologists have attempted to fill in the missing parts with new fossil discoveries, much as a writer might fill in the middle of a novel when the beginning and end are already there.

Two different models—punctuated equilibrium and gradualism—have been developed to explain evolution as observed in the fossil record (Fig. 19-12). The **punctuated equilibrium** model was proposed by paleontologists who question whether the fossil record really is as incomplete as it initially appeared. First advanced by paleontologists Stephen Jay Gould and Niles Eldredge in 1972, the punctuated equilibrium model suggests that the fossil record accurately reflects evolution as it actually occurs. That is, in the history of a species long periods of **stasis** (little or no evolutionary change) are punctuated, or interrupted, by short periods of rapid speciation that are perhaps triggered by changes in the environment, that is, periods of great evolutionary stress. Thus speciation normally proceeds in "spurts." These relatively short periods of active evolution (such as 100,000 years) are followed by long periods (such as 2 million years) of stability.

With punctuated equilibrium, speciation occurs in a relatively short period. Keep in mind, however, that a "short" amount of time for speciation may be thousands of years. Such a span is short when compared with the several million years of a species' existence. Biologists who support the idea of punctuated equilibrium emphasize that sympatric speciation and even allopatric speciation occur in such relatively short periods. Punctuated equilibrium accounts for the abrupt appearance of a new species in the fossil record, with little or no evidence of intermediate forms. Proponents hypothesize that few transitional forms appear in the fossil record, because few transitional forms occurred during speciation.

In contrast, the traditional view of evolution espouses the **gradualism** model, in which evolution proceeds continuously over long periods. Gradualism is rarely observed in the fossil record, because the record is incomplete. (Recall from Chapter 17 that precise conditions are required for fossil formation. Most organisms decompose when they die, leaving no trace of their existence.) Occasionally a complete fossil record of transitional forms is discovered and cited as a strong case for gradualism. The gradualism model maintains that populations slowly diverge from one another by the gradual accumulation of adaptive characteristics within each population. These adaptive characteristics accumulate as a result of different selective pressures encountered in different environments.

PROCESS OF SCIENCE

The abundant fossil evidence of long periods with no change in a species has been used to argue against the gradualism model of evolution. Gradualists, however, maintain that any periods of stasis evident in the fossil record are the result of *stabilizing selection* (see Chapter 18). They also emphasize that stasis in fossils is deceptive, because fossils do not reveal all aspects of evolution. Although fossils display changes in external structure and skeletal structure, other genetic changes—in physiol-

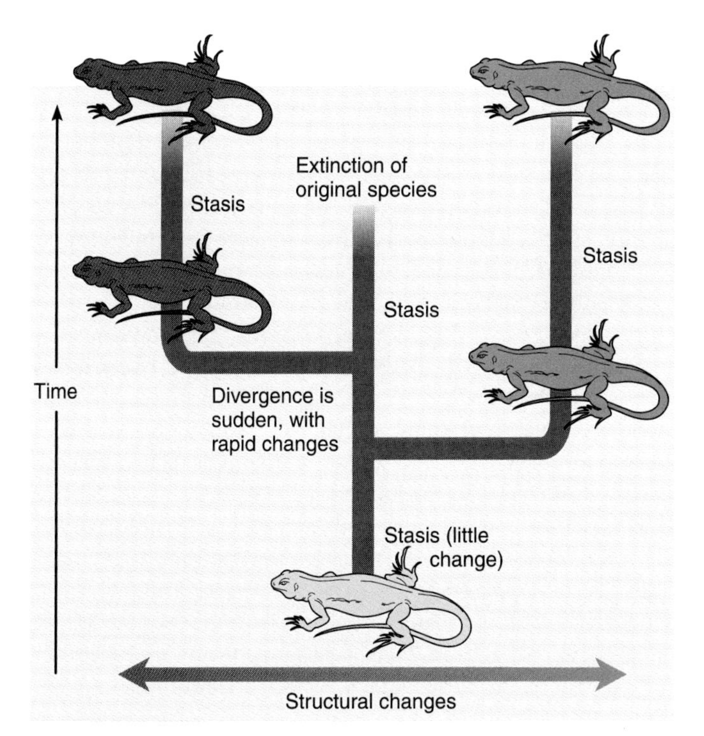

(a) Punctuated equilibrium

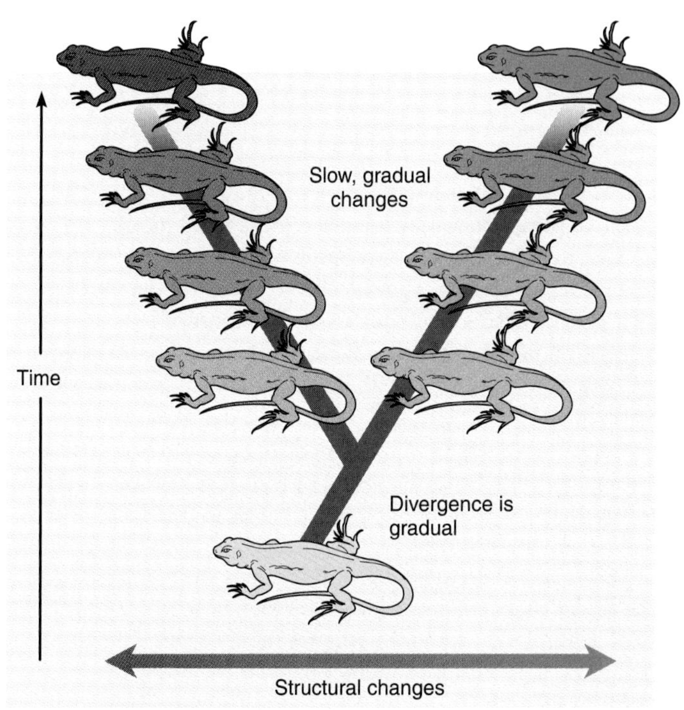

(b) Gradualism

FIGURE **19-12** **Punctuated equilibrium and gradualism.**
In this figure, structural changes in the lizards are represented by changes in skin color. **(a)** In punctuated equilibrium, long periods of stasis are interrupted by short periods of rapid speciation. **(b)** In gradualism, a slow, steady change in species occurs over time.

ogy, internal structure, and behavior—all of which also represent evolution, are not evident. Gradualists recognize rapid evolution only when strong *directional selection* occurs.

Many biologists embrace both models to explain the fossil record; they also contend that the pace of evolution may be abrupt in certain instances and gradual in others and that neither punctuated equilibrium nor gradualism exclusively characterizes the changes associated with evolution. Other biologists do not view the distinction between punctuated equilibrium and gradualism as real. They suggest that genetic changes occur gradually and at a roughly constant pace and that the majority of these mutations do not cause speciation. When the mutations that do cause speciation occur, they are dramatic and produce a pattern consistent with the punctuated equilibrium model.

Review

- If you were in a debate, how would you support the gradualism model?
- How would you support the punctuated equilibrium model?
- Are the gradualism and punctuated equilibrium models mutually exclusive? Explain your answer.

BiologyNow™ Assess your understanding of **the rate of evolutionary change** by taking the pretest on your BiologyNow CD-ROM.

MACROEVOLUTION

Learning Objectives

5 Define *macroevolution.*
6 Discuss macroevolution in the context of novel features, including preadaptations, allometric growth, and paedomorphosis.
7 Discuss the macroevolutionary significance of adaptive radiation and extinction.

Macroevolution is large-scale phenotypic changes in populations that warrant their placement in taxonomic groups at the species level and higher—that is, new species, genera, families, orders, classes, and even phyla, kingdoms, and domains. One concern of macroevolution is to explain evolutionary novelties, which are large phenotypic changes such as the appearance of jointed limbs during the evolution of arthropods (crustaceans, insects, and spiders). These phenotypic changes are so great that the new species possessing them are assigned to different genera or higher taxonomic categories. Studies of macroevolution also seek to discover and explain major changes in species diversity through time, such as occur during *adaptive radiation,* when many species appear, and *mass extinction,* when many species disappear. Thus evolutionary novelties, adaptive radiation, and mass extinction are important aspects of macroevolution.

Evolutionary novelties originate through modifications of pre-existing structures

New designs arise from structures already in existence. A change in the basic pattern of an organism produces something unique, such as wings on insects, flowers on plants, and feathered wings on birds. Usually these evolutionary novelties are variations of some pre-existing structures, called **preadaptations,** that originally fulfilled one role but subsequently changed in a way that was adaptive for a different role. Feathers, which evolved from reptilian scales and may have originally provided thermal insulation in primitive birds and some dinosaurs (see Chapter 20), represent a preadaptation for flight. With gradual modification, feathers evolved to function in flight as well as to fulfill their original thermoregulatory role. (Interestingly, a few feather-footed birds exist; this phenotype is the result of a change in gene regulation that alters scales, normally found on bird feet, into feathers.)

How do such evolutionary novelties originate? Many are probably due to changes during **development,** which is the orderly sequence of events that occurs as an organism grows and matures. *Regulatory genes* exert control over hundreds of other genes during development, and very slight genetic changes in regulatory genes could ultimately cause major structural changes in the organism (see Chapters 16 and 17).

For example, during development most organisms exhibit varied rates of growth for different parts of the body, known as **allometric growth** (from the Greek *allo,* "different," and *metr,* "measure"). The size of the head in human newborns is large in proportion to the rest of the body. As a human grows and matures, its torso, hands, and legs grow more rapidly than the head. Allometric growth is found in many organisms, including the male fiddler crab with its single, oversized claw, and the ocean sunfish with its enlarged tail (Fig. 19-13). If growth rates are altered even slightly, drastic changes in the shape of an organism may result, changes that may or may not be adaptive. For example, allometric growth may help explain the extremely small and relatively useless forelegs of the dinosaur *Tyrannosaurus rex,* compared with those of its ancestors.

Sometimes novel evolutionary changes occur when a species undergoes changes in the *timing* of development, in comparison with its ancestor. Consider, for example, the changes that would occur if juvenile characteristics were retained in the adult stage, a phenomenon known as **paedomorphosis** (from the Greek *paed,* "child," and *morph,* "form"). Adults of some salamander species have external gills and tail fins, features found only in the larval (immature) stages of other salamanders.

Retention of external gills and tail fins throughout life obviously alters the salamander's behavioral and ecological characteristics (Fig. 19-14). Perhaps such salamanders succeeded because they had a selective advantage over "normal" adult salamanders, that is, by remaining aquatic they did not have to compete for food with the terrestrial adult forms of related species. The paedomorphic forms also escaped the typical predators of terrestrial salamanders (although they had other predators in their aquatic environment). Studies suggest that paedomorphosis in salamanders is probably the result of mutations in genes that block the production of hormones that stimulate metamorphic changes. When paedomorphic salamanders receive hormone injections, they develop into adults lacking external gills and tail fins.

Adaptive radiation is the diversification of an ancestral species into many species

Adaptive radiation is the evolutionary diversification of many related species from one or a few ancestral species in a relatively short period. The concept of adaptive zones was developed to help explain why adaptive radiations take place. **Adaptive zones** are new ecological opportunities that were not exploited by an ancestral organism. At the species level, an adaptive zone is essentially identical to one or more similar *ecological niches* (the

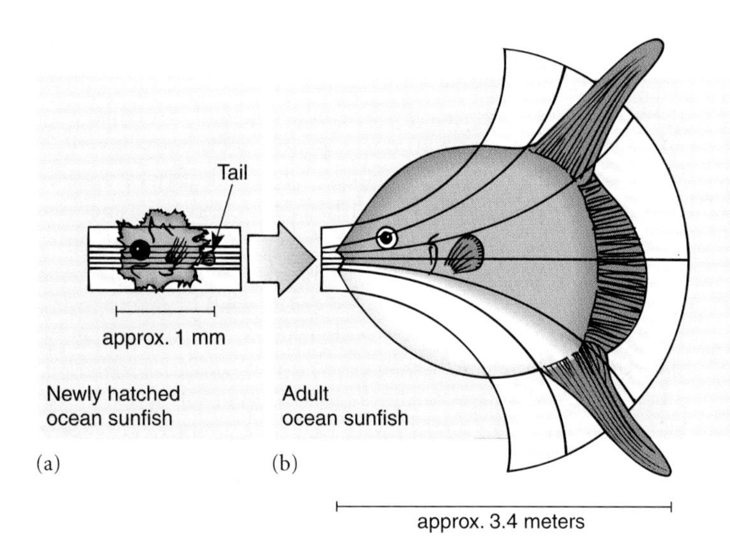

FIGURE **19-13** **Allometric growth in the ocean sunfish.**

The tail end of an ocean sunfish (*Mola mola*) grows faster than the head end, resulting in the unique shape of the adult. **(a)** A newly hatched ocean sunfish, only 1 mm long, has an extremely small tail. **(b)** This allometric transformation is visualized by drawing rectangular coordinate lines through a picture of the juvenile fish and then changing the coordinate lines mathematically. **(c)** An ocean sunfish swims off the coast of southern California. The adult ocean sunfish may reach 4 m (13 ft) and weigh about 1500 kg (3300 lb).

FIGURE **19-14** **Paedomorphosis in a salamander.**

An adult axolotl salamander (*Ambystoma mexicanum*) retains the juvenile characteristics of external gills (feathery structures protruding from the neck) and a tail fin *(not visible)*. Paedomorphosis allows the axolotl to remain permanently aquatic and to reproduce without developing typical adult characteristics.

functional roles of species within a community; see Chapter 52). Examples of adaptive zones include nocturnal flying to catch small insects, grazing on grass while migrating across a savanna, and swimming at the ocean's surface to filter out plankton. When many adaptive zones are empty, as they were in Lake Victoria when it refilled some 12,000 years ago (discussed earlier in the chapter), colonizing species such as the cichlids may rapidly diversify and exploit them.

Because islands have fewer species than do mainland areas of similar size, latitude, and topography, vacant adaptive zones are more common on islands than on continents. Consider the Hawaiian honeycreepers, a group of related birds found on the Hawaiian Islands. When the honeycreeper ancestors reached Hawaii, few other birds were present. The succeeding generations of honeycreepers quickly diversified into many new species and, in the process, occupied the many available adaptive zones that on the mainland are occupied by finches, honeyeaters, treecreepers, and woodpeckers. The diversity of their bills is a particularly good illustration of adaptive radiation (Fig. 19-15). Some honeycreeper bills are curved to extract nectar out of

FIGURE **19-15** **Adaptive radiation in Hawaiian honeycreepers.**

Compare the various beak shapes and methods of obtaining food. Many honeycreeper species are now extinct or nearing extinction as a result of human activities, including the destruction of habitat and the introduction of predators such as rats, dogs, and pigs.

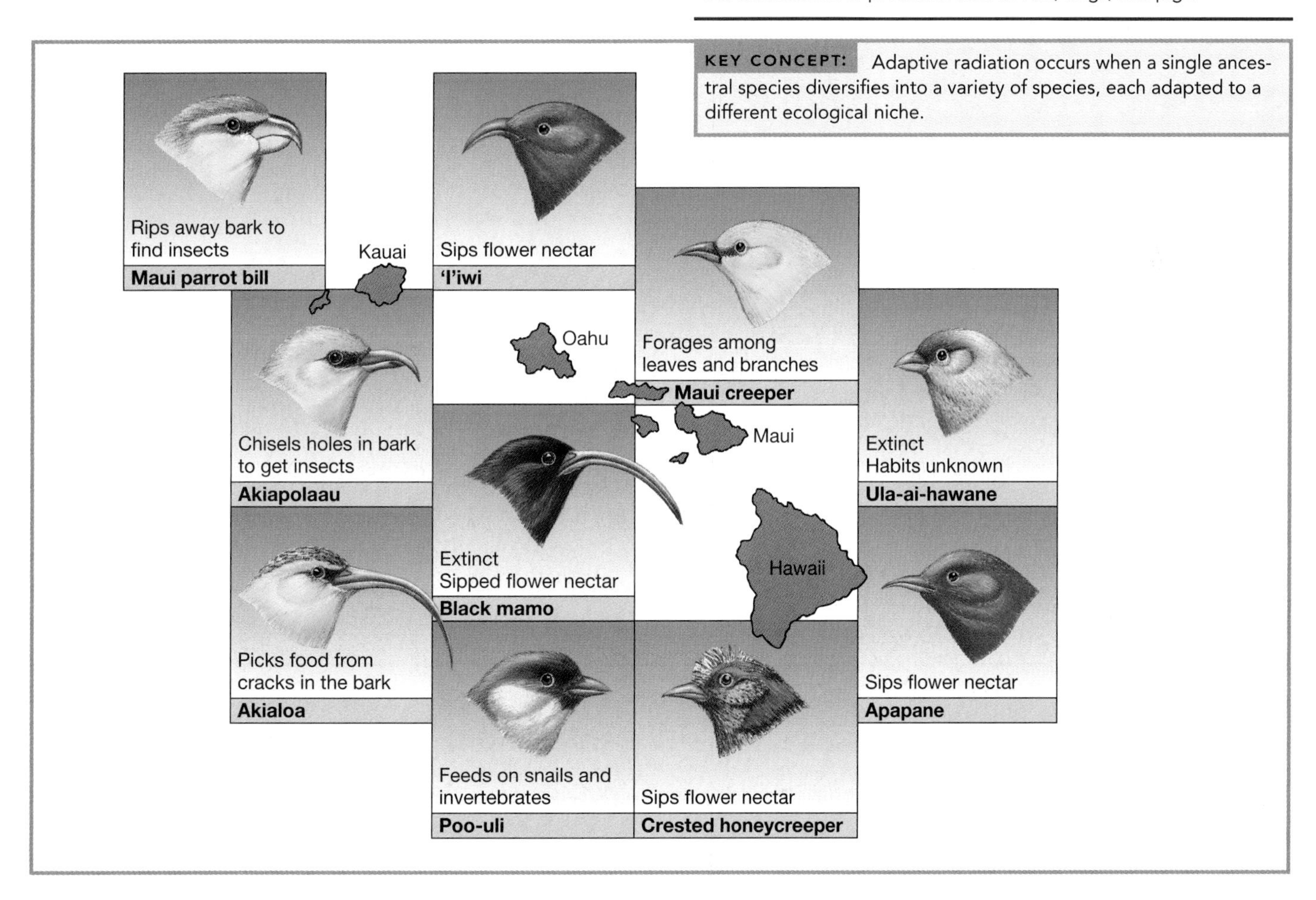

tubular flowers, whereas others are short and thick for ripping away bark in search of insects.

Another example of vacant adaptive zones involves the Hawaiian silverswords, 28 species of closely related plants found only on the Hawaiian Islands. When the silversword ancestor, a California plant related to daisies, reached the Hawaiian Islands, many diverse environments, such as cool, arid mountains; exposed lava flows; dry woodlands; shady, moist forests; and wet bogs were present and more or less unoccupied. The succeeding generations of silverswords quickly diversified in structure and physiology, occupying the many adaptive zones available to them. The diversity in their leaves, which changed during the course of natural selection as different populations adapted to various levels of light and moisture, is a particularly good illustration of adaptive radiation (Fig. 19-16). Leaves of silverswords that are adapted to shady moist forests are large, for example, whereas those of silverswords living in arid areas are small. The leaves of silverswords living on exposed volcanic slopes are covered with dense silvery hairs that may reflect some of the intense ultraviolet radiation off the plant.

Adaptive radiation appears more common during periods of major environmental change, but it is difficult to determine if these changes actually induce adaptive radiation. Possibly major environmental change indirectly affects adaptive radiation by increasing the extinction rate. Extinction produces empty adaptive zones, which provide new opportunities for species that remain. Mammals, for example, had existed as small nocturnal insectivores (insect eaters) for millions of years before undergoing adaptive radiation leading to the modern mammalian orders. This radiation was presumably triggered by the extinction of the dinosaurs. Mammals diversified and exploited a variety of adaptive zones relatively soon after the dinosaurs' demise. Flying bats, running gazelles, burrowing moles, and swimming whales all originated from the small, insect-eating, ancestral mammals.

Extinction is an important aspect of evolution

Extinction, the end of a lineage, occurs when the last individual of a species dies. The loss is permanent, for once a species is extinct it never reappears. Extinctions have occurred continually since the origin of life; by one estimate, only 1 species is alive today for every 2000 that have become extinct. Extinction is the eventual fate of all species, in the same way that death is the eventual fate of all individual organisms.

Although extinction has a negative short-term impact on the number of species, it facilitates evolution over a period of thousands to millions of years. As mentioned previously, when species become extinct their adaptive zones become vacant.

(a)

(b)

(c)

FIGURE **19-16** **Adaptive radiation in Hawaiian silverswords.**
The 28 silversword species, which are classified in three closely related genera, live in a variety of habitats. **(a)** The Haleakala silversword (*Argyroxyphium sandwicense* ssp. *macrocephalum*) is found only in the cinders on the upper slope of Haleakala Crater on the island of Maui. This plant is adapted to low precipitation and high levels of ultraviolet radiation. **(b)** This silversword species (*Wilkesia gymnoxiphium*), which superficially resembles a yucca, is found only along the slopes of Waimea Canyon on the island of Kauai. **(c)** *Daubautia scabra* is a small, herbaceous silversword found in moist to wet environments on several Hawaiian islands. (The fern fronds in the background give an idea of the small size of *D. scabra*.)

Consequently, surviving species are presented with new evolutionary opportunities and may diverge, filling in some of the unoccupied zones. In other words, the extinct species may eventually be replaced by new species.

During the long history of life, extinction appears to have occurred at two different rates. The continuous, low-level extinction of species is sometimes called **background extinction.** In contrast, five or possibly six times during Earth's history, **mass extinctions** of numerous species and higher taxonomic groups have taken place in both terrestrial and marine environments. The most recent mass extinction, which occurred about 65 million years ago (mya), killed off many marine organisms, terrestrial plants, and vertebrates, including the last of the dinosaurs (Fig. 19-17). The time span over which a mass extinction occurred may have been several million years, but that is relatively short compared with the 3.5 billion years or so of Earth's history of life. Each period of mass extinction has been followed by a period of adaptive radiation of some of the surviving groups.

The causes of past episodes of mass extinction are not well understood. Both environmental and biological factors seem to have been involved. Major changes in climate could have adversely affected those plants and animals that lacked the genetic flexibility to adapt. Marine organisms, in particular, are adapted to a steady, unchanging climate. If Earth's temperature were to increase or decrease by just a few degrees overall, many marine species would probably perish.

It is also possible that past mass extinctions were due to changes in the environment induced by catastrophes. If a large comet or small asteroid collided with Earth, the dust ejected into the atmosphere on impact could have blocked much of the sunlight. In addition to disrupting the food chain by killing many plants (and therefore terrestrial animals), this event would have lowered Earth's temperature, leading to the death of many marine organisms. Evidence that the extinction of dinosaurs was caused by an extraterrestrial object's collision with Earth continues to accumulate (see Chapter 20).

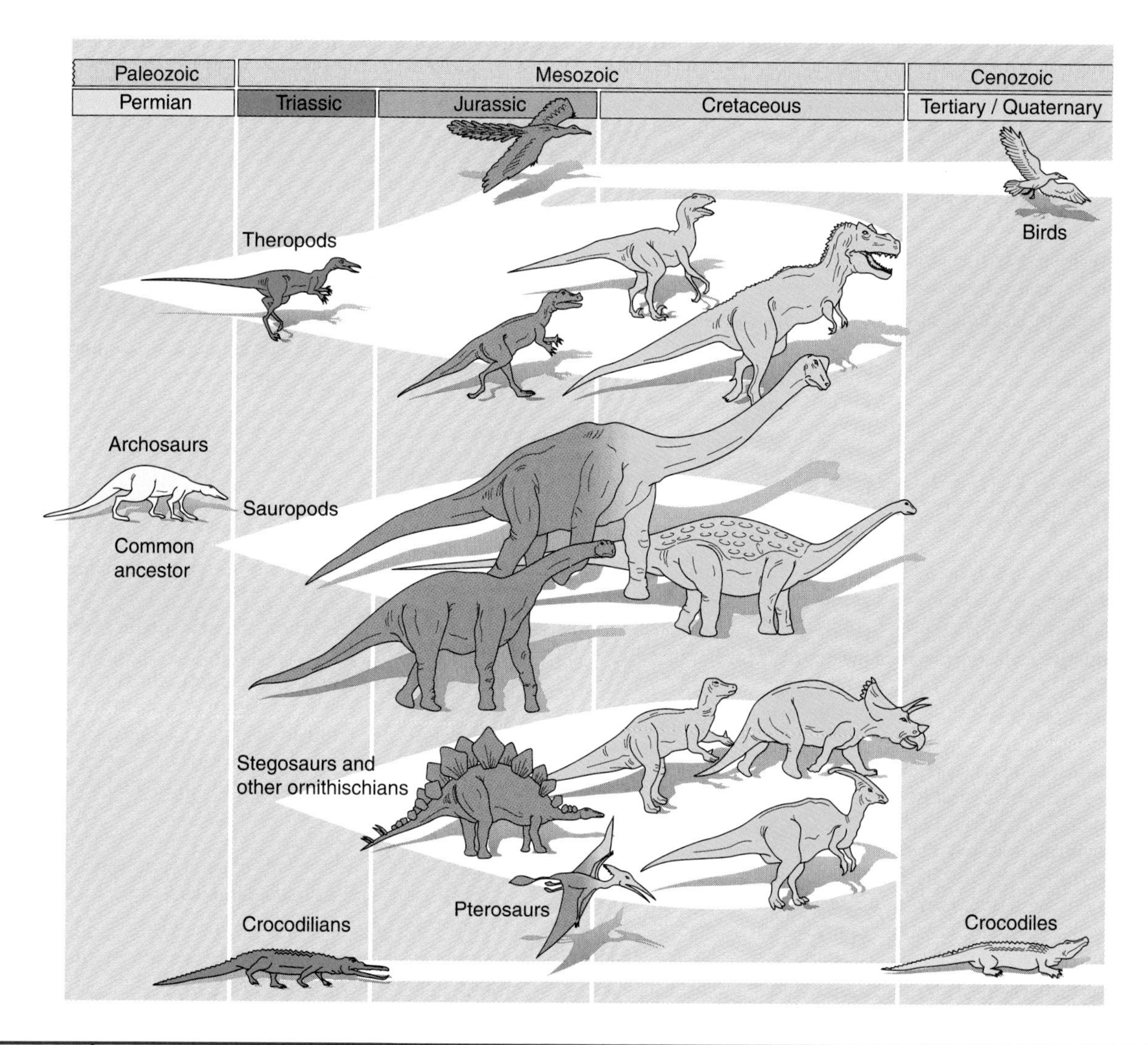

FIGURE **19-17** **Mass extinction of the archosaurs.**

At the end of the Cretaceous period, approximately 65 mya, a mass extinction of many organisms, including the remaining dinosaurs, occurred. (Dinosaurs had already been declining in diversity throughout the latter part of the Cretaceous period.) Most of the archosaurs, one of five main groups of reptiles, became extinct. The only lines to survive were crocodiles and birds, both of which are archosauran descendants.

Biological factors also trigger extinction. Competition among species may lead to the extinction of species that cannot compete effectively. The human species, in particular, has had a profound impact on the rate of extinction. The habitats of many animal and plant species have been altered or destroyed by humans, and habitat destruction can result in a species' extinction. Some biologists think we have entered the greatest mass extinction episode in Earth's history. (Extinction is discussed further in Chapter 55.)

Is microevolution related to speciation and macroevolution?

The concepts presented in Chapters 17 and 18 represent the **modern synthesis,** or *synthetic theory of evolution,* in which mutation provides the genetic variation on which natural selection acts. The modern synthesis combines Darwin's theory with important aspects of genetics. Many aspects of the modern synthesis have been tested and verified at the population and subspecies levels. Many biologists contend that microevolutionary processes (natural selection, mutation, genetic drift, and gene flow) account for the genetic variation within species and for the origin of new species. These biologists also hypothesize that microevolutionary processes explain macroevolutionary patterns.

A considerable body of data from many fields supports the modern synthesis as it relates to speciation and macroevolution. Consider, for example, the evolution of amphibians from fish, which was a major macroevolutionary event in the history of vertebrates. Study of the few known fossil intermediates shows that the transition from aquatic fish to terrestrial amphibian occurred as evolutionary novelties, such as changes in the limbs and skull roof, were added. These novelties accumulated as a succession of small changes over a period of 9 million to 14 million years. This time scale is sufficient to have allowed natural selection and other microevolutionary processes to produce the novel characters.

Although few biologists doubt the role of natural selection and microevolution in generating specific adaptations, some question the *extent* of microevolution's role in the overall pattern of life's history. These biologists ask whether speciation and macroevolution have been dominated by microevolutionary processes or by external, chance events, such as an impact by an asteroid. Chance events do not "care" about adaptive superiority but instead lead to the random extinction or survival of species. In the case of an asteroid impact, for example, certain species may survive because they were "lucky" enough to be in a protected environment at the time of impact. If chance events have been the overriding factor during life's history, then microevolution cannot be the exclusive explanation for the biological diversity we have today (see additional discussion of the role of chance in evolution in Chapter 17).

Review

- Why are evolutionary novelties a topic of scientists studying macroevolution?
- What is a preadaptation? What is paedomorphosis?
- What roles do extinction and adaptive radiation play in macroevolution?

BiologyNow™ Assess your understanding of **macroevolution** by taking the pretest on your BiologyNow CD-ROM.

SUMMARY WITH KEY TERMS

1 Explain the significance of reproductive isolating mechanisms, and distinguish among the different prezygotic and postzygotic barriers.

- **Reproductive isolating mechanisms** restrict gene flow between species.
- **Prezygotic barriers** are reproductive isolating mechanisms that prevent fertilization from taking place. **Temporal isolation** occurs when two species reproduce at different times of the day, season, or year. In **habitat isolation,** two closely related species live and breed in different habitats in the same geographic area. In **behavioral isolation,** distinctive courtship behaviors prevent mating between species. **Mechanical isolation** is due to incompatible structural differences in the reproductive organs of similar species. In **gametic isolation,** gametes from different species are incompatible, owing to molecular and chemical differences.
- **Postzygotic barriers** are reproductive isolating mechanisms that prevent gene flow after fertilization has taken place. **Hybrid inviability** is the death of interspecific embryos during development. **Hybrid sterility** prevents interspecific hybrids that survive to adulthood from reproducing successfully. **Hybrid breakdown** prevents the offspring of hybrids that survive to adulthood and successfully reproduce, from reproducing beyond one or a few generations.

2 Explain the mechanism of allopatric specification, and give an example.

- **Speciation** is the evolution of a new species from an ancestral population. **Allopatric speciation** occurs when one population becomes geographically isolated from the rest of the species and subsequently diverges. Speciation is more likely to occur if the original isolated population is small, because genetic drift is more significant in small populations. Examples of allopatric speciation include Death Valley pupfish, Kaibab squirrels, and Porto Santo rabbits.

3 Explain the mechanisms of sympatric speciation, and give both plant and animal examples.

- **Sympatric speciation** does not require geographic isolation.
- Sympatric speciation in plants results almost exclusively from **allopolyploidy,** in which a **polyploid** individual (one with more than two sets of chromosomes) is a hybrid derived from two species. Two examples of sympatric speciation by allopolyploidy are the kew primroses and hemp nettles.
- Sympatric speciation occurs in animals, such as fruit maggot flies and cichlids, but how often it occurs and under what conditions remain to be determined.

4 Take either side in a debate on the pace of evolution, by representing the opposing views of gradualism and punctuated equilibrium.

- The interpretation of evolution, as observed in the fossil record, is currently being debated.
- According to the **punctuated equilibrium** model, evolution of species proceeds in spurts. Short periods of active speciation are interspersed with long periods of **stasis.**
- According to the **gradualism** model, populations slowly diverge from one another by the accumulation of adaptive characteristics within a population.

5 Define *macroevolution.*

- **Macroevolution** is large-scale phenotypic changes in populations that warrant their placement in taxonomic groups at the species level and higher—that is, new species, genera, families, orders, classes, and even phyla, kingdoms, and domains.

6 Discuss macroevolution in the context of novel features, including preadaptations, allometric growth, and paedomorphosis.

- Macroevolution includes the appearance of evolutionary novelties, which may be due to changes during **development,** the orderly sequence of events that occurs as an organism grows and matures. Slight genetic changes in regulatory genes could cause major structural changes in the organism.
- Evolutionary novelties may originate from **preadaptations,** structures that originally fulfilled one role but changed in a way that was adaptive for a different role. Feathers are an example of a preadaptation.
- Changes in **allometric growth,** varied rates of growth for different parts of the body, result in overall changes in the shape of an organism. Examples include the ocean sunfish and the male fiddler crab.
- **Paedomorphosis,** the retention of juvenile characteristics in the adult, occurs owing to changes in the timing of development. Adult axolotl salamanders, with external gills and tail fins, are an example of paedomorphosis.

7 Discuss the macroevolutionary significance of adaptive radiation and extinction.

- **Adaptive radiation** is the process of diversification of an ancestral species into many new species. **Adaptive zones** are new ecological opportunities that were not exploited by an ancestral organism. When many adaptive zones are empty, colonizing species rapidly diversify and exploit them. Hawaiian honeycreepers and silverswords both underwent adaptive radiation after their ancestors colonized the Hawaiian Islands.
- **Extinction** is the death of a species. When species become extinct, the adaptive zones that they occupied become vacant, allowing other species to evolve and fill those zones. **Background extinction** is the continuous, low-level extinction of species. **Mass extinction** is the extinction of numerous species and higher taxonomic groups in both terrestrial and marine environments.

POST-TEST

1. According to the biological species concept, two populations belong to the same species if (a) their members freely interbreed (b) individuals from the two populations produce fertile offspring (c) their members do not interbreed with individuals of different species (d) a and c are correct (e) a, b, and c are correct
2. The zebrass is an example of (a) a fertile hybrid (b) a sterile hybrid (c) prezygotic barriers (d) a biological species (e) allopolyploidy
3. A prezygotic barrier prevents (a) the union of egg and sperm (b) reproductive success by an interspecific hybrid (c) the development of the zygote into an embryo (d) allopolyploidy from occurring (e) changes in allometric growth
4. The reproductive isolating mechanism in which two closely related species live in the same geographic area but reproduce at different times is (a) temporal isolation (b) behavioral isolation (c) mechanical isolation (d) gametic isolation (e) hybrid inviability
5. Interspecific hybrids, if they survive, are (a) always sterile (b) always fertile (c) usually sterile (d) usually fertile (e) never sterile
6. The process by which populations of one species evolve into distinct species is known as (a) the evolutionary species concept (b) speciation (c) behavioral isolation (d) ploidy (e) hybridization
7. The first step leading to allopatric speciation is (a) hybrid inviability (b) hybrid breakdown (c) adaptive radiation (d) geographic isolation (e) paedomorphosis
8. The pupfish in the Death Valley region are an example of which evolutionary process? (a) background extinction (b) allopatric speciation (c) sympatric speciation (d) allopolyploidy (e) paedomorphosis
9. Sympatric speciation (a) is most common in animals (b) does not require geographic isolation (c) accounts for the evolution of the Hawaiian goose (nene) (d) involves the accumulation of gradual genetic changes (e) usually takes millions of years
10. Which of the following evolutionary processes is associated with allopolyploidy? (a) gradualism (b) allometric growth (c) sympatric speciation (d) mass extinction (e) preadaptation
11. According to the punctuated equilibrium model, (a) populations slowly diverge from one another (b) the evolution of species occurs in spurts interspersed with long periods of stasis (c) evolutionary novelties originate from preadaptations (d) reproductive isolating mechanisms restrict gene flow between species (e) the fossil record, being incomplete, does not accurately reflect evolution as it actually occurred
12. The evolutionary conversion of reptilian scales into feathers is an example of (a) allometric growth (b) paedomorphosis (c) gradualism (d) hybrid breakdown (e) preadaptation
13. Adaptive radiation is common following a period of mass extinction, probably because (a) the survivors of a mass extinction are remarkably well adapted to their environment (b) the unchanging environment following a mass extinction drives the evolutionary process (c) many adaptive zones are empty (d) many ecological niches are filled (e) the environment induces changes in the timing of development for many species

POST-TEST *(continued)*

14. Adaptive radiations do not appear to have ever occurred (a) on isolated islands (b) in birds such as honeycreepers (c) in environments colonized by few species (d) in plants such as silverswords (e) in environments with many existing species
15. The Hawaiian silverswords are an excellent example of which evolutionary process? (a) allometry (b) preadaptation (c) microevolution (d) adaptive radiation (e) extinction

CRITICAL THINKING

1. Why is allopatric speciation more likely to occur if the original isolated population is small?
2. Based on what you have learned about prezygotic and postzygotic isolating mechanisms, which reproductive isolating mechanism(s) would you say is/are probably at work between the Porto Santo rabbit and its mainland relative?
3. Based on what you have learned in the chapter, hypothesize what the common ancestor of the more than 20 species of desert pupfish may have looked like. (*Hint:* The ancestral species lived in one or more large lakes.) How could you test your hypothesis?
4. Could hawthorn and apple maggot flies be considered an example of assortative mating, which was discussed in Chapter 18? Explain your answer.
5. Because mass extinction is a natural process that may facilitate evolution during the period of thousands to millions of years that follow it, should humans be concerned about the current mass extinction that we are causing? Why or why not?

- Visit our Web site at **http://biology.brookscole.com/solomon7** for links to chapter-related resources on the World Wide Web. Additional online materials relating to this chapter can also be found on our Web site.

BIOLOGY NOW RESOURCES

Active Figures

19-1: Reproductive isolation

19-7: Allopatric speciation

Preparing for an exam? Take a diagnostic test on your BiologyNow CD-ROM.

Post-Test Answers

1. e	2. b	3. a	4. a
5. c	6. b	7. d	8. b
9. b	10. c	11. b	12. e
13. c	14. e	15. d	

The Origin and Evolutionary History of Life

Fossil trilobites (*Phacops rana*). These extinct arthropods, which were about 3 cm (1.2 in) long, flourished in the ocean during the Paleozoic era. They ranged from 1 mm to 1 m in length, depending on the species. Note the large, well-developed eyes (visible on either side of the head region).

CHAPTER OUTLINE

- Chemical Evolution on Early Earth
- The First Cells
- The History of Life

The preceding three chapters were concerned with the evolution of organisms, but we have not yet dealt with what many regard as a fundamental question of biological evolution: How did life begin? Although biologists generally accept the hypothesis that life developed from nonliving matter, exactly how this process, called **chemical evolution,** occurred is not certain. Chemical evolution probably involved several stages. Current models suggest that small organic molecules first formed spontaneously and accumulated over time. These molecules may have accumulated rather than being broken down (as occurs today) because environmental conditions were different. The two factors that today break down organic molecules—free oxygen and living organisms—were absent from early Earth.

Large organic macromolecules such as proteins and nucleic acids could have then assembled from the smaller molecules. The macromolecules interacted with one another, combining into more complicated structures that could eventually metabolize and replicate. Natural selection favored macromolecular assemblages with cell-like structures. Their descendants eventually became the first true cells. After the first cells originated, they diverged over several billion years into the rich biological diversity that characterizes our planet today. Photosynthesis, aerobic respiration, and eukaryotic cell structure represent several major advances that evolved during the history of life.

Geologic evidence, in particular the fossil record, tells us much of what we know about the history of life, such as what kinds of organisms existed and where and when they lived. Certain organisms appear in the fossil record, then disappear and are replaced by others. Initially unicellular prokaryotes predominated, followed by unicellular eukaryotes. The first multicellular eukaryotes—soft-bodied animals that did not leave many fossils—appeared in the ocean approximately 630 million years ago (mya). Shelled animals and many other marine invertebrates (animals without backbones) appeared next, as exemplified by trilobites, members of a large group of primitive aquatic arthropods (see photograph). Marine invertebrates were followed by the first vertebrates (animals with backbones). The first fishes with jaws appeared and diversified. Some of these gave rise to amphibians, the first vertebrates with limbs capable of moving about on land. Amphibians also spread and diversified. About 300 mya, amphibians gave rise to reptiles, which diversified and populated the land. Reptiles in turn gave rise independently to birds and to mammals. Plants underwent a comparable evolutionary history and diversification.

In this chapter we survey life over a vast span of time, starting some 3.8 billion years ago (bya) when our planet was relatively young. We examine proposed models about how life began and trace life's long evolutionary history from its beginnings to the present. ■

CHEMICAL EVOLUTION ON EARLY EARTH

Learning Objectives

1. Describe the conditions that geologists think existed on early Earth.
2. Contrast the prebiotic soup hypothesis and the iron-sulfur world hypothesis.

Many biologists speculate that life originated only once and that life's beginnings occurred under environmental conditions quite different from those of today. We must therefore examine the conditions of early Earth to understand the origin of life. Although we will never be certain about the exact conditions that existed when life arose, scientific evidence from many sources provides us with valuable clues that help us formulate plausible scenarios. Study of the origin of life is an active area of scientific research today, and many important contributions are adding to our understanding of how life began.

Astrophysicists and geologists estimate Earth is about 4.6 billion years old. The atmosphere of early Earth apparently included carbon dioxide (CO_2), water vapor (H_2O), carbon monoxide (CO), hydrogen (H_2), and nitrogen (N_2). It may also have contained some ammonia (NH_3), hydrogen sulfide (H_2S), and methane (CH_4), although ultraviolet radiation from the sun probably broke these reduced molecules down rapidly. The early atmosphere probably contained little or no free oxygen (O_2).

Four requirements must have existed for the chemical evolution of life: little or no free oxygen, a source of energy, the availability of chemical building blocks, and time. First, life could have begun only in the absence of free oxygen. Oxygen is quite reactive and would have oxidized the organic molecules that are necessary building blocks in the origin of life. Earth's early atmosphere was probably strongly reducing, which means that any free oxygen would have reacted with other elements to form oxides. Thus oxygen would have been tied up in compounds.

The origin of life would also have required energy to do the work of building biological molecules from simple inorganic chemicals. Early Earth was a place of high energy with violent thunderstorms; widespread volcanic activity; bombardment from meteorites and other extraterrestrial objects; and intense radiation, including ultraviolet radiation from the sun (Fig. 20-1). The young sun probably produced more ultraviolet radiation than it does today, and ancient Earth had no protective ozone layer to filter it.

A third requirement would have been the presence of the chemical building blocks needed for chemical evolution. These included water, dissolved inorganic minerals (present as ions), and the gases present in the early atmosphere. A final requirement for the origin of life was time for molecules to accumulate and react with one another. Earth is approximately 4.6 billion years old, and the earliest traces of life are approximately 3.8 billion years old; therefore, life had several hundred million years to get started.

ACTIVE FIGURE **20-1** **An artist's interpretation of conditions on early Earth.**

Volcanoes erupted, spewing gases that contributed to the atmosphere, and violent thunderstorms triggered torrential rainfall that eroded the land. Meteorites and other extraterrestrial objects continually bombarded Earth, cataclysmically changing the crust, ocean, and atmosphere.

BiologyNow™ Learn more about **conditions on early Earth** by clicking on this figure on your BiologyNow CD-ROM.

Courtesy of Reader's Digest Books. Drawing by H. K. Wimmer.

Organic molecules formed on primitive Earth

PROCESS OF SCIENCE

Because organic molecules are the building materials for organisms, it is reasonable to first consider how they may have originated. Two main models seek to explain how the organic precursors of life originated: The **prebiotic soup hypothesis** proposes that these molecules formed near Earth's surface, whereas the **iron-sulfur world hypothesis** proposes that organic precursors formed at cracks in the ocean's floor.

Organic molecules may have been produced at Earth's surface

The concept that simple organic molecules such as sugars, nucleotide bases, and amino acids could form spontaneously from simpler raw materials was first advanced in the 1920s by two scientists working independently: A.I. Oparin, a Russian biochemist, and J.B.S. Haldane, a Scottish physiologist and geneticist.

Their hypothesis was tested in the 1950s by American biochemists Stanley Miller and Harold Urey, who designed a closed apparatus that simulated conditions that presumably existed on early Earth (Fig. 20-2). They exposed an atmosphere rich in H_2, CH_4, H_2O, and NH_3 to an electrical discharge that simulated lightning. Their analysis of the chemicals produced in a week revealed that amino acids and other organic molecules had formed. Although more recent data suggest that Earth's early atmosphere was not rich in methane or ammonia, similar experiments using different combinations of gases have produced a wide variety of organic molecules that are important in contemporary organisms. These include all 20 amino acids, several sugars, lipids, the nucleotide bases of RNA and DNA, and ATP (when phosphate is present). Thus, before life began, its chemical building blocks were probably accumulating as a necessary step in chemical evolution.

Oparin envisioned that the organic molecules would, over vast spans of time, accumulate in the shallow seas to form a "sea of organic soup." Under such conditions, he envisioned smaller organic molecules (monomers) combining to form larger ones (polymers). Evidence gathered since Oparin's time indicates that organic polymers may have formed and accumulated on rock or clay surfaces rather than in the primordial seas. Clay, which consists of microscopic particles of weathered rock, is particularly intriguing as a possible site for early polymerizations, because it binds organic monomers and contains zinc and iron ions that may have served as catalysts. Laboratory experiments have confirmed that organic polymers form spontaneously from monomers on hot rock or clay surfaces.

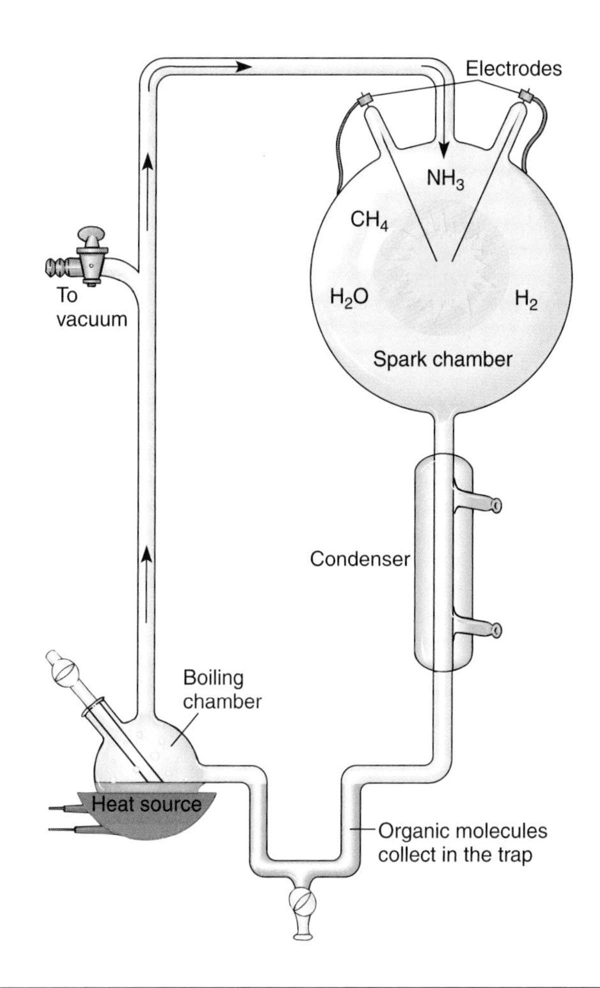

ACTIVE FIGURE **20-2** **Testing the prebiotic soup hypothesis.**

Diagram of the apparatus that Miller and Urey used to simulate the reducing atmosphere of early Earth. An electrical spark was produced in the upper right flask to simulate lightning. The gases present in the flask reacted together, forming a variety of simple organic compounds that accumulated in the trap at the bottom.

BiologyNow™ See the **Miller-Urey experiment** unfold by clicking on this figure on your BiologyNow CD-ROM.

Organic molecules may have formed at hydrothermal vents

In a different scenario of chemical evolution, some biologists hypothesize that early polymerizations leading to the origin of life may have occurred in cracks in the deep ocean floor where hot water, carbon monoxide, and minerals such as sulfides of iron and nickel spew forth—the iron-sulfur world hypothesis. Such **hydrothermal vents** would have been better protected than Earth's surface from the catastrophic effects of meteorite bombardment.

Today these hot springs produce precursors of biological molecules and of energy-rich "food," including the highly reduced compounds, hydrogen sulfide and methane. These chemicals support a diverse community of microorganisms, clams, crabs, tube worms, and other animals (see *Focus On: Life without the Sun* in Chapter 53).

Testing the iron-sulfur world hypothesis at hydrothermal vents is difficult, but laboratory experiments simulating the high pressures and temperatures at the vents have yielded intriguing results. For example, iron and nickel sulfides catalyze reactions between carbon monoxide and hydrogen sulfide, producing acetic acid and other simple organic compounds. Also, experiments show that ammonia, one of the precursors of proteins and nucleic acids, is produced in abundance, suggesting that vents were ammonia-rich environments in the prebiotic world.

Review

- What are the four requirements for chemical evolution, and why is each essential?
- What is the prebiotic soup hypothesis?
- How does the iron-sulfur world hypothesis differ from the prebiotic soup hypothesis?

BiologyNow™ Assess your understanding of **chemical evolution on early Earth,** by taking the pretest on your BiologyNow CD-ROM.

THE FIRST CELLS

Learning Objectives

3 Outline the major steps hypothesized to have occurred in the origin of cells.

4 Explain how the evolution of photosynthetic autotrophs affected both the atmosphere and other organisms.

5 Describe the endosymbiont theory.

After the first polymers formed, could they have assembled spontaneously into more complex structures? Scientists have synthesized several different **protobionts,** which are assemblages of abiotically produced (that is, not produced by organisms) organic polymers. They have recovered protobionts that resemble living cells in several ways, thus providing clues as to how aggregations of complex nonliving molecules took that "giant leap" and became living cells. These protobionts exhibit many functional and structural attributes of living cells. They often divide in half (binary fission) after they have sufficiently "grown." Protobionts maintain an internal chemical environment that is different from the external environment (homeostasis), and some of them show the beginnings of metabolism (catalytic activity). They are highly organized, considering their relatively simple composition.

Microspheres are a type of protobiont formed by adding water to abiotically formed polypeptides (Fig. 20-3). Some microspheres are excitable: They produce an electrical potential across their surfaces, reminiscent of electrochemical gradients in cells. Microspheres can also absorb materials from their surroundings (selective permeability) and respond to changes in osmotic pressure as though membranes enveloped them, even though they contain no lipid.

The study of protobionts shows that relatively simple "pre-cells" have some of the properties of contemporary life. However, it is a major step (or several steps) to go from simple molecular aggregates such as protobionts to living cells. Although we have learned many things about how organic molecules may have formed on primitive Earth, the problem of how pre-cells evolved into living cells remains to be solved. One of the most significant parts of that process was the evolution of molecular reproduction.

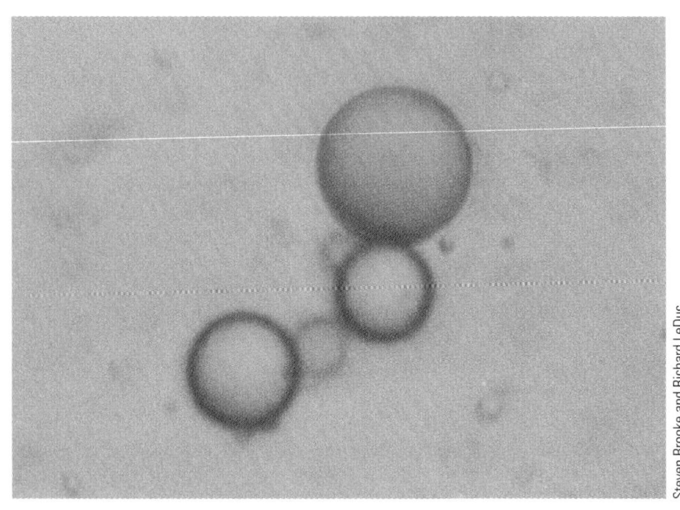
Steven Brooke and Richard LeDuc

2 μm

FIGURE **20-3** **Microspheres.**
These tiny protobionts exhibit some of the properties of life.

Molecular reproduction was a crucial step in the origin of cells

In living cells, genetic information is stored in the nucleic acid DNA, which is transcribed into messenger RNA (mRNA), which in turn is translated into the proper amino acid sequence in proteins. All three macromolecules in the DNA ⟶ RNA ⟶ protein sequence contain precise information, but only DNA and RNA are capable of self-replication, although only in the presence of the proper enzymes. Because both RNA and DNA can form spontaneously on clay in much the same way that other organic polymers do, the question becomes which molecule, DNA or RNA, first appeared in the prebiotic world.

Many scientists have suggested that RNA was the first informational molecule to evolve in the progression toward a self-sustaining, self-reproducing cell and that proteins and DNA came along later. According to a model known as the **RNA world,** the chemistry of prebiotic Earth gave rise to self-replicating RNA molecules that functioned as both enzymes and substrates for their own replication. We represent the replication of RNA in the RNA world scenario as a circular arrow:

One feature of RNA is that it often has catalytic properties; such enzymatic RNAs are called **ribozymes.** Before the evolution of true cells, ribozymes may have catalyzed their own replication in the clays, shallow rock pools, or hydrothermal vents where life originated. When RNA strands are added to a test tube containing RNA nucleotides but no enzymes, the nucleotides combine to form short RNA molecules.

The occurrence of an RNA world early in the history of life can never be proven, but experiments with **in vitro evolution,** also called **directed evolution,** have shown it is feasible. These experiments address an important question about the RNA world, namely, could RNA molecules have catalyzed the many different chemical reactions needed for life? In directed evolution, a large pool of RNA molecules with different sequences are mixed together and selected for their ability to catalyze a single biologically important reaction (Fig. 20-4). Those molecules that have at least some catalytic ability are then amplified

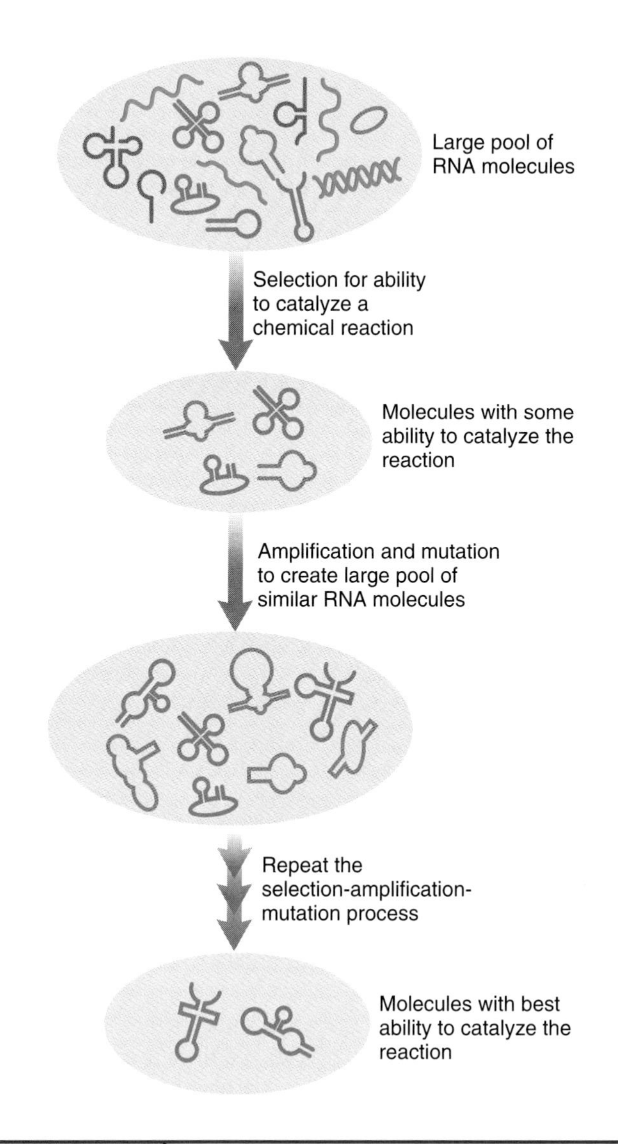

FIGURE **20-4** | **In vitro evolution of RNA molecules.**

RNA molecules are selected from a large pool, based on their ability to catalyze a specific reaction, then amplified and mutated, a process that is then repeated 7 to 20 additional times. The final group of RNA molecules is the most efficient at catalyzing the chemical reaction selected for. Scientists have developed more than two dozen synthetic RNA catalysts by in vitro evolution.

and mutated before being exposed to another round of selection. After this cycle is repeated several times, the RNA molecules at the end of the selection process function efficiently as catalysts for the reaction. In vitro evolution studies have shown that RNA has a large functional repertoire—that is, RNA can catalyze a variety of biologically important reactions.

In the RNA world, biologists hypothesize that ribozymes initially catalyzed protein synthesis and other important biological reactions; only later did protein enzymes catalyze these reactions.

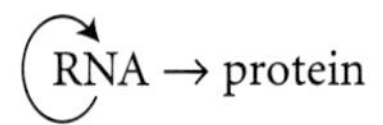

Interestingly, RNA directs protein synthesis by catalyzing peptide bond formation. Some single-stranded RNA molecules fold back on themselves as a result of interactions among the nucleotides composing the RNA strand. Sometimes the conformation (shape) of the folded RNA molecule is such that it weakly binds to an amino acid. If amino acids are held close together by RNA molecules, they may bond together, forming a polypeptide.

We have considered how the evolution of informational molecules may have given rise to RNA and later to proteins. If a self-replicating RNA capable of coding for proteins appeared before DNA, how did DNA, the universal molecule of heredity in cells, become involved? Perhaps RNA made double-stranded copies of itself that eventually evolved into DNA.

$$\text{DNA} \longleftarrow \text{RNA} \longrightarrow \text{protein}$$

The incorporation of DNA into the information transfer system was advantageous because the double helix of DNA is more stable (less reactive) than the single strand of RNA. Such stability in a molecule that stores genetic information would have provided a decided advantage in the prebiotic world (as it does today).

In the DNA/RNA/protein world, then, DNA became the information storage molecule, RNA remained involved in protein synthesis, and protein enzymes catalyzed most cell reactions, including DNA replication, RNA synthesis, and protein synthesis.

$$\circlearrowright\text{DNA} \rightarrow \text{RNA} \rightarrow \text{protein}$$

RNA is still a necessary component of the information transfer system, because DNA is not catalytic. Thus natural selection at the molecular level favored the DNA ⟶ RNA ⟶ protein information sequence. Once DNA was incorporated into this sequence, RNA molecules assumed their present role as an intermediary in the transfer of genetic information.

Several additional steps had to occur before a true living cell could evolve from macromolecular aggregations. For example, the self-replicating genetic code must have arisen extremely early in the prebiotic world because all organisms possess it—but how did it originate? Also, how did a plasma membrane of lipid and protein evolve?

Biological evolution began with the first cells

No one knows exactly when the first cells first appeared on Earth. Nonfossil evidence—isotopic "fingerprints" of organic carbon in ancient rocks in Greenland—indicates that life existed as early as 3.8 bya. **Microfossils,** ancient remains of microscopic life, suggest that cells may have been thriving as long as 3.5 bya, although the original interpretations of the oldest carbon-rich "squiggles" in ancient rocks as microfossils have been recently challenged. The oldest fossil cells that are widely accepted are 2 billion years old.

The earliest cells were prokaryotic. **Stromatolites,** another type of fossil evidence of the earliest cells, are rocklike columns composed of many minute layers of prokaryotic cells (Fig. 20-5). Over time, sediment collects around the cells and mineralizes. Meanwhile, a new layer of living cells grows over the older, dead cells. Fossil stromatolite reefs are found in several places in the

(a)

(b)

FIGURE **20-5** | **Stromatolites.**

(a) These living stromatolites at Hamlin Pool in Western Australia consist of mats of cyanobacteria and minerals such as calcium carbonate. They are several thousand years old. **(b)** Cutaway view of a fossil stromatolite showing the layers of microorganisms and sediments that accumulated over time. This stromatolite, also from Western Australia, is about 3.5 billion years old.

Fred Bavendam/Peter Arnold, Inc.

Biological Photo Service

world, including Great Slave Lake in Canada and the Gunflint Iron Formations along Lake Superior in the United States. Some fossil stromatolites are extremely ancient. One group in Western Australia, for example, is several billion years old. Stromatolite reefs are still living in hot springs and in warm, shallow pools of fresh and salt water.

The first cells were probably heterotrophic

The earliest cells probably obtained the organic molecules they needed from the environment, rather than synthesizing them. These primitive **heterotrophs** probably consumed many types of organic molecules that had formed spontaneously—sugars, nucleotides, and amino acids, to name a few. By fermenting these organic compounds, they obtained the energy they needed to support life. *Fermentation* is, of course, an anaerobic process (performed in the absence of oxygen), and the first cells were almost certainly **anaerobes.**

When the supply of spontaneously generated organic molecules gradually declined, only certain organisms could survive. Mutations had probably already occurred that permitted some cells to obtain energy directly from sunlight, perhaps by using sunlight to make ATP. These cells, which did not require the energy-rich organic compounds that were now in short supply in the environment, had a distinct selective advantage.

Photosynthesis requires not only light energy but also a source of electrons, which are used to reduce CO_2 when organic molecules such as glucose are synthesized (see Chapter 8). Most likely, the first photosynthetic **autotrophs**—organisms that produce their own food from simple raw materials—used the energy of sunlight to split hydrogen-rich molecules such as H_2S, releasing elemental sulfur (not oxygen) in the process. Indeed, the green sulfur bacteria and the purple sulfur bacteria still use H_2S as a hydrogen source for photosynthesis.[1]

The first photosynthetic autotrophs to obtain hydrogen by splitting water were the cyanobacteria. Water was quite abundant on early Earth, as it is today, and the selective advantage that splitting water gave cyanobacteria allowed them to thrive. The process of splitting water released oxygen as a gas (O_2). Initially, the oxygen released during photosynthesis oxidized minerals in the ocean and in Earth's crust, and for a long time oxygen did not begin to accumulate in the atmosphere. Eventually, however, oxygen levels increased in the ocean and the atmosphere.

Scientists estimate the timing of the events just described on the basis of geological and fossil evidence. Fossils from that period, which include rocks containing traces of chlorophyll, as well as the fossil stromatolites discussed earlier, indicate the first photosynthetic organisms appeared approximately 3.1 to 3.5 bya. This evidence suggests heterotrophic forms existed even earlier.

Aerobes appeared after oxygen increased in the atmosphere

By 2 bya, cyanobacteria had produced enough oxygen to begin significantly changing the composition of the atmosphere. The increase in atmospheric oxygen affected life profoundly. The oxygen poisoned *obligate anaerobes* (organisms that can't use oxygen for cellular respiration), and many species undoubtedly perished. Some anaerobes, however, survived in environments where oxygen did not penetrate; others evolved adaptations to neutralize the oxygen so it could not harm them. Some organisms, called **aerobes,** evolved a respiratory pathway that used oxygen to extract more energy from food. Aerobic respiration joined with the existing anaerobic process of glycolysis.

[1] Members of a third group of bacteria, the purple nonsulfur bacteria, use other organic molecules or hydrogen gas as a hydrogen source.

The evolution of organisms that could use oxygen in their metabolism had several consequences. Organisms that respire aerobically gain much more energy from a single molecule of glucose than anaerobes gain by fermentation (see Chapter 7). As a result, the newly evolved aerobic organisms were more efficient and more competitive than anaerobes. Coupled with the poisonous nature of oxygen to many anaerobes, the efficiency of aerobes forced anaerobes into relatively minor roles. Today the vast majority of organisms, including plants, animals, and most fungi, protists, and prokaryotes, use aerobic respiration, whereas only a few bacteria and even fewer protists and fungi are anaerobic.

The evolution of aerobic respiration stabilized both oxygen and carbon dioxide levels in the biosphere. Photosynthetic organisms used carbon dioxide as a source of carbon for synthesizing organic compounds. This raw material would have been depleted from the atmosphere in a relatively brief period without the advent of aerobic respiration, which released carbon dioxide as a waste product from the complete breakdown of organic molecules. Carbon thus started cycling in the biosphere, moving from the nonliving physical environment, to photosynthetic organisms, to heterotrophs that ate the photosynthetic organisms (see Chapter 53). Aerobic respiration released carbon back into the physical environment as carbon dioxide, and the carbon cycle continued. In a similar manner, molecular oxygen was produced by photosynthesis and used during aerobic respiration.

Another significant consequence of photosynthesis occurred in the upper atmosphere, where molecular oxygen reacted to form **ozone,** O_3 (Fig. 20-6). A layer of ozone eventually blanketed Earth, preventing much of the sun's ultraviolet radiation from penetrating to the surface. With the ozone layer's protection from the mutagenic effect of ultraviolet radiation, organisms could live closer to the surface in aquatic environments and eventually move onto land. Because the energy in ultraviolet radiation may have been necessary to form organic molecules, however, their abiotic synthesis decreased.

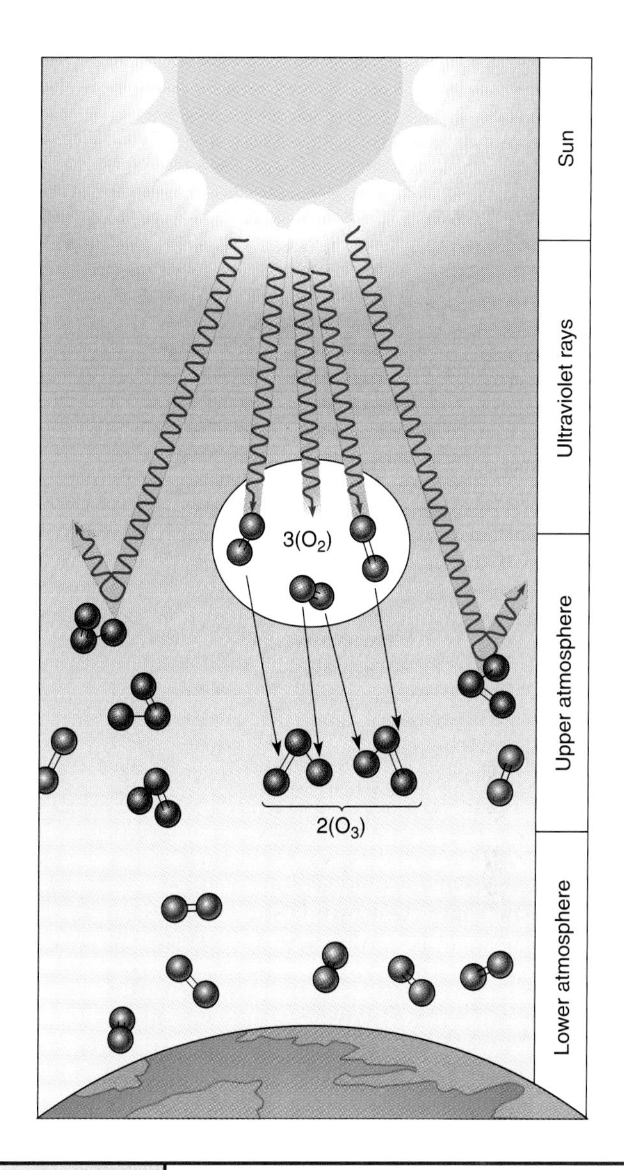

FIGURE **20-6** | **How ozone forms.**

Ozone (O_3) forms in the upper atmosphere when ultraviolet radiation from the sun breaks the double bonds of oxygen molecules.

Eukaryotic cells descended from prokaryotic cells

Eukaryotes may have appeared in the fossil record as early as 1.5 to 1.6 bya, and geochemical evidence suggests that eukaryotes were present much earlier. Steranes, molecules derived from steroids, have been discovered in Australian rocks dated at 2.7 billion years. Because bacteria are not known to produce steroids, the steranes may be biomarkers for eukaryotes. (These ancient rocks lack fossil traces of ancient organisms because the rocks have since been exposed to heat and pressure that would have destroyed any fossilized cells. Steranes, however, are very stable in the presence of heat and pressure.)

Eukaryotes arose from prokaryotes. Recall from Chapter 4 that prokaryotic cells lack nuclear envelopes as well as other membranous organelles such as mitochondria and chloroplasts. The **endosymbiont theory,** advanced by Lynn Margulis, declares that organelles such as mitochondria and chloroplasts may each have originated from mutually advantageous symbiotic relationships between two prokaryotic organisms (Fig. 20-7). Chloroplasts apparently evolved from photosynthetic bacteria (cyanobacteria) that lived inside larger heterotrophic cells, whereas mitochondria presumably evolved from aerobic bacteria (perhaps purple bacteria) that lived inside larger anaerobic cells. Thus early eukaryotic cells were assemblages of formerly free-living prokaryotes.

How did these bacteria come to be **endosymbionts,** which are organisms that live symbiotically inside a host cell? They may originally have been ingested, but not digested, by a host cell. Once incorporated, they could have survived and reproduced along with the host cell so that future generations of the host also contained endosymbionts. The two organisms developed a mutualistic relationship in which each contributed something to the other. Eventually the endosymbiont lost the ability to exist outside its host, and the host cell lost the ability to survive with-

FIGURE **20-7**

The endosymbiont theory.

Chloroplasts and mitochondria of eukaryotic cells are thought to have originated from various bacteria that lived as endosymbionts inside other cells.

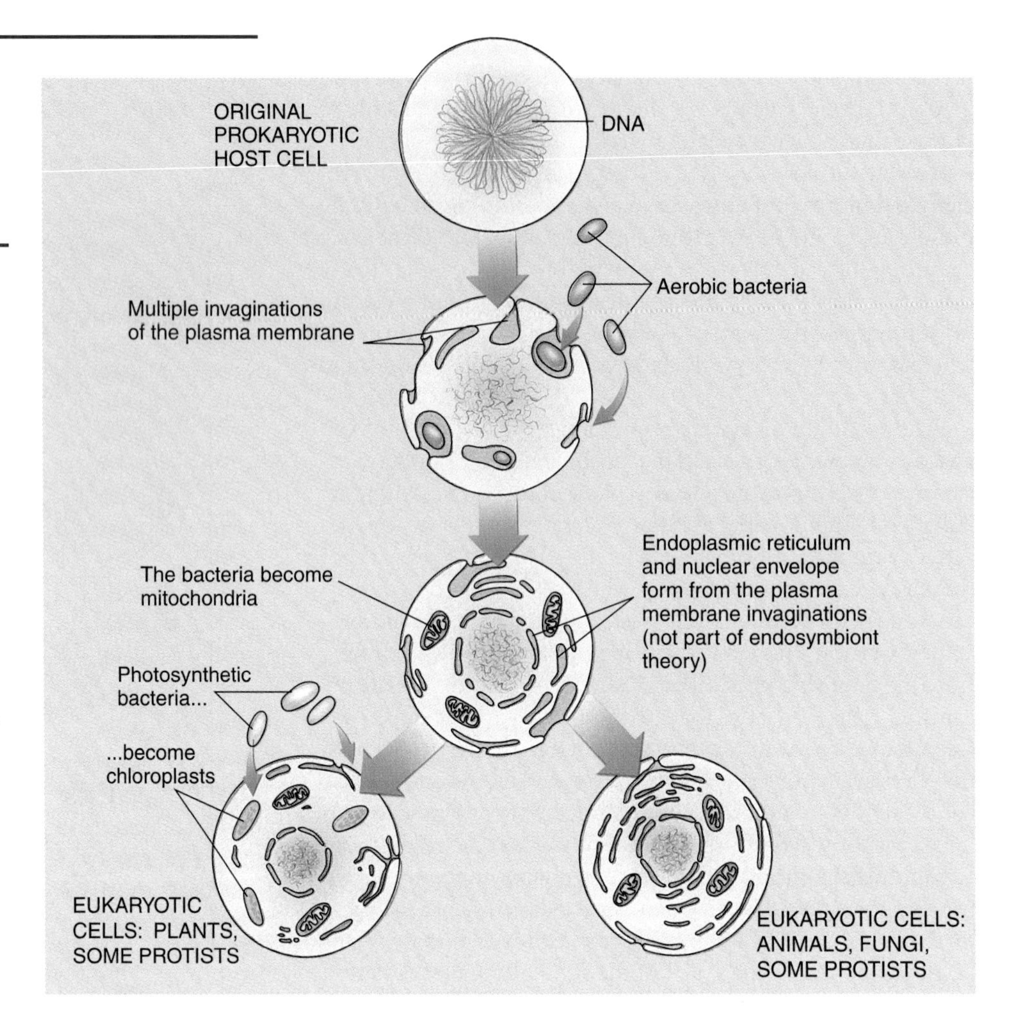

out its endosymbionts. This theory stipulates that each of these partners brought to the relationship something the other lacked. For example, mitochondria provided the ability to carry out the aerobic respiration lacking in the original anaerobic host cell. Chloroplasts provided the ability to use a simple carbon source (CO_2) to produce needed organic molecules. The host cell provided a safe environment and raw materials or nutrients.

The principal evidence in favor of the endosymbiont theory is that mitochondria and chloroplasts possess some (although not all) of their own genetic material and translational components. They have their own DNA (as a circular molecule similar to that of prokaryotes; see Chapter 23) and their own ribosomes (which resemble prokaryotic rather than eukaryotic ribosomes). Mitochondria and chloroplasts also possess some of the machinery for protein synthesis, including tRNA molecules, and conduct protein synthesis on a limited scale independent of the nucleus. Furthermore, it is possible to poison mitochondria and chloroplasts with an antibiotic that affects prokaryotic but not eukaryotic cells. As discussed in Chapter 4, double membranes envelope mitochondria and chloroplasts. The outer membrane apparently developed from the invagination (infolding) of the host cell's plasma membrane, whereas the inner membrane is derived from the endosymbiont's plasma membrane. (The endosymbiont theory is discussed in greater detail in Chapter 24.)

Many endosymbiotic relationships exist today. Many corals have algae living as endosymbionts within their cells (see Fig. 52-11). In the gut of the termite lives a protozoon (*Myxotricha paradoxa*) that in turn has several different endosymbionts, including spirochete bacteria that are attached to the protozoon and function as whiplike flagella, allowing it to move.

The endosymbiont theory does not completely explain the evolution of eukaryotic cells from prokaryotes. It does not explain, for example, how a double membranous envelope came to surround the genetic material in the nucleus.

Review

- What major steps probably occurred in the origin of cells?
- How did the presence of molecular oxygen in the atmosphere affect early life?
- What kinds of evidence support the endosymbiont theory?

BiologyNow™ Assess your understanding of **the first cells** by taking the pretest on your BiologyNow CD-ROM.

THE HISTORY OF LIFE

Learning Objectives

6 List the geological eras in chronological order, and give approximate dates for each.

7 Briefly describe the distinguishing organisms and major biological events of Precambrian time and of the Paleozoic, Mesozoic, and Cenozoic eras.

The sequence of biological, climate, and geological events that make up the history of life is recorded in rocks and fossils. The sediments of Earth's crust consist of five major rock strata (layers), each subdivided into minor strata, lying one on top of the other. Very few places on Earth show all layers, but the strata present typically occur in the correct order, with younger rocks on top of older ones. These sheets of rock formed from the accumulation of mud and sand at the bottoms of the ocean, seas, and lakes. Each layer contains certain characteristic fossils,

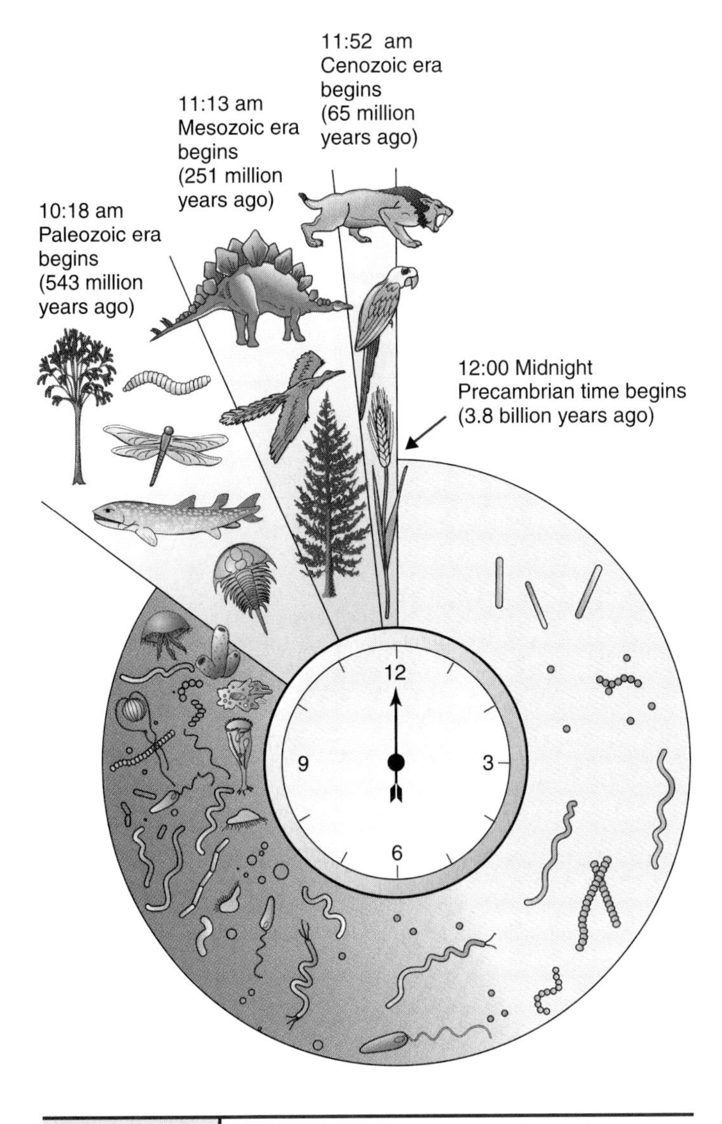

FIGURE **20-8** **An interpretive scale of biological time.**

Because it is difficult to interpret time in millions or billions of years, using a clock helps represent such vast spans of time. Life began 3.8 bya, at 12:00 midnight. More than 10 hours later, at 10:18 A.M., the Paleozoic era began. The beginning of the Mesozoic era, 251 mya, would be at 11:13 A.M. The Cenozoic era, which began 65 mya, would start at 11:52 A.M. The last epoch of the Cenozoic era, the Holocene epoch, began 10,000 years ago, which would be represented by the last 0.1 second before 12:00 noon. Representative life forms for each era are included.

known as **index fossils,** that identify deposits made at about the same time in different parts of the world (see Chapter 17).

Geologists divide Earth's 4.6-billion-year history into units of time based on major geological, climate, and biological events. Relatively little is known about Earth from its beginnings approximately 4.6 bya up to 543 mya, a period known informally as **Precambrian time.** More precisely, that part of Precambrian time from 2.5 bya to 543 mya is known as the **Proterozoic eon.** Beginning about 543 mya, the fossil record of ancient organisms is abundant. This most recent time, from 543 mya to the present, is divided into three **eras** based primarily on organisms that characterized each era (Fig. 20-8 and Table 20-1). Eras are subdivided into **periods,** which in turn are composed of **epochs.**

Precambrian deposits contain fossils of cells and simple animals

Signs of Precambrian life date back as early as 3.8 bya. Not much physical evidence is available, because the rocks of Precambrian time, being extremely ancient, are deeply buried in most parts of the world. Precambrian rocks are exposed in a few places, including the bottom of the Grand Canyon and along the shores of Lake Superior. More than 400 Precambrian rock formations have revealed microfossils.

During Precambrian time, widespread volcanic activity and giant upheavals raised mountains, and the heat, pressure, and churning associated with these movements probably destroyed most of whatever fossils had formed. Some evidence of life still remains as traces of graphite or pure carbon, which may be the transformed remains of primitive life. These remains are especially abundant in what were the ocean and seas of that time. In addition, geologists recovered fossils resembling cyanobacteria from several Precambrian formations. The fossils found in more recent Precambrian rocks show unambiguous examples of some major groups of bacteria, fungi, protists (including multicellular algae), and animals.

One rich source of Precambrian fossil deposits is the Ediacaran Hills (pronounced "ee-dee-ack′uh-ran") in South Australia. **Ediacaran fossils,** the oldest known fossils of multicellular animals, some a meter or more in length, are from very late in Precambrian time—from 600 to 543 mya. Experts interpret some of these fossils as early sponges, jellyfish, and comb jellies, but they have not yet identified all the simple, soft-bodied animals found there and at other late Precambrian sites. The oldest, simplest Ediacaran fossils are from 580-million-year-old rocks in Newfoundland and in the Mackenzie Mountains of northwest Canada.

A diversity of organisms evolved during the Paleozoic era

The **Paleozoic era** began approximately 543 mya and lasted approximately 292 million years. It is divided into six periods: Cambrian, Ordovician, Silurian, Devonian, Carboniferous, and Permian.

Rocks rich in fossils represent the oldest subdivision of the Paleozoic era, the **Cambrian period.** From about 565 mya to 525 mya, evolution was in such high gear, with the sudden appearance of many new animal body plans, that this period is nicknamed the **Cambrian explosion.** Fossils of all contemporary animal phyla are present, along with many bizarre, extinct phyla, in marine sediments. The sea floor was covered with sponges, corals, sea lilies, sea stars, snails, clamlike bivalves, primitive squidlike cephalopods, lamp shells (brachiopods), trilobites (see chapter opening photograph; also see Fig. 29-18).

TABLE 20-1 Some Important Biological Events in Geological Time

Time*	Era	Period	Epoch	Geological/ Climatic Conditions	Plants and Microorganisms	Animals
0.01 (10,000 years ago)	Cenozoic	Quaternary	Holocene	End of last Ice Age; warmer climate; higher sea levels as glaciers melt	Decline of some woody plants; rise of herbaceous plants	Age of *Homo sapiens*
2			Pleistocene	Multiple ice ages; glaciers in Northern Hemisphere	Extinction of some plant species	Extinction of many large mammals at end
5		Tertiary	Pliocene	Uplift and mountain-building; volcanoes; climate much cooler; North and South America join at Isthmus of Panama	Expansion of extensive grasslands and deserts; decline of forests	Many grazing mammals; large carnivorous mammals; human-like primates diversify
24			Miocene	Mountains form; climate drier and cooler	Flowering plants continue to diversify	Great diversity of grazing mammals and songbirds
33			Oligocene	Rise of Alps and Himalayas; most land low; volcanic activity in Rockies; climate cool and dry	Spread of forests; flowering plant communities expand	Apes appear; present mammalian families are represented
55			Eocene	Climate warmer	Flowering plants dominant	Modern mammalian orders appear and diversify; modern bird orders appear
65			Paleocene	Continental seas disappear; climate mild to cool and wet	Semitropical vegetation (flowering plants and conifers) widespread	Primitive mammals diversify rapidly
144	Mesozoic	Cretaceous		Continents separate; most continents low; large inland seas and swamps; climate warm	Rise of flowering plants	Dinosaurs reach peak, then become extinct at end; toothed birds become extinct; primitive mammals
206		Jurassic		Continents low; inland seas; mountains form; continental drift begins; climate mild	Gymnosperms common	Large, specialized dinosaurs; first toothed birds; primitive insectivorous mammals diversify
251		Triassic		Many mountains form; widespread deserts; climate warm and dry	Gymnosperms dominant; ferns common	First dinosaurs; first mammals
290	Paleozoic	Permian		Glaciers; continents rise and merge as Pangaea; climate variable	Conifers diversify; cycads appear	Modern insects appear; mammal-like reptiles; extinction of many Paleozoic invertebrates and vertebrates at end of Permian
354		Carboniferous		Lands low and swampy; climate warm and humid, becoming cooler later	Forests of ferns, club mosses, horsetails, and gymnosperms; mosses and liverworts	First reptiles; spread of ancient amphibians; many insect forms; ancient sharks abundant
408		Devonian		Glaciers; inland seas	Vascular plants diversify and become well established; first forests; gymnosperms appear; bryophytes appear	Many trilobites; fishes with jaws appear and diversify; amphibians appear; wingless insects appear
439		Silurian		Most continents remain covered by seas; climate warm	Algae dominant in aquatic environments; vascular plants appear	Jawless fishes diversify; coral reefs common; terrestrial arthropods
495		Ordovician		Sea covers most continents	Marine algae dominant; fossil spores of terrestrial plants (bryophytes?)	Invertebrates dominant; coral reefs appear; first fishes appear
543		Cambrian		Oldest rocks with abundant fossils; lands low; climate mild and wet	Algae; bacteria and cyanobacteria; fungi	Age of marine invertebrates; modern and extinct animal phyla represented; first chordates

*Time from beginning of period to present (millions of years).

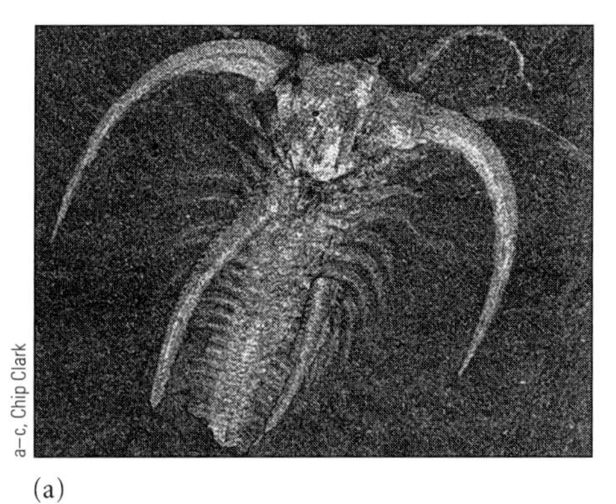
(a)

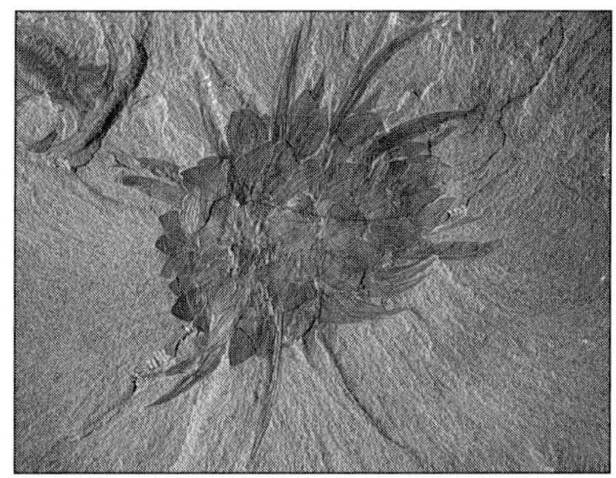
(b)

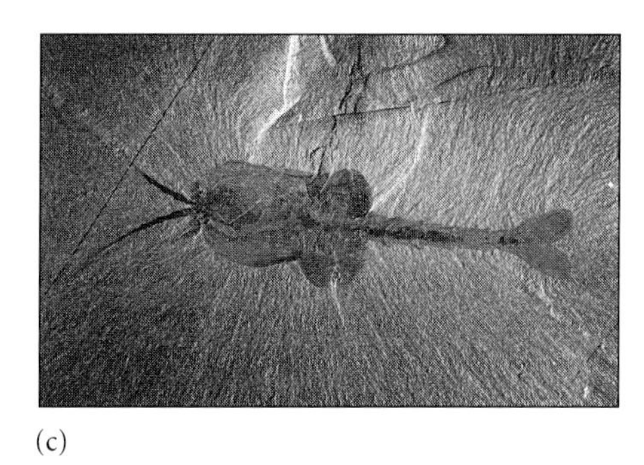
(c)

a–c, Chip Clark

FIGURE **20-9** **Fossils from the Cambrian explosion.**

(a) *Marrella splendens* was a small arthropod. **(b)** *Wiwaxia* was a bristle-covered marine worm that was distantly related to earthworms. It had scaly armor and needlelike spines for protection. **(c)** *Waptia fieldensis* was a crustacean that may have been an ancestor of modern crustaceans such as shrimp. All three of these fossils were discovered in the Burgess Shale in the Canadian Rockies of British Columbia.

In addition, small vertebrates—cartilaginous fishes, first reported in 1999—became established in the marine environment. Scientists have not determined the factors responsible for the Cambrian explosion, a period unmatched in the evolutionary history of life. There is some evidence that oxygen concentrations, which had continued to gradually increase in the atmosphere, passed some critical environmental threshold late in Precambrian time. Scientists who advocate the *oxygen enrichment hypothesis* note that until late in Precambrian time, Earth didn't have enough oxygen to support larger animals. The most important fossil sites that document the Cambrian explosion are the **Chenjiang site** in China (for early Cambrian fossils) and the **Burgess Shale** in British Columbia (for middle Cambrian fossils; Fig. 20-9).

According to geologists, seas gradually flooded the continents during the Cambrian period. In the **Ordovician period,** much land was covered by shallow seas, in which there was another burst of evolutionary diversification, although not as dramatic as the Cambrian explosion. The Ordovician seas were inhabited by giant cephalopods, squidlike animals with straight shells 5 to 7 m (16 to 23 ft) long and 30 cm (12 in) in diameter. Coral reefs first appeared during this period, as did small, jawless, bony-armored fishes called *ostracoderms* (Fig. 20-10). Lacking jaws, these fishes typically had round or slitlike mouth openings that may have sucked in small food particles from the water or scooped up organic debris from the bottom. Ordovician deposits also contain fossil spores of terrestrial (land-dwelling) plants, suggesting the colonization of land had begun.

During the **Silurian period,** jawless fishes diversified considerably, and jawed fishes first appeared. Definitive evidence of two life forms of great biological significance appeared in the Silurian period: terrestrial plants and air-breathing animals. The evolution of plants allowed animals to colonize the land, because plants provided the first terrestrial animals with food and shelter. All air-breathing land animals discovered in Silurian rocks were arthropods—millipedes, spider-like arthropods, and possibly centipedes. From an ecological perspective, the energy flow from plants to animals probably occurred via detritus, which is organic debris from decomposing organisms, rather than directly from living plant material. Millipedes eat plant detritus today, and spiders and centipedes prey on millipedes and other animals.

The **Devonian period** is frequently called the Age of Fishes. This period witnessed the explosive radiation of fishes with jaws, an adaptation that lets a vertebrate chew and bite. Armored *placoderms,* an extinct group of jawed fishes, diversified to exploit varied lifestyles (see Fig. 30-9b). Appearing in Devonian deposits are sharks and the two predominant types of bony fish: lobe-finned fishes and ray-finned

FIGURE **20-10** **Representative ostracoderms.**

(a) *Thelodus,* **(b)** *Pterapsis,* and **(c)** *Jamoytius* are fossil ostracoderms, primitive jawless fishes that appeared in the Devonian period. Ostracoderms ranged from 10 to 50 cm (4 to 20 in) in length.

fishes, which gave rise to the major orders of modern fishes (see Chapter 30). Upper (more recent) Devonian sediments contain fossil remains of salamander-like amphibians (*labyrinthodonts*) that were often quite large, with short necks and heavy, muscular tails (see Fig. 30-15). Wingless insects also originated in the late Devonian period. The early vascular plants diversified during the Devonian period in a burst of evolution that rivaled that of animals during the Cambrian explosion. With the exception of flowering plants, all major plant groups appeared during the Devonian. Forests of ferns, club mosses, horsetails, and seed ferns (an extinct group of ancient plants that had fernlike foliage but reproduced by forming seeds) flourished.

The **Carboniferous period** is named for the great swamp forests whose remains persist today as major coal deposits. Much of the land during this time was covered with low swamps filled with horsetails, club mosses, ferns, seed ferns, and gymnosperms (seed-bearing plants such as conifers) (Fig. 20-11). Amphibians, which underwent an **adaptive radiation** and exploited both aquatic and terrestrial ecosystems, were the dominant terrestrial carnivores of the Carboniferous period. Reptiles first appeared and diverged to form two major lines at this time. One line consisted of mostly small and mid-sized insectivorous (insect-eating) lizards; this line later led to lizards, snakes, crocodiles, dinosaurs, and birds. The other reptilian line led to a diverse group of Permian and early Mesozoic mammal-like reptiles. Two groups of winged insects, cockroaches and dragonflies, appeared in the Carboniferous period.

FIGURE **20-11** | **Reconstruction of a Carboniferous forest.**
Plants of this period included giant ferns, horsetails, and club mosses as well as seed ferns and early gymnosperms.

No. GEO85638c, Field Museum of Natural History, Chicago

Amphibians continued in importance during the **Permian period,** but they were no longer the dominant carnivores in terrestrial ecosystems. During the Permian period, reptiles diversified and dominated both carnivorous and herbivorous terrestrial lifestyles. One important group of mammal-like reptiles, originating in the Permian and extending into the Mesozoic era, were the *therapsids,* a group that included the ancestor of mammals (discussed shortly; also see Fig. 30-21). During the Permian period, seed plants diversified and dominated most plant communities. Cone-bearing conifers were widespread, and cycads (plants with crowns of fernlike leaves and large, seed-containing cones) and ginkgoes (trees with fan-shaped leaves and exposed, fleshy seeds) appeared.

The greatest **mass extinction** of all time occurred at the end of the Paleozoic era, between the Permian and Triassic periods, 251 mya. More than 90% of all existing marine species became extinct at this time, as did more than 70% of the vertebrate genera living on land. There is also evidence of a major extinction of plants. Many causes for the late Permian mass extinction have been suggested. Regardless of cause, evidence suggests the extinction occurred globally in a very compressed period, within a few hundred thousand years. This is extremely short in the geological time scale and suggests some sort of catastrophic event caused the mass extinction.

Dinosaurs and other reptiles dominated the Mesozoic era

The **Mesozoic era** began about 251 mya and lasted some 186 million years. It is divided into the Triassic, Jurassic, and Cretaceous periods. Fossil deposits from the Mesozoic era occur worldwide. Notable sites include the **Yixian formation** in northeastern China, the Solnhofen Limestone in Germany, northwestern Patagonia in Argentina, the Sahara Desert in central Niger, the badlands in South Dakota, and other sites in western North America.

The outstanding feature of the Mesozoic era was the origin, differentiation, and ultimately the extinction of a large variety of reptiles. For this reason, the Mesozoic era is commonly called the Age of Reptiles. Most of the modern orders of insects appeared during the Mesozoic era. Snails and bivalves (clams and their relatives) increased in number and diversity, and sea urchins reached their peak diversity. From a botanical viewpoint, the Mesozoic era was dominated by gymnosperms until the mid-Cretaceous period, when the flowering plants first diversified.

During the **Triassic period,** reptiles underwent an adaptive radiation leading to many groups. On land, the dominant Triassic groups were the mammal-like therapsids, which ranged from small-sized insectivores (insect-eating reptiles) to moderately large herbivores (plant-eating mammals), and a diverse group of *thecodonts,* early "ruling reptiles," that were primarily carnivores (Fig. 20-12a). The thecodont group was ancestral to dinosaurs, flying reptiles, and birds.

FIGURE **20-12** **Representative animals of the Mesozoic era.**

Figures are not drawn to scale. **(a)** This Triassic thecodont, *Euparkia,* was about 150 cm (5 ft) long. **(b)** *Elasmosaurus* was a long-necked plesiosaur. Other plesiosaurs had short necks and superficially resembled seals. **(c)** *Opthalmosaurus* was an ichthyosaur that superficially resembled a shark or porpoise. It was about 3.6 m (12 ft) long. **(d)** *Pteranodon* was a pterosaur from the Cretaceous period with a wingspan of 7 to 9.2 m (23–30 ft), depending on the species. Pterosaur wings were membranes of skin that were supported by an elongated fourth finger bone. **(e)** *Tylosaurus* was a large (about 10 m [33 ft] long) marine lizard (a mosasaur). **(f)** *Giganotosaurus,* whose fossil remains were discovered in Argentina, was the largest (more than 12 m [39 ft] in length) predatory saurischian. **(g)** *Argentinosaurus,* a herbivorous saurischian from Argentina, is the largest known animal to have ever walked on land. **(h)** *Hadrosaurus* was a duck-billed, plant-eating ornithischian. It was 7 to 10 m (23–33 ft) long and had hundreds of cheek teeth (its bill was toothless). **(i)** *Ankylosaurus* was a heavily armored ornithischian. Ankylosaurs were 2 to 6 m (7–20 ft) in length.

In the ocean, several important marine reptile groups, the plesiosaurs and ichthyosaurs, appeared in the Triassic. *Plesiosaurs* were aquatic reptiles with bodies up to 15 m (about 49 ft) long and paddle-like fins (Fig. 20-12b). *Ichthyosaurs,* also aquatic reptiles, had body forms superficially resembling those of sharks or porpoises, with short necks, large dorsal fins, and shark-type tails (Fig. 20-12c). Ichthyosaurs had very large eyes, which may have helped them see at diving depths of 500 m or more.

Pterosaurs, the first flying reptiles, appeared and underwent considerable diversification during the Mesozoic era (Fig. 20-12d). This group produced some quite spectacular forms, most notably the giant *Quetzalcoatlus,* known from fragmentary Cretaceous fossils in Texas to have had a wingspan of 11 to 15 m (36 to 49 ft).

The first mammals to appear in the Triassic period were small insectivores that evolved from the mammal-like therapsids of the Triassic. Mammals diversified into a variety of mostly small, nocturnal insectivores during the remainder of the Mesozoic, with marsupial and placental mammals appearing later in the Mesozoic era.

During the **Jurassic** and **Cretaceous periods,** other important groups—such as crocodiles, lizards, snakes, and birds—appeared, and the dinosaurs diversified dramatically to "inherit the Earth." One group of lizards, the *mosasaurs,* were large, voracious marine predators during the late Cretaceous period (Fig. 20-12e). The mosasaurs, which are now extinct, attained lengths of 10 m (33 ft) or more.

Dinosaurs underwent an impressive radiation throughout the Jurassic and Cretaceous periods. There were two main groups of dinosaurs: the *saurischians,* with pelvic bones similar to those of modern-day lizards, and the *ornithischians,* with pelvic bones similar to those of birds (Fig. 20-13). Some saurischians were fast, bipedal (walking on two feet) forms ranging from those the size of a dog to the ultimate representatives of this group, the gigantic carnivores of the Cretaceous period, *Tyrannosaurus, Giganotosaurus,* and *Carcharodontosaurus* (Figs. 20-12f and 20-14). Other saurischians were huge, quadrupedal (walking on four feet) dinosaurs that ate plants. Some of these were the largest terrestrial animals that have ever lived, including *Argentinosaurus,* with an estimated length of 30 m (98 ft) and an estimated weight of 72 to 90 metric tons (80 to 100 tons) (Fig. 20-12g).

The other group of dinosaurs, the ornithischians, was entirely herbivorous. Although some ornithischians were bipedal, most were quadrupedal. Some had no front teeth and possessed stout, horny, birdlike beaks. In some species these beaks were broad and ducklike, hence the common name, *duck-billed dinosaurs* (Fig. 20-12h). Other ornithischians had great armor plates, possibly as protection against carnivorous saurischians. *Ankylosaurus* had a broad, flat body covered with armor plates (actually bony scales embedded in the skin) and large, laterally projecting spikes (Fig. 20-12i).

Over the past few decades scientists have reconsidered many traditional ideas about dinosaurs and no longer think they were all cold-blooded, slow-moving monsters living in swamps. Recent evidence suggests that at least some dinosaurs were warm-blooded, agile, and able to move extremely fast. Many dinosaurs appear to have had complex social behaviors, including courtship rituals and parental nurturing of their young. Some species lived in social groups and hunted in packs.

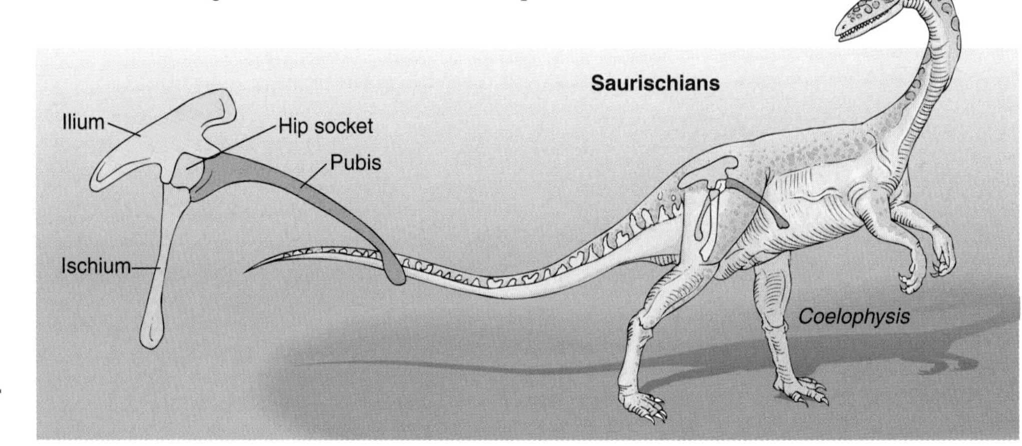

(a)

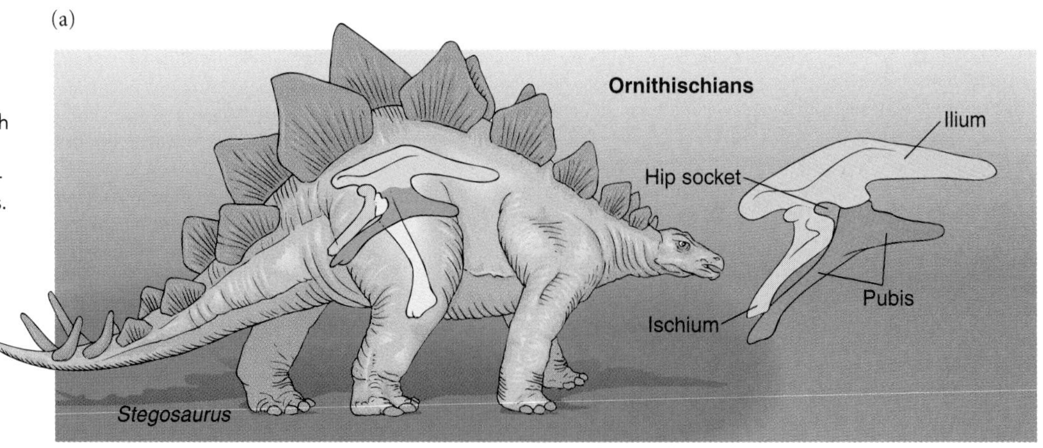

(b)

FIGURE **20-13**

Saurischian and ornithischian dinosaurs.

The two orders of dinosaurs are distinguished primarily by differences in their pelvic bones. (In each dinosaur figure, the pale yellow femur is shown relative to the pelvic bone.) **(a)** The saurischian pelvis. Note the opening (hip socket), a trait possessed by no quadrupedal vertebrates other than dinosaurs. **(b)** The ornithischian pelvis also has the hole in the hip socket but differs from the saurischian pelvis in that it has a backward-directed extension of the pubis.

Paul Sereno. Reprinted with permission of *Discover*

FIGURE **20-14** **Reconstruction of a skull of *Carcharodontosaurus.***

The fossil remains of this fearsome predator were discovered in North Africa. A human skull is added for comparison.

Birds appeared by the late Jurassic period, and most paleontologists think they evolved directly from dinosaurs (see *On the Cutting Edge: The Origin of Flight in Birds*). Excellent bird fossils, many showing the outlines of feathers, are preserved from the Jurassic period. *Archaeopteryx,* the oldest known bird in the fossil record, lived about 150 mya (see Fig. 30-19b). It was about the size of a pigeon and had rather feeble wings that it used to glide rather than actively fly. Although *Archaeopteryx* is considered a bird, it had many reptilian features, including a mouthful of teeth and a long, bony tail. Thousands of well-preserved bird fossils have been found in early Cretaceous deposits in China. These include *Sinornis,* a 135-million-year-old sparrow-sized bird capable of perching, and the magpie-sized *Confuciusornis,* the earliest known bird with a toothless beak. *Confuciusornis* may date back as far as 142 million years.

At the end of the Cretaceous period, 65 mya, dinosaurs, pterosaurs, and many other animals abruptly became extinct (see Fig. 19-17). Many gymnosperms, with the exception of conifers, also perished. Evidence suggests that a catastrophic collision of a large extraterrestrial body with Earth dramatically changed the climate at the end of the Cretaceous period. Part of the evidence is a thin band of dark clay, with a high concentration of iridium, located between Mesozoic and Cenozoic sediments at more than 200 sites around the world. Iridium is rare on Earth but abundant in meteorites, leading many to conclude that Earth was hit by a large extraterrestrial object at that time. (The force of the impact would have driven the iridium into the atmosphere, to be deposited later on the land by precipitation.) The Chicxulub crater, buried under the Yucatan Peninsula in Mexico, is the apparent site of the collision at the close of the Cretaceous period. The impact produced giant tsunamis (tidal waves) that deposited materials from the extraterrestrial body around the perimeter of the Gulf of Mexico, from Alabama to Guatemala. It may have caused global forest fires and giant smoke and dust clouds that lowered temperatures for many years.

Although scientists widely accept that a collision with an extraterrestrial body occurred 65 mya, they have reached no consensus about the effects of such an impact on organisms. The extinction of many marine organisms at or immediately after the time of the impact was probably the result of the environmental upheaval that the collision produced. However, many clam species associated with the mass extinction at the end of the Cretaceous period seem to have gone extinct *before* the impact, suggesting that other factors caused some of the massive extinctions occurring then.

The Cenozoic era is the Age of Mammals

With equal justice the **Cenozoic era** could be called the Age of Mammals, the Age of Birds, the Age of Insects, or the Age of Flowering Plants. This era is marked by the appearance of all these forms in great variety and numbers of species. The Cenozoic era extends from 65 mya to the present. It is subdivided into two periods: the **Tertiary period,** encompassing some 63 million years, and the **Quaternary period,** which covers the last 2 million years. The Tertiary period is subdivided into five epochs, named from earliest to latest: Paleocene, Eocene, Oligocene, Miocene, and Pliocene. The Quaternary period is subdivided into the Pleistocene and Holocene epochs.

Flowering plants, which arose during the Cretaceous period, continued to diversify during the Cenozoic era. During the Paleocene and Eocene epochs, fossils indicate that plant communities ranging from tropical to semitropical extended to relatively high latitudes. Palms, for example, are found in Eocene deposits in Wyoming. Later in the Cenozoic era, there is evidence of more open habitats. Grasslands and savannas spread throughout much of North America during the Miocene epoch, with deserts developing later in the Pliocene and Pleistocene epochs. During the Pleistocene epoch, plant communities changed dynamically in response to the fluctuating climates associated with the multiple advances and retreats of continental glaciers.

During the Eocene epoch, there was an explosive radiation of birds, which acquired adaptations for many different habitats. The jaws and beak of the flightless giant bird *Diatryma,* for example, may have been adapted primarily for crushing and slicing vegetation in Eocene forests, marshes, and grasslands (Fig. 20-15). Other paleontologists hypothesize that these giant birds were carnivores that killed or scavenged mammals and other vertebrates.

During the Paleocene epoch, an explosive radiation of primitive mammals occurred. Most of these were small forest dwellers that are not closely related to modern mammals. During the Eocene epoch, mammals continued to diverge, and all the modern orders first appeared. Again, many of the mammals were small, but there were also some larger herbivores—the *titanotheres,* for example, which got progressively larger during the Eocene epoch (Fig. 20-16a).

During the Oligocene epoch, many modern families of mammals appeared, including the first apes in Africa. Many lineages showed adaptations that suggest a more open type of habitat, such as grassland or savanna. Many mammals were

The Origin of Flight in Birds

PROCESS OF SCIENCE

Hypothesis: Bird flight originated in ancestral animals that lived in trees and glided from treetops.

Method: Evaluate recently discovered fossils of the feathered dinosaur, *Microraptor gui.*

Results: *Microraptor gui* was a small, lightweight dinosaur with feathers on its forelimbs, hindlimbs, and tail. The hindlimb feathers would have made it difficult for *Microraptor* to run on the ground to power flight.

Conclusion: *Microraptor gui* is a significant fossil find that supports the hypothesis that early birds flew by gliding from the treetops.

The evolution of birds is arguably one of the most interesting chapters in Earth's history of life. Given the substantial fossil evidence, most paleontologists have concluded the ancestors of birds were dinosaurs, specifically the *dromaeosaurs,* a group of ground-dwelling, bipedal, carnivorous theropods. Beginning in 1997, paleontologists made several discoveries of fossil dinosaurs with feathers, indicating that feathers appeared before birds. They think that feathers evolved as one or a series of evolutionary novelties, or **preadaptations.** The first feathers may have provided thermal insulation but were subsequently modified for flight (see Chapter 19).

Once dinosaurs and early birds had feathers, how did they fly? Did tree-dwelling animals glide as an intermediate step in bird flight, or did ground-dwelling animals flap their wings and run to provide thrust and lift for a takeoff? The question about how flight originated in birds has intrigued biologists for more than a century. In 1915, for example, American zoologist William Beebe hypothesized that the ancestors of birds were probably tree-dwelling gliders that had feathers on all four limbs.* Because no fossil evidence supported his suggestion, scientists did not consider it seriously at the time.

Almost a century later, in 2003, a group of Chinese paleontologists led by Xing Xu and Zhonghe Zhou of the Institute of Vertebrate Paleontology and Paleoanthropology in Beijing announced the discovery of two nearly complete fossils of the organism that Beebe had hypothesized.† The fossils of a small, feathered dramaeosaur dinosaur were found in Liaoning Province in northeastern China. The dinosaur, *Microraptor gui,* had feathers on both forelimbs and hindlimbs as well as its long tail (see figure). It was small—77 cm in length, including the tail—and appeared to be adapted to life in the trees.

The feathers on *M. gui* were very similar to those of modern-day birds. Downy feathers covered the body, and each limb had about 12 "primary" flight feathers and about 18 shorter, "secondary" feathers. *M. gui*'s flight feathers were asymmetrical, a characteristic associated with flight or gliding in modern birds. The primary and secondary feathers followed a similar pattern on both the forelimbs and hindlimbs, and this pattern resembles that on modern birds. In part because *M. gui's* breastbone was not structured to attach large flight muscles, Xu, Zhou, and colleagues hypothesize that *M. gui* glided rather than flapped its wings. They suggest that stretching out the feathered forelimbs and hindlimbs would make a perfect airfoil for gliding, somewhat like the stretched web of skin in flying squirrels today. However, the flight question may not be resolved until functional anatomists and paleontologists make detailed studies of *M. gui.* Analysis of the shoulder and wing anatomy may clarify whether *M. gui* could power flight or merely glide. Study of the hips may resolve if *M. gui* could rotate its legs to glide.

M. gui, which is about 126 million years old, is not a direct ancestor of birds. Birds had already evolved when *M. gui* existed. The earliest known bird, *Archaeopteryx,* lived about 145 million years ago and therefore predates *M. gui* by about 20 million years. *M. gui* is considered a *basal member*—that is, an earlier evolutionary branch—of the most recent ancestor of *Archaeopteryx* and other birds. Xu and Zhou hypothesize that like *M. gui,* earlier feathered dinosaurs were four-winged organisms that lived and glided in the trees. They suggest that, during the course of bird evolution, the feathered hindlimbs became reduced and eventually lost.

However, an alternative hypothesis is that feathered hindlimbs may have been a failed evolutionary experiment restricted to dromaeosaurs and not important as an intermediate step in bird evolution. Further analysis of *M. gui* and future discoveries of older dromaeosaur fossils may shed some light on the importance of feathered hindlimbs.

Clearly, *M. gui* is a significant discovery that has given paleontologists and biologists much to consider in the evolutionary transition from dinosaurs to birds. Conflicting explanations are at the heart of the scientific process because they stimulate additional research that leads to a clearer picture. However, science rarely, if ever, deals in absolutes. Biologists will continue to study and debate the evolution of flight in birds for many years.

An artist's interpretation of *Microraptor gui.*

© Portia Sloan

*W.H. Beebe, "A Tetrapteryx Stage in the Ancestry of Birds," *Zoologica,* Vol. 2 (1915).

†X. Xu, Z. Zhou, X. Wang, X. Kuang, F. Zhang, and X. Du, "Four-Winged Dinosaurs from China." *Nature,* Vol. 421, 23 Jan. 2003.

larger and had longer legs for running, specialized teeth for chewing coarse vegetation or for preying on animals, and increases in their relative brain sizes. Such adaptations continued to evolve in the Miocene and Pliocene epochs. Human ancestors appeared in Africa during the late Miocene and early Pliocene epochs. The genus *Homo* appeared approximately 2.3 mya. (Primate evolution, including human evolution, is discussed in Chapter 21.)

The Pliocene and Pleistocene epochs witnessed a spectacular North and South American large-mammal fauna, including mastodons, saber-toothed cats, camels, giant ground sloths, giant armadillos, and many other species (Fig. 20-16b–f). However, many of the large mammals became extinct at the end of the Pleistocene. This extinction was possibly due to climate change—the Pleistocene epoch was marked by several ice ages—and/or to the influence of humans, which had spread from Africa to Europe and Asia, and later to North and South America by crossing a land bridge between Siberia and Alaska. Strong archaeological evidence shows that this mass extinction event was concurrent with the appearance of human hunters.

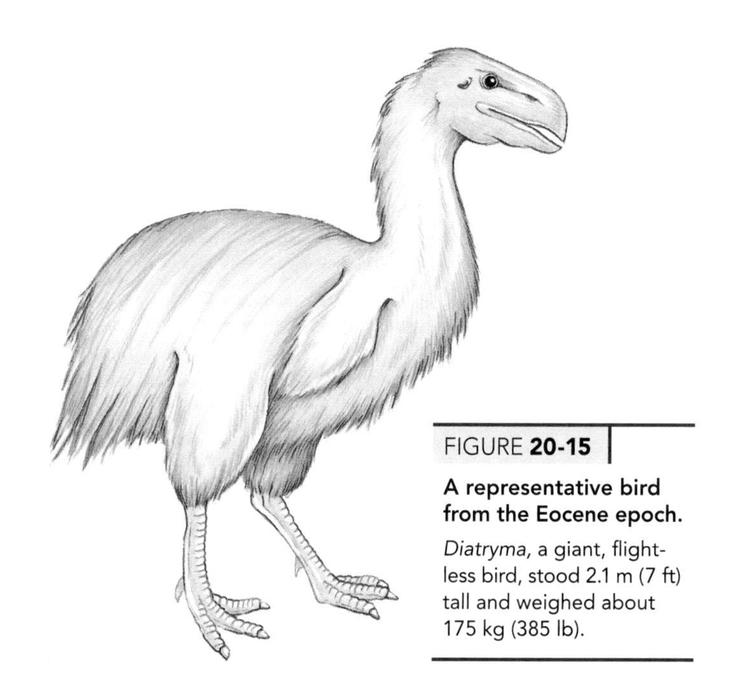

FIGURE **20-15**

A representative bird from the Eocene epoch.

Diatryma, a giant, flightless bird, stood 2.1 m (7 ft) tall and weighed about 175 kg (385 lb).

Review

- What is the correct order of appearance in the fossil record, starting with the earliest: eukaryotic cells, multicellular organisms, prokaryotic cells?
- What is the correct order of appearance in the fossil record, starting with the earliest: reptiles, mammals, amphibians, fish?
- What is the correct order of appearance in the fossil record, starting with the earliest: flowering plants, ferns, gymnosperms?

BiologyNow™ Assess your understanding of **the history of life** by taking the pretest on your BiologyNow CD-ROM.

FIGURE **20-16**

Representative North and South American mammals of the Cenozoic era.

Figures are not drawn to scale. **(a)** *Brontotherium,* a titanothere, was 2.5 m (8 ft) tall at the shoulder. **(b)** The elephant-like mastodon (*Mammut*) was at home in forests, lakes, and rivers. *Mammut* was 2.5 m (8 ft) tall at the shoulder. **(c)** The saber-toothed cat (*Smilodon*), which had short, powerful legs and was about the size of the modern African lion, was found in both North and South America. **(d)** This camel-like mammal (*Macrauchenia*), which was about 3 m (10 ft) tall, probably browsed in forest clearings in South America. **(e)** *Megatherium* was a South American giant ground sloth. As long as 6 m (20 ft), *Megatherium* had 18-cm (7-in) claws that may have stripped bark from trees or stabbed prey. **(f)** The giant armadillo (*Glyptodon*) lived on the pampas of South America. Encased in bony armor, *Glyptodon* was about 2 m (6.5 ft) long.

SUMMARY WITH KEY TERMS

1 Describe the conditions that geologists think existed on early Earth.

- Biologists generally agree that life originated from nonliving matter by **chemical evolution.** Although chemical evolution is very difficult to test experimentally, hypotheses about the origin of life are testable.
- Four requirements for chemical evolution are (1) the absence of oxygen, which would have reacted with and oxidized abiotically produced organic molecules; (2) energy to form organic molecules; (3) chemical building blocks, including water, minerals, and gases present in the atmosphere, to form organic molecules; and (4) sufficient time for molecules to accumulate and react.

2 Contrast the prebiotic soup hypothesis and the iron-sulfur world hypothesis.

- During chemical evolution, small organic molecules formed spontaneously and accumulated.
- The **prebiotic soup hypothesis** proposes that organic molecules formed near Earth's surface in a "sea of organic soup" or on rock or clay surfaces.
- The **iron-sulfur world hypothesis** suggests that organic molecules were produced at **hydrothermal vents,** cracks in the deep ocean floor.

3 Outline the major steps hypothesized to have occurred in the origin of cells.

- After small organic molecules formed and accumulated, macromolecules assembled from the small organic molecules. Macromolecular assemblages called **protobionts** formed from macromolecules. Cells arose from the macromolecular assemblages.
- According to a model known as the **RNA world,** RNA was the first informational molecule to evolve in the progression toward a self-sustaining, self-reproducing cell. Natural selection at the molecular level resulted in the information sequence: DNA ⟶ RNA ⟶ protein.

4 Explain how the evolution of photosynthetic autotrophs affected both the atmosphere and other organisms.

- The first cells were prokaryotic **heterotrophs** that obtained organic molecules from the environment. They were almost certainly **anaerobes.** Later, **autotrophs,** arose—organisms that produce their own organic molecules by photosynthesis.
- The evolution of photosynthesis ultimately changed early life because it generated oxygen, which accumulated in the atmosphere, and permitted the evolution of **aerobes,** organisms that could use oxygen for a more efficient type of cellular respiration.

5 Describe the endosymbiont theory.

- Eukaryotic cells arose from prokaryotic cells. According to the **endosymbiont theory,** certain eukaryotic organelles (mitochondria and chloroplasts) evolved from prokaryotic **endosymbionts** within larger prokaryotic hosts.

6 List the geological eras in chronological order, and give approximate dates for each.

- The Paleozoic era began about 543 mya and ended 251 mya.
- The Mesozoic era began about 251 mya and ended 65 mya.
- The Cenozoic era began about 65 mya and extends to the present.

7 Briefly describe the distinguishing organisms and major biological events of Precambrian time and of the Paleozoic, Mesozoic, and Cenozoic eras.

- During **Precambrian time,** which extended from approximately 3.8 bya up to 543 mya, life began and diverged into different groups of bacteria, protists (including algae), fungi, and simple multicellular animals.
- During the **Paleozoic era,** which lasted approximately 292 million years, all major groups of plants, except for flowering plants, and all animal phyla appeared. Fish and amphibians flourished, and reptiles appeared and diversified. The greatest mass extinction of all time occurred at the end of the Paleozoic era, 251 mya. More than 90% of all existing marine species and more than 70% of land-dwelling vertebrate genera became extinct, as well as many plant species.
- The **Mesozoic era** lasted some 186 million years. During the Mesozoic era, flowering plants appeared and reptiles diversified. Dinosaurs, which descended from early reptiles, dominated. Insects flourished, and birds and early mammals appeared. At the end of the Cretaceous period, 65 mya, a great many animals abruptly became extinct. A collision of Earth with a large extraterrestrial body may have resulted in dramatic climate changes that played a role in this mass extinction episode.
- In the **Cenozoic era,** which extends from 65 mya to the present, flowering plants, birds, insects, and mammals diversified greatly.

POST-TEST

1. Energy, the absence of oxygen, chemical building blocks, and time were the requirements for (a) chemical evolution (b) biological evolution (c) the Cambrian explosion (d) the mass extinction episode at the end of the Cretaceous period (e) directed evolution
2. Protobionts (a) form spontaneously in hydrothermal vents in the ocean floor (b) are heterotrophs that obtain the organic molecules they need from the environment (c) are assemblages of abiotically produced organic polymers that resemble living cells in several ways (d) are autotrophs that use sunlight to split hydrogen sulfide (e) are fossilized mats of cyanobacteria
3. Many scientists think that ______________ was the first information molecule to evolve. (a) DNA (b) RNA (c) a protein (d) an amino acid (e) a lipid
4. The first cells were probably (a) heterotrophs (b) autotrophs (c) anaerobes (d) both a and c (e) both b and c
5. According to the endosymbiont theory, (a) life originated from nonliving matter (b) the pace of evolution quickened at the start of the Cambrian period (c) chloroplasts, mitochondria, and possibly other organelles originated from intimate relationships among prokaryotic organisms (d) banded iron formations reflect the buildup of sufficient oxygen in the atmosphere to oxidize iron at

POST-TEST *(continued)*

Earth's surface (e) the first photosynthetic organisms appeared 3.1 to 3.5 bya

6. Geological time from the first traces of life to the beginning of the Paleozoic era some 543 mya is informally known as (a) the Cenozoic era (b) the Paleozoic era (c) the Mesozoic era (d) Precambrian time (e) the Cambrian period
7. Geologists divide Earth's history, from Precambrian time to the present, into (a) three periods (b) three epochs (c) three eras (d) five periods (e) five eras
8. Ediacaran fossils (a) are the oldest known fossils of multicellular animals (b) come from the Burgess Shale in British Columbia (c) contain remains of large salamander-like organisms (d) are the oldest fossils of early vascular plants (e) contain a high concentration of iridium
9. The correct chronological order of geological eras, starting with the oldest, is (a) Paleozoic, Cenozoic, and Mesozoic (b) Mesozoic, Cenozoic, and Paleozoic (c) Mesozoic, Paleozoic, and Cenozoic (d) Paleozoic, Mesozoic, and Cenozoic (e) Cenozoic, Paleozoic, and Mesozoic
10. The time of greatest evolutionary diversification in the history of life occurred during the (a) Cambrian period (b) Ordovician period (c) Silurian period (d) Carboniferous period (e) Permian period
11. The greatest mass extinction episode in the history of life occurred at what boundary? (a) Pliocene–Pleistocene (b) Permian–Triassic (c) Mesozoic–Cenozoic (d) Cambrian–Ordovician (e) Triassic–Jurassic
12. The Mesozoic era is divided into three periods, which are (a) Cambrian, Ordovician, and Silurian (b) Devonian, Carboniferous, and Permian (c) Triassic, Jurassic, and Cretaceous (d) Cretaceous, Tertiary, and Quaternary (e) Pliocene, Pleistocene, and Holocene
13. The Age of Reptiles corresponds to the (a) Paleozoic era (b) Mesozoic era (c) Cenozoic era (d) Pleistocene epoch (e) Permian period
14. Evidence exists that a catastrophic collision between Earth and a large extraterrestrial body occurred 65 mya, resulting in the extinction of (a) Precambrian worms, mollusks, and soft-bodied arthropods (b) jawless ostracoderms and jawed placoderms (c) dinosaurs, pterosaurs, and many gymnosperm species (d) mastodons, saber-toothed cats, and giant ground sloths (e) ferns, horsetails, and club mosses
15. Flowering plants and mammals diversified and became dominant during the (a) Paleozoic era (b) Mesozoic era (c) Cenozoic era (d) Devonian period (e) Cambrian period

CRITICAL THINKING

1. If you were experimenting on how protobionts evolved into cells and you developed a protobiont that was capable of self-replication, would you consider it a living cell? Why or why not?
2. If living cells were created in a test tube from nonbiological components by chemical processes, would this accomplishment prove that life evolved in a similar manner billions of years ago? Why or why not?
3. Why did the evolution of multicellular organisms such as plants and animals have to be preceded by the evolution of oxygen-producing photosynthesis?

- Visit our Web site at **http://biology.brookscole.com/solomon7** for links to chapter-related resources on the World Wide Web. Additional online materials relating to this chapter can also be found on our Web site.

BIOLOGY NOW RESOURCES

Active Figures

20-1: Early Earth

20-2: Miller-Urey experiment

Preparing for an exam? Take a diagnostic test on your BiologyNow CD-ROM.

Post-Test Answers

1. a	2. c	3. b	4. d
5. c	6. d	7. c	8. a
9. d	10. a	11. b	12. c
13. b	14. c	15. c	

21

The Evolution of Primates

John Reader/Science Photo Library/Photo Researchers, Inc.

Fossilized remains of a Neandertal man. Each of these fossils, which were found in La Chapelle-aux-Saints, France, has been carefully catalogued and studied. Some of the bones are deformed by osteoarthritis.

CHAPTER OUTLINE

Twelve years after Darwin wrote *The Origin of Species by Natural Selection,* he published another controversial book, *The Descent of Man,* which addressed human evolution. In it, Darwin hypothesized that humans and apes share a common ancestry. For nearly a century after Darwin, fossil evidence of human ancestry remained fairly incomplete. However, research over the last several decades, especially in Africa, has yielded a rapidly accumulating set of fossils that gives an increasingly clear answer to the question "Where did we come from?"

Humans and other **primates,** such as lemurs, tarsiers, monkeys, and apes, are mammals that share such traits as flexible hands and feet with five digits, a strong social organization, and front-facing eyes, which permit depth perception. Mammals (class Mammalia) arose from mammal-like reptiles known as *therapsids* more than 200 million years ago (mya), during the Mesozoic era (see Fig. 30-21). These early mammals remained a minor component of life on Earth for almost 150 million years and then rapidly diversified during the Cenozoic era (the last 65 million years). Mammals are **endothermic** (they use metabolic energy to maintain a constant body temperature); produce body hair for such functions as insulation, protective coloration, and waterproofing; and feed their young with milk from mammary glands. Most mammals are **viviparous,** which means that their fertilized eggs develop into young offspring within the female body.

Scientists hypothesize that the three groups of living mammals—the monotremes, marsupials, and placental mammals—all are descended from the same common ancestor. The **monotremes** are mammals, such as the duck-billed platypus, that lay eggs (see Chapter 30). The **marsupials,** such as kangaroos and opossums, give birth to their young in a very underdeveloped condition, and then carry them in an abdominal pouch. **Placental mammals,** the largest and most successful group, possess a **placenta,** an organ that exchanges materials between the mother and the embryo/fetus developing in the uterus. When placental mammals give birth, their young are more developed than the newborn young of marsupials.

Paleontologists hypothesize that the first primates descended from small, shrewlike placental mammals that lived in trees and ate insects. These early primates appeared about 55 mya. Many traits of the 233 living primate species are related to their **arboreal** (tree-dwelling) past. This chapter focuses on humans and their ancestors (see photograph), who differ from most other primates because they did not remain in the trees but instead adapted to a terrestrial way of life.

Fossil evidence has allowed **paleoanthropologists,** scientists who study human evolution, to infer not only the structure but also the habits of early humans and other primates. Teeth and bones are the main fossil evidence studied by paleoanthropologists. Much information can be obtained by studying teeth, which have changed dramatically during the course of primate and human evolution. Because tooth enamel is more mineralized (harder) than bone, teeth are more likely to fossilize. The teeth of each primate species, living or extinct, are distinctive enough to identify the species, approximate age, diet, and even sex of the individual. Consider *Australopithecus robustus,* which lived in southern Africa about 2 mya, at a time when the climate was becoming more arid and the forests were giving way to grasslands. Its large jaws and molars (broad-ridged teeth in the back of the mouths of adult mammals) indicate its diet included tough foods such as roots, tubers, and seeds. ■

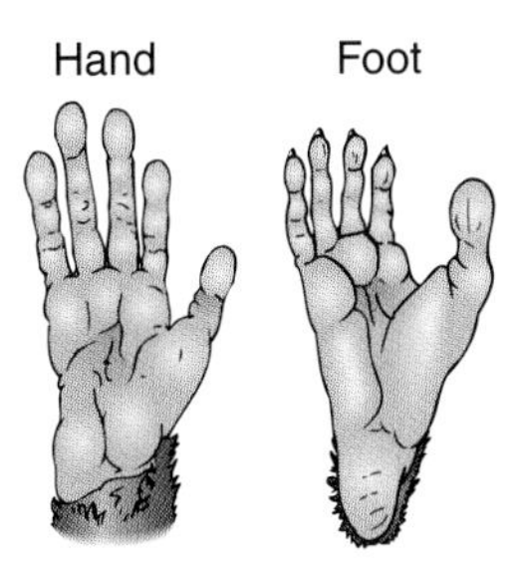

(a) Lemur *(Eulemur mongoz)*

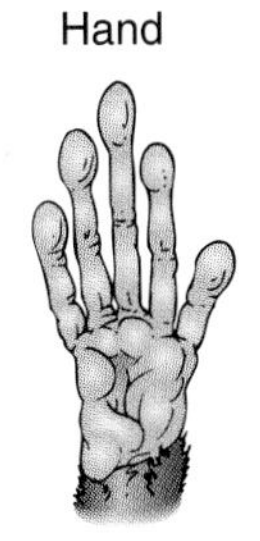

(b) Tarsier *(Tarsius spectrum)*

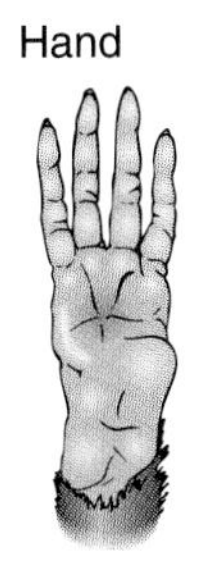
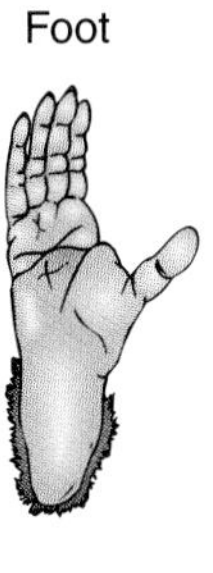
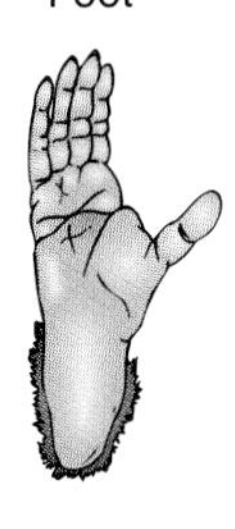

(c) Woolly spider monkey *(Brachyteles arachnoides)*

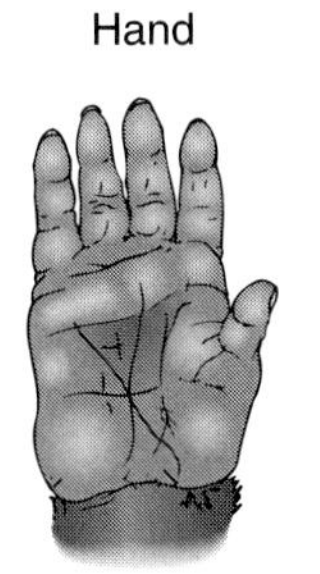
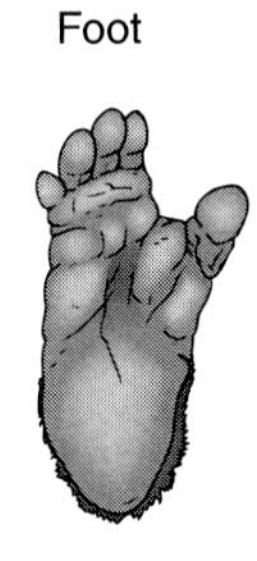

(d) Gorilla *(Gorilla gorilla)*

FIGURE **21-1** **Right hands and feet of selected primates.**
Primates have five grasping digits, and the thumb or big toe is often partially or fully opposable. **(a)** Lemur. **(b)** Tarsier. **(c)** Woolly spider monkey. **(d)** Gorilla. *(Figures not drawn to scale.) (Adapted from A.H. Schultz,* The Life of Primates, *Weidenfeld & Nicholson, London, 1969)*

PRIMATE ADAPTATIONS

Learning Objective

1 Describe the structural adaptations that primates have for life in treetops.

Fossil evidence indicates that the first primates with traits characteristic of modern primates appeared by the early Eocene epoch about 55 mya. These early primates had digits with nails, and their eyes were directed somewhat forward on the head. The climate was mild then, and early primates were widely distributed over much of North America, Europe, and Asia. (Recall from Chapter 17 that North America was still attached to Europe at that time.) As the climate became cooler and drier toward the end of the Eocene epoch, many of these early primates became extinct.

Several novel adaptations evolved in early primates that allowed them to live in trees. One of the most significant primate features is that they have five highly flexible digits: four lateral digits (fingers) plus a partially or fully opposable first digit (thumb and, in many primates, big toe; Fig. 21-1). An **opposable thumb** positions the fingers opposite the thumb, enabling primates to grasp objects such as tree branches with precision. Nails (instead of claws) provide a protective covering for the tips of the digits, and the fleshy pads at the ends of the digits are sensitive to touch. Another arboreal feature is long, slender limbs that rotate freely at the hips and shoulders, giving primates full mobility to climb and search for food in the treetops.

Having eyes located in the front of the head lets primates integrate visual information from both eyes simultaneously; they have *stereoscopic* (three-dimensional) vision. Stereoscopic vision is vital in an arboreal environment, especially for species that leap from branch to branch, because an error in depth perception may cause a fatal fall. In addition to sharp sight, primates hear acutely.

Primates share several other characteristics, including a relatively large brain size. Biologists have suggested that the increased sensory input associated with their sharp vision and greater agility favored the evolution of larger brains. Primates are generally very social and intelligent animals that reach sexual maturity relatively late in life. They typically have long life spans. Females usually bear one offspring at a time; the baby is helpless and requires a long period of nurturing and protection.

Review

- How are primate hands and feet adapted to an arboreal existence?
- Why is the location of primate eyes in front of the head an important adaptation for an arboreal existence?

BiologyNow™ Assess your understanding of **primate adaptations** by taking the pretest on your BiologyNow CD-ROM.

PRIMATE CLASSIFICATION

Learning Objectives

2 List the three suborders of primates, and give representative examples of each.

3 Distinguish among anthropoids, hominoids, and hominids.

4 Describe skeletal and skull differences between apes and hominids.

Now that we have surveyed the general characteristics of primates, let's look at how they are classified. Many biologists currently divide the order Primates into three groups, or suborders (Fig. 21-2). The suborder Prosimii includes lemurs, galagos, and lorises, the suborder Tarsiiformes includes tarsiers, and the suborder Anthropoidea includes **anthropoids** (monkeys, apes, and humans).

All lemurs are restricted to the island of Madagascar off the coast of Africa. Because of extensive habitat destruction and hunting, they are very endangered. Lorises, which live in tropical areas of Southeast Asia and Africa, resemble lemurs in many respects, as do galagos, which live in sub-Saharan Africa. Lemurs, lorises, and galagos have retained several early mammalian features, such as elongated, pointed faces.

Tarsiers are found in rain forests of Indonesia and the Philippines (Fig. 21-3). They are small primates (about the size of a small rat) and are very adept leapers. These nocturnal primates resemble anthropoids in a number of ways, including their shortened snouts and forward-pointing eyes.

Suborder Anthropoidea includes monkeys, apes, and humans

Anthropoid primates arose during the middle Eocene epoch, at least 45 mya. Several different fossil anthropoids have been identified, from Asia and North Africa, and there is a growing consensus about the relationships of these fossil groups to one

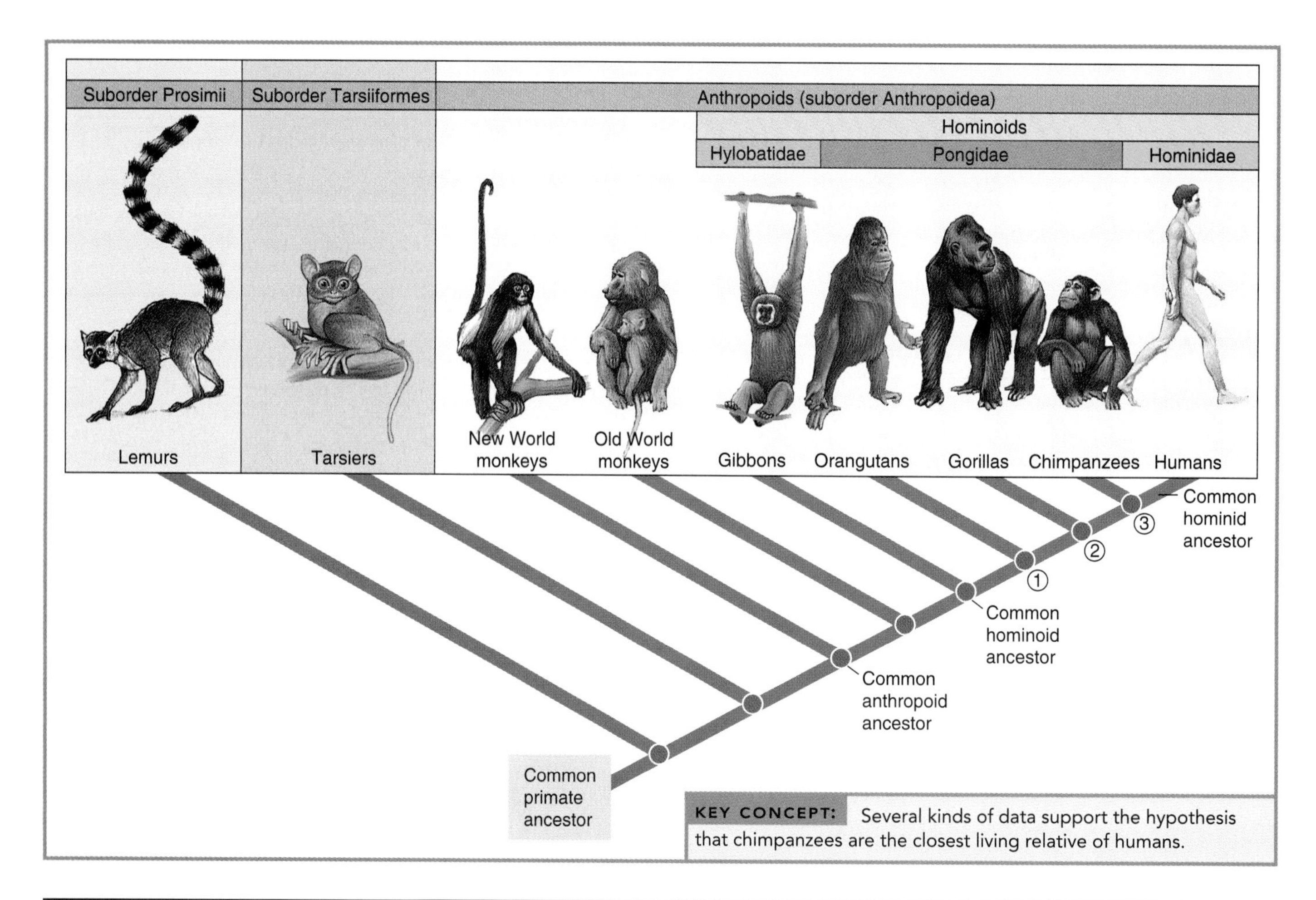

FIGURE **21-2** **Primate evolution.**

This branching diagram, called a cladogram, shows evolutionary relationships among living primates, based on current scientific evidence. The nodes *(circles)* represent branch points where a species splits into two or more lineages. ① The divergence of orangutans from the ape/hominid line occurred some 12 to 16 mya. ② Gorillas separated from the chimpanzee/hominid line an estimated 6 to 8 mya, and ③ the hominid (human) lineage diverged from that of chimpanzees about 4 to 6 mya. *(Figures not drawn to scale.)*

Frans Lanting/Minden Pictures

FIGURE **21-3** **Tarsiers.**

The huge eyes of the tarsier (*Tarsius bancanus*) help it find insects, lizards, and other prey when it hunts at night. When a tarsier sees an insect, it pounces on it and grasps the prey with its hands. Tarsiers live in the rain forests of Indonesia and the Philippines.

another and to living anthropoids. Evidence indicates that anthropoids originated in Africa or Asia. The oldest known anthropoid fossils, such as 42-million-year-old *Eosimias,* have been found in China and Myanmar. Given details about their dentition and the few bones that have been discovered, scientists infer that *Eosimias* and other ancestral anthropoids were small, insect-eating arboreal primates that were active during the day. Once they arose, anthropoids quickly spread throughout Europe, Asia, and Africa and arrived in South America much later. (Paleoanthropologists date the oldest known South American primate, *Branisella,* from Bolivia, at 26 million years.)

One significant difference between anthropoids and other primates is in the size of their brains. The cerebrum, in particular, is more developed in monkeys, apes, and humans, where it functions as a highly complex center for learning, voluntary movement, and interpretation of sensation.

Monkeys are generally diurnal (active during the day) tree dwellers. They tend to eat fruit and leaves, with nuts, seeds, buds, insects, spiders, birds' eggs, and even small vertebrates playing a smaller part in their diets. The two main groups of monkeys, New World monkeys and Old World monkeys, are named for the hemispheres where they diversified. Monkeys in South and Central America are called New World monkeys, whereas monkeys in Africa, Asia, and Europe are called Old World monkeys. New and Old World monkeys have been evolving separately for tens of millions of years.

One of the most important unanswered questions in anthropoid evolution concerns *how* monkeys arrived in South America. Africa and South America had already drifted apart (see Chapter 17), so the ancestors of New World monkeys may have rafted from Africa to South America on floating masses of vegetation. The South Atlantic Ocean would have been about half as wide as it is today, and islands could have provided "stepping stones." Alternatively, the ancestors of New World monkeys may have dispersed from Asia to North America to South America. Once established in the New World, these monkeys rapidly diversified.

New World monkeys are restricted to Central and South America and include marmosets, capuchins, howler monkeys, squirrel monkeys, and spider monkeys. New World monkeys are arboreal, and some have long, slender limbs that permit easy movement in the trees (Fig. 21-4a). A few have **prehensile tails** capable of wrapping around branches and serving as fifth limbs. Some New World monkeys have shorter thumbs, and in certain

FIGURE **21-4** **New and Old World monkeys.**

(a) The white-faced monkey (*Cebus capucinus*) is a New World monkey that has a prehensile tail and a flattened nose with nostrils directed to the side. **(b)** The Anubis baboon (*Papio anubis*) is an Old World monkey native to Africa. Note that its nostrils are directed downward.

C.C. Lockwood/DRK Photo

(a)

S. Meyers/Okapia/Photo Researchers, Inc.

(b)

cases the thumbs are totally absent. Their facial anatomy differs from that of the Old World monkeys; they have flattened noses with the nostrils opening to the side. They live in groups and perform complex social behaviors.

Old World monkeys are distributed in tropical parts of Africa and Asia. In addition to baboons and macaques (pronounced muh-kacks´), the Old World group includes guenons, mangabeys, langurs, and colobus monkeys. Most Old World monkeys are arboreal, although some, such as baboons and macaques, spend much of their time on the ground (Fig. 21-4b). The ground dwellers, which are **quadrupedal** ("four-footed"; they walk on all fours), arose from arboreal monkeys. None of the Old World monkeys has a prehensile tail, and some have extremely short tails. They have a fully opposable thumb, and unlike the New World monkeys, their nostrils are closer together and directed downward. Old World monkeys are intensely social animals.

Many classification schemes place apes and humans in three families

Old World monkeys shared a common ancestor with the **hominoids,** a group composed of apes and **hominids** (humans and their ancestors). A fairly primitive anthropoid, discovered in Egypt, was named *Aegyptopithecus* (Fig. 21-5a). A cat-sized, forest-dwelling arboreal monkey with a few apelike characteristics, *Aegyptopithecus* lived during the Oligocene epoch, approximately 34 mya.

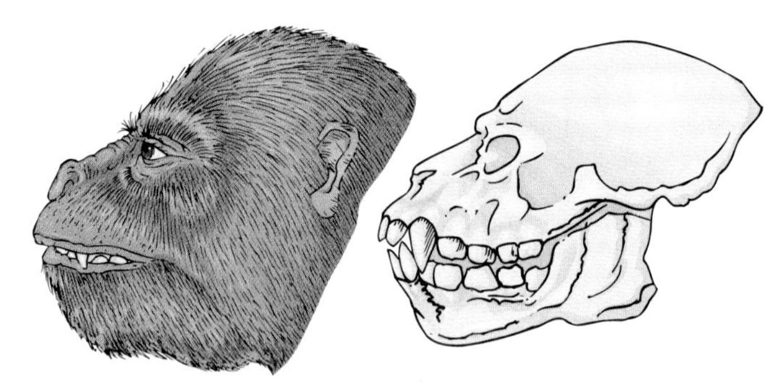

(a) Oligocene anthropoid, *Aegyptopithecus*

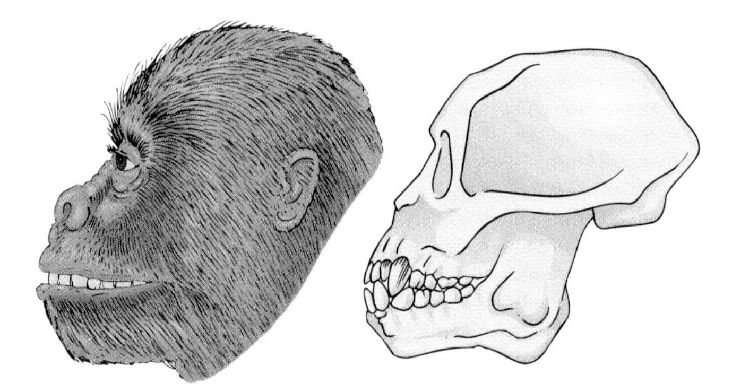

(b) Miocene ape, *Dryopithecus*

FIGURE **21-5** *Aegyptopithecus and Dryopithecus.*
(a) Fossils of *Aegyptopithecus*, a fairly primitive anthropoid, were discovered in Egypt. **(b)** *Dryopithecus*, a more advanced ape, may have given rise to modern hominoids. *(Figures not drawn to scale.)*

During the Miocene epoch, approximately 20 mya, apes and Old World monkeys diverged. Paleoanthropologists discovered the oldest fossils with hominoid features in East Africa, mostly in Kenya. *Proconsul,* for example, appeared early in the Miocene epoch, about 20 mya. It had a larger brain than monkeys, apelike teeth and diet (fruits), but a monkey-like body. At least 30 other early hominoid species lived during the Miocene epoch, but most of them became extinct and were not the common ancestor of modern apes and humans.

Miocene fossils of forest-dwelling, chimpanzee-sized apes called **dryopithecines,** which lived about 15 mya, are of special interest because this hominoid lineage may have given rise to modern apes as well as to the human line (Fig. 21-5b). The dryopithecines, such as *Dryopithecus, Kenyapithecus,* and *Morotopithecus,* were distributed widely across Europe, Africa, and Asia. As the climate gradually cooled and became drier, their range became more limited. These apes had highly modified bodies for swinging through the branches of trees, although there is also evidence that some of them may have left the treetops for the ground as dense forest gradually changed into open woodland. Many questions about the relationships among the various early apes have been generated by the discovery of these and other Miocene hominoids. As future fossil finds are evaluated, they may lead to a rearrangement of ancestors in the hominoid family tree.

Many biologists classify the five genera of hominoids alive today into three families: Gibbons (*Hylobates*) are known as *lesser apes* and are placed in the family Hylobatidae. The family Pongidae includes orangutans (*Pongo*), gorillas (*Gorilla*), and chimpanzees (*Pan*), and the family Hominidae includes humans (*Homo*). Recent molecular evidence indicates a close relationship between humans and the greater apes, particularly chimpanzees, and some scientists now classify them in the same family.

Gibbons are natural acrobats that can **brachiate,** or arm-swing, with their weight supported by one arm at a time (Fig. 21-6a). Orangutans are also tree dwellers, but chimpanzees and especially gorillas have adapted to life on the ground (Fig. 21-6b–d). Gorillas and chimpanzees have retained long arms typical of brachiating primates but use these to assist in quadrupedal walking, sometimes known as **knuckle-walking** because of the way they fold (flex) their digits when moving. Apes, like humans, lack tails. They are generally much larger than monkeys, although gibbons are a notable exception.

Evidence of the close relatedness of orangutans, gorillas, chimps, and humans is abundant at the molecular level. The amino acid sequence of the chimpanzee's hemoglobin is identical to that of the human; hemoglobin molecules of the gorilla and rhesus monkey differ from the human's by 2 and 15 amino acids, respectively. DNA sequence analyses indicate that chimpanzees are likely to be our nearest living relatives among the apes (see Table 17-1). Molecular evidence suggests that orangutans may have diverged from the gorilla, chimpanzee, and hominid lines about 12 to 16 mya. Gorillas may have diverged from the chimpanzee and hominid lines some 6 to 8 mya, whereas chimpanzee and hominid lines probably separated about 4 to 6 mya.

FIGURE **21-6** **Apes.**

(a) A mother white-handed gibbon (*Hylobates lar*) nurses her baby. Gibbons are extremely acrobatic and often move through the trees by brachiation. **(b)** An orangutan (*Pongo pygmaeus*) mother and baby. **(c)** A young lowland gorilla (*Gorilla gorilla*) in knuckle-walking stance. **(d)** A bonobo chimpanzee (*Pan paniscus*) grooms another member of the group. Bonobos are endemic to a single country, the Democratic Republic of Congo (formerly Zaire).

Joe McDonald/Visuals Unlimited

(a)

BIOS/Peter Arnold, Inc.

(b)

Nancy Adams/Tom Stack & Associates

(c)

K. & K. Ammann/Bruce Coleman, Inc.

(d)

Review

- How can you distinguish between anthropoids and hominoids?
- How can you distinguish between hominoids and hominids?
- How do the skulls of apes and humans differ?
- How does an ape skeleton differ from a human skeleton?

BiologyNow™ Assess your understanding of **primate classification** by taking the pretest on your BiologyNow CD-ROM.

HOMINID EVOLUTION

Learning Objectives

5 Compare the following early hominids: *Sahelanthropus tchadensis, Ardipithecus ramidus, Australopithecus anamensis, Australopithecus afarensis,* and *Australopithecus africanus.*

6 Distinguish among the following members of genus *Homo: Homo habilis, Homo ergaster, Homo erectus, Homo heidelbergensis, Homo neanderthalensis,* and *Homo sapiens.*

7 Discuss the current debate over the origin of modern humans, and briefly describe the opposing out-of-Africa and multiregional hypotheses.

PROCESS OF SCIENCE

Scientists have a growing storehouse of hundreds of hominid fossils, which provide useful data about general trends in the body design, appearance, and behavior of ancestral humans. For example, before their brains enlarged early hominids clearly adopted a **bipedal** (two-footed) posture. Despite the wealth of fossil evidence, interpretations of hominid characteristics, taxonomy, and phylogeny continue to be vigorously debated, and new discoveries raise new questions. As in other scientific fields, ideas about hominid evolution are influenced by the different perspectives of the various workers studying it. The lack of scientific consensus regarding certain aspects of hominid evolution is, therefore, an expected part of the scientific process.

Evolutionary changes from the earliest hominids to modern humans are evident in some of the characteristics of the skeleton and skull. Compared with the ape skeleton, the human skeleton shows distinct differences that reflect human's ability to stand erect and walk on two feet (Fig. 21-7). These differences

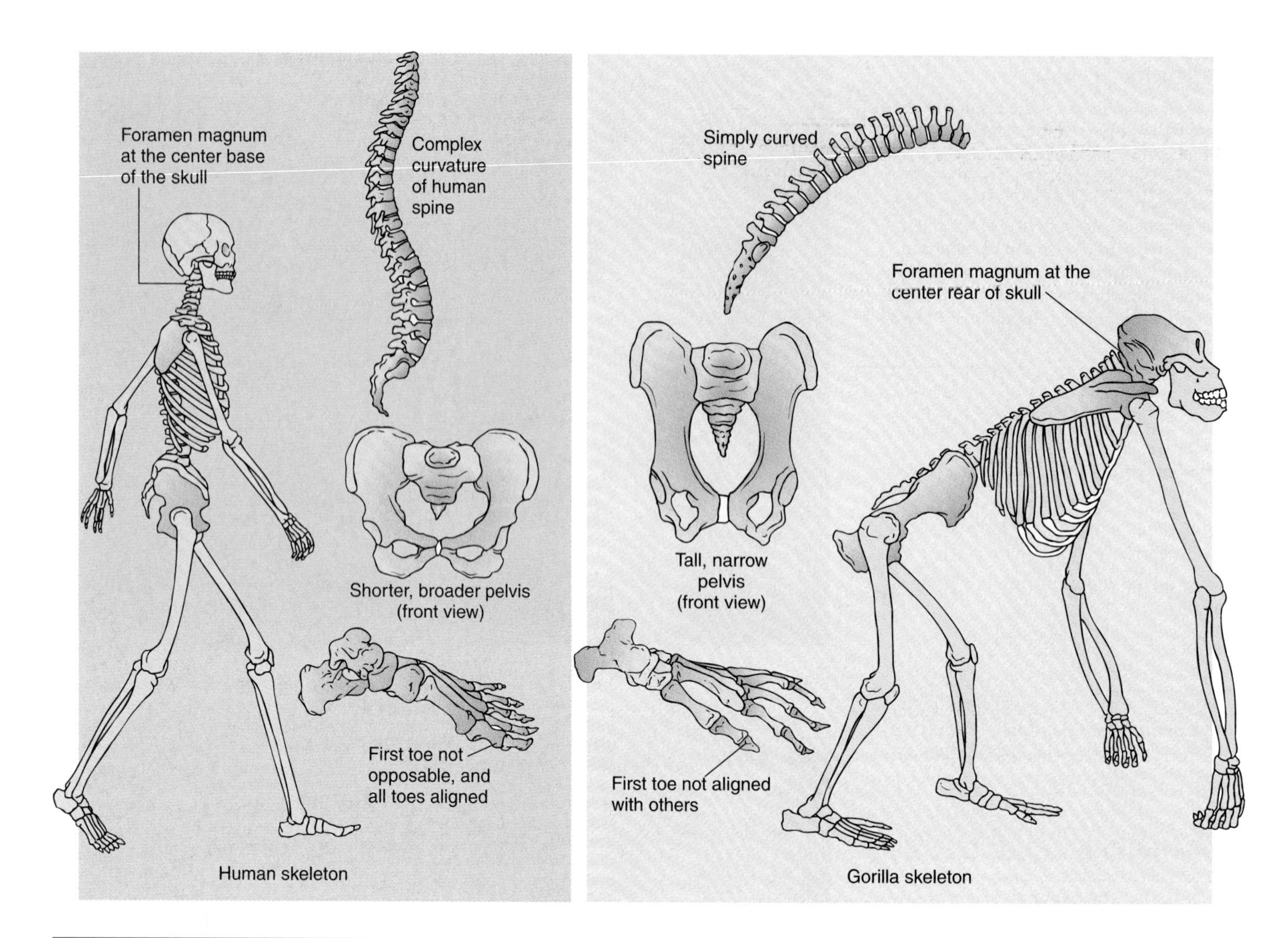

FIGURE **21-7** | **Gorilla and human skeletons.**

When gorilla and human skeletons are compared, the skeletal adaptations for bipedalism in humans become apparent.

also reflect the habitat change for early hominids, from an arboreal existence in the forest to a life spent at least partly on the ground. The curvature of the human spine provides better balance and weight distribution for bipedal locomotion. The human pelvis is shorter and broader than the ape pelvis, allowing better attachment of muscles used for upright walking. In apes the **foramen magnum,** the hole in the base of the skull for the spinal cord, is located in the middle of the rear of the skull. In contrast, the human foramen magnum is centered in the skull base, positioning the head for erect walking. An increase in the length of the legs relative to the arms, and alignment of the big toe with the rest of the toes, further adapted the early hominids for bipedalism.

Another major trend in hominid evolution was an increase in brain size relative to body size (Fig. 21-8). The ape skull has prominent bony ridges above the eye sockets, whereas modern human skulls lack these **supraorbital ridges.** Human faces are flatter than those of apes, and the jaws are different. The arrangement of teeth in the ape jaw is somewhat rectangular, compared with a rounded, or U-shaped, arrangement in humans. Apes have larger front teeth (canines and incisors) than do humans, and their canines are especially large. Gorillas and orangutans also have larger back teeth (premolars and molars) than humans.

We now examine some of the increasing number of generally recognized fossil hominids in the human lineage. As you read the following descriptions of human evolution, keep in mind that much of what is discussed is still open to interpretation and major revision as additional discoveries are made. It is also important to remember that, although we present human evolution in a somewhat linear fashion, from ancient hominids to anatomically and behaviorally modern humans, the human family tree is not a single trunk but has several branches (Fig. 21-9). *Homo sapiens* is the only species of hominid in existence today, but more than one hominid species coexisted at any given time for most of the past 4 million years. In addition, don't make the mistake of thinking that your smaller-brained ancestors were inferior to yourself. Ancestral hominids were evolutionarily successful in that they were well adapted to their environment and survived for millions of years.

The earliest hominid may belong to the genus *Sahelanthropus*

Hominid evolution began in Africa. Although most hominid fossils have been discovered in East Africa, in 2002 French paleon-

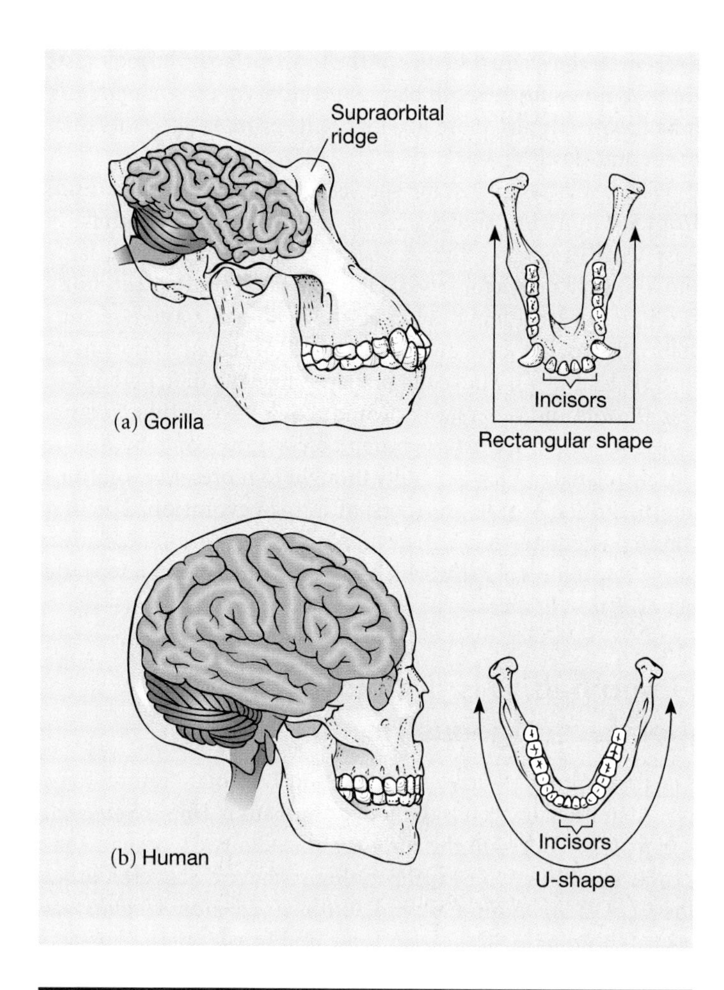

ACTIVE FIGURE **21-8** **Gorilla and human heads.**

(a) The ape skull has a pronounced supraorbital ridge. **(b)** The human skull is flatter in the front and has a pronounced chin. The human brain, particularly the cerebrum *(purple)*, is larger than that of an ape, and the human jaw is structured so that the teeth are arranged in a U shape. Human canines and incisors are also smaller than those of apes.

BiologyNow™ Learn more about **monkey, gorilla, and human skeletons** by clicking on this figure on your BiologyNow CD-ROM.

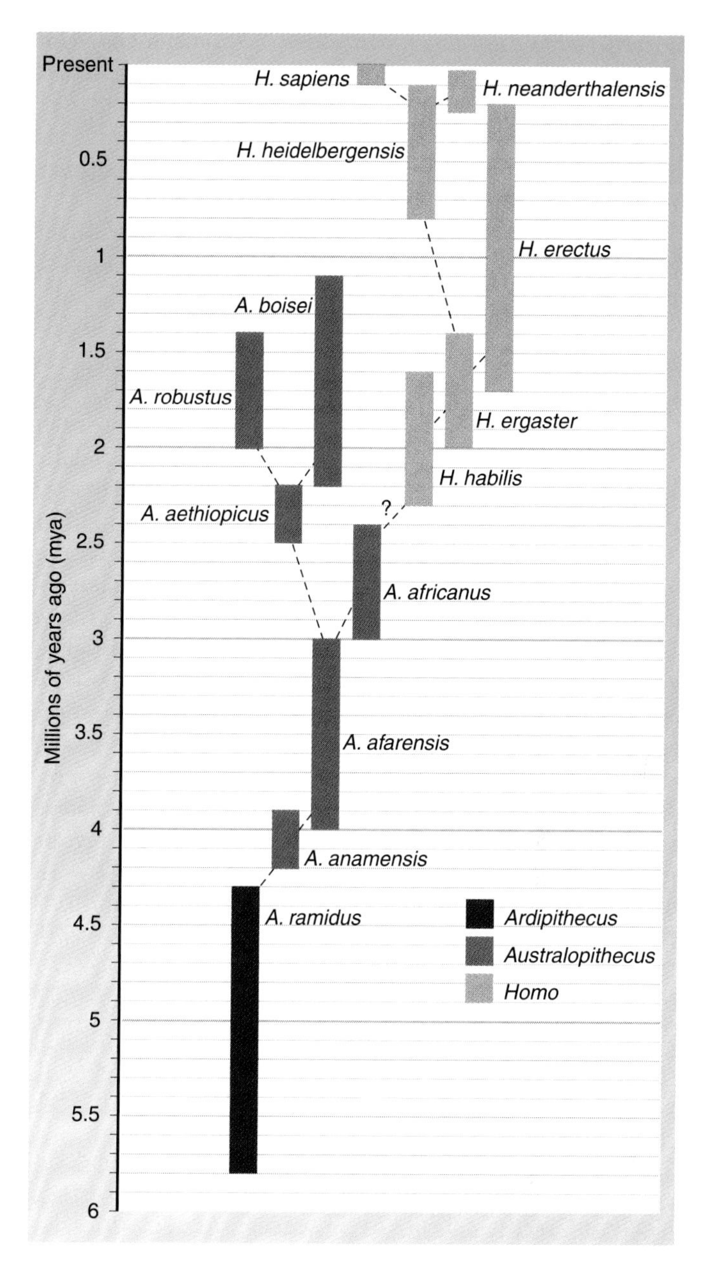

FIGURE **21-9** **One interpretation of hominid evolution.**

Dashed lines show possible evolutionary relationships. Paleoanthropologists do not completely agree about certain specific details of the human family tree and hold many possible interpretations of classification and lines of descent.

tologist Michel Brunet and an international team made a stunning discovery in a dry lakebed in Chad, which is in central Africa. The fossil, which has been reliably dated at 6 to 7 million years old, may be the earliest known hominid. Viewed from the back, the skull of ***Sahelanthropus tchadensis,*** with its small brain case, resembles a chimpanzee. However, viewed from the front the face and teeth have many characteristics of larger-brained human ancestors. Most paleoanthropologists place *Sahelanthropus* close to the base of the human family tree—that is, close to the last common ancestor of hominids and chimpanzees. This discovery is important in its own right, but it is also significant because it shows that early hominids had more variation and lived over a larger area of Africa than had been originally hypothesized.

As with most aspects of human evolution, scientists differ in interpreting *Sahelanthropus.* Some paleoanthropologists hypothesize that *Sahelanthropus* was a forerunner of modern apes, specifically the gorilla, and not one of the earliest humans. This point remains controversial, largely because there are currently no skeletal bones to indicate if *Sahelanthropus* walked upright, a hallmark characteristic of hominids. Future fossil discoveries may clarify this important issue.

Australopithecines are the immediate ancestors of genus *Homo*

After *Sahelanthropus,* the next oldest hominid belongs to the genus *Ardipithecus,* which appeared more than 5 mya. *Ardipithe-*

cus gave rise to *Australopithecus,* a genus that includes several species that lived between 4 and 1 mya. *Ardipithecus* and *Australopithecus* were two genera of early hominids, often referred to as **australopithecines,** or "southern man apes." Both *Ardipithecus* and *Australopithecus* had longer arms, shorter legs, and smaller brains relative to modern humans.

The actual number of australopithecine species for which fossil evidence has been found is under debate. In some cases, differences in the relatively few skeletal fragments could indicate either variation among individuals within a species or evidence of separate species. Most paleoanthropologists recognize at least six species of australopithecines.

To date, two sets of fossils of ***Ardipithecus ramidus*** have been reported, one in 1992 that was dated at 4.4 million years, and the other in 2001 that was dated at 5.2 to 5.8 million years. The specific epithet *ramidus* is derived from a word meaning "root" in the Afar language, spoken in the region of Ethiopia where the fossils were found. Although not as primitive as *Sahelanthropus, Ardipithecus* is close to the "root" of the human family tree.

Hominids that existed between 3.9 and 4.2 mya are assigned to the species ***Australopithecus anamensis,*** first named in 1995 by paleoanthropologist Meave Leakey (1942–) and her coworkers from fossils discovered in East Africa. This hominid species, which has a mixture of apelike and human-like features, presumably arose from *Ardipithecus ramidus.* A comparison of male and female *A. anamensis* body sizes and canine teeth reveals **sexual dimorphism,** marked phenotypic differences between the two sexes of the same species. (The modern-day gorilla is sexually dimorphic.) The back teeth and jaws of *A. anamensis* are larger than those of modern chimpanzees, whereas the front teeth are smaller and more like those of later hominids. A fossil leg bone, the tibia, indicates that *A. anamensis* had an upright posture and was bipedal, although it also may have foraged in the trees. Thus, bipedalism occurred early in human evolution and may have been the first human adaptation.

Australopithecus afarensis, another primitive hominid, appears to have arisen directly from *A. anamensis.* Many fossils of *A. afarensis* skeletal remains have been discovered in Africa, including a remarkably complete skeleton nicknamed Lucy found in Ethiopia in 1974 by a team led by American paleoanthropologist Donald Johanson. Lucy, a small hominid approximately 1.04 m (3 ft, 5 in) is thought to be about 3.2 million years old. In 1978, British paleoanthropologist Mary Leakey (1913–1996) and coworkers discovered beautifully preserved fossil footprints of three *A. afarensis* individuals who walked more than 3.6 mya. In 1994 the first adult skull of *A. afarensis* was found. The skull, characterized by a relatively small brain, pronounced supraorbital ridges, a jutting jaw, and large canine teeth, is an estimated 3 million years old. It is probable that *A. afarensis* did not construct tools or make fires, because no evidence of tools or fire has been found at fossil sites.

Many paleoanthropologists think *A. afarensis* gave rise to several australopithecine species, including ***Australopithecus africanus,*** which may have appeared as early as 3 mya. The first *A. africanus* fossil was discovered in South Africa in 1924, and since then hundreds have been found. This hominid walked erect and possessed hands and teeth that were distinctly human-like. Given the characteristics of the teeth, paleoanthropologists think *A. africanus* ate both plants and animals. Like *A. afarensis,* it had a small brain, more like that of its primate ancestors than of present-day humans.

Three australopithecine species (*Australopithecus robustus* from South Africa, and *A. aethiopicus* and *A. boisei,* both from East Africa) are larger than *A. africanus* and have extremely large molars, very powerful jaws, relatively small brains, and bony skull crests. Most females lacked the skull crests and had substantially smaller jaws, another example of sexual dimorphism in early hominids. The teeth and jaws suggest a diet, perhaps of tough roots and tubers, that would require powerful grinding. These so-called *robust australopithecines* may or may not be closely related but are generally thought to represent evolutionary offshoots, or side branches, of human evolution. The first robust australopithecine, *A. aethiopicus,* appeared about 2.5 mya. Some researchers classify robust australopithecines in a separate genus, *Paranthropus.*

Homo habilis is the oldest member of genus *Homo*

The first hominid to have enough uniquely human features to be placed in the same genus as modern humans is ***Homo habilis.*** It was first discovered in the early 1960s at Olduvai Gorge in Tanzania. Since then paleoanthropologists have discovered other fossils of *H. habilis* in East and South Africa. *Homo habilis* was a small hominid with a larger brain and smaller premolars and molars than the australopithecines. This hominid appeared approximately 2.3 mya and persisted for about 0.75 million years. Fossils of *H. habilis* have been found in numerous areas in Africa. These sites contain primitive tools, stones that had been chipped, cracked, or hammered to make sharp edges for cutting or scraping.[1] *Oldowan* pebble choppers and flakes, for example, were probably used to cut through animal hides to obtain meat and to break bones for their nutritious marrow.

The relationship between the australopithecines and *H. habilis* is not clear. Using physical characteristics of their fossilized skeletons as evidence, many paleoanthropologists have inferred that the australopithecines were ancestors of *H. habilis.* Some researchers do not think *H. habilis* belongs in the genus *Homo,* and they suggest it should be reclassified as *Australopithecus habilis.* Discoveries of additional fossils may help clarify these relationships.

Homo erectus apparently evolved from *Homo habilis*

Investigators found the first fossil evidence of ***Homo erectus*** in Indonesia in the 1890s. Since then, searchers have found numerous fossils of *H. erectus* throughout Africa and Asia. Paleoanthro-

[1] Investigators found the oldest known stone tools in the mid-1990s in Gona, Ethiopia. These ancient tools were made some 2.6 mya, but because searchers have found no hominid remains at the site yet, no one knows who made the tools.

pologists think *Homo erectus* originated in Africa about 1.7 mya and then spread quickly to Europe and Asia. Peking man and Java man, discovered in Asia, were later examples of *H. erectus,* which existed until at least 200,000 years ago; some populations of *H. erectus* may have persisted more recently.

Homo erectus was taller than *H. habilis.* Its brain, which was larger than that of *H. habilis,* got progressively larger during the course of its evolution. Its skull, although larger, did not possess totally modern features, retaining the heavy supraorbital ridge and projecting face that are more characteristic of its ape ancestors (Fig. 21-10). *Homo erectus* is the first hominid to have fewer differences between the sexes.

The increased intelligence associated with an increased brain size enabled these early humans to make more advanced stone tools, known as *Acheulean* tools, including hand axes and other implements that scientists have interpreted as choppers, borers, and scrapers. Their intelligence also allowed these humans to survive in cold areas. *Homo erectus* obtained food by hunting or scavenging and may have worn clothing, built fires, and lived in caves or shelters. Evidence of weapons (spears) has been unearthed at *Homo erectus* sites in Europe.

Ideas regarding *Homo erectus,* like many other aspects of human evolution, are changing with each new fossil discovery. Many scientists hypothesize that the fossils classified as *H. erectus* really represent two species, ***Homo ergaster,*** an earlier African species, and *H. erectus,* a later East Asian offshoot. The best known fossils of *H. ergaster* come from the Lake Turkana region in Kenya. Researchers who support this split speculate that *Homo ergaster* may be the direct ancestor of later humans, whereas *Homo erectus* may be an evolutionary dead end.

Other paleoanthropologists do not think that *H. erectus* should be split into two species. They cite evidence of a 1-million-year-old *H. erectus* fossil that was discovered in Africa in 2002. This fossil shares striking similarities with Asian *H. erectus* fossils. It is a significant discovery because it enables paleoanthropologists to compare *H. erectus* fossils from the same period but from two different places, Africa and Asia. Scientists hope that further analysis of this fossil and of future fossil discoveries will clarify the status of *Homo erectus.*

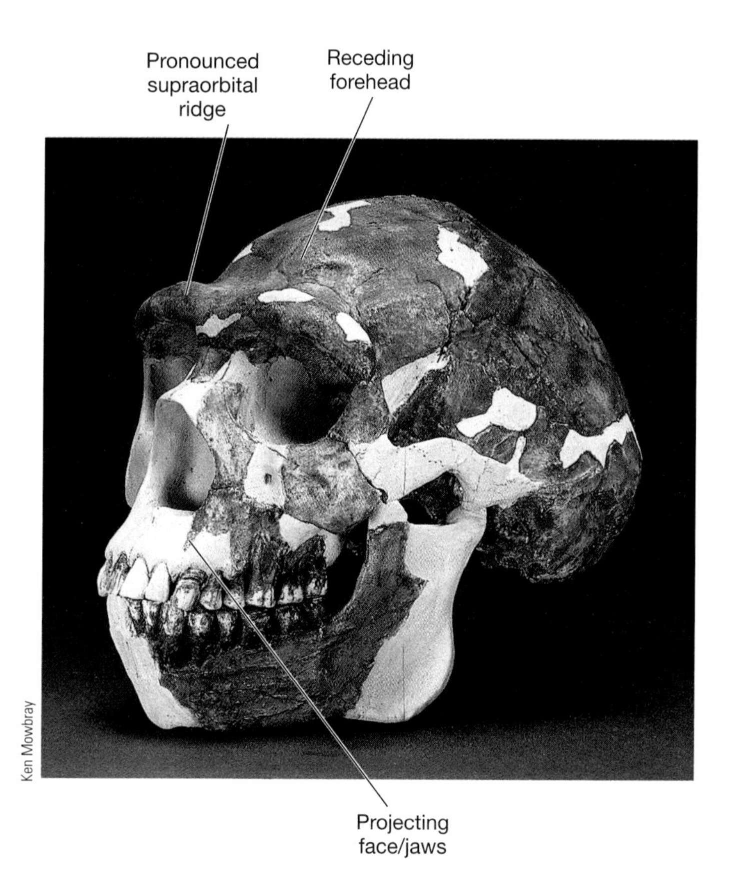

FIGURE **21-10** ***Homo erectus* skull from China.**

The reconstructed portions are white. Note the receding forehead, pronounced supraorbital ridge, and projecting face and jaws.

Archaic *Homo sapiens* appeared about 800,000 years ago

Archaic *Homo sapiens* are regionally diverse descendants of *Homo erectus* or *Homo ergaster* that lived in Africa, Asia, and Europe from about 800,000 to 100,000 years ago. They thus overlapped both *H. erectus* populations and the later appearing Neandertals (discussed in the next section). Some researchers classify archaic *Homo sapiens* as a separate species, ***Homo heidelbergensis.***

Neandertals appeared approximately 230,000 years ago

Fossils of **Neandertals**[2] were first discovered in the Neander Valley in Germany. They lived throughout Europe and western Asia from about 230,000 to 30,000 years ago. These early humans had short, sturdy builds. Their faces projected slightly, their chins and foreheads receded, they had heavy supraorbital ridges and jawbones, and their brains and front teeth were larger than those of modern humans. Their nasal cavities were large, and their cheekbones were receding. Scientists have suggested that the large noses provided larger surface areas in Neandertal sinuses, enabling them to better warm the cold air of Ice Age Eurasia as inhaled air traveled through the head to the lungs.

Scientists have not reached a consensus about whether the Neandertals are a separate species from modern humans. Many think the anatomical differences between Neandertals and modern humans mean that they were separate species, ***Homo neanderthalensis*** and ***Homo sapiens.*** Other scientists disagree and think that Neandertals were a group of *Homo sapiens.*

Neandertal tools, known as *Mousterian* tools, include the oldest known spear points (Fig. 21-11). Neandertal tools were more sophisticated than those of *H. erectus.* Studies of Neandertal sites indicate that they hunted large animals. The existence of skeletons of elderly Neandertals and of Neandertals with healed fractures may demonstrate that they cared for the aged

[2] Neandertal was formerly spelled "Neanderthal." The silent "h" has been dropped in modern German but not in the scientific name.

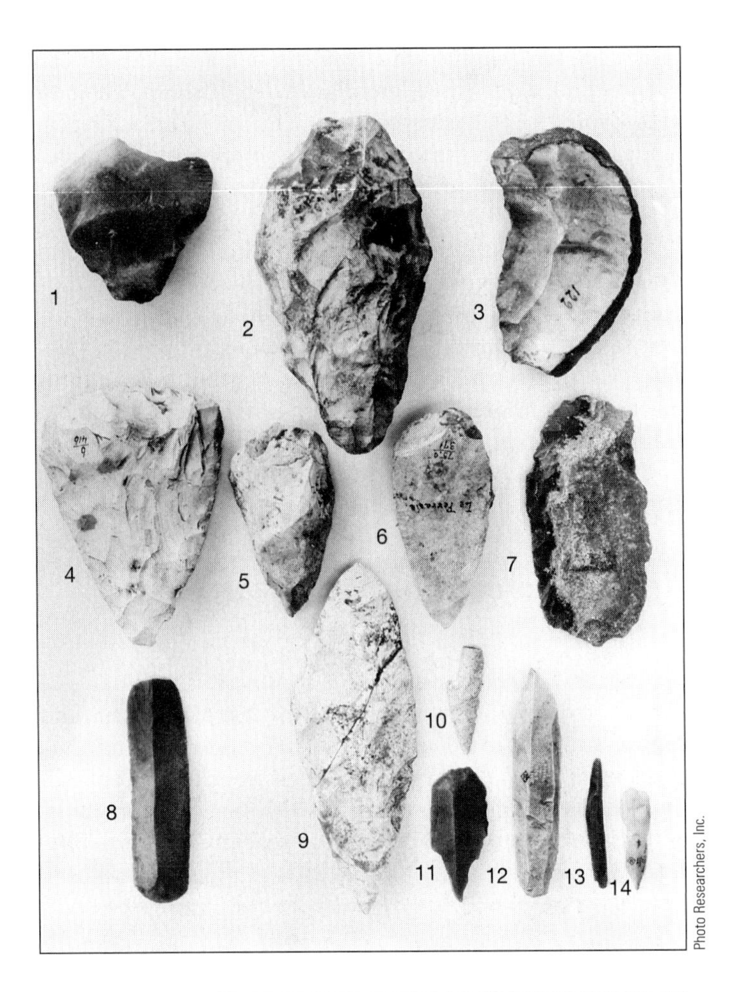

FIGURE **21-11** | **Mousterian tools.**

Paleoanthropologists named Mousterian tools after a Neandertal site in Le Moustier, France. Mousterian tools included a variety of skillfully made stone tools, such as hand axes, flakes, scrapers, borers, and spear points. (*1* to *7* are earlier tools, and *8* to *14* are later tools.)

and the sick, an indication of advanced social cooperation. They apparently had rituals, possibly of religious significance, and sometimes buried their dead.

PROCESS OF SCIENCE

The disappearance of the Neandertals some 30,000 years ago is a mystery that has sparked debate among paleoanthropologists. Other groups of *H. sapiens* with more modern features coexisted for tens of thousands of years with the Neandertals. Perhaps the other humans outcompeted or exterminated the Neandertals, leading to their extinction. It is also possible that the Neandertals interbred with these humans, diluting their features beyond recognition.

Analysis of **mitochondrial DNA (mtDNA)** contributes useful data for such controversies. Each of the several hundred mitochondria within a cell has about 10 copies of a small loop of DNA that codes for transfer RNAs, ribosomal RNAs, and certain respiratory enzymes. MtDNA is only transmitted through the maternal line because eggs, not sperm, contribute mitochondria. Mitochondrial DNA mutates more rapidly than nuclear DNA, so mtDNA is a sensitive indicator of evolution. Researchers extracted the mtDNA from a Neandertal bone and evaluated it. It differs significantly from all modern human mtDNA sequences, although it is more similar to human than to chimpanzee mtDNA. This finding suggests that Neandertals are an evolutionary dead end that did not interbreed with more modern humans.

The question of the relationship between Neandertals and anatomically modern humans remains controversial, however. In 1999 researchers reported the discovery of a 4-year-old child's remains in Portugal. The skeleton, dated at 24,500 years of age, has traits of both modern humans and Neandertals (short lower limb bones). The child lived several thousand years after Neandertals are thought to have disappeared, and the researchers who discovered it view it as an example of mixed ancestry, meaning that Neandertals and anatomically modern humans are members of the same species who interbred freely. Other scientists disagree with their interpretation and think the so-called Neandertal features of the child may reflect normal variation inherent in the human species.

Biologists debate the origin of modern *Homo sapiens*

Homo sapiens with anatomically modern features existed in Africa and the Middle East at least 100,000 years ago. The *H. sapiens* skull lacked a heavy brow ridge and had a distinct chin. By about 30,000 years ago, anatomically modern humans were the only members of genus *Homo* remaining. Paleoanthropologists refer to European populations of these ancient people as **Cro-Magnons.** Their weapons and tools were complex and often made of materials other than stone, including bone, ivory, and wood. They made stone blades that were extremely sharp. Cro-Magnons developed art, including cave paintings, engravings, and sculpture, possibly for ritual purposes. Their sophisticated tools and art indicate they may have possessed language, which they would have used to transmit their culture to younger generations.

PROCESS OF SCIENCE

Two opposing hypotheses currently exist about the origin of these modern humans: the *out-of-Africa* hypothesis and the *multiregional* hypothesis (Fig. 21-12). The out-of-Africa hypothesis holds that modern *H. sapiens* evolved from *H. erectus* in Africa between 200,000 and 100,000 years ago and then migrated to Europe and Asia, displacing the Neandertals and other more primitive humans living there. According to the multiregional hypothesis, modern humans originated, beginning as early as 2 mya, as separately evolving *H. erectus* populations living in several parts of Africa, Asia, and Europe. Each of these populations evolved in its own distinctive way but occasionally met and interbred with other populations, preventing complete reproductive isolation. The variation found today in different geographic populations therefore represents a continuation of this multiregional process.

Data from *Homo* fossils, as well as molecular biology and population genetics studies of modern humans, have been cited in support of both hypotheses, and both have vigorous defenders and strong detractors. Such disagreement is an important

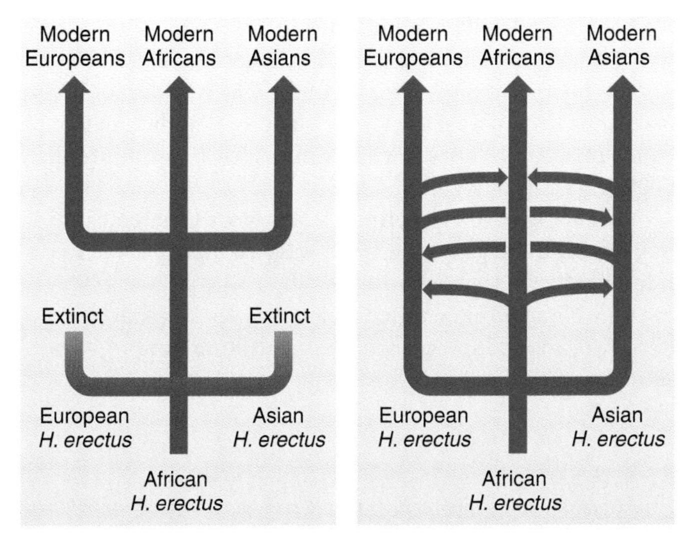

(a) Out-of-Africa hypothesis (b) Multiregional hypothesis

FIGURE **21-12** **Competing hypotheses on the origin of modern humans.**

Scientists agree that *Homo erectus* arose in Africa and migrated to other continents. What happened then is the subject of controversy. **(a)** According to the out-of-Africa hypothesis, all but the African line of *H. erectus* went extinct. In Africa, *H. erectus* evolved into modern humans, which migrated to other continents. **(b)** According to the multiregional hypothesis, modern humans evolved from *H. erectus* populations in different regions of the world. The smaller horizontal arrows in **(b)** represent gene flow (migration and interbreeding) among the different populations.

part of the scientific process, because it stimulates research that may ultimately resolve the issue.

Analysis of DNA provides evidence on the origin of modern humans

Molecular anthropology, the comparison of biological molecules from present-day individuals of regional human populations, provides clues that help scientists unravel the origin of modern humans and trace human migrations. The out-of-Africa hypothesis was originally supported by studies in the late 1980s of mtDNA from various human populations. In 1992 researchers found that the statistical assumptions used in one analysis of mtDNA were wrong, raising questions about the validity of this purported test of the out-of-Africa hypothesis. Multiple subsequent molecular studies comparing DNA sequences from different human groups, however, all have produced essentially the same answer—that modern humans have descended from an early human population that lived in southern Africa.

For example, a 1996 study examined the genetic variation in two stretches of DNA on human chromosome 12 from 1600 people living in 42 different populations around the world (13 African, 2 Middle Eastern, 7 European, 9 Asian, 3 Pacific, and 8 Amerindian populations). The DNA segments varied depending on where the populations lived. Based on the results, scientists divided the present-day human population into three groups: sub-Saharan Africans, northeastern Africans, and non-Africans. (Sub-Saharan Africa includes all countries located south of the Sahara Desert.) The sub-Saharan Africans showed the greatest genetic diversity, whereas the non-African populations were the least diverse. These findings are consistent with the predictions of the out-of-Africa hypothesis for two reasons. First, proponents of the hypothesis expect the sub-Saharan populations to be more diverse because they are older and have had a longer time to accumulate that diversity. Second, the small populations thought to have emigrated from Africa could not have been representative of the total diversity present in the larger African population (recall the discussion in Chapter 18 on the founder effect and genetic drift).

Such research has not disproved the multiregional hypothesis, but it indicates the direction for additional research on other segments of human DNA. A series of recent genetic studies of both mtDNA and nuclear DNA has strengthened the case for Africa as the birthplace of modern humans, and in 1999 scientists announced that the ancestors common to all humans were probably ancient Khoisans, an indigenous group in southern Africa. This conclusion was based on separate analyses of DNA from both mitochondria and the Y chromosome. Both studies indicated that the Khoisan people are the most ancient of all human groups.

Some molecular research has suggested that the out-of-Africa hypothesis may not be as simple as originally envisioned—that is, humans from Africa may not have completely replaced the archaic humans on other continents but may have interbred to some extent with them as they evolved into modern humans. Although many of the human genes that were analyzed have demonstrated an African ancestry, a few appear to have arisen in Asia and to have been introduced at a later time into African populations, probably by migration from Asia to Africa. Thus, certain ancestral human populations in both Africa and Asia may have contributed to the gene pool of modern humans.

Review

- How can you distinguish between *Homo habilis* and *Homo erectus?*
- How do *Homo ergaster* and *Homo erectus* differ?
- How can you distinguish between *Homo erectus* and *Homo heidelbergensis?*
- How do *Homo neanderthalensis* and *Homo sapiens* differ?
- What two hypotheses have been proposed to explain where modern humans originated?

BiologyNow™ Assess your understanding of **hominid evolution** by taking the pretest on your BiologyNow CD-ROM.

CULTURAL EVOLUTION

Learning Objective

8 Describe cultural evolution and its impact on the biosphere.

Genetically speaking, humans are not very different from other primates. At the level of our DNA sequences, we are roughly 98% identical to gorillas and 99% identical to chimpanzees. Our relatively few genetic differences, however, give rise to several important distinguishing features, such as greater intelligence and the ability to capitalize on it through **cultural evolution,** which is the transmission of knowledge from one generation to the next. (Note that researchers increasingly agree that humans are not the only animals to have culture. Chimpanzees have primitive cultures that include tool-using techniques, hunting methods, and social behaviors, all of which vary from one population to another. These cultural traditions are passed from one generation to the next by teaching and imitation; see Chapter 50.)

Human culture is dynamic; it is modified as people obtain new knowledge. Human cultural evolution is generally divided into three stages: (1) the development of hunter-gatherer societies, (2) the development of agriculture, and (3) the Industrial Revolution.

Early humans were hunters and gatherers who relied on what was available in their immediate environment. They were nomadic, and as the resources in a given area were exhausted or as the population increased, they migrated to a different area. These societies required a division of labor and the ability to make tools and weapons, which were needed to kill game, scrape hides, dig up roots and tubers, and cook food. Although scientists are not certain when hunting was incorporated into human society, they do know it declined in importance approximately 15,000 years ago. This may have been due to a decrease in the abundance of large animals, triggered in part by overhunting. A few isolated groups of hunter-gatherer societies, such as the Inuit of the northern polar region and the Mbuti of Africa, have survived into the 21st century.

Development of agriculture resulted in a more dependable food supply

Evidence that humans had begun to cultivate crops approximately 10,000 years ago includes the presence of agricultural tools and plant material at archaeological sites. Agriculture, which involves keeping animals as well as cultivating plants, resulted in a more dependable food supply. Archaeological evidence suggests agriculture arose in several steps. Although there is variation from one site to another, plant cultivation, in combination with hunting, usually occurred first. Animal domestication generally followed later, although in some areas, such as Australia, early humans did not domesticate animals.

Agriculture, in turn, often led to more permanent dwellings, because considerable time was invested in growing crops in one area. Villages and cities often grew up around the farmlands, but the connection between agriculture and the establishment of villages and towns is complicated by certain discoveries. For example, Abu Hureyra in Syria was a village founded *before* agriculture arose. The villagers subsisted on the rich plant life of the area and the migrating herds of gazelle. Once people turned to agriculture, however, they seldom went back to hunting and gathering to obtain food.

Other advances in agriculture include the domestication of animals, which people kept to supply food, milk, and hides. Archaeological evidence indicates that wild goats and sheep were probably the first animals to be domesticated in southwest Turkey, northern Iraq, and Iran. In the Old World, people also used animals to prepare fields for planting. Another major advance in agriculture was irrigation, which began more than 5000 years ago in Egypt.

Producing food agriculturally was more time-consuming than hunting and gathering, but it was also more productive. In hunter-gatherer societies, everyone shares the responsibility for obtaining food. In agricultural societies, fewer people are needed to provide food for everyone. Thus agriculture freed some people to pursue other endeavors, including religion, art, and various crafts.

Cultural evolution has had a profound impact on the biosphere

Cultural evolution has had far-reaching effects on both human society and on other organisms. The Industrial Revolution, which began in the 18th century, drew populations to concentrate in urban areas near centers of manufacturing. Advances in agriculture encouraged urbanization, because fewer and fewer people were needed in rural areas to produce food for everyone. The spread of industrialization increased the demand for natural resources to supply the raw materials for industry.

Cultural evolution has permitted the human population, which reached 6.3 billion in 2003, to expand so dramatically that there are serious questions about Earth's ability to support so many people indefinitely (see Chapter 51). According to the U.N. Food and Agricultural Organization, about 828 million people lack access to the food needed to be healthy and lead productive lives. To further compound the problem, the United Nations projects that 3 billion *additional* people will be added to the world population by the year 2050.

Cultural evolution has resulted in large-scale disruption and degradation of the environment. Tropical rain forests and other natural environments are rapidly being eliminated. Soil, water, and air pollution occur in many places. Since World War II, soil degradation caused by poor agricultural practices, overgrazing, and deforestation has occurred in an area equal to 17% of Earth's total vegetated surface area. Many species cannot adapt to the rapid environmental changes caused by humans and thus are becoming extinct. The decrease in biological diversity from species extinction is alarming (see Chapter 55).

On a positive note, people are aware of the damage we are causing, and have the intelligence to modify behavior to improve these conditions. Education, including the study of biology, may help future generations develop environmental sensitivity, making cultural evolution our salvation rather than our destruction.

Review

- What is cultural evolution?
- How has cultural evolution affected planet Earth?

BiologyNow™ Assess your understanding of **cultural evolution** by taking the pretest on your BiologyNow CD-ROM.

SUMMARY WITH KEY TERMS

1 Describe the structural adaptations that primates have for life in treetops.

- Primates are **placental mammals** that arose from small, **arboreal** (tree dwelling), shrewlike mammals. Primates possess five grasping digits, including an opposable thumb or toe; long, slender limbs that move freely at the hips and shoulders; and eyes located in front of the head.

2 List the three suborders of primates, and give representative examples of each.

- Primates are divided into three suborders. The suborder Prosimii includes lemurs, galagos, and lorises. The suborder Tarsiiformes includes tarsiers. The suborder Anthropoidea includes **anthropoids** (monkeys, apes, and humans).

3 Distinguish among anthropoids, hominoids, and hominids.

- Anthropoids include monkeys, apes, and humans. Early anthropoids branched into two groups: the New and Old World monkeys.
- **Hominoids** include apes and humans. Hominoids arose from the Old World monkey lineage. There are four modern genera of apes: gibbons, orangutans, gorillas, and chimpanzees.
- The **hominid** line consists of humans and their ancestors.

4 Describe skeletal and skull differences between apes and hominids.

- Unlike ape skeletons, hominid skeletons have adaptations that reflect the ability to stand erect and walk on two feet. These adaptations include a complex curvature of the spine; a shorter, broader pelvis; a **foramen magnum** at the base of the skull; and a first toe that is aligned with the other toes.
- The human skull lacks a pronounced **supraorbital ridge,** is flatter in the front, and has a pronounced chin. The human brain is larger, and the jaw is structured so that the teeth are arranged in a U shape.

5 Compare the following early hominids: *Sahelanthropus tchadensis, Ardipithecus ramidus, Australopithecus anamensis, Australopithecus afarensis,* and *Australopithecus africanus.*

- Hominid evolution began in Africa. The earliest known hominids belong to ***Sahelanthropus tchadensis.***
- *Ardipithecus* and *Australopithecus* species are often referred to as **australopithecines.** *Australopithecus* species were **bipedal** (walked on two feet), a hominid feature. ***Ardipithecus ramidus,*** which appeared about 5.2 to 5.8 mya, presumably gave rise to ***Australopithecus anamensis.*** *A. anamensis* may have given rise to another primitive hominid, ***Australopithecus afarensis.*** Many paleoanthropologists think *A. afarensis* gave rise to several australopithecine species, including ***Australopithecus africanus.***
- The genus *Australopithecus* contains the immediate ancestors of the genus *Homo.*

6 Distinguish among the following members of genus *Homo: Homo habilis, Homo ergaster, Homo erectus, Homo heidelbergensis, Homo neanderthalensis,* and *Homo sapiens.*

- ***Homo habilis*** was the earliest known hominid with some of the human features lacking in the australopithecines, including a slightly larger brain. *H. habilis* fashioned crude tools from stone.
- ***Homo erectus*** had a larger brain than *H. habilis;* made more sophisticated tools; and may have worn clothing, built fires, and lived in caves or shelters. Some scientists hypothesize that fossils identified as *Homo erectus* represent two different species, ***Homo ergaster,*** an earlier African species that gave rise to archaic *Homo sapiens,* and *Homo erectus,* a later Asian offshoot that may be an evolutionary dead end.
- **Archaic *Homo sapiens*** lived in Africa, Asia, and Europe from about 800,000 to 100,000 years ago. Some researchers classify archaic *Homo sapiens* as a separate species, ***Homo heidelbergensis.***
- **Neandertals** lived from about 230,000 to 30,000 years ago. Neandertals had short, sturdy builds; receding chins and foreheads; heavy supraorbital ridges and jawbones; larger front teeth; and nasal cavities with unusual triangular bony projections. Many scientists think that Neandertals were a separate species, ***Homo neanderthalensis,*** whereas some scientists think Neandertals were a type of modern human.
- Anatomically modern humans ***(Homo sapiens)*** existed 100,000 years ago. By about 30,000 years ago, anatomically modern humans were the only members of genus *Homo* remaining. European remains of these ancient people are referred to as **Cro-Magnons.**

7 Discuss the current debate over the origin of modern humans, and briefly describe the opposing out-of-Africa and multiregional hypotheses.

- Two hypotheses purport to explain the origin of modern humans. The out-of-Africa hypothesis holds that modern *H. sapiens* arose in Africa and migrated to Europe and Asia, displacing the more primitive humans living there. According to the multiregional hypothesis, modern humans originated as separately evolving populations of *H. erectus* living in several parts of Africa, Asia, and Europe. Each of these populations occasionally met and interbred with other populations, thereby preventing complete reproductive isolation.
- **Molecular anthropology,** the comparison of biological molecules from individuals of regional human populations, generally favors the African origin of modern humans.

8 Describe cultural evolution and its impact on the biosphere.

- **Cultural evolution** is the transmission of knowledge from one generation to the next. Large human brain size makes cultural evolution possible. Two significant advances in cultural evolution were the development of agriculture and the Industrial Revolution.

POST-TEST

1. The first primates evolved about 55 mya from (a) shrewlike monotremes (b) therapsids (c) shrewlike placental mammals (d) tarsiers (e) shrewlike marsupials
2. The anthropoids are more closely related to ____________ than to ____________. (a) tarsiers; lemurs (b) lemurs; monkeys (c) tree shrews; tarsiers (d) lemurs; tarsiers (e) tree shrews; monkeys

POST-TEST *(continued)*

3. Unlike Old World monkeys, some New World monkeys have (a) body hair (b) five grasping digits (c) a well-developed cerebrum (d) a bipedal walk (e) a prehensile tail
4. Apes and humans are collectively called (a) mammals (b) primates (c) anthropoids (d) hominoids (e) hominids
5. With what group do hominoids share the most recent common ancestor? (a) Old World monkeys (b) New World monkeys (c) tarsiers (d) lemurs (e) lorises and galagos
6. The ____________ in humans is centered at the base of the skull, positioning the head for erect walking. (a) supraorbital ridge (b) foramen magnum (c) pelvis (d) bony skull crest (e) femur
7. Scientists collectively call humans and their *immediate* ancestors (a) mammals (b) primates (c) anthropoids (d) hominoids (e) hominids
8. The earliest hominid belongs to the genus (a) *Aegyptopithecus* (b) *Dryopithecus* (c) *Sahelanthropus* (d) *Homo* (e) *Pan*
9. The earliest hominid scientists placed in the genus *Homo* is (a) *H. habilis* (b) *H. ergaster* (c) *H. erectus* (d) *H. heidelbergensis* (e) *H. neanderthalensis*
10. Some scientists now think that fossils identified as *Homo erectus* represent which two different species? (a) *H. habilis* and *H. erectus* (b) *H. ergaster* and *H. erectus* (c) *H. heidelbergensis* and *H. ergaster* (d) *H. neanderthalensis* and *H. erectus* (e) *H. neanderthalensis* and *H. sapiens*
11. Archaic *Homo sapiens* appeared about ____________ years ago. (a) 5 million (b) 800,000 (c) 230,000 (d) 100,000 (e) 5000
12. ____________were an early group of humans with short, sturdy builds and heavy supraorbital ridges that lived throughout Europe and western Asia from about 230,000 to 30,000 years ago. (a) australopithecines (b) dryopithecines (c) archaic *Homo sapiens* (d) Neandertals (e) Cro-Magnons
13. The modern human skull *lacks* (a) small canines (b) a foramen magnum centered in the base of the skull (c) pronounced supraorbital ridges (d) a U-shaped arrangement of teeth on the jaw (e) a large cranium (brain case)
14. The comparison of genetic material from individuals of regional populations of humans, used to help unravel the origin and migration of modern humans, is known as (a) paleoarchaeology (b) cultural anthropology (c) molecular anthropology (d) cytogenetics (e) genetic dimorphism
15. Place the following hominids in chronological order of appearance, beginning with the earliest: *H. habilis, H. sapiens, H. neanderthalensis,* and *H. erectus.* (a) *H. habilis, H. erectus, H. sapiens, H. neanderthalensis* (b) *H. erectus, H. habilis, H. sapiens, H. neanderthalensis* (c) *H. erectus, H. habilis, H. neanderthalensis, H. sapiens* (d) *H. neanderthalensis, H. habilis, H. erectus, H. sapiens* (e) *H. habilis, H. erectus, H. neanderthalensis, H. sapiens*

CRITICAL THINKING

1. What types of as-yet-undiscovered scientific evidence might help explain how monkeys got to South America from the Old World?
2. What was the common ancestor of chimpanzees and humans—a chimpanzee, a human, or neither? Explain your answer.
3. If you were evaluating whether other early humans exterminated the Neandertals, what kinds of archaeological evidence might you look for?
4. The remains of Cro-Magnons have been found in southern Europe alongside reindeer bones, but reindeer currently exist only in northern Europe and Asia. Can you explain the apparent discrepancy?

- Visit our Web site at **http://biology.brookscole.com/solomon7** for links to chapter-related resources on the World Wide Web. Additional online materials relating to this chapter can also be found on our Web site.

BIOLOGY NOW RESOURCES

Active Figure

21-8: Monkey, gorilla, and human skeletons

Preparing for an exam? Take a diagnostic test on your BiologyNow CD-ROM.

Post-Test Answers

1. c	2. a	3. e	4. d
5. a	6. b	7. e	8. c
9. a	10. b	11. b	12. d
13. c	14. c	15. e	

22

Understanding Diversity: Systematics

Pygmy seahorse. This newly described species, generally light orange in color, has been found at depths from 42 to 297 feet.

CHAPTER OUTLINE

About 1.7 million species of living organisms have been described, and about 15,000 new species are discovered each year. The recently discovered pygmy seahorse (*Hippocampus denise*) shown here was described in 2003. Widespread in the western Pacific, this seahorse measures only about 16 mm including its tail (making it one of the smallest fishes). Biologists speculate that at least several million additional species remain to be identified. The total number of living species is estimated to be between 4 million and 100 million. Biologists estimate that to date they have identified less than 10% of bacteria, about 5% of fungi species, only about 2% of nematode (roundworm) species, and less than 20% of insect species.

The variety of living organisms and the variety of ecosystems they form are referred to as **biological diversity,** or **biodiversity.** The totality of life on Earth represents our biological heritage, and the quality of life for all organisms depends on the health and balance of this worldwide web of organisms. For example, we depend on organisms to maintain the life-sustaining composition of gases in the atmosphere, to form soil, break down wastes, recycle nutrients, and provide food for one another. We humans exploit many species for economic benefit. One very anthropocentric (human-centered) example of the practical importance of biodiversity is that more than 40% of the prescriptions that pharmacists dispense in the United States derive from living organisms. Investigators are just learning how to effectively screen organisms for potential drugs. Unfortunately, human activity is seriously reducing biodiversity, and species are becoming extinct more rapidly than researchers can study them (see the discussion on declining biodiversity in Chapter 55).

Alarmed by these extinctions, biologists are mobilizing to more rapidly identify new species and to conserve biodiversity. Many biologists agree that describing and classifying all the surviving species of the world should be one of the great scientific goals of the 21st century. An important step toward that goal is the Catalogue of Life, the work of a collaborative international research team that is setting up a comprehensive global database made available on the Internet. Another project launched in 2002 by the U.N. Environment Programme and the Global Environment Facility focuses on identifying new species in subterranean ecosystems.

To study the diverse life forms that share this planet and to effectively communicate findings, biologists must organize their knowledge. The scientific study of the diversity of organisms and their evolutionary relationships is called **systematics.** An important aspect of systematics is **taxonomy,** the science of naming, describing, and classifying organisms. The term **classification** means arranging organisms into groups based on their similarities, which reflect historical relationships among lineages.

Biologists continuously debate the best methods for inferring the history of life from the data they have, so classifying organisms and establishing their relationships is complex and often controversial. Recently developed molecular approaches to taxonomy are revolutionizing this dynamic science. In 2003, Paul Hebert, of the University of Guelph in Canada, and his colleagues proposed in the British *Proceedings of the Royal Society* that we identify all living things by DNA rather than by their physical structure. Hebert suggested that each organism has a DNA barcode (much like grocery store products) and that biologists can learn to read these barcodes and use them to identify species. DNA is already used to identify bacteria, and Hebert and his colleagues suggest extending this approach to other kinds of organisms.

Systematics proceeds by constantly re-evaluating data, hypotheses, and theoretical constructs. As new data are discovered and old data are reinterpreted, the ideas of systematists change. As a result, our understanding of how organisms are related and the way we classify organisms are continuously being revised.

In this chapter, we explore some approaches and methods used by systematists to infer the evolutionary history of life. In Chapters 23 through 30 we introduce the major groups of organisms that share this planet. We describe how systematists have classified these organisms based on current data from morphology, development, the fossil record, behavior, and molecular biology. ■

BINOMIAL NOMENCLATURE

Learning Objective

1 State at least two justifications for the use of scientific names and classifications of organisms.

Because there are millions of kinds of organisms, scientists need a system for accurately identifying them. Imagine you were going to develop a classification system. How would you use what you already know about living things to assign them to categories? Would you place insects, bats, and birds in one category because they all have wings and fly? And would you, perhaps, place squid, whales, fish, penguins, and Olympic backstroke champions in another category because they all swim? Or would you classify organisms according to a culinary scheme, placing lobsters and tuna in the same part of the menu, perhaps identifying them as "seafood"? Any of these schemes might be valid, depending on your purpose. Similar methods have been used throughout history. For example, St. Augustine in the fourth century classified animals as useful, harmful, or superfluous—to humans. During the Renaissance, scholars began to develop categories based on the characteristics of the organisms themselves.

Of the many classification systems that were developed, the one designed by Carolus Linnaeus in the mid-18th century (described briefly in Chapter 1) has survived with some modification to the present day. As we discuss in the next section, Linnaeus grouped organisms according to their similarities, mainly structural similarities.

Before the mid-18th century, each species had a lengthy descriptive name, sometimes consisting of 10 or more Latin words! Linnaeus simplified scientific classification, developing a **binomial system of nomenclature** in which each species is assigned a unique two-part name.The first part of a scientific name designates the **genus** (pl., *genera*), and the second part is called the **specific epithet.** Note that the specific epithet alone is *not* the species' name. In fact, the same specific epithet can be used as the second name of species in different genera. For example, *Quercus alba* is the scientific species name for the white oak, and *Salix alba* is the species name for the white willow. (*Alba* comes from a Latin word meaning "white.") Thus, both parts of the name must be used to accurately identify the species.

The genus name is always capitalized, whereas the specific epithet is usually not. Both names are underlined or italicized. The genus, or generic, name can be used alone to designate all species in the genus (for example, the genus *Quercus* includes all oak species). However, the specific epithet is never used alone; it must always follow the full or abbreviated genus name (for example, *Quercus alba* or *Q. alba*).

Scientific names are generally derived from Greek or Latin roots or from latinized versions of the names of persons, places, or characteristics. For example, the generic name for the bacterium *Escherichia coli* is based on the name of the scientist, Theodor Escherich, who first described it. The specific epithet *coli* reminds us that *E. coli* lives in the colon (large intestine).

Most areas of biology depend on scientific names and classification. For example, to study the effects of pollution on an aquatic community, biologists must be able to accurately record the relative numbers of each type of organism present and changes in their populations over time. This requires that all investigators identify each species accurately by name.

Scientific names permit biology to be a truly international science. Even though the common names of an organism may vary in different locations and languages, an organism can be universally identified by its scientific name. A researcher in Puerto Rico knows exactly which organisms were used in a study published by a Russian scientist and therefore can repeat or extend the Russian's experiments using the same species.

Review

- What are the key features of the binomial system of nomenclature?
- Why is it important to use scientific, rather than common, names for organisms?

BiologyNow™ Assess your understanding of **binomial nomenclature** by taking the pretest on your BiologyNow CD-ROM.

TAXONOMIC CATEGORIES

Learning Objective

2 Arrange the Linnaean categories in hierarchical fashion, from most inclusive to least inclusive.

Classifying organisms helps biologists organize their knowledge. Linnaeus devised a system for assigning species to a hierarchy of groups. As you move up the hierarchy, each group is more inclusive. When he set up his system, Linnaeus did not have a theory of evolution in mind. Nor did he have any idea of the vast number of extant (living) and extinct organisms that would later be discovered. Yet his system has proved remarkably flexible and adaptable to new biological knowledge and theory. Very few other 18th-century inventions survive today in a form that would still be recognizable to their originators.

The range of taxonomic categories from species to kingdom forms a hierarchy (Fig. 22-1; Table 22-1). Closely related species are assigned to the same genus, and closely related genera are grouped in a single **family.** Families are grouped into **orders,** orders into **classes,** classes into **phyla,** and phyla into **kingdoms** and/or **domains.**

A **taxon** (pl., *taxa*) is a formal grouping of organisms at any given level, such as the species, genus, or phylum. For example, class Mammalia is a taxon that includes many different orders. Each taxon can be separated into subgroupings, for example, subphyla or superclasses. Subphylum Vertebrata, a subgroup of phylum Chordata, is a taxon that contains several classes, including Amphibia and Mammalia.

Some systematists are moving away from the hierarchical Linnaean categories because they do not fit well with recent findings. Also, systematists have experienced some difficulty renaming and reclassifying organisms as quickly as new species are identified and relationships among organisms are adjusted. One group of biologists has proposed a very different classification known as PhyloCode. This system groups organisms into clades; a *clade* is defined as a set of organisms with a common ancestor. Whether (and how soon) biologists will give up conventional nomenclature and the hierarchical system of classification remains to be seen.

Review

- Classify an organism such as the domestic cat using taxa from kingdom to species.

BiologyNow™ Assess your understanding of **taxonomic categories** by taking the pretest on your BiologyNow CD-ROM.

KINGDOMS OR DOMAINS?

Learning Objectives

3 Describe the three domains and six kingdoms of organisms introduced in this chapter, and give the rationales for and against this system of classification.

4 Given its distinguishing characters, classify an organism in the appropriate domain and kingdom.

PROCESS OF SCIENCE

The history of taxonomy at the kingdom level is a good example of the process of science. From the time of Aristotle to the mid-19th century, biologists divided organisms into two kingdoms: **Plantae** and **Animalia.** After the development of microscopes, it became increasingly obvious that many organisms could not be easily assigned to either the plant or the animal kingdom. For example, the unicellular organism *Euglena,* which has been classified at various times in the plant kingdom and in the animal kingdom, carries on photosynthesis in the light but in the dark uses its flagellum to move about in search of food (see Fig. 24-6). In 1866 a German biologist, Ernst Haeckel, proposed that a third kingdom, **Protista,** be established to accommodate bacteria and other microorganisms, such as *Euglena,* that

TABLE 22-1 Classification of Corn

Kingdom	**Plantae** Terrestrial, multicellular, photosynthetic organisms
Phylum	**Anthophyta** Vascular plants with flowers, fruits, and seeds
Class	**Monocotyledones** Monocots: Flowering plants with one seed leaf (cotyledon) and flower parts in threes
Order	**Commelinales** Monocots with reduced flower parts, elongated leaves, and dry 1-seeded fruits
Family	**Poaceae** Grasses with hollow stems; fruit is a grain; and abundant endosperm in seed
Genus	***Zea*** Tall annual grass with separate female and male flowers
Species	***Zea mays*** Only one species in genus—corn

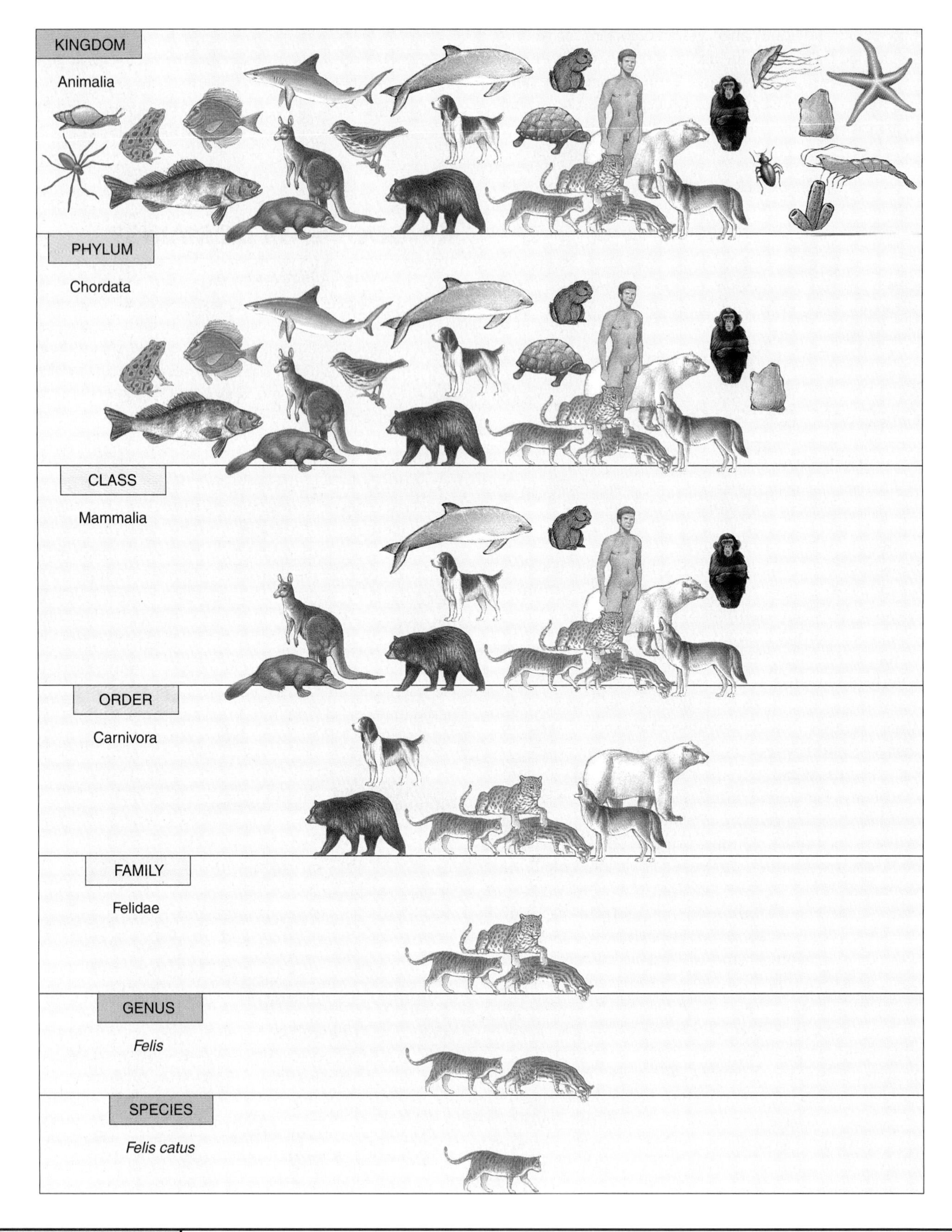

ACTIVE FIGURE **22-1** **The principal categories used in classification.**

The domestic cat (*Felis catus*) is classified here to illustrate the hierarchical organization of our taxonomic system. Each level is more inclusive than the one below it.

BiologyNow™ Explore **the tree of life** by clicking on this figure on your BiologyNow CD-ROM.

did not seem to fit into the plant or animal kingdoms. Today, many biologists place algae (including multicellular forms), protozoa, water molds, and slime molds in kingdom Protista.

In 1937 the French marine biologist Edouard Chatton suggested the term *procariotique* ("before nucleus") to describe bacteria, and the term *eucariotique* ("true nucleus") to describe all other cells. This dichotomy between prokaryotes and eukaryotes is now universally accepted by biologists as a fundamental evolutionary divergence. In the 1960s advances in electron microscopy and biochemical techniques revealed further cell differences that inspired many new proposals for classifying organisms.

In 1969 R.H. Whittaker proposed a five-kingdom classification based mainly on cell structure and how organisms derived nutrition from their environment. Whittaker suggested that the fungi (which include the mushrooms, molds, and yeasts) be removed from the plant kingdom and classified in their own kingdom **Fungi.** After all, fungi are not photosynthetic and must absorb nutrients produced by other organisms. Fungi also differ from plants in the composition of their cell walls, in their body structures, and in their modes of reproduction. Kingdom **Prokaryotae** was established to accommodate the bacteria, which are fundamentally different from all other organisms in that they do not have distinct nuclei and other membranous organelles and do not undergo mitotic division.

In the late 1970s, Carl Woese (pronounced "woes"), of the University of Illinois, and his colleagues began to study the evolutionary relationships among organisms by analyzing their RNA. A component of ribosomes, small subunit ribosomal RNA (SSU rRNA)), specifically the 16S[1] rRNA nucleotide sequence, is highly conserved: The gene that codes SSU rRNA is present in all known organisms, and its function has remained the same through time. Using sequence analysis, Woese used variations in this universal molecule to challenge the long-held view that all prokaryotes were closely related and very similar to one another. He proposed that there are two fundamentally different groups of bacteria, Archaebacteria and Eubacteria and that the prokaryotes account for two of three of the major branches of organisms.

Woese's hypothesis gained support in 1996 when Carol J. Bult of the Institute for Genomic Research in Rockville, Maryland, reported in the journal *Science* that she and her colleagues had sequenced the complete genome of a methane-producing archaebacterium *Methanococcus jannaschii.* When these researchers compared gene sequences with those of two previously sequenced eubacteria, they found that fewer than half of the genes matched. Given this molecular evidence, many biologists divided the prokaryotes into kingdom **Eubacteria** and kingdom **Archaebacteria.**

Most systematists now use a level of classification above the kingdom, called a **domain,** based on fundamental molecular differences among the Eubacteria, Archaebacteria, and eukaryotes. They classify organisms in three domains: **Archaea** (which corresponds to kingdom Archaebacteria), **Eubacteria** (also called Bacteria), and **Eukarya** (eukaryotes; Fig. 22-2). An important character that distinguishes the Archaea from the Eubacteria is the absence of the compound peptidoglycan in the cell walls of the Archaea (discussed in Chapter 23). Gene sequencing indicates the Archaea have a combination of bacteria-like and eukaryote-like genes and appear more closely related to the Eukarya than to the Eubacteria. For example, the Archaea do not have the simple RNA polymerase characteristic of Eubacteria.

[1] The number 16S refers to the sedimentation coefficient (a measure of relative size) when centrifuged.

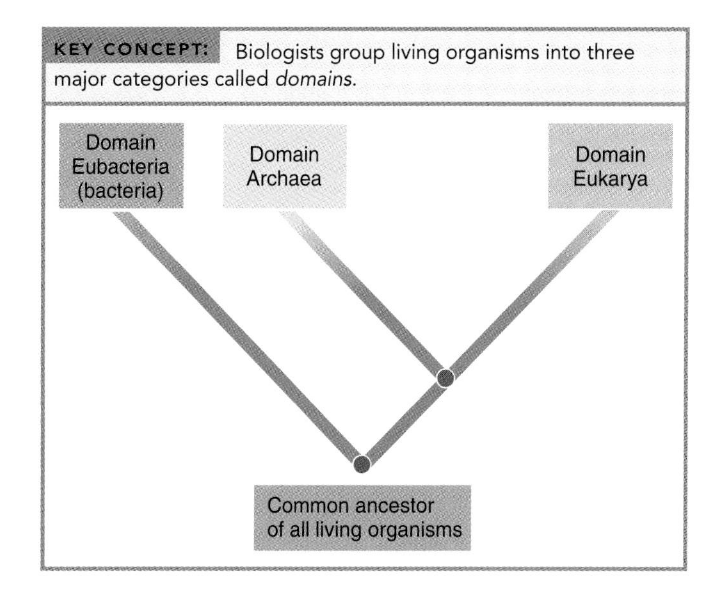

FIGURE **22-2** | **The three domains.**

This branching diagram, called a *cladogram,* illustrates the evolutionary relationships among organisms in the three domains. The branches portray ancestral populations of each group through time. Each node (circle) represents the immediate common ancestor of the groups that branch from it.

Increasingly, many biologists now classify organisms in domains rather than kingdoms. However, some biologists use both kingdoms and domains, classifying organisms into three domains and six kingdoms; we use this classification in the seventh edition of *Biology.* The six kingdoms are Eubacteria, Archaebacteria, Protista, Fungi, Plantae, and Animalia (Fig. 22-3 and Table 22-2).

In recent years, biologists have been surprised to discover that evolution is not linear. During the course of evolution, genes were not only passed down vertically from one generation to the next but were also exchanged laterally (horizontally), for example, by occasional interbreeding between closely related species. Such movement of genes from organisms in one taxon to related organisms in another taxon is called **lateral gene transfer.**

The evolution of systematics reflects the creative and dynamic process of science. Systematists are very responsive to new data, and consequently classification of organisms at all levels is a challenging and continuously changing process. As we discuss in Chapter 24, some biologists are moving toward a 10-kingdom system of classification. An important goal is to base classification on evolutionary change and relationships, so that we can use classification to reconstruct the history of life.

FIGURE **22-3**

The six-kingdom system of classification.

Compare this cladogram to Figure 22-2. Note that kingdoms Protista, Plantae, Animalia, and Fungi can be assigned to domain Eukarya. The Protista are a very diverse group with several major branches. Systematists may eventually divide them into two or more kingdoms. Note that this diagram is greatly simplified. As in Figure 22-2, the nodes represent branch points where a species splits and two or more populations diverge and evolve independently.

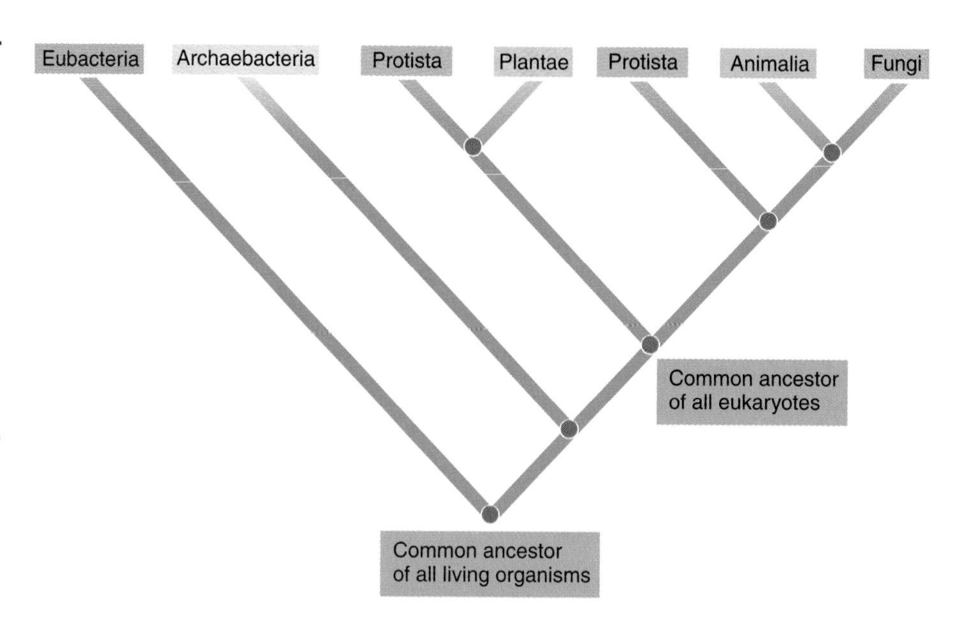

TABLE 22-2 Domain and Kingdoms: Eubacteria, Archaebacteria, Protista, Fungi, Plantae, and Animalia

Domain	Kingdom	Characteristics	Ecological Role and Comments
Eubacteria	Eubacteria	Prokaryotes (lack distinct nuclei and other membranous organelles); unicellular; microscopic; cell walls generally composed of peptidoglycan.	Most are decomposers; some parasitic (and pathogenic): some chemosynthetic autotrophs; some photosynthetic; important in recycling nitrogen and other elements; some used in industrial processes.
Archaea	Archaebacteria	Prokaryotes; unicellular; microscopic; peptidoglycan absent in cell walls; differ biochemically from eubacteria.	Methanogens are anaerobes that inhabit sewage, swamps, and animal digestive tracts; extreme halophiles inhabit salty environments; extreme thermophiles inhabit hot, sometimes acidic environments.
Eukarya	Protista	Eukaryotes; mainly unicellular or simple multicellular. Three informal groups (not taxa) include protozoa; algae; and slime molds and water molds.	Protozoa are an important part of zooplankton. Algae are important producers, especially in marine and freshwater ecosystems; important oxygen source.
	Fungi	Eukaryotes; heterotrophic; absorb nutrients; do not photosynthesize; body composed of threadlike hyphae that form tangled masses that infiltrate food or habitat; cell walls of chitin.	Decomposers; some parasitic (and pathogenic); some used as food; yeast used in making bread and alcoholic beverages; some used to make industrial chemicals or antibiotics; responsible for much spoilage and crop loss.
	Plantae	Eukaryotes; multicellular; photosynthetic; possess multicellular reproductive organs; alternation of generations; cell walls of cellulose.	Terrestrial biosphere depends on plants in their role as primary producers; important source of oxygen in Earth's atmosphere.
	Animalia	Eukaryotes; multicellular heterotrophs; many exhibit tissue differentiation and complex organ systems; most able to move about by muscular contraction; nervous tissue coordinates responses to stimuli.	Consumers; some specialized as herbivores, carnivores, or detritus feeders.

Review

- Contrast the three domains.
- What are the advantages of a "six-kingdom" system over a "two-kingdom" system?
- What types of organisms are especially difficult to assign to a kingdom?
- In which domain and kingdom would you classify each of the following? (a) oak tree (b) amoeba (c) *Escherichia coli* (d) tapeworm (e) black bread mold

BiologyNow™ Assess your understanding of **kingdoms and domains** by taking the pretest on your BiologyNow CD-ROM.

RECONSTRUCTING PHYLOGENY

Learning Objectives

5 Critically summarize the difficulties encountered in choosing taxonomic criteria.

6 Apply the concept of shared derived characters to the classification of organisms.

7 Identify methods of molecular biology now used by taxonomists, and summarize the advantages of molecular taxonomy.

8 Contrast monophyletic, paraphyletic, and polyphyletic taxa.

In Chapter 17 you learned that evolution is the accumulation of inherited changes within populations over time. Recall that a **population** is made up of all the individuals of the same species that live in a particular area. A population has a dimension in space—its geographic range—and also a dimension in time. Each population extends backward in time. Somewhat like branches of a tree, a population may diverge from other populations enough to become a new species (that is, a new tree branch; see Chapters 18 and 19). Species have various degrees of evolutionary relationships with one another, depending on the degree of genetic divergence since their populations branched from a common ancestor.

PROCESS OF SCIENCE

The goal of systematics is to reconstruct **phylogeny** (literally, "production of phyla"), the evolutionary history of a group of organisms from a common ancestor. As they determine evolutionary relationships among and between species and higher taxa, systematists build classifications based on common ancestry. Once phylogenies are established, they help answer other questions in biology. For example, phylogenies help us understand evolutionary patterns that might provide clues to the origin and spread of HIV and other pathogens. Phylogenies also help predict characteristics of newly discovered species. Systematists may classify a new species based on specific characters that it shares with other organisms in a particular grouping. They may then infer that the new species shares other characters with organisms in that group.

As you read the following sections, remember that phylogenies are testable hypotheses. They are supported or falsified by the available data. Thus, systematics is a dynamic science that changes as biologists discover new species and use ever more sophisticated techniques to investigate the evolutionary relationships among organisms.

Homologous structures are important in determining evolutionary relationships

Just how to determine phylogeny and how to group species into higher taxa—genera, families, orders, classes, or phyla—can be difficult decisions. Traditionally, many biologists have based their judgments about the degree of relationship on the extent of similarity among living species and, when available, on the fossil record.

When they evaluate similarities, systematists now consider structural, physiological, developmental, behavioral, and molecular traits. When comparing these similarities, they look for homology in different organisms. Recall from Chapter 17 that **homology** refers to the presence, in two or more species, of a structure derived from a recent common ancestor. For example, the bones in the wings of a bird and a bat are homologous. In contrast, similar structures in two or more species not derived from a recent common ancestor sometimes result when unrelated or distantly related species become adapted to similar environmental conditions. For example, although the wings of a butterfly are adapted for flight, they have a different structure from bird wings, and these animals do not share a common winged ancestor. Sharks and dolphins have similar, but independently derived, body forms because they have become adapted to similar environments (aquatic) and lifestyles (predatory). A characteristic that superficially appears homologous, but is actually independently acquired, is described as exhibiting **homoplasy.**

Shared derived characters provide clues about phylogeny

Organisms sharing many homologous structures are considered closely related, while organisms sharing few homologous characters are presumed less closely related. However, distinguishing between homology and homoplasy is not always straightforward. Therefore, carefully selecting similarities that indicate evolutionary relationships is extremely important.

How does a systematist interpret the significance of these similarities? In making decisions about taxonomic relationships, the systematist first examines the characteristics in the largest group (such as phylum or class) of organisms being studied and interprets them as indicating the most remote common ancestry. These **shared ancestral characters,** or **plesiomorphic characters,** are features that were present in an ancestral species and remain present in all groups descended from that ancestor. For example, the vertebral column, present in all vertebrates, is an ancestral character for study of classes within the subphylum Vertebrata. Studying the presence or absence of the vertebral column does not help us discriminate among various classes of vertebrates. For example, we could not distinguish between amphibians and mammals, because all individuals in these classes have a vertebral column.

Shared derived characters, or **synapomorphic characters,** are traits found in two or more taxa that first appeared in their most recent common ancestor. A feature viewed as a *derived character* in a more inclusive (broader) taxon may also be considered an *ancestral character* in a less inclusive (narrower) taxon. More recent common ancestry is indicated by classification into less and less inclusive taxonomic groups with more and more specific shared derived characters. For example, the three small bones in the middle ear are useful in identifying a branch point between reptiles and mammals. The evolution of this derived character was a unique event, and only mammals have these bones. However, if we compare mammals with one another, the three ear bones are a shared ancestral character, because all mammals have them. Consequently, they have no value in distinguishing among mammalian taxa. Other characters must be used to establish branch points among the mammals.

If we compare dogs, goats, and dolphins (all of which are mammals), we find that dogs and goats have abundant hair whereas dolphins do not. Hair is an ancestral trait in mammals and therefore cannot be used as evidence that dogs and goats share a more recent common ancestor. In contrast, the virtual absence of hair in mature dolphins is a derived character within mammals. When we compare dogs, dolphins, and whales, we find that dolphins and whales share this derived character, providing evidence that these animals evolved from a common ancestor not shared by dogs.

Biologists carefully choose taxonomic criteria

Both fishes and dolphins have streamlined body forms, but this characteristic is homoplastic and does not indicate close evolutionary relationships. The dolphin shares important homologous derived characters (synapomorphies) with mammals such as humans: mammary glands, which produce milk for the young, three small bones in the middle ear, and a muscular diaphragm that helps move air into and out of the lungs. Thus the dolphin is classified as a mammal.

Although dolphins have more shared derived characters in common with humans than with fishes, some characters are shared by all three of these animals. Among these ancestral characters are a dorsal tubular nerve cord and, during embryonic development, a notochord (skeletal rod) and rudimentary gill slits. These shared ancestral characters (plesiomorphies) indicate a common ancestry and serve as a basis for classification. The ancestry is more remote between the dolphin and the fish than between the dolphin and the human. Therefore, although fishes, humans, and dolphins are grouped together in a more inclusive taxon, the phylum Chordata, humans and dolphins are also classified together in class Mammalia, a less inclusive taxon within phylum Chordata, indicating that they are more closely related.

PROCESS OF SCIENCE

Determining which traits best illustrate evolutionary relationships can be challenging. What, for example, are the most important taxonomic characteristics of a bird? We might list feathers, beak, wings, absence of teeth, egg laying, and the fact that they are endotherms (animals that use metabolic heat to maintain a constant body temperature despite variations in surrounding temperature). Some mammals, for example, monotremes such as the duck-billed platypus, have many of these same characteristics: beaks, endothermy, absence of teeth, and egg laying. Yet we do not classify them as birds (Fig. 22-4). No mammal, however, has feathers. Is this trait absolutely diagnostic of birds? According to the conventional taxonomic wisdom, the presence or absence of feathers determines what is and is not a bird. This applies only to modern birds, however. A number of dinosaur fossils indicates that some extinct reptiles had feathers.

Usually organisms are classified on the basis of a combination of traits rather than on any single trait. The significance of these combinations is determined inductively, that is, by integrating and interpretating the data. Such induction is a necessary part of the process of science. Taxonomists may hypothesize, for example, that all birds should have beaks, feathers, no teeth, and so forth. Then they re-examine the living world and observe whether any organisms might reasonably be called birds that do not fit the current definition of "birdness." If not, the definition stands. If too many exceptions emerge, the definition may be modified or abandoned. Sometimes the taxonomist determines that an apparent exception—the bat, for instance—resembles a bird only superficially and should not be considered one. The bat has all the basic characteristics of a mammal, such as hair and mammary glands that produce milk for the young.

Molecular biology provides additional characters

When a new species evolves, it does not always exhibit obvious phenotypic differences relative to closely related species. For example, two distinct species of fruit flies may appear identical. Some of their DNA, proteins, and other molecules, however, are different. Such variations in the structure of specific macromolecules among species, just like differences in anatomical structure, result from mutations. Macromolecules that are functionally similar in two different types of organisms are considered homologous if their subunit sequence is similar.

The science of **molecular systematics** focuses on the use of molecular structure to clarify evolutionary relationships. Advances in molecular biology have provided the tools for biologists to compare the macromolecules of various organisms. DNA, RNA, and amino acid sequencing are used to compare macromolecules. Comparisons of amino acid sequences of proteins and nucleotide sequences of nucleic acids provide systematists with valuable information about the degree of relatedness among organisms. The more subunit sequences of two species correspond, the more closely related they are considered to be. The number of differences in certain DNA or RNA nucleotide sequences or in amino acid sequences in two groups of organisms may reflect how much time has passed since the groups branched from a common ancestor. (This can be true only if the rates of change have remained constant.) Thus specific genes and specific proteins are sometimes used as **molecular clocks** (see Chapter 17).

FIGURE **22-4** | **Is this animal a bird?**

A few mammals, such as the duck-billed platypus, lay eggs, have beaks, and lack teeth. However, the platypus does not have feathers, and it nourishes its young with milk secreted from mammary glands.

Jean Philippe Varin/Jacana/Photo Researchers, Inc.

PROCESS OF SCIENCE

Systematists compare ribosomal RNA structure to help determine phylogenies (evolutionary histories). All known organisms have ribosomes that function in protein synthesis, and certain ribosomal RNA nucleotide sequences have been highly conserved in evolution. Recall that the division of organisms into three domains was based, in large part, on the comparison of ribosomal RNA by Carl Woese and his research team at the University of Illinois. All prokaryotic ribosomes contain three types of RNA, named in order of increasing size: 5S, 16S, and 23S. The 5S and 16S RNAs have been extensively used to determine evolutionary relationships among bacteria because the number of base pairs is manageable and because these molecules are transcribed from highly conserved regions of DNA.

Comparison of ribosomal RNA sequences has also been used to challenge the once widely accepted idea that fungi are closely related to plants. According to ribosomal RNA analysis, fungi are more closely related to animals than to plants; that is, animals and fungi share a more recent common ancestor, perhaps a flagellate protist.

In 1997 molecular taxonomist Robert Wayne, of the University of California at Los Angeles, and his international team of researchers compared mitochondrial DNA sequences from 162 wolves at 27 different geographic locations with those from domestic dogs of 67 different breeds (Fig. 22-5). They reported that dog and wolf nucleotide sequences were similar, differing by no more than 12 substitutions. In contrast, dog sequences differ by at least 20 substitutions from jackal and coyote DNA. Wayne and his colleagues concluded that dogs evolved from wolves. Extending Wayne's work, Peter Savolainen, of the Royal Institute of Technology in Stockholm, and his colleagues compared mitochondrial DNA samples from hundreds of dogs throughout the world. Their findings, reported in 2002, supported Wayne's work, and they further concluded that dogs evolved in a single geographic site of origin, most likely China.

The new tools provided by molecular biology have put systematics on the cutting edge of biological research. Biologists are no longer limited by subjective decisions regarding anatomical similarities among organisms. Comparison of molecular structure is far more precise. For example, until recently the African elephant and the Indian elephant were the only two species of living elephants recognized. Using molecular methods, biologists have demonstrated that there are at least two distinct species of African elephants. Each species has its distinctive DNA barcode.

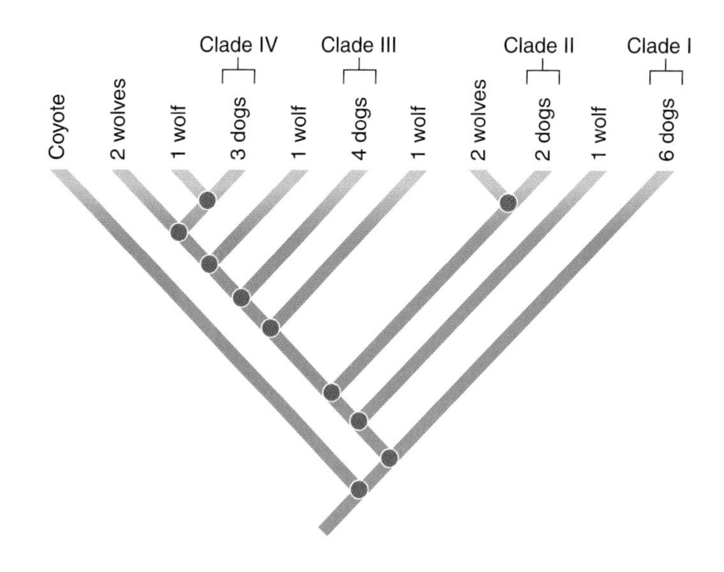

FIGURE **22-5** **Molecular taxonomy.**

Diagram of the relationships of 8 wolves and 15 dogs based on differences in mitochondrial DNA. The researchers compared 1030 base pairs of a segment of mitochondrial DNA. Wolf and dog sequences were very similar, differing by no more than 12 substitutions. Dog and coyote sequences differed by at least 20 substitutions. Four separate clades (lineages) of dogs were identified among various wolf lineages, suggesting multiple origins.

Taxa should reflect evolutionary relationships

Based on molecular data and other taxonomic criteria, systematists recognize three kinds of taxonomic groupings: monophyletic, paraphyletic, and polyphyletic. A **monophyletic taxon** includes an ancestral species and all its descendants (Fig. 22-6a). Mammals, for example, form a monophyletic taxon because all mammals are thought to have evolved from a common ancestral mammal, and all descendants of this ancestor are mammals. Monophyletic taxa are natural groupings, because they represent true evolutionary relationships and include all close relatives.

A **paraphyletic taxon** is a group that contains a common ancestor and some, but not all, of its descendants. As discussed later in this chapter, the class Reptilia is paraphyletic because it does not include all descendants of the most recent common ancestor of reptiles. Birds are thought to share a recent common ancestor with reptiles, but evolutionary systematists assign birds to their own class. Cladistic systematists (discussed later in this chapter) avoid constructing paraphyletic taxa.

A **polyphyletic group,** such as the Protista, consists of several evolutionary lines that do not share the same recent common ancestor (Fig. 22-6b). Biologists might have mistakenly grouped together the members of such a taxon because these organisms share similar (homoplastic) features arising from convergent evolution (see Chapter 17). Systematists avoid constructing polyphyletic taxa, because they are unnatural and misrepresent evolutionary relationships.

Review

- How are shared ancestral characters and shared derived characters different?
- Why don't shared ancestral characters provide evidence for relationships between organisms within a taxon that has those traits? Give an example.
- Of what use to a taxonomist is knowledge of the amino acid sequences of the proteins of various organisms?
- How does a monophyletic taxon differ from a polyphyletic taxon?

BiologyNow™ Assess your understanding of **reconstructing phylogeny** by taking the pretest on your BiologyNow CD-ROM.

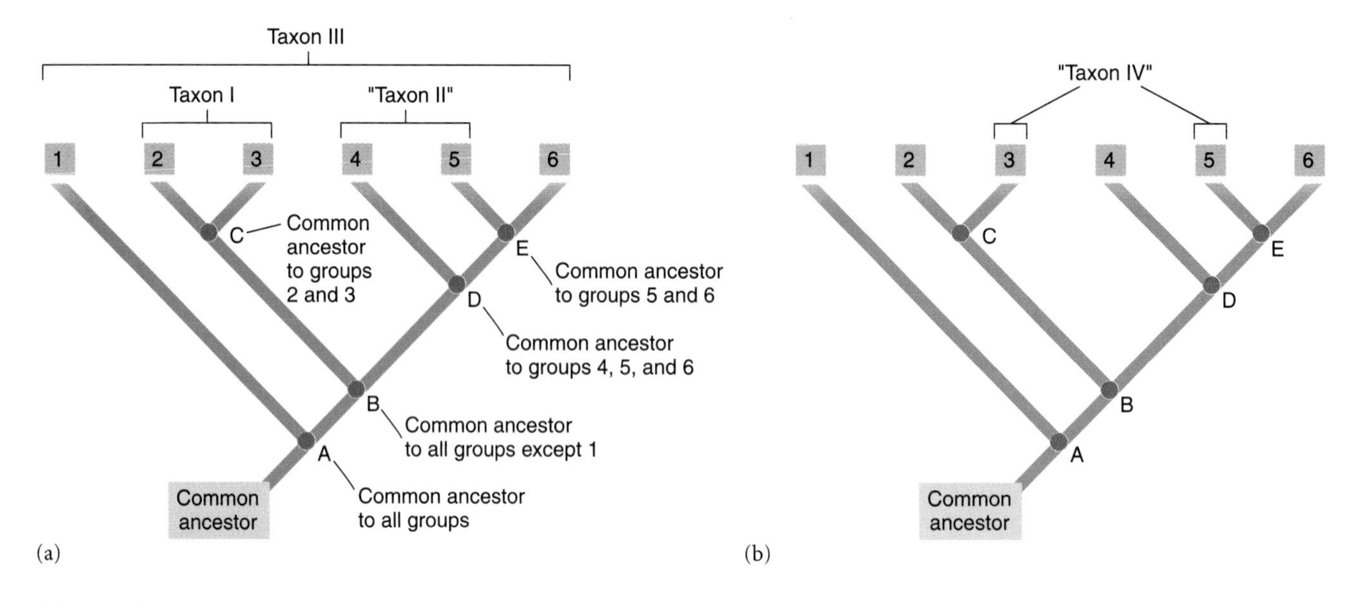

FIGURE **22-6** **Evolutionary relationships.**

In both figures, node A represents the common ancestor of all groups shown. Node C represents the common ancestor for groups 2 and 3. Node D represents the common ancestor for groups 4, 5, and 6. **(a)** Monophyletic and paraphyletic groups. Taxon I and taxon III are monophyletic. Each includes a common ancestor and all its descendants. "Taxon II" is paraphyletic; it does not include all the descendants of a common ancestor. **(b)** "Taxon IV" is polyphyletic; members of this group do not share a most recent common ancestor. (Paraphyletic and polyphyletic "taxa" are indicated with quotation marks because they are not natural evolutionary groupings.)

TWO MAJOR APPROACHES TO SYSTEMATICS

Learning Objective

9 Compare and contrast two approaches to systematics: evolutionary systematics and cladistics (phylogenetic systematics).

In determining the relationships among organisms, systematists use a variety of data and methods. How data are analyzed and interpreted depends on the systematist's approach. **Phenetics,** or **numerical taxonomy,** was an early approach to the quantitative analysis of characters. First gaining attention of systematists in the 1960s, phenetics is based on as many phenotypic similarities (that is, similarities in *appearance*) as possible. In the phenetic approach, computer programs are used to analyze data and group organisms according to the number of shared characteristics. No attempt is made to determine whether their similarities arose from a common ancestor or from convergent evolution. Pheneticists argue that it is not important to try to differentiate between homology and homoplasy, because many more similarities are due to homology than to homoplasy. Thus the number of similarities two organisms have in common reflects the degree of homology.

Some phenetic techniques are currently used in molecular taxonomy. For example, if each amino acid in a protein is considered a character, amino acid sequences of various animals can be determined in the laboratory and compared by computer. The information about differences in amino acid sequences can be used to construct phylogenetic diagrams. Species are placed at relative distances from each other, reflecting the extent of difference in amino acid sequence.

The two major approaches to systematics that are widely used today are **evolutionary systematics** and **cladistics,** also known as **phylogenetic systematics.** Although we discuss them separately here, many modern systematists use aspects of both approaches in their work. Although evolutionary systematists and cladists evaluate phenotypic traits as pheneticists would, they also consider whether similarities arise through common ancestry or convergent evolution. In addition, they consider dissimilar traits that may share a common origin but have diverged over evolutionary time.

Evolutionary systematics allows paraphyletic groups

Evolutionary systematics, sometimes called *classical evolutionary systematics,* uses a system of phylogenetic classification and presents evolutionary relationships in phylogenetic trees. Evolutionary systematists consider both evolutionary branching (as do cladists) and the extent of divergence (structural and other changes) that has occurred in a lineage since it branched from a stem group. They use a combination of shared ancestral characters and shared derived characters to establish evolutionary relationships and build classifications.

Many, perhaps most, of the taxa currently recognized by evolutionary systematists are monophyletic, based on possession of shared derived characters. Cladists also recognize these taxa. For example, proponents of both approaches agree that birds are a monophyletic taxon based on shared derived characters such as feathers. However, evolutionary systematics also recognizes paraphyletic taxa, groups that include some, but not all, subgroups of organisms that share the same most recent common ancestor. Paraphyletic groupings are based on a combination of shared ancestral and shared derived characters.

Evolutionary systematists recognize class Reptilia as a valid group that does not include birds because it includes the common ancestor of all reptiles and it lacks the derived features of birds (such as feathers). Class Reptilia does not include all subgroups, such as birds, that evolved from the ancestral reptile (Fig. 22-7a); evolutionary systematists assign birds to a separate class because they have diverged markedly from the reptiles. In contrast, cladists do not recognize reptiles as a natural grouping containing snakes, lizards, crocodiles, dinosaurs, and turtles because they do not form a monophyletic group, or **clade.** Instead, they recognize a clade that includes birds.

Cladistics emphasizes phylogeny

The cladistic approach to systematics emphasizes common ancestry, rather than phenotypic similarity, as the basis for classification. Cladists base their assessment on shared derived characters that can be structural, behavioral, physiological, or molecular. The characters must be homologous. According to cladistics, dolphins are classified with mammals rather than with fishes, because dolphins and mammals share derived characters not present in fishes, indicating a more recent common ancestor. Cladistics uses shared derived characters to reconstruct phylogenies.

Cladists determine the evolutionary relationships of organisms and express them in branching diagrams called **cladograms.** Each branch on a cladogram represents the divergence, or splitting, of two or more new groups from a common ancestor. Each branch point, referred to as a **node,** represents the immediate common ancestor of the next monophyletic group depicted in the cladogram. Consider the evolutionary grouping of mammals, lizards, snakes, crocodiles, dinosaurs, and birds (Fig. 22-7b). Birds, along with dinosaurs, are thought to share a common ancestor with modern crocodiles and alligators (node D on Fig. 22-7b). Crocodiles, dinosaurs, and birds, then, constitute a monophyletic group, and cladists would classify them in the same clade. Similarly, snakes and lizards form a clade that is the closest group to birds, dinosaurs, and crocodiles. Mammals form an additional clade.

When cladists consider the evolutionary relationships of organisms, they must choose between multiple, competing cladograms. How do they make this choice? The most common criterion is the principle of **parsimony**—they choose the simplest explanation to interpret the data. Parsimony, a guiding principle in many areas of research, is based on the experience that the simplest explanation is most probably the correct one. Applied to choice of cladograms, parsimony requires that the cladogram with the fewest changes in characters (the one with the fewest homoplasies) be accepted as most probable (Fig. 22-8). In actual practice it is often possible to generate several cladograms that are equally parsimonious.

A crucial step in most cladistic analyses is **outgroup analysis,** a method for estimating which attributes are shared derived characters in a given group of taxa. An **outgroup** is a taxon that is considered to have diverged earlier than the taxa under investigation, the **ingroups,** and thus to represent an approximation of the ancestral condition. An ideal outgroup is the closest relative of the group being studied (its sister group) and has not been highly modified since its origin. Systematists argue that such a group is likely to retain the ancestral state for characters being

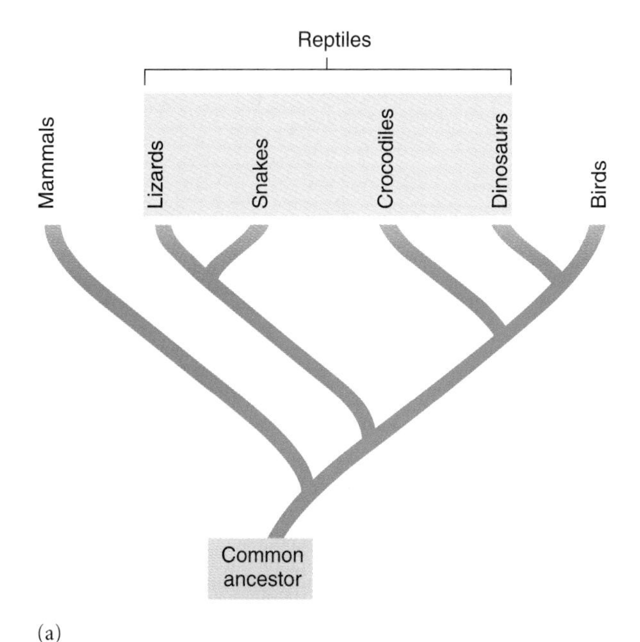

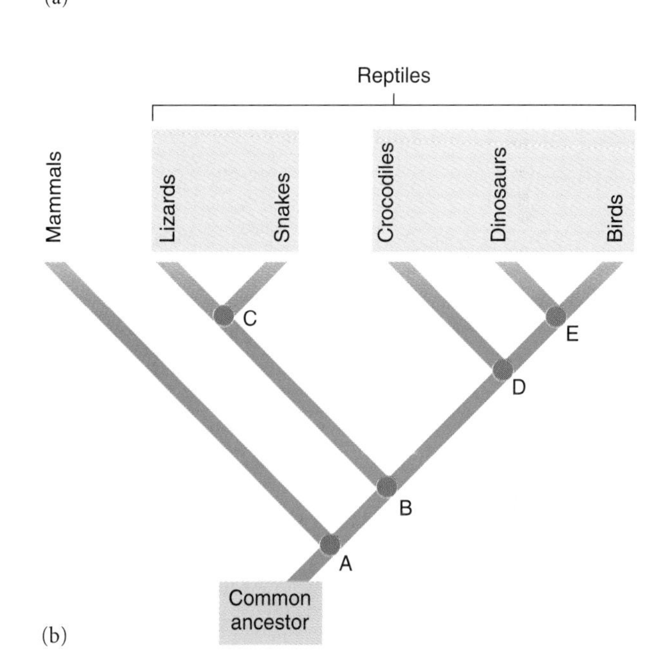

FIGURE **22-7** **Two approaches to the classification of reptiles, birds, and mammals.**

(a) The evolutionary systematics approach considers both common ancestry and extent of divergence that has occurred since two taxa split. The branching points and degrees of difference in the evolution of the major groups of reptiles are shown. Reptiles are a paraphyletic group. Snakes, lizards, and crocodiles are phenotypically most similar, but crocodiles, dinosaurs, and birds are most closely related because they evolved most recently from a common ancestor. **(b)** Cladists classify birds and some reptiles together because they have a recent common ancestor. Node D represents the common ancestor of crocodiles, dinosaurs, and birds.

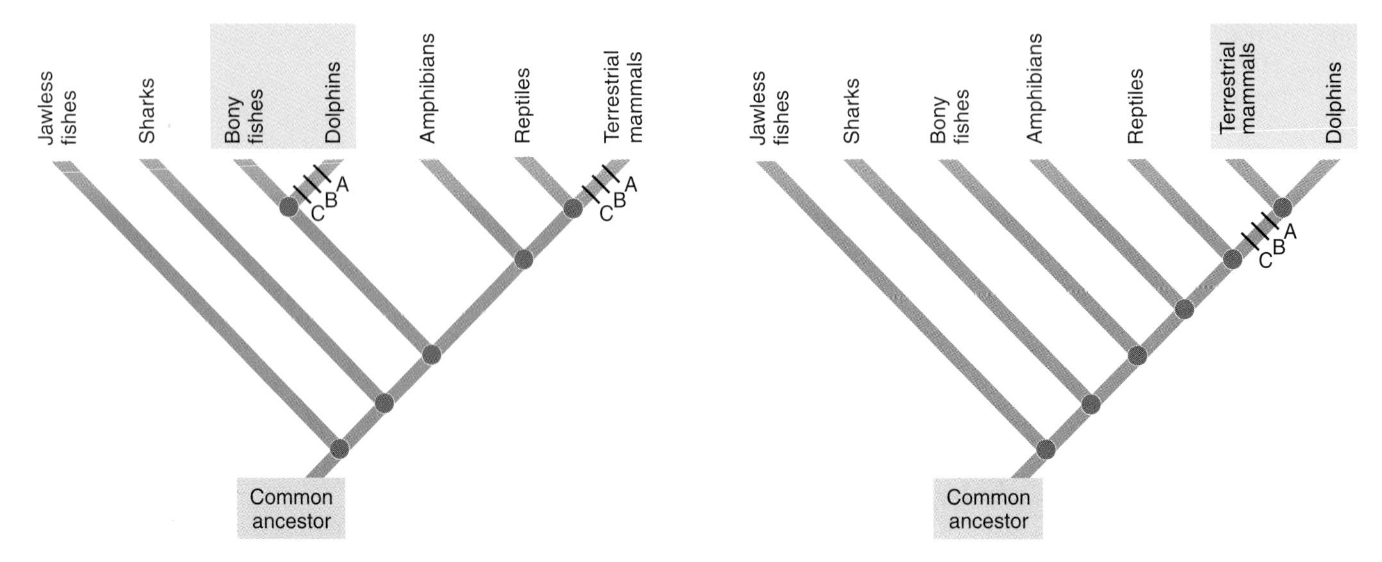

(a) Hypothesis 1: Dolphins and bony fishes are close relatives

(b) Hypothesis 2: Dolphins and terrestrial mammals are close relatives

FIGURE **22-8** | **Using parsimony to choose alternative hypotheses of relationships.**

Are dolphins more closely related to bony fishes or to terrestrial mammals? Three characters are examined: Character A = hair; B = mammary glands; C = middle-ear ossicles. In hypothesis 1, all three characters examined had to evolve independently twice. Hypothesis 2 is chosen because it requires that each character evolve only once.

used in the analysis, allowing them to identify the evolutionary changes leading to derived characters. To help you understand outgroup analysis, we now consider a specific example.

Outgroup analysis is used in building and interpreting cladograms[2]

PROCESS OF SCIENCE

The first step in constructing a cladogram is to select the taxa, which may consist of individuals, species, genera, or other taxonomic levels. Here we use a representative group of eight chordates (Fig. 22-9). The next step is to select the homologous characters to be analyzed. In our example, we use seven characters. For each character, we must define all the different conditions, or states, as they exist in our taxa. For simplicity, we consider our characters to have only two different states: present or absent. Keep in mind that many characters used in cladistics have more than two states. For example, black, brown, yellow, and red may be only a few of the many possible states for the character of hair color.

The last, and often the most difficult, step in preparing the data is to organize the character states into their correct evolutionary order. Outgroup analysis is used. In our example, amphioxus, a small chordate with a fishlike appearance, is the chosen outgroup. It belongs to a taxon that is considered to have diverged earlier than any of the other taxa under investigation but to be closely related to vertebrates; thus amphioxus represents an approximation of the ancestral condition for vertebrates. Therefore, the character state "absent" for a particular character state such as vertebrae is the ancestral condition, and the character state "present" is the derived condition for the characters in Figure 22-9.

A cladogram is constructed by considering shared derived characters

Our objective is to construct a cladogram that requires the fewest number of evolutionary changes in the characters. Recall that in cladistics, taxa are grouped by the presence of shared derived characters. To form a valid monophyletic group, all members must share at least one derived character. Membership in a clade cannot be established by shared ancestral characters.

In our example, notice that all taxa except the outgroup amphioxus possess vertebrae. We can therefore conclude that these seven vertebrate taxa form a valid clade. Next, among the seven vertebrate taxa, note that jaws are present in all groups except for lampreys. Using these data, we construct a preliminary cladogram (Fig. 22-9a).

The base of the cladogram represents the common ancestor for all taxa being analyzed. In Figure 22-9a, node A represents the common chordate ancestor from which the outgroup

FIGURE **22-9** | **Building a cladogram.** ▶

In this example, amphioxus is the chosen outgroup that represents an approximation of the ancestral condition. Refer to the table as you follow the process in (a) through (e).

[2] This discussion of building and interpreting cladograms is based on an essay contributed to *Biology*, 4th ed., by Dr. John Beneski, Department of Biology, West Chester University, West Chester, Pennsylvania.

A = Absent

B = Present

TAXA	Vertebrae (backbones)	Jaws	Tetrapod (4 limbs)	Amniotic egg	Mammary glands	Opposable thumb	Upright posture
	CHARACTERS						
Amphioxus (outgroup)	A	A	A	A	A	A	A
Lamprey	P	A	A	A	A	A	A
Sunfish	P	P	A	A	A	A	A
Newt	P	P	P	A	A	A	A
Lizard	P	P	P	P	A	A	A
Bear	P	P	P	P	P	A	A
Chimpanzee	P	P	P	P	P	P	A
Human	P	P	P	P	P	P	P

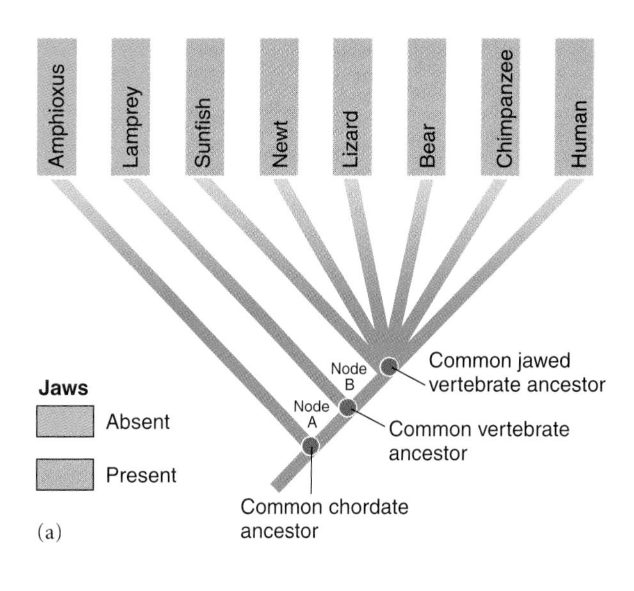

(a)

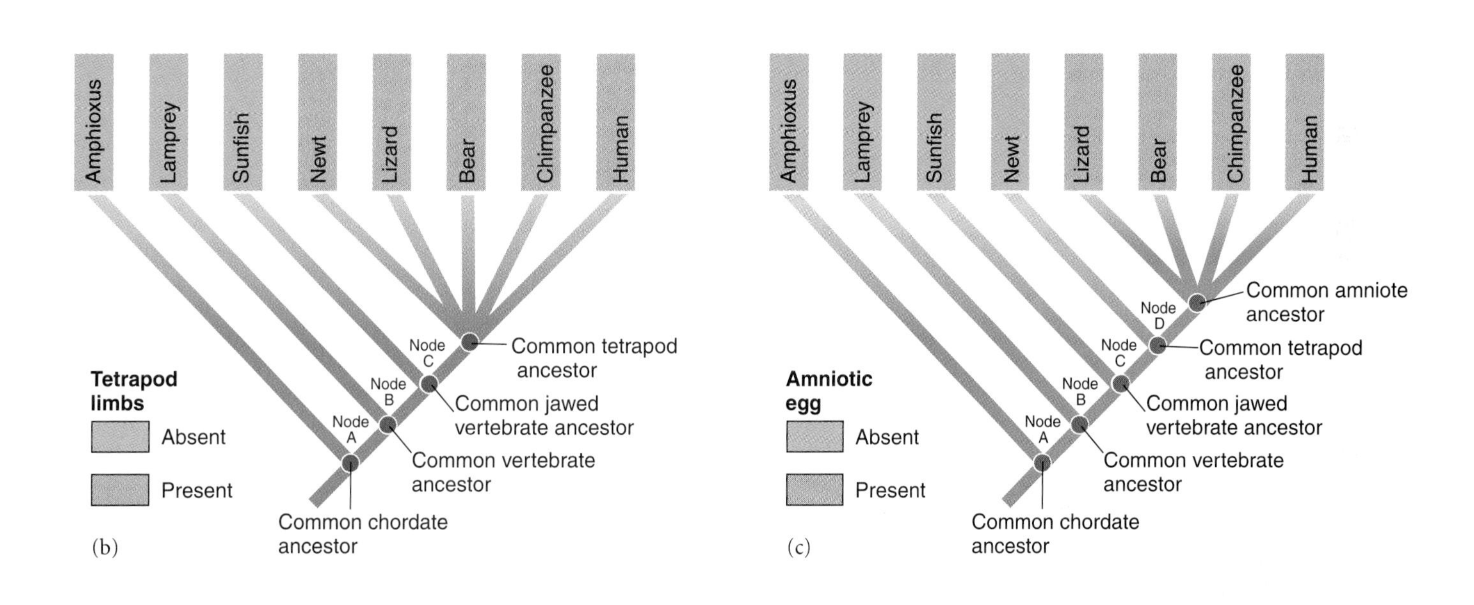

(b)

(c)

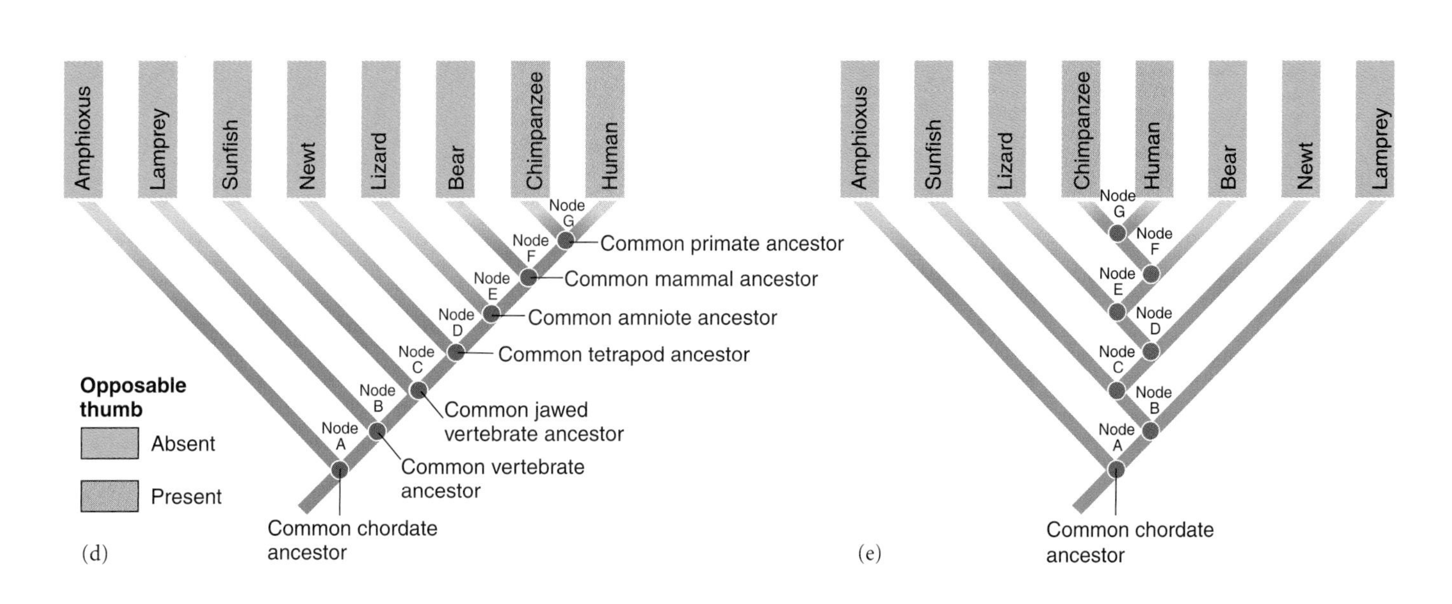

(d)

(e)

(amphioxus) and the seven vertebrate taxa evolved. Similarly, node B represents the common ancestor of the vertebrates and node C, the common ancestor of the jawed vertebrates. Continuing with this procedure, notice that among the six-jawed taxa, all but sunfish are tetrapods (animals with four limbs) (Fig. 22-9b). Among the five tetrapods, all but newts have amniotic eggs (Fig. 22-9c). (In an amniotic egg, the embryo is surrounded by a fluid-filled sac, called an *amnion.*) The branching process is continued, using the data in the table in the upper left-hand corner of the figure, until all clades are established (Fig. 22-9d).

In a cladogram each branch point represents a major evolutionary step

In Figure 22-9d, notice that humans and chimpanzees share a recent common ancestor at node G. This hypothesis is supported by the derived characters (synapomorphies) they share. In the same way, bears are more closely related to the human–chimpanzee clade than to any other clade considered in our example, as indicated by the common ancestor at node F. In comparing the nodes, the order of divergence (branching) is indicated by distance from the base of the diagram. The further a node is located up the cladogram, the more recently the group diverged. In our example, node G represents the most recent divergence, and node A represents the most ancient divergence. Thus, in our example, humans are closely related to chimpanzees (through node G) but more distantly related to bears (through node F). Therefore humans and chimpanzees are assigned to a less inclusive taxon (order Primates) whereas humans, chimpanzees, and bears are assigned to a broader, more inclusive taxon (class Mammalia). In addition, the cladogram reveals that lizards are more closely related to the mammal clade than to newts, sunfish, or any other clade. Can you explain why?

Two important concepts guide interpretation of cladograms. First, the relationships among taxa is determined only by tracing along the branches back to the most recent common ancestor (node) and not by the relative placement of the branches along the horizontal axis. It is possible to represent the same relationships with many different types of branching diagrams. For example, the cladogram in Figure 22-9e is equivalent to the one in Figure 22-9d. (Verify this by comparing the numbered nodes and by checking the relationships described earlier.)

A second key concept is that the cladogram tells us which taxa shared a common ancestor and how recently they shared a common ancestor. The ancestor itself remains unspecified. The cladogram does not establish direct ancestor–descendant relationships among taxa. In other words, a cladogram does not suggest a taxon gave rise to any other taxon.

Many evolutionary questions remain. New hypotheses about how organisms are related suggest new taxonomic schemes. Systematics offers tools to help us test these hypotheses. In Chapters 23 through 30 we explore many evolutionary puzzles and discuss current hypotheses about relationships and classifications of organisms.

Review

- How are the methods of cladists and evolutionary systematists similar?
- Which systematists do not recognize paraphyletic groups?
- What does a node represent in a cladogram?
- What is outgroup analysis?

BiologyNow™ Assess your understanding of **the two major approaches to systematics** by taking the pretest on your BiologyNow CD-ROM.

SUMMARY WITH KEY TERMS

1 State at least two justifications for the use of scientific names and classifications of organisms.

- **Systematics** is the scientific study of the diversity of organisms and their evolutionary relationships. The term **biological diversity,** or **biodiversity,** refers to the variety of living organisms and the ecosystems they form.
- **Taxonomy** is the branch of systematics devoted to naming, describing, and classifying organisms. The process of assigning organisms into groups based on their similarities or relationships is **classification.** Classifications help biologists organize their knowledge.
- Scientific names are important because they allow biologists from different countries with different languages to communicate about organisms. Biologists in distant locations must know with certainty that they are studying the same (or different) organisms.
- Biologists name organisms using the **binomial system of nomenclature** developed by Linnaeus in the mid-18th century. In this system the basic unit of classification is the **species.**
- The name of each species has two parts: the **genus** name followed by the **specific epithet.** For example, the scientific name of the human is *Homo sapiens,* and that of the white oak is *Quercus alba.*

2 Arrange the Linnaean categories in hierarchical fashion, from most inclusive to least inclusive.

- The hierarchical system of classification used in this book includes **domain, kingdom, phylum, class, order, family, genus,** and **species.** Each formal grouping at any given level is a **taxon.**

3 Describe the three domains and six kingdoms of organisms introduced in this chapter, and give the rationales for and against this system of classification.

- The three-domain classification system assigns organisms to domain **Archaea,** domain **Eubacteria,** or domain **Eukarya.**
- The six-kingdom classification recognizes the kingdoms **Archaebacteria, Eubacteria, Protista, Fungi, Plantae,** and **Animalia.**

4 Given its distinguishing characters, classify an organism in the appropriate domain and kingdom.

- Domain Eubacteria (kingdom Eubacteria) is distinguished by the presence of the compound peptidoglycan in cell walls. Peptidoglycan is absent in the cell walls of the Archaea.
- Domain Archaea (kingdom Archaebacteria) has a combination of bacteria-like and eukaryote-like genes and may be more closely related to the Eukarya than to the Eubacteria. For exam-

ple, the Archaea do not have the simple RNA polymerase characteristic of Eubacteria.

- Domain Eukarya consists of the eukaryotes: kingdoms Protista, Fungi, Plantae, and Animalia.
- For characteristics of the six kingdoms, refer to Table 22-2.

5 Critically summarize the difficulties encountered in choosing taxonomic criteria.

- The goal of systematics is to determine evolutionary relationships, or **phylogeny,** based on shared characteristics. Historically, scientists have depended mainly on structural similarities to make decisions about phylogeny.
- **Homology,** the presence in two or more species of a structure derived from a recent common ancestor, implies evolution from a common ancestor. The term **homoplasy** refers to superficially similar characters that are not homologous, because they evolved independently.
- Biologists now use a variety of criteria to determine relationships, including shared derived characters and comparison of DNA, RNA, and protein. Deciding which traits best reflect phylogeny is challenging.

6 Apply the concept of shared derived characters to the classification of organisms.

- **Shared ancestral characters,** or **plesiomorphic characters,** suggest a distant common ancestor. **Shared derived characters,** or **synapomorphic characters,** indicate a more recent common ancestor.

7 Identify methods of molecular biology now used by taxonomists, and summarize advantages of molecular taxonomy.

- **Molecular systematics** uses methods for comparing macromolecules such as nucleic acids and proteins for assessing evolutionary relationships. Comparison of nucleotide sequences in ribosomal RNA has led to important taxonomic decisions regarding domains, kingdoms, and species.

8 Contrast monophyletic, paraphyletic, and polyphyletic taxa.

- A **monophyletic taxon** includes all the descendants of the most recent common ancestor. A **paraphyletic taxon** consists of a common ancestor and some, but not all, of its descendants. The organisms in a **polyphyletic group** evolved from different ancestors.

9 Compare and contrast two approaches to systematics: evolutionary systematics and cladistics (phylogenetic systematics).

- **Evolutionary systematics** considers both evolutionary branching and the extent of divergence. Evolutionary systematics is based on shared ancestral characters as well as shared derived characters.
- **Cladistics,** also called **phylogenetic systematics,** insists that taxa be monophyletic. Each monophyletic taxon, or **clade,** consists of a common ancestor and all its descendants. Cladists use shared derived characters to determine these relationships, and illustrate them with diagrams called **cladograms.**
- Cladists use **outgroup analysis** to determine which characters in a given group of taxa are ancestral and which are derived. An **outgroup** is a taxon that represents the ancestral condition because it diverged earlier than any of the other taxa being investigated.
- Cladists use the principle of **parsimony:** They choose the simplest explanation to interpret the data.
- In interpreting cladograms, the relationships among taxa are determined by tracing along the branches back to the most recent common ancestor (represented by a node on the cladogram). The cladogram indicates which taxa shared a common ancestor and how recently they shared that ancestor.

POST-TEST

1. The science of describing, naming, and classifying organisms is (a) systematics (b) taxonomy (c) cladistics (phylogenetic systematics) (d) phenetics (e) evolutionary systematics
2. Using the binomial system of nomenclature, the scientific name of each species consists of two parts (a) class, specific epithet (b) family, genus (c) genus, specific epithet (d) family, species (e) genus, species
3. The mold that produces penicillin is *Penicillium notatum. Penicillium* is the name of its (a) genus (b) order (c) family (d) species (e) specific epithet
4. Closely related genera may be grouped together in a single (a) phylum (b) domain (c) species (d) family (e) kingdom
5. Related classes are grouped together in the same (a) genus (b) phylum (c) order (d) paraphyletic taxon (e) family
6. In the six-kingdom system, the kingdom that includes the protozoa is (a) Plantae (b) Protista (c) Archaea (d) Eukarya (e) Fungi
7. Decomposers such as molds and mushrooms belong to kingdom (a) Plantae (b) Protista (c) Archaebacteria (d) Eukarya (e) Fungi
8. A taxon that contains a recent common ancestor and all its descendants is (a) polyphyletic (b) paraphyletic (c) monophyletic (d) phyletic (e) plesiomorphic
9. The presence of homologous structures in different organisms suggests that (a) the organisms evolved from a common ancestor (b) convergent evolution has occurred (c) they belong to a polyphyletic group (d) homoplasy has occurred (e) independently acquired characters may evolve when organisms inhabit similar environments
10. The dolphin and the human both have the ability to nurse their young, whereas the less closely related fish does not. The ability to nurse their young is a (a) shared derived character of mammals (b) shared ancestral character of all vertebrates (c) plesiomorphy (d) homologous behavior (e) homoplasy
11. Relative constancy in the rates of DNA and protein evolution permits biologists to use these macromolecules as (a) molecular clocks (b) polymerase chains (c) clades (d) paraphyletic clues (e) outgroups
12. Using DNA as a molecular barcode (a) is effective only with bacteria (b) is possible because of similarity in polymerase chains (c) could cause major mistakes in distinguishing among closely related species (d) is an effective way to identify polyphyletic clades (e) could help taxonomists identify and describe new species
13. The conclusion that fungi are more closely related to animals than to plants was based in large part on comparing (a) molecular

POST-TEST (continued)

clocks (b) polymerase chains (c) nucleotides in DNA (d) ribosomal RNA sequences (e) amino acid sequences

14. Phenetics (a) is a numerical taxonomy based on phenotypic similarities (b) emphasizes common ancestry (c) emphasizes polyphyletic groups (d) focuses on ancestral characters (e) strives to differentiate between homology and homoplasy.
15. Some systematists classify crocodiles and birds in the same taxon because they are monophyletic. These systematists follow which approach? (a) phyletic (b) cladistic (c) evolutionary systematic (d) polyphyletic (e) either cladistic or evolutionary systematic
16. In cladistic analysis (a) ancestral characters are used to reconstruct phylogenies (b) characters must be homoplastic (c) polyphyletic groups are preferred (d) ancestral character analysis is commonly used (e) outgroup analysis is used
17. When cladists use the principle of parsimony, they (a) choose the simplest explanation to interpret the data (b) select multiple hypotheses to explain each relationship (c) do not use outgroups (d) typically use a polyphyletic approach (e) hypothesize that the most complex explanation is most probably the correct one
18. In the illustration, what kind of grouping is represented by the bracketed area?

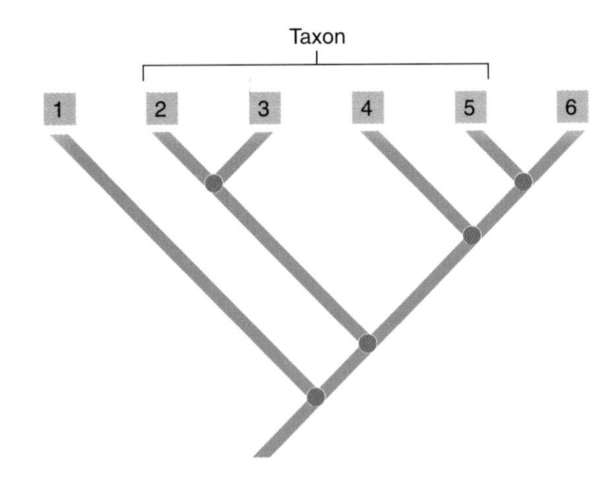

CRITICAL THINKING

1. Are members of a genus similar because they share a common ancestor, or do they belong to the same genus because they are similar? How might your answer vary depending on which approach to systematics you are following?
2. How are new techniques in molecular biology affecting the science of systematics?
3. The TATA-binding protein (TBP) is thought to be necessary for transcription in all eukaryotic cell nuclei. Studies show that archaebacteria, but not eubacteria, have a protein structurally and functionally similar to TBP. What does this similarity suggest regarding the evolution of archaebacteria and eukaryotes? How might knowledge of this similarity affect how systematists classify these organisms?

- Visit our Web site at **http://biology.brookscole.com/solomon7** for links to chapter-related resources on the World Wide Web. Additional online materials relating to this chapter can also be found on our Web site.

BIOLOGY NOW RESOURCES

Active Figure

22-1 The tree of life

Preparing for an exam? Take a diagnostic test on your BiologyNow CD-ROM.

Post-Test Answers

1. b	2. c	3. a	4. d
5. b	6. b	7. e	8. c
9. a	10. a	11. a	12. e
13. d	14. a	15. b	16. e
17. a			

23

Viruses and Prokaryotes

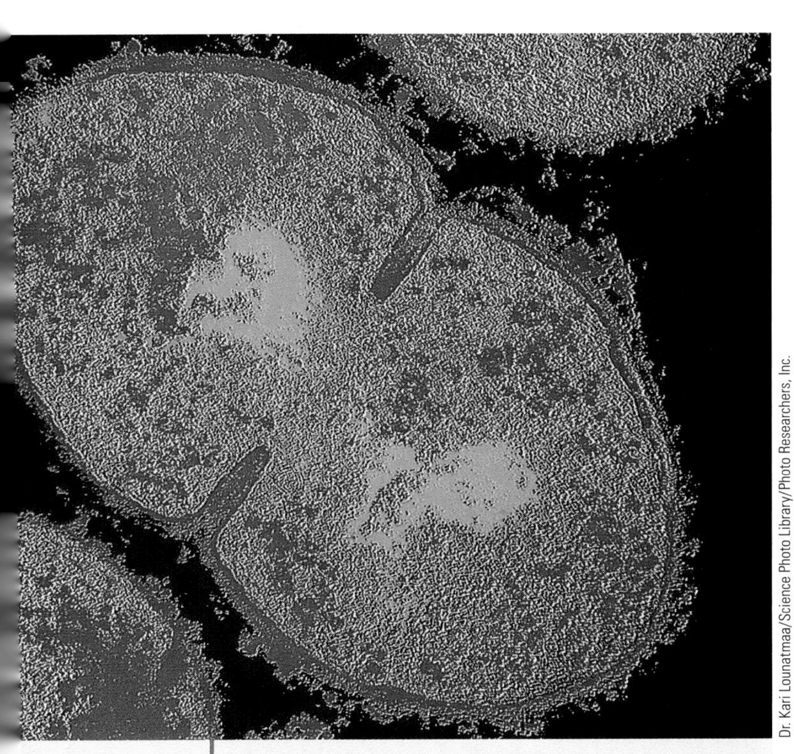

Dr. Kari Lounatmaa/Science Photo Library/Photo Researchers, Inc.

Color-enhanced TEM of a bacterium (*Streptococcus pyogenes*) dividing by binary fission. This bacterium, a pathogen that inhabits the human nose and throat, can cause scarlet fever and inflammation of the heart tissue. The strain shown here is resistant to antibiotics, and infection can be fatal.

CHAPTER OUTLINE

Prokaryotes (bacteria and archaea) have inhabited our planet for more than 3.5 billion years—much longer than eukaryotes, which evolved about 1.5 to 1.6 billion years ago. Although prokaryotes are microscopic, they are so numerous that they, together with the fungi, account for approximately half of Earth's biomass, the mass of living material. In contrast, plants account for about 35% and animals for about 15% of the biomass of our planet.

The Dutch microscopist Anton van Leeuwenhoek discovered bacteria and other microorganisms in 1674 when he looked at a drop of lake water through a glass lens. During the late 1800s, many microorganisms, including bacteria, fungi, and protozoa, were identified as **pathogens,** agents that cause disease (see the adjacent photo).

Although bacteria cause many diseases, including respiratory infections and food poisoning in humans, only a small minority of bacterial species are pathogens. In fact, bacteria play an essential role in the biosphere as decomposers, breaking down organic molecules into their components. Along with fungi, prokaryotes are nature's recyclers. Without these microorganisms, carbon, nitrogen, phosphorus, and sulfur would remain locked up in the wastes and dead bodies of plants and animals, unavailable for the synthesis of new cells and organisms.

Some prokaryotes are producers that carry on photosynthesis. Others convert atmospheric nitrogen to ammonia and then to nitrates, forms that are used by plants (see Fig. 53-7). This conversion enables plants and animals (because they eat plants) to manufacture essential nitrogen-containing compounds such as proteins and nucleic acids.

In contrast to prokaryotes, viruses are acellular. Biologists long considered them nonliving particles, but some biologists now view them as life forms. Viruses contain the nucleic acids necessary to make copies of themselves, and they reproduce by invading living cells and commandeering their metabolic machinery. These tiny, but potent, particles infect cells and produce a wide variety of diseases in plants and animals. Viruses influence many ecological processes, including the recycling of nutrients and the biodiversity of prokaryotes and algae.

This chapter examines the diversity and characteristics of viruses, prokaryotes, and the smaller viroids and prions. They are not a natural group of closely related organisms, and we discuss them in a single chapter only for convenience, not to reflect shared ancestry. ■

VIRUSES

Learning Objectives

1. Describe the structure of a virus, and compare a virus with a free-living cell.
2. Trace the evolutionary origin of viruses according to current hypotheses.
3. Characterize bacteriophages, and contrast a lytic cycle with a lysogenic cycle.
4. Explain how viruses infect animal and plant cells.
5. Describe the reproductive cycle of a retrovirus such as human immunodeficiency virus (HIV).

PROCESS OF SCIENCE

During the late 1800s, botanists searched for the cause of tobacco mosaic disease, which stunts the growth of tobacco plants and gives the infected tobacco leaves a spotted, mosaic appearance. They found they could transmit the disease to healthy plants by daubing their leaves with the sap of diseased plants. In 1892 Dmitrii Ivanowsky, a Russian botanist, showed that the sap was still infective after it had been passed through filters that removed particles the size of all known bacteria. A few years later, in 1898, his work was expanded by Martinus Beijerinck, a Dutch microbiologist. Apparently unaware of Ivanowsky's work, Beijerinck provided independent evidence that the agent that caused tobacco mosaic disease had many characteristics of a living organism. He hypothesized that the infective agent could reproduce only within a living cell, and named it *virus* (the Latin word *virus* means poison).

Early in the 20th century, scientists discovered infective agents, like those responsible for tobacco mosaic disease, that could cause disease in animals or kill bacteria. These pathogens also passed through filters that removed known bacteria and were so small they could not be seen with the light microscope. Curiously, they could not be grown in laboratory cultures unless living cells were present. Microbiologists called viruses that killed bacteria **bacteriophages** ("bacteria eaters"), or **phages.**

A virus particle consists of nucleic acid surrounded by a protein coat

A **virus,** or **virion,** is a tiny, infectious particle consisting of a nucleic acid core (its genetic material) surrounded by a protein coat called a **capsid.** Some viruses are also surrounded by an outer membranous envelope containing proteins, lipids, carbohydrates, and traces of metals. A typical small virus, such as the poliovirus, is about 20 nm in diameter (about the size of a ribosome), whereas a larger virus, such as the poxvirus that causes smallpox, may be 400 nm long and 200 nm wide.

Viruses are acellular and cannot independently perform metabolic activities. They do not have the components necessary to carry on cellular respiration or to synthesize proteins and other molecules. Other living organisms contain both deoxyribonucleic acid (DNA) and ribonucleic acid (RNA), but a virus contains either DNA or RNA, not both. A particular type of virus may have single-stranded (ss) DNA, double-stranded (ds) DNA, ssRNA, or dsRNA.

Viruses reproduce, but only within the complex environment of the living cells they infect. They use their genetic information to force the host cell to replicate the viral nucleic acid and to take over the translational and transcriptional mechanisms of the host cell. The host then synthesizes the capsid and envelope components of the virus.

The organization of protein subunits, called *capsomeres,* that make up the capsid determine the shape of a virus. Viral capsids are generally either helical or polyhedral, or a complex combination of both shapes (Fig. 23-1). Helical viruses, such as the tobacco mosaic virus (TMV), appear as long rods or threads. The capsid is a hollow cylinder made up of proteins that form a groove into which the RNA fits. Polyhedral viruses, such as the adenoviruses (which causes a number of human illnesses, including some respiratory infections), appear somewhat spherical. The T4 phage that infects *Escherichia coli* consists of a polyhedral "head" attached to a helical "tail," a shape commonly found in bacteriophages.

The International Committee on Taxonomy of Viruses classifies viruses

Viruses do not show the characteristics that define living organisms (see Chapter 1). They are acellular, do not carry on metabolic activities, and reproduce only by taking over the reproductive machinery of other cells. For these reasons, viruses cannot be classified in any of the three domains and present a taxonomic challenge to biologists.

Early systems of classifying viruses were based on the host or organ system they infected. Today, viruses are classified by the International Committee on Taxonomy of Viruses (ICTV). This committee groups viruses according to common characteristics such as type of nucleic acid and presence or absence of a capsid. The classification system is not a traditional Linnaean system and does not assign viruses to domains, kingdoms, or phyla.

Viruses may have "escaped" from cells

PROCESS OF SCIENCE

Where did viruses come from? The most widely held hypothesis is that viruses derive from bits of nucleic acid that "escaped"

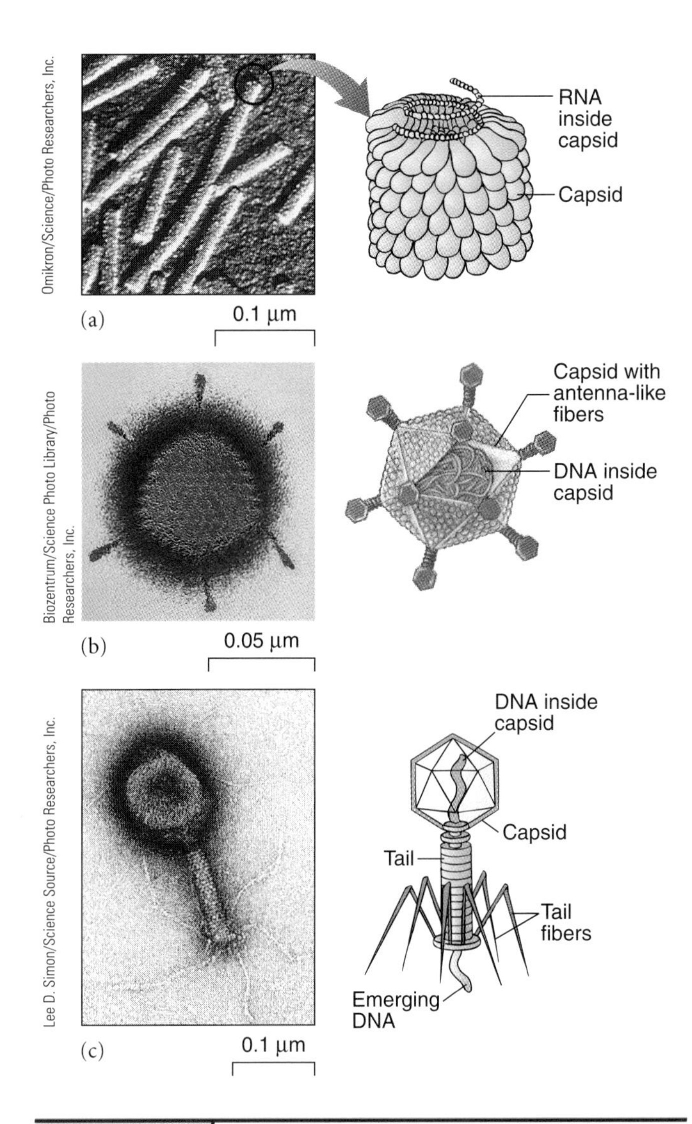

FIGURE **23-1** | **Virus structure.**

(a) TEM of tobacco mosaic virus, a rod-shaped virus with a helical arrangement of capsid proteins. **(b)** Color-enhanced TEM of an adenovirus, which has a capsid composed of 252 subunits (visible as tiny ovals) arranged into a 20-sided polyhedron. Twelve of the subunits have projecting protein spikes that permit the virus to recognize host cells. **(c)** TEM of T4, a bacteriophage that has a polyhedral head and a helical tail.

from cellular organisms. Viruses may have originated as mobile genetic elements such as *transposons* (see Chapter 12) or *plasmids* (small, circular DNA fragments discussed in Chapter 14 and later in this chapter). Such fragments could have moved from one cell and entered another through damaged cell membranes.

According to this *escaped gene hypothesis,* some viruses may trace their origin to animal cells, others to plant cells or to bacterial cells. Their multiple origins may explain why many viruses are species specific; perhaps they infect only those species that are closely related to the organisms from which they originated. This hypothesis is supported by the genetic similarity between some viruses and their host cells—a closer similarity than exists between one type of virus and another.

Another hypothesis suggests that viruses arose early in the history of life, even before the three domains diverged. Evidence for this hypothesis comes from similarities found in the protein structures of some viral capsids and also in genetic similarities between some viruses that infect the Archaea and some that infect the Eubacteria. Molecular biologists studying this hypothesis suggest it is improbable that these similarities evolved independently. This suggests viruses diverged very early. However, viruses are parasites, dependent on their hosts. How could they have existed before their hosts evolved?

Bacteriophages are viruses that attack bacteria

Much of our knowledge of viruses has come from studying bacteriophages, because they can be cultured easily within living bacteria in the laboratory. Bacteriophages are among the most complex viruses (see Fig. 23-1c). Their most common structure consists of a long nucleic acid molecule (usually dsDNA) coiled within a polyhedral head. Many phages have a tail attached to the head. The phage may use fibers extending from the tail to attach to a bacterium.

More than 2000 phages have been identified. They infect a wide variety of organisms. For example, phycoviruses infect algae. Some studies suggest that phages inhibit or control the growth rate of algal blooms.

Before the age of sulfa drugs and antibiotics, phages were used clinically to treat infection. Then in the 1940s, they were abandoned (at least in Western countries) in favor of antibiotics, which were more dependable and easier to use. Now, with the widespread and growing problem of bacterial resistance to antibiotics, these bacteria-killing viruses are once again the focus of research. With new knowledge of phages and sophisticated technology, several research groups are investigating phages to determine which ones kill which species of bacteria. Scientists are genetically engineering phages so that bacteria will be slower to evolve resistance to them. Phages can also be used to improve food safety. For example, certain phages can kill deadly strains of *E. coli* in cattle. These bacteria do not appear to make cattle ill, but can cause illness and death in people who eat infected, undercooked hamburgers.

Lytic reproductive cycles destroy host cells

Two types of reproduction among viruses are lytic and lysogenic cycles. In a **lytic cycle,** the virus lyses (destroys) the host cell. When the virus infects a susceptible host cell, it forces the host to use its metabolic machinery to replicate viral particles. Viruses that have only a lytic cycle are described as **virulent.**

Five steps are typical in lytic viral reproduction (Fig. 23-2):

1. **Attachment (or absorption).** The virus attaches to receptors on the host cell wall.

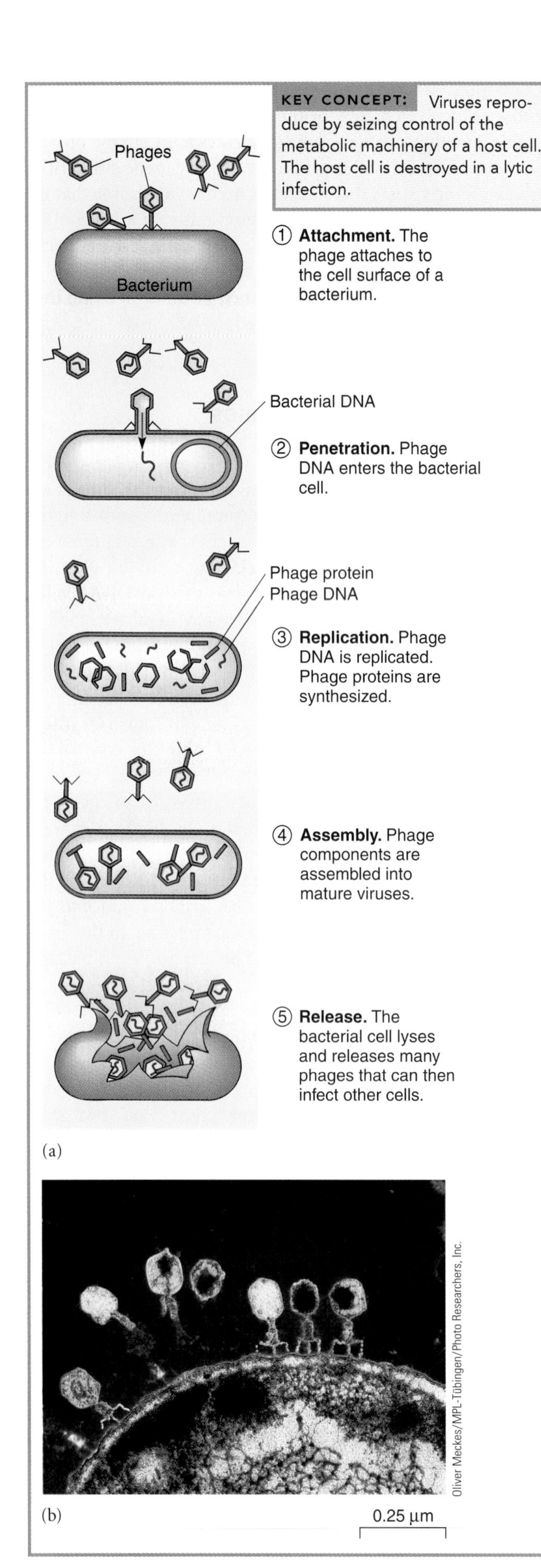

KEY CONCEPT: Viruses reproduce by seizing control of the metabolic machinery of a host cell. The host cell is destroyed in a lytic infection.
Phages
Bacterium
① Attachment. The phage attaches to the cell surface of a bacterium.
Bacterial DNA
② Penetration. Phage DNA enters the bacterial cell.
Phage protein
Phage DNA
③ Replication. Phage DNA is replicated. Phage proteins are synthesized.
④ Assembly. Phage components are assembled into mature viruses.
⑤ Release. The bacterial cell lyses and releases many phages that can then infect other cells.
(a)
(b)
0.25 µm
Oliver Meckes/MPL-Tübingen/Photo Researchers, Inc.

FIGURE **23-2** | **Lytic cycle.** ◀

(a) The sequence of events in a lytic infection. **(b)** Color-enhanced TEM of phages infecting a bacterium, *Escherichia coli.*

2. **Penetration.** The nucleic acid of the virus moves through the plasma membrane and into the cytoplasm of the host cell. The capsid of a phage remains on the outside, but many viruses that infect animal cells enter the host cell intact.
3. **Replication and synthesis.** The viral genome contains all the information necessary to produce new viruses. Once inside, the virus degrades the host cell nucleic acid and uses the host cell to synthesize the necessary components for its replication.
4. **Assembly.** The newly synthesized viral components are assembled into new viruses.
5. **Release.** Assembled viruses are released from the cell. Generally, lytic enzymes, produced by the phage late in the replication process, destroy the host plasma membrane. Phage release typically occurs all at once and results in rapid cell lysis, whereas animal viruses are often released slowly or bud off from the plasma membrane.

The new viruses infect other cells, and the process begins anew. The time required for viral reproduction, from attachment to the bacterium to the release of new viruses, varies from less than 20 minutes to more than an hour.

Temperate viruses integrate their DNA into the host DNA

Temperate viruses do not always destroy their hosts. In a **lysogenic cycle** the viral genome usually becomes integrated into the host bacterial DNA and is then referred to as a **prophage.** When the bacterial DNA replicates, the prophage also replicates (Fig. 23-3). The viral genes that code for viral structural proteins may be repressed indefinitely. Bacterial cells carrying prophages are called *lysogenic cells.* Certain external conditions (such as ultraviolet light and x-rays) cause temperate viruses to revert to a lytic cycle and then destroy their host. Sometimes temperate viruses become lytic spontaneously.

Bacterial cells containing certain temperate viruses may exhibit new properties. This is called **lysogenic conversion.** An interesting example involves the bacterium *Corynebacterium diphtheriae,* which causes diphtheria. Two strains of this species exist, one that produces a toxin (and causes diphtheria) and one that does not. The only difference between these two strains is that the toxin-producing bacteria contain a specific temperate phage. The phage DNA codes for the powerful toxin that causes the symptoms of diphtheria. Similarly, the bacterium *Clostridium botulinum,* which causes botulism, a serious form of food poisoning, is harmless unless it contains certain prophage DNA that induces synthesis of the toxin.

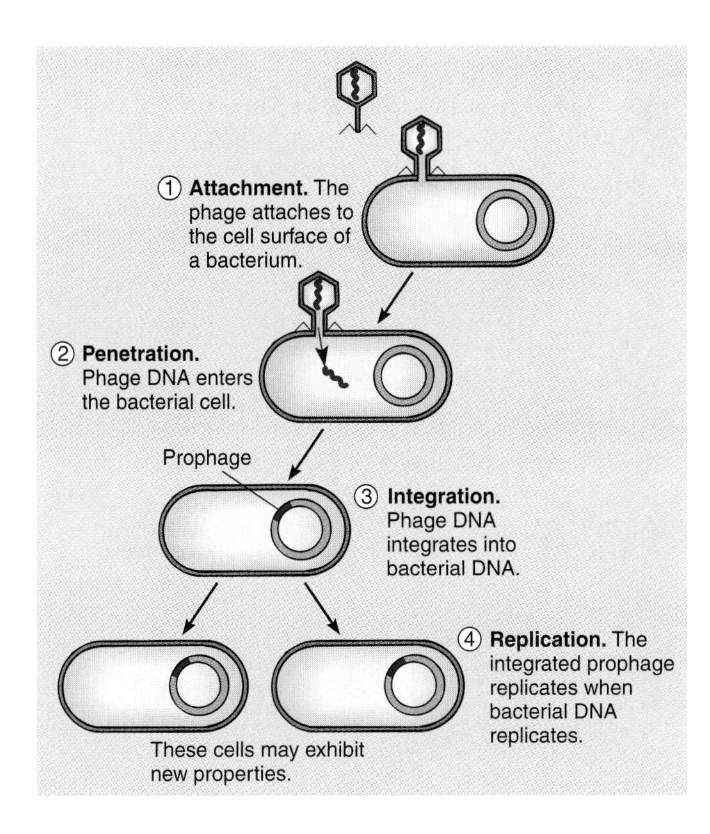
① Attachment. The phage attaches to the cell surface of a bacterium.
② Penetration. Phage DNA enters the bacterial cell.
Prophage
③ Integration. Phage DNA integrates into bacterial DNA.
④ Replication. The integrated prophage replicates when bacterial DNA replicates.
These cells may exhibit new properties.

FIGURE **23-3** | **Lysogenic cycle.**

Temperate phages integrate their nucleic acid into the host cell DNA, making it a lysogenic cell.

Some viruses infect animal cells

Hundreds of different viruses infect humans and other animals. Most viruses cannot survive very long outside a living host cell, so their survival depends on their being transmitted from animal to animal. The type of attachment proteins on the surface of a virus determines what type of cell it can infect. Some viruses, such as the adenoviruses, have fibers that project from the capsid and may help the virus adhere to complementary receptor sites on the host cell. Other viruses, such as those that cause herpes, influenza, and rabies, are surrounded by a lipoprotein envelope with projecting glycoprotein spikes that aid in attachment to a host cell.

Receptor sites typically vary with each species and sometimes with each type of tissue. Thus many human viruses can infect only humans because their attachment proteins combine only with receptor sites found on human cell surfaces. The measles virus and the pox viruses infect many types of human tissue because their attachment proteins combine with receptor sites on a variety of cells. In contrast, polioviruses attach to specific types of cells, such as those that line the digestive tract and motor neurons of the brain and spinal cord.

Viruses have several ways to penetrate animal cells (Fig. 23-4). After attachment to a host-cell receptor, some enveloped viruses fuse with the animal cell's plasma membrane. The viral capsid and nucleic acid are both released into the animal cell. Other viruses enter the host cell by *endocytosis.* In this process, the plasma membrane of the animal cell invaginates to form a membrane-bounded vesicle that contains the virus.

In DNA animal viruses, the synthesis of viral DNA and protein is similar to the processes by which the host cell would normally carry out its own DNA and protein synthesis. In most RNA viruses, RNA replication and transcription take place with the help of an RNA-dependent RNA polymerase. However, **retroviruses** are RNA viruses that have a DNA polymerase called **reverse transcriptase** used to transcribe the RNA genome into a DNA intermediate (Fig. 23-5 on page 441). This DNA becomes integrated into the host DNA by an enzyme also carried by the virus. Copies of the viral RNA are synthesized as the incorporated DNA is transcribed by host RNA polymerases. The human immunodeficiency virus (HIV) that causes acquired immunodeficiency syndrome (AIDS) is a retrovirus. Certain cancer-causing viruses are also retroviruses.

After the viral genes are transcribed, the viral structural proteins are synthesized. The capsid is produced, and new virus particles are assembled. Viruses that do not have an outer envelope exit by cell lysis. The plasma membrane ruptures, releasing many new viral particles. Enveloped viruses obtain their lipoprotein envelopes by picking up a fragment of the host plasma membrane as they leave the infected cell (see Fig. 23-4b and Fig. 23-5).

Viral proteins damage the host cell in a variety of ways. These proteins may alter the permeability of the plasma membrane or may inhibit synthesis of host nucleic acids or proteins. Viruses sometimes damage or kill their host cells by their sheer numbers. A poliovirus can produce 100,000 new viruses within a single host cell!

Animal viruses cause hog cholera, foot-and-mouth disease, canine distemper, swine influenza, and certain types of cancer (such as feline leukemia). Most humans suffer from two to six viral infections each year, including common colds. Viruses cause chickenpox, herpes simplex (one type causes genital herpes), mumps, rubella (german measles), rubeola (measles), rabies, warts, infectious mononucleosis, influenza, viral hepatitis, and AIDS (Table 23-1 on page 442 and *On the Cutting Edge: Emerging and Re-emerging Diseases* on page 443).

Some viruses infect plant cells

Viral diseases are spread among plants by insects such as aphids and leafhoppers as they feed on plant tissues. Because of the thick cell wall, viruses cannot penetrate plant cells unless the cells are damaged. Plant viruses are inherited by way of infected seeds or by asexual propagation. Once a plant is infected, the virus spreads through the plant body by passing through plasmodesmata (cytoplasmic connections) that penetrate the walls between adjacent cells (see Fig. 5-25). The genome of most plant viruses consists of RNA.

Symptoms of viral infection include reduced plant size, and spots, streaks, or mottled patterns on leaves, flowers, or fruits (Fig. 23-6 on page 442). Infected crops almost always produce lower yields. Cures are not known for most viral diseases of plants, so it is common to burn plants that have been infected. Some agricultural scientists are focusing their efforts on pre-

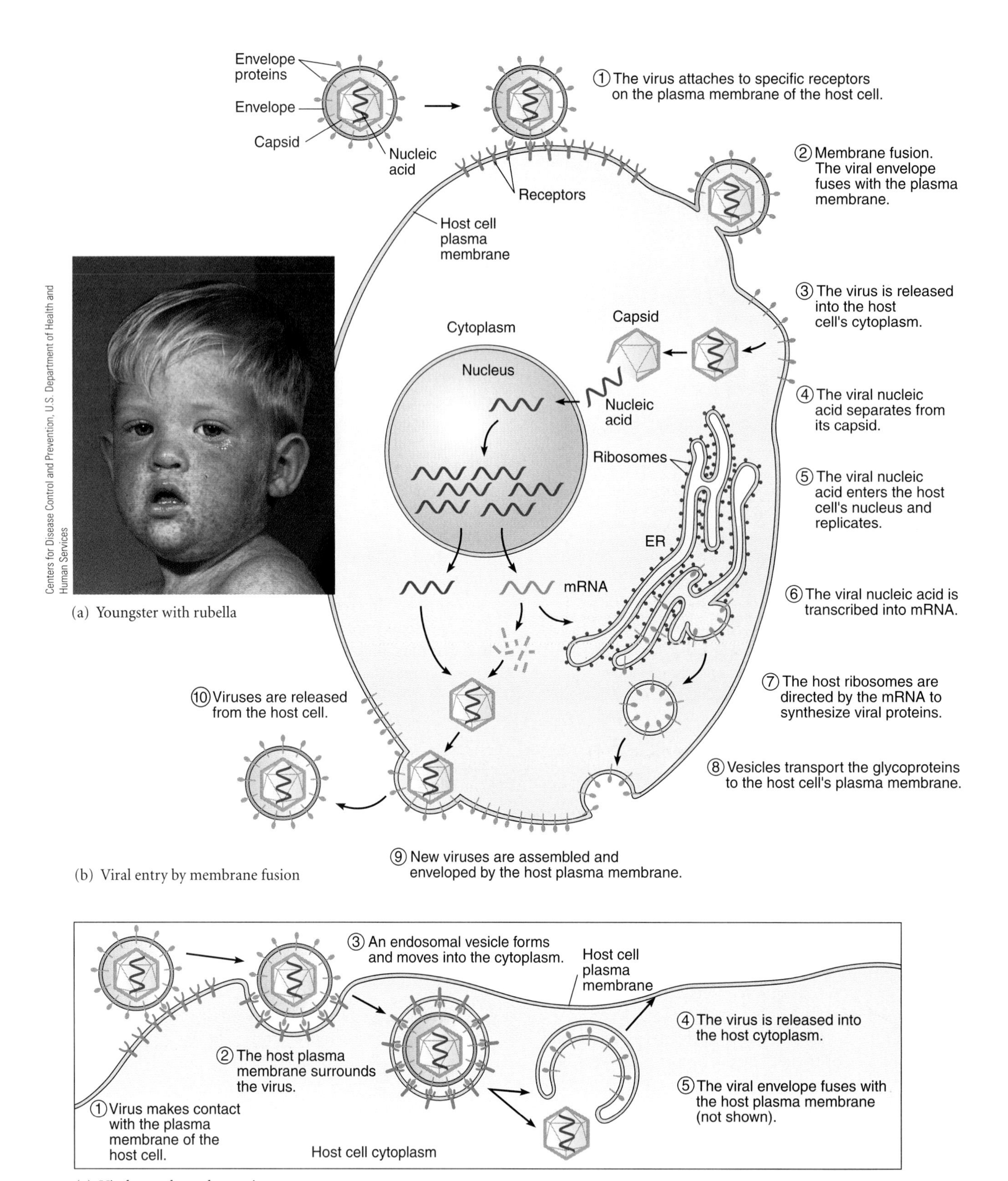

Envelope proteins
Envelope
Capsid
Nucleic acid
Receptors
Host cell plasma membrane
① The virus attaches to specific receptors on the plasma membrane of the host cell.
② Membrane fusion. The viral envelope fuses with the plasma membrane.
③ The virus is released into the host cell's cytoplasm.
Capsid
Cytoplasm
Nucleus
Nucleic acid
④ The viral nucleic acid separates from its capsid.
Ribosomes
⑤ The viral nucleic acid enters the host cell's nucleus and replicates.
ER
mRNA
⑥ The viral nucleic acid is transcribed into mRNA.
⑦ The host ribosomes are directed by the mRNA to synthesize viral proteins.
⑩ Viruses are released from the host cell.
⑧ Vesicles transport the glycoproteins to the host cell's plasma membrane.
⑨ New viruses are assembled and enveloped by the host plasma membrane.
Centers for Disease Control and Prevention, U.S. Department of Health and Human Services
③ An endosomal vesicle forms and moves into the cytoplasm.
Host cell plasma membrane
④ The virus is released into the host cytoplasm.
② The host plasma membrane surrounds the virus.
⑤ The viral envelope fuses with the host plasma membrane (not shown).
① Virus makes contact with the plasma membrane of the host cell.
Host cell cytoplasm

(a) Youngster with rubella

(b) Viral entry by membrane fusion

(c) Viral entry by endocytosis

FIGURE **23-4** **Viruses cause diseases in animals.**

(a) This child's rash is a symptom of rubella (German measles), which is caused by an RNA virus spread by close contact. **(b)** Some viruses enter animal cells by membrane fusion. Replication occurs, and the new viruses are released. **(c)** Some viruses enter the host cell by endocytosis.

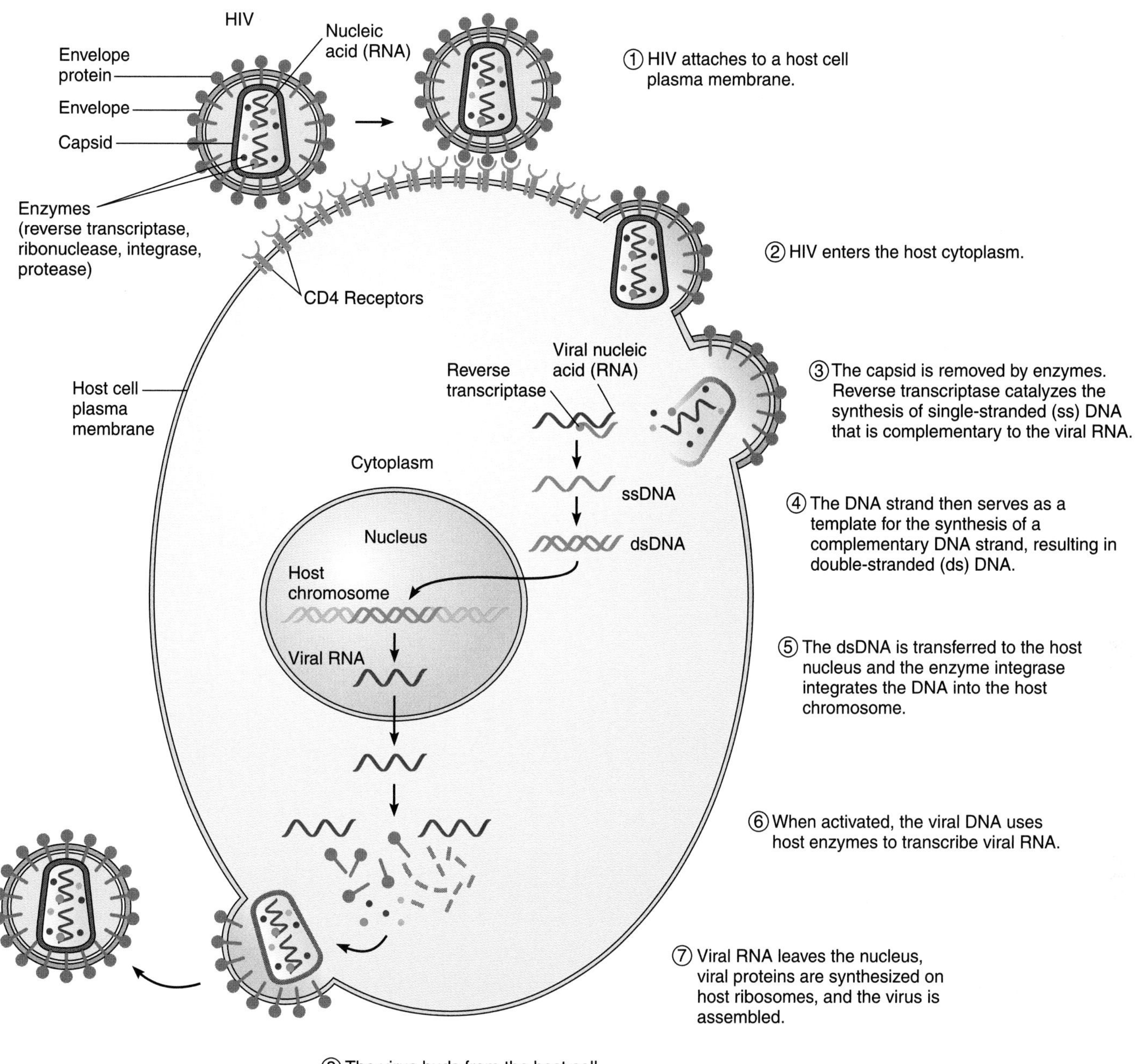
HIV
Envelope protein
Envelope
Capsid
Nucleic acid (RNA)
Enzymes (reverse transcriptase, ribonuclease, integrase, protease)
CD4 Receptors
Host cell plasma membrane
Reverse transcriptase
Viral nucleic acid (RNA)
Cytoplasm
ssDNA
dsDNA
Nucleus
Host chromosome
Viral RNA
① HIV attaches to a host cell plasma membrane.
② HIV enters the host cytoplasm.
③ The capsid is removed by enzymes. Reverse transcriptase catalyzes the synthesis of single-stranded (ss) DNA that is complementary to the viral RNA.
④ The DNA strand then serves as a template for the synthesis of a complementary DNA strand, resulting in double-stranded (ds) DNA.
⑤ The dsDNA is transferred to the host nucleus and the enzyme integrase integrates the DNA into the host chromosome.
⑥ When activated, the viral DNA uses host enzymes to transcribe viral RNA.
⑦ Viral RNA leaves the nucleus, viral proteins are synthesized on host ribosomes, and the virus is assembled.
⑧ The virus buds from the host cell, using the host plasma membrane to make the viral envelope.

ACTIVE FIGURE **23-5** **Life cycle of HIV, the retrovirus that causes AIDS.**

HIV infects T helper cells, specialized cells of the immune system. The virus attaches to protein receptors, known as CD4, on the host plasma membrane. HIV has two identical single-stranded RNA molecules.

BiologyNow™ Watch **the HIV life cycle unfold** by clicking on this figure on your BiologyNow CD-ROM.

venting viral disease by developing virus-resistant strains of important crop plants.

Review

- What characteristics does a virus share with a living cell? What characteristics of life are absent in a virus?
- What are the steps in a lytic cycle? Draw diagrams to illustrate your answer.
- How is a lysogenic cycle different from a lytic cycle?

BiologyNow™ Assess your understanding of **viruses** by taking the pretest on your BiologyNow CD-ROM.

TABLE 23-1 Animal Viruses

Group	Diseases Caused	Characteristics
DNA Viruses		
Poxviruses	Smallpox, cowpox, monkeypox, and economically important diseases of domestic fowl	dsDNA; large, complex viruses; replicates in the cytoplasm of the host cell*
Herpesviruses	Cold sores (herpes simplex virus type 1); genital herpes, a sexually transmitted disease (herpes simplex virus type 2); chickenpox and shingles (herpes varicella-zoster virus); infectious mononucleosis and Burkitt's lymphoma (Epstein-Barr virus)	dsDNA; medium to large, enveloped viruses; replicates in the the host nucleus†
Adenoviruses	Respiratory tract disorders (e.g., sore throat, tonsillitis), conjunctivitis, and gastrointestinal disorders are caused by more than 40 types of adenoviruses in humans; other varieties infect other animals	dsDNA; replicates in the host nucleus
Papovaviruses	Human warts and some degenerative brain diseases; some cancers	dsDNA‡
Parvoviruses	Infections in dogs, swine, arthropods, rodents; gastroenteritis in humans (transmitted by consumption of infected shellfish)	ssDNA; some require a helper virus in order to multiply
RNA Viruses		
Picornaviruses	Polio (poliovirus); hepatitis A (hepatitis A virus); intestinal disorders (enteroviruses); common cold (rhinoviruses); aseptic meningitis (coxsackievirus, echovirus)	ssRNA that can serve as mRNA; diverse group of small viruses
Togaviruses	Rubella (german measles)	ssRNA that can serve as mRNA; large diverse group of medium-sized, enveloped viruses; many transmitted by arthropods
Orthomyxoviruses	Influenza (flu) in humans and other animals	ssRNA that serves as template for mRNA synthesis; medium-sized viruses that often exhibit projecting spikes
Paramyxoviruses	Rubeola (measles) and mumps in humans; distemper in dogs	ssRNA; resemble orthomyxoviruses but somewhat larger
Rhabdoviruses	Rabies	ssRNA
Coronaviruses	Upper respiratory infections, SARS	ssRNA; envelope
Flaviviruses	Yellow fever; West Nile virus; hepatitis C (the most common reason for liver transplants in the United States)	ssRNA
Filoviruses	Hemorrhagic fever, including that caused by the Ebola virus	ssRNA
Bunyaviruses	St Louis encephalitis; hantavirus pulmonary syndrome (caused by Sin Nombre Virus, a hantavirus)	ssRNA; envelope
Reoviruses	Vomiting and diarrhea; encephalitis	dsRNA; no envelope
Retroviruses	AIDS; some types of cancer	RNA viruses that contain reverse transcriptase for transcribing the RNA genome into DNA; two identical molecules of ssRNA

*The vaccinia (cowpox) virus is used to produce genetically engineered vaccines.
†These viruses frequently cause latent infections; some cause tumors.
‡The virus SV40 has been used as a vector to transport genes into cells.

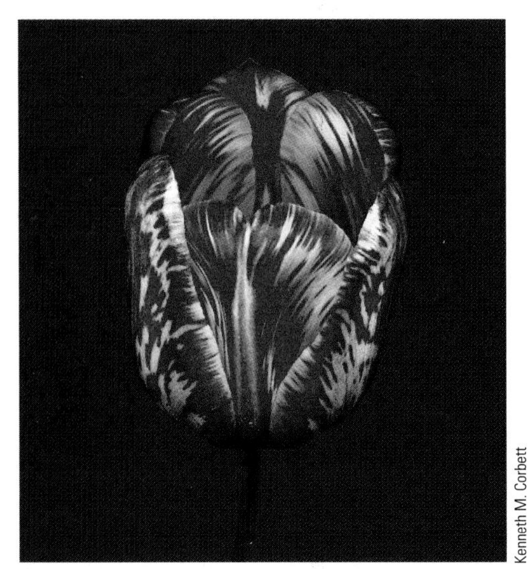
(a)

Kenneth M. Corbett

(b)

Jack M. Bostrack/ Visuals Unlimited

FIGURE 23-6 Plant viruses. **(a)** Virus-streaked tulip. The virus that causes this relatively harmless disease affects pigment formation in the petals. **(b)** Pepper leaves infected with tobacco mosaic virus. The leaf is characteristically mottled with light green areas.

ON THE CUTTING EDGE Emerging and Re-emerging Diseases

Hypothesis: Outbreaks of emerging or re-emerging diseases can be quickly contained to minimize loss of life and to prevent widespread epidemics.

Method: A combination of new technology, increased awareness, and greater cooperation among public health agencies, health care providers, and the media.

Results: Responses to recent outbreaks of SARS have been encouraging.

Conclusion: Not yet known

The development of antibiotics, vaccines, and improved healthcare have made it possible for public health officials to effectively control some diseases. However, many familiar diseases, for example, malaria, tuberculosis, bacterial pneumonias, and smallpox, continue to infect large numbers of people, and sometimes reappear in forms that are resistant to drug treatments. Re-emerging diseases are those that have been almost eradicated, and then strike unpredictably, causing an epidemic, sometimes in a new geographic area. Emerging diseases, those new to the human population—such as AIDS, SARS, ebola, eastern equine encephalitis, and West Nile virus—often appear suddenly. The outbreak of severe acute respiratory syndrome (SARS) in 2003 is a grim reminder that pathogens can strike quickly and spread rapidly.

According to the U.S. Centers for Disease Control and Prevention, more than 200 new, continual, or re-emerging pathogens have the potential to strike globally. Historically, new viral strains have claimed many human lives. For example, in 1918 an influenza epidemic killed more 20 million people throughout the world. Some epidemiologists warn that an influenza epidemic today could kill huge numbers of people, perhaps as many as one third of the human population.

As new technology is developed, researchers are able to more quickly characterize emerging disease organisms and explore treatments and vaccines. For example, hepatitis E, a viral disease that may be re-emerging, was documented as an epidemic in India in 1955, but was not isolated as a separate virus until the development of modern immunologic research tools in the early 1980s. In contrast, severe acute respiratory syndrome (SARS), first recognized as a global threat in mid-March 2003, was successfully contained in less than four months. The SARS virus was isolated and identified as a coronavirus in a matter of a few weeks. This information coupled with public health response helped speed efforts to diagnose and treat the disease, and to reduce transmission, thus preventing a global epidemic of this deadly disease. Still, the World Health Organization cautions that the SARS virus could hide in some animal or environmental reservoir, and resurface when conditions again become favorable for spread to its new human host. Epidemiologists are modeling the SARS epidemic in an effort to answer basic questions such as whether we can move from control of local outbreaks to global eradication of this disease.*

Even at the level of our current knowledge about viruses and epidemiology, just how prepared are we to contain a particularly virulent virus? How well are we containing HIV (the virus that causes AIDS), which continues to claim more than 3 million lives worldwide each year? Since the September 11, 2001 terrorist attacks in the United States, bioterrorism has become a critical concern worldwide. Bioterrorism is the intentional use of microorganisms or toxins derived from living organisms to cause death or disease in humans, animals, or plants on which humans depend. For example, in 2001 *Bacillus anthracis,* the causative agent of anthrax, was intentionally released and disseminated through the U.S. postal system. Terrorists could conceivably initiate epidemics of anthrax, smallpox, plague, and other potentially fatal diseases. The quest for effective treatments and vaccines for these diseases has taken on new urgency.

Whether disease outbreaks occur naturally or intentionally, countries must be prepared to protect their citizens. Are public health agencies prepared to quickly detect and investigate new outbreaks? Are they prepared to help the health care community contain the spread of disease? Are we training physicians, nurses, and other health care providers to recognize, report, and treat infectious diseases? Do we have sufficient supplies of medications and vaccines that would be necessary to contain an epidemic? Are systems of communication effective in reporting and sharing information among health care agencies, providers, and the public?

Human activity, including social factors such as urbanization, global travel, and war, contributes to epidemics of infectious disease. For example, as human populations concentrate in cities, large numbers of people come into close contact, permitting the rapid spread of viruses. Living conditions, including sanitation, nutrition, physical stress, level of health care, and sexual practices, are important factors in the spread of disease. In the United States and other highly developed countries, infectious disease accounts for about 4% to 8% of deaths compared with death rates of 30% to 50% in developing regions.

Changing natural habitats can create the conditions necessary for new pathogens to emerge. For example, cutting down forests can bring disease-carrying insects into contact with humans. Sometimes, even naturally occurring ecological changes can spawn outbreaks of disease. The 1993 outbreak of hantavirus in the Southwestern United States killed more than 50 people. A mild winter coupled with heavy rainfall resulted in a large crop of seeds, which supported a population explosion of field mice. The mice carry the virus. Lessons learned from dealing with emerging viruses and the resurgence of old ones will help us contain future epidemics, but much more research is needed.

*Dye, C., and Gay, N., "Modeling the SARS Epidemic," *Science,* Vol. 300, Issue 5627, 1884–1885, June 20, 2003

VIROIDS AND PRIONS

Learning Objective

6 Compare and contrast viroids and prions.

In 1961, an infective agent in potatoes was discovered that had a short RNA strand but no protein coat. The agent, named a **viroid,** is much smaller than a virus and has no protective protein coat and no associated proteins to assist in duplication. Each viroid consists of a very short strand of naked RNA (only 250–400 nucleotides) that serves as a template copied by host RNA polymerases. Viroids cause a variety of plant diseases. These infective agents are generally found within the host cell nucleus and may interfere with gene regulation. Viroids are extremely hardy and resist heat and ultraviolet radiation because of the condensed folding of their RNA.

PROCESS OF SCIENCE

Stanley Prusiner, professor of neurology and biochemistry at the University of California School of Medicine, San Francisco, began his studies of prions in the early 1970s, motivated by the death of a patient from Creutzfeldt-Jakob disease (CJD, a degenerative brain disease). Prusiner discovered that the infective agent was not affected by radiation (which typically mutates nucleic acids), and he could not find DNA or RNA in the particles. In 1982 he named the infective agent **prion** for "proteinaceous infectious particle." Prusiner's hypothesis that a pathogen could exist and transfer information without nucleic acids went against all accepted biological dogma. However, Prusiner and other researchers continued to study prions, and their findings continued to support Prusiner's hypothesis. In 1997 Prusiner was awarded the Nobel Prize in physiology or medicine for his discovery of prions—a new biological principle of infection.

Prusiner and others have shown that animals have a gene that encodes the prion, a protein consisting of 208 amino acids. This protein is normally harmless and may help process copper. However, it sometimes converts to a different shape, an insoluble variant that accumulates in the brains of patients with **transmissible spongiform encephalopathies (TSEs).** This group of fatal degenerative brain diseases, which have been identified in birds and mammals, are called TSEs because when infected, the brain appears to develop holes, becoming somewhat spongelike. Genetically engineered mice that lack the prion protein gene are immune to TSE infection. According to Prusiner, mutations in the prion protein gene increase the risk that the protein will change shape.

One of the most studied prion diseases is *scrapie* in sheep and goats. When infected, animals lose coordination, become irritable, and itch so severely that they scrape off their wool or hair. Bovine spongiform encephalopathy (BSE), a related prion disease popularly referred to as "mad cow disease," became epidemic in cattle in the United Kingdom in the 1990s. Scientists hypothesized that cattle became infected when they ate feed that had been mixed with sheep offal—brains and other organs—containing prions. After sheep offal was banned as feed, the epidemic was slowly brought under control. This disease led to widespread bans on the export and consumption of British cattle.

More than 120 people in Europe have died from a human variety of BSE, providing evidence that the disease is transmissible from cow to human. The human disease is called vCJD because it is a variant of CJD. The infectious agent has been recovered from infected human neural tissue and appears similar to the prion that causes scrapie in sheep. Human-to-human transmission has been associated with tissue and organ transplants and transfusion with contaminated blood. Recently, *chronic wasting disease,* a disease related to mad cow disease, has spread among deer and elk populations in the United States. Studies are under way to determine whether this disease can infect livestock and humans. Efforts to develop treatments for prion diseases have not yet been successful.

Review

- How do viroids differ from prions?
- How do viroids and prions differ from viruses?

Biology Now™ Assess your understanding of **viroids and prions** by taking the pretest on your BiologyNow CD-ROM.

PROKARYOTES

Learning Objectives

7 Describe the structure and common shapes of prokaryotic cells.

8 Describe asexual reproduction in prokaryotes, and summarize three mechanisms (transformation, conjugation, and transduction) that may lead to genetic recombination.

9 Characterize the metabolic diversity of autotrophic and heterotrophic prokaryotes, including aerobes, facultative anaerobes, and obligate anaerobes.

In contrast to viruses, viroids, and prions, which consist only of nucleic acid and/or protein, **prokaryotes** are cellular organisms. Recall from Chapters 4 and 22 that the cell structure of prokaryotes is fundamentally different from the cells of other living organisms. Microbiologists assign prokaryotes to two domains: **Archaea** and **Eubacteria.**

Most prokaryotic cells are very small. Typically, their diameter ranges from 0.5 to 1.0 μm. Their cell volume is only about one thousandth that of small eukaryotic cells, and their length is only about one tenth. Most prokaryotes are unicellular, but some form colonies or filaments containing specialized cells.

Prokaryotes have several common shapes

Prokaryotes have two basic shapes: spherical and rods. Spherical prokaryotes, known as **cocci** (sing., *coccus*), occur singly in some species and in groups of independent cells in others (Fig. 23-7a). Cells may be grouped in twos *(diplococci),* in long chains *(streptococci),* or in irregular clumps that look like bunches of grapes *(staphylococci).* Rod-shaped prokaryotes, called **bacilli** (sing., *bacillus*), may occur as single rods or as long chains of rods (Fig. 23-7b). Some prokaryotes form spirals. If the spiral-shaped cell is flexible, it is a **spirochete;** if rigid, if it is a **spirillum** (pl., *spirilla*) (Fig. 23-7c). A spirillum shaped like a comma is called a **vibrio.**

Prokaryotic cells lack membrane-enclosed organelles

Prokaryotic cells do not have membrane-enclosed organelles typical of eukaryotic cells. Thus these cells do not have nuclei, mitochondria, chloroplasts, endoplasmic reticula, Golgi complexes, or lysosomes (Fig. 23-8). The dense cytoplasm of the prokaryotic cell contains ribosomes (smaller than those found in eukaryotic cells) and storage granules that hold glycogen,

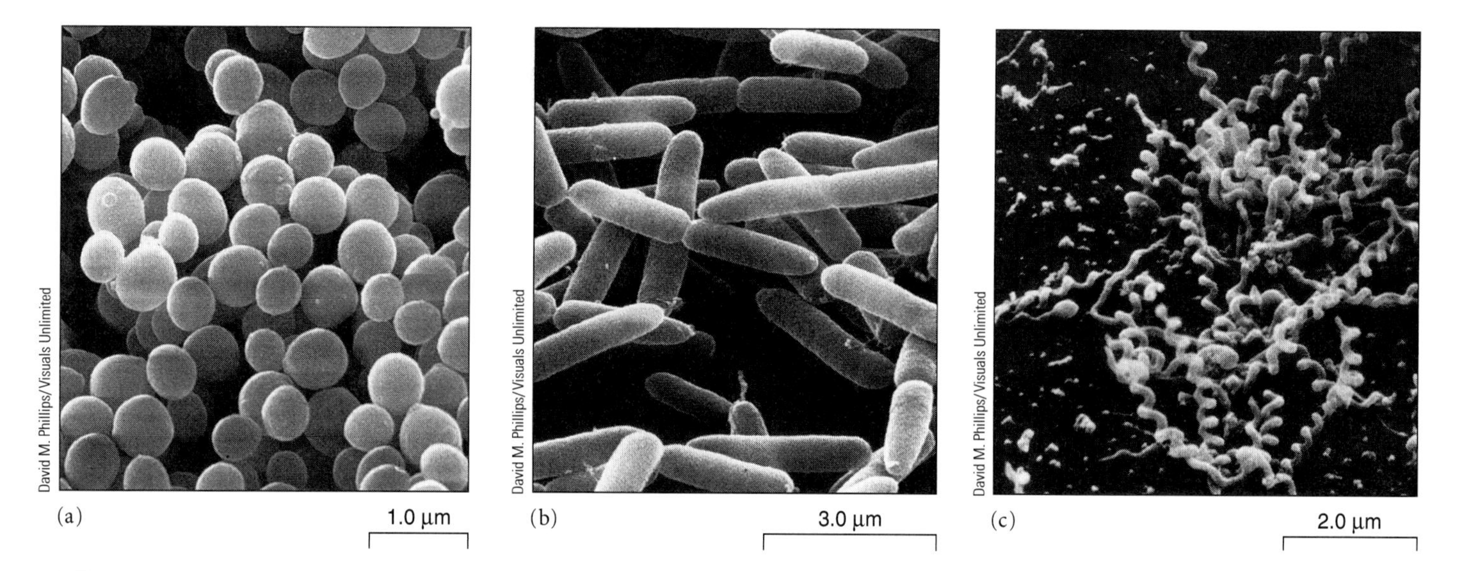

FIGURE **23-7** **Common shapes of prokaryotes.**

(a) SEM of *Micrococcus*, cocci bacteria. **(b)** SEM of *Salmonella*, bacilli bacteria. **(c)** SEM of *Spiroplasma*, spirilla bacteria.

lipid, or phosphate compounds. Enzymes needed for metabolic activities may be located in the cytoplasm. Although the membranous organelles of eukaryotic cells are absent, in some prokaryotic cells the plasma membrane is extensively folded inward. Enzymes needed for cellular respiration and photosynthesis may be associated with the plasma membrane or its folds.

A cell wall typically covers the cell surface

Most prokaryotic cells have a cell wall surrounding the plasma membrane. The cell wall provides a rigid framework that supports the cell, maintains its shape, and keeps it from bursting under hypotonic conditions (see Chapter 5). Most bacteria are adapted to hypotonic surroundings. When wall-less forms of bacteria are produced experimentally, they must be maintained in isotonic solutions to keep them from bursting. However, cell walls are of little help when a bacterium is in a hypertonic environment, as in food preserved by a high sugar or salt content. That is why most bacteria grow poorly in jellies, jams, salted fish, and other foods preserved in these ways.

The eubacterial cell wall includes **peptidoglycan,** a complex polymer that consists of two unusual types of sugars (amino sugars) linked with short polypeptides. The sugars and polypeptides are linked to form a single macromolecule that surrounds the entire plasma membrane.

Differences in bacterial cell wall composition are of great interest to microbiologists and are important clinically. In 1888 the Danish physician Christian Gram developed the Gram staining procedure. Bacteria that absorb and retain crystal violet stain in the laboratory are referred to as **Gram-positive,** whereas those that do not retain the stain when rinsed with alcohol are **Gram-negative.** The cell walls of Gram-positive bacteria are very thick and consist primarily of peptidoglycan. The cell walls of Gram-negative bacteria have two layers: a thin peptidoglycan layer and a thick outer membrane. The outer membrane resembles the plasma membrane but contains polysaccharides bonded to lipids (Fig. 23-9).

Distinguishing between Gram-positive and Gram-negative bacteria is important in treating certain diseases. For example,

FIGURE **23-8** **Structure of a prokaryotic cell.**

This bacillus is a gram-negative bacterium (discussed in text). Note the absence of a nuclear envelope surrounding the bacterial DNA.

KEY CONCEPT: Prokaryotic cells are fundamentally different from eukaryotic cells. The organelles of prokaryotic cells are not enclosed by membranes.

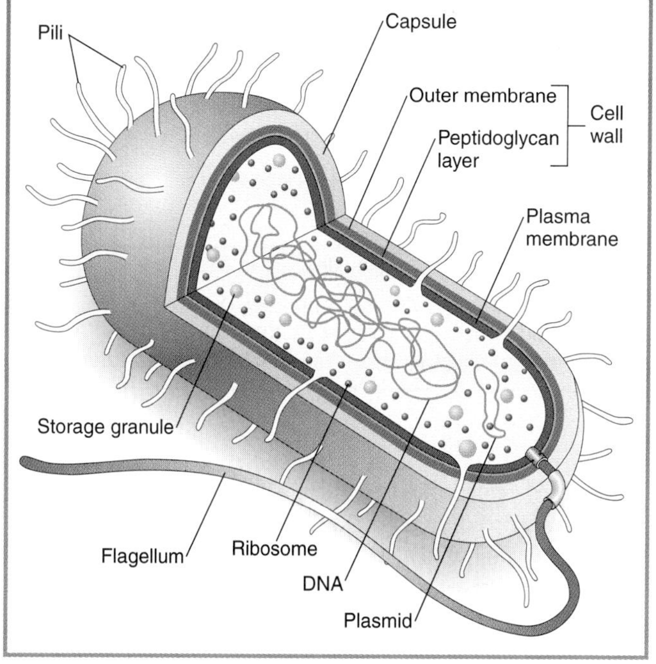

Pili
Capsule
Outer membrane
Peptidoglycan layer
Cell wall
Plasma membrane
Storage granule
Flagellum
Ribosome
DNA
Plasmid

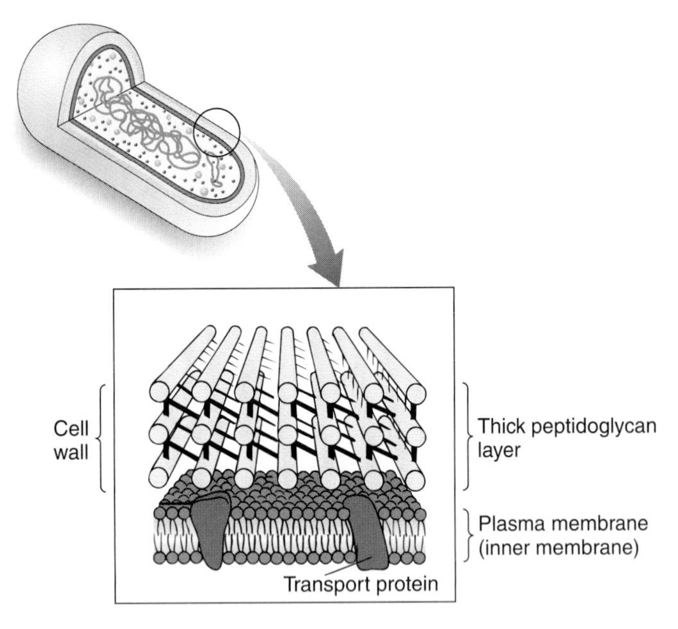

(a) Gram-positive cell wall

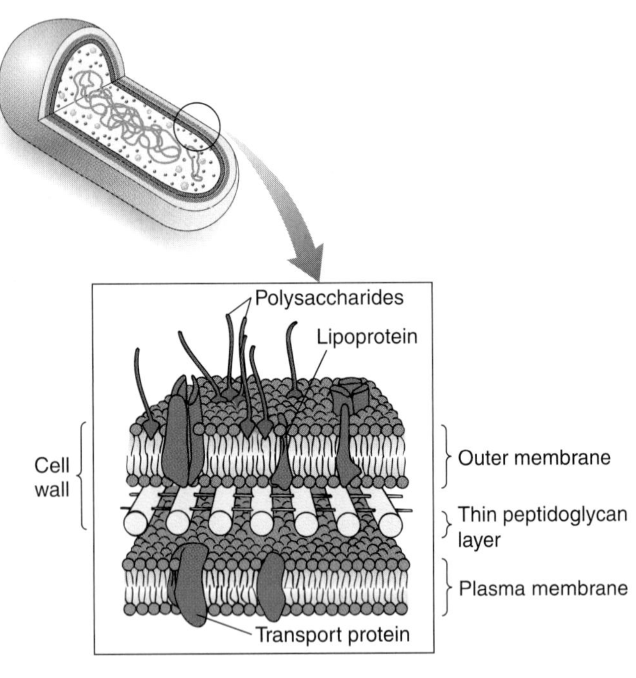

(b) Gram-negative cell wall

FIGURE **23-9** **Bacterial cell walls.**

(a) In the Gram-positive cell wall, a thick layer of peptidoglycan molecules is held together by amino acids. **(b)** In the Gram-negative cell wall, a thin peptidoglycan layer is covered by an outer membrane.

the antibiotic penicillin interferes with peptidoglycan synthesis, ultimately resulting in a fragile cell wall that cannot protect the cell (see Chapter 6, section on drugs that are enzyme inhibitors). Predictably, penicillin works most effectively against Gram-positive bacteria.

Some species of bacteria produce a **capsule** or slime layer that surrounds the cell wall. In free-living species, the capsule may provide the cell with added protection against phagocytosis (engulfment; see Chapter 5) by other microorganisms. In disease-causing bacteria, the capsule may protect against phagocytosis by the host's white blood cells. For example, the ability of *Streptococcus pneumoniae* to cause bacterial pneumonia depends on its capsule. A strain of *S. pneumoniae* that lacks a capsule does not cause the disease. Bacteria also use their capsules to attach to surfaces such as rocks, plant roots, or human teeth (where they cause dental plaque).

Some bacteria have hundreds of hairlike appendages known as **pili** (sing., *pilus*). These protein structures help bacteria adhere to one another or to certain surfaces, such as the cells they infect. Some elongated pili, called *sex pili,* are involved in transmitting DNA between bacteria.

Many types of prokaryotes are motile

Can you imagine trying to swim through molasses? Water has the same relative viscosity to prokaryotes that molasses has to humans. Most motile prokaryotes move by means of rotating **flagella.** The number and location of flagella are important in classifying some bacterial species.

Unlike eukaryotic flagella, prokaryotic flagella do not consist of microtubules (see Chapter 4). A bacterial flagellum consists of three parts: a basal body, a hook, and a single filament (Fig. 23-10). The basal body is a complex structure that anchors the flagellum into the cell wall by disc-shaped plates. The curved hook connects the basal body to the long, hollow filament that extends into the outside environment. The basal body is a motor. The bacterium uses energy from ATP to pump protons out of the cell. Diffusion of these protons back into the cell powers the motor that spins the flagellum, producing a rotary motion that pushes the cell much like a propeller pushes a ship through the water.

Prokaryotes have a circular DNA molecule

The genetic material of a prokaryote lies in the cytoplasm and is not surrounded by a nuclear envelope. In most species, it is contained in a single, circular DNA molecule. If stretched out to its full length, this molecule would be about 1000 times longer than the cell itself. Unlike eukaryotic chromosomes, prokaryote DNA has little protein associated with it.

In addition to the genomic DNA, most bacteria have a small amount of genetic information present as one or more **plasmids,** smaller circular fragments of DNA. Plasmids replicate independently of the genomic DNA (see Chapter 14) or become integrated into it. Bacterial plasmids often have genes that code for catabolic enzymes, for genetic exchange, or for resistance to antibiotics.

Most prokaryotes reproduce by binary fission

Prokaryotes reproduce asexually, generally by **binary fission,** a process in which one cell divides into two similar cells as in the chapter introduction photograph. First the circular DNA repli-

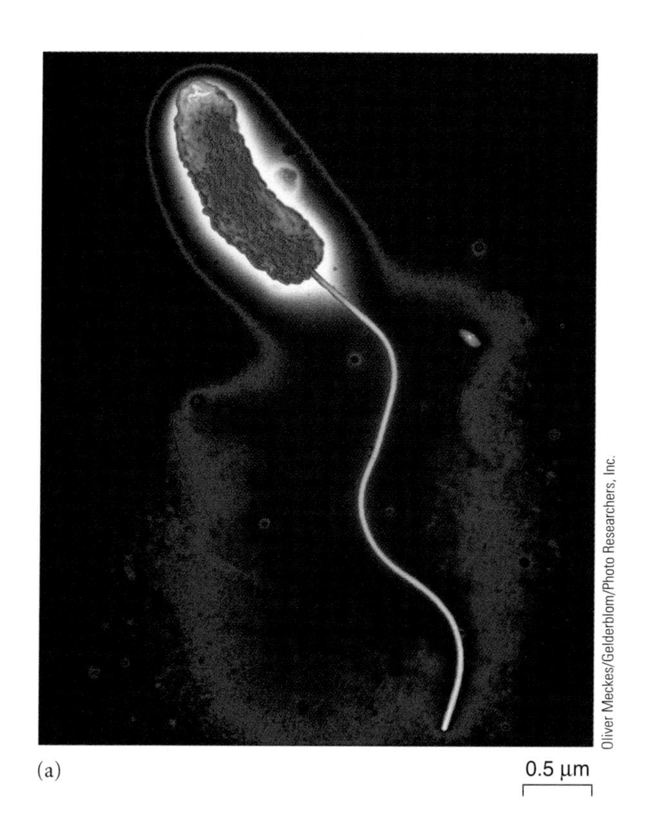

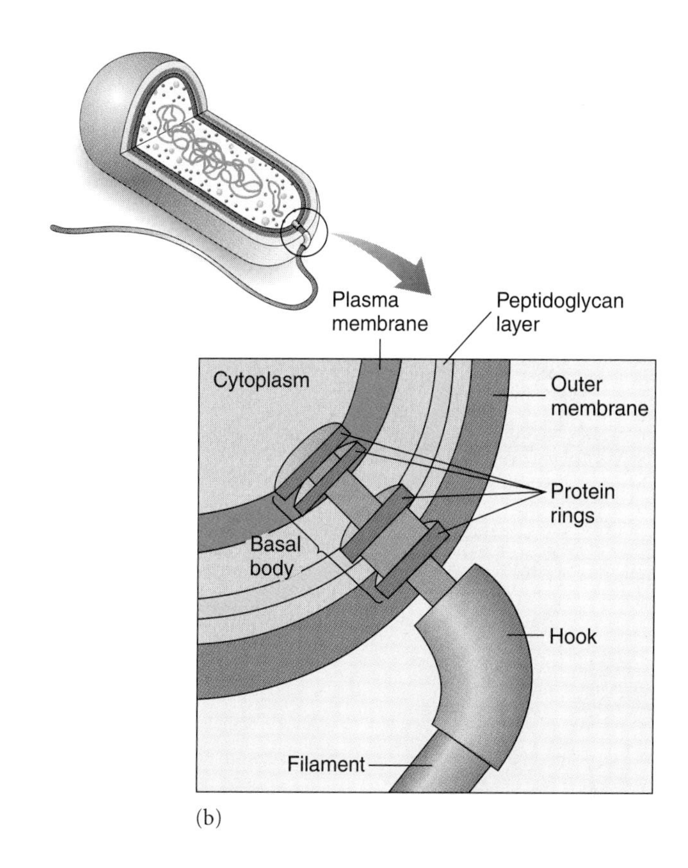

FIGURE **23-10** **Bacterial flagella.**

(a) Color-enhanced TEM of *Vibrio cholerae,* the flagellated bacterium that causes cholera. **(b)** Structure of a bacterial flagellum. The motor is the basal body, which consists of a series of disc-shaped plates that anchor the flagellum to the cell wall and plasma membrane and spin the hook and filament of the flagellum.

cates, and then a transverse wall is formed by an ingrowth of both the plasma membrane and the cell wall.

Binary fission occurs with remarkable speed. Under ideal conditions, some bacterial species divide in less than 20 minutes. At this rate, if nothing interfered, one bacterium could give rise to more than 1 billion bacteria within 10 hours! However, bacteria cannot reproduce at this rate for very long before lack of food or the accumulation of waste products affects their population expansion.

A less common form of asexual reproduction among bacteria is **budding.** In budding a cell develops a bulge, or bud, that enlarges, matures, and eventually separates from the mother cell. A few species of bacteria (actinomycetes) divide by **fragmentation.** Walls develop within the cell, which then separates into several new cells.

Although sexual reproduction involving the fusion of gametes does not occur in prokaryotes, genetic material is sometimes exchanged between individual bacteria. This exchange takes place by three different mechanisms: transformation, transduction, and conjugation.

1. In **transformation,** fragments of DNA released by a cell are taken in by another bacterial cell. Recall that this mechanism was used experimentally to demonstrate that genes are transferred from one bacterium to another and that DNA is the chemical basis of heredity (see Figure 11-1).
2. In a different process of gene transfer, **transduction,** a phage carries bacterial genes from one bacterial cell into another (Fig. 23-11 on the next page).
3. In **conjugation,** two cells of different mating types come together, and genetic material is transferred from one to the other (Fig. 23-12 on the next page). In contrast with transformation and transduction, conjugation involves contact between two cells.

Conjugation has been most extensively studied in the bacterium *Escherichia coli.* In the *E. coli* population there are donor cells (sometimes referred to as male cells) that have plasmids that can be transmitted to recipient (female) cells. A pilus on the donor cell recognizes the recipient cell and makes the first contact. A cytoplasmic bridge forms between the two cells, and DNA is transferred from donor to recipient cell.

Archaea reproduce by binary and multiple fission, budding, and fragmentation. Transduction, transformation, or conjugation has not been observed to date.

Some bacteria form endospores

When the environment becomes unfavorable, for example, when it gets very dry, the cells of some bacteria become dormant.

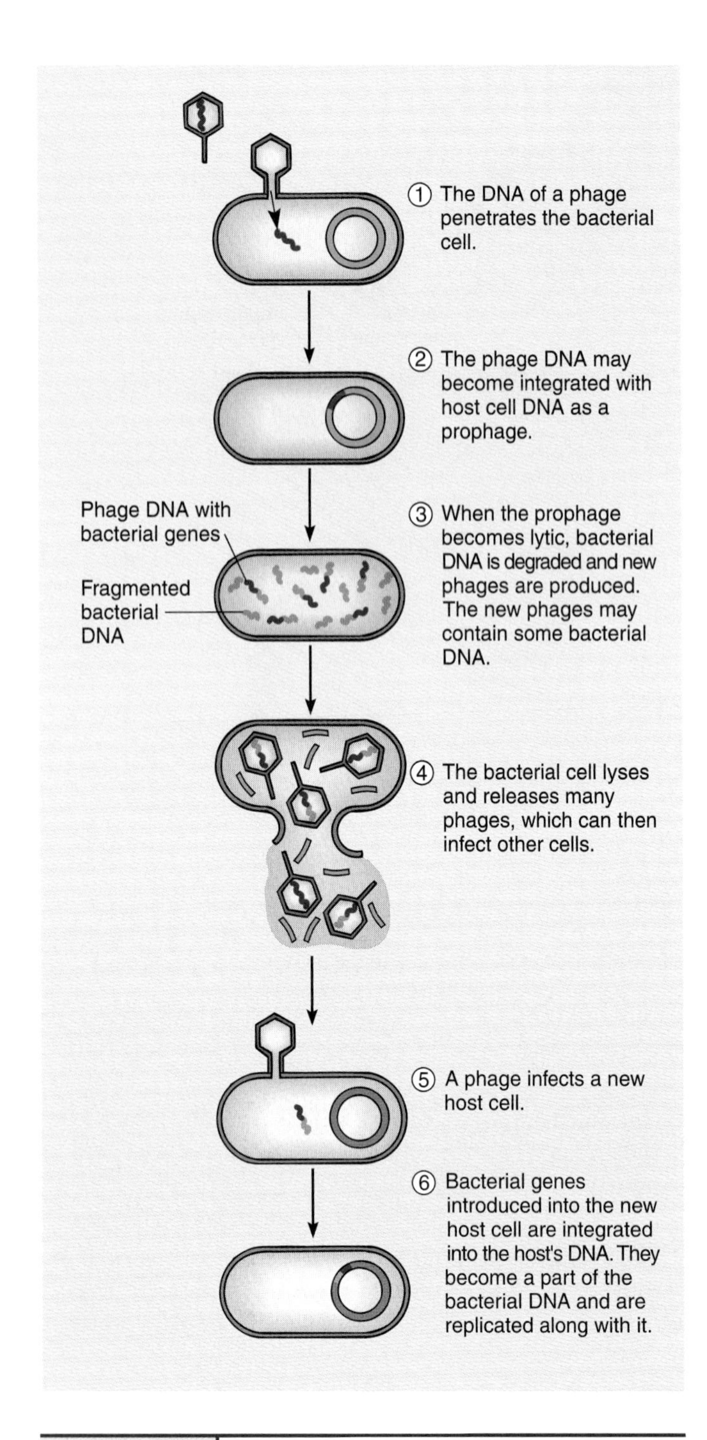

① The DNA of a phage penetrates the bacterial cell.
② The phage DNA may become integrated with host cell DNA as a prophage.
Phage DNA with bacterial genes
Fragmented bacterial DNA
③ When the prophage becomes lytic, bacterial DNA is degraded and new phages are produced. The new phages may contain some bacterial DNA.
④ The bacterial cell lyses and releases many phages, which can then infect other cells.
⑤ A phage infects a new host cell.
⑥ Bacterial genes introduced into the new host cell are integrated into the host's DNA. They become a part of the bacterial DNA and are replicated along with it.

FIGURE **23-11** | **Transduction.**
In this process, a phage transfers bacterial DNA from one bacterium to another.

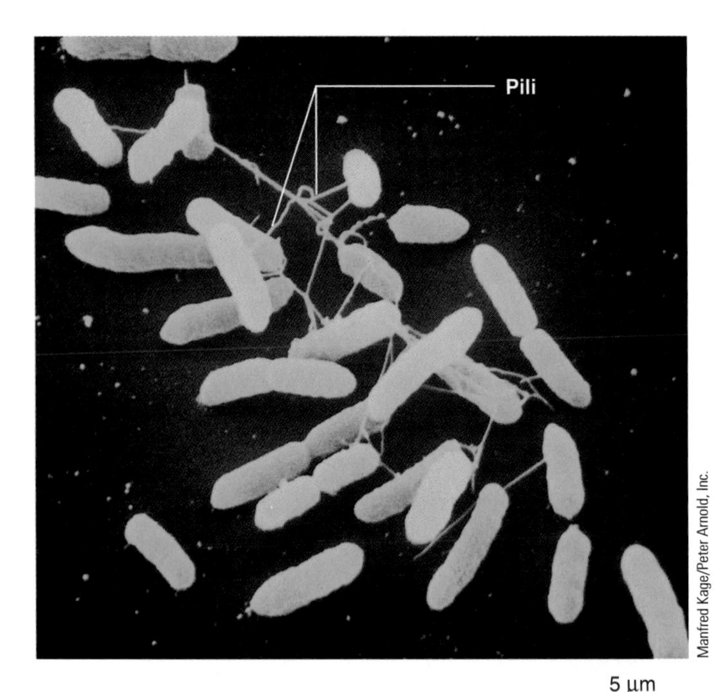

FIGURE **23-12** | **Conjugation.**
Color-enhanced SEM of *Serratia marcescens* bacteria, which are connected by pili. Stimulated by the contact, the cells pull close together and form cytoplasmic bridges between donor and recipient cells *(not shown).*

Cells lose water, shrink slightly, and remain inactive until water is again available. Certain other bacteria form dormant, extremely durable cells called **endospores.** After the endospore forms, the cell wall of the original cell lyses, releasing the endospore. Formation of endospores is not a type of reproduction in bacteria; endospores are not comparable to the reproductive spores of fungi and plants. Only one endospore is formed per original cell, so the total number of individuals does not increase. Endospores have not been found in association with archaea; these prokaryotes produce unique enzymes on the cell surface that protect them from cold, heat, and dessication.

Endospores survive in very dry, hot, or frozen environments, or when food is scarce (Fig. 23-13). Some endospores are so resistant that they can survive an hour or more of boiling or centuries of freezing. When environmental conditions are again suitable for growth, the endospore germinates, forming an active, growing bacterial cell. Several types of bacteria that form endospores are medically important. For example, when surgical instruments are not effectively sterilized, endospores of pathogenic bacteria may survive. Patients may then be exposed to serious diseases such as tetanus (caused by *Clostridium tetani*) and gas gangrene (caused by several other species of *Clostridium*). The endospore of *Bacillus anthracis,* the bacterium that causes anthrax, is so hardy that this pathogen has become a concern as an agent of biological warfare.

Metabolic diversity is evident among prokaryotes

Several methods of obtaining energy and carbon have evolved among the prokaryotes. Most prokaryotes are **heterotrophs** that obtain energy and carbon atoms from the organic compounds of other organisms. The majority are **chemoheterotrophs,** which include free-living decomposers (sometimes called *saprotrophs*) that get their carbon and energy from dead organic

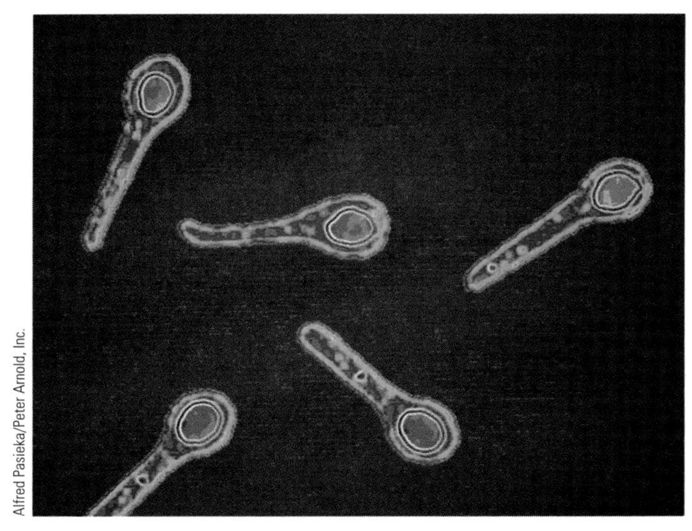

Alfred Pasieka/Peter Arnold, Inc.

FIGURE **23-13** | **Endospores.**

Color-enhanced TEM of *Clostridium tetani,* the bacterium that causes tetanus. Each bacterial cell *(blue)* contains one endospore *(orange),* a resistant, dehydrated cell that develops within the original cell.

matter. Some chemoheterotrophs are pathogens, which obtain their nourishment from living organisms and harm their hosts by causing diseases. Other heterotrophic bacteria benefit their hosts, for example by producing needed vitamins. The purple nonsulfur bacteria are **photoheterotrophs** that get their carbon from other organisms but use chlorophyll and other photosynthetic pigments to trap energy from sunlight.

Some bacteria are **autotrophs** that manufacture their own organic molecules from simple raw materials. **Photoautotrophs** use the energy from sunlight to synthesize organic compounds from carbon dioxide and other inorganic compounds. **Chemoautotrophs** also use carbon dioxide as a carbon source, but they do not use sunlight as their energy source. Instead, they obtain energy by oxidizing inorganic chemical substances such as ammonia (NH_3) and hydrogen sulfide (H_2S).

Whether they are heterotrophs or autotrophs, most bacterial cells are **aerobic** (like animal and plant cells), requiring atmospheric oxygen for cellular respiration. Many are **facultative anaerobes** that use oxygen for cellular respiration if it is available but can also carry on metabolism anaerobically when necessary. Other bacteria are **obligate anaerobes** that carry out energy-yielding metabolism only anaerobically. Certain obligate anaerobes are actually killed by even low concentrations of oxygen. Some bacteria respire with terminal electron acceptors other than oxygen, for example, sulfate (SO_4^{2-}), nitrate (NO_3^-), or iron (Fe^{2+}).

Review

- In what ways do prokaryotic cells differ from eukaryotic cells?
- How do prokaryotes reproduce? What mechanisms result in genetic recombination?
- How do facultative anaerobes differ from obligate anaerobes? How do they differ from aerobes?

BiologyNow™ Assess your understanding of **prokaryotes** by taking the pretest on your BiologyNow CD-ROM.

THE TWO PROKARYOTE DOMAINS

Learning Objectives

10 Compare the three domains: Eubacteria, Archaea, and Eukarya.

11 Distinguish among the three main groups of the Archaea and among several groups of the Eubacteria as described in Table 23-3. Give examples of each group.

Under a microscope, most prokaryotes appear similar in size and form. However, using sequence analysis of small subunit 16S ribosomal RNA (SSU rRNA), Carl Woese and his coworkers demonstrated that there are two fundamentally different groups of prokaryotes (see Chapter 22). Each group has **signature sequences,** regions of SSU rRNA that have unique nucleotide sequences. This is because after they diverged, prokaryote populations diversified and mutations occurred that affect RNA sequences. Based on such analyses, Woese hypothesized that ancient prokaryotes split into two lineages early in the history of life.

Based on Woese's work, microbiologists now classify the modern descendants of these two ancient lines in two domains: the **Archaea** and the **Eubacteria,** also called **Bacteria** (Fig. 23-14). Domain Archaea, which corresponds to kingdom Archaebacteria, includes a group of prokaryotes that produce methane gas from simple carbon sources and two groups that live in extreme environments. Domain Eubacteria, which corresponds to kingdom Eubacteria, includes all other prokaryotes. In this system, all eukaryotes are classified in domain **Eukarya.**

Key characteristics that distinguish the Archaea from Eubacteria are the absence of peptidoglycan in their cell walls and the unique structure of their cell membranes (Table 23-2). Recall

TABLE **23-2** **Comparison of the Three Domains**

Characteristic	Eubacteria	Archaea	Eukarya
Peptidoglycan in cell wall	Present	Absent	Absent
Isoprene units and ether linkages in cell membranes	Absent	Present	Absent
Nuclear envelope	Absent	Absent	Present
Membrane-bounded organelles	Absent	Absent	Present
Simple RNA polymerase	Present	Absent	Absent
70S ribosomes*	Present	Present	Absent
Flagellum (if present) has single filament	Yes	Yes	No
Transcription is sensitive to the antibiotic rifamycin	Yes	No	No

*The number 70S refers to the sedimentation coefficient (a measure of relative size) when centrifuged.

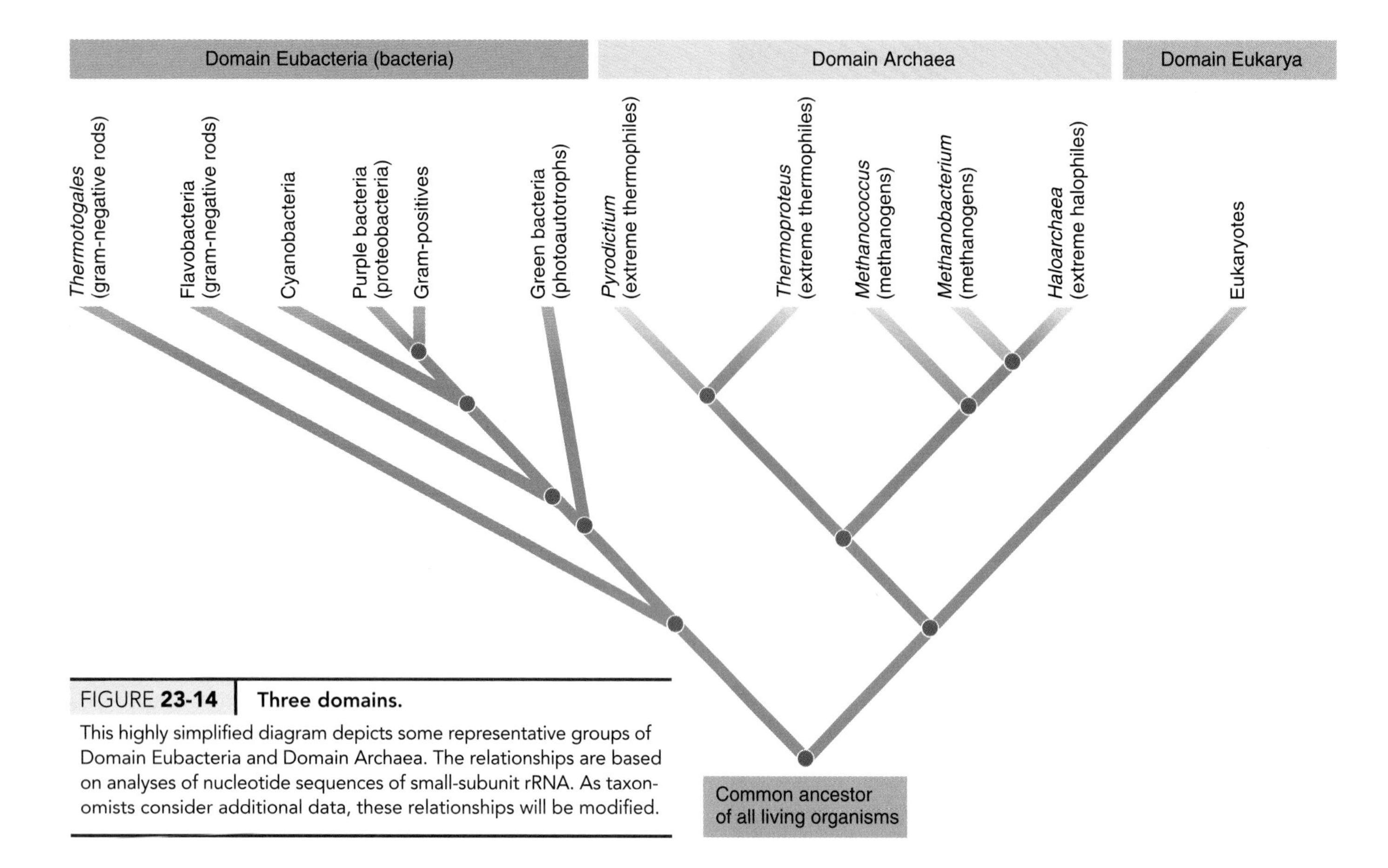

FIGURE **23-14** | **Three domains.**

This highly simplified diagram depicts some representative groups of Domain Eubacteria and Domain Archaea. The relationships are based on analyses of nucleotide sequences of small-subunit rRNA. As taxonomists consider additional data, these relationships will be modified.

that the side chains in the phospholipids of the cell membranes of bacteria and eukaryotes are fatty acids. In contrast, the side chains in the cell membranes of archaea are isoprene units with unique side branches off the main chain. Eubacterial and eukaryotic phospholipids have their fatty acids attached to the glycerol by ester linkages. By contrast, archaeal side chains are bound by ether linkages.

$$R-\overset{\overset{\displaystyle O}{\|}}{C}-O-R \qquad\qquad R-\underset{\underset{\displaystyle H}{|}}{\overset{\overset{\displaystyle H}{|}}{C}}-O-R$$

Ester linkage **Ether linkage**

The absence of a second electronegative oxygen atom makes ether linkages stronger than ester linkages. This unique membrane structure may contribute to the ability of the Archaea to survive and thrive in harsh environments.

In some ways, the Archaea are more like the eukaryotes than like the Eubacteria. For example, they do not have the simple RNA polymerase found in eubacteria. Like eukaryotes, their transcription process is not sensitive to the antibiotic rifamycin. This antibiotic specifically inhibits the type of RNA polymerase found in Eubacteria. The translational mechanisms of archaebacteria are also more like those of eukaryotes.

Prokaryote taxonomy is increasingly based on RNA sequencing. Groups that branched off earlier had more time to accumulate mutations in their SSU rRNA. Their nucleotide sequences are less similar than those of groups that diverged more recently. However, some microbiologists are developing phylogenetic trees based on entire genomes. They argue that there may be 1000 genes that code proteins for every one gene that codes an rRNA. These researchers prefer to consider the proportions of genes (or proteins) that genomes of various groups have in common.

Although prokaryotic taxonomy is controversial and continuously changing, more than 1220 genera and 4000 species of prokaryotes have been classified. The editors of *Bergey's Manual of Systematic Bacteriology,*[1] considered the definitive reference text by microbiologists, have divided prokaryotes into several groups. Several representative groups of prokaryotes are described in Table 23-3 on pages 452–453.

Members of the Archaea survive in harsh environments

Many extant members of the Archaea inhabit environments thought to be similar to conditions on early Earth. *Bergey's Manual* divides the Archaea into two phyla: the Crenarchaeota, which is a small, diverse group that are extremely thermophilic (heat loving), and the Euryarchaeota. A third group, Kararchaeota,

[1] *Bergey's Manual of Systematic Bacteriology,* 2nd ed., Springer, New York, 2001 is the first of five volumes of the second edition.

has been recently named to include uncultured archaea with similar 16S rRNA sequences that inhabit terrestrial hot springs.

Based on their metabolism and ecology, we can identify three main types of archaea: methanogens, extreme halophiles, and extreme thermophiles. The **methanogens** (methane producers) are a large, diverse group that inhabit oxygen-free environments in sewage and swamps and are common in the digestive tracts of humans and other animals. They are strict anaerobes that produce methane gas from simple carbon compounds. The methanogens are important in recycling components of organic products of organisms that inhabit swamps. These archaea produce most of the methane in Earth's atmosphere.

Extreme halophiles are heterotrophs that live in saturated brine solutions such as salt ponds, the Dead Sea, and Great Salt Lake (Fig. 23-15). The extreme halophiles use aerobic respiration to make ATP. However, they also carry out a form of photosynthesis in which they capture the energy of sunlight using a purple pigment (bacteriorhodopsin). (This pigment is very similar to the pigment rhodopsin involved in animal vision.)

Extreme thermophiles normally grow in hot (45° to 110°C), sometimes acidic, environments. One species is found in the hot sulfur springs of Yellowstone Park at temperatures near 60°C and pH values of 1 to 2 (the pH of concentrated sulfuric acid). Others inhabit volcanic areas under the sea. One species, found in hot deep-sea vents on the sea floor, lives at temperatures from 80° to 110°C.

Members of the Archaea also inhabit less extreme conditions. For example, they are abundant in the soil and in the cold ocean surface waters near Antarctica. Microbiologists have suggested that the Archaea may be important in biogeochemical cycles and in marine food chains. The three main metabolic groups of the Archaea are summarized in Table 23-3.

FIGURE **23-15** | **Sea water evaporating ponds.**

These salt ponds near San Francisco Bay are colored pink, orange, and yellow from the large number of extreme halophiles that inhabit them. The colors are a result of pigments (carotenoids) in the cell wall. These bacteria are harmless, and the ponds are used to produce salt commercially.

Helen E. Carr, Biological Photo Service

Eubacteria are the most familiar prokaryotes

The Eubacteria are widely distributed in the environment and are better known to microbiologists than are the Archaea. *Bergey's Manual* divides the Eubacteria into 21 phyla. Several groups of eubacteria are described in Table 23-3.

Review

- What are three main types of archaea based on ecology?
- How do these Eubacteria differ? (a) enterobacteria, (b) cyanobacteria, (c) mycoplasmas.

BiologyNow™ Assess your understanding of **the two prokaryote domains** by taking the pretest on your BiologyNow CD-ROM.

EFFECTS OF PROKARYOTES ON THE ENVIRONMENT

Learning Objective

12 Discuss the ecological roles of prokaryotes, their importance as pathogens, and their commercial importance.

Prokaryotes are of great ecological importance

The metabolically diverse prokaryotes are vital members of the biosphere. Bacteria, especially actinomycetes and myxobacteria, are the most numerous inhabitants of soil. As described in the chapter introduction, prokaryotes are essential **decomposers** (saprotrophs) that break down dead organic matter and wastes. They use the products of decomposition as an energy source. When prokaryotes break down organic compounds, chemical components are recycled. Many nutrients, including nitrogen, oxygen, carbon, phosphorus, sulfur, and certain trace elements, are recycled in this way. (The roles of bacteria in biogeochemical cycles, particularly the nitrogen cycle, are discussed in Chapter 53.)

Nitrogen is constantly removed from the soil by plants and other natural processes, as well as by human activities, such as agriculture. Plant growth depends on the availability of usable nitrogen, so it must be continually added to the soil. Several

TABLE 23-3 Some Major Groups of Prokaryotes

Eubacteria with Gram-Negative Cell Walls

Proteobacteria **A large, very diverse group; not monophyletic and so not included as a clade in Figure 23-14.**

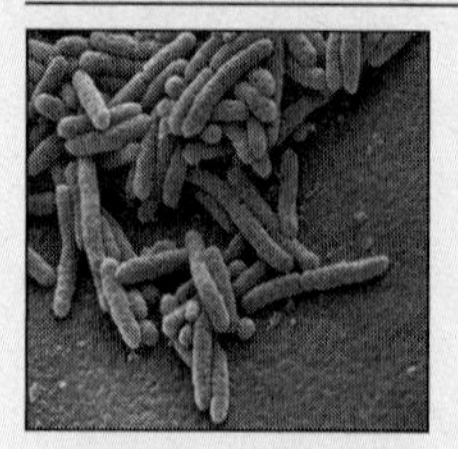

SEM of *Escherichia coli* colony. *(David Scharf/Peter Arnold, Inc.)*

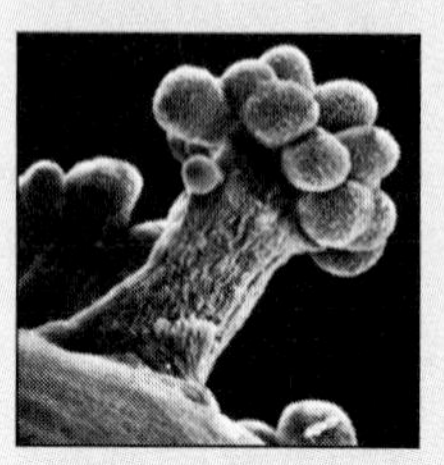

SEM of fruiting body of the myxobacterium *Stigmatella aurantiaca.* Protective resting cells within the fruiting bodies are very resistant to heat and drying. *(From P.L. Grilicone, and J. Pangborn, Journal of Bacteriology 124:1558, 1975)*

Purple bacteria are photoautotrophs that do not produce oxygen.

Enterbacteria include decomposers that live on decaying plant matter, pathogens, and a variety of bacteria that inhabit humans. Although *Escherichia coli* is a normal inhabitant of the animal intestinal tract, certain strains can cause moderate to severe diarrhea. One species of *Salmonella* infects food and produces a toxin that causes a form of food poisoning; another species causes typhoid fever.

Vibrios are mainly marine; some are bioluminescent. *Vibrio cholerae* causes cholera.

Rhizobium species live symbiotically in root nodules of leguminous plants and convert atmospheric nitrogen to a form usable by plants (nitrogen fixation).

Pseudomonads are heterotrophs that produce nonphotosynthetic pigments. They cause disease in plants and animals, including humans.

Azotobacteria inhabit the soil and fix nitrogen under aerobic conditions. They form a resting cell termed a cyst that is resistant to drying.

Rickettsias are very small, rod-shaped bacteria. A few species are pathogenic to humans and other animals and are transmitted by arthropods through bites or contact with their excretions. Among these are typhus (transmitted by fleas and lice) and Rocky Mountain spotted fever (transmitted by ticks).

Myxobacteria (slime bacteria) secrete slime and glide or creep along. When nutrients are exhausted, these bacteria form masses that develop into stalked, multicellular reproductive structures called *fruiting bodies.* During this process, bacterial cells within the fruiting body enter a resting stage. When conditions are favorable, the resting cells become active.

Other medically important Proteobacteria include *Neisseria gonorrhoeae,* which causes gonorrhea, and *Legionella pneumophila,* which causes Legionnaires' disease.

Other Gram-Negative Eubacterial Groups

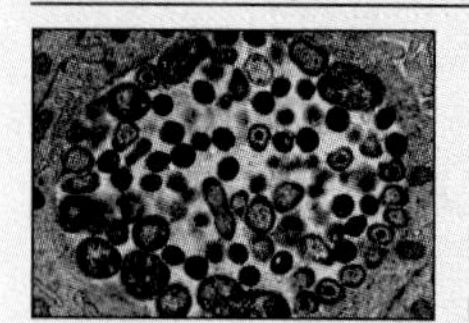

TEM of *Chlamydia trachomatis* in human oviduct cell. *(David M. Phillips/Visuals Unlimited)*

Chlamydias lack peptidoglycan in their cell walls and do not depend on arthropod vectors for transmission. Chlamydias are energy parasites, that is, completely dependent on their host for ATP. Chlamydias infect almost every species of bird and mammal. Trachoma, the leading cause of blindness in the world, is caused by a strain of *Chlamydia;* sexually transmitted chlamydias are the major cause of pelvic inflammatory disease in women.

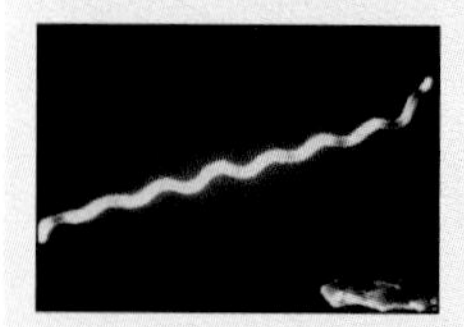

LM of *Treponema pallidum,* the spirochete that causes syphilis. *(Charles W. Stratton/Visuals Unlimited/Science VU)*

Spirochetes are spiral-shaped bacteria with flexible cell walls. They move by means of unique internal flagella called *axial filaments.* Some species are free-living, whereas others form symbiotic associations; a few are parasitic. The spirochete of greatest medical importance is *Treponema pallidum,* which causes syphilis. Lyme disease, a tick-borne disease of humans and some other animals, is caused by a spirochete belonging to the genus *Borrelia.*

Cyanobacteria are gram-negative, photosynthetic bacteria that inhabit ponds, lakes, swimming pools, moist soil, dead logs, and the bark of trees. They contain chlorophyll *a* and use a photosynthetic process similar to that of plants and algae. Many species also fix nitrogen.

---

types of bacteria transform atmospheric nitrogen to forms that can be used by plants (Fig. 23-16 on page 454). Rhizobial bacteria, a type of motile bacteria that inhabit the soil, form mutualistic (mutually beneficial) relationships with the roots of legumes, a large family of plants that include important crops such as beans, peas, and peanuts. The bacteria fix nitrogen for the plant, and the plant provides the bacteria with organic compounds.

Many prokaryotes, such as the cyanobacteria, carry out photosynthesis using water as the electron source and generating oxygen. During this process, they fix huge amounts of carbon dioxide into organic molecules.

Microbiologists are only beginning to unravel the mysteries of prokaryote ecology. For example, the SAR11 clade of proteobacteria are among the most successful organisms on Earth. Although they are one of the most abundant organisms in the Atlantic Ocean, they were not successfully cultured until 2002 and very little is known about their ecological role.

Some types of bacteria may influence the course of evolution of other species. The proteobacteria *Wolbachia* infect many invertebrates, including insects, spiders, crustaceans, and nematodes (roundworms). These bacteria inhabit the cells of their host's reproductive system and have evolved remarkable mech-

TABLE 23-3 Continued

Eubacteria with Gram-Positive Cell Walls

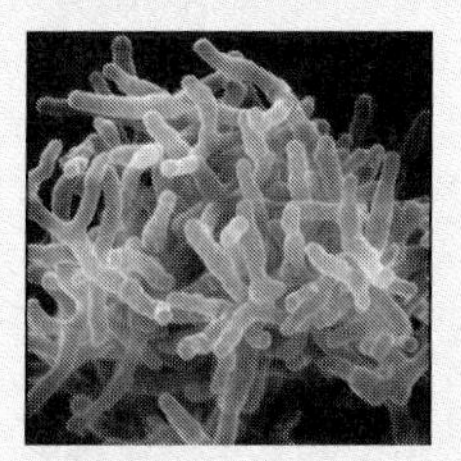

SEM of *Actinomycetes naeslundi*, a soil-dwelling bacterium that forms filamentous colonies. *(David M. Phillips/Visuals Unlimited)*

Actinomycetes superficially resemble fungi. However, they have peptidoglycan in their cell walls, lack nuclear envelopes, and have other prokaryotic characteristics. Actinomycetes decompose organic materials in soil. Most are saprotrophs, and some are anaerobic. Several species of the genus *Streptomyces* produce antibiotics such as streptomycin, erythromycin, chloramphenicol, and the tetracyclines. Some actinomycetes cause serious lung disease or generalized infections in humans and other animals.

Lactic acid bacteria ferment sugar, producing lactic acid as the main end-product. They inhabit decomposing plant material, milk, and other dairy products. The characteristic taste of yogurt, pickles, sauerkraut, and green olives is caused by the action of lactic acid bacteria. These bacteria are also among the normal inhabitants of the human mouth and vagina.

Mycobacteria are slender, irregular rods. They contain a waxy substance in their cell walls. One species causes tuberculosis; another causes leprosy.

Streptococci inhabit the mouth and digestive tract of humans and some other animals. Among the harmful species are those that cause "strep throat," dental caries, a form of pneumonia, scarlet fever, and rheumatic fever.

Staphylococci normally live in the nose and on the skin. They are opportunistic pathogens that cause disease when the immunity of the host is lowered. *Staphylococcus aureus* causes boils and skin infections (some extremely serious) and may infect wounds. Certain strains of *S. aureus* cause a form of food poisoning, and other strains cause toxic shock syndrome.

Clostridia are anaerobic. One species causes tetanus, another causes gas gangrene. *Clostridium botulinum* can cause botulism, an often fatal type of food poisoning.

Eubacteria That Lack Cell Walls

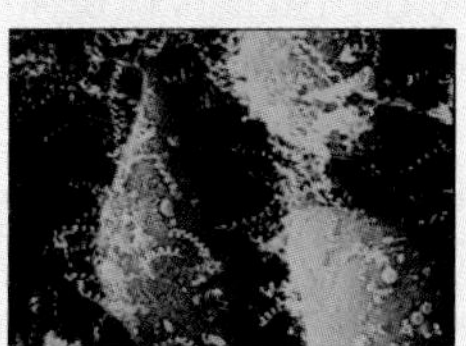

SEM of *Mycoplasma* on fibroblast cells. *(David M. Phillips/Visuals Unlimited)*

Mycoplasmas lack cell walls. These bacteria inhabit soil and sewage; some are parasitic on plants or animals. Some inhabit human mucous membranes but do not generally cause disease; one species causes a mild type of bacterial pneumonia in humans.

Archaea: Prokaryotes with Cell Walls That Have No Peptidoglycan

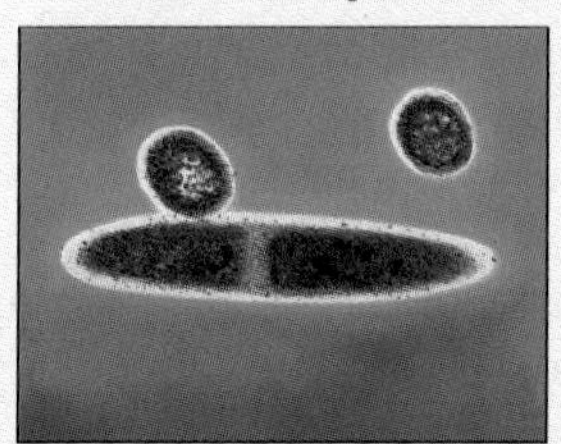

TEM of a methanogen *(Methanospirillum hungatii)* undergoing cell division. Two other archaea are visible in cross-section. When not dividing, these archaea are spiral shaped. *(Dr. Kari Lounatmaa/Science Photo Library/Photo Researchers, Inc.)*

Methanogens are anaerobes that produce methane gas from simple carbon compounds.

Extreme halophiles inhabit saturated salt solutions.

Extreme thermophiles grow at 70°C or higher; some thrive at temperatures greater than 100°C.

anisms that increase the probability of their transmission to new generations of hosts. *Wolbachia* are transmitted from generation to generation in the eggs of their hosts, so male hosts are not useful to them. Consequently, these parasites limit or eradicate males from the population. In some insect species, infected males can reproduce only when they mate with infected females carrying the same *Wolbachia* strain. In other species, these parasites convert male insects into females. Some infected females reproduce parthenogenetically (that is, their eggs develop without fertilization). Because they affect reproduction in their hosts, *Wolbachia* and other reproductive parasites may influence evolutionary divergence and even extinction in some species.

Some prokaryotes cause disease

Some prokaryote species have coevolved with eukaryotes and are interdependent with them. All plants and animals harbor a population of microorganisms that are considered normal microbiota (or microflora)—harmless symbiotic prokaryotes. In fact, the number of bacteria that normally inhabit the human body (approximately 700 trillion cells) exceeds the number of

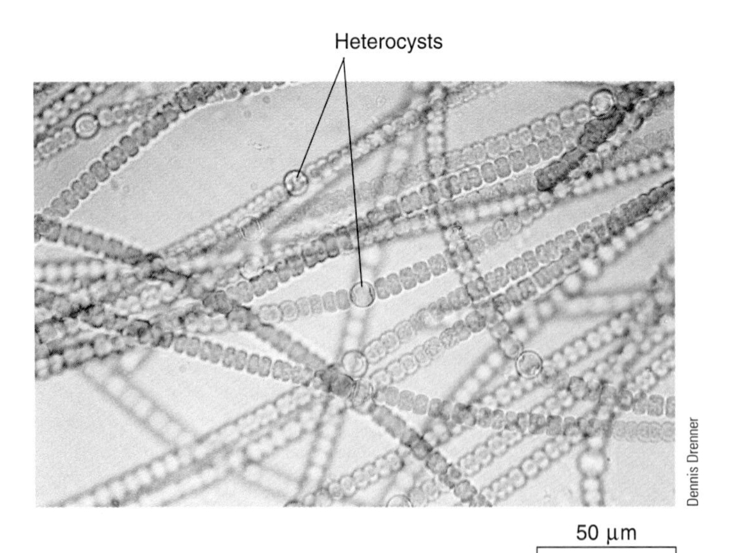

FIGURE **23-16** **LM of *Anabaena*, a filamentous cyanobacterium that fixes nitrogen.**

Nitrogen fixation takes place in the rounded cells, called *heterocysts*.

its own human cells (about 70 trillion cells). Some of these bacteria provide a beneficial service for their host. For example, the presence of certain bacterial populations prevents harmful microorganisms (including other bacteria) from flourishing. Some bacteria that inhabit the human intestine produce vitamin K and some of the B vitamins, which are then absorbed and used.

A small percentage of bacterial species are important pathogens of plants and animals. Some of the normal bacterial (and viral) inhabitants are opportunistic pathogens that cause disease only under certain conditions. For example, when the immune system is compromised, opportunistic bacteria may increase in number and cause disease.

PROCESS OF SCIENCE

The idea that some unknown agent causes disease was debated long before Leeuwenhoek discovered microorganisms with his microscope in the late 1600s. However, not until much later did scientists develop the tools or the methods needed to accurately understand the relationships between bacteria and disease. During the late 19th century several physicians, microbiologists, and chemists working independently laid the foundations for the science of microbiology. French chemist Louis Pasteur disproved the prevailing views of spontaneous generation by demonstrating that sterilization of a sugar and protein culture prevented bacterial growth. Pasteur also developed a rabies vaccine, showing that people can be stimulated to develop immunity to disease.

The German physician Robert Koch was the first to clearly demonstrate that bacteria cause infectious disease. In 1876 he showed that *Bacillus anthracis* causes anthrax. Using a microscope, Koch observed the bacteria in the blood and spleens of dead sheep. When he inoculated mice with the infected sheep blood, he was able to identify *B. anthracis* in the blood of the mice. He also cultured *B. anthracis* and showed that when he injected the bacteria into healthy mice, they developed anthrax.

Koch proposed a set of guidelines, now known as **Koch's postulates,** that are still used to demonstrate that a specific pathogen causes specific disease symptoms: (1) the pathogen must be present in every individual with the disease; (2) a sample of the microorganism taken from the diseased host can be grown in pure culture; (3) when a sample of the pure culture is injected into a healthy host, it causes the same disease; and (4) the microorganism can be recovered from the experimentally infected host.

Pathogenic microorganisms can enter the body in food, dust, droplets, or through wounds. Many diseases are transmitted by insect or animal bites. To cause disease, a pathogen must adhere to a specific cell type, multiply, and produce toxic substances. Adherence and multiplication occurs only when the pathogen competes successfully with the normal microbiota and counteracts the host's defenses against invasion.

Pathogens produce a variety of substances that increase their success. Some bacteria produce **exotoxins,** strong poisons that are either secreted from the cell or leak out when the bacterial cell is destroyed. The toxin, not the presence of the bacteria themselves, is responsible for the disease. As mentioned previously, diphtheria is caused by a Gram-positive bacillus *(Corynebacterium diphtheriae)* that is lysogenized by a phage. The diphtheria toxin kills cells and causes inflammation.

Botulism, a type of food poisoning that can lead to paralysis and sometimes death, results from eating improperly canned food. Botulism is caused by an exotoxin released by the Gram-positive, endospore-forming *Clostridium botulinum.* During the canning process, food must be heated sufficiently to kill any highly heat-resistant endospores that may be present. If not, the endospores can germinate. The resulting bacterial population grows and releases an exotoxin so powerful that 1 g could kill a million humans! Like many exotoxins, the one that causes botulism is inactivated by heating. (Food must be heated to 80°C for 10 minutes, or boiled for 3 to 4 minutes.) The botulism exotoxin, marketed under the trade name Botox, is used in extremely tiny amounts to treat several medical conditions involving spasms (involuntary muscle contractions). Because Botox is a neurotoxin that works by paralyzing muscles, it can also relax facial wrinkles caused by contraction of the underlying muscles. However, its effects last for only three to eight months.

Endotoxins are not secreted by pathogens but are components of the cell walls of most Gram-negative bacteria. These compounds affect the host only when they are released from dead bacteria. Endotoxins bind to the host's macrophages and stimulate them to release substances that cause fever and other symptoms of infection. Unlike exotoxins, which cause specific symptoms, endotoxins appear to affect the entire body. Endotoxins are not destroyed by heating.

Prokaryotes are used in many commercial processes

Many foods and beverages are produced with the help of microbial fermentation. Lactic acid bacteria are used in producing acidophilus milk, yogurt, pickles, olives, and sauerkraut (Fig. 23-17). Several types of bacteria are used to produce cheese. Bacteria

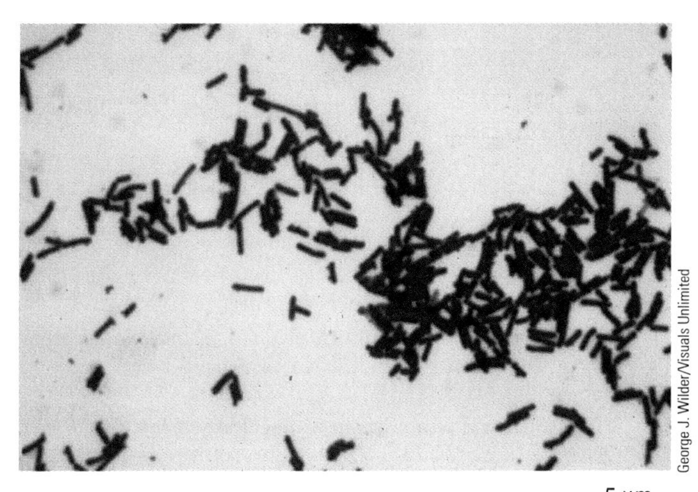

FIGURE **23-17** **LM of lactic acid bacteria *(Lactobacillus acidophilus).***

are involved in making fermented meats such as salami and in the production of vinegar, soy sauce, chocolate, and certain B vitamins (B_{12} and riboflavin). Bacteria are also used in producing citric acid, a compound added to candy and to most soft drinks.

Some microorganisms produce *antibiotics,* compounds that limit competition for nutrients by inhibiting or destroying other microorganisms. By the 1950s antibiotics had become important clinical tools that transformed the treatment of infectious disease. Today about 100 clinically useful antibiotics are available, and literally tons of antibiotics are produced annually. Pharmaceutical companies obtain most antibiotics from three groups of microorganisms: a large group of Gram-positive soil bacteria, the *actinomycetes,* Gram-positive bacteria of the genus *Bacillus,* and molds (eukaryotes belonging to kingdom Fungi).

Because of their prolific reproduction rates, bacteria are ideal "factories" for the production of biomolecules. They have been genetically engineered (see Chapter 14) to produce certain vaccines, human growth hormone, insulin, and many other clinically important compounds. Most of the insulin used to treat diabetics is now derived from genetically engineered bacteria. Researchers are developing genetically engineered bacteria for production of many other medically and agriculturally useful products.

Bacteria are used in sewage treatment and to break down solid wastes in landfills. They are also used in **bioremediation,** a process in which a contaminated site is exposed to microorganisms that break down the toxins, leaving behind harmless metabolic byproducts such as carbon dioxide and chlorides. More than 1000 different species of bacteria and fungi have been used to clean up various forms of pollution, and microbiologists are searching for others. In 2003, a team of German microbiologists reported that they had isolated a species of anaerobic bacteria that has the potential to bioremediate sites contaminated with dioxins, very toxic pollutants that persist in the environment.

Archaea also have achieved economic importance. For example, archaeal enzymes have been added to laundry and industrial detergents to increase their performance at higher temperatures and pH levels. Another enzyme has been useful in the food industry to convert corn starch to dextrins.

Review

- What important roles do prokaryotes play in ecosystems?
- Why is each of Koch's postulates essential?
- What is bioremediation?

BiologyNow™ Assess your understanding of **the effects of prokaryotes on the environment** by taking the pretest on your BiologyNow CD-ROM.

SUMMARY WITH KEY TERMS

1 Describe the structure of a virus, and compare a virus with a free-living cell.

- A **virus,** or **virion,** is a tiny particle consisting of a DNA or RNA genome surrounded by a protein coat, called a **capsid.**
- Viruses are subcellular particles that cannot metabolize on their own. In the past, biologists considered them nonliving particles, but some now view them as life forms.

2 Trace the evolutionary origin of viruses according to current hypotheses.

- Viruses may be bits of nucleic acid that originally "escaped" from animal, plant, or bacterial cells.
- Some biologists hypothesize that viruses must have evolved before the three domains diverged; they think it unlikely that the similarities between viruses that infect the Archaea and those that infect the Eubacteria evolved twice.

3 Characterize bacteriophages, and contrast a lytic cycle with a lysogenic cycle.

- **Phages (bacteriophages)** are viruses that infect bacteria.
- A viral reproductive cycle can be lytic or lysogenic. In a **lytic cycle,** the virus destroys the host cell. The five steps in a lytic cycle are: **attachment** to the host cell; **penetration** of viral nucleic acid into the host cell; **replication** of the viral nucleic acid; **assembly** of newly synthesized components into new viruses; and **release** from the host cell.
- **Temperate** viruses do not always destroy their hosts. In a **lysogenic cycle** the viral genome is replicated along with the host DNA. The nucleic acid of some phages becomes integrated into the bacterial DNA, and is then called a **prophage.** Bacterial cells that carry prophages are **lysogenic** cells. In **lysogenic conversion,** bacterial cells containing certain temperate viruses exhibit new properties.

4 Explain how viruses infect animal and plant cells.

- Viruses enter animal cells by membrane fusion or by endocytosis. Viral nucleic acid is replicated within the host cell, proteins are synthesized, and new viruses are assembled and released from the cell.
- Plant viruses can be spread among plants by insect vectors. Viruses spread through the plant via plasmodesmata.

5 Describe the reproductive cycle of a retrovirus such as human immunodeficiency virus (HIV).

- **Retroviruses,** such as HIV, use **reverse transcriptase** to transcribe their RNA genome into a DNA intermediate that becomes integrated into the host DNA. Copies of the viral RNA are then synthesized.

6 Compare and contrast viroids and prions.

- **Viroids** and **prions** are smaller than viruses. A viroid consists of a short strand of RNA with no protein coat.
- A prion consists only of protein. Prions cause **transmissible spongiform encephalopathies (TSEs).**

7 Describe the structure and common shapes of prokaryotic cells.

- Prokaryotic cells do not have membrane-enclosed organelles such as nuclei and mitochondria.
- Common shapes of bacterial cells include **coccus,** or spherical; **bacillus,** or rod-shaped; and spiral. Spiral bacteria include the **spirillum,** a rigid helix and **spirochete,** a flexible helix. **Vibrios** are comma shaped.
- Most eubacteria have cell walls composed of **peptidoglycan.** The walls of **Gram-positive** bacteria are very thick and consist mainly of peptidoglycan. The cell walls of **Gram-negative** bacteria consist of a thin peptidoglycan layer and an outer membrane resembling the plasma membrane. Some species of eubacteria produce a **capsule** that surrounds the cell wall.
- Some bacteria have **pili,** protein structures that extend from the cell and help bacteria adhere to one another or to certain other surfaces.
- Bacterial **flagella** are structurally different from eukaryotic flagella; each flagellum consists of a basal body, hook, and filament. They produce a rotary motion.

8 Describe asexual reproduction in prokaryotes, and summarize three mechanisms (transformation, conjugation, and transduction) that may lead to genetic recombination.

- The genetic material of a prokaryote typically consists of a circular DNA molecule and one or more **plasmids,** smaller circular fragments of DNA.
- Prokaryotes reproduce asexually by **binary fission** (the cell divides forming two cells), **budding** (a bud forms and separates from the mother cell) or **fragmentation** (walls form inside the cell and it separates into several cells.)
- In bacteria, genetic material may be exchanged by transformation, transduction, or conjugation. In **transformation,** a bacterial cell takes in DNA fragments released by another cell. In **transduction,** a phage carries bacterial DNA from one bacterial cell into another. In **conjugation,** two cells of two different mating types exchange genetic material.

9 Characterize the metabolic diversity of autotrophic and heterotrophic prokaryotes, including aerobes, facultative anaerobes, and obligate anaerobes.

- Most prokaryotes are **heterotrophs** that obtain energy and carbon from other organisms; some are **autotrophs** that make their own organic molecules from simple raw materials.
- The majority of heterotrophic bacteria are free-living **saprotrophs,** that obtain nourishment from dead organic matter.
- Autotrophs may be **photoautotrophs,** which obtain energy from light or **chemoautotrophs,** which obtain energy by oxidizing inorganic chemicals.
- Most bacteria are aerobic; some are **facultative anaerobes** which metabolize anaerobically when necessary; others are **obligate anaerobes,** which can carry on metabolism only anaerobically.

10 Compare the three domains: Eubacteria, Archaea, and Eukarya.

- Prokaryotes are assigned to domain **Eubacteria** and domain **Archaea.**
- Unlike those of the Eubacteria, the cell walls of the Archaea do not have peptidoglycan. The translational mechanisms of the Archaea more closely resemble eukaryotic mechanisms than those of other prokaryotes. Domain Eukarya includes the four kingdoms of eukaryotes.

11 Distinguish among three main groups of the Archaea based on their ecology, and among several groups of the Eubacteria as described in Table 23-3. Give examples of each group.

- Three main groups of the Archaea are methanogens, extreme halophiles, and extreme thermophiles. **Methanogens** produce methane gas from simple carbon compounds. They inhabit anaerobic environments such as marshes, marine sediments, and the digestive tracts of animals. **Extreme halophiles** inhabit saturated salt solutions. **Extreme thermophiles** can inhabit environments at temperatures higher than 100°C.
- See Table 23-3 to review some of the major groups of Eubacteria.

12 Discuss the ecological roles of prokaryotes, their importance as pathogens, and their commercial importance.

- Prokaryotes play essential ecological roles as **decomposers** and are important in recycling nutrients. Some prokaryotes carry out photosynthesis.
- Many bacteria are symbiotic with other organisms; some are important **pathogens** of plants and animals.
- Anton van Leeuwenhoek, Louis Pasteur, and Robert Koch were important pioneers in microbiology. **Koch's postulates** are a set of guidelines used to demonstrate that a specific pathogen causes specific disease symptoms.
- Some pathogenic bacteria produce strong poisons called **exotoxins;** others produce **endotoxins,** poisonous components of their cell walls. Endotoxins are released when bacteria die.

POST-TEST

1. Pathogens (a) are agents that cause disease (b) include most bacteria (c) include most myxobacteria (d) include many autotrophs (e) are cyanobacteria that cause disease
2. The genome of a virus consists of (a) DNA (b) RNA (c) prions (d) DNA and RNA (e) DNA or RNA

POST-TEST (continued)

3. The capsid of a virus consists of (a) protein (b) nucleic acid (c) helical proteins (d) a simple carbohydrate (e) RNA and lipid
4. Viruses that kill host cells are (a) lysogenic (b) lytic (c) viroids (d) prophages (e) temperate
5. The correct sequence in viral reproduction is (a) attachment, penetration, assembly, replication, release (b) penetration, absorption, assembly, replication, release (c) attachment, penetration, replication, assembly, release (d) release, assembly, penetration, replication, attachment (e) attachment, penetration, replication, release, assembly
6. In lysogenic conversion (a) bacterial cells may exhibit new properties (b) the host cell dies (c) prions sometimes convert to viroids (d) reverse transcriptase transcribes DNA into RNA (e) prion proteins convert to infective agents
7. Peptidoglycan is a chemical compound found in the cell walls of (a) most viroids (b) most Archaea (c) all prokaryotes (d) most Eubacteria (e) most Eukarya
8. Retroviruses (a) are DNA viruses (b) are RNA viruses (c) use reverse transcriptase to transcribe DNA into RNA (d) use prion protein to transcribe RNA (e) are DNA viruses that cause AIDS
9. Bacterial flagella (a) are homologous with eukaryotic flagella (b) exhibit a rotary motion (c) consist of a basal body and nine pairs of microtubules (d) are important in transduction (e) are characteristic of Gram-positive bacteria
10. In conjugation (a) two bacterial cells of different mating types come together, and genetic material is transferred from one to another (b) a bacterial cell develops a bulge that enlarges and eventually separates from the mother cell (c) fragments of DNA released by a broken cell are taken in by another bacterial cell (d) a phage carries bacterial genes from one bacterial cell into another (e) walls develop in the cell, which then divides into six new cells
11. Endospores (a) are formed by some viruses (b) are extremely durable cells (c) are comparable to the reproductive spores of fungi and plants (d) cause fever and other symptoms in the host (e) are extremely vulnerable to infection by phages
12. The majority of heterotrophic bacteria are (a) free-living saprotrophs (b) photoautotrophs (c) chemoautotrophs (d) facultative anaerobes (e) obligate anaerobes
13. Which of the following do *not* belong to domain Archaea? (a) bacteria that produce methane from carbon dioxide and hydrogen (b) thermophiles (c) halophiles (d) bacteriophages (e) prokaryotes with cell walls that lack peptidoglycan
14. Which of the following scientists proposed a set of guidelines to demonstrate that a specific pathogen causes specific disease symptoms? (a) Leeuwenhoek (b) Pasteur (c) Koch (d) Prusiner (e) Woese
15. Bacteria that thrive in puncture wounds are likely to be (a) saprotrophs (b) photoautotrophs (c) chemoautotrophs (d) endospores (e) obligate anaerobes
16. Prions (a) consist of a short strand of RNA with no protein coat (b) are deadly viruses that have caused several hundred deaths in Uganda and other parts of Africa (c) appear to consist only of protein; no nucleic acid component has been found (d) are responsible for causing hepatitis C (e) are not transmissible from human to human
17. Bacteria that are autotrophs (a) do not require atmospheric oxygen for cellular respiration (b) manufacture their own organic molecules from simple raw materials (c) must obtain organic compounds from other organisms (d) get their nourishment from dead organisms (e) produce endospores when oxygen levels are too low for active growth
18. Lactic acid bacteria (a) have Gram-negative cell walls (b) fix nitrogen under anaerobic conditions (c) were formerly known as blue-green algae (d) contribute to the unique taste of yogurt, pickles, and sauerkraut (e) secrete slime on which they glide
19. Which group of bacteria contains the Gram-positive, anaerobic bacterium that causes botulism? (a) clostridia (b) actinomycetes (c) enterobacteria (d) spirochetes (e) streptococci

CRITICAL THINKING

1. Historically, biologists thought that viruses, because of their simple structure, evolved before cellular organisms. Later, the hypothesis that viruses escaped from cells gained favor. Recently, microbiologists have again suggested that viruses evolved early in the history of life. Based on what you have learned about viruses, present an argument for one of these hypotheses.
2. Imagine that you discover a new microorganism. After careful study you determine that it should be classified in the domain Archaea. What characteristics might lead you to this classification?
3. How might life be different if all prokaryotes were killed?

- Visit our Web site at **http://biology.brookscole.com/solomon7** for links to chapter-related resources on the World Wide Web. Additional online materials relating to this chapter can also be found on our Web site.

BIOLOGY NOW RESOURCES

Active Figure

23-5 HIV life cycle

Preparing for an exam? Take a diagnostic test on your BiologyNow CD-ROM.

Post-Test Answers

1. a	2. e	3. a	4. b
5. c	6. a	7. d	8. b
9. b	10. a	11. b	12. a
13. d	14. c	15. e	16. c
17. b	18. d	19. a	

24 Protists

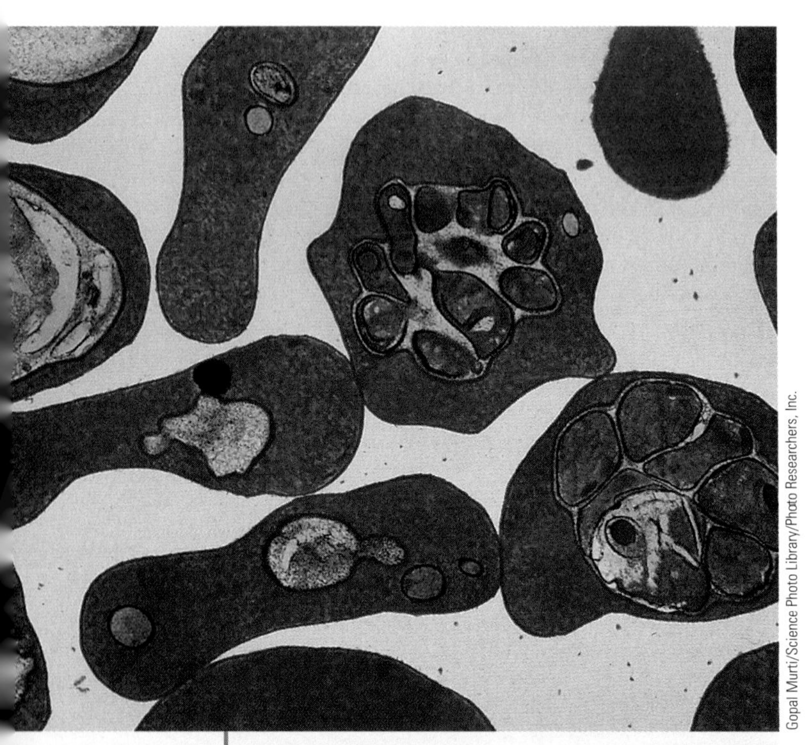

Gopal Murti/Science Photo Library/Photo Researchers, Inc.

Human red blood cells filled with malarial parasites. As shown in this colorized TEM, the parasites multiply inside human red blood cells *(red)*, safe from the body's immune system. Malaria is a tropical disease caused by the protist *Plasmodium* and transmitted to humans by the mosquito *Anopheles gambiae* as it feeds on human blood.

CHAPTER OUTLINE

- **Introduction to the Protists**
- **Evolution of the Eukaryotes**
- **Representative Protists**

In Chapter 22 you learned that the kingdom Protista was first proposed in 1866, to include bacteria and other microorganisms that differ from plants and animals. When Robert Whittaker recommended the five-kingdom system of classification in 1969, only unicellular eukaryotic organisms were placed in kingdom Protista. Since then the boundaries of this kingdom have expanded to include any eukaryote that isn't a fungus, animal, or land plant. More recently, some biologists have proposed eliminating the kingdom Protista by splitting it into several kingdoms based largely on molecular data that have clarified certain evolutionary relationships. Many relationships among protists remain uncertain, however.

Kingdom Protista currently consists of a vast assortment of primarily aquatic eukaryotic organisms whose diverse body forms, types of reproduction, modes of nutrition, and lifestyles make them difficult to characterize. **Protists** are unicellular, colonial, or simple multicellular organisms that have a eukaryotic cell organization. The word *protist,* from the Greek for "the very first," reflects the idea that protists were the first eukaryotes to evolve.

Protists are members of Eukarya, the third domain on the tree of life. Eukaryotic cells are characteristic of protists as well as of complex multicellular organisms belonging to the kingdoms Fungi, Animalia, and Plantae. However, having a eukaryotic cell structure clearly differentiates protists from members of the prokaryotic domains Eubacteria and Archaea. Recall from Chapter 4 that, unlike prokaryotic cells, eukaryotic cells have nuclei and other membrane-enclosed organelles such as mitochondria and plastids, 9 + 2 flagella, and multiple chromosomes in which DNA and proteins form a complex called *chromatin.* Sexual reproduction, meiosis, and mitosis are also characteristic of eukaryotes.

Size varies considerably within the protist kingdom, from microscopic protists to giant kelps, which are brown algae that reach 75 m (about 250 ft) in length. Although most protists are unicellular, some form **colonies,** loosely connected groups of cells; some are **coenocytic,** consisting of a multinucleate mass of cytoplasm; and some are **multicellular,** composed of many cells. Unlike animals, plants, and many fungi, most multicellular protists have relatively simple body forms without specialized tissues.

Because of their huge numbers, members of kingdom Protista are crucial to the natural balance of the living world. Protists are an important source of food for other organisms, and photosynthetic protists supply oxygen to aquatic and terrestrial ecosystems. Certain protists are economically important, and others cause devastating diseases (see photograph).

Biologists currently recognize as many as 60 major groups of protists. Consideration of all protist groups is beyond the scope of this text, but we discuss a number of representative taxa. ■

INTRODUCTION TO THE PROTISTS

Learning Objectives

1. Characterize the features common to the members of kingdom Protista.
2. Discuss in general terms the diversity inherent in the protist kingdom, including means of locomotion, modes of nutrition, interactions with other organisms, habitats, and modes of reproduction.

Size and structural complexity are not the only variable features of protists. During the course of evolutionary history, organisms in kingdom Protista have evolved diversity in their (1) means of locomotion, (2) ways of obtaining nutrients, (3) interactions with other organisms, (4) habitats, and (5) modes of reproduction.

Protists, most of which are motile at some point in their life cycle, have various means of locomotion. They may move by pushing out cytoplasmic extensions (*pseudopodia*) as an amoeba does; by flexing individual cells; by gliding over surfaces; by waving *cilia,* short, hairlike organelles; or by lashing *flagella,* long, whiplike organelles (Fig. 24-1). Some protists have two or more means of locomotion—for example, both flagella and pseudopodia.

Methods of obtaining nutrients differ widely in kingdom Protista. Most algae are autotrophic and photosynthesize as plants do. Some heterotrophic protists obtain their nutrients by absorption, as fungi do, whereas others resemble animals in that they ingest food. Some protists switch their modes of nutrition and are autotrophic at certain times and heterotrophic at others.

Although many protists are free-living, others form stable symbiotic associations with unrelated organisms. These intimate associations range from **mutualism,** a more or less equal partnership where both partners benefit, to **commensalism,** where one partner benefits and the other is unaffected, to **parasitism,** where one partner (the parasite) lives on or in another (the host) and metabolically depends on it (see Chapter 52). Some parasitic protists are important pathogens (disease-causing agents) of plants or animals. Throughout this chapter, we describe specific examples of symbiotic associations involving protists.

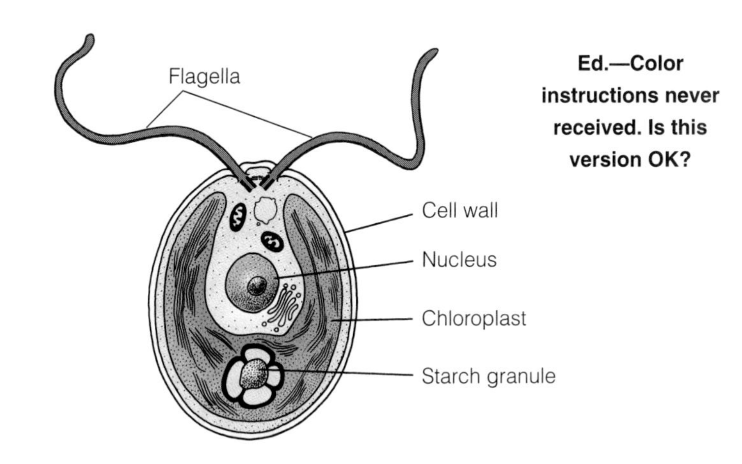

Ed.—Color instructions never received. Is this version OK?

FIGURE **24-1** | **Chlamydomonas, a unicellular protist.**
Chlamydomonas is a motile organism with two flagella and a cup-shaped chloroplast.

Most protists are aquatic and live in the ocean or in freshwater ponds, lakes, and streams. They make up most of the **plankton,** the floating, often microscopic organisms that inhabit surface waters and are the base of the food web in aquatic ecosystems. Other aquatic protists attach to rocks or other surfaces in the water. Even parasitic protists are aquatic, because they live in the watery environments of other organisms' body fluids. Terrestrial protists are restricted to damp places such as soil, cracks in bark, and leaf litter.

Reproduction is varied in kingdom Protista. Almost all protists reproduce asexually, and many also reproduce sexually, often by **syngamy,** the union of gametes. However, most protists do not develop multicellular reproductive organs, nor do they form embryos the way more complex organisms do.

Review

- What are the features of a "typical" protist?
- How do protists vary in their means of obtaining nutrients?
- What are some of the ways protists interact with other organisms?

BiologyNow™ Assess your understanding of **the protists** by taking the pretest on your BiologyNow CD-ROM.

EVOLUTION OF THE EUKARYOTES

Learning Objectives

3. Discuss the endosymbiont theory, and briefly explain some of the evidence that supports it.
4. Describe the kinds of data biologists use to classify eukaryotes.

Biologists hypothesize that protists were the first eukaryotic cells to evolve from ancestral prokaryotes. They may have appeared in the fossil record as early as 1.5 to 1.6 billion years ago. The first eukaryotes were probably *zooflagellates,* heterotrophic unicellular protists that have flagella. However, other than a few protists with hard shells, such as diatoms and foraminiferans, most ancient protists did not leave many fossils, because their bodies were too soft to leave permanent traces. Evolutionary studies of protists focus primarily on molecular and structural comparisons of present-day organisms, which contain many clues about their evolutionary history.

Mitochondria and chloroplasts probably originated from endosymbionts

According to the **endosymbiont theory,** advanced since the 1960s by U.S. biologist Lynn Margulis, certain eukaryotic organelles (particularly mitochondria and chloroplasts) arose from symbiotic relationships between larger cells and smaller prokaryotes that were incorporated and lived within them (see Fig. 20-7). Cell biologists hypothesize mitochondria originated from aerobic eubacteria. Studies of mitochondrial DNA suggest it is a remnant from the mitochondrion's past, when it was an independent organism. Gene sequence data suggest that rickettsias—bacteria that are obligate parasites (see Chapter 23)—are the closest extant relatives of mitochondria.

Chloroplast evolution is more complex, as there were probably several endosymbiotic events. Molecular evidence supports the view that incorporation of an ancient cyanobacterium within a host cell, known as the *primary endosymbiosis,* resulted in the chloroplasts in today's red algae, green algae, and plants. Biologists hypothesize that these chloroplasts, which are enclosed by two external membranes, later provided other nonphotosynthetic eukaryotes with their chloroplasts during *secondary endosymbiosis* (Fig. 24-2).

Secondary endosymbiosis occurred frequently in eukaryote evolution, as evidenced by the presence of additional chloroplast membranes. For example, *three* membranes envelop the chloroplasts of euglenoids and dinoflagellates, and *four* membranes surround the chloroplasts of diatoms, golden algae, and brown algae. Understanding how these membranes originated is an essential aspect of the endosymbiont theory, and many researchers are studying the origin of chloroplasts in different organisms.

Even nonphotosynthetic protists may contain chloroplast relics from secondary endosymbiotic events. Apicomplexans, protists such as *Plasmodium,* which causes malaria, have a nonfunctional chloroplast surrounded by four membranes. This chloroplast, which is probably derived from either a red alga or a green alga, has great medical potential in treating malaria, because some researchers hypothesize that drugs that inhibit chloroplasts also may kill *Plasmodium.*

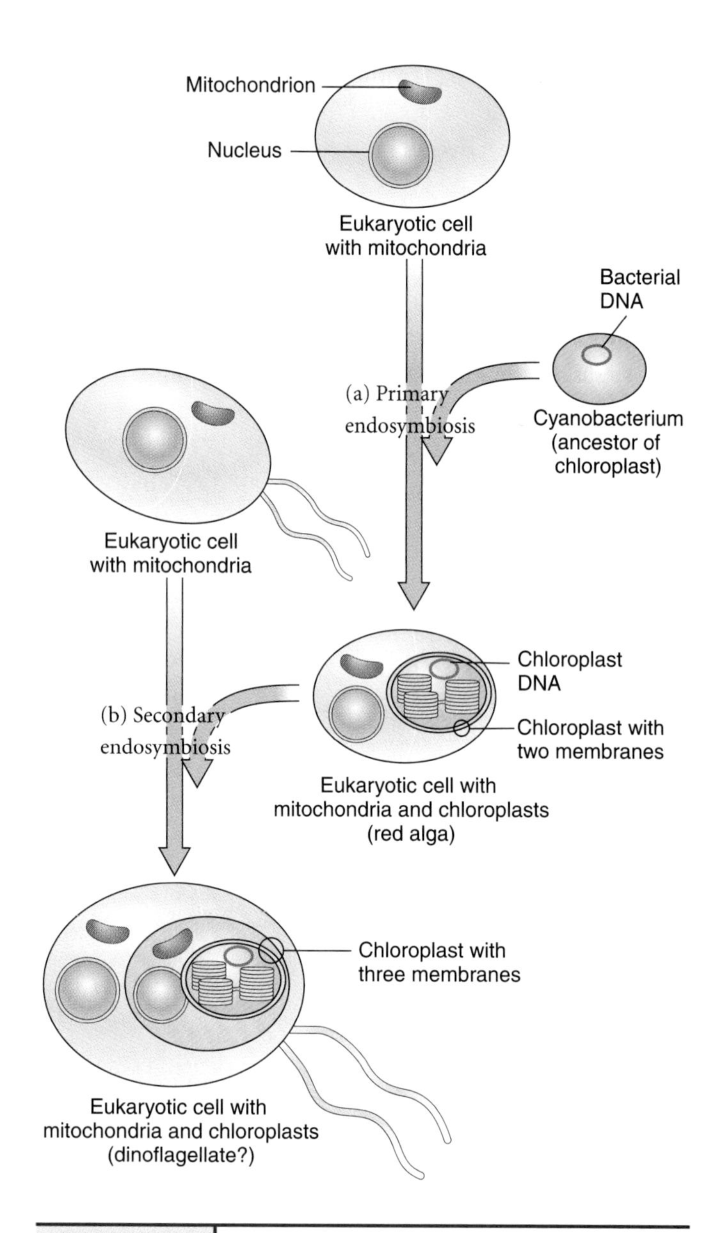

FIGURE **24-2** **Chloroplast evolution by primary and secondary endosymbiosis.**

(a) In primary endosymbiosis, an ancient eukaryotic cell engulfed a cyanobacterium, which survived and evolved into a chloroplast. The ancient cell is depicted as a eukaryotic cell because mitochondria almost certainly evolved before chloroplasts. **(b)** In a secondary endosymbiotic event, a heterotrophic eukaryotic cell (with mitochondria) engulfed a eukaryotic cell with chloroplasts (a red alga is depicted). The red alga survived and evolved into a chloroplast surrounded by three membranes (a dinoflagellate chloroplast is depicted). Other secondary endosymbiotic events resulted in more complex chloroplast membrane structures.

A consensus is slowly emerging in eukaryote classification

PROCESS OF SCIENCE

Scientists re-evaluate evolutionary relationships among the eukaryotes as additional evidence becomes available. Two types of modern research, molecular and ultrastructure, have contributed substantially to scientific understanding of the phylogenetic relationships among protists. Molecular data were initially obtained for the gene that codes for small subunit ribosomal RNA in different eukaryotes (SSU rRNA; see Chapter 23). More recently, biologists compared other nuclear genes, many of which code for proteins, for different protist groups.

Ultrastructure is the fine details of cell structure revealed by electron microscopy. In many cases, ultrastructure data complement molecular data. Electron microscopy reveals similar structural patterns among those protist groups that comparative molecular evidence suggests are **monophyletic**—that is, evolved

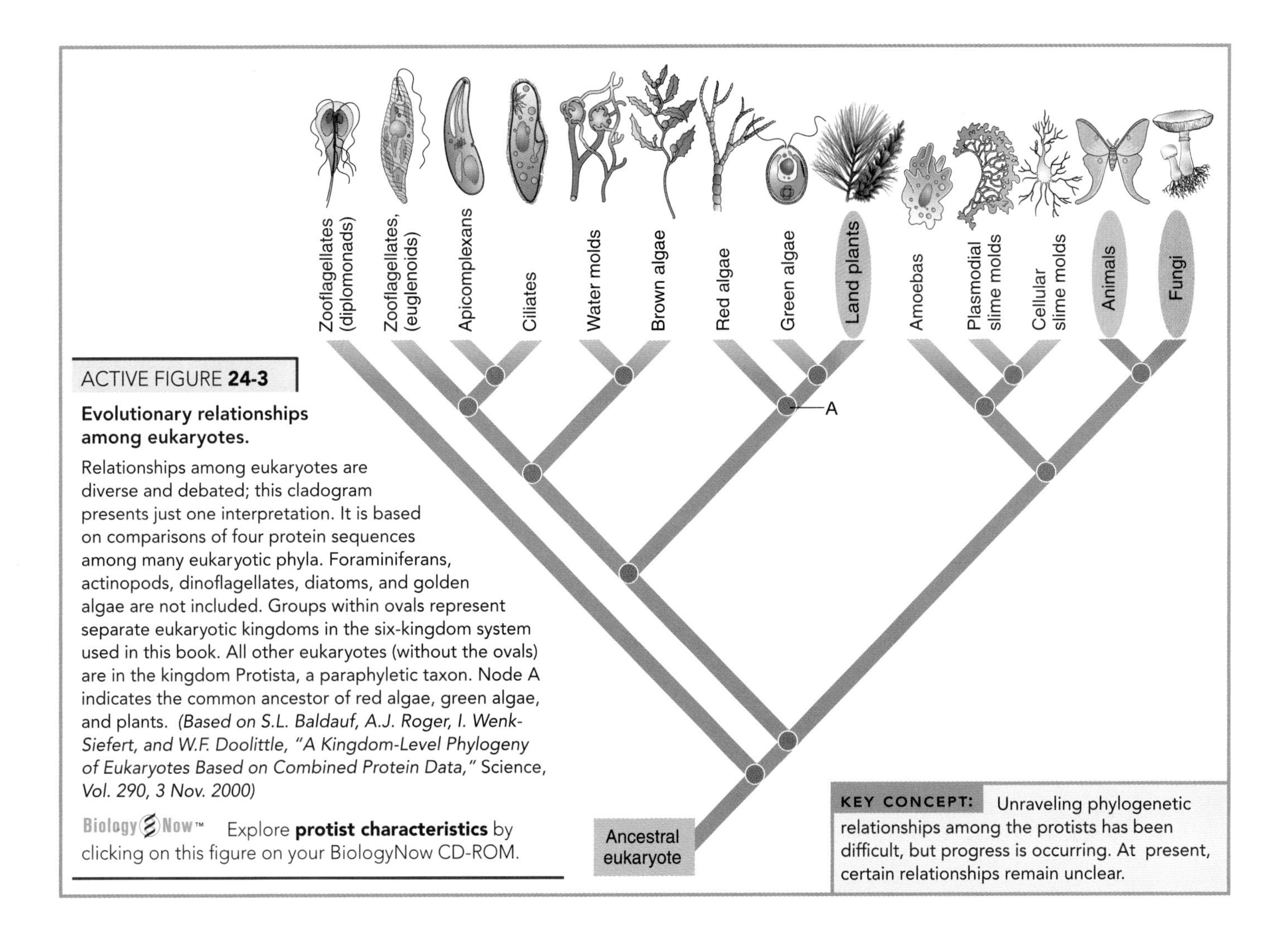

ACTIVE FIGURE **24-3**

Evolutionary relationships among eukaryotes.

Relationships among eukaryotes are diverse and debated; this cladogram presents just one interpretation. It is based on comparisons of four protein sequences among many eukaryotic phyla. Foraminiferans, actinopods, dinoflagellates, diatoms, and golden algae are not included. Groups within ovals represent separate eukaryotic kingdoms in the six-kingdom system used in this book. All other eukaryotes (without the ovals) are in the kingdom Protista, a paraphyletic taxon. Node A indicates the common ancestor of red algae, green algae, and plants. *(Based on S.L. Baldauf, A.J. Roger, I. Wenk-Siefert, and W.F. Doolittle, "A Kingdom-Level Phylogeny of Eukaryotes Based on Combined Protein Data,"* Science, *Vol. 290, 3 Nov. 2000)*

BiologyNow™ Explore **protist characteristics** by clicking on this figure on your BiologyNow CD-ROM.

KEY CONCEPT: Unraveling phylogenetic relationships among the protists has been difficult, but progress is occurring. At present, certain relationships remain unclear.

from a common ancestor (see Chapter 22). For example, molecular and ultrastructure data suggest that water molds, diatoms, golden algae, and brown algae—protist groups that at first glance seem to share few characteristics—are a monophyletic group.

Given the diversity in protist structures and molecules, most biologists regard the protist kingdom as a **paraphyletic group**—that is, the protist kingdom contains some, but not all, of the descendants of a common eukaryote ancestor. Molecular and ultrastructural analyses continue to help biologists clarify relationships among the various protist phyla and among protists and the other eukaryotic kingdoms (Fig. 24-3). Biologists also use these data to develop various proposals that split organisms in kingdom Protista into several kingdoms. One proposal, for example, removes red algae and green algae from the protists and classifies them within the plant kingdom. Figure 24-3 supports such a classification change, because it suggests they share a common ancestor (see node A).

Review

- How does the endosymbiont theory explain the origin of mitochondria and chloroplasts?
- What kinds of evidence support the hypothesis that the protist kingdom is a paraphyletic group?

BiologyNow™ Assess your understanding of **the evolution of the eukaryotes** by taking the pretest on your BiologyNow CD-ROM.

REPRESENTATIVE PROTISTS

Learning Objectives

5 Explain why zooflagellates are no longer classified in a single phylum, and distinguish among diplomonads, euglenoids, and choanoflagellates.

6 Briefly describe and compare the following alveolates: ciliates, dinoflagellates, and apicomplexans.

7 Briefly describe and compare the following heterokonts: water molds, diatoms, golden algae, and brown algae.

8 Describe the foraminiferans and actinopods, and explain why many biologists classify them in the monophyletic group cercozoa.

9 Support the hypothesis that red algae and green algae are a monophyletic group with land plants.

10 Briefly describe and compare the following amoebozoa: amoebas, plasmodial slime molds, and cellular slime molds.

The protists include those organisms traditionally known as protozoa, algae, and "lower fungi." Biologists continue to resolve evolutionary relationships among the protists and between the protists and other eukaryotes (plants, animals, and fungi). As a result of ongoing research, many biologists are developing different ways to classify eukaryotes, including new names for major groups. In the classification scheme we are currently using in this book almost all known eukaryotes are assigned to one of eight major groups: excavates, discicristates, alveolates, heterokonts, cercozoa, plants, amoebozoa, and opisthokonts (Table 24-1). Figure 24-4, shows the groups superimposed over Fig. 24-3. As additional data accumulate, these groups may be elevated to the status of kingdoms, resulting in a 10-kingdom system that replaces the 6-kingdom system currently in use (see Chapter 22). However, it is also possible new data will prompt biologists to abandon or reorganize at least some of the eight groups.

Excavates are anaerobic zooflagellates

Zooflagellates are mostly heterotrophic, unicellular (a few are colonial) organisms with spherical or elongated bodies. They move rapidly by lashing one to many long, whiplike **flagella** which are usually located at the anterior (front) end. Some zooflagellates ingest food by means of a definite "mouth," or *oral groove.* Biologists traditionally classified zooflagellates in a single phylum. However, evidence indicates zooflagellates are polyphyletic, and most biologists have separated them into several groups, such as the excavates and discicristates, based on close evolutionary relationships.

Most **excavates** are endosymbionts and live in anoxic (without oxygen) environments. Unlike virtually all other protists, excavates lack mitochondria, or if they are present, the mitochondria are atypical. Excavates do not carry out aerobic respiration; they obtain energy by glycolysis. These protists are so named because most have a deep, or *excavated,* oral groove.

From an evolutionary perspective, one of the most interesting groups of modern-day excavates is the **diplomonads** (phylum Retortamonada), which retain some characteristics of ancient protists. Diplomonads have one or two nuclei, no mitochondria, no Golgi complex, and up to eight flagella.

Giardia is a parasitic diplomonad. Comparative ribosomal RNA sequencing suggests diplomonads such as *Giardia* may be more closely related to prokaryotes than are any other protists (Fig. 24-5a). Interestingly, *Giardia's* cell structure—it has two haploid nuclei— suggests a partial explanation of how diploid eukaryotes may have arisen from haploid prokaryotes. Some biologists hypothesize that the first eukaryotes had a single haploid nucleus. Most eukaryotes today have a single diploid nucleus formed at some stage in the life cycle when two haploid nuclei fuse. Perhaps the ancestors of *Giardia* represent an intermediate stage in eukaryotic evolution when cells each had two haploid nuclei but fusion had not yet occurred:

Haploid prokaryote ⟶ eukaryote with single haploid nucleus ⟶ eukaryote with two haploid nuclei ⟶ eukaryote with single diploid nucleus

Giardia also lacks mitochondria, which initially suggested the nucleus may have evolved earlier than other membrane-enclosed organelles such as mitochondria. Arguments based on some molecular data dispute this idea, however. *Giardia* contains certain genes that code for proteins associated with mitochondria in other organisms. This information suggests an early eukaryote ancestor of *Giardia* may have possessed mitochondria, which were somehow lost at a later time during its evolutionary history.

Giardia intestinalis causes backpackers' diarrhea, a common infection among campers and hikers, particularly in the mountains of the western United States. *Giardia* is eliminated as a resistant cyst in the feces of many vertebrate animals. These cysts

TABLE 24-1 Major Monophyletic Eukaryote Taxa

Monophyletic Group	Representative Members	Key Characteristics
Excavates	Diplomonads and other zooflagellates	Mitochondria absent or atypical; cells specialized for glycolysis instead of aerobic respiration
Discicristates	Euglenoids and other zooflagellates	Disc-shaped cristae in mitochondria
Alveolates	Ciliates Dinoflagellates Apicomplexans	Alveoli, flattened vesicles just inside the plasma membrane; tubular cristae in mitochondria
Heterokonts	Water molds Diatoms Golden algae Brown algae	No flagella or two flagella, one plain and one with hairs
Cercozoa	Foraminiferans Actinopods	Tests (hard shells) through which cytoplasmic projections extend
Plants	Red algae Green algae Land plants	Chloroplasts surrounded by two membranes (inner and outer membranes)
Amoebozoa	Amoebas Plasmodial slime molds Cellular slime molds	Naked amoebas (no tests) with pseudopods
Opisthokonts	Fungi Choanoflagellates Animals	No flagella or single posterior flagellum on motile cells; flattened cristae in mitochondria

Source: S.L. Baldauf, "The Deep Roots of Eukaryotes," *Science,* Vol. 300, 13 June 2003.

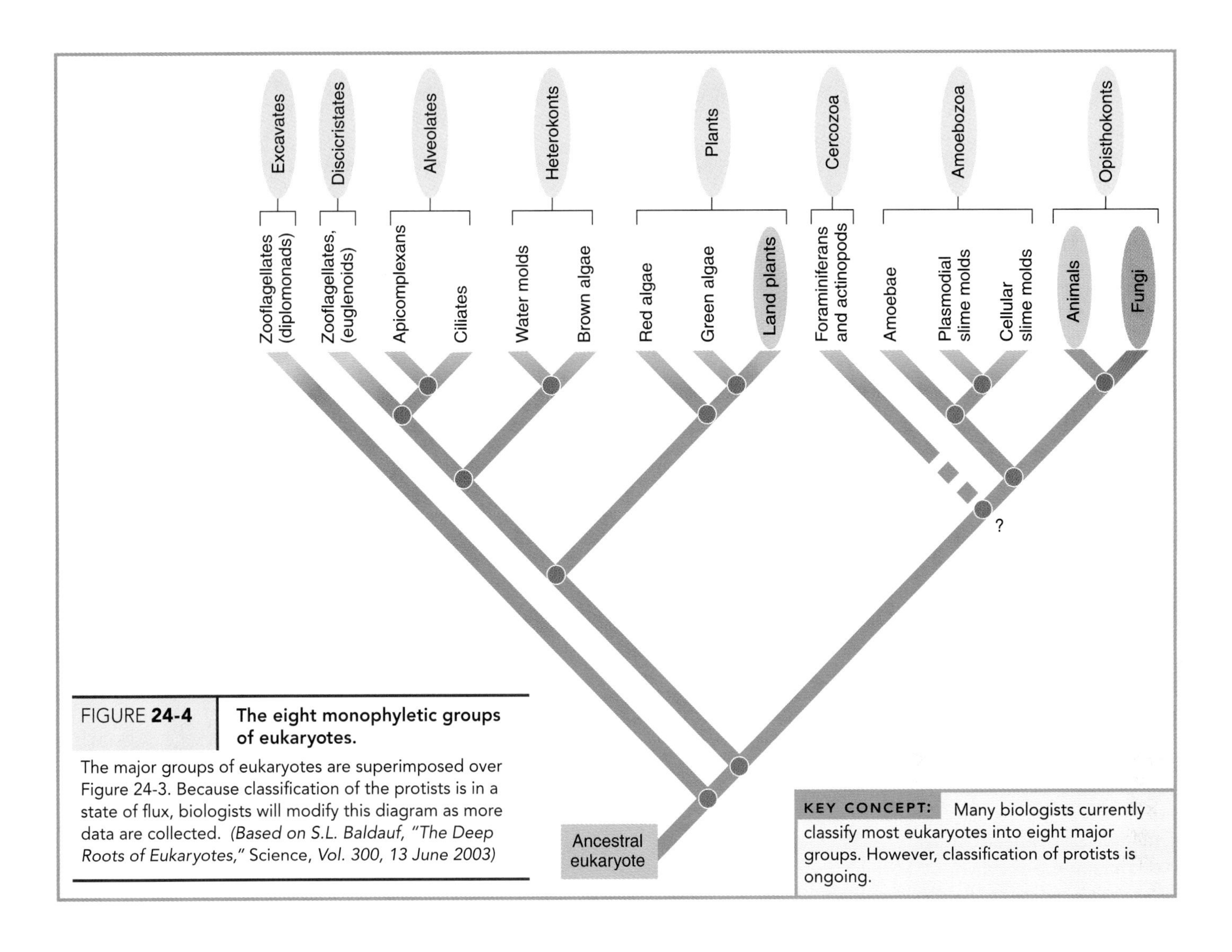

FIGURE **24-4** The eight monophyletic groups of eukaryotes.

The major groups of eukaryotes are superimposed over Figure 24-3. Because classification of the protists is in a state of flux, biologists will modify this diagram as more data are collected. *(Based on S.L. Baldauf, "The Deep Roots of Eukaryotes,"* Science, *Vol. 300, 13 June 2003)*

KEY CONCEPT: Many biologists currently classify most eukaryotes into eight major groups. However, classification of protists is ongoing.

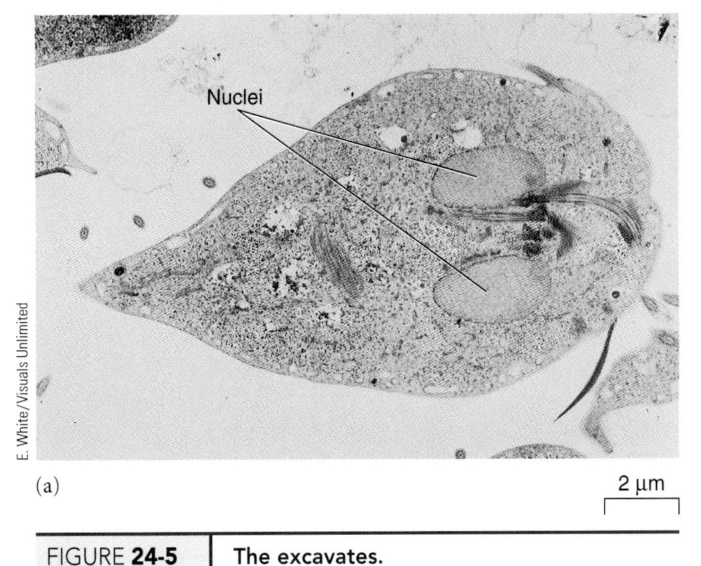

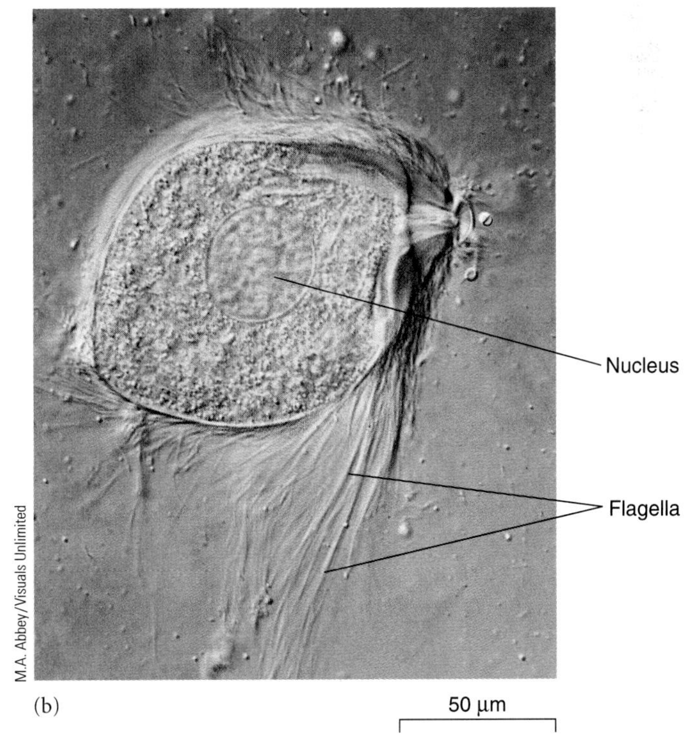

FIGURE **24-5** The excavates.

(a) This colorized TEM of *Giardia intestinalis*, a parasitic diplomonad, reveals two nuclei, suggesting that *Giardia* retains an ancestral condition that was an intermediate stage in the evolution of diploidy in eukaryotic life cycles. **(b)** LM of *Trichonympha*, an excavate that lives in the gut of wood-eating termites and cockroaches. *Trichonympha* has hundreds of flagella.

FIGURE 24-6

The discicristates.

(a) Note the complex cell structure shown in this LM of *Euglena gracilis,* a unicellular, flagellate euglenoid. **(b)** The eyespot may shield a light detector at the base of the long flagellum, thereby helping *Euglena* move to light of an appropriate intensity. *Euglena's* pellicle is flexible and changes shape easily. **(c)** LM of the zooflagellate *Trypanosoma gambiense* that causes sleeping sickness, among red blood cells in a human blood smear.

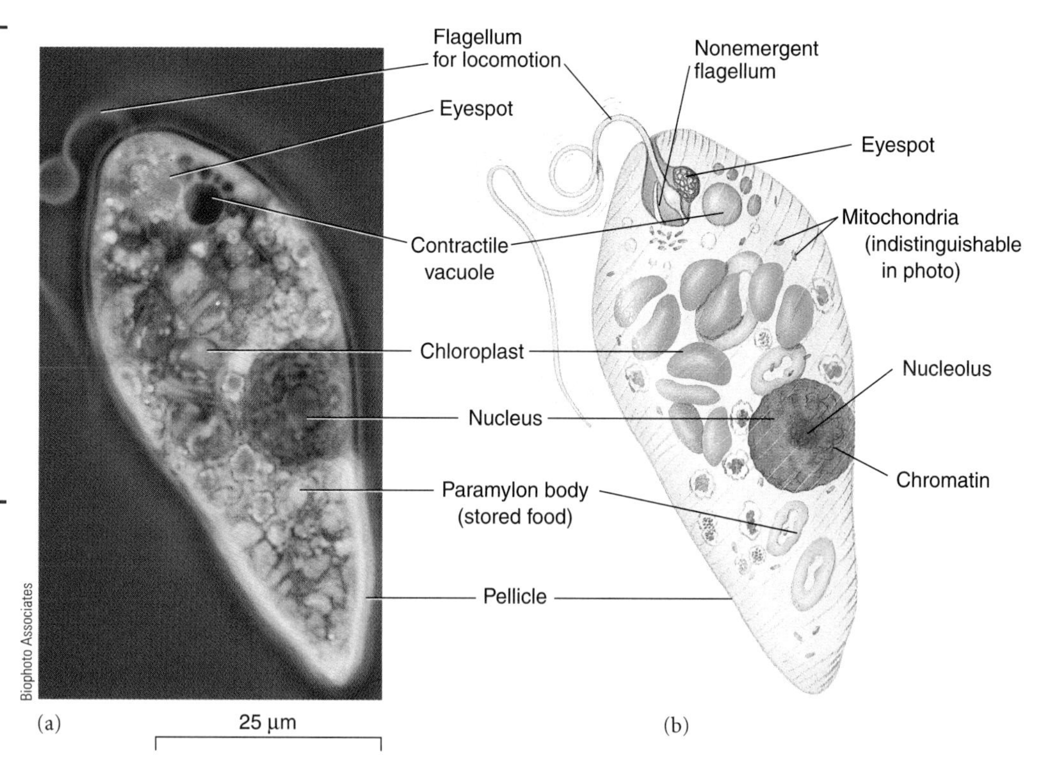

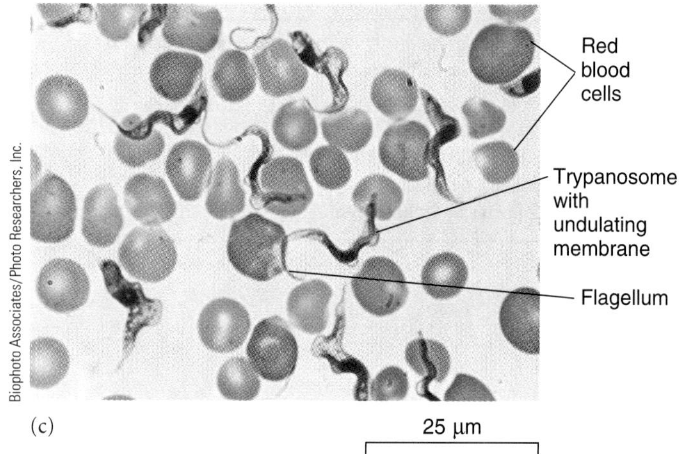

are a common contaminant in mountain streams. Campers and hikers become infected when they drink or rinse dishes in the "clean" mountain water. In a heavy infection, much of the wall of the small intestine is coated with these zooflagellates, which interfere with the absorption of digested nutrients and cause weight loss, abdominal cramps, and diarrhea.

Trichonymphs (phylum Axostylata) are specialized excavates with hundreds of flagella, that live in the guts of termites and wood-eating cockroaches (Fig. 24-5b). These zooflagellates lack mitochondria but have Golgi complexes. Trichonymphs ingest wood chips and rely on endosymbiotic bacteria to digest cellulose in the wood that termites or roaches eat; the insects, trichonymphs, and bacteria obtain their nutrients from this source. This is an excellent example of mutualism.

Discicristates include euglenoids and trypanosomes

Discicristates are zooflagellates named for their *disc*-shaped mitochondrial *cristae.* Discicristates include euglenoids, which are somewhat familiar to most students. Most **euglenoids** (phylum Euglenozoa) are unicellular flagellates, and about one third of them are photosynthetic (Fig. 24-6a,b). They generally have two flagella: one long and whiplike and one so short that it does not extend outside the cell. Some euglenoids, such as *Euglena,* change shape continually as they move through the water, because their **pellicle,** or outer covering, is flexible. Euglenoids reproduce asexually by mitosis; none has been observed to reproduce sexually.

At various times scientists classified euglenoids in the plant kingdom (with the algae) and in the animal kingdom (when protozoa were considered animals). Autotrophic euglenoids have chlorophylls *a* and *b* and yellow and orange **carotenoids,** the same photosynthetic pigments that green algae and plants have. Although the euglenoids have the same pigments as the green algae and plants, they are not closely related to either group, as shown in Fig. 24-3. They store energy reserves as *paramylon,* a polysaccharide. Some photosynthetic euglenoids lose their chlorophyll when grown in the dark, and they obtain their nutrients heterotrophically, by ingesting organic matter. Other species of euglenoids, such as *Peranema,* are always colorless and heterotrophic. Some heterotrophic species absorb organic compounds from the surrounding water, whereas others engulf bacteria and protists by **phagocytosis;** they digest the prey within food vacuoles.

Euglenoids inhabit freshwater ponds and puddles, particularly those with large concentrations of organic material. For that reason, researchers use them as indicator species of organic pollution. Some euglenoids also live in marine waters and mud flats.

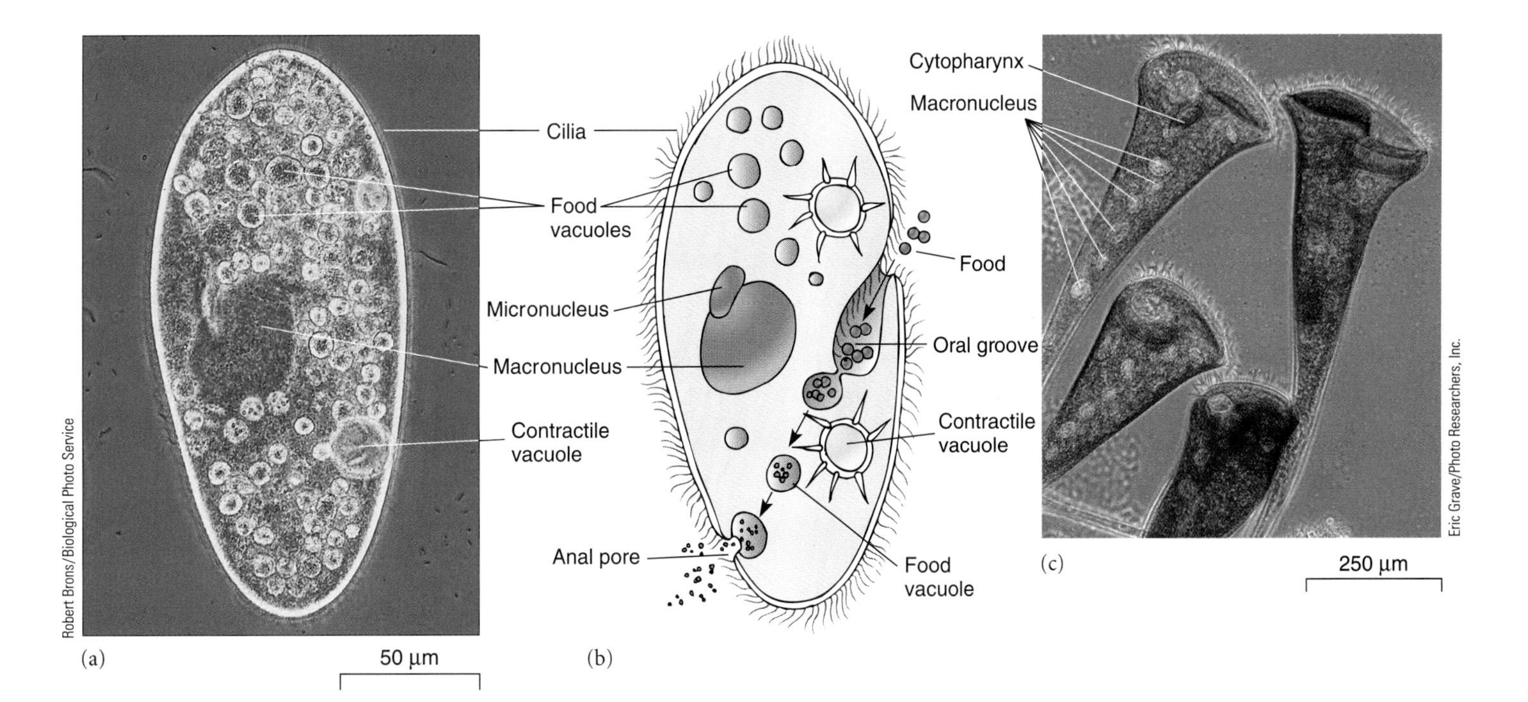

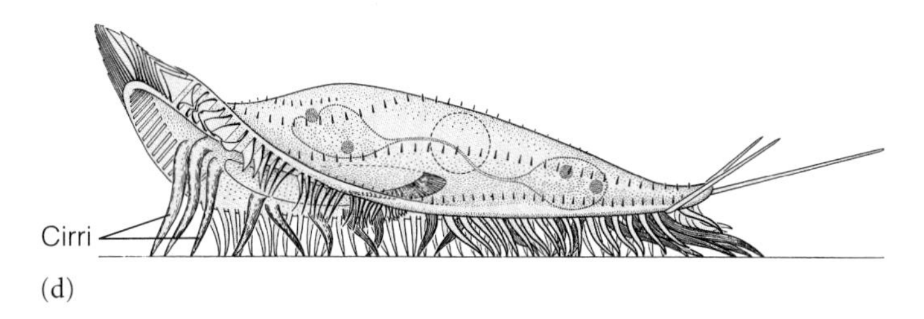

FIGURE **24-7** **Ciliates.**

(a) Note the complex cell structure in this phase contrast LM of *Paramecium,* a freshwater ciliate. Like many ciliates, *Paramecium* has multiple nuclei: a macronucleus and one or more smaller micronuclei. **(b)** Food particles are swept into *Paramecium's* ciliated oral groove and incorporated into food vacuoles. Lysosomes fuse with the food vacuoles, and the food is digested and absorbed; undigested wastes are eliminated through the anal pore. **(c)** LM of *Stentor,* a sessile ciliate. Note the numerous cilia that direct food particles into its funnel-like cytopharynx ("throat"). The elongated macronucleus resembles a string of beads. **(d)** The hypotrich *Euplotes* has stiff ciliary tufts called cirri. *(Source for 24-7(d): By H. Machemer, in K. G. Grell,* Protozoology, *© 1973 Springer-Verlag.)*

Trypanosomes (also in phylum Euglenozoa) are colorless discicristates, many of which are parasitic and cause disease. In vertebrates, including humans, trypanosomes live in the blood. For example, *Trypanosoma brucei* is a human parasite that causes African sleeping sickness (Fig. 24-6c). It is transmitted by the bite of infected tsetse flies. Early symptoms include recurring attacks of fever. Later, when the trypanosomes have invaded the central nervous system, infected people are lethargic and have difficulty speaking or walking; if untreated, African sleeping sickness can cause death. Trypanosomes, like euglenoids, reproduce asexually by mitosis; none has been observed to reproduce sexually.

Alveolates have flattened vesicles under the plasma membrane

The unifying features of protists classified as **alveolates** are similar ribosomal DNA sequences and **alveoli** (sing., *alveolus*), flattened vesicles located just inside the plasma membrane. In some alveolates, the vesicles contain plates of cellulose. Alveolates include the ciliates, dinoflagellates, and apicomplexans.

Ciliates use cilia for locomotion

Ciliates (phylum Alveolata, subphylum Ciliophora) are among the most complex of eukaryotic cells. These unicellular organisms have a flexible outer covering called a *pellicle* that gives them a definite but changeable shape. In *Paramecium,* the surface of the cell is covered with several thousand fine, short, hairlike **cilia** that extend through pores in the pellicle to facilitate movement (Fig. 24-7a,b; also see Fig. 1-1a). The cilia beat with such precise coordination that the organism can back up and turn around as well as move forward. Just under the pellicle, many ciliates have numerous small **trichocysts,** organelles that discharge filaments that may aid in trapping prey.

Not all ciliates are motile. Some sessile forms have stalks, and others, although capable of some swimming, are more likely to remain attached to a rock or other surface at one spot (Fig. 24-7c). Their cilia set up water currents that draw food toward them.

One group of ciliates, the **hypotrichs,** have greatly modified cilia and exhibit an unusual creeping-darting locomotion. Hypotrichs lack cilia over much of the body except on the ventral (lowermost) surface, where they occur in stiff tufts called **cirri** (sing., *cirrus,* from the Latin for "a curl of hair") (Fig. 24-7d). Hypotrichs crawl about by means of these cirri, which beat in a coordinated manner.

Ciliates have a wide range of feeding habits and diets; most ingest bacteria or other tiny protists. Their cilia draw the food into a simple opening in some species and into a funnel-like oral groove in others. A vacuole forms around the food at the end of the opening, and the food is digested. Special organelles called **contractile vacuoles** control water regulation in freshwater ciliates. Being hypertonic to their environment, freshwater ciliates continually take in water by osmosis; the contractile vacuole continually expels excess water.

Ciliates differ from other protists in having two kinds of nuclei: one or more small, diploid **micronuclei** that function in reproduction, and a larger, polyploid **macronucleus** that controls cell metabolism and growth. Most ciliates are capable of a sexual process called **conjugation,** in which two individuals come together and exchange genetic material. Biologists divide each ciliate species into several different mating types. Individuals with different mating types are identical in appearance but genetically different in terms of sexual compatibility. Because there are no known physical differences between the mating types, it is not appropriate to refer to them as "male" and "female."

During conjugation in *Paramecium,* two individuals of compatible mating types press their oral surfaces together (Fig. 24-8). Within each individual the macronucleus disintegrates and the micronucleus undergoes meiosis, forming four haploid nuclei. Three of these degenerate, leaving one, which divides mitotically to form two identical haploid nuclei. One of these remains within the cell, and the other nucleus crosses through the oral region into the other individual. A type of syngamy occurs; that is, the migrant haploid micronucleus fuses with the stationary haploid nucleus in that cell, and the two ciliates separate. The new diploid micronucleus in each cell divides a varying number of times, with one micronucleus developing into a new macronucleus.

A single act of conjugation yields two cross-fertilizations as each cell fertilizes the other. Conjugation results in two "new" cells that are genetically identical to each other but different from what they were before conjugation. Mitosis and cell division need not follow immediately after conjugation. Ciliates usually divide perpendicular to their longitudinal axis.

FIGURE **24-8** | **Conjugation in *Paramecium caudatum*.**

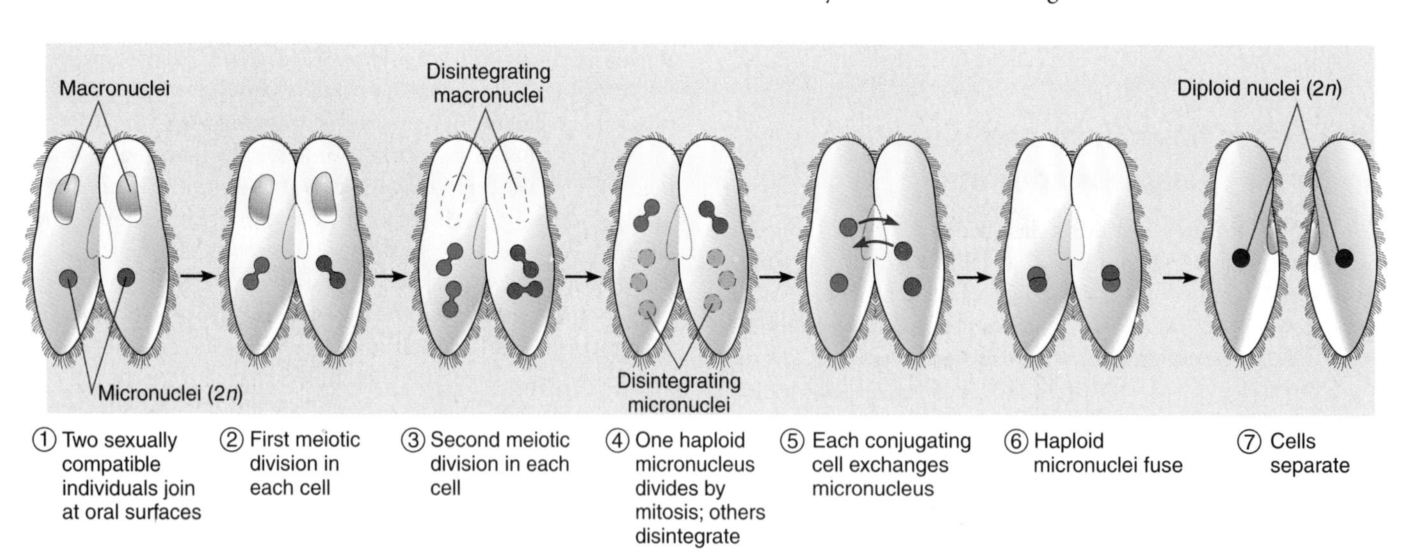

Macronuclei
Disintegrating macronuclei
Diploid nuclei (2n)
Micronuclei (2n)
Disintegrating micronuclei
① Two sexually compatible individuals join at oral surfaces
② First meiotic division in each cell
③ Second meiotic division in each cell
④ One haploid micronucleus divides by mitosis; others disintegrate
⑤ Each conjugating cell exchanges micronucleus
⑥ Haploid micronuclei fuse
⑦ Cells separate

Most dinoflagellates are a part of marine plankton

The group of **dinoflagellates** (phylum Alveolata, subphylum Dinoflagellata) are among the most unusual protists. Most dinoflagellates are unicellular, although a few are colonial. Their alveoli contain interlocking cellulose plates impregnated with silicates (Fig. 24-9a). The typical dinoflagellate has two flagella. One flagellum wraps around a transverse groove in the center of the cell like a belt, and the other lies in a longitudinal groove (perpendicular to the transverse groove), projecting behind the cell. The undulation of these flagella propels the dinoflagellate through the water like a spinning top. Indeed, the name is derived from the Greek *dinos,* meaning "whirling." Many marine dinoflagellates are bioluminescent.

Most dinoflagellates are photosynthetic and possess chlorophylls *a* and *c* and carotenoids, including **fucoxanthin,** a yellow-brown carotenoid (Fig. 24-9b). However, a number are colorless; some of these ingest other microorganisms for food. Dinoflagellates usually store energy reserves as oils or polysaccharides.

Many dinoflagellates are endosymbionts that live in the bodies of marine invertebrates such as mollusks, jellyfish, and corals (see Fig. 52-11). These symbiotic dinoflagellates, called **zooxanthellae,** lack cellulose plates and flagella. Zooxanthellae photosynthesize and provide carbohydrates for their invertebrate partners. Zooxanthellae contribute substantially to the productivity of coral reefs. Other dinoflagellates that are endosymbionts lack pigments and are parasites that live off their hosts.

Reproduction in the dinoflagellates is primarily asexual, by mitosis, although a few species reproduce sexually. The dinoflagellate nucleus is distinctive because the chromosomes are permanently condensed and always evident. Meiosis and mitosis are unusual because the nuclear envelope remains intact throughout cell division and the spindle lies *outside* the nucleus. The chromosomes attach to the nuclear envelope, and the spindle separates the new nuclei from each other.

Ecologically, dinoflagellates are important producers in marine ecosystems. A few dinoflagellates are known to have occa-

FIGURE **24-9** **Dinoflagellates.**

(a) SEM of *Protoperidinium.* Note the cellulose plates that encase the unicellular body and the sutures, or junctions, between adjacent plates. The two flagella (not visible) are located in grooves. **(b)** LM of *Ceratium,* which is photosynthetic (notice the green chlorophyll). Because of the pigments, the cellulose plates are difficult to see in living cells.

sional population explosions, or blooms. These blooms, which frequently color coastal waters orange, red, or brown, are known as **red tides.** The environmental conditions that initiate dinoflagellate blooms are not known, but many experts think human-produced coastal pollution triggers red tides. Some dinoflagellate species that form red tides produce a toxin that attacks the nervous systems of fishes, leading to fish kills. Birds sometimes die when exposed to the toxin by eating the dead fish. Research has also linked red tides to manatee deaths in Florida.

Since 1991, millions of fish with open, bleeding sores have died in estuaries and tributaries of North Carolina and Maryland waters, the result of a population explosion of toxic dinoflagellates that are relatively new to science. The species *Pfiesteria piscicida* was first scientifically named and identified in 1991. *Pfiesteria* and related dinoflagellates may have complex life cycles with about 24 stages, including dormant cysts, nontoxic amoeboid forms, and both nontoxic and toxic flagellate forms. The algal blooms are circumstantially linked to nitrogen and phosphorus pollution in shallow estuaries near hog and chicken operations and municipal sewage treatment plants.

Humans sometimes get paralytic shellfish poisoning by eating filter-feeding mollusks, such as oysters, mussels, or clams, that have fed on certain dinoflagellates. The toxin may be produced by a bacterial endosymbiont living within the dinoflagellate. The poison is a neurotoxin that impairs breathing in humans who eat the contaminated shellfish; death from respiratory failure occasionally occurs. The dinoflagellates do not seem to harm the shellfish.

Apicomplexans are spore-forming parasites of animals

Apicomplexans (phylum Alveolata, subphylum Apicomplexa) are a large group of parasitic, spore-forming protists, some of which cause serious diseases such as malaria in humans. They contain a vestigial plastid and may have evolved from parasitic dinoflagellates living in the intestines of marine invertebrates. Apicomplexans lack specific structures for locomotion (cilia, flagella, or pseudopodia); instead, they move by flexing. Apicomplexans have an *apical complex* of microtubules that attaches the parasite to its host cell; the apical complex is only visible using electron microscopy. At some stage in their life cycle, apicomplexans produce **sporozoites,** small infective agents transmitted to the next host; for this reason, many biologists call these organisms **sporozoa.** Many apicomplexans spend part of their complex life cycle in one host species and part in a different host species.

Malaria is caused by an apicomplexan. According to the World Health Organization, approximately 500 million people currently have malaria, and more than 1 million people, mostly children in developing countries, die from the disease each year. Although for centuries Chinese, Greek, Arabic, and Roman writings described the disease, its causative agent and mode of transmission were not identified until the end of the 19th century. Ronald Ross, an Englishman, received the Nobel Prize in 1902 for his role in elucidating the life cycle. Four species of *Plasmodium* cause malaria in humans. *Plasmodium* sporozoites enter human blood through the bite of an infected female *Anopheles* mosquito (Fig. 24-10). *Plasmodium* first enters liver cells, where it multiplies, and then red blood cells, where it continues to proliferate. When each infected red blood cell bursts, many new parasites are released. The released parasites infect new red blood cells, and the process is repeated. The simultaneous bursting of millions of red cells causes the symptoms of malaria: a chill, followed by high fever caused by toxic substances that are released and affect other organs of the body.

Today malaria is recurring in many countries where it was under control for decades. Formerly effective control methods—antimalarial drugs and pesticides—have lost much of their effectiveness. Chloroquine and several other antimalarial drugs are taken prophylactically to prevent malaria, but *Plasmodium* has evolved resistance to several of these, necessitating

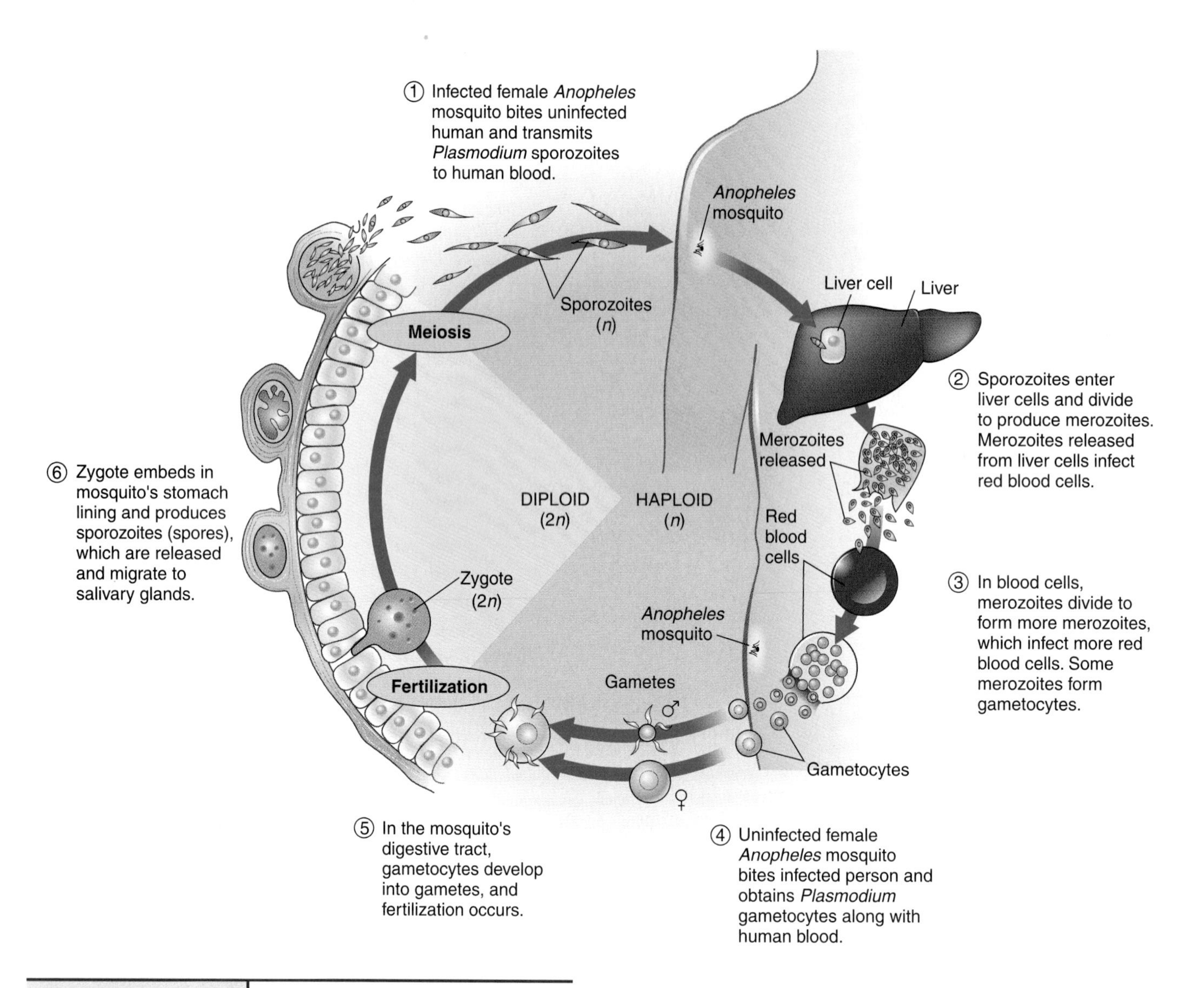

① Infected female Anopheles mosquito bites uninfected human and transmits Plasmodium sporozoites to human blood.
Anopheles mosquito
Liver cell
Liver
Sporozoites (n)
Meiosis
② Sporozoites enter liver cells and divide to produce merozoites. Merozoites released from liver cells infect red blood cells.
Merozoites released
⑥ Zygote embeds in mosquito's stomach lining and produces sporozoites (spores), which are released and migrate to salivary glands.
DIPLOID (2n)
HAPLOID (n)
Red blood cells
Zygote (2n)
③ In blood cells, merozoites divide to form more merozoites, which infect more red blood cells. Some merozoites form gametocytes.
Anopheles mosquito
Fertilization
Gametes
Gametocytes
⑤ In the mosquito's digestive tract, gametocytes develop into gametes, and fertilization occurs.
④ Uninfected female Anopheles mosquito bites infected person and obtains Plasmodium gametocytes along with human blood.

ACTIVE FIGURE **24-10**

The life cycle of *Plasmodium*, the causative agent of malaria.

Biology Now™ Watch **the life-cycle of the malaria parasite** by clicking on this figure on your BiologyNow CD-ROM.

the use of a combination of drugs. Moreover, pesticides are used to control *Plasmodium's* vector (carrier), the mosquito, but mosquitoes have evolved resistance to many pesticides. New antimalarial drugs and several possible vaccines against malaria are currently being tested. In addition, the recent sequencing of the genomes of *Plasmodium falciparum* and the *Anopheles* mosquito has the potential to provide new diagnostics, drugs, and vaccines.

Motile cells of heterokonts are biflagellate

The **heterokonts** include water molds, diatoms, golden algae, and brown algae. At first glance, heterokonts appear too diverse to classify together. The name *heterokont,* derived from the Greek words for "different flagella," provides a clue. Heterokonts have motile cells with two flagella, one of which has tiny hairlike projections off the shaft.

Water molds produce flagellate reproductive cells

Water molds (phylum Oomycota) were once classified as fungi because of their superficial resemblance. Both water molds and fungi have a body, called a **mycelium,** that grows over organic material, digesting it and then absorbing the predigested nutrients. The threadlike **hyphae** that make up the mycelium in water molds are *coenocytic,* meaning that there are no cross walls, and the body consists of a single multinucleate cell. The cell walls of water molds are composed of cellulose (as in plants), chitin (as in fungi), or both.

When food is plentiful and environmental conditions are favorable, water molds reproduce asexually (Fig. 24-11). A hyphal tip swells and a cross wall is formed, separating the hyphal tip from the rest of the mycelium. Within this structure, called a **zoosporangium,** tiny biflagellate zoospores form, each of which develops into a new mycelium. When environmental conditions worsen, water molds initiate sexual reproduction. After fusion

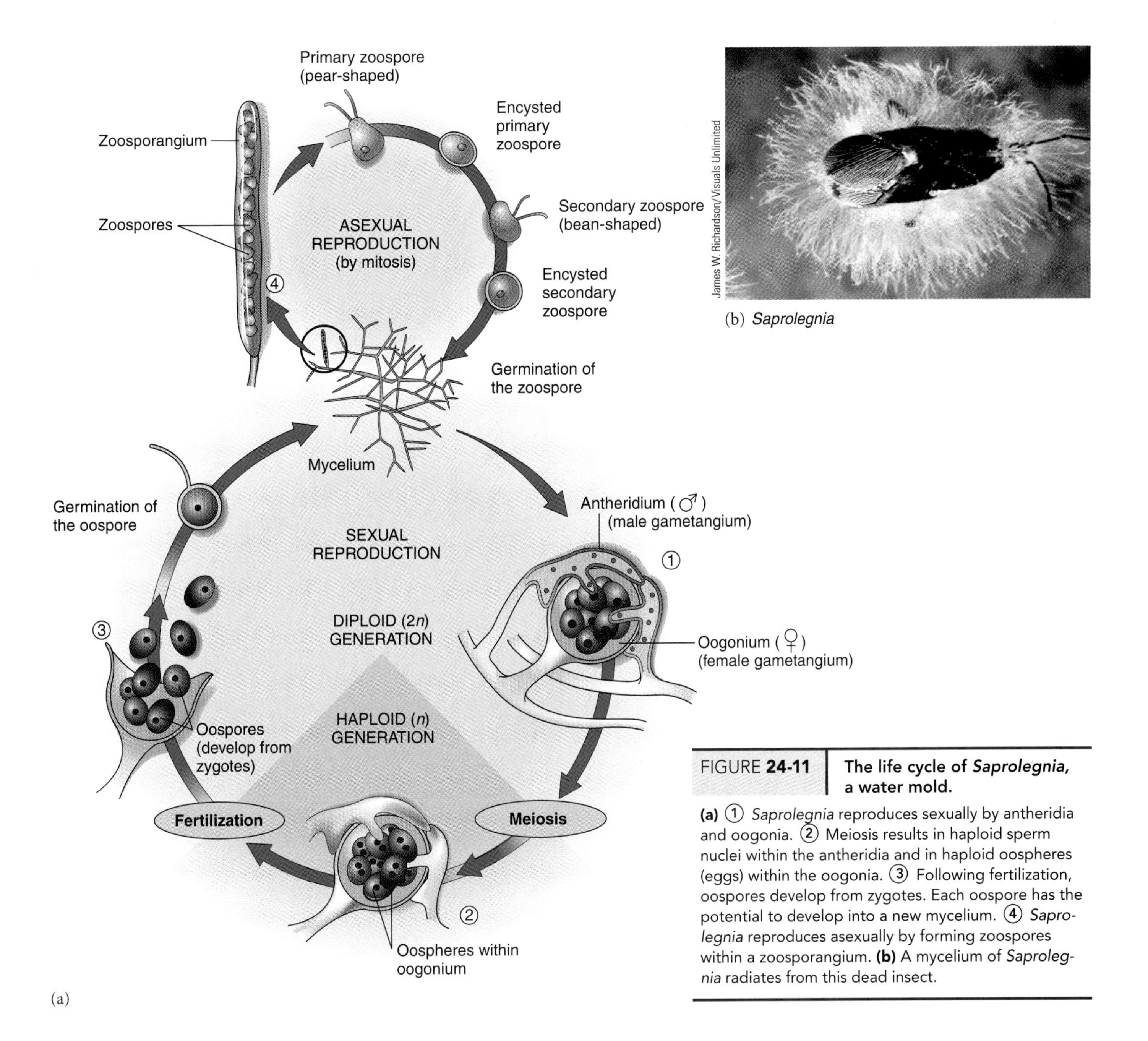

(a)

(b) *Saprolegnia*

FIGURE **24-11** **The life cycle of *Saprolegnia*, a water mold.**

(a) ① *Saprolegnia* reproduces sexually by antheridia and oogonia. ② Meiosis results in haploid sperm nuclei within the antheridia and in haploid oospheres (eggs) within the oogonia. ③ Following fertilization, oospores develop from zygotes. Each oospore has the potential to develop into a new mycelium. ④ *Saprolegnia* reproduces asexually by forming zoospores within a zoosporangium. **(b)** A mycelium of *Saprolegnia* radiates from this dead insect.

of male and female nuclei, thick-walled **oospores** develop from the oospheres. Water molds often spend the winter as oospores.

Some water molds have played infamous roles in human history. For example, the Irish potato famine of the 19th century was precipitated by the water mold *Phytophthora infestans,* which causes late blight of potatoes. (The genus *Phytophthora* is named from the Greek, meaning "plant destruction.") During several rainy, cool summers in Ireland in the 1840s, the water mold multiplied unchecked, causing potato tubers to rot in the fields. Because potatoes were the staple of the Irish peasants' diet, many people starved. Estimates of the number of deaths from the outbreak of the potato famine range from 250,000 to more than 1 million. The famine prompted a mass migration out of Ireland to such countries as the United States.

Late blight is still a problem today. New strains of *P. infestans* have appeared in the United States, Canada, northern Europe, Russia, South America, Japan, South Korea, the Middle East, and Africa. These strains resist the chemicals usually used to control the disease. Because today most people eat a varied diet, late blight does not cause famine, but it costs potato growers millions of dollars annually in lost crops.

A close relative of the late blight water mold, *Phytophthora ramorum,* causes sudden oak death, which is killing oak forests in California and Oregon. Plant pathologists are concerned the disease may spread to midwestern and eastern forests. This particular water mold also attacks redwoods, Douglas firs, bay trees, maples, and several other plant species, but most only have twig and leaf infections, not the rapid death observed in oaks.

Diatoms have shells composed of two parts

Most **diatoms** (phylum Bacillariophyta) are unicellular, although a few exist as colonies. The cell wall of each diatom consists of two shells that overlap where they fit together, much like

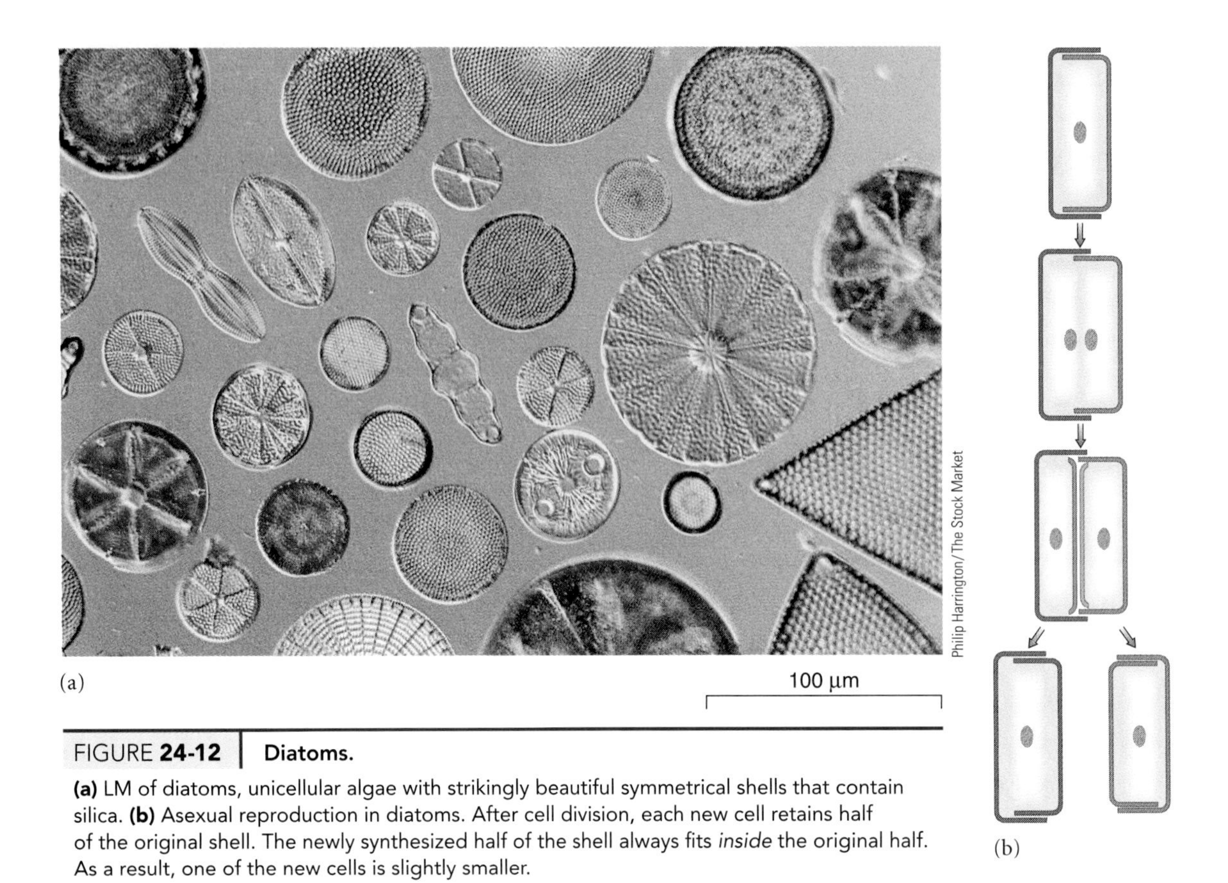

FIGURE **24-12** **Diatoms.**

(a) LM of diatoms, unicellular algae with strikingly beautiful symmetrical shells that contain silica. **(b)** Asexual reproduction in diatoms. After cell division, each new cell retains half of the original shell. The newly synthesized half of the shell always fits *inside* the original half. As a result, one of the new cells is slightly smaller.

a petri dish. Silica is deposited in the shell, and this glasslike material is laid down in intricate patterns (Fig. 24-12a). There are two basic groups of diatoms: those with radial symmetry (wheel-shaped) and those with bilateral symmetry (boat-shaped or needle-shaped). Although some diatoms are part of the floating plankton, others live on rocks and sediments, where they move by gliding. This gliding movement is facilitated by the secretion of a slimy material from a small groove along the shell.

Diatoms contain the photosynthetic pigments chlorophylls *a* and *c* and carotenoids, including fucoxanthin; their pigment composition gives them a yellow or brown color. Energy reserves are stored as oils or the water-soluble carbohydrate *chrysolaminarin,* which is similar to the laminarin stored in brown algae.

Diatoms most often reproduce asexually by mitosis. When a diatom divides, the two halves of its shell separate, and each becomes the larger half of a new diatom shell (Fig. 24-12b). Because the glass shell cannot grow, some diatom cells get progressively smaller with each succeeding generation. When a diatom reaches a fraction of its original size, sexual reproduction occurs, with the production of shell-less gametes. Sexual reproduction restores the diatom to its original size because the resulting *zygote,* a $2n$ cell that results from the fusion of n gametes, grows substantially before producing a new shell.

Diatoms are common in fresh water, but they are especially abundant in relatively cool ocean water. They are major producers in aquatic ecosystems, because of their extremely large numbers. At least one species is toxic and linked to shellfish poisonings, marine mammal strandings, and the deaths of sea lions along the central California coast.

When diatoms die, their shells trickle to the ocean floor and accumulate in layers that eventually become sedimentary rock. After millions of years, geological upheaval exposed some of these deposits on land. Called *diatomaceous earth,* these deposits are mined and used as filtering, insulating, and soundproofing materials. As a filtering agent, diatomaceous earth is used to refine raw sugar and to process vegetable oils. Because of its abrasive properties, diatomaceous earth is a common ingredient in scouring powders and metal polishes; it is no longer added to most toothpastes because it is too abrasive for tooth enamel. The intricately detailed diatom shells are often used to test microscope resolution down to 1 μm.

Most golden algae are flagellate, unicellular organisms

Golden algae (phylum Chrysophyta) are found in both freshwater and marine environments. Most species are biflagellate, unicellular organisms, although some are colonial (Fig. 24-13a). A few lack flagella and are similar to amoebas in appearance except that they contain chloroplasts. Tiny scales of either silica or calcium carbonate may cover the cells. Reproduction in golden algae is primarily asexual and involves the production of flagellate, motile spores called **zoospores.**

Most golden algae are photosynthetic and produce the same pigments as diatoms: chlorophylls *a* and *c* and carotenoids, including fucoxanthin. The pigment composition of golden algae gives them a golden or golden brown color. As in diatoms, energy reserves are stored as oils or carbohydrates. A few species

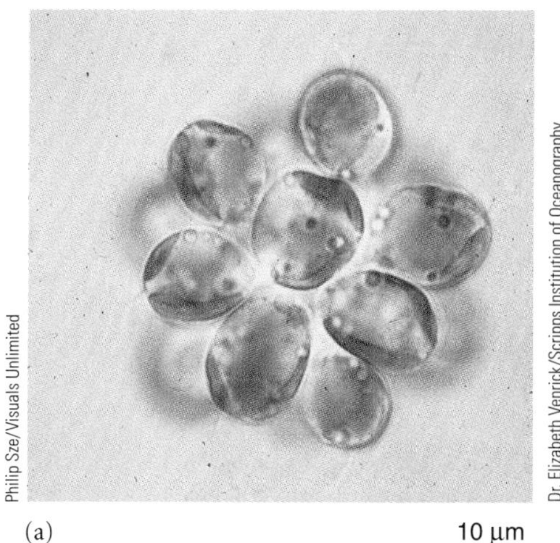

FIGURE **24-13** **Golden algae.**
(a) LM of a colonial golden alga (*Synura*) found in freshwater lakes and ponds.
(b) SEM of a coccolithophorid (*Emiliania huxleyi*). Note the overlapping scales of calcium carbonate.

ingest bacteria and other particles of food. Ecologically, golden algae are important producers in marine environments. They comprise a significant portion of the ocean's **nanoplankton,** extremely minute algae that are major producers because of their great abundance.

Classification of golden algae is controversial. Some biologists lump diatoms and golden algae in a single phylum, whereas others classify them as brown algae. At the other extreme, some biologists divide the golden algae into two phyla by placing many of the marine species, such as **coccolithophorids** (Fig. 24-13b), in a separate phylum.

Brown algae are multicellular seaweeds

Brown algae (phylum Phaeophyta) include the giants of the protist kingdom. They are the largest and most complex of all algae commonly called *seaweeds.* All brown algae are multicellular and range in size from a few centimeters (about an inch) to 75 m (about 250 ft). Their body forms are branched filaments, tufts, fleshy "ropes," or thick, flattened branches. The largest brown algae, called *kelps,* are tough and leathery in appearance. Many kelps have leaflike **blades** in which most photosynthesis occurs, stemlike **stipes,** and rootlike anchoring **holdfasts** (Fig. 24-14a). They often have gas-filled *bladders* that provide buoyancy. (The blades, stipes, and holdfasts of brown algae are not homologous to the leaves, stems, and roots of plants. Brown algae and plants arose from different unicellular ancestors, as shown in Fig. 24-3.)

Brown algae are photosynthetic and have chlorophylls *a* and *c* and carotenoids, including fucoxanthin, in their chloroplasts. The main energy storage reserve in brown algae is a carbohydrate called *laminarin.*

Reproduction is varied and complex in the brown algae. Their reproductive cells, both asexual zoospores and sexual gametes, are usually flagellate. Most have a life cycle that exhibits **alternation of generations,** in which they spend part of their life as haploid organisms and part as diploid organisms (see Fig. 9-16c).

Brown algae are commercially important for several reasons. Their cell walls contain a polysaccharide called *algin* that is harvested from kelps such as *Macrocystis* and used as a thickening and stabilizing agent in ice cream, toothpaste, shaving cream, hair spray, and hand lotion. Brown algae are an important human food, particularly in eastern Asia, and they are rich sources of certain vitamins and minerals such as iodine.

Brown algae are common in cooler marine waters, especially along rocky coastlines, where they live mainly in the intertidal zone or relatively shallow offshore waters. Kelps form extensive underwater "forests," or kelp beds (Fig. 24-14b). They are essential in that ecosystem for two reasons: as important food producers, and as habitat for many marine invertebrates, fish, and mammals. The diversity of life supported by kelp beds rivals that in coral reefs. There is also an extensive population of floating brown algae in a central area of the North Atlantic Ocean called the Sargasso Sea, named for the brown alga *Sargassum.* The Sargasso Sea is not greatly affected by the surface ocean currents rotating around the margins of the North Atlantic, so the floating *Sargassum* remains there.

Cercozoa are amoeboid cells enclosed in shells

The **cercozoa** are amoeboid cells that often have hard outer shells, called **tests,** through which cytoplasmic projections extend. The cytoplasmic projections suggest the name *cercozoa,* from the Greek *cerco,* meaning "rod." Foraminiferans and actinopods are cercozoa.

Foraminiferans extend cytoplasmic projections that form a threadlike interconnected net

Almost all **foraminiferans** (phylum Foraminifera) are marine organisms that produce elaborate tests (Fig. 24-15a). The ocean contains enormous numbers of foraminiferans, which secrete chalky, many-chambered tests with pores through which cytoplasmic projections are extended. The group gets its phylum name from this characteristic, as *foraminifera* is derived from the Latin for "bearing openings." The cytoplasmic projections form a sticky, interconnected net that entangles prey. Many foraminiferans contain unicellular algal endosymbionts (green algae, red algae, or diatoms) that provide food by photosynthe-

(a)

(b)

FIGURE **24-14** **Brown algae.**

(a) *Laminaria* is widely distributed on rocky coastlines of temperate and polar seas. It grows to 2 m (6.5 ft). **(b)** A kelp (*Macrocystis pyrifera*) bed is ecologically important to aquatic organisms, including the sea lion shown here. Photographed off the coast of California.

sis. Many foraminiferan species live on the ocean floor, but others are part of the plankton.

Dead foraminiferans settle on the bottom of the ocean, where their tests form a gray mud that is gradually transformed into chalk. With geological uplifting, these chalk formations become part of the land, like the White Cliffs of Dover in England (Fig. 24-15b). (The White Cliffs of Dover are the remains of a variety of carbonate organisms, not only foraminiferans.) Because foraminiferan tests often appear in rock layers covering oil deposits, geologists exploring for oil look for foraminiferan

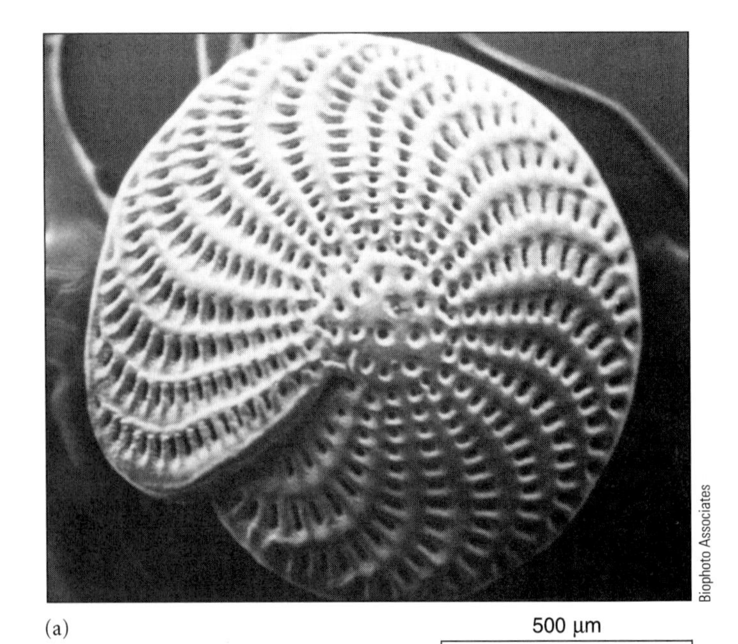

(a)

(b)

Lynn McLaren/Photo Researchers, Inc.

FIGURE **24-15** **Foraminiferans.**

(a) SEM of a foraminiferan test. Note the pores through which cytoplasm extrudes. Most foraminiferans have multichambered shells like this one. **(b)** The White Cliffs of Dover, England, consist largely of the tests of foraminiferans.

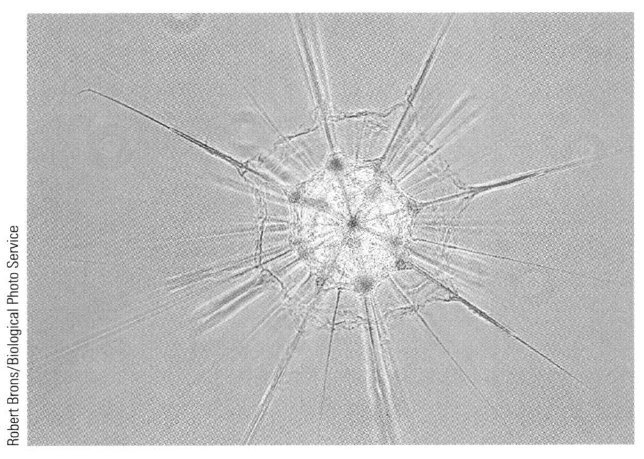

FIGURE **24-16** | **Actinopods.**

LM of an unidentified living actinopod from the Red Sea. Note the many slender axopods that project from the cell. The shell is not visible, because cytoplasm covers it on all sides (the shell is an endoskeleton).

tests in rock strata. Foraminiferans are well preserved in the fossil record, and biologists use some as **index fossils,** markers to help identify sedimentary rock layers (see Chapter 17).

Actinopods project slender axopods

Actinopods (phylum Actinopoda) are mostly marine plankton organisms with long, filamentous cytoplasmic projections called **axopods** that protrude through pores in their shells (Fig. 24-16). A cluster of microtubules strengthens each axopod. Unicellular algae and other prey become entangled in these axopods and are engulfed outside the main body of the actinopod; cytoplasmic streaming carries the prey inside the body. Many actinopods contain algal endosymbionts that provide them with the products of photosynthesis.

Some actinopods secrete elaborate and beautiful glassy shells made of silica. These **radiolarians,** are an important constituent of marine plankton. When radiolarians and other actinopods die, their shells settle and become an ooze (sediment) that can become several meters thick on the ocean floor.

Red algae, green algae, and land plants are collectively classified as plants

In the classification scheme adopted in this text, the monophyletic taxon of **plants** includes red algae and green algae, which are currently in the kingdom Protista, and land plants, which are in a separate kingdom (see Chapters 26 and 27). Biologists classify these together based on the presence of chloroplasts bounded by outer and inner membranes (see Fig. 8-4).

Red algae do not produce motile cells

The vast majority of **red algae** (phylum Rhodophyta) are multicellular organisms, although there are a few unicellular species. The multicellular body form of red algae commonly consists of complex, interwoven filaments that are delicate and feathery (Fig. 24-17a); a few red algae are flattened sheets of cells. Most multicellular red algae attach to rocks or other substrates by a basal holdfast. Reproduction in the red algae is remarkably complex, with an alternation of sexual and asexual stages. Although sexual reproduction is common, no flagellate cells develop during the life history of red algae.

The chloroplasts of red algae contain **phycoerythrin,** a red pigment, and **phycocyanin,** a blue pigment, in addition to chlorophylls *a* and *d* and carotenoids. The red algae store energy reserves as *floridean starch,* a polysaccharide similar to glycogen. They have the same pigment composition as the cyanobacteria, a group of photosynthetic eubacteria (see Chapter 23).

The cell walls of red algae often contain thick, sticky polysaccharides that have commercial value. For example, agar is a polysaccharide humans extract from certain red algae and use as a food thickener and culture medium, a substrate on which to grow microorganisms and propagate some plants, such as orchids. Another polysaccharide extracted from red algae, carrageenan, is a food additive used to stabilize chocolate milk and to provide a thick, creamy texture to ice cream and other soft processed foods. Carrageenan is also used to stabilize paints and cosmetics. Red algae are a source of vitamins (particularly

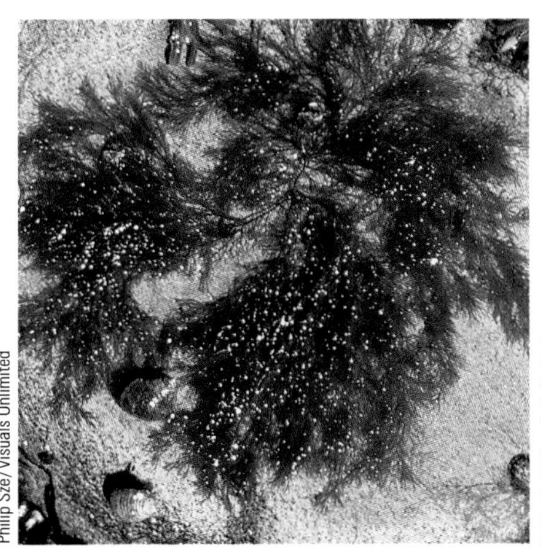

(a)

(c)

FIGURE **24-17** | **Red algae.**

(a) *Polysiphonia,* which is widely distributed throughout the world, has a highly branched body of interwoven filaments. It grows to 30 cm (11.7 in). **(b)** *Bossiella* is a coralline red alga encrusted with calcium carbonate. It lives in the Pacific Ocean, where it grows to 12 cm (4.7 in).

A and C) and minerals for humans, especially in Japan and other East Asian countries, where people eat red algae fresh, dried, or toasted in such traditional foods as sushi and nori.

Red algae primarily live in warm tropical ocean waters, although a few species occur in fresh water and in soil. Some red algae, known as *coralline algae,* incorporate calcium carbonate in their cell walls from the ocean waters (Fig. 24-17b). The hard calcium carbonate may protect coralline algae from the rigors of wave action. These coralline red algae are extremely important in building "coral" reefs—perhaps more crucial than coral animals in this process.

Green algae share many similarities with land plants

Green algae (phylum Chlorophyta) have pigments, energy reserve products, and cell walls that are chemically identical to those of land plants. Green algae are photosynthetic, with chlorophylls *a* and *b* and carotenoids present in chloroplasts of a wide variety of shapes. They store their main energy reserves as starch. Most green algae have cell walls with cellulose, although some lack walls. Because of these and other similarities, biologists generally accept that land plants arose from ancestral green algae (see Fig. 24-3). Using recent molecular and ultrastructure data, some biologists want to reclassify this extremely diverse group in the plant kingdom.

Green algae exhibit a variety of body types, from single cells to colonial forms, to coenocytic algae (multinucleate), to multicellular filaments and sheets (Fig. 24-18; also see Fig. 8-7a). The multicellular forms do not have cells differentiated into tissues, a characteristic that separates them from land plants. Most green algae have, or produce, flagellate cells during their life history, although a few are totally nonmotile.

Reproduction in the green algae is as varied as their body forms, with both sexual and asexual reproduction. Many green algae have life cycles with an alternation of generations. Asexual reproduction is by mitosis and cell division in single cells or by fragmentation in multicellular forms. Many green algae produce spores asexually by mitosis; if these spores have flagella and are motile, they are called *zoospores* (Fig. 24-19).

Sexual reproduction in the green algae involves gamete formation in unicellular **gametangia** (sing., *gametangium*), reproductive structures in which gametes are produced. Green algae have three types of sexual reproduction—isogamous, anisogamous, and oogamous. If the two flagellate gametes that fuse are identical in size and appearance, sexual reproduction is **isogamous** (see Fig. 24-19). **Anisogamous** sexual reproduction involves the fusion of two flagellate gametes of different sizes (Fig. 24-20). Some green algae are **oogamous** and produce a nonmotile egg and a flagellate male gamete. In addition to sexual reproduction by the fusion of gametes, some green algae exchange genetic information by a form of conjugation, in which the genetic material of one cell passes into a recipient cell (Fig. 24-21).

Green algae are found in both aquatic and terrestrial environments. Aquatic green algae primarily inhabit fresh water, although there are many marine species. Terrestrial green algae are restricted to damp soil, cracks in tree bark, and other moist places. Many green algae are symbionts with other organisms; some live as endosymbionts in body cells of invertebrates, and a few grow together with fungi as "dual organisms" called *lichens* (see Chapter 25). Regardless of where they live, green algae are ecologically important as producers.

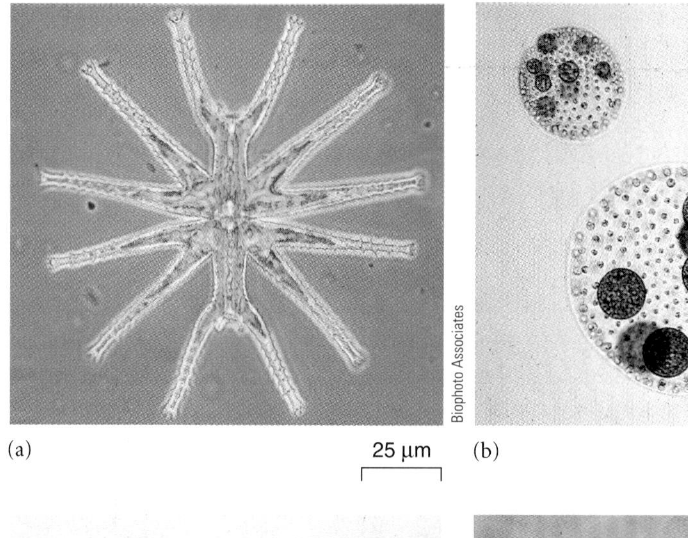

FIGURE **24-18** | **Green algae.**

(a) LM of a widely distributed desmid (*Micrasterias*), a unicellular green alga with mirror-image halves. **(b)** LM of two *Volvox* colonies, each composed of up to 50,000 cells. New colonies are inside the parental colonies, which eventually break apart. **(c)** *Ulva's* thin, sheetlike form suggests its common name, sea lettuce. *Ulva* typically grows to 30 cm (11.7 in). **(d)** *Chara,* a green alga commonly called a stonewort, is closely related to land plants. *Chara* is widely distributed in fresh water, where it grows to about 30 cm (11.7 in).

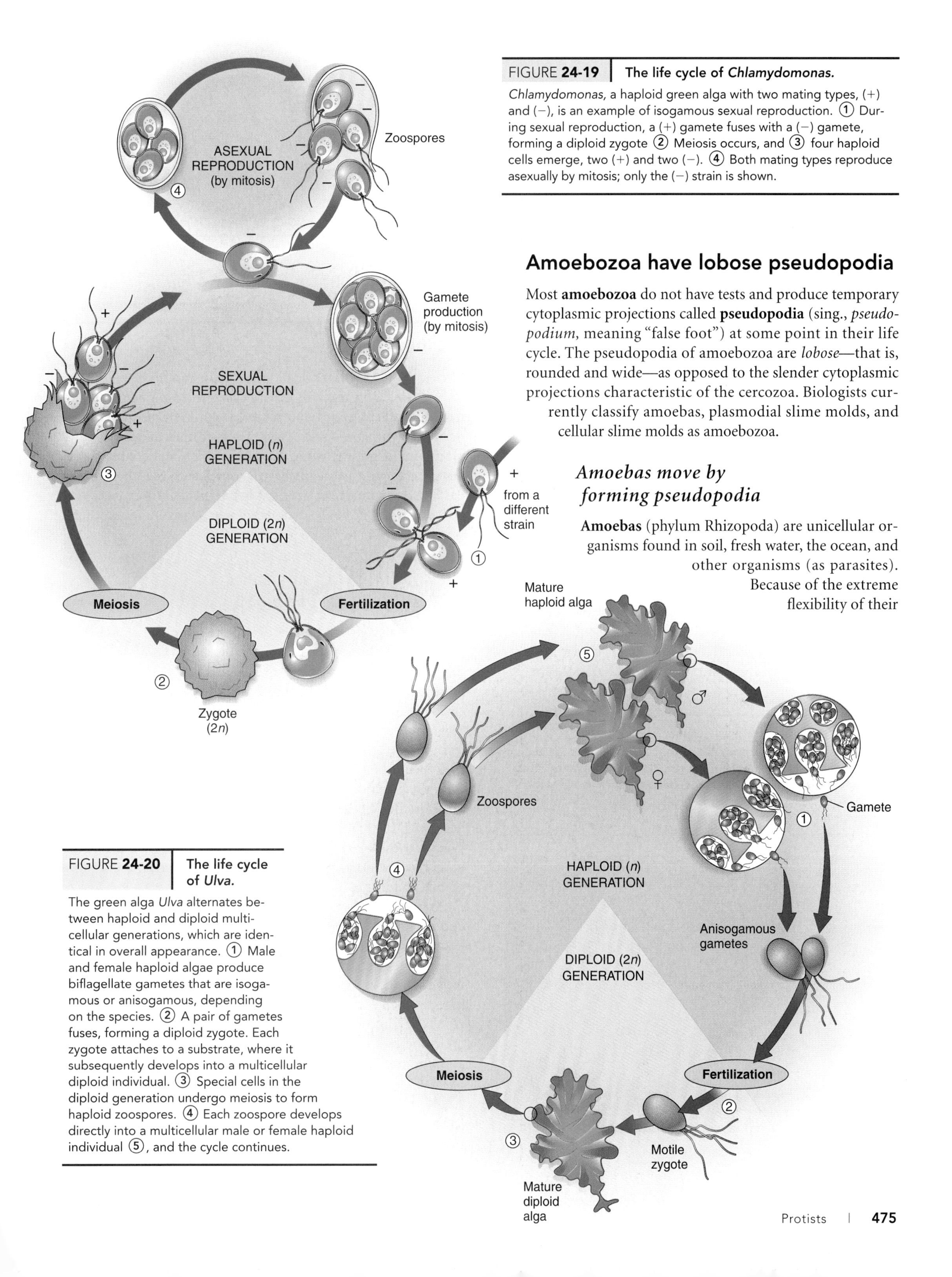

FIGURE 24-19 **The life cycle of *Chlamydomonas*.**

Chlamydomonas, a haploid green alga with two mating types, (+) and (−), is an example of isogamous sexual reproduction. ① During sexual reproduction, a (+) gamete fuses with a (−) gamete, forming a diploid zygote ② Meiosis occurs, and ③ four haploid cells emerge, two (+) and two (−). ④ Both mating types reproduce asexually by mitosis; only the (−) strain is shown.

Amoebozoa have lobose pseudopodia

Most **amoebozoa** do not have tests and produce temporary cytoplasmic projections called **pseudopodia** (sing., *pseudopodium*, meaning "false foot") at some point in their life cycle. The pseudopodia of amoebozoa are *lobose*—that is, rounded and wide—as opposed to the slender cytoplasmic projections characteristic of the cercozoa. Biologists currently classify amoebas, plasmodial slime molds, and cellular slime molds as amoebozoa.

Amoebas move by forming pseudopodia

Amoebas (phylum Rhizopoda) are unicellular organisms found in soil, fresh water, the ocean, and other organisms (as parasites). Because of the extreme flexibility of their

FIGURE 24-20 **The life cycle of *Ulva*.**

The green alga *Ulva* alternates between haploid and diploid multicellular generations, which are identical in overall appearance. ① Male and female haploid algae produce biflagellate gametes that are isogamous or anisogamous, depending on the species. ② A pair of gametes fuses, forming a diploid zygote. Each zygote attaches to a substrate, where it subsequently develops into a multicellular diploid individual. ③ Special cells in the diploid generation undergo meiosis to form haploid zoospores. ④ Each zoospore develops directly into a multicellular male or female haploid individual ⑤, and the cycle continues.

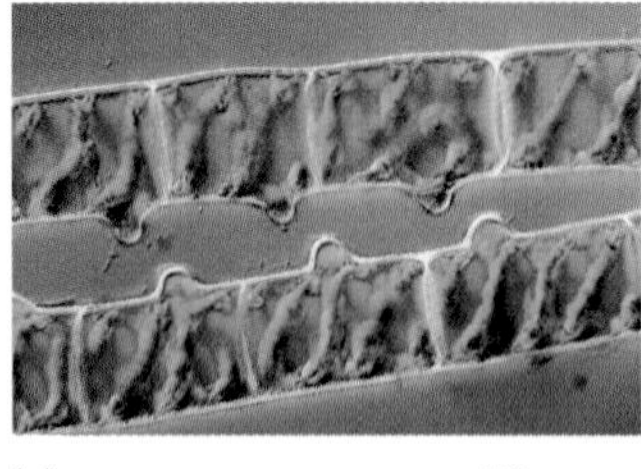
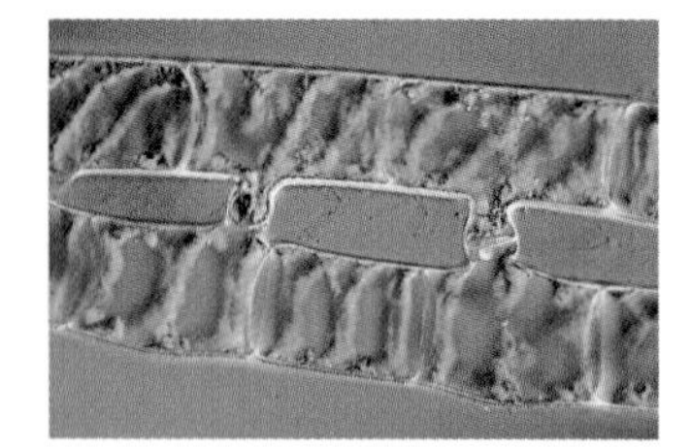
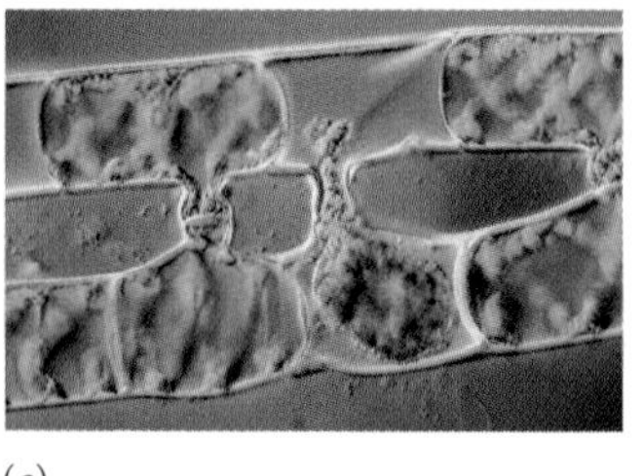
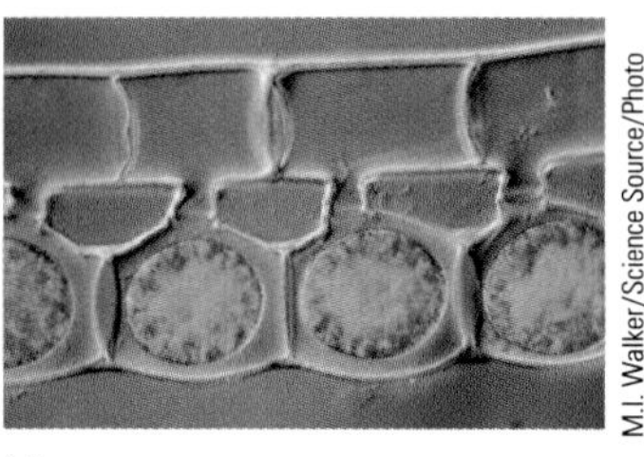

FIGURE **24-21** **Conjugation in *Spirogyra*.**

(a, b) Filaments of two different mating types of the green alga *Spirogyra* align, and conjugation tubes grow between cells of the two haploid filaments. **(c)** The contents of one cell passes into the other through the conjugation tube. **(d)** The two cells fuse, forming a diploid zygote. Following a period of dormancy, the rounded zygote undergoes meiosis, restoring the haploid condition.

outer plasma membrane, many members of this group have an asymmetrical body form and continually change shape as they move. (The word *amoeba* derives from a Greek word meaning "change.") An amoeba moves by pushing out lobose pseudopodia from the surface of the cell. More cytoplasm flows into the pseudopodia, enlarging them until all the cytoplasm has entered and the organism as a whole has moved. Pseudopodia also capture and engulf food by surrounding and forming a vacuole around it (Fig. 24-22). Food particles are digested when the food vacuole fuses with a lysosome containing digestive enzymes. Digested materials are absorbed from the food vacuole, which gradually shrinks as it empties. Amoebas reproduce asexually, splitting into two equal parts after mitotic division of the nucleus; sexual reproduction has not been observed.

Parasitic amoebas include *Entamoeba histolytica,* which causes *amoebic dysentery,* a serious human intestinal disease characterized by severe diarrhea, bloody stools, and ulcers in the intestinal wall. In especially severe cases, the organism spreads from the large intestine and causes abscesses in the liver, lungs, or brain. *Entamoeba histolytica* is transmitted as cysts in contaminated drinking water. A **cyst** is a thick-walled, resistant, resting stage in the life history of some protists. Other amoebas, like *Acanthamoeba,* are usually free-living but produce opportunistic infections such as eye infections in contact lens wearers.

FIGURE **24-22** **Amoeba.**

LM of a giant amoeba (*Chaos carolinense*). This unicellular protist, which moves and feeds by pseudopodia, is surrounding and ingesting a colonial green alga. *Chaos* amoebas are generally scavengers that feed on debris in freshwater habitats, but they ingest living organisms when the opportunity arises.

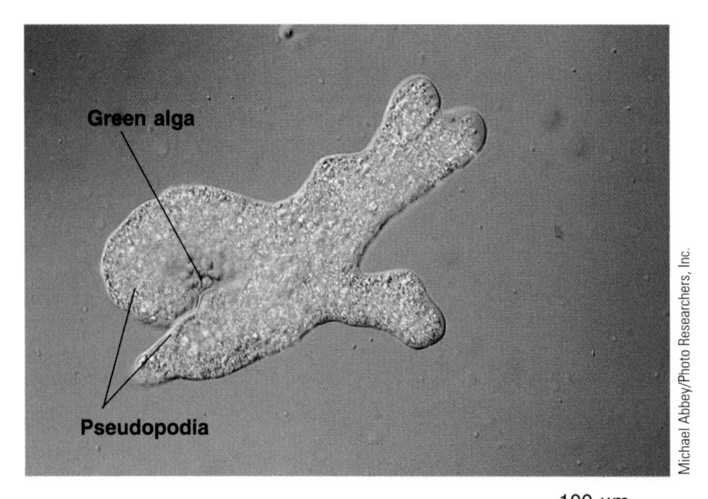

Plasmodial slime molds feed as multinucleate plasmodia

The feeding stage of a **plasmodial slime mold** (phylum Myxomycota) is a **plasmodium,** a multinucleate mass of cytoplasm that grows to 30 cm (1 ft) in diameter (Fig. 24-23a). The plasmodium, which is slimy in appearance, streams over damp, decaying logs and leaf litter, often forming a network of channels that covers a large surface area. As it creeps along, it ingests bacteria, yeasts, spores, and decaying organic matter.

When the food supply dwindles or there is insufficient moisture, the plasmodium crawls to an exposed surface and starts reproducing. Stalked structures of intricate complexity and beauty usually form from the drying plasmodium (Fig. 24-23b). Within these structures, called **sporangia,** meiosis produces haploid spores that are extremely resistant to adverse environmental conditions.

When conditions become favorable, the spores germinate, and a haploid reproductive cell emerges from each. This haploid cell is either a biflagellate *swarm cell* or an amoeboid *myxamoeba,* depending on available moisture; flagellate cells form in wet conditions. Swarm cells and myxamoebas act as gametes, which fuse to form a zygote with a diploid nucleus. The resultant diploid nucleus divides many times by mitosis, but the cytoplasm does not divide, so the result is a multinucleate plasmodium.

The plasmodial slime mold *Physarum polycephalum* is a model organism researchers have used to study many fundamental biological processes, such as growth and differentiation, cytoplasmic streaming, and the function of the cytoskeleton.

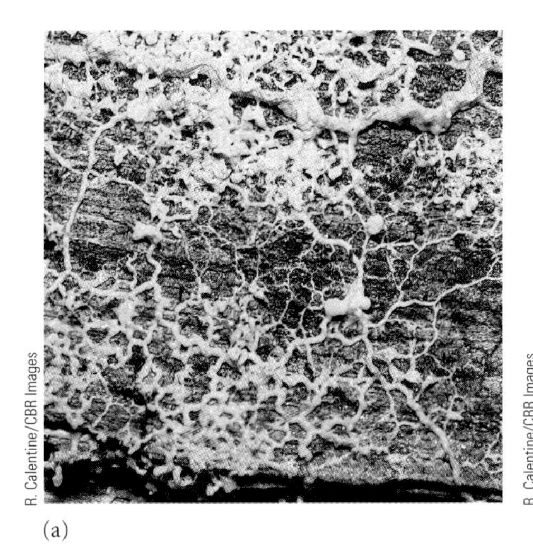

FIGURE **24-23**

The plasmodial slime mold *Physarum polycephalum.*

(a) The brightly pigmented plasmodium, shown on a dead log, feeds on bacteria and other microorganisms. **(b)** The reproductive structures are sporangia on stalks.

Cellular slime molds feed as individual amoeboid cells

The **cellular slime molds** (phylum Dictyostelida) have close affinities with amoebas and plasmodial slime molds. During its feeding stage, each cellular slime mold is an individual amoeboid cell that behaves as a separate, solitary organism. Each cell creeps over rotting logs and soil or swims in fresh water, ingesting bacteria and other particles of food as it goes. Each amoeboid cell has a haploid nucleus and reproduces by mitosis like a true amoeba.

When moisture or food becomes inadequate, certain cells send out a chemical signal, cyclic adenosine monophosphate (cAMP), that causes them to aggregate by the hundreds or thousands (Fig. 24-24a). During this stage the cells creep about for short distances as a single multicellular unit, called a **pseudoplasmodium,** or slug (Fig. 24-24b). Each cell of the slug retains its plasma membrane and individual identity. Eventually the slug settles and reorganizes, forming a stalked fruiting body containing spores (Fig. 24-24c). After being released, each spore opens and a single haploid amoeboid cell—the feeding stage—emerges. The spore-forming reproductive cycle is asexual, although sexual reproduction is observed occasionally. The life cycles of most cellular slime molds lack a flagellate stage.

The cellular slime mold *Dictyostelium discoideum* is a model organism for the study of cell differentiation, cell recognition, and cell motility and adhesion. Its biology has been studied intensively, particularly as it relates to **cell-signaling** molecules, such as cAMP, which are found in many organisms in addition to the cellular slime molds.

Opisthokonts include choanoflagellates, fungi, and animals

The **opisthokonts** are a monophyletic group that includes members of three kingdoms: choanoflagellates (kingdom Protista),

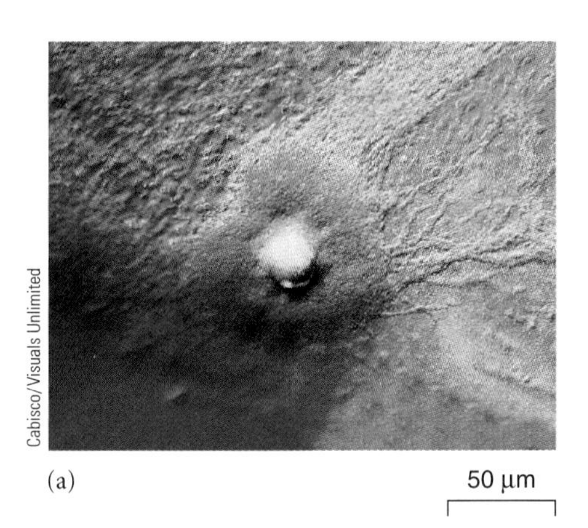

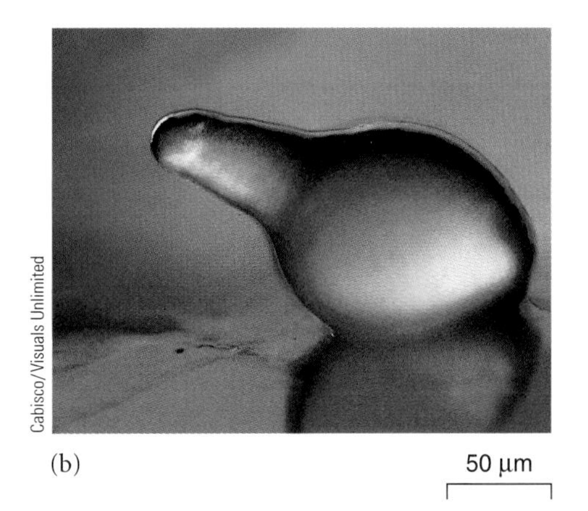

FIGURE **24-24** **The cellular slime mold Dictyostelium discoideum.**

(a) Streams of migrating amoeboid cells aggregate. **(b)** The aggregation organizes into a migrating pseudoplasmodium or slug that eventually forms a fruiting body on a stalk. **(c)** The fruiting body releases spores, each of which opens to liberate an amoeboid cell.

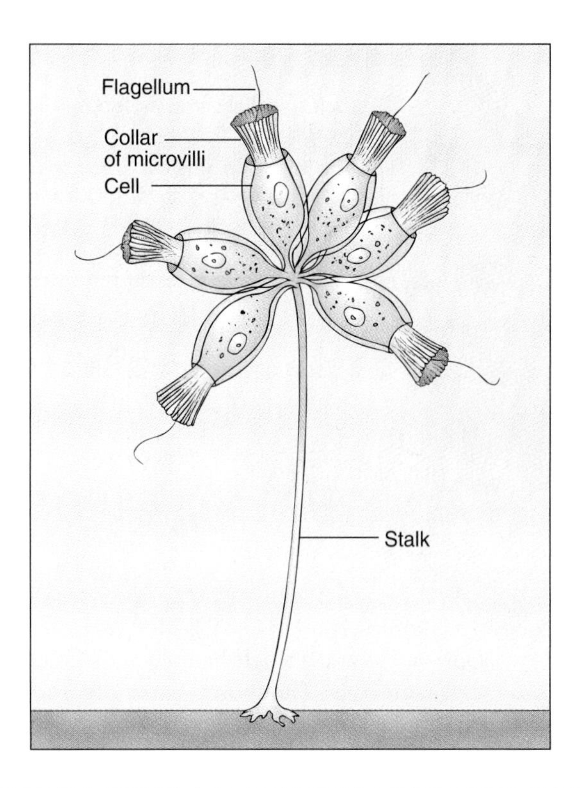

FIGURE **24-25** Choanoflagellate.

Choanoflagellates are free-living zooflagellates that obtain food by waving their flagella, causing water currents to carry bacteria and other small particles of food into the collar of microvilli. A colonial form is shown. Each cell is 5 to 10 μm long, not including the flagellum.

fungi (kingdom Fungi), and animals (kingdom Animalia). We consider fungi and animals in later chapters (see Chapters 25, 28, 29, and 30).

Choanoflagellates (phylum Choanoflagellata), are 600 species of collared zooflagellates. These small, inconspicuous marine and freshwater zooflagellates include both free-swimming and **sessile** species that are permanently attached by a thin stalk to bacteria-rich debris. Their single flagellum is surrounded at the base by a delicate collar of microvilli (Fig. 24-25). They are of special interest because of their striking resemblance to collar cells in sponges (see Fig. 28-6). Other animal phyla, such as flatworms and rotifers, also contain choanoflagellate-like cells. Given structural similarities and comparative ribosomal RNA and DNA sequence data, many biologists hypothesize that choanoflagellates are related to the ancestor of animals—that is, that living choanoflagellates and animals may share a common choanoflagellate-like ancestor.

Review

- What are some unique features of diplomonads?
- What characteristics do ciliates, dinoflagellates, and apicomplexans share?
- What are heterokonts?
- How do cercozoa and amoebozoa differ?
- Why do many biologists classify red algae and green algae with land plants?

BiologyNow™ Assess your understanding of **representative protists** by taking the pretest on your BiologyNow CD-ROM.

SUMMARY WITH KEY TERMS

1 Characterize the features common to the members of kingdom Protista.

- **Protists** are "simple" eukaryotic organisms, most of which are unicellular and live in aquatic environments.
- Protists range in size from microscopic unicellular organisms to **colonies** (loosely connected groups of cells) to **coenocytes** (multinucleate masses of cytoplasm) to **multicellular** organisms (composed of many cells).

2 Discuss in general terms the diversity inherent in the protist kingdom, including means of locomotion, modes of nutrition, interactions with other organisms, habitats, and modes of reproduction.

- Protists have various means of locomotion, including pseudopodia, flagella, and cilia. A few are nonmotile.
- Protists obtain their nutrients autotrophically or heterotrophically.
- Protists are free-living or symbiotic, with symbiotic relationships ranging from **mutualism** to **parasitism.**
- Most protists live in the ocean or in freshwater ponds, lakes, and streams. Parasitic protists live in the body fluids of their hosts.
- Many protists reproduce both sexually, often by **syngamy** (union of gametes), and asexually; others reproduce only asexually.

3 Discuss the endosymbiont theory, and briefly explain some of the evidence that supports it.

- According to the **endosymbiont theory,** mitochondria and chloroplasts arose from symbiotic relationships between larger cells and the smaller prokaryotes that were incorporated and lived within them.
- Mitochondria probably originated from aerobic eubacteria. Rickettsias are the closest living relatives of mitochondria.
- Chloroplasts of red algae, green algae, and plants probably arose in a single primary endosymbiotic event in which a cyanobacterium was incorporated into a cell. Multiple secondary endosymbioses led to chloroplasts in euglenoids, dinoflagellates, diatoms, golden algae, and brown algae, and to the nonfunctional chloroplasts in apicomplexans.

4 Describe the kinds of data biologists use to classify eukaryotes.

- Relationships among protists are determined largely by **ultrastructure,** which is the fine details of cell structure revealed by electron microscopy, and comparative molecular data. Most biologists consider the protist kingdom as a **paraphyletic group** that contains some, but not all, descendants of a common eukaryotic ancestor.

SUMMARY WITH KEY TERMS *(continued)*

5 Explain why zooflagellates are no longer classified in a single phylum, and distinguish among diplomonads, euglenoids, and choanoflagellates.

- **Zooflagellates** are mostly unicellular heterotrophs that move by means of whiplike **flagella.** Evidence indicates that zooflagellates are polyphyletic, and most biologists have separated them into several monophyletic groups.
- **Diplomonads** are **excavates** that have one or two nuclei, no mitochondria, no Golgi complex, and up to eight flagella.
- **Euglenoids** are **discicristates** that are unicellular and flagellate. Some euglenoids are photosynthetic. The discicristate *Trypanosoma* causes African sleeping sickness.
- **Choanoflagellates** are **opisthokonts** related to fungi and animals. A collar of microvilli surrounds their single flagellum at the base.

6 Briefly describe and compare the following alveolates: ciliates, dinoflagellates, and apicomplexans.

- **Ciliates** are **alveolates** that move by hairlike **cilia,** have **micronuclei** (for sexual reproduction) and **macronuclei** (for controlling cell metabolism and growth), and undergo a complex sexual reproduction called **conjugation.**
- **Dinoflagellates** are mostly unicellular, biflagellate, photosynthetic organisms of great ecological importance as major producers in marine ecosystems. Their **alveoli,** flattened vesicles under the plasma membrane, often contain cellulose plates impregnated with silicates. Some dinoflagellates produce toxic blooms known as **red tides.**
- **Apicomplexans** are parasites that produce **sporozoites** and are nonmotile. An apical complex of microtubules attaches the apicomplexan to its host cell. The apicomplexan *Plasmodium* causes malaria.

7 Briefly describe and compare the following heterokonts: water molds, diatoms, golden algae, and brown algae.

- **Water molds** are **heterokonts** that have a coenocytic **mycelium.** They reproduce asexually by forming biflagellate **zoospores** and sexually by forming **oospores.** The water mold *Phytophthora* causes serious plant diseases, such as late blight of potato and sudden oak death.
- **Diatoms,** which are major producers in aquatic ecosystems, are mostly unicellular, with shells containing silica. Some diatoms are part of floating **plankton,** and others live on rocks and sediments where they move by gliding.
- **Golden algae** are mostly unicellular, biflagellate freshwater and marine algae that are of ecological importance as a major component of the ocean's extremely minute **nanoplankton. Coccolithophorids** are golden algae covered by tiny, overlapping scales of calcium carbonate.
- **Brown algae** are multicellular seaweeds that are ecologically important in cooler ocean waters. The largest brown algae (kelps) possess leaflike **blades,** stemlike **stipes,** anchoring **holdfasts,** and gas-filled bladders for buoyancy.

8 Describe the foraminiferans and actinopods, and explain why many biologists classify them in the cercozoa.

- The **cercozoa** are amoeboid cells that often have hard outer shells, called **tests,** through which cytoplasmic projections extend.
- **Foraminiferans** secrete many-chambered tests with pores through which cytoplasmic projections extend to move and obtain food.
- **Actinopods** are mostly marine plankton that obtain food by means of **axopods,** slender cytoplasmic projections that extend through pores in their shells. **Radiolarians** are actinopods with glassy shells.

9 Support the hypothesis that red algae and green algae are a monophyletic group with land plants.

- Red algae, green algae, and land plants, collectively called **plants,** are monophyletic because all have chloroplasts bounded by outer and inner membranes.
- **Red algae,** which are mostly multicellular seaweeds, are ecologically important in warm tropical ocean waters. Red algae that incorporate calcium carbonate in their cell walls are important in reef building.
- **Green algae** exhibit a wide diversity in size, structural complexity, and reproduction. Botanists hypothesize ancestral green algae gave rise to land plants.

10 Briefly describe and compare the following Amoebozoa: amoebas, plasmodial slime molds, and cellular slime molds.

- **Amoebas** move and obtain food by **phagocytosis** using cytoplasmic extensions called **pseudopodia.** The parasitic amoeba *Entamoeba histolytica* causes amoebic dysentery.
- The feeding stage of **plasmodial slime molds** is a multinucleate **plasmodium.** Reproduction is by haploid spores produced within **sporangia.**
- **Cellular slime molds** feed as individual amoeboid cells. They reproduce by aggregating into a **pseudoplasmodium** (slug), then forming asexual spores.

POST-TEST

1. Which of the following is *not* true of the protists? (a) they are unicellular, colonial, coenocytic, or simple multicellular organisms (b) their cilia and flagella have a 9 + 2 arrangement of microtubules (c) they are prokaryotic, like the eubacteria and archaebacteria (d) some are free-living, and some are endosymbionts (e) most are aquatic and live in the ocean or in freshwater ponds
2. Amoebas move and obtain food by means of (a) pseudopodia (b) flagella (c) cilia (d) gametangia (e) trichocysts
3. Foraminiferans (a) are endosymbionts in many marine invertebrates (b) were responsible for the Irish potato famine in the 19th century (c) secrete many-chambered tests with pores through which cytoplasmic extensions project (d) have numerous trichocysts that may aid in trapping and holding prey (e) contain phycocyanin and phycoerythrin, pigments found in no other protist group

POST-TEST (continued)

4. Unicellular protists that are free-living or parasitic, move by means of flagella, and do not photosynthesize are called (a) euglenoids (b) dinoflagellates (c) myxamoebas (d) zooflagellates (e) apicomplexans
5. *Paramecium* and other ciliates often display a sexual phenomenon called (a) oogamy (b) conjugation (c) anisogamy (d) red tide (e) alternation of generations
6. Parasitic alveolates that form spores at some stage in their life belong to which group? (a) actinopods (b) ciliates (c) coccolithophorids (d) apicomplexans (e) dinoflagellates
7. Malaria (a) is transmitted by the bite of a female tsetse fly (b) is caused by a parasitic zooflagellate, *Giardia intestinalis* (c) is a serious form of amoebic dysentery caused by *Entamoeba histolytica* (d) is caused by an apicomplexan that spends part of its life cycle in the *Anopheles* mosquito and part in humans (e) is transmitted when people drink water tainted by a red tide
8. Alveolates characterized by two flagella, one wrapped around the center of the cell like a belt and the other projecting behind the cell, are (a) actinopods (b) ciliates (c) coccolithophorids (d) apicomplexans (e) dinoflagellates
9. Photosynthetic protists with shells composed of two halves that fit together like a petri dish are (a) golden algae (b) diatoms (c) euglenoids (d) brown algae (e) foraminiferans
10. Chlorophyll *a*, chlorophyll *b*, and carotenoids are found in (a) green algae, red algae, and land plants (b) green algae, euglenoids, and land plants (c) brown algae, green algae, and golden algae (d) brown algae, diatoms, and golden algae (e) green algae, euglenoids, and diatoms
11. Which protists have photosynthetic pigments similar to those of the cyanobacteria? (a) golden algae (b) diatoms (c) euglenoids (d) brown algae (e) red algae
12. The multicellular bodies of _______________ are differentiated into blades, stipes, holdfasts, and gas-filled floats. (a) golden algae (b) diatoms (c) euglenoids (d) kelps (e) green algae
13. The feeding stage of plasmodial slime molds is a multinucleate (a) plasmodium (b) pseudoplasmodium (c) pseudopodium (d) gametangium (e) mycelium
14. Cellular slime molds (a) include *Physarum* and *Phytophthora* (b) are more closely related to prokaryotes than are any other protists (c) are responsible for late blight of potatoes, which led to starvation in Ireland in the 1840s (d) have isogamous sexual reproduction (e) form a pseudoplasmodium when cells aggregate in response to cyclic AMP
15. Water molds reproduce asexually by forming _______________, and sexually by forming _______________. (a) oospores; holdfasts (b) zoospores; zooxanthellae (c) zoospores; oospores (d) holdfasts; isogametes (e) oospores; isogametes
16. Label the diagrams. Use Figures 24-7b and 24-6b to check your answers.

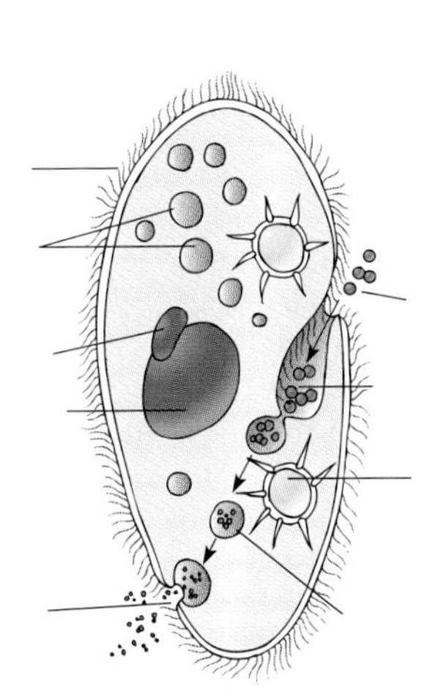

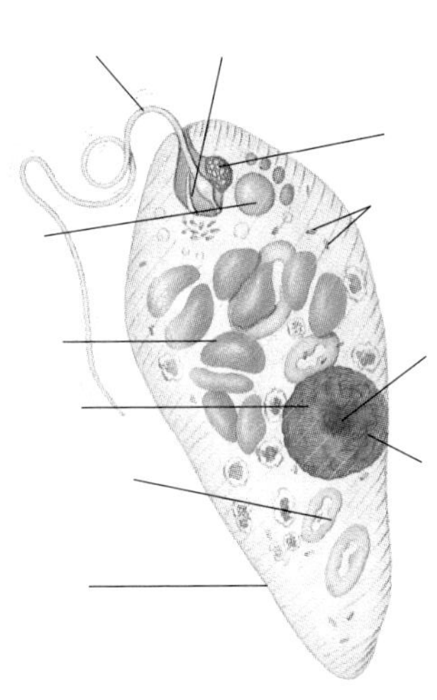

CRITICAL THINKING

1. Given what you have learned about dinoflagellates, where would you place them in the Figure 24-3 cladogram?
2. Where on Figure 24-3 would you place the common ancestor of brown algae and water molds? The common ancestor of animals and fungi? The common ancestor of animals?
3. In this chapter we discussed the hypothesis that choanoflagellates and animals share a common choanoflagellate-like ancestor. If this hypothesis is true, where would you place the choanoflagellate branch on Figure 24-3? Explain.

- Visit our Web site at **http://biology.brookscole.com/solomon7** for links to chapter-related resources on the World Wide Web. Additional online materials relating to this chapter can also be found on our Web site.

BIOLOGY NOW RESOURCES

Active Figures

24-3 Protist characteristics

24-10 Life-cycle of the malaria parasite

Preparing for an exam? Take a diagnostic test on your BiologyNow CD-ROM.

Post-Test Answers

1. c	2. a	3. c	4. d
5. b	6. d	7. d	8. e
9. b	10. b	11. e	12. d
13. a	14. e	15. c	

25

Kingdom Fungi

Jeff Lepore/Photo Researchers, Inc.

Fungal spores. The rounded earthstar (*Geastrum saccatum*) releases a puff of microscopic spores after the sac, which is about 1.3 cm (0.5 in) wide, is hit by a raindrop. This fungus is common in leaf litter under trees throughout North America.

CHAPTER OUTLINE

Mushrooms, morels, and truffles, delights of the gourmet, have a lot in common with baker's yeast, the black mold that forms on stale bread, and the mildew that collects on damp shower curtains. These life forms belong to the kingdom Fungi, a diverse group of more than 81,000 known species, most of which are terrestrial. Although they vary strikingly in size and shape, all **fungi** (sing., *fungus*) are eukaryotes; their cells contain membrane-enclosed nuclei, mitochondria, and other membranous organelles.

Fungi are heterotrophs: They cannot produce their own organic materials from a simple carbon source (carbon dioxide) but instead obtain preformed carbon molecules produced by other organisms. Moreover, fungi do not ingest food as animals do. Instead, they secrete digestive enzymes onto the food source and then absorb the predigested food (as small organic molecules) through their cell walls and plasma membranes.

Most fungi are decomposers; they get their nutrients from dead organic matter. As decomposers, fungi, along with prokaryotes (see Chapter 23), play an important ecological role. Some fungi are parasites that cause disease in animals or plants.

Fungi grow best in moist habitats, but they are found universally wherever organic material is available. They require moisture to grow, and they obtain water from a humid atmosphere as well as from the medium on which they live. When the environment becomes dry, fungi survive by going into a resting stage or by producing spores (see photograph) that resist desiccation (drying out).

Although the optimum pH for most species is about 5.6, various fungi can tolerate and grow in environments where the pH ranges from 2 to 9. Many fungi are less sensitive to high osmotic pressures than are bacteria. As a result, they can grow in concentrated salt solutions, or in sugar solutions such as jelly, that discourage or prevent bacterial growth. Fungi also thrive over a wide temperature range. Even refrigerated food may be invaded by fungi.

Fungi were classified in the plant kingdom until biologists realized they differ greatly from plants. In fact, molecular studies suggest fungi are related more closely to animals than to plants (see Fig. 24-3). Because fungi are distinct from plants, animals, and protists in many ways, they are assigned to a separate kingdom. ■

CHARACTERISTICS OF FUNGI

Learning Objectives

1. Describe the distinguishing characteristics of Kingdom Fungi.
2. Describe the body plan of a fungus.
3. Trace the fate of a fungal spore that lands in a favorable location, such as an overripe peach, and describe conditions that permit fungal growth.

Like the cells of bacteria, certain protists, and plants, fungal cells are enclosed by cell walls during at least some stage in their life cycle. Fungal cell walls, however, have a different chemical composition from cell walls of other organisms. In most fungi, the cell wall consists of complex carbohydrates, including **chitin,** a polymer that consists of subunits of a nitrogen-containing sugar (see Fig. 3-11). Chitin—which is coincidentally a component of the external skeletons of insects and other arthropods—is resistant to breakdown by most microorganisms.

Most fungi have a filamentous body plan

Mycologists (biologists who study fungi) identify two main types of fungi, based on body plan: molds and yeasts. Most fungi are **molds.** The vegetative (nonreproductive) body plan of molds consists of long, branched, threadlike filaments called **hyphae** (sing., *hypha*) (Fig. 25-1a, b). As hyphae elongate, the fungus infiltrates food sources and absorbs nutrients through its very large surface area. Hyphae form a tangled mass or tissue-like aggregation known as a **mycelium** (pl., *mycelia*). The cobweb-like mold sometimes seen on bread is the mycelium of a fungus. What is not seen is the extensive mycelium that grows into the bread.

Some hyphae are **coenocytic;** that is, they are not divided into individual compartments or cells and are instead an elongated, multinucleated giant cell (Fig. 25-1c). Other hyphae are divided by cross walls called **septa** (sing., *septum*) into individual cells containing one or more nuclei (Fig. 25-1d,e). The septa of many septate fungi are perforated by a pore that may be large enough to permit organelles to flow from cell to cell.

Hundreds of fungal species are unicellular, with a round or oval shape. These fungi, called **yeasts,** are widely distributed; they are found in the soil, on leaves, fruits, cured meats, and on and in our bodies. Yeasts are essential in making bread and fermenting alcoholic beverages (discussed later in this chapter). The yeast *Saccharomyces cerevisiae* is an important model eukaryotic organism. Its genome has been sequenced, and scientists are in the process of determining the functions of the 6000 identified proteins encoded by its genes.

Fungi reproduce by spores

Most fungi reproduce by means of microscopic **spores,** which are nonmotile (or, less commonly, motile) reproductive cells dispersed by wind, water, or animals. Spores are usually produced on specialized aerial hyphae or in fruiting structures. This

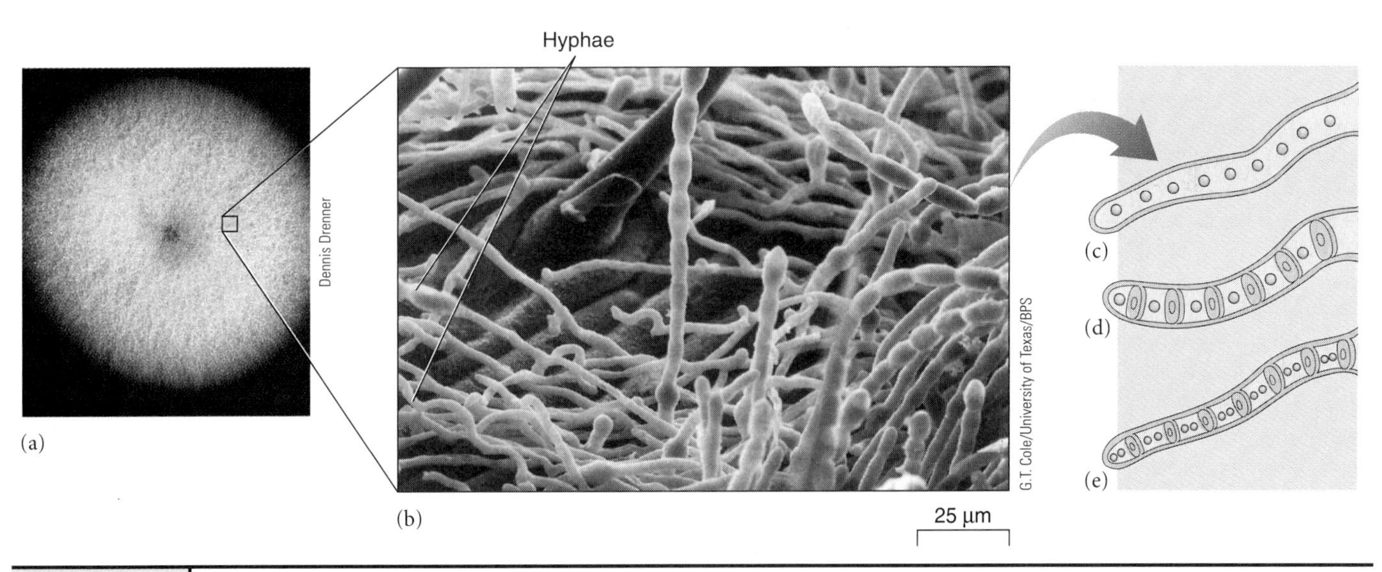

FIGURE **25-1** **Fungus body plan.**

(a) A fungal mycelium, a mass of threadlike hyphae (10 cm [4 in] wide), growing on agar in a culture dish. In nature, fungal mycelia are rarely so symmetric. **(b)** SEM of a mycelium of *Blumeria graminis,* growing on a leaf *(darker area underneath the mycelium).* **(c)** A coenocytic hypha. **(d)** A hypha divided into cells by septa; each cell is monokaryotic (has one nucleus). In some taxa the septa are perforated *(as shown),* permitting cytoplasm to stream from one cell to another. **(e)** A septate hypha in which each cell is dikaryotic (has two genetically distinct nuclei).

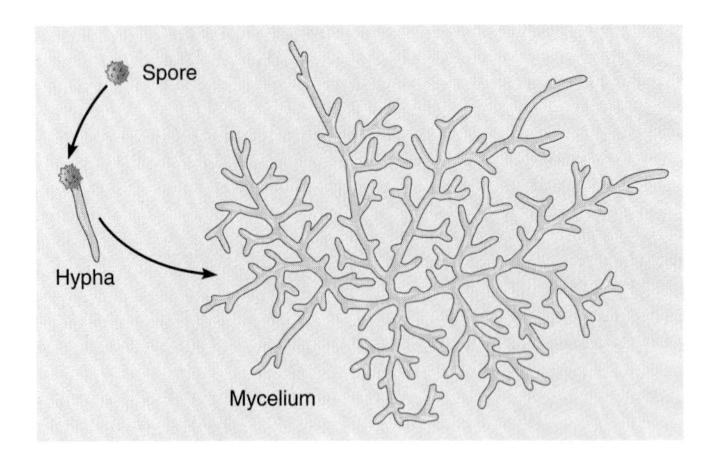

FIGURE **25-2** **Germination of a spore to form a mycelium.** Because the filamentous mycelium spreads throughout its substrate (such as overripe fruit, wood, or soil), it is difficult to estimate the actual size of an individual fungus.

arrangement permits the spores to be more easily dispersed. The aerial hyphae of some fungi form large, complex reproductive structures, called **fruiting bodies,** in which spores are produced. The familiar part of a mushroom is a large fruiting body. We do not normally see the bulk of the fungus, a nearly invisible mycelium buried out of sight in the rotting material or soil on which it grows.

Fungi may produce spores either sexually or asexually. Unlike most animal and plant cells, most fungal cells contain haploid nuclei. In sexual reproduction, the hyphae of two genetically compatible mating types come together, and their cytoplasm and nuclei fuse, forming a diploid cell known as a **zygote.**[1] In some groups, the zygote is the only diploid nucleus. Three types of fungal reproductive structures are gametangia, sporangia, and conidiophores. **Gametangia** are structures in which gametes form. **Sporangia** (sing., *sporangium*) are structures in which spores are produced. **Conidiophores** (from the Greek, meaning "dust-bearers") are specialized hypha that produce asexual spores called **conidia** (sing., *conidium*). The arrangement of conidia on conidiophores varies from species to species.

In two fungal groups—the ascomycetes and basidiomycetes—the hyphae fuse, but the two different nuclei do not fuse immediately; rather, they remain separate within the fungal cytoplasm. Hyphae that contain two genetically distinct, sexually compatible nuclei within each cell are **dikaryotic.** This condition is described as $n + n$ rather than $2n$, because there are two separate haploid nuclei. Hyphae that contain only one nucleus per cell are said to be **monokaryotic.**

When a fungal spore comes into contact with an appropriate food source, perhaps an overripe peach that has fallen to the ground, it germinates and begins to grow (Fig. 25-2). A threadlike hypha emerges from the tiny spore and grows at its tip, branching frequently. Soon a mycelium infiltrates the peach, degrading its complex organic compounds to small organic molecules that the fungus can absorb. Fungi are very efficient at converting nutrients into new cell material. If excessive amounts of nutrients are available, fungi store them, usually as lipid droplets or glycogen, in the mycelium.

[1] In certain fungi (such as *Allomyces*), zygotes are typically formed by the union of two haploid gametes rather than the union of two haploid nuclei in compatible hyphae.

Review

- What characteristics distinguish fungi from other organisms?
- How does the body of a yeast differ from that of a mold?
- What is the function of fruiting bodies?

BiologyNow™ Assess your understanding of **the characteristics of fungi** by taking the pretest on your BiologyNow CD-ROM.

FUNGAL DIVERSITY

Learning Objectives

4 Give arguments to support the hypothesis that chytridiomycetes may have been the earliest fungal group to evolve from the common ancestor of fungi.

5 List distinguishing characteristics, describe a typical life cycle, and give examples of each of the following fungal groups: chytridiomycetes, zygomycetes, ascomycetes, basidiomycetes, and imperfect fungi.

There are few examples of fossilized fungi, and fossil evidence has not helped biologists unravel evolutionary relationships among different fungal groups. Historically, fungi have been classified mainly on the characteristics of their sexual spores and fruiting bodies. Recently, molecular data, such as comparative DNA and RNA sequences, have clarified relationships among fungi.

Biologists assign these diverse organisms to four main phyla: Chytridiomycota, Zygomycota, Ascomycota, and Basidiomycota. (Fig. 25-3 and Table 25-1). As discussed later in this chapter, fungi that do not fit into any of these groups are assigned to a polyphyletic group, Deuteromycota, also referred to as *imperfect fungi.* As biologists learn more about these fungi, they are reassigned. Slime molds and water molds were originally classified as fungi but are now generally considered protists (see Chapter 24).

Chytridiomycetes are the most primitive fungi

Until recently, many biologists considered the **chytridiomycetes,** also called *chytrids,* (phylum Chytridiomycota) to be fungus-like protists similar in many respects to the water molds. However, molecular comparisons, particularly of DNA and RNA sequences, have provided compelling evidence that the approximately 750 species of chytridiomycetes are fungi. The molecular evidence also indicates that chytridiomycetes, which are small, relatively simple aquatic fungi, are the most primitive group of fungi living today. *Primitive* in this context implies that chytridiomycetes were probably the earliest fungal group to evolve from the ancient flagellate protist hypothesized as the common an-

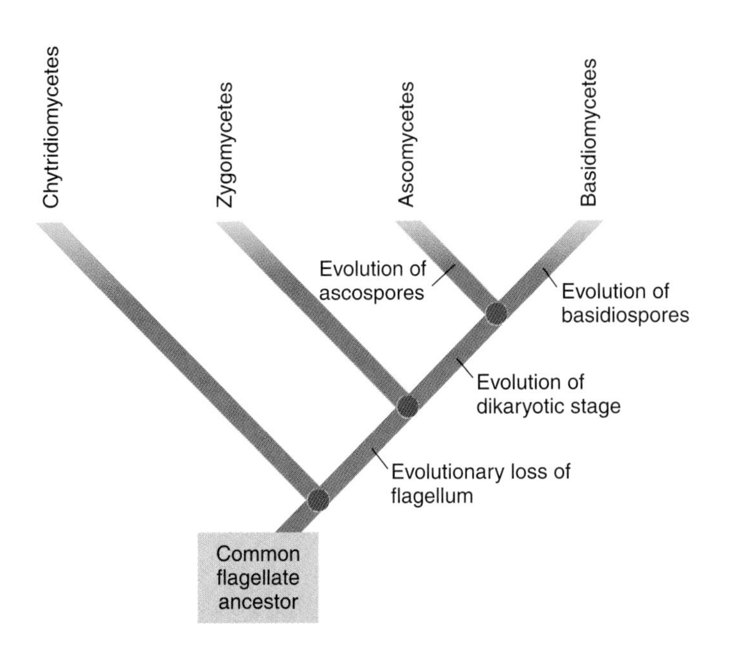

ACTIVE FIGURE **25-3** **Fungal evolution.**

This cladogram shows phylogenetic relationships among living fungi, based on comparisons of ribosomal and nuclear gene sequence data of many species. The chytridiomycetes were the lineage that branched off first during fungal evolution. The deuteromycetes are not shown, because the group is polyphyletic.

BiologyNow™ Learn about **the diversity of fungi** by clicking on this figure on your on your BiologyNow CD-ROM.

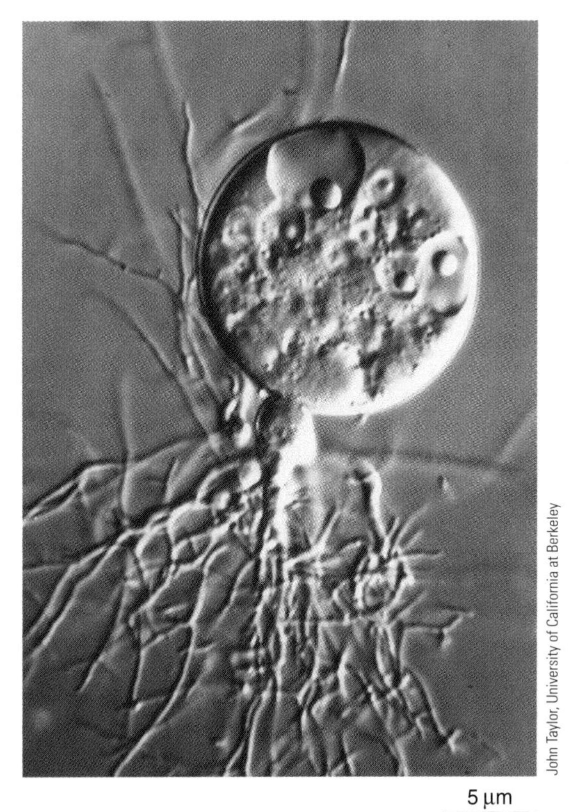

FIGURE **25-4** **Chytrid.**

Nomarski differential interference micrograph of a common chytrid (*Chytridium convervae*). Many chytrids have a microscopic body form consisting of a coenocytic globose (rounded) thallus and branched rhizoids that superficially resemble roots. The rhizoids may anchor the chytrid thallus and absorb predigested food.

cestor of all fungi. Chytridiomycetes, which produce flagellate cells at some stage in their life history, have retained this feature of their protist ancestor. All other fungal phyla are nonflagellate; they apparently lost the ability to produce motile cells at some point in their evolutionary history, perhaps during the transition from aquatic to terrestrial habitats.

The motile cells of chytrids possess a single, posterior flagellum. They reproduce both sexually and asexually. Chytrids are parasites or decomposers found principally in fresh water (Fig. 25-4); however, a few species occur on land (in moist environments) and in marine water. Although the disease was not identified until 1998, a parasitic chytrid may be responsible for declining amphibian populations dating back to the 1970s. Infected frogs have been identified in many parts of the world. (see *Focus On: Declining Amphibian Populations,* in Chapter 55).

Allomyces, a common chytridiomycete, has an unusual life cycle compared to most fungi. It has an **alternation of generations** (common in plants, but rare in fungi), spending part of

TABLE **25-1** **Phyla of Kingdom Fungi**

Phylum	Common Types	Asexual Reproduction	Sexual Reproduction
Chytridiomycota (chytridiomycetes or chytrids)	*Allomyces*	Zoospores	Flagellate gametes in some chytrids
Zygomycota (zygomycetes)	Black bread mold	Nonmotile spores form in a sporangium	Zygospores
Ascomycota (ascomycetes or sac fungi)	Yeasts, powdery mildews, molds, morels, truffles	Conidia pinch off from conidiophores	Ascospores
Basidiomycota (basidiomycetes or club fungi)	Mushrooms, bracket fungi, puffballs, rusts, smuts	Conidia (uncommon)	Basidiospores
Deuteromycota (deuteromycetes or imperfect fungi)	Molds	Conidia	Sexual stage not observed

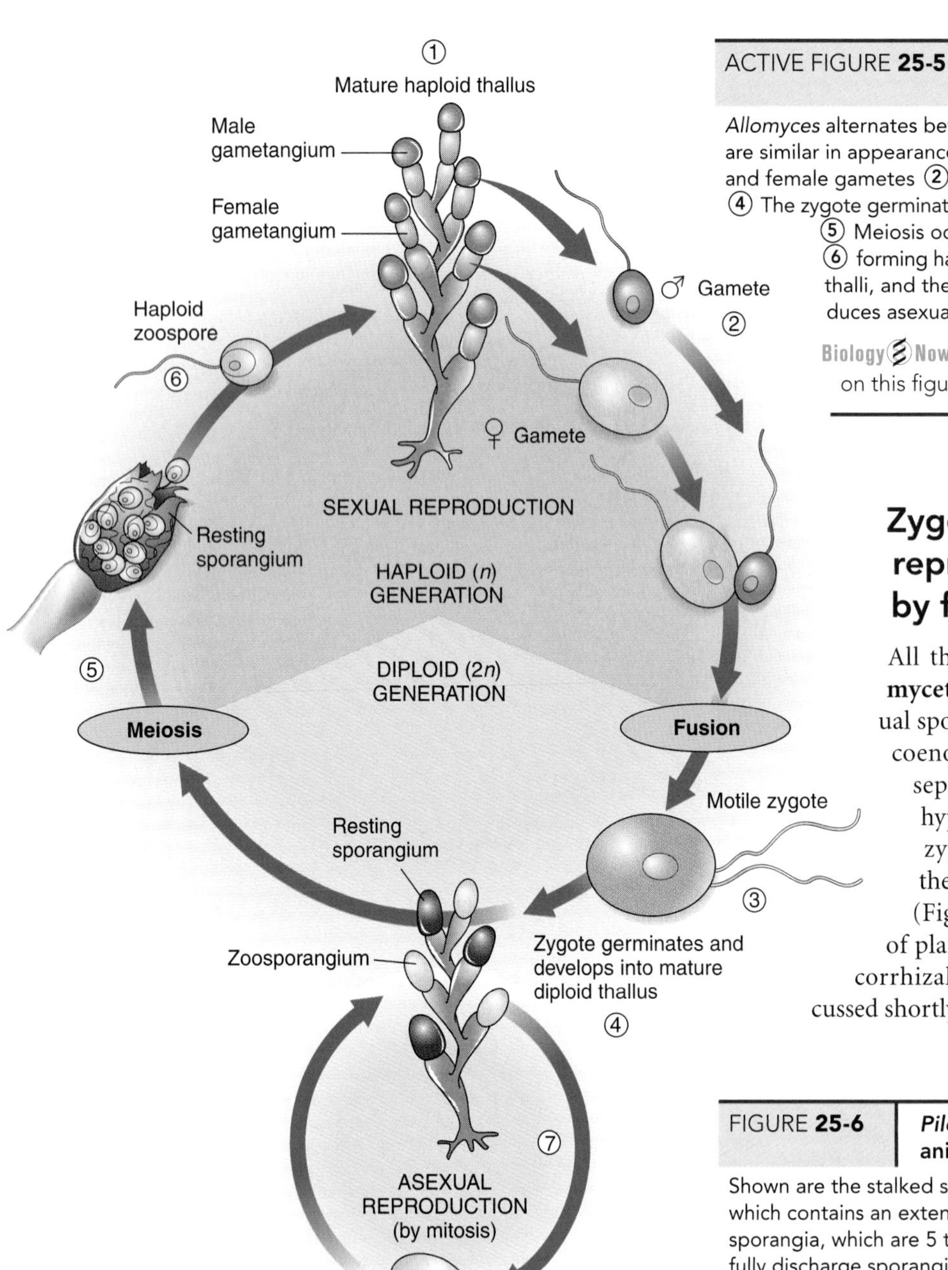

ACTIVE FIGURE **25-5** **Life cycle of *Allomyces arbuscula*, a chytridiomycete.**

Allomyces alternates between haploid and diploid stages, which are similar in appearance. The haploid thallus ① gives rise to male and female gametes ② that fuse, ③ forming a flagellate zygote. ④ The zygote germinates and develops into a diploid thallus. ⑤ Meiosis occurs in the diploid resting sporangia, ⑥ forming haploid zoospores that develop into haploid thalli, and the cycle continues. ⑦ *Allomyces* also reproduces asexually by forming diploid zoospores.

Biology Now™ Explore **fungus life-cycles** by clicking on this figure on your BiologyNow CD-ROM.

its life as a haploid *(n)* **thallus** and part as a diploid (2*n*) thallus (Fig. 25-5). The term *thallus* is used to describe the simple body plan of certain algae, fungi, or plants. The haploid and diploid thalli are similar in appearance and consist of a stout trunklike part with slender branches. At the tips of its branches, the haploid thallus bears male and female gametangia, structures in which gametes form by mitosis. When a flagellate male gamete fuses with a slightly larger, flagellate female gamete, the resulting zygote develops into a diploid thallus.

The diploid thallus bears two kinds of spore cases, zoosporangia and resting sporangia. Zoosporangia produce flagellate diploid **zoospores** that may settle down and develop into new diploid thalli. Meiosis occurs within resting sporangia to form haploid zoospores, each of which has the potential to settle down and develop into a haploid thallus.

Zygomycetes reproduce sexually by forming zygospores

All the approximately 800 species of **zygomycetes** (phylum Zygomycota) produce sexual spores, called **zygospores.** Their hyphae are coenocytic; that is, they lack regularly spaced septa. Septa form, however, to separate the hyphae from reproductive structures. Most zygomycetes are decomposers that live in the soil on decaying plant or animal matter (Fig. 25-6). Some zygomycetes are parasites of plants and animals, and still others are mycorrhizal fungi associated with plant roots (discussed shortly).

FIGURE **25-6** ***Pilobolus,* a zygomycete that grows in animal dung.**

Shown are the stalked sporangia protruding from a pile of dung, which contains an extensive mycelium of the fungus. The stalked sporangia, which are 5 to 10 mm tall, act like shotguns and forcefully discharge sporangia (the black tips) away from the dung, onto nearby grass. When animals such as cattle or horses eat the grass, the spores pass unharmed through the animal's digestive tract to be deposited in a fresh pile of dung.

Cabisco/Visuals Unlimited

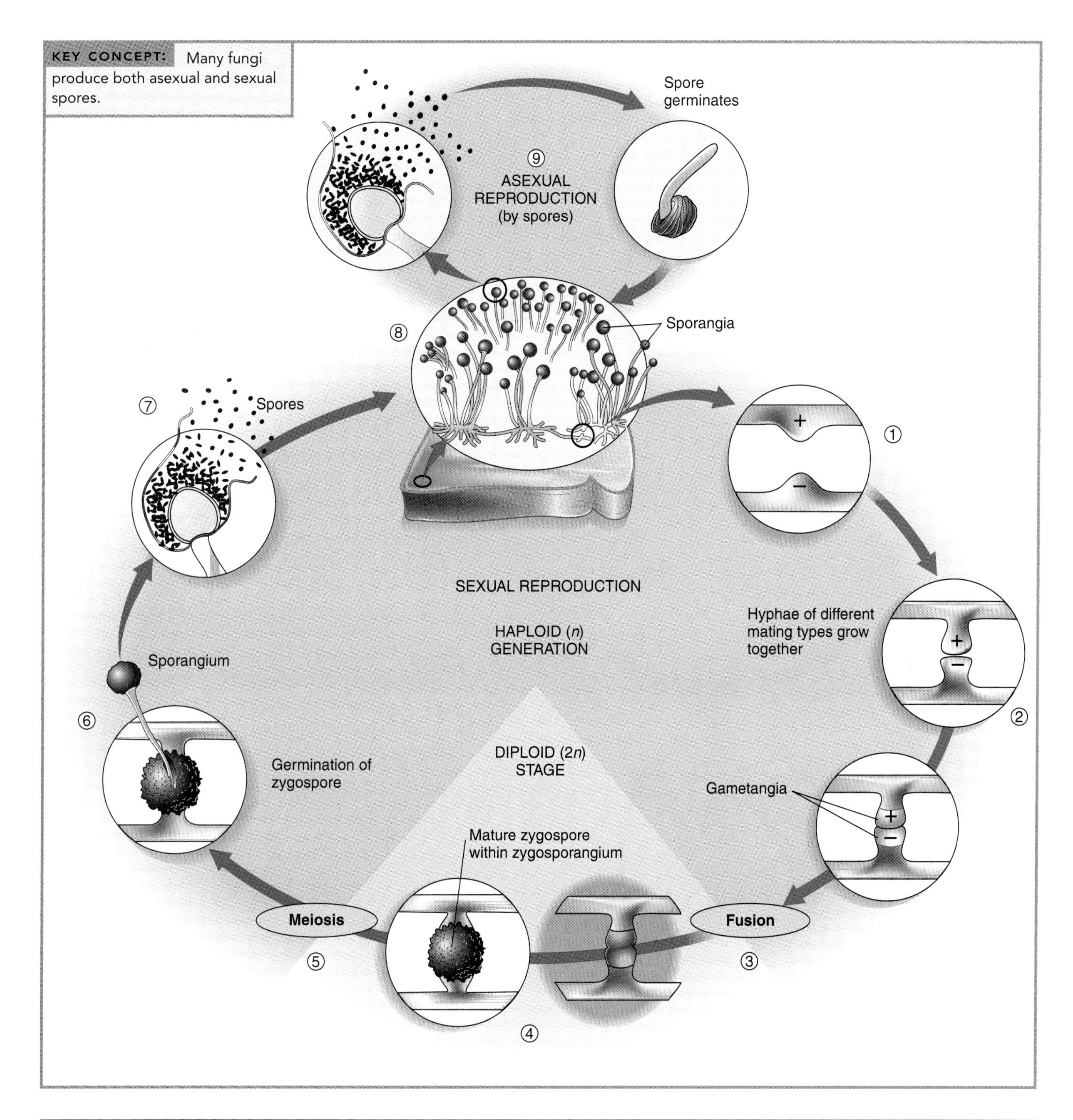

FIGURE **25-7** **Life cycle of the black bread mold (*Rhizopus stolonifer*), a zygomycete.**

① Sexual reproduction takes place only between different mating types, designated (+) and (–). After their hyphae meet, ② they form gametangia. ③ The gametangia unite, and the nuclei fuse. ④ A single zygospore develops within a thick-walled, black zygosporangium. ⑤ Meiosis occurs, the zygospore germinates, and ⑥ the emerging hypha develops a sporangium at its tip. ⑦ Spores are released and may germinate, ⑧ giving rise to new hyphae. ⑨ In asexual reproduction, certain hyphae form sporangia in which clusters of black, asexual, haploid spores develop. When released, they give rise to new hyphae.

One common zygomycete is the black bread mold, *Rhizopus stolonifer*, a decomposer that breaks down bread and other foods (Fig. 25-7). If preservatives are not added, bread left at room temperature often becomes covered with a black, fuzzy growth in a few days. Bread becomes moldy when a spore falls on it and then germinates and grows into a mycelium. Hyphae penetrate

the bread and absorb nutrients. Eventually, certain hyphae grow upward and develop spore sacs called *sporangia* at their tips. Clusters of black asexual spores develop within each sporangium and are released when the delicate sporangium ruptures. The spores give the black bread mold its characteristic color.

Sexual reproduction in the black bread mold occurs when the hyphae of two different mating types, designated as plus (+) and minus (−), grow into contact with one another. The bread mold is **heterothallic,** meaning that an individual fungal hypha is self-sterile and mates only with a hypha of a different mating type. That is, sexual reproduction occurs only between a member of a (+) strain and one of a (−) strain, not between members of two (+) strains or members of two (−) strains. Because there are no physical differences between the two mating types, it is not appropriate to refer to them as "male" and "female."

When hyphae of opposite mating types grow in close proximity, hormones are produced that cause the tips of the hyphae to come together and form gametangia. The gametangia then unite, and the (+) and (−) nuclei fuse to form the zygote, which is a diploid nucleus. A zygospore develops from the zygote, surrounded by a thick protective covering known as a **zygosporangium.** The zygospore may lie dormant for several months and can survive desiccation and extreme temperatures. Meiosis probably occurs at or just before germination of the zygospore. When the zygospore germinates, an aerial hypha develops with a sporangium at the tip. Mitosis within the sporangium produces haploid spores. These spores may be all (+) spores, all (−) spores, or a mixture of (+) and (−) spores. When released, the spores may germinate to form new hyphae. Only the zygote and zygospore of a black bread mold are diploid; all the hyphae and the asexual spores are haploid.

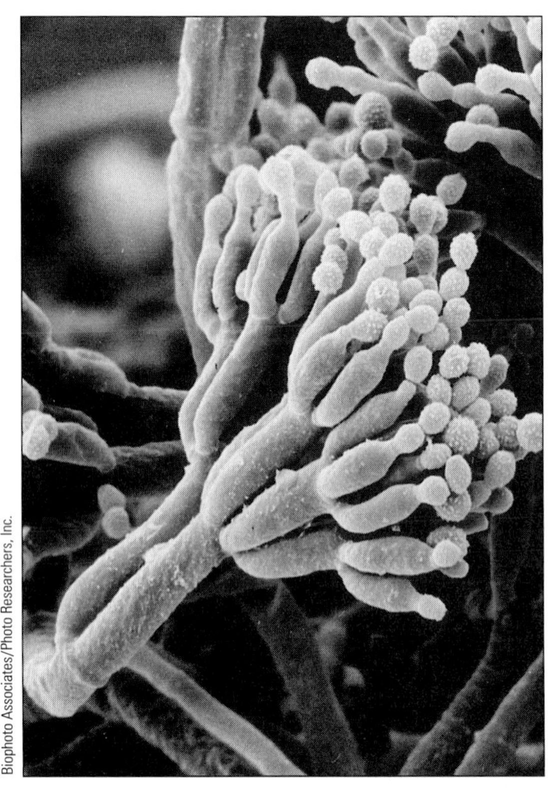

FIGURE **25-8** | **Conidia.**

Conidia are asexual reproductive cells produced by ascomycetes, some basidiomycetes, and most deuteromycetes. Biologists use the arrangement of conidia on conidiophores to identify species of these fungi. Shown is a SEM of *Penicillium* conidiophores, which resemble paintbrushes. Note the conidia pinching off the tips of the "brushes."

Ascomycetes reproduce sexually by forming ascospores

Ascomycetes (phylum Ascomycota) comprise a large group of fungi consisting of about 30,000 described species. Ascomycetes are sometimes referred to as *sac fungi* because their sexual spores are produced in microscopic sacs called **asci** (sing., *ascus*). Their hyphae usually have septa, but these cross walls have pores so that cytoplasm is continuous from one cell compartment to another.

The diverse ascomycetes include most yeasts; the powdery mildews; most of the blue-green, pink, and brown molds that cause food to spoil; decomposer cup fungi; and the edible morels and truffles. Some ascomycetes cause serious plant diseases such as Dutch elm disease, ergot disease on rye, powdery mildew on fruits and ornamental plants, and chestnut blight.

In most ascomycetes, asexual reproduction involves production of spores called *conidia* (sing., *conidium*). Conidia form at the tips of certain specialized hyphae known as *conidiophores* (Fig. 25-8). Production of these spores is a means of rapidly propagating new mycelia when environmental conditions are favorable. Conidia occur in various shapes, sizes, and colors in different species. The color of the conidia produces the characteristic brown, blue-green, pink, or other tints of many of these molds.

Some species of ascomycetes are heterothallic and have different mating strains. Others are **homothallic,** which means they are self-fertile and have the ability to mate with themselves. In both heterothallic and homothallic ascomycetes, sexual reproduction takes place after two gametangia come together and their cytoplasm mingles (Fig. 25-9). Within this fused structure, pairs of haploid nuclei, one from each parent hypha, associate but do not fuse. New hyphae, with dikaryotic cells, develop from the fused structure. The hyphae branch repeatedly until the hyphal tips reach the site where asci will be produced. As the many sac-shaped asci develop, each containing two dissimilar nuclei (one from each parent), they are surrounded by intertwining haploid (monokaryotic) hyphae. These hyphae develop into a fruiting body known as an **ascocarp** (Fig. 25-10a).

Within a cell that develops into an ascus, the two nuclei fuse and form a diploid nucleus, the zygote. The zygote then undergoes meiosis to form four haploid nuclei. One mitotic division of each of the four nuclei usually follows, resulting in eight haploid nuclei. Each haploid nucleus becomes incorporated into a heavy-walled **ascospore;** thus there are usually eight haploid ascospores within the ascus (Fig. 25-10b). The ascospores are usually released through a pore, slit, or hinged lid at the tip of the ascus. Air currents carry individual ascospores, often for long

distances. If one lands in a suitable location, it germinates and forms a new mycelium.

Phylum Ascomycota includes more than 300 species of unicellular yeasts (see Fig. 7-13a). Asexual reproduction of yeasts is mainly by **budding;** in this process a small protuberance (bud) grows and eventually separates from the parent cell. Each bud can grow into a new yeast cell. Some yeasts also reproduce asexually by **binary fission,** in which they undergo mitosis and then divide in half. Yeasts reproduce sexually by forming ascospores.

During sexual reproduction, two haploid yeasts fuse, forming a diploid zygote. The zygote undergoes meiosis, and the resulting haploid nuclei are incorporated into ascospores. These spores remain enclosed for a time within the original cell wall, which corresponds to an ascus.

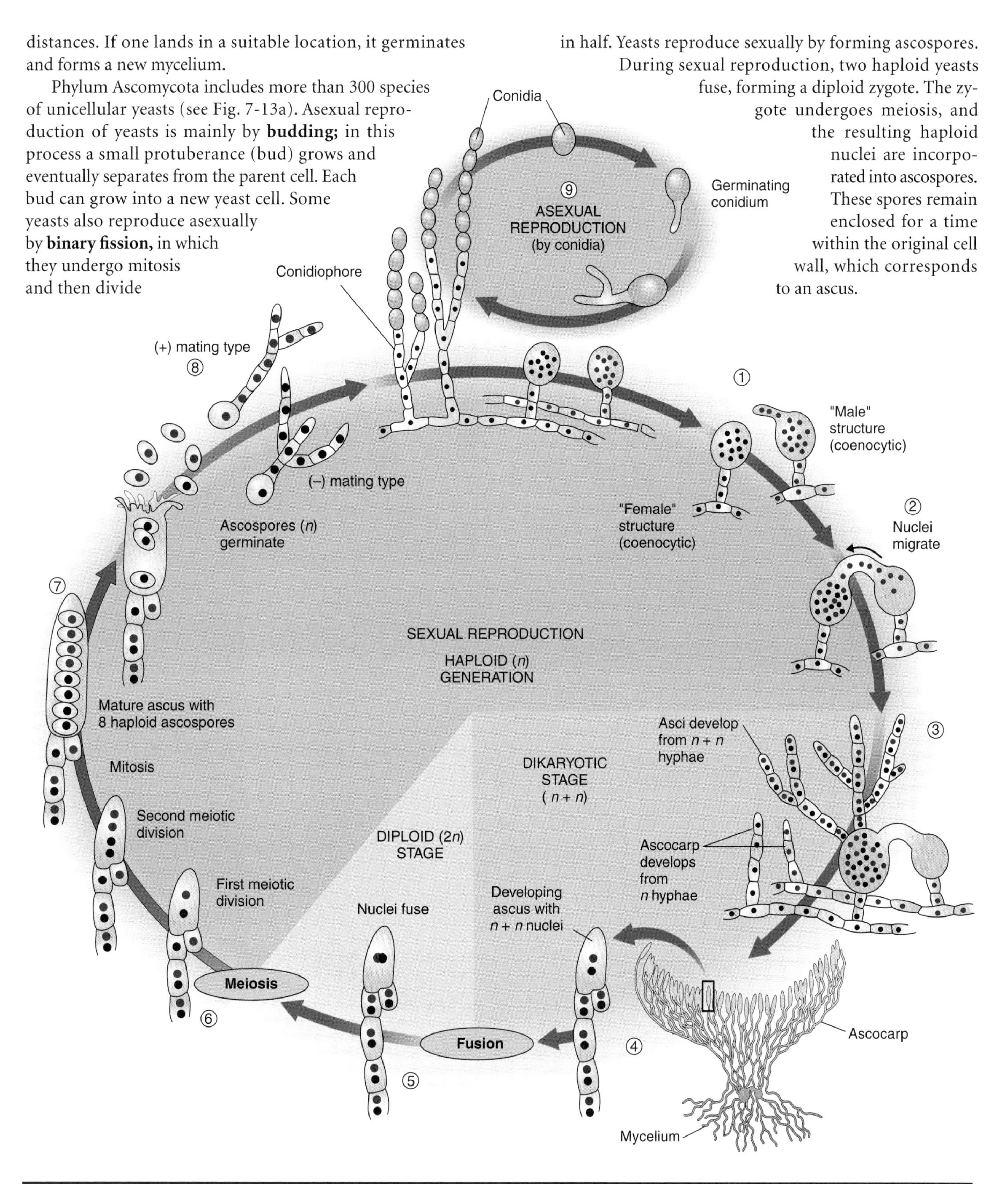

FIGURE **25-9** | **Life cycle of a typical ascomycete.**

① Sexual reproduction involves haploid mycelia of opposite mating types. ② Gametangia of the two different mating types fuse, ③ forming dikaryotic ($n + n$) hyphae from which asci develop. ④ The asci are incorporated into an ascocarp (fruiting body). ⑤ In each ascus, the $n + n$ nuclei fuse forming a diploid nucleus, the zygote. ⑥ Meiosis occurs forming four haploid nuclei. Mitosis then takes place, producing eight haploid nuclei. ⑦ Each becomes incorporated into an ascospore. When ascospores are released, ⑧ they germinate and form new mycelia. ⑨ Asexual reproduction involves the formation of haploid conidia.

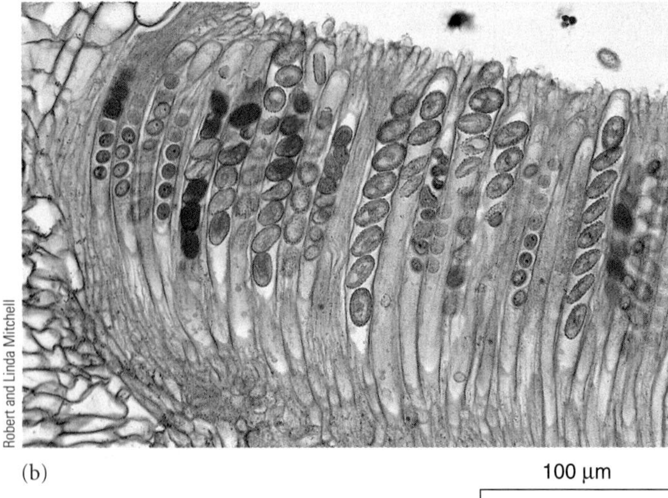

FIGURE **25-10** | **Sexual reproduction in the ascomycetes.**
(a) The ascocarp of the common brown cup (*Peziza badio-confusa*) is shaped like a saucer or bowl and is 3 to 10 cm (1 to 4 in) wide. It is found on damp soil in woods throughout North America. Photographed in Muskegon, Michigan. **(b)** Asci, each containing eight ascospores, line the inner portion of the saucer.

Basidiomycetes reproduce sexually by forming basidiospores

The 25,000 or more species of **basidiomycetes** (phylum Basidiomycota) include the most familiar of the fungi: the mushrooms, bracket fungi, and puffballs (Fig. 25-11). Some destructive plant parasites of important crops, such as wheat rust and corn smut, are also basidiomycetes. Sometimes called *club fungi,* Basidiomycetes derive their name from the fact that they develop microscopic club-shaped **basidia** (sing., *basidium*). Basidia are comparable in function to the asci of ascomycetes. Each basidium is an enlarged hyphal cell, on the tip of which develop four **basidiospores** (Fig. 25-12). Note that basidiospores develop on the *outside* of a basidium, whereas ascospores develop *within* an ascus.

Each individual fungus produces millions of basidiospores, and each basidiospore has the potential to give rise to a new **primary mycelium** (Fig. 25-13). Hyphae of a primary mycelium consist of monokaryotic cells. The mycelium of a basidiomycete such as the commonly cultivated mushroom *Agaricus brunnescens*[2] consists of a mass of white, branching, threadlike hyphae that live mostly below ground. Septa divide the hyphae into cells, but, as in ascomycetes, the perforated septa allow cytoplasmic streaming between cells.

When in the course of its growth a hypha of a primary mycelium encounters another monokaryotic hypha of a different mating type, the two hyphae fuse. As in the ascomycetes, the two haploid nuclei remain separate within each cell. In this way a **secondary mycelium** with dikaryotic hyphae is produced, in which each cell contains two haploid nuclei.

The $n + n$ hyphae of the secondary mycelium grow extensively. When environmental conditions are favorable, the hyphae form compact masses, called *buttons,* along the mycelium. Each button grows into a fruiting body that we call a *mushroom.* A mushroom, which consists of a stalk and a cap, is more formally referred to as a **basidiocarp.** Each basidiocarp consists of intertwined hyphae that are matted together. The lower surface of the cap usually consists of many thin, perpendicular plates called **gills** that radiate from the stalk to the edge of the cap.

Within the young basidia on the gills of the mushroom, haploid nuclei fuse in the dikaryotic cells, forming diploid zygotes. These are the only diploid cells that form during a basidiomycete's life history. Meiosis then takes place, forming four haploid nuclei that move to the outer edge of the basidium. Finger-like extensions of the basidium develop, into which the nuclei and some cytoplasm move; each of these extensions becomes a basidiospore. A septum forms that separates the basidiospore from the rest of the basidium by a delicate stalk that breaks when the basidiospore is forcibly discharged. Asexual reproduction is less common in basidiomycetes than in other groups.

[2] Also called *Agaricus bisporus.*

Imperfect fungi have no known sexual stage

Mycologists have grouped about 25,000 species of fungi as **deuteromycetes** and assigned them to a *form phylum,* Deuteromycota. Members of a form phylum are similar to one another in certain respects but are probably **polyphyletic,** which means they do not share a common ancestor. Fungi classified as deuteromycetes do not have a common ancestor and are grouped together simply as a matter of convenience. Deuteromycetes are also known as **imperfect fungi,** because in most of them no sexual stage has been observed at any point during their life cycle. Some of these fungi have lost the ability to reproduce sexually, whereas others may reproduce sexually only rarely. Should further study reveal a sexual stage, mycologists will reassign the

(a)

FIGURE 25-11 Basidiomycete fruiting bodies.

(a) Basidia line the gills of the Jack-o'lantern mushroom (*Omphalotus olearius*), a poisonous species whose gills produce a greenish glow in the dark. Each cap is about 15 cm (6 in) wide. Photographed at the base of an oak tree in Maryland, the Jack-o'lantern occurs throughout eastern North America and California. **(b)** The elegant stinkhorn (*Phallus ravenelii*) has a foul smell that attracts flies, which help disperse the slimy mass of basidiospores. Fruiting bodies of elegant stinkhorns grow to 18 cm (7 in) tall. Photographed in Pennsylvania. **(c)** Turkey-tail (*Trametes versicolor*) is a common bracket fungus. Bracket fungi grow on both dead and living trees, producing shelf-like fruiting bodies. Basidiospores are produced in pores located underneath each shelf. Photographed in Pennsylvania.

(b)

(c)

sexual state of these species to a different phylum. (By convention, the asexual stage is still identified as a deuteromycete.)

Increasingly, comparing DNA and/or RNA sequences among various species helps biologists find relationships between imperfect fungi and their sexually reproducing relatives. Most imperfect fungi reproduce only by means of conidia and are closely related to the ascomycetes. A few are more closely related to the basidiomycetes.

Review

- Why are the chytridiomycetes considered the earliest fungal group to evolve from the common ancestor of fungi?
- What are the distinguishing characteristics of each of the following fungal groups: (a) zygomycetes (b) ascomycetes (c) basidiomycetes?
- How does the life cycle of a typical basidiomycete differ from that of a typical ascomycete? (Draw diagrams to support your answer.)
- Distinguish among (a) ascocarp, ascus, and ascospore and among (b) basidiocarp, basidium, and basidiospore.
- Why are the deuteromyocetes referred to as imperfect fungi?

BiologyNow™ Assess your understanding of **fungal diversity** by taking the pretest on your BiologyNow CD-ROM.

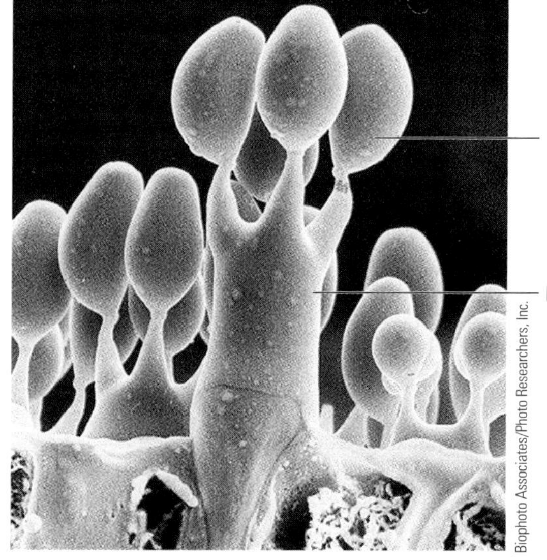

FIGURE 25-12 SEM of a basidium.

Each basidium produces four basidiospores.

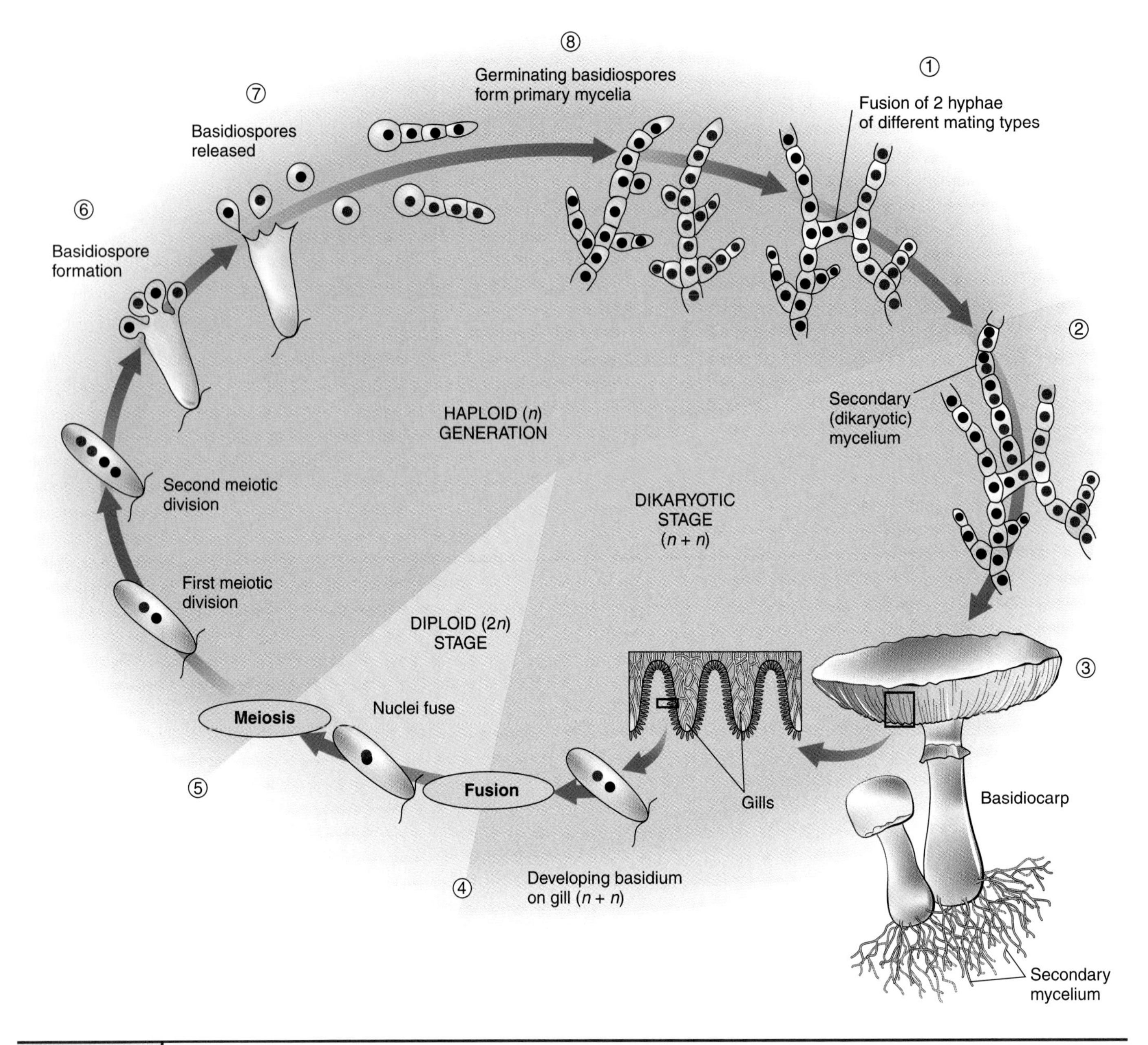

FIGURE 25-13 | Life cycle of a typical basidiomycete.

Sexual reproduction involves ① the fusion of two haploid hyphae of different mating types to form ② a secondary mycelium composed of dikaryotic ($n + n$) hyphae. ③ Fruiting bodies (basidiocarps) periodically develop from the secondary mycelium. ④ Basidia form along the gills of the basidiocarps. In each basidium the $n + n$ nuclei fuse. ⑤ Meiosis occurs, producing ⑥ four basidiospores. ⑦ When basidiospores are released, they germinate and form ⑧ primary mycelia.

LICHENS

Learning Objective

6 Characterize the unique nature of a lichen.

Although a **lichen** looks like a single organism, it is actually a dual organism. A lichen is a symbiotic association between two organisms: a *phototroph* (a photosynthetic organism) and a fungus (Fig. 25-14a). (A symbiotic association is an intimate relationship between organisms of different species.) Almost one fifth of all known fungal species form these symbiotic relationships; about 14,000 kinds of lichens have been described.

The phototrophic component of a lichen is either a green alga, a cyanobacterium, or both. The fungus is most often an ascomycete, although in some tropical lichens the fungal partner is a basidiomycete. Most phototrophic organisms found in lichens also occur as free-living species in nature, but the fungal components are generally found only as a part of the lichen.

In the laboratory, the fungal and phototrophic components of some lichens can be isolated and grown separately in appropriate culture media. The phototroph grows more rapidly when separated, whereas the fungus grows more slowly and requires many complex carbohydrates. Neither organism resembles a lichen in appearance when grown separately. The phototroph

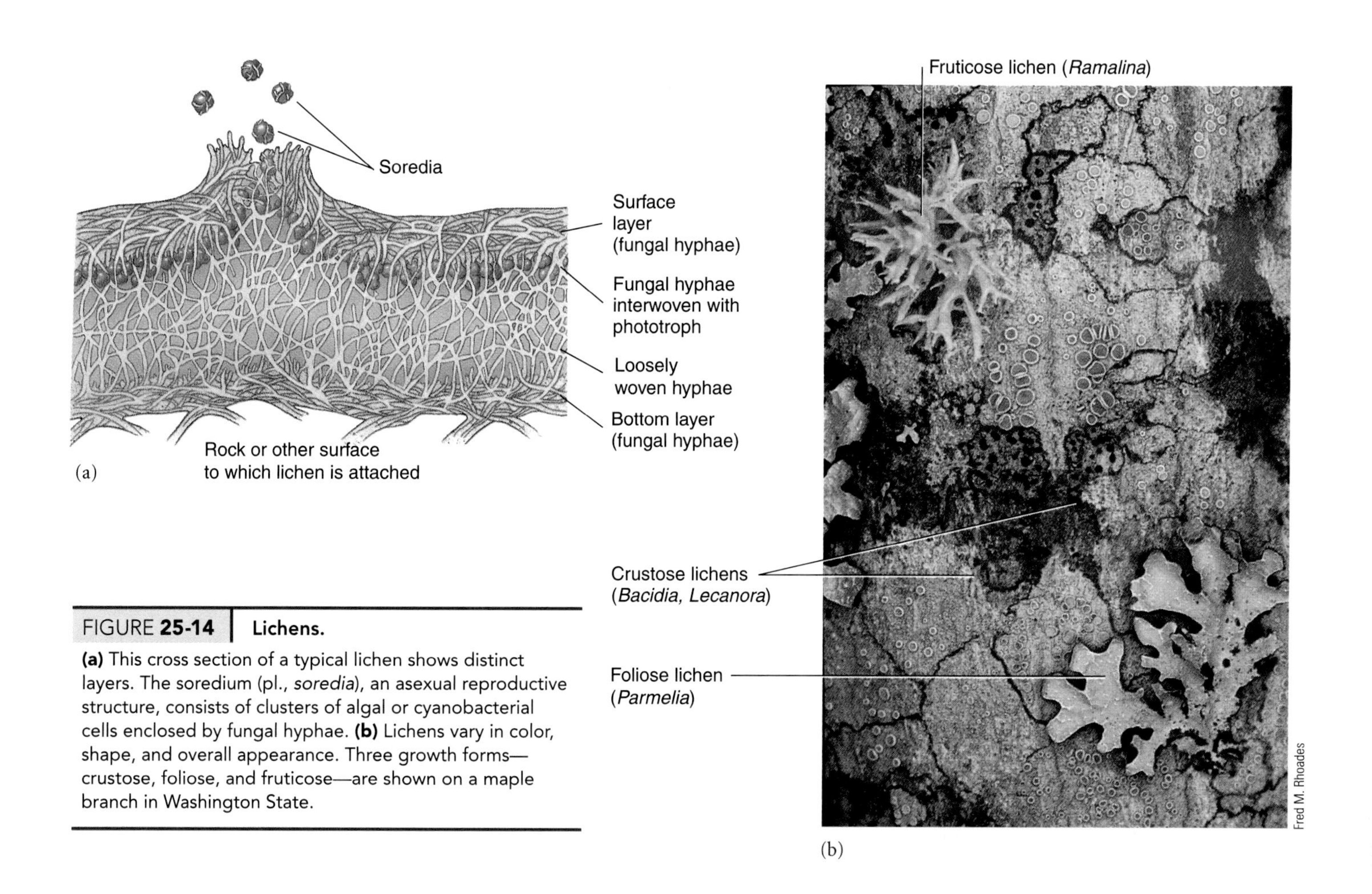

FIGURE **25-14** | **Lichens.**

(a) This cross section of a typical lichen shows distinct layers. The soredium (pl., *soredia*), an asexual reproductive structure, consists of clusters of algal or cyanobacterial cells enclosed by fungal hyphae. **(b)** Lichens vary in color, shape, and overall appearance. Three growth forms—crustose, foliose, and fruticose—are shown on a maple branch in Washington State.

and fungus can be reassembled as a lichen thallus, but only if they are placed in a culture medium under conditions that cannot support either of them independently.

What is the nature of this partnership? The lichen was originally considered a definitive example of **mutualism,** a symbiotic relationship that is beneficial to both species. The phototroph carries on photosynthesis, producing food for both members of the lichen, but it is unclear how the phototroph benefits from the relationship. It has been suggested that the phototroph obtains water and nutrient minerals from the fungus as well as protection against desiccation. More recently some biologists have suggested that the lichen partnership is not really a case of mutualism but one of controlled parasitism of the phototroph by the fungus.

Lichens typically exhibit one of three different growth forms (Fig. 25-14b): *Crustose lichens* are flat and grow tightly against their substrate (the surface they are growing on); *foliose lichens* are also flat, but they have leaflike lobes and are not so tightly pressed to the substrate; and *fruticose lichens* grow erect and have many branches.

Able to tolerate extremes of temperature and moisture, lichens grow in almost all terrestrial environments except polluted cities. They exist farther north than any plants of the arctic region and are equally at home in the steaming equatorial rain forest. They grow on tree bark, leaves, and exposed rock surfaces, from solidified lava to tombstones. In fact, lichens are often the first organisms to inhabit rocky areas. Lichen growth in these areas plays an important role in forming soil from rock because they gradually etch tiny cracks in rock, which eventually breaks into particles of soil. This process sets the stage for further disintegration of the rock by wind and rain.

Reindeer mosses of the arctic region, which serve as the main source of food for migrating herds of caribou, are lichens, not mosses. Some lichens produce colored pigments. One of them, orchil, is used to dye woolens, and another, litmus, is widely used in chemistry laboratories as an acid–base (pH) indicator.

Lichens vary greatly in size. Some are almost invisible, whereas others, like the reindeer mosses, may cover many square kilometers of land with an ankle-deep growth. Growth proceeds slowly; the radius of a lichen may increase by less than 1 mm each year. Some mature lichens are thought to be thousands of years old.

Lichens absorb minerals from the air, rainwater, and the surface on which they grow. They cannot excrete the elements they absorb, and perhaps for this reason they are extremely sensitive to toxic compounds. This sensitivity was first reported in 1866 by a Finnish biologist who observed that lichens growing on tree trunks in Paris were poorly developed or sterile. He deduced that lichens could be used to measure air purity. Today, reduction in lichen growth is used as a sensitive indicator of air pollution, particularly from sulfur dioxide. In 1997, Italian researchers established a relationship between lung cancer and air pollution by comparing the locations of low lichen biodiversity (and therefore of air pollution) with the locations of

lung cancer deaths in young males. The return of lichens to an area indicates an improvement in air quality.

Lichens reproduce mainly by asexual means, usually by fragmentation, a process in which special dispersal units of the lichen, called **soredia,** break off and, if they land on a suitable surface, establish themselves as new lichens. Soredia contain cells of both partners. In some lichens, the fungus produces ascospores, which may be dispersed by wind and find an appropriate algal partner only by chance.

Review

- What are the two components of a lichen?
- What are soredia?

BiologyNow™ Assess your understanding of **lichens** by taking the pretest on your BiologyNow CD-ROM.

ECOLOGICAL IMPORTANCE OF FUNGI

Learning Objectives

7 Explain the ecological significance of fungi as decomposers.

8 Describe the special ecological role of mycorrhizae.

Fungi make important contributions to the ecological balance of our planet. Like bacteria, most fungi are free-living decomposers (saprotrophs) that absorb nutrients from organic wastes and dead organisms. For example, many fungal decomposers degrade cellulose and lignin, the main components of plant cell walls. When fungi degrade wastes and dead organisms, they release water, carbon (as CO_2), and mineral components of organic compounds, and these elements are recycled (see biogeochemical cycles in Chapter 53). Without this continuous decomposition, essential nutrients would remain locked up in huge mounds of animal carcasses, feces, branches, logs, and leaves. The nutrients would be unavailable for use by new generations of organisms, and life would eventually cease.

Although most fungi are decomposers, others form symbiotic relationships of various kinds. Some fungi are **parasites,** organisms that live in or on other organisms and are harmful to their hosts. Parasitic fungi absorb food from the bodies of their hosts.

Mycorrhizae (from Greek words meaning "fungus roots") are mutualistic relationships between fungi and the roots of plants (see Fig. 34-10). Such relationships occur in more than 90% of all plant families. The mycorrhizal fungus decomposes organic material in the soil. It also benefits the plant by increasing its absorptive surface area, enabling it to take in more water and nutrient minerals such as phosphorus. At the same time, the roots supply the fungus with sugars, amino acids, and other organic substances (Fig. 25-15). Scientists have also measured the movement of organic materials from one tree species to another through shared mycorrhizal connections.

The importance of mycorrhizae first became evident when horticulturalists observed that orchids do not grow unless an appropriate fungus lives with them. Similarly, many forest trees such as pines decline and eventually die from mineral deficiencies when transplanted to mineral-rich grassland soils that lack the appropriate mycorrhizal fungi. When forest soil containing the appropriate fungi or their spores is added to the soil around these trees, they quickly resume normal growth.

Review

- What is the ecological importance of fungal decomposers?
- What is the importance of mycorrhizae?

BiologyNow™ Assess your understanding of **the ecological importance of fungi** by taking the pretest on your BiologyNow CD-ROM.

FIGURE 25-15 **Western red cedar (*Thuja plicata*) seedlings and mycorrhizae.**
(a) Control seedlings grown in low phosphorus in the absence of the fungus. **(b)** These seedlings are the same age as the control plants and were grown under conditions identical to the control, except that their roots have formed mycorrhizal associations.

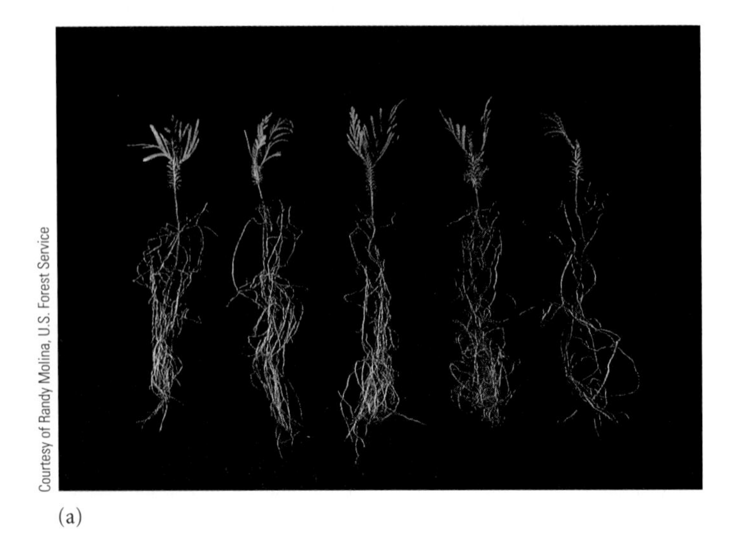

Courtesy of Randy Molina, U.S. Forest Service

(a)

Courtesy of Randy Molina, U.S. Forest Service

(b)

ECONOMIC AND MEDICAL IMPORTANCE OF FUNGI

Learning Objectives

9 Summarize some of the ways humans use fungi.

10 Identify at least three fungal diseases of plants and three of animals.

The same powerful digestive enzymes that fungi use to decompose wastes and dead organisms can also be used to reduce wood, fiber, and food to their basic components with great efficiency. From the human perspective, various fungi cause incalculable damage to stored goods and building materials each year. Bracket fungi, for example, cause enormous losses by decaying wood, both in living trees and in stored lumber.

Fungi are more destructive to plants than any other disease-causing organism. Their activities cost billions of dollars in agricultural damage yearly. Some fungi cause diseases in humans and other animals. Yet fungi are responsible for economic gains as well as losses: People eat them and grow them to make various chemicals.

Fungi provide beverages and food

Humans exploit the capability of yeasts to produce ethyl alcohol and carbon dioxide from sugars such as glucose by fermentation (see Chapter 7) to make wine, beer, and other fermented beverages, and also to make bread. Wine is produced when yeasts ferment fruit sugars, and beer results when yeasts ferment sugars derived from starch in grains (usually barley). During the process of making bread, carbon dioxide produced by yeast becomes trapped in dough as bubbles, causing the dough to rise; this gives leavened bread its light texture. Both the carbon dioxide and the alcohol produced by the yeast escape during baking.

The unique flavor of cheeses such as Roquefort, Brie, Gorgonzola, and Camembert is produced by species of *Penicillium*. *Penicillium roquefortii*, for example, is found in caves near the French village of Roquefort; by French law, only cheeses produced in this area can be called Roquefort cheese. In Roquefort and certain other cheeses, the blue mycelium of the fungus is visible in the cheese.

Aspergillus tamarii and certain other imperfect fungi are used to produce soy sauce by fermenting soybeans with the fungi for three or more months.[3] Soy sauce enriches other foods with more than just its special flavor. It also adds vital amino acids from both the soybeans and the fungi themselves, which supplement a low-protein rice diet.

Among the basidiomycetes are some 200 kinds of edible mushrooms and about 70 species of poisonous ones, sometimes called toadstools. Some edible mushrooms are cultivated commercially. The mushroom *Agaricus brunnescens* is the principal fungal species grown extensively for food. About 30 other mushroom species, such as oyster, shiitake, portobello, and straw mushrooms, are available in supermarkets. Morels, which superficially resemble mushrooms, and truffles, which produce underground fruiting bodies, are ascomycetes (Fig. 25-16). These gourmet delights are now being cultivated as mycorrhizal fungi on the roots of tree seedlings.

Edible and poisonous mushrooms can look very much alike and may even belong to the same genus. There is no simple way to tell them apart; an expert must identify them. Some of the most poisonous mushrooms belong to the genus *Amanita* (Fig. 25-17).

[3] In the United States soy sauce is often made by adding flavoring to salt water rather than by soaking fermented soybeans.

(a)

(b)

FIGURE **25-16** **Edible ascomycetes.**

(a) The yellow morel (*Morchella esculenta*), which grows 6 to 10 cm (2.5 to 4 in) tall, is found throughout North America. Photographed in Michigan. **(b)** The Oregon white truffle (*Tuber gibbosum*), which is found underground near Douglas firs and possibly oak trees in British Columbia and Northern California, is 1 to 5 cm (0.4 to 2 in) wide. People find these subterranean ascocarps with the help of trained dogs or pigs. Here truffles are shown whole and sectioned to show the conspicuous white, marbled tissue.

FIGURE **25-17** | **Poisonous mushrooms.**

The destroying angel (*Amanita virosa*) is an extremely poisonous mushroom that is distinguished, as are other amanitas, by the ring of tissue around its stalk and by the underground cup from which the stalk protrudes. About 50 g (2 oz) of this mushroom can kill an adult man. The destroying angel, which is 7.5 to 20 cm (3 to 8 in) tall, is found in grass or near trees throughout North America.

Toxic species of this genus have been appropriately called such names as "destroying angel" (*Amanita virosa*) and "death cap" (*Amanita phalloides*). Eating a single mushroom of either species can be fatal.

Ingesting certain species of mushrooms causes intoxication and hallucinations. The sacred mushrooms of the Aztecs—*Conocybe* and *Psilocybe*—are still used in religious ceremonies by Native Americans of Central America for their hallucinogenic properties. The chemical ingredient *psilocybin* is responsible for the trances and visions experienced by those who eat these mushrooms. Psilocybin is related to lysergic acid diethylamide (LSD). Ingestion of psychoactive mushrooms is dangerous because negative reactions vary considerably, from mild indigestion, sweating, and heart palpitations, to death. In addition, the possession and use of such mushrooms are illegal in the United States and some other countries.

Fungi produce useful drugs and chemicals

Discovered in 1928 by the British bacteriologist Alexander Fleming, penicillin, produced by the mold *Penicillium notatum,* is still among the most widely used and effective antibiotics (see Chapter 1). Other drugs derived from fungi include the cephalosporin antibiotics (produced by *Cephalosporium*), statins (used to lower blood cholesterol levels), and cyclosporine (used to suppress immune responses in patients who receive organ transplants). Fumagillin is a fungal chemical that inhibits the formation of new blood vessels; because solid tumors need a rich blood supply, fumagillin shows promise as an anticancer agent.

The ascomycete *Claviceps purpurea* infects the flowers of rye plants and other cereals. It produces a structure called an *ergot* where a seed would normally form in the grain head. When livestock eat this grain or when humans eat bread made from ergot-contaminated rye flour, they may be poisoned by the extremely toxic substances in the ergot. However, some ergot compounds are now used clinically in small quantities as drugs to induce labor, to stop uterine bleeding, to treat high blood pressure, and to relieve one type of migraine headache.

Some fungi are grown commercially to produce citric acid and other industrial chemicals. Also, biologists are using recombinant DNA techniques to manipulate yeasts and certain filamentous fungi to produce important biological molecules such as hormones.

Fungi cause many important plant diseases

Fungi are responsible for many serious plant diseases, including epidemic diseases that spread rapidly and often result in complete crop failure. All plants are apparently susceptible to some fungal infection. Damage may be localized in certain tissues or structures of the plant, or the disease may be systemic and spread throughout the entire plant. Fungal infections may cause stunting of plant parts or of the entire plant, may cause growths like warts, or may kill the plant.

A plant often becomes infected after hyphae enter through stomata (pores) in the leaf (Fig. 25-18) or stem or through wounds in the plant body. Alternatively, the fungus may produce *cutinase,* an enzyme that dissolves the waxy cuticle that covers the surface of leaves and stems. After dissolving the cuticle, the fungus easily invades the plant tissues. As the mycelium

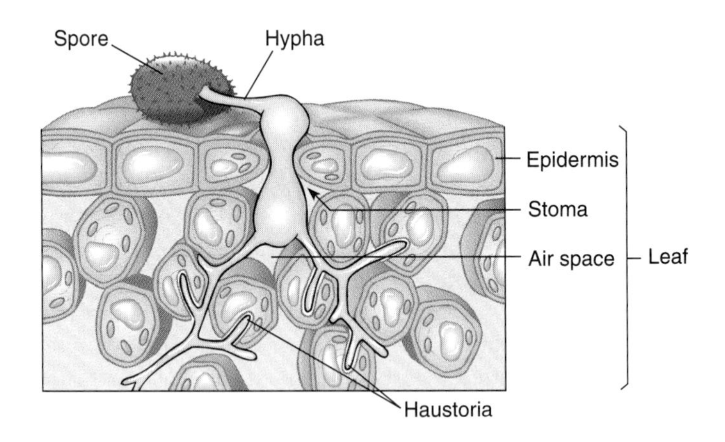

FIGURE **25-18** | **How a fungus parasitizes a plant.**

In this example, the hypha enters the leaf through a stoma. It grows, branching extensively through the internal air spaces, and penetrates plant cells with specialized hyphal extensions called haustoria.

FIGURE **25-19**

Fungi that cause plant diseases.

(a) Brown rot of peaches is caused by *Monilinia fruticola*, an ascomycete. Photographed in Oregon. **(b)** Corn smut on an ear of sweet corn is caused by *Ustilago maydis*, a basidiomycete. Photographed in Pennsylvania.

(a)

Kathy Merrifield/ Photo Researchers, Inc.

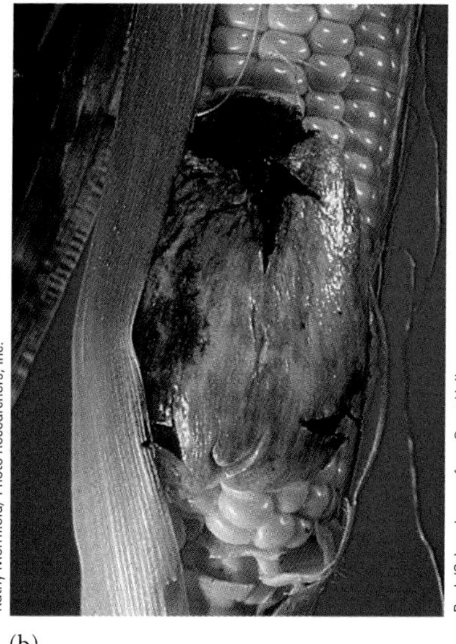

(b)

Runk/Schoenberger, from Grant Heilman

grows, it may remain mainly between the plant cells or it may penetrate the cells. Parasitic fungi often produce special hyphal branches called **haustoria** (sing., *haustorium*) that penetrate the host cells and obtain nourishment from the cytoplasm.

Ascomycetes cause important plant diseases, including powdery mildew, chestnut blight, Dutch elm disease, apple scab, and brown rot, which attacks cherries, peaches, plums, and apricots (Fig. 25-19a). Early in the 20th century the chestnut blight fungus (*Cryphonectria parasitica*) was accidentally imported on diseased Asian chestnuts and quickly attacked native American chestnut trees, which had no resistance to the fungus. By the late 1940s, the blight had killed or damaged several billion mature trees—almost every North American chestnut tree throughout its entire natural range. Plant biologists are working to save the American chestnut tree from extinction. One approach has been to develop a disease-resistant variety by breeding the American chestnut with the Chinese chestnut, a related species that is resistant to chestnut blight. Other biologists are genetically engineering a virus that infects the chestnut blight fungus and reduces its virulence.

Basidiomycetes cause smuts and rusts that attack various plants, for example, corn, wheat, oats, and other grains (Fig. 25-19b). Certain imperfect fungi also cause plant diseases, including verticillium wilt on potatoes, which is caused by the deuteromycete *Verticillium,* and bean anthracnose, which is caused by the deuteromycete *Colletotrichum lindemuthianum.*

Some fungal parasites, such as the stem rust of wheat, have complex life cycles that involve two or more different host plants and the production of several kinds of spores. For example, wheat rust must infect a barberry plant at one stage in its life cycle. Since this fact was discovered, the eradication of barberry plants in wheat-growing regions has reduced infection with wheat rust. Eradication of the barberry has not eliminated wheat rust however, because the fungus overwinters on wheat at the southern end of the Grain Belt and forms asexual spores. During the spring, wind blows these spores for hundreds of kilometers, reinfecting northern areas of the United States and Canada.

Fungi cause certain animal diseases

Some fungi cause superficial infections in which only the skin, hair, or nails are infected. Ringworm and athlete's foot are examples of superficial fungal infections; both are caused by imperfect fungi. Candidiasis, a yeast infection of mucous membranes of the mouth, throat, or vagina, is also caused by an imperfect fungus.

Other fungi infect internal tissues and organs and may spread through many regions of the body. Histoplasmosis, for example, is a serious infection of the lungs caused by inhaling spores of a fungus common in soil contaminated with bird feces. Most people in the eastern and midwestern parts of the United States have been exposed to this fungus at some time, and an estimated 40 million Americans have had mild infections. Fortunately, the infection is usually confined to the lungs and is of short duration, but if the infection spreads through the blood to the heart, brain, or other parts of the body, it can be serious and sometimes fatal.

Most pathogenic fungi are opportunists that cause infections only when the body's immunity is lowered. For example, the deuteromycete *Aspergillus fumigatus* is usually harmless, but causes aspergillosis in people with defective immune systems, such as patients with AIDS. During the course of this disease, the fungus can invade the lungs, heart, brain, kidneys, and other vital organs and can cause death. Other patients at high risk of acquiring life-threatening fungal infections include those with organ transplants or cancer.

Some fungi produce poisonous compounds collectively called **mycotoxins.** A few species of *Aspergillus,* for example, produce potent mycotoxins called *aflatoxins* that harm the liver and are known carcinogens. Foods on which aflatoxin-producing fungi commonly grow include peanuts, pecans, corn, and other grains. Other foods that may contain traces of aflatoxins include animal products such as milk, eggs, and meat (from animals that consumed feed contaminated by aflatoxin). Avoiding aflatoxin in the diet is impossible, but exposure should be minimized as much as possible. Any human food or animal forage

product that has become moldy should be suspected of aflatoxin contamination and discarded.

Fungi contribute to sick building syndrome, a situation in which occupants of a building experience acute health effects linked to the time they spend in a given building. Mold-related insurance claims amount to hundreds of millions of dollars each year. When conditions are moist, molds can grow on carpets, leather, cloth, wood, sheet rock, and insulation, as well as food. Mold spores, fragments, and mold products make their way into the air, and people are exposed through inhalation as well as by skin contact. Exposure to molds and their toxins has been linked to depressed immune function, irritation of the throat and respiratory passageways, infection, and toxicity. The most common responses to mold exposure are allergic reactions that range from mild to severe illnesses, including hayfever, sinusitis, asthma, and dermatitis.

Review

- Some dictionaries erroneously define a morel as a type of mushroom. Why isn't a morel a mushroom?
- What are three important fungal diseases of plants?
- What are three important fungal diseases of humans?

BiologyNow™ Assess your understanding of **the economic and medical importance of fungi** by taking the pretest on your BiologyNow CD-ROM.

SUMMARY WITH KEY TERMS

1 Describe the distinguishing characteristics of kingdom Fungi.

- **Fungi** are eukaryotic heterotrophs that secrete digestive enzymes onto their food source and then absorb the predigested food.
- Most fungi are decomposers, but many form mutualistic relationships with plants, and some are parasites of plants and animals.
- Fungi are characterized by their cell walls containing **chitin.**

2 Describe the body plan of a fungus.

- A fungus may be a unicellular **yeast** or a filamentous, multicellular **mold.** The vegetative body plan of most multicellular fungi consists of long, branched **hyphae** that form a tangled mass called a **mycelium.**
- In the **zygomycetes,** the hyphae are **coenocytic,** that is, they form an elongated, multinuclear cell. In other fungi, perforated **septa** divide the hyphae into individual cells.

3 Trace the fate of a fungal spore that lands in a favorable location, such as an overripe peach, and describe conditions that permit fungal growth.

- Most fungi reproduce both sexually and asexually by means of **spores.**
- When a fungal spore lands in a suitable spot, it germinates. Some hyphae infiltrate the substrate and digest its organic compounds. Spores are produced on aerial hyphae.

4 Give arguments to support the hypothesis that chytridiomycetes may have been the earliest fungal group to have evolved from the common ancestor of fungi.

- **Chytridiomycetes** are simple aquatic fungi that produce flagellate cells at some stage in their life cycle. No other fungi have flagella. Thus chytrids probably evolved from a flagellate protist, the common ancestor of all fungi.

5 List distinguishing characteristics, describe a typical life cycle, and give examples of each of the following fungal groups: chytridiomycetes, zygomycetes, ascomycetes, basidiomycetes, and imperfect fungi.

- *Allomyces,* a common chytridiomycete (chytrid), spends part of its life as a haploid **thallus** and part as a diploid thallus. The haploid thallus bears male and female **gametangia,** structures in which flagellate gametes form. The diploid thallus bears zoosporangia that produce diploid **zoospores** and resting sporangia in which haploid zoospores form by meiosis.
- **Zygomycetes** produce both asexual spores and sexual spores, called **zygospores.** *Pilobolus* and the black bread mold *Rhizopus* are representatives of this group.
- **Ascomycetes** produce asexual spores called **conidia;** sexual spores called **ascospores** are produced in **asci.** Ascomycetes include yeasts, cup fungi, morels, truffles, and pink, brown, and blue-green molds.
- **Basidiomycetes** produce sexual spores called **basidiospores** on the outside of a **basidium;** basidia develop on the surface of **gills** in mushrooms, a type of **basidiocarp.** Basidiomycetes include mushrooms, puffballs, bracket fungi, rusts, and smuts.
- **Imperfect fungi,** or **deuteromycetes,** are a polyphyletic group that lack a sexual stage. Most reproduce asexually by forming conidia. Members of this group include *Aspergillus tamarii* (used to produce soy sauce), some species of *Penicillium,* and fungi that cause certain human infections.

6 Characterize the unique nature of a lichen.

- A **lichen** is a symbiotic combination of a fungus and a phototroph (an alga or cyanobacterium.) Lichens have three main growth forms: crustose, foliose, and fruticose.

7 Explain the ecological significance of fungi as decomposers.

- Most fungi are decomposers that break down organic compounds into simpler nutrients that are recycled.

8 Describe the special ecological role of mycorrhizae.

- **Mycorrhizae** are mutualistic relationships between fungi and the roots of plants. The fungus supplies water and nutrient minerals to the plant, and the plant secretes organic compounds needed by the fungus.

9 Summarize some of the ways humans use fungi.

- Mushrooms, morels, and truffles are foods; yeasts are vital in the production of beer, wine, and bread; certain fungi are used to produce cheeses and soy sauce.
- Fungi are used to make penicillin and other antibiotics; ergot is used to produce certain drugs; other fungi make citric acid and other industrial chemicals.

10 Identify at least three fungal diseases of plants and three of animals.

- Fungi cause many plant diseases, including wheat rust, Dutch elm disease, and chestnut blight; they cause human diseases such as ringworm, athlete's foot, candidiasis, and histoplasmosis.

POST-TEST

1. Most fungi produce cell walls containing the polymer ______________. (a) chitin (b) penicillin (c) cellulose (d) peptidoglycan (e) lysergic acid
2. Which of the following fungi does *not* have a mycelium? (a) black bread mold (b) yeast (c) decomposer cup fungus (d) cultivated mushroom (e) *Penicillium*
3. The condition described as $n + n$ is said to be (a) monokaryotic (b) diploid (c) a primary mycelium (d) coenocytic (e) dikaryotic
4. With the exception of chytridiomycetes, fungi are generally disseminated by (a) water currents (b) fragmentation of hyphae (c) soredia (d) airborne spores (e) flagellate zoospores
5. Which statement is *not* true of the chytridiomycetes? (a) they are simple aquatic fungi (b) they produce motile cells with single, posterior flagella (c) they have both sexual and asexual reproduction (d) the black bread mold is a representative of this group (e) they are the most primitive group of fungi
6. Which statement is *not* true of the zygomycetes? (a) they produce motile cells with single, posterior flagella (b) their sexual spores are called *zygospores* (c) they have both sexual and asexual reproduction (d) the black bread mold is a representative of this group (e) they have coenocytic hyphae
7. Which statement is *not* true of the ascomycetes? (a) their sexual spores are produced in asci (b) their asexual spores are produced in basidia (c) some species in this group are yeasts (d) their asexual spores are called *conidia* (e) some species cause serious plant diseases
8. The ascomycete life cycle typically includes (a) mainly diploid thalli (b) the formation of a thick zygosporangium (c) the production of eight haploid ascospores within an ascus (d) intertwined hyphae to form a basidiocarp (e) the production of ascospores, zoospores, and conidia at different stages
9. Which statement is *not* true of the basidiomycetes? (a) they have a diploid thallus that produces zoospores (b) their sexual spores are called *basidiospores* (c) they produce a secondary mycelium with $n + n$ hyphae (d) mushrooms and bracket fungi are examples of this group (e) this group includes both edible and poisonous species
10. The familiar portion of a mushroom is actually a large fruiting body called a(an) (a) ascocarp (b) basidium (c) basidiocarp (d) gametangium (e) ascus
11. Which statement is *not* true of deuteromycetes? (a) many are ascomycetes that lost the ability to reproduce by forming ascospores (b) they are also known as imperfect fungi (c) they have both sexual and asexual reproduction (d) their asexual spores are called *conidia* (e) some are important plant pathogens
12. A ______________ is a symbiotic association between a phototroph and a fungus. (a) mycorrhizae (b) deuteromycete (c) lichen (d) haustorium (e) saprotroph
13. Which characteristic is true of *all* fungi? (a) saprotrophic (b) parasitic (c) nonflagellate (d) pathogenic (e) heterotrophic
14. Mutualistic relationships between fungi and the roots of plants are called (a) lichens (b) mycorrhizae (c) deuteromycetes (d) haustoria (e) conidiophores
15. *Amanita virosa* (a) is the mushroom commonly cultivated for food (b) is the yeast that ferments wine and beer (c) produces the unique flavor of many cheeses (d) is a highly toxic mushroom (e) has been ingested for its hallucinogenic properties
16. Which of the following describes how a fungus infects a plant? (a) infiltrates leaves with lichens (b) forms relationships by attaching mycorrhizae to stems (c) secretes powerful digestive juices onto the leaves (d) uses haustoria to penetrate stem tissue (e) hyphae enter leaves through a stoma
17. Mycotoxins are (a) released by most lichens (b) typically cause mild allergic reactions (c) can harm the liver (d) reduce the effects of sick building syndrome (e) are produced mainly by chytrids

CRITICAL THINKING

1. How are the life cycles of the marine alga *Ulva* (see Fig. 24-20) and *Allomyces* similar?
2. What measures can you suggest to prevent bread from becoming moldy?
3. Explain the statement: "Mushrooms are like the tips of icebergs." If you do not see toadstools in your lawn, can you conclude that no fungi live there? Why or why not?
4. Biologists have discovered that many mycorrhizal fungi are sensitive to a low pH. What human-caused environmental problem may prove catastrophic for these fungi? How may this problem affect their plant partners?

- Visit our Web site at **http://biology.brookscole.com/solomon7** for links to chapter-related resources on the World Wide Web. Additional online materials relating to this chapter can also be found on our Web site.

BIOLOGY NOW RESOURCES

Active Figures

25-3: The diversity of fungi

25-5: Fungal life-cycles

Preparing for an exam? Take a diagnostic test on your BiologyNow CD-ROM.

Post-Test Answers

1. a	2. b	3. e	4. d
5. d	6. a	7. b	8. c
9. a	10. c	11. c	12. c
13. e	14. b	15. d	16. e
17. c			

26

The Plant Kingdom: Seedless Plants

Sydney Karp/Photo/Nats, Inc.

Moss-covered rocks. Shown is Marron Creek in Snowmass Wilderness, Colorado.

CHAPTER OUTLINE

About 440 million years ago (mya), planet Earth would have seemed most inhospitable because, although life abounded in the oceans, it did not yet exist in abundance on land. The ocean was filled with vast numbers of fish, mollusks, and crustaceans as well as countless microscopic algae, and the water along rocky coastlines was home to large seaweeds. Occasionally, perhaps, an animal would crawl out of the water onto land, but it never stayed there permanently because there was little to eat on land—not a single blade of grass, no fruit, and no seeds.

During the next 30 million years, a time corresponding roughly to the Silurian period of the Paleozoic era in geological time (see Table 20-1), plants appeared in abundance and colonized the land. Where did they come from? Although plants living today exhibit great diversity in size, form, and habitat, botanists hypothesize they all evolved from a common ancestor that was an ancient green alga. Biologists infer this because modern green algae share many biochemical and metabolic traits with modern plants. Both green algae and plants contain the same photosynthetic pigments: chlorophylls *a* and *b* and accessory pigments, the yellow and orange carotenoids, including xanthophylls (yellow pigments) and carotenes (orange pigments). Also, both store excess carbohydrates as starch and have cellulose as a major component of their cell walls. In addition, plants and some green algae share certain details of cell division, including formation of a cell plate during cytokinesis (see Chapter 9).

Recent ultrastructural and molecular data indicate plants probably descended from a group of green algae called **charophytes** or **stoneworts** (see Fig. 24-18d). Molecular comparisons, particularly of DNA and RNA sequences, provide compelling evidence that charophytes are closely allied to plants. These data include comparisons among plants and various green algae, of chloroplast DNA sequences, of certain nuclear genes, and of ribosomal RNA sequences. In each case, the closest match occurs between charophytes and plants, indicating modern charophytes and plants probably share a common charophyte ancestor.

Today the plant kingdom consists of hundreds of thousands of species that live in varied habitats, from frozen Arctic tundra to lush tropical rain forests to harsh deserts to moist stream banks, as shown in the photograph on the previous page. Plants are complex multicellular organisms that range in size from minute, almost microscopic duckweeds to massive giant sequoias, some of the largest organisms that have ever lived. ■

ADAPTATIONS OF PLANTS

Learning Objectives

1. Discuss some environmental challenges of living on land, and describe how several adaptations meet these challenges.
2. Name the protist group from which plants are hypothesized to have descended, and describe supporting evidence.

What are some features of plants that have let them colonize so many different environments? One important difference between plants and algae is that a waxy **cuticle** covers the aerial portion of a plant. Essential for existence on land, the cuticle helps prevent desiccation, or drying out, of plant tissues by evaporation. Most plants that are adapted to moister habitats may have a very thin layer of wax, whereas those adapted to drier environments often have a thick, crusty cuticle. (Many desert plants also have a reduced surface area, particularly of leaves, that minimizes water loss.)

Plants get the carbon they need for photosynthesis from the atmosphere as carbon dioxide (CO_2). To be fixed into organic molecules such as sugar, CO_2 must first diffuse into the chloroplasts that are inside green plant cells. Because a waxy cuticle covers the external surfaces of leaves and stems, however, gas exchange through the cuticle between the atmosphere and the insides of cells is negligible. Tiny pores called **stomata** (sing., *stoma*), which dot the surfaces of leaves and stems of almost all plants, facilitate gas exchange; algae lack stomata.

Most plants have multicellular sex organs called **gametangia** (sing., *gametangium*), whereas the gametangia of algae are unicellular (Fig. 26-1). Each plant gametangium has a layer of sterile (nonreproductive) cells that surrounds and protects the delicate gametes (eggs and sperm cells). In plants, the fertilized egg develops into a multicellular **embryo** (young plant) within the female gametangium. Thus the embryo is protected during its development. In algae, the fertilized egg develops away from its gametangium; in some algae, the gametes are released before fertilization, whereas in others the fertilized egg is released.

FIGURE **26-1** **Generalized reproductive structures of algae and plants.**

(a, b) In algae, gametangia are generally unicellular. When the gametes are released, only the wall of the original cell remains. **(c, d)** In plants, the gametangia are multicellular, but only the inner cells become gametes. The gametes are surrounded by a protective layer of sterile (nonreproductive) cells.

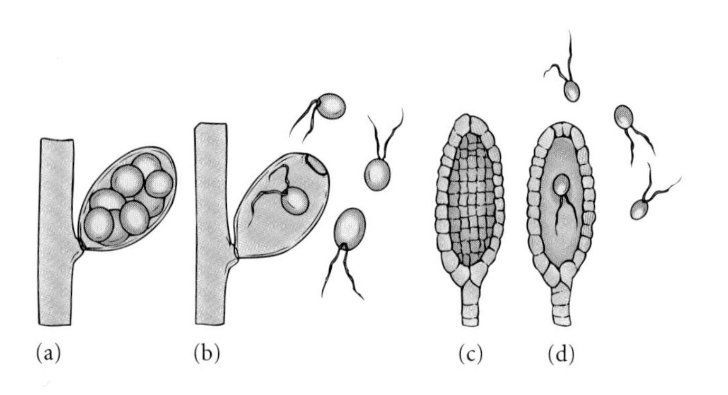

The plant life cycle alternates haploid and diploid generations

Plants have a clearly defined **alternation of generations** in which they spend part of their lives in a multicellular haploid stage and part in a multicellular diploid stage[1] (Fig. 26-2). The haploid portion of the life cycle is called the **gametophyte generation** because it gives rise to haploid gametes by mitosis. When two gametes fuse, the diploid portion of the life cycle, called the **sporophyte generation,** begins. The sporophyte generation produces haploid spores by the process of meiosis; these spores represent the first stage in the gametophyte generation.

[1] For convenience we limit our discussion to plants that are not polyploid, although polyploidy is very common in the plant kingdom. We therefore use the terms *diploid* and $2n$ (and *haploid* and n) interchangeably, although these terms are not actually synonymous.

FIGURE **26-2** **The basic plant life cycle.**

Plants have an alternation of generations, spending part of the cycle in a haploid gametophyte stage and part in a diploid sporophyte stage. Depending on the plant group, the haploid or the diploid stage may be greatly reduced.

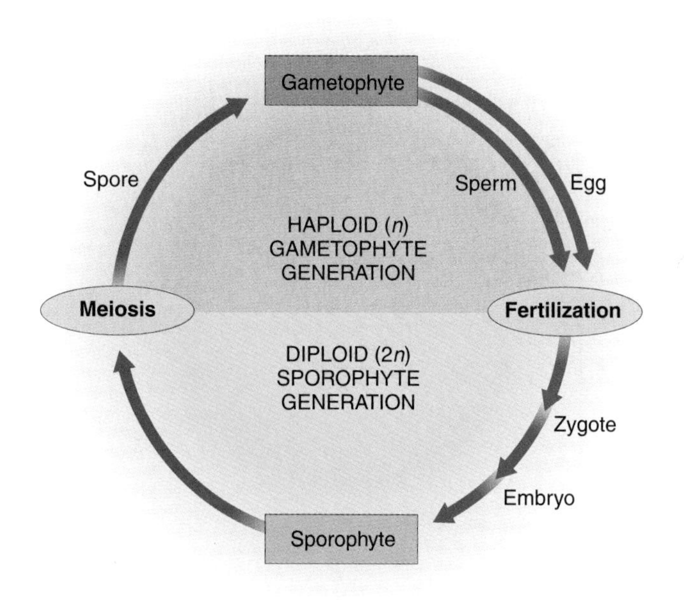

Let's examine alternation of generations more closely. The haploid gametophytes produce **antheridia** (male gametangia), in which sperm cells form, and/or **archegonia** (female gametangia), each bearing a single egg (Fig. 26-3). Sperm cells reach the female gametangium in a variety of ways, and one sperm cell fertilizes the egg to form a **zygote,** or fertilized egg.

The diploid zygote is the first stage in the sporophyte generation. The zygote divides by mitosis and develops into a multicellular embryo, the young sporophyte plant. Embryo development takes place within the archegonium; thus, the embryo is protected as it develops. Eventually the embryo grows into a mature sporophyte plant. The mature sporophyte has special cells called *sporogenous cells* (spore-producing cells, also called *spore mother cells*) that divide by meiosis to form haploid **spores.**

Fertilization of egg by sperm cell ⟶ zygote ⟶ embryo ⟶ mature sporophyte plant ⟶ sporogenous cells ⟶ meiosis ⟶ spores

All plants produce spores by meiosis, in contrast with algae and fungi, which may produce spores by meiosis or mitosis. The spores represent the first stage in the gametophyte generation. Each spore divides by mitosis to produce a multicellular gametophyte, and the cycle continues. Plants therefore alternate between a haploid gametophyte generation and a diploid sporophyte generation.

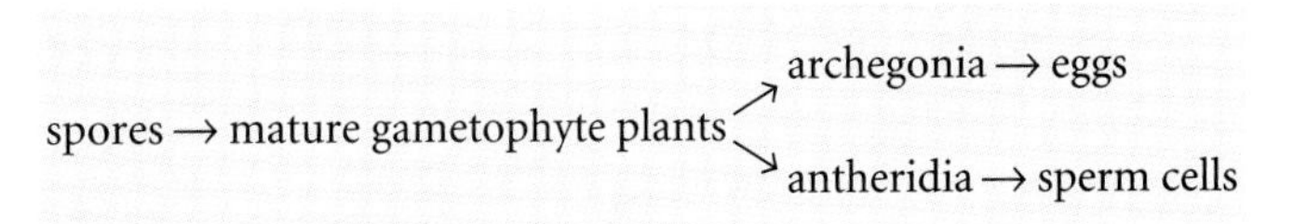

FIGURE **26-3** **Plant gametangia.**

(a) Each antheridium, the male gametangium, produces numerous sperm cells. **(b)** Each archegonium, the female gametangium, produces a single egg. Shown are generalized moss gametangia.

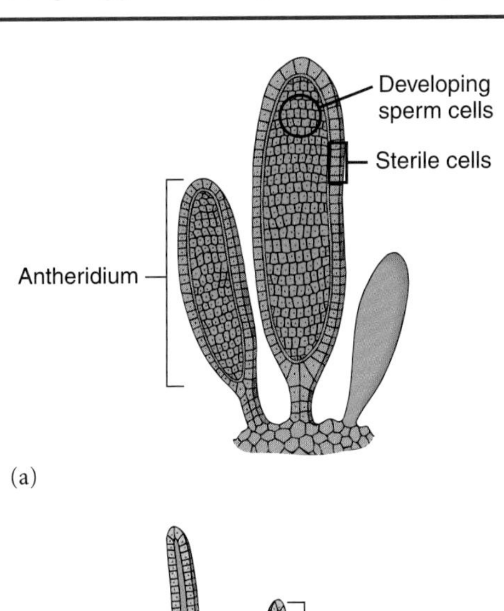

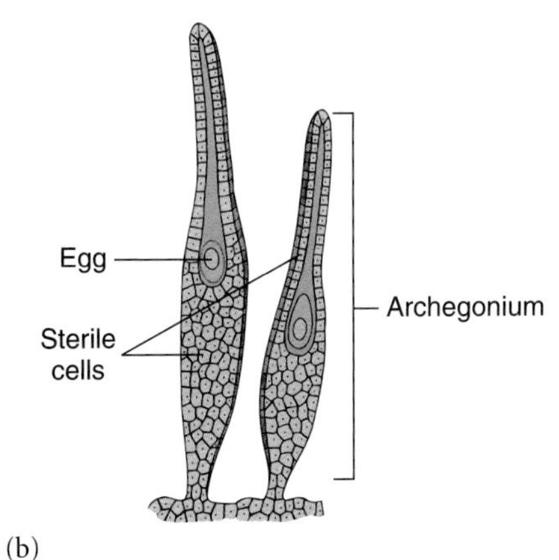

Four major groups of plants evolved

The plant kingdom consists of four major groups: bryophytes, seedless vascular plants, gymnosperms, and flowering plants (Fig. 26-4; see also Table 26-1, which is an overview of living plant phyla discussed in Chapters 26 and 27). The mosses and other bryophytes are small nonvascular plants that lack a specialized vascular, or conducting, system to transport dissolved nutrients, water, and essential minerals throughout the plant body. In the absence of such a system, bryophytes rely on diffusion and osmosis to obtain needed materials. This reliance means bryophytes are restricted in size; if they were much larger, some

ACTIVE FIGURE **26-4** **Plant evolution.**

This cladogram shows hypothetical evolutionary relationships among living plants, based on current evidence. Although the arrangement of nonvascular, seedless vascular, and seed plant groupings is widely recognized, the exact positions of various phyla remain uncertain. The order in which the hornworts, liverworts, and mosses evolved is not yet resolved, and the exact position of the whisk ferns is still unclear.

BiologyNow™ Explore **plant evolution** by clicking on this figure on your BiologyNow CD-ROM.

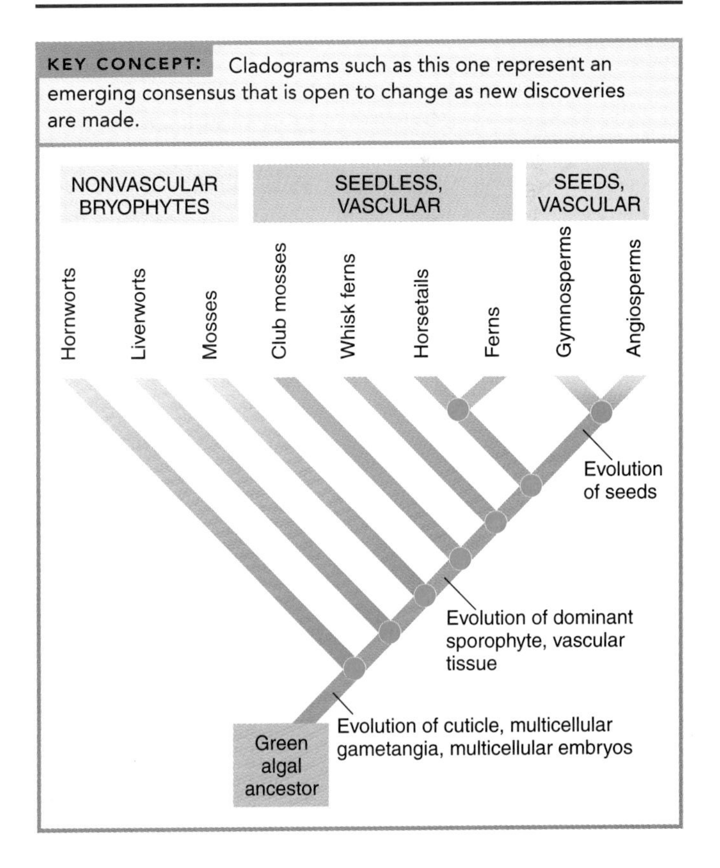

TABLE 26-1 The Plant Kingdom

The Plant Kingdom
Nonvascular plants with a dominant gametophyte generation (bryophytes)
Phylum Bryophyta (mosses)
Phylum Hepaticophyta (liverworts)
Phylum Anthocerotophyta (hornworts)
Vascular plants with a dominant sporophyte generation
Seedless plants
Phylum Polypodiophyta (ferns)
Phylum Psilotophyta (whisk ferns)
Phylum Equisetophyta (horsetails)
Phylum Lycophyta (club mosses)
Seed plants
Plants with naked seeds (gymnosperms)
Phylum Pinophyta (conifers)
Phylum Cycadophyta (cycads)
Phylum Ginkgophyta (ginkgoes)
Phylum Gnetophyta (gnetophytes)
Seeds enclosed within a fruit
Phylum Magnoliophyta (angiosperms or flowering plants)
Class Magnoliopsida (dicots)
Class Liliopsida (monocots)

of their cells could not obtain enough necessary materials. Bryophytes are seedless plants; they do not form seeds, the reproductive structures discussed in Chapter 27. Bryophytes reproduce and disperse primarily via haploid spores. Recent molecular and fossil evidence, discussed later in the chapter, indicates bryophytes may have been some of the earliest plants to colonize land.

The other three groups of plants—ferns, gymnosperms, and flowering plants—have vascular tissues and are thus known as vascular plants. The two vascular tissues are **xylem,** for conducting water and minerals, and **phloem,** for conducting dissolved organic molecules such as sugar. A key step in the evolution of vascular plants was the ability to produce **lignin,** a strengthening polymer in the cell walls of cells that function for support and conduction (see Chapter 31 for a discussion of plant cell wall chemistry, including lignin). The stiffening property of lignin enabled plants to grow tall (which let them maximize light interception). The successful occupation of the land by plants in turn made the evolution of terrestrial animals possible by providing them with both habitat and food.

Ferns and their close relatives, or allies (whisk ferns, horsetails, and club mosses), are seedless vascular plants that, like the bryophytes, reproduce and disperse primarily via spores. Seedless vascular plants arose and diversified during the Silurian and Devonian periods of the Paleozoic era, between 420 and 360 mya (discussed later in the chapter). Ferns and fern allies extend back more than 420 million years and were of considerable importance as Earth's dominant plants in past ages. Fossil evidence indicates that many species of these plants were the size of immense trees. Many ferns and most fern allies are extinct today; a few, mostly small representatives of the ancient groups survive.

The gymnosperms are vascular plants that reproduce by forming seeds (see Chapter 27). Gymnosperms produce seeds borne exposed (unprotected) on a stem or in a cone. Plants with seeds as their primary means of reproduction and dispersal first appeared about 360 mya, at the end of the Devonian period. These early seed plants diversified into many varied species of gymnosperms.

The most recent plant group to appear is the flowering plants, or angiosperms, which arose during the early Cretaceous period of the Mesozoic era about 130 mya. Like gymnosperms, flowering plants reproduce by forming seeds. Flowering plants, however, produce seeds enclosed within a fruit.

Review

- What are the most important environmental challenges that plants face living on land?
- What adaptations do plants have to meet these environmental challenges?
- From which group of protists are plants hypothesized to have descended?
- What is alternation of generations in plants?

BiologyNow™ Assess your understanding of **adaptations of plants** by taking the pretest on your BiologyNow CD-ROM.

BRYOPHYTES

Learning Objectives

3 Summarize the features that distinguish bryophytes from green algae and from other plants.

4 Name and briefly describe the three phyla of bryophytes.

5 Describe the life cycle of mosses, and compare their gametophyte and sporophyte generations.

The **bryophytes** (from the Greek words meaning "moss plant") consist of more than 15,000 species of mosses, liverworts, and hornworts; bryophytes are the only living nonvascular plants (Table 26-2). Because they have no means for extensive internal transport of water, dissolved sugar, and essential nutrient minerals, bryophytes are typically quite small. They generally require a moist environment for active growth and reproduction, but some bryophytes tolerate dry areas.

The bryophytes are divided into three distinct phyla: mosses (phylum Bryophyta), liverworts (phylum Hepaticophyta), and hornworts (phylum Anthocerotophyta). These three groups differ in many ways and may or may not be closely related. They are usually studied together, because they lack vascular tissues and have similar life cycles.

Moss gametophytes are differentiated into "leaves" and "stems"

Mosses (phylum Bryophyta), with about 9000 species, usually live in dense colonies or beds (Fig. 26-5a). Each individual plant has tiny, hairlike absorptive structures called *rhizoids,* and an upright, stemlike structure that bears leaflike blades, each normally consisting of a single layer of undifferentiated cells except at the midrib. Because mosses lack vascular tissues, they do not

TABLE 26-2 A Comparison of Major Groups of Seedless Plants

Plant Group	Dominant Stage of Life Cycle	Representative Genera
Nonvascular; reproduce by spores (bryophytes)		
Mosses (phylum Bryophyta)	Gametophyte: leafy plant	*Polytrichum, Sphagnum, Physcomitrella*
Liverworts (phylum Hepaticophyta)	Gametophyte: thalloid or leafy plant	*Marchantia*
Hornworts (phylum Anthocerotophyta)	Gametophyte: thalloid plant	*Anthoceros*
Vascular; reproduce by spores		
Ferns (phylum Polypodiophyta)	Sporophyte: roots, rhizomes, and leaves (megaphylls)	*Pteridium, Polystichum, Azolla, Platycerium*
Whisk ferns (phylum Psilotophyta)	Sporophyte: rhizomes and erect stems; no true roots or leaves	*Psilotum*
Horsetails (phylum Equisetophyta)	Sporophyte: roots, rhizomes, erect stems, and leaves (reduced megaphylls)	*Equisetum*
Club mosses (phylum Lycophyta)	Sporophyte: roots, rhizomes, erect stems, and leaves (microphylls)	*Lycopodium, Selaginella*

have true roots, stems, or leaves; the moss structures are not homologous to roots, stems, or leaves in vascular plants. Some moss species have water-conducting cells and sugar-conducting cells, although these cells are not as specialized or as effective as the conducting cells of vascular plants.

Alternation of generations is clear in the life cycle of mosses (Fig. 26-6). In ① of the figure, the green moss gametophyte often bears its gametangia at the top of the plant. Many moss species have separate sexes: male plants that bear antheridia and female plants that bear archegonia. Other mosses produce antheridia and archegonia on the same plant.

In ②, fertilization occurs when one of the sperm cells fuses with the egg within the archegonium. Sperm cells, which have flagella, are transported from antheridium to archegonium by flowing water, such as splashing rain droplets. A raindrop lands on the top of a male gametophyte, and sperm cells are released into it from the antheridia. Another raindrop landing on the male plant may splash the sperm-laden droplet into the air and onto the top of a nearby female plant. Alternatively, insects may touch the sperm-laden fluid and inadvertently carry it for considerable distances. Once in a film of water on the female moss, a sperm cell swims into the archegonium, which secretes

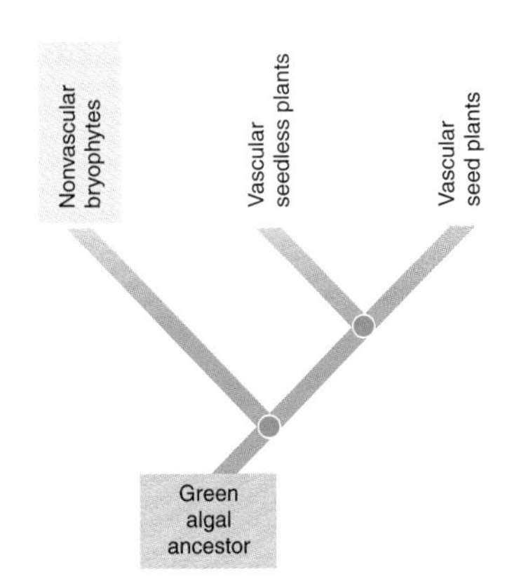

FIGURE 26-5 Mosses, liverworts, and hornworts.

(a) A closeup of haircap moss (*Polytrichum commune*) gametophytes. The haircap moss is a popular ground cover in rock gardens, particularly in Japan. **(b)** Flattened, ribbon-like lobes characterize the gametophyte of the common liverwort (*Marchantia polymorpha*). *Marchantia* grows on moist soil, from the tropics to arctic regions. **(c)** The gametophyte with mature sporophytes (the "horns" projecting out of the gametophyte) of the common hornwort (*Anthoceros natans*).

(a)

(b)

(c)

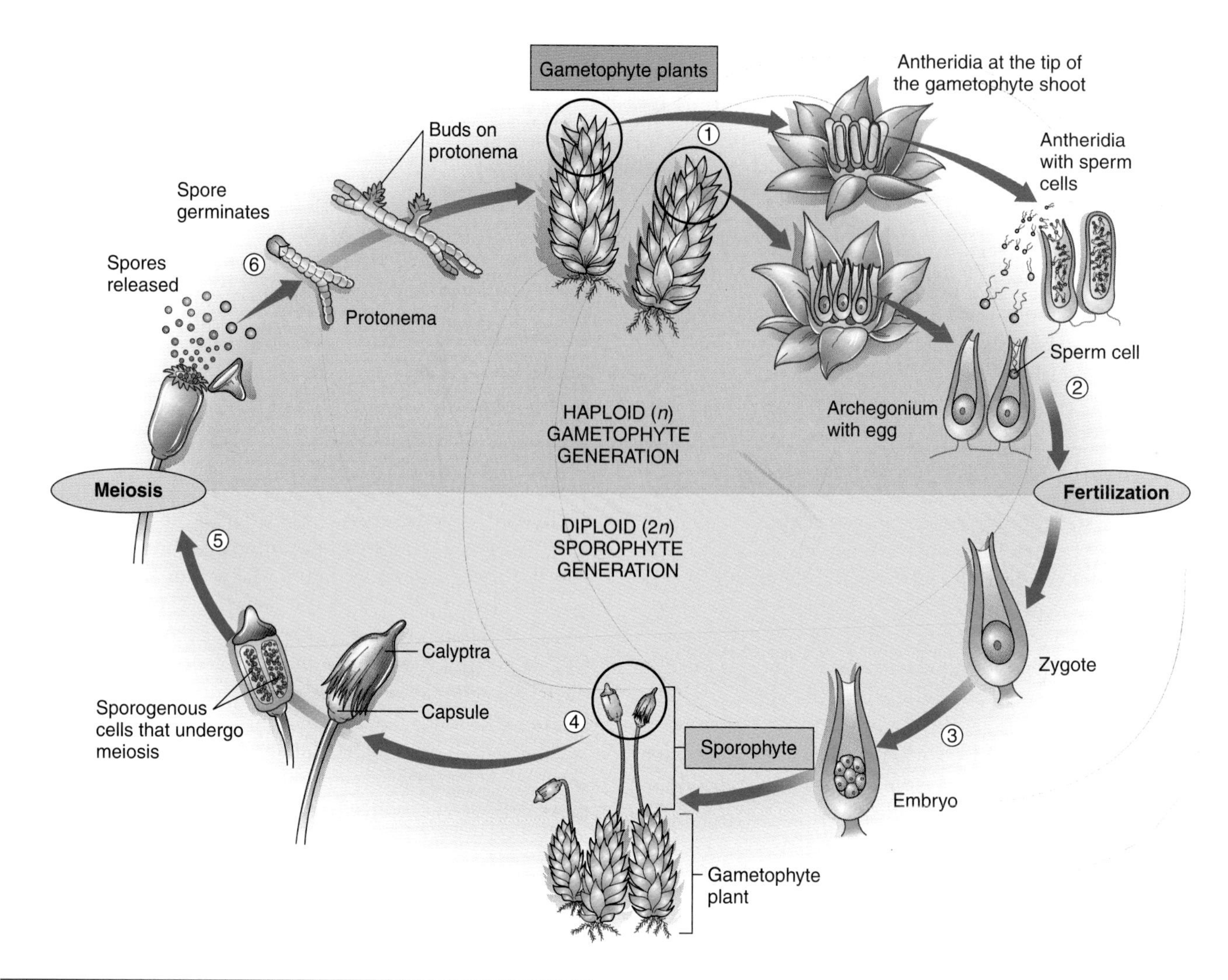

FIGURE **26-6** **The life cycle of mosses.**

The gametophyte generation is dominant in the moss life cycle. After sexual reproduction, the sporophyte grows out of the gametophyte. See text for a detailed description.

chemicals to attract and guide the sperm cells, and fuses with the egg.

In ③, the diploid zygote, formed by fertilization, grows by mitosis into a multicellular embryo that ④ develops into a mature moss sporophyte. This sporophyte grows out of the top of the female gametophyte, remaining attached and nutritionally dependent on the gametophyte throughout its existence (Fig. 26-7). Initially green and photosynthetic, the sporophyte becomes a golden brown at maturity. It consists of three main parts: a *foot,* which anchors the sporophyte to the gametophyte and absorbs minerals and nutrients from it; a *seta,* or stalk; and a *capsule,* which contains sporogenous cells (spore mother cells). The capsule of some species is covered by a caplike structure, the *calyptra,* which is derived from the archegonium.

In ⑤, the sporogenous cells undergo meiosis to form haploid spores. When the spores are mature, the capsule opens and releases the spores, which are then transported by wind or rain. In ⑥, if a moss spore lands in a suitable spot, it germinates and grows into a filament of cells called a **protonema.** The protonema, which superficially resembles a filamentous green alga, forms buds, each of which grows into a green gametophyte, and the life cycle continues.

Biologists consider the haploid gametophyte generation the dominant generation in mosses, because it lives independently of the diploid sporophyte. In contrast, the moss sporophyte is attached to and nutritionally dependent on the gametophyte.

Mosses make up an inconspicuous but significant part of their environment. They play an important role in forming soil. Because they grow tightly packed together in dense colonies, mosses hold soil in place and help prevent erosion.

Commercially, the most important mosses are the peat mosses in the genus *Sphagnum.* One of the distinctive features of *Sphagnum* "leaves" is the presence of many large dead cells that absorb and hold water. This feature makes peat moss particularly beneficial as a soil conditioner. Added to sandy soils, for example, peat moss helps absorb and retain moisture. In some countries, such as Ireland and Scotland, people cut out layers of dead peat moss that have accumulated for hundreds of years from peat bogs, dry them and burn them for fuel.

The name "moss" is often misused to refer to plants that are not truly mosses. For example, reindeer "moss" is a lichen that is a dominant form of vegetation in the Arctic tundra, Spanish "moss" is a flowering plant, and club "moss" (discussed later in this chapter) is a relative of ferns.

FIGURE **26-7** | **Moss sporophytes.**

Each consisting of a foot, seta, and capsule, the sporophytes grow out of the top of the gametophytes. Spores are produced within the capsule. Shown is the haircap moss (*Polytrichum commune*).

Liverwort gametophytes are either thalloid or leafy

Liverworts (phylum Hepaticophyta) consist of about 6000 species of nonvascular plants with a dominant gametophyte generation, but the gametophytes of some liverworts are quite different from those of mosses. Their body form is often a flattened, lobed structure called a **thallus** (pl., *thalli*) that is not differentiated into leaves, stems, or roots. The common liverwort, *Marchantia polymorpha*, is thalloid (see Fig. 26-5b). Liverworts are so named because the lobes of their thalli superficially resemble the lobes of the human liver; *wort* is derived from the Old English word *wyrt,* meaning "plant." On the underside of the liverwort thallus are hairlike rhizoids that anchor the plant to the soil. Other liverworts, known as *leafy liverworts,* superficially resemble mosses, with leaflike blades, "stems," and rhizoids rather than a lobed thallus. As in the mosses, leafy liverwort "leaves" consist of a single layer of undifferentiated cells. Like other bryophytes, liverworts are small, generally inconspicuous plants that are largely restricted to damp environments. Unlike other plants, including mosses and hornworts, liverworts lack stomata (although some liverworts have surface pores thought to be analogous to stomata).

Liverworts reproduce both sexually and asexually (Fig. 26-8). In ①, their sexual reproduction involves production of archegonia and antheridia on the haploid gametophyte. In some liverworts, these gametangia are borne on stalked structures called *archegoniophores,* which bear archegonia, and *antheridiophores,* which bear antheridia. Their life cycle is basically the same as that of mosses, although some of the structures look quite different. In ②, splashing raindrops transport sperm cells to the archengonia, where fertilization takes place. The resulting zygote develops into a multicellular embryo that becomes a mature sporophyte, ③. The liverwort sporophyte, which is usually somewhat spherical, is attached to the gametophyte, as in mosses. In ④, sporogenous cells in the capsule of the sporophyte undergo meiosis, producing haploid spores. In ⑤, each spore has the potential to develop into a green gametophyte, and the cycle continues.

Some liverworts reproduce asexually by forming tiny balls of tissue called **gemmae** (sing., *gemma*), which are borne in a saucer-shaped structure, the gemmae cup, directly on the liverwort thallus. Splashing raindrops and small animals help disperse gemmae. When a gemma lands in a suitable place, it grows into a new liverwort thallus. Liverworts may also reproduce asexually by thallus branching and growth. The individual thallus lobes elongate, and each becomes a separate plant when the older part of the thallus that originally connected the individual lobes dies.

Hornwort gametophytes are inconspicuous thalloid plants

Hornworts (phylum Anthocerotophyta) are a small group of about 100 species of bryophytes whose gametophytes superficially resemble those of the thalloid liverworts. Hornworts live in disturbed habitats such as fallow fields and roadsides.

Hornworts may or may not be closely related to other bryophytes. For example, their cell structure, particularly the presence of a single large chloroplast in each cell, resembles certain algal cells more than plant cells. In contrast, mosses, liverworts, and other plants have many disk-shaped chloroplasts per cell.

In the common hornwort (*Anthoceros natans*), archegonia and antheridia are embedded in the gametophyte thallus rather than on archegoniophores and antheridiophores. After fertilization and development, the needle-like sporophyte projects out of the gametophyte thallus, forming a spike or "horn"—hence the name *hornwort.* A single gametophyte often produces multiple sporophytes (see Fig. 26-5c). Meiosis occurs, forming spores within each **sporangium** (pl., *sporangia*), or spore case. The sporangium splits open from the top to release the spores; each spore can give rise to a new gametophyte thallus. A unique feature of hornworts is that the sporophytes, unlike those of mosses and liverworts, continue to grow from their bases for the remainder of the gametophyte's life.

Bryophytes are used for experimental studies

Botanists use certain bryophytes as experimental models to study many fundamental aspects of plant biology, including genetics, growth and development, plant ecology, plant hormones, and **photoperiodism,** which is plant responses to varying periods

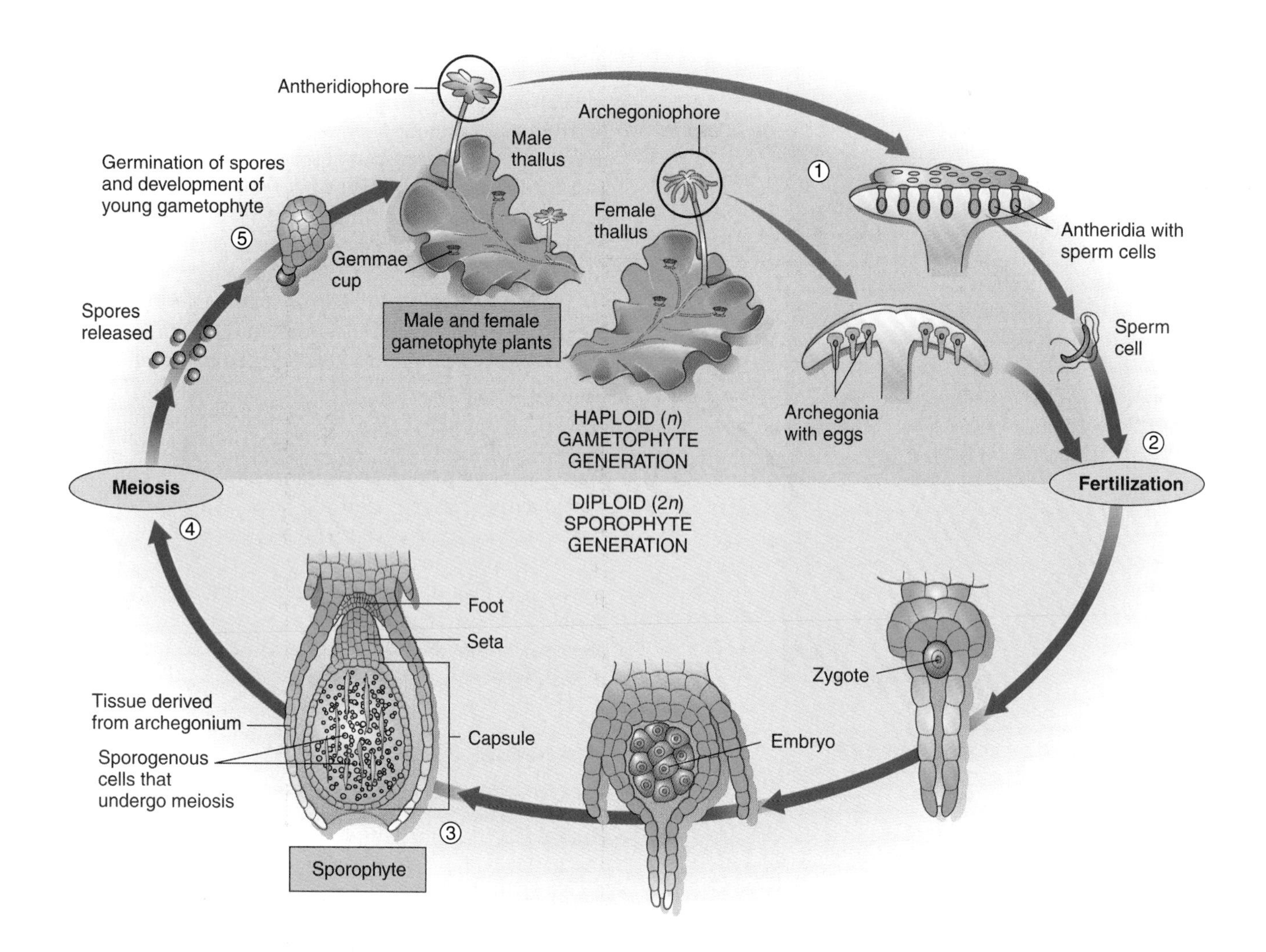

ACTIVE FIGURE **26-8** **The life cycle of the common liverwort (*Marchantia polymorpha*).**

The dominant generation is the gametophyte, represented by separate male and female thalli. The stalked, umbrella-shaped structures are the antheridiophores, with antheridia that produce sperm cells, and the archegoniophores, with archegonia that each bear an egg cell. See text for a detailed description.

BiologyNow™ Watch **the life cycle of the liverwort** by clicking on this figure on your BiologyNow CD-ROM.

FIGURE **26-9** **Mosses as research organisms.**

Researchers inoculated this petri dish with offspring from a genetic cross in the moss *Physcomitrella patens.*They will test the cultures for their phenotypes with respect to vitamin requirements. (The culture medium in which the mosses are currently growing is supplemented with all the vitamins required by the parental strains.)

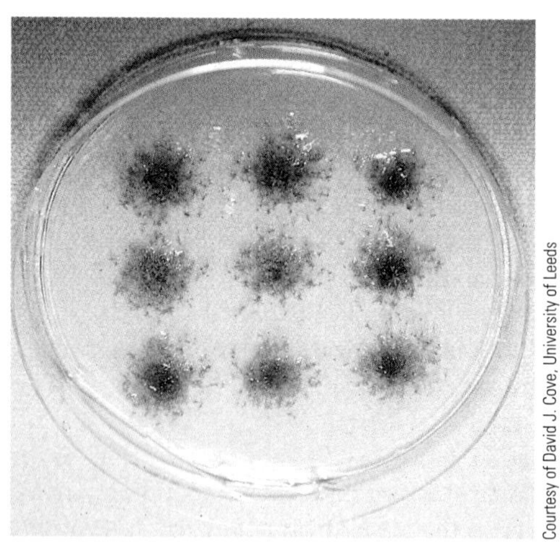

of night and day length (see Chapter 36). As experimental organisms, bryophytes are easy to grow on artificial media and do not require much space, because they are so small (Fig. 26-9).

Bryophyte evolution is based on fossils and on structural and molecular evidence

PROCESS OF SCIENCE

Plants are a **monophyletic group**—that is, all plants probably evolved from a common ancestral green alga. Fossil evidence indicates the bryophytes are ancient plants, probably the first group of plants to arise from the common plant ancestor. The fossil record of ancient bryophytes is incomplete, consisting mostly of spores and small tissue fragments, and can be interpreted in different ways. As a result, it does not provide a definite answer on bryophyte evolution.

The oldest known complete fossil plant, dated at about 400 million years old, was reported in 1995. This fossil, from a coal deposit in the United Kingdom, resembles modern liverworts in many respects, but its spores are virtually identical to those in 460-million-year-old rocks. This discovery suggests liverwort-like plants may have been the earliest plants to colonize land.

Structural and molecular evidence, however, supports the hypothesis that *hornworts* may be the most ancient group of plants alive today. A 1998 cladistic analysis using structural data was the first to strongly support this hypothesis, which was reinforced in 2000 by DNA evidence. The study of nuclear, chloroplast, and mitochondrial genes from numerous plant species, from bryophytes to flowering plants, suggests an early plant ancestor gave rise to two lineages, one from which the hornworts descended and the other from which all other plants descended.

Review

- Which of the following are parts of the gametophyte generation in mosses: antheridia, zygote, embryo, capsule, archegonia, sperm cells, egg cell, spore mother cells, spores, and protonema?
- How are mosses, liverworts, and hornworts similar? How is each group distinctive?
- What adaptations do bryophytes have that algae lack?

BiologyNow™ Assess your understanding of **bryophytes** by taking the pretest on your BiologyNow CD-ROM.

SEEDLESS VASCULAR PLANTS

Learning Objectives

6 Discuss the features that distinguish ferns and other seedless vascular plants from algae and bryophytes.

7 Name and briefly describe the four phyla of seedless vascular plants.

8 Describe the life cycle of ferns, and compare their sporophyte and gametophyte generations.

9 Compare the generalized life cycles of homosporous and heterosporous plants.

About 11,000 species of ferns exist today. Ferns are especially common in temperate woodlands and tropical rain forests, where they are most varied. Three groups of vascular plants—whisk ferns (about 12 species), horsetails (15 species), and club mosses (about 1000 species)—are considered fern allies because their life cycles are similar to those of ferns (Table 26-2). As shown in Figure 26-4, horsetails and ferns are a monophyletic group and the closest living relatives of seed plants.

The most important adaptation found in ferns and their allies, though absent in algae and bryophytes, is specialized vascular tissues—xylem and phloem—for support and conduction. This system of conduction lets vascular plants grow larger than the bryophytes because water, dissolved minerals, and dissolved sugar are transported to all parts of the plant. Although ferns in temperate environments are relatively small, tree ferns in the tropics may grow to heights of 18 m (60 ft). All ferns and fern allies have true stems with vascular tissues, and most also have true roots and leaves.

Botanists have extensively studied the evolution of the leaf as the main organ of photosynthesis. There are two basic types of leaves: microphylls and megaphylls (Fig. 26-10). The **microphyll,** which is usually small and has a single vascular strand, probably evolved from small, projecting extensions of stem tissue *(enations)*. Only one group of living plants, the club mosses, has microphylls. In contrast, **megaphylls** probably evolved from stem branches that gradually filled in with additional tissue *(webbing)* to form most leaves as we know them today. Megaphylls have more than one vascular strand, as we would expect if they evolved from branch systems. Ferns, horsetails, gymnosperms, and flowering plants have mega-

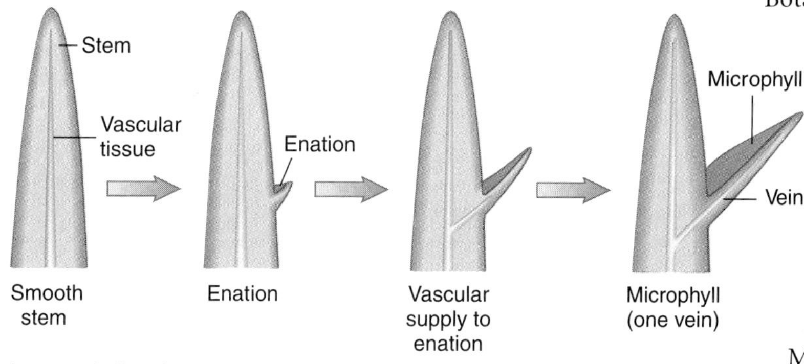

(a) Microphyll evolution

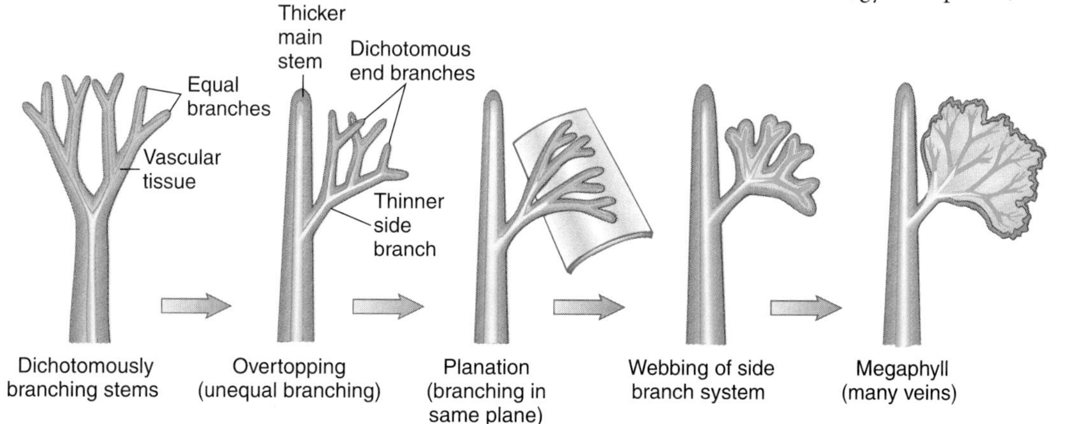

(b) Megaphyll evolution

FIGURE **26-10**

Evolution of microphylls and megaphylls.

Webbing (in **b**) is the evolutionary process in which the spaces between close branches become filled with chlorophyll-containing cells.

phylls. (Note that whisk ferns lack leaves and so are not included in this discussion.)

Recent evidence suggests megaphylls evolved over a 40-million-year period in the late Paleozoic era, in response to a gradual decline in the level of atmospheric CO_2. As CO_2 declined, plants developed a flattened blade with more stomata for gas exchange. (More stomata allowed cells inside the leaf to get enough CO_2.)

Ferns have a dominant sporophyte generation

Most of the 11,000 species of **ferns** (phylum Polypodiophyta) are terrestrial, although a few have adapted to aquatic habitats (Fig. 26-11a, b). Ferns range from the tropics to the Arctic Circle, with most species living in tropical rain forests, where they perch high in the branches of trees (Fig. 26-11c). In temperate regions, ferns commonly inhabit swamps, marshes, moist woodlands, and stream banks. Some species grow in fields, rocky crevices on cliffs or mountains, or even deserts.

The life cycle of ferns involves a clearly defined alternation of generations. The ferns grown as houseplants (such as the Boston fern, maidenhair fern, and staghorn fern) represent the larger, more conspicuous sporophyte generation. As shown in Figure 26-12, number ①, the fern sporophyte consists of a horizontal underground stem, or *rhizome,* that bears leaves, called *fronds,* and true roots. As each young frond first emerges from the ground, it is tightly coiled and resembles the top of a violin, hence the name *fiddlehead* (Fig. 26-13). As fiddleheads grow, they unroll and expand to form fronds. Fern fronds are usually compound (the blade is divided into several leaflets), with the leaflets forming beautifully complex leaves. Fronds, roots, and rhizomes all contain vascular tissues.

Spore production usually occurs in certain areas on the fronds, which develop sporangia. Many species bear the sporangia in clusters, called **sori** (sing., *sorus*). ② Within sporangia, sporogenous cells (spore mother cells) undergo meiosis to form haploid spores. In ③, the sporangia burst open and discharge spores that may germinate and grow by mitosis into mature gametophytes.

In ④, the mature fern gametophyte, which bears no resemblance to the sporophyte, is a tiny (less than half the size of one of your fingernails), green, often heart-shaped structure that grows flat against the ground. Called a **prothallus** (pl., *prothalli*), the fern gametophyte lacks vascular tissues and has tiny, hairlike absorptive rhizoids to anchor it (Fig. 26-14). In ⑤, the prothallus usually produces both archegonia and antheridia on its underside. Each archegonium contains a single egg, whereas numerous sperm cells are produced in each antheridium.

Ferns use water as a transport medium. The flagellate sperm cells swim, usually from a nearby prothallus, to the neck of an archegonium through a thin film of water on the ground

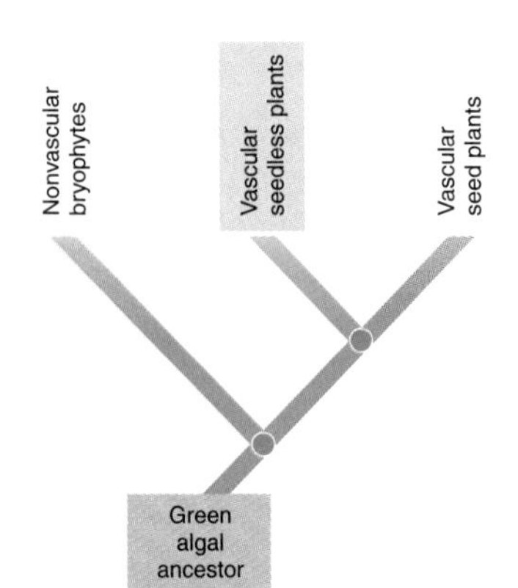

FIGURE **26-11** **Ferns.**

(a) The Christmas fern (*Polystichum acrostichoides*) is green at Christmas, making it a popular decoration. This fern, photographed in the Great Smoky Mountains in Tennessee, has fronds that grow to 0.6 m (2 ft) in length. **(b)** The tiny mosquito fern (*Azolla caroliniana*) is a free-floating aquatic fern that does not resemble "typical" ferns. Although each individual plant grows to only about 1.3 cm (0.5 in) in length, *Azolla* populations sometimes grow so densely across ponds that they reportedly smother mosquito larvae. Photographed in Massachusetts. **(c)** The staghorn fern (*Platycerium bifurcatum*) is native to Australian rain forests and is widely cultivated elsewhere. In nature the staghorn fern is an epiphyte, a plant that grows attached to another organism (in this case, a tree trunk) but derives no nourishment from it. The individual leaves grow to 0.9 m (3 ft) in length.

Ed Reschke

(a)

W. Ormerod/Visuals Unlimited

(b)

Carlyn Iverson

(c)

FIGURE **26-12** **The life cycle of ferns.**

Note the clearly defined alternation of generations between the gametophyte (prothallus) and sporophyte (leafy plant) generations. See text for a detailed description.

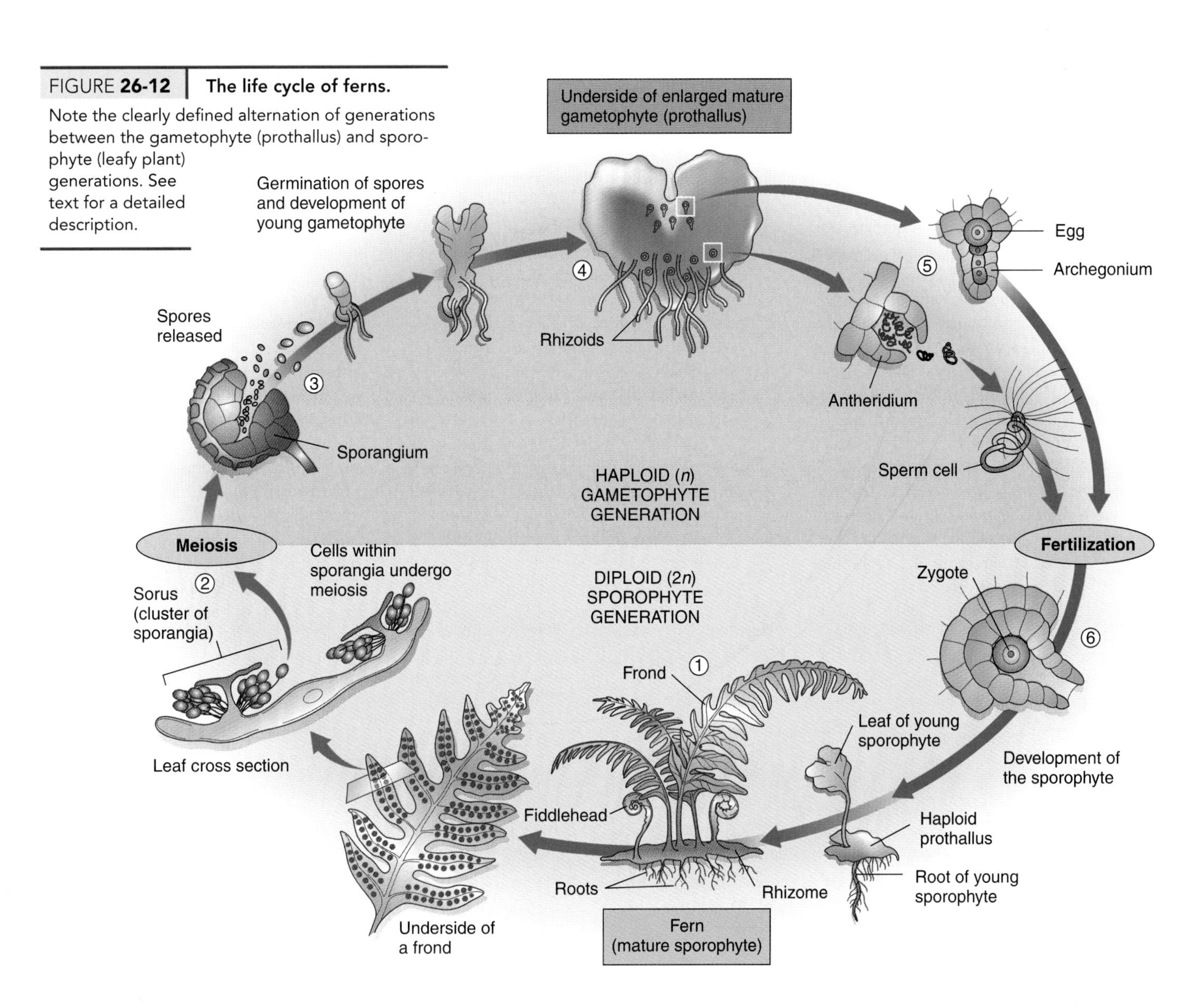

FIGURE **26-13** **Fiddleheads.**

The tightly coiled young fronds, or fiddleheads, are characteristic of ferns.

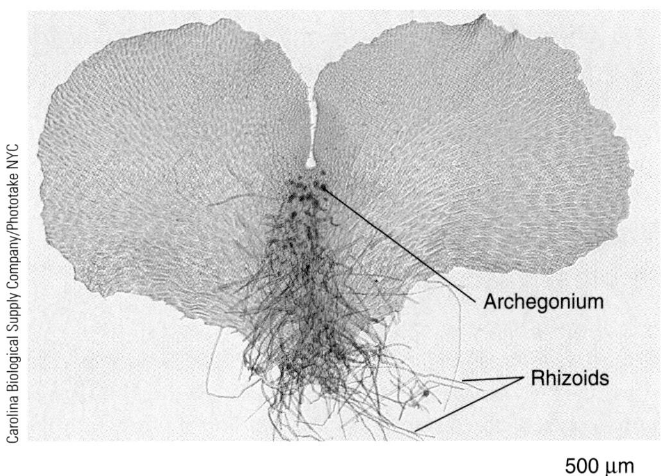

FIGURE **26-14** **Fern prothallus.**

The dark spots near the notch of the "heart" are archegonia; no antheridia are visible. (In many prothalli, archegonia and antheridia mature at different times.) Shown is a prothallus of Virginia chain fern (*Woodwardia virginica*).

FOCUS ON ANCIENT PLANTS AND COAL FORMATION

Industrial society depends on energy from fossil fuels that formed from the remains of ancient organisms. One of the most important fossil fuels is coal, which people burn to produce electricity and to manufacture items of steel and iron. Although mined, coal is not an inorganic mineral like gold or aluminum, but an organic material formed from the remains of ancient vascular plants, particularly those of the Carboniferous period, approximately 300 mya. Five main groups of plants contributed to coal formation. Three were seedless vascular plants: the club mosses, horsetails, and ferns. The other two important groups were seed plants: seed ferns (now extinct) and early gymnosperms.

It is hard to imagine that relatives of the small, relatively inconspicuous club mosses, ferns, and horsetails of today were so significant in forming vast beds of coal. However, many members of these groups that existed during the Carboniferous period were giants compared with their modern counterparts and formed immense forests (see Fig. 20-11).

The climate during the Carboniferous period was warm, moist, and mild. Plants in most locations could grow year-round because of the favorable conditions. Forests of these plants often grew in low-lying, swampy areas that periodically flooded when the sea level rose. As the sea level receded, these plants would re-establish.

When these large plants died or were blown over in storms, they decomposed incompletely, because they were covered by swamp water. (The anaerobic conditions of the water prevented wood-rotting fungi from decomposing the plants, and anaerobic bacteria do not decompose wood rapidly.) Thus, over time the partially decomposed plant material accumulated and consolidated.

Layers of sediment formed over the plant material each time the water level rose and flooded the low-lying swamps. With time, heat and pressure built up in these accumulated layers and converted the plant material to coal and the sediment layers to sedimentary rock. Much later, geological upheavals raised the layers of coal and sedimentary rock. Coal is usually found in seams, underground layers that vary in thickness from 2.5 cm (1 in) to more than 30 m (100 ft).

The various grades of coal (lignite, sub-bituminous, bituminous, and anthracite) formed as a result of the different temperatures and pressures to which the layers were exposed. Coal exposed to high heat and pressure during its formation is drier, is more compact (and therefore harder), and has a higher heating value (that is, a higher energy content).

underneath the prothallus. In ⑥, after one of the sperm cells fertilizes the egg, a diploid zygote grows by mitosis into a multicellular embryo (an immature sporophyte). At this stage, the sporophyte embryo is attached to and dependent on the gametophyte, but as the embryo matures, the prothallus withers and dies, and the sporophyte becomes free-living.

The fern life cycle alternates between the dominant, diploid sporophyte with its rhizome, roots, and fronds, and the haploid gametophyte (prothallus). The sporophyte generation is dominant not only because it is larger than the gametophyte but also because it persists for an extended period (most fern sporophytes are perennials), whereas the gametophyte dies soon after reproducing.

Whisk ferns are the simplest vascular plants

Only about 12 species of **whisk ferns** (phylum Psilotophyta) exist today, and the fossil record contains several extinct species. All are relatively simple in structure and lack true roots and leaves but have vascularized stems. *Psilotum nudum*, a representative whisk fern, has both a horizontal underground rhizome and vertical aerial stems (Fig. 26-15a). Whenever the stem forks or branches, it always divides into two equal halves. Botanists consider this **dichotomous branching** a primitive characteristic. In contrast, when most plant stems branch, one stem is more vigorous and becomes the main trunk.

The upright stems of *Psilotum* are green and are the main organs of photosynthesis. Tiny, round sporangia, borne directly on the erect, aerial stems, contain sporogenous cells that undergo meiosis to form haploid spores. After dispersal, the spores germinate to form haploid prothalli. Because they grow underground, the prothalli of whisk ferns are difficult to study. They are nonphotosynthetic, owing to their subterranean location, and they apparently have a symbiotic relationship with mycorrhizal fungi that provides them with sugar and essential minerals (see Chapter 25).

Most species of whisk ferns are extinct, and the few surviving species live mainly in the tropics and subtropics. Although whisk ferns do not closely resemble ferns in appearance, they are considered fern allies because of similarities in their life cycles.

Botanists have carefully studied whisk ferns in recent years but disagree about how to interpret their structures. Some botanists consider whisk ferns to be surviving representatives of extinct vascular plants (see discussion of rhyniophytes later in the chapter). Other botanists hypothesize that whisk ferns are highly modified relatives of ferns. Recent molecular data, including comparisons of nucleotide sequences of ribosomal RNA, chloroplast DNA, and mitochondrial DNA in living species, support the hypothesis that the whisk ferns are more closely related to ferns than to other seedless vascular plants.

Horsetails have hollow, jointed stems

About 300 mya the **horsetails** (phylum Equisetophyta) were among the dominant plants and grew as large as modern trees (Fig 26-16). Because they contributed to Earth's vast coal deposits, these ancient horsetails are still significant today (see

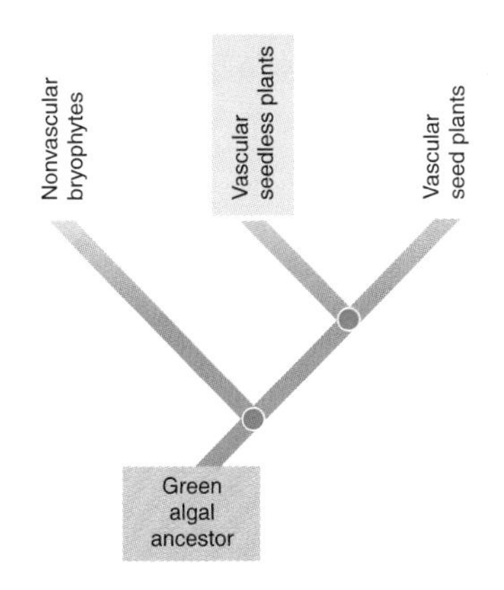

FIGURE **26-15** **Fern allies.**

(a) The sporophyte of *Psilotum nudum,* a whisk fern. The stem is the main organ of photosynthesis in this rootless, leafless, vascular plant. Sporangia, which are initially green but turn yellow as they mature, are borne on short lateral branches directly on the stems. **(b)** *Equisetum telematia,* a horsetail with a wide distribution in Eurasia, Africa, and North America, has unbranched reproductive shoots bearing conelike strobili and separate, highly branched vegetative (nonreproductive) shoots. In some horsetail species, both reproductive and vegetative shoots are unbranched. **(c)** The sporophyte of *Lycopodium,* a club moss, has small, scalelike leaves (microphylls) that are evergreen. Spores are produced in sporangia on reproductive leaves clustered in a conelike strobilus *(as shown)* or, in other species, scattered along the stem.

FIGURE **26-16**

Reconstruction of *Calamites*.

This ancient horsetail was as tall as many modern trees—to 20 m (about 65 ft). *Calamites* had an underground rhizome where roots and aerial shoots originated. *(Redrawn from L. Emberger,* Les Plantes Fossiles, *Masson et Cie, Paris, 1968)*

Focus On: Ancient Plants and Coal Formation on page 510). The few surviving horsetails, about 15 species in the genus *Equisetum,* grow mostly in wet, marshy habitats and are less than 1.3 m (4 ft) tall but extremely distinctive (see Fig. 26-15b). They are widely distributed on every continent except Australia.

Horsetails have true roots, stems (both rhizomes and erect aerial stems), and small leaves. The hollow, jointed stems are impregnated with silica, which gives them a gritty texture. Small leaves, interpreted as reduced megaphylls, are fused in whorls at each node (the area on the stem where leaves attach). The green stem is the main organ of photosynthesis. Horsetails are so named because certain vegetative (nonreproductive) stems have whorls of branches that give the appearance of a bushy horse's tail. In the past horsetails were called "scouring rushes" and were used to scrub pots and pans along stream banks.

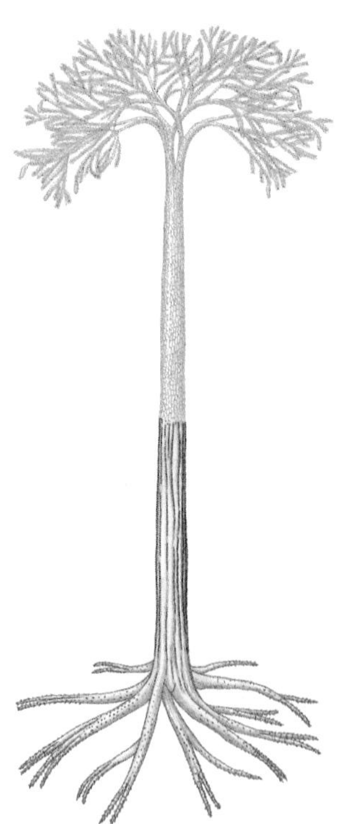

(a)

(b)

FIGURE **26-17** ***Lepidodendron.***

(a) Reconstruction of *Lepidodendron.* This ancient club moss was the size of a large tree—to 40 m (about 130 ft). Numerous fossils of *Lepidodendron* were preserved in coal deposits, particularly in Great Britain and the central United States. **(b)** Closeup of fossil *Lepidodendron* bark, showing scars where leaves were once attached. Each leaf scar is approximately 1.9 cm (0.75 in) wide. *(a, Redrawn from M. Hirmer,* Handbuch der Paläobotanik, *R. Olderbourg, Munich, 1927)*

Each reproductive branch of a horsetail bears a terminal conelike **strobilus** (pl., *strobili*). The strobilus consists of several stalked, umbrella-like structures, each of which bears 5 to 10 sporangia in a circle around a common axis.

The horsetail life cycle is similar in many respects to the fern life cycle. In horsetails, as in ferns, the sporophyte is the conspicuous plant, whereas the gametophyte is a minute, lobed thallus ranging in width from the size of a pinhead to about 1 cm (less than 0.5 in) across. The sporophyte and gametophyte are both photosynthetic and nutritionally independent at maturity. Like ferns, horsetails require water as a medium for flagellate sperm cells to swim to the egg.

Club mosses are small plants with rhizomes and short, erect branches

Like horsetails, **club mosses** (phylum Lycophyta) were important plants millions of years ago, when species that are now extinct often reached great size (Fig. 26-17). These large, treelike plants, like the ancient horsetails, were major contributors to our present-day coal deposits. The 1000 or so species of club mosses living today, such as *Lycopodium* (see Fig. 26-15c), are small (less than 25 cm [10 in] tall), attractive plants common in temperate woodlands. They possess true roots; both rhizomes and erect aerial stems; and small, scale-like leaves (microphylls). Sporangia are borne on reproductive leaves that are either clustered in conelike strobili at the tips of stems or scattered in reproductive areas along the stem. Club mosses are evergreen and often fashioned into Christmas wreaths and other decorations. In some areas overharvesting endangers them.

That common names are sometimes misleading in biology is vividly evident in this group of plants. The most common names for the phylum Lycophyta are "club mosses" and "ground pines," yet these plants are neither mosses nor pines and are most closely allied to the ferns.

Some ferns and club mosses are heterosporous

In the life cycles examined thus far, plants produce only one type of spore as a result of meiosis. This condition, known as **homospory,** is characteristic of bryophytes, horsetails, whisk ferns, and most ferns and club mosses. However, certain ferns and club mosses exhibit **heterospory,** in which they produce two different types of spores: microspores and megaspores. Figure 26-18 illustrates the generalized life cycle of a heterosporous plant.

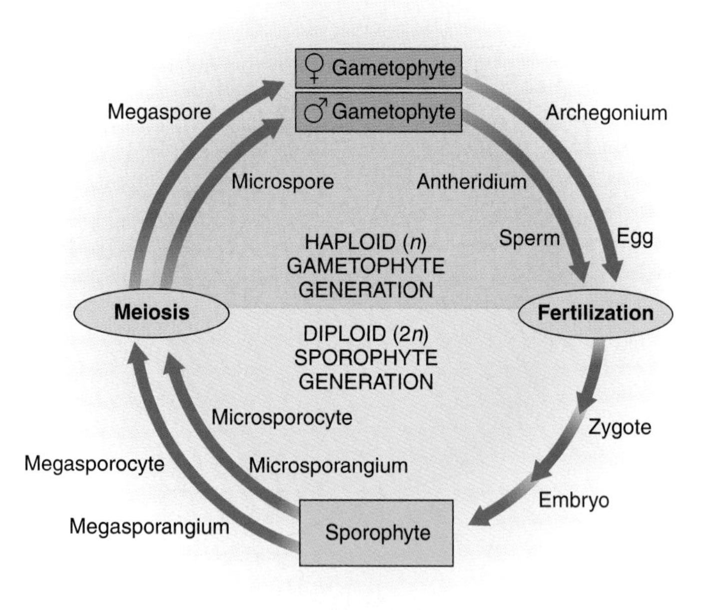

FIGURE **26-18** **The basic life cycle of heterosporous plants.**

Two types of spores, microspores and megaspores, are produced during the life cycle of heterosporous plants.

Spike moss (*Selaginella*), a small, delicate club moss, is an example of a heterosporous plant (Fig. 26-19). Beginning at ①, the sporophyte plant produces sporangia within a conelike strobilus. Each strobilus usually bears two kinds of sporangia: microsporangia and megasporangia. In ②, *microsporangia* are sporangia that produce *microsporocytes* (also called *microspore mother cells*), which undergo meiosis to form microscopic, haploid **microspores.** In ③, each microspore develops into a male gametophyte that produces sperm cells within antheridia. In ④, *megasporangia* in the *Selaginella* strobilus produce *megasporocytes* (also called *megaspore mother cells*). When megasporocytes undergo meiosis, they form haploid **megaspores,** each of which develops into a female gametophyte that produces eggs in archegonia, ⑤. In *Selaginella,* the development of male gametophytes from microspores and of female gametophytes from megaspores occurs within their respective spore walls, using stored food provided by the sporophyte. As a result, the male and female gametophytes are not truly free-living, unlike the gametophytes of other seedless vascular plants. In ⑥, fertilization is followed by the development of a new sporophyte.

Heterospory was a significant development in plant evolution because it was the forerunner of the evolution of seeds. Heterospory characterizes the two most successful groups of plants existing today, the gymnosperms and the flowering plants, both of which produce seeds (see Chapter 27).

Seedless vascular plants are used for experimental studies

Botanists use many seedless vascular plants as experimental models to study certain aspects of plant biology, such as physiology, growth, and development. Ferns and other seedless vascular plants are useful in studying how apical meristems give rise to plant tissues. As discussed in Chapter 31, an **apical meristem** is the area at the tip (apex) of a root or shoot where growth—cell division, elongation, and differentiation—occurs.

FIGURE **26-19** | **The life cycle of spike moss (*Selaginella*).**

Spike moss is heterosporous, producing two types of spores in one strobilus. The megaspores develop into female gametophytes, and the microspores become male gametophytes. See text for a detailed description.

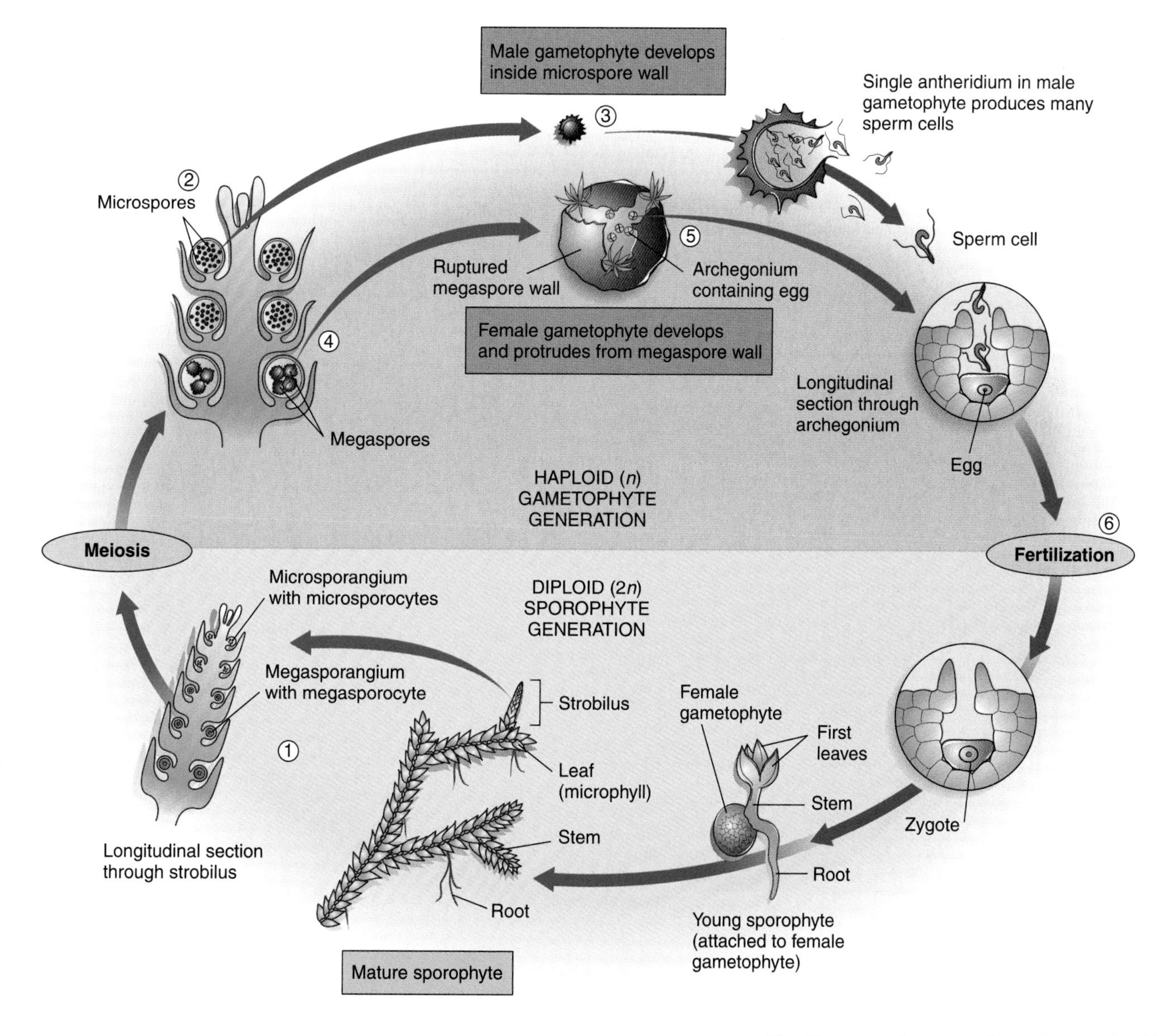

Ferns and other seedless vascular plants have a single large *apical cell* located at the center tip of the apical meristem. This apical cell is the source, by mitosis, of all the cells that eventually make up the root or shoot. The apical cell divides in an orderly fashion, and the smaller daughter cells produced by the apical cell in turn divide, giving rise to different parts of the root or shoot. It is possible to trace mature cells in the root or shoot back to their origin from the single apical cell.

Ferns are interesting research plants for studies in genetics because they are **polyploids** and have multiple sets of chromosomes. (Many ferns have hundreds of chromosomes.) However, gene expression in ferns is exactly what one would expect of a *diploid* plant. Apparently, genes in the extra sets of chromosomes are gradually silenced and therefore not expressed.

Seedless vascular plants arose more than 420 mya

Currently, the oldest known megafossils of early vascular plants are from mid-Silurian (420 mya) deposits in Europe. (Plant *megafossils* are fossilized roots, stems, leaves, and reproductive structures.) Megafossils of several kinds of small, seedless vascular plants were also discovered in Silurian deposits in Bolivia, Australia, and northwestern China. Microscopic spores of early vascular plants appear in the fossil record earlier than megafossils, suggesting that even older megafossils of simple vascular plants may be discovered.

Botanists assign the oldest known vascular plants to phylum Rhyniophyta which, according to the fossil record, arose some 420 mya and became extinct about 380 mya. The rhyniophytes are so named because many fossils of these extinct plants were found in fossil beds near the village of Rhynie, Scotland. *Rhynia gwynne-vaughanii* is an example of an early vascular plant that superficially resembled whisk ferns in that it consisted of leafless upright stems that branched dichotomously from an underground rhizome (Fig. 26-20). *Rhynia* lacked roots, although it had absorptive rhizoids. Sporangia formed at the ends of short branches. The internal structure of its rhizome contained a central core of xylem cells for conducting water and dissolved nutrient minerals.

PROCESS OF SCIENCE

For more than 60 years, botanists considered *Rhynia major*, a plant that grew about 50 cm (20 in) tall and probably lived in marshes, a classic example of a rhyniophyte. Fossils indicate this plant had rhizoids, dichotomously branching rhizomes and upright stems that terminated in sporangia. However, recent microscopic studies of fossil rhizomes indicate that the central core of tissue lacked the xylem cells characteristic of vascular plants. For that reason, *R. major* was reclassified into a new genus, *Aglaophyton*, and is no longer considered a rhyniophyte (Fig. 26-21). Science is an ongoing enterprise, and over time, existing knowledge is re-evaluated in light of newly discovered evidence. *Aglaophyton major* is an excellent example of

FIGURE **26-20** **Reconstruction of *Rhynia gwynne-vaughanii*.**
This leafless plant, one of Earth's earliest vascular plants, is now extinct. It grew about 18 cm (7 in) tall. *(Redrawn from D. Edwards, "Evidence for the Sporophytic Status of the Lower Devonian Plant* Rhynia gwynne-vaughanii," Rev. Palaeobot. Palynol., *Vol. 29, 1980)*

FIGURE **26-21** **Reconstruction of *Aglaophyton major*.**
Recent evidence indicates that this plant, although superficially similar to other early vascular plants, lacked conducting tissues that are characteristic of vascular plants. For that reason, it was reclassified into a new genus and is no longer considered a rhyniophyte. *(Redrawn from J.D. Mauseth,* Botany: An Introduction to Plant Biology, *2nd ed., Saunders College Publishing, Philadelphia, 1995)*

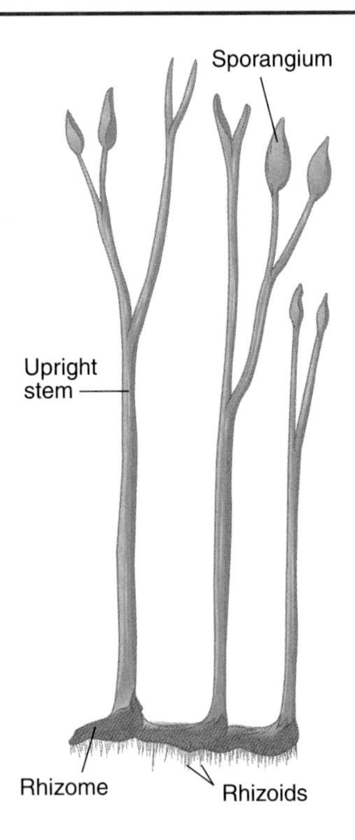

the self-correcting nature of science, which is in a perpetually dynamic state and changes in response to newly available techniques and data.

Review

- What adaptations do ferns have that both algae and bryophytes lack?
- How does one distinguish between megaphylls and microphylls?
- Which of the following are parts of the sporophyte generation in ferns: frond, sperm cells, egg cell, roots, sorus, sporangium, spores, prothallus, rhizome, antheridium, archegonium, and zygote?
- What three groups of plants are known as fern allies?
- How does heterospory modify the life cycle?

BiologyNow™ Assess your understanding of **seedless vascular plants** by taking the pretest on your BiologyNow CD-ROM.

SUMMARY WITH KEY TERMS

1 Discuss some environmental challenges of living on land, and describe how several adaptations meet these challenges.

- The colonization of land by plants required the evolution of many anatomical, physiological, and reproductive adaptations. Plants have a waxy **cuticle** to protect against water loss and **stomata** for gas exchange needed for photosynthesis.
- Plant life cycles have an **alternation of generations** in which they spend part of their life cycle in a haploid **gametophyte generation** and part in a diploid **sporophyte generation.** The gametophyte plant produces gametes by mitosis. During fertilization these gametes fuse to form a **zygote,** the first stage of the sporophyte generation. The zygote develops into a multicellular **embryo** that the gametophyte protects and nourishes. The mature sporophyte plant develops from the embryo and produces sporogenous cells (spore mother cells). These undergo meiosis to form **spores,** the first stage in the gametophyte generation.
- Most plants produce multicellular **gametangia** with a protective jacket of sterile cells surrounding the gametes. **Antheridia** are gametangia that produce sperm cells, and **archegonia** are gametangia that produce eggs.
- Mosses and ferns, although adapted to life on land, have motile sperm cells and require water as a transport medium for fertilization. Ferns and other vascular plants have **xylem** to conduct water and dissolved nutrient minerals, and have **phloem** to conduct dissolved sugar.

2 Name the protist group from which plants are hypothesized to have descended, and describe supporting evidence.

- Recent ultrastructural and molecular data indicate plants probably arose from a group of green algae called **charophytes.** Plants and green algae have similar biochemical characteristics: the same photosynthetic pigments, cell wall components, and carbohydrate storage material. Plants and green algae share similarities in certain fundamental processes such as cell division.

3 Summarize the features that distinguish bryophytes from green algae and from other plants.

- **Mosses** and other **bryophytes** have several adaptations that green algae lack, including a cuticle, stomata, and multicellular gametangia. Unlike other land plants, bryophytes are nonvascular and lack xylem and phloem.
- Bryophytes are the only plants with a dominant gametophyte generation. Their sporophytes remain permanently attached and nutritionally dependent on the gametophyte.

4 Name and briefly describe the three phyla of bryophytes.

- Mosses (phylum Bryophyta) have gametophytes that are green plants that grow from a filamentous **protonema.**
- Many **liverworts** (phylum Hepaticophyta) have gametophytes that are flattened, lobelike **thalli;** others are leafy.
- **Hornworts** (phylum Anthocerotophyta) have thalloid gametophytes.

5 Describe the life cycle of mosses, and compare their gametophyte and sporophyte generations.

- The green moss gametophyte bears archegonia and/or antheridia at the top of the plant. During fertilization, a sperm cell fuses with an egg cell in the archegonium. The zygote grows into an embryo that develops into a moss sporophyte, which is attached to the gametophyte. Meiosis occurs within the capsule of the sporophyte to produce spores. When a spore germinates, it grows into a protonema that forms buds.

6 Discuss the features that distinguish ferns and other seedless vascular plants from algae and bryophytes.

- **Ferns** and fern allies have several adaptations that algae and bryophytes lack, including vascular tissues and a dominant sporophyte generation. As in bryophytes, reproduction in ferns depends on water as a transport medium for their motile sperm cells.

7 Name and briefly describe the four phyla of seedless vascular plants.

- Ferns (phylum Polypodiophyta) are the largest and most diverse group of seedless vascular plants.
- Sporophytes of **whisk ferns** (phylum Psilotophyta) consist of **dichotomously branching** rhizomes and erect stems; they lack true roots and leaves.
- **Horsetail** (phylum Equisetophyta) sporophytes have roots, rhizomes, aerial stems that are hollow and jointed, and leaves that are reduced **megaphylls.**
- Sporophytes of **club mosses** (phylum Lycophyta) consist of roots, rhizomes, erect branches, and leaves that are **microphylls.**

8 Describe the life cycle of ferns, and compare their sporophyte and gametophyte generations.

- Fern sporophytes have roots, rhizomes, and leaves that are megaphylls. Their leaves, or fronds, bear sporangia in clusters called **sori.** Meiosis in sporangia produces haploid spores. The fern gametophyte, called a **prothallus,** develops from a haploid spore and bears both archegonia and antheridia.

9 Compare the generalized life cycles of homosporous and heterosporous plants.

- **Homospory,** the production of one kind of spore, is characteristic of bryophytes, whisk ferns, horsetails, most club mosses, and most ferns. In homospory, spores give rise to gametophyte plants that produce both egg cells and sperm cells.
- **Heterospory,** the production of two kinds of spores (microspores and megaspores), occurs in certain club mosses, certain ferns, and all seed plants. The evolution of heterospory was an essential step in the evolution of seeds. **Microspores** give rise to male gametophytes that produce sperm cells. **Megaspores** give rise to female gametophytes that produce eggs.

POST-TEST

1. The bryophytes (a) include mosses, liverworts, and hornworts (b) include whisk ferns, horsetails, and club mosses (c) are small plants that lack a vascular system (d) both a and c (e) both b and c
2. The waxy layer that covers aerial parts of plants is the (a) cuticle (b) archegonium (c) protonema (d) stoma (e) thallus
3. A strengthening compound found in cell walls of vascular plants is (a) xanthophyll (b) lignin (c) cutin (d) cellulose (e) carotenoid
4. Stomata (a) help prevent desiccation of plant tissues (b) transport water and minerals through plant tissues (c) allow gas exchange for photosynthesis (d) strengthen cell walls (e) produce male gametes
5. The female gametangium, or ______________, produces an egg; the male gametangium, or ______________, produces sperm cells. (a) antheridium; archegonium (b) archegonium; megaphyll (c) megasporangium; antheridium (d) archegonium; antheridium (e) megasporangium; megaphyll
6. Liverworts and hornworts share life cycle similarities with (a) ferns (b) mosses (c) horsetails (d) club mosses (e) whisk ferns
7. The green, gametangia-bearing moss plant (a) is the gametophyte generation (b) is the sporophyte generation (c) is called a protonema (d) contains cells with single large chloroplasts (e) both b and c
8. Seedless vascular plants have ______________ to conduct water and dissolved minerals and ______________ to conduct dissolved sugar. (a) cuticle; xylem (b) phloem; stoma (c) phloem; xylem (d) stoma; cuticle (e) xylem; phloem
9. Whisk ferns, horsetails, and club mosses share life cycle similarities with (a) ferns (b) mosses (c) hornworts (d) liverworts (e) b, c, and d
10. A(an) ______________ is a leaf that arose from a branch system. (a) antheridium (b) microphyll (c) megaphyll (d) sorus (e) microspore
11. These plants have vascularized stems but lack true roots and leaves. (a) mosses (b) club mosses (c) horsetails (d) whisk ferns (e) hornworts
12. These plants have hollow, jointed stems that are impregnated with silica. (a) mosses (b) club mosses (c) horsetails (d) whisk ferns (e) hornworts
13. Spike moss (*Selaginella*) is a (a) homosporous fern (b) homosporous horsetail (c) heterosporous fern (d) heterosporous horsetail (e) heterosporous club moss
14. Which of the following statements about ferns is *not* true? (a) Ferns have motile sperm cells that swim through water to the egg-containing archegonium. (b) Ferns are vascular plants. (c) Ferns are the most economically important group of bryophytes. (d) The fern sporophyte consists of a rhizome, roots, and fronds. (e) The diversity of ferns is greatest in the tropics.
15. Plants probably descended from a group of green algae called (a) rhyniophytes (b) *Calamites* (c) epiphytes (d) charophytes (e) club mosses
16. Which of the following is *not* a characteristic of plants? (a) cuticle (b) unicellular gametangia (c) stomata (d) multicellular embryo (e) alternation of generations
17. In plant life cycles (a) the first products of meiosis are gametes (b) spores are part of the diploid sporophyte generation (c) the embryo gives rise to a zygote (d) the first stage in the diploid sporophyte generation is the zygote (e) the first stage in the haploid gametophyte generation is the prothallus
18. Ferns are (a) seedless, vascular plants (b) vascular plants with seeds (c) seedless, nonvascular plants (d) nonvascular plants with seeds (e) seedless plants with a dominant gametophyte generation
19. Microphylls (a) probably evolved from stem branches that gradually filled in with additional tissues (b) are usually small and have a single vascular strand (c) are characteristics of club mosses (d) both a and c (e) both b and c

CRITICAL THINKING

1. Which group probably colonized the land first, plants or animals? Explain.
2. How may the following trends in plant evolution be adaptive to living on land?
   a. Dependence on water for fertilization ⟶ no need for water as a transport medium
   b. Dominant gametophyte generation ⟶ dominant sporophyte generation
   c. Homospory ⟶ heterospory

- Visit our Web site at **biology.brookscole.com/solomon7** for links to chapter-related resources on the World Wide Web. Additional online materials relating to this chapter can also be found on our Web site.

BIOLOGY NOW RESOURCES

Active Figures

26-4: Plant evolution

26-8: Life cycle of the liverwort

Preparing for an exam? Take a diagnostic test on your BiologyNow CD-ROM.

Post-Test Answers

1. d	2. a	3. b	4. c
5. d	6. b	7. a	8. e
9. a	10. c	11. d	12. c
13. e	14. c	15. d	16. b
17. d	18. a	19. e	

The Plant Kingdom: Seed Plants

Holt Studies Intl./Silvestra Silva/Photo Researchers, Inc.

Guava fruit. Guava (*Psidium guajava*) is a flowering plant, and its seeds are enclosed within a fruit.

CHAPTER OUTLINE

Chapter 26 focused on plants that reproduce by means of *spores,* haploid reproductive units that give rise to gametophytes. Although the most successful and widespread plants also produce spores, their primary means of reproduction and dispersal is by **seeds,** each of which consists of an embryonic sporophyte, nutritive tissue, and a protective coat. Seeds develop from the fertilized egg cell, the female gametophyte, and its associated tissues. The two groups of seed plants, gymnosperms and angiosperms (flowering plants), exhibit the greatest evolutionary complexity in the plant kingdom and are the dominant plants in most terrestrial environments.

Seeds, such as the guava seeds shown in the photograph, are reproductively superior to spores for three main reasons. First, development is further advanced in seeds than in spores. A seed contains a multicellular young plant with embryonic root, stem, and one or more leaves already formed, whereas a spore is a single cell. Second, a seed contains an abundant food supply. After germination, food stored in the seed nourishes the plant embryo until it becomes self-sufficient. Because a spore is a single cell, few food reserves exist for the plant that develops from a spore. Third, a seed is protected by a multicellular **seed coat** that is very thick and hard in some plants, as for example, in lima beans. Like spores, seeds live for extended periods at reduced rates of metabolism, germinating when conditions become favorable.

Seeds and seed plants are intimately connected with the development of human civilization. From prehistoric times, early humans collected and used seeds for food. The food stored in seeds is a concentrated source of proteins, oils, carbohydrates, and vitamins, which are nourishing for humans as well as for germinating plants. Also, seeds are easy to store (if kept dry); this has allowed humans to collect them during times of plenty and save them for times of need. Few other foods are stored as conveniently or for as long. Although flowering plants produce most seeds that humans consume, the seeds of certain gymnosperms—the piñon pine, for example—are edible. They are usually sold as "pine nuts." ■

SEED PLANTS

Learning Objective

1 Compare the features of seeds with those of spores, and discuss the advantages of plants that reproduce primarily by seeds rather than by spores.

Following fertilization in seed plants, an **ovule,** which is a *megasporangium* and its enclosed structures, develops into a seed. Seed plants also have **integuments,** layers of sporophyte tissue that surround and enclose the megasporangium. After fertilization takes place, the seed coat develops from the integuments.

Botanists divide seed plants into two groups based on whether or not an ovary wall surrounds their ovules (an *ovary* is a structure that contains one or more ovules). The two groups of seed plants are the **gymnosperms** and the **angiosperms** (Table 27-1). The word *gymnosperm* is adapted from the Greek for "naked seed." Gymnosperms produce seeds that are totally exposed or borne on the scales of cones. In other words, an ovary wall does not surround the ovules of gymnosperms. Pine, spruce, fir, hemlock, and *Ginkgo* are examples of gymnosperms.

The Greek expression from which the term *angiosperm* is derived translates as "seed enclosed in a vessel or case." Angiosperms are flowering plants that produce their seeds within a fruit (a mature ovary). Thus the ovules of angiosperms are protected. Flowering plants, which are extremely diverse, include corn, oaks, water lilies, cacti, apples, grasses, palms, and buttercups.

Both gymnosperms and flowering plants have vascular tissues: **xylem** for conducting water and dissolved nutrient minerals, and **phloem** for conducting dissolved sugar. Both have life cycles with an **alternation of generations,** that is, they spend part of their lives in the diploid sporophyte stage and part in the haploid gametophyte stage. The sporophyte generation is the dominant stage in each group, and the gametophyte generation is significantly reduced in size and entirely dependent on the sporophyte generation. Unlike the plants we have considered so far (bryophytes, ferns, and most of their allies; see Chapter 26), gymnosperms and flowering plants do not have free-living gametophytes. Instead, the female gametophyte is attached to and nutritionally dependent on the sporophyte generation. All gymnosperms and flowering plants are *heterosporous* and produce two types of spores: microspores and megaspores.

Review

- Why are seeds such a significant evolutionary innovation?
- What is an ovule?

BiologyNow™ Assess your understanding of **seed plants** by taking the pretest on your BiologyNow CD-ROM.

TABLE 27-1 A Comparison of Gymnosperms and Angiosperms

Characteristic	Gymnosperms	Angiosperms
Growth habit	Woody trees and shrubs	Woody or herbaceous
Conducting cells in xylem	Tracheids	Vessel elements and tracheids
Reproductive structures	Cones (usually)	Flowers
Pollen grain transfer	Wind (usually)	Animals or wind
Fertilization	Egg and sperm ⟶ zygote; double fertilization fertilization in gnetophytes	Double fertilization: egg and sperm ⟶ zygote; Two polar nuclei and sperm ⟶ endosperm
Seeds	Exposed or borne on scales of cones	Enclosed within fruit
Number of species	About 760	More than 235,000
Geographical distribution	Worldwide	Worldwide

GYMNOSPERMS

Learning Objectives

2 Trace the steps in the life cycle of a pine, and compare its sporophyte and gametophyte generations.

3 Summarize the features that distinguish gymnosperms from bryophytes and ferns.

4 Name and briefly describe the four phyla of gymnosperms.

The gymnosperms include some of the most interesting members of the plant kingdom, including a number of record holders. For example, a giant sequoia (*Sequoiadendron giganteum*) known as the General Sherman Tree, in Sequoia National Park in California, is one of the world's most massive organisms. It is 82 m (267 ft) tall and has a girth of 23.7 m (77 ft) measured 1.5 m (5 ft) above ground level. Another gymnosperm, a coast redwood (*Sequoia sempervirens*) known as the Mendocino tree, is among the world's tallest trees, measuring 112 m (364 ft) in the year 2000. Botanists using tree ring analysis determined that one of the oldest living trees, a bristlecone pine (*Pinus aristata*) in the White Mountains of California, is 4900 years old!

Gymnosperms are usually classified into four phyla, which represent four different evolutionary lines (Fig. 27-1). Numbering 550 species, the largest phylum of gymnosperms is the phylum Pinophyta, commonly called *conifers.* Two phyla of gymnosperms, the ginkgo (phylum Ginkgophyta) and the cycads (phylum Cycadophyta), represent evolutionary remnants of groups that were more significant in the past. The fourth phylum of gymnosperms, the gnetophytes (phylum Gnetophyta), is a collection of some unusual plants that share certain traits not found in other gymnosperms.

Conifers are woody plants that produce seeds in cones

The **conifers** (phylum Pinophyta), which include pines, spruces, hemlocks, and firs, are the most familiar group of gymnosperms (Fig. 27-2a). These woody trees or shrubs produce annual additions of secondary tissues (wood and bark; see Chapter 33); there are no herbaceous (nonwoody) conifers. The wood *(sec-*

KEY CONCEPT: The arrangement of the phyla shown here may change as future analyses help clarify relationships.

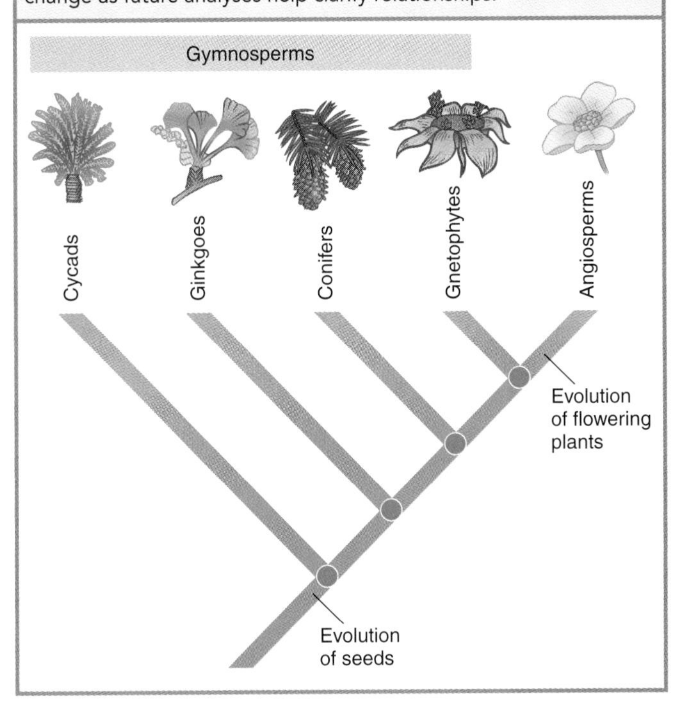

FIGURE **27-1** **Gymnosperm and angiosperm evolution.**
This cladogram shows one hypothesis of phylogenetic relationships among living seed plants. It is based on structural evidence and molecular comparisons. The exact relationships of certain seed plant phyla remain unclear. Some DNA analyses suggest gnetophytes are more closely related to other gymnosperms than to the angiosperms, unlike the depiction shown here.

ondary xylem) consists of **tracheids,** which are long, tapering cells with pits through which water and dissolved nutrient minerals move from one cell to another.

Many conifers produce **resin,** a viscous, clear or translucent substance consisting of several organic compounds that may protect the plant from attack by fungi or insects. The resin collects in resin ducts, tubelike cavities that extend throughout the roots, stems, and leaves. Resin is produced and secreted by cells lining the resin ducts.

Most conifers have leaves called **needles** that are commonly long and narrow, tough, and leathery (Fig. 27-2b). Pines bear clusters of two to five needles, depending on the species. In a few conifers such as American arborvitae, the leaves are scalelike and cover the stem (Fig. 27-2c). Most conifers are evergreen and bear their leaves throughout the year. Only a few, such as the dawn redwood, larch, and bald cypress, are deciduous and shed their needles at the end of each growing season.

Most conifers are **monoecious:** They have separate male and female reproductive parts in different locations on the same plant. These reproductive parts are generally borne in *strobili* (commonly called *cones*), hence the name *conifer,* which means "cone-bearing."

Conifers occupy extensive areas, ranging from the Arctic to the tropics, and are the dominant vegetation in the forested regions of Alaska, Canada, northern Europe, and Siberia. In addition, they are important in the Southern Hemisphere, particularly in wet, mountainous areas of temperate and tropical regions in South America, Australia, New Zealand, and Malaysia. Southwestern China, with more than 60 species of conifers, has the greatest regional diversity of conifer species on Earth. California, New Caledonia (an island east of Australia), southeastern China, and Japan also have considerable diversity of conifer species.

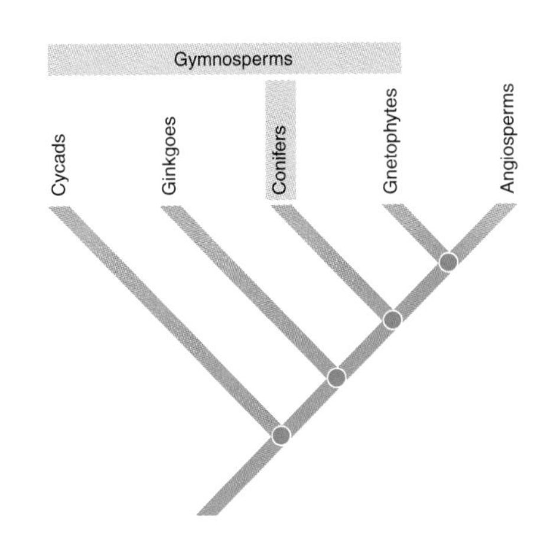

(a)

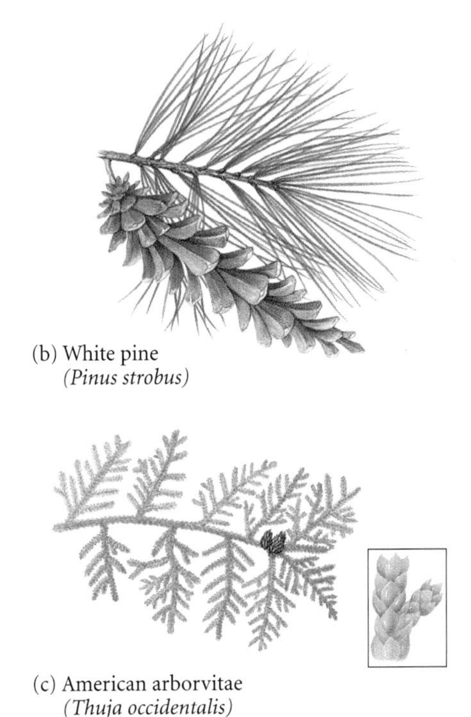

(b) White pine *(Pinus strobus)*

(c) American arborvitae *(Thuja occidentalis)*

FIGURE **27-2** **Conifers.**
(a) The sacred fir (*Abies religiosa*) is native to Mexico, where it grows to 30.5 m (100 ft) or more. **(b, c)** Leaf variation in conifers. **(b)** Needles of white pine (*Pinus strobus*). **(c)** Small, scalelike leaves of American arborvitae (*Thuja occidentalis*).

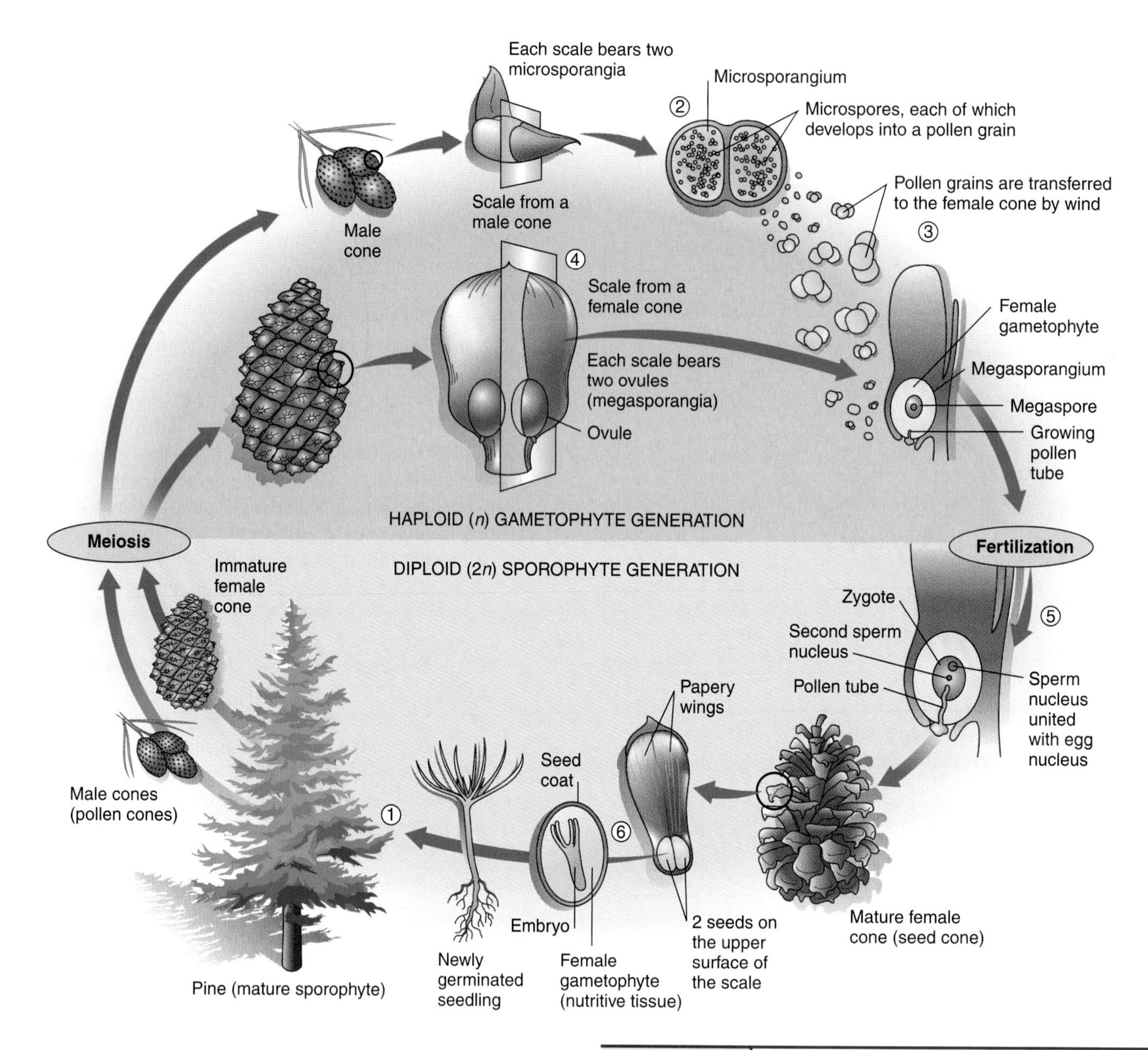

FIGURE **27-3** | **Life cycle of pine.**

One major advantage of gymnosperms over the seedless vascular plants is the production of wind-borne pollen grains. See text for a detailed description.

Ecologically, conifers contribute food and shelter to animals and other organisms, and their roots hold the soil in place and help prevent soil erosion. Humans use conifers for their wood (for building materials as well as paper products), medicine (such as the anticancer drug taxol from the Pacific yew), turpentine, and resins. Because of their attractive appearance, conifers such as firs, spruces, pines, and cedars are grown for landscape design and decorative holiday trees and wreaths.

Pines represent a typical conifer life cycle

The genus *Pinus,* by far the largest genus in the conifers, consists of about 100 species. As shown in ① of Fig. 27-3, a pine tree is a mature sporophyte. Pine is heterosporous and therefore produces microspores and megaspores in separate cones.[1] Male cones, usually 1 cm or less in length, are smaller than female cones and are generally produced on the lower branches each spring (Fig. 27-4). The more familiar, woody female cones, which are on the tree year-round, are usually found on the upper branches of the tree and bear seeds after reproduction. Female cones vary considerably in size. The sugar pine (*Pinus lambertiana*) of California produces the world's longest female cones, which reach lengths of 60 cm (2 ft).

Each male cone, also called a *pollen cone,* consists of **sporophylls,** leaflike structures that bear sporangia on the underside. At the base of each sporophyll are two **microsporangia,** which contain numerous **microsporocytes,** also called *microspore mother cells.* In ② of Fig. 27-3, each microsporocyte undergoes meiosis to form four haploid microspores. **Microspores** then develop into extremely reduced male gametophytes. Each

[1] It may be helpful to review alternation of generations in Chapter 26, including Figure 26-18, which depicts a heterosporous life cycle, before studying the pine life cycle.

Walt Anderson/Visuals Unlimited

FIGURE **27-4** **Male and female cones in lodgepole pine (*Pinus contorta*).**

(Top) Mature female cones have opened to shed their seeds. *(Bottom)* Clusters of male cones produce copious amounts of pollen grains in the spring.

immature male gametophyte, also called a **pollen grain,** consists of four cells, two of which—a *generative cell* and a *tube cell*—are involved in reproduction. The other two cells soon degenerate. Two large air sacs on each pollen grain provide buoyancy for wind dissemination. In ③, male cones shed pollen grains in great numbers, and wind currents carry some to the immature female cones.

Many botanists think the female cones (also called *seed cones*) are modified branch systems. As shown in ④, each cone scale bears two ovules, or **megasporangia,** on its upper surface. Within each megasporangium, meiosis of a **megasporocyte,** or *megaspore mother cell,* produces four haploid **megaspores.** One of these divides mitotically, developing into the female gametophyte, which produces an egg within each of several archegonia. The other three megaspores are nonfunctional and soon degenerate.

When the ovule is ready to receive pollen, it produces a sticky droplet at the opening where the pollen grains land. **Pollination,** the transfer of pollen to the female cones, occurs in the spring during a period of a week or 10 days, after which the pollen cones wither and drop off the tree. One of the many pollen grains that adhere to the sticky female cone grows a **pollen tube,** an outgrowth that digests its way through the megasporangium to the egg within the archegonium. The germinated pollen grain with its pollen tube is the mature male gametophyte. The tube cell, which is involved in the growth of the pollen tube, and the generative cell enter the pollen tube.

The generative cell divides and forms a *stalk cell* and a *body cell;* the body cell later divides and forms two nonmotile sperm cells. When the pollen tube reaches the female gametophyte, it discharges the two sperm cells near the egg. In ⑤, one of these sperm cells fuses with the egg, in the process of **fertilization,** to form a zygote, or fertilized egg, which subsequently grows into a young pine embryo in the seed. The other sperm cell degenerates.

In ⑥ the developing embryo, which consists of an embryonic root and an embryonic shoot with several cotyledons (embryonic leaves), is embedded in haploid female gametophyte tissue that becomes the nutritive tissue in the mature pine seed. A tough, protective seed coat derived from the integuments surrounds the embryo and nutritive tissue. The seed coat forms a thin, papery wing at one end that enables dispersal by air currents.

A long time elapses between the appearance of female cones on a tree and the maturation of seeds in those cones. When pollination occurs during the first spring, the female cone is still immature, and meiosis of the megasporocytes (megaspore mother cells) has not yet occurred. After the megaspore has formed, it takes more than a year for eggs to form within archegonia. Meanwhile, the pollen tube grows slowly through the megasporangium to the archegonia. Fertilization occurs about 15 months after pollination, and the embryo begins to develop. Seed maturation takes several additional months, although some seeds remain within the female cones for several years before being shed.

In the pine life cycle, the sporophyte generation is dominant, and the gametophyte generation is restricted in size to microscopic structures in the cones. Although the female gametophyte produces archegonia, the male gametophyte is so reduced that it does not produce antheridia. The gametophyte generation in pines, as in all seed plants, depends totally on the sporophyte generation for nourishment.

A major adaptation in the pine life cycle is elimination of the need for external water as a sperm transport medium. Instead, air currents carry pine pollen grains to female cones and nonmotile sperm cells accomplish fertilization by moving through a pollen tube to the egg. Pine and other conifers are plants whose reproduction is totally adapted for life on land.

Cycads have seed cones and compound leaves

Cycads (phylum Cycadophyta) were very important during the Triassic period, which began approximately 248 million years ago (mya) and is sometimes referred to as the "Age of Cycads." Most species are now extinct, and the few surviving cycads, about 140 species, are tropical and subtropical plants with stout, trunklike stems and compound leaves that resemble those of palms or tree ferns (Fig. 27-5). Many cycads are endangered, primarily because they are popular as ornamentals and are gathered from the wild and sold to collectors.

Cycad reproduction is similar to that in pines except that cycads are **dioecious** and therefore have seed cones on female plants and pollen cones on male plants. Their seed structure is most like that of the earliest seeds found in the fossil record. Cycads have also retained motile sperm cells, each of which has many hairlike flagella. Motile sperm cells are a vestige retained from the ancestors of cycads, in which sperm cells swam from antheridia to archegonia. In cycads, air or insects carry pollen grains to the female plants and their cones; there the pollen

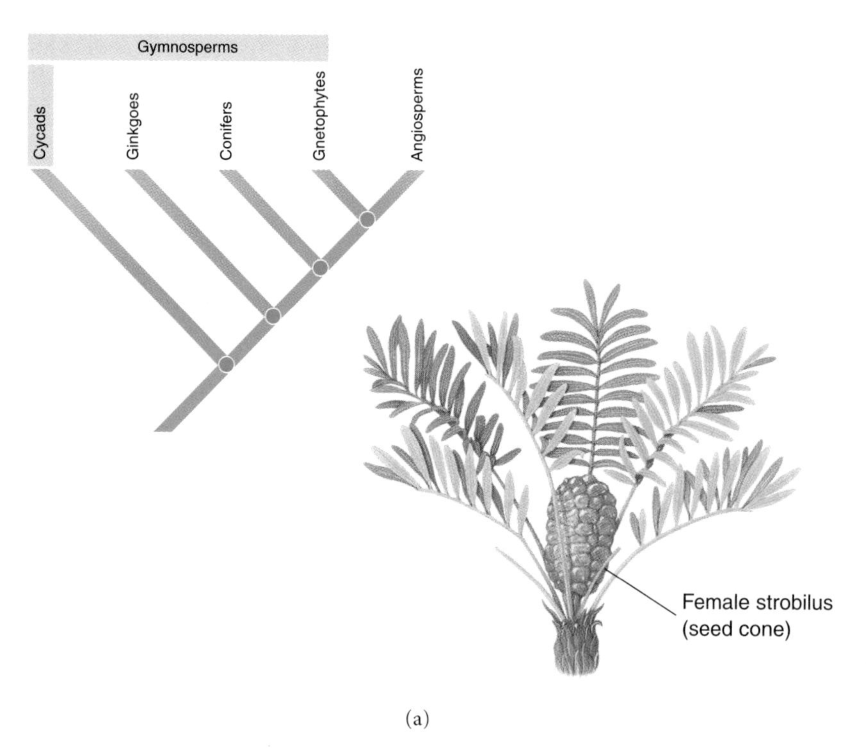

FIGURE **27-5** **Cycads.**

(a) A female coontie (*Zamia integrifolia*) produces seed cones. This plant is the only cycad native to the United States. Like *Zamia,* most cycads are short plants, less than 2 m tall. **(b)** This female cycad (*Encephalartos transvenosus*) in South Africa has a trunk that reaches a height of about 9 m (30 ft) and resembles a palm. Note its immense seed cones, to 0.8 m (30 in) in length.

grain germinates and grows a pollen tube. The sperm cells are released at the top of this tube and swim to get to the egg.

Ginkgo biloba is the only living species in its phylum

Ginkgo (phylum Ginkgophyta) is represented by a single living species, the maidenhair tree, *Ginkgo biloba* (Fig. 27-6). This native of eastern China grew in the wild in only two locations, although people had cultivated it for its edible seeds in China and Japan for centuries. *Ginkgo* is the oldest genus (and species) of living trees. Fossil ginkgoes 200 million years old have been discovered that are similar to the modern ginkgo.

People often plant ginkgo in North America and Europe today, particularly in parks and along city streets, because it is hardy and somewhat resistant to air pollution. Its leaves are deciduous and turn a beautiful yellow before being shed in the fall.

Like cycads, ginkgo is dioecious, with separate male and female trees. It has flagellate sperm cells—an evolutionary vestige that is not required, because ginkgo produces airborne pollen grains. Ginkgo seeds are completely exposed rather than occur-

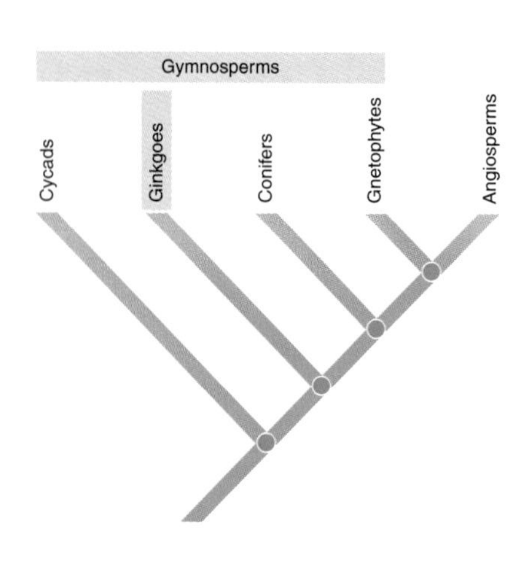

FIGURE **27-6**

Ginkgo, or maidenhair tree (*Gingko biloba*).

Closeup of a branch from a female ginkgo, showing the exposed seeds and the distinctive, fan-shaped leaves.

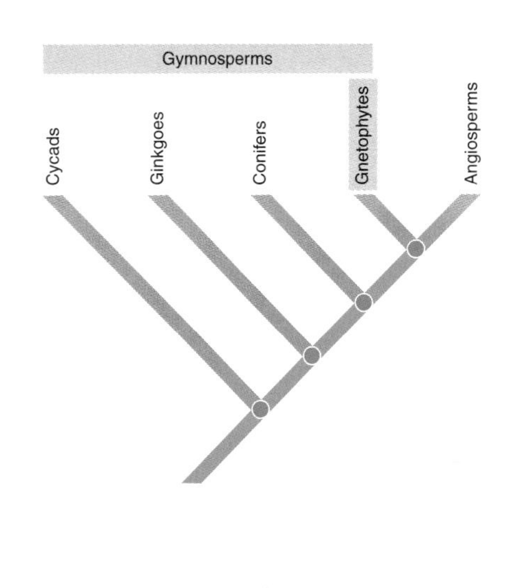

(a)

(b)

(c)

FIGURE **27-7** **Gnetophytes.**

(a) The leaves of *Gnetum gnemon* resemble those of flowering plants. Note the exposed seeds. **(b)** A male joint fir (*Ephedra*) has pollen cones clustered at the nodes. In the 19th century, European pioneers used species native to the American Southwest to make a beverage, Mormon tea. **(c)** *Welwitschia mirabilis* is native to deserts in southwestern Africa. It survives on moisture-laden fogs that drift inland from the ocean. Photographed in the Namib Desert, Namibia.

ring within cones. Male trees are typically planted, because the female trees bear seeds whose fleshy outer coverings give off a foul odor that smells like rancid butter. In China and Japan, where people eat the seeds, the female trees are more common.

Ginkgo has been an important medicinal plant for centuries. Extracts from the leaves may enhance neurological functioning by increasing blood flow to the brain. Several studies are underway to determine if ginkgo improves memory in elderly people.

Gnetophytes include three unusual genera

The **gnetophytes** (phylum Gnetophyta) consist of about 70 species in three diverse genera (*Gnetum, Ephedra,* and *Welwitschia*). Gnetophytes share certain features that make them unique among the gymnosperms. For example, gnetophytes have more efficient water-conducting cells, called *vessel elements,* in their xylem (see Chapter 31). Flowering plants also have vessel elements in their xylem, but no gymnosperms do except the gnetophytes. Also, the cone clusters that some gnetophytes produce resemble flower clusters, and certain details in their life cycles resemble those of flowering plants.

The genus *Gnetum* contains tropical vines, shrubs, and trees with broad leaves that resemble those of flowering plants (Fig. 27-7a). Species in the genus *Ephedra* include many shrubs and vines that grow in deserts and other dry temperate and tropical regions. Some *Ephedra* species resemble horsetails in that they have jointed green stems with tiny leaves (Fig. 27-7b). Commonly called *joint fir, Ephedra* has been used medicinally for centuries. An Asiatic *Ephedra* is the source of ephedrine, which stimulates the heart and raises blood pressure. Ephedrine is sold over the counter in weight control medications and herbal energy-boosters; several deaths have been reported from chronic use or overdose of products containing ephedrine.

The third gnetophyte genus, *Welwitschia,* contains a single species found in deserts of southwestern Africa (Fig. 27-7c). Most of *Welwitschia's* body—a long taproot—grows underground. Its short, wide stem forms a shallow disk, up to 0.9 m (3 ft) in diameter, from which two ribbon-like leaves extend. These two leaves continue to grow from the stem throughout the plant's life, but their ends are usually broken and torn by the wind, giving the appearance of numerous leaves. Each leaf grows to about 2 m (6.5 ft) in length. When *Welwitschia* reproduces, it forms cones around the edge of its disclike stem.

Review

- What is the dominant generation in the pine life cycle? How does pollination occur in gymnosperms?

- What features distinguish gymnosperms from other plants?
- What are the four groups of gymnosperms?
- What features distinguish cycads from ginkgo? From gnetophytes?

BiologyNow™ Assess your understanding of **gymnosperms** by taking the pretest on your BiologyNow CD-ROM.

FLOWERING PLANTS

Learning Objectives

5 Summarize the features that distinguish flowering plants from other plants.

6 Briefly explain the life cycle of a flowering plant, and describe double fertilization.

7 Contrast dicots and monocots, the two classes of flowering plants.

8 Discuss the evolutionary adaptations of flowering plants.

Flowering plants or **angiosperms** (phylum Magnoliophyta) are the most successful plants today, surpassing even the gymnosperms in importance. They have adapted to almost every habitat and, with at least 235,000 species, are Earth's dominant plants. Flowering plants come in a wide variety of sizes and forms, from herbaceous violets to massive eucalyptus trees. Some flowering plants—tulips and roses, for example—have large, conspicuous flowers, whereas others, such as grasses and oaks, produce small and inconspicuous flowers.

Flowering plants are vascular plants that reproduce sexually by forming flowers, and, after a unique double fertilization process that is discussed later, seeds within fruits. The fruit protects the developing seeds and often aids in their dispersal (see Chapter 35). Flowering plants have efficient water-conducting cells called **vessel elements** in their xylem, and efficient sugar-conducting cells called **sieve tube elements** in their phloem (see Chapter 31).

Flowering plants are extremely important to humans because our survival as a species literally depends on them. All our major food crops are flowering plants, including cereal crops such as rice, wheat, corn, and barley. Woody flowering plants such as oak, cherry, and walnut provide valuable lumber. Flowering plants give us fibers such as cotton and linen and medicines such as digitalis and codeine. Products as diverse as rubber, tobacco, coffee, chocolate, and aromatic oils for perfumes come from flowering plants. Economic botany is the subdiscipline of botany that deals with plants of economic importance.

Monocots and dicots are the two classes of flowering plants

Traditionally, phylum Magnoliophyta is divided into two classes: the monocots (class Liliopsida) and the dicots (class Magnoliopsida) (Fig. 27-8). Monocots include palms, grasses, orchids, irises, onions, and lilies. The dicot class includes oaks, roses, mustards, cacti, blueberries, and sunflowers. Dicots are more diverse and include many more species (at least 170,000) than the monocots (at least 65,000). Table 27-2 provides a comparison of some of the general features of the two classes.

FIGURE **27-8** **Flowering plants.**

(a) *Trillium erectum*, like most monocots, has floral parts in threes. Note the three green sepals, three red petals, six stamens, and three stigmas (the compound pistil consists of three fused carpels). **(b)** Most dicots such as this *Tacitus* have their floral parts in fours or fives. Note the five petals, 10 stamens, and five separate pistils. Five sepals are present but barely visible against the background.

John Gerlach/Tom Stack & Associates

(a)

Richard H. Gross

(b)

TABLE 27-2 Distinguishing Features of Dicots and Monocots		
Feature	**Dicots**	**Monocots**
Flower parts	Usually in fours or fives	Usually in threes
Pollen grains	Three furrows or pores	One furrow or pore
Leaf venation	Usually netted	Usually parallel
Vascular bundles in stem cross section	Arranged in a circle (ring)	Usually scattered or more complex arrangement
Roots	Taproot system	Fibrous root system
Seeds	Embryo with two cotyledons	Embryo with one cotyledon
Secondary growth (wood and bark)	Often present	Absent

Monocots are mostly herbaceous plants with long, narrow leaves that have parallel veins (the main leaf veins run parallel to one another). The parts of monocot flowers usually occur in threes or multiples of 3. For example, a flower may have three sepals, three petals, six stamens, and a compound pistil consisting of three fused carpels (these flower parts are discussed shortly). Monocot seeds have a single **cotyledon,** or embryonic seed leaf, and **endosperm,** a nutritive tissue, is usually present in the mature seed.

Dicots are either herbaceous (such as a tomato plant) or woody (such as a hickory tree). Their leaves vary in shape but usually are broader than monocot leaves, with netted veins (branched veins resembling a net). Flower parts usually occur in fours or fives or multiples thereof. Two cotyledons are present in dicot seeds, and endosperm is usually absent in the mature seed, having been absorbed by the two cotyledons.

Flowers are involved in sexual reproduction

Flowers are reproductive shoots usually composed of four parts—sepals, petals, stamens, and carpels—arranged in whorls (circles) on the end of a flower stalk, or **peduncle** (Fig. 27-9). The peduncle may terminate in a single flower or a cluster of flowers known as an **inflorescence.** The tip of the peduncle enlarges to form a **receptacle** that bears some or all of the flower parts.

All four floral parts are important in the reproductive process, but only the stamens (the "male" organs) and carpels (the "female" organs) produce gametes. A flower that has all four parts is **complete,** whereas an **incomplete** flower lacks one or more of these four parts. A flower with both stamens and carpels is **perfect,** whereas an **imperfect** flower has stamens or carpels, but not both.

Sepals, which make up the lowermost and outermost whorl on a floral shoot, are leaflike in appearance and often green (Fig. 27-10a). Sepals cover and protect the other flower parts when the flower is a bud. As the blossom opens, the sepals fold

FIGURE **27-9** **Floral structure.**

This cutaway view of a "typical" flower shows the details of basic floral structure. This flower is both a complete and a perfect flower. Not all flowers have all these structures. (Pistils each consist of one or more carpels. In this example, the pistil consists of two carpels.)

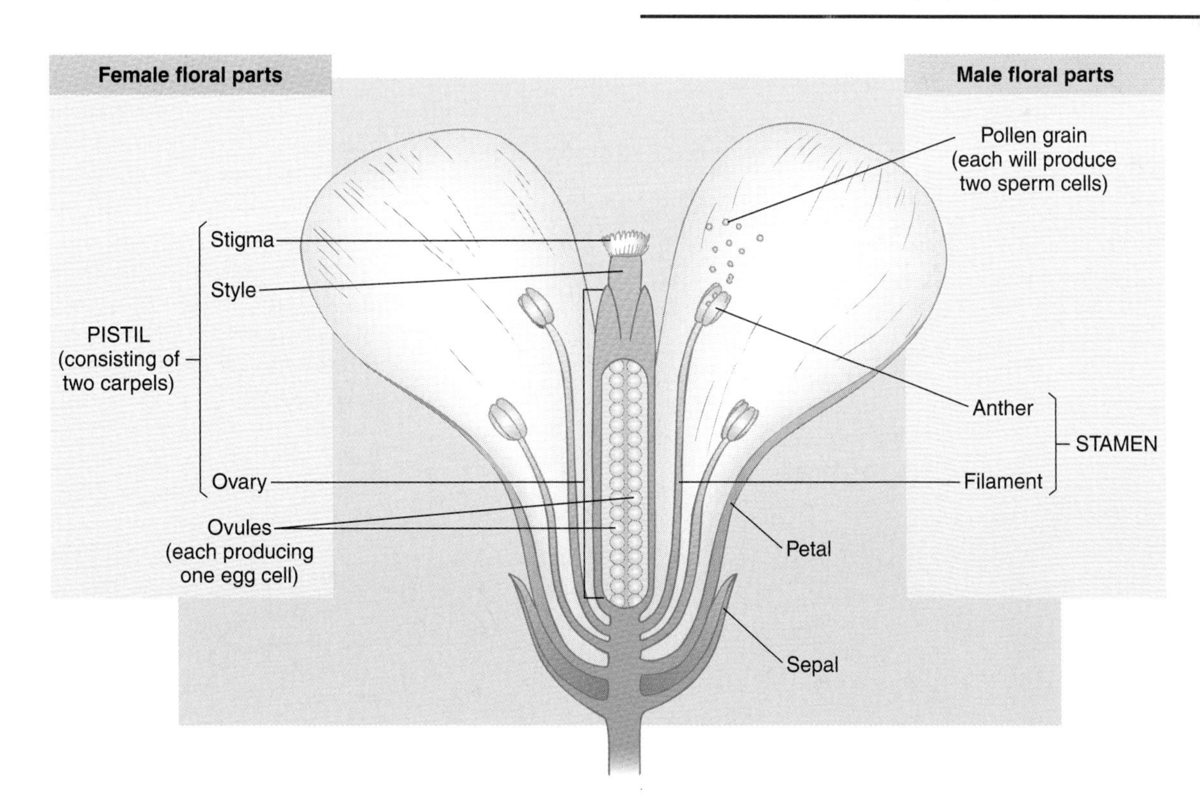

(a)

(b)

FIGURE 27-10 | Parts of a flower.

(a) The leaflike sepals of a rose (*Rosa*) bud enclose and protect the inner flower parts. **(b)** A twinleaf (*Jeffersonia diphylla*) flower has eight yellow stamens. Note the simple pistil with its green ovary in the center of the flower.

back to reveal the more conspicuous petals. The collective term for all the sepals of a flower is the **calyx.**

The whorl just above the sepals consists of **petals,** which are broad, flat, and thin (like sepals and leaves) but vary in shape and are frequently brightly colored. Petals play an important role in attracting animal pollinators to the flower (see Chapter 35). Sometimes petals are fused to form a tube (such as trumpet honeysuckle flowers) or other floral shape (such as snapdragons). The petals of a flower are referred to collectively as the **corolla.**

Just inside the petals is a whorl of **stamens** (Fig. 27-10b). Each stamen is composed of a thin stalk, called a **filament,** and a saclike **anther,** where meiosis occurs to form microspores that develop into pollen grains. Each pollen grain produces two cells surrounded by a tough outer wall. One cell eventually divides to form two male gametes, or sperm cells, and the other produces a pollen tube through which the sperm cells travel to reach the ovule.

In the center of most flowers is one or more closed **carpels,** the "female" reproductive organs. Carpels bear ovules, which, as you may recall, are structures with the potential to develop into seeds. The carpels of a flower are separate or fused into a single structure. The female part of the flower is also called a **pistil** (see Fig. 27-10b). A pistil may consist of a single carpel (a simple pistil) or a group of fused carpels (a compound pistil) (Fig. 27-11). Each pistil generally has three sections: a **stigma,** on which the pollen grain lands; a **style,** a necklike structure

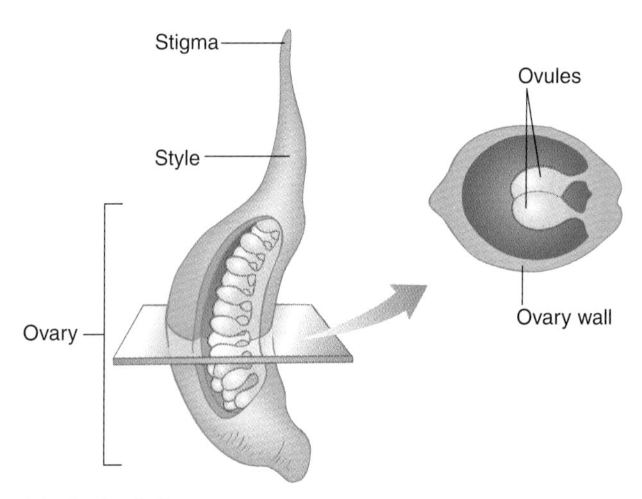

(a) Simple pistil

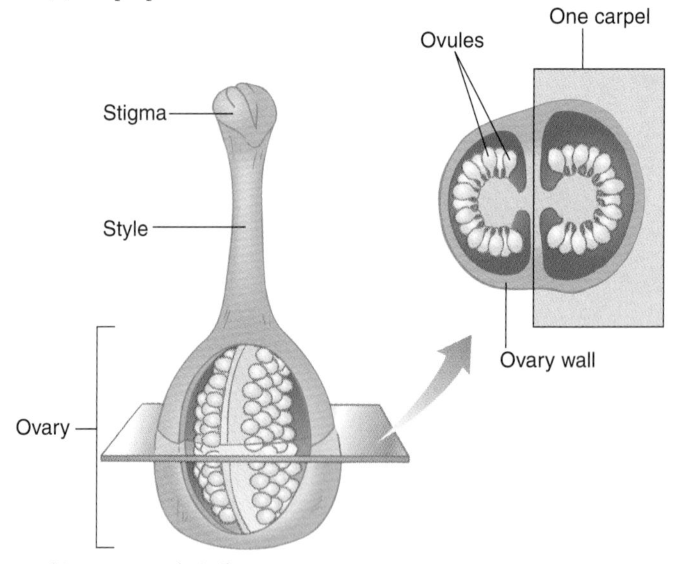

(b) Compound pistil

FIGURE 27-11 | Simple and compound pistils.

(a) This simple pistil consists of a single carpel. **(b)** This compound pistil has two united carpels. In most flowers with single pistils, the pistils are compound, consisting of two or more fused carpels.

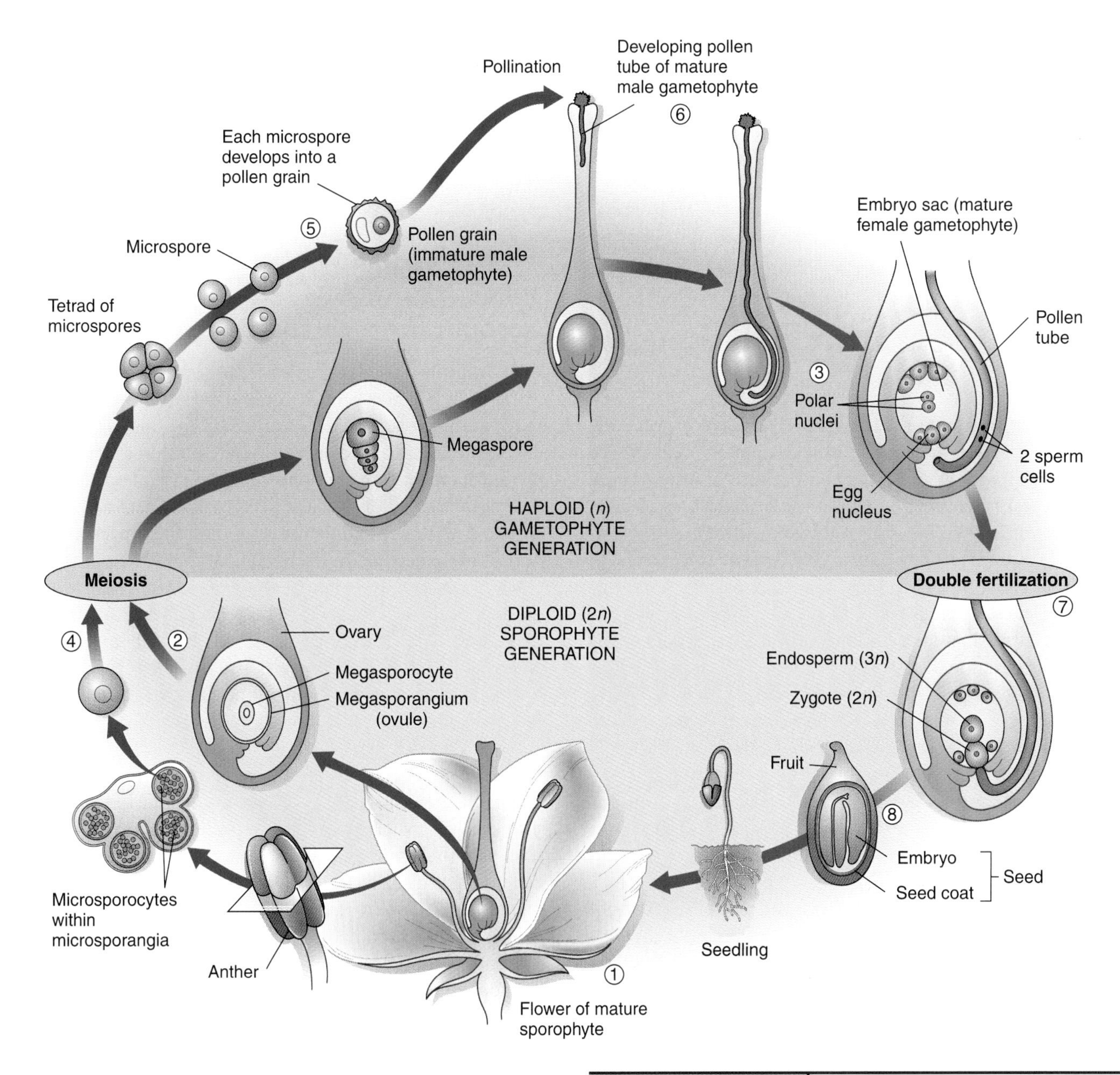

ACTIVE FIGURE **27-12** | **Life cycle of flowering plants.**

A significant feature of the flowering plant life cycle is double fertilization, in which one sperm cell unites with the egg, forming a zygote, and the other sperm cell unites with the two polar nuclei, forming a triploid cell that gives rise to endosperm. See text for a detailed description.

BiologyNow™ Explore **plant life-cycles** by clicking on this figure on your BiologyNow CD-ROM.

through which the pollen tube grows; and an **ovary,** an enlarged structure that contains one or more ovules. Each young ovule contains a female gametophyte that forms one female gamete (an egg), two *polar nuclei,* and several other haploid cells. After fertilization, the ovule develops into a seed and the ovary into a fruit.

The life cycle of flowering plants includes double fertilization

Flowering plants have an alternation of generations in which the sporophyte generation is larger and nutritionally independent. The gametophyte generation in flowering plants is microscopic and nutritionally dependent on the sporophyte. Flowering plants, like gymnosperms and certain other vascular plants, are heterosporous and produce two kinds of spores: microspores and megaspores. As shown in ① of Fig. 27-12, sexual reproduction occurs in the flower.

In ② each young ovule within an ovary contains a megasporocyte (megaspore mother cell) that undergoes meiosis to produce four haploid megaspores. Three of these usually disintegrate, and one divides mitotically and ③ develops into a mature female gametophyte, also called an **embryo sac.** The most widely studied type of embryo sac contains seven cells with eight haploid nuclei. Six of these cells, including the egg cell, contain a single nucleus each, and a central cell has two nuclei, called **polar nuclei.** The egg and the central cell with two polar nuclei are directly involved in fertilization; the other five cells in the embryo sac apparently have no direct role in the fertilization process and disintegrate. As the *synergids* (the two cells closely associated with the egg) disintegrate, however, they release chemicals that may affect the direction of pollen tube growth.

Each pollen sac, or microsporangium, of the anther contains numerous microsporocytes (microspore mother cells) ④, each of which undergoes meiosis to form four haploid microspores. Every microspore develops into ⑤ an immature male gametophyte, also called a *pollen grain.* Pollen grains are extremely small; each consists of two cells: the *tube cell* and the *generative cell.*

The anthers split open and begin to shed pollen. Pollen grains are transferred to the stigma by a variety of agents, including wind, water, insects, and other animal pollinators (see Chapter 35). As shown in ⑥, if compatible with the stigma, the pollen grain germinates; that is, the tube cell forms a pollen tube that grows down the style and into the ovary. The germinated pollen grain with its pollen tube is the mature male gametophyte. Next, the generative cell divides to form two nonmotile sperm cells. The sperm cells move down the pollen tube and are discharged into the embryo sac. Both sperm cells are involved in fertilization.

Something happens during sexual reproduction in flowering plants that does not occur anywhere else in the living world. In ⑦, when the two sperm cells enter the embryo sac, *both* participate in fertilization. One sperm cell fuses with the egg, forming a zygote that grows by mitosis and develops into a multicellular embryo in the seed. The second sperm cell fuses with the two haploid polar nuclei of the central cell to form a triploid ($3n$) cell that grows by mitosis and develops into **endosperm,** a nutrient tissue rich in lipids, proteins, and carbohydrates that nourishes the growing embryo. This fertilization process, which involves two separate nuclear fusions, is called **double fertilization** and is, with two exceptions, unique to flowering plants. (Double fertilization was reported in the gymnosperms *Ephedra nevadensis* [in 1990] and *Gnetum gnemon* [in 1996]. This process differs from double fertilization in flowering plants in that an additional zygote, rather than endosperm, is produced. The second zygote later disintegrates.)

Seeds and fruits develop after fertilization

As a result of double fertilization and subsequent growth and development, each seed contains a young plant embryo and nutritive tissue (the endosperm), both of which are surrounded by a protective seed coat. In monocots the endosperm persists and is the main source of food in the mature seed. In most dicots the endosperm nourishes the developing embryo, which subsequently stores food in its cotyledons.

As a seed develops from an ovule following fertilization, ⑧ the ovary wall surrounding it enlarges dramatically and develops into a **fruit.** In some instances, other tissues associated with the ovary also enlarge to form the fruit. Fruits serve two purposes: to protect the developing seeds from desiccation as they grow and mature and to aid in the dispersal of seeds (see Chapter 35). For example, dandelion fruits have feathery plumes that are lifted and carried by air currents. Once a seed lands in a suitable place, it may germinate and develop into a mature sporophyte that produces flowers, and the life cycle continues as described.

Flowering plants have many adaptations that account for their success

The evolutionary adaptations of flowering plants account for their success in terms of their ecological dominance and their great number of species. Seed production as the primary means of reproduction and dispersal, an adaptation shared with the gymnosperms, is clearly significant and provides a definite advantage over seedless vascular plants. Closed carpels, which give rise to fruits surrounding the seeds, and the process of double fertilization increase the likelihood of reproductive success. The evolution of a variety of interdependencies with many types of insects, birds, and bats, which disperse pollen from one flower to another of the same species, is another reason for angiosperm success. Pollen transfer results in cross-fertilization, which mixes the genetic material and promotes genetic variation among the offspring.

In addition to their highly successful reproduction involving flowers, fruits, and seeds, several distinctive features have contributed to the success of flowering plants. Recall that most flowering plants have very efficient water-conducting vessel elements in their xylem, in addition to tracheids. In contrast, the xylem of almost all seedless vascular plants and gymnosperms consists exclusively of tracheids. Most flowering plants also have efficient carbohydrate-conducting sieve tube elements in their phloem. Vascular plants other than flowering plants and gnetophytes lack vessel elements and sieve tube elements.

The leaves of flowering plants, with their broad, expanded blades, are very efficient at absorbing light for photosynthesis. Abscission (shedding) of these leaves during cold or dry spells reduces water loss and has enabled some flowering plants to expand into habitats that would otherwise be too harsh for survival. The stems and roots of flowering plants are often modified for food or water storage, another feature that helps flowering plants survive in severe environments.

Probably most crucial to the evolutionary success of flowering plants, however, is the overall adaptability of the sporophyte generation. As a group, flowering plants readily adapt to new habitats and changing environments. This adaptability is evident in the great diversity exhibited by the various species of flowering plants. For example, the cactus is remarkably well adapted to desert environments. Its stem stores water; its leaves (spines) have a reduced surface area available for transpiration (loss of water vapor; see Chapter 32) and may also protect against thirsty herbivorous animals; and its thick, waxy cuticle reduces water loss. In contrast, the water lily is well adapted for wet environments, in part because it has air channels that provide adequate oxygen to stems and roots living in oxygen-deficient water and mud.

Studying how flowers evolved provides insights into the evolutionary process

In evolution, new structures or organs often originate by modification of previously existing structures or organs (see Chapter 19). Much evidence supports the classical interpretation that the four organs of a flower—sepals, petals, stamens, and carpels—arose from highly modified leaves. This evidence includes comparisons of the arrangement of vascular tissues in both flowers and leafy stems and of the developmental stages of floral parts and leaves.

Sepals are the most leaflike of the four floral organs, and botanists generally agree that sepals are specialized leaves. Although petals of many flowering plant species are leaflike in appearance, botanists generally view petals as modified stamens that became sterile and leaflike. Cultivated roses and camellias provide evidence supporting this hypothesis; in some varieties the stamens have been transformed into petals, forming showy flowers with large numbers of petals.

PROCESS OF SCIENCE

The remarkably leaflike stamens and carpels of certain tropical trees and other species support the origin of stamens and carpels from leaves or leaflike organs. Consider, for example, the carpel of *Drimys,* a genus of evergreen trees and shrubs native to Southeast Asia, Australia, and South America. This carpel resembles a leaf that is folded inward along the midrib, thereby enclosing the ovules, and joined along the entire length of the leaf's margin (Fig. 27-13).

The fundamental question is whether these leaflike stamens and carpels are early organs that were conserved (retained) during the course of evolution or are highly specialized organs that do not resemble early stamens and carpels. Many botanists who have studied this question have concluded that stamens and carpels are probably derived from leaves. Not all botanists accept the origin of stamens and carpels from highly modified leaves, however. As we noted in Chapter 1, uncertainty and debate are part of the scientific process and scientists can never claim to know a final answer.

FIGURE **27-13** | **Carpel of *Drimys piperita.***

(a) The carpel resembles a folded leaf in which the ovules borne on its upper surface are enclosed. **(b)** A cross section of the carpel, cut along the dashed line in **(a).**

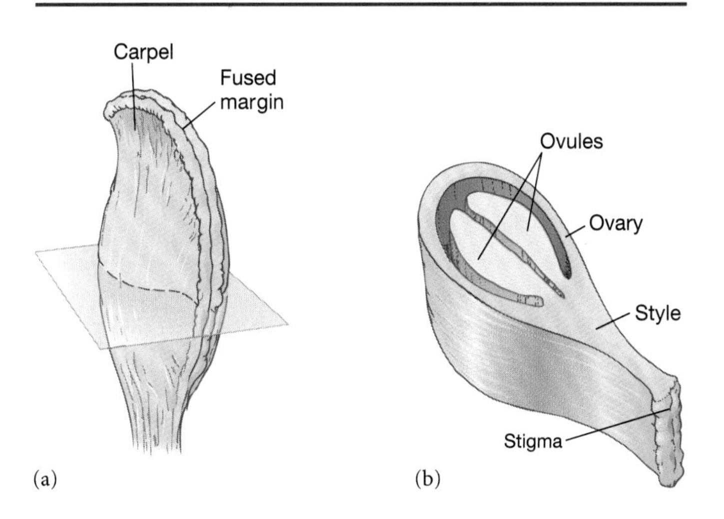

During the course of more than 130 million years of angiosperm evolution, flower structure diversified as floral organs fused together or became reduced in size or number. These changes led to greater complexity in floral structure in some species and to greater simplicity in other species. Interpreting the floral structures of so many different angiosperm species is sometimes difficult, but it is important because correct interpretations are essential to devising a classification scheme that is phylogenetic.

Review

- How do nonreproductive adaptations of flowering plants differ from those of gymnosperms?
- How does the flowering plant life cycle differ from that of the gymnosperms?
- What are the two classes of flowering plants, and how can one distinguish between them?
- How does pollination occur in flowering plants?
- How does fertilization differ in gymnosperms and flowering plants?

BiologyNow™ Assess your understanding of **flowering plants** by taking the pretest on your BiologyNow CD-ROM.

THE EVOLUTION OF SEED PLANTS

Learning Objective

9 Summarize the evolution of gymnosperms from seedless vascular plants, and trace the evolution of flowering plants from gymnosperms.

One group that descended from ancestral seedless vascular plants was the **progymnosperms,** all of which are now extinct. Progymnosperms had two derived features absent in their immediate ancestors: leaves with branching veins (*megaphylls*), and woody tissue (*secondary xylem*) similar to that of modern gymnosperms. Progymnosperms, however, reproduced by spores, not seeds. *Archaeopteris* was a progymnosperm that lived about 370 mya (Fig. 27-14a). In 1999, the discovery and evaluation of 150 fossils of Archaeopteris in southeast Morocco confirmed that this progymnosperm is the earliest known tree with "modern" woody tissue.

Fossils of several progymnosperms with reproductive structures intermediate between those of spore plants and seed plants have been discovered. For example, the evolution of microspores into pollen grains and of megasporangia into ovules (seed-producing structures) can be traced in fossil progymnosperms. Plants producing seeds appeared during the late Devonian period, more than 360 mya. The fossil record indicates different groups of seed plants apparently arose independently several times.

As mentioned previously, fossilized remains of ginkgo are found in 200-million-year-old rocks, and other groups of gymnosperms were well established by 160 to 100 mya. Although the gymnosperms are an ancient group, some questions persist about the exact pathways of gymnosperm evolution. The fossil

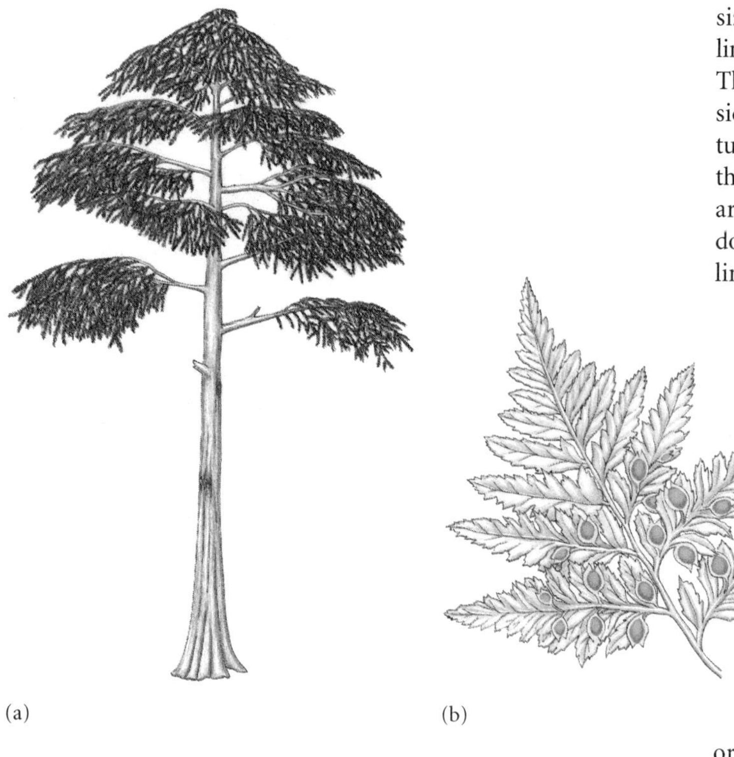

FIGURE **27-14** | **Evolution of seed plants.**

(a) *Archaeopteris,* a progymnosperm that existed about 370 mya, had some features in common with modern seed plants but did not produce seeds. **(b)** *Emplectopteris,* a seed fern, produced seeds on fernlike leaves. Seed ferns existed from about 360 to 250 mya. *(a, Redrawn from C.B. Beck, "Reconstructions of* Archaeopteris *and Further Consideration of Its Phylogenetic Position,"* American Journal of Botany, *Vol. 49, 1962; b, Redrawn from H.N. Andrews,* Ancient Plants and the World They Lived In, *Comstock, New York, 1947)*

record indicates progymnosperms probably gave rise to conifers and to another group of extinct plants called **seed ferns,** which were seed-bearing woody plants with fernlike leaves (Fig. 27-14b). The seed ferns in turn probably gave rise to cycads and possibly ginkgo, as well as to several gymnosperm groups now extinct. The origin of gnetophytes remains unclear, although molecular data indicate they are closely related to conifers.

Flowering plants are the most recent group of plants to evolve. The fossil record, although incomplete, indicates flowering plants probably descended from gymnosperms. By the middle of the Jurassic period, approximately 180 mya, several gymnosperm lines existed with features resembling those of flowering plants. Among other traits, these derived gymnosperms possessed leaves with broad, expanded blades and the first modified seed-bearing leaves, which nearly enclose the ovules. It is also evident beetles were visiting these plants, and biologists have suggested that perhaps this relationship was the beginning of **coevolution,** a mutual adaptation between plants and their animal pollinators (see Chapter 35).

One important task facing paleobotanists (biologists who study fossil plants) is determining which of the ancient gymnosperms are in the direct line of evolution leading to the flowering plants. Given the structural data, most botanists hypothesize flowering plants arose only once, that is, there is only one line of evolution from the gymnosperms to the flowering plants. The gnetophytes are the gymnosperm group some botanists consider the closest living relatives of flowering plants; both structural similarities and certain comparative molecular data support this conclusion. Like angiosperms, gnetophytes have vessels, lack archegonia, have flower-like compound strobili, and undergo double fertilization. However, other molecular data refute a close link between the gnetophytes and flowering plants. It is hoped additional studies will clarify the relationships among the various gymnosperm phyla and the flowering plants.

The oldest definitive trace of flowering plants in the fossil record is of ovules enclosed in tiny pod-like fruits interpreted as carpels in Jurassic and Lower Cretaceous rocks some 125 to 145 million years old (Fig. 27-15). In 1998 Chinese botanists Ge Sun and David Dilcher of the University of Florida discovered the fossil fruits in northeast China. The oldest fossilized flowers are about 118 to 120 million years old.

Based on the fossil record as well as both structural and molecular data of living angiosperms, the first flowering plants were probably small, weedy shrubs or herbaceous plants adapted to disturbed habitats. They may have been fragile plants that were not easily preserved. If so, it explains why there are few fossils of early angiosperms: The environment in which they evolved was repeatedly disturbed and not favorable for their preservation as fossils.

FIGURE **27-15** | **The oldest known fossil angiosperm.**

This fossil of *Archaefructus* shows a carpel-bearing stem. It was discovered in northeast China in 1998.

Botanists hypothesize the rapid diversification of angiosperms did not occur until early flowering plants had invaded lowland regions. By 90 mya, during the Cretaceous period, flowering plants had diversified and had begun to replace gymnosperms as Earth's dominant plants. Fossils of flowering plant leaves, stems, flowers, fruits, and seeds are numerous and diverse. They outnumber fossils of gymnosperms and ferns in late Cretaceous deposits, indicating the rapid success of flowering plants once they appeared (Fig. 27-16). Many angiosperm species apparently arose from changes in chromosome number (see discussion of sympatric evolution in Chapter 19).

Within the angiosperms, current evidence suggests the dicots evolved before the monocots and are therefore an older group. The monocots probably evolved from certain ancient dicots (Fig. 27-17). The cladistic analysis of molecular comparisons suggests the monocots are a monophyletic group but the dicots are not. Instead, the dicots are paraphyletic. Recall from Chapter 22 that a **paraphyletic** group contains a common ancestor and some, but not all, of its descendants. According to molecular evidence such as DNA sequence comparisons, some dicots—the magnolias and laurels—are more closely related to the monocots than to other dicots. Despite these data, many botanists currently recognize the classes Liliopsida and Magnoliopsida for convenience.

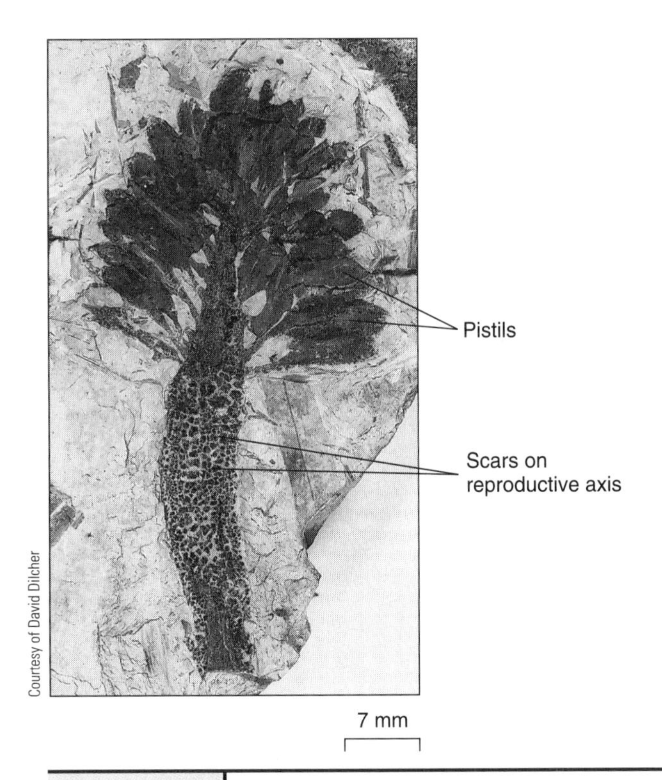

FIGURE **27-16** **Fossil flower.**

The fossilized flower of the extinct plant *Archaeanthus linnenbergeri*, which lived during the Cretaceous period, about 98 to 100 mya. The scars on the reproductive axis (receptacle) may show where stamens, petals, and sepals were originally attached but abscised (fell off). Many spirally arranged pistils were still attached at the time this flower was fossilized.

Review

- What features distinguish progymnosperms from seed ferns?
- What is the oldest known fossil angiosperm?
- Are monocots a monophyletic or paraphyletic group? Why?

Biology Now™ Assess your understanding of **the evolution of seed plants** by taking the pretest on your BiologyNow CD-ROM.

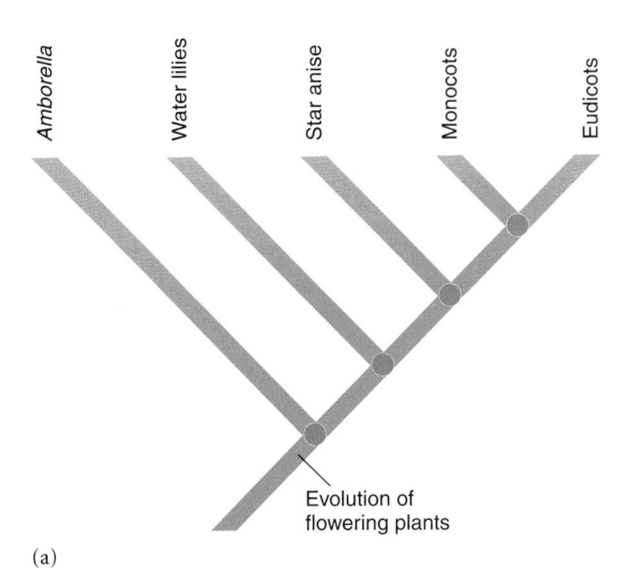

FIGURE **27-17** **Evolution of flowering plants.**

(a) One hypothesis of relationships among the flowering plants, based on fossil and molecular evidence. *Amborella*, water lilies (*Nymphaea*), and star anise (*Illicium verum*) are living dicots whose ancestors apparently branched off the angiosperm family tree early. These early dicot taxa were followed by the monocot branch and then by all other dicots (the Eudicots). **(b)** *Amborella trichopoda*, which is native to New Caledonia, an island in the South Pacific, may be the nearest living relative to the ancestor of all flowering plants. Photographed in the National Tropical Botanical Garden World Collection Greenhouse in Kauai, Hawaii.

SUMMARY WITH KEY TERMS

1 Compare the features of seeds with those of spores, and discuss the advantages of plants that reproduce primarily by seeds rather than by spores.

- **Seeds** are the primary means of reproduction and dispersal of gymnosperms and angiosperms (flowering plants). Seeds are reproductively superior to spores because embryonic development is further advanced in seeds, seeds contain an abundant food supply, and each seed has a protective **seed coat.**

2 Trace the steps in the life cycle of a pine, and compare its sporophyte and gametophyte generations.

- A pine tree is a mature sporophyte; pine gametophytes are extremely small and nutritionally dependent on the sporophyte generation. Pine is heterosporous and produces microspores and megaspores in separate cones.
- Male cones produce **microspores** that develop into **pollen grains** (immature male gametophytes) that are carried by air currents to female cones.
- Female cones produce **megaspores.** One of each four megaspores produced by meiosis develops into a female gametophyte within an **ovule (megasporangium).**
- After **pollination,** the transfer of pollen to the female cones, a **pollen tube** grows through the megasporangium to the egg within the archegonium.
- After **fertilization,** the zygote develops into an embryo encased inside a seed adapted for wind dispersal.

3 Summarize the features that distinguish gymnosperms from bryophytes and ferns.

- **Gymnosperms** are vascular plants with seeds that are totally exposed or borne on the scales of cones. Gymnosperms produce wind-borne **pollen grains,** a feature seedless vascular plants lack.

4 Name and briefly describe the four phyla of gymnosperms.

- **Conifers** (phylum Pinophyta) are the largest phylum of gymnosperms. Conifers are woody plants that bear **needles** (leaves that are usually evergreen) and produce seeds in cones. Most conifers are **monoecious** and have male and female reproductive parts in separate cones on the same plant.
- **Cycads** (phylum Cycadophyta) are palmlike or fernlike in appearance. They are **dioecious**—they have male and female reproductive structures on separate plants—but reproduce with pollen and seeds in conelike structures. There are relatively few living members of this once large phylum.
- *Ginkgo biloba,* the only surviving species in phylum Ginkgophyta, is a deciduous, dioecious tree. The female **ginkgo** produces fleshy seeds directly on branches.
- **Gnetophytes** (phylum Gnetophyta) share a number of traits with angiosperms.

5 Summarize the features that distinguish flowering plants from other plants.

- **Flowering plants,** or **angiosperms** (phylum Magnoliophyta) constitute the phylum of vascular plants that produce flowers and seeds enclosed within a **fruit.** They are the most diverse and most successful group of plants.
- The flower, which may contain **sepals, petals, stamens,** and **carpels,** functions in sexual reproduction. Unlike those of gymnosperms, the ovules of flowering plants are enclosed within an **ovary.** After fertilization, the ovules become seeds, and the ovary develops into a fruit.

6 Briefly explain the life cycle of a flowering plant, and describe double fertilization.

- The sporophyte generation is dominant in flowering plants; gametophytes are extremely reduced in size and nutritionally dependent on the sporophyte generation. Flowering plants are heterosporous and produce microspores and megaspores within the flower.
- Each microspore develops into a pollen grain (immature male gametophyte). One of each four megaspores produced by meiosis develops into an **embryo sac** (female gametophyte). The embryo sac contains seven cells with eight nuclei; the egg cell and the central cell with two **polar nuclei** participate in fertilization.
- **Double fertilization,** which results in the formation of a diploid zygote and triploid **endosperm,** is characteristic of flowering plants.

7 Contrast dicots and monocots, the two classes of flowering plants.

- Most **monocots** (class Liliopsida) have floral parts in multiples of 3, and their seeds each contain one **cotyledon.** The nutritive tissue in their mature seeds is endosperm.
- **Dicots** (class Magnoliopsida) usually have floral parts in multiples of 4 or 5, and their seeds each contain two cotyledons. The nutritive organs in their mature seeds are usually the cotyledons, which have absorbed the nutrients in the endosperm.

8 Discuss the evolutionary adaptations of flowering plants.

- Flowering plants reproduce sexually by forming flowers. After double fertilization, seeds are formed within fruits.
- Flowering plants have efficient water-conducting **vessel elements** in their xylem and efficient carbohydrate-conducting **sieve tube elements** in their phloem.
- Wind, water, insects, or other animals transfer pollen grains in various flowering plants.

9 Summarize the evolution of gymnosperms from seedless vascular plants, and trace the evolution of flowering plants from gymnosperms.

- Seed plants arose from seedless vascular plants.
- **Progymnosperms** were seedless vascular plants that had megaphylls and "modern" woody tissue. Progymnosperms probably gave rise to conifers. Progymnosperms probably gave rise to **seed ferns** as well, which in turn probably gave rise to cycads and possibly ginkgo.
- The evolution of the gnetophytes, particularly their relationship to flowering plants, is unclear.
- Flowering plants probably descended from ancient gymnosperms that had specialized features, such as leaves with broad, expanded blades and closed carpels. Flowering plants probably arose only once. The first flowering plants were probably dicots that were weedy shrubs or small herbaceous plants. *Amborella* is a dicot that may be the nearest living relative to the ancestor of all flowering plants.

POST-TEST

1. Seed plants *lack* which of the following structures? (a) ovules surrounded by integuments (b) microspores and megaspores (c) vascular tissues (d) a large, nutritionally independent sporophyte (e) a large, nutritionally independent gametophyte
2. Conifers, cycads, ginkgo, and gnetophytes are collectively called (a) fern allies (b) gymnosperms (c) angiosperms (d) dicots (e) seedless vascular plants
3. Most conifers are ______________, having male and female reproductive parts at different locations on the same plant. (a) incomplete (b) imperfect (c) monoecious (d) dioecious (e) perfect
4. The immature male gametophytes of pine are called (a) ovules (b) stamens (c) seed cones (d) pollen grains (e) polar nuclei
5. The transfer of pollen grains from the male to the female reproductive structure is known as (a) pollination (b) fertilization (c) embryo sac development (d) seed development (e) fruit development
6. Motile sperm cells are found as vestiges in these two gymnosperm groups: (a) monocots, dicots (b) gnetophytes, conifers (c) gnetophytes, flowering plants (d) cycads, conifers (e) cycads, ginkgo
7. More than ______________ species of flowering plants have been identified. (a) 235 (b) 2350 (c) 23,500 (d) 235,000 (e) 2,350,000
8. This class of flowering plants includes the palms, grasses, and orchids. (a) dicots (b) gnetophytes (c) cycads (d) monocots (e) conifers
9. The pistil has three sections: (a) stigma, style, and anther (b) anther, filament, and ovule (c) stigma, style, and ovary (d) ovary, ovule, and sepal (e) corolla, stamen, and sepal
10. A simple pistil consists of a single (a) calyx (b) carpel (c) ovule (d) filament (e) petal
11. A flower that lacks stamens is both ______________ and ______________. (a) complete; imperfect (b) incomplete; perfect (c) complete; perfect (d) incomplete; imperfect
12. After fertilization, the ______________ develop(s) into a fruit and the ______________ develop(s) into a seed. (a) ovary; ovule (b) polar nuclei; ovule (c) ovary; endosperm (d) ovule; ovary (e) ovule; polar nuclei
13. The female gametophyte in flowering plants is also called the (a) polar nuclei (b) anther (c) embryo sac (d) endosperm (e) sporophyll
14. This living dicot may be the nearest living relative to the ancestor of all flowering plants (a) *Amborella* (b) *Archaeopteris* (c) *Gnetum* (d) water lily (e) *Archaeanthus*
15. 
    This cross section through an ovary reveals that the pistil is (a) simple, with one carpel (b) compound, with two fused carpels (c) compound, with three fused carpels (d) compound, with six separate carpels (e) compound, with six fused carpels

CRITICAL THINKING

1. How are cones and flowers alike? How are they different? (*Hint:* Your answer should consider microspores/megaspores and seeds.)
2. How do the life cycles of seedless plants (see Chapter 26) and seed plants differ? In what fundamental way are they alike?
3. In contrast with the cones of gymnosperms, which are either male or female, most flowers contain both male and female reproductive structures. Explain how bisexual flowers might be advantageous to flowering plants.
4. Contrast the algae, mosses, ferns, gymnosperms, and angiosperms with respect to their dependence on water as a transport medium for reproductive cells. Suggest a hypothesis to explain the differences.

- Visit our Web site at **http://biology.brookscole.com/solomon7** for links to chapter-related resources on the World Wide Web. Additional online materials relating to this chapter can also be found on our Web site.

BIOLOGY NOW RESOURCES

Active Figure

27-12: Plant life-cycles

Preparing for an exam? Take a diagnostic test on your BiologyNow CD-ROM.

Post-Test Answers

1. e	2. b	3. c	4. d
5. a	6. e	7. d	8. d
9. c	10. b	11. d	12. a
13. c	14. a	15. c	

28

The Animal Kingdom: An Introduction to Animal Diversity

Charles V. Angelo/Photo Researchers, Inc.

The tube sponge (*Callyspongia vaginalis*). This animal, which ranges in color from purple to blue to gray, is common on coral reefs in the Caribbean, from Florida to Mexico.

CHAPTER OUTLINE

Biologists have described and named more than a million species of animals, and millions more may remain to be discovered and classified. Although most animal species are readily recognizable as animals, the identity of some others is less obvious. Early naturalists thought sponges were plants, because they did not move from place to place. Some people still mistake certain marine animals, such as sponges and corals, for plants (see photograph). Locomotion is not a requirement for being classified as an animal.

Animal phylogeny has become an exciting and rapidly changing field of study. Biologists have assigned the extant (living) members of **kingdom Animalia** to about 35 phyla, but the relative positions of phyla and classes are a work in progress. As they consider new data, systematists redraw the tree of animal life.

Early animals were soft-bodied forms that left few fossils. The scarcity of fossils has made it difficult to determine the age, rate of divergence, and number of branches of animal groups. Animals belonging to most extant phyla first appeared in the fossil record during a geologically brief 40-million-year span (about 565–525 mya) during the late Precambrian period and early Cambrian period. The rapid appearance of diverse body plans of extant animals and of an amazing variety of extinct animals during this time is known as the **Cambrian explosion** (see Chapter 20). (The term **body plan** refers to the basic structure of the body.) Based on the fossil record, the many major modifications in body plan that occurred during this time account for many of the branches in the animal tree.

Recently, researchers have presented molecular data indicating that animal clades diverged over a much longer period of time during the Proterozoic eon (which preceded the Cambrian period). Analysts of the molecular data estimate that certain groups are about twice as old as the oldest fossils found to date. According to this view, most animal groups diverged several hundred million years before they appear in the fossil record. Later, during the Cambrian explosion, a reorganization of the animal genome led to new body plans. Perhaps fossils of early animals remain to be discovered in Proterozoic rocks.

Systematists are using recently developed molecular tools to clarify the evolutionary relationships among animal groups. For example, biologists historically viewed roundworms and ribbon worms as members of simple groups that appeared early in animal evolution. Analysis of molecular data suggests these worms evolved from more complex animals and then became structurally simpler over time. Given these data, systematists have repositioned these groups.

In this chapter we discuss some of the criteria biologists use to determine evolutionary relationships and classify animals. Then we describe three phyla that diverged early in the evolutionary history of this kingdom. In Chapters 29 and 30 we discuss two major clades of animals: the protostomes and the deuterostomes. Many current hypotheses are presented in these chapters; keep in mind that biologists continually respond to new data by redrawing branches of the animal phylogenetic tree. ■

ANIMAL CHARACTERISTICS

Learning Objective

1 List several characteristics common to most animals.

Animals are so diverse that for almost any definition, we can find exceptions. Still, the following characteristics describe most animals:

1. Animals are multicellular eukaryotes.
2. Animals are **heterotrophs.** As consumers, they depend on producers for their raw materials and energy. Most animals ingest their food first and then digest it inside the body, usually within a digestive system.
3. Cells that make up the animal body are specialized to perform specific functions. In all but the simplest animals, cells are organized to form tissues, and tissues are organized to form organs. In many animals with simple body plans, life processes such as gas exchange, circulation of materials, and waste disposal take place by diffusion directly to and from the environment. In large, complex animals, specialized organ systems and mechanisms have evolved that perform these life processes.
4. Most animals have nervous systems and muscle systems that enable them to respond rapidly to stimuli in their environment.
5. Most animals are capable of locomotion at some time during their life cycle. Some animals (such as sponges and corals) move about as larvae (immature forms) but are **sessile** (firmly attached to the ground or some other surface) as adults (see chapter opening photograph).
6. Most animals are diploid organisms that reproduce sexually, with large, nonmotile eggs and small, flagellated sperm. A haploid sperm unites with a haploid egg, forming a **zygote** (fertilized egg) that undergoes **cleavage,** a series of mitotic cell divisions. During cleavage the zygote develops into a hollow ball of cells called a **blastula.** Cells of the blastula undergo **gastrulation,** a process of forming and segregating specific layers of tissue, called *germ layers.*

Some animals develop directly into adults. Others develop into a **larva,** a sexually immature form that may look very different from the adult. Larvae undergo **metamorphosis,** a developmental process that converts the immature animal into a juvenile form that can then grow into an adult.

Review

- For centuries scientists classified sponges as plants. What criteria do biologists use to classify them as animals?

BiologyNow™ Assess your understanding of **animal characteristics** by taking the pretest on your BiologyNow CD-ROM.

ANIMAL HABITATS

Learning Objective

2 Compare the advantages and disadvantages of life in the ocean, in fresh water, and on land.

Fossil evidence suggests animals evolved during Precambrian time in shallow, marine environments (see Chapter 20). Although animals are now distributed in virtually every environment, members of most animal phyla still inhabit marine environments.

Marine environments offer many advantages. The buoyancy of sea water provides support, and the temperature is relatively stable owing to the large volume of water. The body fluids of most invertebrates have about the same osmotic concentration as sea water, so fluid and salt balance are more easily maintained than in fresh water. **Plankton,** which consists of the mainly microscopic animals and protists that are suspended in water and float with its movement, provides a ready source of food for many aquatic animals.

Life in the ocean also presents some challenges. Although the continuous motion of water brings nutrients to animals and washes their wastes away, they must be able to cope with the water's movements and the currents that could sweep them away. Squids, fishes, and marine mammals are strong swimmers and can usually direct their movements and maintain their location. However, most invertebrates and young vertebrates cannot swim strongly, and they have adapted in many different ways to the tides and currents. Some sessile animals attach permanently to a stable structure such as a rock. Others burrow in the sand and silt that cover the sea bottom. Many invertebrates have adapted by maintaining a small body size and becoming part of the plankton. As they are tossed about, their food supply continues to surround them.

Far fewer kinds of animals make their homes in fresh water than in the ocean. Fresh water is hypotonic to the tissue fluids of animals, so water tends to move into the animal by osmosis. Freshwater species must have mechanisms for removing excess water while retaining salts. This *osmoregulation* requires an expenditure of energy. Fresh water offers a much less constant environment than sea water and generally contains less food. Oxygen content and temperature vary, and turbidity (owing to sediments suspended in the water) and even water volumes fluctuate.

Living on land is even more difficult than living in fresh water and requires many adaptations. Analyzing the fossil record, many biologists hypothesize that the first air-breathing terrestrial animals were scorpion-like arthropods that came ashore in the Silurian period about 450 million years ago (mya). The first vertebrates to inhabit terrestrial environments, the amphibians, did not appear until the latter part of the Devonian period, about 30 million years later.

The chief problem facing all terrestrial organisms is desiccation (drying out). Water is constantly lost by evaporation and is often difficult to replace. A body covering adapted to minimize fluid loss helps solve this problem in many terrestrial animals. Location of the respiratory surface deep within the animal also helps prevent fluid loss. Thus the gills of aquatic animals are typically located externally, but lungs and tracheal tubes of terrestrial animals are internal.

Reproduction on land also poses challenges to protect gametes and the developing offspring from desiccation. Aquatic animals typically shed their gametes in the water, where fertilization occurs. The surrounding water also serves as an effective shock absorber, protecting the delicate embryos as they develop. Some land animals, including most amphibians, return to the water for reproduction, and their larval forms develop in the water.

The evolution of internal fertilization has permitted many terrestrial animals, including earthworms, land snails, insects, reptiles, birds, and mammals, to meet the desiccation challenge. Because they transfer sperm from the body of the male directly into the body of the female by copulation, a watery medium continuously surrounds the sperm. Another important adaptation to reproduction on land is the tough, protective shell that surrounds the eggs of many species. Secreted by the female, this shell protects the developing embryo from drying out. An alternative adaptation for terrestrial reproduction is development of the embryo within the moist body of the mother.

The temperature extremes of terrestrial habitats also present challenges. In later chapters we discuss behavioral and physiological adaptations for maintaining body temperature.

Review

- What are some advantages of marine environments over freshwater and terrestrial habitats?
- What challenges does the terrestrial environment present to animals?

BiologyNow™ Assess your understanding of **animals' habitats** by taking the pretest on your BiologyNow CD-ROM.

RECONSTRUCTING PHYLOGENY

Learning Objectives

3 Critique the use of type of symmetry and type of body cavity to infer relationships among animal phyla.

4 Describe the distinguishing characteristics of the three major clades of coelomate animals: Lophotrochozoa, Ecdysozoa, and Deuterostomia.

Historically, biologists depended on fossils, on similarities in body plan, and on patterns of development to determine evolutionary relationships among various groups of animals. Now, molecular methods are providing additional data that help answer questions about phylogeny. For example, recent molecular analysis indicates the structure of RNA, genes that control development, and many other molecules are very similar among all animal groups that have been studied. Such complex molecules are unlikely to have evolved multiple times. Thus, these data, along with the *principle of parsimony* (see Chapter 22), support the hypothesis that animals evolved only once. They are a monophyletic group.

Biologists generally agree animals evolved from *choanoflagellates* (see Chapter 24). In some colonial flagellates, cells became specialized to perform specific functions such as movement, feeding, or reproduction. As this division of labor evolved, a colony of flagellates reached the level of cooperation and coordination that qualified it to be considered a single organism—the first animal. The choanoflagellates, fungi, and animals are the major groups classified as **opisthokonts.**

Biologists classify animals in two clades: Parazoa and Eumetazoa. The only animals classified as **Parazoa** (*para,* "alongside" and *zoa,* "animals") are sponges (phylum Porifera). The cells of sponges are loosely associated and do not form true tissues. Other animals have true tissues and are classified as **Eumetazoa** (*eu,* "true"; *meta,* "later"; *zoa,* "animals").

Biologists classify animals according to body symmetry

Symmetry refers to the arrangement of body structures in relation to the body axis. Most sponges are not symmetrical, so when cut in half, the two halves are not similar to one another. Most other animals exhibit either radial or bilateral body symmetry. Members of phylum Cnidaria (jellyfish, sea anemones, and their relatives), phylum Ctenophora (comb jellies), and adult echinoderms (sea stars and their relatives) have **radial symmetry.** In radial symmetry, the body has the general form of a wheel or cylinder, and similar structures are regularly arranged as spokes from a central axis (Fig. 28-1a, b). Multiple planes can be drawn through the central axis, each dividing the organism into two mirror images. An animal with radial symmetry receives stimuli equally from all directions in the environment.

Many radially symmetrical animals have modified radial symmetry. For example, sea anemones and ctenophores (comb jellies) actually have **biradial symmetry,** in which parts of the body have become specialized so that only two planes can divide the body into similar halves.

Most animals exhibit **bilateral symmetry,** at least in their larval stages. A bilaterally symmetrical animal can be divided through only one plane (which passes through the midline of the body) to produce roughly equivalent right and left halves that are mirror images (Fig. 28-1c, d). As bilateral symmetry evolved, a corresponding evolutionary trend led toward **cephalization,** the development of a head where sensory structures are concentrated. In many animal groups, concentrations of

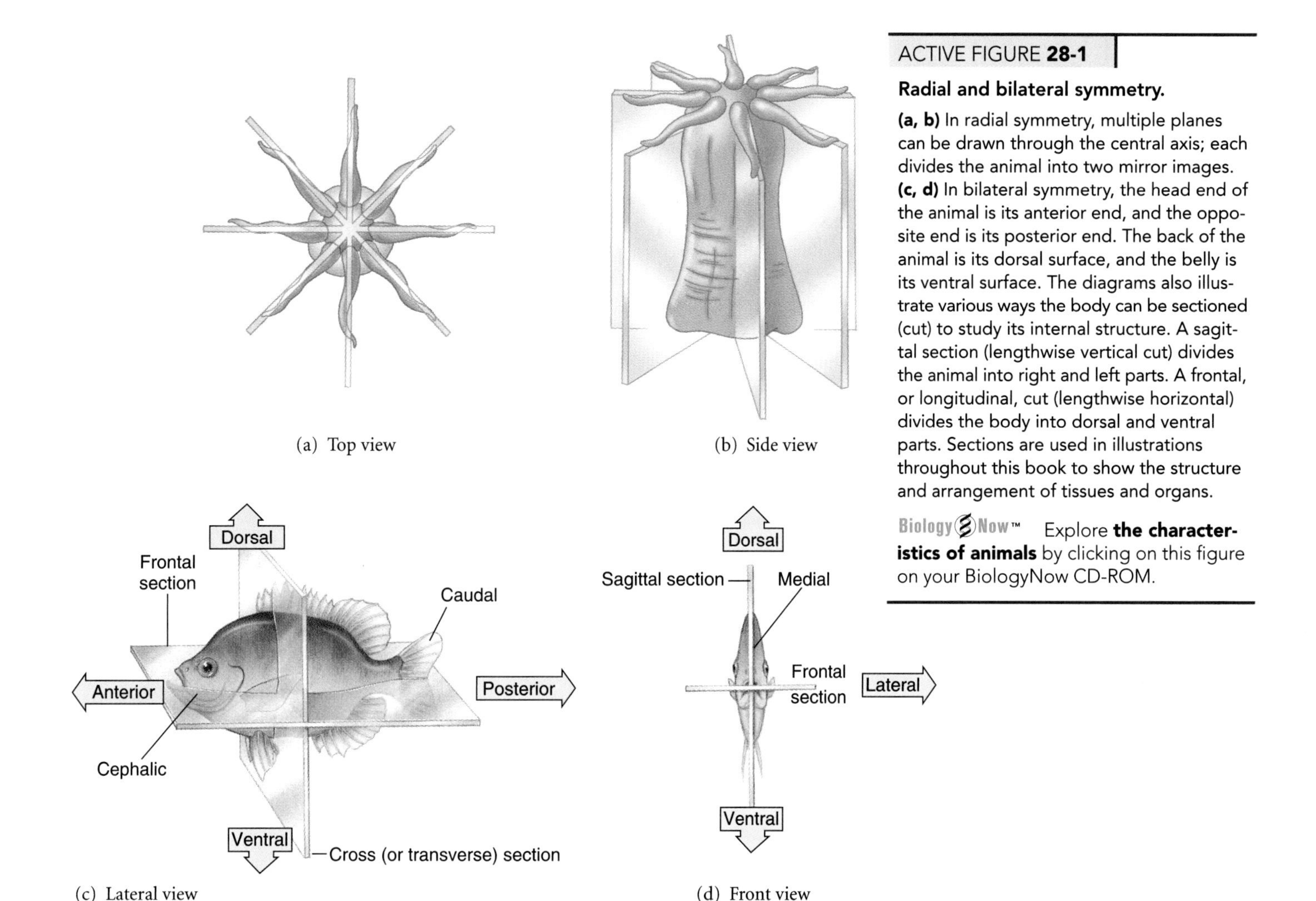

ACTIVE FIGURE **28-1**

Radial and bilateral symmetry.

(a, b) In radial symmetry, multiple planes can be drawn through the central axis; each divides the animal into two mirror images. **(c, d)** In bilateral symmetry, the head end of the animal is its anterior end, and the opposite end is its posterior end. The back of the animal is its dorsal surface, and the belly is its ventral surface. The diagrams also illustrate various ways the body can be sectioned (cut) to study its internal structure. A sagittal section (lengthwise vertical cut) divides the animal into right and left parts. A frontal, or longitudinal, cut (lengthwise horizontal) divides the body into dorsal and ventral parts. Sections are used in illustrations throughout this book to show the structure and arrangement of tissues and organs.

BiologyNow™ Explore **the characteristics of animals** by clicking on this figure on your BiologyNow CD-ROM.

nerve cells in the head form a brain, and a nerve cord extends from the brain toward the rear end of the animal. Bilateral symmetry and cephalization are considered adaptations to locomotion. The head end of the animal meets its environment first and is best equipped to capture food or respond to danger.

To locate body structures in bilaterally symmetrical animals, it is helpful to define some basic terms and directions. The back surface of an animal is its **dorsal** surface; the underside (belly) is its **ventral** surface. **Anterior** (or **cephalic**) means toward the head end of the animal; **posterior,** or **caudal,** means toward the tail end. A structure is said to be **medial** if it is located toward the midline of the body and **lateral** if it is toward one side of the body; for example, the human ear is lateral to the nose. In human anatomy, the term **superior** refers to a structure located above some point of reference, or toward the head end of the body. The term **inferior** is used in human anatomy to mean located below some point of reference, or toward the feet.

A bilaterally symmetrical animal has three axes, each at right angles to the other two: an anterior-posterior axis extending from head to tail, a dorsal–ventral axis extending from back to belly, and a left–right axis extending from side to side. We can distinguish three planes or sections that divide the body into specific parts. A **sagittal plane** divides the body into right and left parts. A sagittal plane passes from anterior to posterior and from dorsal to ventral. A **frontal** plane divides a bilateral body into dorsal and ventral parts. A **transverse section,** or **cross section,** cuts at right angles to the body axis and separates anterior and posterior parts.

Biologists group animals according to type of body cavity

Biologists describe cnidarians and ctenophores as **diploblastic** because they have two tissue layers. These tissues develop from two types of embryonic tissue, or **germ layers.** In all animals, the outer germ layer, or **ectoderm,** gives rise to the outer covering of the body and to nervous tissue. The inner layer, or **endoderm,** forms the lining of the digestive tube and other digestive structures. All other eumetazoa are **triploblastic.** They have a third germ layer, the **mesoderm,** that gives rise to most other body structures, including muscles, skeletal structures, and circulatory system (when present).

Traditionally, biologists have classified triploblastic animals based on the presence and type of **body cavity,** or **coelom** (see′-lum), a fluid-filled space between the body wall and the digestive tube (Fig. 28-2). The flatworms (phylum Platyhelminthes) and ribbon worms (phylum Nemertea) are triploblastic but have a solid body; that is, they have no body cavity. They are called **acoelomates** (*a,* "without" ; *coelom,* "cavity").

FIGURE **28-2**

Three basic body plans in triploblastic animals.

The germ layer from which each tissue was derived is indicated in parentheses. **(a)** An acoelomate animal has no body cavity. **(b)** A pseudocoelomate animal has a body cavity that is not completely lined with mesoderm. **(c)** In a coelomate animal the body cavity, or coelom, is completely lined with tissue that develops from mesoderm. Ectoderm is shown in blue, mesoderm in red, and endoderm in yellow.

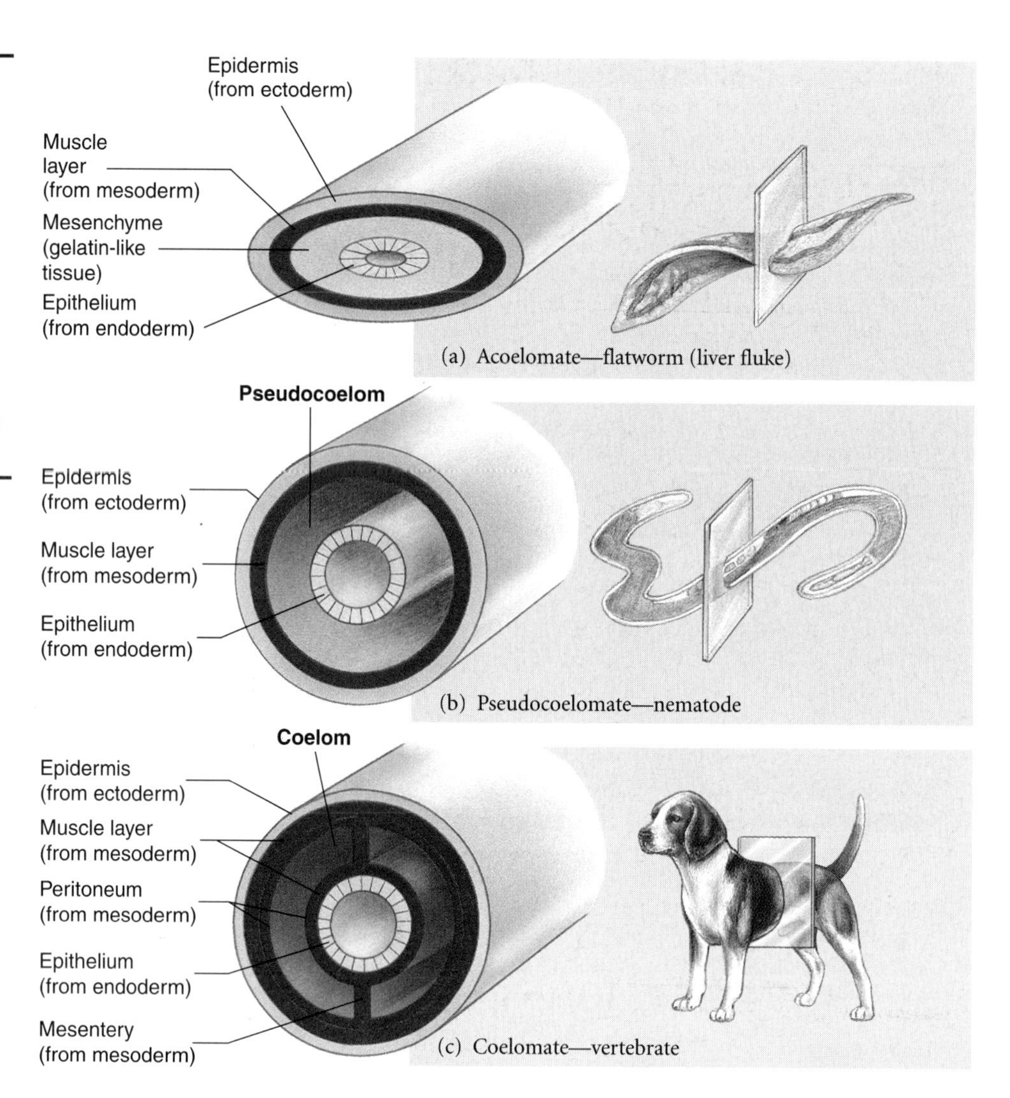

In most animals, the body cavity is completely lined with mesoderm. Such a body cavity is a *true coelom.* Animals with true coeloms are **coelomates.** The coelom is a space that separates the body wall from the digestive tube, or gut, producing a **tube-within-a-tube body plan.** The body wall, which forms the outer tube, is covered with tissue that develops from ectoderm. Tissue derived from endoderm lines the inner digestive tube, which has an opening at each end: the mouth and the anus.

In some (typically, small) animals, the body cavity is not completely lined with mesoderm; it is called a **pseudocoelom** ("false coelom"). Animals with a pseudocoelom, like nematodes (roundworms) and rotifers, are **pseudocoelomates,** and biologists formerly classified them as a separate group. Recent evidence indicates pseudocoelomates are not a monophyletic group and probably evolved through a process of simplification from more than one group of animals with a true coelom.

Coelomate animals form two main groups based on differences in development

Animals with a coelom form two main evolutionary lines: **Protostomia,** which include mollusks, annelids, and arthropods, and **Deuterostomia,** which include the echinoderms (such as sea stars and sea urchins) and chordates (the phylum that includes the vertebrates). Basic differences in their pattern of early development distinguish protostomes from deuterostomes.

One important difference in the development of protostomes and deuterostomes is the pattern of cleavage, the first several cell divisions of the embryo. In many protostomes, the early cell divisions are diagonal to the polar axis (the long axis of the egg), resulting in a somewhat spiral arrangement of cells; any one cell lies between the two cells above or below it (Fig. 28-3). This pattern of division is known as **spiral cleavage.** In **radial cleavage,** characteristic of the deuterostomes, the early divisions are either parallel or at right angles to the polar axis. The resulting cells lie directly above or below one another.

In the protostomes, the developmental fate of each embryonic cell is typically fixed very early. For example, if the first four cells of an annelid embryo are separated, each cell develops into only a fixed quarter of the larva; this is called **determinate cleavage.** In deuterostomes, cleavage is usually **indeterminate.** If the first four cells of a sea star embryo, for instance, are separated, each cell can form a complete, though small, larva. If a few cells are removed from the blastula of an embryo undergoing determinate cleavage, some structure such as a limb, does not develop. In contrast, if a few cells are removed from a blastula undergoing indeterminate cleavage, other cells compensate, and the embryo develops normally.

During gastrulation a group of cells moves inward, forming a sac that becomes the embryonic gut. The opening to the outside is called the **blastopore.** In most protostomes, the blastopore develops into the mouth. The word *protostome* comes from Greek words meaning "first, the mouth." In deuterostomes the blastopore does not give rise to the mouth. Instead, it generally develops into the anus. A second opening that forms later in development gives rise to the mouth. The word *deuterostome* means "second, the mouth."

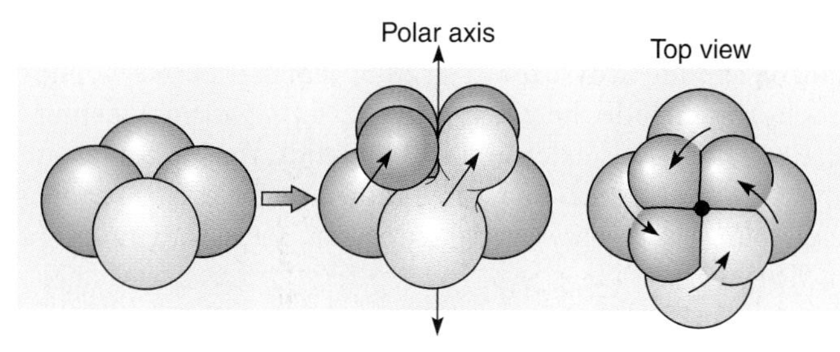

(a) Protostomes are characterized by spiral cleavage.

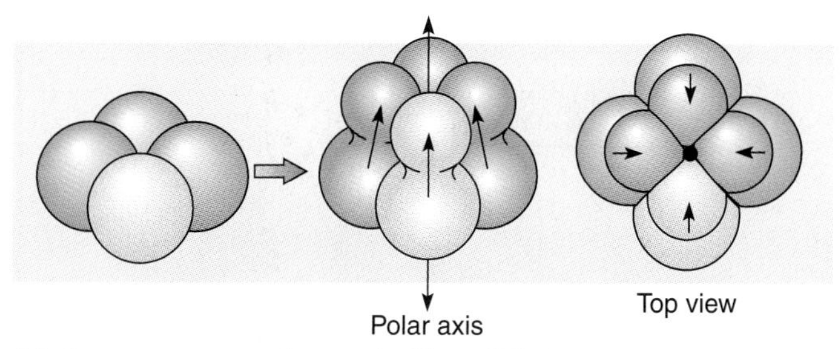

(b) Deuterostomes are characterized by radial cleavage.

FIGURE **28-3** | **Spiral and radial cleavage.**

(a) Spiral cleavage is characteristic of protostomes. Note the spiral arrangement, with the upper cells centered between the lower cells. **(b)** In radial cleavage, characteristic of deuterostome embryos, the early divisions are either parallel to the polar axis or at right angles to it. The cells are stacked, with the upper cells centered directly above the lower cells. The pattern of cleavage can be appreciated by comparing the positions of the purple cells in *(a)* and *(b)*.

Another, though less reliable, difference between protostome and deuterostome development is the manner in which the coelom forms. In most protostomes, the mesoderm splits, and the split widens into a cavity that becomes the coelom (Fig. 28-4). This method of coelom formation is known as **schizocoely.** In deuterostomes, the mesoderm forms as "outpocketings" of the developing gut, a process called **enterocoely.** These outpocketings eventually pinch off and form pouches; the cavity within the pouches becomes the coelom.

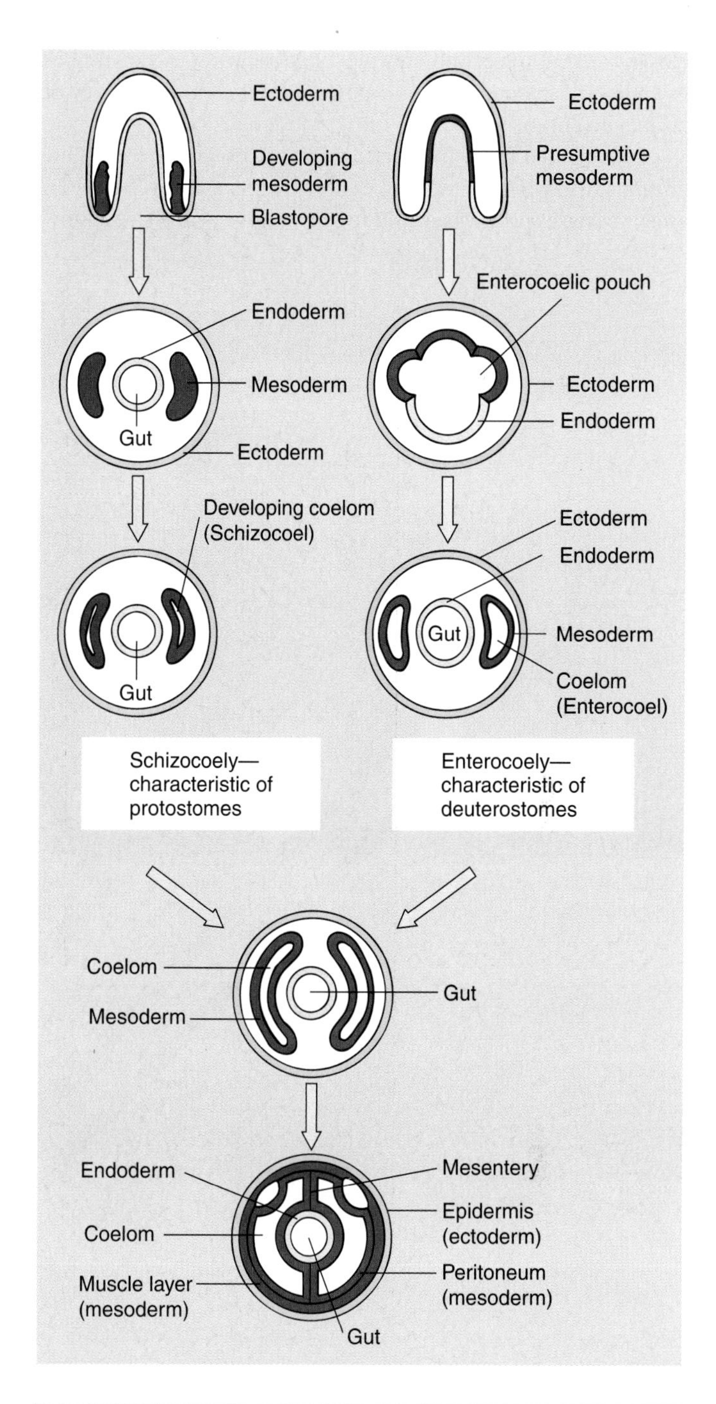

FIGURE **28-4** | **Two types of coelom formation.**

The coelom originates in the embryo from blocks of mesoderm that split off from each side of the embryonic gut. In protostomes the coelom typically forms by the process of schizocoely, in which the mesoderm *(red)* splits. The split widens, forming a cavity that becomes the coelom. In enterocoely, characteristic of deuterostomes, the mesoderm outpockets from the gut, forming pouches. The cavity within these pouches becomes the coelom. Ectoderm is shown in blue, endoderm in yellow. Top row, longitudinal sections of developing embryos; other diagrams are cross sections.

Biologists are using molecular data to rethink animal relationships

PROCESS OF SCIENCE

Using molecular techniques, biologists are exploring similarities in DNA, ribosomal RNA, and *Hox* genes. Recall from Chapter 16 that *Hox* genes are a group of regulatory genes that help control early development. All major animal groups have them except sponges. Investigators suggest all the *Hox* gene groups had evolved by the beginning of the Cambrian period.

Although molecular data have thus far validated much of the phylogeny systematists have based on morphology and development, these data suggest changes in placement of several animal groups. For example, molecular data suggest evolution of animals was not an orderly progression from simple to complex. Until recently, biologists thought the very small (less than 3 mm) marine animals known as **Placozoans** were the earliest animals to evolve. Placozoans have flat, rounded bodies consisting of only a few thousand cells of four types. Biologists formerly classified them along with sponges as parazoa, but molecular data suggest the placozoans evolved later and are more closely related to certain complex invertebrates. According to this view, placozoans became simplified later in the history of animal evolution.

As explained in the last section, biologists classify coelomate animals into protostome and deuterostome groups. Based on an evolving hypothesis of phylogeny, many systematists now subdivide the protostomes into two branches: the **Lophotrochozoa** and the **Ecdysozoa** (Fig. 28-5; Table 28-1). The Lophotrochozoa include the platyhelminthes (flatworms), nemerteans (ribbon worms), mollusks, annelids, the lophophorate phyla (groups that have a ciliated ring of tentacles surrounding the mouth), and rotifers. The Ecdysozoa, animals that molt (a

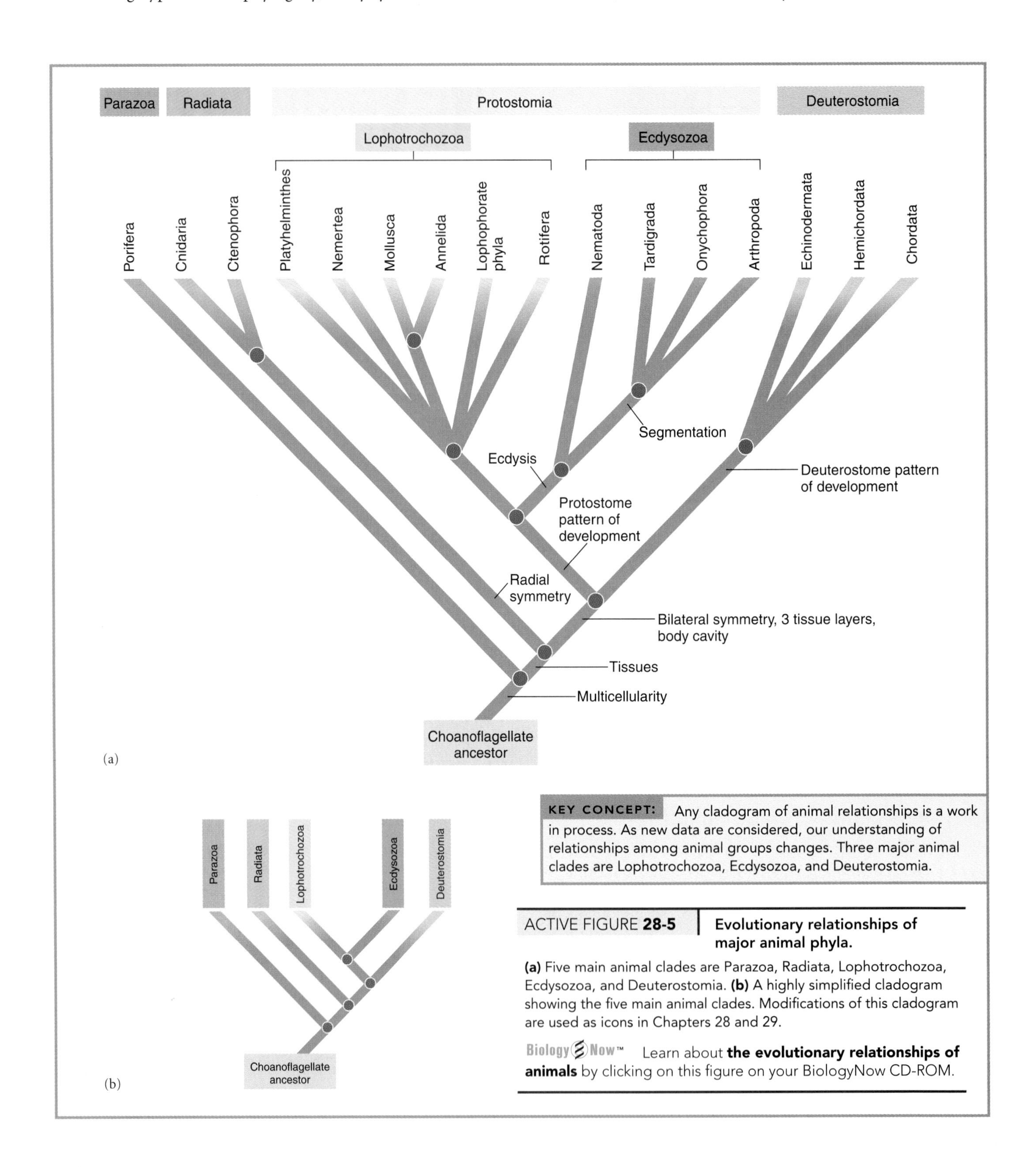

KEY CONCEPT: Any cladogram of animal relationships is a work in process. As new data are considered, our understanding of relationships among animal groups changes. Three major animal clades are Lophotrochozoa, Ecdysozoa, and Deuterostomia.

ACTIVE FIGURE **28-5** | **Evolutionary relationships of major animal phyla.**

(a) Five main animal clades are Parazoa, Radiata, Lophotrochozoa, Ecdysozoa, and Deuterostomia. **(b)** A highly simplified cladogram showing the five main animal clades. Modifications of this cladogram are used as icons in Chapters 28 and 29.

BiologyNow™ Learn about **the evolutionary relationships of animals** by clicking on this figure on your BiologyNow CD-ROM.

TABLE 28-1 Overview of the Animal Kingdom

Major Groups and Phyla	Symmetry	Level of Organization	Digestion	Circulation
Parazoa				
Porifera (sponges)	None	Cells loosely arranged	Intracellular	Diffusion
Radiata				
Cnidarians (hydras, jellyfish, corals)	Radial	Diploblastic; tissues	Gastrovascular cavity with one opening	Diffusion
Ctenophores (comb jellies)	Biradial	Diploblastic; tissues	Gastrovascular cavity with mouth and anal pores	Diffusion
Protostome Coelomates of the Lophotrochozoan Branch				
Platyhelminthes* (flatworms)	Bilateral	Triploblastic; simple organ systems	Gastrovascular cavity with one opening	Diffusion
Nemertea (proboscis worms)	Bilateral	Triploblastic; organ systems	Complete digestive tube**	Blood vessels; no heart
Mollusca (clams, snails, squids)	Bilateral	Triploblastic; organ systems	Complete digestive tube	Open system† (closed in cephalopods)
Annelida (some marine worms, earthworms, leeches)	Bilateral	Triploblastic; organ systems	Complete digestive tube	Closed system
Lophophorates (brachiopods, phoronids, bryozoans)	Bilateral	Triploblastic; organ systems	Complete digestive tube	Closed system
Rotifera (wheel animals)	Bilateral	Triploblastic; organ systems	Complete digestive tube	Closed system
Protostome Coelomates of the Ecdysozoan Branch				
Nematoda (roundworms)	Bilateral	Triploblastic; organ systems	Complete digestive tube	Diffusion
Arthropoda (crustaceans, insects, arachnids)	Bilateral	Triploblastic; organ systems	Complete digestive tube	Open system
Deuterostome Coelomates				
Echinodermata (sea stars, sea urchins)	Embryo bilateral; adult pentaradial	Triploblastic; organ systems	Complete digestive tube	Open system; reduced
Hemichordata	Bilateral	Triploblastic; organ systems	Complete digestive tube	Open system
Chordata (tunicates, lancelets, vertebrates)	Bilateral	Triploblastic; organ systems	Complete digestive tube	Closed system

*Triploblastic, but have solid body (i.e., no body cavity).
**A complete digestive tube has a mouth for food intake and an anus for elimination of wastes.
†A type of circulatory system in which the heart pumps blood through vessels into tissues that have open ends, and the blood bathes the tissues directly.

process called *ecdysis*), include the nematodes and arthropods. Thus this phylogeny has three major clades of coelomate animals: Lophotrochozoa, Ecdysozoa, and Deuterostomia.

Now that we have briefly discussed animal body plans and some of the criteria for determining phylogenetic relationships, we survey representative animal phyla. In the remainder of this chapter, we discuss two groups that appeared early in animal evolution: the Parazoa and the Radiata. The cladogram (phylogenetic tree) shown in Figure 28-5 depicts one current view of the relationships among the major phyla of animals.

Review

- In what way do biologists use the type of body cavity to infer relationships among animal groups?
- How do the three main animal clades differ from one another?

BiologyNow™ Assess your understanding of **reconstructing phylogeny** by taking the pretest on your BiologyNow CD-ROM.

THE PARAZOA

Learning Objective

5 Identify distinguishing characteristics of phylum Porifera.

Sponges, the only members of the Parazoa, diverged early from the lineage leading to other animals. Because they have been evolving independently for so long, they are very different from other animals. Although they are multicellular and can be large, sponges function much like colonial, unicellular protozoa. Most cells of sponges are extremely versatile and can change form and function; the cells do not form true tissues.

Collar cells characterize sponges

Biologists have identified about 10,000 species of sponges and assigned them to phylum **Porifera.** The name *Porifera,* meaning "to have pores," aptly describes the sponges, whose bodies are perforated by tiny holes. Sponges are aquatic, mainly marine, animals that range in size from 1 to 200 cm (0.4–79 in) in height. Many are asymmetrical, but they vary in shape from flat, encrusting growths to balls, cups, fans, or vases. Living sponges may be brightly colored—green, orange, red, yellow, blue, or purple—or they may be white or drab (Fig. 28-6). Some species are inhabited by symbiotic bacteria or algae that give them color.

Sponges most likely evolved from choanoflagellates, protozoa that have a single flagellum surrounded by a collar of microvilli (see Fig. 24-25). Sponges have **collar cells,** or **choanocytes,** flagellate cells that are strikingly similar to the choanoflagellates. Molecular biologists have discovered that choanoflagellates have receptor proteins known as *tyrosine kinases,* characteristic of multicellular animal cells. This finding suggests that some of the mechanisms necessary for cells to interact with one another have evolved in choanoflagellates.

Sponge larvae have flagella and can swim about. Adult sponges attach to some solid object and have long been described as sessile. However, biologists have observed adults of several species moving slowly (about 4 mm per day) by extending footlike appendages.

In a simple sponge, water enters through hundreds of tiny pores *(ostia),* passes into the central cavity, or **spongocoel** (not a digestive cavity), and flows out through the sponge's open end, the **osculum.** In most types of sponges, the body wall is extensively folded, and complicated systems of canals provide increased surface area for food capture.

Although sponges are multicellular, their cells are loosely associated and do not form definite tissues. However, a division of labor exists among the several types of cells that make up the sponge, with certain cells specializing in nutrition, support, contraction, or reproduction. Epidermal cells form the outer layer of the sponge and line the canals. Specialized tubelike cells, called *porocytes,* form the pores of a simple sponge. These cells regulate the diameter of the pores by contracting.

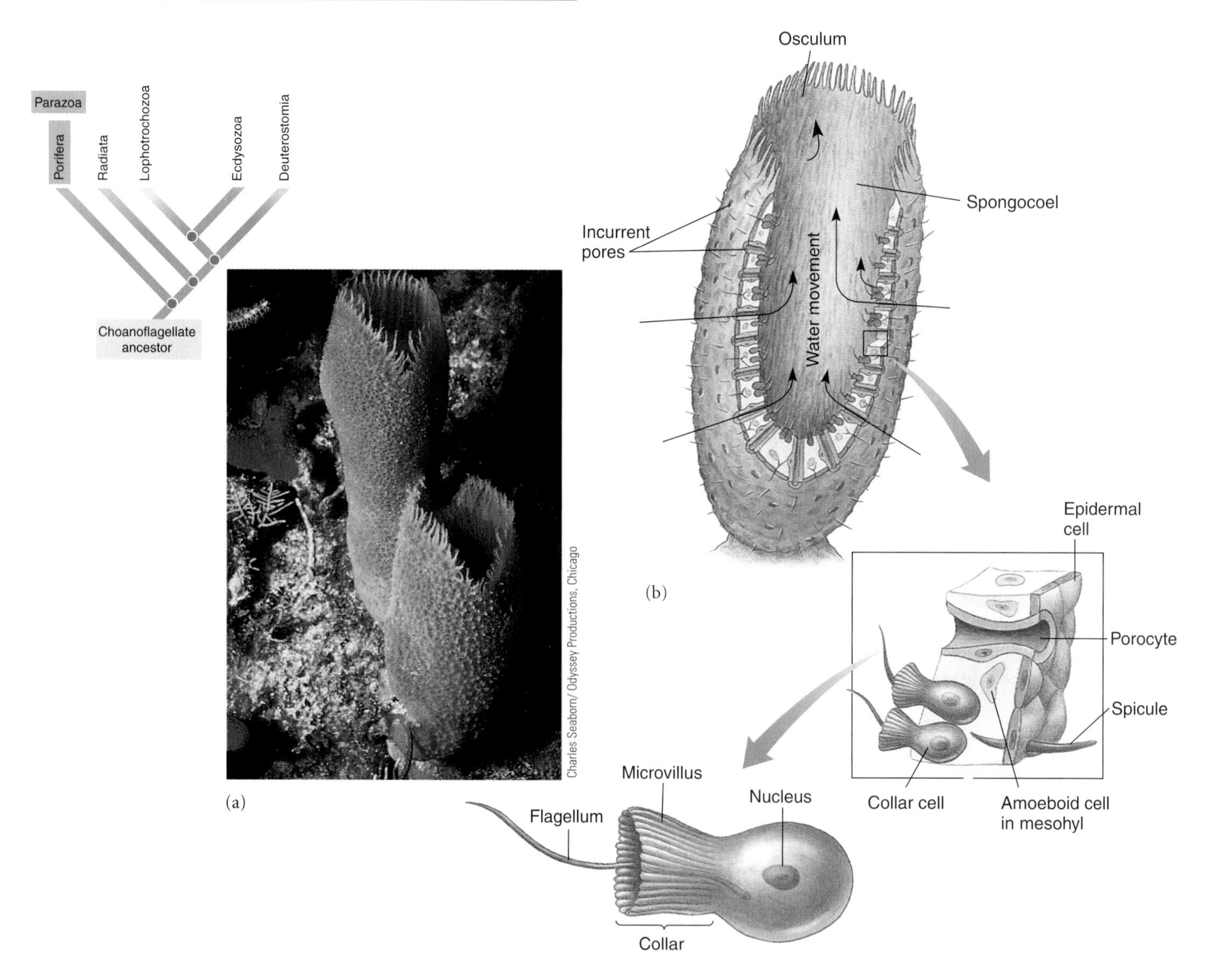

FIGURE **28-6** | **Sponge structure.**

(a) Tube sponges (*Spinosella plicifera*) from the Caribbean, attached to the coral reef substrate. **(b)** A simple sponge cut open to expose its organization. Water drawn through the pores passes through the spongocoel and exits through the osculum. Collar cells trap food particles in the stream of water.

Collar cells make up the inner layer of certain sponges. Each cell is equipped with a tiny collar surrounding the base of the flagellum. The collar is an extension of the plasma membrane and consists of microvilli. These cells create the water current that brings food and oxygen to the cells and carries away carbon dioxide and other wastes. Collar cells also trap and phagocytize food particles. Together, the collar cells of some complex sponges can pump a volume of water equal to the volume of the sponge each minute!

Between the outer and inner cell layers of the sponge body is a gelatin-like layer, the *mesohyl,* supported by slender skeletal spikes, or **spicules.** Amoeboid cells, which wander about in the mesohyl, secrete the spicules, which consist of calcium carbonate, silica, or a fibrous protein material known as *spongin.* Other amoeboid cells in the mesohyl are important in digestion and food transport.

Sponges are *suspension feeders,* adapted for trapping and eating whatever food the water brings to them. As water circulates through the body, food is trapped along the sticky collars of the choanocytes. Food particles are either digested within the collar cell or transferred to an amoeboid cell for digestion and transport of nutrients to epidermal cells. Undigested food passes out through the osculum and is simply eliminated into the water.

Gas exchange and excretion of wastes depend on diffusion into and out of individual cells. Although cells of the sponge are irritable and can react to stimuli, sponges do not have specialized nerve cells and so cannot react as a whole. Behavior appears limited to basic metabolic necessities such as capturing food and regulating the flow of water through the body.

Sponges reproduce both asexually and sexually. In asexual reproduction, a small fragment or bud may break free from the parent sponge and give rise to a new sponge. Such fragments may attach to the parent sponge, forming a colony. Most sponges are **hermaphrodites,** meaning the same individual can produce both eggs and sperm. Some of the amoeboid cells develop into sperm cells, others into egg cells. However, hermaphroditic sponges usually produce eggs and sperm at different times, and they cross-fertilize with other sponges. They release mature sperm into the water, which are taken in by other sponges of the same species.

Fertilization and early development take place within the jelly-like mesohyl. Zygotes develop into flagellate larvae that leave the parent along with the stream of outflowing water. After swimming about for awhile, a larva finds a solid object, attaches to it, and settles down to a sessile life.

Sponges have a remarkable ability to repair themselves when injured and to regenerate lost parts. If the cells of a sponge are separated experimentally, they recognize one another and their place in the whole, and aggregate, forming a complete sponge again.

Review

- Why are choanocytes significant? What is their function in sponges?
- In what ways do the cells of sponges specialize?

BiologyNow™ Assess your understanding of **the Parazoa** by taking the pretest on your BiologyNow CD-ROM.

THE RADIATA

Learning Objectives

6 Identify distinguishing characteristics of phylum Cnidaria, describe three main classes of this phylum, and give examples of animals that belong to each class.

7 Identify distinguishing characteristics of phylum Ctenophora.

Recall that all animals except the sponges are classified as Eumetazoa. The Radiata, animals with radial symmetry, may have been the first eumetazoans to evolve. Biologists currently classify cnidarians and ctenophores as Radiata. However, the relationships of these groups to other animals and to each other is still a matter of debate.

Cnidarians have unique stinging cells

Most of the 10,000 or so species of phylum **Cnidaria** (pronounced "ni-dah′-ree-ah") are marine. The radially symmetrical cnidarian body is organized as a hollow sac with the mouth and surrounding tentacles located at one end. Biologists assign cnidarians to three main classes (Table 28-2). Class **Hydrozoa** includes hydras, and *hydroids,* such as *Obelia* and the Portuguese man-of-war; class **Scyphozoa** includes jellyfish; and class **Anthozoa** includes sea anemones and corals. Some cnidarians live a solitary existence, whereas many others, such as corals, form colonies.

Cnidarians have two body shapes: the polyp and the medusa (Fig. 28-7). The **polyp** form, represented by *Hydra,* typically has a dorsal mouth surrounded by tentacles. In the **medusa** (pl., *medusae*), or jellyfish form, the mouth is located in the lower concave, or *oral,* surface; the convex upper surface is the *aboral* surface. Some cnidarians have the polyp shape during one stage of their life cycle and the medusa form during another stage. The Portuguese man-of-war and some other cnidarians consist of colonies of many individuals, some of which are polyps and others medusae.

Cnidarians get their name from specialized cells, called **cnidocytes** (from a Greek word meaning "sea nettles"), that

TABLE 28-2 Major Classes of Phylum Cnidaria

Class and Representative Animals	Characteristics
Hydrozoa *Hydra, Obelia,* Portuguese man-of-war	Mainly marine, but some freshwater species; alternation of polyp and medusa stages in most species (polyp form only in *Hydra*); some form colonies.
Scyphozoa Jellyfish	Mainly marine; typically inhabit coastal water, free-swimming medusa most prominent form; polyp stage often reduced.
Anthozoa See anemones, corals, sea fans	Marine; solitary or colonial polyps; in most no medusa stage; gastrovascular cavity divided by partitions into chambers, increasing area for digestion; sessile.

contain stinging organelles. Cnidocytes are located mainly in the epidermis, especially on the tentacles. The cnidocytes contain stinging "thread capsules," or **nematocysts** (Fig. 28-8). When stimulated, a nematocyst releases a coiled, hollow thread. Some types of nematocyst threads are sticky. Others are long and coil around prey. A third type bears barbs or spines and can inject a protein toxin that paralyzes prey animals such as small crustaceans. Each cnidocyte has a small, projecting trigger *(cnidocil)* on its outer surface. Stimuli such as touch or chemicals dissolved in the water stimulate the nematocyst to fire its thread.

Cnidarians use their tentacles to capture prey and push it into the mouth. The mouth leads into the **gastrovascular cavity,** where digestion takes place. The mouth is the only opening into the gastrovascular cavity and so must serve for both ingestion of food and expulsion of wastes. Gas exchange and excretion occur by diffusion. The body wall is thin enough that no cell is far from the surface.

More highly organized than the sponge, cnidarians are diploblastic; that is, they have two definite tissue layers. The ectoderm gives rise to the outer **epidermis,** a protective layer covering the body. The endoderm gives rise to the inner **gastrodermis,** which lines the gastrovascular cavity and functions in digestion. These thin layers are separated by a gelatinous, mainly acellular, **mesoglea.**

Cnidarians have nerve cells that form **nerve nets** connecting sensory cells in the body wall to contractile and gland cells. Sense organs—for example, photoreceptors which detect light—are positioned around the edge of the body. An impulse set up by one of them passes in all directions more or less equally. The nerve cells are not organized to form a brain or nerve cord.

Both the epidermis and gastrodermis have cells specialized to contract (however, they are not true muscle cells). Contractile fibers in the epidermal cells run lengthwise, and those in the gastrodermis run circularly. These two sets of contractile cells act on the water-filled gastrovascular cavity, which forms a **hydrostatic skeleton** that supports the body and allows move-

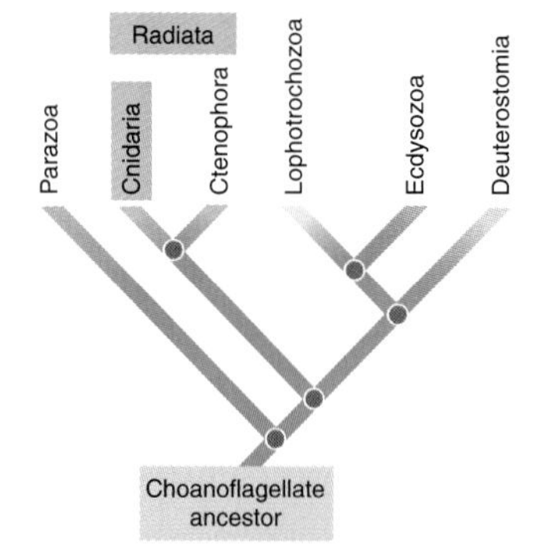

FIGURE **28-7** **Polyp and medusa body forms of cnidarians.**

(a) This hydrozoan (*Gonothyraea loveni*) forms a colony of polyps. The drawing illustrates a single polyp. **(b)** The sea nettle (*Chrysaora fuscescens*), like other jellyfish, uses its tentacles equipped with cnidocytes to capture small animals (zooplankton) suspended in the water and to carry this food to the mouth. **(c)** Coral polyps (*Montastrea cavernosa*) extended for feeding.

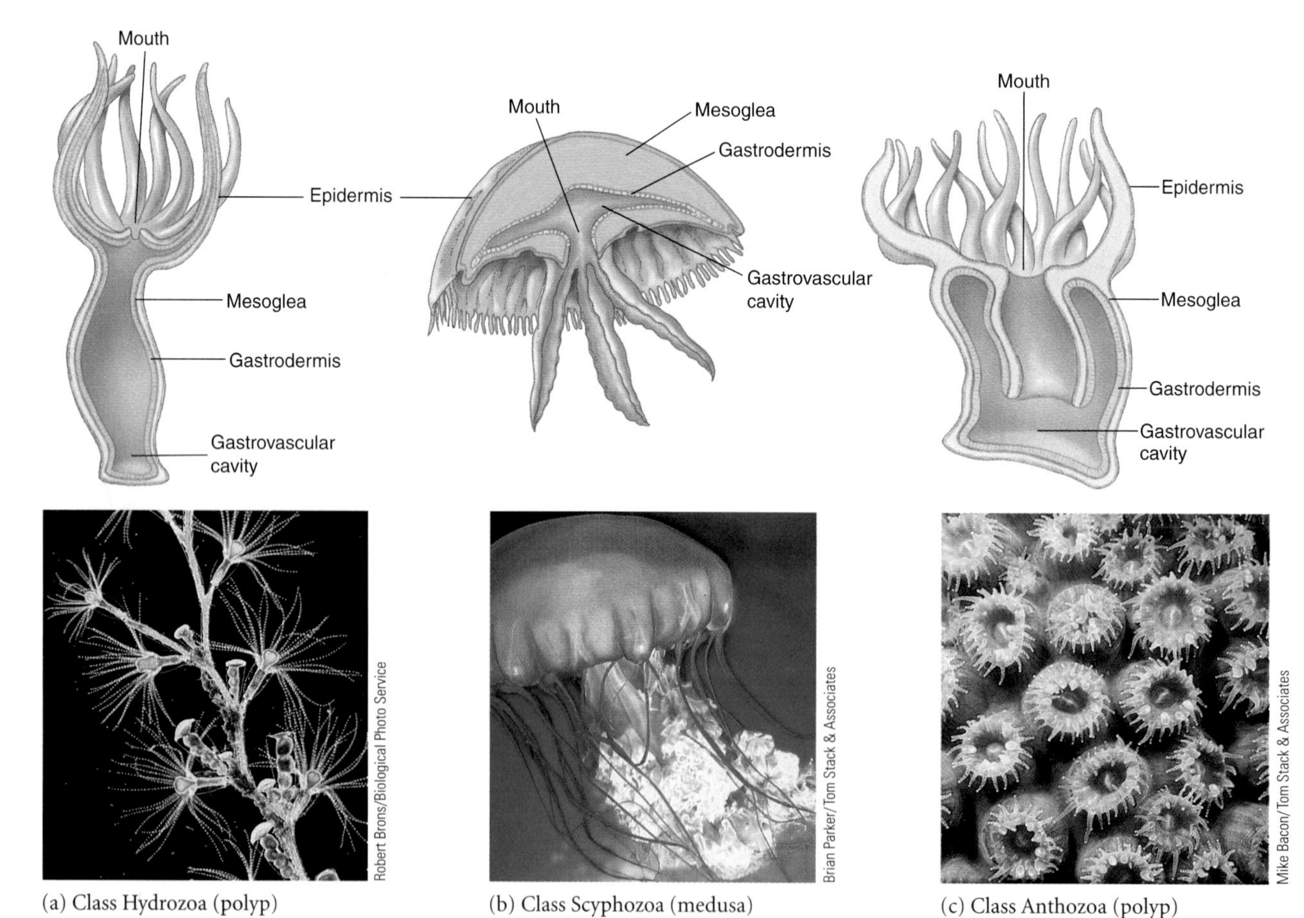

(a) Class Hydrozoa (polyp)

(b) Class Scyphozoa (medusa)

(c) Class Anthozoa (polyp)

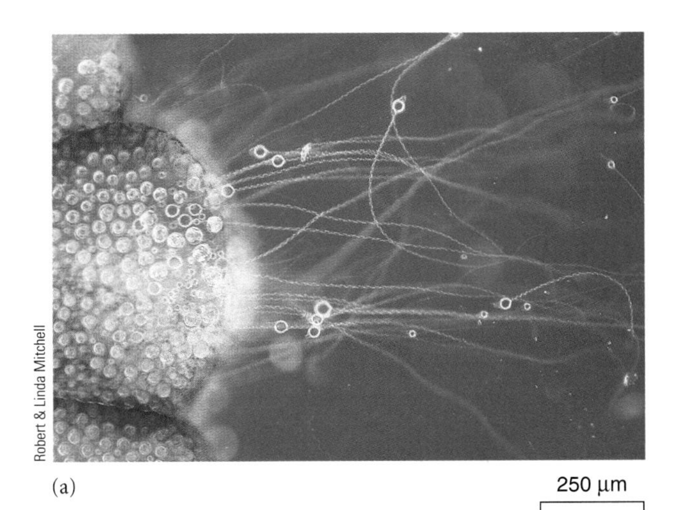

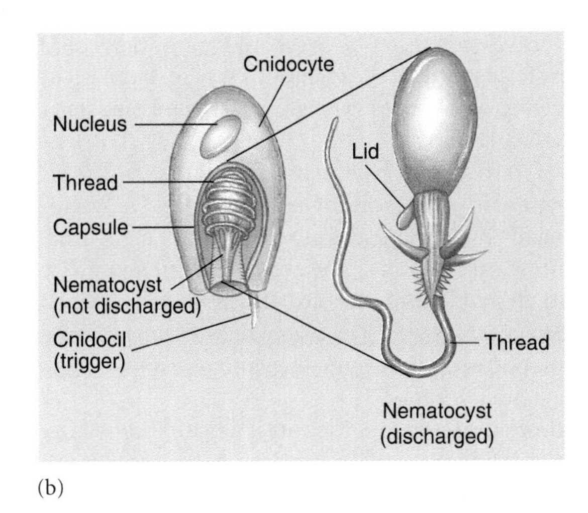

Robert & Linda Mitchell

(a) (b)

FIGURE 28-8 Nematocysts.

When cnidarian stinging cells (cnidocytes) are stimulated, the nematocyst discharges, ejecting a thread that may entangle or penetrate the prey. Some nematocysts secrete a toxic substance that immobilizes the prey. **(a)** LM of discharged nematocysts of a Portuguese man-of-war (*Physalia physalis*). Photographed in the Gulf of Mexico. **(b)** Undischarged and discharged nematocyst. The cnidocil, or trigger, is a mechanoreceptor that discharges the nematocyst when it senses contact with an object.

ment. By contracting one set of contractile cells or the other, the hydra can shorten, lengthen, or bend its body (see Fig. 38-3).

Class Hydrozoa includes solitary and colonial forms

Although not really typical, the cnidarian beginning biology students most often study is the tiny, solitary *Hydra* found in freshwater ponds (Fig. 28-9). To the naked eye *Hydra* looks like a bit of frayed string. Because it has a remarkable ability to regenerate, biologists named *Hydra*

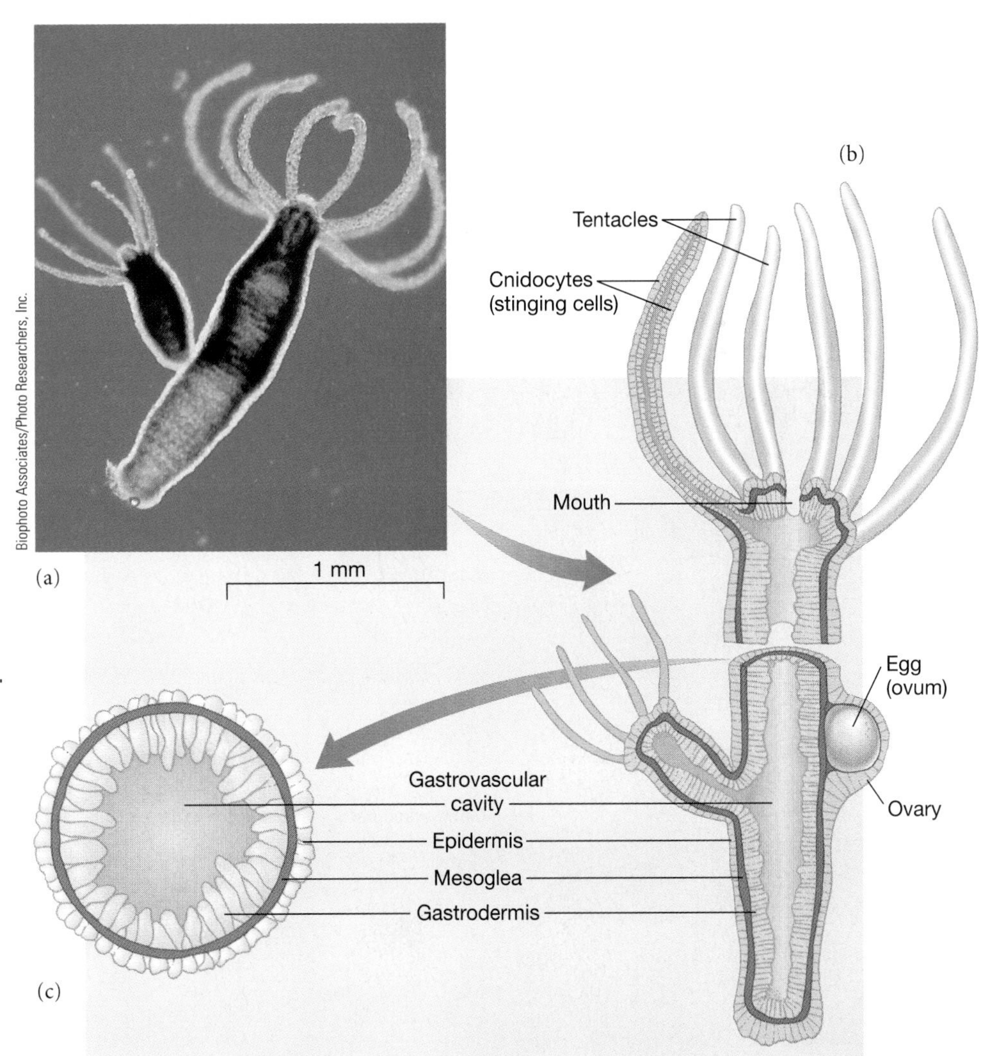

Biophoto Associates/Photo Researchers, Inc.

(a) (b) (c)

FIGURE 28-9

Hydra, a freshwater hydrozoan.

(a) *Hydra viridis* with a large bud. When the bud separates from its parent, it becomes an independent individual. **(b)** Hydra cut longitudinally to show its internal structure. Asexual reproduction by budding is represented on the left; sexual reproduction is represented by the ovary on the right. Male hydras develop testes that produce sperm. **(c)** Cross section through the body of a hydra.

after the multiheaded monster of Greek mythology that could grow two new heads for each head cut off. When *Hydra* is cut into several pieces, each piece may regrow all the missing parts and become a whole animal.

Hydra lives in fresh water and typically attaches to a rock, aquatic plant, or detritus by a disc of cells at its base. Hydras reproduce asexually by budding during periods when environmental conditions are optimal. However, they differentiate as males and females and reproduce sexually in the fall or when pond water becomes stagnant. The zygote may become covered with a shell that protects it through the winter or until conditions become more favorable.

Many hydrozoans form colonies consisting of hundreds or thousands of individuals. A colony begins with a single polyp that reproduces asexually by budding. However, instead of separating from the parent, the bud remains attached and eventually forms additional buds. Several types of individuals may arise in the same colony, some specialized for feeding, some for reproduction, and others for defense.

Some marine cnidarians are remarkable for an alternation of sexual and asexual stages. This alternation of stages differs from the alternation of generations in plants in that both sexual and asexual forms are diploid; only sperm and eggs are haploid. The life cycle of the colonial marine hydrozoan *Obelia* illustrates alternation of sexual and asexual stages (Fig. 28-10).

The medusa stage is dominant among the jellyfish

Among the jellyfish, members of class Scyphozoa, the medusa is the dominant body form. Scyphozoan medusae are generally larger than hydrozoan medusae, and they have a thick, viscous mesoglea that gives firmness to the body. In scyphozoans, the polyp stage is small and inconspicuous or may even be absent. The largest jellyfish, *Cyanea,* may be more than 2 m (6.5 ft) in diameter and have tentacles 30 m (98 ft) long. These orange and blue "monsters," among the largest invertebrates, are dangerous to swimmers in the North Atlantic Ocean.

Anthozoans occur only as polyps

Sea anemones and corals, members of class Anthozoa, have either individual or colonial polyps but no free-swimming medusa stage. The polyp produces eggs and sperm, and the fertilized egg develops into a small, ciliated larva called a **planula.** This larval form may swim to a new location before attaching to develop into a polyp.

Anthozoans differ from hydrozoans in that a series of vertical partitions partially divides the gastrovascular cavity into a number of connected chambers. The partitions increase the surface area for digestion, enabling an anemone to digest an ani-

FIGURE **28-10** ***Obelia,* a marine colonial hydrozoan.**

(a) LM of *Obelia*. Some polyps have tentacles and are specialized for feeding, whereas others are specialized for reproduction. **(b)** Life cycle of *Obelia*. Reproductive polyps give rise asexually to medusae. The free-swimming medusae reproduce sexually, and the zygote develops into a planula larva. The larva develops into a polyp that forms a new colony. The colony grows as new polyps bud and remain attached.

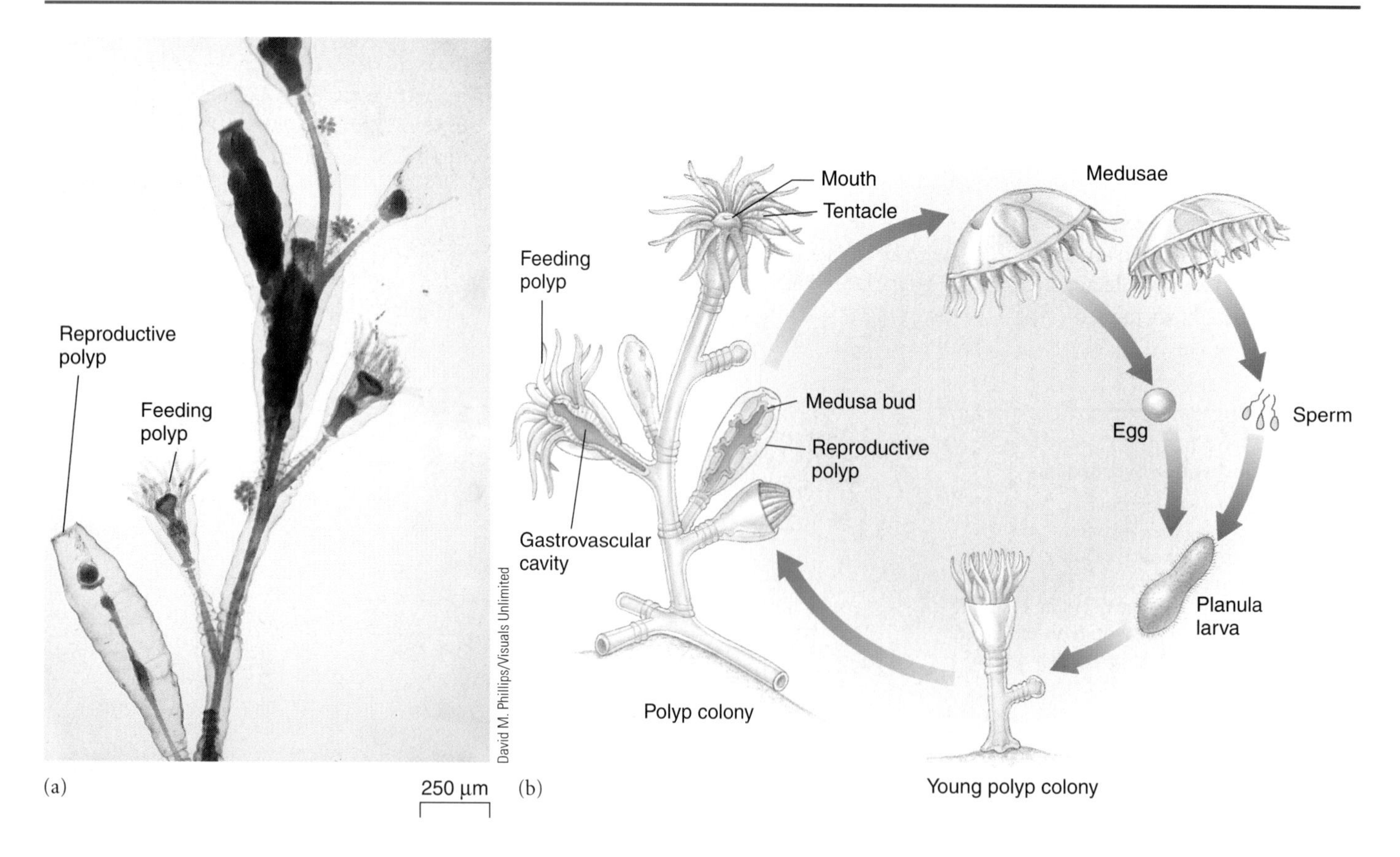

mal as large as a crab. Although corals can capture prey, many tropical species depend for nutrition on photosynthetic algae *(zooxanthellae)* that live within cells lining the coral's digestive cavity (see Chapter 24 and Fig. 52-11). The relationship between coral and zooxanthellae is symbiotic and mutually beneficial. The algae provide the coral with oxygen and with carbon and nitrogen compounds. In exchange, the coral supplies the algae with waste products such as ammonia, from which the algae make nitrogenous compounds for both partners.

In warm, shallow seas, much of the bottom is covered with coral or anemones, most of them brightly colored. Coral reefs consist of colonies of millions of corals, and of certain algae (mainly coralline red algae). Living colonies occur only in the uppermost regions of such reefs, adding their own skeletons to the forming rock. Coral reefs are among the most productive of all ecosystems, rivaling tropical rain forests in species diversity (see Chapter 54). A single reef can serve as home for more than 3000 species of fishes and other marine organisms, and an estimated one fourth of all marine species depend on coral reefs.

During the past several years many coral reefs, especially those in coastal waters, have suffered serious damage from human activities. These include overfishing, mining reefs for building materials, polluting coastal waters with industrial chemicals, and smothering coral with the silt washed downstream from clearcut forests. **Coral bleaching** is the stress-induced loss of the colorful symbiotic algae (zooxanthellae) that inhabit coral cells (Fig. 28-11). Without their algae, coral become malnourished and die. Environmental factors suspected to contribute to coral bleaching include unusually high (associated with global warming) or low temperatures, pollution, changes in salinity, disease, and high doses of ultraviolet radiation (associated with the destruction of the ozone layer).

Photo by Ove Hoegh-Guldberg, Centre for Marine Studies, University of Queensland.

FIGURE **28-11** **Bleached staghorn coral.**
This staghorn coral (*Acropora*) was photographed off Heron Island on the Great Barrier Reef off the coast of Australia.

Comb jellies have adhesive glue cells that trap prey

The 100 or so species of comb jellies, members of phylum **Ctenophora,** all are marine. They are fragile, luminescent animals that may be as small as a pea or larger than a tomato. The outer surface of a ctenophore bears eight rows of cilia, resembling combs (Fig. 28-12). The coordinated beating of the cilia in these combs moves the animal through the water. A sense organ functions in balance and helps the animal orient itself. Some ctenophores have two tentacles. They do not have the stinging nematocysts characteristic of the cnidarians. However, their tentacles are equipped with adhesive glue cells, that trap prey.

Ctenophores are biradially symmetrical, meaning you could obtain equal halves by cutting through the body axis in two different ways. Because of their radial symmetry and feeding tentacles, the ctenophore body plan is somewhat similar to that of a cnidarian medusa. Ctenophores are also like cnidarians in that they consist of two cell layers separated by a thick, jelly-like mesoglea. Because of these similarities, biologists classify ctenophores near the cnidarians. However, their development is

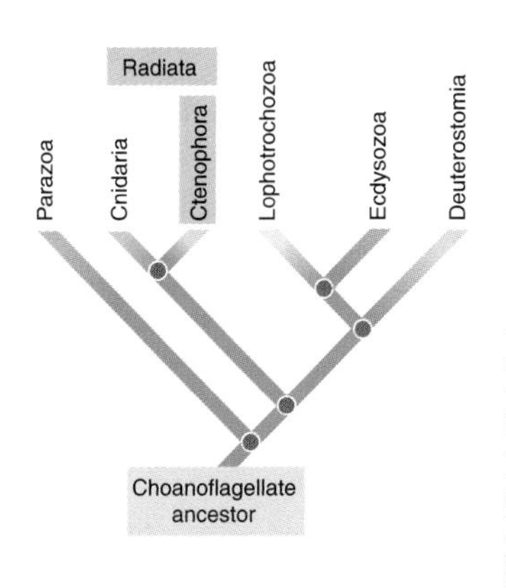

David Wrobel/Biological Photo Service

FIGURE **28-12** **Ctenophore (comb jelly).**
Ctenophores are free-swimming, bioluminescent hermaphrodites capable of self-fertilization. The sea gooseberry (*Pleurobrachia*) has two long tentacles with adhesive cells used to capture prey.

different, and unlike cnidarians, their digestive system has a mouth for food intake at one end and two anal pores for the egestion of water and wastes at the other end. The similarities between ctenophores and cnidarians may be a result of convergent evolution from living in a similar environment, the ocean. Where, then, do the ctenophores belong on our cladogram (see Fig. 28-5)? Systematists will decide as they gather new data.

Review

- In what ways do cnidarians differ from sponges?
- Draw the life cycle of *Obelia.*
- How is a ctenophore like a cnidarian? In what ways is it different?

BiologyNow™ Assess your understanding of **the Radiata** by taking the pretest on your BiologyNow CD-ROM.

SUMMARY WITH KEY TERMS

1 List several characteristics common to most animals.

- Members of **kingdom Animalia** are eukaryotic, multicellular, **heterotrophic** organisms with cells specialized to perform specific functions. Typically, they are capable of locomotion at some time during their life cycle, can respond adaptively to external stimuli, and can reproduce sexually.
- In sexual reproduction, sperm and egg unite to form a **zygote** that undergoes **cleavage.** Multiple cell divisions result in the development of a hollow ball of cells, a **blastula.** The blastula undergoes **gastrulation,** forming embryonic tissues.

2 Compare the advantages and disadvantages of life in the ocean, in fresh water, and on land.

- Marine environments have relatively stable temperatures, provide buoyancy, and provide readily available food. Fluid and salt balance are more easily maintained than in fresh water. Currents and other water movements are a disadvantage.
- Fresh water offers a less constant environment and less food. Because fresh water is hypotonic to tissue fluid, animals must osmoregulate.
- Terrestrial animals must avoid dessication, deal with temperature changes, and protect gametes and embryos.

3 Critique the use of type of symmetry and type of body cavity to infer relationships among animal phyla.

- Many biologists have used the type of **body plan,** the basic structure of the body, to infer evolutionary relationships. For example, they hypothesized that cnidarians and ctenophores were closely related because they share **radial symmetry;** most other animals are **bilaterally symmetrical,** at least in their larval stages. However, analysis of more recent data suggests that cnidarians and ctenophores are not closely related. Adult echinoderms also have radial symmetry and are not related to these groups.
- Biologists have also inferred relationships based on level of tissue development and type of body cavity. Embryonic tissues, called **germ layers,** include the outer **ectoderm** which gives rise to the body covering and the nervous system; the inner, **endoderm,** which lines the gut and other digestive organs; and a middle, **mesoderm,** which gives rise to most other body structures.
- **Triploblastic** animals have traditionally been classified as **acoelomate** (no body cavity), **pseudocoelomate** (body cavity not completely lined with mesoderm), or **coelomate,** those with a true **coelom,** a body cavity completely lined with mesoderm. Recent data indicate pseudocoelomate animals do not form a natural group and probably evolved from coelomate ancestors.

4 Describe the distinguishing characteristics of the three major clades of coelomate animals: Lophotrochozoa, and Ecdysozoa, and Deuterostomia.

- Two major evolutionary branches of coelomates are **Protostomia** (mollusks, annelids, arthropods) and **Deuterostomia** (echinoderms and chordates). In protostomes the blastopore develops into the mouth; in deuterostomes the blastopore typically becomes the anus.
- Based on molecular data, many biologists now subdivide the protostomes into two clades, the **Lophotrochozoa** and the **Ecdysozoa.** The Lophotrochozoa include the platyhelminthes (flat worms), nemerteans (ribbon worms), mollusks, annelids, the lophophorate phyla (groups that have a ciliated ring of tentacles surrounding the mouth), and rotifers. The Ecdysozoa, animals that molt, include the nematodes and arthropods.

5 Identify distinguishing characteristics of phylum Porifera.

- Phylum **Porifera** consists of the sponges, animals characterized by flagellate **collar cells (choanocytes).** Sponges are the only members of the **Parazoa.**
- The sponge body is a sac with tiny openings through which water enters; a central cavity, or **spongocoel;** and an open end, or **osculum,** through which water exits. The cells of sponges are loosely associated; they do not form true tissues.

6 Identify distinguishing characteristics of phylum Cnidaria, describe three main classes of this phylum, and give examples of animals that belong to each class.

- Phylum **Cnidaria** is characterized by radial symmetry, two tissue layers, and **cnidocytes,** cells that contain stinging organelles called **nematocysts.** The **gastrovascular cavity** has a single opening that serves as both mouth and anus. Nerve cells form irregular, nondirectional **nerve nets** that connect sensory cells with contractile and gland cells.
- The life cycle of many cnidarians includes a sessile **polyp** stage and a free-swimming **medusa** stage.
- Phylum Cnidaria includes three main classes. Class **Hydrozoa** (hydras, hydroids, and the Portuguese man-of-war) are typically polyps and may be solitary or colonial. Class **Scyphozoa** (the jellyfish) are generally medusae. Class **Anthozoa** (sea anemones and corals) are polyps and may be solitary or colonial; anthozoans differ from hydrozoans in the organization of the gastrovascular cavity.

7 Identify distinguishing characteristics of phylum Ctenophora.

- Phylum **Ctenophora** consists of the comb jellies, which are fragile, luminescent marine predators with biradial symmetry, and eight rows of cilia that resemble combs. They are diploblastic and have tentacles with adhesive glue cells.

POST-TEST

1. Which of the following is *not* a defining characteristic of animals? (a) heterotrophs (b) multicellular (c) eukaryotic (d) presence of a coelom (e) form a zygote that undergoes cleavage
2. Which of the following is *not* an adaptation to terrestrial living? (a) internal fertilization (b) shell surrounding egg (c) adaptations for maintaining body temperature (d) surface for gas exchange deep in body (e) ability to maintain location
3. The Parazoa (a) are coelomates (b) include cnidarians (c) include all animals with radial symmetry (d) do not form true tissues (e) have a pseudocoelom
4. Cephalization (a) evolved along with bilateral symmetry (b) is the development of a digestive system (c) is characteristic of cnidarians (d) involves a concentration of excretory organs (e) first evolved in deuterostomes
5. The germ layer that gives rise to the outer covering of the body and the nervous system is the (a) gastrodermis (b) ectoderm (c) contractile layer (d) endoderm (e) mesoderm
6. Radial symmetry is characteristic of (a) protostomes (b) acoelomates (c) deuterostomes (d) cnidarians (e) Porifera
7. A true coelom is completely lined with (a) flagella (b) ectoderm (c) a contractile layer (d) endoderm (e) mesoderm
8. Protostomes are characterized by (a) spiral cleavage (b) indeterminate cleavage (c) enterocoely (d) radial symmetry (e) a distinctive body plan that includes a pseudocoelom
9. The evolution of animals (a) followed an orderly progression from simple to complex (b) will be better understood as biologists continue to collect molecular and other types of data (c) has been determined by studying classification (d) began with the placozoans (e) began with their common ancestor, a cnidarian
10. During cleavage, an animal (a) undergoes metamorphosis (b) becomes a larva (c) undergoes a series of mitotic divisions and becomes a blastula (d) becomes diploid (e) reproduces sexually
11. Collar cells (choanocytes) are characteristic of phylum (a) Porifera (b) Cnidaria (c) Coelomata (d) Lophotrochozoa (e) Ecdysozoa
12. Which of the following is an example of a deuterostome? (a) a lophotrochozoan (b) coral (c) chordate (d) planarian (e) a pseudocoelomate
13. Cnidocytes are (a) characteristic of sponges (b) found among cells lining the gastrovascular cavity (c) contain stinging organelles (d) are lined with mesoderm (e) have two main shapes
14. The marine hydrozoan *Obelia* (a) is seen mainly in the medusa form (b) lacks cnidocytes (c) has reproductive polyps that produce eggs and sperm (d) is one of the largest jellyfish known (e) alternates sexual and asexual stages during its life cycle
15. Corals (a) are coelomates (b) form symbiotic relationships with certain algae (c) lack a polyp stage (d) are remarkably resistant to environmental stressors (e) have bilateral symmetry
16. Ctenophores (a) tend to be sessile (b) have a digestive system with only one opening (c) exhibit bilateral symmetry (d) are characterized by adhesive glue cells on their tentacles (e) are free-swimming polyps
17. Label the branches of the diagram. Use Figure 28-5 to check your answers.

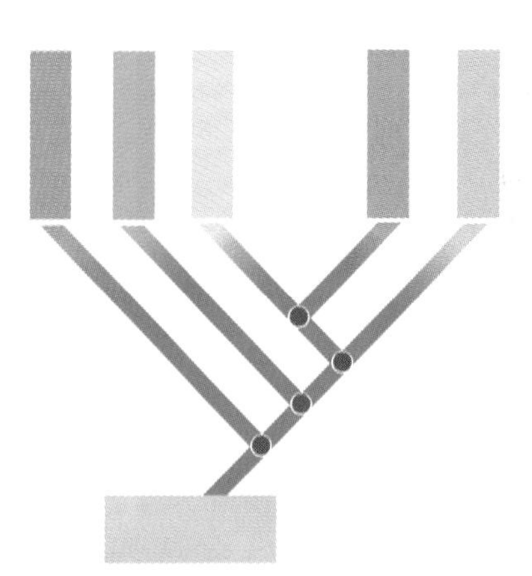

CRITICAL THINKING

1. Imagine you discover a new animal in a rain forest. How would you decide to which phylum it belongs? What are some characteristics that might contribute to your decision?
2. Several international monitoring projects are gathering data to help us understand coral reef destruction. Why is it important to take action to protect coral reefs?

- Visit our Web site at **http://biology.brookscole.com/solomon7** for links to chapter-related resources on the World Wide Web. Additional online materials relating to this chapter can also be found on our Web site.

BIOLOGY NOW RESOURCES

Active Figures

28-1: The characteristics of animals

28-5: The evolutionary relationships of animals

Preparing for an exam? Take a diagnostic test on your BiologyNow CD-ROM.

Post-Test Answers

1. d	2. e	3. d	4. a
5. b	6. d	7. e	8. a
9. b	10. c	11. a	12. c
13. c	14. e	15. b	16. d

29

The Animal Kingdom: The Protostomes

Marty Snyderman/Visuals Unlimited

A bearded fireworm (*Hermodice carunculata*). The fireworm is a segmented worm (an annelid) up to 30 cm (11.8 in) in length. It lives in shallow marine waters—in coral reefs, beds of turtle grass, or under rocks. Fireworm bristles, which are filled with venom, can break off in the skin, causing irritation.

CHAPTER OUTLINE

- **The Lophotrochozoa**
- **The Ecdysozoa**

The **coelom,** one of the most important early animal adaptations, is a fluid-filled body cavity completely lined by mesoderm that lies between the digestive tube and the outer body wall (see Chapter 28). Evolution of the coelom produced a new body design, a **tube-within-a-tube plan.** The digestive tube (the inner tube) is attached to the body wall (the outer tube) at its ends. Typically, the digestive tube has a mouth at one end and an anus for eliminating wastes at the other end. Because the coelom separates the muscles of the body wall from those in the wall of the digestive tract, the digestive tube can move food along independently of body movements.

Animals with a coelom, and to a lesser extent those with a pseudocoelom, have another major advantage over acoelomate animals. Because it is an enclosed compartment (or series of compartments) of fluid under pressure, the coelom can serve as a **hydrostatic skeleton** in which contracting muscles push against a tube of fluid. In many coelomates, such as the fireworm shown here, a hydrostatic skeleton permits a greater range of movement than in acoelomate animals. The evolution of various shapes and divisions of the coelom provided the opportunity for animals to become specialized in swimming, crawling, or walking. Animals that can move quickly to capture food or avoid predators are more likely to survive. The hydrostatic skeleton also shapes the body of soft animals.

Evolution of the coelom provided a space where internal organs could develop. Most coelomate animals have well-developed circulatory, excretory, and nervous systems. Many internal organs are suspended within folds of the tissue lining the coelom and can move independently of the outer body wall. For example, the pumping action of the heart is possible because of the surrounding space the coelom provides. The fluid-filled coelom protects internal organs by cushioning them. The coelom also provides space for the gonads to develop. During the breeding season of many animals, such as birds, the gonads enlarge within the coelom as they fill with ripe gametes.

In some animals, fluid within the coelom helps transport materials such as food, oxygen, and wastes. Cells bathed by the coelomic fluid can exchange materials with it. The cells receive nutrients and oxygen from the coelomic fluid and excrete wastes into it. Some coelomates have excretory structures that remove wastes directly from the coelomic fluid.

Recall from Chapter 28 that animals with a coelom form two major branches: protostomes and deuterostomes, and that the **protostome coelomates** evolved along two branches: the Lophotrochozoa and the Ecdysozoa. These groups are sometimes referred to as *superphyla*. In this chapter we discuss the evolution, relationships, body plans, and life histories of several groups of the Lophotrochozoa and Ecdysozoa. ■

THE LOPHOTROCHOZOA

Learning Objectives

1. Characterize the protostome coelomates, describe their two main evolutionary branches, and give examples of animals assigned to each branch.
2. Identify distinguishing characteristics of phylum Nemertea and phylum Platyhelminthes; describe the main classes of phylum Platyhelminthes, giving examples of animals that belong to each class.
3. Describe the adaptive advantages of cephalization.
4. Describe the distinguishing characteristics of phylum Mollusca, and describe the classes discussed, giving examples.
5. Describe the distinguishing characteristics of phylum Annelida, and describe the three classes of annelids discussed, giving examples.
6. Describe the distinguishing characteristics of the lophophorate phyla.

The **Lophotrochozoa** include the platyhelminthes, nemerteans (ribbon worms), mollusks, annelids, the lophophorate phyla, and the rotifers (see Fig. 28-5). The name Lophotrochozoa comes from the names of the two major animal groups included: the Lophophorata and the Trochozoa. The *Lophophorata* include the lophophorate phyla, which are characterized by a lophophore, a ciliated ring of tentacles surrounding the mouth. The name *Trochozoa* refers to the trochophore larva that characterizes its two major groups—the mollusks and annelids.

Flatworms are bilateral acoelomates

Members of phylum **Platyhelminthes,** the **flatworms,** are flat, elongated, acoelomate animals that are bilaterally symmetrical. These soft-bodied animals left few fossils to provide clues to their evolutionary history. Biologists have historically assigned the 20,000 species to four classes. Class **Turbellaria** consists of the free-living flatworms, including planarians and their relatives. Classes **Trematoda** and **Monogenea** include the flukes, which are either internal or external parasites. Class **Cestoda** includes the tapeworms, which as adults are intestinal parasites of vertebrates (Table 29-1).

PROCESS OF SCIENCE

You may wonder why, if flatworms are acoelomate animals, we classify them as lophotrochozoans, which are coelomate animals. Molecular data, specifically 18S ribosomal RNA sequences, indicate phylum Platyhelminthes is not a monophyletic group. The flatworms known as *acoels*, traditionally classified in one order of class Turbellaria, differ significantly from other flatworms. The acoels may be closely related to the common ancestor that linked radially symmetrical animals with bilateral animals. Other flatworms appear more closely related to coelomates. If so, their body plan became simplified later in their evolutionary history. As biologists consider new data, their assessment of the phylogenetic relationships of flatworms may change.

Flatworms exhibit *bilateral symmetry* and have definite anterior and posterior ends. The beginnings of **cephalization** are evident. A simple brain and paired sense organs are concentrated at the anterior end. An animal with a front end ("head") generally moves forward. With a concentration of sense organs in the part of the body that first meets its environment, the animal can find food or detect an enemy quickly.

Flatworms are triploblastic (have all three germ layers) and have *three definite tissue layers.* In addition to an outer epidermis derived from ectoderm, and an inner endodermis derived from endoderm, the flatworm has a middle tissue layer that develops from mesoderm. Flatworms have no organs for circulation or gas exchange. These functions depend largely on diffusion through the body wall. The activities of some organs are coordinated, and form simple organ systems, for example, the digestive and nervous systems. As in the cnidarians, the digestive system has only one opening, a mouth. The *gastrovascular cavity* is often extensively branched.

Flatworms typically have a simple **nervous system.** The brain consists of two masses of nervous tissue, called **ganglia,** in the head region. In many species, the ganglia connect to two nerve cords that extend the length of the body. This nervous system is

TABLE 29-1 Classes of Phylum Platyhelminthes

Class and Representative Animals	Characteristics
Turbellaria Planarians	Mainly free-living; mainly marine; body covered by ciliated epidermis; typically carnivorous; prey on tiny invertebrates.
Trematoda and Monogenea Flukes	All parasites with a wide range of vertebrate and invertebrate hosts; may require intermediate hosts; adults have suckers for attachment to host.
Cestoda Tapeworms	Parasites of vertebrates; complex life cycle usually with one or two intermediate hosts; larval host may be invertebrate; typically have suckers and sometimes hooks for attachment to host; eggs produced within proglottids, which are shed; no digestive or nervous systems.

sometimes referred to as a "ladder-type nervous system," because a series of nerves connects the cords like the rungs of a ladder.

Parasitic flatworms—flukes and tapeworms—are highly adapted to and modified for their parasitic lifestyle. They have suckers or hooks for holding onto their hosts. The bodies of those that live in digestive tracts resist the digestive enzymes secreted by their hosts. Many have complicated life cycles and produce large numbers of eggs. Other adaptations include the loss of certain structures such as sense organs. Tapeworms have also lost the digestive system.

Planarians are free-living flatworms

Most members of class Turbellaria are free-living flatworms. Most are marine, but many inhabit freshwater habitats, and some are terrestrial. Turbellarian flatworms typically have a muscular pharynx that takes in food; a simple brain, eyespots and other sensory organs in the head; **protonephridia,** structures that function in osmoregulation and metabolic waste disposal (excretion); and complex reproductive organs. **Planarians** are turbellarian flatworms found in ponds and quiet streams throughout the world. The common American planarian *Dugesia* is about 15 mm (0.6 in) long, with what appear to be crossed eyes and flapping "ears" called **auricles** (Fig. 29-1). The auricles actually serve as organs of chemoreception, important in locating food.

Planarians are carnivorous, trapping small animals in a mucous secretion. The digestive system consists of a single opening (the mouth), a tubelike, muscular **pharynx** (the first portion of the digestive tube), and a branched gastrovascular cavity. A planarian projects its pharynx outward through its mouth, using it to suck in prey. The long, highly branched gastrovascular cavity distributes food to all parts of the body, so each cell can receive nutrients by diffusion.

Although excretion also takes place mainly by diffusion, some metabolic wastes are excreted by the *protonephridia.* These structures function primarily in fluid balance, or *osmoregulation.* Protonephridia are blind tubules that end in **flame bulbs,** collecting cells equipped with cilia. The beating of the cilia channels waste through the system of tubules and eventually out of the body through excretory pores. Osmoregulation and protonephridia are discussed further in Chapter 46 (see Fig. 46-3).

Planarians are capable of learning. Memory is not localized within the ganglia but appears to be retained throughout the nervous system. Planarians reproduce either asexually or sexually. In asexual reproduction, an individual constricts in the middle and divides into two planarians. Each regenerates its missing parts. Sexually, these animals are **hermaphrodites.** During the warm months of the year, each is equipped with a complete set of male and female organs. Two planarians come together in copulation and exchange sperm cells so that their eggs are cross-fertilized.

Flukes parasitize other animals

Although their body plan resembles that of the free-living flatworms, the **flukes,** members of classes Trematoda and Monogenea, have structures, such as hooks and suckers, for attachment to the host. Flukes also have extremely complex and prolific reproductive organs.

Flukes that are parasitic in humans include blood flukes, widespread in tropical areas of the world, and liver flukes, common in Asia, particularly in areas where humans use their own feces for fertilizing crops. Blood flukes of the genus *Schistosoma* infect about 200 million people who live in tropical areas. Both blood flukes and liver flukes go through complicated life cycles involving a number of different forms, alternation of sexual and asexual stages, and parasitism on one or more intermediate hosts (such as snails and fishes) (Fig. 29-2). The aquatic snails that serve as intermediate hosts thrive in ponds, rice paddies, and marshy areas that form when dams are built.

Tapeworms inhabit the intestine of vertebrates

Adult members of the more than 5000 different species of class Cestoda live as parasites in the intestine of probably every kind

FIGURE **29-1** | **The common planarian, *Dugesia*.**

(a) Internal structure. **(b)** LM of a living planarian (*Dugesia dorotocephala*). Note the prominent auricles, used to locate food. **(c)** Cross section through a planarian.

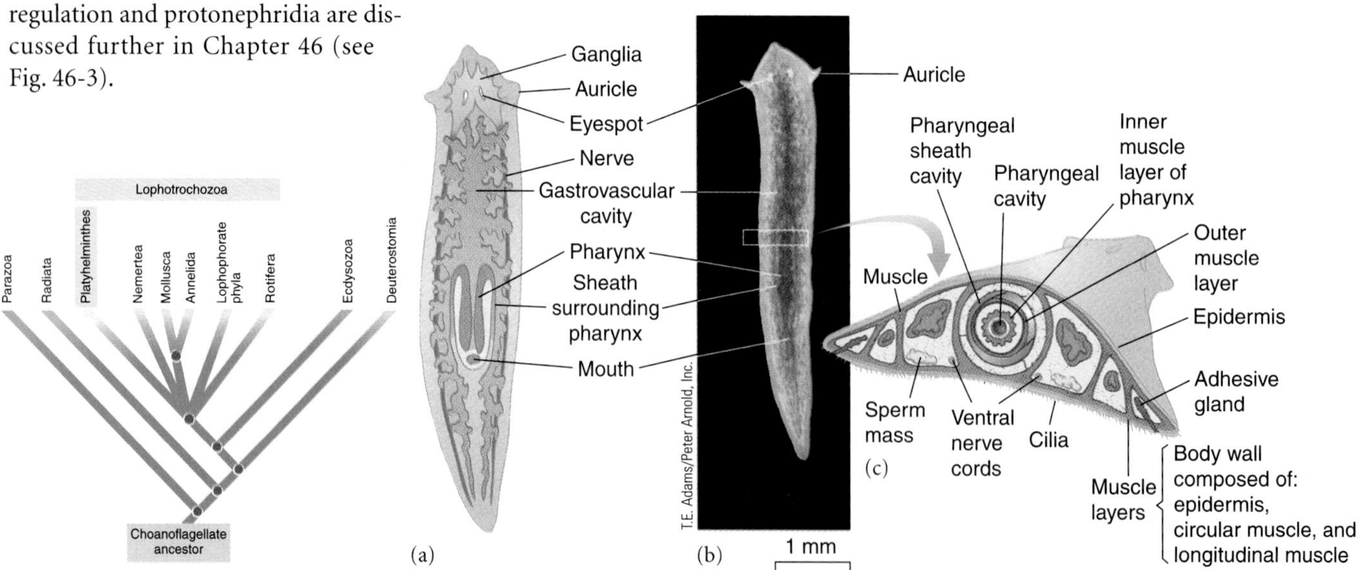

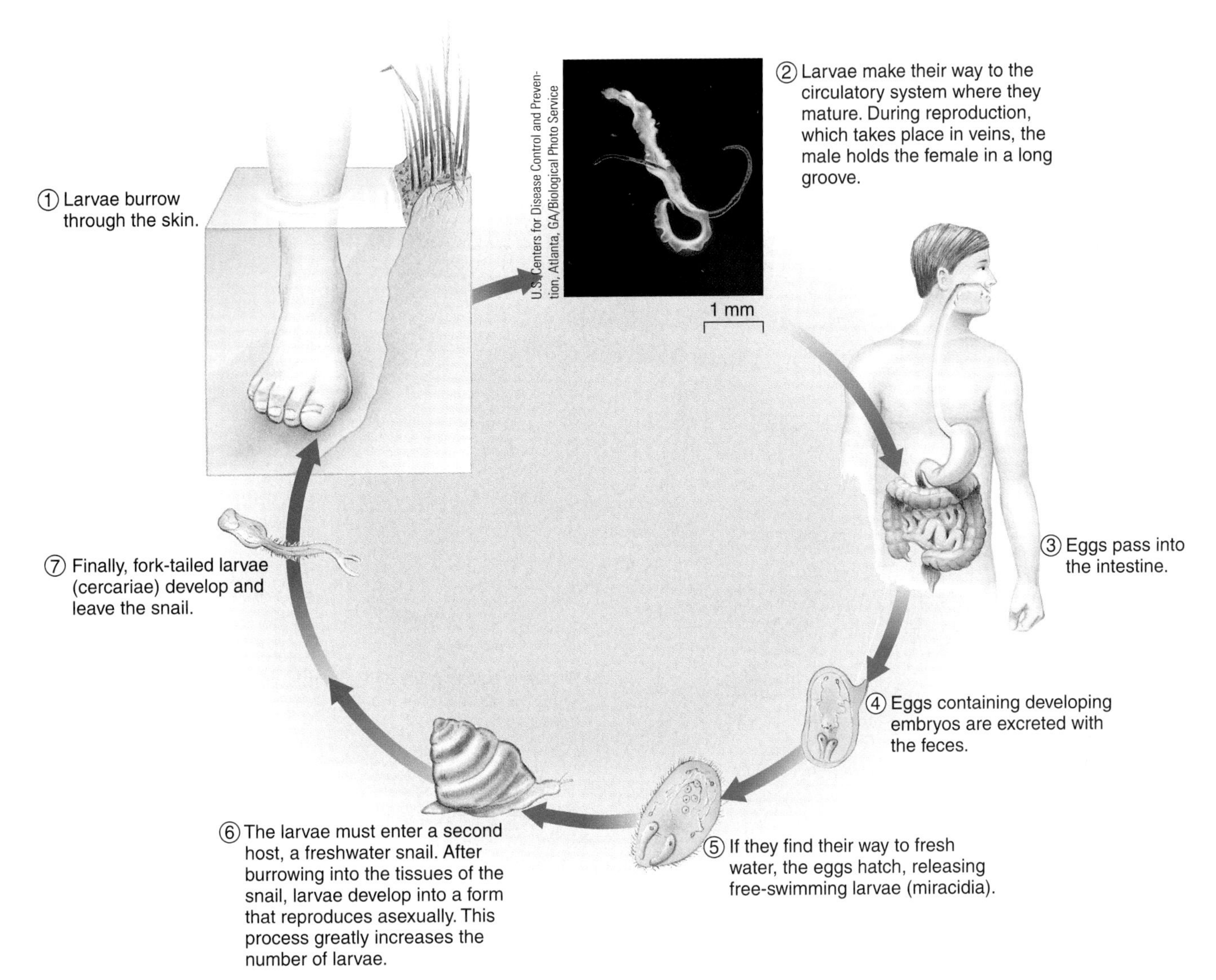

U.S. Centers for Disease Control and Prevention, Atlanta, GA/Biological Photo Service
① Larvae burrow through the skin.
② Larvae make their way to the circulatory system where they mature. During reproduction, which takes place in veins, the male holds the female in a long groove.
1 mm
③ Eggs pass into the intestine.
④ Eggs containing developing embryos are excreted with the feces.
⑤ If they find their way to fresh water, the eggs hatch, releasing free-swimming larvae (miracidia).
⑥ The larvae must enter a second host, a freshwater snail. After burrowing into the tissues of the snail, larvae develop into a form that reproduces asexually. This process greatly increases the number of larvae.
⑦ Finally, fork-tailed larvae (cercariae) develop and leave the snail.

ACTIVE FIGURE **29-2** ▲ **Life cycle of the blood fluke (*Schistosoma*).**

The LM shows an adult male enfolding a smaller female.

BiologyNow™ Watch **the fluke life cycle** by clicking on this figure on your BiologyNow CD-ROM.

of vertebrate, including humans. Tapeworms are long, flat, ribbon-like animals strikingly specialized for their parasitic mode of life. Among their many adaptations are suckers and sometimes hooks on the "head," or **scolex,** that enable the parasite to attach to the host's intestine (Fig. 29-3).

The reproductive adaptations and abilities of tapeworms are extraordinary. The body of the tapeworm consists of a long

FIGURE **29-3** ► **Scolex of the tapeworm.**

The small tapeworm (***Acanthrocirrus retrisrostris***) reaches maturity in the intestine of wading birds that eat barnacles. The color-enhanced SEM shows the piston-like cluster of hooks that can be withdrawn into the head or thrust out and buried in the host's tissue. Beneath the hooks, two of the four powerful suckers are visible.

chain of segments called **proglottids.** Each proglottid is an entire reproductive machine equipped with both male and female reproductive organs and containing up to 100,000 eggs. Because an adult tapeworm may have as many as 2000 segments, its reproductive potential is staggering. A single tapeworm can produce 600 million eggs in a year. Proglottids farthest from the tapeworm's head contain the ripest eggs; these segments are shed from the host's body along with the feces.

The tapeworm has no mouth or digestive system. Digested food from the host is absorbed across the worm's body wall. The tapeworm also lacks well-developed sense organs. Some tapeworms have complex life cycles, spending their larval stage within the body of an intermediate host and their adult life within the body of a different, final host. Figure 29-4 illustrates the life cycle of the beef tapeworm, which can infect humans when they eat undercooked beef containing the larvae.

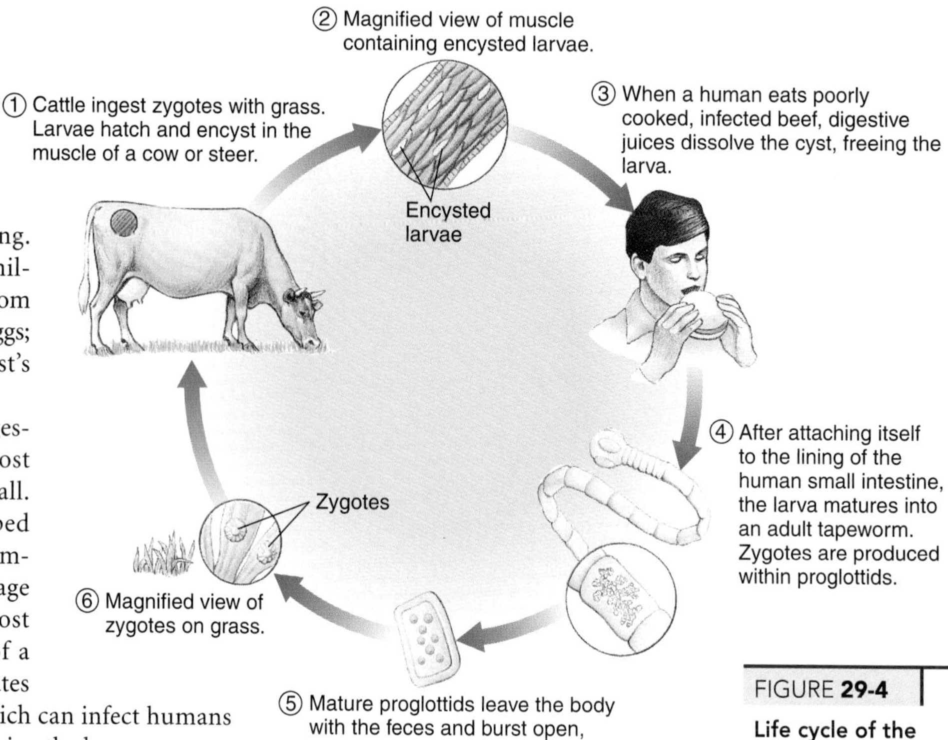

① Cattle ingest zygotes with grass. Larvae hatch and encyst in the muscle of a cow or steer.
② Magnified view of muscle containing encysted larvae.
Encysted larvae
③ When a human eats poorly cooked, infected beef, digestive juices dissolve the cyst, freeing the larva.
④ After attaching itself to the lining of the human small intestine, the larva matures into an adult tapeworm. Zygotes are produced within proglottids.
⑤ Mature proglottids leave the body with the feces and burst open, releasing zygotes.
⑥ Magnified view of zygotes on grass.
Zygotes

FIGURE **29-4**

Life cycle of the beef tapeworm.

Phylum Nemertea is characterized by the proboscis

Phylum **Nemertea,** the ribbon worms or proboscis worms, is a relatively small group (about 1150 species) of free-living animals (Fig. 29-5). Almost all are marine, although a few inhabit fresh water or damp soil. They are carnivorous, feeding on crustaceans and other worms. Nemerteans have long narrow bodies, either cylindrical or flattened, generally ranging in length from 5 cm (2 in) to about 2 m (6.5 ft), although some are much longer. Some are a vivid orange, red, or green, with black or colored stripes.

Their most remarkable organ, the **proboscis,** from which they get one of their common names, *proboscis worms,* is a long, hollow, muscular tube that can be rapidly everted (turned inside out) from the anterior end of the body. The proboscis wraps around prey. In some species it is sharp, and in various species it secretes toxic fluid that immobilizes prey. The proboscis is a *derived character* that distinguishes nemerteans from all other invertebrate groups.

Historically, the nemerteans were of special evolutionary interest because biologists thought they were the earliest animals to have a circulatory system and a tube-within-a-tube body plan. Nemerteans have no heart; blood is circulated by contractions of muscular blood vessels and by movements of the body. Although the nemerteans are functionally acoelomate, the chamber surrounding the proboscis is a true coelomic space, known as a **rhynchocoel,** derived like the coelom in coelomate protostomes (see Chapter 28). The system of blood vessels is probably homologous to the coelom of other coelomate animals. Biologists now view the rhyncocoel as a remnant of the coelom of a coelomate ancestor. Because of its presence and recent molecular evidence, biologists now classify nemerteans with the lophotrochozoans.

Mollusks have a muscular foot, visceral mass, and mantle

Phylum **Mollusca** includes clams, oysters, snails, slugs, octopods, and the largest of all the invertebrates, the giant squid, which averages 9 to 16 m (about 30–53 ft) in length, including its tentacles. More than 50,000 living species and 35,000 fossil species (second only to the arthropods in number) have been described. Representative mollusks are illustrated in Figure 29-6. In this chapter we discuss four of the eight recognized classes and list them in Table 29-2.

Although most mollusks are marine, many snails and clams live in fresh water, and some species of snails and slugs inhabit the land. Mollusks probably evolved early in the history of the protostome clade, soon after the evolution of the coelom but before the origin of the segmented body that is characteristic of annelids. Although mollusks vary widely in outward appearance, most share certain basic characteristics:

1. A soft body, usually covered by a dorsal shell composed mainly of calcium carbonate.
2. A broad, flat, muscular **foot,** located ventrally, which is used for locomotion.
3. The body organs (viscera) concentrated as a **visceral mass** located above the foot.
4. A **mantle,** a thin sheet of tissue that covers the visceral mass and usually contains glands that secrete a shell. The mantle

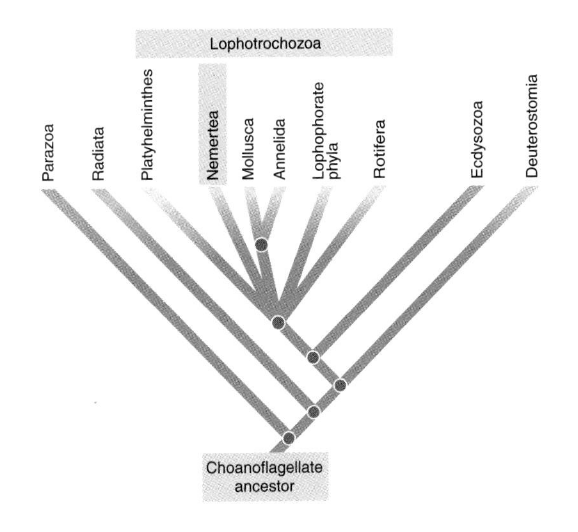

(a)

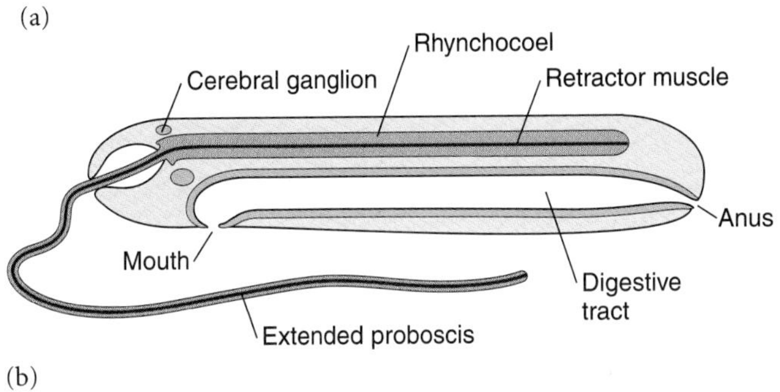

(b)

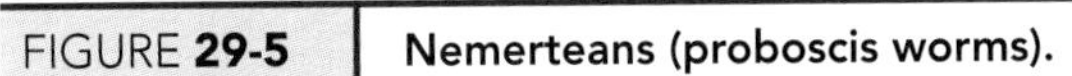

FIGURE 29-5 Nemerteans (proboscis worms).
(a) Proboscis worm (*Lineus*) from the Pacific coast of Panama. **(b)** Lateral view of a typical nemertean. Note the complete digestive tract that extends from mouth to anus, giving this animal a tube-within-a-tube body plan.

generally overhangs the visceral mass, forming a mantle cavity that contains gills and other structures.

5. A rasplike structure called a **radula,** which is a belt of teeth in the mouth region. (The radula is not present in clams or their relatives, which are suspension feeders.)
6. The coelom is generally reduced to small compartments around certain organs, including the heart and excretory organs *(metanephridia)*. The main body cavity is typically a **hemocoel,** a space containing blood (see following discussion of open circulatory system).

Mollusks have all organ systems typical of complex animals. The digestive system is a tube, often coiled, consisting of a mouth, buccal cavity (mouth cavity), esophagus, stomach, intestine, and anus. The mollusk radula, located within the buccal cavity, projects out of the mouth and is used to scrape particles of food from the surface of rocks or the ocean floor. Sometimes the radula is used to drill a hole in the shell of a prey animal or to tear off pieces of a plant.

Most mollusks have an **open circulatory system** in which the blood, called **hemolymph,** bathes the tissues directly. The heart pumps blood into a single blood vessel, the aorta, which may branch into other vessels. Eventually blood flows into a network of large spaces called *sinuses,* where the tissues are bathed directly; this network makes up the hemocoel, or blood cavity. From the sinuses, blood drains into vessels that conduct it to the gills, where it is recharged with oxygen. After passing through the gills, the blood returns to the heart. Thus, blood flow in a mollusk follows the pattern:

Heart ⟶ aorta ⟶ smaller blood vessels ⟶ blood sinuses (hemocoel) ⟶ blood vessels to gills ⟶ heart

In open circulatory systems, blood pressure tends to be low, and tissues are not very efficiently oxygenated. Because most mollusks are slow-moving animals with low metabolic rates, this type of circulatory system is adequate. The active cephalopods (the class that includes the squids and octopods), have a **closed circulatory system,** in which blood flows through a complete circuit of blood vessels.

Most marine mollusks pass through one or more larval stages. The first larval stage is typically a **trochophore larva,** a free-swimming, top-shaped larva with two bands of cilia around its middle (Fig. 29-7). In many mollusks (gastropods

TABLE 29-2 Major Classes of Phylum Mollusca

Class and Representative Animals	Characteristics
Polyplacophora Chitons	Primitive marine animals with shell consisting of eight separate transverse plates; head reduced; broad foot used for locomotion.
Gastropoda Snails, slugs, nudibranchs	Marine, freshwater, or terrestrial; coiled shell in many species; torsion of visceral mass; well-developed head with tentacles and eyes.
Bivalvia Clams, oysters, mussels	Marine or freshwater; body laterally compressed; two-part shell hinged dorsally; hatchet-shaped foot; suspension feeders.
Cephalopoda Squids, octopods	Marine; predatory; foot modified into tentacles, usually bearing suckers; well-developed eyes; closed circulatory system.

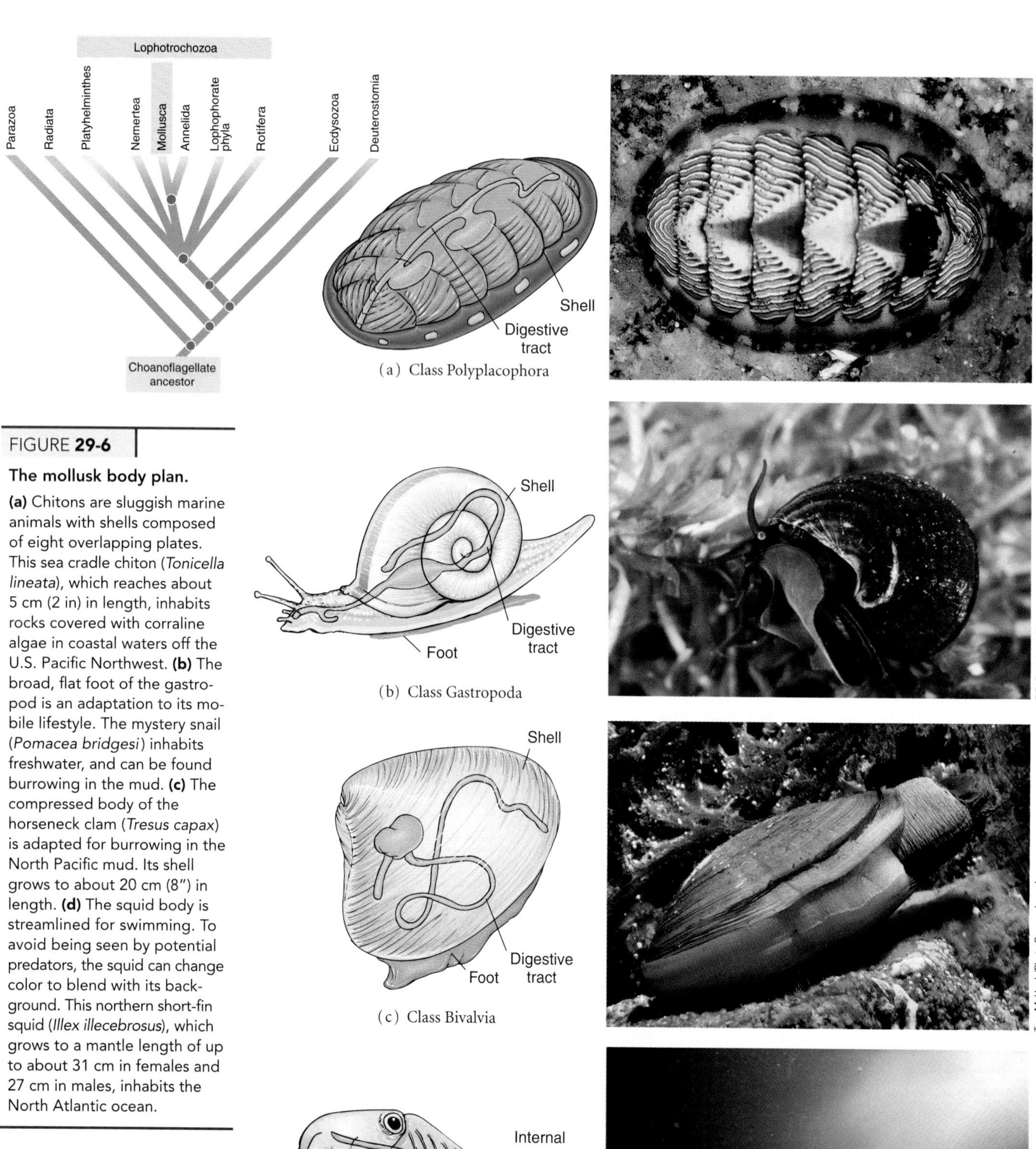

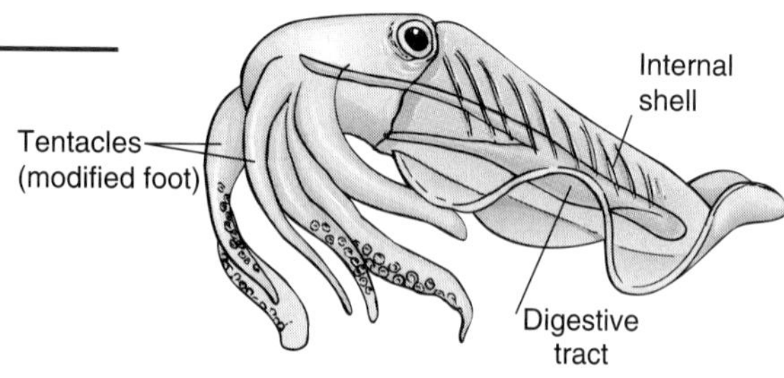

FIGURE **29-6**

The mollusk body plan.

(a) Chitons are sluggish marine animals with shells composed of eight overlapping plates. This sea cradle chiton (*Tonicella lineata*), which reaches about 5 cm (2 in) in length, inhabits rocks covered with corraline algae in coastal waters off the U.S. Pacific Northwest. **(b)** The broad, flat foot of the gastropod is an adaptation to its mobile lifestyle. The mystery snail (*Pomacea bridgesi*) inhabits freshwater, and can be found burrowing in the mud. **(c)** The compressed body of the horseneck clam (*Tresus capax*) is adapted for burrowing in the North Pacific mud. Its shell grows to about 20 cm (8") in length. **(d)** The squid body is streamlined for swimming. To avoid being seen by potential predators, the squid can change color to blend with its background. This northern short-fin squid (*Illex illecebrosus*), which grows to a mantle length of up to about 31 cm in females and 27 cm in males, inhabits the North Atlantic ocean.

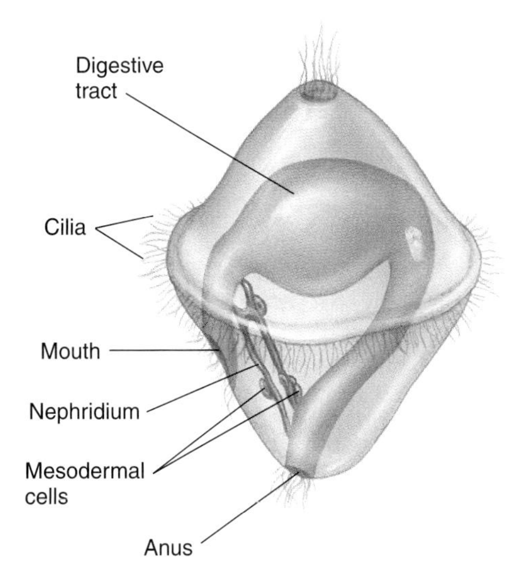

FIGURE **29-7** **A trochophore larva.**

The first larval stage of a marine mollusk, the trochophore larva, is also characteristic of annelids. Just above the mouth a characteristic band of ciliated cells functions as a swimming organ and, in some species, collects suspended food particles.

such as snails and bivalves such as clams), the trochophore larva develops into a **veliger larva,** which has a shell, foot, and mantle. The veliger larva is unique to the mollusks.

Chitons may be similar to ancestral mollusks

Class **Polyplacophora** (meaning "many plates") consists of the **chitons,** sluggish marine animals with flattened bodies (see Fig. 29-6a). Their most distinctive feature is a shell composed of eight separate but overlapping dorsal plates. The head is reduced, and there are no eyes or tentacles.

The chiton inhabits rocky intertidal zones, using its broad, flat foot to move and to hold firmly onto rocks. By pressing its mantle against the substratum and lifting the inner edge of the mantle, the chiton can produce a partial vacuum. The resulting suction lets the animal adhere powerfully to its perch. Using its radula for grazing, the chiton scrapes algae and other small organisms off rocks and shells.

Gastropods are the largest group of mollusks

Class **Gastropoda,** which includes snails, slugs, and their relatives, is the largest and most diverse group of mollusks (see Fig. 29-6b). In fact, with more than 40,000 extant species, gastropods comprise the second largest class in the animal kingdom, second only to insects. Most gastropods inhabit marine waters, but others make their homes in brackish or fresh water, or on land. We think of snails as having a single, spirally coiled shell into which they can withdraw the body, and many do. However, other gastropods, such as limpets, have shells like flattened dunce caps. Still others, such as garden slugs and the beautiful marine slugs known as **nudibranchs,** have no shell at all (Fig. 29-8).

Bruce Watkins/Animals Animals

FIGURE **29-8** **Nudibranch.**

(a) A nudibranch (*Chromodoris porterae*) feeds on a lightbulb tunicate (*Clavelina huntsmani*).

Many gastropods have a well-developed head with tentacles. Two simple eyes may be located on stalks that extend from the head. The gastropod uses its broad, flat foot for creeping. Most land snails do not have gills. Instead, the mantle is highly vascularized and functions as a lung. Biologists describe these garden snails and slugs as **pulmonate** ("having a lung").

Torsion, a twisting of the visceral mass, is a unique feature of gastropods. This twisting is unrelated to the coiling of the shell. As the bilateral larva develops, one side of the visceral mass grows more rapidly than the other side. This uneven growth rotates the visceral mass. The visceral mass and mantle twist permanently up to 180 degrees relative to the head. As a result, the digestive tract becomes somewhat U-shaped, and the anus comes to lie above the head and gill (Fig. 29-9). Subsequent growth is dorsal and usually in a spiral coil. Torsion limits space in the body, and typically the gill, metanephridium (excretory organ), and gonad are absent on one side. Some biologists hypothesize that torsion is an adaptation that protects the head by allowing it to enter the shell first during withdrawal from potential predators. Without torsion, the foot would be withdrawn first.

FIGURE **29-9** **Embryonic torsion in a gastropod.**

As the bilateral larva develops, the visceral mass twists 180 degrees relative to the head. The digestive tube coils, and the anus becomes relocated near the mouth.

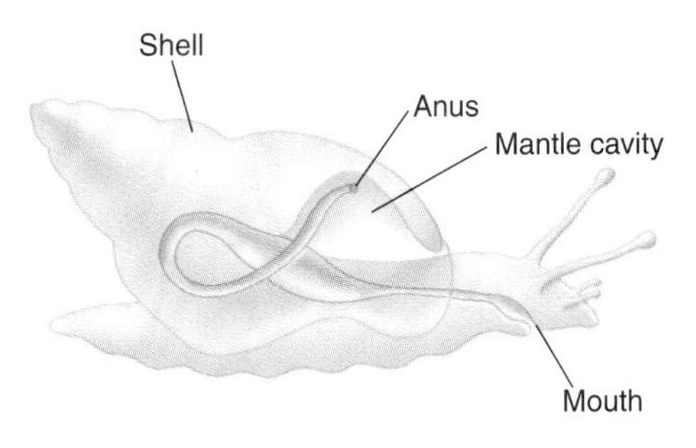

Bivalves typically burrow in the mud

Class **Bivalvia** includes the clams, oysters, mussels, scallops and their relatives (see Fig. 29-6c). Typically, the soft body of members of class Bivalvia is laterally compressed and completely enclosed by a two-part shell that hinges dorsally and opens ventrally (Fig. 29-10). This arrangement allows the hatchet-shaped foot to protrude ventrally for locomotion and for burrowing in mud. The two parts, or **valves,** of the shell are connected by an elastic ligament. Stretching of the ligament opens the shell. Large, strong adductor muscles attached to the shell permit the animal to close its shell.

The inner, pearly layer of the bivalve shell is made of calcium carbonate secreted in thin sheets by the epithelial cells of the mantle. Known as *mother-of-pearl,* this material is valued for making jewelry and buttons. Should a bit of foreign matter lodge between the shell and the mantle, the epithelial cells covering the mantle secrete concentric layers of calcium carbonate around the intruding particle. Many species of oysters and clams form pearls in this way.

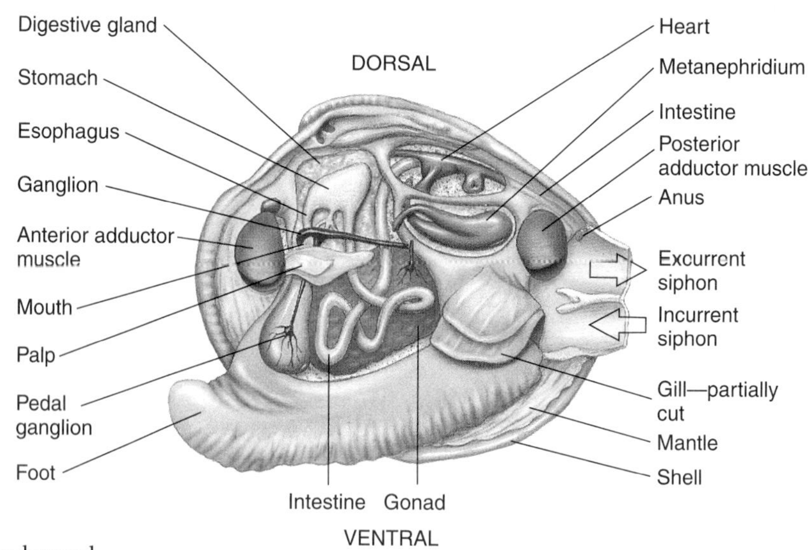

FIGURE **29-10** | **Internal anatomy of a clam.**
The two shells of a bivalve hinge dorsally and open ventrally.

Some bivalves, such as oysters, attach permanently to the substrate. Others burrow slowly through rock or wood, seeking protected dwellings. The shipworm, *Teredo,* a clam that damages dock pilings and other marine installations, is just looking for a home. A few bivalves, such as scallops, swim rapidly by clapping their two shells together with the contraction of a large adductor muscle (the part of the scallop that humans eat).

Clams and oysters are suspension feeders that trap food particles in sea water. They take water in through an extension of the mantle called the **incurrent siphon.** Water leaves by way of an **excurrent siphon.** As the water passes over the gills, mucus secreted by the gills traps food particles in the water. Cilia move the food to the mouth. An oyster can filter more than 30 L (about 32 quarts) of water per hour. As suspension feeders, bivalves have no radula, and indeed they are the only group of mollusks that lack this structure.

Cephalopods are active, predatory animals

In contrast with most other mollusks, members of the class **Cephalopoda** (meaning "head-foot") are fast-swimming predatory animals (see Fig. 29-6d). The cephalopod mouth is surrounded by tentacles, or arms: 8 in octopods, 10 in squids, and as many as 90 in the chambered nautilus. The large head has well-developed eyes that form images. Although they develop differently, the eyes are structurally similar to vertebrate eyes and function in much the same way. The octopus has no shell, and the shell of the squid is greatly reduced and is located inside the body.

Nautilus has a coiled shell consisting of many chambers built up over time. Each year the animal lives in the newest and largest chamber of the series. *Nautilus* secretes a mixture of gases similar to air into the other chambers. By regulating the amount of gas in the chambers, the animal controls its depth in the ocean.

The tentacles of squids, octopods, and cuttlefish are covered with suckers for seizing and holding prey. In addition to a radula, the mouth has two strong, horny beaks used to kill prey and tear it to bits. The mantle is thick and muscular and fitted with a funnel-like **siphon.** By filling the cavity with water and ejecting it through the siphon, the cephalopod achieves forceful jet propulsion. Squids and cuttlefish have streamlined bodies adapted for efficient swimming.

In addition to speed, the cephalopod has two other adaptations that facilitate escape from its predators, which include certain whales and moray eels. One is its ability to confuse the enemy by rapidly changing colors. By expanding and contracting *chromatophores,* cells in the skin that contain pigment granules, the cephalopod can display an impressive variety of mottled colors. Another defense mechanism is its **ink sac,** which produces a thick, black liquid the animal releases in a dark cloud when alarmed. While its enemy pauses, temporarily blinded and confused, the cephalopod escapes. The ink inactivates the chemical receptors of some predators, making them incapable of detecting their prey.

The octopus feeds on crabs and other arthropods, catching and killing them with a poisonous secretion of its salivary glands. During the day the octopus usually hides among the rocks; in the evening it emerges to hunt for food. Its motion is incredibly fluid, giving little hint of the considerable strength in its eight arms.

Small octopods survive well in aquariums and have been studied extensively. Because they are relatively intelligent and can make associations among stimuli, researchers have used them as models for studying learning and memory. Their highly adaptable behavior more closely resembles that of the vertebrates than the more stereotypic patterns of behavior seen in other invertebrates (see Chapter 50).

tends beneath the digestive tract to the posterior end of the body. In each segment along the nerve cord is a pair of fused segmental ganglia. Nerves extend laterally from the segmental ganglia to the muscles and other structures of that segment. The segmental ganglia coordinate muscle contraction of the body wall, allowing the worm to creep along.

Like other oligochaetes, earthworms are hermaphroditic. During copulation, two worms, headed in opposite directions, press their ventral surfaces together. These surfaces become glued together by the thick mucous secretions of each worm's *clitellum,* a thickened ring of epidermis. Sperm are then exchanged and stored in the *seminal receptacles* (small sacs) of the other worm. A few days later each clitellum secretes a membranous cocoon containing a sticky fluid. As the cocoon is slipped forward, eggs are laid in it. Sperm are added as the cocoon passes over the openings of the seminal receptacles. As the cocoon slips free over the worm's head, its openings constrict so that a spindle-shaped capsule is formed. The fertilized eggs develop into tiny worms within this capsule. This complex reproductive pattern is an adaptation to terrestrial life; the cocoon protects the delicate gametes and young worms from drying out.

Many leeches are blood-sucking parasites

Most leeches, members of class **Hirudinea,** inhabit fresh water, but some live in the sea or in moist areas on land. About 75% of the known species of leeches are blood-sucking parasites. Some leeches are nonparasitic predators that capture small invertebrates such as earthworms and snails. Leeches differ from other annelids in having neither setae nor parapodia.

Leeches have muscular suckers at both anterior and posterior ends. Most parasitic leeches attach themselves to a vertebrate host, bite through the skin, and suck out a quantity of blood, which is stored in pouches in the digestive tract. *Hirudin,* an anticoagulant secreted by glands in the crop, ensures leeches a full meal of blood. In 30 minutes, a leech can suck out as much as ten times its own weight in blood. Some leeches feed only about twice each year because they digest their food slowly over several months.

Leeches have been used since ancient times for drawing blood from areas swollen by poisonous stings and bites. During the 19th century they were widely used to remove "bad blood," thought to be the cause of many diseases. Leeches are used in modern medicine to remove excess fluid and blood that accumulates within body tissues as a result of injury, disease, or surgery (Fig. 29-13). The leech attaches its sucker near the site of injury, makes an incision, and secretes hirudin. The hirudin prevents the blood from clotting and dissolves already existing clots.

The lophophorate phyla are distinguished by a ciliated ring of tentacles

Biologists are still debating the evolutionary position of the **lophophorate phyla**—Brachiopoda, Phoronida, and Bryozoa. With a few exceptions, lophophorates are marine animals adapted for life on the ocean floor. The **lophophore,** a ciliated ring of tentacles that surrounds the mouth, is specialized for capturing suspended particles in the water.

Brachiopods, or lampshells, are solitary marine animals that inhabit cold water. They superficially resemble clams and other bivalve mollusks because the body is enclosed between two shells (Fig. 29-14a). However, they differ from bivalve mol-

FIGURE **29-13** **Leeches.**

(a) LM of a blood-sucking leech (*Helobdella stagnalis*) that feeds on mammals. The digestive tract *(dark area)* of its swollen body is filled with ingested blood. **(b)** Medicinal leeches (*Hirudo medicinalis*) are used here to treat hematoma, an accumulation of blood within tissues that results from injury or disease.

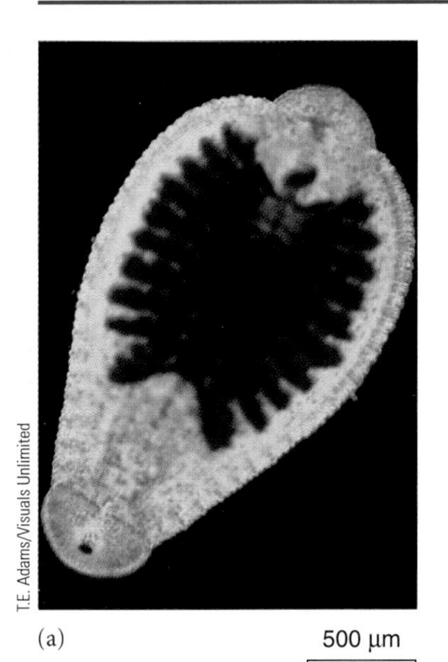

T.E. Adams/Visuals Unlimited

(a) 500 μm

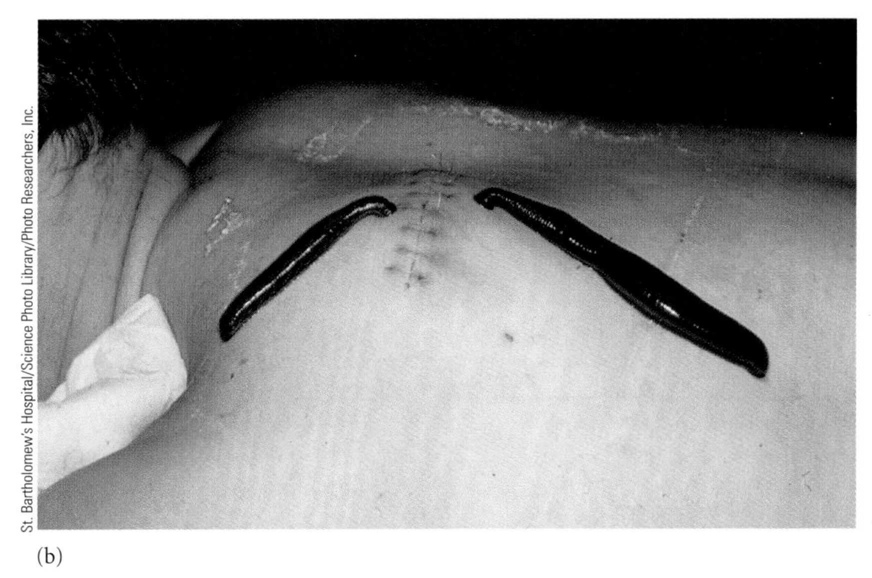

St. Bartholomew's Hospital/Science Photo Library/Photo Researchers, Inc.

(b)

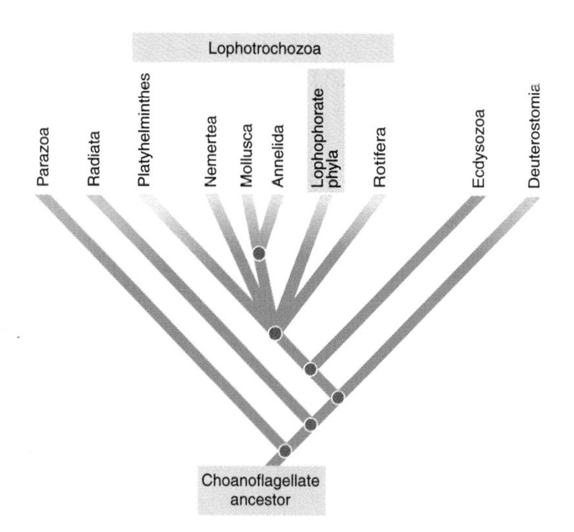

(a) Phylum Brachiopoda

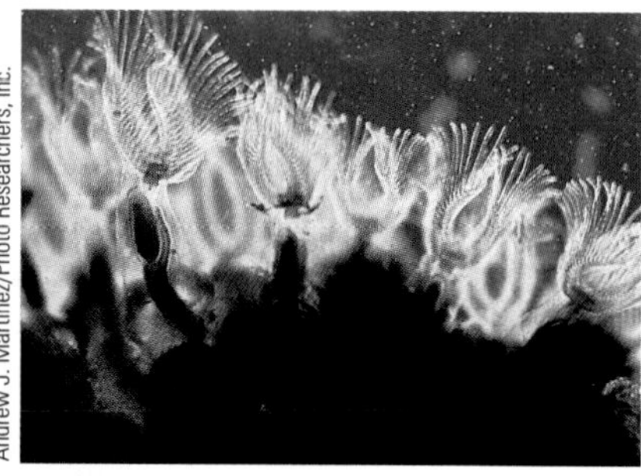

(b) Phylum Bryozoa

FIGURE **29-14** | **Representative lophophorates.**

(a) Northern lamp shells (*Terebratulina septentrionalis*), like other brachiopods, superficially resemble clams. **(b)** Like many bryozoans, this freshwater species (*Cristatella mucedo*) forms colonies by asexual budding.

lusks in that the shells are dorsal and ventral, rather than lateral; each shell is symmetrical about the midline, and the two shell valves are typically of unequal size. Brachiopods attach to the substrate by a long stalk. Cilia on the lophophore beat, bringing water laden with food into the slightly opened shell. Although there are now only about 350 living species, brachiopods were abundant and diversified during the Paleozoic era. Paleobiologists have described about 30,000 fossil species.

Only about 12 extant species of **phoronids** are known. Some phoronids secrete tubes of sediment and live in them, extending their lophophores from their tubes for feeding. **Bryozoa,** or "moss animals," form sessile colonies by asexual budding (Fig. 29-14b). Their colonies, which can consist of millions of individuals, may appear plantlike. Other bryozoan colonies have the appearance of coral. About 5000 extant species are known.

Rotifers have a crown of cilia

Among the less familiar invertebrates are the "wheel animals" of phylum **Rotifera.** Although no larger than many protozoa, these aquatic, microscopic animals are multicellular. More than 1800 species have been described. Most inhabit fresh water, but some live in marine environments or damp soil. Rotifers were formerly classified as pseudocoelomates, but many biologists now hypothesize that rotifers evolved from animals with a coelom. Because they have a cuticle, some systematists classify them as ecdysozoans related to nematodes and arthropods. However, other systematists classify rotifers with the lophotrochozoans.

Rotifers have a characteristic crown of cilia on their anterior end. The cilia beat rapidly during swimming and feeding, giving the appearance of a spinning wheel (Fig. 29-15). Rotifers

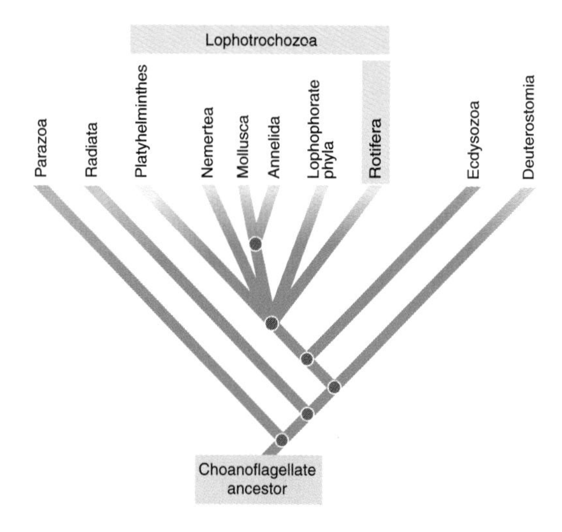

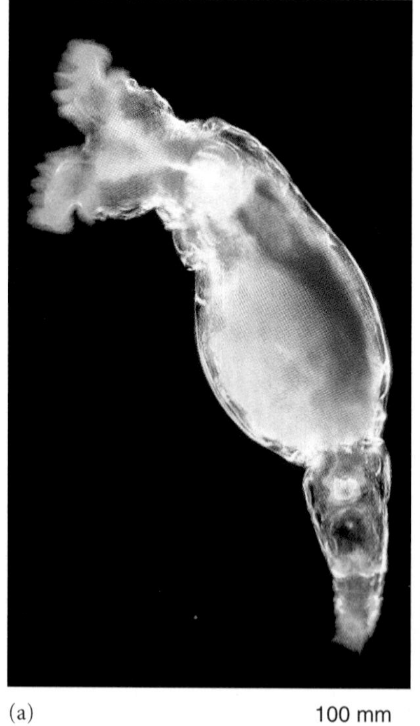

(a)

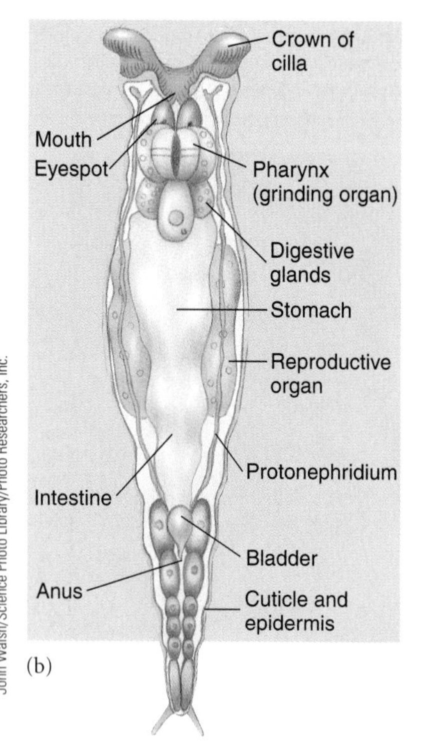

(b)

FIGURE **29-15** | **Rotifers (wheel animals).**

(a) LM of an Antarctic rotifer (*Philodina gregaria*) that survives the winter by forming a cyst. It reproduces in great numbers, sometimes coloring the water red. **(b)** Longitudinal section showing rotifer anatomy. The motion of its cilia draws particles of food such as algae into the mouth.

feed on tiny organisms suspended in the stream of water drawn into the mouth by the cilia. A muscular organ posterior to the mouth grinds the food. The complete digestive tract ends in an anus through which wastes are eliminated.

Rotifers have a nervous system with a "brain" and sense organs, including eyespots. Protonephridia with flame cells remove excess water from the body and may also excrete metabolic wastes.

Biologists have discovered that, like some other animals with a pseudocoelom, rotifers are "cell constant." Each member of a given species has exactly the same number of cells. Indeed, each part of the body has a precisely fixed number of cells arranged in a characteristic pattern. Cell division does not take place after embryonic development, and mitosis cannot be induced; growth and repair are not possible. One of the challenging problems of biological research is discovering the difference between such nondividing cells and the dividing cells of other animals. Do you think rotifers could develop cancer?

Review

- What are some advantages of bilateral symmetry and cephalization?
- On what basis have biologists now classified nemerteans as lophotrochozoans?
- How does the lifestyle of a gastropod differ from that of a cephalopod? Identify adaptations that have evolved in each for its particular lifestyle.
- What are two distinguishing characteristics for each of the following: (a) mollusks (b) annelids (c) nematodes?

BiologyNow™ Assess your understanding of **the lophotrochozoa** by taking the pretest on your BiologyNow CD-ROM.

THE ECDYSOZOA

Learning Objectives

7 Describe the distinguishing characteristics of phylum Nematoda.

8 Describe the distinguishing characteristics of phylum Arthropoda; distinguish among the subphyla and classes of arthropods, giving examples of animals that belong to each group.

9 Discuss factors that have contributed to the great biological success of insects.

The **Ecdysozoa** include the nematodes (roundworms) and arthropods, animals characterized by a **cuticle,** a noncellular body covering secreted by the epidermis. The name Ecdysozoa refers to the process of **ecdysis,** or **molting,** characteristic of animals in this group. During ecdysis, an animal sheds its cuticle. The taxonomic validity of the Ecdysozoa has been supported by many types of data, most notably by a 2001 study that analyzed more than 300 ribosomal RNA genes and 138 morphologic characters.

Roundworms are of great ecological importance

Members of phylum **Nematoda,** the **roundworms,** play key ecological roles as decomposers and predators of smaller organisms. Many soil nematodes eat bacteria. Nematodes are numerous and widely distributed in soil and in marine and freshwater sediments. Nearly 20,000 species have been identified, and many remain to be discovered. The nematodes rival the insects in number of individuals. A spadeful of soil may contain more than a million of these mainly microscopic worms, which thrash around, coiling and uncoiling.

The elongated, cylindrical, threadlike nematode body is pointed at both ends and covered with a tough, flexible *cuticle* (Fig. 29-16). Secreted by the underlying epidermis, the thick cuticle gives the nematode body shape and offers some protection. The epidermis is unusual in that it does not consist of distinct cells.

Nematodes have a fluid-filled pseudocoelom that serves as a hydrostatic skeleton. It transmits the force of muscle contraction to the enclosed fluid. Movement of fluid in the pseudocoelom is also important in nutrient transport and distribution. Like ribbon worms, nematodes exhibit bilateral symmetry, a complete digestive tract, three definite tissue layers, and definite organ systems; however, they lack specific circulatory structures. The sexes are usually separate, and the male is generally smaller than the female.

Traditionally, biologists classified nematodes as a single phylum based on the pseudocoelom, wormlike body structure, and other characters such as absence of a well-defined head. However, molecular evidence suggests that nematodes are not a monophyletic group and that they evolved from more complex

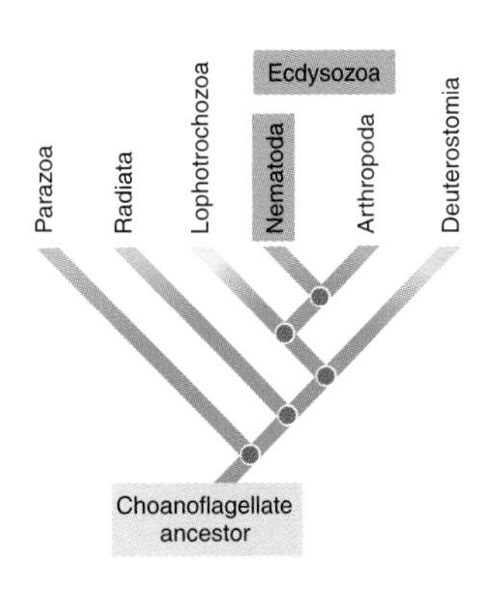

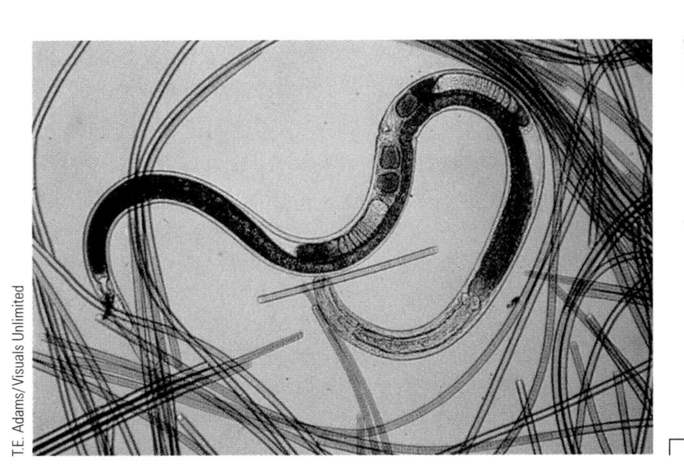

FIGURE **29-16**

LM of a free-living nematode.

This unidentified aquatic nematode is shown among the cyanobacterium *Oscillatoria,* which it eats.

animals by a simplification in body plan. Systematists now classify nematodes with the Ecdysozoa. Like arthropods and other ecdysozoans, the nematode sheds its cuticle periodically.

Caenorhabditis elegans, a free-living soil nematode, is an important model research organism for biologists studying the genetic control of development (see Chapter 16). Although most nematodes are free-living, others are important parasites in plants and animals. More than 50 species of roundworms are human parasites, including *Ascaris,* hookworms, pinworms, and the trichina worm.

The common intestinal parasite ***Ascaris*** is a white worm about 25 cm (10 in) long. *Ascaris* spends its adult life in the human intestine, where it ingests partly digested food (Fig. 29-17). Like most parasites, it devotes a great deal of effort to reproduction to ensure survival of its species. A mature female may produce as many as 200,000 eggs a day.

Hookworms hook onto the lining of the intestine and suck blood, potentially causing serious tissue damage and blood loss. **Pinworms** are the most common worms found in children. The tiny pinworm eggs are often ingested by eating with hands contaminated with them. The **trichina worm** lives inside a variety of animals, including pigs, rats, and bears. Humans typically become infected by eating undercooked, infected meat. Larvae encyst in skeletal muscle.

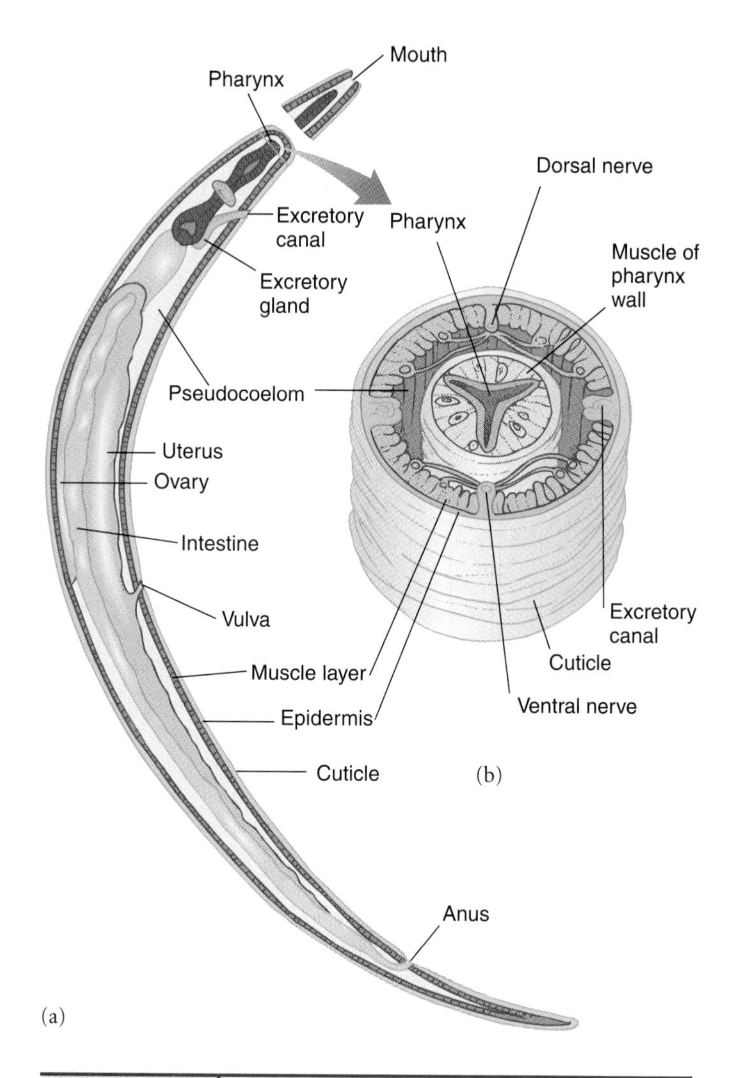

FIGURE **29-17** | **The roundworm *Ascaris.***

(a) Longitudinal section. Note the complete digestive tract that extends from mouth to anus. **(b)** Cross section through *Ascaris* shows the tube-within-a-tube body plan. The protective cuticle that covers the body helps the animal resist drying.

Arthropods are characterized by jointed appendages and an exoskeleton of chitin

Phylum **Arthropoda,** the most biologically successful group of animals, includes more than 1 million species, and biologists predict millions more will be identified. More than 80% of all known animals are arthropods! They are more diverse and live in a greater range of habitats than do the members of any other animal phylum. The following adaptations have greatly contributed to their success:

1. The arthropod body, like that of the annelid, is segmented. *Segmentation* is important from an evolutionary perspective because it provides the opportunity for specialization of body regions. In arthropods groups of segments are specialized, by virtue of their shape, muscles, or the appendages they bear, to perform particular functions.
2. A hard **exoskeleton,** composed of chitin and protein, covers the entire body and appendages. The exoskeleton serves as a coat of armor protecting against predators and helps prevent excessive loss of moisture. It also supports the underlying soft tissues. Arthropods move effectively because distinct muscles attach to the inner surface of the exoskeleton and operate the joints of the body and appendages. A disadvantage of the exoskeleton is that it is nonliving, and the arthropod periodically outgrows it. *Molting* is the process of shedding an old exoskeleton and growing a larger one. The shed exoskeleton represents a net metabolic loss, and molting also leaves the arthropod temporarily vulnerable to predators.
3. **Paired, jointed appendages,** from which this phylum gets its name (*arthropod* means "jointed foot"), are modified for many different functions. They serve as swimming paddles, walking legs, mouth parts for capturing and manipulating food, sensory structures, or organs for transferring sperm.
4. The nervous system, which resembles that of the annelids, consists of a "brain" (cerebral ganglia) and ventral nerve cord with ganglia. In some arthropods, successive ganglia may fuse together. Arthropods have a variety of very effective sense organs. Many have organs of hearing, and **antennae** that sense taste and touch. Most insects and many crustaceans have **compound eyes** composed of many light-sensitive units called **ommatidia** (see Figure 41-13). The compound eye can form an image and is especially adapted for detecting movement.

Arthropods have an *open circulatory system.* A dorsal, tubular heart pumps hemolymph into a dorsal artery, which may branch into smaller arteries. From the arteries, hemolymph flows into large spaces that collectively make up the hemocoel. Eventually hemolymph re-enters the heart through openings, called *ostia,* in its walls.

The exoskeleton presents a barrier to diffusion of oxygen and carbon dioxide through the body wall, necessitating the evolution of specialized respiratory systems for gas exchange. Most aquatic arthropods have gills that function in gas exchange, whereas many terrestrial forms have a system of internal branching air tubes called **tracheae,** or **tracheal tubes.** Other terrestrial arthropods have platelike **book lungs.**

The coelom is small and filled chiefly by the organs of the reproductive system. The arthropod digestive system is a tube similar to that of the earthworm. Excretory structures vary somewhat from class to class.

Arthropod evolution and classification are controversial

PROCESS OF SCIENCE

Two groups of animals thought to be closely related to arthropods are the onychophorans and the tardigrades. The wormlike members of phylum **Onychophora,** or velvet worms, most likely branched early from the arthropod line. They have paired appendages, although they are not jointed. As in arthropods, their jaws are derived from appendages. Some biologists place the tardigrades, or water bears, close to the arthropods (see Fig. 2-12). Assigned to phylum **Tardigrada,** these tiny animals have unbranched, clawed legs. Like arthropods, onychophorans and tardigrades have a thick cuticle and must molt in order to grow. The body is segmented in all three groups, and all share similarities in the circulatory system.

Arthropod fossils have been identified that date back to the Precambrian era. Early in their evolutionary history (more than 540 mya), arthropods diverged into several groups. Arthropods evolved rapidly during the Cambrian explosion, as evidenced by many diverse fossils in the Burgess Shale and other ancient sites.

We discuss five main groups: based on emerging molecular and other data: the extinct trilobites, and the extant Myriapoda, Chelicerata, Crustacea, and Hexapoda (Table 29-4). Some systematists consider each of these groups a phylum; others view them as subphyla or classes. In this edition, we classify these groups as subphyla. However, as you study the following sections, bear in mind that arthropod systematics is a work in progress. Both the relationship of arthropods to other protostomes and the relationships among arthropods continue to be topics of lively debate.

TABLE 29-4 Extant Arthropod Subphyla

Subphylum and Selected Classes	Body Divisions	Appendages: Antennae, Mouthparts	Appendages: Legs	Gas Exchange	Development	Main Habitat
Myriapoda Class Chilopoda (centipedes) Class Diplopoda (millipedes)	Head with segmented body	Uniramous. Antennae: 1 pair Mouth parts: mandibles; Maxillae	Chilopods: 1 pair/segment Diplopods: usually 2 pairs/ segment	Tracheae	Direct	Terrestrial
Chelicerata Class Merostoma (horseshoe crabs) Class Arachnida (spiders, scorpions, ticks, mites)	Cephalothorax and abdomen	Uniramous. Antennae: none Mouthparts: chelicerae; pedipalps	Merostomes: 5 pairs of walking legs Arachnids: 4 pairs on cephalothorax	Merostomes: gills Arachnids: book lungs or tracheae	Direct, except mites and ticks	Merostomes: marine Arachnids: mainly terrestrial
Crustacea Class Malacostraca (lobsters, crabs, shrimp, isopods) Class Cirripedia (barnacles) Class Copepoda (copepods)	Head, thorax, abdomen	Biramous. Antennae: 2 pairs Mouthparts: mandibles, 2 pairs of maxillae (for handling food)	Typically 1 pair/segment	Gills	Usually larval stages (nauplius larva)	Marine or freshwater; a few are terrestrial
Hexapoda Class Insecta (bees, ants, grasshoppers, roaches, flies, beetles)	Head, thorax, and abdomen	Uniramous. Antennae: 1 pair Mouthparts: mandibles, maxillae	3 pairs on thorax	Tracheae	Typically, larval stages; most with complete metamorphosis	Mainly terrestrial

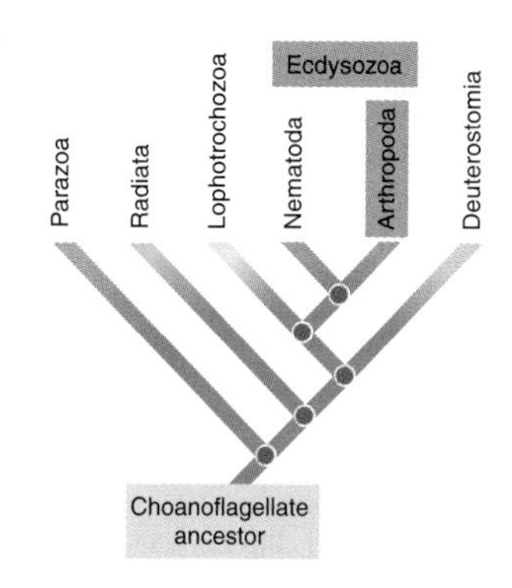

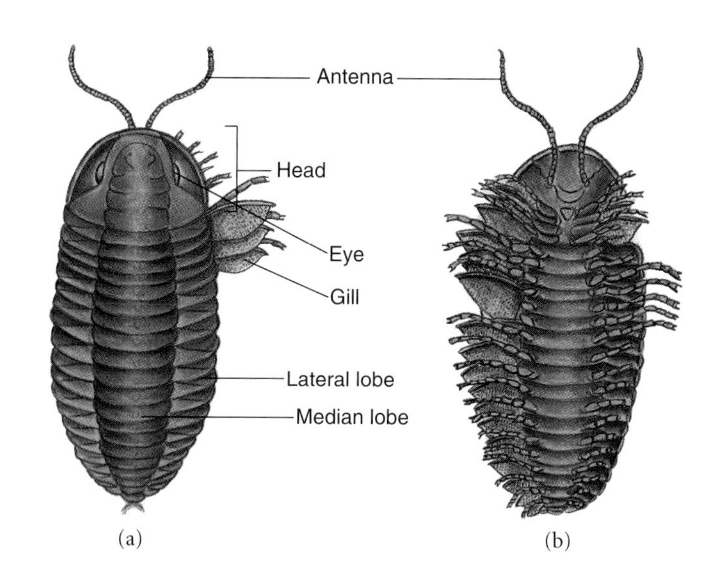

FIGURE **29-18** **Trilobites.**

Biologists consider these extinct marine arthropods the most primitive members of the phylum. **(a)** Dorsal view of a trilobite. **(b)** Ventral view.

Trilobites were early arthropods

Among the earliest arthropods to evolve, **trilobites** inhabited shallow Paleozoic seas more than 500 million years ago. These arthropods, which have been extinct for about 250 million years, lived on the sea bottom and dug into the mud. Most ranged from 3 to 10 cm, but a few reached almost 1 m in length.

Covered by a hard, segmented exoskeleton, the trilobite body was a flattened oval divided into three parts: an anterior head bearing a pair of antennae and a pair of compound eyes; a thorax; and a posterior abdomen (Fig. 29-18; also see introduction to Chapter 20). At right angles to these divisions, two dorsal grooves extended the length of the animal, dividing the body into a median lobe and two lateral lobes. (The name *trilobite* derives from this division of the body into these three longitudinal parts.) Each segment had a pair of segmented **biramous appendages,** appendages with two jointed branches extending from their base. In trilobites, each appendage consisted of an inner walking leg and an outer branch with gills.

Subphylum Myriapoda includes centipedes and millipedes

Members of subphylum **Myriapoda** are characterized by **uniramous** (unbranched) **appendages,** jawlike **mandibles,** and a single pair of antennae. Centipedes (class **Chilopoda**) and millipedes (class **Diplopoda**) are terrestrial and are typically found beneath stones or wood in the soil in both temperate and tropical regions. These animals are similar in having a head and an elongated trunk with many segments, each bearing uniramous legs (Fig. 29-19).

Centipedes ("hundred-legged") have one pair of legs on each segment behind the head. Most centipedes do not have enough legs to merit their name; typically they have 30 or so, although a few species have 100 or more. The legs of centipedes are long, enabling them to run rapidly. Centipedes are carnivorous and feed on other animals, mostly insects. Larger centipedes will eat snakes, mice, and frogs. With their poison claws located just behind the head on the first trunk segment, centipedes capture and kill their prey.

Millipedes ("thousand-legged") have two pairs of legs on most body segments. Diplopods are not as agile as chilopods, and most species can crawl only slowly over the ground. However, they can powerfully force their way through earth and rotting wood. Millipedes are generally herbivorous and feed on both living and dead vegetation.

(a)

Dwight Kuhn

(b)

John R. MacGregor/Peter Arnold, Inc.

FIGURE **29-19** **Myriapods.**

(a) Centipede (*Lithobius*), a member of class Chilopoda. Centipedes have one pair of uniramous appendages per segment. **(b)** Millipede (*Diplopoda pachydesmus*), a member of class Diplopoda. Millipedes have two pairs of uniramous appendages per segment.

(a)

(b)

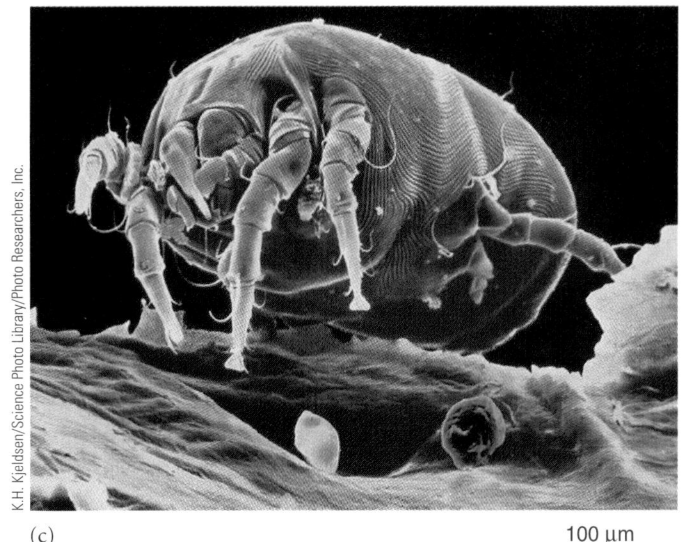

(c)

FIGURE **29-20** | **Chelicerates.**

(a) The only living merostomes are a few closely related species of horseshoe crabs. Seasonally, horseshoe crabs (*Limulus polyphemus*) return to beaches for mating. In this photograph, males are competing for a female. **(b)** A female black widow spider (*Latrodectus mactans*) rests on her web. The venom of the black widow spider is a neurotoxin. **(c)** Color-enhanced SEM of a house dust mite (*Dermatophagoides*), a common inhabitant of homes. This mite is associated with house dust allergies.

Chelicerates have no antennae

Subphylum **Chelicerata** includes the merostomes (horseshoe crabs) and the arachnids. The chelicerate body consists of a *cephalothorax* (fused head and thorax) and an abdomen. Chelicerates are the only arthropods without antennae. They have no chewing mandibles. Instead, the first pair of appendages, located immediately anterior to the mouth, are the **chelicerae** (sing. *chelicera*), fanglike feeding appendages. The second pair of appendages are the **pedipalps.** The chelicerae and pedipalps are modified to perform different functions in various groups, including manipulation of food, locomotion, defense, and copulation. The four pairs of legs on the cephalothorax are specialized for walking.

Almost all the **merostomes** are extinct. Only the horseshoe crabs have survived, essentially unchanged for more than 350 million years. *Limulus polyphemus,* the species common along the shore of North America, is horseshoe-shaped (Fig. 29-20a). Its long, spikelike tail is used in locomotion, not for defense or offense. Horseshoe crabs feed on mollusks, worms, and other invertebrates that they find on the sandy ocean floor.

Arachnids include spiders, scorpions, ticks, harvestmen (daddy long-legs), and mites (Fig. 29-20b, c). Most of the 65,000 or so named species are carnivorous and prey on insects and other small arthropods. The arachnid body consists of a cephalothorax and abdomen, with six pairs of jointed appendages. In spiders, the first pair, the chelicerae, are fanglike structures used to penetrate prey. Some spider species use the chelicerae to inject poison into their prey. Spiders use the second pair of appendages, the pedipalps, to hold and chew food. Many arachnid species have pedipalps modified as sense organs for tasting food or for reproduction (for sperm transfer and courtship displays). Scorpions use their very large pedipalps as pincers for capturing prey. Chelicerates use the remaining four pairs of appendages for walking.

Typically, spiders have 8 eyes arranged in two rows of 4 each along the anterior dorsal edge of the cephalothorax. The eyes detect movement and locate objects, and in some species form a relatively sharp image.

Gas exchange in arachnids takes place by either tracheal tubes, book lungs, or both. A **book lung** consists of 15 to 20 plates, like pages of a book, that contain tiny blood vessels. Air enters the body through abdominal slits and circulates between the plates. As air passes over the blood vessels in the plates, oxygen diffuses into the blood and carbon dioxide diffuses out of the blood into the air. As many as four pairs of book lungs provide an extensive surface area for gas exchange.

The spider has unique glands in its abdomen that secrete silk, an elastic protein that is spun into fibers by organs called *spinnerets.* The silk is liquid as it emerges from the spinnerets but hardens after it leaves the body. Spiders use silk to build nests, to encase their eggs in a cocoon, and, in some species, to trap prey in a web. Many spiders lay down a silken dragline that serves as a safety line and also as a means of communication between members of a species. From a dragline another spider can determine the sex and maturity level of the spinner.

Although all spiders, as well as scorpions, have poison glands useful in capturing prey, only a few produce poison toxic to humans. The most widely distributed poisonous spider in the United States is the black widow (*Latrodectus mactans;* Fig. 29-20b). Its poison is a neurotoxin that interferes with transmission of messages from nerves to muscles. The brown recluse spider (*Loxosceles reclusa*) is more slender than the black widow and has a violin-shaped dorsal stripe on its cephalothorax. Its venom destroys the tissues surrounding the bite and, like the venom of the black widow, can occasionally be fatal. Although painful, spider bites cause fewer than five fatalities per year in the United States.

Although some mites contribute to soil fertility by breaking down organic matter, many are serious nuisances. They eat crops, infest livestock and pets, and inhabit our own bodies. Certain mites cause mange in dogs and other domestic animals. Chiggers (red bugs), the larval form of red mites, attach themselves to the skin and secrete an irritating digestive fluid that may cause itchy red welts.

Larger than mites, ticks are parasites on dogs, deer, and many other animals. They transmit diseases such as Rocky Mountain spotted fever, Texas cattle fever, relapsing fever, and Lyme disease (see Fig. 52-1).

Crustaceans are vital members of marine food webs

Subphylum **Crustacea** includes lobsters, crabs, shrimp, barnacles, and their relatives (Fig. 29-21). Paleobiologists have dated crustacean fossils back to the Cambrian period (more than 505 million years ago). Recent studies suggest crustaceans form a clade with members of subphylum Hexapoda.

Many crustaceans are primary consumers of algae and detritus. Countless billions of microscopic crustaceans are part of marine zooplankton, the free-floating, mainly microscopic organisms in the upper layers of the ocean. These crustaceans are food for many fishes and other marine animals, such as certain baleen whales.

A distinctive feature of crustaceans is the *nauplius larva*, which is often the first stage after hatching. This larva has only the most anterior three pairs of appendages. Crustaceans are also characterized by mandibles, biramous appendages, and two pairs of antennae. Their antennae serve as sensory organs for touch and taste. Most adult crustaceans have compound eyes, and many crustaceans have **statocysts,** sense organs that detect gravity (see Fig. 41-4).

The hard **mandibles,** used for biting and grinding food, are the third pair of appendages; they lie on each side of the ventral mouth. Posterior to the mandibles are two pairs of appendages, the first and second **maxillae,** used for manipulating and holding food. Several other pairs of appendages are present. Some are modified for walking. Others may be specialized for swimming, sperm transmission, carrying eggs and young, or sensation.

Crustaceans are the only group of arthropods that are primarily aquatic, and so, not surprisingly, they typically have gills for gas exchange. Two large glands located in the head excrete metabolic wastes and regulate salt balance. Most crustaceans have separate sexes. During copulation, the male uses specialized appendages to transfer sperm into the female. Newly hatched animals may resemble adults, or they may pass by successive molts through a series of larval stages before they develop the adult body.

Barnacles, the only sessile crustaceans, differ markedly in external anatomy from other members of the subphylum. They are marine suspension feeders that secrete complex limestone cups within which they live (see Fig. 29-21a). The larvae of barnacles are free-swimming forms that go through several molts. They eventually become sessile and develop into the adult form. The 19th-century naturalist Louis Agassiz described the barnacle as "nothing more than a little shrimplike animal standing on its head in a limestone house and kicking food into its mouth." The bane of marine boaters, barnacles can proliferate on ship

FIGURE **29-21**

Crustaceans.

(a) Goose barnacles (*Lepas fascicularis*) hanging from driftwood. These stalked barnacles occur in large numbers on intertidal rocks along the West Coast of the United States. **(b)** This sponge crab (*Criptodromia octodenta*), photographed in Australia, wears a living sponge on its back as camouflage.

(a)

Hal Harrison/Grant Heilman

(b)

Fred Bavendam/Peter Arnold, Inc.

bottoms in great numbers. Their presence can reduce the speed of a ship by more than 30%.

Isopods are mainly tiny (5–15 mm in length) marine crustaceans that inhabit the ocean floor. However, this group includes some familiar terrestrial animals: pillbugs and sowbugs. The mainly microscopic *copepods* are the largest group of crustaceans. Marine copepods are the most numerous component of zooplankton.

The largest order of crustaceans, *Decapoda,* contains more than 10,000 species of lobsters, crayfish, crabs, and shrimp. Most decapods are marine, but a few, such as the crayfish, certain shrimp, and a few crabs, live in fresh water. The crustaceans in general and the decapods in particular show striking specialization and differentiation of parts in the various regions of the animal. In the lobster, the appendages in the different parts of the body differ markedly in form and function (Fig. 29-22).

The five segments of the lobster's head and the eight segments of its thorax are fused into a cephalothorax, which is covered on the top and sides by a shield, the **carapace,** composed of chitin impregnated with calcium salts. The two pairs of antennae serve as chemoreceptors and tactile sense organs. The mandibles are short and heavy, with opposing surfaces used in grinding and biting food. Behind the mandibles are two pairs of accessory feeding appendages: the first and second maxillae. The appendages of the first three segments of the thorax are the maxillipeds, which aid in chopping up food and passing it to the mouth. The fourth segment of the thorax has a pair of large **chelipeds,** or pinching claws. The last four thoracic segments bear walking legs.

The appendages of the first abdominal segment are part of the reproductive system and function in the male as sperm-transferring structures. The following four abdominal segments bear paired **swimmerets,** small paddle-like structures used by some decapods for swimming and by the females of all species for holding eggs. Each branch of the sixth abdominal appendages consists of a large flattened structure. Together with the flattened posterior end of the abdomen, they form a tail fan used for swimming backward.

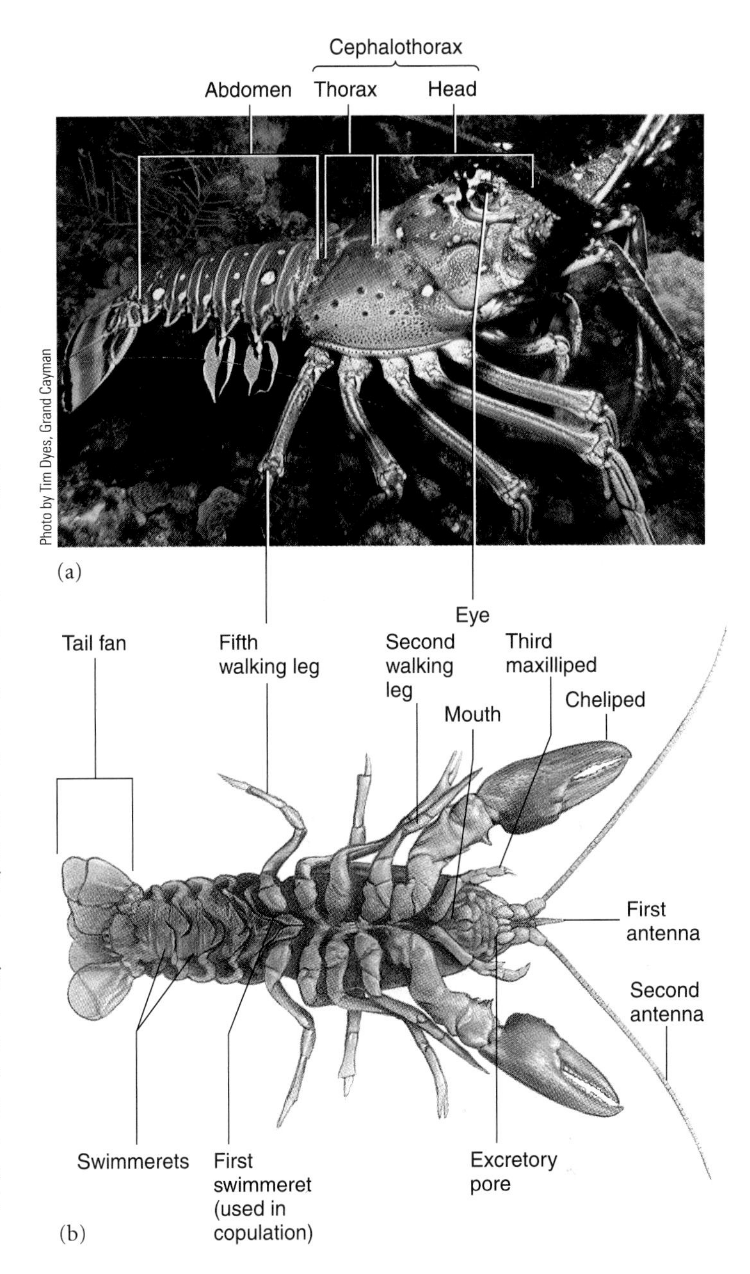

FIGURE **29-22** **Anatomy of the lobster.**

(a) Like other decapods, the spiny lobster (*Panulirus argus*) has five pairs of walking legs. The first pair of walking legs is modified as chelipeds *(large claws).* Photographed in Grand Cayman. **(b)** Ventral view of a lobster. Note the variety of specialized appendages.

Insects belong to subphylum Hexapoda

With more than 1 million described species (and perhaps millions more not yet identified), class **Insecta** of subphylum **Hexapoda,** is the most successful group of animals on our planet in terms of diversity, geographic distribution, number of species, and number of individuals (Table 29-5). What they lack in size, insects make up in sheer numbers. If we could weigh all the insects in the world, their weight would exceed that of all the remaining terrestrial animals. Although primarily terrestrial, some insect species live in fresh water, a few are truly marine, and others inhabit the shore between the tides.

The earliest fossil hexapods—primitive, wingless species—date back to the Devonian period more than 360 million years ago (mya). Hexapod fossils from the Carboniferous period more than 286 mya include both wingless and primitive winged species. Cockroaches, mayflies, and cicadas are among the insects that have survived relatively unchanged from the Carboniferous period to the present day. Until recently biologists thought insects made up a monophyletic taxon and were closely related to myriapods (centipedes and millipedes). Recent molecular and developmental studies suggest insects are not a monophyletic group and are more closely related to crustaceans. Molecular data also suggest a group of wingless, soil hexapods known as *springtails* may have evolved independently of the insects.

We describe an insect as an **articulated** (jointed), **tracheated** (having tracheal tubes for gas exchange) **hexapod** (having six feet). The insect body consists of three distinct parts: head, tho-

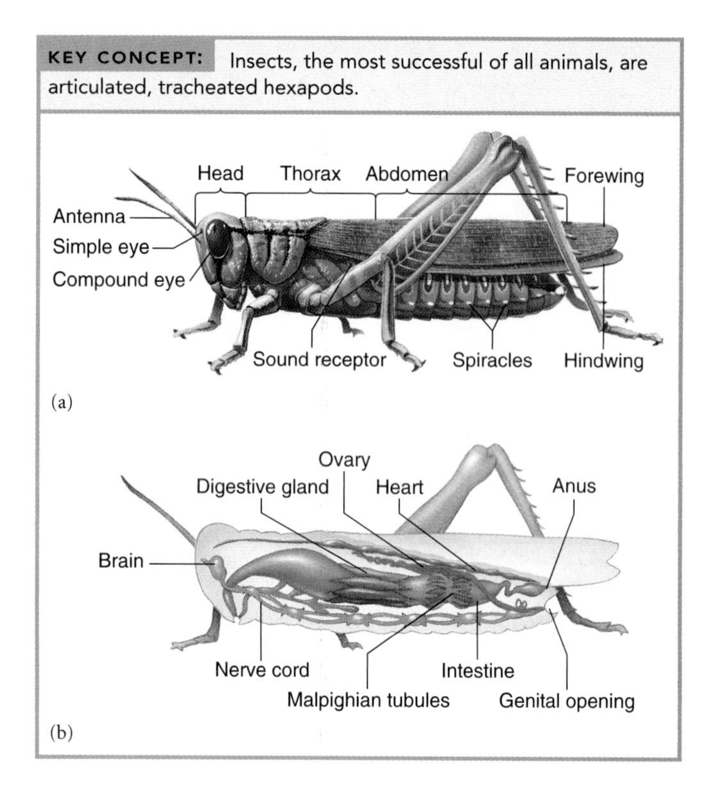

ACTIVE FIGURE **29-23** | **Anatomy of the grasshopper.**

(a) External structure. Note the three pairs of segmented legs. **(b)** Internal anatomy.

BiologyNow™ See **the butterfly life-cycle** in action by clicking on this figure on your BiologyNow CD-ROM.

rax, and abdomen (Fig. 29-23). Their uniramous (unbranched) appendages include three pairs of legs that extend from the adult thorax, and, in many orders, one or two pairs of wings. One pair of antennae protrudes from the head, and the sense organs include both simple and compound eyes. Their complex mouthparts, which include mandibles and maxillae, are adapted for piercing, chewing, sucking, or lapping.

The tracheal system of insects has made an important contribution to their diversity. Air enters the tracheal tubes, or tracheae, through **spiracles,** tiny openings in the body wall. Oxygen passes directly to the internal organs. This effective oxygen delivery system permits insects to have the high metabolic rate necessary for activities such as flight, even though they have an open circulatory system.

Excretion is accomplished by two or more **Malpighian tubules,** which receive metabolic wastes from the blood, concentrate the wastes, and discharge them into the intestine. Unique to terrestrial arthropods, Malpighian tubules also perform the extremely important function of conserving water (see Fig. 46-4).

The sexes are separate, and fertilization takes place internally. Several molts occur during development. The immature stages between molts are called **instars.** Primitive insects with no wings, such as silverfish, have direct, or simple, development: The young hatch as juveniles that resemble the adult form. Other insects, such as grasshoppers and cockroaches, undergo **incomplete metamorphosis** in which the egg gives rise to a larva that resembles the adult in many ways but lacks functional wings and reproductive structures. The larva goes through a series of molts during which it becomes more like the adult.

Most insects, including bees, butterflies, and fleas, undergo **complete metamorphosis** with four distinct stages in the life cycle: **egg, larva, pupa,** and **adult.** The wormlike larva does not look at all like the adult. For example, caterpillars have an entirely different body form from butterflies. Typically, an insect spends most of its life as a larva. Eventually the larva stops feeding, molts, and enters a pupal stage, usually within a protective cocoon or underground burrow. The pupa does not feed and typically cannot defend itself. Energy reserves stored during the larval stage are spent remodeling its body. When it emerges as an adult, it is equipped with functional wings and reproductive organs.

Certain species of bees, ants, and termites exist as colonies or societies made up of several different types of individuals, each adapted for some particular function. The members of some insect societies communicate with each other by "dances" and by means of **pheromones,** substances secreted to the external environment. Social insects and their communication are discussed in Chapter 50.

Adaptations that contribute to the biological success of insects. What are the secrets of insect success? The tough exoskeleton protects insects from predators and helps prevent water loss. Segmentation allows mobility and flexibility, and permits regional specialization of the body. For example, cephalization and highly developed sense organs are important in both offense and defense. The insect body plan has been modified and specialized in so many ways that insects are adapted to a remarkable number of lifestyles. The jointed appendages are specialized for various types of feeding, sensory functions, and different types of locomotion (walking, jumping, swimming). For example, grasshopper mouthparts are adapted for biting and chewing leaves. Moth and butterfly mouthparts are adapted for sucking nectar from flowers, and mosquito mouthparts are adapted for sucking blood. Aphids and leafhoppers have mouthparts specialized to pierce plants and feed on plant juices.

Another adaptation that contributes to insect success is the ability to fly. Unlike other terrestrial invertebrates, which creep slowly along on or under the ground, many insects fly rapidly through the air. Their wings and small size facilitate their wide distribution and, in many instances, their immediate survival. For example, when a pond dries up, adult aquatic insects can fly to another habitat.

The reproductive capacity of insects is amazing. Under ideal conditions, the fruit fly *Drosophila* can produce 25 generations in one year! Insect eggs are protected by a thick membrane, and several eggs may, in addition, be enclosed in a protective egg case. By dividing the insect life cycle into different stages, metamorphosis reduces intraspecific competition; larval forms have different lifestyles so they do not compete with adults for food or habitats.

Insects have many interesting adaptations for offense and defense (see Fig. 6-8). We all are familiar with the stingers of bees and wasps, which are specialized egg-laying structures (ovi-

TABLE 29-5 Some Orders of Insects

Order, Number of Species, Some Characteristics, and Representative Adult

Insects with Incomplete Metamorphosis (egg ⟶ larva ⟶ adult)

Order	Characteristics	Representative Adult
Homoptera (33,000)	Aphids, Leafhoppers, Cicadas, Scale insects 2 pairs membranous wings; piercing-sucking mouthparts form beak; parasites of plants; vectors of plant diseases	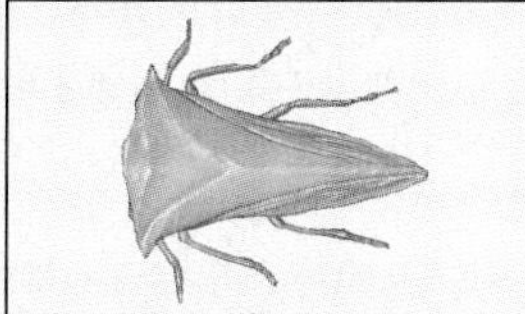 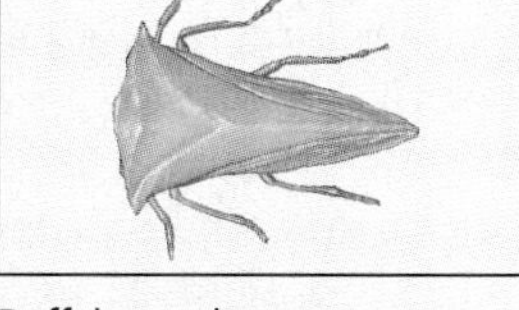 Buffalo treehopper *Stictocephala bubalus*
Odonata (6000)	Dragonflies, Damselflies Two pairs of long membranous wings; chewing mouthparts; large, compound eyes; active predators; larvae very different from adult	Damselfly *Ischnura*
Orthoptera (20,000)	 Grasshoppers, Crickets Forewings leathery, hindwings membranous; chewing mouthparts; most herbivorous, some cause crop damage; some predatory	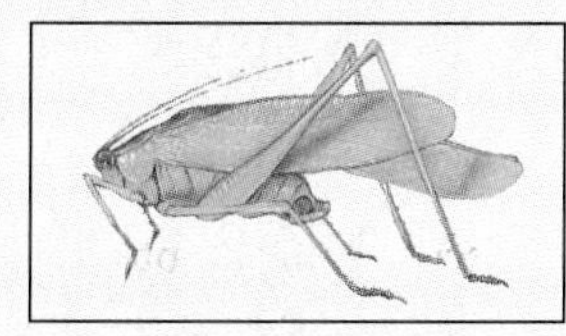 Fork-tailed bush katydid *Scudderia furcata*
Blattaria (3700)	Cockroaches When wings present, forewings leathery, hindwings membranous; chewing mouthparts; legs adapted for running	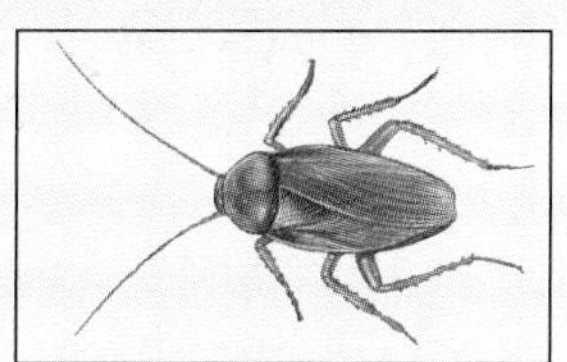 American cockroach *Periplaneta americana*
Isoptera (2000)	Termites 2 pairs of wings, or none; wings shed by sexual forms after mating; chewing mouthparts; social insects, form large colonies; eat wood	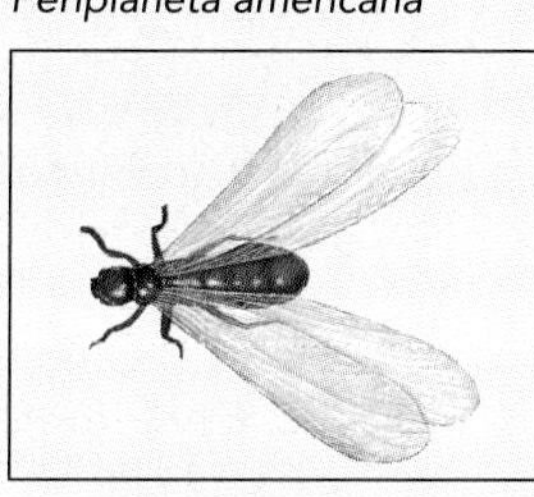 Eastern subterranean termite *Reticulitermes flavipes*
Anoplura (500)	Sucking lice No wings; piercing-sucking mouthparts; ectoparasites of mammals; head and body louse and crab louse are human parasites; vectors of typhus fever	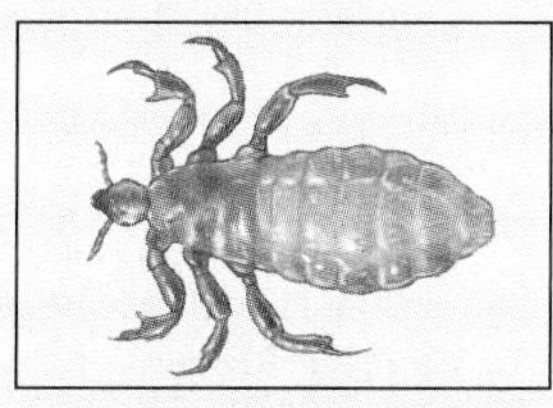 Human head and body louse *Pediculus humanus*
Hemiptera (true bugs) (62,000)	Chinch bugs, Bed bugs, Water striders Hindwings membranous; forewings smaller; piercing-sucking mouthparts form beak; most herbivorous; some parasitic.	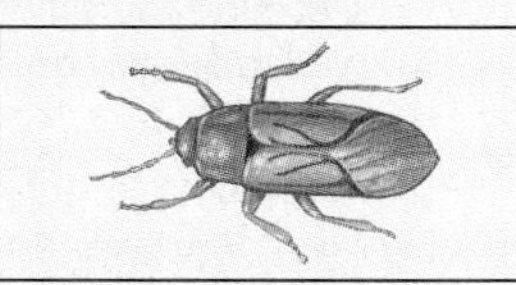 Chinch bug *Blissus leucopterus*

Insects with Complete Metamorphosis (egg ⟶ larva ⟶ pupa ⟶ adult)

Order	Characteristics	Representative Adult
Lepidoptera (140,000)	Moths, Butterflies Usually 2 pairs of membranous, colorful, scaled wings; sucking mouthparts; larvae are wormlike caterpillars that eat plants; adults suck flower nectar; important pollinators	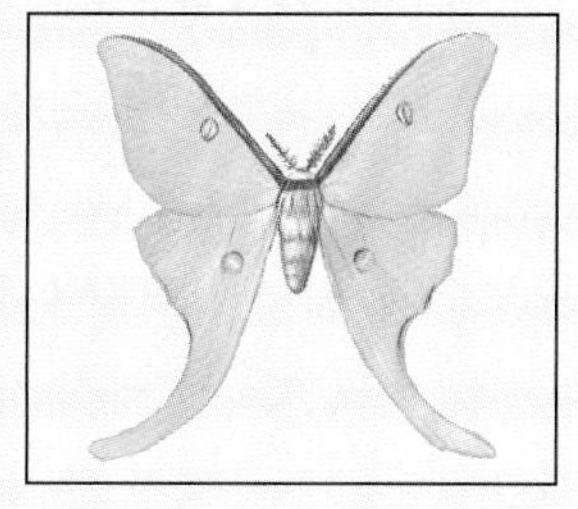 Luna moth *Aetias luna*
Diptera (150,000)	Houseflies, Mosquitos Only forewings functional in flying; hindwings small, knoblike balancing organs (halteres); mouthparts usually adapted for sucking (and piercing in some); larvae are maggots that damage domestic animals or food; adults transmit diseases such as sleeping sickness, yellow fever, or malaria	 Deerfly *Chrysops vittatus*
Siphonaptera (1750)	Fleas No wings; piercing sucking mouthparts; legs adapted for clinging and jumping; parasites on birds and mammals; vectors of bubonic plague and typhus	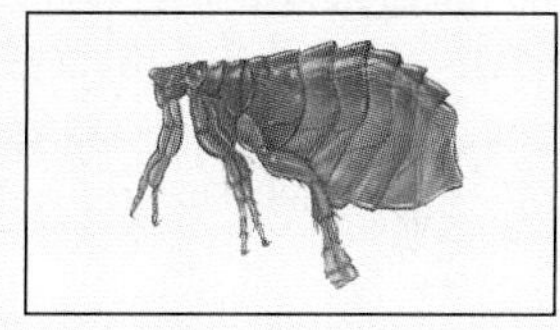 Dog Flea *Ctenocephalides canis*
Coleoptera (350,000)	Beetles, Weevils Forewings modified as protective coverings for membranous hindwings (which are sometimes absent); chewing mouthparts; largest order of insects; most herbivorous; some aquatic	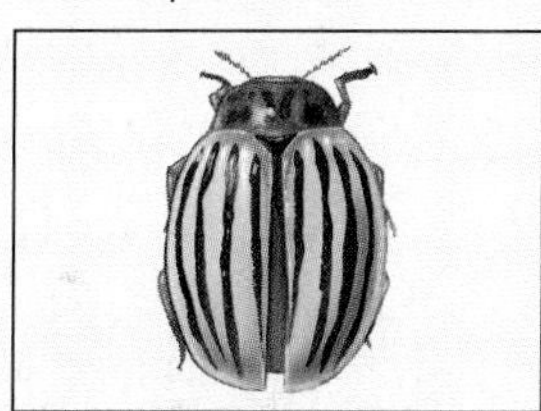 Colorado potato beetle *Leptinotarsa decemlineata*
Hymenoptera (190,000)	Ants, Bees, Wasps Usually 2 pairs of membranous wings; mouthparts may be modified for sucking or lapping nectar; many are social insects; some sting; important pollinators	 Bald-faced hornet *Vespula maculata*

Insects with No Metamorphosis (egg ⟶ immature form ⟶ adult)

Order	Characteristics	Representative Adult
Thysanura (320)	Silverfish No wings; biting-chewing mouthparts; 2–3 "tails" extend from posterior tip of abdomen; inhabit dead leaves; eat starch in books	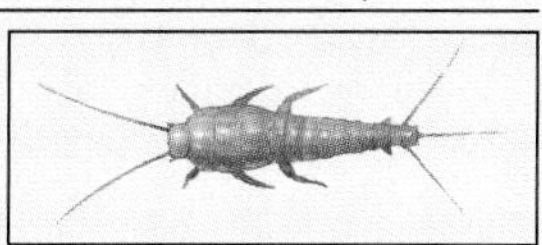 Silverfish *Lepisma saccharina*

positors). Many insects have **cryptic coloration,** blending into the background of their habitat; some look like dead twigs or leaves. Others mimic poisonous insects, deriving protection from this resemblance. Still others "play dead" by remaining motionless.

Insects communicate by tactile, auditory, visual, or chemical signals. For example, certain ant species use pheromones to mark trails and to warn of danger. Other insects use pheromones to attract a mate.

Impact of insects on humans. Not all insects compete with humans for food or cause us to scratch, swell up, or recoil from their presence. Bees, wasps, beetles, and many other insects pollinate flowers of crops and fruit trees. Some insects destroy other insects that are harmful to humans. For example, dragonflies eat mosquitos. And some organic farmers and home gardeners buy ladybird beetles, which are adept at ridding plants of aphids and other insect pests. Insects are important members of many food webs. Many birds, mammals, amphibians, reptiles, and some fishes depend on insects for food. Many beetles and the larvae (maggots) of flies are detritus feeders—they break down dead plants and animals and their wastes, permitting nutrients to be recycled.

Many insect products are useful to humans. Bees produce honey as well as beeswax, which we use to make candles, lubricants, and other products. Shellac is made from lac, a substance given off by certain scale insects that feed on the sap of trees. And the labor of silkworms provides us with beautiful fabric.

On the negative side, insects destroy billions of dollars worth of crops each year. Termites destroy buildings, and moths damage clothing. Fire ants not only inflict painful stings but cause farmers serious economic loss because of their large mounds, which damage mowers and other farm equipment.

Blood-sucking flies, screw worms, lice, fleas, and other insects annoy and transmit disease in humans and domestic animals. Mosquitos are vectors of malaria (see Fig. 24-10) and yellow fever. Body lice transmit the typhus rickettsia, and houseflies sometimes transmit typhoid fever and dysentery. Tsetse flies transmit African sleeping sickness, and fleas may be vectors of bubonic plague.

Review

- What function does the pseudocoelom have in nematodes?
- Why is segmentation an important characteristic of the arthropod body?
- How do myriapods differ from chelicerates?
- What are the functions of each of these appendages in a lobster? maxillae; chelipeds; swimmerets?
- Define an insect. What are four adaptations that have contributed to insect success?

BiologyNow™ Assess your understanding of **the ecdysozoa** by taking the pretest on your BiologyNow CD-ROM.

SUMMARY WITH KEY TERMS

1 Characterize the protostome coelomates, describe their two main evolutionary branches, and give examples of animals assigned to each branch.

- The **protostome coelomates** are characterized by spiral cleavage, determinate cleavage, and the development of the mouth from the blastopore. The **coelom** is a fluid-filled cavity completely lined with mesoderm. The evolution of the coelom permitted many innovations, including the **tube-within-a-tube body plan** and the **hydrostatic skeleton.** The coelom provides space for internal organs and for gonads to develop. It helps transport materials and protects internal organs.
- The protostome coelomates include two main branches: the Lophotrochozoa and the Ecdysozoa. The **Lophotrochozoa** include the platyhelminthes, nemerteans, mollusks, annelids, the lophophorate phyla, and the rotifers. The **Ecdysozoa** include the nematodes (roundworms), and arthropods.

2 Identify distinguishing characteristics of phylum Nemertea and phylum Platyhelminthes; describe the main classes of phylum Platyhelminthes, giving examples of animals that belong to each class.

- Phylum **Nemertea** (ribbon worms) is a lophotrochozoan group characterized by the **proboscis,** a muscular tube used in capturing food and in defense. The coelom is reduced; it consists of the **rhynchocoel,** a space surrounding the proboscis. Nemerteans have a tube-within-a-tube body plan, a complete digestive tract with mouth and anus, and a circulatory system.
- Phylum **Platyhelminthes,** the flatworms, are acoelomate animals with bilateral symmetry, **cephalization,** three definite tissue layers, and well-developed organs. Many flatworms are **hermaphrodites;** a single animal produces both sperm and eggs.
- Flatworms have a ladder-type nervous system, typically consisting of sense organs and a simple brain consisting of two **ganglia** connected to two nerve cords that extend the length of the body. They have **protonephridia,** organs that function in osmoregulation and disposal of metabolic wastes.
- Phylum Platyhelminthes includes four classes: Class **Turbellaria** is comprised of free-living flatworms, including **planarians.** Classes **Trematoda** and **Monogenea** include the parasitic **flukes,** and the parasitic tapeworms comprise class **Cestoda.** The parasitic flukes and tapeworms typically have suckers or hooks for holding onto their hosts; they have complicated life cycles with intermediate hosts and produce large numbers of eggs.

3 Describe the adaptive advantages of cephalization.

- Flatworms show the beginnings of cephalization, the evolution of a head with the concentration of sense organs and nerve cells (a simple brain) at the anterior end. Cephalization increases the effectiveness of a bilateral animal to find food and detect enemies.

4 Describe the distinguishing characteristics of phylum Mollusca, and describe the classes discussed, giving examples.

- Members of phylum **Mollusca** are soft-bodied animals usually covered by a shell. They have a ventral **foot** for locomotion and a **mantle** that covers the **visceral mass,** a concentration of body organs.

- Mollusks have an **open circulatory system** with the exception of cephalopods, which have a **closed circulatory system.** Mollusks have paired excretory tubules called metanephridia. A rasplike **radula** functions as a scraper in feeding in all groups except the bivalves, which are suspension feeders. Typically, marine mollusks have a free-swimming, ciliated **trochophore larva.**
- Class **Polyplacophora** includes the sluggish marine **chitons,** which have shells that consist of eight overlapping plates.
- Class **Gastropoda,** the largest group of mollusks, includes the snails, slugs, and their relatives. In gastropods, the body undergoes **torsion,** a twisting of the visceral mass. The shell (when present) is coiled. Torsion is not related to shell coiling.
- Class **Bivalvia** includes the aquatic clams, scallops, and oysters; a two-part shell, hinged dorsally, encloses the bodies of these suspension feeders.
- Class **Cephalopoda** includes the squids, octopods, and *Nautilus.* Cephalopods are active, predatory swimmers. Tentacles surround the mouth, located in the large head.

5 Describe the distinguishing characteristics of phylum Annelida, and describe the three classes of annelids discussed, giving examples.

- Phylum **Annelida,** the segmented worms, includes many aquatic worms, earthworms, and leeches. Annelids have conspicuously long bodies with **segmentation** both internally and externally; their large compartmentalized coelom serves as a hydrostatic skeleton.
- Class **Polychaeta** consists of marine worms characterized by appendages called **parapodia,** which are used for locomotion and gas exchange. The parapodia bear many bristle-like structures called **setae.**
- Class **Oligochaeta,** comprised of the earthworms, is characterized by a few short setae per segment. The body is divided into more than 100 segments separated internally by **septa.**
- Class **Hirudinea,** the leeches, is characterized by the absence of setae and appendages. Parasitic leeches are equipped with suckers for holding onto their host.

6 Describe the distinguishing characteristics of the lophophorate phyla.

- The **lophophorate phyla,** marine animals that have a lophophore, include the **brachiopods, phoronids,** and **bryozoa.** The **lophophore,** a ciliated ring of tentacles surrounding the mouth, is specialized for capturing suspended particles in the water.

7 Describe the distinguishing characteristics of phylum Nematoda.

- Phylum **Nematoda,** the **roundworms,** are pseudocoelomates with bilateral symmetry, three tissue layers, and a complete digestive tract. The nematode body is covered by a tough **cuticle** that helps prevent desiccation. Parasitic nematodes in humans include ***Ascaris,*** **hookworms, trichina worms,** and **pinworms.**

8 Describe the distinguishing characteristics of phylum Arthropoda; distinguish among the subphyla and classes of arthropods, giving examples of animals that belong to each group.

- Phylum **Arthropoda** is composed of segmented animals with **paired, jointed appendages** and an armor-like **exoskeleton** of chitin. **Molting** is necessary for the arthropod to grow.
- Arthropods have an open circulatory system with a dorsal heart that pumps **hemolymph.**
- Aquatic forms have gills; terrestrial forms have either **tracheae** or **book lungs.**
- The **trilobites** are extinct marine arthropods covered by a hard, segmented shell.
- Subphylum **Myriapoda** includes class **Chilopoda,** the centipedes, and class **Diplopoda,** the millipedes; members of this subphylum have unbranched appendages and a single pair of antennae.
- Subphylum **Chelicerata** includes the **merostomes** (horseshoe crabs) and the **arachnids** (spiders, mites, and their relatives). The chelicerate body consists of a cephalothorax and abdomen; there are six pairs of jointed appendages, of which four pairs serve as legs.
- The first pair of appendages are **chelicerae,** the second pair are **pedipalps.** These appendages are adapted for manipulation of food, locomotion, defense, or copulation. Chelicerates have no antennae and no mandibles.
- Subphylum **Crustacea** includes lobsters, crabs, shrimp, pillbugs, and barnacles. The body consists of a cephalothorax and abdomen; they typically have five pairs of walking legs.
- Crustaceans have two pairs of **antennae** that sense taste and touch. The third appendages are **mandibles** used for chewing. Two pairs of **maxillae,** posterior to the mandibles, manipulate and hold food.
- Subphylum **Hexapoda** includes class **Insecta.** Insects have unbranched appendages and a single pair of antennae. An insect is an **articulated, tracheated hexapod;** its body consists of head, thorax, and abdomen. Insects have tracheae for gas exchange and **Malpighian tubules** for excretion.

9 Discuss factors that have contributed to the great biological success of insects.

- The biological success of the insects results from many adaptations, including the versatile exoskeleton, segmentation, specialized jointed appendages, ability to fly, highly developed sense organs, **metamorphosis** (which reduces intraspecific competition), effective reproductive strategies, effective mechanisms for defense and offense, and the ability to communicate.

POST-TEST

1. Which of the following is associated with the evolution of the coelom? (a) radial symmetry (b) tube-within-a-tube body plan (c) incomplete metamorphosis (d) bilateral symmetry (e) development of ganglia
2. Cephalization (a) evolved along with bilateral symmetry (b) refers to the development of a digestive system (c) is associated with radial symmetry (d) involves a concentration of excretory organs (e) first evolved in arthropods
3. Rudimentary cephalization, protonephridia, and auricles characterize some numbers of phylum (a) Arthropoda (b) Cnidaria (c) Platyhelminthes (d) Mollusca (e) Crustacea

4. Trochophore larvae are characteristic of (a) Arthropoda (b) Cnidaria (c) Platyhelminthes (d) Mollusca (e) Crustacea
5. Tapeworms are classified in phylum (a) Porifera (b) Cnidaria (c) Platyhelminthes (d) Ctenophora (e) Coelomata
6. Which of the following is *not* an adaptation of parasitic life? (a) production of a few well-protected eggs (b) hooks (c) suckers (d) reduced digestive system (e) intermediate host
7. The hydrostatic skeleton (a) results in torsion (b) permits contracting muscles to recover quickly (c) permits a greater range of movement than possible in acoelomate animals (d) is a unique characteristic of annelids (e) permits ecdysis
8. Which of the following is *not* characteristic of nemerteans (proboscis worms)? (a) large coelom (b) tube-within-a-tube body plan (c) complete digestive tube (d) muscular tube for capturing food (e) circulatory system
9. Which of the following characteristics is associated with phylum Mollusca? (a) mandibles (b) mantle (c) pedipalps (d) chelipeds (e) setae
10. Which of the following belong to phylum Mollusca? (a) gastropods and crustaceans (b) oligochaetes and polychaetes (c) chelicerates and bryozoans (d) crustaceans and nemerteans (e) gastropods and cephalopods
11. Which of the following belongs to Ecdysozoa? (a) mollusks (b) annelids (c) nematodes (d) nemerteans (e) lophophorates
12. Which of the following is *not* a nematode? (a) hookworm (b) trichina worm (c) *Ascaris* (d) leech (e) pinworm
13. Which of the following is *not* true of the lophophorates? (a) this group includes the bryozoa (b) some have shells and look somewhat like mollusks (c) characterized by a ciliated ring of tentacles at anterior end (d) mainly marine animals (e) some of its members are articulated, tracheated hexapods
14. Torsion in mollusks (a) is characteristic of bivalves (b) is a twisting of the visceral mass (c) involves coiling of the molluscan shell (d) begins in the adult stage (e) depends on action of parapodia
15. Trilobites (a) were early mollusks (b) are members of phylum Onychophora (c) are characterized by parapodia and setae (d) were early arthropods (e) are an evolutionary link between annelids and arthropods
16. Which of the following belong to subphylum Hexapoda? (a) insects (b) horseshoe crabs (c) crustaceans (d) centipedes (e) spiders
17. The correct sequence in insect complete metamorphosis is (a) egg ⟶ immature form ⟶ adult (b) egg ⟶ trochophore larva ⟶ veliger larva ⟶ adult (c) egg ⟶ pupa ⟶ larva ⟶ adult (d) egg ⟶ larva ⟶ pupa ⟶ adult (e) adult ⟶ larva ⟶ egg ⟶ pupa
18. Which of the following is characteristic of insects? (a) biramous appendages (b) mandibles (c) chelicerae (d) eight legs (e) two pairs of antennae
19. Which of the following is *not* characteristic of arthropods? (a) exoskeleton (b) trochophore larva (c) paired, jointed appendages (d) chitin (e) segmentation
20. Spiders are characterized by (a) mandibles and maxillae (b) six pairs of legs on the abdomen (c) one pair of antennae (d) biramous appendages (e) chelicerae and pedipalps

CRITICAL THINKING

1. Discuss the idea that every evolutionary adaptation has both advantages and disadvantages, using each of the following characteristics as an example: (a) cephalization (b) the arthropod exoskeleton (c) segmentation with specialization
2. Hypothesize why oysters secrete calcium carbonate layers around foreign particles.
3. Insects that undergo complete metamorphosis outnumber those that do not by more than ten to one. Hypothesize an explanation.

- Visit our Web site at **http://biology.brookscole.com/solomon7** for links to chapter-related resources on the World Wide Web. Additional online materials relating to this chapter can also be found on our Web site.

BIOLOGY NOW RESOURCES

Active Figures

29-2: The fluke life-cycle

29-23: The butterfly life-cycle

Preparing for an exam? Take a diagnostic test on your BiologyNow CD-ROM.

Post-Test Answers

1. b	2. a	3. c	4. d
5. c	6. a	7. c	8. a
9. b	10. e	11. c	12. d
13. e	14. b	15. d	16. a
17. d	18. b	19. b	20. e

30

The Animal Kingdom: The Deuterostomes

© Ed Robinson/Tom Stack & Associates

Representative deuterostomes This marine habitat photographed in Hawaii includes an echinoderm known as a pencil urchin (*Heterocentrotrus mammilatus*), and chordates, represented by fishes, the clown wrasse (*Coris gaimard*) with its yellow tail and the Moorish idol (*Zanchus cornutus*).

CHAPTER OUTLINE

What does a sea star have in common with a fish, frog, hawk, or human? You may think it strange to group the **echinoderms**—the sea stars, sea urchins, and sand dollars—with the **chordates,** the phylum to which we humans and other **vertebrates** (animals with a backbone) belong. However, even though these animals look and behave very differently, evidence suggests that echinoderms and chordates share a common ancestor and are closely related. They are both **deuterostomes,** animals that make up a major evolutionary branch of the animal kingdom.

Biologists also include the **hemichordates,** a small group of wormlike marine animals, with the deuterostomes. The most familiar of the hemichordates are the acorn worms, animals that live buried in mud or sand. The hemichordates have a characteristic ring of cilia surrounding the mouth. A distinguishing derived character is the three-part body: a proboscis, collar, and trunk.

In this chapter we focus on echinoderms, such as the pencil urchin, and chordates, represented in the photograph by the fishes. The largest chordate subphylum is Vertebrata, which includes the animals with which we are most familiar—fishes, amphibians, reptiles, birds, and mammals.

Deuterostomes evolved from a common ancestor more than 550 mya (during the Precambrian). These animals share several derived characters (evolutionary novelties absent in the common ancestor of deuterostomes; see Chapter 22). Their defining characters are similarities in their patterns of development (see Chapter 28). Deuterostomes are characterized by radial, rather than spiral, cleavage. Their cleavage is indeterminate, which means the fate of their cells is fixed later in development than is the case in protostomes. In deuterostomes, the blastopore becomes the anus (or is located near the future site of the anus), and the mouth develops from a second opening at the anterior end of the embryo. Pharyngeal slits are also characteristic of deuterostomes, but the extant echinoderms have lost this character.

Fossil evidence of the first chordates, and perhaps also the first vertebrates, has been found in rocks more than 550 million years old from the Cambrian period. However, the interpretation of these fossils is controversial. Investigators disagree about which fossil or group represents the deuterostome ancestor of the chordate lineage. They also disagree about which group represents the ancestral chordate that gave rise to the vertebrates. Another issue is how the invertebrate chordates are related to the vertebrate groups. Investigators are studying recently discovered fossils and applying molecular techniques to these problems. ■

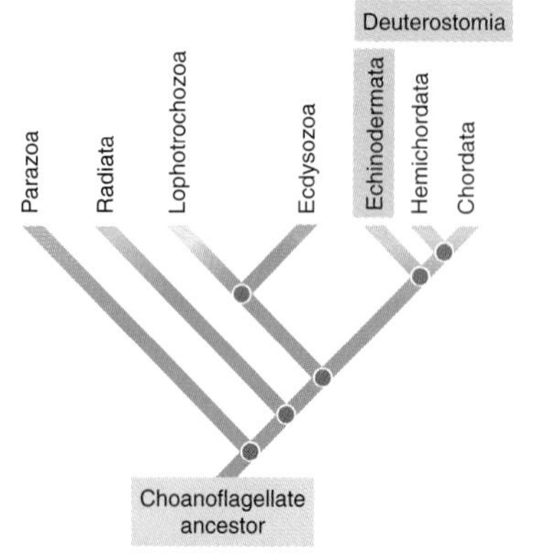

(a) Crinoidea

FIGURE 30-1 | Echinoderms.

(a) Feather stars use their slender, jointed appendages to cling to the surface of a rock or coral reef. They can creep and often move away to escape predators. **(b)** Orange and red sea star (*Fromia monilis*) on bubble coral. **(c)** Daisy brittle star (*Ophiopholis aculeata*), photographed in Muscongus Bay, Maine. **(d)** With its flattened, circular body, the sand dollar (*Dendraster excentricus*) is adapted for burrowing on the ocean floor. **(e)** A sea cucumber (*Thelonota*), raises its body to spawn.

(b) Asteroidea

(c) Ophiuroidea

(d) Echinoidea

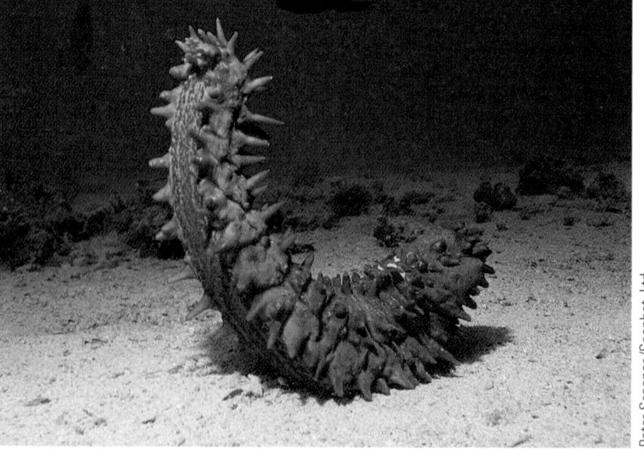

(e) Holothuroidea

ECHINODERMS

Learning Objectives

1. Identify the three main branches of deuterostomes.
2. List three shared, derived characters of echinoderms, and describe the main classes of echinoderms.

All members of phylum **Echinodermata** inhabit marine environments. They are found in the ocean at all depths. About 7000 living and more than 13,000 extinct species have been identified. The echinoderms probably evolved during the early Cambrian period and achieved maximum diversity by the middle of the Paleozoic era, about 400 million years ago (mya). By the beginning of the Mesozoic era 248 mya, they had declined, leaving six main groups that have survived to the present day (Fig. 30-1): class Crinoidea, sea lilies and feather stars; class Asteroidea, sea stars; class Ophiuroidea, basket stars and brittle stars; class Echinoidea, sea urchins and sand dollars; class Holothuroidea, sea cucumbers; and class Concentricycloidea, sea daisies.

The echinoderms are in many ways unique in the animal kingdom. Echinoderm larvae are bilaterally symmetrical, ciliated, and free-swimming. However, during development the body form reorganizes, and the adult exhibits **pentaradial symmetry,** in which the body is arranged in five parts around a central axis. Echinoderms have an **endoskeleton,** an internal skeleton, that is covered by a thin, ciliated epidermis. The endoskeleton consists of $CaCO_3$ plates and spines. The name *Echinodermata,* meaning "spiny-skinned," was inspired by the spines that project outward from the endoskeleton. Some groups have pincer-like, modified spines called **pedicellariae** on the body surface (Fig. 30-2). These structures, unique to the echinoderms, keep the surface of the animal free of debris.

The most unique derived character of echinoderms is the hydraulic **water vascular system,** a network of fluid-filled canals that functions in locomotion, feeding, and gas exchange. Branches of the water vascular system lead to numerous tiny **tube feet** that extend when filled with fluid. Each tube foot receives fluid from the main system of canals. A rounded muscular sac, or **ampulla,** at the base of the foot, stores fluid and is used to operate the tube foot. A valve separates each tube foot from other parts of the system. When the valve shuts, the ampulla contracts, forcing fluid into the tube foot so that it extends. At the bottom of the foot is a suction-type structure that presses against and adheres to whatever surface the tube foot is on.

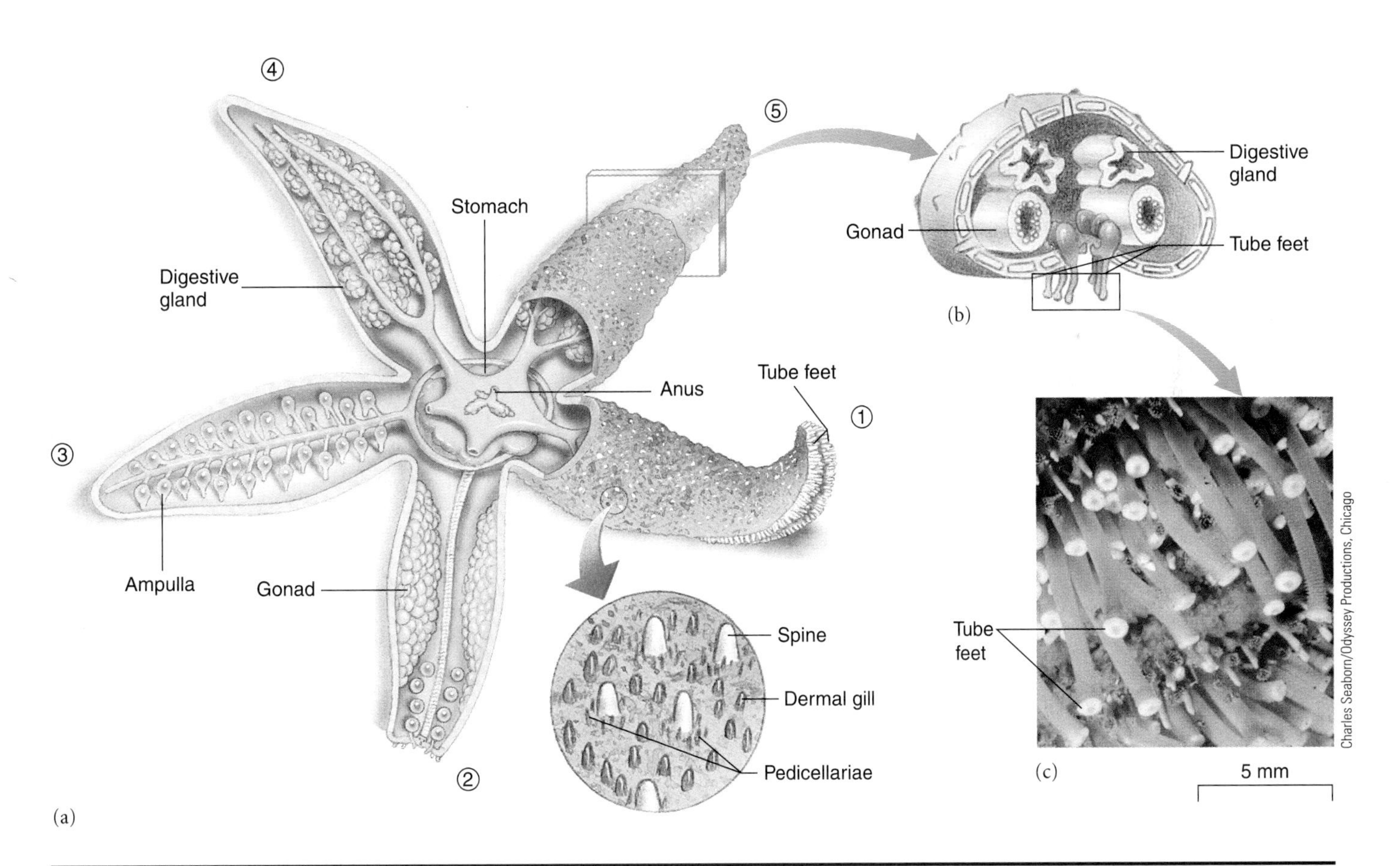

FIGURE 30-2 Body plan of a sea star.

(a) A sea star viewed from above, with its arms in various stages of dissection. Similar structures are present in each arm. The two-part stomach is in the central disc with the anus on the aboral *(upper)* surface and the mouth beneath on the oral surface. ① Upper surface with magnified detail. The end is turned up to show the tube feet on the lower surface. ② The arm is dissected to show well-developed gonads. ③ Upper body and digestive glands have been removed, exposing the ampullae of some of the hundreds of tube feet *(magnified view).* ④ Other organs have been removed to show the digestive glands. ⑤ Upper surface. **(b)** Cross section through arm and tube feet. **(c)** LM of tube feet of a sea star.

Echinoderms have a well-developed coelom, and the coelomic fluid transports materials. Although its structure varies in different groups, the complete digestive system is the most prominent body system. A variety of respiratory structures are found in the various classes. No excretory organs are present. The nervous system is simple, generally consisting of a nerve ring with nerves that extend out from it. Echinoderms have no brain. The sexes are usually separate, and eggs and sperm are generally released into the water, where fertilization takes place.

Members of class Crinoidea are suspension feeders

Class **Crinoidea,** the oldest class of living echinoderms, includes the feather stars and the sea lilies (see Fig. 30-1a). Although a great many extinct crinoids are known, there are relatively few living species. The feather stars are motile crinoids, although they often remain in the same location for long periods. Sea lilies are sessile and remain attached to the ocean floor by a stalk.

Crinoids remove suspended food from the water. In all other echinoderms, the mouth is located on the underside of the disk toward the substratum, but in crinoids the **oral surface** (mouth surface) is on the upper side of the disk. A number of branched, feathery arms also extend upward. Along the feathery arms are numerous tube feet shaped like small tentacles and coated with mucus that traps microscopic organisms.

Many members of class Asteroidea capture prey

Sea stars, or starfish, are members of class **Asteroidea** (see Fig. 30-1b). Their bodies consist of a central disk from which extend 5 to more than 20 arms, or rays (see Fig. 30-2). The undersurface of each arm has hundreds of pairs of tube feet. The mouth lies in the center of the underside of the disk. The endoskeleton consists of a series of calcareous plates that permit some movement in the arms. Delicate dermal gills, small extensions of the body wall, carry on gas exchange.

Most sea stars are carnivorous predators and scavengers that feed on crustaceans, mollusks, annelids, and even other echinoderms. Occasionally they catch small fish. The sea star's water vascular system does not permit rapid movement, so its prey are usually slow-moving or stationary animals such as clams.

To attack a clam or other bivalve mollusk, the sea star mounts it and assumes a humped position as it straddles the edge opposite the hinge. Then, holding itself in position with its tube feet, the sea star slides its thin, flexible stomach out through its mouth and between the closed, or slightly gaping, valves (shell parts) of the clam. The sea star secretes enzymes that digest the soft parts of the clam to the consistency of a thick soup while the clam is still in its own shell. The partly digested meal passes into the sea star body, where it is further digested by enzymes secreted from glands located in each arm.

The circulatory system in sea stars is poorly developed and probably of little help in circulating materials. Instead, this function is assumed by the coelomic fluid, which fills the large coelom and bathes the internal tissues. Metabolic wastes pass to the outside by diffusion across the tube feet and dermal gills. The nervous system consists of a ring of nervous tissue encircling the mouth and a nerve extending from this ring into each arm.

Class Ophiuroidea is the largest class of echinoderms

Basket stars and brittle stars (serpent stars), members of class **Ophiuroidea,** are the largest group of echinoderms, both in number of species and in number of individuals (see Fig. 30-1c). These animals resemble sea stars in that their bodies consist of a central disk with arms, but the arms are long and slender and more sharply set off from the central disk. Ophiuroids can move more rapidly than sea stars, using their arms to perform rowing or even swimming movements. Their tube feet lack suckers and are not used in locomotion; they are used to collect and handle food and may also serve a sensory function, perhaps that of taste.

Members of class Echinoidea have movable spines

Sea urchins and sand dollars, the animals of class **Echinoidea,** have no arms (see Fig. 30-1d and chapter opening photograph). Their skeletal plates are flattened and fused to form a solid shell called a **test.** The flattened body of the sand dollar is adapted for burrowing in the sand, where it feeds on tiny organic particles. Sand dollars have smaller spines than do sea urchins.

The sea urchin body is covered with spines that in some species can penetrate flesh and are difficult to remove. So threatening are these spines that swimmers on tropical beaches are often cautioned to wear shoes when venturing offshore, where these living pincushions may live in abundance. Sea urchins use their tube feet for locomotion. They also push themselves along with their movable spines. Many sea urchins graze on algae, scraping the sea floor with their calcareous teeth.

Members of class Holothuroidea are elongated, sluggish animals

Sea cucumbers, members of class **Holothuroidea,** are appropriately named, for some species are about the size and shape of a cucumber. The elongated body is a flexible, muscular sac (see Fig. 30-1e). The mouth is usually surrounded by a circle of tentacles that are modified tube feet. The endoskeleton consists of microscopic plates embedded in the body wall. More highly developed than that of other echinoderms, the circulatory system functions to transport oxygen and perhaps nutrients as well.

Sea cucumbers are sluggish animals that usually live on the bottom of the sea, sometimes burrowing in the mud. Some graze with their tentacles, whereas others stretch their branched tentacles out in the water and wait for dinner to float by. Algae and other morsels are trapped in mucus along the tentacles.

An interesting habit of some sea cucumbers is evisceration, in which the digestive tract, respiratory structures, and gonads are ejected from the body, usually when environmental conditions are unfavorable. When conditions improve, the lost parts are regenerated. Even more curious is the fact that when certain sea cucumbers are irritated or attacked, they direct their rear end toward the enemy and shoot red tubules out of their anus! These unusual weapons are sticky, and the attacking animal may become hopelessly entangled. Some of the tubules release a toxic substance.

Members of class Concentricycloidea have a unique water vascular system

Biologists established class **Concentricycloidea** to accommodate sea daisies, a group of interesting echinoderms first discovered in 1986. Sea daisies inhabit bacteria-rich wood sunk in deep water and apparently ingest bacteria through their body surface. Their small, disk-shaped, flat bodies (less than 1 cm in diameter) have no arms or mouth.

Review

- What are three derived characters of echinoderms? Describe each.
- How do sea stars differ from crinoids? From sea urchins?

BiologyNow™ Assess your understanding of **echinoderms** by taking the pretest on your BiologyNow CD-ROM.

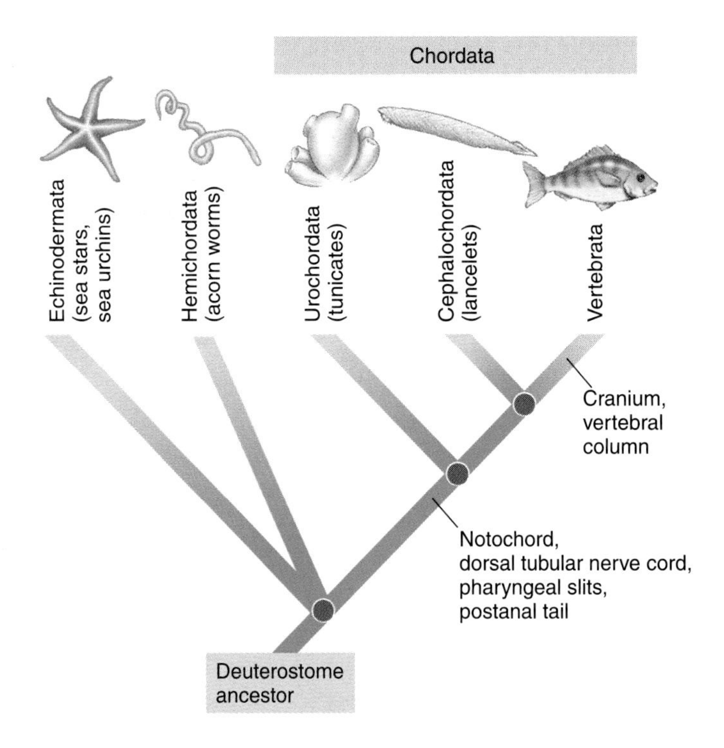

FIGURE **30-3** **Evolutionary relationships of the chordates.**

This diagram shows possible phylogenetic relationships among deuterostomes, including chordates (tan region), based on morphological and DNA data.

CHORDATE CHARACTERS

Learning Objective

3 Describe four shared derived characters of chordates.

Biologists currently divide phylum **Chordata** into three subphyla: Urochordata, which consists of marine animals called *tunicates;* Cephalochordata, marine animals called *lancelets;* and Vertebrata, animals with backbones. Of these extant chordate groups, the urochordates were the earliest to evolve (Fig. 30-3).

Chordates are deuterostome coelomates with bilateral symmetry, a tube-within-a-tube body plan, and three well-developed germ layers (Fig. 30-4). Typically, they have an endoskeleton and a closed circulatory system with a ventral heart. Chordates have segmented bodies, but specialization is so pronounced that the basic segmentation of the body plan may not be apparent. For example, in vertebrates, although segmentation of the body is not obvious, muscles and nerves are segmentally organized. Chordates are compared to echinoderms in Table 30-1.

Four shared derived characters that distinguish the chordates are the notochord; the dorsal, tubular nerve cord; gill slits; and a postanal tail.

1. All chordates have a **notochord** during some time in their life cycle. The notochord is a dorsal longitudinal rod that is firm, but flexible, and supports the body.
2. At some time in their life cycle, chordates have a **dorsal, tubular nerve cord.** The chordate nerve cord differs from the nerve cord of most other animals in that it is located dorsally rather than ventrally, is hollow rather than solid, and is single rather than double.
3. Chordates have **pharyngeal slits** (also called *pharyngeal gill slits*) during some time in their life cycle. In the embryo, a series of alternating branchial (gill) arches and grooves develop in the body wall in the pharyngeal (throat) region. Pharyngeal pouches extend laterally from the anterior portion of the digestive tract toward the grooves.

FIGURE **30-4** **Generalized chordate body plan.**

Note the notochord; dorsal, tubular nerve cord; pharyngeal (gill) slits; and postanal tail.

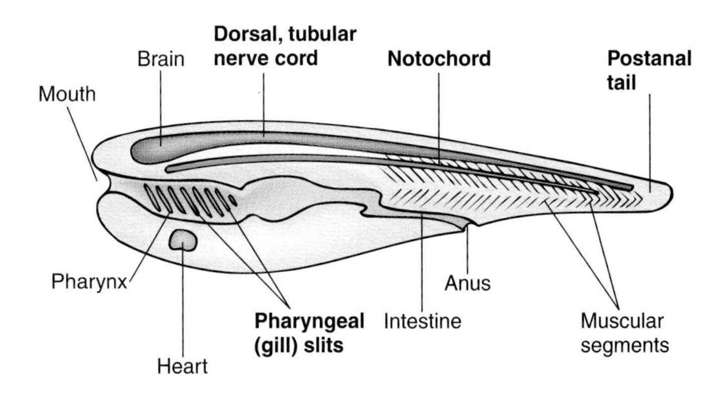

Early chordates, like many living chordates, were probably suspension feeders. The arrangement of pharyngeal pouches and slits permitted them to take water in through the mouth, concentrate small particles of food in the gut, and let the water escape from the body through the slits. The pharyngeal slits and the arches supporting them became modified for gas exchange in fishes and some other aquatic vertebrates. Early in vertebrate evolution, anterior pharyngeal arches evolved into jaws.

4. Chordates have a larva or embryo with a muscular **postanal tail,** an appendage that extends posterior to the anus.

Review

- What is a notochord?
- What is the significance of pharyngeal slits?

BiologyNow™ Assess your understanding of **chordate characters** by taking the pretest on your BiologyNow CD-ROM.

INVERTEBRATE CHORDATES

Learning Objectives

4 Describe the invertebrate chordate subphyla.

5 Discuss the evolution of chordates.

The invertebrate chordates include tunicates and lancelets. Biologists assign the tunicates to subphylum Urochordata, and the lancelets to subphylum Cephalochordata.

Tunicates are common marine animals

The **tunicates,** which make up subphylum **Urochordata,** include the sea squirts, or **ascidians,** and their relatives. Larval tunicates have typical chordate characteristics and superficially resemble tadpoles. The expanded body has a pharynx with slits, and the long muscular tail contains a notochord and a dorsal, tubular nerve cord. Some tunicates *(appendicularians)* are common members of the zooplankton that retain their chordate features and ability to swim.

Most tunicates are ascidians, commonly known as *sea squirts* (class Ascidiacea). A sea squirt larva swims for a time, then attaches itself to the sea bottom and loses its tail, notochord, and much of its nervous system. Adult sea squirts are barrel-shaped, sessile marine animals unlike other chordates. Indeed, they are often mistaken for sponges or cnidarians (Fig. 30-5). Only the pharyngeal slits and the structure of its larva indicate that the sea squirt is a chordate.

Adult tunicates develop a protective covering, or **tunic,** that may be soft and transparent or quite leathery. Curiously, the tunic consists of a carbohydrate much like cellulose. The tunic has two openings: the incurrent siphon, through which water and food enter, and the excurrent siphon, through which water, waste products, and gametes pass to the outside. Sea squirts get their name from their practice of forcefully expelling a stream of water from the excurrent siphon when irritated.

Tunicates are suspension feeders that remove plankton from the stream of water passing through the pharynx. Food particles are trapped in mucus secreted by cells of the *endostyle,* a groove that extends the length of the pharynx. Ciliated cells of the pharynx move the stream of food-laden mucus into the esophagus. Much of the water entering the pharynx passes out through the pharyngeal slits into an **atrium** (chamber) and is discharged through the excurrent siphon.

Some species of tunicates form large colonies in which members share a common tunic and excurrent siphon. Colonial forms often reproduce asexually by budding. Sexual forms are usually hermaphroditic.

Lancelets may be closely related to vertebrates

Most species of subphylum Cephalochordata belong to the genus *Branchiostoma,* which consists of animals commonly known as **lancelets,** or **amphioxus.** Lancelets are translucent, fish-shaped animals, 5 to 10 cm long and pointed at both ends. They are widely distributed in shallow seas, either swimming freely or burrowing in the sand near the low-tide line. In some parts of the world, lancelets are an important source of food. One Chinese fishery reports an annual catch of 35 tons (about 1 billion lancelets).

Chordate characteristics are highly developed in lancelets. The notochord extends from the anterior tip ("head"; hence the name *Cephalochordata*) to the posterior tip. Many pairs of pharyngeal slits are evident in the large pharyngeal region, and a dorsal, tubular nerve cord extends the entire length of the animal (Fig. 30-6). Although superficially similar to fishes, lancelets have a far simpler body plan. They do not have paired fins, jaws, sense organs, a heart, or a well-defined head or brain.

Like the tunicates, lancelets use their cilia to draw a current of water into the mouth and then strain out microscopic organisms. Food particles are trapped in mucus in the pharynx and are then carried back to the intestine.

Water passes through the pharyngeal slits into the atrium, a chamber with a ventral opening (the *atriopore*) just anterior to the anus. Metabolic wastes are excreted by segmentally arranged, ciliated *protonephridia* that open into the atrium. In contrast with other invertebrates, the blood flows anteriorly in the ventral vessel and posteriorly in the dorsal vessel. This circulatory pattern is similar to that of fishes.

Recent molecular and morphological findings suggest that lancelets are the vertebrates' closest living relative. One similarity is the lancelet nerve cord, which has specialized regions that correspond to the vertebrate forebrain and midbrain.

Chordate phylogeny is controversial

PROCESS OF SCIENCE

No clear fossil record of chordate ancestors exists, but evidence suggests they were small, soft-bodied animals. **Yunnanozoans** are a distinctive group of well-preserved fossils that occur in **Chengjiang,** an early Cambrian fossil site in China. At this site,

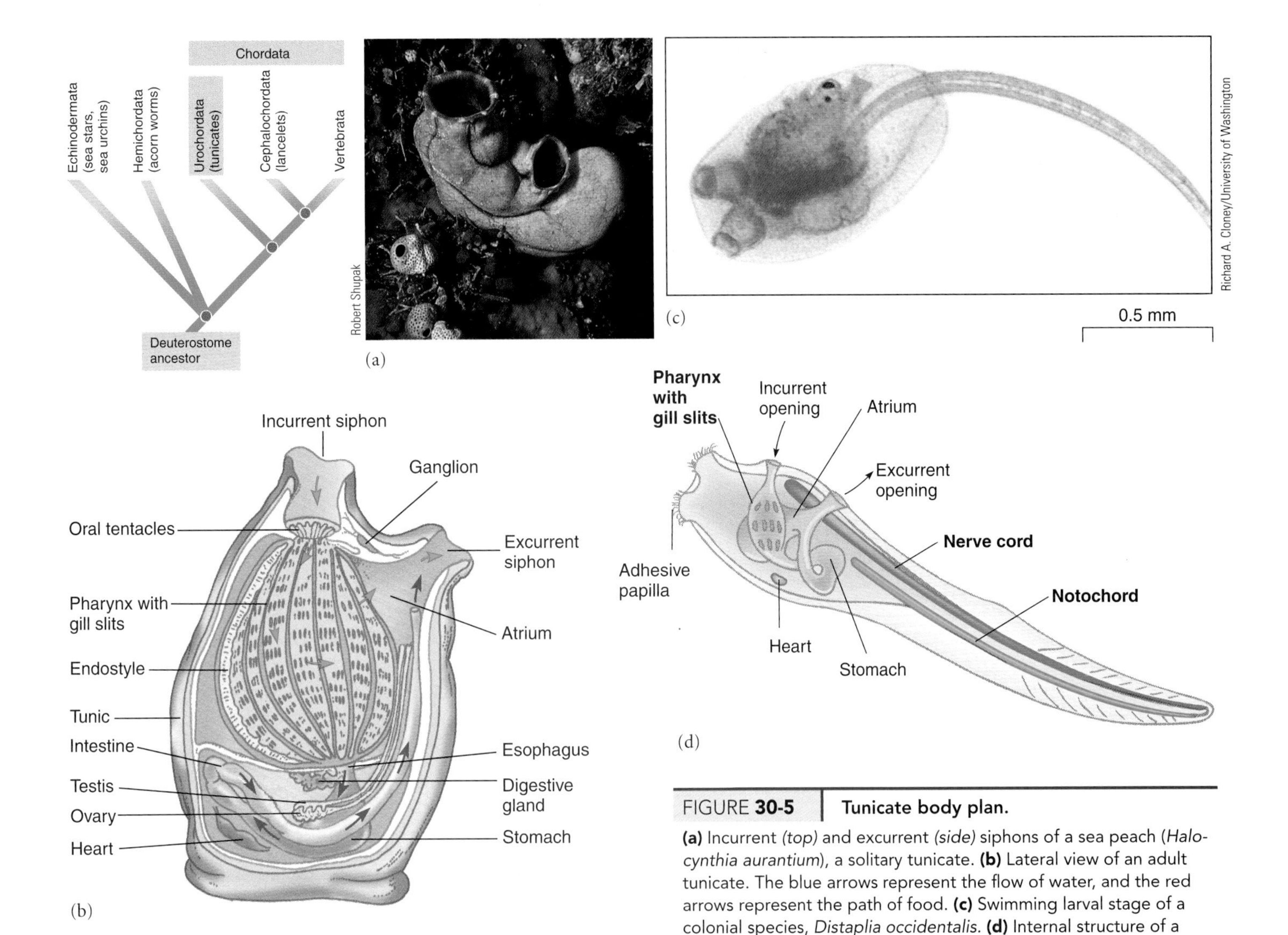

FIGURE **30-5** **Tunicate body plan.**

(a) Incurrent *(top)* and excurrent *(side)* siphons of a sea peach (*Halocynthia aurantium*), a solitary tunicate. **(b)** Lateral view of an adult tunicate. The blue arrows represent the flow of water, and the red arrows represent the path of food. **(c)** Swimming larval stage of a colonial species, *Distaplia occidentalis.* **(d)** Internal structure of a larval tunicate *(lateral view).*

fine-grained rocks about 530 million years old have preserved soft-bodied animals in great detail. One yunnanozoan, named ***Haikouella,*** was about 3 cm long and had a nerve cord and notochord. Some biologists interpret a species of *Haikouella,* reported in 2003, as an early deuterostome, but not as a chordate. Other scientists disagree, and instead, hypothesize that the yunnanozoans were vertebrate-like chordates. Biologists need additional data to settle this controversy.

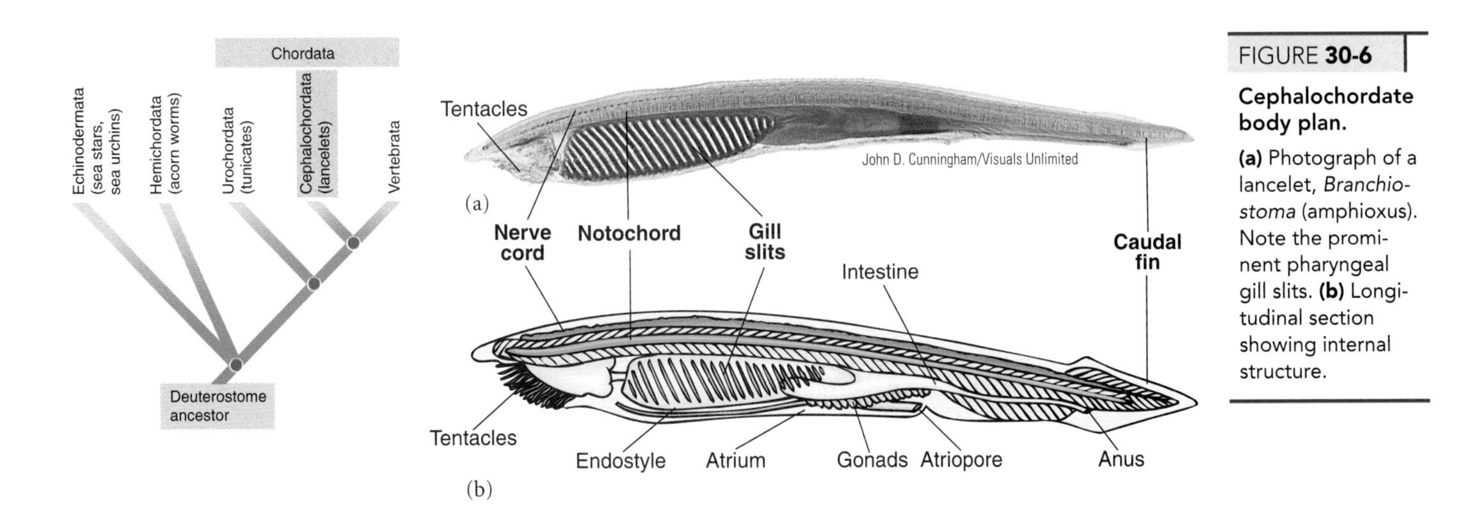

FIGURE **30-6**

Cephalochordate body plan.

(a) Photograph of a lancelet, *Branchiostoma* (amphioxus). Note the prominent pharyngeal gill slits. **(b)** Longitudinal section showing internal structure.

The rocks of the Burgess Shale of British Columbia, Canada, which date back to the Cambrian period, have also been a rich source of fossils. Among these fossils, biologists discovered a lancelet-like animal they named ***Pikaia.*** This animal had a notochord with muscles attached to it allowing for locomotion. Biologists interpret the fossil as an early cephalochordate.

Based on morphological and molecular evidence, the hemichordates are probably closely related to the echinoderms. Many systematists consider subphyla cephalochordata and Vertebrata to be **sister taxa,** groups that diverged from a recent common ancestor. That ancestor may have resembled a tunicate larva.

Recently, scientists sequenced part of the genome of the sea squirt *Ciona intestinalis,* the most studied ascidian tunicate. Based on their findings, these scientists reported their insights into chordate and vertebrate evolution. They found that *Ciona* has most vertebrate gene families in simplified form. For example, *Ciona* has only a single copy of each gene family involved in cell signaling and regulation of development, whereas vertebrates have two or more copies of these gene families. Interestingly, *Ciona* has genes for some vertebrate structures and processes, even though it does not express these genes. These data suggest the sea squirt genetic makeup corresponds to that of the chordate ancestor.

Review

- How are the main derived chordate characters evident in a tunicate larva, an adult tunicate, and in a lancelet?
- How does sequencing animal genomes help clarify relationships among chordate groups?

BiologyNow™ Assess your understanding of **the invertebrate chordates** by taking the pretest on your BiologyNow CD-ROM.

INTRODUCING THE VERTEBRATES

Learning Objective

6 Describe four shared derived characters of vertebrates.

With about 48,000 species, the **vertebrates,** members of subphylum **Vertebrata,** are less diverse and much less numerous than the insects but rival them in their adaptations to an enormous variety of lifestyles. In addition to the basic chordate characteristics, vertebrates have a number of shared derived characters not found in other groups.

The vertebral column is a key vertebrate character

The vertebrates are distinguished from other chordates in having a backbone, or **vertebral column,** that forms the skeletal axis of the body. This flexible support develops around the notochord, and in most species it largely replaces the notochord during embryonic development. The vertebral column consists of cartilaginous or bony segments called **vertebrae.** Dorsal projections of the vertebrae enclose the nerve cord along its length. Anterior to the vertebral column, a cartilaginous or bony **cranium,** or braincase, encloses and protects the brain, the enlarged anterior end of the nerve cord.

The cranium and vertebral column are part of the **endoskeleton.** In contrast with the nonliving exoskeleton of many invertebrates, the vertebrate endoskeleton is a living tissue that grows with the animal. In most vertebrates, the skeleton is mainly bone, a tissue that contains fibers made of the protein collagen. The hard matrix of bone consists of the compound hydroxyapatite, composed mainly of calcium phosphate.

Many characters common to vertebrates have been derived from a group of cells called **neural crest cells.** These cells, found only in vertebrates, appear early in development and migrate to various parts of the embryo. Neural crest cells give rise to or influence the development of many structures, including nerves, head muscles, cranium, and jaws.

Recall that invertebrates show an evolutionary trend toward **cephalization,** the concentration of nerve cells and sense organs in a definite head. Vertebrate evolution is characterized by *pronounced* cephalization. The brain has become larger and more elaborate, and its various regions have specialized to perform different functions. Ten or 12 pairs of **cranial nerves** emerge from the brain and extend to various organs of the body. Vertebrates have well-developed sense organs concentrated in the head: eyes; ears that serve as organs of balance and, in some vertebrates, for hearing as well; and organs of smell and taste.

Most vertebrates have two pairs of appendages. The fins of fishes are appendages that stabilize them in the water. Paired pectoral and pelvic fins are also used in steering. Biologists hypothesize that jointed appendages that facilitated locomotion on land evolved from the lobed (divided) fins of lungfishes.

Vertebrates have a closed circulatory system with a ventral heart and blood containing hemoglobin. The complete digestive tract has specialized regions and large digestive glands (the liver and pancreas). Several complex **endocrine glands,** which are ductless glands, secrete hormones. Paired kidneys regulate fluid balance. The sexes are typically separate.

Certain aspects of vertebrate phylogeny are still unclear

The study of evolutionary relationships of vertebrates is an important focus of research. Biologists consider the vertebrates and lancelets as sister groups that diverged from a common ancestor. The earliest "vertebrates" to evolve were **agnathans** (*a,* "without"; *gnathos,* "jaw"): the hagfishes, followed by the lampreys. Biologists assign hagfishes to class **Myxini** and lampreys to class **Cephalaspidomorphi.** However, a problem hinders this classification because the hagfishes do not quite qualify as vertebrates. Although they have many vertebrate characters, including a cranium, hagfishes have no trace of vertebrae. The notochord is their only axial support. Based on the cranium, many biologists now use the term **Craniata** to designate a clade that includes the vertebrates plus the hagfishes. They view the Myxini as the sister group of all other craniates.

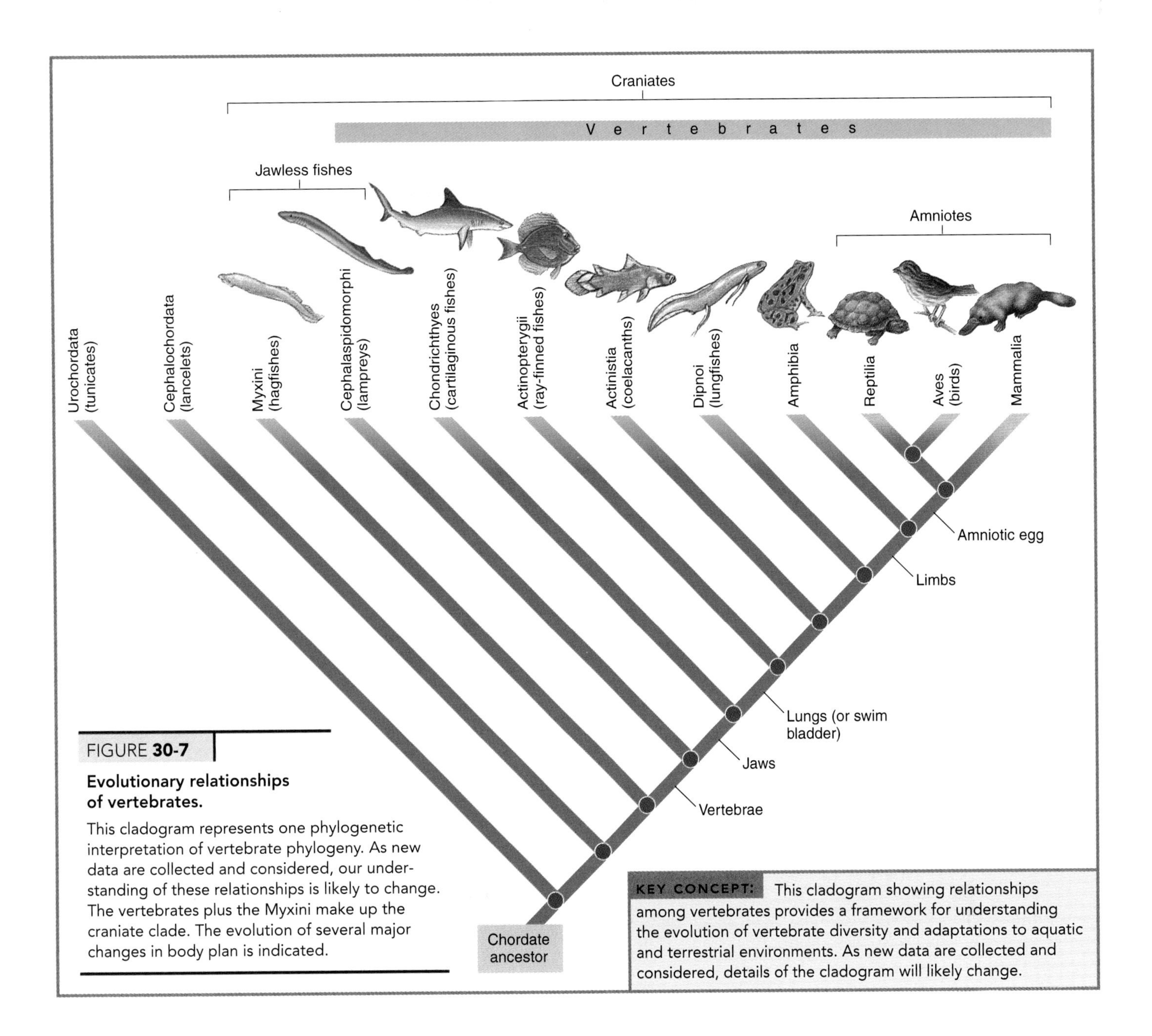

FIGURE **30-7**

Evolutionary relationships of vertebrates.

This cladogram represents one phylogenetic interpretation of vertebrate phylogeny. As new data are collected and considered, our understanding of these relationships is likely to change. The vertebrates plus the Myxini make up the craniate clade. The evolution of several major changes in body plan is indicated.

KEY CONCEPT: This cladogram showing relationships among vertebrates provides a framework for understanding the evolution of vertebrate diversity and adaptations to aquatic and terrestrial environments. As new data are collected and considered, details of the cladogram will likely change.

If we exclude the Myxini, the extant vertebrates are currently assigned to 9 classes: 5 classes of fishes and 4 of tetrapods (four-limbed vertebrates) (Fig. 30-7 and Table 30-2). Some biologists give these groups superclass status: superclass **Pisces** (fishes) and superclass **Tetrapoda.** In this classification scheme, the fishes include the lampreys and four classes of jawed fishes: Chondrichthyes, Actinopterygii, Actinistia, and Dipnoi.

The four-limbed land vertebrates, or tetrapods, include class Amphibia, the frogs, toads, and salamanders; class Reptilia, the lizards, snakes, turtles, and alligators; class Aves, the birds; and class Mammalia, the mammals. As we discuss later in the chapter, many cladists classify birds with reptiles, rather than assigning them to a separate class. The reptiles, birds, and mammals form a clade known as *amniotes,* animals fully adapted to life on land.

One approach to understanding the phylogenetic relationships of various taxa is to compare gene groups, such as *Hox* genes, of different animals. Recall that *Hox* genes are important in determining pattern development in embryos (see Chapter 16). Specifically, *Hox* genes determine the fate of cells in the anterior–posterior axis. These genes occur in clusters on specific chromosomes. All invertebrates and early chordates that have been studied have one *Hox* gene cluster. Investigators have suggested that duplication of *Hox* clusters may have contributed to the evolution of vertebrate body plans. Amphibians, reptiles, birds, and mammals have four *Hox* gene clusters, each containing about 10 different genes. Interestingly, the zebrafish and its close relatives have seven *Hox* clusters. We discuss the evolutionary implications of this fact later in the chapter.

Review

- What derived characters distinguish the vertebrates from the rest of the chordates?

Biology Now™ Assess your understanding of **the vertebrates** by taking the pretest on your BiologyNow CD-ROM.

TABLE 30-1 Extant Vertebrate Classes

Class	Examples	Characteristics
Cephalaspidomorphi	Lampreys	Jawless, freshwater and marine fishes with skeleton of cartilage; complete cranium and rudimentary vertebrae; hatch as small larvae
Chondrichthyes	Sharks, rays, skates, chimeras	Jawed marine and freshwater fishes with skeleton of cartilage; notochord replaced by vertebrae in adult; gills; placoid scales; two pairs of fins; oviparous, ovoviparous, or viviparous (a few species); well-developed sense organs (including lateral line system)
Actinopterygii (ray-finned fishes)	Perch, salmon, tuna, trout	Bony, marine and freshwater fishes; gills; swim bladder; generally oviparous
Actinistia (lobe-finned fishes)	Coelacanths	Bony fishes; marine, nocturnal predators on fish; seven-lobed fins
Dipnoi (lungfishes)	Lungfishes	Bony freshwater fishes; four similarly sized limbs, which are similar in structure and position to those of tetrapods
Amphibia	Salamanders, frogs and toads, caecilians	Tetrapods; aquatic larva typically undergoes metamorphosis into terrestrial adult; gas exchange through lungs and/or moist skin; heart consists of two atria and single ventricle; systemic and pulmonary circulation
Reptilia Cladists divide this group into anapsid (turtles), diapsids (snakes, lizards, crocodiles, alligators), and birds.*	Turtles, lizards, snakes, alligators	Amniotes with horny scales; adapted for reproduction on land (internal fertilization, leathery shell, amnion); lungs; ventricles of heart partly divided
Aves	Robins, pelicans, eagles, ducks, penguins, ostriches	Amniotes with feathers; anterior limbs modified as wings; compact, streamlined body; four-chambered heart; endothermic; vocal calls and complex songs
Mammalia	Monotremes (Holotheria), marsupials (Metatheria), placental mammals (Eutheria)	Amniotes with hair; females nourish young with mammary glands; differentiation of teeth; 3 middle ear bones; diaphragm; four-chambered heart; endothermic; highly developed nervous system

*This classification of amniotes is illustrated in Figure 30-17.

JAWLESS FISHES

Learning Objective

7 Distinguish among the major groups of jawless fishes.

Some of the earliest known vertebrates, collectively referred to as **ostracoderms,** consisted of several groups of small, armored, jawless fishes that lived on the bottom and strained their food from the water (see Fig. 20-10). Their heads were protected from predators by thick bony plates, and their trunks and tails were covered with thick scales. Most ostracoderms lacked fins. Fragments of ostracoderm scales have been found in rocks from the Cambrian period, but most ostracoderm fossils are from the Ordovician and Silurian periods. They became extinct by the end of the Devonian period.

Like ostracoderms, present-day hagfishes and lampreys have neither jaws nor paired fins. They are eel-shaped animals, up to 1 m long. Their smooth skin lacks scales, and they are supported by a cartilaginous skeleton and well-developed notochord.

Hagfishes, assigned to class Myxini and no longer considered vertebrates, are marine scavengers. They burrow for worms and other invertebrates or prey on dead and disabled fishes. As a defense mechanism, hagfishes secrete large amounts of sticky slime.

Lampreys are jawless vertebrates assigned to class Cephalaspidomorphi. Some spend their adult lives in the ocean and return to fresh water to reproduce. Many species of adult lampreys are parasites on other fishes (Fig. 30-8). Adult parasitic lampreys have a circular sucking disk around the mouth, which lies on the ventral side of the anterior end of the body. Using this disk to attach to a fish, the lamprey bores through the skin of its host with horny (made of keratin rather than bone) teeth on the disk and tongue. Then the lamprey injects an anticoagulant into its host and sucks out blood and soft tissues.

As in hagfish, the notochord persists throughout life and is not replaced by a vertebral column. However, lampreys have rudiments of vertebrae, cartilaginous segments called *neural arches* that extend dorsally around the spinal cord.

Review

- How do lampreys and hagfishes resemble ostracoderms?
- How do hagfishes differ from other fishes?

BiologyNow™ Assess your understanding of **jawless fishes** by taking the pretest on your BiologyNow CD-ROM.

FISHES AND AMPHIBIANS

Learning Objective

8 Trace the evolution of jawed fishes and early tetrapods, and identify major taxa of jawed fishes and amphibians.

Fossil evidence suggests that jaws and paired fins evolved during the late Silurian and Devonian periods. Two early groups of jawed fishes, now extinct, were the **acanthodians,** armored fishes

FIGURE **30-8** | **Lampreys.**

(a) Three lampreys are attached to a carp by their suction-cup mouths. Note the absence of jaws and paired fins. **(b)** Suction-cup mouth of adult lamprey (*Entosphenus japonicus*). Note the rasplike teeth.

courtesy of Dr. Kiyoko Uehara

Tom Stack/Tom Stack &Associates

(a)

(b)

with paired spines and pectoral and pelvic fins, and **placoderms,** armored fishes with paired fins (Fig. 30-9).

The evolution of jaws from a portion of the gill arch skeleton and the development of fins enabled fishes to change from suspension-feeding bottom dwellers to active predators. This change afforded many new opportunities for finding food. The

FIGURE **30-9** | **Early jawed fishes.**

Acanthodians and placoderms flourished in the Devonian period. **(a)** *Climatius*, a spiny-skinned acanthodian with large fin spines and five pairs of accessory fins between the pectoral and pelvic pairs. *Climatius* was a small fish that reached a length of 8 cm (3 in). **(b)** *Dinichthys*, a giant placoderm that grew to a length of 8 m (26 ft). (Most placoderms were only about 20 cm, or 8 in, long.) Its head and thorax were covered by bony armor, but the rest of the body and tail were naked.

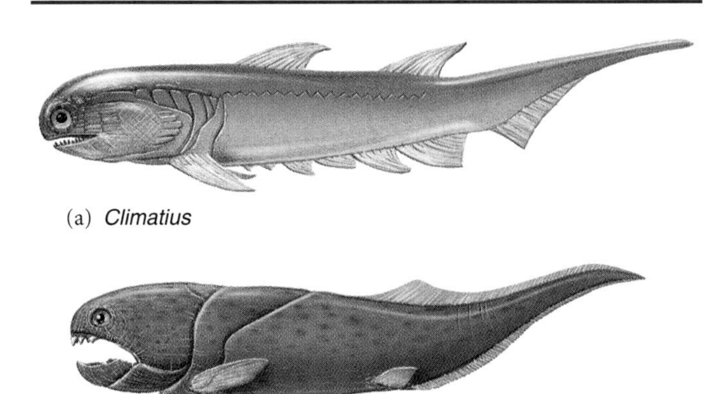

success of the jawed vertebrates may have contributed to the extinction of the ostracoderms.

Members of class Chondrichthyes are cartilaginous fishes

Members of class **Chondrichthyes,** the cartilaginous fishes, evolved as successful marine forms in the Devonian period. Class Chondricthyes, which is considered monophyletic, includes the sharks, rays, and skates (Fig. 30-10). Most species are ocean dwellers, but a few have invaded fresh water. Except for the whales, the sharks are the largest living vertebrates. Some whale sharks (*Rhincodon*) exceed 15 m (49 ft) in length, making them the largest fish.

Most rays and skates are sluggish, flattened creatures that live partly buried in the sand. Their enormous pectoral fins propel them along the bottom, where they feed on mussels and clams. The sting ray has a whiplike tail with a barbed spine at its base that can inflict a painful wound. The electric ray has electric organs on either side of the head. These modified muscles can discharge enough electric current (up to 2500 watts) to stun fairly large fishes, as well as human swimmers.

The chondrichthyes retain their cartilaginous embryonic skeleton. Although this skeleton is not replaced by bone, a deposit of calcium salts may strengthen it. All chondrichthyes have paired jaws and two pairs of fins. The skin contains **placoid scales.** Each scale is a toothlike structure consisting of an outer layer of enamel and an inner layer of dentine (Fig. 30-11). The lining of the mouth contains larger, but essentially similar, scales that serve as teeth. The teeth of other vertebrates are homologous with these scales. Shark teeth are embedded in the flesh and not attached to the jawbones; new teeth develop continuously in rows behind the functional teeth and migrate forward to replace any that are lost.

The shark body is adapted for swimming. Lift is provided by body shape and fins and by swimming swiftly. The shark stores a great deal of oil in its large liver (which may account

(a)

(b)

FIGURE **30-10** | **Cartilaginous fishes.**

(a) Blue-spotted sting ray (*Taeniura lymma*). Sting rays typically feed on shellfish and bottom-dwelling fishes. **(b)** The great white shark (*Carcharodon carcharias*), photographed in Australia, is considered the most dangerous shark to humans. This shark is actually white only on its ventral aspect; the rest of the body is brownish-gray or bluish-gray.

for up to 30% of the body weight). Fats and oils decrease the overall density of fishes and contribute to buoyancy. Even so, the shark body is denser than water, so sharks tend to sink unless they are actively swimming.

Most sharks are streamlined predators that swim actively and catch and eat other fishes as well as crustaceans and mollusks. The largest sharks and rays, like the largest whales, are suspension feeders that strain plankton from the water. They gulp water through the mouth. Then, as the water passes through the pharynx and out the gill slits, food particles are trapped in a sievelike structure.

Predatory sharks are attracted to blood, so a wounded swimmer or a skin diver towing speared fish is a target. However, despite the common portrayal of sharks in books and films as monstrous enemies, most do not go out of their way to attack humans. In fact, of the approximately 350 known shark species, fewer than 30 are known to attack humans.

The shark has a complex brain, and a spinal cord protected by vertebrae. Their well-developed sense organs very effectively locate prey in the water. Sharks may detect other animals electrically before sensing them by sight or smell. **Electroreceptors** on the shark's head sense weak electric currents generated by the muscular activity of animals. The **lateral line organ,** found in all fishes, is a groove along each side of the body with many tiny openings to the outside. Sensory cells in the lateral line organ are sensitive to waves and other motion in the water, alerting the shark to the presence of predator or prey (see Fig. 41-5).

Cartilaginous fishes have no lungs. They have five to seven pairs of gills. A current of water enters the mouth and passes over the gills and out the gill slits, constantly providing the fish with a fresh supply of dissolved oxygen. Sharks that actively swim depend on their motion to enhance gas exchange. Sharks that spend time on the ocean floor, and rays and skates, use muscles of the jaw and pharynx to pump water over their gills.

The digestive tract of sharks consists of the mouth cavity; a long pharynx leading to the stomach; a short, straight intestine; and a **cloaca,** which opens on the underside of the body and is characteristic of many vertebrates. The liver and pancreas discharge digestive juices into the intestine. The cloaca receives digestive wastes, as well as metabolic wastes from the urinary system. In females, the cloaca also serves as a reproductive organ.

The sexes are separate, and fertilization is internal. In the mature male, each pelvic fin has a slender, grooved section, known as a **clasper,** used to transfer sperm into the female's cloaca. The

FIGURE **30-11** | **Sharks.**

(a) Internal structure of a shark. **(b)** Structure of a placoid scale.

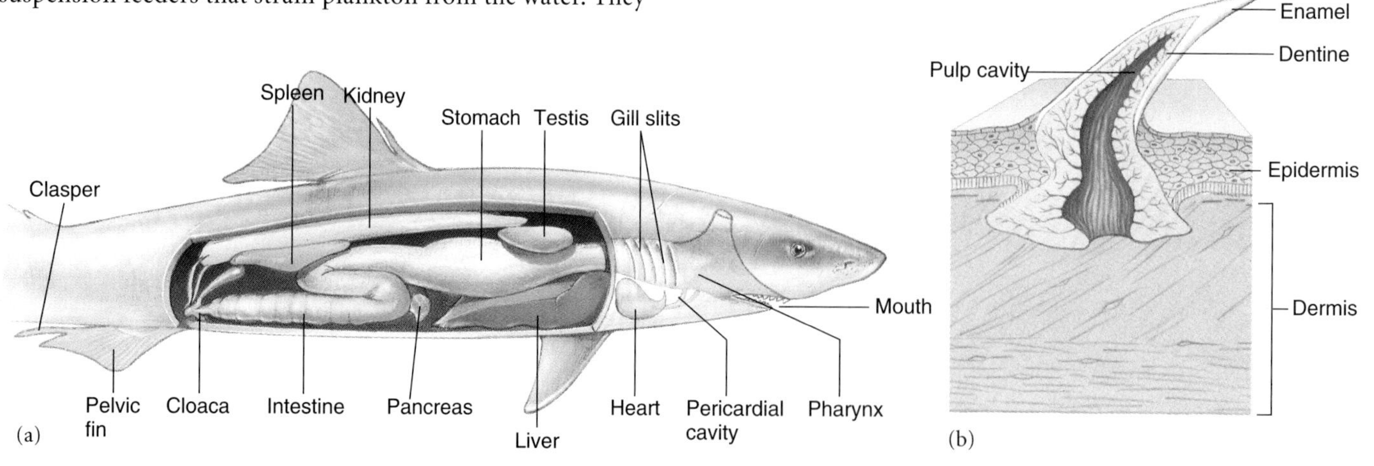

eggs are fertilized in the upper part of the female's oviducts. Part of the oviduct is modified as a shell gland, which secretes a protective coat around the egg.

Skates and some species of sharks are **oviparous;** that is, they lay eggs. Many species of sharks, however, are **ovoviviparous,** meaning their young are enclosed in eggs and incubated within the mother's body. During development, the young depend on stored yolk for their nourishment rather than on transfer of materials from the mother. The young are born after hatching from the eggs.

A few species of sharks are **viviparous.** Not only do the embryos develop within the uterus, but much of their nourishment is delivered to them by the mother's blood. Nutrients are transferred between the blood vessels in the lining of the uterus and the yolk sac surrounding each embryo.

The ray-finned fishes gave rise to modern bony fishes

Although bony fishes appear earlier in the fossil record than cartilaginous fishes, both groups may have evolved about the same time, during the Devonian period. The two groups share many characteristics (such as continuous tooth replacement), but they also differ in important ways.

Most bony fishes are characterized by a bony skeleton with many vertebrae. Bone has advantages over cartilage: It provides excellent support and serves as a very effective storage site for calcium. Most species have flexible median and paired fins, supported by long rays made of cartilage or bone. Overlapping, bony dermal scales cover the body. A lateral bony flap, the **operculum,** extends posteriorly from the head and protects the gills.

Unlike most sharks, bony fishes are oviparous. They lay an impressive number of eggs and fertilize them externally. The ocean sunfish, for example, lays more than 300 million eggs! Of course, most of the eggs and young become food for other animals. The probability of survival is increased by certain behavioral adaptations. For example, many species of fishes build nests for their eggs and protect them.

During the Devonian period, the bony fishes diverged into two major groups: the **Sarcopterygii** and the **ray-finned fishes,** class **Actinopterygii.** Fossils of the earliest sarcopterygians date back to the Devonian about 400 mya. Lungs and fleshy, **lobed fins** characterize these fishes. The flexible fins had a central appendage with many bones and muscles and could support the body. Early sarcopterygians evolved along two separate lines: the **lungfishes** (class **Dipnoi**) and **lobe-finned fishes** (class **Actinistia**). Three genera of lungfishes survive today in the rivers of tropical Africa, Australia, and South America.

The ray-finned fishes underwent two important adaptive radiations. The first gave rise during the late Paleozoic era to a group of fishes that are now mostly extinct. The second radiation began during the early Mesozoic era and gave rise to the modern bony fishes. The ancestors of the modern bony fishes had lungs, which were retained by the lobe-finned fishes and lungfishes. In the ray-finned fishes, however, the lungs became modified as a **swim bladder,** an air sac that helps regulate buoyancy (Fig. 30-12). Bones and muscles are heavier than water, and without the swim bladder the fish would sink. By regulating gas exchange between its blood and swim bladder, a fish can control the amount of gas in the swim bladder, changing the overall density of its body. This ability allows a bony fish, in contrast to a shark, to hover at a given depth of water without much muscular effort.

There are more species of bony fishes than any other group of vertebrates. Biologists have identified about 49,000 living species of freshwater and saltwater bony fishes, of many shapes and colors (Fig. 30-13). Bony fishes range in size from the Philippine goby, which is only about 1 cm (0.4 in) long, to the ocean sunfish (or *Mola;* see Fig. 19-13c), which may reach 4 m and weigh about 1500 kg (about 3300 lb).

The diversity of bony fishes may have resulted from the action of *Hox* genes. Early chordates have only one *Hox* cluster, but tetrapods have four clusters. The zebrafish has seven *Hox* clusters. During the radiation of ray-finned fishes, entire chromosomes duplicated, resulting in one or more additional clusters of *Hox* genes. These genes could have provided the genetic material for the evolution of the diverse species of ray-finned fish.

Did descendants of the lobe-finned fishes or lungfishes move onto the land?

PROCESS OF SCIENCE

Biologists thought the lobe-finned fishes were extinct by the end of the Paleozoic era, so in 1938 the scientific community

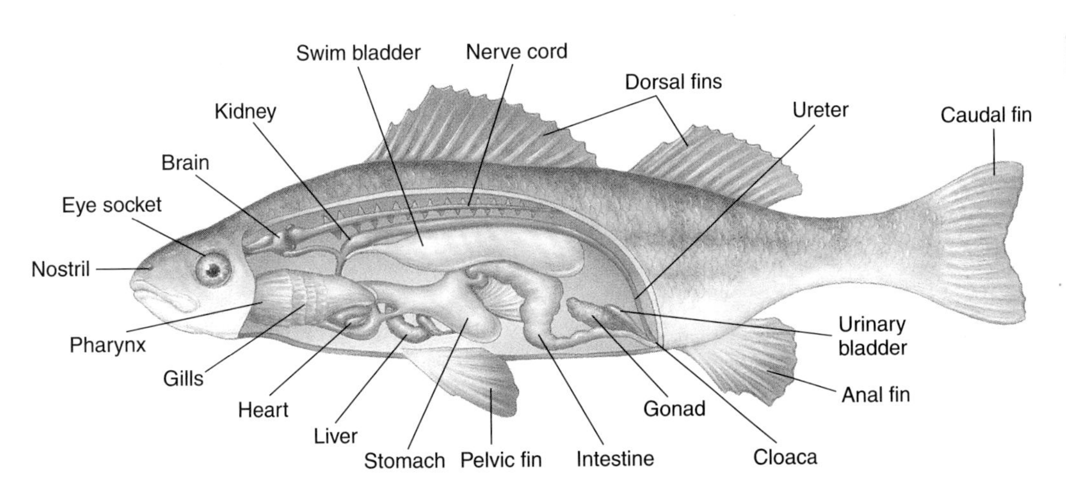

FIGURE 30-12 **Perch, a representative bony fish.**
The swim bladder is a hydrostatic organ that enables the fish to change the density of its body and remain stationary at a given depth. Pectoral *(not shown)* and pelvic fins are paired.

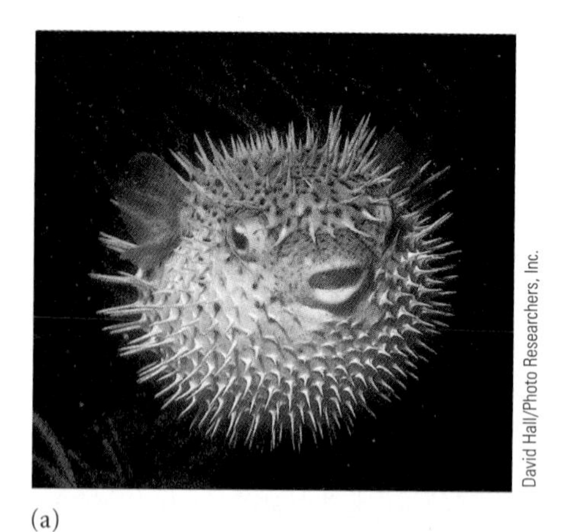

David Hall/Photo Researchers, Inc.

(a)

Jeffrey L. Rotman/Peter Arnold, Inc.

(b)

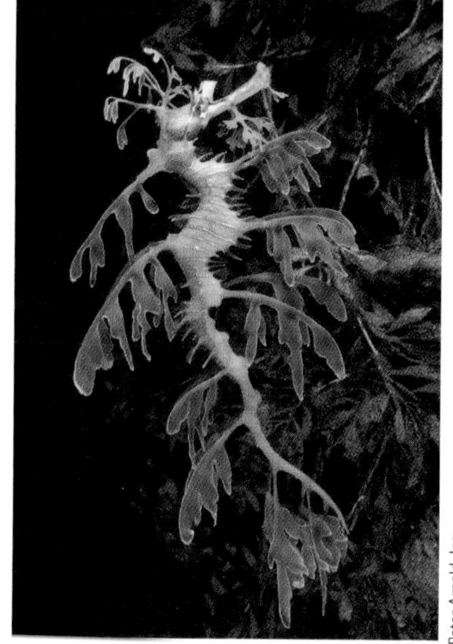

Peter Arnold, Inc.

(c)

FIGURE **30-13** | **Modern bony fishes.**

(a) The porcupinefish (*Diodon hystrix*) swallows air or water to inflate its body, a strategy to discourage potential predators. Photographed in the Virgin Islands. **(b)** The parrotfish (*Scarus gibbus*) feeds on coral, grinds it in its digestive tract, and extracts the coralline algae. The fish eliminates a fine white sand. These fishes contribute to white sand beaches in many parts of the world. Parrotfish begin life as females and later become males. **(c)** The leaflike extensions of the body wall of this Australian leafy sea dragon (*Phyllopteryx taeniolatus*) camouflage it in surrounding kelp.

was very excited when a commercial fisherman caught one off the coast of South Africa. Since that time more than 200 specimens of these giant "living fossils" have been found in the deep waters off the east coast of Africa near the Comoro Islands (Fig. 30-14). These fishes, known as **coelacanths,** measure nearly 2 m (about 6 ft) in length. They are nocturnal predators on other fishes. Coelacanths have been discovered recently off the coasts of Indonesia and Kenya.

The pieces seemed to fit together perfectly. The coelacanths were surely the living fossil representatives of the group that gave rise to the land vertebrates. This was the prevailing hypothesis until the 1980s, when a few investigators began to question it because of morphological data. The external nasal openings of fossil and living lungfish suggested they, rather than the coelacanths, might be the ancestors of the land vertebrates. Other morphological similarities between lungfishes and land vertebrates are tooth enamel and four similarly sized limbs that have a similar structure and position. (Recall that the fins of lungfishes are also lobed.) During the 1990s, the lungfish hypothesis gained support from comparisons of mitochondrial DNA. Current data suggest lungfishes are probably the ancestors of tetrapods. Future sequencing of nuclear DNA may support or refute this hypothesis.

One hypothesis that explains the origin of tetrapod limbs is as follows. During the frequent seasonal droughts in the Devonian period, swamps became stagnant or even dried up completely. Fishes with lobed fins would have had a tremendous advantage for survival under those conditions. They were strikingly preadapted for moving onto the land. They had lungs for breathing air, and their sturdy, fleshy fins may have allowed them to "walk" along in shallow water. These fins could support the fish's weight, enabling it to emerge onto dry land and make its way to another pond or stream.

Some biologists question this account. A competing view holds that limbs may have evolved in a fully aquatic environ-

FIGURE **30-14** | **Diver swimming with coelacanth.**

For many years, biologists thought that ancestors of this lobe-finned fish (*Latimeria chalumnae*) gave rise to the amphibians. Living coelacanths are difficult to observe because they inhabit deep ocean waters.

Mark Erdmann

ment. Animals with more developed limbs could move along in shallow water or creep through dense aquatic vegetation more efficiently than their lobe-finned ancestors. Whichever view of the evolution of tetrapods is correct, natural selection favored those individuals adapted for making their way on land.

The ability to move about, however awkwardly, on dry land gave animals access to new food sources. Terrestrial plants were already established, and terrestrial insects and arachnids were rapidly evolving. A vertebrate that could survive on land had less competition for food. Laying eggs on land, away from the many ocean predators, also increased their chances for successful reproduction. The early vertebrate experience on land was so successful that their lineage gave rise to the amphibians and, later, the reptiles, birds, and mammals.

Amphibians were the first successful land vertebrates

The first successful **tetrapods,** or land vertebrates, were the **labyrinthodonts** (Fig. 30-15), clumsy animals with short necks and heavy, muscular tails. They had limbs strong enough to support the weight of their bodies on land. The largest labyrinthodonts were the size of crocodiles. They flourished during the late Paleozoic and early Mesozoic eras, then became extinct. Biologists hypothesize that the labyrinthodonts gave rise to frogs and salamanders.

Biologists classify modern amphibians (class **Amphibia**) in three orders: Order **Urodela** ("visible tail") includes salamanders, mudpuppies, and newts, all animals with long tails; order **Anura** ("no tail") is made up of frogs and toads, with legs adapted for hopping; and order **Apoda** ("no feet") contains the wormlike caecilians (Fig. 30-16). Although some adult amphibians are quite successful as land animals and live in dry environments, most return to the water to reproduce. Eggs and sperm are generally released in the water.

Amphibians undergo a complex change, or **metamorphosis,** from larva to adult. The embryos of most frogs and toads develop into larvae called **tadpoles.** These larvae have tails and gills, and most feed on aquatic plants. After a time, the tadpole undergoes metamorphosis, a process regulated by hormones secreted by the *thyroid gland.* During metamorphosis, gills and gill slits disappear, the tail is resorbed, and limbs emerge. The digestive tract shortens, and food preference shifts from plant material to a carnivorous diet; the mouth widens; a tongue develops; the tympanic membrane (ear drum) and eyelids appear; and the eye lens changes shape. Many biochemical changes also accompany the transformation from a completely aquatic life to an amphibious one.

Several salamanders, such as the mudpuppy *Necturus,* do not undergo complete metamorphosis; they retain many larval characteristics even when sexually mature adults. Recall from Chapter 19 that this is an example of **paedomorphosis** (see Fig. 19-14). This type of development permits these salamanders to remain aquatic rather than having to compete on land.

The coloration of amphibians may conceal them in their habitat or may be very bright and striking. Many of the brightly colored species are poisonous (Fig. 30-16a). Their distinctive

FIGURE **30-15** **An artist's conception of labyrinthodonts.**

These early amphibians were large (some more than 5 m long) predators that roamed Earth about 150 mya.

Photograph by Logan, courtesy of Department of Library Services, American Museum of Natural History

ACTIVE FIGURE **30-16** **Modern amphibians.**

(a) This red dart frog (*Dendrobates pumilio*) is a poison arrow frog. **(b)** The red spotted newt (*Notophthalmus viridescens*) is mainly aquatic but spends one to three years as a pre-adult in a moist terrestrial environment.

BiologyNow™ See **how frogs reproduce** by clicking on this figure on your BiologyNow CD-ROM.

Geralc and Buff Corsi/Tom Stack

(a)

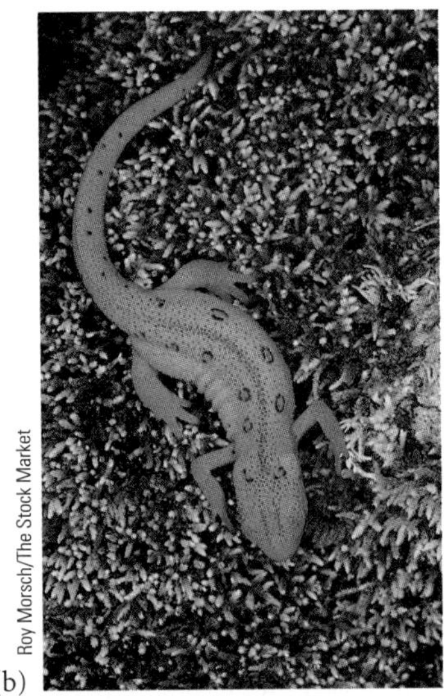

Roy Morsch/The Stock Market

(b)

colors warn predators they are not encountering an ordinary amphibian. Some frogs camouflage themselves by changing color.

Adult amphibians do not depend solely on their primitive lungs for the exchange of respiratory gases. Their moist, glandular skin, which lacks scales and is plentifully supplied with blood vessels, also serves as a respiratory surface. The numerous mucous glands within the skin help keep the body surface moist, which is important in gas exchange. The mucus also makes the animal slippery, facilitating its escape from predators. Some amphibians have glands in their skin that secrete poisonous substances harmful to predators.

The amphibian heart is divided into three chambers: Two **atria** receive blood, and a single **ventricle** pumps it into the arteries. A double circuit of blood vessels keeps oxygen-rich and oxygen-poor blood partially separate. Blood passes through the **systemic circulation** to the various tissues and organs of the body. Then, after returning to the heart, it is directed through the **pulmonary circulation** to the lungs and skin, where it is recharged with oxygen. The oxygen-rich blood returns to the heart to be pumped out into the systemic circulation again. We discuss the comparative anatomy of the heart and circulation of various vertebrate classes in Chapter 42.

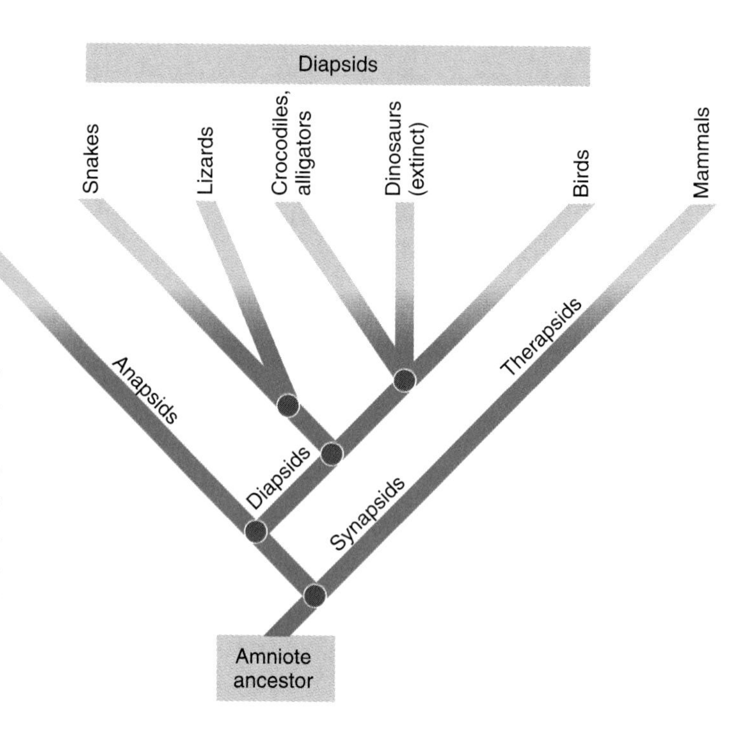

FIGURE **30-17** | **Phylogeny of the amniotes.**
This cladogram shows proposed evolutionary relationships among reptiles, birds, and mammals. Some of these relationships, for example, classification of the turtle as an anapsid, are currently being studied and may be reinterpreted.

Review

- From an evolutionary perspective, what is the significance of each of the following? (a) placoderms (b) lungfishes (c) labyrinthodonts?
- What changes allow amphibians to move from aquatic life as a tadpole to terrestrial life as an adult?
- Support the hypothesis that lungfishes gave rise to tetrapods.

BiologyNow™ Assess your understanding of **fishes and amphibians** by taking the pretest on your BiologyNow CD-ROM.

AMNIOTES

Learning Objectives

9 Describe three vertebrate adaptations to terrestrial life.
10 Describe the reptiles and birds, and give an argument for including the birds in the reptile clade.
11 Contrast monotremes, marsupials, and placental mammals, and give examples of animals that belong to each group.

The evolution of reptiles from ancestral amphibians required many adaptations that allowed them to be completely terrestrial. Evolution of the **amniotic egg** was an extremely important event because it let terrestrial vertebrates complete their life cycles on land. This egg contains an **amnion,** a membrane that forms a fluid-filled sac around the embryo. The amnion provides the embryo with its own private "pond," permitting independence from a watery external environment. In addition to keeping the embryo moist, the amniotic fluid serves as a shock absorber, cushioning the developing embryo.

The evolution of the amniotic egg is so important to the success of terrestrial vertebrates—reptiles, birds, and mammals—that biologists refer to these animals as **amniotes.** Amniotes are a monophyletic group, because they have a common ancestor that was itself an amniote (Fig. 30-17).

In addition to the amnion, the amniotic egg has three other extraembryonic (not part of the developing body itself) membranes: yolk sac, chorion, and allantois. These membranes protect the developing embryo, store nutrients *(yolk sac),* carry on gas exchange *(chorion and allantois),* and store wastes *(allantois).* We discuss development of the extraembryonic membranes in Chapter 49. Although most extant mammals do not lay eggs, their embryos have an amnion and other extraembryonic membranes.

Another important adaptation to terrestrial life is a body covering that minimizes water loss. However, such a covering severely decreases gas exchange across the body surface. Amniotes have efficient lungs and circulatory systems for exchange of oxygen and carbon dioxide. Amniotes also have physiological mechanisms for conserving water. For example, to avoid water loss during excretion of metabolic wastes, much of the fluid filtered from the blood by the kidneys is reabsorbed in the kidney tubules and urinary bladder.

Our understanding of amniote phylogeny is changing

Until recently, biology instructors taught that the three classes of amniotes were Reptilia, Aves, and Mammalia. However, cladistic analysis indicates class Reptilia is not a monophyletic group. It is paraphyletic because it includes some, but not all, of its descendants (see Fig. 22-7). Our discussion of amniotes reflects the cladistic view, although we describe modern birds separately for convenience.

Biologists hypothesize that the earliest amniotes resembled lizards. By the late Carboniferous period, about 290 mya, amniotes had undergone an impressive adaptive radiation and diverged into two branches. One of these branches, the **synapsids,** gave rise to the mammals. The second branch, the **sauropsids,** gave rise to all the other reptiles and to the birds. The sauropsids divided to form two sub-branches: **anapsids** and **diapsids.** The turtles may be the only extant anapsids. The diapsids include the snakes, lizards, crocodilians, and many extinct reptiles, including the dinosaurs. Based on cladistic analysis, biologists also classify the birds as diapsids.

A second great amniote adaptive radiation occurred during the Mesozoic era that ended about 65 mya. In fact, the Mesozoic era is known as the Age of Reptiles. During that time reptiles were the dominant terrestrial animals (see Chapter 20). They had radiated into an impressive variety of ecological lifestyles (see Fig. 20-12). Some could fly, others became marine, and many filled terrestrial habitats. Two major groups were the **dinosaurs** and the **pterosaurs,** the flying reptiles.

Some of the dinosaurs were among the largest land animals to have ever walked on Earth (see Fig. 20-14). Some dinosaurs apparently traveled in social groups and took care of their young. Fossil evidence supports the hypothesis that at least some dinosaurs were **endothermic,** meaning they used metabolic energy to maintain a constant body temperature despite changes in the temperature of the environment. An advantage of endothermy is that it allows animals to be more active.

The reptiles were the dominant land animals for almost 200 million years. Then, toward the end of the Mesozoic era, many, including all the dinosaurs and pterosaurs, disappeared from the fossil record. In fact, more than half of all animal species became extinct at that time (see Chapter 20). As a result of this mass extinction, there are many more extinct reptiles than living species.

Reptiles have many terrestrial adaptations

Many reptilian characters are adaptations to terrestrial life. The female reptile secretes a protective leathery shell around the egg, which helps prevent the developing embryo from drying out. Sperm cannot penetrate this shell, and internal fertilization takes place within the body of the female before the shell is added. In this process, the male uses a copulatory organ (penis) to transfer sperm into the female reproductive tract. An amnion surrounds the embryo, as it develops within the protective shell.

The hard, dry, horny scales that retard drying are another adaptation to life on land. This scaly protective armor, which also protects the reptile from predators, is shed periodically. The dry reptilian skin does not allow effective gas exchange. Reptilian lungs are better developed than the saclike lungs of amphibians. Divided into many chambers, the reptilian lung provides a greatly increased surface area for gas exchange. Most reptiles have a three-chambered heart that is more efficient than the amphibian heart. The single ventricle has a partition, though incomplete, that separates oxygen-rich and oxygen-poor blood, facilitating oxygenation of body tissues. In crocodiles the partition is complete, and the heart has four chambers.

Like fishes and amphibians, reptiles generally lack metabolic mechanisms for regulating body temperature. They are **ectothermic,** meaning that their body temperature fluctuates with the temperature of the surrounding environment. Some reptiles have behavioral adaptations that let them maintain a body temperature higher than that of their environment. For example, you may have observed a lizard basking in the sun, which raises its body temperature and so increases its metabolic rate. This permits the lizard to hunt actively for food. When the body of a reptile is cold, its metabolic rate is low and it tends to be sluggish. Ectothermy may explain why reptiles are more successful in warm than in cold climates.

Many reptiles are carnivores (animals that eat meat). Their paired limbs, usually with five toes, are well adapted for running and climbing, and their well-developed sense organs enable them to locate prey.

Ignoring birds for the moment, biologists assign the extant reptiles to three main groups. Some consider these groups as classes, whereas others rank them as orders. The anapsids gave rise to the turtles, terrapins, and tortoises. These animals can be grouped in order Chelonia. Order Squamata includes lizards and snakes, and order Crocodilia contains crocodiles, alligators, caimans, and gavials (Fig. 30-18).

Members of order **Chelonia** are enclosed in a protective shell made of bony plates overlaid by horny scales. Some terrestrial species can withdraw their heads and legs completely into their shells. The size of adult turtles ranges from about 8 cm (3 in) long to that of the great marine species, which measure more than 2 m (6.5 ft) long and may weigh 450 kg (almost 1000 lb). Biologists usually refer to the land species as *tortoises,* whereas the aquatic forms are called *turtles;* freshwater types are sometimes called *terrapins.*

Lizards and snakes, assigned to order **Squamata,** are the most common of the modern reptiles. These animals have rows of scales that overlap like shingles on a roof, forming a continuous, flexible armor that is shed periodically. Lizards range in size from certain geckos, which weigh as little as 1 g, to the Komodo dragon of Indonesia, which may weigh 100 kg (220 lb). Their body sizes and shapes vary greatly. Some lizards, like the glass snake (it's really a lizard), are legless.

Snakes are characterized by a flexible, loosely jointed jaw structure that lets them swallow animals larger than the diameter of their own jaws. Snakes lack legs, and their bodies are elongated. (Remember that although not all the tetrapods have four limbs, all evolved from four-limbed ancestors.) Their eyes, covered by a transparent scale, do not have movable eyelids. Also absent are an external ear opening, a tympanic membrane (ear drum), and a middle ear cavity.

The snake uses its forked tongue as an accessory sensory organ for touch and smell. Chemicals from the ground or air adhere to the tongue. The snake rubs the tip of its tongue across the opening of a sense organ in the roof of the mouth that de-

Carlyn Iverson

(a) Order Chelonia

BIOS/Peter Arnold, Inc.

(b) Order Squamata

Frans Lanting/Minden Pictures

(c) Order Crocodilia

FIGURE **30-18** | **Modern reptiles.**

(a) The paddlelike appendages of this green turtle (*Chelonia mydas*) are adapted for swimming. **(b)** This basilisk lizard (*Basiliscus plumifrons*), photographed in Costa Rica, thermoregulates by basking in the sun. **(c)** A Nile crocodile (*Crocodilus niloticus*) emerges from its leathery egg.

tects odors. Pit vipers and some boas also have a prominent **pit organ** on each side of the head that enables them to detect heat from endothermic prey (see Fig. 41-1). These snakes use their pit organs to locate and capture small nocturnal mammals.

Some snakes, such as king snakes, pythons, and boa constrictors, kill their prey by rapidly wrapping themselves around the animal and squeezing so it cannot breathe. Others have fangs, which are hollow teeth connected to venom glands. When the snake bites, it pumps venom through the fangs into the prey. Some snake venoms cause the breakdown of red blood cells; others, such as that of the coral snake, are neurotoxins, which interfere with nerve function. Venomous snakes of the United States include rattlesnakes, copperheads, cottonmouths, and coral snakes. All these except the coral snakes are pit vipers.

Order **Crocodilia** includes three groups: (1) the crocodiles of Africa, Asia, and America; (2) the alligators of the southern United States and China, plus the caimans of Central America; and (3) the gavials of South Asia. Most species live in swamps, in rivers, or along sea coasts, feeding on various kinds of animals. Crocodiles are the largest living reptiles; some exceed 7 m (23 ft) in length. The cranial skeleton is adapted for aquatic life. The crocodile can be distinguished from the alligator or caiman by its long, slender snout and by the large fourth tooth on the bottom jaw that is visible when the mouth is closed.

Are birds really dinosaurs?

Cladistic analyses indicate that birds that evolved from saurischian dinosaurs (see Fig. 20-13a), specifically carnivorous theropods. Many theropods were long-tailed animals that moved about on two feet and had forelimbs with three clawed fingers. Many biologists now view birds as living dinosaurs!

Birds are the only extant animals with feathers. However, recent molecular and fossil evidence indicates that feathers evolved in dinosaurs before birds evolved. Biologists have suggested several hypotheses to explain the early functions of feathers. They may have been an adaptation that conserved body heat. Insulating feathers may have contributed to the evolution of endothermy (the ability to maintain a constant body temperature), permitting animals to be more active. Feathers may have been important in courtship rituals, or in camouflage.

Chinese, American, and Canadian paleontologists have discovered, in the Yixian formation in China, an amazing array of fossils from the early Cretaceous period (124 to 128 mya). They have identified a number of theropod dinosaur fossils that demonstrate the evolution of feathers from very simple, tubular structures to complex modern forms. One feathered dinosaur discovered there is *Caudipteryx,* which was about the size of a turkey. *Caudipteryx* has well-preserved imprints of complex feathers on its tail and forelimbs, and its bones show many bird characteristics (Fig. 30-19a). Analyzing fossil evidence and recent molecular data, biologists conclude that feather evolution took place in terrestrial, bipedal dinosaurs before the evolution of birds or flight.

In 2003, a team of paleontologists led by Xing Xu and Zhonghe Zhou of the Institute of Vertebrate Paleontology and Paleoanthropology in Beijing reported a new species of feathered dinosaur (*Microraptor gui*) with asymmetrical feathers on its forelimbs and hindlimbs. Modern birds use asymmetrical feathers for flight. These small dinosaurs, which lived about 126 mya, probably coexisted with early birds. Their small size and body structure supports the hypothesis that flying evolved from gliding; bird ancestors climbed trees and used their feathered limbs for gliding (see *On the Cutting Edge: Origin of Flight in Birds,* in Chapter 20).

Although the bones of birds are fragile and disintegrate quickly, a few fossils of early birds have been found. The first birds looked very much like reptiles. They had teeth (which

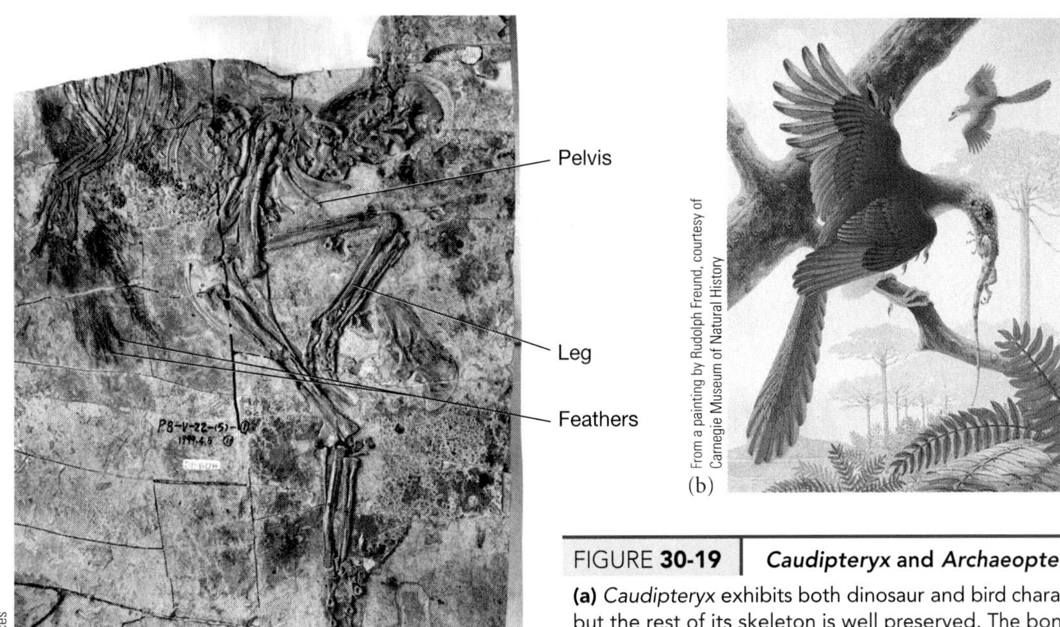

FIGURE **30-19** *Caudipteryx* and *Archaeopteryx*.

(a) *Caudipteryx* exhibits both dinosaur and bird characteristics. This fossil lacks a head, but the rest of its skeleton is well preserved. The bones in the foot and the shape and orientation of the pelvis are similar to those of the theropods, but the fossil also has bird characters, including impressions of feathers. **(b)** This reconstruction represents the view that *Archaeopteryx* was a climbing animal that had at least some ability to use its wings and feathers for gliding. Another view is that *Archaeopteryx* remained mainly on the ground, using its wings to trap small insects and its feathers for insulation.

modern birds lack), a long tail, and bones with thick walls. Unlike reptiles, their jaws were elongated into beaks, and they had feathers and wings.

One of the earliest known birds, *Archaeopteryx* (meaning "ancient wing") was about the size of a pigeon. Several specimens of this genus have been found in the Jurassic limestone of Bavaria, which was laid down about 145 mya. Unlike extant birds, its jawbones were armed with reptilian-type teeth, and it had a long reptilian tail covered with feathers (Fig. 30-19b). Each *Archaeopteryx* wing had three claw-bearing digits. Like modern birds, *Archaeopteryx* had wings, feathers, reduced digits, and a **furcula,** or "wishbone." Its feathers were very similar to those of modern birds. Biologists do not think that *Archaeopteryx* gave rise to modern birds but, rather, that it shared a common ancestor with other birds.

Cretaceous rocks have yielded fossils of other early birds. *Hesperornis,* which lived in North America, was a toothed, aquatic diving bird with rudimentary wings and broad, webbed feet for swimming. *Ichthyornis* was a toothed, flying bird about the size of a sea gull. From the Tertiary period onward, the fossil record of birds shows an absence of teeth and progressive changes leading to the modern birds.

Modern birds are adapted for flight

About 9000 species of extant birds have been described, and they have been classified into about 23 monophyletic groups. Birds inhabit a wide variety of habitats and can be found on all continents, most islands, and even the open sea. Birds are a diverse group (Fig. 30-20). The largest living birds are the ostriches of Africa, which may be up to 2 m (6.5 ft) tall and weigh 136 kg (300 lb), and the great condors of the Americas, with wingspans of up to 3 m (10 ft). The smallest known bird is Helena's hummingbird of Cuba, with a length of less than 6 cm (2 in) and a weight of less than 4 g (0.14 oz).

Like their reptilian ancestors, modern birds lay eggs and have reptilian-type scales on their legs. Birds have evolved remarkable specializations for flight. Their feathers are very light, yet flexible and strong, and present a flat surface to the air. Feathers also protect the body and decrease water and heat loss. The anterior limbs of birds are wings, usually adapted for flight. The posterior limbs are modified for walking, swimming, or perching.

In addition to feathers and wings, birds have many other adaptations for flight. Their bodies are compact and streamlined, and the fusion of many bones gives them the rigidity needed for flying. Their bones are strong but very light. Many are hollow, containing large air spaces. The avian jaw is light and instead of teeth, birds have a light, horny beak. The breastbone is broad and keeled, for the attachment of the large flight muscles.

Birds have efficient lungs with thin-walled extensions, called **air sacs,** that occupy spaces between the internal organs and within certain bones. They have a unique "one-way" flow of air through their respiratory system (discussed in Chapter 44). Birds have a four-chambered heart and a double circuit of blood flow. Blood delivers oxygen to the tissues and then recharges with oxygen in the lungs before the heart pumps it into the systemic circulation again. The very effective respiratory and circulatory systems provide the cells with enough oxygen to permit a high metabolic rate, which is necessary for the tremendous muscular activity that flying requires. Some of the heat a bird gen-

Frans Lanting/Minden Pictures

(a)

Gary Meszaros/Dembinsky Photo Associates

(b)

Smithsonian, Vol. 31, Feb. 2001/Tom Blagden

(c)

FIGURE **30-20** **Modern birds.**

Although they are diverse, most birds are highly adapted for flight, and their basic structure is similar. **(a)** A male peacock (*Pavo cristatus*) impressively displaying his feathers. **(b)** This Atlantic puffin (*Fratercula arctica*) has caught several fish. **(c)** Two fledgling wood storks (*Mycteria americana*), about 8 to 9 weeks old, preen their feathers atop their nest in South Carolina's White Hall rookery.

erates by metabolic activities helps it maintain a constant body temperature. Birds are endothermic, which permits them to remain active in cold climates.

Birds excrete nitrogenous wastes mainly as semisolid uric acid. Because birds typically do not have a urinary bladder, these solid wastes are delivered into the cloaca. They leave the body with the feces, which are dropped frequently. This adaptive mechanism helps maintain a light body weight.

Birds have become adapted to a variety of environments, and various species have very different types of beaks, feet, wings, tails, and behavioral patterns. Not all birds fly. Some, such as penguins, have small, flipper-like wings used in swimming. Ratites are a group of mainly large, flightless birds, including the ostriches, cassowarys, and kiwis. These birds have only vestigial wings, but they have well-developed legs, which they use for running.

Bills are specifically adapted for the type of food the bird eats. Although all birds must eat frequently (because they have a high metabolic rate and typically do not store much fat), the choice of food varies widely among species. Most birds eat energy-rich foods such as seeds, fruits, worms, mollusks, or arthropods. Warblers and some other species eat mainly insects. Owls and hawks eat rodents, rabbits, and other small mammals. Vultures feed on dead animals. Pelicans, gulls, terns, and kingfishers catch fish. Some hawks catch snakes and lizards.

An interesting feature of the bird digestive system is the **crop,** an expanded, saclike portion of the digestive tract below the esophagus, in which food is temporarily stored. The stomach is divided into a **proventriculus,** which secretes gastric juices, and a thick, muscular **gizzard,** which grinds food. The bird swallows small bits of gravel that act as "teeth" in the gizzard, mechanically breaking down food.

Birds have a well-developed nervous system, with a brain that is proportionately larger than that of reptiles. Birds rely heavily on vision, and their eyes are relatively larger than those of other vertebrates. Hearing is also well developed.

In striking contrast with the relatively silent reptiles, birds are very vocal. Most have short, simple *calls* that signal danger or influence feeding, flocking, or interaction between parent and young. *Songs* are usually more complex than calls and are performed mainly by males. Songs are related to attracting and keeping a mate, and to claiming and defending a territory.

One of the most fascinating aspects of bird behavior is the annual migration that many species make. Some birds, such as the golden plover and Arctic tern, fly from Alaska to Patagonia, South America, and back each year, covering perhaps 40,250 km (25,000 mi) en route. Migration and navigation are discussed in Chapter 50.

Many birds have beautiful and striking colors. The color is due partly to pigments deposited during the development of the feathers and partly to reflection and refraction of light of certain wavelengths. Many birds, especially females, are protectively colored by their plumage. During the breeding season the male often assumes brighter colors to help in attracting a mate.

Mammals are characterized by hair and mammary glands

Derived characters that distinguish mammals (class Mammalia) include hair, which insulates and protects the body; **mammary glands,** which produce milk for the young; differentiation of teeth into incisors, canines, premolars, and molars; and three middle-ear bones (malleus, incus, and stapes) that conduct vibrations.

A muscular **diaphragm** helps move air into and out of the lungs. Like birds, mammals are *endotherms,* but biologists think mammals evolved endothermy independently of birds. Maintaining a constant body temperature is enhanced by the covering of insulating hair, by the four-chambered heart, and by separate pulmonary and systemic circulations. Red blood cells without nuclei serve as efficient oxygen transporters.

Contributing significantly to the success of the mammals, the complex nervous system is more highly developed than in any other group of animals. The cerebrum is especially large and complex, with an outer gray region called the **cerebral cortex.**

Fertilization is always internal, and, except for the primitive monotremes that lay eggs, mammals are viviparous. Most mammals develop a **placenta,** an organ of exchange between developing embryo and mother, through which the embryo receives its nourishment and oxygen and rids its blood of wastes. By carrying their developing young internally, mammals avoid the hazards of having their eggs consumed by predators. By nourishing the young and caring for them, the parents offer both protection and an "education" on how to obtain food and avoid being eaten.

The limbs of mammals are variously adapted for walking, running, climbing, swimming, burrowing, or flying. In four-legged mammals, the limbs are more directly under the body than they are in extant reptiles, which contributes to speed and agility.

Early mammals were small, endothermic animals

Mammals descended from **therapsids,** a group of synapsid reptiles, during the Triassic period some 200 mya. The therapsids were doglike carnivores with differentiated teeth and legs adapted for running (Fig. 30-21). The fossil record indicates that early mammals were small, about the size of a mouse or shrew. Some may have been endothermic, and some may have had fur.

How did mammals manage to coexist with reptiles during the 160 million years or so that reptiles ruled Earth? Many adaptations permitted early mammals to compete for a place on this planet. Perhaps one of the most important was their skill at being inconspicuous. They were **arboreal** (tree-dwelling) and **nocturnal** (active at night), searching for food (mainly insects and plant material, and perhaps reptile eggs) at night while the reptiles were inactive. Although their eyes may have been small, early mammals had well-developed sense organs for hearing and smell.

As many reptiles became extinct, the mammals adapted to their abandoned niches (lifestyles). During this time, the flowering plants, including many trees, underwent an adaptive radiation, providing new habitats, sources of food, and protection from predators. Larger forms and numerous varieties of mammals evolved. During the early Cenozoic era (more than 55 mya), the mammals underwent an adaptive radiation, becoming widely distributed and adapted to an impressive variety of ecological lifestyles (see Fig. 20-16).

Painting by John C. Germann, Department of Library Services, American Museum of Natural History

FIGURE **30-21** **Therapsid (*Lycaenops*).**
The therapsids were synapsid reptiles. *Lycaenops,* a carnivore about the size of a coyote, lived in South Africa during the late Permian period.

Modern mammals are assigned to three subclasses By the end of the Cretaceous period, three main groups of mammals had evolved (Fig. 30-22). Today, mammals inhabit virtually every corner of Earth; they live on land, in fresh and salt water, and in the air. They range in size from the tiny pigmy

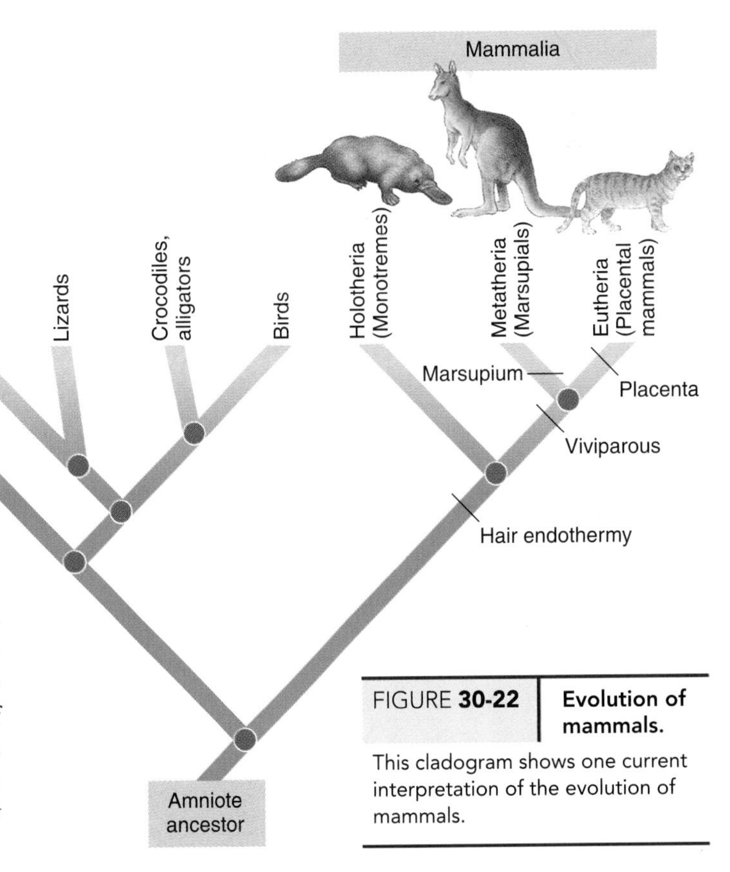

FIGURE **30-22** **Evolution of mammals.**
This cladogram shows one current interpretation of the evolution of mammals.

FIGURE **30-23** **Spiny anteater.**

The spiny anteater (*Tachyglossus aculeatus*) is an egg-laying monotreme.

shrew, weighing about 2.5 g (less than 0.1 oz), to the blue whale, which may weigh up to 136,000 kg (150 tons) and which biologists think is one of the largest animals that has ever lived.

Modern mammals are classified in three subclasses: **Holotheria** includes the egg-laying mammals, or monotremes; **Metatheria** includes the marsupials, or pouched mammals; and **Eutheria** includes the placental mammals.

The **monotremes** are the only living order of subclass Holotheria. One genus includes the duck-billed platypus (*Ornithorhynchus*) and a second, the spiny anteaters or echidnas (*Tachyglossus*) (Fig. 30-23; see also Figs. 1-16 and 22-4). These animals live in Australia and Tasmania; two of the three spiny anteater species also live in New Guinea. The females lay eggs that they carry in a pouch on the abdomen or keep warm in a nest. When

(a)

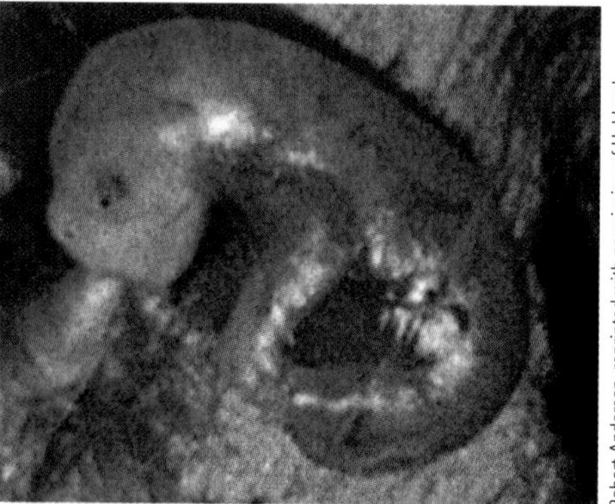

(b)

FIGURE **30-24** **Marsupials.**

(a) Eastern gray kangaroo (*Macropus giganteus*) with young, known as a joey. The kangaroo is native to Australia. **(b)** A kangaroo soon after birth. Marsupials are born in an embryonic state and continue to develop in the safety of the mother's marsupium (pouch).

FIGURE **30-25** **Convergent evolution in placental and marsupial mammals.**

For each mammal in one group, there is a counterpart in the other group. Their similarities include both lifestyles and structural features.

the young hatch, they lap up milk secreted by their mother's mammary glands. Unlike other mammals, nipples are absent.

With their long beaks and long, sticky tongues, spiny anteaters capture ants and termites The duck-billed platypus, which lives in burrows along river banks, preys on freshwater invertebrates. It has webbed feet and a flat, beaver-like tail, which helps it swim.

Marsupials include pouched mammals such as kangaroos and opossums (Fig. 30-24a). Embryos begin their development in the mother's uterus, where they are nourished by fluid and yolk. After a few weeks, still in a very undeveloped stage, the young are born. In many species, the young crawl to the **marsupium** (pouch), where they complete their development. The young marsupial attaches its mouth to a mammary gland nipple and is nourished by its mother's milk (Fig. 30-24b).

At one time, marsupials probably inhabited much of the world, but placental mammals largely outcompeted them. Now marsupials live mainly in Australia and in Central and South America. The only marsupial species that ranges into the United States is the Virginia opossum (*Didelphis virginiana*). Australia became geographically isolated from the rest of the world before placental mammals migrated there, and the marsupials remained the dominant mammals (see discussion of continental drift in Chapter 17). Australian marsupials underwent adaptive radiation, paralleling the evolution of placental mammals elsewhere. Thus, marsupials that live in Australia and adjacent islands correspond to North American placental wolves, bears, rats, moles, flying squirrels, and even cats (Fig. 30-25).

Most familiar to us are **placental mammals,** characterized by development of a placenta, an organ of exchange between the developing embryo and its mother (Fig. 30-26). The placenta forms from both embryonic membranes and the maternal uterine wall. In it the blood vessels of the embryo come very close to the blood vessels of the mother, so materials can be exchanged by diffusion. (The two circulations do not normally mix.) The placenta enables the young to remain within the mother's body until embryonic development is complete.

Placental mammals are born at a more mature stage than marsupials. Indeed, among some species the young can walk around and begin to interact with other members of the group within a few minutes after birth. Biologists classify extant placental mammals into about 16 orders. Table 30-3 gives a brief summary of some of these orders. Remember that aquatic mammals, such as dolphins, whales, and seals, evolved from terrestrial ancestors.

Review

- What vertebrate adaptations are important for terrestrial life?
- What are some arguments for classifying birds and reptiles in a single clade?
- What are the three main groups of mammals? How do they differ?
- Which amniotes are endothermic? What are some advantages of endothermy?

BiologyNow™ Assess your understanding of **amniotes** by taking the pretest on your BiologyNow CD-ROM.

FIGURE **30-26** **Placental mammals.**

(a) Wildebeest (*Connochaetes taurinus*), photographed in Tanzania. The dominant plains antelope in many areas of eastern and southern Africa, wildebeest migrate long distances during the dry season in search of food and water. **(b)** Humpback whales (*Megaptera novaeangliae*) exhibiting a rare double breach. **(c)** Polar bears (*Ursus maritimus*), the largest living land carnivores, feed mainly on seals.

James D. Watt/Animals Animals

(b)

McMurray Photography

(a)

Michio Hoshino/Minden Pictures

(c)

TABLE 30-2 Some Orders of Living Placental Mammals

Order, Representative Members, and Some Characteristics

Insectivora (moles, hedgehogs, shrews)

Nocturnal; eat insects; considered most primitive placental mammals. Shrew is smallest living mammal; some weigh less than 5 g.

African hedgehog

Chiroptera (bats)

Adapted for flying; fold of skin extends from elongated fingers to body and legs, forming wing. Guided in flight by type of biological sonar; they emit high-frequency squeaks and are guided by echoes from obstructions. Eat insects and fruit, or suck blood of other animals.

Bat

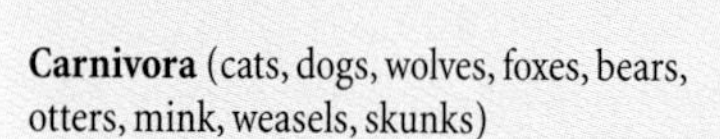

Carnivora (cats, dogs, wolves, foxes, bears, otters, mink, weasels, skunks)

Carnivores with sharp, pointed canine teeth and molars for shearing. Keen sense of smell; complex social interactions. Among fastest, strongest, and smartest animals.

Wolf

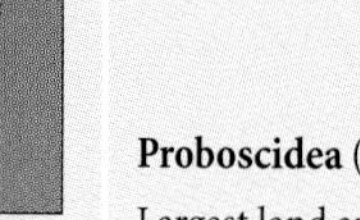

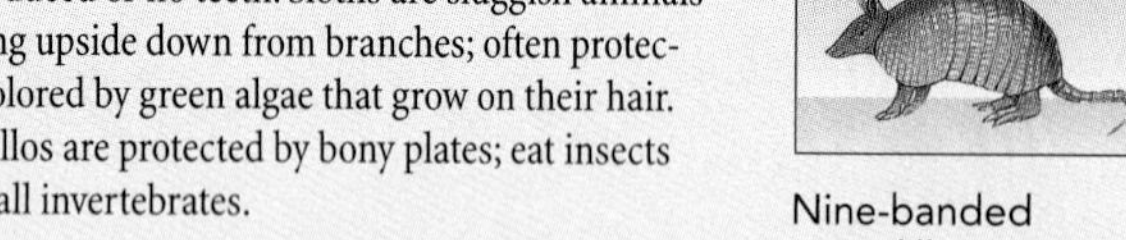

Xenarthra (sloths, anteaters, armadillos)

Teeth reduced or no teeth. Sloths are sluggish animals that hang upside down from branches; often protectively colored by green algae that grow on their hair. Armadillos are protected by bony plates; eat insects and small invertebrates.

Nine-banded armadillo

Rodentia (squirrels, beavers, rats, mice, hamsters, porcupines, guinea pigs)

Gnawing animals with chisel-like incisors. As they gnaw, teeth are worn down, and so must grow continually.

Flying squirrel

Lagomorpha (rabbits, hares, pikas)

Like rodents, have chisel-like incisors. Typically have long hind legs adapted for jumping. Many have long ears.

Pika

Primates lemurs, monkeys, apes, humans)

Highly developed brains and eyes. Nails instead of claws. Opposable thumb. Eyes directed forward. Omnivores. Most species arboreal. (Primate evolution discussed in Chapter 21.)

Ring-tailed lemur

Perissodactyla (horses, zebras, tapirs, rhinoceroses)

Herbivores. Hoofed with odd number of digits per foot; one or three toes. Teeth adapted for chewing. Usually large animals with long legs.

Tapir

Artiodactyla (cattle, sheep, pigs, deer, giraffes)

Hoofed with even numbers of digits per foot; most have two toes, some have four. Most have antlers or horns. Herbivores; most are ruminants, chew a cud and have series of stomachs in which bacteria that digest cellulose.

American elk

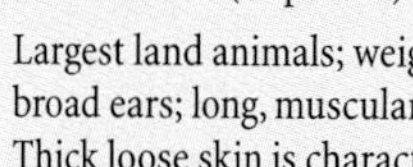

Proboscidea (elephants)

Largest land animals; weigh up to 7 tons; large head; broad ears; long, muscular, flexible trunk (proboscis). Thick loose skin is characteristic. Two upper incisors are elongated as tusks. Includes extinct mastodons and wooly mammoths.

African elephant

Sirenia (sea cows, manatees)

Herbivorous, aquatic mammals with finlike forelimbs and no hind limbs. They are probably the basis for most tales about mermaids.

Manatee

Cetacea (whales, dolphins, porpoises)

Adapted for aquatic life with fish-shaped body and broad, paddle-like forelimbs (flippers). Posterior limbs absent. Many have thick layer of blubber under skin. Some are suspension feeders. Mate and bear young in the water. Very intelligent.

Humpback whale

Pinnipedia (seals, sea lions, walruses)

Marine; limbs adapted as flippers for swimming. Carnivores; eat fish.

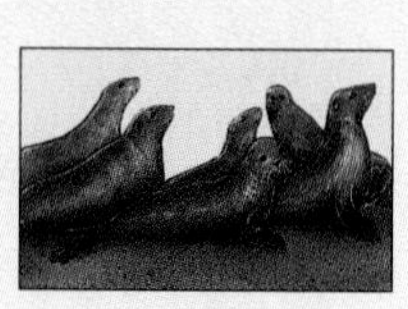

Sea lions

SUMMARY WITH KEY TERMS

1 Identify the main branches of deuterostomes.

- **Deuterostomes** include echinoderms, hemichordates, and chordates. **Hemichordates** (acorn worms) are marine deuterostomes with a three-part body, including proboscis, collar, and trunk.

2 List three shared derived characters of echinoderms; and describe the main classes of echinoderms.

- Phylum **Echinodermata** includes marine animals with a spiny "skin," **water vascular system, tube feet,** and **endoskeleton.** The larvae have bilateral symmetry; most of the adults exhibit **pentaradial symmetry.**
- Class **Crinoidea** includes sea lilies and feather stars. The **oral surface** of crinoids is turned upward; some crinoids are sessile.
- Class **Asteroidea** consists of the sea stars. They have a central disk with five or more arms. They use tube feet for locomotion.
- Class **Ophiuroidea** includes the brittle stars, which resemble sea stars but have longer, more slender arms that are set off more distinctly from the central disk. They use their arms for locomotion. Their tube feet lack suckers and are not used in locomotion.
- Class **Echinoidea** includes the sea urchins and sand dollars. Echinoids lack arms; they have a solid shell and are covered with spines.
- Class **Holothuroidea** consists of sea cucumbers, animals with elongated flexible bodies. The mouth is surrounded by a circle of modified tube feet that serve as tentacles.

3 Describe four shared derived characters of chordates

- Phylum **Chordata** has three subphyla: Urochordata, Cephalochordata, and Vertebrata. At some time in its life cycle, a chordate has a flexible, supporting **notochord;** a **dorsal, tubular nerve cord; pharyngeal (gill) slits;** and a **muscular postanal tail.**

4 Describe the invertebrate chordate subphyla.

- The **tunicates,** which belong to subphylum **Urochordata,** are suspension-feeding, marine animals with **tunics.** Larvae have typical chordate characteristics and are free-swimming. Adults of most groups are sessile suspension feeders.
- Subphylum **Cephalochordata** consists of the **lancelets,** small, segmented, fishlike animals that exhibit chordate characteristics.

5 Discuss the evolution of chordates.

- The urochordates (tunicates) were probably the first chordates to evolve. Subphyla Cephalochordata and **Vertebrata** to be **sister taxa,** groups that diverged from a recent common ancestor.

6 Describe four shared derived characters of vertebrates.

- The vertebrates have a **vertebral column** that forms the chief skeletal axis of the body, and a braincase called a **cranium. Neural crest cells** give rise to or influence the development of many structures, including the cranium and jaws. Vertebrates have pronounced cephalization, a complex brain, and muscles attached to the **endoskeleton** for movement.

7 Distinguish among the major groups of jawless fishes.

- Some of the earliest known vertebrates, called **ostracoderms,** were jawless fishes that are now extinct.
- The extant jawless fishes, or **agnathans,** are the hagfishes, assigned to class **Cephalaspidomorphi** and the lampreys, class **Myxini.** Systematists no longer consider hagfishes vertebrates, because they have no trace of vertebrae. They classify the vertebrates plus the hagfishes as **craniates** (Craniata).

8 Trace the evolution of jawed fishes and early tetrapods, and identify major taxa of jawed fishes and amphibians.

- Class **Chondrichthyes,** the cartilaginous fishes, includes the sharks, rays, and skates. These fishes have jaws, two pairs of fins, and **placoid scales.** Skates and some species of sharks are **oviparous,** that is, they lay eggs. Many species of sharks are **ovoviparous;** their young are enclosed by eggs that are incubated in the mother's body. A few shark species are **viviparous;** the young develop in the mother's uterus and are nourished by transfer of nutrients from the mother's blood.
- The bony fishes are assigned to three classes: **Actinopterygii,** ray-finned fishes; **Actinistia,** coelancanths; and **Dipnoi,** lungfishes.
- During the Devonian, bony fishes gave rise to two evolutionary lines: the Actinopterygii, or **ray-finned fishes,** and the **Sarcopterygii,** or **lobe-finned fishes.**
- The ray-finned fishes gave rise to the modern bony fishes. In these fishes, the lungs have been modified as a **swim bladder,** an air sac for regulating buoyancy.
- The sarcopterygii probably gave rise to the **lungfishes** (class **Dipnoi**) and **coelacanths** (class **Actinistia**).
- Both coelancanths and lungfishes were apparently preadapted for life on land. Biologists think the lungfish gave rise to the **tetrapods,** the land vertebrates.
- The first successful tetrapods were the now-extinct **labyrinthodonts,** which may be the ancestors of frogs and salamanders.
- Class **Amphibia** includes salamanders, frogs and toads, and wormlike caecilians. Most amphibians return to the water to reproduce. Frog embryos develop into **tadpoles,** which undergo **metamorphosis** to become adults.
- Amphibians use their moist skin as well as lungs for gas exchange. They have a three-chambered heart and **systemic** and **pulmonary circulations.**

9 Describe three vertebrate adaptations to terrestrial life.

- Terrestrial vertebrates, or **amniotes,** include reptiles, birds, and mammals. The evolution of the **amniotic egg** with its shell and amnion was an important adaptation for life on land. The **amnion** is a membrane that forms a fluid-filled sac around the embryo. Amniotes have a body covering that retards water loss. They also have physiological mechanisms that conserve water.

10 Describe the reptiles and birds, and give an argument for including the birds in the reptile clade.

- Class **Reptilia,** as traditionally defined, is paraphyletic. It includes dinosaurs, turtles, lizards, snakes, and alligators. Many biologists now include the birds in the dinosaur clade.
- In reptiles, fertilization is internal and most reptiles secrete a leathery protective shell around the egg. The embryo develops protective membranes, including an amnion that keeps it moist.
- Reptiles have dry skin with horny scales, lungs with many chambers, and a three-chambered heart with some separation of oxygen-rich and oxygen-poor blood.
- Paleontologists have discovered feathered dinosaurs, and many biologists now consider the birds as feathered dinosaurs; they classify the birds and most of the reptiles as **diapsids.**
- Birds have many adaptations for powered flight, including feathers, wings, and light, hollow bones containing air spaces. Birds have a four-chambered heart, very efficient lungs, a high metabolic rate, and a constant body temperature; they excrete solid metabolic wastes (uric acid). They have a well-developed nervous system and excellent vision and hearing.

SUMMARY WITH KEY TERMS *(continued)*

11 Contrast monotremes, marsupials, and placental mammals, and give examples of animals that belong to each group.

- **Mammals** have hair, **mammary glands,** differentiated teeth, and three middle-ear bones. They maintain a constant body temperature and have a highly developed nervous system and a muscular **diaphragm.**
- **Monotremes,** which belong to subclass **Holotheria,** include the duck-billed platypus and spiny anteaters. Monotremes lay eggs.
- **Marsupials** (subclass **Metatheria**) include pouched mammals, such as kangaroos and opossums. The young are born in an embryonic stage and complete their development in their mother's **marsupium,** where they are nourished with milk from the mammary glands.
- **Placental mammals** (subclass **Eutheria**) are characterized by the **placenta,** an organ of exchange that develops between the embryo and the mother.

POST-TEST

1. Which of the following is *not* a deuterostome? (a) echinoderm (b) chordate (c) hemichordate (d) arthropod (e) hagfish
2. Which of the following belongs to subphylum Vertebrata? (a) lophophorate (b) lamprey (c) lancelet (d) tunicate (e) crinoid
3. Which of the following is/are found in tunicates? (a) notochord (b) tube feet (c) anal gill slits (d) two pairs of appendages (e) vertebral column
4. Which of the following was an early jawed fish? (a) placoderm (b) crinoid (c) therapsid (d) labyrinthodont (e) ostracoderm.
5. Arrange the following animals to reflect an accepted sequence of evolution: 1. lungfish 2. reptile 3. amphibian 4. bird. (a) 3, 1, 2, 4, (b) 1, 3, 2, 4 (c) 2, 1, 3, 4 (d) 1, 2, 3, 4
6. Which of the following characteristics is *not* associated with sea stars? (a) tube feet (b) water vascular system (c) central disk with five or more arms (d) spiny skeleton (e) notochord
7. Which of the following characteristics is associated with amphibians? (a) mantle (b) placoid scales (c) three-chambered heart (d) swim bladder (e) amnion
8. Which of the following is *not* characteristic of birds? (a) feathers (b) ectothermic (c) amnion (d) high metabolic rate (e) reptilian-like scales on legs
9. Which of the following is *not* true of the duck-billed platypus? (a) classified as a marsupial (b) lays eggs (c) embryo has a notochord (d) embryo has pharyngeal slits (e) is a predator
10. Which of the following is *not* a placental mammal? (a) shrew (b) human (c) dolphin (d) spiny anteater (e) bat
11. Which of the following is *true* of mammals? (a) all have placentas (b) evolved from saurischian dinosaurs (c) three middle-ear bones (d) most are ovoviviparous (e) evolved during the Devonian period
12. The vertebral column (a) is characteristic of all chordates (b) forms the skeletal axis of the chordate body (c) may consist of cartilaginous or bony segments (d) is well developed in lancelets (e) three of the preceding answers are correct
13. A shark is characterized by (a) placoid scales (b) bony skeleton (c) water vascular system (d) amnion (e) muscular diaphragm
14. Reptiles (a) have dry, scaly skin (b) are amniotes (c) gave rise to the birds and mammals (d) answers a, b, and c are correct (e) answers a and c only are correct
15. Amniotes (a) include amphibians, reptiles, birds, and mammals (b) are especially well adapted for fresh water (c) typically have physiological mechanisms for conserving water (d) typically excrete ammonia (e) gave rise to the amphibians
16. Which of the following is *not* true? (a) feathers evolved in dinosaurs (b) birds are endothermic (c) birds excrete uric acid (d) birds have an amniotic egg (e) *Archaeopteryx* was the ancestor of feathered dinosaurs

CRITICAL THINKING

1. Discuss the relationships among the echinoderms, hemichordates, and chordates, describing shared derived characters that support grouping them together as deuterostomes.
2. Some paleontologists consider monotremes to be therapsid reptiles rather than mammals. Give arguments for and against this position.
3. Which are more specialized, birds or mammals? Explain your answer.
4. Imagine you discover an interesting new animal. You find it has a dorsal, tubular nerve cord, a cranium, moist skin, and a heart with two atria and a ventricle. How would you classify the animal? Explain each step in your decision.

- Visit our Web site at **http://biology.brookscole.com/solomon7** for links to chapter-related resources on the World Wide Web. Additional online materials relating to this chapter can also be found on our Web site.

BIOLOGY NOW RESOURCES

Active Figures

30-17: How frogs reproduce

Preparing for an exam? Take a diagnostic test on your BiologyNow CD-ROM.

Post-Test Answers

1. d	2. b	3. a	4. a
5. b	6. e	7. c	8. b
9. a	10. d	11. c	12. c
13. a	14. d	15. c	16. e

Plant Structure, Growth, and Differentiation

Runk/Schoenberger from Grant Heilman

Oak seedling. A massive oak tree may grow from this young basket oak (*Quercus prinus*) seedling.

CHAPTER OUTLINE

About 90% of the approximately 262,000 species of plants are flowering plants. These vascular plants are characterized by flowers, double fertilization, endosperm, and seeds enclosed within fruits (see Chapter 27). Because flowering plants are the largest, most successful group of plants, they are the focus of much of this chapter and of Chapters 32 to 36. Let us begin our study of the structure and functions of plants by examining a familiar species, the oak, and its fruit, the acorn.

Squirrels often temporarily store acorns in holes in the ground to provide winter food. Many of these are never retrieved, and so a new oak often begins its life after a squirrel has buried an acorn in the soil. The seed within the acorn first absorbs water from the surrounding soil. Then germination occurs as a root emerges and works its way down into the soil, absorbing additional water and anchoring the young plant. A miniature shoot begins to grow upward and breaks through the soil. At this point the young oak, shown in the photograph, is a seedling—it has just emerged from the seed and is still dependent on the food supply stored within the seed.

The stem continues to elongate, and small leaves develop, expand, and begin to photosynthesize. The young plant is now independently established: It is anchored in the ground, absorbs water and dissolved nutrient minerals from the soil, and uses energy from organic molecules, such as glucose, that it produces by photosynthesis.

Smaller roots branch off the original root, and the young tree grows taller, always by growth at the tips of its branches. The roots likewise elongate at their tips. As it ages, the tree also increases in girth (thickness) by forming additional wood and bark. This growth occurs along the sides of the more mature stems and roots. Thus, the oak progressively accumulates more wood and bark, more stem and root tissues, and more leaves.

Growth and expansion of both the root and shoot systems continue throughout the life of the oak. As in all plants, growth is flexible and dynamic, enabling the oak to respond to its environment. For example, the young oak may grow very slowly for several years, particularly if shaded by mature trees. When conditions change, perhaps when an older tree nearby dies and falls to the ground, permitting direct sunlight to reach the young tree, it is poised for rapid growth.

After several years of growth, the oak tree becomes reproductively mature. Oaks reproduce by forming separate male and female flowers in spring. The minute male flowers, more conspicuous than the female flowers, are found together on slender, drooping, caterpillar-shaped catkins (clusters); in contrast, the tiny female flowers are often solitary. After pollen from a male flower is blown by the wind to a female flower, a sperm cell fertilizes the egg, and an acorn develops, partially enclosed by a scaly cup. First green in color, the acorn fruit turns a deep brown as it matures. Within the acorn is a seed containing an embryo of a miniature oak, along with a supply of stored food. Thus the cycle of life repeats itself, as squirrels busily collect these acorns and cache them for later retrieval.

PROCESS OF SCIENCE

In this chapter we examine the external structure of the flowering plant body; the organization of its cells, tissues, and tissue systems; and its basic growth patterns. Information of this type, which is based on observation and description, rarely receives as much attention as experimental science. However, observation and description are crucial aspects of the scientific process. ■

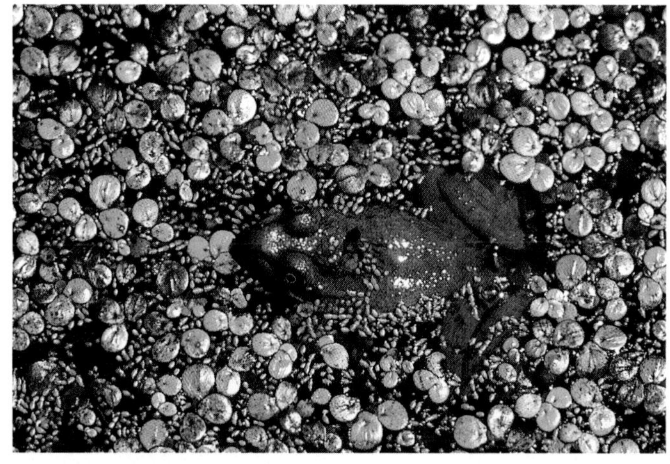

(a)

Carlyn Iverson

(b)

Joyce Photographers/Photo Researchers, Inc.

FIGURE **31-1** | **Size variation in plants.**

(a) Duckweeds (*Spirodela*), the larger green plants in the photograph, are tiny floating aquatic plants about 1 cm (3⁄8 in) across. If you look closely at the frog's body, you will see minute green dots. Each is a water-meal (*Wolffia*) plant. These tiny floating herbs, about 1.5 mm (1⁄16 in) wide, are the smallest known flowering plants.
(b) This giant gum tree (*Eucalyptus regnans*) was photographed in Bushy Park, Australia. Although most giant gums grow to 75 m (246 ft), some specimens have been measured at heights of 100 m (328 ft) and are the world's tallest flowering plants.

PLANT STRUCTURE AND LIFE SPAN

Learning Objectives

1. Distinguish between herbaceous and woody plants.
2. Discuss the differences among annuals, biennials, and perennials, and give an example of each.
3. Contrast two different life history strategies in plants.

Plants range in size from the minute floating water-meal (*Wolffia microscopica*), the smallest flowering plant known, to Australian gum trees (*Eucalyptus*), some of Earth's tallest trees (Fig. 31-1). The 235,000 or so species of flowering plants that live in and are adapted to the many environments offered by our planet represent remarkable variety. Yet all of these—from desert cacti with enormously swollen stems, to cattails partly submerged in marshes, to orchids growing in the uppermost tree branches of lush tropical rain forests—are recognizable as plants. Almost all plants have the same basic body plan, which consists of roots, stems, and leaves.

Plants are either herbaceous or woody. *Herbaceous* plants are nonwoody. In temperate climates, the aerial parts (stems and leaves) of herbaceous plants die back to the ground at the end of the growing season. In contrast, the aerial parts of *woody* plants (trees and shrubs) persist. Botanically speaking, woody plants produce hard, lignified secondary tissues (the cell walls of secondary tissues contain lignin), and herbaceous plants do not. Lignin and the production of secondary tissues are discussed later in the chapter.

Annuals are herbaceous plants (such as corn, geranium, and marigold) that grow, reproduce, and die in one year or less. Other herbaceous plants (such as carrot, Queen Anne's lace, cabbage, and foxglove) are **biennials;** they take two years to complete their life cycles before dying. During their first season biennials produce extra carbohydrates, which they store and use during their second year when they typically form flowers and reproduce. **Perennials** are herbaceous and woody plants that have the potential to live for more than two years. In temperate climates, the aerial (aboveground) stems of herbaceous perennials such as iris, rhubarb, onion, and asparagus die back each winter. Their underground parts (roots and underground stems) become dormant during the winter and send out new growth each spring. (In **dormancy,** an organism reduces its metabolic state to a minimum level to survive unfavorable conditions.) Similarly, in certain tropical climates with pronounced wet and dry seasons, the aerial parts of herbaceous perennials die back and the underground parts become dormant during the unfavorable dry season. Other tropical plants, such as orchids, are herbaceous perennials that grow year-round.

All woody plants are perennials, and some of them live for hundreds or even thousands of years. In temperate climates, the aboveground stems of woody plants become dormant during the winter. Many temperate woody perennials are **deciduous;** that is, they shed their leaves before winter and produce new stems with new leaves the following spring. Other woody perennials are **evergreen** and shed their leaves over a long period, so that some leaves are always present. Because they have permanent woody stems that are the starting points for new growth the following season, many trees attain massive sizes.

Plants have different life history strategies

Plants exhibit a variety of patterns of growth, reproduction, and longevity. Woody perennials often live for hundreds of years, whereas some herbaceous annuals may live for only a few weeks or months. Such characteristic features of an organism's life cycle, particularly as they relate to its survival and successful reproduction, are known as **life history strategies.** Biologists try to understand the relative advantages of each life history strategy, including any tradeoffs involved in the allocation of various resources. It appears that in some environments a longer life span is advantageous, whereas in other environments a shorter life span increases a species' chances for reproductive success—that is, for long-term survival.

When an environment is relatively favorable, it is filled with plants competing for available space. Because such an environment is so crowded, it has few open spots in which new plants can become established. When a plant dies, the empty area is quickly filled by another plant, but not necessarily by the same species as before. Thus an adult perennial survives well, but young plants, whether perennials or annuals, do not. A plant with a long life span thrives in this type of environment because it can occupy a piece of soil and continue to produce seeds for many years. In a tropical rain forest, for example, competition prevents most young plants from becoming established, and woody perennials predominate.

In a relatively unfavorable environment, many possible sites are usually available. This type of environment is not crowded, and young plants usually do not have to compete against large, fully established plants. Here, smaller, short-lived plants have the reproductive advantage. These plants are opportunists: They grow and mature quickly during the brief periods when environmental conditions are most favorable. As a result, all their resources are directed into producing as many seeds as possible before dying. In deserts following a rainy spell, for example, annuals are more prevalent than woody perennials.

Thus each species has its own characteristic life history strategy, with some plants adapted to variable environments and others adapted to stable environments. The longer life span characteristic of woody perennials is just one of several successful life history plans. We return to life history strategies in our discussion of population ecology (see Chapter 51).

Review

- Consider a common plant such as a marigold. Is it herbaceous or woody? An annual, biennial, or perennial? Explain your answers.
- Consider a common plant such as a carrot. Is it herbaceous or woody? An annual, biennial, or perennial? Explain.
- Consider a common plant such as an oak. Is it herbaceous or woody? An annual, biennial, or perennial? Explain.

BiologyNow™ Review your understanding of **plant structure and life span** by taking the pretest on your BiologyNow CD-ROM.

THE PLANT BODY

Learning Objectives

4. Discuss the functions of various parts of the vascular plant body, including the nutrient- and water-absorbing root system and the photosynthesizing shoot system.
5. Describe the structure and functions of the ground tissue system (parenchyma tissue, collenchyma tissue, and sclerenchyma tissue).
6. Describe the structure and functions of the vascular tissue system (xylem and phloem).
7. Describe the structure and functions of the dermal tissue system (epidermis and periderm).

The plant body of flowering plants (and other vascular plants) is usually organized into a root system and a shoot system (Fig. 31-2). The **root system** is generally underground. The aboveground portion, the **shoot system,** generally consists of a vertical stem that bears leaves, and, in flowering plants, flowers and fruits that contain seeds.

Each plant typically grows in two different environments: the dark, moist soil and the illuminated, relatively dry air. Plants usually have both roots and shoots because they require resources from both environments. Thus, roots branch extensively through the soil, forming a network that anchors the plant firmly in place and absorbs water and dissolved nutrient minerals from the soil. Leaves, the flattened organs of photosynthesis, are attached more or less regularly on the stem, where they absorb the sun's light and atmospheric CO_2 used in photosynthesis.

The plant body consists of cells and tissues

As in other organisms, the basic structural and functional unit of plants is the cell. During the course of evolution, plants have developed a diversity of cell types, each specialized for particular functions.

Like animal cells, plant cells are organized into tissues. A **tissue** is a group of cells that form a structural and functional unit. *Simple tissues* are composed of only one kind of cell, whereas *complex tissues* have two or more kinds of cells.

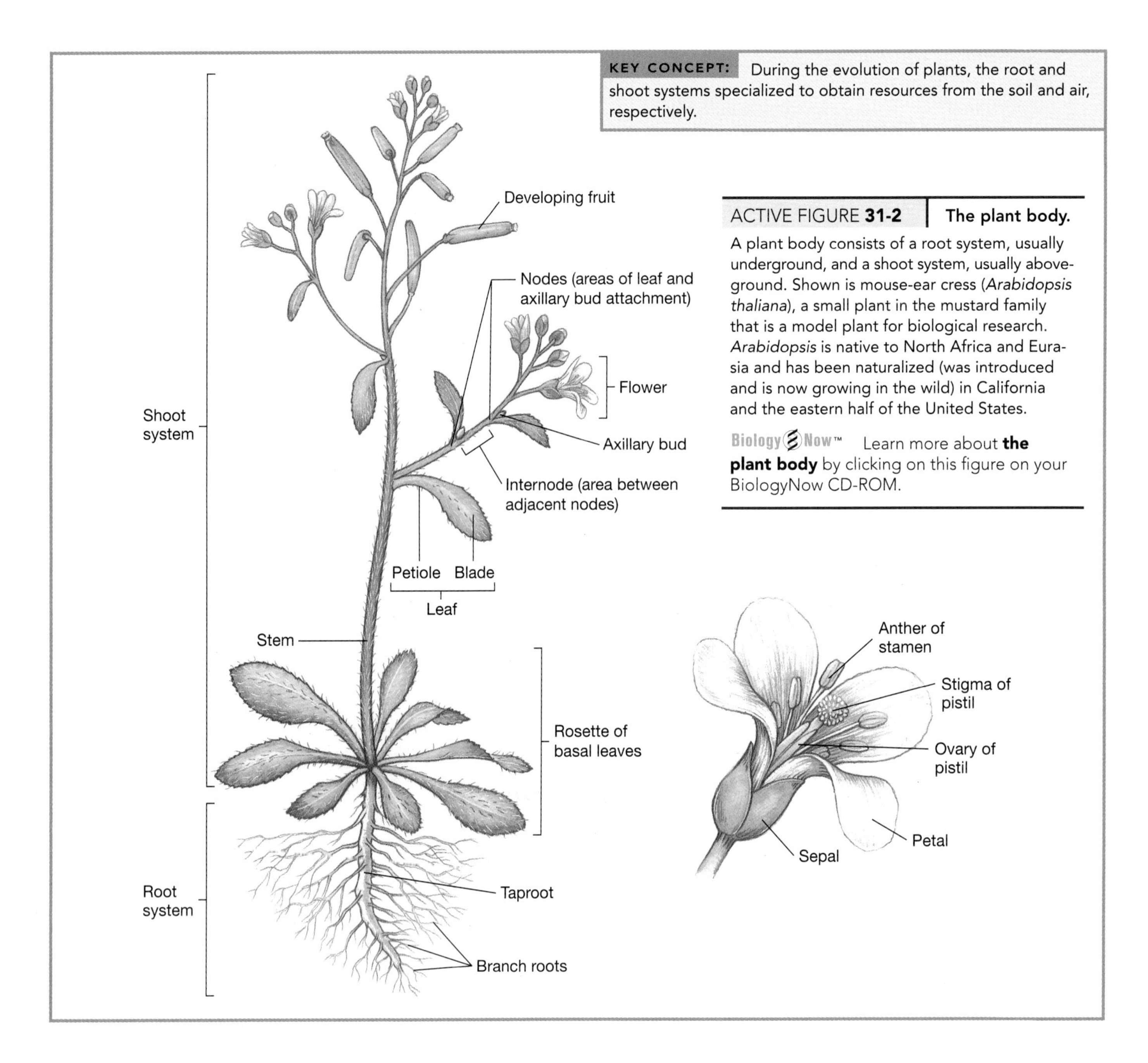

ACTIVE FIGURE **31-2** | **The plant body.**

A plant body consists of a root system, usually underground, and a shoot system, usually above-ground. Shown is mouse-ear cress (*Arabidopsis thaliana*), a small plant in the mustard family that is a model plant for biological research. *Arabidopsis* is native to North Africa and Eurasia and has been naturalized (was introduced and is now growing in the wild) in California and the eastern half of the United States.

BiologyNow™ Learn more about **the plant body** by clicking on this figure on your BiologyNow CD-ROM.

In vascular plants, tissues are organized into three tissue systems, each of which extends throughout the plant body (Fig. 31-3). Each tissue system contains two or more kinds of tissues (Table 31-1). Most of the plant body is composed of the **ground tissue system,** which has a variety of functions, including photosynthesis, storage, and support. The **vascular tissue system,** an intricate conducting system that extends throughout the plant body, is responsible for conduction of various substances, including water, dissolved nutrient minerals, and food (dissolved sugar). The vascular tissue system also functions in strengthening and supporting the plant. The **dermal tissue system** provides a covering for the plant body.

Roots, stems, leaves, flower parts, and fruits are **organs,** because each is composed of all three tissue systems. The tissue systems of different plant organs form an interconnected network throughout the plant. For example, the vascular tissue system of a leaf is continuous with the vascular tissue system of the stem to which it is attached, and the vascular tissue system of the stem is continuous with the vascular tissue system of the root.

The ground tissue system is composed of three simple tissues

The bulk of an herbaceous plant is its ground tissue system, which is composed of three tissues: parenchyma, collenchyma, and sclerenchyma (Table 31-2). These tissues can be distinguished by their cell wall structures. Recall that plant cells are surrounded by a cell wall that provides structural support (see Chapter 4). A growing plant cell secretes a thin **primary cell wall,** which stretches and expands as the cell increases in size. After the cell stops growing, it sometimes secretes a thick, strong **secondary cell wall,** which is deposited *inside* the primary cell

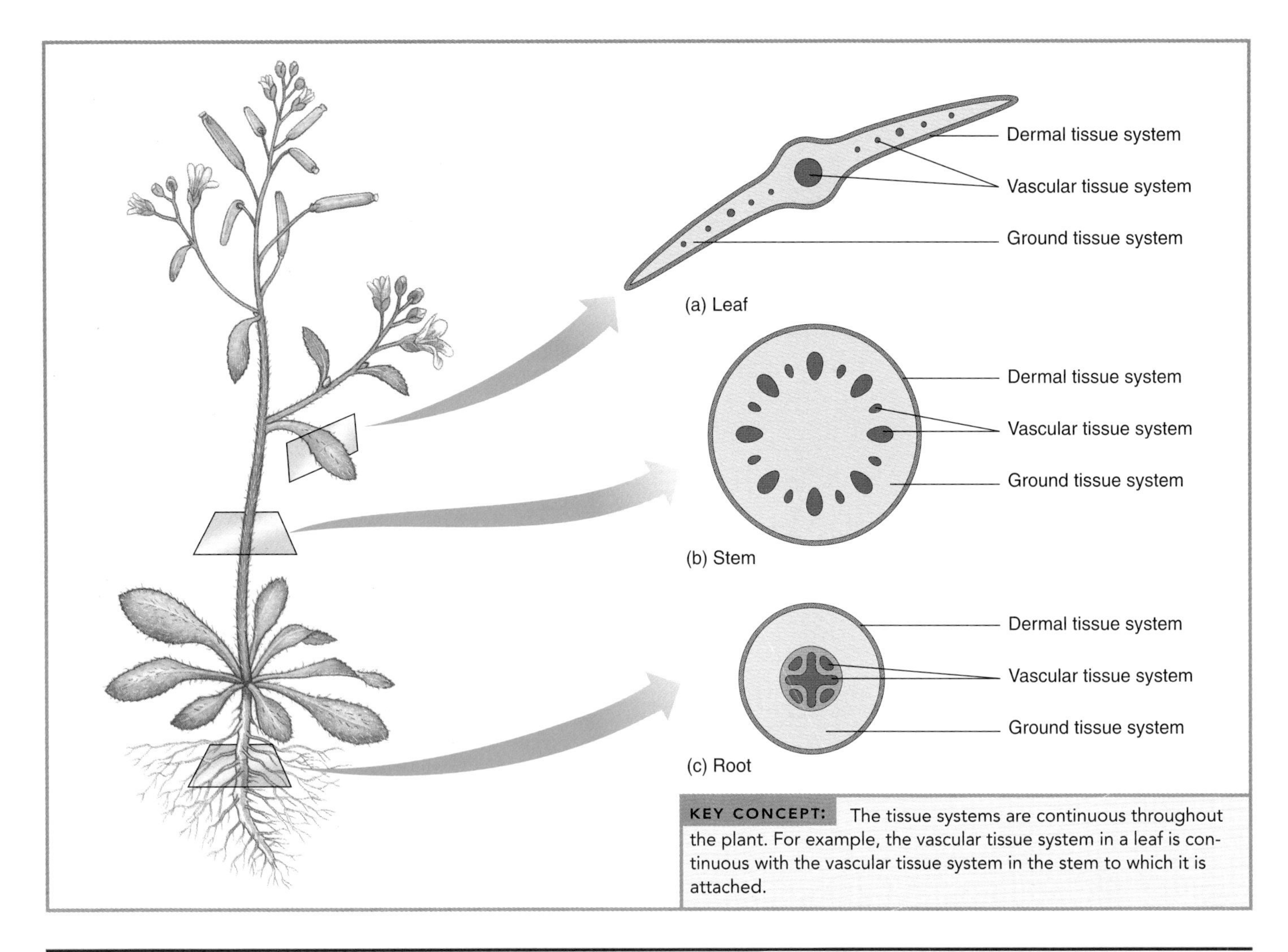

KEY CONCEPT: The tissue systems are continuous throughout the plant. For example, the vascular tissue system in a leaf is continuous with the vascular tissue system in the stem to which it is attached.

FIGURE **31-3** | **The three tissue systems in the plant body.**

This figure shows the distribution of the ground tissue system, vascular tissue system, and dermal tissue system in the (a) leaves, (b) stems, and (c) roots of a herbaceous dicot such as *Arabidopsis*.

TABLE **31-1** **Tissue Systems, Tissues, and Cell Types of Flowering Plants**

Tissue System	Tissue	Cell Types	Main Functions of Tissue
Ground tissue system	Parenchyma tissue	Parenchyma cells	Storage, secretion, photosynthesis
	Collenchyma tissue	Collenchyma cells	Support
	Sclerenchyma tissue	Sclerenchyma cells (sclereids or fibers)	Support, strength
Vascular tissue system	Xylem	Tracheids	Conduction of water and nutrient minerals, support
		Vessel elements	Conduction of water and nutrient minerals, support
		Xylem parenchyma cells	Storage
		Fibers (sclerenchyma cells)	Support, strength
	Phloem	Sieve tube elements	Conduction of sugar in solution, support
		Companion cells	May control functioning of sieve tube elements
		Phloem parenchyma cells	Storage, loading sugar into sieve tube elements
		Fibers (sclerenchyma cells)	Support, strength
Dermal tissue system	Epidermis	Epidermal cells	Protective covering over surface of plant body
		Guard cells	Regulate stomata
		Trichomes	Variable functions
	Periderm	Cork cells	Protective covering over surface of plant body
		Cork cambium cells	Meristematic (form new cells)
		Cork parenchyma cells	Storage

TABLE 31-2 Cell Types in the Ground Tissue System

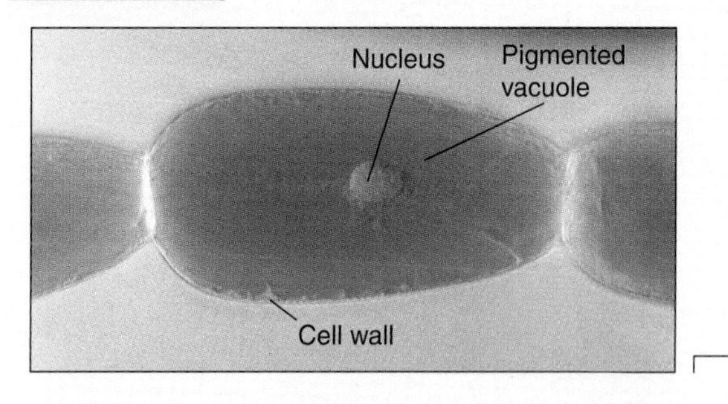

Parenchyma cell

Description
Living, actively metabolizing; thin primary cell walls

Functions
Storage; secretion, and/or photosynthesis

Location and comments
Throughout the plant body; shown is an LM of a stamen hair of a spiderwort (*Tradescantia virginiana*) flower; note the large pigmented vacuole; the nucleus is not inside the vacuole but is lying on *top* of the vacuole
(Phil Gates/Biological Photo Service)

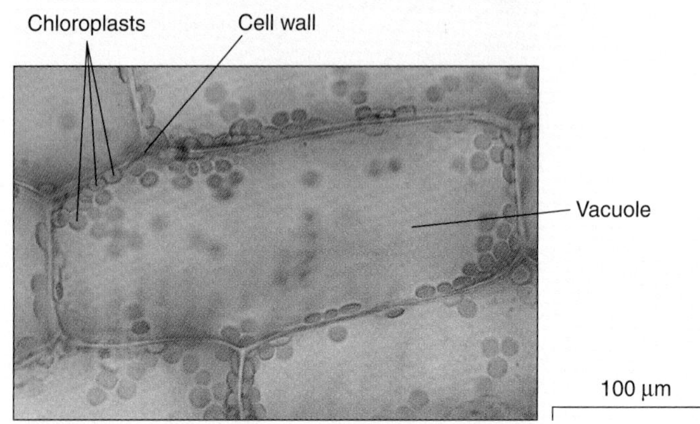

Parenchyma cell

Description
Living, actively metabolizing; thin primary cell walls

Functions
Storage; secretion; photosynthesis

Location and comments
Throughout the plant body; shown is an LM of leaf cells from an aquatic plant (*Elodea);* note the many chloroplasts in the thin layer of cytoplasm surrounding the large, transparent vacuole
(Dennis Drenner)

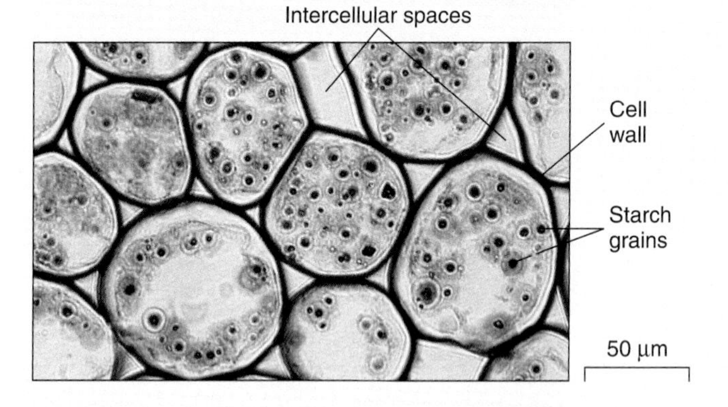

Parenchyma cell

Description
Living, actively metabolizing; thin primary cell walls

Functions
Storage; secretion; photosynthesis

Location and comments
Throughout the plant body; shown is an LM of a cross section of part of a buttercup (*Ranunculus*) root; note the starch grains filling the cells
(Dennis Drenner)

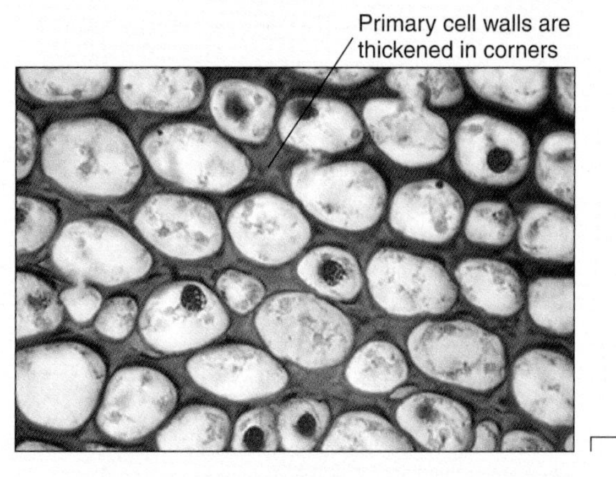

Collenchyma cell

Description
Living; unevenly thickened primary cell walls

Function
Elastic support

Location and comments
Just under stem epidermis; shown is an LM of a cross section of an elderberry (*Sambucus*) stem; note the unevenly thickened cell walls that are especially thick in the corners, making the cell contents assume a spherical shape
(Ed Reschke)

wall—that is, between the primary cell wall and the plasma membrane (see Fig. 4-28).

Plant cell walls have several important roles in plants. Cell walls are involved in growth; the ability of primary cell walls to expand allows cells to increase in size. Scientists are increasingly aware that **signal transduction** takes place in plant cell walls, because many carbohydrate and protein molecules in plant cell walls communicate with other molecules both inside

TABLE 31-2 Continued

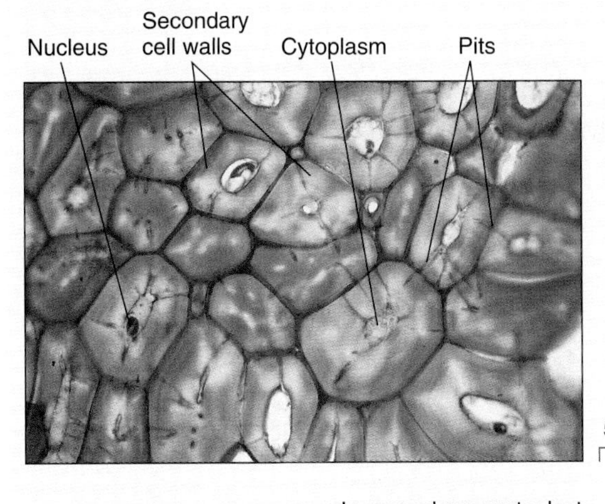

Sclereid (Sclerenchyma cell)

Description
May be living or dead at maturity; thick secondary cell walls; lacks secondary wall at pits

Functions
Strength; sclereid-rich tissue is hard and inflexible

Location and comments
Shells of walnuts and coconuts; pits of cherries and peaches; shown is an LM of living sclereids in a section through a cherry (*Prunus avium*) pit
(James Mauseth, University of Texas)

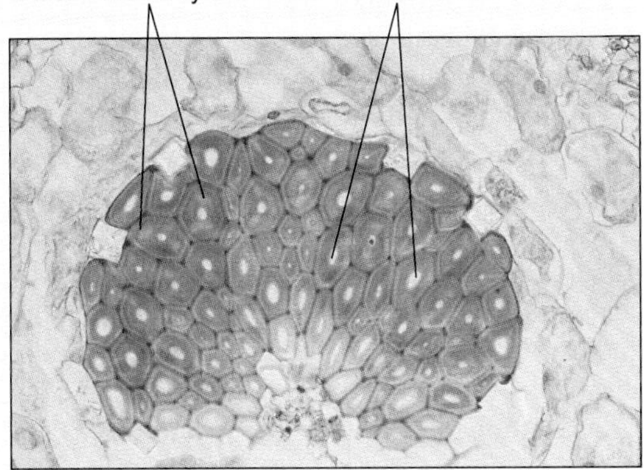

Fiber (Sclerenchyma cell)

Description
Often dead at maturity; thick secondary cell walls; fewer pits than in sclereids

Functions
Support; provides strength

Location and comments
Throughout the plant body; common in stems and certain leaves; shown is an LM of a cross section through a clump of fibers from a century plant (*Agave*) leaf
(James Mauseth, University of Texas)

and outside the cell. The cell wall is also the first line of defense against disease-causing organisms.

Parenchyma cells have thin primary walls

Parenchyma tissue, a simple tissue composed of parenchyma cells, is found throughout the plant body and is the most common type of cell and tissue (Fig. 31-4a). The soft parts of a plant, such as the edible part of an apple or a potato, consist largely of parenchyma.

Parenchyma cells perform several important functions, such as photosynthesis, storage, and secretion. Parenchyma cells that function in photosynthesis contain green chloroplasts, whereas nonphotosynthetic parenchyma cells lack chloroplasts and are often colorless. Materials stored in parenchyma cells include starch grains, oil droplets, water, and salts, which are sometimes visible as crystals. Resins, tannins, hormones, enzymes, and sugary nectar are examples of substances that may be secreted by parenchyma cells. The various functions of parenchyma require that they be living, metabolizing cells.

Parenchyma cells have the ability to *differentiate* into other kinds of cells, particularly when a plant has been injured. If xylem (water-conducting) cells are severed, for example, adjacent parenchyma cells may divide and differentiate into new xylem cells within a few days. (Recall from Chapter 16 that it is possible to induce certain plant cells to become the equivalent of embryonic cells that can then differentiate into specialized cells.)

Collenchyma cells have unevenly thickened primary walls

Collenchyma tissue, a simple plant tissue composed of collenchyma cells, is an extremely flexible structural tissue that provides much of the support in soft, nonwoody plant organs (Fig. 31-4b). Support is crucial for plants, in part because it lets them grow upward to compete with other plants for available sunlight in a plant-crowded area. Plants lack the bony skeletal system typical of many animals; instead, individual cells, including collenchyma cells, support the plant body.

Collenchyma cells, which are usually elongated, are alive at maturity. Their primary cell walls are unevenly thickened and are especially thick in the corners. Collenchyma is not found uniformly throughout the plant and often occurs as long strands near stem surfaces and along leaf veins. The "strings" in a celery stalk, for example, consist of collenchyma tissue.

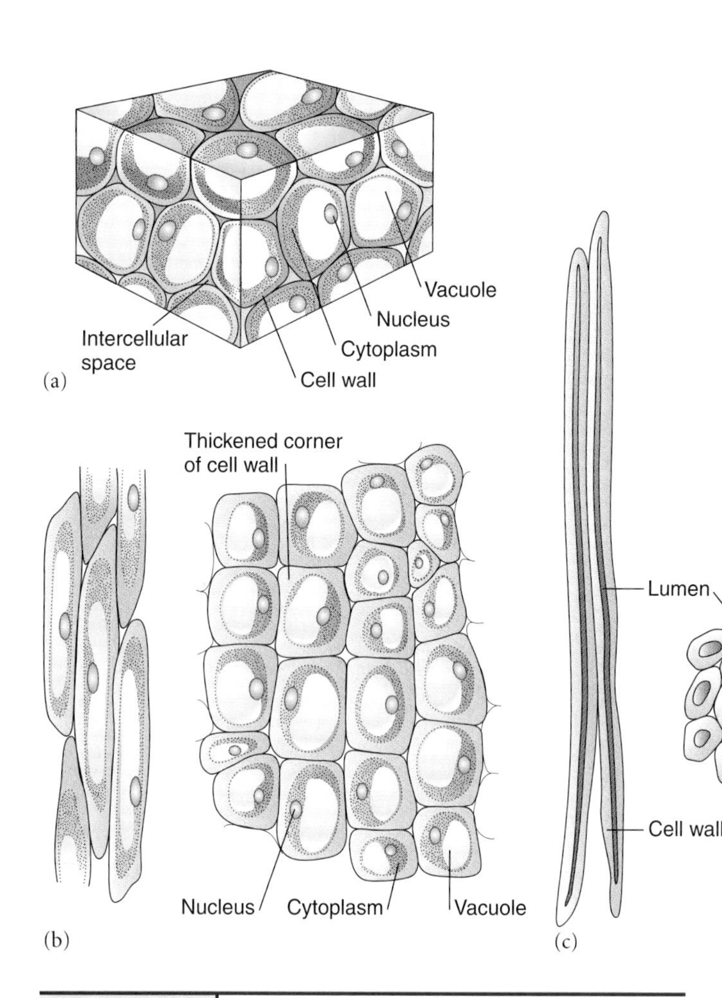

FIGURE **31-4** **Cell types: parenchyma, collenchyma, and sclerenchyma.**

(a) 3-D view of parenchyma cells. Parenchyma cells vary in size and structure, depending on their various functions within the plant body. **(b)** Collenchyma cells in longitudinal section *(left)* and cross section. Note the elongated cells, evident in longitudinal section, and the unevenly thickened cell walls, evident in cross section. **(c)** Sclerenchyma cells (fibers) in longitudinal section *(left)* and cross section. Mature fibers are often dead at functional maturity and therefore lack nuclei and cytoplasm; the lumen is the space formerly occupied by the living cell.

Sclerenchyma cells have both primary walls and thick secondary walls

Another simple plant tissue specialized for structural support is **sclerenchyma** tissue, whose cells have both primary and secondary cell walls. The root of the word *sclerenchyma* is derived from a Greek word *(sclero)* meaning "hard." The secondary cell walls of sclerenchyma cells become strong and hard because of extreme thickening. Thus sclerenchyma cells are unable to stretch or elongate. At functional maturity, when sclerenchyma tissue is providing support for the plant body, its cells are often dead.

Sclerenchyma tissue may be located in several areas of the plant body. Two types of sclerenchyma cells are sclereids and fibers. **Sclereids** are cells of variable shape common in the shells of nuts and in the pits of stone fruits such as cherries and peaches. Pears owe their slightly gritty texture to the presence of clusters of sclereids. **Fibers,** which are long, tapered cells that often occur in patches or clumps, are particularly abundant in the wood, inner bark, and leaf ribs (veins) of flowering plants (Fig. 31-4c).

Cells of the three simple tissues vary in their cell wall chemistry

Parenchyma, collenchyma, and sclerenchyma cells can be distinguished by the chemistry of their cell walls. Cell walls may contain cellulose, hemicelluloses, pectin, and lignin. **Cellulose,** the most abundant polymer in the world, accounts for about 40% to 60% of the dry weight of plant cell walls. As discussed in Chapter 3, cellulose is a **polysaccharide** composed of glucose units joined by β-1,4 bonds. Each cellulose molecule consists of thousands of glucose subunits joined to form a flat, ribbon-like chain. From 40 to 70 of these chains lie parallel to one another and connect by hydrogen bonding to form a **cellulose microfibril,** a strong, tiny strand visible under the electron microscope (see Fig. 3-10a).

Cellulose microfibrils are cemented together by a matrix of hemicelluloses and pectins. **Hemicelluloses** are a group of polysaccharides that are more soluble than cellulose. Hemicelluloses vary in their chemical composition from one species to another. Some hemicelluloses are composed of *xyloglucan,* which consists of a backbone of β-1,4-glucose molecules to which side chains of *xylose,* a five-carbon sugar, are attached. Despite their name, the chemical structure of the hemicelluloses is significantly different from that of cellulose. **Pectin,** another cementing polysaccharide, is less variable in its monomer composition than are the hemicelluloses. Pectin monomer units are *α-galacturonic acid,* a six-carbon molecule that is a derivative of glucose.

An important component of secondary plant cell walls, particularly those of wood, is **lignin.** Comprising as much as 35% of the dry weight of the secondary cell wall, lignin is an extremely complex strengthening polymer composed of monomers derived from certain amino acids. The chemical structure of lignin has not been completely determined, because it is hard to remove from cellulose and other cell wall materials to which it is covalently bound.

Having examined four main components in plant cell walls, we can now generalize about the cell chemistry of parenchyma, collenchyma, and sclerenchyma cells. The thin primary cell walls of parenchyma cells contain predominantly cellulose, although they also contain hemicelluloses and pectin. Both parenchyma and collenchyma cells have primary cell walls, but their walls are chemically distinct because the thickened areas of collenchyma walls contain large quantities of pectin in addition to cellulose and hemicelluloses. The thick secondary walls of sclerenchyma cells are chemically different because they are rich in lignin, in addition to cellulose, hemicelluloses, and pectin.

The vascular tissue system consists of two complex tissues

The vascular tissue system, which is embedded in the ground tissue, transports needed materials throughout the plant via two complex tissues: xylem and phloem (Table 31-3). Both xylem and

TABLE 31-3 Selected Cell Types in the Vascular Tissue System

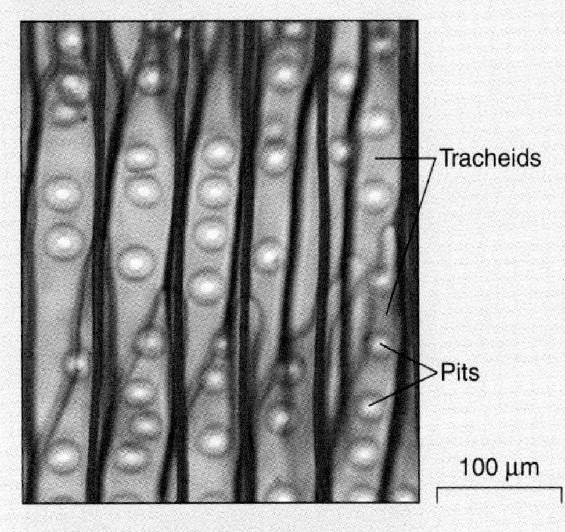

Tracheid

Description
Dead at maturity; lacks secondary wall at pits

Functions
Conduction of water and nutrient minerals, support

Location and comments
Occurs in clumps in xylem throughout plant body; shown is an LM of a longitudinal section of tracheids from white pine (*Pinus strobus*) wood
(John D. Cunningham/Visuals Unlimited)

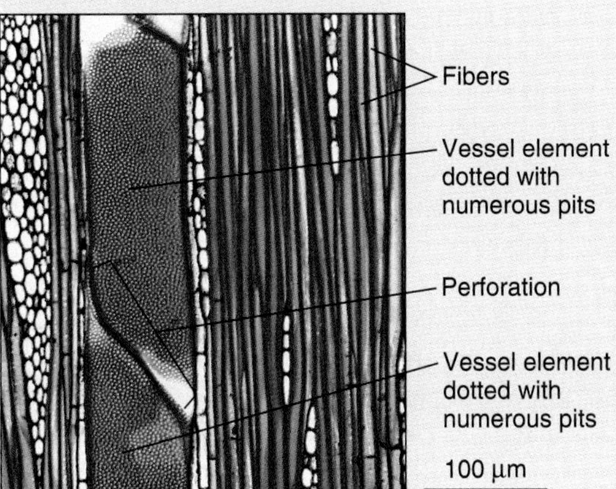

Vessel element

Description
Dead at maturity; end walls have perforations; lacks secondary wall at pits

Functions
Conduction of water and nutrient minerals, support

Location and comments
Xylem throughout plant body; vessel elements are more efficient than tracheids in conduction; shown is an LM of a longitudinal section of two vessel elements from an unknown woody dicot
(James Mauseth, University of Texas)

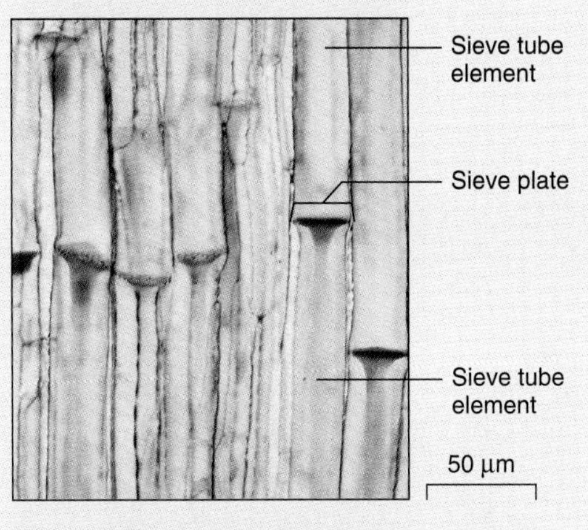

Sieve tube element

Description
Living but lacks nucleus and other organelles at maturity; end walls are sieve plates

Function
Conduction of sugar in solution

Location and comments
Phloem throughout plant body; shown is an LM of a longitudinal section through a clump of sieve tube elements in a squash (*Cucurbita*) petiole (leaf stalk)
(James Mauseth, University of Texas)

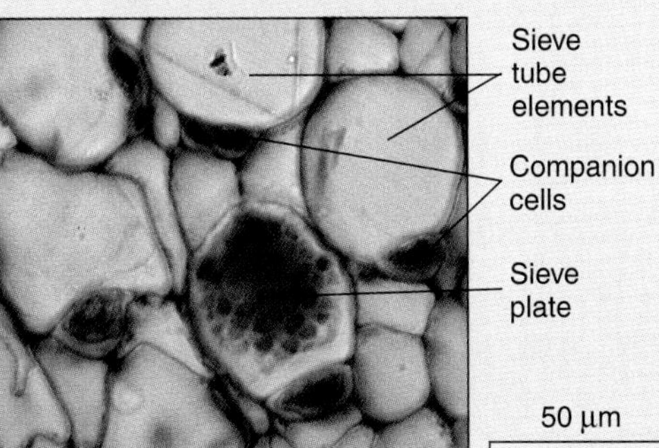

Companion cell

Description
Living; has cytoplasmic connections with sieve tube element

Function
Assists in moving sugars into and out of sieve tube element

Location and comments
Phloem throughout plant body; shown is an LM of phloem from a squash (*Cucurbita*) petiole (leaf stalk) in cross section
(J. Robert Waaland/Biological Photo Service)

phloem are continuous throughout the plant body. (Chapter 33 discusses the mechanisms of transport in xylem and phloem.)

The conducting cells in xylem are tracheids and vessel elements

Xylem conducts water and dissolved nutrient minerals from the roots to the stems and leaves and provides structural support. In flowering plants, xylem is a complex tissue composed of four different cell types: tracheids, vessel elements, parenchyma cells, and fibers. Two of the four cell types found in xylem, the **tracheids** and **vessel elements,** conduct water and dissolved nutrient minerals. In addition to these cells, xylem also contains parenchyma cells, known as *xylem parenchyma,* that perform storage functions, and fibers that provide support.

Tracheids and vessel elements are highly specialized for conduction. When mature, both cell types are dead and therefore hollow; only their cell walls remain. Tracheids, the chief water-conducting cells in gymnosperms (such as pine) and seedless vascular plants (such as ferns), are long, tapering cells located in patches or clumps (Fig. 31-5a). Water is conducted upward, from roots to shoots, passing from one tracheid into another through *pits,* thin areas in the tracheids' cell walls where a secondary wall did not form.

In addition to a relatively few tracheids, flowering plants possess extremely efficient water-conducting cells called *vessel elements* (Fig. 31-5b). The cell diameters of vessel elements are usually greater than those of tracheids. Vessel elements are hollow, but unlike tracheids, the end walls have holes, known as *perforations,* or are entirely dissolved away. Vessel elements are stacked one on top of the other, and water is conducted readily from one vessel element into the next. A stack of vessel elements, called a **vessel,** resembles a miniature water pipe. Vessel elements also have pits in their side walls that permit the lateral transport (sideways movement) of water from one vessel to another.

Sieve tube elements are the conducting cells of phloem

Phloem conducts food materials, that is, carbohydrates formed in photosynthesis, throughout the plant and provides structural support. In flowering plants, phloem is a complex tissue composed of four different cell types: sieve tube elements, companion cells, fibers, and phloem parenchyma cells (Fig. 31-5c, d). Fibers are frequently extensive in the phloem of herbaceous plants, providing additional structural support for the plant body.

Food materials are conducted in *solution,* that is, dissolved in water, through the **sieve tube elements,** which are among the most specialized cells. Sieve tube elements are joined end-on-end to form long **sieve tubes.** The cells' end walls, called *sieve plates,* have a series of holes through which cytoplasm extends from one sieve tube element into the next. Sieve tube elements are living at maturity, but many of their organelles, including the nucleus, vacuole, mitochondria, and ribosomes, disintegrate or shrink as they mature.

Sieve tube elements are among the few eukaryotic cells that can function without nuclei. These cells typically live for less than a year. There are, however, notable exceptions: Certain palms have sieve tube elements that remain alive for approximately 100 years!

Adjacent to each sieve tube element is a **companion cell** that assists in the functioning of the sieve tube element. The companion cell is a living cell, complete with a nucleus. This nucleus is thought to direct the activities of both the companion cell and sieve tube element. Numerous **plasmodesmata**—cytoplasmic connections through which cytoplasm extends from one cell to another (see Chapter 5)—occur between a companion cell and its adjoining sieve tube element. The companion cell plays an essential role in moving sugar into the sieve tube elements for transport to other parts of the plant.

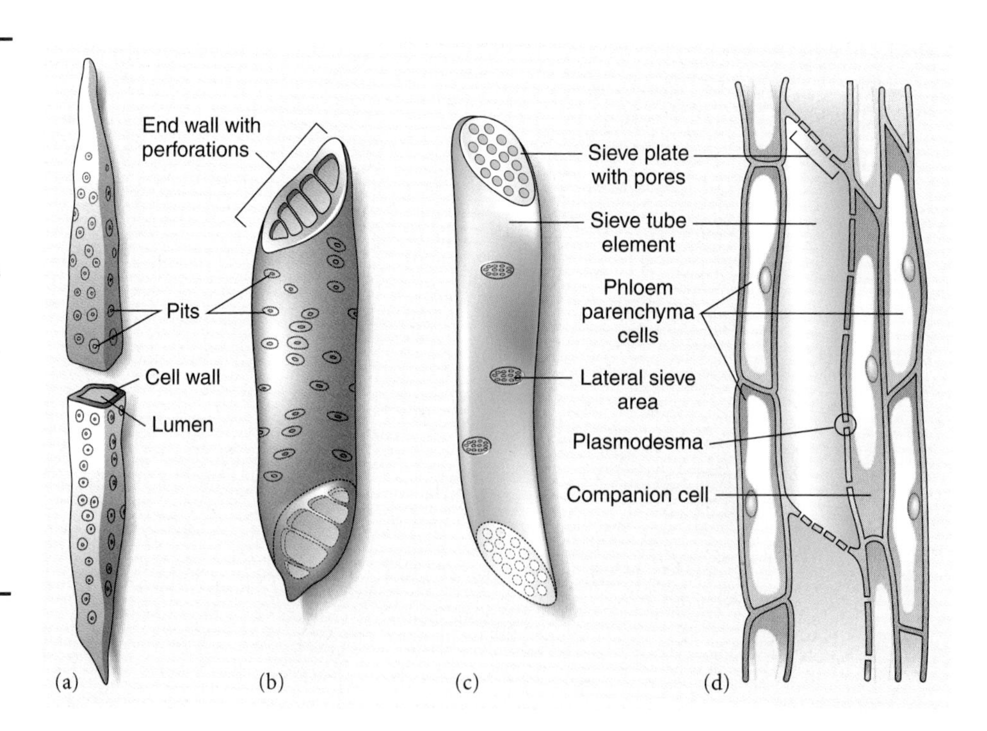

FIGURE **31-5**

Longitudinal views of cell types in xylem and phloem tissues.

(a) 3-D view of a tracheid (xylem cell). The tracheid is opened to reveal the cell's appearance in cross section. At maturity, tracheids are usually dead. **(b)** 3-D view of a vessel element (xylem cell). The end walls of vessel elements have a variety of openings, or perforations, depending on the species. Vessel elements are joined end-to-end to produce minute "water pipes" that run the length of the plant, from roots to leaves and other shoot parts. **(c)** 3-D view of a sieve tube element, showing the sieve plate. **(d)** Longitudinal section of phloem tissue, showing sieve tube elements, companion cells, and phloem parenchyma cells. Fiber cells are not shown.

The dermal tissue system consists of two complex tissues

The dermal tissue system, the epidermis and periderm, provides a protective covering over plant parts (Table 31-4). In herbaceous plants, the dermal tissue system is a single layer of cells called the epidermis. Woody plants initially produce an epidermis, but it splits apart as the plant increases in girth owing to the production of additional woody tissues inside the epidermis. Periderm, a tissue several to many cell layers thick, forms under the epidermis to provide a new protective covering as the epidermis is destroyed. Periderm, which replaces the epidermis in the stems and roots of older woody plants, composes the outer bark.

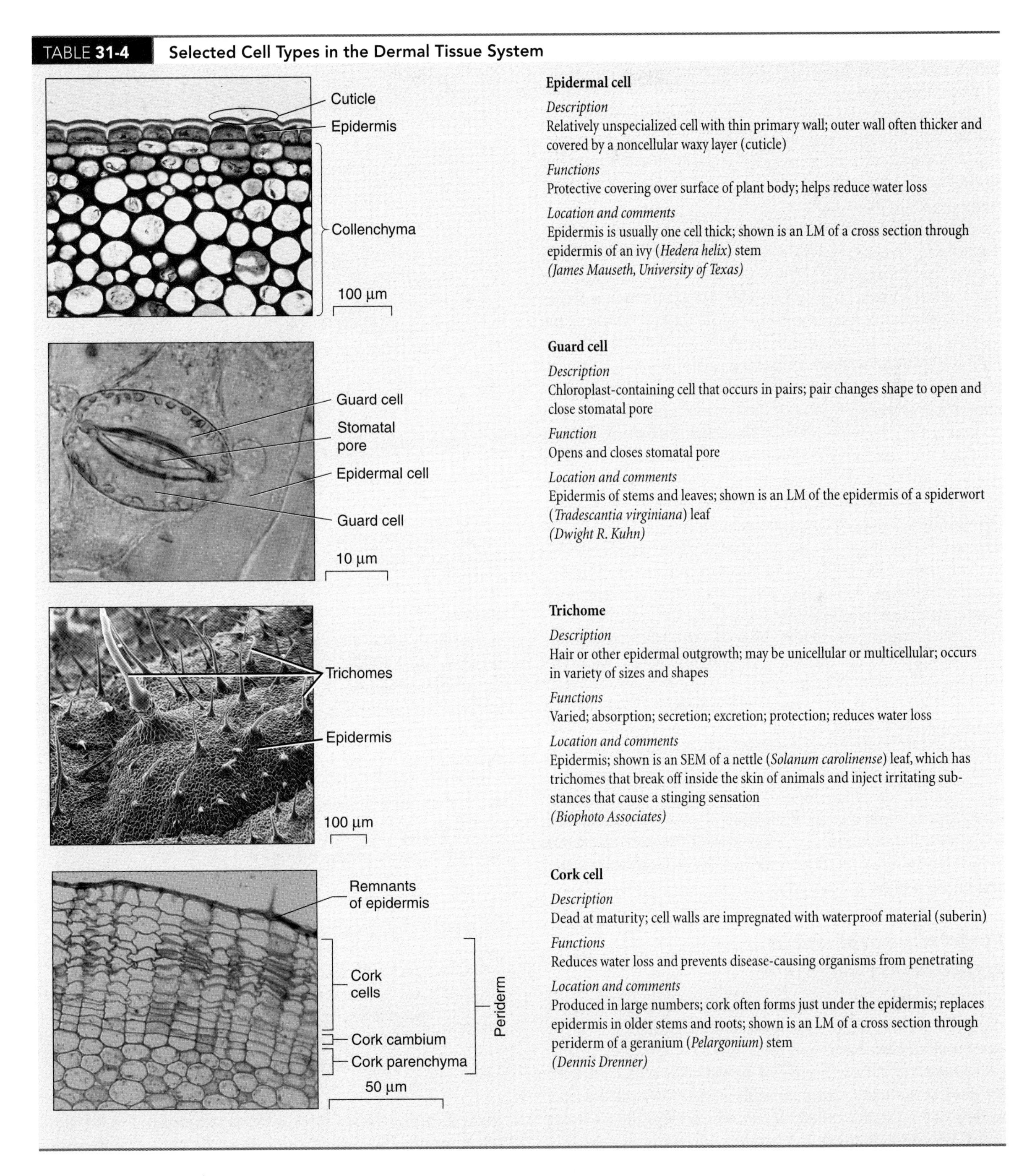

TABLE 31-4 Selected Cell Types in the Dermal Tissue System

Epidermal cell

Description
Relatively unspecialized cell with thin primary wall; outer wall often thicker and covered by a noncellular waxy layer (cuticle)

Functions
Protective covering over surface of plant body; helps reduce water loss

Location and comments
Epidermis is usually one cell thick; shown is an LM of a cross section through epidermis of an ivy (*Hedera helix*) stem
(James Mauseth, University of Texas)

Guard cell

Description
Chloroplast-containing cell that occurs in pairs; pair changes shape to open and close stomatal pore

Function
Opens and closes stomatal pore

Location and comments
Epidermis of stems and leaves; shown is an LM of the epidermis of a spiderwort (*Tradescantia virginiana*) leaf
(Dwight R. Kuhn)

Trichome

Description
Hair or other epidermal outgrowth; may be unicellular or multicellular; occurs in variety of sizes and shapes

Functions
Varied; absorption; secretion; excretion; protection; reduces water loss

Location and comments
Epidermis; shown is an SEM of a nettle (*Solanum carolinense*) leaf, which has trichomes that break off inside the skin of animals and inject irritating substances that cause a stinging sensation
(Biophoto Associates)

Cork cell

Description
Dead at maturity; cell walls are impregnated with waterproof material (suberin)

Functions
Reduces water loss and prevents disease-causing organisms from penetrating

Location and comments
Produced in large numbers; cork often forms just under the epidermis; replaces epidermis in older stems and roots; shown is an LM of a cross section through periderm of a geranium (*Pelargonium*) stem
(Dennis Drenner)

Epidermis is the outermost layer of a herbaceous plant

The **epidermis** is a complex tissue composed primarily of relatively unspecialized living cells. Dispersed among these cells are more specialized guard cells and outgrowths called trichomes (discussed shortly). In most plants, the epidermis consists of a single layer of flattened cells (Fig. 31-6). Epidermal cells generally contain no chloroplasts and are therefore transparent, allowing light to penetrate into the interior tissues of stems and leaves. In both stems and leaves, photosynthetic tissues lie *beneath* the epidermis.

An important requirement of the aerial parts (stems, leaves, flowers, and fruits) of a plant is the ability to control water loss. Epidermal cells of aerial parts secrete a waxy **cuticle** over the surface of their exterior walls; this waxy layer greatly restricts water loss from plant surfaces.

Although the cuticle is extremely efficient at preventing most water loss through epidermal cells, it also slows the diffusion of carbon dioxide, needed for photosynthesis, from the atmosphere into the leaf or stem. The diffusion of carbon dioxide is facilitated by **stomata** (singular, *stoma*). Stomata are minute pores in the epidermis, surrounded by two cells called **guard cells** (see Figure 32-5). Many gases, including carbon dioxide, oxygen, and water vapor, pass through the stomata by diffusion. Stomata are generally open during the day when photosynthesis is occurring, and the water loss that also takes place when stomata are open provides some evaporative cooling. During the night, stomata usually close. During drought conditions, the need to conserve water overrides the need to cool the leaves and exchange gases. Thus, during a drought, the stomata close in the daytime. Stomata are discussed in greater detail in Chapter 32.

The epidermis also contains special outgrowths, or hairs, called **trichomes,** which occur in many sizes and shapes and have a variety of functions (see Figure 32-4). Plants that tolerate salty environments such as the seashore often have specialized trichomes on their leaves to remove excess salt that accumulates in the plant. The presence of trichomes on the aerial parts of desert plants may increase the reflection of light off the plants, thereby keeping the internal tissues cooler and decreasing water loss. Other trichomes have a protective function. For example, the trichomes on stinging nettle leaves and stems contain irritating substances that may discourage herbivorous animals from eating the plant. **Root hairs** are simple, unbranched trichomes that increase the surface area of the root epidermis (which comes into contact with the soil) for more effective water and nutrient mineral absorption.

Epidermis is replaced by periderm in woody plants

As a woody plant begins to increase in girth, its epidermis is sloughed off and replaced by **periderm.** Periderm forms the outer bark of older stems and roots. It is a complex tissue composed mainly of cork cells and cork parenchyma cells. **Cork cells** are dead at maturity, and their walls are heavily coated with a waterproof substance called *suberin,* which helps reduce water loss. **Cork parenchyma** cells function primarily in storage.

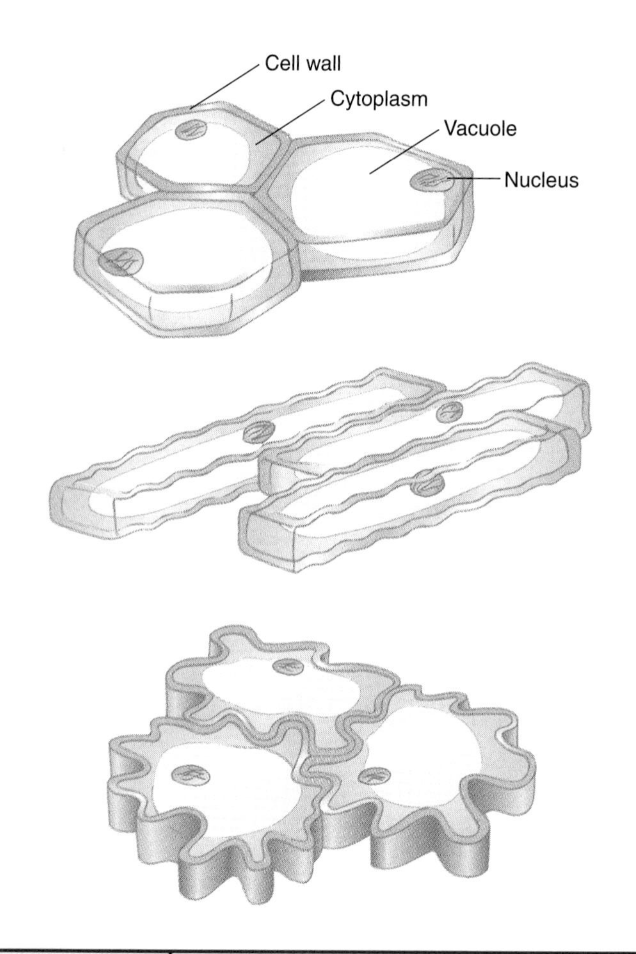

FIGURE **31-6** | **Cell types: epidermal cells.**
3-D views of epidermal cells from three different plants. Note that, regardless of the cell shape, epidermal cells fit tightly together.

Review

- What are some of the functions of roots? Of shoots?
- What are the three tissue systems in plants? Describe the functions of each.
- How do parenchyma, collenchyma, and sclerenchyma tissues differ in cellular structure and function?
- What are the functions of xylem and phloem? Describe the conducting cells that occur in each of these complex tissues.
- How are epidermis and periderm alike? How are they different?

BiologyNow™ Assess your understanding of **the plant body** by taking the pretest on your BiologyNow CD-ROM.

PLANT MERISTEMS

Learning Objectives

8 Discuss what is meant by growth in plants, and state how it differs from growth in animals.

9 Distinguish between primary and secondary growth.

10 Distinguish between apical meristems and lateral meristems.

Plant growth involves three different processes: cell division, cell elongation (the lengthening of a cell), and cell differentiation. Cell division is an essential part of plant growth that results in an increase in the number of cells. These new cells elongate as the cytoplasm grows and the vacuole fills with water, which exerts pressure on the cell wall and causes it to expand. In an onion root cell, the vacuole increases in size by some 30 to 150 times during elongation. Plant cells also differentiate, or specialize, into the various cell types just discussed. These cell types constitute the mature plant body and perform the various functions required in a multicellular organism. Although differentiation does not contribute to an increase in size, it is an important aspect of growth and development because it is essential for tissue formation.

One difference between plants and animals is the *location* of growth. When a young animal is growing, all parts of its body grow, although not necessarily at the same rate. However, when plants grow, their cells divide only in specific areas, called **meristems,** which are composed of cells whose primary function is the formation of new cells by mitotic division. Meristematic cells do not differentiate. Instead, they retain the ability to divide by mitosis, a trait that many differentiated cells lose. The persistence of mitotically active meristems means that plants, unlike most animals, can grow throughout their entire life span.

Two kinds of meristematic growth may occur in plants. **Primary growth** is an increase in stem and root length. All plants exhibit primary growth, which produces the entire plant body in herbaceous plants and the young, soft shoots and roots in woody trees and shrubs. **Secondary growth** is an increase in the girth of a plant. For the most part, only gymnosperms and woody dicots have secondary growth.[1] Tissues produced by secondary growth comprise the wood and bark, which make up most of the bulk of trees and shrubs. A few annuals—geranium and sunflower, for example—have limited secondary growth despite the fact that they lack obvious wood and bark tissues.

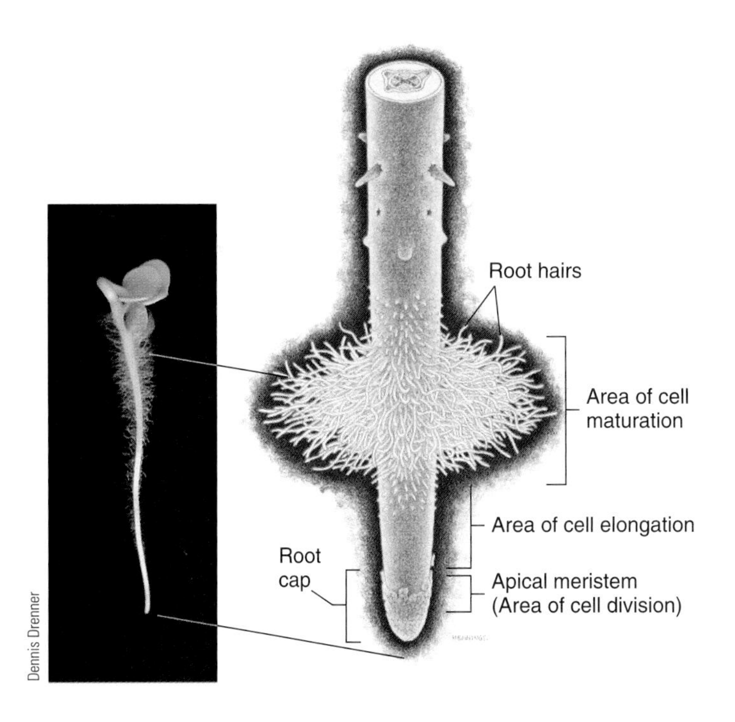

FIGURE **31-7** | **A root tip.**
The root apical meristem (where cells divide and thus increase in number) is protected by a root cap. Farther from the tip is an area of cell elongation, where cells enlarge and begin to differentiate. The area of cell maturation has fully mature, differentiated cells. Note the young root hairs in this area and on the root of a young radish (*Raphanus sativus*) seedling, which is approximately 5 cm (2 in) long *(left).*

Primary growth takes place at apical meristems

Primary growth occurs as a result of the activity of **apical meristems,** areas located at the tips of roots and shoots, including within the buds of stems. (**Buds** are dormant embryonic shoots that eventually develop into branches.)

Primary growth is evident when a root tip (Fig. 31-7) is examined. A protective layer of cells called a **root cap** covers the root tip. Directly behind the root cap is the root apical meristem, which consists of meristematic cells. Meristematic cells, which are quite small and cube-shaped, remain small because they are continually dividing. (In meristems, as daughter cells begin to enlarge, one or both will divide again.)

Farther from the root tip, just behind the area of cell division, is an area of cell elongation where the cells have been displaced from the meristem. Here the cells are no longer dividing but instead are growing longer, pushing the root tip ahead of them, deeper into the soil. Some differentiation also begins to occur in the area of cell elongation, and immature tissues such as differentiating xylem and phloem, become evident. The tissue systems continue to develop and differentiate into primary tissues (epidermis, xylem, phloem, and ground tissues) of the adult plant. Farther from the tip, behind the area of cell elongation, most cells have completely differentiated and are fully mature. Root hairs are evident in this area.

A shoot apex (such as the terminal bud) is quite different in appearance from a root tip (Fig. 31-8). A dome of minute, regularly arranged meristematic cells—the shoot apical meristem—is located in each shoot apex. **Leaf primordia** (developing leaves) and **bud primordia** (developing buds) arise from the shoot apical meristem. The tiny leaf primordia cover and protect the shoot apical meristem. As the cells formed by the shoot apical meristem elongate, the shoot apical meristem is pushed upward. Subsequent cell divisions produce additional stem tissue and cause new leaf and bud primordia to appear. Farther from the stem tip, the immature cells differentiate into the three tissue systems of the mature plant body.

Secondary growth takes place at lateral meristems

In addition to primary growth, trees and shrubs have secondary growth. These plants increase in length by primary growth and

[1] Recall from Chapter 27 that, on the basis of structural features, flowering plants are divided into two groups, informally called dicots and monocots. Oak, sycamore, ash, cherry, apple, and maple are examples of woody dicots, whereas bean, daisy, and snapdragon are herbaceous dicots. Palm, corn, bluegrass, lily, and tulip are monocots.

ACTIVE FIGURE **31-8**

A shoot apex.

LM of a longitudinal section through a shoot apex of *Coleus,* showing the shoot apical meristem, leaf primordia, and bud primordia.

BiologyNow™ See **shoot development in action** by clicking on this figure on your BiologyNow CD-ROM.

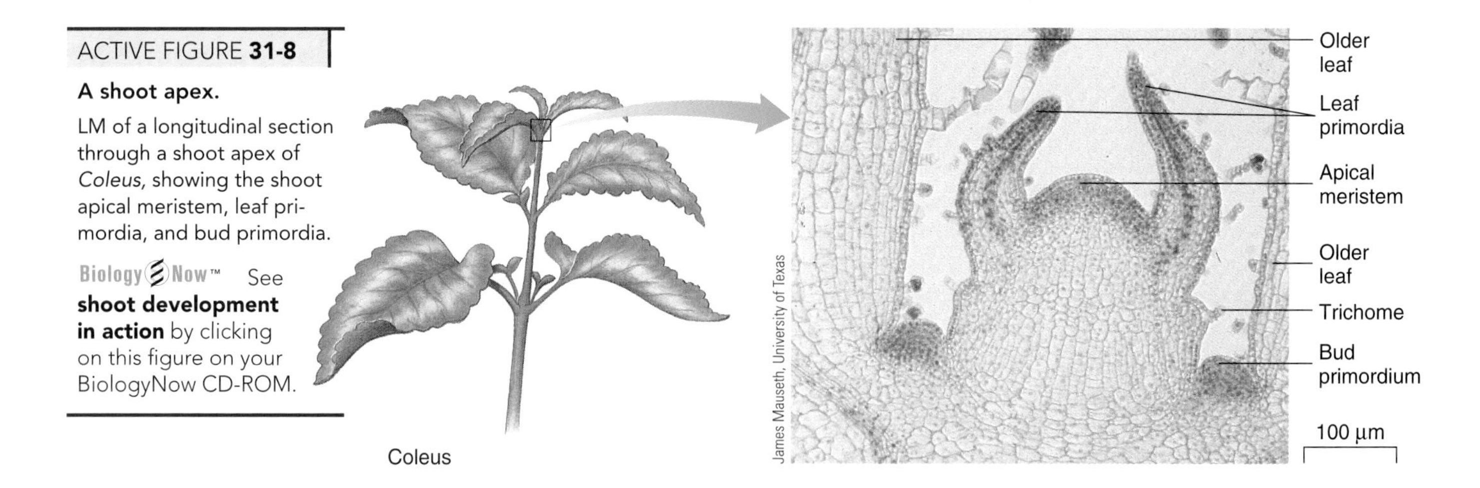

increase in girth by secondary growth. This increase in girth, which occurs in areas that are no longer elongating, is due to cell divisions that take place in **lateral meristems,** areas extending along the entire length of the stems and roots, except at the tips. Two lateral meristems, the vascular cambium and the cork cambium, are responsible for secondary growth, which is the formation of secondary tissues: secondary xylem, secondary phloem, and periderm (Fig. 31-9).

The **vascular cambium** is a layer of meristematic cells that forms a long, thin, continuous cylinder within the stem and root. It is located between the wood and bark of a woody plant. Cells of the vascular cambium divide, adding more cells to the wood (secondary xylem) and inner bark (secondary phloem).

The **cork cambium** is composed of a thin cylinder or irregular arrangement of meristematic cells and is located in the outer bark. Cells of the cork cambium divide and form **cork cells** toward the outside and one or more underlying layers of **cork parenchyma** cells that function in storage. Collectively, cork cells, cork cambium, and cork parenchyma make up the periderm (discussed previously).

We are now ready to give a more precise definition of bark. **Bark,** the outermost covering over woody stems and roots, consists of all plant tissues located outside the vascular cambium. Bark has two regions, a living inner bark composed of secondary phloem and a mostly dead outer bark composed of periderm. A more comprehensive discussion of secondary growth is given in Chapters 33 and 34.

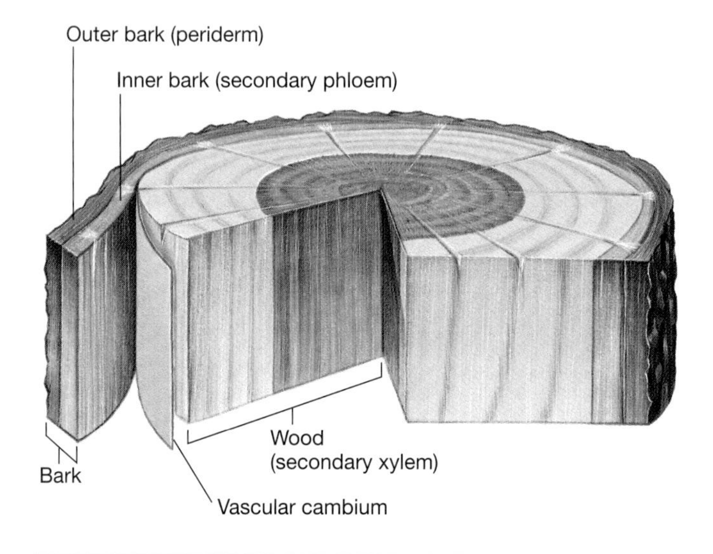

ACTIVE FIGURE **31-9** | **Secondary growth.**

The vascular cambium, a thin layer of cells sandwiched between the wood and bark, produces the secondary vascular tissues: the wood, which is secondary xylem, and the inner bark, which is secondary phloem. The cork cambium produces the periderm, the outer bark tissue that replaces the epidermis in a plant with secondary growth.

BiologyNow™ Experiment with **tree-ring development** by clicking on this figure on your BiologyNow CD-ROM.

Review

- What is the role of plant meristems? Does animal growth involve meristems?
- What is primary growth? Secondary growth?
- What are apical meristems? Lateral meristems?

BiologyNow™ Assess your understanding of **plant meristems** by taking the pretest on your BiologyNow CD-ROM.

SUMMARY WITH KEY TERMS

1 Distinguish between herbaceous and woody plants.

- Plants are either herbaceous (nonwoody) or woody. In temperate climates, the aerial parts of herbaceous plants die back, whereas the aerial parts of woody plants persist.

2 Discuss the differences among annuals, biennials, and perennials, and give an example of each.

- **Annuals** such as corn, geranium, and marigold are herbaceous plants that grow, reproduce, and die in one year or less.

- **Biennials** such as carrot and Queen Anne's lace take two years to complete their life cycles before dying.
- **Perennials** such as asparagus and oak trees are herbaceous and woody plants that have the potential to live for more than two years.

3 Contrast two different life history strategies in plants.

- Plants have a variety of life history strategies to ensure their successful reproduction and survival.
- Long-lived trees thrive in a tropical rain forest where competition prevents most small, short-lived plants from becoming established. Small, short-lived plants thrive in a relatively unfavorable environment such as a desert following a rainy spell.

4 Discuss the functions of various parts of the vascular plant body, including the nutrient- and water-absorbing root system and the photosynthesizing shoot system.

- The vascular plant body typically consists of a root system and a shoot system. The **root system** is generally underground and obtains water and dissolved nutrient minerals for the plant. Roots also anchor the plant firmly in place.
- The **shoot system** is generally aerial and obtains sunlight and exchanges gases such as carbon dioxide, oxygen, and water vapor. The shoot system consists of a vertical stem that bears leaves (the main **organs** of photosynthesis) and reproductive structures (flowers and fruits in flowering plants). **Buds,** undeveloped embryonic shoots, develop on stems.
- Although separate organs (roots, stems, leaves, flower parts, and fruits) exist in the plant, tissue systems are integrated throughout the plant body, providing continuity from organ to organ. The plant body is composed of three tissue systems: ground, vascular, and dermal.

5 Describe the ground tissue system (parenchyma tissue, collenchyma tissue, and sclerenchyma tissue).

- The **ground tissue system** consists of three tissues with a variety of functions. **Parenchyma** tissue is composed of living parenchyma cells that have thin **primary cell walls.** Functions of parenchyma tissue include photosynthesis, storage, and secretion.
- **Collenchyma** tissue consists of collenchyma cells with unevenly thickened primary cell walls. This tissue provides flexible structural support.
- **Sclerenchyma** tissue is composed of sclerenchyma cells—**sclereids** or **fibers**—that have both primary cell walls and **secondary cell walls.** Sclerenchyma cells are often dead at maturity, but provide structural support.

6 Describe the structure and function of the vascular tissue system (xylem and phloem).

- The **vascular tissue system** conducts materials throughout the plant body and provides strength and support.
- **Xylem** is a complex tissue that conducts water and dissolved nutrient minerals. The actual conducting cells of xylem are **tracheids** and **vessel elements.**
- **Phloem** is a complex tissue that conducts sugar in solution. **Sieve tube elements** are the conducting cells of phloem; they are assisted by **companion cells.**

7 Describe the dermal tissue system (epidermis and periderm).

- The **dermal tissue system** is the outer protective covering of the plant body.
- The **epidermis** is a complex tissue that covers the herbaceous plant body. The epidermis that covers aerial parts secretes a waxy **cuticle** that reduces water loss. **Stomata** permit gas exchange between the interior of the shoot system and the surrounding atmosphere. **Trichomes** are outgrowths, or hairs, that occur in many sizes and shapes and have a variety of functions.
- The **periderm** is a complex tissue that covers the woody parts of the plant body in woody plants.

8 Discuss what is meant by growth in plants, and state how it differs from growth in animals.

- Growth in plants, unlike animals, is localized in specific regions, called **meristems,** and involves three processes: cell division, cell elongation, and cell differentiation.

9 Distinguish between primary and secondary growth.

- **Primary growth** is an increase in stem or root length. Primary growth occurs in all plants.
- **Secondary growth** is an increase in stem or root girth (thickness). Secondary growth is localized, typically occurring in long cylinders of meristematic cells throughout the length of older stems and roots.

10 Distinguish between apical meristems and lateral meristems.

- Primary growth results from the activity of **apical meristems** that are localized at the tips of roots and shoots and within the buds of stems.
- The two **lateral meristems** responsible for secondary growth are the **vascular cambium** and the **cork cambium.**

POST-TEST

1. Plants that complete their life cycles in one year are called ______________; those that complete them in two years are ______________; and those that live year after year are ______________. (a) annuals; perennials; biennials (b) biennials; annuals; perennials (c) annuals; biennials; perennials (d) perennials; annuals; biennials (e) perennials; biennials; annuals
2. Which of the following plant life history strategies would be successful in a relatively favorable environment such as a tropical rain forest? (a) long life span with flowers and seeds produced each year (b) long life span with flowers and seeds produced only when the plant is very young (c) short life span with flowers and seeds produced each year (d) short life span with flowers and seeds produced only when the plant is very young (e) very short life span with flowering at end of life
3. Most of the plant body consists of the ______________ tissue system. (a) ground (b) vascular (c) periderm (d) dermal (e) cortex
4. The cell walls of parenchyma cells: (a) contain large quantities of pectin in the thickened corners (b) are rich in lignin but do not contain hemicelluloses and pectin (c) are predominantly cellulose, although they also contain hemicelluloses and pectin (d) contain cellulose, hemicelluloses, and lignin in approximately equal amounts (e) contain hemicelluloses, pectin, and lignin but no cellulose

POST-TEST *(continued)*

5. Which tissue system provides a covering for the plant body? (a) ground (b) vascular (c) periderm (d) dermal (e) cortex
6. Storage, secretion, and photosynthesis are the functions of (a) collenchyma (b) vessel elements (c) lateral meristems (d) sclerenchyma (e) parenchyma
7. The two simple tissues that are specialized for support are (a) parenchyma and collenchyma (b) collenchyma and sclerenchyma (c) sclerenchyma and parenchyma (d) parenchyma and xylem (e) xylem and phloem
8. Sclereids and fibers are examples of which plant tissue? (a) parenchyma (b) collenchyma (c) sclerenchyma (d) xylem (e) epidermis
9. Which of the following statements about the vascular tissue system is *not* true? (a) Xylem and phloem are continuous throughout the plant body. (b) Xylem not only conducts water and dissolved nutrient minerals, but also provides support. (c) Four different cell types occur in phloem: sieve tube elements, companion cells, tracheids, and vessel elements. (d) Sieve tube elements lack nuclei. (e) Vessel elements are hollow, and their end walls have perforations or are entirely dissolved away.
10. Conduction of water and nutrient minerals in xylem occurs in vessel elements and (a) sieve tube elements (b) tracheids (c) collenchyma (d) cork cells (e) phloem
11. Conduction of sugar in solution in the sieve tube elements is aided by (a) cork cells (b) sclerenchyma (c) parenchyma (d) guard cells (e) companion cells
12. The outer tissue that covers plants with primary growth is ______________, whereas plants with secondary growth are covered by ______________. (a) cuticle; cork parenchyma (b) periderm; phloem (c) epidermis; periderm (d) epidermis; collenchyma (e) cellulose; lignin
13. The noncellular waxy layer secreted by the epidermis over its aerial surface is called (a) lignin (b) cuticle (c) periderm (d) cellulose (e) trichome
14. Minute pores known as ______________ dot the surface of the epidermis of leaves and stems; each pore is bordered by two ______________. (a) stomata; guard cells (b) stomata; fibers (c) sieve tube elements; companion cells (d) sclereids; guard cells (e) cuticle; guard cells
15. Localized areas within the plant body where cell divisions occur are known as (a) organs (b) fibers (c) meristems (d) cork parenchyma (e) stomata
16. Primary growth, an increase in the length of a plant, occurs at the (a) cork cambium (b) apical meristem (c) vascular cambium (d) lateral meristem (e) periderm
17. The two lateral meristems responsible for secondary growth are the (a) cork cambium and apical meristem (b) apical meristem and cork parenchyma (c) vascular cambium and apical meristem (d) vascular cambium and cork cambium (e) cork cambium and cork parenchyma

CRITICAL THINKING

1. Grasses have a special meristem situated at the base of the leaves. Relate this information to what you know about the growth of grass after you mow the lawn.
2. A couple carved a heart with their initials into a tree trunk 4 feet above ground level; the tree was 25 feet tall at the time. Twenty years later the tree was 50 feet tall. How far above the ground were the initials? Explain your answer.
3. Sclerenchyma in plants is the functional equivalent of bone in humans (both sclerenchyma and bone provide support). However, sclerenchyma is dead, whereas bone is living tissue. What are some of the advantages of a plant having dead support cells? Can you think of any disadvantages?

- Visit our Web site at **http://biology.brookscole.com/solomon7** for links to chapter-related resources on the World Wide Web. Additional online materials relating to this chapter can also be found on our Web site.

BIOLOGY NOW RESOURCES

Active Figures

31-2: Plant body
31-8: Shoot development
31-9: Tree-ring development

Preparing for an exam? Take a diagnostic test on your BiologyNow CD-ROM.

Post-Test Answers

1. c	2. a	3. a	4. c
5. d	6. e	7. b	8. c
9. c	10. b	11. e	12. c
13. b	14. a	15. c	16. b
17. d			

Leaf Structure and Function

Leaves. Note how the red maple (*Acer rubrum*) leaves are arranged to efficiently capture light.

CHAPTER OUTLINE

Suppose for a moment that you are taking a course in engineering and have been asked to design an efficient solar collector capable of converting the radiant energy it collects into chemical energy of organic compounds. Where would you start? It might be helpful to check library books and periodicals to see how solar collectors/energy converters have been designed in the past. In this instance, it would also be wise to ask a biologist, or even a biology student like yourself, whether anything comparable exists in nature. The answer, of course, is yes. Plants have organs that are effective solar collectors and energy converters; they are called *leaves.*

Plants allocate many resources to the production of leaves. According to John E. Dale, a University of Edinburgh botanist who specializes in leaf development, a large maple tree annually produces 46.5 m^2 (500 ft^2) of leaves, which weigh more than 113 kg (250 lb). The metabolic cost of producing so many leaves is high, but leaves are essential to the tree's survival.

Leaves gather the sunlight necessary for **photosynthesis,** the biological process that converts radiant energy into the chemical energy of carbohydrate molecules (see Chapter 8). Plants then use these molecules as starting materials to synthesize all other organic compounds (such as starch and cellulose, amino acids, lipids, and nucleic acids) and as fuel to provide energy for their metabolic processes. During a single summer, the leaves of the red maple shown here will "fix" about 454 kg (1000 lb) of carbon dioxide into sugars and other organic compounds.

Leaves are the main photosynthetic organs of most plants, and the structure of a leaf is superbly adapted for its primary function of photosynthesis (see Chapter 26 for a discussion of how leaves evolved). Most leaves are thin and flat, a shape that allows maximum absorption of light energy and the efficient internal diffusion of gases such as CO_2 and O_2. As a result of their ordered arrangement on the stem, leaves efficiently catch the sun's rays. The leaves of plants form an intricate green mosaic, bathed in sunlight and atmospheric gases.

Leaves possess several features that help reduce or control water loss, the most important of which is a thin, transparent layer of wax that covers the leaf surface. Structural adaptations are compromises between competing needs, and some features that optimize photosynthesis actually *promote* water loss. Plants have minute pores that allow an exchange of gases for photosynthesis, but these tiny openings also allow water to escape into the atmosphere as water vapor. Thus, leaf structure represents a tradeoff between photosynthesis and water conservation. ■

LEAF FORM AND STRUCTURE

Learning Objectives

1. Discuss variation in leaf form, including simple versus compound leaves, leaf arrangement on the stem, and venation patterns.
2. Describe the major tissues of the leaf (epidermis, photosynthetic ground tissue, xylem, and phloem), and label them on a diagram of a leaf cross section.
3. Compare leaf anatomy in dicots and monocots.
4. Relate leaf structure to its function of photosynthesis.

Foliage leaves are the most variable of plant organs, so much so that plant biologists developed specific terminology to describe their shapes, margins (edges), vein patterns, and the way they attach to stems. Because each leaf is characteristic of the species on which it grows, many plants can be identified by their leaves alone. Leaves may be round, needle-like, scalelike, cylindrical, heart-shaped, fan-shaped, or thin and narrow. They vary in size from those of the raffia palm (*Raphia ruffia*), whose leaves often grow more than 20 m (65 ft) long, to those of water-meal (*Wolffia*), whose leaves are so small that 16 of them laid end-to-end measure 2.5 cm (1 in) (see Fig. 31-1a).

The broad, flat portion of a leaf is the **blade;** the stalk that attaches the blade to the stem is the **petiole.** Some leaves also have **stipules,** which are leaflike outgrowths usually present in pairs at the base of the petiole (Fig. 32-1). Some leaves do not have petioles or stipules.

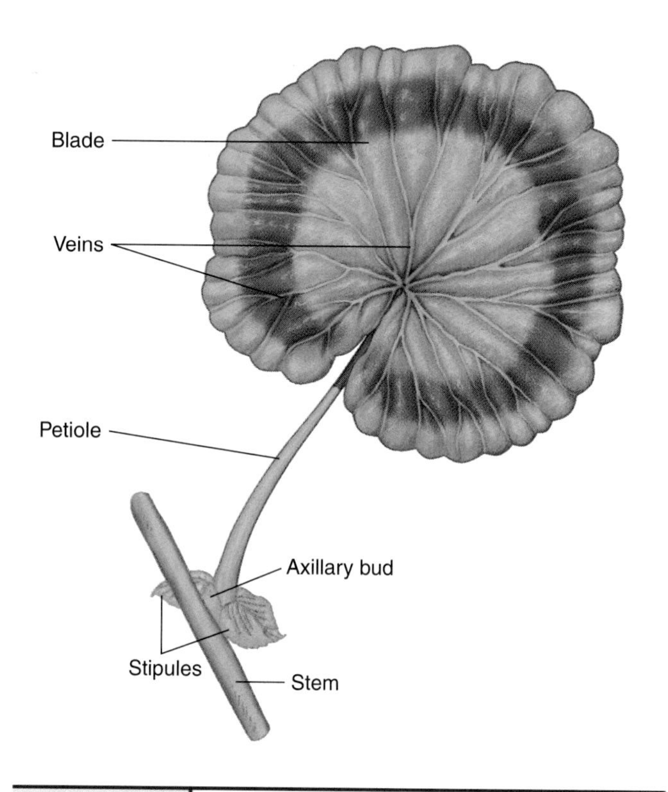

FIGURE **32-1** **Parts of a leaf.**
A geranium leaf consists of a blade, a petiole, and two stipules at the base of the leaf. Note the axillary bud in the leaf axil.

Leaves may be *simple* (having a single blade) or *compound* (having a blade divided into two or more leaflets) (Fig. 32-2a). Sometimes it is difficult to tell whether a plant has formed one compound leaf or a small stem bearing several simple leaves. One easy way to determine if a plant has simple or compound leaves is to look for axillary buds, so called because each develops in a leaf *axil* (the angle between the stem and petiole). Axillary buds form at the base of a leaf, whether it is simple or compound. However, axillary buds never develop at the base of leaflets. Also, the leaflets of a compound leaf lie in a single plane (you can lay a compound leaf flat on a table), whereas simple leaves usually are not arranged in one plane on a stem.

Leaves are arranged on a stem in one of three possible ways (Fig. 32-2b). Plants such as beeches and walnuts have an *alternate leaf arrangement,* with one leaf at each **node,** the area of the stem where one or more leaves are attached. In an *opposite leaf arrangement,* as occurs in maples and ashes, two leaves grow at each node. In a *whorled leaf arrangement,* as in catalpa trees, three or more leaves grow at each node.

Leaf blades may possess *parallel venation,* in which the primary **veins**—strands of vascular tissue—run approximately parallel to one another (generally characteristic of monocots), or *netted venation,* in which veins are branched in such a way that they resemble a net (generally characteristic of dicots; Fig. 32-2c).[1] Netted veins can be *pinnately netted,* in which major veins branch off along the entire length of the **midvein** (main or central vein of a leaf), or *palmately netted,* in which several major veins radiate out from one point.

Leaf structure consists of an epidermis, photosynthetic ground tissue, and vascular tissue

The leaf is a complex organ composed of several tissues organized to optimize photosynthesis (Fig. 32-3). The leaf blade has upper and lower surfaces that are covered by an epidermal layer. The **upper epidermis** covers the upper surface, and the **lower epidermis** covers the lower surface. Most cells in these layers lack chloroplasts and are relatively transparent. One interesting

[1] Recall that flowering plants, the focus of this chapter, are divided into two groups, informally called *dicots* and *monocots* (see Chapter 27). Dicots include well-known plants such as beans, petunias, oaks, cherry trees, roses, and snapdragons; examples of monocots include corn, lilies, grasses, palms, tulips, orchids, and bananas.

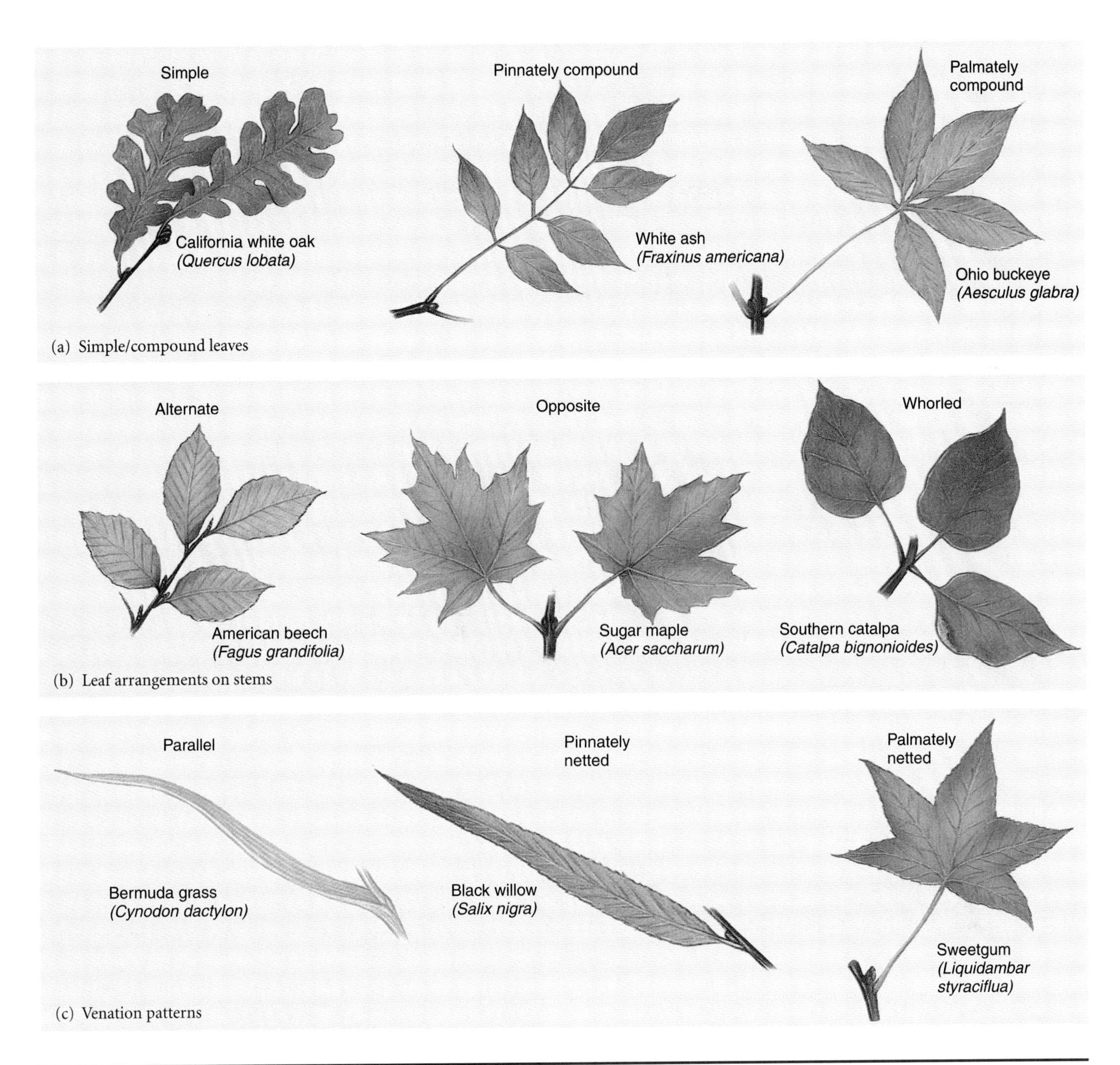

FIGURE **32-2** | **Leaf morphology.**

(a) Simple, pinnately compound, and palmately compound leaves. **(b)** Leaf arrangement may be alternate, opposite, or whorled. **(c)** Venation patterns include parallel, pinnately netted, and palmately netted. All leaves shown are woody dicot trees from North America, except bermuda grass, which is a herbaceous monocot native to Europe and Asia.

feature of leaf epidermal cells is that the cell wall facing toward the outside environment is somewhat thicker than the cell wall facing inward. This extra thickness may provide the plant with additional protection against injury or water loss.

Because leaves have such a large surface area exposed to the atmosphere, water loss by evaporation from the leaf's surface is unavoidable. However, epidermal cells secrete a waxy layer, the **cuticle,** that reduces water loss from their exterior walls (see Table 31-4). The cuticle, which consists primarily of a waxy substance called **cutin,** varies in thickness in different plants, in part owing to environmental conditions. As one might expect, the leaves of plants adapted to hot, dry climates have extremely thick cuticles. Furthermore, a leaf's exposed (and warmer) upper epidermis generally has a thicker cuticle than its shaded (and cooler) lower epidermis.

The epidermis of many leaves is covered with various hair-like structures called **trichomes** (see Table 31-4). Some leaves, such as those of the popular cultivated plant called lamb's ear, have so many trichomes that they feel fuzzy (Fig. 32-4). Trichomes have several functions. Trichomes of some plants help

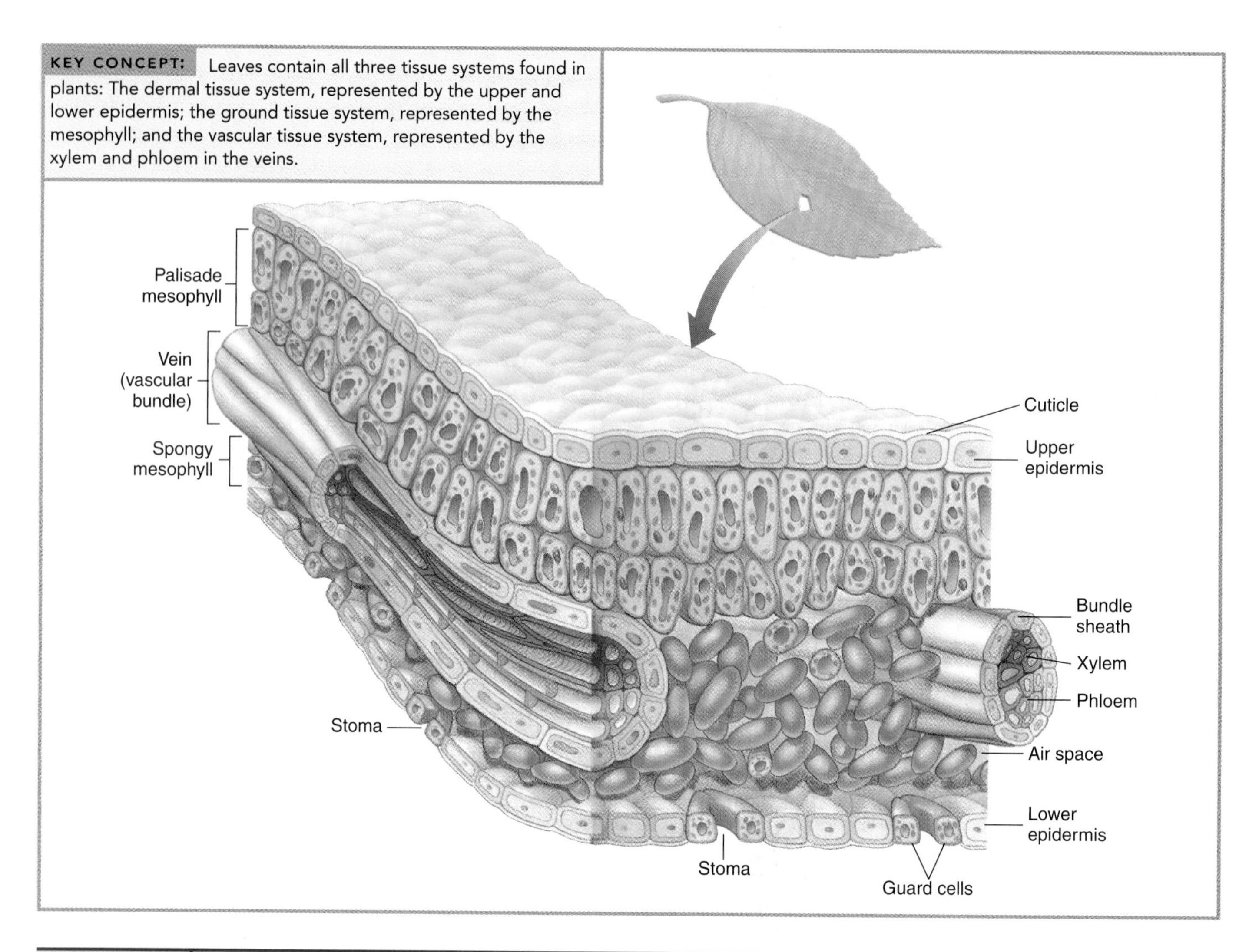

FIGURE **32-3** **Tissues in a typical leaf blade.**
An upper epidermis and lower epidermis cover the blade. The photosynthetic ground tissue, called *mesophyll,* is often arranged into palisade and spongy layers. Veins branch throughout the mesophyll.

reduce water loss from the leaf surface by retaining a layer of moist air next to the leaf and by reflecting sunlight, thereby protecting the plant from overheating. Some trichomes secrete stinging irritants for deterring **herbivores,** animals that feed on plants. In addition, a leaf covered with trichomes is difficult for an insect to walk over or eat. Other trichomes excrete excess salts absorbed from a salty soil.

The leaf epidermis contains minute openings, or **stomata** (sing., *stoma*), for gas exchange. Each stoma is flanked by two specialized epidermal **guard cells** (Fig. 32-5). The guard cells are responsible for opening and closing the stoma. Guard cells are usually the only epidermal cells with chloroplasts.

Stomata are especially numerous on the lower epidermis of horizontally oriented leaves (an average of about 100 stomata per mm^2), and in many species are located *only* on the lower surface. The lower epidermis of apple (*Malus sylvestris*) leaves, for example, has almost 400 stomata per mm^2, whereas the upper epidermis has none. This adaptation reduces water loss because stomata on the lower epidermis are shielded from direct sunlight and are therefore cooler than those on the upper epidermis. In

FIGURE **32-4** **Trichomes.**
Lamb's ear (*Stachys byzantina*) a popular garden plant, has so many trichomes (the tiny white hairs covering the leaf's surface) that the leaves look and feel like velvet. Each leaf grows to 10 cm (4 in) long.

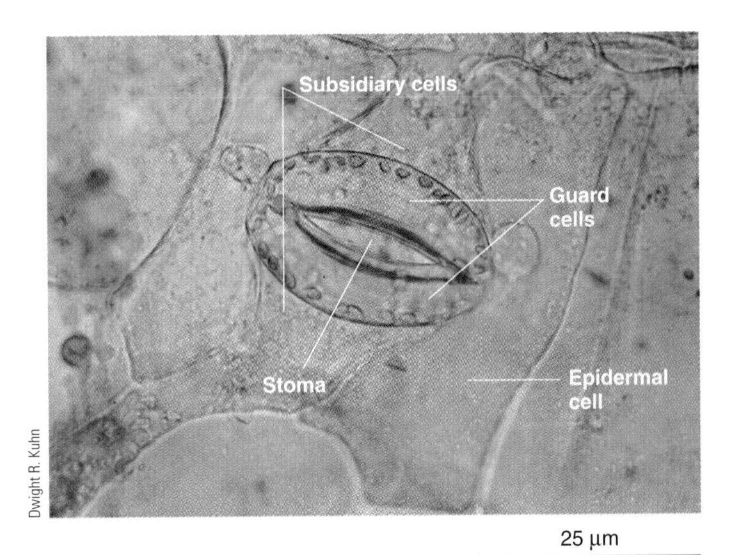

FIGURE **32-5** **Stoma.**

LM of an open stoma from the leaf epidermis of inch plant (*Zebrina pendula*). Note the chloroplasts present in the guard cells and the thicker inner wall of each guard cell.

contrast, floating leaves of aquatic plants such as water lilies have stomata only on the upper epidermis.

Guard cells are associated with special epidermal cells called **subsidiary cells** that are often structurally different from other epidermal cells. Subsidiary cells provide a reservoir of water and ions that move into and out of the guard cells as they change shape during stomatal opening and closing (discussed later in this chapter).

The photosynthetic ground tissue of the leaf, called the **mesophyll** (from the Greek *meso,* "the middle of," and *phyll,* "leaf"), is sandwiched between the upper epidermis and the lower epidermis. Mesophyll cells, which are parenchyma cells (see Chapter 31) packed with chloroplasts, are loosely arranged, with many air spaces between them that facilitate gas exchange. These intercellular air spaces account for as much as 70% of the leaf's volume.

In many plants, the mesophyll is divided into two sublayers. Toward the upper epidermis, the columnar cells are stacked closely together in a layer called **palisade mesophyll.** In the lower portion, the cells are more loosely and irregularly arranged, in a layer called **spongy mesophyll.** The two layers have different functions. Palisade mesophyll is the main site of photosynthesis in the leaf. Photosynthesis also occurs in the spongy mesophyll, but the primary function of the spongy mesophyll is to allow diffusion of gases, particularly CO_2, within the leaf.

Palisade mesophyll may be further organized into one, two, three, or even more layers of cells. The presence of additional layers of palisade mesophyll is at least partly an adaptation to environmental conditions. Leaves exposed to direct sunlight contain more layers of palisade mesophyll than do shaded leaves on the same plant. In direct sunlight, the light is strong enough to effectively penetrate multiple layers of palisade mesophyll, allowing all layers to photosynthesize efficiently.

The veins, or **vascular bundles,** of a leaf extend through the mesophyll. Branching is extensive, and no mesophyll cell is more than two or three cells away from a vein. Therefore, the slow process of diffusion does not limit the movement of needed resources between mesophyll cells and veins. Each vein contains two types of vascular tissue: xylem and phloem (see Chapter 31). **Xylem,** which conducts water and dissolved nutrient minerals, is usually located in the upper part of a vein, toward the upper epidermis, whereas **phloem,** which conducts dissolved sugars, is usually confined to the lower part of a vein.

One or more layers of nonvascular cells surround the larger veins and make up the **bundle sheath.** Bundle sheaths are composed of parenchyma or sclerenchyma cells (see Chapter 31). Frequently the bundle sheath has support columns, called **bundle sheath extensions,** that extend through the mesophyll from the upper epidermis to the lower epidermis (Fig. 32-6). Bundle sheath

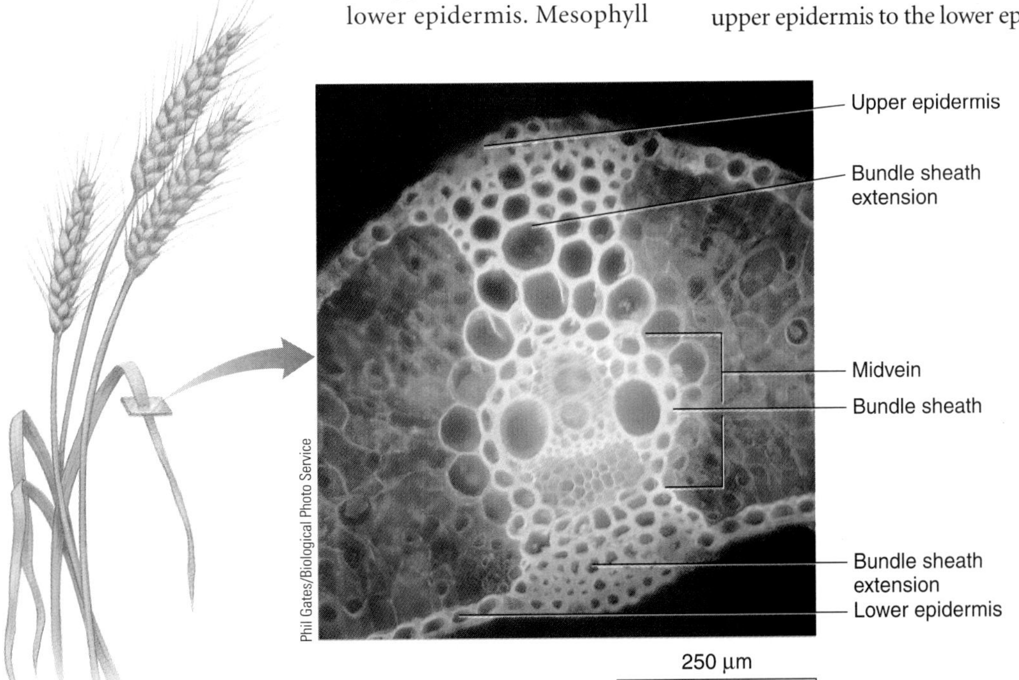

FIGURE **32-6**

Bundle sheath extension.

LM of a wheat (*Triticum aestivum*) midvein in cross section. Note the bundle sheath extensions to both the upper epidermis and lower epidermis. Photographed using fluorescence microscopy.

extensions may be composed of parenchyma, collenchyma, or sclerenchyma cells.

Leaf structure differs in dicots and monocots

A dicot leaf is usually composed of a broad, flattened blade and a petiole. As mentioned previously, dicot leaves typically have netted venation. In contrast, monocot leaves often lack a petiole; they are narrow and the base of the leaf often wraps around the stem, forming a sheath. Parallel venation is characteristic of monocot leaves.

Dicots and certain monocots also differ in internal leaf anatomy (Fig. 32-7). Although most dicots and monocots have both palisade and spongy layers, some monocots (corn and other grasses) do not have mesophyll differentiated into distinct palisade and spongy layers. Because dicots have netted veins, a cross section of a dicot blade often shows veins in both cross-sectional and lengthwise views. In a cross section of a monocot leaf, in contrast, the parallel venation pattern produces evenly spaced veins, all of which appear in cross section.

Differences between the guard cells in dicot and certain monocot leaves also occur (Fig. 32-8). The guard cells of dicots and many monocots are shaped like kidney beans. Other monocot leaves (those of grasses, reeds, and sedges) have guard cells shaped like dumbbells. These structural differences affect how the cells swell or shrink to open or close the stoma.

Leaf structure is related to function

How is leaf structure related to its primary function of photosynthesis? The epidermis of a leaf is relatively transparent and allows light to penetrate to the interior of the leaf where the photosynthetic ground tissue, the mesophyll, is located. Stomata, which dot the leaf surfaces, permit the exchange of gases between the atmosphere and the leaf's internal tissues. Carbon dioxide, a raw material of photosynthesis, diffuses into the leaf through stomata, and the oxygen produced during photosynthesis diffuses rapidly out of the leaf through stomata. Stomata also permit other gases, including air pollutants, to enter the leaf (see *Focus On: Air Pollution and Leaves*).

Water required for photosynthesis is obtained from the soil and transported in the xylem to the leaf, where it diffuses into the mesophyll and moistens the surfaces of mesophyll cells. The loose arrangement of the mesophyll cells, with air spaces between cells, allows for rapid diffusion of carbon dioxide to the mesophyll cell surfaces; there it dissolves in a film of water before diffusing into the cells.

The veins not only supply the photosynthetic ground tissue with water and minerals (from the roots, by way of the xylem) but also carry (in the phloem) dissolved sugar produced during photosynthesis to all parts of the plant. Bundle sheaths and bundle sheath extensions associated with the veins provide additional support to prevent the leaf, which is structurally weak because of the large amount of air space in the mesophyll, from collapsing under its own weight.

Leaves are adapted to help a plant survive in its environment

Leaf structure reflects the environment to which a particular plant is adapted. Although

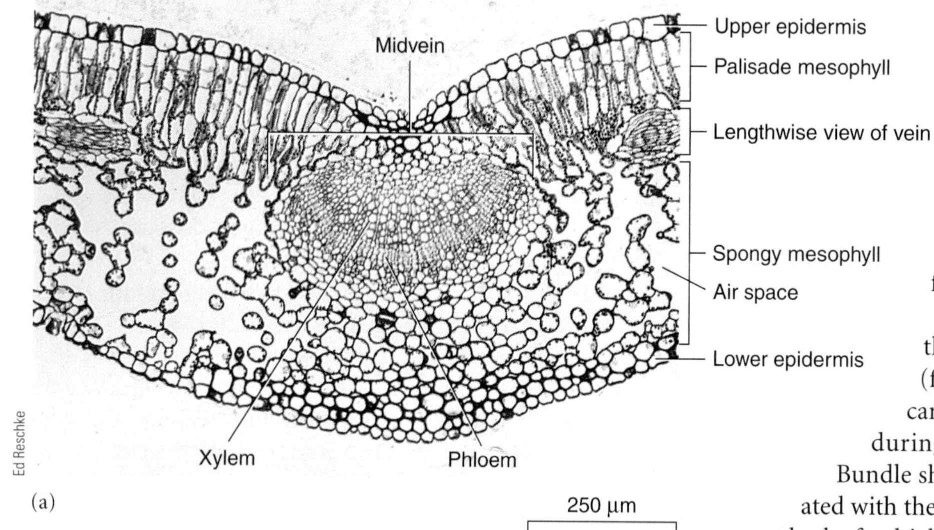

FIGURE 32-7 Dicot and monocot leaf cross sections.
(a) LM of part of a leaf cross section of privet (*Ligustrum vulgare*), a dicot. The mesophyll has distinct palisade and spongy sections. **(b)** LM of part of a leaf cross section of corn (*Zea mays*), a monocot. Corn leaves lack distinct regions of palisade and spongy mesophyll.

FIGURE **32-8** | **Variation in guard cells.**

(a) Guard cells of dicots and many monocots are bean-shaped. **(b)** Some monocot guard cells are narrow in the center and thicker at each end. Guard cells are associated with special epidermal cells called subsidiary cells.

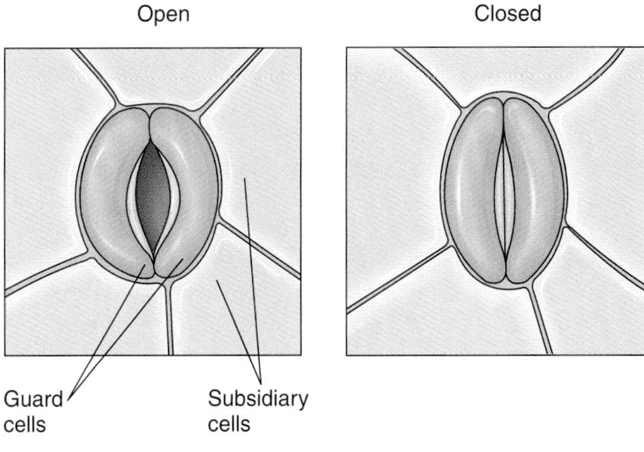

(a) Dicots and some monocots

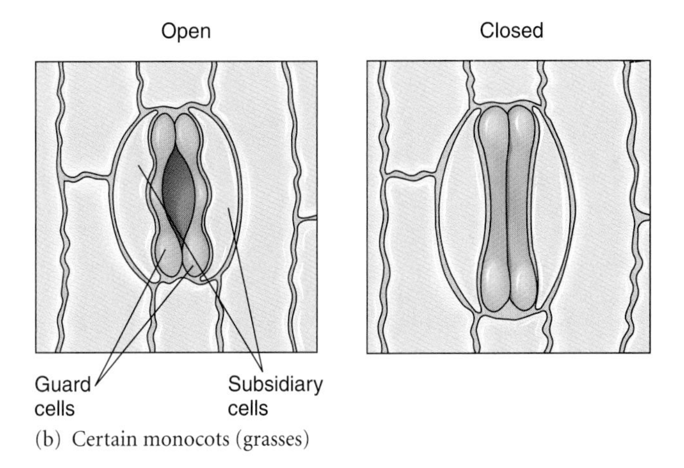

(b) Certain monocots (grasses)

both aquatic plants and those adapted to dry conditions perform photosynthesis and have the same basic leaf anatomy, their leaves are modified to enable them to survive different environmental conditions. The leaves of water lilies have petioles long enough to allow the blade to float on the water's surface. Large air spaces in the mesophyll make the floating blade buoyant. The petioles and other submerged parts have an internal system of air ducts; oxygen moves through these ducts from the floating leaves to the underwater roots and stems, which live in a poorly aerated environment.

Conifers are an important group of woody trees and shrubs that includes pine, spruce, fir, redwood, and cedar.[2] Most conifers are evergreen, which means they produce and lose leaves throughout the year rather than during certain seasons. Conifers dominate a large portion of Earth's land area, particularly in northern forests and mountains. The leaves of most conifers are waxy needles. Their needles have structural adaptations that help them survive winter, the driest part of the year. (Winter is arid even in areas of heavy snows because roots cannot absorb water from soil when the soil temperature is very low.) Indeed, many of the structural features of needles are also found in many desert plants.

Figure 32-9 shows a cross section of a pine needle. Note that the needle is somewhat thickened rather than thin and blade-like. The needle's relative thickness, which results in less surface area exposed to the air, reduces water loss. Other features that help conserve water include the thick, waxy cuticle and sunken stomata; these permit gas exchange while minimizing water loss. Thus needles help conifers tolerate the dry (low relative humid-

[2] As discussed in Chapter 27, conifers are gymnosperms, one of the two groups of seed plants (the other group is the flowering plants, or angiosperms). Unlike flowering plants, whose seeds are enclosed in fruits, conifers bear "naked" seeds on the scales of female cones.

FIGURE **32-9** | **LM of a pine needle in cross section.**

The thick, waxy cuticle and sunken stomata are two structural adaptations that enable pine (*Pinus*) to retain its needles throughout the winter.

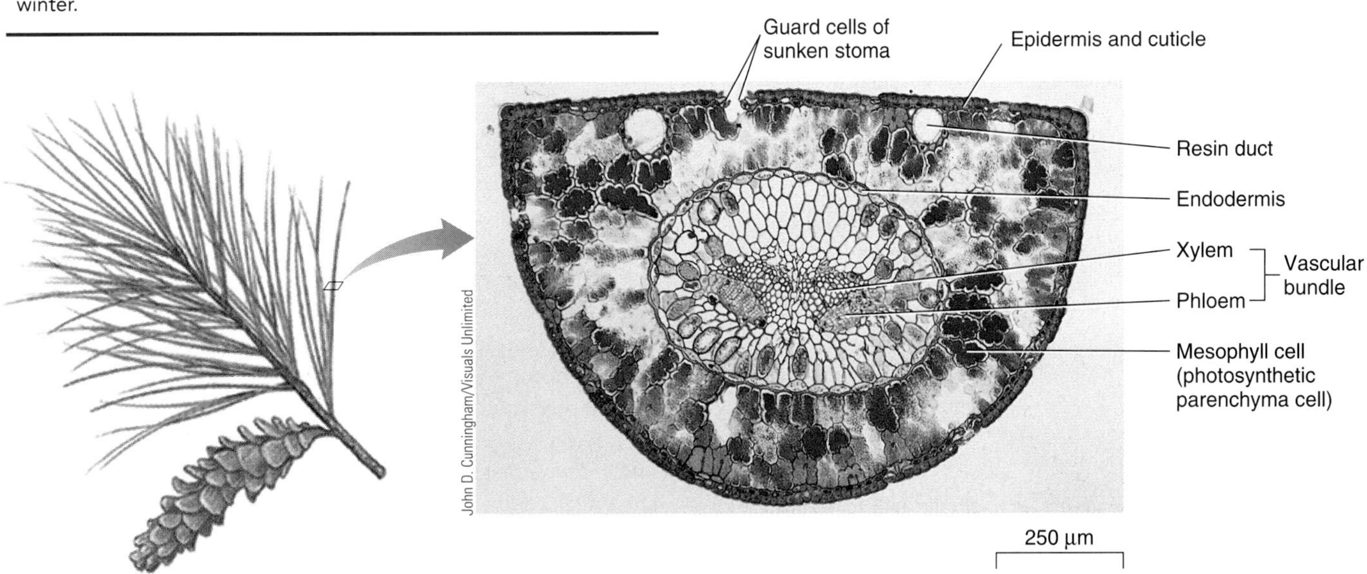

FOCUS ON AIR POLLUTION AND LEAVES

The air we breathe is often dirty and contaminated with many pollutants, particularly in urban areas. Air pollution consists of gases, liquids, or solids present in the atmosphere in levels high enough to harm humans and other organisms as well as nonliving materials. Although air pollutants can come from natural sources, as, for example, a lightning-caused fire or a volcanic eruption, human activities are a major cause of global air pollution. Motor vehicles and industry are the two main human sources of air pollution.

All parts of a plant can be damaged by air pollution, but leaves are particularly susceptible because of their structure and function. The thin blade provides a large surface area that comes into contact with the surrounding air. The thousands of stomatal pores that dot the epidermis and allow gas exchange with the atmosphere also permit pollutants to diffuse into the leaf. Just as lungs, the organs of gas exchange in humans and other terrestrial vertebrates are affected by air pollution, so too the leaves of a plant are most affected.

Many studies have shown that high levels of most forms of air pollution reduce the overall productivity of crop plants. The worst pollutant in terms of yield loss is ozone, a toxic gas produced when sunlight catalyzes a reaction between pollutants emitted by motor vehicles and industries. Ozone has been observed to reduce soybean yields by as much as 35%, and the average yield reduction from exposure to ozone is 15%. Ozone inhibits photosynthesis because it damages the mesophyll cells, probably by altering the permeability of their cell membranes. Exposure to low levels of air pollution often causes a decline in photosynthesis without other symptoms of injury. Lesions on leaves and other obvious symptoms appear at much higher levels of air pollution.

When air pollution is combined with other environmental stressors (such as low winter temperatures, prolonged droughts, insects, and bacterial, fungal, and viral diseases), it causes plants to decline and die. More than half of the red spruce trees in the mountains of the northeastern United States have died since the mid-1970s, and sugar maples in eastern Canada and the United States are also dying. Many still-living trees are exhibiting symptoms of **forest decline,** characterized by gradual deterioration and eventual death of trees. The general symptoms of forest decline are reduced vigor and growth, but some plants exhibit specific symptoms, such as yellowing of needles in conifers. Forest decline is more pronounced at higher elevations, possibly because most trees growing at high elevations are at the limit of their normal range and are therefore more susceptible to wind and low temperatures.

Many factors can interact to decrease the health of trees, and no single factor accounts for the recent instances of forest decline. Several air pollutants have been implicated, including acid rain, ozone, and toxic heavy metals such as lead, cadmium, and copper. Power plants, ore smelters, refineries, and motor vehicles produce these pollutants. Insects and weather factors such as drought and severe winters may also be important. To complicate matters further, the actual causes of forest decline may vary from one tree species to another and from one location to another. Thus forest decline appears to result from the combination of multiple stressors. When one or more stressors weaken a tree, then an additional stressor may be enough to cause its death.

ity) winds that occur during winter. With the warming of spring, soil water again becomes available, and the needles quickly resume photosynthesis.

Review

- Sketch two shoots, one with simple leaves, palmate venation, and alternate leaf arrangement; and the second with palmately compound leaves, pinnate venation, and opposite leaf arrangement.
- How are leaves adapted to conserve water?
- What is the photosynthetic ground tissue of a leaf called? What are its two sublayers?
- What are the two types of vascular tissue in a vascular bundle? Which vascular tissue is usually located on the upper part of the vascular bundle?
- What are some of the anatomical differences between dicot and monocot leaves?
- How is the leaf organized to deliver the raw materials and remove the products of photosynthesis?

BiologyNow™ Assess your understanding of **leaf form and structure** by taking the pretest on your BiologyNow CD-ROM.

STOMATAL OPENING AND CLOSING

Learning Objectives

5 Explain the role of blue light in the opening of stomata.

6 Outline the physiological changes that accompany stomatal opening and closing.

Stomata are adjustable pores that are usually open during the day when carbon dioxide is required for photosynthesis and closed at night when photosynthesis is shut down (see the section on CAM photosynthesis in Chapter 8 for an interesting exception). The opening and closing of stomata are controlled by changes in the shape of the two guard cells that surround each pore. When water moves into guard cells from surrounding cells, they become turgid (swollen) and bend, producing a pore. When water leaves the guard cells, they become flaccid (limp) and collapse against one another, closing the pore. What causes water to move into and out of guard cells?

Blue light triggers stomatal opening

Data from numerous experiments and observations are beginning to explain the complexities of stomatal movements. Let us begin with stomatal opening, which occurs when the plant detects light from the rising sun. You already know that light is a form of energy; plants absorb light and convert it to chemical energy in the process of photosynthesis. However, light is also an important *environmental signal* for plants—that is, light provides plants with information about their environment that they use to modify various activities at the molecular and cellular levels.

In stomatal opening and several other plant responses, **blue light,** which has wavelengths from 400 to 500 nm, is an environmental signal. Any plant response to light must involve a **pigment,** a molecule that absorbs the light before the induction of a particular biological response. Data such as the responses of stomata to different colors of light suggest the pigment involved in stomatal opening and closing is yellow (yellow pigments strongly absorb blue light). The yellow pigment is thought to be located in the guard cells, probably in their plasma membranes.

In step ① of Figure 32-10, blue light, which is a component of sunlight, triggers the activation of an enzyme, **ATP synthase,** located in the guard cell plasma membrane. Blue light also triggers the synthesis of malic acid and the **hydrolysis** (splitting) of starch, about which we say more shortly.

In step ② ATP synthase actively pumps protons (H^+) out of the guard cells. The H^+ that are pumped are formed when malic acid produced in the guard cells ionizes to form H^+ and negatively charged malate ions. The enzyme ATP synthase may be familiar to you, because ATP synthase is also involved in pumping protons in mitochondria (for aerobic respiration; see Chapter 7) and in chloroplasts (for photosynthesis; see Chapter 8). As ATP synthase in guard cells pumps protons out of the guard cell plasma membranes, an **electrochemical gradient**—that is, a charge and concentration difference—forms on the two sides of the guard cell plasma membrane.

In step ③ the resulting electrochemical gradient of H^+ drives the facilitated diffusion of large numbers of potassium ions *into* guard cells. This movement occurs through **voltage-activated potassium channels,** which open when a certain voltage (difference in charge between the two sides of the guard cell plasma membrane) is attained. This movement of potassium ions has been experimentally measured by **patch clamp techniques** (see Figure 5-16). (Interestingly, in animals, voltage-activated ion channels are found in the plasma membranes of nerve cells and are involved in transmitting neural impulses; see Chapter 39.) As shown in step ④, chloride ions are also taken into the guard cells through ion channels in the guard cell plasma membrane. The negatively charged chloride and malate ions help to electrically balance the positively charged potassium ions.

The potassium, chloride, and malate ions accumulate in the vacuoles of the guard cells, increasing the solute concentration in the vacuoles. You may recall from the discussion of osmosis in Chapter 5 that when a cell has a solute concentration greater than that of surrounding cells, water flows *into* the cell. Thus in step ⑤, water enters the guard cells from surrounding epidermal cells by osmosis. The increased turgidity of the guard cells changes their shape, because the thickened inner cell walls do not expand as much as the outer walls, causing the stoma to open.

Blue light activates ATP synthase ⟶ proton pump moves H^+ out of guard cells ⟶ K^+ and Cl^- move into guard cells through voltage-activated ion channels ⟶ water diffuses by osmosis into guard cells ⟶ guard cells change shape and stoma opens

In the late afternoon or early evening, stomata close, but not by an exact reversal of the opening process. Recent studies have demonstrated that the concentration of potassium ions in guard cells slowly decreases during the day. However, the guard-cell concentration of sucrose, another osmotically active substance, increases during the day—maintaining the open pore—and then

ACTIVE FIGURE **32-10** | **Mechanism of stomatal opening.**

The accumulation of osmotically active ions (K^+ and Cl^-) in the guard cells is driven by a proton (H^+) gradient.

BiologyNow™ Watch **stomata** in action by clicking on this figure on your BiologyNow CD-ROM.

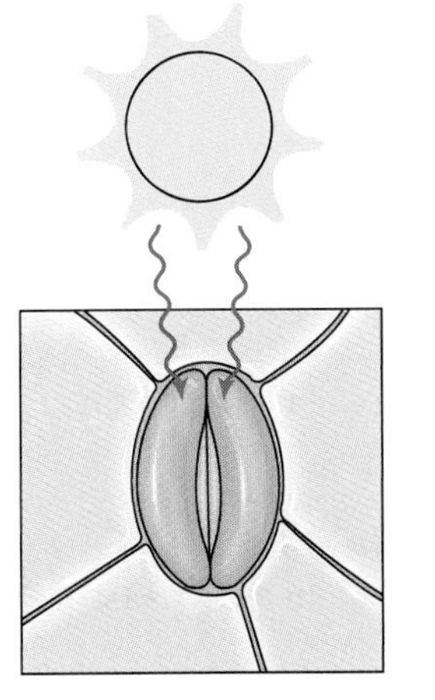

① Blue light activates ATP synthase.

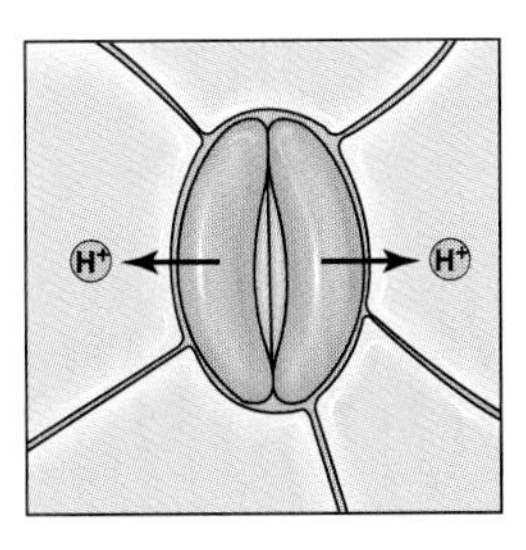

② Protons are pumped out of guard cells, forming electrochemical gradient.

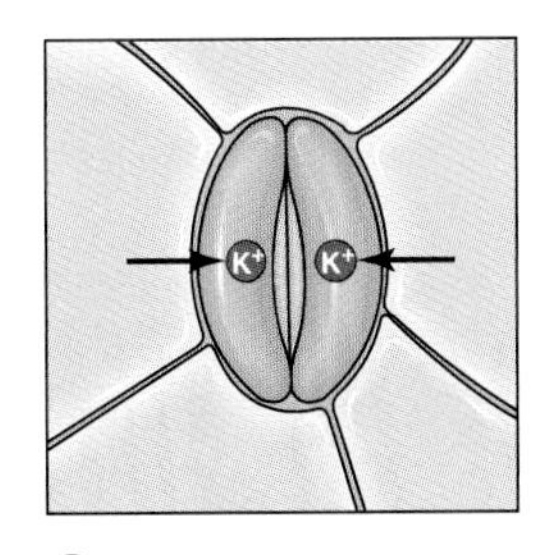

③ Potassium ions enter guard cells through voltage-activated ion channels.

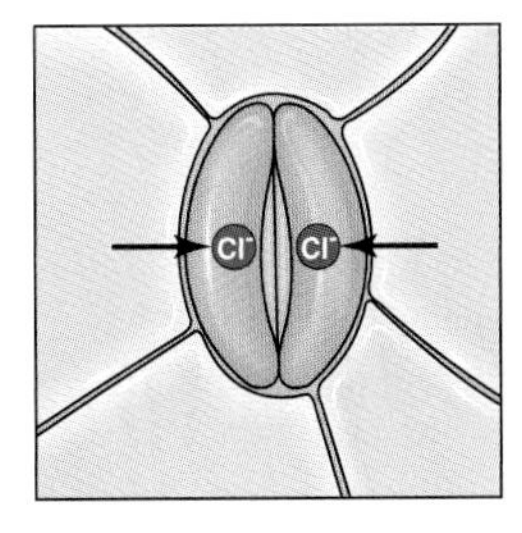

④ Chloride ions also enter guard cells through ion channels.

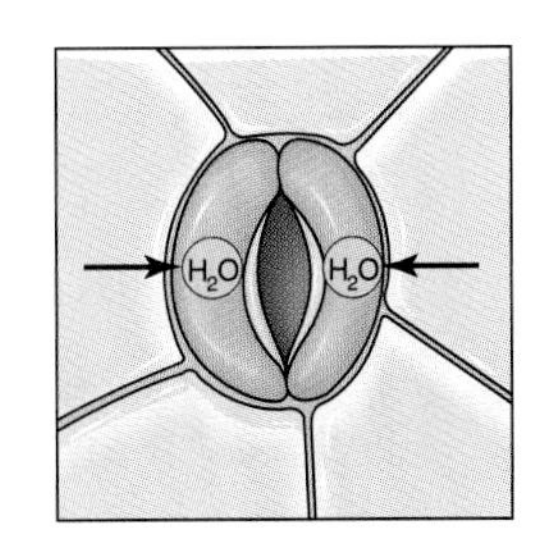

⑤ Water enters guard cells by osmosis, and stoma opens.

slowly decreases as evening approaches. This sucrose comes from the hydrolysis of the polysaccharide, starch, which is stored in the guard cell chloroplasts. As evening approaches, the sucrose concentration in the guard cells declines as sucrose is converted back to starch (which is osmotically inactive), water leaves by osmosis, the guard cells lose their turgidity, and the pore closes.

To summarize, different mechanisms appear to regulate the opening and closing of stomata. The uptake of potassium and chloride ions is mainly associated with stomatal opening, and the declining concentration of sucrose is mainly associated with stomatal closing. In Chapter 36, we discuss other blue light responses in plants.

Other factors also affect stomatal opening and closing

Although light and darkness trigger the opening and closing of stomata, other environmental factors are also involved, including carbon dioxide concentration. A low concentration of CO_2 in the leaf induces stomata to open even in the dark. The effects of light and CO_2 concentration on stomatal opening are interrelated. Photosynthesis, which occurs in the presence of light, reduces the internal concentration of CO_2 in the leaf, triggering stomatal opening. Another environmental factor that affects stomatal opening and closing is dehydration (water stress). During a prolonged drought, stomata remain closed even during the day. Stomatal opening and closing are under hormonal control, particularly by the plant hormone *abscisic acid* (see Chapter 36).

The opening and closing of stomata also appear to be regulated by an internal biological clock that in some way measures time. For example, after plants are placed in continual darkness, their stomata continue to open and close at more or less the same time each day. Such biological rhythms that follow an approximate 24-hour cycle are known as **circadian rhythms.** Other examples of circadian rhythms are provided in Chapter 36.

Review

- How does blue light trigger stomatal opening?
- What physiological changes occur in guard cells during stomatal opening? During stomatal closing?

BiologyNow™ Assess your understanding of **stomatal opening and closing** by taking the pretest on your BiologyNow CD-ROM.

TRANSPIRATION AND GUTTATION

Learning Objectives

7 Discuss transpiration and its effects on plants.

8 Distinguish between transpiration and guttation.

Despite leaf adaptations such as the cuticle, approximately 99% of the water that a plant absorbs from the soil is lost by evaporation from the leaves and, to a lesser extent, the stems. Loss of water vapor by evaporation from aerial plant parts is called **transpiration.**

The cuticle is extremely effective in reducing water loss by transpiration. It is estimated that only 1% to 3% of the water lost from a plant passes directly through the cuticle. Most transpiration occurs through open stomata. The numerous stomatal pores that are so effective in gas exchange for photosynthesis also provide openings through which water vapor escapes. In addition, the loose arrangement of the spongy mesophyll cells provides a large surface area within the leaf from which water can evaporate.

Several environmental factors influence the rate of transpiration. More water is lost from plant surfaces at higher temperatures. Light increases the transpiration rate, in part because it triggers stomatal opening and in part because it increases the leaf's temperature. Wind and dry air increase transpiration, but humid air *decreases* transpiration because the air is already saturated, or nearly so, with water vapor.

Although transpiration may seem wasteful, particularly to farmers in arid lands, it is an essential process that has adaptive value. Transpiration is responsible for water movement in plants, and without it water would not reach the leaves from the soil (see Chapter 33, discussion of the tension-cohesion model). The large amount of water that plants lose by transpiration may provide some additional benefits. Transpiration, like sweating in humans, cools the leaves and stems. When water passes from a liquid state to a vapor, it absorbs a great deal of heat. When the water molecules leave the plant as water vapor, they carry this heat with them. Thus the cooling effect of transpiration may prevent the plant from overheating, particularly in direct sunlight.

A second benefit of transpiration is that it distributes essential minerals throughout the plant. The water a plant transpires is initially absorbed from the soil, where it is present, not as pure water, but as a dilute solution of dissolved mineral salts. The water and dissolved nutrient minerals are then transported in the xylem throughout the plant, including its leaves. Water moves from the plant to the atmosphere during transpiration, but minerals remain in plant tissues. Many of these minerals are required for the plant's growth. It has been suggested that transpiration enables a plant to take in sufficient water to provide enough essential minerals, and that plants cannot satisfy their mineral requirements if the transpiration rate is not high enough.

There is no doubt, however, that under certain circumstances excessive transpiration can be harmful to a plant. On hot summer days, plants frequently lose more water by transpiration than they can take in from the soil. Their cells experience a loss of turgor, and the plant wilts (Fig. 32-11). If a plant is able to recover overnight, because of the combination of negligible transpiration (recall that stomata are closed) and absorption of water from the soil, the plant is said to have experienced *temporary wilting.* Most plants recover from temporary wilting with no ill effects. In cases of prolonged drought, however, the soil may not contain sufficient moisture to permit recovery from wilting. A plant that cannot recover is said to be *permanently wilted* and will die.

Transpiration is an important part of the **hydrologic cycle** (see Chapter 53), in which water cycles from the ocean and land to the atmosphere, and then back to the ocean and land. As a result of transpiration, water evaporates from leaves and stems to form clouds in the atmosphere. Thus, transpiration eventu-

(a)

(b)

FIGURE **32-11** | **Temporary wilting.**

Shown are the same squash (*Cucurbita pepo*) leaves **(a)** in the late afternoon of a hot day, after the leaves have wilted owing to water loss, and **(b)** the following morning, after water in the leaves has been replenished from the soil. Note that wilting helps reduce the surface area from which transpiration occurs. During the night while transpiration is negligible, the plants recover by absorbing water from the soil.

ally results in precipitation. As you might expect, forest trees release substantial amounts of moisture into the air by transpiration. Researchers have determined that at least half the rain that falls in the Amazon rainforest basin is recycled again and again by transpiration and precipitation.

Some plants exude liquid water

Many leaves have special structures through which liquid water is literally forced out. This loss of liquid water, known as **guttation,** occurs when transpiration is negligible and available soil moisture is high. Guttation typically occurs at night because the stomata are closed, but water continues to move into the roots by osmosis. People sometimes think erroneously that the early morning water droplets on leaf margins are dew rather than guttation (Fig. 32-12). Unlike dew, which condenses from cool night air, guttation droplets come from within the plant. (The mechanism for guttation is discussed in Chapter 33.)

Review

- What is transpiration? How is leaf structure related to transpiration?
- How do environmental factors (sunlight, temperature, humidity, and wind) influence the rate of transpiration?
- How does guttation differ from transpiration?

Biology Now™ Assess your understanding of **transpiration and guttation** by taking the pretest on your BiologyNow CD-ROM.

FIGURE **32-12** | **Guttation.**

Shown is a compound leaf of strawberry (*Fragaria*). Many people mistake guttation for early morning dew.

LEAF ABSCISSION

Learning Objective

9 Define leaf *abscission,* explain why it occurs, and describe the physiological and anatomical changes that precede it.

All trees shed leaves. Many conifers shed their needles in small numbers year-round. The leaves of deciduous plants turn color and **abscise,** or fall off, once a year—as winter approaches in temperate climates or at the beginning of the dry period in tropical climates with pronounced wet and dry seasons. In temperate forests, most woody plants with broad leaves shed their leaves to survive the low temperatures of winter. During winter, the plant's metabolism, including its photosynthetic machinery, slows down or halts temporarily.

Another reason for abscission is related to a plant's water requirements, which become critical during the physiological drought of winter. As mentioned previously, as the ground chills, absorption of water by the roots is inhibited. When the ground freezes, *no* absorption occurs. If the broad leaves were to stay on the plant during the winter, the plant would continue to lose water by transpiration but would be unable to replace it with water absorbed from the soil.

Leaf abscission is a complex process that involves many physiological changes, all initiated and orchestrated by changing levels of plant hormones, particularly **ethylene** (see Chapter 36). Briefly, the process is this: As autumn approaches, sugars, amino acids, and many essential minerals (such as nitrogen, phosphorus, and possibly potassium) are mobilized and transported from the leaves to other plant parts. Chlorophyll breaks down, allowing the **carotenoids** (orange carotenes and yellow xanthophylls), some of the accessory pigments in the chloroplasts of leaf cells, to become evident (see Fig. 3-14b). Recall from Chapter 8 that accessory pigments are always present in the leaf but are masked by the green of the chlorophyll. In addition, red water-soluble pigments called **anthocyanins** may accumulate in the vacuoles of epidermal leaf cells in some species; anthocyanins may protect leaves against damage by ultraviolet radiation. The various combinations of carotenoids and anthocyanins are responsible for the brilliant colors found in autumn landscapes in temperate climates.

In many leaves, abscission occurs at an abscission zone near the base of the petiole

The area where a petiole detaches from the stem is a structurally distinct area called the **abscission zone.** Composed primarily of thin-walled parenchyma cells, it is anatomically weak because it contains few fibers (Fig. 32-13). A protective layer of cork cells develops on the stem side of the abscission zone. These cells have a waxy, waterproof material impregnated in their walls. Enzymes then dissolve the **middle lamella** (the "cement" that holds the primary cell walls of adjacent cells together) in the abscission zone (see Fig. 4-28). Once this process is completed, nothing holds the leaf to the stem but a few xylem cells. A sudden breeze is enough to make the final break, and the leaf detaches. The protective layer of cork remains, sealing off the area and forming a leaf scar.

Although the structure of abscission zones and the physiological changes associated with abscission are well known, the genes responsible for the development of an abscission zone have been investigated only recently. In 2000, researchers from Clemson University and the University of Iowa reported in the journal *Nature* that they had isolated a gene involved in the development of an abscission zone on tomato flowers and fruits. (Like leaves, flowers and fruits have abscission zones and are also shed.) Tomato plants with a mutated version of the gene did not develop normal abscission zones on their flower stalks.

Review

- Why do many woody plants living in temperate zones lose their leaves in autumn?
- What physiological changes occur during leaf abscission?
- What is the abscission zone?

BiologyNow™ Assess your understanding of **leaf abscission** by taking the pretest on your BiologyNow CD-ROM.

MODIFIED LEAVES

Learning Objective

10 List at least four examples of modified leaves, and give the function of each.

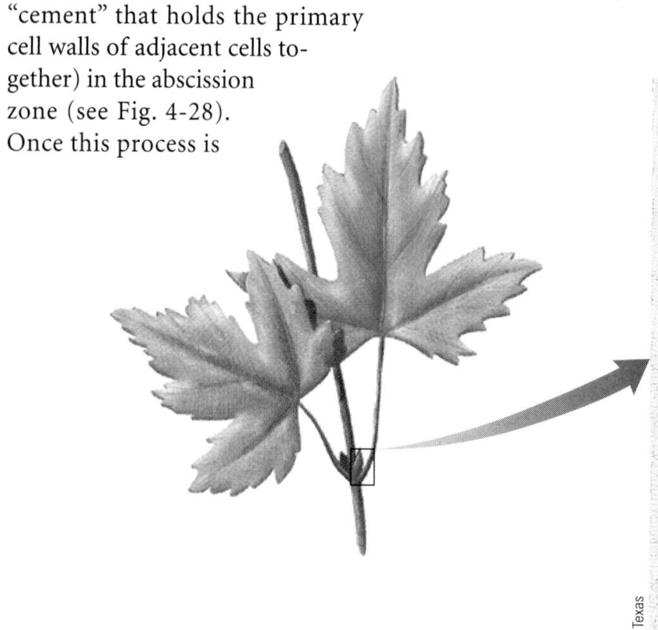

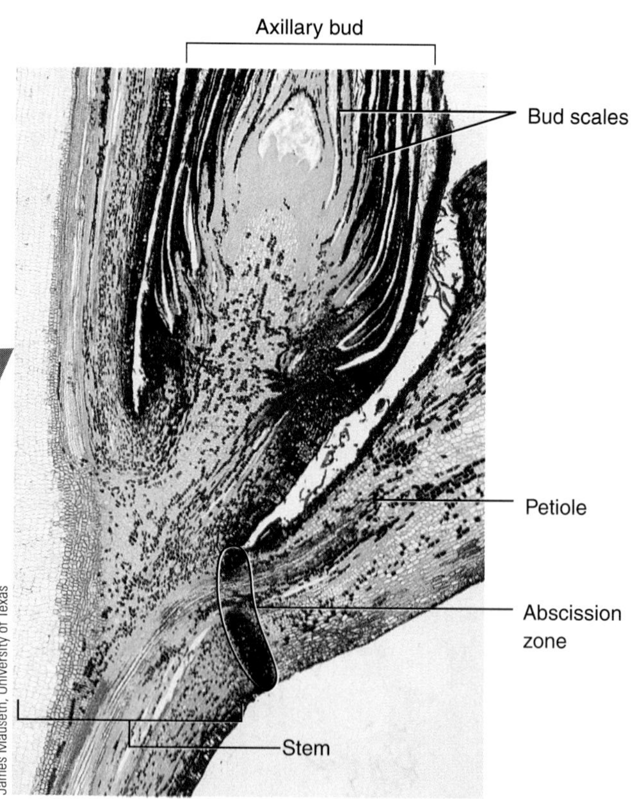

FIGURE **32-13** Abscission zone.

This LM of a longitudinal section through a silver maple (*Acer saccharinum*) branch shows the abscission zone at the base of a petiole. An axillary bud with its protective bud scales is also evident above the petiole.

Although photosynthesis is the main function of leaves, certain leaves have special modifications for other functions. Some plants have leaves specialized for deterring herbivores. **Spines,** modified leaves that are hard and pointed, are found on many desert plants such as cacti (Fig. 32-14a). In the cactus, the main organ of photosynthesis is the stem rather than the leaf. Spines discourage animals from eating the succulent stem tissue.

Vines are climbing plants whose stems cannot support their own weight, so they often possess **tendrils** that help keep the vine attached to the structure on which it is growing. The tendrils of many vines, such as peas, cucumbers, and squash, are specialized leaves (Fig. 32-14b). However, some tendrils, such as those of ivy, Virginia creeper, and grape, are specialized stems.

The winter buds of a dormant woody plant are covered by **bud scales,** modified leaves that protect the delicate meristematic tissue of the bud from injury, freezing, or drying out (Fig. 32-14c).

Leaves may also be modified for storage of water or food. For example, a **bulb** is a short underground stem to which large, fleshy leaves are attached (Fig. 32-14d). Onions and tulips form bulbs. Many plants adapted to arid conditions, such as jade

FIGURE **32-14** **Leaf modifications.**

(a) The leaves of this pincushion cactus (*Mammilaria*) are modified as spines for deterring herbivores. Photographed in Arizona. **(b)** Tendrils of bur cucumber (*Echinocystis lobata*) are modified leaves. Tendrils, which wind around objects and aid vines in climbing, may be modified leaves or stems. **(c)** A terminal bud and two axillary buds of a maple (*Acer*) twig have overlapping bud scales to protect buds. **(d)** The leaves of bulbs such as the onion (*Allium cepa*) are fleshy for storage of food materials and water. **(e)** Some plants have thick, succulent leaves modified for water storage as well as photosynthesis. The succulent leaves of the string-of-beads plant (*Senecio rowleyanus*) are spherical to minimize surface area, thereby conserving water.

plant, medicinal aloe, and string-of-beads, have succulent leaves for water storage (Fig. 32-14e). These leaves are usually green and also function in photosynthesis.

Some leaves are modified for sexual reproduction (see discussion of flower evolution in Chapter 27, including Fig. 27-13) or asexual reproduction (see Fig. 35-17).

Modified leaves of insectivorous plants capture insects

Insectivorous plants are plants that capture insects. Most insectivorous plants grow in poor soil that is deficient in certain essential minerals, particularly nitrogen. These plants meet some of their mineral requirements by digesting insects and other small animals. The leaves of insectivorous plants are adapted to attract, capture, and digest their animal prey.

Some insectivorous plants have passive traps. The leaves of a pitcher plant, for example, are shaped so that rainwater collects and forms a reservoir that also contains acid secreted by the plant (Fig. 32-15). Some pitchers are quite large; in the tropics pitcher plants may be large enough to hold 1 L (approximately 1 qt) or more of liquid. An insect attracted by the odor or nectar of the pitcher may lean over the edge and fall in. Although it may make repeated attempts to escape, the insect is prevented from crawling out by the slippery sides and the rows of stiff hairs that point downward around the lip of the pitcher. The insect eventually drowns, and part of its body eventually disintegrates and is absorbed.

Although most insects are killed in pitcher plants, the larvae of several insects (certain fly, midge, and mosquito species), as well as a large community of microorganisms, actually live inside the pitchers. These insect species obtain their food from the insect carcasses, and the pitcher plant digests what remains. It is not known how these insects survive the acidic environment inside the pitcher.

The Venus flytrap is an insectivorous plant with active traps. Its leaf blades resemble tiny bear traps (see Fig. 1-3). Each side of the leaf blade contains three small, stiff hairs. If an insect alights and brushes against two of the hairs, or against the same hair twice in quick succession, the trap springs shut with amazing rapidity. The spines along the margins of the blades fit closely together to prevent the insect from escaping. Digestive glands on the surface of the trap secrete enzymes in response to the insect pressing against them. After the insect has died and been digested, the trap reopens and the indigestible remains fall out.

Bill Lea/Dembinsky Photo Associates

FIGURE **32-15** | **A common pitcher plant.**
This species (*Sarracenia purpurea*), whose pitchers grow to 30.5 cm (12 in), is widely distributed in acidic bogs and marshes in eastern North America. Young pitchers are green but turn red as they age. Note the dead beetle in the "pitcher."

Review

- What are the primary functions of each of the following modified leaves: spines, tendrils, and bud scales?
- What are the functions of bulbs? Of succulent leaves?
- What are some of the specialized features of the leaves of insectivorous plants?

BiologyNow™ Assess your understanding of **modified leaves** by taking the pretest on your BiologyNow CD-ROM.

SUMMARY WITH KEY TERMS

1 Discuss variation in leaf form, including simple versus compound leaves, leaf arrangement on the stem, and venation patterns.

- Leaves typically consist of a broad, flat **blade** and a stalklike **petiole.** Some leaves also have small, leaflike outgrowths from the base called **stipules.**
- Leaves may be simple (having a single blade) or compound (having a blade divided into two or more leaflets).
- Leaf arrangement on a stem may be alternate (one leaf at each **node**), opposite (two leaves at each node), or whorled (three or more leaves at each node).
- Leaves may have parallel or netted venation. Netted venation may be pinnately netted, with several major veins radiating from one point, or palmately netted, with veins branching along the entire length of the midvein.

2 Describe the major tissues of the leaf (epidermis, photosynthetic ground tissue, xylem, and phloem), and label them on a diagram of a leaf cross section.

- Upper and lower surfaces of the leaf blade are covered by an **epidermis.** A waxy **cuticle** coats the epidermis, enabling the plant to survive the dry conditions of a terrestrial existence.

SUMMARY WITH KEY TERMS (continued)

- **Stomata** are small pores in the epidermis that permit gas exchange needed for photosynthesis. Each pore is surrounded by two **guard cells** that are often associated with special epidermal cells called **subsidiary cells.** Subsidiary cells provide a reservoir of water and ions that move into and out of the guard cells as they change shape during stomatal opening and closing.
- **Mesophyll** consists of photosynthetic parenchyma cells. Mesophyll is divided into **palisade mesophyll,** which functions primarily for photosynthesis, and **spongy mesophyll,** which functions primarily for gas exchange.
- Leaf **veins** have **xylem** to conduct water and essential minerals to the leaf and **phloem** to conduct sugar produced by photosynthesis to the rest of the plant.

3 Compare leaf anatomy in dicots and monocots.

- Monocot leaves have parallel venation, whereas dicot leaves have netted venation.
- Some monocots (corn and other grasses) do not have mesophyll differentiated into distinct palisade and spongy layers.
- Some monocots (grasses, reeds, and sedges) have guard cells shaped like dumbbells, unlike the more common bean-shaped guard cells.

4 Relate leaf structure to its function of photosynthesis.

- Leaf structure is adapted for its primary function of **photosynthesis.** Most leaves have a broad, flattened blade that is quite efficient in collecting the sun's radiant energy.
- Stomata generally open during the day for gas exchange needed during photosynthesis and close at night to conserve water when photosynthesis is not occurring.
- The transparent epidermis allows light to penetrate into the middle of the leaf, where photosynthesis occurs.
- Air spaces in mesophyll tissue permit the rapid diffusion of carbon dioxide and water into, and oxygen out of, mesophyll cells.

5 Explain the role of blue light in the opening of stomata.

- **Blue light,** which is a component of sunlight, triggers the activation of **ATP synthase** located in the guard cell plasma membrane. Blue light also triggers the synthesis of malic acid and the **hydrolysis** of starch.

6 Outline the physiological changes that accompany stomatal opening and closing.

- ATP synthase pumps protons (H^+) out of the guard cells. The protons are produced when malic acid ionizes. As protons leave the guard cells, an **electrochemical gradient** (a charge and concentration difference) forms on the two sides of the guard cell plasma membrane.
- The electrochemical gradient drives the uptake of potassium ions through **voltage-activated potassium channels** into the guard cells. Chloride ions are also taken into the guard cells through ion channels. These osmotically active ions increase the solute concentration in the guard cell vacuoles.
- The resulting osmotic movement of water into the guard cells causes them to become turgid, forming a pore.
- As the day progresses, potassium ions slowly leave the guard cells, and starch is hydrolyzed to sucrose, which increases in concentration in the guard cells. Stomata close when water leaves the guard cells as a result of a decline in the concentration of sucrose, an osmotically active solute. The sucrose is converted to starch, which is osmotically inactive.
- Several environmental factors affect stomatal opening and closing, including light or darkness, CO_2 concentration, water stress, and the plant's circadian rhythm.

7 Discuss transpiration and its effects on plants.

- **Transpiration** is the loss of water vapor from aerial parts of plants. Transpiration occurs primarily through the stomata.
- The rate of transpiration is affected by environmental factors such as temperature, wind, and relative humidity.
- Transpiration appears to be both beneficial and harmful to the plant—that is, transpiration represents a tradeoff between the CO_2 requirement for photosynthesis and the need for water conservation.

8 Distinguish between transpiration and guttation.

- **Guttation,** the release of liquid water from leaves of some plants, occurs through special structures when transpiration is negligible and available soil moisture is high. In contrast, transpiration is the loss of water vapor and occurs primarily through the stomata.

9 Define leaf *abscission*, explain why it occurs, and describe the physiological and anatomical changes that precede it.

- Leaf **abscission** is the loss of leaves that often occurs as winter approaches in temperate climates or at the beginning of the dry period in tropical climates with wet and dry seasons.
- Abscission is a complex process involving physiological and anatomic changes that occur prior to leaf fall. An **abscission zone** develops where the petiole detaches from the stem. Sugars, amino acids, and many essential minerals are transported from the leaves to other plant parts. Chlorophyll breaks down, and **carotenoids** and **anthocyanins** become evident.

10 List at least four examples of modified leaves, and give the function of each.

- **Spines** are leaves adapted to deter herbivores.
- Some **tendrils** are leaves modified for grasping and holding onto other structures (to support weak stems).
- **Bud scales** are leaves modified to protect delicate meristematic tissue or dormant buds.
- **Bulbs** are short underground stems with fleshy leaves specialized for storage.
- Many plants adapted to arid conditions have succulent leaves for water storage.
- Insectivorous plants have leaves modified to trap insects.

POST-TEST

1. Plants with an alternate leaf arrangement have (a) blades divided into two or more leaflets (b) major veins that radiate out from one point (c) one leaf at each node (d) major veins branching off along the entire length of the midvein (e) two leaves at each node
2. The photosynthetic ground tissue in the middle of the leaf is called (a) cutin (b) mesophyll (c) the abscission zone (d) subsidiary cells (e) palisade and spongy stomata

POST-TEST *(continued)*

3. The primary function of the spongy mesophyll is (a) reduction of water loss from the leaf surface (b) changing the shape of the guard cells (c) support to prevent the leaf from collapsing under its own weight (d) diffusion of gases within the leaf (e) deterrence of herbivores
4. Gas exchange occurs through microscopic pores formed by two (a) subsidiary cells (b) abscission cells (c) mesophyll cells (d) guard cells (e) stipules
5. Most stomata are usually located in the ______________ of the leaf. (a) upper epidermis (b) lower epidermis (c) cuticle (d) spongy mesophyll (e) palisade mesophyll
6. The thin, noncellular layer of wax secreted by the epidermis of leaves is the (a) stoma (b) subsidiary cell (c) trichome (d) bundle sheath (e) cuticle
7. The ______________ encircles a vein. (a) palisade mesophyll (b) guard cell (c) bundle sheath (d) blade (e) cuticle
8. The ______________ of a leaf vein transports water and dissolved nutrient minerals, whereas the ______________ transports sugars produced by the leaf during photosynthesis. (a) xylem; phloem (b) xylem; bundle sheath (c) phloem; xylem (d) phloem; vein (e) vascular bundle; bundle sheath
9. Which of the following is *not* an adaptation of pine needles to conserve water? (a) less surface area exposed to the air than thin-bladed leaves (b) a relatively thick cuticle (c) sunken stomata (d) netted veins instead of parallel veins (e) both c and d are not adaptations of pine needles
10. Most of the water that a plant absorbs from the soil is lost by the process of (a) guttation (b) circadian rhythm (c) abscission (d) transpiration (e) photosynthesis
11. When transpiration is negligible, plants such as grasses exude excess water by (a) guttation (b) circadian rhythm (c) abscission (d) pumping H^+ out of and K^+ into guard cells (e) photosynthesis
12. At sunrise, the accumulation in the guard cells of the osmotically active substance, ________, causes an inflow of water and the opening of the pore. (a) protons (b) starch (c) ATP synthase (d) sucrose (e) potassium ions
13. Stomatal opening is most pronounced in response to ________ light. (a) green (b) yellow (c) blue (d) ultraviolet (e) infrared
14. The seasonal detachment of leaves is known as (a) forest decline (b) transpiration (c) abscission (d) guttation (e) dormancy
15. Anatomically, the abscission zone where a petiole detaches from a stem consists of (a) thin-walled parenchyma cells with few fibers (b) thick-walled cork parenchyma cells (c) clusters of fibers and collenchyma strands (d) hard and pointed stipules (e) epidermal cells with sunken stomata
16. Modified leaves that enable a stem to climb are called ______________ whereas modified leaves that cover the winter buds of a dormant woody plant are called ______________ (a) spines; bud scales (b) bud scales; tendrils (c) tendrils; bud scales (d) tendrils; spines (e) insectivorous leaves; spines
17. Leaves have a tradeoff between photosynthesis and transpiration that results from (a) numerous stomatal pores that provide both gas exchange for photosynthesis and openings through which water vapor escapes (b) secretion of a waxy layer, the cuticle, that reduces water loss (c) blue light triggering an influx of potassium ions (K^+) into the guard cells (d) abscission of leaves of deciduous plants as winter approaches in temperate climates (e) stomata being closed at night, although water continues to move into the roots by osmosis

CRITICAL THINKING

1. Suppose you are asked to observe a micrograph of a leaf cross section and distinguish between the upper and lower epidermis. How would you make this decision?
2. Given that (a) xylem is located toward the upper epidermis in leaf veins while phloem is toward the lower epidermis, and (b) the vascular tissue of a leaf is continuous with that of the stem, suggest one possible arrangement of vascular tissues in the stem that might account for the arrangement of vascular tissue in the leaf.
3. What might be some of the advantages of a plant having a few very large leaves? What might some disadvantages be? What might be some advantages of having many very small leaves? What disadvantages might this entail? How would your answer differ along a moisture gradient, from a humid environment to a desert?
4. Briefly explain why additional research on the molecular mechanism of stomatal closure might be of future use in agriculture.

- Visit our Web site at **http://biology.brookscole.com/solomon7** for links to chapter-related resources on the World Wide Web. Additional online materials relating to this chapter can also be found on our Web site.

BIOLOGY NOW RESOURCES

Active Figure

32-10: Stomata

Preparing for an exam? Take a diagnostic test on your BiologyNow CD-ROM.

Post-Test Answers

1. c	2. b	3. d	4. d
5. b	6. e	7. c	8. a
9. d	10. d	11. a	12. e
13. c	14. c	15. a	16. c
17. a			

33

Stems and Plant Transport

Michael Fairchild/Peter Arnold, Inc.

Baobab tree. Baobabs (*Adansonia digitata*), which are native to Africa, Madagascar, and Australia, store large volumes of water and starch in their massive trunks. Baobab trees are relatively short (growing to 18 m, or 60 ft), but their trunks can be as much as 9 m, or 30 ft, in diameter. Photographed in Namibia, with children around the tree as a size reference.

CHAPTER OUTLINE

A vegetative (not sexually reproductive) vascular plant has three parts: roots, leaves, and stems. As discussed in Chapter 31, roots serve to anchor the plant and absorb materials from the soil, whereas leaves are primarily for photosynthesis, converting radiant energy into the chemical energy of carbohydrate molecules. Stems, the focus of this chapter, link a plant's roots to its leaves and are usually located above ground, although many plants have underground stems. Stems exhibit varied forms, ranging from ropelike vines to massive tree trunks. They can be either herbaceous (consisting of soft, nonwoody tissues) or woody (with extensive hard tissues of wood and bark).

Stems perform three main functions in plants. First, stems of most species support leaves and reproductive structures. The upright position of most stems and the arrangement of the leaves on them allow each leaf to absorb light for use in photosynthesis. Reproductive structures (flowers and fruits) are located on stems in areas accessible to insects, birds, and air currents, which transfer pollen from flower to flower and help disperse seeds and fruits.

Second, stems provide internal transport. They conduct water and dissolved nutrient minerals from the roots, where these materials are absorbed from the soil, to leaves and other plant parts. Stems also conduct the sugar produced in leaves by photosynthesis to roots and other parts of the plant. Remember, however, that stems are not the only plant organs that conduct materials. The vascular system is continuous throughout all parts of a plant, and conduction occurs in roots, stems, leaves, and reproductive structures.

Third, stems produce new living tissue. They continue to grow throughout a plant's life, producing buds that develop into stems with new leaves and/or reproductive structures. In addition to the main functions of support, conduction, and production of new stem tissues, stems of some species are modified for asexual reproduction (see Chapter 35) or, if green, to manufacture sugar by photosynthesis. Also, some stems are specialized to store starch *(see photograph)*. ■

EXTERNAL STEM STRUCTURE IN WOODY TWIGS

Learning Objective

1 Describe the external features of a woody twig.

Although stems exhibit great variation in structure and growth, they all have **buds,** which are embryonic shoots. A **terminal bud** is the embryonic shoot located at the tip of a stem. The dormant (not actively growing) apical meristem of a terminal bud is covered and protected by an outer layer of **bud scales,** which are modified leaves (see Fig. 32-14c). **Axillary buds,** also called **lateral buds,** are located in the axils of a plant's leaves (see Fig. 32-1). An axil is the upper angle between a leaf and the stem to which it is attached. When terminal and axillary buds grow, they form branches that bear leaves and/or flowers. The area on a stem where each leaf is attached is called a **node,** and the region between two successive nodes is an **internode.**

A woody twig of a deciduous tree that has shed its leaves can be used to demonstrate certain structural features of the stem (Fig. 33-1). Bud scales cover the terminal bud and protect its delicate apical meristem during dormancy. When the bud resumes growth, the bud scales covering the terminal bud fall off, leaving **bud scale scars** on the stem where they were attached. Because temperate-zone woody plants form terminal buds at the end of each year's growing season, the number of sets of bud scale scars on a twig indicates its age. A **leaf scar** shows where each leaf was attached on the stem; the pattern of leaf scars can be used to determine leaf arrangement on a stem—alternate, opposite, or whorled (see Fig. 32-2b). The vascular (conducting) tissue that extends from the stem out into the leaf forms **bundle scars** within a leaf scar. Axillary buds may be found above the leaf scars. Also, the bark of a woody twig has **lenticels,** sites of loosely arranged cells that allow oxygen to diffuse into the interior of the woody stem. Lenticels look like tiny specks on the bark of a twig.

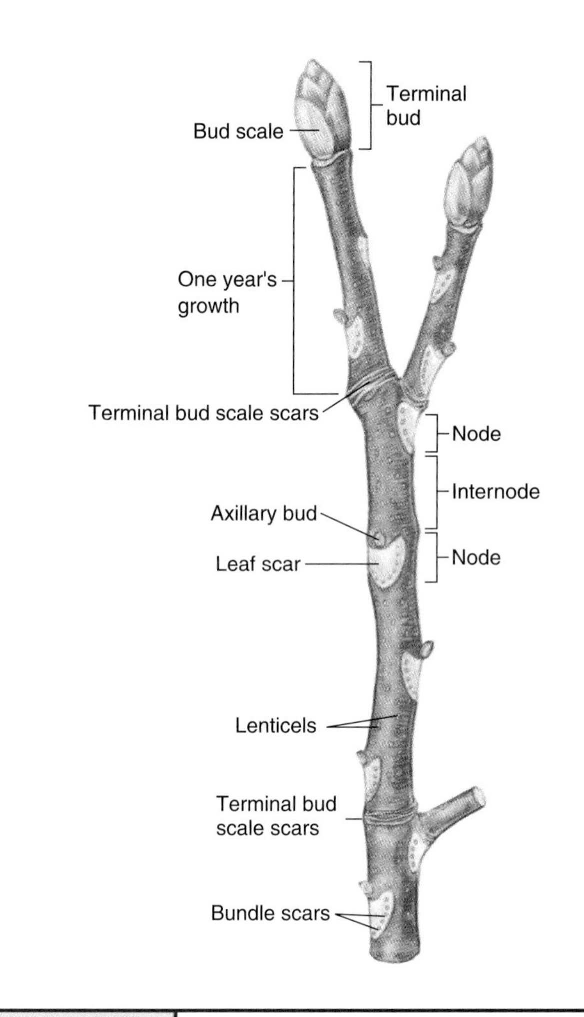

FIGURE **33-1** **External structure of a woody twig in its winter condition.**

The age of a woody twig can be determined by the number of sets of bud scale scars (do not count side branches). How old is this twig?

Review

- What is the difference between terminal and axillary buds?
- What is the function of bud scales? Of lenticels?
- How can you tell the age of a woody twig?

BiologyNow™ Assess your understanding of **external stem structure in woody plants** by taking the pretest on your BiologyNow CD-ROM.

STEM GROWTH AND STRUCTURE

Learning Objectives

2 Label cross sections of herbaceous dicot and monocot stems, and describe the functions of each tissue.

3 Name the two lateral meristems, and describe the tissues that arise from each.

4 Outline the transition from primary growth to secondary growth in a woody stem.

You may recall from Chapter 31 that plants have two different types of growth. **Primary growth** is an increase in the length of a plant and occurs at **apical meristems** located at the tips of roots and shoots and also within the buds of stems. **Secondary growth** is an increase in the girth (thickness) of a plant owing to the activity of **lateral meristems** located within stems and roots. The new tissues formed by the lateral meristems are called *secondary tissues* to distinguish them from *primary tissues* produced by apical meristems.

All plants have primary growth; some plants have both primary and secondary growth. Stems with only primary growth are herbaceous, whereas those with both primary and secondary growth are woody.[1] A woody plant increases in length by primary growth at the tips of its stems and roots, while its older stems and roots farther back from the tips increase in girth by secondary growth. In other words, at the same time that second-

[1] Certain herbaceous stems, such as geranium and sunflower, also have a limited amount of secondary growth.

ary growth is adding wood and bark, thereby causing the stem to thicken, primary growth is increasing the length of the stem.

Herbaceous dicot and monocot stems differ in internal structure

Although considerable structural variation exists in stems, they all possess an outer protective covering (epidermis or periderm), one or more types of ground tissue, and vascular tissues (xylem and phloem). Let us first consider the structure of herbaceous dicot stems and then of monocot stems.

Vascular bundles of herbaceous dicot stems are arranged in a circle in cross section

A young sunflower stem is a representative herbaceous dicot stem that exhibits primary growth (Fig. 33-2). Its outer covering, the **epidermis,** provides protection in herbaceous stems, as it does in leaves and herbaceous roots (see Table 31-4 and Fig. 31-6). The **cuticle,** a waxy layer of *cutin,* covers the stem epidermis and reduces water loss from the stem surface. **Stomata** permit gas exchange. (Recall from Chapter 32 that a cuticle and stomata are also associated with the leaf epidermis.)

Inside the epidermis is the **cortex,** a cylinder of ground tissue that may contain parenchyma, collenchyma, and sclerenchyma cells (see Table 31-2 and Fig. 31-4). As might be expected from the various types of cells that it contains, the cortex in herbaceous dicot stems can have several functions, such as photosynthesis, storage, and support. If a stem is green, photosynthesis occurs in chloroplasts of cortical parenchyma cells. Parenchyma in the cortex also stores starch (in amyloplasts) and crystals (in vacuoles). Collenchyma and sclerenchyma in the cortex confer strength and structural support for the stem.

The vascular tissues provide conduction and support. In herbaceous dicot stems, the vascular tissues are located in bundles that, when viewed in cross section, are arranged in a circle. However, viewed lengthwise, these bundles extend as long strands throughout the length of a stem and are continuous with vascular tissues of both roots and leaves.

Each vascular bundle contains both **xylem,** which transports water and dissolved nutrient minerals from roots to leaves, and **phloem,** which transports dissolved sugar (see Table 31-3 and Fig. 31-5). Xylem is located on the inner side of the vascular bundle, and phloem is found toward the outside. Sandwiched between xylem and phloem in some herbaceous stems is a single layer of cells called the *vascular cambium,* a lateral meristem responsible for secondary growth (discussed later).

Because most stems support the aerial plant body, they are much stronger than roots. The thick walls of tracheids and vessel elements in xylem help support the plant. Fibers also occur in both xylem and phloem, although they are usually more extensive in phloem. These fibers add considerable strength to the herbaceous stem. In sunflowers and certain other herbaceous dicot stems, phloem contains a cluster of fibers toward the outside of the vascular bundle, called a **phloem fiber cap,** that helps strengthen the stem. The phloem fiber cap is not present in all herbaceous dicot stems.

The **pith** is a ground tissue at the center of the herbaceous dicot stem that consists of large, thin-walled parenchyma cells

FIGURE **33-2** | **LMs of a herbaceous dicot stem.**

(a) Cross section of a sunflower (*Helianthus annuus*) stem. Note the vascular bundles arranged in a circle around a central core of pith. **(b)** Closeup of two vascular bundles. In each bundle, xylem is located toward the stem's interior, and phloem toward the exterior. Each vascular bundle is "capped" by a batch of fibers for additional support.

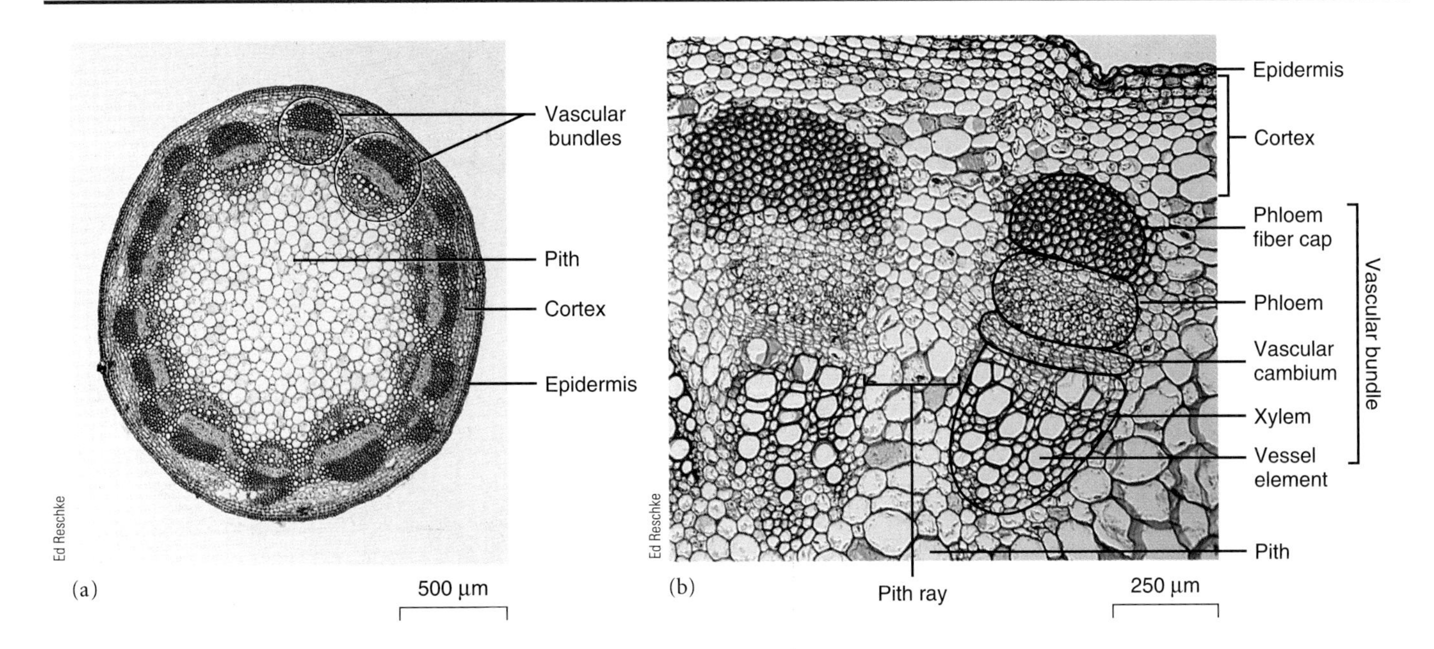

that function primarily in storage. Because of the arrangement of the vascular tissues in bundles, there is no distinct separation of cortex and pith between the vascular bundles. The areas of parenchyma between the vascular bundles are often referred to as **pith rays.**

Vascular bundles are scattered throughout monocot stems

An epidermis with its waxy cuticle covers monocot stems such as the herbaceous stem of corn. As in herbaceous dicot stems, the vascular tissues run in strands throughout the length of a stem. In cross section the vascular bundles contain xylem toward the inside and phloem toward the outside. In contrast with herbaceous dicots, however, vascular bundles of monocots are not arranged in a circle but instead are scattered throughout the stem (Fig. 33-3). Each vascular bundle is enclosed in a bundle sheath of supporting sclerenchyma cells. The monocot stem does not have distinct areas of cortex and pith. The ground tissue in which the vascular tissues are embedded performs the same functions as cortex and pith in herbaceous dicot stems.

Monocot stems do not possess lateral meristems (vascular cambium and cork cambium) that give rise to secondary growth. Monocots have primary growth only and do not produce wood and bark. Although some treelike monocots such as palms attain considerable size, they do so by a modified form of primary growth in which parenchyma cells divide and enlarge. Stems of some monocots such as bamboo and palm contain a great deal of sclerenchyma tissue, which makes them hard and woodlike in appearance.

Woody plants have stems with secondary growth

Woody plants undergo secondary growth, an increase in the girth of stems and roots. Secondary growth occurs as a result of the activity of two lateral meristems: vascular cambium and cork cambium. Among flowering plants, only woody dicots (such as apple, hickory, and maple) have secondary growth. Cone-bearing gymnosperms (such as pine, juniper, and spruce) also have secondary growth.

Cells in the **vascular cambium** divide and produce two conducting and supporting tissues: secondary xylem (wood) to replace primary xylem, and secondary phloem (inner bark) to replace primary phloem. Primary xylem and primary phloem are not able to transport materials indefinitely and so are replaced in plants that have extended life spans.

Cells of the outer lateral meristem, called **cork cambium,** divide and produce cork cells and cork parenchyma. Cork cambium and the tissues it produces are collectively referred to as **periderm** (outer bark), which functions as a replacement for the epidermis (see the last LM in Table 31-4).

Vascular cambium gives rise to secondary xylem and secondary phloem

Primary tissues in woody dicot stems are organized like those in herbaceous dicot stems, with the vascular cambium a thin layer of cells sandwiched between xylem and phloem in the vascular bundles. Once secondary growth begins, the internal structure of a stem changes considerably (Fig. 33-4). Although vas-

FIGURE **33-3** | **LMs of a monocot stem.**

(a) Cross section of a corn (*Zea mays*) stem shows vascular bundles scattered throughout ground tissue. **(b)** Closeup of a vascular bundle. The air space is the site where the first xylem elements were formed and later disintegrated. The entire bundle is enclosed in a bundle sheath of sclerenchyma for additional support.

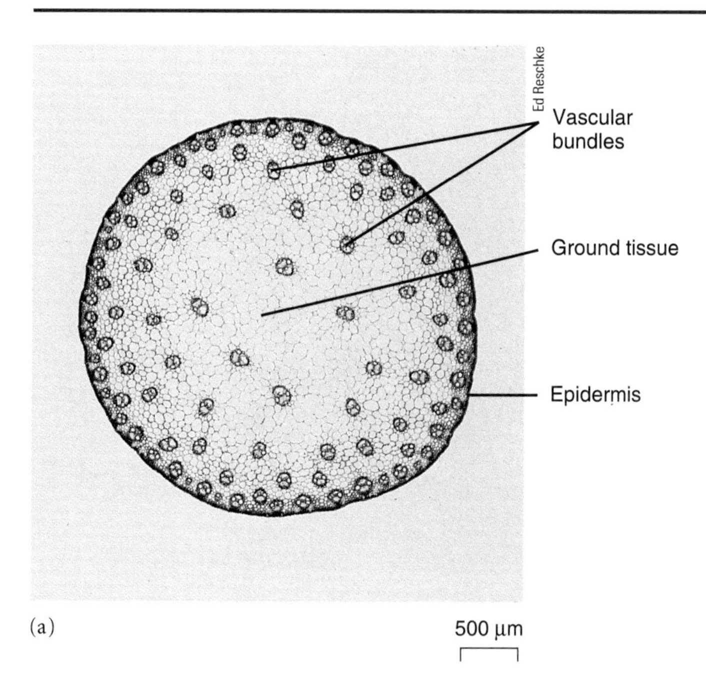

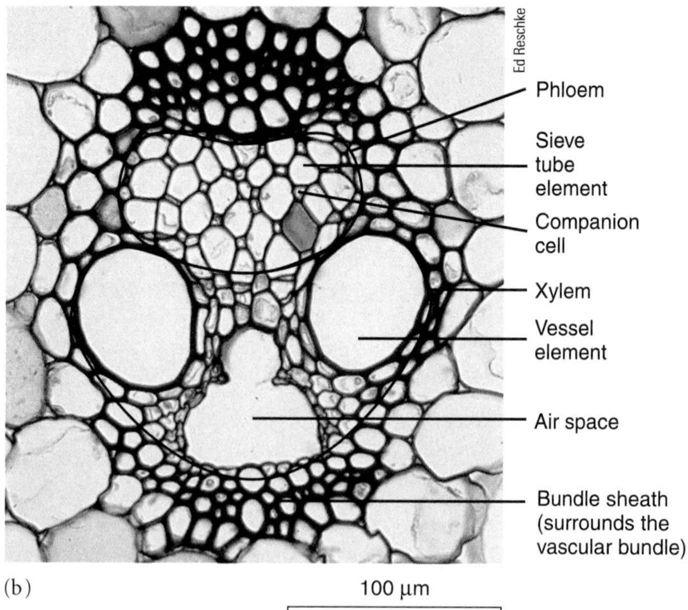

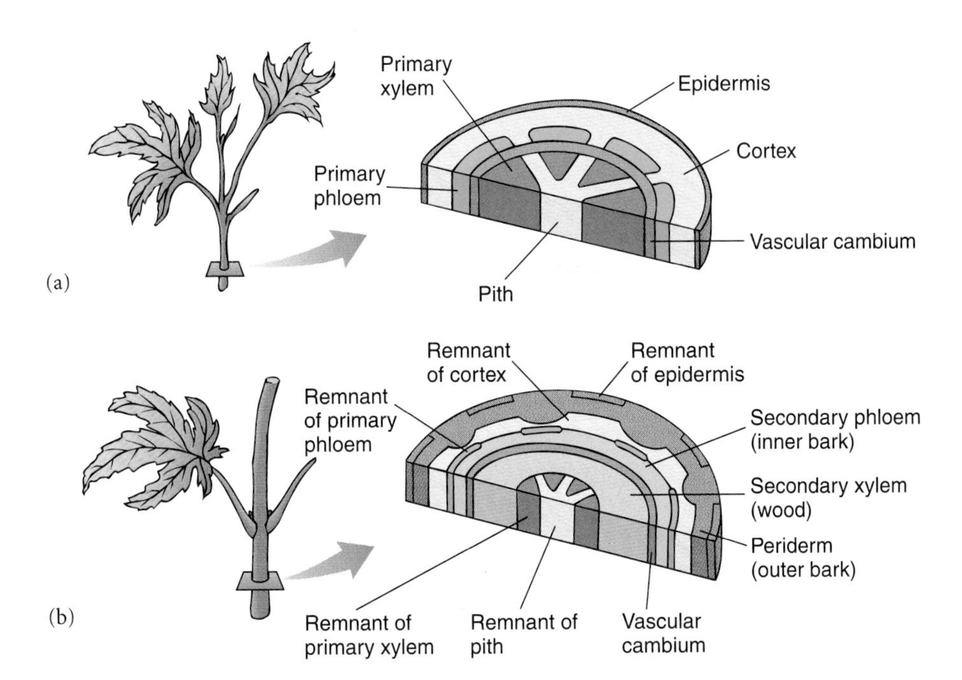

FIGURE **33-4**

Development of secondary growth.

Vascular cambium and the tissues it produces are shown in cross section; cork cambium is not depicted. **(a)** At the onset of secondary growth, vascular cambium arises in the parenchyma between the vascular bundles (that is, in the pith rays), forming a cylinder of meristematic tissue (blue circle in cross section). **(b)** Vascular cambium begins to divide, forming secondary xylem on the inside and secondary phloem on the outside. **(c)** A young woody stem. Vascular cambium produces significantly more secondary xylem than secondary phloem. (The figures change in scale owing to space limitations; pith and primary xylem are actually the same size in all three diagrams, but the change in scale makes those tissues appear to shrink from part a to part b to part c.)

cular cambium is not initially a continuous cylinder of cells (because the vascular bundles are separated by pith rays), it becomes continuous when production of secondary tissues begins. This continuity develops because certain parenchyma cells in each pith ray retain the ability to divide. These cells connect to vascular cambium cells in each vascular bundle, forming a complete ring of vascular cambium.

Cells in the vascular cambium divide and produce daughter cells in two directions. The cells formed from the dividing vascular cambium are located either *inside* the ring of vascular cambium (to become secondary xylem, or wood), or *outside* it (to become secondary phloem, or inner bark) (Fig. 33-5). When a cell in the vascular cambium divides tangentially (inward or outward), one daughter cell remains meristematic; that is, it remains part of the vascular cambium. The other cell may divide again several times, but eventually it stops dividing and develops into mature secondary tissue. Thus vascular cambium is a thin layer of cells sandwiched between the wood and inner bark, the two tissues it produces (Fig. 33-6).

As the stem increases in circumference, the number of cells in the vascular cambium also increases. This occurs by an occasional radial division of a vascular cambium cell, at right angles to its normal direction of division. In this case, both daughter cells remain meristematic.

What happens to the original primary tissues of a stem once secondary growth develops? As a stem increases in thickness, the orientation of the original primary tissues changes. For example, secondary xylem and secondary phloem are laid down between the primary xylem and primary phloem within each vascular bundle. Therefore, as vascular cambium forms secondary tissues, the primary xylem and primary phloem in each vascular bundle become separated from one another (Fig. 33-7). The primary tissues located outside the cylinder of secondary growth (that is, primary phloem, cortex, and epidermis) are subjected to the mechanical pressures produced by secondary growth and are gradually crushed or torn apart and sloughed off.

Secondary tissues replace the primary tissues in function. Secondary xylem conducts water and dissolved nutrient minerals from roots to leaves in the woody plant. It contains the same types of cells found in primary xylem: water-conducting tracheids and vessel elements, in addition to xylem parenchyma cells and fibers. The arrangement of the different cell types in secondary xylem produces the distinctive wood characteristics of each species.

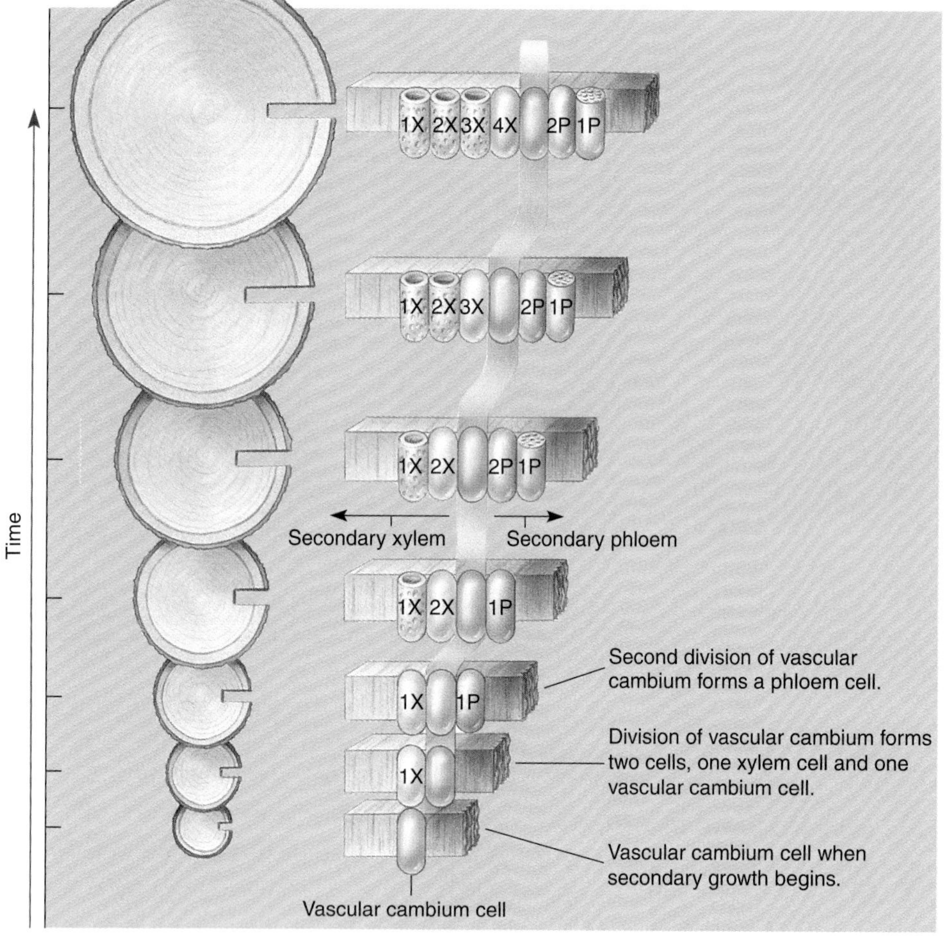

Time

1X 2X 3X 4X 2P 1P

1X 2X 3X 2P 1P

1X 2X 2P 1P

Secondary xylem

Secondary phloem

1X 2X 1P

1X 1P

Second division of vascular cambium forms a phloem cell.

1X

Division of vascular cambium forms two cells, one xylem cell and one vascular cambium cell.

Vascular cambium cell when secondary growth begins.

Vascular cambium cell

FIGURE **33-5**

Development of secondary xylem and secondary phloem.

To study the figure, which shows a radial view of a dividing vascular cambium cell, start at the bottom and move up. Note that vascular cambium *(the blue cell)* divides in two directions, forming secondary xylem (X) to the inside and secondary phloem (P) to the outside. These cells, which are numbered in the order in which they are produced, differentiate to form the mature cell types associated with xylem and phloem. As secondary xylem accumulates, vascular cambium "moves" outward, and the woody stem increases in diameter.

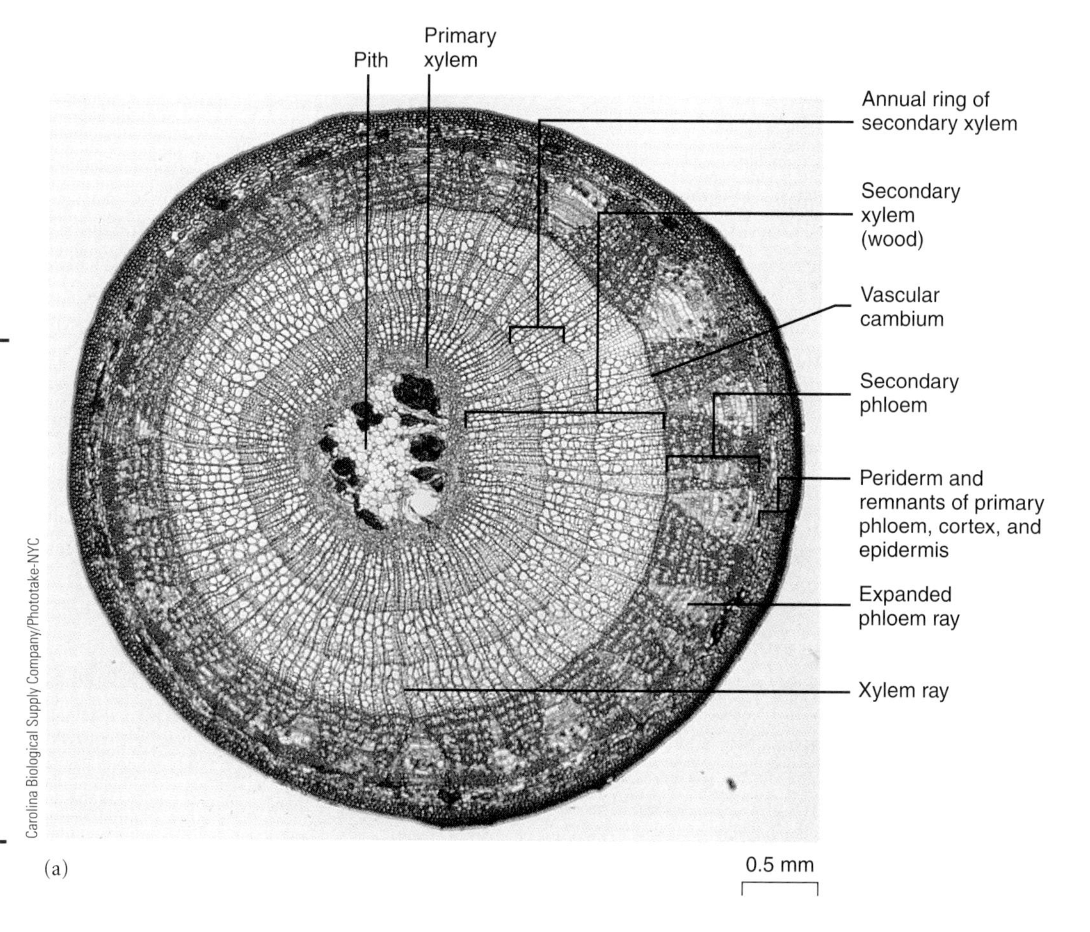

FIGURE **33-6**

Three-year-old stem in cross section.

(a) LM of entire cross section of basswood (*Tilia americana*) stem. **(b)** Sketch of a pie-shaped segment of the cross section (on the facing page). It may be easier to study the parts labeled on the sketch and then locate them on the micrograph. Note the location of the vascular cambium between the secondary xylem (wood) and secondary phloem (inner bark). The primary phloem is not labeled, because it is crushed beyond recognition.

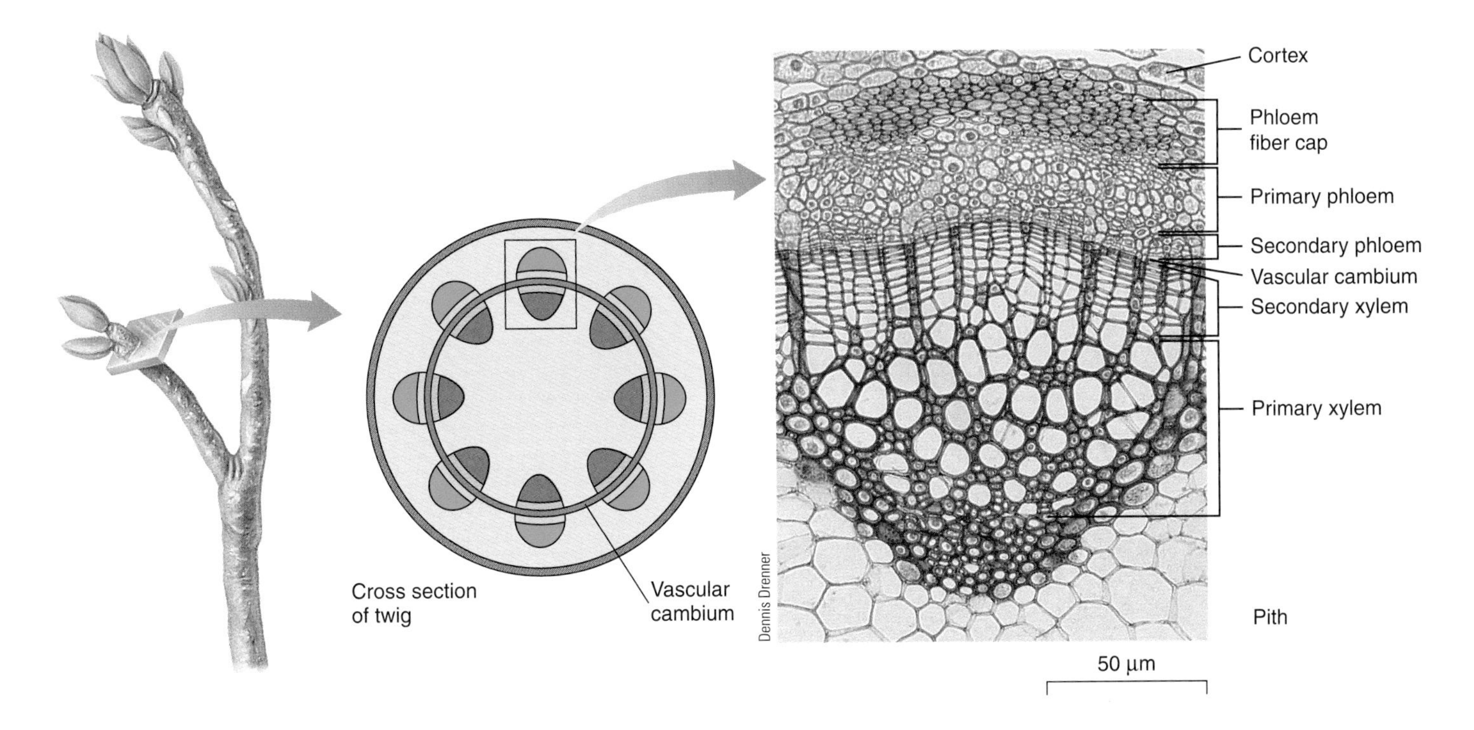

FIGURE **33-7** **LM of part of a young *Magnolia* stem in cross section.**

Note that secondary growth is splitting the vascular bundle apart.

Secondary phloem conducts dissolved sugar from its place of manufacture (leaves) to a place of use and storage (such as roots). The same types of cells found in primary phloem (sieve tube elements, companion cells, phloem parenchyma cells, and fibers) are also found in secondary phloem, although secondary phloem usually has more fibers than primary phloem.

Secondary xylem and secondary phloem transport water, minerals, and sugar vertically throughout the woody plant body. However, materials must also move horizontally (laterally). Lateral movement occurs through **rays,** which are chains of parenchyma cells that radiate out from the center of the woody stem or root (see Fig. 33-6). The vascular cambium forms rays, which are often continuous from the secondary xylem to the secondary phloem. Water and dissolved nutrient minerals are transported laterally through rays, from the secondary xylem to the secondary phloem. Likewise, rays form pathways for the lateral transport of dissolved sugar, from the secondary phloem to the secondary xylem, and of waste products to the center, or heart, of the tree (discussed later).

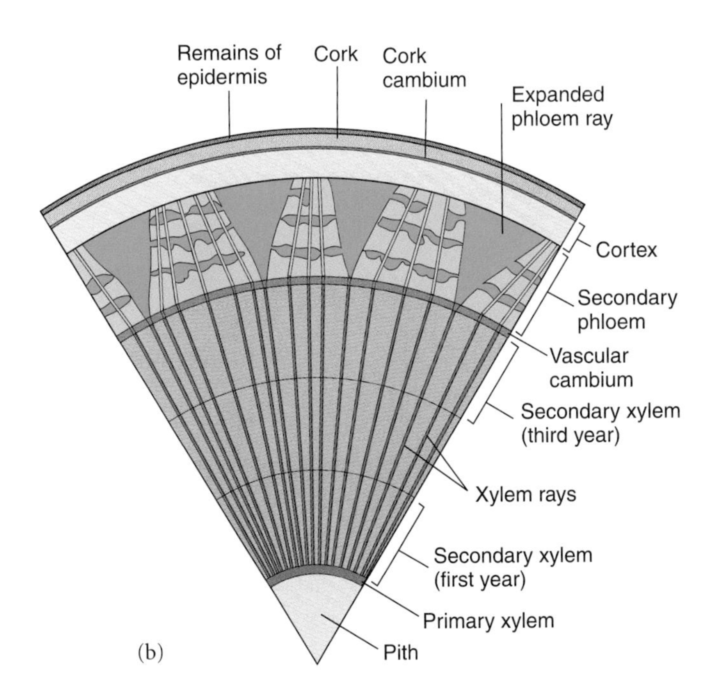

Cork cambium produces periderm

Cork cambium, which usually arises from parenchyma cells in the outer cortex, produces **periderm,** the functional replacement for the epidermis. Cork cambium is either a continuous cylinder of dividing cells (similar to vascular cambium) or a series of overlapping arcs of meristematic cells that form from parenchyma cells in successively deeper layers of the cortex and, eventually, secondary phloem. Variation in cork cambia and their rates of division explain why the outer bark of some tree species is fissured (as in bur oak), rough and shaggy (shagbark hickory), scaly (Norway pine), or smooth and peeling (paper birch).

As is true of vascular cambium, cork cambium divides to form new tissues in two directions—to its inside and its outside. Cork cells, formed to the outside of cork cambium, are dead at maturity and have walls that contain layers of *suberin* and waxes, making them waterproof. These cork cells protect the woody stem against mechanical injury, mild fires, attacks by insects and fungi, temperature extremes, and water loss. To its inside, cork cambium sometimes forms cork parenchyma cells that store water and starch granules. Cork parenchyma is only one to several cells thick, much thinner than the cork cell layer.

Cork cells are impermeable to water and gases, yet the living internal cells of the woody stem require oxygen and must

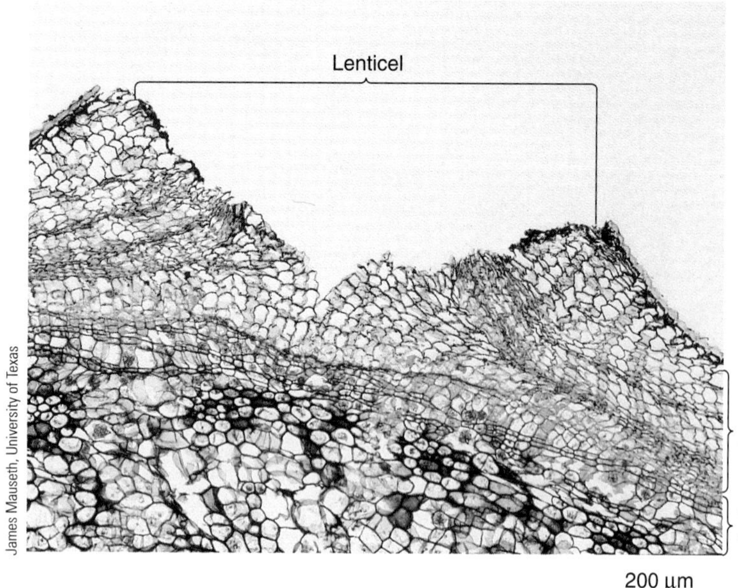

FIGURE **33-8** **LM of stem periderm, showing a lenticel.**

The epidermis has ruptured owing to the proliferation of loosely arranged cork cells in the lenticel. From the bark of a calico flower (*Aristolochia elegans*) stem.

be able to exchange gases with the surrounding atmosphere. As a stem thickens from secondary growth, the epidermis, including stomata that exchange gases for the herbaceous stem, dies. Stomata are replaced by lenticels, which permit gas exchange through the periderm (Fig. 33-8).

Common terms associated with wood are based on plant structure

If you have ever examined different types of lumber, you may have noticed that some trees have wood with two different colors (Fig. 33-9). The functional secondary xylem, that is, the part that conducts water and dissolved nutrient minerals, is the *sapwood,* a thin layer of younger, lighter-colored wood that is closest to the bark. *Heartwood,* the older wood in the center of the tree, is typically a brownish red. A microscopic examination of heartwood reveals that its vessels and tracheids are plugged with pigments, tannins, gums, resins, and other materials. Therefore, heartwood no longer functions in conduction but instead functions as a storage site for waste products. Heartwood is denser than sapwood and therefore provides structural support for trees. Some evidence suggests heartwood is also more resistant to decay.

Almost everyone has heard of hardwood and softwood. Botanically speaking, *hardwood* is the wood of flowering plants and *softwood* is the wood of conifers (cone-bearing gymnosperms). The wood of pine and other conifers typically lacks fibers (with their thick secondary cell walls) and vessel elements; the conducting cells in gymnosperms are tracheids. These cell differences generally make conifer wood softer than the wood of flowering plants, although there is a substantial variation from one species to another. The balsa tree, for example, is a flowering plant whose extremely light, soft wood is used to fashion airplane models.

Woody plants that grow in temperate climates where there is a growing period (during spring and summer) and a dormant period (during winter) exhibit *annual rings,* concentric circles found in cross sections of wood. To determine the age of a woody stem in the temperate zone, simply count the annual rings. In the tropics, environmental conditions, particularly seasonal or year-round precipitation patterns, determine the presence or absence of rings, so rings are not a reliable method of determining the ages of most tropical trees.

Examination of annual rings with a magnifying lens reveals no actual "ring," or line, separating one year's growth from the next. The appearance of a ring in cross section is due to differences in cell size and cell wall thickness between secondary xylem formed at the end of the preceding year's growth and that formed at the beginning of the following year's growth. In the

FIGURE **33-9** **Heartwood and sapwood.**

The wood of older trees consists of a dense, central heartwood and an outer layer of sapwood. The sapwood is the functioning xylem that conducts water and dissolved nutrient minerals. The annual rings in the heartwood are very conspicuous.

spring, when water is plentiful, wood formed by vascular cambium has large-diameter conducting cells (tracheids and vessel elements) and few fibers and is appropriately called *springwood* or *early wood.* As summer progresses and water becomes less plentiful, the wood formed, known as *summerwood* or *late wood,* has narrower conducting cells and many fibers. It is this difference in cell size between the summerwood of one year and the springwood of the following year that gives the appearance of rings (Fig. 33-10). A great deal of information about climate in past times can be learned from the study of annual rings of ancient trees (see *Focus On: Tree Ring Analysis and Climate Change*).

As a woody stem increases in girth over the years, the branches that it bears grow along with it as long as they are alive. If a branch dies, it no longer continues to grow with the stem. In time, as the stem increases in girth, it surrounds the base of the dead branch. The basal portion of an embedded dead branch is called a *knot.* It is possible for a knot to contain bark as well as wood. The presence of knots in wood reduces its commercial value, except for ornamental purposes.

FIGURE 33-10 LM of a portion of a basswood (*Tilia americana*) stem cross section.

One annual ring, or growth increment, is shown. Note the differences in cell size between the vessel elements of springwood and summerwood. The pink cells in the secondary phloem are fibers.

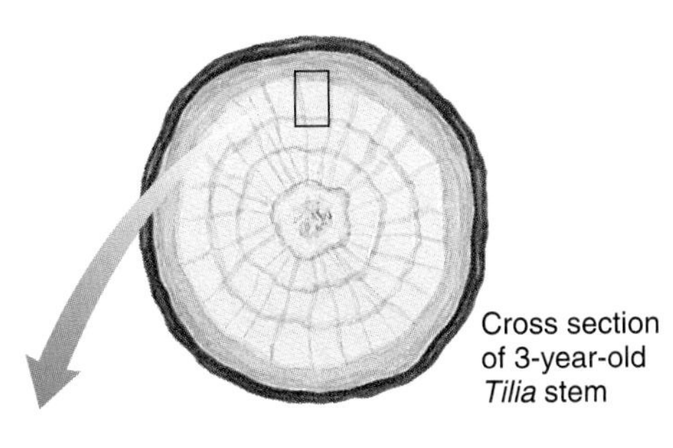

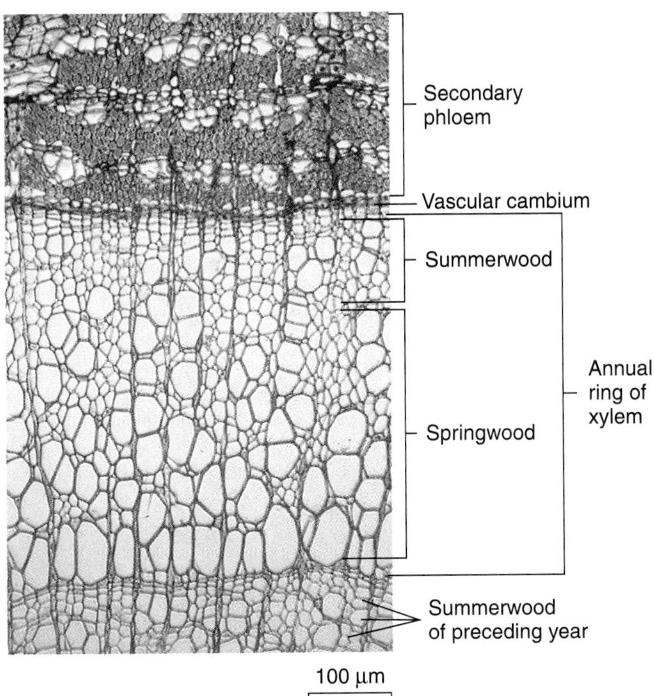

Review

- How does the arrangement of vascular tissue in a stem cross section differ in dicots and monocots?
- What is the difference between vascular cambium and cork cambium? What tissues arise from vascular cambium? From cork cambium?
- What happens to the primary tissues of a stem when secondary growth occurs?
- How do growth rings form in woody stems?
- When a strip of bark is peeled off a tree branch, what tissues are usually removed?

BiologyNow™ Assess your understanding of **stem growth and structure** by taking the pretest on your BiologyNow CD-ROM.

TRANSPORT IN THE PLANT BODY

Learning Objectives

5 Describe the pathway of water movement in plants.
6 Define *water potential.*
7 Explain the roles of tension-cohesion and root pressure as mechanisms responsible for the rise of water and dissolved nutrient minerals in xylem.
8 Describe the pathway of sugar translocation in plants.
9 Discuss the pressure-flow hypothesis of sugar translocation in phloem.

Now that we have discussed stem structure and primary and secondary growth, we examine internal transport in the vascular system of the plant. Roots obtain water and dissolved nutrient minerals from the soil. Once inside roots, these materials are transported upward to stems, leaves, flowers, fruits, and seeds. Furthermore, sugar molecules manufactured in leaves by photosynthesis are transported dissolved in water throughout the plant, including into the subterranean roots. Water and dissolved nutrient minerals are transported from roots to other parts of the plant in xylem, whereas dissolved sugar is **translocated** in phloem.

Xylem transport and phloem translocation do not resemble the movement of materials in animals, because in plants nothing *circulates* in a system of vessels. Water and minerals, transported in xylem, travel in one direction only (upward), whereas translocation of dissolved sugar may occur upward or downward in separate phloem cells. In addition, xylem transport and phloem translocation differ from internal circulation in animals because movement in both xylem and phloem is driven largely by natural physical processes rather than by a pumping organ, or heart.

How, exactly, do materials travel in the continuous system of the plant's vascular tissues? We first examine water and its movement through the plant, and later we discuss the translocation of dissolved sugar.

In temperate climates, counting the number of annual rings determines the age of a tree. The size of each ring varies depending on local weather conditions, including precipitation and temperature. Sometimes the variation in tree rings can be attributed to a single environmental factor, and similar patterns appear in the rings of different tree species over a large geographic area. For example, trees in the U.S. Southwest have similar ring patterns, caused by variations in the amount of annual precipitation. Years with adequate precipitation produce wider rings of growth, whereas years of drought produce much narrower rings.

It is possible to study ring sequences that go back several thousand years. First, a *master chronology,* a complete sample of rings dating back as far as possible, is developed (see figure). To obtain a sample of rings, a small core of wood is bored out of the trunk of an old living tree. The oldest rings (those toward the center of the tree) are matched with the youngest rings (those toward the outside) of an older tree or even an old piece of wood from a house. A master chronology of the area is obtained by using successively older and older sections of wood, even those found in prehistoric dwellings, and by overlapping their matching ring sequences. The longest master chronology is of bristlecone pines in the western United States; it goes back almost 9000 years.

Dendrochronology, the study of both visible and microscopic details of tree rings, has been used extensively in several fields. Tree ring analysis has been extremely useful in dating prehistoric sites of Native Americans in the Southwest. For example, the Cliff Palace in the Mesa Verde National Park dates back to the year 1073. Tree ring analysis indicates that an extended drought forced the original inhabitants to abandon their homes. Tree ring analysis is also useful in other disciplines, including ecology (to study changes in a forest community over time), environmental science (to study the effects of air pollution on tree growth), and geology (to date earthquakes and volcanic eruptions).

Climatologists are increasingly using tree ring data to study past climate patterns. Annual ring widths of certain tree species that grow at high elevations are sensitive to yearly temperature variations; the rings of these trees are wider in warm years and narrower in cool years. Studying tree rings across long time sequences helps researchers determine the natural pattern of global temperature fluctuations. This information is particularly important because of concerns about the human influence on global climate. Scientists generally agree Earth has warmed in recent decades, and there is little doubt that human production of "greenhouse gases" such as carbon dioxide has contributed to this warming (see Chapter 55).

Researchers are not sure how much of the recent warming is the result of human influence, as opposed to natural climate variability. Tree ring analysis may help answer this vital question. Scientists in nine European countries are cooperating in a massive tree ring analysis to construct an annual history of temperatures across northern Europe and Asia since the end of the last Ice Age, about 10,000 years ago. Local diving clubs are obtaining samples of well-preserved logs that are thousands of years old from river and lake sediments. Scientists will use these and other samples, including living trees, to piece together a 10,000-year master chronology. When that is accomplished (sometime before the year 2010), it should be possible to determine conclusively if today's climate patterns are distinctly different from the natural climate patterns of the past.

Water and minerals are transported in xylem

Water initially moves horizontally into roots from the soil, passing through several tissues until it reaches xylem. Once the water moves into the tracheids and vessel elements of root xylem, it travels upward through a continuous network of these hollow, dead cells from root to stem to leaf. Dissolved nutrient minerals are carried along passively in the water. The plant does not expend any energy of its own to transport water, which moves as a result of natural physical processes. The transport of xylem sap is the most rapid of any movement of materials in plants (Table 33-1).

How does water move to the tops of plants? It is either pushed up from the bottom of the plant or pulled up to the top of the plant. Although plants use both mechanisms, current evidence indicates most water is transported through xylem by being *pulled* to the top of the plant.

TABLE 33-1 Xylem and Phloem Transport Rates in Selected Plants

Plant	Maximum Rate in Xylem (cm\min)	Maximum Rate in Phloem (cm\min)
Conifer	2	0.8
Woody dicot	73	2
Herbaceous dicot/monocot	100	2.8–11
Herbaceous vine	250	1.2

Note: Xylem and phloem rates are from different plants within each general group and should be used for comparative purposes only.

Source: Adapted form J.D. Mauseth, *Botany: An Introduction to Plant Biology,* 2nd ed., Philadelphia, Saunders College Publishing, 1995.

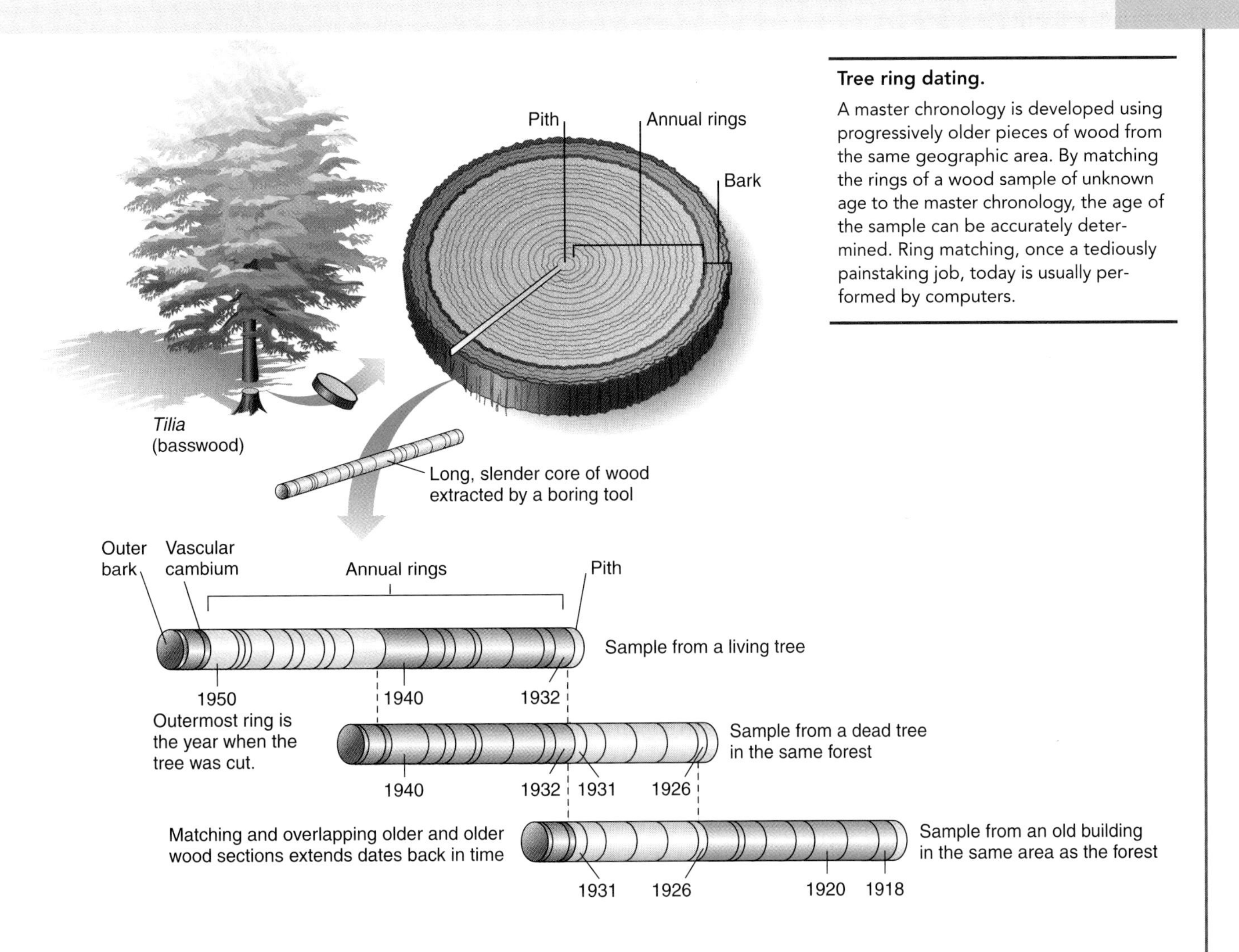

Tree ring dating.

A master chronology is developed using progressively older pieces of wood from the same geographic area. By matching the rings of a wood sample of unknown age to the master chronology, the age of the sample can be accurately determined. Ring matching, once a tediously painstaking job, today is usually performed by computers.

Water movement can be explained by a difference in water potential

To understand how water moves, it is helpful to introduce **water potential,** which is defined as the **free energy** (see Chapter 6) of water. Water potential is important in plant physiology because it is a measure of a cell's ability to absorb water by **osmosis** (see Chapter 5). Water potential also provides a measure of water's tendency to evaporate from cells.

The water potential of pure water is conventionally set at 0 megapascals (MPa), because it cannot be measured directly. (A megapascal is a unit of pressure equal to about 10 atmospheres, or 145.1 pounds per square inch.) However, botanists can measure differences in the free energy of water molecules in different situations. When solutes dissolve in water, the free energy of water decreases. Solutes induce **hydration,** in which water molecules surround ions and polar molecules, keeping them in solution by preventing them from coming together (see Fig. 2-10). The association of water molecules with hydrated molecules and ions reduces the motion of water molecules, decreasing their free energy. Thus *dissolved solutes lower the water potential to a negative number. Water moves from a region of higher (less negative) water potential to a region of lower (more negative) water potential.*

The water potential of the soil varies, depending on how much water it contains. When a soil is extremely dry, its water potential is very low (very negative). When a soil is moister, its water potential is higher, although it is still a negative number because dissolved nutrient minerals are present in dilute concentrations.

The water potential in root cells is also negative owing to the presence of dissolved solutes. Roots contain more dissolved materials than does soil water, unless the soil is extremely dry. This means that *under normal conditions the water potential of the root is more negative than the water potential of the soil.* Thus water moves by osmosis from the soil into the root.

The tension-cohesion model pulls water up a stem

According to the **tension-cohesion model,** also known as the **transpiration-cohesion model,** water is pulled up the plant as a result of a *tension* produced at the top of the plant (Fig. 33-11). This tension, which resembles that produced when drinking a liquid through a straw, is caused by the evaporative pull of transpiration. Recall from Chapter 32 that **transpiration** is the evaporation of water vapor from plants. Most water loss from transpiration takes place through stomata, the numerous microscopic pores present on leaf and stem surfaces. The tension extends from leaves, where most transpiration occurs, down the stems and into the roots. It draws water up stem xylem to leaf cells that have lost water as a result of transpiration, and pulls water from root xylem into stem xylem. As water is pulled upward, additional water from the soil is drawn into the roots. Thus the pathway of water movement is as follows:

Soil ⟶ root tissues (epidermis, cortex, and so forth) ⟶ root xylem ⟶ stem xylem ⟶ leaf xylem ⟶ leaf mesophyll ⟶ atmosphere

This upward pulling of water is only possible as long as there is an unbroken column of water in xylem throughout the plant. Water forms an unbroken column in xylem because of the *cohesiveness* of water molecules. Recall from Chapter 2 that water molecules are **cohesive,** that is, strongly attracted to one another, because of **hydrogen bonding.** In addition, the **adhesion** of water to the walls of xylem cells, also the result of hydrogen bonding, is an important factor in maintaining an unbroken column of water. Thus the cohesive and adhesive properties of water enable it to form a continuous column that can be pulled up through the xylem.

The movement of water in xylem by the tension-cohesion mechanism can be explained in terms of water potential. The atmosphere has an extremely negative water potential. For example, air with a relative humidity of 50% has a water potential of −100 MPa; even moist air at a relative humidity of 90% has a negative water potential of −13 MPa. Thus *there is a water potential gradient from the least negative (the soil) up through the plant to the most negative (the atmosphere).* This gradient literally pulls the water from the soil up through the plant.

Is the tension-cohesion model powerful enough to explain the rise of water in the tallest plants? Plant biologists have calculated that the tension produced by transpiration is strong enough

KEY CONCEPT: Transpiration is the driving force of the tension-cohesion model.

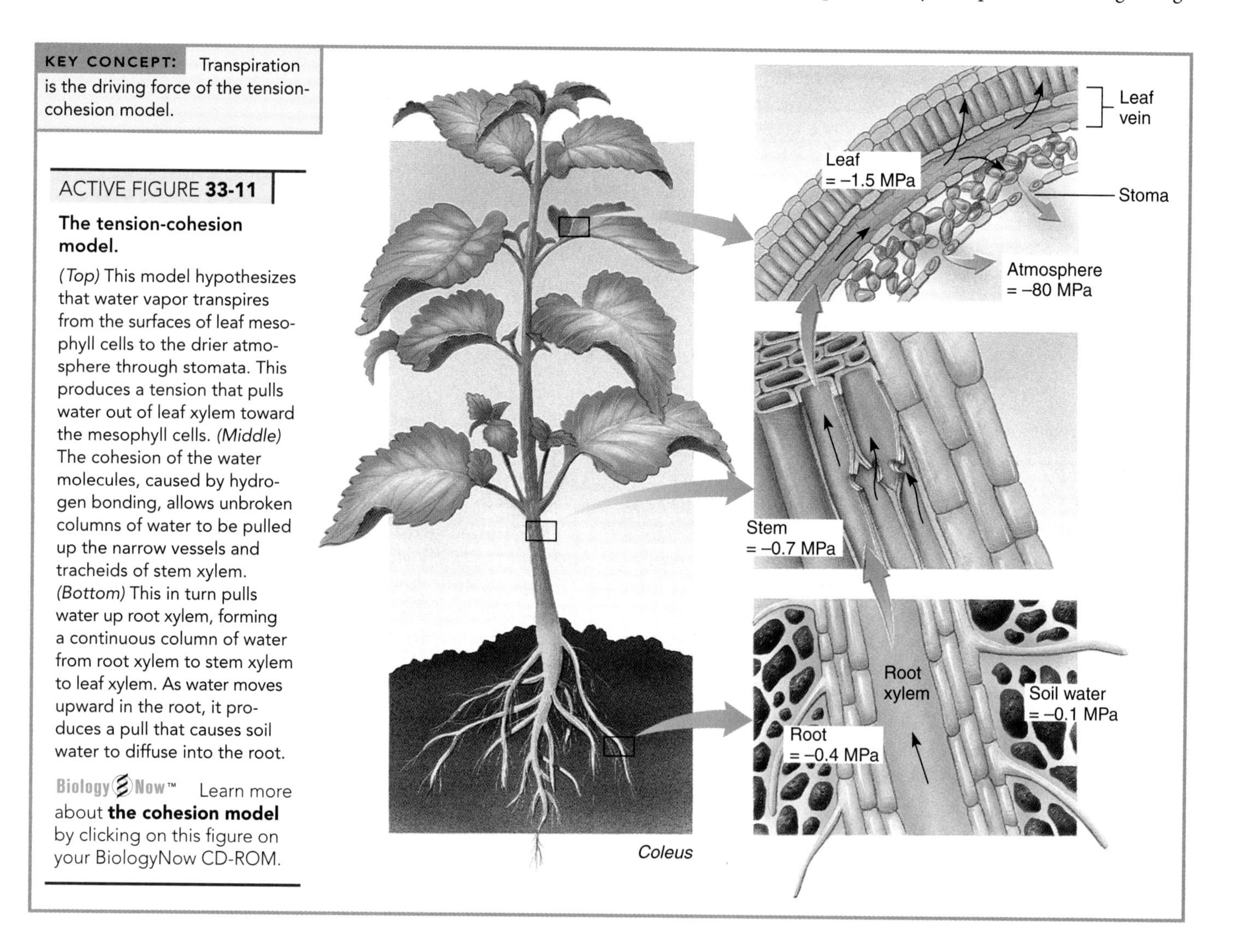

ACTIVE FIGURE **33-11**

The tension-cohesion model.

(Top) This model hypothesizes that water vapor transpires from the surfaces of leaf mesophyll cells to the drier atmosphere through stomata. This produces a tension that pulls water out of leaf xylem toward the mesophyll cells. *(Middle)* The cohesion of the water molecules, caused by hydrogen bonding, allows unbroken columns of water to be pulled up the narrow vessels and tracheids of stem xylem. *(Bottom)* This in turn pulls water up root xylem, forming a continuous column of water from root xylem to stem xylem to leaf xylem. As water moves upward in the root, it produces a pull that causes soil water to diffuse into the root.

BiologyNow™ Learn more about **the cohesion model** by clicking on this figure on your BiologyNow CD-ROM.

to pull water upward 150 m (500 ft) in tubes the diameter of xylem vessels. Because the tallest trees are about 117 m (375 ft), the tension-cohesion model easily accounts for the transport of water. Currently, most botanists consider the tension-cohesion model to be the dominant mechanism of xylem transport in most plants.

Although the tension-cohesion model was first proposed toward the end of the 19th century, conclusive experimental evidence to support this mechanism was first obtained in 1995 by two research groups working independently. Both groups demonstrated that large negative pressures exist in xylem and that the water potential gradients in root, stem, and leaf xylem are adequate to explain the observed movement of water.

Root pressure pushes water from the root up a stem

In the less important mechanism for water transport, known as **root pressure,** water that moves into roots from the soil is *pushed* up through xylem toward the top of the plant. Root pressure occurs because nutrient mineral ions that are actively absorbed from the soil are pumped into the xylem, decreasing its water potential. Water then moves into xylem cells from surrounding root cells. In turn, water moves into roots by osmosis because of the difference in water potential between the soil and root cells. The accumulation of water in root tissues produces a positive pressure (as high as +0.2 MPa) that forces the water up through the xylem.

Guttation, a phenomenon in which liquid water is forced out through special openings in the leaves (see Chapter 32), results from root pressure. However, root pressure is not strong enough to explain the rise of water to the tops of coastal redwoods and other tall trees. Root pressure exerts an influence in smaller plants, particularly in the spring when the soil is quite wet, but it clearly does not cause water to rise 100 m (330 ft) or more in the tallest plants. Furthermore, root pressure does not occur to any appreciable extent in summer (when water is often not plentiful in soil), yet the movement of water is greatest during hot summer days.

Sugar in solution is translocated in phloem

The sugar produced during photosynthesis is converted into sucrose (common table sugar), a disaccharide composed of one molecule of glucose and one of fructose (see Fig. 3-8b), before being loaded into phloem and translocated to the rest of the plant. Sucrose is the predominant photosynthetic product carried in phloem. Phloem sap also contains much smaller amounts of other materials, such as amino acids, organic acids, proteins, hormones, certain minerals, and sometimes disease-causing plant viruses. Translocation of phloem sap is not as rapid as xylem transport (see Table 33-1).

Fluid within phloem tissue moves both upward and downward. Sucrose is translocated in individual sieve tubes from a *source,* an area of excess sugar supply (usually a leaf), to a *sink,* an area of storage (as insoluble starch) or of sugar use such as roots, apical meristems, fruits, and seeds.

The pressure-flow hypothesis explains translocation in phloem

Current experimental evidence supports the translocation of dissolved sugar in phloem by the **pressure-flow hypothesis,** which was first proposed in 1930 by the German scientist Ernst Münch. The pressure-flow hypothesis states that solutes (such as dissolved sugars) move in phloem by means of a pressure gradient—that is, a difference in pressure. The pressure gradient exists between the source, where the sugar is loaded into phloem, and the sink, where the sugar is removed from phloem.

At the source, the dissolved sucrose is moved from a leaf's mesophyll cells, where it was manufactured, into the companion cell, which loads it into the sieve tube elements of phloem. This loading occurs by active transport, a process that requires adenosine triphosphate (ATP) (Fig. 33-12). The ATP supplies energy to pump protons out of the sieve tube elements, producing a proton gradient that drives the uptake of sugar through specific channels by the **cotransport** of protons back into the sieve tube elements (an example of a linked cotransport system, discussed in Chapter 5). The sugar therefore accumulates in the sieve tube element. The increase in dissolved sugars in the sieve tube element at the source—a concentration that is two to three times greater than in surrounding cells—decreases (makes more negative) the water potential of that cell. As a result, water moves by osmosis from the xylem cells into the sieve tubes, increasing the **turgor pressure** (hydrostatic pressure) inside them. Thus, phloem loading at the source is as follows:

> Proton pump moves H^+ out of sieve tube element ⟶ sugar is actively transported into sieve tube element ⟶ water diffuses from xylem into sieve tube element ⟶ turgor pressure increases within sieve tube

At its destination (the sink), sugar is unloaded by various mechanisms, both active and passive, from the sieve tube elements. With the loss of sugar, the water potential in the sieve tube elements at the sink increases (becomes less negative). Therefore, water moves out of the sieve tubes by osmosis and into surrounding cells where the water potential is more negative. Most of this water diffuses back to the xylem to be transported upward. This water movement decreases the turgor pressure inside the sieve tubes at the sink. Thus phloem unloading at the sink proceeds as follows:

> Sugar is transported out of sieve tube element ⟶ water diffuses out of sieve tube element and into xylem ⟶ turgor pressure decreases within sieve tube

The pressure-flow hypothesis explains the movement of dissolved sugar in phloem by means of a pressure gradient. The difference in sugar concentrations between the source and the sink causes translocation in phloem, as water and dissolved sugar flow along the pressure gradient. This pressure gradient

ACTIVE FIGURE **33-12** | **The pressure-flow hypothesis.**

Sugar is actively loaded into the sieve tube element at the source. As a result, water diffuses from the xylem into the sieve tube element. At the sink, the sugar is actively or passively unloaded, and water diffuses from the sieve tube element into the xylem. The pressure gradient within the sieve tube, from source to sink, causes translocation from the area of higher turgor pressure (the source) to the area of lower turgor pressure (the sink).

BiologyNow™ Learn more about **how sugar travels** by clicking on this figure on your BiologyNow CD-ROM.

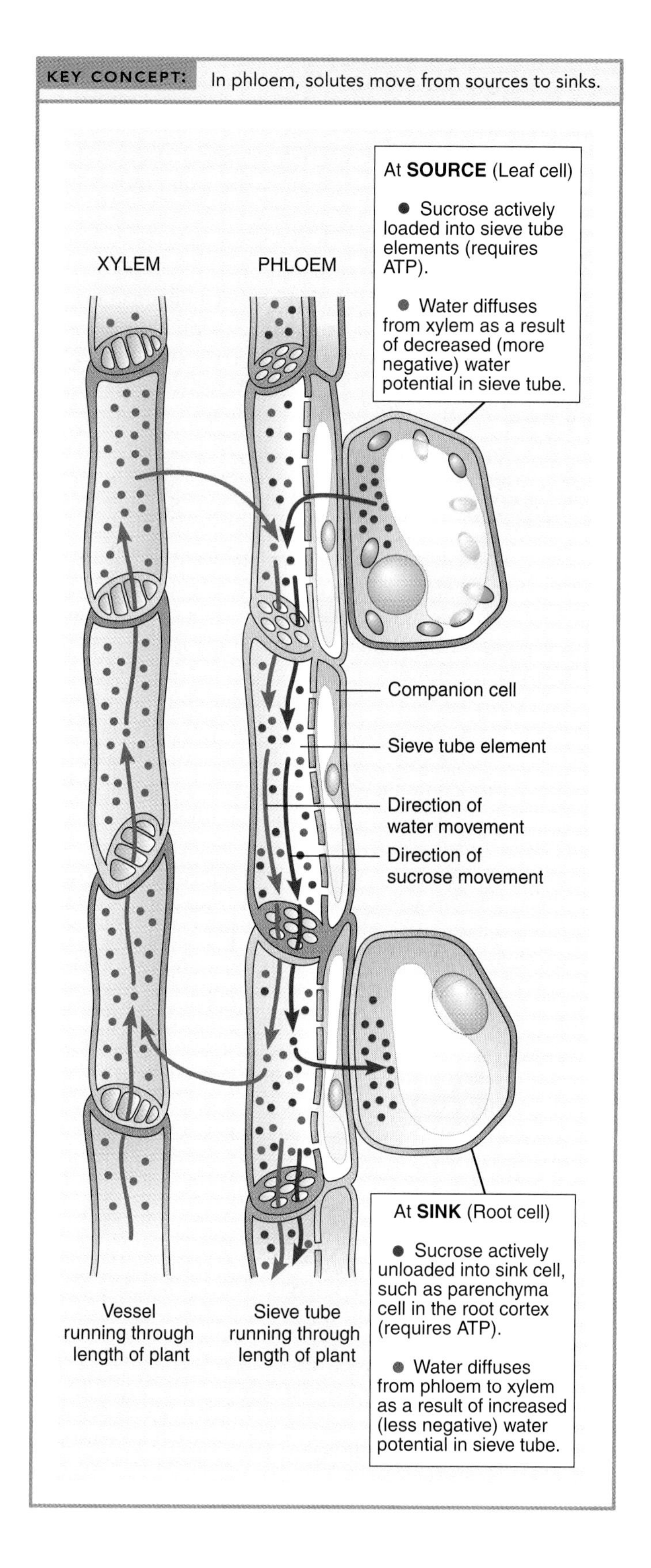
KEY CONCEPT: In phloem, solutes move from sources to sinks.
At SOURCE (Leaf cell)
Sucrose actively loaded into sieve tube elements (requires ATP).
Water diffuses from xylem as a result of decreased (more negative) water potential in sieve tube.
XYLEM
PHLOEM
Companion cell
Sieve tube element
Direction of water movement
Direction of sucrose movement
At SINK (Root cell)
Sucrose actively unloaded into sink cell, such as parenchyma cell in the root cortex (requires ATP).
Water diffuses from phloem to xylem as a result of increased (less negative) water potential in sieve tube.
Vessel running through length of plant
Sieve tube running through length of plant

pushes the sugar solution through phloem much as water is forced through a hose.

The actual translocation of dissolved sugar in phloem does not require metabolic energy. However, the loading of sugar at the source and the active unloading of sugar at the sink require energy derived from ATP to move the sugar across cell membranes by active transport.

PROCESS OF SCIENCE

Although the pressure-flow hypothesis adequately explains current data on phloem translocation, much remains to be learned about this complex process. Phloem translocation is difficult to study in plants. Because phloem cells are under pressure, cutting into phloem to observe it releases the pressure and causes the contents of the sieve tube elements (the phloem sap) to exude and mix with the contents of other severed cells that are also unavoidably cut. In the 1950s botanists developed a unique research tool to avoid contaminating the phloem sap: aphids, which are small insects that insert their mouthparts into phloem sieve tubes for feeding (Fig. 33-13). The pressure in the punctured phloem drives the sugar solution through the aphid's mouthpart and into its digestive system. When the aphid's mouthpart is severed from its body by a laser beam, the sugar solution continues to flow through the mouthpart at a rate proportional to the pressure in phloem. This rate can be measured, and the effects on phloem transport of different environmental conditions—varying light intensities, darkness, and mineral deficiencies, for example—can be ascertained.

The identity and proportions of translocated substances can also be determined using severed aphid mouthparts. This technique has verified that in most plant species the sugar sucrose is the primary carbohydrate transported in phloem; however, some species transport other sugars, such as raffinose, or sugar alcohols, such as sorbitol.

Review

- How does the direction of transport differ in xylem and phloem?
- What is water potential? How is the movement of water related to water potential?
- How does the tension-cohesion model explain the rise of water in the tallest trees?
- How does the pressure-flow hypothesis explain sugar movement in phloem? In your answer, make sure you include the activities at source and sink.

BiologyNow™ Assess your understanding of **transport in the plant body,** by taking the pretest on your BiologyNow CD-ROM.

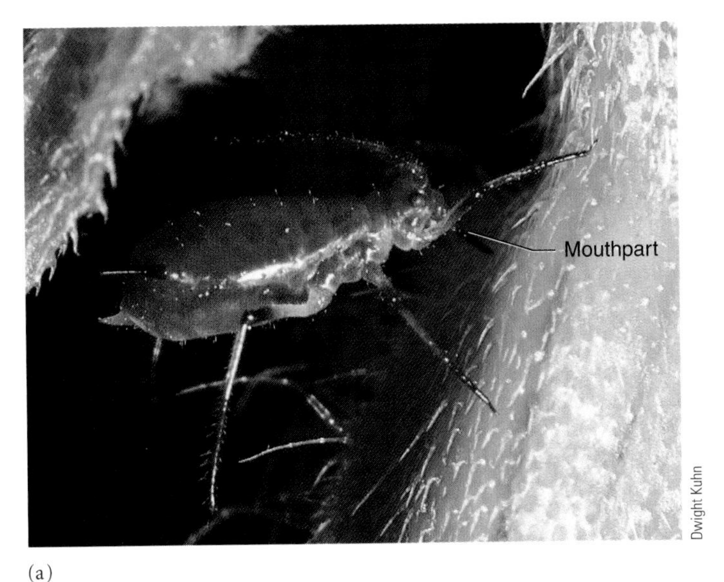

(a)

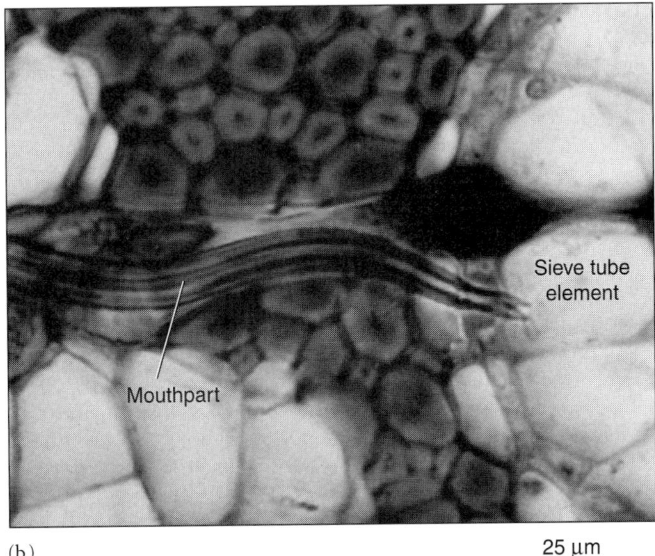

(b) 25 μm

FIGURE **33-13** **Aphids are used to study translocation in phloem.**

(a) Mature aphid, a tiny insect about 3 to 6 mm in length, feeding on a stem. **(b)** LM of phloem cells, showing a sieve tube element that has been penetrated by the aphid mouthpart. *(M.H. Zimmerman,* Science, *Vol. 133, pp. 73–79 [Fig. 4], 13 Jan. 1961. Copyright 2005 by the American Association for the Advancement of Science)*

SUMMARY WITH KEY TERMS

1 Describe the external features of a woody twig.

- Woody twigs demonstrate the external structure of stems. **Buds** are undeveloped embryonic shoots. A **terminal bud** is located at the tip of a stem, whereas **axillary buds (lateral buds)** are located in leaf axils. A dormant bud is covered and protected by **bud scales.** When the bud resumes growth, bud scales covering the bud fall off, leaving **bud scale scars.**
- The area on a stem where each leaf is attached is called a **node,** and the region of a stem between two successive nodes is an **internode.** A **leaf scar** shows where each leaf was attached to the stem. **Bundle scars** are the areas within a leaf scar where the vascular tissue extended from the stem to the leaf.
- **Lenticels** are sites of loosely arranged cells that allow oxygen to diffuse into the interior of a woody stem.

2 Label cross sections of herbaceous dicot and monocot stems, and describe the functions of each tissue.

- Herbaceous stems possess an epidermis, vascular tissue, and either ground tissue or cortex and pith. The **epidermis** is a protective layer covered by a water-conserving **cuticle. Stomata** permit gas exchange. **Xylem** conducts water and dissolved nutrient minerals, and **phloem** conducts dissolved sugar. The **cortex, pith,** and **ground tissue** function primarily for storage.
- Although all herbaceous stems have the same basic tissues, their arrangement varies considerably. Herbaceous dicot stems have the vascular bundles arranged in a circle (in cross section) and have a distinct cortex and pith. Monocot stems have vascular bundles scattered in ground tissue.

3 Name the two lateral meristems, and describe the tissues that arise from each.

- **Vascular cambium** is the lateral meristem that produces secondary xylem (wood) and secondary phloem (inner bark).
- **Cork cambium** produces **periderm,** which consists of cork parenchyma and cork cells. Cork cells are the functional replacement for epidermis in a woody stem. Cork parenchyma functions primarily for storage in a woody stem.

4 Outline the transition from primary growth to secondary growth in a woody stem.

- **Secondary growth** (the production of the secondary tissues, wood and bark) occurs in some flowering plants (woody dicots) and in all cone-bearing gymnosperms.
- During secondary growth, the vascular cambium, which develops between the primary xylem and the primary phloem, divides in two directions to form secondary xylem (to the inside) and secondary phloem (to the outside). The primary xylem and primary phloem in the original vascular bundles become separated as secondary growth proceeds.

5 Describe the pathway of water movement in plants.

- Water and dissolved nutrient minerals move from the soil into root tissues (epidermis, cortex, and so forth). Once in root xylem, water and dissolved minerals move upward, from root xylem to stem xylem to leaf xylem. Much of the water entering the leaf exits leaf veins and passes into the atmosphere.

6 Define *water potential.*

- **Water potential** is a measure of the **free energy** of water. Pure water has a water potential of 0 megapascals, whereas water with dissolved solutes has a negative water potential.

- Water moves from an area of higher (less negative) water potential to an area of lower (more negative) water potential.

7 Explain the roles of tension-cohesion and root pressure as mechanisms responsible for the rise of water and dissolved nutrient minerals in xylem.

- The **tension-cohesion model** explains the rise of water in even the largest plants. The evaporative pull of **transpiration** causes tension at the top of the plant. This tension is the result of a water potential gradient that ranges from the slightly negative water potentials in the soil and roots to the very negative water potentials in the atmosphere. As a result of the **cohesive** and **adhesive** properties of water, the column of water pulled up through the plant remains unbroken.
- **Root pressure,** caused by the movement of water into roots from the soil as a result of the active absorption of nutrient mineral ions from the soil, helps explain the rise of water in smaller plants, particularly when the soil is wet. Root pressure pushes water up through xylem.

8 Describe the pathway of sugar translocation in plants.

- Dissolved sugar is **translocated** upward or downward in phloem, from a source (an area of excess sugar, usually a leaf) to a sink (an area of storage or of sugar use, such as roots, apical meristems, fruits, and seeds). Sucrose is the predominant sugar translocated in phloem.

9 Discuss the pressure-flow hypothesis of sugar translocation in phloem.

- The movement of materials in phloem is explained by the **pressure-flow hypothesis.**
- Companion cells actively load sugar into the sieve tubes at the source; ATP is required for this process. The ATP supplies energy to pump protons out of the sieve tube elements. The proton gradient drives the uptake of sugar by the cotransport of protons back into the sieve tube elements. Sugar therefore accumulates in the sieve tube element, causing the movement of water into the sieve tubes by **osmosis.**
- Companion cells actively (requiring ATP) and passively (not requiring ATP) unload sugar from the sieve tubes at the sink. As a result, water leaves the sieve tubes by osmosis, decreasing the **turgor pressure** (hydrostatic pressure) inside the sieve tubes.
- The flow of materials between source and sink is driven by the turgor pressure gradient produced by water entering phloem at the source and water leaving phloem at the sink.

POST-TEST

1. The three main functions of stems are (a) support, conduction, and photosynthesis (b) support, anchorage in soil, and production of new living tissues (c) conduction, production of new living tissues, and sexual reproduction (d) conduction, asexual reproduction, and sexual reproduction (e) support, conduction, and production of new living tissues
2. All stems have undeveloped embryonic shoots called (a) lenticels (b) buds (c) vines (d) phloem fiber caps (e) periderm
3. Axillary buds are located (a) at the tips of stems (b) in unusual places, such as on roots (c) in the region between two successive nodes (d) in the upper angle between a leaf and the stem to which it is attached (e) within the loosely arranged cells of lenticels
4. The tissue in monocot stems in which the vascular tissues are embedded is (a) cork cambium (b) cortex (c) ground tissue (d) pith (e) phloem
5. The protective outer layer of cells covering herbaceous stems is the (a) periderm (b) cork cambium (c) lateral meristem (d) epidermis (e) bud scale
6. Ground tissue in monocot stems performs the same functions as ______________ and ______________ in herbaceous dicot stems (a) phloem; xylem (b) cork cambium; vascular cambium (c) epidermis; periderm (d) primary xylem; secondary xylem (e) cortex; pith
7. Which of the following statements is *false?* (a) Primary growth is an increase in the length of a plant (b) Primary growth occurs at both apical and lateral meristems (c) All plants have primary growth (d) Herbaceous stems have primary growth, whereas woody stems have both primary and secondary growth (e) Buds are embryonic shoots that contain apical meristems
8. The two lateral meristems responsible for secondary growth are (a) phloem and xylem (b) cork cambium and vascular cambium (c) epidermis and periderm (d) primary xylem and secondary xylem (e) cortex and pith
9. Cork cambium and the tissues it produces are collectively called (a) periderm (b) lenticels (c) cortex (d) epidermis (e) wood
10. Horizontal movement of materials in woody plants occurs in (a) bud scales (b) cortex (c) rays (d) lenticels (e) pith rays
11. The older wood in the center of a tree trunk is commonly called (a) hardwood (b) softwood (c) sapwood (d) heartwood (e) cork
12. Each annual ring in a section of wood represents one year's growth of (a) primary xylem (b) secondary xylem (c) primary xylem or secondary xylem in alternate years (d) primary phloem (e) secondary phloem
13. Water potential is (a) the formation of a proton gradient across a cell membrane (b) the transport of a watery solution of sugar in phloem (c) the transport of water in both xylem and phloem (d) the removal of sucrose at the sink, causing water to move out of the sieve tubes (e) the free energy of water in a particular situation
14. Which of the following is a mechanism of water movement in xylem that does *not* generate sufficient force to explain the rise of water to the tops of the tallest trees? (a) pressure-flow hypothesis (b) tension-cohesion (c) root pressure (d) active transport of potassium ions into guard cells (e) transpiration
15. Which of the following is a mechanism of water movement in xylem that combines the evaporative pull of transpiration with the cohesive and adhesive properties of water? (a) pressure-flow (b) tension-cohesion (c) root pressure (d) active transport of potassium ions into guard cells (e) guttation
16. Which of the following is a mechanism of phloem transport in which dissolved sugar is moved by means of a pressure gradient

POST-TEST (continued)

that exists between the source and the sink? (a) pressure-flow (b) tension-cohesion (c) root pressure (d) active transport of potassium ions into guard cells (e) guttation

17. How does increasing solute concentration affect water potential? (a) water potential becomes more positive (b) water potential becomes more negative (c) water potential becomes more positive under certain conditions and more negative under other conditions (d) water potential is not affected by solute concentration (e) water potential is always zero when solutes are dissolved in water

18. Label the various tissues, give at least one function for each tissue, and identify the stem as a herbaceous dicot, monocot, or woody plant. Use Figure 33-2 to check your answers.

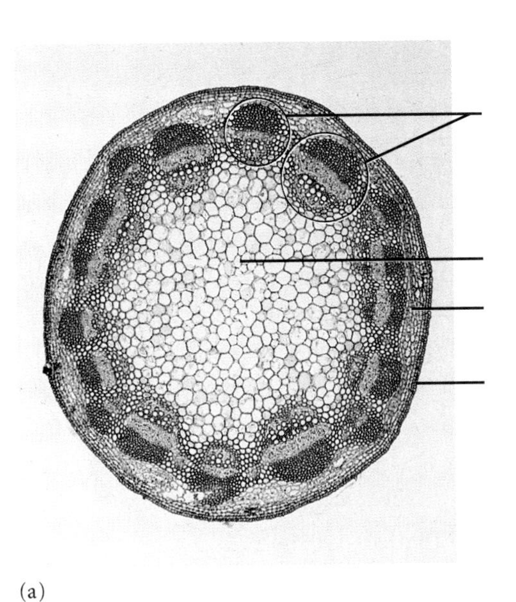
(a)

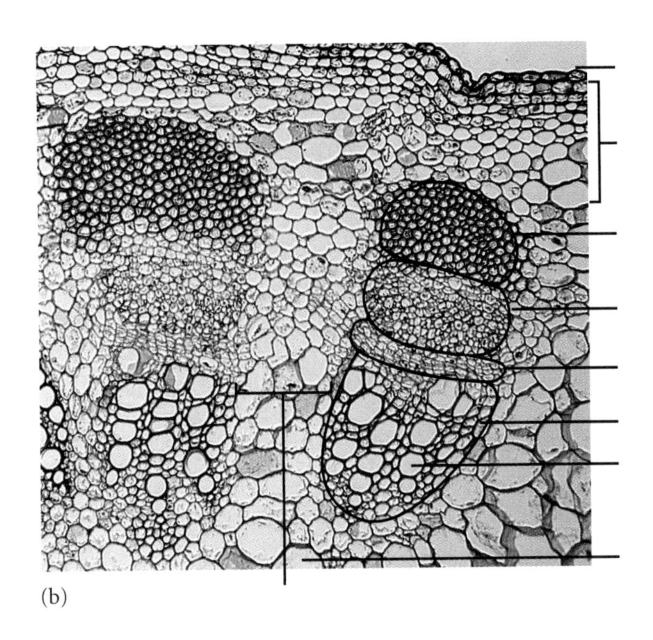
(b)

CRITICAL THINKING

1. When secondary growth is initiated, certain cells become meristematic and begin to divide. Could a mature tracheid ever do this? A sieve tube element? Why or why not?
2. Why does the wood of many tropical trees lack annual rings? Why does the wood of other tropical trees possess annual rings?
3. Why is hardwood more desirable than softwood for making furniture? Explain your answer, based on the structural differences between hardwood and softwood.
4. Why should you cut off a few inches from the stem ends of cut flowers before placing them in water? Base your answer on what you have learned about the tension-cohesion model.

■ Visit our Web site at **http://biology.brookscole.com/solomon7** for links to chapter-related resources on the World Wide Web. Additional online materials relating to this chapter can also be found on our Web site.

BIOLOGY NOW RESOURCES

Active Figures

33-11: The tension-cohesion model

33-12: How sugar travels in a plant

Preparing for an exam? Take a diagnostic test on your BiologyNow CD-ROM.

Post-Test Answers

1. e	2. b	3. d	4. c
5. d	6. e	7. b	8. b
9. a	10. c	11. d	12. b
13. e	14. c	15. b	16. a
17. b			

34 Roots and Mineral Nutrition

R. Calentine/Visuals Unlimited

Storage roots. Carrots (*Daucus carota*) are biennials that live for two years. During the first year's growth, food is stored in the fleshy root system; during the second year, the shoot elongates and produces flowers. Carrots and other root crops are important sources of human food.

CHAPTER OUTLINE

- **Root Structure and Function**
- **Root Associations with Other Organisms**
- **The Soil Environment**

In Chapters 32 and 33 we discussed the aerial vegetative structures of a plant: the leaves and stems. In this chapter we turn to the third major vegetative organ: the roots. Branching underground root systems are often more extensive than a plant's aerial parts. The roots of a corn plant, for example, may grow to a depth of 2.5 m (about 8 ft) and spread outward 1.2 m (4 ft) from the stem. Desert-dwelling tamarisk (*Tamarix*) trees reportedly have roots that grow to a depth of 50 m (163 ft) to tap underground water. The total root length, not counting root hairs, of a four-month-old rye (*Secale cereale*) plant was found to exceed 500 km (310 mi)! The extent of a plant's root depth and spread varies considerably among different species and even among different individuals in the same species. Soil conditions, discussed in this chapter, greatly affect the extent of root growth.

Because roots are usually underground and out of sight, people do not always appreciate the important functions they perform. First, as anyone who has ever pulled weeds can attest, roots anchor a plant securely in the soil. A plant needs a solid foundation from which to grow. Firm anchorage is essential to a plant's survival so that the stem remains upright, enabling leaves to absorb sunlight effectively.

Second, roots absorb water and dissolved nutrient minerals such as nitrates, phosphates, and sulfates, which are necessary for synthesizing important organic molecules. These dissolved nutrient minerals are then transported throughout the plant in the xylem.

Storage is the third main function performed by many roots. Carrots (see photograph), sweet potatoes, cassava, and other root crops are important sources of human food. Surplus sugars produced in the leaves by photosynthesis are transported in the phloem to the roots for storage (usually as starch or sucrose) until needed. Carrot roots have extensive phloem for this purpose. Although roots use some photosynthetic products for their own respiratory needs, most are stored and later transported out of the roots for use by the plant. Both *taproots* (carrots, beets, radishes, and turnips) and *fibrous roots* (sweet potatoes) may be modified for storage. Plants with storage taproots are often **biennials** (see Chapter 31) that, as part of the strategy to survive winter, store their food reserves in the root during the first year's growth and use these reserves to reproduce during the second year's growth. Other plants, particularly those living in arid regions, possess storage roots adapted to store water.

In certain species, roots are modified for functions other than anchorage, absorption, conduction, and storage. Roots specialized to perform uncommon functions are discussed later in this chapter. ■

ROOT STRUCTURE AND FUNCTION

Learning Objectives

1. Distinguish between taproot and fibrous root systems.
2. Label cross sections of a primary dicot root and a monocot root, and describe the functions of each tissue.
3. Trace the pathway of water and nutrient mineral ions from the soil through the various root tissues, and distinguish between the symplast and apoplast.
4. Discuss the structure of roots with secondary growth.
5. Describe at least four roots that are modified to perform uncommon functions.

Two types of root systems, a taproot system and a fibrous root system, occur in plants (Fig. 34-1). A **taproot** system consists of one main root that formed from the seedling's enlarging **radicle,** or embryonic root. Many lateral roots of various sizes branch out of a taproot. Taproots are characteristic of many dicots and gymnosperms. A dandelion is a good example of a common herbaceous plant with a taproot system. A few trees, such as hickory, retain their taproots, which become quite massive as the plants age. Most trees, however, have taproots when young and later develop large, shallow lateral roots from which other roots branch off and grow downward.

A **fibrous root** system has several to many roots of similar size developing from the end of the stem, with lateral roots of various sizes branching off these roots. Fibrous root systems form in plants that have a short-lived embryonic root. The roots first originate from the base of the embryonic root and later from stem tissue. The main roots of a fibrous root system do not arise from pre-existing roots, but rather from the stem; such roots are called **adventitious roots.** Adventitious organs occur in an unusual location, such as roots that develop on a stem, or buds that develop on roots. Onions, crabgrass, and other monocots have fibrous root systems.

Taproot and fibrous root systems are adapted to obtain water in different sections of the soil. Taproot systems often extend down into the soil to obtain water located deep underground, whereas fibrous root systems, which are located relatively close to the soil surface, are adapted to obtain rainwater from a larger area as it drains into the soil.

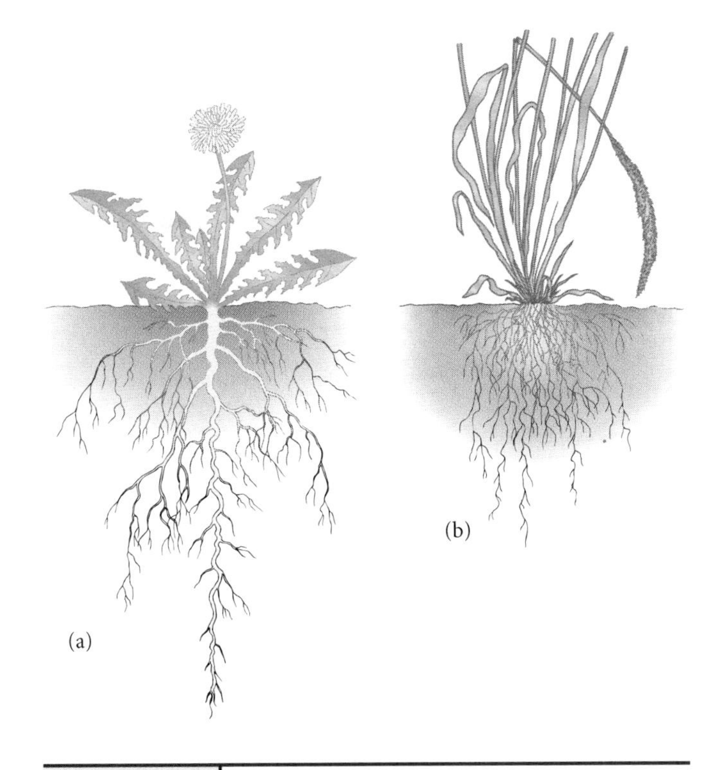

FIGURE **34-1** **Root systems.**

(a) A taproot system develops from the embryonic root in the seed. **(b)** The roots of a fibrous root system are adventitious and develop from stem tissue.

Roots have root caps and root hairs

Because of the need to adapt to the soil environment instead of the atmospheric environment, roots have several structures, such as root caps and root hairs, that shoots lack. Although stems and leaves have various types of hairs, they are distinct from root hairs in structure and function.

Each root tip is covered by a **root cap,** a protective, thimble-like layer many cells thick that covers the delicate root **apical meristem** (Fig. 34-2a; also see Fig. 31-7). As the root grows, pushing its way through the soil, cells of the root cap are sloughed off by the frictional resistance of the soil particles and replaced by new cells formed by the root apical meristem. The root cap cells secrete lubricating polysaccharides that reduce friction as the root passes through the soil. The root cap also appears to be involved in orienting the root so that it grows downward (see discussion of gravitropism in Chapter 36). When a root cap is removed, the root apical meristem grows a new cap. However, until the root cap has regenerated the root grows randomly rather than in the direction of gravity.

Root hairs are short-lived tubular extensions of epidermal cells located just behind the growing root tip. Root hairs continually form in the area of cell maturation closest to the root tip to replace those that are dying off at the more mature end of the root hair zone (Fig. 34-2b; also see Fig. 31-7). Each root hair is short (typically less than 1 cm, or 0.4 in, in length), but they are quite numerous. Root hairs greatly increase the absorptive

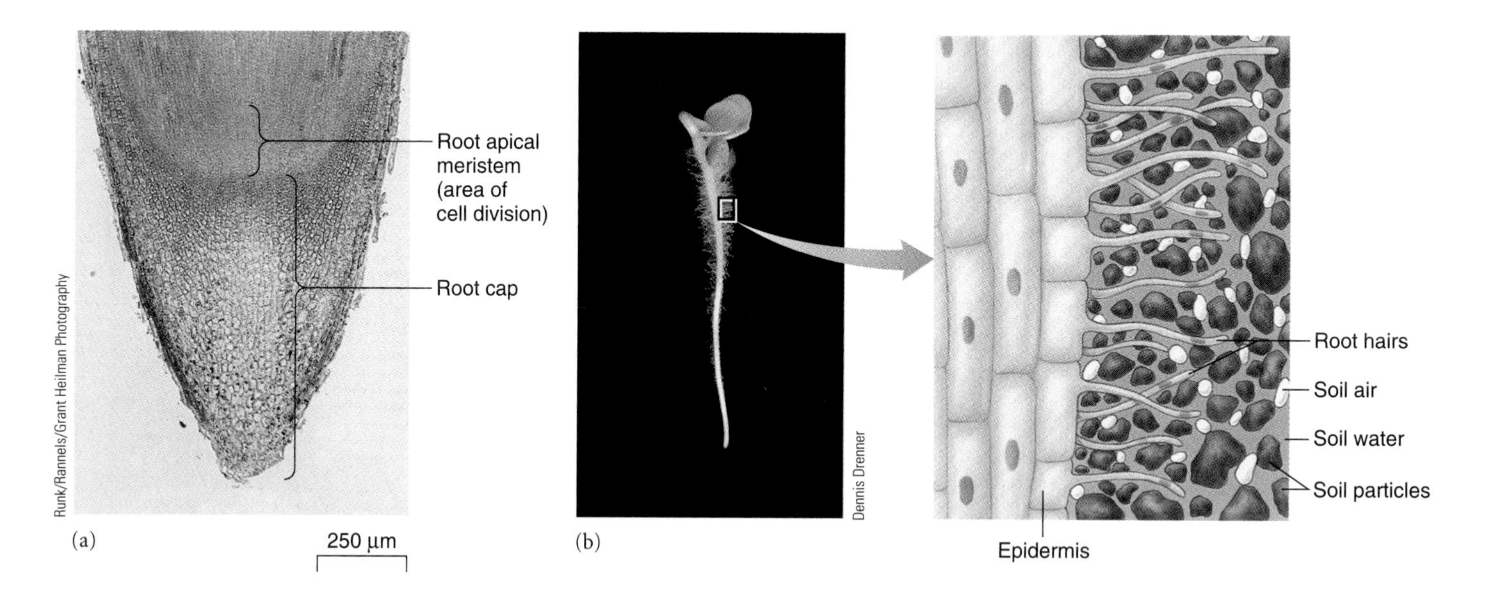

FIGURE **34-2** | **Structures unique to roots.**

(a) LM of an oak (*Quercus* sp.) root tip showing its root cap. The root apical meristem is protected by the root cap. **(b)** Root hairs on a radish seedling. Each delicate hair is an extension of a single cell of the root epidermis. Root hairs increase the surface area of the root in contact with the soil. The seedling is approximately 5 cm (2 in) long.

capacity of roots by increasing their surface area in contact with moist soil. Soil particles are coated with a microscopically thin layer of water in which nutrient minerals are dissolved. The root hairs establish an intimate contact with soil particles, which allows absorption of much of the water and nutrient minerals.

Unlike stems, roots lack nodes and internodes and do not usually produce leaves or buds. Although herbaceous roots have certain primary tissues (such as epidermis, xylem, phloem, cortex, and pith) found in herbaceous stems, these tissues are arranged quite differently.

The arrangement of vascular tissues distinguishes the roots of herbaceous dicots and monocots

Although considerable variation exists in herbaceous dicot and monocot roots, they all have an outer protective covering (epidermis), a cortex for storage of starch and other organic molecules, and vascular tissues for conduction. Let us first consider the structure of herbaceous dicot roots.

In most herbaceous dicot roots, the central core of vascular tissue lacks pith

The buttercup root is a representative dicot root with primary growth (Fig. 34-3). Like other parts of this herbaceous dicot, a single layer of protective tissue, the **epidermis,** covers its roots. The root hairs are a modification of the root epidermis that enables it to absorb more water from the soil. The root epidermis does not secrete a thick, waxy cuticle in the region of root hairs, because this layer would impede the absorption of water from the soil. Both the lack of a cuticle and the presence of root hairs increase absorption. (*How* water moves from the soil into the root was explained in Chapter 33.)

Most of the water that enters the root moves along the cell walls rather than entering the cells. One of the major components of cell walls is cellulose, which absorbs water like a sponge. An example of the absorptive properties of cellulose is found in cotton balls, which are almost pure cellulose.

The **cortex,** which is primarily composed of loosely packed **parenchyma** cells, comprises the bulk of a herbaceous dicot root. Roots usually lack supporting **collenchyma** cells, probably because the soil supports the root, although roots may develop some **sclerenchyma** (another supporting tissue; see Chapter 31) as they age. The primary function of the root cortex is storage. A microscopic examination of the parenchyma cells that form the cortex often reveals numerous amyloplasts (see Fig. 34-3b and Fig. 3-9a), which store starch. Starch, an insoluble carbohydrate composed of glucose subunits, is the most common form of stored energy in plants. When used at a later time, these reserves provide energy for such activities as growth following winter or cell replacement following an injury.

The large intercellular (between-cell) spaces, a common feature of the root cortex, provide a pathway for water uptake and allow for aeration of the root. The oxygen that root cells need for aerobic respiration diffuses from air spaces in the soil into the intercellular spaces of the cortex and from there into the cells of the root.

The inner layer of the cortex, the **endodermis,** regulates the movement of nutrient minerals that enter the xylem in the root's center. Structurally, the endodermis differs from the rest of the cortex. Endodermal cells fit snugly against each other, and each has a special bandlike region, called a **Casparian strip** (Fig. 34-4),

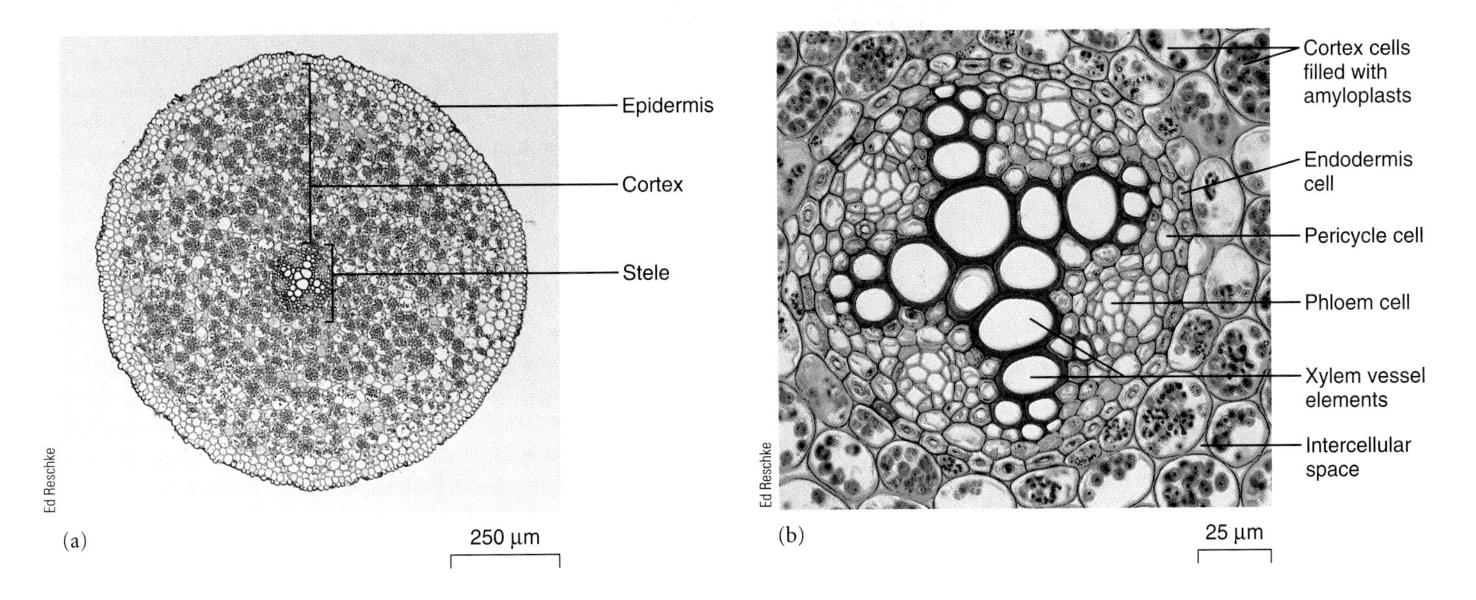

FIGURE **34-3** **LMs of cross sections of a herbaceous dicot root.**

Shown is a buttercup (*Ranunculus*) root. **(a)** Cortex comprises the bulk of herbaceous dicot roots. Note the X-shaped xylem in the center of the root. **(b)** Closeup of the root's stele. Surrounding the solid core of vascular tissues is a single layer of pericycle, which is meristematic in growing roots.

on its radial (side) and transverse (upper and lower) walls. If you compare the endodermis to a cylinder constructed of bricks, endodermal cells correspond to the bricks, and the Casparian strips correspond to the mortar between them. Casparian strips contain *suberin,* a fatty material that is waterproof. (Recall from Chapter 33 that suberin is also the waterproof material in cork cell walls.)

The water and dissolved nutrient minerals that enter the root cortex from the epidermis move in solution along two pathways: the symplast and apoplast. The **symplast** is the con-

FIGURE **34-4** **Endodermis and nutrient mineral uptake.**

Note the Casparian strip around the radial and transverse walls that prevents water and dissolved nutrient minerals from passing into the stele along endodermal cell walls. To reach the vascular tissues, water and dissolved nutrient minerals must pass through the plasma membranes of endodermal cells.

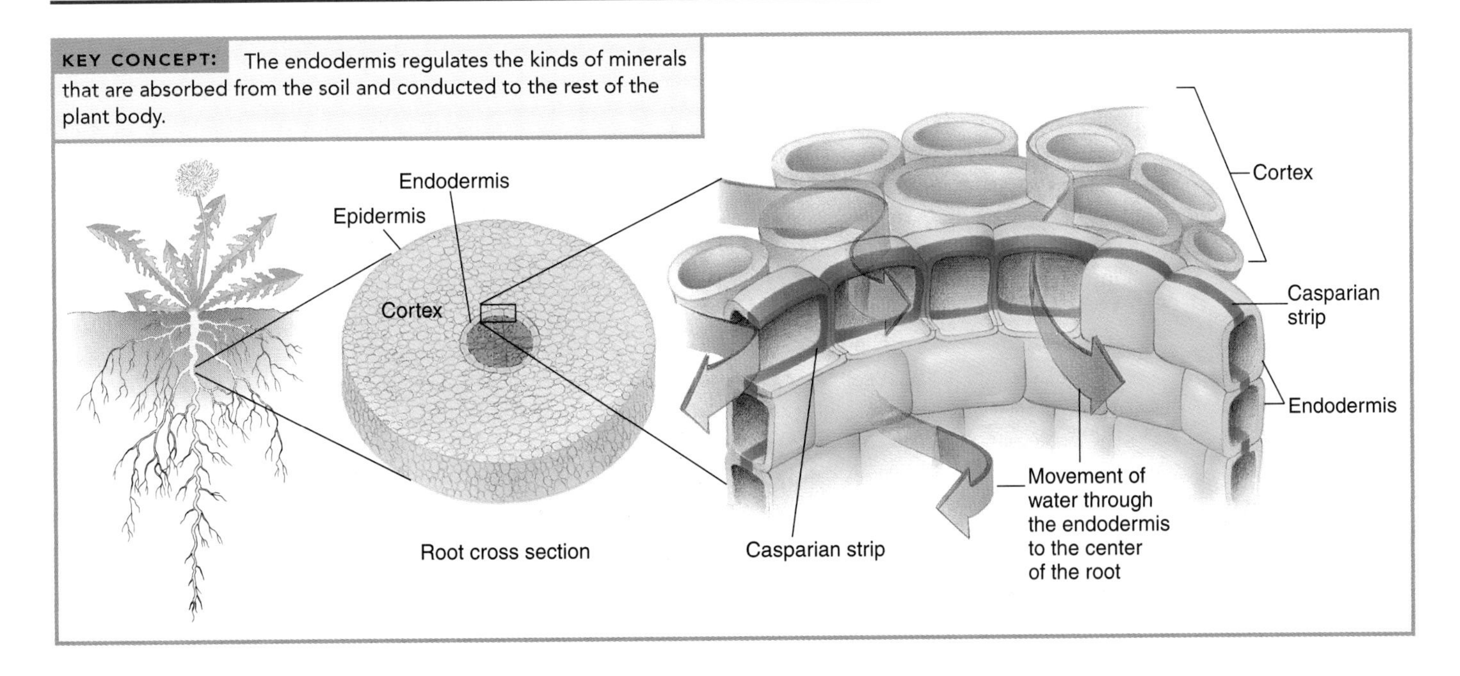

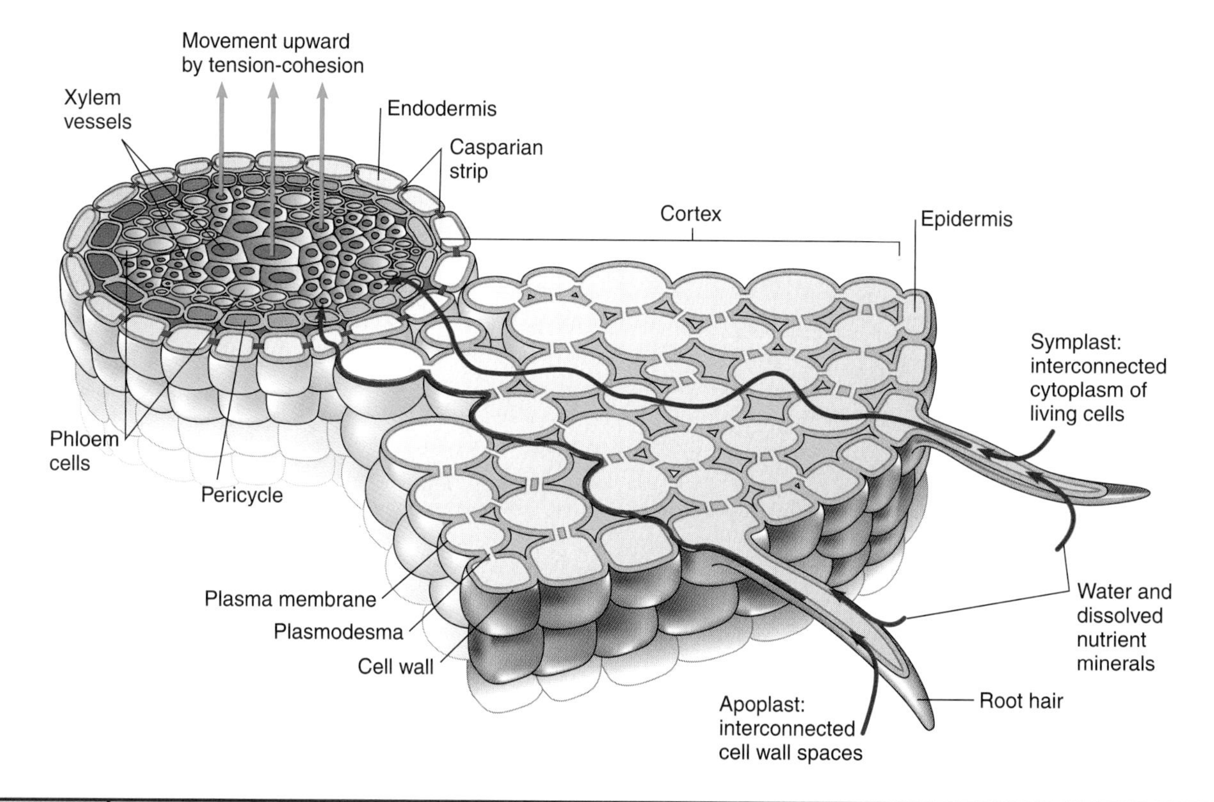

FIGURE **34-5** | **Pathways of water and dissolved nutrient minerals in the root.**

Water and dissolved nutrient minerals travel from cell to cell along the interconnected porous cell walls (the apoplast) or from one cell's cytoplasm to another through plasmodesmata (the symplast). On reaching the endodermis, water and nutrient minerals can only continue to move into the root's center if they pass through a plasma membrane and enter the cytoplasm of an endodermal cell. The Casparian strip blocks the passage of water and nutrient minerals along the cell walls between adjoining endodermal cells.

tinuum of living cytoplasm, which is connected from one cell to the next by cytoplasmic bridges called **plasmodesmata** (Fig. 34-5 and Fig. 5-25). Some dissolved mineral ions move from the epidermis through the cortex via the symplast. The **apoplast** consists of the interconnected porous cell walls of a plant, along which water and nutrient mineral ions move freely. The water and mineral ions can diffuse across the cortex without ever entering a living cell.

Until the endodermis is reached, most of the water and dissolved nutrient minerals have traveled along the apoplast and therefore have not passed through a plasma membrane or entered the cytoplasm of a root cell. However, the waterproof Casparian strip on the radial and transverse walls of the endodermal cells prevents water and nutrient minerals from continuing to move passively along the cell walls. For substances to pass further into the root interior, they must move from the cell walls into the cytoplasm of the endodermal cells. Water enters by osmosis, whereas nutrient minerals enter the endodermal cells by passing through carrier proteins in their plasma membranes. Even though dissolved nutrient minerals pass through the epidermis and cortex to reach the endodermis, it is the endodermis that is responsible for controlling the movement of nutrient minerals from the soil into the vascular tissue of the root (and from there to the rest of the plant body).

Dissolved nutrient mineral ions are actively transported through carrier proteins in the plasma membranes of endodermal cells (see Chapter 5). In **carrier-mediated active transport,** the nutrient mineral ions move *against* their concentration gradient—that is, from an area of *low* concentration of that mineral in the soil solution to an area of *high* concentration in the plant's cells. One of many reasons why root cells require sugar and oxygen for aerobic respiration is that this active transport requires the expenditure of cellular energy, usually in the form of ATP. From the endodermis, water and nutrient mineral ions enter the root xylem (botanists do not know precisely how this is done) and are conducted to the rest of the plant.

At the center of a dicot primary root is the **stele** or **vascular cylinder,** a central cylinder of vascular tissues (see Fig. 34-3a). The outermost layer of the stele is the **pericycle,** which is just inside the endodermis. The pericycle consists of a single layer of parenchyma cells that gives rise to multicellular lateral roots, also called *branch roots* (Fig. 34-6). Lateral roots originate when cells in a portion of the pericycle start dividing. As it grows, the lateral root pushes through several layers of root tissue (endodermis, cortex, and epidermis) before entering the soil. Each lateral root has all the structures and features—root cap, root hairs, epidermis, cortex, endodermis, pericycle, xylem, and phloem—of the larger root from which it emerges. In addition to pro-

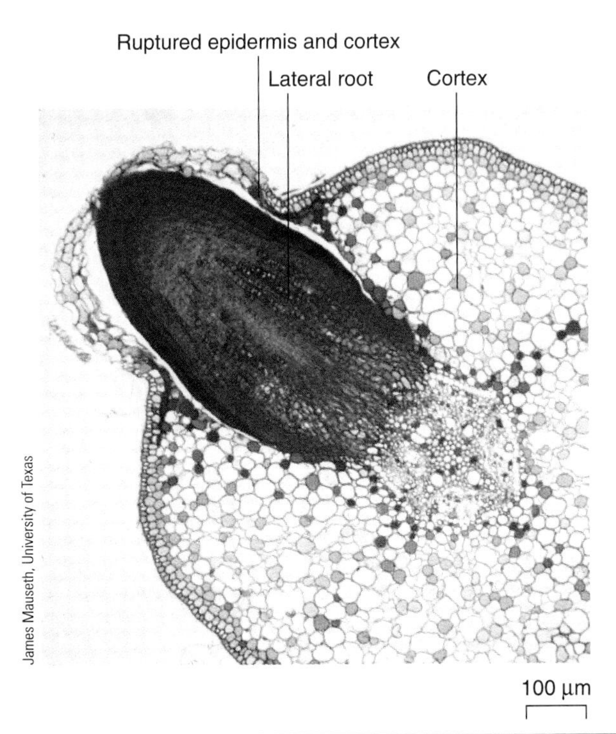

FIGURE **34-6** | **LM of a lateral root.**
Lateral roots originate at the pericycle.

ducing lateral roots, the pericycle is involved in forming the lateral meristems that produce secondary growth in woody roots (discussed later in the chapter).

Xylem, the centermost tissue of the stele, often has two, three, four, or more extensions, or "xylem arms" (see Fig. 34-3b). **Phloem** is located in patches between the xylem arms. The xylem and phloem of the root have the same functions and kinds of cells as in the rest of the plant: Water and dissolved nutrient minerals are conducted in **tracheids** and **vessel elements** of xylem, and dissolved sugar (sucrose) is conducted in **sieve tube elements** of phloem.

After passing through the endodermal cells, water enters the root xylem, often at one of the xylem arms. Up to this point the pathway of water has been horizontal from the soil into the center of the root:

Root hair/epidermis ⟶ cortex ⟶ endodermis ⟶ pericycle ⟶ root xylem

Once water enters the xylem, it is transported upward through root xylem into stem xylem and from there to the rest of the plant.

One direction of phloem conduction is from the leaves, where sugar is made by photosynthesis, to the root, where sugar is used for the growth and maintenance of root tissues or stored, usually as starch. Another direction of phloem conduction is from the root, where sugar is stored as starch, to other parts of the plant, where sugar is used for growth and maintenance of tissues. The *vascular cambium,* which gives rise to secondary tissues in woody plants, is sandwiched between the xylem and phloem. Because it has an inner core of vascular tissue, the primary dicot root lacks **pith,** a ground tissue found in the centers of many stems and roots.

Xylem does not form the central tissue in some monocot roots

Monocot roots vary considerably in internal structure, compared with dicot roots. Starting at the outside of some monocot roots, there is epidermis, then cortex, endodermis, and pericycle (Fig. 34-7). Unlike the xylem in herbaceous dicot roots, the xylem in a monocot root does not form a solid cylinder in the center. Instead, the phloem and xylem are in separate alternating bundles arranged around the central pith, which consists of parenchyma cells.

Because virtually no monocots have secondary growth, no vascular cambium exists in monocot roots. Despite their lack of secondary growth, long-lived monocots, such as palms, may have thickened roots produced by a modified form of primary growth in which parenchyma cells in the cortex divide and enlarge.

Woody plants have roots with secondary growth

Plants that produce stems with secondary growth also produce roots with secondary growth. Recall from Chapter 33 that these plants—gymnosperms and woody dicots—have primary growth

FIGURE **34-7** | **LM of a cross section of a monocot root.**
Shown is a greenbriar (*Smilax*) root. As in herbaceous dicot roots, the cortex of a monocot root is extensive.

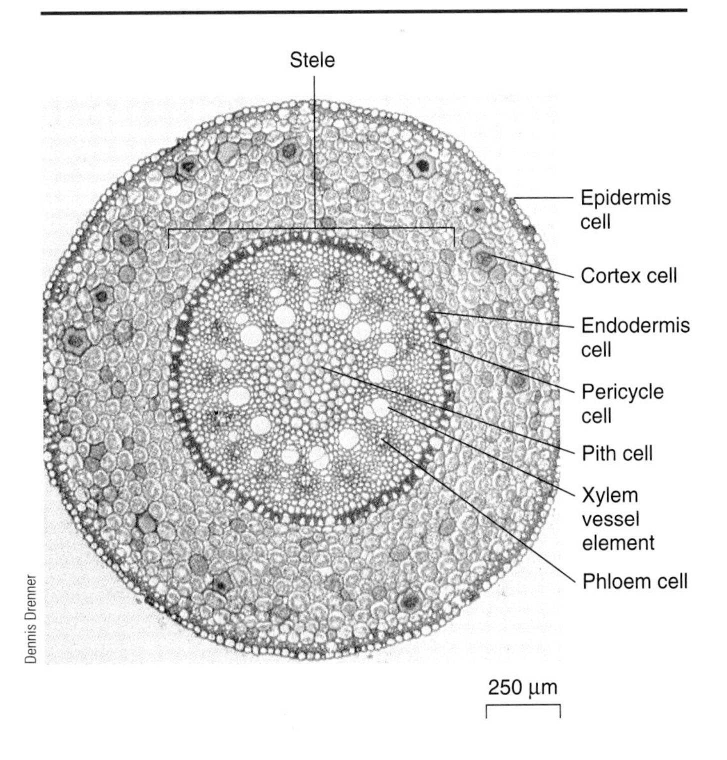

at apical meristems and secondary growth at lateral meristems. The production of secondary tissues occurs some distance back from the root tips and results from the activity of the same two lateral meristems found in woody stems: the vascular cambium and the cork cambium. Major roots of trees are often massive and have both wood and bark. In temperate climates, the wood of both roots and stems exhibits annual rings in cross section.

Before secondary growth starts in a root, the **vascular cambium** is sandwiched between the primary xylem and the primary phloem (Fig. 34-8a). At the onset of secondary growth, the vascular cambium extends out to the pericycle, which develops into vascular cambium opposite the xylem arms. As a result, the pericycle links the separate sections of vascular cambium so that the vascular cambium becomes a continuous, noncircular loop of cells in cross section (Fig. 34-8b). As the vascular cambium divides to produce secondary tissues, it eventually forms a cylinder of vascular cambium that continues to divide, producing secondary xylem (wood) to the inside and secondary phloem (inner bark) to the outside (Fig. 34-8c,d). The root increases in girth (thickness), and the vascular cambium continues to move outward.

The epidermis, cortex, endodermis, and primary phloem are gradually torn apart as the root increases in girth. The root epidermis is replaced by **periderm,** composed of cork cells and cork parenchyma, both produced by the **cork cambium** (the last figure in Table 31-4 shows a LM of periderm). The cork cambium in the root initially arises from regions in the pericycle.

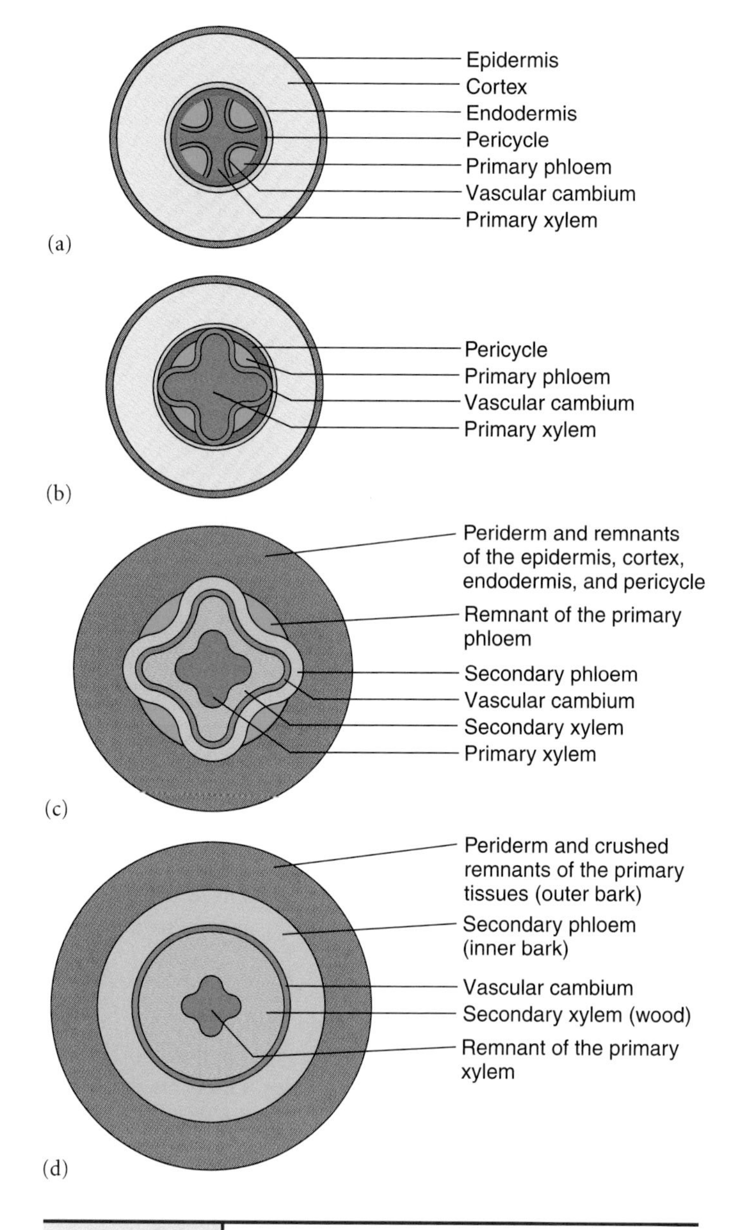

FIGURE **34-8** **Development of secondary vascular tissues in a primary root.**

(a) The tissues in a primary root. **(b)** At the onset of secondary growth, the vascular cambium extends out to the pericycle, forming a continuous, noncircular loop. **(c)** The vascular cambium produces secondary xylem to its inside and secondary phloem to its outside. **(d)** Over time, the ring of vascular cambium gradually becomes circular. As the vascular cambium continues to divide, the epidermis, cortex, and primary phloem located in the outer bark are torn apart. (The figures are not drawn to scale because of space limitations; the primary xylem is actually the same size in all four diagrams, but differences in scale make the xylem appear different sizes.)

Some roots are specialized for unusual functions

Adventitious roots often arise from the nodes of stems. Many aerial adventitious roots are adapted for functions other than anchorage, absorption, conduction, or storage. **Prop roots** are adventitious roots that develop from branches or a vertical stem and that grow downward into the soil to help support the plant in an upright position (Fig. 34-9a). Prop roots are more common in monocots than in dicots. Corn and sorghum, both monocots, are herbaceous plants that produce prop roots. Many tropical and subtropical dicot trees, such as red mangrove, and banyan also produce prop roots.

The roots of many tropical rainforest trees are shallow and concentrated near the surface in a mat only a few centimeters (an inch or so) thick. The root mat catches and absorbs almost all nutrient minerals released from leaves by decomposition. Swollen bases or braces called **buttress roots** hold the trees upright and aid in the extensive distribution of the shallow roots (Fig. 34-9b).

In swampy or tidal environments where the soil is flooded or waterlogged, some roots grow upward until they are above the high-tide level. Even though roots live in the soil, they still require oxygen for aerobic respiration. Flooded soils are depleted of oxygen, so these aerial "breathing" roots, known as **pneumatophores,** may assist in getting oxygen to the submerged roots (Fig. 34-9c). Pneumatophores, which also help anchor the plant, have a well-developed system of internal air spaces that is continuous with the submerged parts of the root, presumably allowing gas exchange. Black mangrove, white mangrove, and bald cypress are examples of plants with pneumatophores.

Climbing plants and **epiphytes,** which are plants that grow attached to other plants, have aerial roots that anchor the plant

FIGURE 34-9 | Specialized roots.

(a) Prop roots are adventitious roots that arise near the base of the stem and provide additional support. *Pandanus* has an elaborate set of aerial prop roots. Photographed in Kauai, Hawaii. **(b)** Tropical rainforest trees typically possess buttress roots that support them in the shallow, often wet soil. Shown are buttress roots on Australian banyan (*Ficus macrophylla*). Photographed at Selby Gardens in Sarasota, Florida. **(c)** White mangrove (*Laguncularia racemosa*) produces pneumatophores, shown protruding from the wet mud in the foreground. Pneumatophores may provide oxygen for roots buried in anaerobic (oxygen-deficient) soil. Photographed in Isla del Carmen, Mexico. **(d)** The moth orchid (*Phalaenopsis* hybrid) has photosynthetic aerial roots. **(e)** Plants that produce corms or bulbs often have contractile roots. During successive seasons, contractile roots pull the corm or bulb deeper into the soil.

to the bark, branch, or other surface on which they grow. Some epiphytes have aerial roots specialized for functions other than anchorage. Certain epiphytic orchids, for example, have photosynthetic roots (Fig. 34-9d). Epiphytic roots may absorb moisture as well. Some parasitic epiphytes such as mistletoe have modified roots that penetrate the host plant tissues and absorb water. Another plant that starts its life as an epiphyte is the strangler fig, which produces long roots that eventually reach the ground and anchor the plant (now a tree rather than an epiphyte) in the soil. The tree on which the strangler fig originally grew is often killed as the strangler fig grows around it, competing with it for light and other resources and crushing its secondary phloem.

Plants that produce corms or bulbs (underground stems or buds specialized for asexual reproduction; see Chapter 35) often have wiry **contractile roots** in addition to their "normal" roots. The contractile roots grow into the soil and then contract (the cortical cells shorten or totally collapse), thus pulling the corm or bulb deeper into the soil (Fig. 34-9e). Contractile roots are necessary for corms because each succeeding year's growth is *on top of* the preceding year's growth. As a result, corms tend to move upward in the soil over time. Without contractile roots they would eventually be exposed at the soil's surface. Contractile roots are more common in monocots, but certain dicots and ferns also possess them.

Review

- What are the advantages of a taproot system? Of a fibrous root system?
- If you were examining a cross section of a primary root of a flowering plant, how would you determine whether it was a dicot or a monocot?
- What is the symplast? The apoplast?
- How does a herbaceous dicot root develop secondary tissues?
- What are the functions of each of the following? (a) prop roots, (b) buttress roots, (c) pneumatophores, and (d) contractile roots

BiologyNow™ Assess your understanding of **root structure and function** by taking the pretest on your BiologyNow CD-ROM.

ROOT ASSOCIATIONS WITH OTHER ORGANISMS

Learning Objective

6 List and describe two mutualistic relationships between roots and other organisms.

The roots of most plant species form **mutualistic**—that is, mutually beneficial—relationships with certain soil fungi (see Chapters 25 and 52). These subterranean associations, known as **mycorrhizae,** permit the transfer of materials (such as sugars) from roots to the fungus. At the same time, essential nutrient minerals such as phosphorus move from fungus to the roots of the host plant. The threadlike body of the fungal partner extends into the soil, extracting nutrient minerals well beyond the reach of the plant's roots. In some mycorrhizae, the fungal mycelium encircles the root like a sheath; in others, the fungus penetrates root cells (Fig. 34-10). The relationship is mutually beneficial because when mycorrhizae are not present, neither the fungus nor the plant grows as well (see Fig. 25-15). Recent evidence indicates that the hyphal network of mycorrhizae simultaneously interconnects different plant species in the community and that carbon compounds may flow from one plant to another through their mutual fungal partner.

Certain nitrogen-fixing bacteria, collectively called *rhizobia,* form associations with the roots of leguminous plants—clover, peas, and soybeans, for example. **Nodules** (swellings) that house millions of the rhizobia develop on the roots (see Fig. 53-8). As with mycorrhizae, the association between nitrogen-fixing bacteria and the roots of plants is mutually beneficial. Bacteria receive photosynthetic products from plants while helping them meet their nitrogen requirements by producing ammonia (NH_3) from atmospheric nitrogen.

For several decades, biologists have studied the molecular basis of associations between plants and rhizobia. The initial contact between the two partners in this association involves **cell signaling,** that is, communication between cells. Some signal molecules produced by the bacteria are nodulation factors that cause the roots of leguminous plants to produce nodules.

Roots also form associations with organisms other than fungi and bacteria. As roots of certain trees grow through the soil, they sometimes encounter roots of other trees of the same or different species. When this occurs, they may grow together by secondary growth to form a natural **graft.** Because their vascular tissues are connected in the graft, dissolved sugars and other materials such as hormones pass between the two trees; disease organisms can also be transmitted in this way. Root grafts have been observed in more than 160 tree species.

Review

- How are mycorrhizae and root nodules similar? How are they different?
- What are root grafts?

BiologyNow™ Assess your understanding of **root associations with other organisms** by taking the pretest on your BiologyNow CD-ROM.

THE SOIL ENVIRONMENT

Learning Objectives

7 Describe the roles of weathering, organisms, climate, and topography in soil formation.

8 List the four components of soil, and give the ecological significance of each.

9 Describe how roots absorb positively charged nutrient mineral ions by the process of cation exchange.

10 Distinguish between macronutrients and micronutrients.

11 Explain the impacts of mineral depletion and soil erosion on plant growth.

We now examine the soil environment in which most roots live. Soil is a relatively thin layer of Earth's crust that has been modified by the natural actions of weather, wind, water, and organisms. It is easy to take soil for granted. We walk on and over it throughout our lives but rarely stop to think about how impor-

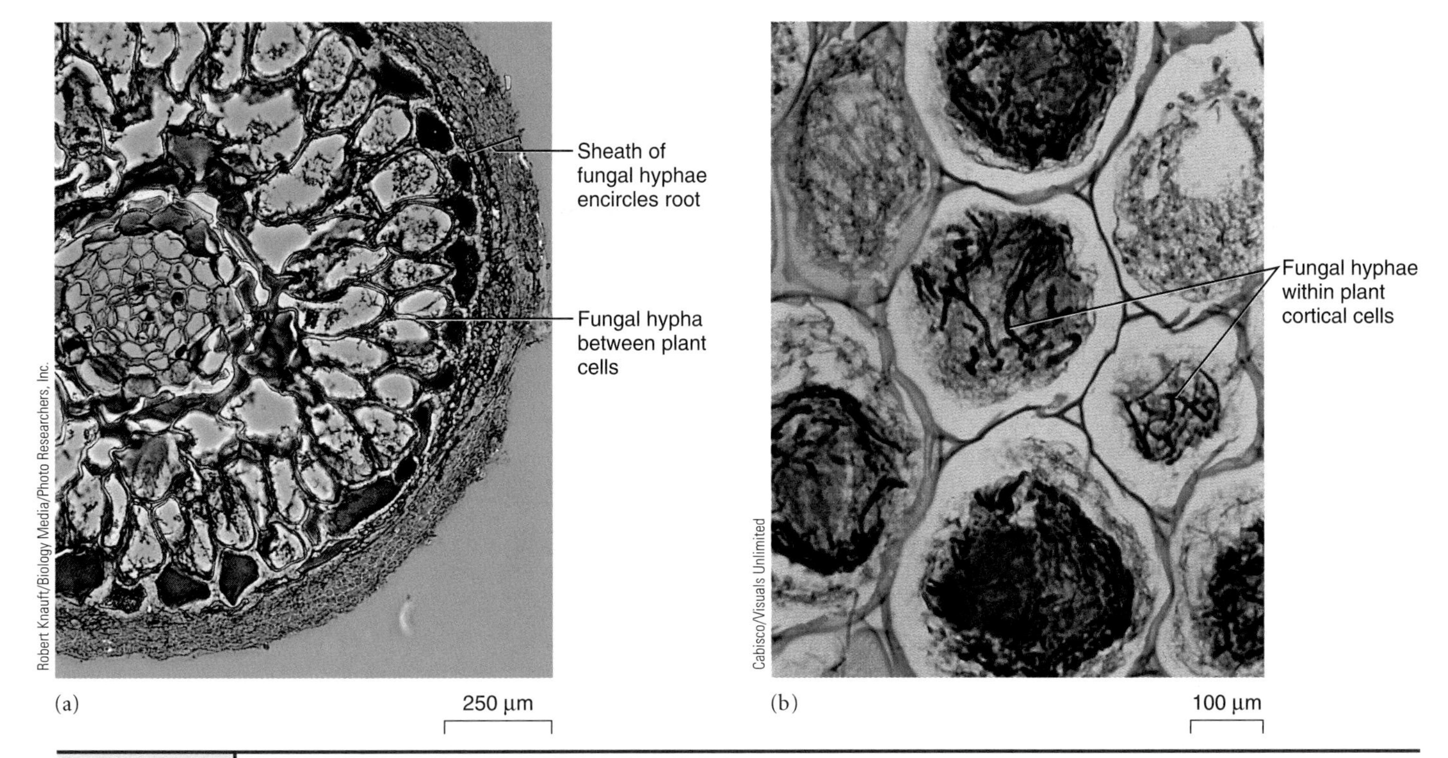

FIGURE **34-10** | **Mycorrhizae.**

Mycorrhizae enhance plant growth by providing soil nutrients to the roots. **(a)** LM of ectomycorrhizae, fungal associations that form a sheath around the root. The fungal hyphae penetrate the root between cortical cells but do not enter the cells. **(b)** LM of endomycorrhizae, fungal associations in which the fungal hyphae penetrate root cells of the cortex and form branched haustoria (absorbing organs) within the cells to aid in delivering and receiving nutrients. Roots of most vascular plant species are colonized by endomycorrhizae.

tant it is to our survival. Vast numbers and kinds of organisms colonize soil and depend on it for shelter and food. Most plants anchor themselves in soil, and from it they receive water and essential nutrient minerals. Most elements essential for plant growth are obtained directly from the soil. Most plants cannot survive on their own without soil, and because we depend on plants for our food, humans could not exist without soil either.

Most soils are formed from rock (called *parent material*) that is gradually broken down, or fragmented, into smaller and smaller particles by biological, chemical, and physical **weathering processes.** Two important factors that work together in the weathering of rock are climate and organisms. When plant roots and other organisms living in the soil respire, they produce carbon dioxide (CO_2), which diffuses into the soil and reacts with soil water to form carbonic acid (H_2CO_3). Soil organisms such as lichens[1] also produce other kinds of acids. These acids etch tiny cracks, or fissures, in the rock surface; water then seeps into these cracks. If the parent material is located in a temperate climate, the alternate freezing and thawing of the water during winter cause the cracks to enlarge, breaking off small pieces of rock. Small plants can then become established and send their roots into the larger cracks, fracturing the rock further.

Topography, a region's surface features, such as the presence or absence of mountains and valleys, is also involved in soil formation. Steep slopes often have little or no soil on them, because the soil and rock are continually transported down the slopes by gravity. Runoff from precipitation tends to amplify erosion on steep slopes. Moderate slopes, in contrast, may encourage the formation of deep soils.

The disintegration of solid rock into finer and finer mineral particles and the accumulation of organic material (discussed in the next section) in the soil take an extremely long time, sometimes thousands of years. Soil forms constantly as the weathering of parent material beneath already formed soil continues to add new soil.

Soil is composed of inorganic minerals, organic matter, air, and water

Four distinct components comprise soil—inorganic mineral particles (which make up about 45% of a typical soil), organic matter (about 5%), water (about 25%), and air (about 25%). The plants, animals, fungi, and microorganisms that inhabit soil interact with it, and nutrient minerals are continually cycled from the soil to organisms, which use them in their biological processes. When the organisms die, bacteria and other soil organisms decompose the remains, returning the nutrient minerals to the soil.

The inorganic mineral particles, which come from weathered rock, provide anchorage and essential nutrient minerals for plants, as well as pore space for water and air. Because different rocks consist of different minerals, soils vary in mineral

[1] A lichen is a dual organism composed of a fungus and a phototroph (photosynthetic organism). See Chapter 25 for further discussion of lichens.

composition and chemical properties. Rocks rich in aluminum form acidic soils, for example, whereas rocks that contain silicates of magnesium and iron form soils that may be deficient in calcium, nitrogen, and phosphorus. Also, soils formed from the same kind of parent material may not develop in the same way, because other factors, such as weather, topography, and organisms, differ.

The texture, or structural characteristic, of a soil is determined by the percentages (by weight) of the different-sized inorganic mineral particles—sand, silt, and clay—in it. The size assignments for sand, silt, and clay give soil scientists a way to classify soil texture. Particles larger than 2 mm in diameter, called gravel or stones, are not considered soil particles, because they do not have any direct value to plants. The largest soil particles, are called *sand* (0.02 to 2 mm in diameter), the medium-sized particles are called *silt* (0.002 to 0.02 mm in diameter), and the smallest particles are called *clay* (less than 0.002 mm in diameter). Sand particles are large enough to be seen easily with the eye, and silt particles (about the size of flour particles) are barely visible. Most individual clay particles are too small to be seen with an ordinary light microscope; they can be seen only under an electron microscope.

A soil's texture affects many of that soil's properties, in turn influencing plant growth. The clay component of a soil is particularly important in determining many of its characteristics because clay particles have the greatest surface area of all soil particles. If the surface areas of about 450 g (1 lb) of clay particles were laid out side by side, they would occupy 1 hectare (2.5 acres).

Each clay particle has predominantly negative electrical charges on its outer surface that attract and reversibly bind **cations,** which are positively charged mineral ions such as potassium (K^+) and magnesium (Mg^{2+}). Because many cations are essential for plant growth, cation absorption to soil particles is an important aspect of soil fertility. Roots secrete protons (hydrogen ions, H^+), which are exchanged for other positively charged mineral ions absorbed to the surface of soil particles in a process known as **cation exchange.** These "freed" ions and the water that forms a film around the soil particles are absorbed by the plant's roots (Fig. 34-11).

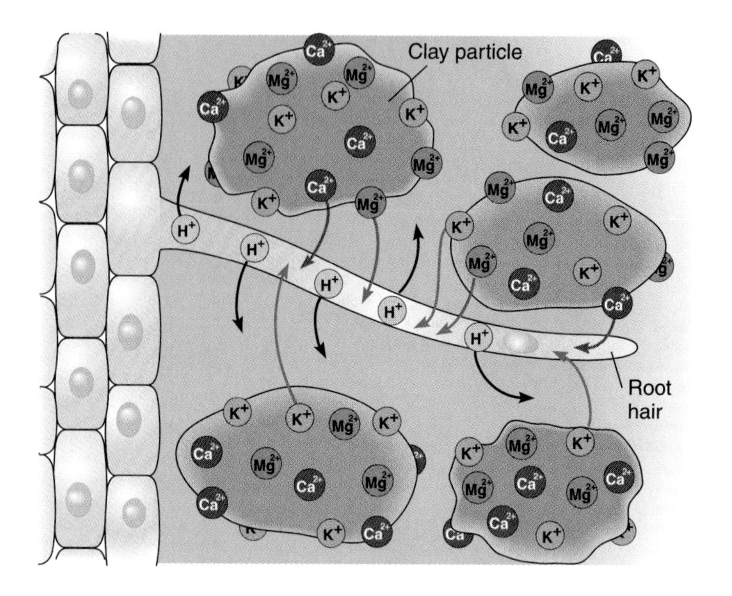

FIGURE **34-11** **Cation exchange.**

Negatively charged clay particles bind to positively charged nutrient mineral cations, holding them in the soil. Roots pump out protons (H^+), which are exchanged for the cations, facilitating their absorption.

In contrast, **anions,** which are negatively charged mineral ions, are repelled by the negative surface charges of clay particles and tend to remain in solution. Anions such as nitrate (NO_3^-) are often washed out of the root zone by water moving through the soil.

Soil always contains a mixture of different-sized particles, but the proportions vary from one soil to another. A *loam,* which is an ideal agricultural soil, has an optimum combination of different soil particle sizes: It contains approximately 40% each of sand and silt and about 20% of clay. Generally, the larger particles provide structural support, aeration, and permeability to the soil, whereas the smaller particles bind together into aggregates, or clumps, and hold nutrient minerals and water. Soils with larger proportions of sand are not as desirable for most plants, because they do not hold water and nutrient mineral ions well. Plants grown in such soils are more susceptible to drought and nutrient mineral deficiencies. Soils with larger proportions of clay are also not desirable for most plants because they provide poor drainage and often do not provide enough oxygen. Clay soils used in agriculture tend to get compacted, which reduces the number of soil spaces that can be filled by water and air.

A soil's organic matter consists of the wastes and remains of soil organisms

The organic matter in soil is composed of litter (dead leaves and branches on the soil's surface); droppings (animal dung); and the dead remains of plants, animals, and microorganisms in various stages of decomposition. Organic matter is decomposed by microorganisms, particularly bacteria and fungi, that inhabit the soil. During decomposition, essential nutrient mineral ions are released into the soil, where they may be bound by soil particles or absorbed by plant roots. Organic matter increases the soil's water-holding capacity by acting much like a sponge. For this reason gardeners often add organic matter to soils, especially sandy soils, which are naturally low in organic matter.

The partly decayed organic portion of the soil is referred to as **humus**. Humus, which is not a single chemical compound but a mix of many organic compounds, binds nutrient mineral ions and holds water. On average, humus persists in agricultural soil for about 20 years. Certain components of humus may persist in the soil for hundreds of years. Although humus is somewhat resistant to decay, a succession of microorganisms gradually reduces it to carbon dioxide, water, and nutrient minerals.

About 50% of soil volume is composed of pore spaces

Soil has numerous pore spaces of different sizes around and among the soil particles. Pore spaces occupy roughly 50% of a soil's volume and are filled with varying proportions of air and water. Both soil air and soil water are necessary to produce a moist but aerated soil that sustains plants and other soil-dwelling organisms. Water is usually held in the smaller pores, whereas air is found in the larger pores. After a prolonged rain, almost all the pore spaces may be filled with water, but water drains rapidly from the larger pore spaces, drawing air from the atmosphere into those spaces.

Soil air contains the same gases as atmospheric air, although they are usually present in different proportions. As a result of respiration by soil organisms, there is less oxygen and more carbon dioxide in soil air than in atmospheric air. (Recall from Chapter 7 that aerobic respiration uses oxygen and produces carbon dioxide.) Among the important gases in soil are oxygen (O_2), required by soil organisms for aerobic respiration; nitrogen (N_2), used by nitrogen-fixing bacteria; and carbon dioxide (CO_2), involved in soil weathering.

Soil water originates as precipitation, which drains downward, or as groundwater (water stored in porous underground rock), which rises upward from the water table (the uppermost level of groundwater). Soil water contains low concentrations of dissolved nutrient minerals that enter the roots of plants when they absorb water. Water not bound to soil particles or absorbed by roots percolates (moves down) through soil, carrying dissolved nutrient minerals with it. The removal of dissolved materials from soil by percolating water is called **leaching.** The deposition of leached material in the lower layers of soil is known as **illuviation.** Iron and aluminum compounds, humus, and clay are some illuvial materials that can gather in the subsurface portion of the soil. Some substances completely leach out of the soil because they are so soluble that they migrate down into the groundwater. It is also possible for water that is moving *upward* through the soil to carry dissolved materials with it.

The organisms living in the soil form a complex ecosystem

A single teaspoon of fertile agricultural soil may contain millions of microorganisms such as bacteria, fungi, algae, protozoa, and microscopic nematodes and other worms. Many other organisms also colonize the soil ecosystem, including plant roots, earthworms, insects, moles, snakes, and groundhogs (Fig. 34-12). Most numerous in soil are bacteria, which number in the hundreds of millions per gram of soil. Scientists have identified about 170,000 species of soil organisms, but thousands remain to be identified. Little is known about the roles of soil organisms, in part because it is hard to study their activities under natural conditions.

Worms are some of the most important organisms living in soil. Earthworms, probably one of the most familiar soil inhabitants, ingest soil and obtain energy and raw materials by digesting humus. *Castings,* bits of soil that have passed through the gut of an earthworm, are deposited on or near the soil surface. In this way, nutrient minerals from deeper layers are brought to upper layers. Earthworm tunnels serve to aerate the soil, and the worms' waste products and corpses add organic material to the soil.

Ants live in the soil in enormous numbers, constructing tunnels and chambers that help aerate it. Members of soil-dwelling ant colonies forage on the surface for bits of food, which they carry back to their nests. Not all this food is eaten, however, and its eventual decomposition helps increase the organic matter in the soil. Many ants are also indispensable in plant reproduction because they bury seeds in the soil.

Soil pH affects soil characteristics and plant growth

As discussed in Chapter 2, acidity is measured using the pH scale, which extends from 0 (extremely acidic) through 7 (neutral)

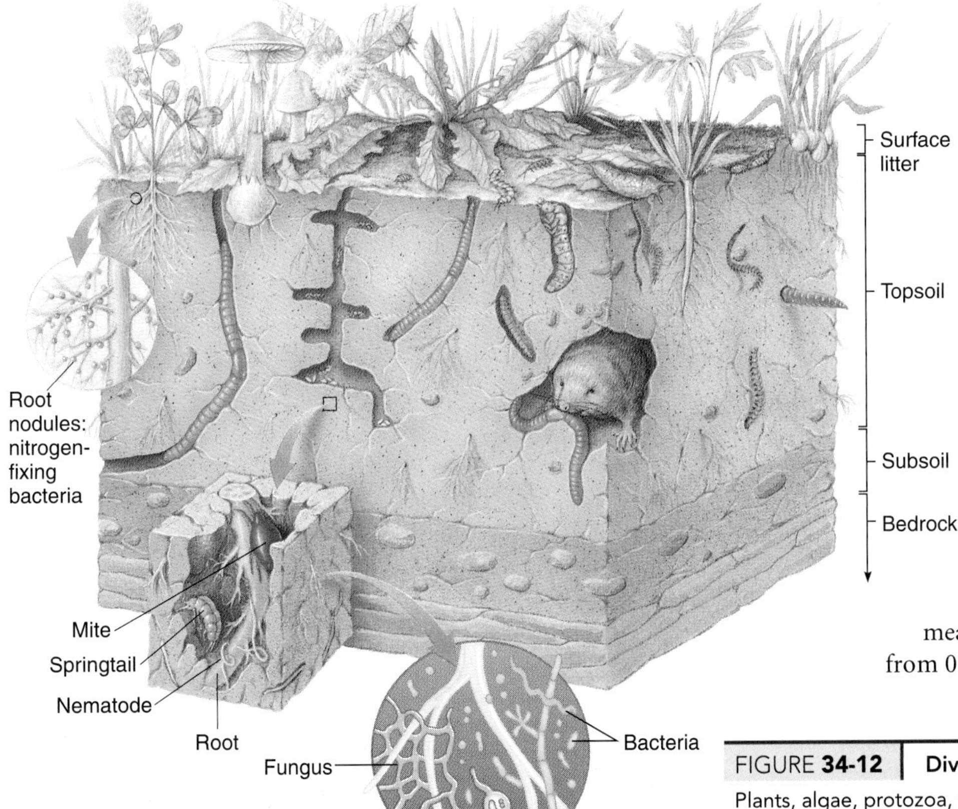

FIGURE **34-12** | **Diversity of life in fertile soil.**

Plants, algae, protozoa, fungi, bacteria, earthworms, flatworms, roundworms, insects, spiders and mites, and burrowing animals such as moles and groundhogs live in soil.

to 14 (extremely alkaline). The pH of most soils ranges from 4 to 8, but some soils are outside this range. The soil of the Pygmy Forest in Mendocino County, California, is extremely acidic (pH 2.8–3.9). At the other extreme, certain soils in Death Valley, California, have a pH of 10.5.

Plants are affected by soil pH, partly because the solubility of certain nutrient minerals varies with differences in pH. Soluble nutrient minerals can be absorbed by the plant, whereas insoluble forms cannot. At a low pH, for example, aluminum and manganese in soil water are more soluble and are sometimes absorbed by the roots in toxic concentrations. At a higher pH, certain nutrient mineral salts essential for plant growth, such as calcium phosphate, become less soluble and thus less available to plants.

Soil pH also affects the leaching of nutrient minerals. An acidic soil has less ability to bind positively charged ions to it because the soil particles also bind the abundant protons (Fig. 34-13). As a result, certain nutrient mineral ions essential for plant growth, such as potassium (K^+), are leached more readily from acidic soil. The optimum soil pH for most plant growth is 6.0 to 7.0, because most essential nutrient minerals are available to plants in that pH range.

Soil pH affects plants, and plants and other organisms in turn influence soil pH. The decomposition of humus and the cellular respiration of soil organisms and roots decrease the pH of the soil.

Acid precipitation, a type of air pollution in which sulfuric and nitric acids produced by human activities fall to the ground as acid rain, sleet, snow, or fog, can seriously decrease soil pH. Acid precipitation is one of several factors implicated in **forest decline,** the gradual deterioration, and often death, of trees that has been observed in many European and North American forests. Forest decline may be partly the result of soil changes caused by acid precipitation. In central European forests that have experienced forest decline, for example, a strong correlation exists between forest damage and soil chemistry altered by acid precipitation.

FIGURE **34-13** | **How acid alters soil chemistry.**

(a) In normal soil, positively charged nutrient mineral ions are attracted to the negatively charged soil particles. **(b)** In acidified soil, protons displace the cations. Aluminum ions released from inorganic mineral particles when the soil becomes acidified also adhere to soil particles.

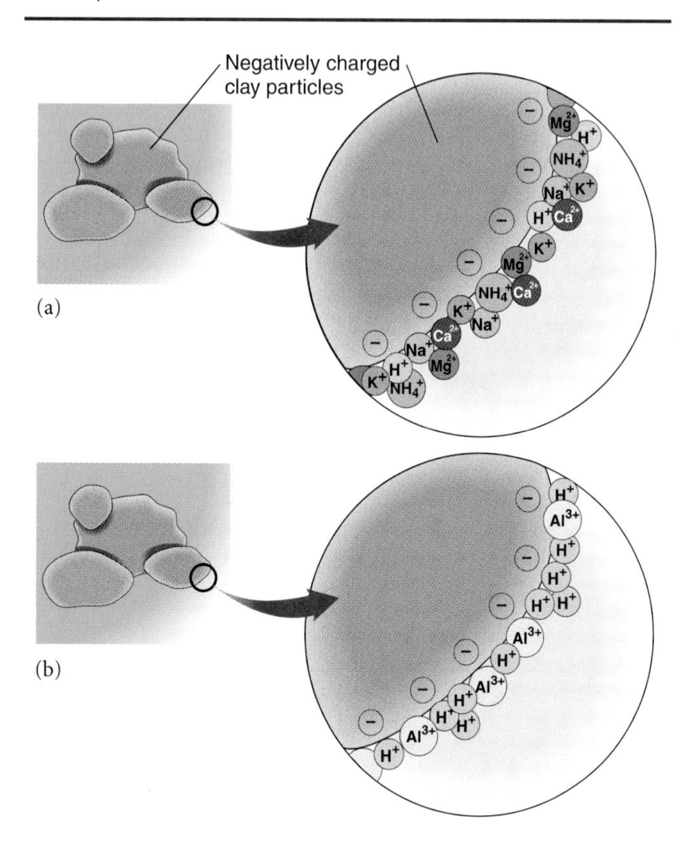

Soil provides most of the minerals found in plants

More than 90 naturally occurring elements exist on Earth, and more than 60 of these, including elements as common as carbon and as rare as gold, have been found in plant tissues. Not all these elements are essential for plant growth, however.

Nineteen elements have been found essential for most, if not all, plants (Table 34-1). Ten of these are required in fairly large quantities (greater than 0.05% dry weight) and are therefore known as **macronutrients.** These include carbon, hydrogen, oxygen, nitrogen, potassium, calcium, magnesium, phosphorus, sulfur, and silicon. The remaining nine **micronutrients** are needed in trace amounts (less than 0.05% dry weight) for normal plant growth and development. These include chlorine, iron, boron, manganese, sodium, zinc, copper, nickel, and molybdenum.

Four of the 19 elements—carbon, oxygen, hydrogen, and nitrogen—come directly or indirectly from soil water or from gases in the atmosphere. Carbon is obtained from carbon dioxide in the atmosphere during photosynthesis. Oxygen is obtained from atmospheric oxygen (O_2) and water (H_2O). Water also supplies hydrogen to the plant. Plants absorb their nitrogen from the soil as ions of nitrogen salts, but the nitrogen in nitrogen salts ultimately comes from atmospheric nitrogen. The remaining 15 essential elements are obtained from the soil as dissolved nutrient mineral ions. Their ultimate source is the parent material from which the soil was formed.

Let us examine the main functions of essential elements. Carbon, hydrogen, and oxygen are found as part of the structure of all biologically important molecules, including lipids, carbohydrates, nucleic acids, and proteins. Nitrogen is part of proteins, nucleic acids, and chlorophyll.

Potassium, which plants use in fairly substantial amounts, is not found in a specific organic compound in plant cells. Instead, it remains as free K^+ and plays a key physiological role in maintaining the turgidity of cells because it is osmotically active. The presence of K^+ in cytoplasm causes water to pass through the plasma membrane into the cell by osmosis. Potassium is also involved in the opening and closing of stomata (see Chapter 32).

Calcium plays a key structural role as a component of the middle lamella (the cementing layer between cell walls of adja-

TABLE 34-1 Functions of Elements Required by Most Plants

Element	Taken Up As	Major Functions
Macronutrients		
Carbon	CO_2	Component of carbohydrate, lipid, protein, and nucleic acid molecules
Hydrogen	H_2O	Component of carbohydrate, lipid, protein, and nucleic acid molecules
Oxygen	CO_2, H_2O	Component of carbohydrate, lipid, protein, and nucleic acid molecules
Nitrogen	NO_3^-, NH_4^+	Component of proteins, nucleic acids, chlorophyll, certain coenzymes
Potassium	K^+	Osmotic and ionic balance; opening and closing of stomata; enzyme activator (for 40+ enzymes)
Calcium	Ca^{2+}	In cell walls; involved in membrane permeability; enzyme activator; second messenger
Magnesium	Mg^{2+}	In chlorophyll; enzyme activator in carbohydrate metabolism
Phosphorus	HPO_4^{2-}, $H_2PO_4^-$	In nucleic acids, phospholipids, ATP (energy transfer compound)
Sulphur	SO_4^{2-}	In certain amino acids and vitamins
Silicon	SiO_3^{2-}	In cell walls
Micronutrients		
Chlorine	Cl^-	Ionic balance; involved in photosynthesis
Iron	Fe^{2+}, Fe^{3+}	In enzymes and electron transport molecules involved in photosynthesis, respiration and nitrogen fixation
Boron	$H_2BO_3^-$	In cell walls; involved in nucleic acid metabolism and in cell growth
Manganese	Mn^{2+}	In enzymes involved in respiration and nitrogen metabolism; required for photosynthesis
Sodium	Na^+	Involved in photosynthesis; substitutes for potassium in osmotic and ionic balance
Zinc	Zn^{2+}	In enzymes involved in respiration and nitrogen metabolism
Copper	Cu^+, Cu^{2+}	In enzymes involved in photosynthesis
Nickel	Ni^{2+}	In enzymes (urease) involved in nitrogen metabolism
Molybdenum	MoO_4^{2-}	In enzymes involved in nitrogen metabolism

cent plant cells). Calcium ions (Ca^{2+}) are also important in a number of physiological roles in plants, such as altering membrane permeability and as **second messengers** in various cell signaling responses (see Chapter 5).

Magnesium is critical for plants because it is part of the chlorophyll molecule. Phosphorus is a component of nucleic acids, phospholipids (an essential part of cell membranes), and energy transfer molecules such as ATP. Sulfur is essential because it is found in certain amino acids and vitamins. Many plants accumulate silicon in their cell walls and intercellular spaces. Silicon enhances the growth and fertility of some species and may help reinforce cell walls.

Chlorine and sodium are micronutrients that help maintain turgor of cells. In addition to this osmotic role, chloride (Cl^-) and sodium (Na^+) ions are essential for photosynthesis. Six of the micronutrients (iron, manganese, zinc, copper, nickel, and molybdenum) are associated with various plant enzymes, often as **cofactors,** and are involved in certain enzymatic reactions. Boron, present in cell walls, is also involved in nucleic acid metabolism and in cell growth.

How do biologists determine whether an element is essential?

It is impossible to conduct mineral nutrition experiments by growing plants in soil, because soil is too complex and contains too many elements. Therefore, nutritional studies require special methods. One of the most useful techniques to test whether an element is essential is **hydroponics,** which is the growing of plants in aerated water to which nutrient mineral salts have been added. Hydroponics also has commercial applications in addition to its scientific use (Fig. 34-14).

If biologists suspect a particular element is essential for plant growth, they grow plants in a nutrient solution that contains all known essential elements *except* the one in question. If plants grown in the absence of that element can't develop normally or complete their life cycle, the element may be essential. Additional criteria are used to confirm whether an element is essential. For example, it must be demonstrated that the element has a direct effect on the plant's metabolism and that the element is essential for a wide variety of plants.

Soil can be damaged by human mismanagement

Soil is a valuable natural resource on which humans depend for food. Many human activities generate or aggravate soil problems, including nutrient mineral depletion, soil erosion, and accumulation of salt.

Nutrient mineral depletion may occur in soils that are farmed

In a natural ecosystem, the essential nutrient minerals removed from the soil by plants are returned when the plants or the ani-

Hank Morgan/Photo Researchers, Inc.

ACTIVE FIGURE **34-14** | **Hydroponically grown lettuce.**

Lettuce is one of several hydroponic crops grown commercially. The white boards support the plants, which are not anchored in soil. A chemically defined liquid solution of nutrient minerals trickles over their roots, which also have access to atmospheric oxygen.

Biology Now™ Diagnose a sick plant to **learn more about mineral nutrition** by clicking on this figure on your BiologyNow CD-ROM.

mals that eat them die and are decomposed. An agricultural system disrupts this pattern of nutrient cycling when the crops are harvested. Plant material containing nutrient minerals is removed from the nutrient cycle, and the harvested crops fail to decay and release their nutrients back to the soil. Over time, farmed soil inevitably loses its fertility, that is, its ability to produce abundant crops. Homeowners often mow their lawns and remove the clippings, similarly preventing decomposition and cycling of nutrient minerals that were in the grass blades.

The essential material (water, sunlight, or some essential element) that is in shortest supply usually limits plant growth. This phenomenon is sometimes called the concept of **limiting resources.** The three elements that are most often limiting resources for plants are nitrogen, phosphorus, and potassium. To sustain the productivity of agricultural soils, fertilizers are periodically added to depleted soils to replace the nutrient minerals that limit plant growth.

The two main types of fertilizers are organic and commercial inorganic. *Organic fertilizers* include such natural materials as cow manure, crop residues, bone meal, and compost. Green manure, a special type of organic fertilizer, is actually a crop that is planted and deliberately plowed into the soil to decompose instead of being harvested. Frequently the plants grown for green manure have nitrogen-fixing bacteria living in root nodules, thereby increasing the amount of nitrogen in the soil. Organic fertilizers are complex, and their exact compositions vary. The nutrient minerals in organic fertilizers become available to plants only as the organic material decomposes. For that reason, organic fertilizers are slow-acting and long-lasting.

Commercial inorganic fertilizers are manufactured from chemical compounds, and their exact compositions are known. Because they are soluble, they are immediately available to plants. Commercial inorganic fertilizers are available in the soil for only a short period (relative to organic fertilizers), because they quickly leach away, often into groundwater or surface runoff, polluting the water. Most commercial inorganic fertilizers contain the three elements (nitrogen, phosphorus, and potassium) that are usually the limiting resources in plant growth. The numbers on fertilizer bags (such as 10–20–20) tell the percentage concentrations of each of these three elements (N, P, and K).

Soil erosion is the loss of soil from the land

Water, wind, ice, and other agents cause **soil erosion,** the wearing away or removal of soil from the land. Water and wind are particularly effective in moving soil from one place to another. Rainfall loosens soil particles that can then be transported away by moving water (Fig. 34-15). Wind loosens soil and blows it away, particularly if the soil is barren and dry. Soil erosion is a natural process that can be greatly accelerated by human activities.

Soil erosion is a national and international problem that does not make the headlines very often. The U.S. Department of Agriculture estimates that approximately 4.2 billion metric tons (4.6 billion tons) of topsoil are lost each year from U.S. croplands and rangelands as a result of soil erosion. Approximately one fifth of U.S. cropland is vulnerable to soil erosion damage. Erosion causes a loss in soil fertility because essential nutrient minerals and organic matter that are a part of the soil are removed. As a result of these losses, the productivity of eroded agricultural soils declines, and more fertilizer must be used to replace the nutrients lost to erosion.

Humans often accelerate soil erosion through poor soil management practices. Agriculture is not the only culprit, because the removal of natural plant communities during surface mining, unsound logging practices (such as clearcutting large areas

FIGURE **34-15** | **Soil erosion in an open field.**

Removal of plant cover exposes bare soil to erosion in heavy rains. Precipitation can cause gullies to enlarge rapidly, because they provide channels for the runoff of water.

USDA/Natural Resources Conservation Service

of forest), and the construction of roads and buildings also accelerate erosion.

Soil erosion has an impact on other natural resources. Sediment that enters streams, rivers, and lakes degrades water quality and fish habitats. Pesticides and fertilizer residues in sediment may add pollutants to the water. Also, when forests are removed within the watershed of a hydroelectric power facility, accelerated soil erosion fills the reservoir behind the dam with sediment much faster than usual. This process reduces electricity production at that facility.

Sufficient plant cover reduces the amount of soil erosion: Leaves and stems cushion the impact of rainfall, and roots help hold the soil in place. Although soil erosion is a natural process, abundant plant cover makes it negligible in many natural ecosystems.

Salt accumulates in soil that is improperly irrigated

Although irrigation improves the agricultural productivity of arid and semiarid lands, it sometimes causes salt to accumulate in the soil, a process called **salinization.** In a natural scenario, as a result of precipitation runoff rivers carry dissolved salts away. Irrigation water, however, normally soaks into the soil and does not run off the land into rivers, so when water evaporates, the salt remains behind and accumulates in the soil. Salty soil results in a decline in productivity and, in extreme cases, renders the soil completely unfit for crop production.

Most plants cannot obtain all the water they need from salty soil, because a water balance problem exists: Water moves by osmosis *out* of plant roots and into the salty soil. Obviously, most plants cannot survive under these conditions (see Fig. 5-13). Plant species that thrive in saline soils have special adaptations that enable them to tolerate the high amount of salt. Black mangroves, for example, excrete excess salt through their leaves.

Most crops, unless they have been genetically selected to tolerate high salt, are not productive in saline soil. Research on the molecular aspects of salt tolerance suggests that mutations in membrane transport proteins confer salt tolerance. A salt-tolerant variety of *Arabidopsis* was genetically engineered to overexpress a single gene that codes for a sodium transport protein in the vacuolar membrane. These plants are able to store large quantities of sodium in their vacuoles, thereby tolerating saline soils. Other salt-resistant plants exclude sodium from entering the cell through the plasma membrane.

Review

- What are the four components of soil, and how is each important to plants?
- What is cation exchange?
- How do weathering processes convert rock into soil?
- What are macronutrients and micronutrients?
- How are nutrient minerals lost from the soil?

BiologyNow™ Assess your understanding of **the soil environment** by taking the pretest on your BiologyNow CD-ROM.

SUMMARY WITH KEY TERMS

1 Distinguish between taproot and fibrous root systems.

- A **taproot** system has one main root (formed from the radicle), from which many lateral roots extend.
- A **fibrous root** system has several to many **adventitious roots** of the same size developing from the end of the stem. Lateral roots branch from these adventitious roots.

2 Label cross sections of a primary dicot root and a monocot root, and describe the functions of each tissue.

- Primary roots have an epidermis, ground tissues (cortex and, in certain roots, pith), and vascular tissues (xylem and phloem). Each root tip is covered by a **root cap,** a protective layer that covers the delicate root **apical meristem** and may orient the root so that it grows downward.
- **Epidermis** protects the root. **Root hairs,** short-lived extensions of epidermal cells, increase the surface area of the root in contact with soil. Root hairs aid in absorption of water and dissolved nutrient minerals.
- **Cortex** consists of parenchyma cells that often store starch. **Endodermis,** the innermost layer of the cortex, regulates the movement of nutrient minerals into the root xylem. Cells of the endodermis have a **Casparian strip** around their radial and transverse walls that is impermeable to water and dissolved nutrient minerals. Nutrient minerals are actively transported through carrier proteins in the plasma membranes of endodermal cells. Pericycle, xylem, and phloem collectively make up the root's **stele,** or **vascular cylinder.**
- **Pericycle** gives rise to lateral roots and lateral meristems. **Xylem** conducts water and dissolved nutrient minerals; **phloem** conducts dissolved sugar.
- Xylem of a herbaceous dicot root forms a solid core in the center of the root. In contrast, the center of a monocot root often consists of **pith** surrounded by a ring of alternating bundles of xylem and phloem. Monocot roots lack a **vascular cambium** and therefore do not have secondary growth.

3 Trace the pathway of water and nutrient mineral ions from the soil through the various root tissues, and distinguish between the symplast and apoplast.

- As water and dissolved nutrient mineral ions move from the soil into the root, they pass through the following tissues: root hair/epidermis ⟶ cortex ⟶ endodermis ⟶ pericycle ⟶ root xylem.
- The water and dissolved nutrient minerals move through the epidermis and cortex along one of two pathways, the **apoplast** (along the interconnected porous cell walls) or the **symplast** (from one cell's cytoplasm to the next through **plasmodesmata**).

4 Discuss the structure of roots with secondary growth.

- Roots of gymnosperms and woody dicots develop secondary tissues (wood and bark). The production of secondary tissues is the result of the activity of two lateral meristems, the vascular cambium and cork cambium. The vascular cambium produces secondary xylem (wood) and secondary phloem (inner bark). The **cork cambium** produces **periderm** (outer bark).

SUMMARY WITH KEY TERMS *(continued)*

5 Describe at least four roots that are modified to perform uncommon functions.

- **Prop roots** develop from branches or from a vertical stem and grow downward into the soil to help support certain plants in an upright position. **Buttress roots** are swollen bases or braces that support certain tropical rainforest trees that have shallow root systems.
- **Pneumatophores** are aerial "breathing" roots that may assist in getting oxygen to submerged roots.
- Certain epiphytes have roots modified to photosynthesize, absorb moisture, or, if parasitic, penetrate host tissues.
- Corms and bulbs often have **contractile roots** that grow into the soil and then contract, thereby pulling the corm or bulb deeper into the soil.

6 List and describe two mutualistic relationships between roots and other organisms.

- **Mycorrhizae** are mutually beneficial associations between roots and soil fungi.
- Root **nodules** are swellings that develop on roots of leguminous plants and house millions of rhizobia (nitrogen-fixing bacteria).
- A root **graft** is a natural connection between the roots of trees belonging to the same or different species.

7 Describe the roles of weathering, organisms, climate, and topography in soil formation.

- Factors influencing soil formation include parent material (type of rock), climate, organisms, the passage of time, and topography. Most soils are formed from parent material that is broken into smaller and smaller particles by **weathering processes.** Climate and organisms work together in weathering rock.
- Soil organisms such as plants, algae, fungi, worms, insects, spiders, and bacteria are important not only in forming soil, but also in cycling nutrient minerals.
- Topography, a region's surface features, affects soil formation. Steep slopes have little or no soil on them, whereas moderate slopes often have deep soils.

8 List the four components of soil, and give the ecological significance of each.

- Soil is composed of inorganic minerals, organic matter, air, and water. Inorganic minerals provide anchorage and essential nutrient minerals for plants. Organic matter increases the soil's water-holding capacity. As it decomposes, organic matter releases essential nutrient minerals into the soil. Soil air provides oxygen for soil organisms to use during aerobic respiration. Soil water provides water and dissolved nutrient minerals to plants and other organisms.

9 Describe how roots absorb positively charged nutrient mineral ions by the process of cation exchange.

- In the soil, **cations,** positively charged nutrient mineral ions, are attracted and reversibly bound to clay particles, which have predominantly negative charges on their outer surfaces. In **cation exchange,** roots secrete protons (H^+), which are exchanged for other positively charged mineral ions, freeing them into the soil water to be absorbed by roots.

10 Distinguish between macronutrients and micronutrients.

- Plants require 19 essential elements for normal growth. Ten elements are **macronutrients:** carbon, hydrogen, oxygen, nitrogen, potassium, calcium, magnesium, phosphorus, sulfur, and silicon. Macronutrients are required in fairly large quantities.
- Nine elements are **micronutrients:** chlorine, iron, boron, manganese, sodium, zinc, copper, nickel, and molybdenum. Micronutrients are needed in trace amounts.

11 Explain the impacts of mineral depletion and soil erosion on plant growth.

- Mineral depletion may occur in soils that are farmed because the natural pattern of nutrient cycling is disrupted when crops are harvested (and not allowed to decompose into the soil).
- **Soil erosion** is the removal of soil from the land by the actions of agents such as water and wind. Erosion causes a loss in soil fertility, because essential nutrient minerals and organic matter that are a part of the soil are removed.

POST-TEST

1. One main root, formed from the enlarging embryonic root, with many smaller lateral roots coming out of it is a(an) (a) fibrous root system (b) adventitious root system (c) taproot system (d) contractile root system (e) prop root system
2. Roots produced at unusual places on the plant are (a) fibrous (b) adventitious (c) taproots (d) contractile (e) mycorrhizae
3. Root hairs (a) cover and protect the delicate root apical meristem (b) increase the absorptive capacity of roots (c) secrete a waxy cuticle (d) orient the root so it grows downward (e) store excess sugars produced in the leaves
4. Plants with corms often have ____________ roots that pull the bulb deeper into the ground. (a) fibrous (b) adventitious (c) tap (d) contractile (e) prop
5. Certain plants adapted to flooded soil produce aerial "breathing" roots known as (a) fibrous roots (b) pneumatophores (c) mycorrhizae (d) contractile roots (e) prop roots
6. Unlike stems, roots produce (a) nodes and internodes (b) root caps and internodes (c) axillary buds and root hairs (d) terminal buds and axillary buds (e) root caps and root hairs
7. The waterproof region around the radial and transverse walls of endodermal cells is the (a) Casparian strip (b) pericycle (c) apoplast (d) symplast (e) pneumatophore
8. The apoplast is (a) a layer of cells that surrounds the vascular region in roots (b) the layer of cells just inside the endodermis (c) a system of interconnected plant cell walls through which water moves (d) the central cylinder of the root that comprises the vascular tissues (e) a continuum of cytoplasm of many cells, all connected by plasmodesmata
9. Plants obtain positively charged nutrient mineral ions from clay particles in the soil by cation exchange, in which (a) roots passively absorb the positively charged mineral ions they require (b) mineral ions flow freely along porous cell walls (c) roots se-

POST-TEST (continued)

crete protons, which free other positively charged mineral ions to be absorbed by roots (d) the Casparian strip effectively blocks the passage of water and nutrient mineral ions along the endodermal cell wall (e) a well-developed system of internal air spaces in the root allows both gas exchange and cation exchange

10. The cell layer from which lateral roots originate is the (a) epidermis (b) cortex (c) endodermis (d) pericycle (e) vascular cambium
11. The center of a herbaceous dicot root is composed of ___________, whereas the center of a monocot root is composed of ___________ (a) pith; cortex (b) xylem; phloem (c) phloem; xylem (d) xylem; pith (e) pith; xylem
12. Mutually beneficial associations between certain soil fungi and the roots of most plant species are called (a) mycorrhizae (b) pneumatophores (c) nodules (d) rhizobia (e) humus
13. Which of the following statements about soil is correct? (a) pore spaces are always filled with about 50% air and 50% water (b) a single teaspoon of fertile agricultural soil may contain up to several hundred living microorganisms (c) the texture of a soil is determined by the soil's pH (d) a soil's organic matter includes litter, droppings, and the dead remains of plants, animals, and microorganisms (e) soil formation is unaffected by a region's climate or topography
14. The technique of growing plants in aerated water containing dissolved nutrient mineral salts is known as (a) hydration (b) hydroponics (c) hydrophilic (d) hydrostatic (e) hydrolysis
15. Carbon, hydrogen, oxygen, nitrogen, potassium, calcium, magnesium, phosphorus, sulfur, and silicon are collectively known as (a) micronutrients (b) microvilli (c) micronuclei (d) macronuclei (e) macronutrients
16. In roots of woody plants (a) xylem does not form the central tissue of the root (b) the cortex comprises the bulk of the root (c) the vascular cambium gives rise to secondary xylem and secondary phloem (d) the pericycle gives rise to the apical meristem (e) secondary growth occurs despite the lack of a vascular cambium
17. Corn, sorghum, red mangrove, and banyan are plants that have (a) prop roots (b) buttress roots (c) pneumatophores (d) contractile roots (e) root nodules

CRITICAL THINKING

1. A mesquite root is found penetrating a mine shaft about 46 m (150 ft) below the surface of the soil. How could you determine *when* the root first grew into the shaft? (*Hint:* Mesquite is a woody plant.)
2. A barrel cactus that is 60 cm tall and 30 cm in diameter has roots more than 3 m long. However, all the plant's roots are found in the soil at a depth of 5 to 15 cm. What possible adaptive value does such a shallow root system confer on a desert plant?
3. How would you distinguish between a root hair and a small lateral root?
4. You are given a plant part that was found growing in the soil and are asked to determine whether it is a root or an underground stem. How would you identify the structure without a microscope? With a microscope?
5. How would you design an experiment to determine whether gold is essential for plant growth? What would you use for an experimental control?
6. Why does overwatering a plant often kill it?
7. Explain why, once secondary growth has occurred, that portion of the root is no longer involved in absorption. Where does absorption of water and dissolved nutrient minerals occur in plants that have roots with secondary growth?

- Visit our Web site at **http://biology.brookscole.com/solomon7** for links to chapter-related resources on the World Wide Web. Additional online materials relating to this chapter can also be found on our Web site.

BIOLOGY NOW RESOURCES

Active Figure

34-14: Mineral nutrition

Preparing for an exam? Take a diagnostic test on your BiologyNow CD-ROM.

Post-Test Answers

1. c	2. b	3. b	4. d
5. b	6. e	7. a	8. c
9. c	10. d	11. d	12. a
13. d	14. b	15. e	16. c
17. a			

35

Reproduction in Flowering Plants

Skip Moody/Dembinsky Photo Associates

Painted trillium. This plant, (*Trillium undulatum*), which grows to 50 cm (20 in), inhabits bogs and moist, acid woodlands in southern Canada and the northeastern United States, as well as the mountains south to Georgia.

CHAPTER OUTLINE

- **The Flowering Plant Life Cycle**
- **Pollination**
- **Fertilization and Seed/Fruit Development**
- **Asexual Reproduction in Flowering Plants**
- **A Comparison of Sexual and Asexual Reproduction**

In Chapter 31 you learned that flowering plants, which include at least 235,000 species, are the largest, most successful group of plants. One reason for the success of flowering plants, or **angiosperms,** is that they can reproduce both sexually and asexually. You may have admired flowers for their fragrances as well as their appealing colors and varied shapes (see photograph). The biological function of flowers, however, is sexual reproduction. Their colors, shapes, and fragrances are adaptations that increase the likelihood that pollen grains, which produce sperm cells, will be carried from plant to plant.

As in most organisms, sexual reproduction in plants includes meiosis and the fusion of reproductive cells—egg and sperm cells, collectively called **gametes.** The union of gametes, which is called *fertilization,* occurs within the flower's ovary. After fertilization, flowering plants produce seeds inside fruits.

The offspring of parents that reproduce sexually, show considerable genetic variation. They may resemble one of the parent plants, both of the parents, or neither of the parents. Sexual reproduction offers the advantage of new gene combinations, not found in either parent, that may make an individual plant better suited to its environment. These new gene combinations largely result from the **independent assortment** of chromosomes that occurs during meiosis before the production of both egg and sperm cells. (How this variation occurs, that is, the details of independent assortment and genetic recombination, is discussed in Chapters 9 and 10.)

Many flowering plants also reproduce asexually. Asexual reproduction doesn't usually involve the formation of flowers, seeds, and fruits. Instead, offspring generally form when a vegetative part of an existing plant expands, grows, and then becomes separated from the rest of the plant, often by the death of tissues. This part then forms a complete, independent plant.

Flowering plants have many methods of asexual reproduction, most of which involve modified vegetative organs such as stems, roots, and leaves. In particular, many modified stems, such as rhizomes, tubers, corms, and stolons, reproduce asexually. Because asexual reproduction requires only one parent, and no meiosis or fusion of gametes occurs, the offspring of asexual reproduction are virtually genetically identical to each other and to the parent plant from which they came.[1]

In this chapter we look at various aspects of both sexual and asexual reproduction in flowering plants, including floral adaptations that are important in pollination, seed and fruit structure and dispersal, and several kinds of asexual reproduction. We conclude with a discussion of the evolutionary advantages and disadvantages of sexual and asexual reproduction. ■

THE FLOWERING PLANT LIFE CYCLE

Learning Objectives

1. Label the parts of a flower on a diagram, and describe the functions of each part.
2. Identify where eggs and pollen grains are formed within the flower.

In Chapters 26 and 27 you learned that angiosperms and other plants have a cyclic **alternation of generations** in which they spend a portion of their life cycle in a multicellular haploid stage and a portion in a multicellular diploid stage. The haploid portion, called the **gametophyte generation,** gives rise to gametes by mitosis. When two gametes fuse during **fertilization,** the diploid portion of the life cycle, called the **sporophyte generation,** begins. The sporophyte generation produces haploid spores by meiosis. Each spore has the potential to give rise to a gametophyte plant, and the cycle continues.

In flowering plants the diploid sporophyte generation is larger and nutritionally independent. The haploid gametophyte generation, which is located in the flower, is microscopic and nutritionally dependent on the sporophyte. We revisit alternation of generations in flowering plants later in the chapter, after our discussion of the role of flowers as reproductive structures. It may be helpful for you to review Figure 27-12, which shows the main stages in the flowering plant life cycle.

Flowers are involved in sexual reproduction

How does a plant "know" it is time to start forming flowers? Correct timing is crucial to ensure reproductive success when a plant switches from vegetative to reproductive development. What happens, for example, if a plant flowers so late in the season that it does not have enough time to set seed before winter? A variety of environmental cues, such as temperature and day length, ensure proper timing, and different species are adapted to respond to distinctly different environmental cues. When environmental conditions induce flowering, *floral meristem identity genes* are activated, and the shoot apical meristem undergoes a transition from vegetative growth to reproductive growth—that is, to produce flowers. In Chapter 36 we say more about the initiation of flowering.

Flowers are reproductive shoots, usually consisting of four kinds of organs—sepals, petals, stamens, and carpels—arranged in whorls (circles) on the end of a flower stalk (Fig. 35-1; also see Fig. 27-10). In flowers with all four organs, the normal order of whorls from the flower's periphery to the center (or from the flower's base upward) is

Sepals ⟶ petals ⟶ stamens ⟶ carpels

The tip of the stalk enlarges to form a **receptacle** on which some or all of the flower parts are borne. All four floral parts are important in the reproductive process, but only the stamens (the "male" organs) and carpels (the "female" organs) participate directly in sexual reproduction—sepals and petals are sterile.

Sepals, which constitute the outermost and lowest whorl on a floral shoot, cover and protect the flower parts when the flower is a bud. Sepals are leaflike in shape and form and are often green. Some sepals, such as those in lily flowers, resemble petals. As the blossom opens from a bud, the sepals fold back to reveal the more conspicuous petals. The collective term for all the sepals of a flower is **calyx.**

The whorl just inside and above the sepals consists of **petals,** which are broad, flat, and thin (like sepals and leaves) but tremendously varied in shape and frequently brightly colored, attracting pollinators. As discussed later, petals play an important role in ensuring that sexual reproduction will occur. Sometimes petals fuse to form a tube or other floral shape. The collective term for all the petals of a flower is **corolla.**

Just inside and above the petals are the **stamens,** the male reproductive organs. Each stamen has a thin stalk, called a **filament,** at the top of which is an **anther,** a saclike structure in which **pollen grains** form. For sexual reproduction to occur, pollen grains must be transferred from the anther to the female reproductive structure (the carpel), usually of another flower of the same species. At first, each pollen grain consists of two cells surrounded by a tough outer wall. One cell, the **generative cell,** divides mitotically to form two nonflagellated male gametes, known as *sperm cells.* The other cell, the **tube cell,** produces a **pollen tube,** through which the sperm cells travel to reach the ovule.

In the center or top of most flowers are one or more **carpels,** the female reproductive organs. Carpels bear **ovules,** which are structures with the potential to develop into seeds. The carpels of a flower may be separate or fused together into a single structure. The female part of the flower, often called a **pistil,** may be a single carpel (a *simple pistil*) or a group of fused carpels (a *compound pistil*) (see Fig. 27-11). Each pistil has three sections: a **stigma,** on which the pollen grains land; a **style,** a necklike structure through which the pollen tube grows; and an **ovary,** a juglike structure that contains one or more ovules and can develop into a fruit.

[1] Although offspring of asexual reproduction are generally considered genetically uniform, somatic mutations can result in some variability among asexually derived offspring.

(a)

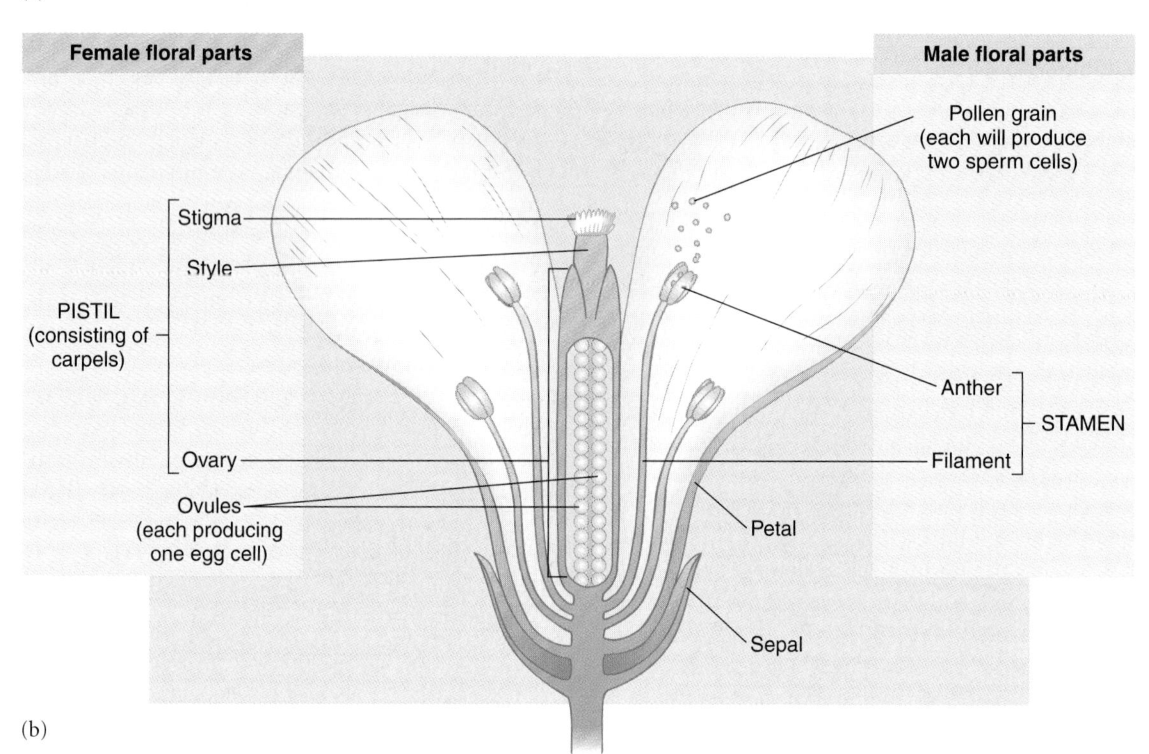

(b)

FIGURE **35-1** **Floral structure.**

(a) An *Arabidopsis thaliana* flower. **(b)** Cutaway view of an *Arabidopsis* flower. Each flower has four sepals (two are shown), four petals (two are shown), six stamens, and one pistil. Four of the stamens are long, and two are short (two long and two short are shown). Pollen grains develop within sacs in the anthers. In *Arabidopsis,* the compound pistil consists of two carpels that each contain numerous ovules.

Female gametophytes are produced in the ovary, male gametophytes in the anther

Before we proceed, it may be helpful to relate the stages in alternation of generations to floral structure. As discussed in Chapters 26 and 27, angiosperms and certain other plants are heterosporous and produce two kinds of spores: megaspores and microspores (Fig. 35-2).

Each young ovule within an ovary contains a diploid cell, the *megasporocyte,* which undergoes meiosis to produce four haploid **megaspores.** Three of these usually disintegrate, and the fourth, the functional megaspore, divides mitotically to produce a multicellular female gametophyte, also called an **embryo sac.** The embryo sac, which is embedded in the ovule, typically contains seven cells with eight haploid nuclei. Six of these cells, including the egg cell (the female gamete), contain a single nucleus each; a large central cell has two nuclei, called **polar nuclei.** The egg and both polar nuclei participate directly in fertilization.

Pollen sacs within the anther contain numerous diploid cells called *microsporocytes,* each of which undergoes meiosis to produce four haploid cells called **microspores.** Each microspore divides mitotically to produce an immature male gametophyte, also called a pollen grain, that consists of two cells, the tube cell and the generative cell. The pollen grain becomes mature when its generative cell divides to form two nonmotile sperm cells.

Review

- How do petals differ from sepals? How are they similar?
- How do stamens differ from carpels? How are they similar?
- What are female gametophytes, and where are they formed?
- What are male gametophytes, and where are they formed?

Biology Now™ Assess your understanding of **the flowering plant life cycle** by taking the pretest on your BiologyNow CD-ROM.

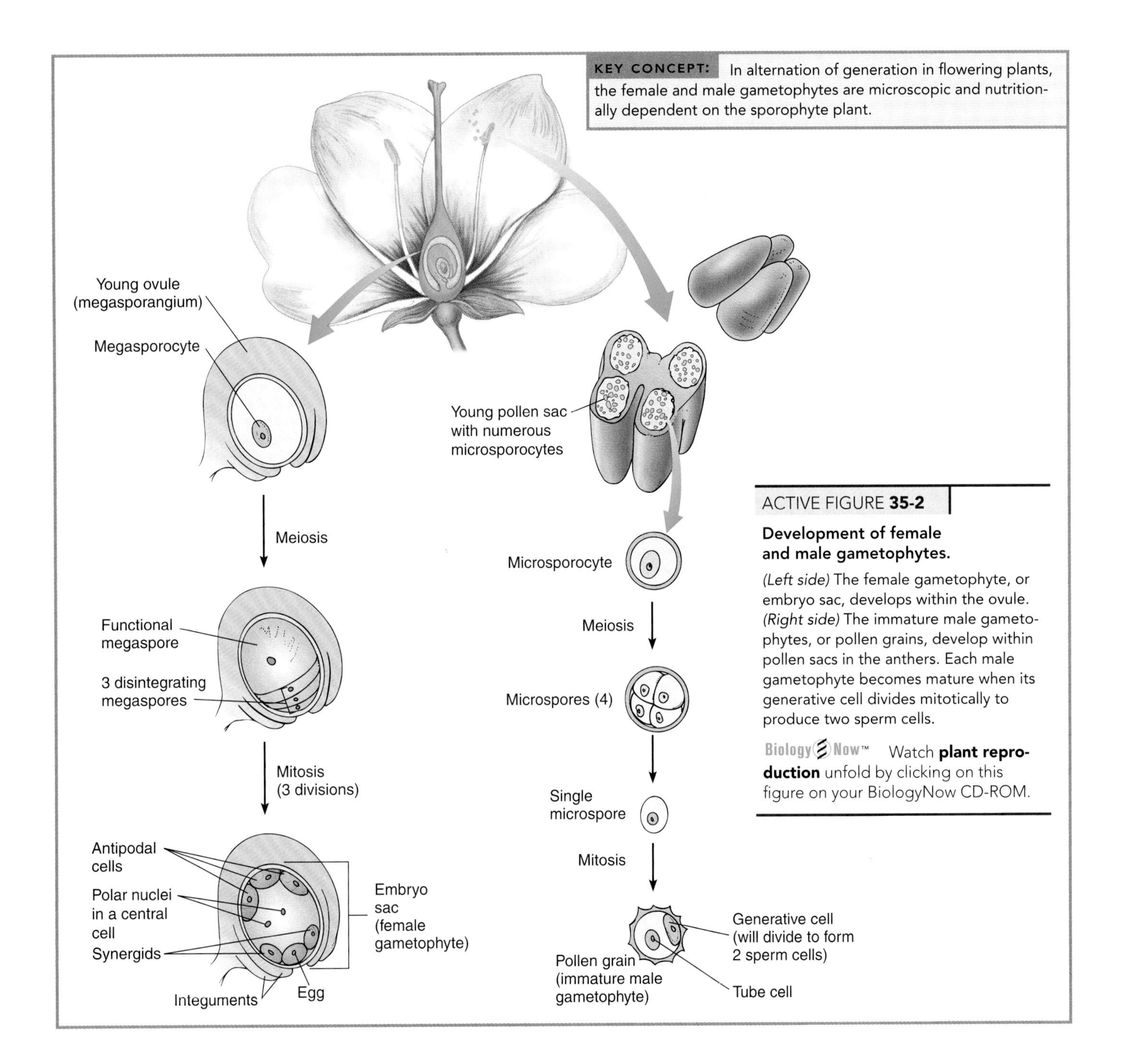

ACTIVE FIGURE **35-2**

Development of female and male gametophytes.

(Left side) The female gametophyte, or embryo sac, develops within the ovule. *(Right side)* The immature male gametophytes, or pollen grains, develop within pollen sacs in the anthers. Each male gametophyte becomes mature when its generative cell divides mitotically to produce two sperm cells.

BiologyNow™ Watch **plant reproduction** unfold by clicking on this figure on your BiologyNow CD-ROM.

POLLINATION

Learning Objectives

3 Compare the evolutionary adaptations that characterize flowers pollinated in different ways (by insects, birds, bats, and wind).

4 Define *coevolution,* and give examples of ways that plants and their animal pollinators have affected one another's evolution.

Before fertilization can occur, pollen grains must travel from the anther (where they form) to the stigma. The transfer of pollen grains from anther to stigma is known as **pollination.** Plants are *self-pollinated* if pollination occurs within the same flower or a different flower on the same individual plant. When pollen grains are transferred to a flower on another individual of the same species, we say the plant is *cross-pollinated.* Flowering plants accomplish pollination in a variety of ways. Beetles, bees, flies, butterflies, moths, wasps, and other insects pollinate many flowers. Other animals such as birds, bats, snails, and small nonflying mammals (rodents, primates, and marsupials) also pollinate plants. Wind is an agent of pollination for certain flowers, whereas a few aquatic flowers are pollinated by water.

Many plant species have mechanisms to prevent self-pollination

In plant sexual reproduction, the two gametes that unite to form a zygote may be from the same parent or from two different parents. The combination of gametes from two different parents increases the variation in offspring, and this variation may confer a selective advantage. Some offspring, for example, may be able to survive environmental changes better than either parent can.

Plants have evolved a variety of mechanisms that prevent self-pollination and thus prevent **inbreeding,** which is the mating of genetically similar individuals. Inbreeding can increase the concentration of deleterious genes in the offspring. Some species have separate male and female individuals; the male plants have staminate flowers that lack carpels, and the female plants have pistillate flowers that lack stamens. Other species have flowers with both stamens and pistils, but the pollen is shed from a given flower either before or after the time that the stigma of that flower is receptive to pollen.

Many species have genes for **self-incompatibility,** a genetic condition in which the pollen is ineffective in fertilizing the same flower or other flowers on the same plant. In other words, an individual plant can identify and reject its own pollen. Genes for self-incompatibility usually inhibit the growth of the pollen tube in the stigma and style, thereby preventing sperm cell delivery to the ovules. Self-incompatibility, which is more common in wild species than in cultivated plants, ensures that reproduction occurs only if the pollen comes from a genetically different individual.

The molecular basis of self-incompatibility in *Arabidopsis* and related plants is an area of active scientific interest. It appears that self-incompatibility in these plants is the ancestral (normal) condition. Self-fertile individuals contain mutations in one or more of the genes involved in self-incompatibility.

Flowering plants and their animal pollinators have coevolved

Animal pollinators and the plants they pollinate have had such close, interdependent relationships over time that they have affected the evolution of certain physical and behavioral features in each other. The term **coevolution** describes such reciprocal adaptation, in which two species interact so closely that they become increasingly adapted to one another as they each undergo evolutionary change by *natural selection.* Let's examine some of the features of flowers and their pollinators that may be the products of coevolution.

Flowers pollinated by animals have various features to attract their pollinators, including showy petals (a visual attractant) and scent (an olfactory attractant). One reward for the animal pollinator is food. Some flowers produce nectar, a sugary solution, in special floral glands called *nectaries.* Pollinators use nectar as an energy-rich food. Pollen grains are also a protein-rich food for many animals. As they move from flower to flower searching for food, pollinators inadvertently carry along pollen grains on their body parts, helping the plants reproduce sexually.

Biologists estimate that insects pollinate about 70% of all flowering plant species. Bees are particularly important as pollinators of crop plants. Crops pollinated by bees provide about 30% of human food. Plants pollinated by insects often have blue or yellow petals. The insect eye does not see color the same way the human eye does. Most insects see well in the violet, blue, and yellow range of visible light but do not perceive red as a distinct color. So flowers pollinated by insects are not usually red. Insects can also see in the ultraviolet range, wavelengths that are invisible to the human eye. Insects see ultraviolet radiation as a color called *bee's purple.* Many flowers have dramatic uv markings that may or may not be visible to humans but that direct insects to the center of the flower where the pollen grains and nectar are (Fig. 35-3).

Insects have a well-developed sense of smell, and many insect-pollinated flowers have a strong scent that may be pleasant or foul to humans. The carrion plant, for example, is pollinated by flies and smells like the rotting flesh in which flies lay their eggs. As flies move from one reeking flower to another looking for a place to lay their eggs, they transfer pollen grains.

Birds such as hummingbirds are important pollinators (Fig. 35-4a). Flowers pollinated by birds are usually red, orange, or yellow, because birds see well in this range of visible light. Because birds do not have a strong sense of smell, bird-pollinated flowers usually lack a scent.

Bats, which feed at night and do not see well, are important pollinators, particularly in the tropics, where they are most abun-

(a)

(b)

FIGURE **35-3** **Ultraviolet markings on insect-pollinated flowers.**

(a) This flower as seen by the human eye is solid yellow. **(b)** The same flower viewed under ultraviolet radiation provides clues about how the insect eye perceives it. The light blue portions of the petals appear purple to a bee, whereas the dark blue inner parts appear yellow. These differences in color draw attention to the center of the flower, where the pollen grains and nectar are located.

Dan Dempster/Dembinsky Photo Associates

(a)

Merlin Tuttle/Bat Conservation International/Photo Researchers, Inc.

(b)

FIGURE **35-4** | **Animal pollinators.**

(a) A ruby-throated hummingbird (*Archilochus colubris*) obtains nectar from a trumpet vine flower (*Campis radicans*). The pollen grains on the bird's feathers are carried to the next plant. **(b)** A lesser long-nosed bat (*Leptonycteris curasoae*) obtains nectar from a cardon cactus flower (*Pachycereus*) and transfers pollen as it moves from flower to flower. Bats pollinate several hundred species of plants.

dant (Fig. 35-4b). Bat-pollinated flowers bloom at night and have dull white petals and a strong scent, usually of fermented fruit. Nectar-feeding bats are attracted to the flowers by the scent; they lap up the nectar with their long, extendible tongues. As they move from flower to flower, they transfer pollen grains. At least one bat-pollinated flower (the tropical vine *Mucana holtonii*) has evolved an unusual adaptation to encourage pollination by bats. When the pollen in a given flower is mature, a concave petal lifts up. The petal bounces the echo from the bat's echolocating calls back to the bat, helping it find the flower.

During the time when plants were coevolving specialized features such as petals, scent, and nectar to attract pollinators, animal pollinators also coevolved. Specialized body parts and behaviors adapted animals to aid pollination as they obtain nectar and pollen grains as a reward. For example, coevolution has selected for bumblebees' hairy bodies, which catch and hold the sticky pollen grains for transport from one flower to another.

Coevolution may also have led to the long, curved beaks of the 'i'iwi, one of the Hawaiian honeycreepers, that inserts its beak into tubular flowers to obtain nectar (see Figs. 1-11b and 19-15). The long, tubular corolla of the flowers that 'i'iwis visit probably also came about through coevolution. During the 20th century, some of the tubular flower species (such as lobelias) became rare, largely as a result of grazing by nonnative cows and feral goats, and about 25% of lobelia species have become extinct. The 'i'iwi now feeds largely on the flowers of the 'ohi'a tree, which lacks petals, and the 'i'iwi bill appears to be slowly adapting to this change in feeding preference. Comparison of the bills of 'i'iwi museum specimens collected in 1902 with the bills of live birds captured in the 1990s showed that 'i'iwi bills were about 3% shorter in the 1990s than in 1902.

Animal behavior has also coevolved, sometimes in bizarre and complex ways. The flowers of certain orchids (*Ophrys*) resemble female bees in coloring and shape. The unpollinated flowers also secrete a scent similar to that produced by female bees, and the males are irresistibly attracted to it. The resemblance between *Ophrys* flowers and female bees is so strong that male bees mount the flowers and try to copulate with them. During this pseudocopulation, a pollen sac usually attaches to the back of the bee. When the frustrated bee departs and tries to copulate with another orchid flower, it transfers pollen grains to that flower. Interestingly, once a flower has been pollinated, it emits a scent like that released by female bees that have already mated. Male bees have no interest in visiting flowers that have been pollinated (just like they lose interest in female bees that have already been inseminated).

Some flowering plants depend on wind to disperse pollen

Some flowering plants, such as grasses, ragweed, maples, and oaks, are pollinated by wind. Wind-pollinated plants produce many small, inconspicuous flowers (Fig. 35-5). They don't produce large, colorful petals, scent, or nectar. Some have large, feathery stigmas, presumably to trap wind-borne pollen grains. Because wind pollination is a hit-or-miss affair, the likelihood of a particular pollen grain landing on a stigma of the same species of flower is slim. Wind-pollinated plants produce large quantities of pollen grains, which increase the likelihood that some pollen grains will land on the appropriate stigma.

Review

- A flower has yellow petals, a pleasant scent, and sugary nectar. How is it probably pollinated?
- A flower has small, inconspicuous flowers without petals, a scent or nectar. How is it probably pollinated?
- What is coevolution?

BiologyNow™ Assess your understanding of **pollination** by taking the pretest on your BiologyNow CD-ROM.

Dr. Jeremy Burgess/Science Photo Library/Photo Researchers, Inc.

FIGURE **35-5** | **Wind pollination.**

Each cluster of male oak flowers (*Quercus*) dangles from a tree branch and sheds a shower of pollen when the wind blows. These flowers lack petals.

FERTILIZATION AND SEED/FRUIT DEVELOPMENT

Learning Objectives

5 Distinguish between pollination and fertilization.

6 Trace the stages of embryo development in flowering plants, and list and define the main parts of seeds.

7 Explain the relationships among the following: ovules, ovaries, seeds, and fruits.

8 Distinguish among simple, aggregate, multiple, and accessory fruits; give examples of each type; and cite several different methods of seed and fruit dispersal.

Once pollen grains have been transferred from anther to stigma, the tube cell, one of the two cells in the pollen grain, grows a thin pollen tube down through the style and into an ovule in the ovary. How does the pollen tube "know" where to grow? Scientists found that molecular signals from the embryo sac guide the growing pollen tube toward the ovule. Once a pollen tube penetrates the ovule, the attracting signals cease. As a result, only one pollen tube enters each embryo sac. The second cell within the pollen grain divides to form two male gametes (the sperm cells), which move down the pollen tube and enter the ovule (Fig. 35-6).

A unique double fertilization process occurs in flowering plants

The egg within the ovule unites with one of the sperm cells, forming a zygote (fertilized egg) that develops into an embryonic plant contained in the future seed. The two polar nuclei in the central cell of the ovule fuse with the second sperm cell to form the first cell of the triploid ($3n$) **endosperm,** the tissue with nutritive and hormonal functions that surrounds the developing embryonic plant in the seed. This process, in which two separate cell fusions occur, is called **double fertilization.** It is, with two exceptions, unique to flowering plants. (A type of double fertilization has been reported in two gymnosperm species, *Ephedra nevadensis* and *Gnetum gnemon.*)

After double fertilization, the ovule develops into a *seed,* and the surrounding ovary develops into a *fruit.* (We discuss double fertilization and the flowering plant life cycle in greater detail in Chapter 27.)

Embryonic development in seeds is orderly and predictable

Flowering plants package a young plant embryo, complete with stored nutrients, in a compact **seed,** which develops from the ovule after fertilization. The nutrients in seeds are not only used by *germinating* plant embryos but also eaten by animals, including humans (see *Focus On: Seed Banks*). Development of the embryo and endosperm following fertilization is possible because of the constant flow of nutrients into the developing seed from the parent plant.

Cell divisions of the fertilized egg to form a multicellular embryo proceed in several ways in flowering plants. We next describe dicot embryonic development; monocot embryonic development is similar in the early stages.

The two cells (basal cell and apical cell) that are formed as a result of the first division of the fertilized egg establish *polarity,* or direction, in the embryo. The large **basal cell** (located toward the outside of the ovule) typically develops into a **suspensor,** an embryonic tissue that anchors the developing embryo and aids in nutrient uptake from the endosperm. The **apical cell** (toward the inside of the ovule) develops into the plant embryo. Scientists are currently studying molecular differences between the apical cell and basal cell to determine the initial molecular signals that are involved in establishing polarity.

Initially, the apical cell divides to form a short cluster of cells, called a *proembryo* (Fig. 35-7 on page 677). As cell division continues, a small sphere of cells, often called a *globular embryo,* develops. Cells begin to develop into specialized tissues during this stage. When the dicot embryo starts to develop its two cotyledons (seed leaves), it has two lobes and resembles a heart; this is often called the *heart stage.* During the *torpedo stage,* the embryo continues to grow as the cotyledons elongate. As the embryo enlarges, it often curves back on itself and crushes the suspensor beyond recognition.

The mature seed contains an embryonic plant and storage materials

A mature seed contains an embryonic plant and food (stored in the cotyledons or endosperm), surrounded by a tough, protective **seed coat** derived from the **integuments,** which are

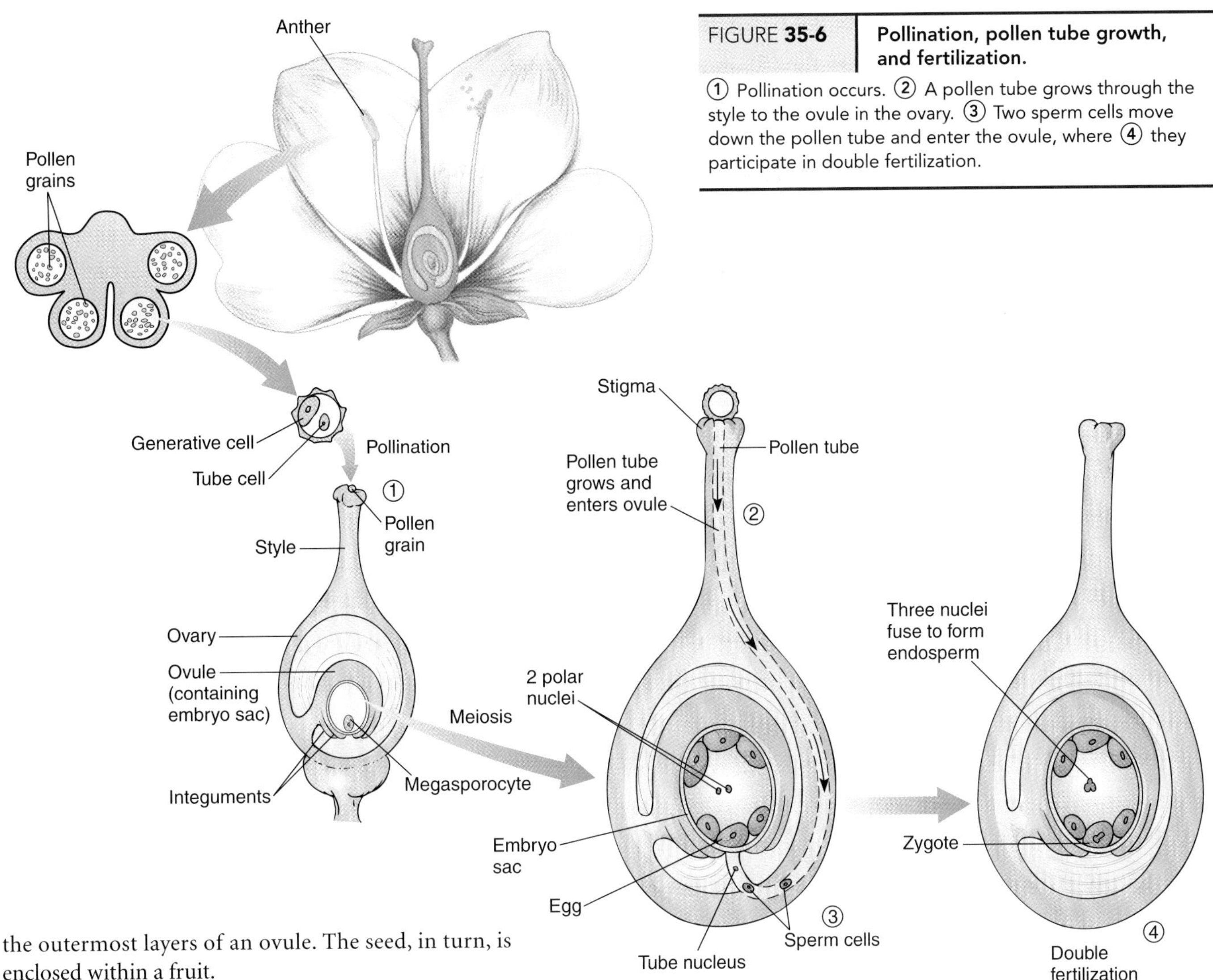

FIGURE **35-6** **Pollination, pollen tube growth, and fertilization.**

① Pollination occurs. ② A pollen tube grows through the style to the ovule in the ovary. ③ Two sperm cells move down the pollen tube and enter the ovule, where ④ they participate in double fertilization.

the outermost layers of an ovule. The seed, in turn, is enclosed within a fruit.

The mature embryo within the seed consists of a short embryonic root, or **radicle;** an embryonic shoot; and one or two seed leaves, or **cotyledons.** Monocots have a single cotyledon, whereas dicots have two (Fig. 35-8 on page 677). The short portion of the embryonic shoot connecting the radicle to one or two cotyledons is the **hypocotyl.** The shoot apex, or terminal bud, located above the point of attachment of the cotyledon(s), is the **plumule.** After the radicle, hypocotyl, cotyledon(s), and plumule have formed, the young plant's development is arrested, usually by desiccation or dormancy. When conditions are right for continuing the developmental program, the seed **germinates,** or sprouts, and the embryo resumes growth.

Because the embryonic plant is nonphotosynthetic, it must be nourished during germination until it becomes photosynthetic and therefore self-sufficient. The cotyledons of many plants function as storage organs and become large, thick, and fleshy as they absorb the food reserves (starches, oils, and proteins) initially produced as endosperm. Seeds that store nutrients in cotyledons have little or no endosperm at maturity. Examples of such seeds are peas, beans, squashes, sunflowers, and peanuts. Other plants, wheat and corn, for example, have thin cotyledons that function primarily to help the young plant digest and absorb food stored in the endosperm. (See Chapter 36 for a discussion of seed germination and early growth.)

Seed size is related to survival in a particular environment

In flowering plants seed size varies considerably, from the microscopic, dustlike seeds of orchids to the giant seeds of the double coconut (*Lodoicea maldivica*), which weigh as much as 27 kg (almost 60 lb). Despite this variation among different species, seed size is a remarkably constant trait within a species.

Assuming that a given plant species invests a fixed amount of its energy in reproduction, is it more advantageous to produce a large number of small seeds, or a few big ones? Observing the seed sizes that predominate in different environments, biologists have suggested that in some environments, a smaller seed seems advantageous, whereas in others a larger seed may be better.

Seed size for each plant species probably represents a tradeoff between the requirements for dispersal and for successful establishment of seedlings. For example, plants that grow in widely scattered open sites (such as old fields) usually produce smaller seeds, perhaps because they can be more easily dispersed over large areas than can larger seeds. However, wide dispersal is probably less important for plants adapted to densely

To preserve older, more diverse varieties of plants, many countries are collecting plant **germplasm,** which is any plant material used in breeding. Germplasm includes seeds, plants, and plant tissues of traditional crop varieties. The International Plant Genetics Resource Institute in Rome, Italy, is the scientific organization that oversees plant germplasm collections worldwide. More than 100 seed collections, called *seed banks,* exist around the world and collectively hold more than 3 million samples of thousands of different kinds of plants (see figure). The U.S. National Plant Germplasm System in Fort Collins, Colorado, stores seeds of about 250,000 different species and varieties.

Most seed banks help preserve the genetic variation within different varieties of crops and their wild relatives. Farmers typically stop planting local varieties when newer, improved varieties become available. The newer varieties have desirable genetic characteristics such as a greater yield, but the local, discarded varieties also contain valuable genes. Each local variety's characteristic combination of genes gives it distinctive nutritional value, size, color, flavor, resistance to disease, and adaptability to different climates and soil types. Maintaining the genetic diversity present in local crop varieties and their wild relatives helps preserve genes we may need in the future. The gene combinations of local varieties are potentially valuable to agricultural breeders because they can be transferred to other varieties, either by traditional breeding methods or by genetic engineering.

Seed banks offer the advantage of storing a large amount of live plant genetic material in a very small space. Seeds stored in seed banks are safe from habitat destruction, climate changes, disease, predators, and general neglect. In some instances, seeds from seed banks have even been used to reintroduce a plant species that has become extinct in the wild.

Courtesy of Nordiska Genbanken, Alnarp, Svenge

Seed storage. These moisture-proof bottles, vials, and packets of seeds are stored at very low temperatures in the seed bank in Svalbard, Norway.

There are some disadvantages to seed banks. Seeds don't stay alive forever and must be periodically germinated so that new seeds can be collected. Growing, harvesting, and returning seeds to storage is the most expensive aspect of storing plant material in seed banks. According to the U.N. Food and Agricultural Organization, many countries establish seed banks but do not provide the funds to pay for periodic replanting.

Many types of plants, such as avocados and coconuts, cannot be stored as seeds. The seeds of these plants do not tolerate being dried out, which is necessary before they are sealed in moisture-proof containers for storage at −18°C (−0.4°F). If 20% or more moisture remains in a frozen seed, it will probably die. Some seeds cannot be stored successfully because they only stay viable for a short time—a few months or even just a few days. Cryopreservation in liquid nitrogen at −160°C, or −256°F, is a new method being developed for certain kinds of seeds. Seeds stored at this temperature survive for longer periods than seeds stored at warmer temperatures.

Perhaps the most important disadvantage of seed banks is that plants stored in this manner remain stagnant in an evolutionary sense. Removed from their natural habitats, they are no longer subject to the forces of natural selection. As a result, they may be less fit for survival when they are reintroduced into the wild.

Despite their shortcomings, seed banks are increasingly viewed as an important way to safeguard seeds for future generations. For example, the Millennium Seed Bank Project at the Royal Botanic Gardens (Kew Gardens) in Great Britain is currently collecting and storing seeds from 10% of the world's plant species, including all species native to Great Britain.

vegetated areas such as forests. These plants generally produce larger seeds with an ample food reserve that may confer a greater likelihood of successfully establishing seedlings in a shaded environment. The stored energy may allow the young seedling to grow tall enough to reach adequate sunlight for photosynthesis.

Other ecological factors are associated with seed size. Larger seeds are typical of many plants that live in arid habitats, possibly because the food stored in a large seed lets a young seedling establish an extensive root system quickly, enabling it to survive the dry climate. Island plants also produce larger seeds than similar species on the nearby mainland. In this case, biologists hypothesize that large seeds are less likely to be widely dispersed and therefore less likely to fall into the ocean, where chances of survival are slim. Seeds that remain dormant in the soil for a long period tend to be smaller; the exact reason for this tendency is not known.

Despite many associations of seed size with specific environments, the general principles concerning the adaptive advantages of large seeds versus small seeds remain to be determined.

Fruits are mature, ripened ovaries

After double fertilization takes place within the ovule, the ovule develops into a seed (just described), and the ovary surround-

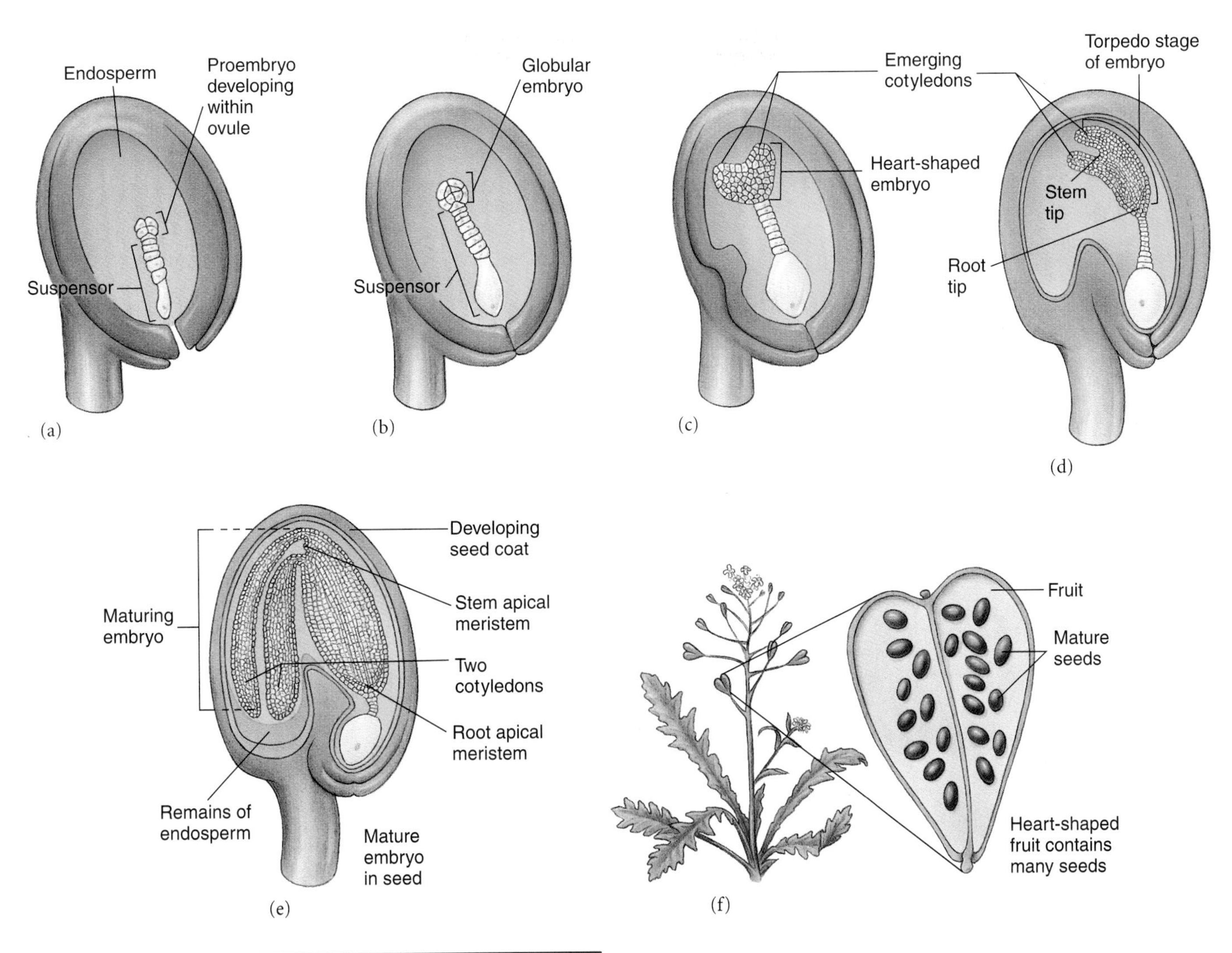

FIGURE 35-7 | Embryonic development.

(a) The proembryo in shepherd's purse (*Capsella bursa-pastoris*) is the earliest multicellular stage of the embryo (the ovule is shown apart from the ovary). **(b)** As cell division continues, the embryo becomes a ball of cells, called the globular stage. **(c)** As the two cotyledons begin to emerge, the embryo is shaped like a heart. **(d)** The cotyledons continue to elongate, forming the torpedo stage. **(e)** A maturing embryo within the seed. The food originally stored in the endosperm has been almost completely depleted during embryonic growth and development, and most of the food for the embryonic plant is stored in its cotyledons. **(f)** A longitudinal section through a heart-shaped fruit of shepherd's purse reveals numerous tiny seeds, each containing a mature embryo. Each seed developed from an ovule.

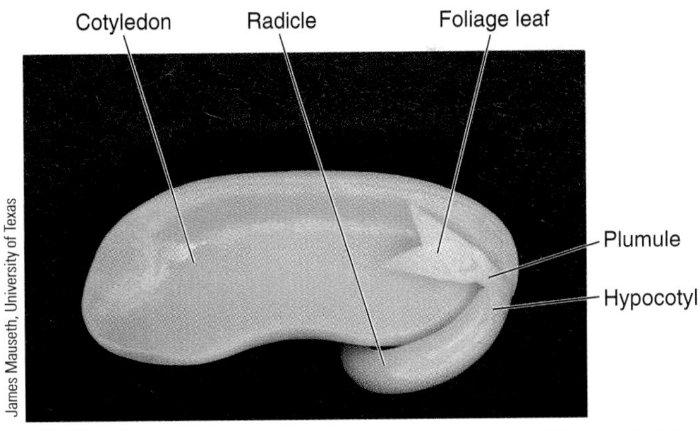

FIGURE 35-8 | Seed structure.

A bean seed has been dissected—its seed coat and one of its two cotyledons were removed—to show the radicle, hypocotyl, remaining cotyledon, and plumule with its foliage leaf at the shoot apex. This particular seed had begun germinating, so its radicle is larger than in an ungerminated seed.

ing it develops into a **fruit.** For example, a pea pod is a fruit, and the peas within it are seeds. A fruit may contain one or more seeds; some orchid fruits contain several thousand to a few million seeds! Fruits provide protection for the enclosed seeds and sometimes aid in their dispersal.

There are several types of fruits; their differences result from variations in the structure or arrangement of the flowers from which they were formed. The four basic types of fruits—simple fruits, aggregate fruits, multiple fruits, and accessory fruits—are summarized in Figure 35-9.

Most fruits are simple fruits. A **simple fruit** develops from a single pistil, which may consist of a single carpel or several fused carpels. At maturity, simple fruits may be fleshy or dry. Two examples of simple, fleshy fruits are berries and drupes. A **berry**

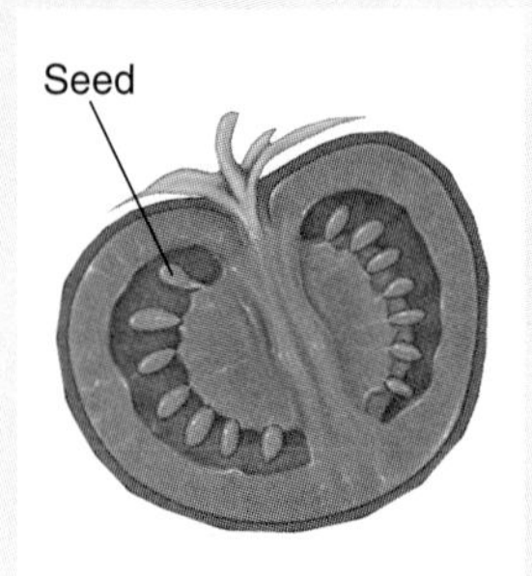

Berry (simple fruit)
A simple, fleshy fruit in which the fruit wall is soft throughout.

Tomato *(Lycopersicon lycopersicum)*

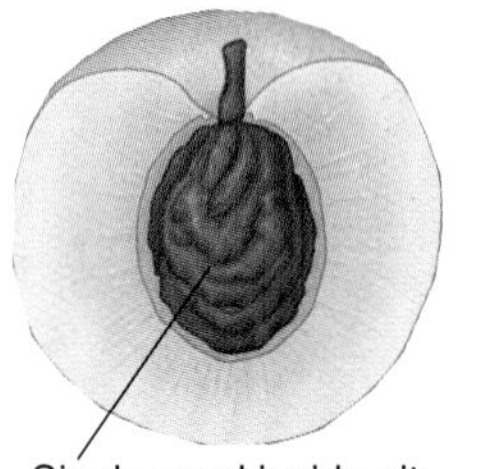
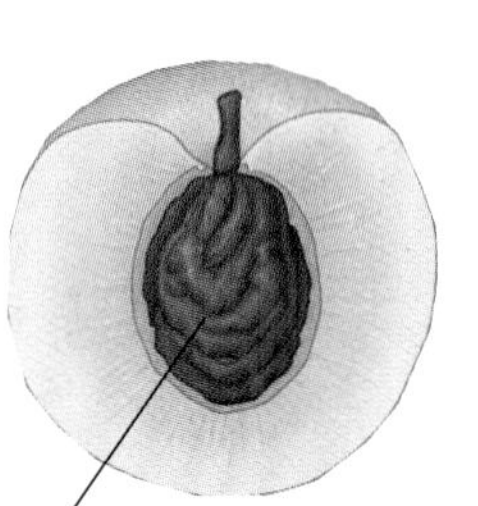

Drupe (simple fruit)
A simple, fleshy fruit in which the inner wall of the fruit is hard and stony (the pit).

Peach *(Prunus persica)*

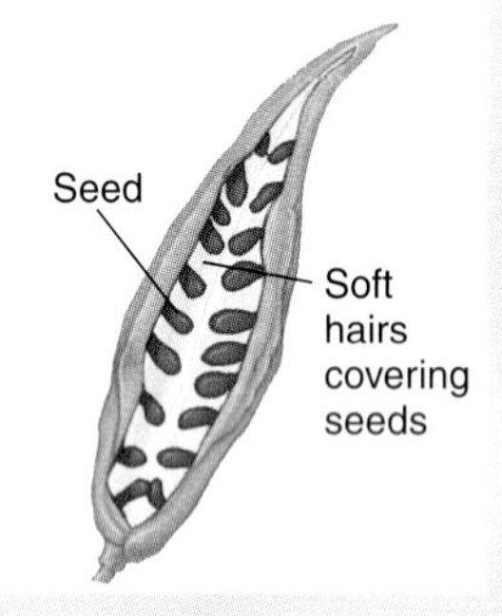

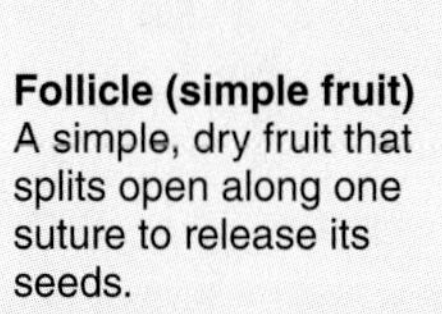

Follicle (simple fruit)
A simple, dry fruit that splits open along one suture to release its seeds.

Milkweed *(Asclepias syriaca)*

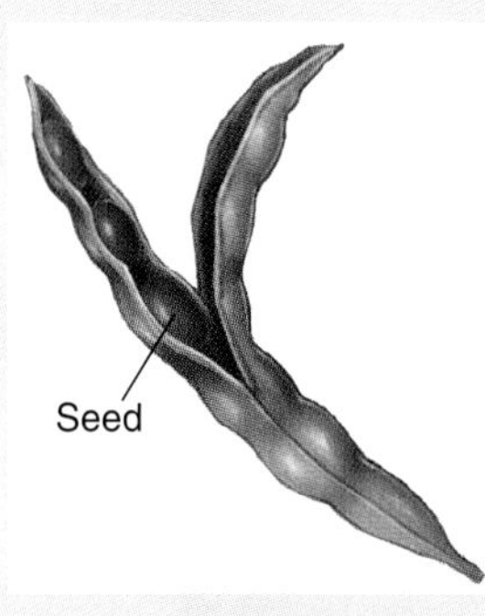

Legume (simple fruit)
A simple, dry fruit that splits open along two sutures to release its seeds.

Green bean *(Phaseolus vulgaris)*

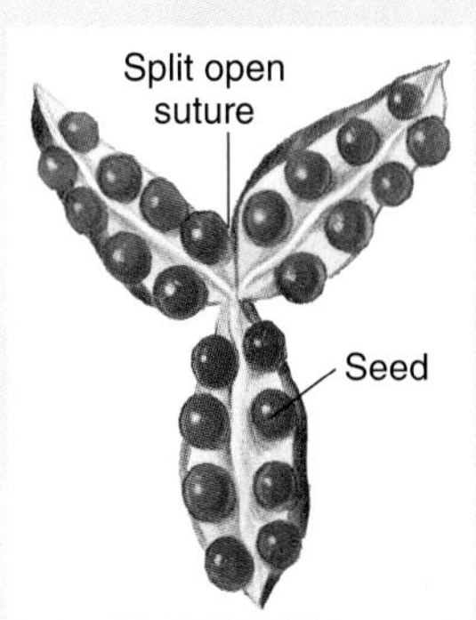

Capsule (simple fruit)
A simple, dry fruit that splits open along three or more sutures or pores to release its seeds.

Iris *(Iris)*

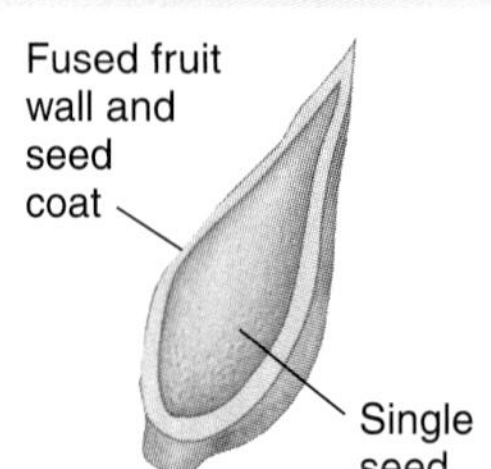
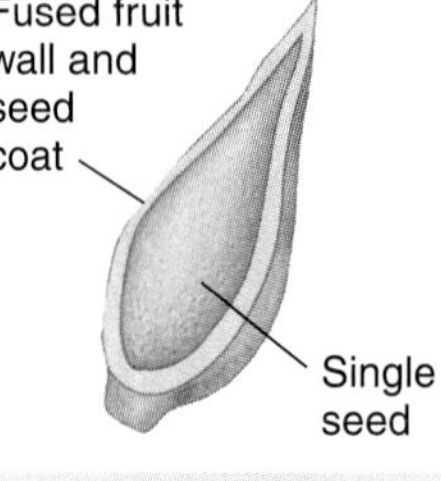

Caryopsis (simple fruit)
A simple, dry fruit in which the fruit wall is fused to the seed coat.

Wheat *(Triticum)*

Achene (simple fruit)
A simple, dry fruit in which the fruit wall is separate from the seed coat.

Sunflower *(Helianthus annuus)*

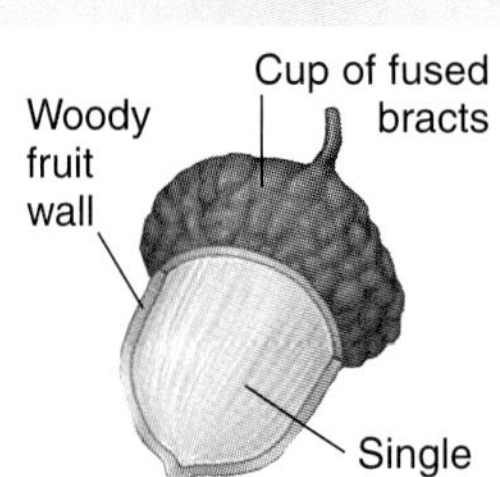
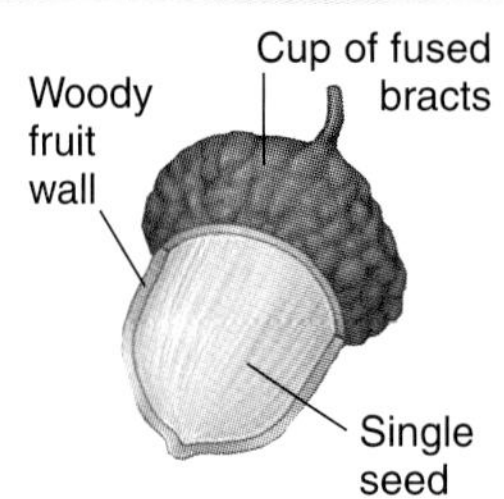

Nut (simple fruit)
A simple, dry fruit that has a stony wall, is usually large, and does not split open at maturity.

Oak *(Quercus)*

Aggregate fruit
A fruit that develops from a single flower with several to many pistils (i.e., carpels are not fused into a single pistil).

Blackberry *(Rubus)*

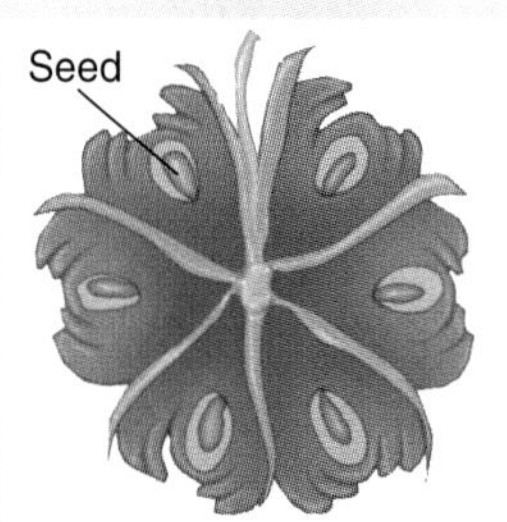

Multiple fruit
A fruit that develops from the ovaries of a group of flowers.

Mulberry *(Morus)*

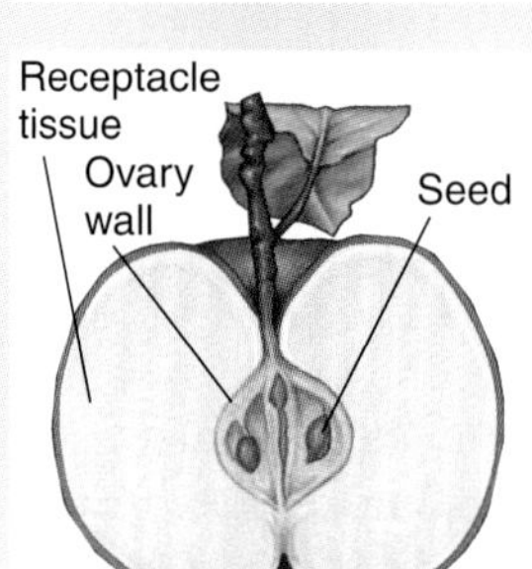

Accessory fruit
A fruit composed primarily of tissue (such as the receptacle) other than ovary tissue.

Apple *(Malus sylvestris)*

ACTIVE FIGURE **35-9** | **Fruit types.**

Fruits are botanically classified into four groups—simple, aggregate, multiple, and accessory fruits—based on their structure and mechanism of seed dispersal.

BiologyNow™ Learn how fruit types relate to the environment by clicking on this figure on your BiologyNow CD-ROM.

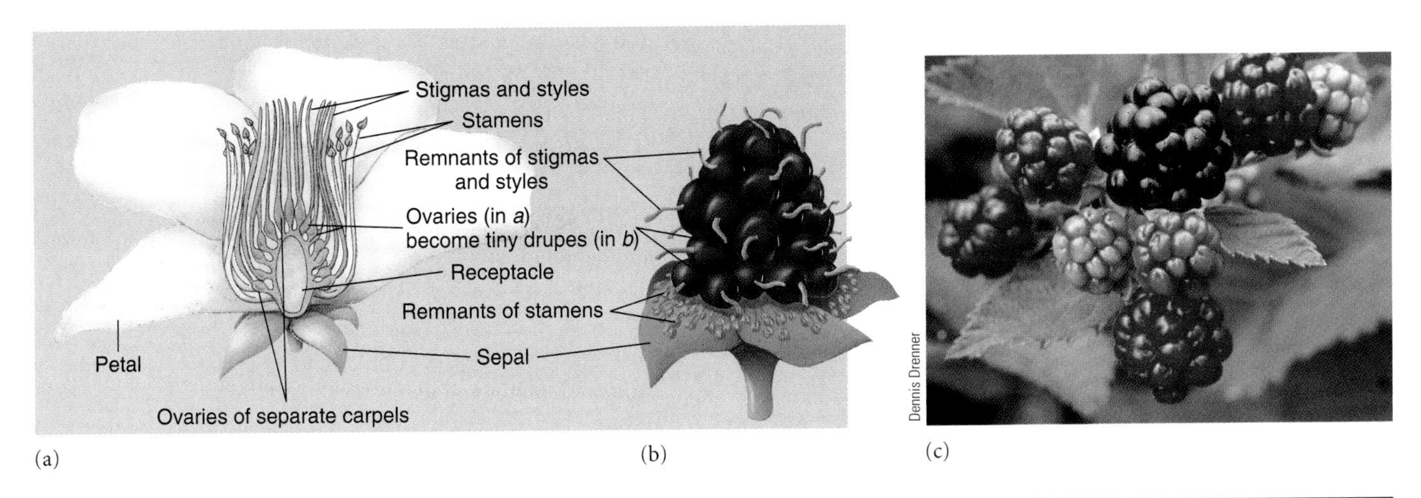

FIGURE **35-10** **An aggregate fruit.**

(a) Cutaway view of a blackberry (*Rubus*) flower, showing the many separate carpels in the center of the flower. **(b)** A developing blackberry fruit is an aggregate of tiny drupes. The little "hairs" on the blackberry are remnants of stigmas and styles. **(c)** Developing blackberry fruits at various stages of maturity.

is a fleshy fruit that has soft tissues throughout and contains few to many seeds; a blueberry is a berry, as are grapes, cranberries, bananas, and tomatoes. Many so-called berries do not fit the botanical definition. Strawberries, raspberries, and mulberries, for example, are not berries; these three fruits are discussed shortly.

A **drupe** is a simple, fleshy or fibrous fruit that contains a hard, stony pit surrounding a single seed. Examples of drupes include peaches, plums, olives, avocados, and almonds. The almond shell is actually the stony pit, which remains after the rest of the fruit has been removed.

Many simple fruits are dry at maturity; some of these split open, usually along seams, called *sutures,* to release their seeds. A milkweed pod is an example of a **follicle,** a simple, dry fruit that splits open along one suture to release its seeds. A **legume** is a simple, dry fruit that splits open along two sutures. Pea pods are legumes, as are green beans, although both are generally harvested before the fruit has dried out and split open. Pea seeds are usually removed from the fruit and consumed, whereas in green beans the entire fruit and seeds are eaten. A **capsule** is a simple, dry fruit that splits open along multiple sutures or pores. Iris, poppy, and cotton fruits are capsules.

Other simple, dry fruits, such as **caryopses** (sing., *caryopsis*), or **grains,** don't split open at maturity. Each caryopsis contains a single seed. Because the seed coat is fused to the fruit wall, a caryopsis looks like a seed rather than a fruit. Kernels of corn and wheat are fruits of this type.

An **achene** is similar to a caryopsis in that it is simple and dry, does not split open at maturity, and contains a single seed. However, the seed coat of an achene is not fused to the fruit wall. Instead, the single seed is attached to the fruit wall at one point only, permitting an achene to be separated from its seed. The sunflower fruit is an example of an achene. One can peel off the fruit wall (the shell) to reveal the sunflower seed within.

Nuts are simple, dry fruits that have a stony wall and do not split open at maturity. Unlike achenes, nuts are usually large and are often derived from a compound pistil. Examples of nuts include chestnuts, acorns, and hazelnuts. Many so-called nuts do not fit the botanical definition. Peanuts and brazil nuts, for example, are seeds, not nuts.

Aggregate fruits are a second main type of fruit. An **aggregate fruit** is formed from a single flower that contains several to many separate (free) carpels (Fig. 35-10). After fertilization, each ovary from each individual carpel enlarges. As they enlarge, the ovaries may fuse together to form a single fruit. Raspberries, blackberries, and magnolia fruits are examples of aggregate fruits.

A third type is the **multiple fruit,** formed from the ovaries of many flowers that grow in proximity on a common floral stalk. The ovary from each flower fuses with nearby ovaries as it develops and enlarges after fertilization. Pineapples, figs, and mulberries are multiple fruits (Fig. 35-11).

Accessory fruits are the fourth type. They differ from other fruits in that plant tissues in addition to ovary tissue make up the fruit. For example, the edible portion of a strawberry is the red, fleshy receptacle. Apples and pears are also accessory fruits; the outer part of each of these fruits is an enlarged *floral tube,* consisting of receptacle tissue along with portions of the calyx, that surrounds the ovary (Fig. 35-12).

Seed dispersal is highly varied

Wind, animals, water, and explosive dehiscence disperse the various seeds and fruits of flowering plants. Effective methods of seed dispersal have made it possible for certain plants to expand their geographic range. In some cases, the seed is the actual agent of dispersal, whereas in others the fruit performs this role. In tumbleweeds, such as russian thistle, the entire plant is the agent of dispersal because it detaches and blows across the ground, scattering seeds as it bumps along. Tumbleweeds are lightweight and are sometimes blown many kilometers by the wind.

Wind is responsible for seed dispersal in many plants. Plants such as maple trees have winged fruits adapted for wind dispersal. Light, feathery plumes enable other seeds or fruits to be transported by wind, often for considerable distances. Both dan-

FIGURE **35-11** | **A multiple fruit.**

Mulberry (*Morus*) is formed from the ovaries of many flowers that fused to become a multiple fruit. Mulberry flowers are imperfect and contain either stamens or pistils. Also shown is an inflorescence of female flowers from which the mulberry fruit develops.

delion fruits and milkweed seeds (Fig. 35-13a) have this type of adaptation.

Some plants have special structures that aid in dispersal of their seeds and fruits by animals. The spines and barbs of burdock burs and similar fruits catch in animal fur and fall off as the animal moves about (Fig. 35-13b). Fleshy, edible fruits are also adapted for animal dispersal. As the animal eats these fruits, it either discards or swallows the seeds. Many seeds that are swallowed have thick seed coats and are not digested, but instead pass through the digestive tract and are deposited in the animal's feces some distance from the parent plant. In fact, some seeds do not germinate unless they have passed through an animal's digestive tract; the animal's digestive juices probably aid germination by helping break down the seed coat. Some edible fruits apparently contain chemicals that function as laxatives that speed seeds through an animal's digestive tract; the less time these seeds spend in the gut, the more likely they are to germinate.

Animals such as squirrels and many bird species also help disperse acorns and other fruits and seeds by burying them for winter use. Many buried seeds are never used by the animal and germinate the following spring. Ants collect the seeds of many plants and take them underground to their nests. Ants disperse and bury seeds for hundreds of different plant species in almost every terrestrial environment, from northern coniferous forests to tropical rain forests to deserts.

Both ants and flowering plants benefit from their association. The ants ensure the reproductive success of the plants whose seeds they bury, and the plants supply food to the ants. A seed that is collected and taken underground by ants often contains a special structure called an *elaiosome,* or *oil body,* that protrudes from the seed (Fig. 35-14). Elaiosomes are a nutritious food for ants, which carry seeds underground before removing the elaiosome. Once an elaiosome is removed from a seed, the ants discard the undamaged seed in an underground refuse pile, which happens to be rich in organic material (such

FIGURE **35-12** | **An accessory fruit.**

(a) Note the floral tube surrounding the ovary in the pear (*Pyrus communis*) flower. This tube becomes the major edible portion of the pear. **(b)** Longitudinal section through a pear showing the fruit tissue, which derives from both the floral tube and the ovary.

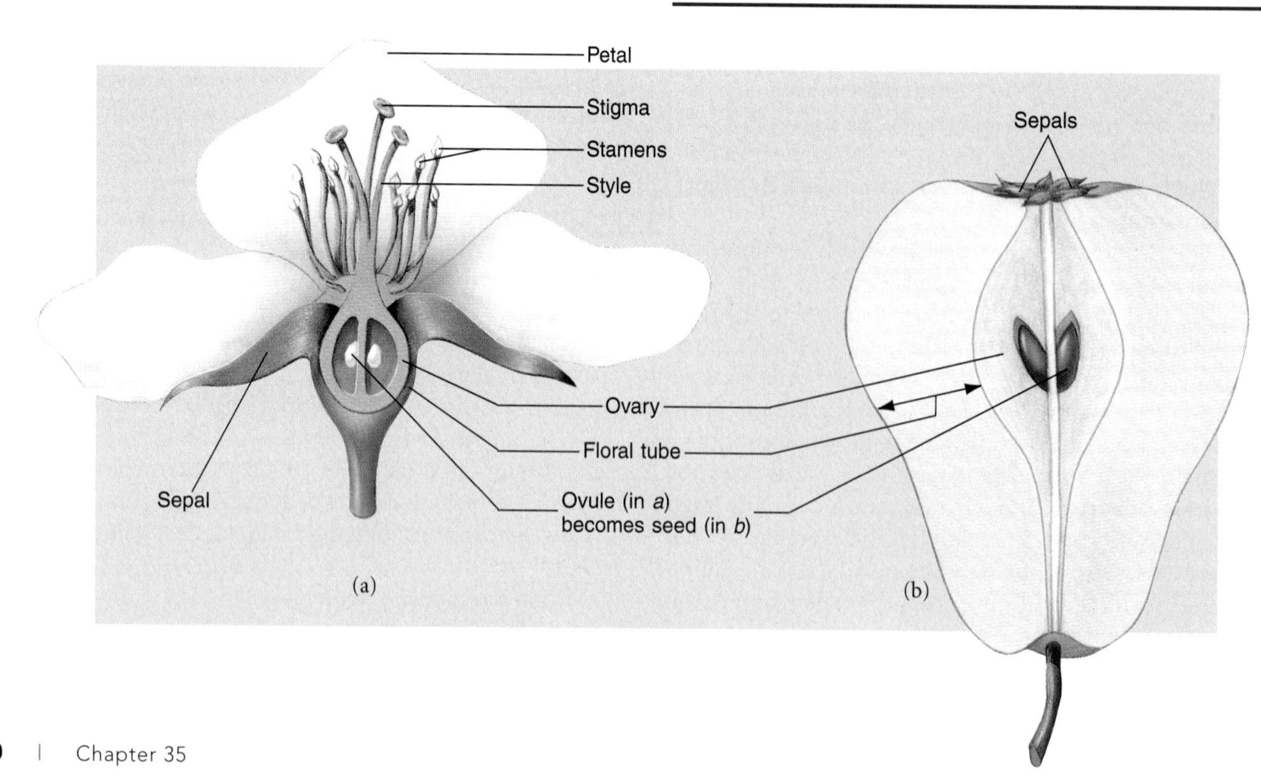

John Serrao/Visuals Unlimited

(a)

DPA /Dembinsky Photo Associates

(b)

FIGURE **35-13** **Seed (and fruit) dispersal.**

(a) The feathery plumes of milkweed (*Asclepias syriaca*) seeds make them buoyant for dispersal by wind. **(b)** Burdock (*Arctium minus*) burs (the hooked fruits) are carried away from the parent plant after sticking to bird feathers, mammal fur, or human clothing.

as ant droppings and dead ants) and contains the minerals required by young seedlings. Thus ants not only bury the seeds away from animals that might eat them but also place the seeds in rich soil that is ideal for germination and seedling growth.

The coconut is an example of a fruit adapted for dispersal by water. The coconut has air spaces that make it buoyant and capable of being carried by ocean currents for thousands of kilometers. When it washes ashore, the seed may germinate and grow into a coconut palm tree.

FIGURE **35-14** **Dispersal of seeds by ants.**

The brown part of each bloodroot (*Sanguinaria canadensis*) seed is the seed proper, and the white part is the elaiosome, or oil body. The seeds have been placed along the midvein of an oak leaf to indicate scale.

Marion Lobstein

Some seeds are dispersed neither by wind, animals, nor water. Such seeds are found in fruits that use *explosive dehiscence,* in which the fruit bursts open suddenly and quite often violently, forcibly discharging its seeds (Fig. 35-15). Pressures due to differences in **turgor** or to drying out cause these fruits to burst open. The fruits of plants such as touch-me-not and bitter cress split open so explosively that seeds are scattered a meter or more.

Review

- What is the difference between pollination and fertilization? Which process occurs first?
- What are the main stages of dicot embryonic development?

FIGURE **35-15** **Explosive dehiscence.**

(a) An intact fruit of bitter cress (*Cardamine pratensis*) before it has opened. **(b)** The fruit splits open with explosive force, flinging the seeds some distance from the plant.

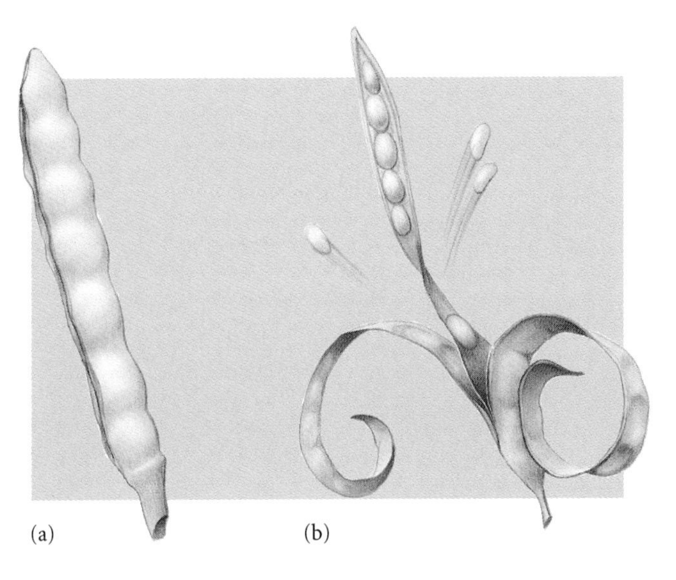

- What is a fruit?
- How are the following fruits distinguished: simple, aggregate, multiple, and accessory fruits?
- What are some characteristics of animal-dispersed seeds and fruits?

BiologyNow™ Assess your understanding of **fertilization and seed/fruit development** by taking the pretest on your BiologyNow CD-ROM.

ASEXUAL REPRODUCTION IN FLOWERING PLANTS

Learning Objectives

9 Explain how the following structures may be used to propagate plants asexually: rhizomes, tubers, stolons, corms, bulbs, plantlets, and suckers.

10 Define *apomixis,* and explain how it occurs.

Flowering plants have many kinds of asexual reproduction, several of which involve modified stems: rhizomes, tubers, bulbs, corms, and stolons. A **rhizome** is a horizontal underground stem that may or may not be fleshy. Fleshiness indicates the rhizome is used for storing food materials such as starch (Fig. 35-16a). Although rhizomes resemble roots, they are really stems, as indicated by the presence of scalelike leaves, buds, nodes, and internodes. (Roots have none of these features.) Rhizomes frequently branch in different directions. Over time, the old portion of the rhizome dies, and the two branches eventually separate to become distinct plants. Irises, bamboos, ginger, and many grasses are examples of plants that reproduce asexually by forming rhizomes.

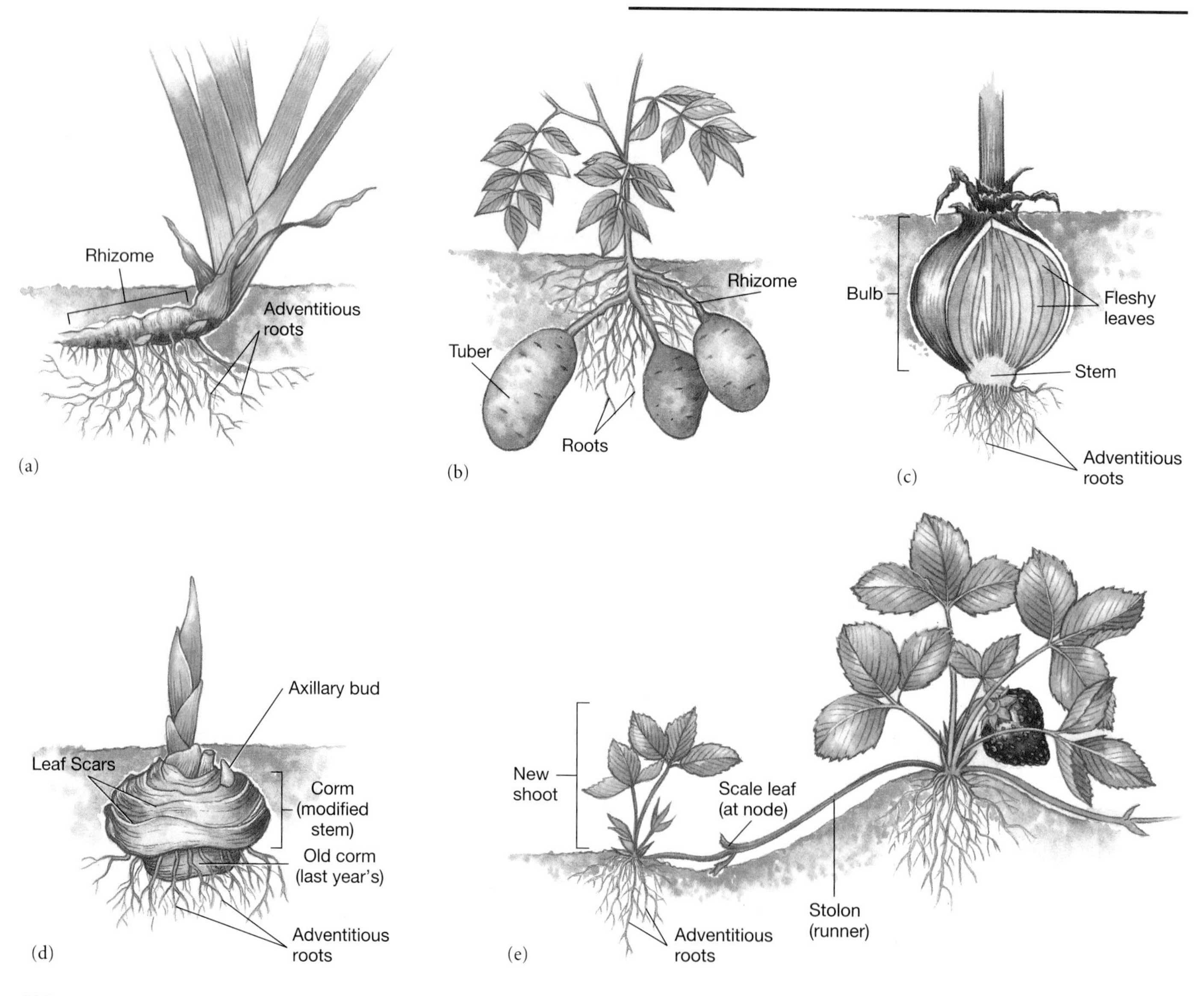

FIGURE 35-16 Modified Stems.
(a) Irises have horizontal underground stems called rhizomes. New aerial shoots arise from buds that develop on the rhizome. **(b)** Potato plants form rhizomes, which enlarge into tubers (the potatoes) at the ends. **(c)** A bulb is a short underground stem to which overlapping, fleshy leaves are attached; most of the bulb consists of leaves. **(d)** A corm is an underground stem that is almost entirely stem tissue surrounded by a few papery scales. **(e)** Strawberries reproduce asexually by forming stolons, or runners. New plants (shoots and roots) are produced at every other node.

Some rhizomes produce greatly thickened ends called **tubers,** which are fleshy underground stems enlarged for food storage. When the attachment between a tuber and its parent plant breaks, often as a result of the death of the parent plant, the tuber grows into a separate plant. Potatoes and elephant's ear (*Caladium*) are examples of plants that produce tubers (Fig. 35-16b). The "eyes" of a potato are axillary buds, evidence that the tuber is an underground stem rather than a storage root such as sweet potatoes or carrots.

A **bulb** is a modified underground bud in which fleshy storage leaves are attached to a short stem (Fig. 35-16c). A bulb is globose (round) and covered by paper-like bulb scales, which are modified leaves. It frequently forms axillary buds that develop into small daughter bulbs (bulblets). These new bulbs are initially attached to the parent bulb, but when the parent bulb dies and rots away, each daughter bulb can become established as a separate plant. Lilies, tulips, onions, and daffodils are some plants that form bulbs.

A **corm** is a very short, erect underground stem that superficially resembles a bulb (Fig. 35-16d). Unlike the bulb, whose food is stored in underground leaves, the corm's storage organ is a thickened underground stem covered by papery scales (modified leaves). Axillary buds frequently give rise to new corms; the death of the parent corm separates these daughter corms, which then become established as separate plants. Familiar garden plants that produce corms include crocus, gladiolus, and cyclamen.

Stolons, or **runners,** are horizontal, aboveground stems that grow along the surface and have long internodes (Fig. 35-16e). Buds develop along the stolon, and each bud gives rise to a new shoot that roots in the ground. When the stolon dies, the daughter plants live separately. The strawberry plant produces stolons.

Some plants form detachable **plantlets** (small plants) in notches along their leaf margins. *Kalanchoe,* whose common name is "mother of thousands," has meristematic tissue that gives rise to an individual plantlet at each notch in the leaf (Fig. 35-17). When these plantlets reach a certain size, they drop to the ground, root, and grow.

Some plants reproduce asexually by producing **suckers,** aboveground shoots that develop from adventitious buds on roots (Fig. 35-18). Each sucker grows additional roots and becomes an independent plant when the parent plant dies. Examples of plants that form suckers include black locust, pear, apple, cherry, blackberry, and aspen. A quaking aspen (*Populus tremuloides*) colony in the Wasatch Mountains of Utah contains at least 47,000 tree trunks formed from suckers that can be traced back to a single individual; this massive "organism" occupies almost 43 hectares (106 acres). Some weeds, such as field bindweed, produce many suckers. These plants are difficult to control, because pulling the plant out of the soil seldom removes all the roots, which can grow as deep as 3 m (10 ft). In fact, in response to wounding, the roots produce additional suckers, which can be a considerable nuisance to humans.

Apomixis is the production of seeds without the sexual process

Sometimes flowering plants produce embryos in seeds without meiosis and the fusion of gametes. This asexual process is **apomixis.** For example, an embryo may develop from a diploid cell in the ovule rather than from a diploid zygote that forms from the union of two haploid gametes. Seed production by apomixis is a form of asexual reproduction. Because there is no fusion of gametes, the embryo is virtually genetically identical to the maternal genotype. However, the advantage of apomixis over other methods of asexual reproduction is that the seeds

FIGURE **35-17** | **Plantlets.**
The "mother of thousands" (*Kalanchoe*) produces detachable plantlets along the margins of its leaves. The young plantlets drop off and root in the ground.

Jerome Wexler/Photo Researchers, Inc.

FIGURE **35-18** | **Suckers in various stages of development.**
All the suckers are connected to one another through the root system.

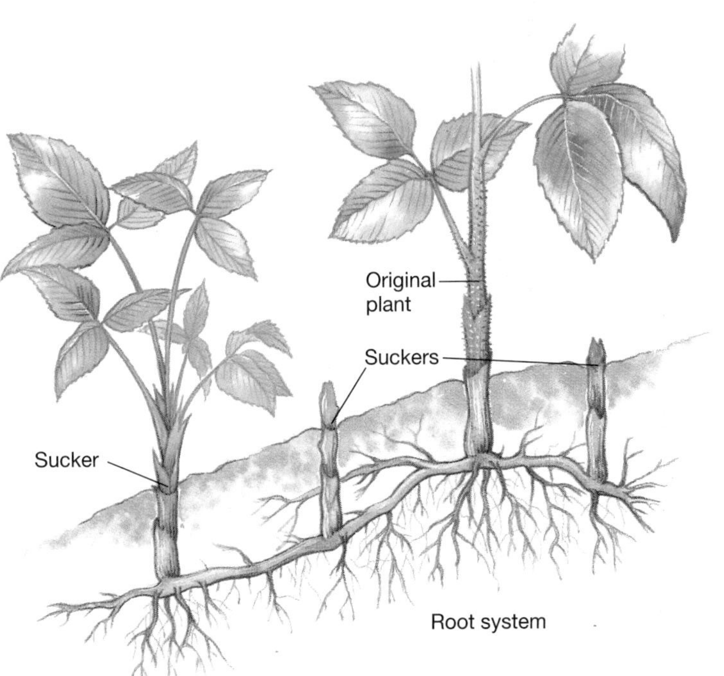

and fruits produced by apomixis can be dispersed by methods associated with sexual reproduction. Apomixis occurs in various species of more than 40 angiosperm families. Examples of plants that can reproduce by apomixis include dandelions, citrus trees, mango, blackberries, garlic, and certain grasses.

Plant biologists are actively engaged in identifying the genes involved in asexual reproduction by apomixis. If researchers can transfer these genes to crop plants that possess superior traits such as high yield, they may be able to develop apomictic crop plants whose traits do not get lost or diluted during the genetic shuffling of sexual reproduction.

Review

- How are rhizomes and tubers involved in asexual reproduction?
- How are corms and bulbs similar? How are they different?
- What is apomixis? How does the ability to reproduce by apomixis help a plant survive?

BiologyNow™ Assess your understanding of **asexual reproduction in flowering plants** by taking the pretest on your BiologyNow CD-ROM.

A COMPARISON OF SEXUAL AND ASEXUAL REPRODUCTION

Learning Objective

11 State the differences between sexual and asexual reproduction, and discuss the evolutionary advantages and disadvantages of each.

Sexual and asexual reproduction are suited for different environmental circumstances. As you know, sexual reproduction results in offspring that are genetically different from the parents; that is, the parental genotypes are not preserved in the offspring. This genetic diversity of offspring may be selectively advantageous, particularly in an unstable, or changing, environment. (We define *environment* broadly to include all external conditions, both living and nonliving, that affect an organism.) If a plant species that reproduces sexually (and is therefore genetically diverse) is exposed to increasing annual temperatures as a result of global warming, for example, some of the individuals may be more fit than either the parents or other offspring. The genetic diversity from sexual reproduction may also let the individuals of a species exploit new environments, thereby expanding their range.

You have seen that asexual reproduction results in offspring that are virtually genetically identical to the parent; that is, the parental genotype is preserved. Assuming the parent is well adapted to its environment (has a favorable combination of alleles), this genetic similarity may be selectively advantageous if the environment remains stable (unchanging) for several generations. None of the offspring of asexual reproduction is more fit than the parent, but neither is any of them less fit.

Despite the apparent advantages of asexual reproduction, most plant species whose reproduction is primarily asexual occasionally reproduce sexually. Even in what appears a stable environment, plants are exposed to changing selective pressures, such as changes in the number and kinds of predators and parasites, the availability of food, competition from other species, and climate. Sexual reproduction permits species whose reproduction is primarily asexual to increase their genetic variability so at least some individuals are adapted to the changing selective pressures in a stable environment.

Sexual reproduction has some disadvantages

Although the genetic diversity produced by sexual reproduction is advantageous to a species' long-term survival, sexual reproduction is a "costly" form of reproduction. Sexual reproduction requires both male and female gametes, which must meet for reproduction to occur. The many adaptations of flowers for different modes of pollination represent one cost of sexual reproduction.

Sexual reproduction produces some individuals with genotypes that are well adapted to the environment, but it also produces some individuals that are less well adapted. Therefore, sexual reproduction is usually accompanied by high death rates among offspring, particularly when selective pressures are strong. As discussed in Chapter 17, however, this aspect of sexual reproduction is an important part of evolution by natural selection.

Every biological process involves tradeoffs, and sexual reproduction is no exception. Although sexual reproduction has its costs, the adaptive advantages of sexual reproduction clearly outweigh any disadvantages.

Review

- How does sexual reproduction in plants differ from asexual reproduction?
- What is the adaptive advantage of sexual reproduction? Why is sexual reproduction considered a "costly" form of reproduction?

BiologyNow™ Assess your understanding of **sexual and asexual reproduction** by taking the pretest on your BiologyNow CD-ROM.

SUMMARY WITH KEY TERMS

1 Label the parts of a flower on a diagram, and describe the functions of each part (see Figure 35-1).

- Sexual reproduction occurs in the flower. A flower may contain sepals, petals, stamens, and carpels (pistils).
- **Sepals** cover and protect the flower parts when the flower is a bud. **Petals** play an important role in attracting animal pollinators to the flower.
- **Stamens** produce pollen grains. Each stamen consists of a thin stalk (the **filament**) attached to a saclike structure (the **anther**).

- The **carpel** is the female reproductive unit. A **pistil** may consist of a single carpel or a group of fused carpels. Each pistil has three sections: a **stigma,** on which the pollen grains land; a **style,** through which the pollen tube grows; and an **ovary** that contains one or more **ovules.**

2 Tell where eggs and pollen grains are formed within the flower.

- Pollen forms within pollen sacs in the anther. Each **pollen grain** contains two cells. One generates two sperm cells, and the other produces a **pollen tube** through which the sperm cells reach the ovule.
- An egg and two **polar nuclei,** along with several other nuclei, are formed in the ovule. Both egg and polar nuclei participate directly in fertilization.

3 Compare the evolutionary adaptations that characterize flowers pollinated in different ways (by insects, birds, bats, and wind).

- Flowers pollinated by insects are often yellow or blue and have a scent.
- Bird-pollinated flowers are often yellow, orange, or red and do not have a strong scent.
- Bat-pollinated flowers often have dusky white petals and are scented.
- Plants pollinated by wind often have smaller petals or lack petals altogether and do not produce a scent or nectar; wind-pollinated flowers make large amounts of pollen.

4 Define *coevolution,* and give examples of ways that plants and their animal pollinators have affected one another's evolution.

- **Coevolution** is reciprocal adaptation caused by two different species (such as flowering plants and their animal pollinators) forming an interdependent relationship and affecting the course of one another's evolution. For example, certain flowers have evolved large, showy petals and scent, whereas bees have evolved hairy bodies that catch and hold sticky pollen grains.

5 Distinguish between pollination and fertilization.

- **Pollination** is the transfer of pollen grains from anther to stigma. After pollination, **fertilization,** the fusion of gametes, occurs.
- Flowering plants have **double fertilization.** In the ovule, the egg fuses with one sperm cell, forming a zygote (fertilized egg) that eventually develops into a multicellular embryo in the **seed.** The two polar nuclei fuse with the second sperm cell, forming a triploid nutritive tissue called **endosperm.**

6 Trace the stages of embryo development in flowering plants, and list and define the main parts of seeds.

- A dicot embryo develops in the seed in an orderly fashion, from proembryo to globular embryo to the heart stage to the torpedo stage. The mature flowering plant embryo consists of a **radicle,** a **hypocotyl,** a **plumule,** and **cotyledons** (one in monocots or two in dicots).
- A mature seed contains both a young embryo and nutritive tissue (stored in the **endosperm** or cotyledons) for use during germination.

7 Explain the relationships among the following: ovules, ovaries, seeds, and fruits.

- Ovules are structures with the potential to develop into seeds, whereas ovaries are structures with the potential to develop into fruits. Seeds are enclosed within **fruits,** which are mature, ripened ovaries.

8 Distinguish among simple, aggregate, multiple, and accessory fruits; give examples of each type; and cite several different methods of seed and fruit dispersal.

- **Simple fruits** develop from a single pistil that consists of one carpel or several fused carpels. Some simple fruits (**berries, drupes**) are fleshy at maturity, whereas others (**follicles, legumes, capsules, caryopses, achenes, nuts**) are dry.
- **Aggregate fruits** develop from a single flower with many separate ovaries. **Multiple fruits** develop from the ovaries of many flowers growing in close proximity on a common axis. In **accessory fruits,** the major part of the fruit consists of tissue other than ovary tissue.
- Seeds and fruits are adapted for various means of dispersal, including animals, wind, water, and explosive dehiscence.

9 Explain how the following structures may be used to propagate plants asexually: rhizomes, tubers, stolons, corms, bulbs, plantlets, and suckers.

- Rhizomes, tubers, bulbs, corms, and stolons are stems specialized for asexual reproduction. A **rhizome** is a horizontal underground stem. A **tuber** is a fleshy underground stem enlarged for food storage. A **bulb** is a modified underground bud with fleshy storage leaves attached to a short stem. A **corm** is a short, erect underground stem covered by papery scales. A **stolon** is a horizontal aboveground stem with long internodes.
- Some leaves have meristematic tissue along their margins and give rise to detachable **plantlets.**
- Roots may develop adventitious buds that develop into **suckers.** Suckers produce additional roots and may give rise to new plants.

10 Define *apomixis,* and explain how it occurs.

- **Apomixis** is the production of seeds and fruits without sexual reproduction.

11 State the differences between sexual and asexual reproduction, and discuss the evolutionary advantages and disadvantages of each.

- Sexual reproduction involves the union of two gametes; the offspring produced by sexual reproduction are genetically variable. Asexual reproduction involves the formation of offspring without the fusion of gametes; the offspring are virtually genetically identical to the single parent.
- The parental genotypes are not preserved in the offspring of sexual reproduction. Genetic diversity among offspring produced by sexual reproduction may be selectively advantageous, particularly in an unstable, or changing, environment. Genetic diversity may also let individuals exploit new environments. However, sexual reproduction is costly because both male and female gametes must be produced and must meet.
- The parental genotype is preserved in asexual reproduction. Genetic similarity may be selectively advantageous if the environment remains stable (unchanging) for several generations. In asexual reproduction all individuals can produce offspring. Despite the apparent advantages of asexual reproduction, most plant species whose reproduction is primarily asexual occasionally reproduce sexually, thereby increasing their genetic variability.

POST-TEST

1. In flowering plants, the ____________ is/are large (multicellular) and nutritionally independent. (a) gametes (b) microspores (c) megaspores (d) mature gametophyte (e) mature sporophyte
2. The normal order of whorls from the flower's periphery to the center is (a) sepals ⟶ petals ⟶ carpels ⟶ stamens (b) stamens ⟶ carpels ⟶ sepals ⟶ petals (c) sepals ⟶ petals ⟶ stamens ⟶ carpels (d) petals ⟶ carpels ⟶ stamens ⟶ sepals (e) carpels ⟶ stamens ⟶ petals ⟶ sepals
3. The pistil consists of (a) stigma, style, and stamen (b) anther and filament (c) sepal and petal (d) stigma, style, and ovary (e) radicle, hypocotyl, and plumule
4. The petals of a flower are collectively called a(an) (a) calyx (b) capsule (c) carpel (d) cotyledon (e) corolla
5. The transfer of pollen grains from anther to stigma is (a) fertilization (b) double fertilization (c) pollination (d) germination (e) apomixis
6. The observation that insects with long mouthparts, pollinate long, tubular flowers, whereas insects with short mouthparts, pollinate short flowers, is explained by (a) coevolution (b) germination (c) double fertilization (d) apomixis (e) explosive dehiscence
7. The process of ____________ in flowering plants involves one sperm cell fusing with an egg cell and one sperm cell fusing with two polar nuclei. (a) coevolution (b) germination (c) double fertilization (d) apomixis (e) pollination
8. The nutritive tissue in the seeds of flowering plants that is formed from the union of a sperm cell and two polar nuclei is called the (a) plumule (b) endosperm (c) cotyledon (d) hypocotyl (e) radicle
9. The ____________ is a multicellular structure that anchors the embryo and aids in nutrient uptake from the endosperm (a) proembryo (b) ovule (c) suspensor (d) cotyledon (e) pollen tube
10. After fertilization the ovule develops into a ____________, and the ovary into a ____________ (a) fruit; seed (b) seed; fruit (c) calyx; corolla (d) corolla; calyx (e) follicle; legume
11. In plants that lack endosperm in their mature seeds, the cotyledons function (a) to enclose and protect the seed (b) to aid in seed dispersal (c) as an absorptive embryonic root (d) to store food reserves (e) to attach the embryo within the ovule
12. ____________ fruits develop from many ovaries of a single flower, whereas ____________ fruits develop from the ovaries of many separate flowers (a) multiple; accessory (b) simple; accessory (c) aggregate; multiple (d) accessory; aggregate (e) simple; multiple
13. Apples, strawberries, and pears are examples of what kind of fruit? (a) accessory (b) simple (c) multiple (d) aggregate (e) legume
14. A horizontal, underground stem that may or may not be fleshy and that is often specialized for asexual reproduction is called a (a) stolon (b) bulb (c) corm (d) rhizome (e) tuber
15. Place the following events in the correct order: (1) pollen tube grows into ovule (2) insect lands on flower to drink nectar (3) embryo develops within the seed (4) fertilization occurs (5) pollen carried by insect contacts stigma (a) 2-5-1-4-3 (b) 1-4-2-5-3 (c) 3-2-5-1-4 (d) 5-1-3-4-2 (e) 2-5-4-3-1
16. In flowering plant reproduction, the multicellular female gametophyte is also called a(an) (a) pollen grain (b) embryo sac (c) generative cell (d) endosperm (e) hypocotyl
17. A simple, dry fruit that splits open along one suture to release its seeds is a(an) (a) berry (b) legume (c) achene (d) caryopsis (e) follicle

CRITICAL THINKING

1. Draw pictures to show the kinds of flowers that form simple, aggregate, multiple, and accessory fruits.
2. Is seed dispersal by ants an example of coevolution? Why or why not?
3. Using what you have learned in this chapter, speculate whether it is more likely that offspring of asexual reproduction develop in close proximity to or widely dispersed from the parent plant. Explain your reasoning. How could you design an experiment to test your hypothesis?
4. Which type of reproduction, sexual or asexual, might be more beneficial in the following circumstances, and why? (a) a perennial (plant that lives more than two years) in a stable environment, (b) an annual (plant that lives one year) in a rapidly changing environment, (c) a plant adapted to an extremely narrow climate range.
5. Using what you have learned in this chapter, offer an explanation of why telephone poles and wires strung across grassy fields or plains often have tree seedlings growing under them.

- Visit our Web site at **http://biology.brookscole.com/solomon7** for links to chapter-related resources on the World Wide Web. Additional online materials relating to this chapter can also be found on our Web site.

BIOLOGY NOW RESOURCES

Active Figures

35-2: Reproduction in flowering plants

35-9: How fruit types relate to environmental factors

Preparing for an exam? Take a diagnostic test on your BiologyNow CD-ROM.

Post-Test Answers

1. e	2. c	3. d	4. e
5. c	6. a	7. c	8. b
9. c	10. b	11. d	12. c
13. a	14. d	15. a	16. b
17. e			

Plant Growth and Development

Black-eyed Susan. This plant (*Rudbeckia hirta*), which grows to 0.9 m (3 ft), produces flowers in response to the shortening nights of spring and early summer.

CHAPTER OUTLINE

As you learned in Chapter 16, the ultimate control of plant growth and *development,* which includes all the changes that take place during the entire life of an individual, is genetic. If the genes required for development of a particular trait, such as the shape of a leaf, the color of a flower, or the type of root system, are not present, that characteristic does not develop. When a particular gene is present, its *expression,* that is, how it exhibits itself as an observable feature of an organism, is determined by a variety of factors, including signals from other genes and from the environment. The location of a cell in the young plant body has a profound effect on gene expression during development. Experiments suggest that chemical signals from adjacent cells help the cell "perceive" its location within the plant body. Each cell's spatial environment helps determine what that cell ultimately becomes.

Environmental cues, such as changing day length and variations in precipitation and temperature, also exert an important influence on gene expression, as they do on all aspects of plant growth and development. The initiation of sexual reproduction is often under environmental control, particularly in temperate latitudes, and plants switch from vegetative growth (in which their leaves, stems, and roots grow) to reproductive growth after receiving the appropriate signals from the environment. Such control is important for the plant's survival because the timing of sexual reproduction is critical for reproductive success: All flowering plants in temperate climates must flower and form seeds before the onset of winter induces **dormancy,** a temporary state of reduced physiological activity. Many flowering plants are sensitive to changes in the relative amounts of daylight and darkness that accompany the changing seasons, and these plants flower in response to those changes (see photograph). Other plants have temperature requirements that induce sexual reproduction. Plants, then, continually perceive information from the environment and use this information to help regulate normal growth and development.

Growth and development are controlled by plant *hormones,* organic compounds that are present in very low concentrations in plant tissues and that act as chemical signals between cells. **Germination**—the process of a seed sprouting—and the growth of young seedlings into mature plants are aspects of growth and development. A plant's responses to changes in various conditions in its environment, including temperature, light, gravity, and touch, are other aspects of growth and development. We consider all of these topics in this chapter. ■

FIGURE **36-1** **Imbibition.**

Pinto bean (*Phaseolus vulgaris*) seeds *(left)* before imbibition and *(right)* after. Dry seeds imbibe water before they germinate.

Marion Lobstein

GERMINATION AND EARLY GROWTH

Learning Objectives

1 Discuss genetic and environmental factors that affect plant growth and development.

2 Summarize the influence of internal and environmental factors on the germination of seeds.

In Chapter 35 you learned how pollination and fertilization are followed by seed and fruit development in flowering plants. Each seed develops from an ovule and contains an embryonic plant and food to provide nourishment for the embryo during germination.

Seed germination requires favorable environmental conditions

Within a given species, the precise requirements for seed germination represent evolutionary adaptations that protect the young seedlings from adverse environmental conditions. Environmental cues, such as the presence of water and oxygen, proper temperature, and sometimes the presence of light penetrating the soil surface, influence whether or not a seed germinates.

No seed germinates unless it has absorbed water. The embryo in a mature seed is dehydrated, and a watery environment is necessary for active metabolism. When a seed germinates, its metabolic machinery is turned on, and numerous materials are synthesized and others degraded. Therefore, water is an absolute requirement for germination. The absorption of water by a dry seed is **imbibition.** As a seed imbibes water, it often swells to several times its original, dry size (Fig. 36-1). Cells imbibe water by adhesion of water onto and into materials such as cellulose, pectin, and starches within the seed. Water molecules are attracted to and bound to these materials by *adhesion,* the attraction between unlike materials (see Chapter 2).

Germination and subsequent growth require a great deal of energy. Because young plants obtain this energy by converting the energy of food molecules stored in the seed's endosperm or cotyledons to ATP during aerobic respiration, much oxygen is usually needed during germination. Some plants, such as rice, carry out *alcohol fermentation* (see Fig. 7-13b) during the early stages of germination and seedling growth. Fermentation enables rice plants to grow and become established in flooded soil, an environment that would suffocate most young plants.

Temperature is another environmental factor that affects germination. Each species has an optimal, or ideal, temperature at which the germination percentage is highest. For most plants, the optimal germination temperature is between 25°C and 30°C (77°F and 86°F). Some seeds, such as those of apples, require prolonged exposure to low temperatures before their seeds break dormancy and germinate. Some of the environmental factors needed for germination help ensure the survival of the young plant. The requirement of a prolonged low-temperature period ensures that seeds adapted to temperate climates germinate in the spring rather than the fall.

Some plants, especially those with tiny seeds, such as lettuce, require light for germination. A light requirement ensures that a tiny seed germinates only if it is close to the soil surface. If such a seed were to germinate several centimeters below the soil surface, it might not have enough food reserves to grow to the surface. But if this light-dependent seed remains dormant until the soil is disturbed and it is brought to the surface, it has a much greater likelihood of survival.

Some seeds do not germinate immediately

A mature seed is often dormant and may not germinate immediately even if growing conditions are ideal. In certain seeds, internal factors, which are under genetic control, prevent germination even when all external conditions are favorable. Many seeds are dormant because certain chemicals are present or absent or because the seed coat restricts germination. For example, the seeds of many desert plants contain high concentrations of *abscisic acid* (discussed later in this chapter), which inhibits germination under unfavorable conditions. Abscisic acid is washed out only when rainfall is sufficient to support the plant's growth after the seed germinates.

Some seeds, such as certain legumes, have extremely hard, thick seed coats that prevent water and oxygen from entering, thereby inducing dormancy. *Scarification,* the process of scratching or scarring the seed coat (physically with a knife or chemically with an acid) before sowing it, induces germination in these plants. Two examples of how scarification occurs in nature are when these seeds pass through the digestive tracts of animals or when the seed coats are partially digested by soil bacteria.

Dicots and monocots exhibit characteristic patterns of early growth

Once conditions are right for germination, the first part of the plant to emerge from the seed is the **radicle,** or embryonic root. As the root grows and forces its way through the soil, it encounters considerable friction from soil particles. A root cap protects the delicate apical meristem of the root tip (see Chapter 34). The shoot is next to emerge from the seed. (Recall from Chapter 31 that a *shoot* is a collective term for a vertical stem with its leaves and reproductive structures.) Stem tips are not protected by a structure comparable to a root cap, but plants have ways to protect the delicate stem tip as it grows through the soil to the surface. The stem of a bean seedling (a dicot), for instance, curves over to form a hook so that the stem tip and cotyledons are actually *pulled up* through the soil (Fig. 36-2a). Corn and other grasses (monocots) have a special sheath of cells called a **coleoptile** that surrounds and protects the young shoot (Fig. 36-2b). First, the coleoptile pushes up through the soil, and then the leaves and stem grow through the tip of the coleoptile.

Certain parts of a plant grow throughout its life. This **indeterminate growth,** the ability to grow indefinitely, is characteristic of stems and roots, both of which arise from apical meristems. Hypothetically, stems and roots could continue growing forever. In contrast, many leaves and flowers have **determinate growth;** that is, they stop growing after reaching a certain size. The size of leaves and flowers with determinate growth varies from species to species and from individual to individual, depending on the plant's genetic programming and on environmental conditions, such as availability of sunlight, water, and essential nutrient minerals.

FIGURE **36-2** | **Germination and seedling growth.**

(a) Common bean, a dicot. The hook in the stem protects the delicate stem tip as it grows through the soil. Once the shoot emerges from the soil, the hook straightens. **(b)** Corn, a monocot. The coleoptile, a sheath of cells, emerges first from the soil. The shoot and leaves grow through the middle of the coleoptile. *(From* Biology, Unity and Diversity of Life, *9th edition by Cecie Starr and Ralph Taggart, p. 547. Art by Raychel Ciemma.)*

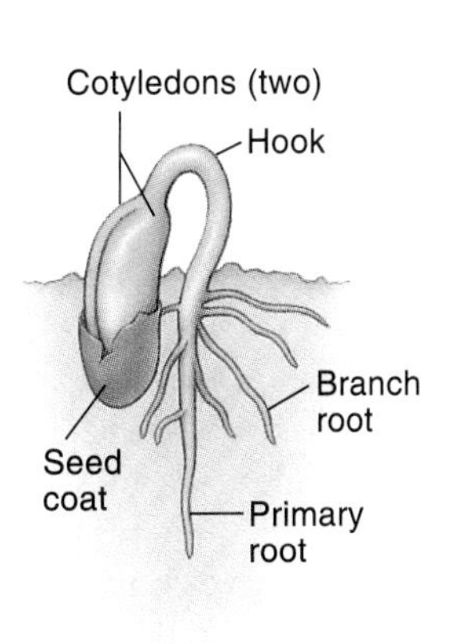

(a)

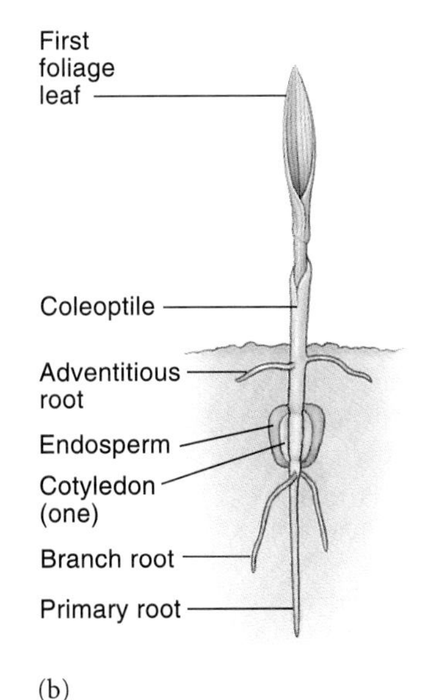

(b)

Review

- What factors influence the germination of seeds?
- Why are plant growth and development so sensitive to environmental cues?

BiologyNow™ Assess your understanding of **germination and early growth** by taking the pretest on your BiologyNow CD-ROM.

LIGHT SIGNALS AND PLANT DEVELOPMENT

Learning Objectives

3 Explain how varying amounts of light and darkness induce flowering, and describe the role of phytochrome in flowering, including a brief discussion of phytochrome signal transduction.

4 Define *circadian rhythm,* and give an example.

5 Distinguish between phytochrome and cryptochrome.

Photoperiodism is any response of a plant to the relative lengths of daylight and darkness. Initiation of flowering at the shoot apical meristem is one of several physiological activities that are photoperiodic in many plants. Plants are classified into four main groups—short-day, long-day, intermediate-day, and day-neutral—on the basis of how photoperiodism affects their transition from vegetative growth to flowering.

Short-day plants (also called **long-night plants**) flower when the night length is equal to or greater than some critical period (Fig. 36-3a). There are two kinds of short-day plants: qualitative and quantitative. In *qualitative short-day plants,* flowering occurs only in short days, whereas in *quantitative short-day plants,* flowering is accelerated by short days. The initiation of flowering in short-day plants is not due to the shorter period of daylight but to the long, uninterrupted period of darkness. The minimum critical night length varies considerably from one plant species to another but falls between 12 and 14 hours for many. Examples of short-day plants are florist's chrysanthemum, cocklebur, and poinsettia, which typically flower in later summer or fall. Poinsettias, for example, typically initiate flower buds in early October in the Northern Hemisphere and flower about 8 to 10 weeks later; hence their traditional association with Christmas. Short-day plants detect the lengthening nights of late summer or fall, and they flower at that time.

Long-day plants (also called **short-night plants**) flower when the night length is equal to or less than some critical period (Fig. 36-3b). In *qualitative long-day plants,* flowering occurs only in long days, whereas in *quantitative long-day plants,* long days accelerate flowering. Plants such as spinach, black-eyed Susan (see Chapter introduction photo) and the model research plant *Arabidopsis thaliana* flower in late spring or summer and are long-day plants. These plants detect the shortening nights of spring and early summer, and they flower at that time.

Intermediate-day plants do not flower when night length is either too long or too short (Fig. 36-3c). Sugarcane and coleus

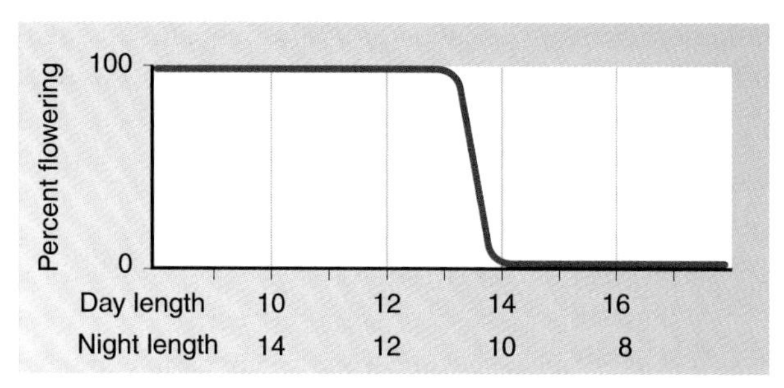

(a) Short-day plant

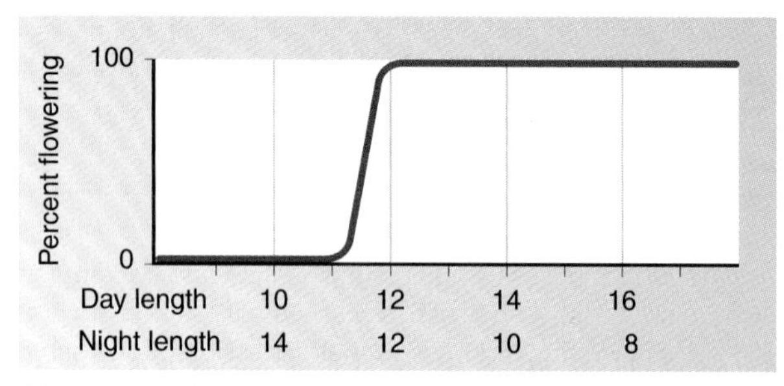

(b) Long-day plant

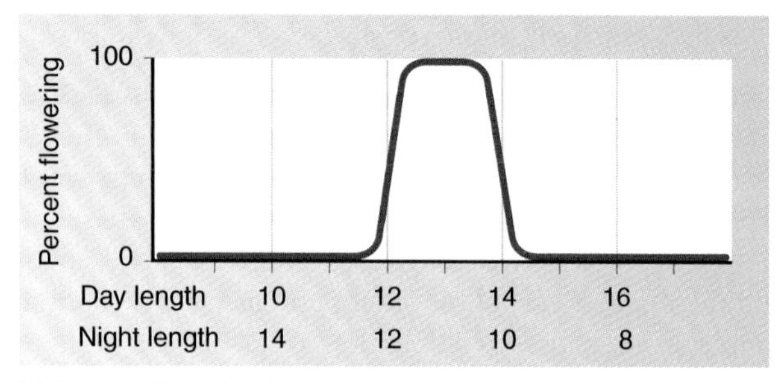

(c) Intermediate-day plant

FIGURE **36-3**

Generalized photoperiodic responses in short-day, long-day, and intermediate-day plants.

(a) Short-day plants flower when the night length is equal to or exceeds a certain critical length in a 24-hour period. **(b)** Long-day plants flower when the night length is equal to or less than a certain critical length in a 24-hour period. The critical length varies among different species; short-day plants may require a shorter night length than long-day plants, as shown. **(c)** Intermediate-day plants have a narrow night length requirement.

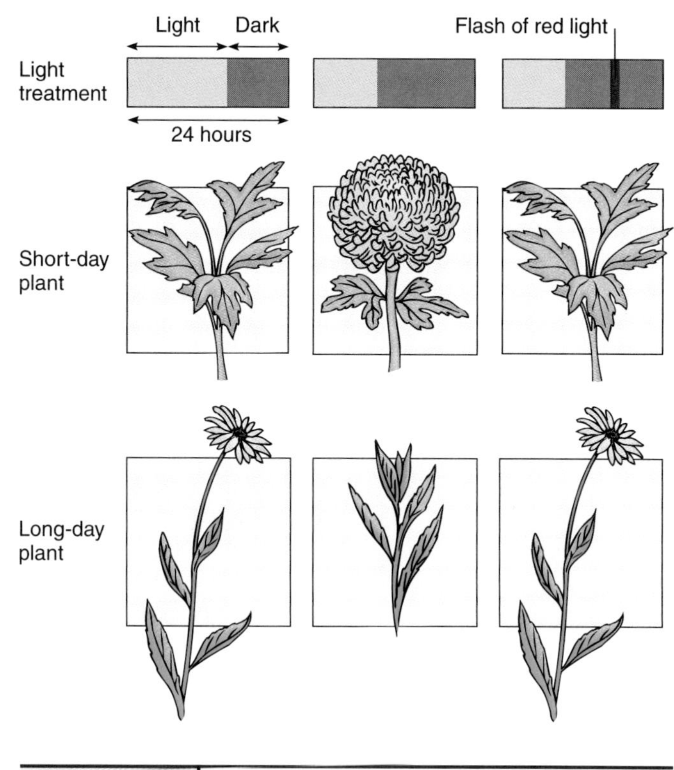

FIGURE **36-4** **Photoperiodic responses of short-day and long-day plants.**

A short-day plant flowers when it is grown under long-night conditions *(top middle)*, but it does not flower when exposed to a long night interrupted with a brief flash of red light *(top right)*. A long-day plant does not flower when grown under long-night conditions *(bottom middle)* unless the long night is interrupted with a brief flash of red light *(bottom right)*.

are intermediate-day plants. These plants flower when exposed to days and nights of indeterminate length.

Some plants, called **day-neutral plants,** do not initiate flowering in response to seasonal changes in the period of daylight and darkness but instead respond to some other type of stimulus, external or internal. Cucumber, sunflower, corn, and onion are examples of day-neutral plants. Many of these plants originated in the tropics where day length does not vary appreciably during the year. (In contrast, short-day, long-day, and intermediate-day plants are temperate, or mid-latitude, species.)

Plant biologists have experimented with the effects of various light regimens on flowering. Figure 36-4 shows how flowering in long-day and short-day plants is affected by different light treatments, including a short-day/long-night regimen with a *night break,* a short burst of light in the middle of the night.

Phytochrome detects day length

For a plant or any organism to have a biological response to light, it must contain a light-sensitive substance, called a *photoreceptor,* to absorb the light. The main photoreceptor for photoperiodism and many other light-initiated plant responses (such as germination and seedling establishment) is **phytochrome,** a family of about 5 blue-green pigment proteins, each of which is coded for by a different gene. A mixture of phytochrome proteins is present in cells of all vascular plants examined so far. For example, five members of the phytochrome family, designated phyA, phyB, phyC, phyD, and phyE, occur in *Arabidopsis thaliana.*

Much current knowledge of phytochrome in *Arabidopsis* is based on various mutant plants that do not express a specific phytochrome gene, such as the gene that codes for phyA. By studying the physiological response of plants that do not produce an individual phytochrome, biologists have concluded that the individual forms of phytochrome have both unique and overlapping functions. PhyB appears to exert its influence at all stages of the plant life cycle, whereas the other forms of phytochrome have narrower functions at specific stages in the life cycle.

PhyA and phyB may have antagonistic (opposite) effects on flowering. In long-day plants, phyB may inhibit flowering and phyA may induce flowering. Flowering occurs more rapidly in long-day plants with mutations in the gene that codes for phyB; such mutations reduce or eliminate the production of phyB. In contrast, flowering is delayed or prevented in long-day plants with mutations in the gene that codes for phyA.

Each member of the phytochrome family exists in two forms and readily converts from one form to the other after absorption of light of specific wavelengths. One form, designated **Pr** (for *r*ed-absorbing *p*hytochrome), strongly absorbs light with a relatively short red wavelength (660 nm). In the process, the shape of the molecule changes to the second form of phytochrome, **Pfr,** so designated because it absorbs *f*ar-*r*ed light, which is light with a relatively long red wavelength (730 nm) (Fig. 36-5). When Pfr absorbs far-red light, it reverts back to the original form, Pr. Pfr is the active form of phytochrome, triggering or inhibiting physiological responses such as flowering.

What does a pigment that absorbs red light and far-red light have to do with daylight and darkness? Sunlight consists of various amounts of the entire spectrum of visible light, in addition to ultraviolet and infrared radiation. Because sunlight contains more red than far-red light, however, when a plant is exposed to sunlight, the level of Pfr increases. During the night, the level of Pfr slowly decreases as Pfr, which is less stable than Pr, is degraded.

The importance of phytochrome to plants cannot be overemphasized. Timing of day length and darkness is the most reliable way for plants to measure the change from one season to the next. This measurement, which synchronizes the stages of plant development, is crucial for survival, particularly in environments where the climate has an annual pattern of favorable and unfavorable seasons.

FIGURE **36-5** | **Phytochrome.**

This pigment occurs in two forms, designated Pr and Pfr, and readily converts from one form to the other. Red light (660 nm) converts Pr to Pfr, and far-red light (730 nm) converts Pfr to Pr.

KEY CONCEPT: Phytochrome is a unique photoreversible pigment that undergoes changes in conformation on exposure to light of different wavelengths.

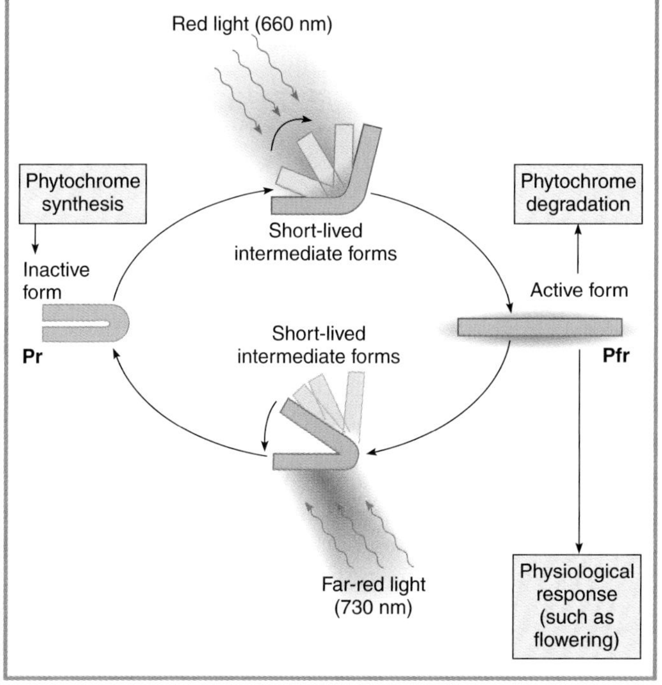

Competition for sunlight among shade-avoiding plants involves phytochrome

Plants sense the proximity of nearby plants, which are potential competitors, and react by changing the way they grow and develop. Many plants, from small herbs to large trees, compete for light, a response known as **shade avoidance,** in which plants tend to grow taller when closely surrounded by other plants. If successful, the shade-avoiding plant projects its new growth into direct sunlight, increasing its chances of survival.

Since the 1970s, botanists have recognized the environmental factor that triggers shade avoidance: Plants perceive changes in the ratio of red to far-red light that result from the presence of nearby plants. The leaves of neighboring plants absorb much more red light than far-red light. (Recall from Chapter 8 that the green pigment chlorophyll used in photosynthesis strongly absorbs red light.) In a densely plant-populated area, the ratio of red light to far-red light (r/fr) decreases. The greater relative amount of far-red light triggers a series of responses that cause the shade-avoiding plant, which is adapted to full light environments, to grow taller or flower earlier. Figure 36-6 shows an interesting application of additional (reflected) far-red light without the reduction in incoming sunlight that occurs in shade avoidance.

When a plant is using many of its resources for stem elongation, it has fewer resources for new leaves and branches, storage tissues, or reproductive tissues. However, for a shade-avoiding plant that is shaded by its neighbors, a rapid increase in stem length is advantageous, because once this plant is taller than its neighbors it obtains a larger share of unfiltered sunlight.

Phytochrome is involved in other responses to light, including germination

Phytochrome is involved in the light requirement that some seeds have for germination. Seeds with a light requirement must be exposed to light containing red wavelengths. Exposure to red light converts Pr to Pfr, and germination occurs. Many temperate forest species with small seeds require light for germination. (Larger seeds generally do not have a light requirement.) This adaptation enables the seeds to germinate at the optimal time. During early spring, sunlight, including red light, penetrates the bare branches of overlying deciduous trees and reaches the soil between the trees. As spring temperatures warm, the seeds on the soil absorb red light and germinate. During their early growth, the newly germinated seedlings do not have to compete with the taller trees for sunlight.

Cary Wolinsky

FIGURE **36-6** **Colored mulches and plant growth processes.**

The red mulch reflects far-red light from the ground to the tomato plants. The plants, absorbing the extra far-red light, react as if nearby plants are present. Their aboveground growth is greater, and their fruits ripen earlier. Manipulating mulch colors also modifies yield, flavor, and nutrient content.

Other physiological functions under the influence of phytochrome include sleep movements in leaves (discussed later); shoot dormancy; leaf abscission (see Chapter 32); and pigment formation in flowers, fruits, and leaves.

Phytochrome acts by signal transduction

PROCESS OF SCIENCE

Some phytochrome-induced responses are very rapid and short-term, whereas others are slower and long-term. The rapid responses probably involve reversible changes in the properties of membranes, altering cell ionic balances. Red light, for example, causes potassium ion (K^+) channels to open in cell membranes involved in plant movements caused by changes in turgor (discussed later). Exposure to far-red light causes the K^+ channels to close. Slower phytochrome-induced responses involve the selective regulation of transcription of numerous genes. For example, phytochrome activates transcription of the gene for the small subunit of *Rubisco,* an enzyme involved in photosynthesis (see Chapter 8).

Each phytochrome molecule consists of a protein attached to a light-absorbing photoreceptor. Biologists hypothesize the absorption of light by the photoreceptor portion of phytochrome elicits a change in the shape of the larger protein component. This change in turn triggers one or more signal transduction pathways. In **signal transduction,** a receptor converts an extracellular signal into an intracellular signal that causes some change in the cell. Signal transduction amplifies the original signal and ultimately results in a physiological or developmental response, such as promotion or inhibition of flowering. (It may be helpful to review the general discussion of signal transduction in Chapter 5.)

Some biologists have focused on elucidating phytochrome signal transduction. One important research tool is mutant plants that respond as if they were exposed to a particular light stimulus even when they are not. Research indicates that the active form of phytochrome moves from the cytoplasm into the nucleus, where it affects gene expression by activating a **transcription factor** (see Chapter 13). When activated, the transcription factor, which binds to the promoter of a gene, either turns on or represses transcription that leads to protein synthesis. (Recall from Chapters 12 and 13 that the **promoter** is the nucleotide sequence in DNA to which RNA polymerase attaches to begin transcription.) Biologists have identified **phytochrome-interacting factor (PIF-3),** a transcription factor involved with phytochrome signaling.

The pathway in phytochrome signal transduction with PIF-3 is elegantly simple (Fig. 36-7). In step ①, inactive phytochrome (Pr) in the cytoplasm absorbs red light and is converted into the active form, Pfr, which ② moves into the nucleus. ③ There, phytochrome binds to the transcription factor PIF3 (which is already bound to the promoter) and ④ activates (or represses) the transcription of light-responsive genes. The signal transduction pathway, as currently understood, is summarized as follows:

Red light ⟶ conversion of Pr to Pfr ⟶ movement of Pfr to nucleus ⟶ formation of Pfr-PIF3 complex bound to promoter region ⟶ light-responsive gene switched on (or off)

The signal transduction pathway is shut down by far-red light, which is absorbed by the Pfr in the Pfr-PIF3 complex in the nucleus. When Pfr is converted to Pr, it dissociates from PIF3.

The signal transduction pathway just described is not the end of the story. There are probably other non-PIF3 pathways by which phytochrome regulates light-responsive gene expression. Some of the molecules that may be involved in other phytochrome-mediated signaling pathways include Ca^{2+}, calmodulin, G proteins, and cyclic guanosine monophosphate (cGMP); these molecules are also implicated in signal transduction pathways in animals (see Chapter 47).

Light influences circadian rhythms

Almost all organisms, including plants, animals, fungi, eukaryotic microorganisms, and many prokaryotes, appear to have an internal timer, or biological clock, that approximates a 24-hour cycle, the time it takes for Earth to rotate around its own axis. These internal cycles, known as **circadian rhythms** (from the Latin *circum,* "around," and *diurn,* "daily"), help the organism detect the time of day. In contrast, photoperiodism enables a plant to detect the time of year.

One example of a circadian rhythm in plants is the opening and closing of stomata, independent of light and darkness. Plants

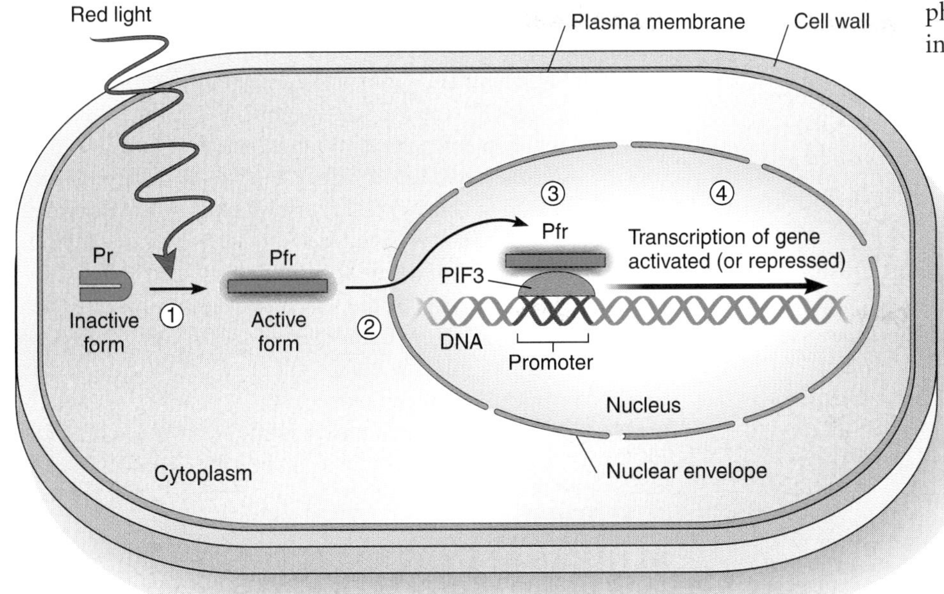

FIGURE **36-7** **Phytochrome signal transduction.**
The numbered steps are explained in the text.

placed in continual darkness for extended periods continue to open and close their stomata on an approximate 24-hour cycle. **Sleep movements** observed in the common bean and other plants are another example of a circadian rhythm (Fig. 36-8). During the day, bean leaves are horizontal, possibly for optimal light absorption, but at night the leaves fold down or up, a movement that orients them perpendicular to their daytime position. The biological significance of sleep movements is unknown at this time.

When constant environmental conditions are maintained, circadian rhythms repeat every 20 to 30 hours, at least for several days. In nature, the rising and setting of the sun reset the biological clock so that the cycle repeats every 24 hours.

For many plants, two photoreceptors—the red light–absorbing phytochrome and the blue light–absorbing **cryptochrome**—are implicated in resetting the biological clock. Certain amino acid sequences of the protein portion of phytochrome are homologous with amino acid sequences of clock proteins in fruit flies, fungi, mammals, and bacteria; this molecular evidence strongly supports the circadian clock role of phytochrome. The evidence for cryptochrome as a clock protein is also convincing. First discovered in plants, cryptochrome counterparts are found in the fruit fly and mouse biological clock proteins. Possibly both photoreceptors are involved in resetting the biological clock in plants: Researchers have evidence that phytochrome and cryptochrome sometimes interact to regulate similar responses.

Why do plants and other organisms exhibit circadian rhythms? Predictable environmental changes, such as sunrise and sunset, occur during the course of each 24-hour period. These predictable changes may be important to an individual organism, causing it to change its physiological activities or its behavior (in the case of animals). It is thought that circadian rhythms help an organism to synchronize repeated daily activities so that they occur at the appropriate time each day. If, for example, an insect-pollinated flower does not open at the time of day that pollinating insects are foraging for food, reproduction will be unsuccessful. Likewise, if a firefly flashes its light on and off at the wrong time of day, it will not find a mate.

Review

- What is phytochrome? What are two roles of phytochrome?
- What is signal transduction?
- In what process is cryptochrome involved?

BiologyNow™ Assess your understanding of **light signals and plant development** by taking the pretest on your BiologyNow CD-ROM.

FIGURE **36-8** **Sleep movements.**
(a) Leaf position at noon in a bean (*Phaseolus vulgaris*) seedling.
(b) Leaf position at midnight.

Dennis Drenner
(a)

Dennis Drenner
(b)

NASTIC MOVEMENTS AND TROPISMS

Learning Objectives

6 Distinguish between a nastic movement and a tropism.

7 Describe phototropism, gravitropism, thigmotropism, and heliotropism.

You've just seen that some plants have sleep movements. Plants exhibit other kinds of movements in response to environmental stimuli such as light, touch, and gravity. Plants may respond quickly (by nastic movements) or slowly (by tropisms) to adjust the position of their organs. Most of these movements are the result of cell growth.

Changes in turgor induce nastic movements

The sensitive plant (*Mimosa pudica*) dramatically folds its leaves and droops in response to touch (or to an electrical, chemical, or thermal stimulus) (Fig. 36-9). The response, which typically occurs within a few seconds, spreads throughout the plant even if only one leaflet is initially stimulated. When a sensitive plant is touched, an electrical impulse moves down the leaf to special cells housed in structures called **pulvini** (sing., *pulvinus*) at the base of each leaflet, each cluster of leaflets, and each petiole. Each pulvinus is a somewhat swollen joint that acts as a hinge. When the electrical signal reaches cells in the pulvinus, it induces a chemical signal that increases membrane permeability to certain ions. A loss of turgor occurs in certain pulvinus cells as potassium ions (K^+) and chloride ions (Cl^-) exit through the now permeable plasma membrane, causing water to leave the cells by osmosis (see Chapter 5 for a discussion of turgor pressure). The sudden change in turgor causes the leaf movement. Such **nastic movements** occur in response to external stimuli, but the direction of movement is predetermined and is independent of the direction of the stimulus. Nastic movements are temporary and reversible. The movement of K^+, Cl^-, and water back into the pulvinus cells causes the plant part to return to its original position, although recovery takes several to many minutes longer than the original movement.

A tropism is directional growth in response to an external stimulus

A plant may respond to an external stimulus, such as light, gravity, or touch, by directional growth (that is, the direction of growth depends on the direction of the stimulus). Such a directional growth response, called a **tropism,** results in a change in the position of a plant part. Tropisms are irreversible and may be positive or negative, depending on whether the plant grows toward the stimulus (a positive tropism) or away from it (a negative tropism). Tropisms are under hormonal control, which is discussed later in this chapter.

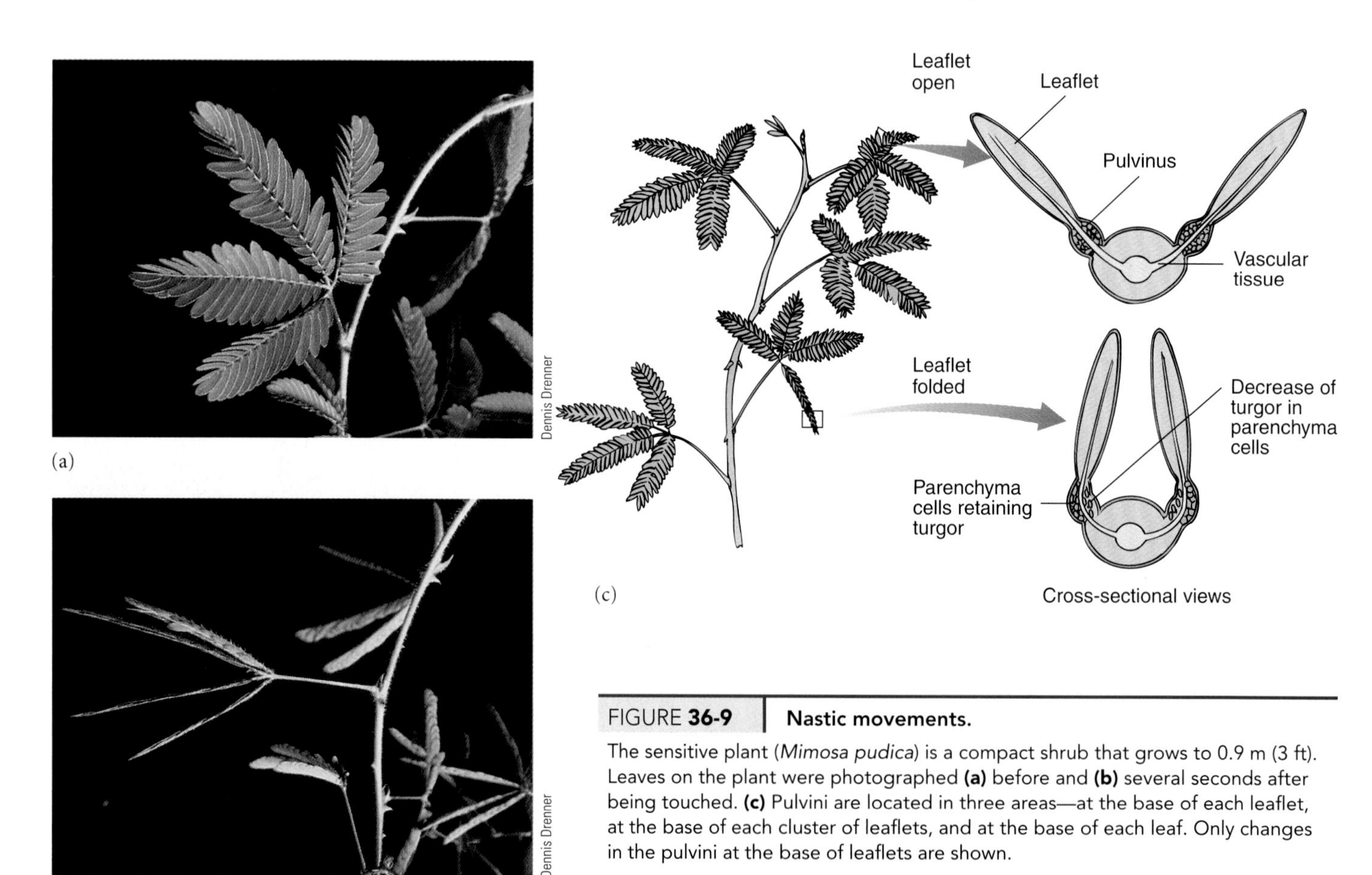

FIGURE **36-9** **Nastic movements.**

The sensitive plant (*Mimosa pudica*) is a compact shrub that grows to 0.9 m (3 ft). Leaves on the plant were photographed **(a)** before and **(b)** several seconds after being touched. **(c)** Pulvini are located in three areas—at the base of each leaflet, at the base of each cluster of leaflets, and at the base of each leaf. Only changes in the pulvini at the base of leaflets are shown.

FIGURE **36-10** | **Phototropism.**

Stems of corn (*Zea mays*) seedlings grow in the direction of light and therefore exhibit positive phototropism. The bending is caused by greater elongation on the shaded side of a stem than on the lighted side.

Phototropism is the directional growth of a plant caused by light (Fig. 36-10). Most growing shoot tips exhibit positive phototropism by bending (growing) toward light, something you may have observed if you place houseplants near a sunny window. This growth response increases the likelihood that stems and leaves receive adequate light for photosynthesis. The bending response of phototropism is triggered by blue light with wavelengths less than 500 nm. (You may recall from Chapter 32 that blue light also induces stomata to open.)

The photoreceptor that absorbs blue light and triggers the phototropic response is thought to be a yellow pigment. Traditionally, two families of yellow pigments—flavins and carotenoids—were leading candidates for the blue light photoreceptor. In 1998, researchers convincingly demonstrated in *Arabidopsis* that a flavoprotein (a flavin attached to a protein molecule) is the photoreceptor for phototropism. In 2002 the photoreceptor was named **phototropin.** Knowledge of the signal transduction pathway by which blue light triggers phototropism and other responses in plants is the focus of current research. There is evidence that phototropin becomes phosphorylated—that is, a phosphate group is added—in response to blue light. Thus, phosphorylation may be an early step in the blue light–signaling pathway.

Growth in response to the direction of gravity is called **gravitropism.** Most stem tips exhibit negative gravitropism by growing away from the center of Earth, whereas most root tips exhibit positive gravitropism (Fig. 36-11). The root cap is the site of gravity perception in roots; when the root cap is removed, the root continues to grow, but it loses any ability to perceive gravity. Special cells in the root cap possess starch-containing **amyloplasts** that collect toward the bottoms of the cells in response to gravity, and these amyloplasts may initiate at least some of the gravitropic response. If the root is put in a different position, as when a potted plant is laid on its side, the amyloplasts tumble to a new position, always settling in the direction of gravity. The gravitropic response (bending) occurs shortly thereafter and involves the hormone *auxin* (discussed later in the chapter). Despite the movement of amyloplasts in response to gravity, their role in gravitropism is questioned. A mutant *Arabidopsis* plant that lacks amyloplasts in its root cap cells still responds gravitropically when placed on its side, indicating that roots do not necessarily need amyloplasts to respond to gravity. Ongoing research may clarify how roots perceive gravity.

Thigmotropism is growth in response to a mechanical stimulus, such as contact with a solid object. The twining or curling growth of tendrils or stems, which helps attach a climbing plant such as a vine to some type of support, is an example of thigmotropism (see Fig. 32-14b).

Heliotropism, also called **solar tracking,** is the ability of leaves or flowers of certain plants, such as sunflower, soybean, and cot-

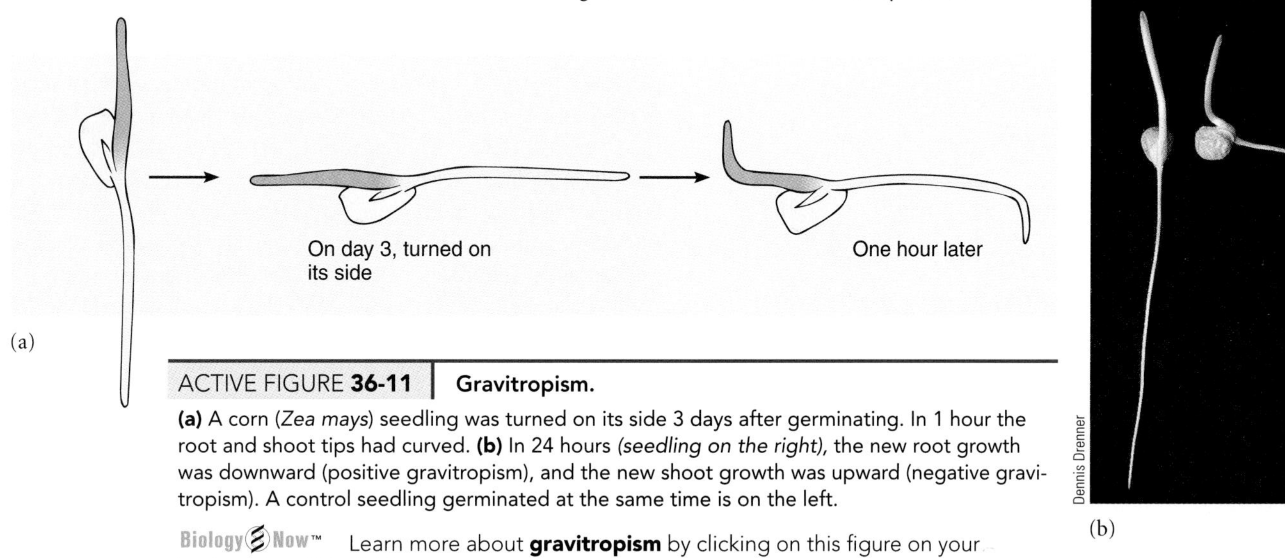

ACTIVE FIGURE **36-11** | **Gravitropism.**

(a) A corn (*Zea mays*) seedling was turned on its side 3 days after germinating. In 1 hour the root and shoot tips had curved. **(b)** In 24 hours *(seedling on the right),* the new root growth was downward (positive gravitropism), and the new shoot growth was upward (negative gravitropism). A control seedling germinated at the same time is on the left.

Biology Now™ Learn more about **gravitropism** by clicking on this figure on your BiologyNow CD-ROM.

FIGURE 36-12

Heliotropism.

Note that the leaves are oriented so they are perpendicular to the sun's rays throughout the day, thereby maximizing the amount of light absorbed.

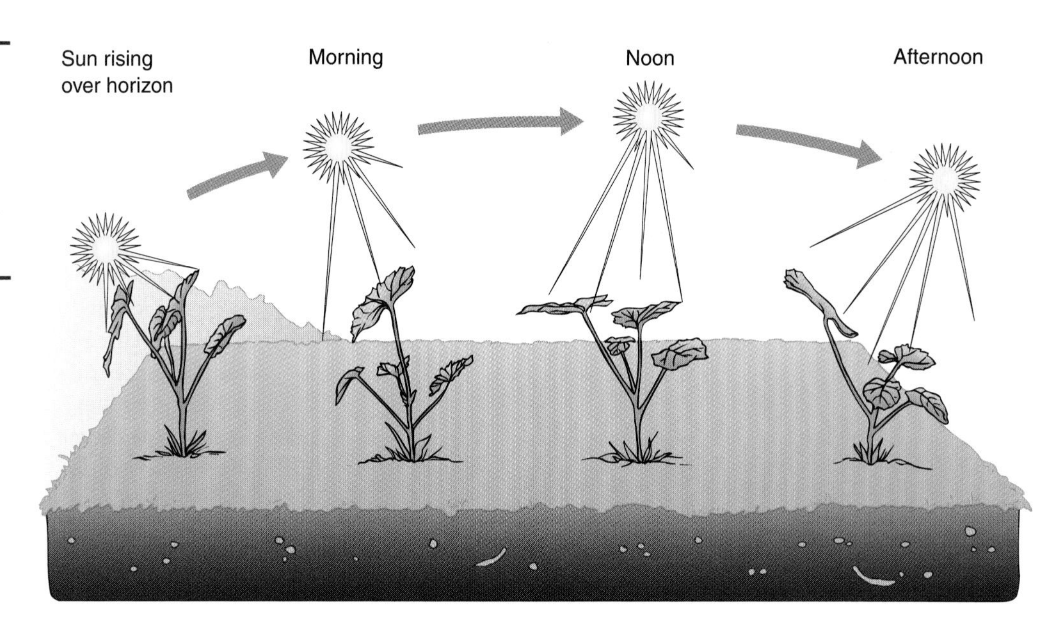

ton, to follow the sun's movement across the sky (Fig. 36-12). Frequently, the leaves of such plants arrange themselves so that they are perpendicular to the sun's rays, regardless of the time of day or the sun's position in the sky. This positioning allows for maximal light absorption. Many solar trackers have pulvini at the bases of their petioles. Changes in turgor in the cells of the pulvinus help position the leaf in its optimum orientation relative to the sun. Like phototropism, heliotropism is triggered by blue light.

Review

- How do nastic movements and tropisms differ?
- What is phototropism? How does blue light trigger the phototropic response?
- How do gravitropism, thigmotropism, and heliotropism differ?

BiologyNow™ Assess your understanding of **nastic movements and tropisms** by taking the pretest on your BiologyNow CD-ROM.

PLANT HORMONES AND DEVELOPMENT

Learning Objectives

8 Describe a general mechanism of action for plant hormones.

9 List several ways each of these hormones affects plant growth and development: auxins, gibberellins, cytokinins, ethylene, and abscisic acid.

10 Summarize the activities of these plant hormones and hormone-like signaling molecules: brassinolides, salicylic acid, systemin, oligosaccharins, and jasmonates.

A plant **hormone** is an organic compound that acts as a chemical signal eliciting a variety of responses that regulate growth and development. The study of plant hormones is challenging because hormones are effective in extremely small amounts (less than 10^{-6} moles per liter). In addition, the effects of different plant hormones overlap, and it is difficult to determine which hormone, if any, is the primary cause of a particular response. Plant hormones may also stimulate a response at one concentration and inhibit that same response at a different concentration.

For many years, biologists studied five major classes of plant hormones: auxins, gibberellins, cytokinins, ethylene, and abscisic acid. More recently, researchers have uncovered compelling evidence for a variety of signaling molecules, such as brassinolide, salicylic acid, systemin, oligosaccharins, and jasmonic acid. (Table 36-1 summarizes the 10 plant hormones and hormone-like signaling molecules discussed in this chapter.)

Plant hormones act by signal transduction

PROCESS OF SCIENCE

Researchers have used molecular genetic techniques to better understand the biology of plant hormones. *Arabidopsis* mutants are particularly useful. For example, different mutants have defects in hormone synthesis, hormone transport, signal reception, or signal transduction. These mutants enable plant biologists to identify and clone genes involved in these aspects of hormone biology. Studying the mutant phenotypes helps plant biologists establish connections between the mutant genes and specific physiological activities involved in growth and development.

Plant and animal hormones are similar in their basic mechanism of action. Like animal hormones, plant hormones bind to specific receptor proteins on the plasma membrane or in the target cells (Fig. 36-13). As shown in step ①, each receptor has a 3-D shape that binds only with one kind of hormone molecule. ② This binding activates a G protein in the plasma membrane (see Chapter 5). The G protein triggers the production of a **second messenger,** an intracellular signaling molecule that affects the function of the cell. Ions such as Ca^{2+} serve as second messengers in many plant cells. Cyclic AMP (see Fig. 3-25),

TABLE 36-1 Plant Hormones and Signaling Molecules

Hormone	Site of Production	Principal Actions
Auxins (e.g., IAA) 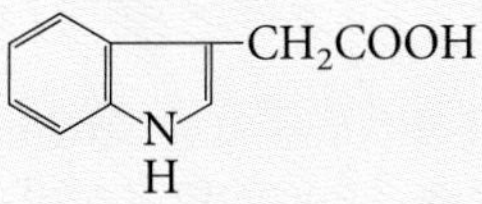	Shoot apical meristem, young leaves, seeds	Stem elongation, apical dominance, root initiation, fruit development
Gibberellins (e.g., GA_3) 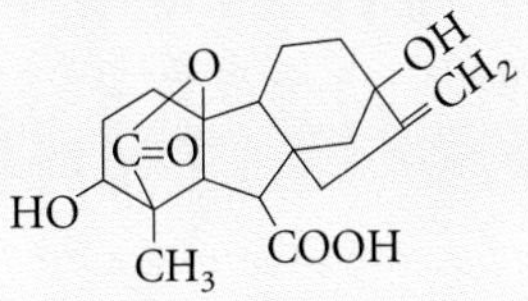	Young leaves and shoot apical meristems, embryo in seed	Seed germination, stem elongation, flowering, fruit development
Cytokinins (e.g., Zeatin) 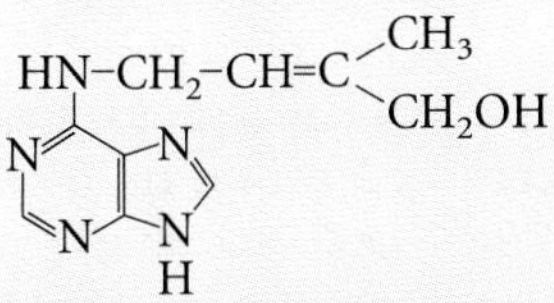	Roots	Cell division, delay of leaf senescence, inhibition of apical dominance, flower development, embryo development, seed germination
Ethylene 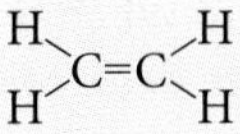	Stem nodes, ripening fruit, damaged or senescing tissue	Fruit ripening, responses to environmental stressors, seed germination, maintenance of apical hook on seedlings, root initiation, senescence and abscission in leaves and flowers
Abscisic acid 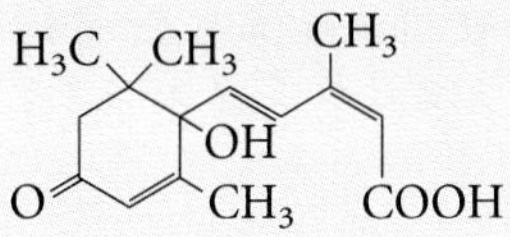	Almost all cells that contain plastids (leaves, stems, roots)	Seed dormancy, responses to water stress
Brassinosteroids (e.g., Brassinolide) 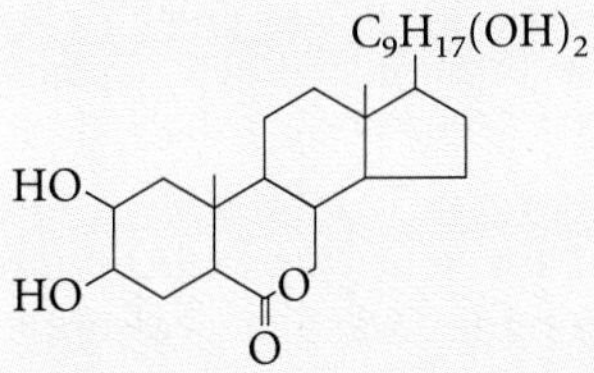	Unknown	Light-mediated gene expression, cell division, stem elongation, flower development, leaf senescence
Salicylic acid 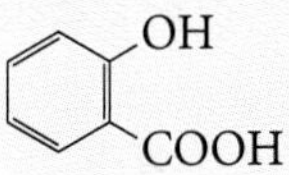	Wound (site of infection)	Resistance to disease organisms
Systemin* 	Wound (site of herbivore or pathogen attack)	Initiation of defenses against predators (herbivores) or disease organisms
Oligosaccharins* 	Unknown	May function in normal cell growth and development; defense responses to disease organisms
Jasmonates (e.g., Jasmonic acid) 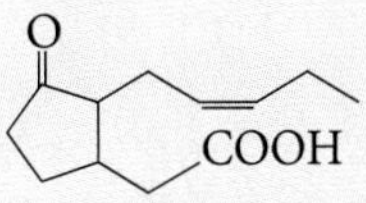	Leaves? Probably many tissues	Initiation of defenses against predators or disease organisms

*Key (A) = amino acid; Sug = sugar.

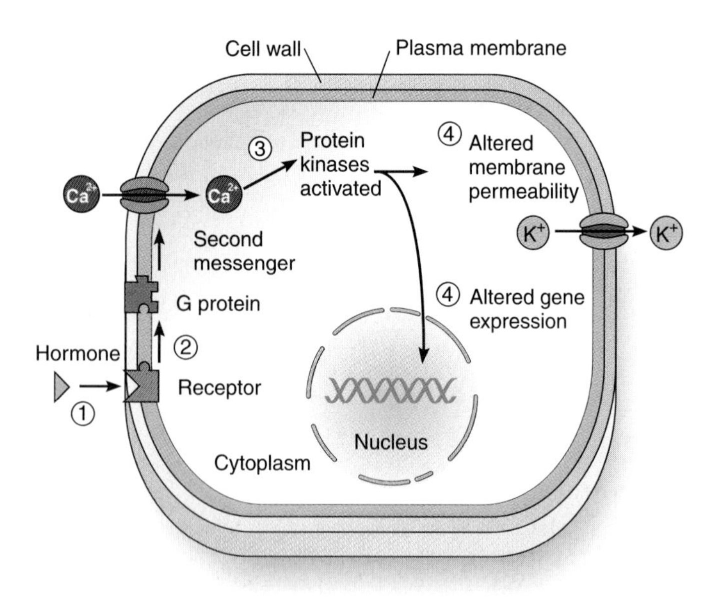

FIGURE **36-13** **General mechanism of action of plant hormones.**

This diagram includes aspects of signal reception, signal transduction, and altered cell activity (responses) that have been demonstrated for various plant hormones. The numbered steps are explained in the text.

an important second messenger in a variety of prokaryotes and eukaryotes, may also be a signaling molecule for at least one plant hormone (auxin).

As shown in ③, once the concentration of Ca^{2+} or other second messenger increases in the cell, it may bind to proteins and activate or inactivate certain enzymes, such as protein kinases. The enzymes can phosphorylate proteins, which alter cell activity in some way, ④ such as altered membrane permeability and/or altered gene expression (that is, altered transcription and/or translation). (Chapter 47 discusses second messengers as they relate to animal hormones.)

Auxins promote cell elongation

PROCESS OF SCIENCE

Charles Darwin, the British naturalist best known for developing the theory of natural selection to explain evolution, also provided the first evidence for the existence of auxins. The experiments that Darwin and his son Francis performed in the 1870s involved positive phototropism, the directional growth of plants toward light. The plants they used were newly germinated canary grass seedlings. As in all grasses, the first part of a canary grass seedling to emerge from the soil is the coleoptile, a protective sheath that encircles the stem. When coleoptiles are exposed to light from only one direction, they bend toward the light. The bending occurs below the tip of the coleoptile.

The Darwins tried to influence this bending in several ways (Fig. 36-14). For example, they covered the tip of the coleoptile as soon as it emerged from the soil. When they covered that part of the coleoptile above where the bend would be expected to occur, the plants did not bend. On other plants, they removed the coleoptile tip and found bending did not occur. When the bottom of the coleoptile where the curvature would occur was shielded from the light, the coleoptile bent toward light. From these experiments, the Darwins concluded that "some influence is transmitted from the upper to the lower part, causing it to bend."

In the 1920s Frits Went, a young Dutch scientist, isolated the phototropic hormone from oat coleoptiles. He removed the coleoptile tips and placed them on tiny blocks of agar for a period of time. When he put one of these agar blocks squarely on a decapitated coleoptile, normal growth resumed. When he placed one of these agar blocks to one side of the tip of a decapitated coleoptile in the dark, bending occurred (Fig. 36-15). This indicated the substance had diffused from the coleoptile tip into the agar and, later, from the agar into the decapitated coleoptile.

ACTIVE FIGURE **36-14** **The Darwins' phototropism experiments.**

(a) Some canary grass coleoptiles were uncovered, some were covered only at the tip, some had the tip removed, and some were covered everywhere but at the tip. The covers were impervious to light. **(b)** After exposure to light coming from one direction, the uncovered plants and the plants with uncovered tips grew toward the light. The plants with tips covered or removed did not bend toward light.

BiologyNow™ See **the Darwin's Experiments** in action by clicking on this figure on your BiologyNow CD-ROM.

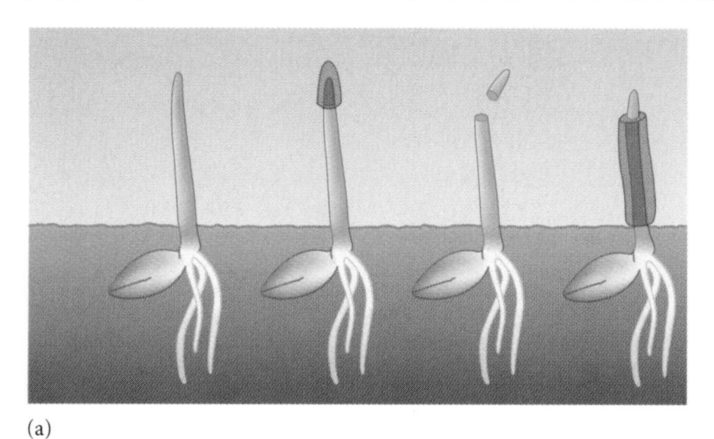

(a)

(b)

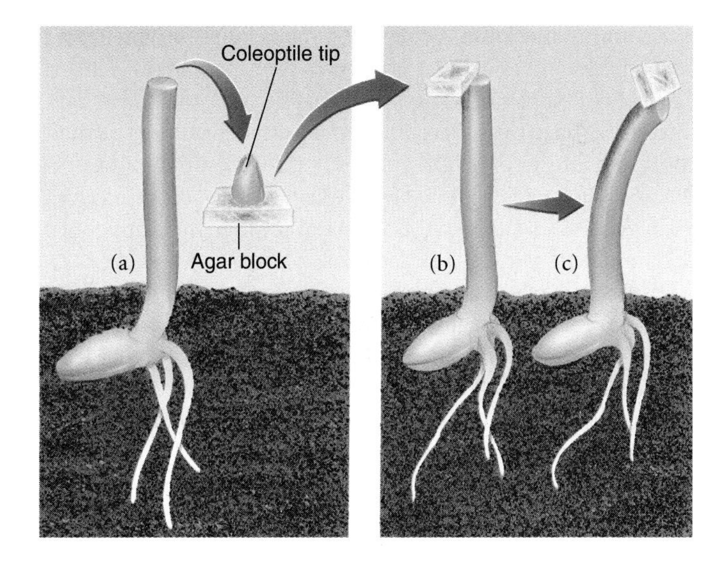

FIGURE **36-15** **Isolating auxin from coleoptiles.**

(a) Coleoptile tips were placed on agar blocks for a period. **(b)** The agar block was transferred to a decapitated coleoptile. It was placed off-center, and the coleoptile was left in darkness. **(c)** The coleoptile bent. This indicated that a chemical moved from the original coleoptile tip to the agar block and from there to one side of the decapitated coleoptile, causing that side to elongate.

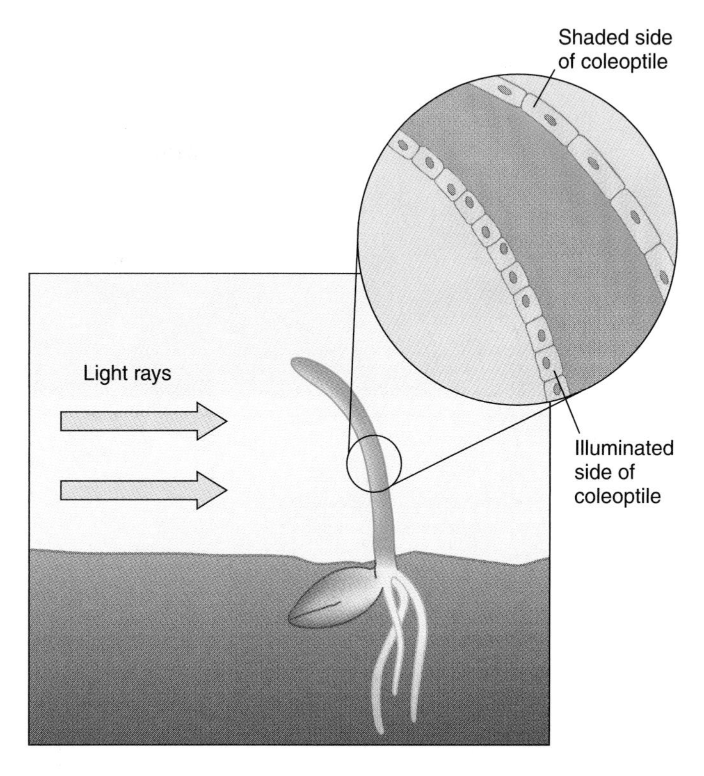

FIGURE **36-16** **Phototropism and the unequal distribution of auxin.**

Auxin travels down the side of the stem or coleoptile *away* from the light, causing cells on the shaded side to elongate. Therefore, the stem or coleoptile bends toward light.

Went named this substance **auxin** (from the Greek *aux,* "enlarge" or "increase"). The purification and elucidation of the primary auxin's chemical structure were accomplished in the mid-1930s by a research team led by U.S. biologist Kenneth Thimann at the California Institute of Technology.

Auxin is a group of several natural (and artificial) plant hormones; the most common and physiologically important auxin is **indoleacetic acid (IAA).** The movement of auxin in the plant is said to be *polar,* or unidirectional. Auxin moves downward along the shoot–root axis from its site of production, usually the shoot apical meristem. Young leaves and seeds are also sites of auxin production.

Auxin's most characteristic action is promotion of cell elongation in stems and coleoptiles. This effect, apparently exerted by acidification of cell walls, increases their plasticity, enabling them to expand under the force of the cell's internal turgor pressure. Auxin's effect on cell elongation also provides an explanation for phototropism. When a plant is exposed to a light from only one direction, some of the auxin migrates laterally to the shaded side of the stem before moving down the stem by polar transport. Because of the greater auxin concentration on the shaded side of the stem, the cells there elongate more than the cells on the light side, and the stem bends toward the light (Fig. 36-16). Auxin is also involved in gravitropism, thigmotropism, and possibly heliotropism.

Auxin exerts other effects on plants. For example, some plants tend to branch out very little when they grow. Growth in these plants occurs almost exclusively from the apical meristem rather than from axillary buds, which do not develop as long as the terminal bud is present. Such plants are said to exhibit **apical dominance,** the inhibition of axillary bud growth by the apical meristem. In plants with strong apical dominance, auxin produced in the apical meristem inhibits axillary buds near the apical meristem from developing into actively growing shoots. When the apical meristem is pinched off, the auxin source is removed, and axillary buds grow to form branches. Apical dominance is often quickly re-established, however, as one branch begins to inhibit the growth of others. Recent evidence indicates that other hormones (ethylene and cytokinin, both discussed later) are also involved in apical dominance. As with other physiological activities, the changing ratios of these hormones may be the factor responsible for apical dominance.

Auxin produced by developing seeds stimulates the development of the fruit. When auxin is applied to certain flowers in which fertilization has not occurred (and, therefore, in which seeds are not developing), the ovary enlarges and develops into a seedless fruit. Seedless tomatoes are produced in this manner.[1] Auxin is not the only hormone involved in fruit development, however.

Some manufactured, or synthetic, auxins have structures similar to IAA. The synthetic auxin naphthalene acetic acid stimulates root development on stem cuttings and is used for asexual

[1] Not all seedless fruits are produced by treatment with auxin. In Thompson seedless grapes, fertilization occurs but the embryos abort, and therefore the seeds fail to develop. Thompson seedless grapes are sprayed with the hormone gibberellin to increase berry size.

Joe Eakes, Color Advantage/Visuals Unlimited

FIGURE **36-17** **Auxin and root development on stem cuttings.**

(Left) Many adventitious roots developed on a honeysuckle (*Lonicera fragrantissima*) cutting placed in a solution with a high concentration of synthetic auxin. *(Middle)* Fewer roots developed in a lower auxin concentration. *(Right)* The cutting placed in water (no auxin) served as a control and did not form roots in the same period.

propagation, particularly of woody plants with horticultural importance (Fig. 36-17). The synthetic auxins 2,4-D and 2,4,5-T are used as selective herbicides (weed killers). These compounds kill plants with broad leaves but, for reasons not completely understood, do not kill grasses. Both herbicides are similar in structure to IAA and disrupt the plants' normal growth processes. Because many of the world's most important crops are grasses (such as wheat, corn, and rice), 2,4-D and 2,4,5-T can be used to kill broadleaf weeds that compete with these crops. The use of 2,4,5-T is no longer allowed in the United States, however, because of its association with dioxins, a group of mildly to very toxic compounds formed as byproducts during the manufacture of 2,4,5-T.

Gibberellins promote stem elongation

In the 1920s, a Japanese biologist was studying a disease of rice in which the young rice seedlings grew extremely tall and spindly, fell over, and died. The cause of the disease, dubbed the "foolish seedling" disease, was a fungus (*Gibberella fujikuroi*) that produces a chemical substance named **gibberellin.** Not until after World War II did scientists in Europe and North America learn of the exciting work done by the Japanese. During the 1950s and 1960s, studies in the United States and Great Britain showed that gibberellins are produced by healthy plants as well as by the fungus that causes foolish seedling disease.

Gibberellins are hormones involved in many normal plant functions. The symptoms of foolish seedling disease are caused by an abnormally high gibberellin concentration in the plant tissue (because both plant and fungus produce gibberellin). Currently, more than 110 naturally occurring gibberellins are known, although many are probably inactive precursors; there are no synthetic gibberellins.

As in foolish seedling disease, gibberellins promote stem elongation in many plants. When gibberellin is applied to a plant, particularly certain dwarf varieties, this elongation may be spectacular. Some corn and pea plants that are dwarfs as a result of one or more mutations grow to a normal height when treated with gibberellin (Fig. 36-18a). Short-stemmed, high-yielding varieties of wheat are short-stemmed because they have a reduced response to gibberellin. These varieties put less of their resources into stem height and more resources into grain production. Gibberellins are also involved in **bolting,** the rapid elongation of a floral stalk that occurs naturally in many plants when they initiate flowering (Fig. 36-18b).

Gibberellins cause stem elongation by stimulating cells to divide as well as elongate. The actual mechanism of cell elongation differs from that caused by auxin, however. Recall that IAA-induced cell elongation involves the acidification of the cell wall. In gibberellin-induced cell elongation, cell wall acidification does not occur; instead, gibberellin increases the mechanical extensibility of a cell wall.

Gibberellins affect several reproductive processes in plants. They stimulate flowering, particularly in long-day plants. In addition, they substitute for the low temperature that biennials require before they begin flowering. If gibberellins are applied to biennials during their first year of growth, flowering occurs without exposure to a period of low temperature. Gibberellins, like auxin, affect the development of fruits. Agriculturalists apply gibberellins to several varieties of grapes to produce larger fruits.

Gibberellins are also involved in seed germination in many plants. In a classic experiment involving barley seed germination, researchers showed that the release of gibberellin from the embryo triggers the synthesis of α-amylase, an enzyme that digests starch in the endosperm. As a result, glucose becomes available for absorption by the embryo. Although enzymes mobilize starch reserves in many types of seeds, gibberellin control of seed enzymes appears restricted to cereals and other grasses. In addition to mobilizing food reserves in newly germinated grass seeds, application of gibberellins substitutes for low-temperature or light requirements for germination in seeds such as lettuce, oats, and tobacco.

Cytokinins promote cell division

During the 1940s and 1950s, researchers were trying to find substances that might induce plant cells to divide in **tissue culture,** a technique in which cells are isolated from plants and grown in a nutrient medium (see *Focus On: Cell and Tissue Culture*). They discovered that cells would not divide without a substance found in coconut milk. Because coconut milk has a complex chemical composition, investigators did not chemically identify the division-inducing substance for some time. Finally, researchers isolated an active substance from a different source, aged DNA from herring sperm. They called it **cytokinin** because it induces cytokinesis, or cytoplasmic division. In 1963 researchers identified the first natural plant cytokinin, zeatin, in corn. Since then similar molecules have been identified in other plants. Biologists have also synthesized several cytokinins. Cytokinins are structurally similar to adenine, a purine base that is part of DNA and RNA molecules.

Cytokinins promote cell division and differentiation of young, relatively unspecialized cells into mature, more specialized cells in intact plants. They are a required ingredient in any plant tissue culture medium and must be present for cells to di-

Courtesy of B.O. Phinney, University of California, Los Angeles

(a)

Robert E. Lyons

(b)

FIGURE **36-18** **Gibberellin and stem elongation.**

(a) An experiment testing the effects of gibberellin on normal and dwarf corn (*Zea mays*) plants shows that dwarf plants respond to gibberellin much more dramatically than normal plants. (This dwarf variety is a mutant with a single recessive gene that impairs gibberellin biosynthesis.) From left to right: dwarf, untreated; dwarf, treated with gibberellin; normal, treated with gibberellin; normal, untreated. Normal corn grows to 4.5 m (15 ft). **(b)** *(left)* Like many biennials, Indian blanket (*Gaillardia pulchella*) grows as a rosette, which is a circular cluster of leaves close to the ground, during its first year. *(right)* It then bolts when it initiates flowering in the second year. The plant in bloom grows to 0.6 m (2 ft).

vide. In tissue culture, cytokinins interact with auxin during the formation of plant organs such as roots and stems (Fig. 36-19). For example, in tobacco tissue culture a high ratio of auxin to cytokinin induces root formation, whereas a low ratio of auxin to cytokinin induces shoot formation.

Cytokinins and auxin also interact in the control of apical dominance. Here their relationship is antagonistic: Auxin inhibits the growth of axillary buds, and cytokinin promotes their growth.

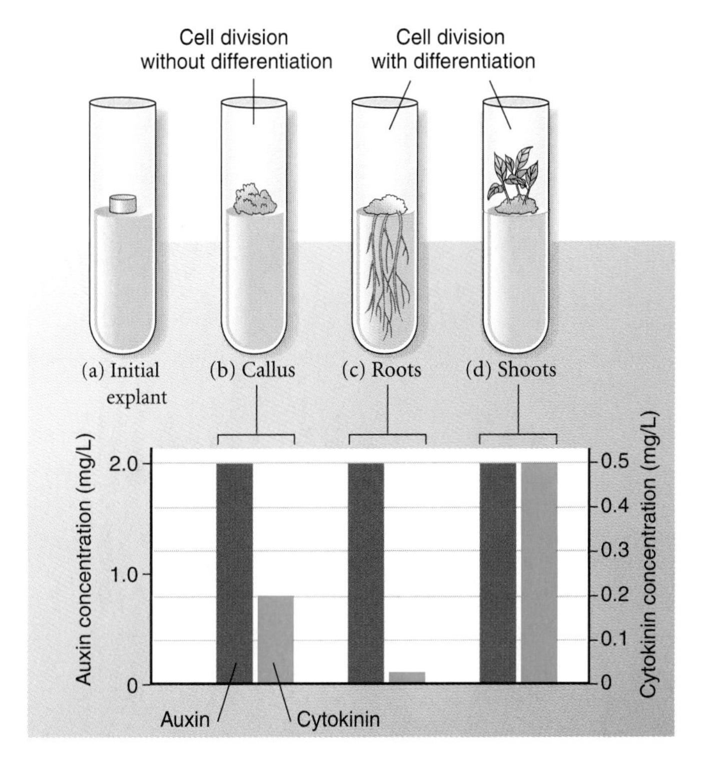

FIGURE **36-19** **Auxin-cytokinin interactions in tissue culture.**

Varying amounts of auxin and cytokinin in the culture media produce different growth responses. **(a)** The initial explant is a small piece of sterile tissue from the pith of a tobacco stem, which is placed on nutrient agar. **(b)** Nutrient agar containing 2.0 mg/L of auxin and 0.2 mg/L of cytokinin causes cells to divide and form a clump of undifferentiated tissue called a callus. **(c)** Agar with a high ratio of auxin (2.0 mg/L) to cytokinin (0.02 mg/L) stimulates root growth. **(d)** Agar with a lower ratio of auxin (2.0 mg/L) to cytokinin (0.5 mg/L) stimulates shoot growth.

Cells can be isolated from certain plants and grown in a chemically defined, sterile nutrient medium. In initial experiments with such cultures, plant cells could be kept alive, but they did not divide. Researchers later discovered that adding certain natural materials such as the liquid endosperm of coconut, also known as coconut milk, induced cells to divide in culture. By the late 1950s, plant cells from a variety of sources could be cultured successfully, dividing to produce a mass of disorganized, relatively undifferentiated cells, or **callus.**

In 1958, F.C. Steward, a plant physiologist at Cornell University, succeeded in generating an entire carrot plant from a single callus cell derived from a carrot root (see Fig. 16-2). This demonstrated conclusively that each plant cell contains a genetic blueprint for all features of an entire organism. His work also showed that an entire plant can be grown from a single cell, provided the proper genes are expressed at the appropriate times.

Since Steward's pioneering work, biologists have successfully cultured many plants using a variety of cell sources. Plants have been regenerated from different tissues, organ explants (excised organs or parts such as root apical meristems), and single cells. Cell and tissue culture techniques help answer many fundamental questions involving growth and development in plants. These techniques also have great practical potential. Using tissue culture, it is possible to regenerate many genetically identical plants from the cells of a single, genetically superior plant. Many kinds of plants—from orchids and African violets to coastal redwoods—have been cultured in this way.

It is also possible to alter the genetic composition of a cell while it is in culture and then have these changes expressed in the whole plant following regeneration. Thus tissue culture provides a valuable tool to genetic engineers who wish to introduce desirable new traits, such as better nutritional properties, into crop species such as rice (see Fig. 14-16b).

One effect of cytokinins on plant cells is to delay the aging process. Plant cells, like all living cells, go through a natural aging process known as **senescence.** Senescence is accelerated in cells of plant parts that are cut, such as flower stems. Botanists think plants must have a continual supply of cytokinins from the roots. Cut stems, of course, lose their source of cytokinins and therefore age rapidly. When cytokinins are sprayed on leaves of a cut stem of many species, they remain green, whereas unsprayed leaves turn yellow and die.

Despite their involvement in regulating many aspects of plant growth and development, cytokinins currently have few commercial applications other than plant tissue culture. However, in 1995, molecular biologists at the University of Wisconsin combined a promoter from a gene activated during normal senescence with a gene that encodes an enzyme involved in cytokinin synthesis. The leaves of transgenic tobacco plants that contained this recombinant DNA produced more cytokinin and therefore lived longer and continued to photosynthesize (Fig. 36-20). This molecular technology has the potential to increase the longevity and productivity of certain crops.

Courtesy of Dr. Richard M. Amasino, University of Wisconsin

FIGURE **36-20** **Cytokinin synthesis and delay of senescence.**

The tobacco (*Nicotiana tabacum*) plant on the left was genetically engineered to produce additional cytokinin as it aged, whereas the tobacco plant of the same age on the right served as a control. Note the extensive senescence and death of older leaves on the control plant. Depending on the variety, tobacco grows 0.9 to 3 m (3 to 10 ft) tall.

Ethylene promotes abscission and fruit ripening

During the early 20th century, scientists observed that the gas **ethylene** (C_2H_4) has several effects on plant growth, but not until 1934 did they demonstrate that plants produce ethylene. This natural hormone influences many plant processes. Ethylene inhibits cell elongation, promotes seed germination, promotes apical dominance, and is involved in plant responses to wounding or invasion by disease-causing microorganisms.

Ethylene also has a major role in many aspects of senescence, including fruit ripening. As a fruit ripens, it produces ethylene, which triggers an acceleration of the ripening process. This induces the fruit to produce more ethylene, which further accelerates ripening. The expression "one rotten apple spoils the barrel" is true. A rotten apple is one that is overripe and produces large amounts of ethylene, which diffuses and triggers the ripening process in nearby apples. Humans use ethylene commercially to uniformly ripen bananas and tomatoes. These fruits are picked while green and shipped to their destination, where they are exposed to ethylene before they are delivered to grocery stores (Fig. 36-21).

Plants growing in a natural environment encounter rain, hail, wind, and contact with passing animals. All these mechanical stressors alter their growth and development, making them shorter and stockier than plants grown in a greenhouse. Ethylene regulates such developmental responses to mechanical stimuli, known as **thigmomorphogenesis.** Plants that are mechanically disturbed produce additional ethylene, which in turn inhibits stem elongation and enhances cell wall thickening in supporting cells such as collenchyma and sclerenchyma. These changes are adaptive because shorter, thicker stems are less likely to be damaged by mechanical stressors.

Ethylene is involved in leaf abscission, which is actually influenced by two antagonistic hormones, ethylene and auxin. As a leaf ages (when autumn approaches, for deciduous trees in temperate climates), the level of auxin in the leaf decreases. Concurrently, cells in the abscission layer at the base of the petiole (where the leaf will break away from the stem) begin producing ethylene.

FIGURE **36-21** | **Ethylene and fruit ripening.**

Both boxes of tomatoes were picked at the same time, while green. The tomatoes in the box on the right were exposed to an atmosphere containing 100 ppm of ethylene for three days.

Illustration Services

Abscisic acid promotes seed dormancy

In 1963 two different research teams discovered **abscisic acid.** Despite its name, abscisic acid is involved in seed dormancy and in a plant's response to stress; abscisic acid does not induce abscission in most plants. As an environmental stress hormone, abscisic acid particularly promotes changes in plant tissues that are water stressed. (Recall that ethylene also affects plant responses to certain stressors, such as mechanical stressors and wounding.)

Botanists best understand the effect of abscisic acid on plants suffering from water stress. The level of abscisic acid increases dramatically in the leaves of plants exposed to severe drought conditions. The high level of abscisic acid in the leaves activates a signal transduction process that leads to the closing of stomata. The closing of stomata saves the water that the plant would normally lose by transpiration, thereby increasing the plant's likelihood of survival. As knowledge of abscisic acid signaling in guard cells increases, botanists hope to use this information to engineer crops and horticultural plants that are resistant to drought.

The onset of winter is also a type of stress on plants. A winter adaptation that involves abscisic acid is dormancy in seeds. Many seeds have high levels of abscisic acid in their tissues and do not germinate until the abscisic acid washes out. In a corn mutant unable to synthesize abscisic acid, the seeds germinate as soon as the embryos are mature, even while attached to the ear (Fig. 36-22).

FIGURE **36-22** | **Abscisic acid and seed germination.**

In a corn (*Zea mays*) mutant that does not produce abscisic acid, some of the kernels have germinated while still on the ear, producing roots *(arrows)*.

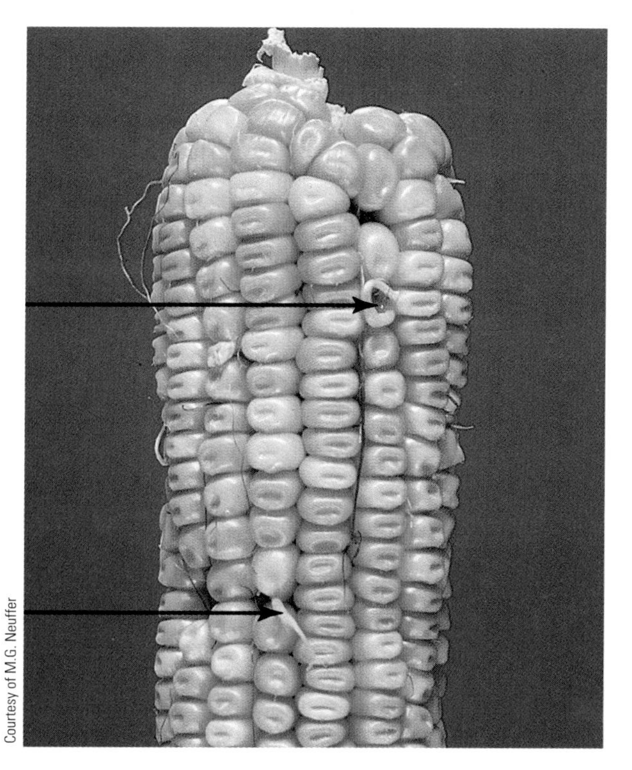

Courtesy of M.G. Neuffer

Abscisic acid is not the only hormone involved in seed dormancy. For example, addition of gibberellin reverses the effects of dormancy. In seeds, the level of abscisic acid decreases during the winter, and the level of gibberellin increases. Cytokinins are also implicated in breaking dormancy. Once again, you see that a single physiological activity such as seed dormancy may be controlled in plants by the interaction of several hormones. The plant's actual response may result from changing ratios of hormones rather than the effect of each individual hormone.

Additional signaling molecules affect growth and development, including plant defenses

Biologists continue to discover new plant hormones and hormone-like signaling molecules . Many of these signaling molecules are involved in defensive responses of plants to disease organisms and insects. Here we briefly consider five groups—the brassinolides, salicylic acid, systemin, oligosaccharins, and jasmonates.

Brassinolides are steroids

Although steroid hormones have crucial roles in animals, their roles in plants are unclear. The **brassinolides** are a group of steroids that may function as plant hormones. Brassinolides appear to be involved in several aspects of growth and development. *Arabidopsis* mutants that cannot synthesize brassinolides are dwarf plants with reduced fertility. Researchers reverse this defect by applying brassinolides. Studies of these mutants suggest that the brassinolides are involved in developmental processes such as cell division and cell elongation.

Salicylic acid is a phenolic compound

For centuries people chewed willow (*Salix*) bark to treat headaches and other types of pain. **Salicylic acid** was first extracted from willow bark and is chemically related to aspirin (acetylsalicylic acid). More recently, biologists have shown that salicylic acid helps plants defend against insect pests and pathogens such as viruses. When a plant is under attack, the concentration of salicylic acid increases, and it spreads systemically throughout the plant. It is thought that salicylic acid binds to a cell receptor, triggering a signal transduction pathway that switches on genes. Presumably, these genes code for proteins that fight infection and promote wound healing.

Plants also use salicylic acid to signal nearby plants. Tobacco plants infected with tobacco mosaic virus release into the air a volatile form of salicylic acid, known as *methyl salicylate,* or *oil of wintergreen.* When nearby healthy plants receive the airborne chemical signal, they begin synthesizing antiviral proteins that enhance their resistance to the virus.

Systemin is a small polypeptide

Although many animal hormones are polypeptides, the first plant polypeptide with hormonal properties was not isolated until 1991. This 18-amino-acid polypeptide, **systemin,** is transported systemically throughout the plant in response to wounding by insects. Systemin may stimulate natural defense mechanisms at extremely low concentrations, as low as one part per trillion. Systemin may trigger the plant to activate genes that produce *protease inhibitors,* molecules that disrupt insect digestion, curbing leaf damage done by caterpillars and other herbivorous (plant-eating) insects. The discovery of systemin in tomato leaves prompted a flurry of research in search of polypeptide regulators and resulted in the discovery of additional polypeptides in other plants.

Oligosaccharins are composed of sugar residues

Oligosaccharins are carbohydrate fragments that consist of short, branched chains of sugar molecules. They are present in extremely small quantities in cells and active at much lower concentrations (100 to 1000 times lower) than hormones such as auxin. Different oligosaccharins may have distinct functions. Some trigger the production of **phytoalexins** (from the Greek *phyto,* "plant," and *alexi,* "to ward off"), antimicrobial compounds that limit the spread of plant pathogens such as fungi. Other oligosaccharins inhibit flowering and induce vegetative growth.

Jasmonates are derivatives of fatty acids

Jasmonates affect several plant processes, such as pollen development, root growth, fruit ripening, and senescence. These lipid-derived plant hormones, which are structurally similar to *prostaglandins* in animals, are also produced in response to the presence of insect pests and disease-causing organisms. Jasmonates trigger the production of enzymes that confer an increased resistance against herbivorous insects (see *On the Cutting Edge: Herbivore Defense Against Plant-Produced Signaling Molecules*). For example, caterpillar-infested tomato plants release volatile chemicals into the air that attract natural caterpillar enemies, such as parasitic wasps that lay their eggs on the caterpillar's bodies. When the eggs hatch, the larvae consume, and ultimately kill, the host insects. In one study, treating plants with jasmonic acid increased parasitism on caterpillar pests twofold over control fields in which the plants did not have jasmonic acid applications. Jasmonates may be of practical value in controlling certain insect pests without the use of chemical pesticides.

Unidentified plant hormones remain to be discovered

PROCESS OF SCIENCE

Experiments in which different tobacco species are grafted together indicate that unidentified flower-promoting and flower-inhibiting substances may exist. *Nicotiana silvestris* is a long-day tobacco plant, and a variety of *Nicotiana tabacum* is a day-neutral tobacco plant. When a long-day tobacco is grafted to a day-neutral tobacco and exposed to short nights, both plants flower (Fig. 36-23). The day-neutral tobacco plant flowers sooner than

ON THE CUTTING EDGE: Herbivore Defense Against Plant-Produced Signaling Molecules

PROCESS OF SCIENCE

Hypothesis: Some plant-eating insects "eavesdrop" on plant defensive signals and respond by activating their own defenses.

Methods: Researchers fed corn earworms diets of plant signaling molecules (either jasmonate or salicylate) and evaluated their responses.

Results: Corn earworms with either jasmonate or salicylic acid in their diets rapidly produced enzymes to counter plant defensive toxins.

Conclusion: Corn earworms use jasmonate and salicylate to activate their production of enzymes that detoxify plant-produced chemicals.

TABLE 36-A Effects of Jasmonate and Salicylate on Mortality in Corn Earworms

Pre-exposure to signaling molecule	Diet treatment	Percent mortality (after 3 days)
No (control)	Control diet—no toxin	0.0
No (control)	Xanthotoxin	8.9
No (control)	Celery leaves	33.3
Jasmonate	Control diet—no toxin	0.0
Jasmonate	Xanthotoxin	0.0
Jasmonate	Celery leaves	3.0
Salicylate	Control diet—no toxin	3.3
Salicylate	Xanthotoxin	0.0
Salicylate	Celery leaves	7.0

Plants protect themselves from being eaten by producing an array of defensive chemicals that are unpalatable or even toxic to herbivores. However, some herbivores have evolved countermeasures, including the production of enzymes that break down, or detoxify, the plant's defensive chemicals. Insects that respond in this way can eat a variety of host plants. The corn earworm (*Helicoverpa zea),* for example, has more than 100 known host plants; in addition to corn, it eats beans, peas, peppers, potatoes, squash, tomatoes, and many other agriculturally important plants.

Graduate student Xianchun Li and professors Mary Schuler and May Berenbaum from the University of Illinois decided to study the highly successful corn earworm's ability to eavesdrop on its host plant's defenses.* They reared 270 caterpillars and divided them into three different feeding groups. One group of 90 caterpillars ate a diet that contained a small amount of the plant signaling molecule *jasmonate,* another group of 90 ate a diet containing the signaling molecule *salicylate,* and the third group (the control) ate a diet without either signaling molecule.

After 12 hours of feeding, each caterpillar group was subdivided into three groups, each containing 30 caterpillars. One of these subgroups was fed a diet containing *xanthotoxin,* a plant toxin produced in response to jasmonate signaling. The second subgroup was fed a diet of celery leaves; celery is known to produce toxins in response to feeding damage. The third subgroup served as a control and was fed a diet without toxins. The biologists monitored the caterpillars' growth and development, for example, by measuring differences in mortality (see table).

Examine the percent mortality after 3 days: Caterpillars exposed to jasmonate and then fed xanthotoxin had no mortality, compared to a death rate of 8.9% in the control. Caterpillars exposed to jasmonate and then fed celery leaves experienced 3.0% mortality, compared to 33.3% in the contol. The results are similar for caterpillars exposed to salicylate. Exposure to jasmonate or salicylate reduced mortality, presumably because the jasmonate or salicylate activated caterpillar genes coding for enzymes that detoxify plant toxins.

To test the hypothesis that jasmonate activates catepillar genes coding for enzymes that detoxify plant poisons, the researchers used molecular techniques to examine gene expression in the caterpillars. Several cytochrome P450 genes are known to code for enzymes that detoxify plant defensive chemicals in the caterpillar's digestive system. The biologists demonstrated that, after exposure to plant signaling molecules, four of the caterpillar's cytochrome P450 genes are rapidly expressed, either before or at around the same time that plants are synthesizing defensive molecules. (Other researchers demonstrated that, in celery, defensive compounds begin to accumulate 24 hours after the plant is exposed to jasmonate. The compounds reach their maximal concentration in plant tissues about 4 to 6 days after exposure.)

In summary, this research demonstrates that corn earworms detect the signaling molecules jasmonate and salicylate, which plants produce in response to feeding injuries. Jasmonate and salicylate trigger plants to produce defensive toxins, but they also activate genes in corn earworms that code for enzymes that detoxify the plant toxins.

Some researchers have suggested that crop plants could be genetically engineered to produce airborne chemical signals such as jasmonate and salicylate, thereby reducing the number of herbivores feeding on them (see *On the Cutting Edge: New Possibilities for Environmentally Friendly Pest-Control Strategies,* in Chapter 1). However, the results of Li, Schuler, and Berenbaum indicate that the strategy of using signaling molecules to control some insect pests may not produce the desired outcome.

*Li, Xianchun, M.A. Schuler, and M.R. Berenbaum, "Jasmonate and Salicylate Induce Expression of Herbivore Cytochrome P450 Genes," *Nature,* Vol. 419, 17 Oct. 2002.

it normally would. Biologists think that a flower-promoting substance, **florigen,** may be induced in the long-day plant and transported to the day-neutral plant through the graft union, causing the day-neutral tobacco plant to flower sooner than expected. An intact plant may produce florigen in the leaves and transport it in the phloem to the shoot apical meristem. There, it induces a transition from vegetative to reproductive development—that is, to a meristem that produces flowers.

When a botanist grafts a long-day tobacco to a day-neutral tobacco and exposes them to long nights, neither plant flowers. As long as these conditions continue, the day-neutral plants do not flower even when they would normally do so. In this case,

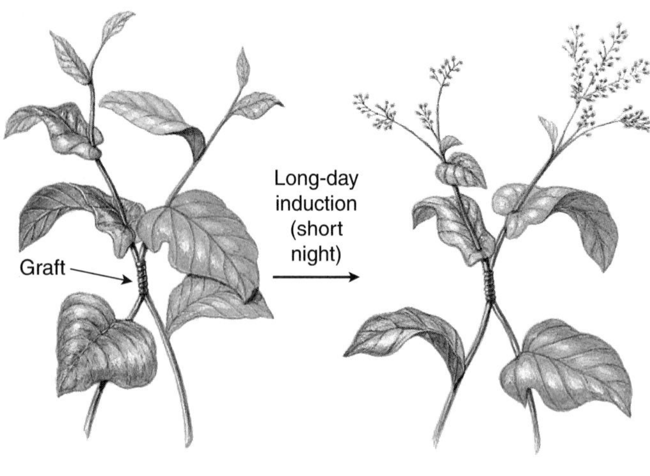

FIGURE 36-23 | Evidence for the existence of a flower-promoting substance.

When a long-day tobacco plant (*Nicotiana silvestris*) is grafted to a day-neutral tobacco plant (*N. tabacum*) and both plants are exposed to a long-day/short-night regimen, they both flower. The day-neutral plant flowers sooner than it normally would, presumably because a flower-promoting substance passes from the long-day plant to the day-neutral one through the graft.

the long-day tobacco may produce a flower-inhibiting substance that is transported to the day-neutral tobacco through the graft union. This substance prevents the day-neutral tobacco from flowering. Biologists have not yet isolated and chemically characterized substances that are clearly identifiable as flower promoters (florigen) or flower inhibitors.

Review

- How are signal reception, signal transduction, and altered cell activity part of the general mechanism of action for plant hormones?
- How is auxin involved in phototropism?
- Which hormones are involved in each of the following physiological processes: (a) seed germination, (b) stem elongation, (c) fruit ripening, (d) leaf abscission, (e) seed dormancy?
- How does salicylic acid help plants defend against insects and viruses?

BiologyNow™ Assess your understanding of **plant hormones and development** by taking the pretest on your BiologyNow CD-ROM.

SUMMARY WITH KEY TERMS

1 Discuss genetic and environmental factors that affect plant growth and development.

- Plant growth and development are controlled by internal factors, such as a cell's location in the plant body, and by environmental factors, such as changing day length. The location of a cell in the young plant body affects gene expression during development by causing some genes in that cell to be turned off and others to be turned on.

2 Summarize the influence of internal and environmental factors on the germination of seeds.

- **Germination** is the process of seed sprouting. Internal factors affecting whether a seed germinates include the maturity of the embryo, the presence or absence of chemical inhibitors, and the presence or absence of hard, thick seed coats.
- External environmental factors that may affect germination include requirements for oxygen, water, temperature, and light. For example, before germinating, dry seeds absorb water by **imbibition.**

3 Explain how varying amounts of light and darkness induce flowering, and describe the role of phytochrome in flowering, including a brief discussion of phytochrome signal transduction.

- **Photoperiodism** is any response of plants to the duration and timing of light and dark. In many plants, flowering is a photoperiodic response; some are **short-day plants,** some are **long-day plants,** and others are **intermediate-day plants.** In **day-neutral plants,** flowering is not affected by photoperiod.
- The photoreceptor in photoperiodism is **phytochrome,** a family of about five blue-green pigments. Each type of phytochrome has two forms, **Pr** and **Pfr,** named by the wavelength of light they absorb. Pfr is the active form, triggering or inhibiting physiological responses such as flowering, **shade avoidance,** and a light requirement for germination. Each type of phytochrome may have both unique and overlapping physiological functions. PhyB seems to exert its influence at all stages of the plant life cycle.
- Some phytochrome-induced responses are rapid and short term (such as changes in membrane properties); others are slower and long term (such as the regulation of gene transcription).
- The absorption of light by phytochrome triggers one or more signal transduction pathways. In **signal transduction,** a receptor converts an extracellular signal into an intracellular signal that causes some change in the cell.

4 Define *circadian rhythm,* and give an example.

- A **circadian rhythm** is a regular period in the growth or activities of a plant or other organism that approximates the 24-hour day and is reset by the rising and setting of the sun. Two plant examples of circadian rhythms are the opening and closing of stomata and **sleep movements.**

5 Distinguish between phytochrome and cryptochrome.

- Both phytochrome and **cryptochrome** are photoreceptors that sometimes interact to regulate similar responses, such as resetting the biological clock. Phytochrome strongly absorbs red light, whereas cryptochrome absorbs blue light.

6 Distinguish between a nastic movement and a tropism.

- Nastic movements and tropisms are the two kinds of plant movements that occur in response to external stimuli. **Nastic movements,** which are temporary and reversible, occur in response to external stimuli, but the direction of movement is independent of the direction of the stimulus.

SUMMARY WITH KEY TERMS *(continued)*

- **Tropisms** are directional growth responses (i.e., the direction of growth is dependent on the direction of the stimulus).

7 Describe phototropism, gravitropism, thigmotropism, and heliotropism.

- **Phototropism** is plant growth in response to the direction of light. **Gravitropism** is plant growth in response to the influence of gravity. **Thigmotropism** is plant growth in response to contact with a solid object. **Heliotropism** is the ability of leaves or flowers to track the sun across the sky.

8 Describe a general mechanism of action for plant hormones.

- Plants produce and respond to **hormones,** organic compounds that act as highly specific chemical signals, eliciting a variety of responses that regulate growth and development. Hormones are effective in extremely small concentrations. The functions of some plant hormones overlap, and many physiological activities of plants are regulated by the interactions of several hormones rather than a single hormone.
- Plant hormones bind to specific receptor proteins in or on target cells. This binding may trigger the production of a **second messenger** such as Ca^{2+}. The second messenger may, in turn, bind to and activate or inactivate certain enzymes. This activation or inactivation may lead to altered membrane permeability and/or altered gene expression.

9 List several ways each of these hormones affects plant growth and development: auxins, gibberellins, cytokinins, ethylene, and abscisic acid.

- **Auxin** is involved in cell elongation; tropisms; **apical dominance,** the inhibition of axillary buds by the apical meristem; and fruit development. Auxin also stimulates root development on stem cuttings. Some synthetic auxins (2,4-D and 2,4,5-T) are selective herbicides.
- **Gibberellins** are involved in stem elongation, flowering, and germination.
- **Cytokinins** promote cell division and differentiation; delay **senescence,** the natural aging process; and interact with auxin and ethylene in apical dominance. Cytokinins induce cell division in **tissue culture,** a technique in which cells are isolated from plants and grown in a nutrient medium.
- **Ethylene** plays a role in ripening fruits; apical dominance; leaf abscission; wound response; **thigmomorphogenesis,** a developmental response to mechanical stressors such as wind; and senescence.
- **Abscisic acid** is an environmental stress hormone involved in stomatal closure caused by water stress and in seed dormancy.

10 Summarize the activities of these plant hormones and hormone-like signaling molecules: brassinolides, salicylic acid, systemin, oligosaccharins, and jasmonates.

- Plant steroids known as **brassinolides** are probably involved in several aspects of plant growth and development, such as cell division and cell elongation.
- **Salicylic acid** helps defend plants against pathogens and insect pests. It may bind to a cell receptor, thereby switching on genes that fight infection and promote wound healing. A volatile form of salicylic acid may serve as an airborne chemical signal from virus-infected plants to healthy ones.
- **Systemin,** a plant polypeptide with hormonal properties, stimulates a natural defense mechanism in which the plant produces protease inhibitors, molecules that disrupt insect digestion.
- **Oligosaccharins,** short, branched chains of sugar molecules, have various functions: they inhibit flowering and stimulate vegetative growth. Oligosaccharins bind to membrane receptors and alter gene expression.
- **Jasmonates** affect several plant processes, such as pollen development, root growth, fruit ripening, and senescence. They are also produced in response to the presence of insect pests and disease-causing organisms.

POST-TEST

1. A plant's response to the relative amounts of daylight and darkness is (a) apical dominance (b) bolting (c) gravitropism (d) photoperiodism (e) phototropism
2. Pfr, the active state of ______________, forms when red light is absorbed. (a) photosystem I (b) phytochrome (c) far-red light (d) phototropin (e) cryptochrome
3. Which of the following represents the correct order in the phytochrome signal transduction pathway? (1) red light (2) light-responsive gene is switched on (or off) (3) movement of Pfr to nucleus (4) conversion of Pr to Pfr (5) formation of PFr-PIF3 complex that is bound to promoter region (a) 1-3-5-4-2 (b) 1-5-3-2-4 (c) 1-2-3-4-5 (d) 1-4-3-2-5 (e) 1-4-3-5-2
4. Which of the following statements about phytochrome is *incorrect?* (a) Phytochrome is the main photoreceptor for photoperiodism. (b) Phytochrome is a family of about five blue-green pigment proteins, each coded by a different gene. (c) Most of our knowledge of phytochrome function is based on mutant corn (*Zea mays*) plants. (d) Each member of the phytochrome family exists in two forms: Pr and Pfr. (e) Phytochrome helps plants sense the presence of nearby plants with which they must compete for light.
5. Which photoreceptor(s) is/are implicated in resetting the biological clock? (a) phytochrome only (b) cryptochrome only (c) gibberellin only (d) cryptochrome and gibberellin (e) phytochrome and cryptochrome
6. In ______________, leaves or other plant organs track the sun's movement across the sky. (a) heliotropism (b) thigmotropism (c) phototropism (d) bolting (e) photoperiodism
7. The orientation of the growth of a plant according to the direction of light is called ______________, whereas the twining of tendrils is an example of ______________. (a) heliotropism; gravitropism (b) photoperiodism; nastic movements (c) phototropism; gravitropism (d) phototropism; thigmotropism (e) photoperiodism; thigmotropism
8. When you prune shrubs to make them "bushier"—that is, to prevent apical dominance—you are affecting the distribution and action of which plant hormone? (a) auxin (b) jasmonic acid (c) systemin (d) ethylene (e) abscisic acid
9. A synthetic ______________ known as 2,4-D is used as a selective herbicide. (a) auxin (b) gibberellin (c) cytokinin (d) ethylene (e) abscisic acid

POST-TEST *(continued)*

10. Research on a fungal disease of rice provided the first clues about the plant hormone (a) auxin (b) gibberellin (c) cytokinin (d) ethylene (e) abscisic acid
11. This plant hormone interacts with auxin during the formation of plant organs in tissue culture. (a) florigen (b) gibberellin (c) cytokinin (d) ethylene (e) abscisic acid
12. The plant hormone ______________ delays senescence, whereas the plant hormone ______________ promotes senescence. (a) cytokinin; auxin (b) auxin; cytokinin (c) cytokinin; ethylene (d) abscisic acid; ethylene (e) gibberellin; auxin
13. The stress hormone that helps plants respond to drought is (a) auxin (b) gibberellin (c) cytokinin (d) ethylene (e) abscisic acid
14. This hormone promotes seed dormancy. (a) auxin (b) gibberellin (c) cytokinin (d) ethylene (e) abscisic acid
15. Which signaling molecule triggers the release of volatile substances that attract parasitic wasps to plant-eating caterpillars? (a) brassinolide (b) jasmonic acid (c) systemin (d) salicylic acid (e) oligosaccharin

CRITICAL THINKING

1. Predict whether flowering in a short-day plant with a minimum critical night length of 14 hours would be expected to occur in the following situations. Explain each answer. (a) The plant is exposed to 15 hours of daylight and 9 hours of darkness. (b) The plant is exposed to 9 hours of daylight and 15 hours of darkness. (c) The plant is exposed to 9 hours of daylight and 15 hours of darkness, with a 10-minute exposure to red light in the middle of the night.
2. If you transplanted the short-day plant discussed in question 1 to the tropics, would it flower? Explain your answer.
3. Many solar-tracking flowers live in cool arctic or alpine environments. The temperature of certain heliotropic flowers is up to 14°F warmer than that of the surrounding air. Insects are drawn to these warm flowers. Given these observations, suggest a possible adaptive explanation for heliotropism in such environments.
4. What biological advantages are conferred on a plant whose stems are positively phototropic and whose roots are positively gravitropic?

- Visit our Web site at **http://biology.brookscole.com/solomon7** for links to chapter-related resources on the World Wide Web. Additional online materials relating to this Chapter can also be found on our Web site.

BIOLOGY NOW RESOURCES

Active Figures

36-11: Gravitropism

36-14: Darwin's Experiments with coleoptiles

Preparing for an exam? Take a diagnostic test on your BiologyNow CD-ROM.

Post-Test Answers

1. d	2. b	3. e	4. c
5. e	6. a	7. d	8. a
9. a	10. b	11. c	12. c
13. e	14. e	15. b	

37

The Animal Body: Introduction to Structure and Function

Fritz Polking/Dembinsky Photo Associates

Larger body size does not mean bigger cells. The cells of the Yacare caiman (*Caiman yacare*) and the flambeau butterfly (*Dryas julia*) on its head are all about the same size. The caiman is larger because its genes specify that its body should consist of a larger number of cells.

CHAPTER OUTLINE

- **Tissues**
- **Organs and Organ Systems**
- **Regulating Body Temperature**

As you read in Chapters 28 through 30, scientists have identified more than a million animal species. In those chapters, we introduced you to dramatically diverse animal groups with radically different body structures. Yet despite their differences, animal groups share many characteristics. One of these characteristics is their relatively large size.

Why *are* most animals larger than bacteria, protists, and fungi? Many biologists think the answer is related to *ecological niches,* which are the functional roles of a species within a community. By the time animals evolved, bacteria, protists, and fungi already occupied most available ecological niches. For new species to succeed, they had to displace other organisms from a niche or adapt to a new one. Success in a new niche required a new body plan, and new body plans often involved larger size. Large size gave more opportunity for capturing food. Predators are typically larger than their prey.

To grow larger than their bacterial and protist competitors, animals had to be multicellular. Recall that the size of a single cell is limited by the ratio of its surface area (plasma membrane) to its volume (see Chapter 4). The plasma membrane needs to be large enough relative to the cell's volume to permit passage of materials into and out of the cell, so that the conditions necessary for life can be maintained. In a multicellular animal, each cell has a large enough surface-to-volume ratio to effectively regulate its internal environment. Individual cells live and die and are replaced while the organism continues to maintain itself and thrive. The number of cells, not their individual sizes, is usually responsible for the size of an animal.

In unicellular organisms, such as bacteria and many protists, the single cell carries on all the activities necessary for life. Recall that unicellular and small, flat organisms depend on diffusion for many life processes, including gas exchange and disposal of metabolic wastes. One reason they can be small is that they do not have complex organ systems. They move about without muscles, coordinate their activities without a nervous system, and process food without a digestive system.

Consider how different the caiman and the butterfly are, not only in size but also in body form and lifestyle (see photograph). In a multicellular organism, cells specialize to perform specific tasks. A **tissue** consists of a group of closely associated, similar cells that carry out specific functions.

Tissues associate to form **organs** such as the heart or stomach. Groups of tissues and organs form the **organ systems** of a complex **organism.** In a large, complex animal, billions of cells may be organized to form tissues, organs, and organ systems.

Cells, tissues, organs, and organ systems work together to maintain appropriate conditions in the body. The tendency to maintain a relatively constant internal environment is called **homeostasis,** and the processes that accomplish the task are **homeostatic mechanisms.** For example, in mammals, nervous, endocrine, and circulatory systems work together to regulate body temperature.

In this chapter we focus on the types and functions of tissues characteristic of animals. We then introduce the organ systems of a complex animal and discuss the important concept of homeostasis, using regulation of body temperature as an example. In the remaining chapters of this unit, we discuss how animals carry out life processes and how the organ systems of a complex animal work together to maintain homeostasis of the organism as a whole. ■

TISSUES

Learning Objectives

1. Compare the structure and function of the four main kinds of animal tissues: epithelial, connective, muscle, and nervous tissues.
2. Describe the main types of epithelial tissue, and state their functions.
3. Compare the main types of connective tissue, and summarize their functions.
4. Contrast the three types of muscle tissue and their functions.
5. Relate the structure of the neuron to its function.

Biologists classify animal tissues as epithelial, connective, muscle, or nervous tissue. Classification of tissues depends on their structure and origin. Each kind of tissue (and subtissue) is composed of cells with characteristic sizes, shapes, and arrangements, and each type of tissue is specialized to perform a specific function or group of functions. For example, some tissues are specialized to transport materials, whereas others contract, enabling the animal to move. Still others secrete hormones that regulate metabolic processes. As we discuss each tissue type, notice the relationship between its form and its function.

Epithelial tissues cover the body and line its cavities

Epithelial tissue (also called **epithelium**) consists of cells fitted tightly together to form a continuous layer, or sheet, of cells. One surface of the sheet is typically exposed because it covers the body (outer layer of the skin) or lines a cavity, such as the lumen (cavity) of the intestine. The other surface of an epithelial layer attaches to the underlying tissue by a noncellular **basement membrane** consisting of tiny fibers and nonliving polysaccharide material that the epithelial cells produce.

Epithelial tissue forms the outer layer of the skin and the linings of the digestive, respiratory, excretory, and reproductive tracts. As a result, everything that enters or leaves the body must cross at least one layer of epithelium. Food taken into the mouth and swallowed is not really "inside" the body until it is absorbed through the epithelium of the gut and enters the blood. To a large extent, the permeabilities of the various epithelial tissues regulate the exchange of substances between the different parts of the body, as well as between the animal and the external environment.

Epithelial tissues perform many functions, including protection, absorption, secretion, and sensation. The epithelial layer of the skin, the **epidermis,** covers the entire body and protects it from mechanical injury, chemicals, bacteria, and fluid loss. The epithelial tissue lining the digestive tract absorbs nutrients and water into the body. Some epithelial cells form **glands** that secrete cell products such as hormones, enzymes, or sweat. Other epithelial cells are sensory receptors that receive information from the environment. For example, epithelial cells in taste buds and in the nose specialize as chemical receptors.

Table 37-1 illustrates the main types of epithelial tissue, indicates their locations in the body, and describes their functions. You can distinguish three types of epithelial cells on the basis of shape. **Squamous** epithelial cells are thin, flat cells shaped like flagstones. **Cuboidal** epithelial cells are short cylinders that from the side appear cube-shaped, like dice. Actually, each cuboidal cell is typically hexagonal in cross section, making it an eight-sided polyhedron. **Columnar** epithelial cells look like columns or cylinders when viewed from the side. The nucleus is usually located near the base of the cell. Viewed from above or in cross section, these cells often appear hexagonal. On its free surface, a columnar epithelial cell may have cilia that beat in a coordinated way, moving materials over the tissue surface. Most of the upper respiratory tract is lined with ciliated columnar epithelium that moves particles of dust and other foreign material away from the lungs.

Epithelial tissue is also classified by the number of layers. **Simple epithelium** is composed of one layer of cells. It is usually located where substances are secreted, excreted, or absorbed, or where materials diffuse between compartments. For example, simple squamous epithelium lines the air sacs in the lungs. The structure of this thin tissue is wonderfully suited to permit diffusion of gases in and out of air sacs.

Stratified epithelium, which has two or more layers, is found where protection is required. Stratified squamous epithelium, which makes up the outer layer of your skin, continuously regenerates as it is sloughed off during normal wear and tear.

The cells of **pseudostratified epithelium** falsely appear layered. Although all its cells rest on a basement membrane, not every cell extends to the exposed surface of the tissue. This arrangement gives the impression of two or more cell layers. Some of the respiratory passageways are lined with pseudostratified epithelium equipped with cilia.

The cells lining blood and lymph vessels, called **endothelium,** have a different embryonic origin from "true" epithelium. However, these cells are structurally similar to squamous epithelial cells and can be included in that category.

A gland consists of one or more epithelial cells specialized to produce and secrete a product such as sweat, milk, mucus, wax, saliva, hormones, or enzymes (Fig. 37-1). Epithelial tissue lining the cavities and passageways of the body typically has some specialized mucus-secreting cells called **goblet cells.** The mucus lubricates these surfaces, offers protection, and facilitates the movement of materials.

Glands are classified as exocrine or endocrine. **Exocrine glands,** like goblet cells and sweat glands, secrete their products onto a free epithelial surface, typically through a duct (tube). **Endocrine glands** lack ducts. These glands release their products, called **hormones,** into the **interstitial fluid** (tissue fluid) or blood; hormones are typically transported by the circulatory system. (Endocrine glands are discussed in Chapter 47.)

An **epithelial membrane** consists of a sheet of epithelial tissue and a layer of underlying connective tissue. Types of epithelial membranes include mucous membranes and serous membranes. A **mucous membrane,** or mucosa, lines a body cavity that opens to the outside of the body, such as the digestive or respiratory tract. The epithelial layer secretes mucus that lubricates the tissue and protects it from drying.

A **serous membrane** lines a body cavity that does not open to the outside of the body. It consists of simple squamous epithelium over a thin layer of loose connective tissue. This type of membrane secretes fluid into the cavity it lines. Examples of serous membranes are the pleural membranes lining the pleural cavities around the lungs and the pericardial membranes lining the pericardial cavity around the heart.

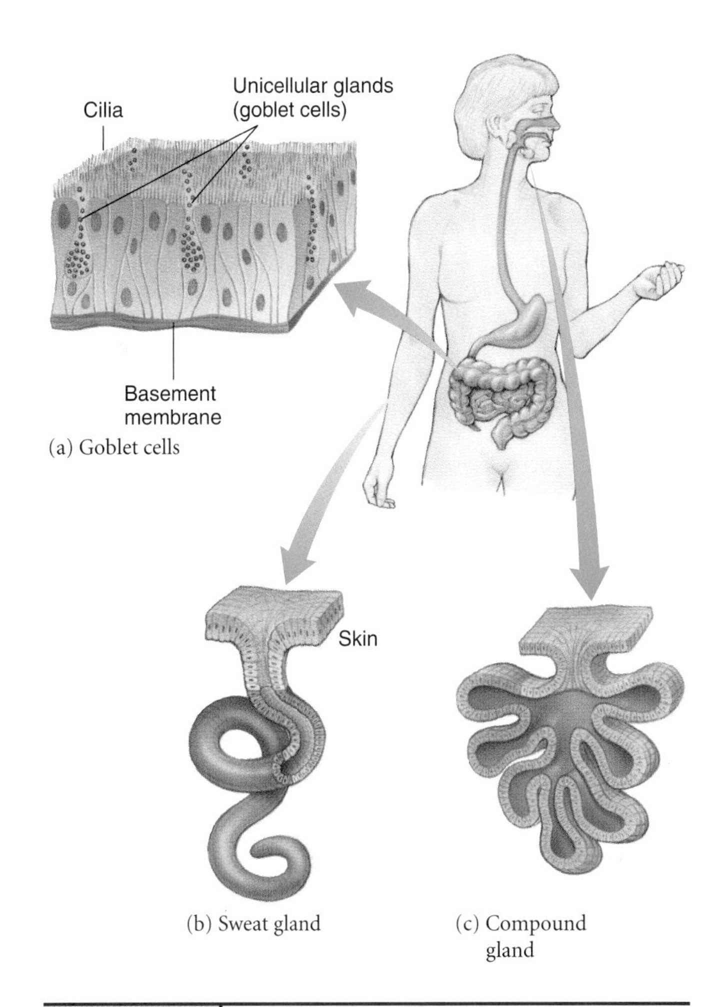

FIGURE **37-1** **Glands.**

A gland consists of one or more epithelial cells. **(a)** Goblet cells are unicellular glands that secrete mucus. **(b)** Sweat glands are simple glands consisting of coiled tubes. Their walls are constructed of simple cuboidal epithelium. **(c)** Compound glands, such as the parotid salivary glands, have branched ducts.

Connective tissues support other body structures

Almost every organ in the body has a framework of **connective tissue** that supports and cushions it. Typically, connective tissues contain relatively few cells. Its cells are embedded in an extensive **intercellular substance** consisting of threadlike, microscopic **fibers** scattered throughout a **matrix,** a thin gel of polysaccharides that the cells secrete. The nature and function of each kind of connective tissue are determined in part by the structure and properties of the intercellular substance.

Connective tissue typically contains three types of fibers: collagen, elastic, and reticular. **Collagen fibers,** the most numerous type, are made of **collagens,** a group of fibrous proteins found in all animals (see Fig. 3-22b). Collagens are the most abundant proteins in mammals, accounting for about 25% of their total protein mass. Collagen is very tough (meat is tough because of its collagen content). The tensile strength (ability to stretch without tearing) of collagen fibers is comparable to that of steel. Collagen fibers are wavy and flexible, allowing them to remain intact when tissue is stretched.

Elastic fibers branch and fuse to form networks. They can be stretched by a force and then (like a stretched rubber band) return to their original size and shape when the force is removed. Elastic fibers, composed of the protein elastin, are an important component of structures that must stretch.

Reticular fibers are very thin, branched fibers that form delicate networks joining connective tissues to neighboring tissues. Reticular fibers consist of collagen and some glycoprotein.

The cells of different kinds of connective tissues differ in their shapes and structures and in the kinds of fibers and matrices they secrete. **Fibroblasts** are connective tissue cells that produce the fibers, as well as the protein and carbohydrate complexes, of the matrix. Fibroblasts release protein components that become arranged to form the characteristic fibers. These cells are especially active in developing tissue and are important in

TABLE 37-1 Epithelial Tissues

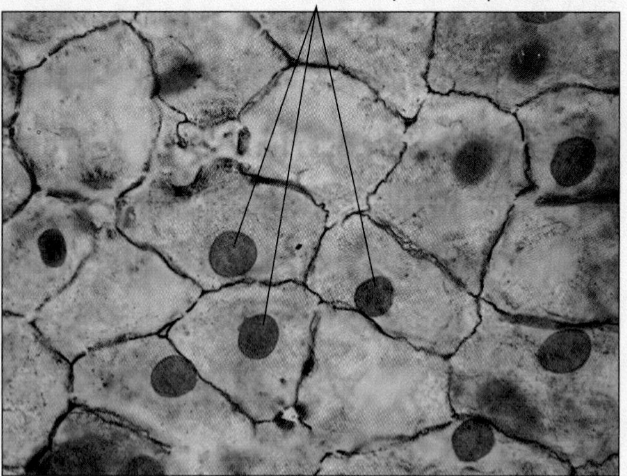

LM of simple squamous epithelium.

Simple Squamous Epithelium

Main Locations
Air sacs of lungs; lining of blood vessels

Functions
Passage of materials where little or no protection is needed and where diffusion is major form of transport

Description and Comments
Cells are flat and arranged as single layer

25 μm

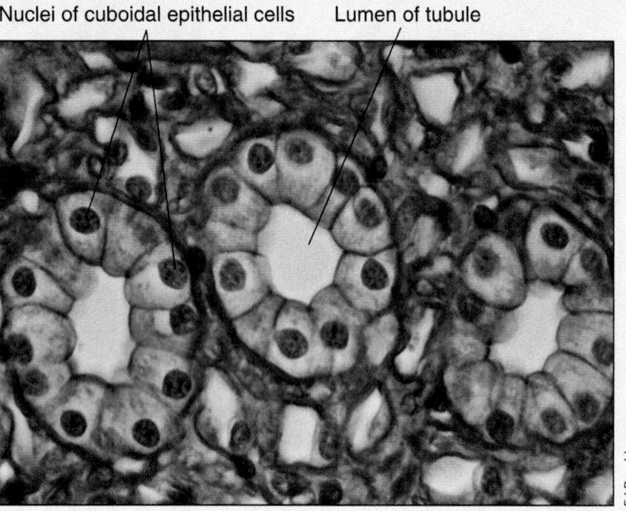

LM of simple cuboidal epithelium.

Simple Cuboidal Epithelium

Main Locations
Linings of kidney tubules; gland ducts

Functions
Secretion and absorption

Description and Comments
Single layer of cells; LM shows cross section through tubules; from the side each cell looks like a short cylinder; some have microvilli for absorption

25 μm

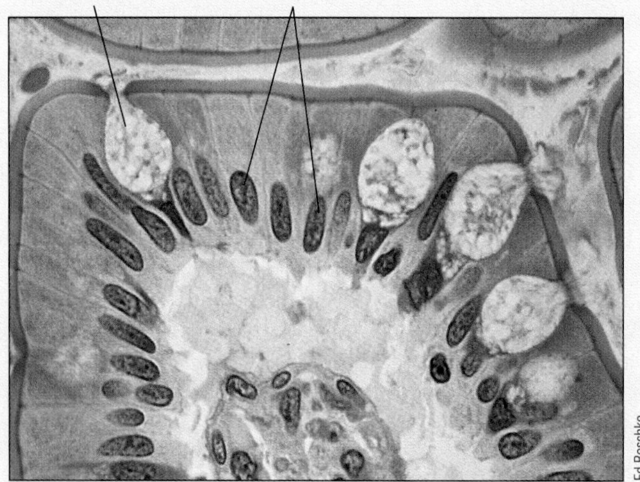

LM of simple columnar epithelium.

Simple Columnar Epithelium

Main Locations
Linings of much of digestive tract and upper part of respiratory tract

Functions
Secretion, especially of mucus; absorption; protection; movement of layer of mucus

Description and Comments
Single layer of columnar cells; sometimes with enclosed secretory vesicles (in goblet cells); highly developed Golgi complex; often ciliated

25 μm

TABLE 37-1 Epithelial Tissues (continued)

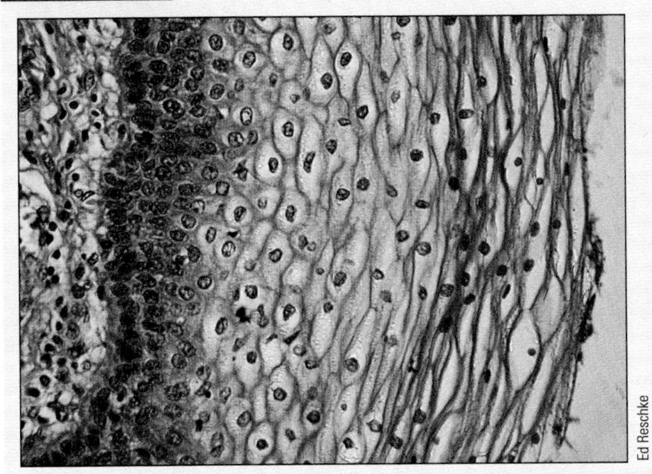

LM of stratified squamous epithelium.

Stratified Squamous Epithelium

Main Locations
Skin; mouth lining; vaginal lining

Functions
Protection only; little or no absorption or transit of materials; outer layer continuously sloughed off and replaced from below

Description and Comments
Several layers of cells, with only the lower ones columnar and metabolically active; division of lower cells causes older ones to be pushed upward toward surface, becoming flatter as they move.

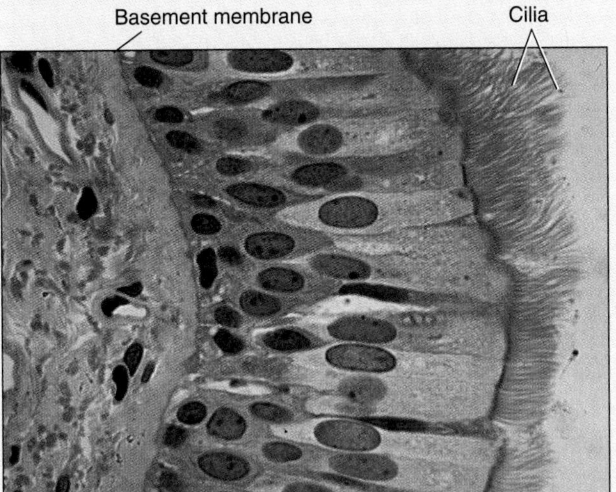

LM of pseudostratified columnar epithelium, ciliated.

Pseudostratified Epithelium

Main Locations
Some respiratory passages; ducts of many glands

Functions
Secretion; protection; movement of mucus

Description and Comments
Ciliated, mucus-secreting, or with microvilli; comparable in many ways to columnar epithelium except that not all cells are the same height. Thus, though all cells contact the same basement membrane, the tissue appears stratified

healing wounds. As tissues mature, the number of fibroblasts decreases and they become less active. **Macrophages,** the body's scavenger cells, commonly wander through connective tissues, cleaning up cell debris and phagocytizing foreign matter, including bacteria.

Some of the main types of connective tissue are (1) loose and dense connective tissues; (2) elastic connective tissue; (3) reticular connective tissue; (4) adipose tissue; (5) cartilage; (6) bone; and (7) blood, lymph, and tissues that produce blood cells. These tissues vary widely in their structural details and in the functions they perform (Table 37-2).

Loose connective tissue is the most widely distributed connective tissue in the vertebrate body. This thin filling between body parts serves as a reservoir for fluid and salts. Nerves, blood vessels, and muscles are wrapped in this tissue. Together with adipose tissue, loose connective tissue forms the subcutaneous (below the skin) layer that attaches skin to the muscles and other structures beneath. Loose connective tissue consists of fibers running in all directions through a semifluid matrix. Its flexibility permits the parts it connects to move.

Dense connective tissue, found in the dermis (lower layer) of the skin, is very strong, though somewhat less flexible than loose connective tissue. Collagen fibers predominate. **Tendons,** the cords that connect muscles to bones, and **ligaments,** the cables that connect bones to one another, consist of dense connective tissue in which collagen bundles are arranged in a definite pattern.

Elastic connective tissue consists mainly of bundles of parallel elastic fibers. This tissue is found in structures that must expand and then return to their original size, such as lung tissue and the walls of large arteries.

Reticular connective tissue is composed mainly of interlacing reticular fibers. It forms a supporting internal framework in many organs, including the liver, spleen, and lymph nodes.

The cells of **adipose tissue** store fat and release it when fuel is needed for cellular respiration. Adipose tissue is found in the subcutaneous layer and in tissue that cushions internal organs.

TABLE 37-2 Connective Tissues

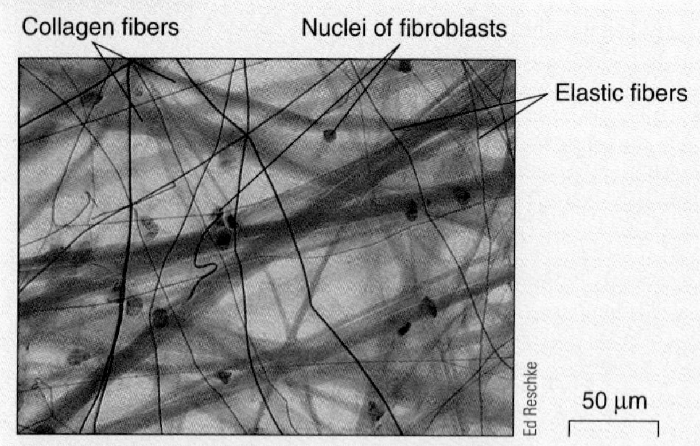

LM of loose connective tissue.

Loose Connective Tissue

Main Locations
Everywhere that support must be combined with elasticity, such as subcutaneous tissue (the layer of tissue beneath the dermis of the skin)

Functions
Support; reservoir for fluid and salts

Description and Comments
Fibers produced by fibroblast cells embedded in semifluid matrix and mixed with miscellaneous other cells

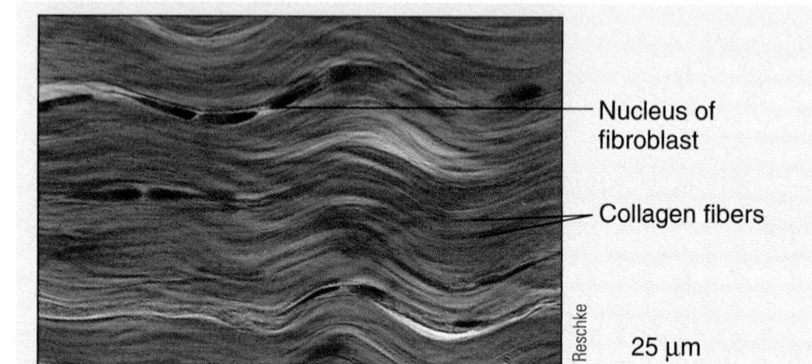

LM of dense connective tissue.

Dense Connective Tissue

Main Locations
Tendons; many ligaments; dermis of skin

Functions
Support; transmission of mechanical forces

Description and Comments
Collagen fibers may be regularly or irregularly arranged

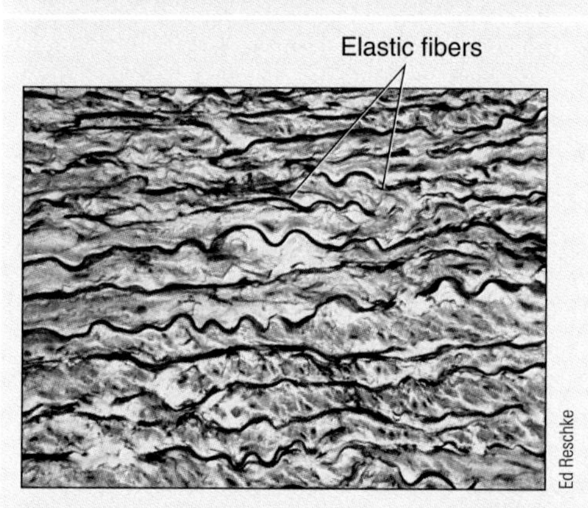

LM of elastic connective tissue.

Elastic Connective Tissue

Main Locations
Structures that must both expand and return to their original size, such as lung tissue and large arteries

Function
Confers elasticity

Description and Comments
Branching elastic fibers interspersed with fibroblasts

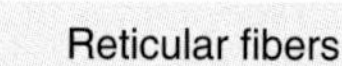

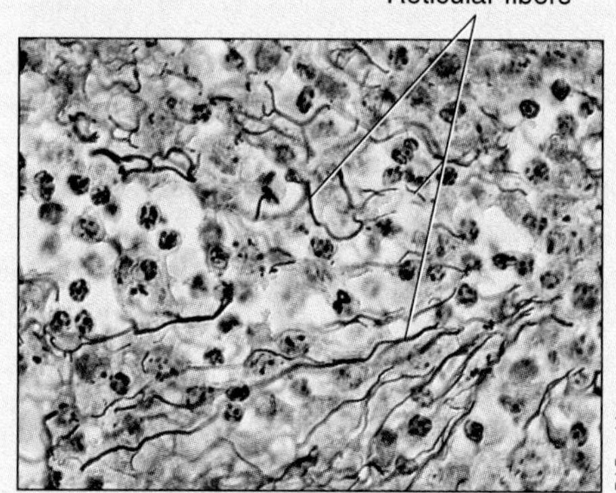

LM of reticular connective tissue.

Reticular Connective Tissue

Main Locations
Framework of liver; lymph nodes; spleen

Function
Support

Description and Comments
Consists of cells dispersed in plasma (a fluid interlacing fiber)

TABLE 37-2 Connective Tissues (continued)

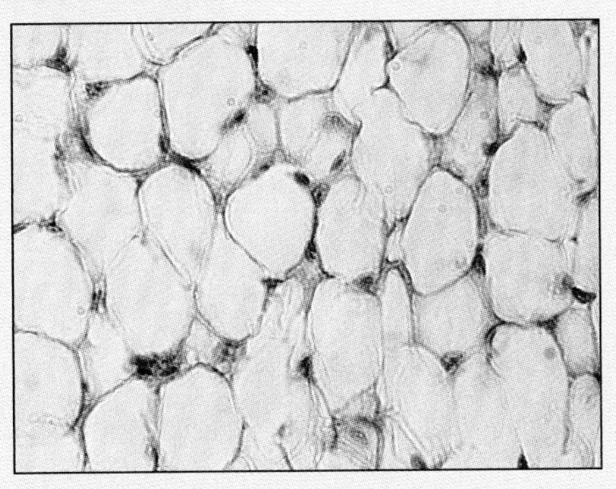

LM of adipose tissue.

Adipose Tissue

Main Locations
Subcutaneous layer; pads around certain internal organs

Functions
Food storage; insulation; support of such organs as mammary glands, kidneys

Description and Comments
Fat cells are star-shaped at first; fat droplets accumulate until typical ring-shaped cells are produced

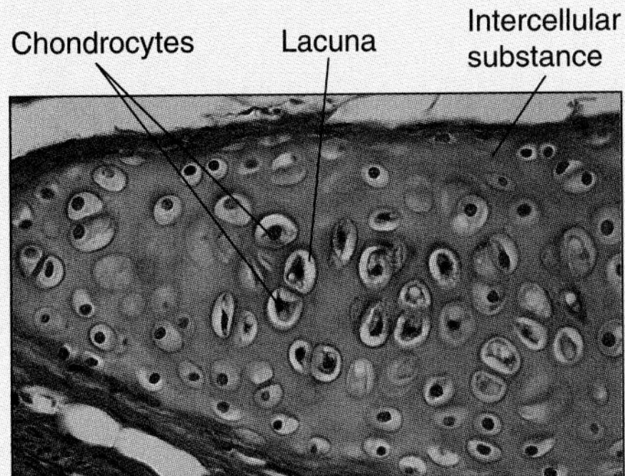

LM of cartilage.

Cartilage

Main Locations
Supporting skeletons in sharks and rays; ends of bones in mammals and some other vertebrates; supporting rings in walls of some respiratory tubes; tip of nose; external ear

Function
Flexible support

Description and Comments
Cells (chondrocytes) separated from one another by intercellular substance; cells occupy lacunae

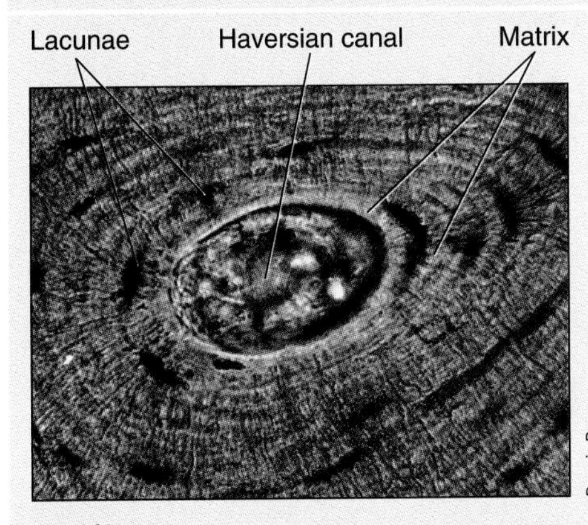

LM of bone.

Bone

Main Locations
Forms skeletal structure in most vertebrates

Functions
Support and protection of internal organs; calcium reservoir; skeletal muscles attach to bones

Description and Comments
Osteocytes in lacunae; in compact bone, lacunae arranged in concentric circles surrounding Haversian canals

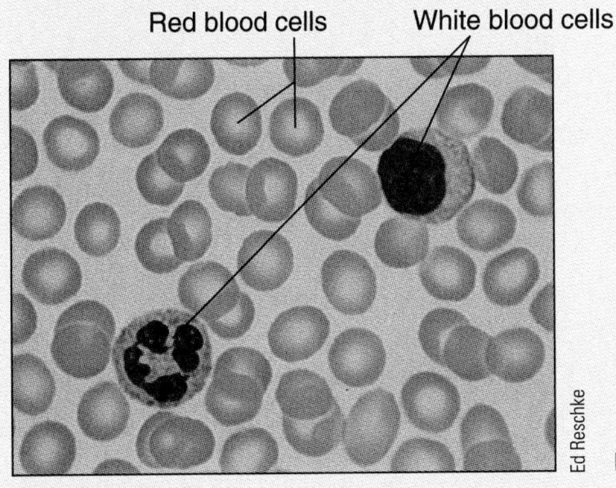

LM of blood.

Blood

Main Locations
Within heart and blood vessels of circulatory system

Functions
Transports oxygen, nutrients, wastes, and other materials

Description and Comments
Consists of cells dispersed in fluid intercellular substance

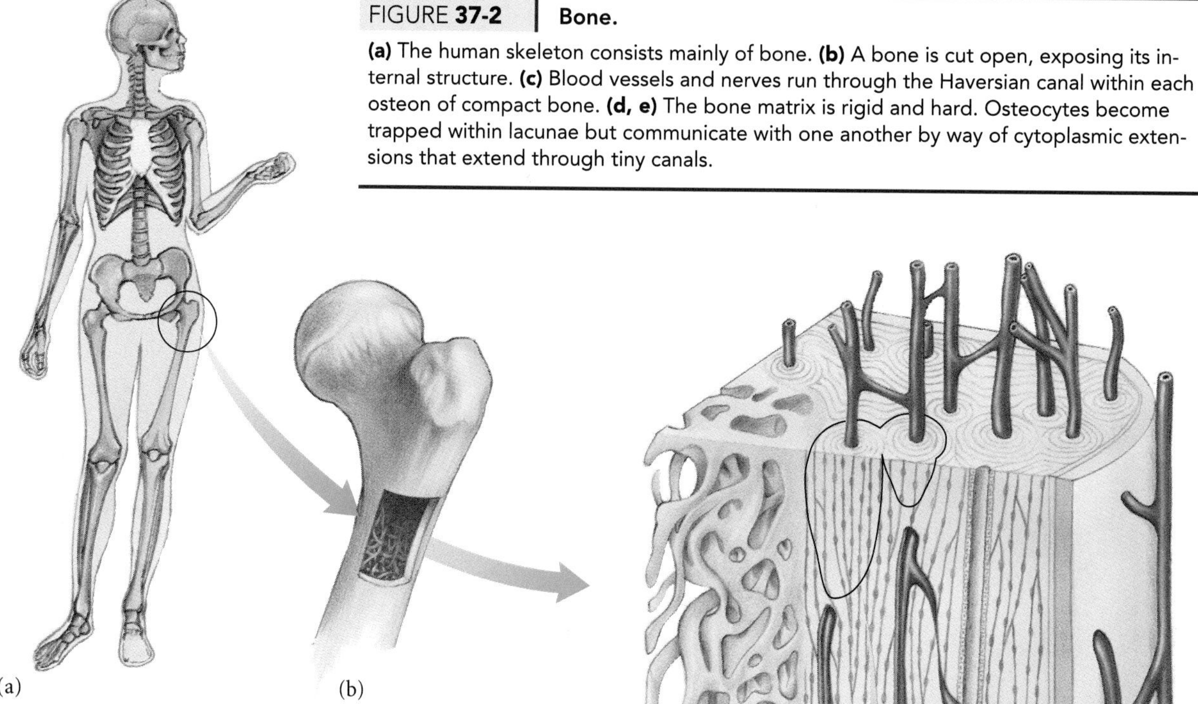

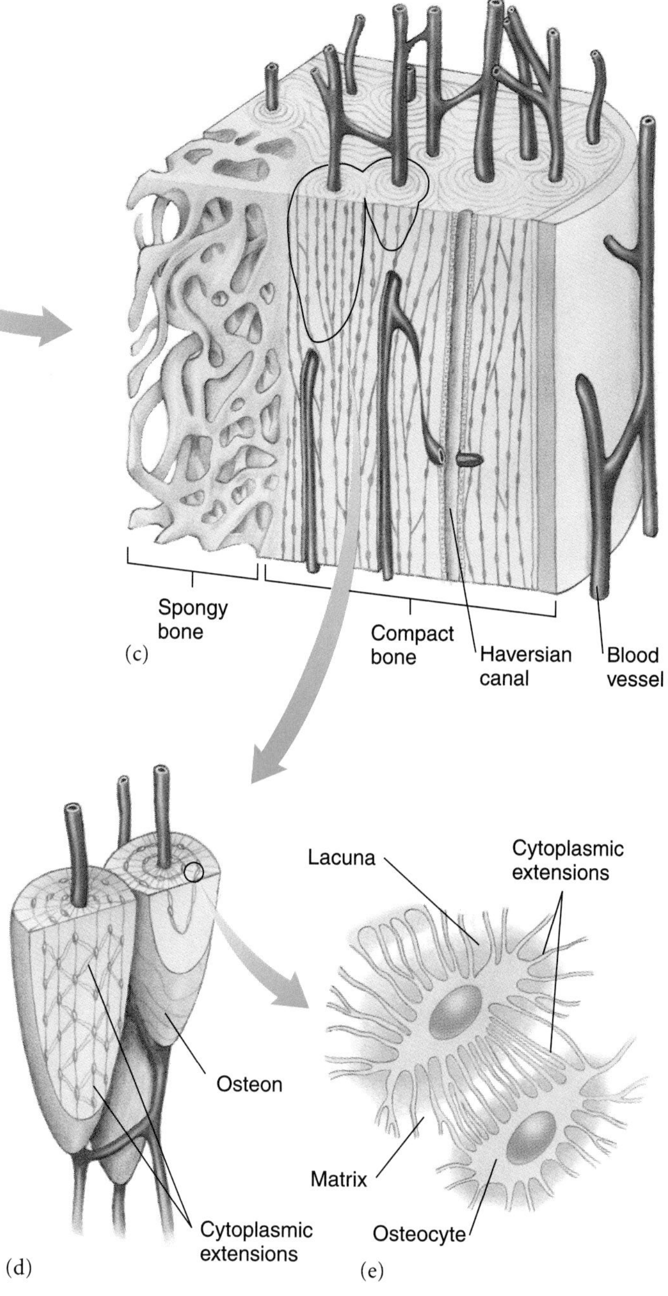

FIGURE **37-2** **Bone.**

(a) The human skeleton consists mainly of bone. **(b)** A bone is cut open, exposing its internal structure. **(c)** Blood vessels and nerves run through the Haversian canal within each osteon of compact bone. **(d, e)** The bone matrix is rigid and hard. Osteocytes become trapped within lacunae but communicate with one another by way of cytoplasmic extensions that extend through tiny canals.

The supporting skeleton of a vertebrate is made of cartilage, or of both cartilage and bone. Recall that **cartilage** is the supporting skeleton in the embryonic stages of all vertebrates. In most vertebrates, bone replaces cartilage during development. However, cartilage remains in some supporting structures. For example, in humans cartilage is found in the external ear, the supporting rings in the walls of the respiratory passageways, the tip of the nose, the ends of some bones, and the discs that serve as cushions between the vertebrae.

Cartilage is firm yet elastic. Its cells, called **chondrocytes,** secrete a hard, rubbery matrix around themselves and also secrete collagen fibers, which become embedded in the matrix and strengthen it. Chondrocytes eventually come to lie, singly or in groups of two or four, in small cavities in the matrix called **lacunae.** These cells remain alive and are nourished by nutrients and oxygen that diffuse through the matrix. Cartilage tissue lacks nerves, lymph vessels, and blood vessels.

Bone, the main vertebrate skeletal tissue, is like cartilage in that it consists mostly of matrix material containing lacunae. The bone cells, called **osteocytes,** secrete and maintain the matrix (Fig. 37-2). Unlike cartilage, however, bone is a highly vascular tissue, with a substantial blood supply. Osteocytes communicate with one another and with capillaries by tiny channels (canaliculi) that contain long cytoplasmic extensions of the osteocytes.

A typical bone has an outer layer of *compact bone* surrounding a filling of *spongy bone.* Compact bone consists of spindle-shaped units called **osteons.** Within each osteon, osteocytes are arranged in concentric layers called *lamellae,* which are formed by the matrix. In turn, the lamellae surround central microscopic channels known as **Haversian canals,** through which capillaries and nerves pass.

Bones are amazingly light and strong. Calcium salts of bone render the matrix very hard, and collagen prevents the bony matrix from being overly brittle. Most bones have a large central **marrow cavity** that contains a spongy tissue called *marrow.* Yellow marrow consists mainly of fat. Red marrow is the connective tissue in which blood cells are produced. Chapter 38 discusses bone in more detail.

Blood and **lymph** are circulating tissues that help other parts of the body communicate and interact. Like other connective tissues, they consist of specialized cells dispersed in an intercellular substance.

TABLE 37-3 Muscle Tissues

	Skeletal	Cardiac	Smooth
Location	Attached to skeleton	Walls of heart	Walls of stomach, intestines, etc.
Type of Control	Voluntary	Involuntary	Involuntary
Shape of Fibers	Elongated, cylindrical, blunt ends	Elongated, cylindrical, fibers that branch and fuse	Elongated, spindle-shaped, pointed ends
Striations	Present	Present	Absent
Number of Nuclei per Fiber	Many	One or two	One
Position of Nuclei	Peripheral	Central	Central
Speed of Contraction	Most rapid	Intermediate (varies)	Slowest
Resistance to Fatigue (with repetitive contraction)	Least	Intermediate	Greatest

Nuclei Striations

Skeletal muscle fibers

Nuclei Intercalated discs

Cardiac muscle fibers

Nuclei

Smooth muscle fibers

In mammals, blood consists of **red blood cells, white blood cells,** and **platelets** all suspended within **plasma,** the liquid, noncellular part of the blood. In humans and other vertebrates, red blood cells contain the respiratory pigment which transports oxygen. White blood cells defend the body against disease-causing microorganisms (see Chapter 43). Platelets, small fragments broken off from large cells in the bone marrow, play a key role in blood clotting. Plasma consists of water, proteins, salts, and a variety of soluble chemical messengers such as hormones that it transports from one part of the body to another. We discuss blood in Chapter 42.

Muscle tissue is specialized to contract

Most animals move by contracting the long, cylindrical or spindle-shaped cells of **muscle tissue.** Each muscle cell is called a **muscle fiber,** because of its length. A muscle fiber contains many thin, longitudinal, parallel contractile units called **myofibrils.** Two proteins, **myosin** and **actin,** are the chief components of myofibrils and play a key role in contraction of muscle fibers.

Some invertebrates have skeletal and smooth muscle. Vertebrates have three types of muscle tissue: skeletal, cardiac, and smooth (Table 37-3). **Skeletal muscle** makes up the large muscle masses attached to the bones of the body. Skeletal muscle fibers are very long and each fiber has many nuclei. The nuclei of skeletal muscle fibers are also unusual in their position. They lie just under the plasma membrane, which frees the entire central part of the skeletal muscle fiber for the myofibrils. This adaptation appears to increase the efficiency of contraction. Whereas skeletal muscle fibers are generally under voluntary control, you do not normally contract your cardiac and smooth muscle fibers at will.

Light microscopy shows that both skeletal and cardiac fibers have alternating light and dark transverse stripes, or **striations,** that change their relative sizes during contraction. Striated muscle fibers contract rapidly but cannot remain contracted for a long period. They must relax and rest momentarily before contracting again.

Cardiac muscle is the main tissue of the heart. The fibers of cardiac muscle join end-to-end, and they branch and rejoin, forming complex networks. One or two nuclei lie within each fiber. A characteristic feature of cardiac muscle tissue is the presence of *intercalated discs,* specialized junctions where the fibers join.

Smooth muscle occurs in the walls of the digestive tract, uterus, blood vessels, and many other internal organs. Each spindle-shaped fiber contains a single, central nucleus. (Muscle contraction is discussed in Chapter 38.)

Nervous tissue controls muscles and glands

Nervous tissue consists of neurons and glial cells. **Neurons** are specialized for receiving and transmitting signals. **Glial cells** support and nourish the neurons (Fig. 37-3).

A typical neuron has a **cell body** containing the nucleus, and two types of cytoplasmic extensions (see Chapter 39). **Dendrites** are cytoplasmic extensions specialized for receiving signals and transmitting them to the cell body. The single **axon** transmits signals, called *nerve impulses,* away from the cell body. Axons range in length from 1 or 2 mm to more than a meter. Those extending from the spinal cord down the arm or leg in a human, for example, may be a meter or more in length.

Certain neurons receive signals from the external or internal environment and transmit them to the spinal cord and brain. Other neurons relay, process, or store information. Still others transmit signals from the brain and spinal cord to the muscles and glands. Neurons communicate at junctions called **synapses.**

A neoplasm ("new growth"), or **tumor,** is an abnormal mass of cells. A neoplasm may be benign or malignant (cancerous). A **benign** ("kind") tumor tends to grow slowly, and its cells stay together. Because benign tumors form masses with distinct borders, they can usually be removed surgically.

A **malignant** ("wicked") **neoplasm,** or **cancer,** usually grows much more rapidly and invasively than a benign tumor. In fact, two basic defects in behavior that characterize most cancer cells are rapid multiplication and abnormal relations with neighboring cells. Unlike normal cells, which respect one another's boundaries and form tissues in an orderly, organized manner, cancer cells grow helter-skelter on one another and infiltrate normal tissues. They apparently no longer receive or respond appropriately to signals from surrounding cells; communication is lacking (see figure).

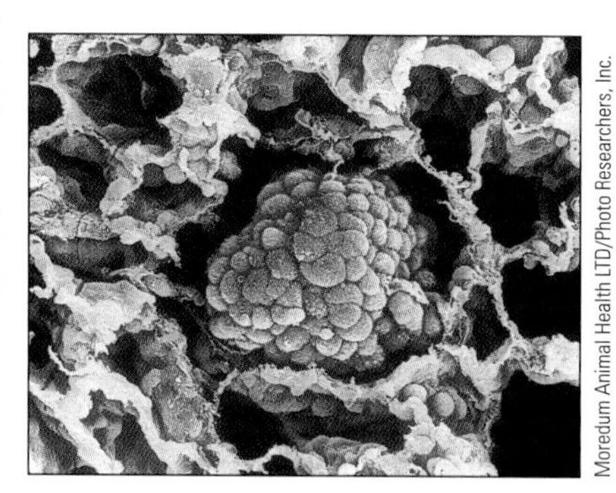

Moredum Animal Health LTD/Photo Researchers, Inc.

Lung cancer. Cancer cells in this color-enhanced SEM of human lung tissue multiply rapidly and invade normal tissues, interfering with normal function. A mass of malignant cells (pink) occupies the air sac in the center, and cancer cells that have separated from the main tumor can be seen in other air sacs. Microvilli on the surface of the cancer cells give them a fuzzy appearance.

Earlier in this book (Chapter 16), you learned that cancer results from abnormal expression of specific genes critical for cell division. When a cancer cell multiplies, all the cells derived from it are also abnormal. Unlike the cells of benign tumors, cancer cells do not retain normal structural features. Cancers that develop from connective tissues or muscle are referred to as **sarcomas.** Those that originate in epithelial tissue are called **carcinomas.** Most human cancers are epithelial.

Death from cancer results from **metastasis,** a migration of cancer cells through blood or lymph channels to distant parts of the body. Once there, cancer cells multiply, forming new malignant neoplasms that interfere with the normal functions of the tissues being invaded. Cancer often spreads so rapidly and extensively that surgeons cannot locate or remove all the malignant masses.

Studies suggest that many neoplasms grow to several millimeters in diameter and then enter a dormant stage, which may last for months or even years. Some researchers hypothesize that eventually cells of the neoplasm release a chemical substance that stimulates nearby blood vessels to develop new capillaries that grow into the abnormal mass of cells. Nourished by its new blood supply, the neoplasm begins to grow rapidly. Newly formed blood vessels have leaky walls and so are an important route for metastasis. Malignant cells enter the blood through these walls and are transported to new sites.

Cancer is the second greatest cause of death in the United States. One in three people in the United States gets cancer at some time in his or her life. Currently, the key to survival is early diagnosis and treatment with some combination of surgery, hormonal treatment, radiation therapy, and drugs that suppress mitosis, such as chemotherapy. Cancer is actually a large family of closely related diseases (there are hundreds of distinct varieties), so treatment must be tailored to the particular type of cancer.

Alleles of some genes appear to affect an individual's level of tolerance to carcinogens (cancer-producing agents). More than 80% of cancer cases are thought to be triggered by carcinogens in the environment. You can decrease risk of developing cancer by following these recommendations:

1. Do not smoke or use tobacco. Smoking is responsible for more than 80% of lung cancer cases and it increases the risk for many other cancers.
2. Avoid prolonged exposure to the sun. When in the sun, use sunscreen or sun block. Exposure to the sun is responsible for almost all the 400,000 cases of skin cancer reported each year in the United States.
3. Eat a healthy diet, including fruit and vegetables. Although research results are not clear, it appears prudent to increase the fiber content of your diet and avoid high-fat, smoked, salt-cured, and nitrite-cured foods.
4. Avoid unnecessary exposure to x-rays.
5. Women should examine their breasts each month, have regular mammograms, and obtain annual Papanicolaou's (Pap) tests.
6. Men should regularly examine their testes and should have prostate examinations yearly after age 50. They should also have the prostate-specific antigen (PSA) blood test. These routine tests detect cancer at an early, more treatable stage.
7. Beginning at age 50, both men and women should be screened for colorectal cancer. Detection and removal of polyps can prevent cancer.

A **nerve** consists of a great many neurons bound together by connective tissue.

In this chapter, we have focused on normal tissues. For a discussion of some abnormal tissues, see *Focus On: Unwelcome Tissues: Cancers.*

Review

- What are the main differences in structure and function between epithelial tissue and connective tissue?
- What type of tissue lines the air sacs of the lungs? How is its structure adapted to its function?

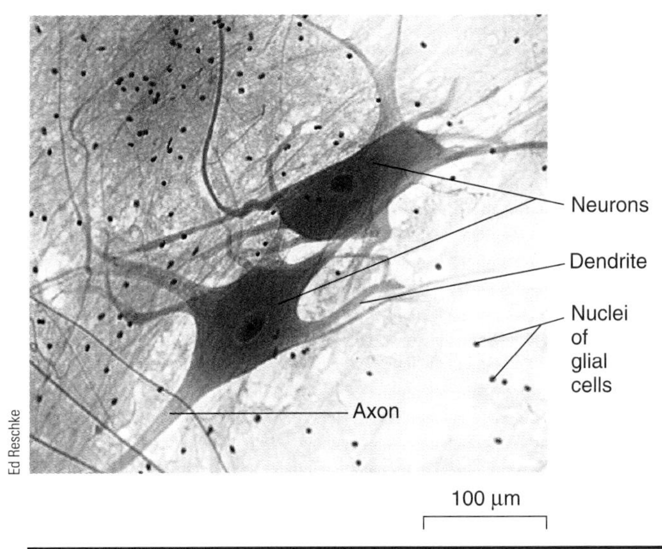

FIGURE **37-3** **LM of nervous tissue.**
Neurons transmit impulses. Glial cells support and nourish neurons.

- How is the structure of adipose tissue adapted to its function? Where is it found in the body?
- What are some differences between the three types of muscle tissue? How is the structure of each type adapted for its function?
- How is the neuron uniquely adapted for its function?

BiologyNow™ Assess your understanding of **tissues** by taking the pretest on your BiologyNow CD-ROM.

ORGANS AND ORGAN SYSTEMS

Learning Objectives

6 Briefly describe the organ systems of a complex animal.
7 Define *homeostasis,* and contrast negative and positive feedback mechanisms.
8 Summarize the homeostatic actions of each organ system.

Although an animal organ may be composed mainly of one type of tissue, other types are needed to support, protect, provide a blood supply, and transmit information. For example, the heart is mainly cardiac muscle tissue, but it is lined with endothelium and contains blood vessels made of endothelium, smooth muscle, and connective tissue. The heart also has nerves that transmit information and help regulate the rate and strength of its contractions.

An organized group of tissues and organs that together perform a specialized set of functions make up an organ system. We can identify 10 major organ systems that work together to make up a complex animal: **integumentary, skeletal, muscular, digestive, circulatory, respiratory, urinary, nervous, endocrine,** and **reproductive systems.** Table 37-4 and Figure 37-4 summarize their principal organs and functions.

The body maintains homeostasis

To survive and function, the organism must carefully regulate the composition of the fluids that bathe its cells. It must maintain the appropriate concentration of nutrients, oxygen, and other

TABLE **37-4** The Mammalian Organ Systems and Their Functions

System	Components	Functions	Homeostatic Ability
Integumentary	Skin, hair, nails, sweat glands	Covers and protects body	Sweat glands help control body temperature; as a barrier, the skin helps maintain a steady state
Skeletal	Bones, cartilage, ligaments	Supports and protects body; provides for movement and locomotion; stores calcium	Helps maintain constant calcium level in blood
Muscular	Skeletal muscle; cardiac muscle; smooth muscle	Moves parts of skeleton; provides locomotion; moves internal materials	Ensures vital functions requiring movement, e.g., cardiac muscle circulates the blood
Digestive	Mouth, esophagus, stomach, intestine, liver, pancreas, salivary glands	Ingests and digests food; absorbs nutrients into blood	Maintains adequate supplies of fuel molecules and building materials
Circulatory	Heart, blood vessels, blood; lymph and lymph structures (lymphatic system is a subsystem of the circulatory system)	Transports materials from one part of body to another; defends body against disease organisms	Transports oxygen, nutrients, hormones, wastes; maintains water and salt balance of tissues
Respiratory	Lungs, trachea, and other air passageways	Exchanges gases between blood and external environment	Maintains adequate blood oxygen content and helps regulate blood pH; removes carbon dioxide from the blood
Urinary	Kidney, bladder, and associated ducts	Excretes metabolic wastes; removes excessive substances from blood	Helps regulate volume and composition of blood and body fluids
Nervous	Nerves and sense organs; brain and spinal cord	Receives stimuli from external and internal environment; conducts impulses; integrates activities of other systems	Principal regulatory system
Endocrine	Ductless glands (e.g., pituitary, adrenal, thyroid) and tissues that secrete hormones	Regulates blood chemistry and many body functions	Regulates metabolic activities and blood levels of various substances
Reproductive	Testes, ovaries, and associated structures	Sexual reproduction	Maintains sexual characteristics

ACTIVE FIGURE **37-4** The ten principal organ systems of the human body.

BiologyNow™ Watch **body systems** work together by clicking on this figure on your BiologyNow CD-ROM.

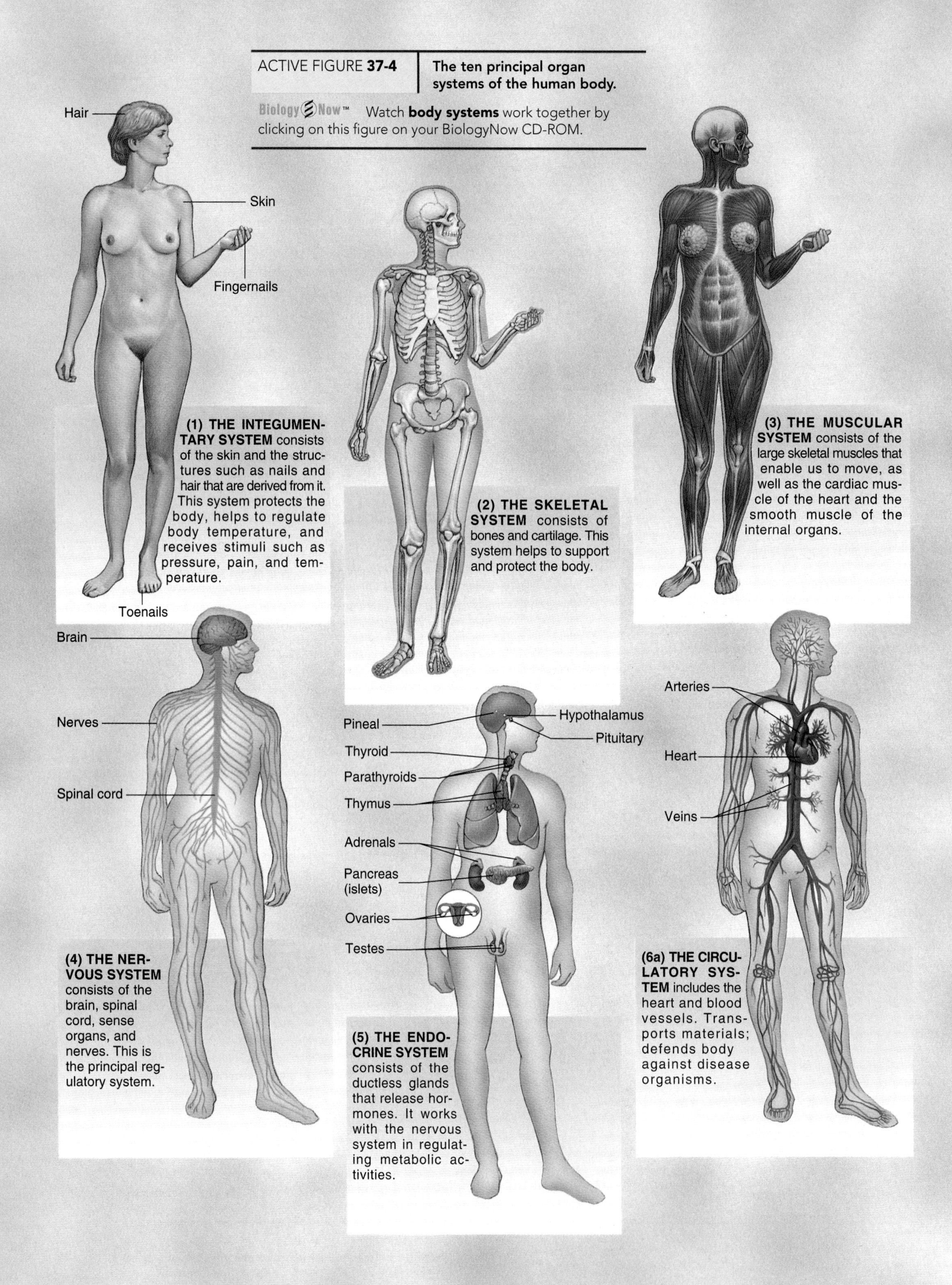

(1) THE INTEGUMENTARY SYSTEM consists of the skin and the structures such as nails and hair that are derived from it. This system protects the body, helps to regulate body temperature, and receives stimuli such as pressure, pain, and temperature.

(2) THE SKELETAL SYSTEM consists of bones and cartilage. This system helps to support and protect the body.

(3) THE MUSCULAR SYSTEM consists of the large skeletal muscles that enable us to move, as well as the cardiac muscle of the heart and the smooth muscle of the internal organs.

(4) THE NERVOUS SYSTEM consists of the brain, spinal cord, sense organs, and nerves. This is the principal regulatory system.

(5) THE ENDOCRINE SYSTEM consists of the ductless glands that release hormones. It works with the nervous system in regulating metabolic activities.

(6a) THE CIRCULATORY SYSTEM includes the heart and blood vessels. Transports materials; defends body against disease organisms.

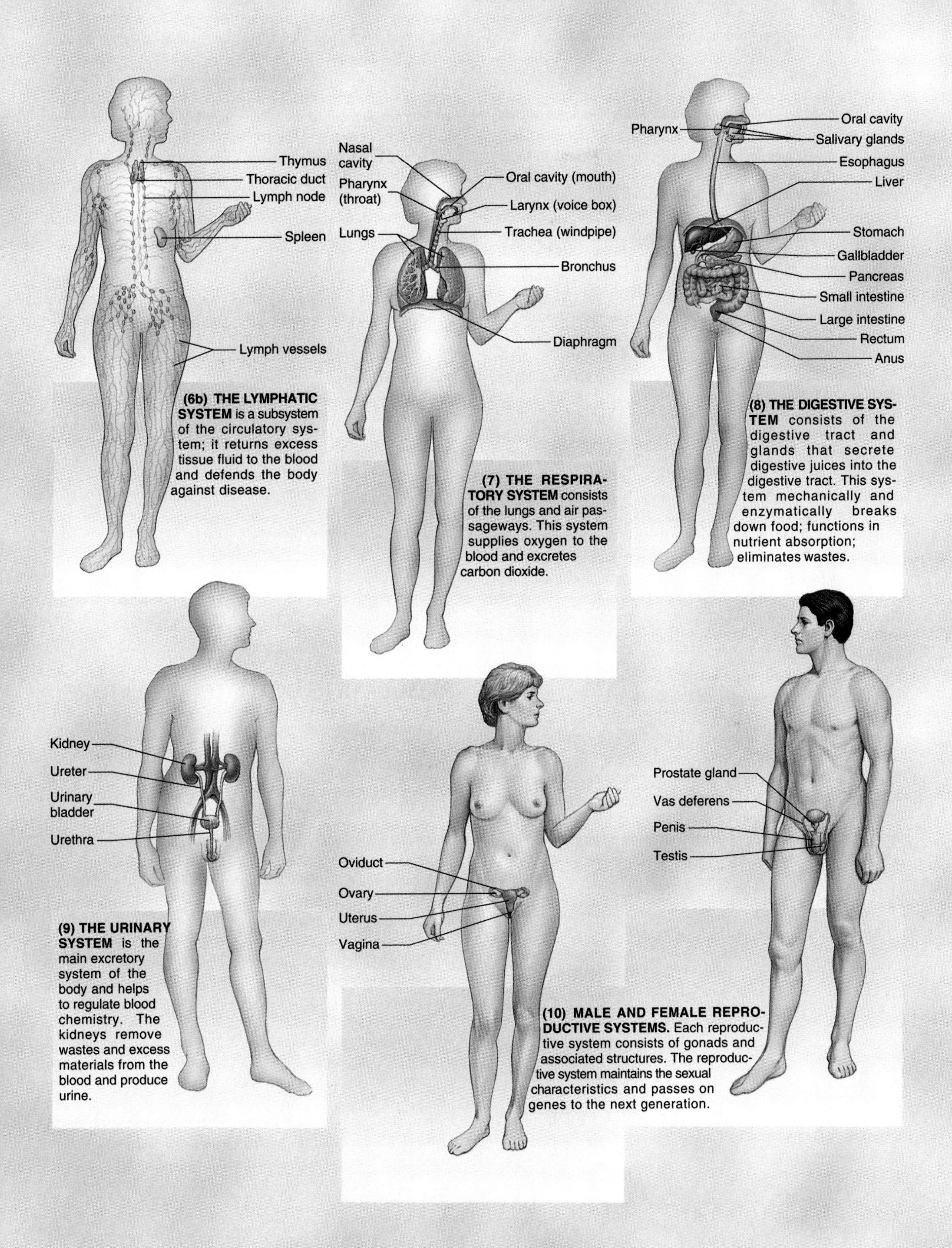
Thymus
Thoracic duct
Lymph node
Spleen
Lymph vessels
(6b) THE LYMPHATIC SYSTEM is a subsystem of the circulatory system; it returns excess tissue fluid to the blood and defends the body against disease.
Nasal cavity
Pharynx (throat)
Lungs
Oral cavity (mouth)
Larynx (voice box)
Trachea (windpipe)
Bronchus
Diaphragm
(7) THE RESPIRATORY SYSTEM consists of the lungs and air passageways. This system supplies oxygen to the blood and excretes carbon dioxide.
Pharynx
Oral cavity
Salivary glands
Esophagus
Liver
Stomach
Gallbladder
Pancreas
Small intestine
Large intestine
Rectum
Anus
(8) THE DIGESTIVE SYSTEM consists of the digestive tract and glands that secrete digestive juices into the digestive tract. This system mechanically and enzymatically breaks down food; functions in nutrient absorption; eliminates wastes.
Kidney
Ureter
Urinary bladder
Urethra
(9) THE URINARY SYSTEM is the main excretory system of the body and helps to regulate blood chemistry. The kidneys remove wastes and excess materials from the blood and produce urine.
Oviduct
Ovary
Uterus
Vagina
Prostate gland
Vas deferens
Penis
Testis
(10) MALE AND FEMALE REPRODUCTIVE SYSTEMS. Each reproductive system consists of gonads and associated structures. The reproductive system maintains the sexual characteristics and passes on genes to the next generation.

gases, ions, and compounds needed for metabolism at all times. In addition, it must maintain internal temperature and pressure within relatively narrow limits.

Cells, tissues, organs, and organ systems work together to maintain appropriate conditions in the body. The tendency to maintain a balanced internal environment is called **homeostasis,** and the control processes that maintain these conditions are **homeostatic mechanisms.** Homeostasis is a basic concept in physiology. First coined by U.S. physiologist Walter Cannon, the word *homeostasis* is derived from the Greek *homoios,* meaning "same," and *stasis,* "standing." Actually, the internal environment never really stays the same. It is a dynamic equilibrium in which conditions are maintained within narrow limits.

Stressors, changes in the internal or external environment that affect normal conditions within the body, continuously challenge homeostasis. An internal condition that moves out of its homeostatic range (either too high or too low) causes **stress.** An organism functions effectively because very precise homeostatic mechanisms interact continuously to manage stress.

How do homeostatic mechanisms work? Many are **feedback systems,** sometimes called *biofeedback systems.* Such a system consists of a cycle of events in which information about a change (such as a change in temperature) is fed back into the system so that the regulator (the temperature-regulating center in the brain) can control the process (temperature regulation). The desired condition (normal body temperature) is referred to as the **set point.** When body temperature becomes too high or too low, the change serves as input, triggering the regulator to counteract the change. The regulator activates mechanisms that bring the system back to the set point. The return to normal temperature signals the temperature-regulating center to "shut off" the mechanism.

In this type of feedback system, the response counteracts the inappropriate change, restoring the steady state. This is a **negative feedback mechanism,** because the response of the regulator is *opposite* (negative) to the output (Fig. 37-5). Most homeostatic mechanisms in the body are negative feedback systems. When some condition varies too far from the steady state (either too high or too low), a control system using negative feedback brings the condition back to the steady state.

A few **positive feedback mechanisms** also operate in the body. In these systems a deviation from the steady state sets off a series of changes that intensify (rather than reverse) the changes. Although many positive feedback mechanisms are beneficial, they do not maintain homeostasis. A positive feedback cycle operates during the birth of a baby. As the baby's head pushes against the opening of the uterus (cervix), a reflex action causes the uterus to contract. The contraction forces the head against the cervix again, resulting in another contraction, and the positive feedback cycle repeats again and again until the baby is born. Some positive feedback sequences, such as those that deepen circulatory shock following severe hemorrhage, can disrupt steady states and even lead to death.

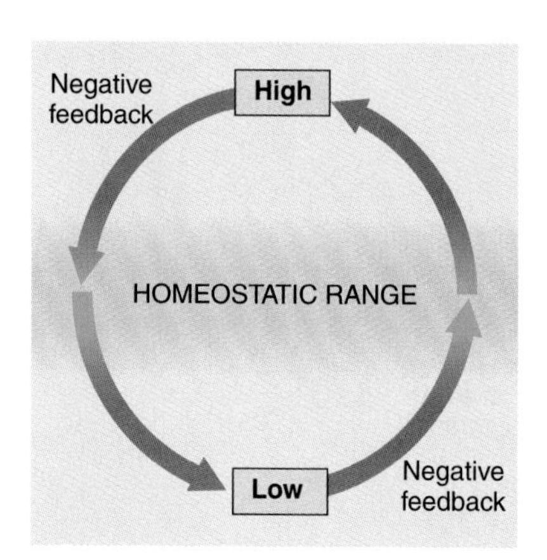

FIGURE **37-5**

Negative feedback mechanisms.

In negative feedback, the response of the regulator is opposite to the output. Regulatory mechanisms bring conditions back to a homeostatic range.

Homeostatic mechanisms maintain the internal environment within the physiological limits that support life. All organ systems participate in these regulatory processes, but the nervous and endocrine systems play major roles in controlling them. As you continue your study of animal processes, you will learn many ways in which organ systems interact to maintain the *steady state* (normal homeostatic state) of the organism. In the next section, we discuss some homeostatic mechanisms that help regulate body temperature.

Review

- What are the main functions of each organ system?
- What is the basic difference between negative and positive feedback mechanisms?
- How do each of the following organ systems help maintain homeostasis? (a) respiratory (b) urinary (c) endocrine? (You can consult Table 37-4 for help.)

BiologyNow™ Assess your understanding of **organs and organ systems** by taking the pretest on your BiologyNow CD-ROM.

REGULATING BODY TEMPERATURE

Learning Objectives

9 Describe advantages and disadvantages of ectothermy.

10 Describe advantages and disadvantages of endothermy, and describe strategies endotherms use to survive in extreme temperature conditions.

Many animals have elaborate homeostatic mechanisms for regulating body temperature. Some are physiological, others are structural or behavioral. **Thermoregulation** is the ability to maintain body temperature within certain limits, even when the temperature of the environment changes a lot.

Animals produce heat as a byproduct of metabolic activities. Body temperature is determined by the rate at which heat is produced and the rate at which heat is lost to, or gained from, the outside environment.

The strategies for maintaining body temperature that are available to an animal may restrict the type of environment it can inhabit. Each animal species has an optimal environmental temperature range. Some animals, such as snowshoe hares, snowy owls, and weasels, can survive in cold arctic regions. Others, such as the Cape ground squirrel, are adapted to hot tropical climates. Although some animals can survive at temperature extremes, most survive only within moderate temperature ranges.

Ectotherms absorb heat from their surroundings

In most animals, body temperature is determined mainly by the changing temperature of the environment. Such animals are **ectotherms.** Most of the heat for their thermoregulation comes from the sun. An advantage of ectothermy is that such animals use very little energy to maintain a high metabolic rate. An ectotherm's metabolic rate tends to change with the weather. As a result, ectotherms have a much lower daily energy expenditure than do endotherms. They survive on less food and convert more of the energy in their food to growth and reproduction. One disadvantage of ectothermy is that daily and seasonal temperature conditions may limit activity.

Many ectotherms use behavioral strategies to adjust body temperature. For example, lizards may keep warm by burrowing in the soil at night. During the day, lizards take in heat by basking in the sun, orienting their bodies to expose the maximum surface area to the sun's rays (Fig. 37-6a). Other behavioral strategies for regulating temperature include hibernation and migration.

Endotherms derive heat from metabolic processes

Birds and mammals, as well as some species of fish (such as tuna) and insects, are **endotherms,** which means they have homeostatic mechanisms that maintain body temperature despite changes in the external temperature. Important benefits of endothermy include constant body temperature and increased rate of enzyme activity. Endotherms can be active even in low winter temperatures. However, endotherms must pay the high energy cost of thermoregulation during times when they are inactive. You must maintain your body temperature even when you are asleep.

Endotherms have structural adaptations for maintaining body temperature. For example, the insulating feathers of birds, hair of mammals, insulating layers of fat in birds and mammals, and hairs or fur of chitin on some insects are structures that reduce heat loss from the body. Birds and mammals also have behavioral adaptations for maintaining body temperature. For example, the Cape ground squirrel positions its tail to shade its body from the direct rays of the sun. Elephants spray themselves with cool water. We humans put on warm coats and install furnaces in our homes.

Some insects use a combination of structural, behavioral, and physiological mechanisms to regulate body temperature. The "furry" body of the moth helps conserve body heat. When a moth prepares for flight, it contracts its flight muscles with little movement of its wings. The metabolic heat generated enables the moth to sustain the intense metabolic activity needed for flight.

Endotherms have a variety of physiologic mechanisms for maintaining temperature homeostasis. They regulate heat production and regulate heat exchange with the environment. Most of their body heat comes from their own metabolic processes (see *Focus On: Electron Transport and Heat,* Chapter 7).

Breck P. Kent/Animals Animals

(a)

Frans Lanting/ Minden Pictures

(b)

FIGURE **37-6** **Behavioral adaptations for thermoregulation.**

(a) This marine iguana (*Amblyrhynchus cristatus*) is an ectotherm that increases its body temperature by sunning itself. **(b)** The body temperature of this baby sooty tern (*Sterna fuscata*) of Hawaii, an endotherm, is reduced as heat leaves the body through its open mouth.

In mammals, receptors located in the hypothalamus of the brain and in the spinal cord regulate temperature. Heat from metabolic activities can be increased either directly, or indirectly by the action of hormones (such as thyroid hormones) that increase metabolic rate. Heat production is increased by contracting muscles, and in cold weather many animals shiver.

When body temperature rises, birds and many mammals pant and some mammals sweat (Fig. 37-6b). These processes provide fluid for **evaporation,** the conversion of a liquid, such as sweat, to water vapor. Recall from Chapter 2 that when molecules enter the vapor phase, they take their heat energy with them. Heat transfers from the body to the surroundings, resulting in evaporative cooling.

In the human body, about 2.5 million sweat glands secrete sweat, and its evaporation from the skin surface lowers body temperature. Constriction and dilation of capillaries in the skin are also homeostatic mechanisms for regulating body temperature.

In humans, information about body temperature is sent to the temperature-regulating center in the hypothalamus (Fig. 37-7). When your body temperature rises above normal, your hypothalamus sends messages by way of nerves to the sweat glands, increasing sweat secretion. At the same time the hypothalamus sends messages to smooth muscle in the walls of blood vessels in the skin, causing them to dilate. More blood circulates through the skin, bringing heat to the body surface. The skin acts as a heat radiator, allowing heat to radiate from the body surface into the environment. These homeostatic mechanisms help return body temperature to normal.

When body temperature decreases below normal, blood vessels in the skin constrict so less heat is brought to the body surface. Hormonal messages increase heat production by body tissues. Nerves signal muscles to shiver, or let you move muscles voluntarily to increase body temperature.

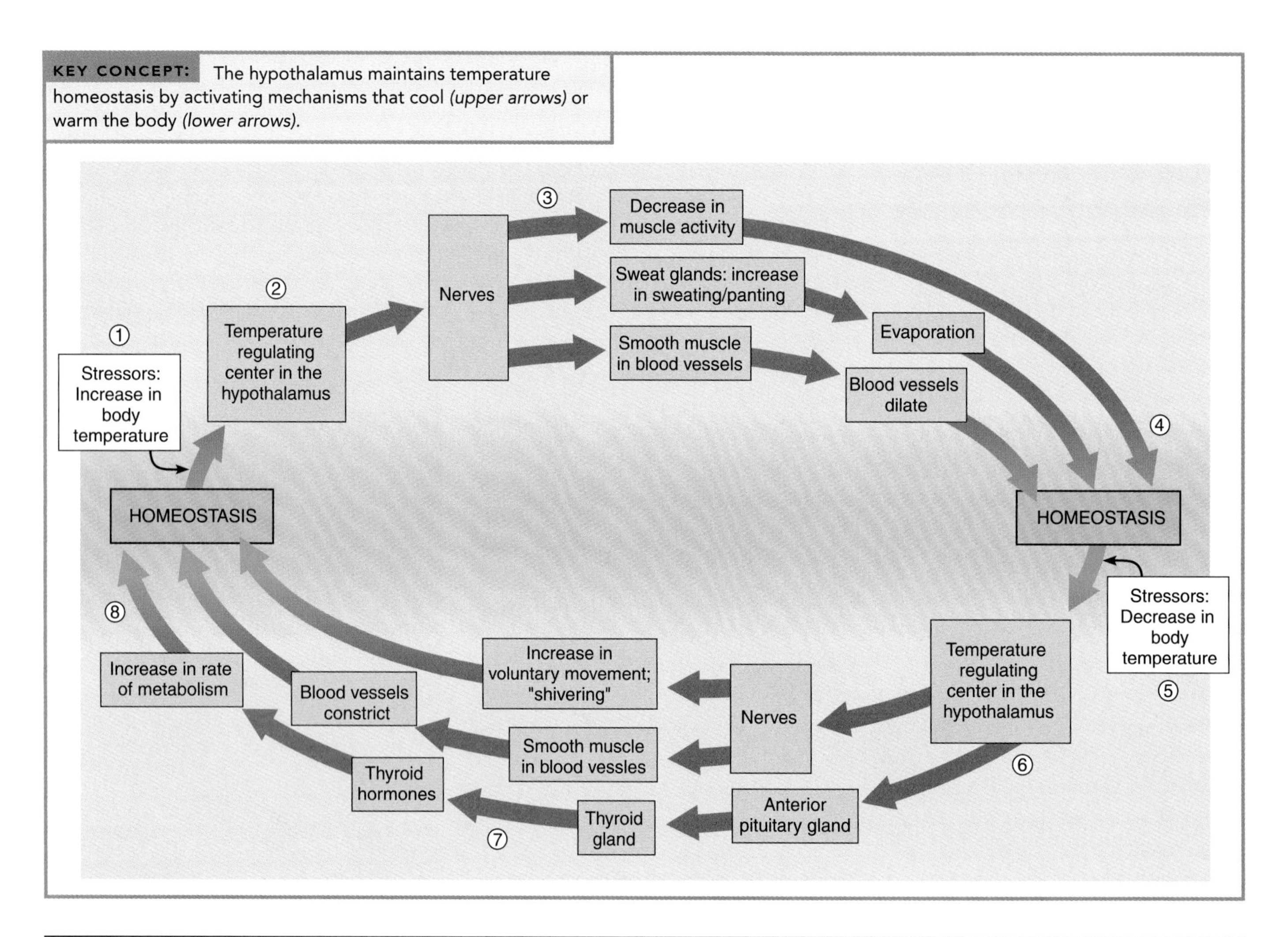

ACTIVE FIGURE **37-7** **Regulation of temperature in the human body.**

① When stressors increase body temperature above homeostatic levels, ② the temperature-regulating center in the hypothalamus ③ activates homeostatic mechanisms that ④ restore normal body temperature. ⑤ When stressors decrease body temperature, ⑥ the temperature-regulating center ⑦ activates mechanisms that ⑧ increase temperature, again restoring homeostasis.

BiologyNow™ Explore **temperature regulation** in humans and other animals by clicking on this figure on your BiologyNow CD-ROM.

Many animals respond physiologically to changes in environmental temperature

Animals adjust to seasonal changes, a process called **acclimatization.** A familiar example is the thickening of a dog's coat in the winter. As water temperature decreases during fall and winter, a trout's enzyme systems decrease their level of activity, allowing the trout to remain active at the lowest metabolic cost.

When stressed by cold, many animals sink into **torpor,** a decrease in body temperature below normal levels (an adaptive hypothermia). The decrease in body temperature is brought about by a decrease in the temperature set point in the hypothalamus. In a state of torpor, metabolism may be dramatically depressed. Heart and respiratory rates decrease, and animals are less responsive to external stimulation.

Hibernation is long-term torpor in response to winter cold and scarcity of food. **Estivation** is a state of torpor caused by lack of food or water during periods of high temperature. During estivation some mammals retreat to their burrows. The cactus mouse enters a state of hibernation during the winter in response to cold and scarcity of food. In summer, it estivates in response to lack of either food or water.

Review

- How do ectotherms adjust body temperature? What are some costs and benefits of ectothermy?
- What are some costs and benefits of endothermy?
- How is body temperture regulated in humans? Draw a diagram to illustrate your answer.
- What is acclimatization? Give an example.
- How do hibernation and estivation differ?

BiologyNow™ Assess your understanding of **regulating body temperature** by taking the pretest on your BiologyNow CD-ROM.

SUMMARY WITH KEY TERMS

1 Compare the structure and function of the four main kinds of animal tissues: epithelial, connective, muscle, and nervous tissues.

- A **tissue** consists of a group of similarly specialized cells that associate to perform one or more functions.
- **Epithelial tissue (epithelium)** forms a continuous layer, or sheet, of cells covering a body surface or lining a body cavity. Epithelial tissue functions in protection, absorption, secretion, or sensation.
- In contrast, **connective tissue** consists of relatively few cells separated by **intercellular substance,** which consists of **fibers** scattered through a **matrix.** The intercellular substance contains **collagen fibers, elastic fibers,** and/or **reticular fibers.** Connective tissue contains specialized cells such as **fibroblasts** and **macrophages.** Connective tissue joins other tissues of the body, supports the body and its organs, and protects underlying organs.
- **Muscle tissue** consists of cells specialized to contract. Each cell is an elongated muscle fiber containing many contractile units called **myofibrils.**
- **Nervous tissue** consists of elongated cells called **neurons,** specialized for transmitting impulses, and **glial cells,** which support and nourish the neurons.

2 Describe the main types of epithelial tissue, and state their functions.

- Epithelial cells may be **squamous, cuboidal,** or **columnar** in shape. Epithelial tissue may be **simple, stratified,** or **pseudostratified.** Simple squamous epithelium lines blood vessels and the air sacs in the lungs; it permits exchange of materials by diffusion. Simple cuboidal and columnar epithelia line passageways and are specialized for secretion and absorption. Stratified squamous epithelium forms the outer layer of the skin and lines passageways into the body; it provides protection. Pseudostratified epithelium also lines passageways and protects underlying tissues.
- Some epithelial tissue is specialized to form **glands. Goblet cells** are unicellular glands that secrete mucus. Goblet cells are **exocrine glands** that secrete their product through a duct onto an exposed epithelial surface. In contrast, **endocrine glands** release hormones into the **interstitial fluid** or blood.
- An **epithelial membrane** consists of a sheet of epithelial tissue and a layer of underlying connective tissue. A **mucous membrane** lines a cavity that opens to the outside of the body. A **serous membrane** lines a body cavity that does not open to the outside.

3 Compare the main types of connective tissue, and summarize their functions.

- Some main types of connective tissue are **loose connective tissue, dense connective tissue, elastic connective tissue, reticular connective tissue, adipose tissue, cartilage, bone,** and **blood** (summarized in Table 37-2). Loose connective tissue lies in the subcutaneous tissue and between many body parts. It consists of fibers in a semifluid matrix.
- Cartilage cells, called **chondrocytes,** lie in **lacunae,** small cavities in the cartilage matrix. **Osteocytes** secrete and maintain the matrix of bone. Compact bone consists of spindle-shaped units called **osteons.** Each osteon has a central blood vessel that runs through a **Haversian canal,** surrounded by lamellae, concentric layers containing osteocytes.

4 Contrast the three types of muscle tissue and their functions.

- **Skeletal muscle** is **striated** and under voluntary control. Each elongated, cylindrical **muscle fiber** has several nuclei.
- **Cardiac muscle** is striated; its contraction is involuntary. Its elongated, cylindrical fibers branch and fuse; each fiber has one or two central nuclei.
- **Smooth muscle** contracts involuntarily. Its elongated, spindle-shaped fibers lack striations. Each fiber has a single central nucleus. Smooth muscle is responsible for movement of food through the digestive tract and for movement of other body organs.

5 Relate the structure of the neuron to its function.

- The elongated neurons are adapted for receiving and transmitting information. **Dendrites** receive signals and transmit them to the **cell body.** The **axon** transmits signals away from the cell body to other neurons or to a muscle or gland. A **synapse** is a junction between neurons.

6 Briefly describe the organ systems of a complex animal.

- Tissues and **organs** work together, forming **organ systems.** In complex animals, 10 principal organ systems work together,

SUMMARY WITH KEY TERMS *(continued)*

making up the living **organism.** These include the **integumentary, skeletal, muscular, digestive, circulatory, respiratory, urinary, nervous, endocrine,** and **reproductive systems** (see Table 37-4).

7 Define *homeostasis,* and contrast negative and positive feedback mechanisms.

- **Homeostasis** is the body's automatic tendency to maintain a balanced internal environment, or steady state. This steady state is actually a dynamic equilibrium maintained by **negative feedback** systems in which the regulators respond to counteract the changes caused by **stressors.**

8 Summarize the homeostatic actions of each organ system.

- See Table 37-4

9 Describe advantages and disadvantages of ectothermy.

- Animals have structural, behavioral, and physiologic strategies for **thermoregulation.** In **ectotherms,** body temperature depends to a large extent on the temperature of the environment.
- An advantage of ectothermy is that very little energy is used to maintain a high metabolic rate. As a result, ectotherms can survive on less food. A disadvantage is that activity may be limited by daily and seasonal temperature conditions.

10 Describe advantages and disadvantages of endothermy, and describe strategies endotherms use to survive in extreme temperature conditions.

- **Endotherms** have homeostatic mechanisms for regulating body temperature within a narrow range.
- Advantages include constant body temperature and increased rate of enzyme activity. Many endotherms can carry out their activities even in low winter temperatures. A disadvantage of endothermy is its high energy cost.
- **Acclimatization** is the process of adjustment to seasonal changes. **Torpor** is an adaptive hypothermia; body temperature is maintained below normal levels. **Hibernation** is long-term torpor in response to winter cold.
- **Estivation** is torpor caused by lack of food or water during summer heat.

POST-TEST

1. In a multicellular animal, (a) organ systems are always present (b) cells specialize to perform specific functions (c) small size provides more opportunity for capturing food (d) organelles are more numerous (e) organelles carry on more functions
2. A group of closely associated cells that carry out specific functions is a(an) (a) colony (b) tissue (c) organ system (d) organelle (e) homeostatic mechanism
3. Which tissue consists of cells that fit tightly together to form a sheet of cells? (a) muscle (b) nervous tissue (c) connective tissue (d) epithelial tissue (e) reticular tissue
4. Epithelial cells that produce and secrete a product into a duct form a(an) (a) endocrine gland (b) exocrine gland (c) pseudostratified membrane (d) endocrine or exocrine gland (e) serous membrane
5. A serous membrane (a) lines a body cavity that does not open to the outside of the body (b) consists of loose connective tissue (c) has a framework of reticular connective tissue (d) covers the body (e) lines the digestive and respiratory systems
6. The most numerous fibers in connective tissue are (a) reticular (b) myofibrils (c) collagen (d) elastic (e) glial
7. Dense connective tissue is likely to be found in (a) the outer layer of skin (b) tendons (c) bone (d) the air sacs of lungs (e) the framework of the lymph nodes
8. Osteocytes are most likely to be found in (a) the outer layer of skin (b) bone (c) cartilage (d) lungs (e) blood
9. Tissue that contains fibroblasts and a great deal of intercellular substance is (a) connective tissue (b) muscle tissue (c) nervous tissue (d) pseudostratified epithelium (e) adipose tissue
10. Tissue that contracts and is striated and involuntary is (a) connective tissue (b) smooth muscle (c) skeletal muscle (d) pseudostratified (e) cardiac muscle
11. Glial cells are most characteristic of (a) muscle that is striated and voluntary (b) cardiac muscle tissue (c) nervous tissue (d) stratified epithelium (e) loose connective tissue
12. Tissue that has a hard, rubbery matrix and lacks blood vessels is (a) integumentary tissue (b) bone (c) nervous tissue (d) stratified epithelium (e) cartilage
13. The contractile elements in muscle tissue are (a) myofibrils (b) elastic (c) collagen (d) lacunae (e) stress proteins
14. Which system consists of glands that secrete hormones? (a) integumentary (b) skeletal (c) nervous (d) endocrine (e) exocrine
15. Which system has the homeostatic function of helping to maintain constant calcium level in the blood? (a) integumentary (b) skeletal (c) nervous (d) reproductive (e) exocrine
16. Which system has the homeostatic function of helping to regulate volume and composition of blood and body fluids? (a) integumentary (b) muscular (c) reproductive (d) urinary (e) exocrine
17. Many homeostatic functions are maintained by (a) negative feedback systems (b) positive feedback systems (c) set points (d) stressors (e) exocrine glands
18. An ectotherm (a) may use behavioral strategies to help adjust body temperature (b) has a variety of homeostatic mechanisms to precisely regulate body temperature (c) depends on sensors in the hypothalamus to regulate temperature (d) has a higher rate of enzyme activity than a typical endotherm (e) must expend more energy on thermoregulation than an endotherm
19. The state of torpor caused in some animals by lack of food or water during periods of high temperature is known as (a) positive feedback (b) hibernation (c) endothermy (d) ectothermy (e) estivation

CRITICAL THINKING

1. Imagine that all the epithelium in a complex animal, such as a human, suddenly disappeared. What effects might this have on the body and its ability to function?
2. What would connective tissue be like if it had no intercellular substance? What effect would the absence of intercellular substance have on the body?
3. A high concentration of carbon dioxide in the blood and interstitial fluid results in more rapid breathing. Explain this observation in terms of homeostasis.
4. What are some advantages of multicellularity?

- Visit our Web site at **http://biology.brookscole.com/solomon7** for links to chapter-related resources on the World Wide Web. Additional online materials relating to this chapter can also be found on our Web site.

BIOLOGY NOW RESOURCES

Active Figures

37-4: Body systems

37-7: Temperature regulation

Preparing for an exam? Take a diagnostic test on your BiologyNow CD-ROM.

Post-Test Answers

1. b	2. b	3. d	4. b
5. a	6. c	7. b	8. b
9. a	10. e	11. c	12. e
13. a	14. d	15. b	16. d
17. a	18. a	19. e	

38

Protection, Support, and Movement

Stephen Dalton

Grasshopper in mid-jump The grasshopper (*Tropidacris dux*) uses powerful muscles in its hind legs to catapult itself into the air. Insect muscles attach to the exoskeleton.

CHAPTER OUTLINE

In Chapter 37 we described animal tissues, organs, and systems, laying the foundation for the more detailed discussions in this and later chapters. Here we focus on epithelial coverings, skeletal systems, and muscle—structures that are closely interrelated in function. In our survey of the animal kingdom (see Chapters 28 to 30), we described many adaptations involving these structures. For example, we discussed how the tough insect exoskeleton contributes to the great biological success of these animals. In this chapter, we compare these systems in several animal groups and then focus on their structures and functions in mammals.

In addition to protecting underlying tissues, the **epithelial coverings** of animals may be specialized to exchange gases, excrete wastes, regulate temperature, or secrete substances such as poison or mucus. Most animals are also protected by a **skeletal system.** Whether it is a fluid-filled compartment, a shell, or a system of bones, the skeletal system provides support for the body and typically protects internal organs. In vertebrates, bones store calcium and are important in maintaining homeostatic levels of calcium in the blood.

The skeletal system and **muscular system** work together to produce movement. Muscle tissue is specialized to contract. Although a few animals remain rooted to one spot, sweeping their surroundings with tentacles, **locomotion**—movement from place to place in the environment—is characteristic of most animals. In fact, for most animals dependable, rapid, and responsive movement is key to finding food and escaping from predators. Animals creep, walk, run, jump, swim, or fly (see photograph). Muscles responsible for locomotion anchor to the skeleton, which gives them something firm on which to act. In most vertebrates, bones serve as levers that transmit the force necessary to move various parts of the body.

Muscle powers both locomotion and the physiological actions necessary to maintain homeostasis. Many animals have internal circulating fluids, pumped by hearts and contained by hollow vessels that maintain their pressure with gentle squeezing. Most have digestive systems that push food along with peristaltic contractions. In complex animals, effective movement depends on the cooperation and interactions of several body systems, including the muscular, skeletal, nervous, circulatory, respiratory, and endocrine systems. ■

EPITHELIAL COVERINGS

Learning Objectives

1. Compare the structure and functions of the external epithelium of invertebrates and vertebrates.
2. Relate the structure of vertebrate skin to its functions.

Epithelial tissue covers all external and internal surfaces of the animal body. The structure and functions of the epithelial covering are adapted to the animal's environment and lifestyle. In both invertebrates and vertebrates, epithelial coverings protect the body. These coverings may also be specialized for secretion or gas exchange, and typically contain receptors that receive sensory signals from the environment.

Invertebrate epithelium may function in secretion or gas exchange

In many invertebrates, the outer epithelium is specialized to secrete protective or supportive layers of nonliving material. In insects and many other animals, the secreted material forms an outer covering called a **cuticle.** Many animals, including corals and mollusks, secrete a shell made mainly of calcium carbonate. In addition to its protective function, the external epithelium of invertebrates may be specialized for secretion or gas exchange. Epithelial cells may also be modified as sensory cells that are selectively sensitive to light, chemical stimuli, or mechanical stimuli such as contact with an object, or pressure.

In many species, the epithelium contains secretory cells that secrete lubricants or adhesives. Such cells may release odorous secretions used for communication or for marking trails. In other species, the cells produce poisonous secretions used for offense or defense. The earthworm epithelium secretes a lubricating mucus that promotes efficient diffusion of gases across the body wall and also reduces friction during movement through the soil.

Vertebrate skin functions in protection and temperature regulation

The **integumentary system** of vertebrates includes the skin and structures that develop from it. In many fishes, in the African ant-eating pangolin (a mammal; see Fig. 17-13c), and in some reptiles, skin has developed into a set of scales formidable enough to be considered armor. Even human skin has considerable strength.

Derivatives of skin differ considerably among vertebrates. Fish have bony or toothlike scales. Amphibians have naked skin covered with mucus, and some species are equipped with poison glands. Reptiles have epidermal scales, mammals have hair, and birds have feathers that provide even more effective insulation than fur. The feathers of birds and the fur of mammals help maintain body temperature. Skin and its derivatives are often brilliantly colored in connection with courtship rituals, territorial displays, and other kinds of communication.

In mammals, structures derived from skin include claws (modified as fingernails and toenails in primates), hair, sweat glands, oil (sebaceous) glands, and several types of sensory receptors that give us the ability to feel pressure, temperature, and pain. The skin of mammals also contains mammary glands, specialized in females for secretion of milk.

In humans and other mammals, oil glands empty via short ducts into hair follicles. A **hair follicle** is the part of a hair below the skin surface, together with the surrounding epithelial tissue. Oil glands secrete **sebum,** a complex mixture of fats and waxes that inhibits the growth of harmful bacteria.

Although the skin varies among species, the basic structure of skin is the same in all vertebrates. The outer layer of skin, the **epidermis,** is a waterproof protective barrier. The epidermis consists of several strata, or sublayers. The deepest is **stratum basale,** and the most superficial is **stratum corneum** (Fig. 38-1). Pigment cells in stratum basale and in the dermis produce **melanin,** a pigment that contributes to the color of the skin. In stratum basale, cells divide and are pushed outward as other cells are produced below them. The epidermal cells mature as they move toward the skin surface.

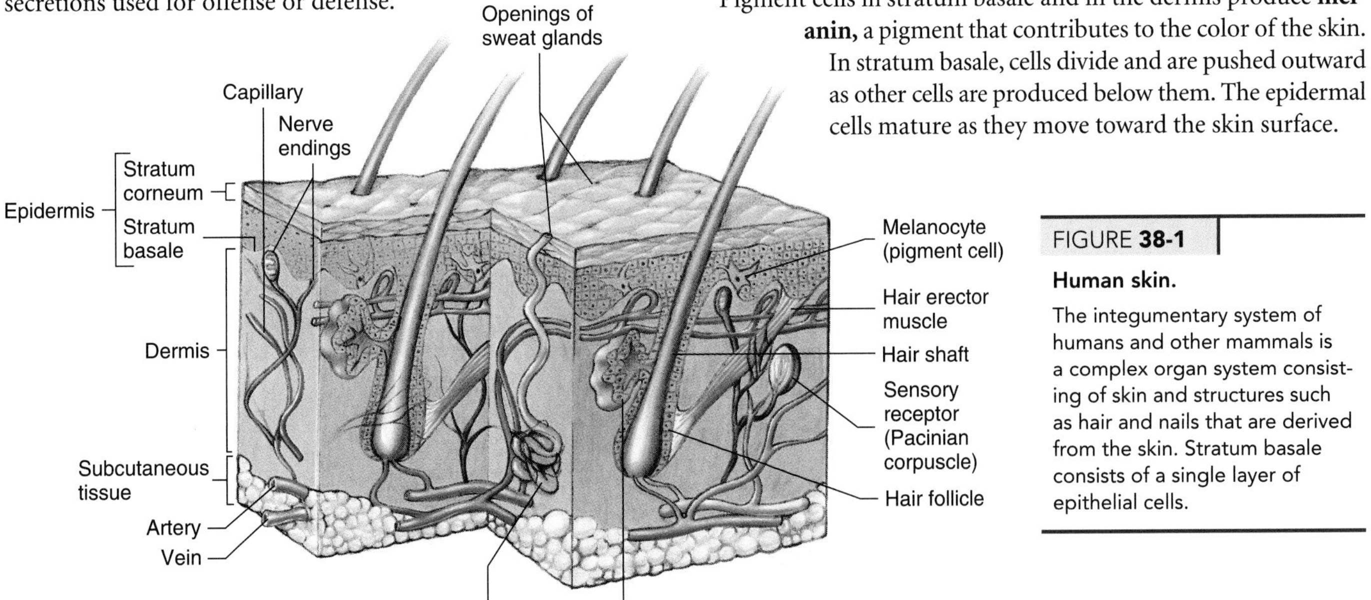

FIGURE **38-1** **Human skin.** The integumentary system of humans and other mammals is a complex organ system consisting of skin and structures such as hair and nails that are derived from the skin. Stratum basale consists of a single layer of epithelial cells.

As they move toward the body surface, epidermal cells manufacture **keratin,** an elaborately coiled protein that gives the skin considerable mechanical strength and flexibility. Keratin is insoluble and serves as a diffusion barrier for the body surface. As epidermal cells move through stratum corneum, they die. When they reach the outer surface of the skin, they wear off and must be continuously replaced.

Beneath the epidermis lies the **dermis** (see Fig. 38-1), a dense, fibrous connective tissue made up mainly of collagen fibers. Collagen imparts strength and flexibility to the skin. The dermis also contains blood vessels that nourish the skin and sensory receptors for touch, pain, and temperature. Hair follicles and, in most mammals, sweat glands lie embedded in the dermis. In birds and mammals, the dermis rests on a layer of **subcutaneous tissue** composed mainly of adipose tissue, that insulates the body from outside temperature extremes.

In humans, exposure to ultraviolet (UV) radiation, the short, invisible rays from the sun, causes the epidermis to thicken and stimulates pigment cells in the skin to produce melanin at an increased rate. Melanin is an important protective screen against the sun because it absorbs some of the harmful UV rays. An increase in melanin causes the skin to darken. The sun tan so prized by sun worshipers is actually a sign that the skin has been exposed to too much UV radiation. When the melanin cannot absorb all the UV rays, the skin becomes inflamed, or sunburned. Because dark-skinned people have more melanin, they suffer less sunburn, wrinkling, and skin cancer, although they are still at risk. UV radiation damages DNA, causing mutations that lead to malignant transformation (see Chapter 16).

Most cases of skin cancer are caused by excessive, chronic exposure to UV radiation. The incidence of *malignant melanoma,* the most lethal type of skin cancer, is increasing faster than any other type of cancer.

Review

- How are the external epithelia of invertebrates and vertebrates alike? How are they different?
- How is the structure of vertebrate skin related to its specific functions?
- What properties does keratin give human skin?
- What is the function of melanin?

BiologyNow™ Assess your understanding of **epithelial coverings** by taking the pretest on your BiologyNow CD-ROM.

SKELETAL SYSTEMS

Learning Objectives

3 Compare the structure and functions of different types of skeletal systems, including the hydrostatic skeleton, exoskeleton, and endoskeleton.

4 Describe the main divisions of the vertebrate skeleton and the bones that make up each division.

5 Describe the structure of a typical long bone, and differentiate between endochondral and intramembranous bone development.

6 Compare the main types of vertebrate joints.

In addition to an epithelial covering, many animals are protected by a hard skeleton that forms the framework of the body. The skeleton also functions in locomotion. In complex animals, muscles typically act on hard structures such as chitin or bone. These skeletal structures transmit and transform mechanical forces generated by muscle contraction into the variety of motions that animals make. Three types of skeletons are hydrostatic skeletons, exoskeletons, and endoskeletons.

In hydrostatic skeletons, body fluids transmit force

Many soft-bodied invertebrates have **hydrostatic skeletons** made of fluid-filled body compartments. Imagine an elongated balloon full of water. If you pull on it, it lengthens and becomes thinner. It does the same if you squeeze it. Conversely, if you push the ends toward the center, it shortens and thickens. Many animals, including cnidarians and annelids, have a hydrostatic skeleton that works something like a balloon filled with water. Fluid in a closed compartment of the body is held under pressure. When muscles in the compartment wall contract, they push against the tube of fluid. Because fluids cannot be compressed, the force is transmitted through the fluid, changing the shape and movement of the body. Most cnidarians, flatworms, annelids, and roundworms have hydrostatic skeletons.

In *Hydra* and other cnidarians, cells of the two body layers are capable of contracting. The contractile cells in the outer epidermal layer lie longitudinally, whereas the contractile cells of the inner layer (the gastrodermis) are arranged circularly around the central body axis (Fig. 38-2). The two groups of cells work in **antagonistic** fashion: What one can do, the other can undo. When the epidermal (longitudinal) layer contracts, the hydra shortens. Because of the fluid in the gastrovascular cavity, force is transmitted so that the hydra thickens as well. In contrast, when the inner (circular) layer contracts, the hydra thins, and its fluid contents force it to lengthen.

Mechanically, the hydra is a bag of fluid. The fluid acts as a hydrostatic skeleton because it transmits force when the contractile cells contract against it. (Although technically not a closed compartment, the gastrovascular cavity functions as a hydrostatic skeleton because its opening is small.) Hydrostatic skeletons permit only crude mass movements of the body or its appendages. Delicate movements are difficult, because force tends to be transmitted equally in all directions throughout the entire fluid-filled body of the animal. For example, it isn't easy for the hydra to thicken one part of its body while thinning another.

The more sophisticated hydrostatic skeleton of the annelid worm lets it move more flexibly. An earthworm's body consists of a series of segments divided by transverse partitions, or septa (see Fig. 29-12). The septa isolate portions of the body cavity and its contained coelomic fluid, permitting the hydrostatic skeletons of each segment to be largely independent of one another. Thus contraction of the circular muscle in the elongating anterior end doesn't interfere with the action of the longitudinal muscle in the segments of the posterior end.

You can find some examples of hydrostatic skeletons even in complex invertebrates equipped with shells or endoskele-

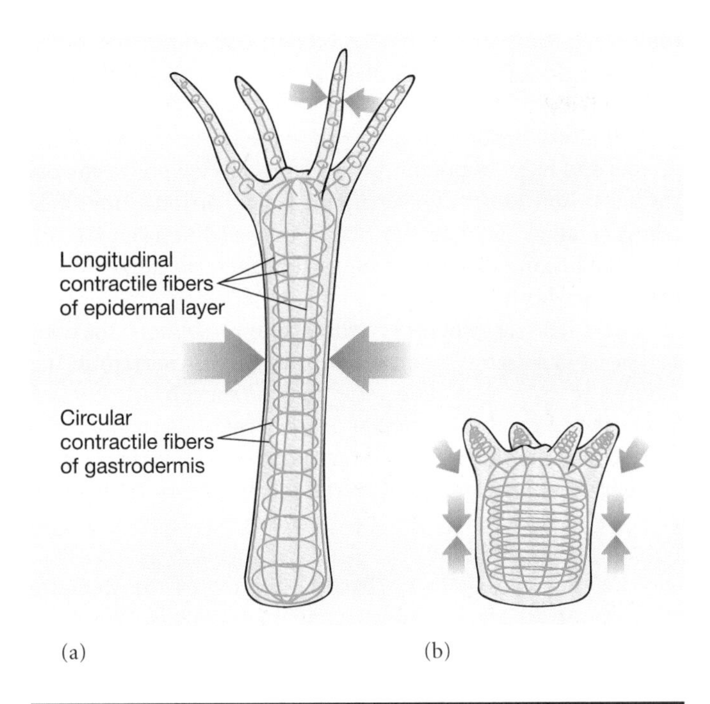

FIGURE **38-2** **Hydrostatic skeleton.**

The longitudinally arranged contractile cells of *Hydra* are antagonistic to the cells arranged in circles around the body axis. **(a)** Contraction of the circular contractile fibers elongates the body. **(b)** Contraction of the longitudinal fibers shortens the body.

tons, and in vertebrates with endoskeletons of cartilage or bone. Sea stars and sea urchins move their tube feet by an ingenious version of the hydrostatic skeleton (see Chapter 30). And even the human penis becomes erect and stiff because of the turgidity of pressurized blood in its internal spaces.

Mollusks and arthropods have nonliving exoskeletons

In most animals, the skeleton is a lifeless shell, or **exoskeleton,** deposited atop the epidermis (cells of the outer epithelium). In mollusks, the exoskeleton is a calcium carbonate shell secreted by the mantle, a thin sheet of epithelial tissue that extends from the body wall. The exoskeleton provides protection, a retreat used in emergencies, with the bulk of the naked, tasty body exposed at other times.

Exoskeletons of arthropods serve not only to protect but also to transmit forces. In this respect they are comparable to the skeletons of vertebrates. The arthropod exoskeleton is a nonliving cuticle that contains the polysaccharide **chitin.** Although it is a continuous, one-piece sheath covering the entire body, the arthropod exoskeleton varies greatly in thickness and flexibility. Large, thick, inflexible plates are separated from one another by thin, flexible joints arranged segmentally. Enough joints are provided to make the arthropod's body as flexible as those of many vertebrates. The arthropod exoskeleton is adapted to a vast variety of lifestyles, and parts of it are modified to function as specialized tools or weapons.

A disadvantage of the rigid arthropod exoskeleton is that to accommodate growth, an arthropod must **molt,** that is, shed its exoskeleton and replace it with a new, larger one (Fig. 38-3). During molting the animal is weak and vulnerable to predators.

Internal skeletons are capable of growth

You are probably most familiar with the **endoskeleton** of echinoderms and chordates. This living internal skeleton consists of plates or shafts of calcium-impregnated tissue (such as cartilage or bone). Composed of living tissue, the endoskeleton grows along with the animal as a whole. The echinoderm endoskeleton consists of spicules and plates of calcium salts embedded in the body wall, beneath an epidermis that covers the body. This endoskeleton forms what amounts to an internal shell that provides support and protection (Fig. 38-4). Many echinoderm endoskeletons bear spines that project to the outer surface.

The internal vertebrate skeleton provides support and protection, and transmits muscle forces. Members of class Chondrichthyes (sharks and rays) have skeletons of cartilage, but in most vertebrates the skeleton consists mainly of bone. Many bones form systems of levers that transmit muscle forces.

FIGURE **38-3** **Molting.**

A greengrocer cicada (*Magicicada*) requires 13 years to mature. It then emerges from the soil, climbs a tree, and molts prior to reproducing.

Judy Davidson/Science Photo Library/Photo Researchers, Inc.

FIGURE **38-4** **The echinoderm endoskeleton.**

The endoskeleton provides support and protection. As in other echinoderms, the endoskeleton of the red sea star (*Formia milleporella*) is composed of spicules and plates of nonliving calcium salts embedded in the body wall. Photographed in Indonesia.

The vertebrate skeleton has two main divisions

The two main divisions of the vertebrate skeleton are the axial and appendicular skeletons. The **axial skeleton,** located along the central axis of the body, consists of the skull, vertebral column, ribs, and sternum (breastbone). The **appendicular skeleton** consists of the bones of the limbs (arms and legs) plus the bones making up the pectoral (shoulder) girdle and most of the pelvic (hip) girdle; these girdles connect the limbs to the axial skeleton (Fig. 38-5).

The **skull,** the bony framework of the head, consists of the cranial and facial bones. In the human, 8 cranial bones enclose the brain, and 14 bones make up the facial portion of the skull. Several cranial bones that are single in the adult human result from the fusion of 2 or more bones that are separate in the embryo or newborn.

The vertebrate spine, or **vertebral column,** supports the body and bears its weight. In humans it consists of 24 **vertebrae** and 2 bones composed of fused vertebrae, the **sacrum** and **coccyx.** The vertebral column consists of the **cervical** (neck) region, with 7 vertebrae; the **thoracic** (chest) region, with 12 vertebrae; the **lumbar** (back) region, with 5 vertebrae; the **sacral** (pelvic) region, with 5 fused vertebrae; and the **coccygeal** region, also composed of fused vertebrae.

The **rib cage** is a bony "basket" formed by the **sternum** (breastbone), thoracic vertebrae, and, in mammals, 12 pairs of ribs. The rib cage protects the internal organs of the chest, including the heart and lungs. It also supports the chest wall, preventing it from collapsing as the diaphragm contracts with each breath. Each pair of ribs is attached dorsally to a separate vertebra. Of your 12 pairs of ribs, the first 7 are attached ventrally to the sternum; the next 3 are attached indirectly by cartilages; and the last 2, the "floating ribs," have no attachments to the sternum.

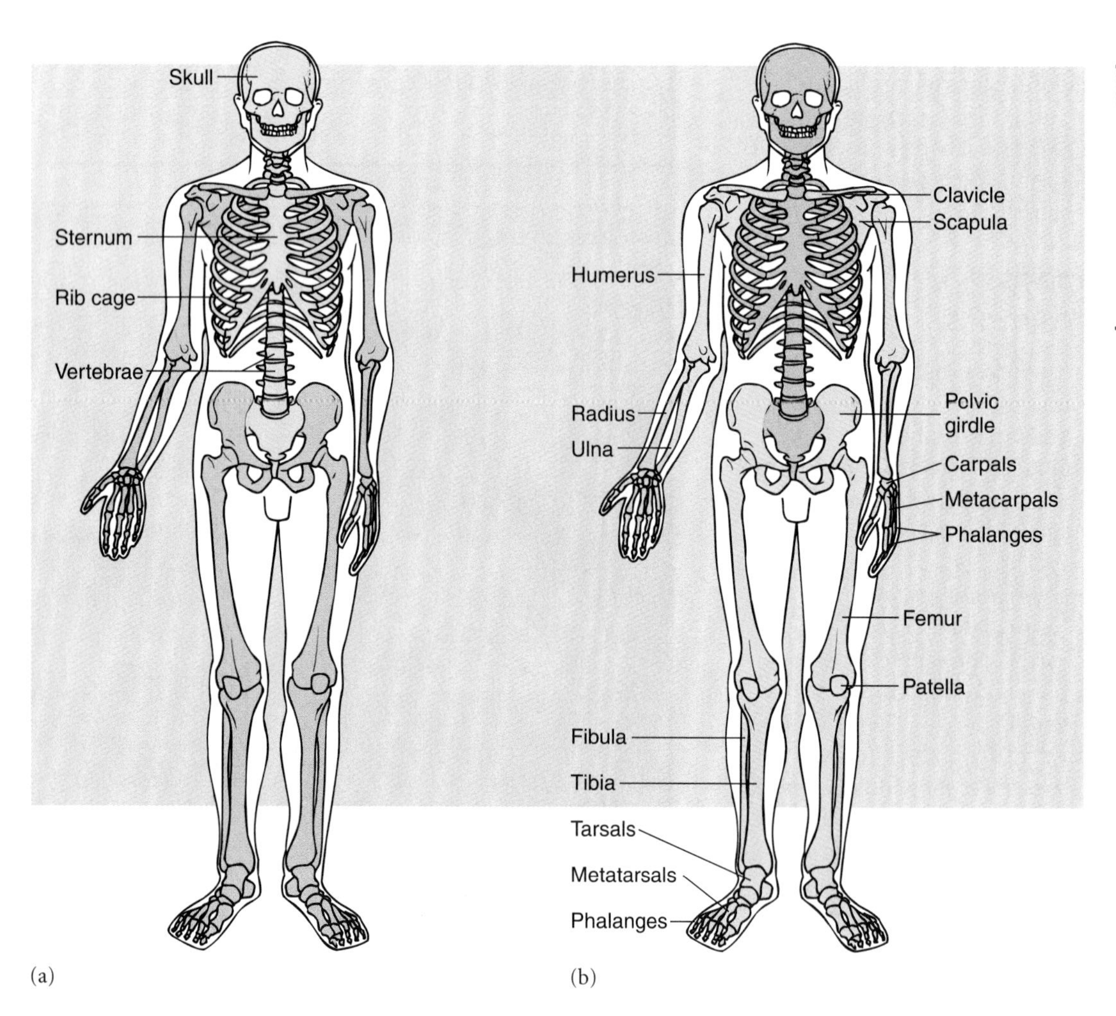

FIGURE **38-5** **The human skeletal system.**

(a) Bones of the axial skeleton *(tan)*, anterior view. **(b)** Bones of the appendicular skeleton *(tan)*, anterior view.

The **pectoral girdle** consists of two collarbones, or **clavicles,** and two shoulder blades, or **scapulas.** The **pelvic girdle** consists of a pair of large bones, each composed of three fused hipbones. Whereas the pelvic girdle is securely fused to the vertebral column, the pectoral girdle is loosely and flexibly attached to it by muscles.

Each human limb consists of 30 bones and terminates in five **digits,** the fingers and toes. The more specialized appendages of other tetrapods may be characterized by four digits (as in the pig), three (as in the rhinoceros), two (as in the camel), or one (as in the horse).

A typical long bone consists of compact and spongy bone

The radius, one of the two bones of the forearm, is a typical long bone (Fig. 38-6). Its numerous muscle attachments are arranged in such a way that the bone rotates about its long axis and also operates as a lever, amplifying the motion generated by the muscles. By themselves, muscles cannot shorten enough to produce large movements of the body parts to which they are attached.

Like other bones, the radius is covered by a connective tissue membrane, the **periosteum,** to which muscle tendons and ligaments attach. The periosteum can produce new layers of bone, thus increasing the bone's diameter. The main shaft of a long bone is its **diaphysis;** each expanded end is an **epiphysis.** In children, a disk of cartilage, the **metaphysis,** lies between the epiphyses and the diaphysis. Metaphyses are growth centers that disappear at maturity, becoming vague **epiphyseal lines.** Long bones have a central cavity that contains **bone marrow.** Yellow marrow consists mainly of a fatty connective tissue; the red marrow in certain bones produces blood cells.

The radius has a thin outer shell of **compact bone,** which is very dense and hard. Compact bone lies primarily near the surfaces of a bone, where it provides great strength. Recall from Chapter 37 that compact bone consists of interlocking spindle-shaped units called **osteons,** or **Haversian systems** (see Fig. 37-2). Within an osteon, **osteocytes** (bone cells) lie in small cavities called **lacunae** (sing., *lacuna*). The lacunae are arranged in concentric circles around central **Haversian canals.** Blood vessels that nourish the bone tissue pass through the Haversian canals. Osteocytes are connected by threadlike extensions of their cytoplasm that extend through narrow channels called *canaliculi.*

Interior to the thin shell of compact bone is a filling of **spongy bone,** which provides mechanical strength. Spongy bone consists of a network of thin strands of bone. Its spaces are filled with bone marrow.

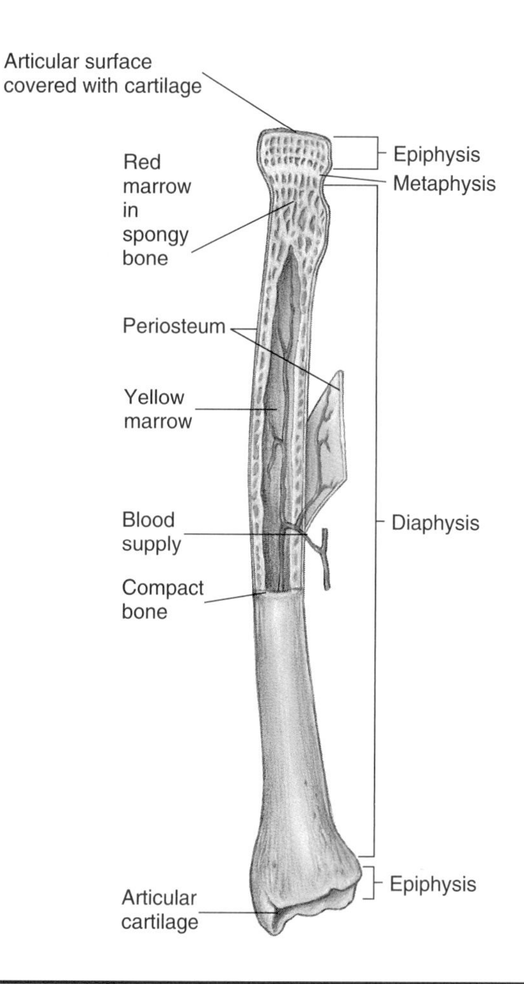

FIGURE **38-6** **A typical long bone.**
A long bone has a thin shell of compact bone and a filling of spongy bone that contains marrow.

Bones are remodeled throughout life

During fetal development, bones form in two ways. Long bones, such as the radius, develop from cartilage templates in a process called **endochondral bone development.** A bone begins to ossify in its diaphysis, and secondary sites of bone production develop in the epiphyses. The part of the bone between the ossified regions can grow. Eventually the ossified regions fuse. In contrast, other bones, including the flat outer bones of the skull, develop from a noncartilage, connective tissue scaffold. This process is known as **intramembranous bone development.**

Osteoblasts are bone-building cells. They secrete the protein collagen, which forms the strong fibers of bone. The compound hydroxyapatite, composed mainly of calcium phosphate, is present in the interstitial fluid. It automatically crystallizes around the collagen fibers, forming the hard matrix of bone. As the matrix forms around the osteoblasts, they become isolated within the lacunae. The trapped osteoblasts are then called *osteocytes.*

Bones are modeled during growth and remodeled continuously throughout life in response to physical stresses and other changing demands. As muscles develop in response to physical activity, the bones to which they are attached thicken and become stronger. As a bone grows, tissue is removed from its interior, especially from the walls of the marrow cavity.

Osteoclasts are large, multinucleated cells that break down (resorb) bone. The osteoclasts move about, secreting hydrogen ions that dissolve the crystals, and enzymes that digest the collagen. Osteoclasts and osteoblasts are synergistic; together they shape bones. The remodeling process is extensive—the adult human skeleton is completely replaced every 10 years!

In *osteoporosis,* the most common progressive degenerative bone disease, bone resorption takes place more rapidly than bone formation. Patients lose so much bone mass that their bones become fragile, which greatly increases their risk for fracture. Certain drugs inhibit osteoclast-mediated bone resorption. Researchers are also testing drugs that promote osteoblast activity, with the goal of promoting bone formation.

Joints are junctions between bones

Joints, or articulations, are junction sites of two or more bones. Joints facilitate flexibility and movement. At the joint, the outer surface of each bone consists of articular cartilage. One way to classify joints is according to the degree of movement they allow. The *sutures* between bones of the human skull are **immovable joints.** In a suture, bones are held together by a thin layer of dense fibrous connective tissue, which may be replaced by bone in the adult. **Slightly movable joints,** found between vertebrae, are made of cartilage and help absorb shock.

Most joints are **freely movable joints.** Each is enclosed by a joint capsule of connective tissue and lined with a membrane that secretes a lubricant called **synovial fluid.** This viscous fluid reduces friction during movement and also absorbs shock. The joint capsule is typically reinforced by **ligaments,** bands of fibrous connective tissue that connect bones and limit movement at the joint.

With time and use, joints wear down. In *osteoarthritis,* a common joint disorder, cartilage repair does not keep up with degeneration, and the articular cartilage wears out. In *rheumatoid arthritis,* an autoimmune disease, the synovial membrane thickens and becomes inflamed. Synovial fluid accumulates, causing pressure, pain, stiffness, and progressive deformity, leading to loss of function.

Review

- What functions does a hydrostatic skeleton perform?
- How do the septa in the annelid worm contribute to the effectiveness of its hydrostatic skeleton?
- What are some disadvantages of an exoskeleton? Some advantages?
- What are the two main divisions of the human skeleton? Describe each division.
- How do osteoblasts and osteoclasts remodel bone?

BiologyNow™ Assess your understanding of **skeletal systems** by taking the pretest on your BiologyNow CD-ROM.

MUSCLE CONTRACTION

Learning Objectives

7 Relate the structure and function of insect flight muscles.

8 Describe the structure of skeletal muscle.

9 List, in sequence, the events that take place during muscle contraction.

10 Compare the roles of glycogen, creatine phosphate, and ATP in providing energy for muscle contraction.

11 Describe how muscles work, including the antagonistic action of skeletal muscles.

12 Compare the structures and functions of slow and fast muscle fibers.

Throughout most of the animal kingdom, muscle tissue generates the mechanical forces and motion necessary for locomotion, manipulation of objects, circulation of blood, movement of food through the digestive tract, and many other life processes. Animals with very simple body plans do not have muscle tissue, but all eukaryotic cells do contain the contractile protein **actin.** The major component of microfilaments (see Fig. 4-25), actin is important in many cell processes, including amoeboid movement and attachment of cells to surfaces. In most cells, actin is functionally associated with the contractile protein **myosin.** Actin and myosin are most highly organized in **muscle** cells.

Invertebrate muscle varies among groups

Earlier in the chapter, we discussed the contractile cells of the hydrostatic skeleton of *Hydra* and other cnidarians. In flatworms and most other animal groups, muscle is a specialized tissue organized into definite layers, or straplike bands. Some invertebrate phyla have skeletal and smooth muscle. Bivalve mollusks, such as clams, have two sets of muscles for opening and closing the shell. Smooth muscle, which is capable of slow, sustained contraction, keeps the two shells tightly closed for long periods, even days or weeks at a time. Striated muscle, which contracts rapidly, is used to swim and to shut the shell quickly when the mollusk is threatened. Arthropod muscles are striated, even in the walls of the digestive tract.

PROCESS OF SCIENCE

Insect flight muscles contract more rapidly than any other known muscle, up to 1000 contractions per second. Not surprisingly, insect flight muscles in action have the highest known metabolic rate of any muscle tissue. As you might imagine, these muscles are structurally well adapted to their function. They contain more mitochondria than any known variety of muscle! Insect muscles are also elaborately infiltrated with *tracheae,* tiny air-filled tubes that carry oxygen directly to each fiber.

Insects were the first animals to evolve flight, an adaptation that has contributed to their impressive biological success (see Chapter 29). Just how insects fly has been an aerodynamic mystery, and scientists are working to understand their remarkable ability to maneuver. Somehow insects create lift that is 20 or more times their body weight. A central question has been how their flapping wings generate enough force to keep them airborne. Insect flight involves much more than just flapping the wings up and down. The flapping motion of the insect wing changes direction and speed, and upstrokes alternate with downstrokes at very high rates. At each shift of stroke, the wing rotates about its long axis and tilts to just the right angle for the new direction of motion.

In many flying insects, the striated flight muscles attach not directly to the wings but to the flexible portions of the exoskeleton that articulate with the wings. Each contraction of the muscles "dimples" the exoskeleton in association with a downstroke and sometimes, depending on the exact arrangement of the muscles, on the upstroke as well. When the dimple springs back into resting position, the muscles attached to it stretch. The stretching immediately initiates another contraction, and the cycle repeats.

The deformation of the exoskeleton is transmitted as a force to the wings, which beat so fast we may perceive the sounds as a musical tone. In the common blowfly, for instance, the wings may beat at 120 cycles *per second*. Yet in the same blowfly, the neurons that innervate those furiously contracting flight muscles are delivering impulses to them at the astonishingly low frequency of 3 per second. The mechanical properties of the musculoskeletal arrangement provide the stimuli for contraction, by stretching the muscle fibers at a high frequency, but the nerve impulses are needed to maintain contraction.

Researchers are using a variety of approaches to unravel the mysteries of insect flight. Some groups of investigators are using computer simulations, and one group has designed a fly robot to use as a model. Another group is painting the wings of honeybees with a dye that responds to changes in air pressure. Under UV light, the dye phosphoresces bright red. Oxygen in the air diminishes the red glow, so regions on the wings that are affected by the highest air pressure phosphoresce least (more oxygen molecules are present in denser air). The researchers can map out the forces acting on the wings by recording the intensity of the red glow. Because a bee's wing-beat cycle takes place in 5 milliseconds, mapping these forces is a major challenge.

A vertebrate muscle may consist of thousands of muscle fibers

In vertebrates, skeletal muscle is the most abundant tissue in the body. Its elongated cells, or **muscle fibers,** are organized in bundles wrapped by connective tissue. The biceps in your arm, for example, consists of thousands of individual muscle fibers and their connective tissue coverings.

Each striated muscle fiber is a long, cylindrical cell with many nuclei (Fig. 38-7). The plasma membrane, known as the **sarcolemma** in a muscle fiber, has multiple inward extensions that form a set of **T tubules** (transverse tubules). The cytoplasm of a muscle fiber is called **sarcoplasm,** and the endoplasmic reticulum is called the **sarcoplasmic reticulum.**

Threadlike structures called **myofibrils** run lengthwise through the muscle fiber. They consist of even smaller structures, the **myofilaments** or simply **filaments.** There are two types of myofilaments: myosin and actin filaments. **Myosin filaments** are thick, consisting mainly of the protein myosin. The thin **actin filaments** consist mostly of the protein actin; they also contain the proteins **tropomyosin** and the **troponin complex** that regulate the actin filament's interaction with myosin filaments.

Myosin and actin filaments are organized into repeating units called **sarcomeres,** the basic units of muscle contraction. Hundreds of sarcomeres connected end-to-end make up a myofibril. Sarcomeres are joined at their ends by an interweaving of filaments called the **Z line.** Each sarcomere consists of overlapping myosin and actin filaments. The filaments overlap lengthwise in the muscle fibers, producing the pattern of transverse bands or striations characteristic of striated muscle (Figs. 38-7 and 38-8). The bands are designated by the letters *A, H,* and *I.*

The **I bands,** which consist of actin filaments, lie at both ends of the sarcomere, immediately adjacent to the Z line. The **A band** is the wide, dark region of overlapping myosin and actin filaments. Within the A band is a narrow, light area, the **H zone,** made up exclusively of myosin filaments; the actin filaments do not extend into this region.

Contraction occurs when actin and myosin filaments slide past each other

Muscle contraction occurs when the sarcomeres, and thus the muscle fibers, shorten. This theory of muscle contraction, known as the **sliding filament model,** was developed in the 1950s by two British biologists, Hugh Huxley and Andrew Huxley. We now know that the muscle shortens as the actin and myosin filaments slide past each other, increasing their overlap. You might think of an extension ladder. The overall ladder length changes as the ends get closer or farther apart, but the length of each ladder section stays the same. The I band and H zone decrease in length, but neither the actin nor myosin filaments themselves shorten.

In a **motor unit,** a motor neuron (a nerve cell that stimulates muscle) is functionally connected with an average of about 150 muscle fibers (Fig. 38-9). Each junction of a motor neuron with a muscle fiber is called a **neuromuscular junction.** When a motor neuron transmits a message, it releases the neurotransmitter **acetylcholine** into the synaptic cleft (a small space) between the motor neuron and each muscle fiber. Acetylcholine binds with receptors on each muscle fiber, causing **depolarization,** a change in the distribution of electrical charge across its sarcolemma. Depolarization can cause an electrical signal, or **action potential,** to be generated in the muscle fiber.

In a muscle fiber, an action potential is a wave of depolarization that travels along the sarcolemma and into the system of T-tubule membranes. Depolarization of the T tubules opens calcium channels in the sarcoplasmic reticulum, releasing stored calcium ions into the myofibrils. Calcium ions bind to a protein in the troponin complex on the actin filaments, which changes the shape of the troponin. This change results in the troponin pushing the tropomyosin away from the active sites on the actin filament (Fig. 38-10 on page 738). These active sites, also called *myosin-binding sites,* are now exposed.

One end of each myosin molecule is folded into two globular structures called *heads.* The rounded heads of the myosin molecules extend away from the body of the myosin filament. Each myosin molecule also has a long tail that joins other myosin tails to form the body of the thick filament. ATP is bound to the myosin when the muscle fiber is at rest (not contracting). Myosin is an adenosine triphosphatase (ATPase), an enzyme that splits ATP, forming ADP and inorganic phosphate (P_i). Myosin converts the chemical energy of ATP into the mechanical energy of sliding filaments.

According to the current model of muscle contraction, the ADP and P_i initially remain attached to the myosin head. The myosin head (with ADP and P_i still bound to it) is in an ener-

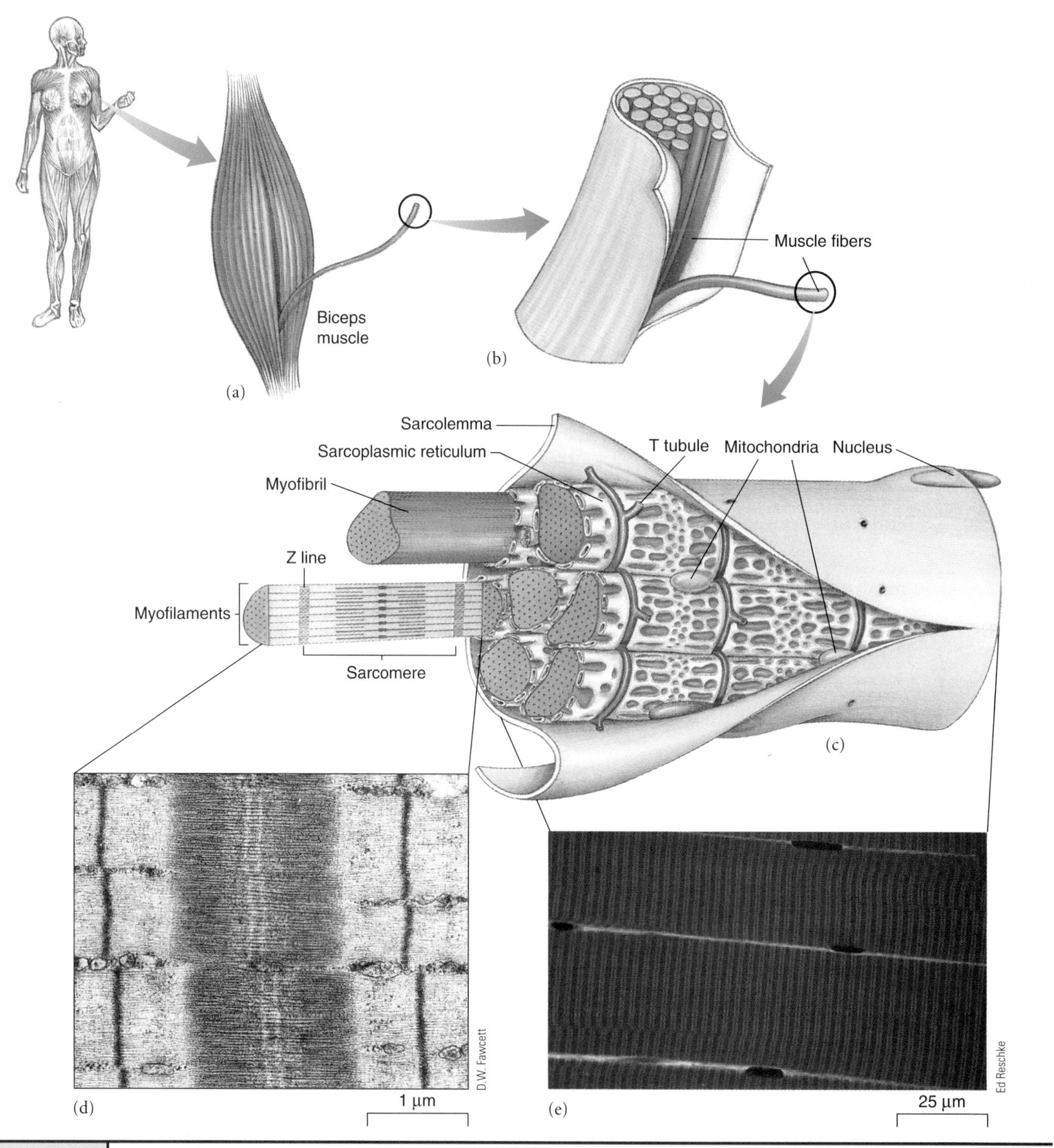

FIGURE **38-7** **Muscle structure.**

(a) A muscle such as the biceps in the arm consists of many fascicles (bundles) of muscle fibers. **(b)** A fascicle wrapped in a connective tissue covering. **(c)** Part of a muscle fiber showing the structure of myofibrils. The Z lines mark the ends of the sarcomeres. **(d)** TEM of striated muscle. **(e)** LM showing striations.

gized state; it is "cocked." The myosin head binds to an exposed active site on the actin filament, forming a **cross bridge** linking the myosin and actin filaments. The inorganic phosphate (P_i) is then released, which triggers a conformational change in the myosin head. The myosin head bends about 45 degrees, in a flexing motion. This movement, the *power stroke,* pulls the actin filament closer to the center of the sarcomere. During the power stroke, the ADP is released.

A new ATP must bind to the myosin head before the myosin can detach from the actin. If sufficient calcium ions are pres-

FIGURE **38-8** | **Muscle contraction.**

Myofibrils are threadlike structures that contain actin and myosin filaments. **(a)** Cross section of a myofibril shows the arrangement of actin and myosin filaments. **(b)** Part of a muscle fiber showing the location of the filaments. **(c)** The regular pattern of overlapping filaments gives skeletal and cardiac muscle their striated appearance. **(d)** During contraction actin filaments slide toward each other, increasing the amount of overlap between actin and myosin filaments. **(e)** As the sarcomeres become shorter, the muscle fiber shortens. Note that the I band and H zone decrease in length as the filaments slide past each other. The filaments themselves do not become shorter. (The I band consists of actin filaments of two adjacent sarcomeres.)

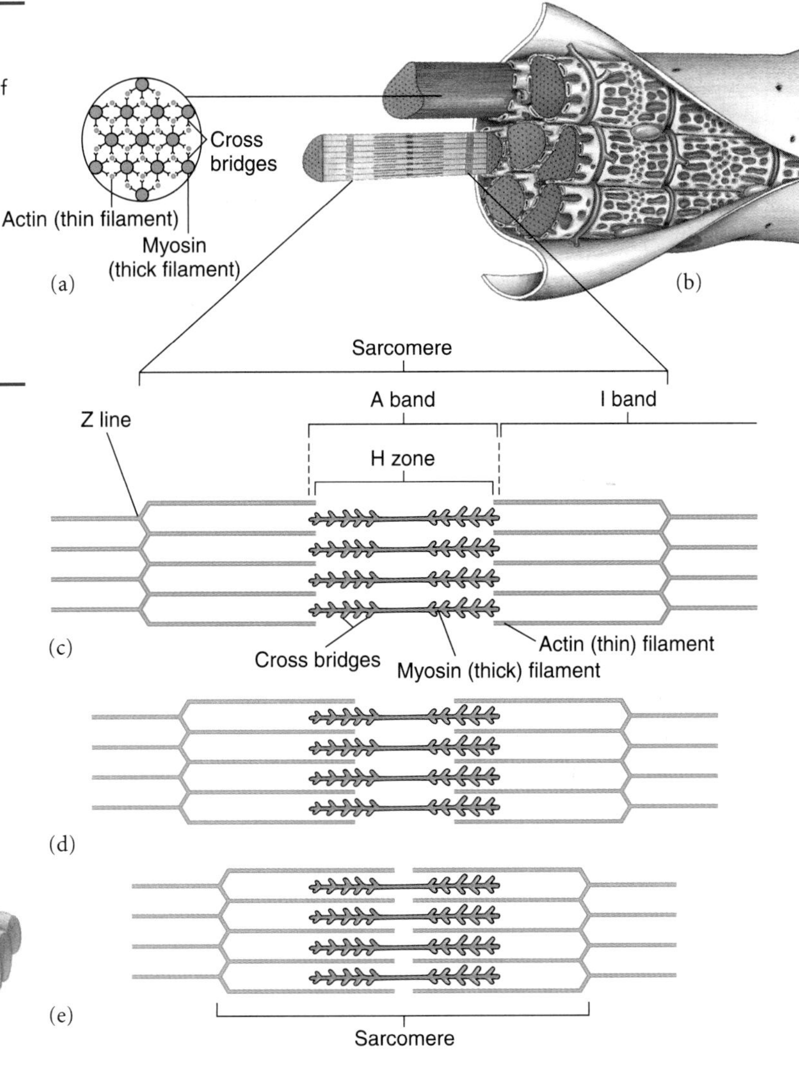

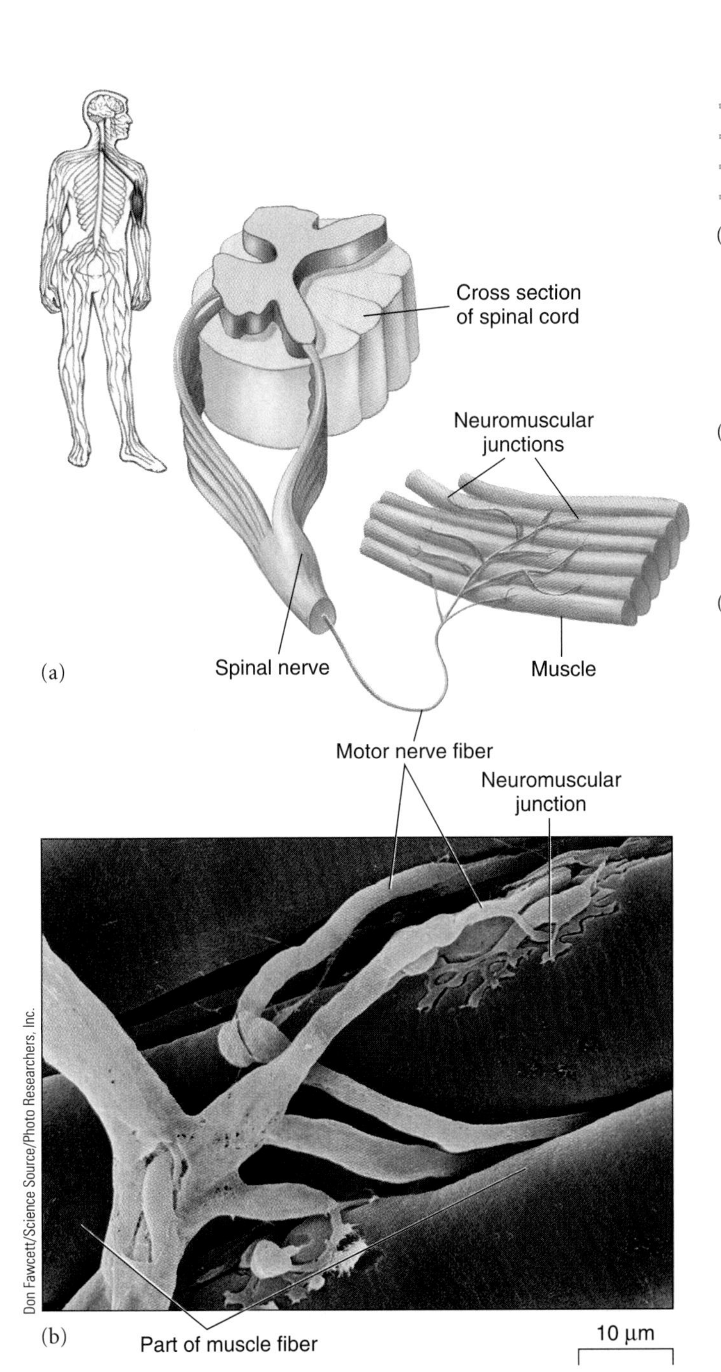

ent, the cycle begins anew. Energized once again, myosin heads contact a second set of active sites on the actin filament; this next set is further down the molecule, closer to the end of the sarcomere. The process repeats with a third set, and so on. This series of stepping motions pulls the actin filaments toward the center of the sarcomere.

Each time the myosin heads attach, move 45 degrees, detach, and then reattach farther along the actin filament, the muscle shortens. One way to visualize this process is to imagine the myosin heads engaging "hand-over-hand" with the actin filaments. When many sarcomeres contract simultaneously,

FIGURE **38-9** | **A motor unit.**

A motor unit consists of a motor nerve fiber and muscle fibers that it innervates. A motor unit typically includes about 150 muscle fibers, but some units have less than a dozen fibers, whereas others have several hundred. **(a)** The motor unit illustrated here shows only a single motor nerve fiber. **(b)** SEM of some of the fibers in a motor unit. Note how the neuron branches to innervate all muscle fibers in the motor unit.

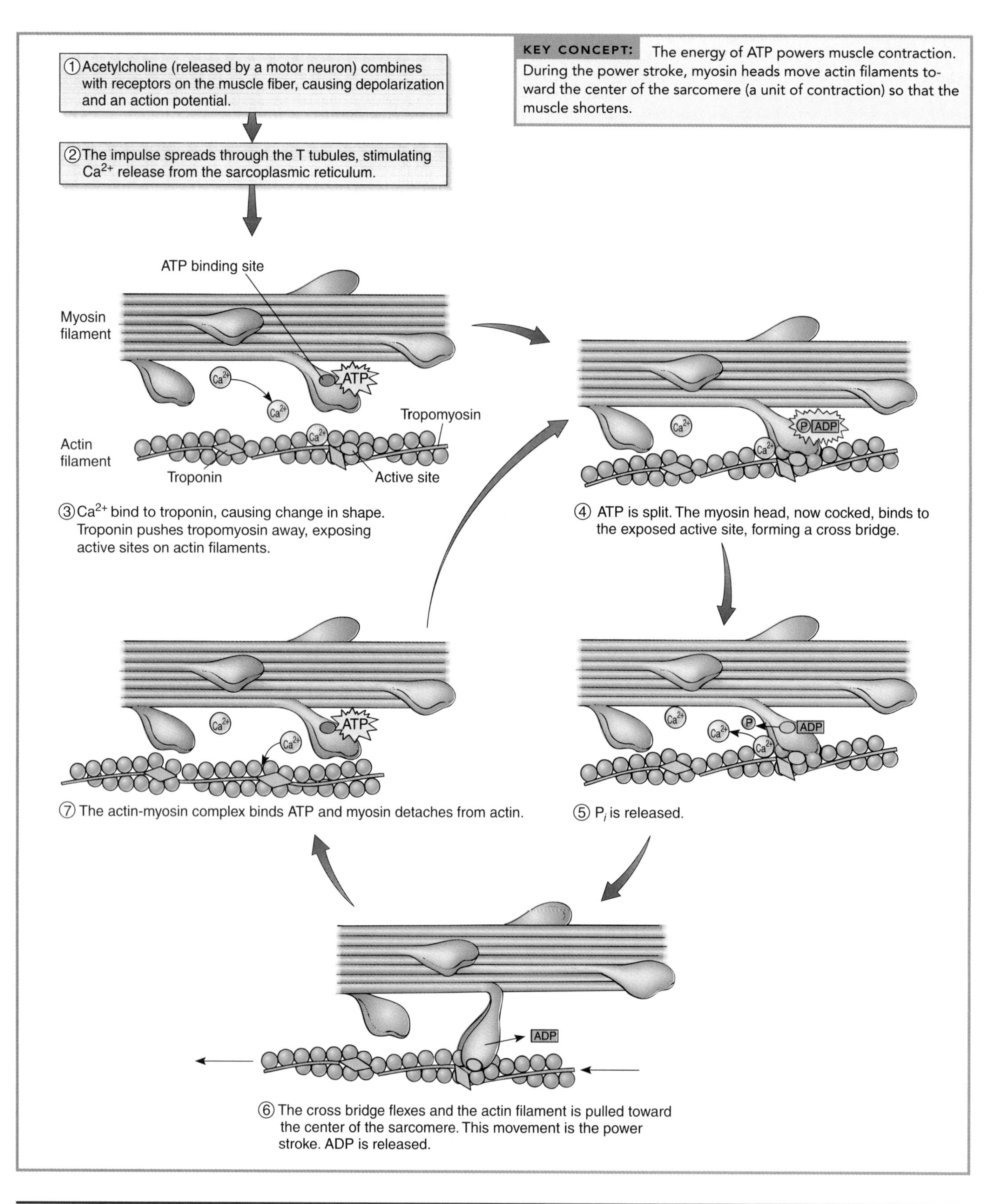

KEY CONCEPT: The energy of ATP powers muscle contraction. During the power stroke, myosin heads move actin filaments toward the center of the sarcomere (a unit of contraction) so that the muscle shortens.
① Acetylcholine (released by a motor neuron) combines with receptors on the muscle fiber, causing depolarization and an action potential.
② The impulse spreads through the T tubules, stimulating Ca^{2+} release from the sarcoplasmic reticulum.
ATP binding site
Myosin filament
ATP
Tropomyosin
Actin filament
Troponin
Active site
③ Ca^{2+} bind to troponin, causing change in shape. Troponin pushes tropomyosin away, exposing active sites on actin filaments.
P ADP
④ ATP is split. The myosin head, now cocked, binds to the exposed active site, forming a cross bridge.
⑤ P_i is released.
⑦ The actin-myosin complex binds ATP and myosin detaches from actin.
ADP
⑥ The cross bridge flexes and the actin filament is pulled toward the center of the sarcomere. This movement is the power stroke. ADP is released.

ACTIVE FIGURE **38-10** **Model of muscle contraction.**

Contraction results when actin filaments slide toward the center of individual sarcomeres of a myofibril. After Step ⑦, the cycle repeats from step ④.

BiologyNow™ Watch **the steps of muscle contraction** by clicking on this figure on your BiologyNow CD-ROM.

they contract the muscle as a whole. The sequence of events in muscle contraction can be summarized as follows (circled numbers correspond to steps in Fig. 38-10):

① Acetylcholine (released from motor neuron) combines with receptors on muscle fiber causing depolarization and an action potential
② the action potential spreads through T tubules, triggering Ca^{2+} release from sarcoplasmic reticulum
③ When troponin binds Ca^{2+} it undergoes a conformational change that causes active sites on actin filaments to be exposed
④ ATP (attached to myosin) is split and the energized myosin head is cocked; it binds to active site on actin filament, forming a cross bridge
⑤ P_i is released from myosin head, triggering the power stroke
⑥ the power stroke occurs as the actin filament is pulled toward center of sarcomere; and ADP is released
⑦ the myosin head binds ATP and detaches from actin If sufficient Ca^{2+} is present, the sequence repeats from step ④.

When impulses from the motor neuron cease, muscle fibers return to their resting state. The enzyme acetylcholinesterase in the synaptic cleft inactivates acetylcholine. Calcium ions are pumped back into the sarcoplasmic reticulum by active transport, a process that requires ATP. Without calcium ions, the tropomyosin-troponin complex once again covers active sites on the actin filaments. The actin filaments slide back to their original position, and the muscle relaxes. This entire series of events happens in milliseconds.

Even when you are not moving, your muscles are in a state of partial contraction known as **muscle tone.** At any given moment some muscle fibers are contracted, stimulated by messages from motor neurons. Muscle tone is an unconscious process that keeps muscles prepared for action. When the motor nerve to a muscle is cut, the muscle becomes limp (completely relaxed, or flaccid) and eventually atrophies (decreases in size).

ATP powers muscle contraction

Muscle cells often perform strenuously and need large amounts of energy. As noted, the immediate source of energy for muscle contraction is ATP. Energy stored in ATP molecules powers the repeated cocking, attachment of the myosin heads to the actin filaments, flexion, and release of the myosin heads. Note that energy is needed not only for the pull exerted by the cross bridges but also for their release from each active site. *Rigor mortis,* the temporary but very marked muscular rigidity that appears after death, results from ATP depletion following the end of cellular respiration. Rigor mortis does not persist indefinitely, because the entire contractile apparatus of the muscles eventually decomposes, restoring pliability. The phenomenon is temperature dependent, so, given the prevailing temperature, a medical examiner can estimate the time of death of a cadaver from its degree of rigor mortis.

ATP molecules can provide energy for only a few seconds of strenuous activity. Fortunately, muscle cells have a backup energy storage compound, called **creatine phosphate,** that can be stockpiled. The energy stored in creatine phosphate is transferred to ATP as needed. But during vigorous exercise the supply of creatine phosphate is quickly depleted. Muscle cells must replenish their supplies of these energy-rich compounds.

Chemical energy is stored in muscle fibers in **glycogen,** a large polysaccharide formed from hundreds of glucose molecules. Glycogen is degraded, yielding glucose, which then breaks down in cellular respiration. When sufficient oxygen is available, enough energy is captured from glucose to produce needed quantities of ATP and creatine phosphate.

During a burst of strenuous exercise, the circulatory system cannot deliver enough oxygen to keep up with the demand of the rapidly metabolizing muscle cells. This results in an **oxygen debt.** Under these conditions, muscle cells break down fuel molecules anaerobically (without oxygen) for short periods. Lactic acid fermentation is a method of generating ATP anaerobically, but not in great quantity (see Fig. 7-3). ATP depletion results in weaker contractions and muscle fatigue. Accumulation of the waste product **lactic acid** also contributes to muscle fatigue. Well-conditioned athletes develop the ability to tolerate the high levels of lactic acid generated during high-performance activity. The period of rapid breathing that generally follows strenuous exercise pays back the oxygen debt by consuming lactic acid.

Skeletal muscle action depends on muscle pairs that work antagonistically

Skeletal muscles produce movements by pulling on **tendons,** tough cords of connective tissue that anchor muscles to bone. Tendons then pull on bones. Skeletal muscles, or their tendons, pass across a joint and are attached to the bones on each side of the joint. When the muscle contracts, it pulls one bone toward or away from the bone with which it articulates.

Because muscles can only contract, they can only pull; they cannot push. Muscles act **antagonistically** to one another, which means that the movement produced by one can be reversed by another. The biceps muscle, for example, flexes (bends) your arm, whereas contraction of the triceps muscle extends it once again (Fig. 38-11). Thus the biceps and triceps work antagonistically.

The muscle that contracts to produce a particular action is known as the **agonist.** The muscle that produces the opposite movement is the **antagonist.** When the agonist is contracting, the antagonist is relaxed. Generally, movements are accomplished by groups of muscles working together, so several agonists and several antagonists may take part in any action. Note that muscles that are agonists in one movement may serve as antagonists in another. The superficial skeletal muscles of the human body are shown in Figure 38-12.

Muscle fibers may be specialized for slow or quick responses.

One component of myosin, the heavy chain, exists in three different forms: Types I, IIa, and IIx. The three types of myosin

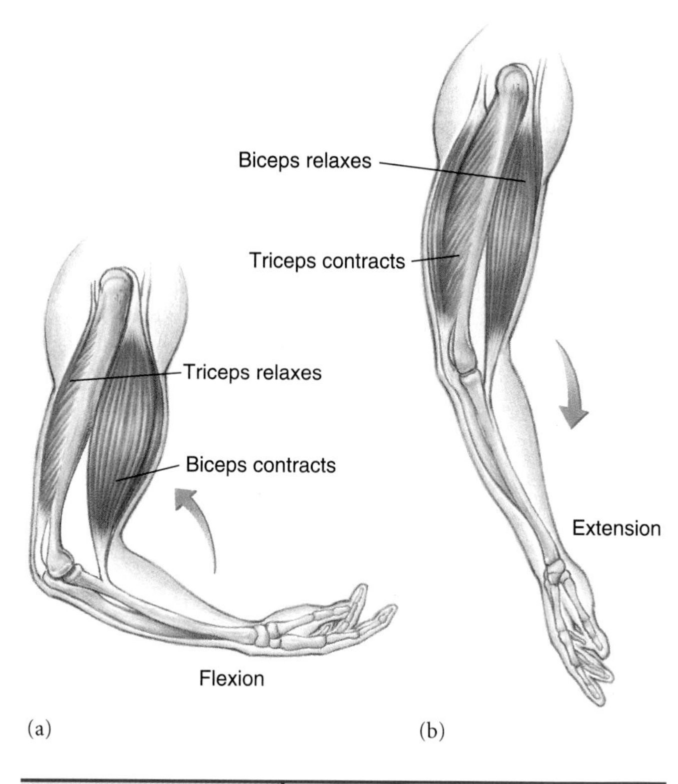

ACTIVE FIGURE **38-11** | **Muscle action.**
The biceps and triceps muscles function antagonistically.

BiologyNow™ Watch **a muscle in action** by clicking on this figure on your BiologyNow CD-ROM.

apparently evolved early and have been conserved throughout the animal kingdom. Muscle fibers that contain mainly Type I myosin are called Type I fibers, or **slow fibers.** These fibers are specialized for endurance activities such as swimming, long-distance running, or maintaining posture. Slow fibers require a steady supply of oxygen. They derive most of their energy from aerobic metabolism and are rich in mitochondria and capillaries. Slow fibers are also called **red fibers,** because they are rich in **myoglobin,** a red pigment similar to hemoglobin in red blood cells. Myoglobin, like hemoglobin, stores oxygen. Myoglobin enhances rapid diffusion of oxygen from blood into muscles during strenuous muscle exertion.

Muscle fibers that contain Type IIa and IIx myosin are known as **fast fibers,** or **white fibers.** (Type IIx fibers are the fastest.) White fibers generate a great deal of power and carry out rapid movements but can sustain that activity for only a short time. They are important in activities such as sprinting and weight lifting. White fibers have few mitochondria and must obtain most of their energy from glycolysis. When their glycogen supply is depleted, they fatigue rapidly. People who are sedentary have higher numbers of Type IIx fibers than do physically fit individuals. With physical training, Type IIx fibers change to Type IIa fibers.

Entire muscles may be specialized for quick or slow responses. In chickens, for instance, the white breast muscles are efficient for quick responses, perhaps because a short flight is an escape mechanism for chickens. However, they walk about on the ground all day; the dark (red) meat of the leg and thigh is muscle specialized for more sustained activity. Birds that fly have red breast muscles specialized to support sustained activity.

The proportions of slow and fast fibers vary from person to person and from muscle to muscle in the same person. Although the relative proportions of the two types of fibers appear genetically influenced, appropriate training can change the proportions. Someone whose leg and thigh muscles contain a high proportion of fast fibers can, with proper training, become a good sprinter. An athlete with a greater proportion of slow fibers may be better suited to marathon activities.

Smooth, cardiac, and skeletal muscle respond in specific ways

The three types of vertebrate muscle—skeletal, smooth, and cardiac—are compared in Chapter 37, Table 37-3. Each type of muscle is specialized for particular types of responses. **Smooth muscle** is not attached to bones but instead forms tubes that squeeze like the muscle tissue in the body wall of the earthworm. Smooth muscle often contracts in response to simple stretching, and its contraction tends to be sustained. It is well adapted to performing such tasks as regulating blood pressure by sustained contraction of the walls of the arterioles. Although smooth muscle contracts slowly, it shortens much more than striated muscle does; it squeezes superlatively.

Smooth muscle is not striated because its actin and myosin filaments are not organized into myofibrils or into sarcomeres. The fibers of smooth muscle tissue function as a unit because they are connected by *gap junctions* (see Chapter 5). Gap junctions permit electrical signals to pass rapidly from fiber to fiber. Although smooth muscle contraction is basically similar to contraction of skeletal muscle (occurring by a sliding filament mechanism), the cross bridges in smooth muscle remain in the attached state longer. This translates into less ATP being required to maintain a high level of force.

Cardiac muscle contracts and relaxes in alternating rhythm, propelling blood with each contraction. Sustained contraction of cardiac muscle would be disastrous! Like smooth muscle fibers, cardiac muscle fibers are electrically coupled by gap junctions (the intercalated discs). Cardiac muscle produces its own signals for contraction (discussed in Chapter 42).

Skeletal muscle, when stimulated by a single brief electrical stimulus, contracts with a single quick contraction called a **simple twitch.** Typically, skeletal muscle receives a series of separate stimuli timed very close together. These do not produce a series of simple twitches, however, but a single, smooth, sustained contraction called **tetanus.** Depending on the identity and number of our muscle cells tetanically contracting, we might thread a needle, rock a baby, or run a mile.

Review

- Describe a skeletal muscle fiber. What are its two types of filaments?
- What events occur when a muscle fiber contracts (begin with release of acetylcholine and include cross-bridge action)?

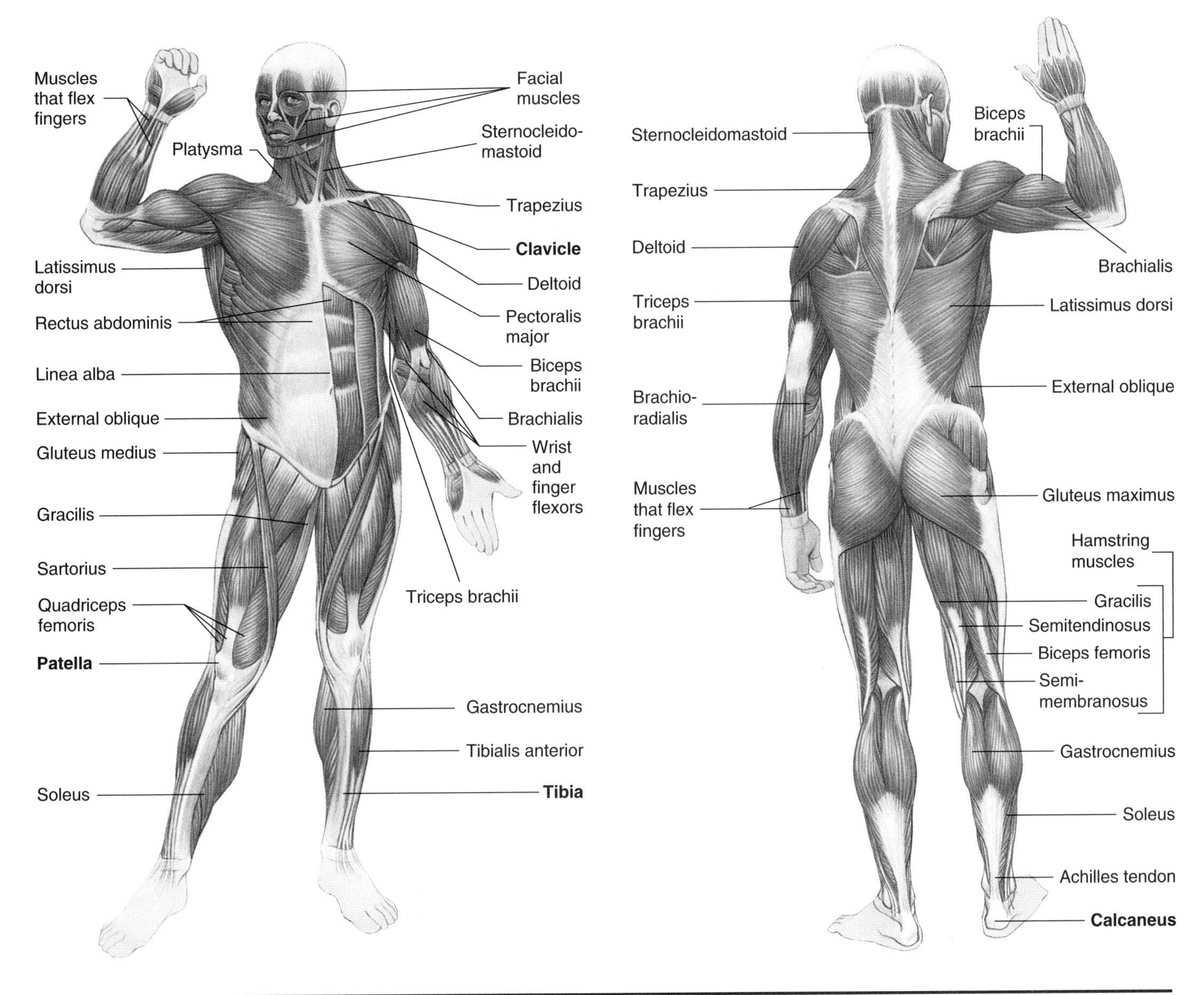

FIGURE **38-12** **Some superficial muscles of the human body.**
(a) Anterior view. **(b)** Posterior view. The labels in boldface are bones.

- What is the role of ATP in muscle contraction? What are the functions of creatine phosphate and glycogen?

BiologyNow™ Assess your understanding of **muscle contraction** by taking the pretest on your BiologyNow CD-ROM.

SUMMARY WITH KEY TERMS

1 Compare the structure and functions of the external epithelium of invertebrates and vertebrates.

- **Epithelial coverings** in both invertebrates and vertebrates protect underlying tissues, and may be specialized for sensory or respiratory functions. The outer epithelium may be specialized to secrete lubricants or adhesives, or odorous or poisonous substances.
- In many invertebrates, the outer epithelium is specialized to secrete a protective **cuticle** or shell.
- The **integumentary system** of vertebrates includes the skin and structures that develop from it. Mammalian skin includes hair, claws or nails, sweat glands, oil glands, and sensory receptors.

2 Relate the structure of vertebrate skin to its functions.

- The feathers of birds and the hair of mammals help maintain a constant body temperature.
- Mammalian skin is well adapted to protect the body from the wear and tear that occurs as it interacts with the outer environ-

ment. **Stratum corneum,** the most superficial layer of the epidermis, consists of dead cells. These cells are filled with **keratin,** which gives mechanical strength to the skin and reduces water loss. Cells in **stratum basale** of the **epidermis** divide and are pushed upward toward the skin surface. These cells mature, flatten, produce keratin, and eventually die and slough off.

- The **dermis** consists of dense, fibrous connective tissue. In birds and mammals, the dermis rests on a layer of **subcutaneous tissue** composed largely of insulating fat.

3 Compare the different types of skeletal systems, including the hydrostatic skeleton, exoskeleton, and endoskeleton.

- The **skeletal system** supports and protects the body, and transmits mechanical forces generated by muscles.
- Many soft-bodied invertebrates, including cnidarians, flatworms, and annelids, have a **hydrostatic skeleton** in which fluid in a closed body compartment is used to transmit forces generated by contractile cells or muscle.
- **Exoskeletons** are characteristic of mollusks and arthropods. The arthropod skeleton, composed partly of **chitin,** is jointed for flexibility, and adapted for many lifestyles. This nonliving skeleton does not grow, making it necessary for arthropods to **molt** periodically.
- The **endoskeletons** of echinoderms and chordates consist of living tissue and therefore can grow.

4 Describe the main divisions of the vertebrate skeleton and the bones that make up each division.

- The vertebrate skeleton consists of an **axial skeleton** and an **appendicular skeleton.**
- The axial skeleton consists of **skull, vertebral column, ribs,** and **sternum.**
- The appendicular skeleton consists of bones of the limbs, **pectoral girdle,** and **pelvic girdle.**

5 Describe the structure of a typical long bone, and differentiate between endochondral and intramembranous bone development.

- A typical long bone consists of a thin outer shell of **compact bone** surrounding the inner **spongy bone** and a central cavity that contains **bone marrow.**
- Long bones develop from cartilage templates during **endochondral bone formation.** Other bones, such as the flat bones of the skull, develop from a noncartilage connective tissue model by **intramembranous bone development.**
- **Osteoblasts,** cells that produce bone, and **osteoclasts,** cells that break down bone, work together to shape and remodel bone.

6 Compare the main types of vertebrate joints.

- **Joints** are junctions of two or more bones. **Ligaments** are connective tissue bands that connect bones and limit movement at the joint.
- The sutures of the skull are **immovable joints.** Joints between vertebrae are **slightly movable joints.**
- A **freely movable joint** is enclosed by a joint capsule lined with a membrane that secretes **synovial fluid.**

7 Relate the structure and function of insect flight muscles.

- Large numbers of mitochondria and tracheae (air tubes) present in insect flight muscles support the high metabolic rate required for flight.

8 Describe the macroscopic and microscopic structures of skeletal muscle.

- A **muscular system** is found in most invertebrate phyla and in vertebrates. As muscle tissue contracts (shortens), it moves body parts by pulling on them. Three types of muscle are **skeletal, smooth,** and **cardiac muscle.**
- A skeletal muscle such as the biceps is an organ made up of hundreds of **muscle fibers.** Each fiber consists of threadlike **myofibrils** composed of smaller **myofilaments,** or simply **filaments.**
- The striations of skeletal muscle fibers reflect the overlapping of their **actin filaments** and **myosin filaments.** A **sarcomere** is a contractile unit of actin (thin) and myosin (thick) filaments.

9 List, in sequence, the events that take place during muscle contraction.

- **Acetylcholine** released by a motor neuron combines with receptors on the surface of a muscle fiber. This may cause **depolarization** of the sarcolemma and transmission of an **action potential.**
- The action potential spreads through the **T tubules,** releasing calcium ions from the **sarcoplasmic reticulum.**
- Calcium ions bind to **troponin** in the actin filaments, causing the troponin to change shape. Troponin pushes **tropomyosin** away from the active sites on the actin filaments.
- ATP binds to myosin; ATP is split, putting the myosin head in a high-energy state (it is "cocked"). Energized myosin heads bind to the exposed active sites on the actin filaments, forming **cross bridges** linking the myosin and actin filaments.
- After myosin attaches to the actin filament, phosphate is released, flexing the cross bridge. The actin filament is pulled toward the center of the sarcomere. This is the power stroke. ADP is released.
- The myosin head binds a new ATP, which lets the myosin head detach from the actin. As long as the calcium ion concentration remains elevated, the new ATP is split, and the sequence repeats. The myosin reattaches to new active sites so that the filaments are pulled past one another, and the muscle continues to shorten.

10 Compare the roles of glycogen, creatine phosphate, and ATP in providing energy for muscle contraction.

- During muscle contraction, the myosin filaments pull the actin filaments toward the center of the myofibril. This movement requires energy.
- ATP is the immediate source of energy for muscle contraction. The energy from ATP hydrolysis provides the energy to "cock" the myosin.
- Muscle tissue has an intermediate energy storage compound, **creatine phosphate.**
- **Glycogen** is the fuel stored in muscle fibers.

11 Describe how muscles work, including the antagonistic action of skeletal muscles.

- Muscles pull on **tendons,** connective tissue cords that attach muscles to bones. When a muscle contracts, it pulls a bone toward or away from the bone with which it articulates. **Muscle tone** is the state of partial contraction characteristic of muscles.
- Muscles act **antagonistically** to one another. The muscle that produces a particular action is the **agonist;** the **antagonist** produces the opposite movement.
- When activated by a brief electrical stimulus, skeletal muscle responds with a **simple twitch.** Typically skeletal muscle is stim-

SUMMARY WITH KEY TERMS *(continued)*

ulated with a series of separate stimuli timed close together and responds with a smooth, sustained contraction called **tetanus.**

12 Compare the structures and functions of slow and fast muscle fibers.

- **Slow fibers,** or **red fibers,** are rich in mitochondria and **myoglobin.** They are specialized for endurance activities.
- **Fast fibers,** or **white fibers,** are specialized for rapid response. Using ATP from glycolysis, they can generate a great deal of power for a brief period.

POST-TEST

1. The vertebrate skin consists of (a) outer epidermis, inner hypodermis (b) outer epidermis, inner endoskeleton (c) outer endodermis, inner epidermis (d) outer epidermis, inner dermis (e) outer subcutaneous layer, inner dermis
2. Cells actively divide in (a) stratum basale (b) stratum corneum (c) stratum dermis (d) the layer with cells that contain keratin (e) two of the preceding answers are correct
3. An endoskeleton (a) is typically composed of dead tissue (b) is characterized by fluid in a closed compartment (c) is typical of echinoderms (d) is typical of arthropods (e) requires the animal to molt
4. Which of the following is *not* part of the axial skeleton? (a) skull (b) vertebral column (c) pelvic girdle (d) rib cage (e) sternum
5. The thin outer shell of a long bone is made of (a) compact bone (b) spongy bone (c) epiphyses (d) cancellous bone (e) mainly chondrocytes
6. Which of the following connects bones to one another? (a) tendons (b) ligaments (c) osteoclasts (d) synovial membranes (e) smooth fibers
7. In endochondral bone formation (a) osteoclasts produce bone (b) joints connect fibers (c) the skeleton consists of cartilage (d) bones develop from cartilage templates (e) bones form in noncartilage connective tissue
8. All animals have (a) muscles (b) actin (c) bones (d) endoskeleton or exoskeleton (e) dermis
9. An energy storage compound that can be stockpiled in muscle cells for short-term use is (a) creatine phosphate (b) ADP (c) troponin (d) myosin (e) myoglobin
10. Myosin binds to actin, forming a cross bridge. What happens next? (a) acetylcholine is released (b) calcium ions stimulate process that leads to exposure of active sites (c) filaments slide past each other, and the muscle fiber shortens (d) myosin is activated (e) P_i is released, and the cross bridge flexes
11. Calcium ions are released from the sarcoplasmic reticulum. What happens next? (a) acetylcholine is released (b) active sites on the actin filaments are exposed (c) filaments slide past each other, and the muscle fiber shortens (d) myosin is activated (e) P_i is released, and the cross bridge flexes
12. When skeletal muscle is stimulated by a series of closely timed separate stimuli (a) it responds with a smooth, sustained contraction called *tetanus* (b) a simple twitch occurs (c) white fibers respond (d) red fibers respond (e) muscle tone occurs
13. Glycogen is (a) produced by actin (b) depleted within 1 second of strenuous activity (c) a form of long-term energy storage (d) synthesized when cross bridges form (e) depleted before creatine phosphate reserves are used
14. Slow fibers (a) are also called *white fibers* (b) do not depend on ATP (c) have few mitochondria (d) obtain most of their energy from glycolysis (e) are rich in myoglobin
15. Insect flight muscles (a) work best at low temperature (b) have a very high metabolic rate (c) do not create much lift (d) evolved after bird wings (e) are always attached directly to the wing and make up the bulk of the wing
16. Which of the following pairs of systems function together most closely? (a) integumentary and skeletal (b) integumentary and digestive (c) muscular and epithelial (d) skeletal and muscular (e) epithelial and muscular

CRITICAL THINKING

1. Compare the arthropod exoskeleton with the vertebrate endoskeleton. What are some advantages and some disadvantages of each type of skeleton?
2. What are some examples of hydrostatic support in plants?
3. Why is it important that a muscle be able to shift functions, sometimes acting as an agonist and at other times acting as an antagonist?
4. Summarize the functional relationship between skeletal and muscle tissues.

- Visit our Web site at **http://biology.brookscole.com/solomon7** for links to chapter-related resources on the World Wide Web. Additional online materials relating to this chapter can also be found on our Web site.

BIOLOGY NOW RESOURCES

Active Figures

38-10: Muscle action

38-11: Some superficial muscles of the human body

Preparing for an exam? Take a diagnostic test on your BiologyNow CD-ROM.

Post-Test Answers

1. d	2. a	3. c	4. c
5. a	6. b	7. d	8. b
9. a	10. e	11. b	12. a
13. c	14. e	15. b	16. d

39 Neural Signaling

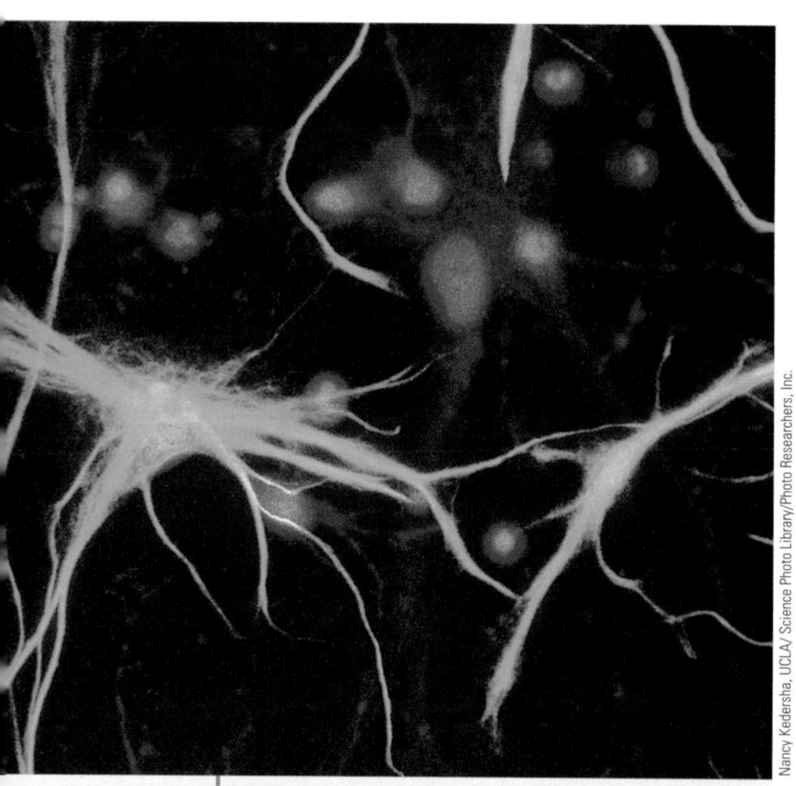

Nancy Kedersha, UCLA/ Science Photo Library/Photo Researchers, Inc.

Immunofluorescence LM of several astrocytes *(green)* from mammalian spinal cord tissue. The smaller cells *(red)* are neurons. Their cell bodies appear pink. The blue dots are the nuclei of other glial cells.

CHAPTER OUTLINE

An organism's ability to survive and to maintain homeostasis depends largely on how effectively it detects and responds to **stimuli**—changes in the environment. Stimuli within the body include internal signals such as hunger or lowered blood pressure. Stimuli from the outside world include changes in temperature, or light, an odor, or movement that may indicate the presence of a predator or of prey. In all animals except the sponges, responses to stimuli depend on *cell signaling* by networks of nerve cells, or **neurons.** These cells are specialized for transmitting **neural signals,** which are electrical signals and chemical messages.

In most animals, neurons and supporting cells are organized as a **nervous system** that, like a computer, takes in information, integrates it, and responds. Just how animals respond to stimuli depends on how their neurons are organized and connected to one another. A single neuron in the vertebrate brain may be functionally connected to thousands of other neurons. The endocrine system works with the nervous system to regulate many behaviors and physiological processes. The endocrine system generally provides relatively slow and long-lasting regulation, whereas the nervous system typically permits more rapid, but brief, responses.

Neurobiology is one of the most exciting areas of biological research. Many investigators are studying *neurotransmitters,* the chemical messengers used by neurons to signal other neurons, and the *receptors* that bind with the neurotransmitters. Another active area of research is the role of *glial cells* in the nervous system. These cells support and protect the neurons and have many regulatory functions. The glial cells shown in the micrograph are *astrocytes.* These glial cells provide glucose for neurons and also help regulate the composition of the extracellular fluid in the brain and spinal cord. Neurobiologists recently demonstrated that astrocytes induce and stabilize synapses (connections between neurons) in the brain. Although astrocytes can generate weak electrical signals, they communicate with one another and with neurons mainly with chemical signals. Based on recent findings, some researchers hypothesize that astrocytes may participate in information signaling in the brain by coordinating activity among neurons.

Astrocytes play a role in development by guiding neurons to appropriate locations in the body. Neurobiologists have shown that certain populations of astrocytes function as stem cells in the brain and spinal cord. These cells can give rise to neurons, additional astrocytes, and certain other glial cells. Taken out of their normal environment in the adult mouse brain, astrocytes can give rise to cells of all germ layers (that is, the embryonic tissue layers: ectoderm, mesoderm, and endoderm). Human astrocytes may some day be used to produce specific types of cells needed for treating various medical conditions. ■

INFORMATION FLOW THROUGH THE NERVOUS SYSTEM

Learning Objective

1 Trace the flow of information through the nervous system and describe the four processes involved in neural signaling: reception, transmission, integration, and action by effectors.

Thousands of stimuli constantly bombard an animal, and its survival depends on identifying these stimuli and responding appropriately. Appropriate response to a stimulus depends on communication by neurons, or **neural signaling.** In most animals, neural signaling involves four processes: reception, transmission, integration, and action by effectors (muscles or glands). **Reception,** the process of detecting a stimulus, is the job of the neurons and of specialized sensory receptors such as those in the skin, eyes, and ears. **Transmission** is the process of sending messages along a neuron, from one neuron to another or from a neuron to a muscle or gland. In vertebrates, a neural message is transmitted from a receptor to the **central nervous system (CNS),** which consists of the brain and spinal cord. Neurons that transmit information to the CNS are called **afferent** (meaning "to carry toward") **neurons,** or **sensory neurons.**

Afferent neurons generally transmit information to **interneurons,** or association neurons, in the CNS. Most neurons, perhaps 99%, are interneurons. Their function is to integrate input and output. **Integration** involves sorting and interpreting incoming sensory information and determining the appropriate response. Neural messages are transmitted from the CNS by **efferent** (meaning "to carry away") **neurons,** or **motor neurons,** to **effectors** (muscles and glands). The **action by effectors** is the actual response to the stimulus (Fig. 39-1). Sensory receptors and afferent and efferent neurons are part of the **peripheral nervous system (PNS).** In summary, information flows through the nervous system in the following sequence:

Reception by sensory receptor ⟶ transmission by afferent neuron ⟶ integration by interneurons in CNS ⟶ transmission by efferent neuron ⟶ action by effectors

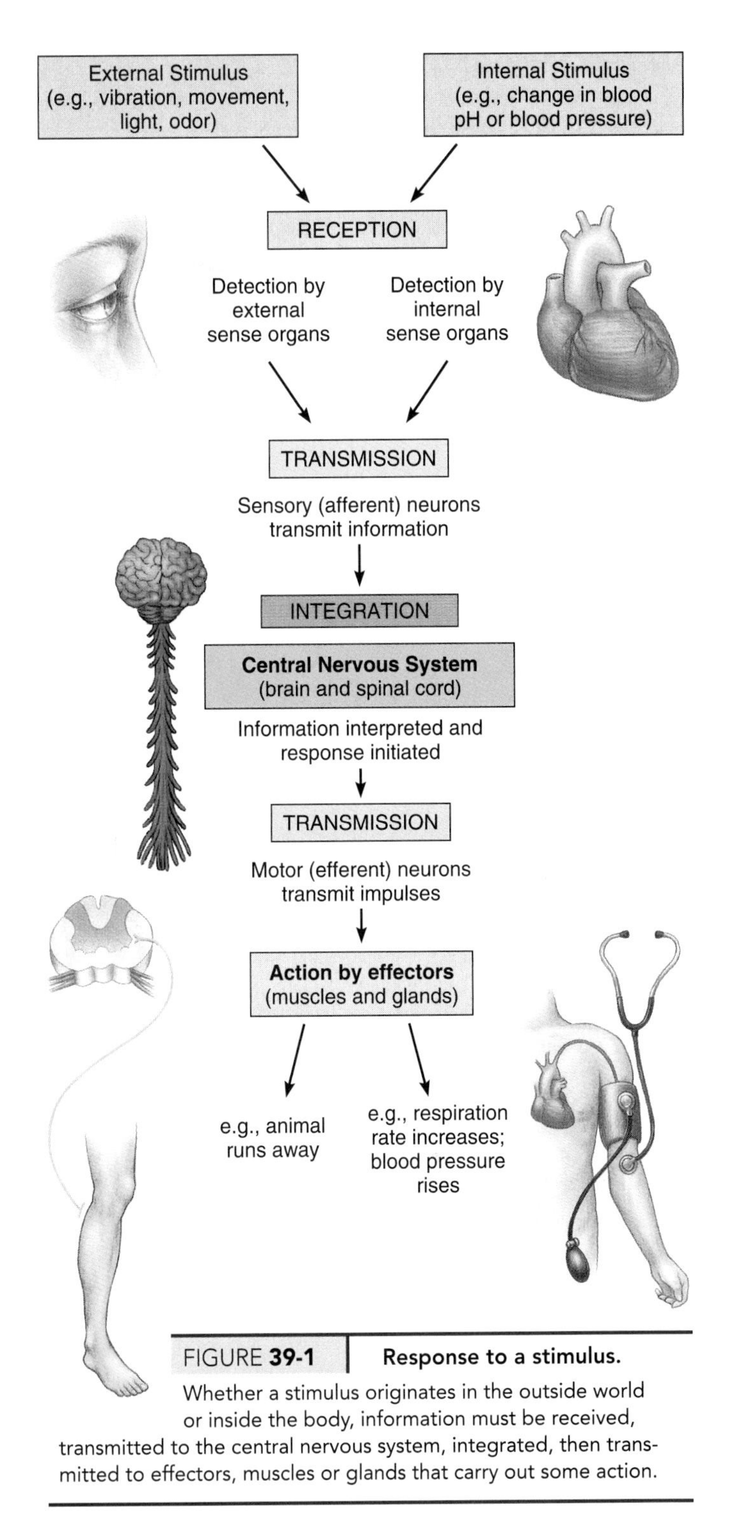

FIGURE **39-1** **Response to a stimulus.** Whether a stimulus originates in the outside world or inside the body, information must be received, transmitted to the central nervous system, integrated, then transmitted to effectors, muscles or glands that carry out some action.

Review

- Imagine you are swimming and you suddenly spot a shark fin moving in your direction. What processes must take place within your nervous system before you can make your escape?
- What happens after neural signals are integrated in the CNS?

BiologyNow™ Assess your understanding of **information flow through the nervous system** by taking the pretest on your BiologyNow CD-ROM.

NEURONS AND GLIAL CELLS

Learning Objectives

2 Describe the functions of each type of vertebrate glial cell.

3 Describe the structure of a typical neuron, and give the function of each of its parts.

Two types of cells unique to the nervous system are neurons and glial cells. Neurons are specialized to receive and send information. Glial cells support and protect the neurons.

Glial cells provide metabolic and structural support

Glial cells collectively make up the **neuroglia** (literally, "nerve glue"). The human brain has 10 times as many glial cells as neurons. Vertebrates have three main types of glial cells in the CNS: microglia, astrocytes, and oligodendrocytes. **Microglia** are phagocytic cells that remove cell debris. They are found near blood vessels in the nervous system. Recent findings suggest that microglia can respond to signals from neurons and may be important in mediating responses in injury or disease.

Astrocytes are star-shaped glial cells that provide neurons with glucose (see chapter opening photograph). Astrocytes help regulate the composition of the extracellular fluid in the CNS by removing potassium ions and excess neurotransmitters. Some astrocytes position the ends of their long **processes** (cytoplasmic extensions) on blood vessels in the brain. In response, endothelial cells lining the blood vessels form *tight junctions* (Chapter 5) that prevent many substances in the blood from entering the brain tissue. This protective wall is the *blood–brain barrier.* Neurobiologists have recently shown that astrocytes are functionally connected throughout the brain and they communicate through gap junctions and with signaling molecules.

Oligodendrocytes are glial cells that envelop neurons in the CNS, forming insulating sheaths around them. This covering consists of **myelin,** a white, fatty substance found in the plasma membrane of the glial cell. Because it is an excellent electrical insulator, its presence speeds transmission of neural impulses. In the neurological disease **multiple sclerosis,** which affects more than 400,000 people in North America alone, patches of myelin deteriorate at irregular intervals along axons in the CNS and are replaced by scar tissue. This damage interferes with conduction of neural impulses, and the victim suffers loss of coordination, tremor, and partial or complete paralysis of parts of the body. Evidence suggests that multiple sclerosis is an *autoimmune disease,* in which the body attacks its own tissue, in this case glial cells (see Chapter 43 for a discussion of autoimmune diseases).

In vertebrates, **Schwann cells,** another type of glial cell, are located outside the CNS. Schwann cells form myelin sheaths around some axons (discussed later).

A typical neuron consists of a cell body, dendrites, and an axon

The **neuron** is highly specialized to receive stimuli and to produce and transmit electrical signals called **nerve impulses,** or **action potentials.** The neuron is distinguished from all other cells by its long processes. Examine the structure of a common type of neuron, the multipolar neuron in Figure 39-2.

The largest portion of the neuron, the **cell body,** contains the nucleus, the bulk of the cytoplasm, and most of the organelles. Typically, two types of cytoplasmic extensions project from the cell body of a multipolar neuron. Numerous dendrites extend from one end, and a long, single axon projects from the opposite end. **Dendrites** are typically short, highly branched processes specialized to receive stimuli and send signals to the cell body. The cell body integrates incoming signals.

Although microscopic in diameter, an **axon** may be 1 m (over 1 yd) or more in length and may divide, forming branches called **axon collaterals.** The axon conducts nerve impulses away from the cell body to another neuron, or to a muscle or gland. At its

FIGURE **39-2** **Structure of a multipolar neuron.**
The cell body contains most of the organelles. Many dendrites and a single axon extend from the cell body. Schwann cells form a myelin sheath around the axon.

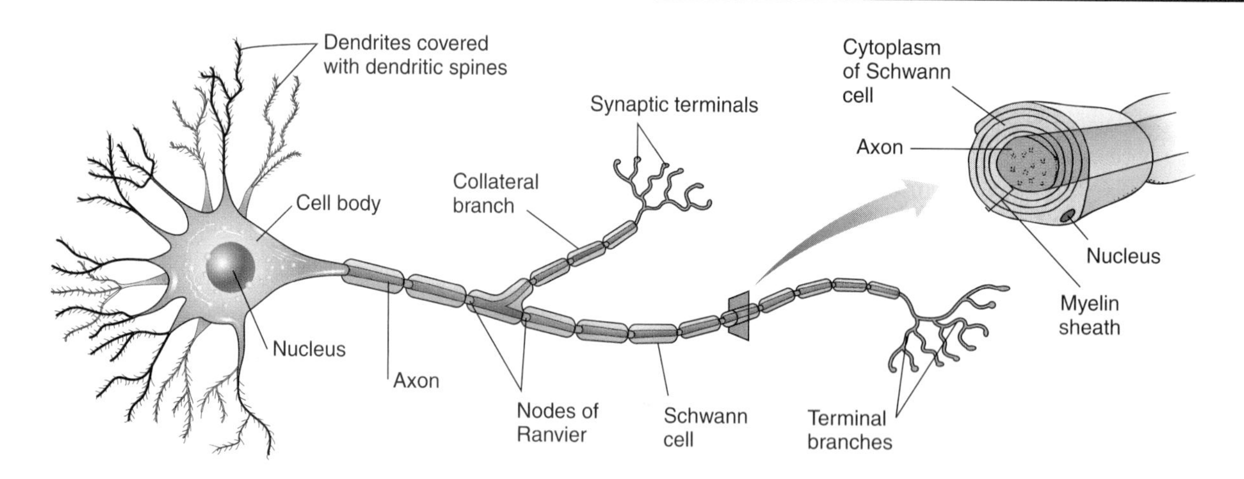

end the axon divides, forming many **terminal branches** that end in **synaptic terminals.** The synaptic terminals release **neurotransmitters,** chemicals that transmit signals from one neuron to another, or from neuron to effector. The junction between a synaptic terminal and another neuron (or effector) is called a **synapse.** Typically, a small space exists between the membranes of these two cells.

In vertebrates, the axons of many neurons outside the CNS are surrounded by a series of Schwann cells that form an insulating covering, the **myelin sheath.** Gaps in the myelin sheath, called **nodes of Ranvier,** occur between successive Schwann cells. At these points the axon is not insulated with myelin. Axons more than 2 μm in diameter have myelin sheaths and are described as *myelinated.* Those of smaller diameter are generally unmyelinated.

A **nerve** consists of hundreds or even thousands of axons wrapped together in connective tissue (Fig. 39-3). We can compare a nerve to a telephone cable. The individual axons correspond to the wires that run through the cable, and the sheaths and connective tissue coverings correspond to the insulation. Within the CNS, bundles of axons are called **tracts** or **pathways** rather than nerves. Outside the CNS, the cell bodies of neurons are usually grouped together in masses called **ganglia** (sing., *ganglion*). Inside the CNS, collections of cell bodies are generally called *nuclei* rather than *ganglia.*

Review

- What are the functions of (a) astrocytes and (b) microglia?
- What are the functions of (a) dendrites (b) synaptic terminals (c) the myelin sheath?
- How is a neuron different from a nerve?

BiologyNow™ Assess your understanding of **neurons and glial cells** by taking the pretest on your BiologyNow CD-ROM.

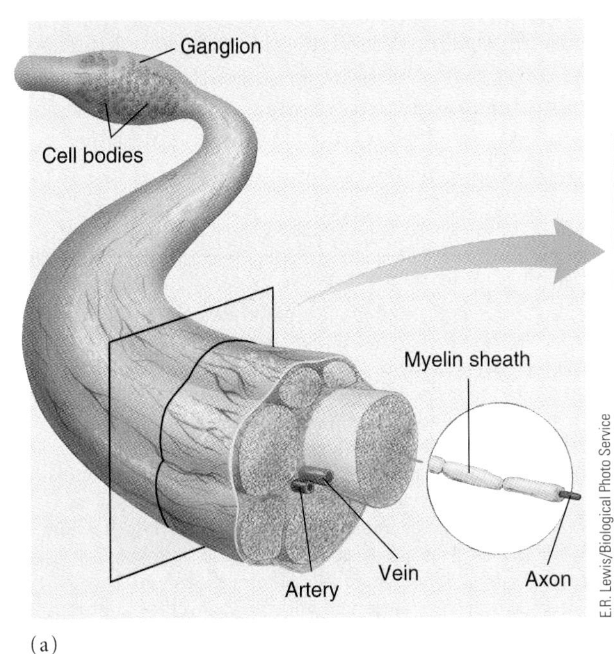

(a)

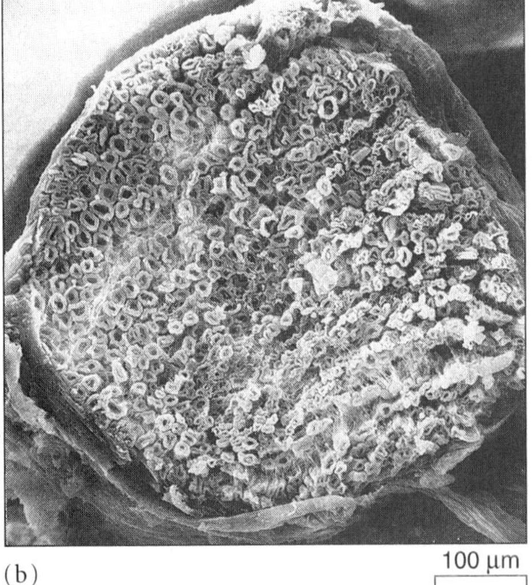

(b)

FIGURE **39-3**

Structure of a nerve.

(a) A nerve consists of bundles of axons held together by connective tissue. The cell bodies belonging to the axons of a nerve are grouped together in a ganglion. **(b)** SEM showing a cross section through a myelinated afferent nerve of a bullfrog.

TRANSMITTING INFORMATION ALONG THE NEURON

Learning Objectives

4 Explain how the neuron develops and maintains a resting potential.

5 Compare a graded potential with an action potential, describing the production and transmission of each.

6 Contrast continuous conduction with saltatory conduction.

Most animal cells have a difference in electrical charge across the plasma membrane—a more negative electrical charge inside the cell compared with the electrical charge of the extracellular fluid outside. The plasma membrane is said to be electrically **polarized,** meaning that one side, or pole, has a different charge from the other side. When electrical charges are separated in this way, a potential energy difference exists across the membrane. This difference in electrical charge across the plasma membrane gives rise to an electrical voltage gradient. The voltage measured across the plasma membrane is called the **membrane potential.** If the charges are permitted to come together, they have the ability to do work. Thus the cell can be thought of as a biological battery. In excitable cells, such as neurons and muscle cells, the membrane potential can change rapidly, and such changes can transmit signals to other cells.

The neuron membrane has a resting potential

The membrane potential in a resting (not excited) neuron or muscle cell is its **resting potential.** The resting potential is generally expressed in units called *millivolts* (mV). (A millivolt equals one thousandth of a volt.) Voltage is the force that causes charged particles to flow between two points. Like other cells that can produce electrical signals, the neuron has a resting potential of about 70 mV. By convention this is expressed as −70 mV because the cytosol close to the plasma membrane is negatively charged relative to the extracellular fluid (Fig. 39-4).

Biologists can measure the potential across the membrane by placing one electrode inside the cell and a second electrode outside the cell, and connecting them through a very sensitive voltmeter or oscilloscope. If you place both electrodes on the outside surface of the neuron, no potential difference between them is registered. (All points on the same side of the membrane have the same charge.) However, once one of the electrodes penetrates the cell, the voltage changes from zero to approximately −70 mV.

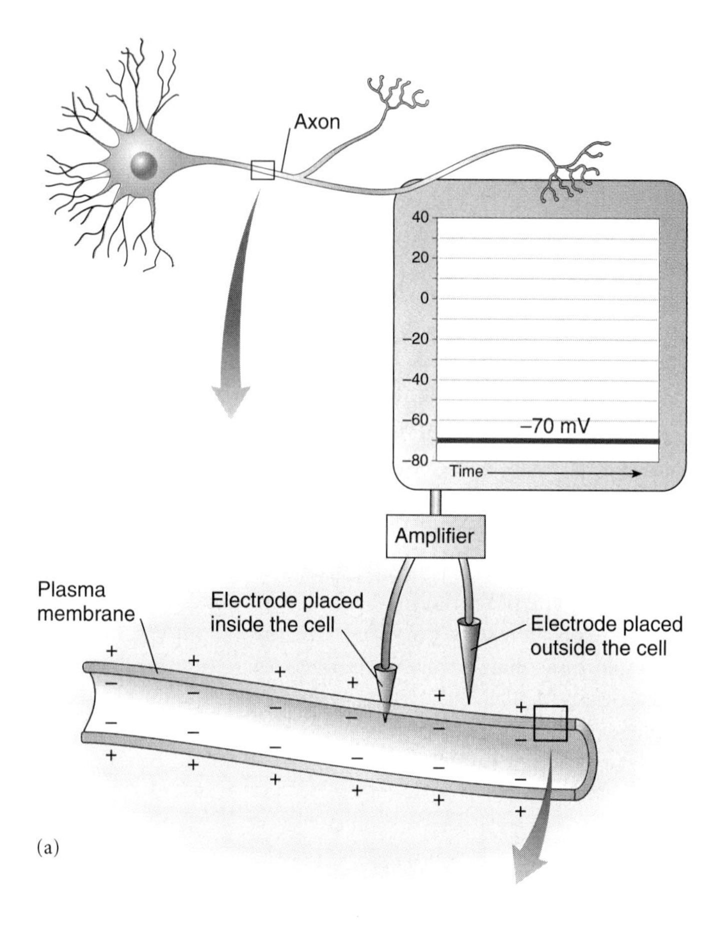

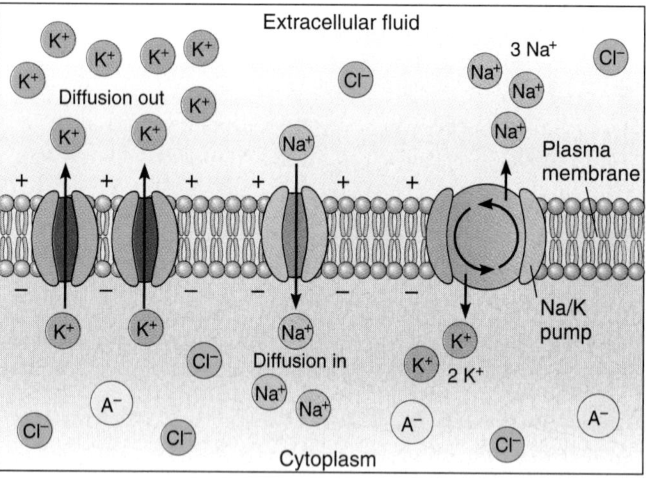

Two main factors determine the magnitude of the membrane potential: (1) differences in the concentrations of specific ions inside the cell compared with the extracellular fluid, and (2) selective permeability of the plasma membrane to these ions. The distribution of ions inside neurons and in the extracellular fluid surrounding them is like that of most other cells in the body. The potassium ion (K^+) concentration is about 10 times greater inside than outside the cell. In contrast, the sodium ion (Na^+) concentration is about 10 times greater outside than inside. This asymmetric distribution of ions across the plasma membrane at rest is brought about by the action of selective ion channels and ion pumps. In vertebrate neurons (and skeletal muscle fibers), the resting membrane potential depends mainly on the diffusion of ions down their concentration gradients.

Ions cross the plasma membrane by diffusion through ion channels that are formed by membrane proteins. Net movement of ions occurs from an area of higher concentration of that ion to one of lower concentration (see discussion of diffusion in Chapter 5). Typically, these channels allow only specific types of ions to pass. Neurons have three types of ion channels: *passive ion channels, voltage-activated channels,* and *chemically activated ion channels.* **Passive ion channels** permit the passage of specific ions such as Na^+, K^+, Cl^-, and Ca^{2+} (see Fig. 39-4). Unlike voltage-activated and chemically activated ion channels, passive ion channels are not controlled by gates.

Potassium channels are the most common type of passive ion channel in the plasma membrane, and cells are more permeable to potassium than to other ions. In fact, in the resting neuron, the plasma membrane is up to 100 times more permeable to K^+ than to Na^+. Sodium ions pumped out of the neuron cannot easily pass back into the cell, but K^+ pumped into the neuron easily diffuse out.

Potassium ions leak out through passive ion channels following their concentration gradient. As these positively charged ions diffuse out of the neuron, they increase the positive charge in the extracellular fluid outside the cell relative to the charge inside the cell. The resulting change in the electrical gradient across the membrane influences the flow of ions. This electrical gradient forces some of the positively charged potassium ions back into the cell.

FIGURE **39-4** | **Resting potential.**

(a) As shown in this segment of an axon of a resting (nonconducting) neuron, the axon is negatively charged compared with the surrounding extracellular fluid. The difference in electrical charge across the plasma membrane is measured by placing an electrode just inside the neuron and a second electrode in the extracellular fluid just outside the plasma membrane. **(b)** The membrane is less permeable to sodium ions than to potassium ions. Potassium ions diffuse out along their concentration gradient and are largely responsible for the voltage across the membrane. Sodium-potassium pumps in the plasma membrane actively pump sodium ions out of the cell and pump potassium ions in. The plasma membrane is permeable to negatively charged chloride ions, and they contribute slightly to the negative charge. Large anions (A^-) such as proteins also contribute to the negative charge.

The membrane potential at which the flux (flow) of K^+ inward (due to the electrical gradient) equals the flux of K^+ outward (because of the concentration gradient) is the **equilibrium potential** for K^+. The equilibrium potential for any particular ion is a steady state in which opposing chemical and electrical fluxes are equal and there is no net movement of the ion. The equilibrium potential for K^+ in the typical neuron is -80 mV. The equilibrium potential for Na^+ is $+40$ mV; it differs from the equilibrium potential of K^+ because of the difference in concentration of Na^+ across the membrane (concentration is high outside and low inside the cell). Because the membrane is much more permeable to K^+ than to Na^+, the resting potential of the neuron is closer to the potassium equilibrium potential than to the equilibrium potential of sodium. (Remember, the resting potential of the neuron is about -70 mV.)

Although the resting potential is primarily established by K^+, and less so by Na^+, chloride ions also contribute slightly because the plasma membrane is permeable to negatively charged chloride ions. These ions accumulate in the cytosol close to the plasma membrane. Also contributing to the negative charge in the cytosol are negative charges on large molecules such as proteins. These large anions cannot cross the plasma membrane.

Ion pumping maintains the gradients that determine the resting potential. The neuron plasma membrane has very efficient **sodium-potassium pumps** that actively transport sodium ions out of the cell and potassium ions into the cell (see Fig. 5-15). Both sodium and potassium ions are pumped against their concentration and electrical gradients, and these pumps require ATP. For every three sodium ions pumped out of the cell, two potassium ions are pumped in. Thus, more positive ions are pumped out than in. The sodium-potassium pumps maintain a higher concentration of K^+ inside the cell than outside, and a higher concentration of Na^+ outside than inside.

In summary, the development of the resting potential depends mainly on the diffusion of K^+ out of the cell. However, once the neuron reaches a steady state, Na^+ diffusion into the cell is greater than K^+ diffusion out of the cell. This situation is offset by the sodium-potassium pumps that pump three Na^+ out of the cell for every two K^+ pumped in.

Graded local signals vary in magnitude

Neurons are excitable cells. They can respond to stimuli and convert stimuli into nerve impulses. An electrical, chemical, or mechanical stimulus may alter the resting potential by increasing the membrane's permeability to sodium ions.

When a stimulus causes the membrane potential to become less negative (closer to zero) than the resting potential, the membrane is **depolarized.** Because depolarization brings a neuron closer to transmitting a neural impulse, it is described as *excitatory.* In contrast, when the membrane potential becomes more negative than the resting potential, the membrane is **hyperpolarized.** Hyperpolarization is *inhibitory;* it decreases the ability of the neuron to generate a neural impulse.

A stimulus may change the potential in a relatively small region of the plasma membrane. Such a **graded potential** is a local response that functions as a signal only over a very short distance, because it fades out within a few millimeters of its point of origin. A graded potential varies in magnitude; that is, the potential charge varies depending on the strength of the stimulus applied.

An action potential is generated by an influx of Na^+ and an efflux of K^+

If a stimulus is strong enough, a neuron fires a nerve impulse, or **action potential.** An action potential is an electrical excitation that travels rapidly down the axon into the synaptic terminals. All cells can generate graded potentials, but only neurons, muscle cells, and a few other cell types (certain cells of the endocrine and immune systems) can generate action potentials.

In a series of experiments in the 1940s, pioneering English researchers Alan Hodgkin and Andrew Huxley inserted electrodes into large axons found in squids. By measuring voltage changes as they varied ion concentrations, they showed that passage of Na^+ ions into the neuron and K^+ ions out of the neuron resulted in an action potential. In 1963 Hodgkin and Huxley received the Nobel Prize in Medicine for this research.

We now know that specific **voltage-activated ion channels** (also called **voltage-gated ion channels**) in the plasma membrane of the axon and cell body are regulated by changes in voltage. These channels have been studied using the **patch clamp technique** (see Fig. 5-16). Using this technique, biologists can measure currents across a very small segment of the plasma membrane rather than across the entire membrane. The patch of membrane measured is so small that it may contain a single channel.

In investigating membrane channels and how they function, biologists have used certain toxins that affect the nervous system. For example, the poison tetrodotoxin (TTX) binds to voltage-activated sodium channels, blocking the passage of Na^+. Using TTX and other toxins, researchers were able to identify the channel protein. TTX was first isolated from the Japanese pufferfish, which is eaten as a delicacy (Fig. 39-5). In small amounts TTX tingles the taste buds, but in a larger dose it can

FIGURE **39-5** | **Japanese pufferfish.**

A poison, tetrodotoxin, isolated from this fish (*Fugu rubripes*) binds to voltage-activated sodium ion channels, allowing neurobiologists to study them.

Jeffrey L. Rotman/Peter Arnold, Inc.

prevent breathing. (The Japanese government operates a certification program for pufferfish chefs!)

Investigators have established that voltage-activated ion channels have charged regions that act as gates. Voltage-activated Na^+ channels have two gates—an *activation gate* and an *inactivation gate* (see Fig. 39-6a). Voltage-activated K^+ channels have a single gate (see Fig. 39-6b).

An action potential is generated when the voltage reaches a certain critical point known as the **threshold level.** The membrane of most neurons can depolarize by about 15 mV, that is, to a potential of about −55 mV, without initiating an action potential. However, when depolarization is greater than −55 mV, the threshold level is reached, and an action potential is generated. The neuron membrane quickly reaches zero potential and even **overshoots** to +35 mV or more as a momentary reversal in polarity takes place. The sharp rise and fall of the action potential is called a *spike.* Figure 39-7 illustrates an action potential recorded by placing one electrode inside an axon and one just outside.

At resting potentials, Na^+ activation gates are closed, and Na^+ cannot pass through them into the neuron (Fig. 39-7b). When the voltage reaches the threshold level, the protein making up the channel changes shape, opening the activation gate. When the gate opens, Na^+ flow through the channel and the inside of the neuron becomes positively charged relative to the extracellular fluid outside the plasma membrane; an action potential may be generated. Inactivation gates that are open in the resting neuron close slowly in response to depolarization. Sodium ions can pass through the channel only during the brief period when both activation and inactivation gates are open.

The generation of an action potential depends on a **positive feedback mechanism.** (Recall from Chapter 37 that in a positive feedback mechanism, a change in some condition triggers a response that intensifies the change.) As just discussed, when a neuron is depolarized, voltage-activated sodium channels open, increasing the membrane permeability to Na^+. Sodium ions diffuse into the cell, moving from an area of higher concentration to an area of lower concentration. As Na^+ flows into the cell, the cytosol becomes positively charged relative to the extracellular fluid. This charge further depolarizes the neuron so that more voltage-gated sodium channels open, further increasing membrane permeability to sodium.

When, after a certain period, inactivation sodium gates close, the membrane again becomes impermeable to sodium. This inactivation begins the process of **repolarization,** during which the membrane potential returns to its resting level. Voltage-gated potassium ion channels slowly open in response to depolarization. When potassium ion channels open, potassium leaks out of the neuron (Fig. 39-7b, ③). This decrease in intracellular K^+ returns the interior of the membrane to its relatively negative state, repolarizing the membrane. Potassium channels remain open until the resting potential has been restored (Fig. 39-7b, ④). As a wave of depolarization moves down the membrane of the neuron, the normal polarized state is quickly re-established behind it. The entire process—depolarization and then repolarization—can take place in less than 1 millisecond.

During the millisecond or so in which it is depolarized, the axon membrane is in an **absolute refractory period:** It cannot transmit another action potential no matter how great a stimulus is applied. This is because the voltage-activated Na^+ channels are inactivated. Until their gates are reset, they cannot be reopened. When enough Na^+ channel gates have been reset, the neuron enters a **relative refractory period** that lasts for a few more milliseconds. During this period, the axon can transmit impulses, but the threshold is higher (more negative). Even with the limits imposed by their refractory periods, most neurons can transmit several hundred impulses per second.

In summary, neural action proceeds as follows:

Resting state ⟶ stimulus causes depolarization ⟶ action potential ⟶ repolarization and return to resting state

Local anesthetics such as procaine (novocaine), lidocaine (Xylocaine), and also cocaine, bind to voltage-activated sodium ion channels and block them. The channels are not able to open in response to depolarization, and the neuron cannot transmit an impulse through the anesthetized region. Signals do not reach the brain, so pain is not experienced.

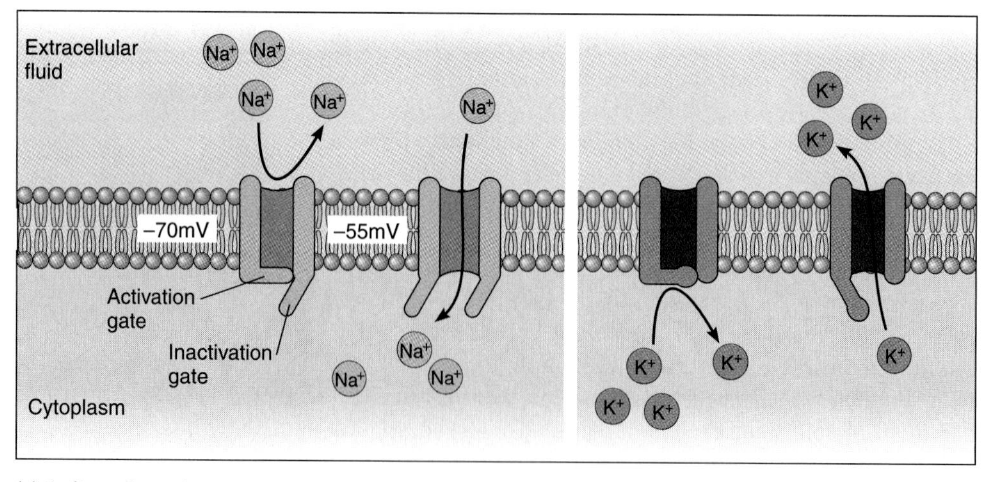

(a) Sodium channels

(b) Potassium channels

FIGURE **39-6**

Voltage-activated ion channels.

(a) In the resting state, voltage-activated Na^+ channels are closed. When the voltage reaches threshold level, activation gates open, allowing Na^+ to pass into the cell. After a certain amount of time elapses, activation gates close, blocking the channels. Inactivation gates are open when a neuron is in the resting state; they close slowly in response to depolarization. The ion channel is open only during the brief period when both activation and inactivation gates are open. **(b)** Potassium channels have activation gates that open slowly in response to depolarization. These channels have no inactivation gates.

The action potential is an all-or-none response

Any stimulus too weak to depolarize the plasma membrane to threshold level cannot fire the neuron. It merely sets up a local signal that fades and dies within a few millimeters from the point of stimulus. Only a stimulus strong enough to depolarize the membrane to its critical threshold level results in transmission of an impulse along the axon. An action potential is an **all-or-none response,** because either it occurs or it does not. No variation exists in the strength of a single impulse.

If the all-or-none law is valid, how can we explain differences in the levels of intensity of sensations? After all, we have no difficulty distinguishing between the pain of a severe toothache and that of a minor cut on the arm. We can explain this apparent inconsistency by the fact that intensity of sensation depends on the *number of neurons stimulated and on their frequency of discharge.* For example, suppose you burn your hand. If a large area is burned, more pain receptors are stimulated and more neurons become depolarized. Also, each neuron transmits more action potentials per unit of time.

An action potential is self-propagating

Once begun, the action potential is self-propagating. During depolarization, the affected area of the membrane is more positive relative to adjacent regions where the membrane is still at resting potential. The difference in potentials between active and resting membrane regions causes ions to flow between them (an electrical current). The flow of Na^+ into the adjacent region induces the opening of voltage-activated sodium channels in one

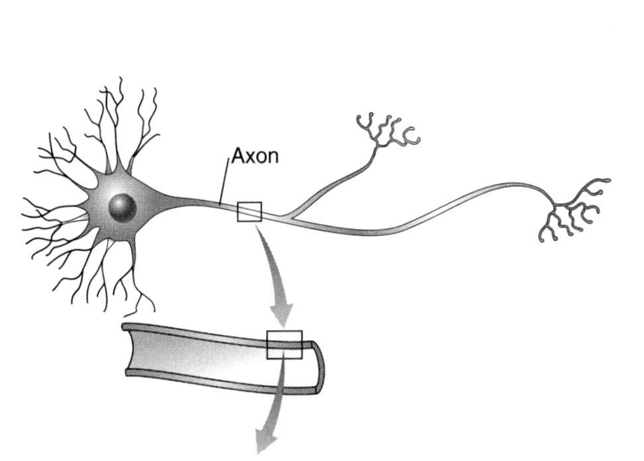

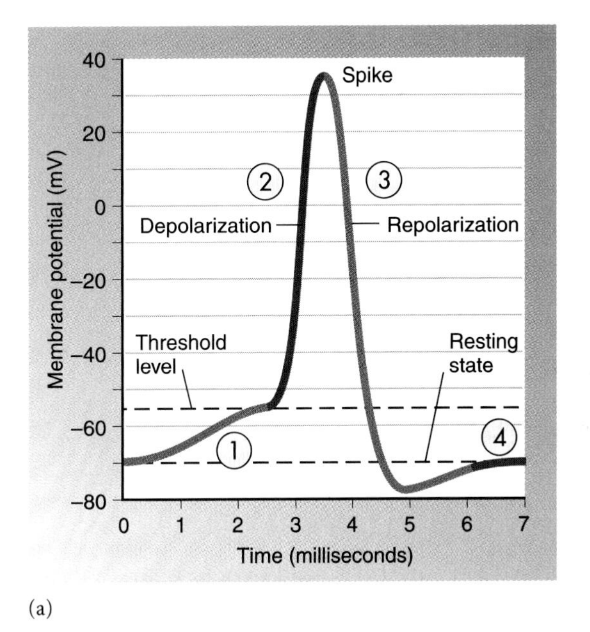

(a)

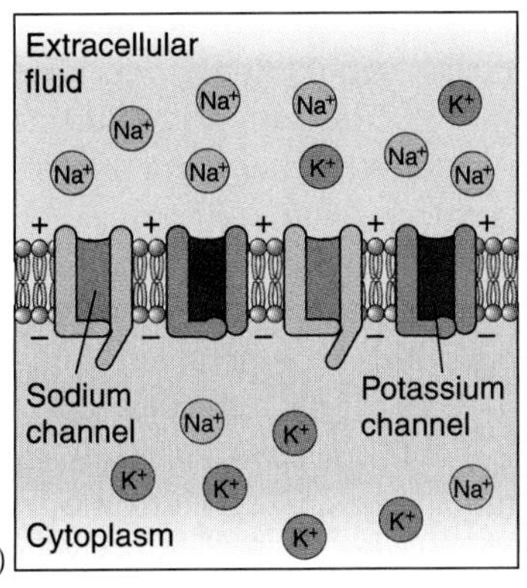

① Resting state.
Voltage-activated Na^+ and K^+ channels are closed.

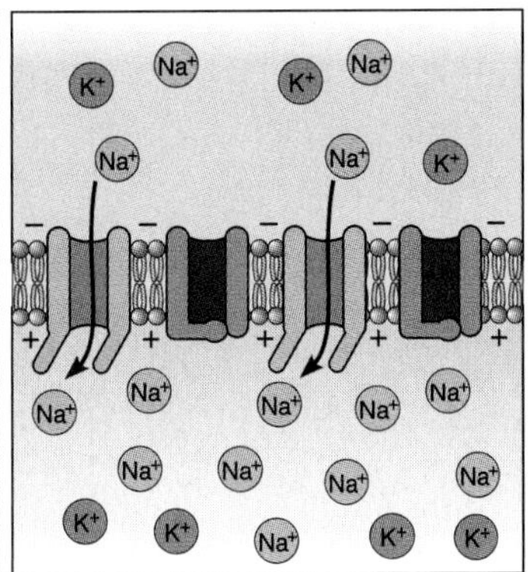

② Depolarization.
Voltage-activated Na^+ channels open. Na^+ ions enter cell; inside of neuron becomes positive relative to outside.

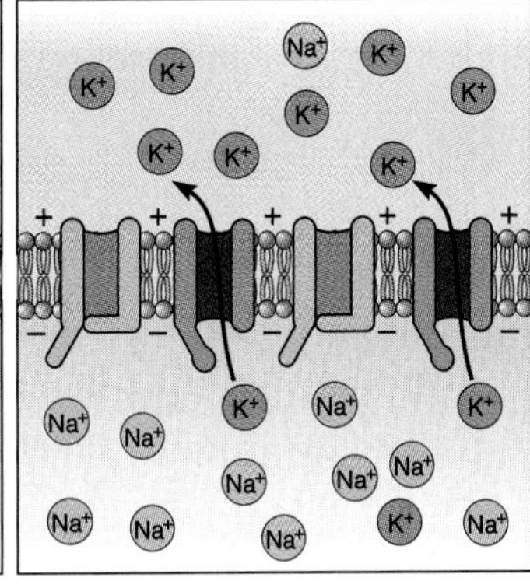

③ Repolarization.
Voltage-activated Na^+ channels close; K^+ channels are open; K^+ moves out of cell, restoring negative charge to inside of cell.

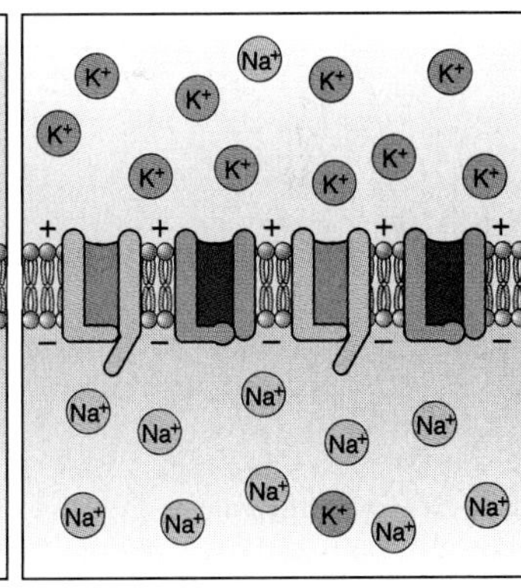

④ Return to resting state.
Voltage-activated Na^+ and K^+ channels close.

(b)

ACTIVE FIGURE **39-7** | **Action of voltage-activated ion channels.**

(a) Action potential. When the axon depolarizes to about −55 mV, an action potential is generated. (The numerical values vary for different nerve cells.) **(b)** The state of ion channels in the plasma membrane determines the state of the neuron. ① Resting state. ② Depolarization. At threshold, voltage-activated Na^+ channels open and Na^+ enters the neuron, causing further depolarization. An action potential is generated. K^+ channels slowly begin to open. ③ Repolarization. ④ Resting conditions are restored.

BiologyNow™ Learn more about **the action potential** by clicking on this figure on your BiologyNow CD-ROM.

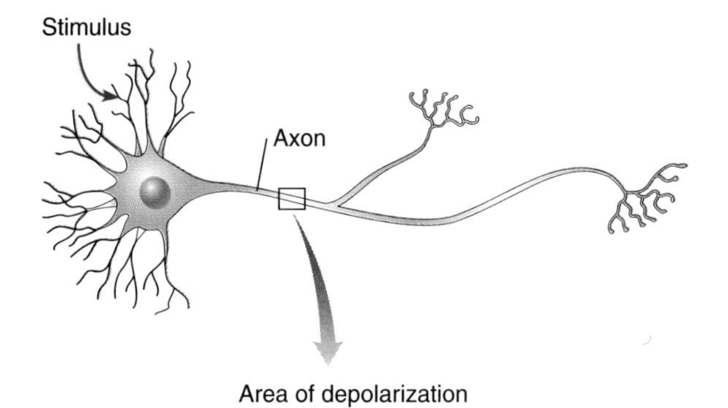

ACTIVE FIGURE **39-8** **Transmission of an action potential along the axon.**

The dendrites (or cell body) of a neuron are stimulated enough to depolarize the membrane to its firing level. ① An action potential is transmitted as a wave of depolarization that travels down the axon. At the region of depolarization, sodium ions diffuse into the cell. ② As the action potential progresses along the axon, repolarization occurs quickly behind it.

BiologyNow™ Watch **transmission of an action potential** by clicking on this figure on your BiologyNow CD-ROM.

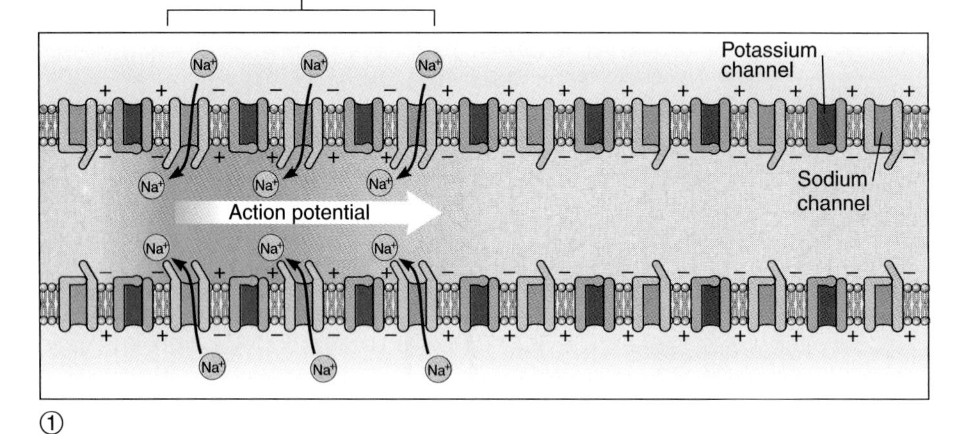

area of the membrane (Fig. 39-8). This action causes the next adjacent voltage-activated ion channels to open, permitting sodium ions to enter in that area, and the process is repeated like a chain reaction until the end of the axon is reached. Thus an action potential is a **wave of depolarization** that travels down the length of the axon. It is self-propagating and moves at a velocity that is constant for each type of neuron. Conduction of a neural impulse is sometimes compared with burning a trail of gunpowder. Once the gunpowder is ignited at one end of the trail, the flame moves steadily to the other end by igniting the powder particles ahead of it.

Compared with the flow of electrons in electrical wiring or the speed of light, a nerve impulse travels rather slowly. Most axons transmit impulses at about 1 to 10 m per second. The smooth, progressive impulse transmission just described, called **continuous conduction,** occurs in unmyelinated neurons. In unmyelinated axons, the speed of transmission is proportional to the diameter of the axon. Axons with larger diameters transmit faster than do those with smaller diameters, because an axon with a large diameter presents less internal resistance to the ions flowing along its length. Squids and certain other invertebrates have giant axons, up to 1 mm in diameter, that let them respond rapidly when escaping from predators.

Myelinated neurons transmit impulses rapidly

In vertebrates, another strategy has evolved that speeds transmission—myelinated neurons. Myelin acts as an effective electrical insulator around the axon except at the nodes of Ranvier, which are not myelinated. The axon plasma membrane makes direct contact with the surrounding extracellular fluid only at the nodes, and voltage-activated sodium and potassium ion channels are concentrated there. Ion movement across the membrane occurs only at the nodes. The ion activity at the active node results in diffusion of ions along the axon that depolarize the next node; the action potential appears to jump from one node of Ranvier to the next (Fig. 39-9). This type of impulse transmission is known as **saltatory conduction** (from the Latin word *saltus,* which means "to leap").

In myelinated neurons, the distance between successive nodes of Ranvier affects speed of transmission: When nodes are farther apart, less of the axon must depolarize, and the axon conducts the impulse faster. Using saltatory conduction, a myelinated axon can conduct an impulse up to 50 times faster than the fastest unmyelinated axon. Saltatory conduction has another advantage over continuous conduction: It requires less energy. Ions move across the plasma membrane only at the nodes, so fewer sodium and potassium ions are displaced. As a result, the

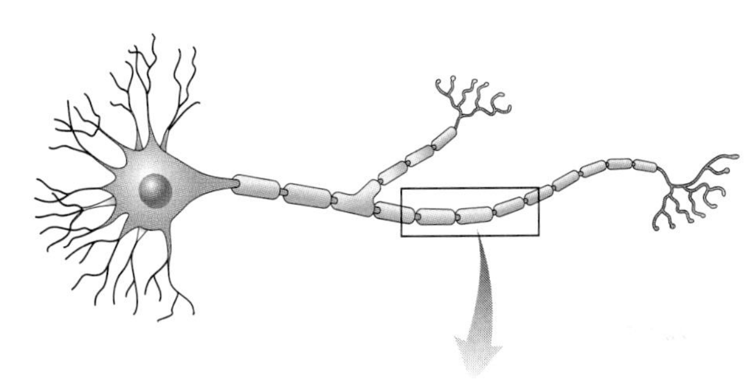

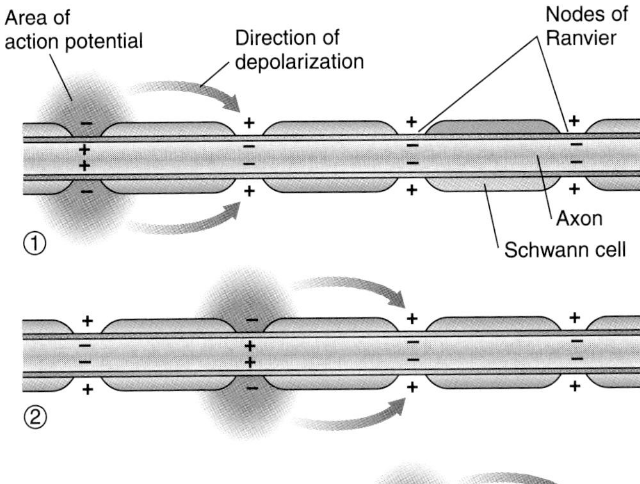

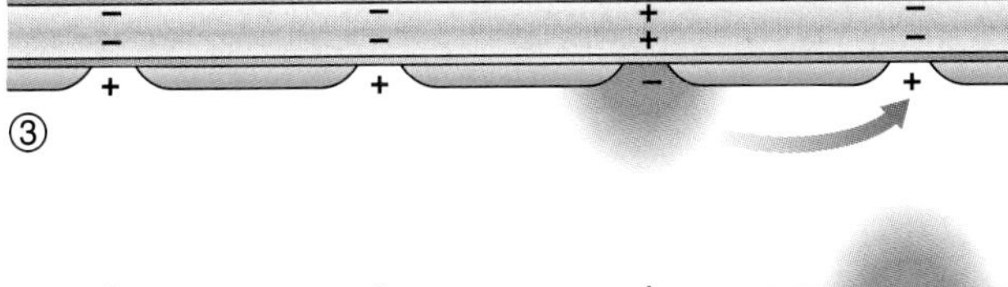

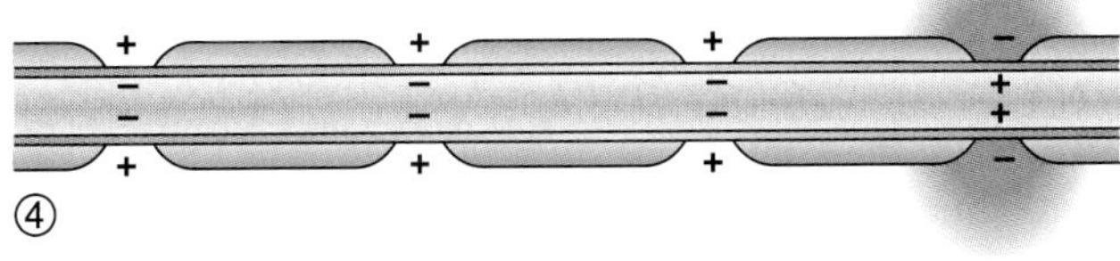

FIGURE **39-9** | **Saltatory conduction.**

In a myelinated axon, action potentials leap from one node of Ranvier to the next.

sodium-potassium pumps do not expend as much ATP to reestablish resting conditions each time an impulse is conducted.

Review

- How do ion channels and sodium-potassium pumps contribute to the resting potential?
- What sequence of events occurs when threshold level is exceeded in a neuron?
- How do voltage-activated ion channels work?
- How does an action potential differ from a graded potential?
- What are the benefits of saltatory conduction compared with continuous conduction?

BiologyNow™ Assess your understanding of **transmitting information along the neuron** by taking the pretest on your BiologyNow CD-ROM.

NEURAL SIGNALING ACROSS SYNAPSES

Learning Objectives

7 Describe the actions of the neurotransmitters identified in the chapter.

8 Trace the events that take place in synaptic transmission; draw diagrams to support your description.

9 Compare excitatory and inhibitory signals and their effects.

A **synapse** is a junction between two neurons or between a neuron and an effector, such as between a neuron and a muscle cell. A neuron that *terminates* at a specific synapse is called a **presynaptic neuron,** whereas a neuron that *begins* at that synapse is a **postsynaptic neuron.** Note that these terms are relative to a specific synapse. A neuron that is postsynaptic with respect to one synapse may be presynaptic to the next synapse in the sequence.

Signals across synapses can be electrical or chemical

Based on how presynaptic and postsynaptic neurons communicate, two types of synapses have been identified: electrical synapses and chemical synapses. In **electrical synapses,** the presynaptic and postsynaptic neurons occur very close together (within 2 nm of one another) and form *gap junctions* (see Chapter 5). The interiors of the two cells are physically connected by a protein channel. Electrical synapses let ions pass from one cell to another, permitting an impulse to be directly and rapidly transmitted from a presynaptic to a postsynaptic neuron. The escape responses of many animals involve electrical synapses. For example, the "tail-flick" escape response of the crayfish involves giant neurons in the nerve cord that form electrical synapses with large motor neurons. The motor neurons signal muscles in the abdomen to contract.

The majority of synapses are **chemical synapses.** Presynaptic and postsynaptic neurons are separated by a space, the **synaptic cleft,** about 20 nm wide (less than one millionth of an inch). Because depolarization is a property of the plasma membrane, when an action potential reaches the end of the axon it cannot jump the gap. The electrical signal must be converted into a chemical one. Neurotransmitters are the chemical messengers that conduct the neural signal across the synapse and bind to chemically activated ion channels in the membrane of the postsynaptic neuron. When a postsynaptic neuron reaches threshold depolarization, it transmits an action potential.

Neurons use neurotransmitters to signal other cells

More than 60 different chemical compounds are now known or suspected to function as neurotransmitters, chemical messengers used by neurons to signal other neurons or to signal muscle or gland cells (Table 39-1).

TABLE 39-1 Selected Neurotransmitters

Substance	Source	Actions
Acetylcholine	Some neurons in the brain and certain neurons in the autonomic nervous system; motor neurons	Excitatory effect on skeletal muscle; inhibitory effect on cardiac muscle
Biogenic Amines		
Norepinephrine	Neurons in both the central and autonomic nervous systems	Excitatory or inhibitory effect; may be involved in dreaming during REM sleep; level in brain affects mood
Dopamine	Central nervous system (CNS)	Helps maintain balance between inhibition and excitation of neurons; important in motor functions; level in brain affects mood
Serotonin (5-hydroxytryptamine, 5-HT)	CNS	Excitatory effect on pathways that control muscle action; inhibitory effect on pathways involved in sensations; also functions in regulation of food intake and affects mood
Amino Acids		
Glutamate	Brain (major excitatory neurotransmitter in the brain)	Functions in memory and learning; *In vertebrates:* excitatory effect on pathways in the brain; *In invertebrates:* excitatory effect at neuromuscular junctions
Aspartate	CNS	Excitatory effect on CNS pathways; functions in memory and learning
Glycine	Inhibitory interneurons in CNS	Inhibitory effect on CNS pathways
Gamma-aminobutyric acid (GABA)	Inhibitory interneurons in CNS	*In vertebrates;* inhibitory effect on CNS pathways *In invertebrates;* inhibitory effect at neuromuscular junctions
Neuropeptides		
Endorphins (opioids)	Brain	Pain regulation
Enkephalins (opioids)	Brain	Pain regulation
Substance P	Sensory neurons	Pain regulation
Gaseous Neurotransmitter		
Nitric oxide (NO)	Most neurons?	Retrograde messenger; transmits signals in opposite direction from other neurotransmitters, i.e., from postsynaptic to presynaptic neuron

Acetylcholine is a low-molecular-weight neurotransmitter that is released from motor neurons and triggers muscle contraction. Acetylcholine is also released by some neurons in the brain and in the autonomic nervous system (Focus on Alzheimer's disease; also see Chapter 40). Cells that release this neurotransmitter are called **cholinergic neurons.**

Neurons that release **norepinephrine** are called **adrenergic neurons.** Norepinephrine and the neurotransmitter **dopamine** are **catecholamines.** The catecholamines and the neurotransmitter **serotonin** belong to a class of compounds called **biogenic amines.** Biogenic amines affect mood, and their imbalance has been linked to several disorders, including major depression, attention deficit disorder (ADD), and schizophrenia.

The major excitatory neurotransmitter in the brain is the amino acid **glutamate.** One type of glutamate receptor is the target of several mind-altering drugs, including phencyclidine ("angel dust").

The amino acid **glycine** and a modified amino acid, **gamma-aminobutyric acid (GABA),** are neurotransmitters that inhibit neurons in the brain and spinal cord. Drugs that reduce anxiety enhance the actions of GABA by permitting lower concentrations of GABA to open Cl^- channels. These drugs include benzodiazepines such as diazepam (Valium) and alprazolam (Xanax), and barbiturates, such as phenobarbital.

Opiate drugs, for example, morphine and codeine, are powerful analgesics, drugs that relieve pain without causing loss of consciousness. The body has its own endogenous opioids **enkephalin** and **endorphins,** such as **beta-endorphin.** These neuropeptides bind to opioid receptors (the same receptors to which opiate drugs bind) and block pain signals. Opioids modulate the effect of other neurotransmitters. (Chapter 41 discusses pain perception.)

Much recent research has focused on the signaling molecule **nitric oxide (NO).** NO is a gas that acts as a retrograde messenger at some synapses. It transmits information from the postsynaptic neuron to the presynaptic neuron, the opposite direction of other neurotransmitters.

Neurotransmitters bind with receptors on postsynaptic cells

Neurotransmitters are stored in the synaptic terminals within small membrane-bounded sacs called **synaptic vesicles** (Fig. 39-10a). Each time an action potential reaches a synaptic terminal, voltage-gated calcium channels open. Calcium ions from the extracellular fluid then flow into the synaptic terminal. The Ca^{2+} ions cause synaptic vesicles to fuse with the presynaptic membrane and release neurotransmitter molecules into the synaptic cleft by exocytosis (Fig. 39-10b).

Neurotransmitter molecules diffuse across the synaptic cleft and combine with specific receptors on the dendrites or cell bodies of postsynaptic neurons (or on the plasma membranes of

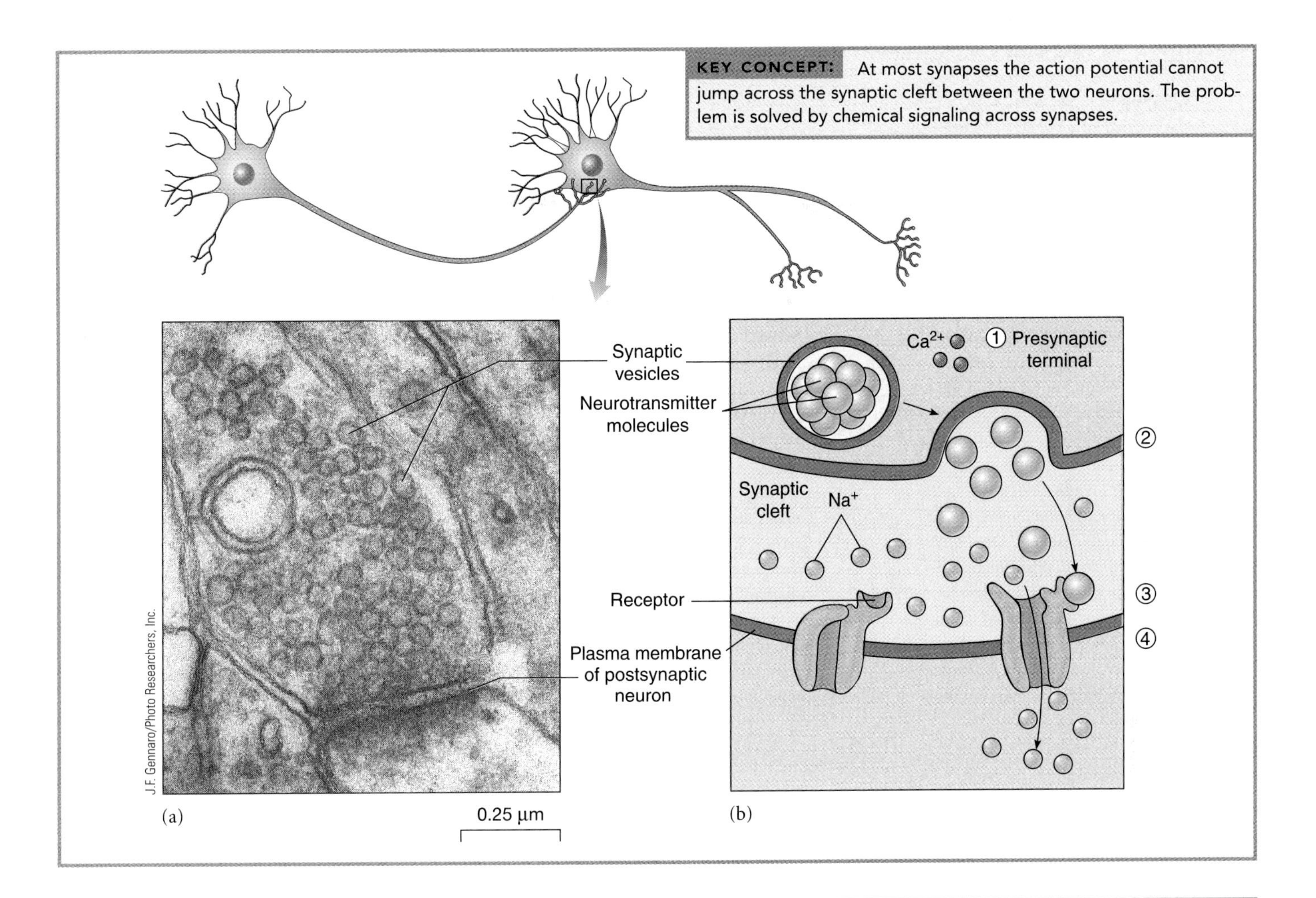

FIGURE **39-10** **Synaptic transmission.**

(a) The TEM shows synaptic terminals filled with synaptic vesicles. **(b)** How a neural impulse is transmitted across a synapse. ① When an action potential reaches the synaptic terminals at the end of a presynaptic neuron, calcium ions enter the synaptic terminal from the extracellular fluid. ② The calcium ions cause the synaptic vesicles to fuse with the plasma membrane and release a neurotransmitter into the synaptic cleft. ③ The neurotransmitter diffuses across the synaptic cleft and combines with receptors in the membrane of the postsynaptic neuron. ④ The receptors cause ion channels to open or close, resulting in either depolarization or hyperpolarization. When depolarization reaches threshold level, an action potential is generated in the postsynaptic neuron.

effector cells). Many neurotransmitter receptors are chemically activated ion channels known as **ligand-gated ion channels.** When the neurotransmitter, the **ligand,** binds with the receptor, the ion channel opens. The acetylcholine receptor, for example, is an ion channel that permits the passage of Na^+ and K^+.

Some neurotransmitters, such as serotonin, operate by a different mechanism. They work indirectly through a **second messenger.** Binding of the neurotransmitter with a receptor activates a **G protein.** The G protein then activates an enzyme, such as adenylyl cyclase, in the postsynaptic membrane. Adenylyl cyclase converts ATP to **cyclic AMP (cAMP),** which acts as a second messenger (see Fig. 3-25 and Chapters 5 and 47). Cyclic AMP activates a kinase that phosphorylates a protein, which then closes K^+ channels.

If a postsynaptic neuron is to repolarize quickly, any excess neurotransmitter in the synaptic cleft must be removed. Some neurotransmitters are inactivated by enzymes. For example, excess acetylcholine is degraded into its chemical components, choline and acetate, by the enzyme acetylcholinesterase. Other neurotransmitters, such as the biogenic amines, are actively transported back into the synaptic terminals, a process known as **reuptake.** These neurotransmitters are repackaged in vesicles and recycled.

Many drugs inhibit the reuptake of neurotransmitters. For example, some antidepressants are selective serotonin reuptake inhibitors, or SSRIs (such as fluoxetine [Prozac]) that selectively inhibit the reuptake of serotonin in the brain. This action concentrates serotonin in the synaptic cleft, elevating mood. Some antidepressants (and cocaine) also inhibit the reuptake of dopamine.

Neurotransmitters and their receptors can send excitatory or inhibitory signals

A postsynaptic neuron may have receptors for several types of neurotransmitters. Some of its receptors may be excitatory, and others may be inhibitory. Depending on the type of postsynap-

In **Alzheimer's disease (AD),** a progressive, degenerative brain disorder, neural signaling is disrupted and neurons are damaged and die prematurely. Alzheimer's disease is the leading cause of senile dementia, which is the loss of memory, judgment, and the ability to reason that may be associated with aging. This disease afflicts more than 4 million people in the United States alone.

The development of AD may be a lifelong process. One startling study demonstrated that a person's writing early in life can predict the disease. Young adults whose writings had a lower density of ideas were more likely to develop the disease. One approach to detecting development of AD is the use of positron emission tomography (PET) scans to study glucose uptake in the brain. Low glucose uptake in certain areas of the brain indicates that neurons in these areas may be damaged.

In AD, neurons are damaged in certain parts of the brain, including the cerebral cortex and hippocampus, areas that are important in thinking and remembering. Neurons that secrete the neurotransmitter acetylcholine are especially affected. Investigators have demonstrated that two of the abnormalities that develop in brain tissue as we age, amyloid plaques and neurofibrillary tangles, are especially characteristic of AD. Understanding the biochemistry and genetic basis of both plaques and tangles may provide clues to the causes and cures of AD.

Amyloid plaques were first identified in autopsied brains in 1906 by the German physician Alois Alzheimer. These plaques are extracellular deposits of a peptide called β-amyloid. Neurons near these plaques appear swollen and misshapen. Microglia are typically present, suggesting an attempt to remove the damaged cells or the plaques themselves.

Researchers have found that β-amyloid is cut from a protein called β-amyloid precursor protein, or β-APP, a large transmembrane protein coded by a gene on chromosome 21. Normal β-amyloid molecules have 40 amino acids. However, about 10% of the peptides produced have two additional amino acids. This longer peptide appears to form the plaques. Investigators have identified mutations in specific genes that result in the production of abnormal β-amyloid.

Just how the longer β-amyloid leads to AD is not known. The abnormal peptide appears to disrupt calcium regulation, killing neurons in the brain. The peptide may also damage mitochondria, releasing harmful free radicals. In 2002 Dennis Selkoe, of Harvard Medical School, reported that AD begins with subtle changes in synaptic function in the hippocampus, a part of the brain important in memory. An early symptom is the loss of the ability to encode new memories. Selkoe hypothesizes that the

tic receptor with which it combines, the same neurotransmitter can have different effects. For example, acetylcholine has an excitatory effect on skeletal muscle and an inhibitory effect on cardiac muscle.

When neurotransmitter molecules bind to receptors, they directly or indirectly cause ion channels to open or close, thus changing the permeability of the postsynaptic membrane to certain ions. The resulting redistribution of ions changes the electrical potential of the membrane, and the membrane may depolarize. If sufficiently intense, a local depolarization can set off an action potential. For example, when neurotransmitter molecules combine with receptors that open sodium channels, the resulting influx of sodium ions partially depolarizes the membrane. A change in membrane potential that brings the neuron closer to firing is called an **excitatory postsynaptic potential (EPSP)**. Imagine that sufficient sodium ions enter to change the membrane potential, from −70 mV to −60 mV. The membrane would be only −5 mV away from threshold. Under such conditions, an additional relatively weak stimulus, causing a small influx of Na^+, can cause the neuron to fire.

Some neurotransmitter-receptor combinations hyperpolarize the postsynaptic membrane. The membrane potential becomes more negative. Because such an action takes the neuron farther away from the firing level, a potential change in this direction is called an **inhibitory postsynaptic potential (IPSP)**. For example, if the membrane potential changes from −70 mV to −80 mV, the membrane is farther away from threshold, and a stronger stimulus is required to fire the neuron.

Like an EPSP, an IPSP can be produced in several ways. Two types of GABA receptors are known to produce IPSPs. Binding of GABA to a $GABA_B$ receptor opens K^+ channels. As K^+ diffuses out, the neuron becomes more negative, hyperpolarizing the membrane. Activated $GABA_A$ receptors produce IPSPs by opening Cl^- channels. In this case, the influx of negative ions hyperpolarizes the membrane. Although the mechanisms are different, the inhibiting effect of GABA is the same in both cases.

In summary, when an action potential reaches the synaptic terminals at the end of a presynaptic neuron, the following sequence of events takes place:

Calcium ions enter synaptic terminal ⟶ synaptic vesicles release neurotransmitter into the synaptic cleft ⟶ neurotransmitter diffuses across synaptic cleft and combines with receptors in membrane of the postsynaptic neuron ⟶ ion channels open or close ⟶ depolarization or hyperpolarization of postsynaptic membrane

Review

- What are some functions of biogenic amines?
- What are the functions and mechanisms of action of the following substances: (a) acetylcholine (b) acetylcholinesterase; (c) norepinephrine (d) GABA?
- How are EPSPs produced? IPSPs?

BiologyNow™ Assess your understanding of **neural signaling across synapses** by taking the pretest on your BiologyNow CD-ROM.

loss of synaptic function is caused by abnormal β-amyloid.

Another piece of the AD puzzle may be **neurofibrillary tangles,** which form in the neuron cytoplasm. A cytoskeletal protein called **tau** normally associates with the protein tubulin, which forms microtubules. Mutations in the *tau* gene may interfere with the way the tau protein binds to microtubules, resulting in abnormal microtubules that cannot effectively transport materials through the axons. Abnormal tau proteins accumulate in the cytoplasm, forming fibrous deposits that make up the neurofibrillary tangles. These tangles interfere with neural signaling.

Recent findings indicate that the mechanisms that cause amyloid plaques and neurofibrillary tangles are connected. In 2001 two groups of investigators working with transgenic mice independently demonstrated that β-amyloid deposits in the brain are responsible for the formation of neurofibrillary tangles in areas of the brain known to be affected in Alzheimer's disease.

Several drugs designed to slow the progress of Alzheimer's disease are now in clinical trials. One of the challenges of developing a drug has been getting the medication through the blood–brain barrier (see Chapter 40). Two approaches to developing a treatment for AD are blocking formation of β-amyloid and dissolving the abnormal β-amyloid deposits that are already present.

A promising strategy is to stimulate the immune system to destroy β-amyloid deposits. In 2000 Dale Schenk and his colleagues at Elan Pharmaceuticals demonstrated that when they immunized genetically engineered mice with a vaccine made of β-amyloid, the mice developed antibodies against the protein and destroyed the β-amyloid plaques. In 2001 investigators demonstrated that this vaccine also reduces memory loss and enhances learning in aging mice. Vaccinated mice developed fewer plaques and performed significantly better on memory tests. Researchers were optimistic when human clinical trials began. Unfortunately, these trials were halted when about 5% of the patients developed a dangerous brain inflammation. Studies are currently underway to develop a safer vaccine.

People with high cholesterol levels are at greater risk for developing AD because cholesterol apparently promotes β-amyloid production. Clinical trials are in progress to test the use of statins, drugs that lower cholesterol levels in the blood. Other approaches include the use of anti-inflammatory drugs and of antioxidants (such as vitamin C or E). The drug memantine, used to treat AD in Germany, is in clinical trials in the United States. Memantine blocks the action of the neurotransmitter glutamate, which appears to be overproduced in the brains of AD patients.

NEURAL INTEGRATION

Learning Objective

10 Describe how a postsynaptic neuron integrates incoming stimuli and "decides" whether or not to fire.

Each EPSP or IPSP is a graded potential that varies in magnitude depending on the strength of the stimulus applied. One EPSP is usually too weak to trigger an action potential by itself. Its effect is below threshold level. Even though subthreshold EPSPs do not produce an action potential, they do affect the membrane potential. EPSPs may add together in a process known as **summation.**

Temporal summation occurs when repeated stimuli cause new EPSPs to develop before previous EPSPs have decayed. The summation of several EPSPs may bring the neuron to the critical firing level. Spatial summation can also bring the postsynaptic neuron to the threshold level. **Spatial summation** occurs when several closely spaced synaptic terminals release neurotransmitter simultaneously. The postsynaptic neuron is stimulated at several places at once.

Neural integration is the process of summing, or integrating, incoming signals. Each neuron may synapse with hundreds of other neurons. Indeed, thousands of synaptic terminals of presynaptic neurons may cover as much as 40% of a postsynaptic neuron's dendritic surface and cell body. It is the job of the dendrites and the cell body of every neuron to integrate the messages that continually bombard them.

EPSPs and IPSPs are produced continually in postsynaptic neurons, and IPSPs cancel the effects of some EPSPs. It is important to remember that each EPSP and IPSP is not an all-or-none response. Rather, each is a graded response that does not travel like an action potential but may be added to or subtracted from other EPSPs and IPSPs. As the postsynaptic neuron membrane continuously updates its molecular tabulations, the neuron may be inhibited or brought to threshold level. This mechanism integrates hundreds of signals (EPSPs and IPSPs) before an action potential is actually transmitted along the axon of a postsynaptic neuron. Local responses permit the neuron and the entire nervous system a far greater range of response than would be possible if every EPSP generated an action potential.

Where does neural integration take place? Every neuron sorts through (on a molecular level) the hundreds and thousands of bits of information simultaneously bombarding it. In vertebrates more than 90% of the neurons in the body are located in the CNS, so most neural integration takes place there, within the brain and spinal cord. These neurons are responsible for making most of the "decisions." In the next chapter, the brain and spinal cord are examined in some detail.

Review

- How do temporal and spatial summation differ?
- What is neural integration?
- Where does most neural integration take place in vertebrates?

BiologyNow™ Assess your understanding of **neural integration** by taking the pretest on your BiologyNow CD-ROM.

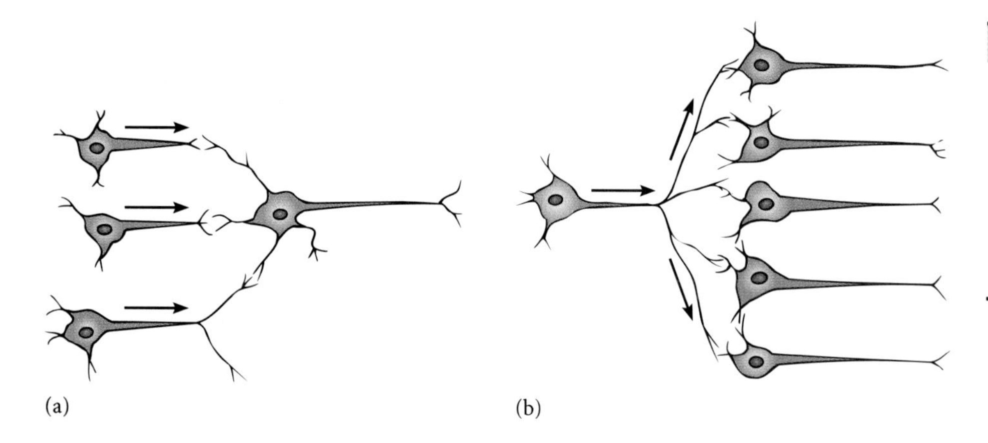

FIGURE **39-11** **Neural circuits.**

(a) Convergence of neural input permits one neuron to receive signals from many sources. Several presynaptic neurons synapse with one postsynaptic neuron. **(b)** Divergence of neural output allows one neuron to communicate with many others. A single presynaptic neuron synapses with many postsynaptic neurons.

NEURAL CIRCUITS

Learning Objective

11 Distinguish among convergence, divergence, and reverberation, and explain why each is important.

The CNS contains millions of neurons, but it is not just a tangled mass of nerve cells. Its neurons are organized into separate **neural networks,** and within each network the neurons are arranged in specific pathways, or **neural circuits.** Although each network has some special characteristics, the neural circuits in all the networks share many organizational features. For example, convergence and divergence are probably characteristic of them all.

In **convergence,** a single neuron is controlled by converging signals from two or more presynaptic neurons (Fig. 39-11a). An interneuron in the spinal cord, for instance, may receive converging information from sensory neurons entering the cord, from neurons bringing information from various parts of the brain, and from neurons coming from different levels of the spinal cord. Information from all these sources must be integrated before an action potential is transmitted and an appropriate motor neuron stimulated. Convergence is an important mechanism by which the CNS integrates the information that impinges on it from various sources.

In **divergence,** a single presynaptic neuron stimulates many postsynaptic neurons (Fig. 39-11b). Each presynaptic neuron may branch and synapse with thousands of different postsynaptic neurons. For example, a single neuron transmitting an impulse from the motor area of the brain may synapse with hundreds of interneurons in the spinal cord, and each of these may diverge in turn, so that hundreds of muscle fibers are stimulated.

Another important type of neural circuit is the **reverberating circuit,** a neural pathway arranged so that an axon collateral synapses with an interneuron (Fig. 39-12). The interneuron synapses with a neuron in a sequence that can send new impulses through the circuit. This is an example of positive feedback. New impulses can be generated again and again until the synapses fatigue (from depletion of neurotransmitter) or until stopped by some sort of inhibition. Researchers think reverberating circuits are important in rhythmic breathing, mental alertness, and short-term memory.

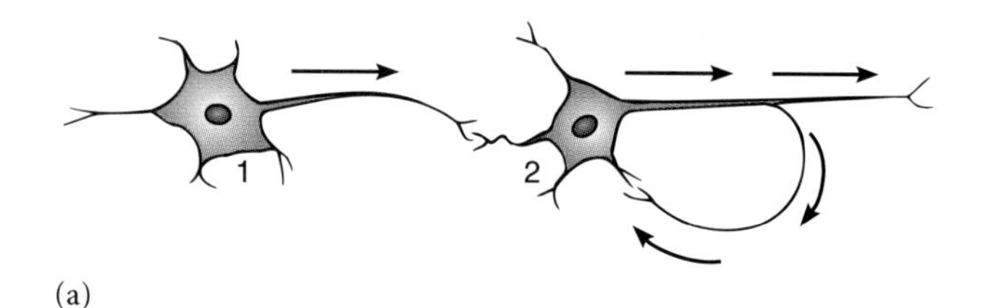

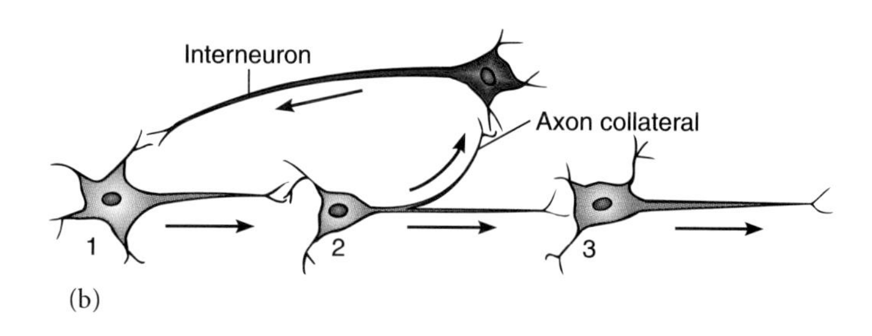

FIGURE **39-12** **Reverberating circuits.**

(a) A simple reverberating circuit in which an axon collateral of the second neuron turns back on its own dendrites, so the neuron continues to stimulate itself. **(b)** In this neural circuit an axon collateral of the second neuron synapses with an interneuron. The interneuron synapses with the first neuron in the sequence. New impulses are triggered again and again in the first neuron, causing reverberation.

Review

- How does convergence differ from divergence?
- What is the role of positive feedback in a reverberating circuit?

BiologyNow™ Assess your understanding of **neural circuits** by taking the pretest on your BiologyNow CD-ROM.

SUMMARY WITH KEY TERMS

1 Trace the flow of information through the nervous system, and describe the processes involved in neural signaling: reception, transmission, integration, and action by effectors.

- Neural signaling involves (a) **reception** of information by a sensory receptor (b) **transmission** by an **afferent neuron** to the CNS (c) **integration** by interneurons in the **central nervous**

system (CNS) (d) transmission by an **efferent neuron** to other neurons or to an effector (e) action by **effectors,** the muscles and glands. Sensory receptors and neurons outside the CNS make up the **peripheral nervous system (PNS).**

2 Describe the functions of each type of vertebrate glial cell.

- **Glial cells** support and nourish neurons. **Microglia** are phagocytic cells. Some **astrocytes** are phagocytic, and others help regulate the composition of the extracellular fluid in the CNS. Astrocytes may also induce and stabilize synapses.
- **Oligodendrocytes** are glial cells that form myelin sheaths around axons in the CNS; **Schwann cells** form sheaths around axons in the PNS.

3 Describe the structure of a typical neuron, and give the function of each of its parts.

- **Neurons** are specialized to receive stimuli and transmit electrical signals. In a typical neuron a **cell body** contains the nucleus and most of the organelles. Many branched **dendrites** extend from the cell body; they are specialized to receive stimuli and send signals to the cell body.
- The single long **axon** extends from the cell body and forms branches called **axon collaterals.** The axon transmits signals into its **terminal branches** that end in **synaptic terminals.** Many axons are surrounded by an insulating **myelin sheath. Nodes of Ranvier** are gaps in the sheath between successive Schwann cells. A **nerve** consists of several hundred axons wrapped in connective tissue; a **ganglion** is a mass of neuron cell bodies.

4 Explain how the neuron develops and maintains a resting potential.

- Neurons use electrical signals to transmit information. In a resting neuron, one that is not transmitting an impulse, the inner surface of the plasma membrane is negatively charged compared with the outside; the membrane is **polarized.**
- The **resting potential** is the potential difference of about −70 mV that exists across the membrane. The magnitude of the resting potential is determined by (1) differences in concentrations of specific ions (mainly Na^+ and K^+) inside the cell relative to the extracellular fluid, and (2) selective permeability of the plasma membrane to these ions.
- Ions pass through specific **passive ion channels.** K^+ leaks out more readily than Na^+ can leak in. Cl^- ions accumulate along the inner surface of the plasma membrane. Large anions such as proteins that cannot cross the plasma membrane contribute negative charges.
- The gradients that determine the resting potential are maintained by ATP-requiring **sodium-potassium pumps** that continuously transport three sodium ions out of the neuron for every two potassium ions transported in.

5 Compare a graded potential with an action potential, describing the production and transmission of each.

- If a stimulus causes the membrane potential to become less negative, the membrane becomes **depolarized.** If the membrane potential becomes more negative than the resting potential, the membrane is **hyperpolarized.**
- A **graded potential** is a local response that varies in magnitude depending on the strength of the applied stimulus. A graded potential fades out within a few mm of its point of origin.
- If voltage across the membrane declines to a critical point, called the **threshold level,** the **voltage-activated ion channels** open, allowing Na^+ to flow into the neuron, and an **action potential** is generated. The action potential is a **wave of depolarization** that moves down the axon.
- The action potential conforms to an **all-or-none response,** no variation exists in the strength of a single impulse. The membrane potential either exceeds threshold level, leading to transmission of an action potential, or it does not. Once begun, an action potential is self-propagating.
- As the action potential moves down the axon, **repolarization** occurs behind it. During depolarization, the axon enters an **absolute refractory period,** a time when it cannot transmit another action potential. When enough gates controlling Na^+ channels have been reset, the neuron enters a **relative refractory period,** a time when the threshold is higher.

6 Contrast continuous conduction with saltatory conduction.

- **Continuous conduction** takes place in unmyelinated neurons; it involves the entire axon plasma membrane.
- **Saltatory conduction,** which is more rapid than continuous conduction, takes place in myelinated neurons. Depolarization skips along the axon from one **node of Ranvier** to the next—sites where the axon is not covered by myelin and where Na^+ channels are concentrated.

7 Describe the actions of the neurotransmitters identified in the chapter.

- The junction between two neurons or between a neuron and effector is a **synapse.** Although there are electrical synapses, most synapses are chemical. Synaptic transmission generally depends on release of a **neurotransmitter** from **synaptic vesicles** in the synaptic terminals of a **presynaptic neuron.**
- **Acetylcholine** triggers contraction of skeletal muscle.
- The **biogenic amines** include **norepinephrine, serotonin,** and **dopamine.** These neurotransmitters play important roles in regulating mood. Dopamine is also important in motor function.
- Several amino acids function as neurotransmitters including **glutamate,** the main excitatory neurotransmitter in the brain and **GABA**, a widespread inhibitory neurotransmitter.
- The neuropeptides include the **endorphins** (for example, **beta-endorphin**) and the **enkephalins.** These neuropeptides are opioids.
- **Nitric oxide (NO)** is a gaseous neurotransmitter that transmits signals from the postsynaptic neuron to the presynaptic neuron, the opposite direction from other neurotransmitters.

8 Trace the events that take place in synaptic transmission; draw diagrams to support your description.

- Calcium ions cause synaptic vesicles to fuse with the presynaptic membrane and release neurotransmitter into the **synaptic cleft.** A neurotransmitter diffuses across the synaptic cleft and combines with specific receptors on a postsynaptic neuron.
- Many neurotransmitter receptors are proteins that form **ligand-gated ion channels.** Others work through a **second messenger** such as **cAMP.**

9 Compare excitatory and inhibitory signals and their effects.

- Binding of a neurotransmitter to a receptor can cause either an **excitatory postsynaptic potential (EPSP)** or an **inhibitory postsynaptic potential (IPSP),** depending on the type of receptor.
- EPSPs bring the neuron closer to firing. IPSPs move the neuron farther away from its firing level.

10 Describe how a postsynaptic neuron integrates incoming stimuli and "decides" whether or not to fire.

- Each EPSP or IPSP is a graded potential that varies in magnitude depending on the strength of the stimulus applied.
- The mechanism of neural integration is **summation,** the process of adding and subtracting incoming signals. By summation of several EPSPs, the neuron may be brought to critical firing level.
- **Temporal summation** occurs when repeated stimuli cause new EPSPs to develop before previous EPSPs have decayed. **Spatial summation** occurs when several closely spaced synaptic terminals release neurotransmitter simultaneously, stimulating the postsynaptic neuron at several different places.

11 Distinguish among convergence, divergence, and reverberation, and explain why each is important.

- Complex **neural circuits** are possible because of associations such as convergence and divergence. In **convergence,** a single neuron is controlled by converging signals from two or more presynaptic neurons. Convergence permits the CNS to integrate incoming information from various sources. In **divergence,** a single presynaptic neuron stimulates many postsynaptic neurons, allowing widespread effect.
- **Reverberating circuits** depend on positive feedback. New impulses can be generated again and again until the synapses fatigue. Reverberating circuits are important in rhythmic breathing, mental alertness, and short-term memory.

POST-TEST

1. Summing incoming neural signals is part of (a) reception (b) transmission (c) integration (d) action by effectors (e) afferent neuron transmission
2. Which of the following are phagocytic cells that remove debris from tissue in the CNS? (a) Schwann cells (b) axons (c) oligodendrocytes (d) effectors (e) microglia
3. Most of the neuron cytoplasm is located in the (a) cell body (b) axon (c) dendrites (d) synaptic terminals (e) nodes
4. Action potentials are transmitted to synaptic terminals by the (a) ganglia (b) axon (c) dendrites (d) cell body (e) nodes
5. In responding to a stimulus, a motor neuron signals a(an) (a) efferent neuron (b) receptor (c) afferent neuron (d) interneuron (e) effector
6. The myelin sheath is produced around axons outside the CNS by the (a) axon (b) neuron cell body (c) dendrites (d) Schwann cells (e) synaptic terminals
7. Neurotransmitters are released by the (a) axon (b) neuron cell body (c) dendrites (d) astrocytes (e) synaptic terminals
8. Which of the following does *not* contribute to the resting potential of a neuron? (a) sodium-potassium pumps (b) ion channels (c) differences in concentration of ions across the membrane (d) differences in permeability to certain ions (e) graded potentials
9. Which of the following occurs first when voltage reaches the threshold level? (a) gates of certain voltage-activated ion channels open (b) K^+ channels close (c) the membrane hyperpolarizes (d) neurotransmitter is released (e) neurotransmitter reuptake takes place at the synapse
10. Saltatory conduction (a) requires more energy than continuous conduction (b) occurs in unmyelinated neurons (c) occurs when the action potential jumps from one node of Ranvier to the next (d) slows transmission of an impulse (e) depends on the presence of GABA
11. Acetylcholine (a) is a biogenic amine (b) is recycled (c) is released by motor neurons and by some neurons in the brain (d) is deactivated by G proteins (e) binds to opioid receptors
12. Neurotransmitter receptors often (a) are voltage-activated ion channels (b) permit influx of chloride ions, leading to depolarization of the membrane (c) work through a second messenger (d) inhibit reuptake of the neurotransmitter (e) are passive ion channels
13. IPSPs (a) excite presynaptic neurons (b) excite postsynaptic neurons (c) cancel the effects of some EPSPs (d) release large amounts of neurotransmitters (e) are released by postsynaptic cell bodies
14. A presynaptic neuron in the cerebrum synapses with hundreds of other neurons. This is an example of (a) convergence (b) divergence (c) summation (d) a reverberating circuit (e) a graded potential
15. During a relative refractory period (a) IPSPs are generated (b) opioid neurotransmitters are released (c) voltage-activated sodium channels are inactivated (d) an axon can transmit impulses but the threshold is higher (more negative) (e) an axon cannot transmit an action potential
16. Label the diagram. Use Figure 39-2 to check your answers.

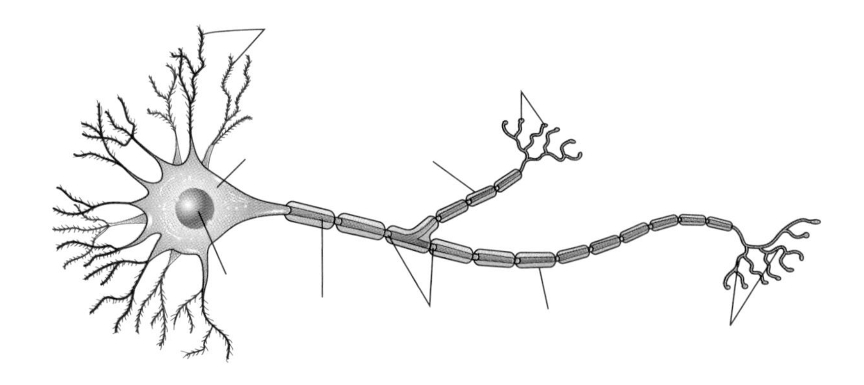

CRITICAL THINKING

1. Discuss several specific ways in which the nervous system helps maintain homeostasis.
2. Stimulant drugs such as amphetamines increase the activity of the nervous system. Describe two mechanisms involving synaptic transmission that could explain the action of such drugs.
3. Develop a hypothesis to explain the fact that acetylcholine has an excitatory effect on skeletal muscle but an inhibitory effect on cardiac muscle.
4. Investigators have genetically engineered cells to produce acetylcholine, dopamine, GABA, and other neurotransmitters. In what ways might these cells be useful in neurobiological research? How might they be useful in clinical medicine?

■ Visit our Web site at **http://biology.brookscole.com/solomon7** for links to chapter-related resources on the World Wide Web. Additional online materials relating to this chapter can also be found on our Web site.

BIOLOGY NOW RESOURCES

Active Figures

39-7: The action potential

39-8: Transmission of an action potential

Preparing for an exam? Take a diagnostic test on your BiologyNow CD-ROM.

Post-Test Answers

1. c	2. e	3. a	4. b
5. e	6. d	7. e	8. e
9. a	10. c	11. c	12. c
13. c	14. b	15. d	

40

Neural Regulation

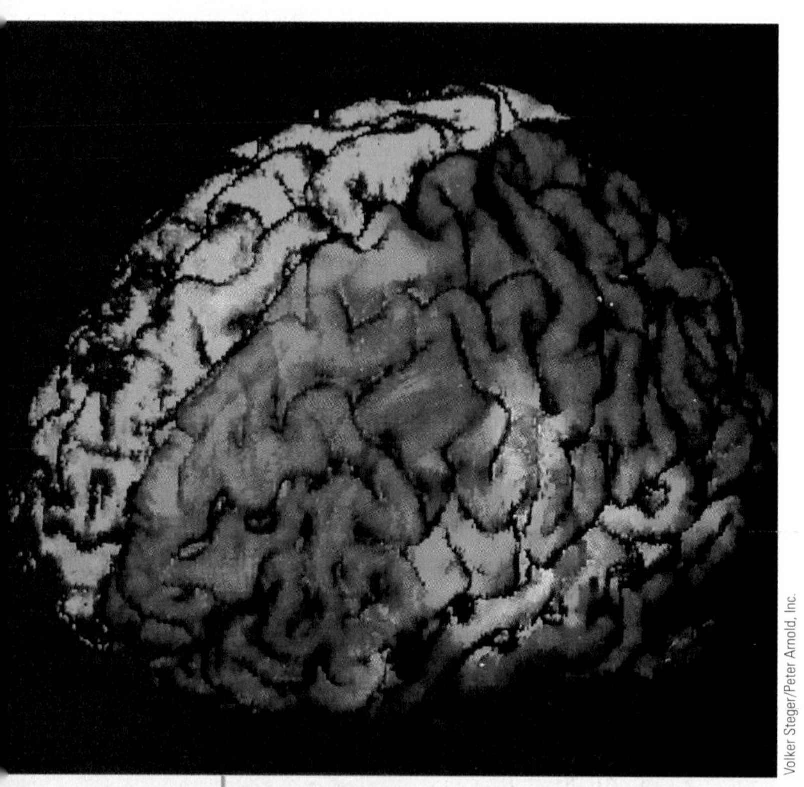

Volker Steger/Peter Arnold, Inc.

Functional magnetic resonance imaging (fMRI) of the brain. In this image, red indicates areas of greatest activation during speech; yellow indicates medium activation.

CHAPTER OUTLINE

- **Invertebrate Nervous Systems**
- **Organization of the Vertebrate Nervous System**
- **Evolution of the Vertebrate Brain**
- **The Human Central Nervous System**
- **The Peripheral Nervous System**
- **Effects of Drugs on the Nervous System**

A frog flicks out its tongue with lightning speed to capture a fly, a rabbit escapes from a predator, and you learn biology—perhaps not that quickly. In a complex nervous system, millions of neurons work together to produce effective responses to stimuli in the external environment. The nervous system also regulates heart rate, breathing, and hundreds of other internal activities Researchers have focused extensively on just how neurons interact, and we are only beginning to understand the mechanisms that permit the complex functions of the nervous system.

Neurobiologists have investigated two major aspects of neural function: hard-wiring of the nervous system and neural plasticity. The term **hard-wiring** refers to how neurons signal one another, how they connect, and how they carry out basic functions such as regulating heart rate, blood pressure, respiration, and sleep–wake cycles. Hard-wiring was the major focus of neurobiologists for decades.

Eventually, neurobiologists began to focus on their observation that even animals with very simple nervous systems can learn to repeat behaviors associated with reward and to avoid behaviors that cause pain. Such changes in behavior are possible because of **neural plasticity,** the ability of the nervous system to change in response to experience. Such change modifies structure and function. Many areas of the brain once thought hard-wired are now known to be flexible and capable of change. Familiar examples of neural plasticity in human motor skills include learning to walk, ride a bicycle, or catch a baseball. At first you were probably clumsy, but with practice your performance became smoother and more precise. Similarly, the ability to learn languages, solve problems, and perform scientific experiments depends on neural plasticity.

What are the mechanisms by which neurons signal one another to learn diverse activities such as skilled movement and studying biology? Canadian psychologist Donald O. Hebb proposed in 1949 that when two neurons connected by a *synapse* are active simultaneously, the synapse is somehow strengthened.

In 1973, British researchers T.V.P. Bliss and T.J. Lomo found experimental evidence for Hebb's hypothesis. Investigating rabbit brains, these researchers found that when neurons in the hippocampus (a region that converts information into memories) were stimulated by a series of high-frequency electrical stimuli, functional changes occurred at synapses.

Because neurons signal one another at synapses, these junctions between neurons are currently the focus of intense research. Neural plasticity may, in fact, be **synaptic plasticity,** because changes that occur during learning and remembering appear to take place at synapses. Working on a molecular level, neurobiologists are studying neurotransmitter action on membrane receptors. They are asking questions more quickly than they can find answers. How is synaptic activity depressed, and how is it enhanced? Do neurons change the contacts they make with one another? Do they make connections with new neurons? What mechanisms are activated by neurotransmitter-receptor binding? These mechanisms are known to include complex intracellular signaling systems that involve second messengers, gene activation, and protein synthesis.

Neurobiologists use a variety of methods to study the mechanisms of neural function. For example, improved imaging methods have revolutionized the study of the brain. Functional magnetic resonance imaging (fMRI) has provided investigators with a window through which to observe brain function. Functional MRI allows neurobiologists to study the responses of neural networks in the brain while an individual is actually performing a task, such as speaking (see figure). During the performance of the task, an area of the brain becomes active, and flow of oxygenated blood to that area increases. Functional MRI detects changes that take place in response to a very brief stimulus. For example, a visual stimulus lasting only 30 milliseconds stimulates brain activation that can be detected by fMRI.

In this chapter we compare various animal nervous systems. We then examine the structure and function of the vertebrate nervous system, with emphasis on the function of the human brain. We explore some of the frontiers of neurobiology such as the mechanisms involved in information processing. ■

INVERTEBRATE NERVOUS SYSTEMS

Learning Objectives

1. Contrast nerve nets and radial nervous systems with bilateral nervous systems.
2. Identify trends in the evolution of the invertebrate nervous system.

An animal's lifestyle is closely related to the organization and complexity of its nervous system. *Hydra* and other cnidarians have a **nerve net** consisting of interconnected neurons with no central control organ. Electrical signals are sent from neuron to neuron in more than one direction (Fig. 40-1). Responses may involve large parts of the body. An advantage of the nerve net is that the cnidarian responds effectively to predator or prey approaching from any direction. *Hydra* responds by discharging nematocysts (stinging structures) and coordinating the movements of its tentacles to capture food. Some cnidarians have two or more nerve nets. In some jellyfish, a slow-transmission nerve net coordinates movement of the tentacles, and a second nerve net, which is faster, coordinates swimming.

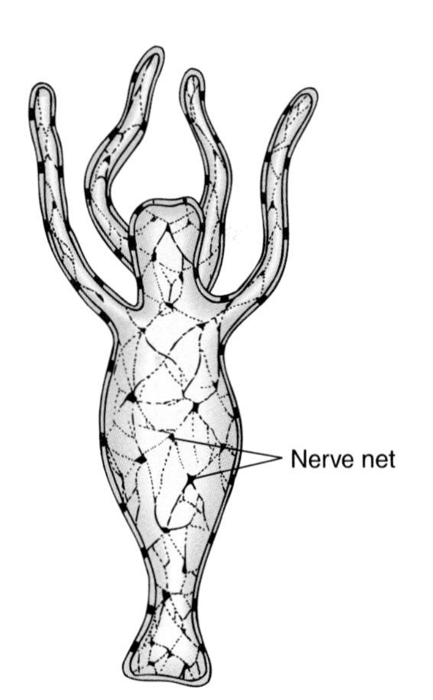

FIGURE **40-1**

***Hydra's* nerve net.**

Hydra and other cnidarians have a network of neurons with no central control organ.

The **radial nervous system** of the sea star and other echinoderms (see Chapter 30) is a modified nerve net. This system shows some degree of selective organization of neurons into more than a diffuse network. It consists of a nerve ring around the mouth, from which a large radial nerve extends into each arm. Branches of these nerves, which form a network somewhat similar to the nerve net of *Hydra,* coordinate the animal's movement.

Bilaterally symmetrical animals have a **bilateral nervous system.** We can identify the following trends in the evolution of nervous systems:

1. *Increased number of nerve cells.*
2. *Concentration of nerve cells,* forming masses of tissue that become **ganglia** and **brain,** and thick cords of tissue that become **nerve cords** and **nerves.**
3. *Specialization of function.* For example, transmission of nerve impulses in one direction requires both **afferent nerves,** which conduct impulses toward a central nervous system (CNS), and **efferent nerves,** which transmit impulses away from the CNS to the **effectors** (muscles and glands). Certain parts of the CNS are typically specialized to perform specific functions, and distinct structural and functional regions can be identified.
4. *An increased number of interneurons and more complex synaptic contacts* permit greater integration of incoming

messages, provide a greater range of responses, and allow more precision in responses.

5. ***Cephalization,*** *or formation of a head.* A bilaterally symmetrical animal generally moves in a forward direction. Concentration of sense organs at the front end of the body lets the animal detect an enemy quickly enough to escape or to see or smell food in time to capture it. Response can be rapid if short pathways link sense organs to decision-making nerve cells nearby. Therefore, nerve cells are typically concentrated in the head region and organized to form ganglia or a brain.

In planarian flatworms, the head region contains concentrations of nerve cells called **cerebral ganglia** (Fig. 40-2). These serve as a primitive brain and have some control over the rest of the nervous system. Typically, two solid ventral longitudinal nerve cords extend from the ganglia to the posterior end of the body. Transverse nerves connect the two nerve cords and connect the brain with the eyespots. This arrangement is called a "ladder-type" nervous system.

Annelids and arthropods typically have a ventral nerve cord (Fig. 40-3). The cell bodies of many of the neurons are massed into ganglia. Afferent and efferent neurons lie in lateral nerves that link the ganglia with muscles and other body structures. If an earthworm's brain is removed, the animal can move almost as well as before. However, when it bumps into an obstacle, it persists in a futile effort to move forward rather than turning aside. To respond adaptively to environmental change, the earthworm needs its brain.

The cerebral ganglia of some arthropods differ from those of annelids in that they have specific functional regions. These areas are specialized for integrating information transmitted to the ganglia from sense organs.

Mollusks with inactive lifestyles have relatively simple nervous systems with little cephalization and very simple sense organs. In contrast, the active cephalopod mollusks (squids and octopods) have complex nervous systems with neurons concentrated in a central region. Ganglia are massed in a ring surrounding the esophagus, making up a brain that contains about 168 million nerve cells. This complex nervous system, which includes well-developed sense organs, is an adaptation that correlates with the active, predatory lifestyle of these animals. With its highly developed nervous system, the octopus is capable of considerable learning and can be taught quite complex tasks. In fact, cephalopods are considered the most intelligent invertebrates.

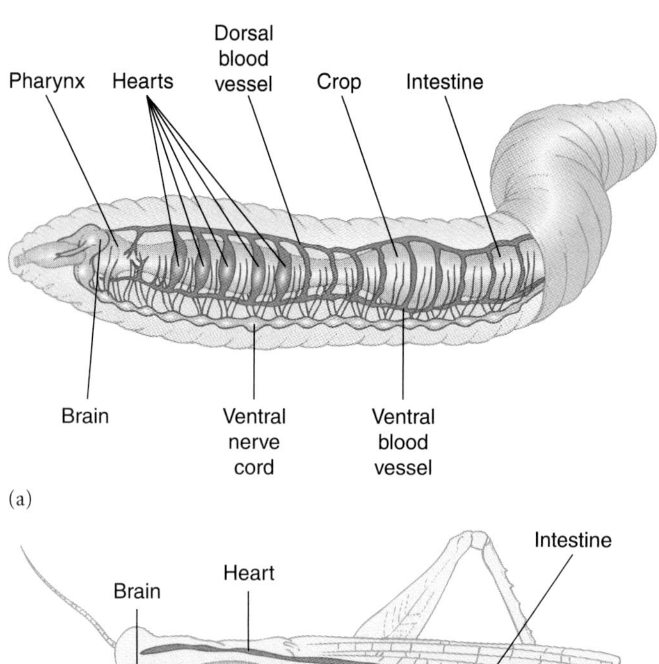

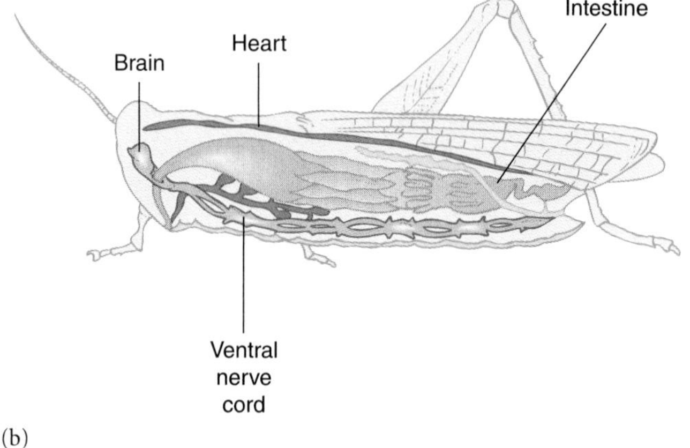

FIGURE **40-3** **The annelid and arthropod nervous systems.**

(a) Like other annelids, the earthworm nervous system includes a dorsal anterior brain and one or more ventral nerve cords. Cell bodies of the neurons are located in ganglia that are connected by the ventral nerve cord. **(b)** In the insect nervous system, the brain is connected to a ventral nerve cord. The brain is more specialized than in annelids.

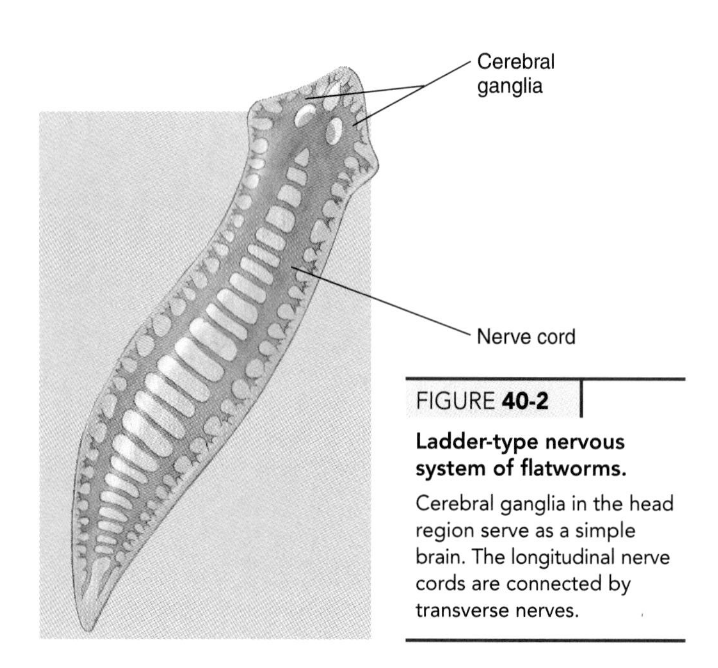

FIGURE **40-2**

Ladder-type nervous system of flatworms.

Cerebral ganglia in the head region serve as a simple brain. The longitudinal nerve cords are connected by transverse nerves.

Review

- What are some trends in the evolution of nervous systems?
- What are some advantages of cephalization?

BiologyNow™ Assess your understanding of **invertebrate nervous systems** by taking the pretest on your BiologyNow CD-ROM.

ORGANIZATION OF THE VERTEBRATE NERVOUS SYSTEM

Learning Objective

3 Describe the two main divisions of the vertebrate nervous system.

An animal's range of possible responses depends in large part on the number of its neurons and how they are organized in the nervous system. As animal groups evolved, nervous systems became increasingly complex. The vertebrate nervous system has two main divisions: the **central nervous system (CNS)** and the **peripheral nervous system (PNS)** (Table 40-1). The CNS consists of a complex brain that is continuous with the dorsal tubular **spinal cord.** Serving as central control, these organs integrate incoming information and determine appropriate responses.

The PNS is made up of the sensory receptors (for example, tactile, auditory, and visual receptors) and the nerves, which are the communication lines. Various parts of the body are linked to the brain by cranial nerves and to the spinal cord by spinal nerves. Afferent neurons in these nerves continuously inform the CNS of changing conditions. Then efferent neurons transmit the "decisions" of the CNS to appropriate muscles and glands, which make the adjustments needed to maintain homeostasis.

For convenience the PNS is subdivided into somatic and autonomic divisions. Most of the receptors and nerves concerned with changes in the external environment are somatic. Those that regulate the internal environment are autonomic. Both divisions have afferent nerves, which transmit messages from receptors to the CNS, and efferent nerves, which transmit information back from the CNS to the structures that respond. The autonomic division has two kinds of efferent pathways: sympathetic and parasympathetic nerves.

TABLE 40-1 Divisions of the Vertebrate Nervous System

Central Nervous System (CNS)
- Brain
- Spinal cord

Peripheral Nervous System (PNS)
- Somatic division
  1. Receptors
  2. Afferent (sensory) nerves—transmit information from receptors to CNS
  3. Efferent (motor) nerves—transmit information from CNS to skeletal muscles
- Autonomic division
  1. Receptors
  2. Afferent (sensory) nerves—transmit information from receptors in internal organs to CNS
  3. Efferent nerves—transmit information from CNS to glands and involuntary muscle in organs
     a. Sympathetic nerves—generally stimulate activity that results in mobilization of energy (e.g., speeds heartbeat)
     b. Parasympathetic nerves—action results in energy conservation or restoration (e.g., slows heart but speeds digestion)

Review

- What are the main components of the CNS?
- What are the main components of the PNS?

BiologyNow™ Assess your understanding of **the organization of the vertebrate nervous system** by taking the pretest on your BiologyNow CD-ROM.

EVOLUTION OF THE VERTEBRATE BRAIN

Learning Objective

4 Trace the development of the principal vertebrate brain regions from the forebrain, midbrain, and hindbrain, and compare the brains of fishes, amphibians, reptiles, birds, and mammals.

All vertebrates, from fishes to mammals, have the same basic brain structure. Different parts of the brain are specialized in the various vertebrate classes, and the evolutionary trend is toward increasing complexity, especially of the cerebrum and cerebellum.

In the early vertebrate embryo, the brain and spinal cord differentiate from a single tube of tissue, the **neural tube.** Anteriorly, the tube expands and develops into the brain. Posteriorly, the tube becomes the spinal cord. Brain and spinal cord remain continuous, and their cavities communicate. As the brain begins to differentiate, three bulges become visible: the hindbrain, midbrain, and forebrain (Fig. 40-4).

FIGURE 40-4 Early development of the vertebrate nervous system.

Early in the development of the vertebrate embryo, the anterior end of the neural tube differentiates into the forebrain, midbrain, and hindbrain. These primary divisions subdivide and eventually give rise to specific structures of the adult brain (see Table 40-2).

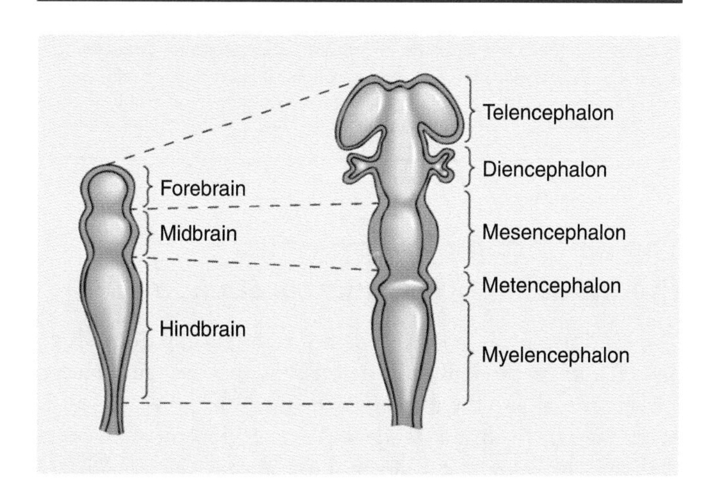

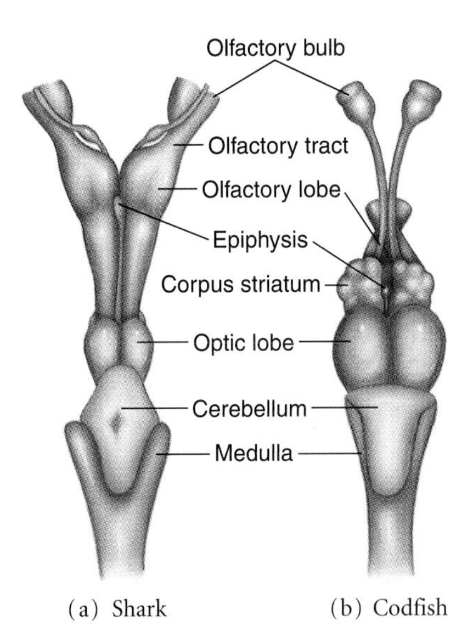

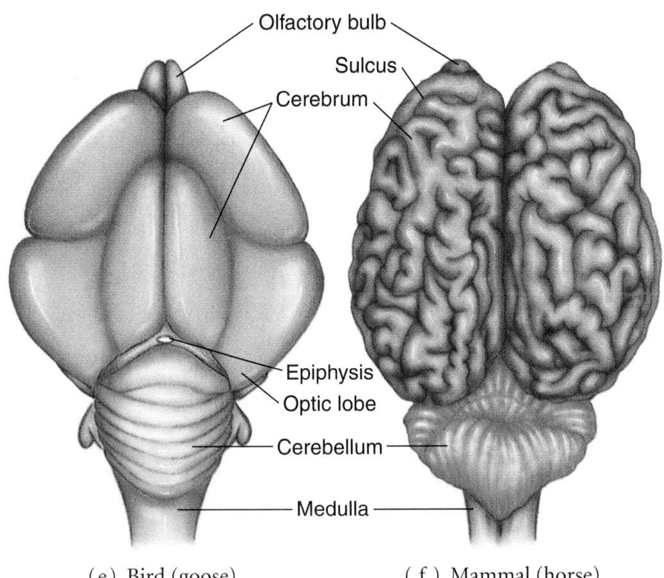

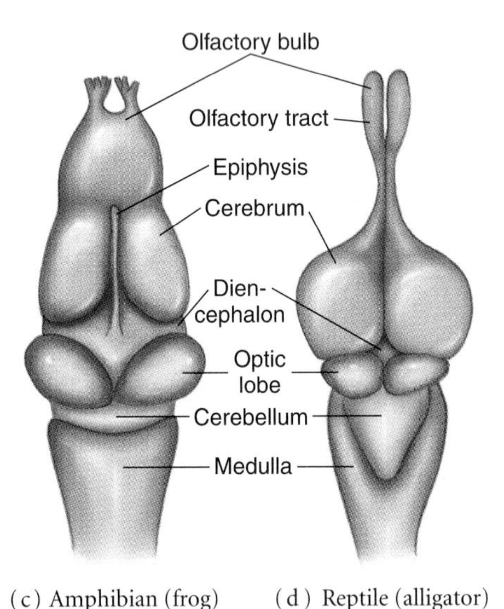

FIGURE **40-5** **Evolution of the vertebrate brain.**
Comparison of the brains of members of six vertebrate classes reveals basic similarities and evolutionary trends. (Brains are not drawn to scale.) Note that different parts of the brain are specialized in the various groups. The large olfactory lobes in the shark brain **(a)** are essential to this predator's highly developed sense of smell. **(b** through **f)** During the course of evolution, the cerebrum and cerebellum have become larger and more complex. In mammals **(f),** the cerebrum is the most prominent part of the brain; the cerebral cortex, the thin outer layer of the cerebrum, is highly convoluted (folded), which greatly increases its surface area.

The hindbrain develops into the medulla, pons, and cerebellum

The **hindbrain** subdivides to form the **metencephalon,** which gives rise to the **cerebellum** and **pons,** and the **myelencephalon,** which gives rise to the **medulla.** The medulla, pons, and midbrain make up the **brain stem,** the elongated portion of the brain that looks like a stalk holding up the cerebrum.

The medulla, the most posterior part of the brain, is continuous with the spinal cord. Its cavity, the **fourth ventricle,** is continuous with the central canal of the spinal cord and with a channel that runs through the midbrain. The walls of the medulla are thick and made up largely of *nerve tracts* (bundles of axons) that connect the spinal cord with various parts of the brain. In complex vertebrates, the medulla contains vital centers that regulate respiration, heartbeat, and blood pressure. Other reflex centers in the medulla regulate activities such as swallowing, coughing, and vomiting.

The size and shape of the cerebellum vary among the vertebrate classes (Fig. 40-5). Development of the cerebellum in different animals is roughly correlated with the extent and complexity of muscular activity, reflecting the principle that the relative size of a brain part correlates with its importance to the behavior of the species. In some fishes, birds, and mammals, the cerebellum is highly developed, whereas it tends to be small in jawless fishes, amphibians, and reptiles. The cerebellum coordinates muscle activity and is responsible for muscle tone, posture, and equilibrium.

Injury or removal of the cerebellum results in impaired muscle coordination. A bird without a cerebellum cannot fly, and its wings thrash about jerkily. When the human cerebellum is injured by a blow or by disease, muscular movements are uncoordinated. Any activity requiring delicate coordination, such as threading a needle, is very difficult, if not impossible.

In mammals, a large mass of fibers forms the pons, a bridge connecting the spinal cord and medulla with upper parts of the brain. The pons contains centers that help regulate respiration and nuclei that relay impulses from the cerebrum to the cerebellum. A **nucleus** is a group of cell bodies within the CNS.

The midbrain is prominent in fishes and amphibians

In fishes and amphibians, the **midbrain,** or **mesencephalon,** is the most prominent part of the brain, serving as the main association area. It receives incoming sensory information, integrates it, and sends decisions to appropriate motor nerves. The dorsal portion of the midbrain is differentiated to some extent. For example, the optic lobes are specialized for visual interpretations.

In reptiles, birds, and mammals, many functions of the optic lobes are assumed by the cerebrum, which develops from the forebrain. In mammals, the midbrain consists of the **superior colliculi,** centers for visual reflexes such as pupil constriction, and the **inferior colliculi,** centers for certain auditory reflexes. The mammalian midbrain also contains a center (the *red nucleus*) that helps maintain muscle tone and posture.

The forebrain gives rise to the thalamus, hypothalamus, and cerebrum

As indicated in Table 40-2, the **forebrain** subdivides to form the **telencephalon** and **diencephalon.** The telencephalon gives rise to the **cerebrum,** and the diencephalon to the **thalamus** and **hypothalamus.** The *lateral ventricles* (also called the first and second ventricles) lie within the cerebrum. Each lateral ventricle connects with the third ventricle (within the diencephalon) by way of a channel.

In all vertebrate classes, the thalamus is a relay center for motor and sensory messages. In mammals, all sensory messages except those from the olfactory receptors are delivered to the thalamus before they are relayed to the sensory areas of the cerebrum.

The hypothalamus, which lies below the thalamus, forms the floor of the third ventricle. The hypothalamus is a major coordinating center for regulating autonomic and somatic responses. It integrates incoming information and provides input to centers in the medulla and spinal cord that regulate activities such as heart rate, respiration, and digestive system function. In reptiles, birds, and mammals, the hypothalamus controls body temperature. It also contains olfactory centers, regulates appetite and water balance, and is important in emotional and sexual responses. As discussed in Chapter 47, the hypothalamus links the nervous and endocrine systems and produces certain hormones.

The telencephalon gives rise to the cerebrum and, in most vertebrate groups, the **olfactory bulbs.** These structures are important in the chemical sense of smell—the dominant sense in most aquatic and terrestrial vertebrates. In fact, much of brain development in vertebrates appears to focus on integrating olfactory information. In fish and amphibians, the cerebrum is almost entirely devoted to this function.

Birds are an exception among the vertebrates in that their sense of smell is generally poor. A part of their cerebrum, the *corpus striatum,* however, is highly developed. This structure controls eating, flying, singing, and other complex action patterns.

In most vertebrates, the cerebrum is divided into right and left **cerebral hemispheres.** Most of the cerebrum is made of **white matter,** which consists mainly of myelinated axons that connect various parts of the brain. In mammals and most reptiles, a layer of **gray matter,** the **cerebral cortex,** makes up the outer portion of the cerebrum. Gray matter contains cell bodies and dendrites.

Certain reptiles and all mammals have a type of cerebral cortex, the **neocortex,** not found in less complex vertebrates. The neocortex serves as an association area—a region that links sensory and motor functions and is responsible for higher functions such as learning. The neocortex is very extensive in mammals, making up the bulk of the cerebrum. In humans, about 90% of the cerebral cortex is neocortex, consisting of six distinct cell layers.

In mammals, the cerebrum is the most prominent part of the brain. During embryonic development, it expands and grows backward, covering many other brain structures. The cerebrum is responsible for many functions performed by other parts of the brain in other vertebrates. In addition, the mammalian brain has many complex association functions that reptiles, amphibians, and fishes lack.

In small or simple mammals, the cerebral cortex may be smooth. However, in large complex mammals, the surface area is greatly expanded by numerous folds called **convolutions.** The furrows between them are called **sulci** (sing., *sulcus*) if shallow and **fissures** if deep. The number of folds (not the size of the brain) has been associated with complexity of brain function.

Review

- What structures of the vertebrate brain derive from the embryonic forebrain?
- How does the mammalian brain differ from the brain of an amphibian?

BiologyNow™ Assess your understanding of **the evolution of the vertebrate brain** by taking the pretest on your BiologyNow CD-ROM.

TABLE 40-2 Differentiation of CNS Structures

Early Embryonic Divisions	Subdivisions	Derivatives in Adult	Cavity
Brain			
Forebrain	Telencephalon	Cerebrum	Lateral ventricles (first and second ventricles)
	Diencephalon	Thalamus, hypothalamus, epiphysis (pineal body)	Third ventricle
Midbrain	Mesencephalon	Optic lobes in fish and amphibians; superior and inferior colliculi	Cerebral aqueduct
Hindbrain	Metencephalon	Cerebellum, pons	
	Myelencephalon	Medulla	Fourth ventricle
Spinal Cord		Spinal Cord	Central canal

THE HUMAN CENTRAL NERVOUS SYSTEM

Learning Objectives

5 Describe the structure and functions of the human spinal cord.

6 Describe the structure and functions of the human cerebrum.

7 Describe the sleep–wake cycle, and contrast rapid-eye-movement (REM) sleep and non-REM sleep.

8 Describe the actions of the limbic system.

9 Summarize how the brain processes information.

The soft, fragile human brain and spinal cord are well protected. Encased within bone, they are covered by three layers of connective tissue, the **meninges.** The three meningeal layers are the tough, outer **dura mater,** the middle **arachnoid,** and the thin, vascular **pia mater,** which adheres closely to the tissue of the brain and spinal cord (Fig. 40-6). *Meningitis* is a disease in which these coverings become infected and inflamed.

The space between the arachnoid and the pia mater is the subarachnoid space, which contains **cerebrospinal fluid (CSF).** This fluid is produced by special networks of capillaries, collectively called the **choroid plexus,** that extend from the pia mater into the ventricles. CSF is a shock-absorbing fluid that cushions the brain and spinal cord against mechanical injury. This fluid also serves as a medium for exchange of nutrients and waste products between the blood and brain. CSF circulates down through the ventricles and passes into the subarachnoid space. It is then reabsorbed into large blood sinuses within the dura mater.

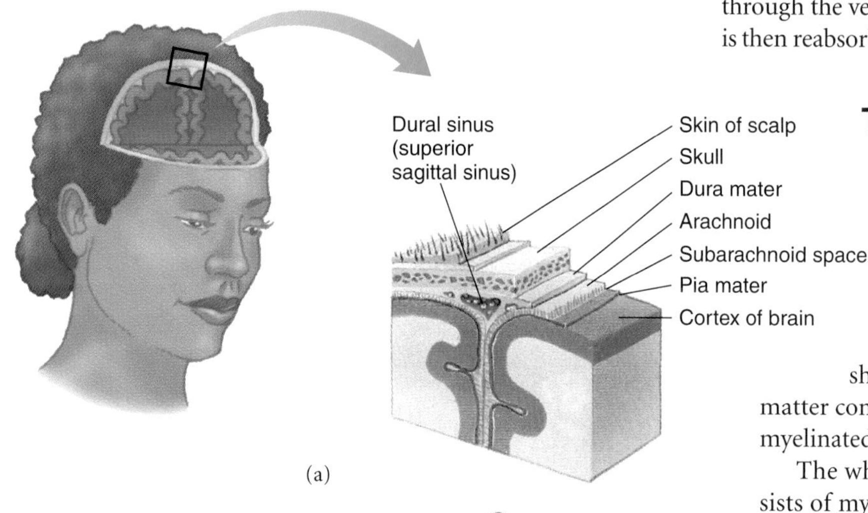

The spinal cord transmits impulses to and from the brain

The tubular **spinal cord** extends from the base of the brain to the level of the second lumbar vertebra. A cross-section view through the spinal cord reveals a small **central canal** surrounded by an area of gray matter shaped a bit like the letter H (Fig. 40-7). The gray matter consists of large masses of cell bodies, dendrites, unmyelinated axons, and glial cells.

The white matter, found outside the gray matter, consists of myelinated axons arranged in bundles called **tracts** or **pathways. Ascending tracts** conduct impulses up the cord to the brain. For example, the spinothalamic tracts in the anterior and lateral columns of the white matter conduct pain and temperature information from sensory neurons in the skin. The pyramidal tracts are **descending tracts** that convey impulses from the cerebrum to motor nerves at various levels in the cord. We describe the spinal nerves later in this chapter.

FIGURE **40-6** **Protection of the brain and spinal cord.**

The CNS is well protected by the skull and meninges and by cerebrospinal fluid (CSF). **(a)** Frontal section through the superior part of the brain. Note the large dural sinus, a blood sinus, between two layers of the dura mater. Blood leaving the brain flows into such sinuses and then circulates to the large jugular veins in the neck. **(b)** Sagittal section. The CSF, which cushions the brain and spinal cord, is produced by the choroid plexi in the walls of the ventricles. This fluid circulates through the ventricles and subarachnoid space. It is continuously produced and continuously reabsorbed into the blood of the dural sinuses.

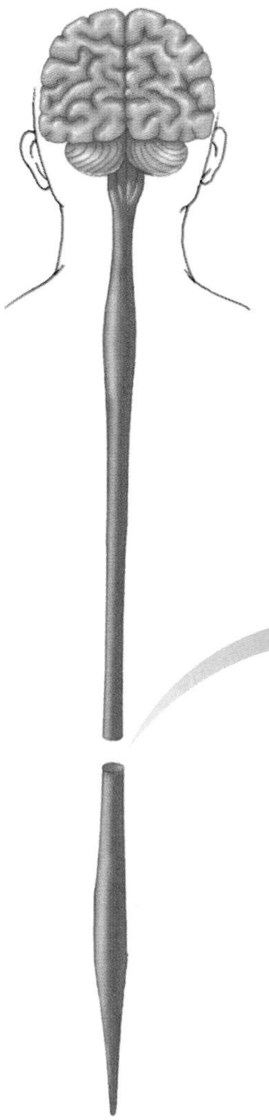

In addition to transmitting impulses to and from the brain, the spinal cord controls many reflex activities. A **reflex action** is a relatively simple, involuntary motor response to a stimulus. Although most reflex actions are much more complex, let us consider a **withdrawal reflex,** in which a neural circuit consisting of three types of neurons carries out a response to a stimulus (Fig. 40-8). Suppose you touch a flame. Almost instantly, and even before you become consciously aware of what has happened, you jerk your hand away. In this brief instant a sensory neuron transmits a message from pain receptors in the skin to the spinal cord. Within the spinal cord, a sensory neuron transmits the signal to an interneuron. The interneuron integrates the information and signals an appropriate motor neuron, which conducts it to groups of muscles. The muscles respond by contracting and moving the limb toward the midline of the body, moving the hand away from the flame. Actually, many neurons

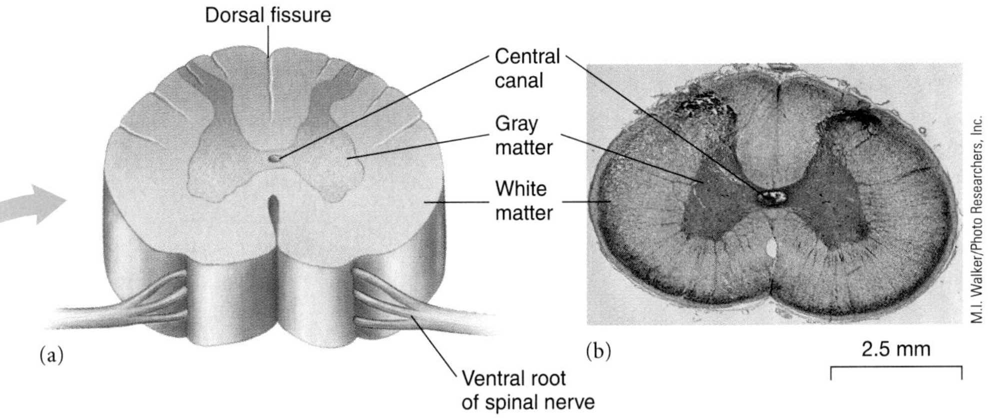

FIGURE **40-7** | **Structure of the spinal cord.**

The spinal cord consists of gray matter and white matter. **(a)** Cross section through the spinal cord. **(b)** LM of a cross section through the spinal cord.

KEY CONCEPT: A reflex action is a coordinated, involuntary motor response to a stimulus.

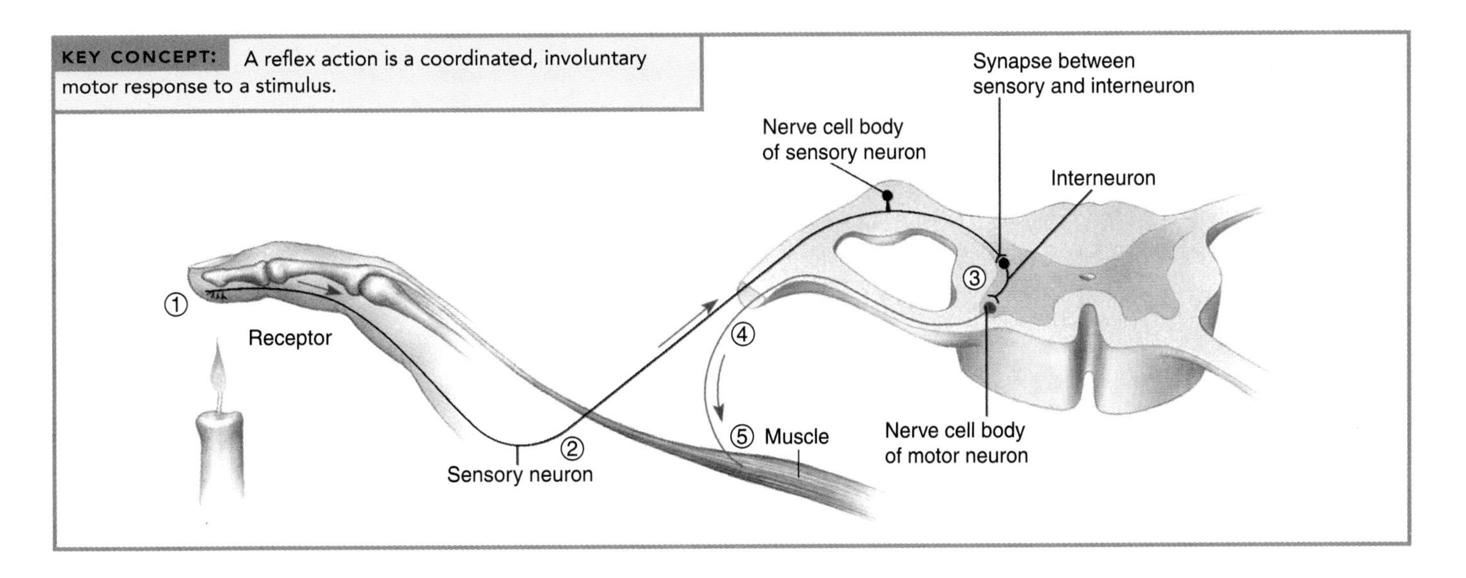

FIGURE **40-8** | **Withdrawal reflex.**

① A sensory receptor signals a sensory neuron that ② transmits impulses to the CNS. ③ An interneuron integrates the information. ④ Then an appropriate motor neuron *(shown in red)* transmits impulses to the muscles. ⑤ The muscles contract, moving the hand away from the flame (the actual response).

in sensory, association, and motor nerves participate in such a reflex action. Generally we are not even consciously aware all these responding muscles exist. In summary, signals are transmitted through a withdrawal reflex in the following sequence:

Reception by sensory receptor in skin ⟶ sensory neuron transmits signal ⟶ CNS interneuron integrates information ⟶ motor neuron transmits signal ⟶ muscle contracts

At the same time the reflex pathway is activated, a message is sent up the spinal cord to the conscious areas of the brain. As you pull back your hand from the flame, you become aware of what has happened and feel the pain. This awareness, however, is not part of the reflex response.

Contrary to long-accepted dogma, the spinal cord has plasticity and is capable of training. For example, sensory feedback from walking exercise may increase the strength of neural connections in the spinal cord. If descending connections with the brain are intact, some victims of spinal cord injury (some 200,000 patients in the United States alone) could potentially develop at least some limited ability to walk again.

The most prominent part of the human brain is the cerebrum

The structure and functions of the main parts of the human brain are summarized in Table 40-3, and the brain is illustrated in Figures 40-9 and 40-10. As in other mammals, the human cerebral cortex consists of right and left *cerebral hemispheres* and is functionally divided into three areas: (1) **sensory areas** that receive incoming signals from the sense organs; (2) **motor areas** that control voluntary movement; and (3) **association areas** that link the sensory and motor areas and are responsible for thought, learning, language, memory, judgment, and personality.

Investigators have mapped the cerebral cortex, locating the areas responsible for different functions. The **occipital lobes** contain the visual centers. Stimulation of these areas, even by a blow on the back of the head, causes the sensation of light; their removal causes blindness. The centers for hearing are located in the **temporal lobes** of the brain above the ear; stimulation by a blow causes a sensation of noise. Removal of both auditory areas causes deafness. Removal of one does not cause deafness in

TABLE 40-3 The Human Brain

Structure	Description	Function
Brain stem		
Medulla	Continuous with spinal cord; primarily made up of nerves passing from spinal cord to rest of brain	Contains vital centers (clusters of neuron cell bodies) that control heartbeat, respiration, and blood pressure; contains centers that control swallowing, coughing, vomiting
Pons	Forms bulge on anterior surface of brain stem	Connects various parts of brain with one another; contains respiratory and sleep centers
Midbrain	Just above pons	Center for visual and auditory reflexes (e.g., pupil reflex, blinking, adjusting ear to volume of sound)
Thalamus	At top of brain stem	Main sensory relay center for conducting information between spinal cord and cerebrum; neurons in thalamus sort and interpret all incoming sensory information (except olfaction) before relaying messages to appropriate neurons in cerebrum
Hypothalamus	Just below thalamus; pituitary gland is connected to hypothalamus by stalk of neural tissue	Contains centers for control of body temperature, appetite, fat metabolism, and certain emotions; regulates pituitary gland
Cerebellum	Second largest division of brain	Reflex center for muscular coordination and refinement of movements
Cerebrum	Largest, most prominent part of human brain; longitudinal fissure divides cerebrum into right and left hemispheres, each divided into lobes: frontal, parietal, temporal, and occipital lobes	Center of intellect, memory, consciousness, and language; also controls sensation and motor functions
Cerebral cortex (outer gray matter)	Arranged into convolutions (folds) that increase surface area; functionally, cerebral cortex is divided into	
	1. Motor areas	Control movement of voluntary muscles.
	2. Sensory areas	Receive incoming information from eyes, ears, pressure and touch receptors, etc.
	3. Association areas	Sites of intellect, memory, language, and emotion; interpret incoming sensory information
White matter	Consists of myelinated axons of neurons that connect various regions of brain; these axons are arranged into bundles (tracts)	Connects the following: 1. Neurons within same hemisphere 2. Right and left hemispheres 3. Cerebrum with other parts of brain and spinal cord

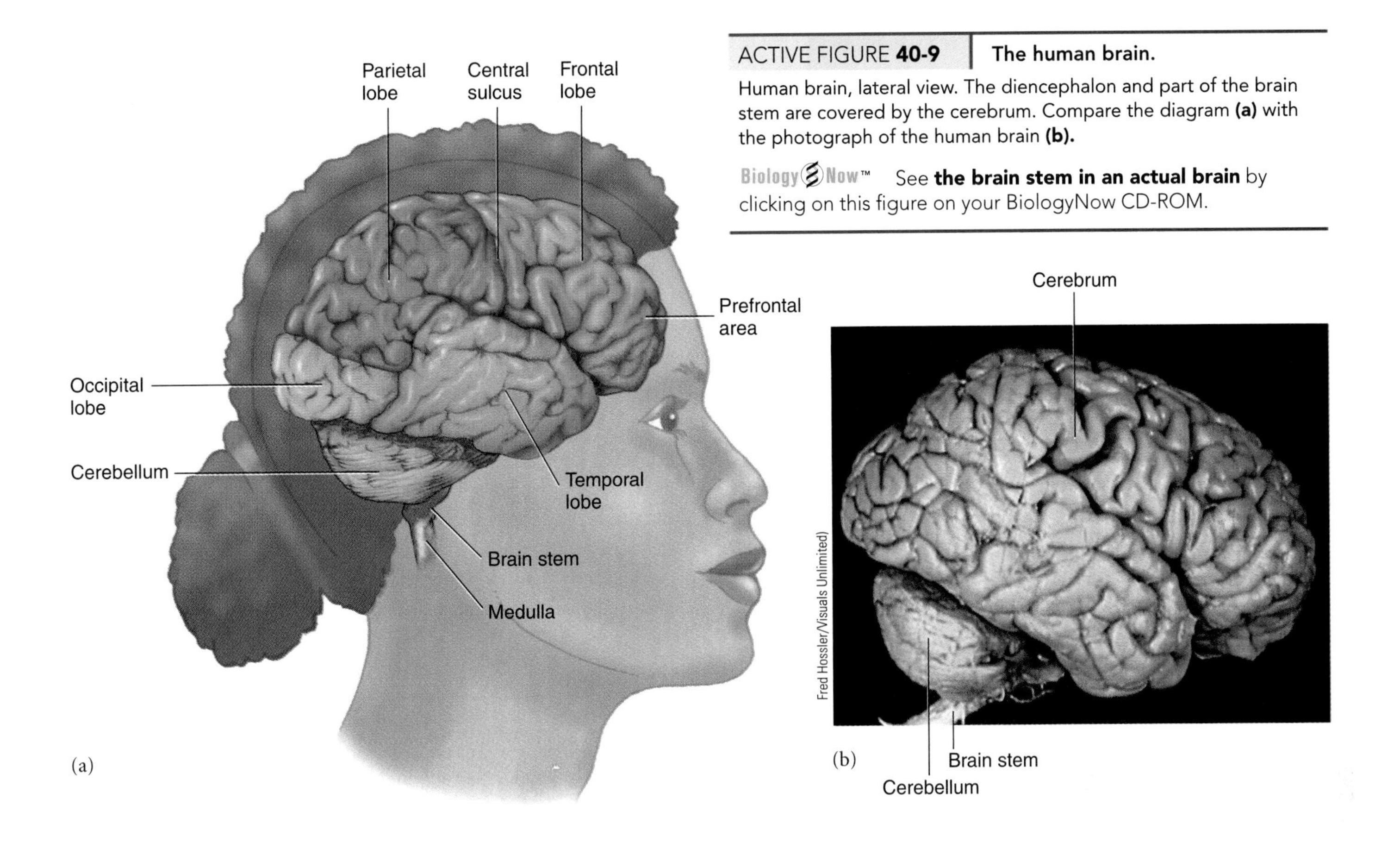

ACTIVE FIGURE 40-9 The human brain.

Human brain, lateral view. The diencephalon and part of the brain stem are covered by the cerebrum. Compare the diagram **(a)** with the photograph of the human brain **(b).**

BiologyNow™ See **the brain stem in an actual brain** by clicking on this figure on your BiologyNow CD-ROM.

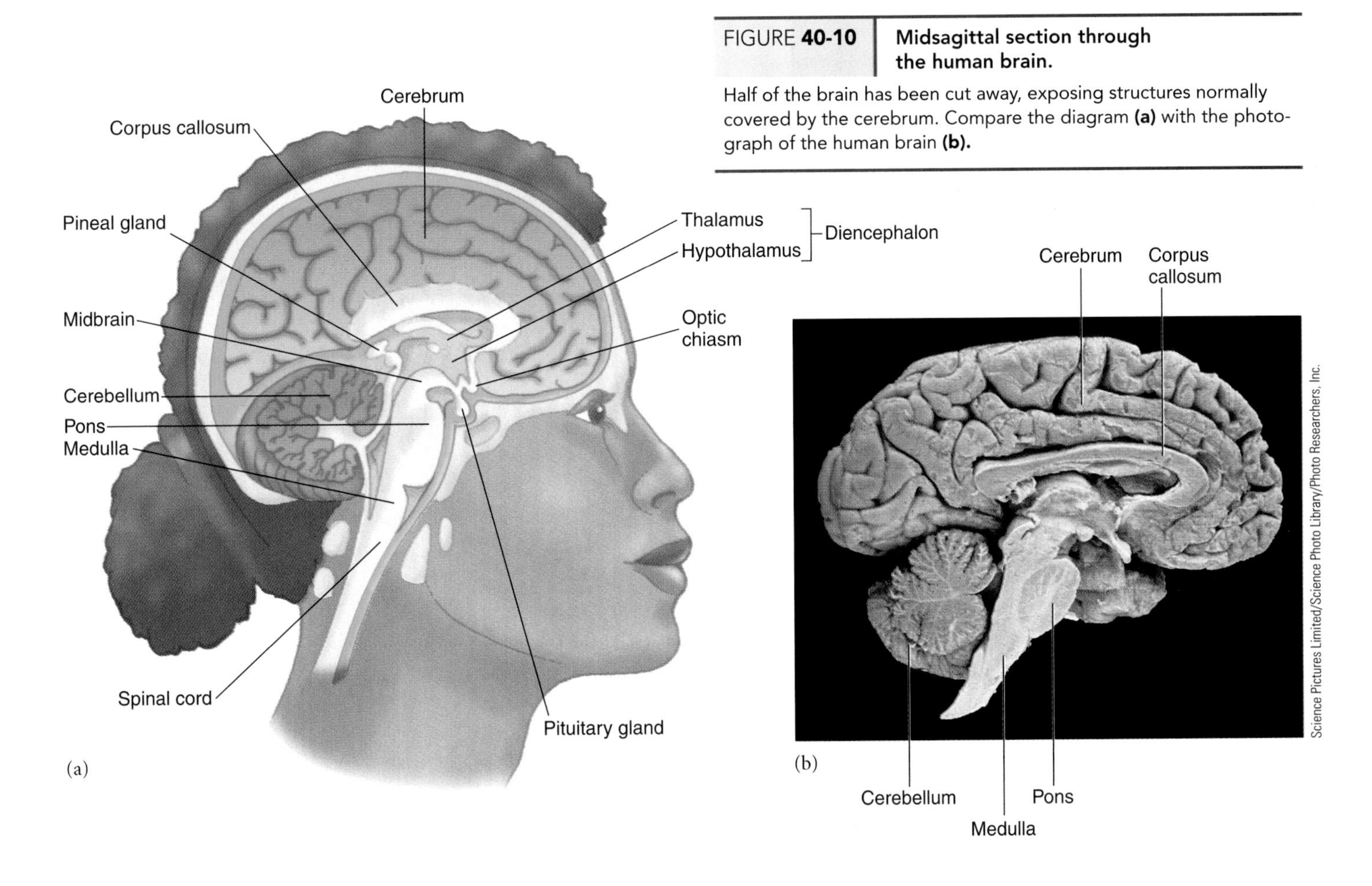

FIGURE 40-10 Midsagittal section through the human brain.

Half of the brain has been cut away, exposing structures normally covered by the cerebrum. Compare the diagram **(a)** with the photograph of the human brain **(b).**

one ear but rather produces a decrease in the auditory acuity of both ears.

A groove called the **central sulcus** crosses the top of each hemisphere from medial to lateral edge. This groove partially separates the **frontal lobes** from the **parietal lobes.** The frontal lobes have important motor and association areas. The **prefrontal cortex** is an association area in each frontal lobe that is crucial in evaluation, judgment, planning, and organizing responses. The **general motor areas** in the frontal lobes control the skeletal muscles. The **primary sensory areas** in the parietal lobes receive information regarding heat, cold, touch, and pressure from sense organs in the skin.

The size of the motor area in the brain for any given part of the body is proportional not to the amount of muscle, but to the complexity of movement involved. Predictably, areas that control the hands and face are relatively large (Fig. 40-11). A similar relationship exists between the sensory area and the sensitivity of the region of the skin from which it receives impulses. Neural fibers in the brain cross so that one side of the brain controls the opposite side of the body. As a result of another "reversal," the uppermost part of the cortex controls the lower limbs of the body.

When all the areas of known function are plotted, they cover almost all of the rat's cortex, a large part of the dog's, a moderate amount of the monkey's, and only a small part of the total surface of the human cortex. The remaining cortical areas are the association areas. Somehow the association regions integrate all the diverse impulses reaching the brain into a meaningful unit so that an appropriate response is made. When disease or accident destroys the functioning of one or more association areas, the ability to recognize certain kinds of symbols may be lost. For example, the names of objects may be forgotten, although their functions are remembered and understood.

The white matter of the cerebrum lies beneath the cerebral cortex. Nerve fibers of the white matter connect the cortical areas with one another and with other parts of the nervous system. A large band of white matter, the **corpus callosum,** connects the right and left hemispheres (see Fig. 40-10).

Deep within the white matter of the cerebrum lie the **basal ganglia,** paired groups of nuclei (gray matter). These nuclei play

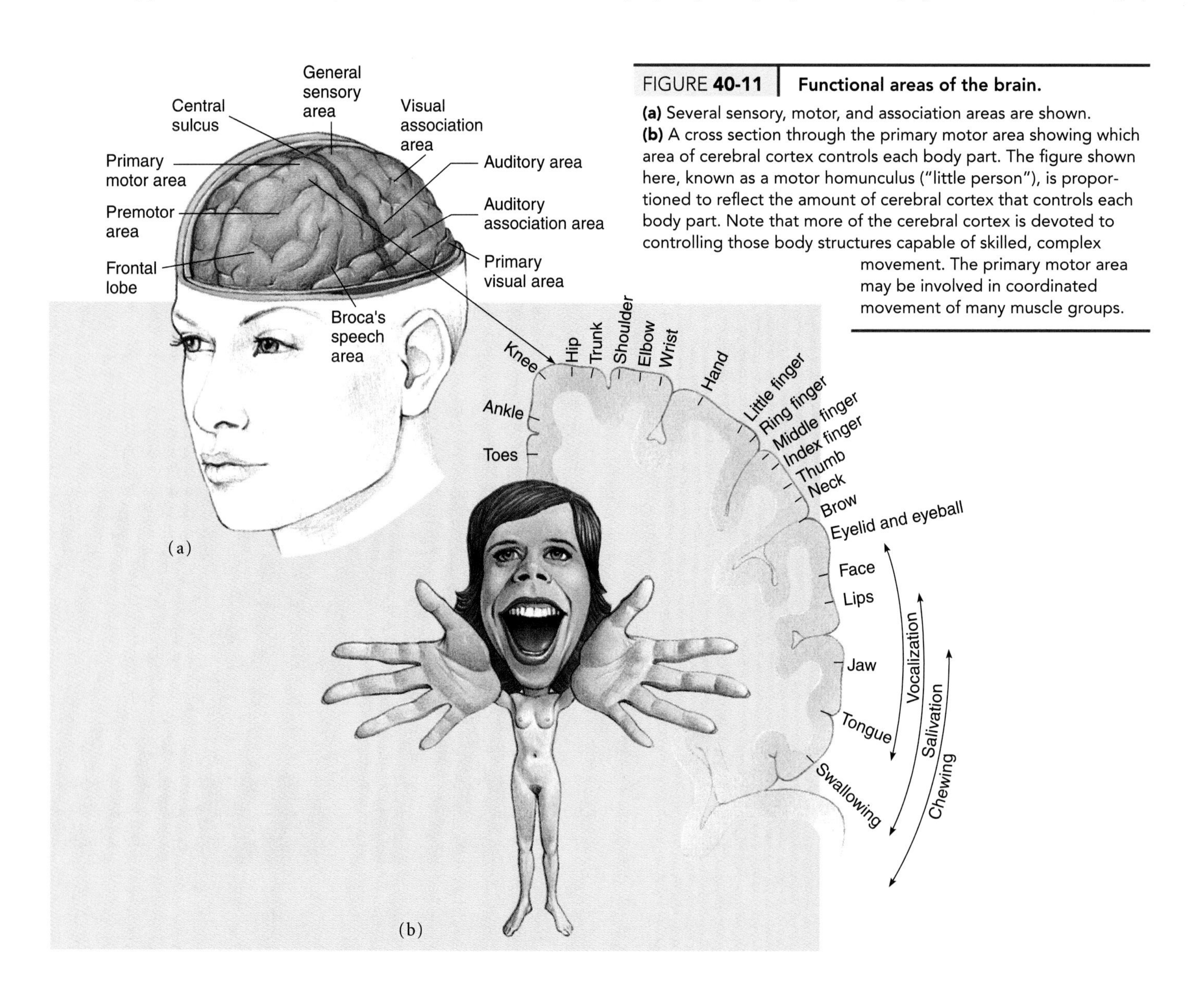

FIGURE **40-11** | **Functional areas of the brain.**
(a) Several sensory, motor, and association areas are shown. **(b)** A cross section through the primary motor area showing which area of cerebral cortex controls each body part. The figure shown here, known as a motor homunculus ("little person"), is proportioned to reflect the amount of cerebral cortex that controls each body part. Note that more of the cerebral cortex is devoted to controlling those body structures capable of skilled, complex movement. The primary motor area may be involved in coordinated movement of many muscle groups.

an important role in coordinating movement. The basal ganglia send signals to the **substantia nigra** in the midbrain, and they receive inputs back from the substantia nigra. The neurons that project (extend) to the basal ganglia produce **dopamine,** a neurotransmitter essential for sensory-motor coordination. Dopamine helps balance inhibition and excitation of neurons involved in motor function. Other neurons from the substantia nigra signal nuclei in the thalamus, which in turn relay information to the motor cortex. The substantia nigra neurons that signal the thalamus release the neurotransmitter **gamma-aminobutyric acid (GABA).** Just how these areas work together to coordinate motor function is not yet fully understood.

Brain activity cycles in a sleep–wake pattern

Brain activity can be studied by measuring and recording the electrical potentials, or "brain waves," generated by thousands of active neurons in various parts of the brain. This electrical activity can be recorded by a device known as an *electroencephalograph.* To obtain a recording of this electrical activity, called an **electroencephalogram (EEG),** electrodes are taped to different parts of the scalp, and the activity of the cerebral cortex is measured. The EEG shows the brain is continuously active. The most regular indication of activity appears to be **alpha waves,** which occur rhythmically at the rate of about 10 per second (Fig. 40-12).

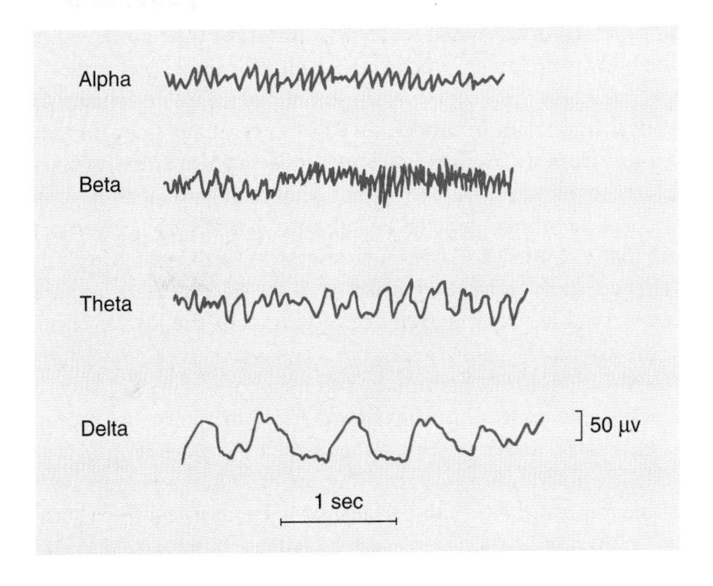

FIGURE **40-12** **EEGs showing electrical activity in the brain.**

The regular waves characteristic of the relaxed state are called *alpha waves.* During mental activity, rapid, low-amplitude beta waves are dominant. Recordings made in various stages of sleep show theta and delta waves, which have a lower frequency and greater amplitude.

Alpha waves are generated mainly from the visual areas in the occipital lobes when the person being tested is resting quietly with eyes closed. When the eyes are opened, alpha waves disappear and are replaced by more rapid, irregular waves. As you are reading this biology text, your brain is emitting **beta waves.** These waves have a fast-frequency rhythm that represent heightened mental activity such as information processing. During sleep, the brain emits waves with a lower frequency and a higher amplitude. The slow, large waves associated with certain stages of sleep are called **delta waves** and **theta waves.** During dreaming, flurries of irregular waves occur.

Certain brain diseases change the pattern of brain waves. Individuals with epilepsy, for example, exhibit a distinctive, recognizable, abnormal wave pattern. The location of a brain tumor or the site of brain damage caused by a blow to the head can sometimes be determined by noting the part of the brain that shows abnormal waves.

The **reticular activating system (RAS)** is a complex neural pathway within the brain stem and thalamus. The RAS receives messages from neurons in the spinal cord and from many other parts of the nervous system and communicates with the cerebral cortex by complex neural circuits. Components of the RAS help regulate **consciousness**—a state of awareness, or of selective attention. When certain neurons of the RAS bombard the cerebral cortex with stimuli, you feel alert and can focus on specific thoughts. If the RAS is severely damaged, the victim may pass into a deep, permanent coma.

Sleep is an alteration of consciousness during which there is decreased electrical activity of the cerebral cortex and from which a person can be aroused. Two main stages of sleep are recognized: **non-REM** and **REM.** *REM* is an acronym for *rapid eye movement.* During non-REM sleep, sometimes called normal sleep, metabolic rate decreases, breathing slows, and blood pressure decreases. Slower-frequency, higher-amplitude waves (theta and delta waves) are characteristic of non-REM sleep. This electrical activity is thought to be generated spontaneously by the cerebral cortex when it is not driven by impulses from other parts of the brain.

Every 90 minutes or so, a sleeping person enters the REM stage for a time. During this stage (about one fourth of total sleep time), the eyes move about rapidly beneath the closed but fluttering lids. Brain waves change to a desynchronized pattern. Sleep researchers claim everyone dreams, especially during REM sleep. Positron emission tomography (PET) scans of sleeping subjects indicate that during REM sleep, blood flow in the frontal lobes is reduced. In contrast, blood flow increases in areas that produce visual scenes and emotion. During REM sleep, neurons in the RAS release the neurotransmitter **norepinephrine,** which stimulates heightened activity in certain regions of the brain.

The hypothalamus and brain stem regulate the sleep-wake cycle. Using rats, researchers have demonstrated that stimulation of the *preoptic nucleus* of the hypothalamus results in non-REM sleep. The preoptic nucleus is located near the **suprachiasmatic nucleus,** the most important of the body's biological clocks. The suprachiasmatic nucleus receives information about the duration of light and dark from the retina of the eye and apparently transmits this information to the preoptic nucleus. The fatigue and decreased physical and mental performance referred to as jet lag occur when we fly to a different time zone where the body is no longer synchronized with the light–dark cycle.

The *raphe nuclei* in the brain stem (lower pons and medulla) appear important in producing REM sleep. Many neurons that project from the raphe nuclei release **serotonin,** a neurotransmitter involved in sleep. After many hours of activity, the sleep–wake cycle may be affected by fatigue of the RAS. Sleep centers are then activated, and their neurons release serotonin. After sufficient rest, the inhibitory neurons of the sleep centers become less excitable, and the excitatory neurons of the RAS become more excitable.

Although birds, mammals, and many other animals have a sleep–wake cycle, neurophysiologists do not know why sleep is necessary. Recent research suggests sleep may be important in learning and in processing memories. When a person stays awake for unusually long periods, fatigue and irritability result, and even routine tasks aren't performed well. Both non-REM and REM sleep are necessary. In sleep deprivation experiments with human volunteers, lack of REM sleep makes the participants anxious and irritable. When they are permitted to sleep normally again, they spend more time than usual in the REM stage for a while. Many types of drugs, such as amphetamines, alter sleep patterns and affect the amount of REM sleep. Certain drugs that induce sleep may increase total sleeping time but decrease REM time. When a person stops taking such a drug, several weeks may be required before normal sleep patterns are re-established.

The limbic system affects emotional aspects of behavior

The **limbic system** includes parts of the cerebrum (parts of the frontal lobe, temporal lobe, hippocampus, and amygdala), parts of the thalamus, and hypothalamus, several nuclei in the midbrain, and the neural pathways that connect these structures. The limbic system affects the emotional aspects of behavior, evaluates rewards, and is important in motivation. This system also plays a role in sexual behavior, biological rhythms, and autonomic responses.

The **hippocampus** functions in the formation and retrieval of verbal and emotional memories. It lets us place our experience in categories and store it along with similar memories. The **amygdala** evaluates the emotional aspects of experience and when it perceives a threat, it signals danger. The amygdala becomes hypersensitive to possible danger following a traumatic experience (see *On the Cutting Edge: The Neurophysiology of Traumatic Experience.*)

Researchers first discovered that the limbic system is part of an important motivational system in the 1950s when they implanted electrodes in certain areas of the brains of laboratory animals. They found that a rat may press, as many as 15,000 times per hour, a lever that stimulates a reward circuit in the brain. The rat forgoes food and water in favor of self-stimulation until it drops from exhaustion. This reward circuit, the **mesolimbic dopamine pathway,** includes two important areas in the midbrain, the substantia nigra and adjacent ventral tegmental area. These areas, which extend into behavioral control centers in the limbic system, contain the largest group of neurons that release dopamine in the brain. The mesolimbic dopamine pathway (important in motivation and experiencing emotion) enables an individual to experience pleasure in response to certain events or substances.

Dopamine neurons are activated by novel stimuli associated with reward or pleasure. These neurons signal us about surprising events or stimuli that predict rewards. Dopamine may motivate us to act when something important is happening. These mechanisms may help us understand attention deficit disorder (ADD) and schizophrenia, disorders associated with excessive amounts of dopamine and other neurotransmitters in the brain. People with these disorders have difficulty filtering out sensory stimuli. The presence of excess dopamine may drive them to divert their attention to so many sensory stimuli that they have difficulty focusing on the most important stimuli.

Learning involves the storage of information and its retrieval

Learning is the process by which we acquire information as a result of experience. Laboratory experiments have shown that members of every animal phylum can learn. For learning to occur, we must be able to remember what we experience. **Memory** is the process by which information is encoded, stored, and retrieved. **Implicit memory** is unconscious memory for perceptual and motor skills, such as riding a bicycle. **Explicit memory** involves factual knowledge of people, places, or objects, and requires conscious recall of the information.

Information processing involves short- and long-term memory

At any moment you are bombarded with thousands of bits of sensory information. At this very moment, your eyes are receiving information about the words on this page, the objects around you, and the intensity of the light in the room. At the same time you may be hearing a variety of sounds—music, your friends talking in the next room, the hum of an air conditioner. Your olfactory epithelium may sense cologne or the smell of coffee. Maybe you are eating as you read. Sensory receptors in your hands may be receiving information regarding the weight and position of your book. Most sensory stimuli are not important to remember and so are filtered out.

When we select information to remember, we must process it. Information processing involves both **short-term memory,** which lasts only minutes, and **long-term memory,** which may last for many years. Typically, we can hold only about seven chunks of information (a chunk is some unit such as a word, syllable, or number) at a time in short-term memory. Short-term memory lets us recall information for a few minutes. Usually when we look up a phone number, for example, we remember it only long enough to dial. If we need the same number an

The Neurophysiology of Traumatic Experience

Hypothesis: Traumatic experience can cause changes in the brain that affect information processing.

Method: Neuroimaging studies were performed on the brain structure and function of individuals who had survived traumatic experiences.

Results: Both structural and functional changes were identified, including increased activation of the amygdala and decreased activity in the anterior cingulate cortex and in Broca's area.

Conclusion: Traumatic experience causes changes in the brain that may mediate responses to reminders of the experience.

Terrorist acts, such as the 9/11 terrorist attack on the World Trade Center and the recent war in Iraq, have heightened our awareness of the effects of traumatic experiences. A *traumatic experience* is an event (or events) that causes intense fear, helplessness, or horror and that overwhelms normal coping and defense mechanisms. Whether you are a combat veteran, a survivor of childhood abuse, or the victim of a robbery, rape, a hurricane, or domestic violence, you may have experienced the effects of trauma.

We can process moderately disturbing experiences more easily than traumatic experiences. If you were one of the millions of people who were glued to their television sets on 9/11 or watched as people were wounded in Iraq, you may have been disturbed by the events you witnessed. You probably processed what you saw and heard by thinking and talking about it, perhaps even writing about it. You may have dreamed about your experience. As the brain actively reviews and sorts out an experience, we make sense of what happened and store the memory along with other memories of more ordinary past events. The emotional intensity decreases, and the memory of the uncomfortable experience fades in importance.

Traumatic events are more difficult to process. If you are one of the survivors who actually ran for your life as the World Trade Center fell, or if you are a combat veteran, your body's response to the danger is more intense. Years later, your brain may still be secreting abnormally high amounts of norepinephrine and other neurotransmitters, and you may remain on red alert for possible danger.

How does the brain respond in a traumatic experience, and how does the experience affect the brain? When danger threatens, the amygdala sends messages to both the hypothalamus (which signals the autonomic system and the endocrine system) and the neocortex (which allows us to be conscious of our experience). The amygdala is programmed to remember the smells, sounds, and sensations that are part of the experience. Until the memories of the experience are fully processed, similar smells, sounds, and sensations remind us of the traumatic event and trigger the body to prepare for danger. The individual experiences fear and anxiety. Some trauma survivors develop Posttraumatic Stress Disorder (PTSD), a condition in which they experience (1) intrusive thoughts, images, sensory experiences, memories, and dreams; (2) an urge to avoid reminders of the traumatic event; and (3) physiological hyperarousal, a condition in which the body is on red alert, continuously scanning the environment for potential danger.

Because they are so overwhelming, traumatic memories are very difficult to process. They cause us so much discomfort and anxiety that we tend to avoid them rather than intentionally focus on them and sort them out. As a result, traumatic memories seem to stay "stuck" in the limbic system, and when triggered, the experience is replayed with its original emotional intensity (a flashback).

Several neuroimaging studies have shown an association between prolonged trauma (for example, severe child abuse) and long-term changes in the brain. These changes include significantly smaller total brain and cerebral volumes, decreased right–left hemisphere integration, decreased size of the corpus callosum, EEG abnormalities, and decreased volume of the hippocampus and amygdala.*

Neuroimaging studies have also demonstrated changes in brain function, including increased amygdalar response and decreased cortical response. Scott Rauch, of Massachusetts General Hospital and Harvard Medical School, and his colleagues used functional magnetic resonance imaging (fMRI) to measure amygdalar response to masked, fearful face targets.† They found exaggerated amygdalar responses in subjects with PTSD. These investigators suggested these intense responses occurred in the absence of the normal mediating influence of the medial prefrontal cortex.

Using fMRI, Lisa Shin of Harvard Medical School, and her colleagues reported findings that support the hypothesis of the Rauch group.‡ When these researchers presented subjects with reminders of their traumatic experience, they observed decreased activity in the rostral anterior cingulate cortex (ACC). This area, part of the medial prefrontal cortex, normally inhibits the amygdala so that it does not respond too strongly. When the ACC does not function normally, cerebral blood flow to the amygdala increases (and can be seen on neuroimages). The amygdala overresponds, and the individual experiences physiological hyperarousal, distress, and anxiety, which may lead to intense emotional reactions that are rooted in the traumatic experience.

Broca's area, a region in the posterior part of the left frontal cortex, is also affected by trauma. Broca's area is critically important in generating words and thus in expressing language. Researchers have shown that when subjects are exposed to accounts of their traumatic experiences, activity in Broca's area decreases. This deactivation appears to be the physiologic basis for the difficulty that trauma survivors have describing their experience in words. Trauma survivors have described flashbacks of their experience as being in a state of "speechless terror." Without words, it is difficult to process and resolve a traumatic experience.

Given these and other studies, we can conclude that traumatic experience affects the brain and that many trauma survivors respond strongly to triggers that remind them of their experience. The exaggerated responses of the amygdala to harmless stimuli perceived as threatening is dissociated from cortical (medial prefrontal) activation. Responses generated by the limbic system are rooted in emotion rather than in rationality and judgment.

*G. Villareal, & C.Y. King, "Brain Imaging in Posttraumatic Stress Disorder," *Seminars in Clinical Neuropsychiatry*, Vol. 6, No. 2 (April 2001).

†S.L. Rauch, P.J. Whalen, L.M. Shin, S.C. McInerney, M.L. Macklin, N.B. Lasko, S.P. Orr, & R.K. Pitman, "Exaggerated Amygdala Response to Masked Facial Stimuli in Posttraumatic Stress Disorder: A Functional MRI Study," *Biological Psychiatry*, Vol. 47, 2000.

‡L.M. Shin, P.J. Whalen, R.K. Pitman, G. Bush, M.L. Macklin, N.B. Lasko, S.P. Orr, S.C. McInerney, & S. L. Rauch, "An fMRI Study of Anterior Cingulate Function in Posttraumatic Stress Disorder," *Biological Psychiatry*, Vol. 50, 2000.

hour later, most of us have to look it up again. One hypothesis of short-term memory suggests that reverberating circuits are involved (see Fig. 39-12). A memory circuit may continue to reverberate for several minutes until it fatigues or until new signals interfere with the old.

When we begin to process information, we *encode* it. We recognize patterns and meaningfully associate the stimuli to past experience or knowledge. We integrate the new information with other knowledge already stored in the brain. When information is selected for long-term storage, the brain rehearses the material and then stores it in long-term memory in association with similar memories. Several minutes are required for the brain to *consolidate* a new memory. Consolidation depends on the hippocampus and involves expression of genes and protein synthesis. If a person suffers a brain concussion or undergoes electroconvulsive therapy, memory of what happened immediately before the incident may be completely lost. This is known as *retrograde amnesia.* When the hippocampus is damaged, short-term memory is unimpaired and the patient can still recall information stored in the past. However, new short-term memories can no longer be converted to long-term memories. The hippocampus temporarily holds new information and may integrate various aspects of an experience, including odors, sounds, and other information. Consolidation allows memories to be *stored* for long periods.

Where are memories stored? Some forms of learning take place in association areas within lower brain regions, such as the thalamus. When large areas of the mammalian cerebral cortex are destroyed, information is lost somewhat in proportion to the extent of lost tissue. However, no specific area can be labeled the "memory bank." Rather, memories seem to be stored within many areas of the brain. For example, visual memories may be stored in the visual centers of the occipital lobes, and auditory memories may be stored in the temporal lobes. Researchers think memories are integrated in many areas of the brain, including the amygdala and the hippocampus, association areas of the cerebral cortex, and the thalamus and hypothalamus. **Wernicke's area** in the temporal lobe has been identified as a very important association area for complex thought processes. This area is an important center for language function involved in recognizing and interpreting words. Neurons within the association areas form highly complex pathways that permit complicated reverberation.

Retrieval of information stored in long-term memory is of considerable interest—especially to students! The challenge is to find information when you need it. When you seem to forget something, the problem may be that you have not searched effectively for the memory. Information retrieval can be improved by careful encoding, for example, forming strong associations between items.

Neurophysiological changes occur during learning

Short-term memory involves changes in the neurotransmitter receptors of postsynaptic neurons. These changes strengthen synaptic connections. The receptors are linked by second messengers (for example, cyclic AMP) to ion channels.

In long-term memory, gene expression and protein synthesis take place. This process involves slower, but longer lasting, changes in synaptic connections. Long-term memory depends on activated receptors linked to G proteins. Cyclic AMP acts as a second messenger. A high level of cyclic AMP activates a protein kinase that enters the nucleus, leading to gene activation and protein synthesis. In this process, protein kinase phosphorylates a transcription factor known as **CREB** (for *cyclic AMP response element binding protein*). CREB then turns on the transcription process of certain genes. CREB has been shown to be a signaling molecule in the memory pathway in many animals, including fruit flies and mice. The molecules and processes involved in learning and memory have been highly conserved during evolution.

In some types of learning, changes take place in presynaptic terminals or postsynaptic neurons that permanently enhance or inhibit the transmission of impulses. In some cases, specific neurons may become more sensitive to neurotransmitter. Unraveling the mechanisms that change the strength of synaptic connections appears important in understanding learning. Neurons typically transmit action potentials in bursts, and the amount of neurotransmitter released by each action potential may increase or decrease.

Recall from the beginning of the chapter that repeated electrical stimulation of neurons causes a functional change at synapses. When a presynaptic neuron continues to transmit action potentials at a high rate for a minute or longer, there is a long-lasting, *increased* response of postsynaptic neurons. This increased strength of the synaptic connections is known as **long-term potentiation (LTP).** (Potentiation means to strengthen or make more potent.)

Low-frequency stimulation of neurons results in a long-lasting *decrease* in the strength of their synaptic connections, called **long-term depression (LTD),** also known as synaptic depression. Induction of LTP and LTD require activation of *NMDA receptors,* which are present on the plasma membranes of postsynaptic neurons. These receptors control the passage of calcium ions into neurons. Information storage and forgetting appear to depend on the strengthening and weakening of synaptic connections brought about by LTP and LTD.

Experience affects development and learning

Many studies have demonstrated neural plasticity, the ability to change in response to environmental stimuli, in rats and other laboratory animals exposed to enriched environments. In contrast with rats housed in standard cages and provided with the basic necessities, those exposed to enriched environments are given toys and other stimulating objects, as well as the opportunity to socially interact with other rats. Animals reared in a complex environment exhibit increased synaptic contacts and process and remember information more quickly than animals not given such advantages. Mice exposed to enriched environments develop significantly greater numbers of neurons in the hippocampus and learn mazes significantly faster than controls.

Early environmental stimulation can also enhance the development of motor areas in the brain. For example, the brains of rats encouraged to exercise become slightly heavier than those of control animals. Characteristic changes occur within the cerebellum, including the development of larger dendrites.

Apparently, during early life certain critical or sensitive periods of nervous system development occur that are influenced by environmental stimuli. For example, when the eyes of young mice first open, neurons in the visual cortex develop large numbers of dendritic spines (structures on which synaptic contact takes place). If the animals are kept in the dark and deprived of visual stimuli, fewer dendritic spines form. If the mice are exposed to light later in life, some new dendritic spines form, but never as many as develop in a mouse reared in a normal environment.

Studies linking the development of the brain with environmental experience indicate that early stimulation is important for the sensory, motor, and intellectual development of children. Such studies have led to the rapidly expanding educational toy market and to widespread acceptance of early education programs. Researchers have also suggested that continuing environmental stimulation is needed to maintain the status of the cerebral cortex in later life.

Review

- How is the human CNS protected? What is the function of the cerebrospinal fluid?
- What are two main functions of the vertebrate spinal cord?
- What are the functions of each of the major lobes of the cerebrum?
- What is the function of the reticular activating system?
- How is information processed? How might short-term memory be adaptive?

BiologyNow™ Assess your understanding of **the human central nervous system** by taking the pretest on your BiologyNow CD-ROM.

THE PERIPHERAL NERVOUS SYSTEM

Learning Objectives

10 Describe the organization of the peripheral nervous system, comparing the somatic and autonomic divisions.

11 Contrast the sympathetic and parasympathetic divisions of the autonomic system, giving examples of the effects of these systems on specific organs such as the heart and intestine.

The peripheral nervous system (PNS) consists of the sensory receptors, the nerves that link these receptors with the CNS, and the nerves that link the CNS with the effectors (muscles and glands). The somatic division of the PNS helps the body respond to changes in the external environment. The nerves and receptors that maintain homeostasis despite internal changes make up the autonomic division.

The somatic division helps the body adjust to the external environment

The **somatic division** of the PNS includes the receptors that react to changes in the external environment, the sensory neurons that inform the CNS of those changes, and the motor neurons that adjust the positions of the skeletal muscles that help maintain the body's posture and balance. In mammals, 12 pairs of **cranial nerves** emerge from the brain. They transmit information regarding the senses of smell, sight, hearing, and taste from the special sensory receptors as well as information from the general sensory receptors, especially in the head region. For example, cranial nerve II, the optic nerve, transmits signals from the retina of the eye to the brain. The cranial nerves also bring orders from the CNS to the voluntary muscles that control movements of the eyes, face, mouth, tongue, pharynx, and larynx. Cranial nerve VII, the facial nerve, for example, transmits signals to the muscles used in facial expression and to the salivary glands.

In humans, 31 pairs of **spinal nerves** emerge from the spinal cord. Named for the general region of the vertebral column from which they originate, they comprise 8 pairs of cervical, 12 pairs of thoracic, 5 pairs of lumbar, 5 pairs of sacral, and 1 pair of coccygeal spinal nerves.

Each spinal nerve has a dorsal root and a ventral root (Fig. 40-13). The dorsal root consists of afferent fibers that transmit information from the sensory receptors to the spinal cord. Just before the dorsal root joins with the cord, it is marked by a swelling, the spinal ganglion, which consists of the cell bodies of the sensory neurons. The ventral root is a grouping of the efferent fibers leaving the cord en route to the muscles and glands. Cell bodies of the motor neurons occur within the gray matter of the cord.

Peripheral to the junction of the dorsal and ventral roots, each spinal nerve divides into branches. The autonomic branch innervates the viscera. The dorsal branch serves the skin and muscles of the back. The ventral branch serves the skin and muscles of the sides and ventral part of the body. The ventral branches of several spinal nerves form tangled networks called *plexi* (sing., *plexus*). Within a plexus, the fibers of a spinal nerve may separate and then regroup with fibers that originated in other spinal nerves. Thus nerves emerging from a plexus consist of neurons from several different spinal nerves.

The autonomic division regulates the internal environment

The **autonomic division** helps maintain homeostasis in the internal environment. For instance, it regulates the heart rate and helps maintain a constant body temperature. The autonomic system works automatically and without voluntary input. Its effectors are smooth muscle, cardiac muscle, and glands. Like the somatic system, it is functionally organized into reflex pathways. Receptors within the viscera relay information via afferent nerves to the CNS. The information is integrated at various

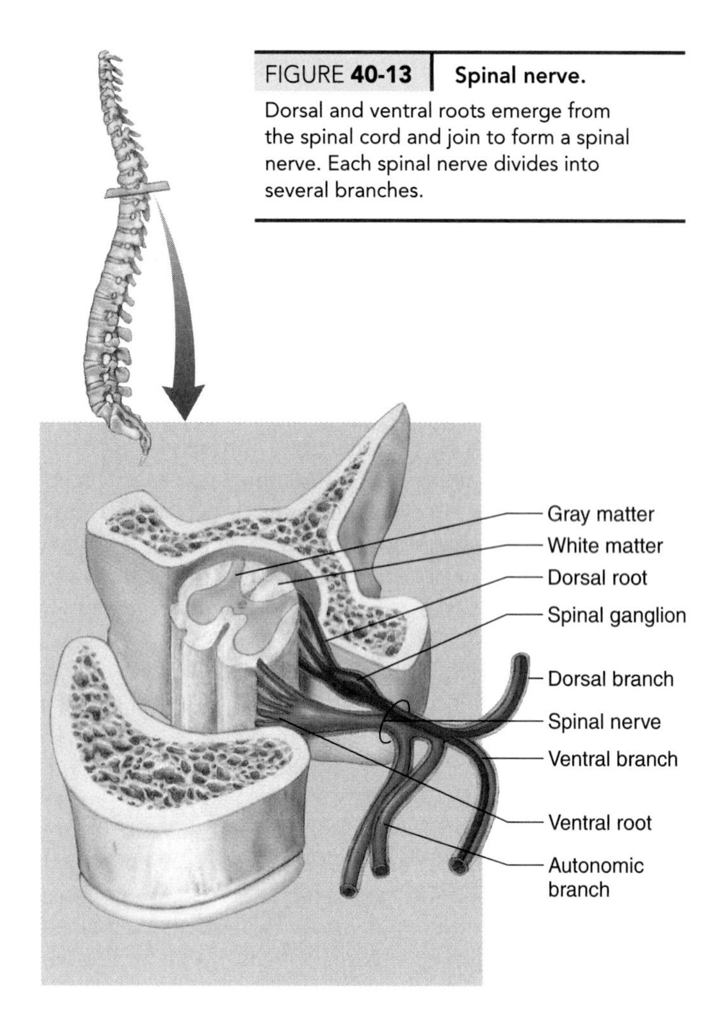

FIGURE **40-13** **Spinal nerve.**
Dorsal and ventral roots emerge from the spinal cord and join to form a spinal nerve. Each spinal nerve divides into several branches.

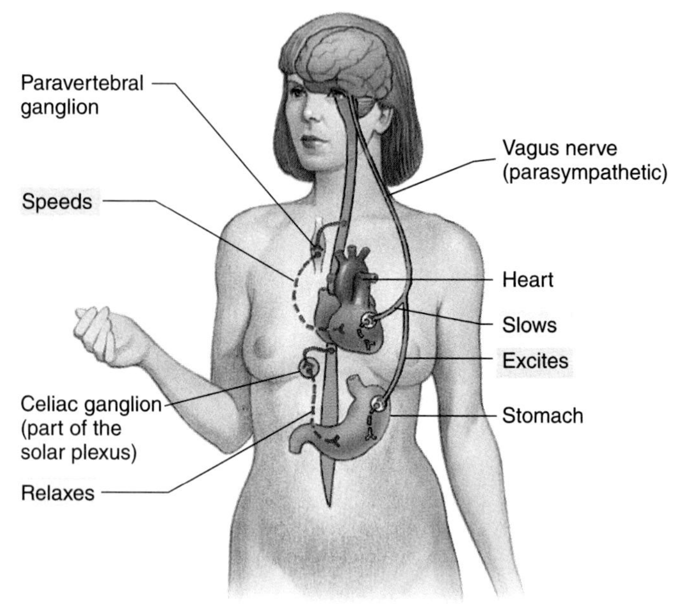

FIGURE **40-14** **Dual innervation of the heart and stomach.**
A sympathetic nerve increases the heart rate, whereas the vagus nerve, a parasympathetic nerve, slows the heartbeat. In contrast, sympathetic nerves slow contractions of the stomach and intestine while the vagus nerve stimulates contractions of the digestive organs. Sympathetic nerves are shown in red; postganglionic fibers are shown as dotted lines.

levels. Then the decision is transmitted along efferent nerves to the appropriate muscles or glands.

Afferent and efferent neurons of the autonomic division lie within cranial or spinal nerves. For example, afferent fibers of cranial nerve X, the vagus nerve, transmit signals from many organs of the chest and upper abdomen to the CNS. Its efferent fibers transmit signals from the brain to the heart, stomach, small intestine, and several other organs.

The efferent portion of the autonomic division is subdivided into **sympathetic** and **parasympathetic systems.** Many organs are innervated by both types of nerves. In general, sympathetic and parasympathetic systems have *opposite* effects (Figs. 40-14 and 40-15). For example, the heart rate is speeded up by messages from sympathetic nerve fibers and slowed by impulses from its parasympathetic nerve fibers. In many cases sympathetic nerves operate to stimulate organs and to mobilize energy, especially in response to stress, whereas the parasympathetic nerves influence organs to conserve and restore energy, particularly during quiet, calm activities.

Instead of using a single efferent neuron, as in the somatic system, the autonomic system uses a relay of two neurons between the CNS and the effector. The first neuron, called the **preganglionic neuron,** has a cell body and dendrites within the CNS. Its axon, part of a peripheral nerve, ends by synapsing with a **postganglionic neuron.** The dendrites and cell body of the postganglionic neuron are in a ganglion outside the CNS. Its axon terminates near or on the effector. The sympathetic ganglia are paired, and a chain of them, the **paravertebral sympathetic ganglion chain,** runs on each side of the spinal cord from the neck to the abdomen. Some sympathetic preganglionic neurons do not end in these ganglia but instead pass on to **collateral ganglia** in the abdomen, close to the aorta and its major branches. Parasympathetic preganglionic neurons synapse with postganglionic neurons in **terminal ganglia** near or within the walls of the organs they innervate.

The sympathetic and parasympathetic systems also differ in the neurotransmitters they release at the synapse with the effector. Both preganglionic and postganglionic parasympathetic neurons secrete acetylcholine. Sympathetic postganglionic neurons release norepinephrine (although preganglionic sympathetic neurons secrete acetylcholine).

The autonomic division got its name from the original belief that it was independent of the CNS, that is, autonomous. Neurobiologists have shown that this is not so and that the hypothalamus and many other parts of the CNS help regulate the autonomic system. Although the autonomic system usually functions automatically, its activities can be consciously influenced. Biofeedback provides a person with visual or auditory evidence concerning the status of an autonomic body function. For example, a tone may be sounded when blood pressure reaches a desirable level. Using such techniques, subjects have

FIGURE **40-15** **The sympathetic and parasympathetic systems.**

Complex as it appears, this diagram has been greatly simplified. Red lines represent sympathetic nerves, black lines represent parasympathetic nerves, and dotted lines represent postganglionic nerves.

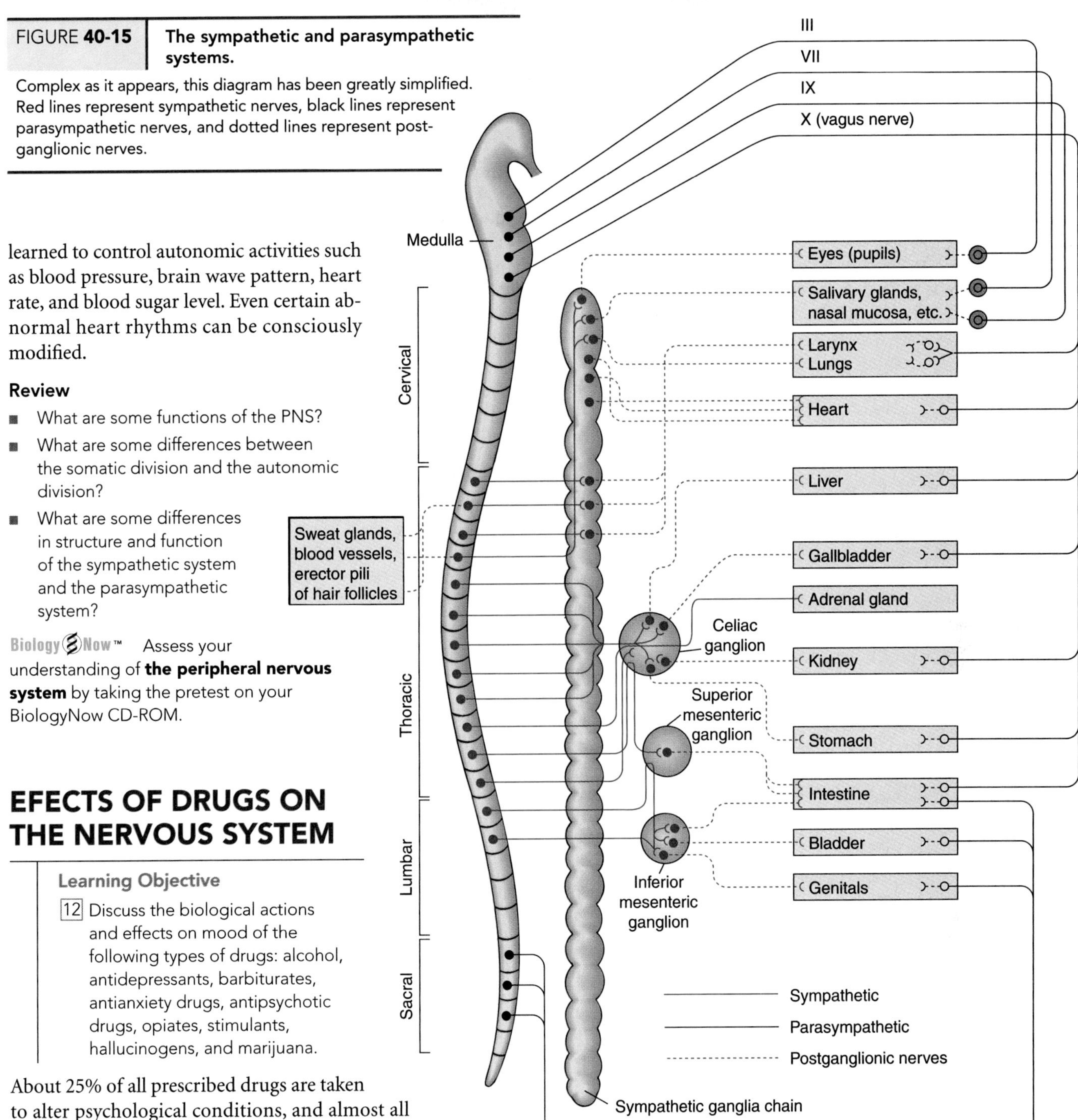

learned to control autonomic activities such as blood pressure, brain wave pattern, heart rate, and blood sugar level. Even certain abnormal heart rhythms can be consciously modified.

Review

- What are some functions of the PNS?
- What are some differences between the somatic division and the autonomic division?
- What are some differences in structure and function of the sympathetic system and the parasympathetic system?

BiologyNow™ Assess your understanding of **the peripheral nervous system** by taking the pretest on your BiologyNow CD-ROM.

EFECTS OF DRUGS ON THE NERVOUS SYSTEM

Learning Objective

12 Discuss the biological actions and effects on mood of the following types of drugs: alcohol, antidepressants, barbiturates, antianxiety drugs, antipsychotic drugs, opiates, stimulants, hallucinogens, and marijuana.

About 25% of all prescribed drugs are taken to alter psychological conditions, and almost all the commonly abused drugs affect mood. Many act by changing the levels of neurotransmitters within the brain. In particular, levels of norepinephrine, serotonin, and dopamine influence mood.

According to the Substance Abuse and Mental Health Services Administration, an estimated 15.9 million individuals in the United States were current users of illicit drugs in 2001, meaning they used an illicit drug at least once during the 30 days before the interview. More than 9 million children live with a parent who is dependent on alcohol and/or illicit drugs. Table 40-4 lists several commonly used and abused drugs and gives their effects. Also see *Focus On: Alcohol: The Most Abused of All Drugs.*

Habitual use of almost any mood-altering drug can result in **psychological dependence,** in which the user becomes emotionally dependent on the drug. When deprived of it, the user craves the feeling of euphoria (well-being) the drug induces. Some drugs induce **tolerance** after several days or weeks. Response to the drug therefore decreases, and greater amounts are required to obtain the desired effect. Tolerance often occurs because the liver cells are stimulated to produce more of the enzymes that metabolize and inactivate the drug. It can also occur when the number of postsynaptic receptors that bind with the drug decrease (a mechanism known as *downregulation*).

TABLE 40-4 Effects of Some Commonly Used Drugs

Name of Drug	Effect on Mood	Actions on Body	Side Effects/Dangers Associated with Abuse
Antidepressants			
Tricyclic antidepressants (e.g., Elavil, Anafranil)	Elevate mood; relieve depression; also used to treat obsessive-compulsive behavior	Most block reuptake of biogenic amines (especially norepinephrine), increasing their concentration in synapses	Sedation, weight gain, sexual dysfunction
Selective serotonin reuptake inhibitors (SSRIs) (e.g., Prozac, Zoloft, Lexapro)	Relieve depression; used to treat obsessive-compulsive behavior	Block serotonin reuptake	Nausea, headache, insomnia, anxiety
MAO inhibitors (e.g., Parnate, Nardil)	Relieve depression	Block enzymatic breakdown of biogenic amines, increasing their concentration in synapses	Liver toxicity, excessive CNS stimulation; overdose may affect blood pressure, cause hallucinations
Anti-anxiety Drugs			
Benzodiazepines (e.g., Xanax, Valium, Librium)	Sedation; induce sleep	Bind to GABA receptors on post-synaptic neuron; increase effectiveness of GABA in opening chloride channels, causing hyperpolarization; cause relaxation of skeletal muscles	Drowsiness, confusion; psychological dependence, addiction; effects additive with alcohol
Barbiturates "Downers" (e.g., phenobarbital)	Sedation; induce sleep	Bind to receptor adjacent to GABA receptor on chloride channels; chloride channels open so more chloride ions move in, causing hyperpolarization	Drowsiness, confusion; psychological dependence, tolerance, addiction; severe CNS depression, resulting in coma and death (especially lethal in combination with alcohol)
Antipsychotic Medications			
Phenothiazines (e.g., Thorazine, Mellaril, Stelazine)	Relieve symptoms of schizophrenia; reduce impulsive and aggressive behavior	Block dopamine receptors	Muscle spasms, shuffling gait, and other symptoms of Parkinson's disease
Newer antipsychotic medications (e.g., Abilify, Risperdal)	Similar to phenothiazines, but also increase motivation	Block dopamine receptors and other neurotransmitter receptors	Fewer side effects than with the phenothiazines
Narcotic Analgesics			
Opiates (e.g., morphine, codeine, Demerol, heroin)	Euphoria; sedation; relieve pain	Mimic actions of endorphins; bind to opiate receptors	Depress respiration; constrict pupils; impair coordination; tolerance, psychological dependence, addiction; convulsions, death from overdose
Cocaine	Brief, intense high with euphoria, followed by fatigue and depression	Stimulates release and inhibits reuptake of norepinephrine and dopamine, leading to CNS stimulation followed by depression; autonomic stimulation; dilation of pupils; local anesthesia	Mental impairment, convulsions, hallucinations, unconsciousness; death from overdose
Marijuana	Euphoria	Impairs coordination; impairs depth perception and alters sense of timing; impairs short-term memory (probably by decreasing acetylcholine levels in the hippocampus); inflames eyes; causes peripheral vasodilation	In large doses, sensory distortions, hallucinations, evidence of lowered sperm counts and testosterone levels

Drug addiction is a serious societal problem that involves compulsive use of a drug despite negative health consequences and negative effects on the ability to function socially and occupationally. Use of some drugs, such as heroin, tobacco, alcohol, and barbiturates, may also result in physical dependence, in which physiological changes occur that make the user dependent on the drug. For example, when heroin or alcohol is withheld, the addict suffers characteristic withdrawal symptoms. Physical addiction can occur when a drug, for example, morphine, has components similar to substances that body cells normally manufacture on their own. Some highly addictive drugs such as crack cocaine and methamphetamine do not cause serious withdrawal symptoms.

TABLE 40-4 Effects of Some Commonly Used Drugs (continued)

Name of Drug	Effect on Mood	Actions on Body	Side Effects/Dangers Associated with Abuse
Amphetamines			
"Uppers," "pep pills," "crystal meth" (e.g., Dexedrine); methamphetamine is a popular club drug	Euphoria, stimulation, hyperactivity	Stimulate release and block reuptake of dopamine and norepinephrine; enhance flow of impulses in RAS, leading to increased heart rate, blood pressure; pupil dilation	Tolerance, physical dependence; hallucinations; increased blood pressure; psychotic episodes; death from overdose
MDMA (3,4-methylenedioxy-methamphetamine) "ecstasy" (most popular of club drugs)	Stimulant, sense of pleasure, self-confidence, feelings of closeness to others; mild hallucinogenic	CNS stimulant; damages serotonin pathways in rat and monkey brains	Hyperthermia (excessive body heat); overdose causes rapid heartbeat, high blood pressure, panic attacks, and seizures; long-term neurological changes; cognitive impairment
Rohypnol (flunitrazepam) "Date rape" drug, "roofies," "roach"	Sedative-hypnotic	Muscle relaxation; amnesia	Physical and psychological dependence; withdrawal may cause seizures; may be lethal when mixed with alcohol and/or other depressants
PCP (phencyclidine) "Angel dust," "ozone"	Feelings of strength, power, invulnerability, and a numbing effect on the mind	Increase in blood pressure and heart rate; shallow respiration; flushing, profuse sweating; numbness of extremities; muscular incoordination; changes in body awareness, similar to those associated with alcohol intoxication	Psychological dependence, craving; in adolescents, interferes with hormones related to growth and development; may interfere with learning; causes hallucinations, delusions, disordered thinking; high doses cause drop in blood pressure and heart rate, seizures, coma, and death; chronic use causes memory loss, difficulties with speech and thinking, and depression
LSD (lysergic acid diethylamide)	Overexcitation; sensory distortions, hallucinations	Alters levels of transmitters in brain; potent CNS stimulator; dilates pupils, sometimes unequally; increases heart rate; raises blood pressure	Irrational behavior
Methaqualone (e.g., Quaalude, Sopor)	Hypnotic	Depresses CNS; depresses certain spinal reflexes	Tolerance, physical dependence; convulsions, death
Caffeine	Increases mental alertness; decreases fatigue and drowsiness	Acts on cerebral cortex; relaxes smooth muscle; stimulates cardiac and skeletal muscle; increases urine volume	Very large doses stimulate centers in the medulla (may slow the heart); toxic doses may cause convulsions
Nicotine	Lessens psychological tension	Stimulates sympathetic nervous system; increases dopamine in mesolimbic pathway	Tolerance, physical dependence; stimulates development of athero-sclerosis by stimulating synthesis of lipid in arterial wall
Alcohol	Euphoria, relaxation, release of inhibitions	Depresses CNS; impairs vision, coordination, judgement; lengthens reaction time	Physical dependence; damage to pancreas and liver (e.g., cirrhosis), brain damage

The neurophysiological mechanisms for drug addiction involve the *mesolimbic dopamine pathway.* Neurons in this pathway signal neurons in an area of the limbic system (the nucleus accumbens). Facilitation of neurons in the mesolimbic dopamine pathway has been demonstrated in nicotine, heroin, amphetamine, and cocaine addiction. For example, amphetamines stimulate dopamine release. Cocaine blocks the reuptake of dopamine into presynaptic neurons, prolonging the stimulation. Amphetamines or cocaine can increase dopamine concentration in the limbic system by up to 1000 times normal amounts! Prolonged drug use alters gene expression and changes neuron structure.

Review

- How does each of the following drugs affect the CNS: (a) alcohol (b) antipsychotic drugs (c) antidepressants (d) amphetamines (e) MDMA ("ecstasy")?

BiologyNow™ Assess your understanding of **the effects of drugs on your nervous system** by taking the pretest on your BiologyNow CD-ROM.

FOCUS ON ALCOHOL: THE MOST ABUSED OF ALL DRUGS

After tobacco, alcohol is the leading cause of premature death in the United States. It is linked to more than 100,000 deaths every year and costs our society more than $185 billion annually. According to the National Institute on Alcohol Abuse and Alcoholism, in 2002 an estimated 14 million Americans—1 in every 13 adults—abused alcohol or were alcoholic. More than half of men and women in the United States report that one or more of their close relatives have a drinking problem. Alcohol abuse is not limited to adults. About 4.6 million adolescents, or nearly one of every three high school students, experience negative consequences from alcohol use, including difficulty with parents, poor performance at school, and trouble with the law.

Alcohol abuse results in physiological, psychological, and social impairment for the abuser and has serious negative consequences for family, friends, and society. Alcohol abuse has been linked to the following:

- More than 50% of all traffic fatalities
- More than 50% of violent crimes, and more than 60% of child abuse and spouse abuse cases
- More than 50% of suicides
- More than 15,000 babies born each year with serious birth defects because their mothers drank alcohol during pregnancy; **fetal alcohol syndrome** is the leading cause of preventable mental retardation in the United States
- Increased risk for certain cancers, especially those of the liver, esophagus, throat, and larynx

A single drink, that is, 12 oz of beer, 5 oz of wine, or 1.5 oz of 80-proof liquor, results in a blood alcohol concentration of approximately 20 mg/dL (milligrams per deciliter). This represents about 0.5 oz of pure alcohol in the blood. Alcohol accumulates in the blood because absorption occurs more rapidly than do oxidation and excretion. Every cell in the body can take in alcohol from the blood. At first the drinker may feel stimulated, but alcohol actually depresses the CNS.

As blood alcohol concentration rises, information processing, judgment, memory, sensory perception, and motor coordination all become progressively impaired. Depression and drowsiness generally occur. Contrary to popular belief, alcohol decreases sexual performance. Some individuals become loud, angry, or violent. A blood alcohol concentration of 80 mg/dL (or 0.08 gm 100 mL) legally defines driving while intoxicated (DWI). A 170-pound man typically reaches this level by drinking four drinks in an hour on an empty stomach.

Alcohol metabolism occurs at a fixed rate in the liver so only time, not coffee, decreases its effects. Because alcohol inhibits water reabsorption in the kidneys, more fluid is excreted as urine than is consumed. This results in dehydration that, together with low blood sugar level, may cause the stupor produced by excessive drinking.

In chronic drinkers, cells of the CNS adapt to the presence of the drug. This causes *tolerance* (more and more alcohol is needed to experience the same effect) and physical dependence. Abrupt withdrawal can result in sleep disturbances, severe anxiety, tremors, seizures, hallucinations, and psychoses.

The *A1* allele of the D2 dopamine receptor gene increases the risk for alcoholism. Individuals with this allele have fewer D2 dopamine receptors in their brain. Alcohol, as well as cocaine, heroin, amphetamines, and nicotine, enhances dopamine activity in the *mesolimbic dopamine pathway* of the brain. Individuals with the *A1* allele may compensate for their low dopamine levels by using alcohol and other drugs. However, the mechanisms underlying alcohol addiction are far more complex. Evidence suggests that many neurotransmitters are involved, including glutamate, serotonin, GABA, and opioid peptides.

Treatment for alcohol problems includes various forms of psychotherapy, including relapse prevention therapy, in which individuals are encouraged not to consider lapses from abstinence as failure. The group support offered by Alcoholics Anonymous (AA) has proved effective for many struggling with alcohol abuse.

SUMMARY WITH KEY TERMS

1 Contrast nerve nets and radial nervous systems with bilateral nervous systems.

- Nerve nets and radial nervous systems are typical of radially symmetrical invertebrates. Bilateral nervous systems are characteristic of bilaterally symmetrical animals.
- Cnidarians have a **nerve net** of nerve cells scattered throughout the body; they have no central control organ. Echinoderms typically have a **radial nervous system** consisting of a nerve ring and nerves that extend to various parts of the body.
- In a **bilateral nervous system,** nerve cells concentrate to form **nerves, nerve cords, ganglia,** and **brain;** typically, sense organs are concentrated in the head region. An increased number of association neurons and more complex synaptic contacts permit a wide range of responses.
- In planarian flatworms the bilateral nervous system includes **cerebral ganglia** and, typically, two solid ventral nerve cords connected by transverse nerves. Annelids and arthropods typically have a ventral nerve cord and numerous ganglia. The cerebral ganglia of arthropods have specialized regions. Octopods and other cephalopod mollusks have complex nervous systems with neurons concentrated in a central region.

2 Identify trends in the evolution of the invertebrate nervous system.

- Trends in nervous system evolution include increased numbers and concentration of nerve cells, specialization of function, increased number of association neurons; more complex synaptic contacts and **cephalization,** formation of a head.

3 Describe the two main divisions of the vertebrate nervous system.

- The vertebrate nervous system consists of the **central nervous system (CNS)** and **peripheral nervous system (PNS).** The CNS consists of the brain and dorsal, tubular **spinal cord.** The PNS consists of sensory receptors and nerves.

4 Trace the development of the principal vertebrate brain regions from the forebrain, midbrain, and hindbrain, and compare the brains of fishes, amphibians, reptiles, birds, and mammals.

- In the vertebrate embryo, the brain and spinal cord arise from the **neural tube.** The anterior end of the tube differentiates into **forebrain, midbrain,** and **hindbrain.**
- The hindbrain subdivides into the **metencephalon** and **myelencephalon.** The myelencephalon develops into the **medulla,** which contains vital centers and other reflex centers. The **fourth ventricle,** the cavity of the medulla, communicates with the central canal of the spinal cord. The metencephalon gives rise to the **cerebellum,** which is responsible for muscle tone, posture, and equilibrium, and to the **pons,** which connects various parts of the brain.
- The midbrain is the largest part of the brain in fishes and amphibians. It is their main association area, linking sensory input and motor output. In reptiles, birds, and mammals, the midbrain serves as a center for visual and auditory reflexes.
- The medulla, pons, and midbrain make up the **brain stem.**
- The forebrain differentiates to form the **diencephalon** and **telencephalon.** The diencephalon develops into the thalamus and hypothalamus. The **thalamus** is a relay center for motor and sensory information. The **hypothalamus** controls autonomic functions; links nervous and endocrine systems; controls temperature, appetite, and fluid balance; and is involved in some emotional and sexual responses.
- The telencephalon develops into the cerebrum and **olfactory bulbs.** In most vertebrates the **cerebrum** is divided into right and left **hemispheres.** In fishes and amphibians, the cerebrum mainly integrates incoming sensory information.
- In birds, the corpus striatum controls complex behavior patterns, such as flying and singing.
- In mammals, the **neocortex** accounts for a large part of the **cerebral cortex,** the **gray matter** of the brain. The cerebrum has complex association functions.

5 Describe the structure and functions of the human spinal cord.

- The human brain and spinal cord are protected by bone and three **meninges**—the **dura mater, arachnoid,** and **pia mater;** brain and spinal cord are cushioned by **cerebrospinal fluid** (**CSF**).
- The spinal cord transmits impulses to and from the brain and controls many **reflex actions.** The spinal cord consists of **ascending tracts,** which transmit information to the brain, and **descending tracts,** which transmit information from the brain. Its gray matter contains nuclei that serve as reflex centers.
- A **withdrawal reflex** involves sensory receptors; sensory neurons, interneurons, and motor neurons; and effectors, such as muscles.

6 Describe the structure and functions of the human cerebrum.

- The human cerebral cortex consists of gray matter, which forms folds or **convolutions.** Deep furrows between the folds are called **fissures.**
- The cerebrum is functionally divided into **sensory areas** that receive incoming sensory information; **motor areas** that control voluntary movement; and **association areas** that link sensory and motor areas and are responsible for learning, language, thought, and judgment.
- The cerebrum consists of lobes including the **frontal lobes, parietal lobes, temporal lobes,** and **occipital lobes.**
- The **white matter** of the cerebrum lies beneath the cerebral cortex. The **corpus callosum,** a large band of white matter, connects right and left hemispheres. The **basal ganglia,** a cluster of **nuclei** within the white matter, are important centers for motor function.

7 Describe the sleep–wake cycle, and contrast rapid-eye-movement (REM) sleep and non-REM sleep.

- Brain activity cycles in a sleep–wake pattern that is regulated by the hypothalamus and brain stem.
- **Alpha wave** patterns are characteristic of relaxed states. **Beta wave** patterns accompany heightened mental activity; and slower-frequency, higher-amplitude **theta** and **delta waves,** are characteristic of non-REM sleep.
- The **reticular activating system** (**RAS**) is an arousal system; its neurons filter sensory input, selecting which information is transmitted to the cerebrum.
- Electrical activity of the cerebral cortex and metabolic rate slows during **non-REM sleep,** whereas dreaming characterizes **REM sleep.**
- The body's main biological clock—the **suprachiasmatic nucleus**—receives information about light and dark and apparently transmits it to other nuclei that regulate sleep. When sleep centers are activated, they release **serotonin.**

8 Describe the actions of the limbic system.

- The **limbic system** of the brain affects the emotional aspects of behavior, motivation, sexual behavior, autonomic responses, and biological rhythms. The **hippocampus** is important in categorizing information and consolidating memories. The **amygdala** evaluates incoming information and signals danger.

9 Summarize how the brain processes information.

- **Learning** is the process by which we acquire information as a result of experience. **Memory** is the process by which information is encoded, stored, and retrieved.
- **Implicit memory** is unconscious memory for perceptual and motor skills. **Explicit memory** is factual memory of people, places, or objects.
- **Short-term memory** lets us recall information, such as a telephone number, for a few minutes. Information can be transferred from short-term memory to **long-term memory.**
- **Synaptic plasticity** is the ability of the nervous system to modify synapses during learning and remembering. Long-term memory storage involves gene activation and long-term functional changes at synapses. Known as **long-term potentiation (LTP),** such changes involve increased sensitivity to an action potential by a postsynaptic neuron. **Long-term depression (LTD)** is a long-lasting decrease in the strength of synaptic connections.
- The physical structure and chemistry of the brain can be altered by environmental experience.

10 Describe the organization of the peripheral nervous system, comparing the somatic and autonomic divisions.

SUMMARY WITH KEY TERMS *(continued)*

- The PNS consists of sensory receptors and nerves, including the **cranial nerves** and **spinal nerves** and their branches.
- The **somatic division** of the PNS responds to changes in the external environment. The **autonomic division** regulates the internal activities of the body.

11 Contrast the sympathetic and parasympathetic divisions of the autonomic system, giving examples of the effects of these systems on specific organs such as the heart and intestine.

- The **sympathetic system** permits the body to respond to stressful situations. The **parasympathetic system** influences organs to conserve and restore energy.
- Many organs are innervated by sympathetic and parasympathetic nerves, which function in an opposite way. For example, the sympathetic system increases heart rate, whereas the parasympathetic system decreases heart rate.

12 Discuss the biological actions and effects on mood of the following types of drugs: alcohol, antidepressants, barbiturates, antianxiety drugs, antipsychotic drugs, opiates, stimulants, hallucinogens, and marijuana.

- Many drugs alter mood by increasing or decreasing the concentrations of specific neurotransmitters within the brain. See Table 40-4 for a summary of these drugs.
- Habitual use of mood-altering drugs can result in **psychological dependence** or **drug addiction.** Many drugs induce **tolerance,** in which the body's response to the drug decreases so that greater amounts are needed to obtain the desired effect.

POST-TEST

1. A radially symmetrical animal is likely to have (a) a forebrain (b) a nerve net (c) cerebral ganglia (d) a ventral nerve cord (e) cerebral ganglia and a nerve net
2. In vertebrate embryos the brain develops from the (a) spinal cord (b) sympathetic nervous system (c) parasympathetic nervous system (d) neural tube (e) forebrain
3. Which part of the brain maintains posture, muscle tone, and equilibrium? (a) cerebrum (b) medulla (c) cerebellum (d) neocortex (e) thalamus
4. Which part of the brain controls autonomic functions and regulates body temperature? (a) cerebrum (b) hypothalamus (c) cerebellum (d) pons (e) thalamus
5. In a withdrawal reflex, following reception, a signal is transmitted by (a) a motor neuron to an association neuron in the CNS (b) an association neuron in the CNS to an afferent neuron (c) an afferent neuron in the CNS to a motor neuron (d) a sensory neuron to an interneuron in the CNS (e) a sensory neuron to an interneuron in the PNS
6. Association areas in the human brain are concentrated in the (a) cerebral cortex (b) medulla (c) ventricle (d) hippocampus (e) meninges
7. The human brain is protected by (a) meninges, cerebrospinal fluid, and skull bones (b) meninges and skull bones only (c) dura mater and fourth ventricle (d) pia mater and skull bones (e) arachnoid, pia mater, cerebrospinal fluid, and ganglia
8. Which of the following is *not* a function of the spinal cord? (a) controls many reflex actions (b) transmits information to the brain (c) transmits information from the brain (d) regulates sleep–wake cycles (e) controls the withdrawal reflex
9. The most prominent part of the amphibian brain is the (a) midbrain (b) medulla (c) cerebellum (d) neocortex (e) cerebrum
10. The visual centers are located in the (a) parietal lobes (b) thalamus (c) occipital lobes (d) limbic lobes (e) frontal lobes
11. As you are answering these questions, what is the predominant type of wave your brain is emitting? (a) REM (b) alpha (c) theta (d) delta (e) beta
12. Implicit memory is (a) short-term memory (b) long-term memory (c) factual knowledge of people, places, or objects (d) unconscious memory for perceptual or motor skills (e) learning that depends on long-term depression (LTD)
13. Long-term potentiation (LTP) (a) is mainly the responsibility of cranial nerve X (b) is a long-lasting increase in synaptic strength (c) is associated with short-term memory (d) is a long-lasting decrease in synaptic strength (e) occurs mainly during REM sleep
14. The heart rate is slowed by (a) sympathetic nerves (b) parasympathetic nerves (c) corpus callosum (d) both sympathetic and parasympathetic nerves (e) the hippocampus working together with sympathetic nerves
15. After taking a mood-altering drug for several weeks, a patient notices it no longer works as effectively. This is an example of (a) psychological dependence (b) withdrawal (c) addiction (d) tolerance (e) neurotransmitter increases
16. Amphetamines (a) facilitate neurons in the mesolimbic dopamine pathway (b) decrease the amount of dopamine secreted (c) do not result in tolerance (d) are sometimes referred to as "downers" (e) are CNS depressants
17. Label the diagram. Use Figure 40-10a to check your answers.

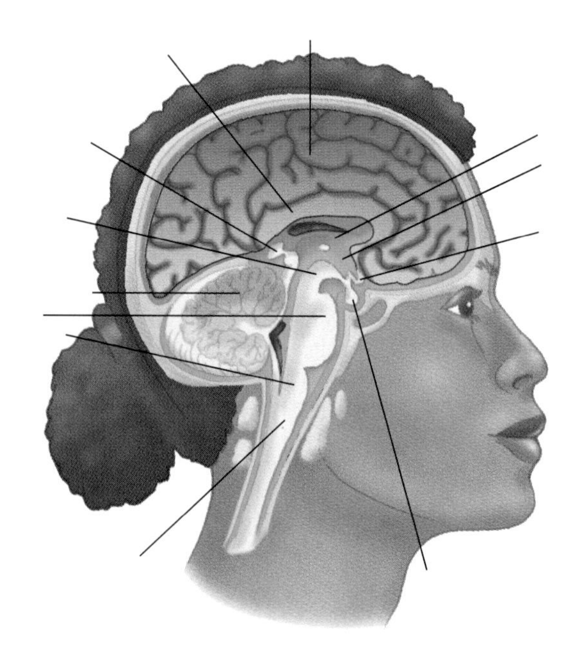

CRITICAL THINKING

1. What general trends can you identify in the evolution of the vertebrate brain?
2. Imagine you have just become a parent. What kind of things can you do to enhance the development of your child's intellectual abilities?
3. Hypothesize a possible relationship between inheritance and intelligence based on gene activation during information processing.
4. What is the adaptive function of the limbic system when you are in danger?

- Visit our Web site at **http://biology.brookscole.com/solomon7** for links to chapter-related resources on the World Wide Web. Additional online materials relating to this chapter can also be found on our Web site.

BIOLOGY NOW RESOURCES

Active Figure

40-9: The human brain

Preparing for an exam? Take a diagnostic test on your BiologyNow CD-ROM.

Post-Test Answers:

1. b	2. d	3. c	4. b
5. d	6. a	7. a	8. d
9. a	10. c	11. e	12. d
13. b	14. b	15. d	16. a

41

Sensory Reception

Merlin D. Tuttle/Bat Conservation International/Photo Researchers, Inc.

Bats are guided by echoes. This California leaf-nosed bat (*Macrotus californicus*) has located an insect.

CHAPTER OUTLINE

- Types of Sensory Receptors
- How Sensory Receptors Function
- Mechanoreceptors
- Chemoreceptors
- Photoreceptors

Have you heard dolphins make clicking sounds? Humans can hear only a limited number of the wide range of sound frequencies that dolphins emit. Bats, dolphins, and a few other vertebrates detect distant objects by **echolocation,** sometimes called biosonar. Echolocation works something like radar but uses sound rather than radio waves. For example, a bat emits high-pitched sounds that bounce off objects in its path and echo back. By rapidly responding to the echo, the bat skillfully avoids obstacles and captures prey (see photograph). Using echolocation, bats and dolphins determine the size, shape, texture, and density of an object, as well as its location. The ability to detect sounds emitted by some foraging bats has evolved in certain moths and other insect prey. Some species of bats have adapted, in turn, by emitting very high-frequency sounds that their prey cannot detect.

Sensory receptors are structures that detect information about changes in the internal or external environment. By informing an animal about its environment, sensory receptors are the link between the animal and the outside world. The kinds of sensory receptors an animal has determine just how it perceives its surroundings. We humans live in a world of rich colors, numerous shapes, and varied sounds. But we cannot hear the high-pitched whistles that are audible to dogs and cats, or the ultrasonic echoes by which bats navigate. Nor do we ordinarily recognize our friends by their distinctive odors! And although vision is our dominant and most refined sense, we are blind to the ultraviolet (UV) hues that light up the world for insects.

Sensory receptors consist of specialized neuron endings or specialized cells in close contact with neurons. These receptors *transduce* or convert the energy of the stimulus to electrical signals, the information currency of the nervous system. The transduction mechanisms that couple the stimulus with the opening or closing of ion channels in the plasma membrane of sensory receptors is the focus of current research. When a sensory receptor is depolarized, an action potential may be initiated in a sensory neuron. As we discussed in Chapter 40, **sensory neurons,** also known as *afferent neurons,* transmit information from receptors to the central nervous system (CNS).

Sensory receptors, along with other types of cells, make up complex **sense organs,** such as eyes, ears, nose, and taste buds. A human taste bud, for example, consists of modified epithelial cells that detect chemicals dissolved in saliva. In addition to the five senses of sight, hearing, smell, taste, and touch, neurobiologists recognize balance as a sense. They view touch as a compound sense that allows us to detect pressure, pain, and temperature. In this chapter, we also consider sensory receptors that enable us to sense muscle tension and joint position. ■

TYPES OF SENSORY RECEPTORS

Learning Objective

1 Distinguish among five kinds of sensory receptors according to the types of energy they transduce.

Animals receive information about their environment in various energy forms. Sensory receptors transduce these types of energy into graded potentials that can result in action potentials (see Chapter 39). **Mechanoreceptors** transduce mechanical energy—touch, pressure, gravity, stretching, and movement (Table 41-1). These receptors convert mechanical forces directly into electrical signals. **Chemoreceptors** transduce certain chemical compounds, and **photoreceptors** transduce light energy.

Thermoreceptors respond to heat and cold. Mosquitoes, ticks, and other blood-sucking arthropods use thermoreception in their search for an endothermic host. Some have temperature receptors on their antennae that are sensitive to changes of less than 0.5°C. At least two types of snakes—pit vipers and boas—use thermoreceptors to locate their prey (Fig. 41-1). In mammals, which are endothermic, free nerve endings and specialized receptors in the skin and tongue detect temperature changes in the outside environment. Thermoreceptors in the

FIGURE **41-1** **Thermoreception.**
The pit organ of this bamboo viper (*Trimeresurus stejnegeri*) is a sense organ located between each eye and nostril. The pit organ of many snakes can detect the heat from an endothermic animal up to a distance of 1 to 2 m.

TABLE **41-1** **Classification of Receptors by Type of Energy They Transduce**

Type of Receptor	Examples	Effective Stimuli
Mechanoreceptors	Tactile receptors Pacinian corpuscles; Merkel discs; Ruffini corpuscles; Meissner corpuscles	Touch, pressure
	Nociceptors (pain receptors)	Strong touch, pressure, temperature extreme
	Proprioceptors Muscle spindles Golgi tendon organs Joint receptors	Movement, body position Muscle contraction Stretch of a tendon Movement in ligaments
	Statocysts in invertebrates (hair cells)	Gravity
	Lateral line organs in fish	Waves, currents in water
	Vestibular apparatus Saccule and utricle (hair cells) Semicircular canals (hair cells)	 Gravity; linear acceleration Angular acceleration
	Organ of Corti of the cochlea (hair cells)	Pressure waves (sound)
Chemoreceptors	Taste buds; olfactory epithelium	Specific chemical compounds
Thermoreceptors	Temperature receptors in blood-sucking insects and ticks; pit organs in pit vipers; nerve endings and receptors in skin and tongues of many animals	Heat
Electroreceptors	Organs in skin of some fishes	Electrical currents in water
Photoreceptors	Eyespots; ommatidia of arthropods; rods and cones in retinas of vertebrates	Light energy

hypothalamus detect internal changes in temperature, and receive and integrate information from thermoreceptors on the body surface. The hypothalamus then initiates homeostatic mechanisms that ensure a constant body temperature.

Electroreceptors sense differences in electrical potential. Some predatory species of sharks, rays, and bony fishes detect electrical fields generated in the water by the muscle activity of their prey. Some electroreceptors are sensitive enough to detect Earth's magnetic field.

Several groups of fishes have electric organs, specialized muscle or nerve cells that produce external electrical fields. In species that produce a weak electric current, such organs may help in orientation. This is particularly useful in murky water, where visibility and olfaction are poor. Electroreception also appears important in communication, for example, in recognizing a potential mate. Males and females have different frequencies of electric discharge. A few fishes, such as electric eels or electric rays, have electric organs in their heads capable of delivering powerful shocks that stun prey or predators.

Each kind of sensory receptor is especially sensitive to one particular form of energy. The photoreceptors of the human eye are stimulated by an extremely faint beam of light, and taste receptors are stimulated by a minute amount of a chemical compound.

Sensory receptors can also be classified according to the source of the stimuli to which they respond. **Exteroceptors** receive stimuli from the outside environment, enabling an animal to know and explore the world, search for food, find and attract a mate, recognize friends, and detect enemies. **Interoceptors** are sensory receptors within body organs that detect changes in pH, osmotic pressure, body temperature, and the chemical composition of the blood. You are not usually aware of messages sent to the CNS by these receptors as they work continuously to maintain homeostasis. You become aware of their activity when they signal certain internal conditions such as thirst, hunger, nausea, pain, and orgasm.

Review

- Name the five kinds of sensory receptors. What is the specific function of each?

BiologyNow™ Assess your understanding of **types of sensory receptors** by taking the pretest on your BiologyNow CD-ROM.

HOW SENSORY RECEPTORS FUNCTION

Learning Objective

2 Describe how a sensory receptor functions; define *energy transduction, receptor potential,* and *sensory adaptation* as part of your answer.

Sensory receptors produce receptor potentials. They (1) absorb a small amount of energy from some stimulus; (2) convert the energy of the stimulus into electrical energy, the process known as **energy transduction;** and (3) produce a **receptor potential,** a depolarization or hyperpolarization of the membrane. With minor variations, this is how all receptors operate.

The function of a receptor potential is to generate action potentials that transmit information to the CNS. However, as we discuss later, each receptor potential does not directly trigger an action potential. When unstimulated, a sensory receptor maintains a resting potential, that is, a difference in charge between the inside and the outside of the cell. When the receptor cell is stimulated and its membrane potential changes, specific types of ion channels in its plasma membrane open or close, altering the permeability of the plasma membrane to various ions. A change in ion distribution can cause a change in voltage across the membrane. If the difference in charge increases, the receptor becomes hyperpolarized. If the potential decreases, the receptor becomes depolarized. The change in membrane potential is the receptor potential.

Like an excitatory postsynaptic potential, or EPSP (see Chapter 39), the receptor potential is a **graded response** in which the magnitude of change depends on the energy of the stimulus. A graded response spreads relatively slowly down the dendrite, fading as it goes. In contrast, according to the *all-or-none law,* the amplitude (size) of each action potential does not vary and has no relation to the magnitude of the stimulus.

The specialized region of the receptor plasma membrane that transduces energy does not generate action potentials. Current generated by receptor potentials flows to a region along the axon where an action potential can be generated. If the receptor membrane is part of a separate cell, receptor potentials stimulate the release of a neurotransmitter, which flows across the synapse and binds to sites on the sensory neuron. When the sensory neuron becomes sufficiently depolarized to reach its threshold level, an action potential is generated. The action potential travels along the axon of the sensory neuron to the CNS. In summary,

Stimulus (such as light energy) ⟶ transduction into electrical energy ⟶ receptor potential ⟶ action potential

Sensation depends on transmission of a coded message

How do you know whether you are seeing a blue sky, tasting a doughnut, or hearing a note played on a piano? All action potentials are qualitatively the same. Light of the wavelength 450 nm (blue), sugar molecules (sweet), and sound waves of 440 hertz (Hz) (*A* above middle *C*) all cause transmission of similar action potentials. Our ability to differentiate stimuli depends on both the sensory receptor itself and on the brain. We can distinguish the color of a blue sky from the scent of cologne; a sweet taste from a light breeze; or the sound of a piano from the heat of the sun because cells of each sensory receptor are connected to specific neurons in particular parts of the brain. Because a receptor normally responds to only one type of stimulus, (for example, light), the brain interprets a message arriving from a particular receptor as meaning that a certain type of stimulus occurred (such as a flash of color).

Sensation takes place in the brain. Interpretation of the message and the type of sensation depends on which association neurons receive the message. The rods and cones of the eye do not see. When stimulated, they send a message to the brain that interprets the signals and translates them into a rainbow, an elephant, or a child. Artificial (for example, electrical) stimulation of brain centers can also result in sensation. In contrast, many sensory messages never give rise to sensations at all. For example, certain chemoreceptors sense internal changes in the body but never stir our consciousness.

When stimulated, a sensory receptor initiates what might be considered a coded message, composed of action potentials transmitted by sensory neurons. This coded message is later decoded in the brain. Impulses from the sensory receptor may differ in the (1) total number of sensory neurons transmitting the signal, (2) specific neurons transmitting action potentials and their targets, (3) total number of action potentials transmitted by a given neuron, and (4) frequency of the action potentials transmitted by a given fiber. The intensity of a stimulus is coded by the frequency of action potentials fired by sensory neurons during the stimulus. A strong stimulus results in a greater depolarization of the receptor membrane, causing the sensory neuron to fire action potentials with greater frequency than would a weak stimulus. This variation is possible because as a graded response, the receptor potential can vary in magnitude.

The difference in sound intensity between the gentle rustling of leaves and a clap of thunder depends on the number of neurons transmitting action potentials, as well as the frequency of the action potentials transmitted by each neuron. Just how the sensory receptor initiates different codes and how the brain analyzes and interprets them to produce various sensations are not yet completely understood.

Sensory receptors adapt to stimuli

Many sensory receptors do not continue to respond at the initial rate, even if the stimulus continues at the same intensity. With time, the response to a continued, constant stimulus decreases. This decrease in frequency of action potentials in a sensory neuron, even though the stimulus is maintained, is called **sensory adaptation.** Sensory adaptation occurs for two reasons. First, during a sustained stimulus, the receptor sensitivity decreases and produces a smaller receptor potential (resulting in a lower frequency of action potentials in the sensory neurons). Second, changes take place at synapses in the neural pathway activated by the receptor. For example, the release of neurotransmitter from a presynaptic terminal may decrease in response to a series of action potentials.

Some receptors, such as those for pain or cold, adapt so slowly that they continue to trigger action potentials as long as the stimulus persists. Other receptors adapt rapidly, permitting an animal to ignore persistent unpleasant or unimportant stimuli. For example, when you first pull on a pair of tight jeans, your pressure receptors let you know that you are being squished, and you may feel uncomfortable. Soon, though, these receptors adapt, and you hardly notice the sensation of the tight fit. In the same way, people quickly adapt to odors that at first seem to assault their senses. Sensory adaptation enables an animal to discriminate between unimportant background stimuli that can be ignored and new or important stimuli that require attention.

Review

- What is a receptor potential?
- What is the function of sensory adaptation?

BiologyNow™ Assess your understanding of **how sensory receptors function,** by taking the pretest on your BiologyNow CD-ROM.

MECHANORECEPTORS

Learning Objectives

3 Describe the following mechanoreceptors: tactile receptors, nociceptors, proprioceptors, statocysts, hair cells, and lateral line organs.

4 Compare the function of the saccule and utricle with that of the semicircular canals in maintaining equilibrium.

5 Trace the path taken by sound waves through the structures of the ear, and explain how the organ of Corti functions as an auditory receptor.

Mechanoreceptors are activated when they change shape as a result of being mechanically pushed or pulled. They transduce mechanical energy, permitting animals to feel, hear, and maintain balance. These receptors provide information about the shape, texture, weight, and topographic relations of objects in the external environment. Some mechanoreceptors are extremely sensitive. For example, alligators have pressure receptors on their faces that can detect the movement in the surrounding water caused by a single falling drop of water.

Certain mechanoreceptors enable an organism to maintain its body position with respect to gravity (for us, head up and feet down; for a dog, dorsal side up and ventral side down; for a tree sloth, ventral side up and dorsal side down). When displaced from its normal position, the animal quickly adjusts its body to reassume normal orientation. In complex animals, receptors continually send information to the CNS regarding the position and movements of the body. Mechanoreceptors also provide information about the operation of internal organs. For example, they inform us about the presence of food in the stomach, feces in the rectum, urine in the bladder, and a fetus in the uterus.

Touch receptors are located in the skin

In many invertebrates as well as vertebrates, **tactile** (touch) **receptors** lie at the base of a hair or bristle. Tactile hairs may sense the body's orientation in space with respect to gravity. They may also detect air and water vibrations and contact with other objects. The tactile receptor is stimulated indirectly when the hair is bent or displaced. A receptor potential develops, and a few action potentials may be generated. This type of receptor responds only when the hair is moving. Even though the hair may be maintained in a displaced position, the receptor is not stimulated unless there is motion.

The simplest mechanoreceptors are free nerve endings in the skin. They detect touch, pressure, and pain when stimulated

FIGURE 41-2

Sensory reception in human skin.

(a) This diagrammatic section through the human skin illustrates several mechanoreceptors that respond to touch and pressure. Also shown are free nerve endings that respond to pain. **(b)** Pacinian corpuscle, a deep pressure receptor. **(c)** The mechanical forces these receptors sense cause Na^+ channels to open, depolarizing the axon.

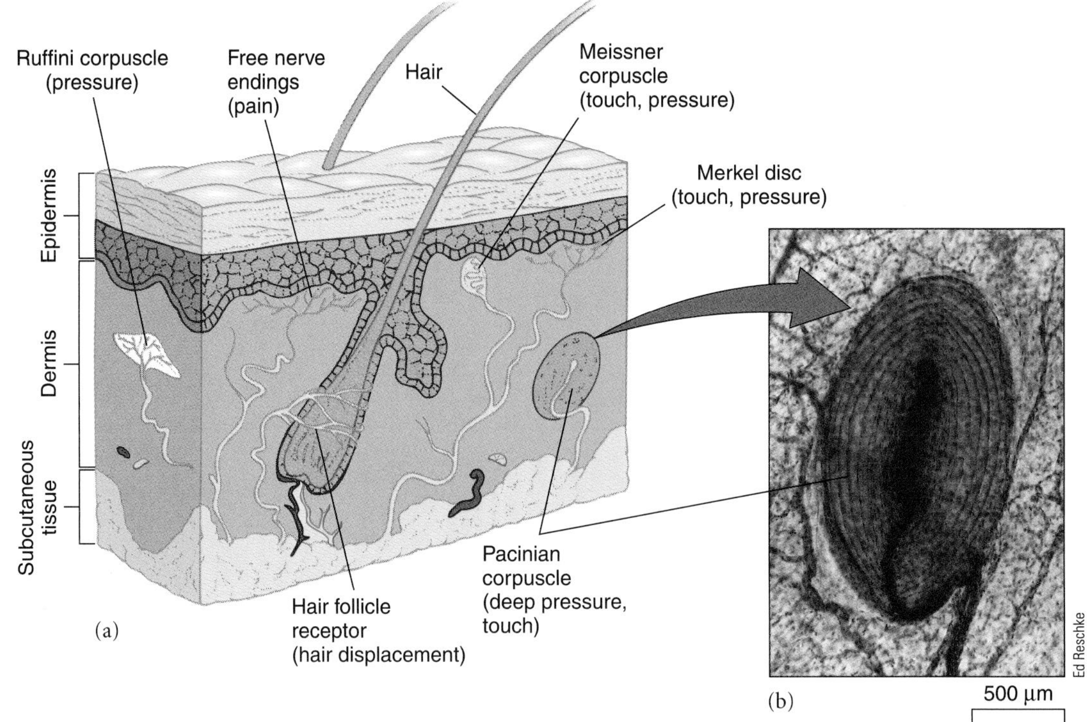

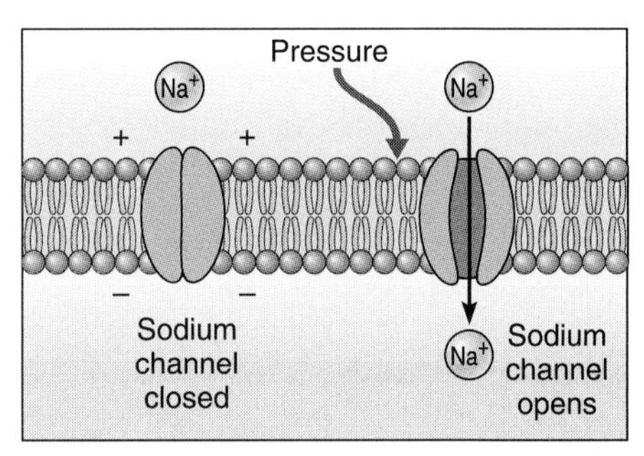

(c) Membrane of the Pacinian corpuscle

by objects that contact the body surface. Thousands of specialized tactile receptors are located in mammalian skin (Fig. 41-2). **Merkel discs** sense touch and pressure. They adapt slowly, permitting us to know that an object continues to touch the skin. Three types of mechanoreceptors in the skin have encapsulated endings: Meissner corpuscles, Ruffini corpuscles, and Pacinian corpuscles. **Meissner corpuscles** are sensitive to light touch and pressure and adapt quickly to a sustained stimulus. **Ruffini corpuscles,** which adapt very slowly, inform us of heavy and continuous touch and pressure.

The **Pacinian corpuscle** is sensitive to deep pressure that causes rapid movement of the tissues. It is especially sensitive to stimuli that vibrate. The Pacinian corpuscle consists of a neuron ending surrounded by concentric connective tissue layers interspersed with fluid. Compression causes displacement of the layers, which stimulates the axon. Even though the displacement is maintained under steady compression, the receptor potential rapidly falls to zero, and action potentials cease—an excellent example of sensory adaptation. The Pacinian corpuscle is stimulated only when there is movement of the tissue.

In humans, pain receptors, called **nociceptors** (from the Latin *nocere,* to injure), are free nerve endings (dendrites) of certain sensory neurons found in almost every tissue. Several types of nociceptors have been identified. Mechanical nociceptors respond to strong tactile stimuli such as penetration by sharp objects or pinching. Thermal nociceptors respond to temperature extremes (temperatures above 45°C or below 5°C). Other nociceptors respond to a variety of stimuli, including certain chemicals.

When stimulated, nociceptors transmit signals through sensory neurons to interneurons in the spinal cord. The sensory neurons release the neurotransmitter *glutamate,* as well as several neuropeptides, including **substance P,** that enhance and prolong the actions of glutamate. The interneuron transmits the message to the opposite side of the spinal cord and then upward to the thalamus, where pain perception begins. From the thalamus, impulses are sent into the parietal lobes and several other cortical regions, including areas of the limbic system where the emotional aspects of the pain are processed. When pain signals reach the cerebrum, we become fully aware of the pain and can evaluate the situation.

Proprioceptors help coordinate muscle movement

Proprioceptors help animals maintain postural relations—the position of one part of the body with respect to another. Located within muscles, tendons, and joints, proprioceptors continually respond to tension and movement. With their continuous input, an animal can perceive the positions of its arms, legs, head, and other body parts, along with the orientation of its body as a whole. This information is essential for all forms of locomotion and for all coordinated and skilled movements, from spinning a cocoon to completing a reverse one-and-a-half dive with twist. With the help of proprioceptors, we carry out activities such as dressing or playing the piano even in the dark.

Vertebrates have three main types of proprioceptors: **muscle spindles,** which detect muscle movement (Fig. 41-3); **Golgi tendon organs,** which respond to tension in contracting muscles and in the tendons that attach muscle to bone; and **joint receptors,** which detect movement in ligaments. Impulses from the proprioceptors help coordinate the contractions of the several distinct muscles involved in a single movement. Without such receptors, complicated, skilled acts would be impossible. Proprioceptors are also important in maintaining balance.

Proprioceptors are probably more numerous and more continually active than any of the other sensory receptors, although we are less aware of them than of most of the others. As long as the stimulus is present, receptor potentials are maintained (although not at constant magnitude) and action potentials continue to be generated. Proprioceptors continuously send signals to inform the brain about position.

The mammalian muscle spindle, one of the more versatile stretch receptors, helps maintain muscle tone. It is a bundle of specialized muscle fibers, with a central region encircled by sensory neuron endings. These neurons continue to transmit signals for a prolonged period, in proportion to the degree of stretch.

Many invertebrates have gravity receptors called statocysts

Many invertebrates—from jellyfish to crayfish—have gravity receptors called **statocysts.** A statocyst is basically an infolding of the epidermis lined with receptor cells, called sensory hair cells, equipped with sensory hairs (not true hairs) (Fig. 41-4). The cavity contains one or more **statoliths,** tiny granules of loose sand grains or calcium carbonate, held together by an adhesive material secreted by cells of the statocyst. Normally the particles are pulled downward by gravity and stimulate the sensory cells. When the animal changes position, the position of the statolith also changes, stimulating different sensory cells. Each sensory cell responds maximally when the animal is at a particular position with respect to gravity. The mechanical displacement results in receptor potentials and action potentials that inform the CNS of the change in position. By "knowing" which sensory cells are firing, the animal knows where "down" is and can correct any abnormal orientation.

In a classic experiment, the function of the statocyst was demonstrated by substituting iron filings for sand grains in the statocysts of crayfish. The force of gravity was overcome by holding magnets above the animals. The iron filings were attracted upward toward the magnets, and the crayfish began to swim upside down in response to the new information provided by their gravity receptors.

Hair cells are characterized by stereocilia

Vertebrate hair cells are mechanoreceptors that detect movement. They are the sensory receptors in the lateral line of fish that detect water movements. Hair cells help maintain position

FIGURE **41-3** | **Proprioceptors.**

Muscle spindles detect muscle movement. A muscle spindle consists of an elongated bundle of specialized muscle fibers. Golgi tendon organs respond to tension in muscles and their associated tendons.

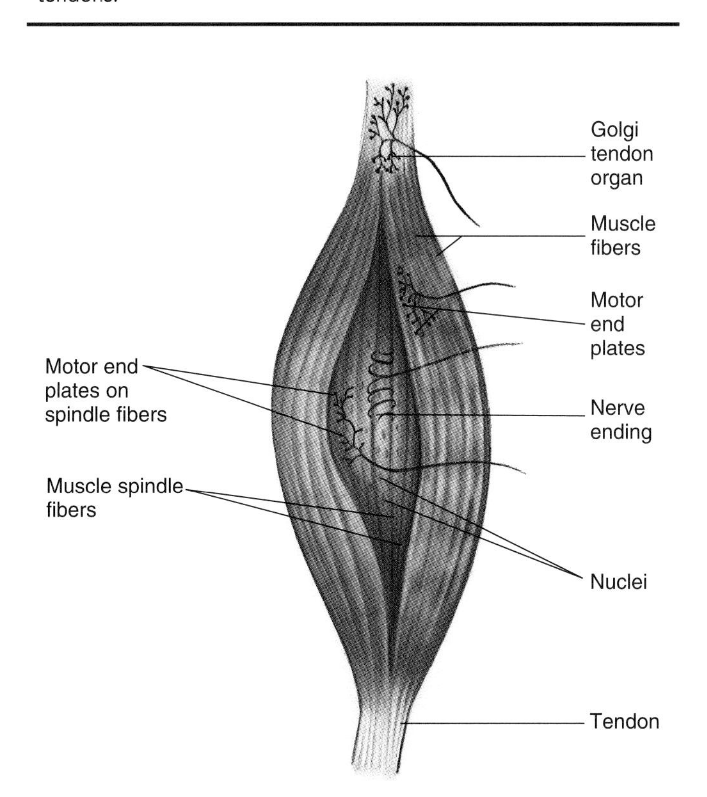

FIGURE **41-4** | **A statocyst.**

Many invertebrates have statocysts, receptors that sense gravitational force and provide information about orientation of the body with respect to gravity.

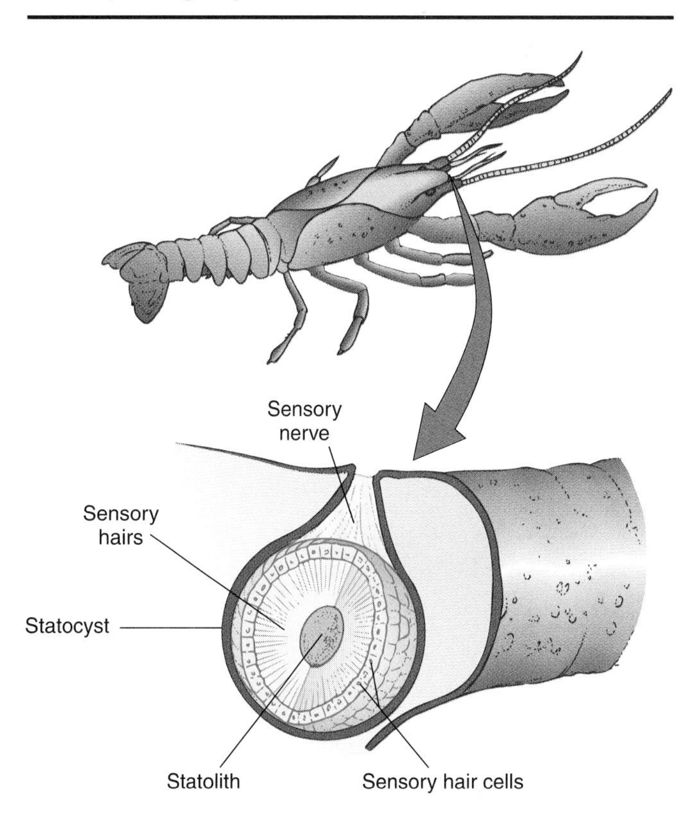

and equilibrium, and are important in hearing. The surface of a vertebrate hair cell typically has a single long **kinocilium** and a number of shorter **stereocilia,** hairlike projections that increase in length from one side of the hair cell to the other. A kinocilium is a true cilium with a 9 + 2 arrangement of microtubules (see Chapter 4). In contrast, stereocilia are not really cilia. They are microvilli that contain actin filaments.

Mechanical stimulation of the stereocilia causes voltage changes. Mechanical stimulation in one direction causes depolarization and release of neurotransmitter, whereas mechanical stimulation in the opposite direction results in hyperpolarization. However, hair cells have no axons and do not produce their own action potentials. Rather, they release neurotransmitters that depolarize associated neurons.

Lateral line organs supplement vision in fish

Lateral line organs of fish and aquatic amphibians detect vibrations in the water. They inform the animal of obstacles in its way and of moving objects such as prey, enemies, and others in its school. Typically the lateral line organ is a long canal running the length of the body and continuing into the head (Fig. 41-5). The canal is lined with sensory hair cells. The tips of the stereocilia are enclosed by a **cupula,** a mass of gelatinous material secreted by the hair cells.

Pressure waves, currents, and even slight movement in the water cause vibrations in the lateral line organ. The water moves the cupula, bending the stereocilia. This changes the membrane potential of the hair cell which may then release neurotransmitter. If a sensory neuron is sufficiently stimulated, an action potential may be dispatched to the CNS.

The vestibular apparatus maintains equilibrium

Probably, when you think of the ear you think of hearing. However, in vertebrates its main function is to help maintain equilibrium. Many vertebrates do not have outer or middle ears, but all have inner ears (Fig. 41-6). The mammalian **inner ear** contains organs of equilibrium equipped with hair cells that sense the position of the body with respect to gravity.

The inner ear is a complicated group of interconnected canals and sacs, called the **labyrinth,** which includes a membranous labyrinth that fits inside a bony labyrinth. In mammals the membranous labyrinth has two saclike chambers—the **saccule** and the **utricle**—and three **semicircular canals.** Collectively, the saccule, utricle, and semicircular canals are called the **vestibular apparatus** (see Fig. 41-6). Destruction of these organs leads to a considerable loss of the sense of equilibrium. A pigeon whose vestibular apparatus has been destroyed cannot fly but in time can relearn how to maintain equilibrium using visual stimuli.

The saccule and utricle house gravity detectors in the form of small calcium carbonate ear stones called **otoliths** (Fig. 41-7). The sensory cells of the saccule and utricle are hair cells similar to those of the lateral line organ. The stereocilia projecting from the hair cells are covered by a gelatinous cupula in which the otoliths are embedded. The hair cells in the saccule and utricle lie in different planes.

Normally, the pull of gravity causes the otoliths to press against the stereocilia, stimulating them to initiate impulses. Sensory neurons at the bases of the hair cells transmit these signals to the brain. When the head is tilted or in linear acceleration (increasing speed when the body is moving in a straight line), otoliths press on the stereocilia of different cells, deflecting them. Deflection of the stereocilia toward the kinocilium depolarizes the hair cell. Deflection in the opposite direction hyperpolarizes the hair cell. Thus, depending on the direction of movement, the hair cells release more or less neurotransmitter. The brain interprets the neural messages so the animal is aware of its full body position relative to the ground regardless of how its head is positioned.

Information about turning movements, referred to as angular acceleration, is provided by the three semicircular canals. Each canal, a hollow ring connected with the utricle, lies at right angles to the other two and is filled with fluid called **endolymph.**

FIGURE **41-5** **Lateral line organ.**

(a) The lateral line organ is a canal that extends the length of the body. **(b)** The canal has numerous openings to the outside environment. **(c)** The receptor cells are hair cells. Each hair cell has stereocilia and a kinocilium that are covered with a gelatinous cupula. The hair cells respond to waves, currents, or other disturbances in the water, informing the fish of obstacles or moving objects.

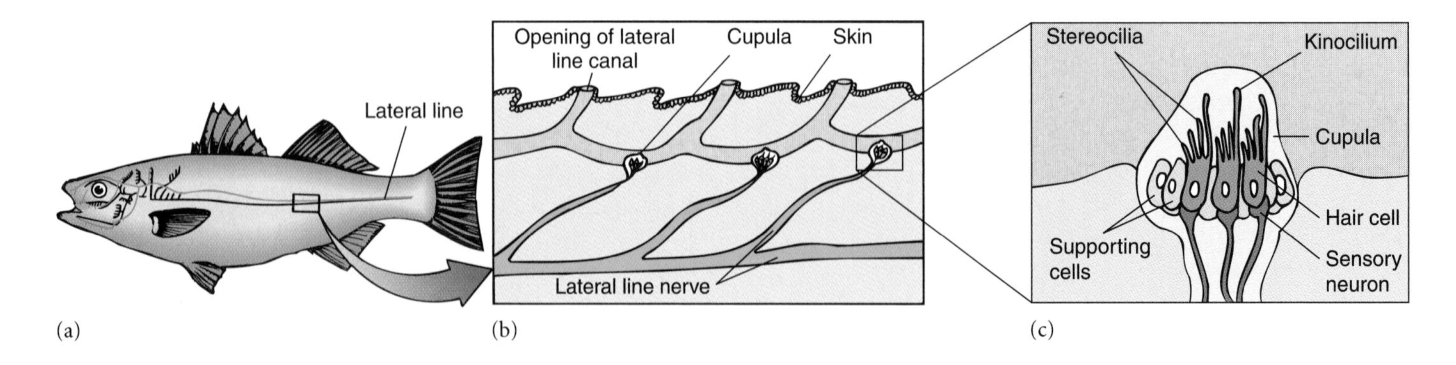

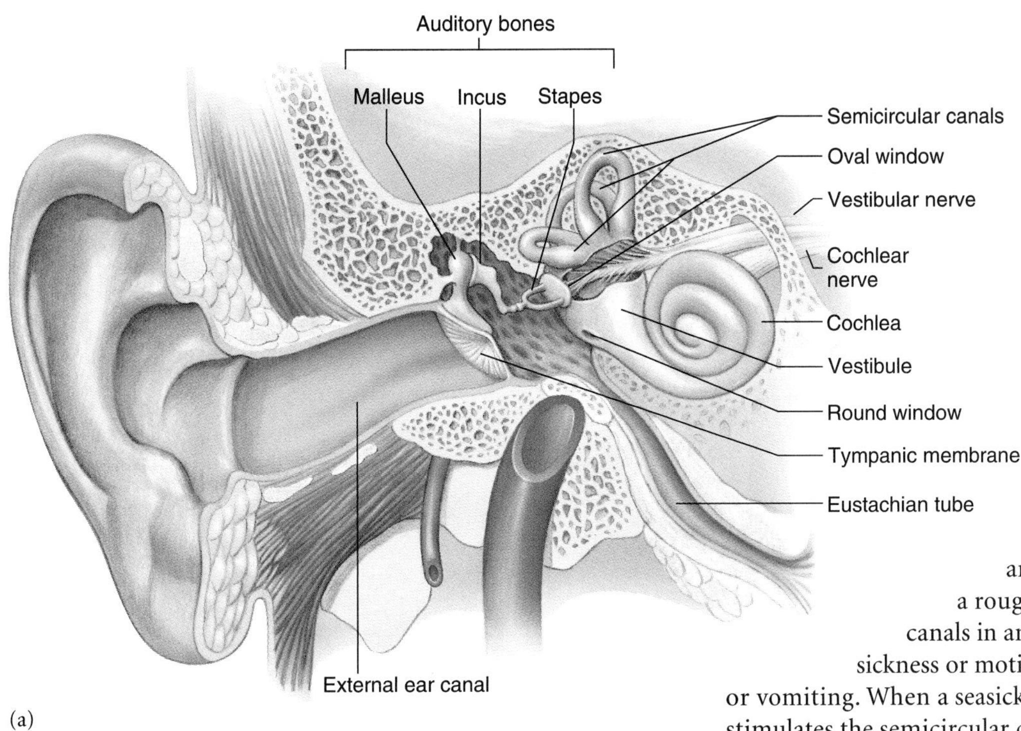

(a)

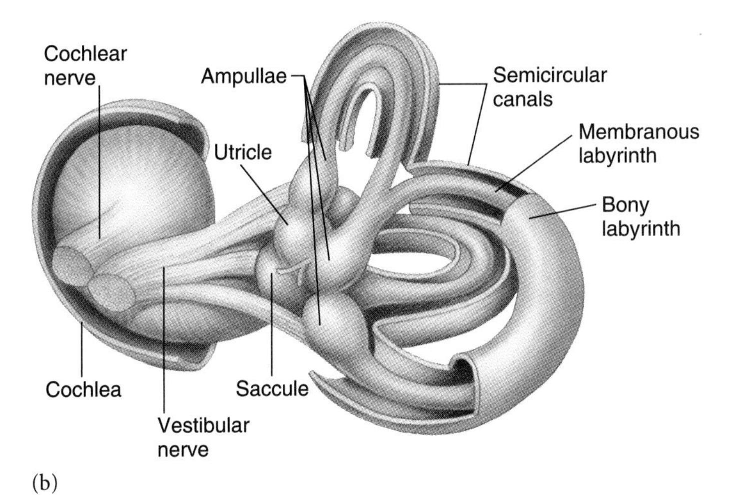

(b)

ACTIVE FIGURE **41-6**

The human ear.

(a) Human ear, anterior view. **(b)** The inner ear comprises the saccule, utricle, and semicircular canals, collectively referred to as the vestibular apparatus. The utricle and saccule are best seen in the posterior view shown here.

Biology Now™ Watch the **human ear** in action by clicking on this figure on your BiologyNow CD-ROM.

At one of the openings of each canal into the utricle is a small, bulblike enlargement, the **ampulla.** Within each ampulla lies a clump of hair cells called a **crista** (pl., *cristae*), similar to the groups of hair cells in the utricle and saccule. No otoliths are present. The stereocilia of the hair cells of the cristae are stimulated by movements of the endolymph in the canals (Fig. 41-8).

When the head is turned, there is a lag in the movement of the fluid within the canals. The stereocilia move in relation to the fluid and are stimulated by its flow. This stimulation produces not only the consciousness of rotation but also certain reflex movements in response to it. These reflexes cause the eyes and head to move in a direction opposite that of the original rotation. Because the three canals are in three different planes, head movement in any direction stimulates fluid movement in at least one of the canals.

We humans are used to movements in the horizontal plane but not to vertical movements (parallel to the long axis of the upright body). The motion of an elevator or of a ship pitching in a rough sea stimulates the semicircular canals in an unusual way and may cause seasickness or motion sickness, with resultant nausea or vomiting. When a seasick person lies down, the movement stimulates the semicircular canals in a more familiar way, and nausea is less likely to occur.

Auditory receptors are located in the cochlea

Many arthropods and most vertebrates have sound receptors, but for many of them hearing does not seem to be a sensory priority. It is important in tetrapods, however, and both birds and mammals have a highly developed sense of hearing. Their auditory receptors, located in the **cochlea** of the inner ear, contain mechanoreceptor hair cells that detect pressure waves.

The cochlea is a spiral tube that resembles a snail's shell (Fig. 41-9). If the cochlea were uncoiled, you would see that it consists of three canals separated from each other by thin membranes and coming almost to a point at the apex. Two of these canals, or ducts, the **vestibular canal** and the **tympanic canal,** connect at the apex of the cochlea and are filled with a fluid known as **perilymph.** The middle canal, the **cochlear duct,** is filled with endolymph and contains the auditory organ, the **organ of Corti.**

Each organ of Corti contains about 18,000 hair cells, in rows that extend the length of the coiled cochlea. Each hair cell has stereocilia that extend into the cochlear duct. The hair cells rest on the **basilar membrane,** which separates the cochlear duct from the tympanic canal. Overhanging and in contact with the hair cells of the organ of Corti is another membrane, the **tectorial membrane.**

In terrestrial vertebrates sound waves in the air are transformed into pressure waves in the cochlear fluid. In the human ear, for example, sound waves pass through the **external auditory canal** and vibrate the **tympanic membrane,** or eardrum (the membrane separating the outer ear and the middle ear). The vibrations are transmitted across the middle ear by three

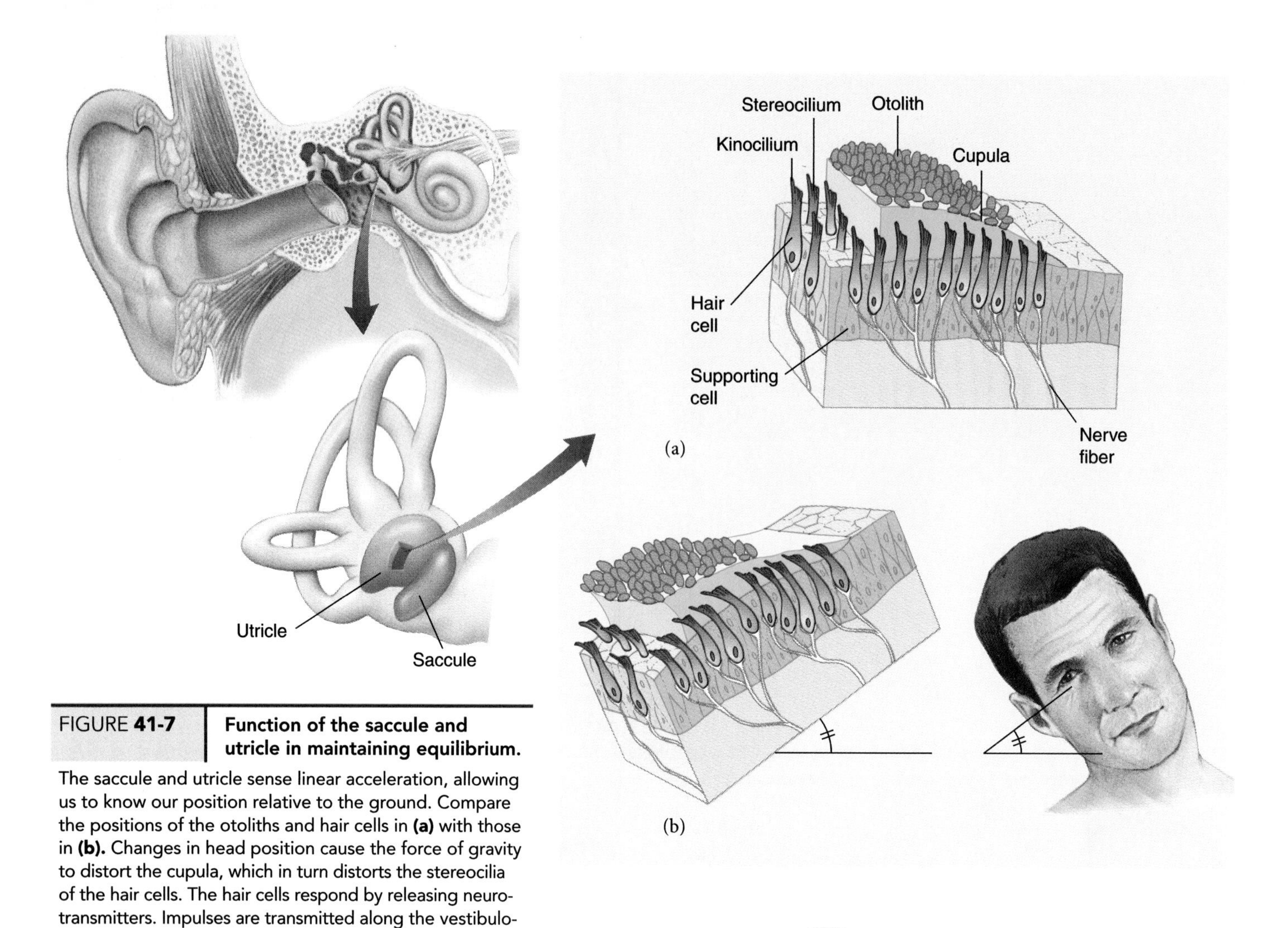

FIGURE **41-7** **Function of the saccule and utricle in maintaining equilibrium.**

The saccule and utricle sense linear acceleration, allowing us to know our position relative to the ground. Compare the positions of the otoliths and hair cells in **(a)** with those in **(b).** Changes in head position cause the force of gravity to distort the cupula, which in turn distorts the stereocilia of the hair cells. The hair cells respond by releasing neurotransmitters. Impulses are transmitted along the vestibulocochlear nerve to the brain.

tiny bones—the **malleus, incus,** and **stapes** (or hammer, anvil, and stirrup, so called because of their shapes). The malleus is in contact with the eardrum, and the stapes is in contact with a thin, membranous region of the cochlea, called the **oval window.** These bones act as three interconnected levers that amplify the vibrations. A very small movement in the malleus causes a larger movement in the incus and a very large movement in the stapes. The vibrations pass through the oval window to the perilymph in the vestibular canal. If sound waves were conducted directly from air to the oval window, much energy would be lost. The middle ear functions to couple sound waves in the air with the pressure waves conducted through the fluid in the cochlea.

Because liquids cannot be compressed, the oval window could not cause movement of the fluid in the vestibular canal if there were no escape valve for the pressure.

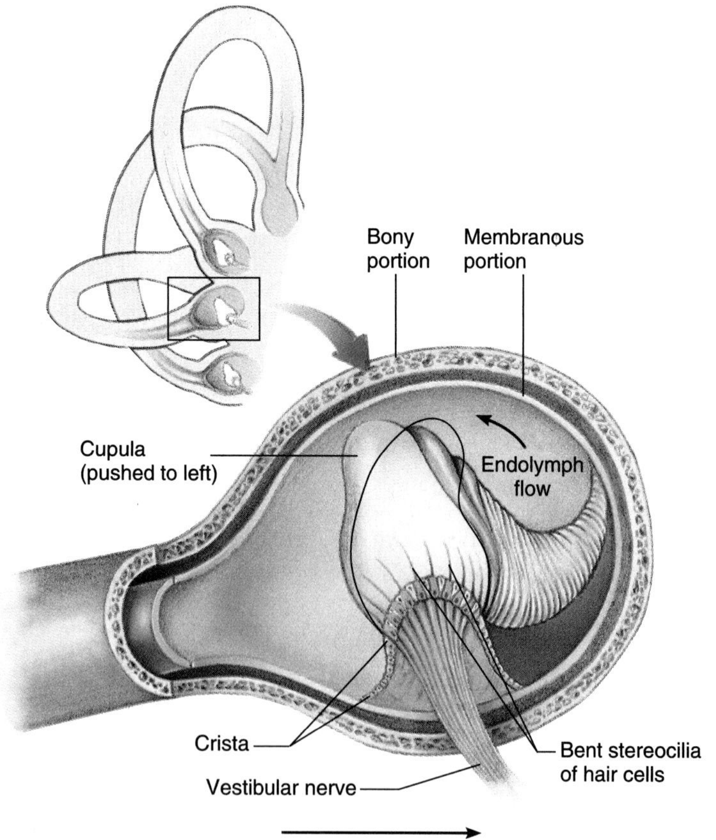

FIGURE **41-8** **Semicircular canals and equilibrium.**

When the head changes its rate of rotation, endolymph within the ampulla of the semicircular canal distorts the cupula. The stereocilia of the hair cells bend, increasing the frequency of action potentials in sensory neurons. Information is transmitted to the brain via the vestibular nerve.

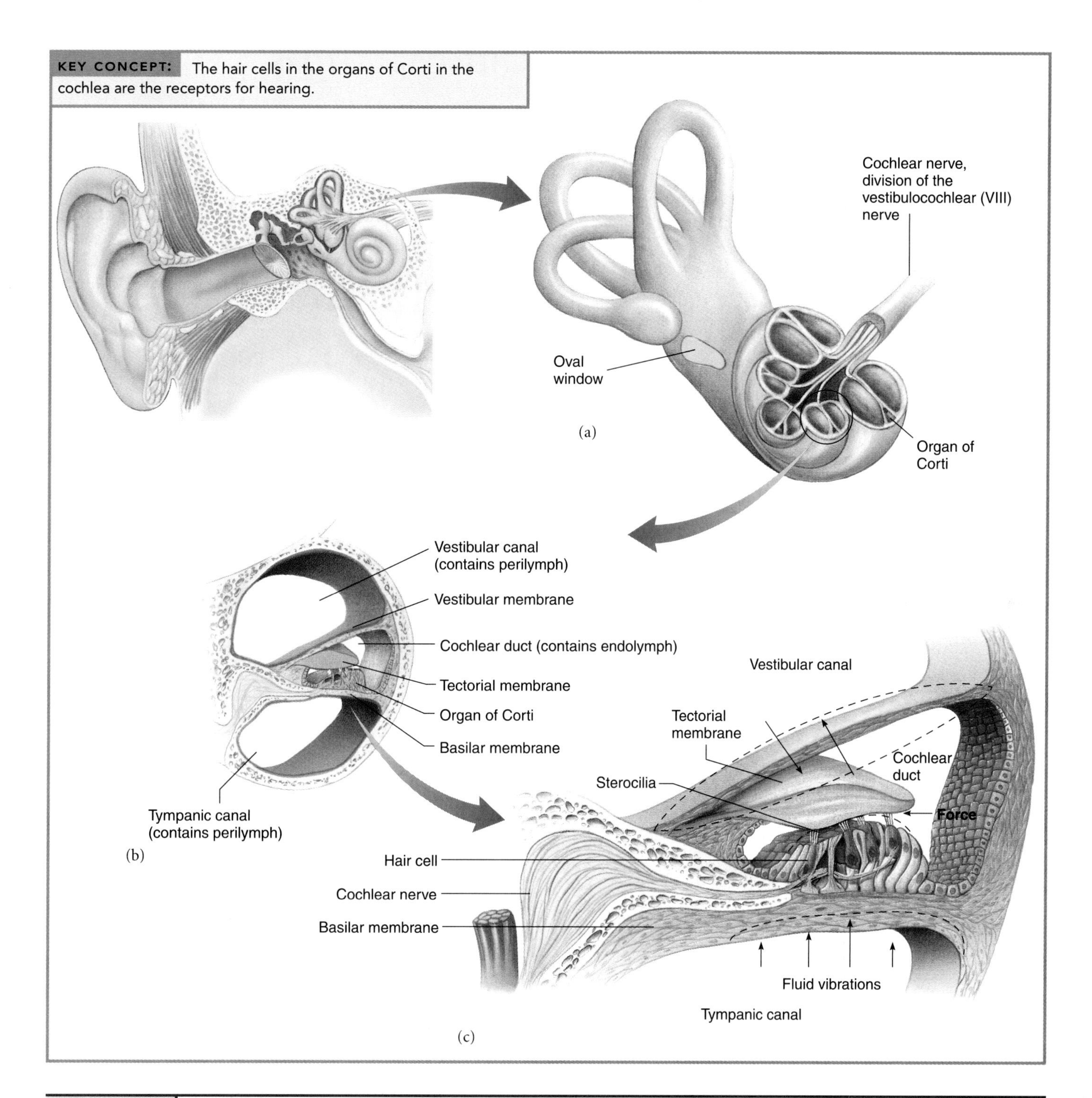

FIGURE **41-9** **The cochlea, organ of hearing.**

(a) Cross section through the cochlea showing the cochlear nerve. **(b)** Enlarged view of the organ of Corti resting on the basilar membrane and covered by the tectorial membrane. **(c)** How the organ of Corti works. Vibrations transmitted by the malleus, incus, and stapes set the fluid in the tympanic canal in motion. These vibrations *(dotted lines)* are transmitted to the basilar membrane, and, as the membrane vibrates, hair cells of the organ of Corti rub against the overlying tectorial membrane. This stimulation depolarizes the hair cells, prompting action potentials in the sensory neurons of the cochlear nerve.

This valve is provided by the membranous **round window** at the end of the tympanic canal. The pressure wave presses on the membranes separating the three ducts, is transmitted to the tympanic canal, and makes the round window bulge. The movements of the basilar membrane produced by these pulsations cause the stereocilia of the organ of Corti to rub against the overlying tectorial membrane. When stereocilia are bent by this contact, ion channels in the plasma membrane of the hair cells open. Calcium ions move into the hair cell, causing the release of the neurotransmitter *glutamate,* which binds to receptors on sensory neurons that synapse on each hair cell. Glutamate binds to receptors on sensory neurons that synapse on each hair cell, leading to depolarization of sensory neurons. Axons of the sensory neurons join to form the **cochlear nerve,** a component of the vestibulocochlear nerve (cranial nerve VIII; also referred to as the *auditory nerve*).

We can summarize the sequence of events involved in hearing as follows:

Sound waves enter external auditory canal ⟶ tympanic membrane vibrates ⟶ malleus, incus, and stapes amplify vibrations ⟶ oval window vibrates ⟶ vibrations are conducted through fluid ⟶ basilar membrane vibrates ⟶ hair cells in organ of Corti are stimulated ⟶ cochlear nerve transmits impulses to brain

Sounds differ in pitch, loudness, and tone quality. **Pitch** depends on frequency of sound waves, or number of vibrations per second, and is expressed as hertz (Hz). Low-frequency vibrations result in the sensation of low pitch, whereas high-frequency vibrations result in the sensation of high pitch. Sounds of a given frequency set up resonance waves in the cochlear fluid that cause a particular section of the basilar membrane to vibrate. High frequencies are detected by hair cells located near the base of the cochlea, whereas low frequencies are sensed by hair cells near the apex of the cochlea. The brain infers the pitch of a sound from the particular hair cells that are stimulated.

Loud sounds cause resonance waves of greater amplitude (height). The hair cells are more intensely stimulated, and the cochlear nerve then transmits a greater number of impulses per second. Variations in quality of sound, such as when an oboe, a cornet, and a violin play the same note, depend on the number and kinds of overtones, or harmonics, produced. These stimulate different hair cells in addition to the main stimulation common to all three instruments. Thus differences in tone quality are recognized in the pattern of the hair cells stimulated.

Your ear typically registers sound frequencies between about 20 and 20,000 Hz. (Dogs and some other animals can hear sounds of much higher frequencies.) The human ear is most sensitive to sounds between 1000 and 4000 Hz. Comparing the energy of audible sound waves with the energy of visible light waves, your ear is 10 times more sensitive than your eye.

Deafness may be caused by injury to, or malformation of, either the sound-transmitting mechanism of the outer, middle, or inner ear, or the sound-perceiving mechanism of the inner ear. Efferent neurons from the brain stem protect the hair cells of the organ of Corti by dampening their response. Even so, exposure to high-intensity sound, such as heavily amplified music, damages hair cells in the organ of Corti.

Review

- Which sensory receptors permit you to perform actions such as getting dressed or finding your way into bed in the dark?
- How do hair cells work?
- What are otoliths, and what is their role in maintaining equilibrium?
- What is the sequence of events involved in hearing?

BiologyNow™ Assess your understanding of **mechanoreceptors** by taking the pretest on your BiologyNow CD-ROM.

CHEMORECEPTORS

Learning Objective

6 Compare the structure and function of the receptors of taste and smell.

Two highly sensitive chemoreceptive systems are the senses of **taste** (gustation) and **smell** (olfaction). These chemical senses, found throughout the animal kingdom, allow animals to detect chemical substances in food, water, and air. Evaluating such chemical cues allows animals to find food and mates and to avoid predators. Chemoreception is also used by members of the same species to communicate information about food sources, potential mates, and danger.

For terrestrial vertebrates, the sense of smell involves gaseous substances that reach olfactory receptors through the air. The sense of taste involves materials dissolved in water (or saliva) in the mouth. For aquatic animals, these distinctions blur. Does a catfish smell or taste the water?

Taste buds detect dissolved food molecules

The ability to discriminate among tastes has survival value. For example, foods with a high caloric value often taste sweet, whereas poisons are generally bitter. The organs of taste in insects are sensory hairs. Mammals sense taste with **taste buds** located in the mouth. In humans, taste buds lie mainly in tiny elevations, or papillae, on the tongue. Each of the thousands of taste buds is an oval epithelial capsule containing about 100 taste receptor cells interspersed with supporting cells (Fig. 41-10).

The plasma membrane at the tip of each taste receptor cell has microvilli that extend into a taste pore on the surface of the tongue, where they are bathed in saliva. The taste receptors detect chemical substances dissolved in saliva. Certain molecules, such as those perceived as sweet, activate a signal transduction process involving a G protein (see Fig. 41-10c). Adenylyl cyclase activity increases, elevating cyclic AMP levels. A protein kinase is activated that phosphorylates and closes K^+ channels. This decrease in K^+ permeability sets up a depolarizing receptor potential. Action potentials are then generated in sensory neurons that synapse with the taste receptor cell. One sensory neuron can innervate several taste buds.

Traditionally, four basic tastes have been recognized: sweet, sour, salty, and bitter. In 2000, physiologists reported a fifth taste, glutamate. Although the idea of a fifth taste is still somewhat controversial, it was first suggested almost a century ago by the Japanese physiologist Kikunae Ikeda. In 1908, Ikeda identified glutamate as the compound that triggers the taste he called umami, responsible for the savoriness of certain aged cheeses, soy sauce, anchovy, and some seafoods.

Flavor depends on the four or five basic tastes in combination with smell, texture, and temperature. Smell affects flavor because odors pass from the mouth to the nasal chamber. No doubt you have observed that when your nose is congested, food seems to have little "taste." The taste buds are not affected, but the blockage of nasal passages severely reduces the participation of olfactory reception in the composite sensation of flavor.

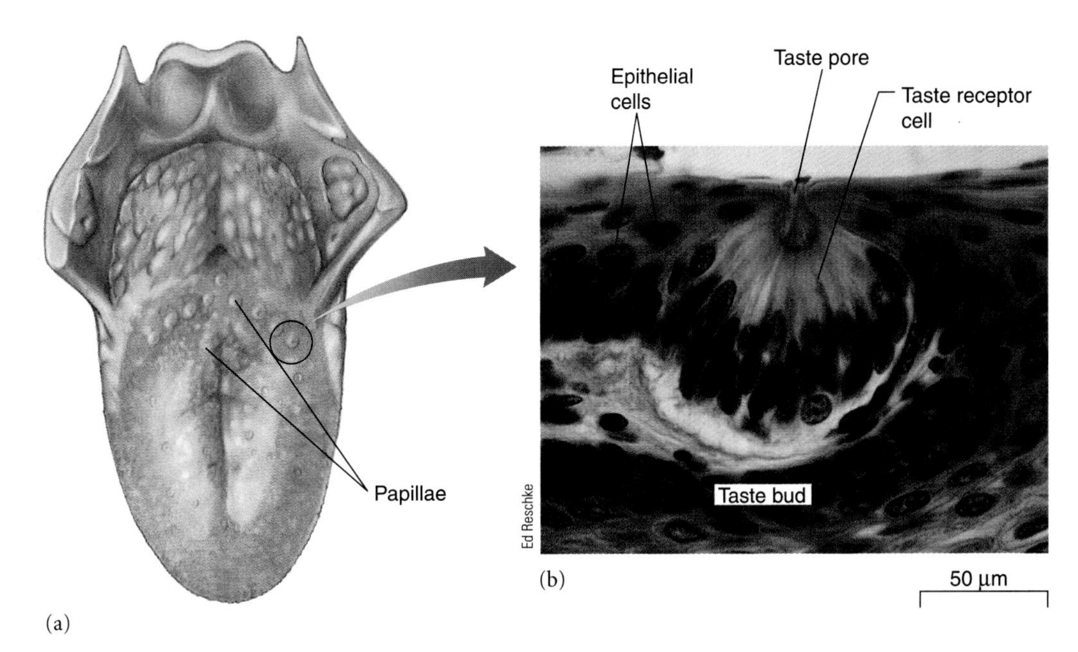

FIGURE **41-10** | **Taste buds.**

(a) The taste buds are located in the papillae, which cover the surface of the tongue. **(b)** LM of a taste bud: an epithelial capsule containing several taste receptor cells. A single taste receptor cell may respond to more than one category of taste. **(c)** A sugar molecule activates a signal transduction process. (1) The sugar binds with a receptor in the plasma membrane of a taste receptor cell. (2) A G protein is activated, and (3) a GTP-coupled subunit of the G protein activates adenylyl cyclase. (4) ATP is converted to cyclic AMP (cAMP). (5) Cyclic AMP activates a protein kinase. (6) Action of the protein kinase closes K^+ channels. The decrease in K^+ permeability results in an action potential in a sensory neuron that synapses with the taste receptor cell.

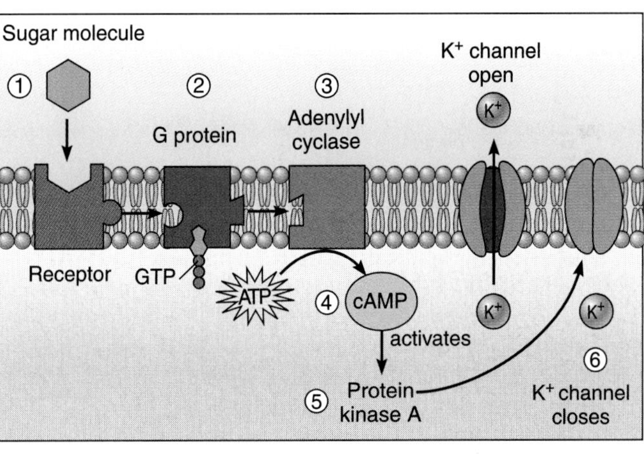

PROCESS OF SCIENCE

Awareness of the genetic component of taste can be traced to 1931 when Arthur L. Fox, a chemist at the DuPont Company synthesized a compound called *phenylthiocarbamide (PTC)*. Some PTC blew into the air and was inhaled by a colleague, who experienced it as very bitter. Because Fox himself could not taste it, he became intrigued and asked other people to taste it. Fox found that about 25% of people were nontasters. Everyone else experienced PTC as bitter. Further research showed that the ability to taste PTC is inherited as a dominant trait.

In the 1970s Linda Bartoshuk of Yale University continued taste research using a chemical compound called PROP (6-*n*-propylthiouracil). She found some tasters were especially sensitive to the bitter taste of PROP. Subsequent studies suggest that about 25% of the U.S. population are supertasters, 50% are regular tasters, and 25% are nontasters.

In 1997, Adam Drewnowski at the University of Michigan and his team reported that supertasters avoid broccoli, brussels sprouts, cabbage, and many other vegetables and fruits that taste bitter. These foods contain flavonoids and other compounds that are thought to protect against cancer. Continuing investigation of taste preferences and their nutritional consequences may lead to more effective approaches to improving nutrition and to a better understanding of the relative importance of genetics and learning in food selection.

The olfactory epithelium is responsible for the sense of smell

Most invertebrates depend on **olfaction,** the detection of odors, as their main sensory modality. Recent findings indicate that the mechanisms for chemoreception have been highly conserved throughout the animal kingdom.

In terrestrial vertebrates, olfaction occurs in the nasal epithelium. In humans, the **olfactory epithelium** is found in the roof of the nasal cavity (Fig. 41-11). It contains about 100 million olfactory receptor cells with ciliated tips. The long, nonmotile cilia extend into a layer of mucus on the epithelial surface of the nasal passageway. Receptor molecules on the cilia bind with compounds dissolved in the mucus. The other end of each olfactory receptor cell is an axon that projects directly to the brain. These axons make up the **olfactory nerve** (the first cranial nerve), which extends to the olfactory bulb in the brain.

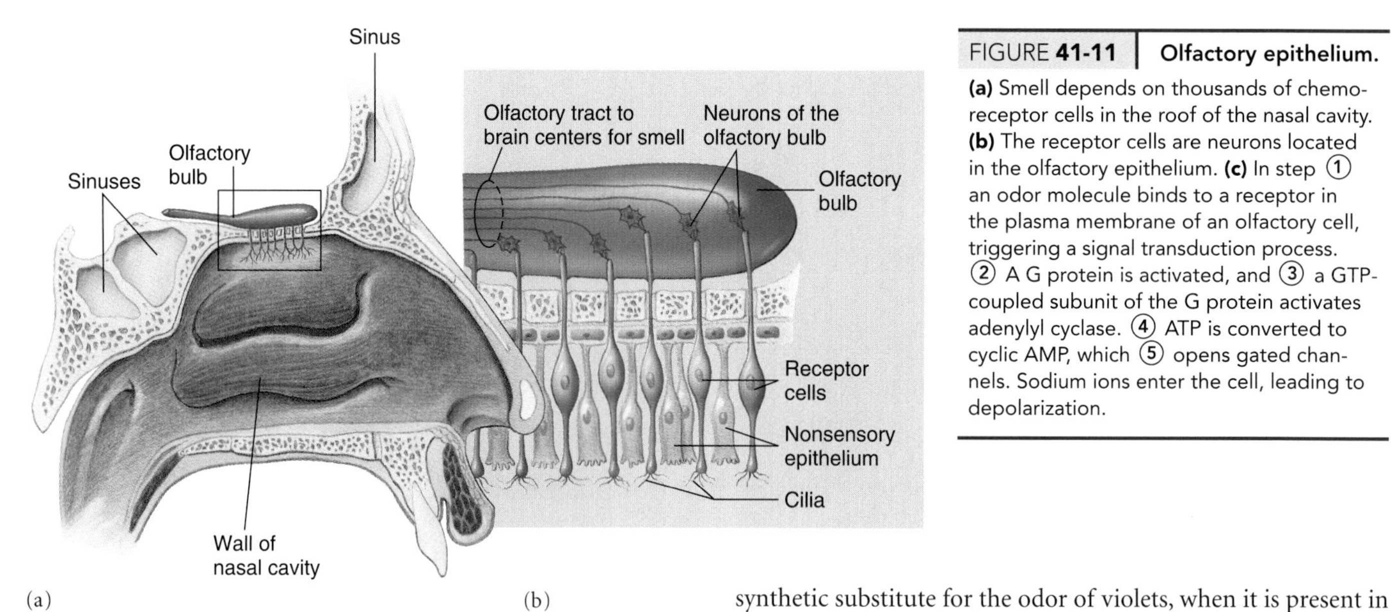

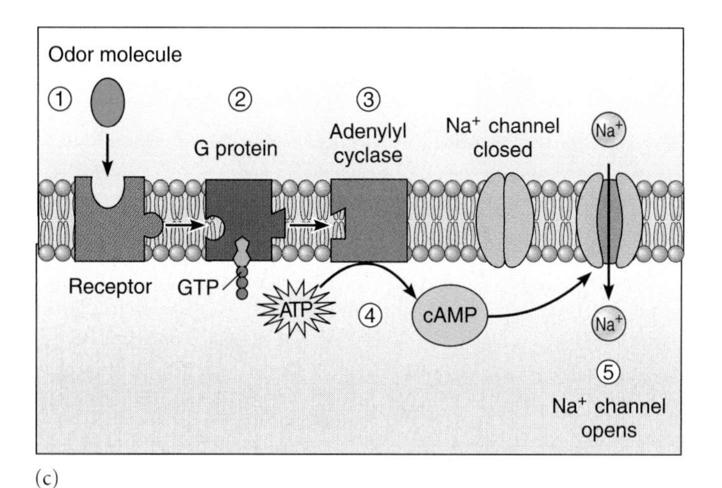

FIGURE 41-11 Olfactory epithelium. **(a)** Smell depends on thousands of chemoreceptor cells in the roof of the nasal cavity. **(b)** The receptor cells are neurons located in the olfactory epithelium. **(c)** In step ① an odor molecule binds to a receptor in the plasma membrane of an olfactory cell, triggering a signal transduction process. ② A G protein is activated, and ③ a GTP-coupled subunit of the G protein activates adenylyl cyclase. ④ ATP is converted to cyclic AMP, which ⑤ opens gated channels. Sodium ions enter the cell, leading to depolarization.

From there information is transmitted to the **olfactory cortex,** which is in the limbic system, a part of the brain also associated with emotional behavior. (Odors are often associated with feelings and memories.)

When a molecule binds with a receptor on the cilia of an olfactory receptor cell, a signal transduction process is initiated (see Fig. 41-11c). A G protein is activated, leading to the synthesis of cyclic AMP, which opens gated channels in the plasma membrane. These channels permit Na^+ and other cations to enter the cell, causing depolarization, which is the receptor potential. The number of odorous molecules determines the intensity, which in turn determines the magnitude of the receptor potential.

Humans can detect at least seven main groups of odors: camphor, musk, floral, peppermint, ethereal, pungent, and putrid. About 1000 genes code for 1000 types of olfactory receptors. Each odor consists of several component chemical groups, and each type of receptor may bind with a particular component. The combination of receptors activated determines what odor you perceive. You are capable of perceiving about 10,000 scents.

The olfactory receptors respond to remarkably small amounts of a substance. For example, most people can detect ionone, the synthetic substitute for the odor of violets, when it is present in a concentration of only one part to more than 30 billion parts of air. Smell is perhaps the sense that adapts most quickly. The olfactory receptors adapt about 50% in the first second or so after stimulation, so even offensively odorous air may seem odorless after only a few minutes.

Animals within many species communicate with one another by releasing **pheromones,** small volatile molecules that are secreted into the environment. For example, female moths release pheromones that attract males, and dogs and wolves use pheromones to mark territory. Mammals have specialized chemoreceptor cells that detect pheromones. These cells make up the **vomeronasal organ** located in the epithelium of the nose (separate from the main olfactory epithelium). The vomeronasal sensory neurons send signals to areas of the hypothalamus, which serves as a link with the endocrine system. (We discuss pheromones further in Chapter 50.)

Review

- How are the signal transduction processes of taste buds and olfactory receptor cells alike? How are they different?

BiologyNow™ Assess your understanding of **chemoreceptors** by taking the pretest on your BiologyNow CD-ROM.

PHOTORECEPTORS

Learning Objectives

7 Contrast simple eyes, compound eyes, and vertebrate eyes.

8 Describe the functions of each structure of the vertebrate eye.

9 Describe the events that take place in human vision; compare the two types of photoreceptors, and describe the signal transduction pathway as part of your explanation.

Most animals have photoreceptors that use pigments to absorb light energy. **Rhodopsins** are the photopigments in the eyes of cephalopod mollusks, arthropods, and vertebrates. Light energy striking a light-sensitive receptor cell containing these pigments triggers chemical changes in the pigment molecules that result in receptor potentials.

Invertebrate photoreceptors include eyespots, simple eyes, and compound eyes

The simplest light-sensitive structures in animals are found in certain cnidarians and flatworms (Fig. 41-12). They are **eyespots,** called **ocelli,** that detect light but do not form images. Eyespots are often bowl-shaped clusters of light-sensitive cells within the epidermis. They may detect the direction of the light source and distinguish light intensity.

Effective image formation requires a more complex **eye,** usually with a **lens.** A lens is a structure that concentrates light on a group of photoreceptors. Vision also requires a brain that can interpret the action potentials generated by the photoreceptors. The brain integrates information about movement, brightness, location, position, and shape of the visual stimulus.

Two fundamentally different types of eyes evolved: the camera eye of vertebrates and some mollusks (squids and octopods), and the compound eye of arthropods.

The **compound eyes** of crustaceans and insects differ structurally and functionally from vertebrate eyes (Fig. 41-13). The surface of a compound eye appears faceted, which means "having many faces," like a diamond. Each **facet** is the convex *cornea* of one of the eye's visual units, called **ommatidia** (sing., *ommatidium*). The number of ommatidia varies with the species. For example, each eye of certain crustaceans has only 20 ommatidia, whereas the dragonfly eye has as many as 28,000.

The optical part of each ommatidium includes a biconvex lens and a **crystalline cone.** These structures focus light onto photoreceptor cells called **retinular cells.** These cells have a light-sensitive membrane made up of microvilli containing rhodopsin. The membranes of adjacent retinular cells may fuse, forming a rod-shaped **rhabdome** that is sensitive to light.

Compound eyes do not perceive form well. Although the lens system of each ommatidium is adequate to focus a small inverted image, little evidence suggests the animal actually perceives them as images. However, all the ommatidia together produce a composite image, or **mosaic** picture. Each ommatidium, in gathering a point of light from a narrow sector of the visual field, is in fact sampling a mean intensity from that sector. All these points of light taken together form a mosaic picture (Fig. 41-13d).

Arthropod eyes usually adapt to different intensities of light. A sheath of pigmented cells envelops each ommatidium, and screening pigments are present in cells called iris cells and in retinular cells. In nocturnal and crepuscular (active at dusk) insects and many crustaceans, pigment migrates proximally and distally within each pigmented cell. When the pigment is in the proximal position, each ommatidium is shielded from its neighbor, and only light entering directly along its axis can stimulate the receptors. When the pigment is in the distal position, light striking at any angle can pass through several ommatidia and stimulate many retinal units. As a result, sensitivity is increased in dim light, and the eye is protected from excessive stimulation in bright light. Pigment migration is under neural control in insects and under hormonal control in crustaceans. In some species it follows a daily rhythm.

Although the compound eye can form only coarse images, it compensates by following flickers to higher frequencies. Flies can detect up to about 265 flickers per second. In contrast, the human eye can detect only 45 to 53 flickers per second; for us, flickering lights fuse above these values, so we see light provided by an ordinary bulb as steady and the movement in motion pictures as smooth. To an insect, both room lighting and motion pictures must flicker horribly. The insect's high critical flicker fusion threshold permits immediate detection of even slight movement by prey or enemy. The compound eye is an important adaptation to the arthropod's way of life.

Compound eyes differ from our eyes in another respect. They are sensitive to wavelengths of light in the range from red to ultraviolet (UV). Accordingly, an insect can see UV light well, and its world of color is very different from ours. Because they reflect UV light to various degrees, flowers that appear identically colored to humans may appear strikingly dissimilar to insects (see Fig. 35-3).

FIGURE **41-12** | **Eyespots.**

The simplest light-sensitive structures are the ocelli, or eyespots, found in certain invertebrates. **(a)** A planarian worm showing eyespots. **(b)** Structure of the eyespot of a planarian worm.

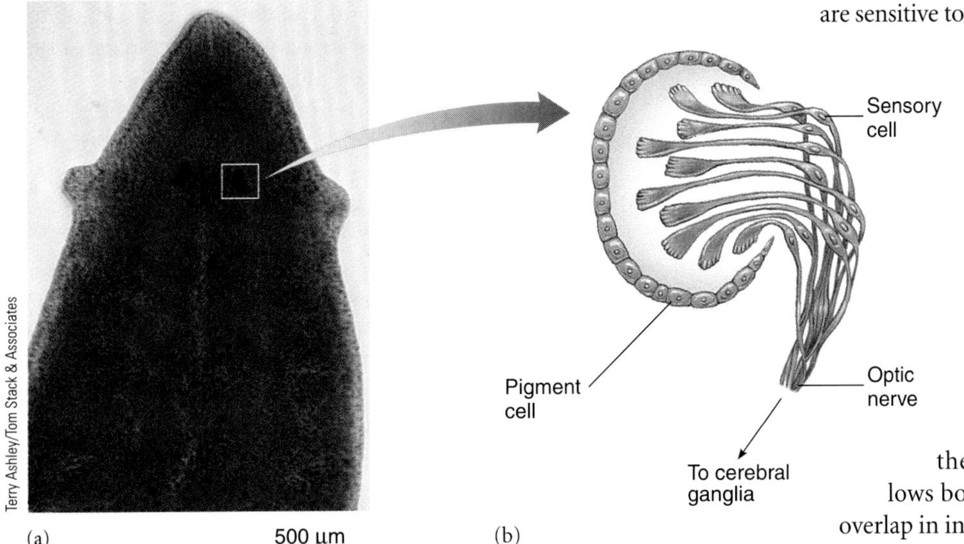

Vertebrate eyes form sharp images

The position of the eyes in the front of the head of primates and certain birds allows both eyes to focus on the same object. The overlap in information they receive results in the same visual information striking the two retinas (light-sensitive

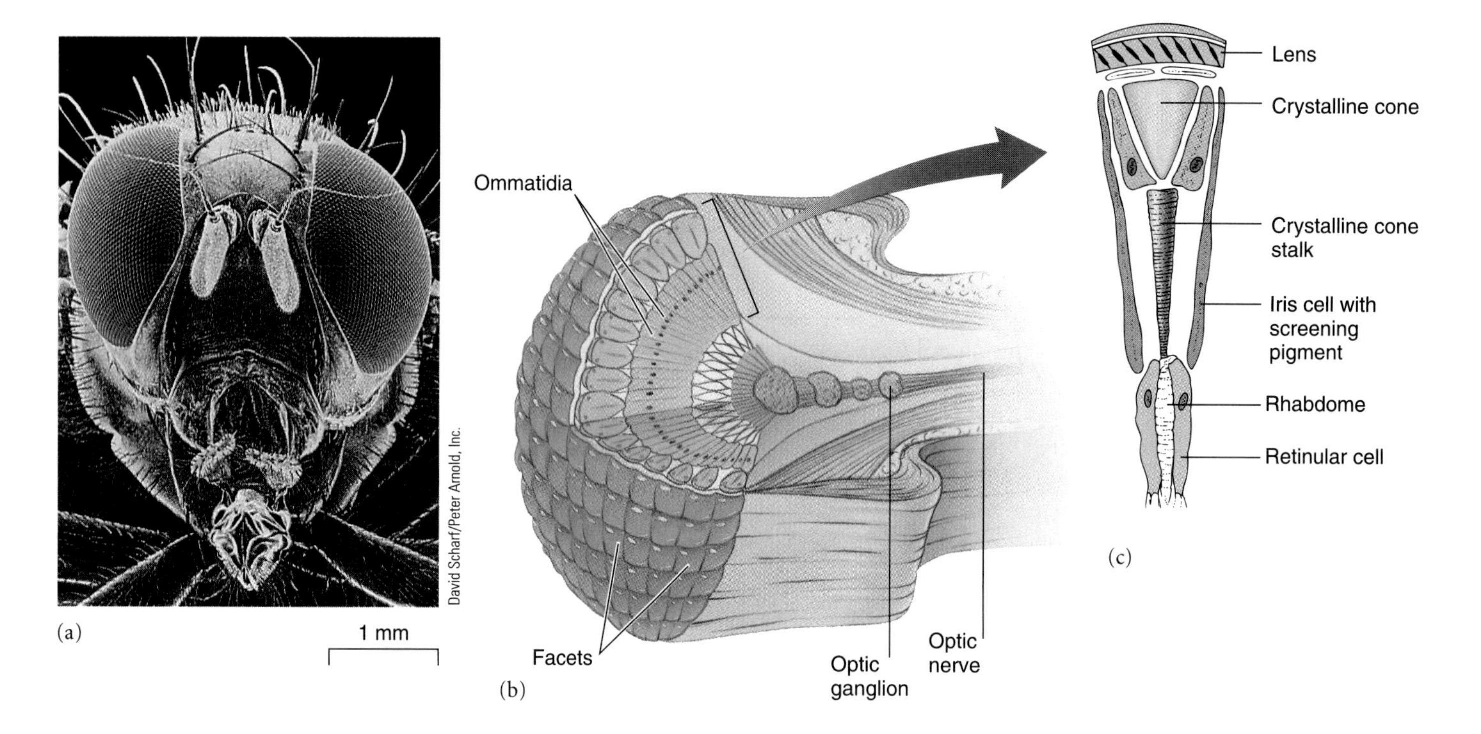

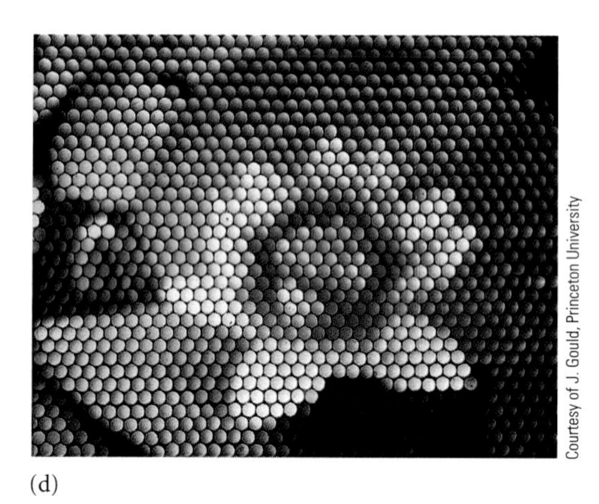

FIGURE **41-13** **Compound eyes.**

(a) SEM of the Mediterranean fruit fly (*Ceratitis capitata*) showing its prominent compound eyes. **(b)** Structure of the compound eye showing several ommatidia. The eye registers changes in light and shade, permitting the animal to detect movement. **(c)** Structure of an ommatidium. The rhabdome is the light-sensitive core of the ommatidium. **(d)** A bee eye view of a flower. This photo was taken using an optical device to achieve an approximate simulation of how the bee might see this flower. However, the bee would see UV light rather than red, and circles would be vertically elongated ellipses.

areas) at the same time. This **binocular vision** is important in judging distance and in depth perception.

The vertebrate eye can be compared to a camera. An adjustable lens can be focused for different distances, and a diaphragm, called the **iris,** regulates the size of the light opening, called the **pupil** (Fig. 41-14). The **retina** corresponds to the light-sensitive film used in a camera. Outside the retina is the **choroid layer,** a sheet of cells filled with black pigment that absorbs extra light, preventing light from being reflected back into the photoreceptors that would cause blurring of images. (Cameras are also black on the inside.) The choroid is rich in blood vessels that supply the retina.

The outer coat of the eyeball, called the **sclera,** is a tough, opaque, curved sheet of connective tissue that protects the inner structures and helps maintain the rigidity of the eyeball. On the front surface of the eye, this sheet becomes the thinner, transparent **cornea,** through which light enters. The cornea serves as a fixed lens that focuses light.

The lens of the eye is a transparent, elastic ball just behind the iris. It bends the light rays coming in and brings them to a focus on the retina. The lens is aided by the curved surface of the cornea and by the refractive properties (ability to bend light rays) of the liquids inside the eyeball. The **anterior cavity** between the cornea and the lens is filled with a watery substance, the **aqueous fluid.** The larger **posterior cavity** between the lens and the retina is filled with a more viscous fluid, the **vitreous body.** Both fluids are important in maintaining the shape of the eyeball by providing an internal fluid pressure.

At its anterior margin, the choroid is thick and projects medially into the eyeball to form the **ciliary body,** which consists of ciliary processes and the ciliary muscle. The ciliary processes are glandlike folds that project toward the lens and secrete the aqueous fluid.

We focus a camera by changing the distance between the lens and the film. The eye has the power of **accommodation,** the ability to change focus for near or far vision by changing the shape of the lens (Fig. 41-15). This is accomplished by the **ciliary muscle,** a part of the ciliary body. To focus on close objects, the ciliary muscle contracts, making the elastic lens rounder. To focus on more distant objects, the ciliary muscle relaxes and the lens assumes a flattened (ovoid) shape.

The amount of light entering the eye is regulated by the iris, a ring of smooth muscle that appears as blue, green, gray, or

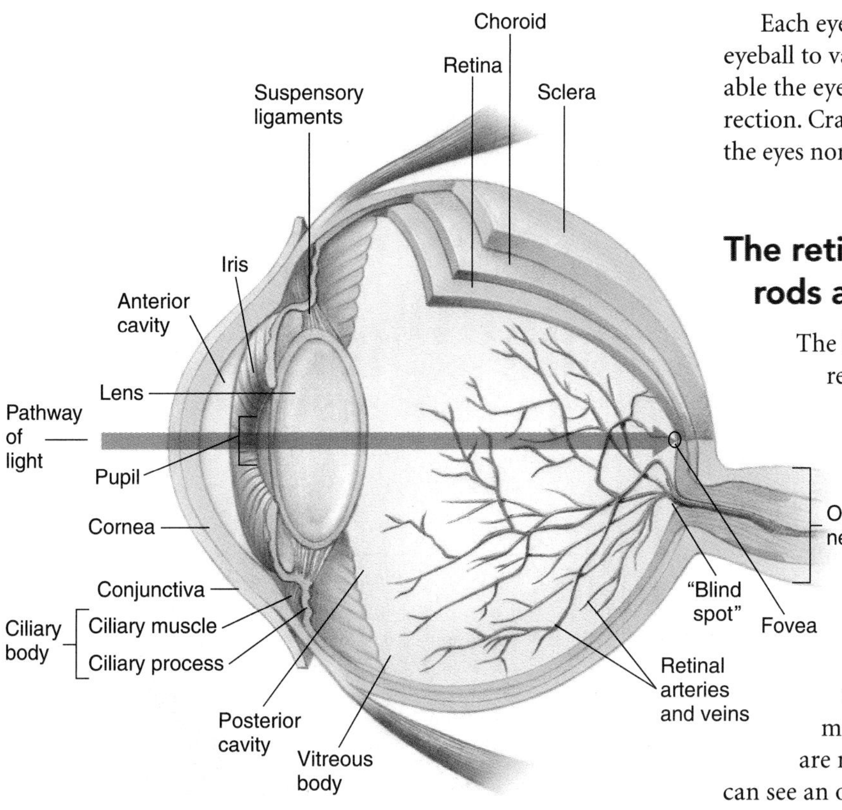

FIGURE **41-14** | **Structure of the human eye.**

Light passes through the human eye to photoreceptor cells in the retina. In this lateral view, the eye is shown partly sectioned along the sagittal plane to show its internal structures.

brown depending on the amount and nature of pigment present. The iris consists of two mutually antagonistic sets of muscle fibers. One set is arranged circularly and contracts to *decrease* the size of the pupil. The other is arranged radially and contracts to *increase* the size of the pupil.

ACTIVE FIGURE **41-15** | **Accomodation.**

How the eye changes focus for **(a)** near and **(b)** distant vision.

BiologyNow™ See how **vision disorders** change the eye by clicking on this figure on your BiologyNow CD-ROM.

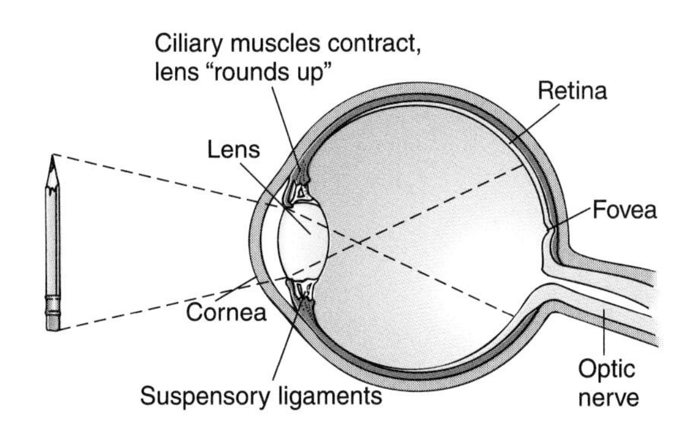

(a) Near vision

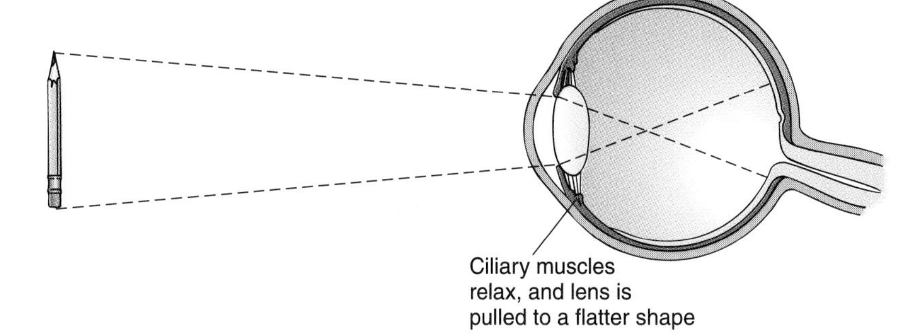

(b) Distant vision

Each eye has six muscles that extend from the surface of the eyeball to various points in the bony socket. These muscles enable the eye as a whole to move and be oriented in a given direction. Cranial nerves innervate the muscles in such a way that the eyes normally move together and focus on the same area.

The retina contains light-sensitive rods and cones

The light-sensitive structure in the vertebrate eye is the retina, which lines the posterior two thirds of the eyeball, covering the choroid. The retina, which is composed of 10 layers, contains the photoreceptor cells called, according to shape, **rods** and **cones.** In both rod and cone cells, infoldings of the plasma membrane form stacks of membranous discs that contain the photopigments. The discs greatly increase the light-absorbing surface. The human eye has about 125 million rods and 6.5 million cones. Rods function in dim light, letting us detect shape and movement. They are not sensitive to colors. Because the rods are more numerous in the periphery of the retina, you can see an object better in dim light if you look slightly to one side of it (allowing the image to fall on the rods).

Cones respond to light at higher levels of intensity, for example daylight, and they allow us to perceive fine detail. Cones are responsible for color vision; they are differentially sensitive to different wavelengths (colors) of light. In primates and certain hunting birds (such as eagles), the cones are most concentrated in the **fovea,** a small depressed area in the center of the retina. The fovea is the region of sharpest vision because it has the greatest density of receptor cells and because the retina is thinner there.

Light must pass through several layers of connecting neurons in the retina to reach the rods and cones (Fig. 41-16). This arrangement lets the rods and cones contact a layer of pigmented epithelium that provides retinal, a component of rhodopsin. The retina has five main types of neurons: (1) Photoreceptors (rods and cones) synapse on (2) **bipolar cells,** which make synaptic contact with (3) **ganglion cells.** Two types of lateral interneurons are the (4) **horizontal cells,** which receive information from the photoreceptor cells and send it to bipolar cells, and (5) **amacrine cells,** which receive messages from the bipolar cells and send signals back to the bipolar cells or to ganglion cells (Fig. 41-17). Interestingly, researchers recently reported that one group of ganglion cells project to the suprachiasmatic nucleus, the body's main biological clock. These ganglion cells contain a light-sensitive pigment (melanopsin) that is important in nonvisual responses to light.

The axons of the ganglion cells extend across the surface of the retina and unite to form the **optic nerve.** The area where the optic nerve passes out of the eyeball, the **optic disk,** is known as the "blind spot"; because it lacks rods and cones, images falling on it cannot be perceived. A simplified summary of the visual pathway follows:

> Light passes through cornea ⟶ through aqueous fluid ⟶ through lens ⟶ through vitreous body ⟶ image forms on photoreceptor cells in retina ⟶ signal bipolar cells ⟶ signal ganglion cells ⟶ optic nerve transmits signals to thalamus ⟶ integration by visual areas of cerebral cortex

FIGURE **41-16** | **Organization of the retina.**

(a) The elaborate interconnections among the various layers of neurons in the retina let them interact and influence one another. The rods and cones are the photoreceptor cells. **(b)** Rods *(red elongated structures)* and two cones *(shorter, thicker, yellow structures)* are seen in this SEM. The elongated rods permit you to see shape and movement, whereas the shorter cones allow you to view the world in color.

KEY CONCEPT: The rods and cones in the retina are the photoreceptors in vertebrates.

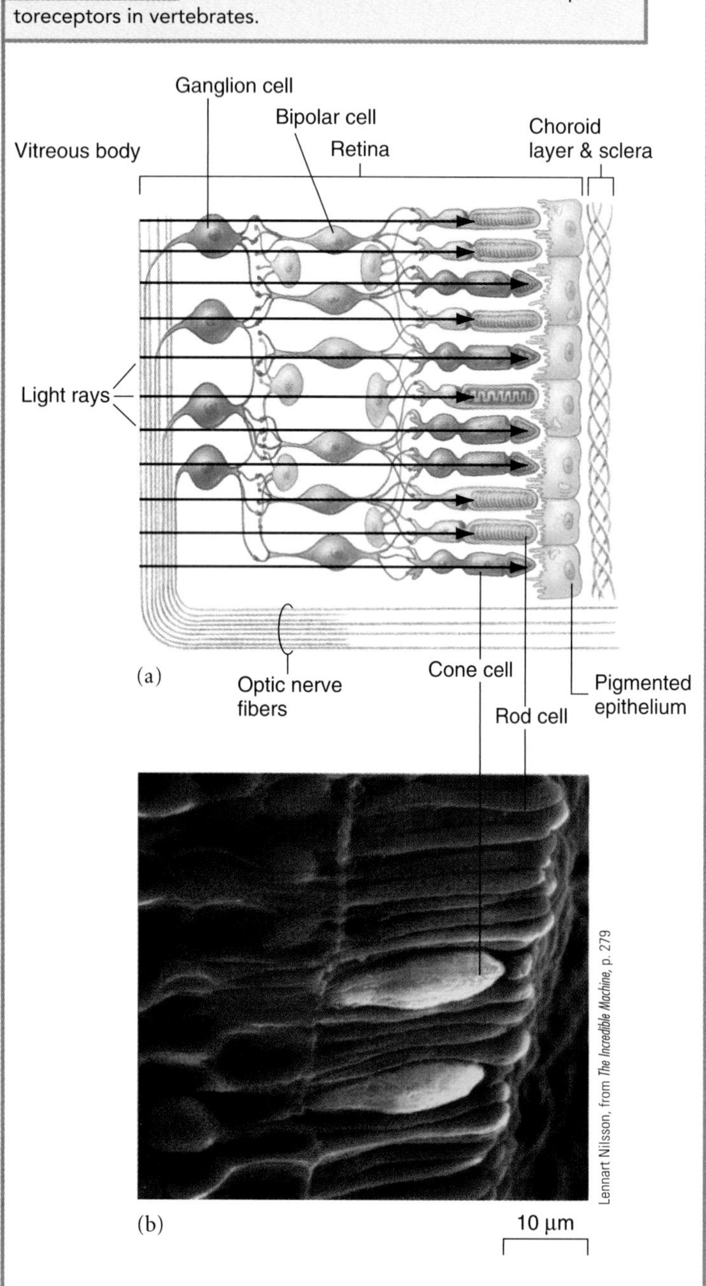

FIGURE **41-17** | **Neural pathway in the retina.**

The photoreceptor cells (rods and cones) are located in the back of the retina. They synapse with bipolar and horizontal cells. The bipolar cells synapse with ganglion cells and amacrine cells. Axons of ganglion cells make up the optic nerve.

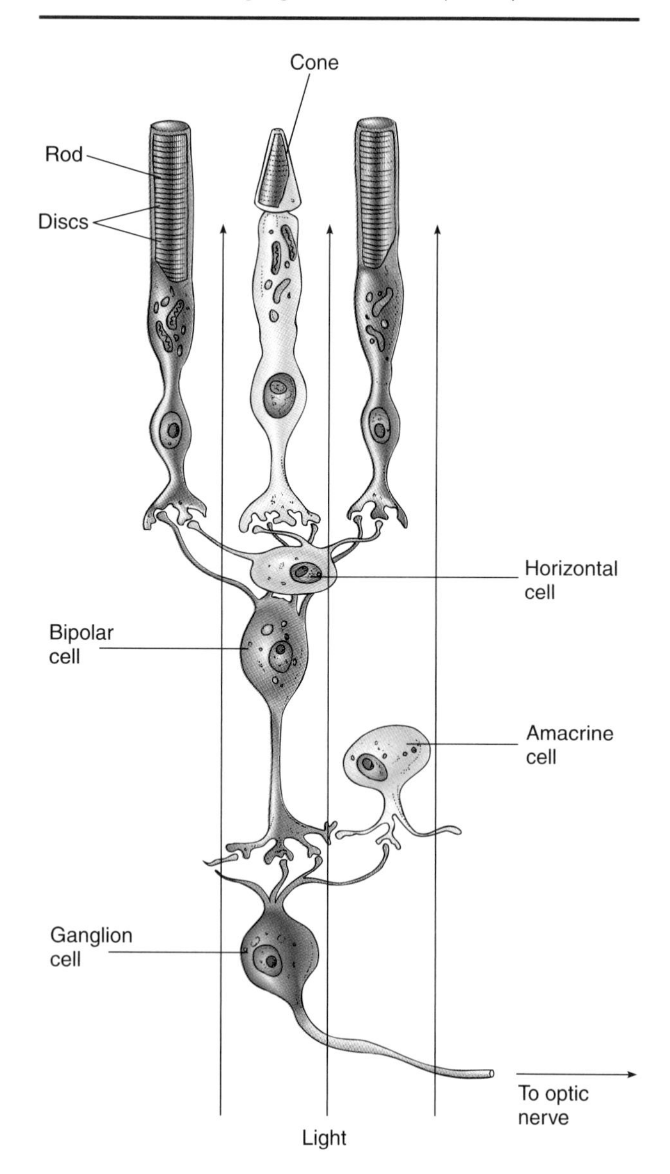

A chemical change in rhodopsin leads to the response of a rod to light

How is light converted into the neural signals that transmit information about environmental stimuli into pictures in the brain? Rod cells are so sensitive to light that they can respond to a single photon. Rhodopsin in the rod cells and some very closely related photopigments in the cone cells are responsible for the ability to see. Rhodopsin consists of *opsin,* a large protein that is chemically joined with *retinal,* an aldehyde of vitamin A (see Fig. 3-14). Two isomers of retinal exist: the *cis* form, which is folded, and the *trans* form, which is straight.

When it is dark, opsin binds to retinal in the *cis* form. Cyclic GMP (cyclic guanosine monophosphate, a molecule similar to cyclic AMP) opens nonspecific channels that permit passage of Na^+ and other cations into the rod cell (Fig. 41-18). This depolarizes the rod cell, which releases the neurotransmitter glutamate. The glutamate hyperpolarizes the membrane of the bipolar cell so it does not transmit messages. Note that the photoreceptor differs from other neurons in that the ion channels in its membrane are normally open; it is depolarized and continually releases neurotransmitter. The steady flow of ions into the cell when it is dark is called the *dark current.* Another unusual characteristic of photoreceptors, and also bipolar cells, is that they do not produce action potentials. Their release of neurotransmitter is graded, regulated by the extent of depolarization.

When light strikes rhodopsin, light is transduced. This process can be considered in three stages:

1. Light activates rhodopsin. Light transforms *cis*-retinal to *trans*-retinal. This causes rhodopsin to change shape and to break down into its components, opsin and retinal. This is the light-dependent process in vision.
2. Rhodopsin is part of a signal transduction pathway. When it changes shape, it binds with a G protein called **transducin.** Transducin activates an esterase that hydrolyzes cyclic GMP (cGMP) to GMP, reducing the concentration of cGMP.
3. When the concentration of cGMP in the rod decreases, its Na^+ channels begin to close. Fewer cations pass into the rod cell, and it becomes more negative (hyperpolarized). The rod cell then releases less glutamate. Thus light decreases the number of neural signals from the rod cells.

The release of glutamate normally hyperpolarizes the membrane of the bipolar cell, so a decrease in glutamate release results in its depolarization. The depolarized bipolar cell increases its release of a neurotransmitter, which typically stimulates the ganglion cell.

When you move from bright daylight to a dark room, or from a dimly lit room to bright light, you experience "blindness" for a few moments while your eyes adapt. In dim light, vision depends on the rods. However, exposure to bright light may completely activate rhodopsin in the rods. Dark adaptation occurs as enzymes restore rhodopsin to a form in which it can respond to light.

Color vision depends on three different types of cones

Many invertebrates and at least some animals in each vertebrate class have color vision. Color perception depends on cones. Most mammals have two types of cones, but humans and other primates have three types: blue, green, and red cones. Each contains a slightly different photopigment. Although the retinal portion of the pigment molecule is the same as in rhodopsin, the opsin protein differs slightly in each type of photoreceptor. Each type of cone responds to light within a considerable range of wavelengths but is named for the ability of its pigment to absorb a particular wavelength more strongly than other cones. For example, red light can be absorbed by all three types of cones, but those cones most sensitive to red act as red receptors. By comparing the relative responses of the three types of cones,

Discs

Rod

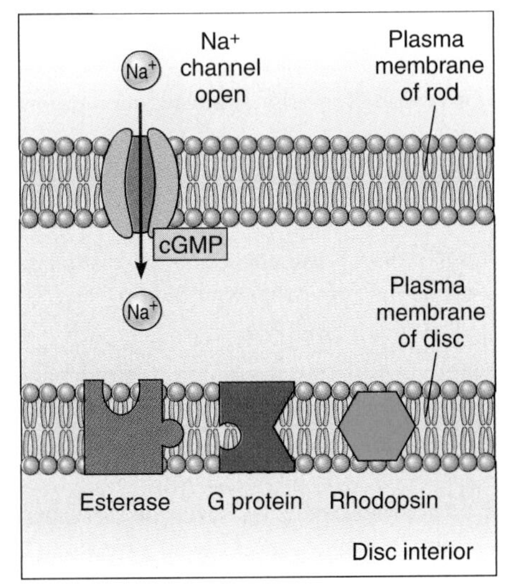

(a) In the dark, the rod cell is depolarized

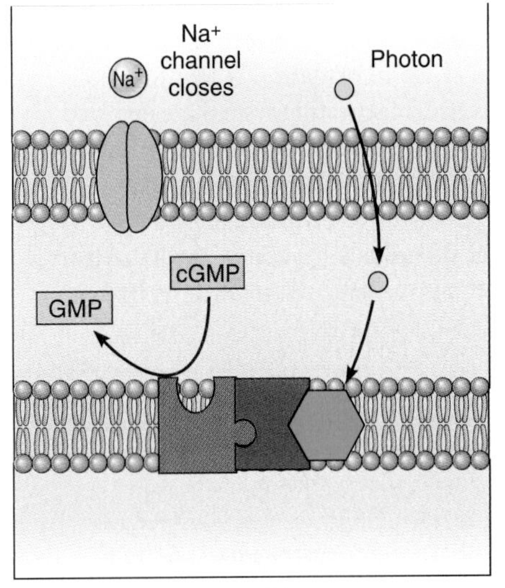

(b) In the light, the rod cell becomes hyperpolarized

FIGURE **41-18**

The biochemistry of vision.

(a) In the dark, sodium channels are open and the rod cell is depolarized. It releases the neurotransmitter glutamate. **(b)** Light makes rhodopsin change shape, activating a signal transduction pathway. The activation of a G protein leads to hyperpolarization and a decrease in signals from the rod cell. The discs are infoldings of the plasma membrane.

the brain can detect colors of intermediate wavelengths. Color blindness occurs when there is a deficiency of one or more of the three types of cones. This is usually an inherited X-linked condition (see Chapter 10, Fig. 10-16).

Integration of visual information begins in the retina

The size, intensity, and location of light stimuli determine initial processing in the retina. Rods and cones send signals directly to bipolar cells, and bipolar cells send signals directly to ganglion cells. However, as illustrated in Figure 41-17, photoreceptor cells can also transmit signals to horizontal cells, and bipolar cells can transmit signals to amacrine cells. Ganglion cells are inhibited by amacrine cells and can be inhibited indirectly by horizontal cells, by their action on bipolar cells. Thus the horizontal and amacrine cells integrate signals laterally.

Bipolar, horizontal, and amacrine cells combine signals from several photoreceptors. Each ganglion cell has a **receptive field,** a specific group of photoreceptors that light must strike for the ganglion cell to be stimulated. Ganglion cells are activated or inhibited, depending on which photoreceptor cells have been stimulated.

The pattern of neuron firing in the retina is very important. The signals sent to ganglion cells depend on spatial patterns and timing of the light striking the retina. Ganglion cells receive signals about specific types of visual stimuli such as color, brightness, and motion and transmit this information about the characteristics of a visual image. Ganglion cells produce action potentials, in contrast with rods and cones and most other neurons of the retina, which produce only graded potentials.

The retina begins the construction of visual images, but those images are interpreted in the brain. Axons of more than a million ganglion cells form the **optic nerves** that transmit information to the brain by complex, encoded signals. The optic nerves cross in the floor of the hypothalamus, forming an X-shaped structure, the **optic chiasm** (Fig. 41-19). Some of the axons of the optic nerves cross over and then extend to the opposite side of the brain. Axons of the optic nerves end in the **lateral geniculate nuclei** of the thalamus. From there, neurons send signals to the **primary visual cortex** in the occipital lobe of the cerebrum. The lateral geniculate nucleus controls which information is sent to the visual cortex. Neurons in the reticular activating system are involved in this integration. Information is transmitted from the primary visual cortex to other cortical areas for further processing.

Different types of ganglion cells receive information about different aspects of a visual stimulus. As a result, information from each point in the retina is transmitted in parallel by several different types of ganglion cells. These cells project to different kinds of neurons in the lateral geniculate nucleus and primary visual cortex.

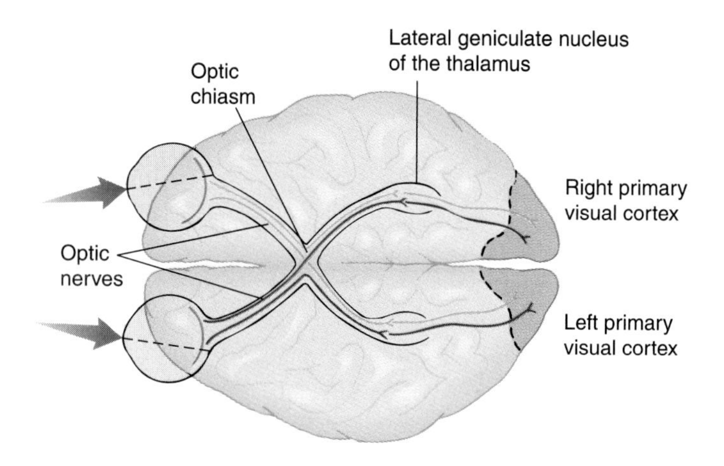

FIGURE **41-19** **Neural pathway for transmission of visual information.**

Axons of ganglion cells form the optic nerves. The optic nerves cross, forming the optic chiasm. Many optic nerve fibers end in the lateral geniculate nuclei of the thalamus. From there, signals are sent to the visual cortex.

Neurobiologists have not yet discovered all the mechanisms by which the brain makes sense out of the visual information it receives. We know that a large part of the association areas of the cerebrum are involved in integrating visual input. The neurons of the visual cortex are organized as a map of the external visual field. They are also organized into columns. Each column of neurons receives signals originating from light entering either the right or left eye. Columns are also organized by orientation of light. Neurons in a column receive signals originally triggered by light with the same orientation. However, these signals originate from different locations in the retina.

Review

- How does the insect's compound eye compare with the vertebrate eye?
- What happens when light strikes rhodopsin? (Include a description of the signal transduction pathway in your answer.)
- What is the sequence of neural signaling in the retina?
- What is meant by the statement "Vision happens mainly in the brain?"

BiologyNow™ Assess your understanding of **photoreceptors** by taking the pretest on your BiologyNow CD-ROM.

SUMMARY WITH KEY TERMS

1 Distinguish among five kinds of sensory receptors according to the types of energy they transduce.

- **Sensory receptors** are neuron endings or specialized receptor cells in close contact with neurons. **Sense organs** consist of sensory receptors and accessory cells.
- **Mechanoreceptors** transduce mechanical energy, permitting animals to feel, hear, and maintain balance.
- **Chemoreceptors** transduce certain kinds of chemical compounds, allowing taste and olfaction.
- **Thermoreceptors** transduce thermal energy. In endothermic animals, thermoreceptors provide cues about body temperature; some invertebrates use thermoreceptors to locate endothermic prey.
- **Electroreceptors,** used by predatory fishes to detect prey, respond to electrical stimuli.

- **Photoreceptors** transduce light energy and serve as the sensory receptors in eyespots and eyes.

2 Describe how a sensory receptor functions; define *energy transduction*, receptor potential, and *sensory adaptation* as part of your answer.

- Receptor cells absorb energy, transduce that energy—**energy transduction**—into electrical energy, and produce **receptor potentials,** which are depolarizations or hyperpolarizations of the membrane. Receptor potentials are **graded responses.**
- The ability to experience sensation depends on sensory receptors transmitting coded signals and on the brain interpreting those signals.
- **Sensory adaptation,** the decrease in frequency of action potentials in a sensory neuron even when the stimlulus is maintained, decreases response to that stimulus.

3 Describe the following mechanoreceptors: tactile receptors, nociceptors, proprioceptors, statocysts, hair cells, and lateral line organs.

- Mechanoreceptors respond to touch, pressure, gravity, stretch, or movement. They are activated when they change shape as a result of being mechanically pushed or pulled.
- **Tactile receptors** in the skin respond to mechanical displacement of hairs or to displacement of the receptor cells themselves. For example, the **Pacinian corpuscle** is a tactile receptor that responds to touch and pressure.
- **Nociceptors** are pain receptors; they are free nerve endings of certain sensory neurons that respond to strong tactile stimuli, temperature extremes, and certain chemicals.
- **Proprioceptors** enable the animal to perceive orientation of the body and the positions of its parts. **Muscle spindles, Golgi tendon organs,** and **joint receptors** are proprioceptors that continually respond to tension and movement.
- **Statocysts** are gravity receptors found in many invertebrates.
- Vertebrate **hair cells** detect movement; they are found in the lateral line of fishes, the vestibular apparatus, semicircular canals, and cochlea. Each hair cell has a single **kinocilium,** which is a true cilium, and **stereocilia,** which are microvilli that contain actin filaments.
- **Lateral line organs** supplement vision in fish and some amphibians by informing the animal of moving objects or objects in its path.

4 Compare the function of the saccule and utricle with that of the semicircular canals in maintaining equilibrium.

- The vertebrate **inner ear** consists of a **labyrinth** of fluid-filled chambers and canals that help maintain equilibrium. The **vestibular apparatus** in the upper part of the labyrinth consists of the **saccule, utricle,** and **semicircular canals.**
- The saccule and utricle contain **otoliths** that change position when the head is tilted or when the body is moving in a straight line. The otoliths stimulate hair cells that send signals to the brain, enabling the animal to perceive the direction of gravity.
- The semicircular canals inform the brain about turning movements. Clumps of hair cells, called **cristae,** are located within each bulblike enlargement, an **ampulla. Cristae** are stimulated by movements of the **endolymph,** a fluid that fills each canal.

5 Trace the path taken by sound waves through the structures of the ear, and explain how the organ of Corti functions as an auditory receptor.

- In birds and mammals, the **organ of Corti** within the **cochlea** contains auditory receptors.
- Sound waves pass through the **external auditory canal** and cause the **tympanic membrane** (eardrum) to vibrate. The ear bones—**malleus, incus,** and **stapes**—transmit and amplify the vibrations through the middle ear.
- Vibrations pass through the **oval window** to fluid within the vestibular duct. Pressure waves press on the membranes that separate the three ducts of the cochlea.
- The bulging of the **round window** serves as an escape valve for the pressure. The pressure waves cause movements of the **basilar membrane.** These movements stimulate the hair cells of the organ of Corti by rubbing them against the overlying **tectorial membrane.**
- Neural impulses are initiated in the dendrites of neurons that lie at the base of each hair cell, and are transmitted by the **cochlear nerve** to the brain.

6 Compare the structure and function of the receptors of taste and smell.

- The receptors for taste and smell are chemoreceptors. Taste receptor cells are specialized epithelial cells in **taste buds.** The **olfactory epithelium** contains specialized olfactory cells with axons that extend to the brain as fibers of the **olfactory nerves.**
- When a molecule binds with a receptor on a taste receptor cell or an olfactory receptor cell, a signal transduction process involving a G protein is initiated.

7 Contrast simple eyes, compound eyes, and vertebrate eyes.

- **Eyespots,** or **ocelli,** found in cnidarians and flatworms, detect light but do not form images.
- The **compound eye** of insects and crustaceans consists of visual units called **ommatidia,** which collectively produce a **mosaic** image. Each ommatidium has a transparent **lens** and a **crystalline cone** that focus light onto receptor cells called **retinular cells.**
- The vertebrate eye can be compared to a camera.

8 Describe the functions of each structure of the vertebrate eye.

- In the human eye, light enters through the **cornea,** is focused by the lens, and produces an image on the **retina.** The **iris** regulates the amount of light that can enter.

9 Describe the events that take place in human vision; compare the two types of photoreceptors and describe the signal transduction pathway as part of your explanation.

- The retina contains the photoreceptor cells: **rods,** which function in dim light and form images in black and white; and **cones,** which function in bright light and permit color vision.
- The retina also contains **bipolar cells** that send signals to **ganglion cells.** Two types of lateral interneurons integrate information: **Horizontal cells** receive signals from the rods and cones and send signals to bipolar cells, and **amacrine cells** receive signals from bipolar cells and send signals back to bipolar cells and to ganglion cells.
- When the environment is dark, ion channels in the plasma membranes of rod cells are open and the cells are depolarized; they release glutamate, which hyperpolarizes the membranes of bipolar cells so they do not send signals.
- When light strikes the photopigment **rhodopsin** in the rod cells, its retinal portion changes shape and initiates a signal transduc-

SUMMARY WITH KEY TERMS *(continued)*

tion process that involves **transducin.** This G protein activates an esterase that hydrolyzes cGMP, reducing its concentration. As a result, ion channels close and the membrane becomes hyperpolarized. The rod cells release less glutamate, and fewer signals are transmitted. As a result, bipolar cells become depolarized, and release neurotransmitter that stimulates ganglion cells.

- Axons of the ganglion cells make up the **optic nerves.** The optic nerves transmit information to the **lateral geniculate nuclei** in the thalamus. From there neurons project to the **primary visual cortex** and then to integration centers in the cerebral cortex.

POST-TEST

1. Which of the following is *not* true of a sensory receptor? (a) detects a stimulus in the environment (b) converts energy of the stimulus into electrical energy (c) produces a receptor potential (d) produces a graded response (e) interprets sensory stimuli
2. A sensory receptor absorbs energy from some stimulus. The next step is (a) release of neurotransmitter (b) transmission of an action potential (c) energy transduction (d) transmission of a receptor potential (e) sensory adaptation
3. Interoceptors (a) help maintain homeostasis (b) are located in the skin (c) are photoreceptors (d) adapt rapidly (e) convert absorbed energy directly into action potentials
4. Which of the following are *not* correctly matched? (a) mechanoreceptors—touch, pressure (b) electroreceptors—voltage (c) photoreceptors—light (d) chemoreceptors—gravity (e) nociceptors—extreme temperature
5. A Pacinian corpuscle (a) is a mechanoreceptor (b) responds to heat (c) responds to pressure (d) answers a, b, and c are correct (e) answers a and c only
6. Which of the following is *not* located in the vertebrate inner ear? (a) vestibular apparatus (b) cochlea (c) malleus (d) organ of Corti (e) utricle
7. The auditory receptors are located in the (a) utricles (b) organs of Corti (c) vestibular apparati (d) saccule (e) semicircular canals
8. Which of the following is/are associated with informing the brain about turning movements? (a) semicircular canals (b) saccule (c) lymph (d) otoliths (e) utricle
9. Which of the following receptors do *not* have hair cells? (a) lateral line organ (b) saccule (c) utricle (d) semicircular canals (e) rods
10. In the process of hearing, the basilar membrane vibrates. Which event occurs next? (a) tympanic membrane vibrates (b) bones in middle ear amplify and conduct vibrations (c) cochlear nerve transmits impulses to organ of Corti (d) hair cells in organ of Corti are stimulated (e) vibrations conducted to chemoreceptors
11. In olfaction, (a) pheromones stimulate the taste buds (b) the number of odorous molecules determines the magnitude of the receptor potential (c) chemical signals are converted directly into electrical signals (d) a G protein is activated and leads to the closing of gated channels (e) the concentration of cyclic AMP increases, leading to hyperpolarization of the receptor
12. The vomeronasal organ (a) detects pheromones (b) is located at the back of the tongue (c) sends signals directly to the thalamus (d) releases pheromones (e) detects signals from predators
13. Mechanical stimulation of stereocilia (a) produces action potentials (b) inhibits glutamate secretion (c) causes voltage changes (d) degrades opiates (e) causes rhodopsin release
14. In the human visual pathway, after light passes through the cornea, it (a) stimulates ganglion cells (b) passes through the lens (c) sends signals through the optic nerve (d) depolarizes horizontal cells (e) hyperpolarizes rod cells
15. Cones (a) are most concentrated in the ganglion area (b) are more numerous than rods (c) are responsible for vision in bright light (d) are found in all animals with photoreceptors (e) are a type of bipolar cell
16. Rhodopsin (a) is concentrated in bipolar cells (b) binds with a G protein (c) changes shape when the membrane of the cone is depolarized (d) sends signals to the lateral geniculate nuclei (e) activates substance P

CRITICAL THINKING

1. If all neurons transmit the same type of message, how do you know the difference between sound and light? How are you able to distinguish between an intense pain and a mild one? How are these discriminations adaptive?
2. Connoisseurs can recognize many varieties of cheese or wine by "tasting." How can this be, when there are only a few types of taste receptors? How might it be adaptive for organisms to have only four or five basic tastes?
3. Design a synaptic mechanism by which the same neurotransmitter has an excitatory effect on one bipolar cell and an inhibitory effect on another.

- Visit our Web site at **http://biology.brookscole.com/solomon7** for links to chapter-related resources on the World Wide Web. Additional online materials relating to this chapter can also be found on our Web site.

BIOLOGY NOW RESOURCES

Active Figures

41-6: The human ear

41-15: Vision disorders

Preparing for an exam? Take a diagnostic test on your BiologyNow CD-ROM.

Post-Test Answers

1. e	2. c	3. a	4. d
5. e	6 c	7. b	8. a
9. e	10. d	11. b	12. a
13. c	14. b	15. c	16. b

Internal Transport

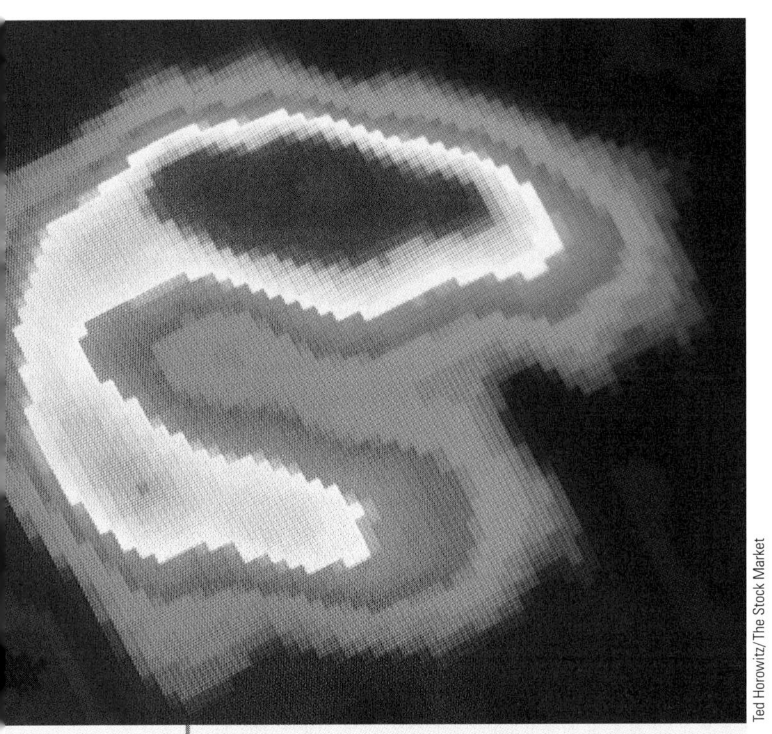

Ted Horowitz/The Stock Market

PET scan visualizing the left ventricle of the human heart. The scan measures metabolic activity based on glucose levels in the cardiac muscle tissue, providing information about the condition of the heart. (The color progression from least to most active is blue, green, yellow, red.)

CHAPTER OUTLINE

Most cells need a continuous supply of nutrients and oxygen and the removal of waste products. In very small animals, these metabolic needs are met by simple **diffusion,** the net movement of particles from a region of higher concentration to a region of lower concentration, resulting from random motion. A molecule can diffuse 1 micrometer (μm) in less than 1 millisecond (msec), so diffusion is adequate over microscopic distances. In invertebrates that are only a few cells thick, diffusion is an effective mechanism for distributing materials to and from their cells. No specialized circulatory structures are present in sponges, cnidarians (such as jellyfish), ctenophores (comb jellies), flatworms, or nematodes (roundworms). As in all animals, the fluid between the cells, called **interstitial fluid,** or tissue fluid, bathes the cells and provides a medium for diffusion of oxygen, nutrients, and wastes.

How much time diffusion requires increases with the square of the distance over which diffusion occurs. A cell that is 10 μm away from its oxygen (or nutrient) supply can receive oxygen by diffusion in about 50 msec, but a cell that is 1000 μm (1 mm) away from its oxygen supply would have to wait several minutes and couldn't survive if it had to depend on diffusion alone. In animals that are many cells thick, specialized **circulatory systems** transport oxygen, nutrients, hormones, and other materials to the interstitial fluid surrounding all the cells and remove metabolic wastes. A circulatory system reduces the diffusion distance that needed materials must travel. In most animals, a circulatory system interacts with every other organ system in the body.

A circulatory system typically has the following: (1) **blood,** a connective tissue consisting of cells and cell fragments dispersed in fluid known as *plasma;* (2) a pumping organ, generally a **heart;** and (3) a system of **blood vessels** or spaces through which blood circulates.

The human circulatory system, known as the **cardiovascular system,** is the focus of extensive research because cardiovascular disease is the leading cause of death in the United States and most other industrial societies. The positron emission tomography (PET) scan shown here visualizes the left ventricle of the human heart. A major risk factor for cardiovascular disease is elevated levels of cholesterol and low-density lipoprotein (LDL) in the blood. In contrast, high-density lipoprotein (HDL) seems to play a protective role, removing excess cholesterol from the blood and tissues. Cells in the liver and certain other organs bind with the HDL, remove the cholesterol, and use it in synthesizing needed compounds.

Researchers have identified a gene in mice that codes for an HDL receptor. When investigators knock out this gene, cholesterol levels more than double. As new receptors and mechanisms involved in lipid transport and metabolism are discovered, they serve as targets for new drugs and other treatments for cardiovascular disease. ■

TYPES OF CIRCULATORY SYSTEMS

Learning Objective

1 Compare and contrast internal transport in animals with no circulatory system, those with an open circulatory system, and those with a closed circulatory system.

As indicated, many small, aquatic invertebrates have no circulatory system. In cnidarians, the central gastrovascular cavity serves as a circulatory organ as well as a digestive organ (Fig. 42-1a).

FIGURE **42-1** **Invertebrates with no circulatory system.**

(a) In *Hydra* and other cnidarians, oxygen and nutrients circulate through the gastrovascular cavity and come in contact with the inner layer of body cells. These materials diffuse the short distance to the outer layer of cells. **(b)** In planarian flatworms, the branched intestine circulates nutrients and oxygen within close proximity to the body cells.

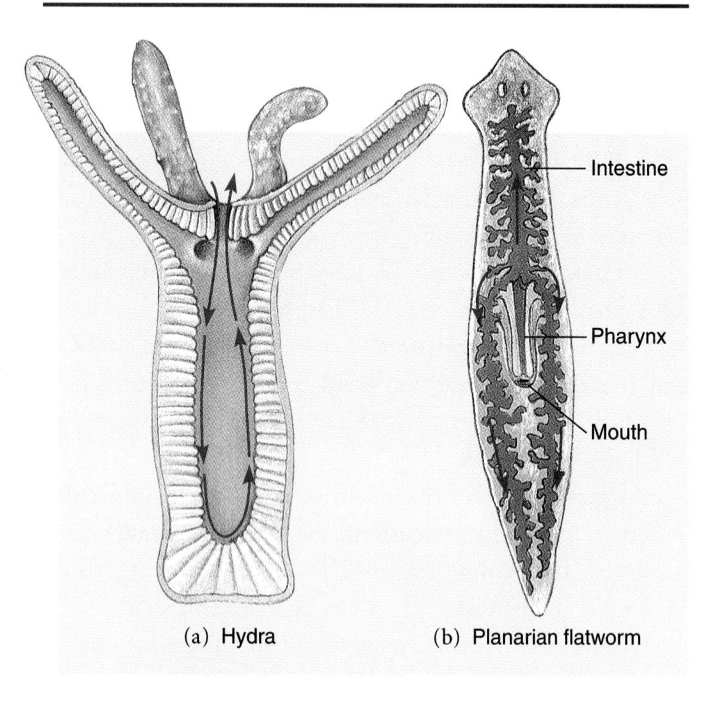

As the animal stretches and contracts, body movements stir up the contents of the gastrovascular cavity and help distribute nutrients.

The flattened body of the flatworm permits effective gas exchange by diffusion (Fig. 42-1b). Its branched intestine brings nutrients close to all the cells. As in cnidarians, circulation is aided by contractions of the body wall muscles, which agitate the intestinal fluid and the tissue fluid. The branching excretory system of planarians provides for internal transport of metabolic wastes that are then expelled from the body.

Fluid in the body cavity of nematodes and other pseudocoelomate animals helps circulate materials. Nutrients, oxygen, and wastes dissolve in this fluid and diffuse through it to and from the cells. Body movements of the animal move the fluid, distributing these materials.

Many invertebrates have an open circulatory system

Arthropods and most mollusks have an **open circulatory system,** in which the heart pumps blood into vessels that have open ends. Their blood and interstitial fluid are indistinguishable and are collectively referred to as **hemolymph.** Hemolymph spills out of the open ends of the blood vessels, filling large spaces, called *sinuses.* The sinuses make up the **hemocoel** (blood cavity), which is not part of the coelom. (In arthropods and mollusks, the coelom is reduced.) The hemolymph bathes the cells of the body directly. Blood re-enters the circulatory system through openings in the heart (in arthropods) or through open-ended vessels that lead to the gills (in mollusks).

In the open circulatory system of most mollusks, the heart typically has three chambers (Fig. 42-2a). The two atria receive hemolymph from the gills. Then the single ventricle pumps oxygen-rich hemolymph into blood vessels that conduct it into the large sinuses of the hemocoel. After bathing the body cells, the hemolymph passes into vessels that lead to the gills, where it is recharged with oxygen. The hemolymph then returns to the heart.

Some mollusks, as well as arthropods, have a hemolymph pigment, **hemocyanin,** containing copper that binds with oxygen. When oxygenated, hemocyanin is blue and imparts a bluish color to the hemolymph of these animals (the original blue bloods!).

In arthropods, a tubular heart pumps hemolymph into blood vessels (arteries) that deliver it to the sinuses of the hemocoel (Fig. 42-2b). Hemolymph then circulates through the hemocoel, eventually returning to the pericardial cavity surrounding the heart. Hemolymph enters the heart through tiny openings (*ostia*) equipped with valves to prevent backflow. The rate of hemolymph circulation increases when the insect moves. Thus, when an animal is active and most needs nutrients for fuel, its own movement ensures effective circulation.

An open circulatory system cannot provide enough oxygen to maintain the active lifestyle of insects. Indeed, insect hemolymph mainly distributes nutrients and hormones. Oxygen diffuses directly to the cells through a system of air tubes (tracheae) that make up the respiratory system (see Figure 44-2). In cray-

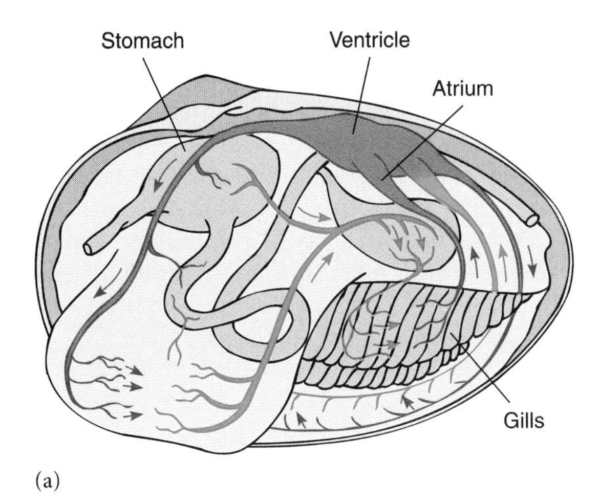

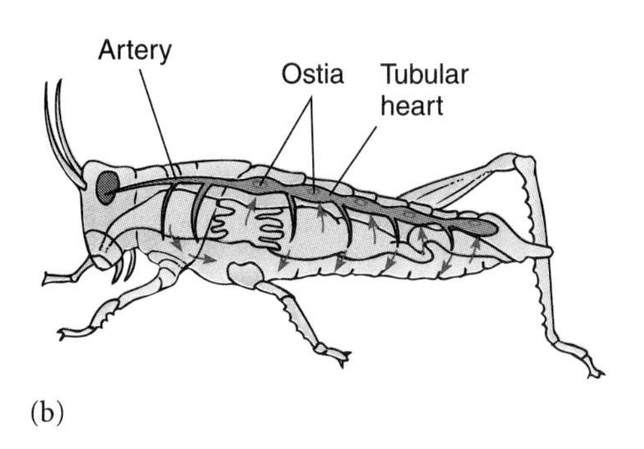

FIGURE **42-2** **Open circulatory systems.**

In mollusks and arthropods, a heart pumps the blood into arteries that end in sinuses of the hemocoel. Hemolymph circulates through the hemocoel. **(a)** In most mollusks, hemolymph enters vessels that conduct it to the gills, where it is recharged with oxygen. Blood then returns to the heart. **(b)** In arthropods, a tubular heart pumps hemolymph into arteries that deliver it to the sinuses of the hemocoel. After circulating, hemolymph re-enters the heart through ostia in the heart wall.

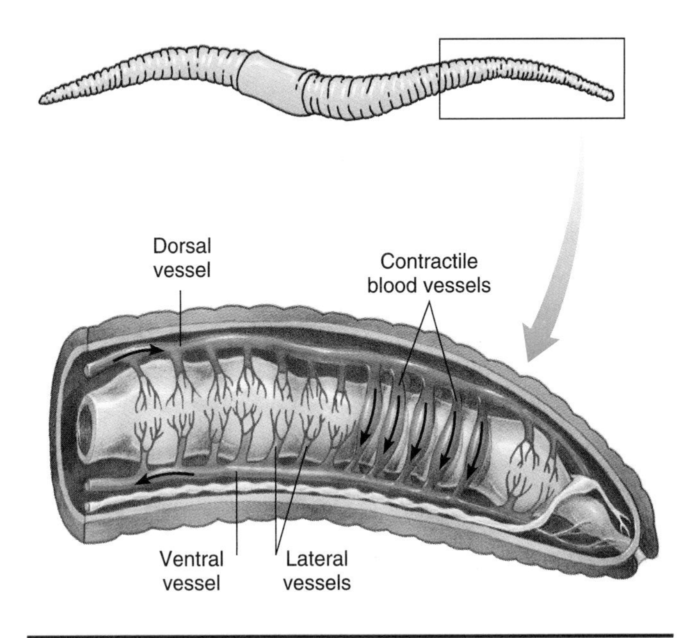

FIGURE **42-3** **Closed circulatory system of the earthworm.**

Blood circulates through a continuous system of blood vessels. Five pairs of contractile blood vessels deliver blood from the dorsal vessel to the ventral vessel.

fish and other crustaceans, gas exchange takes place as hemolymph circulates through the gills.

Some invertebrates have a closed circulatory system

Annelids, some mollusks (cephalopods), and echinoderms have a **closed circulatory system.** In them, blood flows through a continuous circuit of blood vessels. The walls of the smallest blood vessels, the **capillaries,** are thin enough to permit diffusion of gases, nutrients, and wastes between blood in the vessels and the interstitial fluid that bathes the cells.

A rudimentary closed circulatory system characterizes proboscis worms (phylum Nemertea). This system consists of a complete network of blood vessels but no heart. Blood flow depends on movements of the animal and on contractions in the walls of large blood vessels.

Earthworms and other annelids have a complex, closed circulatory system (Fig. 42-3). Two main blood vessels extend lengthwise in the body. The ventral vessel conducts blood posteriorly, and the dorsal vessel conducts blood anteriorly. Dorsal and ventral vessels are connected by lateral vessels in every segment. Branches of the lateral vessels deliver blood to the surface, where it is oxygenated. In the anterior part of the worm, five pairs of contractile blood vessels (sometimes called "hearts") connect dorsal and ventral vessels. Contractions of these paired vessels and of the dorsal vessel, as well as contraction of the body wall muscles, circulate the blood. Earthworms have **hemoglobin,** the same red pigment that transports oxygen in vertebrate blood. However, their hemoglobin is not contained within red blood cells but is dissolved in the blood plasma.

Although other mollusks have an open circulatory system, the fast-moving cephalopods, such as the squid and octopus, require a more efficient means of internal transport. They have a closed system made even more effective by accessory "hearts" at the base of the gills, which speed passage of blood through the gills.

All vertebrates have a closed circulatory system

The circulatory system is basically similar in all vertebrates, from fishes, frogs, and reptiles to birds and mammals. The system consists of heart, blood vessels, blood, lymph, lymph vessels, and associated organs such as the thymus, spleen, and liver. All vertebrates have a ventral, muscular heart that pumps blood into a closed system of blood vessels. Capillaries, the tiniest blood vessels, have very thin walls that permit exchange of materials between blood and interstitial fluid.

The vertebrate circulatory system performs several functions:

1. Transports nutrients from the digestive system and from storage depots to each cell
2. Transports oxygen from respiratory structures (gills or lungs) to the cells
3. Transports metabolic wastes from each cell to organs that excrete them
4. Transports hormones from endocrine glands to target tissues
5. Helps maintain fluid balance
6. Defends the body against invading microorganisms
7. Helps distribute metabolic heat within the body, which helps maintain a constant body temperature in endothermic animals
8. Helps maintain appropriate pH

Review

- How are nutrients and oxygen transported to the body cells in a hydra, flatworm, earthworm, insect, and frog?
- How does an open circulatory system differ from a closed circulatory system?
- What are five functions of the vertebrate circulatory system?

BiologyNow™ Assess your understanding of **types of circulatory systems** by taking the pretest on your BiologyNow CD-ROM.

VERTEBRATE BLOOD

Learning Objectives

2 Compare the structure and function of red blood cells, white blood cells, and platelets.

3 Summarize the sequence of events involved in blood clotting.

In vertebrates, blood consists of a pale yellowish fluid called **plasma,** in which red blood cells, white blood cells, and platelets are suspended (Fig. 42-4; Table 42-1). In humans the total circulating blood volume is about 8% of the body weight—5.6 L (6 qt) in a 70-kg (154-lb) person. About 55% of the blood volume is plasma. The remaining 45% is made up of blood cells and platelets. Because cells and platelets are heavier than plasma, they can be separated from it by centrifugation. Plasma does not separate from blood cells in the body, because the blood is constantly mixed as it circulates in the blood vessels.

Plasma is the fluid component of blood

Plasma consists of water (about 92%), proteins (about 7%), salts, and a variety of materials being transported, such as dissolved gases, nutrients, wastes, and hormones. Plasma is in dynamic equilibrium with the interstitial fluid bathing the cells and with the intracellular fluid. As blood passes through the capillaries, substances continuously move into and out of the plasma. Changes in its composition signal one or more organs of the body to restore homeostasis.

Plasma contains several kinds of **plasma proteins,** each with specific properties and functions: **fibrinogen;** alpha, beta, and gamma **globulins;** and **albumin.** Fibrinogen is one of the proteins involved in the clotting process. When the proteins involved in blood clotting have been removed from the plasma, the remaining liquid is called **serum.** Alpha globulins include certain hormones and proteins that transport hormones; prothrombin, a protein involved in blood clotting; and high-density lipoproteins (HDL), which transport fats and cholesterol. Beta globulins include other lipoproteins that transport fats and cholesterol, as well as proteins that transport certain vitamins and minerals. The **gamma globulin** fraction contains many types of antibodies that provide immunity to diseases such as measles and infectious hepatitis. Purified human gamma globulin is

TABLE **42-1** Cell Components of Blood

	Normal Range	Function	Pathology
Red Blood Cells (RBCs)	Male: 4.2–5.4 million/μL Female: 3.6–5.0 million/μL	Oxygen transport; carbon dioxide transport	Too few: anemia Too many: polycythemia
Platelets	150,000–400,000/μL	Essential for clotting	Clotting malfunctions; bleeding; easy bruising
White Blood Cells (WBCs)	5000–10,000/μL		
Neutrophils	About 60% of WBCs	Phagocytosis	Too many: may be due to bacterial infection, inflammation, myelogenous leukemia
Eosinophils	1%–3% of WBCs	Play some role in allergic response	Too many: may result from allergic reaction, parasitic infestation
Basophils	1% of WBCs	May play role in prevention of inappropriate clotting	
Lymphocytes	25%–35% of WBCs	Produce antibodies; destroy foreign cells	Atypical lymphocytes present in infectious mononucleosis; too many may be due to lymphocytic leukemia, certain viral infections
Monocytes	6% of WBCs	Differentiate to form macrophages	May increase in monocytic leukemia and fungal infections

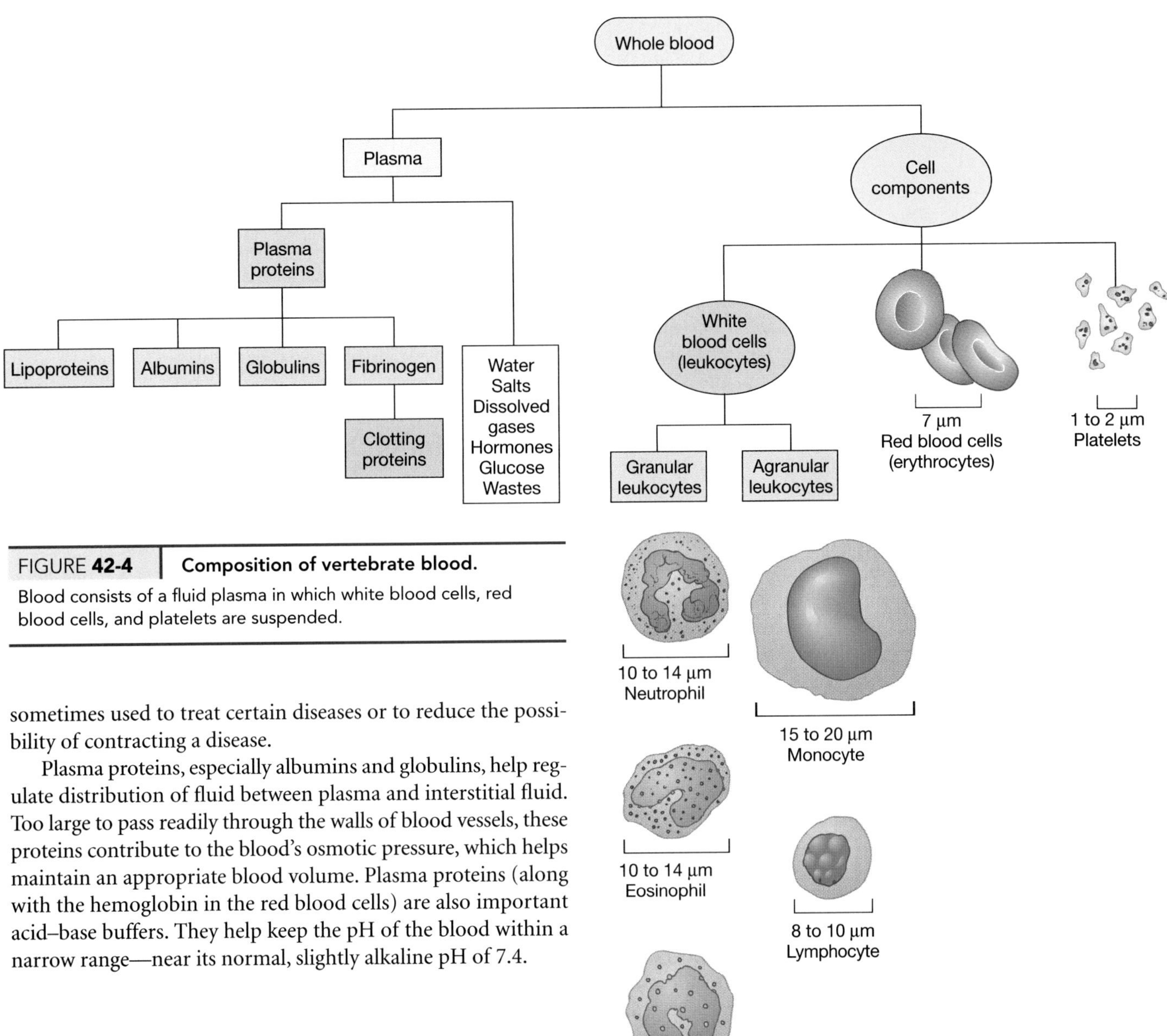

FIGURE **42-4** **Composition of vertebrate blood.**
Blood consists of a fluid plasma in which white blood cells, red blood cells, and platelets are suspended.

sometimes used to treat certain diseases or to reduce the possibility of contracting a disease.

Plasma proteins, especially albumins and globulins, help regulate distribution of fluid between plasma and interstitial fluid. Too large to pass readily through the walls of blood vessels, these proteins contribute to the blood's osmotic pressure, which helps maintain an appropriate blood volume. Plasma proteins (along with the hemoglobin in the red blood cells) are also important acid–base buffers. They help keep the pH of the blood within a narrow range—near its normal, slightly alkaline pH of 7.4.

Red blood cells transport oxygen

Erythrocytes, also called **red blood cells (RBCs),** are highly specialized for transporting oxygen. In most vertebrates except mammals, circulating RBCs have nuclei. For example, birds have large, oval, nucleated RBCs. In mammals, the nucleus is ejected from the RBC as the cell develops. Each mammalian RBC is a flexible, biconcave disc, 7 to 8 μm in diameter and 1 to 2 μm thick. An internal elastic framework maintains the disc shape and permits the cell to bend and twist as it passes through blood vessels even smaller than its own diameter. Its biconcave shape provides a high ratio of surface area to volume, allowing efficient diffusion of oxygen and carbon dioxide into and out of the cell. In an adult human, about 30 trillion RBCs circulate in the blood, approximately 5 million per μL.

Erythrocytes are produced within the red bone marrow of certain bones: vertebrae, ribs, breastbone, skull bones, and long bones. As an RBC develops, it produces great quantities of hemoglobin, the oxygen-transporting pigment that gives vertebrate blood its red color. (Oxygen transport is discussed in Chapter 44.) The life span of a human RBC is about 120 days. As blood circulates through the liver and spleen, phagocytic cells remove worn-out RBCs from the circulation. These RBCs are then disassembled, and some of their components are recycled. In the human body, more than 2.4 million RBCs are destroyed every second, so an equal number must be produced in the bone marrow to replace them. Red blood cell production is regulated by the hormone **erythropoietin,** which the kidneys release in response to a decrease in oxygen.

Anemia is a deficiency in hemoglobin (often accompanied by a decrease in the number of RBCs). When hemoglobin is insufficient, the amount of oxygen transported is inadequate to supply the body's needs. An anemic person may complain of feeling weak and may become easily fatigued. Three general causes of anemia are (1) loss of blood from hemorrhage or internal bleeding; (2) decreased production of hemoglobin or red blood cells as in iron-deficiency anemia or pernicious anemia (which can be caused by vitamin B_{12} deficiency); and (3) increased rate of RBC destruction—the **hemolytic anemias,** such as sickle cell anemia (see Chapter 15).

White blood cells defend the body against disease organisms

The **leukocytes,** or **white blood cells (WBCs),** are specialized to defend the body against harmful bacteria and other microorganisms. Leukocytes are amoeba-like cells capable of independent movement. Some types routinely slip through the walls of blood vessels and enter the tissues. Human blood contains five kinds of leukocytes, classified as either granular or agranular (see Fig. 42-4). Both types are manufactured in the red bone marrow.

The **granular leukocytes** are characterized by large, lobed nuclei and distinctive granules in their cytoplasm. The three varieties of granular leukocytes are the neutrophils, eosinophils, and basophils. **Neutrophils,** the principal phagocytic cells in the blood, are especially adept at seeking out and ingesting bacteria. They also phagocytize dead cells, a cleanup task especially demanding after injury or infection. Most granules in neutrophils contain enzymes that digest ingested material.

Eosinophils have large granules that stain bright red with eosin, an acidic dye. The lysosomes of these WBCs contain enzymes such as oxidases and peroxidases, suggesting that they function in detoxifying foreign proteins and other substances. Eosinophils increase in number during allergic reactions and during parasitic (for example, tapeworm) infestations. **Basophils** exhibit deep blue granules when stained with basic dyes. Like eosinophils, these cells play a role in allergic reactions. Basophils do not contain lysosomes. Granules in their cytoplasm contain **histamine,** a substance that dilates blood vessels and makes capillaries more permeable. Basophils release histamine in injured tissues and in allergic responses. Other basophil granules contain **heparin,** an anticoagulant that helps prevent blood from clotting inappropriately within the blood vessels.

Agranular leukocytes lack large, distinctive granules, and their nuclei are rounded or kidney-shaped. Two types of agranular leukocytes are lymphocytes and monocytes. Some **lymphocytes** are specialized to produce antibodies, whereas others directly attack foreign invaders such as bacteria or viruses. Just how they manage these feats is discussed in Chapter 43.

Monocytes are the largest WBCs, reaching 20 μm in diameter. After circulating in the blood for about 24 hours, a monocyte leaves the circulation and completes its development in the tissues. The monocyte greatly enlarges and becomes a **macrophage,** a giant scavenger cell. All macrophages found in the tissues develop this way. Macrophages voraciously engulf bacteria, dead cells, and debris.

Human blood normally has about 7000 WBCs per μL of blood (only 1 for every 700 RBCs). During bacterial infections the number may rise sharply, so a WBC count is useful in diagnosis. The proportion of each kind of WBC is determined by a differential WBC count. The normal distribution of leukocytes is indicated in Table 42-1.

Leukemia is a form of cancer in which any one of the various kinds of WBCs multiplies rapidly within the bone marrow. Many of these cells do not mature, and their large numbers crowd out developing RBCs and platelets, leading to anemia and impaired clotting. A common cause of death from leukemia is internal hemorrhaging, especially in the brain. Another frequent cause of death is infection; although the WBC count may rise dramatically, the cells are immature and abnormal and cannot defend the body against disease organisms.

Platelets function in blood clotting

In most vertebrates other than mammals, the blood contains small, oval cells called **thrombocytes,** which have nuclei. In mammals, thrombocytes are tiny spherical or disc-shaped bits of cytoplasm that lack nuclei. They are usually called **platelets.** About 300,000 platelets per μL are present in human blood. Platelets are pinched off from very large cells in the bone marrow. Thus, a platelet is not a whole cell but a fragment of cytoplasm enclosed by a membrane.

When a blood vessel is cut, it constricts, reducing blood loss. Platelets stick to the rough, cut edges of the vessel, physically patching the break. As platelets begin to gather, they release substances that attract other platelets. The platelets become sticky and adhere to collagen fibers in the blood vessel wall. Within about 5 minutes after injury, they form a platelet plug, or temporary clot.

At the same time the temporary clot forms, a stronger, more permanent clot begins to develop. More than 30 different chemical substances interact in this very complex process. The series of reactions that leads to clotting is triggered when one of the clotting factors in the blood is activated by contact with the injured tissue. In **hemophilia,** one clotting factor is absent, as a result of an inherited genetic mutation (see Chapter 15).

Prothrombin, a plasma protein manufactured in the liver, requires vitamin K for its production. In the presence of clotting factors, calcium ions, and compounds released from platelets, prothrombin is converted to **thrombin.** Then thrombin catalyzes the conversion of the soluble plasma protein **fibrinogen** to an insoluble protein, **fibrin.** Once formed, fibrin polymerizes, producing long threads that stick to the damaged surface of the blood vessel and form the webbing of the clot. These threads trap blood cells and platelets, which help strengthen the clot. The clotting process is summarized in Figure 42-5.

Review

- What are the functions of the main groups of plasma proteins?
- What is the function of red blood cells? Of neutrophils?
- What are the major steps in blood clotting?

BiologyNow™ Assess your understanding of **vertebrate blood** by taking the pretest on your BiologyNow CD-ROM.

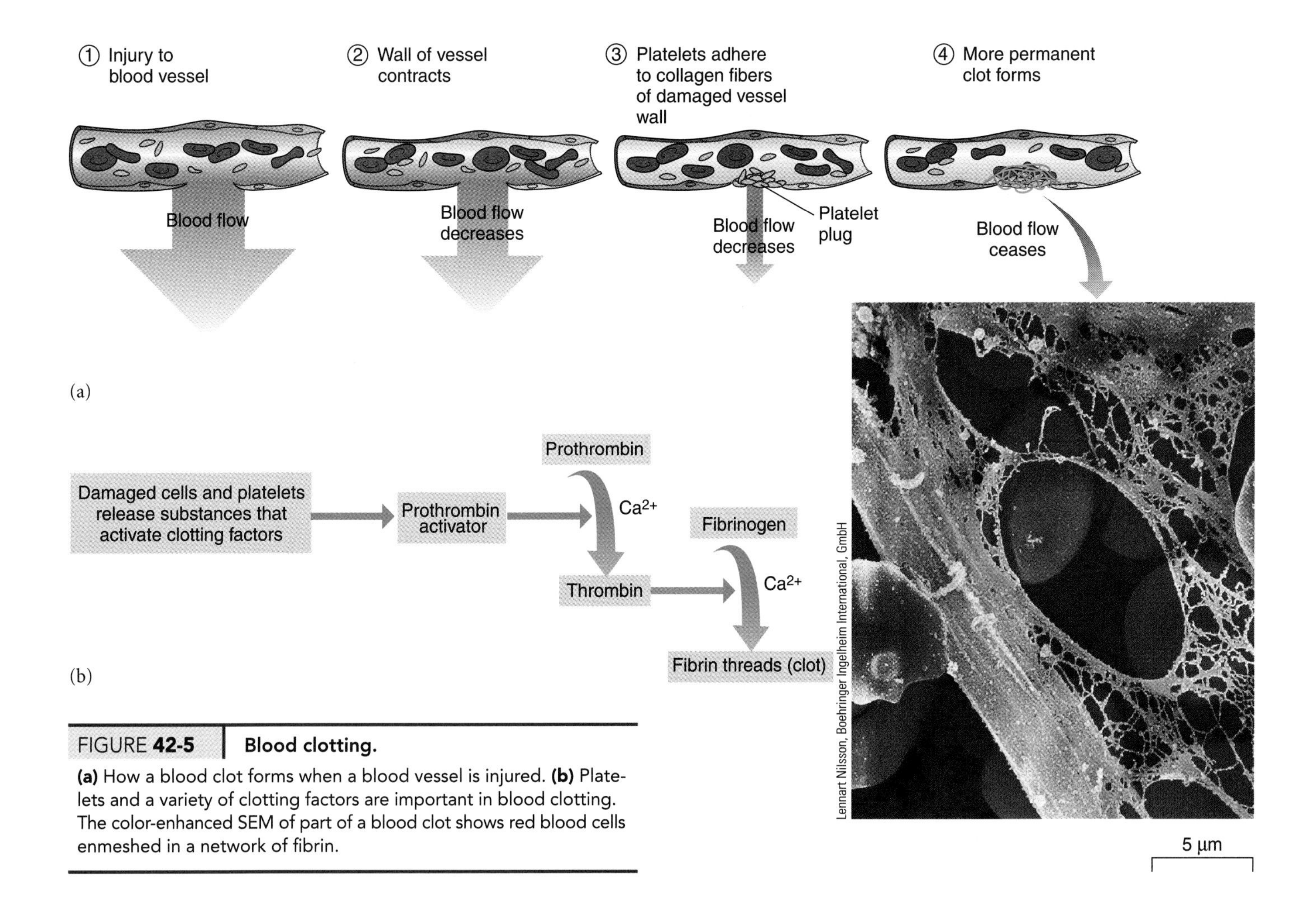

FIGURE **42-5** **Blood clotting.**

(a) How a blood clot forms when a blood vessel is injured. **(b)** Platelets and a variety of clotting factors are important in blood clotting. The color-enhanced SEM of part of a blood clot shows red blood cells enmeshed in a network of fibrin.

VERTEBRATE BLOOD VESSELS

Learning Objective

4 Compare the structure and function of different types of blood vessels, including arteries, arterioles, capillaries, and veins.

The vertebrate circulatory system includes three main types of blood vessels: arteries, capillaries, and veins (Fig. 42-6). An **artery** carries blood away from a heart chamber, toward other tissues. When an artery enters an organ, it divides into many smaller branches called **arterioles.** The arterioles deliver blood into the microscopic capillaries. After blood circulates through an organ, capillaries merge to form **veins** that channel the blood back toward the heart.

The wall of an artery or vein has three layers (Fig. 42-6b). The innermost layer, which lines the blood vessel, consists mainly of **endothelium,** a tissue that resembles squamous epithelium (see Chapter 37). The middle layer is connective tissue and smooth muscle cells, and the outer coat is connective tissue rich in elastic and collagen fibers.

The thick walls of arteries and veins prevent gases and nutrients from passing through. Materials are exchanged between the blood and interstitial fluid bathing the cells through the capillary walls, which are only one cell thick. Capillary networks in the body are so extensive that at least one of these tiny vessels is located close to every cell in the body. The total length of all capillaries in the body has been estimated to be over 96,000 km!

Smooth muscle in the arteriole wall can constrict (**vasoconstriction**) or relax (**vasodilation),** changing the radius of the arteriole. Such changes help maintain appropriate blood pressure and help control the volume of blood passing to a particular tissue. Changes in blood flow are regulated by the nervous system in response to the metabolic needs of the tissue, as well as by the demands of the body as a whole. For example, when a tissue is metabolizing rapidly, it needs more nutrients and oxygen. During exercise, arterioles within skeletal muscles dilate, increasing by more than 10-fold the amount of blood flowing to these muscle tissues.

If all your blood vessels dilated at the same time, you would not have enough blood to fill them completely. Normally your liver, kidneys, and brain receive the lion's share of blood. However, if an emergency suddenly occurred requiring rapid action, your blood would be rerouted quickly in favor of heart and muscles. At such a time the digestive system and kidneys can do with less blood, because they are not crucial in responding to the crisis.

The small vessels that directly link arterioles with venules (small veins) are **metarterioles.** The true capillaries branch off from the metarterioles and then rejoin them (Fig. 42-7). True capillaries also interconnect with one another. Wherever a capillary branches from a metarteriole, a smooth muscle cell called a *precapillary sphincter* is present. Precapillary sphincters open and close continuously, directing blood first to one and then to another section of tissue. These sphincters (along with the smooth muscle in the walls of arteries and arterioles) regulate the blood supply to each organ and its subdivisions.

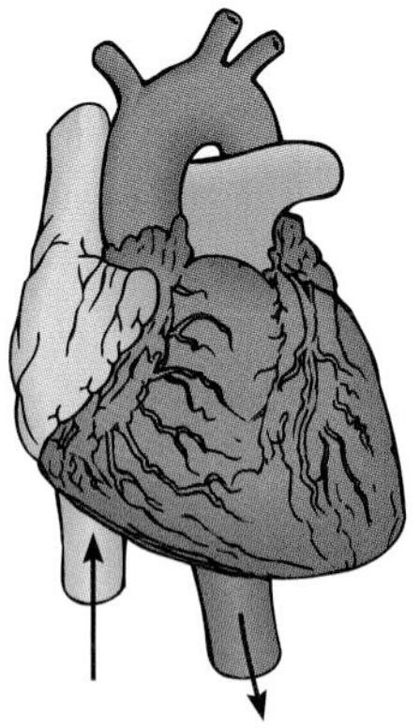

Review

- Compare the functions of arteries, capillaries, and veins.
- How do arterioles function in maintaining homeostasis?

BiologyNow™ Assess your understanding of **vertebrate blood vessels** by taking the pretest on your BiologyNow CD-ROM.

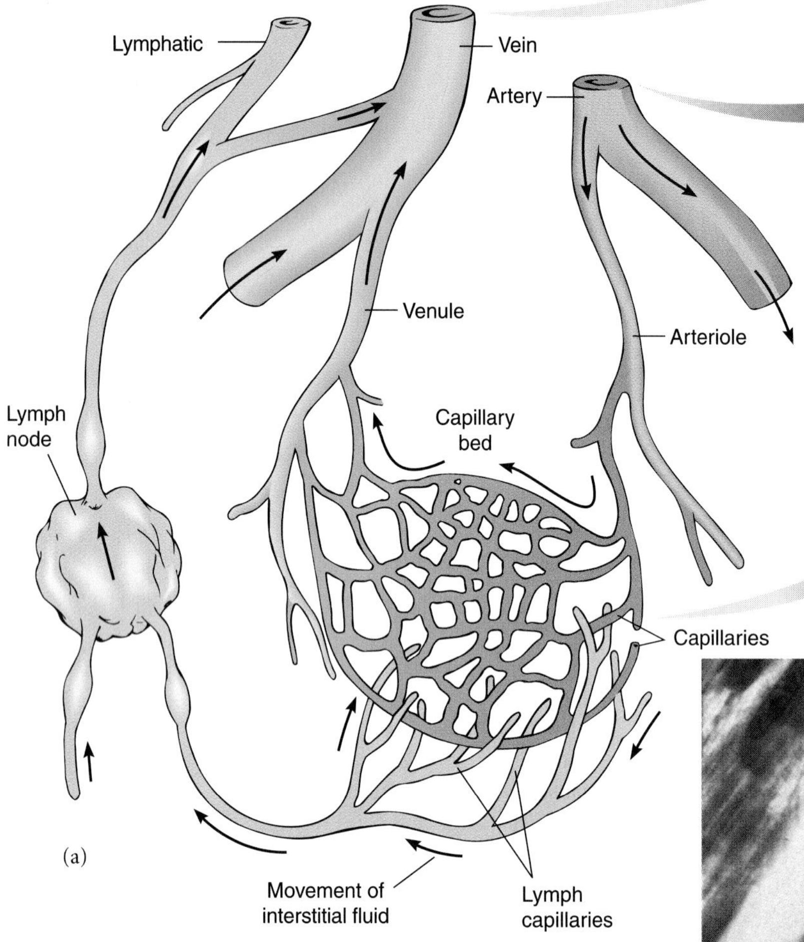

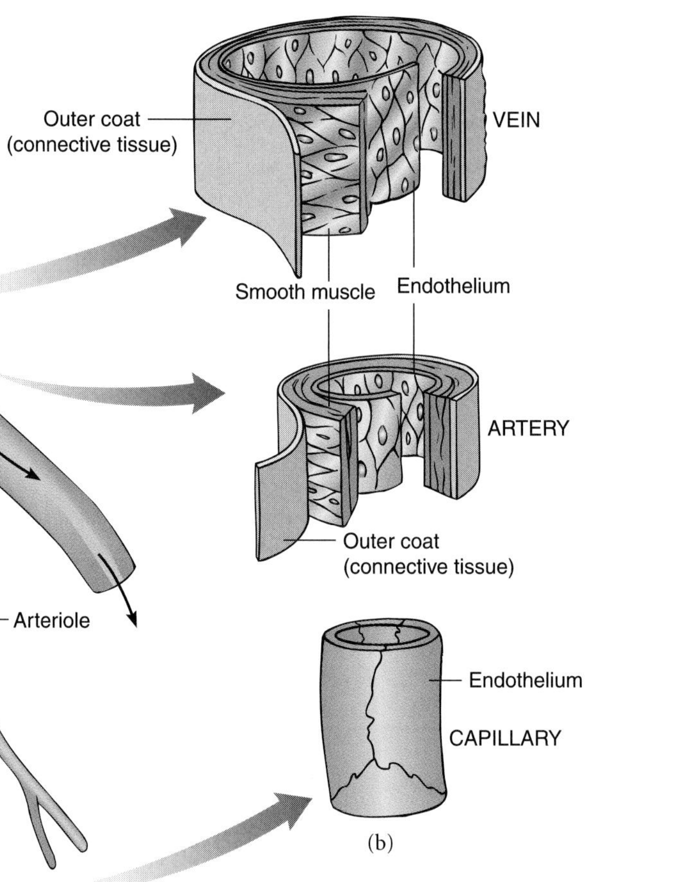

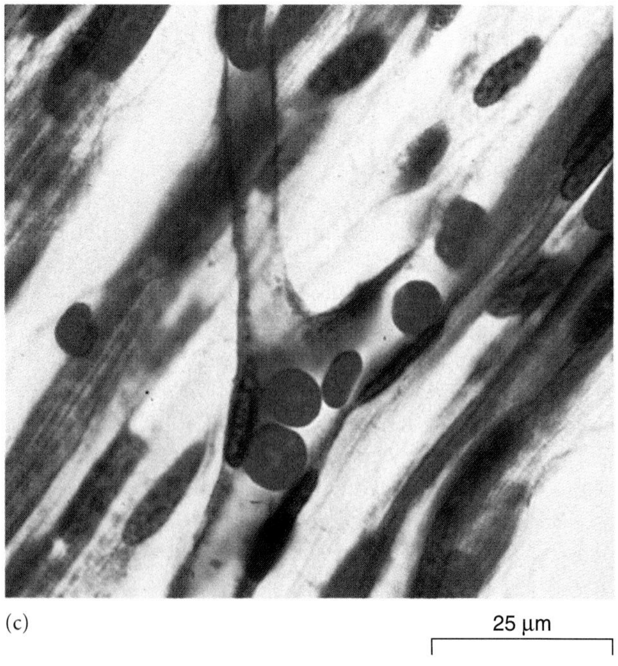

Lennart Nilsson, Boehringer Ingelheim International, GmbH

FIGURE **42-6** **Blood flow through blood and lymphatic vessels.**
(a) The heart pumps blood into arteries. Blood then flows through arterioles, capillaries, and veins, which return it to the heart. Some plasma leaves the capillaries and becomes interstitial fluid. Lymphatic vessels return excess interstitial fluid to the blood by way of ducts that lead into large veins in the shoulder region. Blood vessels with oxygen-rich blood are shown in red. Those with oxygen-poor blood are shown in blue. **(b)** Comparison of the walls of an artery, vein, and capillary. All three vessels are lined with endothelium. The capillary wall is only one cell thick, allowing exchange of materials. **(c)** LM of red blood cells passing through capillaries almost in single file.

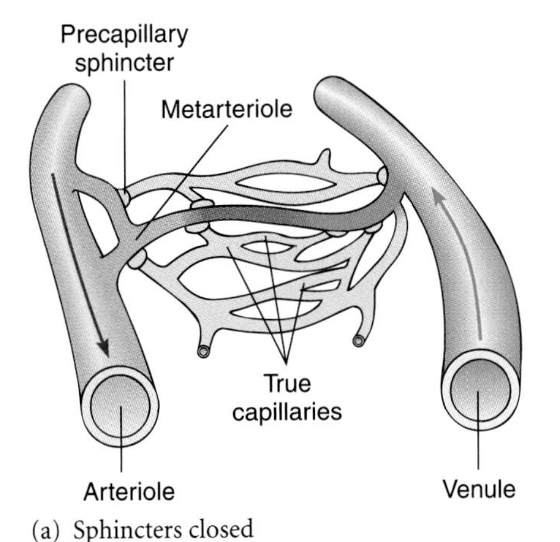

(a) Sphincters closed

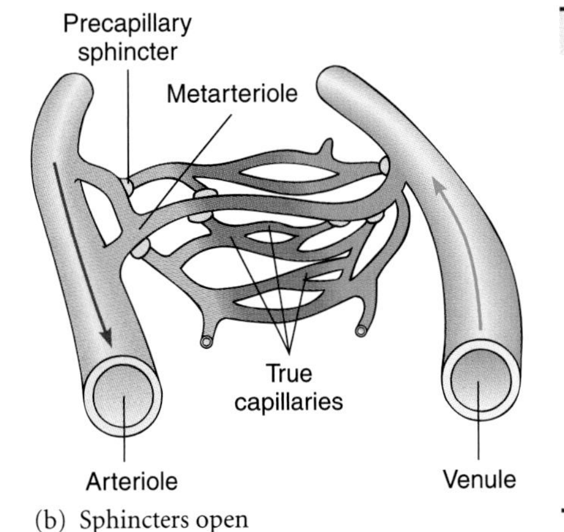

(b) Sphincters open

FIGURE **42-7**

Blood flow through a capillary network.

As a tissue becomes active, the pattern of blood flow through its capillary networks changes. **(a)** When a tissue is inactive, only its metarterioles are open. **(b)** When the tissue becomes active, decreased oxygen tension in the tissue relaxes the precapillary sphincters, and the capillaries open. This increases the blood supply and thus the delivery of nutrients and oxygen to the active tissue.

EVOLUTION OF THE VERTEBRATE CARDIOVASCULAR SYSTEM

Learning Objective

5 Trace the evolution of the vertebrate cardiovascular system from fish to mammal.

The vertebrate cardiovascular system became modified in the course of evolution, as the site of gas exchange shifted from gills to lungs and as certain vertebrates became active, endothermic animals with higher metabolic rates. The vertebrate heart has one or two **atria** (sing., *atrium*), chambers that receive blood returning from the tissues, and one or two **ventricles** that pump blood into the arteries (Fig. 42-8). Some vertebrate classes have additional chambers.

The fish heart has one atrium and one ventricle. The atrium pumps blood into the ventricle, which pumps blood into a single circuit of blood vessels. Blood is oxygenated as it passes through capillaries in the gills. After blood circulates through the gill capillaries, its pressure is low, so blood passes very slowly to the other organs. The fish's swimming movements facilitate circulation. Blood returning to the heart has a low oxygen content. A thin-walled *sinus venosus* receives blood returning from the tissues and pumps it into the atrium.

In amphibians, blood flows through a double circuit: the **pulmonary circulation** and the **systemic circulation.** Oxygen-rich and oxygen-poor blood are kept somewhat separate. The amphibian heart has two atria and one ventricle. A *sinus venosus* collects oxygen-poor blood returning from the veins and pumps it into the right atrium. Blood returning from the lungs passes directly into the left atrium. Both atria pump into the single ventricle, but oxygen-poor blood is pumped out of the ventricle before oxygen-rich blood enters it. Blood passes into an artery, the *conus arteriosus,* equipped with a fold that helps keep the blood separate. Much of the oxygen-poor blood is directed into the pulmonary circulation, which delivers it to the lungs and skin where it is recharged with oxygen. The systemic circulation delivers oxygen-rich blood into arteries that conduct it to the various tissues of the body.

Most reptiles also have a double circuit of blood flow, made more efficient by a wall that partly divides the ventricle. Mixing of oxygen-rich and oxygen-poor blood is minimized by the timing of contractions of the left and right sides of the heart and by pressure differences. In crocodilians (crocodiles and alligators), the wall between the ventricles is complete, so the heart consists of two atria and two ventricles. Thus a four-chambered heart first evolved among the reptiles.

Unlike birds and mammals, amphibians and reptiles do not ventilate their lungs continuously. Therefore, it would be inefficient to pump blood through the pulmonary circulation continuously. The shunts between the two sides of the heart allow blood to be distributed to the lungs as needed.

In all birds and mammals (and in crocodilians), the septum (wall) between the ventricles is complete. Biologists hypothesize that the completely divided heart evolved twice during the course of vertebrate evolution; first in the crocodilian-bird clade and then independently in mammals The interventricular septum prevents oxygen-rich blood in the left chamber from mixing with oxygen-poor blood in the right chamber. The conus arteriosus has split and become the base of the **aorta** (the largest artery) and the pulmonary artery. No sinus venosus is present as a separate chamber, but a vestige remains as the *sinoatrial node* (the pacemaker).

Complete separation of the right and left sides of the heart requires blood to pass through the heart twice each time it tours the body. The complete double circuit allows birds and mammals to maintain higher blood pressures in the systemic circulation and modest pressures in the pulmonary circulation. This system delivers materials to the tissues rapidly and efficiently. Because their blood contains more oxygen per unit volume and circulates more rapidly than in other vertebrates, the tissues of birds and mammals receive more oxygen. As a result, these animals can maintain a higher metabolic rate and a constant, high body temperature even in cold surroundings.

The pattern of blood circulation in birds and mammals can be summarized as follows:

Veins (conduct blood from organs) ⟶ right atrium ⟶right ventricle ⟶ pulmonary arteries ⟶ capillaries in the lungs ⟶ pulmonary veins ⟶ left atrium ⟶ left ventricle ⟶ aorta ⟶ arteries (conduct blood to organs) ⟶ arterioles ⟶ capillaries

KEY CONCEPT: The adaptations of the cardiovascular system reflect the evolution of vertebrates as they became terrestrial, active, endothermic animals.

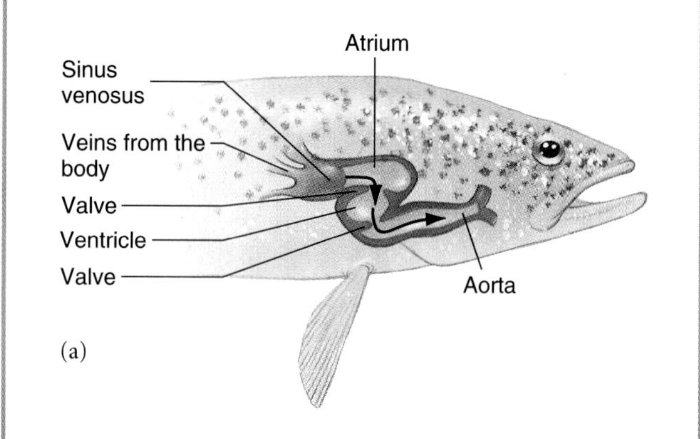

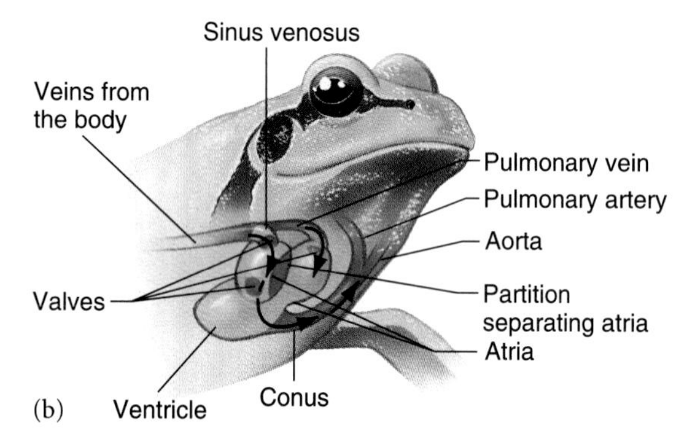

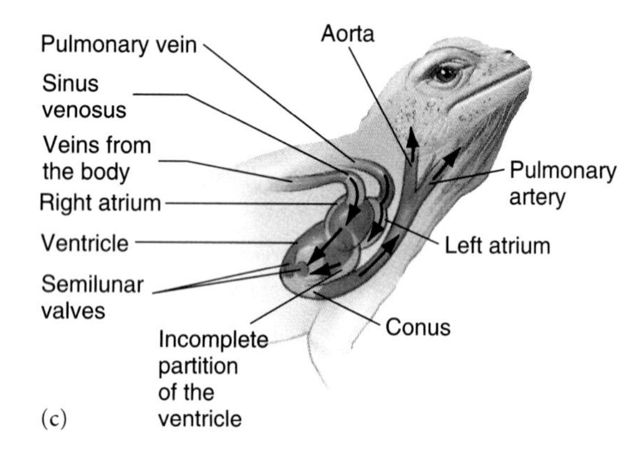

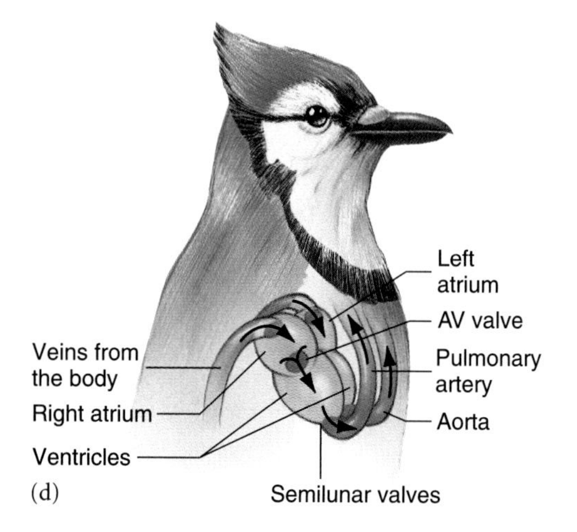

FIGURE **42-8** **Evolution of the vertebrate cardiovascular system.**

(a) The single atrium and ventricle of the fish heart are part of a single circuit of blood flow. **(b)** In amphibians blood flows through a double circuit and the heart consists of two atria and one ventricle. **(c)** The reptilian heart has two atria and two ventricles; in all but the crocodiles and alligators, the wall separating the ventricles is incomplete so blood from the right and left chambers mixes to some extent. **(d)** Birds and mammals have two atria and two ventricles, and blood rich in oxygen is kept completely separate from oxygen-poor blood.

Review

- What are some of the major adaptations that occurred in the evolution of the vertebrate cardiovascular system?

BiologyNow™ Assess your understanding of the **evolution of the vertebrate cardiovascular system** by taking the pretest on your BiologyNow CD-ROM.

THE HUMAN HEART

Learning Objectives

6 Describe the structure and function of the human heart. (Include the heart's conduction system in your answer.)

7 Trace the events of the cardiac cycle, and relate normal heart sounds to these events.

8 Define *cardiac output,* describe how it is regulated, and identify factors that affect it.

Not much bigger than a fist and weighing less than a pound, the human **heart** is a remarkable organ that beats about 2.5 billion times in an average lifetime, pumping about 300 million L (80 million gal) of blood. To meet the body's changing needs, the heart can vary its output from 5 to more than 20 L of blood per minute.

Your heart is a hollow, muscular organ located in the chest cavity directly under the breastbone. Enclosing it is a tough connective tissue sac, the **pericardium.** A smooth layer of endothelium covers the inner surface of the pericardium and the outer surface of the heart. Between these two surfaces is a small **pericardial cavity** filled with fluid, which reduces friction to a minimum as the heart beats. The wall of the heart is mainly cardiac muscle attached to a framework of collagen fibers.

The right atrium and ventricle are separated from the left atrium and ventricle by a wall, or **septum.** Between the atria the wall is known as the **interatrial septum;** between the ventricles, it is the **interventricular septum.** A shallow depression, the **fossa ovalis,** on the interatrial septum marks the place where an opening, the **foramen ovale,** was located in the fetal heart. In the fetus, the foramen ovale lets the blood flow directly from right to left atrium, so very little passes to the nonfunctional lungs. At the upper surface of each atrium lies a small muscular pouch, the **auricle.**

To keep blood from flowing backward, the heart has valves that close automatically (Fig. 42-9). The valve between the right atrium and right ventricle is called the **right atrioventricular (AV) valve,** or **tricuspid valve.** The **left AV valve** (between

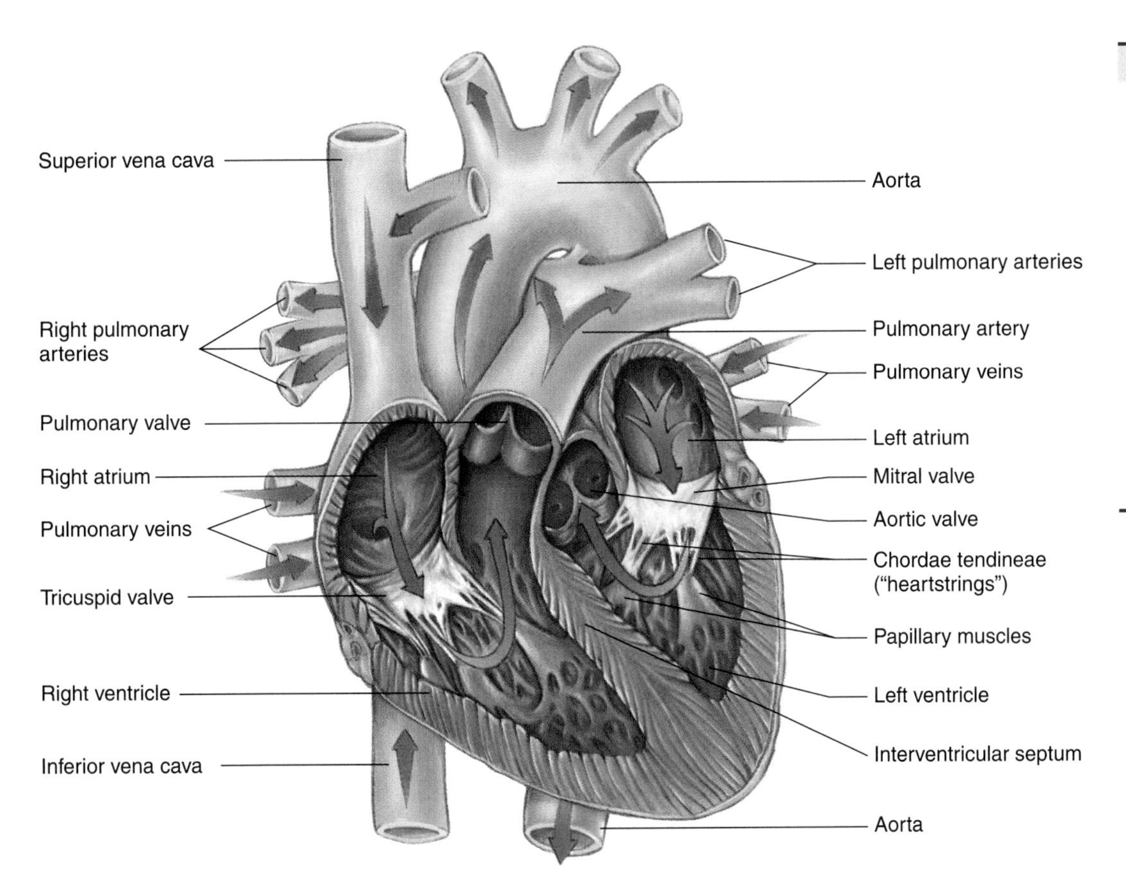

ACTIVE FIGURE **42-9**

Section through the human heart showing the valves.

Note the right and left atria, which receive blood, and the right and left ventricles, which pump blood into the arteries. Arrows indicate the pattern of blood flow.

BiologyNow™ Learn more about **heart anatomy** by clicking on this figure on your BiologyNow CD-ROM.

the left atrium and left ventricle) is the **mitral valve,** or **bicuspid valve.** The AV valves are held in place by stout cords, or "heart-strings," the **chordae tendineae.** These cords attach the valves to the papillary muscles that project from the walls of the ventricles.

When blood returning from the tissues fills the atria, blood pressure on the AV valves forces them to open into the ventricles, which then fill with blood. As the ventricles contract, blood is forced back against the AV valves, pushing them closed. Contraction of the papillary muscles and tensing of the chordae tendineae prevent the valves from opening backward into the atria. These valves are like swinging doors that open in only one direction.

Semilunar valves (named for their flaps, which are shaped like half-moons) guard the exits from the heart. The semilunar valve between the left ventricle and the aorta is the **aortic valve,** and the one between the right ventricle and the pulmonary artery is the **pulmonary valve.** When blood passes out of the ventricles, the flaps of the semilunar valves are pushed aside and offer no resistance to blood flow. But when the ventricles are relaxing and filling with blood from the atria, blood pressure in the arteries is higher than in the ventricles. Blood then fills the pouches of the valves, stretching them across the artery so blood cannot flow back into the ventricle.

Each heartbeat is initiated by a pacemaker

Horror films sometimes feature a scene in which a heart cut out of a human body continues to beat. Scriptwriters of these tales actually have some factual basis for their gruesome fantasies, because a heart carefully removed from the body does continue to beat for many hours if kept in a nutritive, oxygenated fluid. This is possible because the contractions of cardiac muscle begin within the muscle itself and can occur independently of any nerve supply (Fig. 42-10).

At their ends cardiac muscle cells are joined by dense bands called **intercalated discs** (Fig. 42-10b, c). Each disc is a type of *gap junction* (see Chapter 5) in which two cells connect through pores. This type of junction is of great physiological importance, because it offers very little resistance to the passage of an action potential. Ions move easily through the gap junctions, allowing the entire atrial (or ventricular) muscle mass to contract as one giant cell.

Compared with skeletal muscle that has action potentials typically lasting 1 to 2 msec, the action potentials of cardiac muscle are much longer, several hundred milliseconds. Voltage-activated calcium ion channels open during depolarization of cardiac muscle fibers. Entry of Ca^{2+} contributes to the longer depolarization time. Another factor is a type of potassium channel that stays open when the cell is at its resting potential but closes during depolarization, lengthening depolarization time by decreasing the permeability of the membrane to K^+. From patch-clamp experiments (see Chapter 5) on isolated cardiac muscle fibers, investigators have found that spontaneous contraction results from the combination of a slow decrease in potassium permeability and a slow increase in sodium and calcium permeability.

A specialized conduction system ensures that the heart beats in a regular and effective rhythm. Each beat is initiated by the pacemaker, also called the **sinoatrial (SA) node.** The SA node is a small mass of specialized cardiac muscle in the posterior wall of the right atrium near the opening of a large vein, the superior vena cava. The action potential in the SA node is triggered mainly by the opening of Ca^{2+} channels. Ends of the SA node fibers fuse with surrounding ordinary atrial muscle fibers so each action potential spreads through both atria, producing atrial contraction.

One group of atrial muscle fibers conducts the action potential directly to the **atrioventricular (AV) node,** located in the right atrium along the lower part of the septum. Here transmission is delayed briefly, letting the atria finish contracting before the ventricles begin to contract. From the AV node the action potential spreads into specialized muscle fibers called **Purkinje fibers.** These large fibers make up the **AV bundle.** The AV bundle divides, sending branches into each ventricle. When an impulse reaches the ends of the Purkinje fibers, it spreads through the ordinary cardiac muscle fibers of the ventricles.

SA node ⟶ atrial muscle fibers (atria contract) ⟶ AV node ⟶ AV bundle (Purkinje fibers) ⟶right and left branches (Purkinje fibers) ⟶ ventricular muscle fibers (ventricles contract)

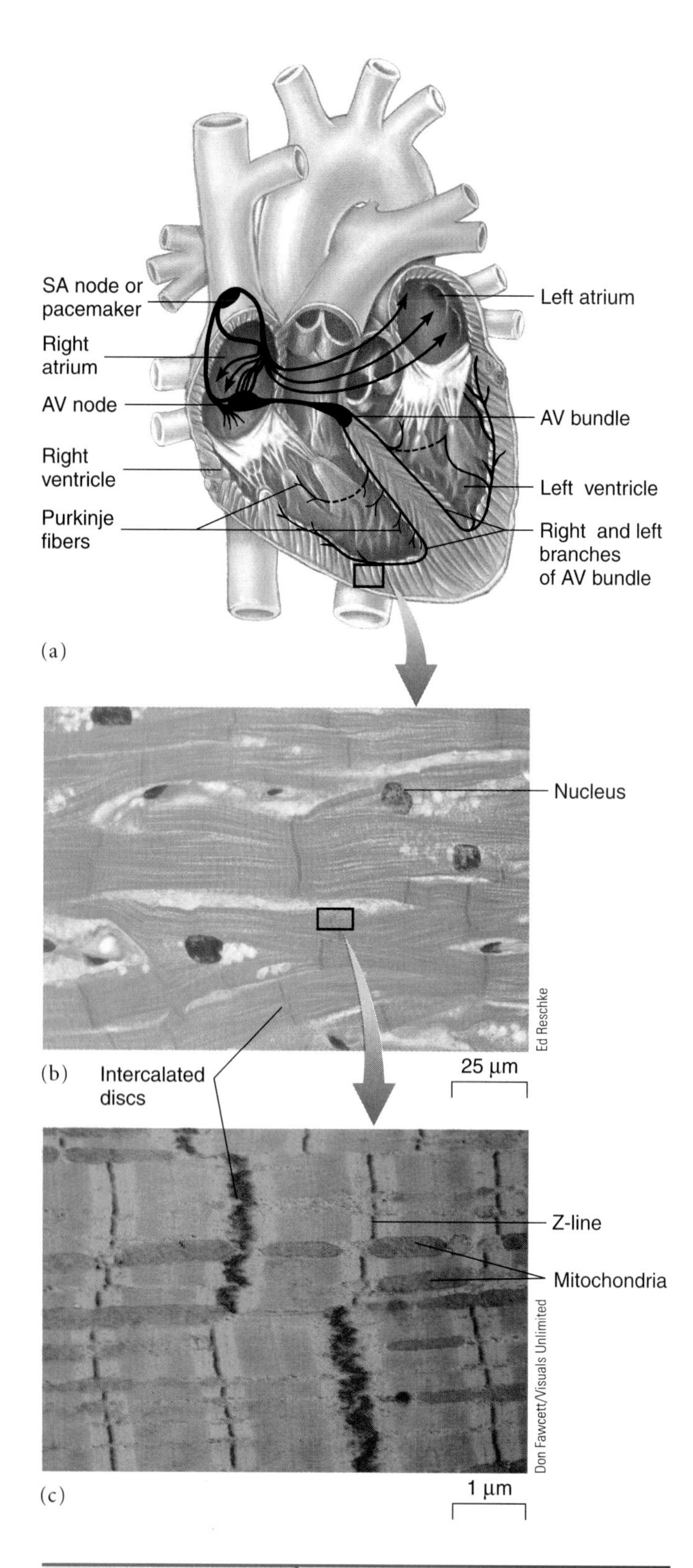

ACTIVE FIGURE **42-10** **Conduction system of the heart.**

(a) The sinoatrial (SA) node initiates each heartbeat. The action potential spreads through the muscle fibers of the atria, producing atrial contraction. Transmission is briefly delayed at the atrioventricular (AV) node before the action potential spreads through specialized muscle fibers into the ventricles. **(b)** LM of cardiac muscle. **(c)** TEM of cardiac muscle.

BiologyNow™ See **cardiac conduction** in action by clicking on this figure on your BiologyNow CD-ROM.

As each wave of contraction spreads through the heart, electrical currents flow into the tissues surrounding the heart and onto the body surface. By placing electrodes on the body surface on opposite sides of the heart, the electrical activity can be amplified and recorded either by an oscilloscope or an electrocardiograph. The graph produced is called an **electrocardiogram (ECG or EKG).** Abnormalities in the EKG indicate disorders in the heart or its rhythm. For example, in **heart block** impulse transmission is delayed or blocked at some point in the conduction system. **Artificial pacemakers** can help patients with severe heart block. The pacemaker is implanted beneath the skin, and its electrodes connected to the heart. This device provides continuous rhythmic impulses that avoid the block and drive the heartbeat.

Each minute the heart beats about 70 times. One complete heartbeat takes about 0.8 second and is referred to as a **cardiac cycle.** That portion of the cycle in which contraction occurs is known as **systole;** the period of relaxation is **diastole.** Figure 42-11 shows the sequence of events that occur during one cardiac cycle.

You can measure your heart rate by placing a finger over the radial artery in your wrist or the carotid artery in your neck and counting the pulsations for 1 minute. **Arterial pulse** is the alternate expansion and recoil of an artery. Each time the left ventricle pumps blood into the aorta, the elastic wall of the aorta expands to accommodate the blood. This expansion moves in a wave down the aorta and the arteries that branch from it. When this pressure wave passes, the elastic arterial wall snaps back to its normal size.

When you listen to the heartbeat with a stethoscope, you can hear two main heart sounds, "lub-dup," which repeat rhythmically. These sounds result from the heart valves closing. When the valves close, they cause turbulence in blood flow that sets up vibrations in the walls of the heart chambers. The first heart sound, "lub," is low-pitched, not very loud, and fairly long-

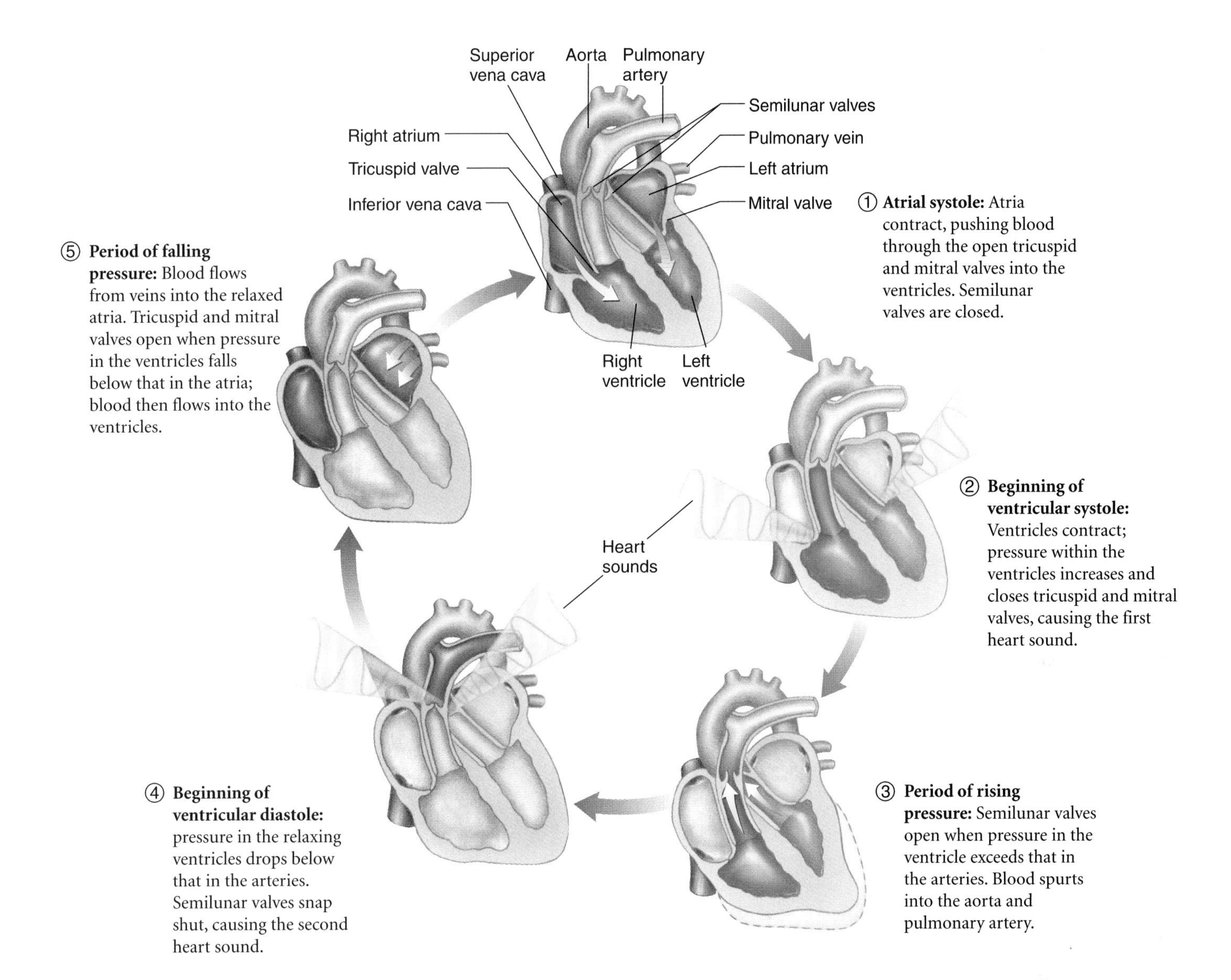
Superior vena cava
Aorta
Pulmonary artery
Semilunar valves
Pulmonary vein
Left atrium
Mitral valve
Right atrium
Tricuspid valve
Inferior vena cava
Right ventricle
Left ventricle
① Atrial systole: Atria contract, pushing blood through the open tricuspid and mitral valves into the ventricles. Semilunar valves are closed.
② Beginning of ventricular systole: Ventricles contract; pressure within the ventricles increases and closes tricuspid and mitral valves, causing the first heart sound.
③ Period of rising pressure: Semilunar valves open when pressure in the ventricle exceeds that in the arteries. Blood spurts into the aorta and pulmonary artery.
④ Beginning of ventricular diastole: pressure in the relaxing ventricles drops below that in the arteries. Semilunar valves snap shut, causing the second heart sound.
⑤ Period of falling pressure: Blood flows from veins into the relaxed atria. Tricuspid and mitral valves open when pressure in the ventricles falls below that in the atria; blood then flows into the ventricles.
Heart sounds

FIGURE **42-11** The cardiac cycle.

The cycle comprises contraction of both atria followed by both ventricles. White arrows indicate the direction of blood flow; dotted lines indicate the change in size as contraction occurs.

lasting. It is caused mainly by the closing of the AV (mitral and tricuspid) valves and marks the beginning of ventricular systole. The "lub" sound is quickly followed by the higher-pitched, louder, sharper, and shorter "dup" sound. Heard almost as a quick snap, the "dup" marks the closing of the semilunar valves and the beginning of ventricular diastole.

The quality of these sounds tells a discerning physician much about the state of the valves. For example, when a valve doesn't close tightly blood may flow backward, causing a **heart murmur.** When the semilunar valves are injured, a soft, hissing noise ("lub-shhh") is heard in place of the normal sound. Valve deformities are sometimes present at birth, or they may result from certain diseases such as rheumatic fever or syphilis. Diseased or deformed valves can be surgically repaired or replaced with artificial valves.

The nervous system regulates heart rate

Although the heart can beat independently, its rate is, in fact, carefully regulated by the nervous system and endocrine system (Fig. 42-12). Sensory receptors in the walls of certain blood vessels and heart chambers are sensitive to changes in blood pressure. When stimulated, they send messages to **cardiac centers** in the medulla of the brain. These cardiac centers govern two sets of autonomic nerves that pass to the SA node: parasympathetic and sympathetic nerves.

Parasympathetic and sympathetic nerves have opposite effects on heart rate (see Fig. 40-15). Parasympathetic nerves release the neurotransmitter *acetylcholine,* which slows the heart. Acetylcholine slows the rate of depolarization by increasing the membrane's permeability to potassium (Fig. 42-13a). Sympathetic nerves release norepinephrine, which speeds the heart rate and increases the strength of contraction. Norepinephrine stimulates calcium ion channel opening during depolarization (Fig. 42-13b). Both norepinephrine and acetylcholine act indirectly on ion channels. They activate a *signal transduction* process involving a G protein. Norepinephrine binds to *beta-adrenergic receptors,* one of the two main types of adren-

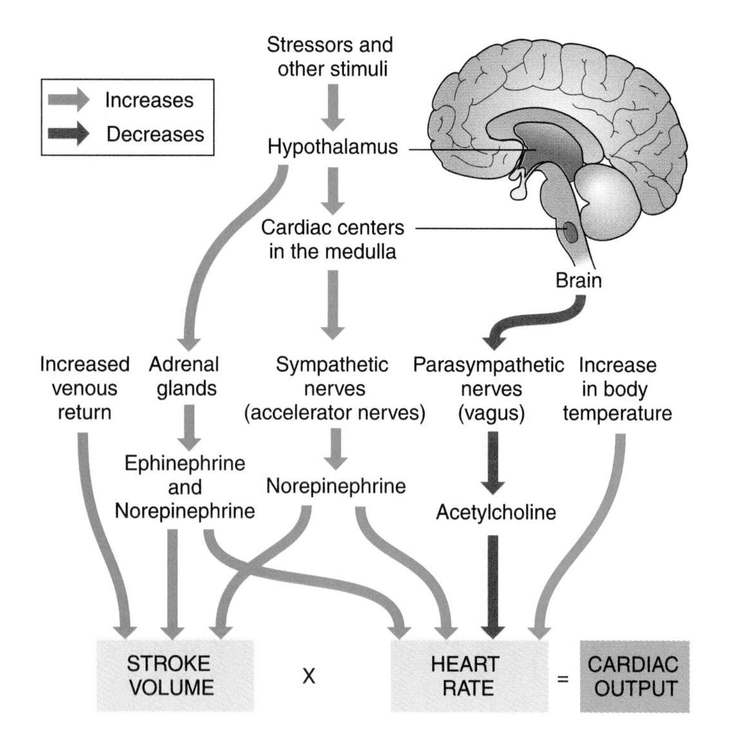

FIGURE **42-12** | **Some factors that influence cardiac output.**

ergic receptors. These receptors are targeted by *beta blockers,* drugs that block the actions of norepinephrine on the heart and are used clinically in treating hypertension and other types of heart disease.

In response to physical and emotional stressors, the adrenal glands release epinephrine and norepinephrine, which speed the heart. An elevated body temperature also increases heart rate. During fever, the heart may beat more than 100 times per minute. As you might expect, heart rate decreases when body temperature is lowered. This is why physicians may deliberately lower a patient's temperature during heart surgery.

Stroke volume depends on venous return

The volume of blood one ventricle pumps during one beat is the **stroke volume.** Stroke volume depends mainly on **venous return,** the amount of blood the veins deliver to the heart. According to **Starling's law of the heart,** if the veins deliver more blood to the heart, the heart pumps more blood (within physiological limits). When extra amounts of blood fill the heart chambers, the cardiac muscle fibers stretch more and contract with greater force, pumping a larger volume of blood into the arteries. Norepinephrine released by sympathetic nerves and

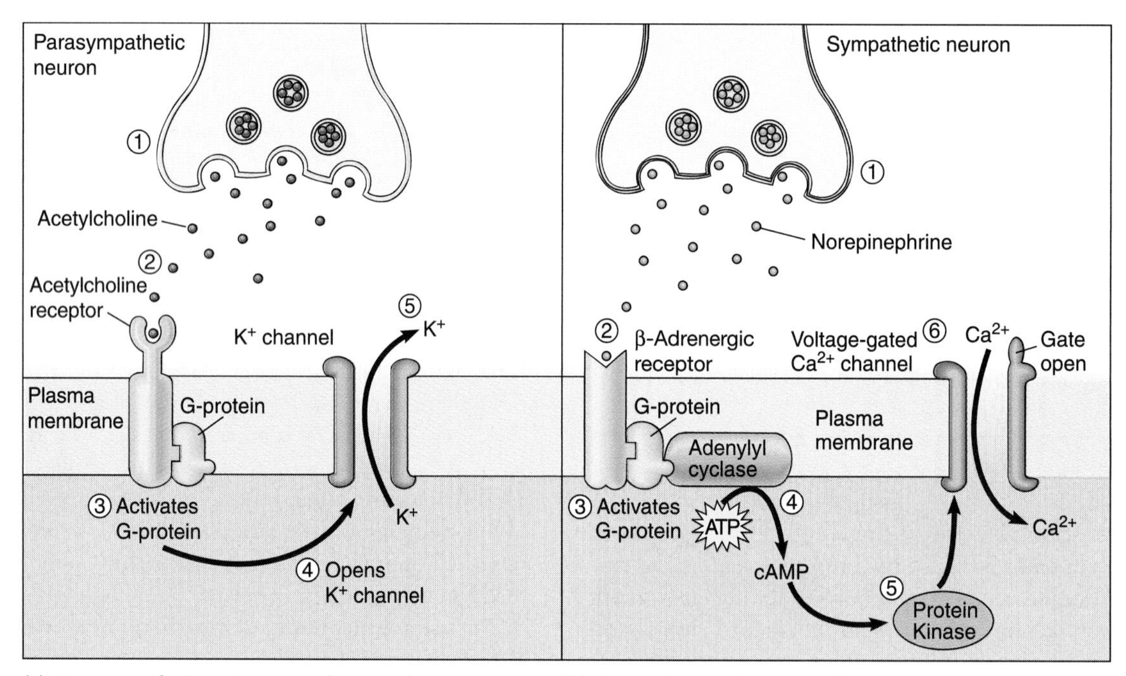

(a) Parasympathetic action on cardiac muscle

(b) Sympathetic action on cardiac muscle

FIGURE **42-13** | **Actions of sympathetic and parasympathetic neurons on cardiac muscle cells.**

Neurotransmitters released by sympathetic and parasympathetic neurons initiate a signal transduction process. **(a)** ① Parasympathetic neurons release acetylcholine which ② binds with receptors on the plasma membrane of cardiac muscle. ③ The receptor activates a G protein, which ④ binds with a K^+ channel, causing the channel to open. ⑤ K^+ leaves the cell, causing the membrane to become hyperpolarized. The rate of action potentials slows.

(b) ① Sympathetic neurons release norepinephrine, which ② binds with receptors on the plasma membrane of cardiac muscle. ③ The receptor then activates a G protein, which ④ activates the enzyme (adenylyl cyclase) that converts ATP to cyclic AMP. ⑤ Cyclic AMP activates a protein kinase. ⑥ The protein kinase phosphorylates calcium ion channels so that they open more easily when the neuron is depolarized. Action potentials occur more rapidly.

epinephrine released by the adrenal glands during stress also increase the force of contraction of cardiac muscle fibers.

Cardiac output varies with the body's need

By multiplying the stroke volume by the number of times the left ventricle beats per minute, we can compute the **cardiac output (CO).** The CO is the volume of blood pumped by the left ventricle into the aorta in 1 minute. For example, in a resting adult the heart may beat about 72 times per minute and pump about 70 mL of blood with each contraction.

$$\begin{aligned} CO &= \text{stroke volume} \times \text{heart rate (number of ventricular contractions per minute)} \\ &= 70 \text{ mL/stroke} \times 72 \text{ strokes/min} \\ &= 5040 \text{ mL/min (about 5 L/min)} \end{aligned}$$

Cardiac output varies with changes in either stroke volume or heart rate (see Fig. 42-12). When stroke volume increases, CO increases. The CO varies dramatically with the changing needs of the body. During stress or heavy exercise, the normal heart can increase its CO fourfold to fivefold, so that it pumps 20 to 30 L of blood per minute.

Review

- What factors influence cardiac output?
- What sequence of events occurs during cardiac conduction?
- How is the heart regulated? (Include a description of the actions of acetylcholine and norepinephrine.)

BiologyNow™ Assess your understanding of **the human heart** by taking the pretest on your BiologyNow CD-ROM.

BLOOD PRESSURE

Learning Objective

9 Identify factors that determine and regulate blood pressure, and compare blood pressure in different types of blood vessels.

Blood pressure is the force exerted by the blood against the inner walls of the blood vessels. It is determined by CO, blood volume, and the resistance to blood flow (Fig. 42-14a). When CO increases, blood flow increases, causing a rise in blood pressure. When CO decreases, blood flow decreases, causing a fall in blood pressure. If the volume of blood is reduced by hemorrhage or by chronic bleeding, the blood pressure drops. In contrast, an increase in blood volume results in an increase in blood pressure. For example, a high dietary intake of salt causes water retention. This increases blood volume and raises blood pressure.

Blood flow is impeded by resistance; when the resistance to flow increases, blood pressure rises. **Peripheral resistance** is the resistance to blood flow caused by blood viscosity and by friction between blood and the blood vessel wall. In the blood of a healthy person, viscosity remains fairly constant and is only a minor influence on changes in blood pressure. More important is the friction between blood and the blood vessel wall. The length and diameter of a blood vessel determine the amount of surface area in contact with the blood. Blood vessel length does not change, but the diameter, especially of an arteriole, does. A small change in blood vessel diameter causes a big change in blood pressure. Constriction of blood vessels raises blood pressure; dilation lowers blood pressure.

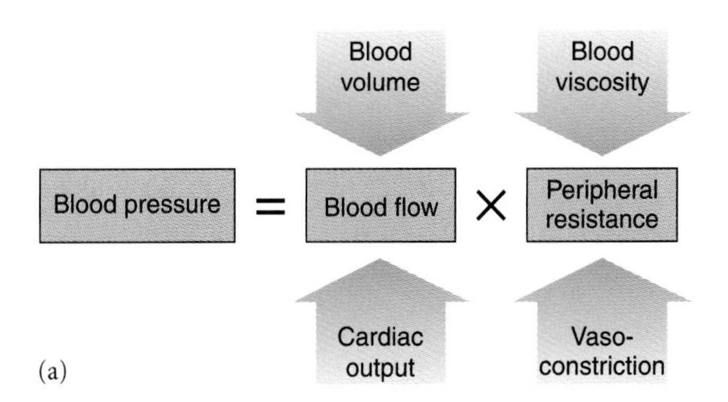

(a)

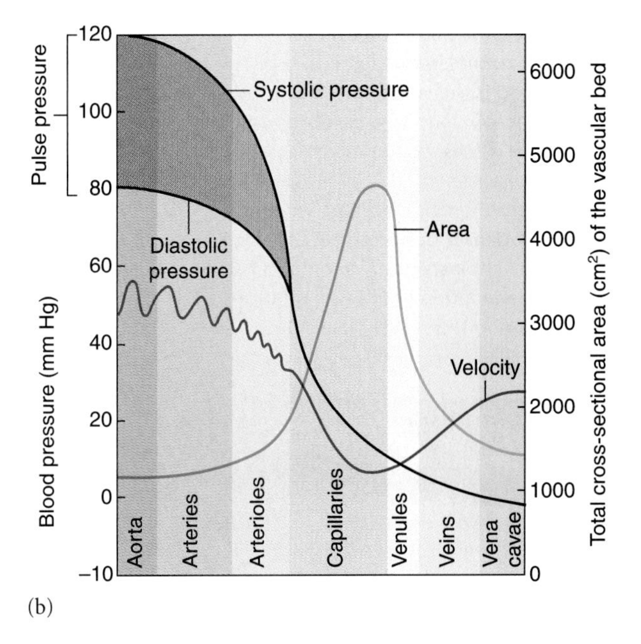

(b)

ACTIVE FIGURE **42-14** **Blood pressure.**

(a) Blood pressure depends on blood flow and resistance to that flow. A variety of factors affect blood flow and resistance. **(b)** Blood pressure in different types of blood vessels. Systolic and diastolic variations in arterial blood pressures are shown. Note that the venous pressure drops below zero (below atmospheric pressure) near the heart.

BiologyNow™ Learn more about **determining blood pressure** by clicking on this figure on your BiologyNow CD-ROM.

Blood pressure in arteries rises during systole and falls during diastole. In 2003 the National Institutes of Health revised its guidelines for prevention and treatment of **hypertension,** or high blood pressure. It included a new guideline for normal blood pressure, defining it as a systolic pressure of less than 120 and a diastolic pressure of less than 80. An example of a normal blood pressure, measured in the upper arm with a sphygmomanometer, is 110/73 mm of mercury, abbreviated mm Hg. Systolic pressure is indicated by the first number, diastolic by

Cardiovascular disease is the number one cause of death in the United States and in most other industrial countries. Most often death results from some complication of **atherosclerosis,** a disease in which the arteries are damaged, inflamed, and narrow as a result of lipid deposits in their walls. Although it can affect almost any artery, the disease most often develops in the aorta and in the coronary and cerebral arteries. When it occurs in the cerebral arteries, less blood is delivered to the brain. This condition can lead to a **cerebrovascular accident (CVA),** commonly referred to as a stroke.

Atherosclerosis develops when excess low-density lipoprotein (LDL), known as "bad" cholesterol, in the blood triggers inflammation in the wall of arteries. (Inflammation, discussed in Chapter 43, is one of the body's most important responses to infection or injury.) The LDLs become oxidized and adhere to the wall of the blood vessel. They stimulate a complex series of events that promote inflammation. Underlying smooth muscle fibers then migrate to the inner lining of the artery and produce components of the extracellular matrix. The muscle fibers and extracellular matrix components form a fibrous cap of tissue over the injured area of the artery.

As atherosclerotic plaque develops, arteries lose their ability to stretch when filled with blood, and they gradually become occluded (blocked), as shown in the figure. The occluded vessels deliver less blood to the tissues, which then become **ischemic** (lacking in blood). Now the tissue is deprived of an adequate oxygen and nutrient supply.

In **ischemic heart disease,** the increased need for oxygen during exercise or emotional stress results in the pain characteristic of **angina pectoris.** Nitroglycerin pills or a nitroglycerin patch dilate veins, decreasing venous return. Cardiac output is lowered so the heart is working less hard and requires less oxygen. Nitroglycerin also dilates the coronary arteries slightly, allowing more blood to reach the heart muscle. *Calcium ion channel blockers* are drugs that slow the heart by inhibiting the passage of calcium into cardiac muscle fibers. They also dilate the coronary arteries.

Myocardial infarction (MI), or heart attack, is a very serious, often fatal, consequence of cardiovascular disease. MI often results from a sudden decrease in coronary blood supply. The part of

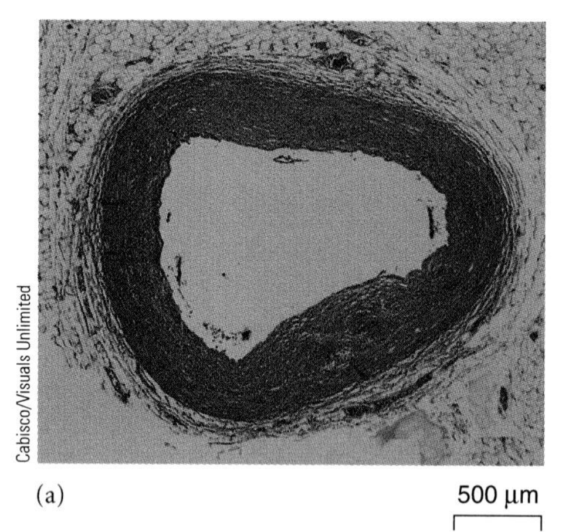

(a)

Cabisco/Visuals Unlimited

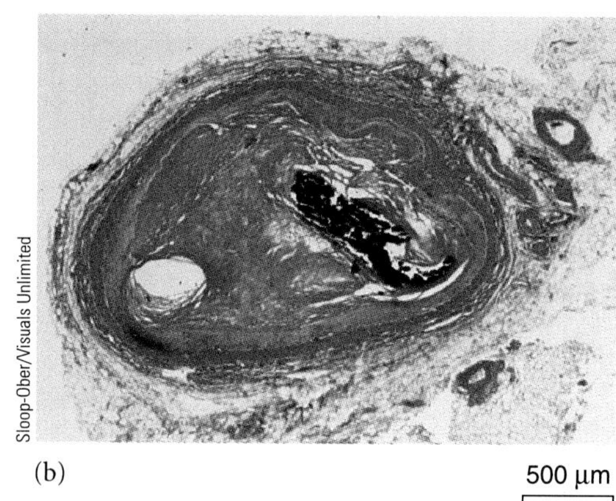

(b)

Sloop-Ober/Visuals Unlimited

Progression of atherosclerosis. LMs of cross sections through two arteries showing changes that take place in atherosclerosis. **(a)** Normal coronary artery. **(b)** This artery is almost completely blocked with atherosclerotic plaque.

the second. If you have a blood pressure of 120 to 139 systolic over 80 to 89 diastolic, you are considered *prehypertensive,* and you need to modify your lifestyle to prevent cardiovascular disease and stroke. Lifestyle changes that will reduce your risk include: exercise, lose excess weight, follow a heart-healthy diet, reduce salt intake, limit alcohol, and don't smoke.

If your systolic pressure consistently measures 140 mm Hg or higher or your diastolic pressure measures 90 mm Hg or higher, you have hypertension. Hypertension is a risk factor for atherosclerosis and other cardiovascular disease (see *Focus On: Cardiovascular Disease*). In hypertension, there is usually increased vascular resistance, especially in the arterioles and small arteries. The heart's workload increases because it must pump against this greater resistance. If this condition persists, the left ventricle enlarges and may deteriorate in function. Heredity, aging, and ethnicity contribute to the development of hypertension.

Blood pressure is highest in arteries

As you might guess, blood pressure is greatest in the large arteries, decreasing as blood flows away from the heart and through the smaller arteries and capillaries (Fig. 42-14b). By the time blood enters the veins, its pressure is very low, even approaching zero. Flow rate can be maintained in veins at low pressure, because they are low-resistance vessels. Their diameter is larger than that of corresponding arteries, and their walls have little smooth muscle. Flow of blood through veins depends on several factors, including skeletal muscle movement, which compresses veins. Most veins larger than 2 mm (0.08 in) in diameter that conduct blood against the force of gravity are equipped with valves to prevent backflow (Fig. 42-15). Such valves usually consist of two cusps formed by inward extensions of the vein wall.

cardiac muscle deprived of oxygen dies within a few minutes and is then referred to as an **infarct.** MI is the leading cause of death and disability in the United States. The sudden decrease in blood supply most often occurs when inflammation causes the fibrous cap over the atherosclerotic plaque to break open. Platelets adhere to the roughened arterial wall and initiate clotting.

A **thrombus,** a clot that forms within a blood vessel or within the heart, can block a sizable branch of a coronary artery. Blood flow to a portion of heart muscle is impeded or completely halted. When such blockage, referred to as a *coronary occlusion,* prevents blood flow to a large region of cardiac muscle, the heart may stop beating; that is, **cardiac arrest** occurs, and death can follow within moments. If only a small part of the heart is affected, however, the heart may continue to function. Cells in the region deprived of oxygen die and are replaced by scar tissue.

An episode of ischemia may trigger a fatal arrhythmia such as **ventricular fibrillation,** a condition in which the ventricles contract very rapidly without actually pumping blood. The pulse may stop, and blood pressure may fall precipitously. Ventricular fibrillation has been linked with about 65% of cardiac arrests. The only effective treatment for fibrillation is **defibrillation** with electric shock. The shock appears to depolarize every muscle fiber in the heart so that its timing mechanism can reset.

Patients with progressive cardiovascular disease can be treated with *coronary bypass surgery* in which veins from another location in the patient's body are grafted around occluded coronary arteries. The newly positioned blood vessels restore adequate blood flow to the affected area. Another procedure, *coronary angioplasty,* involves inserting a small balloon into an occluded coronary artery. Inflating the balloon breaks up the plaque in the arterial wall, widening the vessel.

Approaches to preventing and treating cardiovascular disease include use of anti-inflammatory agents and statins, medications that lower cholesterol level. A molecular marker of inflammation called C-reactive protein (CRP) is used to assess the risk of heart disease. Several major modifiable risk factors for cardiovascular disease have been identified:

1. **Elevated LDL-cholesterol levels** in the blood, associated with diets rich in total calories, total fats, saturated fats, and cholesterol.
2. **Hypertension.** The higher the blood pressure, the greater the risk. The hormone angiotensin II, which at abnormally high levels is thought to contribute to hypertension, also stimulates inflammation.
3. **Cigarette smoking.** The risk of developing atherosclerosis is two to six times greater in smokers than in nonsmokers and is directly proportional to the number of cigarettes smoked daily. Components of cigarette smoke cause oxidants to form, which may contribute to the inflammation by increasing oxidation of the altered LDL in the arterial wall.
4. **Diabetes mellitus,** an endocrine disorder in which glucose is not metabolized normally. The body shifts to fat metabolism, and there is a marked increase in circulating lipids, leading to atherosclerosis.
5. **Physical inactivity.** One in four adults in the United States has a sedentary lifestyle and does not engage in regular exercise. Physical inactivity contributes to more than one third of the approximately 500,000 deaths related to heart disease in the United States each year. Exercise reduces the concentrations of triacylglycerols (triglycerides) and cholesterol in the blood (these lipids have been associated with heart disease). At the same time, exercise increases the concentration of high-density lipoproteins (HDLs), which protect against heart disease.
6. **Obesity** can affect cholesterol levels and increase the risk of hypertension and diabetes.

The risk of developing cardiovascular disease also increases with age. Other probable risk factors currently being studied are hereditary predisposition, hormone levels, stress and behavior patterns, and dietary factors.

When a person stands perfectly still for a long time, as when a soldier stands at attention or a store clerk stands at a cash register, blood tends to pool in the veins. When fully distended, veins can accept no more blood from the capillaries. Pressure in the capillaries increases, and large amounts of plasma are forced out of the circulation through the thin capillary walls. Within just a few minutes, as much as 20% of the blood volume is lost from the circulation—with drastic effect. Arterial blood pressure

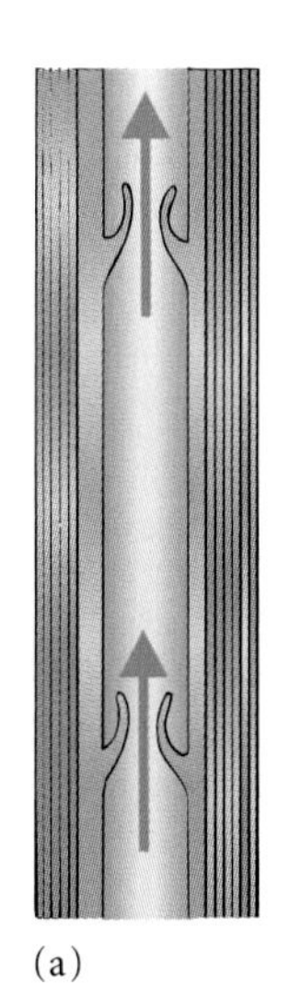
(a)

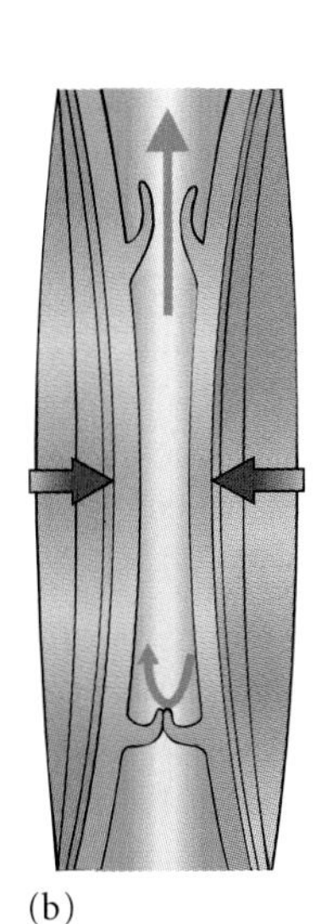
(b)

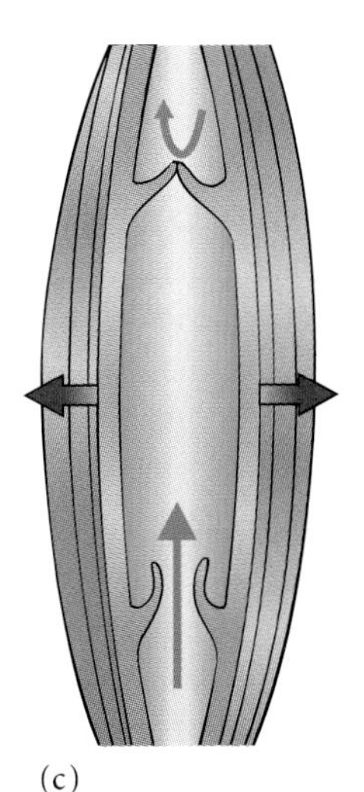
(c)

FIGURE **42-15** | **Venous blood flow.**

Contraction of skeletal muscles helps move blood through the veins. **(a)** Resting condition. **(b)** Muscles contract and bulge, compressing veins and forcing blood toward the heart. The lower valve prevents backflow. **(c)** Muscles relax, and the vein expands and fills with blood from below. The upper valve prevents backflow.

falls dramatically, reducing blood flow to the brain. Sometimes the resulting lack of oxygen in the brain causes fainting, a protective response. Lying in a prone position increases blood supply to the brain. In fact, lifting a person who has fainted to an upright position can result in circulatory shock and even death.

Blood pressure is carefully regulated

Each time you get up from a horizontal position, your blood pressure changes. Several complex mechanisms interact to maintain normal blood pressure so that you do not faint when you get out of bed each morning or change position during the day. When blood pressure falls, sympathetic nerves to the blood vessels stimulate vasoconstriction, causing the pressure to rise again.

The **baroreceptors** present in the walls of certain arteries and in the heart wall are sensitive to changes in blood pressure. When an increase in blood pressure stretches the baroreceptors, messages are sent to the cardiac and vasomotor centers in the medulla of the brain. The cardiac center stimulates parasympathetic nerves that slow the heart, lowering blood pressure. The vasomotor center inhibits sympathetic nerves that constrict arterioles; this action causes vasodilation, which also lowers blood pressure. These neural reflexes continuously work in a complementary way to maintain blood pressure within normal limits.

Hormones are also involved in regulating blood pressure. In response to low blood pressure, the kidneys release **renin,** which activates the **renin-angiotensin-aldosterone pathway** (discussed in Chapter 46). Renin acts on a plasma protein (angiotensinogen), triggering a cascade of reactions that produces the hormone **angiotensin II,** a powerful vasoconstrictor. Angiotensin II also acts indirectly to maintain blood pressure by increasing the synthesis and release of the hormone **aldosterone** by the adrenal glands. Aldosterone increases retention of sodium ions by the kidneys, resulting in greater fluid retention and increased blood volume.

Review

- What is peripheral resistance? How does it affect blood pressure?
- Blood pressure is low in capillaries. How does this help retain fluid in the circulation?

BiologyNow™ Assess your understanding of **blood pressure** by taking the pretest on your BiologyNow CD-ROM.

KEY CONCEPT: The left side of the heart pumps oxygen-rich blood into the systemic circulation. The right side of the heart receives oxygen-poor blood and pumps it into the pulmonary circulation.

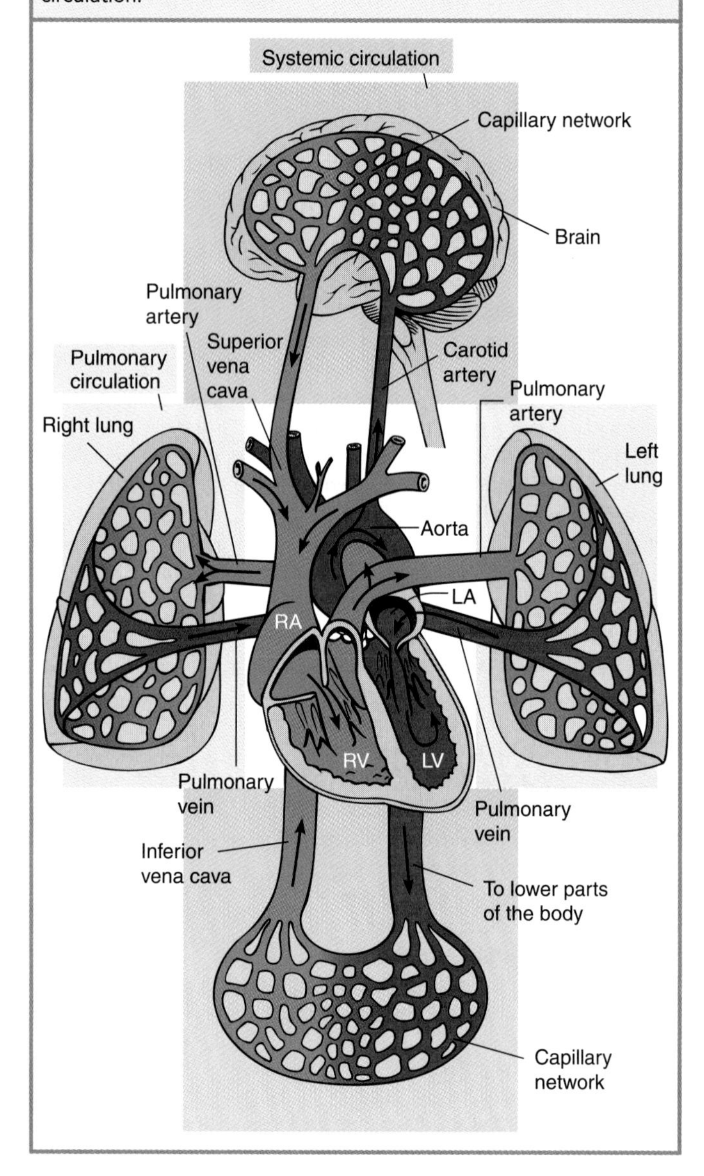

FIGURE **42-16** **Systemic and pulmonary circulation.**
In this highly simplified diagram, red represents oxygen-rich blood; blue represents oxygen-poor blood. Green screen highlights systemic circulation. Yellow screen highlights pulmonary circulation.

THE PATTERN OF CIRCULATION

Learning Objective

10 Trace a drop of blood through the pulmonary and systemic circulations, naming in sequence each structure through which it passes.

Most vertebrates other than fish have a double circuit of blood vessels: (1) the pulmonary circulation connects the heart and lungs; and (2) the systemic circulation connects the heart with all the body tissues. You can trace this general pattern of circulation in Figure 42-16.

The pulmonary circulation oxygenates the blood

Blood from the tissues returns to the right atrium of the heart. This oxygen-poor blood, loaded with carbon dioxide, is pumped by the right ventricle into the pulmonary circulation. As it emerges from the heart, the large pulmonary trunk branches to form the two **pulmonary arteries,** one going to each lung. These are the only arteries that carry oxygen-poor blood.

In the lungs the pulmonary arteries branch into smaller and smaller vessels, which finally give rise to extensive networks of pulmonary capillaries that surround the air sacs of the lungs. As blood circulates through the pulmonary capillaries, carbon dioxide diffuses out of the blood and into the air sacs. Oxygen from the air sacs diffuses into the blood so that by the time it enters the **pulmonary veins** leading back to the left atrium of the heart, the blood is charged with oxygen. Pulmonary veins are the only veins in the body that carry blood rich in oxygen.

In summary, blood flows through the pulmonary circulation in the following sequence:

Right atrium ⟶ right ventricle ⟶ pulmonary arteries ⟶ pulmonary capillaries (in lungs) ⟶ pulmonary veins ⟶ left atrium

The systemic circulation delivers blood to the tissues

Blood entering the systemic circulation is pumped by the left ventricle into the **aorta,** the largest artery. Arteries that branch off from the aorta conduct blood to all regions of the body. Some of the principal branches include the **coronary arteries** to the heart wall itself, the **carotid arteries** to the brain, the **subclavian arteries** to the shoulder region, the **mesenteric artery** to the intestine, the **renal arteries** to the kidneys, and the **iliac arteries** to the legs (Fig. 42-17). Each of these arteries gives rise to smaller and smaller vessels, somewhat like branches of a tree that divide until they form tiny twigs. Eventually blood flows into the capillary network within each tissue or organ.

Blood returning from the capillary networks within the brain passes through the **jugular veins.** Blood from the shoulders and arms drains into the **subclavian veins.** These veins and others returning blood from the upper portion of the body merge to form a very large vein that empties blood into the right atrium. In humans this vein is called the **superior vena cava.** The **renal veins** from the kidneys, **iliac veins** from the lower limbs, **hepatic veins** from the liver, and other veins from the lower portion of the body return blood to the **inferior vena cava,** which delivers blood to the right atrium.

As an example of blood circulation through the systemic circuit, let us trace a drop of blood from the heart to the right leg and back to the heart:

Left atrium ⟶ left ventricle ⟶ aorta ⟶ right common iliac artery ⟶ smaller arteries in leg ⟶ capillaries in leg ⟶ small veins in leg ⟶ common iliac vein ⟶ inferior vena cava ⟶ right atrium

In most fishes, amphibians, and reptiles, the heart muscle is spongy and receives oxygen directly from the blood as it passes through the heart chambers. However, in birds and mammals

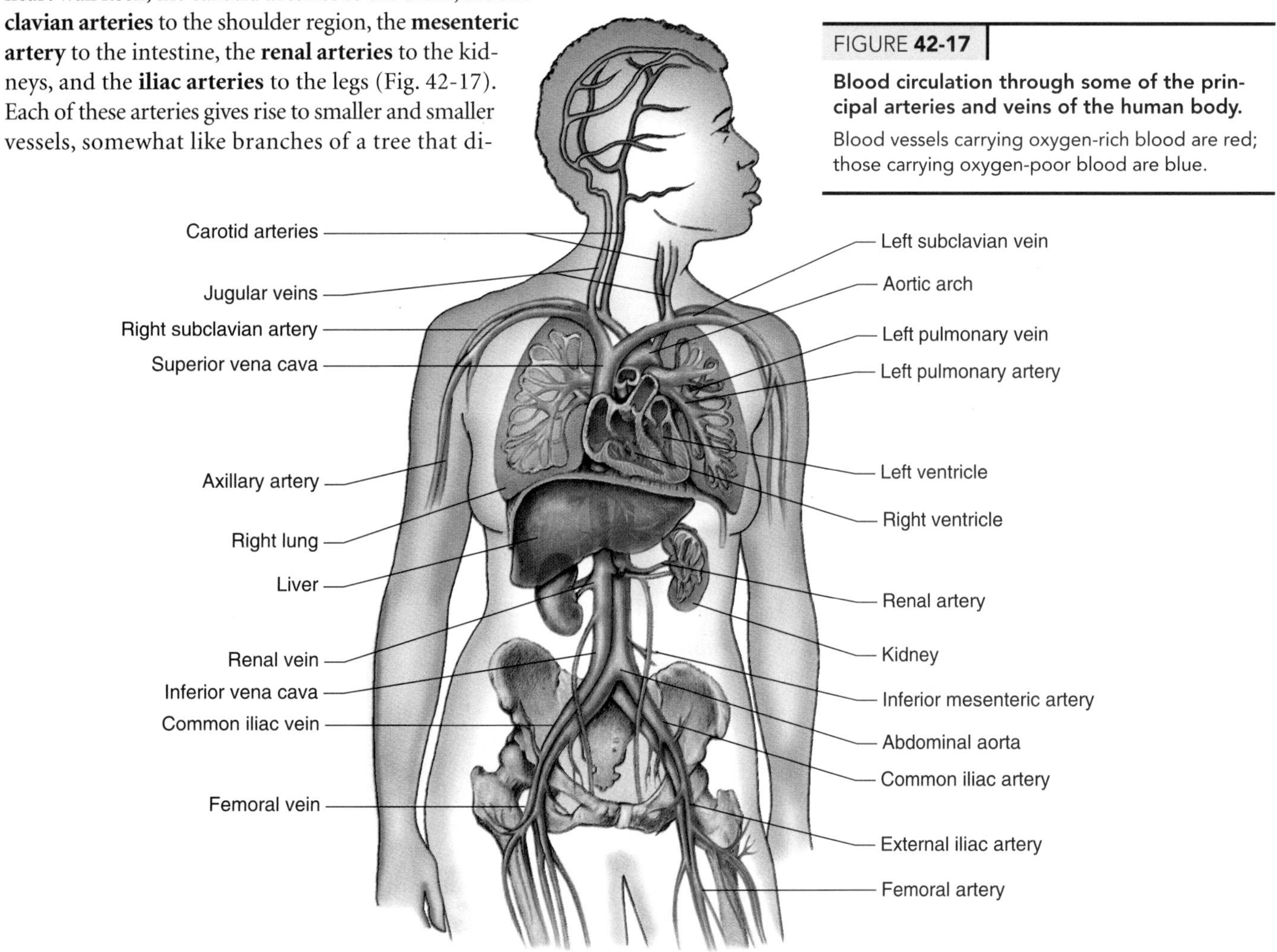

FIGURE **42-17**

Blood circulation through some of the principal arteries and veins of the human body.

Blood vessels carrying oxygen-rich blood are red; those carrying oxygen-poor blood are blue.

the walls of the heart are too thick for nutrients and oxygen to diffuse through them to reach all the muscle fibers. Instead, the cardiac muscle has its own system of blood vessels. In humans, **coronary arteries** give rise to a network of capillaries within the heart wall. The **coronary veins** join to form a large vein, the **coronary sinus** that empties directly into the right atrium.

Blood almost always travels from artery to capillary to vein to the heart. An exception occurs in the **hepatic portal system.** Instead of leading directly back to the heart (as most veins do), the hepatic portal vein delivers nutrients from the intestine to the liver. Within the liver, the hepatic portal vein gives rise to an extensive network of tiny blood sinuses. As blood courses through the hepatic sinuses, liver cells remove nutrients and store them. Eventually liver sinuses merge to form hepatic veins, which deliver blood to the inferior vena cava.

Review

- What sequence of blood vessels and heart chambers would a red blood cell pass through on its way (a) from the inferior vena cava to the aorta; and (b) from the renal vein to the renal artery?
- What is the function of the hepatic portal system? How does its sequence of blood vessels differ from that in most other circulatory routes?

BiologyNow™ Assess your understanding of **the pattern of circulation** by taking the pretest on your BiologyNow CD-ROM.

THE LYMPHATIC SYSTEM

Learning Objective

11 Describe the structure and functions of the lymphatic system.

In addition to the blood circulatory system, vertebrates have an accessory circulatory system, the **lymphatic system** (Fig. 42-18), which has three important functions: (1) to collect and return interstitial fluid to the blood; (2) to launch immune responses that defend the body against disease organisms; and (3) to absorb lipids from the digestive tract. In this section we focus on the first function. We discuss immunity in Chapter 43, and lipid absorption in Chapter 45.

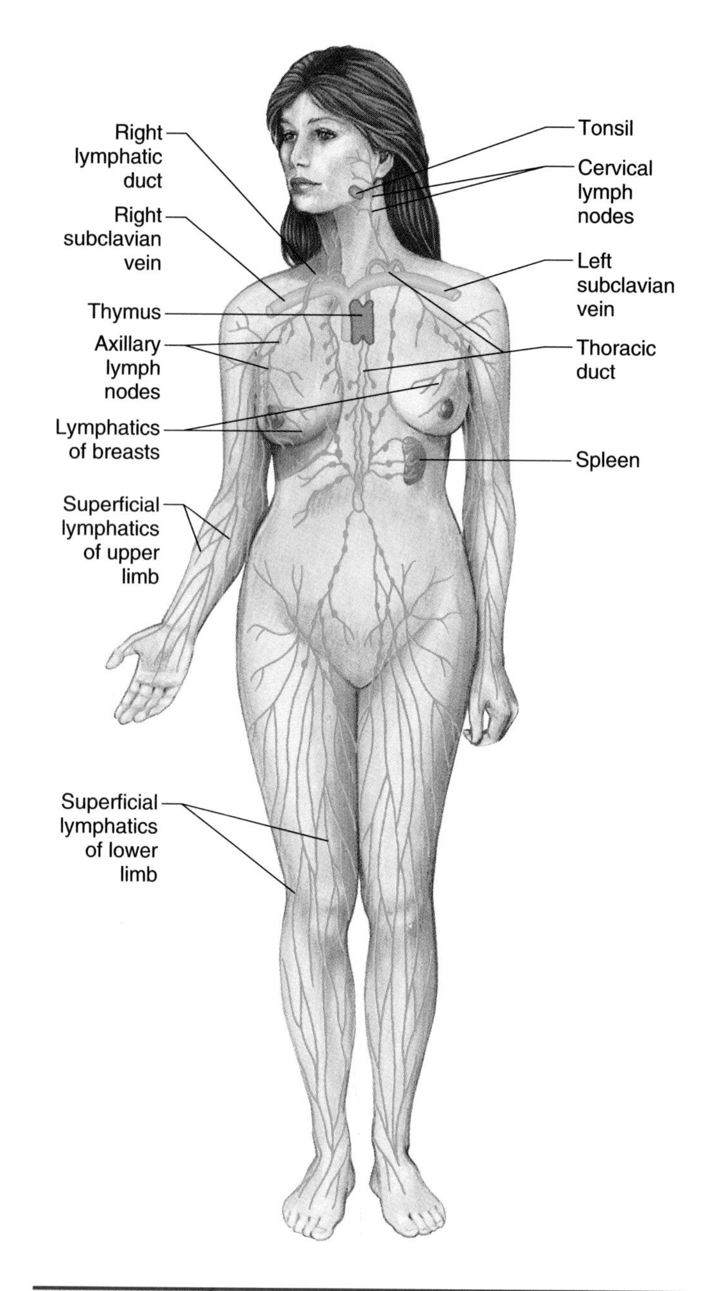

FIGURE **42-18** **Human lymphatic system.**
Note that while the lymphatic vessels extend into most tissues of the body, the lymph nodes are clustered in certain regions. The right lymphatic duct drains lymph from the upper right quadrant of the body. The thoracic duct drains lymph from other regions of the body.

The lymphatic system consists of lymphatic vessels and lymph tissue

The lymphatic system consists of (1) an extensive network of **lymphatic vessels,** or simply **lymphatics,** that conduct **lymph,** the clear, watery fluid formed from interstitial fluid, and (2) **lymph tissue,** a type of connective tissue with large numbers of lymphocytes. Lymph tissue is organized into small masses of tissue called **lymph nodes** and **lymph nodules.** The tonsils, thymus gland, and spleen, which consist mainly of lymph tissue, are also part of the lymphatic system.

Tiny "dead-end" capillaries of the lymphatic system extend into almost all body tissues (Fig. 42-19). Lymph capillaries join to form larger lymphatics (which you can think of as lymph veins). There are no lymph arteries.

Interstitial fluid enters lymph capillaries, becoming lymph. The lymph is conveyed into lymphatics, which at certain locations empty into lymph nodes. As lymph circulates through the lymph nodes, phagocytes filter out bacteria and other harmful materials. The lymph then flows into lymphatics that conduct it away from the lymph node. Lymphatics from all over the body conduct lymph toward the shoulder region. These vessels join the circulatory system at the base of the subclavian veins by way of ducts: the **thoracic duct** on the left side and the **right lymphatic duct** on the right.

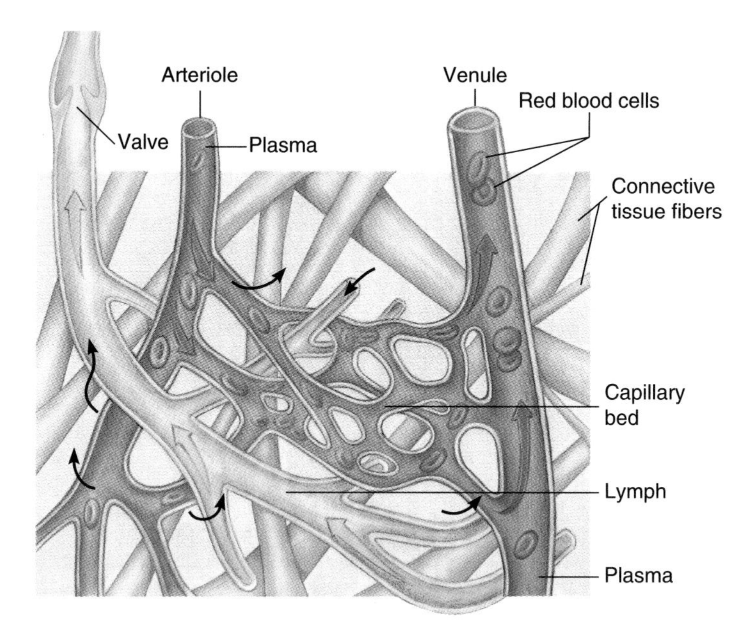

FIGURE **42-19** | **Lymph capillaries.**

Lymph capillaries drain excess interstitial fluid from the tissues. Note that blood capillaries are connected to vessels at both ends, whereas lymph capillaries, shown in green, are "dead-end streets." The arrows indicate direction of flow.

Tonsils are masses of lymph tissue under the lining of the oral cavity and throat. (When enlarged, the pharyngeal tonsils in back of the nose are called **adenoids.**) Tonsils help protect the respiratory system from infection by destroying bacteria and other foreign matter that enter the body through the mouth or nose. Unfortunately, tonsils are sometimes overcome by invading bacteria and become the site of frequent infection themselves.

Some nonmammalian vertebrates, such as frogs, have lymph "hearts," which pulsate and squeeze lymph along. However, in mammals the walls of lymphatic vessels themselves pulsate. Valves within the lymphatics prevent the lymph from flowing backward. When muscles contract or when arteries pulsate, pressure on the lymphatic vessels increases lymph flow. The rate at which lymph flows is slow and variable, and the total lymph flow is about 100 mL per hour—far slower than the 5 L per min of blood flowing in the vascular system.

The lymphatic system plays an important role in fluid homeostasis

When blood enters a capillary network, it is under rather high pressure, so some plasma is forced out of the capillaries and into the tissues. Once it leaves the blood vessels, this fluid is called **interstitial fluid,** or tissue fluid. It contains no red blood cells or platelets and only a few white blood cells. Its protein content is about one fourth that found in plasma, because proteins are too large to pass easily through capillary walls. However, smaller molecules dissolved in the plasma do pass out with the fluid leaving the blood vessels. Thus interstitial fluid contains glucose, amino acids, other nutrients, and oxygen, as well as a variety of salts. This nourishing fluid bathes all the cells.

The main force *(filtration pressure)* pushing plasma out of the blood is hydrostatic pressure, that is, the blood pressure against the capillary wall (Fig. 42-20). The osmotic pressure of the interstitial fluid adds to the filtration pressure. The principal opposing force is the osmotic pressure of the blood which restrains fluid loss from the capillary.

At the venous ends of the capillaries the blood pressure is much lower, and the osmotic pressure of the blood draws fluid back into the capillary. However, not as much fluid is absorbed back into the circulation as is filtered out. Furthermore, protein does not return effectively into the venous capillaries and instead tends to accumulate in the interstitial fluid. These potential problems are so serious that without the lymphatic system, fluid balance in the body would be significantly disturbed within a few hours, and death would occur within about 24 hours. The lymphatic system preserves fluid balance by collecting some of

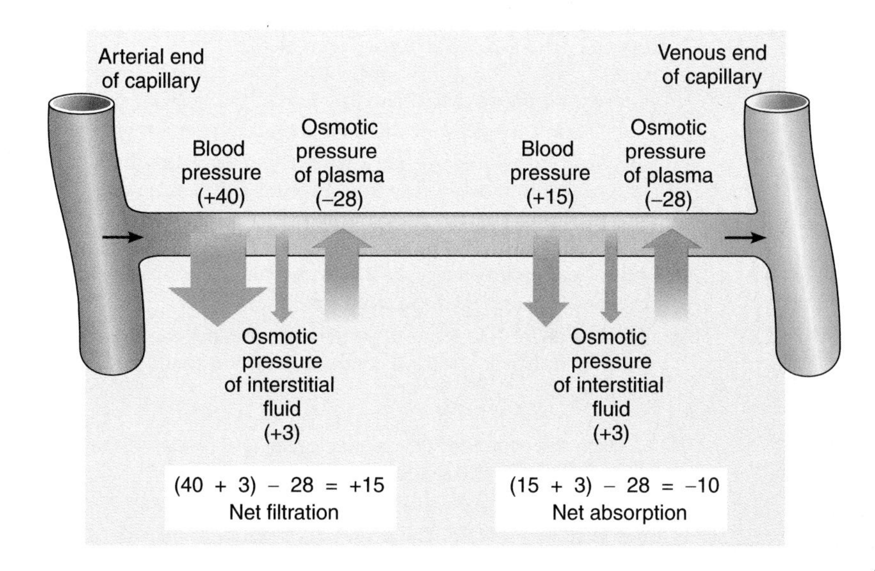

FIGURE **42-20** | **Fluid movement between blood and interstitial fluid.**

Blood pressure and osmotic pressures are responsible for fluid movement, and, thus, exchange of dissolved materials, between blood and interstitial fluid. At the arterial end of a capillary, blood pressure forces plasma out of the capillary. The osmotic pressure of the blood is an opposing force acting to draw fluid into the blood. Osmotic pressure of the interstitial fluid contributes to the net filtration pressure but does not change much between arterial and venous ends of the capillary. At the venous end of the capillary, fluid enters the blood because blood pressure is much lower. However, more fluid leaves the blood than returns, and the lymphatic system collects the excess interstitial fluid. The numbers given are hypothetical and represent millimeters of mercury. Net filtration is the total pressure moving fluid out of the capillary. Net absorption is the total pressure drawing fluid into the capillary.

the interstitial fluid (about 10%), including the protein that accumulates in it, and returning it to the circulation.

The walls of the lymph capillaries consist of endothelial cells that overlap slightly. When interstitial fluid accumulates, it presses against these cells, pushing them inward like tiny swinging doors that swing in only one direction. As fluid accumulates within the lymph capillary, these cell doors are pushed closed.

Obstruction of the lymphatic vessels causes **edema,** swelling from excessive accumulation of interstitial fluid. Lymphatic vessels can be blocked as a result of injury, inflammation, surgery, or parasitic infection. For example, when a breast is removed (mastectomy) because of cancer, lymph nodes in the underarm region may also be removed, to prevent the spread of cancer cells. The disrupted lymph circulation makes the patient's arm swell. New lymphatic vessels develop within a few months, and the swelling slowly subsides.

Review

- What are the relationships among plasma, interstitial fluid, and lymph?
- How does the lymphatic system help maintain fluid balance?

BiologyNow™ Assess your understanding of **the lymphatic system** by taking the pretest on your BiologyNow CD-ROM.

SUMMARY WITH KEY TERMS

1 Compare and contrast internal transport in animals with no circulatory system, those with an open circulatory system, and those with a closed circulatory system.

- Small, simple invertebrates, such as sponges, cnidarians, and flatworms, depend on **diffusion** for internal transport. Larger animals require a specialized **circulatory system,** which typically consists of **blood,** a **heart,** and a system of blood vessels or spaces through which blood circulates. In all animals, **interstitial fluid,** the tissue fluid between cells, brings oxygen and nutrients into contact with cells.
- Arthropods and most mollusks have an **open circulatory system** in which blood flows into a **hemocoel,** bathing the tissues directly.
- Some invertebrates and all vertebrates have a **closed circulatory system** in which blood flows through a continuous circuit of blood vessels.
- The vertebrate circulatory system consists of a muscular heart that pumps blood into a system of **arteries, capillaries,** and **veins.** This system transports nutrients, oxygen, wastes, and hormones; helps maintain fluid balance, appropriate pH, and body temperature; and defends the body against disease.

2 Compare the structure and function of plasma, red blood cells, white blood cells, and platelets.

- **Plasma** consists of water, salts, substances in transport, and **plasma proteins,** including albumins, globulins, and fibrinogen.
- **Red blood cells,** also called **erythrocytes,** transport oxygen and carbon dioxide. Red blood cells produce large quantities of **hemoglobin,** a red pigment that binds with oxygen.
- **White blood cells,** also called **leukocytes,** defend the body against disease organisms. **Lymphocytes** and **monocytes** are agranular white blood cells; **neutrophils, eosinophils,** and **basophils** are granular white blood cells.
- **Platelets** patch damaged blood vessels and release substances essential for blood clotting.

3 Summarize the sequence of events involved in blood clotting.

- Damaged cells and platelets release substances that activate clotting factors. Prothrombin is converted to **thrombin,** which catalyzes the conversion of **fibrinogen** to an insoluble protein, **fibrin.** Fibrin forms long threads that form the webbing of the clot.

4 Compare the structure and function of different types of blood vessels, including arteries, arterioles, capillaries, and veins.

- Arteries carry blood away from the heart; veins return blood to the heart.
- **Arterioles** constrict (**vasoconstriction**) and dilate (**vasodilation**) to regulate blood pressure and distribution of blood to the tissues.
- Capillaries are the thin-walled exchange vessels through which blood and tissues exchange materials.

5 Trace the evolution of the vertebrate heart from fish to mammal.

- The vertebrate heart has one or two **atria,** which receive blood, and one or two **ventricles,** which pump blood into the arteries.
- The fish heart consists of a single atrium and ventricle that are part of a single circuit of blood flow.
- In terrestrial vertebrates, complex circulatory systems separate oxygen-rich from oxygen-poor blood; this allows the higher metabolic rate needed to support an active terrestrial lifestyle.
- Amphibians have two atria and a ventricle, and blood flows through a double circuit so that oxygen-rich blood is partly separated from oxygen-poor blood. Most reptiles have a wall that partly divides the ventricles, minimizing the mixing of oxygen-rich and oxygen-poor blood.
- The four-chambered hearts of birds and mammals separate oxygen-rich blood from oxygen-poor blood.

6 Describe the structure and function of the human heart. (Include the heart's conduction system in your answer.)

- The heart is enclosed by a **pericardium** and has valves that prevent backflow of blood. The valve between right atrium and ventricle is the **right atrioventricular (AV) valve,** or **tricuspid valve.** The valve between left atrium and ventricle is the **mitral valve. Semilunar valves** guard the exits from the heart. Cardiac muscle fibers are joined by **intercalated discs.**
- The **sinoatrial (SA) node,** or **pacemaker,** initiates each heartbeat. A specialized electrical conduction system coordinates heartbeats.

7 Trace the events of the cardiac cycle, and relate normal heart sounds to these events.

- One complete heartbeat makes up a **cardiac cycle.** Contraction occurs during **systole.** The period of relaxation is **diastole.** At the beginning of ventricular systole, the closing of the AV valves, makes a low-pitched "lub" sound. The closing of the semilunar valves, the beginning of ventricular diastole, makes a short, loud, sharp "dup" sound.

8 Define *cardiac output,* describe how it is regulated, and identify factors that affect it.

- **Cardiac output (CO)** equals **stroke volume** times heart rate. Stroke volume depends on **venous return** and on neural messages and hormones, especially epinephrine and norepinephrine.
- According to **Starling's law of the heart,** the more blood delivered to the heart by the veins, the more blood the heart pumps.
- Heart rate is regulated mainly by the nervous system and is influenced by hormones and body temperature.

9 Identify factors that determine and regulate blood pressure, and compare blood pressure in different types of blood vessels.

- **Blood pressure** is the force blood exerts against the inner walls of the blood vessel. Blood pressure is greatest in the arteries and decreases as blood flows through the capillaries.
- Blood pressure depends on cardiac output, blood volume, and resistance to blood flow. **Peripheral resistance** is the resistance to blood flow caused by blood viscosity and by friction between blood and blood vessel wall.
- **Baroreceptors** sensitive to blood pressure changes send messages to the cardiac and vasomotor centers in the medulla of the brain. When informed of an increase in blood pressure, the cardiac center stimulates parasympathetic nerves that slow heart rate, and the vasomotor center inhibits sympathetic nerves that constrict blood vessels. These actions reduce blood pressure.
- **Angiotensin II** is a hormone that raises blood pressure. **Aldosterone** helps regulate salt excretion, which affects blood volume and blood pressure.

10 Trace a drop of blood through the pulmonary and systemic circulations, naming in sequence each structure through which it passes.

- The **pulmonary circulation** connects heart and lungs; the **systemic circulation** connects the heart and the tissues.
- In the pulmonary circulation, the right ventricle pumps blood into the **pulmonary arteries,** one going to each lung. Blood circulates through pulmonary capillaries in the lung and then is conducted to the left atrium by a **pulmonary vein.**
- In the systemic circulation, the left ventricle pumps blood into the **aorta,** which branches into arteries leading to the body organs. After flowing through capillary networks within various organs, blood flows into veins that conduct it to the **superior vena cava** or **inferior vena cava** which returns blood to the right atrium.
- The **coronary arteries** supply the heart muscle with blood.
- The **hepatic portal system** circulates nutrient-rich blood through the liver.

11 Describe the structure and functions of the lymphatic system.

- The **lymphatic system** collects interstitial fluid, and returns it to the blood. It plays an important role in homeostasis of fluids. The lymph system also defends the body against disease and absorbs lipids from the digestive tract.
- **Lymphatic vessels** conduct **lymph,** a clear fluid formed from interstitial fluid, to the **thoracic duct** and **right lymphatic duct** in the shoulder region; these ducts return lymph to the blood circulatory system.
- **Lymph nodes** are small masses of tissue that filter bacteria and harmful materials out of lymph.

POST-TEST

1. An open circulatory system (a) is found in flatworms (b) typically includes a hemocoel (c) has a continuous circuit of vessels with openings in the capillaries (d) is characteristic of vertebrates (e) is typically found in animals with a two-chambered heart
2. Which of the following is *not* a function of the vertebrate circulatory system? (a) helps maintain appropriate pH (b) transports nutrients, oxygen, and metabolic wastes (c) helps maintain fluid balance (d) produces hemocyanin (e) provides internal defense
3. Lipoproteins (a) are mainly transported in granular leukocytes (b) transport cholesterol (c) have been linked to clotting disorders (d) are associated with platelets (e) are stored in red blood cells
4. Which of the following are most closely associated with oxygen transport? (a) red blood cells (b) platelets (c) neutrophils (d) basophils (e) lymphocytes
5. Which of the following are most closely associated with blood clotting? (a) red blood cells (b) platelets (c) neutrophils (d) basophils (e) lymphocytes
6. In blood clotting (a) thrombin ⟶ prothrombin; fibrinogen ⟶ fibrin (b) prothrombin ⟶ thrombin; fibrin ⟶ fibrinogen (c) prothrombin ⟶ thrombin; fibrinogen ⟶ fibrin (d) clotting factors ⟶ platelets; thrombin ⟶ fibrinogen (e) prothrombin ⟶ thrombin; fibrinogen ⟶ platelets
7. Blood vessels that carry blood away from the heart are (a) arteries (b) sinuses (c) veins (d) capillaries (e) arterioles and venules
8. Arterioles (a) help regulate blood pressure (b) help regulate distribution of blood to the tissues (c) deliver blood to arteries (d) answers a, b, and c are correct (e) answers a and b only
9. Which choice most accurately describes one sequence of blood flow? (a) right atrium ⟶ right ventricle ⟶ pulmonary artery (b) right atrium ⟶ left atrium ⟶ left ventricle ⟶ aorta (c) left atrium ⟶ left ventricle ⟶ pulmonary artery (d) left ventricle ⟶ left atrium ⟶ aorta (e) right atrium ⟶ right ventricle ⟶ aorta
10. Which choice most accurately describes one sequence of blood flow? (a) pulmonary vein ⟶ pulmonary artery ⟶ right atrium (b) pulmonary artery ⟶ left atrium ⟶ left ventricle (c) pulmonary artery ⟶ pulmonary capillaries ⟶ pulmonary vein ⟶ left atrium (d) left ventricle ⟶ aorta ⟶ pulmonary artery (e) pulmonary artery ⟶ pulmonary capillaries ⟶ pulmonary vein ⟶ right atrium

POST-TEST *(continued)*

11. A cardiac cycle (a) consists of one ventricular heartbeat (b) includes a systole (c) equals stroke volume times heart rate (d) includes a diastole (e) includes a systole and a diastole
12. Blood pressure is determined by (a) cardiac output (b) peripheral resistance (c) blood volume (d) answers a, b, and c are correct (e) answers b and c only
13. Lymph forms from (a) interstitial fluid (b) blood serum (c) plasma combined with protein (d) fluid released by lymph nodes (e) angiotensins
14. The valve between the right atrium and right ventricle is the (a) mitral valve (b) semilunar valve (c) tricuspid valve (d) pulmonary valve (e) aortic valve
15. Norepinephrine (a) slows heart rate (b) is released in cardiac muscle by parasympathetic nerves (c) causes potassium channels in cardiac muscle to open (d) decreases stroke volume (e) causes calcium channels in cardiac muscle to open
16. Baroreceptors (a) stimulate renin release (b) activate the rennin-angiotensin-aldosterone pathway (c) stimulate sympathetic nerves (d) are stimulated by increased blood pressure (e) send messages to cardiac centers that increase blood pressure
17. Atherosclerosis (a) is associated with thickening of arteries and veins (b) is associated with high concentrations of low-density lipoprotein (c) can lead to ischemic heart disease (d) answers a, b, and c are correct (e) answers b and c only
18. Label the diagram. See Figure 42-9 to check your answers.

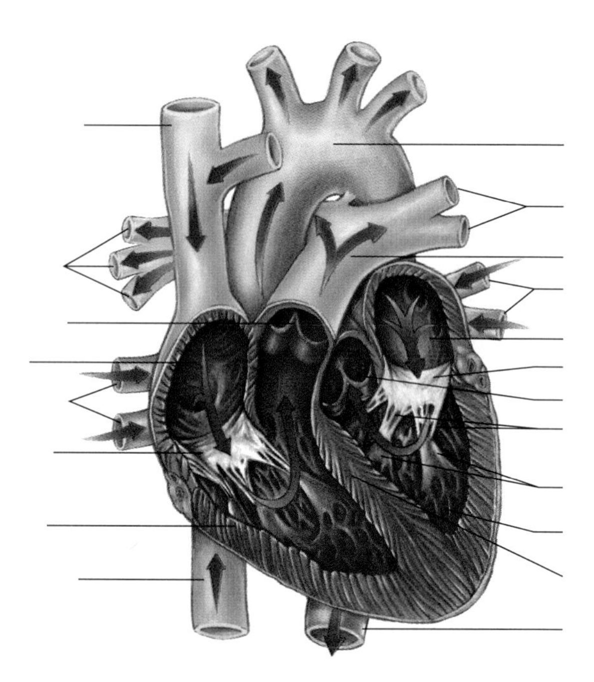

CRITICAL THINKING

1. How is the heart of the fish specifically adapted to its lifestyle?
2. When the nerves to the heart are cut, the heart rate increases to about 100 contractions per minute. What does this indicate about the regulation of the heart rate?
3. List five modifiable risk factors associated with the development of cardiovascular disease. Describe the disease process in atherosclerosis, and explain the association between atherosclerosis and ischemic heart disease. What happens in myocardial infarction?

- Visit our Web site at **http://biology.brookscole.com/solomon7** for links to chapter-related resources on the World Wide Web. Additional online materials relating to this chapter can also be found on our Web site.

BIOLOGY NOW RESOURCES

Active Figures

42-9: Anatomy of the heart
42-10: Conduction system of the heart
42-14: Blood pressure

Preparing for an exam? Take a diagnostic test on your BiologyNow CD-ROM.

Post-Test Answers

1. b	2. d	3. b	4. a
5. b	6. c	7. a	8. e
9. a	10. c	11. e	12. d
13. a	14. c	15. e	16. d
17. d			

43

The Immune System: Internal Defense

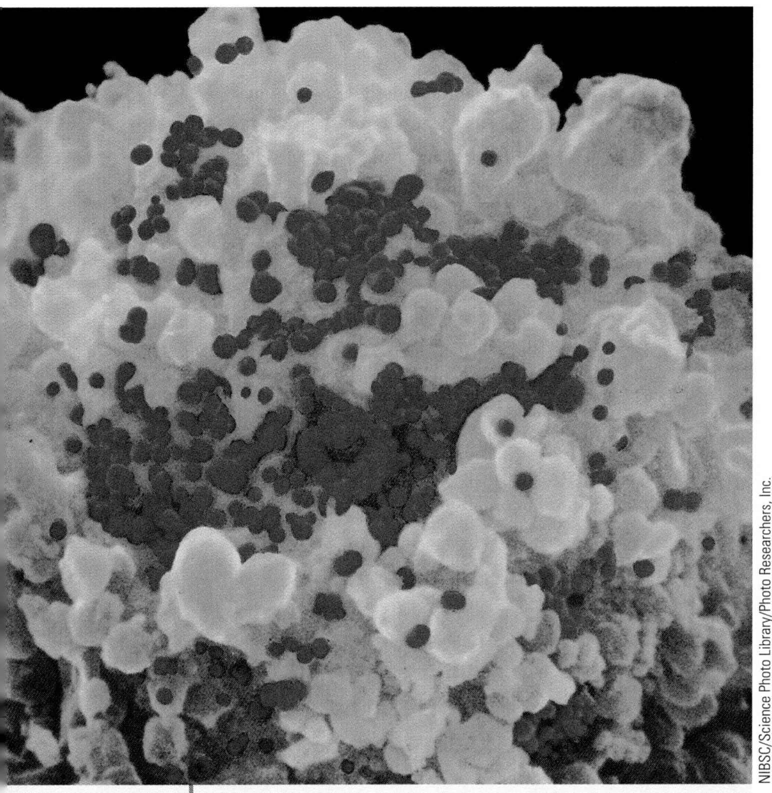

NIBSC/Science Photo Library/Photo Researchers, Inc.

A T cell infected with HIV. In this colorized SEM, an immune system T cell *(green)* is infected by HIV viruses *(red)*.

CHAPTER OUTLINE

The **immune system,** our internal defense system, protects the body against disease-causing organisms and certain toxins. Disease-causing organisms, or **pathogens,** include certain viruses, bacteria, fungi, and protozoa. Pathogens enter the body with air, food, and water; during copulation; and through wounds in the skin. The immune system recognizes pathogens and toxins and responds to eliminate them. Derived from the Latin for "safe," the word *immune* refers to the early observation that when a person recovered from smallpox and other serious infection, they were safe from contracting the same illnesses again. **Immunology,** the study of internal defense systems of humans and other animals, is one of the most rapidly changing, challenging, and exciting fields of biomedical research today.

For more than 50 years, immunologists based their work on the hypothesis that internal defense depends on the animal's ability to distinguish between *self* and *nonself.* Such recognition is possible because each individual is biochemically unique. Cells have surface proteins different from those on the cells of other species or even other members of the same species. An animal's immune system recognizes its own cells and can identify those of other organisms as foreign. Thus when a pathogen invades an animal, its distinctive macromolecules stimulate the animal's defensive responses. A single bacterium may have from 10 to more than 1000 distinct macromolecules on its surface.

Immunologists are aware of several limitations of the self–nonself hypothesis. For example, the immune system does not typically respond to foreign molecules that are harmless. In 1994, Polly Matzinger, of the U. S. National Institutes of Health, proposed the **danger model,** which hypothesizes that the immune system does more than distinguish between self and nonself. It responds to danger signals from injured tissues, such as proteins released when cell membranes are damaged.

Most immunologists now agree that internal defense relies on a combination of factors, including the ability to identify foreign molecules and to respond to chemical clues from injured tissues. The immune system is a collection of many types of cells and of tissues scattered throughout the body. Immune responses require communication among cells, or **cell signaling.** Cells of the immune system communicate directly by means of their surface molecules and indirectly by releasing messenger molecules. Understanding the complex signaling systems of the immune system is a major focus of research.

Sometimes pathogens overcome the body's internal defenses, resulting in disease. Some diseases, as well as certain genetic mutations, prevent or compromise immune function. HIV, the retrovirus that causes AIDS, infects T cells, an important component of the immune system (see photograph). The immune system may overfunction, as in allergic reactions, or it may respond in ways that are clinically important, such as in Rh incompatibility or the destruction of the cells of organ transplants. In about 5% of adults in highly developed countries, certain immune responses are directed against self tissues, resulting in autoimmune disease.

Among the greatest accomplishments of immunologists are the development of vaccines that prevent disease, and techniques for successful tissue and organ transplantation. Sophisticated research tools, such as gene transfer, have enabled immunologists to expand their knowledge of the cells and molecules that interact to generate immune responses, and to develop new approaches to the prevention and treatment of disease. Much has been learned, and many challenges lie ahead. ■

NONSPECIFIC AND SPECIFIC IMMUNITY: AN OVERVIEW

Learning Objectives

1. Distinguish between nonspecific and specific immune responses.
2. Compare, in general terms, the immune responses of invertebrates and vertebrates.

An **immune response** is the process of recognizing foreign or dangerous macromolecules and responding to eliminate them. Two main types of immune responses protect the body: nonspecific and specific. **Nonspecific immune responses,** or *innate immunity,* provide general protection against pathogens, parasites, some toxins and drugs, and cancer cells. Nonspecific immune responses prevent most pathogens from entering the body and rapidly destroy those that do penetrate the outer defenses. For example, the cuticle or skin provides a physical barrier to pathogens that come in contact with an animal's body. Phagocytosis, another nonspecific defense, destroys bacteria that invade the body. Some of the molecules important in nonspecific immune responses recognize and attack certain pathogen-associated molecular patterns, which are shared by whole groups of viruses, bacteria, or fungi.

Specific immune responses, also referred to as *adaptive* or *acquired immunity,* are highly specific. Any molecule that cells of the immune system specifically recognize as foreign is called an **antigen.** Proteins are the most powerful antigens, but some polysaccharides and lipids can be antigenic. **Antibodies** are highly specific proteins that recognize and bind to specific antigens. Specific immune responses are directed toward particular antigens and typically include the production of antibodies. In complex animals, an important characteristic of specific immune responses is **immunological memory,** the capacity to respond more effectively the second time foreign molecules invade the body.

Invertebrates launch nonspecific immune responses

All invertebrate species that researchers have studied demonstrate the ability to distinguish between their own cells and those of other species. For example, sponge cells have specific glycoproteins on their surfaces that enable them to recognize their own species. When cells of two different species are mixed together, they reassort according to species. When two different species of sponges are forced to grow in contact with each other, tissue is destroyed along the region of contact. Cnidarians (such as corals), annelids (such as earthworms), arthropods (such as insects), and echinoderms (such as sea stars) reject tissue grafted from other animals, even from the same species.

Invertebrates have very efficient nonspecific immune mechanisms. For example, many invertebrates (cnidarians, annelids, and mollusks) are covered by mucus that traps and kills pathogens. Tough external skeletons, such as shells or cuticles, shield the body of many invertebrates. Most invertebrate coelomates (animals with a coelom) have amoeba-like **phagocytes** that engulf and destroy bacteria and other foreign matter. In mollusks, substances in the hemolymph (blood) enhance phagocytosis.

Researchers have identified **antimicrobial peptides** in all eukaryotes (including plants) that have been studied, suggesting an early common origin of these molecules. More than 800 of these peptides that inactivate or kill pathogens have been described! When researchers inject an antigen into an insect, as many as 15 antimicrobial peptides are produced within a few hours. Antimicrobial peptides are very effective because of their small size (a dozen or fewer amino acids), which facilitates their rapid production and diffusion.

What stimulates an animal to produce antimicrobial peptides and other immune defenses? Animal cells have receptors that recognize certain types of pathogen molecules, and then signal the cell to produce antimicrobial peptides. One important family of these signaling receptors, the *Toll group,* is a focus of current research. Immunologists first identified the Toll group in the fruit fly *Drosophila.* Toll receptors recognize some common molecular features of classes of pathogens called **pathogen-associated molecular patterns,** or **PAMPs.** Examples of PAMPs include the double-stranded RNA of certain viruses and peptidoglycan in Gram-positive bacteria.

Certain cnidarians, arthropods, some echinoderms and simple chordates (such as tunicates) appear to remember antigens for a short period. As mentioned earlier, immunological memory enables the body to respond more effectively when it encounters the same pathogens again. Although certain invertebrates demonstrate some specificity and memory, their immune responses are primarily nonspecific. Echinoderms and tunicates are the simplest animals known to have differentiated white blood cells that perform limited immune functions.

Vertebrates launch nonspecific and specific immune responses

A specialized lymphatic system, including **lymphocytes,** white blood cells specialized to carry out immune responses, evolved in the vertebrates (see Chapter 42). The lymphatic system performs the sophisticated specific immune responses of vertebrates. In the discussion that follows, we focus on the human immune system, with references to immune mechanisms of other animals (Fig. 43-1).

Review

- What are two key ways in which specific immune responses differ from nonspecific immune responses?
- How do vertebrate immune responses differ from invertebrate responses?

BiologyNow™ Assess your understanding of **nonspecific and specific immunity,** in general by taking the pretest on your BiologyNow CD-ROM.

NONSPECIFIC IMMUNE RESPONSES

Learning Objective

3 Describe nonspecific immune responses, including physical and chemical barriers; soluble molecules such as cytokines and the proteins that make up the complement system; phagocytes; natural killer cells; and the inflammatory response.

Physical and chemical barriers prevent most pathogens from entering the body. An animal's first line of defense is its outer covering—skin, cuticle, shell, or chitin—which blocks the entry of pathogens. Microorganisms that enter with food are usually destroyed by the acid secretions and enzymes of the stomach, which are effective chemical barriers.

Mucous membranes of the respiratory and reproductive tracts protect passageways that open to the outside of the body. For example, pathogens that enter the body with inhaled air may be filtered out by hairs in the nose or trapped in the sticky mucous lining of the respiratory tract. Once trapped, they are destroyed by phagocytes. Mucus contains a substance called *mucin,* which chemically destroys invaders. **Lysozyme,** an enzyme found in many tissues, as well as in tears and other body fluids, attacks the cell walls of many Gram-positive bacteria.

When pathogens breach the body's outer barriers, their chemical properties activate other nonspecific defense responses, and most invaders are quickly eliminated. In addition to the front-line physical and chemical barriers, nonspecific immune responses involve soluble molecules, phagocytes, and natural killer (NK) cells. The activation and concentration of these molecules and cells produce inflammation, the major nonspecific immune response.

Soluble molecules mediate immune responses

Cells of the immune system secrete a remarkable number of regulatory and antimicrobial peptides and proteins. Two major

ACTIVE FIGURE **43-1** **Overview of human immune responses.**

Nonspecific immune responses include physical and chemical barriers that prevent the entrance of pathogens; soluble molecules; phagocytosis; and inflammation. During inflammation, certain cells release cytokines that activate specific immune responses: cell-mediated and antibody-mediated immunity.

BiologyNow™ Watch **the immune system fight a cold** by clicking on this figure on your BiologyNow CD-ROM.

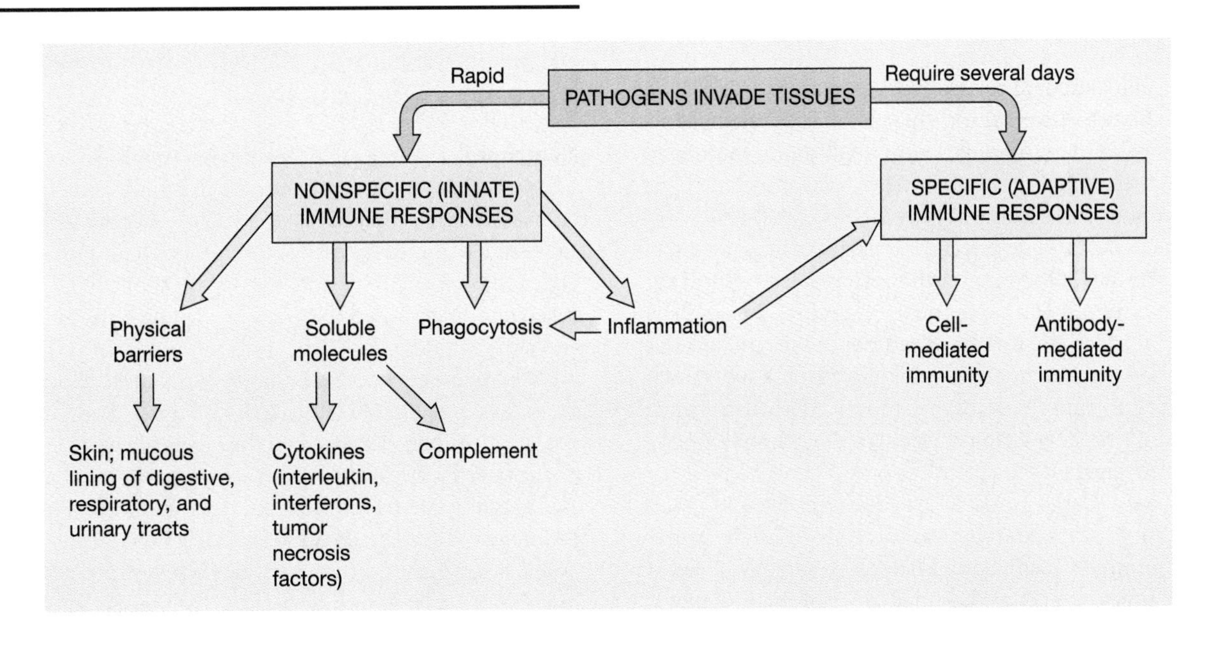

groups of soluble molecules in nonspecific defense responses are cytokines and complement.

Cytokines are important signaling molecules

Cytokines are a large group of peptides and proteins (most are glycoproteins) that serve as signals and perform regulatory functions during both nonspecific and specific immune responses. Cytokines help regulate the intensity and duration of immune responses, and they are also important in regulating many other biological processes, such as cell growth, repair, and cell activation. Cytokines can act as *autocrine agents,* affecting the very cells that produce them, or as *paracrine agents* regulating the activity of nearby cells (see Chapter 47). Some cytokines circulate in the blood and affect distant tissues. Biologists name the types of cytokines for their function or origin. We discuss four groups: interferons, interleukins, chemokines, and tumor necrosis factors.

When infected by viruses or other intracellular parasites (some types of bacteria, fungi, and protozoa), cells respond by secreting cytokines called **interferons.** Type I interferons are produced by either macrophages, which are large phagocytic cells, or fibroblasts, cells that produce the fibers of connective tissues. Type I interferons inhibit viral replication and also activate natural killer cells that have antiviral actions. Viruses produced in cells exposed to Type I interferons are not very effective at infecting other cells.

Type II interferons, produced as a specific immune response, also exhibit anti-viral activity, but in addition, they enhance the activities of other immune cells. For example, Type II interferons stimulate macrophages to destroy tumor cells and host cells that have been infected by viruses.

Since their discovery in 1957, interferons have been the focus of a great deal of research. Pharmaceutical companies now use recombinant DNA techniques to produce large quantities of some interferons. The U.S. Food and Drug Administration (FDA) has approved interferons for treating several diseases, including hepatitis B and hepatitis C, genital warts, a type of leukemia, a type of multiple sclerosis, and AIDS-related Kaposi's sarcoma. Researchers are testing interferons in clinical trials for the treatment of HIV infection and several types of cancer.

Interleukins are a diverse group of cytokines secreted mainly by macrophages and lymphocytes. Interleukins are numbered according to their order of discovery. They regulate interactions between lymphocytes and other cells of the body, and some interleukins have widespread effects. For example, during infection *interleukin-1 (IL-1)* can reset the body's thermostat in the hypothalamus, resulting in **fever** and its symptoms. Just as there are overlaps between the functions of nonspecific and specific immune responses, the actions of cytokines of these subsystems also overlap. For example, cytokines produced by nonspecific cells such as macrophages can activate lymphocytes involved in specific immune responses.

Chemokines, a large group of cytokines, are signaling molecules that attract, activate, and direct the movement of various cells of the immune system. Some chemokines are produced in response to infection and are key mediators of the inflammatory response.

Tumor necrosis factors (TNFs) are cytokines secreted by macrophages and by lymphocytes called *T cells.* TNF stimulates immune cells to initiate an inflammatory response. TNF also kills tumor cells, offering promise in terms of immunotherapy for cancer patients. Sometimes infection by Gram-negative bacteria, such as *Salmonella typhi,* results in the release of large amounts of TNF and other cytokines. This results in a cascade of reactions leading to *septic shock,* a potentially fatal condition that may involve high fever and malfunctioning of the circulatory system. Thus cytokines can sometimes have harmful effects.

Complement leads to the destruction of pathogens

Cytokines produced by phagocytes can activate the *complement system.* **Complement,** so named because it "complements" the action of other defensive responses, consists of more than 20 proteins present in plasma and other body fluids. Similarities in complement proteins in many species, including horseshoe crabs, sea urchins, tunicates, and mammals, suggest these molecules evolved millions of years ago and have been conserved.

Normally, complement proteins are inactive until the body is exposed to an antigen. Certain pathogens activate the complement system directly. In other cases, the binding of an antigen and antibody stimulate activation. Complement activation involves a cascade of reactions; each component acts on the next in the series. Proteins of the complement system then work to destroy pathogens.

Complement proteins are activated against many antigens, and their actions are nonspecific: (1) They lyse viruses, bacteria, and other cells; (2) they coat pathogens, making them less "slippery" so that phagocytes (macrophages and neutrophils) phagocytose them more easily; (3) they attract white blood cells to the site of infection (a process called *chemotaxis*); and (4) they bind to specific receptors on cells of the immune system, stimulating specific actions, such as secreting regulatory molecules, and enhancing the inflammatory response.

Phagocytes and natural killer cells destroy pathogens

Neutrophils, the most common type of white blood cell, (see Chapter 42) and macrophages are the main phagocytes in the body. Recall that **phagocytosis** is a type of endocytosis in which cells engulf microorganisms, foreign matter, or other cells (see Fig. 5-18). A neutrophil can phagocytize about 20 bacteria before it becomes inactivated (perhaps by leaking lysosomal enzymes) and dies.

Macrophages are large phagocytes that develop from nongranular white blood cells called *monocytes.* A macrophage can phagocytize about 100 bacteria during its life span. Some macrophages patrol the body's tissues, phagocytizing damaged cells or foreign matter (including bacteria) and, when appropriate, they release antiviral agents (Fig. 43-2). Others stay in one place and destroy bacteria that pass by. For example, air sacs in the lungs contain large numbers of macrophages that destroy foreign matter entering with inhaled air.

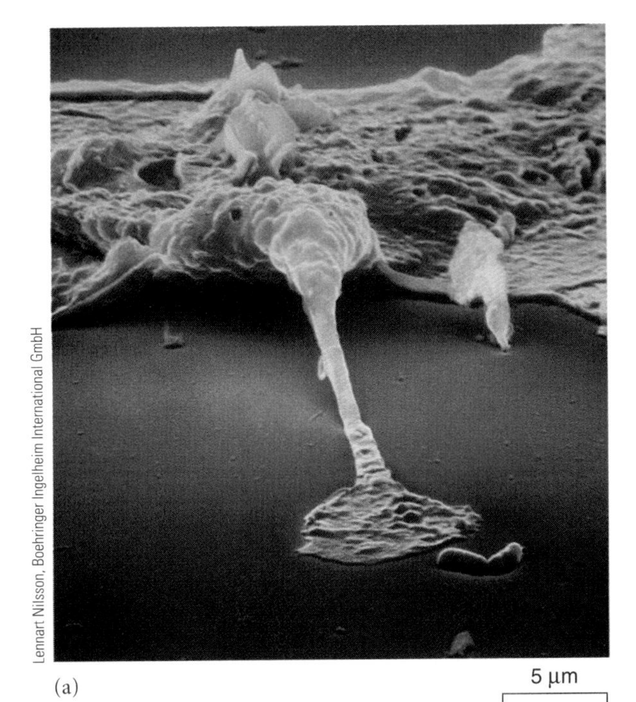

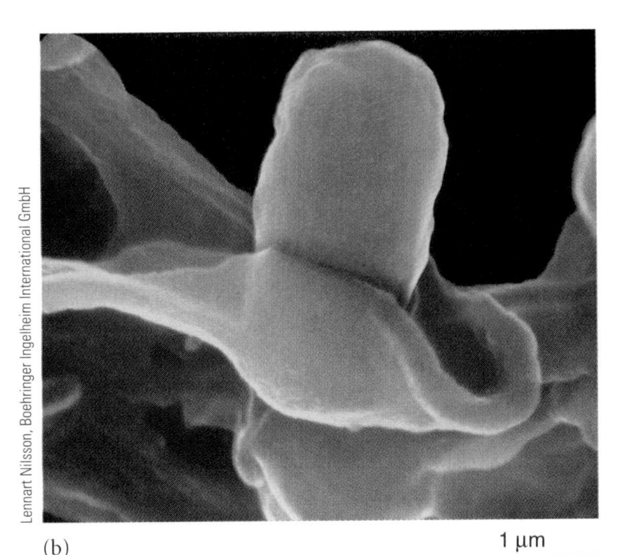

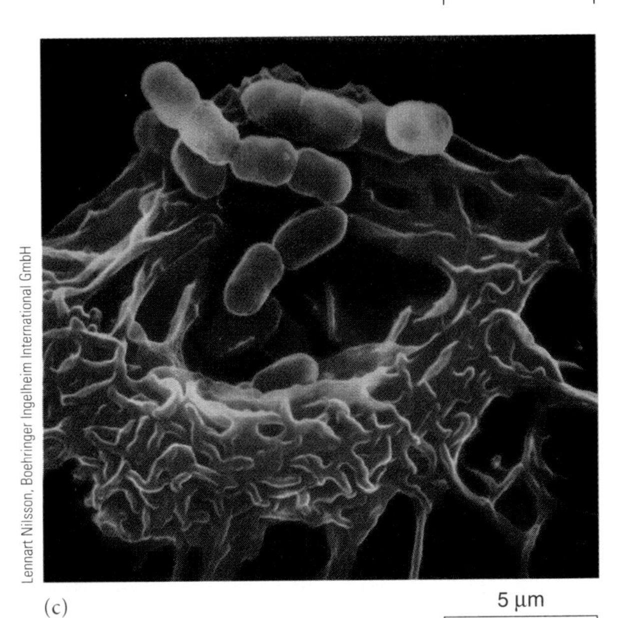

FIGURE **43-2** | **Phagocytosis.**

These colorized SEMs show macrophages, which are efficient warriors. **(a)** A macrophage extends a pseudopod toward an invading *E. coli* bacterium that is already multiplying. **(b)** The bacterium is trapped within the engulfing pseudopod. **(c)** The macrophage takes in the trapped bacteria along with its own plasma membrane. The macrophage plasma membrane will seal over the bacteria, and powerful lysosomal enzymes will destroy them.

Vertebrate macrophages have **Toll-like receptors** (a family of receptors related to the Toll receptors of insects) that recognize certain PAMPs and respond by producing cytokines e.g. TNF. For example, when stimulated by a lipopolysaccharide found on Gram-negative bacteria, macrophages release molecules that enhance the inflammatory response.

Can bacteria counteract the phagocyte's attack? Many strategies have evolved in bacteria that protect them from their hosts. *Streptococcus pneumoniae,* for example, has cell walls or capsules that resist the action of the phagocyte's lysosomal enzymes. Other bacteria release enzymes that destroy the lysosomal membranes. The powerful lysosomal enzymes then spill out into the cytoplasm and may destroy the phagocyte. *Listeria monocytogenes* bacteria escape destruction by surviving within phagocytic cells.

Natural killer (NK) cells are large, granular lymphocytes that originate in the bone marrow. They account for about 10% of circulating lymphocytes. NK cells are active against tumor cells and cells infected with some types of viruses. They destroy target cells by both nonspecific and specific (antibody-mediated) processes. NK cells release cytokines, as well as perforins and granzymes, enzymes that destroy target cells. **Perforins** cause pores to form in the plasma membrane of the target cell, allowing granzymes to enter the cell. **Granzymes** then activate a cascade of reactions that cause the cell to self-destruct by **apoptosis,** programmed cell death. Several cytokines stimulate NK cell activity. When NK cell levels are high, resistance to certain cancers may be strengthened. Psychological stressors are thought to decrease NK cell activity and thus enhance tumor growth.

Inflammation is a protective response

The **inflammatory response,** or *inflammation,* begins immediately after pathogen invasion or physical injury (Fig. 43-3). Tissue injury activates a clotting factor in the blood plasma that turns on three different interconnecting molecular cascades in the plasma, including the clotting cascade. The reactions generate molecules—for example, the peptide *bradykinin*—that mediate the inflammatory process. These plasma mediators dilate blood vessels and increase capillary permeability. Inflammation also activates the complement system.

Large numbers of neutrophils migrate into the inflamed tissue within a few hours. These cells secrete cytokines, including chemokines, that are important mediators of the inflammatory response. Macrophages and **mast cells** stationed in the tissues also respond rapidly to damaged tissue or infection. Mast cells release **histamine,** cytokines, and other compounds that dilate blood vessels in the affected area and increase capillary perme-

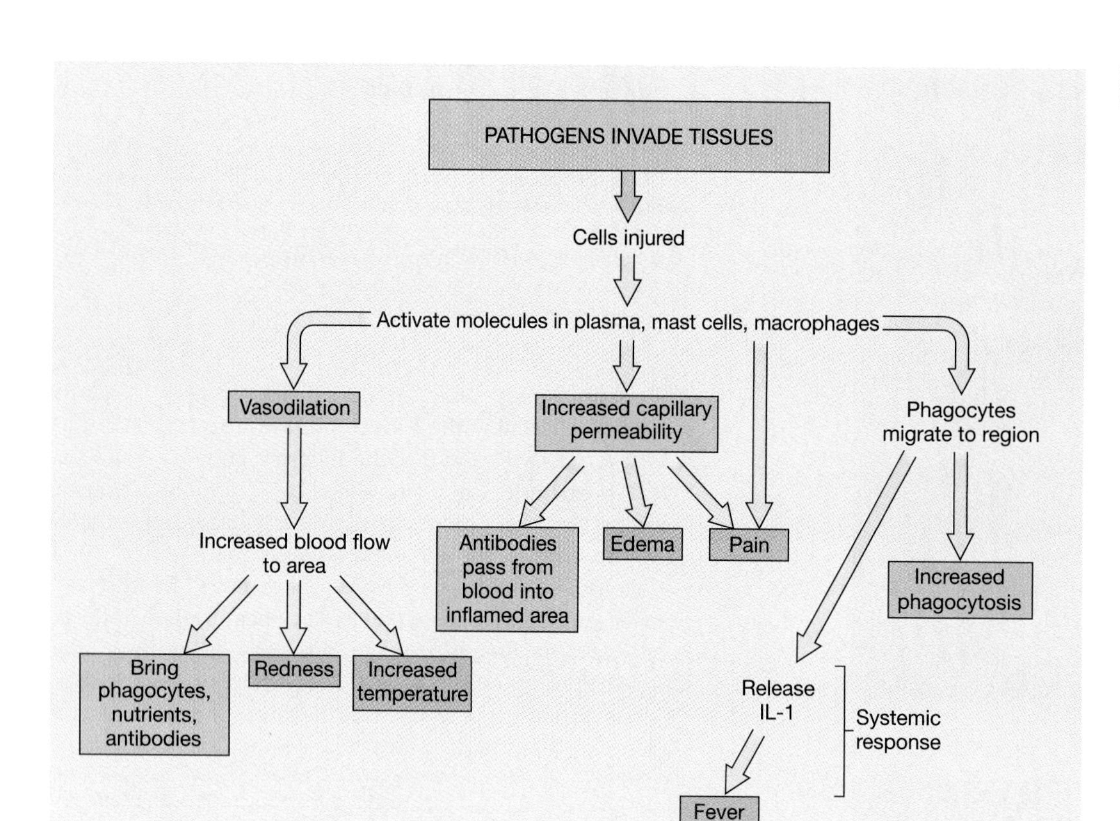

FIGURE **43-3**

The inflammatory response.

ability. Mast cells and macrophages release signaling molecules that attract and activate additional neutrophils. In turn, neutrophils send signals that attract lymphocytes.

The clinical characteristics of inflammation are heat, redness, edema (swelling), and pain. The inflammatory response includes three main processes.

1. *Vasodilation.* Expanded blood vessel diameter brings more blood to the area. The increased blood flow warms the skin and reddens skin that contains little pigment.
2. *Increased capillary permeability.* Fluid and antibodies leave the circulation and enter the tissues. As the volume of interstitial fluid increases, *edema* occurs. The edema, along with the action of certain enzymes in the plasma, cause the pain that characterizes inflammation.
3. *Increased phagocytosis.* Increased blood flow brings large numbers of neutrophils and other phagocytic cells to the infected region. The phagocytes migrate out of the capillaries and into the infected tissues. One of the main functions of inflammation is increased phagocytosis.

Although the inflammatory response begins as a local response, sometimes the entire body becomes involved. Fever, a common clinical symptom of widespread inflammation, helps the body fight infection. Elevated body temperature increases phagocytosis and interferes with the growth and replication of microorganisms. Fever breaks down lysosomes, destroying cells infected by viruses. Fever also promotes the activity of certain lymphocytes. A short-term, low fever speeds recovery.

Review

- What are the main groups of cytokines? What is the function of each group?
- What processes are involved in the inflammatory response? What types of cells help mediate the inflammatory response?

BiologyNow™ Assess your understanding of **nonspecific immune responses** by taking the pretest on your BiologyNow CD-ROM.

SPECIFIC IMMUNE RESPONSES

Learning Objectives

4 Distinguish between cell-mediated and antibody-mediated immunity.

5 Describe the principal cells of the immune system, and summarize the function of the major histocompatibility complex.

While nonspecific immune responses are destroying pathogens and preventing the spread of infection, the body is mobilizing its specific immune responses. It takes several days to activate specific immune responses, but once in gear, these mechanisms are extremely effective. As discussed earlier, specific immune responses are precisely targeted to destroy specific antigens, and they have immunological memory. Two types of specific immune responses are cell-mediated and antibody-mediated immunity.

Many types of cells are involved in specific immune responses

The primary types of cells of the immune system are shown in Figure 43-4. Two key cell types that participate in specific immune responses are lymphocytes and antigen-presenting cells. Phagocytes and many other cells important in nonspecific immune responses also participate in specific responses. **Eosinophils**

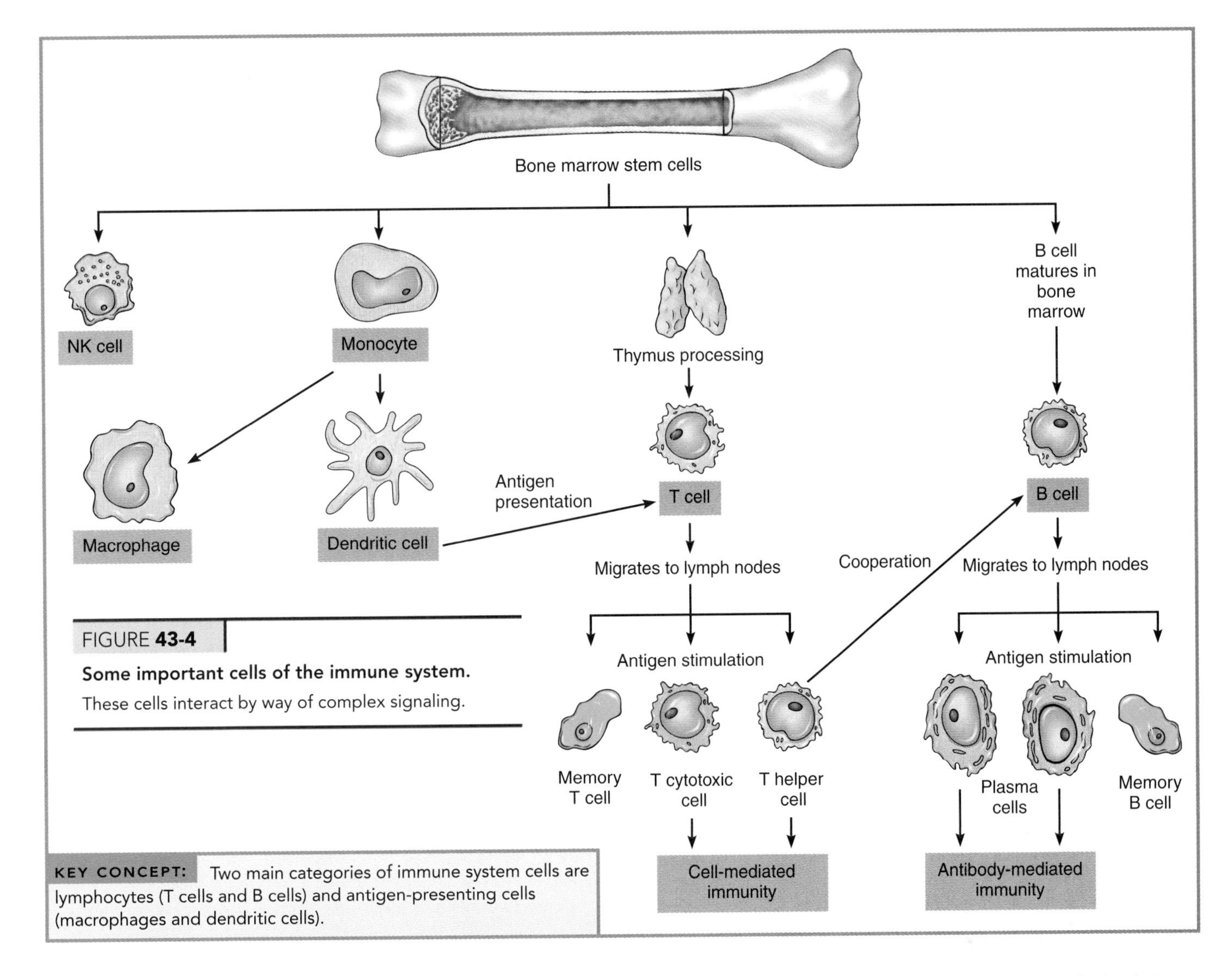

FIGURE **43-4**

Some important cells of the immune system.
These cells interact by way of complex signaling.

KEY CONCEPT: Two main categories of immune system cells are lymphocytes (T cells and B cells) and antigen-presenting cells (macrophages and dendritic cells).

are white blood cells that release toxins from their granules to destroy parasitic worms.

Lymphocytes are the principal warriors in specific immune responses

Three main types of lymphocytes are T lymphocytes, B lymphocytes, and NK cells. As already discussed, NK cells kill virally infected cells and tumor cells. **T lymphocytes,** or **T cells,** are responsible for **cell-mediated immunity.** T cells are the body's cellular soldiers. They travel to the site of infection and attack body cells infected by invading pathogens, as well as foreign cells (such as those introduced in tissue grafts or organ transplants). T cells also destroy cells altered by mutation (cancer cells).

B lymphocytes, or **B cells,** are responsible for **antibody-mediated immunity.** B cells mature into **plasma cells,** which produce specific antibodies. Antibodies bind to specific antigens, neutralizing them or marking them for destruction. Antibody-mediated immunity is one of the body's chief chemical defense strategies.

All lymphocytes develop from stem cells in the bone marrow. B cells complete their development in the adult bone marrow (from which the *B* in their name is derived). T cells mature in the **thymus gland** (from which the *T* in their name derives). Large numbers of mature lymphocytes reside in the secondary lymph organs, including the spleen, lymph nodes, tonsils, and other lymph tissues strategically positioned throughout the body.

Although T and B cells are similar in appearance when viewed with a light microscope, sophisticated techniques such as fluorescence microscopy demonstrate that these cells can be differentiated by their unique cell surface macromolecules. T and B cells have different functions and life histories, and tend to locate in (or "home" to) separate regions of the spleen, lymph nodes, and other lymph tissues.

Millions of B cells are produced in the bone marrow daily. Each B cell is genetically programmed to encode a glycoprotein receptor that binds with a specific type of antigen. When a B cell comes into contact with an antigen that binds to its receptors, it becomes activated. B-cell activation is a complex process that typically requires the participation of a particular type of T cell. Once activated, a B cell divides rapidly, forming a clone of identical cells. The cloned B cells differentiate into plasma cells, which produce antibody, a soluble form of the cell's glycoprotein receptor molecule that can be secreted. A plasma cell can produce more than 10 million molecules of antibody per hour! The antibody binds to the antigen that originally activated the B cells. Some activated

B cells do not become plasma cells, but instead become long-living **memory B cells** that continue to produce small amounts of antibody after the body overcomes an infection.

On their way from the bone marrow to the lymph tissues, T cells are processed in the **thymus gland.** The thymus makes T cells *immunocompetent,* capable of immunological response. As T cells move through the thymus, they divide many times and develop specific surface proteins with distinctive receptor sites. Only the T cells that have specific receptors are selected to divide; this is a form of *positive selection.*

T cells that react to self-antigens undergo apoptosis. This is a form of *negative selection.* Immunologists estimate that more than 90% of developing T cells are negatively selected. The remaining T cells differentiate and leave the thymus to take up residence in other lymph tissues or to launch immune responses in infected tissues. By selecting only appropriate T cells, the thymus gland ensures that T cells can distinguish between the body's own molecules and foreign antigens.

Most T cells in the thymus differentiate just before birth and during the first few months of postnatal life. If the thymus is removed before this processing takes place, an animal does not develop cell-mediated immunity. If the thymus is removed after that time, cell-mediated immunity is less seriously impaired.

T cells are distinguished by the **T-cell receptor (TCR)** which recognizes specific antigens. Two main populations of T cells develop in the thymus: T cytotoxic and T helper cells. **T cytotoxic (T_C) cells** are also known as *CD8 cells* because they have a glycoprotein designated CD8 on their plasma membrane surface. Known less formally as *killer T cells,* T_C cells recognize and destroy cells with foreign antigens on their surfaces. Among their target cells are virus-infected cells, cancer cells, and foreign tissue grafts.

T helper (T_H) cells, also known as *CD4 cells,* have a surface glycoprotein designated CD4. T_H cells are regulatory cells. They secrete cytokines that activate B cells and macrophages. After an infection, both cytotoxic and helper **memory T cells** remain in the body.

Antigen-presenting cells activate T helper cells.

Macrophages, dendritic cells, and B cells function as "professional" **antigen-presenting cells (APCs)** that display foreign antigens as well as their own surface proteins. APCs are inactive until their pattern recognition receptors recognize PAMPs on pathogens. Once activated, an APC ingests the pathogen and lysosomal enzymes degrade most, but not all, of the bacterial antigens. The APC displays fragments of the foreign antigens on its cell surface in association with a type of self-molecule (class II MHC, discussed later). The activated APC expresses additional signaling molecules, called **co-stimulatory molecules,** along with the displayed antigen. APCs present displayed antigens to T cells.

Macrophages secrete more than 100 different compounds, including cytokines and enzymes that destroy bacteria. Macrophages stimulated by bacteria secrete interleukins that activate B cells and certain T cells. Interleukins also promote nonspecific immune responses, such as fever.

Although biologists first identified **dendritic cells** in 1868 and then rediscovered them in 1973, their small numbers made them hard to study until sophisticated techniques became available in the 1990s. Their name derives from their long cytoplasmic extensions known as *dendrites* (not to be confused with the dendrites of neurons). Like macrophages, they develop from monocytes. Dendritic cells are strategically stationed in all tissues of the body that come into contact with the environment—the skin and the linings of the digestive, respiratory, urinary, and vaginal passageways.

When pathogens infect a tissue, dendritic cells are activated by PAMPs and capture the pathogens (or their products) by phagocytosis or by receptor-mediated endocytosis (see Chapter 5). Guided by chemokines expressed in the lymph nodes and lymph vessels, dendritic cells make their way to the lymph nodes. As they migrate, they mature. Mature dendritic cells break down antigens and display the fragments on their cell surface. Co-stimulatory molecules, along with the displayed antigen, attract and activate specific T cells capable of responding to the antigen. Thus dendritic cells are specialized to process, transport, and present antigens to T cells.

The major histocompatibility complex is responsible for recognition of self

The ability of the vertebrate immune system to distinguish self from nonself depends largely on a cluster of closely linked genes known as the **major histocompatibility complex (MHC).** In humans, the MHC is also called the **HLA (human leukocyte antigen) complex.** The MHC genes are polymorphic (variable). Within the population there are multiple alleles, more than 40 for each locus. As a result, the cell surface proteins for which they code are generally different in each individual. With so many possible combinations, no two people, except identical twins, are likely to have all the same MHC proteins on their cells. The more closely related two individuals are, the more MHC genes they have in common. Thus, a person's MHC proteins serve as a biochemical "fingerprint."

The MHC genes encode **MHC antigens,** or self-antigens, that differ in chemical structure, function, and tissue distribution. Class I MHC genes encode glycoproteins expressed on the surface of most nucleated cells. These antigens bind with foreign antigens from viruses or other pathogens within the cell, forming a foreign antigen–class I MHC glycoprotein complex. The cell displays this complex on its surface, and presents it to T_C cells. Thus any infected cell can function as an antigen-presenting cell and can activate certain T_C cells.

Class II MHC genes encode glycoproteins expressed primarily on "professional" APCs—dendritic cells, macrophages, and B cells. These MHC antigens combine with foreign antigen from bacteria, and the cell presents the complex to T_H cells. Class III MHC genes encode many secreted proteins, including components of the complement system and TNFs.

Review

- How are cell-mediated immunity and antibody-mediated immunity different?
- What is the function of antigen-presenting cells? Describe two types.

- What are two main types of T cells? What are their functions?

BiologyNow™ Assess your understanding of **specific immune responses** by taking the pretest on your BiologyNow CD-ROM.

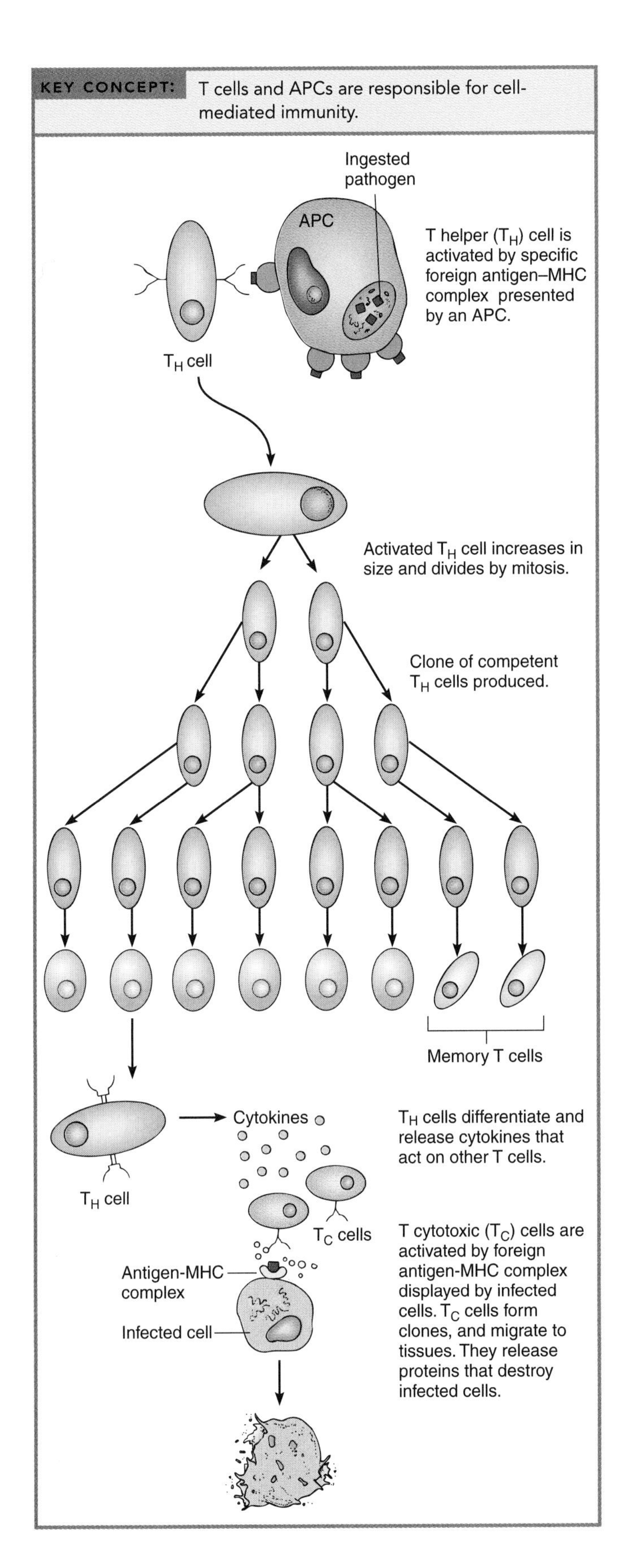

KEY CONCEPT: T cells and APCs are responsible for cell-mediated immunity.
Ingested pathogen
APC
T helper (T_H) cell is activated by specific foreign antigen–MHC complex presented by an APC.
T_H cell
Activated T_H cell increases in size and divides by mitosis.
Clone of competent T_H cells produced.
Memory T cells
Cytokines
T_H cells differentiate and release cytokines that act on other T cells.
T_H cell
T_C cells
Antigen-MHC complex
Infected cell
T cytotoxic (T_C) cells are activated by foreign antigen-MHC complex displayed by infected cells. T_C cells form clones, and migrate to tissues. They release proteins that destroy infected cells.

CELL-MEDIATED IMMUNITY

Learning Objective

6 Describe the sequence of events in cell-mediated immunity.

T cells and APCs (mainly dendritic cells and macrophages) are responsible for cell-mediated immunity (Fig. 43-5). T cells destroy cells infected with viruses and cells that have been altered in some way, such as cancer cells. They also destroy the cells of foreign grafts such as a transplanted kidney. There are thousands of different populations of T cells. Each T_C cell expresses the CD8 molecule on its surface, as well as more than 50,000 identical T cell receptors (TCRs) that bind to one specific type of antigen.

How do T cells know which cells to target? T cells do not recognize antigens unless they are presented properly. When a virus (or other pathogen) infects a cell, some of the viral protein is broken down to peptides and displayed with class I MHC molecules on the cell surface. Only T_C cells with receptors that bind to the specific antigen–MHC class I complex become activated. Generally, fewer than 1 in 10,000 T cells have the same antigen specificity and can respond. T_C cell activation requires at least two signals in addition to the presented antigen: a costimulatory signal and an interleukin signal.

Once activated, a T_C cell increases in size and gives rise to a clone. Activated T_C cells are the effector cells that make up the cell infantry. They leave the lymph nodes and make their way to the infected area, where they destroy target cells within seconds after contact. When a T_C cell combines with antigen on the surface of the target cell, it destroys it in much the same way that NK cells destroy their target cells. The T_C secretes perforins and granzymes that perforate the plasma membrane of the target cell and induce it to kill itself by apoptosis. After releasing cytotoxic substances, the T cell disengages itself from its victim cell and seeks out a new target. This sequence is summarized as follows:

Virus invades body cell ⟶ infected cell displays foreign antigen–MHC class I antigen complex ⟶ specific T_C cell activated by this complex ⟶ clone of T_C cells produced ⟶ T_C cells migrate to area of infection ⟶ T_C cells release proteins that stimulate destruction of target cells

T_H cells are activated by the foreign antigen–MHC class II antigen complex displayed on the surface of APCs. Once activated, a T_H cell gives rise to a clone of T_H cells. Some T_H cells attract macrophages to the site of infection and promote the destruction of intracellular pathogens. Others function mainly in antibody-mediated immunity.

FIGURE **43-5** **Cell-mediated immunity.**
When activated by a foreign antigen–MHC complex presented by an APC, and by cytokines, a T helper (T_H) cell divides, giving rise to a clone of cells. T cytotoxic (T_C) cells are activated by antigens presented on the surfaces of infected cells. T_C cells also form a clone. Both types of T cells migrate to the site of infection, and T_C cells release proteins that destroy invading pathogens.

Review

- How do T cytotoxic cells function in cell-mediated immunity?
- How do T helper cells function in cell-mediated immunity?

BiologyNow™ Assess your understanding of **cell-mediated immunity** by taking the pretest on your BiologyNow CD-ROM.

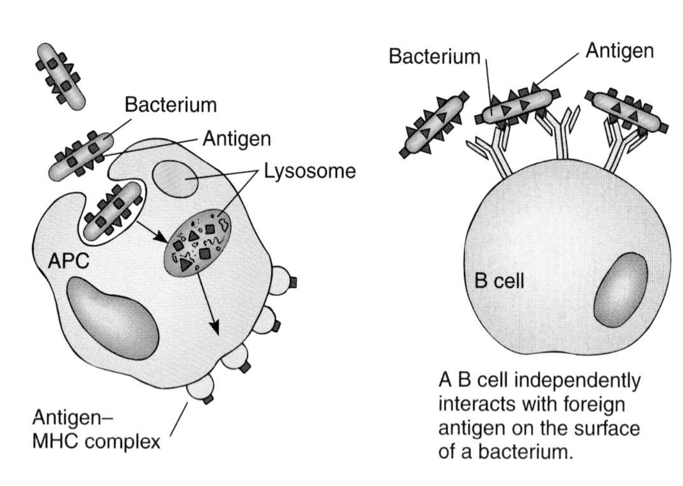

(a) An APC degrades antigen and displays it in combination with MHC class II.

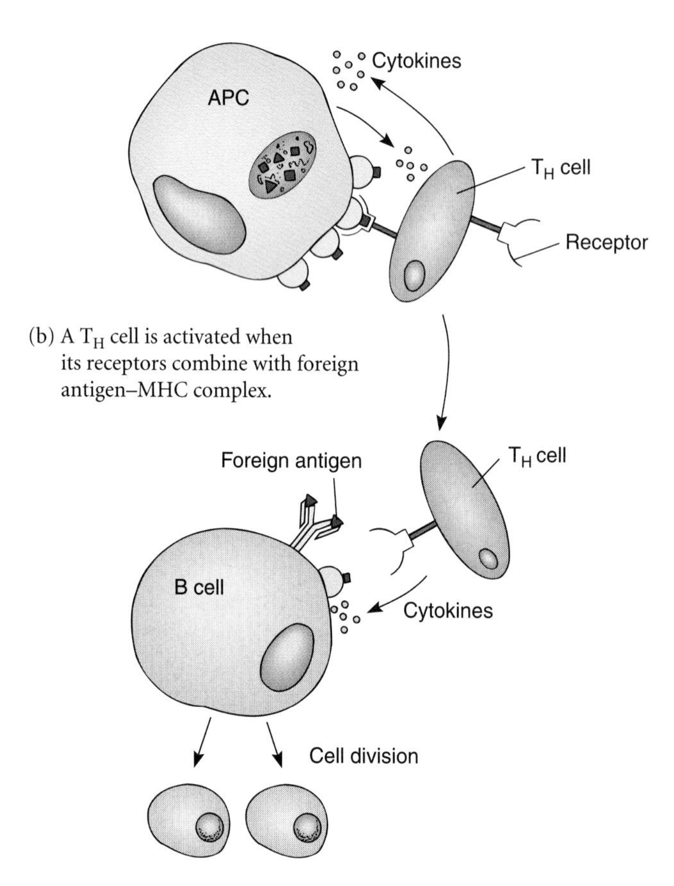

(b) A T_H cell is activated when its receptors combine with foreign antigen–MHC complex.

(c) The activated T_H cell recognizes antigen–MHC complex on a B cell. The T_H cell secretes cytokines that activate the B cell.

ANTIBODY-MEDIATED IMMUNITY

Learning Objectives

7. Summarize the sequence of events in antibody-mediated immunity, including the effects of antigen-antibody complexes on pathogens.
8. Describe the basic structure and function of an antibody, and explain the basis of antibody diversity.

B cells are responsible for antibody-mediated immunity (also called *humoral immunity*). A given B cell can produce many copies of one specific antibody. Antibody molecules serve as cell surface receptors that combine with antigens. Only a B cell displaying a matching receptor on its surface can bind a particular antigen. Inside the B cell, the antigen is degraded into peptide fragments. The B cell then displays the peptide fragments together with MHC protein class II on its surface.

In most cases, the activation of B cells is a complex process that involves APCs and T_H cells (Fig. 43-6). T_H cells stimulate B cells to divide and produce antibodies. However, the T_H cell itself must first be activated. T_H cells do not recognize an antigen that is presented alone. The antigen must be presented as part of a foreign antigen–class II MHC complex on the surface of an APC. When an APC displaying a foreign antigen–MHC complex contacts a T_H cell with complementary T cell receptors, a complicated interaction occurs. Multiple chemical signals are sent back and forth between cells. For example, the APC secretes interleukins, such as IL-1, that activate T_H cells.

B cells serve as APCs to T cells. An activated T_H cell binds with the foreign antigen–MHC complex on the B cell. The T_H cell then releases interleukins, which, together with antigen, activate the B cell.

Once activated, a B cell increases in size, and then divides by mitosis, giving rise to a clone of identical cells (Fig. 43-7). This cell division in response to a specific antigen is known as *clonal selection* (discussed later). The cloned B cells mature into plasma cells that secrete antibodies specific to the antigen that activated the original B cell. It is important to remember that the specificity of the clone is determined *before* the B cell encounters the antigen.

Unlike T cells, most plasma cells do not leave the lymph nodes. Only the antibodies they secrete pass out of the lymph tissues and make their way via the lymph and blood to the infected area. This sequence can be summarized as follows:

FIGURE **43-6** | **B-cell activation.**

(a) An antigen-presenting cell (APC) takes in a bacterium, breaks it down, and presents foreign antigens on its surface in combination with MHC class II antigens. B cells can combine with foreign antigens on infected cells, but they are typically not active until stimulated by T helper (T_H) cells. **(b)** T_H cells are activated when their receptors combine with foreign antigen–MHC antigen complexes on an APC. Complex signaling takes place via cytokines secreted by both cells. **(c)** The activated T_H cell secretes cytokines that activate a B cell that has interacted with foreign antigen. The B cell then divides, forming a clone of identical B cells.

Pathogen invades body ⟶ APC phagocytizes pathogen ⟶ foreign antigen–MHC complex displayed on APC surface ⟶ T_H cell binds with foreign antigen–MHC complex ⟶ activated T_H cell interacts with a B cell that displays the same antigen ⟶ B cell activated ⟶ clone of B cells ⟶ B cells differentiate, becoming plasma cells ⟶ plasma cells secrete antibodies ⟶ antibodies form complexes with pathogen ⟶ destroy pathogen

KEY CONCEPT: B cells are responsible for antibody-mediated immunity.

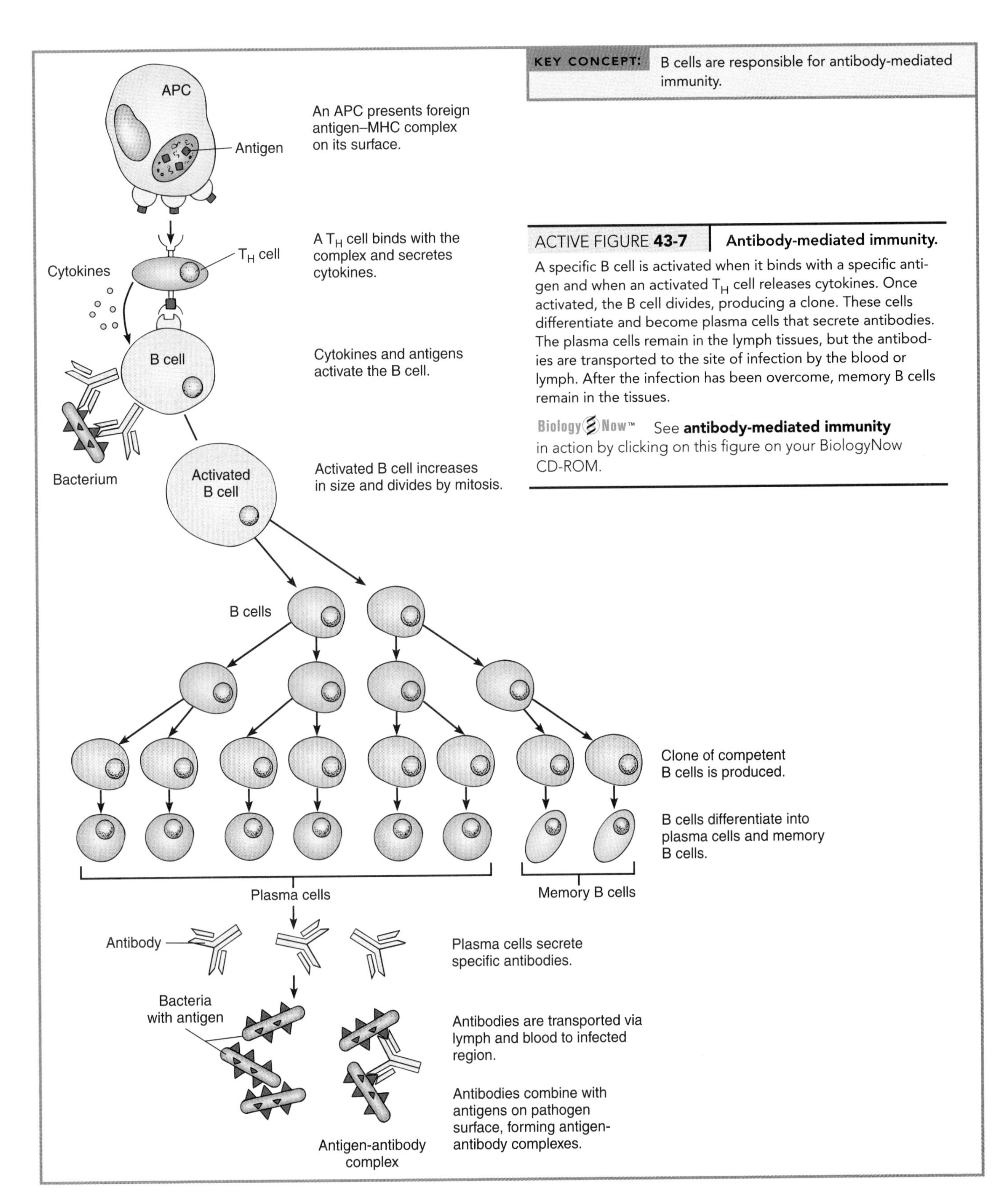

APC
Antigen
An APC presents foreign antigen–MHC complex on its surface.
T_H cell
Cytokines
A T_H cell binds with the complex and secretes cytokines.
B cell
Cytokines and antigens activate the B cell.
Bacterium
Activated B cell
Activated B cell increases in size and divides by mitosis.
B cells
Clone of competent B cells is produced.
B cells differentiate into plasma cells and memory B cells.
Plasma cells
Memory B cells
Antibody
Plasma cells secrete specific antibodies.
Bacteria with antigen
Antibodies are transported via lymph and blood to infected region.
Antibodies combine with antigens on pathogen surface, forming antigen-antibody complexes.
Antigen-antibody complex

ACTIVE FIGURE **43-7** **Antibody-mediated immunity.**

A specific B cell is activated when it binds with a specific antigen and when an activated T_H cell releases cytokines. Once activated, the B cell divides, producing a clone. These cells differentiate and become plasma cells that secrete antibodies. The plasma cells remain in the lymph tissues, but the antibodies are transported to the site of infection by the blood or lymph. After the infection has been overcome, memory B cells remain in the tissues.

BiologyNow™ See **antibody-mediated immunity** in action by clicking on this figure on your BiologyNow CD-ROM.

Some activated B cells do not differentiate into plasma cells but instead become memory B cells (discussed in a later section).

A typical antibody consists of four polypeptide chains

An antibody molecule, also called **immunoglobulin (Ig),** has two main functions: It combines with antigen, and it activates processes that destroy the antigen that binds to it. For example, an antibody may stimulate phagocytosis. Note that an antibody does not destroy an antigen directly; rather, it *labels* the antigen for destruction.

PROCESS OF SCIENCE

The basic structure of the immunoglobulin molecule was clarified by Rodney Porter, of the University of Oxford in England, and Gerald Edelman, of Rockefeller University in New York, during the 1960s. Porter used the plant enzyme papain, a protease, to split Ig molecules into fragments. Based on his findings, Porter developed a working model of the structure of the Ig molecule and was the first to suggest that it is Y-shaped. These researchers then constructed an accurate model of the antibody molecule. Porter and Edelman won the 1972 Nobel Prize in Medicine for their contributions.

Two fragments of the antibody molecule bind antigen and are referred to as *Fab fragments* (*Fab* stands for *antigen-binding fragment*). The fragment that interacts with cells of the immune system is the *Fc fragment* (*c* indicates that this fragment crystallizes during cold storage). Many cells of the immune system have Fc receptors.

A typical antibody is a Y-shaped molecule in which the two arms of the Y (the Fab portions) bind with antigen (Fig. 43-8). This shape enables the antibody to combine with two antigen molecules and allows the formation of **antigen-antibody complexes.** While the arms of the Y bind to antigen, the tail of the Y, the Fc portion, interacts with cells of the immune system, such as phagocytes, or binds with molecules of the complement system.

The antibody molecule consists of four polypeptide chains: two identical long chains called *heavy chains,* and two identical short chains called *light chains.* Each chain has a constant region and a variable region. In the **constant (C) region,** of the heavy chains, the amino acid sequence is constant within a particular immunoglobulin class. You can think of the C region as the handle portion of a door key. Like the pattern of bumps and notches at the part of a key that slides into a lock, the **variable (V) region,** has a unique amino acid sequence. The variable region of the immunoglobulin extends outward from the B cell, whereas the constant region anchors the molecule to the B cell.

At its variable regions, the antibody folds three-dimensionally, assuming a shape that enables it to combine with a specific antigen. When they meet, antigen and antibody fit together somewhat like a lock and key. They must fit in just the right way for the antibody to be effective (Fig. 43-9). A given antibody can bind with different strengths, or *affinities,* to different antigens. In the course of an immune response, higher-affinity antibodies are generated.

In an antigen, specific sequences of amino acids make up an **antigenic determinant,** or *epitope.* These sequences give part of the antigen molecule a specific shape that is recognized by an antibody or T cell receptor. Usually, an antigen has many different antigenic determinants on its surface; some have hundreds.

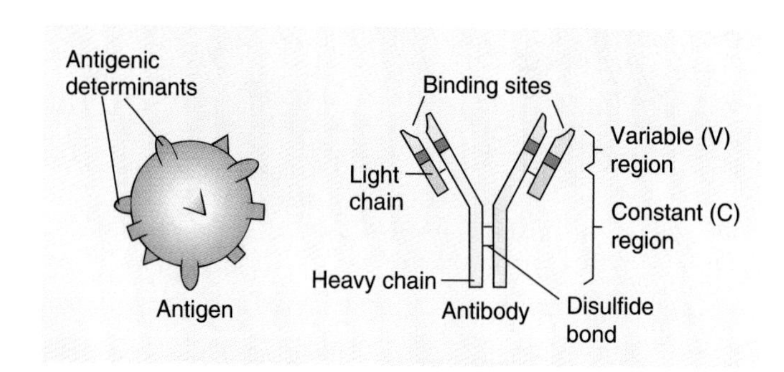

(a)

Fc
Fab
Antigen-antibody complex
Inactivates pathogen
Increases phagocytosis
Destroys pathogen
Activates complement
Inactivates toxins
Coats surface of pathogen
Stimulates mast cells to release histamine
Attracts phagocytes
Increases phagocytosis
Increases inflammation

(b)

FIGURE **43-8** **Antibody structure and function.**

Antibodies combine with antigens, forming antigen-antibody complexes. **(a)** The antibody molecule is composed of two light chains and two heavy chains, joined by disulfide bonds. The constant (C) and variable (V) regions of the chains are labeled. **(b)** The Fab portion of the antibody binds to antigen. The Fc portion of the molecule binds to a cell of the immune system. Antigen-antibody complexes directly inactivate pathogens and increase phagocytosis. They also activate the complement system.

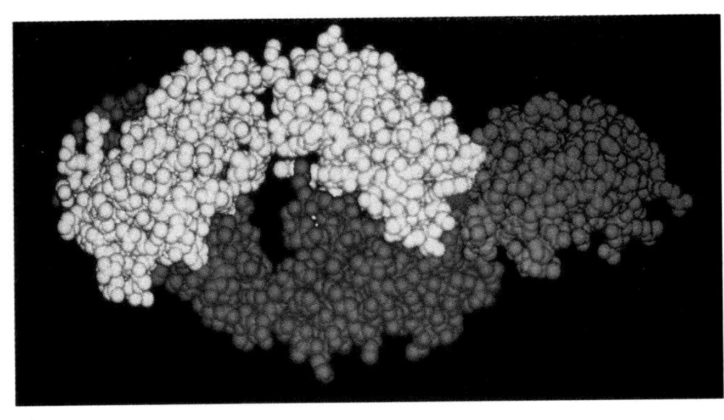

(a)

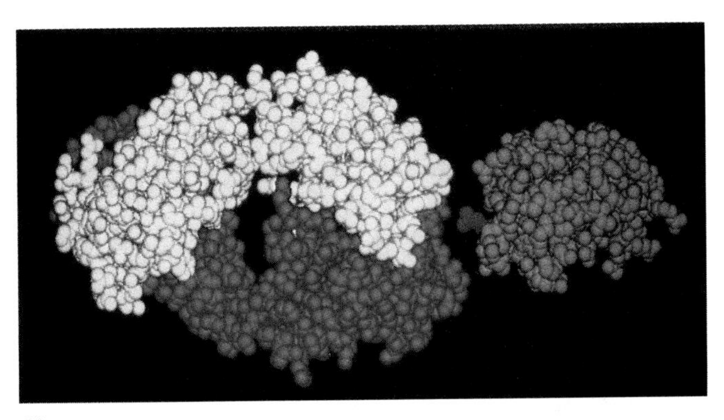

(b)

FIGURE **43-9** | **An antigen-antibody complex.**

The components of an antigen-antibody complex fit together as shown in this computer simulation of their molecular structure. The antigen lysozyme is shown in green, the heavy chain of the antibody is shown in blue, and the light chain is shown in yellow. **(a)** A portion of the antigenic determinant, shown in red, fits into a groove in the antibody molecule. **(b)** The antigen-antibody complex has been pulled apart to show its structure.

Antigenic determinants often differ from one another, so several different kinds of antibodies can combine with a single antigen.

Antibodies are grouped in five classes

Antibodies are grouped in five classes, defined by unique amino acid sequences in the C regions of the heavy chains. Using the abbreviation *Ig* for *immunoglobulin,* the classes are designated IgG, IgM, IgA, IgD, and IgE. In humans, about 75% of the circulating antibodies belong to the **IgG** class; these are part of the gamma globulin fraction of the plasma. IgG and **IgM** interact with macrophages and activate the complement system. They defend against many pathogens carried in the blood, including bacteria, viruses, and some fungi. **IgA,** present in mucus, tears, saliva, and breast milk, prevents viruses and bacteria from attaching to epithelial surfaces. This immunoglobulin, which defends against inhaled or ingested pathogens, is strategically secreted into the respiratory, digestive, urinary, and reproductive tracts.

IgD has a low concentration (less than 1%) in plasma. Along with IgM, it is an important immunoglobulin on the B-cell surface. IgD helps activate B cells following antigen binding. **IgE,** which has an even lower plasma concentration, binds to mast cells, which contain potent signaling molecules such as histamine. When an antigen binds to IgE on a mast cell, these molecules are released. Histamine triggers many allergy symptoms, including inflammation. IgE is also responsible for an immune response to invading parasitic worms.

Antigen–antibody binding activates other defenses

Antibodies mark a pathogen as foreign by combining with an antigen on its surface. Generally, several antibodies bind with several antigens, creating a mass of clumped antigen-antibody complexes. The combination of antigen and antibody activates several defensive responses:

(1) The antigen-antibody complex may inactivate the pathogen or its toxin. For example, when an antibody attaches to its surface, a virus may lose the ability to attach to a host cell. Snake venom antitoxin contains antibodies that neutralize the toxins that enter the body with poisonous snake bites.

(2) The antigen-antibody complex stimulates phagocytic cells to ingest the pathogen.

(3) Antibodies of the IgG and IgM groups work mainly through the complement system. When antibodies combine with a specific antigen on a pathogen, complement proteins destroy the pathogen. IgG molecules have an Fc fragment that binds Fc receptors, expressed on most immune cells. When an antibody molecule is bound to a pathogen and the Fc portion of the antibody binds with an Fc receptor on a phagocyte, the pathogen is more easily destroyed.

The immune system responds to millions of different antigens

PROCESS OF SCIENCE

How can the immune system recognize every possible antigen, even those produced by newly mutated viruses never before encountered during the evolution of our species?

In the 1950s, the Danish researcher Niels Jerne of the Basel Institute for Immunology in Basel, Switzerland, David Talmage of the University of Chicago, and Frank Macfarlane Burnet of the Walter and Eliza Hall Institute for Medical Research in Melbourne, Australia, developed the **clonal selection** hypothesis, which proposes that before a lymphocyte ever encounters an antigen, the lymphocyte has specific receptors for that antigen on its surface. (Jerne and Burnet later won Nobel prizes in Medicine for their work in immunology). When an antigen binds to a matching receptor on a lymphocyte, it activates the lymphocyte, which then gives rise to a clone of cells with identical receptors. A major problem with this hypothesis was its suggestion that our cells must contain millions of separate antibody genes, each coding for an antibody with a different specificity. Although each human cell has a large amount of DNA, it is not enough to provide a different gene to code for each of the millions of possible specific antibody molecules.

In 1965 W. J. Dreyer of the Institute of Biochemistry at the University of Zurich, Switzerland, and Joe Claude Bennett of the University of Alabama at Birmingham, suggested that the C region and the V region of an immunoglobulin are encoded by two separate genes. Their hypothesis was at first rejected by many biologists because it contradicted the prevailing hypothesis that one gene codes for one polypeptide. The technology needed to test Dreyer and Bennett's hypothesis was not immediately available, and it was not until 1976 that Susumu Tonegawa, of the Massachusetts Institute of Technology, and his colleagues demonstrated that separate genes encode the V and C regions of immunoglobulins. Tonegawa further showed that three separate families of genes code for immunoglobulins and that each gene family contains a large number of DNA segments that code for V regions. Recombination of these DNA segments during the differentiation of B cells is responsible for antibody diversity. Tonegawa was awarded the 1987 Nobel Prize in Physiology or Medicine for his work, which transformed the emerging science of immunology.

We now understand that in undifferentiated B cells, gene segments are present for a number of different V regions, for one or more junction (J) regions, and for one or more different C regions (Fig. 43-10). Rearrangement of these DNA segments produces an enormous number of potential combinations! Millions of different types of B (and T) cells are produced. By chance, one of those cells may produce just the right antibody to destroy a pathogen that invades the body. To appreciate the capacity for antibody diversity, consider the diverse combinations of things you create in your everyday life. A familiar example is making an ice cream sundae. Imagine the varied combinations that are possible using 10 flavors of ice cream, six types of sauce, and 15 kinds of toppings.

Researchers have identified additional sources of antibody diversity. For example, the DNA of the mature B cells that codes for the regions of immunoglobulins mutates very readily. These somatic mutations produce genes that code for slightly different antibodies.

We have used antibody diversity here as an example, but similar genetic mechanisms account for the diversity of T cell receptors. Although we may actually use only a relatively few types of antibodies or T cells in a lifetime, the remarkable diversity of the immune system prepares it to attack most potentially harmful antigens that may invade the body.

Monoclonal antibodies are highly specific

PROCESS OF SCIENCE

Before 1975, the only method for obtaining antibodies for medicine or research was immunizing animals and collecting their

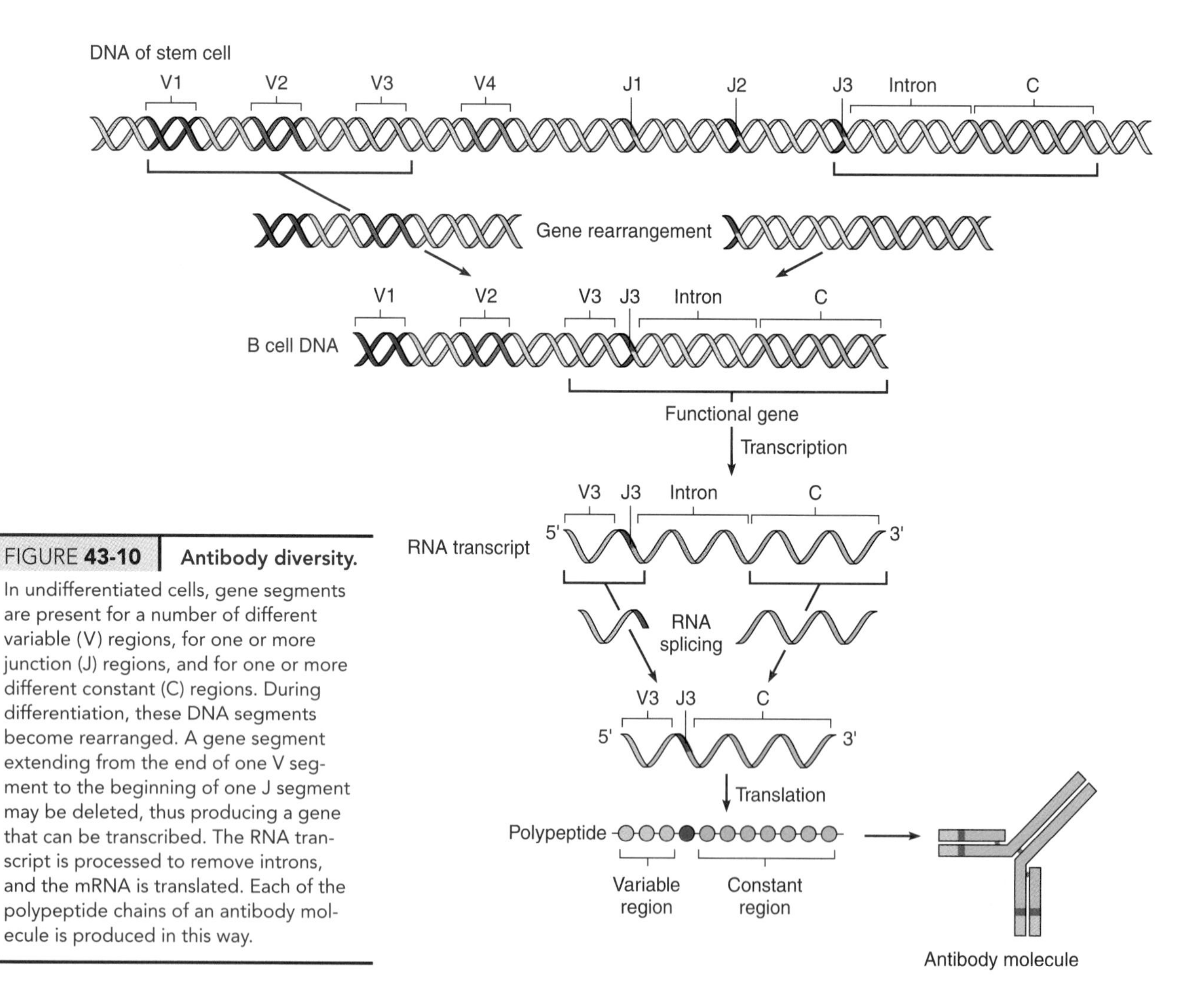

FIGURE 43-10 | Antibody diversity.
In undifferentiated cells, gene segments are present for a number of different variable (V) regions, for one or more junction (J) regions, and for one or more different constant (C) regions. During differentiation, these DNA segments become rearranged. A gene segment extending from the end of one V segment to the beginning of one J segment may be deleted, thus producing a gene that can be transcribed. The RNA transcript is processed to remove introns, and the mRNA is translated. Each of the polypeptide chains of an antibody molecule is produced in this way.

blood. Then Cesar Milstein and Georges Kohler at the Laboratory of Molecular Biology in Cambridge, England, developed **monoclonal antibodies,** identical antibodies produced by cells cloned from a single cell.

One way to produce monoclonal antibodies in the laboratory is to inject mice with the antigen of interest, for example, an antigen from a particular bacterium. After the mice have produced antibodies to the antigen, their B cells are collected. However, these normal cells survive in culture for only a few generations, limiting the amount of antibody that can be produced. In contrast, cancer cells live and divide in tissue culture indefinitely. The B cells are suspended in a culture medium together with lymphoma cells from other mice. (Lymphoma is cancer of lymphocytes.) Then the B cells and lymphoma cells are induced to fuse. They form hybrid cells, known as **hybridomas,** which have properties of the two "parent" cells. Hybridomas can be cultured indefinitely (a cancer cell property) and continue to secrete antibodies (a B cell property).

Researchers select hybrid cells that are manufacturing the specific antibody needed and then clone them in a separate cell culture. Cells of this clone secrete large amounts of the particular antibody—thus the name *monoclonal antibodies.* Each type of monoclonal antibody is specific for a single antigenic determinant.

Immunologists also use recombinant DNA technology to produce monoclonal antibodies. Because of their purity and specificity, monoclonal antibodies have proved to be invaluable tools in modern biology. For example, a researcher may want to detect a specific molecule present in very small amounts in a mixture. The reaction of that molecule (the antigen) with a specific monoclonal antibody makes its presence known. Monoclonal antibodies are used in similar ways in various diagnostic tests. For example, home pregnancy tests make use of a monoclonal antibody that is specific for human chorionic gonadotropin (hCG), a hormone produced by a developing human embryo (see Chapters 48 and 49). Several monoclonal antibodies are being used clinically to treat cancer. For example, the drug herceptin is a monoclonal antibody used in treating a form of breast cancer.

Review

- How do antibodies recognize antigens?
- What are three ways that antigen-antibody complexes affect pathogens?

BiologyNow™ Assess your understanding of **antibody-mediated immunity** by taking the pretest on your BiologyNow CD-ROM.

IMMUNOLOGICAL MEMORY

Learning Objectives

9. Describe the basis of immunological memory; contrast secondary and primary immune responses.
10. Compare active and passive immunity, giving examples of each.

Memory B and memory T cells are responsible for long-term immunity. Following an immune response, memory T cells group strategically in many nonlymphatic tissues, including the lung, liver, kidney, and intestine. Many infections occur at such sites, and T cells stationed in those locations can respond quickly. In response to antigen, memory T cells rapidly become T_C cells that produce substances that kill invading cells.

Memory B cells have a "survival gene" that prevents apoptosis, the programmed cell death that is the eventual fate of plasma cells. Memory B cells continue to live and produce small amounts of antibody long after the body has overcome an infection. This circulating antibody is part of the body's arsenal of chemical weapons. If the same pathogen enters the body again, the antibody immediately targets it for destruction. At the same time, specific memory cells are stimulated to divide, producing new clones of plasma cells that produce the same antibody. The presence of circulating antibodies is used clinically to detect previous exposure to specific pathogens. For example, some HIV screening tests measure antibody to the virus that causes AIDS.

A secondary immune response is more effective than a primary response

The first exposure to an antigen stimulates a **primary immune response.** An infection, or an antigen injected into an animal causes specific antibodies to appear in the blood plasma in 3 to 14 days. After antigen injection, there is a brief *latent period* during which the antigen is recognized and appropriate lymphocytes begin to form clones. A *logarithmic phase* follows, during which the antibody concentration rises rapidly for several days

FIGURE **43-11** **Immunological memory.**

Antigen 1 was injected at day 0, and the immune response was assessed by measuring antibody levels to the antigen. At week 4, the primary immune response had subsided. Antigen 1 was injected again, together with a new protein, antigen 2. The secondary immune response was greater and more rapid than the primary response. It was also specific to antigen 1. A primary immune response was made to the newly encountered antigen 2.

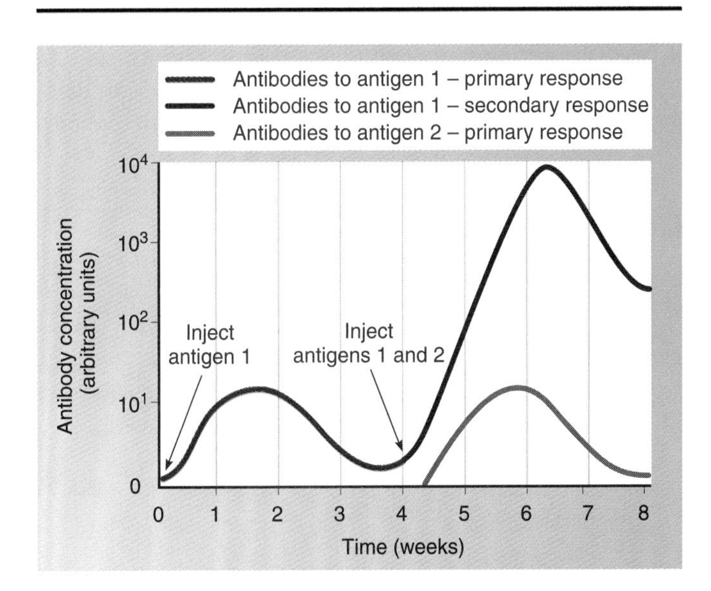

until it reaches a peak (Fig. 43-11). IgM is the principal antibody synthesized during the primary immune response. Finally, there is a *decline phase,* during which the antibody concentration decreases to a very low level.

A second exposure to (for example, by injection) of the same antigen, even years later, results in a **secondary immune response.** Because memory B cells bearing antibodies to that antigen (and also memory T cells) persist for many years, the secondary immune response is much more rapid than the primary immune response, with a shorter latent period. Much less antigen is necessary to stimulate a secondary immune response than a primary response, and more antibodies are produced. In addition, the affinity of antibodies is generally much higher. The predominant antibody in a secondary immune response is IgG.

The body's ability to launch a rapid, effective response during a second encounter with an antigen explains why we do not usually suffer from the same infectious disease several times. For example, most people contract measles or chickenpox only once. When exposed a second time, the immune system responds quickly, destroying the pathogens before they have time to multiply and cause symptoms of the disease. Booster shots of vaccine are given to elicit a secondary immune response, thus reinforcing immunological memory.

You may wonder, then, how a person can get influenza (the flu) or a cold more than once. The reason is that there are many varieties of these viral diseases, each caused by a virus with slightly different antigens. For example, more than 100 different viruses cause the common cold, and new varieties of cold and flu viruses evolve continuously by mutation (a survival mechanism for them), which may result in changes in their surface antigens. Even a slight change may prevent recognition by memory cells. Because the immune system is so specific, each different antigen is treated by the body as a new immunological challenge.

Immunization induces active immunity

We have been considering **active immunity,** immunity that develops following exposure to antigens. Active immunity can be naturally or artificially induced (Table 43-1). If someone with chickenpox sneezes near you and you contract the disease, you develop active immunity naturally. Active immunity can also be artificially induced by **immunization,** that is, by exposure to a *vaccine.* When an effective vaccine is introduced into the body, the immune system actively develops clones of cells, produces antibodies, and develops memory cells.

PROCESS OF SCIENCE

The first vaccine was prepared in 1796 by British physician Edward Jenner against vaccinia, the cowpox virus. The term *vaccination* was thus derived. Jenner's vaccine provided humans with immunity against the deadly disease smallpox. Jenner had no knowledge of microorganisms or of immunology, and it remained for French chemist Louis Pasteur to begin to develop scientific methods for preparing vaccines 100 years later. Pasteur showed that inoculations with preparations of attenuated (weakened) pathogens could be used to develop immunity against the virulent (infectious) form of the pathogen. However, not until 20th century advances in immunology—for example, Burnet's clonal selection theory in 1957 and the discovery of T and B cells in 1965—did scientists gain a modern understanding of vaccines. Effective vaccination stimulates the body to launch an immune response against the antigens contained in the vaccine. Memory cells develop, and future encounters with the same pathogen are dealt with rapidly.

Microbiologists prepare effective vaccines in a number of ways. They can attenuate a pathogen so it loses its ability to cause disease. When pathogens are cultured for long periods in nonhuman cells, mutations adapt the pathogen to the nonhuman host so that they no longer cause disease in humans. This is how the Sabin polio vaccine and the measles vaccine are produced.

Whooping cough and typhoid fever vaccines are made from killed pathogens that still have the necessary antigens to stimulate an immune response. Tetanus and botulism vaccines are made from toxins secreted by the respective pathogens. The toxin is altered so that it can no longer destroy tissues, but its antigenic determinants are still intact.

Most vaccines consist of the entire pathogen, attenuated or killed, or of a protein from the pathogen. Researchers are investigating several approaches that would reduce potential side effects. For example, they are developing **DNA vaccines** (or RNA vaccines), made from a part of the pathogen's genetic material. The DNA of the pathogen is altered so that it transfers genes that specify antigens. When injected into a patient, the altered DNA is taken up by cells and makes its way to the nucleus. The encoded antigens are manufactured and stimulate both cell-mediated and antibody-mediated immunity. Several DNA vaccines, including vaccines to prevent and treat HIV infection, are in clinical trials.

TABLE **43-1** Active and Passive Immunity

Type of Immunity	When Developed	Development of Memory Cells	Duration of Immunity
Active			
Naturally induced	After pathogens enter the body through natural encounters (e.g., a person with measles sneezes on you)	Yes	Many years
Artificially induced	After immunization with a vaccine	Yes	Many years
Passive			
Naturally induced	After transfer of antibodies from mother to developing baby	No	Few months
Artificially induced	After injection with gamma globulin	No	Few months

Passive immunity is borrowed immunity

To provide **passive immunity,** physicians inject people with antibodies actively produced by another organism. The serum or gamma globulin that contains these antibodies is obtained from humans or other animals. Nonhuman serums are less desirable, because nonhuman proteins can act as antigens, stimulating an immune response that may result in serum sickness.

Passive immunity is borrowed immunity. Its effects are temporary whether it is artificially or naturally induced (see Table 43-1). Physicians use artificially induced passive immunity to boost the body's defenses temporarily against a particular disease. For example, when exposed to hepatitis A, a form of viral hepatitis spread through contaminated food or water, people at risk of infection are injected with gamma globulin containing antibodies to the hepatitis pathogen. However, gamma globulin injections offer protection for only a few weeks. Because the body has not actively launched an immune response, it has no memory cells and cannot produce antibodies to the pathogen. Once the injected antibodies are broken down, the immunity disappears.

Pregnant women confer naturally induced passive immunity on their developing babies by manufacturing antibodies for them. These maternal antibodies, of the IgG class, pass through the placenta (the organ of exchange between mother and developing fetus) and provide the fetus and newborn infant with a defense system until its own immune system matures. Babies who are breast-fed continue to receive immunoglobulins, particularly IgA, in their milk.

Review

- A playmate in kindergarten exposes John and Jack to measles. John has been immunized against measles, but Jack has not received the measles vaccine. How are their immune responses different? Five years later, John and Jack are playing together when Judy, who has just been diagnosed with measles, sneezes on both of them. Compare their immune responses.
- Why is passive immunity temporary?

BiologyNow™ Assess your understanding of **immunological memory** by taking the pretest on your BiologyNow CD-ROM.

THE IMMUNE SYSTEM AND DISEASE

Learning Objective

11 Describe the body's response to cancer cells and HIV.

The immune system is generally very effective in defending the body against invading pathogens. However, many factors can impair immune function, including genetic mutations, malnutrition, sleep deprivation, pre-existing disease, stress, and virulent pathogens.

Cancer cells evade the immune system

Cancer is now the second leading cause of death in Western countries. An estimated one person in three in the United States will develop cancer, and one in five will die of cancer. We have discussed cancer in several previous chapters, including Chapters 16 and 37. Here we focus on the interaction between cancer cells and cells of the immune system, and on new treatment strategies based on immune mechanisms.

Cancer cells are body cells that have been transformed in a way that alters their normal growth-regulating mechanisms. Every day, a few normal cells may be transformed into precancer cells in each of us in response to the sun's UV rays, x-rays, certain viruses, chemical carcinogens in the environment, and yet unknown factors. How does the immune system respond? And why is immune response sometimes ineffective?

Human T cells recognize two main groups of cancer cell antigens: tumor-specific antigens and tumor-associated antigens. *Tumor-specific antigens* are unique to cancer cells. They are induced by certain viruses, and by chemical and physical carcinogens, such as ultraviolet light. Most cancer antigens are *tumor-associated antigens,* which are not unique to cancer cells; normal cells express them during fetal development. Some tumor-associated antigens are not expressed after birth, whereas others are expressed at very low levels. High levels stimulate the immune system. Researchers have shown that oncogenes encode tumor-associated antigens (see Chapter 16).

Antigens on cancer cells induce both cell-mediated and antibody-mediated immune responses. NK cells and macrophages destroy cancer cells. Macrophages produce cytokines, including TNFs (tumor necrosis factors) that inhibit tumor growth.

Dendritic cells present antigens to T cells, stimulating them to produce interferons, which have an antitumor effect. T_C cells attack cancer cells directly, and also produce interleukins, which attract and activate macrophages and NK cells (Fig. 43-12).

Sometimes cancer cells evade the immune system and multiply unchecked. Some types of cancer cells can block T_C cells. Others dramatically decrease their expression of class I MHC molecules. Recall that T_C cells only recognize antigen associated with class I MHC, so a low level of these molecules prevents tumor destruction. Some cancer cells do not produce costimulatory molecules needed to activate T_C cells.

Conventional cancer treatments such as chemotherapy and radiation destroy normal cells as well as cancer cells. To treat cancer effectively, treatment strategies must be more specific, or even customized to the particular cancer. For example, cancer researchers are genetically engineering cancer cells to secrete cytokines that would stimulate the patient's immune response.

A few new cancer drugs, such as Herceptin, are monoclonal antibodies. Herceptin binds with a growth-factor receptor that is present in excessive numbers on the cells of about 30% of metastatic breast cancers. Herceptin blocks the growth factors that would stimulate proliferation of the cells.

More than 50 *angiogenesis inhibitors* are currently being tested. These drugs inhibit the development of blood vessels tumors need, and although these inhibitors do not cure cancer, they slow its growth.

Researchers are developing vaccines that stimulate a cell-mediated immune response tumor-associated antigens. These vaccines differ from traditional vaccines that are used to destroy pathogens *before* they cause disease. Several research groups are experimenting with cancer vaccines made from dendritic cells from a cancer patient mixed with peptides from the patient's

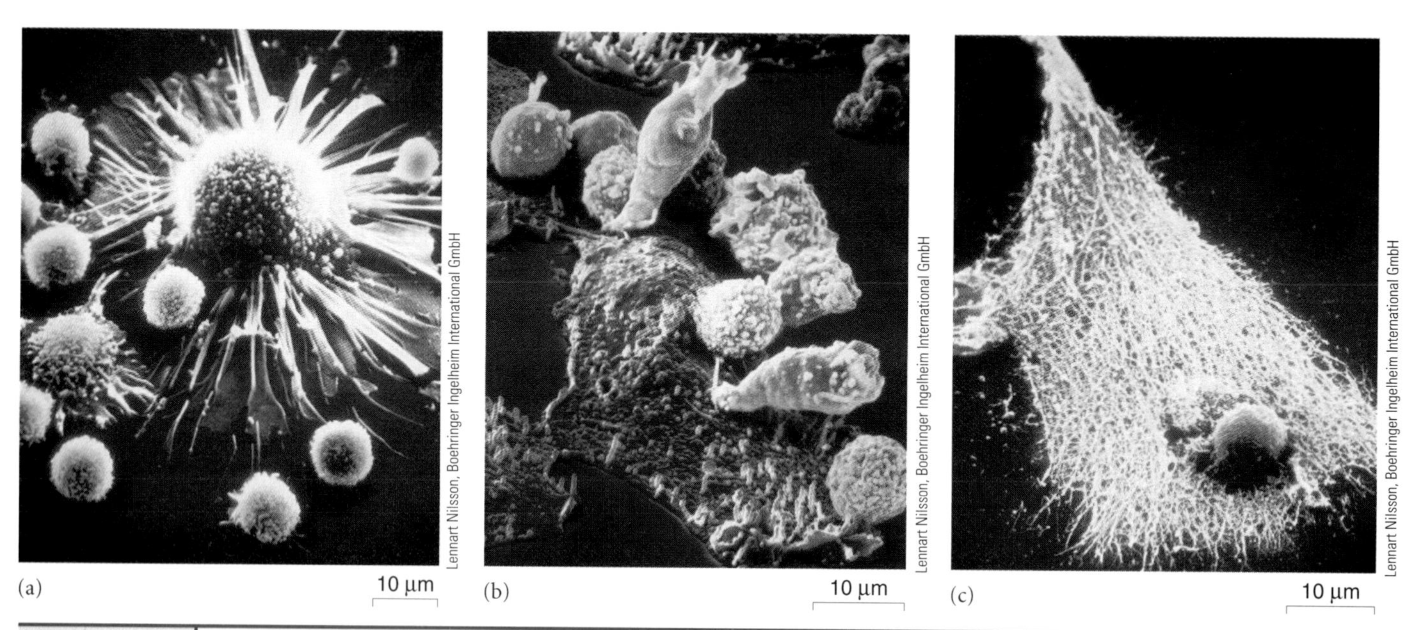

FIGURE **43-12** | **Colorized SEMs showing cancer cell destruction.**

(a) An army of T_C cells surrounds a large cancer cell. The T_C cells recognize the cancer cell as nonself because it displays altered or unique antigens on its surface. **(b)** Some T_C cells elongate as they chemically attack the cancer cell, breaking down its plasma membrane. **(c)** The cancer cell has been destroyed. Only a collapsed fibrous cytoskeleton remains.

cancer cells. The hypothesis is that the dendritic cells will become sensitized to the cancer peptides and become more effective in recognizing the cancer cells in the patient.

DNA microarrays detect patterns of gene expression in cancer cells. The microarray is a slide or chip that the technician can dot with DNA from thousands of genes (see Fig. 15-3). These genes are used as probes to determine which genes are active in cancer cells. Researchers use microarrays to identify cancer subtypes, information that can help identify and develop the most effective treatments.

Immunodeficiency disease can be inherited or acquired

The absence or failure of some component of the immune system can result in **immunodeficiency disease,** a condition that increases susceptibility to infection. Inherited immunodeficiencies have an estimated incidence of 1 per 10,000 births. *Severe combined immunodeficiency syndromes (SCIDs)* are X-linked and autosomal recessive disorders that profoundly affect both cell-mediated immunity and antibody-mediated immunity, resulting in multiple infections. Babies born with SCID typically die by 2 years of age unless they are maintained in a protective bubble until they can be effectively treated, typically with bone marrow transplants.

In *DiGeorge syndrome,* the thymus is reduced or absent and the patient is deficient in T cells. Children born with this disorder are prone to serious viral infections. Treatment involves transplanting bone marrow or fetal thymus tissue.

Researchers use several animal models to study immunodeficiency. For example, the nude mouse (a hairless mutant mouse) does not develop a functional thymus. Because they are deficient in mature T cells, nude mice are used to study the effects of compromised cell-mediated immunity, as well as possible treatments. *Stem cell research* and advances in genetic engineering suggest new approaches to treating immunodeficiency (see Chapters 15 and 16).

Worldwide, the leading cause of acquired immunodeficiency in children is protein malnutrition. Lack of protein decreases T-cell numbers, and decreases the ability to manufacture antibodies, increasing the risk of contracting opportunistic infections. Another important cause of acquired immunodeficiency is chemotherapy administered to cancer patients.

HIV is the major cause of acquired immunodeficiency

The **human immunodeficiency virus (HIV)** was first isolated in 1983 and was shown to be the cause of **acquired immunodeficiency syndrome (AIDS)** in 1984. HIV, a retrovirus, has been studied more than any other virus. (Recall from Chapter 23 that a retrovirus is an RNA virus that uses its RNA as a template to make DNA with the help of reverse transcriptase.) Several different strains of the virus are known; HIV-1 is the most virulent form in humans.

The AIDS pandemic has claimed more than 26 million lives. It kills more than 3 million people every year, making it the fourth highest cause of death globally. Epidemiologists estimate that more than 38.6 million adults and 3.2 million children worldwide are now infected with HIV. The World Health Organization and the Joint United Nations Programme on HIV/AIDS predict that if the pandemic proceeds at its current alarming rate, there will be 45 million new infections by 2010 and nearly 70 million deaths by 2020. In 2001, the General Assembly of the

United Nations passed a declaration calling AIDS a "global emergency" and outlining measures for mobilizing world resources to combat this disease. The declaration, signed by 189 countries, views AIDS as more than a medical issue, stating that it is a political, economic, and human rights threat.

HIV is transmitted mainly during sexual intercourse with an infected person, or by direct exposure to infected blood or blood products. The virus is not spread by casual contact. People do not contract HIV by hugging, casual kissing, or using the same bathroom facilities. Friends and family members who have contact with AIDS patients are not more likely to get the virus.

Most individuals currently infected in the United States are men who engage in homosexual or bisexual behavior and people who use intravenous drugs. New infections are increasing most rapidly among women and teenagers who contract the virus through heterosexual contact. Heterosexual contact with infected individuals accounts for more than one third of HIV infections in women and for an increasing number of cases in both men and women. Use of a latex condom during sexual intercourse provides some protection against HIV transmission, and use of a spermicide containing nonoxynol-9 may provide additional protection.

An estimated 10% of AIDS patients are children born to infected mothers, but recent drug treatment protocols have reduced the rate of HIV transmission between mother and fetus. Mothers infected with HIV can also transmit the virus to their babies by breast-feeding. Effective blood-screening procedures have been developed to safeguard blood bank supplies, markedly reducing the risk of infection by blood transfusion in highly developed countries.

When HIV infects the body, an immune response is launched. Dendritic cells in the mucous membranes appear to be the first cells that HIV targets. When inflammation from a sexually transmitted disease or other infection is present in the mucosa of the cervix or rectum, large numbers of dendritic cells are present, increasing the risk of HIV infection. Dendritic cells transport HIV from their point of entry to the lymph nodes.

HIV has a protein (gp120) on its outer envelope that attaches to CD4 on the surface of T_H cells (CD4 T cells), its main target (Fig. 43-13; also see chapter opening figure). The virus also binds to a second receptor that normally binds chemokines. HIV enters the T_H cell and destroys it. Over time HIV causes a dramatic decrease in the T_H cell population, which severely impairs the body's ability to resist infection. T_C cells appear to be the main cells that attack HIV, limiting viral replication and delaying the progress of the disease. Too often, the virus eventually wins the battle.

Although most individuals have no symptoms, about 15% of infected individuals experience mild flulike symptoms (fever

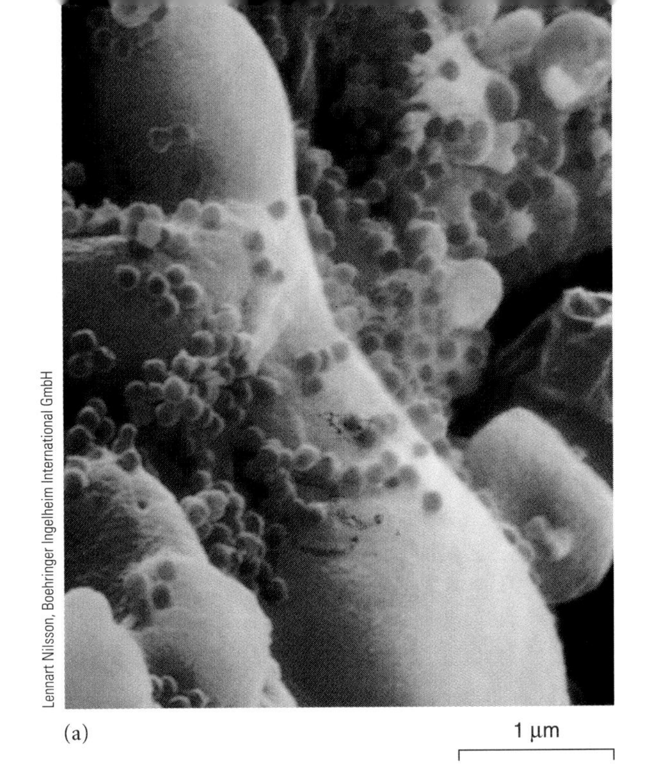

(a)

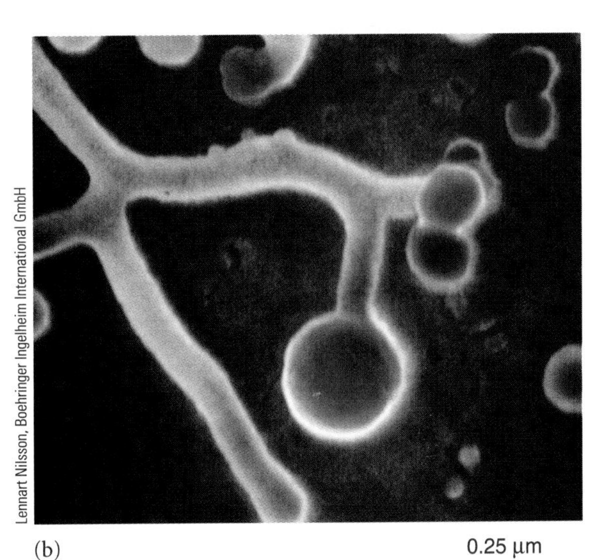

(b)

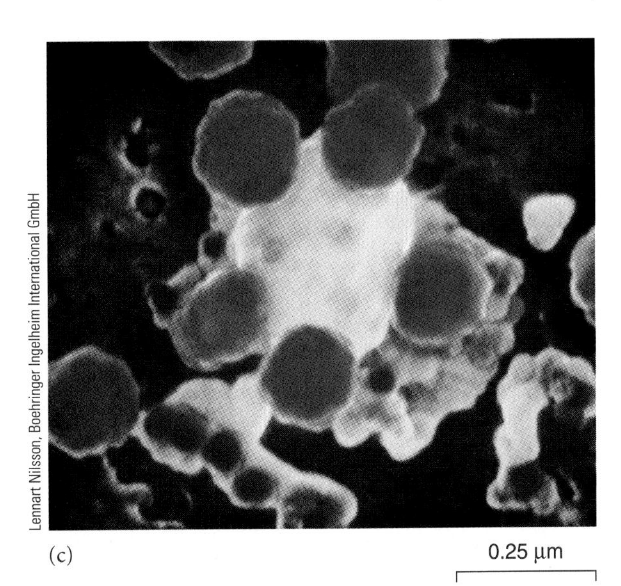

(c)

FIGURE **43-13** **Colorized SEMs of HIV infecting a T helper cell.**

(a) HIV particles *(blue)* attack a T_H cell *(light green)*. **(b)** HIV particles budding from the ends of the branched microvilli of a T_H cell. **(c)** An even higher magnification of viral particles budding from a bleb (a cytoplasmic extension broader than a microvillus).

and aching muscles) for a week or so. Some cells infected by HIV are not destroyed, and the virus may continue to replicate slowly for many years. After a time, a progression of symptoms occurs, including swollen lymph glands, night sweats, fever, and weight loss (Fig. 43-14).

In about one third of AIDS patients, the virus infects the nervous system, causing *AIDS dementia complex.* These patients exhibit progressive cognitive, motor, and behavioral dysfunction that typically ends in coma and death. As a consequence of immunosuppression, many AIDS patients develop and die from serious opportunistic infections or rare forms of cancer, such as Kaposi's sarcoma, an endothelial cell tumor that causes purplish spots on the skin.

FIGURE **43-14** | **The course of HIV infection.**

Exposure to semen or blood that contains HIV can lead to HIV infection and AIDS. Although some exposed individuals apparently do not become infected, the risk increases with multiple exposures. Many factors determine whether a person exposed to HIV will develop AIDS.

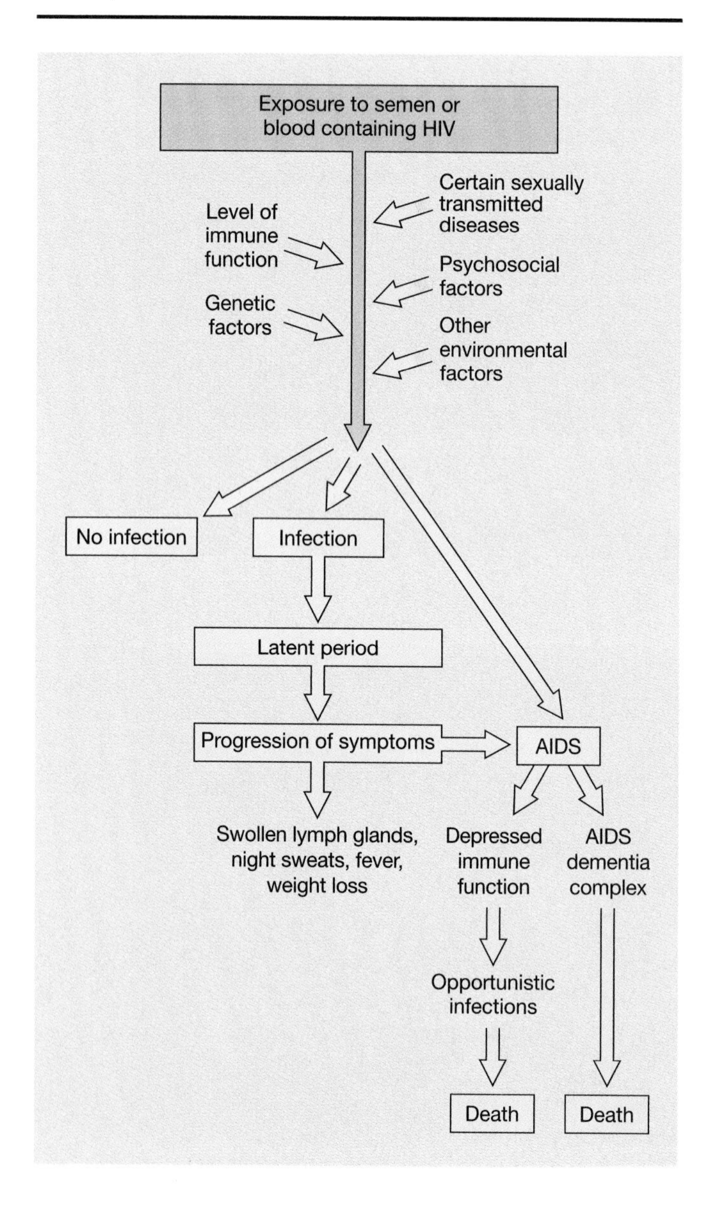

Researchers throughout the world are searching for drugs that will successfully combat the AIDS virus. Because HIV often infects the central nervous system, an effective drug must cross the blood–brain barrier. **AZT** (azidothymidine), the first drug developed to treat HIV infection, can prolong the period prior to the onset of AIDS symptoms. AZT blocks HIV replication by inhibiting the action of reverse transcriptase, the enzyme the retrovirus uses to synthesize DNA. Without producing DNA, the virus cannot incorporate itself into the host cell's DNA. Unfortunately, the reverse transcriptase HIV uses to synthesize DNA makes many mistakes: About 1 in every 2000 nucleotides it incorporates is incorrect. By mutating in this way, viral strains resistant to AZT have evolved.

Protease inhibitors block the viral enzyme protease, resulting in viral copies that cannot infect new cells. Protease inhibitors are used in combination with AZT and other reverse transcriptase inhibitors. Triple-combination treatment has been very effective for many AIDS patients, preventing opportunistic infections and prolonging life. In fact, combination treatment has led to a decline in AIDS incidence and mortality in the United States. Unfortunately, people in many parts of the world cannot afford these expensive drugs, and in poor areas, the AIDS epidemic continues to grow. In addition, the emergence of drug-resistant HIV viruses has been a serious problem.

Worldwide containment of the AIDS epidemic will require an effective vaccine that prevents the spread of HIV. Developing such a vaccine remains a daunting challenge for immunologists. More than three dozen vaccines have been developed and tested clinically, but none has proved effective. An effective vaccine would have to counteract HIV's destruction of key cells needed to mount an immune response. Also, because mutations arise at a very high rate, new viral strains with new antigens evolve quickly. A vaccine would not be effective against new antigens and would quickly become obsolete. Other obstacles to the development of a vaccine include the absence of an effective animal model for AIDS, and the ethical and practical difficulties associated with using human volunteers to test the vaccine.

While immunologists work to develop a successful vaccine and more effective drugs to treat infected patients, massive educational programs have been launched to slow the spread of HIV. Educating the public that having multiple sexual partners increases the risk of AIDS, and teaching sexually active individuals the importance of "safe sex" may help contain the pandemic. Some have suggested that public health facilities should provide free condoms to those who are sexually active and free sterile hypodermic needles to those addicted to intravenous drugs. The cost of these measures would be far less than the cost of medical care for increasing numbers of AIDS patients and the toll in human suffering.

Review

- How does the body defend itself against cancer?
- Why is it difficult to develop a vaccine against HIV?

BiologyNow™ Assess your understanding of **the immune system and disease** by taking the pretest on your BiologyNow CD-ROM.

HARMFUL IMMUNE RESPONSES

Learning Objective

12 Summarize the immunological basis of graft rejection, and describe the events that occur during hypersensitivity reactions, including Rh incompatibility, allergic reactions, and autoimmune diseases.

The immune system's ability to distinguish self from nonself sometimes interferes with clinical interventions. For example, efforts to save a patient's life with a blood transfusion or organ transplant can be disastrous if physicians ignore the effects of immune function. Sometimes the immune system malfunctions. **Hypersensitivity** is an exaggerated, damaging immune response to an antigen that is normally harmless, as occurs in an allergic reaction. Inappropriately directed or ineffective immune responses also result in disease states.

Graft rejection is an immune response against transplanted tissue

Skin can be successfully transplanted from one part of the same body to another or from one identical twin to another. However, when skin is taken from one person and grafted onto the body of a nontwin, it is rejected and sloughs off. Why? Recall that tissues from the same individual or from identical twins have identical MHC alleles and thus the same MHC antigens. Such tissues are compatible.

Because there are many alleles for each of the MHC genes, it is difficult to find identical matches. When a tissue or organ is taken from a donor and transplanted to the body of a nontwin host, several of the MHC antigens are likely to be different. The host's immune system regards the graft as foreign and launches an immune response called **graft rejection.** In the first stage of rejection, the sensitization stage, T_H cells and T_C cells recognize the antigens. In the second stage, the effector stage, T_C cells, the complement system, and cytokines secreted by T_H cells attack the transplanted tissue and can destroy it within a week.

Tissue from donor transplanted into body of recipient $\longrightarrow$ T_C cells recognize MHC antigens on transplant cells as foreign $\longrightarrow$ T_C cells launch immune response (graft rejection) $\longrightarrow$ T_C cells destroy transplant cells

Before transplants are performed, tissues from the patient and from potential donors must be typed and matched as closely as possible. Cell typing is somewhat similar to blood typing but is more complex. If all the MHC antigens are matched, the graft has about a 95% chance of surviving the first year. Unfortunately, not many people are lucky enough to have an identical twin to supply spare parts, so perfect matches are difficult to find. Furthermore, some organs such as the heart cannot be spared. Most organs to be transplanted, therefore, are removed from unrelated donors, often from patients who have just died.

To prevent graft rejection in less compatible matches, physicians use anti-rejection drugs and/or radiation to suppress the immune system. Unfortunately, these treatments make the transplant patient more vulnerable to pneumonia or other infections, and increase the risk of certain types of cancer. If the patient survives the first few months, immunosuppressant drug dosages are reduced. Researchers are developing specific immunosuppression techniques, monoclonal antibodies that will target the specific lymphocytes that cause graft rejection.

Within the United States alone, thousands of patients are in need of organ transplants. Because the number of human donors does not meet this need, investigators are developing techniques for transplanting animal tissues and organs to humans. Challenges include risks of transmitting animal diseases to humans, and graft rejection. Researchers are developing methods for genetically engineering pigs and other animals so that they don't produce antigens that stimulate immune responses in human recipients. These animals could then be cloned and used as donors of hearts, kidneys, and other organs. Effective artificial organs are also being developed.

The body has a few immunologically privileged locations where foreign tissue is accepted. For example, corneal transplants are highly successful because the cornea has almost no associated blood or lymphatic vessels and is thus out of reach of most lymphocytes. Furthermore, antigens in the corneal graft probably would not find their way into the circulatory system and therefore could not stimulate an immune response.

Rh incompatibility can result in hypersensitivity

Named for the rhesus monkeys in whose blood it was first found, the Rh system consists of more than 40 kinds of Rh antigens, each referred to as an **Rh factor.** By far the most important of these factors is *antigen D.* About 85% of U.S. residents who are of western European descent are Rh-positive. This means they have antigen D on the surfaces of their red blood cells (in addition to the antigens of the ABO system and other blood group antigens). The 15% or so of this population who are Rh-negative have no antigen D. Unlike the situation discussed in Chapter 10 for the ABO blood group, Rh-negative individuals do not naturally produce antibodies against antigen D (anti-D). However, they produce anti-D antibodies if they are exposed to Rh-positive blood. The allele coding for antigen D is dominant to the allele for the absence of antigen D. Hence, Rh-negative people are homozygous recessive, and Rh-positive people are heterozygous or homozygous dominant.

Although several kinds of maternal-fetal blood type incompatibilities are known, **Rh incompatibility** is the most serious (Fig. 43-15). If a woman is Rh-negative and the father of the fetus she is carrying is Rh-positive, the fetus may also be Rh-positive, having inherited the D allele from the father. Ordinarily no mixing of maternal and fetal blood occurs; nutrients, oxygen, and other substances are exchanged between these two circulatory systems across the placenta. However, late in pregnancy or during the birth process, a small quantity of blood from the fetus may pass through a defect in the placenta.

The fetus's red blood cells, which bear antigen D, activate the mother's immune system, stimulating B cells to produce

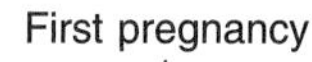
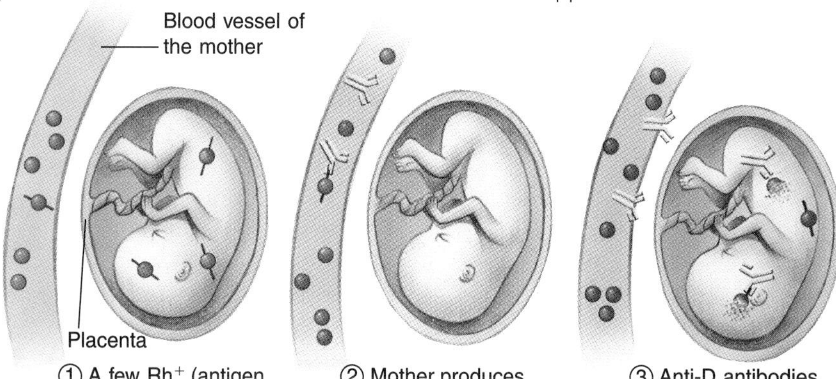

FIGURE **43-15** | **Rh incompatibility.**

When an Rh-negative woman produces Rh-positive offspring, her immune system can be sensitized. In subsequent pregnancies she may produce anti-D antibodies that cross the placenta and destroy fetal red blood cells.

antibodies to antigen D. If the woman becomes pregnant again, she produces anti-D antibodies that cross the placenta and enter the fetal blood. They combine with antigen D molecules on the surface of the fetal red blood cells, causing hemolysis (cells rupture and release hemoglobin into the circulation). The ability to transport oxygen is reduced, and breakdown products of the released hemoglobin damage organs, including the brain. In extreme cases of this disease, known as **erythroblastosis fetalis,** so many fetal red blood cells are destroyed that the fetus may die.

When Rh-incompatibility problems are suspected, fetal blood can be exchanged by transfusion before birth, but this is a risky procedure. More commonly, Rh-negative women are treated just after childbirth (or at termination of pregnancy by miscarriage or abortion) with a preparation of anti-D antibodies known as RhoGAM. These antibodies clear the Rh-positive fetal red blood cells from the mother's blood very quickly, minimizing the chance for her own white blood cells to become sensitized, and develop memory cells. The antibodies are also soon eliminated from her body. As a result, if she becomes pregnant again, her blood does not contain the anti-D that could harm her baby.

Allergic reactions are directed against ordinary environmental antigens

About 20% of the U.S. population is plagued by an allergic disorder such as allergic asthma or hayfever. A predisposition toward these disorders appears to be inherited. In **allergic reactions,** hypersensitivity results in the manufacture of antibodies against mild antigens, called **allergens,** that normally do not stimulate an immune response. Common environmental agents such as house-dust mites, roach feces, pet dander, and pollen can trigger allergic reactions in some individuals. Allergic reactions are referred to as *Type I hypersensitivity.* In many kinds of allergic reactions, distinctive IgE immunoglobulins are produced.

Let's examine a common allergic reaction, a hayfever response to ragweed pollen (Fig. 43-16). Step ① Exposure to pollen causes *sensitization.* Macrophages degrade the allergen and present fragments of it to T cells. The activated T cells then stimulate B cells to become plasma cells and produce pollen-specific IgE. ② These antibodies attach to receptors on mast cells. Each IgE molecule attaches to a mast cell receptor by its C region end, leaving the V region end free to combine with the ragweed pollen allergen.

③ When a sensitized, allergic person inhales the microscopic pollen, ④ allergen molecules rapidly attach to the IgE on sensitized mast cells. This binding of allergen with IgE antibody ⑤ stimulates the mast cell to release granules filled with histamine and other molecules that cause inflammation. ⑥ Blood vessels dilate and capillaries become more permeable, leading to edema and redness. Such responses cause the nasal passages to become swollen and irritated. A runny nose, sneezing, watery eyes, and a general feeling of discomfort are common.

A prolonged allergic response occurs when neutrophils are attracted by chemical compounds released by the mast cells. Neutrophils migrate to the inflamed area, and release compounds that prolong the allergic reaction.

In **allergic asthma,** an allergen-IgE response occurs in the bronchioles of the lungs. Mast cells release substances that cause smooth muscle to contract, and the airways in the lungs sometimes constrict for several hours, making breathing difficult. An antibody to IgE is being clinically tested as a treatment for allergic asthma.

Certain foods or drugs act as allergens in some people, causing a reaction in the walls of the digestive tract that leads to discomfort and diarrhea. The allergen may be absorbed and cause mast cells to release granules elsewhere in the body. When the allergen-IgE reaction takes place in the skin, the histamine released by mast cells causes the swollen red welts we know as **hives.**

Systemic anaphylaxis is a dangerous allergic reaction that can occur when a person develops an allergy to a specific drug such as penicillin, to compounds in the venom injected by a stinging insect, or even to certain foods. Within minutes after the substance enters the body, a widespread allergic reaction takes place. Mast cells release large amounts of histamine and other compounds into the circulation, causing extreme vasodilation and capillary permeability. So much plasma may be lost from the

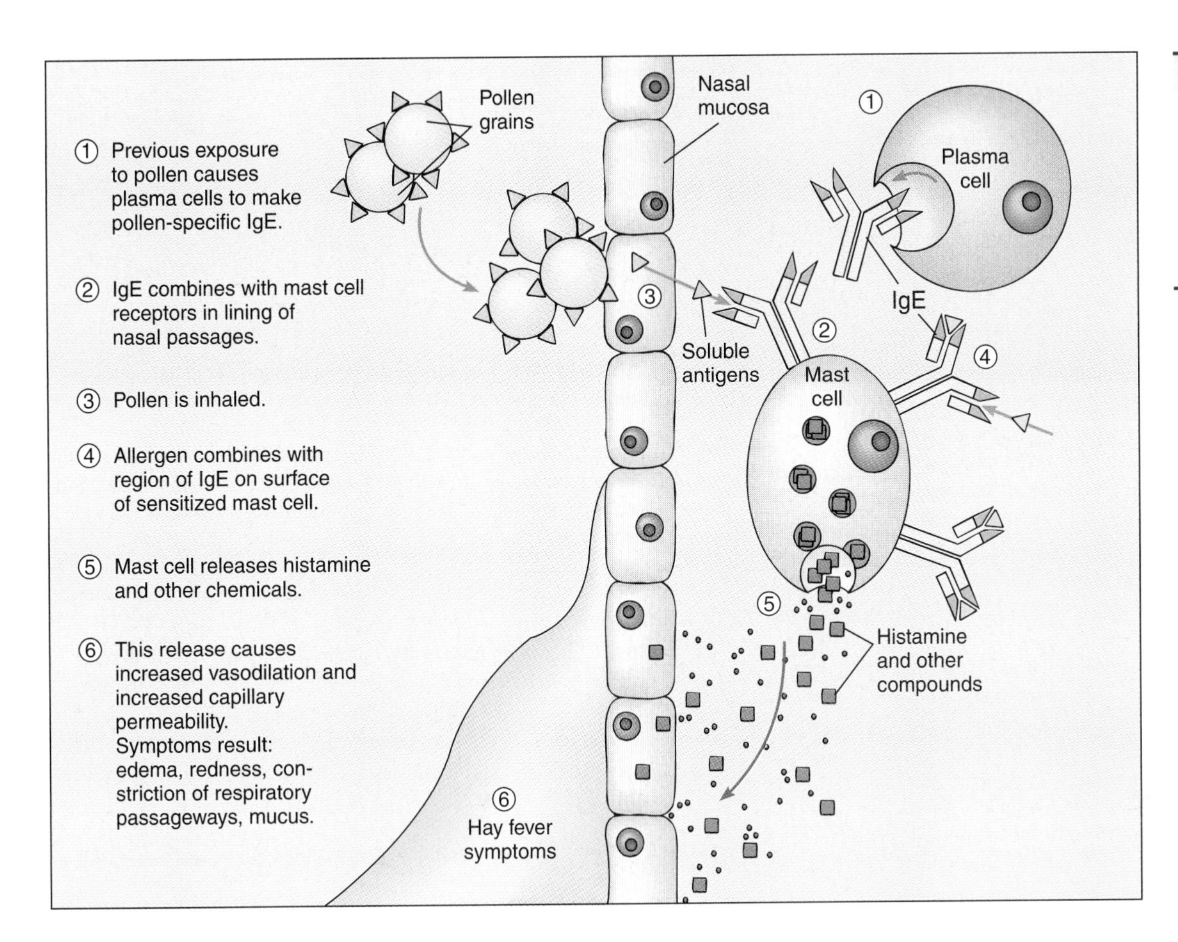

① Previous exposure to pollen causes plasma cells to make pollen-specific IgE.
② IgE combines with mast cell receptors in lining of nasal passages.
③ Pollen is inhaled.
④ Allergen combines with region of IgE on surface of sensitized mast cell.
⑤ Mast cell releases histamine and other chemicals.
⑥ This release causes increased vasodilation and increased capillary permeability. Symptoms result: edema, redness, constriction of respiratory passageways, mucus.
Pollen grains
Nasal mucosa
Plasma cell
IgE
Soluble antigens
Mast cell
Histamine and other compounds
Hay fever symptoms

FIGURE **43-16**

An allergic reaction.

Pollen causes a common type of allergic reaction in many people.

blood that circulatory shock and death can occur within a few minutes.

The symptoms of allergic reactions are often treated with **antihistamines,** drugs that block the effects of histamine. These drugs compete for the same receptors on cells targeted by histamine. When the antihistamine combines with the receptor, it prevents the histamine from binding and thus prevents its harmful effects. Antihistamines are useful clinically in relieving the symptoms of hives and hayfever. They are not completely effective, because mast cells release molecules other than histamine that also cause allergic symptoms.

In an autoimmune disease, the body attacks its own tissues

During their development, complex mechanisms establish self-tolerance (self-recognition) so that lymphocytes do not attack tissues of their own body. However, some lymphocytes remain that have the potential to be *autoreactive,* that is, to launch an immune response against self tissues. Such autoreactivity can lead to a form of hypersensitivity known as *autoimmunity,* or **autoimmune disease,** in which T cells react immunologically against the body's own cells. Some of the diseases that result from such failure in self tolerance are rheumatoid arthritis, systemic lupus erythematosus (SLE), insulin-dependent diabetes, scleroderma, and multiple sclerosis.

In autoimmune diseases, antibodies and T cells attack the body's own tissues. In **rheumatoid arthritis,** T cells in the area of inflammation produce IL-15, an interleukin that promotes inflammation, and generates an immune complex. In **multiple sclerosis,** antibodies attack the glial cells that produce the myelin sheath surrounding neurons in the brain and spinal cord. MRI scans show the absence of myelin sheaths around axons that are normally myelinated.

Genetic risk factors are important in autoimmune diseases. Mutations that cause abnormal apoptosis or inadequate clearing of dead cells may result in danger signals that activate the immune system. Another important fact is that viral or bacterial infection often precedes the onset of an autoimmune disease. Some pathogens have evolved a tactic known as *molecular mimicry.* They trick the body by producing molecules that look like self molecules. For example, an adenovirus that causes respiratory and intestinal illness produces a peptide that mimics myelin protein. When the body launches responses to the adenovirus peptide, it may also begin to attack the similar self molecule, myelin.

To maintain homeostasis and health, the numerous types of cells, regulatory and signaling molecules, and the complex actions of the immune system must be carefully regulated. Specific T cells and B cells proliferate and respond to challenges by pathogens, but after an effective immune response the expanded populations of T and B cells must be decreased to normal numbers. Cells that are no longer needed die by apoptosis. Overactivation of critical components can result in inflammation and autoimmune disease. Underactivation would allow pathogens to cause serious disease. Either situation can lead to death.

Review

- What is the immunological basis for graft rejection?
- What immunological events take place in a common type of allergic reaction such as hayfever?
- What is an autoimmune disease? Give two examples.

Biology Now™ Assess your understanding of **harmful immune responses** by taking the pretest on your BiologyNow CD-ROM.

SUMMARY WITH KEY TERMS

1 Distinguish between nonspecific and specific immune responses.

- The body defends itself against **pathogens,** toxins, and other harmful agents. **Immunology** is the study of internal defensive responses. An **immune response** is the process of recognizing foreign or dangerous macromolecules and responding to eliminate them.
- **Nonspecific immune responses** provide general and immediate protection against pathogens, some toxins and drugs, and cancer cells.
- **Specific immune responses** are highly specific and include immunological memory. An **antigen** is a molecule specifically recognized as foreign or dangerous by cells of the immune system. **Antibodies** are highly specific proteins that recognize and bind to specific antigens.

2 Compare, in general terms, the immune responses of invertebrates and vertebrates.

- Invertebrates depend on nonspecific immune responses such as physical barriers (cuticle, skin, mucous membranes), **phagocytosis,** and **antimicrobial peptides,** soluble molecules that destroy pathogens.
- Vertebrates use both nonspecific and specific immune responses.

3 Describe nonspecific immune responses, including physical and chemical barriers; soluble molecules such as cytokines and the proteins that make up the complement system; phagocytes; natural killer cells; and the inflammatory response.

- Vertebrate nonspecific immune responses include physical barriers, such as the skin and the mucous linings of the respiratory and digestive tracts. When pathogens break through these first-line defenses, other nonspecific defenses are activated.
- Soluble molecules important in immune responses include antimicrobial peptides, regulatory peptides, and proteins that destroy pathogens.
- **Cytokines** are signaling proteins that regulate interactions between cells. Important groups are interferons, interleukins, chemokines, and tumor necrosis factors. **Interferons** inhibit viral replication and activate natural killer cells. **Interleukins** help regulate interactions between lymphocytes and other cells of the body; some have widespread effects. **Chemokines** attract, activate, and direct the movement of certain cells of the immune system. **Tumor necrosis factors (TNFs)** kill tumor cells and stimulate immune cells to initiate an inflammatory response.
- **Complement** proteins lyse the cell wall of pathogens, coat pathogens, enhancing phagocytosis, and attract white blood cells to the site of infection. These actions enhance the inflammatory response.
- **Phagocytes,** including **neutrophils** and **macrophages,** destroy bacteria. **Natural killer cells (NK cells)** destroy cells infected with viruses and foreign or altered cells such as tumor cells.
- When pathogens invade tissues, they trigger an **inflammatory response,** which includes three main processes: vasodilation (increased blood vessel diameter); increased capillary permeability, which allows fluid and antibodies to leave the circulation and enter the tissues; and increased phagocytosis. In response to tissue injury, several types of molecules in the plasma that mediate inflammation are activated. **Mast cells** release **histamine** and other compounds that cause vasodilation and increased capillary permeability.

4 Distinguish between cell-mediated and antibody-mediated immunity.

- In **cell-mediated immunity,** specific T cells are activated; these cells release proteins that destroy cells infected with viruses or other intracellular pathogens.
- In **antibody-mediated immunity,** specific B cells are activated; they multiply and differentiate into plasma cells, which produce antibodies.

5 Describe the principal cells of the immune system, and summarize the function of the major histocompatibility complex.

- Two main types of cells important in specific immune responses are lymphocytes and antigen-presenting cells.
- Lymphocytes develop from stem cells in the bone marrow. **T cells** are lymphocytes responsible for cell-mediated immunity. The **thymus gland** confers immunocompetence on T cells by making them capable of distinguishing between self and nonself. Three types of T cells are **T cytotoxic cells (T_C cells), T helper cells (T_H),** and **memory T cells.** T cells are distinguished by their **T-cell receptors (TCRs).**
- **B cells** are lymphocytes responsible for antibody-mediated immunity. B cells differentiate into **plasma cells,** which produce antibodies. Some activated B cells become **memory B cells,** which continue to produce antibodies after an infection has been overcome.
- **Antigen-presenting cells (APCs)** display foreign antigens as well as their own surface proteins. Dendritic cells, macrophages, and B cells are important APCs. **Dendritic cells** are located in the skin and other tissues of the body that interact with the environment. They are specialized to process, transport, and present antigens.
- Immune responses depend on the **major histocompatibility complex (MHC),** a group of genes that encode MHC proteins. Class I MHC genes encode self antigens, glycoproteins expressed on the surface of most nucleated cells. Class II MHC genes encode glycoproteins expressed on APCs of the immune system. Class III MHC genes encode components of the complement system and TNFs.

6 Describe the sequence of events in cell-mediated immunity.

- In cell-mediated immunity, specific T cells are activated by a foreign antigen–MHC complex on the surface of an infected cell. A co-stimulatory signal and interleukins are also required.
- Activated T_C cells multiply, giving rise to a clone. These cells migrate to the site of infection and destroy pathogen-infected cells.

SUMMARY WITH KEY TERMS (continued)

- Activated T_H cells give rise to a clone of T_H cells, which secrete cytokines that activate B cells and macrophages.

7 Summarize the sequence of events in antibody-mediated immunity, including the effects of antigen-antibody complexes on pathogens.

- In antibody-mediated immunity, B cells are activated when they combine with antigen. Activation requires an APC (such as a dendritic cell or macrophage) that has a foreign antigen–MHC complex displayed on its surface; a T_H cell that secretes interleukins is also needed.
- Activated B cells multiply, giving rise to clones of cells. The cloned cells differentiate, forming plasma cells.
- Plasma cells produce specific antibodies, also called **immunoglobulins (Ig),** in response to the specific antigens that activated them.
- An antibody combines with a specific antigen to form an **antigen-antibody complex,** which may inactivate the pathogen, stimulate phagocytosis, or activate the complement system.

8 Describe the basic structure and function of an antibody, and explain the basis of antibody diversity.

- A typical antibody is Y-shaped; the two arms combine with antigen. An antibody molecule consists of four polypeptide chains: two identical heavy chains and two shorter light chains. Each chain has a **constant (C) region** and a **variable (V) region.**
- Rearrangement of DNA segments during the differentiation of B cells is the main factor responsible for antibody diversity; millions of different types of B (and T) cells are produced.

9 Describe the basis of immunological memory; contrast secondary and primary immune responses.

- After an infection, memory B and memory T cells remain in the body. These cells are responsible for long-term immunity.
- The first exposure to an antigen stimulates a **primary immune response.** A second exposure to the same antigen evokes a **secondary immune response,** which is more rapid and more intense than the primary response.

10 Compare active and passive immunity, giving examples of each.

- **Active immunity** develops as a result of exposure to antigens; it may occur naturally after recovery from a disease or can be artificially induced by immunization with a vaccine.
- **Passive immunity** is a temporary condition that develops when an individual receives antibodies produced by another person or animal.

11 Describe the body's response to cancer cells and HIV.

- NK cells, macrophages, T cells, and other cells of the immune system recognize antigens on cancer cells and launch an immune response against them. Cancer cells evade the immune system by blocking T_C directly or by decreasing the class I MHC molecules on T_C cells.
- **Acquired immunodeficiency syndrome (AIDS)** is caused by the **human immunodeficiency virus (HIV),** a retrovirus. HIV destroys T helper cells, severely impairing immunity and placing the patient at risk for opportunistic infections.

12 Summarize the immunological basis of graft rejection, and describe the events that occur during hypersensitivity reactions, including Rh incompatibility, allergic reactions, and autoimmune diseases.

- Transplanted tissues have MHC antigens that stimulate **graft rejection,** an immune response in which T cells destroy the transplant.
- When an Rh-negative woman gives birth to an Rh-positive baby, she may develop anti-D antibodies. **Rh incompatibility** can then occur in future pregnancies.
- In an **allergic reaction,** an **allergen** stimulates the production of IgE, which combines with receptors on mast cells. The mast cells release histamine and other molecules that cause inflammation and other symptoms of allergy. **Systemic anaphylaxis** is a rapid, widespread allergic reaction that can lead to death.
- In **autoimmune diseases,** the body reacts immunologically against its own tissues.

POST-TEST

1. A molecule recognized as foreign by cells of the immune system is a (an) (a) antibody (b) antigen (c) immunoglobulin (d) interferon (e) cytokine
2. Nonspecific (innate) immune responses include (a) inflammation (b) antigen-antibody complexes (c) immunoglobulin action (d) complement and memory T cells (e) interferon and memory B cells
3. Invertebrate defense responses include (a) phagocytosis (b) antimicrobial peptides (c) ability to distinguish between self and nonself (d) answers a, b, and c are correct (e) answers a and c only
4. Cytokines (a) are regulatory nucleic acids (b) prevent the inflammatory response (c) include interferons and interleukins (d) are immunoglobulins (e) include interleukins and complement proteins
5. Which of the following is *not* an action of complement? (a) enhances phagocytosis (b) enhances inflammatory response (c) coats pathogens (d) lyses viruses release (e) stimulates allergen release
6. Which of the following cells are antigen-presenting cells? (a) NK cells and monocytes (b) macrophages and plasma cells (c) dendritic cells and macrophages (d) mast cells and B cells (e) memory T cells and memory B cells
7. Which of the following cells are especially adept at destroying tumor cells? (a) NK cells (b) plasma cells (c) neutrophils (d) B cytotoxic cells (e) mast cells
8. Which of the following cells become immunologically competent after processing in the thymus gland? (a) NK cells (b) T cells (c) macrophages (d) B cells (e) plasma cells
9. Cells that have a surface marker called CD4 are (a) NK cells (b) T cytotoxic cells (c) T helper cells (d) B cells (e) plasma cells
10. The major histocompatibility complex (MHC) (a) encodes a group of cell surface proteins (b) encodes certain antibodies (c) is important mainly in allergic reactions (d) inhibits complement release from macrophages (e) consists of Y-shaped molecules

POST-TEST *(continued)*

11. Which sequence most accurately describes antibody-mediated immunity? (1) B cell divides and gives rise to clone (2) antibodies produced (3) cells differentiate, forming plasma cells (4) activated helper T cell interacts with B cell displaying same antigen complex (5) B cell activated (a) 1 ⟶ 2 ⟶ 3 ⟶ 4 ⟶ 5 (b) 3 ⟶ 2 ⟶ 1 ⟶ 4 ⟶ 5 (c) 4 ⟶ 5 ⟶ 3 ⟶ 4 ⟶ 1 (d) 4 ⟶ 5 ⟶ 1 ⟶ 3 ⟶ 2 (e) 4 ⟶ 3 ⟶ 1 ⟶ 2 ⟶ 5
12. A typical antibody (a) is activated by APCs (b) has four identical heavy chains and four identical light chains (c) has IgG and IgD components (d) suppresses phagocyte actions (e) has a Y shape
13. Immunoglobulin A (a) combines with mast cells in allergic reactions (b) combines with NK cells (c) prevents pathogens from attaching to epithelial surfaces (d) is found mainly on B-cell surfaces (e) is found mainly on T cells
14. When a person is exposed to the same antigen a second time, the response is (a) called a *secondary immune response* (b) more rapid (c) mediated by dendritic cells (d) answers a, b, and c are correct (e) answers a and b only
15. Graft rejection (a) is an example of passive immunity (b) occurs in mild form after immunization (c) generates monoclonal antibodies (d) does not occur when tissue is transplanted from one identical twin to the other (e) is initiated by the thymus gland
16. In an allergic reaction (a) the body is immunodeficient (b) an allergen binds with IgE (c) T helper cells release histamine (d) allergen stimulates graft rejection (e) mast cells are deactivated
17. HIV (a) is a retrovirus (b) destroys T cytotoxic cells (c) is attacked mainly by B cells (d) answers a, b, and c are correct (e) none of the preceding answers is correct

CRITICAL THINKING

1. Specificity, diversity, and memory are key features of the immune system. Giving specific examples, explain how each of these features is important.
2. Macrophages can be selectively destroyed in the body by the administration of a certain chemical. What would be the effects of such a loss of macrophages? Which do you think would have a greater effect on the immune system, loss of macrophages or loss of B cells?
3. What are the advantages of having MHC antigens? Disadvantages? What do you think would be the consequences of not having them?
4. Imagine that you are a researcher developing new HIV treatments. What approaches might you take? What public policy decisions would you recommend that might help slow the spread of AIDS while new treatments or vaccines are being developed?

- Visit our Web site at **http://biology.brookscole.com/solomon7** for links to chapter-related resources on the World Wide Web. Additional online materials relating to this chapter can also be found on our Web site.

BIOLOGY NOW RESOURCES

Active Figures

43-1: Overview of human immune responses

43-7: Antibody-mediated immunity.

Preparing for an exam? Take a diagnostic test on your BiologyNow CD-ROM.

Post-Test Answers

1. b	2. a	3. d	4. c
5. e	6. c	7. a	8. b
9. c	10. a	11. d	12. e
13. c	14. e	15. d	16. b
17. a			

44

Gas Exchange

Skip Moody/Dembinsky Photo Associates

The white-tailed deer (*Odocoileus virginanus*) and other terrestrial vertebrates ventilate their lungs by breathing air.

CHAPTER OUTLINE

- **Adaptations for Gas Exchange in Air or Water**
- **Types of Respiratory Surfaces**
- **The Mammalian Respiratory System**
- **Breathing Polluted Air**

Most animal cells require a continuous supply of oxygen for cellular respiration. Some cells, such as mammalian brain cells, may be damaged beyond repair if their oxygen supply is cut off for only a few minutes. Animal cells must also rid themselves of carbon dioxide. The exchange of gases between an organism and its environment is known as **respiration.** Two phases of respiration are organismic and cellular respiration. During **organismic respiration,** oxygen from the environment is taken up by the animal and delivered to its individual cells. At the same time, carbon dioxide generated during cellular respiration is excreted into the environment. In **aerobic cellular respiration,** which takes place in mitochondria, oxygen is necessary for the citric acid cycle to proceed. Oxygen serves as the final electron acceptor in the mitochondrial electron transport chain (see Chapter 7). Carbon dioxide is a metabolic waste product of cellular respiration.

In small, aquatic organisms such as sponges, hydras, and flatworms, gas exchange occurs entirely by simple **diffusion,** the passive movement of particles (atoms, ions, or molecules) from a region of higher concentration to a region of lower concentration, that is, down a concentration gradient. Most cells are in direct contact with the environment. Dissolved oxygen from the surrounding water diffuses into the cells, whereas carbon dioxide diffuses out of the cells and into the water. No specialized respiratory structures are needed.

Oxygen diffuses through tissues slowly, however. In an animal more than about 1 mm thick, oxygen cannot diffuse quickly enough through layers of cells to support life. Specialized respiratory structures such as gills or lungs are required to deliver oxygen to the cells or to a transport system and to facilitate excretion of carbon dioxide. Such respiratory systems, working with circulatory systems, provide the efficient intake and transport of oxygen necessary to support high metabolic rates.

If the air or water supplying oxygen to the cells can be continuously renewed, more oxygen will be available. For this reason animals carry on **ventilation;** that is, they actively move air or water over their respiratory surfaces. Sponges do this with flagella, setting up a current of water through the channels of their bodies. Most fishes gulp water, which then passes over their gills. Terrestrial vertebrates have lungs and breathe air (see photograph); the diaphragm and other muscles move air in and out of the lungs. ■

ADAPTATIONS FOR GAS EXCHANGE IN AIR OR WATER

Learning Objective

1 Compare the advantages and disadvantages of gas exchange in air with those of gas exchange in water.

Gills are adapted for gas exchange in water, and the tracheal tubes of insects and lungs of vertebrates are respiratory structures adapted for gas exchange in air. Respiratory surfaces must be kept moist to prevent drying out, and oxygen and carbon dioxide are dissolved in the fluid that bathes the cells of these surfaces. Whether an animal makes its home on land or water, gas exchange takes place across a moist surface.

Animals that carry out gas exchange in water require no special mechanisms for maintaining moisture. In contrast, animals that respire in air struggle continuously with water loss. Adaptations have evolved that keep respiratory surfaces moist and minimize desiccation. For example, the lungs of air-breathing vertebrates are located deep within the body, not exposed like gills. Air is humidified and brought to body temperature as it passes through the upper respiratory passageways, and expired air must again pass through these airways (providing opportunity for retaining water) before leaving the body. These adaptations protect the lungs from the drying and cooling effects of air.

Gas exchange in air has certain advantages over gas exchange in water. Compared with water, air contains a much higher concentration of molecular oxygen. In addition, oxygen diffuses much faster through air than through water. Another advantage is that less energy is needed to move air than to move water over a gas exchange surface. This is because air is less dense and less viscous.

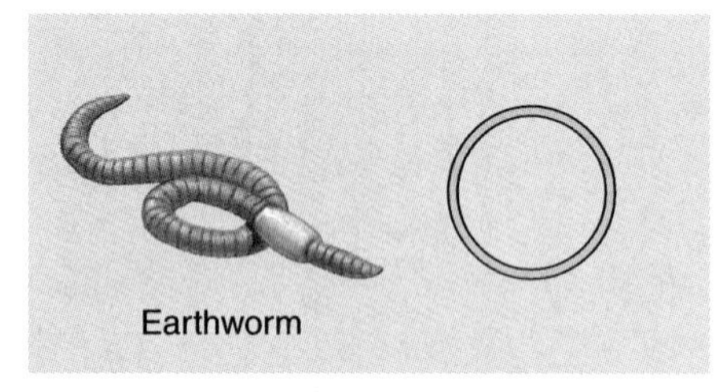

(a) Gas exchange across body surface

(b) Tracheal tubes

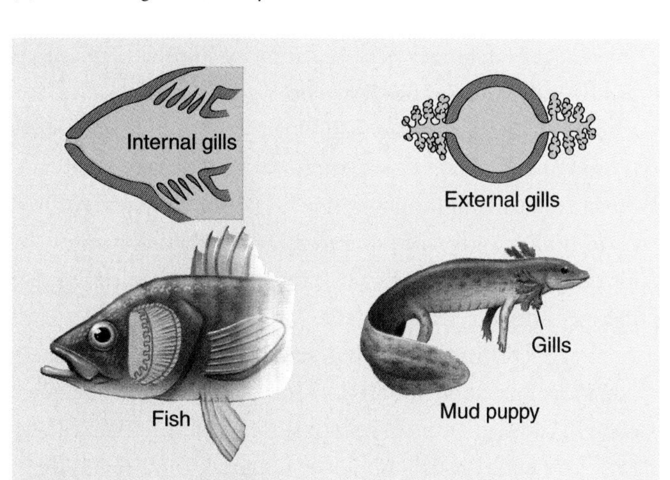

(c) Gills

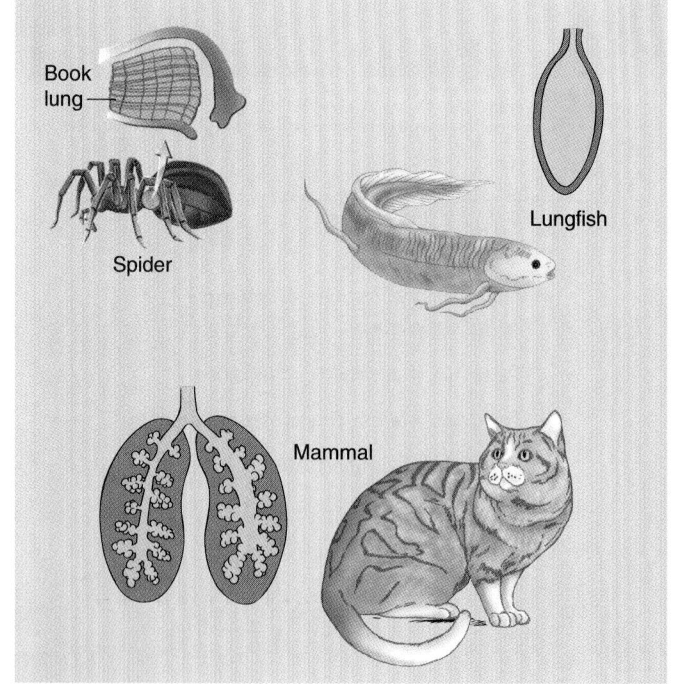

(d) Lungs

FIGURE **44-1** **Adaptations for gas exchange.**

(a) Some animals exchange gases through the body surface. **(b)** Insects and some other arthropods exchange gases through a system of tracheal tubes, or tracheae. **(c)** Gills can be external or internal. **(d)** Lungs are adaptations for terrestrial gas exchange.

Review

- What are some advantages of gas exchange in air over gas exchange in water?

BiologyNow™ Assess your understanding of **adaptations for gas exchange in air or water** by taking the pretest on your BiologyNow CD-ROM.

TYPES OF RESPIRATORY SURFACES

Learning Objective

2 Describe the following adaptations for gas exchange: body surface, tracheal tubes, gills, and lungs.

Not only must respiratory structures be moist, but they must also have thin walls through which diffusion can easily occur. Respiratory structures are generally richly supplied with blood vessels to facilitate transport and exchange of respiratory gases. Four main types of respiratory surfaces have evolved in animals: the animal's own body surface, tracheal tubes, gills, and lungs (Fig. 44-1). Some animals use a combination of these adaptations.

The body surface may be adapted for gas exchange

Gas exchange occurs through the entire body surface in many animals, including nudibranch mollusks, most annelids, and some amphibians. All these animals are small, with a high surface-to-volume ratio. They also have a low metabolic rate that requires smaller quantities of oxygen per cell. In aquatic animals, the body surface is kept moist by the surrounding water. In terrestrial animals, the body secretes fluids that keep its surface moist. Many animals that exchange gases across the body surface also have gills or lungs.

Tracheal tube systems of arthropods deliver air directly to the cells

The relatively inefficient open circulatory system of arthropods cannot supply these active animals with enough oxygen. They need a specialized respiratory system. In insects and some other arthropods (such as chilopods, diplopods, some mites, and some spiders), the respiratory system is a network of **tracheal tubes,** also called **tracheae** (Fig. 44-2). Air enters the tracheal tubes through a series of up to 20 tiny openings called **spiracles** along the body surface. In some insects, especially large, active ones, muscles help ventilate the tracheae by pumping air in and out of the spiracles. For example, the grasshopper draws air in through the first four pairs of spiracles when the abdomen expands. Then the abdomen contracts, forcing air out through the last six pairs of spiracles.

Once inside the body, the air passes through a system of branching tracheal tubes, which extend to all parts of the animal. The tracheal tubes terminate in microscopic, fluid-filled tracheoles. Gases are exchanged between this fluid and the body cells. The tracheal system supplies enough oxygen to support the high metabolic rates characteristic of many insects.

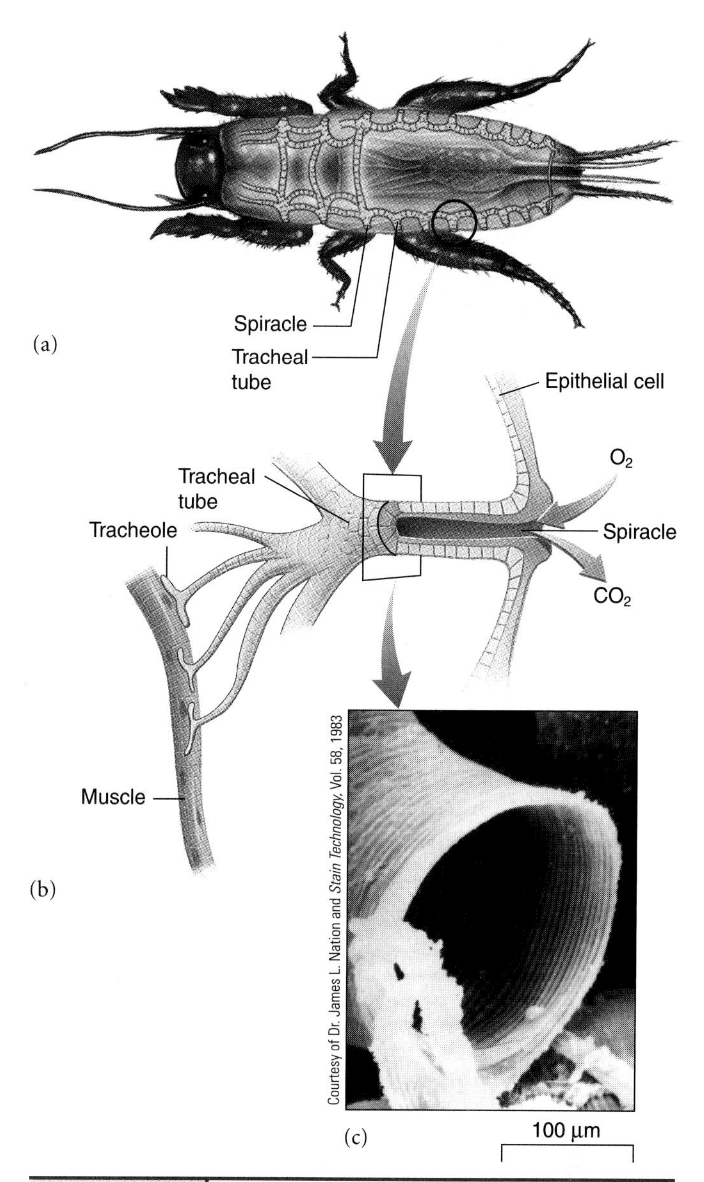

FIGURE **44-2** **Tracheal tubes.**

(a) Air enters the system of tracheal tubes through openings called spiracles. **(b)** Air passes through a system of branching tracheal tubes that conduct oxygen to all cells of the insect. **(c)** SEM of a mole cricket trachea. The wrinkles are a part of a long spiral that wraps around the tube, strengthening the tracheal wall somewhat as a spring strengthens the plastic hoses of many vacuum cleaners. The tracheal wall is composed of chitin.

Gills of aquatic animals are respiratory surfaces

Found mainly in aquatic animals, **gills** are moist, thin structures that extend from the body surface. They are supported by the buoyancy of water but tend to collapse in air. In many animals, the outer surface of the gills is exposed to water, whereas the inner side is in close contact with networks of blood vessels.

Sea stars and sea urchins have **dermal gills** that project from the body wall. Their ciliated epidermal cells ventilate the gills by beating a stream of water over them. Gases are exchanged between the water and the coelomic fluid inside the body by diffusion through the gills.

Various types of gills are found in some annelids, aquatic mollusks, crustaceans, fishes, and amphibians. Mollusk gills are folded, providing a large surface for respiration. In clams and other bivalve mollusks and in simple chordates, gills may also be adapted for trapping and sorting food. The rhythmic beating of cilia draws water over the gill area, and food is filtered out of the water while gases are exchanged. In mollusks, gas exchange also takes place through the mantle.

In chordates, gills are usually internal. A series of slits perforates the pharynx, and the gills lie along the edges of these gill slits (see Fig. 30-4). In bony fishes, the fragile gills are protected by an external bony plate, the **operculum.** In some fish, movements of the jaw and operculum help pump water rich in oxygen through the mouth and across the gills. The water leaves through the gill slits.

Each gill in the bony fish consists of many **filaments,** which provide an extensive surface for gas exchange (Fig. 44-3). The filaments extend out into the water, which continuously flows over them. A capillary network delivers blood to the gill filaments, facilitating diffusion of oxygen and carbon dioxide between blood and water. The impressive efficiency of this system is possible because blood flows in a direction opposite to the movement of the water. This arrangement, called a **countercurrent**

FIGURE **44-3** | **Gills in bony fish.**

(a) The gills lie under a bony plate, the operculum, which has been removed in this side view. The gills form the lateral wall of the pharyngeal cavity. **(b)** Each gill consists of a cartilaginous gill arch to which two rows of leaflike gill filaments attach. As water flows past the gill filaments, blood circulates within them. **(c)** Each gill filament has many smaller extensions rich in capillaries. Blood entering the capillaries is deficient in oxygen. The blood flows through the capillaries in a direction opposite to that taken by the water. This countercurrent exchange system efficiently charges the blood with oxygen. **(d)** If the system were concurrent, that is, if blood flowed through the capillaries in the same direction as the flow of the water, much less of the oxygen dissolved in the water could diffuse into the blood. **(e)** Gills of the salmon.

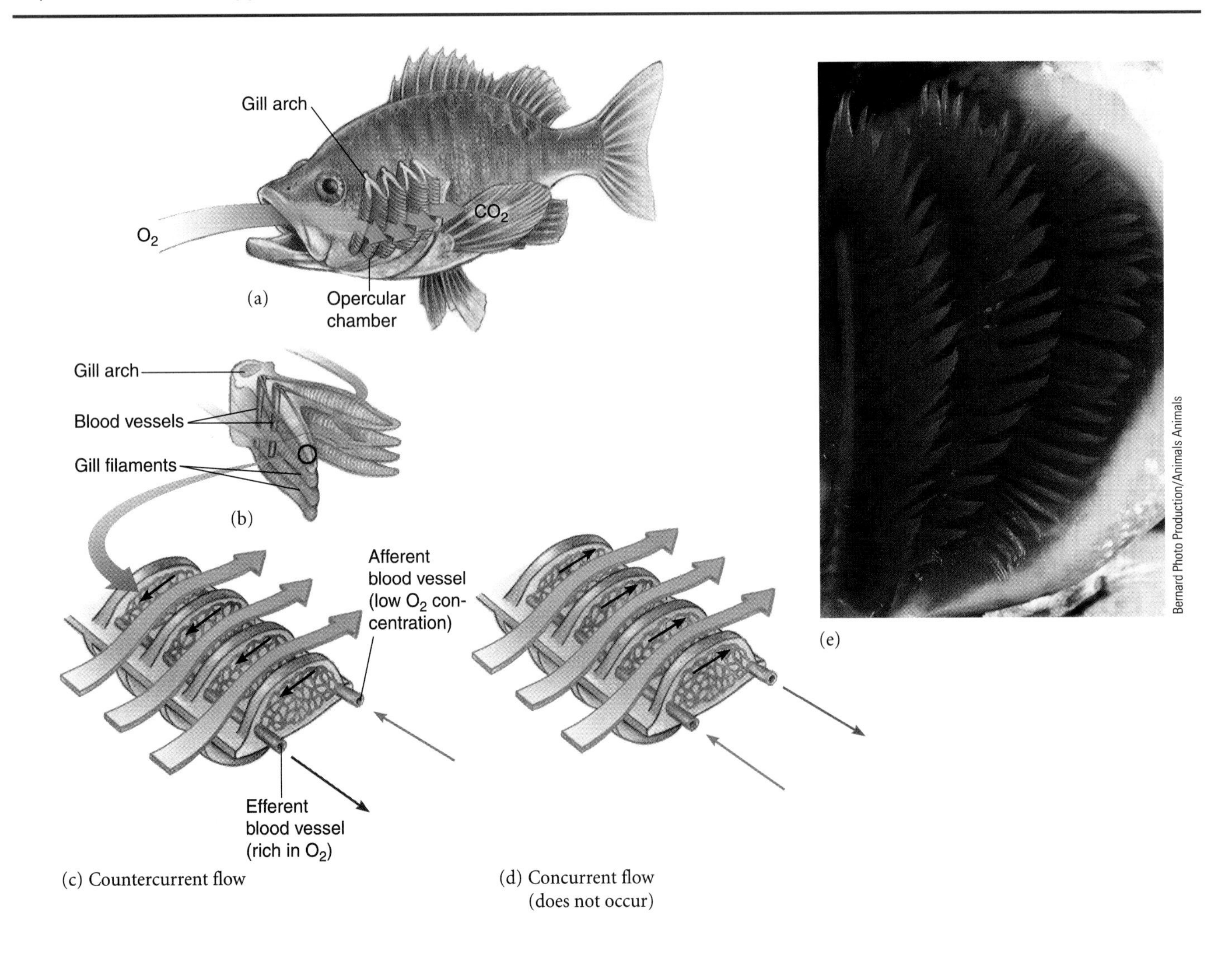

exchange system, maximizes the difference in oxygen concentration between blood and water throughout the area where the two remain in contact (Fig. 44-3c).

If blood and water flowed in the *same* direction, that is, *concurrent exchange,* the difference between the oxygen concentrations in blood (low) and water (high) would be very large initially and very small at the end. The oxygen concentration in the water would decrease as that in the blood increased. When the concentrations in the two fluids became equal, an equilibrium would be reached, and the net diffusion of oxygen would stop. Only about 50% of the oxygen dissolved in the water could diffuse into the blood.

In the countercurrent exchange system, however, blood low in oxygen comes in contact with water that is partly depleted of oxygen. Then, as the blood flowing through the capillaries becomes more and more rich in oxygen, it comes in contact with water with a progressively higher concentration of oxygen. Thus, all along the capillaries, the diffusion gradient favors passage of oxygen from the water into the gill. A high rate of diffusion is maintained, ensuring that a very high percentage (more than 80%) of the available oxygen in the water diffuses into the blood.

Oxygen and carbon dioxide do not interfere with one another's diffusion, and they simultaneously diffuse in opposite directions. This is because oxygen is more concentrated outside the gills than within, but carbon dioxide is more concentrated inside the gills than outside. Thus, the same countercurrent exchange mechanism that ensures efficient inflow of oxygen also results in equally efficient outflow of carbon dioxide.

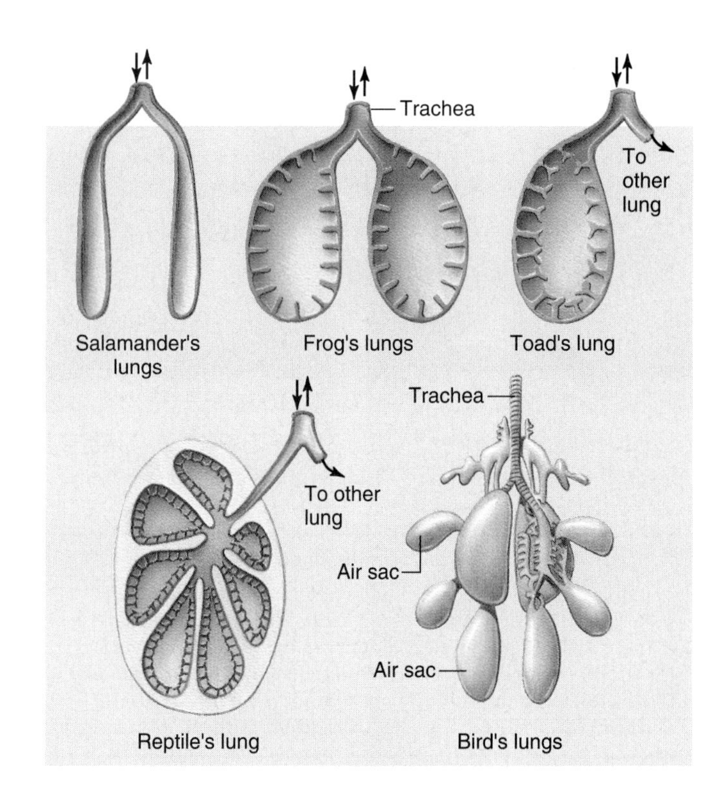

FIGURE **44-4** | **Comparison of vertebrate lungs.**

The surface area of the lung has increased during vertebrate evolution. Salamander lungs are simple sacs. Other amphibians and reptiles have lungs with small ridges or folds that help increase surface area. Birds have an elaborate system of lungs and air sacs. Mammalian lungs have millions of alveoli that increase the surface available for gas exchange (see Fig. 44-7).

Terrestrial vertebrates exchange gases through lungs

Lungs are respiratory structures that develop as ingrowths of the body surface or from the wall of a body cavity such as the pharynx (throat region). For example, the **book lungs** of spiders are enclosed in an inpocketing of the abdominal wall. These lungs consist of a series of thin, parallel plates of tissue (like the pages of a book) filled with hemolymph (see Fig. 44-1d). The plates of tissue are separated by air spaces that receive oxygen from the outside environment through a spiracle. A different type of lung evolved in land snails and slugs. (These terrestrial mollusks lack gills.) Gas exchange takes place through a lung, that is a vascularized region of the mantle.

Fossil evidence suggests that early lobe-finned fishes had lungs somewhat similar to those of modern lungfish. The Australian lungfish can use either its gills or its lungs, depending on the conditions in its environment. The gills of African and South American lungfishes degenerate with age, so adults must rise to the surface and exchange gases entirely through their lungs.

Remains of early lobe-finned fishes occur extensively in the fossil record. Those found in Devonian strata are thought to be similar to the ancestors of amphibians, and many amphibian fossils have been found in adjacent ancient strata. Some paleontologists hypothesize that lungs evolved as an adaptation to the periodic droughts in Devonian times and that all early bony fishes may have had lungs or lunglike structures. Most modern bony fishes have no lungs, but nearly all have homologous **swim bladders** (see Chapter 30). By adjusting the amount of gas in its swim bladder, the fish can control its buoyancy.

Some amphibians do not have lungs. Among plethodontid (lungless) salamanders, for example, all gas exchange takes place in the pharynx or across the thin, wet skin. However, even though they depend mainly on their body surface for gas exchange, most amphibians have lungs (Fig. 44-4). The lungs of salamanders are two long, simple sacs richly supplied by capillaries. Frogs and toads have ridges containing connective tissue on the inside of the lungs, somewhat increasing the respiratory surface.

The lungs of most reptiles are rather simple sacs, with only some folding of the wall to increase the surface for gas exchange. The gas exchange is not very efficient and does not supply enough oxygen to sustain long periods of activity. In some lizards, turtles, and crocodiles, the lungs are somewhat more complex, with subdivisions that give them a spongy texture.

Birds have the most efficient respiratory system of any living vertebrate. Very active, endothermic animals with high metabolic rates, birds require large amounts of oxygen to sustain flight and other activities. Their small, bright red lungs have extensions (usually nine) called **air sacs,** which reach into all parts of the body and even connect with air spaces in some of the bones (Fig. 44-5). The air sacs act as bellows, drawing air into the sys-

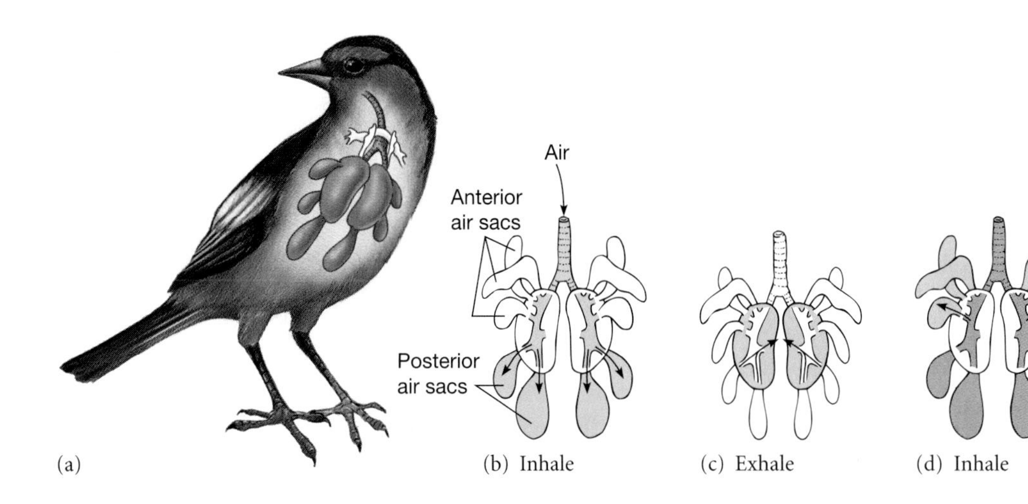

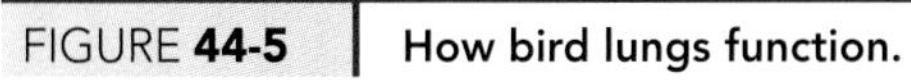

FIGURE **44-5** **How bird lungs function.**

The bird's breathing process requires two cycles of inhalation and exhalation to support a one-way flow of air through the lungs. **(a)** The bird respiratory system includes lungs and extensions called air sacs. **(b)** As the bird inhales, new air flows into the posterior air sacs *(blue)* and partly into the lungs (not shown). **(c)** As the bird exhales, this air is forced into the lungs. **(d)** At the second inhalation, most of the air from the first breath moves into the anterior air sacs and partly into the lungs (not shown). Air from the second inhalation flows into the posterior air sacs *(pink)*. **(e)** At the second exhalation, most of the air from the first inhalation leaves the body, and air from the second inhalation flows into the lungs.

tem. Collapse of the air sacs during exhalation forces air out. Gas exchange does not take place across the walls of the air sacs.

The high-performance bird respiratory system is arranged so that air flows in one direction through the lungs and is renewed during a two-cycle process. Air entering the body passes into the posterior air sacs and the part of the lungs closest to these air sacs. When the bird exhales, that air flows into the lungs. At the second breath, the air flows from the lungs to the anterior air sacs. Finally, at the second exhalation, the air leaves the body as another breath of air enters the lungs. Thus, a bird gets fresh air across its lungs through both inhalation and exhalation.

Bird lungs have tiny, thin-walled tubes, the **parabronchi,** which are open at both ends. Gas exchange takes place across the walls of these tubes. Chickens and other weak-flying birds have about 400 parabronchi per lung compared with pigeons and other strong-flying birds, which have about 1800 parabronchi per lung. The direction of blood flow in the lungs is opposite that of air flow through the parabronchi. This arrangement, similar in principle to the countercurrent exchange in the gills of fishes, increases the amount of oxygen that enters the blood. However, in birds the capillaries are oriented at right angles to the parabronchi rather than along their length. For this reason, this arrangement is referred to as *crosscurrent,* rather than as countercurrent.

Review

- Why are specialized respiratory structures necessary in a tadpole but not in a flatworm?
- How does gas exchange differ among the following animals: (a) earthworm (b) grasshopper (c) fish (d) bird?
- How does the countercurrent exchange system increase the efficiency of gas exchange between a fish's gills and blood?

BiologyNow™ Assess your understanding of **types of respiratory surfaces** by taking the pretest on your BiologyNow CD-ROM.

THE MAMMALIAN RESPIRATORY SYSTEM

Learning Objectives

3 Trace the passage of oxygen through the human respiratory system from nostrils to alveoli.

4 Summarize the mechanics and the regulation of breathing in humans, and describe gas exchange in the lungs and tissues.

5 Explain the role of hemoglobin in oxygen transport, and identify factors that determine and influence the oxygen–hemoglobin dissociation curve.

6 Summarize the mechanisms by which carbon dioxide is transported in the blood.

7 Describe the physiological effects of hyperventilation, and of sudden decompression when a diver surfaces too quickly from deep water.

The respiratory system of mammals consists of the lungs and a series of tubes through which air passes on its journey from the nostrils to the lungs and back (Fig. 44-6). The complex lungs have an enormous surface area. In the following sections we focus on the human respiratory system.

The airway conducts air into the lungs

A breath of air enters the body through the **nostrils** and flows through the **nasal cavities.** As air passes through the nose, it is filtered, moistened, and brought to body temperature. The nasal cavities are lined with a moist, ciliated epithelium rich in blood vessels. Inhaled dirt, bacteria, and other foreign particles are trapped in the stream of mucus produced by cells within the epithelium and pushed along toward the throat by the cilia. In this way, foreign particles are delivered to the digestive system,

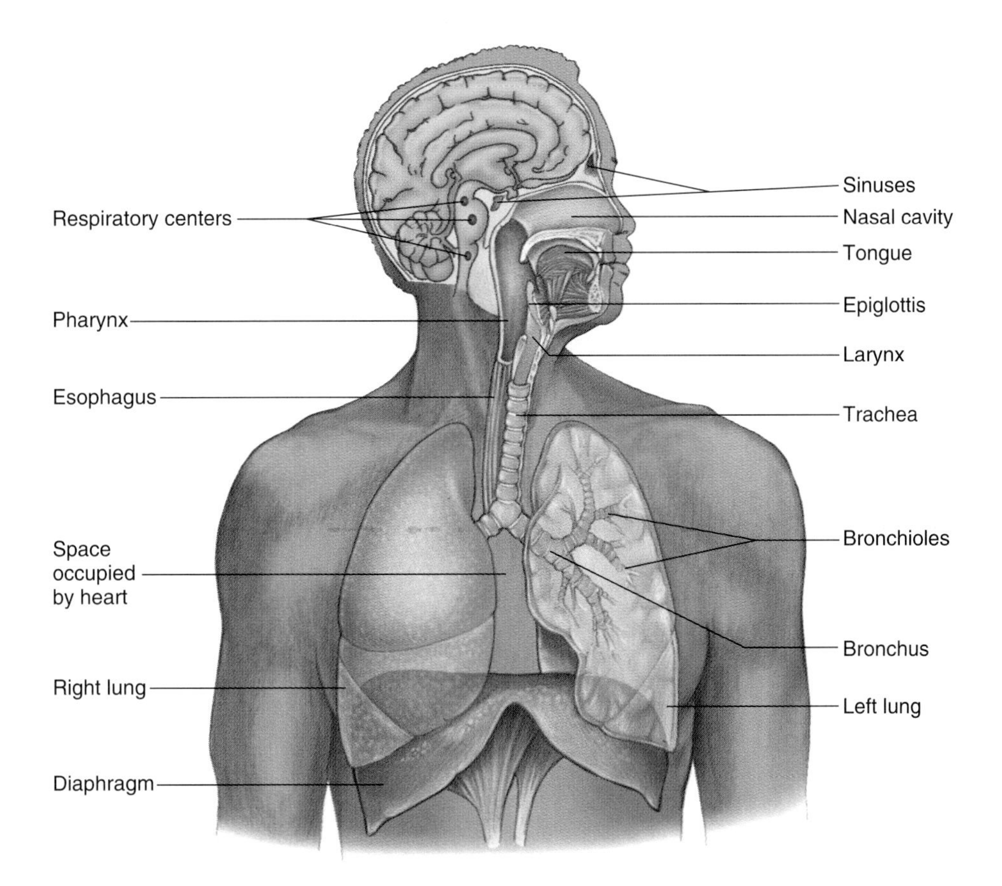

ACTIVE FIGURE **44-6** | **The human respiratory system.**

The muscular diaphragm forms the floor of the thoracic cavity. The internal view of one lung illustrates a portion of its extensive system of air passageways. The respiratory centers in the brain regulate the rate of respiration.

BiologyNow™ Learn more about **the human respiratory system** by clicking on this figure on your BiologyNow CD-ROM.

which can more effectively dispose of such materials than can the delicate lungs. A person normally swallows more than a pint of nasal mucus each day, and even more during an infection or allergic reaction.

The back of the nasal cavities is continuous with the throat region, or **pharynx.** Air finds its way into the pharynx whether one breathes through the nose or mouth. An opening in the floor of the pharynx leads into the **larynx,** sometimes called the "Adam's apple." Because the larynx contains the vocal cords, it is also referred to as the voice box. Cartilage embedded in its wall prevents the larynx from collapsing and makes it hard to the touch when felt through the neck.

During swallowing, a flap of tissue called the **epiglottis** automatically closes off the larynx so that food and liquid enter the esophagus rather than the lower airway. If this mechanism fails and foreign matter enters the sensitive larynx, a cough reflex expells the material. Despite these mechanisms, choking sometimes occurs.

From the larynx, air passes into the **trachea,** or windpipe, which is kept from collapsing by rings of cartilage in its wall. The trachea divides into two branches, the **bronchi** (sing., *bronchus*), each of which connects to a lung. Both trachea and bronchi are lined by a mucous membrane containing ciliated cells that traps medium-sized particles that escape the cleansing mechanisms of nose and larynx. Mucus containing these particles is constantly beaten upward by the cilia to the pharynx, where it is periodically swallowed. This mechanism, functioning as a cilia-propelled elevator of mucus, helps keep foreign material out of the lungs.

Gas exchange occurs in the alveoli of the lungs

The lungs are large, paired, spongy organs occupying the *thoracic* (chest) *cavity.* The right lung is divided into three lobes, the left lung into two lobes. Each lung is covered with a **pleural membrane,** which forms a continuous sac that encloses the lung and becomes the lining of the thoracic cavity. The *pleural cavity* is the space between the pleural membranes. A film of fluid in the pleural cavity provides lubrication between the lungs and the chest wall.

Because the lung consists largely of air tubes and elastic tissue, it is a spongy, elastic organ with a very large internal surface area for gas exchange. Inside the lungs the bronchi branch, becoming smaller and more numerous. These branches give rise to more than 1 million tiny **bronchioles** in each lung. Each bronchiole ends in a cluster of tiny air sacs, the **alveoli** (sing., *alveolus*) (Fig. 44-7). Each human lung contains more than 300 million alveoli, and thus has an internal surface area the approximate size of a tennis court. Each alveolus is lined by an extremely thin, single layer of epithelial cells. Gases diffuse freely through the wall of the alveolus and into the capillaries that surround it. Only two thin cell layers, the epithelia of the alveolar wall and the capillary wall, separate the air in the alveolus from the blood.

In summary, the sequence of structures through which air passes after it enters the body is as follows:

Nostrils ⟶ nasal cavities ⟶ pharynx ⟶ larynx ⟶ trachea ⟶ bronchi ⟶ bronchioles ⟶ alveoli

Ventilation is accomplished by breathing

Breathing is the mechanical process of moving air from the environment into the lungs and of expelling air from the lungs. Inhaling air is called **inhalation** or *inspiration;* exhaling air is **exhalation** or *expiration.* The thoracic cavity is closed so no air can enter except through the trachea. (When the chest wall is punctured, for example, by a fractured rib or gunshot wound, air enters the pleural space and the lung collapses.)

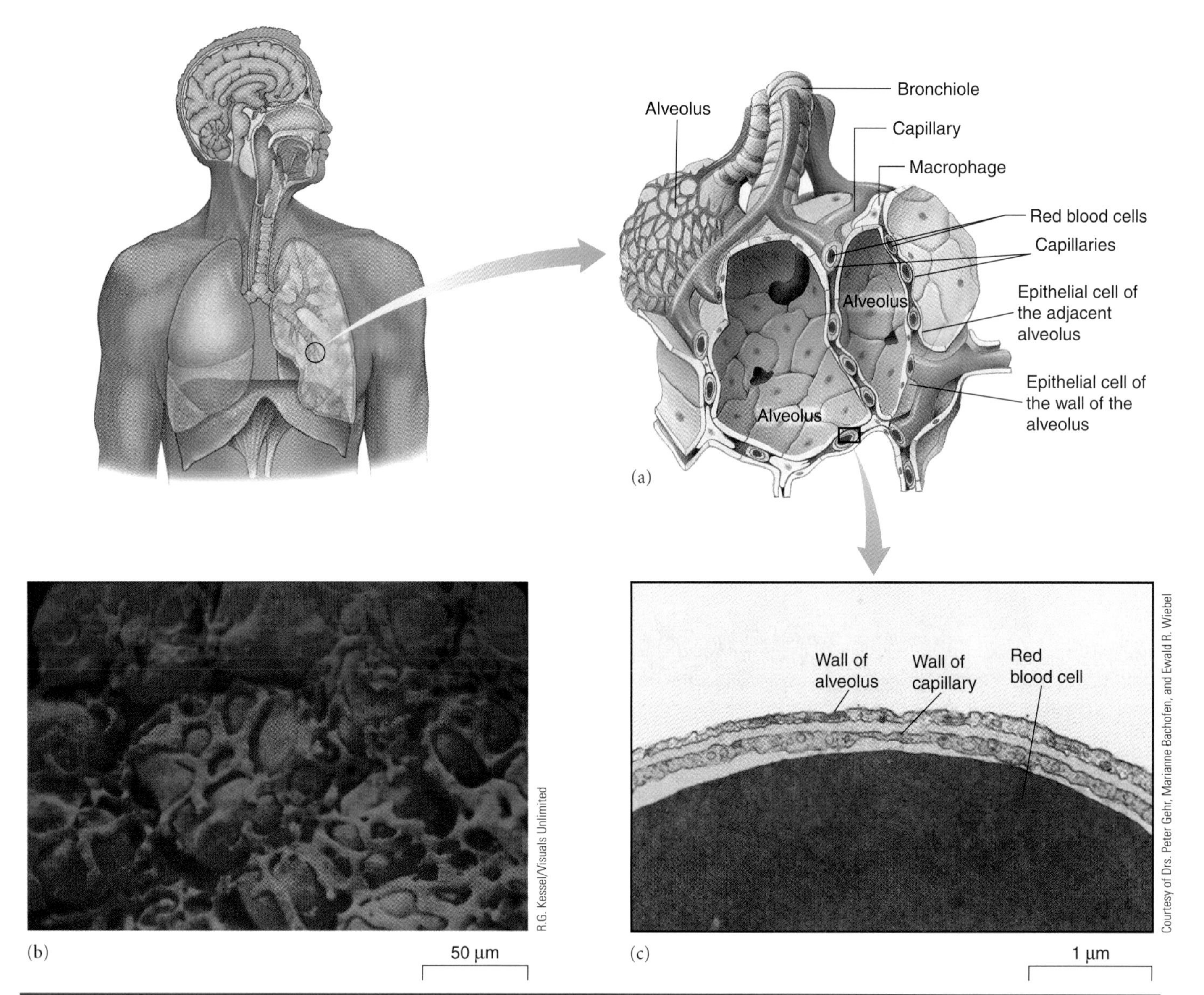

FIGURE **44-7** **Structure of alveoli.**

(a) Gas exchange takes place across the thin wall of the alveolus. Extensive capillary networks lie between the walls of the alveoli. **(b)** Color-enhanced SEM showing the capillary network surrounding a portion of several alveoli. **(c)** Color-enhanced TEM of a portion of a capillary and the wall of an alveolus. The dark structure extending through the capillary is part of a red blood cell. The wall of the alveolus is visible just above the wall of the capillary. Notice the very short distance oxygen must diffuse to get from the air within the alveolus to the red blood cells that transport it to the body tissues.

During inhalation, the volume of the thoracic cavity is increased by the contraction of the **diaphragm,** the dome-shaped muscle that forms its floor. When the diaphragm contracts, it moves downward, increasing the volume of the thoracic cavity (Fig. 44-8). During forced inhalation, when a large volume of air is inhaled, the *external intercostal muscles* contract as well. This action moves the ribs upward, also increasing the volume of the thoracic cavity.

Because the lungs adhere to the walls of the thoracic cavity, when the volume of the thoracic cavity increases the space within each lung also increases. The air in the lungs now has more space in which to move about, and the air pressure in the lungs falls by 2 or 3 millimeters of mercury (mm Hg) below the air pressure outside the body. As a result of this pressure difference, air from the outside rushes in through the respiratory passageways and fills the lungs until the two pressures are equal once again.

Exhalation occurs when the diaphragm relaxes. The volume of the thoracic cavity decreases, raising the pressure in the lungs to 2 to 3 mm Hg above atmospheric pressure. The millions of distended air sacs partially deflate, expelling the inhaled air. The pressure returns to normal, and the lung is ready for another inhalation. Thus in inhalation the millions of alveoli fill with air like so many tiny balloons. Then, during exhalation, the air rushes out of the alveoli, partially deflating them. During deep or forced breathing, the *external intercostal muscles* also contract, pulling the ribcage upward and outward. This action further

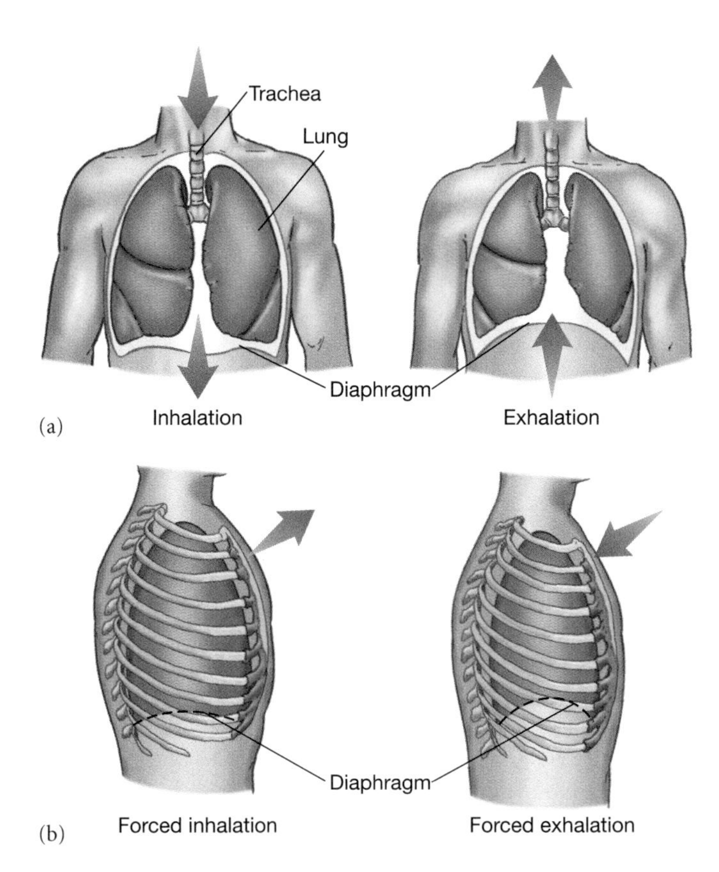

ACTIVE FIGURE **44-8** **Mechanics of breathing.**

(a) Changes in position of the diaphragm in exhalation and inhalation change the volume of the thoracic cavity. During inhalation, the diaphragm contracts, increasing the volume of the thoracic cavity. When the volume of the thoracic cavity increases, air moves into the lungs. During exhalation, the diaphragm relaxes, decreasing the volume of the thoracic cavity. **(b)** During forced inhalation, the external intercostal muscles contract, pulling the rib cage upward and outward. This increases the front-to-back dimension of the chest and correspondingly increases the volume of the thoracic cavity. During forced exhalation, the intercostal muscles contract, pulling the rib cage downward and inward, which decreases the volume of the thoracic cavity.

BiologyNow™ See **the breathing mechanisms in action** by clicking on this figure on your BiologyNow CD-ROM.

increases the volume of the thoracic cavity. During forced exhalation, muscles of the abdominal wall and the *internal intercostal muscles* contract, pushing the diaphragm up and the ribs down. This action decreases the volume of the thoracic cavity.

Some of the work in stretching the thorax and the lungs is necessary to stretch elastic connective tissue. Work is also required to overcome the cohesive force of water molecules associated with the respiratory membrane. The forces between the water molecules produce a surface tension that resists stretching. The work of breathing is reduced by **pulmonary surfactant,** a detergent-like phospholipid mixture secreted by specialized epithelial cells in the lining of the alveoli. Pulmonary surfactant intersperses between the water molecules, reducing their cohesive force. This action markedly reduces the surface tension of the water, prevents the alveoli from collapsing, and reduces the energy required to stretch the lungs. Premature infants often cannot produce enough surfactant and thus suffer from respiratory distress syndrome. In these infants, high surface tension makes it difficult to inflate the lungs, and the alveoli collapse during expiration. Breathing is labored. These infants may be placed on respirators that help them breathe.

The quantity of respired air can be measured

The amount of air moved into and out of the lungs with each normal resting breath is called the **tidal volume.** The normal tidal volume is about 500 mL. The **vital capacity** is the maximum amount of air a person can exhale after filling the lungs to the maximum extent. Vital capacity is greater than tidal volume because the lungs are not completely emptied of stale air and filled with fresh air with each normal resting breath. The volume of air that remains in the lungs at the end of a normal expiration is the **residual capacity.**

Gas exchange takes place in the alveoli

The respiratory system delivers oxygen to the alveoli, but if oxygen remained in the lungs all the other body cells would soon die. The vital link between alveolus and body cell is the circulatory system. Each alveolus serves as a tiny depot from which oxygen diffuses into blood brought close to the alveolar air by capillaries (Fig. 44-9).

Oxygen molecules efficiently pass by simple diffusion from the alveoli, where they are more concentrated, into the blood in the pulmonary capillaries, where they are less concentrated. At the same time, carbon dioxide moves from the blood, where it is more concentrated, to the alveoli, where it is less concentrated. Each gas diffuses through the single layer of cells lining the alveoli and the single layer of cells lining the capillaries.

Cellular respiration results in the continuous use of oxygen and production of carbon dioxide. Inhaled (atmospheric) air contains about 20.9% oxygen, but exhaled (alveolar) air contains only 14% oxygen. Because carbon dioxide is produced during cellular respiration, exhaled air contains 100 times more (5.6%) carbon dioxide than inhaled air (.04% carbon dioxide).

The concentration of oxygen in the cells is lower than in the capillaries entering the tissues, and the concentration of carbon dioxide is higher in the cells than in the capillaries. As blood circulates through capillaries of a tissue such as brain or muscle, oxygen moves by simple diffusion from the blood to the cells, and carbon dioxide moves from the cells into the blood.

The factor that determines the direction and rate of diffusion is the pressure or tension of the particular gas. According to **Dalton's law of partial pressures,** in a mixture of gases the total pressure of the mixture is the sum of the pressures of the individual gases. Each gas exerts, independently of the others, a **partial pressure**—the same pressure it would exert if it were present alone. At sea level, the barometric pressure (the pressure of Earth's atmosphere) typically supports a column of mer-

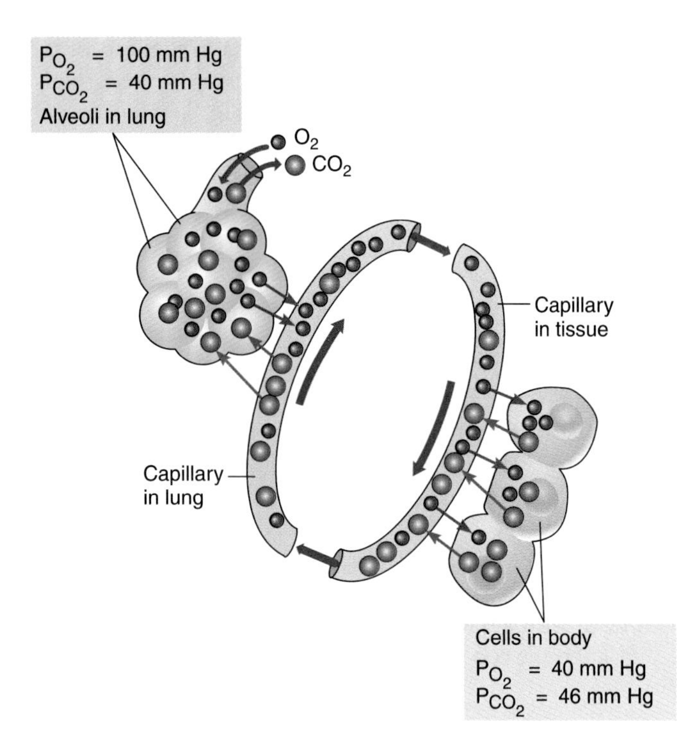

FIGURE **44-9** **Gas exchange in the lungs and tissues.**
The concentration of oxygen is greater in the alveoli than in the pulmonary capillaries, so oxygen diffuses from the alveoli into the blood. Carbon dioxide is more concentrated in the blood than in the alveoli, so it diffuses out of the capillaries and into the alveoli. In the tissues, oxygen is more concentrated in the blood than in the body cells; it diffuses out of the capillaries into the cells. Carbon dioxide is more concentrated in the cells, so it diffuses out of the cells and moves into the blood. Note the differences in partial pressures of oxygen and carbon dioxide before and after gases are exchanged in the tissues.

cury 760 mm high. Because oxygen makes up about 21% of the atmosphere, oxygen's share of that pressure is $0.21 \times 760 = 160$ mm Hg. Thus 160 mm Hg is the partial pressure of atmospheric O_2, abbreviated Po_2. In contrast, the partial pressure of atmospheric CO_2 is 0.3 mm Hg, abbreviated P_{CO_2}.

Fick's law of diffusion explains that the amount of oxygen or carbon dioxide that diffuses across the membrane of an alveolus depends on the differences in partial pressure on the two sides of the membrane and also on the surface area of the membrane. The gas diffuses faster if the difference in pressure or the surface area increases.

Gas exchange takes place in the tissues

The partial pressure of oxygen in arterial blood is about 100 mm Hg. The P_{O_2} in the tissues is still lower, ranging from 0 to 40 mm Hg. Consequently, oxygen diffuses out of the capillaries and into the tissues. Not all the oxygen leaves the blood, however. The blood passes through the tissue capillaries too rapidly for equilibrium to be reached. As a result, the partial pressure of oxygen in venous blood returning to the lungs is about 40 mm Hg. Thus, exhaled air has had only part of its oxygen removed—a good thing for those in need of mouth-to-mouth resuscitation!

Respiratory pigments increase capacity for oxygen transport

Complex animals have **respiratory pigments** that combine reversibly with oxygen and greatly increase the capacity of blood to transport it. **Hemocyanins** are copper-containing proteins dispersed in the hemolymph of many species of mollusks and arthropods. Without oxygen, these pigments are colorless. When oxygen combines with the copper, hemocyanins are blue.

Hemoglobin and **myoglobin** are the most common respiratory pigments in animals. Myoglobin is a form of hemoglobin found in muscle fibers. Hemoglobin is the only type of pigment found in the blood of vertebrates. It is also present in many invertebrate species, including annelids, nematodes, mollusks, and arthropods. In some of these animals, the hemoglobin is dispersed in the plasma rather than confined to blood cells.

In humans and other mammals, inhaled oxygen diffuses out of the alveoli and enters the pulmonary capillaries. Plasma in equilibrium with alveolar air can take up only 0.25 mL of oxygen per 100 mL. However, oxygen diffuses into red blood cells (RBCs) and combines with hemoglobin. The properties of hemoglobin permit whole blood to carry some 20 mL of oxygen per 100 mL. Hemoglobin transports almost 99% of the oxygen. The rest is dissolved in the plasma.

The term hemoglobin is actually a general name for a group of related compounds, all of which consist of an iron-porphyrin, or heme, group bound to a protein known as a *globin.* The protein portion of the molecule varies in size, amino acid composition, and physical properties among various species. When combined with oxygen, hemoglobin is bright red; without oxygen, it appears dark red, imparting a purplish color to venous blood.

The protein portion of hemoglobin is composed of four peptide chains, typically two α and two β chains, each attached to a heme (iron-porphyrin) ring (see Fig. 3-22a). An iron atom is bound in the center of each heme ring. Hemoglobin has the remarkable property of forming a weak chemical bond with oxygen. An oxygen molecule can attach to the iron atom in each heme. In the lung (or gill), oxygen diffuses into the red blood cells and combines with hemoglobin (Hb) to form **oxyhemoglobin (HbO_2).** Because the chemical bond formed between the oxygen and the hemoglobin is weak, the reaction is readily reversible. As blood circulates through tissues where the oxygen concentration is low, the reaction proceeds to the left. Hemoglobin releases oxygen, which diffuses out of the blood and into the tissue cells.

$$Hb + O_2 \rightleftharpoons HbO_2$$

The maximum amount of oxygen that hemoglobin can transport is called the **oxygen-carrying capacity.** The actual amount of oxygen bound to hemoglobin is the **oxygen content.** The ratio of O_2 content to O_2 capacity is the **percent O_2 saturation** of the hemoglobin. The percent saturation is highest in the pulmonary capillaries, where the concentration of oxygen is greatest. In the capillaries of the tissues, where there is less oxygen, the oxyhemoglobin dissociates, releasing oxygen. There, the percent saturation of hemoglobin is correspondingly lower. The **oxygen–hemoglobin dissociation curve** shown in Figure 44-10a illustrates this relationship. As oxygen concentration increases,

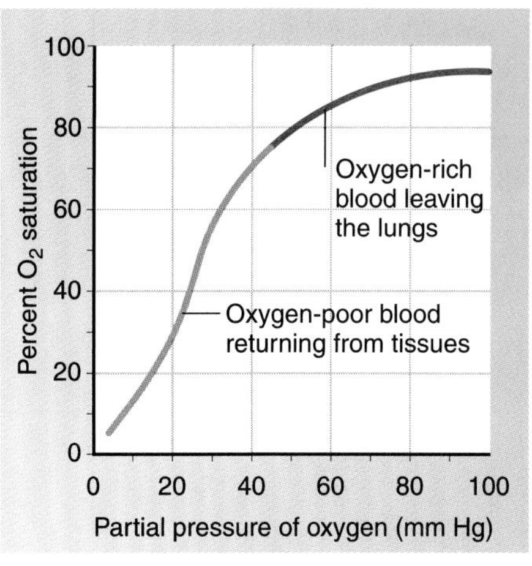

(a) Normal oxygen–hemoglobin dissociation curve

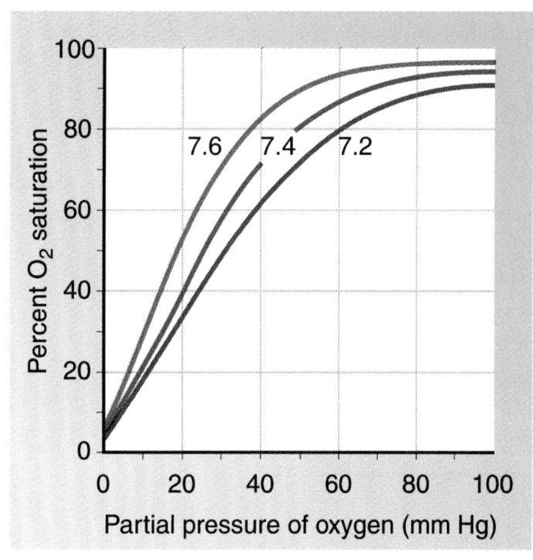

(b) Effect of pH

FIGURE **44-10** **Oxygen–hemoglobin dissociation curves.**
(a) Normal curve showing the relationship between the partial pressure of oxygen and the percent oxygen saturation of hemoglobin. Oxygen binds to hemoglobin in the lungs and unloads from hemoglobin in the tissues. **(b)** Effect of pH on the oxygen–hemoglobin curve (Bohr effect). The normal pH of human blood is 7.4. Find the location on the horizontal axis where the partial pressure of oxygen is 40, and follow the line up through the curves. Notice how the saturation of hemoglobin with oxygen differs among the three curves, even though the partial pressure of oxygen is the same. Oxygen loading increases at higher pH (7.6). At lower pH (7.2), hemoglobin unloads more oxygen. For example, during muscular activity the pH decreases, and more oxygen is unloaded and available for the muscle cells.

there is a progressive increase in the percentage of hemoglobin that is combined with oxygen. The ability of oxygen to combine with hemoglobin and be released from oxyhemoglobin is influenced by several factors in addition to percent O_2 saturation; these include pH, carbon dioxide concentration, and temperature.

Carbon dioxide formed in respiring tissue reacts with water in the plasma to form carbonic acid, H_2CO_3. In this way any increase in the carbon dioxide concentration also increases the acidity (lowers the pH) of the blood. Oxyhemoglobin unloads oxygen more readily in an acidic environment than in an environment with normal pH. Displacement of the oxygen–hemoglobin dissociation curve by a change in pH is known as the **Bohr effect** (Fig. 44-10b). Lactic acid released from active muscles also lowers blood pH and has a similar effect on the oxygen–hemoglobin dissociation curve—more oxygen is unloaded and available for the muscle cells.

Some carbon dioxide is transported by the hemoglobin molecule. Although it attaches to the hemoglobin molecule in a different way and at a different site from oxygen, the attachment of a carbon dioxide molecule releases an oxygen molecule from the hemoglobin. The effect of carbon dioxide concentration on the oxygen–hemoglobin dissociation curve is important. In the capillaries of the lungs (or gills in fishes), carbon dioxide concentration is relatively low and oxygen concentration is high, so oxygen combines with a very high percentage of hemoglobin. In the capillaries of the tissues, carbon dioxide concentration is high and oxygen concentration is low, so hemoglobin readily unloads oxygen. The oxygen and carbon dioxide bound to hemoglobin in red blood cells does not contribute to the partial pressures that govern diffusion. The loading and unloading of gases onto hemoglobin depend on partial pressures of the plasma and interstitial fluid (which is determined by the gases dissolved in these fluids).

Carbon dioxide is transported mainly as bicarbonate ions

Blood transports carbon dioxide in three forms. About 10% of carbon dioxide dissolves in plasma. Another 30% enters the RBCs and combines with hemoglobin. Because the bond between the hemoglobin and carbon dioxide is very weak, the reaction is readily reversible. Most carbon dioxide (about 60%) moves through plasma as **bicarbonate ions** (HCO_3^-).

In plasma, carbon dioxide slowly combines with water to form carbonic acid. This reaction proceeds much more rapidly inside RBCs owing to the action of the enzyme **carbonic anhydrase** (Fig. 44-11). Carbonic acid dissociates, forming hydrogen ions and bicarbonate ions.

$$\underset{\text{Carbon dioxide}}{CO_2} + \underset{\text{Water}}{H_2O} \xrightarrow{\text{Carbonic anhydrase}} \underset{\text{Carbonic acid}}{H_2CO_3} \longrightarrow \underset{\text{Hydrogen ion}}{H^+} + \underset{\text{Bicarbonate ion}}{HCO_3^-}$$

Most hydrogen ions released from carbonic acid combine with hemoglobin, which is a very effective buffer. Many of the bicarbonate ions diffuse into the plasma. The action of carbonic anhydrase in the RBCs maintains a diffusion gradient for carbon dioxide to move into and then out of RBCs. As the negatively charged bicarbonate ions move out of RBCs, chloride ions (Cl^-) in the plasma diffuse into the RBCs to replace them, a process known as the **chloride shift.** In the alveolar capillaries, CO_2 diffuses out of the plasma and into the alveoli. As the CO_2 concentration decreases, the reaction sequence just described reverses.

Any condition (such as emphysema) that interferes with the removal of carbon dioxide by the lungs can lead to *respiratory acidosis.* Carbon dioxide is produced more rapidly than it is excreted by the lungs. As a result, the concentration of carbonic acid in the blood increases. When blood pH falls below 7.0, the central nervous system becomes depressed. Such depression disorients the individual, and untreated respiratory acidosis can cause coma and death.

Breathing is regulated by respiratory centers in the brain

Breathing is a rhythmic, involuntary process regulated by **respiratory centers** in the brain stem (see Fig. 44-6). Groups of neurons in the medulla regulate the basic rhythm of breathing.

FIGURE **44-11** | **Carbon dioxide transport.**

(a) In the tissues, carbon dioxide diffuses from cells into the plasma. Most of the carbon dioxide enters red blood cells, where the enzyme carbonic anhydrase rapidly converts it to carbonic acid. Carbonic acid dissociates, forming bicarbonate and hydrogen ions. As bicarbonate ions move out into the plasma, chloride ions replace them. Hemoglobin combines with hydrogen ions, preventing a decrease in pH. **(b)** In the lungs, carbon dioxide diffuses out of the plasma and into the alveoli, and these processes are reversed. Bicarbonate ions diffuse from the plasma into the red blood cells. H^+ released from hemoglobin combines with bicarbonate ions, forming carbonic acid. Carbon dioxide produced from the carbonic acid diffuses out of the blood and into the alveoli.

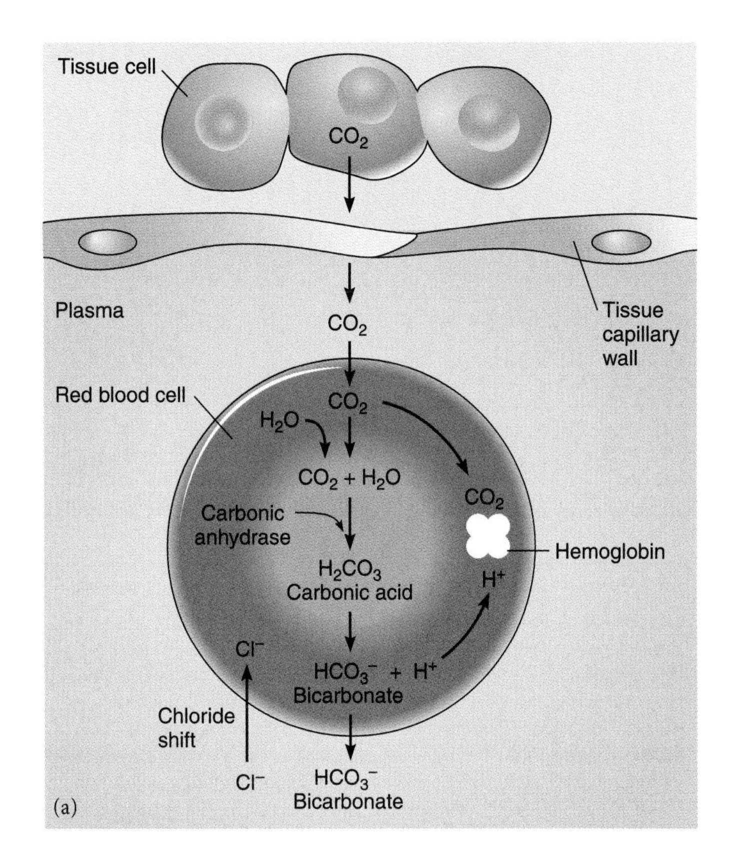

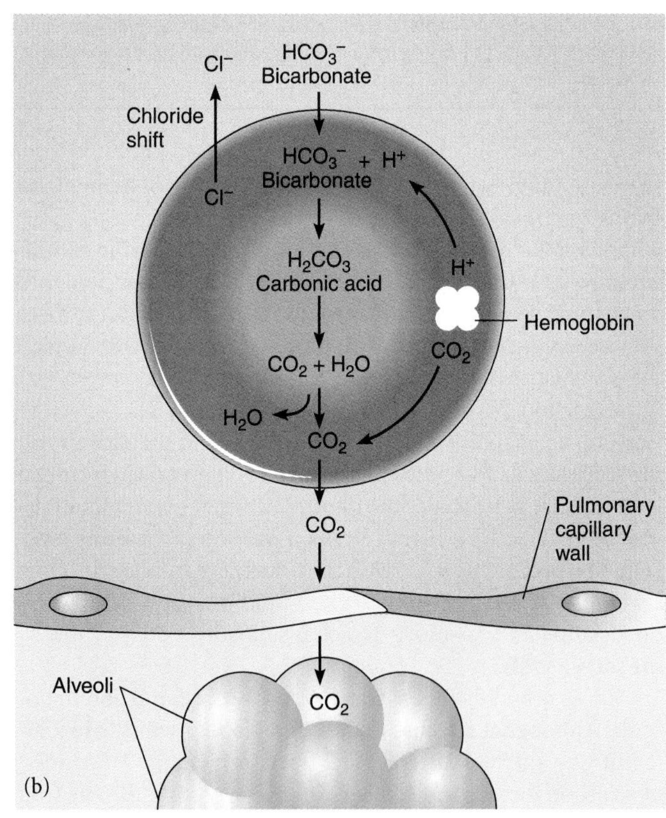

These neurons send a burst of impulses to the diaphragm and external intercostal muscles, causing them to contract. After several seconds, these neurons become inactive, the muscles relax, and exhalation occurs. Respiratory centers in the pons help control the transition from inspiration to exhalation. These centers can stimulate or inhibit the respiratory centers in the medulla. The cycle of activity and inactivity repeats itself so that at rest you breathe about 14 times per minute. Overdose of certain medications such as barbiturates depresses the respiratory centers and may lead to respiratory failure.

The basic rhythm of respiration changes in response to needs of the body. When you are playing a fast game of tennis, you need more oxygen than when you're studying biology. During exercise the rate of aerobic cellular respiration increases, producing more carbon dioxide. Your body must dispose of this carbon dioxide through increased ventilation.

Carbon dioxide concentration is the most important chemical stimulus for regulating the rate of respiration. Specialized **chemoreceptors** in the medulla and in the walls of the aorta and carotid arteries are sensitive to changes in arterial carbon dioxide concentration. When stimulated, they send impulses to the respiratory centers, which increase breathing rate.

The chemoreceptors in the walls of the aorta, called *aortic bodies,* and those in the walls of the carotid arteries, the *carotid bodies,* are sensitive to changes in hydrogen ion concentration and oxygen concentration, as well as to carbon dioxide levels. Recall that an increase in carbon dioxide concentration increases hydrogen ions from carbonic acid, lowering blood pH. Even a slight decrease in pH stimulates these chemoreceptors, leading to faster breathing. As the lungs remove carbon dioxide, the hydrogen ion concentration in the blood and other body fluids decreases, and homeostasis is re-established.

Interestingly, oxygen concentration generally does not play an important role in regulating respiration. Only if the partial pressure of oxygen falls markedly do the chemoreceptors in the aorta and carotid arteries become stimulated to send messages to the respiratory centers.

Although breathing is involuntary, the action of the respiratory centers can be consciously influenced for a short time by stimulating or inhibiting them. For example, you can inhibit respiration by holding your breath. You can't hold your breath indefinitely, however, because eventually you feel a strong urge to breathe. Even if you could ignore this urge, you would eventually pass out, and would resume breathing.

People who have stopped breathing because of drowning, smoke inhalation, electric shock, or cardiac arrest can sometimes be sustained by mouth-to-mouth resuscitation until their own breathing reflexes return. **Cardiopulmonary resuscitation (CPR)** is a method for aiding victims who have suffered respiratory and cardiac arrest. CPR must be started immediately, because irreversible brain damage occurs within about 4 minutes of oxygen deprivation. A number of organizations offer training in CPR.

Hyperventilation reduces carbon dioxide concentration

Underwater swimmers and some Asian pearl divers voluntarily **hyperventilate** before going under water. By making a series of deep inhalations and exhalations, they "blow off" CO_2, markedly reducing the carbon dioxide content of the alveolar air and of the blood. As a result, they can last longer before the urge to breathe becomes irresistible.

When hyperventilation continues for a long period, dizziness and sometimes unconsciousness may occur. This is because a certain concentration of carbon dioxide is needed in the blood to maintain normal blood pressure. (This mechanism operates by way of the vasoconstrictor center in the brain, which maintains the muscle tone of blood vessel walls.) Furthermore, if divers hold their breath too long, the low concentration of oxygen may result in unconsciousness and drowning.

High flying or deep diving can disrupt homeostasis

The barometric pressure decreases at progressively higher altitudes. Because the concentration of oxygen in the air remains at 21%, the partial pressure of oxygen decreases along with the barometric pressure. At an altitude of 6,000 m (19,500 ft), the barometric pressure is about 350 mm Hg, the partial pressure of oxygen is about 75 mm Hg, and the hemoglobin in arterial blood is about 70% saturated with oxygen. At 10,000 m (33,000 ft) the barometric pressure is about 225 mm Hg, the partial pressure of oxygen is 50 mm Hg, and arterial oxygen saturation is only 20%. Thus getting sufficient oxygen from the air becomes an ever-increasing problem at higher altitudes.

When a person moves to a high altitude, the body adjusts over time by producing a greater number of RBCs. In a person breathing pure oxygen at 10,000 m, the oxygen would have a partial pressure of 225 mm Hg and the hemoglobin would be almost fully saturated with oxygen. Above 13,000 m, however, barometric pressure is so low that even breathing pure oxygen does not permit complete oxygen saturation of arterial hemoglobin.

A person becomes unconscious when the arterial oxygen saturation falls to between 40% and 50%. This level is reached at about 7000 m (23,000 ft) when the person is breathing air, or 14,500 m (47,100 ft) when pure oxygen is used. All high-flying jets have airtight cabins pressurized to the equivalent of the barometric pressure at an altitude of about 2000 m.

Shallow breathing, which occurs in many respiratory diseases, causes **hypoxia,** a deficiency of oxygen. Even *rapid,* shallow breathing results in hypoxia, because the tidal volume includes the air in the respiratory passages. When we breathe shallowly, we don't clear out the stale air in the airway and ventilate the lung. Hypoxia causes drowsiness, mental fatigue, headache, and sometimes euphoria. The ability to think and make judgments is impaired, as is the ability to perform tasks requiring coordination. If a jet flying at 11,700 m (over 38,000 ft) suddenly decompressed, the pilot would lose consciousness in about 30 seconds and become comatose in about 1 minute.

In addition to the problems of hypoxia, a rapid decrease in barometric pressure causes **decompression sickness** (commonly known as the "bends" because those suffering from it bend over in pain). Whenever the barometric pressure drops below the total pressure of all gases dissolved in the blood and other body fluids, the dissolved gases tend to come out of solution and form gas bubbles. A familiar example occurs each time you uncap a bottle of soda, reducing pressure in the bottle. Carbon dioxide is released from solution and bubbles out into the air. In the body, nitrogen has a low solubility in blood and tissues. When it comes out of solution, the bubbles formed may damage tissues and block capillaries, interfering with blood flow. The clinical effects of decompression sickness are pain, dizziness, paralysis, unconsciousness, and even death.

Decompression sickness is more common in scuba diving than in high-altitude flying. As a diver descends, the surrounding pressure increases tremendously—1 atmosphere (the atmospheric pressure at sea level, which equals 760 mm Hg) for each 10 m. To prevent lung collapse, a diver must be supplied with air under pressure, exposing the lungs to very high alveolar gas pressures.

At sea level an adult human has about 1 L of nitrogen dissolved in the body, with about half in the fat and half in the body fluids. After a diver's body has been saturated with nitrogen at a depth of 100 m (325 ft), the body fluids contain about 10 L of nitrogen. To prevent this nitrogen from rapidly bubbling out of solution and causing decompression sickness, the diver must rise to the surface gradually, with stops at certain levels on the way up. This allows nitrogen to be expelled slowly through the lungs.

Some mammals are adapted for diving

Some air-breathing mammals can spend rather long periods in the ocean depths without coming up for air. Dolphins, whales, seals, and beavers have structural and physiological adaptations that allow them to dive for food or to elude their enemies (Fig. 44-12). With their streamlined bodies and forelimbs modified as fins or flippers, diving mammals perform impressive aquatic feats. The Weddell seal can swim under the ice at a depth of 596 m (1955 ft) for more than an hour without coming up for air. The enormous northern elephant seal, which measures about 5 m (16 to 18 ft) in length and weighs 2 to 4 tons, can plunge even deeper. A female elephant seal can dive to more than 1500 m (almost 5000 ft) and stay beneath the surface for more than an hour.

Turtles and birds that dive depend on oxygen stored in their lungs. Diving mammals do not take in and store extra air before a dive. In fact, seals exhale before they dive. With less air in their lungs, they are less buoyant. Their lungs collapse at about 50 to 70 m into their dive and then reinflate as they ascend. This means their lungs do not function for most of the dive. These adaptations are thought to reduce the chance of decompression sickness, because with less air in the lungs there is less nitrogen in the blood to dissolve during the dive.

Flip Nicklin/Minden Pictures

FIGURE **44-12** | **Deep diver.**

Elephant seals (*Mirounga angustirostris*) may be the deepest divers on Earth. Researchers have recorded their dives at depths of more than 5,000 feet.

Physiologic adaptations, including ways to distribute and store oxygen, permit some mammals to dive deeply and remain under water for long periods. Seals have about twice the volume of blood, relative to their body weight, as nondiving mammals. Diving mammals also have high concentrations of myoglobin, which stores oxygen in muscles. These animals have up to 10 times more myoglobin than do terrestrial mammals. The very large spleen typical of many diving mammals stores oxygen-rich red blood cells. Under pressure (during a dive), the spleen is squeezed and releases these red blood cells into the circulation.

Diving mammals markedly reduce the energy expended in deep (more than 200 m) diving by gliding. Filmed video sequences of diving seals and whales show that they glide most of the way down. Gliding is possible because the animal's buoyancy decreases as the lungs gradually collapse, reducing the amount of air in the lungs.

When a mammal dives to its limit, physiological mechanisms known collectively as the **diving reflex** are activated. Metabolic rate decreases by about 20%, which conserves oxygen. Breathing stops, and bradycardia (slowing of the heart rate) occurs. The heart rate may decrease to one tenth of the normal rate, reducing the body's consumption of oxygen and energy. Blood is redistributed; skin, muscles, digestive organs, and other internal organs can survive with less oxygen and receive less blood while an animal is submerged.

The diving reflex is present to some extent in humans, where it may act as a protective mechanism during birth, when an infant may be deprived of oxygen for several minutes. Cases of near-drownings, especially of young children, have been documented in which the victim was submerged for as long as 45 minutes in very cold water before being rescued and resuscitated. Many of these survivors showed no brain damage. The shock of icy water slows heart rate, increases blood pressure, and shunts blood to internal organs of the body that most need oxygen (blood flow in arms and legs decreases). Metabolic rate decreases, so less oxygen is required.

Review

- What is the sequence of inhaled air flow through the respiratory structures in a mammal?
- What is the function of respiratory pigments? How do they work?
- Why does alveolar air differ in composition from atmospheric air? Explain.
- What physiological mechanisms bring about an increase in rate and depth of breathing during exercise? Why is such an increase necessary?

BiologyNow™ Assess your understanding of **the mammalian respiratory system** by taking the pretest on your BiologyNow CD-ROM.

BREATHING POLLUTED AIR

Learning Objective

8 Describe the defense mechanisms that protect the lungs, and describe the effects of polluted air on the respiratory system.

Several defense mechanisms protect the delicate lungs from the harmful substances we breathe (Fig. 44-13). The hair in the

FIGURE **44-13** | **Urban air pollution.**

Industry spews tons of pollutants into the atmosphere. Harmful gases, such as sulfur dioxide and nitrogen oxides, contribute to acid rain.

David Nunuk/Photo Researchers, Inc.

Smoking is the single most preventable cause of death in the United States. More than 440,000 adults die each year of tobacco-related causes. Tobacco smoke is by far the most important risk factor for lung cancer, the most common lethal cancer in the United States and in the world. Tobacco smoke is also an important risk factor for cardiovascular disease and a number of other diseases. Still, an estimated 26% of men and 21% of women ages 18 and over continue to smoke. Nonsmokers are also affected by tobacco smoke, which is a "portable" air pollutant. This secondhand smoke has been linked to death from lung cancer of about 3000 nonsmokers each year in the United States. The direct medical costs of tobacco use amount to more than $75 billion dollars annually

Nicotine is highly addictive. Like morphine, amphetamines, and cocaine, nicotine increases dopamine concentration and activates cells in the *nucleus accumbens,* an area at the base of the forebrain. This area helps integrate emotion, and dopamine may facilitate learning an association between the pleasurable effects of the drug and other stimuli such as the smell of smoke. In 2001, Eric Nestler of the University of Texas Southwestern Medical Center reported in *Science* that nicotine, like cocaine and other abused drugs, causes long-lasting changes in the brain. This researcher pointed out that cell signaling and other mechanisms underlying drug addiction are very similar to those of learning and memory. Investigators are searching for drugs that block the dopamine release caused by nicotine and other addictive drugs.

Here are some facts about smoking:

- The life of a 30-year-old who smokes 15 cigarettes a day is shortened by an average of more than five years.
- If you smoke more than one pack per day, you are about 20 times more likely to develop lung cancer than is a nonsmoker. According to the American Cancer Society, cigarette smoking causes more than 75% of all lung cancer deaths.
- If you smoke, you double your chances of dying from cardiovascular disease.
- If you smoke, you are 20 times more likely to develop chronic bronchitis and emphysema than is a nonsmoker.
- If you smoke, you have about 5% less oxygen circulating in your blood (because carbon monoxide binds to hemoglobin) than does a nonsmoker.
- If you smoke when you are pregnant, your baby will weigh about 6 ounces less at birth, and there is double the risk of miscarriage, stillbirth, and infant death.
- Infants whose parents smoke have double the risk of contracting pneumonia or bronchitis in their first year of life.
- Secondhand smoke causes up to 300,000 lung infections (such as pneumonia and bronchitis) in infants and young children each year.
- When smokers quit smoking, their risk of dying from chronic obstructive pulmonary disease, cardiovascular disease, or cancer gradually decreases. (Precise changes in risk depend on the number of years the person smoked, the number of cigarettes smoked per day, the age of starting to smoke, and the number of years since quitting.)
- Nicotine replacement with gum, patches, or nasal spray has been shown to be effective as an aid to smoking cessation, especially used with behavioral therapy. More recently certain antidepressants have been used to reduce the craving for nicotine.
- Almost every American who takes up smoking is a teenager—5000 every day, almost 2 million every year. Of children who begin smoking, 10% start by the fourth grade, and nearly two thirds start by the tenth grade.

nostrils, the ciliated mucous lining in the nose and pharynx, and the cilia-mucus elevator of the trachea and bronchi trap foreign particles in inspired air. One of the body's most rapid defense responses to breathing dirty air is **bronchial constriction.** In this process, the bronchial tubes narrow, increasing the chance that inhaled particles will land on the sticky mucous lining. Unfortunately, bronchial constriction narrows the airway so less air can reach the lungs, decreasing the amount of oxygen available to body cells. Fifteen puffs on a cigarette during a 5-minute period increases airway resistance as much as threefold, and this added resistance to breathing lasts more than 30 minutes. Chain-smokers and those who breathe heavily polluted air are in a state of chronic bronchial constriction.

Neither the smallest bronchioles nor the alveoli are equipped with mucus or ciliated cells. Foreign particles that get through other respiratory defenses and find their way into the alveoli may be engulfed by macrophages. The macrophages may then accumulate in the lymph tissue of the lungs. Lung tissues of chronic smokers and those who work in dirty industries burning fossil fuel, contain large blackened areas where carbon particles have been deposited (Fig. 44-14).

Continued insult to the respiratory system results in disease. Chronic bronchitis, pulmonary emphysema, and lung cancer have been linked to cigarette smoking and breathing polluted air. More than 75% of patients with **chronic bronchitis** have a history of heavy cigarette smoking (see *Focus On: The Effects of Smoking*). Victims of chronic bronchitis often develop **pulmonary emphysema,** a disease also most common in cigarette smokers. In this disorder, alveoli lose their elasticity, and walls between adjacent alveoli are destroyed. The surface area of the lung is so reduced that gas exchange is seriously impaired. Air is not expelled effectively, and stale air accumulates in the lungs. The emphysema victim struggles for every breath, and still the body does not get enough oxygen. To compensate, the right ven-

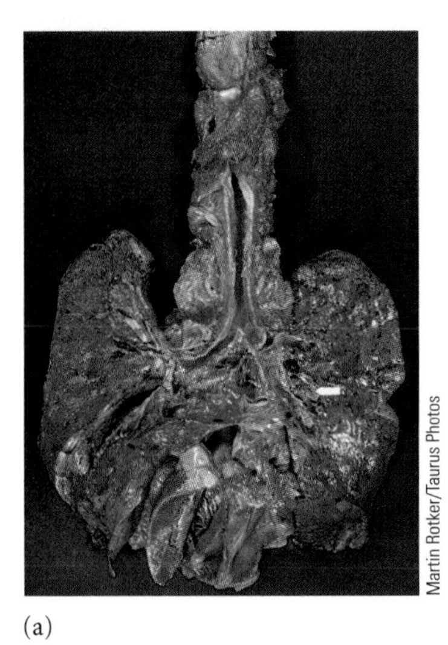

(a)

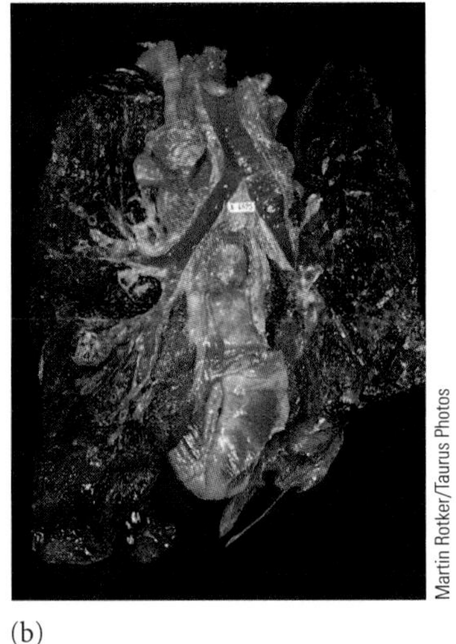

(b)

FIGURE **44-14** | **Effects of cigarette smoking.**

(a) Lungs and major bronchi of a nonsmoker. **(b)** Lungs and heart of a cigarette smoker. The dark spots in the lung tissue are particles of carbon, tar, and other substances that passed through the respiratory defenses and lodged in the lungs.

Martin Rotker/Taurus Photos

tricle of the heart pumps harder and becomes enlarged. Emphysema patients frequently die of heart failure.

Most patients with **chronic obstructive pulmonary disease (COPD),** a condition characterized by obstructed airflow, have both chronic bronchitis and emphysema. Asthma also contributes to COPD. People with asthma respond to inhaled stimuli with exaggerated bronchial constriction, and the airway is typically inflamed.

Cigarette smoking is also the main cause of **lung cancer.** More than 69 of the 4800 chemical compounds in tobacco smoke cause cancer in humans and animals. These carcinogenic substances irritate the cells lining the respiratory passages and alter their metabolic balance. Normal cells are transformed into cancer cells, which may multiply rapidly and invade surrounding tissues.

Review

- What mechanisms does the human respiratory system have for getting rid of inhaled dirt?
- What happens when so much dirty air is inhaled that these mechanisms cannot function effectively?

BiologyNow™ Assess your understanding of **breathing polluted air** by taking the pretest on your BiologyNow CD-ROM.

SUMMARY WITH KEY TERMS

1 Compare the advantages and disadvantages of gas exchange in air with those of gas exchange in water.

- Air contains a higher concentration of molecular oxygen than does water, and oxygen diffuses more rapidly through air than through water. Air is less dense and less viscous than water, so less energy is needed to move air over a gas exchange surface.
- Terrestrial animals have adaptations that protect their respiratory surfaces from drying.

2 Describe the following adaptations for gas exchange: body surface, tracheal tubes, gills, and lungs.

- Small aquatic animals exchange gases by diffusion, requiring no specialized respiratory structures.
- Some invertebrates, including most annelids, and a few vertebrates (many amphibians) exchange gases across the body surface.
- In insects and some other arthropods, air enters a network of **tracheal tubes,** or **tracheae,** through openings, called **spiracles,** along the body surface. Tracheal tubes branch and extend to all regions of the body.
- **Gills** are moist, thin projections of the body surface found mainly in aquatic animals. In chordates, gills are usually internal, located along the edges of the gill slits. In bony fishes an **operculum** protects the gills; A **countercurrent exchange system** maximizes diffusion of oxygen into the blood and diffusion of carbon dioxide out of blood.
- Animals carry on **ventilation,** the process of actively moving air or water over respiratory surfaces. Terrestrial vertebrates have **lungs** and some means of ventilating them. Most fishes do not have lungs but have a homologous swim bladder that permits the fish to control its buoyancy. Amphibians and reptiles have lungs with only some ridges or folds that increase surface area.
- In birds, the lungs have extensions, called **air sacs,** that act as bellows, drawing air into the system. Two cycles of inhalation and exhalation support a one-way flow of air through the lungs. Air flows from the outside into the posterior air sacs, to the lung, through the anterior air sacs, and then out of the body. Gas exchange takes place through the walls of the **parabronchi** in the lungs. A crosscurrent arrangement, in which blood flow is at right angles to the parabronchi, increases the amount of oxygen that enters the blood.

3 Trace the passage of oxygen through the human respiratory system from nostrils to alveoli.

- The mammalian respiratory system includes the lungs and a system of airways. A breath of air passes in sequence through the **nostrils, nasal cavities, pharynx, larynx, trachea, bronchi, bronchioles,** and **alveoli.** Each lung occupies a pleural cavity and is covered with a **pleural membrane.**

4 Summarize the mechanics and the regulation of breathing in humans, and describe gas exchange in the lungs and tissues.

- During breathing, the **diaphragm** contracts, expanding the chest cavity. The membranous walls of the lungs move outward along with the chest walls, lowering pressure within the lungs. Air from outside the body rushes in through the air passageways and fills the lungs until the pressure equals atmospheric pressure.
- **Tidal volume** is the amount of air moved into and out of the lungs with each normal breath. **Vital capacity** is the maximum volume that can be exhaled after the lungs fill to the maximum extent. The volume of air that remains in the lungs at the end of a normal expiration is the **residual capacity.**
- **Respiratory centers** in the medulla and pons regulate respiration. These centers are stimulated by **chemoreceptors** sensitive to an increase in carbon dioxide concentration. They also respond to an increase in hydrogen ions and to very low oxygen concentration.

SUMMARY WITH KEY TERMS *(continued)*

- Oxygen and carbon dioxide are exchanged between alveoli and blood by diffusion. The pressure of a particular gas determines its direction and rate of diffusion.
- **Dalton's law of partial pressures** explains that in a mixture of gases, the total pressure of the mixture is the sum of the pressures of the individual gases. Thus each gas in a mixture of gases exerts a **partial pressure,** the same pressure it would exert if it were present alone. The partial pressure of atmospheric oxygen, P_{O_2}, is 160 mm Hg at sea level.
- According to **Fick's law of diffusion,** the greater the difference in pressure on the two sides of a membrane and the larger the surface area, the faster the gas diffuses across the membrane.

5 Explain the role of hemoglobin in oxygen transport, and identify factors that determine and influence the oxygen–hemoglobin dissociation curve.

- **Hemoglobin** is the respiratory pigment in the blood of vertebrates. Almost 99% of the oxygen in human blood is transported as **oxyhemoglobin (HbO_2).**
- The maximum amount of oxygen that can be transported by hemoglobin is the **oxygen-carrying capacity.** The actual amount of oxygen bound to hemoglobin is the **oxygen content.** The **percent O_2 saturation,** the ratio of oxygen content to oxygen-carrying capacity, is highest in pulmonary capillaries, where oxygen concentration is greatest.
- The **oxygen–hemoglobin dissociation curve** shows that as oxygen concentration increases, there is a progressive increase in the amount of hemoglobin that combines with oxygen. The curve is affected by pH, temperature, and CO_2 concentration.
- Owing to lowered pH caused by carbonic acid, oxyhemoglobin dissociates more readily as carbon dioxide concentration increases. This is the **Bohr effect.**

6 Summarize the mechanisms by which carbon dioxide is transported in the blood.

- About 60% of the carbon dioxide in the blood is transported as bicarbonate ions. About 30% combines with hemoglobin, and another 10% is dissolved in plasma.
- Carbon dioxide combines with water to form carbonic acid; the reaction is catalyzed by **carbonic anhydrase.** Carbonic acid dissociates, forming bicarbonate ions (HCO_3^-) and hydrogen ions (H^+).
- Hemoglobin combines with H^+, buffering the blood. Many bicarbonate ions diffuse into the plasma and are replaced by Cl^- ions; this exchange is known as the **chloride shift.**

7 Describe the physiological effects of hyperventilation and of sudden decompression when a diver surfaces too quickly from deep water.

- **Hyperventilation** reduces the concentration of carbon dioxide in the alveolar air and in the blood. A certain carbon dioxide concentration in the blood is needed to maintain normal blood pressure.
- As altitude increases, barometric pressure (the pressure of Earth's atmosphere) falls, and less oxygen enters the blood. This situation can lead to **hypoxia,** or oxygen deficiency, which can lead to loss of consciousness and death. In addition to hypoxia, rapid decrease in barometric pressure can cause **decompression sickness,** especially common among divers who ascend too rapidly from a dive.
- Diving mammals have high concentrations of **myoglobin,** a pigment that stores oxygen, in their muscles. The **diving reflex,** a group of physiological mechanisms (including a decrease in metabolic rate) is activated when a mammal dives to its limit.

8 Describe the defense mechanisms that protect the lungs, and describe the effects of polluted air on the respiratory system.

- The ciliated mucous lining of the nose, pharynx, trachea, and bronchi trap inhaled particles.
- Inhaling polluted air results in **bronchial constriction,** increased mucous secretion, damage to ciliated cells, and coughing. Breathing polluted air or inhaling cigarette smoke can cause **chronic bronchitis, pulmonary emphysema,** and **lung cancer.**

POST-TEST

1. Breathing is an example of (a) countercurrent exchange (b) cellular respiration (c) ventilation (d) diffusion (e) gas exchange through the body surface
2. Which of the following is a benefit of gas exchange in air compared with water? (a) higher concentration of molecular oxygen (b) oxygen diffuses more slowly in air (c) no energy required for ventilation (d) moist respiratory surface not needed (e) air is denser than water
3. Which of the following adaptations for gas exchange is most characteristic of insects? (a) lungs (b) tracheal tubes (tracheae) (c) parabronchi (d) air sacs (e) dermal gills
4. Which of the following are accurately matched? (a) bony fish—operculum (b) insect—alveoli (c) bird—spiracles (d) mammal—gill filaments (e) shark—dermal gills
5. Tracheal tubes (tracheae) (a) are typically found in mollusks (b) are highly vascular (c) branch and extend to all the cells (d) are characteristic of many mammals (e) end in book lungs
6. The most efficient vertebrate respiratory system is that of (a) amphibians (b) birds (c) reptiles (d) mammals (e) humans
7. In a bird, the correct sequence for a breath of air is (a) anterior air sacs ⟶ posterior air sacs ⟶ lung (b) posterior air sacs ⟶ lung ⟶ anterior air sacs (c) parabronchi ⟶ posterior air sacs ⟶ anterior air sacs (d) posterior air sacs ⟶ alveoli ⟶ anterior air sacs (e) posterior air sacs ⟶ capillaries ⟶ cells
8. Respiratory pigments (a) combine reversibly with oxygen (b) are found only in vertebrates (c) all have a heme (porphyrin) group that combines with oxygen (d) diffuse into the air sacs (e) attach to the alveolar wall
9. Which sequence most accurately describes the sequence of air flow in the human respiratory system? (a) pharynx ⟶ bronchus ⟶ trachea ⟶ alveolus (b) pharynx ⟶ parabronchi ⟶ alveoli ⟶ bronchioles (c) bronchus ⟶ trachea ⟶ larynx ⟶ lung (d) larynx ⟶ trachea ⟶ bronchus ⟶ bronchiole (e) trachea ⟶ larynx ⟶ bronchus ⟶ alveolus

POST-TEST *(continued)*

10. The amount of air moved in and out of the lungs with each normal resting breath is the (a) vital capacity (b) residual capacity (c) vital volume (d) partial pressure (e) tidal volume
11. The greater the difference in pressure and the larger the surface area, the faster a gas will diffuse. This is explained by (a) Dalton's law of partial pressure (b) Fick's law of diffusion (c) the percent saturation (d) the Bohr effect (e) the oxygen–hemoglobin dissociation curve
12. Oxygen in the blood is transported mainly (a) in combination with hemoglobin (b) as bicarbonate ions (c) as carbonic acid (d) dissolved in plasma (e) combined with carbon dioxide
13. The concentration of which of the following substances is most important in regulating the rate of respiration? (a) chloride ions (b) oxygen (c) bicarbonate ions (d) nitrogen (e) carbon dioxide
14. When a diver ascends too rapidly (a) bronchial constriction occurs (b) a diving reflex is activated (c) nitrogen rapidly bubbles out of solution in the body fluids (d) nitrogen hypoxia occurs (e) carbon dioxide bubbles damage the alveoli
15. Which of the following is *not* true of the diving reflex? (a) breathing stops (b) the heart slows (c) less blood is distributed to the muscles (d) metabolic rate increases by about 20% (e) energy consumption decreases
16. Pulmonary emphysema (a) results from chronic obstructive pulmonary disease (b) is uncommon in cigarette smokers (c) results from bronchial constriction (d) is sometimes referred to as the Bohr effect (e) is characterized by loss of elasticity of the alveolar walls
17. Label the figure. Use Fig. 44-6 to check your answers.

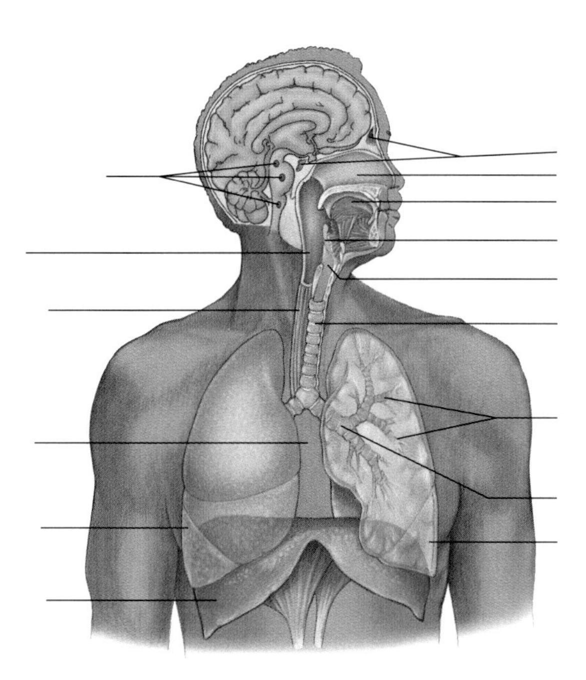

CRITICAL THINKING

1. What problems would be faced by a terrestrial animal having gills instead of lungs?
2. Under what conditions might it be advantageous for a fish to have lungs as well as gills? What function do the "lungs" of modern fishes serve?
3. Aquatic mammals such as whales and dolphins use lungs rather than gills for gas exchange. Propose a hypothesis to explain this.
4. What are the advantages of having millions of alveoli rather than a pair of simple, balloon-like lungs?

- Visit our Web site at **http://biology.brookscole.com/solomon7** for links to chapter-related resources on the World Wide Web. Additional online materials relating to this chapter can also be found on our Web site.

BIOLOGY NOW RESOURCES

Active Figures

44-6: The human respiratory system

44-8: Mechanics of breathing

Preparing for an exam? Take a diagnostic test on your BiologyNow CD-ROM.

Post-Test Answers

1. c	2. a	3. b	4. a
5. c	6. b	7. b	8. a
9. d	10. e	11. b	12 a
13. e	14. c	15. d	16. e

45

Processing Food and Nutrition

McMurray Photography

Obtaining nutrients. Spotted hyenas (*Crocuta crocuta*) and several species of birds of prey gathered around a dead elephant. Photographed in East Africa.

CHAPTER OUTLINE

- **Nutritional Styles and Adaptations**
- **The Vertebrate Digestive System**
- **Required Nutrients**
- **Energy Metabolism**

Sponges and baleen whales feed on food particles suspended in the water. Giraffes eat leaves, and antelope graze on grasses. Frogs, lions, and some insects capture other animals. Although their choices of food and feeding mechanisms are diverse, all animals are **heterotrophs,** organisms that must obtain their energy and nourishment from the organic molecules manufactured by other organisms. What they eat, how they obtain food, and how they process and use food are the focus of this chapter.

Nutrients are substances in food that are used as energy sources to run the systems of the body, as ingredients to make compounds for metabolic processes, and as building blocks in the growth and repair of tissues. Obtaining nutrients is of such vital importance that both individual organisms and ecosystems are structured around the central theme of **nutrition,** the process of taking in and using food. An organism's body plan and its lifestyle are adapted to its particular mode of obtaining food. For example, spotted hyenas, which are formidable predators as well as scavengers, have massive jaws that let them consume an entire elephant carcass, including bones and hide (see photograph). Few nutrients are wasted.

Nutrition has also been an important force in human evolution. Through natural selection, the human diet became more varied than the diets of other primates. Humans also became very efficient at obtaining food. Their higher-quality diet supported the evolution of a larger, more complex brain.

With only slight variations, all animals require the same basic nutrients: minerals, vitamins, carbohydrates, lipids, and proteins. Carbohydrates, lipids, and proteins are used as energy sources. Eating too much of any of these nutrients can result in weight gain, whereas eating too few nutrients or an unbalanced diet can result in malnutrition and death. **Malnutrition,** or poor nutritional status, results from dietary intake that is either below or above required needs. In human populations, both undernutrition (particularly protein deficiency) and obesity (which results from overnutrition) are serious health problems.

Food processing and nutrition are active areas of research. For example, researchers investigating the regulation of digestion are discovering a number of peptide messengers in the digestive tract. Nutritionists are studying a variety of plant compounds that may be important in maintaining health. And the food industry continues to search for new fat and sugar substitutes. ■

NUTRITIONAL STYLES AND ADAPTATIONS

Learning Objective

1 Describe food processing, including ingestion, digestion, absorption, and egestion or elimination, and compare the digestive system of a cnidarian (such as *Hydra*) with that of an earthworm or vertebrate.

Feeding is the selection, acquisition, and ingestion of food. **Ingestion** is the process of taking food into the digestive cavity. In many animals, including vertebrates, ingestion includes taking food into the mouth and swallowing it. Most animals have a specialized digestive system that processes the food they eat. The process of breaking down food is called **digestion.** Because animals eat the macromolecules tailor-made by and for other organisms, they must break down these molecules and refashion them for their own needs. For example, a hyena cannot incorporate the proteins and other complex organic compounds from an elephant carcass directly into its own cells. It must *mechanically digest* its food, and then *chemically digest* it by enzymatic hydrolysis (see Chapter 3). During digestion, complex organic compounds are degraded into smaller molecular components. For example, proteins are broken down into their component amino acids.

Amino acids and other nutrients pass through the lining of the digestive tract and into the blood by **absorption.** Then the circulatory system transports them to the body cells. In the cells they are used to synthesize proteins and other complex organic compounds that the animal needs. Food that is not digested and absorbed is discharged from the body. Biologists call this process **egestion** in simple animals and **elimination** in more complex animals.

Animals are adapted to their mode of nutrition

We classify animals as herbivores, carnivores, or omnivores on the basis of the type of food they typically eat (Fig. 45-1). Animals that feed directly on producers are **herbivores,** or primary consumers. Animals cannot digest the cellulose of plant cell walls, and many adaptations have evolved for extracting nutrients from the plant material they eat. Many herbivores, including termites, cows, and horses, have a symbiotic relationship with bacteria that inhabit their digestive tracts. For example, cud-chewing ruminants (cattle, sheep, deer, giraffes), have a stomach divided into four chambers. Bacteria living in the first two chambers digest cellulose, splitting some of it into sugars, which are then used by the host and the bacteria themselves. The bacteria produce fatty acids during their metabolism, some of which are absorbed by the animal and serve as an important energy source. Food that is not sufficiently chewed, called cud, is regurgitated into the animal's mouth and chewed again.

Most of what a herbivore eats is not efficiently digested and is eliminated from the body, almost unchanged, as waste. For this reason herbivores must eat large quantities of food to obtain the nourishment they need. Many herbivores—grasshoppers, locusts, elephants, and cattle, for example—spend a major part of their lives eating.

Herbivores are sometimes eaten by flesh-eating **carnivores,** which may also eat one another. Many carnivores (secondary and higher-level consumers in ecosystems) are predators, adapted for capturing and killing prey. Some carnivores seize their victims and swallow them alive and whole (Fig. 45-1d). Others paralyze, crush, or shred their prey before ingesting it. Carnivorous mammals have well-developed canine teeth for stabbing their prey during combat. The digestive juice of the stomach breaks down proteins, and because meat is more easily digested than plant food, their digestive tracts are shorter than those of herbivores.

Omnivores, such as bears and humans, consume both plants and animals. Earthworms ingest large amounts of soil containing both animal and plant material. The blue whale, the largest animal, is a filter feeder that strains out tiny algae and animals as it swims. Omnivores often have adaptations that help them distinguish among a wide range of smells and tastes and thereby select a variety of foods.

Animals can also be classified according to the mechanisms they use to feed. Many omnivores are **suspension feeders** that remove suspended food particles from the water. Some animals expose a sticky, mucus-coated surface to flowing water; suspended particles adhere to the surface. For example, some echinoderms have tentacles coated with mucus. Others, like bivalve mollusks, filter water. Baleen whales use rows of hard plates (baleen) suspended from the roof of the mouth to filter small crustaceans.

Some animals feed on fluids by piercing and sucking. Mosquitos have highly adapted structures for piercing skin and sucking blood. Birds that feed on pollen and nectar have long bills and tongues. The shape, size, and curve of the beak may be specialized for feeding on a particular type of flower (see Fig. 19-15). Bats that feed on nectar have a long tongue and reduced dentition (number of teeth).

Many animals, including carnivores, ingest large pieces of food. Adaptations for this type of feeding include claws, fangs, poison glands, tentacles, and teeth. The beaks of birds and the teeth of many vertebrates are specialized for cutting, tearing, or chewing food.

Some invertebrates have a digestive cavity with a single opening

The simplest invertebrates, sponges, obtain food by filtering microscopic organisms from the surrounding water. Individual

FIGURE **45-1** **Adaptations for obtaining and processing food.**
(a) The impressively long "snout" of the herbivorous acorn weevil (*Curculio*) is adapted both for feeding and for making a hole in the acorn through which it deposits an egg. When it has hatched, the larva feeds on the contents of the acorn seed. **(b)** The herbivorous giant panda's (*Ailuropoda melanoleuca*) large, flat teeth and well-developed jaws and jaw muscles are adaptations for grinding high-fiber plant food. **(c)** The mouth *(on left)* of the carnivorous long-nose butterfly fish (*Forcipiger longirostris*) is adapted for extracting small worms and crustaceans from tight spots in coral reefs. **(d)** This carnivorous snake (*Dromicus*) is strangling a lava lizard (*Tropidurus*).

cells phagocytize the food particles, and digestion is *intracellular* within food vacuoles. Wastes are egested into the water that continuously circulates through the sponge body.

Most animals have a digestive cavity. Digestion within a cavity is more efficient than intracellular digestion because digestive enzymes can be released into one confined space, and less surface area is required. Cnidarians (such as hydras and jellyfish) and flatworms have a **gastrovascular cavity,** a central digestive cavity with a single opening. Cnidarians capture small aquatic animals with the help of their stinging cells and tentacles (Fig. 45-2a). The mouth opens into the gastrovascular cavity. Cells lining this digestive cavity secrete enzymes that break down proteins. Digestion continues *intracellularly* within food vacuoles, and digested nutrients diffuse into other cells. Body contractions promote egestion of undigested food particles through the mouth.

Free-living flatworms begin to digest their prey even before ingesting it. They extend their pharynx out through their mouth and secrete digestive enzymes onto the prey (Fig. 45-2b). When ingested, the food enters the branched gastrovascular cavity, where enzymes continue to digest it. Partly digested food frag-

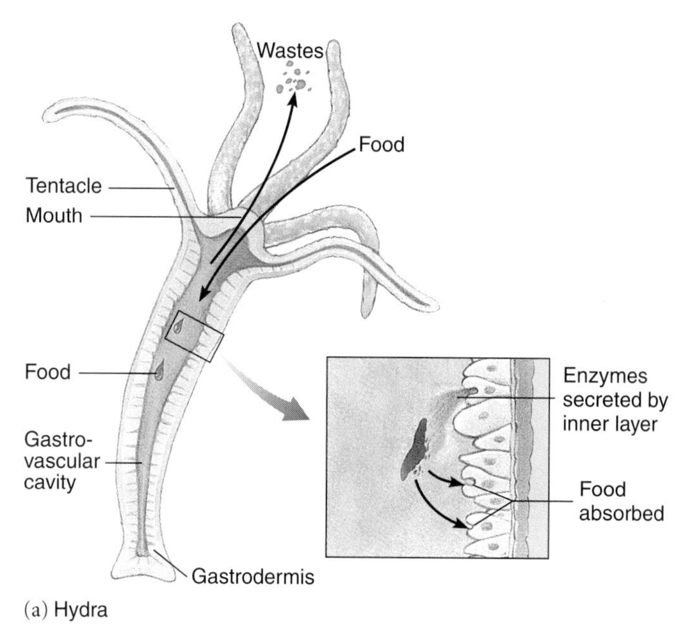

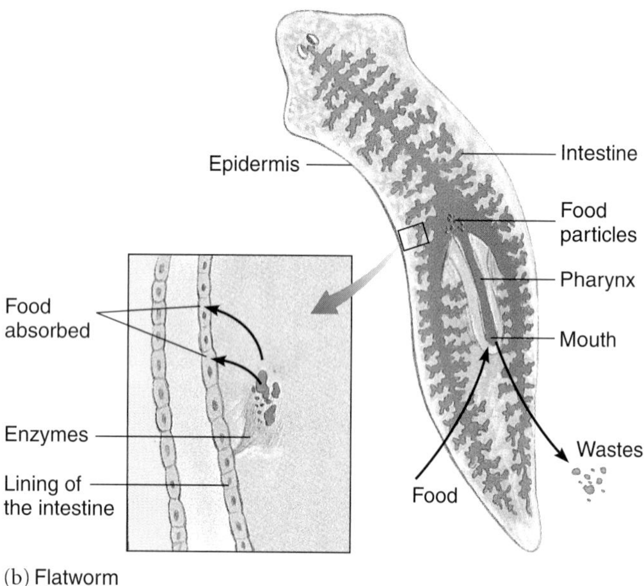

FIGURE **45-2** **Simple invertebrate digestive systems.**
(a) Hydras and **(b)** flatworms (planarians) have digestive tracts with a single opening that serves as both mouth and anus.

ments are then phagocytized by cells lining the gastrovascular cavity, and digestion is completed within food vacuoles. As in cnidarians, the flatworm digestive cavity has only one opening, so undigested wastes are egested through the mouth.

Most animal digestive systems have two openings

Most invertebrates, and all vertebrates, have a tube-within-a-tube body plan. The body wall forms the outer tube. The inner tube is a digestive tract with two openings, sometimes referred to as a complete digestive system (Fig. 45-3). Food enters through

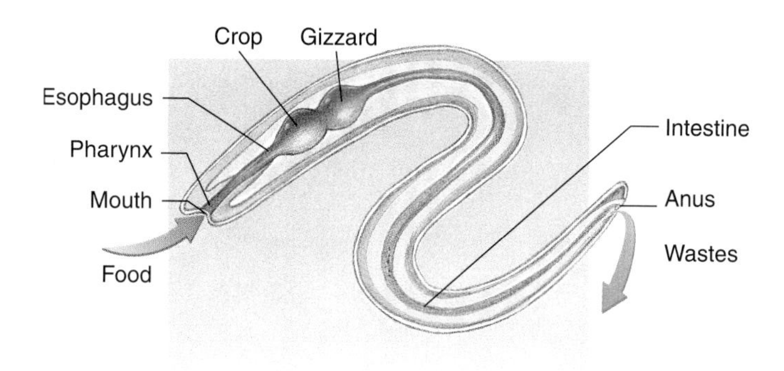

FIGURE **45-3** **Digestive tract with two openings.**
The earthworm, like most complex animals, has a complete digestive tract extending from mouth to anus. Various regions of the digestive tract are specialized to perform different food processing functions.

the mouth, and undigested food is eliminated through the anus. **Motility** is the mixing and propulsive movements of the digestive tract. The propulsive activity characteristic of most regions of the digestive tract is **peristalsis,** waves of muscular contraction that push the food in one direction. More food can be taken in while previously eaten food is being digested and absorbed farther down the digestive tract. In a digestive tract with two openings, various regions of the tube are adapted to perform specific functions.

Review

- How have carnivores adapted to their mode of nutrition?
- How does food processing differ in earthworms and flatworms?

BiologyNow™ Assess your understanding of **nutritional styles and adaptations,** by taking the pretest on your BiologyNow CD-ROM.

THE VERTEBRATE DIGESTIVE SYSTEM

Learning Objectives

2 Trace the pathway traveled by an ingested meal in the human digestive system, describing the structure and function of each organ involved.
3 Describe the step-by-step digestion of carbohydrate, protein, and lipid.
4 Describe the structural adaptations that increase the surface area of the digestive tract.
5 Compare lipid absorption with absorption of other nutrients.

Various regions of the vertebrate digestive tract are specialized to perform specific functions (Fig. 45-4). Food passes in sequence through the following specialized regions:

Mouth ⟶ pharynx (throat) ⟶ esophagus ⟶ stomach ⟶ small intestine ⟶ large intestine ⟶ anus

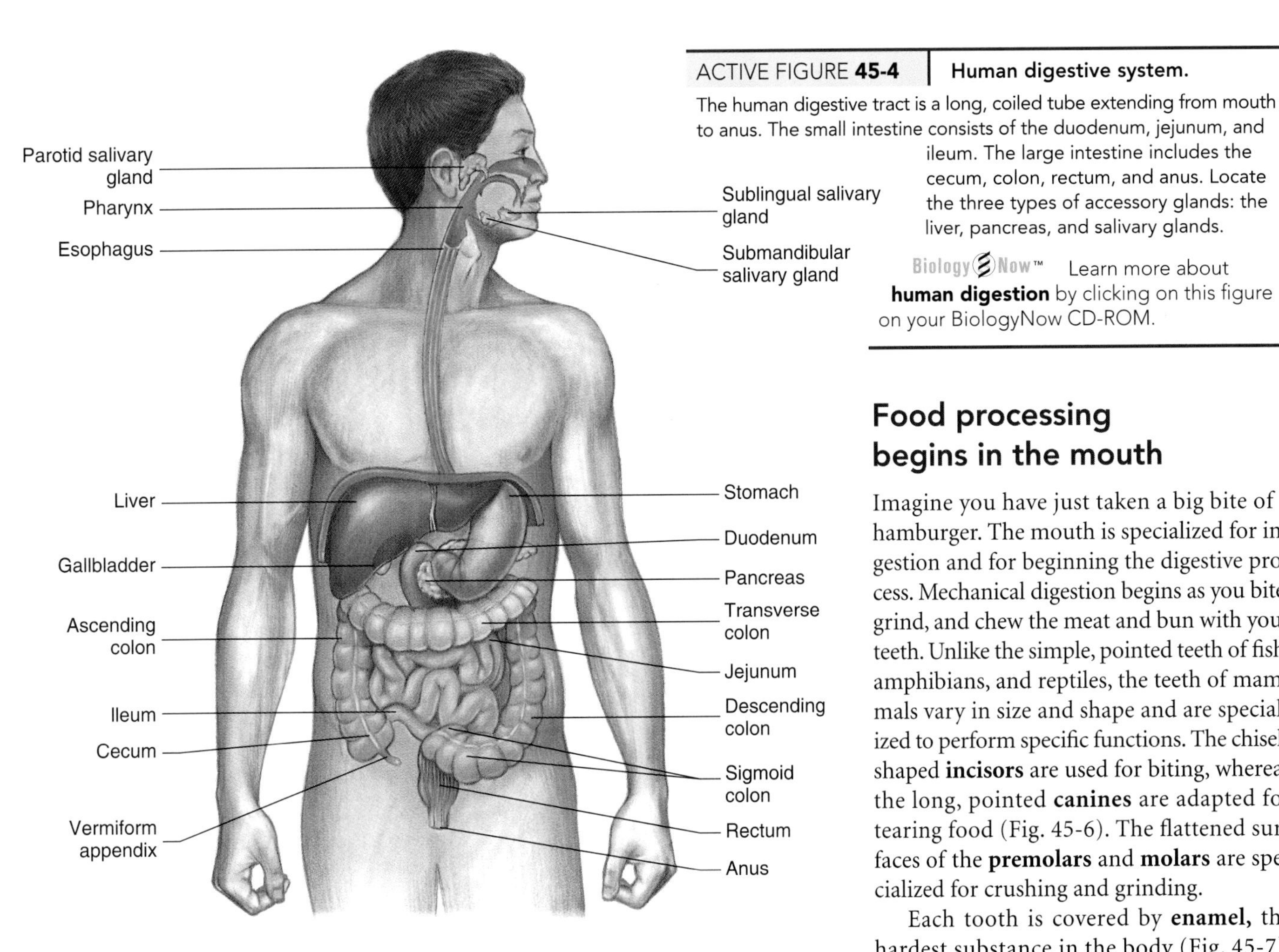

ACTIVE FIGURE **45-4** | **Human digestive system.**

The human digestive tract is a long, coiled tube extending from mouth to anus. The small intestine consists of the duodenum, jejunum, and ileum. The large intestine includes the cecum, colon, rectum, and anus. Locate the three types of accessory glands: the liver, pancreas, and salivary glands.

BiologyNow™ Learn more about **human digestion** by clicking on this figure on your BiologyNow CD-ROM.

Vertebrates have accessory glands that secrete digestive juices into the digestive tract. These include the liver, the pancreas, and, in terrestrial vertebrates, the salivary glands.

The wall of the digestive tract has four layers. Although various regions differ somewhat in structure, the layers are basically similar throughout the digestive tract (Fig. 45-5). The **mucosa,** a layer of epithelial tissue and underlying connective tissue, lines the *lumen* (inner space) of the digestive tract. In the stomach and intestine, the mucosa is greatly folded to increase the secreting and absorbing surface. Surrounding the mucosa is the **submucosa,** a connective tissue layer rich in blood vessels, lymphatic vessels, and nerves.

A **muscle layer,** consisting of two sublayers of smooth muscle, surrounds the submucosa. In the inner sublayer, the muscle fibers are arranged circularly around the digestive tube. In the outer sublayer the muscle fibers are arranged longitudinally. Below the level of the diaphragm, the outer connective tissue coat of the digestive tract is called the **visceral peritoneum.** By various folds it is connected to the **parietal peritoneum,** a sheet of connective tissue that lines the walls of the abdominal and pelvic cavities. The visceral and parietal peritonea enclose part of the coelom called the **peritoneal cavity.** Inflammation of the peritoneum, called **peritonitis,** can be very serious, because infection can spread along the peritoneum to most of the abdominal organs.

Food processing begins in the mouth

Imagine you have just taken a big bite of a hamburger. The mouth is specialized for ingestion and for beginning the digestive process. Mechanical digestion begins as you bite, grind, and chew the meat and bun with your teeth. Unlike the simple, pointed teeth of fish, amphibians, and reptiles, the teeth of mammals vary in size and shape and are specialized to perform specific functions. The chisel-shaped **incisors** are used for biting, whereas the long, pointed **canines** are adapted for tearing food (Fig. 45-6). The flattened surfaces of the **premolars** and **molars** are specialized for crushing and grinding.

Each tooth is covered by **enamel,** the hardest substance in the body (Fig. 45-7). Most of the tooth consists of **dentin,** which resembles bone in composition and hardness. Beneath the dentin is the **pulp cavity,** a soft connective tissue containing blood vessels, lymph vessels, and nerves.

FIGURE **45-5** | **Wall of the digestive tract.**

From inside out, the layers of the wall are the mucosa, submucosa, muscle layer, and visceral peritoneum.

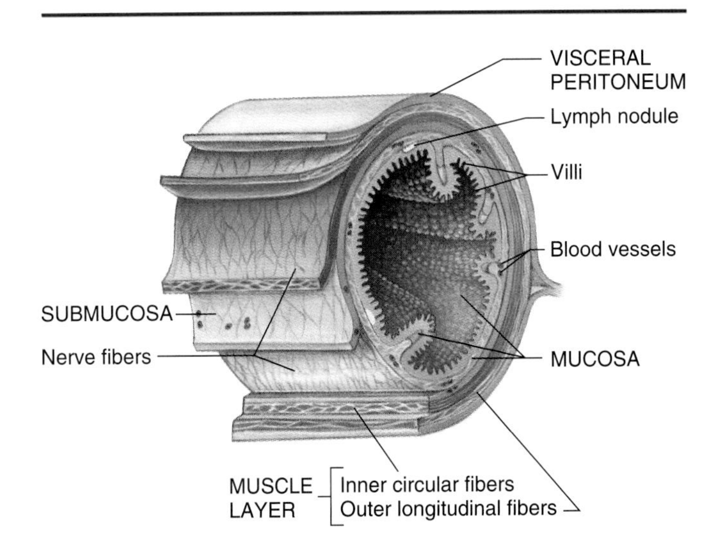

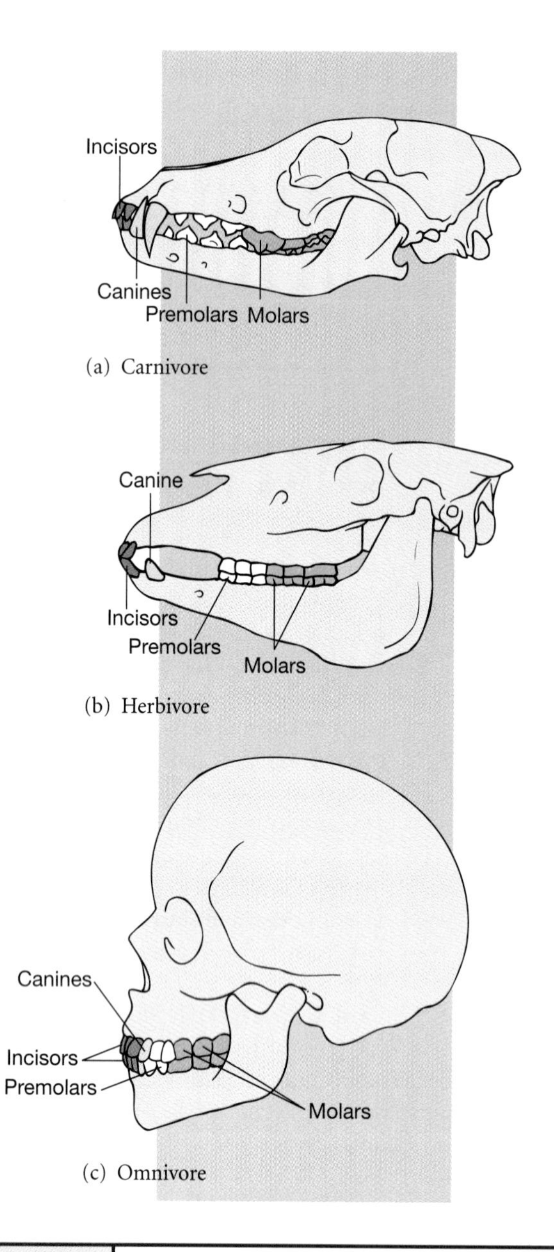

FIGURE **45-6** **Teeth and diet.**

(a) Skull of a coyote showing the pointed incisors and canines, adaptations for ripping flesh. **(b)** In contrast, herbivores, such as horses, have incisors (and sometimes canines) adapted for cutting off bits of vegetation. Canines are absent in some herbivores. The broad, ridged surfaces of the molars are adapted for grinding plant material. **(c)** The teeth of omnivores, such as humans, are adapted for chewing a variety of foods.

While food is being mechanically disassembled by the teeth, it is moistened by saliva. Some of the food molecules dissolve, enabling you to taste the food. Recall from Chapter 41 that taste buds are located on the tongue and other surfaces of the mouth. Three pairs of **salivary glands** secrete about a liter of saliva into the mouth cavity each day. Saliva contains an enzyme, **salivary amylase,** which begins the chemical digestion of starch into sugar.

The pharynx and esophagus conduct food to the stomach

After being chewed and fashioned into a lump called a **bolus,** the bite of food is swallowed—moved through the **pharynx** into the **esophagus.** The pharynx, or throat, is a muscular tube that serves as the hallway of both the respiratory system and the digestive system. During swallowing, a small flap of tissue, the **epiglottis,** closes the opening to the airway.

Waves of peristalsis sweep the bolus through the pharynx and esophagus toward the stomach (Fig. 45-8). Circular muscle fibers in the wall of the esophagus contract around the top of the bolus, pushing it downward. Almost at the same time, longitudinal muscles around the bottom of the bolus and below it contract, shortening the tube.

When the body is in an upright position, gravity helps move the food through the esophagus, but gravity is not essential. Astronauts eat in its absence, and even if you are standing on your head, food will reach your stomach.

Food is mechanically and enzymatically digested in the stomach

The entrance to the large, muscular **stomach** is normally closed by a ring of muscle at the lower end of the esophagus. When a peristaltic wave passes down the esophagus, the muscle relaxes, permitting the bolus to enter the stomach (Fig. 45-9). When empty, the stomach is collapsed and shaped almost like a hot dog. Folds of the stomach wall, called **rugae,** give the inner lining a wrinkled appearance. As food enters, the rugae gradually smooth out, expanding the capacity of the stomach to more than a liter.

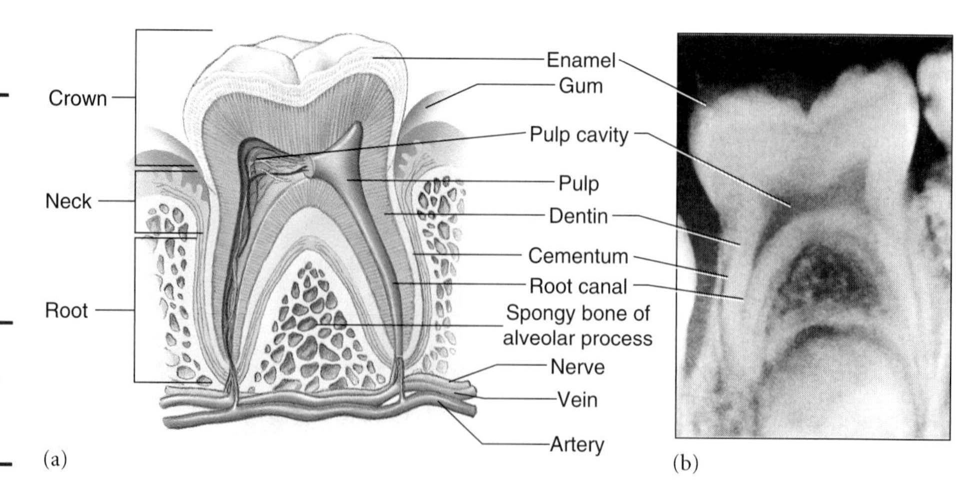

FIGURE **45-7** **Tooth structure.**

(a) Sagittal section through a human lower molar showing the crown, neck, and root. **(b)** X-ray of a healthy tooth.

FIGURE **45-8** **Peristalsis.**

Food is moved through the digestive tract by waves of muscular contractions known as *peristalsis.* ① A bolus is moved down through the esophagus by peristaltic contractions. ② When the sphincter (ring of muscle) at the entrance of the stomach opens, food enters the stomach.

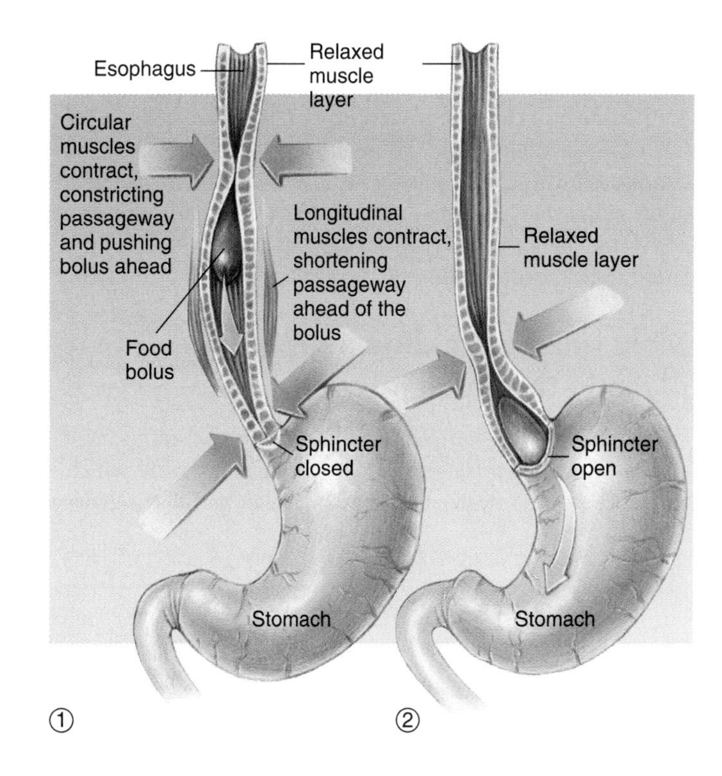

The stomach is lined with a simple, columnar epithelium that secretes large amounts of mucus. Tiny pits mark the entrances to the millions of **gastric glands,** which extend deep into the stomach wall. **Parietal cells** in the gastric glands secrete hydrochloric acid and *intrinsic factor,* a substance needed for adequate absorption of vitamin B_{12}. **Chief cells** in the gastric glands secrete **pepsinogen,** an inactive enzyme precursor. When pepsinogen comes in contact with the acidic gastric juice in the stomach, it is converted to **pepsin,** the main digestive enzyme of the stomach. Pepsin hydrolyzes proteins, converting them to short polypeptides.

Several protective mechanisms prevent the gastric juice from digesting the wall of the stomach. Cells of the gastric mucosa secrete an alkaline mucus that coats the stomach wall and neutralizes the acidity of the gastric juice along the lining. In addition, the epithelial cells of the lining fit tightly together, preventing gastric juice from leaking between them and into the tissue beneath. If some of the epithelial cells become damaged, they are quickly replaced. In fact, about a half million of these cells are shed and replaced every minute!

Sometimes, these protective mechanisms malfunction and part of the stomach lining is digested, leaving an open sore, or **peptic ulcer.** Such ulcers often occur in the duodenum and sometimes in the lower part of the esophagus. The bacterium *Helicobacter pylori* has been implicated as a causative factor in ulcers. *H. pylori* infects the mucus-secreting cells of the stomach lining, decreasing the protective mucus, which can lead to peptic ulcers or cancer. *H. pylori* infection responds to antibiotic therapy.

What changes occur in a bite of hamburger during its three- to four-hour stay in the stomach? The stomach churns and chemically degrades the

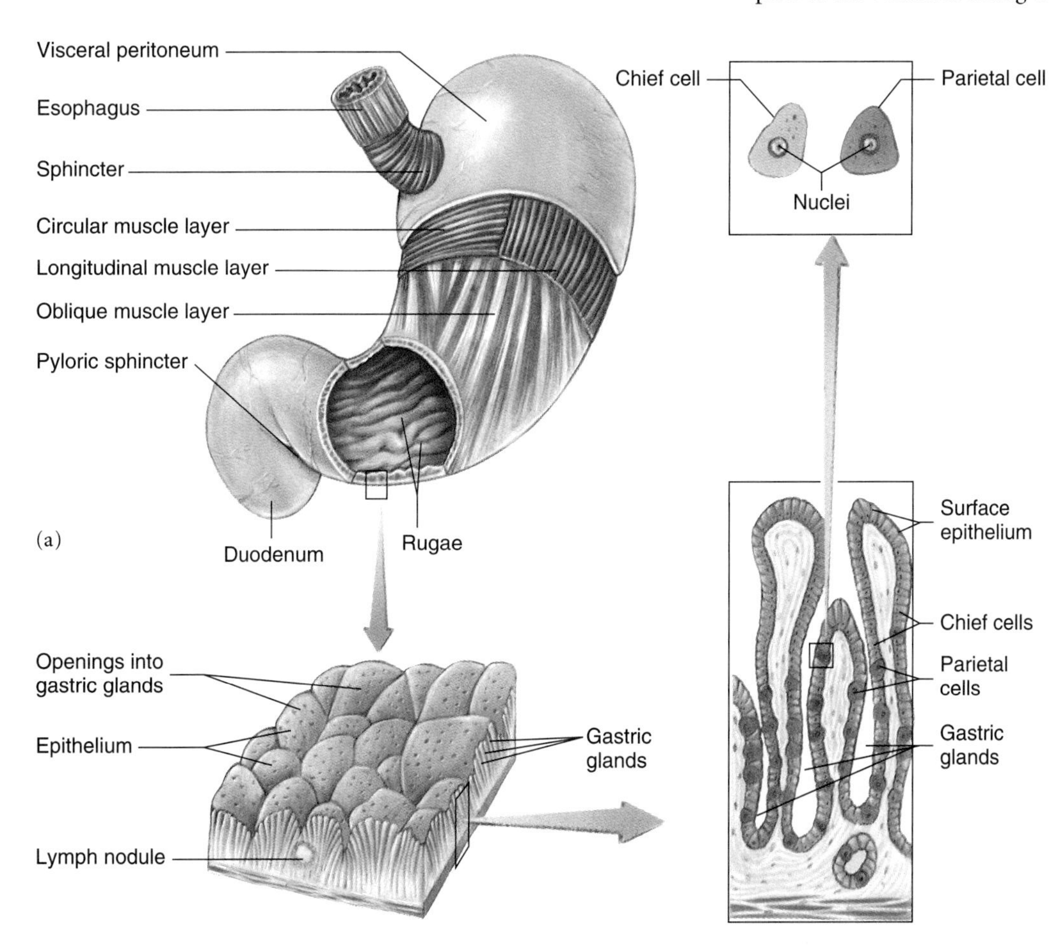

FIGURE **45-9**

Structure of the stomach.

From the esophagus, food enters the stomach, where it is mechanically and enzymatically digested. **(a)** The wall of the stomach has been progressively removed to show muscle layers and rugae. **(b)** Stomach lining and gastric glands.

food so that it assumes the consistency of a thick soup; this partially digested food is called **chyme.** Protein digestion then degrades much of the hamburger protein to polypeptides. Digestion of the starch in the bun to small polysaccharides and maltose continues until salivary amylase is inactivated by the acidic pH of the stomach. Over a period of several hours, peristaltic waves release the chyme in spurts through the stomach exit, the **pylorus,** and into the small intestine.

Most enzymatic digestion takes place in the small intestine

Digestion of food is completed in the **small intestine,** and nutrients are absorbed through its wall. The small intestine, which is 5 to 6 m (about 17 ft) in length, has three regions: the **duodenum, jejunum,** and **ileum.** Most chemical digestion takes place in the duodenum, the first portion of the small intestine, not in the stomach. Bile from the liver and enzymes from the pancreas are released into the duodenum and act on the chyme. Then enzymes produced by the epithelial cells lining the duodenum catalyze the final steps in the digestion of the major types of nutrients.

The lining of the small intestine appears velvety because of its millions of tiny finger-like projections, the intestinal **villi** (sing., *villus*) (Fig. 45-10). The villi increase the surface area of the small intestine for digestion and absorption of nutrients. The intestinal surface is further expanded by thousands of **microvilli,** folds of cytoplasm on the exposed surface of the simple columnar epithelial cells of the villi. About 600 microvilli protrude from the surface of each cell, giving the epithelial lining a fuzzy appearance when viewed with the electron microscope.

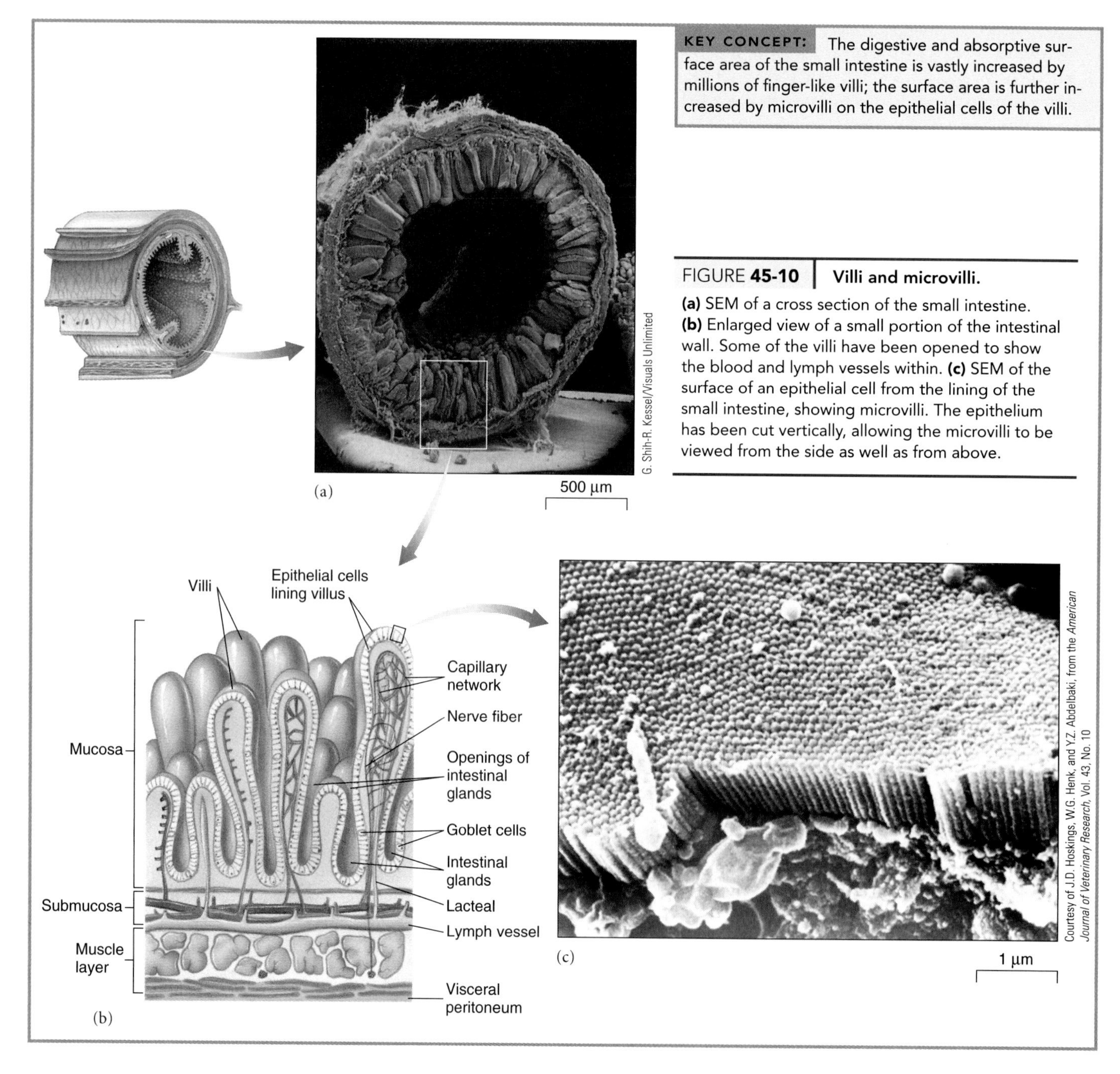

KEY CONCEPT: The digestive and absorptive surface area of the small intestine is vastly increased by millions of finger-like villi; the surface area is further increased by microvilli on the epithelial cells of the villi.

FIGURE **45-10** **Villi and microvilli.**

(a) SEM of a cross section of the small intestine. **(b)** Enlarged view of a small portion of the intestinal wall. Some of the villi have been opened to show the blood and lymph vessels within. **(c)** SEM of the surface of an epithelial cell from the lining of the small intestine, showing microvilli. The epithelium has been cut vertically, allowing the microvilli to be viewed from the side as well as from above.

If the intestinal lining were smooth, like the inside of a water pipe, food would zip right through the intestine, and many valuable nutrients would not be absorbed. Folds in the wall of the intestine, the villi, and microvilli together increase the surface area of the small intestine by about 600 times. If we could unfold and spread out the lining of the small intestine of an adult human, its surface would approximate the size of a tennis court.

The liver secretes bile

The **liver,** the largest internal organ and also one of the most complex organs in the body, lies in the upper right part of the abdomen just under the diaphragm (Fig. 45-11). A single liver cell can carry on more than 500 separate, specialized metabolic activities. The liver:

1. Secretes **bile,** which is important in the mechanical digestion of fats
2. Helps maintain homeostasis by removing or adding nutrients to the blood
3. Converts excess glucose to glycogen and stores it
4. Converts excess amino acids to fatty acids and urea
5. Stores iron and certain vitamins
6. Detoxifies alcohol and other drugs and poisons

Bile consists of water, bile salts, bile pigments, cholesterol, salts, and lecithin (a phospholipid). The pear-shaped gallbladder stores and concentrates the bile, and releases it into the duodenum as needed. Bile mechanically digests fats by a detergent-like action (discussed in a later section). Because it contains no digestive enzymes, bile does not enzymatically digest food.

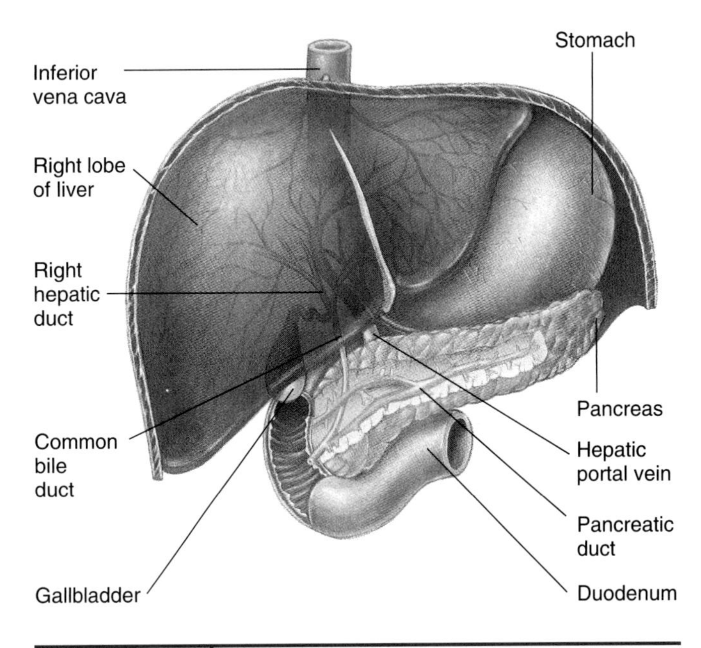

FIGURE **45-11** | **The liver and pancreas.**
The gallbladder stores bile from the liver. Note the ducts that conduct bile to the gallbladder and the duodenum. The stomach has been displaced to expose the pancreas.

The pancreas secretes digestive enzymes

The **pancreas** is an elongated gland that secretes both digestive enzymes and hormones that help regulate the level of glucose in the blood. Among its enzymes are **trypsin** and **chymotrypsin,** which digest polypeptides to dipeptides; **pancreatic lipase,** which degrades fats; **pancreatic amylase,** which breaks down almost all types of carbohydrates, except cellulose, to disaccharides; and **ribonuclease** and **deoxyribonuclease,** which split the nucleic acids ribonucleic acid (RNA) and deoxyribonucleic acid (DNA) to free nucleotides.

Nutrients are digested as they move through the digestive tract

Chyme moves through the digestive tract by peristalsis, mixing contractions, and motions of the villi. Nutrients in the chyme come into contact with enzymes that digest them (Table 45-1).

TABLE **45-1** **Summary of Digestion**

	Carbohydrates	Proteins	Lipids
Mouth	Polysaccharides ↓ Salivary amylase Maltose and small polysaccharides		
Stomach	Action continues until acidic pH inactivates salivary amylase	Protein ↓ Pepsin Short polypeptides	
Small intestine	Undigested polysaccharides ↓ Pancreatic amylase Maltose and other disaccharides ↓ Maltase, sucrase, lactase Monosaccharides	Polypeptides ↓ Trypsin, chymotrypsin Small polypeptides and peptides ↓ Carboxypeptidase, Peptidases, Dipeptidases Amino acids	Glob of fat ↓ Bile salts Emulsified fat droplets ↓ Pancreatic lipase Fatty acids and glycerol

Carbohydrates are digested to monosaccharides

Polysaccharides, such as starch and glycogen, are important components of the food ingested by humans and most other animals. The glucose units of these large molecules are connected by glycosidic bonds linking carbon 4 (or 6) of one glucose molecule with carbon 1 of the adjacent glucose molecule. These bonds are hydrolyzed by **amylases** that digest polysaccharides to the disaccharide maltose. Although amylase can split the α-glycosidic linkages present in starch and glycogen, it cannot split the β-glycosidic linkages present in cellulose (see Figs. 3-9 and 3-10).

Amylase cannot split the bond between the two glucose units of maltose. Enzymes produced by the cells lining the small intestine break down disaccharides such as maltose to monosaccharides. **Maltase,** for example, splits maltose into two glucose molecules (see Fig. 3-8a). Hydrolysis occurs while the disaccharides are being absorbed through the epithelium of the small intestine.

Proteins are digested to amino acids

Several kinds of proteolytic enzymes are secreted into the digestive tract. Each breaks peptide bonds at one or more specific locations in a polypeptide chain. Trypsin, secreted in an inactive form by the pancreas, is activated by an enzyme called *enterokinase.* The trypsin then activates chymotrypsin and carboxypeptidase, as well as additional trypsin. Pepsin, trypsin, and chymotrypsin break certain internal peptide bonds of proteins and polypeptides. Carboxypeptidase removes amino acids with free terminal carboxyl groups from the end of polypeptide chains. **Dipeptidases** released by the duodenum then split the small peptides to amino acids.

Fats are digested to fatty acids and monoacylglycerols

Lipids are usually ingested as large masses of triacylglycerols (also called triglycerides). They are digested mainly within the duodenum by pancreatic lipase. Like many other proteins, lipase is water-soluble, but its substrates are not. Thus the enzyme can attack only the fat molecules at the surface of a mass of fat. Bile salts act like detergents, reducing the surface tension of fats. Their action, called **emulsification,** breaks large masses of fat into smaller droplets. Emulsification greatly increases the surface area of fat exposed to the action of pancreatic lipase and so increases the rate of lipid digestion.

Conditions in the intestine are usually not optimal for the complete hydrolysis of lipids to glycerol and fatty acids. Consequently, the products of lipid digestion include monoacylglycerols (monoglycerides) and diacylglycerols (diglycerides) as well as glycerol and fatty acids. Undigested triacylglycerols remain as well, and some of these are absorbed without digestion.

Nerves and hormones regulate digestion

Most digestive enzymes are produced only when food is present in the digestive tract. Salivary gland secretion is controlled entirely by the nervous system, but secretion of other digestive juices is regulated by both nerves and hormones. The wall of the digestive tract contains dense networks of neurons. This so-called *enteric nervous system* continues to regulate many motor and secretory activities of the digestive system even if sympathetic and parasympathetic nerves to these organs are cut. Many neuropeptides present in the brain are also released by neurons in the digestive tract and help regulate digestion. For example, *substance P* stimulates smooth muscle contraction of the digestive tract, and *enkephalin* inhibits it.

Several hormones, including **gastrin, secretin, cholecystokinin (CCK),** and **gastric inhibitory peptide (GIP),** help regulate the digestive system (Table 45-2). All these hormones are polypeptides secreted by endocrine cells in the mucosa of certain regions of the digestive tract. Investigators are studying several other messenger peptides that are important in regulating digestive activity.

As an example of the regulation of the digestive system, consider the secretion of gastric juice. Seeing, smelling, tasting, or even thinking about food causes the brain to send neural signals to the gastric glands in the stomach, stimulating them to secrete. In addition, when food distends the stomach, stretch receptors send neural messages to the medulla. The medulla then sends messages to endocrine cells in the stomach wall that secrete the

TABLE 45-2 Some Hormones That Regulate Digestion

Hormone	Source	Target Tissue	Actions	Factors That Stimulate Release
Gastrin	Stomach (mucosa)	Stomach (gastric glands)	Stimulates gastric glands to secrete pepsinogen	Distention of the stomach by food; certain substances such as partially digested proteins and caffeine
Secretin	Duodenum (mucosa)	Pancreas	Signals secretion of sodium bicarbonate	Acidic chyme acting on mucosa of duodenum
		Liver	Stimulates bile secretion	
Cholecystokinin (CCK)	Duodenum (mucosa)	Pancreas	Stimulates release of digestive enzymes	Presence of fatty acids and partially digested proteins in duodenum
		Gallbladder	Stimulates emptying of bile	
Gastric inhibitory peptide (GIP)	Duodenum (mucosa)	Stomach	Decreases stomach churning, thus slowing emptying	Presence of fatty acids or glucose in duodenum

hormone gastrin. Gastrin is absorbed into the blood; it stimulates the stomach to release gastric juice and also stimulates gastric emptying and intestinal motility.

Absorption takes place mainly through the villi of the small intestine

Only a few substances—water, simple sugars, salts, alcohol, and certain drugs—are small enough to be absorbed through the stomach wall. Absorption of nutrients is primarily the job of the intestinal villi. As illustrated in Figure 45-10, the wall of a villus is a single layer of epithelial cells. Inside each villus is a network of capillaries and a central lymph vessel, called a **lacteal.**

To reach the blood (or lymph), a nutrient molecule must pass through an epithelial cell lining the intestine and through a cell lining a blood or lymph vessel. Absorption occurs by a combination of simple diffusion, facilitated diffusion, and active transport. Because glucose and amino acids cannot diffuse through the intestinal lining, they must be absorbed by active transport. Absorption of these nutrients is coupled with the active transport of sodium (see Chapter 5). Fructose is absorbed by facilitated diffusion.

Amino acids and glucose are transported directly to the liver by the **hepatic portal vein.** In the liver this vein divides into a vast network of tiny blood sinuses, vessels similar to capillaries. As the nutrient-rich blood moves slowly through the liver, nutrients and certain toxic substances are removed from the circulation.

The products of lipid digestion are absorbed by a different process and different route. After free fatty acids and monoacylglycerols enter an epithelial cell in the intestinal lining, they are reassembled as triacylglycerols in the smooth endoplasmic reticulum. The triacylglycerols, along with absorbed cholesterol and phospholipids, are packaged into protein-covered fat droplets, called **chylomicrons.** After they are released into the interstitial fluid, chylomicrons enter the lacteal (lymph vessel) of the villus. Chylomicrons are transported in the lymph to the subclavian veins, where the lymph and its contents enter the blood. About 90% of absorbed fat enters the blood circulation in this indirect way. The rest, mainly short-chain fatty acids such as those in butter, are absorbed directly into the blood. After a meal rich in fats, the great number of chylomicrons in the blood may give the plasma a turbid, milky appearance for a few hours.

Most of the nutrients in the chyme are absorbed by the time it reaches the end of the small intestine. What is left (mainly waste) passes through a sphincter, the **ileocecal valve,** into the large intestine.

The large intestine eliminates waste

Undigested material, such as the cellulose of plant foods, along with unabsorbed chyme, passes into the **large intestine** (see Fig. 45-4). Although only about 1.3 m (about 4 ft) long, this part of the digestive tract is referred to as "large" because its diameter is greater than that of the small intestine. The small intestine joins the large intestine about 7 cm (2.8 in) from the end of the large intestine, forming a blind pouch, the **cecum.** The **vermiform appendix** projects from the end of the cecum. (Appendicitis is an inflammation of the appendix.) Herbivores such as rabbits have a large, functional cecum that holds food while bacteria digest its cellulose. In humans, the functions of the cecum and appendix are not known, and these structures are generally considered vestigial organs.

From the cecum to the **rectum** (the last portion of the large intestine), the large intestine is known as the **colon.** The regions of the large intestine are the cecum; ascending colon; transverse colon; descending colon; sigmoid colon; rectum; and **anus,** the opening for the elimination of wastes.

As chyme passes slowly through the large intestine, water and sodium are absorbed from it, and it gradually assumes the consistency of normal feces. Bacteria inhabiting the large intestine are nourished by the last remnants of the meal and benefit their host by producing vitamin K and certain B vitamins that can be absorbed and used.

A distinction should be made between elimination and excretion. *Elimination* is the process of getting rid of digestive wastes—materials that have not been absorbed from the digestive tract and did not participate in metabolic activities. In contrast, *excretion* is the process of getting rid of metabolic wastes, which in mammals is mainly the function of the kidneys and lungs. The large intestine, however, does excrete bile pigments.

When chyme passes through the intestine too rapidly, **defecation** (expulsion of feces) becomes more frequent, and the feces are watery. This condition, called *diarrhea,* may be caused by anxiety, certain foods, or disease organisms that irritate the intestinal lining. Prolonged diarrhea results in loss of water and salts and leads to dehydration—a serious condition, especially in infants. *Constipation* results when chyme passes through the intestine too slowly. Because more water than usual is removed from the chyme, the feces may be hard and dry. Constipation is often caused by a diet deficient in fiber.

In Western countries, colorectal cancer (cancer of the colon and rectum) is the third most common cancer (Fig. 45-12). Factors contributing to colon cancer are not yet understood, but high consumption of red meat increases the risk, perhaps from carcinogens produced during cooking.

Review

- Imagine you have just taken a bite of steak. List in sequence the structures through which it passes in its journey through the digestive system. What happens in each structure?
- What is the function of villi?
- What are chylomicrons?

BiologyNow™ Assess your understanding of **the vertebrate digestive system** by taking the pretest on your BiologyNow CD-ROM.

REQUIRED NUTRIENTS

Learning Objectives

6 Summarize the nutritional requirements for dietary carbohydrates, lipids, and proteins, and trace the fate of glucose, lipids, and amino acids after their absorption.

7 Describe the nutritional functions of vitamins, minerals, and phytochemicals.

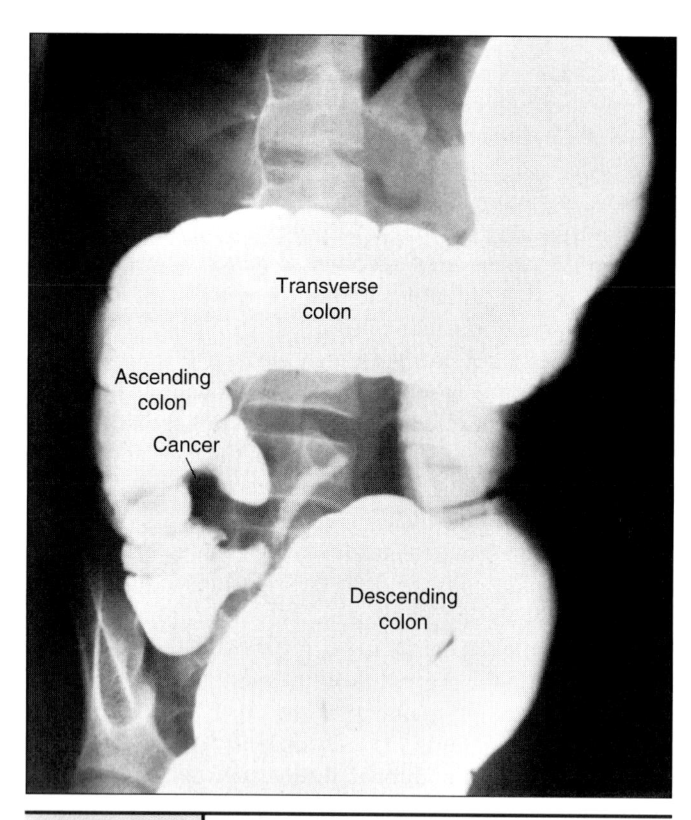

FIGURE **45-12** | **Colon cancer.**

In this radiographic view of the large intestine, cancer is evident as a mass that projects into the lumen of the colon. The large intestine has been filled with a suspension of barium sulfate, which makes irregularities in the wall visible.

All animals require carbohydrates, lipids, proteins, vitamins, and minerals. In addition, nutritionists are investigating plant compounds, called *phytochemicals,* that appear important in nutrition. Although not considered a nutrient in a strict sense, water is a necessary dietary component. Enough fluid must be ingested to replace fluid lost in urine, sweat, feces, and breath.

Adequate amounts of essential nutrients are necessary for metabolic processes. Nutritionists are re-examining the *relative* amounts of various types of nutrients recommended for a healthy diet. Many long-held beliefs about nutrients and diet appear to be flawed. For example, high-carbohydrate diets are not necessarily healthy and not all fat is bad. The Institute of Medicine of the National Academies recently suggested that 20% to 35% of one's calories should come from carbohydrates, 45% to 65% from fat, and 10% to 35% from protein.

Recall from Chapter 1 that **metabolism** includes all the chemical processes that take place in the body. Metabolic processes include anabolism and catabolism. **Anabolism** includes the synthetic aspects of metabolism such as the production of proteins and nucleic acids. **Catabolism** includes breakdown processes such as hydrolysis. Nutritionists measure the energy value of food in kilocalories, or simply Calories. A Calorie, spelled with a capital *C,* is a **kilocalorie (kcal),** defined as the amount of heat required to raise the temperature of a kilogram of water 1 degree C.

Carbohydrates provide energy

Sugars and starches are important sources of energy in the human diet. Most carbohydrates are ingested in the form of starch and cellulose, both polysaccharides composed of long chains of glucose subunits. (You may want to review the discussion of carbohydrates in Chapter 3.) Nutritionists refer to polysaccharides as **complex carbohydrates.** Foods rich in complex carbohydrates include rice, potatoes, corn, and other cereal grains.

When we eat an excess of carbohydrate-rich food, the liver cells become fully packed with glycogen and convert excess glucose to fatty acids and glycerol. Liver cells convert these compounds to triacylglycerols and send them to the fat depots of the body for storage.

Refined carbohydrates, such as white bread and white rice, are unhealthy because the refining process removes fiber and many vitamins and minerals. The refining process also produces a form of starch that the digestive system rapidly breaks down to glucose. The resulting rapid increase in glucose concentration in the blood stimulates the pancreas to release a large amount of insulin, the hormone that stimulates the liver and muscles to remove glucose from the blood. Insulin lowers blood glucose level. When glucose and insulin levels are high, triglyceride levels increase and high-density lipoprotein (HDL; the good cholesterol) concentration decreases. These metabolic events can lead to cardiovascular disease and also increase the risk of type 2 diabetes (discussed in Chapter 47).

Dietary intake of fiber decreases cholesterol concentration in the blood and is associated with a lower risk for cardiovascular disease and diabetes. **Fiber** is mainly a complex mixture of cellulose and other indigestible carbohydrates. We obtain dietary fiber by eating fruits, vegetables, and whole grains. The U.S. diet is low in fiber because of low intake of fruits and vegetables and use of refined flour. Increasing fiber in the diet has a variety of health benefits. Fiber also stimulates the feeling of being satisfied (satiety) after food intake and thus is useful in treating obesity.

Lipids provide energy and are used to make biological molecules

Cells use ingested lipids as an energy source and to make a variety of lipid compounds, such as components of cell membranes, steroid hormones and bile salts. We ingest about 98% of our dietary lipids in the form of triacylglycerols (triglycerides). (Recall from Chapter 3 that a triacylglycerol is a glycerol molecule chemically combined with three fatty acids; see Fig. 3-12b.) Triacylglycerols may be saturated, that is, fully loaded with hydrogen atoms, or their fatty acids may be monounsaturated (containing one double bond) or polyunsaturated (containing two or more double bonds). Three polyunsaturated fatty acids (linoleic, linolenic, and arachidonic acids) are essential fatty acids that humans must obtain from food. Given these and sufficient nonlipid nutrients, the body can make all the lipid compounds (including fats, cholesterol, phospholipids, and prostaglandins) that it needs. The average U.S. diet provides far more cholesterol than the recommended daily maximum of 300 mg. High-cholesterol sources include egg yolks, butter, and meat.

Lipid transport and metabolism are complex. Recall that chylomicrons transport lipids from the intestine to the liver and to other tissues. As chylomicrons circulate in the blood, the enzyme **lipoprotein lipase** breaks down triacylglycerols. The fatty acids and glycerol can then be taken up by the cells. What is left of the chylomicron, a remnant made up mainly of cholesterol and protein, is taken up by the liver.

Liver cells repackage cholesterol and triacylglycerols. These lipids are bound to proteins and transported as large molecular complexes, called **lipoproteins.** Some plasma cholesterol is transported by **high-density lipoproteins,** or **HDLs** ("good" cholesterol), but most is transported by **low-density lipoproteins,** or **LDLs** ("bad" cholesterol). LDLs deliver cholesterol to the cells. For cells to take in LDLs, a protein (apolipoprotein B) on the LDL surface must bind with a protein **LDL receptor** on the plasma membrane (Fig. 5-20). After binding, the LDL enters the cell and its cholesterol and other components are used. When cholesterol levels are high, HDLs collect the excess cholesterol and transport it to the liver. HDLs decrease risk for cardiovascular disease

When needed, stored fats are hydrolyzed to fatty acids and released into the blood. Before cells can use these fatty acids as fuel, they are broken down into smaller compounds and combined with coenzyme A to form molecules of acetyl coenzyme A (acetyl CoA; Fig. 45-13). Acetyl CoA enters the citric acid cycle (see Chapter 7). The conversion of fatty acids to acetyl CoA is accomplished in the liver by a process known as *β-oxidation.*

For transport to the cells, acetyl coenzyme A is converted into one of three types of *ketone bodies* (four-carbon ketones). Normally, the level of ketone bodies in the blood is low, but in certain abnormal conditions, such as starvation and diabetes mellitus, fat metabolism is tremendously increased. Ketone bodies are then produced so rapidly that their level in the blood becomes excessive, which may make the blood too acidic. Such disrupted pH balance can lead to death. What is the relationship between fat and cholesterol intake and cardiovascular disease?

Lipids play a key role in the development of atherosclerosis, a progressive disease in which the arteries become clogged with fatty material (see Chapter 42). The type of fat consumed, as well as other dietary and lifestyle factors, is important. A healthy proportion of HDL to LDL can be promoted by a regular exercise program, healthy diet, appropriate body weight (obesity raises LDL and triacylglycerol levels), and by not smoking cigarettes.

In general, animal foods are rich in both saturated fats and cholesterol, whereas most plant foods contain unsaturated fats and no cholesterol. Butter contains mainly saturated fats. Commonly used polyunsaturated vegetable oils are corn, soy, cottonseed, and safflower oils. Monounsaturated fats raise HDL level, as does stearic acid, the saturated fat in chocolate. Olive, canola, and peanut oils contain large amounts of monounsaturated fats. A few plant oils, including palm oil and coconut oil, are high in saturated fats. **Omega-3 fatty acids** (found in fish and some plant oils) decrease LDL levels and play other protective roles in decreasing the risk for coronary heart disease.

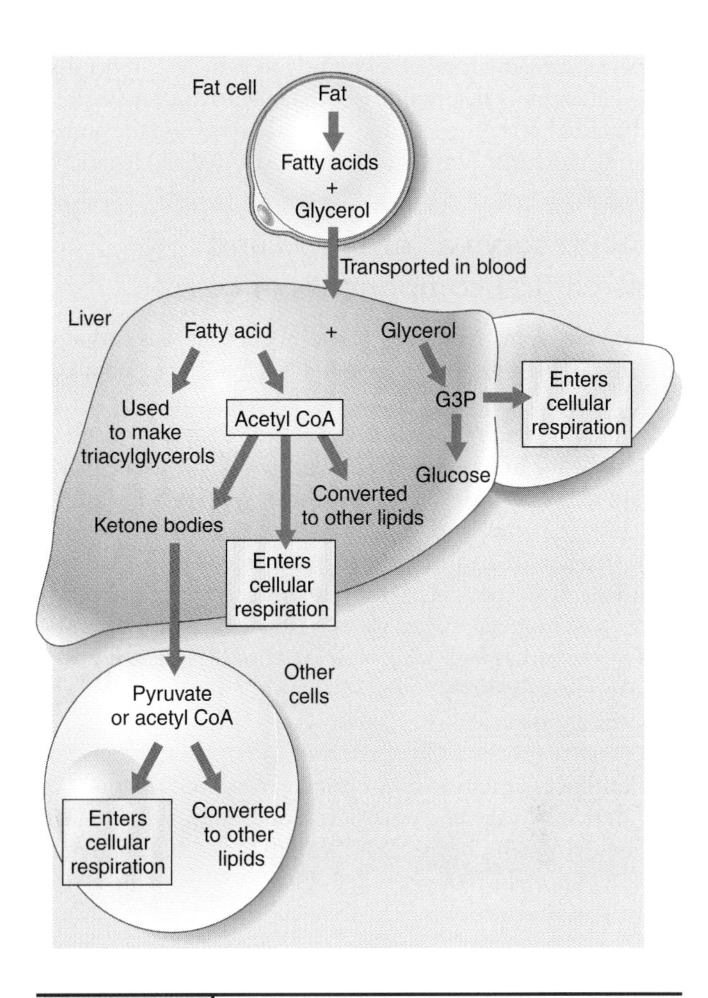

FIGURE **45-13** | **How the body uses fat.**

The liver converts glycerol and fatty acids to compounds that are used as fuel in cellular respiration. Recall from Chapter 7 that G3P is glyceraldehyde-3-phosphate.

In about one third of the U.S. population, a diet high in saturated fats and cholesterol raises the blood cholesterol level by as much as 25%. Because polyunsaturated fats decrease the blood cholesterol level, many people now cook with vegetable oils rather than with butter and lard, drink skim milk rather than whole milk, and eat lowfat ice cream. Some use margarine instead of butter. However, during margarine production vegetable oils are partially hydrogenated (some of the carbons accept hydrogen to become fully saturated). During hydrogenation, some double bonds are transformed from a *cis* arrangement to a *trans* arrangement (see Fig. 3-3b), forming ***trans* fatty acids;** the harder the margarine, the higher the *trans* fatty acid content. Health concerns have been raised because trans fatty acids increase LDL cholesterol in the blood and also lower HDL. Many processed foods, including cookies, crackers, and potato chips, contain *trans* fatty acids.

If you have good genes, control your weight, and exercise regularly, your body may be able to process more unhealthy fat. It is important to remember that some fats, for example, *trans*-fatty acids and saturated fats are unhealthy, but others, such as omega-3 fatty acids and monounsaturated fat, are healthy. The

Inuits in Greenland eat diets extremely high in fat, but because their dietary fat is rich in omega-3 fatty acids, they have a low incidence of heart disease. These fatty acids decrease the probability of ventricular fibrillation, which can cause sudden death.

Proteins serve as enzymes and as structural components of cells

Proteins are essential building blocks of cells, serve as enzymes, and are also used to make compounds such as hemoglobin and myosin. Protein consumption is an index of a country's (or an individual's) economic status, because high-quality protein tends to be the most expensive and least available of the nutrients. In many parts of the world, protein poverty is one of the most pressing health problems.

Ingested proteins are degraded in the digestive tract to small peptides and amino acids. Of the 20 or so amino acids important in nutrition, approximately 9 (10 in children) cannot be synthesized by humans at all, or at least not in sufficient quantities to meet the body's needs. The diet must provide these **essential amino acids** (see Chapter 3).

Complete proteins, those that contain the most appropriate distribution of amino acids for human nutrition, are found in fish, meat, nuts, eggs, and milk. Some foods, such as gelatin and legumes (soybeans, beans, peas, peanuts) contain a high proportion of protein. However, they either do not contain all the essential amino acids, or they do not contain them in proper nutritional proportions. Most plant proteins are deficient in one or more essential amino acids. The healthiest sources of protein are fish, chicken, nuts, and legumes.

Amino acids circulating in the blood are taken up by cells and used for the synthesis of proteins. The liver removes excess amino acids from the circulation. Liver cells deaminate amino acids, that is, remove the amino group (Fig. 45-14). During deamination, ammonia forms from the amino group. Ammonia, which is toxic at high concentrations, is converted to urea and excreted from the body. The remaining carbon chain of the amino acid (called a *keto acid*) may be converted into carbohydrate or lipid and used as fuel or stored. Thus even people who eat high-protein diets can gain weight if they eat too much.

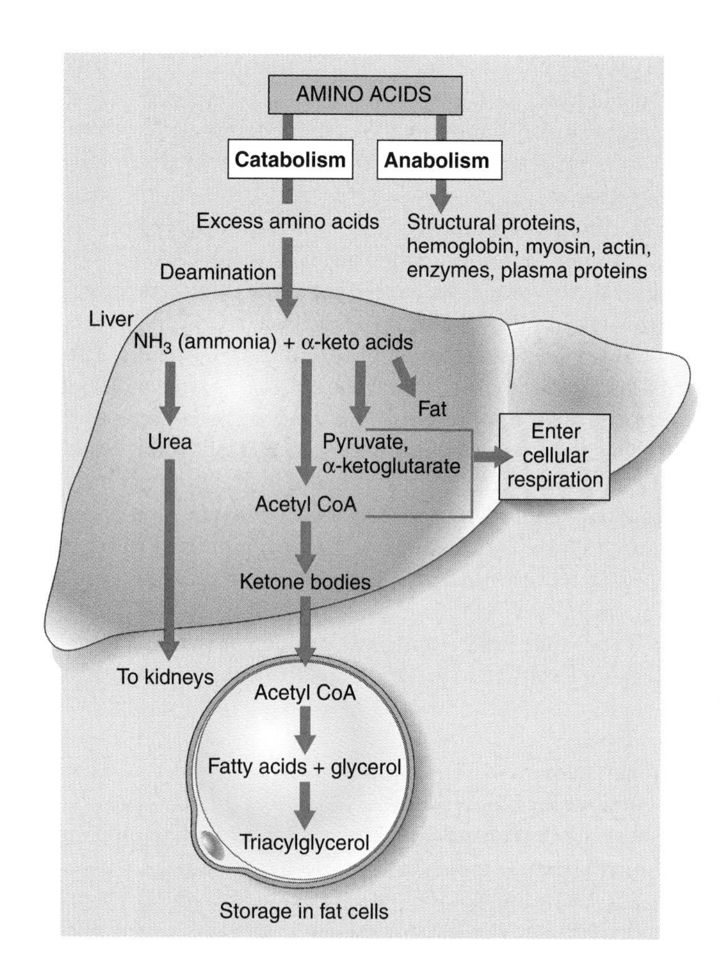

FIGURE **45-14** **How the body uses protein.**
The liver plays a central role in protein metabolism. Deamination of amino acids and conversion of the amino groups to urea take place there. In addition, many proteins are synthesized in the liver. Note that you can gain weight on a high-protein diet, because excess amino acids can be converted to fat.

Vitamins are organic compounds essential for normal metabolism

Vitamins are organic compounds required in the diet in relatively small amounts for normal biochemical functioning. Many are components of coenzymes (see Chapter 6). Nutritionists divide vitamins into two main groups. **Fat-soluble vitamins** include vitamins A, D, E, and K. **Water-soluble vitamins** are the B and C vitamins. Fruit and vegetables are rich sources of vitamins. Table 45-3 provides the sources, functions, and consequences of deficiency for most of the vitamins.

Health professionals debate the advisability of taking large amounts of certain specific vitamins, such as vitamin C to prevent colds or vitamin E to protect against vascular disease. Some studies suggest that vitamin A (found in yellow and green vegetables) and vitamin C (found in citrus fruit and tomatoes) help protect against certain forms of cancer. We do not yet understand all the biochemical roles played by vitamins or the interactions among various vitamins and other nutrients. We do know that large overdoses of vitamins, like vitamin deficiency, can be harmful. Moderate overdoses of the B and C vitamins are excreted in the urine, but surpluses of the fat-soluble vitamins are not easily excreted and can accumulate to harmful levels.

Minerals are inorganic nutrients

Minerals are inorganic nutrients ingested in the form of salts dissolved in food and water (Table 45-4). We need certain minerals, including sodium, chloride, potassium, calcium, phosphorus, magnesium, and sulfur, in amounts of 100 mg or more daily. Several others, such as iron, copper, iodide, fluoride, and selenium, are **trace elements,** minerals that we require in amounts of less than 100 mg per day.

Minerals are necessary components of body tissues and fluids. Salt content (about 0.9% in plasma) is vital in maintaining

TABLE 45-3 The Vitamins

Vitamins and U.S. RDA*	Actions	Effect of Deficiency	Sources
Fat soluble			
Vitamin A, retinol 5000 IU†	Converted to retinal; essential for normal vision; essential for normal growth and differentiation of cells; reproduction; immunity	Growth retardation; night blindness; worldwide 250 million children are at risk of blindness from vitamin A deficiency	Liver, fortified milk, yellow and green vegetables such as carrots and broccoli
Vitamin D, calciferol 400 IU	Promotes calcium and phosphorus absorption from digestive tract; essential to normal growth and maintenance of bone	Weak bones; bone deformities; rickets in children, osteomalacia in adults	Fish oils, egg yolk, fortified milk, butter, margarine
Vitamin E, tocopherols 30 IU	Antioxidant; protects unsaturated fatty acids and cell membranes	Increased catabolism of unsaturated fatty acids, so that not enough are available for maintenance of cell membranes; prevention of normal growth; nerve damage	Vegetable oils, nuts; leafy greens
Vitamin K, about 80 mcg‡	Synthesis of blood-clotting proteins	Prolonged blood clotting time	Normally supplied by intestinal bacteria; leafy greens, legumes
Water soluble			
Vitamin C, ascorbic acid 60 mg	Collagen synthesis; antioxidant; needed for synthesis of some hormones and neurotransmitters; important in immune function	Scurvy (wounds heal very slowly and scars become weak and split open; capillaries become fragile; bone does not grow or heal properly)	Citrus fruit, strawberries, tomatoes leafy vegetables, cabbage
B-complex vitamins			
Vitamin B_1, thiamine 1.5 mg	Active form is a coenzyme in many enzyme systems; important in carbohydrate and amino acid metabolism	Beriberi (weakened heart muscle, enlarged right side of heart, nervous system and digestive tract disorders); common in alcoholics	Liver, yeast, whole and enriched grains, meat, green leafy vegetables
Vitamin B_2, riboflavin 1.7 mg	Used to make coenzymes (e.g., FAD) essential in cellular respiration	Dermatitis, inflammation and cracking at corners of mouth; confusion	Liver, milk, eggs, green leafy vegetables, enriched grains
Niacin 20 mg	Component of important coenzymes (NAD^+ and $NADP^+$); essential to cellular respiration	Pellagra (dermatitis, diarrhea, mental symptoms, muscular weakness, fatigue)	Liver, chicken, tuna, milk, green leafy vegetables, enriched grains
Vitamin B_6, pyridoxine 2 mg	Derivative is coenzyme in many reactions in amino acid metabolism	Dermatitis, digestive tract disturbances; convulsions	Meat, whole grains, legumes, green leafy vegetables
Pantothenic acid 10 mg	Constituent of coenzyme A (important in cellular metabolism)	Deficiency extremely rare	Meat, whole grains, legumes
Folate 400 mcg	Coenzyme needed for nucleic acid synthesis and for maturation of red blood cells	A type of anemia; certain birth defects; increased risk of cardiovascular disease; deficiency in alcoholics, smokers, and pregnant women	Liver, legumes, dark green leafy vegetables, orange juice
Biotin 30 mcg	Coenzyme important in metabolism	—	Produced by intestinal bacteria; liver; chocolate, egg yolk
Vitamin B_{12} 6 mcg	Coenzyme important in metabolism; contains cobalt	A type of anemia	Liver, meat, fish

*RDA is the recommended dietary allowance, established by the Food and Nutrition Board of the National Research Council, to maintain good nutrition for healthy adults.

†International Unit: the amount that produces a specific biological effect and is internationally accepted as a measure of the activity of the substance.

‡mcg = micrograms.

the fluid balance of the body, and because salts are lost from the body daily in sweat, urine, and feces, they must be replaced by dietary intake. Sodium chloride (common table salt) is the salt needed in largest quantity in blood and other body fluids. A deficiency results in dehydration. Iron is the mineral most often deficient in the diet. In fact, iron deficiency is one of the most widespread nutritional problems in the world.

Antioxidants protect against oxidants

Normal cell processes that require oxygen produce **oxidants,** highly reactive molecules such as free radicals, peroxides, and superoxides. **Free radicals,** molecules or ions with one or more unpaired electrons, are also generated by ionizing radiation, to-

TABLE 45-4 Some Important Minerals and Their Functions

Mineral	Functions	Sources; Comments
Calcium	Main mineral of bones and teeth; essential for normal blood clotting, muscle function, nerve function, and regulation of cell activities	Milk and other dairy products, fish, green leafy vegetables; bones serve as calcium reservoir
Phosphorus	Performs more functions than any other mineral; structural component of bone; component of ATP, DNA, RNA, and phospholipids	Meat, dairy products, cereals
Sulfur	Component of many proteins and vitamins	High-protein foods such as meat, fish, legumes, nuts
Potassium	Principal positive ion within cells; important in muscle contraction and nerve function	Fruit, vegetables, grains
Sodium	Principal positive ion in interstitial fluid; important in fluid balance; neural transmission	Many foods, table salt; too much ingested in average American diet; excessive amounts may contribute to high blood pressure
Chloride	Principal negative ion of interstitial fluid; important in fluid balance and in acid–base balance	Many foods; table salt
Magnesium	Needed for normal muscle and nerve function	Nuts; whole grains; green, leafy vegetables
Copper	Component of several enzymes; essential for hemoglobin synthesis	Liver, eggs, fish, whole wheat flour, beans
Iodide	Component of thyroid hormones (hormones that increase metabolic rate); deficiency results in goiter (abnormal enlargement of thyroid gland)	Seafood, iodized salt, vegetables grown in iodine-rich soils
Manganese	Activates many enzymes	Whole-grain cereals, egg yolks, green vegetables; poorly absorbed from intestine
Iron	Component of hemoglobin, myoglobin, important respiratory enzymes (cytochromes), and other enzymes essential to oxygen transport and cellular respiration; deficiency results in anemia and may impair cognitive function	Mineral most likely to be deficient in diet. Good sources: meat (especially liver), fish, nuts, egg yolk, legumes, dried fruit
Fluoride	Component of bones and teeth; makes teeth resistant to decay; excess causes tooth mottling	Fish; in areas where it does not occur naturally, fluoride may be added to municipal water supplies (fluoridation)
Zinc	Cofactor for at least 70 enzymes; helps regulate synthesis of certain proteins; needed for growth and repair of tissues; deficiency may impair cognitive function	Meat, fish, milk, yogurt, grains, vegetables
Selenium	Antioxidant (part of a peroxidase that breaks down peroxides)	Seafood, eggs, meat, garlic, mushrooms, whole grains

bacco smoke, and other forms of air pollution. Oxidants damage DNA, proteins, and unsaturated fatty acids by snatching electrons. Damage to DNA causes mutations that lead to cancer, and injury to unsaturated fatty acids can damage cell membranes. Free radicals are thought to contribute to atherosclerosis by causing oxidation of LDL cholesterol. Oxidative damage to the body over many years contributes to the aging process.

Cells have **antioxidants** that destroy free radicals and other reactive molecules. Antioxidants in tissues include certain enzymes, for example, catalase, and peroxidase. Their action requires minerals such as selenium, zinc, manganese, and copper. Certain vitamins—vitamin C, vitamin E, and β-carotene—have strong antioxidant activity. Vitamins A and E protect cell membranes from free radicals. In addition, a variety of phytochemicals are potent antioxidants (discussed in next section).

Many antioxidants have more than one action. For example, vitamin E also helps prevent selenium deficiency. Nutritionists recommend that we increase the antioxidants in our diet by eating fruits, vegetables, and other foods high in antioxidants. We do not yet know whether antioxidant supplements are useful.

Phytochemicals play important roles in maintaining health

Diets rich in fruits and vegetables lower the incidence of heart disease and of certain types of cancer. Studies comparing diets in various countries suggest that intake of fruits and vegetables may be more important than differences in dietary fat. Yet, nutritionists estimate that in the United States only about 1 in 11 people eats the daily recommended three to five servings of vegetables and two to three of fruit. A diet that includes all the essential nutrients does not provide the same health benefits as one rich in fruit and vegetables. The missing ingredients appear to be **phytochemicals,** plant compounds that promote health. Some phytochemicals function as antioxidants. In Asian countries where diets are low in fat and high in soy and green tea, the incidence of breast, prostate, and colorectal cancer is low.

Nutritionists are just beginning to intensively investigate phytochemicals. Among the important classes of phytochemicals are the carotenoids, allium compounds, flavonoids, indoles, and isocyanates. Except for carotenoids, yellow-orange pigments that the body can convert into vitamin A, phytochemicals have not been established as essential nutrients. Among the carotenoids are the *lycopenes*, powerful antioxidants that are responsible for the red color of tomatoes.

Review

- How do complex carbohydrates affect the body?
- What is the function of glucose? Of essential amino acids?
- What is the function of high-density lipoproteins (HDLs)?
- What measures can you take to lower your risk for heart disease?

BiologyNow™ Assess your understanding of **required nutrients** by taking the pretest on your BiologyNow CD-ROM.

ENERGY METABOLISM

Learning Objectives

8 Contrast basal metabolic rate with total metabolic rate; write the basic energy equation for maintaining body weight, and describe the consequences of altering it in either direction.

9 In general terms, describe the effects of malnutrition, including both undernutrition and overnutrition.

10 Summarize current hypotheses about the regulation of food intake and energy homeostasis. (Include the roles of leptin and neuropeptide Y.)

The amount of energy liberated by the body per unit time is a measure of its **metabolic rate.** Much of the energy expended by the body is ultimately converted to heat. Metabolic rate may be expressed either in kilocalories of heat energy expended per day or as a percentage above or below a standard normal level.

The **basal metabolic rate (BMR)** is the rate at which the body releases heat as a result of breaking down fuel molecules. BMR is the body's basic cost of living, that is, the rate of energy used during resting conditions. This energy is required to maintain body functions such as heart contraction, breathing, and kidney function. A person's **total metabolic rate** is the sum of his or her BMR and the energy used to carry on all daily activities. For example, a laborer has a greater total metabolic rate than does an account executive whose job requirements do not include a substantial amount of physical activity and who does not exercise regularly.

An average-sized person who does not exercise and who sits at a desk all day expends about 2000 kcal daily. If the food the individual eats each day also contains about 2000 kcal, the body will be in a state of energy balance; that is, energy input will equal energy output. This is an extremely important concept, because body weight remains constant when

$$\text{Energy input} = \text{energy output}$$

When energy output is greater than energy input, stored fat is burned and body weight decreases. People gain weight when they take in more energy in food than they expend in daily activity, in other words, when

$$\text{Energy input} > \text{energy output}$$

Undernutrition can cause serious health problems

Millions of people do not have enough to eat, or do not eat a balanced diet. Individuals suffering from undernutrition are weak, easily fatigued, and highly susceptible to infection. Iron, calcium, and vitamin A are commonly deficient nutrients, but essential amino acids are the ones most often deficient in the diet. Millions of people suffer from poor health and lowered resistance to disease because of protein deficiency. Children's physical and mental development are retarded when these essential building blocks of cells are not provided in the diet. Because their bodies cannot manufacture antibodies (which are proteins) and cells needed to fight infection, common childhood diseases, such as measles, whooping cough, and chickenpox, are often fatal in children suffering from protein malnutrition.

In young children, severe protein malnutrition results in the condition known as **kwashiorkor.** This Ashanti (West African) word means "first-second." It refers to the situation in which a first child is displaced from its mother's breast when a younger sibling is born. The older child is then given a diet of starchy cereal or cassava that is deficient in protein. Growth becomes stunted, muscles are wasted, and edema develops (as displayed by a swollen belly); the child becomes apathetic and anemic, with an impaired metabolism (Fig. 45-15). Without essential

FIGURE **45-15** **Protein deficiency.**
Millions of children suffer from kwashiorkor, a disease caused by severe protein deficiency. Note the characteristic swollen belly, which results from fluid imbalance.

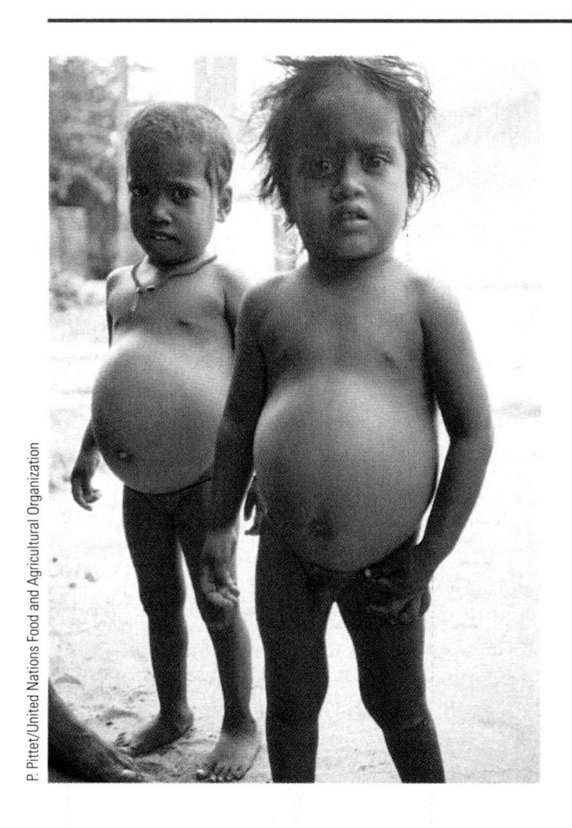

P. Pittet/United Nations Food and Agricultural Organization

amino acids, digestive enzymes cannot be manufactured, so what little protein is ingested cannot be digested.

Obesity is a serious nutritional problem

Obesity, the excess accumulation of body fat, is a serious form of malnutrition that has become a problem of epidemic proportions in affluent societies. The World Health Organization considers obesity among the top 19 global health problems. Approximately 30% of U.S. adults are obese and another 35% are overweight. An estimated 20% of U.S. children and adolescents are overweight. Obesity contributes to about 300,000 deaths annually in the United States and is the second leading preventable cause of death (second only to smoking). It is a major risk factor for heart disease, diabetes mellitus, osteoarthritis, and certain types of cancer, including breast and colon cancers. In the well-known Framingham study of more than 2000 men, those who were 20% overweight had a significantly higher mortality rate from all causes.

Body mass index (BMI), now used worldwide as a measure of body size, is an index of weight in relation to height. It is calculated by dividing the square of the weight (kg^2) by height (m). The English equivalent is 4.89 times the weight (lb) divided by the square of the height (ft^2). A person is considered obese if the BMI is 30 or more. Each of us appears to have a **set point,** or steady state, around which body weight is regulated. When BMI decreases below an individual's set point, energy-conserving mechanisms are activated and energy expenditure decreases.

Obesity can result from an increase in the size of fat cells or from an increase in the number of fat cells, or both. The number of fat cells in the adult is apparently determined mainly by the amount of fat stored during infancy and childhood. When we are overfed early in life, abnormally large numbers of fat cells are formed. Later in life, these fat cells may be fully stocked with excess lipids or may shrink in size, but they are always there. People with such increased numbers of fat cells are thought to be at greater risk for obesity than are those with normal numbers.

PROCESS OF SCIENCE

In their search for the causes of obesity, biologists are focusing on the genetic aspects of obesity and on signaling pathways. This research has its roots in the 1950s when a spontaneous recessive mutation in a strain of laboratory mice, produced mutant mice that were grossly obese. The increased adipose tissue in these mice is part of a syndrome that parallels morbid obesity in humans, a condition in which an individual's body weight is 100 lb (45.5 kg) or more above normal. In 1994, Jeffrey M. Friedman's research team at Rockefeller University isolated the (*ob* locus) that, when mutated, was responsible for the obese phenotype. Mice with the mutated allele apparently lacked some weight-regulating substance.

In 1995, three different research teams reported in *Science* that injections of the normal *ob* gene product, later named **leptin,** resulted in weight loss by obese and nonobese mice. When researchers injected leptin into grossly obese mice, their appetites decreased and their energy use increased, resulting in weight loss, mainly from loss of body fat. Friedman's team fed the same diet to obese mice that were injected with leptin and

Courtesy of John Sholtis/ Rockefeller University, New York City

ACTIVE FIGURE **45-16** **Mutant *Ob* mice before *(left)* and after *(right)* treatment with leptin.**

BiologyNow™ Learn more about **leptins and body weight** by clicking on this figure on your BiologyNow CD-ROM.

to a control group of obese mice that did not receive leptin. The mice in the treated group lost 50% more weight than the untreated animals (Fig. 45-16). Researchers have discovered that leptin is a hormone produced in adipose tissue. It signals centers in the brain about the status of energy stores. The brain then adjusts feeding behavior and energy metabolism. An *ob* gene has been identified in humans, indicating that this gene has been conserved during evolution. Unfortunately, most obese humans are resistant to the actions of leptin.

During the past few years, researchers have identified several gene mutations and signaling molecules that help regulate food intake and energy homeostasis. These signaling molecules form complex regulatory pathways involving genetic, neural, and endocrine mechanisms. Leptin and insulin are long-term regulators that stabilize the body's fat stores. During negative energy balance, the body's adipose tissue decreases, depressing both leptin and insulin secretion. When leptin levels and food intake are low, as during dieting or starvation, the hypothalamus increases secretion of the neurotransmitter **neuropeptide Y (NPY).** This neuropeptide increases appetite and slows metabolism, actions that help restore energy homeostasis.

Short-term appetite regulators signal us to eat or to stop eating. For example, when the stomach is empty it secretes the hormone *ghrelin,* which stimulates appetite. As the stomach fills, the gastrointestinal tract releases the hormone cholecystokinin and *PYY* (a peptide), which signal that we are full. The blood level of PYY rises after eating and remains high between meals. Investigators are testing this hormone for use as an appetite suppressant.

Although an estimated 40% to 70% of the factors involved in obesity are inherited, there is plasticity in the system that regulates body weight. High-calorie diets, overeating, and under-exercising lead to obesity, and psychosocial factors influence these behaviors. A sedentary lifestyle combined with the ready availability of high-energy foods contributes to the problem.

One researcher cleverly described the roots of the obesity problem as the combination of "computer chips and potato chips." For every 9.3 kcal of excess food taken into the body, about 1 g of fat is stored. (An excess of about 140 kcal, less than a typical candy bar, per day for a month results in a 1-lb weight gain.)

As researchers discover the genetic and biochemical mechanisms that regulate energy metabolism, they provide potential targets for the development of pharmacological treatments for obesity. The following are among the many targets for drug action that are the focus of research: (1) Reduce appetite so that food intake is decreased; (2) block absorption of fat; (3) uncouple metabolism from energy production so that food energy is dissipated as heat (see *Focus On: Electron Transport and Heat* in Chapter 7); and (4) block receptors for molecules that send signals leading to increased energy storage.

Review

- Write an equation to describe energy balance, and explain what happens when the equation is altered in either direction.
- What is body mass index (BMI)?
- What is the function of leptin?

BiologyNow™ Assess your understanding of **energy metabolism** by taking the pretest on your BiologyNow CD-ROM.

SUMMARY WITH KEY TERMS

1 Describe food processing, including ingestion, digestion, absorption, and egestion or elimination, and compare the digestive system of a cnidarian (such as *Hydra*) with that of an earthworm or vertebrate.

- **Nutrition** is the process of taking in and using food. Animals are **heterotrophs;** they must obtain their **nutrients** from the organic molecules manufactured by other organisms.
- **Feeding** is the selection, acquisition, and **ingestion** of food. The process of breaking down food mechanically and chemically is **digestion.** Nutrients pass through the lining of the digestive tract and into the blood by **absorption.** Food that is not digested and absorbed is discharged from the body by **egestion** or **elimination.**
- In cnidarians and flatworms, food is digested in the **gastrovascular cavity.** This cavity has only one opening that serves as both mouth and anus.
- In more complex invertebrates and in all vertebrates, the digestive tract is a complete tube with an opening at each end. Digestion takes place as food passes through the tube.
- Various parts of the digestive tract are specialized to perform specific functions. In vertebrates, food passes in sequence through the mouth, pharynx, esophagus, stomach, small intestine, large intestine, and anus.

2 Trace the pathway traveled by an ingested meal in the human digestive system, describing the structure and function of each organ involved.

- Mechanical and enzymatic digestion of carbohydrates begin in the mouth. Mammalian teeth include **incisors** for biting, **canines** for tearing food, and **premolars** and **molars** for crushing and grinding. Three pairs of **salivary glands** secrete saliva, a fluid containing the enzyme **salivary amylase** that digests starch.
- As food is swallowed, it is propelled through the **pharynx** and **esophagus.** A **bolus** of food is moved along through the digestive tract by **peristalsis,** waves of muscular contraction that push food along. The mixing and propulsive movements of the digestive tract are referred to as **motility.**
- In the **stomach,** food is mechanically digested by vigorous churning, and proteins are enzymatically digested by the action of **pepsin** in the gastric juice. **Rugae** are folds in the stomach wall that expand as the stomach fills with food. **Gastric glands** secrete hydrochloric acid and **pepsinogen,** the precursor of pepsin.
- After several hours, a soup of partly digested food, called **chyme,** leaves the stomach through the **pylorus** and enters the **small intestine** in spurts. Most enzymatic digestion takes place in the **duodenum,** which produces several digestive enzymes and also receives secretions from the liver and pancreas.
- The **liver** produces **bile,** which emulsifies fats.
- The **pancreas** releases enzymes that digest protein, lipid, and carbohydrate, as well as RNA and DNA. **Trypsin** and **chymotrypsin** digest polypeptides to dipeptides. **Pancreatic lipase** degrades fats, and **pancreatic amylase** digests complex carbohydrates.
- The **large intestine,** which consists of the **cecum, colon, rectum,** and **anus,** is responsible for eliminating undigested wastes. It also incubates bacteria that produce vitamin K and certain B vitamins.

3 Describe the step-by-step digestion of carbohydrate, protein, and lipid.

- Nutrients in chyme are enzymatically digested as they move through the digestive tract.
- Polysaccharides are digested to the disaccharide maltose by salivary and pancreatic amylases. Maltase in the small intestine splits maltose into glucose, the main product of carbohydrate digestion.
- Proteins are split by pepsin in the stomach and by proteolytic enzymes in the pancreatic juice. Dipeptidases then split the peptides and dipeptides produced. The end products of protein digestion are amino acids.
- Lipids are emulsified by bile salts and then hydrolyzed by pancreatic lipase.

4 Describe the structural adaptations that increase the surface area of the digestive tract.

- The surface area of the small intestine is greatly expanded by folds in its wall, by the intestinal **villi,** and by **microvilli.**

5 Compare lipid absorption with absorption of other nutrients.

- Nutrients are absorbed through the thin walls of the intestinal villi. The **hepatic portal vein** transports amino acids and glucose to the liver.
- Fatty acids and monacylglycerols enter epithelial cells in the intestinal lining where they are reassembled into triacylglycerols. They are packaged into **chylomicrons,** droplets that also contain cholesterol and phospholipids and are covered by a protein coat. The lymphatic system transports chylomicrons to the blood circulation.

6 Summarize the nutritional requirements for dietary carbohydrates, lipids, and proteins, and trace the fate of glucose, lipids, and amino acids after their absorption.

- For a balanced diet, humans and other animals require carbohydrates, lipids, proteins, vitamins, and minerals.
- Most carbohydrates are ingested in the form of polysaccharides—starch and cellulose. Polysaccharides are referred to as **complex carbohydrates. Fiber** is mainly a mixture of cellulose and other indigestible carbohydrates. Carbohydrates are used mainly as an energy source. Glucose concentration in the blood is carefully regulated. Excess glucose is stored as glycogen and can also be converted to fat.
- Lipids are used as an energy source, as components of cell membranes, and to synthesize steroid hormones and other lipid substances. Most lipids are ingested in the form of triacylglycerols. Fatty acids are converted to molecules of acetyl coenzyme A which enter the citric acid cycle. Excess fatty acids are converted to triacylglycerol and stored as fat.
- Lipids are transported as large molecular complexes called **lipoproteins. Low-density lipoproteins (LDLs)** deliver cholesterol to the cells. **High-density lipoproteins (HDLs)** collect excess cholesterol and transport it to the liver.
- Proteins serve as enzymes and are essential structural components of cells. The best distribution of **essential amino acids** is found in the complete proteins of animal foods. Excess amino acids are deaminated by liver cells. Amino groups are converted to urea and excreted in urine; the remaining keto acids are converted to carbohydrate and used as fuel or converted to lipid and stored in fat cells.

7 Describe the nutritional functions of vitamins, minerals, and phytochemicals.

- **Vitamins** are organic compounds required in small amounts for many biochemical processes. Many serve as components of coenzymes. **Fat-soluble vitamins** include vitamins A, D, E, and K. **Water-soluble vitamins** are the B and C vitamins.
- **Minerals** are inorganic nutrients ingested as salts dissolved in food and water. **Trace elements** are minerals required in amounts less than 100 mg per day.
- **Phytochemicals** are plant compounds that promote health. Some phytochemicals are **antioxidants** that destroy **oxidants,** which are free radicals and other reactive molecules that damage DNA by snatching electrons.

8 Contrast basal metabolic rate with total metabolic rate; write the basic energy equation for maintaining body weight, and describe the consequences of altering it in either direction.

- **Basal metabolic rate (BMR)** is the body's cost of metabolic living. **Total metabolic rate** is the BMR plus the energy used to carry on daily activities.
- When energy (kilocalories) input equals energy output, body weight remains constant. When energy output is greater than energy input, body weight decreases. When energy input exceeds energy output, body weight increases.

9 In general terms, describe the effects of malnutrition, including both undernutrition and overnutrition.

- Millions of people suffer from undernutrition. Essential amino acids are the nutrients most often deficient in the diet.
- **Obesity** is a serious nutritional problem in which an excess amount of fat accumulates in adipose tissues. A person gains weight by taking in more energy, in the form of kilocalories, than is expended in activity.

10 Summarize current hypotheses about the regulation of food intake and energy homeostasis. (Include the roles of leptin and neuropeptide Y.)

- Researchers are identifying signaling molecules that make up the complex pathways that regulate food intake and energy metabolism. The hormone **leptin** is produced by fat cells in proportion to body fat mass; it signals the brain about the status of energy stores. **Neuropeptide Y (NPY),** a neurotransmitter produced in the hypothalamus, increases appetite and slows metabolism when leptin levels and food intake are low.
- Short-term appetite regulators signal us to start and stop eating.

POST-TEST

1. The process of taking in and using food is (a) nutrition (b) chemical digestion (c) egestion (d) ingestion (e) absorption
2. Animals that feed mainly on producers are (a) herbivores (b) secondary consumers (c) animals with gastrovascular cavities (d) carnivores (e) secondary consumers and carnivores
3. Teeth adapted for crushing and grinding are (a) incisors (b) canines (c) premolars and molars (d) incisors and premolars (e) canines and molars
4. The layer of tissue that lines the lumen of the digestive tract is the (a) muscle layer (b) visceral peritoneum (c) parietal peritoneum (d) mucosa (e) submucosa
5. Which of the following are accessory digestive glands? (a) salivary glands (b) pancreas (c) liver (d) answers a, b, and c are correct (e) answers a and b only
6. Which of the following is the correct sequence? (a) pharynx ⟶ stomach ⟶ esophagus (b) stomach ⟶ large intestine ⟶ small intestine (c) esophagus ⟶ pharynx ⟶ stomach (d) cecum ⟶ colon ⟶ duodenum (e) pharynx ⟶ esophagus ⟶ stomach
7. Amylase is produced by the (a) liver and pancreas (b) stomach and pancreas (c) colon and salivary glands (d) liver and pancreas (e) pancreas and salivary glands
8. Pepsin is produced by the (a) liver (b) stomach (c) pancreas (d) duodenum (e) salivary glands
9. Which sequence most accurately describes the digestion of protein? (a) polypeptide ⟶ monoacylglycerol ⟶ amino acids (b) polypeptide ⟶ dipeptides ⟶ amino acids (c) protein ⟶ amylose ⟶ amino acid (d) protein ⟶ emulsified peptide ⟶ fatty acids and glycerol (e) protein ⟶ triacylglycerol ⟶ fatty acids and glycerol
10. The surface area of the small intestine is increased by (a) folds in its wall (b) villi (c) microvilli (d) a, b, and c are correct (e) a and b only

POST-TEST *(continued)*

11. Lipids are transported from the intestine to the cells of the body by (a) chylomicrons (b) neuropeptide Y (c) rugae (d) villi (e) leptin
12. Most vitamins are (a) inorganic compounds (b) components of coenzymes (c) used as fuel (d) electrolytes (e) required by herbivores and carnivores but not omnivores
13. When energy input is greater than energy output (a) weight loss occurs (b) weight remains stable (c) weight gain occurs (d) leptin prevents weight gain (e) the *ob* gene is activated
14. A hormone that stimulates gastric glands to secrete pepsinogen is (a) secretin (b) gastric inhibitory peptide (c) gastrin (d) cholecystokinin (CCK) (e) substance P
15. Neuropeptide Y (NPY) (a) increases appetite (b) increases metabolism (c) is encoded by the mutant *ob* gene (d) is a hormone produced by the pancreas (e) decreases bile production
16. Phytochemicals (a) are harmful to the body (b) cause high cholesterol levels (c) inhibit gastrin (d) are mainly used as energy sources (e) contain antioxidants
17. Label the diagram. Use Figure 45-4 to check your answers.

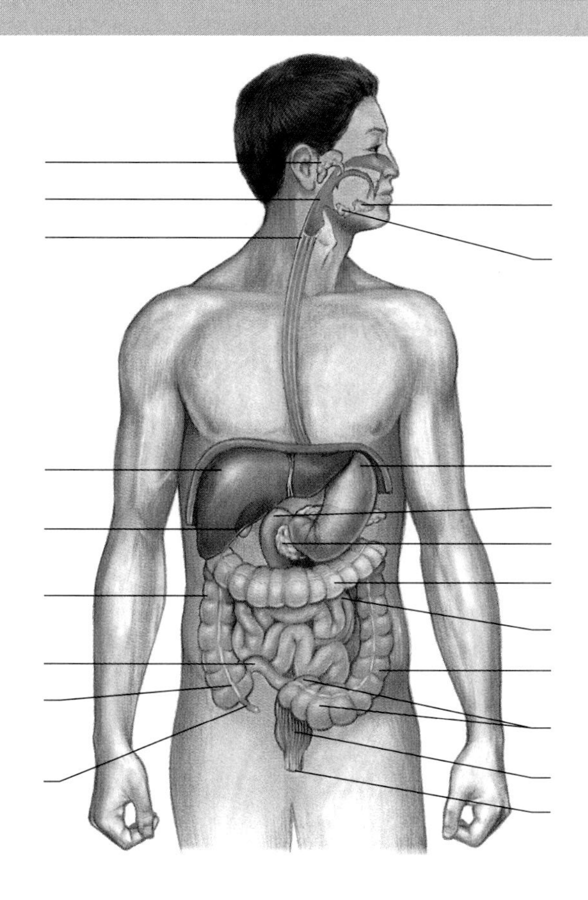

CRITICAL THINKING

1. If you were presented with an unfamiliar animal and asked to determine its nutritional lifestyle, how would you do so?
2. Design an experiment to test the hypothesis that the B vitamin pyridoxine is an essential nutrient in mice.
3. Why are proteolytic enzymes produced in an inactive form?
4. Investigators are unraveling the biochemical pathways that help regulate body weight. How might their work lead to a cure for obesity?

- Visit our Web site at **http://biology.brookscole.com/solomon7** for links to chapter-related resources on the World Wide Web. Additional online materials relating to this chapter can also be found on our Web site.

BIOLOGY NOW RESOURCES

Active Figures

45-4: Human digestive system
45-16: Leptins and body weight

Preparing for an exam? Take a diagnostic test on your BiologyNow CD-ROM.

Post-Test Answers

1. a	2. a	3. c	4. d
5. d	6. e	7. e	8. b
9. b	10. d	11. a	12. b
13. c	14. c	15. a	16. e

46

Osmoregulation and Disposal of Metabolic Wastes

McMurray Photography

Most terrestrial animals inhabit areas near water sources. Zebra, wildebeest, and a variety of birds gather to drink at a lake in Ngorongoro Crater in Tanzania.

CHAPTER OUTLINE

Water, the most abundant molecule both in the cell and on Earth, shapes life and the distribution of organisms on our planet. Water is the medium in which most metabolic reactions take place, and the osmotic and ionic composition of an animal's body fluids must be maintained within homeostatic limits. Natural selection has resulted in the evolution of a variety of homeostatic mechanisms that regulate the volume and composition of fluids in the internal environment. In this chapter we discuss some of these mechanisms.

Many small animals live in the ocean and get their food and oxygen directly from the seawater that surrounds them; they release waste products into the surrounding seawater. In larger, more complex animals and in most terrestrial animals, the extracellular fluid serves as an internal sea. **Extracellular fluid** includes the fluid outside of the cells, such as interstitial fluid (fluid between cells) and blood (or hemolymph). In vertebrates, blood plasma, which is mainly water, transports nutrients, gases, waste products, and other materials throughout the body. Interstitial fluid forms from the blood plasma and bathes all the cells. Excess water evaporates from the body surface or is excreted by specialized structures.

Terrestrial animals have a continuous need to conserve water. Its loss from the body must be carefully regulated, and water lost must be replaced. Water is taken in with food and drink and is also produced in metabolic reactions. Like the animals in the photograph, most animals need a dependable source of water with which to replenish their body fluids, and so they often live near water sources.

Two processes that maintain homeostasis of fluids in animals are osmoregulation and excretion of metabolic wastes. **Osmoregulation** is the active regulation of osmotic pressure of body fluids to keep them from becoming too dilute or too concentrated. **Excretion** is the process of ridding the body of metabolic wastes. Efficient **excretory systems** have evolved that function in osmoregulation and in disposal of metabolic wastes, excess water and ions, and harmful substances. As we discuss, hormones are important signaling molecules in these regulatory processes.

Excretory systems maintain homeostasis by selectively adjusting the concentrations of salts and other substances in blood and other body fluids. Typically, an excretory system collects fluid, generally from the blood or interstitial fluid. It then adjusts the composition of this fluid by selectively returning needed substances to the body fluid. Finally, the body releases the adjusted excretory product containing excess or potentially toxic substances. ■

METABOLIC WASTE PRODUCTS

Learning Objective

1 Define *osmoregulation* and *excretion* and contrast the advantages and disadvantages of excreting ammonia, uric acid, or urea.

Metabolic wastes must be excreted so that they do not accumulate and reach concentrations that would disrupt homeostasis. The principal metabolic waste products in most animals are water, carbon dioxide, and **nitrogenous wastes,** those that contain nitrogen. Carbon dioxide is excreted mainly by respiratory structures (see Chapter 44). Water is also lost from respiratory surfaces in terrestrial animals. Excretory organs, such as kidneys, remove and excrete most of the water and nitrogenous wastes.

Nitrogenous wastes include ammonia, uric acid, and urea. Recall that amino acids and nucleic acids contain nitrogen. During the metabolism of amino acids, the nitrogen-containing amino group is removed (in a process known as deamination) and converted to **ammonia** (Fig. 46-1). However, ammonia is highly toxic. Some aquatic animals excrete it into the surrounding water before it can build up to toxic concentrations in their tissues. A few terrestrial animals, including some terrestrial snails and wood lice, vent it directly into the air. But many animals, humans included, convert ammonia to some less toxic nitrogenous waste such as uric acid or urea.

Uric acid is produced both from ammonia and by the breakdown of nucleotides from nucleic acids. Uric acid is insoluble in water and forms crystals that are excreted as a crystalline paste with little fluid loss. This is an important water-conserving adaptation in many terrestrial animals, including insects, certain reptiles, and birds. Also, because uric acid is not toxic and can be safely stored, its excretion is an adaptive advantage for species whose young begin their development enclosed in eggs.

Urea is the principal nitrogenous waste product of amphibians and mammals. It is produced in the liver from ammonia. Urea is synthesized from ammonia and carbon dioxide by a sequence of reactions known as the **urea cycle.** Like the formation of uric acid, these reactions require specific enzymes and the input of energy by the cells. Urea has the advantage of being far less toxic than ammonia and can accumulate in higher concentrations without causing tissue damage; thus, it can be excreted in more concentrated form. Because urea is highly soluble, however, it is dissolved in water, and more water is needed to excrete urea than to excrete uric acid.

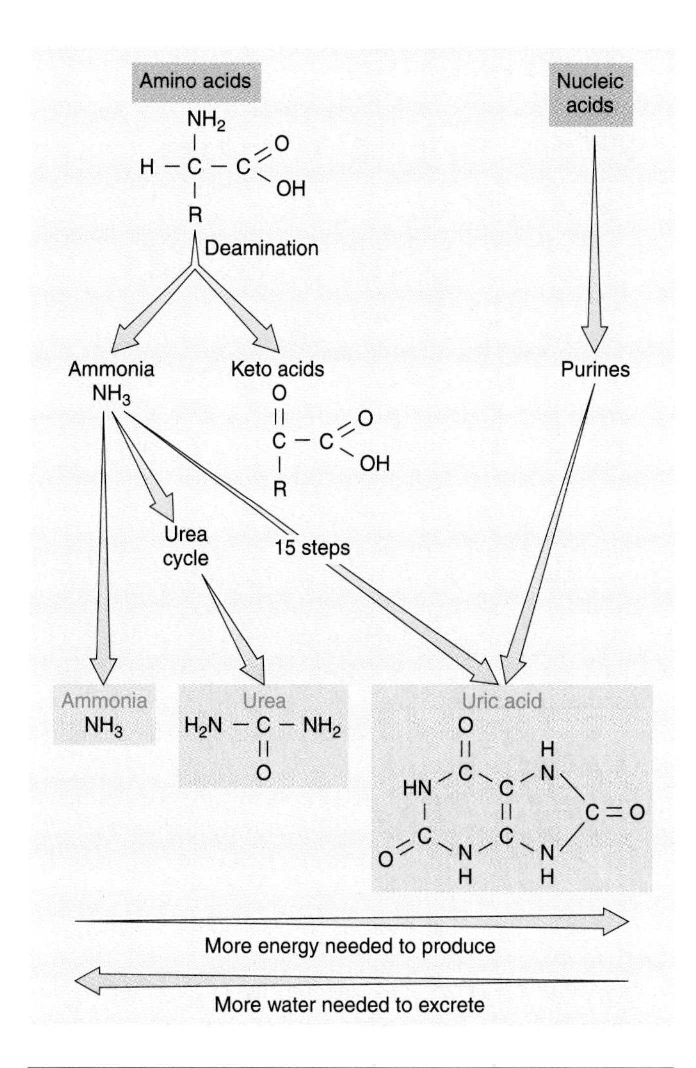

FIGURE **46-1** **Formation of nitrogenous wastes.**

Deamination of amino acids and metabolism of nucleic acids produce nitrogenous wastes. Ammonia is the first metabolic product of deamination. Many aquatic animals excrete ammonia, but terrestrial animals convert it to the less toxic urea and uric acid. Most amphibians and mammals convert ammonia to urea via the urea cycle. Insects, many reptiles, and birds convert ammonia to uric acid. Energy is required to convert ammonia to urea and uric acid, but less water is required to excrete these wastes.

Review

- What are the principal types of nitrogenous wastes? Give examples of animals that excrete each type.
- What are the advantages of excreting nitrogenous wastes in the form of uric acid?
- What is an advantage of excreting nitrogenous wastes in the form of urea? What is a disadvantage?

BiologyNow™ Assess your understanding of **metabolic waste products** by taking the pretest on your BiologyNow CD-ROM.

OSMOREGULATION AND METABOLIC WASTE DISPOSAL IN INVERTEBRATES

Learning Objective

2 Compare osmoconformers and osmoregulators, and describe protonephridia, metanephridia, and Malpighian tubules.

The ocean is a stable environment, and its salt concentration does not vary much. The body fluids of most marine invertebrates are in osmotic equilibrium with the surrounding seawater. These animals are known as **osmoconformers,** because the concentration of their body fluids varies along with changes in the seawater. Even so, many osmoconforming marine animals regulate some ions in their body fluids.

Coastal habitats, such as estuaries that contain brackish water, are much less stable environments than is the open ocean. Salt concentrations change frequently with shifting tides. Many invertebrates (as well as vertebrates) that inhabit these environments are **osmoregulators,** which maintain an optimal salt concentration in their tissues regardless of changes in the salt concentration of their surroundings.

In a coastal environment where fresh water enters the ocean, the water may have a lower salt concentration than the body fluids of the animal. Water osmotically moves into the body, and salt diffuses out. An animal adapted to this environment has excretory structures that remove the excess water. Certain polychaete worms and the blue crab are among the animals that can be osmoconformers or osmoregulators depending on environmental conditions.

Dehydration is a constant threat to terrestrial animals. Because their fluid concentration is higher than that of the air around them, they tend to lose water by evaporation from both the body surface and respiratory surfaces and may also lose water as wastes are excreted. As animals moved onto the land, natural selection favored the evolution of structures and processes that conserve water.

Nephridial organs are specialized for osmoregulation and/or excretion

Marine sponges and cnidarians need no specialized excretory structures, and they expend little or no energy to excrete wastes. Their wastes pass by diffusion from their cells to the external environment and are washed away by water currents. When water stagnates and currents do not wash away wastes, aquatic environments, such as coral reefs, are damaged by their accumulation.

Nephridial organs, or nephridia, are osmoregulatory structures that evolved in many invertebrates, including flatworms, nemerteans, rotifers, annelids, mollusks, and lancelets. Each nephridial organ consists of simple or branching tubes that typically open to the outside of the body through excretory pores, called *nephridiopores.* Two types of nephridial organs are protonephridia and metanephridia.

In flatworms and nemerteans, metabolic wastes pass through the body surface by diffusion, but these animals also have **protonephridia** (Fig. 46-2). These nephridial organs are composed of tubules with no internal openings. Their enlarged blind ends consist of **flame cells** with brushes of cilia, so named because their constant motion reminded early biologists of flickering flames. The flame cells lie in the interstitial fluid that bathes the body cells. Fluid enters the flame cells, and the beating of the cilia propels the fluid through the tubules. Excess fluid leaves the body through nephridiopores.

Most annelids and mollusks have more complex nephridial organs called **metanephridia** (Fig. 46-3). Each segment of an earthworm has a pair of metanephridia. A metanephridium is a tubule open at both ends. The inner end opens into the coelom as a ciliated funnel, and the outer end opens to the outside through a nephridiopore. Around each tubule is a network of capillaries. Fluid from the coelom passes into the tubule, bringing with it whatever it contains—glucose, salts, or wastes. As fluid moves through the tubule, needed materials (such as water and glucose) are removed from the fluid by the tubule and are reabsorbed by the capillaries, leaving the wastes behind. In this way, urine is produced that contains concentrated wastes.

Malpighian tubules conserve water

The excretory system of insects and spiders consists of several hundred **Malpighian tubules** (Fig. 46-4). Malpighian tubules are

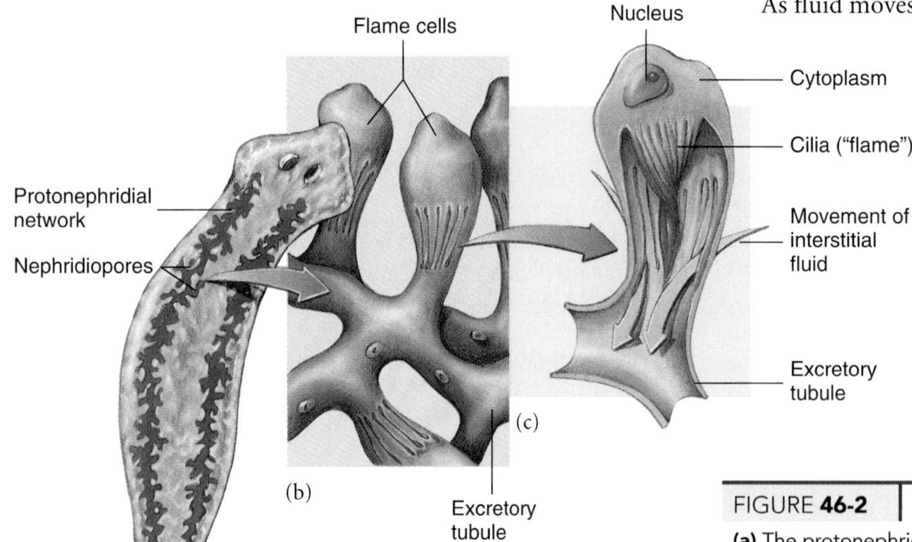

FIGURE **46-2** **Protonephridia of a flatworm.**

(a) The protonephridia of a planarian, which function mainly in osmoregulation, form a system of branching tubules. **(b)** Interstitial fluid enters flame cells and passes through a series of tubules. Excess fluid leaves the body through nephridiopores. **(c)** A single flame cell.

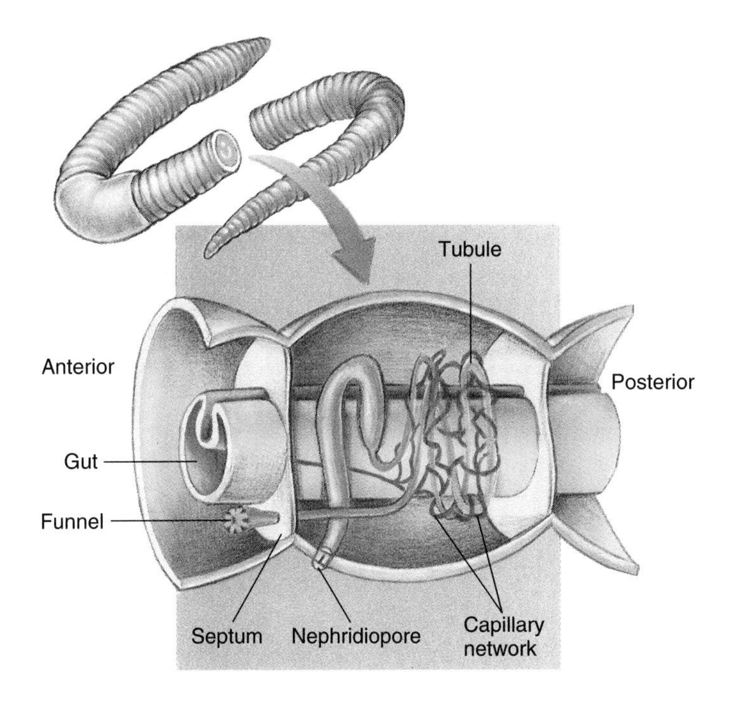

FIGURE **46-3** | **Metanephridium of an earthworm.**
Each metanephridium consists of a ciliated funnel opening into the coelom, a coiled tubule, and a nephridiopore opening to the outside. This 3-D internal view shows parts of three segments.

slender extensions of the gut wall. Their blind ends lie in the hemocoel (blood cavity) and are bathed in hemolymph. Cells of the tubule wall actively transport uric acid, potassium ions, and

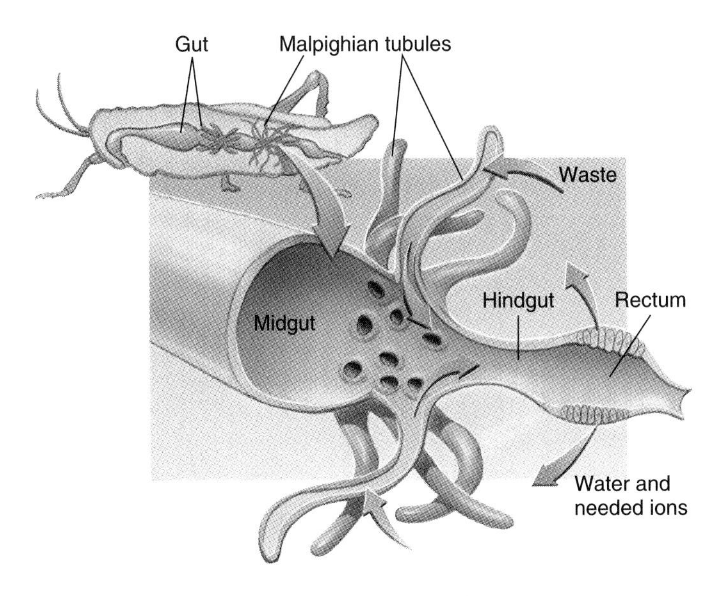

FIGURE **46-4** | **Malpighian tubules of an insect.**
The slender Malpighian tubules have blind ends that extend into the hemocoel. Their cells transfer uric acid and some ions from the hemolymph to the cavity of the tubule. Water follows by diffusion. The wastes are discharged into the gut. The epithelium lining the rectum (part of the hindgut) actively reabsorbs most of the water and needed ions.

some other substances from the hemolymph into the tubule lumen. Other solutes and water follow by diffusion. The Malpighian tubules empty into the gut. Water, some salts, and other solutes are reabsorbed into the hemolymph by a specialized epithelium in the rectum.

Uric acid, the major waste product, is excreted as a semidry paste with a minimum of water. Because Malpighian tubules effectively conserve body fluids, they have contributed to the success of insects in terrestrial environments.

Review

- How are nephridial organs and Malpighian tubules alike? How are they different?

BiologyNow™ Assess your understanding of **osmoregulation and metabolic waste disposal in invertebrates** by taking the pretest on your BiologyNow CD-ROM.

OSMOREGULATION AND METABOLIC WASTE DISPOSAL IN VERTEBRATES

Learning Objectives

3 Compare adaptations to the challenges of osmoregulation in the following animals: freshwater fishes, marine bony fishes, sharks, marine mammals, and terrestrial vertebrates.

4 Relate the function of the vertebrate kidney to the success of vertebrates in a wide variety of habitats.

Vertebrates live successfully in a wide range of habitats—in fresh water, the ocean, tidal regions, and on land, even in extreme environments such as deserts. In response to the requirements of these diverse environments, adaptations have evolved for regulating salt and water content and for excreting wastes. The main osmoregulatory and excretory organ in most vertebrates is the **kidney.** The kidneys excrete most nitrogenous wastes and help maintain fluid balance by adjusting the salt and water content of the urine. The skin, lungs or gills, and digestive system also help maintain fluid balance and dispose of metabolic wastes.

Freshwater vertebrates must rid themselves of excess water

As fishes began to move into freshwater habitats about 460 million years ago (mya), there was strong selection for the evolution of adaptations for effective osmoregulation. Evolution of body fluids more dilute than seawater is one of their chief adaptations. However, the salt concentration of their body fluids is still higher than that of the fresh water that surrounds them, and so they are hypertonic to their watery environment. As a result, water passes into them osmotically, and they are in constant danger of becoming waterlogged.

Freshwater fishes are covered by scales and a mucous secretion that retard the passage of water into the body. However, water enters through the gills. Some water leaves the body through the gills, and the kidneys of these fishes are adapted to excrete large amounts of dilute urine (Fig. 46-5a). The kidneys have large *glomeruli* (capillary clusters that filter the blood and produce urine).

Water entry is only part of the challenge of osmoregulation in freshwater fishes. These animals also tend to lose salts by diffusion through the gills into the surrounding water. To compensate, special gill cells have evolved that actively transport salts (mainly sodium chloride) from the water into the body.

The gills excrete most of the nitrogenous wastes, but the kidneys are also important excretory organs. Ammonia is the main nitrogenous waste of freshwater fishes. About 10% of nitrogenous wastes are excreted as urea.

Most amphibians are at least semiaquatic, and their mechanisms of osmoregulation are similar to those of freshwater fishes. They, too, produce large amounts of dilute urine. In its urine and through its skin, a frog can lose an amount of water equivalent to one third of its body weight in one day. Active transport of salt inward by special cells in the skin compensates for loss of salt through skin and in urine.

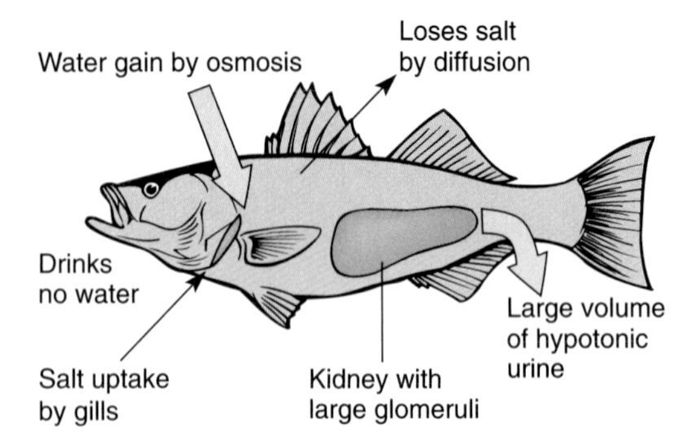

(a) Freshwater fish

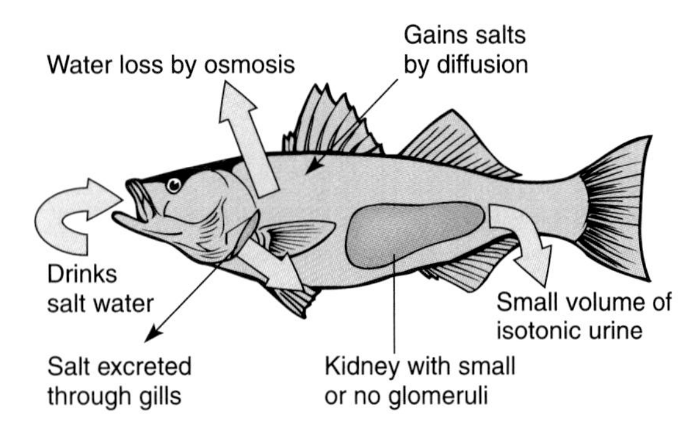

(b) Marine bony fish

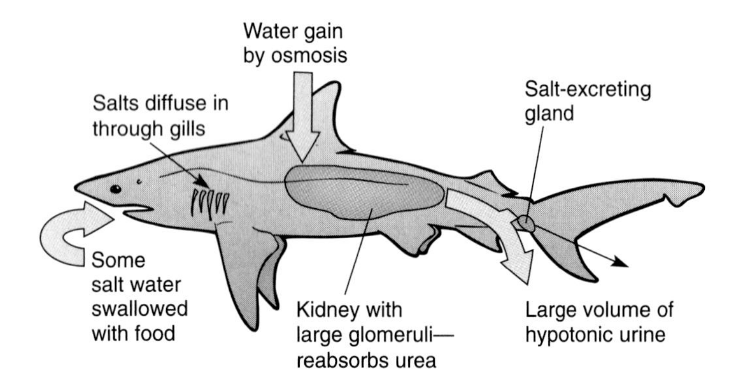

(c) Cartilaginous fish (shark)

FIGURE **46-5** **Osmoregulation in fishes.**

(a) Freshwater fishes live in a hypotonic medium, so water continuously enters the body by osmosis, and salts diffuse out. These fishes excrete large quantities of dilute urine and actively transport salts in through the gills. **(b)** Marine fishes live in a hypertonic medium and therefore lose water by osmosis. They gain salts from the seawater they drink and by diffusion. To compensate, the fish drinks water, excretes the salt, and produces a small volume of urine. **(c)** In contrast, the shark kidney reabsorbs urea in high enough concentration that its tissues become hypertonic to the surrounding medium. As a result, water enters the shark by osmosis. A large quantity of dilute urine is excreted.

Marine vertebrates must replace lost fluid

Freshwater fishes adapted very successfully to their aquatic habitats. Thus, when some freshwater fishes returned to the sea about 200 mya, their blood and body fluids were less salty than (hypotonic to) their surroundings. They tended to lose water osmotically and to take in salt. To compensate for fluid loss, many marine bony fishes drink seawater (Fig. 46-5b). They retain the water and excrete salt by the action of specialized cells in their gills. Very little urine is excreted by the kidneys; the kidneys have only small (or no) glomeruli.

Marine cartilaginous fishes (sharks and rays) have different osmoregulatory adaptations that allow them to tolerate the salt concentrations of their environment. These animals accumulate and tolerate urea (Fig. 46-5c). Their tissues are adapted to function at concentrations of urea that would be toxic to most other animals. The high urea concentration makes the body fluids slightly hypertonic to seawater, resulting in a net inflow of water. Their well-developed kidneys excrete a large volume of urine. Excess salt is excreted by the kidneys and, in many species, by a rectal salt gland.

The heads of certain reptiles and marine birds have salt glands that excrete salt that enters the body with ingested seawater or in salty food. Salt glands are usually inactive; they function only in response to osmotic stress.

Whales, dolphins, and other marine mammals ingest seawater along with their food. Their kidneys produce a concentrated urine, much saltier than seawater. This is an important physiological adaptation, especially for marine carnivores. The high-protein diet of these animals results in the production of large amounts of urea, which must be excreted in urine without losing much water.

Terrestrial vertebrates must conserve water

Adult amphibians generally inhabit moist environments. They excrete urea, and reabsorb some water from the urinary bladder. Amniotes (reptiles, birds, and mammals) have more effective adaptations for life on land. Their skin minimizes water loss by evaporation, and many amniotes excrete uric acid, which requires very little water.

Because they are endothermic (maintain a constant body temperature) and have a high rate of metabolism, birds and mammals produce a relatively large volume of nitrogenous wastes. They have numerous adaptations, including very efficient kidneys, for conserving water. An extreme example is the desert-dwelling kangaroo rat, which obtains most of its water from its own metabolism (recall from Chapter 7 that aerobic respiration produces water). Its kidneys are so efficient that it loses little fluid as urine.

Birds conserve water by excreting nitrogen as uric acid and by efficiently reabsorbing water from the cloaca and intestine. Mammals excrete urea. Their kidneys produce a very concentrated (hypertonic) urine.

In terrestrial vertebrates, the lungs, skin, and digestive system are important in osmoregulation and waste disposal (Fig. 46-6). Most carbon dioxide is excreted by the lungs. In birds and mammals, some water is lost from the body as water vapor in exhaled air. Although primarily concerned with the regulation of body temperature, the sweat glands of humans and some other mammals excrete 5% to 10% of all metabolic wastes.

The liver produces both urea and uric acid, which are transported by the blood to the kidneys. Most of the bile pigments produced by the breakdown of red blood cells are normally excreted by the liver into the intestine. From the intestine they pass out of the body with the feces.

Review

- What type of osmoregulatory challenge is faced by marine fishes? By freshwater fishes? What adaptations have evolved that meet these challenges?
- What adaptations have evolved in birds and mammals to meet the challenges of osmoregulation and metabolic waste disposal?

BiologyNow™ Assess your understanding of **osmoregulation and metabolic waste disposal in vertebrates** by taking the pretest on your BiologyNow CD-ROM.

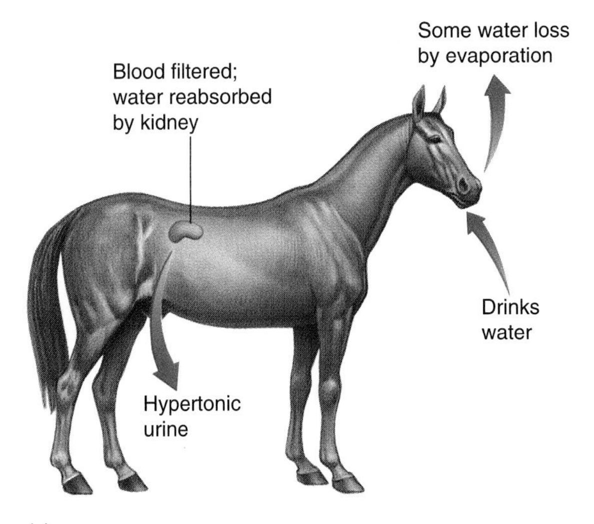

(a)

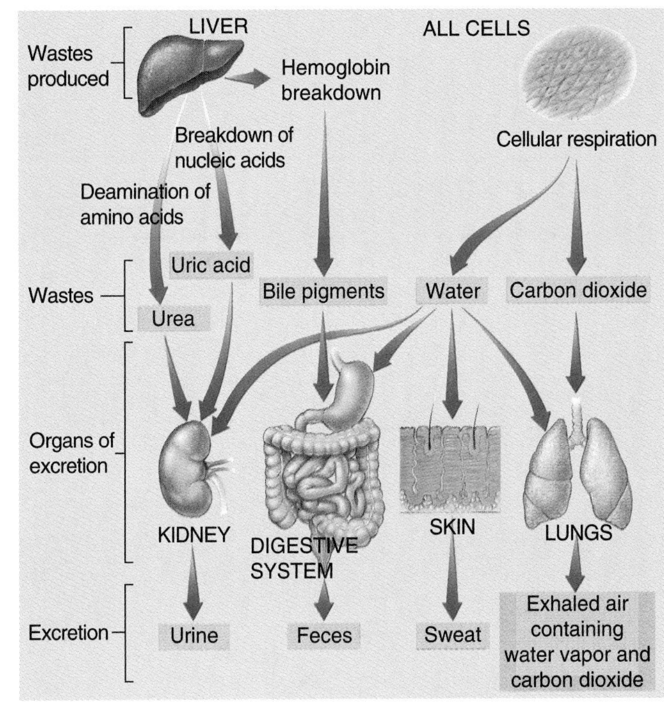

(b)

FIGURE **46-6** **Excretory organs in terrestrial vertebrates.**
(a) The vertebrate kidney conserves water by reabsorbing it. **(b)** Disposal of metabolic wastes in humans and other terrestrial mammals. To conserve water, a small amount of hypertonic urine is produced. Nitrogenous wastes are produced by the liver and transported to the kidneys. All cells produce carbon dioxide and some water during cellular respiration.

THE URINARY SYSTEM

Learning Objectives

5 Describe (or label on a diagram) the organs of the mammalian urinary system, and give the functions of each.

6 Describe (or label on a diagram) the structures of a nephron (including associated blood vessels), and give the functions of each structure.

7 Trace a drop of filtrate from Bowman's capsule to its release from the body as urine.

8 Describe the hormonal regulation of fluid balance by antidiuretic hormone (ADH), aldosterone, angiotensin II, and atrial natriuretic peptide (ANP).

The mammalian **urinary system** consists of the kidneys, the urinary bladder, and associated ducts. The overall structure of the human urinary system is shown in Figure 46-7. Located just below the diaphragm in the "small of the back," the kidneys look like a pair of giant, dark-red lima beans, each about the size of a fist. Each kidney is covered by a connective tissue capsule (Fig. 46-8). The outer portion of the kidney is called the **renal cortex;** the inner portion is the **renal medulla.** The renal medulla contains 8 to 10 cone-shaped structures called **renal pyramids.** The tip of each pyramid is a **renal papilla.** Each papilla has several pores, the openings of **collecting ducts.**

As urine is produced, it flows from collecting ducts through a renal papilla and into the **renal pelvis,** a funnel-shaped chamber. Urine then flows into one of the paired **ureters,** ducts that

FIGURE **46-7** **The human urinary system.**

The kidneys produce urine, which is conveyed by the ureters to the urinary bladder for temporary storage. The urethra then conducts urine from the bladder to the outside of the body.

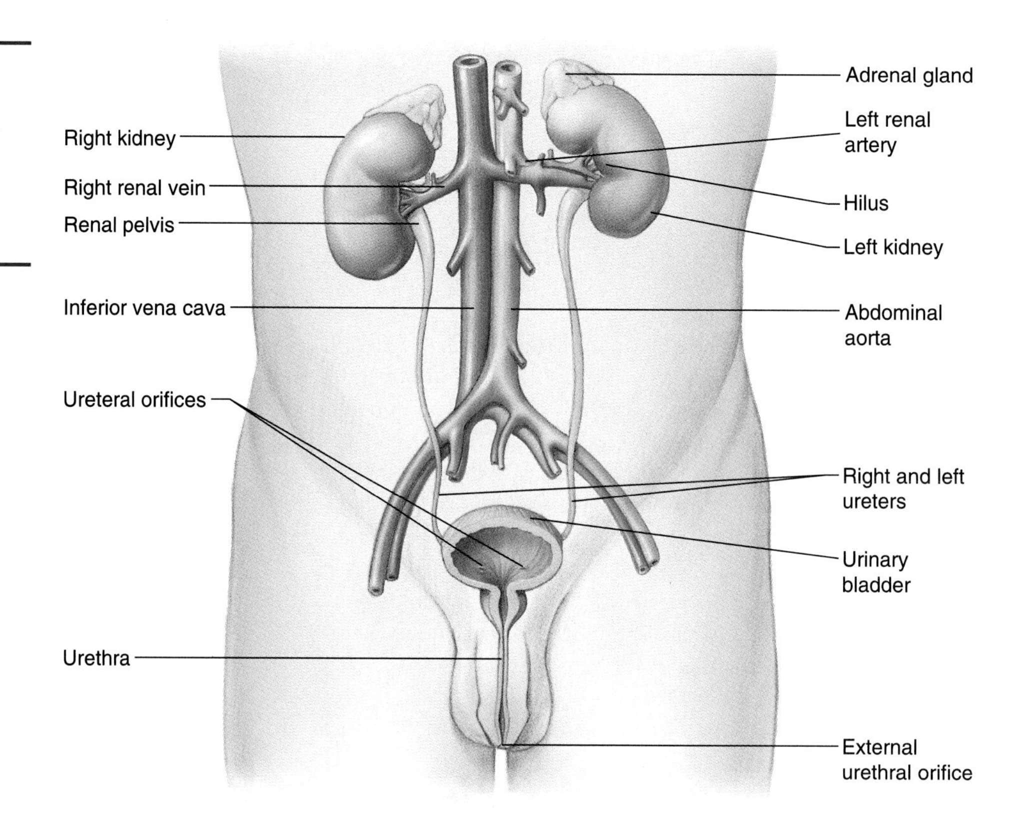

ACTIVE FIGURE **46-8** **Structure of the kidney.**

(a) The kidney is covered by a fibrous capsule. The outer region of the kidney is the cortex; the inner region is the medulla. When urine is produced, it flows into the renal pelvis and leaves the kidney through the ureter. **(b)** Longitudinal section showing the location of the two main types of nephrons: the juxtamedullary nephron and the cortical nephron.

BiologyNow™ Explore **kidney structure** by clicking on this figure on your BiologyNow CD-ROM.

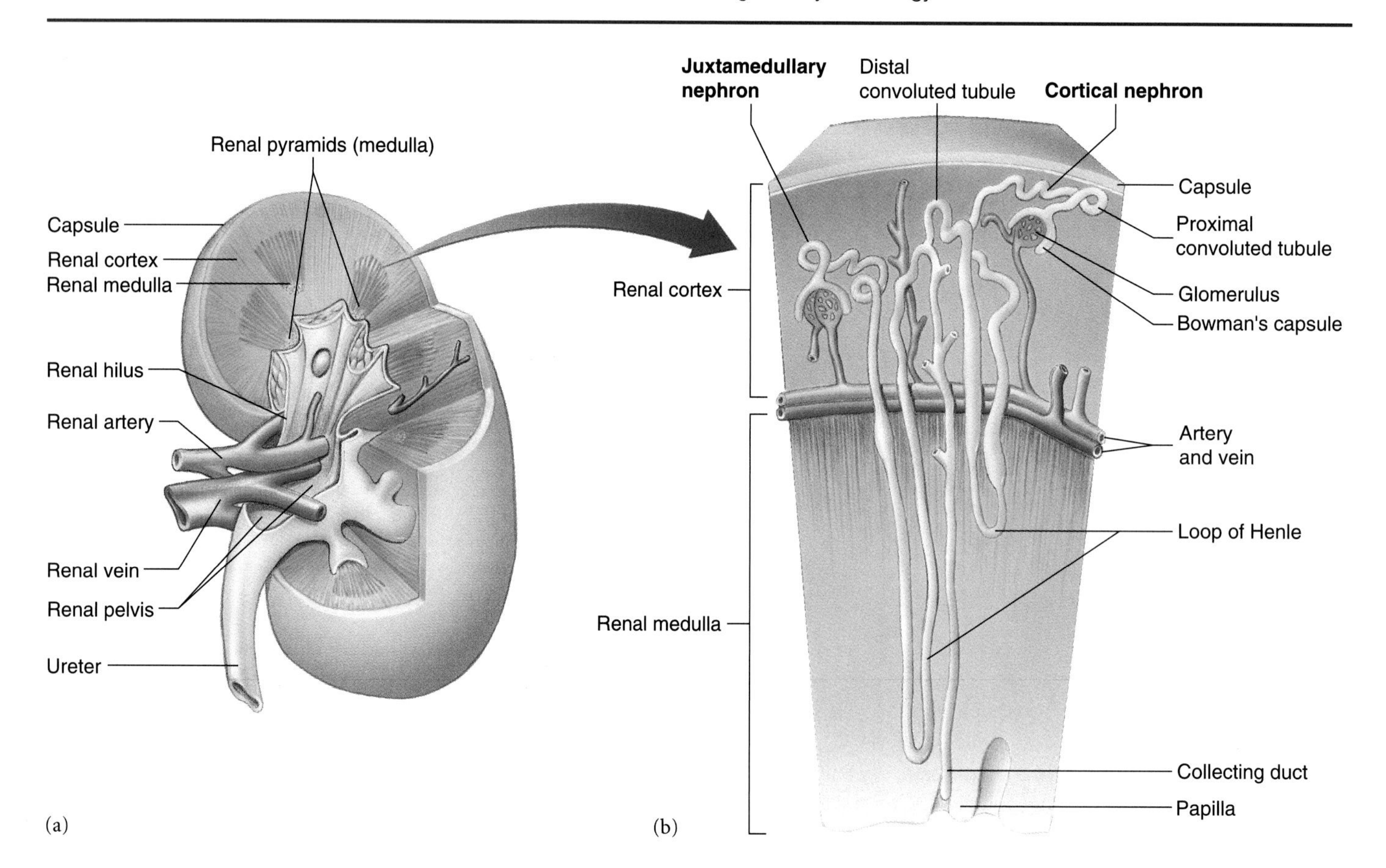

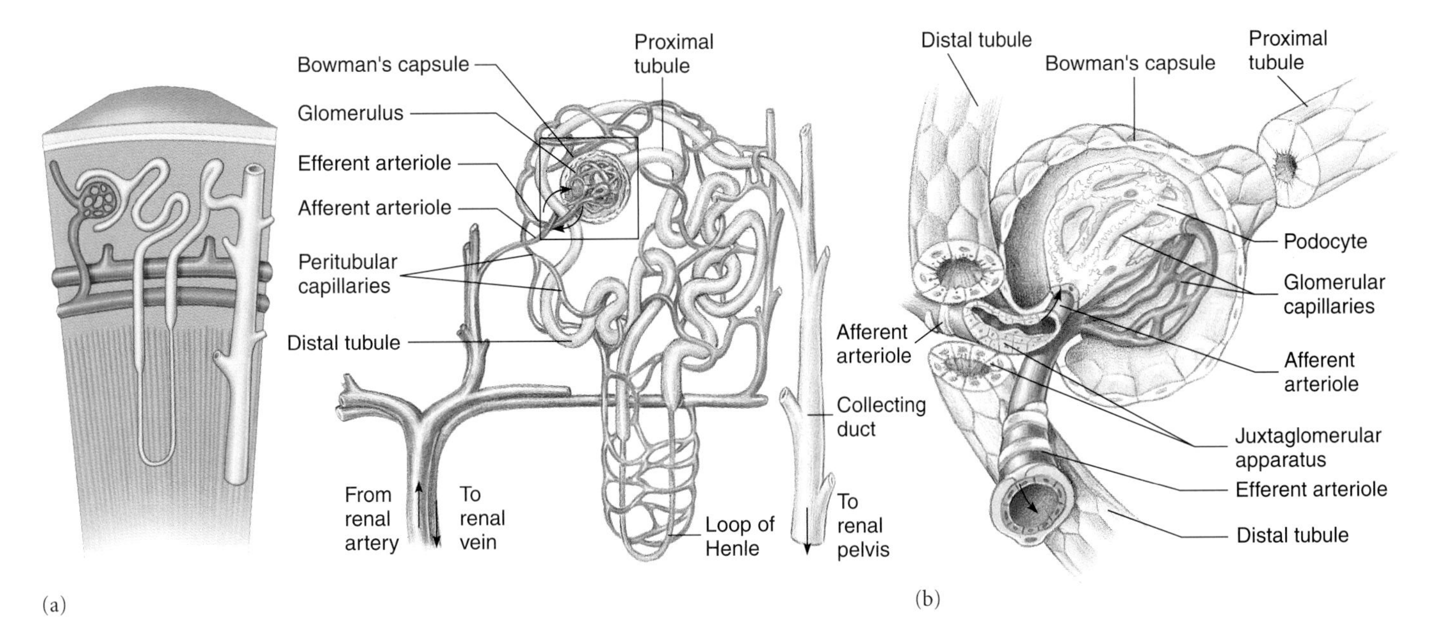

connect the kidney with the **urinary bladder.** The urinary bladder is a remarkable organ capable of holding (with practice) up to 800 mL (about a pint and a half) of urine. Emptying the bladder changes it from the size of a small melon to that of a pecan. This feat is made possible by the smooth muscle and specialized epithelium of the bladder wall, which is capable of great shrinkage and stretching.

During **urination,** urine is released from the bladder and flows through the **urethra,** a duct leading to the outside of the body. In the male, the urethra is lengthy and passes through the penis. Semen, as well as urine, passes through the male urethra. In the female, the urethra is short and transports only urine. Its opening to the outside is just above the opening of the vagina. The length of the male urethra discourages bacterial invasions of the bladder. This length difference helps explain why bladder infections are more common in females than in males. In summary, urine flows through the following structures:

Kidney (through renal pelvis) ⟶ ureter ⟶ urinary bladder ⟶ urethra

The nephron is the functional unit of the kidney

The kidneys produce the enzyme *renin* which helps regulate fluid balance and blood pressure (discussed later in this chapter). The kidneys also produce at least two hormones: *erythropoietin,* which stimulates red blood cell production, and *1,25-dihydroxyvitamin* D_3, which stimulates calcium absorption by the intestine. The principal function of the kidneys is to help maintain homeostasis by regulating fluid balance and excreting metabolic wastes.

Each kidney has more than one million functional units called **nephrons.** A nephron consists of a cuplike **Bowman's capsule** connected to a long, partially coiled **renal tubule** (Fig. 46-9). Positioned within Bowman's capsule is a cluster of capillaries called a **glomerulus.**

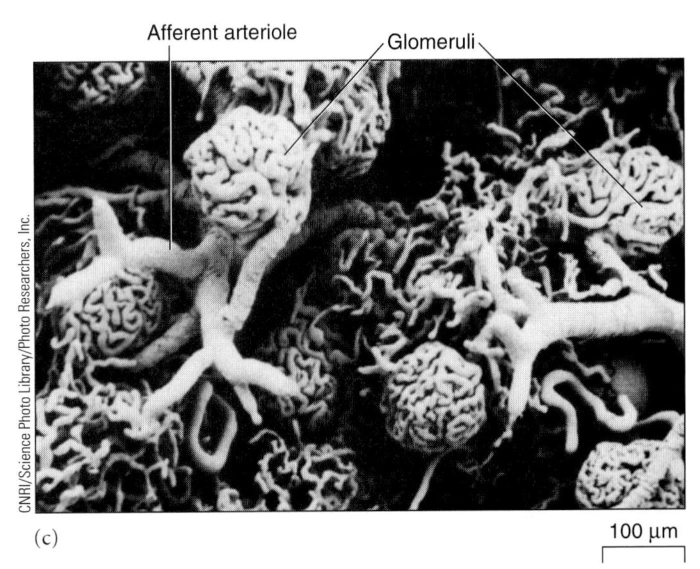

FIGURE **46-9** **Nephron structure.**

(a) The location and the basic structure of a nephron. Urine forms by filtration of the blood in the glomeruli and by adjustment of the filtrate as it passes through the tubules that drain the glomeruli. **(b)** Detailed view of Bowman's capsule. Note that the distal tubule is adjacent to the afferent and efferent arterioles. The juxtaglomerular apparatus (discussed later in the chapter) is a small group of cells located in the walls of the tubule and arterioles. **(c)** Low-power SEM of a portion of the kidney cortex. The tissue was treated to remove some structures and to show glomeruli and associated blood vessels.

Three main regions of the renal tubule are the **proximal convoluted tubule,** which conducts the filtrate from Bowman's capsule; the **loop of Henle,** an elongated, hairpin-shaped portion; and the **distal convoluted tubule,** which conducts the filtrate to a collecting duct. Thus, filtrate passes through the following structures:

Bowman's capsule ⟶ proximal convoluted tubule ⟶ loop of Henle ⟶ distal convoluted tubule ⟶ collecting duct

The human kidney has two types of nephrons: the more numerous (85%) cortical nephrons and the more internal juxtamedullary nephrons (see Fig. 46-8b). **Cortical nephrons** have relatively small glomeruli and are located almost entirely within the cortex or outer medulla. **Juxtamedullary nephrons** have large glomeruli, and their very long loops of Henle extend deep into the medulla. The loop of Henle consists of a *descending limb* that receives filtrate from the proximal convoluted tubule and an *ascending limb,* through which the filtrate passes on its way to the distal convoluted tubule. The juxtamedullary nephrons contribute to the ability of the mammalian kidney to concentrate urine. Excretion of urine that is hypertonic to body fluids is an important mechanism for conserving water.

Blood is delivered to the kidney by the **renal artery.** Small branches of the renal artery give rise to **afferent arterioles.** (Afferent means "to carry toward.") An afferent arteriole conducts blood into the capillaries that make up each glomerulus. As blood flows through the glomerulus, some of the plasma is forced into Bowman's capsule.

You may recall that in a typical circulatory route, capillaries deliver blood into veins. Circulation in the kidneys is an exception, because blood flowing from the glomerular capillaries next passes into an **efferent arteriole.** Each efferent arteriole conducts blood *away* from a glomerulus. The efferent arteriole delivers blood to a second capillary network, the **peritubular capillaries** surrounding the renal tubule.

As blood flows through the first set of capillaries, those of the glomerulus, it is filtered. The peritubular capillaries receive materials returned to the blood by the renal tubule. Blood from the peritubular capillaries enters small veins that eventually lead to the **renal vein.** In summary, blood circulates through the kidney in the following sequence:

Renal artery ⟶ afferent arterioles ⟶ capillaries of glomerulus ⟶ efferent arterioles ⟶ peritubular capillaries ⟶ small veins ⟶ renal vein

Urine is produced by filtration, reabsorption, and secretion

Urine, the watery discharge of the urinary system, is produced by a combination of three processes: filtration, reabsorption, and tubular secretion (Fig. 46-10).

Filtration is not selective with regard to ions and small molecules

Blood flows through the glomerular capillaries under high pressure, forcing more than 10% of the plasma out of the capillaries and into Bowman's capsule. **Filtration** is similar to the mechanism whereby interstitial fluid is formed as blood flows through other capillary networks in the body. However, blood flow through glomerular capillaries is at much higher pressure, so more plasma is filtered in the kidney.

Several factors contribute to filtration. First, the hydrostatic blood pressure in the glomerular capillaries is higher than in other capillaries. This high pressure is mainly due to the high resistance to outflow presented by the efferent arteriole, which is smaller in diameter than the afferent arteriole (see Fig. 46-9b).

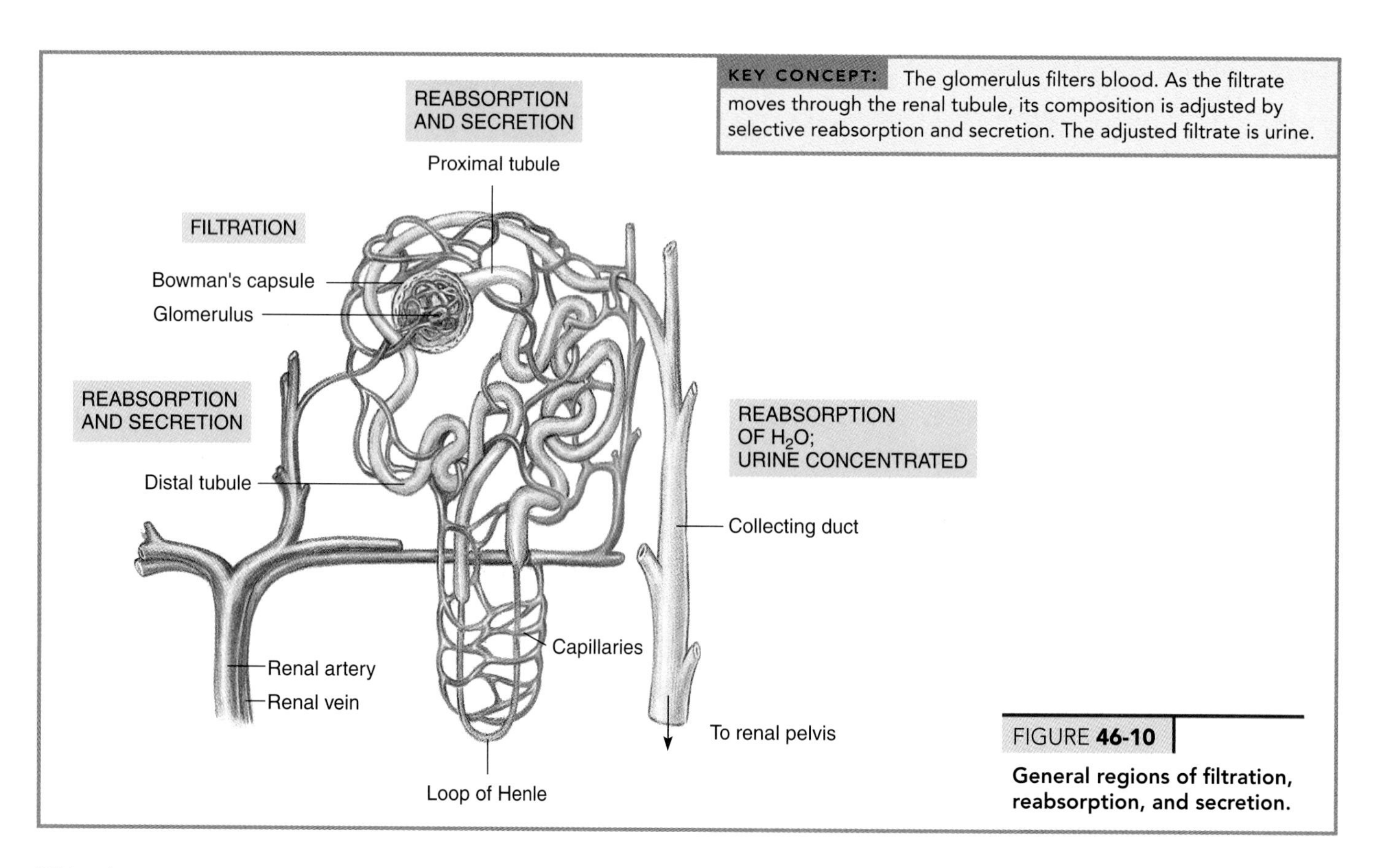

KEY CONCEPT: The glomerulus filters blood. As the filtrate moves through the renal tubule, its composition is adjusted by selective reabsorption and secretion. The adjusted filtrate is urine.

FIGURE **46-10**

General regions of filtration, reabsorption, and secretion.

A second factor contributing to the large amount of **glomerular filtrate** is the large surface area for filtration provided by the highly coiled glomerular capillaries. A third factor is the great permeability of the glomerular capillaries. Numerous small pores between the endothelial cells that form their walls make the glomerular capillaries more porous than typical capillaries.

The wall of Bowman's capsule in contact with the capillaries consists of specialized epithelial cells called **podocytes.** These cells have numerous cytoplasmic extensions called *foot processes* that cover most of the surfaces of the glomerular capillaries (Fig. 46-11). Foot processes of adjacent podocytes are separated by narrow gaps called **filtration slits.** The porous walls of the glomerular capillaries and the filtration slits of the podocytes form a **filtration membrane** that permits fluid and small solutes dissolved in the plasma, such as glucose, amino acids, sodium, potassium, chloride, bicarbonate, other salts, and urea, to pass through and become part of the filtrate. This filtration membrane holds back blood cells, platelets, and most of the plasma proteins.

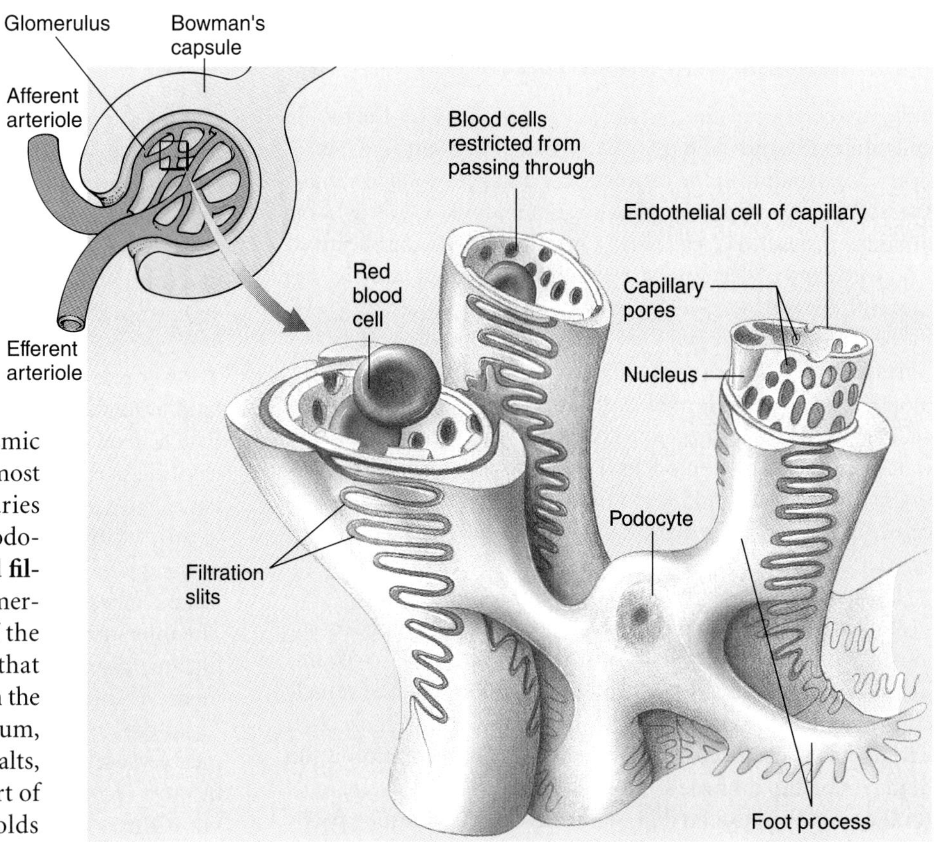

FIGURE **46-11** **Filtration membrane of the kidney.**
The porous walls of the glomerular capillaries and the filtration slits between podocytes form a filtration membrane that is highly permeable to water and small molecules but restricts the passage of blood cells and large molecules.

The total volume of blood passing through the kidneys is about 1200 mL per minute, or about one fourth of the entire cardiac output. As plasma passes through the glomerulus, it loses more than 10% of its volume to the glomerular filtrate. The normal glomerular filtration rate amounts to about 180 L (about 45 gal) each 24 hours. This is four and a half times the amount of fluid in the entire body! Common sense tells us that urine could not be excreted at that rate. Within a few moments, dehydration would become life threatening.

Reabsorption is highly selective

The threat to homeostasis posed by the vast amounts of fluid filtered by the kidneys is avoided by **reabsorption.** The renal tubules reabsorb about 99% of the filtrate into the blood, leaving only about 1.5 L to be excreted as urine during a 24-hour period. Reabsorption permits precise regulation of blood chemistry by the kidneys. Wastes, excess salts, and other materials remain in the filtrate and are excreted in the urine, whereas needed substances such as glucose and amino acids are returned to the blood. Each day the tubules reabsorb more than 178 L of water, 1200 g (2.6 lb) of salt, and about 250 g (0.5 lb) of glucose. Most of this, of course, is reabsorbed many times over.

The simple epithelial cells lining the renal tubule are well adapted for reabsorbing materials. Their abundant microvilli increase the surface area for reabsorption. These epithelial cells contain numerous mitochondria that provide the energy for running the cell pumps that actively transport materials.

Most (about 65%) of the filtrate is reabsorbed as it passes through the proximal convoluted tubule. Glucose, amino acids, vitamins, and other substances of nutritional value are entirely reabsorbed there. Many ions, including sodium, chloride, bicarbonate, and potassium, are partially reabsorbed. Some of these ions are actively transported; others are reabsorbed by diffusion. Reabsorption continues as the filtrate passes through the loop of Henle and the distal convoluted tubule. Then the filtrate is further concentrated as it passes through the collecting duct that leads to the renal pelvis.

Normally, substances that are useful to the body, such as glucose or amino acids, are reabsorbed from the renal tubules. If the concentration of a particular substance in the blood is high, however, the tubules may not be able to reabsorb it all. The maximum rate at which a substance can be reabsorbed is called its **tubular transport maximum (Tm).** When that rate is reached, the binding sites are saturated on the membrane proteins that transport the substance. For example, the tubular load of glucose is about 125 mg per minute, and almost all of it is normally reabsorbed. However, in a person with uncontrolled diabetes mellitus, the concentration of glucose in the blood exceeds its Tm (glucose concentration in excess of 320 mg per minute). The excess glucose cannot be reabsorbed and is excreted in the urine (glucosuria), a symptom of this disorder (see Chapter 47).

Some substances are actively secreted from the blood into the filtrate

Tubular **secretion** is the passage of substances across the tubule epithelium in a direction opposite to that of reabsorption. Secretion occurs mainly in the region of the distal convoluted tubule. Potassium, hydrogen ions, and ammonium ions, as well as some organic ions such as creatinine (a metabolic waste), are secreted into the filtrate. Certain drugs, such as penicillin, are also removed from the blood by secretion.

Secretion of hydrogen ions by the collecting ducts is an important homeostatic mechanism for regulating the pH of the blood. Carbon dioxide, which diffuses from the blood into the cells of the distal tubules and collecting ducts, combines with water, producing carbonic acid. This acid then dissociates, forming hydrogen ions and bicarbonate ions. When the blood becomes too acidic, more hydrogen ions are secreted into the urine:

$$CO_2 + H_2O \rightleftharpoons H_2CO_3 \rightleftharpoons H^+ + HCO_3^-$$

Potassium secretion is another important homeostatic mechanism. When potassium concentration is too high, nerve impulses are not effectively transmitted, and the strength of muscle contraction decreases. The heart rhythm becomes irregular, and cardiac arrest can occur. When potassium concentration is too high, potassium ions are secreted from the blood into the renal tubules and then excreted in the urine. Secretion results partly from a direct effect of the potassium ions on the tubules. In addition, the adrenal cortex increases its output of the hormone aldosterone, which further stimulates secretion of potassium.

Urine becomes concentrated as it passes through the renal tubule

We can survive with limited fluid intake because the kidneys can produce highly concentrated urine—more than four times as concentrated as blood. The osmolarity of human blood is about 300 milliosmols per liter (mOsm/L).[1] The kidneys can produce urine with an osmolarity of about 1400 mOsm/L.

As the filtrate passes through various regions of the renal tubule, salt (NaCl) is reabsorbed and a salt concentration gradient is established (Figs. 46-12 and 46-13). The gradient is used to produce a concentrated urine.

When filtrate flows from Bowman's capsule to the proximal tubule, its osmolarity is the same as that of blood (about 300 mOsm/L). The proximal tubule reabsorbs water and salt. Sodium ions are actively transported out of the proximal tubule, and chloride follows passively. As salt passes into the interstitial fluid, water follows osmotically.

The loop of Henle is specialized to highly concentrate sodium chloride in the medulla. It maintains a highly hypertonic interstitial fluid in the medulla near the bottom of the loop, which in turn lets the kidneys produce a concentrated urine. The walls of the descending limb of the loop of Henle are relatively permeable to water but relatively impermeable to sodium and urea. The interstitial fluid has a high concentration of Na^+, so as the filtrate passes down the loop of Henle water moves out by osmosis. This concentrates the filtrate inside the loop of Henle.

At the turn of the loop of Henle, the walls become more permeable to salt and less permeable to water. As the concentrated filtrate begins to move up the ascending limb (the thin region), salt diffuses out into the interstitial fluid. This contributes to the high salt concentration of the interstitial fluid in the medulla surrounding the loop of Henle. Further along the ascending limb (the thick region), sodium is actively transported out of the tubule.

[1] An osmol is a unit of osmotic pressure equal to the molarity of the solution divided by the number of particles produced when the solute dissolves. For example, glucose dissolves to give only one kind of particle, but NaCl produces two. A milliosmol is 1/1000 of an osmol.

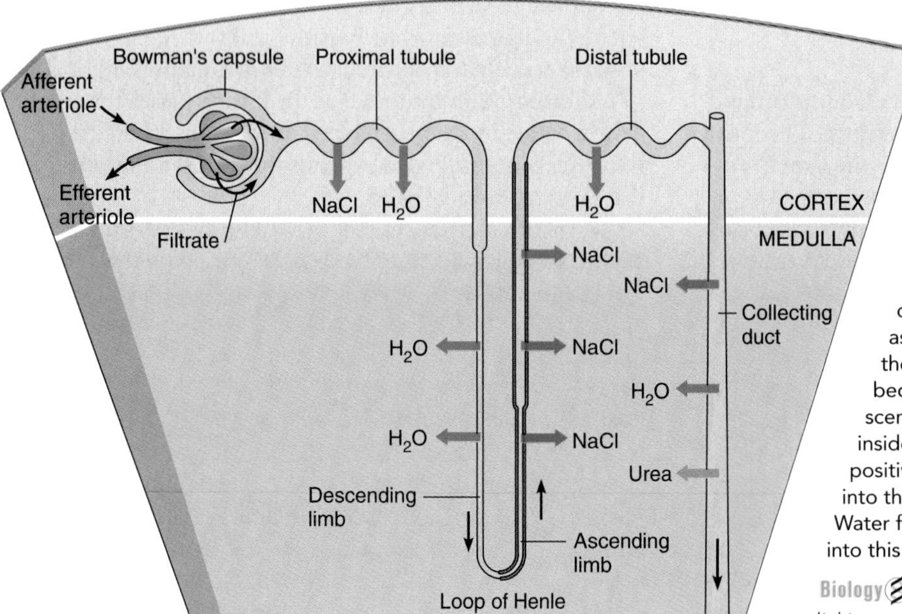

ACTIVE FIGURE **46-12**

Movement of water, ions, and urea through the renal tubule and collecting duct.

Water passes out of the descending limb of the loop of Henle, leaving a more concentrated filtrate inside. The heavy outline along the ascending limb indicates that this region is relatively impermeable to water. NaCl diffuses out from the lower (thin) part of the ascending limb. In the upper (thick) part of the ascending limb, NaCl is actively transported into the interstitial fluid. The saltier the interstitial fluid becomes, the more water moves out of the descending limb. This leaves a concentrated filtrate inside, so more salt passes out. Note that this is a positive feedback system. Some urea also moves out into the interstitial fluid through the collecting ducts. Water from the collecting ducts moves out osmotically into this hypertonic interstitial fluid.

BiologyNow™ Watch **renal processes in action** by clicking on this figure on your BiologyNow CD-ROM.

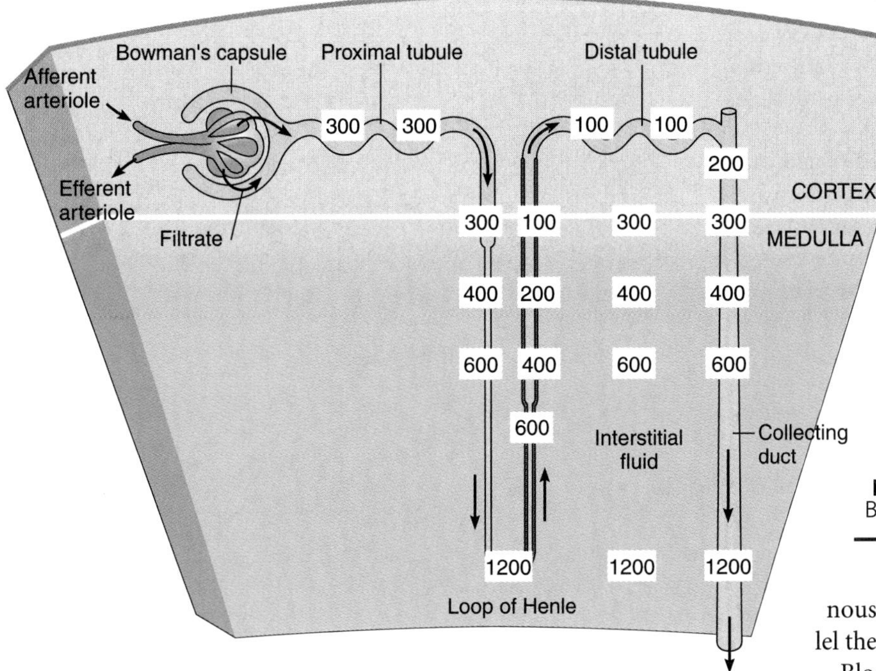

ACTIVE FIGURE **46-13**

Concentration of the filtrate as it moves through the nephron.

This figure shows the relative concentration of ions, mainly Na^+ and Cl^-, during formation of a very concentrated urine. Numbers indicate the concentration of salt in the filtrate expressed in milliosmols per liter. The more hypertonic a solution is, the higher its osmotic pressure. The very hypertonic interstitial fluid near the renal pelvis draws water osmotically from the filtrate in the collecting ducts. The heavy outline along the ascending loop indicates this region is relatively impermeable to water.

BiologyNow™ Learn more about **filtrate processing** by clicking on this figure on your BiologyNow CD-ROM.

Because water passes out of the descending limb of the loop of Henle, the filtrate at the bottom of the loop has a high salt concentration. However, because salt (but not water) is removed in the ascending limb, by the time the filtrate moves into the distal tubule its osmolarity may be the same or even lower than that of blood.

As it passes through the distal tubule, the filtrate may become even more dilute. The distal tubule is relatively impermeable to water but actively transports salt out into the interstitial fluid. The filtrate passes from the renal tubule into a larger collecting duct that eventually empties into the renal pelvis.

Note that there is a *counterflow* of fluid through the two limbs of the loop of Henle. Filtrate passing down through the descending limb is flowing in a direction opposite the filtrate moving upward through the ascending limb. The filtrate becomes concentrated as it moves down the descending limb and diluted as it moves up the ascending limb. This countercurrent mechanism helps maintain a high salt concentration in the interstitial fluid of the medulla. The hypertonic interstitial fluid draws water osmotically from the filtrate in the collecting ducts.

The inner medullary collecting ducts are permeable to urea, allowing the concentrated urea in the filtrate to diffuse out into the interstitial fluid. This urea contributes to the high solute concentration of the inner medulla and so helps in the process of concentrating urine.

The collecting ducts pass through the zone of very salty interstitial fluid. As the filtrate moves through the collecting duct, water passes osmotically into the interstitial fluid, where it is collected by capillaries. Sufficient water can leave the collecting ducts to produce highly concentrated urine. A hypertonic urine conserves water.

Some of the water that diffuses from the filtrate into the interstitial fluid is removed by the **vasa recta,** long, straight capillaries that extend from the efferent arterioles of the juxtamedullary nephrons. The vasa recta extend deep into the medulla, only to negotiate a hairpin curve and return to the cortical venous drainage of the kidney. These capillaries parallel the renal tubules.

Blood flows in opposite directions in the ascending and descending regions of the vasa recta, just as filtrate flows in opposite directions in the ascending and descending limbs of the loop of Henle. As a consequence of this countercurrent flow, much of the salt and urea that enter the blood through the descending region of the vasa recta leave again from the ascending region. As a result, the solute concentration of the blood leaving the vasa recta is only slightly higher than that of the blood entering. This mechanism helps maintain the high solute concentration of the interstitial fluid in the renal medulla.

Urine consists of water, nitrogenous wastes, and salts

By the time the filtrate reaches the renal pelvis, its composition has been precisely adjusted. The adjusted filtrate, called *urine,* consists of approximately 96% water, 2.5% nitrogenous wastes (mainly urea), 1.5% salts, and traces of other substances, such as bile pigments, that may contribute to the characteristic color and odor. Healthy urine is sterile and has been used to wash battlefield wounds when clean water was not available. However, when exposed to bacterial action, urine swiftly decomposes, forming ammonia and other products. Ammonia produces the diaper rash of infants.

The composition of urine yields many clues to body function and malfunction. **Urinalysis,** the physical, chemical, and microscopic examination of urine, is a very important diagnostic tool that has been used to monitor diabetes mellitus and many other disorders. Urinalysis is also extensively used in drug testing, because breakdown products of some drugs can be identified in the urine for several days or weeks after the drugs are taken.

Kidney function is regulated by hormones

Several hormones regulate urine volume and concentration, and additional hormones that help maintain fluid homeostasis remain to be identified (Table 46-1). The amount of urine produced depends on the body's need to retain or rid itself of water. We have seen that salt reabsorption in the loops of Henle establishes a very salty interstitial fluid that draws water osmotically from the collecting ducts.

When fluid intake is low, the body begins to dehydrate, causing the blood volume to decrease. As blood volume decreases, the concentration of salts dissolved in the blood becomes greater, causing an increase in osmotic pressure. Certain receptors in the hypothalamus are sensitive to this osmotic change and stimulate the posterior lobe of the pituitary to release **antidiuretic hormone (ADH).** This hormone is actually produced in the hypothalamus, but it is stored in the posterior pituitary and released as needed. A thirst center in the hypothalamus also responds to dehydration, stimulating an increase in fluid intake.

ADH regulates the permeability of the collecting ducts to water. When the body needs to conserve water, the posterior pituitary gland increases its release of ADH (Fig. 46-14). This hormone makes the collecting ducts more permeable to water so that more water is reabsorbed and a small volume of concentrated urine is produced. ADH acts on aquaporin-2, a membrane protein that forms gated water channels in the wall of the collecting ducts (see Chapter 5). These channels allow water to pass rapidly through the plasma membrane.

When you drink a large volume of water, your blood becomes diluted and its osmotic pressure falls. Release of ADH by the pituitary gland decreases, lessening the amount of water reabsorbed from the collecting ducts. The kidneys produce a large volume of dilute urine.

Occasionally the pituitary gland malfunctions and does not produce enough ADH. The resulting condition is called *diabetes insipidus* (not to be confused with the more common disorder, diabetes mellitus). Diabetes insipidus also results from an acquired insensitivity of the kidney to ADH. In diabetes insipidus, water is not efficiently reabsorbed from the ducts, so the body produces a large volume of urine. A person with severe diabetes insipidus may excrete up to 25 quarts of urine each day, a serious water loss. The affected individual becomes dehydrated and must

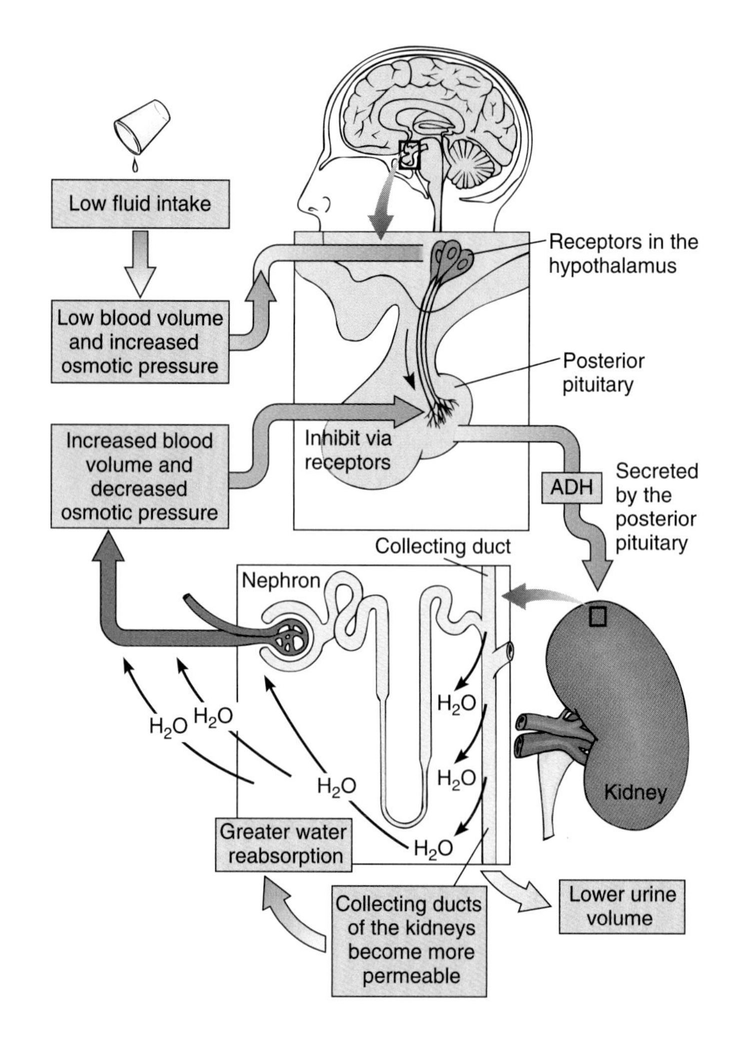

FIGURE **46-14** **Regulation of urine volume by antidiuretic hormone (ADH).**

When the body is dehydrated, the hormone ADH increases the permeability of the collecting ducts to water. More water is reabsorbed, and only a small volume of concentrated urine is produced.

TABLE **46-1** **Hormonal Control of Kidney Function**

Hormone	Source	Target Tissue	Actions	Factors that Stimulate Release
Antidiuretic hormone (ADH)	Produced in hypothalamus; released by posterior pituitary gland	Collecting ducts	Increases permeability of collecting ducts to water, increasing reabsorption and decreasing water excretion	Low fluid intake decreases blood volume and increases osmotic pressure of blood; receptors in hypothalamus stimulate posterior pituitary
Aldosterone	Adrenal glands (cortex)	Distal tubules and collecting ducts	Increases sodium reabsorption	Angiotensin II (when blood pressure decreases)
Angiotensin II	Produced from angiotensin I	Blood vessels and adrenal glands	Constricts blood vessels, which raises blood pressure; stimulates aldosterone secretion	Decrease in blood pressure causes renin secretion; renin catalyzes conversion of angiotensinogen to angiotensin I, which is then converted to angiotensin II by ACE
Atrial natriuretic peptide (ANP)	Atrium of heart	Afferent arterioles; collecting ducts	Dilates afferent arterioles; inhibits sodium reabsorption by collecting ducts; inhibits aldosterone secretion	Stretching of atria due to increased blood volume

drink almost continually to offset fluid loss. Injections of ADH or use of an ADH nasal spray can often control diabetes insipidus.

Sodium is the most abundant extracellular ion, accounting for about 90% of all positive ions in the extracellular fluid. The actions of several hormones precisely regulate sodium concentration. **Aldosterone,** which is secreted by the cortex of the adrenal glands, stimulates the distal tubules and collecting ducts to increase sodium reabsorption. When researchers remove the adrenal glands of experimental animals, too much sodium is excreted, leading to serious depletion of the extracellular fluid.

Aldosterone secretion is stimulated by a decrease in blood pressure (which is caused by a decrease in volume of blood and interstitial fluid). When blood pressure falls, cells of the **juxtaglomerular apparatus** secrete the enzyme **renin** and activate the **renin-angiotensin-aldosterone pathway.** The juxtaglomerular apparatus is a small group of cells in the region where the renal tubule contacts the afferent and efferent arterioles (see Fig. 46-9b). Renin acts on a plasma protein (angiotensinogen), converting it to angiotensin I. *Angiotensin-converting enzyme (ACE)* converts angiotensin I into its active form, **angiotensin II,** a peptide hormone that stimulates aldosterone secretion. ACE is produced by the endothelial cells of capillary walls, especially by those in the lung.

Angiotensin II not only increases the synthesis and release of aldosterone but also raises blood pressure directly by constricting blood vessels. Angiotensin II also stimulates sodium reabsorption by the proximal convoluted tubules and may stimulate the posterior pituitary to release ADH. All these actions help increase extracellular fluid volume and raise blood pressure. In individuals with hypertension, *ACE inhibitors* are sometimes used to block the production of angiotensin II. We can summarize the renin-angiotensin-aldosterone pathway as follows:

Blood volume decreases ⟶ blood pressure decreases ⟶ cells of juxtaglomerular apparatus secrete renin ⟶ angiotensinogen ⟶ angiotensin I ⟶ angiotensin II ⟶ constricts blood vessels and stimulates aldosterone secretion ⟶ aldosterone increases sodium reabsorption ⟶ blood pressure increases

Atrial natriuretic peptide (ANP), a hormone produced by the heart, increases sodium excretion and decreases blood pressure. ANP is stored in granules in atrial muscle cells. When blood volume increases, the atria stretch and respond by releasing ANP into the circulation. ANP dilates afferent arterioles, thereby increasing glomerular filtration rate. It inhibits sodium reabsorption by the collecting ducts directly and also indirectly by inhibiting aldosterone secretion. ANP also reduces plasma aldosterone concentration by inhibiting renin release. These actions of ANP raise urine output and lower blood volume and blood pressure.

Blood volume increases ⟶ blood pressure increases ⟶ atria of heart stretched ⟶ atria release ANP ⟶ directly inhibits sodium reabsorption and inhibits aldosterone secretion (which also inhibits sodium reabsorption) ⟶ larger urine volume ⟶ blood volume decreases ⟶ blood pressure decreases

The renin-angiotensin system and ANP work antagonistically in regulating fluid balance, salt (electrolyte) balance, and blood pressure.

Review

- Which structure(s) in the mammalian body is/are associated with each of the following: (a) urea formation (b) urine formation (c) temporary storage of urine (d) conduction of urine out of the body?
- Which part(s) of the nephron is/are associated with the following? (a) filtration (b) reabsorption (c) secretion
- Through what sequence of structures does a drop of filtrate pass as it moves from Bowman's capsule to the urinary bladder?
- Through what sequence of blood vessels does a drop of blood pass as it is conducted to and from a nephron?
- How is urine volume regulated? Explain.
- What are the actions of the renin-angiotensin-aldosterone pathway?

BiologyNow™ Assess your understanding of **the urinary system** by taking the pretest on your BiologyNow CD-ROM.

SUMMARY WITH KEY TERMS

1 Define *osmoregulation* and *excretion,* and contrast the advantages and disadvantages of excreting ammonia, uric acid, or urea.

- **Osmoregulation** is the active regulation of osmotic pressure of body fluids so that homeostasis is maintained. **Excretion** is the process of ridding the body of metabolic wastes. **Excretory systems** help maintain homeostasis by regulating the concentration of body fluids through osmoregulation and excretion of metabolic wastes.
- The principal waste products of animal metabolism are water; carbon dioxide; and **nitrogenous wastes,** including ammonia, urea, and uric acid. **Ammonia** is toxic and is excreted mainly by aquatic animals.
- **Urea** and **uric acid** are far less toxic than ammonia, but their synthesis requires energy. Urea excretion requires water. Uric acid can be excreted as a semisolid paste, a water-conserving adaptation.

2 Compare osmoconformers and osmoregulators, and describe protonephridia, metanephridia, and Malpighian tubules.

- Most marine invertebrates are **osmoconformers**—the salt concentration of their body fluids varies with changes in the seawater. Some marine invertebrates, especially those inhabiting coastal habitats, are **osmoregulators** that maintain an optimal salt concentration despite changes in salinity of their surroundings.
- **Nephridial organs,** which include protonephridia and metanephridia, function in osmoregulation and waste disposal. **Protonephridia,** found in flatworms and nemerteans, are tubules with no internal openings. Interstitial fluid enters their blind

ends, which consist of **flame cells,** cells with brushes of cilia. Beating of the cilia propels fluid through the tubules; excess fluid leaves through nephridiopores.

- Most annelids and mollusks have excretory tubules called **metanephridia,** which are open at both ends. As fluid from the coelom moves through the tubule, needed materials are reabsorbed by capillaries. Urine, containing wastes, exits the body through nephridiopores.
- **Malpighian tubules,** extensions of the insect gut wall, have blind ends that lie in the hemocoel. Cells of the tubule actively transport uric acid and some other substances from the hemolymph into the tubule, and water follows by diffusion. The contents of the tubule pass into the gut, and water and some solutes are reabsorbed in the rectum. Malpighian tubules effectively conserve water and have contributed to the success of insects as terrestrial animals.

3 Compare the adaptations to the challenges of osmoregulation in the following animals: freshwater fishes, marine bony fishes, sharks, marine mammals, and terrestrial vertebrates.

- Freshwater fishes take in water osmotically; they excrete a large volume of dilute urine.
- Marine bony fishes lose water osmotically. They compensate by drinking seawater and excreting salt through their gills; they produce only a small volume of urine.
- Sharks and other marine cartilaginous fishes retain large amounts of urea, allowing them to take in water osmotically through the gills. They excrete a large volume of urine.
- Marine mammals ingest seawater with their food. They produce a concentrated urine.
- Terrestrial vertebrates must conserve water. Endotherms have a high metabolic rate and produce a large volume of nitrogenous wastes. They have numerous adaptations for conserving water, including efficient kidneys.

4 Relate the function of the vertebrate kidney to the success of vertebrates in a wide variety of habitats.

- The vertebrate **kidney** functions in excretion and osmoregulation and is vital in maintaining homeostasis. Its structure and function are adapted to the lifestyle of the animal.

5 Describe (or label on a diagram) the organs of the mammalian urinary system, and give the functions of each.

- The kidney is the key organ of the **urinary system,** the principal excretory system in humans and other vertebrates. In mammals, the kidneys produce urine, which passes through the **ureters** to the **urinary bladder** for storage. During urination, the urine is released from the body through the **urethra.**
- The outer portion of each kidney is the **renal cortex;** the inner portion is the **renal medulla.** The renal medulla contains 8 to 10 **renal pyramids.** The tip of each pyramid is a **renal papilla.** As urine is produced, it flows into **collecting ducts,** which empty through a renal papilla into a funnel-shaped chamber, the **renal pelvis.** Each kidney has more than a million functional units called *nephrons.*

6 Describe (or label on a diagram) the structures of a nephron (including associated blood vessels), and give the functions of each structure.

- Each **nephron** consists of a cluster of capillaries, called a **glomerulus,** surrounded by a **Bowman's capsule** that opens into a long, coiled **renal tubule.** The renal tubule consists of a **proximal convoluted tubule, loop of Henle,** and **distal convoluted tubule.**
- **Cortical nephrons,** located almost entirely within the cortex or outer medulla, have small glomeruli. **Juxtamedullary nephrons** have large glomeruli and long loops of Henle that extend deep into the medulla. These nephrons are important in concentrating urine.
- Blood flows from small branches of the **renal artery** to **afferent arterioles** and then to glomerular capillaries. Blood then flows into an **efferent arteriole** that delivers blood into a second set of capillaries, the **peritubular capillaries** that surround the renal tubule. Blood leaves the kidney through the **renal vein.**

7 Trace a drop of filtrate from Bowman's capsule to its release from the body as urine.

- Urine formation is accomplished by the **filtration** of plasma, **reabsorption** of needed materials, and **secretion** of a few substances such as potassium and hydrogen ions into the renal tubule.
- Plasma filters through the glomerular capillaries and into Bowman's capsule. The permeable walls of the capillaries and **filtration slits** between **podocytes,** specialized epithelial cells that make up the inner wall of Bowman's capsule, serve as a **filtration membrane.** Filtration is nonselective with regard to small molecules; glucose and other needed materials, as well as metabolic wastes, become part of the filtrate.
- About 99% of the filtrate is reabsorbed from the renal tubules into the blood. Reabsorption is a highly selective process that returns usable materials to the blood but leaves wastes and excesses of other substances to be excreted in the urine. The maximum rate at which a substance can be reabsorbed is its **tubular transport maximum (Tm).**
- In secretion, hydrogen ions, certain other ions, and some drugs are actively transported into the renal tubule to become part of the urine.
- Production of a concentrated urine depends on a high salt and urea concentration in the interstitial fluid of the kidney medulla. The interstitial fluid in the medulla has a concentration gradient in which the salt is most concentrated around the bottom of the loop of Henle. This gradient is maintained, in part, by salt reabsorption from various parts of the renal tubule. A counterflow of fluid through the two limbs of the loop of Henle concentrates filtrate as it moves down the descending loop and dilutes it as it moves up the ascending loop.
- Water is drawn by osmosis from the filtrate as it passes through the collecting ducts. This concentrates urine in the collecting ducts.
- Some of the water that diffuses from the filtrate into the interstitial fluid is removed by the **vasa recta,** a system of capillaries that extend from the efferent arterioles.
- **Urine** is a watery solution of nitrogenous wastes, excess salts, and other substances not needed by the body.

8 Describe the hormonal regulation of fluid balance by antidiuretic hormone (ADH), aldosterone, angiotensin II, and atrial natriuretic peptide (ANP).

- Urine volume is regulated by **antidiuretic hormone (ADH),** which is released by the posterior lobe of the pituitary gland in

SUMMARY WITH KEY TERMS *(continued)*

response to an increase in osmotic concentration of the blood (caused by dehydration). ADH increases the permeability of the collecting ducts to water. As a result, more water is reabsorbed and only a small volume of urine is produced.

- Aldosterone and atrial natriuretic peptide work antagonistically. When blood pressure decreases, cells of the **juxtaglomerular apparatus** secrete the enzyme **renin,** which activates a pathway leading to production of **angiotensin II.** This hormone stimulates aldosterone release. **Aldosterone** increases sodium reabsorption, raising blood pressure.
- When blood pressure increases, **atrial natriuretic peptide (ANP)** increases sodium excretion and inhibits aldosterone secretion. These actions increase urine output and lower blood pressure.

POST-TEST

1. The process that maintains homeostasis of body fluids, keeping them from becoming too dilute or too concentrated is called (a) excretion (b) elimination (c) osmoregulation (d) glomerular filtration (e) tubular secretion
2. The main nitrogenous waste product of insects and birds is (a) urea (b) uric acid (c) ammonia (d) carbon dioxide (e) nitrate
3. The main nitrogenous waste product of amphibians and mammals is (a) urea (b) uric acid (c) ammonia (d) carbon dioxide (e) nitrate
4. Osmoconformers (a) maintain an optimal salt concentration despite fluctuations in salt concentration of their surroundings (b) include many animals that inhabit coastal habitats (c) experience variation in concentration of their body fluids along with changes in salinity of the seawater (d) typically have cells in their gills that remove salts from the surrounding water (e) answers (a) and (b) are correct
5. Which of the following is *not* a correct pair? (a) protonephridia/flatworm (b) metanephridia/annelid (c) flame cell/flatworm (d) Malpighian tubule/ mollusk (e) kidney/vertebrate
6. To compensate for fluid loss, many marine bony fishes (a) accumulate urea (b) have glands that excrete glucose (c) eat a low-protein diet (d) excrete a large volume of hypertonic urine (e) drink seawater
7. Arrange the following structures into an accurate sequence through which urine passes. (1) urethra (2) urinary bladder (3) kidney (4) ureter (a) 4, 3, 2, 1 (b) 3, 4, 2, 1 (c) 1, 2, 3, 4 (d) 4, 2, 1, 3 (e) 3, 1, 2, 4
8. Arrange the following structures into an accurate sequence through which filtrate passes. (1) proximal convoluted tubule (2) loop of Henle (3) collecting duct (4) distal convoluted tubule (5) Bowman's capsule (a) 5, 4, 3, 2, 1 (b) 3, 4, 2, 5, 1 (c) 1, 5, 2, 3, 4 (d) 5, 4, 2, 3, 1 (e) 5, 1, 2, 4, 3
9. The afferent arteriole delivers blood to the (a) renal artery (b) efferent arteriole (c) renal vein (d) capillaries of the glomerulus (e) peritubular capillaries
10. Which of the following does *not* contribute to the process of filtration? (a) high hydrostatic blood pressure in glomerular capillaries (b) large surface area for filtration (c) permeability of glomerular capillaries (d) active transport by epithelial cells lining renal tubules (e) podocytes
11. Tubular transport maximum is (a) the maximum concentration of a substance in the plasma that can be reabsorbed by the kidney (b) the most rapid rate at which urine can be transported through the ureter (c) the maximum rate at which a substance can be reabsorbed from the filtrate in the renal tubules (d) the maximum rate at which a substance can pass through the loop of Henle (e) the maximum rate at which a substance can be secreted into the filtrate
12. Which of the following does *not* contribute to the high salt concentration in the interstitial fluid in the medulla of the kidney? (a) reabsorption of salt from various regions of Bowman's capsule (b) diffusion of salt from the ascending limb of the loop of Henle (c) active transport of sodium from the upper part of the ascending limb (d) counterflow of fluid through the two limbs of the loop of Henle (e) diffusion of urea out of the collecting duct
13. Which of the following is *not* normally present in urine? (a) urea (b) glucose (c) salts (d) water (e) traces of bile pigments
14. Which is *not* true of ADH? (a) released by posterior lobe of the pituitary gland (b) increases water reabsorption (c) secretion increases when osmotic pressure in body increases (d) increases urine volume (e) secretion decreases when you drink a lot of water
15. Aldosterone (a) is released by the posterior pituitary gland (b) decreases sodium reabsorption (c) secretion is stimulated by an increase in blood pressure (d) is an enzyme that converts angiotensin into angiotensin II (e) secretion increases in response to angiotensin II
16. Label the diagram. Use Figure 46-9a to check your answers.

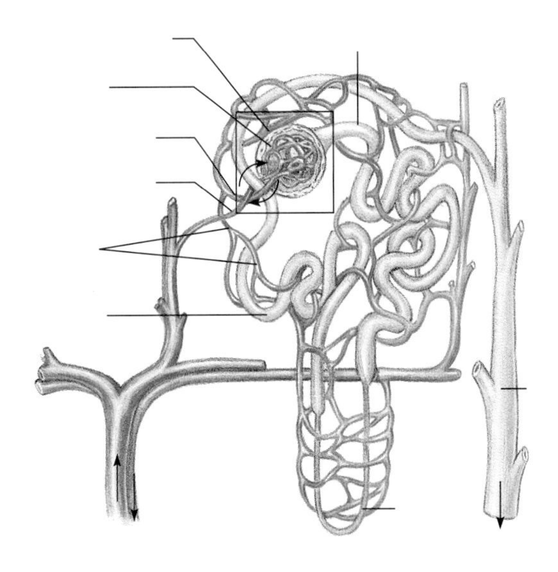

CRITICAL THINKING

1. The number of protonephridia in a planarian is related to the salinity of its environment. Planaria inhabiting slightly salty water develop fewer protonephridia, but the number quickly increases when the concentration of salt in the environment is lowered. Explain why.
2. What types of osmoregulatory challenges do humans experience? Explain. What mechanisms do we have to meet these challenges?
3. Why is glucose normally not present in urine? Why is it present in the urine of individuals with diabetes mellitus? Why do you suppose diabetics experience an increased output of urine?
4. The kangaroo rat's diet consists of dry seeds, and it drinks no water. Speculate about the adaptations this animal needs in order to survive.

- Visit our Web site at **http://biology.brookscole.com/solomon7** for links to chapter-related resources on the World Wide Web. Additional online materials relating to this chapter can also be found on our Web site.

BIOLOGY NOW RESOURCES

Active Figures

46-8: Renal anatomy
46-12: Basic renal process
46-13: Filtrate Processing

Preparing for an exam? Take a diagnostic test on your BiologyNow CD-ROM.

Post-Test Answers

1. c	2. b	3. a	4. c
5. d	6. e	7. b	8. e
9. d	10. d	11. c	12. a
13. b	14. d	15. e	

47

Endocrine Regulation

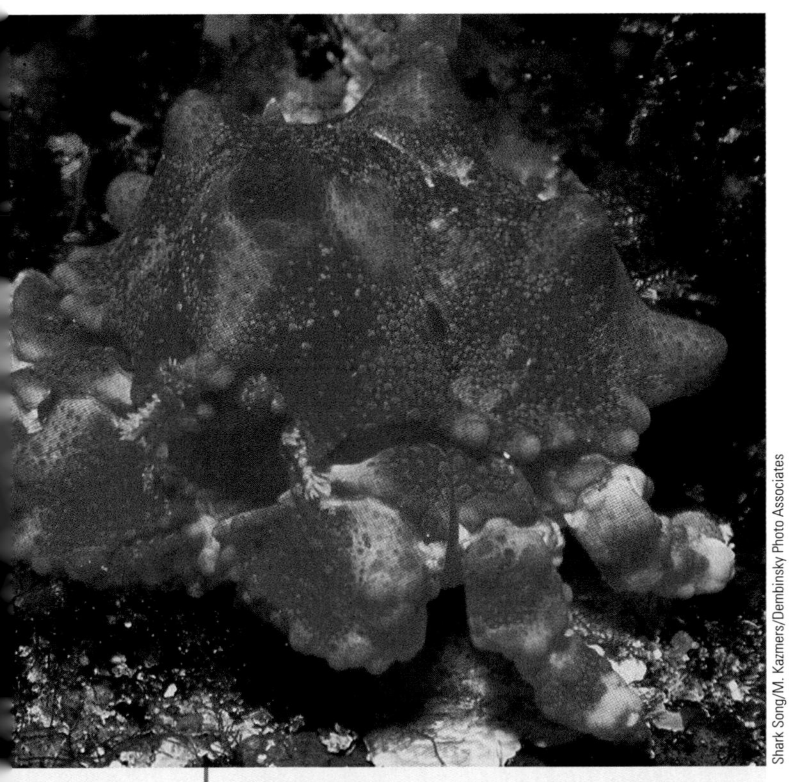
Shark Song/M. Kazmers/Dembinsky Photo Associates

This juvenile Puget Sound king crab (*Lopholithodes mandtii*) changes color to blend with its background. Caused by changes in the distribution of pigment in its cells, color changes are regulated by hormones.

CHAPTER OUTLINE

A caterpillar becomes a butterfly. A crustacean changes color to blend with its background (see photograph). A young girl develops into a woman. An adult copes with chronic stress. These physiological processes and many other adjustments of metabolism, fluid balance, growth and development, and reproduction are regulated by the **endocrine system.** This system works closely with the nervous system to maintain homeostatis, the steady state of the body. **Endocrinology,** the study of endocrine activity, is a very active and exciting field of biomedical research.

The endocrine system is a diverse collection of cells, tissues, and organs, including specialized **endocrine glands** that produce and secrete **hormones,** chemical messengers responsible for the regulation of many body processes. The term *hormone* is derived from a Greek word meaning "to excite." Hormones excite, or stimulate, changes in specific tissues.

Endocrine glands have no ducts; they secrete hormones directly into the interstitial fluid or blood. Hormones are typically transported by the blood and produce a characteristic response only after they reach target cells and bind with specific receptors. Target cells, the cells influenced by a particular hormone, may be in another endocrine gland or in an entirely different type of organ, such as a bone or the kidney. Target cells may be located far from the endocrine gland. For example, the vertebrate thyroid gland secretes hormones that stimulate metabolism in tissues throughout the body. Several types of hormones may be involved in regulating the metabolic activities of a particular type of cell. In fact, many hormones produce a synergistic effect in which the presence of one hormone enhances the effects of another.

Endocrinology has its roots in experiments performed by German physiologist A.A. Berthold in the 1840s. Berthold removed the testes from young roosters and observed that their combs (a male secondary sex characteristic) did not grow as large as those in normal roosters. He then transplanted testes into some of the birds and observed that the combs grew to normal size.

Berthold's methods became a model for subsequent studies in endocrinology and are still used by researchers today. To test the hypothesis that a particular tissue is endocrine, an investigator removes the tissue. Does removal of the tissue produce deficiency symptoms in the experimental animal? Investigators then replace the suspected endocrine tissue by transplanting similar tissue from another animal. They typically transplant the new tissue to a different location in the body to determine whether the effects depend on a signal that moves throughout the body in the blood. As with Berthold's roosters, the changes induced by removing the tissue should be reversed by replacing it.

Endocrinologists extract the suspected compound from the endocrine tissue of one animal and inject it into an experimental animal from which the tissue producing the compound has been removed. Deficiency symptoms should be relieved by replacing the suspected hormone. Researchers then isolate the active compound and determine its chemical structure. Finally, the compound is synthesized in the laboratory and injected into experimental animals. If its effects are those predicted, the researchers have data to support their hypothesis.

Using such procedures, endocrinologists have identified about 10 discrete endocrine glands. More recently investigators have identified specialized cells in the digestive tract, heart, kidneys, and many other parts of the body that also release hormones. In addition, some neurons release hormones. As a result of these discoveries, the scope of endocrinology has been broadened to include the production and actions of chemical messengers produced by a wide variety of organs, tissues, and cells.

One active focus of research is the study of the mechanisms of hormone action, which includes characterizing receptors and identifying the molecules involved in *signal transduction.* Some endocrinologists now use a reverse strategy for discovering new hormones and signaling pathways within the cell. They focus on "orphan" nuclear receptors, those for which ligands (the molecules that bind with them) are not yet known. Some of these "orphan" receptors are receptors for hormones that have not yet been identified. Using this strategy, researchers have identified intracellular signaling pathways for steroids, fatty acids, and several other compounds.

In this chapter we discuss the actions of a variety of hormones. We also examine how overproduction or deficiency of various hormones interferes with normal functioning. ■

CELL SIGNALING

Learning Objectives

1. Describe three main types of chemical signals used by cells, and compare types of endocrine signaling.
2. Identify four main chemical groups to which hormones are assigned, and give two examples for each group.

Cells signal one another with neurotransmitters, hormones, and local regulators. Some chemical compounds function as all three of these types of signals. Thus a neuron, endocrine gland, or some other cell type all may secrete the same chemical message. However, the same message can have different meanings for various target cells. We discussed neurotransmitters in Chapters 39 through 41. In this chapter we focus on hormones and local regulators, compounds that act on nearby cells.

Pheromones are chemical messengers that animals produce for communication with other animals of the same species. Because pheromones are generally produced by exocrine glands and do not regulate metabolic activities within the animal that produces them, most biologists do not classify them as hormones. Their role in regulating behavior is discussed in Chapter 50.

In classical endocrine signaling, hormones are secreted by endocrine glands

Endocrine glands differ from **exocrine glands** (such as sweat glands and gastric glands) that release their secretions into ducts. Endocrine glands have no ducts, and secrete their hormones into the surrounding interstitial fluid or blood. Typically, hormones diffuse into capillaries and are transported by the blood to target cells (Fig. 47-1a). Biologists have discovered that in addition to the discrete classical endocrine glands, specialized cells in many tissues and organs (such as the kidneys and heart) also release hormones or hormone-like substances.

The complexity of animal physiology challenges simplistic definitions. As new chemical signals and their modes of action have been discovered, the traditional definition of a hormone as a substance secreted by an endocrine gland and transported by the blood has become inadequate.

Neurohormones are transported in the blood

Certain neurons, known as **neuroendocrine cells,** are an important link between the nervous and endocrine systems. Neuroendocrine cells produce **neurohormones** that are transported down axons and released into the interstitial fluid. They typically diffuse into capillaries and are transported by the blood (Fig. 47-1b). Invertebrate endocrine systems are largely neuroendocrine. In vertebrates, the hypothalamus produces several neurohormones that link the nervous system with the pituitary gland, an endocrine gland that secretes several hormones.

Many endocrinologists include some local regulators as hormones

A **local regulator** is a signaling molecule that diffuses through the interstitial fluid and acts on nearby cells. Certain chemical compounds that are indisputably hormones because they are typically transported by the blood, under some conditions act as local regulators. In **autocrine regulation,** a hormone, or other

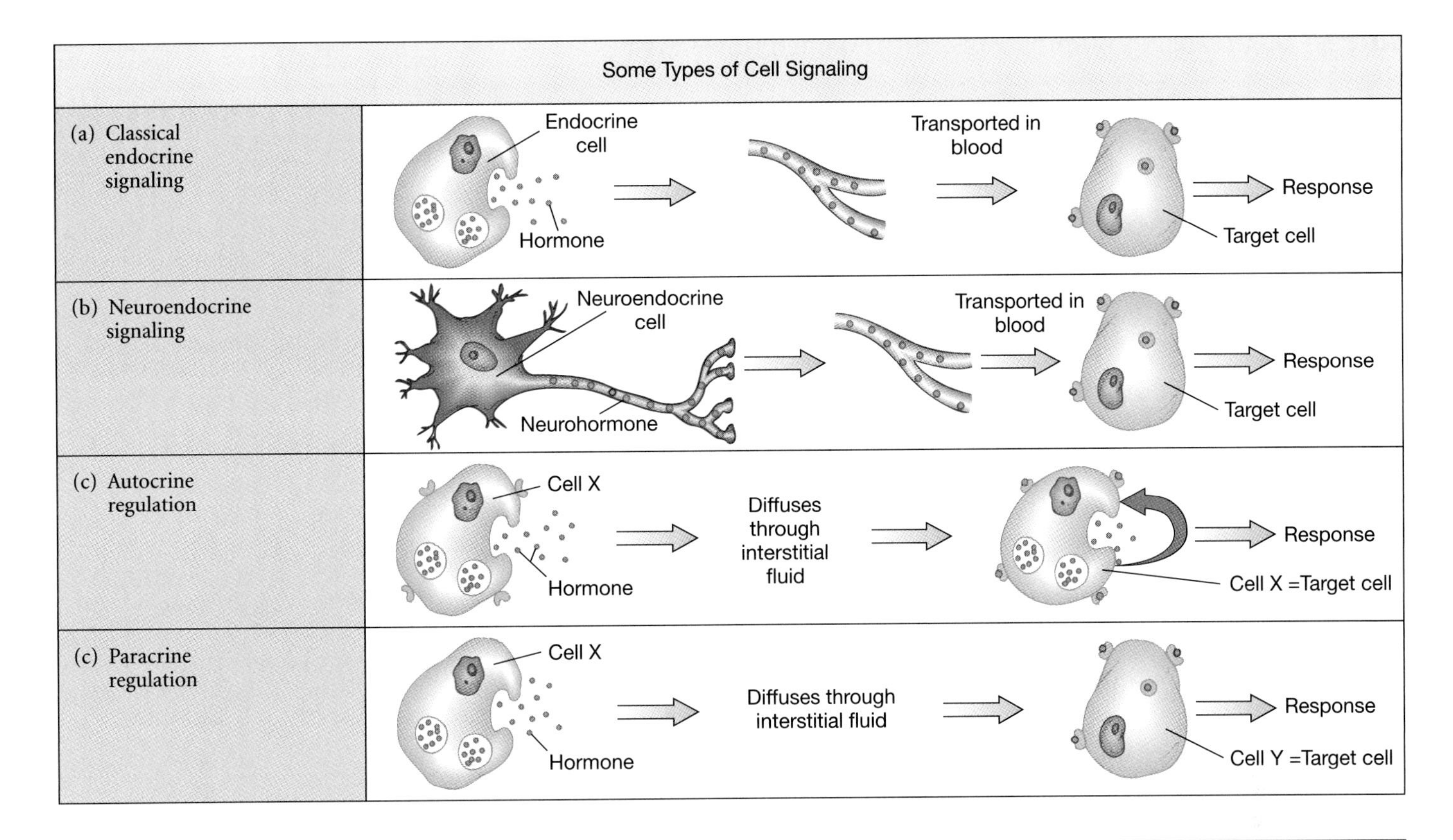

FIGURE **47-1** **Some types of endocrine signaling.**

(a) In classical endocrine signaling, endocrine cells release hormones that are transported to target cells by the blood. **(b)** In neuroendocrine signaling, neurons release neurohormones, which are transported by blood or diffuse through interstitial fluid. **(c)** In autocrine regulation, a hormone acts on the very cells that produce it. **(d)** In paracrine regulation, hormones diffuse through the interstitial fluid and act on nearby target cells.

regulator, acts on the very cells that produce it (Fig. 47-1c). For example, the female hormone estrogen, which functions as a classical hormone, may also exert an autocrine effect that stimulates additional estrogen secretion. Estrogen can also act on nearby cells, a type of local regulation known as **paracrine regulation** (Fig. 47-1d).

Local regulators include local chemical mediators such as histamine, growth factors, and prostaglandins. Recall from Chapter 43 that **histamine** is stored in mast cells and is released in response to allergic reactions, injury, or infection. Histamine causes blood vessels to dilate and capillaries to become more permeable. More than 50 **growth factors,** typically peptides, stimulate cell division and normal development in specific types of cells. Growth factors have autocrine or paracrine effects. **Nitric oxide** (NO), another local regulator, is produced by many types of cells, including those lining blood vessels. It relaxes nearby smooth muscle fibers, dilating the blood vessel.

Prostaglandins are modified fatty acids released continuously by the cells of most tissues. Biologists have grouped them into nine different classes. Although present in very small quantities, these local regulators affect a wide range of body processes. Prostaglandins are paracrine regulators that act on cells in their immediate vicinity. They modify cyclic adenosine monophosphate (cAMP) levels (discussed later in the chapter) and interact with other hormones to regulate various metabolic activities.

The major prostaglandin target is smooth muscle. Some prostaglandins stimulate smooth muscle to contract, whereas others cause relaxation. Thus some reduce blood pressure, whereas others raise it. Prostaglandins synthesized in the temperature-regulating center of the hypothalamus cause fever. In fact, nonsteroidal anti-inflammatory drugs (NSAIDs) such as aspirin and ibuprofen reduce fever and decrease inflammation by inhibiting prostaglandin synthesis.

Because prostaglandins are involved in the regulation of so many metabolic processes, they have great potential for a variety of clinical uses. Physicians use them to induce labor in pregnant women, to induce abortion, and to promote the healing of ulcers in the stomach and duodenum. Prostaglandins may someday be used to treat a wide variety of illnesses, including asthma, arthritis, kidney disease, certain cardiovascular disorders, and some forms of cancer.

Hormones are assigned to four chemical groups

Although hormones are chemically diverse, they generally belong to one of four different chemical groups: (1) fatty acid derivatives, (2) steroids, (3) amino acid derivatives, or (4) peptides or proteins (Fig. 47-2).

FOCUS ON ANABOLIC STEROIDS AND OTHER ABUSED HORMONES

An estimated 1 million people in the United States, half of them adolescents, abuse anabolic steroids, a group of synthetic hormones. Anabolic steroids are synthetic androgens (male reproductive hormones) that were developed in the 1930s to prevent muscle atrophy in patients with diseases that prevented them from moving about. In the 1950s, anabolic steroids became popular with professional athletes, who used them to increase muscle mass, physical strength, endurance, and aggressiveness. In truth, their athletic performance was probably enhanced, at least in part, by drug-induced euphoria and increased enthusiasm for training.

As with other hormones, the concentration of steroids circulating in the body is precisely regulated, so use of anabolic steroids interferes with normal physiological processes. Even during short-term use and at relatively low doses, anabolic steroids have a significant effect on mood and behavior. At higher doses, users experience disturbed thought processes, forgetfulness, and confusion, and often find themselves easily distracted. The term "steroid rage" refers to the mood swings, unpredictable anger, increased aggressiveness, and irrational behavior exhibited by many users.

Anabolic steroids elevate blood pressure, damage the liver, and increase low-density lipoprotein (LDL) concentration, raising the risk of cardiovascular disease. In adolescents, these steroids cause severe acne and stunt growth by prematurely closing the growth plates in bones. Abuse of these hormones reduces sexual function and can shrink the testes, leading to sterility. These drugs remain in the body for a long time. Their metabolites (breakdown products) can be detected in the urine for up to six months.

When their serious side effects became known in the 1960s, anabolic steroid use became controversial, and in 1973 the Olympic Committee banned their use. They are now prohibited worldwide by amateur and professional sports organizations. However, according to the U.S. Drug Enforcement Administration, a multimillion-dollar black market exists for these synthetic hormones. They are both injected and taken in pill form. Steroid abusers who share needles or use nonsterile techniques when they inject steroids are at risk for contracting hepatitis, HIV, and other serious infections.

The typical anabolic steroid user is a male (95%) athlete (65%), most often a football player, weight lifter, or wrestler. Surprisingly, though, about 10% of male high school students have used anabolic steroids and about one third of these students are not even on a high school team. These adolescents use the hormone only to change their physical appearance—to pump up their muscles ("bulk up")—and increase endurance. Many anabolic steroid users have difficulty realistically perceiving their body images. They remain unhappy even after dramatic increases in muscle mass.

People also abuse other hormones, including human growth hormone (GH) and erythropoietin. Human growth hormone, like anabolic steroids, helps build muscle mass but also causes acromegaly. Erythropoietin increases the concentration of red blood cells. Although increased oxygen transport enhances the performance of an endurance athlete up to 10%, abnormally high concentrations of red blood cells can cause serious cardiovascular problems. Erythropoietin abuse has caused the deaths of several athletes.

Prostaglandins and the juvenile hormones of insects are **fatty acid derivatives** (Fig. 47-2a). Prostaglandins are synthesized from arachidonic acid, a 20-carbon fatty acid. A prostaglandin has a five-carbon ring in its structure.

The molting hormone of insects is a **steroid hormone** (Fig. 47-2b). In vertebrates, the adrenal cortex, testis, ovary, and placenta secrete steroid hormones synthesized from cholesterol. Examples of steroid hormones are cortisol, secreted by the adrenal cortex, the male hormone testosterone secreted by the testis, and the female hormones progesterone and estrogens produced by the ovary. Athletes (and others) sometimes abuse synthetic hormones known as anabolic steroids (see *Focus On: Anabolic Steroids and Other Abused Hormones*).

Chemically, the simplest hormones are **amino acid derivatives** (Fig. 47-2c). The thyroid hormones (T_3 and T_4) are synthesized from the amino acid tyrosine and iodide. Epinephrine (also known as adrenaline) and norepinephrine (also known as noradrenaline), produced by the medulla of the adrenal gland, are also derived from tyrosine. Melatonin is synthesized from the amino acid tryptophan.

The water-soluble **peptide hormones** are the largest hormone group. (Endocrinologists include protein hormones in this group.) **Neuropeptides** are a large group of signaling molecules produced by neurons. Oxytocin and antidiuretic hormone (ADH), produced by neuroendocrine cells in the hypothalamus, are short neuropeptides, each composed of nine amino acids (Fig. 47-2d). Seven of the amino acids are identical in the two hormones, but the actions of these hormones are quite different. The hormones glucagon (see Fig. 3-19), secretin, adrenocorticotropic hormone (ACTH), and calcitonin are somewhat longer peptides consisting of about 30 amino acids.

Insulin is a small protein consisting of two peptide chains joined by disulfide bonds. Growth hormone, thyroid-stimulating hormone, and the gonadotropic hormones, all secreted by the anterior lobe of the pituitary gland, are large proteins with molecular weights of 25,000 or more.

Review

- How has the definition of the term *hormone* changed in recent years? Explain your answer.
- How are neuroendocrine and paracrine signaling different?
- What are the four major chemical groups of hormones?

BiologyNow™ Assess your understanding of **cell signaling** by taking the pretest on your BiologyNow CD-ROM.

FIGURE **47-2**

Major chemical groups of hormones.

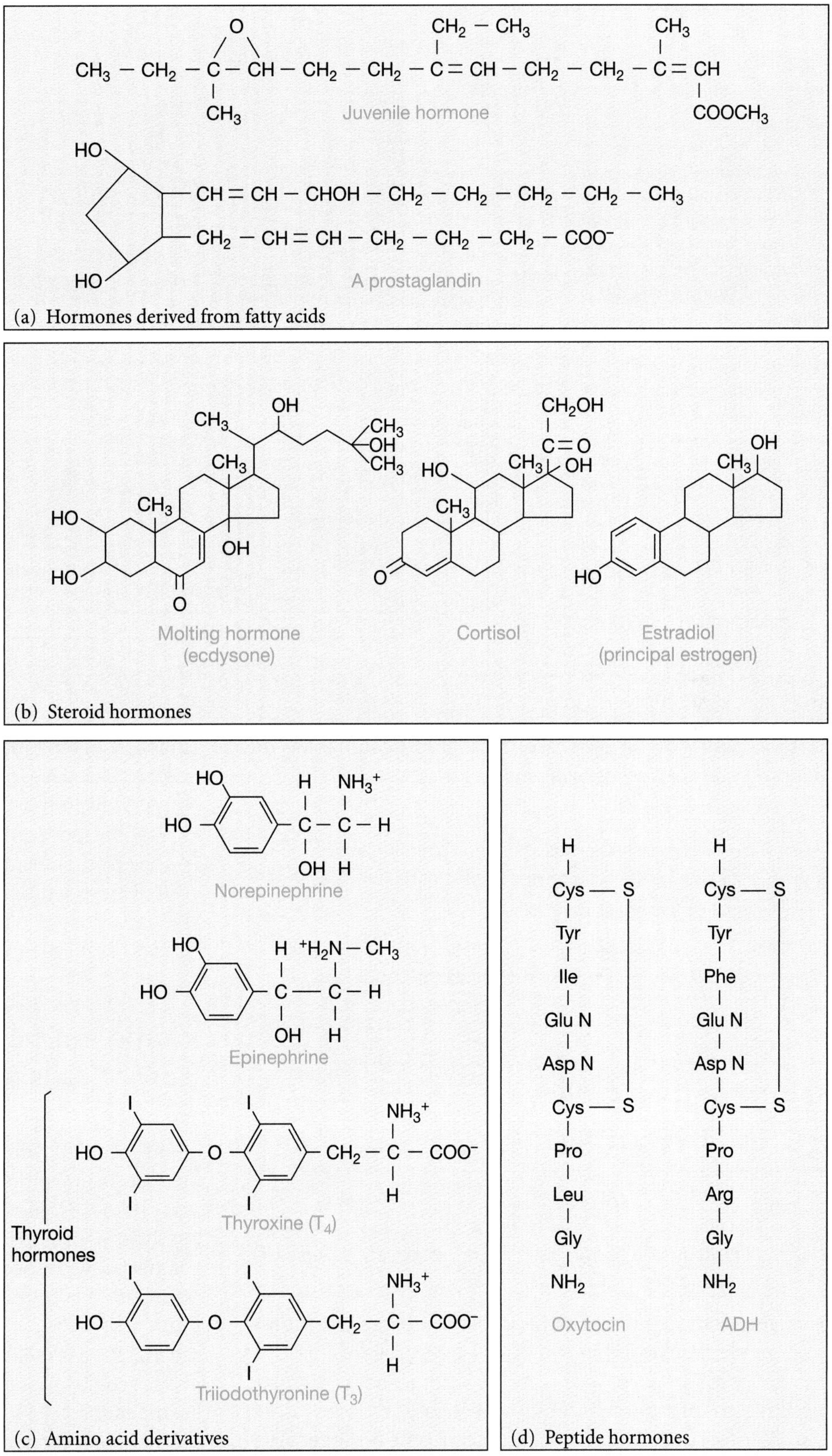

Juvenile hormone

A prostaglandin

(a) Hormones derived from fatty acids

Molting hormone (ecdysone)

Cortisol

Estradiol (principal estrogen)

(b) Steroid hormones

Norepinephrine

Epinephrine

Thyroid hormones

Thyroxine (T_4)

Triiodothyronine (T_3)

(c) Amino acid derivatives

Oxytocin

ADH

(d) Peptide hormones

REGULATION OF HORMONE SECRETION

Learning Objective

3 Summarize the regulation of endocrine glands by negative feedback mechanisms.

Although present in minute amounts, more than 50 different hormones may be circulating in the blood of a vertebrate at any time. Steroid hormones and thyroid hormones are transported bound to plasma proteins. Peptide hormones are water soluble and are transported dissolved in the plasma. Hormone molecules continuously move out of the circulation, and bind with target cells. They are removed from the blood by the liver, which inactivates some hormones, and by the kidneys, which excrete them.

Hormone secretion is typically regulated by **negative feedback mechanisms,** in which a hormone is released in response to some change in a steady state and triggers a response that counteracts the changed condition, restoring homeostasis (see Chapter 37). Information about the concentration of hormone in the blood or interstitial fluid is fed back to the gland, which then responds to restore homeostasis. The parathyroid glands, located in the neck of tetrapod vertebrates, provide a good example of how negative feedback works (Fig. 47-3).

The parathyroid glands regulate the calcium concentration of the blood. When the calcium concentration is not within homeostatic limits, nerves and muscles cannot function properly. For example, when too few calcium ions are present, neurons fire spontaneously, causing muscle spasms. When calcium concentration varies too far from the steady state (either too high or too low), negative feedback mechanisms restore homeostasis.

A decrease in the calcium concentration in the plasma signals the parathyroid glands to release more parathyroid hormone. This hormone increases the concentration of calcium in the blood. (The details of parathyroid action are discussed later in this chapter.) When the calcium concentration rises above normal limits, the parathyroid glands slow their output of hormone. Both responses are negative feedback mechanisms. An increase in calcium concentration results in decreased release of parathyroid hormone, whereas a decrease in calcium leads to increased

FIGURE **47-3**

Regulation by negative feedback.

When calcium concentration exceeds its normal range in the blood, the parathyroid glands are inhibited and slow their release of parathyroid hormone. When the blood calcium concentration falls below its normal range, the parathyroid glands release more parathyroid hormone. This hormone acts on target tissues that increase the calcium concentration in the blood, thus restoring homeostasis.

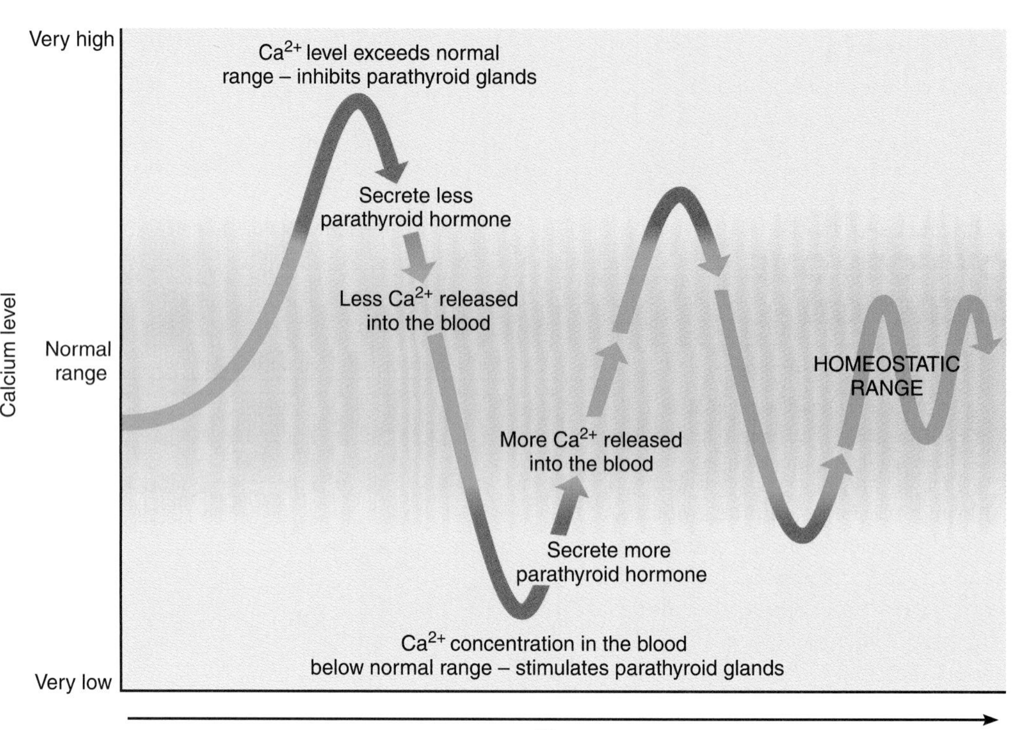

hormone secretion. In each case, the response counteracts the inappropriate change, restoring the steady state.

Review

- How does the body respond when the concentration of calcium in the blood is below the homeostatic level?
- How does negative feedback work?

BiologyNow™ Assess your understanding of the **regulation of hormone secretion,** by taking the pretest on your BiologyNow CD-ROM.

MECHANISMS OF HORMONE ACTION

Learning Objective

4 Compare the mechanisms of action of steroid and protein-type hormones; include the role of second messengers, such as cyclic AMP.

A hormone present in the extracellular fluid, remains "unnoticed" until it reaches its target cells. The **target cells** of a particular hormone have receptors that recognize and bind it. Hormone receptors are large proteins or glycoproteins. The receptor site is similar to a lock, and the hormones are like different keys. Only the hormone that fits the specific receptor can influence the metabolic machinery of the cell. Receptors are responsible for the specificity of the endocrine system.

Receptors are continuously synthesized and degraded. Their numbers are increased or decreased by **receptor up-regulation** and **receptor down-regulation.** For example, when the concentration of the hormone insulin is too high for an extended period, cells decrease the number of their insulin receptors. This down-regulation dampens insulin's ability to stimulate cells to take in glucose. Down-regulation suppresses the sensitivity of target cells to the hormone. Up-regulation occurs in response to low hormone concentrations. A greater number of receptors on the plasma membrane amplifies the hormone's effect on the cell. Receptor up-regulation and down-regulation is triggered by signals to genes that code for the receptors, as well as by several other mechanisms.

Some hormones enter target cells and activate genes

Steroid hormones and thyroid hormones are relatively small, lipid-soluble (hydrophobic) molecules that easily pass through the plasma membrane of the target cell. Specific protein receptors in the cytoplasm or in the nucleus bind with the hormone to form a hormone-receptor complex (Fig. 47-4). This complex combines with specific acceptor sites on the nuclear DNA, and turns specific genes on or off, activating or repressing transcription of messenger RNA molecules that code for specific proteins. These proteins produce the changes in structure or metabolic activity responsible for the hormone's action. Interestingly, some target cells have plasma membrane receptors, as well as intracellular receptors, for steroid hormones.

Many hormones bind to cell-surface receptors

Because peptide hormones are hydrophilic and not soluble in the lipid layer of the plasma membrane, they do not enter the target cell. Instead, they bind to a specific cell-surface receptor in

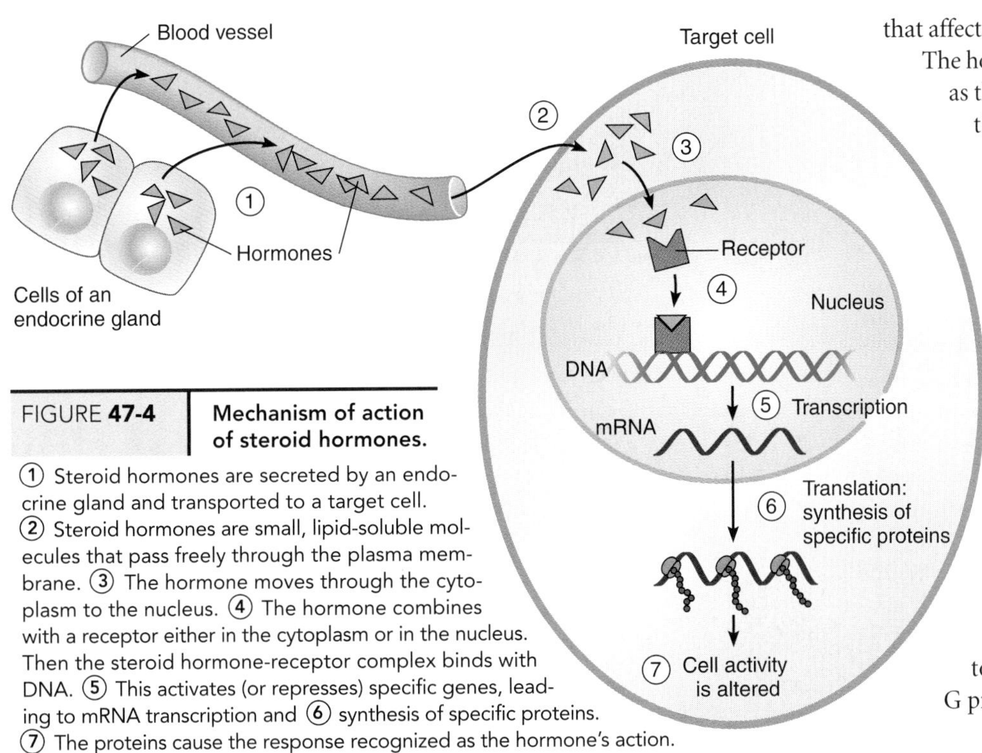

FIGURE **47-4** Mechanism of action of steroid hormones.

① Steroid hormones are secreted by an endocrine gland and transported to a target cell. ② Steroid hormones are small, lipid-soluble molecules that pass freely through the plasma membrane. ③ The hormone moves through the cytoplasm to the nucleus. ④ The hormone combines with a receptor either in the cytoplasm or in the nucleus. Then the steroid hormone-receptor complex binds with DNA. ⑤ This activates (or represses) specific genes, leading to mRNA transcription and ⑥ synthesis of specific proteins. ⑦ The proteins cause the response recognized as the hormone's action.

the plasma membrane. Two main types of cell-surface receptors are G protein–linked receptors and enzyme-linked receptors.

Cell signaling through G protein–linked receptors

G protein–linked receptors are transmembrane proteins that loop back and forth through the plasma membrane seven times. These receptors initiate **signal transduction;** that is, they convert an extracellular hormone signal into an intracellular signal that affects some intracellular process (Fig. 47-5). The hormone does not enter the cell. It serves as the first messenger and relays information to a **second messenger,** or intracellular signal. The second messenger then signals effector molecules that carry out the action. Thus, many hormones activate a series of molecular events that comprise a *signaling cascade.*

G protein–linked receptors activate **G proteins,** a group of integral regulatory proteins (Fig. 47-6). The G indicates that they bind **guanosine triphosphate (GTP),** which, like ATP, is an important molecule in energy reactions. When the system is inactive, G protein binds to **guanosine diphosphate (GDP),** which is similar to ADP, the hydrolyzed form of ATP. G proteins shuttle a signal between the re-

FIGURE **47-5** Overview of peptide hormone action.

① Peptide hormones bind with receptors on the plasma membrane of a target cell. The receptor is a signal transducer that converts the hormone signal to an intracellular signal. ② The signal is relayed by a second messenger. Typically, a sequence of several signaling molecules relays the message. ③ Some cell activity is altered.

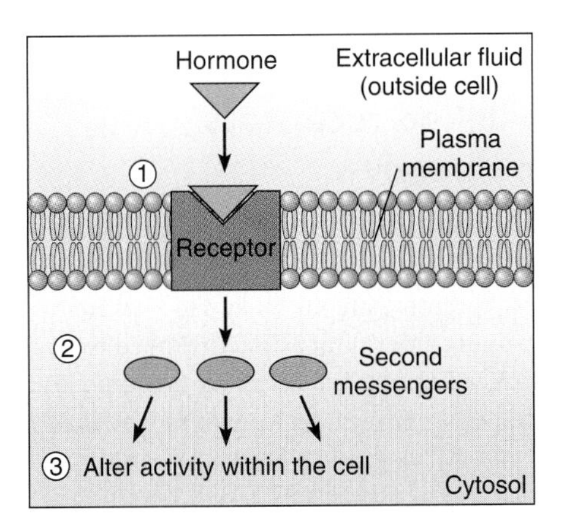

FIGURE **47-6**

Mechanism of action of hormones that use G protein–linked receptors and second messengers.

① A hormone binds with a G protein–linked receptor in the plasma membrane of a target cell, ② activating a G protein (GDP on the G protein is replaced by GTP). ③ The G protein activates adenylyl cyclase. ④ ATP is converted to cAMP, a second messenger, which ⑤ activates protein kinase A. ⑥ This enzyme phosphorylates proteins ⑦ altering their function. (GTP, guanosine triphosphate; cAMP, cyclic AMP).

KEY CONCEPT: Many hormones signal target cells by way of a G protein–linked receptor. Cyclic AMP (or some other second messenger) relays the signal as part of a cascade of reactions that changes some cell process (the hormone action).

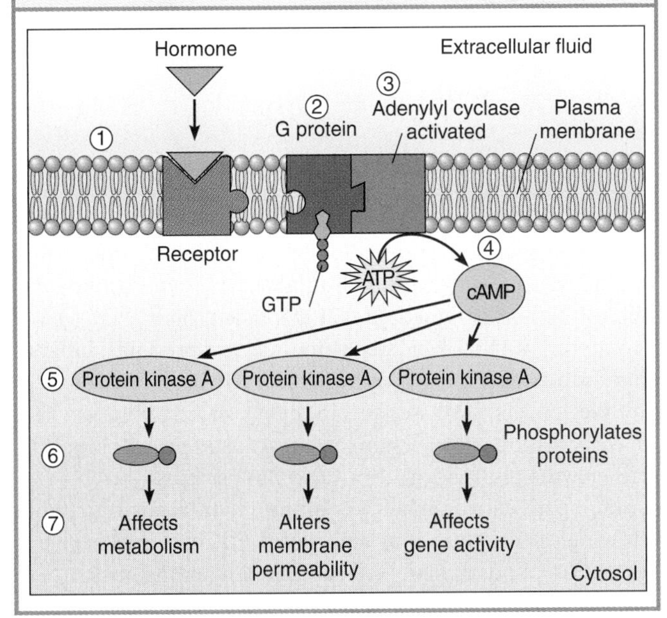

ceptor and a second messenger. Hundreds of different G protein–linked receptors have been identified.

Cyclic AMP is a common second messenger

When a hormone-bound receptor binds to it, a stimulatory G protein releases GDP and replaces it with GTP. The G protein undergoes a conformational change (change in shape), allowing it to bind with and activate **adenylyl cyclase,** an enzyme on the cytoplasmic side of the plasma membrane. Adenylyl cyclase catalyzes the conversion of ATP to cAMP. Note that a G protein couples adenylyl cyclase action to the hormone-receptor complex (Fig. 47-6). One type of G protein, G_s, stimulates adenylyl cyclase, and another, G_i, inhibits it.

Once activated, adenylyl cyclase catalyzes the conversion of ATP to **cyclic AMP (cAMP).** By coupling the hormone-receptor complex to an enzyme that generates a signal, G proteins amplify hormone effects and many second-messenger molecules are rapidly produced. In the 1960s, Earl Sutherland identified cAMP as a second messenger, and for his pioneering work he received the 1971 Nobel Prize in Medicine. Cyclic AMP is a common second messenger in prokaryotic and animal cells.

Cyclic AMP activates **protein kinase A,** an enzyme that phosphorylates (adds a phosphate group to) certain specific proteins. When a protein is phosphorylated, its function is altered, and it triggers a chain of reactions leading to some metabolic change. The substrates for protein kinase A are different in various cell types, so the effect of the enzyme varies. In skeletal muscle cells, protein kinase A triggers the breakdown of glycogen to glucose, whereas in cells of the hypothalamus the same enzyme activates the gene that encodes a growth-inhibiting hormone.

Any increase in cAMP is temporary. Cyclic AMP is rapidly inactivated by **phosphodiesterases,** which convert it to AMP. Thus the concentration of cAMP depends on the activity of both adenylyl cyclase, which produces it, and of phosphodiesterase, which breaks it down.

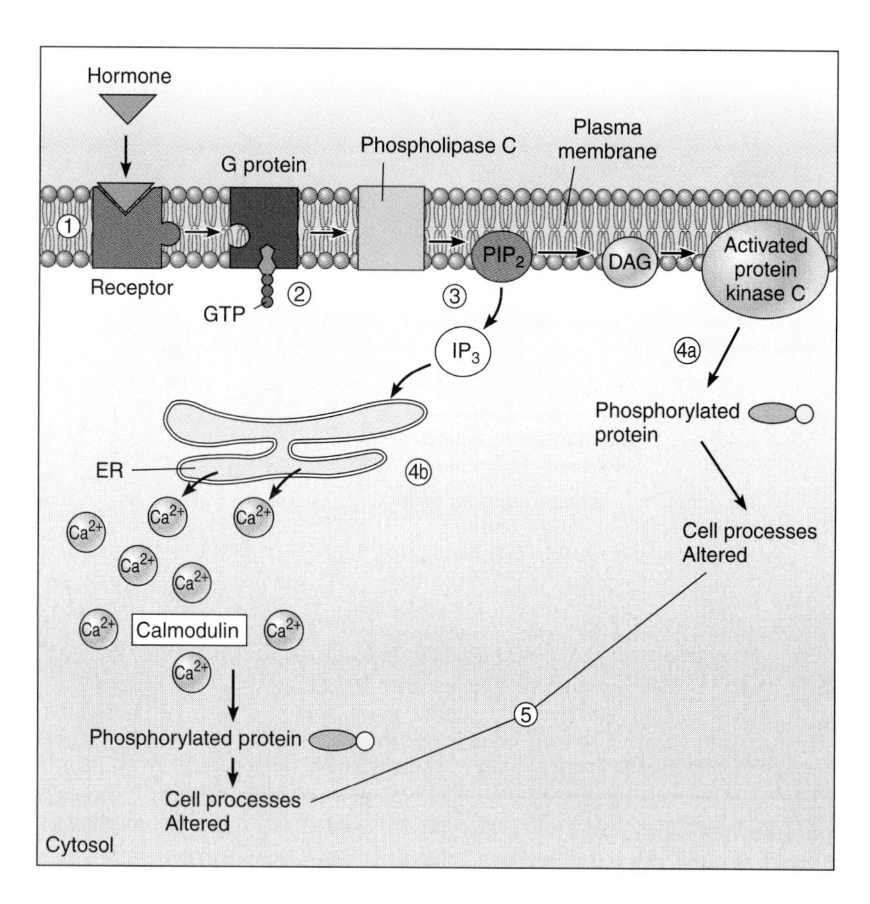

FIGURE **47-7** **Phospholipid products as second messengers.**

① A hormone binds with a cell-surface receptor, which ② activates a G protein that activates phospholipase C. ③ This enzyme splits PIP_2 into two products, DAG and IP_3. ④ⓐ DAG activates protein kinase C, which phosphorylates proteins. ④ⓑ IP_3 opens calcium channels in the ER. Calcium ions combine with calmodulin, which activates proteins. ⑤ Cell processes are altered by the actions of both DAG and IP_3. (GTP, guanosine trisphosphate; PIP_2, phosphotidylinositol 4,5-bisphosphate; DAG, diacylglycerol; IP_3, inositol trisphosphate).

Phospholipid products and calcium ions can act as second messengers

Certain hormone-receptor complexes activate a G protein that then activates the membrane-bound enzyme phospholipase C (Fig. 47-7). This enzyme splits a membrane protein, PIP_2 (phosphotidylinositol 4,5-bisphosphate), into two products, **inositol trisphosphate (IP_3)** and **diacylglycerol (DAG);** both act as second messengers. DAG remains in the plasma membrane where (in combination with calcium ions) it activates **protein kinase C.** This enzyme phosphorylates a variety of proteins.

IP_3 opens calcium channels in the endoplasmic reticulum, releasing calcium ions into the cytosol. Calcium ions have important functions in many cell processes, including muscle contraction, neural signaling, microtubule disassembly, blood clotting, and activation of some enzymes. However, calcium ions do not act alone; they exert their effects by binding to certain proteins such as **calmodulin,** which is found in all eukaryotic cells. The calmodulin molecules change shape and then activate certain proteins, altering their activity.

Enzyme-linked receptors function directly

Enzyme-linked receptors are transmembrane proteins with a hormone-binding site outside the cell and an enzyme site inside the cell. These receptors are not linked to G proteins. They function directly as enzymes or are directly linked to enzymes. Most enzyme-linked receptors are **receptor tyrosine kinases.** These receptors bind signal molecules known as *growth factors,* including insulin and nerve growth factor. When the growth factor binds to the receptor, the receptor is activated and phosphorylates the amino acid tyrosine in specific cell proteins.

Hormone signals are amplified

Although hormones are present in very small amounts, they effectively regulate many physiological processes. This is in large part the result of **signal amplification,** an increase in signal strength. For example, a single hormone molecule activates up to 100 G proteins. Each adenylyl cyclase molecule produces many cAMP molecules. In turn, each cAMP activates a protein kinase that phosphorylates many protein molecules. Thus, through a cascade of signals and reactions, a single hormone molecule activates many proteins.

Review

- What are the roles of receptors and second messengers in hormone action?
- How are the mechanisms of action of steroid and peptide hormones different?
- What is the mechanism of action of a hormone that uses cyclic AMP as a second messenger?

BiologyNow™ Assess your understanding of the **mechanisms of hormone action** by taking the pretest on your BiologyNow CD-ROM.

INVERTEBRATE NEUROENDOCRINE SYSTEMS

Learning Objective

5 Identify four functions of hormones in invertebrates, and describe the hormonal regulation of insect development.

Among invertebrates, hormones are secreted mainly by neurons rather than by endocrine glands. These **neurohormones** regulate regeneration in hydras, flatworms, and annelids; color changes in crustaceans; and growth, development, metabolic rate, gamete production, and reproduction, including reproductive behavior in many other groups. Trends in the evolution of invertebrate endocrine systems include a larger number of both neurohormones and hormones secreted by endocrine glands, and a greater role of hormones in physiological processes.

Insects have endocrine glands as well as neuroendocrine cells. Their various hormones and neurohormones interact with one another to regulate metabolism, growth, and development, including molting and metamorphosis. Hormonal control of development in insects varies among species. Generally, some environmental factor (such as temperature change) activates neuroendocrine cells in the brain. These cells then produce a neurohormone referred to as **brain hormone (BH),** which is transported down axons and stored in the paired **corpora cardiaca** (sing., *corpus cardiacum;* Fig. 47-8). When released from the corpora cardiaca, BH stimulates the **prothoracic glands,** endocrine glands in the prothorax, to produce **molting hormone (MH),** also called **ecdysone.** Molting hormone, a steroid hormone, stimulates growth and molting.

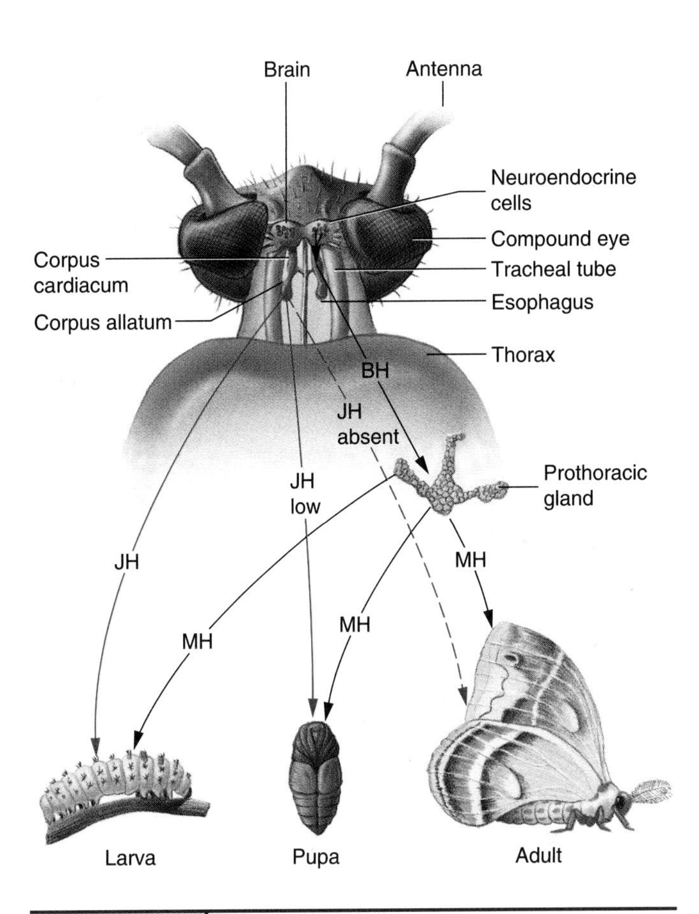

FIGURE **47-8** **Regulation of growth and molting in insects.**

The dissection of an insect head shows the brain, paired corpora allata, and corpora cardiaca. Neuroendocrine cells secrete brain hormone (BH), which is stored in the corpora cardiaca. When released, BH stimulates the prothoracic glands to secrete molting hormone (MH), which stimulates growth and molting. In the immature insect, the corpora allata secrete juvenile hormone (JH), which suppresses metamorphosis at each larval molt. Metamorphosis to the adult form occurs when molting hormone acts in the absence of juvenile hormone.

In the immature insect, paired endocrine glands called **corpora allata** (sing., *corpus allatum*) secrete **juvenile hormone (JH).** This hormone suppresses metamorphosis at each larval molt so that the insect increases in size while remaining in its immature state; after the molt, the insect is still in a larval stage. When the concentration of JH decreases, metamorphosis occurs, and the insect is transformed into a pupa (see Chapter 29). In the absence of JH, the pupa molts and becomes an adult. The nervous system regulates the secretory activity of the corpora allata, and the amount of JH decreases with successive molts.

Review

- What are four actions of hormones in invertebrates?
- What is the function of juvenile hormone in insects?

BiologyNow™ Assess your understanding of **invertebrate neuroendocrine systems** by taking the pretest on your BiologyNow CD-ROM.

THE VERTEBRATE ENDOCRINE SYSTEM

Learning Objectives

6 Identify the principal vertebrate endocrine glands, and describe the actions of their hormones.

7 Describe the mechanisms by which the hypothalamus and pituitary gland integrate many regulatory functions; describe the actions of the hypothalamic and pituitary hormones.

8 Describe the actions of the thyroid and parathyroid hormones, their regulation, and the effects of malfunction.

9 Contrast the actions of insulin and glucagon, and describe the disorders associated with impaired function of these hormones or their receptors.

10 Describe the actions of the adrenal hormones, including their role in helping the body cope with stress.

Vertebrate hormones regulate such diverse activities as growth, development, reproduction, metabolic rate, fluid balance, blood homeostasis, and coping with stress. Most vertebrates have similar endocrine glands, although the actions of some hormones may be different in various groups. Table 47-1 gives the sources, target tissues, and principal physiological actions of some of the major vertebrate hormones. Many of these hormones are regulated by the hypothalamus and pituitary gland. The principal human endocrine glands are illustrated in Figure 47-9.

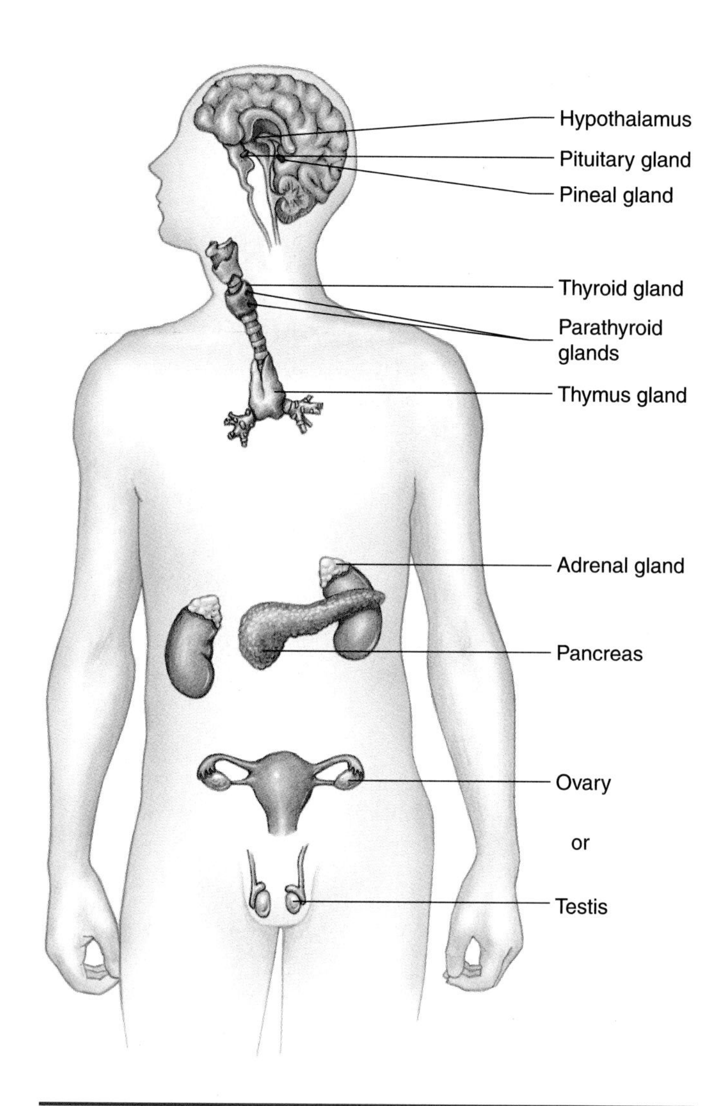

FIGURE **47-9** **The classical human endocrine glands.** The hormones produced by these glands and their actions are discussed in this chapter.

Homeostasis depends on normal concentrations of hormones

When a disorder or disease process affects an endocrine gland, the rate of secretion may become abnormal. If **hyposecretion** (abnormally reduced output) occurs, target cells are deprived of needed stimulation. If **hypersecretion** (abnormally increased output) occurs, the target cells may be overstimulated. In some endocrine disorders, an appropriate amount of hormone is secreted, but the target cell receptors do not function properly. As a result, the target cells cannot respond to the hormone. Any of these abnormalities leads to loss of homeostasis, resulting in predictable metabolic malfunctions and clinical symptoms (Table 47-2).

The hypothalamus regulates the pituitary gland

Most endocrine activity is controlled directly or indirectly by the **hypothalamus,** which links the nervous and endocrine systems both anatomically and physiologically. The **pituitary gland** is connected to the hypothalamus by the pituitary stalk. In response to input from other areas of the brain and from hormones in the blood, neurons of the hypothalamus secrete neurohormones that regulate specific physiological processes.

Because its secretions control the activities of several other endocrine glands, biologists consider the pituitary gland to be the master gland of the body. Although it is only the size of a large pea and weighs only about 0.5 g (0.02 oz), the pituitary gland produces and secretes at least seven peptide hormones that exert far-reaching influence over body activities. The human pituitary gland consists of two main parts: the **anterior lobe** and the **posterior lobe.** In some vertebrates, an intermediate lobe is present.

The posterior lobe of the pituitary gland releases hormones produced by the hypothalamus

The **posterior lobe** of the pituitary gland develops from brain tissue. This neuroendocrine organ *secretes* two peptide hormones, **oxytocin** and **antidiuretic hormone,** or **ADH** (also known as *vasopressin*). (The function of ADH in regulating water reabsorption in the kidneys was discussed in Chapter 46.) These

TABLE 47-1	Some Endocrine Glands and Their Hormones*		
Gland	**Hormone**	**Target Tissue**	**Principal Actions**
Hypothalamus	Releasing and inhibiting hormones	Anterior lobe of pituitary	Regulate secretion of hormones by the anterior pituitary
Posterior pituitary (storage and release of hormones produced by hypothalamus)	Oxytocin	Uterus Mammary glands	Stimulates contraction Stimulates ejection of milk into ducts
	Antidiuretic hormone (ADH)	Kidneys (collecting ducts)	Stimulates reabsorption of water
Anterior pituitary	Growth hormone (GH)	General	Stimulates growth of skeleton and muscle
	Prolactin	Mammary glands	Stimulates milk production
	Thyroid-stimulating hormone (TSH)	Thyroid gland	Stimulates secretion of thyroid hormones
	Adrenocorticotropic hormone (ACTH)	Adrenal cortex	Stimulates secretion of adrenal cortical hormones
	Gonadotropic hormones* (follicle-stimulating hormone [FSH]; luteinizing hormone [LH])	Gonads	Stimulate gonad function and growth
Thyroid gland	Thyroxine (T_4) and triiodothyronine (T_3)	General	Stimulate metabolic rate; regulate energy metabolism
	Calcitonin	Bone	Lowers blood-calcium level
Parathyroid glands	Parathyroid hormone	Bone, kidneys digestive tract	Regulates blood-calcium level
Pancreas	Insulin	General	Lowers blood glucose concentration
	Glucagon	Liver, adipose tissue	Raises blood glucose concentration
Adrenal Medulla	Epinephrine and norepinephrine	Muscle; blood vessels; liver; adipose tissue	Help body cope with stress; increase metabolic rate; raise blood glucose level; increase heart rate and blood pressure
Adrenal Cortex	Mineralocorticoids	Kidney tubules	Maintain sodium and potassium balance
	Glucocorticoids	General	Help body cope with long-term stress; raise blood-glucose level
Pineal gland	Melatonin	Hypothalamus	Important in biological rhythms
Ovary†	Estrogens	General; uterus	Develop and maintain sex characteristics in female; stimulate growth of uterine lining
	Progesterone	Uterus; breast	Stimulates development of uterine lining
Testis†	Testosterone	General; reproductive structures	Develops and maintains sex characteristics in males; promotes spermatogenesis

*Discussed in Chapter 48.

†For more detailed description see Tables 48-1 and 48-2.

TABLE 47-2	Consequences of Endocrine Malfunction	
Hormone	**Hyposecretion**	**Hypersecretion**
Growth hormone	Pituitary dwarfism	Gigantism if malfunction occurs in childhood; acromegaly in adult
Thyroid hormones	Cretinism (in children); myxedema, a condition of prounounced adult hypothyroidism; dietary iodine deficiency leads to hyposecretion and goiter	Hyperthyroidism; increased metabolic rate, nervousness, irritability; goiter, can be caused by Grave's disease
Parathyroid hormone	Spontaneous discharge of nerves; spasms; tetany; death	Weak, brittle bones; kidney stones
Insulin	Diabetes mellitus	Hypoglycemia
Hormones of adrenal cortex	Addison's disease	Cushing's disease

hormones are actually *produced* by neuroendocrine cells in two distinct areas of the hypothalamus. Enclosed within vesicles, oxytocin and ADH are transported down the axons of these neuroendocrine cells into the posterior lobe of the pituitary gland (Fig. 47-10). The vesicles are stored in the axon terminals until the neuron is stimulated. Then the hormones are released and diffuse into surrounding capillaries.

Oxytocin concentration in the blood rises toward the end of pregnancy, stimulating the strong contractions of the uterus needed to expel a baby. Oxytocin is sometimes administered clinically to initiate or speed labor. After birth, when an infant sucks at its mother's breast, sensory neurons signal the pituitary to release oxytocin. The hormone stimulates contraction of smooth muscle cells surrounding the milk glands so that milk is let down into the ducts, from which it can be sucked by the infant. Because oxytocin also stimulates the uterus to contract, breast-feeding promotes recovery of the uterus to nonpregnant size.

The anterior lobe of the pituitary gland regulates growth and other endocrine glands

The **anterior lobe** of the pituitary develops from epithelial cells rather than neural cells. The anterior lobe functions like a classical endocrine gland: it receives signals by way of the blood and releases its hormones into the blood. The hypothalamus produces several **releasing hormones** and **inhibiting hormones** that regulate the production and secretion of specific hormones by the anterior pituitary gland. These neurohormones enter capillaries and pass through special portal veins that connect the hypothalamus with the anterior lobe of the pituitary. (These portal veins, like the hepatic portal vein, do not deliver blood to a larger vein directly but connect two sets of capillaries.) Within the anterior lobe of the pituitary, the portal veins divide into a second set of capillaries. The releasing and inhibiting hormones pass through the walls of these capillaries into the tissue of the anterior lobe.

The anterior lobe secretes prolactin, growth hormone, and several **tropic hormones** that stimulate other endocrine glands (Fig. 47-11). **Prolactin** stimulates the cells of the mammary glands to produce milk in a nursing mother.

Growth hormone stimulates protein synthesis

Small children measure themselves periodically against their parents, eagerly awaiting that time when they, too, will be "big."

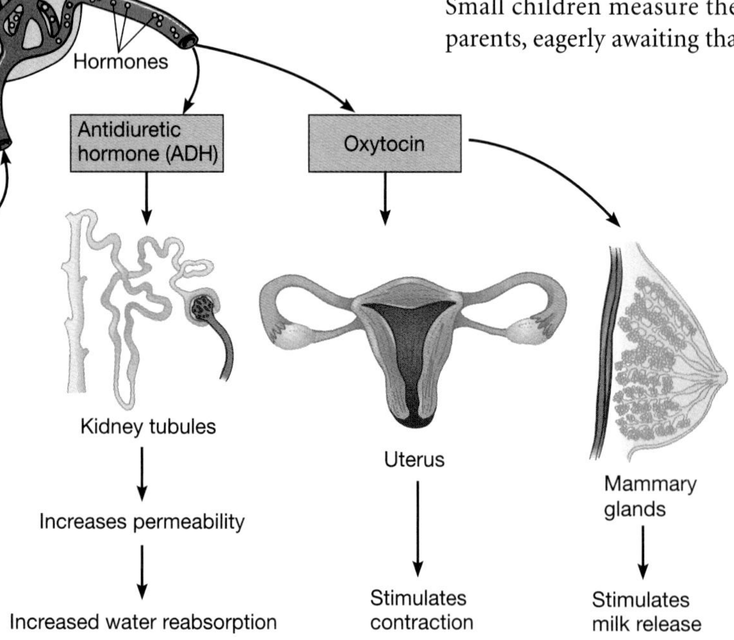

FIGURE **47-10**

The relationship between the hypothalamus and posterior lobe of the pituitary gland.

Neuroendocrine cells of the hypothalamus manufacture hormones that are secreted by the posterior lobe of the pituitary. The axons of these neurons extend down into the posterior lobe of the pituitary. Their hormones are packaged in vesicles that are transported through the axons and stored in their ends. When needed, the hormones are secreted, enter the blood, and are transported by the circulatory system.

Whether someone is tall or short depends on many factors, including genes, diet, and hormonal balance. **Growth hormone** (**GH;** also called *somatotropin*) is referred to as an **anabolic hormone** because it promotes tissue growth. Many of the effects of GH on skeletal growth are indirect. GH stimulates liver cells (and cells of many other tissues) to produce peptides called **insulin-like growth factors (IGFs).** These growth factors (1) promote the linear growth of the skeleton by stimulating cartilage formation in growth areas of bones, and (2) stimulate general tissue growth and increase in organ size by promoting protein synthesis and other anabolic processes.

In adults as well as in growing children, GH is secreted in pulses throughout the day. The hypothalamus regulates GH secretion by secreting a **growth hormone–releasing hormone (GHRH)** and a **growth hormone–inhibiting hormone (GHIH;** also called somatostatin). A high level of GH in the blood signals the hypothalamus to secrete GHIH and the pituitary slows its release of GH. A low level of GH in the blood stimulates the hypothalamus to secrete GHRH, which stimulates the pituitary gland to release more GH. Many other factors, including blood sugar level, amino acid concentration, and stress, influence GH secretion.

Research supports the age-old notions that to grow, children need plenty of sleep, a proper diet, and regular exercise. Secretion of GH increases during exercise, probably because rapid metabolism by muscle cells lowers the blood glucose level. GH is secreted about 1 hour after the onset of deep sleep and in a series of pulses 2 to 4 hours after a meal.

Emotional support is also necessary for proper growth. Growth is retarded in children who are deprived of cuddling, playing, and other forms of nurture, even when their needs for food and shelter are met. In extreme cases, childhood stress can produce a form of retarded development known as *psychosocial dwarfism*. Some emotionally deprived children exhibit abnormal sleep patterns, which may be the basis for decreased secretion of GH.

Other hormones also influence growth. Thyroid hormones appear necessary for normal GH secretion and function and for normal tissue response to IGFs. Sex hormones must be present for the adolescent growth spurt to occur. However, the presence

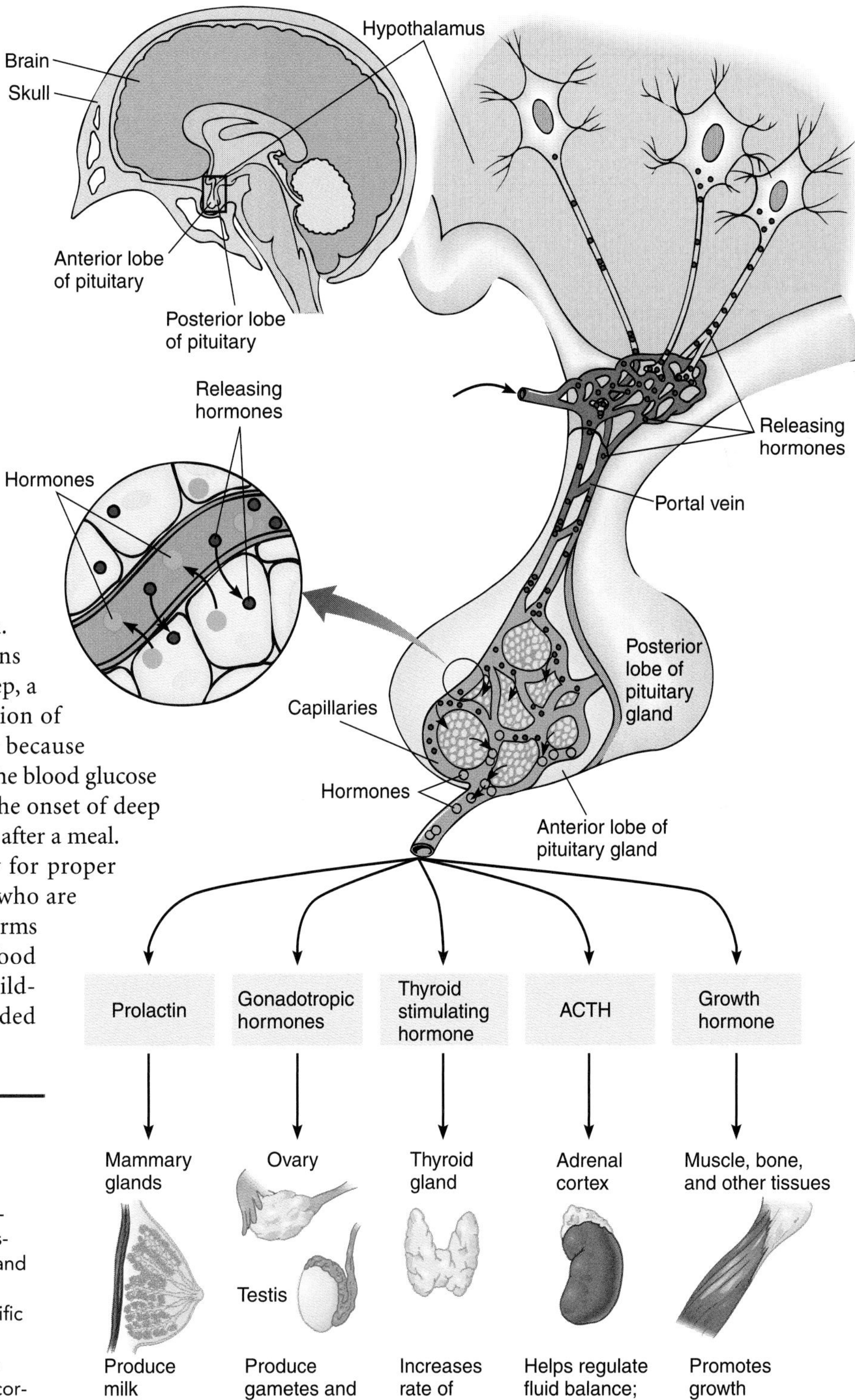

FIGURE **47-11**

The hypothalamus regulates the anterior lobe of the pituitary gland.

The hypothalamus secretes several specific releasing and inhibiting hormones that are transported to the anterior lobe of the pituitary gland by way of portal veins. These regulatory hormones stimulate or inhibit the release of specific hormones by cells of the anterior lobe. The anterior lobe secretes hormones that act on a wide variety of target tissues. (ACTH, adrenocorticotropic hormone)

of sex hormones eventually causes the growth centers within the long bones to ossify, so that further increase in height is impossible even when GH is present.

Inappropriate amounts of growth hormone secretion result in abnormal growth

Have you ever wondered why some people fail to grow normally? Deficiency in GH during childhood is called **pituitary dwarfism** because it is usually caused by insufficient production of GH by the pituitary gland. However, dwarfism also results when the liver produces insufficient IGF or when target tissues do not respond to IGF. Although miniature, a pituitary dwarf has normal intelligence and is usually well proportioned. If the growth centers in the long bones are still active when this condition is diagnosed, it can often be effectively treated by human GH injection. GH is now synthesized by recombinant DNA technology.

A person grows abnormally tall when the anterior pituitary secretes excessive amounts of GH during childhood. This condition is called **gigantism.** If hypersecretion of GH occurs during adulthood, the individual cannot grow taller. Instead, connective tissue thickens, and bones in the hands, feet, and face may increase in diameter. This condition is known as **acromegaly** (acromegaly means "large extremities").

Thyroid hormones increase metabolic rate

The **thyroid gland** is located in the neck region, in front of the trachea and below the larynx. Two of the **thyroid hormones—thyroxine,** also known as $\mathbf{T_4}$, and **triiodothyronine,** or $\mathbf{T_3}$, are synthesized from the amino acid tyrosine and from iodine. Thyroxine has four iodine atoms attached to each molecule; T_3 has three. In most target tissues, T_4 is converted to T_3, the more active form (see Fig. 47-2c). We describe the actions of calcitonin, another hormone secreted by the thyroid gland, when we discuss the parathyroid glands.

In vertebrates, thyroid hormones are essential for normal growth and development, and they increase the rate of metabolism in most body tissues. After T_3 binds with its receptor in the nucleus of a target cell, the T_3 receptor complex induces or suppresses synthesis of specific enzymes and other proteins. Thyroid hormones also help regulate the synthesis of proteins necessary for cell differentiation. For example, tadpoles cannot develop into adult frogs without thyroxine.

Thyroid secretion is regulated by negative feedback mechanisms

The regulation of thyroid hormone secretion depends mainly on a negative feedback loop between the anterior pituitary and the thyroid gland (Fig. 47-12). When the concentration of thyroid hormones in the blood rises above normal, the anterior pituitary secretes less **thyroid-stimulating hormone (TSH):**

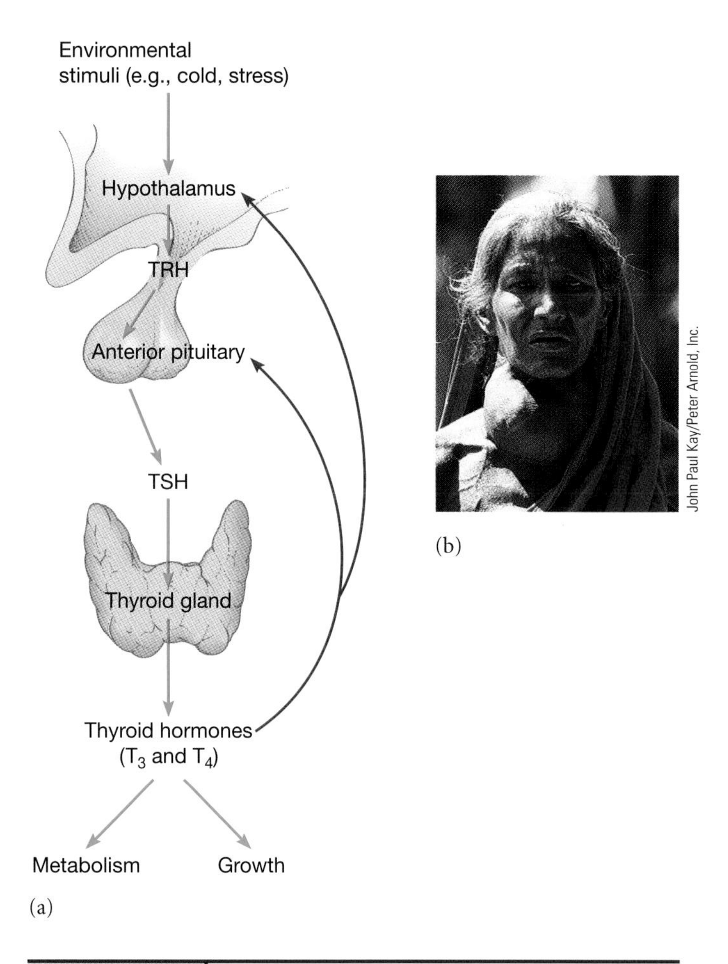

FIGURE **47-12** **Regulation of thyroid hormone secretion by negative feedback.**

(a) An increase in concentration of thyroid hormones above normal levels signals the anterior pituitary and hypothalamus to limit thyroid-stimulating hormone (TSH) production. Thus, thyroid hormones limit their own production by negative feedback. Green arrows indicate stimulation; red arrows indicate inhibition. (TRH, TSH-releasing hormone) **(b)** When iodine is deficient in the diet, the thyroid gland cannot produce its hormones. TSH secretion increases causing goiter, an enlargement of the thyroid gland.

Thyroid hormone concentration high ⟶ anterior pituitary secretes less TSH ⟶ thyroid gland secretes less hormone ⟶ homeostasis

Too much thyroid hormone in the blood also affects the hypothalamus, inhibiting secretion of a **TSH-releasing hormone, TRH.** However, the hypothalamus probably exerts its regulatory effects mainly in certain stressful situations, such as extreme weather change. Exposure to very cold weather may stimulate the hypothalamus to increase secretion of TRH. This action raises body temperature through increased metabolic heat production. Increased body temperature has a negative feedback effect, limiting further secretion of thyroid hormones.

When the concentration of thyroid hormones decreases, the pituitary secretes more TSH. TSH acts by way of cAMP to promote synthesis and secretion of thyroid hormones and also to promote increased size of the thyroid gland itself. The effect of TSH can be summarized as follows:

Thyroid hormone concentration low ⟶ anterior pituitary secretes more TSH ⟶ thyroid gland secretes more hormone ⟶ homeostasis

Malfunction of the thyroid gland leads to specific disorders

Extreme hypothyroidism during infancy and childhood results in low metabolic rate and can lead to **cretinism,** a condition of retarded mental and physical development. When hypothyroidism is diagnosed early enough and treated with thyroid hormones, cretinism can be prevented.

An adult who feels like sleeping all the time, has little energy, and is mentally slow or confused may be suffering from hypothyroidism. When there is almost no thyroid function, the basal metabolic rate is reduced by about 40% and the patient develops **myxedema,** characterized by a slowing down of physical and mental activity. Hypothyroidism is treated by oral administration of the missing hormone.

Hyperthyroidism does not cause abnormal growth but does increase metabolic rate by 60% or even more. This increase in metabolism results in the rapid use of nutrients, causing the individual to be hungry and to eat more. But this is not sufficient to meet the demands of the rapidly metabolizing cells, so people with this condition often lose weight. They also tend to be nervous, irritable, and emotionally unstable. The most common form of hyperthyroidism is **Grave's disease,** an autoimmune disease. Abnormal antibodies bind to TSH receptors, activating them. This leads to increased production of thyroid hormones.

An abnormal enlargement of the thyroid gland, or **goiter,** may be associated with either hypersecretion or hyposecretion (see Figure 47-12). In Grave's disease, the abnormal antibodies that activate TSH receptors can cause the development of a goiter. Hyposecretion can be caused by dietary iodine deficiency. Without iodine the gland cannot make thyroid hormones, so their concentration in the blood decreases. In compensation, the anterior pituitary secretes large amounts of TSH, which stimulates growth of the thyroid gland, sometimes to gigantic proportions. However, enlargement of the gland does not increase production of the hormones, because the needed ingredient is still missing. Seafood is a rich source of iodine, and iodine is also added to table salt as a nutritional supplement. In fact, thanks to iodized salt, goiter is no longer common in the United States and other highly developed countries. In other parts of the world, however, hundreds of thousands still suffer from this easily preventable disorder.

The parathyroid glands regulate calcium concentration

The **parathyroid glands** are typically embedded in the connective tissue surrounding the thyroid gland. These glands secrete **parathyroid hormone (PTH),** a polypeptide that helps regulate the calcium level of the blood and interstitial fluid (Fig. 47-13a). Parathyroid hormone acts by way of a G protein–linked receptor and cAMP to stimulate calcium release from bones and to increase calcium reabsorption from the kidney tubules. It also activates vitamin D, which then increases the amount of calcium absorbed from the intestine.

Calcitonin, a peptide hormone secreted by the thyroid gland, works antagonistically to parathyroid hormone (Fig. 47-13b). When the concentration of calcium rises above homeostatic levels, calcitonin is released and rapidly inhibits removal of calcium from bone.

The islets of the pancreas regulate glucose concentration

In addition to secreting digestive enzymes (see Chapter 45), the pancreas is an important endocrine gland. Its hormones, insulin and glucagon, are secreted by cells that occur in little clusters, the **islets of Langerhans,** throughout the pancreas (Fig. 47-14). About 1 million islets are present in the human pancreas. They consist mainly of **beta cells,** which secrete insulin, and **alpha cells,** which secrete glucagon.

Insulin lowers the concentration of glucose in the blood

Insulin is an anabolic hormone that promotes storage of fuel molecules. It stimulates cells of many tissues, including liver, muscle, and fat cells, to take up glucose from the blood by facilitated diffusion (see Chapter 5). Once glucose enters muscle cells, it is either used immediately as fuel or stored as glycogen. Insulin also inhibits liver cells from releasing glucose. Thus insulin activity *lowers* the glucose level in the blood.

Insulin helps regulate fat and protein metabolism. It reduces the use of fatty acids as fuel and instead stimulates their storage in adipose tissue. Insulin has an anabolic effect on protein metabolism, resulting in a net increase in protein. It promotes protein synthesis by increasing the number of amino acid transporters in the plasma membrane, thus stimulating the transport of certain amino acids into cells. Insulin also promotes transcription and translation.

Glucagon raises the concentration of glucose in the blood

The effects of **glucagon** are opposite to those of insulin. Its main action is to raise blood glucose level. It does this by stimulating liver cells to convert glycogen to glucose, a process known as *glycogenolysis.* Glucagon also stimulates *gluconeogenesis,* the production of glucose from other metabolites. Glucagon mobilizes fatty acids and amino acids as well as glucose.

Insulin and glucagon secretion are regulated by glucose concentration

The concentration of glucose in the blood directly controls insulin and glucagon secretion (Fig. 47-15). After a meal, when the blood-glucose level rises as a result of intestinal absorption,

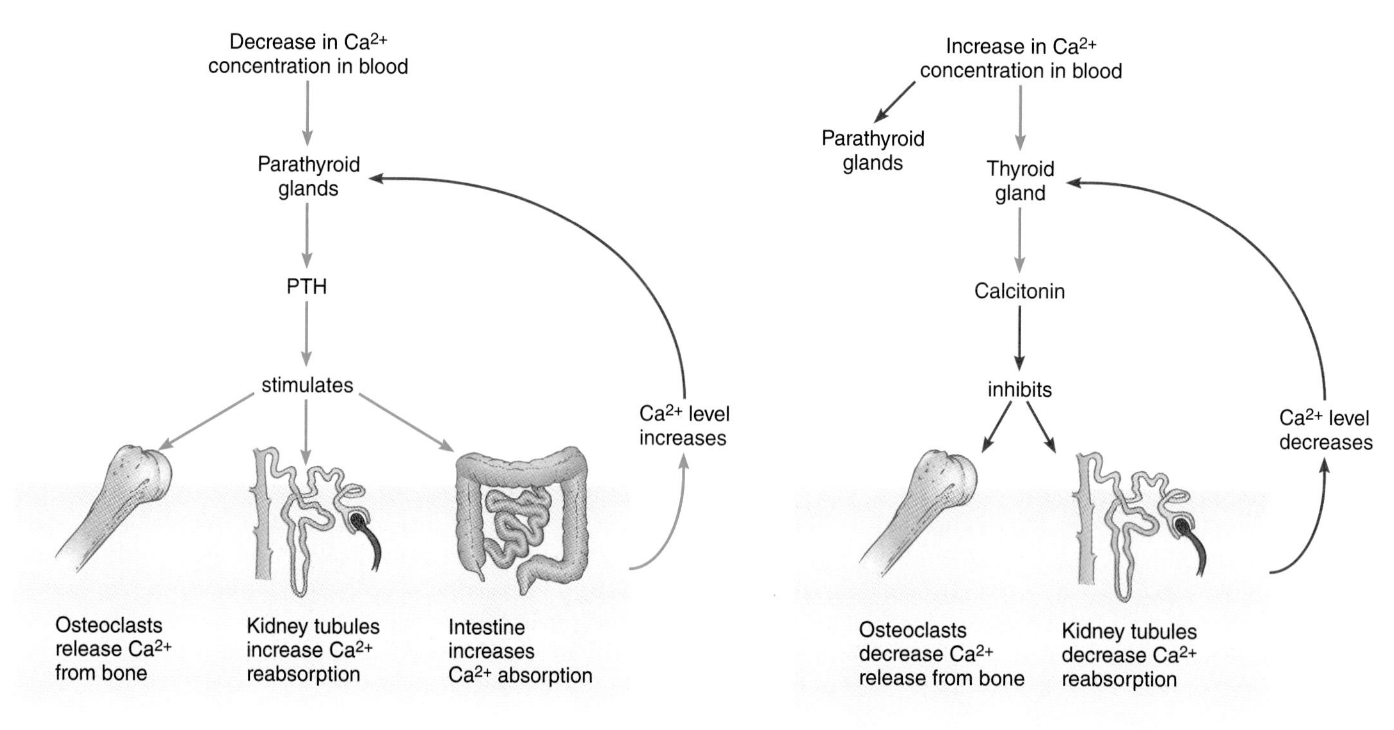

(a) Calcium concentration too low

(b) Calcium concentration too high

FIGURE **47-13** **Regulation of calcium homeostasis by parathyroid hormone (PTH) and calcitonin.** **(a)** When calcium concentration decreases below the normal range, the parathyroid glands secrete parathyroid hormone (PTH). This hormone stimulates homeostatic mechanisms that restore appropriate calcium concentration. **(b)** When calcium concentration increases above normal limits, the parathyroid glands are inhibited. The thyroid gland secretes calcitonin, which inhibits Ca^{2+} release from bone and decreases Ca^{2+} reabsorption in the kidneys. Green arrows indicate stimulation; red arrows indicate inhibition.

beta cells are stimulated to increase insulin secretion. Then, as cells remove glucose from the blood, decreasing its concentration, insulin secretion decreases.

High blood glucose concentration ⟶ beta cells secrete insulin ⟶ blood glucose concentration decreases ⟶ homeostasis

When you have not eaten for several hours, the concentration of glucose in your blood begins to fall. When it decreases from its normal fasting level of about 90 mg of glucose per 100 mL of blood to about 70 mg of glucose, the alpha cells of the islets increase their secretion of glucagon. Glucose is mobilized from storage in the liver cells, and blood glucose concentration returns to normal:

Low blood glucose concentration ⟶ alpha cells secrete glucagon ⟶ blood glucose concentration increases ⟶ homeostasis

Alpha cells respond to the glucose concentration within their own cytoplasm, which reflects the blood glucose level. When blood glucose level is high, there is generally a high level of glucose within the alpha cells, and glucagon secretion is inhibited.

Note that these are negative feedback mechanisms and that insulin and glucagon work antagonistically to keep blood glucose concentration within normal limits. When the glucose level rises, insulin release brings it back to normal; when the glucose concentration falls, glucagon acts to raise it again. The insulin-glucagon system is a powerful, fast-acting mechanism for keeping blood glucose level within normal limits. Why do you think it is important to maintain a constant blood glucose level? Recall that brain cells ordinarily are unable to use other nutrients as fuel and so depend on a continuous supply of glucose. As we will discuss, several other hormones also affect blood glucose concentration.

Diabetes mellitus is a serious disorder of carbohydrate metabolism

Diabetes mellitus, the most common endocrine disorder, is a serious health problem, and the World Health Organization reports that its global prevalence is increasing. An estimated 150 million people worldwide have this disorder, and the number is expected to reach more than 220 million by the year 2010. Diabetes, a leading cause of premature death, is a major cause of cardiovascular disease, blindness, nerve disease, kidney disorders, and gangrene of the limbs. (Note that diabetes mellitus is an entirely different disorder from diabetes insipidus, which is discussed in Chapter 46.)

Diabetes mellitus is actually a group of related disorders characterized by high blood glucose concentrations. Two main

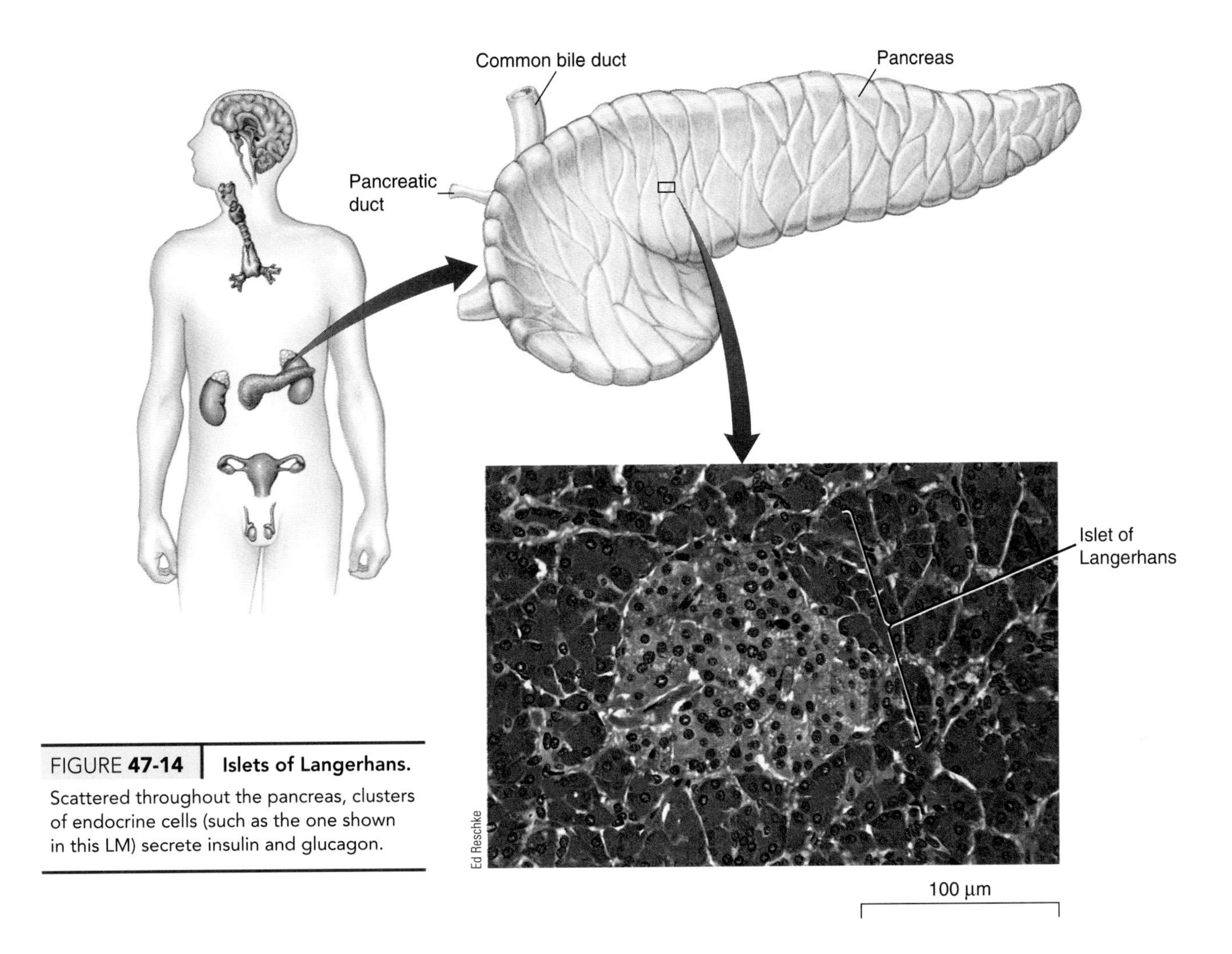

FIGURE **47-14** **Islets of Langerhans.**

Scattered throughout the pancreas, clusters of endocrine cells (such as the one shown in this LM) secrete insulin and glucagon.

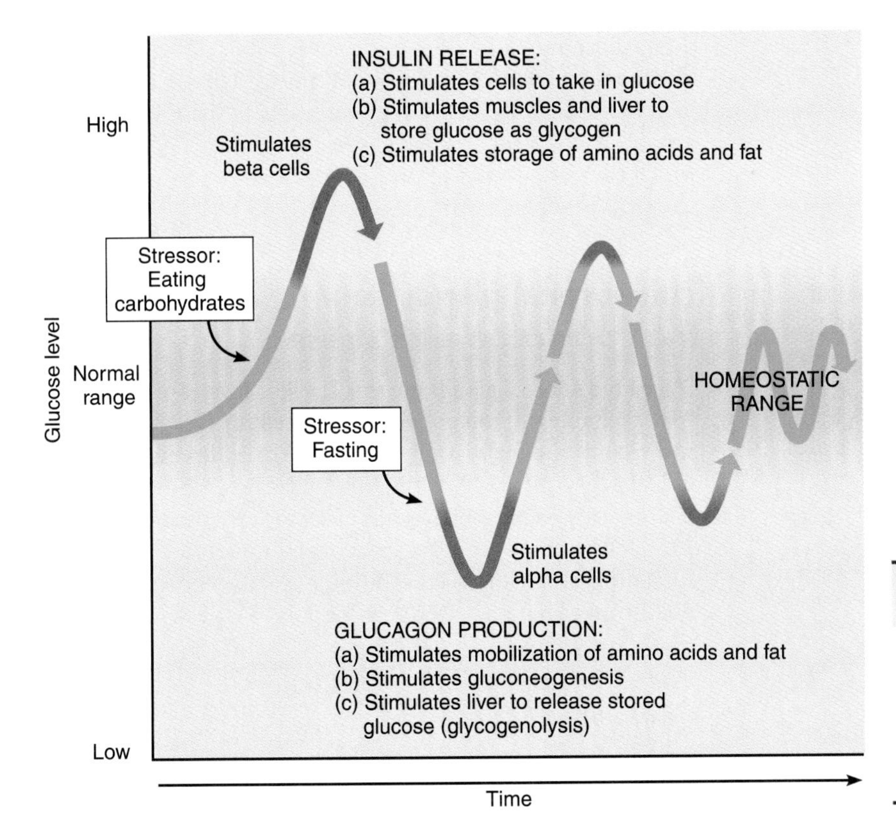

types of diabetes are type 1 and type 2. *Type 1 diabetes* is an autoimmune disease in which antibodies mark the beta cells for destruction. T cells destroy the beta cells, resulting in insulin deficiency. Insulin injections are needed to correct the carbohydrate imbalance that results. This disorder usually develops before age 30, often during childhood. Type 1 diabetes is caused by a combination of genetic predisposition and environmental factors, possibly including infection by a virus.

More than 90% of diabetics have *type 2 diabetes.* This disorder develops gradually, usually in overweight people. In type 2 diabetes, normal (or greater than normal) concentrations of insulin are present in the blood, but insulin recep-

ACTIVE FIGURE **47-15** **Regulation of glucose concentration.**

Insulin and glucagon work antagonistically to regulate blood glucose levels.

BiologyNow™ Learn more about **the hormonal response to stress** by clicking on this figure on your BiologyNow CD-ROM.

tors on target cells do not bind it, a condition known as **insulin resistance.** Biologists do not fully understand the causes of insulin resistance, but do know there is a strong genetic component. Even if identical twins are raised separately in very different environments, when one identical twin develops insulin resistance the other twin almost always does also. Investigators have also reported an association between increased fat tissue and insulin resistance.

Many type 2 diabetics can keep their blood glucose levels within normal range by diet management, weight loss, and regular exercise. When this treatment approach is not effective, they are treated with oral drugs that stimulate insulin secretion and promote its actions. Approximately one third of type 2 diabetics eventually need insulin injections.

Recently, a research team at Yale University found that a high percentage of obese children (25% of children under age 10 and 21% between ages 11 and 18) have impaired glucose tolerance, an early warning sign of diabetes. Impaired glucose tolerance, which affects more than 200 million people worldwide, is defined as hyperglycemia (a blood glucose concentration that is abnormally high but lower than the concentration that is diagnostic for diabetes) following ingestion of a large amount of glucose (a glucose load).

Similar metabolic disturbances occur in both types of diabetes mellitus: disruption of carbohydrate, fat, and protein metabolism, and electrolyte imbalance.

1. Decreased use of glucose. Because the cells of diabetics cannot take up glucose from the blood, it accumulates there, causing hyperglycemia. Instead of the normal fasting concentration of about 90 mg per 100 mL, the level exceeds 140 mg per 100 mL and may reach concentrations of more than 500 mg per mL. When glucose concentration exceeds its **tubular transport maximum (Tm),** the maximum rate at which it can be reabsorbed, glucose is excreted in the urine.
2. Increased fat mobilization. Despite the large quantities of glucose in the blood, insulin-dependent cells in a diabetic can take in only about 25% of the glucose they require for fuel. Cells turn to fat and protein for energy. The absence of insulin promotes the mobilization of fat stores, providing nutrients for cellular respiration. But unfortunately, the blood lipid level may reach five times the normal level, leading to the development of atherosclerosis (see Chapter 42).

   Increased fat metabolism increases the formation of ketone bodies. These compounds build up in the blood, causing *ketoacidosis,* a condition in which the body fluids and blood become too acidic. When the ketone level in the blood rises, ketones appear in the urine, another clinical indication of diabetes mellitus. If severe, ketoacidosis can lead to coma and death.
3. Increased protein use. Lack of insulin also results in increased protein breakdown relative to protein synthesis, so the untreated diabetic becomes thin and emaciated.
4. Electrolyte imbalance. When ketone bodies and glucose are excreted in the urine, water follows by osmosis; as a result, urine volume increases. The resulting dehydration causes thirst and challenges electrolyte homeostasis. When ketones are excreted, they also take sodium, potassium, and some other cations with them, contributing to electrolyte imbalance.

The dramatic increase in type 2 diabetes is a public health problem resulting from a sedentary lifestyle and an increasing prevalence of obesity. Researchers are looking for causes and cures of diabetes. Some research groups are focusing on education and lifestyle changes, whereas others are looking for the genes that contribute to diabetes or for signaling molecules involved in glucose metabolism.

Researchers have linked a number of hormones produced by adipose tissue to diabetes. These include leptin (obese people produce large amounts but are resistant to its effects), resistin (antagonizes the action of insulin), and adiponectin (promotes insulin's effects, but obese people produce low amounts). Several research teams are investigating the role of fatty acids. In obese individuals, fatty acids build up in muscle fibers, the principal insulin target that removes glucose from the blood. Fatty acids appear to interfere with the signal pathway used by insulin.

In hypoglycemia the blood glucose concentration is too low

Low blood glucose concentration, or *hypoglycemia,* sometimes occurs in people who later develop diabetes. It may be an overreaction by the islets to glucose challenge; that is, too much insulin is secreted in response to carbohydrate ingestion. About 3 hours after a meal, the blood sugar concentration falls below normal, making the individual feel very drowsy. If this reaction is severe enough, the patient may become uncoordinated or even unconscious.

Serious hypoglycemia can develop if diabetics receive injections of too much insulin or if the islets, because of a tumor, secrete too much insulin. The blood glucose concentration may then fall drastically, depriving brain cells of their needed fuel supply. These events can lead to **insulin shock,** a condition in which the patient may appear drunk or may become unconscious, suffer convulsions, and even die.

The adrenal glands help the body cope with stress

The paired **adrenal glands** are small, yellow masses of tissue that lie in contact with the upper ends of the kidneys (Fig. 47-16). Each gland consists of a central portion, the **adrenal medulla,** and a larger outer section, the **adrenal cortex.** Although joined anatomically, the adrenal medulla and cortex develop from different types of tissue in the embryo and function as distinctly different glands. Both secrete hormones that help regulate metabolism, and both help the body respond to stress.

The adrenal medulla initiates an alarm reaction

The adrenal medulla is a neuroendocrine gland that is coupled to the sympathetic nervous system. It develops from neural tis-

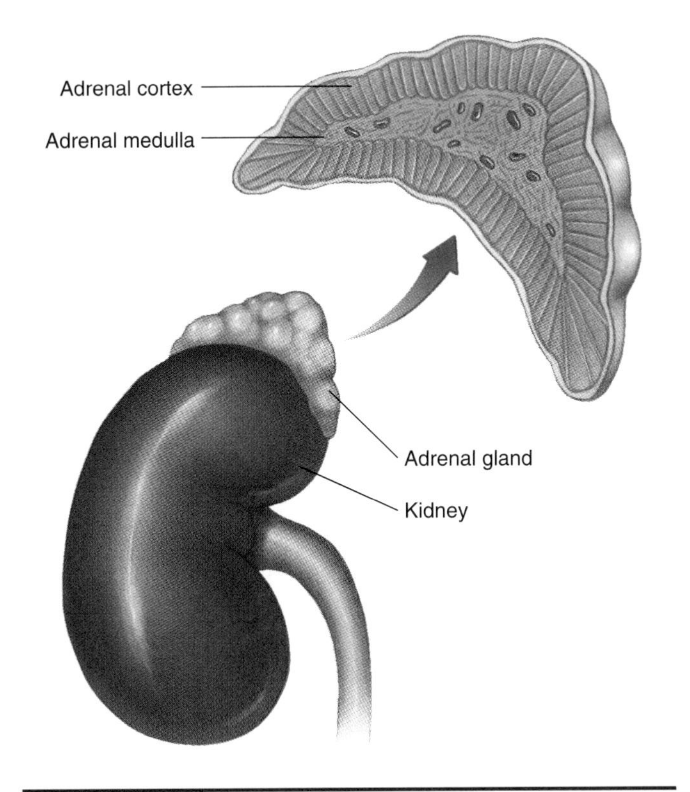

FIGURE **47-16** | **Adrenal gland.**

The adrenal glands are located above the kidneys. Each gland consists of a central adrenal medulla surrounded by an adrenal cortex.

sue, and its secretion is controlled by sympathetic nerves. The adrenal medulla secretes **epinephrine** and **norepinephrine.** Chemically, these hormones are very similar; they belong to the chemical group known as **catecholamines.** Recall that norepinephrine also serves as a neurotransmitter released by sympathetic neurons and by some neurons in the central nervous system. Most of the hormone output of the adrenal medulla is epinephrine.

Under normal conditions, the adrenal medulla continuously secretes small amounts of both epinephrine and norepinephrine. Their secretion is under neural control. When anxiety is aroused, the brain sends messages via sympathetic nerves to the adrenal medulla. Sympathetic neurons release acetylcholine, which triggers the adrenal medulla to release larger amounts of epinephrine and norepinephrine.

During a stressful situation, adrenal medullary hormones initiate an *alarm reaction* enabling you to think more quickly, fight harder, or run faster than usual. Metabolic rate increases by as much as 100%. Blood is rerouted to those organs essential for emergency action (Fig. 47-17). Blood vessels going to the brain, muscles, and heart are dilated, whereas those to the skin and kidneys are constricted. Constriction of blood vessels serving the skin has the added advantage of decreasing blood loss in case of hemorrhage (and explains the sudden paling that comes with fear or rage). At the same time, the heart beats faster. Thresholds in the reticular activating system of the brain are lowered, so you become more alert (see Chapter 40). Strength of muscle contraction increases. The adrenal medullary hormones also raise fatty acid and glucose levels in the blood, ensuring needed fuel for extra energy.

The adrenal cortex helps the body deal with chronic stress

The adrenal cortex synthesizes steroid hormones from cholesterol (which in turn is made in the liver from acetyl coenzyme A.) Although the adrenal cortex produces more than 30 types of steroids, it secretes only three types of hormones in significant amounts: sex hormone precursors, mineralocorticoids, and glucocorticoids.

In both sexes, the adrenal cortex secretes sex hormone precursors, such as androgens (masculinizing hormones). Certain tissues convert sex hormone precursors to **testosterone,** the principal male sex hormone, and **estradiol,** the principal female sex hormone. In males androgen production by the adrenal cortex is not significant, because the testes produce testosterone. However, in females the adrenal cortex secretes the androgen used to produce most of the testosterone circulating in the blood.

The principal **mineralocorticoid** is **aldosterone.** Recall from Chapter 46 that aldosterone helps regulate fluid balance by regulating salt balance. In response to aldosterone, the kidneys reabsorb more sodium and excrete more potassium. As a result of increased sodium, extracellular fluid volume increases, which results in greater blood volume and elevated blood pressure.

When the adrenal glands do not produce enough aldosterone, large amounts of sodium are excreted in the urine. Water leaves the body with the sodium (because of osmotic pressure), and the blood volume may be so markedly reduced that the patient dies of low blood pressure.

Cortisol, also called *hydrocortisone,* accounts for about 95% of the **glucocorticoid** activity of the human adrenal cortex. Cortisol helps ensure adequate fuel supplies for the cells when the body is under stress. Its principal action is to stimulate gluconeogenesis in liver cells. Cortisol helps provide nutrients for glucose production by stimulating amino acid transport into liver cells (see Fig. 47-17). It also promotes mobilization of fats so that glycerol from triacylglycerol molecues is available for conversion to glucose. These actions ensure that glucose and glycogen are produced in the liver, and the concentration of glucose in the blood rises. Thus the adrenal cortex provides an important backup system for the adrenal medulla, ensuring glucose supplies when the body is under stress and in need of extra energy.

During stress, the brain and the adrenal glands work together to help the body cope effectively (see Chapter 40, *On the Cutting Edge: The Neurophysiology of Traumatic Experience*). Almost any type of stress stimulates the hypothalamus to secrete **corticotropin-releasing factor (CRF)** which stimulates the anterior pituitary to secrete **adrenocorticotropic hormone (ACTH).** ACTH regulates both glucocorticoid and aldosterone secretion. ACTH is so potent that it can result in a 20-fold increase in cortisol secretion within minutes. When the body is not under stress, high levels of cortisol in the blood inhibit both CRF secretion by the hypothalamus and ACTH secretion by the pituitary.

Destruction of the adrenal cortex and the resulting decrease in aldosterone and cortisol secretion cause **Addison's disease.**

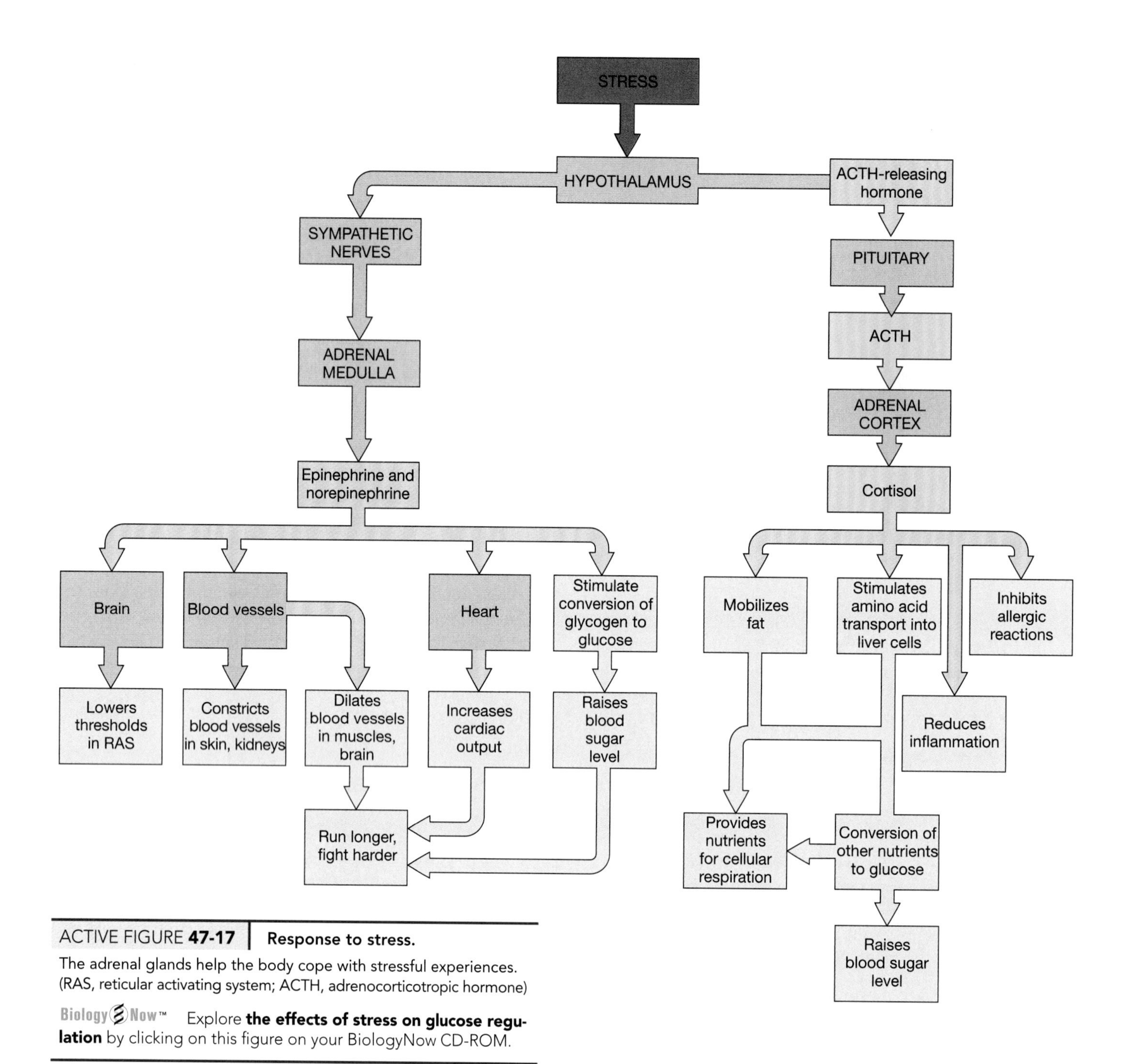

ACTIVE FIGURE **47-17** | **Response to stress.**

The adrenal glands help the body cope with stressful experiences. (RAS, reticular activating system; ACTH, adrenocorticotropic hormone)

Biology Now™ Explore **the effects of stress on glucose regulation** by clicking on this figure on your BiologyNow CD-ROM.

Reduction in cortisol prevents the body from regulating the blood glucose concentration because it cannot synthesize enough glucose. The cortisol-deficient patient also loses the ability to cope with stress. If cortisol levels are significantly depressed, even the stress of mild infections can cause death.

Glucocorticoids are used clinically to reduce inflammation in allergic reactions, infections, arthritis, and certain types of cancer. These hormones inhibit production of prostaglandins (which are mediators of inflammation). Glucocorticoids also reduce inflammation by decreasing the capillary permeability, thereby reducing swelling. In addition, they reduce the effects of histamine and so are used to treat allergic symptoms.

When used in large amounts over long periods, glucocorticoids can cause serious side effects. Although they help stabilize lysosome membranes so that they do not destroy tissues with their potent enzymes, the ability of lysosomes to degrade foreign molecules is also decreased. Glucocorticoids decrease interleukin-1 production, blocking cell-mediated immunity and reducing the patient's ability to fight infections. Other side effects include ulcers, hypertension, diabetes mellitus, and atherosclerosis.

Abnormally large amounts of glucocorticoids, whether due to disease or drugs, can result in **Cushing's disease.** In this condition, fat is mobilized from the lower part of the body and deposited about the trunk. Edema gives the patient's face a full-moon appearance. Blood glucose level rises to as much as 50% above normal, causing *adrenal diabetes.* If this condition persists for several months, the beta cells in the pancreas may "burn out" from secreting excessive amounts of insulin, leading to permanent diabetes mellitus. Reduction in protein synthesis causes weakness and decreases immune responses, so the untreated patient may die of infection.

Many other hormones are known

Many other tissues of the body secrete hormones. The **pineal gland,** located in the brain, produces **melatonin,** which is derived from the amino acid tryptophan, and influences biological rhythms and the onset of sexual maturity. In humans, melatonin facilitates the onset of sleep. Exposure to light suppresses melatonin secretion.

Several hormones secreted by the digestive tract and by adipose tissue regulate digestive processes (see Chapter 45). The **thymus gland** produces **thymosin,** a hormone that plays a role in immune responses. **Atrial natriuretic factor (ANF),** secreted by the atrium of the heart, promotes sodium excretion and lowers blood pressure (see Chapter 46). In Chapter 48 we discuss the principal reproductive hormones.

Review

- Why is the hypothalamus considered the link between the nervous and endocrine systems? What is the role of the anterior lobe of the pituitary? Of the posterior lobe?
- How are thyroid hormone secretion and parathyroid hormone secretion regulated?
- What are the antagonistic actions of insulin and glucagon in regulating blood glucose level? What is insulin resistance?
- What are the actions of epinephrine and norepinephrine? How is the adrenal medulla regulated?
- How do the adrenal glands help the body respond to stress?

BiologyNow™ Assess your understanding of **the vertebrate endocrine system** by taking the pretest on your BiologyNow CD-ROM.

SUMMARY WITH KEY TERMS

1 Describe three main types of chemical signals used by cells, and compare types of endocrine signaling.

- The **endocrine system** consists of endocrine glands, cells, and tissues that secrete **hormones.** Hormones are an important type of chemical signal by which cells communicate with one another. Classically, hormones were defined as chemical messengers secreted by discrete **endocrine glands,** glands that lack ducts. Hormones are secreted into the interstitial fluid and typically are transported by the blood. They bind with receptors on or in specific **target cells.**
- **Neuroendocrine cells** are neurons that secrete **neurohormones.** These hormones are transported down axons and then secreted and transported by the blood.
- Endocrinologists have broadened the definition of hormones to include some **local regulators,** signaling molecules that diffuse through the interstitial fluid and act on nearby cells. These include **growth factors,** peptides that stimulate cell division and development, and **prostaglandins,** a group of local hormones that help regulate many metabolic processes.
- In **autocrine regulation,** a hormone (or other signal molecule) is secreted into the interstitial fluid and then acts on the very cell that produced it. In **paracrine regulation,** a hormone (or other signal molecule) diffuses through interstitial fluid and acts on nearby target cells.

2 Identify four main chemical groups to which hormones are assigned, and give two examples for each group.

- Hormones can be grouped as fatty acid derivatives, steroids, amino acid derivatives, or peptides and proteins.
- Prostaglandins and the juvenile hormone of insects are **fatty acid derivatives.**
- Hormones secreted by the adrenal cortex, ovary, and testis, as well as the molting hormone of insects are **steroid hormones.**
- Thyroid hormones and epinephrine are **amino acid derivatives.**
- Antidiuretic hormone (ADH) and glucagon are examples of **peptide hormones.** Insulin is a small protein.

3 Summarize the regulation of endocrine glands by negative feedback mechanisms.

- Hormone secretion is typically regulated by **negative feedback mechanisms,** in which a hormone is released in response to some change in a steady state and triggers a response that counteracts the changed condition; this process restores homeostasis.

4 Compare the mechanisms of action of steroid and protein-type hormones; include the role of second messengers, such as cyclic AMP.

- Steroid hormones and thyroid hormones are hydrophobic molecules that pass through the plasma membrane and combine with receptors within the target cell; the hormone-receptor complex may activate or repress transcription of messenger RNA coding for specific proteins.
- Peptide hormones are hydrophilic and do not enter target cells. They combine with receptors on the plasma membrane of target cells. Many hormones bind to **G protein–linked receptors** that act via **signal transduction.** An extracellular hormone signal is transduced into an intracellular signal by the receptor.
- Most peptide hormones are first messengers that carry out their actions by way of **second messengers,** such as **cyclic AMP (cAMP)** or calcium ions. The G protein–linked receptor activates a **G protein.** The G protein either stimulates or inhibits an enzyme that affects the second messenger. For example, G proteins stimulate or inhibit **adenylyl cyclase,** the enzyme that catalyzes the conversion of ATP to cAMP.
- Many second messengers stimulate the activity of **protein kinases,** enzymes that phosphorylate specific proteins that affect the activity of the cell.
- **Inositol trisphosphate (IP_3)** and **diacylglycerol (DAG)** are second messengers that increase calcium concentration and activate enzymes. Calcium ions bind with **calmodulin,** which activates certain enzymes.
- **Receptor tyrosine kinases** are **enzyme-linked receptors** that bind growth factors, including insulin and nerve growth factor.
- **Signal amplification** occurs as each hormone-receptor complex stimulates the production of many second messenger molecules. Second messengers, in turn, activate protein kinase molecules that activate many protein molecules.

5 Identify four functions of hormones in invertebrates, and describe the hormonal regulation of insect development.

- Many invertebrate hormones are **neurohormones** secreted by neurons rather than by endocrine glands. Invertebrate hormones and neurohormones help regulate metabolism, growth and development, regeneration, molting, metamorphosis, reproduction, and behavior.
- Hormones control development in insects. When stimulated by some environmental factor, neuroendocrine cells in the insect brain secrete **brain hormone (BH).** BH stimulates the prothoracic glands to produce **molting hormone,** which stimulates growth and molting.
- In the immature insect, the **corpora allata** secrete **juvenile hormone (JH),** which suppresses metamorphosis at each larval molt. The amount of JH decreases with successive molts.

6 Identify the principal vertebrate endocrine glands, and describe the actions of their hormones.

- Vertebrates hormones regulate growth and development, reproduction, salt and fluid balance, many aspects of metabolism, and behavior.
- Endocrine disorders can result from **hyposecretion** (abnormally reduced output) or **hypersecretion** (abnormally increased output) of hormones.
- Consult Table 47-1 for a description of the main endocrine glands and their actions.

7 Describe the mechanisms by which the hypothalamus and pituitary gland integrate many regulatory functions; describe the actions of the hypothalamic and pituitary hormones.

- Nervous and endocrine regulation are integrated in the **hypothalamus,** which regulates the activity of the **pituitary gland.**
- The neurohormones **oxytocin** and **antidiuretic hormone (ADH)** are produced by the hypothalamus and released by the **posterior lobe** of the pituitary. Oxytocin stimulates contraction of the uterus and stimulates ejection of milk by the mammary glands. ADH stimulates reabsorption of water by the kidney tubules.
- The hypothalamus secretes **releasing hormones** and **inhibiting hormones** that regulate the hormone output of the **anterior lobe** of the pituitary gland.
- The anterior lobe of the pituitary gland secretes growth hormone, prolactin, and several **tropic hormones** that stimulate other endocrine glands. **Prolactin** stimulates the mammary glands to produce milk.
- **Growth hormone (GH)** is an **anabolic hormone** that stimulates body growth by promoting protein synthesis. GH stimulates the liver to produce **insulin-like growth factors (IGFs),** which promote skeletal growth and general tissue growth.

8 Describe the actions of the thyroid and parathyroid hormones, their regulation, and the effects of malfunction.

- The **thyroid gland** secretes **thyroid hormones: thyroxine,** or T_4, and **triiodothyronine,** or T_3. Thyroid hormones stimulate the rate of metabolism.
- Regulation of thyroid secretion depends mainly on a negative feedback system between the anterior pituitary gland and the thyroid gland.
- Hyposecretion of thyroxine during childhood may lead to **cretinism;** during adulthood it may result in **myxedema. Goiter,** an abnormal enlargement of the thyroid gland, is associated with both hyposecretion and hypersecretion. The most common cause of hyperthyroidism is **Grave's disease,** an autoimmune disease.
- The **parathyroid glands** secrete **parathyroid hormone (PTH),** which regulates the calcium level in the blood. Parathyroid hormone increases calcium concentration by stimulating calcium release from bones, increasing calcium reabsorption by kidney tubules, and increasing calcium reabsorption from the intestine.
- **Calcitonin,** secreted by the thyroid gland, acts antagonistically to parathyroid hormone.

9 Contrast the actions of insulin and glucagon, and describe the disorders associated with impaired function of these hormones or their receptors.

- The **islets of Langerhans** in the pancreas secrete **insulin** and **glucagon.** Insulin and glucagon secretion are regulated directly by glucose concentration.
- Insulin stimulates cells to take up glucose from the blood and so lowers blood glucose concentration.
- Glucagon raises blood glucose concentration by stimulating conversion of glycogen to glucose (glycogenolysis) and by stimulating production of glucose from other nutrients (gluconeogenesis).
- In **diabetes mellitus,** either insulin deficiency or **insulin resistance** results in decreased use of glucose, increased fat mobilization, increased protein use, and electrolyte imbalance.

10 Describe the actions of the adrenal hormones, including their role in helping the body cope with stress.

- The **adrenal glands** secrete hormones that help the body cope with stress.
- The **adrenal medulla** secretes **epinephrine** and **norepinephrine,** hormones that help the body respond to danger by increasing the heart rate, metabolic rate, and the strength of muscle contraction. These hormones also reroute blood to organs needed for fight or flight.
- The **adrenal cortex** secretes sex hormones; **mineralocorticoids,** such as **aldosterone;** and **glucocorticoids,** such as **cortisol.** Aldosterone increases the rate of sodium reabsorption and potassium excretion by the kidneys. Cortisol promotes gluconeogenesis.
- During stress, the adrenal cortex ensures adequate fuel supplies for the rapidly metabolizing cells.

POST-TEST

1. Which of the following is *not* true of endocrine glands? (a) they secrete hormones (b) they have ducts (c) their product is typically transported by the blood (d) they are typically regulated by negative feedback (e) when removed from an experimental animal, the animal exhibits symptoms of deficiency

POST-TEST *(continued)*

2. A cell secretes a product that diffuses through the interstitial fluid and acts on nearby cells. This is an example of (a) neuroendocrine secretion (b) autocrine regulation (c) paracrine regulation (d) classical endocrine control (e) peptide hormone function
3. Paracrine regulators that are derived from fatty acids and are found in many different organs are (a) prostaglandins (b) thyroid hormones (c) growth factors (d) anabolic steroids (e) G proteins
4. Which of the following is/are true of steroid hormones? (a) hydrophilic (b) secreted by the posterior pituitary (c) typically work through G proteins and cyclic AMP (d) typically bind with receptor in nucleus and affect transcription (e) answers (a) and (c) are correct
5. Which of the following is *not* a correct pair? (a) neurohormone; brain hormone (b) calcium; calmodulin (c) posterior lobe of pituitary; releasing hormone (d) anterior lobe of pituitary; growth hormone (e) thyroid hyposecretion; cretinism
6. Which of the following is *not* a second messenger? (a) hormone-receptor complex (b) calcium ions (c) inositol trisphosphate (IP_3) (d) cyclic AMP (e) diacylglycerol (DAG)
7. Growth hormone (a) is regulated mainly by calcium level (b) stimulates the liver to produce insulin-like growth factors (c) is a catabolic hormone (d) stimulates metabolic rate (e) signals the hypothalamus to produce a releasing hormone
8. Arrange the following events into an appropriate sequence. (1) high thyroid hormone concentration (2) anterior pituitary inhibited (3) homeostasis (4) lower level of thyroid-stimulating hormone (5) thyroid gland secretes less thyroid hormone (a) 1, 2, 4, 5, 3 (b) 5, 4, 3, 2, 1 (c) 1, 2, 5, 4, 3 (d) 4, 5, 2, 3, 1 (e) 1, 4, 2, 5, 3
9. Arrange the following events into an appropriate sequence. (1) blood glucose concentration increases (2) alpha cells in islets stimulated (3) homeostasis (4) low blood-glucose concentration (5) glucagon secretion increases (a) 1, 2, 3, 5, 4 (b) 5, 4, 2, 1, 3 (c) 1, 2, 5, 4, 3 (d) 4, 2, 5, 1, 3 (e) 4, 5, 1, 2, 3
10. Parathyroid hormone (a) increases glucose level in blood (b) helps body cope with stress (c) increases permeability of kidney tubules to water (d) promotes uptake of amino acids (e) increases calcium concentration in blood
11. An action of cortisol is (a) decreases glucose level in blood (b) helps body cope with stress (c) increases permeability of kidney tubules to water (d) promotes uptake of amino acids (e) increases calcium concentration in blood
12. Which of the following is *not* a correct pair? (a) thyroid gland; calcitonin (b) islets of Langerhans; glucagon (c) posterior lobe of pituitary; oxytocin (d) anterior lobe of pituitary; cortisol (e) adrenal medulla; epinephrine
13. Which of the following occurs in diabetes mellitus? (a) decreased use of glucose (b) decreased fat metabolism (c) decreased protein use (d) increased concentration of thyroid-releasing hormone (e) answers b and c are correct
14. Insulin resistance is associated with (a) low insulin secretion by the islets of Langerhans (b) type 1 diabetes (c) impaired function of receptors on target cells (d) hypoglycemia (e) hypersecretion of glucagon
15. Aldosterone (a) is released by posterior pituitary (b) is an androgen (c) secretion is stimulated by an increase in thyroid-stimulating hormone (d) is an enzyme that converts epinephrine to norepinephrine (e) increases sodium reabsorption
16. Which of the following is an action of epinephrine and norepinephrine? (a) decreases glucose use (b) increases cardiac output (c) constricts blood vessels in brain (d) reduces inflammation (e) mobilizes fat

CRITICAL THINKING

1. How do receptors impart specificity within the endocrine system? What might be some advantages of having complex mechanisms for hormone action (such as second messengers)?
2. Why do you think it is important to maintain a constant blood glucose level? Several hormones discussed in this chapter affect carbohydrate metabolism. Why is it important to have more than one? How do they interact?
3. An injection of too much insulin may cause a diabetic to go into insulin shock, in which the patient may appear drunk or may become unconscious, suffer convulsions, and even die. From what you know about the actions of insulin, explain the physiological causes of insulin shock.

- Visit our Web site at **http://biology.brookscole.com/solomon7** for links to chapter-related resources on the World Wide Web. Additional online materials relating to this chapter can also be found on our Web site.

BIOLOGY NOW RESOURCES

Active Figures

47-15: Pancreatic hormones

47-17: Glucose regulation and stress

Preparing for an exam? Take a diagnostic test on your BiologyNow CD-ROM.

Post-Test Answers

1. b	2. c	3. a	4. d
5. c	6. a	7. b	8. a
9. d	10. e	11. b	12. d
13. a	14. c	15. e	16. b

48 Reproduction

Bruce Watkins/Animals Animals

Mating nudibranchs (*Chromodoris*). Photographed in Indonesia.

CHAPTER OUTLINE

- Asexual and Sexual Reproduction
- Human Reproduction: The Male
- Human Reproduction: The Female
- Sexual Response
- Fertilization and Early Development
- The Birth Process
- Birth Control Methods
- Sexually Transmitted Diseases

The ability to reproduce and perpetuate its species is a basic characteristic of living things. The survival of each species requires that its members produce new individuals to replace those that die. The instructions for accomplishing this process are encoded in each organism's nucleic acids.

Many invertebrates, including sponges, cnidarians, and some rotifers, flatworms and annelids, can reproduce asexually. In **asexual reproduction,** a single parent gives rise to offspring that are genetically identical (except for mutations) to the parent. Asexual reproduction is an adaptation of sessile animals that cannot move about to search for mates. For animals that do move about, this method of reproduction is advantageous when the population density is low and mates are not readily available. Some animals reproduce asexually under some conditions and sexually at other times.

Sexual reproduction in animals involves the production and fusion of two types of **gametes**—sperm and eggs. Typically, as exemplified by the mating nudibranchs in the photograph, two different individuals are required. A male parent contributes **sperm** and a female parent contributes an egg, or **ovum** (pl., *ova*). The sperm provides genes coding for some of the male parent's traits, and the egg contributes genes coding for some of the female parent's traits. When sperm and egg unite, a **zygote,** or fertilized egg, forms. The zygote develops into a new animal, similar to both parents but not identical to either. Sexual reproduction typically involves remarkably complex structural, functional, and behavioral processes. In vertebrates, hormones secreted by the hypothalamus, pituitary gland, and gonads regulate these processes. This chapter summarizes some major features of animal reproduction and then focuses on human reproduction. ■

ASEXUAL AND SEXUAL REPRODUCTION

Learning Objective

1. Compare the advantages of asexual and sexual reproduction; describe each mode of reproduction, giving specific examples.

To paraphrase American sociobiologist E. O. Wilson, an animal is really the animal genes' way of making more copies of themselves. How they accomplish this depends on the relative benefits (and costs) of the reproductive strategies available in any given situation. Many variations of both asexual and sexual reproduction have evolved.

Asexual reproduction is an efficient strategy

In asexual reproduction, a single parent may split, bud, or fragment to give rise to two or more offspring. Except for mutations, the offspring have hereditary traits identical to those of the parent. Sponges and cnidarians are among the animals that can reproduce by **budding.** A small part of the parent's body separates from the rest and develops into a new individual (Fig. 48-1). Sometimes the buds remain attached and become more or less independent members of a colony.

Oyster farmers learned long ago that when they tried to kill sea stars by chopping them in half and throwing the pieces back into the sea, the number of sea stars preying on the oyster bed doubled! In some flatworms, nemerteans, and annelids, this ability to regenerate is part of a method of reproduction known as **fragmentation.** The body of the parent breaks into several pieces; each piece regenerates the missing parts and develops into a whole animal.

Parthenogenesis ("virgin development") is a form of asexual reproduction in which an unfertilized egg develops into an adult animal. The adult is typically haploid. Parthenogenesis is common among insects (especially honeybees and wasps) and other arthropods; it also occurs among some other invertebrate and vertebrate groups, including some species of nematodes, gastropods, fishes, amphibians, and reptiles.

Although a few species appear to reproduce solely by parthenogenesis, in most species episodes of parthenogenesis alternate with periods of sexual reproduction. Parthenogenesis may occur for several generations, followed at some point by sexual reproduction in which males develop, produce sperm, and mate with the females to fertilize their eggs. In some species, parthenogenesis is a means of rapidly producing individuals when conditions are favorable.

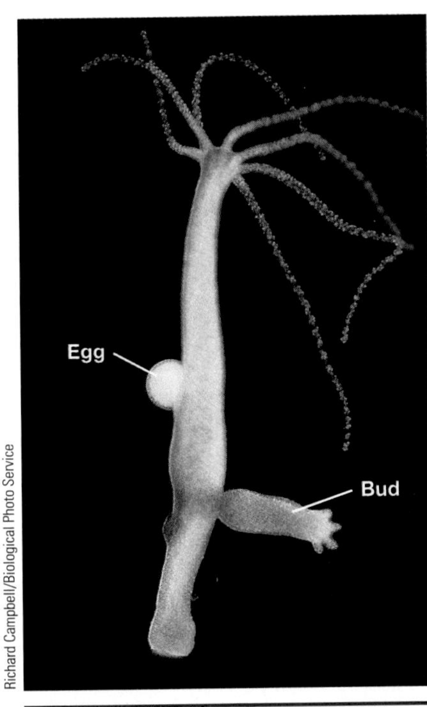

FIGURE **48-1** **Asexual reproduction by budding.**
A part of *Hydra's* body grows outward, then separates and develops into a new individual. The region of the parent body that buds is not specialized exclusively for reproduction. The *Hydra* shown here is also reproducing sexually, as evidenced by the egg *(left).*

Sexual reproduction is the most common type of animal reproduction

Most animals reproduce sexually by fusion of sperm and egg. The egg is typically large and nonmotile, with a store of nutrients that supports the development of the embryo. The sperm is usually small and motile, and adapted to propel itself by beating its long, whiplike flagellum.

Many aquatic animals practice **external fertilization** in which the gametes meet outside the body (Fig. 48-2a). Mating partners usually release eggs and sperm into the water simultaneously. Gametes live only for a short time, and many are lost in the water; some are eaten by predators. However, so many gametes are released that sufficient numbers of sperm and egg cells meet to perpetuate the species.

In **internal fertilization,** matters are left less to chance. The male generally delivers sperm cells directly into the body of the female. Her moist tissues provide the watery medium required for the movement of sperm, and the gametes fuse inside the body. Most terrestrial animals, sharks, and aquatic reptiles, birds, and mammals practice internal fertilization (Fig. 48-2b).

Hermaphroditism is a form of sexual reproduction in which a single individual produces both eggs and sperm. A few hermaphrodites, such as the tapeworm, are capable of self-fertilization. Earthworms are more typical hermaphrodites. Two animals copulate, and mutual cross-fertilization occurs, with each inseminating the other. In some hermaphroditic species, self-fertilization is prevented by the development of testes and ovaries at different times.

PROCESS OF SCIENCE

Asexual reproduction is the fastest and most efficient way to reproduce, allowing an animal to pass all of its genes to its offspring. Compared to asexual reproduction, sexual reproduction is more expensive in terms of energy expenditure, because the animal must produce gametes and find mates. It is also less effi-

(a)

Carmela Leszczynski/Animals Animals

(b)

Fritz Polking/Dembinsky Photo Associates

FIGURE 48-2 | **External and internal fertilization.**
(a) Like many aquatic animals, these spawning frogs *(Rana temporaria),* practice external fertilization. The female lays a mass of eggs, while the male mounts her and simultaneously deposits his sperm in the water. **(b)** Internal fertilization is practiced by some fishes, aquatic reptiles, birds, and mammals and by most terrestrial animals, such as these lions (*Panthera leo*).

cient, because two cells, rather than just one, are required to make a new organism. Why then do most animals reproduce sexually?

Many biologists argue that sexual reproduction has the biological advantage of promoting genetic variety among the members of a species. Each offspring is the product of a particular combination of genes contributed by both parents, rather than a genetic copy of a single individual. By combining inherited traits of two parents, sexual reproduction gives rise to offspring that may be better able to survive than either parent. Also, because the offspring are diploid, they have a backup copy of their genes in case one copy gets damaged by mutation.

Although they generally agree that sexual reproduction has some selective advantage, biologists do not agree on the details, and they are exploring two major possibilities:

1. *Does sexual reproduction permit beneficial mutations from each parent to come together in offspring that can reproduce and spread these mutations through the population?* For example, certain beneficial mutations permit animals to protect themselves from predators and to resist parasites. Sexual reproduction would enable such mutations to spread through the population.
2. *Does sexual reproduction remove harmful mutations?* Mutations occur constantly, and most mutations are harmful. When animals reproduce asexually, all the offspring inherit all the harmful mutations. As mutations accumulate in a population, individuals carry a bigger and bigger load of harmful genes. In contrast, when animals with different mutations mate, offspring inherit varying numbers and combinations of mutations. Offspring that inherit too many harmful mutations are selected against. They may not live to reproduce, so their harmful mutations are removed from the population.

Wayne Getz, an applied mathematician at the University of California, Berkeley, developed a mathematical model that predicts whether asexual or sexual reproduction will exist within a population. According to his model, clones of animals produced by asexual reproduction would be favored in an unchanging environment. The model further predicts that sexual reproduction will be adaptive in an unstable, changing environment and that both types of reproduction can coexist under conditions of moderate change.

Some biologists think the Getz model is too narrow, because it does not consider the benefit of sexual reproduction in both passing on beneficial mutations *and* ridding the genome of harmful mutations. Competing hypotheses are an expected part of the scientific process because scientific discovery is rarely a straightforward, linear sequence of question–answer, question–answer. In fact, as scientific knowledge expands, many creative hypotheses are discarded as dead ends.

Review

- How would you distinguish among budding, fragmentation, and parthenogenesis?
- What are the advantages and disadvantages of asexual reproduction compared to sexual reproduction?

BiologyNow™ Assess your understanding of **asexual and sexual reproduction** by taking the pretest on your BiologyNow CD-ROM.

HUMAN REPRODUCTION: THE MALE

Learning Objectives

2 Describe the structure and function of each organ of the human male reproductive system.

3 Trace the passage of sperm cells through the human male reproductive system from their origin in the seminiferous tubules to their expulsion from the body in the semen. (Include a description of spermatogenesis.)

4 Describe the endocrine regulation of reproduction in the human male.

The human male, like other male mammals, has the reproductive role of producing sperm cells and delivering them into the female reproductive tract. When a sperm combines with an egg, it contributes its genes and determines the sex of the offspring.

The testes produce gametes and hormones

In humans and other vertebrates, **spermatogenesis,** the process of sperm cell production, occurs in the paired male gonads, or **testes** (sing., *testis*) (Figure 48-3a). Spermatogenesis takes place within a vast tangle of hollow tubules, the **seminiferous tubules,** within each testis (Fig. 48-3b). Spermatogenesis begins with undifferentiated cells, the **spermatogonia** in the walls of the tubules (Fig. 48-4).

The spermatogonia, which are diploid cells, divide by mitosis, producing more spermatogonia. Some enlarge and become **primary spermatocytes,** which undergo **meiosis,** producing haploid gametes. (You may want to review the discussion of meiosis in Chapter 9.) In many animals, gamete production occurs only in the spring or fall, but humans have no special breeding season. In the human adult male, spermatogenesis proceeds continuously, and millions of sperm are produced each day.

Each primary spermatocyte undergoes a first meiotic division, producing **secondary spermatocytes** (Fig. 48-5). During the second meiotic division, each of the two secondary spermatocytes gives rise to two **spermatids.** Four spermatids are produced from the original primary spermatocyte. Each haploid spermatid differentiates into a mature sperm. The sequence is as follows:

Spermatogonium (diploid) ⟶ primary spermatocyte (diploid) ⟶ two secondary spermatocytes (haploid) ⟶ four spermatids (haploid) ⟶ four mature sperm (haploid)

Each mature sperm consists of a head, midpiece, and flagellum (Fig. 48-6). The head consists almost entirely of the nucleus. Part of the nucleus is covered by the **acrosome,** a vesicle that differentiates from the Golgi complex. The acrosome contains proteins, including enzymes that help the sperm penetrate the egg.

Mitochondria, located in the midpiece of the sperm, provide the energy for movement of the flagellum. The sperm flagellum has the typical eukaryotic 9 + 2 arrangement of microtubules. During its development, most of the sperm's cytoplasm is discarded and is phagocytized by the large, nutritive **Sertoli cells** that ring the fluid-filled lumen of the seminiferous tubule.

Sertoli cells secrete hormones and other signaling molecules and have several other important functions. Sertoli cells are joined to one another by tight junctions (see Chapter 5) and

ACTIVE FIGURE **48-3** **Male reproductive system.**

(a) The scrotum, penis, and pelvic region of the human male are shown in sagittal section to illustrate their internal structure. **(b)** The testis, epididymis, and spermatic cord are shown partly dissected and exposed. The testis is shown in sagittal section to illustrate the arrangement of the seminiferous tubules.

BiologyNow™ Learn more about **the male reproductive system** by clicking on this figure on your BiologyNow CD-ROM.

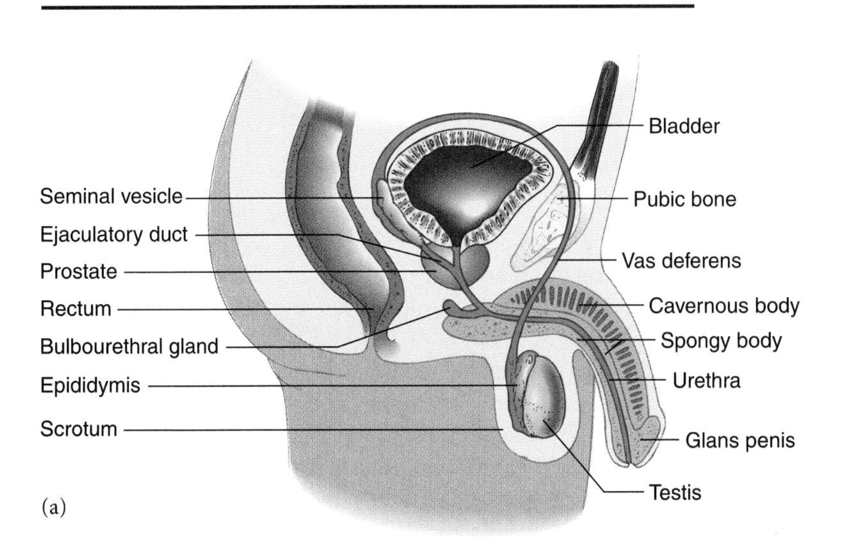

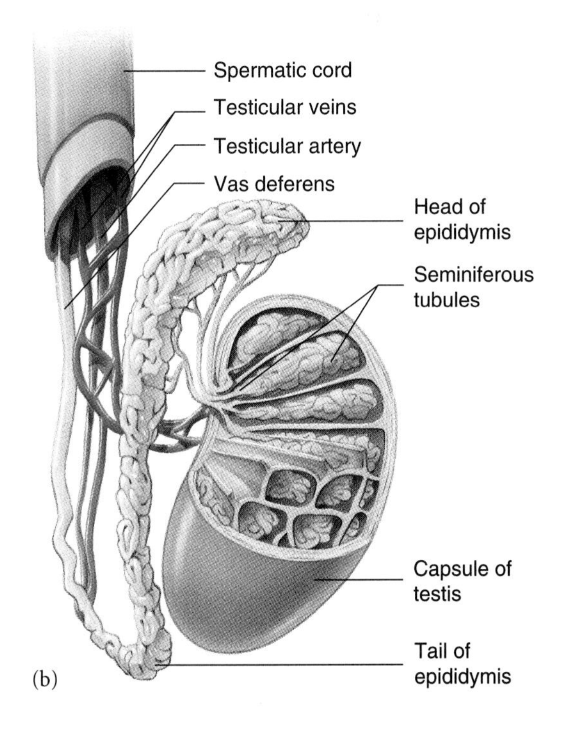

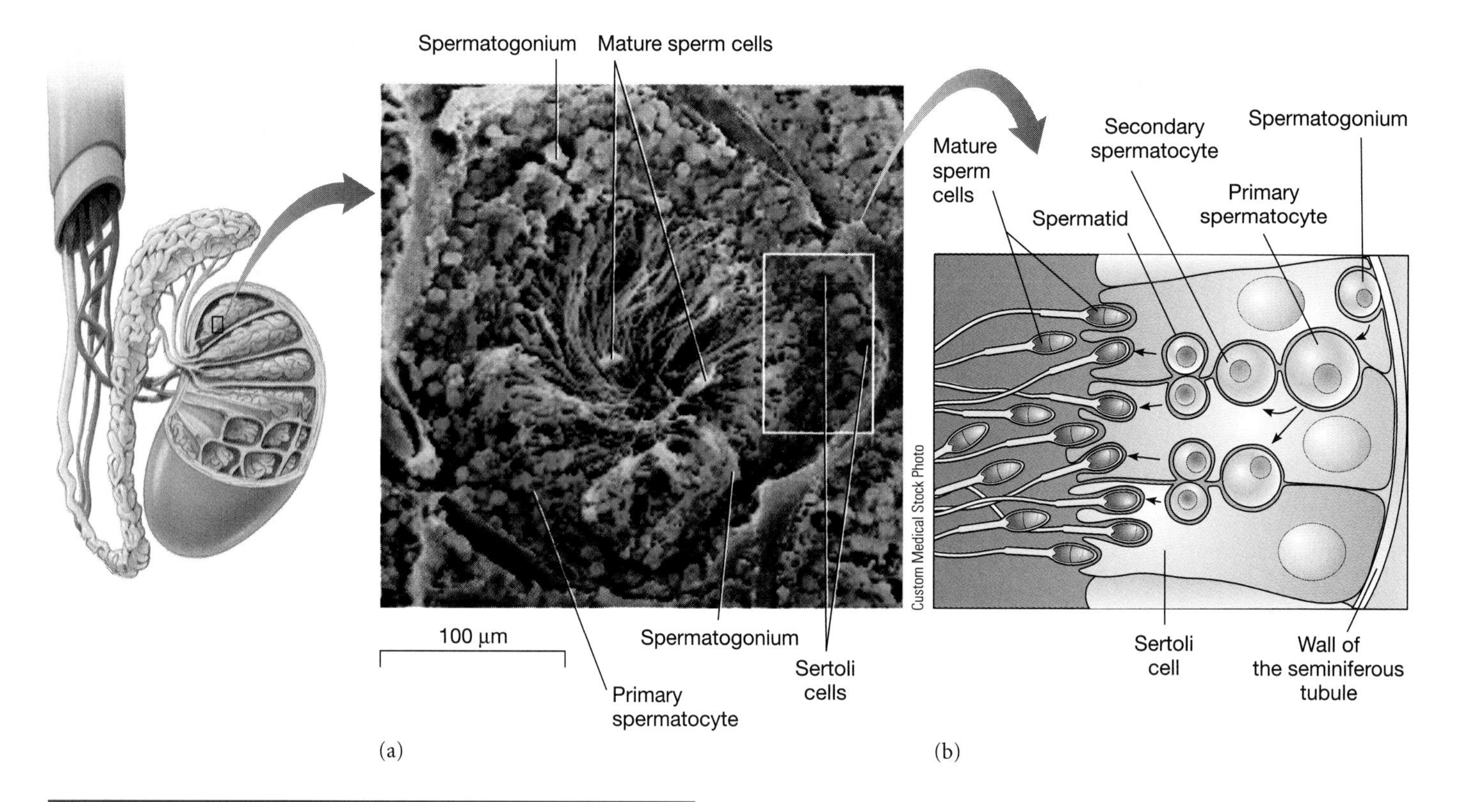

FIGURE **48-4** **Spermatogenesis in the seminiferous tubules.**

(a) Color-enhanced SEM of a transverse section through a seminiferous tubule. **(b)** Sperm cells can be seen in various stages of development between the large, nutritive Sertoli cells.

together form a blood–testis barrier. This barrier prevents the entrance into the tubule of harmful substances that could interfere with spermatogenesis. It also stops sperm from passing out of the tubule and into the blood, where they could stimulate an immune response. The tight junctions between Sertoli cells also form compartments that separate sperm cells in various stages of development.

Human sperm cells cannot develop at body temperature. Although the testes develop within the abdominal cavity of the male embryo, about two months before birth they descend into the **scrotum,** a skin-covered sac suspended from the groin. The scrotum serves as a cooling unit, maintaining sperm below body temperature. In rare cases, the testes do not descend. If this condition is not corrected surgically, or with hormone treatment, the seminiferous tubules eventually degenerate and the male becomes **sterile,** unable to produce offspring.

The scrotum is an outpocketing of the pelvic cavity and is connected to it by the **inguinal canals.** As they descend, the testes pull their blood vessels, nerves, and conducting tubes after them. The inguinal region is a weak place in the abdominal wall. Straining the abdominal muscles by lifting heavy objects sometimes tears the inguinal tissue. A loop of intestine can then bulge into the scrotum through the tear, a condition known as an *inguinal hernia.*

A series of ducts store and transport sperm

Sperm cells leave the seminiferous tubules of each testis and pass into a larger coiled tube, the **epididymis.** There sperm finish maturing and are stored. During ejaculation, sperm pass from each epididymis into a sperm duct, the **vas deferens** (pl., *vasa deferentia*). The vas deferens extends from the scrotum through the inguinal canal and into the pelvic cavity.

Each vas deferens empties into a short **ejaculatory duct,** which passes through the prostate gland and then opens into the single **urethra.** The urethra, which at different times conducts urine and semen, passes through the penis to the outside of the body. Thus the sperm pass in sequence through the following:

Seminiferous tubules ⟶ epididymis ⟶ vas deferens ⟶ ejaculatory duct ⟶ urethra ⟶ release from body

The accessory glands produce the fluid portion of semen

As sperm travel through the conducting tubes, they mix with secretions from three types of accessory glands. Approximately 3 mL of **semen** is ejaculated during sexual climax. Semen consists of about 200 million sperm cells suspended in the secretions of these glands.

The paired **seminal vesicles** secrete a nutritive fluid rich in fructose and prostaglandins into the vasa deferentia (see Fig. 48-3). Nutrients in this secretion provide energy for the sperm after they are ejaculated. Prostaglandins stimulate contractions of the uterus, which help move sperm up the female re-

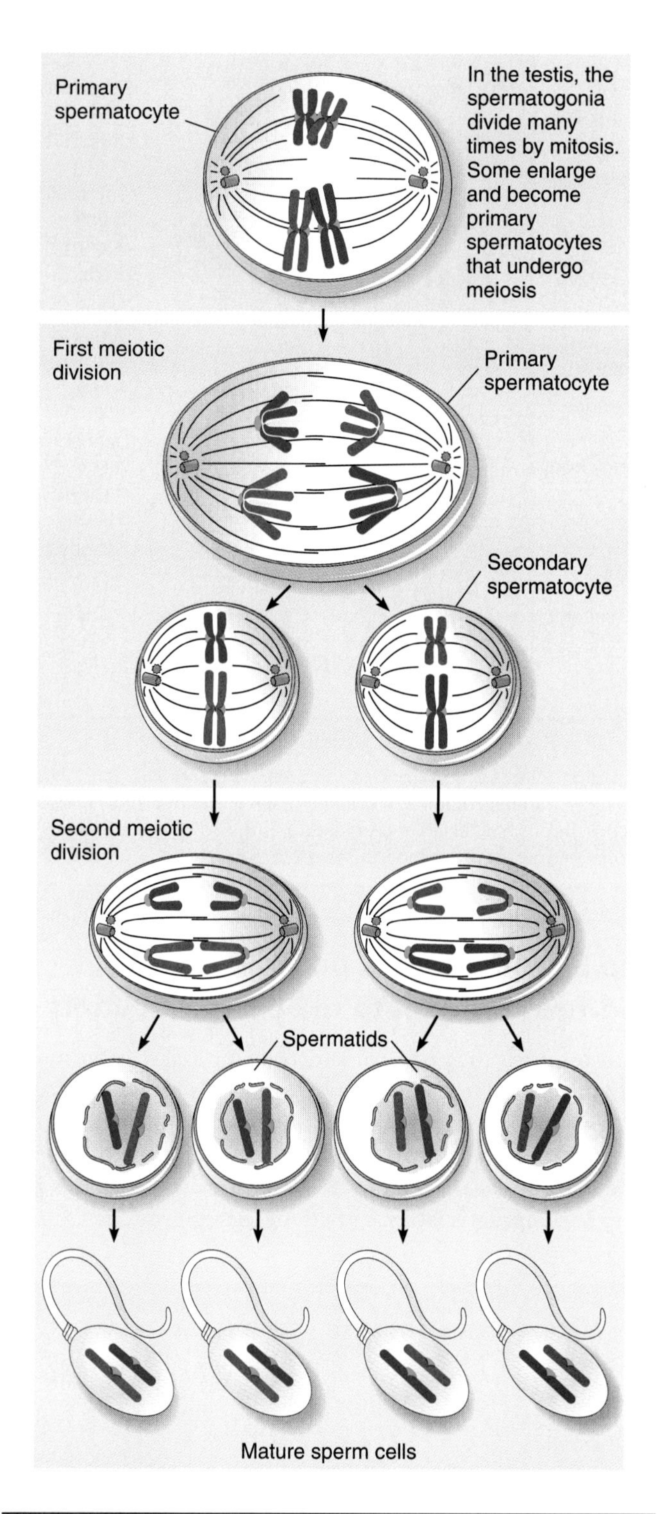

FIGURE **48-5** **Spermatogenesis.**

Spermatogonia divide by mitosis. Some enlarge and become primary spermatocytes, which undergo meiosis. Each primary spermatocyte gives rise to four spermatids. The spermatids differentiate, becoming mature sperm cells. Note that in this example, four chromosomes are present in the primary spermatocyte ($2n$) and that meiosis produces the haploid (n) number (2) in the secondary spermatocytes, spermatids, and mature sperm cells.

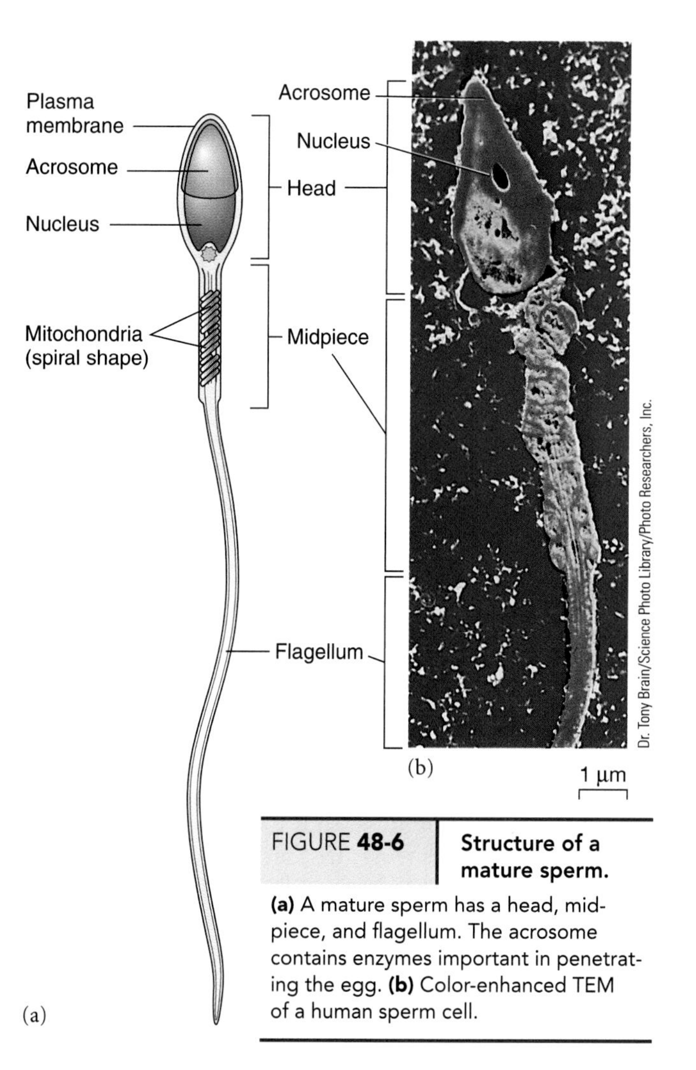

FIGURE **48-6** **Structure of a mature sperm.**

(a) A mature sperm has a head, midpiece, and flagellum. The acrosome contains enzymes important in penetrating the egg. **(b)** Color-enhanced TEM of a human sperm cell.

productive tract. The single **prostate gland** secretes an alkaline fluid containing calcium, citric acid, and enzymes. The prostatic fluid may be important in neutralizing the acidic environment of the vagina and in increasing sperm cell motility. During sexual arousal, the paired **bulbourethral glands,** located on each side of the urethra, release a mucous secretion. This fluid lubricates the penis, facilitating its penetration into the vagina.

A major cause of male infertility is insufficient sperm production. When sperm counts drop below 35 million per mL of semen, fertility is impaired, and males with a sperm count lower than 20 million/mL are usually considered sterile. When a couple's attempts to produce a child are unsuccessful, a sperm count and analysis may be performed in a clinical laboratory. Sometimes semen is found to contain large numbers of abnormal sperm or, occasionally, no sperm at all.

In the United States in the 1970s, an average healthy young man produced about 100 million sperm per mL of semen. Today that average has dropped to about 60 million. Although the cause of this decrease is not known, low sperm counts have been linked to a variety of environmental factors, including chronic marijuana use, alcohol abuse, and cigarette smoking. Studies also show that men who smoke tobacco are more likely than nonsmokers to produce abnormal sperm. Exposure to industrial and environmental toxins such as DDT and PCBs (polychlorinated biphenyls) may contribute to low sperm count and sterility. The use of anabolic steroids by athletes to accelerate muscle development can cause sterility in both males and females (see *Focus On: Anabolic Steroids and Other Abused Hormones* in Chapter 47).

The penis transfers sperm to the female

The **penis** is an erectile copulatory organ that delivers sperm into the female reproductive tract. It is a long shaft that enlarges to form an expanded tip, the **glans.** Part of the loose-fitting skin of the penis folds down and covers the proximal portion of the glans, forming a cuff called the **prepuce,** or foreskin. The foreskin is removed during circumcision (a procedure commonly performed on male babies either for hygienic or religious reasons).

Under the skin, the penis consists of three parallel columns of **erectile tissue:** two **cavernous bodies** and one **spongy body** (Fig. 48-7). The spongy body surrounds the portion of the urethra that passes through the penis. Erectile tissue contains numerous blood vessels. When the male is sexually stimulated, parasympathetic neurons dilate the arteries in the penis. Blood fills the blood vessels of the erectile tissue, causing the tissue to swell. This compresses veins that conduct blood away from the penis, slowing the outflow of blood. Thus, more blood enters the penis than can leave, further engorging the erectile tissue with blood. **Penile erection** occurs; the penis increases in length, diameter, and firmness. Although the human penis contains no bone, penis bones do occur in some other mammals, such as bats, rodents, and some primates. Erectile dysfunction, the chronic inability to sustain an erection, prevents effective sexual intercourse. Associated with a variety of physical causes and psychological issues, this common disorder is now treated with sildenafil (Viagra), which blocks the action of an enzyme that breaks down cyclic GMP (a second messenger).

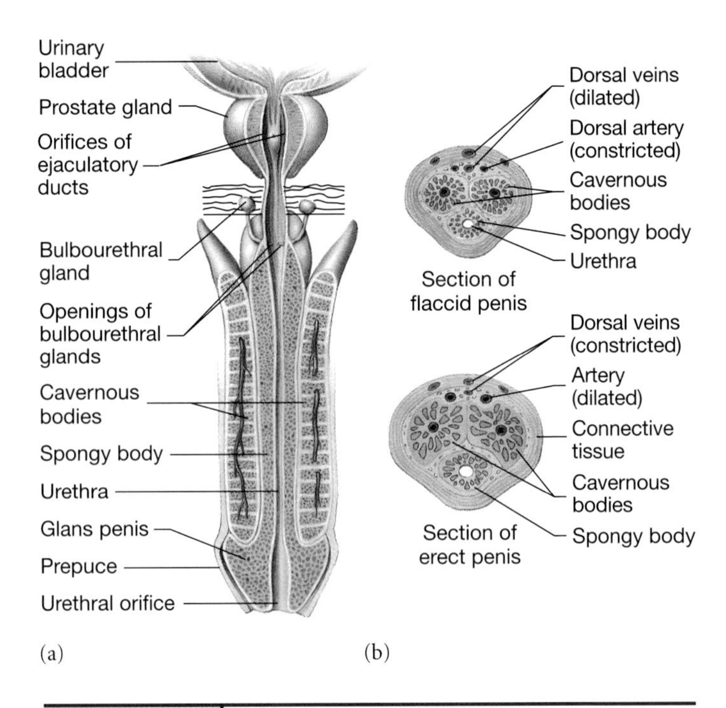

FIGURE **48-7** | **Internal structure of the penis.**
(a) Longitudinal section through the prostate gland and penis. Note the three parallel columns of erectile tissue in the penis. **(b)** Cross section through a flaccid and an erect penis, showing the erectile tissues engorged with blood in the erect penis.

The hypothalamus, pituitary, and testes regulate male reproduction

Testosterone is the principal **androgen,** or male sex hormone. It is a steroid produced by the **interstitial cells** between the seminiferous tubules in the testes. Testosterone directly affects muscle and bone, and stimulates the adolescent growth spurt in males at about age 13 (Table 48-1). It produces the male's **primary sex characteristics:** growth of the reproductive organs

TABLE **48-1** **Principal Male Reproductive Hormones**

Endocrine Gland and Hormones	Principal Target Tissue	Principal Actions
Hypothalamus		
Gonadotropin-releasing hormone (GnRH)	Anterior pituitary	Stimulates release of FSH and LH
Anterior Pituitary		
Follicle-stimulating hormone (FSH)	Testes	Stimulates development of seminiferous tubules; stimulates spermatogenesis
Luteinizing hormone (LH)	Testes	Stimulates interstitial cells to secrete testosterone
Testes		
Testosterone	General	*Before birth:* stimulates development of primary sex organs and descent of testes into scrotum *At puberty:* responsible for growth spurt; stimulates development of reproductive structures and secondary sex characteristics *In adult:* maintains secondary sex characteristics; stimulates spermatogenesis
Inhibin	Anterior pituitary	Inhibits FSH secretion

and spermatogenesis. Testosterone also stimulates the development of the **secondary sex characteristics** at puberty, including growth of facial and body hair, muscle development, and the increase in vocal cord length and thickness that causes the voice to deepen. Testosterone is necessary for normal sex drive.

In some of its target tissues, testosterone is converted to other steroids. Interestingly, in brain cells testosterone is converted to *estradiol,* the principal female sex hormone. The implications of this transformation are not yet understood.

When a boy is about 10 years old the hypothalamus begins to secrete **gonadotropin-releasing hormone (GnRH).** This hormone stimulates the anterior pituitary to secrete the gonadotropic hormones **follicle-stimulating hormone (FSH)** and **luteinizing hormone (LH).** Both FSH and LH are glycoproteins that use cyclic AMP as a second messenger (see Chapter 47). FSH stimulates Sertoli cells to secrete **androgen-binding protein (ABP)** and other signaling molecules that are necessary for spermatogenesis (Fig. 48-8a). LH stimulates the interstitial cells to secrete **testosterone.**

FSH, LH, and testosterone all directly or indirectly stimulate testosterone secretion and spermatogenesis. A high concentration of testosterone in the testes is required for spermatogenesis. Testosterone and FSH stimulate Sertoli cells to produce ABP, which binds to testosterone and concentrates it in the tubules. Interestingly, developing sperm cells appear to lack receptors for sex steroids, so just how testosterone acts on them is not known.

Reproductive hormone concentrations are regulated by negative feedback mechanisms. Testosterone inhibits mainly LH secretion. It acts on the hypothalamus, decreasing its secretion of GnRH, which results in decreased FSH and LH secretion by the pituitary. Testosterone also directly inhibits the anterior lobe of the pituitary by blocking the normal actions of GnRH on LH synthesis and release.

FSH secretion is inhibited mainly by **inhibin,** a peptide hormone secreted by Sertoli cells (Fig. 48-8b). FSH itself stimulates inhibin secretion. The endocrine regulation of reproductive function is extremely complex, and it is likely that other hormones and signaling molecules will be identified.

Insufficient testosterone results in sterility. If a male is **castrated,** that is, the testes are removed, before puberty, he is deprived of testosterone and becomes a eunuch. He retains childlike sexual organs and does not develop secondary sexual characteristics. If castration occurs after puberty, increased secretion of male hormones by the adrenal glands helps maintain masculinity.

Review

- Explain the physiological basis for erection of the penis.
- Compare the functions of ovaries and testes.

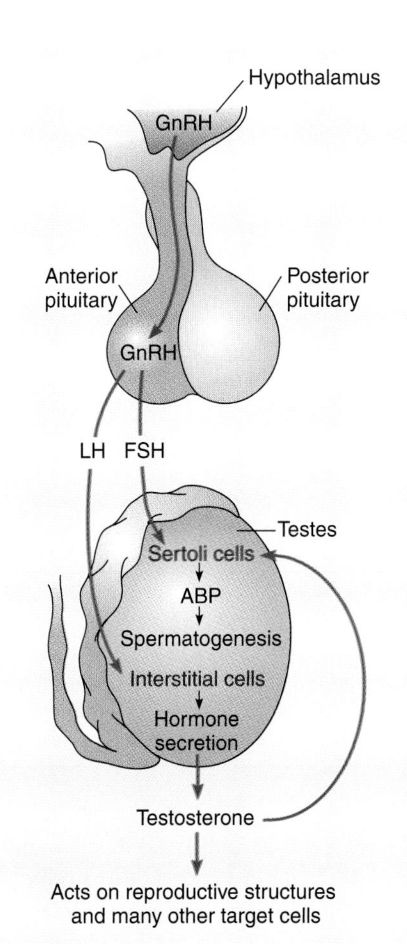

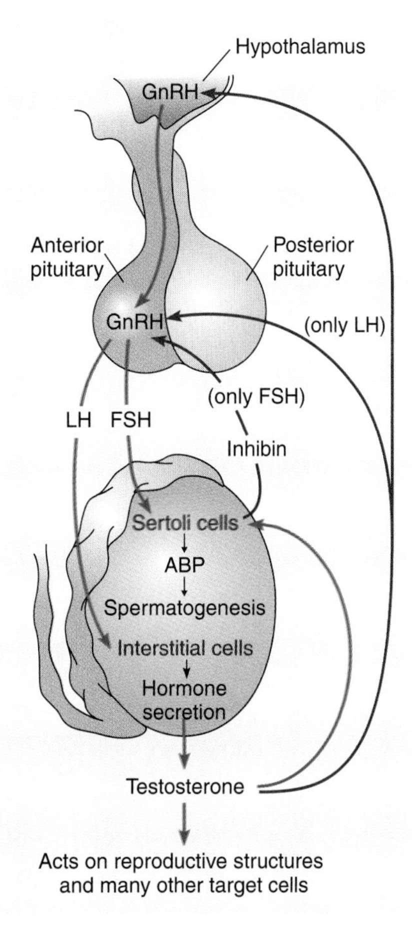

FIGURE **48-8**

Regulation of reproduction in the male.

The hypothalamus, anterior pituitary, and testes interact to regulate reproductive function. **(a)** Overview of hormone action. The hypothalamus secretes gonadotropin-releasing hormone (GnRH), which stimulates the anterior pituitary to secrete follicle-stimulating (FSH) and luteinizing hormone (LH). LH stimulates the interstitial cells in the testes to secrete testosterone. FSH acts on the Sertoli cells, stimulating them to secrete androgen-binding protein (ABP) and other signaling molecules necessary for spermatogenesis. **(b)** Several negative-feedback systems regulate hormone level. Testosterone mainly inhibits LH production. It does this by decreasing GnRH secretion by the hypothalamus and inhibiting LH secretion by the pituitary. Inhibin inhibits FSH secretion. Red arrows indicate inhibition.

- Trace the passage of sperm from a seminiferous tubule through the male reproductive system until it leaves the male body during ejaculation.
- What are the actions of testosterone? Give an overview of the endocrine regulation of male reproduction.

BiologyNow™ Assess your understanding of **male human reproduction** by taking the pretest on your BiologyNow CD-ROM.

HUMAN REPRODUCTION: THE FEMALE

Learning Objectives

5 Describe the structure and function of each organ of the human female reproductive system.

6 Trace the development of a human ovum (egg) and its passage through the female reproductive system until it is fertilized.

7 Describe the endocrine regulation of reproduction in the human female, and identify the important events of the menstrual cycle, such as ovulation and menstruation.

The female reproductive system produces oocytes (immature gametes), receives the penis and sperm released from it during sexual intercourse, houses and nourishes the embryo during prenatal development, gives birth, and produces milk for the young (lactation). These processes are regulated and coordinated by the interaction of hormones secreted by the hypothalamus, pituitary gland, and ovaries.

The ovaries produce gametes and sex hormones

Like the male gonads, the female gonads, or **ovaries,** produce both gametes and sex hormones. About the size and shape of large almonds, the ovaries lie close to the lateral walls of the pelvic cavity and are held in position by several connective tissue ligaments (Figures 48-9). Internally, the ovary consists mainly of connective tissue containing scattered ova in various stages of maturation.

The process of ovum production, called **oogenesis,** begins in the ovaries. Before birth, hundreds of thousands of **oogonia** are present in the ovaries. All of a female's oogonia form during embryonic development. No new oogonia are formed after birth. During prenatal development, the oogonia increase in size and become **primary oocytes.** By the time of birth, they are in the prophase of the first meiotic division. At this stage, they enter a resting phase that lasts throughout childhood and into adult life.

A primary oocyte and the **granulosa cells** surrounding it together make up a **follicle** (Fig. 48-10). The granulosa cells are connected by tight junctions that form a protective barrier around the oocyte. With the onset of puberty, a few follicles begin to mature each month in response to FSH secreted by the anterior pituitary gland. As a follicle grows, the granulosa cells proliferate, forming several layers. Connective tissue cells surrounding the granulosa differentiate, forming a layer of **theca cells.**

As the follicle matures, the primary oocyte completes its first meiotic division. The two haploid cells produced differ in size (Fig. 48-11). The smaller one, the first **polar body,** may later divide, forming two polar bodies, but these eventually disinte-

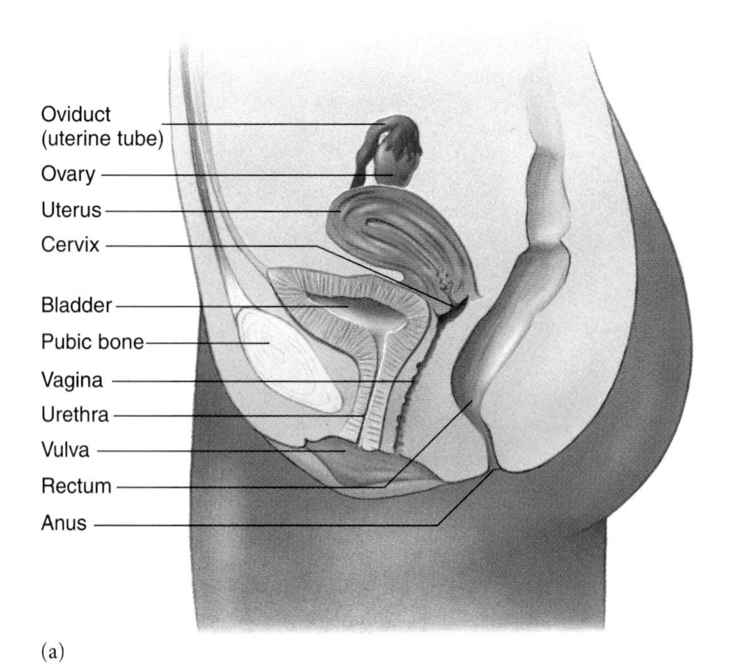

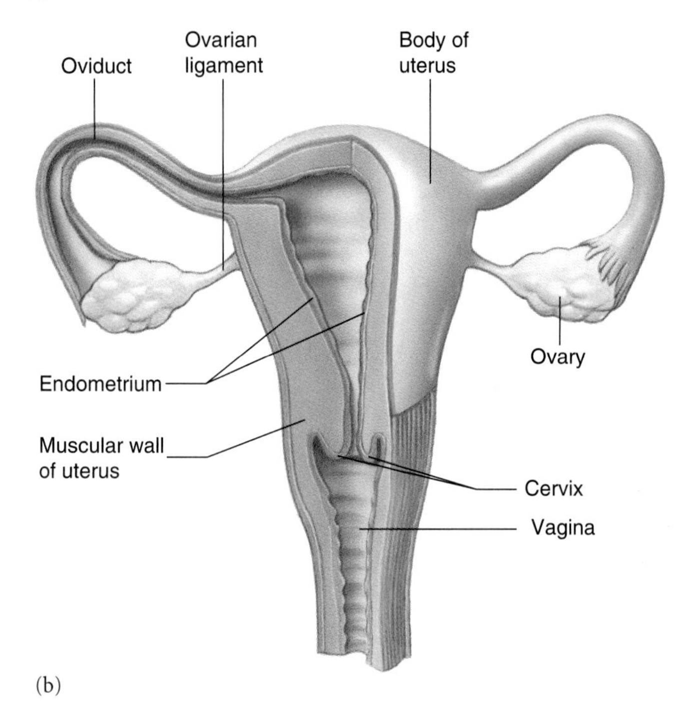

FIGURE 48-9 Female reproductive system.
(a) Midsagittal section through the female pelvis. Note the position of the uterus relative to the vagina. **(b)** In this anterior view of the female reproductive system, some organs have been cut open to expose their internal structure. Connective tissue ligaments anchor the reproductive organs in place.

grate. The larger cell, the **secondary oocyte,** proceeds to the second meiotic division but remains in metaphase II until it is fertilized. When meiosis continues, the second meiotic division gives rise to a single ovum and a second polar body. The polar bodies are small and apparently dispose of unneeded chromosomes with a minimal amount of cytoplasm. The sequence is as follows:

Oogonium (diploid) ⟶ primary oocyte (diploid) ⟶
secondary oocyte + first polar body (both haploid) ⟶ ovum
+ second polar body (both haploid)

Recall that in the male, each primary spermatocyte gives rise to four functional sperm cells. In contrast, each primary oocyte generates only one ovum.

As an oocyte develops, it becomes separated from its surrounding follicle cells by a layer of glycoproteins called the **zona pellucida.** As the follicle develops, follicle cells secrete fluid, which collects in the antrum (space) between them (see Fig. 48-10). The follicle cells also secrete **estrogens,** female sex hormones. The principal estrogen is **estradiol** (see Fig. 47-2b). Typically, only one follicle fully matures each month. Several others may develop for awhile and then deteriorate by apoptosis.

As a follicle matures, it moves closer to the surface of the ovary, eventually resembling a fluid-filled bulge on the ovarian surface. Follicle cells secrete proteolytic enzymes that break down a small area of the ovary wall. During **ovulation,** the secondary oocyte ejects through the ovary wall and into the pelvic cavity. The portion of the follicle that remains in the ovary develops into the **corpus luteum,** a temporary endocrine gland that secretes estrogen and **progesterone.**

The oviducts transport the secondary oocyte

Almost immediately after ovulation, the secondary oocyte is swept into the funnel-shaped opening of the **oviduct,** or **uterine tube** (also called the *fallopian tube*). Action of cilia on the epithelial lining of the oviduct both sweeps the secondary oocyte into the oviduct and moves it along toward the uterus. Fertilization takes place within the oviduct. If fertilization does not occur, the secondary oocyte degenerates there.

Scarring of the oviducts (for example, by pelvic inflammatory disease, a sexually transmitted disease) can block the tubes so that the fertilized ovum cannot pass to the uterus. Sometimes partial constriction of the oviduct results in *tubal pregnancy,* in which the embryo begins to develop in the wall of the oviduct because it cannot progress to the uterus. Oviducts are not adapted to bear the burden of a developing embryo; thus the oviduct and the embryo it contains must be surgically re-

Zona pellucida
Granulosa cells
Theca
Secondary oocyte
Antrum
500 μm
Biophoto Associates
(a)

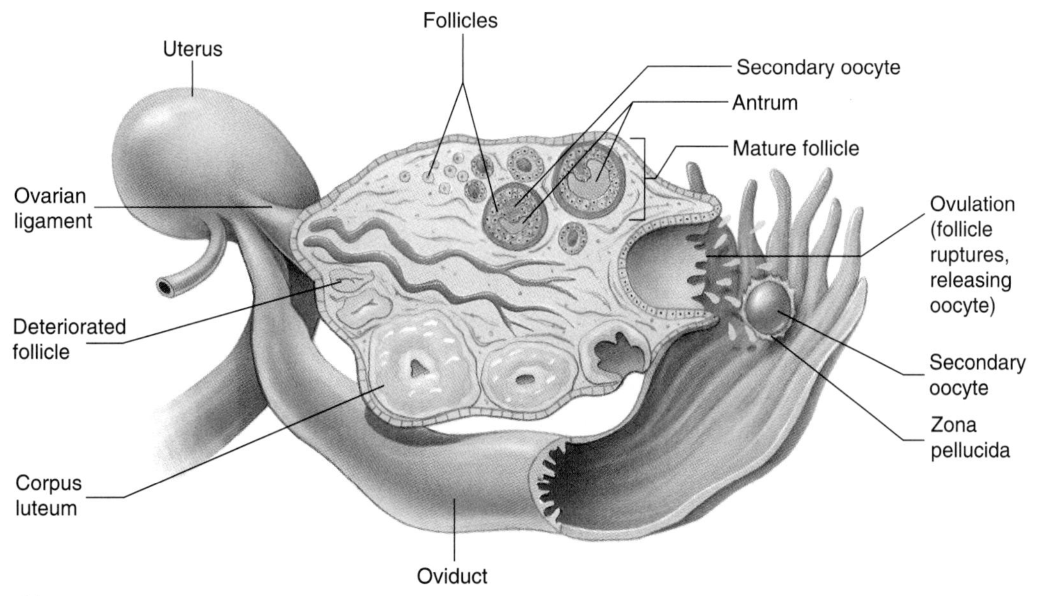

(b)

FIGURE **48-10**

Development of follicles in the ovary.

(a) LM of a developing follicle. The secondary oocyte is surrounded by the zona pellucida (a layer of glycoproteins) and by granulosa cells. Connective tissue cells surrounding the granulosa cells form a layer of theca cells. **(b)** Follicles in various stages of development are scattered throughout the ovary. This is a composite drawing; these stages would not all be present at the same time.

FIGURE **48-11** | **Oogenesis.**

Before birth, oogonia (not shown) divide many times by mitosis. Some oogonia differentiate to become primary oocytes that undergo meiosis. Only one functional ovum is produced from each primary oocyte. The other cells produced are polar bodies that degenerate. The second meiotic division is completed after fertilization. Note that in this example, four chromosomes are present in the primary oocyte (2*n*) and that meiosis produces the haploid (*n*) number (2) in the polar bodies, secondary oocyte, and mature ovum.

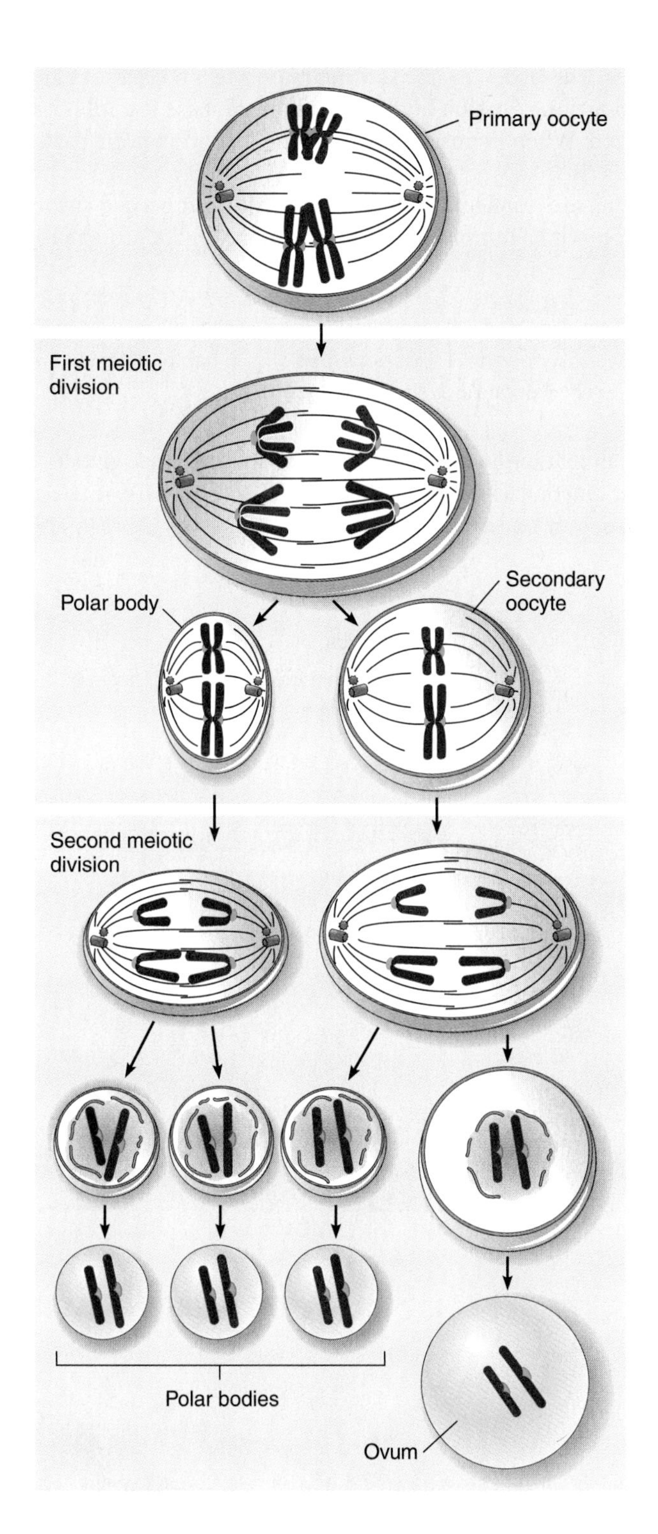

moved before it ruptures and endangers the life of the mother. Women with blocked oviducts may be infertile.

The uterus incubates the embryo

The oviducts open into the upper corners of the pear-shaped **uterus** (see Fig. 48-9b). About the size of a fist, the uterus (or womb) occupies a central position in the pelvic cavity. It has thick walls of smooth muscle and an epithelial lining, the **endometrium,** that thickens each month in preparation for possible pregnancy. Up to 15% of women (more than 5 million women in the United States alone) are affected by **endometriosis,** a painful disorder in which fragments of the endometrium migrate to other areas such as the oviducts or ovaries. Like pelvic inflammatory disease, endometriosis causes scarring that can lead to infertility.

If a secondary oocyte is fertilized, the tiny embryo enters the uterus and implants in the endometrium. As it grows and develops, it is sustained by nutrients and oxygen delivered by surrounding maternal blood vessels. If fertilization does not occur during the monthly cycle, the endometrium sloughs off and is discharged in the process known as **menstruation.**

The lower portion of the uterus, called the **cervix,** extends slightly into the vagina. The cervix is a common site of cancer in women. Detection is usually possible by the routine Papanicolaou test (Pap smear) in which a few cells are scraped from the cervix during a regular gynecological examination and studied microscopically. When cervical cancer is detected at very early stages of malignancy, the chances that the patient can be cured are good.

The vagina receives sperm

The **vagina** is an elastic, muscular tube that extends from the uterus to the exterior of the body. The vagina serves as a receptacle for sperm during sexual intercourse and as part of the birth canal (see Fig. 48-9).

The vulva are external genital structures

The female external genitalia, collectively known as the **vulva,** include several structures. Liplike folds, the **labia minora,** surround the vaginal and urethral openings (Fig. 48-12). The area enclosed by the labia minora is the **vestibule** of the vagina. Vestibular glands secrete a lubricating mucus into the vestibule. The **hymen** is a thin ring of tissue that forms a border around the entrance to the vagina.

Anteriorly, the labia minora merge to form the prepuce of the **clitoris,** a small erectile structure comparable to the male glans penis. Like the penis, the clitoris contains erectile tissue that becomes engorged with blood during sexual excitement. Rich in nerve endings, the clitoris is highly sensitive to touch, pressure, and temperature and serves as a center of sexual sensation in the female.

External to the delicate labia minora are the thicker **labia majora.** The **mons pubis** is the mound of fatty tissue just above

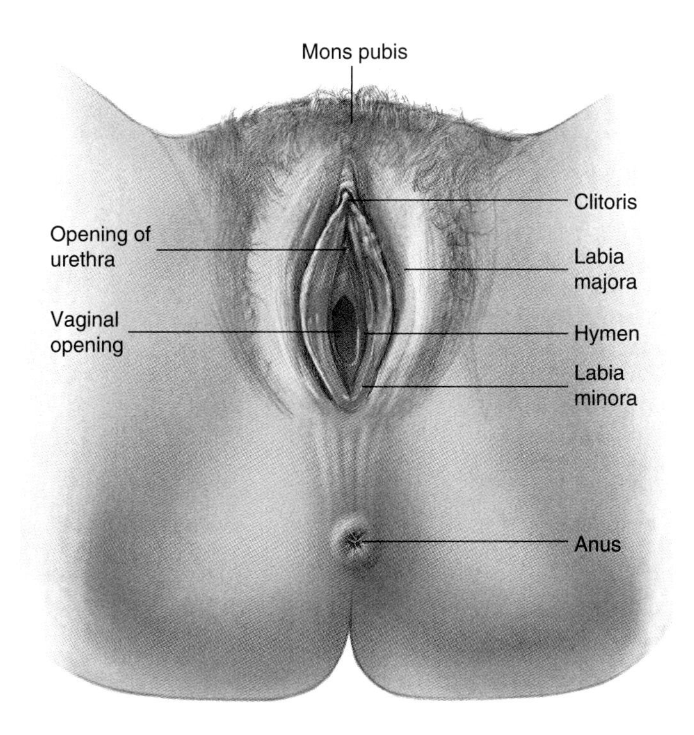

FIGURE **48-12** **External female genital structures.**
Collectively, the structures shown (excluding the anus) are referred to as the *vulva.*

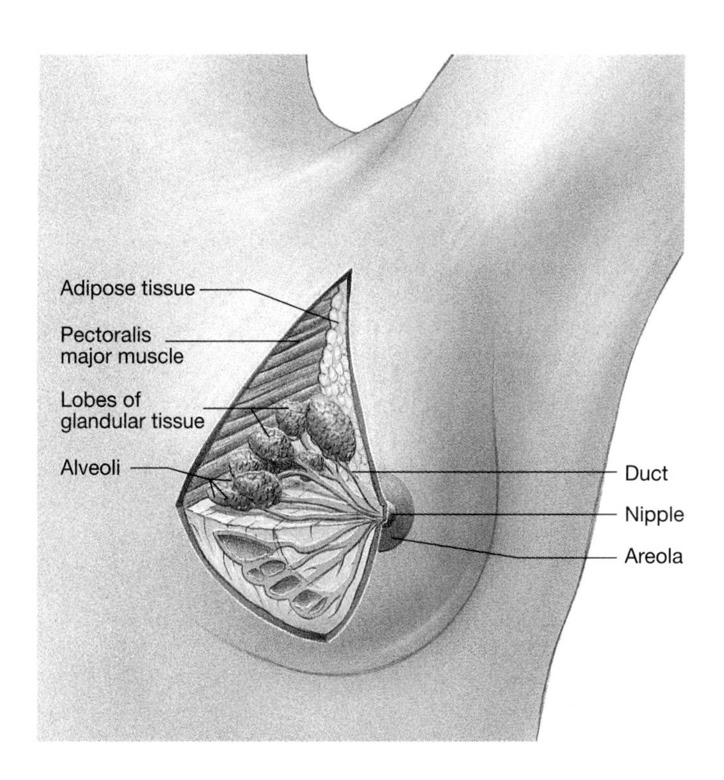

FIGURE **48-13** **Structure of the mature female breast.**
The breast contains lobes of glandular tissue. The lobes consist of alveoli, clusters of gland cells.

the clitoris at the junction of the thighs and torso. At puberty the mons pubis and labia majora become covered by coarse pubic hair.

The breasts function in lactation

Each breast consists of 15 to 20 lobes of glandular tissue. The amount of adipose tissue around these lobes determines the size of the breasts and accounts for their softness. Gland cells are arranged in grapelike clusters called **alveoli** (Fig. 48-13). Ducts from each cluster join to form a single duct from each lobe, producing 15 to 20 tiny openings on the surface of each nipple. The breasts are a common site of cancer in women (see *Focus On: Breast Cancer*).

Lactation is the production of milk for nourishing the young. During pregnancy, high concentrations of the female reproductive hormones, estrogen and progesterone, stimulate the breasts to increase in size. For the first couple of days after childbirth, the mammary glands produce a fluid called **colostrum,** which contains protein and lactose but little fat. After birth the hormone **prolactin,** secreted by the anterior pituitary, stimulates milk production. When a baby suckles, the posterior pituitary releases **oxytocin,** which stimulates ejection of milk from the alveoli into the ducts.

Breast-feeding promotes recovery of the uterus because oxytocin released during breast-feeding stimulates the uterus to contract to nonpregnant size. Breast-feeding offers advantages to the baby as well. It promotes a close bond between mother and child and provides milk tailored to the nutritional needs of the human infant. Breast milk contains antibodies, and breast-fed infants have a lower incidence of diarrhea, ear and respiratory infections, and hospital admissions than do formula-fed babies.

The hypothalamus, pituitary, and ovaries interact to regulate female reproduction

As in the male, regulation of female reproduction is complex, involving many hormones and other signal molecules. In the male, concentration of sex hormones is kept fairly constant. In contrast, concentrations of female sex hormones change in a monthly cycle.

Table 48-2 lists the actions of the principal female reproductive hormones. Like testosterone in the male, estrogens are responsible for growth of the sex organs at puberty, for body growth, and for the development of secondary sexual characteristics. In the female, these include development of the breasts, broadening of the pelvis, and the characteristic development and distribution of muscle and fat responsible for the female body shape.

Hormones of the hypothalamus, anterior pituitary, and ovaries regulate the **menstrual cycle,** the monthly sequence of events that prepares the body for possible pregnancy. (The term *menstrual cycle* is sometimes used more narrowly to refer to the changes that occur in the uterus; we use the term here to include both the ovarian and uterine cycles.) The menstrual cycle runs its course every month from puberty until menopause oc-

Focus On Breast Cancer

Other than skin cancer, breast cancer is the most common type of cancer among women. It is a leading cause of cancer deaths in women, second only to lung cancer. The causes of breast cancer are not known, but there appears to be a higher risk in women with a family history of the disease. An estimated 10% of breast cancers are familial, and about half of these patients have mutations in a tumor suppressor gene, either *BRCA1* or *BRCA2*. When the normal protein product of the *BRCA1* gene is phosphorylated by a specific protein kinase, it interacts with the normal protein product of the *BRCA2* gene and certain other compounds to repair DNA damage.

Investigators are studying a variety of suspected risk factors for breast cancer, including a high-fat diet, obesity, exposure to radiation, and exposure to certain chemicals. Smoking cigarettes increases a woman's risk of dying from breast cancer by at least 25%. Women who smoke two packs or more of cigarettes a day have a 75% greater risk.

About 50% of breast cancers begin in the upper, outer quadrant of the breast. As a malignant tumor grows, it may adhere to the deep tissue of the chest wall. Sometimes it extends to the skin, causing dimpling. Eventually the cancer spreads to the lymphatic system. About two thirds of breast cancers have metastasized (spread) to the lymph nodes by the time they are first diagnosed. When diagnosis and treatment begin early, 86% of patients survive for five years, and 65% survive for 20 years or longer. Untreated patients have a five-year survival rate of only 20%.

Mastectomy (surgical removal of the breast) and radiation treatment are common methods of treating breast cancer. Lumpectomy (surgical removal of only the affected portion of the breast) in conjunction with radiation treatment appears to be as effective as mastectomy in some cases. Chemotherapy is useful in preventing metastasis, especially in premenopausal patients. A recent development in cancer treatment is the use of *biological response modifiers,* which include substances such as interferons, interleukins, and monoclonal antibodies (see Chapter 43).

About one third of breast cancers are estrogen dependent; that is, their growth depends on circulating estrogens. Removing the ovaries in patients with these tumors relieves the symptoms and may cause remission of the disease for months or even years. Developing pharmacological agents that antagonize the action of estrogen receptors has been an important approach in the treatment of breast cancer. One challenge has been to inhibit estrogen in the breast while at the same time retaining its beneficial effects on other parts of the body.

According to the American Cancer Society, when breast cancer is confined to the breast, the five-year survival rate is almost 100%. Because early detection of breast cancer greatly increases the chances of cure and survival, campaigns have been launched to educate women on the importance of self-examination. Mammography, a soft-tissue radiological study of the breast, is helpful in detecting very small lesions that might not be identified by routine examination. In mammography, lesions show on an x-ray plate as areas of increased density (see figure).

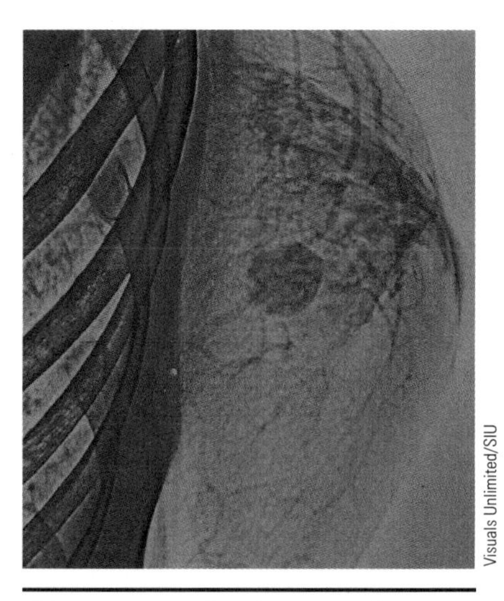
Visuals Unlimited/SIU

Mammogram showing area of breast cancer. Note the extensive vascularization.

curs at about age 50. Although wide variations exist, a typical menstrual cycle is 28 days long (Fig. 48-14). The first two weeks of the menstrual cycle are the **preovulatory phase.** The first day of the cycle is marked by the onset of menstruation, the monthly discharge through the vagina of blood and tissue from the endometrium. Ovulation occurs on about the 14th day of the cycle. The third and fourth weeks of the menstrual cycle are the **postovulatory phase.**

During the menstrual phase (the first five days of the preovulatory phase), gonadotropin-releasing hormone (GnRH) is released from the hypothalamus. GnRH stimulates the anterior pituitary to release follicle-stimulating hormone (FSH) and luteinizing hormone (LH) (Fig. 48-15a). During the preovulatory phase, FSH stimulates a few follicles to begin to develop, and stimulates the granulosa cells to multiply and produce estrogen. Some of the estrogen diffuses into the blood, but estrogen also has an autocrine action (see Chapter 47) on the granulosa cells that produce it and a paracrine effect on nearby granulosa cells. Estrogen stimulates the granulosa cells to multiply (which increases estrogen production).

The amount of estrogen secreted by the granulosa cells is enhanced by the action of LH on the theca cells. LH stimulates the theca cells to proliferate and produce androgens that diffuse into the granulosa cells, where they are converted to estrogen. Estrogen stimulates growth of the endometrium, which thickens and develops new blood vessels and glands.

After the first week of the menstrual cycle, only one follicle continues to develop. Its granulosa cells become sensitive to LH as well as to FSH. This dominant follicle now secretes enough estrogen to cause a rise in the concentration of estrogen in the blood.

TABLE 48-2 Principal Female Reproductive Hormones

Endocrine Gland and Hormones	Principal Target Tissue	Principal Actions
Hypothalamus		
Gonadotropin-releasing hormone (GnRH)	Anterior pituitary	Stimulates release of FSH and LH
Anterior Pituitary		
Follicle-stimulating hormone (FSH)	Ovary	Stimulates development of follicles and secretion of estrogen
Luteinizing hormone (LH)	Ovary	Stimulates ovulation and development of corpus luteum
Prolactin	Breast	Stimulates milk production (after breast has been prepared by estrogen and progesterone)
Posterior Pituitary		
Oxytocin	Uterus	Stimulates contraction and stimulates prostaglandin release
	Mammary glands	Stimulates ejection of milk into ducts
Ovaries		
Estrogen (estradiol)	General	Stimulates growth of sex organs at puberty and development of secondary sex characteristics
	Reproductive structures	Induces maturation; stimulates monthly preparation of endometrium for pregnancy; makes cervical mucus thinner and more alkaline
Progesterone	Uterus	Completes preparation of endometrium for pregnancy
Inhibin	Anterior pituitary	Inhibits secretion of FSH

KEY CONCEPT: The events that take place within the ovary and uterus are precisely coordinated by hormones released by the hypothalamus, anterior pituitary, and ovary.

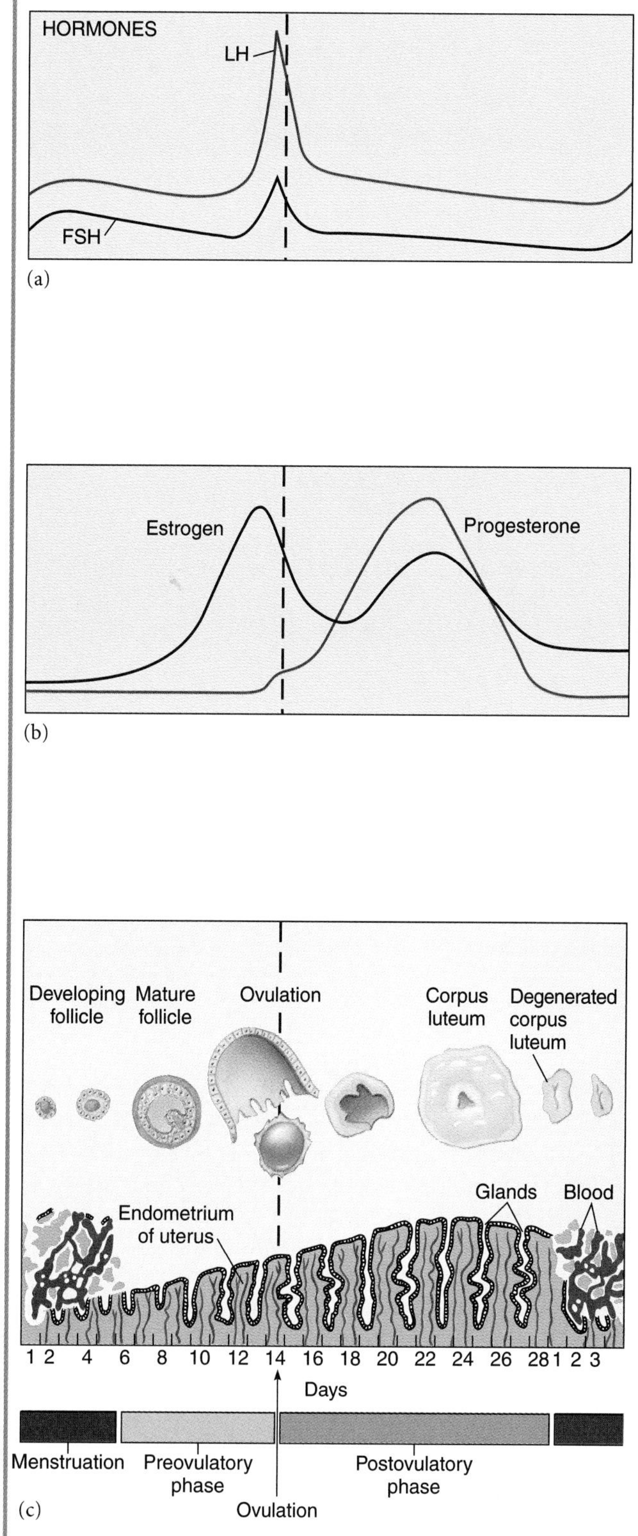

ACTIVE FIGURE 48-14 Endocrine regulation of the menstrual cycle.

When fertilization does not occur, the menstrual cycle repeats about every 28 days. Note that estrogen concentration is highest during the preovulatory phase, whereas progesterone concentration is highest during the postovulatory phase. FSH, follicle-stimulating hormone; LH, luteinizing hormone.

Biology Now™ Explore **hormonal control of the menstrual cycle** by clicking on this figure on your BiologyNow CD-ROM.

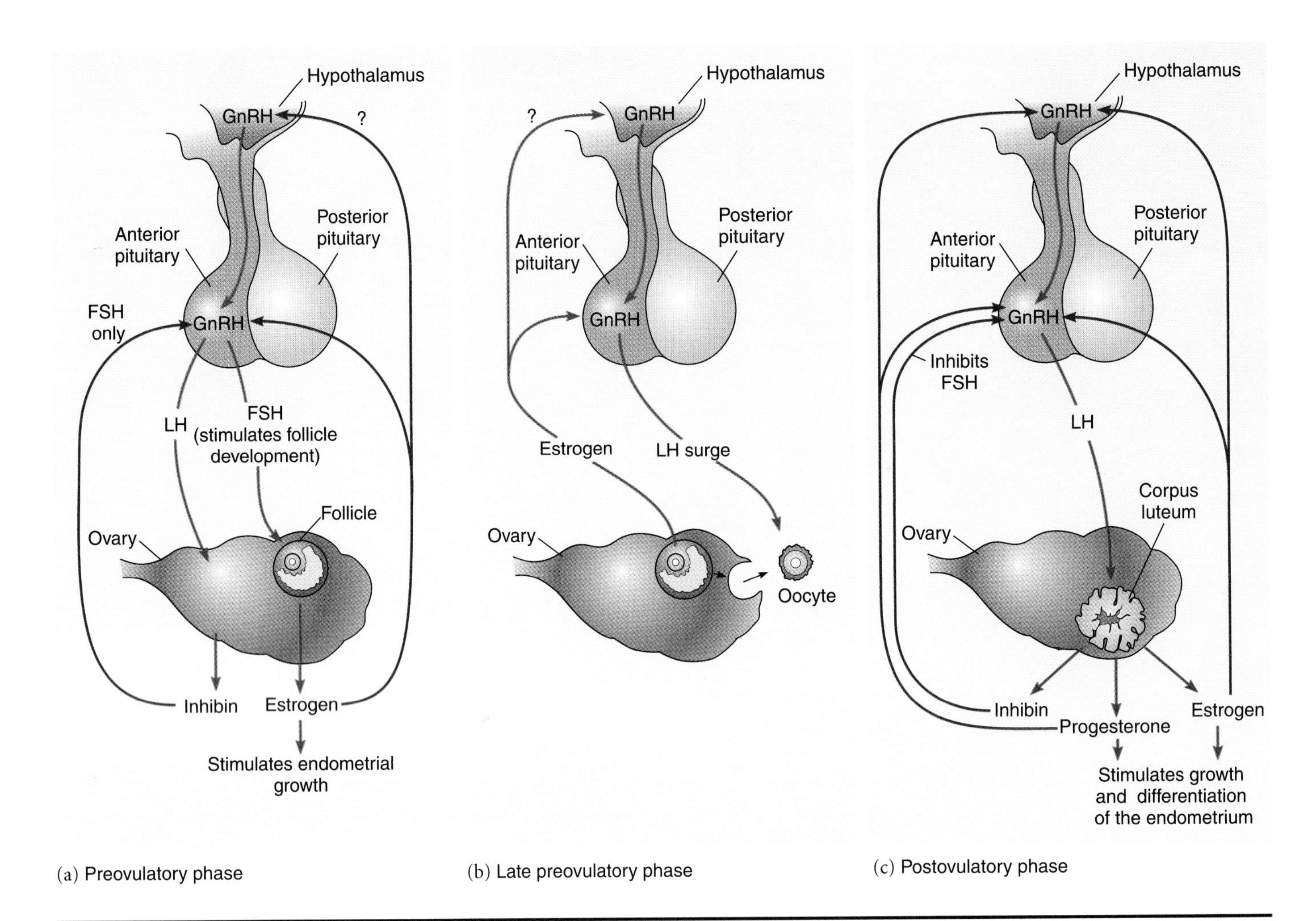

(a) Preovulatory phase (b) Late preovulatory phase (c) Postovulatory phase

FIGURE **48-15** **Feedback mechanisms in endocrine regulation of female reproduction.** Green arrows indicate stimulation, red arrows indicate inhibition. **(a)** Hormone interactions during the preovulatory phase. Estrogen has a negative feedback effect on the pituitary, and possibly the hypothalamus. **(b)** During the late preovulatory phase, estrogen has a positive feedback effect on the pituitary and hypothalamus. FSH and a surge of LH stimulate ovulation. **(c)** During the postovulatory phase, estrogen again has a negative feedback effect on the pituitary and hypothalamus.

Although still at relatively low concentration, estrogen inhibits secretion of FSH and LH from the pituitary and may also act on the hypothalamus, decreasing secretion of GnRH. In addition, the granulosa cells secrete inhibin, a hormone that inhibits mainly FSH secretion. As a result of these negative feedback signals, FSH (and to a lesser extent, LH) concentration decreases.

As its concentration in the blood peaks during the late preovulatory phase, estrogen signals the anterior pituitary to secrete LH (Fig. 48-15b). This is a positive feedback mechanism. The surge of LH secreted at the middle of the menstrual cycle stimulates the final maturation of the follicle, and stimulates ovulation.

After the secondary oocyte has been ejected from the ovary, the postovulatory phase begins. LH stimulates development of the corpus luteum, which secretes a large amount of progesterone and estrogen, and also inhibin (Fig. 48-15c). These hormones stimulate the uterus to continue its preparation for pregnancy. Progesterone stimulates tiny glands in the endometrium to secrete a fluid rich in nutrients.

During the postovulatory phase, the high concentration of progesterone in the blood, along with estrogen, inhibits secretion of GnRH, FSH, and LH. Progesterone is thought to act mainly on the hypothalamus. Inhibin acts on the pituitary to further inhibit FSH secretion. Thus during the postovulatory phase, FSH and LH concentrations are low and no new follicles develop.

If the secondary oocyte is not fertilized, the corpus luteum begins to degenerate after about eight days. Although the mechanism responsible for corpus luteum degeneration is not completely understood, a decrease in LH may be a factor. In addition, the corpus luteum may become less sensitive to LH.

When the corpus luteum stops secreting progesterone and estrogen, the concentrations of these hormones in the blood fall markedly. As a result, small arteries in the endometrium constrict, reducing the oxygen supply. Menstruation, which marks the beginning of a new cycle, begins as cells die and damaged arteries rupture and bleed. The low levels of estrogen and progesterone are insufficient to inhibit the anterior pituitary, and secretion of FSH and LH increases once again.

Review

- How is oogenesis different from spermatogenesis? Why are so many sperm produced in the male and so few ova produced in the female?
- What is the function of the corpus luteum? What is the fate of the corpus luteum when the ovum is not fertilized?
- What are specific actions of FSH and LH in the female? Of estrogens? Of progesterone?

BiologyNow™ Assess your understanding of **female human reproduction** by taking the pretest on your BiologyNow CD-ROM.

SEXUAL RESPONSE

Learning Objective

8 Describe the physiological changes that occur during sexual response.

During copulation, also called **coitus** or sexual intercourse in humans, the male deposits semen into the upper end of the vagina. The complex structures of the male and female reproductive systems, and the physiological, endocrine, and psychological processes associated with sexual activity, are adaptations that promote fertilization of the secondary oocyte, and development of the resulting embryo.

Sexual stimulation results in two basic physiological responses: (1) vasocongestion, the concentration of blood in reproductive structures, as well as in other areas of the body, and (2) increased muscle tension. Sexual response includes four phases: sexual excitement, plateau, orgasm, and resolution. The *desire* to have sexual activity may be motivated by fantasies or thoughts about sex. This anticipation can lead to (physical) sexual excitement and a sense of sexual pleasure. Physiologically, the **sexual excitement phase** involves vasocongestion and increased muscle tension. Before the penis can enter the vagina and function in coitus, it must be erect. Penile erection is the first male response to sexual excitement. In the female, vasocongestion occurs in the vagina, clitoris and breasts, and the vaginal epithelium secretes a sticky lubricant. Vaginal lubrication is the female's first response to effective sexual stimulation. During the excitement phase, the vagina lengthens and expands in preparation for receiving the penis.

If erotic stimulation continues, sexual excitement heightens to the **plateau phase**. Vasocongestion and muscle tension increase markedly. In both sexes, muscle tension, heart rate, blood pressure, and respiration rate all increase. Sexual intercourse is usually initiated during the plateau phase. The penis creates friction as it is moved inward and outward in the vagina, in actions referred to as pelvic thrusts. Physical and psychological sensations resulting from this friction (and from the emotional intimacy experienced) may lead to **orgasm**, the climax of sexual excitement. In the female, stimulation of the clitoris heightens the sexual excitement that leads to orgasm.

Although it lasts only a few seconds, orgasm is the phase of maximum sexual tension and its release. In both sexes, orgasm is marked by rhythmic contractions of the muscles of the pelvic floor and reproductive structures. These muscular contractions continue at about 0.8-second intervals for several seconds. After the first few contractions, their intensity decreases, and they become less regular and less frequent. Heart rate and respiration more than double, and blood pressure rises markedly, just before and during orgasm. Musculoskeletal contractions occur throughout the body. In the male, orgasm is marked by the ejaculation of semen from the penis. No fluid ejaculation accompanies orgasm in the female. Orgasm is followed by the **resolution phase**, a state of well-being during which the body is restored to its unstimulated state.

Review

- What physiological changes take place during the sexual response cycle?

BiologyNow™ Assess your understanding of **sexual response** by taking the pretest on your BiologyNow CD-ROM.

FERTILIZATION AND EARLY DEVELOPMENT

Learning Objective

9 Describe the processes of human fertilization and early development, and summarize the actions of hormones that regulate pregnancy.

Fertilization is the fusion of sperm and egg. Fertilization and the subsequent establishment of pregnancy together are referred to as **conception.** After ejaculation into the female reproductive tract, sperm remain alive and retain their ability to fertilize an ovum for only a few days. The ovum remains fertile for about 24 hours after ovulation. Thus conception is most probable when intercourse takes place on the day of ovulation or the five days preceding ovulation. In a very regular 28-day menstrual cycle, sexual intercourse on days 12 to 16 is most likely to result in fertilization. However, many women do not have regular menstrual cycles, and many factors cause irregular cycles even in women who are generally regular.

When conditions in the vagina and cervix are favorable, sperm begin to arrive at the site of fertilization in the upper oviduct within a few minutes after ejaculation. At the time of ovulation, when estrogen concentration is high, the cervical mucus has a thin consistency that permits passage of sperm from the vagina into the uterus. After ovulation, when progesterone concentration rises, the cervical mucus becomes thick and sticky, blocking entrance of sperm (as well as bacteria that might harm the developing embryo). Once sperm enter the uterus, contractions of the uterine wall help transport them. The sperm's own motility is important, especially in approaching and fertilizing the ovum.

When a sperm encounters an egg, openings develop in the sperm acrosome, allowing enzymes to digest a path through the zona pellucida surrounding the secondary oocyte. As soon as one sperm enters the secondary oocyte, changes occur that prevent the entrance of other sperm As the fertilizing sperm enters, it usually loses its flagellum (Fig. 48-16). Sperm entry stimulates the secondary oocyte to complete its second meiotic

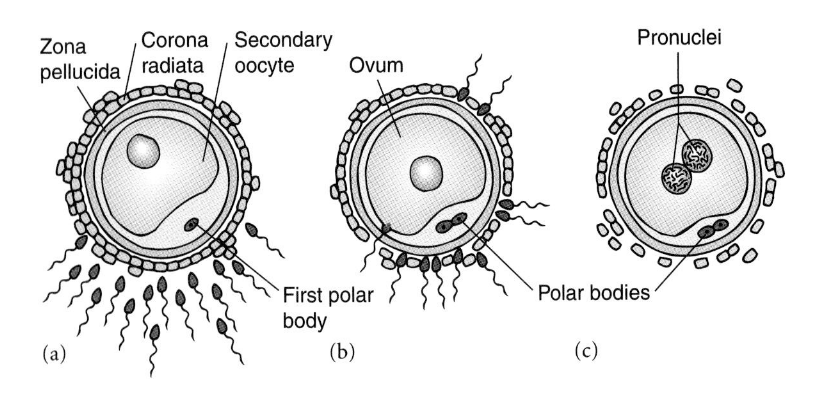

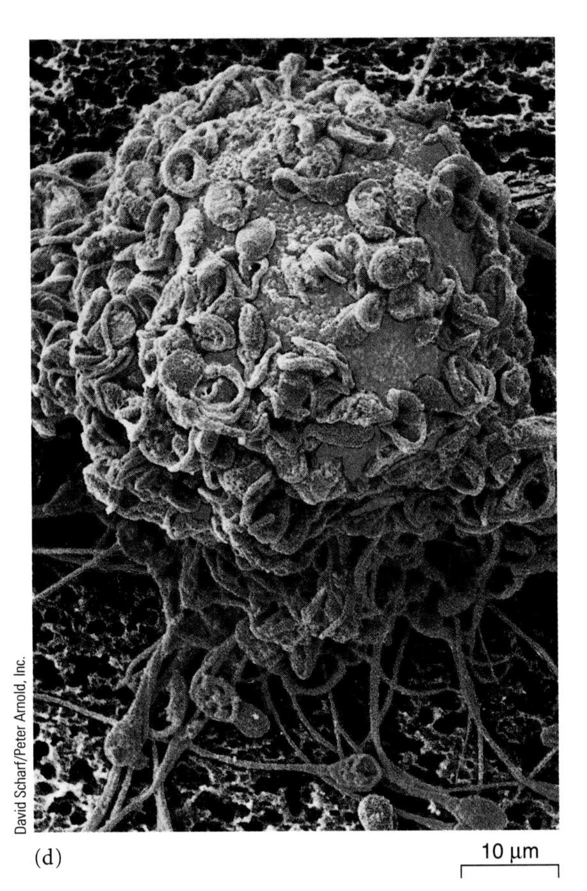

FIGURE 48-16 Fertilization.

(a) Each sperm releases a small amount of an enzyme that helps disperse the layer of follicle cells (corona radiata) surrounding the secondary oocyte. **(b)** After a sperm cell enters, the secondary oocyte completes its second meiotic division, producing an ovum and a polar body. **(c)** Pronuclei of sperm and ovum combine, producing a zygote with the diploid number of chromosomes. **(d)** Color-enhanced SEM of human sperm cells surrounding a test ovum. Sperm are being tested for viability.

division. The head of the haploid sperm then swells to form the **male pronucleus** and fuses with the **female pronucleus,** forming the diploid nucleus of the zygote. The process of fertilization is described in more detail in Chapter 49. (Also see *Focus On: Novel Origins.*)

If only one sperm is needed to fertilize a secondary oocyte, why are millions ejaculated? Many die as a result of unfavorable pH or phagocytosis by leukocytes and macrophages in the female tract. Only a few hundred succeed in traversing the correct oviduct and reaching the vicinity of the secondary oocyte.

If the secondary oocyte is fertilized, development begins as the embryo is slowly moved to the uterus by the cilia lining the oviduct. When it enters the uterus, the embryo consists of a ball of about 32 cells. After floating free in the uterus for about three days, the embryo begins to implant in the thick endometrium on about the seventh day after fertilization (Fig. 48-17). (Development is discussed in Chapter 49.)

Membranes that develop around the embryo secrete **human chorionic gonadotropin (hCG),** a

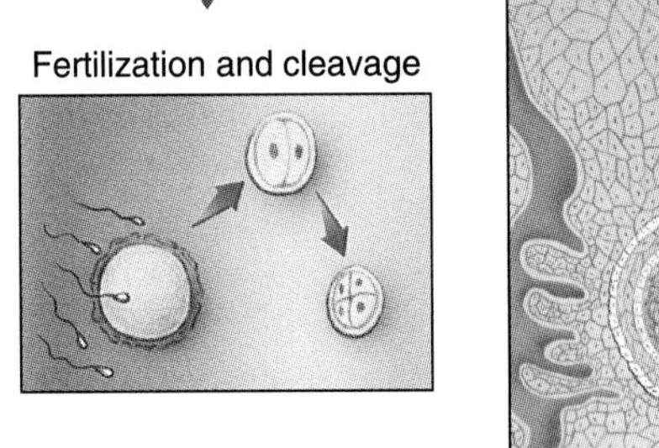

FIGURE 48-17 Events following fertilization.

The ovarian and uterine cycles are interrupted when pregnancy occurs. The corpus luteum does not degenerate, and menstruation does not take place. Instead, the wall of the uterus thickens even more, permitting the embryo to implant and develop within it. (Cleavage is the early series of mitotic divisions that converts the zygote to a multicellular embryo.)

About 16% of married couples in the United States are affected by infertility, the inability of a couple to achieve conception after using no contraception for at least one year. About 30% of cases involve both male and female factors. Male infertility is often attributed to low sperm count. Among the common causes of female infertility are failure to ovulate, production of infertile eggs (common in older women), and oviduct scarring (often caused by pelvic inflammatory disease) that blocks the passage of the secondary oocyte to the uterus. Women with blocked oviducts can usually produce ova and incubate an embryo normally but need clinical assistance in getting the ovum from the ovary to the uterus.

In the United States, more than 3 million infertile couples consult health care professionals each year. Some are helped with conventional treatment, for example, with hormone therapy that regulates ovulation or with fertility drugs. But more than 40,000 couples need more sophisticated clinical help and turn to high-tech assisted reproductive techniques that have been developed through reproduction and human embryo research. At present these techniques are expensive and the success rate is low, less than one third.

The most common assisted reproductive procedure is **artificial insemination,** in which a catheter is used to inject sperm directly into the cervix or uterus. Transfer into the uterus is called **intrauterine insemination (IUI).** More than 600,000 IUI procedures are performed annually with a success rate of about 10%. Artificial insemination is indicated when the male partner of a couple desiring a child is sterile or carries a genetic defect. If the male is infertile because of a low sperm count, his sperm can be concentrated, or alternatively, sperm from a donor can be used. Although the sperm donor usually remains anonymous to the couple, his genetic qualifications are screened by physicians.

With **in vitro fertilization (IVF),** a woman takes a fertility drug that induces ovulation of several ova. The ova are removed and fertilized with sperm. Embryos are screened for chromosome or gene abnormalities, and healthy ones are transferred into the uterus through the cervix. This procedure was first used in England in 1978 to help a couple who had tried unsuccessfully for several years to have a child. Since that time, thousands of "test-tube babies" have been conceived in this way and born to previously infertile women. The success rate of in vitro fertilization is about 31%.

In **gamete intrafallopian transfer (GIFT),** a laparoscope (a fiber-optic instrument) is used to guide the transfer of ova and sperm into a woman's oviduct through a small incision in her abdomen. More than 4000 of these expensive procedures are performed each year with a success rate of about 28%. In **zygote intrafallopian transfer (ZIFT),** ova are fertilized in the laboratory, and a laparoscope is used to guide the transfer of the resulting zygotes into the oviduct. In these procedures, the patient's own ova may be used. However, if she does not produce fertile ova, they can be contributed by a donor **(oocyte donation).**

Another novel procedure is **host mothering.** An embryo is removed from its natural mother and implanted into a female substitute. The foster mother can support the developing embryo either until birth or temporarily until it is implanted again into the original mother or another host. This technique has proved useful to animal breeders. For example, embryos from prize sheep can be temporarily implanted into rabbits for easy shipping by air and then implanted into a host ewe, perhaps one of inferior quality. Host mothering allows an animal with superior genetic traits to produce more offspring than would be naturally possible. This procedure is also used to increase the populations of certain endangered species (see figure).

Technology is available to freeze the gametes or embryos of many species, including humans, and then transplant them into their donors or into host mothers. Freezing eggs may become popular with young women not yet ready to become parents, but who want to preserve young eggs with lower risk for chromosome abnormality, and reimplant them at a later time.

Courtesy of Betsy Dresser

Newborn bongo with its surrogate mother, an eland. As a young embryo, the bongo was transplanted into the eland's uterus, where it implanted and developed. Bongos are a rare and elusive species inhabiting dense forests in Africa. The larger and more common elands inhabit open areas.

hormone that signals the mother's corpus luteum to continue to function. (The presence of hCG in urine or blood is used as an early pregnancy test.) Concentrations of estrogen and progesterone remain high throughout pregnancy. During the first two months, the corpus luteum secretes almost all the estrogen and progesterone necessary to maintain the pregnancy. During this time, membranes surrounding the embryo, together with uterine tissue, form the **placenta,** the organ of exchange between the mother and developing embryo. As the corpus luteum slows its secretion and deteriorates after about three months, the placenta takes over and secretes large amounts of estrogen and progesterone.

Estrogen and progesterone are necessary to maintain pregnancy. Estrogen stimulates development of the uterine wall, including the muscle needed to expel the fetus during delivery. Progesterone inhibits uterine contractions so that the fetus is not expelled too soon. Because these hormones also inhibit FSH and LH, new follicles do not develop and the menstrual cycle stops during pregnancy.

Review

- What is the function of hCG?
- What are the actions of estrogen and progesterone in maintaining pregnancy?

BiologyNow™ Assess your understanding of **fertilization and early development** by taking the pretest on your BiologyNow CD-ROM.

THE BIRTH PROCESS

Learning Objective

10 Summarize the birth process.

A normal human pregnancy is about 38 weeks long, counting from the day of conception, or about 40 weeks from the first day of the last menstrual cycle. The mechanisms that terminate pregnancy and initiate the birth process, called **parturition,** are not fully understood.

At the end of pregnancy, the stretching of the uterine muscle by the growing fetus combined with the effects of increased estrogen concentration and oxytocin produce strong uterine contractions. A long series of involuntary contractions of the uterus are experienced as **labor,** which begins when uterine contractions occur every 10 to 15 minutes. Labor can be divided into three stages. During the first stage, which typically lasts about 12 hours, the contractions of the uterus move the fetus toward the cervix, causing the cervix to *dilate* (open) to a maximum diameter of 10 cm (4 in). The cervix also becomes *effaced;* that is, it thins out so that the fetal head can pass through. During the first stage of labor the amnion (the membrane that forms a fluid-filled sac around the embryo/fetus) usually ruptures, releasing about a liter of amniotic fluid, which flows out through the vagina.

During the second stage, which normally lasts between 20 minutes and an hour, the fetus passes through the cervix and the vagina and is born, or "delivered" (Fig. 48-18). With each uterine contraction, the woman bears down so that the fetus is expelled by the combined forces of uterine contractions and contractions of abdominal wall muscles.

At birth, the baby is still connected to the placenta by the umbilical cord. Contractions of the uterus squeeze much of the fetal blood from the placenta into the infant. The cord is tied and cut, separating the child from the mother. (The stump of the cord gradually shrivels until nothing remains but the scar, the **navel.**)

During the third stage of labor, which lasts 10 or 15 minutes after the birth, the placenta and the fetal membranes are loosened from the lining of the uterus by another series of contractions and expelled. At this stage they are collectively called the **afterbirth,** because these tissues are expelled from the vagina after the baby has been delivered.

During labor an obstetrician may administer oxytocin or prostaglandins to increase the contractions of the uterus or may assist with special forceps or a vacuum device. In some women, the opening between the pelvic bones is too small to permit the passage of the baby vaginally. In this situation, or if the baby's position prevents normal delivery, the obstetrician may perform a **cesarean section,** an operation in which the baby is delivered through an incision made in the abdominal and uterine walls.

Review

- What is the role of estrogen in initiating the birth process? Of oxytocin?

BiologyNow™ Assess your understanding of **the birth process** by taking the pretest on your BiologyNow CD-ROM.

BIRTH CONTROL METHODS

Learning Objective

11 Compare the modes of action, effectiveness, advantages, and disadvantages of the methods of birth control discussed; include sterilization and emergency contraception.

When a fertile, heterosexually active woman uses no form of birth control, her chances of becoming pregnant during the course of a year are about 90%. Any method for deliberately separating sexual intercourse from reproduction is **contraception** (literally, "against conception"). Although many couples worldwide agree it is best to have babies by choice rather than by chance, most do not have contraceptives available to them. Worldwide about 133 million births occur each year. Population experts estimate about 33 million (one fourth) of these are unplanned. If we add the estimated 46 million induced abortions performed annually, the total number of unplanned pregnancies is about 79 million each year.

Teen pregnancy is a serious problem. Nearly 1 million teenagers become pregnant each year in the United States, and almost half a million give birth. Thousands of these girls are aged 14 or younger. Many sexually active teenagers do not consistently use birth control. Teens often lack the knowledge and means of protecting themselves from unwanted pregnancy.

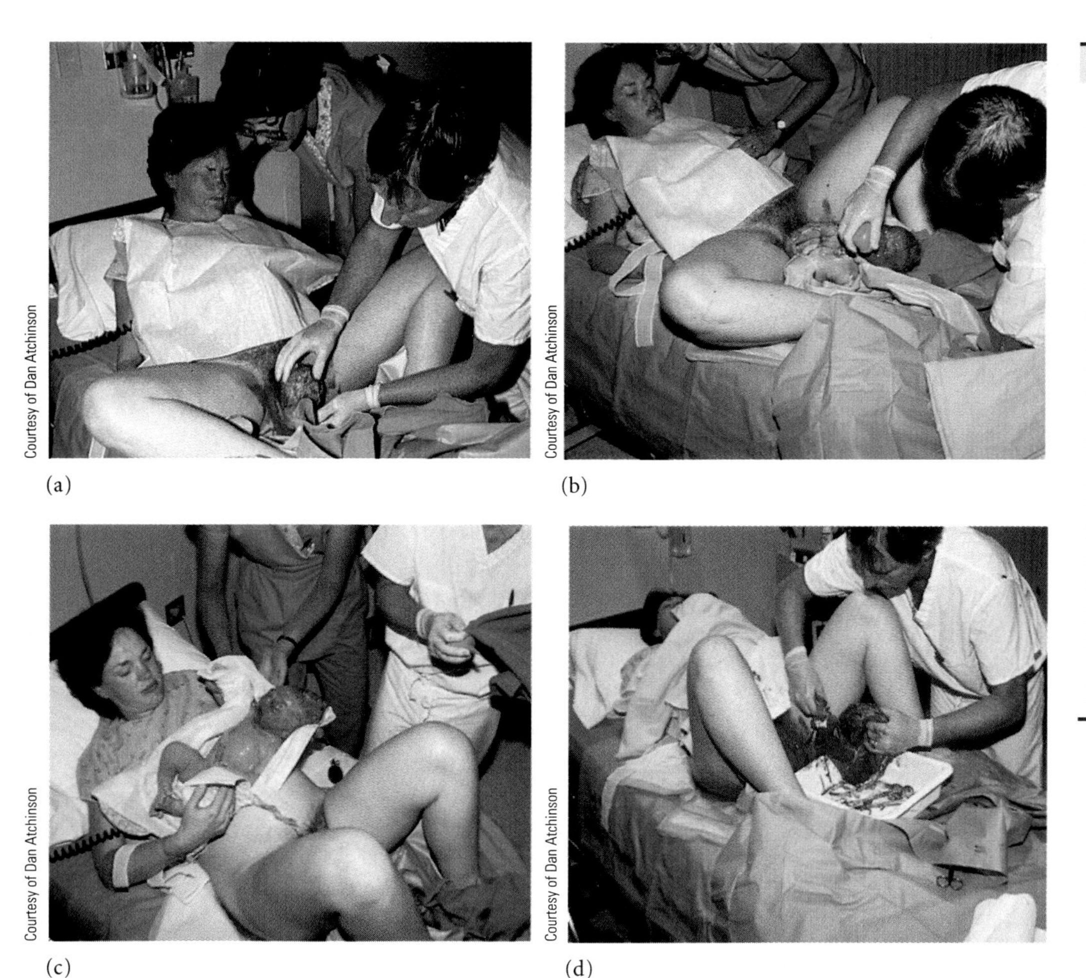

FIGURE **48-18** **Parturition.** In about 95% of all human births, the baby descends through the cervix and vagina in the head-down position. **(a)** The mother bears down hard with her abdominal muscles, helping to push the baby out. When the head fully appears, the physician or midwife gently grasps it and guides the baby's entrance into the outside world. **(b)** Once the head has emerged, the rest of the body usually follows readily. The physician gently aspirates the mouth and pharynx to clear the upper airway of amniotic fluid, mucus, or blood. At this time the newborn takes its first breath. **(c)** The baby, still attached to the placenta by its umbilical cord, is presented to its mother. **(d)** During the third stage of labor, the placenta is delivered.

Since ancient times, humans have searched for effective contraceptive methods. Scientists have developed a variety of contraceptives with a high percentage of reliability, but the ideal contraceptive has not yet been developed. Among the issues that must be considered in developing contraceptives are safety, effectiveness, cost, convenience, and ease of use. Risks of cancer, birth defects in case the method fails and the woman becomes pregnant, permanent infertility, and side effects such as menstrual abnormalities must be minimized.

Researchers predict that the contraceptive methods of the 21st century will control regulatory peptides and the genes that code for them. For example, the genes that code for the pituitary hormones that stimulate the ovaries to release estrogen could be turned off selectively, or other hormone signals could be interrupted. Other approaches being studied are molecular interruption of fertilization and contraceptive vaccines. Researchers are testing a sugar molecule that causes reversible sterility in mice by impairing sperm motility. Within 20 years, current methods of contraception will likely be replaced by more sophisticated molecular methods.

Some of the more common methods of birth control are described in the following paragraphs and in Table 48-3 (see also Fig. 48-19). Oral contraceptives, IUDs, and female sterilization account for more than two-thirds of all contraception practiced worldwide.

Most hormone contraceptives prevent ovulation

Oral contraceptives, contraceptive patches and injectable contraceptives are hormone contraceptives. More than 80 million women worldwide (more than 8 million in the United States alone) use oral contraceptives. When taken correctly, oral contraceptives are about 99.7% effective in preventing pregnancy. They are also used to regulate menstrual cycles. The most common preparations are combinations of progestin and synthetic estrogen. (Natural hormones are destroyed by the liver almost immediately, but synthetic ones are chemically modified during production so that they can be absorbed effectively and metabolized slowly.) In a typical regimen, a woman takes one pill each day for about three weeks. Then for one week she takes a sugar pill that allows menstruation to occur because of the withdrawal of the hormones. A new oral contraceptive (Seasonale) works on a 91-day regimen, reducing menstruation to four times per year.

Oral contraceptives prevent ovulation. When postovulatory levels of ovarian hormones are maintained in the blood, the body is tricked into responding as though conception had occurred. The pituitary gland is inhibited and does not produce the surge of LH that stimulates ovulation.

Studies suggest that women over the age of 35 who smoke or have other risk factors, such as untreated hypertension, should

TABLE **48-3** Selected Contraceptive Methods

Method	Failure Rate*	Mode of Action	Advantages	Disadvantages
Oral contraceptives	0.3; 5	Inhibit ovulation; may affect endometrium and cervical mucus and prevent implantation	Highly effective; regulate menstrual cycle	Minor discomfort in some women; should not be used by women over age 35 who smoke
Injectable contraceptives (Depo-Provera, Lunelle)	About 1	Inhibit ovulation	Effective; long-lasting	Irregular menstrual bleeding; fertility may not return for 6–12 months after contraceptive is discontinued
Intrauterine device (IUD)	1; 1	Probably stimulates inflammatory response and prevents implantation	Provides continuous protection; highly effective for several years	Cramps; increased menstrual flow; increased risk of pelvic inflammatory disease and infertility; not recommended for women who have not had a child
Spermicides foams, jellies, creams	3; 20	Chemically kill sperm	No known side effects; can be used with a condom or diaphragm to improve efficacy	Messy; must be applied before intercourse
Contraceptive diaphragm (with jelly)†	3; 14	Diaphragm mechanically blocks entrance to cervix; jelly is spermicidal	No side effects	Must be inserted prior to intercourse and left in place for several hours afterward
Condom	2.6; 14	Mechanically prevents sperm from entering vagina	No side effects; some protection against STDs, including HIV	Slightly decreased sensation for male; could break
Rhythm‡	13; 19	Abstinence during fertile period	No known side effects	Not very reliable
Withdrawal (coitus interruptus)	9; 22	Male withdraws penis from vagina prior to ejaculation	No side effects	Not reliable; sperm in fluid secreted before ejaculation may be sufficient for conception
Sterilization				
Tubal ligation	0.04	Prevents ovum from leaving uterine tube	Most reliable method	Requires surgery; considered permanent
Vasectomy	0.15	Prevents sperm from leaving vas deferens	Most reliable method	Requires surgery; considered permanent
Chance (no contraception)	About 90			

*Lower figure is the failure rate of method; higher figure is rate of method failure plus failure of the user to apply the method correctly. Based on number of failures per 100 women who use method per year in the United States.

†Failure rate is lower when diaphragm is used with spermicides.

‡There are several variations of rhythm method. For those who use the calendar method alone, the failure is about 35. However, if body temperature is taken daily and careful records are kept (temperature rises after ovulation), failure rate can be reduced. When women use some method to determine time of ovulation and have intercourse *only* more than 48 hours *after* ovulation, failure rate can be reduced to about 7.

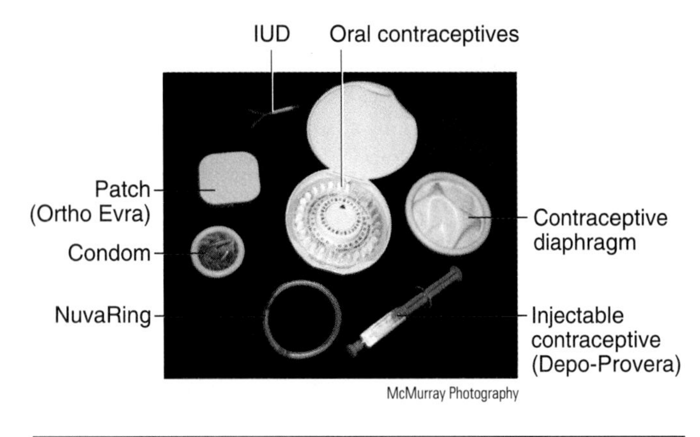

FIGURE **48-19** **Some commonly used contraceptives.** The only common male method of contraception is the condom. (Norplant is an injectable hormonal contraceptive.)

not take oral contraceptives. Women in this category who take oral contraceptives have an increased risk of death from stroke and myocardial infarction. Low-dose oral contraceptive pills appear safe for nonsmokers up to the time of menopause. Oral contraceptives are linked to deaths in about 3 per 100,000 users. This compares favorably with the death rate of about 9 per 100,000 pregnancies.

The birth control patch (Ortho Evra) delivers estrogen and progestin through the skin. The patch is applied weekly for three weeks and removed for the fourth week. During the patch-free week, menstruation usually occurs. Another approach is a thin, flexible plastic ring (NuvaRing) that women can flatten like a rubber band and insert into the vagina once a month. The ring releases progestin and estrogen in the amounts present in low-dose birth control pills.

Injectable progestin (Depo-Provera) is injected intramuscularly every three months. It prevents ovulation by suppressing anterior pituitary function; it also thickens the cervical mucus (which makes it more difficult for the sperm to reach the egg).

Intrauterine devices are widely used

The **intrauterine device (IUD)** is the most widely used method of reversible contraception in the world, used by an estimated 90 million women. However, in the United States only a small percentage of women use them. The IUD is a small device that is inserted into the uterus by a medical professional. IUDs in current use have been shown to be safe, and are about 99% effective. Disadvantages of the IUD include uterine cramping and bleeding in a small percentage of women.

One IUD presently being used in the United States is the Copper T380 (ParaGard) which can be left in place for up to 10 years. A newer IUD (Mirena), a small, plastic T-shaped device, releases a very small amount of progestin onto the inner wall of the uterus. This IUD can be left in place for up to five years and causes fewer side effects.

The IUD's mode of action is not well understood. The copper or hormone slowly released from an IUD apparently interferes with embryo implantation. In addition, white blood cells mobilized in response to the foreign body in the uterus may produce substances toxic to sperm. Mirena causes thickening of the cervical mucus, preventing passage of sperm into the uterus.

Other common contraceptive methods include the diaphragm and condom

The **contraceptive diaphragm** mechanically blocks the passage of sperm from the vagina into the cervix. It is covered with spermicidal jelly or cream and inserted just prior to sexual intercourse.

The **condom** is also a mechanical method of birth control. The only contraceptive device currently sold for men, the condom provides a barrier that contains the semen so that sperm cannot enter the female tract. The latex condom is the only contraceptive that provides some protection against infection with human immunodeficiency virus (HIV) and other sexually transmitted diseases (STDs). The female condom is a strong, soft, polyurethane sheath inserted in the vagina before sexual intercourse; it provides protection against both pregnancy and sexually transmitted infections.

Emergency contraception is available

Emergency contraception is after-the-fact contraception for rape victims and others who have had unprotected intercourse. If physicians advocated and women used emergency contraception, an estimated 2 million unwanted pregnancies and thousands of abortions could be prevented. Insertion of a copper IUD within a week of unprotected intercourse is more than 99% effective in preventing pregnancy. High doses of oral hormone contraceptives change the endometrium so that the embryo cannot implant in the uterine wall. Taken up to 72 hours after unprotected intercourse, they are about 75% effective in preventing pregnancy.

Sterilization renders an individual incapable of producing offspring

Sterilization is the only method of contraception not affected by human error. Worldwide, female sterilization is the most popular contraceptive method, and accounts for one third of all contraceptive use.

Male sterilization is performed by vasectomy

An estimated 1 million **vasectomies** are performed each year in the United States. With the use of a local anesthetic, a small incision is made on each side of the scrotum. Then each vas deferens is cut and its ends sealed so that they cannot grow back together (Fig. 48-20a). Because testosterone secretion and transport are not affected, vasectomy does not affect masculinity. Sperm continue to be produced, though at a much slower rate, and are destroyed by macrophages in the testes. No change in amount of semen ejaculated is noticed, because sperm account for very little of the semen volume.

By surgically reuniting the ends of the vasa deferentia, surgeons can successfully reverse male sterilization in about 70% of attempts. Apparently, some sterilized men eventually develop antibodies against their own sperm and remain sterile even after their vasectomies have been surgically reversed. Therefore, fertility rates are low (about 30%) for men who undergo vasectomy reversal after 10 or more years.

An alternative to vasectomy reversal is the storage of frozen sperm in sperm banks. If the male should decide to father another child after he has been sterilized, he simply "withdraws" his sperm to artificially inseminate his mate. Sperm banks have

FIGURE **48-20** | **Sterilization.**

(a) In vasectomy, the vas deferens (sperm duct) on each side is cut and cauterized. **(b)** In tubal ligation, each oviduct is cut and cauterized so that ovum and sperm can no longer meet.

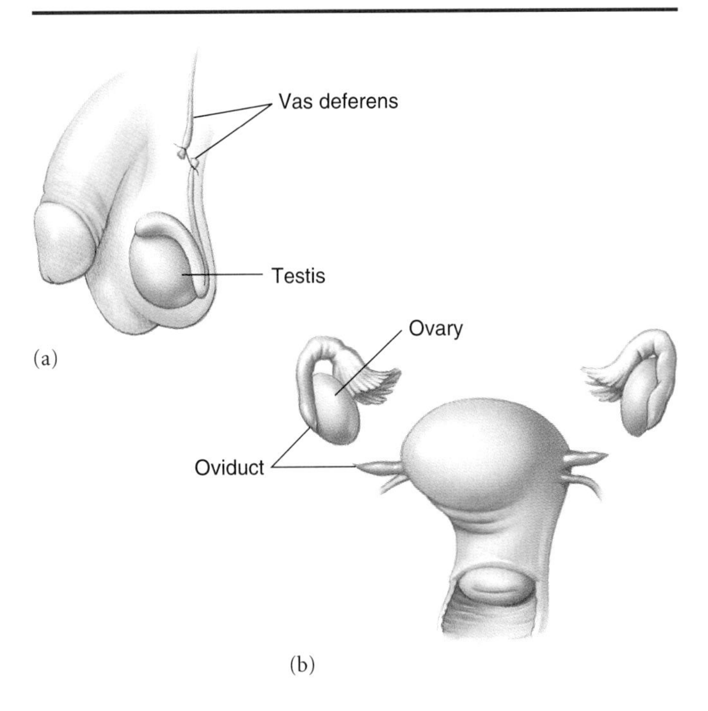

TABLE 48-4 Some Common Sexually Transmitted Diseases

Disease and Causative Organism	Course of Disease	Treatment
Chlamydia (*Chlamydia trachomatis,* a bacterium)	Discharge and burning with urination or asymptomatic; men 15–30 years old with multiple sex partners are most at risk	Antibiotics
Gonorrhea (*Neisseria gonorrhoeae,* a gonococcus bacterium)	Bacterial toxin may produce redness and swelling at infection site; symptoms in males are painful urination and discharge of pus from penis; in about 60% of infected women no symptoms initially; can spread to epididymis (in males) or uterine tubes and ovaries (in females), causing sterility; can cause widespread infection; damage to heart valves, meninges (outer coverings of brain and spinal cord), and joints	Antibiotics
Syphilis (*Treponema pallidum,* a spirochete bacterium)	Bacteria enter body through defect in skin; primary chancre (small, painless ulcer) at site of initial infection; highly infectious at this stage; secondary stage, widespread rash and influenza-like symptoms; scaly lesions may occur that are highly infectious; latent stage that follows can last 20 years; eventually, lesions called *gummae* may form that damage liver, brain, bone, or spleen; death results in 5–10% of untreated cases	Penicillin
Genital herpes (herpes simplex type 2 virus)	Tiny, painful blisters on genitals; may develop into ulcers; influenza-like symptoms may occur; recurs periodically; threat to fetus or newborn infant	No effective cure; some drugs may shorten outbreaks or reduce severity; medication can be taken to prevent outbreak
Human papillomavirus (HPV; about 30 strains of HPV can infect reproductive organs)	Some strains cause abnormal Pap smears and have been linked with cervical cancer; other strains cause genital warts	Immune modifiers; interferon
Trichomoniasis (a protozoon)	Itching, discharge, soreness; can be contracted from dirty toilet seats and towels; may be asymptomatic in males	Metronidazole (an antibiotic)
"Yeast" infections (genital candidiasis; *Candidiasis albicans*)	Irritation, soreness, discharge; especially common in females; rare in males; can be acquired nonsexually	Antifungal drugs
Pelvic inflammatory disease (PID); primarily caused by chlamydia or gonorrhea)	Infection of reproductive organs and pelvic cavity; may lead to sterility (> 15% of cases)	Antibiotics, surgical removal of affected organs
Acquired immunodeficiency syndrome (AIDS)* (caused by human immunodeficiency virus, HIV)	Influenza-like symptoms: swollen lymph glands, fever, night sweats, weight loss; decreased immunity, leading to pneumonia, rare forms of cancer	No effective cure; a variety of drugs reduce symptoms and prolong life

*AIDS was discussed in Chapter 43.

been established throughout the United States. Not much is known yet about the effects of long-term sperm storage, and there may be an increased risk of genetic defects.

Female sterilization is by tubal ligation

In the United States about 26% of never-married women between the ages of 15 and 44 have chosen **tubal ligation** as a means of birth control. This procedure is performed under general anesthesia by making a small abdominal incision and using a laparoscope to locate and cut the oviducts (Fig. 48-20b). The tubes are then cauterized. As in the male, hormone balance and sexual performance are not affected by sterilization.

Abortions can be spontaneous or induced

Abortion is termination of pregnancy that results in the death of the embryo or fetus. **Spontaneous abortions** (commonly referred to as miscarriages) occur without intervention. Embryos that are spontaneously aborted are frequently abnormal. **Induced abortions** are performed deliberately either for therapeutic reasons or as a means of birth control. **Therapeutic abortions** are performed to protect the mother's life or when there is reason to suspect that the embryo is grossly abnormal. Worldwide, about as many unplanned as planned pregnancies occur each year, and more than half of the unplanned pregnancies end in abortion. An estimated 46 million induced abortions are performed each year (more than a million in the United States).

Most first-trimester abortions (those performed during the first three months of pregnancy) are performed by a suction method or by administering drugs that interrupt the pregnancy and induce labor. In the suction method of abortion, the cervix is dilated, and a suction aspirator is inserted into the uterus. The embryo and other products of conception are evacuated.

The drug *RU-486* (mifepristone) binds competitively with progesterone receptors in the uterus but does not activate them. Because the normal actions of progesterone are blocked, the endometrium breaks down and uterine contractions occur. These conditions prevent implantation of an embryo. The action of methotrexate, a drug used to treat cancer, is similar to that of RU-486. Both of these drugs are used to interrupt pregnancy. Prostaglandins may then be administered to induce the uterine contractions that expel the embryo.

After the first trimester, abortions are performed mainly because of fetal defects or maternal illness. The method most commonly used is dilation and evacuation. The cervix is dilated, forceps are used to remove the fetus, and suction is used to aspirate the remaining contents of the uterus. Drugs such as prostaglandins are being increasingly used to induce labor in second-trimester abortions.

When abortions are performed by skilled medical personnel, the mortality rate in the United States is less than 1 per 100,000. In countries where abortion is illegal, the death rate is as high as 700 per 100,000. These statistics can be contrasted with the death rate from pregnancy and childbirth of about 9 per 100,000. The World Health Organization estimates that one third of all maternal deaths result from poorly done illegal abortions.

Review

- What are three types of hormonal contraception?
- What is emergency contraception?

BiologyNow™ Assess your understanding of **birth control methods** by taking the pretest on your BiologyNow CD-ROM.

SEXUALLY TRANSMITTED DISEASES

Learning Objective

12 Identify common sexually transmitted diseases, and describe their symptoms, effects, and treatments.

Sexually transmitted diseases (STDs), also called venereal diseases (VDs), are, next to the common cold, the most prevalent communicable diseases in the world. The World Health Organization has estimated that more than 250 million people are infected each year with **gonorrhea,** and more than 50 million are infected with **syphilis.**

In the United States, more than 65 million people are currently living with an incurable STD, and more than 12 million people (including 3 million teens) are diagnosed with new STD infections each year. One in three sexually active young people will acquire an STD by age 24. One out of four women and one out of five men are infected with **genital herpes.** However, the most common STD in the United States is **chlamydia;** it is also the most frequently reported infectious disease. Chlamydia is the most common cause of **pelvic inflammatory disease (PID),** an infection of the female reproductive organs that often leads to sterility.

Some strains of human papillomavirus (HPV) cause abnormal Pap smears and have been linked with cervical cancer. Other strains cause genital warts. Some common STDs are described in Table 48-4. (HIV was discussed in Chapter 43.)

Review

- What is the most common STD in the United States?
- What is PID?

BiologyNow™ Assess your understanding of **sexually transmitted diseases** by taking the pretest on your BiologyNow CD-ROM.

SUMMARY WITH KEY TERMS

1 Compare the advantages of asexual and sexual reproduction; describe each mode of reproduction, giving specific examples.

- In **asexual reproduction,** a single parent endows its offspring with a set of genes identical to its own (except for mutations). Asexual reproduction is energy efficient, and most successful in a stable environment.
- In **budding,** a part of the parent's body grows and separates from the rest of the body. In **fragmentation,** the parent's body may break into several pieces; each piece can develop into a new animal. In **parthenogenesis,** an unfertilized egg develops into an adult.
- In **sexual reproduction,** offspring are produced by the fusion of two types of **gametes,** ovum (egg) and **sperm.** When sperm and ovum fuse, a fertilized egg, or **zygote,** forms. Sexual reproduction promotes genetic variety and is especially adaptive in an unstable, changing environment.
- In **external fertilization,** mating partners typically release eggs and sperm into the water simultaneously. In **internal fertilization,** the male delivers sperm into the female's body.
- In **hermaphroditism,** a single individual produces both eggs and sperm.

2 Describe the structure and function of each organ of the human male reproductive system.

- The human male reproductive system includes the testes, which produce sperm and testosterone; a series of conducting ducts; accessory glands; and the penis.
- The **testes,** housed in the **scrotum,** contain the **seminiferous tubules,** where **spermatogenesis** (sperm production) takes place. The **interstitial cells** in the testes secrete testosterone. **Sertoli cells** produce signaling molecules and a fluid that nourishes sperm cells.
- Sperm complete their maturation and are stored in the **epididymis** and **vas deferens.** During ejaculation, sperm pass from the vas deferens to the **ejaculatory duct** and then into the **urethra,** which passes through the penis.
- Each ejaculate of semen contains about 200 million sperm suspended in the secretions of the **seminal vesicles** and **prostate gland.** The **bulbourethral glands** release a mucous secretion.
- The penis consists of three columns of **erectile tissue,** two **cavernous bodies** and one **spongy body** that surrounds the urethra. When its erectile tissue becomes engorged with blood, the penis becomes erect.

3 Trace the passage of sperm cells through the human male reproductive system from their origin in the seminiferous tubules to their expulsion from the body in the semen. (Include a description of spermatogenesis.)

- Spermatogenesis takes place in the seminiferous tubules of the testes. **Spermatogonia** divide by mitosis; some differentiate and become **primary spermatocytes,** which undergo **meiosis.** The first meiotic division produces two **secondary spermatocytes.** In the second meiotic division, *each* secondary spermatocyte yields two **spermatids.** Each spermatid differentiates to form a mature sperm. The head of a sperm consists of the nucleus and a cap, or **acrosome,** containing enzymes that help the sperm penetrate the egg.
- Sperm pass in sequence through the seminiferous tubules of the testis ⟶ epididymis ⟶ vas deferens ⟶ ejaculatory duct ⟶ urethra.

4 Describe the endocrine regulation of reproduction in the human male.

- Testosterone establishes and maintains male **primary sex characteristics** and **secondary sex characteristics.**
- Endocrine regulation of male reproduction involves the hypothalamus, pituitary gland, and testes. The hypothalamus secretes **gonadotropin-releasing hormone (GnRH),** which stimulates the anterior pituitary gland to secrete the gonadotropic hormones: **follicle-stimulating hormone (FSH)** and **luteinizing hormone (LH).**
- FSH, LH, and testosterone directly or indirectly stimulate sperm production. LH stimulates the interstitial cells of the testes to produce testosterone. FSH stimulates the Sertoli cells to produce (1) **androgen-binding protein (ABP),** which binds to testosterone and concentrates it; and (2) **inhibin,** a hormone that inhibits FSH secretion.

5 Describe the structure and function of each organ of the human female reproductive system.

- The **ovaries** produce gametes and the steroid hormones **estrogen** and **progesterone.** Fertilization takes place in the **oviducts.**
- The **uterus** serves as an incubator for the developing embryo. The epithelial lining of the uterus, the **endometrium,** thickens each month in preparation for possible pregnancy. The lower part of the uterus, the **cervix,** extends into the vagina.
- The **vagina** receives the penis during sexual intercourse and is the lower part of the birth canal. The **vulva** include the **labia majora, labia minora, vestibule** of the vagina, **clitoris**, and **mons pubis.**
- The breasts function in **lactation,** production of milk for the young. Each breast consists of 15 to 20 lobes of glandular tissue. Gland cells are arranged in **alveoli.** The hormone **prolactin** stimulates milk production; **oxytocin** stimulates ejection of milk from the alveoli into the ducts from which it can be sucked.

6 Trace the development of a human ovum (egg) and its passage through the female reproductive system until it is fertilized.

- **Oogenesis** takes place in the ovaries. **Oogonia** differentiate into **primary oocytes.** A primary oocyte and the **granulosa cells** surrounding it make up a **follicle.**
- As the follicle grows, connective tissue cells surrounding the granulosa cells form a layer of **theca cells.** As the follicle matures, the primary oocyte undergoes the first meiotic division, giving rise to a **secondary oocyte** and a **polar body.**
- During **ovulation,** the secondary oocyte is ejected from the ovary and enters one of the paired **oviducts (uterine tubes),** where it may be fertilized. The part of the follicle remaining in the ovary develops into a **corpus luteum,** a temporary endocrine gland.

7 Describe the endocrine regulation of reproduction in the human female, and identify the important events of the menstrual cycle, such as ovulation and menstruation.

- Endocrine regulation of female reproduction involves the hypothalamus, pituitary gland, and ovaries. The first day of the **menstrual cycle** is marked by the beginning of menstrual bleeding. Ovulation occurs at about day 14 in a typical 28-day menstrual cycle.
- During the **preovulatory phase, gonadotropin-releasing hormone (GnRH)** from the hypothalamus stimulates the anterior lobe of the pituitary to secrete follicle-stimulating hormone (FSH) and luteinizing hormone (LH). FSH stimulates follicle development and stimulates the granulosa cells to produce estrogen. LH stimulates theca cells to multiply and produce androgens, which are converted to estrogen. Estrogen is responsible for primary and secondary female sex characteristics.
- Estrogen stimulates development of the endometrium. After the first week only one follicle continues to develop. Although at relatively low concentration, estrogen inhibits FSH and LH secretion. Granulosa cells produce inhibin, which also inhibits FSH secretion.
- During the late preovulatory phase, estrogen concentration peaks and signals the anterior pituitary to secrete LH. LH stimulates final maturation of the follicle and stimulates ovulation.
- During the **postovulatory phase,** LH promotes development of the corpus luteum. The corpus luteum secretes progesterone and estrogen, which stimulate final preparation of the uterus for pregnancy. During the postovulatory phase, progesterone, along with estrogen, inhibits secretion of GnRH, FSH, and LH.
- If fertilization does not occur, the corpus luteum degenerates, concentrations of estrogen and progesterone in the blood fall, and menstruation occurs.

8 Describe the physiological changes that occur during sexual response.

- Vasocongestion and increased muscle tension are physiological responses to sexual stimulation. The phases of sexual response include **sexual excitement, plateau, orgasm,** and **resolution.**

9 Describe the processes of human fertilization and early development, and summarize the actions of hormones that regulate pregnancy.

- Human **fertilization** is the fusion of secondary oocyte and sperm to form a zygote. Fertilization and establishment of pregnancy together are called **conception.**
- If the secondary oocyte is fertilized, development begins and the embryo implants in the uterus. Membranes that develop around the embryo secrete **human chorionic gonadotropin (hCG),** a hormone that maintains the corpus luteum. During the first two to three months of pregnancy, the corpus luteum secretes the large amounts of estrogen and progesterone needed to maintain pregnancy. After three months, the **placenta,** the organ of exchange between mother and embryo, assumes this function.

10 Summarize the birth process.

- Several hormones, including estrogen, oxytocin, and prostaglandins, regulate **parturition,** the birth process. Labor can be divided into three stages; the baby is delivered during the second stage.

11 Compare the modes of action, effectiveness, advantages, and disadvantages of the methods of birth control discussed; include sterilization and emergency contraception.

- Effective methods of **contraception** include hormonal methods, such as oral contraceptives and injectable of progestin; **intrauterine devices; condoms; contraceptive diaphragms;** and **sterilization** (**vasectomy** or **tubal ligation**); see Table 48-3. Emergency contraception can be used to prevent unwanted pregnancy after rape or unprotected intercourse.
- **Spontaneous abortions** (miscarriages) occur without intervention. **Induced abortions** include **therapeutic abortions,** performed to maintain the mother's health or when the embryo is thought to be grossly abnormal, and abortions performed as a means of birth control.

SUMMARY WITH KEY TERMS *(continued)*

12 Identify common sexually transmitted diseases, and describe their symptoms, effects, and treatments.

- Among the common types of **sexually transmitted diseases (STDs)** are **chlamydia, gonorrhea, syphilis, pelvic inflammatory disease (PID), genital herpes,** and HIV. See Table 48-4.

POST-TEST

1. Which of the following is *not* an example of asexual reproduction? (a) budding (b) external fertilization (c) fragmentation (d) parthenogenesis (e) regeneration to form a new individual from a fragment
2. Hermaphroditism (a) is a form of asexual reproduction (b) occurs when an unfertilized egg develops into an adult animal (c) involves cross-fertilization between two animals (d) typically involves self-fertilization (e) typically involves only one animal
3. The seminiferous tubules (a) are the site of spermatogenesis (b) produce most of the seminal fluid (c) empty directly into the vas deferens (d) are located within the cavernous body (e) receive fluid from the bulbourethral glands
4. Arrange the following stages into the correct sequence. (1) spermatogonium (2) spermatid (3) primary spermatocyte (4) secondary spermatocyte (5) sperm (a) 2, 1, 3, 4, 5 (b) 5, 1, 3, 4, 2 (c) 4, 3, 1, 2, 5 (d) 1, 4, 3, 2, 5 (e) 1, 3, 4, 2, 5
5. Which sequence best describes the passage of sperm? (1) seminiferous tubules (2) vas deferens (3) epididymis (4) ejaculatory duct (5) urethra (a) 3, 1, 2, 4, 5 (b) 1, 3, 2, 4, 5 (c) 5, 4, 2, 3, 1 (d) 1, 3, 4, 2, 5 (e) 3, 1, 4, 2, 5
6. Which of the following characteristics is *not* associated with testosterone? (a) maintains secondary sex characteristics (b) responsible for primary sex characteristics (c) principal androgen (d) protein hormone (e) necessary for spermatogenesis
7. Androgen-binding protein (a) is secreted by Sertoli cells (b) stimulates estrogen production (c) inhibits secretion of FSH (d) inhibits spermatogenesis (e) two of the preceding answers are correct
8. Which of the following cells is haploid? (a) primary oocyte (b) oogonium (c) secondary oocyte (d) corpus luteum (e) follicle cell
9. The corpus luteum (a) is surrounded by ova (b) degenerates if fertilization occurs (c) develops in the preovulatory phase (d) is maintained by prostaglandins (e) serves as a temporary endocrine gland
10. After ovulation the secondary oocyte enters the (a) ovary (b) corpus luteum (c) cervix (d) oviduct (e) vagina
11. The endometrium (a) is the muscle layer of the uterus (b) is thickest during the preovulatory phase (c) is the site of embryo implantation (d) lines the vagina (e) is directly affected by FSH
12. The hormone that reaches its highest level during the postovulatory phase is (a) progesterone (b) estrogen (c) FSH (d) LH (e) testosterone
13. Uterine contraction is strongly stimulated by (a) progesterone (b) FSH (c) LH (d) inhibin (e) oxytocin
14. Which of the following is *not* a hormonal type of contraception? (a) Depo-Provera (b) contraceptive diaphragm (c) injectable progestin (d) birth control patch (e) oral contraceptives
15. Pelvic inflammatory disease is commonly caused by (a) syphilis (b) gonorrhea (c) genital herpes (d) HIV (e) chlamydia

CRITICAL THINKING

1. Contrast the biological advantages of hermaphroditism that involves cross-fertilization with hermaphroditism that involves self-fertilization.
2. What would happen if ovulation occurred but no corpus luteum developed?
3. Why do you think a variety of effective methods of male contraception have not yet been developed? If you were a researcher, what are some approaches you might take to developing a method of male contraception?

- Visit our Web site at **http://biology.brookscole.com/solomon7** for links to chapter-related resources on the World Wide Web. Additional online materials relating to this chapter can also be found on our Web site.

BIOLOGY NOW RESOURCES

Active Figures

48-3: Male reproductive system

48-14: Female menstrual cycle

Preparing for an exam? Take a diagnostic test on your BiologyNow CD-ROM.

Post-Test Answers

1. b	2. c	3. a	4. e
5. b	6. d	7. a	8. c
9. e	10. d	11. c	12. a
13. e	14. b	15. e	

49

Animal Development

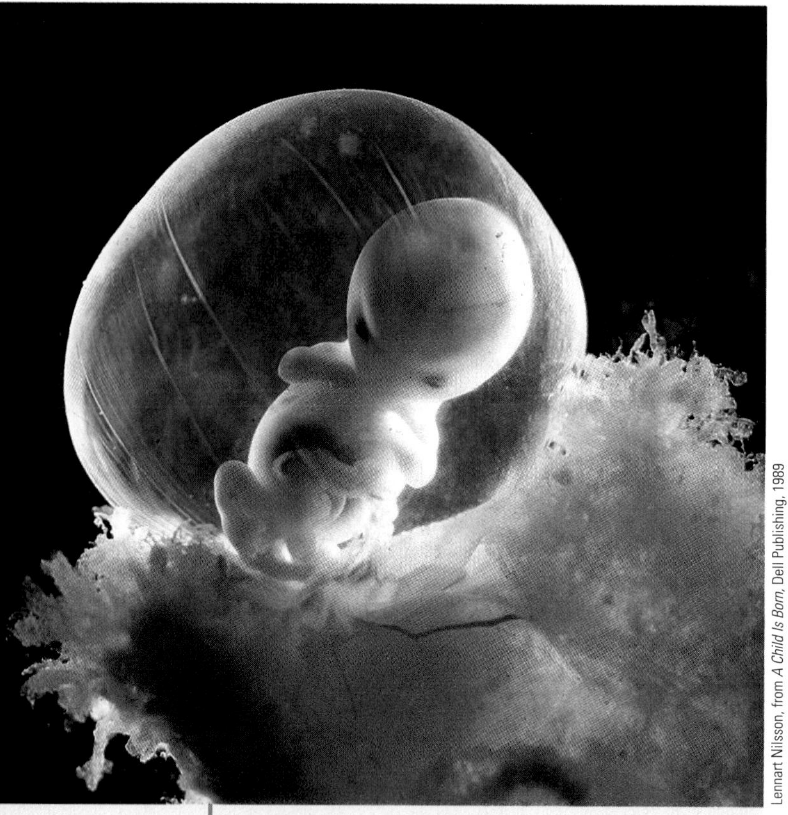

Lennart Nilsson, from *A Child Is Born*, Dell Publishing, 1989

Human embryo in its seventh week of development. The embryo is 2 cm (0.8 in) long.

CHAPTER OUTLINE

Development includes all the changes that take place during the entire life of an individual. In this chapter, however, we focus mainly on the fertilization of the egg to form a zygote, and the subsequent development of the young animal before birth or hatching. Just how does a microscopic, unicellular zygote give rise to the bones, muscles, brain, and other structures of a complex animal? These are derived from a balanced combination of several fundamental processes: cell division and growth, cell determination and cell differentiation, and pattern formation and morphogenesis.

The zygote divides by mitosis, forming an **embryo** that subsequently undergoes an orderly sequence of cell divisions. In animals, growth (that is, increase in mass) occurs primarily by an increase in the number of cells but it also occurs by an increase in cell size, as in fat cells. Although the very early embryo usually does not grow, later cell divisions typically contribute to growth. However, by itself cell division, which is governed by a genetic program that interacts with environmental signals, would produce only a formless heap of similar cells.

As embryonic development proceeds, certain cells become biochemically and structurally specialized to carry out specific functions, through a process known as **cell differentiation.** Through our discussion of the genetic control of development in Chapter 16, you learned how cell differentiation occurs through **cell determination,** a series of molecular events in which the activities of certain genes are altered in ways that cause a cell to progressively commit to a particular differentiation pathway. This process proceeds even though there may not be any immediate changes in the cell's morphology. You also evaluated evidence supporting the principle of **nuclear equivalence,** which states that in most cases neither cell determination nor cell differentiation entails a loss of genetic information from the cell nucleus. That is, the nuclei of virtually all differentiated cells of an animal contain the same genetic information present in the zygote, but each cell type *expresses* a different subset of that information. In fact, nuclear equivalence is the principle that underlies the cloning of organisms (see Chapter 16).

Cell differentiation is therefore an expression of changes in the activity of specific genes, which in turn are influenced by a variety of factors inside and outside the cell. This **differential gene expression** is responsible for variations in chemistry, behavior, and structure among cells. Through this process, an embryo develops into an organism of more than 200 types of cells, each specialized to perform specific functions. However, not all cells differentiate. Some, known as **stem cells,** remain in a relatively undifferentiated state and retain the ability to give rise to various cell types (see Chapter 16).

Cell differentiation by itself does not explain development. The differentiated cells must become progressively organized, shaping the intricate pattern of tissues and organs that characterizes a multicellular animal. This development of form, known as **morphogenesis,** proceeds through the process of **pattern formation.** Pattern formation is a series of steps requiring signaling between cells, changes in the shapes of certain cells, precise cell migrations, interactions with the extracellular matrix, and even **apoptosis** (programmed cell death) of some cells.

The photograph shows a human embryo, as well as part of the placenta (large, fluffy mass in the lower right-hand corner). In this chapter, you have an opportunity to compare and contrast the sequences of developmental events in several different animals in addition to humans. Researchers have chosen these model organisms because they have certain desirable characteristics. For example, some embryos are particularly easy to observe because they are transparent. As you read, note the intimate interrelationships among the basic developmental processes, as well as the fundamental similarities among the early developmental events in the animals featured. ■

FERTILIZATION

Learning Objectives

1. Describe the four processes involved in fertilization.
2. Describe fertilization in echinoderms, and point out some ways in which mammalian fertilization differs.

In Chapter 48 you studied **spermatogenesis** and **oogenesis,** the processes by which meiosis leads to the formation of haploid cells, which differentiate as sperm and eggs, respectively. In **fertilization** a usually flagellated, motile sperm fuses with a much larger, immotile ovum to produce a **zygote,** or fertilized egg. Fertilization has two important genetic consequences: Restoration of the diploid chromosome number and, in mammals and many other animals, determination of the sex of the offspring (see Chapter 10). Fertilization also has profound physiological effects, because it activates the egg, initiating reactions that permit development.

Fertilization involves four events, some of which may occur simultaneously and which do not necessarily follow the same temporal sequence in all animals: (1) The sperm contacts the egg and recognition occurs; (2) the sperm or sperm nucleus enters the egg; (3) the egg becomes activated, and certain developmental changes begin; and (4) the sperm and egg nuclei fuse. Unless otherwise stated, our discussion applies to sea urchins and other echinoderms such as sea stars, which have been studied intensively because they produce large numbers of gametes and because fertilization is external.

The first step in fertilization involves contact and recognition

Although eggs are immotile, they are active participants in fertilization. An egg is surrounded not only by a plasma membrane but also by one or more external coverings that are important in fertilization. For example, as discussed in Chapter 48, a mammalian egg is enclosed by a thick, noncellular, **zona pellucida,** which is surrounded by a layer of granulosa cells derived from the follicle in which the egg developed.

The egg coverings not only facilitate fertilization by sperm of the same species but also bar interspecific fertilization, a particularly important function in species that practice external fertilization. External to the plasma membrane of a sea urchin egg are two acellular layers that interact with sperm: a very thin **vitelline envelope** and, outside of this, a thick glycoprotein layer called the **jelly coat.** Sea urchin sperm become motile when released into seawater, and their motility increases when they reach the vicinity of a sea urchin egg. Motility improves the probability that sperm will encounter the egg, but motility alone may not be sufficient to ensure actual contact with the egg. In sea urchins, as well as certain fish and some cnidarians, the egg or one of its coverings secretes a chemotactic substance that attracts sperm from the same species. However, for a great many species researchers have not found a specific chemical that attracts the sperm to the egg.

When a sea urchin sperm contacts the jelly coat, it undergoes an **acrosome reaction,** in which the membranes surrounding the acrosome (the cap at the head of the sperm, see Fig. 48-6) fuse, and pores in the membrane enlarge. Calcium ions from the seawater move into the acrosome, which swells and begins to disorganize. The acrosome then releases proteolytic enzymes that digest a path through the jelly coat to the vitelline envelope of the egg. If the sperm and egg are of the same species, **bindin,** a species-specific binding protein located on the acrosome, adheres to species-specific bindin receptors located on the egg's vitelline envelope.

Before a mammalian sperm participates in fertilization, it first undergoes **capacitation,** a maturation process in the female reproductive tract. During capacitation, which in humans can take several hours, sperm become increasingly motile and capable of undergoing an acrosome reaction when they encounter an egg. Evidence indicates that mammalian fertilization requires interaction between sperm and species-specific glycoproteins in the egg's zona pellucida.

Sperm entry is regulated

In sea urchins, once contact occurs between acrosome and vitelline envelope, enzymes dissolve a bit of the vitelline envelope in the area of the sperm head. The plasma membrane of the egg is covered with microvilli, several of which elongate to surround the head of the sperm. As they do so, the plasma membranes of sperm and egg fuse and a **fertilization cone** is formed that contracts to draw the sperm into the egg (Fig. 49-1).

Fertilization of the egg by more than one sperm, a condition known as **polyspermy,** results in an offspring with extra sets of chromosomes, which is usually lethal. Two reactions, known as the *fast* and *slow blocks* to polyspermy, prevent this. In the fast block to polyspermy the egg plasma membrane depolarizes, preventing its fusion with additional sperm. An unfertilized egg is polarized, that is, the cytoplasm is negatively charged relative to the outside. However, within a few seconds after sperm fusion, ion channels in the plasma membrane open, permitting positively charged calcium ions to diffuse across the membrane and depolarize the egg.

The slow block to polyspermy, the **cortical reaction,** requires about one to several minutes to complete, but it is a complete block. Binding of the sperm to receptors on the vitelline envelope activates one or more *signal transduction* pathways (see Chapter 5) in the egg. These events cause calcium ions stored in the egg endoplasmic reticulum to be released into the cytosol. This rise in the level of intracellular calcium causes thousands of cortical granules, membrane-bounded vesicles in the egg cortex (region close to the plasma membrane), to release enzymes, various proteins, and other substances by exocytosis into the space between the plasma membrane and the vitelline envelope (Fig. 49-2). Some of the enzymes dissolve the protein linking the two membranes, allowing the space to enlarge as additional substances released by the cortical granules raise the osmotic pressure, causing an influx of water from the surroundings. Thus the vitelline envelope becomes elevated away from the plasma membrane and forms the fertilization envelope, a hardened covering that prevents entry of additional sperm.

In mammals, a fertilization envelope does not form, but the enzymes released during exocytosis of the cortical granules alter the sperm receptors on the egg's zona pellucida so that no additional sperm bind to them.

Not all species have these types of blocks to polyspermy. In some, such as salamanders, several sperm usually enter the egg, but only one sperm nucleus fuses with the egg nucleus, and the rest degenerate.

FIGURE **49-1** **Fertilization.**

In this TEM, a fertilization cone (FC) forms as a sperm enters a sea urchin egg. SN, sperm nucleus; SM, sperm mitochondrion; SF, sperm flagellum.

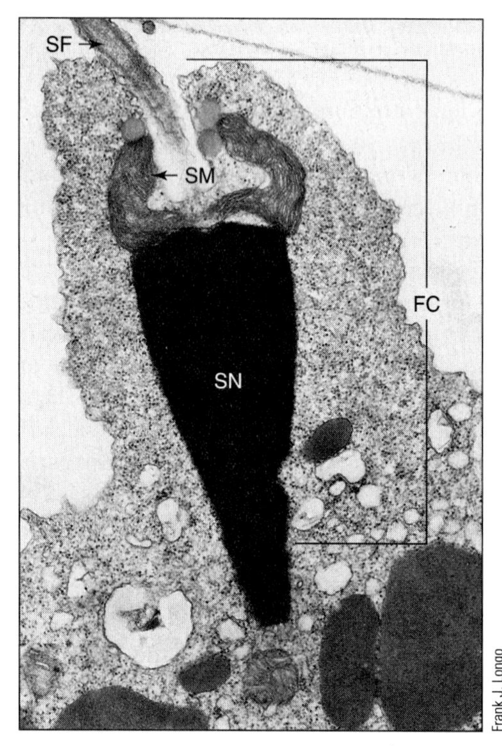

Fertilization activates the egg

Release of calcium ions into the egg cytoplasm does more than stimulate the cortical reaction; it also triggers the *activation program,* a series of metabolic changes within the egg. Aerobic respiration increases, certain maternal enzymes and other proteins become active, and within a few minutes after sperm entry, a burst of protein synthesis occurs. In addition, the egg nucleus is stimulated to complete meiosis. (Recall from Chapter 48 that in most animal species, including mammals, at the time of fertilization an egg is actually a secondary oocyte, arrested early in the second meiotic division.)

FIGURE **49-2** **The cortical reaction.**

Three cortical granules are undergoing exocytosis in this TEM of the plasma membrane of a sea urchin egg shortly after fertilization. This release of the contents of the cortical granules initiates the slow block to polyspermy. *(From E. Anderson. Reprinted from* The Journal of Cell Biology, *Vol. 37, 1968, by copyright permission of the Rockefeller University Press)*

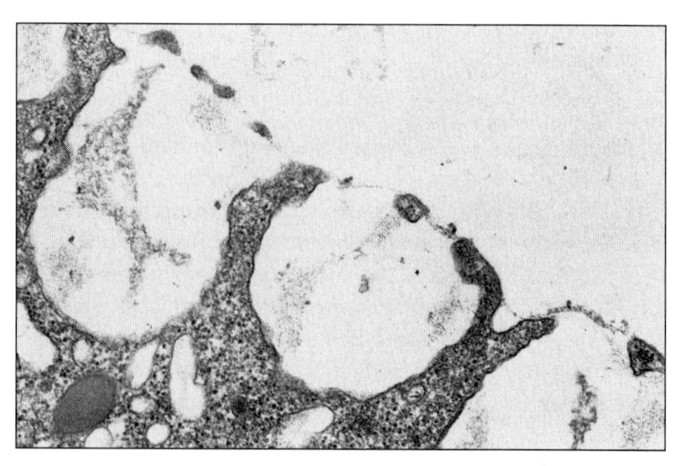

In some species, an egg can be artificially activated without sperm penetration by swabbing it with blood and pricking the plasma membrane with a needle, by calcium injection, or by certain other treatments. These haploid eggs go through some developmental stages *parthenogenetically,* that is, without fertilization (see Chapter 48), but they cannot complete development. Special mechanisms of egg activation have evolved in species that are naturally parthenogenetic.

Sperm and egg pronuclei fuse, restoring the diploid state

After the sperm nucleus enters the egg, researchers think it is guided toward the egg nucleus by a system of microtubules that forms within the egg. The sperm nucleus swells, forming the **male pronucleus.** The nucleus formed during completion of meiosis in the egg becomes the **female pronucleus.** (Recall from Figure 48-11 that the other nucleus formed during meiosis II in the egg becomes the second polar body.) The haploid male and female pronuclei then fuse to form the diploid nucleus of the zygote, and DNA synthesis occurs in preparation for the first cell division.

Review

- What mechanisms ensure fertilization by only one sperm of the same species?
- How do the mechanisms of fertilization ensure control of both quality (fertilization by a sperm of the same species) and quantity (fertilization by only one sperm)?

BiologyNow™ Assess your understanding of **fertilization** by taking the pretest on your BiologyNow CD-ROM.

CLEAVAGE

Learning Objectives

3 Trace the generalized pattern of early development of the embryo from zygote through early cleavage and formation of the morula and blastula.

4 Contrast early development, including cleavage in the echinoderm (or in amphioxus), the amphibian, and the bird, paying particular attention to the importance of the amount and distribution of yolk.

Despite its simple appearance, the zygote is **totipotent,** that is, it gives rise to all the cell types of the new individual. Because the ovum is very large compared with the sperm, the bulk of the zygote cytoplasm and organelles comes from the ovum. However, the sperm and ovum usually contribute equal numbers of chromosomes.

Shortly after fertilization, the zygote undergoes **cleavage,** a series of rapid mitotic divisions with no period of growth during each cell cycle. For this reason, although the cell number increases, the embryo does not increase in size. The zygote initially divides to form a 2-celled embryo. Then each of these cells undergoes mitosis and divides, bringing the number of cells to 4. Repeated divisions increase the number of cells, called **blastomeres,** that make up the embryo. At about the 32-cell stage the embryo is a solid ball of blastomeres called a **morula.** Eventually, anywhere from 64 to several hundred blastomeres form the **blastula,** which is usually a hollow ball with a fluid-filled cavity, the blastocoel.

The pattern of cleavage is affected by yolk

Many animal eggs contain **yolk,** a mixture of proteins, phospholipids, and fats that serves as food for the developing embryo. The amount and distribution of yolk vary among different animal groups, depending on the needs of the embryo. Mammalian eggs have very little yolk, because the embryo obtains maternal nutritional support throughout most of its development, whereas the eggs of birds and reptiles must contain sufficient yolk to sustain the embryo until hatching. Echinoderm eggs typically need only enough yolk to nourish the embryo until it becomes a tiny larva capable of obtaining its own food.

Most invertebrates and simple chordates have **isolecithal** eggs with relatively small amounts of yolk uniformly distributed through the cytoplasm. Isolecithal eggs divide completely **(holoblastic cleavage).** Cleavage of these eggs is radial or spiral. Radial cleavage is characteristic of *deuterostomes* such as chordates and echinoderms; spiral cleavage is common in the embryos of *protostomes* such as annelids and mollusks (see Fig. 28-3).

In **radial cleavage,** the first division is vertical and splits the egg into two equal cells. The second cleavage division, also vertical, is at right angles to the first division and separates the two cells into four equal cells. The third division is horizontal, at right angles to the other two, and separates the four cells into eight cells: four above and four below the third line of cleavage. This pattern of radial cleavage occurs in echinoderms and in the cephalochordate amphioxus (Figs. 49-3a–e and 49-4).

In **spiral cleavage,** after the first two divisions, the plane of cytokinesis is diagonal to the polar axis (Fig. 49-5). This results in a spiral arrangement of cells, with each cell located above and between the two underlying cells. This pattern is typical of annelids and mollusks.

Many vertebrate eggs are **telolecithal,** meaning they have large amounts of yolk concentrated at one end of the cell, known as the **vegetal pole.** The opposite, more metabolically active, pole is the **animal pole.** The eggs of amphibians are moderately telolecithal (mesolecithal). Although cleavage is radial and holoblastic, the divisions in the vegetal hemisphere are slowed by the presence of the inert yolk. As a result, the blastula consists of many small cells in the animal hemisphere, and fewer but larger cells in the vegetal hemisphere (Fig. 49-6). The blastocoel is displaced toward the animal pole.

The telolecithal eggs of reptiles and birds have very large amounts of yolk at the vegetal pole and only a small amount of cytoplasm concentrated at the animal pole. The yolk of such eggs never cleaves. Cell division is restricted to the **blastodisc,** the small disc of cytoplasm at the animal pole (Fig. 49-7); this type of cleavage is termed **meroblastic.** In birds and some reptiles, the blastomeres form two layers, separated by the blastocoel cavity: an upper *epiblast* and below that a thin layer of flat cells, the *hypoblast.*

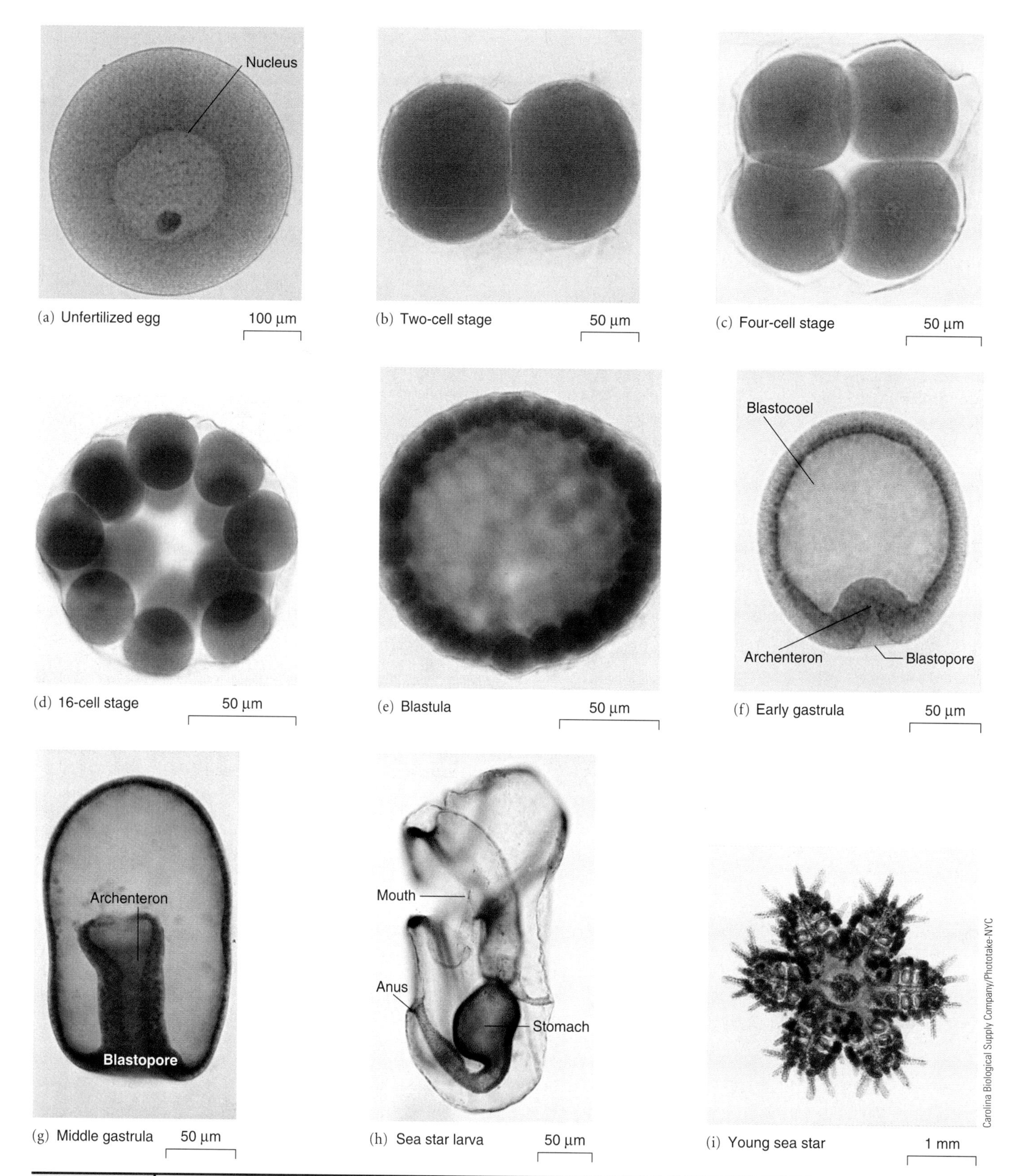

FIGURE **49-3** | **LMs showing sea star development.**

(a) The isolecithal egg has a small amount of uniformly distributed yolk. **(b–e)** The cleavage pattern is radial and holoblastic (the entire egg becomes partitioned into cells). **(f, g)** The three germ layers form during gastrulation. The blastopore is the opening into the developing gut cavity, the archenteron. The rudiments of organs are evident in the sea star larva **(h)** and the young sea star **(i).** All views are side views with the animal pole at the top, except **(c)** and **(i),** which are top views. Note that the sea star larva is bilaterally symmetrical, but differential growth produces a radially symmetrical young sea star.

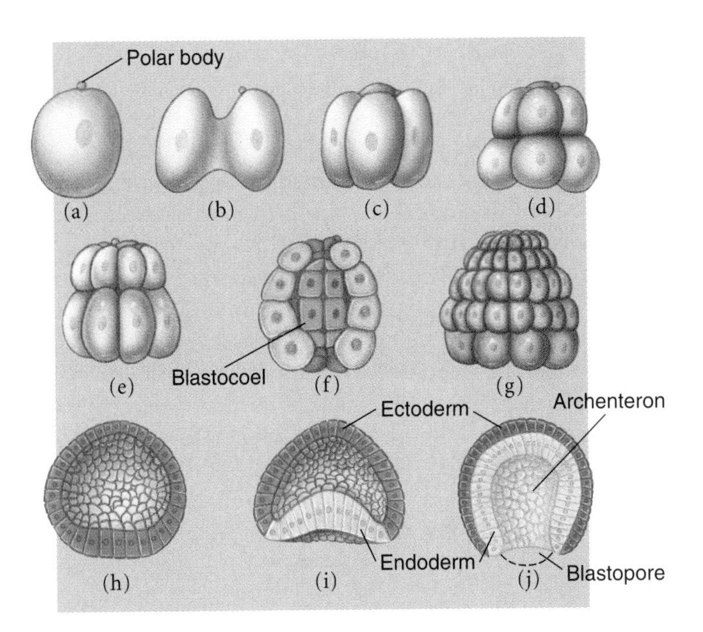

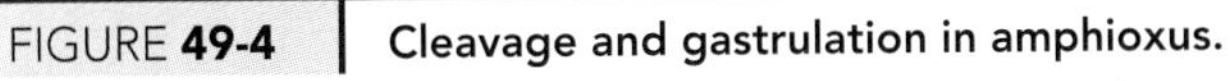

FIGURE **49-4** | **Cleavage and gastrulation in amphioxus.**

As in the sea star, cleavage is holoblastic and radial. The embryos are shown from the side. **(a)** Mature egg with polar body. **(b–e)** The 2-, 4-, 8-, and 16-cell stages. **(f)** Embryo cut open to show the blastocoel. **(g)** Blastula. **(h)** Blastula cut open. **(i)** Early gastrula showing beginning of invagination at vegetal pole. **(j)** Late gastrula. Invagination is completed, and the blastopore has formed.

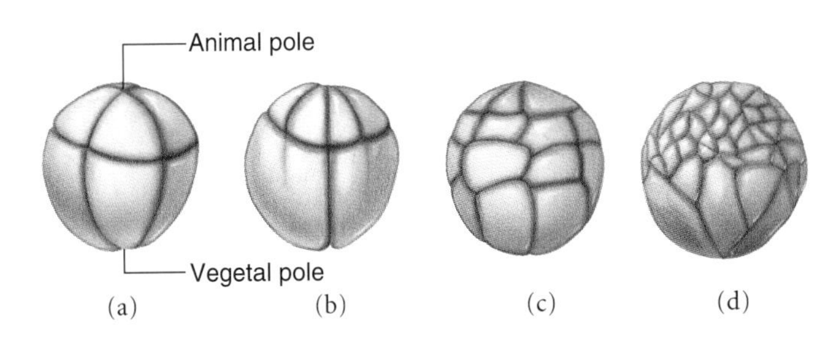

ACTIVE FIGURE **49-6** | **Cleavage pattern in a frog egg.**

(a–d) Although cleavage is holoblastic, the large amount of yolk concentrated in the vegetal hemisphere slows cleavage. As a result, fewer cells develop at the vegetal hemisphere than at the animal hemisphere. These embryos are shown from the side.

BiologyNow™ Learn more about **cleavage in a frog egg** by clicking on this figure on your BiologyNow CD-ROM.

Cleavage may distribute developmental determinants

The pattern of cleavage in a particular species depends not only on yolk but on other factors as well. Recall from Chapter 16 that some organisms have relatively rigid developmental patterns, known as **mosaic development.** This is largely a consequence of the unequal distribution of important materials in the cyto-

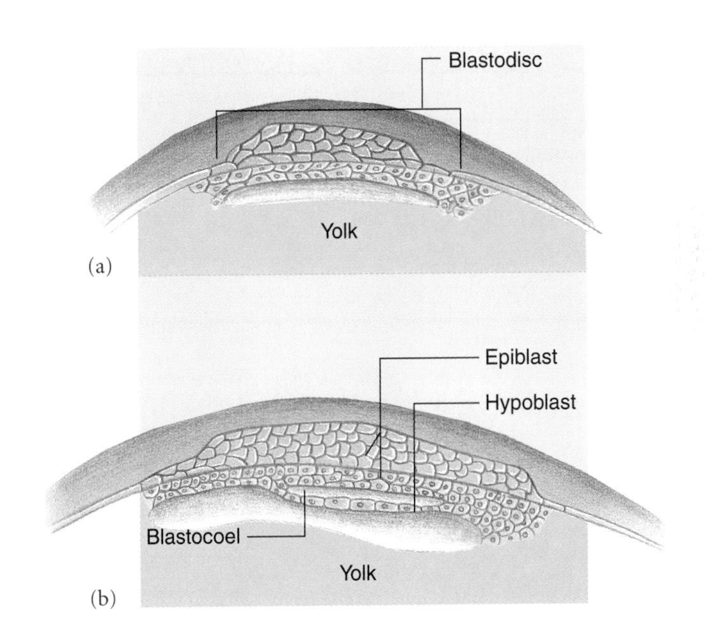

FIGURE **49-7** | **Cleavage in a bird embryo.**

Cleavage is meroblastic; that is, it is restricted to the blastodisc, a small disk of cytoplasm on the upper surface of the egg yolk. **(a)** Early blastodisc formation. This cutaway view shows cells on the blastodisc surface, as well as in the interior. **(b)** The blastodisc splits into two tissue layers, an upper epiblast and a lower hypoblast, separated by the blastocoel.

FIGURE **49-5** | **Spiral cleavage in an annelid embryo.**

(a–f) Top views of the animal pole. The successive cleavage divisions occur in a spiral pattern as illustrated. **(g)** A typical trochophore larva. The upper half of the trochophore develops into the extreme anterior end of the adult worm; all the rest of the adult body develops from the lower half.

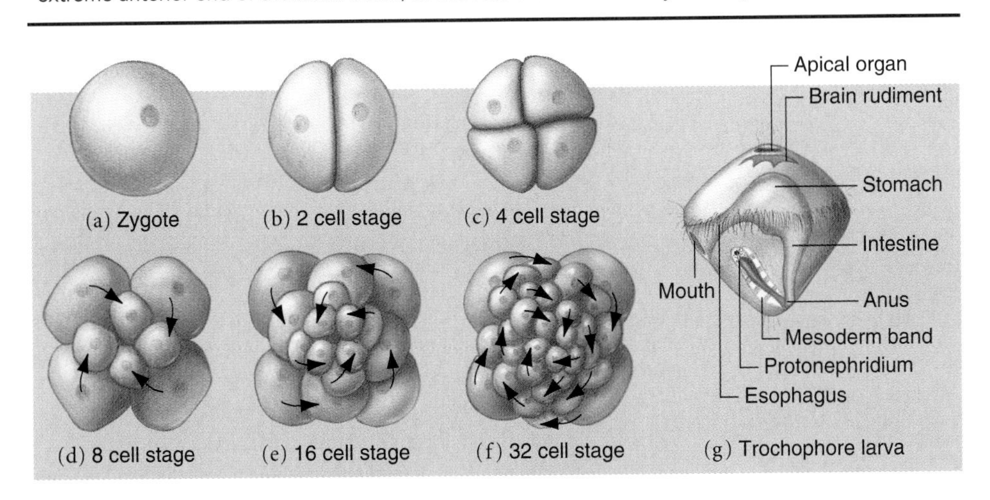

plasm of the zygote. Because the zygote cytoplasm is not homogeneous, cytoplasmic developmental determinants portioned out to each new cell during cleavage may be different. Such differences help determine the course of development. At the other extreme are mammals, which have zygotes with very homogenous cytoplasm. They exhibit highly **regulative development,** in which individual cells produced by the cleavage divisions are equivalent, allowing the embryo to develop as a self-regulating whole. Developmental patterns of most animals fall somewhere on a continuum between these two extremes.

In some species the distribution of developmental determinants in the unfertilized egg is maintained in the zygote. In others sperm penetration initiates rearrangement of the cytoplasm. For example, fertilization in the amphibian egg causes a shift of some of the cortical cytoplasm. These cytoplasmic movements can be easily followed because the egg cortex contains dark pigment granules. As a result of this rearrangement, a crescent-shaped region of underlying lighter-colored (gray) cytoplasm becomes evident directly opposite the point on the cell where the sperm penetrated the egg (Fig. 49-8a). This **gray crescent** region is thought to contain growth factors and other developmental determinants. The first cleavage bisects the gray crescent, distributing half to each of the first two blastomeres; in this way the position of the gray crescent establishes the future right and left halves of the embryo. As cleavage continues, the gray crescent material becomes partitioned into certain blastomeres. Those that contain parts of the gray crescent eventually develop into the dorsal region of the embryo.

Experiments have confirmed the importance of determinants in the gray crescent to development. If the first two frog blastomeres are separated experimentally, each develops into a complete tadpole (Fig. 49-8b). When the plane of the first division is altered so that the gray crescent is completely absent from one of the cells, that cell does not develop normally (Fig. 49-8c).

FIGURE **49-8** **Cytoplasmic determinants in frog development.**

(a) The position of the gray crescent in the frog zygote determines the main axes of the body. **(b)** The first division of the zygote partitions the gray crescent into the two daughter cells. If these cells are separated, each develops into a tadpole. **(c)** If the plane of cleavage is changed experimentally so that only one cell receives the gray crescent, only that cell develops into a tadpole.

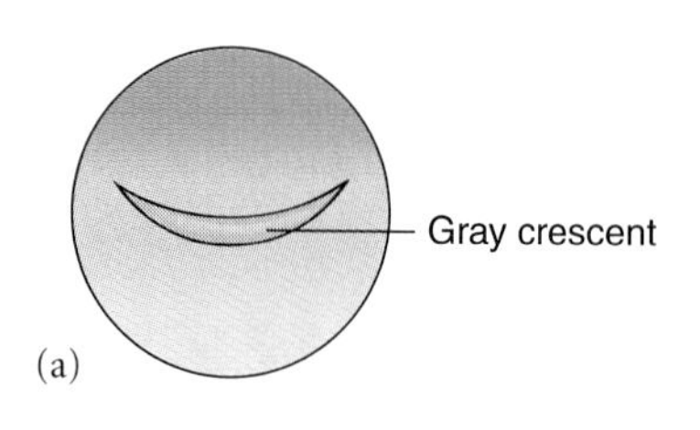

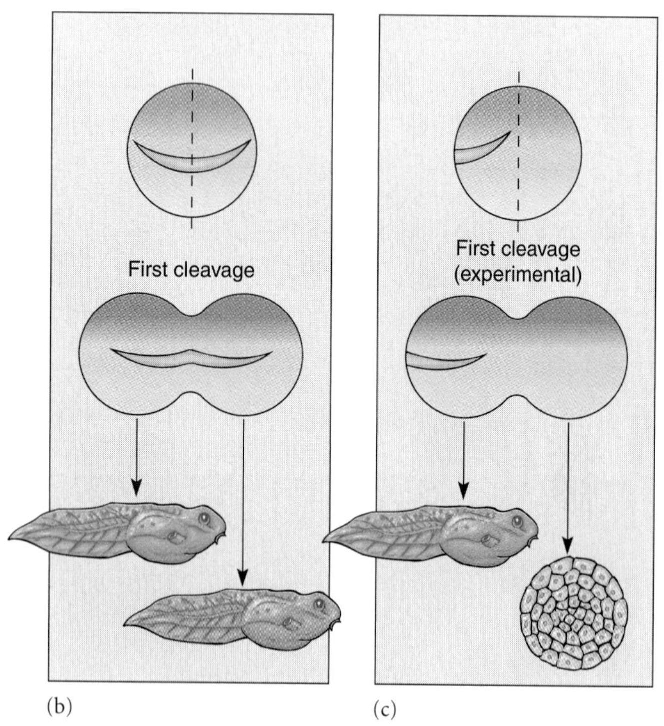

Cleavage provides building blocks for development

Cleavage partitions the zygote into many small cells that serve as the basic building blocks for subsequent development. At the end of cleavage, the small cells that make up the blastula begin to move about with relative ease, arranging themselves into the patterns necessary for further development. Surface proteins are important in helping cells "recognize" one another and therefore in determining which ones adhere to form tissues.

Review

- What is radial cleavage? Spiral cleavage?
- What type of cleavage is characteristic of telolecithal eggs of reptiles and birds?
- How does the cytoplasm influence early development?

BiologyNow™ Assess your understanding of **cleavage** by taking the pretest on your BiologyNow CD-ROM.

GASTRULATION

Learning Objectives

5 Identify the significance of gastrulation in the developmental process, and compare gastrulation in the echinoderm (or in amphioxus), the amphibian, and the bird.

The process by which the blastula becomes a three-layered embryo, or **gastrula,** is called **gastrulation.** Thus, early development proceeds through the following stages:

Zygote ⟶ early cleavage stages ⟶ morula ⟶ blastula ⟶ gastrula

During gastrulation, the embryo begins to approximate its body plan as cells arrange themselves into three distinct **germ layers,** or embryonic tissue layers: the outermost layer, the **ectoderm;** the innermost, the **endoderm;** and the **mesoderm,** which develops between them. Additional cell divisions take place during gastrulation, and the germ layers become established through a combination of processes. Many cells lose their old cell-to-cell contacts and establish new ones through cell recognition and adhesion processes involving complex interactions among the integrins and other plasma membrane proteins, and the extracellular matrix (see Chapters 4 and 5). Many cells undergo cytoskeletal changes, particularly alterations in the distribution of actin microfilaments; these changes in the internal architecture of the cells allow them to change shape and/or undergo specific, directional, amoeboid movements. As a result of these movements, many cells ultimately take up new positions in the interior of the embryo.

Each of the germ layers develops into specific parts of the embryo. As you study the illustrations, keep in mind that a conventional color code accepted by developmental biologists is used to depict the germ layers: endoderm, yellow; mesoderm, red or pink; and ectoderm, blue.

The pattern of gastrulation is affected by the amount of yolk

The simple type of gastrulation that occurs in echinoderms and in amphioxus is illustrated in Figures 49-3f, g and 49-4i, j, respectively. Gastrulation begins when a group of cells at the vegetal pole undergoes a series of changes in shape, causing that part of the blastula wall to first flatten and then bend inward (invaginate). The invaginated wall eventually meets the opposite wall, obliterating the blastocoel.

You can roughly demonstrate this type of gastrulation by pushing inward on the wall of a partly deflated rubber ball until it rests against the opposite wall. In a similar way the embryo is converted into a double-walled, cup-shaped structure. The new internal wall lines the **archenteron,** the newly formed cavity of the developing gut. The opening of the archenteron to the exterior, the **blastopore,** is the site of the future anus in deuterostomes.

This simple type of gastrulation cannot occur in the amphibian embryo, because the large yolk-laden cells in the vegetal half of the blastula obstruct any inward movement at the vegetal pole. Instead, cells from the animal pole move down the embryo surface and when they reach the region derived from the gray crescent they move into the interior. This inward movement is accomplished as the cells change shape, first becoming flask- or bottle-shaped (so that most of their mass is actually below the surface) and then sinking into the interior as they lose their remaining connections with other cells on the surface. This spot on the embryo's surface, referred to as the **dorsal lip of the blastopore,** is marked by a dimple, shaped like a C lying on its side (Fig. 49-9). As the process continues, the blastopore becomes ring-shaped as cells lateral, and then ventral, to its dorsal lip become involved in similar movements. The yolk-filled cells fill the space enclosed by the lips of the blastopore, forming the yolk plug.

The archenteron forms and is lined on all sides by cells that have moved in from the surface. At first, the archenteron is a narrow slit, but it gradually expands at the anterior end of the embryo, causing the blastocoel to progressively shrink and eventually to disappear.

Although the details differ somewhat, gastrulation in birds is basically similar to amphibian gastrulation. Cells that make up the upper layer (the epiblast) migrate toward the midline to form a thickened region known as the **primitive streak,** which elongates and narrows as it develops. At its center is a narrow furrow, the **primitive groove.** The primitive streak is a dynamic structure. Its cells constantly change as they migrate in from the epiblast, sink inward at the primitive groove, then move out laterally and anteriorly in the interior (Fig. 49-10). The primitive groove is the functional equivalent of the blastopore of the echinoderm, amphioxus, and amphibian embryos. However, a bird embryo contains no cavity homologous to the archenteron.

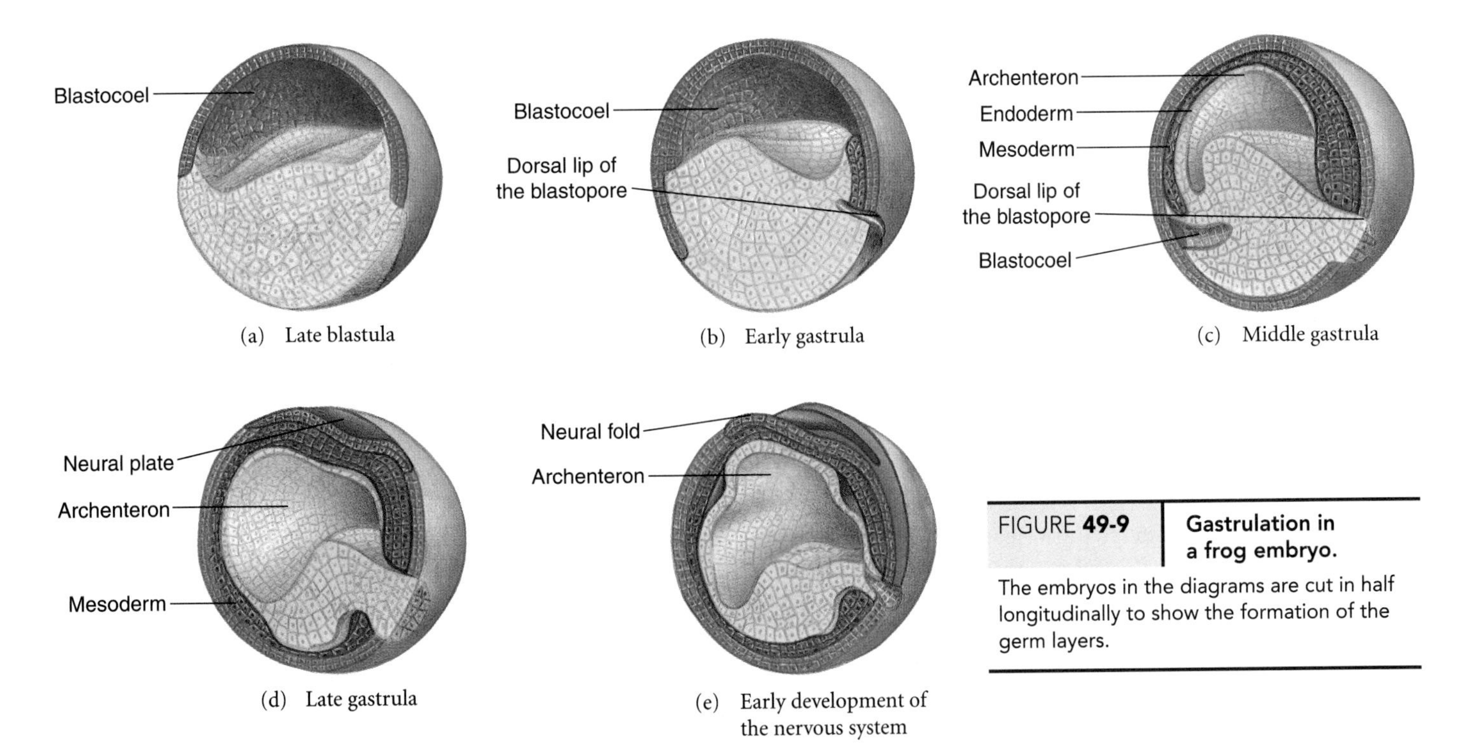

FIGURE 49-9 Gastrulation in a frog embryo.

The embryos in the diagrams are cut in half longitudinally to show the formation of the germ layers.

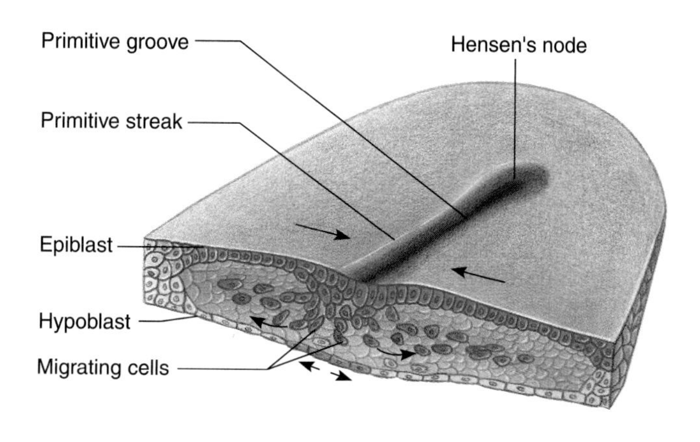

FIGURE **49-10** **Gastrulation in birds.**

A three-layered embryo forms as cells move toward the primitive streak, move inward at the primitive groove, and migrate laterally and forward in the interior.

At the anterior end of the primitive streak, cells destined to form the **notochord** (supporting rod of mesodermal, cartilage-like cells that serves as the flexible skeletal axis in all chordate embryos) concentrate in a thickened knot known as **Hensen's node.** These cells sink into the interior and then move anteriorly just beneath the epiblast, forming a narrow extension from the node. Cells that will form other types of mesoderm move laterally and anteriorly from the primitive streak, between the epiblast (which becomes ectoderm) and the hypoblast. Other cells form the endoderm, displacing the hypoblast cells and causing them to move laterally. These displaced cells will form part of the extraembryonic membranes, discussed in a later section.

Review

- What is the archenteron? Is an archenteron formed in sea stars? In amphibians? In birds?
- How does the amount and distribution of yolk influence gastrulation?

BiologyNow™ Assess your understanding of **gastrulation** by taking the pretest on your BiologyNow CD-ROM.

ORGANOGENESIS

Learning Objectives

6 Define organogenesis, and summarize the fate of each of the germ layers.

7 Trace the early development of the vertebrate nervous system.

Gastrulation leads to **organogenesis,** or organ formation, in which the cells of the three germ layers continue the processes of pattern formation that lead to the formation of specific structures. The ectoderm eventually forms the outer layer of the skin and gives rise to the nervous system and sense organs. Tissues that eventually line the digestive tract, and organs that develop as outgrowths of the digestive tract (including the liver, pancreas, and lungs) are all of endodermal origin. Skeletal tissue, muscle, and the circulatory, excretory, and reproductive systems all are derived from mesoderm (Table 49-1; see Fig. 16-1).

TABLE **49-1** **Fate of the Germ Layers Formed at Gastrulation**

Germ layer	Fate
Ectoderm	Nervous system and sense organs
	Outer layer of skin (epidermis) and its associated structures (nails, hair, etc.)
	Pituitary gland
Mesoderm	Notochord
	Skeleton (bone and cartilage)
	Muscles
	Circulatory system
	Excretory system
	Reproductive system
	Inner layer of skin (dermis)
	Outer layers of digestive tube and of structures that develop from it, such as part of respiratory system
Endoderm	Lining of digestive tube and of structures that develop from it, such as lining of the respiratory system

The notochord, brain, and spinal cord are among the first organs to develop in the early vertebrate embryo (Fig. 49-11; see also Fig. 49-9d and e). First the notochord, which is mesodermal tissue, grows forward along the length of the embryo as a cylindrical rod of cells. The notochord eventually is replaced by the vertebral column, although notochord remnants will remain in the cartilage discs between the vertebrae.

The developing notochord has yet another crucial function. Numerous experiments in which researchers have transplanted portions of the notochord mesoderm to other locations in the embryo show that the notochord causes the overlying ectoderm to thicken and differentiate to form the precursor of the central nervous system, the **neural plate.** Such phenomena, in which certain cells stimulate or otherwise influence the differentiation of their neighbors, are examples of **induction** (see Chapter 16).

Repeated testing supports the hypothesis that induction of the neural plate cells does not require direct cell-to-cell contact with the developing notochord cells. Researchers therefore think that this induction involves the diffusion of some type of signal molecule, although they haven't yet identified the chemical signal itself. The induction of the neural plate by the notochord is a good example of the importance of a cell's position in relation to other cells. That position is often critical in determining the fate of a cell because it determines the cell's exposure to substances released from other cells.

The cells of the neural plate undergo changes in shape that cause the central cells of the neural plate to move downward and form a depression called the **neural groove;** the cells flanking the groove on each side form **neural folds.** Continued changes in cell shape bring the folds closer together until they meet and fuse, forming the **neural tube.** In this process, the hollow neural tube comes to lie beneath the surface. The ectoderm overlying it will form the outer layer of skin. The anterior portion of the

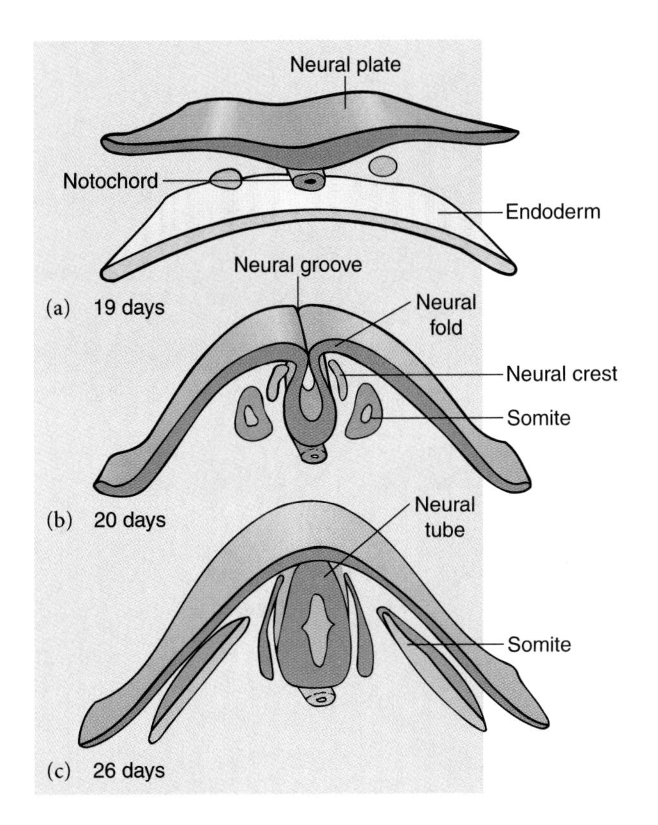

ACTIVE FIGURE **49-11** **Development of the human nervous system.**

(a) Approximately 19 days. The neural plate has indented to form a shallow groove flanked by neural folds. **(b)** Approximately 20 days. The neural folds approach one another. The neural crest cells, derived from ectoderm, will migrate in the embryo and give rise to sensory neurons; the somites are blocks of mesoderm that become the vertebrae and other segmented body parts. The endoderm is not shown. **(c)** Approximately 26 days. The neural folds have joined to form the hollow neural tube, which gives rise to the brain at the anterior end of the embryo and the spinal cord posteriorly.

BiologyNow™ Explore **nervous system development in a frog** by clicking on this figure on your BiologyNow CD-ROM.

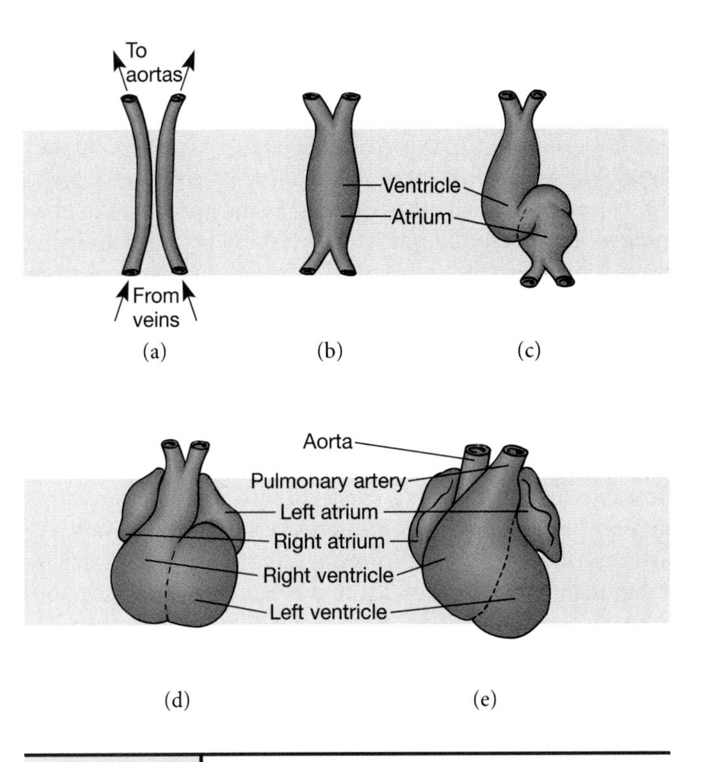

FIGURE **49-12** **Formation of the heart in birds and mammals.**

These ventral views of successive stages show that the heart forms from the fusion of two blood vessels and that at first the end that receives blood from the veins is at the bottom. The developing heart twists to carry the atrium to a position above the ventricle, and the chambers then divide to form the four-chambered heart.

neural tube grows and differentiates into the brain; the rest of the tube develops into the spinal cord. The brain and the spinal cord remain hollow even in the adult; the ventricles of the brain and central canal of the spinal cord are persistent reminders of their embryonic origins.

Various motor nerves grow out of the developing brain and spinal cord, but sensory nerves have a separate origin. When the neural folds fuse to form the neural tube, bits of nervous tissue known as the **neural crest** arise from ectoderm on each side of the tube. Neural crest cells migrate downward from their original position and form the dorsal root ganglia of the spinal nerves and the postganglionic sympathetic neurons. From sensory cells in the dorsal root ganglia, dendrites grow out to the sense organs, and axons grow into the spinal cord. Other neural crest cells migrate to various locations in the embryo. They give rise to parts of certain sense organs, nearly all pigment-forming cells in the body, and parts of the head. Some head derivatives of the neural crest are the cartilage and bone of the skull, face, and jaw; connective tissue in the eye and face dermis; sensory nerves and other parts of the peripheral nervous system; and ganglia of the autonomic nervous system.

As the nervous system develops, other organs also begin to take shape. Blocks of mesoderm known as **somites** form on either side of the neural tube. These will give rise to the vertebrae, muscles, and other components of the body axis. In addition to the skeleton and muscles, mesodermal structures include the kidneys and reproductive structures, as well as the circulatory organs. In fact, the heart and blood vessels are among the first structures to form, and they must function while still developing.

The development of the four-chambered heart of birds and mammals is a good example of the origin of a complex organ (Fig. 49-12; see Chapter 42). The heart originates through the fusion of paired blood vessels. Venous blood enters the single atrium; passes into the single ventricle, which is located above the atrium; and is then pumped into the embryo. During subsequent development, the atrium undergoes a process of torsion that brings it to a position above the ventricle, and partitions form that divide the atrium and the ventricle into right and left chambers.

The digestive tract is first formed as a separate foregut and hindgut as the body wall grows and folds, cutting them off from the yolk sac as two simple tubes. As the embryo grows, these tubes, which are lined with endoderm, grow and become greatly elongated. The liver, pancreas, and trachea originate as hollow, tubular outgrowths from the gut. As the trachea grows downward, it gives rise to the paired lung buds, which develop into lungs. The most anterior part of the foregut becomes the pharynx. A series of small outpocketings of the pharynx, the **pharyngeal pouches,** bud out laterally. These pouches meet the **branchial grooves,** a corresponding set of inpocketings from the overlying ectoderm. The arches of tissue formed between the grooves are called **branchial arches.** They contain the rudimentary skeletal, neural, and vascular elements of the face, jaws, and neck.

In fishes and some amphibians, the pharyngeal pouches and branchial grooves meet and form a continuous passage from the pharynx to the outside; these gill slits function as respiratory organs. In terrestrial vertebrates, each branchial groove remains separated from the corresponding pharyngeal pouch by a thin membrane of tissue, and these structures give rise to organs more appropriate for life on land. For example, the middle ear cavity and its connection to the pharynx, the eustachian tube, derive from a pharyngeal pouch.

Review

- How do cell division, growth, cell differentiation, and morphogenesis interact in the formation of the neural tube?
- What is the developmental significance of the neural crest cells?
- What adult structures develop from each germ layer?

BiologyNow™ Assess your understanding of **organogenesis** by taking the pretest on your BiologyNow CD-ROM.

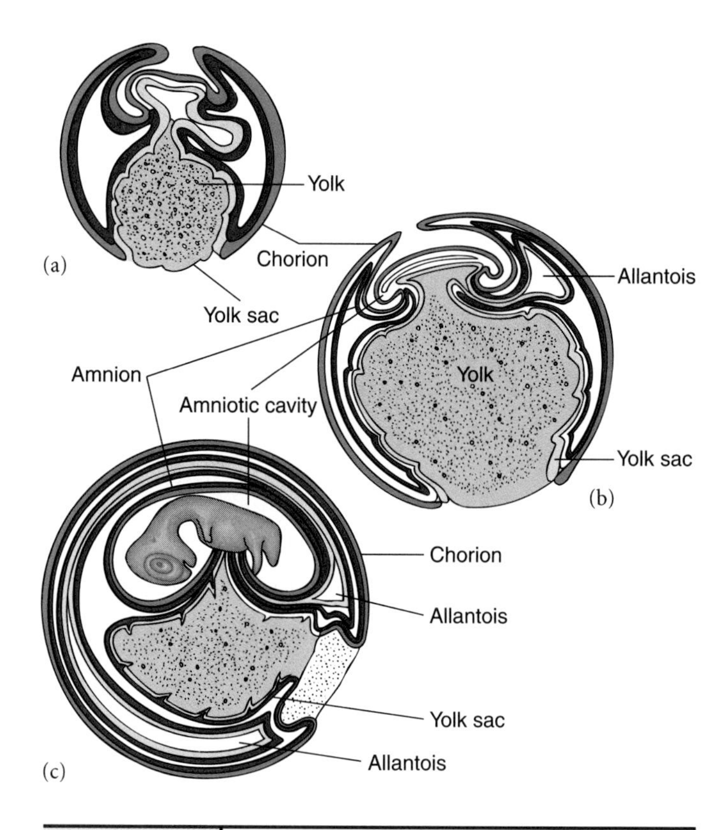

FIGURE **49-13** **Extraembryonic membranes.**
The formation of the extraembryonic membranes of the chick is illustrated at **(a)** 4 days, **(b)** 5 days, and **(c)** 9 days of development. Each of the membranes develops from a combination of two germ layers. The chorion and amnion form from lateral folds of the ectoderm and mesoderm that extend over the embryo and fuse. The allantois and the yolk sac develop from endoderm and mesoderm.

EXTRAEMBRYONIC MEMBRANES

Learning Objectives

8 Give the origins and functions of the chorion, amnion, allantois, and yolk sac.

In terrestrial vertebrates, the three germ layers also give rise to four **extraembryonic membranes:** the chorion, amnion, allantois, and yolk sac (Fig. 49-13). Although they develop from the germ layers, these membranes are not part of the embryo itself and are discarded at hatching or birth. The extraembryonic membranes are adaptations to the challenges of embryonic development on land. They protect the embryo, prevent it from drying out, and help in obtaining food and oxygen and eliminating wastes.

The outermost membrane, the **chorion,** encloses the entire embryo. Lying underneath the egg shell in birds and reptiles, it functions as the major organ of gas exchange. As you will see in the next section, its functions have been extended in humans and most other mammals.

Like the chorion, the **amnion** also encloses the entire embryo. The amniotic cavity, the space between the embryo and the amnion, becomes filled with amniotic fluid secreted by the membrane. Recall from Chapter 30 that terrestrial vertebrates are known as *amniotes,* because their embryos develop within this pool of fluid. The amniotic fluid prevents the embryo from drying out and permits it a certain freedom of motion. The fluid also serves as a protective cushion that absorbs shocks and prevents the amniotic membrane from sticking to the embryo. In current medical practice a sample of amniotic fluid is obtained through a procedure known as *amniocentesis* and analyzed for biochemical or chromosomal abnormalities that indicate a possible birth defect (see Fig. 15-11).

The **allantois** is an outgrowth of the developing digestive tract. In reptiles and birds, it stores nitrogenous wastes. In humans, the allantois is small and nonfunctional, except that its blood vessels contribute to the formation of umbilical vessels joining the embryo to the placenta.

In vertebrates with yolk-rich eggs, the **yolk sac** encloses the yolk, slowly digests it, and makes it available to the embryo. A yolk sac, connected to the embryo by a yolk stalk, develops even in mammalian embryos that have little or no yolk. Its walls serve as temporary centers for the formation of blood cells.

Review

- What are the roles of the chorion and amnion in terrestrial vertebrate embryos?
- How do the functions of the allantois and yolk sac differ between birds and mammals?

BiologyNow™ Assess your understanding of **extraembryonic membranes** by taking the pretest on your BiologyNow CD-ROM.

HUMAN DEVELOPMENT

Learning Objectives

9 Describe the general course of early human development, including fertilization, the fates of the trophoblast and inner cell mass, implantation, and the role of the placenta.

10 Contrast postnatal with prenatal life, describing several changes that occur at or shortly after birth that allow the neonate to live independently.

11 Describe some anatomical and physiological changes that occur with aging.

The human **gestation period,** the duration of pregnancy, averages 266 days (38 weeks, or about 9 months) from the time of fertilization to the birth of the baby (Table 49-2). We will use this system of timing when discussing the events of early development. Because the time of fertilization is not easily marked, obstetricians usually time the pregnancy by counting from the date of onset of the mother's last menstrual period; by this calculation, an average pregnancy is 280 days (40 weeks).

Fertilization occurs in the oviduct, and within 24 hours the human zygote has divided to become a two-celled embryo (Fig. 49-14). Cleavage continues as the embryo is propelled along the oviduct by ciliary action.

When the embryo enters the uterus on about the fifth day of development, the *zona pellucida* (its surrounding coat) is dissolved. During the next few days, the embryo floats free in the uterine cavity, nourished by a nutritive fluid secreted by the glands of the uterus. Its cells arrange themselves, forming a blastula, which in mammals is called a **blastocyst.** The outer layer of cells, the **trophoblast,** eventually forms the chorion and amnion that surround the embryo. A little cluster of cells, the **inner cell mass,** projects into the cavity of the blastocyst. The inner cell mass gives rise to the embryo proper.

On about the seventh day of development the embryo begins the process of **implantation,** in which it embeds in the endometrium of the uterus (Fig. 49-15). The trophoblast cells in contact with the endometrium secrete enzymes that erode an area just large enough to accommodate the tiny embryo. As the embryo slowly works its way into the underlying connective and vascular tissues, the opening in the endometrium repairs itself. All further development of the embryo takes place *within* the endometrium.

TABLE 49-2 Some Important Developmental Events in the Human Embryo

Time from Fertilization	Event
24 hours	Embryo reaches two-cell stage
3 days	Morula reaches uterus
7 days	Blastocyst begins to implant
2.5 weeks	Notochord and neural plate are formed; tissue that will give rise to heart is differentiating; blood cells are forming in yolk sac and chorion
3.5 weeks	Neural tube forming; primordial eye and ear visible; pharyngeal pouches forming; liver bud differentiating; respiratory system and thyroid gland just beginning to develop; heart tubes fuse, bend, and begin to beat; blood vessels are laid down
4 weeks	Limb buds appear; three primary divisions of brain forming
2 months	Muscles differentiating; embryo capable of movement; gonad distinguishable as testis or ovary; bones begin to ossify; cerebral cortex differentiating; principal blood vessels assume final positions
3 months	Sex can be determined by external inspection; notochord degenerates; lymph glands develop
4 months	Face begins to look human; lobes of cerebrum differentiate; eyes, ears, and nose look more "normal"
Third trimester	A covering of downy hair covers the fetus, then later is shed; neuron myelination begins; tremendous growth of body
266 days (from conception)	Birth

The placenta is an organ of exchange

In placental mammals, the **placenta** is the organ of exchange between mother and embryo (see Fig. 49-15). The placenta provides nutrients and oxygen for the fetus and removes wastes, which the mother then excretes. In addition, the placenta is an endocrine organ that secretes estrogens and progesterone to maintain pregnancy. The placenta develops from both the chorion of the embryo and the uterine tissue of the mother. In

FIGURE 49-14 Cleavage in a human embryo.

(a) Male and female pronuclei prior to fusion. **(b)** Two-cell stage. **(c)** Eight-cell stage. **(d)** Cleavage continues, giving rise to a morula. *(Lennart Nilsson, from* Being Born, *1992, pp. 14, 15, 17. The Putnam Publishing Group)*

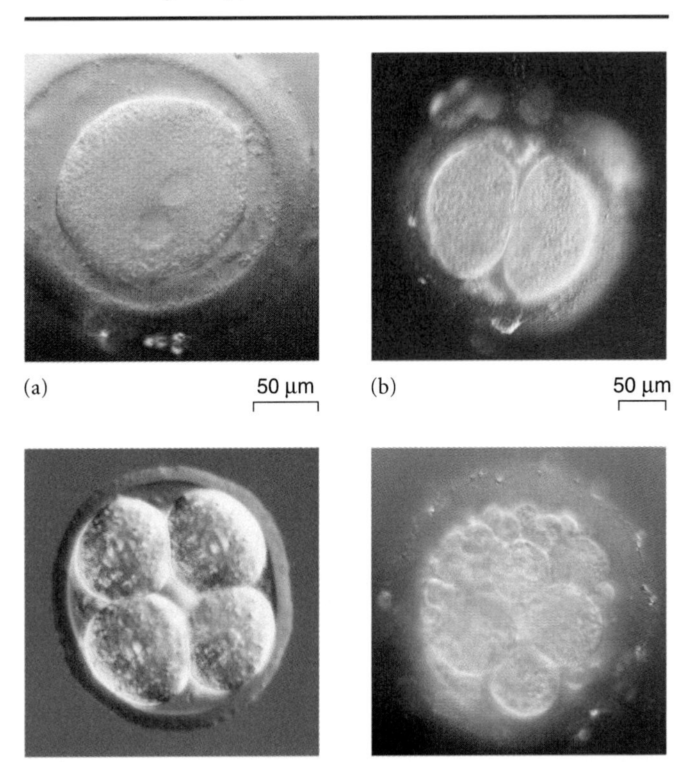

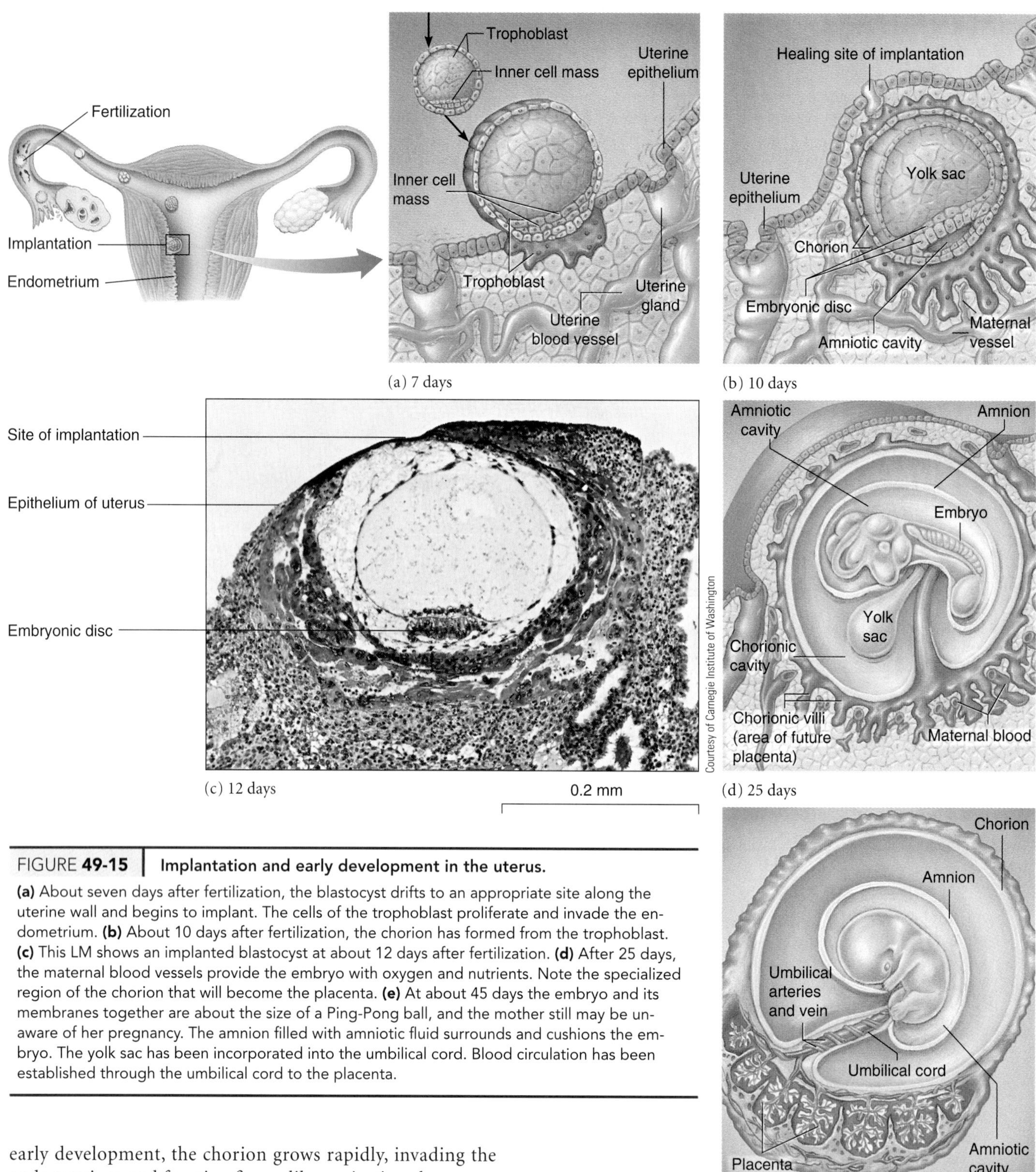

(a) 7 days

(b) 10 days

(c) 12 days

(d) 25 days

(e) 45 days

FIGURE **49-15** | **Implantation and early development in the uterus.**

(a) About seven days after fertilization, the blastocyst drifts to an appropriate site along the uterine wall and begins to implant. The cells of the trophoblast proliferate and invade the endometrium. **(b)** About 10 days after fertilization, the chorion has formed from the trophoblast. **(c)** This LM shows an implanted blastocyst at about 12 days after fertilization. **(d)** After 25 days, the maternal blood vessels provide the embryo with oxygen and nutrients. Note the specialized region of the chorion that will become the placenta. **(e)** At about 45 days the embryo and its membranes together are about the size of a Ping-Pong ball, and the mother still may be unaware of her pregnancy. The amnion filled with amniotic fluid surrounds and cushions the embryo. The yolk sac has been incorporated into the umbilical cord. Blood circulation has been established through the umbilical cord to the placenta.

early development, the chorion grows rapidly, invading the endometrium and forming finger-like projections known as *chorionic villi.* The chorionic villus sampling technique allows prenatal detection of certain genetic disorders (see Fig. 15-12). The villi become vascularized (infiltrated with blood vessels) as the embryonic circulation develops.

As the human embryo grows, the **umbilical cord** develops and connects the embryo to the placenta (see Fig. 49-15). The umbilical cord contains the two umbilical arteries and the umbilical vein. The umbilical arteries connect the embryo to a vast network of capillaries developing within the chorionic villi. Blood from the villi returns to the embryo through the umbilical vein.

The placenta consists of the portion of the chorion that develops villi, together with the underlying uterine tissue that contains maternal capillaries and small pools of maternal blood. The fetal blood in the capillaries of the chorionic villi comes in close contact with the mother's blood in the tissues between the villi. However, they are always separated by a membrane through which substances may diffuse or be actively transported. Maternal and fetal blood do not normally mix in the placenta or any other place.

The placenta produces several hormones. From the time the embryo first begins to implant itself, its trophoblastic cells release **human chorionic gonadotropin (hCG),** which signals the corpus luteum (see Chapter 48) that pregnancy has begun. In response, the corpus luteum increases in size and releases large amounts of progesterone and estrogens. These hormones stimulate continued development of the endometrium and placenta.

Without hCG, the corpus luteum would degenerate and the embryo would abort and be flushed out with the menstrual flow. The woman would probably not even know she had been pregnant. If the corpus luteum is removed before about the 11th week of pregnancy, the embryo spontaneously aborts. After that time, the placenta itself produces enough progesterone and estrogens to maintain pregnancy.

Organ development begins during the first trimester

Gastrulation occurs during the second and third weeks of development. Then the notochord begins to form and induces formation of the neural plate. The neural tube develops, and the forebrain, midbrain, and hindbrain are evident by the fifth week of development. A week or so later the forebrain begins to grow outward, forming the rudiments of the cerebral hemispheres.

The heart begins to develop, and after 3.5 weeks begins to beat spontaneously. Pharyngeal pouches, branchial grooves, and branchial arches form in the region of the developing pharynx. In the floor of the pharynx, a tube of cells grows downward to form the primordial trachea, which gives rise to the lung buds. The digestive system also gives rise to outgrowths that develop into the liver, gallbladder, and pancreas. A thin tail becomes evident but does not grow as rapidly as the rest of the body, and so becomes inconspicuous by the end of the second month. Near the end of the fourth week, the limb buds begin to differentiate; these eventually give rise to arms and legs.

All the organs continue to develop during the second month (Fig. 49-16). Muscles develop, and the embryo becomes capable of movement. The brain begins to send impulses that regulate the functions of some organs, and a few simple reflexes are evident. After the first two months of development, the embryo is referred to as a **fetus** (Fig. 49-17).

By the end of the **first trimester** (the first three months of development), the fetus is about 56 mm (2.2 in) long and weighs about 14 g (0.5 oz). Although small, it looks human. The external genital structures have differentiated, indicating the sex of the fetus. Ears and eyes approach their final positions. Some of the skeleton becomes distinct, and the developing vertebral column has replaced the notochord. The fetus performs breathing movements, pumping amniotic fluid into and out of its lungs, and even makes sucking motions.

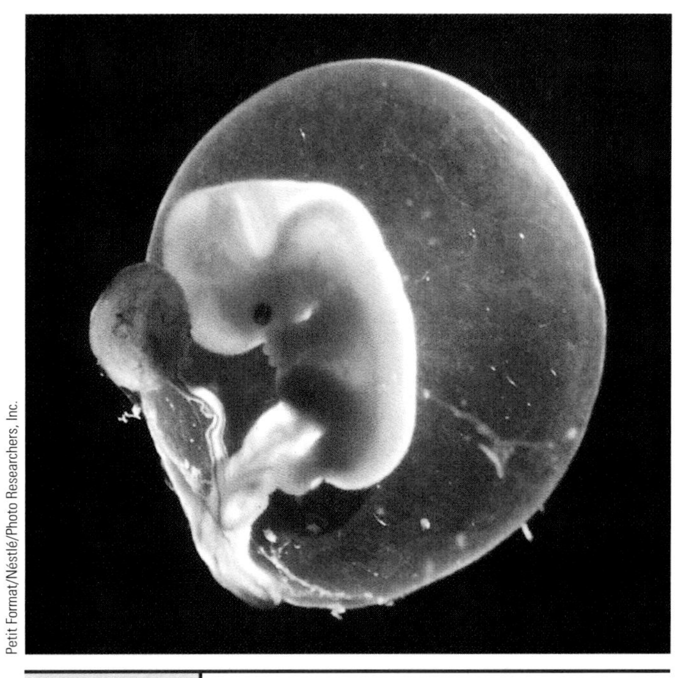

Petit Format/Néstlé/Photo Researchers, Inc.

FIGURE **49-16** | **The second month of development.**
The amnion is prominent as a transparent fluid-filled sac surrounding this 5.5-week-old human embryo, which is 1 cm (0.4 in) long. Note the limb buds and the eyes. The balloon-like object at the left, connected to the embryo by a stalk, is the yolk sac. This embryo is in a period of very rapid development, as you can see by a comparison with the chapter opening photograph of a human embryo in its seventh week.

FIGURE **49-17** | **Human fetus at 10 weeks.**
Note the position of the fetus within the uterine wall.

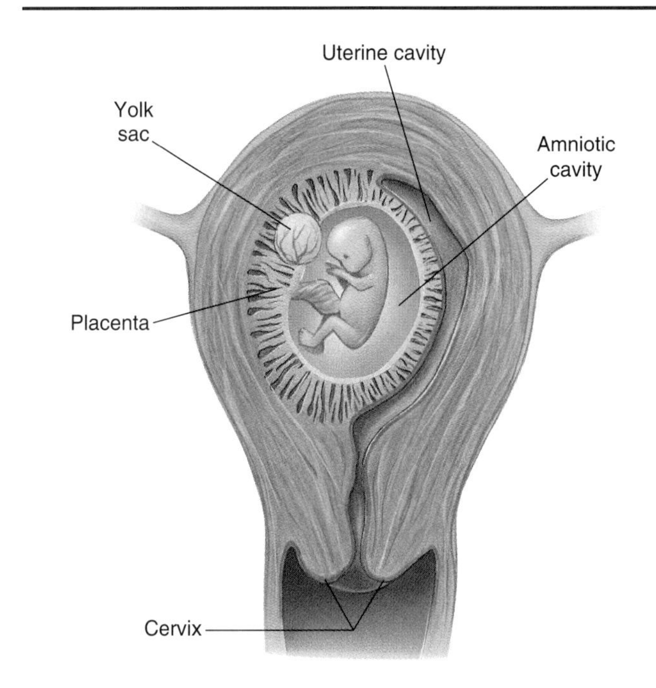

Development continues during the second and third trimesters

During the second trimester (months 4 through 6), the fetal heart can be heard with a stethoscope. The fetus moves freely within the amniotic cavity, and during the fifth month the mother usually becomes aware of relatively weak fetal movements ("quickening").

The fetus grows rapidly during the final trimester (months 7 through 9), and final differentiation of tissues and organs occurs. If born at 24 weeks (out of 40 weeks), the fetus has only about a 50% chance of surviving, even with the best of medical care. Its brain is not yet sufficiently developed to sustain vital functions such as rhythmic breathing, and the kidneys and lungs are immature.

During the seventh month the cerebrum grows rapidly and develops convolutions. The grasping and sucking reflexes are evident, and the fetus may suck its thumb. Any infant born before 37 weeks of gestation is considered premature. If born after 30 weeks, however, the baby has a good chance of surviving. At birth the average full-term baby weighs about 3000 g (6.6 lb) and measures about 52 cm (20 in) in total length.

More than one mechanism can lead to a multiple birth

Occasionally the cells of the two-cell embryo separate, and each cell develops into a complete organism. Or sometimes the inner cell mass subdivides, forming two groups of cells, each of which develops independently. Because these cells have identical sets of genes, the individuals formed are exactly alike—**monozygotic,** or **identical, twins.** Very rarely, the two inner cell masses do not completely separate and so give rise to **conjoined twins,** who are physically attached and usually share one or more body parts.

Dizygotic twins, also called **fraternal twins,** develop when two eggs are ovulated, and each is fertilized by a different sperm. Each zygote has its own distinctive genetic endowment, so the individuals produced are not identical. They may not even be of the same sex. Similarly, triplets (and other multiple births) may be either identical or fraternal.

Before fertility-inducing agents became available in the United States, twins were born once in about 80 births (about 30% of twins are monozygotic), triplets once in 80^2 (or 1 in 6400), and quadruplets once in 80^3 (or 1 in 512,000). However, the Centers for Disease Control and Prevention reported that between 1980 and 1995 the rate of twin births increased by about 30%, and the rate of triplet births by over 240%. This increase was attributed to widespread use of drugs to improve fertility, which increases the chance of dizygotic twinning.

A family history of twinning increases the probability of having dizygotic twins. However, giving birth to monozygotic twins is probably not influenced by heredity, age of the mother, or other known factors. An estimated two thirds of multiple pregnancies end in the birth of a single baby; the other embryo(s) may be absorbed within the first 10 weeks of pregnancy, or spontaneously aborted. Ultrasound imaging techniques are used to produce a type of image known as a **sonogram,** which can provide valuable data regarding the presence of multiple embryos. This is medically important, because multiple births are associated with higher infant mortality, which is largely a consequence of an increased risk of prematurity and low birth weight.

Environmental factors affect the embryo

We all know that the growth and development of babies are influenced by the food they eat, the air they breathe, the disease organisms that infect them, and the chemicals or drugs to which they are exposed. **Prenatal** development is also affected by these environmental influences. Life before birth is even more sensitive to environmental changes than it is for the fully formed baby. Although there is no direct mixing of maternal and fetal blood, diffusion and various other mechanisms allow many substances—nutrients, drugs, pathogens, and gases—to travel across the placenta.

Table 49-3 describes some environmental influences on development. Some of these are **teratogens,** drugs or other substances that interfere with morphogenesis, causing malformations. Many factors, such as smoking, alcohol use, and poor nutrition contribute to low birth weight, a condition responsible for a great number of infant deaths.

About 5% of newborns (more than 150,000 babies per year) in the United States have a defect of clinical significance. Such birth defects account for about 22% of deaths among newborns. Birth defects may be caused by genetic or environmental factors, or a combination of the two. Genetic factors were discussed in Chapter 15. In this section, we examine some environmental conditions that affect the well-being of the embryo.

Timing is important. Each developing structure has a critical period during which it is most susceptible to unfavorable conditions. Generally, this critical period occurs early in the structure's development, when interference with cell movements or divisions may prevent formation of normal shape or size, resulting in permanent malformation. Because most structures form during the first three months of embryonic life, the embryo is most susceptible to environmental factors during this early period. During much of this time, the mother may not even realize she is pregnant and so may not take special precautions to minimize potential dangers.

Physicians diagnose some defects while the embryo is in the uterus. In some cases, treatment is possible before birth. Amniocentesis and chorionic villus sampling, discussed in Chapter 15, are techniques used to detect certain defects. Ultrasound imaging techniques discussed previously help diagnose defects and determine the position of the fetus. Unlike imaging techniques that use radiation, the available evidence indicates that ultrasound is harmless to the fetus. New methods currently under development include special types of MRI and high-resolution three-dimensional ultrasound imaging (Fig. 49-18).

The neonate must adapt to its new environment

Important changes take place after birth. During prenatal life, the fetus received both food and oxygen from the mother

TABLE 49-3 Environmental Influences on the Embryo

Factor	Example and Effect	Comment
Nutrition	Severe protein malnutrition doubles number of defects, fewer brain cells are produced, and learning ability may be permanently affected; deficiency of folic acid (a vitamin) linked to CNS defects such as spina bifida (open spine), low birth weight	Growth rate mainly determined by rate of net protein synthesis by embryo's cells
Medications	Many medications, even aspirin, affect development of the fetus	
Excessive vitamins	Vitamin D essential, but excessive amounts may result in a form of mental retardation; an excess of vitamins A and K may also be harmful	Vitamin supplements are normally prescribed for pregnant women, but only the recommended dosage should be taken
Thalidomide	Thalidomide, marketed in Europe as a mild sedative, was responsible for serious malformations in more than 7000 babies born in the late 1950s in 20 countries; principal defect was phocomelia, a condition in which babies are born with extremely short limbs, often with no fingers or toes	This teratogenic drug interferes with the development of blood vessels and nerves; most hazardous when taken during fourth to fifth weeks, when limbs are developing; thalidomide has been approved for the treatment of leprosy and certain cancers (e.g., myeloma), and clinical trials are being conducted to test its usefulness in the treatment of other cancers and some of the complications of AIDS*
Accutane (isotretinoin)	Accutane, a synthetic derivitive of vitamin A, is used for the treatment of cystic acne. A woman who takes Accutane during pregnancy has a 1 in 5 chance of having a child with serious malformations of the brain (causing mental retardation), head and face, thymus gland (interfering with immune function) or heart	Although Accutane is marketed with severe restrictions to prevent pregnant women from taking this teratogenic drug, several children are born each year with birth defects attributable to Accutane exposure
Pathogens		
Rubella	Rubella (german measles) virus crosses placenta and infects embryo; interferes with normal metabolism and cell movements; causes syndrome that involves blinding cataracts, deafness, heart malformations, and mental retardation; risk is greatest (about 50%) when rubella is contracted during the first month of pregnancy; risk declines with each succeeding month	Rubella epidemic in the United States in 1963–1965 resulted in about 20,000 fetal deaths and 30,000 infants born with serious defects; immunization is available but must be administered several months before conception
HIV*	HIV can be transmitted from mother to baby before birth, during birth, or by breastfeeding	See discussion of AIDS in Chapter 43
Syphilis	Syphilis is transmitted to fetus in about 40% of infected women; fetus may die or be born with defects and congenital syphilis	Pregnant women are routinely tested for syphilis during prenatal examinations; most cases can be safely treated with antibiotics
Ionizing radiation	When a pregnant woman is subjected to x-rays or other forms of radiation, infant has a higher risk of birth defects and leukemia	Radiation was one of the earliest causes of birth defects to be recognized
Recreational and/or abused substances		
Alcohol	When a woman drinks heavily during pregnancy, the baby may be born with fetal alcohol syndrome (FAS), which includes certain physical deformities, and mental and physical retardation; low birth weight and structural abnormalities have been associated with as little as two drinks a day; some cases of hyperactivity and learning disabilities have occurred	Fetal alcohol syndrome is the leading cause of preventable mental retardation in the United States
Cigarette smoking	Cigarette smoking reduces the amount of oxygen available to the fetus because some of the maternal hemoglobin is combined with carbon monoxide; may slow growth and can cause subtle forms of damage	Mothers who smoke deliver babies with lower-than-average birth weights and have a higher incidence of spontaneous abortions, stillbirths, and neonatal deaths; studies also indicate a possible relationship between maternal smoking and slower intellectual development in offspring
Cocaine	Constricts fetal arteries, resulting in retarded development and low birth weight; severe cases may be mentally retarded, have heart defects and other medical problems	Cocaine users frequently abuse alcohol as well, so it is difficult to separate the effects
Heroin	High rates of mortality and prematurity; low birth weight	Infants that survive are born addicted and must be treated for weeks or months

*HIV, human immunodeficiency virus; AIDS, acquired immunodeficiency syndrome.

through the placenta. Now the newborn's own digestive and respiratory systems must function. Correlated with these changes are several major changes in the circulatory system.

Normally, the **neonate** (newborn infant) begins to breathe within a few seconds of birth and cries within half a minute. If anesthetics have been given to the mother, however, the fetus

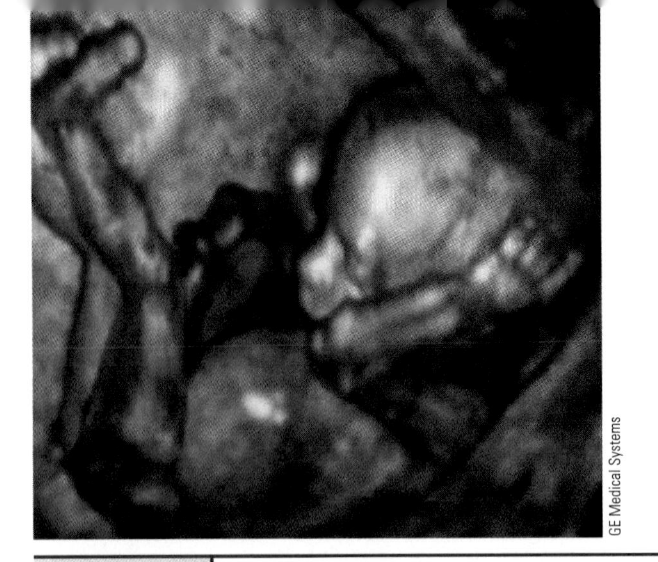

FIGURE **49-18** **Three-dimensional ultrasound image of a human fetus at 12 to 20 weeks gestation.**

Note the enhanced soft tissue detail.

may also be anesthetized, and its breathing and other activities may be depressed. Some infants may not begin breathing until several minutes have passed. This is one reason why many women request childbirth methods that minimize the use of medication.

Researchers think the neonate's first breath is initiated by the accumulation of carbon dioxide in the blood after the umbilical cord is cut. Carbon dioxide stimulates the respiratory centers in the medulla. The resulting expansion of the lungs enlarges its blood vessels (which in the uterus were partially collapsed). Blood from the right ventricle flows in increasing amounts through these larger pulmonary vessels. (During fetal life, blood bypasses the lungs in two ways: by flowing through an opening, the *foramen ovale,* which shunts blood from the right atrium to the left atrium, and by flowing through an arterial duct connecting the pulmonary artery and aorta. Both of these routes close off after birth.)

Aging is not a uniform process

You have examined briefly the development of the embryo and fetus, the birth process, and the adjustments required of the neonate. The human life cycle then proceeds through the stages of infant, child, adolescent, young adult, middle-aged adult, and elderly adult.

Development encompasses any biological change within an organism over time, including the changes commonly called **aging.** Changes during the aging process decrease function in the older organism. The declining capacities of the various systems in the human body, although most apparent in the elderly, may begin much earlier in life.

The aging process is far from uniform among different individuals or in various parts of the body. The body systems generally decline at different times and rates. On average, between the ages of 30 and 75 a man loses 64% of his taste buds, 44% of the glomeruli in his kidneys, and 37% of the axons in his spinal nerves. His nerve impulses are propagated at a 10% slower rate, the blood supply to his brain is 20% less, his glomerular filtration rate has decreased 31%, and the vital capacity of his lungs has declined 44%. However, the human body has considerable functional capacity in reserve, so bodily functions are usually adequate. Furthermore, there is evidence that some of these declines can be significantly lessened by modifying the lifestyle (such as diet and exercise). Women also undergo body system declines in function, although on average they live about 8 years longer than men.

Although marked improvements in medicine and public health have led to longer life expectancies, there has been no corresponding increase in the *maximum* life expectancy. Relatively little is known about the aging process itself; this is now an active field of investigation, and researchers are gaining considerable insight through genetic studies on model organisms such as the nematode worm, *C. elegans;* the fruit fly, *Drosophila melanogaster;* and the laboratory mouse, *Mus* (see Chapter 16).

Homeostatic response to stress decreases during aging

Research findings support the idea that most aging occurs because a combination of inheritance and environmental stress makes the individual less able to respond to additional stressors. One major question is whether a genetic program has evolved to cause aging, or if genetic involvement in the process is more circumstantial. Most available evidence favors the latter view.

Some genetically programmed developmental events do seem related to the aging process. Cells that normally stop dividing when they differentiate appear more subject to the changes of aging than cells that continue to divide throughout life. Furthermore, researchers have hypothesized that certain parts of the body begin to malfunction because a genetic program stops key cells from dividing and replenishing themselves. In one model known as **cell aging,** when grown in culture normal human cells eventually lose their ability to divide . Furthermore, cells taken from an older person divide fewer times than those from a younger person. Cell aging appears related to the fact that most normal human somatic cells are genetically programmed to lose the ability to produce active telomerase, an enzyme that replicates the DNA of the end caps (telomeres) of the chromosomes (see Chapter 11).

Genetically programmed cell death, *apoptosis,* is an essential developmental mechanism that may play a role in aging (see Chapters 4 and 16). However, researchers think most aging is not a direct consequence of the genetic program leading to apoptosis; instead, mistakes in the *control* of apoptosis may lead to certain degenerative conditions, such as Alzheimer's disease, or to some of the cell deaths that occur following a heart attack or stroke.

Researchers are investigating genes that affect how the body is maintained and repaired in the face of various stressors. Like other life processes, aging may be accelerated by certain environmental influences and may vary because of inherited differences among individuals. Experimental evidence suggests that aging, at least in rats and certain other mammals, can be delayed by severe caloric restriction. Evidence also indicates that premature aging can be precipitated by hormonal changes; by various malfunctions of the immune system, including autoimmune responses; by accumulation of specific waste products within the cells; by changes in the molecular structure of macromolecules such as collagen; and by damage to DNA from continued exposure to cosmic radiation and x-rays.

Review

- How does a human blastocyst form and implant?
- How does the placenta form, and what is its function?
- What adaptations must the neonate make immediately after birth?
- During what trimester do each of the following occur? Cerebrum develops convolutions; limb buds develop; mother feels fetal movements.
- What changes take place during the aging process?
- How are model organisms contributing to the study of aging?

BiologyNow™ Assess your understanding of **human development** by taking the pretest on your BiologyNow CD-ROM.

SUMMARY WITH KEY TERMS

1 Describe the four processes involved in fertilization.

- Contact and recognition occur between noncellular egg coverings and sperm.
- Sperm entry is regulated to prevent interspecific fertilization and **polyspermy,** which is fertilization of the egg by more than one sperm.
- Fertilization activates the egg, triggering the events of early development.
- Sperm and egg pronuclei fuse and initiate DNA synthesis.

2 Describe fertilization in echinoderms, and point out some ways in which mammalian fertilization differs.

- The coverings of echinoderm eggs are the **vitelline envelope** and the **jelly coat;** a **zona pellucida** encloses the mammalian egg.
- On contact, a sperm undergoes an **acrosome reaction,** which facilitates penetration of the egg coverings; in mammals the reaction is preceded by **capacitation,** a maturation process that results in the ability of a sperm to fertilize an egg.
- Sea urchin fertilization is followed by a fast block to polyspermy (depolarization of the plasma membrane) and a slow block to polyspermy (the **cortical reaction**); changes in the zona pellucida prevent polyspermy in mammals.

3 Trace the generalized pattern of early development of the embryo from zygote through early cleavage and formation of the morula and blastula.

- The zygote undergoes **cleavage,** a series of rapid cell divisions without a growth phase. The main effect of cleavage is to partition the zygote into many small cells **(blastomeres).**
- Cleavage forms a solid ball of cells (the **morula**) and then usually a hollow ball of cells (the **blastula**).

4 Contrast early development, including cleavage in the echinoderm (or in amphioxus), the amphibian, and the bird, paying particular attention to the importance of the amount and distribution of yolk.

- The **isolecithal** eggs of most invertebrates and simple chordates have evenly distributed **yolk.** They undergo **holoblastic cleavage,** which involves division of the entire egg.
- In the moderately **telolecithal** eggs of amphibians, a concentration of yolk at the **vegetal pole** slows cleavage so that only a few large cells form there, compared to a large number of smaller cells at the **animal pole.**
- The highly telolecithal eggs of reptiles and birds, with a large concentration of yolk at one end, undergo **meroblastic cleavage,** which is restricted to the **blastodisc.**
- Animals whose zygotes have relatively homogenous cytoplasm exhibit **regulative development,** in which the embryo develops as a self-regulating whole. The relatively rigid developmental patterns of some animals (**mosaic development**) is a consequence of the unequal distribution of cytoplasmic components.

The **gray crescent** of an amphibian zygote determines the body axis of the embryo.

5 Identify the significance of gastrulation in the developmental process, and compare gastrulation in the echinoderm (or in amphioxus), the amphibian, and the bird.

- In **gastrulation,** the basic body plan is laid down as three **germ layers** form: the outer **ectoderm,** the inner **endoderm,** and the **mesoderm** between them.
- The **archenteron,** the forerunner of the digestive tube, forms in some groups; its opening to the exterior is the **blastopore.**
- In gastrulation in the sea star and in amphioxus, cells from the blastula wall invaginate and eventually meet the opposite wall, forming the archenteron.
- In the amphibian, invagination at the vegetal pole is obstructed by large, yolk-laden cells; instead, cells from the animal pole move down over the yolk-rich cells and invaginate, forming the **dorsal lip of the blastopore.**
- In the bird, invagination occurs at the **primitive streak,** and no archenteron forms.

6 Define *organogenesis,* and summarize the fate of each of the germ layers.

- Organogenesis is the process of organ formation.
- *Ectoderm* becomes nervous system, sense organs, and outer layer of skin (epidermis).
- *Mesoderm* becomes notochord, skeleton, muscles, circulatory system, and inner layer of skin (dermis).
- *Endoderm* becomes lining of digestive tube.

7 Trace the early development of the vertebrate nervous system.

- The developing **notochord** is responsible for **induction,** causing the ectoderm to differentiate and form the central nervous system.
- The brain and spinal cord develop from the **neural tube.**
- The **chorion** and **amnion** derive from ectoderm and mesoderm; the **allantois** and **yolk sac** from endoderm and mesoderm.

8 Give the origins and functions of the chorion, amnion, allantois, and yolk sac.

- The **chorion** is used in gas exchange.
- The **amnion** is a fluid-filled sac that surrounds the embryo and keeps it moist; it also acts as a shock absorber.
- The **allantois** stores nitrogenous wastes.
- The **yolk sac** makes food available to the embryo.

9 Describe the general course of early human development, including fertilization, the fates of the trophoblast and inner cell mass, implantation, and the role of the placenta.

- Fertilization occurs in the oviduct.
- Cleavage takes place as the embryo is moved down the oviduct.

SUMMARY WITH KEY TERMS *(continued)*

- In the uterus, the embryo develops into a **blastocyst** consisting of an outer **trophoblast,** which gives rise to the chorion and amnion, and an **inner cell mass,** which becomes the embryo proper. The blastocyst undergoes **implantation** in the endometrium.
- The **umbilical cord** connects the embryo to the **placenta,** the organ of exchange between the maternal and fetal circulation. The placenta derives from the embryonic chorion and maternal tissue.

10 Contrast postnatal with prenatal life, describing several changes that occur at or shortly after birth that allow the neonate to live independently.

- Human **prenatal** development requires 266 days from the time of fertilization; organogenesis begins during the **first trimester.**
- After the first two months of development, the embryo is referred to as a **fetus.**
- Growth and refinement of the organs continue in the second and third trimesters.
- The **neonate** (newborn) must undergo rapid adaptations, especially changes in the respiratory and digestive systems.

11 Describe some anatomical and physiological changes that occur with aging.

- The **aging** process is marked by a decrease in homeostatic response to stress.
- All body systems decline with age, but not at the same rate.

POST-TEST

1. The main function of the acrosome reaction is to (a) activate the egg (b) improve sperm motility (c) prevent interspecific fertilization (d) facilitate penetration of the egg coverings by the sperm (e) cause fusion of the sperm and egg pronuclei
2. The fast block to polyspermy in sea urchins (a) is a depolarization of the egg plasma membrane (b) requires exocytosis of the cortical granules (c) includes the elevation of the fertilization envelope (d) involves the hardening of the jelly coat (e) is a complete block
3. Place the following events of sea urchin fertilization in the proper sequence. ① fusion of egg and sperm pronuclei ② DNA synthesis ③ increased protein synthesis ④ release of calcium ions into the egg cytoplasm (a) 4, 3, 1, 2 (b) 3, 2, 4, 1 (c) 2, 3, 1, 4 (d) 1, 2, 3, 4 (e) 4, 1, 2, 3
4. The cleavage divisions of a sea urchin embryo (a) occur in a spiral pattern (b) do not include DNA synthesis (c) do not include cytokinesis (d) are holoblastic (e) do not occur at the vegetal pole
5. Meroblastic cleavage is typical of embryos formed from ______ eggs. (a) moderately telolecithal (b) highly telolecithal (c) isolecithal (d) b and c (e) a, b, and c
6. The primitive groove of the bird embryo is the functional equivalent of the ______ in the amphibian embryo. (a) yolk plug (b) archenteron (c) blastocoel (d) gray crescent (e) blastopore
7. Which of the following are mismatched? (a) endoderm; lining of the digestive tube (b) ectoderm; circulatory system (c) mesoderm; notochord (d) mesoderm; reproductive system (e) ectoderm; sense organs
8. Which of the following has three germ layers? (a) morula (b) gastrula (c) blastula (d) blastocyst (e) trophoblast
9. An unidentified substance (or substances) released from the developing notochord causes the overlying ectoderm to form the neural plate. This phenomenon is known as (a) activation (b) determination (c) induction (d) implantation (e) mosaic development
10. Which of the following consists of both fetal and maternal tissues? (a) umbilical cord (b) placenta (c) amnion (d) allantois (e) yolk sac
11. The embryo proper of a mammal develops from the (a) trophoblast (b) umbilical cord (c) inner cell mass (d) entire blastocyst (e) yolk sac
12. Which of the following statements about vertebrate organogenesis is *not* true? (a) The notochord, brain, and spinal cord are among the first organs to develop in the early embryo. (b) The developing notochord causes the overlying ectoderm to differentiate into the neural plate. (c) The neural folds meet and fuse, forming the four-chambered heart. (d) Some neural crest cells differentiate into neurons. (e) Blocks of mesoderm known as *somites* form on either side of the neural tube.
13. On about the seventh day of development, the human embryo (a) implants in the wall of the uterus (b) has a fully developed placenta for obtaining nutrients and oxygen (c) releases human chorionic gonadotropin (d) both a and c (e) a, b, and c

CRITICAL THINKING

1. What is the adaptive value of developing a placenta?
2. For almost 200 years, scientists debated whether an egg or sperm cell contains a completely formed, miniature human (preformation) or if structures develop gradually from a formless zygote (epigenesis). Relate these views to current concepts of development.
3. Not all teratogenic medications are banned by the U.S. Food and Drug Administration. Why?

- Visit our Website at **http://biology.brookscole.com/solomon7** for links to chapter-related resources on the WorldWide Web. Additional online materials relating to this chapter can also be found on our Web site.

BIOLOGY NOW RESOURCES

Active Figures

49-6: Cleavage pattern in a frog egg

49-11: Development of the nervous system

Preparing for an exam? Take a diagnostic test on your BiologyNow CD-ROM.

Post-Test Answers

1. d	2. a	3. a	4. d
5. b	6. e	7. b	8. b
9. c	10. b	11. c	12. c
13. d			

Animal Behavior

This wasp (*Philanthus triangulum*) is digging a burrow.

CHAPTER OUTLINE

Suppose your professor gave you a hypodermic syringe full of poison and told you to find a particular type of insect, one that you had never seen before and that was armed with active defenses. You then had to inject the ganglia of your prey's nervous system (about which you had been taught nothing) with just enough poison to paralyze but not kill it. You would have difficulty accomplishing these tasks—but a solitary wasp no larger than the first joint of your thumb does it all with elegance and surgical precision, without instruction.

The wasp *Philanthus* captures a bee (or beetle), stings it, and places the paralyzed insect in a burrow excavated in the sand (see photograph). She then lays an egg on her prey, which is devoured alive by the larva that hatches from that egg. From time to time, *Philanthus* returns to her hidden nest to reprovision it, until autumn when the larva becomes a hibernating pupa. Her offspring will repeat this behavior, precisely executing each step without ever having seen it done.

An animal's **behavior** is what it *does* and how it does it, usually in response to stimuli in its environment. A dog may wag its tail, a bird may sing, a butterfly may release a volatile sex attractant. Behavior is just as diverse as biological structure and just as characteristic of a given species as its anatomy or physiology. Like its morphology and physiology, an animal's behavior is the product of natural selection on phenotypes and indirectly on the genotypes that code for those phenotypes. Thus an animal's repertoire of behavior is a set of adaptations that equip it for survival in a particular environment.

The capacity for behavior is inherited, but much inherited behavior can be modified by experience. **Learning** involves persistent changes in behavior that result from experience. In considering complex behaviors such as the reproductive behavior of *Philanthus,* we might wonder *why* she behaves as she does, and we might also be interested in *how* she accomplishes her task. Early investigators of animal behavior focused on *how* questions.

These questions address **proximate causes,** immediate causes such as the genetic, developmental, and physiological processes that permit the animal to carry out the particular behavior.

More recently, biologists have added the *why* perspective, asking questions that address the **ultimate causes.** These questions, which have evolutionary explanations, ask *why* a particular proximate cause has evolved. Ultimate considerations address costs and benefits of behavior patterns. When studying ultimate causes, we may ask what the adaptive value of a particular behavior might be. An understanding of behavior requires consideration of both proximate and ultimate causes. ■

UNDERSTANDING BEHAVIOR

Learning Objective

1 Apply cost–benefit analysis and the concept of ultimate cause to the process of deciding whether a particular behavior is adaptive.

Whether biologists study behavior in an animal's natural environment or in the laboratory, they must consider that what an animal does cannot be isolated from the way in which it lives. **Behavioral ecology** is the study of behavior in natural environments from an evolutionary perspective. For more than two decades, behavioral ecology has been the main approach of biologists who study animal behavior. Before this approach emerged, the study of animal behavior was called *ethology,* and this term is sometimes still used to refer to the overall study of animal behavior.

Behavioral ecologists use **cost–benefit analysis** to understand specific behaviors. A behavior may help an animal obtain food or water, protect itself, reproduce, or acquire and maintain territory in which to live. The benefits typically contribute to **direct fitness,** which is an individual's reproductive success, measured by the number of viable offspring. Reproduction is, of course, a key to evolutionary success. Behaviors also involve costs. For example, while a parent is off hunting for food for its offspring the young it has left alone may be killed by predators. If the benefits are greater than the costs, the behavior is adaptive.

The ultimate cause of a behavior is therefore to increase the probability that the genes of the individual animal will be passed to future generations. Certain responses may lead to the death of the individual while increasing the chance that copies of its genes will survive through the enhanced production or survival of its offspring or other relatives. In this chapter, we consider how an animal's behavior contributes to its reproductive success and to the survival of its species.

Review

- In what ways are the behaviors of *Philanthus,* the sand wasp, adaptive?

BiologyNow™ Assess your **understanding of the adaptiveness of most behavior** by taking the pretest on your BiologyNow CD ROM.

INTERACTION OF GENES AND ENVIRONMENT

Learning Objective

2 Describe the interactions of heredity, environment, and maturation in animal behavior.

Early biologists debated about nature versus nurture, that is, the relative importance of genes compared with environmental experience. They defined **innate behavior** (inborn behavior, popularly referred to as *instinct*) as genetically programmed, and **learned behavior** as behavior that has been modified in response to environmental experience. More recently, behavioral ecologists have recognized that no true dichotomy exists. All behavior has a genetic basis. Even the *capacity* for learned behavior is inherited. However, behavior is modified by the environment in which an animal lives; it is a product of the interaction between genetic capacity and environmental influences. Thus behavior begins with an inherited framework that experience can modify.

We can think of a range of behaviors from the more rigidly genetically programmed types, through those that, although they have a genetic component, are extensively developed through experience. The wasp *Philanthus,* discussed in the chapter introduction, efficiently carries out a complex, largely genetically programmed, sequence of behaviors. How to dig the burrow, how to cover it, how to kill the bees—these behaviors are genetically determined. Yet some of her behavior is learned. There is no way her ability to locate the burrow could be genetically programmed. Because a burrow can be dug only in a suitable spot, its location must be learned *after* it is dug. When *Philanthus* covers a nest with sand, she takes precise bearings on the location of the burrow by circling the area a few times before flying off again to hunt.

The Dutch ethologist Niko Tinbergen studied this behavior of *Philanthus.* He surrounded the wasp's burrow with a circle of pine cones as potential landmarks (Fig. 50-1). Before she returned with another bee, Tinbergen rearranged or removed the pine cones. Without them, the wasp could not find her burrow. When Tinbergen moved the pine cones to an area where there was no burrow, the female wasp responded as though the burrow were there. When the investigator completely removed them, the female seemed very confused. Only when Tinbergen restored the cones to their original location could the wasp find her burrow. When he substituted a ring of stones for the cones, the wasp responded as though the nest were in the center of the stones. Thus, *Philanthus* responds to the *arrangement* of the cones, rather than to the cones themselves. Tinbergen's findings demonstrate that for *Philanthus,* landmark learning is critical for nest locating.

Studies of fruit fly courtship and mating have provided interesting examples of interaction between genes and behavior. A ritual consisting of a complex sequence of steps, almost like a dance, must occur before mating takes place. This courtship ritual involves an exchange of visual, auditory, tactile, and chemical signals between the male and female. J.B. Hall at Brandeis University and his colleagues have identified more than a dozen

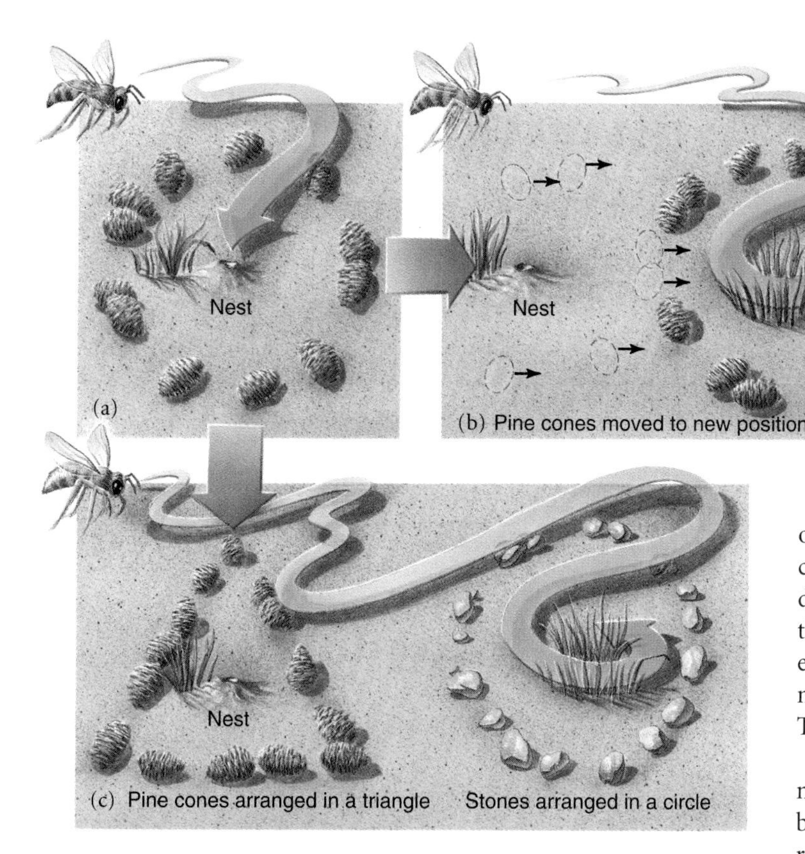

FIGURE **50-1** **Niko Tinbergen's sand wasp experiment.** Although the ability of *Philanthus* to learn is quite limited, it is adequate for most natural situations. When the ring of pine cones is moved from position **(a)** to position **(b)**, *Philanthus* behaves as if her nest were still located at the center of the ring because she learned its position in relation to the cones. **(c)** The wasp responds to the arrangement of the cones, rather than the cones themselves, as shown by the substitution of a ring of stones for cones. *(After N. Tinbergen,* Curious Naturalists, *1958, Doubleday, Garden City, New York)*

genes controlling these actions, suggesting that courtship behavior is largely inherited and preprogrammed. Nevertheless, the fruit fly has the capacity to learn from experience, a capacity that, of course, is also inherited.

The interaction between genes and environment has been studied in many vertebrates. Several species of the lovebird *(Agapornis)* differ not only in appearance but in behavior. One species uses its bill to transport small pieces of bark for building a nest. Another species tucks nest-building materials under its rump feathers. Hybrid birds attempt to tuck material under their feathers, then try to carry it in their beaks, repeating the pattern several times. Eventually most of the birds carry the material in their bills, but it takes them up to three years to perfect this behavior, and most continue to make futile attempts to tuck material into their feathers. Thus the method of transporting materials is inherited but somewhat flexible. (In nature hybrids would likely experience dramatically reduced fitness because of the long delay in breeding onset.)

Behavior depends on physiological readiness

Although behavior involves all body systems, it is influenced mainly by the nervous and endocrine systems. The capacity for behavior depends on the genetic characteristics that govern the development and functions of these systems. Before an animal can exhibit any pattern of behavior, it must be physiologically ready to produce the behavior. For example, breeding behavior does not ordinarily occur among birds or most mammals unless certain concentrations of sex hormones are present in their blood. A human baby cannot walk until its muscles and neurons are sufficiently developed. These states of physiological readiness are themselves produced by a continuous interaction with the environment. The level of sex hormones in a bird's blood may be determined by seasonal variations in day length. The baby's muscles develop with experience, as well as age.

Several factors influence the development of song in male white-crowned sparrows (generally only male songbirds sing a complex song). These birds show considerable regional variation in their song. During early development, young sparrows normally hear adult males sing the distinctive song of their population. Days 10 to 50 after hatching are a critical period for learning the song. When he is several months old, a young male sparrow "practices" the song over several weeks until he eventually sings in the local "dialect."

In laboratory experiments, birds kept in isolation and deprived of the acoustic experience of hearing the song of mature males eventually sing a very poorly developed but recognizable white-crowned sparrow song. When a young white-crowned sparrow is permitted to interact socially with a strawberry finch (which belongs to a different genus of birds), it learns the song of the finch, rather than its own species-specific song. This occurs even if the white-crowned sparrow can hear the song of other sparrows but does not interact with them socially. From these experiments, investigators have concluded that although the white-crowned sparrow is hatched equipped with a rough genetic pattern of its song, social and acoustic stimuli are both important in developing its ability to sing its specific song.

Many behavior patterns depend on motor programs

Many behaviors that we think of as automatic depend on coordinated sequences of muscle actions called **motor programs.** Some motor programs—for example, walking in newborn gazelles—seem mainly innate. Others, such as walking in human infants, have a greater learned component.

A classic example of a motor program in vertebrates is egg rolling in the European graylag goose (Fig. 50-2). Once acti-

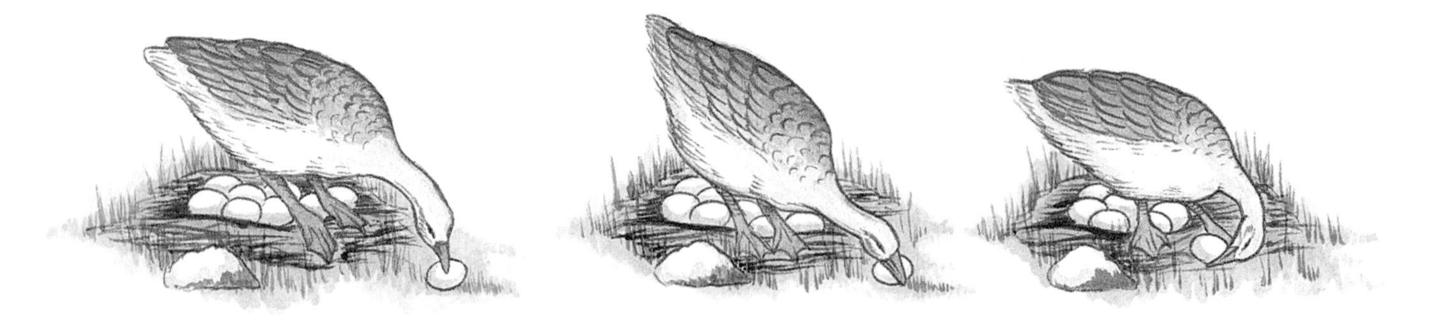

FIGURE **50-2** **Egg-rolling behavior in the European graylag goose.**

This behavior is a fixed action pattern (FAP). The goose reaches out by extending her neck and uses her bill to pull the egg back into the nest. If the investigator quickly removes the egg while the goose is in the process of reaching for it or pulling it back, she continues the FAP to completion, as though pulling the now absent egg back to the nest.

vated by a simple sensory stimulus, egg-rolling behavior continues to completion regardless of sensory feedback. There is little flexibility. Ethologists called this behavior a **fixed action pattern (FAP).** An FAP can be elicited by a **sign stimulus,** or **releaser,** a simple signal that triggers a specific behavioral response. A wooden egg is a sign stimulus that elicits egg-rolling behavior in the graylag goose. Another classic example of a sign stimulus is the red stripe on the ventral surface of a male stickleback fish. The red stripe triggers aggressive behavior by a male whose territory is being invaded. Tinbergen found that crude models painted with a red belly were more likely to be attacked than more realistic models lacking the red belly (Fig. 50-3).

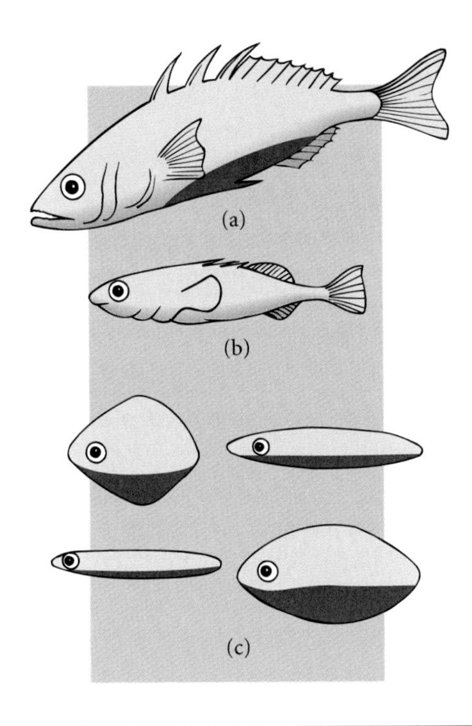

FIGURE **50-3** **A sign stimulus triggers a fixed action pattern.**

(a) A male stickleback fish will not attack **(b)** a realistic model of another male stickleback if it lacks a red belly, but **(c)** it will attack another model, however unrealistic, that has a red "belly." The aggressive behavior is triggered by the red sign stimulus rather than by recognition based on a combination of features.

Review

- Give an example showing behavior capacity is inherited and is modified by learning.
- How does physiological readiness affect innate behavior? How does it affect learned behavior?

BiologyNow™ Assess your understanding of **the interaction between genes and environment** by taking the pretest on your BiologyNow CD-ROM.

LEARNING FROM EXPERIENCE: BIOLOGICAL RHYTHMS AND MIGRATION

Learning Objective

3 Discuss the adaptive significance of habituation, imprinting, classical conditioning, operant conditioning, and insight learning.

Recall that learning is a persistent change in behavior, caused by experience. The capacity to learn appropriate responses to new situations is adaptive, enabling animals to survive as their environment changes. Learning abilities are biased; information most important to survival appears most easily learned. The same rat that may have taken a dozen trials to learn the artificial task of pushing a lever to get a reward, learns from one experience to avoid a food that has made it ill. People who poison rats to get rid of them can readily appreciate the adaptive value of this learning ability. Such quick learning in response to an unpleasant experience forms the basis of **warning coloration,** which is found in many poisonous insects and brilliantly colored, but distasteful, bird eggs. Once made ill by such a meal, predators quickly learn to avoid them. In the following sections, we consider several types of learning including habituation, imprinting, classical conditioning, operant conditioning, and insight learning.

An animal habituates to irrelevant stimuli

Habituation is a type of learning in which an animal learns to ignore a repeated, irrelevant stimulus, that is, one that neither rewards nor punishes. Pigeons gathered in a city park learn by repeated harmless encounters that humans are not dangerous to them and behave accordingly. This behavior benefits them. A pigeon intolerant of people might waste energy by flying away each time a human approached and might not get enough to eat. Many African animals habituate to humans on photo safari and to the vans that transport them (Fig. 50-4). Urban humans habituate to the noise of traffic. In fact, many urban dwellers report that they do not sleep well when they visit a quiet rural area.

Imprinting occurs during an early critical period

Anyone who has watched a mother duck with her ducklings must have wondered how she can "keep track" of such a horde of almost identical little creatures tumbling about in the grass, let alone distinguish them from those of another duck (Fig. 50-5). Although she can recognize her offspring to an extent, basically they have the responsibility of keeping track of her. The survival of a duckling requires that it quickly learn to discriminate its mother (care provider) from others.

Imprinting, a type of social learning based on early experience, has been studied in some mammals as well as birds. It occurs during a **critical period,** usually within a few hours or days after birth (or hatching). Konrad Lorenz, an Austrian physician and early ethologist, discovered that a newly hatched bird imprints on the first moving object it sees—even a human or an inanimate object such as a colored sphere or light. Although the process of imprinting is genetically determined, the bird *learns* to respond to a particular animal or object.

J.H. Dick/VIREO

FIGURE **50-5** | **Imprinting.**
Parent–offspring bonds generally form during an early critical period. These goslings have imprinted on their mother, an upland goose (*Chloephaga picta*). Photographed in the Falkland Islands.

FIGURE **50-4** | **Habituation.**
After repeated safe encounters with vans transporting humans on photo safari, many animals, including giraffes, zebras, and lions in the Serengeti, learn to ignore them. Elephants typically ignore the vans unless the driver provokes them by moving too close. In that event an elephant may challenge and even charge the van. Photographed in Tanzania.

McMurray Photography

Among many types of birds, especially ducks and geese, the older embryos can exchange calls with their nest mates and parents through the porous eggshell. When they hatch, at least one parent is normally on hand, emitting the characteristic sounds with which the hatchlings are already familiar. If the parent moves, the chicks follow. This movement plus the sounds produce imprinting. During a brief critical period after hatching, the chicks learn the appearance of the parent.

Imprinting in some mammals depends on scent. Baby shrews, for example, become imprinted on the odor of their mother (or any female nursing them). In many species, the mother also learns to distinguish her offspring during a critical period. The mother in some species of hoofed mammals, such as sheep, will accept her offspring for only a few hours after its birth. If they are kept apart past that time, the young are rejected. Normally, the mother learns to distinguish her own offspring from those of others by olfactory cues.

In classical conditioning, a reflex becomes associated with a new stimulus

In **classical conditioning,** an association is formed between some normal body function and a new stimulus. If you have observed dog or cat behavior, you know that the sound of a can opener at dinner time can captivate a pet's attention. Early in

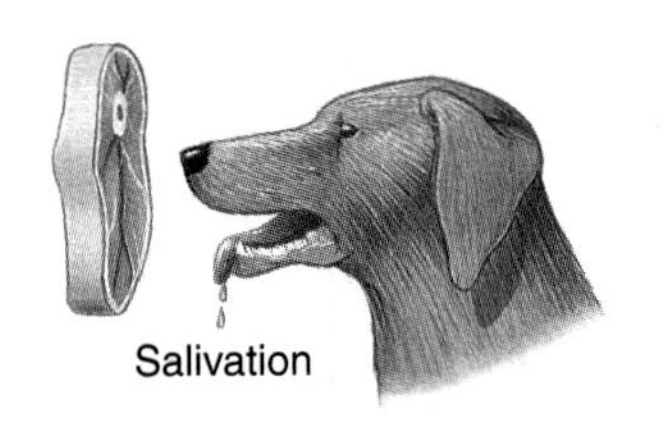
Salivation
When presented with food (the unconditioned stimulus) the dog begins to salivate.

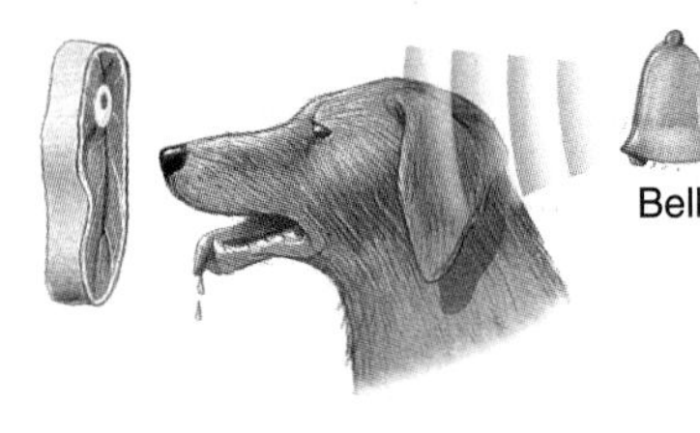
Bell
A bell (the conditioned stimulus) is rung whenever food is given to the dog. This is repeated a number of times so that an association between the food and the bell is formed.

Eventually the dog salivates at the sound of the bell alone.

ACTIVE FIGURE **50-6** | **Classical conditioning.**

Pavlov's experiment demonstrated that, through classical conditioning, dogs learn to substitute a new stimulus (the conditioned stimulus) for one that was physiologically meaningful (the unconditioned stimulus).

BiologyNow™ Learn more about **classical conditioning and other types of learning** by clicking on this figure on your BiologyNow CD-ROM. (Explore Life 26.1 Animal Behavior)

the 20th century Ivan Pavlov, a Russian physiologist, discovered that if he rang a bell just before he fed a dog, the dog formed an association between the sound of the bell and the food. Eventually (Fig. 50-6), the dog salivated even when the bell was rung in the absence of food. Pavlov called the physiologically meaningful stimulus (food, in this case) the *unconditioned stimulus.* The normally irrelevant stimulus (the bell) that became a substitute for it was the *conditioned stimulus.* Because a dog does not normally salivate at the sound of a bell, the association was clearly learned. It could also be forgotten. If the bell no longer signaled food, the dog eventually stopped responding to it. Pavlov called this latter process **extinction.**

In operant conditioning, spontaneous behavior is reinforced

In **operant conditioning,** the animal must do something to gain a reward (positive reinforcement) or avoid punishment. Operant conditioning has been studied in many animals, including flatworms, insects, spiders, birds, and mammals. In a typical laboratory experiment, a rat is placed in a cage containing a movable bar. When random actions of the rat result in pressing the bar, a pellet of food rolls down a chute to the rat. Thus the rat is positively reinforced for pressing the bar. Eventually the rat learns the association and presses the bar to obtain food.

In negative reinforcement, removal of a stimulus increases the probability that a behavior will occur. For example, a rat may be subjected to an unpleasant stimulus, such as a low-level electric shock. When the animal presses a bar, this negative reinforcer is removed.

Many variations of these techniques have been developed. A pigeon may be trained to peck at a lighted circle to obtain food, a chimpanzee may learn to perform some task to get tokens that can be exchanged for food, or children may learn to stay quietly in their seats at school to obtain the teacher's praise. Operant conditioning is probably the way animals learn to perform complex tasks such as perfecting feeding skills.

Operant conditioning plays a role in the development of some behaviors that appear genetically programmed. An example is the feeding behavior of gull chicks. Herring gull chicks peck the beaks of the parents, which stimulates the parents to regurgitate partially digested food for them. The chicks are attracted by two stimuli: the general appearance of the parent's beak with its elongated shape and distinctive red spot, and its downward movement as the parent lowers its head. Like the rat's chance pressing of the bar, their initial exploratory pecking behavior is sufficiently functional to get the chicks their first meal, but they waste a lot of energy in pecking. Some pecks are off target and are therefore not rewarded. However, pecking behavior becomes more efficient with practice. Thus a behavior

FIGURE **50-7** | **Operant conditioning.**

With the experience of being positively reinforced for success, the pelican chick learns to be more accurate in begging for food from a parent.

H. Cruickshank/VIREO

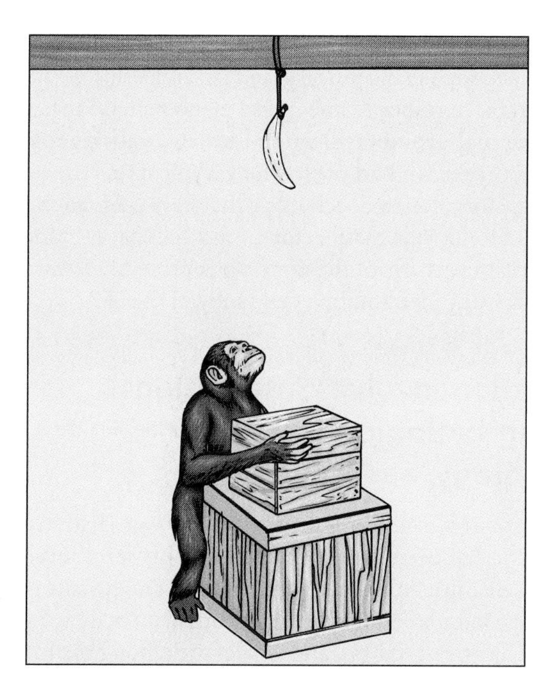

FIGURE **50-8** **Insight learning.**
Confronted with the problem of reaching food hanging from the ceiling, a chimpanzee may stack boxes until it can climb and reach the food. What former experience might the chimp be applying to this new situation?

that might appear to be entirely genetic is improved by learning (Fig. 50-7).

Insight learning uses recalled events to solve new problems

Perhaps the most complex learning is **insight learning,** which is the ability to adapt past experiences that may involve different stimuli to solve a new problem. A dog can be placed in a blind alley that it must circumvent to reach a reward. The difficulty lies in the fact that the animal must move *away* from the reward to get to it. Typically, the dog flings itself at the barrier nearest the food. Eventually, by trial-and-error, the frustrated dog may find its way around the barrier and reach the reward.

In contrast with the dog, a chimpanzee placed in similar situations is likely to make new associations between tasks it has learned previously to solve the problem (Fig. 50-8). Primates are especially skilled at insight learning, but some other mammals and a few birds also seem to have this ability to some degree.

Play may be practice behavior

Perhaps you have watched a kitten pouncing on a dead leaf or practicing a carnivore neck bite or a hind-claw disemboweling stroke on a littermate without causing injury. Many animals, especially young birds and mammals, appear to practice adult

FIGURE **50-9** **Cheetah cubs playing.**
Play may serve as a means of practicing behavior that will be used in earnest later in life, possibly in hunting, fighting for territory, or competing for mates. Photographed in Kenya.

patterns of behavior while they play (Fig. 50-9). They may improve their ability to escape, kill prey, or perform sexual behavior. Play may be an example of operant conditioning in action.

Some investigators have suggested that young animals play just to have "fun." Dolphins seem to play just for pleasure. For example, bottlenose dolphins swirl water with their fins, then blow bubbles to produce rings and helices of air. Dolphins develop more complex play behavior over time. Hypotheses for the ultimate causes of play behavior include exercise, learning to coordinate movements, and learning social skills. Of course, there could be some other explanation, yet unknown.

Review

- How is imprinting adaptive?
- How is operant conditioning adaptive?

BiologyNow™ Assess your understanding of **learning from experience** by taking the pretest on your BiologyNow CD-ROM.

BIOLOGICAL RHYTHMS AND MIGRATION

Learning Objectives

4 Give examples of biological rhythms, and describe some of the mechanisms responsible for them.

5 Analyze costs and benefits of migrations, and distinguish between directional orientation and navigation.

Biologists have identified many types of biological rhythms, including daily, monthly, and annual rhythms. Among many animals, physiological cycles such as body temperature fluctuations and hormone secretion are rhythmic. Human body temperature, for example, follows a typical daily curve. Biological rhythms control many behaviors, including activity, sleep, feeding, drinking, and migration.

Biological rhythms affect behavior

The behavior of many animals, like the activities of many plants (see Chapter 36), is organized around **circadian** (meaning "approximately one day") **rhythms,** which are daily (24-hour) cycles of activity. **Diurnal** animals, such as honeybees and pigeons, are most active during the day. Most bats, moths, and cats are **nocturnal** animals, most active during the hours of darkness. **Crepuscular** animals, like many mosquitoes and fiddler crabs, are busiest at dawn or dusk, or both. Generally, there are ecological reasons for these patterns. If an animal's food is most plentiful in the early morning, for example, its cycle of activity must be regulated so that it becomes active shortly before dawn.

Some biological rhythms of animals reflect the **lunar** (moon) **cycle.** The most striking rhythms are those in marine organisms that are attuned to changes in tides and phases of the moon. For instance, a combination of tidal, lunar, and annual rhythms governs the reproductive behavior of the grunion, a small fish that lives off the Pacific coast of North America. Grunions swarm from April through June on those three or four nights when the highest tides of the year occur. At precisely the high point of the tide, the fish squirm onto the beach and deposit eggs and sperm in the sand. They return to the sea in the next wave. By the time the next tide reaches that portion of the beach 15 days later, the young fish have hatched in the damp sand and are ready to enter the sea. This synchronization may help protect fish eggs from aquatic predators.

An animal's metabolic processes and behavior are typically synchronized with the cyclic changes in its external environment. Its behavior anticipates these regular changes. The little fiddler crabs of marine beaches often emerge from their burrows at low tide to engage in social activities such as territorial disputes. To avoid being washed away, they must return to their burrows before the tide returns. How do the crabs "know" that high tide is about to occur? One might guess that the crabs recognize clues present in the seashore. However, when the crabs are isolated in the laboratory away from any known stimulus that could relate to time and tide, their characteristic behavioral rhythms persist.

Many biological rhythms are regulated by *internal* timing mechanisms that serve as **biological clocks.** As illustrated by the fiddler crabs, these timing mechanisms do not simply respond to environmental cues, but are capable of sustaining biological rhythms independently. Such internal clocks have been identified in almost every eukaryote, as well as some bacteria. Molecular biologists have demonstrated that genes control biological clocks. In *Drosophila,* seven genes produce clock proteins that seem to interact in feedback loops in many cell types both inside and outside the nervous system. Researchers have identified similar genes and proteins in many animal groups, including mammals. The principal clock is located in specific areas of the central nervous system.

In mammals, the master clock is located in the **suprachiasmatic nucleus (SCN)** in the hypothalamus. This SCN clock generates approximately 24-hour cycles even without input from the environment. However, the SCN-generated cycles are normally adjusted every day based on visual input received from the retina. Signals from the retina reflecting changes in light intensity adjust internal circadian rhythms to light–dark cycles in the environment. The SCN sends rhythmic signals, in the form of neuropeptides, to the **pineal gland,** an endocrine gland located in the brain. In response, the pineal gland secretes melatonin, a hormone that promotes sleep in humans. Clock genes in the SCN are turned on and off by the very proteins they encode, setting up complex feedback loops that have a 24-hour cycle.

In addition to the master clock, most cells have timing mechanisms that use many of the same clock proteins. These peripheral clocks may function independently of the SCN.

Migration involves interactions among biological rhythms, physiology, and environment

Ruby-throated hummingbirds cross the vast distance of the Gulf of Mexico twice each year, and the sooty tern travels across the entire South Atlantic from Africa to reach its tiny island breeding grounds south of Florida. Birds, butterflies, fishes, sea turtles, wildebeest, zebras, and whales are among the many animals that travel long distances. Behavioral ecologists define **migration** as a periodic long-distance travel from one location to another. Many migrations involve astonishing feats of endurance and navigation.

Why do animals migrate? Ultimate causes of migration apparently involve the advantages of moving from an area that seasonally becomes less hospitable to a region more likely to support reproduction or survival. Seasonal changes include shifts in climate, availability of food resources, and safe nesting sites. For example, as winter approaches, many birds migrate to a warmer region.

An interesting example of migration is the annual journey of millions of monarch butterflies (*Danaus plexippus*) from Canada and the continental United States to Mexico, a journey of 2500 km (about 1500 mi) for some. Their dramatic annual migration appears related to the availability of milkweed plants on which females lay their eggs. On hatching, the larvae (caterpillars) feed on the leaves of the milkweed plant. In cold regions these plants die in late autumn and grow again with the warmer weather of spring. The wintering destinations of monarchs also may be determined by temperature and humidity. Possible benefits include the opportunity to winter in an area that offers an abundant food supply, nonfreezing temperatures, and moist air. Thus, the investment made by migrating monarch butterflies in their long journey increases their chances of surviving the winter.

The benefits of migration are not without costs in time, energy, and even survival. Many weeks may be spent each year on energy-demanding journeys. Some animals may become lost or die along the way from fatigue or predation. When in unfamiliar areas, migrating individuals are often at greater risk from predators. In recent years, human activities have interfered with migrations of many kinds of animals. For example, after millions of years of biological success, survival of sea turtles is threatened by fishing and shrimping nets in which they become entangled. As a result, thousands of migrating turtles drown each year. Sea turtles also die as a result of ingesting floating plastic bags they mistake for jellyfish.

Proximate causes of migration include signals from the environment that trigger physiological responses leading to migration. In migratory birds, for example, the pineal gland senses changes in day length and then releases hormones that cause restless behavior. The birds show an increased readiness to fly, and to fly for longer periods.

How do migrating animals find their way? The term **directional orientation** refers to travel in a specific direction. To travel in a straight line toward a destination requires a sense of direction, or **compass sense.** Many animals use the sun to orient themselves. Because the sun appears to move across the sky each day, an animal must have a sense of time. Biological clocks seem to sense time and regulate circadian rhythms.

Navigation is more complex, requiring both compass sense and **map sense,** which is an "awareness" of location. Navigation involves the use of cues to change direction when necessary to reach a specific destination. When navigating, an animal must integrate information about distance and time, as well as direction.

DNA tests confirm that loggerhead sea turtles that hatch on Florida beaches along the Atlantic swim hundreds of miles across the ocean to the Mediterranean Sea, an area rich in food. Several years later, those that survive to become adults mate, and the females navigate back, often to the same beach, to lay their eggs. Their journey requires both compass and map sense.

Marine biologist Kenneth Lohmann, of the University of North Carolina, fitted hatchling turtles with harnesses connected to a swivel arm in the center of a large tank (Fig. 50-10). A computer connected to the swivel arm recorded the turtles' swimming movements. Manipulating light and magnetic fields, Lohmann demonstrated that both these environmental cues are important in turtle migration. The turtles swam toward the east until Lohmann reversed the magnetic field. The turtles reversed their direction to swim toward the new "magnetic east," which was now actually west. Recent work suggests that young turtles use wave direction to help set magnetic direction preference. Some biologists suspect that turtles also use their sense of smell to guide them, particularly to guide the females to the very same beach to lay their eggs. Similarly, adult salmon use the unique odors of different streams to help find their way back to the same stream from which they hatched.

Birds and some other animals that navigate by day rely on the position of the sun (our local star); those that travel at night use the stars to guide them. When birds see the star patterns of the night sky, they seek to fly in the appropriate migration direction for their species, even when they have had no opportunity to learn this feat from other birds.

Studies by Stephen Emlen of Cornell University showed that young indigo buntings learn the constellations, using the position of the North Star as their reference point. (Other stars in the Northern Hemisphere appear to rotate about the North Star.) When Emlen rearranged the constellations in a planetarium sky, birds learned the altered patterns of stars and later attempted to fly in a direction consistent with the artificial pattern. These and other studies suggested that birds have a genetic ability to learn constellation patterns and use them to orient themselves during migration.

Investigators have long observed that some species of birds navigate even when the sky is overcast and they cannot see the stars. They hypothesized that these birds use Earth's magnetic field to navigate. Studies of garden warblers indicate that, as these birds make their way from central Europe to Africa each winter, they navigate both by the stars and by Earth's magnetic field. The warblers use stars to determine the general direction of travel, then use magnetic information to refine and correct their course. In addition to birds and sea turtles, honeybees, some fishes, and some other animals are sensitive to Earth's magnetic field and use it as a guide.

Kenneth Lohmann

FIGURE **50-10** | **Navigation by light and magnetic field.**
To study turtle navigation, researcher Kenneth Lohmann harnessed leatherback turtles (*Dermochelys coriacea*) such as this hatchling and wired them to a computer that recorded their swimming direction.

Review

- Why is it adaptive for some species to be diurnal but for others to be nocturnal or crepuscular?
- What is the difference between directional orientation and navigation?

BiologyNow™ Assess your understanding of **biological rhythms and migration** by taking the pretest on your BiologyNow CD-ROM.

FORAGING BEHAVIOR

Learning Objective

6 Discuss the hypothesis that optimal foraging behavior is adaptive.

Feeding behavior, or **foraging,** involves locating and selecting food, as well as food gathering and food capture. Some behavioral ecologists study the costs and benefits of searching for and selecting certain types of food, as well as the mechanisms used to locate prey. For example, many camouflage strategies have evolved that make potential prey difficult to detect. As predators experience repeated success in locating a particular prey species, they are thought to develop a *search image,* a constellation of cues that help them identify hidden prey.

Why do grizzly bears spend hours digging Arctic ground squirrels out of burrows while ignoring larger prey such as caribou? It is more energy efficient to dig for squirrels because the bears' efforts most probably will be rewarded, whereas caribou are more likely to escape, leaving the bears hungry. This is an example of **optimal foraging,** the most efficient way for an animal to obtain food. When animals maximize energy obtained per unit of foraging time, they may maximize their reproductive success. Many factors, such as avoiding predators while foraging, must be considered in determining efficient or optimal strategies.

In habitats where the most preferred food items are abundant and an animal does not have to travel far to obtain them, animals can afford to be very selective. In contrast, the optimal strategy in poor habitats, where it takes longer to find the best food items, is to select less preferred items. Animals may learn to forage efficiently through operant conditioning, that is, by randomly trying various strategies and selecting the one associated with the most rewards (and the fewest hunger pangs at the end of the day).

Foraging is also affected by the forager's risk of predation. U.K. biologist Guy Cowlishaw studied foraging behavior in a population of baboons in Namibia. He found that the baboons spent more time foraging in a habitat in which food was relatively scarce than in an area with more abundant food that had a high risk of predation by lions and leopards.

Lions live, and often forage, in social units called *prides.* A pride typically consists of a group of related adult females, their cubs, and unrelated males. Biologist Craig Packer, of the University of Minnesota, and his research team have studied lion behavior in Serengeti National Park, Tanzania. Packer has radio-collared at least one female in each of 21 prides and has tracked them for several years. During the season when prey is abundant, lions hunt wildebeest, gazelle, and zebra that have migrated into the area. When these herds migrate out of the area during the dry season, these prey become scarce, and lions feed mainly on warthog and Cape buffalo.

Packer has reported that when prey is abundant, the size of the foraging group has little effect on daily food intake. Whether a lion hunts individually or in small (two to four females) or large (five to seven) groups, food is plentiful and is captured by individuals or groups of any size. During the season when prey is scarce, however, lions are more successful if they hunt alone or in large groups (Fig. 50-11). Nevertheless, Packer's data from radio-collared lions indicated that females typically forage in as large a group as they can, even if the group consists of three or four females, an approach that decreases foraging efficiency. Thus selection pressure seems to be stronger to protect cubs and defend territory against larger prides than to forage maximally (alone).

FIGURE **50-11** | **Optimal foraging and group size in lions.**

During times of prey scarcity, optimal foraging is related to group size: Lions hunt most successfully alone or in large groups of five to seven. Observations of group size suggest that the benefits of optimal defense often outweigh the benefits of optimal foraging. *(Based on data of Craig Packer: C. Packer, D. Scheel, and A. E. Pusey, "Why Lions Form Groups: Food Is Not Enough,"* American Naturalist, *Vol. 136, Jul. 1990)*

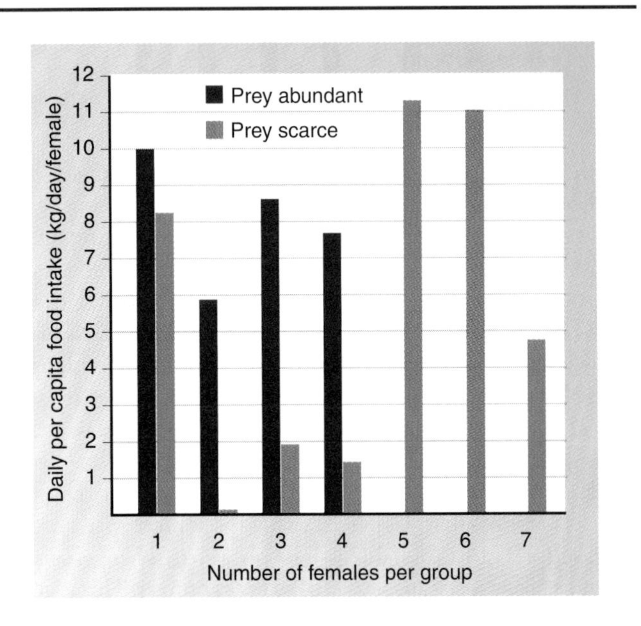

Review

- What is optimal foraging?
- How is optimal foraging adaptive?

Biology Now™ Assess your understanding of **foraging behavior** by taking the pretest on your BiologyNow CD-ROM.

SOCIAL BEHAVIOR

Learning Objectives

7 Analyze social behavior in terms of costs and benefits, and describe modes of animal communication.

8 Describe the concept of a dominance hierarchy, and discuss its adaptive significance and social function.

9 Distinguish between home range and territory, and describe costs and benefits of territoriality.

10 Contrast a society of social insects with a vertebrate society, and give examples of cultural variation in vertebrate populations.

The mere presence of more than one individual does not mean a behavior is social. Many factors of the physical environment bring animals together in **aggregations,** but whatever interaction they experience may be circumstantial. A light shining in the dark is a stimulus that draws large numbers of moths, and the high humidity under a log attracts wood lice. Although these aggregations may have adaptive value, they are not truly social, because the animals are not responding to one another.

We can define **social behavior** as the interaction of two or more animals, usually of the same species. Many animals benefit from living in groups. By cooperation and division of labor, some insects construct elaborate nests and raise young by mass-production methods. Schools of fishes tend to confuse predators, and so individuals within the school may be less vulnerable to predators than a solitary fish would be (see Fig. 51-1b). Zebras can more effectively protect themselves when they are in groups (Fig. 50-12). Zebra stripes appear to be a visual antipredator adaptation. When viewed from a distance, the stripes tend to visually break up the form of the animal so that individuals cannot be distinguished. A herd of zebras confuses predators, whereas a solitary animal is easy prey for lions, cheetahs, or spotted hyenas.

McMurray Photography

FIGURE 50-12 **Social behavior in zebras.**
The social unit consists of a fairly stable group of females with young and a dominant stallion. The stripes of a group of zebras confuse predators. Photographed in East Africa.

R. Lindholm/Visuals Unlimited

(a)

April Ottey, Chimpanzee and Human Communication Institute, Central Washington University

(b)

FIGURE 50-13 **Animal communication.**
(a) A male spring peeper frog *(Hyla crucifer)* calling to locate a mate. **(b)** Communicating with language. Chimpanzee Tatu *(top)* is signing "food" to Washoe *(below)*, the first chimp to learn American Sign Language from a human. Washoe then taught other chimps how to sign.

Social foraging is an adaptive strategy used routinely by many animal species. A pack of wolves has greater success in hunting than an individual wolf would have. Among some birds of prey, such as certain hawks, group hunting locates prey more quickly.

Social behavior offers benefits that increase the chances of perpetuating the genes that produce such behavior. However, social behavior also has certain costs. Living together means increased competition for food and habitats, increased risk of attracting predators, and increased risk of transmitting disease.

Communication is necessary for social behavior

One animal can influence the behavior of another only if the two of them can exchange mutually recognizable signals (Fig. 50-13). **Communication** is most evident when one animal performs an act that changes the behavior of another. Communication may be important in finding food, as in the elaborate dances of honeybees. Animals may communicate to hold a group together, warn of danger, signal social status, indicate willingness to accept or provide care, identify members of the same species, or indicate sexual maturity or readiness.

Animals use auditory, visual, tactile, chemical, and electrical signals to transmit information to one another. In many bird species, territorial males announce their presence and willingness to interact socially by singing. Along with their songs, many birds present visual displays. Some birds respond to a particular sound by matching it. Among nonhuman mammals, only bottlenose dolphins are known to match sounds in communicating. These dolphins respond to the whistle of a member of its own species by imitating and emitting the same sound. Researchers suggest that vocal matching was important in the evolution of human language.

Orcas (formerly known as killer whales) communicate by sounds and songs. Members of a given pod (social group) have an average of 12 different calls, which vary in pitch and duration and appear to reflect their "emotional" state. Certain fishes (gymnotids) use electric pulses for navigation and communication, including territorial threat, in a fashion similar to bird vocalization. As Harvard sociobiologist Edward O. Wilson has said, "The fish, in effect, sing electrical songs."

Some animals communicate by scent

Dogs and wolves mark territory by frequent urination. Antelopes, deer, and cats rub facial gland secretions on conspicuous objects in their vicinity and urinate on the ground. **Pheromones** are chemical signals secreted into the environment that convey information between members of a species. These small volatile molecules provide a simple, widespread means of communication. Animals use pheromones to communicate danger, ownership of territory, and availability for mating. For example, female moths release pheromones that attract males. Ants mark their trails with pheromones.

Most pheromones elicit a very specific, immediate, but transitory type of behavior. Others trigger hormonal activities that result in slow but long-lasting responses. Some pheromones may act in both ways. An advantage to pheromone communication is that relatively little energy is needed to synthesize the simple, but distinctive, organic compounds involved. Members of the same species have receptors that fit the molecular configuration of the pheromone; other species usually ignore it or do not detect it at all. Pheromones are effective in the dark, they can pass around obstacles, and they last for several hours or longer. Major disadvantages of pheromone communication are slow transmission and limited information content. Some animals compensate for the latter disadvantage by secreting different pheromones with different meanings.

Pheromones are important in attracting the opposite sex and in sex recognition in many species. Many female insects produce pheromones that attract males, and these chemical signals govern the reproduction of many social insects. Humans have taken advantage of some sex-attractant pheromones to help control pests such as gypsy moths by luring the males to traps baited with synthetic versions of female gypsy-moth pheromones. In honeybees, certain fatty acids are mixed with hydrocarbons from wax glands, and this mixture is transferred onto worker bees when they touch the comb. This mixture serves as a pheromone that identifies all the bees that belong to a particular hive. Should a bee from another hive attempt to enter, guard bees sting and may even kill the foreigner.

In vertebrates, pheromones affect sexual cycles and reproductive behavior, including choice of a mate, and may play a role in defending territory. Among some mammals, an ovulating female (one physiologically ready to mate) releases pheromones as part of her vaginal secretion. When males detect these chemical odors, their sexual interest increases. When the odor of a male mouse is introduced among a group of females, the reproductive cycles of the female mice synchronize. In some species of mice, the odor of a strange male, a sign of high population density, causes a newly impregnated female to abort.

The extent to which humans respond to pheromones is the focus of research. One interesting finding suggests that an unconsciously perceived body odor can synchronize the menstrual cycles of women who associate closely (for instance, college roommates or cellmates in prison). A recent study of the effects of human steroids used in commercial fragrances, such as perfumes, suggests that such compounds may act as subtle signals that modulate mood and behavior.

As discussed in Chapter 41, mammals detect pheromones with specialized chemoreceptor cells that make up the **vomeronasal organ** in the epithelium of the nose. The vomeronasal sensory neurons signal the amygdala and hypothalamus, brain structures that regulate emotional responses and certain endocrine processes. When neurons of the vomeronasal system are damaged in virgin mice, they do not mate. Biologists have identified about 100 genes that code for pheromone receptors in the mouse and rat. These receptors initiate **signal transduction** processes that involve G proteins.

Dominance hierarchies are social rankings

In the spring, female paper-wasps awaken from hibernation and begin to build a nest together. During the early course of construction, a series of squabbles among the females takes place in which the combatants bite one another's bodies or legs. Finally, one of the wasps emerges as dominant. After that, she is rarely challenged. This queen wasp spends more and more time tending the nest and less and less time out foraging for herself. She takes the food she needs from the others as they return.

The queen then begins to take an interest in raising a family—her family. Because she is almost always in the nest, she can prevent other wasps from laying eggs in the brood cells by rushing at them, jaws agape. Because of her supreme dominance, the queen can bite any other wasp without serious risk of retaliation.

The other wasps of the nest are further organized into a **dominance hierarchy,** a ranking of social status in which each wasp has more status than the wasps that are lower in rank. Wasps lower in the hierarchy are subordinate to those above them, as follows:

Queen ⟶ Wasp A ⟶ Wasp B ⟶ . . . Wasp M ⟶ Wasp N

Once a dominance hierarchy is established, little or no time is wasted in fighting. When challenged, subordinate wasps take submissive poses that, in turn, inhibit the queen's aggressive behavior. Consequently, few or no colony members are lost through wounds sustained in fighting one another. This ensures greater reproductive success for the colony.

In many species, males and females have separate dominance systems. For example, female, as well as male, chimpanzees establish dominance hierarchies. Some females apparently become dominant through aggressive behavior, whereas others achieve high rank by virtue of their mother's status. Dominant females are more successful reproductively than those with lower social status. In many monogamous animals, especially birds, the female acquires the dominance status of her mate by virtue of their relationship and the male's willingness to defend his mate.

Like many fishes and some invertebrates, certain coral reef fishes (labrids) are capable of sex reversal. The largest, most dominant individual of a group is always male, and the remaining fish within his territory are all female. If the male dies or is removed, the most dominant female becomes the new male. Should any harm come to him, the next-ranking female undergoes sex reversal, takes charge, and protects the group territory. Other fishes exhibit the reverse behavior, in which the most dominant fish is always female. In such species, size is less important for aggressive defense than for maximal egg production.

In establishing dominance, males expend energy in posturing, roaring, leaping about, or sometimes fighting fiercely. These behaviors appear to be a test of male quality. The male with the greatest endurance will likely gain dominance and will have the greatest opportunity to mate and to perpetuate his genes. In many animals, social dominance is a function of aggressiveness (Fig. 50-14). Tremendous energy is often needed for the fights engaged in by some birds and many mammals, including ba-

KEY CONCEPT: In many social groups, animals signal other members to establish and maintain dominance.

Gerald Lacz/Peter Arnold, Inc.

FIGURE **50-14** **Communicating dominance.**

This baboon (*Papio*) bares his teeth and screams in an unmistakable show of aggression, a signal that allows him to establish and maintain dominance.

boons, rams, and elephant seals. Establishing dominance is often strenuous and dangerous.

Sex hormones, particularly the male reproductive hormone testosterone, increases aggressiveness. The female hormone estrogen sometimes reduces dominance. Among chickens, the rooster is the most dominant. If a hen receives testosterone injections, her place in the dominance hierarchy shifts upward. When male rhesus monkeys are dominant, their testosterone levels are much higher than when they have been defeated. Not only can testosterone increase dominance, but dominance may even increase testosterone production. It is not always easy to determine cause and effect. (Studies suggest that the effects of testosterone on human mood and behavior may be different from that in other vertebrates.)

Many animals defend a territory

Most animals have a **home range,** a geographic area they seldom leave (Fig. 50-15). Because the animal has the opportunity to become familiar with everything in that range, it has an advantage over its competitors, predators, and prey in negotiating the terrain and finding food. Some, but not all, animals exhibit **territoriality.** They defend a **territory,** a portion of the home range, often against other individuals of their species and sometimes against individuals of other species. Many species are territorial for only part of the year, often during the breeding season, but others are territorial throughout the year. Territoriality has been positively correlated with the availability of needed resources that occur in small areas that can be defended.

Territoriality is easily studied in birds. Typically, the male chooses a territory at the beginning of the breeding season. This behavior results from high sex hormone concentrations in his blood. The males of adjacent territories fight until territorial boundaries become established. Generally, the dominance of a male is directly associated with how close he is to the center of his territory. Close to home he acts like a lion, but when invading some other bird's territory, he may behave more like a lamb. Sometimes males respect a neutral line, an area between their territories in which neither is dominant.

Bird songs announce the territory and often substitute for fighting. Songs also announce to eligible females that a propertied male resides in the territory. Typically, male birds take up a conspicuous station, sing, and sometimes display striking patterns of coloration or aerial acrobatics to their neighbors, their rivals, and sometimes their mates.

The costs of territoriality include the time and energy expended in staking out and defending a territory and the risks

McMurray Photography

FIGURE **50-15**

Home range.

The Cape buffalo (*Syncerus caffer*) lives in herds of several hundred animals. Each herd, such as this one photographed in the Serengeti, has a fairly constant home range. Buffalo protect herd members, especially calves.

involved in fighting for it. Benefits often include exclusive rights to food within the territory and greater reproductive success. The males of a lion pride father all the cubs born within that pride. Their ability to defend their territory from invasion by strange males ensures their reproductive success. Among many species, animals that fail to establish territories fail to reproduce. Territoriality also tends to reduce conflict among members of the same species and ensures efficient use of environmental resources by encouraging individuals to spread throughout a habitat.

Usually, territorial behavior is related to the specific lifestyle of the animal and to whatever aspect of its environment is most critical to its reproductive success. For instance, sea birds may range over hundreds of square kilometers of open water but exhibit territorial behavior only at crowded nesting sites on an island. The nesting sites are their scarcest resource and the one for which competition is keenest.

Treat Davidson/Photo Researchers, Inc.

FIGURE **50-16** | **Maintaining a complex social structure.** Numerous worker honeybees surround their queen *(center)*. Workers constantly lick secretions from the queen bee. These secretions, transmitted throughout the hive, suppress the activity of the workers' ovaries.

Some species that engage in social behavior form societies

A **society** is an actively cooperating group of individuals belonging to the same species and often closely related. A hive of bees, a flock of birds, a pack of wolves, and a school of fish are examples of societies. Some societies are loosely organized, whereas others have complex structures. Characteristics of a highly organized society include cooperation and division of labor among animals of different sexes, age groups, or castes. A complex system of communication reinforces the organization of the society. The members tend to remain together and to resist attempts by outsiders to enter the group.

Social insects form elaborate societies

Many insects, such as tent caterpillars, which spin communal nests, cooperate socially. However, the most elaborate insect societies are formed by bees, ants, wasps, and termites. (The first three of these all belong, not coincidentally, to the same order, Hymenoptera.) Insect societies are held together by a complex system of sign stimuli that are keyed to social interaction, and their behaviors tend to be quite rigid.

The social organization of honeybees has been studied more extensively than that of any other social insect. Instructions for the honeybee society are inherited and preprogrammed, and the size and structure of the bee's nervous system permit only a limited range of behavioral variation. Yet these insects are not automatons. Within those limits, the complex bee society can respond with some flexibility to food and other stimuli in the environment.

A honeybee society generally consists of a single adult queen, up to 80,000 worker bees (all female), and, at certain times, a few males called drones that fertilize newly developed queens. The queen, the only female in the hive capable of reproduction, deposits about 1000 fertilized eggs per day in the wax cells of a comb.

The composition of a bee society is generally controlled by a pheromone secreted by the queen. It inhibits the workers from raising a new queen and prevents development of ovaries in the workers (Fig. 50-16). If the queen dies, or if the colony becomes so large that the inhibiting effect of the pheromone dissipates, the workers begin to feed some larvae special food that promotes their development into new queens.

The queen bee stores sperm cells from previous matings in a seminal receptacle. If she releases sperm to fertilize an egg as it is laid, the resulting offspring is female; otherwise, it is male. Thus, males develop parthenogenesis from unfertilized eggs and are haploid. Because a drone is haploid, each of his sperm cells has *all* his chromosomes; that is, meiosis does not occur during sperm production. The queen bee stores this sperm throughout her lifetime and uses it to produce worker bees.

The worker bees of a hive are more closely related to one another than would be sisters born of a diploid father. Indeed, they have up to three quarters of their genes in common. (Assuming the queen bred with a single drone, they share half of the queen's chromosomes and all of the drone's.) As a consequence, they are more closely related to one another than they would be to their own offspring, if they could have any. (A worker bee's offspring would have only half of its genes in common with its worker mother.) New queens are also their sisters. Worker bees are therefore more likely to pass on copies of their genes to the next generation by raising these individuals than they would if they were to produce their own offspring.

Division of labor among worker bees is mostly determined by age. The youngest worker bees serve as nurses that nourish larval bees. After about a week as nurse bees, workers begin to produce wax and build and maintain the wax cells. Older workers are foragers, bringing home nectar and pollen. Most worker bees die at the ripe old age of 42 days.

The most sophisticated known mode of communication among bees (and among other nonmammals) is a stereotyped series of body movements called a *dance*. The dance re-enacts, in miniature, the bee's flight to the food. When a honeybee scout locates a rich source of nectar, it apparently observes the angles among the food, hive, and the sun. The bee then communicates the direction and distance of the food source relative to the hive (Fig. 50-17). If the food supply is within about 50 m (about 55 yd), the scout performs a **round dance,** which generally excites the other bees and prompts them to fly short distances in all directions from the hive until they find the nectar. If the food is distant, however, the scout performs a **waggle dance,** which follows a figure-eight pattern.

In the 1940s the German zoologist Karl von Frisch pioneered studies in bee communication. He found that the waggle dance conveys information about both distance and direction. The bee typically performs the dance in the hive on the vertical surface of the comb. During the straight run part of the figure eight, the number and frequency of the waggles indicate the distance. The orientation of the movements indicates the direction of the food source in reference to the position of the sun. For example, if the bee dances straight up, the nectar is located directly toward the sun. If the bee dances 40 degrees to the left of the vertical surface of the comb, the nectar is located 40 degrees to the left of a line between the hive and the sun.

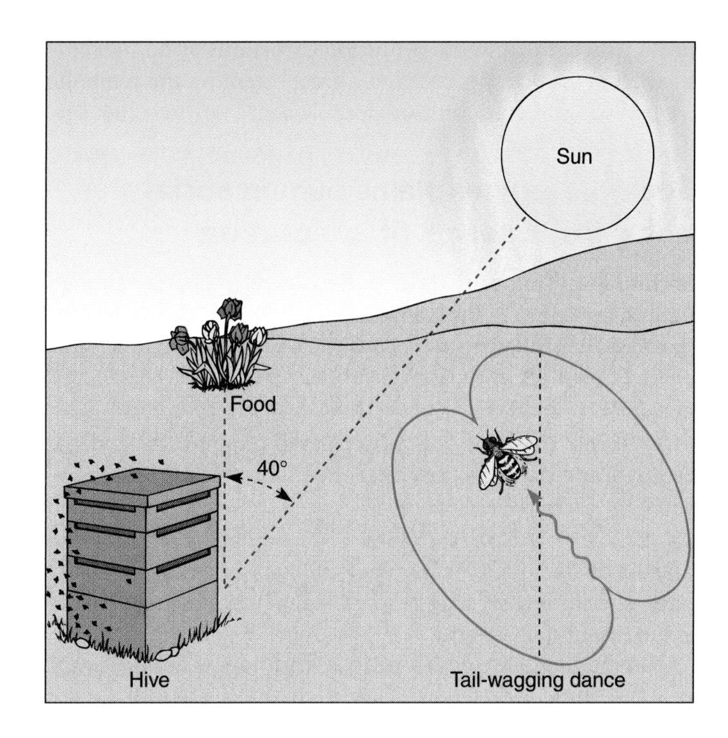

ACTIVE FIGURE **50-17** | **The honeybee waggle dance.**

The scout is waggling upward, indicating that the food is toward the sun, and inclined 40 degrees to the left, which reveals the angle of the food source relative to the sun. The dance takes place inside the hive.

BiologyNow™ See **honeybee dances in action** by clicking on this figure on your BiologyNow CD-ROM. (Non-Starr Media: honeybee_dances. dcr)

Vertebrate societies tend to be relatively flexible

Vertebrate societies usually have nothing comparable to the physically and behaviorally specialized castes of honeybees or ants. An exception is the naked mole rat of southern Africa, a rodent that has a social structure closely resembling that of the social insects. Although most vertebrate societies seem simpler than insect societies, they are also more flexible. Vertebrate societies share a great range and plasticity of potential behaviors and can effectively modify behavior to meet environmental challenges.

PROCESS OF SCIENCE

The behavioral plasticity of vertebrates makes possible the transmission of culture in some bird and mammal species. **Culture** is behavior common to a population, learned from other members of the group, and transmitted from one generation to another. Culture is not inherited genetically. It is maintained by social learning—for example, through imitation or teaching. Whether culture is present in nonhumans is controversial among behavioral ecologists. Researchers base their conclusions on careful observation and deductive reasoning. When they cannot account for differences in behavior among populations by genetic or environmental factors, they conclude that the behaviors are being transmitted culturally.

About 17 behaviors that may be considered cultural have been described among cetaceans (whales, dolphins, porpoises). For example, female orcas teach their offspring to hunt seals according to the custom of their particular group. Orcas also have local dialects that have been documented for at least six generations, suggesting that they are passed on from parent to offspring. Separate dialects are maintained even when various populations of orcas interact socially. Bottlenose dolphins are well known for their ability to learn by imitation. Behavioral ecologists have documented foraging practices that appear to be cultural—transmitted by social learning.

Recent studies of nonhuman primates demonstrate that some groups develop local customs and teach them to their offspring. Using data provided by researchers at seven chimpanzee research sites, animal behaviorists identified 39 chimpanzee behaviors that they considered cultural variations. These behaviors included various ways of using tools, courtship rituals, and grooming techniques. Each of these local customs was learned from other chimpanzees in the population, and customs varied among different populations. For example, chimpanzees on the western side of the Sassandra-N'Zo River use stone hammers to crack coula nuts. Researchers have photographed mother chimpanzees teaching this behavior to their offspring. Just a few miles away on the eastern side of the river, chimpanzees do not crack nuts even though they are available. When a chimp migrates and joins a new population, she can transmit knowledge learned in her previous culture. In this way, customs spread from one population to another.

As researchers continue to study cultural variation among chimpanzee populations, learning ever more about our closest relatives, it is important to remember that the capacity to learn such behaviors is the product of natural selection. The behaviors that are maintained are adaptive for these animals in their environment. Studying culture in nonhuman primates may provide insights into the evolutionary origins of human culture. Human society is based to a great extent on the symbolic transmission of culture through spoken and written language.

Sociobiology explains human social behavior in terms of adaptation

Sociobiology focuses on the evolution of social behavior through natural selection. In his landmark book, *Sociobiology: The New Synthesis*, published in 1975, Edward O. Wilson combined principles of population genetics, evolution, and animal behavior to present a comprehensive view of the evolution of social behavior. Wilson's work influenced the development of behavioral ecology, and many of the concepts discussed in this chapter, such as paternal investment in care of the young (discussed shortly), are based on contributions made by sociobiologists. Like Darwin and many other biologists of the past, Wilson and other sociobiologists suggest that human behavior can be studied in evolutionary terms.

Sociobiology is controversial, at least in part because of its possible ethical implications. This approach is sometimes viewed as denying that human behavior is flexible enough to permit substantial improvements in the quality of our social lives. Yet sociobiologists agree with their critics that human behavior is flexible, and at least some of the debate focuses on the degree to which human behavior is genetically determined and the extent to which it can be modified.

As sociobiologists acknowledge, people can, through culture, change their way of life far more profoundly in a few years than could a hive of bees or a troop of baboons in hundreds of generations of genetic evolution. This capacity to make changes is indeed genetically determined, and that is a great gift. How we use it and what we accomplish with it are not a gift but a responsibility on which our own well-being and the well-being of other species depend.

Review

- What distinguishes an organized society from a mere aggregation of organisms?
- What determines an animal's place in a dominance hierarchy? What costs and benefits accrue to dominant individuals in a dominance hierarchy? To subordinate individuals?
- What is territoriality? What possible functions does it serve?
- How does the dance of bees compare with human language?

BiologyNow™ Assess your understanding of **social behavior** by taking the pretest on your BiologyNow CD-ROM.

SEXUAL SELECTION

Learning Objective

11 Define *sexual selection*, and describe different types of mating systems and approaches to parental care.

In many species, individuals actively compete for mates. Typically in such populations an abundance of males compete for a limited number of receptive females. For a male, reproductive success depends on how many females he can impregnate. Females may have the opportunity to select a sexual partner from among several males. For a female, reproductive success depends on how many eggs she can produce during her reproductive lifetime, on the quality of the sperm that fertilize them, and on the survival of her offspring to reproductive age. Recall that an animal's reproductive success is a measure of its *direct fitness.* **Sexual selection,** a type of natural selection, occurs when individuals vary in their ability to compete for mates. Sexual selection results in the reproductive advantage that some individuals have over others of the same sex and species.

Animals seek quality mates

In many species, success of a male in dominance encounters with other males indicates his quality to the female, and she allows the victorious male to court her. Several studies confirm that males ranking higher in a dominance hierarchy mate more frequently than males that rank lower. However, exceptions demonstrate the complexity of reproductive behavior among some species. For example, among baboons with a definite dominance hierarchy, males lower in the hierarchy copulated with females as frequently as males of higher status. Investigators observed, though, that dominant males copulated more frequently with females who were in *estrus,* their fertile period. In addition, some lower-ranking males develop alternative strategies for attracting females. For instance, a male may win a female's interest by protecting her baby, even though it is not his own.

Females of some species select their mates based on ornamental displays. Male fishes are often brightly colored (see Fig. 19-11). Among many bird species, males exhibit bright colors and dramatic plumage, and male deer display elaborate antlers. Expression of ornamental traits may give the female important information about the male's physical condition and ability to fight. The size of the antlers of red deer, for example, may indicate combat effectiveness, as well as proper nutrition and good health. Female lions are attracted to males with thick, dark manes, an indicator of good nutrition and plenty of testosterone (which regulates growth of hair and melanin production.) Thus in some species ornamentation signals good genes.

Among crickets and many other insect species, a courting male offers a gift of food to a prospective mate. Studies show that the larger or higher quality the food offering, the better the chances the male will be accepted.

Among some species of insects, birds, and bats, males gather in a small display area called a **lek,** where they compete for females. When a receptive female appears, males may excitedly display themselves and compete for her attention. Among some species, the female selects a male based on his location in the lek rather than on his appearance. The dominant male may occupy a central position and be chosen by most of the females. The female selects a male, mates with him, and then leaves the lek. The male remains to woo other females.

Courtship rituals ensure that the male is indeed a male and is a member of the same species. Rituals provide the female further opportunity to evaluate him. In some species, courtship may also be necessary as a signal to trigger nest building or ovulation. Courtship rituals can last seconds, hours, or days and often involve a series of fixed action patterns (Fig. 50-18). The first

(a)

(b)

FIGURE **50-18** | **Courtship rituals.**

(a) The male great frigatebird (*Frigata minor*) inflates his red throat sac in display as part of a courtship ritual. Photographed on Christmas Island in the Pacific. **(b)** Egrets (*Egretta rufescens*) performing a mating dance.

display by the male releases a counterbehavior by the female. This, in turn, releases additional male behavior, and so on until the pair is physiologically ready for copulation. Specific cues enable courtship rituals to function as reproductive isolating mechanisms among species (see Figure 19-2).

An extreme courtship ritual has been described for redback spiders, which are closely related to the black widow spider. During copulation, the small male spider positions himself above his larger mate's jaws. During 65% of matings, the finale is that the female eats her suitor. The apparent explanation for this behavior is that the risk-taking male is able to copulate for a longer period and thus fertilize more eggs than noncannibalized males and that the female is more likely to reject additional mates.

Sexual selection favors polygynous mating systems

In most species, males make little parental investment in their offspring, apart from providing sperm. Males ensure reproductive success by impregnating many females, increasing the probability that their genes will be propagated in multiple offspring. Thus sexual selection often favors **polygyny,** a mating system in which males fertilize the eggs of many females during a breeding season.

In the mating system known as **polyandry,** a female mates with several males. Benefits may include receiving gifts from several males or enlisting several males' help in caring for the young. Data collected at Jane Goodall's research center at Gombe National Park in Tanzania indicate that female chimpanzees practice polyandry. In addition to mating with males of their own group, when females are most fertile, many slip off to neighboring communities and copulate with less familiar males. Perhaps these sexual rendezvous protect against inbreeding. Investigators also hypothesize that mating with many males provides insurance against infanticide. Males do not aggress against infants of mothers with whom they have copulated. Recent studies indicate that females with multiple mates are more fertile and produce more offspring. Males in polyandrous species may accept their status because **mate guarding** may be costly, ineffective, or both. Females may range over a wide area, easily escaping male guarding behavior.

Polyandry and polygyny sometimes occur in the same species. After mating, a female giant water bug attaches a clutch of eggs to her mate's back (Fig. 50-19a). He cares for the eggs until they hatch. She then may mate with a different male and glue the new clutch of eggs to his back. However, if the male has more space for eggs, he may mate with another female.

Apparently, in most species it is not certain who fathered the offspring. However, raising some other male's offspring is a genetic disadvantage, so males in some species may compromise mate chasing in favor of mate guarding. The male guards his partner after copulation to ensure that she does not copulate with another male. Mate-guarding behavior is likely to occur when the female is receptive and has eggs that might be fertilized by another male. For example, dominant male African elephants guard a female only during the phase of estrus when she is most likely to have a fertile egg. Before that time, or later in estrus after mating has already occurred, younger male elephants of lesser social status can copulate with the female. A high cost of mate guarding is the loss of opportunity for a dominant male to mate with other females.

Sexual selection favors males that inseminate many females and produce many offspring. Perhaps for this reason, **monogamy,** a mating system in which a male mates with only one female during a breeding season, is not common. Less than 10% of mammals are monogamous.

Researchers long thought monogamy was common among birds, because many species form **pair bonds,** stable relationships that ensure cooperative behavior in mating and the rearing of the young. However, genetic evidence shows that some offspring are fathered by males other than the one caring for them. For example, DNA tests show that among eastern bluebirds, 15% to 20% of the chicks are fathered by other males.

The extra-pair male contributes new genes, and the female produces offspring with greater genetic variability, increasing their chances for survival. Some researchers distinguish between social monogamy, in which animals form a pair bond, and genetic (sexual) monogamy.

Monogamy does occur in some species, typically when males are needed to protect and feed the young. For example, the California mouse is genetically, as well as socially, monogamous. The offspring need their parents' body heat to survive, and the parents take turns keeping them warm.

Some animals care for their young

Most animals do not invest time and energy in caring for their offspring, because the costs of parenting are high, including production of fewer offspring and the risks taken in protecting the young from predators. The benefit of investing in parental care is the greater probability that each offspring will survive. Natural selection has favored parental care in species in which the female produces few young or reproduces only once during a breeding season. Parental care is an important part of successful reproduction in some invertebrates, including some cnidarians (jellyfish), rotifers, mollusks, and arthropods (crustaceans, insects, spiders, and scorpions), and caring for the young is common among vertebrates (Fig. 50-19).

Females of many vertebrate animals produce relatively few, large eggs. Because of the time and energy invested in producing eggs and carrying the developing embryo, the female has more to lose than the male if the young do not develop. Thus females are more likely than males to brood eggs and young, and usually females invest more in parental care. Parental care is especially skewed toward the female in mammals because females provide milk to nourish their young. Investing time and effort in care of the young is usually less advantageous to a male (assuming the female can handle the job by herself), because time spent in parenting is time lost from inseminating other females.

David M. Dennis/Animals, Animals

(a)

Dominique Braud/Dembinsky Photo Associates

(b)

FIGURE **50-19** | **Caring for the young.**

Parental investment in offspring increases the probability that the young will survive. **(a)** Giant water bug (*Belostomatidae*) male carrying eggs on his back. **(b)** Adult robins invest energy in feeding their young. **(c)** A female crocodile transports hatchlings in her mouth from their nest site to Lake St. Lucia in South Africa. **(d)** Female cheetah (*Acinonyx jubatus*) stands watch as her cubs eat the Thompson's gazelle she has hunted and killed for them. Photographed in Masi Mara, East Africa.

M. Reardon/Photo Researchers, Inc.

(c)

McMurray Photography

(d)

In some situations, a male benefits by helping to rear his own young or even those of a genetic relative. Receptive females may be scarce, breeding territories may be difficult to establish or guard, and gathering sufficient food may require more effort than one parent can provide. In some habitats, the young need protection against predators or cannibalistic males of the same species. Among wolves and many other carnivores, for example, males defend the young and the food supply.

Among many species of fish, the male cares for the young. The ultimate cause for this behavior appears to be that the parenting costs to the males are less than they would be for the females. A male fish must have a territory to attract a female. While guarding his territory, the male guards the fertilized eggs as well. Having eggs also appears to make the male more attractive to potential mates. In contrast, when a female is caring for her eggs or young, she is not able to feed optimally. As a result, her body size is smaller and her fertility is reduced.

Review

- What is sexual selection? What are some factors that influence mate choice?
- Under what conditions does sexual selection favor polygynous (or polyandrous) mating systems? Why is monogamy uncommon?
- What are some advantages of courtship rituals?
- Why do more females than males invest in caring for their offspring? Explain in terms of costs, benefits, and fitness.

BiologyNow™ Assess your understanding of **sexual selection** by taking the pretest on your BiologyNow CD-ROM.

HELPING BEHAVIOR

Learning Objective

12 Describe helping behavior, and relate the concepts of inclusive fitness and kin selection to altruistic behavior.

From the sociobiological perspective, an animal and its adaptations, including its behavior, ensure that its genes make more copies of themselves. The cells and tissues of the body support the functions of the reproductive system, and the reproductive system transmits genetic information to succeeding generations. If an animal's principal evolutionary mandate is to perpetuate its genes, we must wonder why some animals spend time and energy in helping others.

We have discussed examples of **cooperative behavior,** also called mutualism, such as group hunting in which each animal in the group benefits. In a type of cooperative behavior known as **reciprocal altruism,** one animal helps another with no immediate benefit. However, at some later time the animal that was helped repays the debt. If not, the animal will probably not be helped again.

Altruistic behavior can be explained by inclusive fitness

In **altruistic behavior** one individual behaves in a way that seems to benefit others rather than itself, with no potential payoff. However, how could true altruistic behavior be adaptive if the animal's response decreases its reproductive success relative to individuals that do not exhibit the altruistic behavior? In 1964 English biologist William D. Hamilton offered a plausible explanation. He suggested that evolution does not distinguish between genes transmitted directly from parent to offspring and those transmitted indirectly through close relatives.

According to Hamilton, natural selection favors animals that help a relative because the relative's offspring carries some of the helper's alleles. **Inclusive fitness** is the sum of direct fitness (which can be measured by the number of alleles an animal perpetuates in its offspring) plus indirect fitness (the alleles it helps perpetuate in the offspring of its kin).

We can measure the potential genetic gain from helping relatives by the **coefficient of relatedness,** the probability that two individuals inherit the same uncommon allele from a recent common ancestor. The coefficient of relatedness for two animals that are not related is 0, whereas between a parent and its offspring this coefficient would be 0.5, indicating that half of their genes are the same. The coefficient of relatedness for siblings is also 0.5; for first cousins it is 0.125. The higher the coefficient of relatedness, the more likely an animal will help a relative. DNA fingerprinting of lions in the Serengeti National Park and in the Ngorongoro Crater in Tanzania has confirmed that lion brothers that cooperate in prides have a greater probability of perpetuating their genes than those that go off on their own. This is true even if the lion perpetuates his genes only by proxy, through his brother. **Kin selection,** also called *indirect selection,* is a form of natural selection that increases inclusive fitness through the breeding success of close relatives.

The bee society provides an interesting example of kin selection. Recall that worker bees do not reproduce, but they nourish larval bees. The worker bees have up to three quarters of their genes in common with the larval bees and so are more likely to pass on copies of their genes to the next generation by raising these individuals than they would by producing their own offspring.

Among some birds (such as Florida jays), nonreproducing individuals aid in the rearing of the young. Nests tended by these additional helpers as well as parents produce more young than do nests with the same number of eggs overseen only by parents. The nonreproducing helpers are close relatives who increase their own biological success by ensuring the successful, though indirect, perpetuation of their genes. Helpers may be prevented from producing their own offspring by limiting factors such as a shortage of mates or territories.

Cooperative behavior may have alternative explanations

PROCESS OF SCIENCE

Some behavioral ecologists are questioning the role of kin selection in cooperative behavior. One of the time-honored examples of kin selection has been sentinel behavior, keeping watch for predators and warning other members of the group when danger threatens. Until recently, biologists thought that sentinels were at greater risk for predation, yet selflessly engaged in risky behavior that benefited their relatives. This explanation appears valid for some species, including prairie dogs and ground squirrels (Fig. 50-20). However, recent studies suggest an opposing hypothesis for guarding behavior in other species.

U.S. biologist Peter A. Bednekoff developed a model to explain how guarding could result from selfish behavior. Bednekoff observed that for some species no evidence suggests sentinels are at higher risk than others in their group. In fact, the opposite may be true and sentinels may actually have an advantage. By detecting predators first, the sentinels can more easily escape.

Studies of sentinel behavior in the African mongoose support Bednekoff's model. During 2000 hours of researcher observation, no sentinel was killed. In fact, investigators concluded that guarding is a selfish behavior in African mongoose populations, because sentinels were able to escape quickly into their nearby burrows. This research is an important reminder that it is risky for biologists to generalize the meanings of complex social behavior from one species to another. Many explanations are possible. Recent findings suggest that helpers often derive direct benefits in safety and eventual reproductive success. Among some species, for example, younger animals who help are permitted to remain in the social group and derive the benefits of group protection. These helpers are also positioned to take over territories and mates when older members of the group die.

FIGURE **50-20** | **Kin selection in prairie dogs.**
Low-ranking members of this social rodent group act as sentries, risking their own lives by exposing themselves outside their burrows. However, by protecting their siblings, they ensure that the genes shared in common will be perpetuated in the population.

Review

- How do sociobiologists use kin selection to explain the evolution of altruistic behavior?
- What other hypotheses may account for cooperative behavior?

BiologyNow™ Assess your understanding of **helping behavior** by taking the pretest on your BiologyNow CD-ROM.

SUMMARY WITH KEY TERMS

1 Apply cost–benefit analysis and the concept of ultimate cause to the process of deciding whether a particular behavior is adaptive.

- **Behavioral ecology** is the scientific study of behavior in natural environments from an evolutionary perspective. **Behavior** is what an animal does and how it does it, usually in response to stimuli in its environment.
- **Proximate causes** of behavior are immediate causes such as genetic, developmental, and physiological processes that permit the animal to carry out a specific behavior. Proximate causes answer *how* questions. **Ultimate causes** are the evolutionary explanations for *why* a certain behavior occurs.
- We can use **cost–benefit analysis** to determine whether a behavior is adaptive. Benefits contribute to **direct fitness,** the animal's reproductive success measured by the number of viable offspring. When benefits outweigh costs, the behavior is adaptive.

2 Describe the interactions of heredity, environment, and maturation in animal behavior.

- Behavior results from the interaction of genes (**innate behavior**) and environmental factors. The capacity for behavior is inherited, but behavior is modified in response to environmental experience; this is **learned behavior.**
- An organism must be mature—physiologically ready to produce a given behavior—before it can perform that pattern of behavior. Walking and many other behaviors that we view as automatic are **motor programs,** coordinated sequences of muscle actions.
- A **fixed action pattern (FAP)** is an automatic behavior that, once activated by a simple sensory stimulus, continues to completion regardless of sensory feedback. An FAP can be triggered by a specific unlearned **sign stimulus,** or **releaser.**

3 Discuss the adaptive significance of habituation, imprinting, classical conditioning, operant conditioning, and insight learning.

- **Habituation** is a type of learning in which an animal learns to ignore a repeated, irrelevant stimulus, allowing it to focus on finding food and other life activities.
- **Imprinting** establishes a parent–offspring bond during a critical period early in development, ensuring that the offspring recognizes the mother.
- In **classical conditioning,** an association is formed between some normal body function and a new stimulus. This type of learning allows an animal to make an association between two stimuli.
- In **operant conditioning,** an animal learns a behavior to receive positive reinforcement or to avoid punishment. This type of learning is important in many natural situations, including young herring gulls perfecting their pecking behavior to obtain food.
- **Insight learning** is the ability to adapt past experiences that may involve different stimuli to solve a new problem.
- Play may give a young animal a chance to learn and practice adult patterns of behavior.

4 Give examples of biological rhythms, and describe some of the mechanisms responsible for them.

- It is adaptive for an organism's metabolic processes and behavior to be synchronized with cyclical changes in the environment. In many species, physiological processes and activity follow **circadian rhythms,** which are daily cycles of activity.
- **Diurnal** animals are most active during the day, **nocturnal** animals at night, and **crepuscular** animals at dawn and/or dusk. Some biological rhythms reflect the **lunar cycle** or the changes in tides due to changes in the phase of the moon.
- Many biological rhythms are regulated by internal timing mechanisms that serve as **biological clocks.** In mammals the master clock is located in the **suprachiasmatic nucleus (SCN)** in the hypothalamus.

5 Analyze costs and benefits of migrations, and distinguish between directional orientation and navigation.

- **Migration** is periodic long-distance travel from one location to another. Proximate causes of migration include physiological responses that are triggered by environmental changes. Ultimate causes of migration include the benefit of moving away from an area that seasonally becomes too cold, too dry, or depleted of food to a more hospitable area. Costs of migration include time, energy, and greater risk of predation.
- **Directional orientation** is travel in a specific direction and requires **compass sense,** a sense of direction. Many migrating animals use the sun to orient themselves.
- **Navigation** requires both compass sense and **map sense,** an awareness of location. Birds that navigate at night use the stars as guides. Birds and many other animals also use Earth's magnetic field to navigate.

6 Discuss the hypothesis that optimal foraging behavior is adaptive.

- **Optimal foraging,** the most efficient strategy for an animal to get food, often enhances reproductive success.

7 Analyze social behavior in terms of costs and benefits, and describe modes of animal communication.

- **Social behavior** is adaptive interaction, usually among members of the same species. Many animal societies are characterized by a means of communication, cooperation, division of labor, and a tendency to stay together. Benefits include cooperative foraging or hunting and defense from predators. Costs include increased competition for food and habitats and increased risks of attracting predators and transmitting disease.
- Animal **communication** involves the exchange of mutually recognizable signals, which can be auditory, visual, electrical, or chemical. **Pheromones** are chemical signals that convey information between members of a species.

8 Describe the concept of a dominance hierarchy, giving at least one example, and discuss its adaptive significance and social function.

- A **dominance hierarchy** is a ranking of status within a group in which more dominant members are accorded benefits (such as food or mates) by subordinates, often without overt aggressive behavior.

9 Distinguish between home range and territory, and describe costs and benefits of territoriality.

- Animals often inhabit a **home range,** a geographic area that they seldom leave but do not necessarily defend. A defended area within a home range is called a **territory,** and the defensive behavior is **territoriality.** Some animals, such as lions, engage in group territoriality. Costs of territoriality include time and energy expended in staking out and defending a territory, and risks in fighting for it. Benefits include rights to food in the territory and reduction in conflict among members of a population.

10 Contrast a society of social insects with a vertebrate society, and give examples of cultural variation in vertebrate populations.

- A **society** is a group of individuals of the same species that may work together in an adaptive manner. Insect societies tend to be rigid, with the role of the individual narrowly defined. Insect division of labor is mainly determined by age.
- Vertebrate societies are more flexible than insect societies. Some bird and mammal species develop **culture,** learned behavior common to a population that is transmitted from one generation to the next. Symbolic transmission of culture is important in human societies.
- Examples of cultural variation include local dialects in orcas, and use of tools, distinctive courtship rituals, and grooming techniques in chimpanzee populations.

11 Define *sexual selection,* and describe different types of mating systems and parental care.

- **Sexual selection,** a type of natural selection, occurs when individuals vary in their ability to compete for mates. Individuals with reproductive advantages are selected over others of the same sex and species.
- Mate choice may be influenced by dominance, gifts, ornaments, and courtship displays. Males of some species gather in a **lek,** a small display area where they compete for females. **Courtship rituals** ensure that the male is a member of the same species and permit the female to assess the quality of the male.
- Sexual selection often favors **polygyny,** a mating system in which a male mates with many females. In **polyandry,** a female mates with several males. **Monogamy,** mating with a single partner during a breeding season, is less common.
- Among many species, males engage in **mate guarding,** especially when the female is most fertile, to prevent other males from fertilizing her eggs.
- A **pair bond** is a stable relationship between a male and a female that may involve cooperative behavior in mating and in rearing the young.
- Parental care increases the probability that offspring will survive. A high investment in parenting is typically less advantageous to the male than to the female.

12 Describe helping behavior, and relate the concepts of inclusive fitness and kin selection to altruistic behavior.

- Some animals engage in **cooperative behavior** (such as in hunting), that benefits all. In a type of cooperative behavior known as **reciprocal altruism,** the helper does not immediately benefit but is helped later by the animal it helped.
- In **altruistic behavior,** one individual appears to behave in a way that benefits others rather than itself.
- **Inclusive fitness** is the sum of an individual's direct fitness and indirect fitness (number of offspring of kin). This concept suggests that natural selection favors animals that help a relative because this is an indirect way to perpetuate some of the helper's own alleles.
- **Kin selection** is a type of natural selection that increases inclusive fitness through successful reproduction of close relatives.

POST-TEST

1. The contemporary study of behavior in natural environments from the point of view of adaptation is (a) ecology (b) ethology (c) behavioral ecology (d) evolutionary ecology (e) behavioral science.
2. To understand *why* the graylag goose "rolls" a nonexistent egg toward her nest, you might explore (a) proximate causes (b) ultimate causes (c) pheromones (d) hormonal factors (e) its social learning
3. The responses of an organism to signals from its environment are its (a) behavior (b) culture (c) ultimate behavior (d) releasers (e) motor programs
4. A fixed action pattern (FAP) is elicited by a (a) motor program (b) sign stimulus (c) releaser (d) answers a, b, and c are correct (e) only answers b and c are correct
5. A form of learning in which a young animal forms a strong attachment to a moving object (usually its parent) within a few hours of birth is (a) classical conditioning (b) operant conditioning (c) imprinting (d) insight learning (e) parental investment
6. An animal learns to ignore a repeated, irrelevant stimulus. This is (a) classical conditioning (b) operant conditioning (c) imprinting (d) insight learning (e) habituation
7. Salivation by a student when the noon bell rings is an example of (a) classical conditioning (b) operant conditioning (c) imprinting (d) insight learning (e) habituation
8. Both compass sense and map sense are necessary for (a) navigation (b) migration (c) directional orientation (d) both navigation and migration (e) the action of biological clocks
9. In optimal foraging (a) animals always hunt in social groups (b) an animal obtains food in the most efficient way (c) animals can rarely afford to be selective (d) groups of five to seven animals are most successful (e) animals are typically most successful when they hunt alone
10. Chemical signals that convey information between members of a species are (a) pheromones (b) hormones (c) neurotransmitters (d) leks (e) neuropeptides
11. The benefits of territoriality include (a) rights to defend a home range (b) increased reproductive success (c) energy investment in staking out and defending the area (d) monogamy (e) pair bonding
12. Sexual selection (a) is a form of natural selection (b) occurs when animals are very similar in their ability to compete for mates (c) results in animals that have lower direct fitness (d) occurs mainly among animals that practice polygyny (e) occurs mainly among animals that practice polyandry
13. Mate choice is least likely influenced by (a) ornamental displays such as antlers (b) dominance (c) competition in a lek (d) courtship behavior (e) fidelity of the male
14. The round dance of the bee (a) indicates that food is close to the hive (b) communicates direction (c) results in bees flying long distances in all directions (d) indicates that food is distant from the hive (e) indicates both direction and height of food
15. Behavior that appears to have no payoff, that is, an individual appears to act to benefit others rather than itself is known as (a) mutualism (b) altruistic behavior (c) reciprocal altruism (d) inclusive fitness (e) helping behavior
16. Inclusive fitness (a) considers only the genes an animal transmits to its own offspring (b) suggests that natural selection favors animals that help a relative (c) explains territoriality (d) is a form of social behavior (e) is a form of reciprocal altruism
17. Kin selection (a) increases inclusive fitness (b) is a way of perpetuating genes of nonrelatives (c) accounts for some forms of migration (d) typically involves mate guarding (e) involves ornamental displays and use of a lek
18. Culture (a) is common among invertebrate groups (b) does not vary within the same species (c) is a shared derived character of humans (d) has a strong genetic component in primates (e) is behavior learned from other members of the group and shared by members of a population

CRITICAL THINKING

1. What might be the adaptive value of sea turtle migration? Consider how this adaptive value has changed if, as a result of human activities, migration now puts sea turtles at greater risk than if they restricted their habitat to a single location. Discuss the possible evolutionary mechanisms by which the behavior of these species may (or may not) adapt to these environmental pressures. What conservation efforts should we take to ensure successful migration?
2. Behavioral ecologists have demonstrated that young tiger salamanders can become cannibals and devour other salamanders. Investigators observed that the cannibal salamanders preferred eating unrelated salamanders over their own cousins and preferred eating cousins over their siblings. Explain this behavior based on what you have learned in this chapter.
3. How is the society of a social insect different from human society? What are some similarities between the transmission of information by heredity and by culture? What are some differences?

- Visit our Web site at **http://biology.brookscole.com/solomon7** for links to chapter-related resources on the World Wide Web. Additional online materials relating to this chapter can also be found on our Web site.

BIOLOGY NOW RESOURCES

Active Figures

50-6: Classical conditioning and other types of learning

50-17: Honeybee dances

Preparing for an exam? Take a diagnostic test on your BiologyNow CD-ROM.

Post-Test Answers

1. c	2. b	3. a	4. e
5. c	6. e	7. a	8. a
9. b	10. a	11. b	12. a
13. e	14. a	15. b	16. b
17. a	18. e		

Introduction to Ecology: Population Ecology

A population of Mexican poppies blooms in the desert after the winter rains. Mexican poppies (*Eschsolzia mexicana*), which thrive on gravelly desert slopes, grow to a height of 40 cm (16 in.).

CHAPTER OUTLINE

The science of **ecology** is the study of how living organisms and the physical environment interact in an immense and complicated web of relationships. Biologists call the interactions among organisms **biotic factors,** and those between organisms and their nonliving, physical environment, **abiotic factors.** These include precipitation, temperature, pH, wind, and chemical nutrients. Ecologists formulate hypotheses to explain such phenomena as the distribution and abundance of life on Earth, the ecological role of specific species, the interactions among species in communities, and the importance of ecosystems in maintaining the health of the biosphere. They then test these hypotheses.

The focus of ecology can be local or global, specific or generalized, depending on what questions the scientist is asking and trying to answer. Ecology is the broadest field in biology, with explicit links to evolution and every other biological discipline. Its universality also encompasses subjects that are not traditionally part of biology. Earth science, geology, chemistry, oceanography, climatology, and meteorology are extremely important to ecology, especially when ecologists examine the abiotic environment of planet Earth. Because humans are part of Earth's complex web of life, all our activities, including economics and politics, have profound ecological implications. **Environmental science,** a scientific discipline with ties to ecology, focuses on how humans interact with the environment.

As you learned in Chapter 1, most ecologists are interested in the levels of biological organization including and above the level of the individual organism: population, community, ecosystem, landscape, and biosphere. Each level has its own characteristic composition, structure, and functioning.

An individual belongs to a **population,** a group consisting of members of the same species that live together in a prescribed area at the same time. The boundaries of the area are defined by the ecologist performing a particular study. A population ecologist might study a population of microorganisms, animals, or plants, like the Mexican poppies in the photograph, to see how individuals within it live and interact with each other, with other species in their community, and with their physical environment.

In this chapter we begin our study of ecological principles by focusing on the study of populations as functioning systems, ending with a discussion of the human population. Subsequent chapters examine the interactions among different populations within communities (Chapter 52), the dynamic exchanges between communities and their physical environments (Chapter 53), the characteristics of Earth's major biological ecosystems (Chapter 54), and some of the impacts of humans on the biosphere (Chapter 55). ■

FEATURES OF POPULATIONS

Learning Objective

1 Define *population density* and *dispersion,* and describe the main types of population dispersion.

Populations exhibit characteristics distinctive from those of the individuals of which they are composed. Some features discussed in this chapter that characterize populations are population density, population dispersion, birth and death rates, growth rates, survivorship, and age structure.

Although communities consist of all the populations of all the different species that live together within an area, populations have properties that communities lack. Populations, for example, share a common gene pool (see Chapter 18). As a result, allele frequency changes resulting from natural selection occur in populations. Natural selection, therefore, acts directly to produce adaptive changes in populations and only indirectly affects the community level.

Population ecology considers both the number of individuals of a particular species that are found in an area and the dynamics of the population. **Population dynamics** is the study of changes in populations—how and why those numbers increase or decrease over time. Population ecologists try to determine the processes common to all populations. They study how a population interacts with its environment, such as how individuals in a population compete for food or other resources, and how predation, disease, and other environmental pressures affect the population. Population growth, whether of bacteria, maples, or giraffes, cannot increase indefinitely because of such environmental pressures.

Additional aspects of populations that interest biologists are their reproductive success or failure (extinction), their evolution, their genetics, and how they affect the normal functioning of communities and ecosystems. Biologists in applied disciplines, such as forestry, agronomy (crop science), and wildlife management, must understand population ecology to manage populations of economic importance, such as forests, field crops, game animals, or fishes. Understanding the population dynamics of endangered and threatened species plays a key role in efforts to prevent their slide to extinction. Knowledge of population ecology helps in efforts to prevent the increase of pest populations to levels that cause significant economic or health impacts.

Density and dispersion are important features of populations

The concept of population size is meaningful only when the boundaries of that population are defined. Consider, for example, the difference between 1000 mice in 100 hectares (250 acres) and 1000 mice in 1 hectare (2.5 acres). Often a population is too large to study in its entirety. Researchers examined such a population by sampling a part of it and then expressing the population in terms of density. Examples include the number of dandelions per square meter of lawn, the number of water fleas per liter of pond water, or the number of cabbage aphids per square centimeter of cabbage leaf. **Population density,** then, is the number of individuals of a species per unit of area or volume at a given time.

Different environments vary in the population density of any species they can support. This density may also vary in a single habitat from season to season or year to year. For example, red grouse are ground-dwelling game birds whose populations are managed for hunting. Consider two red grouse populations in the treeless moors of northwest Scotland, at locations only 2.5 km (1.5 mi) apart. At one location the population density remained stationary during a three-year period, but at the other site it almost doubled in the first two years and then declined to its initial density in the third year. The reason was likely a difference in habitat. Researchers had experimentally burned the area where the population density increased initially and then decreased. Young heather (*Calluna vulgaris*) shoots produced after the burn provided nutritious food for the red grouse. So population density may be determined in large part by biotic or abiotic factors in the environment that are external to the individuals in the population.

The individuals in a population often exhibit characteristic patterns of **dispersion,** or spacing, relative to each other. Individuals may be spaced in a clumped, uniform, or random dispersion (Fig. 51-1a). Perhaps the most common spacing is **clumped dispersion,** also called **aggregated distribution** or **patchiness,** which occurs when individuals are concentrated in specific parts of the habitat. Clumped dispersion often results from the patchy distribution of resources in the environment. It also occurs among animals, owing to the presence of family groups and pairs, and among plants, owing to limited seed dispersal or asexual reproduction. An entire grove of aspen trees, for example, may originate asexually from a single plant. Clumped dispersion may sometimes be advantageous, because social animals derive many benefits from their association. Many fish species, for example, associate in dense schools for at least part of their life cycle, possibly because schooling may reduce the risk of predation for any particular individual (Fig. 51-1b). The many pairs of eyes of schooling fish tend to detect predators more effectively than a single pair of eyes of a single fish. When threatened, schooling fish clump together more closely, making it difficult for a predator to single out an individual.

Uniform dispersion occurs when individuals are more evenly spaced than would be expected from a random occupation of a given habitat. A nesting colony of seabirds, in which the birds are nesting in a relatively homogeneous environment and place their nests at a more or less equal distance from each other, is an example of uniform dispersion (Fig. 51-1c). What might this spacing pattern tell us? In this case, uniform dispersion may occur as a result of nesting territoriality. Aggressive interactions among the nesting birds as they peck at one another from their nests cause each pair to place its nest just beyond the reach of nearby nesting birds. Uniform dispersion also occurs when competition among individuals is severe, when plant roots or leaves that have been shed produce toxic substances that inhibit the growth of nearby plants, or when animals establish feeding or mating territories.

Random dispersion occurs when individuals in a population are spaced throughout an area in a manner that is unrelated to the presence of others. Of the three major types of dispersion, random dispersion is least common and hardest to observe in nature, leading some ecologists to question its existence. Trees of the same species, for example, sometimes appear to be distributed randomly in a tropical rain forest. However, an international team of 13 ecologists studied 6 tropical forest plots that were 25 to 52 hectares (62 to 128 acres) in area and reported that most of the 1000 tree species observed were clumped and not randomly dispersed. (Ecologists determine both clumped and uniform dispersion by statistically testing for differences from an assumed random distribution.) Random dispersion may occur infrequently, because important environmental factors affecting dispersion usually do not occur at random. Flour beetle larvae in a container of flour are randomly dispersed, but their environment (flour) is unusually homogeneous.

Some populations have different spacing patterns at different ages. Competition for sunlight among same-aged sand pine in a Florida scrub community resulted in a change over time from either random or clumped dispersion when the plants were young, to uniform dispersion when the plants were old. Sand pine is a fire-adapted plant with cones that do not release their seeds until they have been exposed to high temperatures (45°C to 50°C or higher). As a result of seed dispersal and soil conditions following a fire, the seedlings grow back in dense stands that exhibit random or slightly clumped dispersion. Over time, however, many of the more crowded trees tend to die from shading or competition, resulting in uniform dispersion of the surviving trees (Table 51-1).

FIGURE **51-1** **Dispersion of individuals within a population.**

(a) Diagram of the spacing of individuals in clumped, uniform, and random dispersion. **(b)** The schooling behavior of certain fish species is an example of clumped dispersion. Shown are bluestripe snappers (*Lutjanus kasmira*), photographed in Hawaii. This introduced fish, which grows to 30 cm (12 in), may be displacing native fish species in Hawaiian waters. **(c)** Cape gannets (*Morus capensis*) nesting on the coast of South Africa. The birds space their nests more or less evenly and so are an example of uniform dispersion.

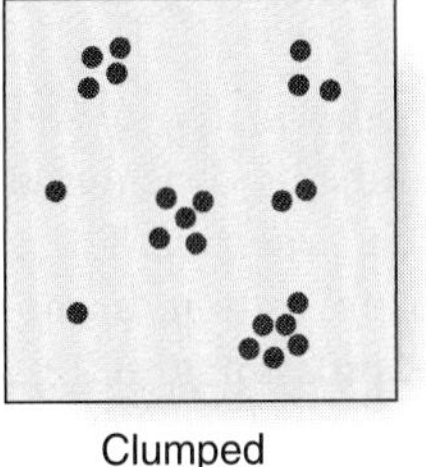

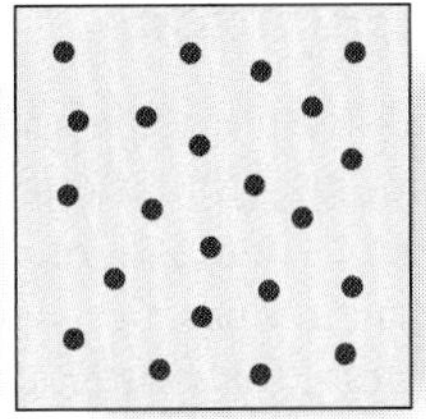

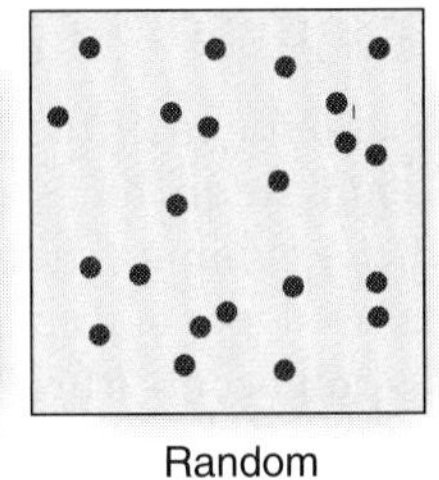

Andrew G. Wood/Photo Researchers, Inc.

Robert Hernandez/Photo Researchers, Inc.

(c)

Review

- What is the difference between population density and disperson?
- What are some biological advantages of a clumped dispersion? What are the disadvantages?

BiologyNow™ Assess your understanding of **population density and dispersion** by taking the pretest on your BiologyNow CD-ROM.

TABLE **51-1** **Dispersion in a Sand Pine Population in Florida**

Tree Trunks Examined*	Density (per m²)	Dispersion
All (alive and dead)	0.16	Random
Alive only	0.08	Uniform

*Adapted from Laessle, A.M. "Spacing and Competition in Natural Stands of Sand Pine." *Ecology*, Vol. 46, 1965, pp. 65–72. Data were collected 51 years after a fire.

CHANGES IN POPULATION SIZE

Learning Objectives

2 Explain the four factors (natality, mortality, immigration, and emigration) that produce changes in population size, and solve simple problems involving these changes.

3 Define *intrinsic rate of increase* and *carrying capacity*, and explain the differences between J-shaped and S-shaped growth curves.

PROCESS OF SCIENCE

One goal of science is to discover common patterns among separate observations. As mentioned previously, population ecologists wish to understand general processes that are shared by many different populations, so they develop mathematical models based on equations that describe the dynamics of a single population. Population models are not perfect representations of a population, but models help illuminate complex processes. Moreover, mathematical modeling enhances the scientific process by providing a framework with which experimental population studies can be compared. We can test a model and see how it fits or does not fit with existing data. Data that are inconsistent with the model are particularly useful, because they demand that we ask how the natural system differs from the mathematical model that we developed to explain it. As more knowledge accumulates from observations and experiments, the model is refined and made more precise.

Population size, whether of sunflowers, elephants, or humans, changes over time. On a global scale, this change is ultimately caused by two factors, expressed on a per capita (that is, per individual) basis: **natality,** the average per capita birth rate, and **mortality,** the average per capita death rate. In humans the birth rate is usually expressed as the number of births per 1000 people per year and the death rate as the number of deaths per 1000 people per year.

To determine the rate of change in population size, we must also take into account the time interval involved, that is, the change in time. To express change in equations, we employ the Greek letter delta (Δ). In equation (1), ΔN is the change in the number of individuals in the population, Δt the change in time, N the number of individuals in the existing population, b the natality, and d the mortality.

$$(1)\ \Delta N/\Delta t = N(b - d)$$

The **growth rate (*r*),** or rate of change (increase or decrease) of a population on a per capita basis is the birth rate minus the death rate:

$$(2)\ r = b - d$$

As an example, consider a hypothetical human population of 10,000 in which there are 200 births per year (that is, by convention, 20 births per 1000 people) and 100 deaths per year (10 deaths per 1000 people):

$$r = 20/1000 - 10/1000 = 0.02 - 0.01 = 0.01, \text{ or } 1\% \text{ per year}$$

A modification of equation (1) tells us the rate at which the population is growing at a particular instant in time, that is, its instantaneous growth rate (dN/dt). (The symbols dN and dt are the mathematical differentials of N and t, respectively; they are not products, nor should the d in dN or dt be confused with the death rate, d.) Using differential calculus, this growth rate can be expressed thus:

$$(3)\ dN/dt = rN$$

where N is the number of individuals in the existing population, t the time, and r the per capita growth rate.

Because $r = b - d$, if individuals in the population are born faster than they die, r is a positive value, and population size increases. If individuals in the population die faster than they are born, r is a negative value, and population size decreases. If r is equal to zero, births and deaths match, and population size is stationary despite continued reproduction and death.

Dispersal affects the growth rate in some populations

In addition to birth and death rates, **dispersal,** which is movement of individuals among populations, must be considered when examining changes in populations on a *local* scale. There are two types of dispersal: immigration and emigration. **Immigration** occurs when individuals enter a population and thus increase its size. **Emigration** occurs when individuals leave a population and thus decrease its size. The growth rate of a local population must take into account birth rate (b), death rate (d), immigration rate (i), and emigration rate (e) on a per capita basis. The per capita growth rate equals the birth rate minus the death rate, plus the immigration rate minus the emigration rate:

$$(4)\ r = (b - d) + (i - e)$$

For example, the growth rate of a human population of 10,000 that has 200 births (by convention, 20 per 1000), 100 deaths (10 per 1000), 10 immigrants (1 per 1000), and 100 emigrants (10 per 1000) in a given year would be calculated as follows:

$$r = (20/1000 - 10/1000) + (1/1000 - 10/1000)$$
$$r = (0.020 - 0.010) + (0.001 - 0.010)$$
$$= 0.010 - 0.009 = 0.001, \text{ or } 0.1\% \text{ per year}$$

Each population has a characteristic intrinsic rate of increase

The maximum rate at which a population of a given species could increase under ideal conditions, when resources are abundant and its population density is low, is known as its **intrinsic rate of increase (r_{max}).** Different species have different intrinsic rates of increase. A particular species' intrinsic rate of increase is influenced by several factors. These include the age at which reproduction begins, the fraction of the **life span** (duration of the individual's life) during which the individual is capable of reproducing, the number of reproductive periods per lifetime, and the number of offspring the individual is capable of producing during each period of reproduction. These factors, which we discuss in greater detail later in the chapter, determine whether a particular species has a large or small intrinsic rate of increase.

Generally, large species such as blue whales and elephants have the smallest intrinsic rates of increase, whereas microorganisms have the greatest intrinsic rates of increase. Under ideal conditions (an environment with unlimited resources), certain bacteria can reproduce by binary fission every 20 minutes. At this rate of growth, a single bacterium would increase to a population of more than 1 billion in just 10 hours!

If we plot the population size versus time, under optimal conditions, the graph has a J shape that is characteristic of **exponential population growth,** which is the accelerating population growth rate that occurs when optimal conditions allow a constant per capita growth rate (Fig. 51-2). When a population grows exponentially, the larger that population gets, the faster it grows.

Regardless of which species we are considering, whenever a population is growing at its intrinsic rate of increase, population size plotted versus time gives a curve of the same shape. The only variable is time. It may take longer for an elephant population than for a bacterial population to reach a certain size (because elephants do not reproduce as rapidly as bacteria), but both populations will always increase exponentially as long as their per capita growth rates remain constant.

No population can increase exponentially indefinitely

Certain populations may grow exponentially for brief periods. Exponential growth has been experimentally demonstrated in certain insects and in bacterial and protistan cultures (by continually supplying nutrients and removing waste products). However, organisms cannot reproduce indefinitely at their intrinsic rate of increase, because the environment sets limits. These include such unfavorable environmental conditions as the limited availability of food, water, shelter, and other essential resources (resulting in increased competition) as well as limits imposed by disease and predation.

FIGURE **51-2** **Exponential population growth.**

When bacteria divide every 20 minutes, their numbers (expressed in millions) increase exponentially. The curve of exponential population growth has a characteristic J shape. The ideal conditions under which bacteria or other organisms reproduce exponentially rarely occur in nature, and when these conditions do occur, they are of short duration.

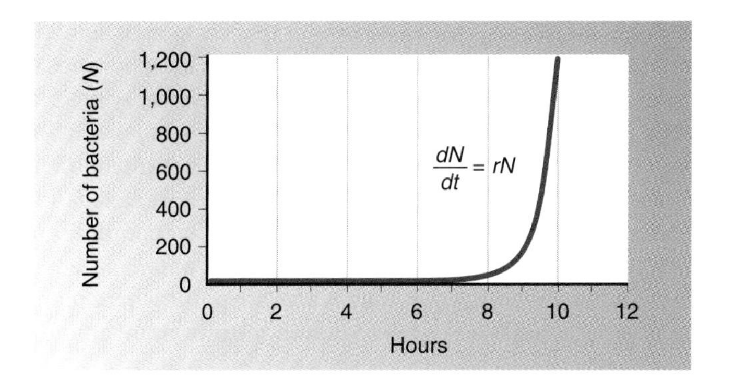

Using the earlier example, bacteria in nature would never be able to reproduce unchecked for an indefinite period, because they would run out of food and living space, and poisonous wastes would accumulate in their vicinity. With crowding, bacteria would also become more susceptible to parasites (high population densities facilitate the spread of infectious organisms such as viruses among individuals) and predators (high population densities increase the likelihood of a predator catching an individual). As the environment deteriorated, their birth rate (b) would decline and their death rate (d) would increase. Conditions might worsen to a point where d would exceed b, and the population would decrease. The number of individuals in a population, then, is controlled by the ability of the environment to support it. As the number of individuals in a population (N) increases, environmental limits act to control population growth.

Over longer periods, the rate of population growth may decrease to nearly zero. This leveling out occurs at or near the limits of the environment to support the population. The **carrying capacity (*K*)** represents the largest population that can be maintained for an indefinite period by a particular environment, assuming there are no changes in that environment. In nature, the carrying capacity is dynamic and changes in response to environmental changes. An extended drought, for example, could decrease the amount of vegetation growing in an area, and this change, in turn, would lower the carrying capacity for deer and other herbivores in that environment.

When a population regulated by environmental limits is graphed over longer periods, the curve has a characteristic S shape. The curve shows the population's initial exponential increase (note the curve's J shape at the start, when environmental limits are few), followed by a leveling out as the carrying capacity of the environment is approached (Fig. 51-3). The S-shaped growth curve, also called **logistic population growth,** can be modeled by a modified growth equation called a logistic equation. The logistic model of population growth was developed

FIGURE **51-3** **Carrying capacity and logistic population growth.**

In many laboratory studies, exponential population growth slows as the carrying capacity (K) of the environment is approached. The logistic model of population growth, when graphed, has a characteristic S-shaped curve.

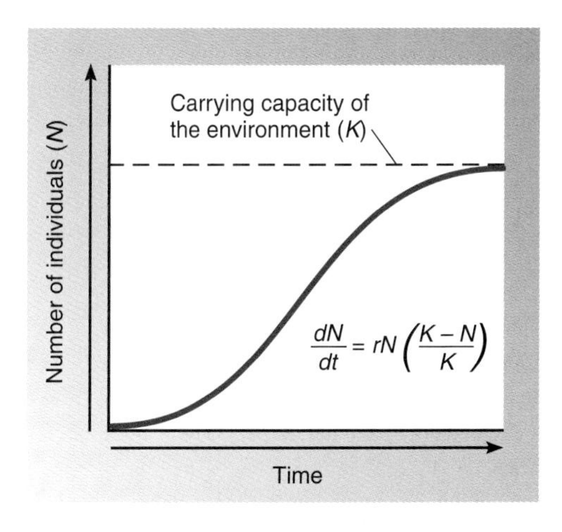

to explain population growth in continually breeding populations. Similar models exist for populations that have specific breeding seasons. The logistic model describes a population increasing from a small number of individuals to a larger number of individuals that are ultimately limited by the environment. The logistic equation, then, takes into account the carrying capacity of the environment:

$$(5)\ dN/dt = rN[(K - N)/K]$$

Note that part of the equation is the same as equation (3). The added element $[(K - N)/K]$ reflects a decline in growth as a population size approaches its carrying capacity. When the number of organisms (N) is small, the rate of population growth is unchecked by the environment, because the expression $[(K - N)/K]$ has a value of almost 1. But as the population (N) begins to approach the carrying capacity (K), the growth rate declines because the value of $[(K - N)/K]$ approaches zero.

Although the S curve is an oversimplification of how most populations change over time, it does appear to fit some populations that have been studied in the laboratory, as well as a few that have been studied in nature. For example, Georgyi F. Gause, a Russian ecologist who conducted experiments during the 1930s, grew a population of a single species, *Paramecium caudatum,* in a test tube. He supplied a limited amount of food (bacteria) daily and replenished the growth medium occasionally to eliminate the accumulation of metabolic wastes. Under these conditions, the population of *P. caudatum* increased exponentially at first, but then its growth rate declined to zero, and the population size leveled off (see Fig. 52-4, middle graph).

A population rarely stabilizes at K (carrying capacity) but may temporarily rise higher than K. It will then drop back to, or below, the carrying capacity. Sometimes a population that overshoots K will experience a *population crash,* an abrupt decline from high to low population density. Such an abrupt change is commonly observed in bacterial cultures, zooplankton, and other populations whose resources have been exhausted.

The carrying capacity for reindeer, which live in cold northern habitats, is determined largely by the availability of winter forage. In 1910 humans introduced a small herd of 26 reindeer onto one of the Pribilof Islands of Alaska. The herd's population increased exponentially for about 25 years until there were approximately 2000 reindeer, many more than the island could support, particularly in winter. The reindeer overgrazed the vegetation until the plant life was almost wiped out. Then, in slightly over a decade, as reindeer died from starvation, the number of reindeer plunged to eight, one third the size of the original introduced population. Recovery of subarctic and arctic vegetation after overgrazing by reindeer can take 15 to 20 years, during which time the carrying capacity for reindeer is greatly reduced.

Review

- What effect does each of the following have on population size? (a) natality; (b) mortality; (c) immigration; (d) emigration
- How does a J-shaped population growth curve differ from an S-shaped curve in terms of intrinsic rate of increase and carrying capacity?
- What would be the main difference between graphs representing the long-term growth of two populations of bacteria cultured in a test tubes, one in which the nutrient medium is replenished, and the other in which it is not replenished?

BiologyNow™ Assess your understanding of **changes in population size** by taking the pretest on your BiologyNow CD-ROM.

FACTORS INFLUENCING POPULATION SIZE

Learning Objective

4 Contrast the influences of density-dependent and density-independent factors on population size, and give examples of each.

Certain natural mechanisms influence population size. Factors that affect population size fall into two categories: density-dependent factors and density-independent factors. These two sets of factors vary in importance from one species to another and, in most cases, probably interact simultaneously in complex ways to determine the size of a population.

Density-dependent factors regulate population size

Sometimes the influence of an environmental factor on the individuals in a population varies with the density or crowding of that population. If a change in population density alters how an environmental factor affects that population, then the environmental factor is said to be a **density-dependent factor.**

As population density increases, density-dependent factors tend to slow population growth by causing an increase in death rate and/or a decrease in birth rate. The effect of these density-dependent factors on population growth increases as the population density increases; that is, density-dependent factors affect a larger proportion, not just a larger number, of the population. Density-dependent factors can also affect population growth when population density declines, by decreasing the death rate and/or increasing the birth rate. Density-dependent factors, then, tend to regulate a population at a relatively constant size that is near the carrying capacity of the environment. (Keep in mind, however, that the carrying capacity of the environment frequently changes.) Density-dependent factors are an excellent example of a **negative feedback system** (Fig. 51-4; also see Chapter 37).

Predation, disease, and competition are examples of density-dependent factors. As the density of a population increases, predators are more likely to find an individual of a given prey species. When population density is high, the members of a population encounter one another more frequently, and the chance of their transmitting parasites and infectious disease organisms increases. As population density increases, so does competition for resources such as living space, food, cover, water, minerals, and sunlight; eventually, it may reach the point where many members of a population fail to obtain the minimum amount

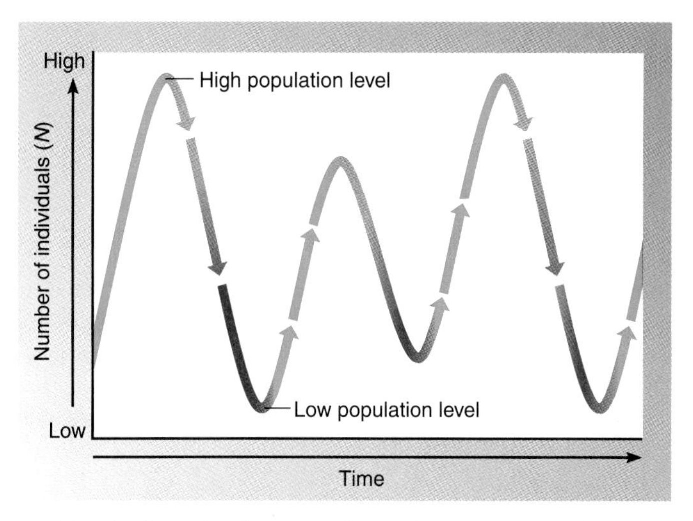

Density-dependent factors are increasingly severe: Population peaks and begins to decline.

Density-dependent factors are increasingly relaxed: Population bottoms out and begins to increase.

ACTIVE FIGURE **51-4** **Density-dependent factors and negative feedback.**

When the number of individuals in a population increases, density-dependent factors cause a decline in the population. When the number of individuals in a population decreases, a relaxation of density-dependent factors allows the population to increase.

BiologyNow™ Learn more about **population growth curves** by clicking on this figure on your BiologyNow CD-ROM.

Tom McHugh/Photo Researchers, Inc.

FIGURE **51-5** **Lemming**

The brown lemming (*Lemmus trimucronatus*) lives in the arctic tundra. Although lemming populations have been studied for decades, much about the cyclic nature of lemming population oscillations and their effects on the rest of the tundra ecosystem is still not well understood. This species ranges from Alaska eastward to the Hudson Bay.

of whatever resource is in shortest supply. At higher population densities, density-dependent factors raise the death rate and/or lower the birth rate, inhibiting further population growth. The opposite effect occurs when the density of a population decreases. Predators are less likely to encounter individual prey, parasites and infectious diseases are less likely to be transmitted from one host to another, and competition among members of the population for resources such as living space and food declines.

Density-dependent factors may explain what makes certain populations fluctuate cyclically over time

Lemmings are small, stumpy-tailed rodents that are found in colder regions of the Northern Hemisphere (Fig. 51-5). They are herbivores that feed on sedges and grasses in the arctic tundra. It has long been known that lemming populations have a three- to four-year cyclical oscillation that is often described as "boom or bust." That is, the population increases dramatically and then crashes; the population peaks may be 100 times as high as the low points in the population cycle. Many other populations, such as snowshoe hares and red grouse, also exhibit cyclic fluctuations.

What is the driving force behind these fluctuations? Several hypotheses have been proposed to explain the cyclical periodicity of lemming and other boom-or-bust populations; many hypotheses involve density-dependent factors. One possibility is that as a prey population becomes more dense it overwhelms its food supply, resulting in population decline. In the lemming example, researchers have recently studied the shape of several lemming population curves during their oscillations; the data suggest that lemming populations crash because they overgraze the plants, not because predators eat them.

Another explanation is that the population density of predators, such as long-tailed jaegers (birds that eat lemmings), increases in response to the increasing density of prey. Few jaegers breed when the lemming population is low, whereas when the lemming population is high, most jaegers breed, and the number of eggs per clutch is greater than usual. As more predators consume the abundant prey, the prey population declines. Later, with fewer prey in an area, the population of predators declines (some disperse out of the area, and fewer offspring are produced).

In 2003 researchers at the University of Helsinki, Finland, reported the results of a long-term study of collared lemmings in Greenland. They designed a model based on their data that indicates the four-year fluctuations in lemming populations are not affected by availability of food or living space. Instead, predation by the stoat, a member of the weasel family that preys almost exclusively on lemmings, drives the population cycle. Three other lemming predators—the snow owl, the arctic fox, and the long-tailed skua—stabilize the population cycle that the stoat establishes.

Parasites may also interact with their hosts to cause regular cyclic fluctuations. Detailed studies of red grouse have shown that even managed populations in wildlife preserves may have significant cyclic oscillations. Reproduction in red grouse is related to the density of parasitic nematodes (roundworms) living in adult intestines. Fewer birds breed successfully when adults are infected with worms; thus a high density of worms leads to a population crash. Hypothesizing that red grouse populations fluctuate in response to the parasites, ecologists at the University of Stirling in the United Kingdom recently reduced or eliminated population fluctuations in several red grouse populations. They accomplished this by catching and orally treating the birds with a chemical that causes worms to be ejected from their bodies.

Competition is an important density-dependent factor

Competition is an interaction among two or more individuals that attempt to use the same essential resource, such as food, water, sunlight, or living space, that is in limited supply. The use of the resource by one of the individuals reduces the availability of that resource for other individuals. Competition occurs both within a given population (**intraspecific competition**) and among populations of different species (**interspecific competition**). We consider the effects of intraspecific competition here; interspecific competition is discussed in Chapter 52.

Individuals of the same species compete for a resource in limited supply by interference competition or by exploitation competition. In **interference competition,** also called **contest competition,** certain dominant individuals obtain an adequate supply of the limited resource at the expense of other individuals in the population; that is, the dominant individuals actively interfere with other individuals' access to resources. In **exploitation competition,** also called **scramble competition,** all the individuals in a population "share" the limited resource more or less equally so that at high population densities none of them obtains an adequate amount. The populations of species in which exploitation competition operates often oscillate over time, and there is always a risk that the population size will drop to zero. In contrast, those species in which interference competition operates experience a relatively small drop in population size, caused by the death of individuals that are unable to compete successfully.

Intraspecific competition among red grouse involves interference competition. When red grouse populations are small, the birds are less aggressive, and most young birds establish a feeding territory (an area defended against other members of the same species). However, when the population is large, establishing a territory is difficult because there are more birds than there are territories, and the birds are much more aggressive. Those birds without territories often die from predation or starvation. Thus, birds with territories use a larger share of the limited resource (the territory with its associated food and cover), whereas birds without territories cannot compete successfully.

The moose population on Isle Royale, Michigan, the largest island in Lake Superior, provides a vivid example of exploitation competition that is similar to the reindeer population on the Pribilof Islands (discussed earlier). Isle Royale differs from most islands in that large mammals can walk to it when the lake freezes over in winter. The minimum distance to be walked is 24 km (15 mi), however, so this movement has happened infrequently. Around 1900, a small herd of moose wandered across the ice of frozen Lake Superior and reached the island for the first time. By 1934 the moose population on the island had increased to about 3000 and had consumed almost all the edible vegetation. In the absence of this food resource, there was massive starvation in 1934. More than 60 years later, in 1996, a similar die-off claimed 80% of the moose after they had again increased to a high density. Thus, exploitation competition for scarce resources can result in dramatic population oscillations.

The effects of density-dependent factors are difficult to assess in nature

Most studies of density dependence have been conducted in laboratory settings where all density-dependent (and density-independent) factors except one are controlled experimentally. But populations in natural settings are exposed to a complex set of variables that continually change. As a result, in natural communities it is difficult to evaluate the relative effects of different density-dependent factors.

Ecologists from the University of California at Davis noted that few spiders occur on tropical islands inhabited by lizards, whereas more spiders and more species of spiders are found on lizard-free islands. Deciding to study these observations experimentally, David Spiller and Thomas Schoener staked out plots of vegetation (mainly seagrape shrubs) and enclosed some of them with lizard-proof screens (Fig. 51-6a). Some of the plots were emptied of all lizards; each control enclosure had approximately nine lizards. Nine web-building spider species were observed in the enclosures. Spiders were counted approximately 30 times from 1989 to 1994 (Fig. 51-6b). During the 4.5 years of observations reported here, spider population densities were higher in the lizard-free enclosures than in enclosures with lizards. Moreover, the enclosures without lizards had more species of spiders. Therefore, we might conclude that lizards control spider populations.

But even this relatively simple experiment may be explained by a combination of two density-dependent factors, predation (lizards eat spiders) and interspecific competition (lizards compete with spiders for insect prey—that is, both spiders and lizards eat insects). In this experiment, the effects of the two density-dependent factors in determining spider population size cannot be evaluated separately. Additional lines of evidence support the actions of both competition and predation at these sites.

Density-independent factors are generally abiotic

Any environmental factor that affects the size of a population but is not influenced by changes in population density is called a **density-independent factor.** Such factors are typically abiotic. Random weather events that reduce population size serve as density-independent factors. These often affect population density in unpredictable ways. A killing frost, severe blizzard, or hurricane, for example, may cause extreme and irregular reductions in a vulnerable population, regardless of its size, and thus may be considered largely density-independent.

Consider a density-independent factor that influences mosquito populations in arctic environments. These insects produce several generations per summer and achieve high population densities by the end of the season. A shortage of food does not seem to be a limiting factor for mosquitoes, nor is there any shortage of ponds in which to breed. What puts a stop to the skyrocketing mosquito population is winter. Not a single adult mosquito survives winter, and the entire population must grow afresh the next summer from the few eggs and hibernating larvae that survive. Thus severe winter weather is a density-independent factor that affects arctic mosquito populations.

Density-independent and density-dependent factors are often interrelated. Social animals, for example, often resist dangerous weather conditions by collective behavior, as in the case of sheep huddling together in a snowstorm. In this case it appears that the greater the population density, the better their ability to resist the environmental stress of a density-independent event (such as a snowstorm).

Review

- What are three examples of density-dependent factors that affect population growth? What are three density-independent factors?

BiologyNow™ Assess your understanding of **factors influencing population size,** by taking the pretest on your BiologyNow CD-ROM.

(a)

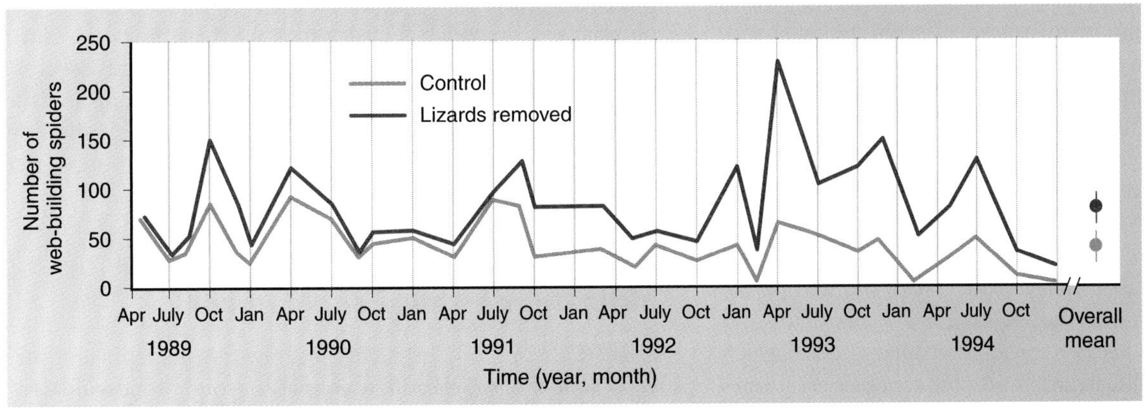

(b)

FIGURE **51-6** **Interaction of density-dependent factors.**

(a) Shown is one of the enclosures that Spiller and Schoener used for their field experiment. Spiders and insects can pass freely into and out of the enclosures, but lizards cannot. **(b)** Graph of the mean number of web-building spiders at each count in enclosures with lizards (control, *blue*) and enclosures with lizards removed *(red)*. An earlier experiment indicated that the enclosures themselves have no effect on the number of web-building spiders. *(b, From D.A Spiller and T.W. Schoener, "Lizards Reduce Spider Species Richness by Excluding Rare Species,"* Ecology, *Vol. 79, No. 2. Copyright © 1998 Ecological Society of America. Reprinted with permission.)*

LIFE HISTORY TRAITS

Learning Objectives

5 Contrast semelparous and iteroparous reproduction.

6 Distinguish among species exhibiting an *r* strategy, those with a *K* strategy, and those that do not easily fit either category.

7 Describe Type I, Type II, and Type III survivorship curves, and explain how life tables and survivorship curves indicate mortality and survival.

Each species is uniquely suited to its lifestyle. Many years pass before a young magnolia tree flowers and produces seeds, whereas a poppy plant grows from seed, flowers, and dies in a single season. A mating pair of black-browed albatrosses produces a single chick every year, but a mating pair of gray-headed albatrosses produces a single chick biennially (every other year).

Species that expend their energy in a single, immense reproductive effort are said to be **semelparous.** Most insects and invertebrates, many plants, and some species of fish exhibit semelparity. Pacific salmon, for example, hatch in fresh water and swim to the ocean, where they live until they mature. Adult salmon swim from the ocean back into the same rivers or streams in which they hatched to spawn (reproduce). After they spawn, the salmon die.

Agaves are semelparous plants that are common in arid tropical and semitropical areas. The thick, fleshy leaves of the agave plant are crowded into a rosette at the base of the stem. Commonly called the century plant because it was mistakenly thought to flower only once in a century, agaves can flower after they are 10 years old or so, after which the entire plant dies (Fig. 51-7).

Many species are **iteroparous** and exhibit repeated reproductive cycles—that is, reproduction during several breeding seasons—throughout their lifetimes. Iteroparity is common in most vertebrates, perennial herbaceous plants, shrubs, and trees.

FIGURE **51-7** **Semelparity.**

Agaves flower once and then die. The agave is a succulent with sword-shaped leaves arranged as a rosette around a short stem. Shown is *Agave shawii*, whose leaves grow to 61 cm (2 ft). Note the floral stalk, which can grow more than 3 m (10 ft) tall.

The timing of reproduction, earlier or later in life, is a crucial aspect of iteroparity and involves tradeoffs. On the one hand, reproducing earlier in life may mean a reduced likelihood of survival (because an individual is expending energy toward reproduction instead of its own growth), which reduces the potential for later reproduction. On the other hand, reproducing later in life means the individual has less time for additional reproductive events.

Ecologists try to understand the adaptive consequences of various **life history traits,** such as semelparity and iteroparity. Adaptations such as reproductive rate, age at maturity, and **fecundity** (potential capacity to produce offspring), all of which are a part of a species' life history traits, influence an organism's survival and reproduction. The ability of an individual to reproduce successfully, thereby making a genetic contribution to future generations of a population, is called its **fitness** (recall the discussions of fitness in Chapters 18 and 50).

Although many different life histories exist, some ecologists recognize two extremes: *r*-selected species and *K*-selected species. Keep in mind, as you read the following descriptions of *r* selection and *K* selection, that these concepts, although useful, oversimplify most life histories. Species tend to possess a combination of *r*-selected and *K*-selected traits, as well as traits that cannot be classified as either *r*-selected or *K*-selected. In addition, some populations within a species may exhibit the characteristics of *r*-selection, whereas other populations in different environments may assume *K*-selected traits.

Populations described by the concept of ***r* selection** have traits that contribute to a high population growth rate. Recall that *r* designates the per capita growth rate. Because such organisms have a high *r*, biologists call them ***r* strategists** or ***r*-selected species.** Small body size, early maturity, short life span, large broods, and little or no parental care are typical of many *r* strategists, which are usually opportunists found in variable, temporary, or unpredictable environments where the probability of long-term survival is low. Some of the best examples of *r* strategists are insects such as mosquitoes and common weeds such as the dandelion.

In populations described by the concept of ***K* selection,** traits maximize the chance of surviving in an environment where the number of individuals (N) is near the carrying capacity (K) of the environment. These organisms, called ***K* strategists** or ***K*-selected species,** do not produce large numbers of offspring. They characteristically have long life spans with slow development, late reproduction, large body size, and a low reproductive rate. *K* strategists tend to be found in relatively constant or stable environments, where they have a high competitive ability. Redwood trees are classified as *K* strategists. Animals that are *K* strategists typically invest in parental care of their young.

Tawny owls (*Strix aluco*), for example, are *K* strategists that pair-bond for life, with both members of a pair living and hunting in adjacent, well-defined territories. Their reproduction is regulated in accordance with the resources, especially the food supply, present in their territories. In an average year, 30% of the birds do not breed at all. If food supplies are more limited than initially indicated, many of those that do breed fail to incubate their eggs. Rarely do the owls lay the maximum number of eggs that they are physiologically capable of laying, and breeding is often delayed until late in the season, when the rodent populations on which they depend have become large. Thus, the behavior of tawny owls ensures better reproductive success of the individual and leads to a stable population at or near the carrying capacity of the environment. Starvation, an indication that the tawny owl population has exceeded the carrying capacity, rarely occurs.

Life tables and survivorship curves indicate mortality and survival

A **life table** can be constructed to show the mortality and survival data of a population or **cohort,** a group of individuals of the same age, at different times during their life span. Insurance companies were the first to use life tables, to calculate the relationship between a client's age and the likelihood of the client surviving to pay enough insurance premiums to cover the cost of the policy. Ecologists construct such tables for animals and plants based on data that rely on a variety of population sampling methods and age determination techniques.

Table 51-2 shows a life table for a cohort of 530 gray squirrels. The first two columns show the units of age (years) and the

TABLE 51-2 Life Table for a Cohort of 530 Gray Squirrels (*Sciurus carolinensis*)

Age Interval (Years)	Number Alive at Beginning of Age Interval	Proportion Alive at Beginning of Age Interval	Proportion Dying During Age Interval	Death Rate for Age Interval
0–1	530	1.000	0.747	0.747
1–2	134	0.253	0.147	0.581
2–3	56	0.106	0.032	0.302
3–4	39	0.074	0.031	0.418
4–5	23	0.043	0.021	0.488
5–6	12	0.022	0.013	0.591
6–7	5	0.009	0.006	0.666
7–8	2	0.003	0.003	1.000
8–9	0	0.000	0.000	—

Source: Adapted from Table 13.1, p. 150, in Smith, R.L. and T.M. Smith, *Elements of Ecology,* 4th ed., San Francisco, Benjamin/Cummings Science Publishing, 1998.

number of individuals in the cohort that were alive at the beginning of each age interval (the actual data collected in the field by the ecologist). The values for the third column (the proportion alive at the beginning of each age interval) are calculated by dividing each number in column 2 by 530, the number of squirrels in the original cohort. The values in the fourth column (the proportion dying during each age interval) are calculated using the values in the third column—by subtracting the number of survivors at the beginning of the next interval from those alive at the beginning of the current interval. For example, the proportion dying during interval 0–1 years $= 1.000 - 0.253 = 0.747$. The last column, the death rate for each age interval, is calculated by dividing the proportion dying during the age interval (column 4) by the proportion alive at the beginning of the age interval (column 3). For example, the death rate for the age interval 1–2 years is $0.147 \div 0.253 = 0.581$.

Survivorship is the probability that a given individual in a population or cohort will survive to a particular age. Plotting the logarithm (base 10) of the number of surviving individuals against age, from birth to the maximum age reached by any individual, produces a **survivorship curve.** Figure 51-8 shows the three main survivorship curves that ecologists recognize.

In Type I survivorship, as exemplified by bison and humans, the young and those at reproductive age have a high probability of surviving. The probability of survival decreases more rapidly with increasing age; mortality is concentrated later in life. Figure 51-9 shows a survivorship curve for a natural population of

FIGURE 51-8 Survivorship curves.

These curves represent the ideal survivorships of species in which mortality is greatest in old age (Type I), spread evenly across all age groups (Type II), and greatest among the young (Type III). The survivorship of most organisms can be compared to these curves.

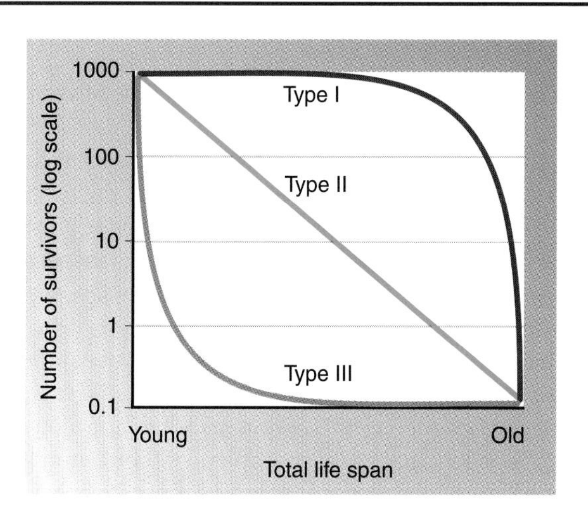

FIGURE 51-9 Survivorship curve for a Drummond phlox population.

Drummond phlox (*Phlox drummondii*) has a Type I survivorship after germination of the seeds. Bars above the graph indicate the various stages in the Drummond phlox life history. Data were collected at Nixon, Texas, in 1974 and 1975. Survivorship on the Y-axis begins at 0.296 instead of 1.00 because the study took into account death during the seed dormancy period prior to germination *(not shown). [From W.J. Leverich and D.A. Levin, "Age-Specific Survivorship and Reproduction in* Phlox drummondii," *American Naturalist, 113:6 (1979), p. 1148. Copyright © 1979 University of Chicago Press. Reprinted with permission.]*

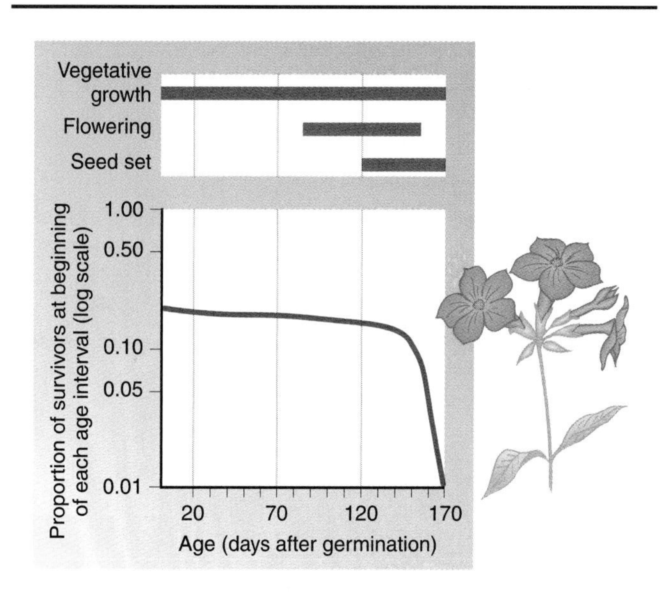

Drummond phlox, an annual native to East Texas that became widely distributed in the southeastern United States after it escaped from cultivation. Because most Drummond phlox seedlings survive to reproduce, after germination the plant exhibits a Type I survivorship that is typical of annuals.

In Type III survivorship, the probability of mortality is greatest early in life, and those individuals that avoid early death subsequently have a high probability of survival, that is, the probability of survival increases with increasing age. Type III survivorship is characteristic of oysters; young oysters have three free-swimming larval stages before settling down and secreting a shell. These larvae are vulnerable to predation, and few survive to adulthood.

In Type II survivorship, which is intermediate between Types I and III, the probability of survival does not change with age. The probability of death is equally likely across all age groups, resulting in a linear decline in survivorship. This constancy probably results from essentially random events that cause death with little age bias. Although this relationship between age and survivorship is rare, some lizards have Type II survivorship.

The three survivorship curves are generalizations, and few populations exactly fit one of the three. Some species have one type of survivorship curve early in life and another type as adults. Herring gulls, for example, start out with a Type III survivorship curve but develop a Type II curve as adults (Fig. 51-10). The survivorship curve shown in this figure is characteristic of birds in general. Note that most death occurs almost immediately after hatching, despite the protection and care given to the chicks by the parent bird. Herring gull chicks die from predation or attack by other herring gulls, inclement weather, infectious diseases, or starvation following death of the parent. Once the chicks become independent, their survivorship increases dramatically, and death occurs at about the same rate throughout their remaining lives. Few or no herring gulls die from the degenerative diseases of "old age" that cause death in most humans.

FIGURE **51-10** **Survivorship curve for a herring gull population.**

Herring gulls have Type III survivorship as chicks and Type II survivorship as adults. Data were collected from Kent Island, Maine, during a 5-year period in the 1930s; baby gulls were banded to establish identity. The very slight increase just prior to 20 years of age is due to sampling error. *[Paynter, R. A. Jr., A New Attempt to Construct Lifetables for Kent Island Herring Gulls,* Bull. Mus. Comp. Zool., *133(11), 489–528, 1966]*

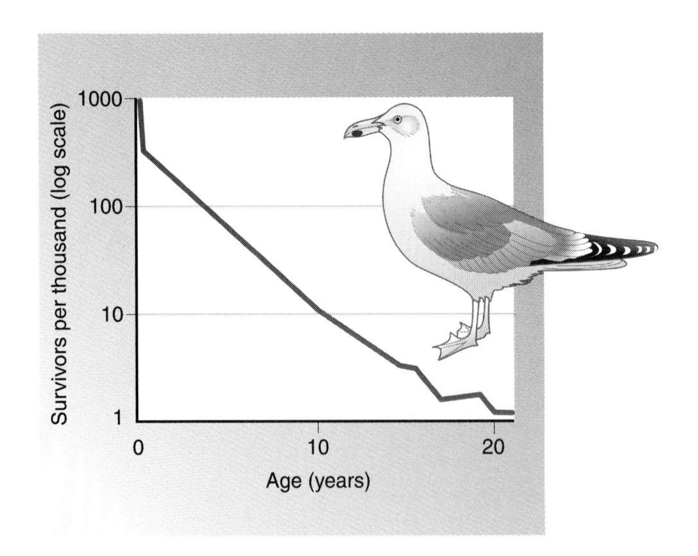

Review

- What are the advantages of semelparity? Of iteroparity? Are there disadvantages?
- Why is parental care of young a common characteristic of *K* strategists?
- Do all survivorship curves neatly fit the Type I, II, or III models? Explain.

METAPOPULATIONS

Learning Objective

8 Define *metapopulation*, and distinguish between source habitats and sink habitats.

The natural environment is a heterogeneous **landscape** consisting of interacting ecosystems that provide a variety of habitat patches. Landscapes, which are typically several to many square kilometers in area, cover larger land areas than individual ecosystems. Consider a forest, for example. The forest landscape is a mosaic of different elevations, temperatures, levels of precipitation, soil moisture, soil types, and other properties. Because each species has its own habitat requirements, this heterogeneity in physical properties is reflected in the different organisms that occupy the various patches in the landscape (Fig. 51-11). Some species occur in very narrow habitat ranges, whereas others have wider habitat distributions.

Population ecologists have discovered that many species are not distributed as one large population across the landscape. Instead, many species exist as a series of local populations distributed in distinct habitat patches. Each local population has its own characteristic demographic features, such as birth, death, emigration, and immigration rates. A population that is divided into several local populations among which individuals occasionally disperse (emigrate and immigrate) is known as a **metapopulation.** For example, note the various local populations of red oak on the mountain slope in Figure 51-11b.

The spatial distribution of a species occurs because different habitats vary in suitability, from unacceptable to preferred. The preferred sites are more productive habitats that increase the likelihood of survival and reproductive success for the individuals living there. Good habitats, called **source habitats,** are areas where local reproductive success is greater than local mortality. **Source populations** generally have greater population densities than populations at less suitable sites, and surplus individuals in the source habitat disperse and find another habitat in which to settle and reproduce.

Individuals living in lower-quality habitats may suffer death or, if they survive, poor reproductive success. Lower-quality habitats, called **sink habitats,** are areas where local reproductive success is less than local mortality. Without immigration from other areas, a **sink population** declines until extinction

(a)

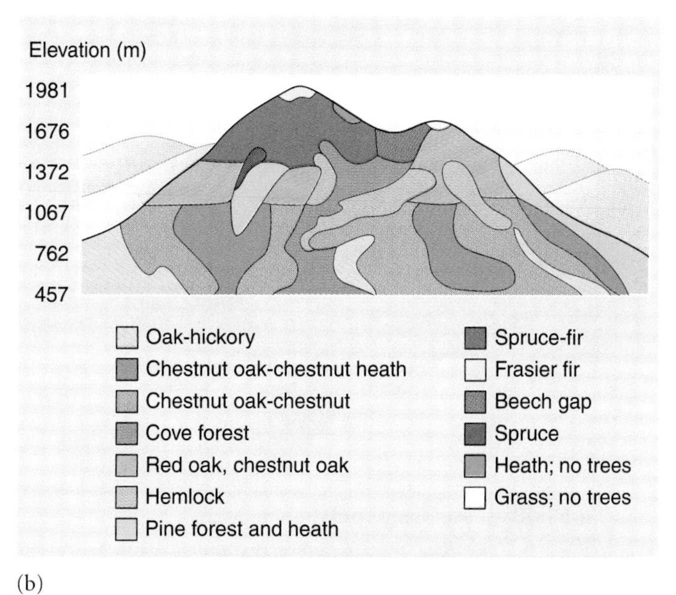

(b)

FIGURE **51-11** | **The mosaic nature of landscapes.**

(a) This early spring view of Deep Creek Valley in the Great Smoky Mountains National Park gives some idea of the heterogeneity of the landscape. During the summer, when all the vegetation is a deep green, the landscape appears homogeneous. **(b)** An evaluation of the distribution of vegetation on a typical west-facing slope in the Great Smoky Mountains National Park reveals that the landscape consists of patches. Chestnut oak is a species of oak (*Quercus prinus*); chestnut heath are areas within chestnut oak forest where the trees are widely scattered and the slopes underneath are covered by a thick growth of laurel (*Kalmia*) shrubs; cove forest is a mixed stand of deciduous trees. *(b, Adapted from Whittaker, R.H. "Vegetation of the Great Smoky Mountains."* Ecological Monographs, *Vol. 26, 1956)*

occurs. If a local population becomes extinct, individuals from a source habitat may recolonize the vacant habitat at a later time. Source and sink habitats, then, are linked to one another by dispersal (Fig. 51-12).

FIGURE **51-12** | **Source and sink populations in a hypothetical metapopulation.**

The local populations in the habitat patches shown here collectively make up the metapopulation. Source habitats provide individuals that emigrate and colonize sink habitats. Studied over time, the metapopulation is shown to exist as a shifting pattern of occupied and vacant habitat patches.

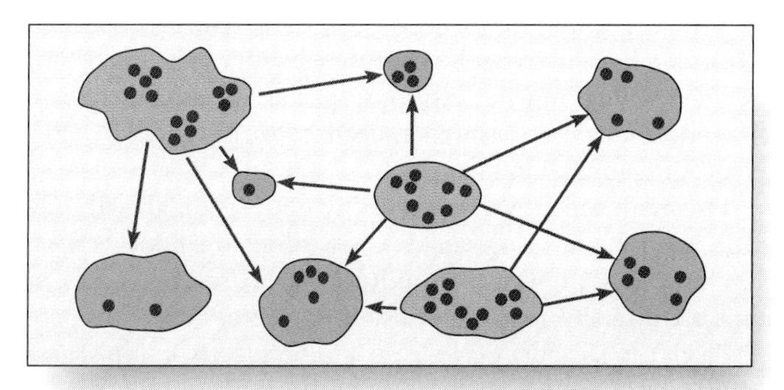

Metapopulations are becoming more common as humans alter the landscape, fragmenting existing habitats to accommodate homes and factories, agricultural fields, and logging. As a result, the concept of metapopulations, particularly as it relates to endangered and threatened species, has become an important area of study in conservation biology (see Chapter 55).

Review

- What is a metapopulation?
- How do you distinguish between a source habitat and a sink habitat?

BiologyNow™ Assess your understanding of **metapopulations** by taking the pretest on your BiologyNow CD-ROM.

HUMAN POPULATIONS

Learning Objectives

9 Summarize the history of human population growth.

10 Explain how highly developed and developing countries differ in population characteristics such as infant mortality rate, total fertility rate, replacement-level fertility, and age structure.

11 Distinguish between people overpopulation and consumption overpopulation.

Now that we have examined some of the basic concepts of population ecology, we can apply those concepts to the human population. Examine Figure 51-13, which shows the world increase in the human population since the development of agriculture

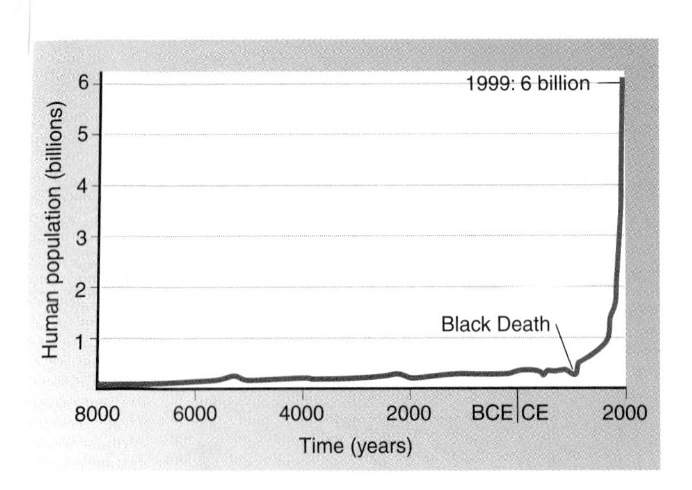

FIGURE **51-13** **Human population growth.**

During the last 1000 years, the human population has been increasing nearly exponentially. Population experts predict that the population will level out during the 21st century, forming the S curve observed in other species. (Black Death refers to a devastating disease, probably bubonic plague, that decimated Europe and Asia in the 14th century.)

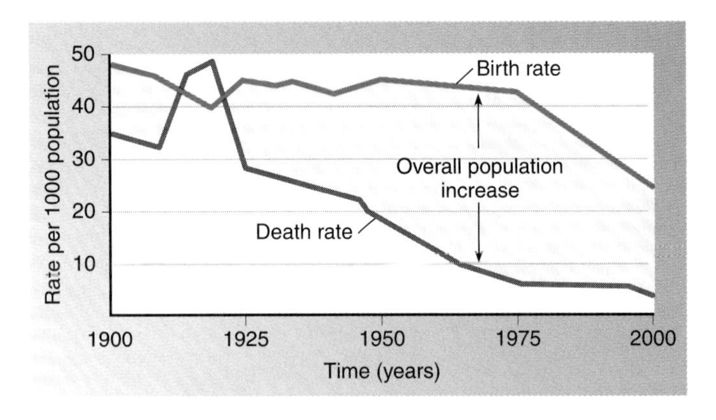

ACTIVE FIGURE **51-14** **Birth and death rates in Mexico, 1900 to 2000.**

Both birth and death rates declined during the 20th century, but because the death rate declined much more than the birth rate, Mexico has experienced a high growth rate. (The high death rate prior to 1920 was caused by the Mexican Revolution.) *(Population Reference Bureau)*

BiologyNow™ Explore **human populations and the future** by clicking on this figure on your BiologyNow CD-ROM.

approximately 10,000 years ago. Now look back at Figure 51-2 and compare the two curves. The characteristic J curve of exponential population growth shown in Figure 51-13 reflects the decreasing amount of time it has taken to add each additional billion people to our numbers. It took thousands of years for the human population to reach 1 billion, a milestone that took place around 1800. It took 130 years to reach 2 billion (in 1930), 30 years to reach 3 billion (in 1960), 15 years to reach 4 billion (in 1975), 12 years to reach 5 billion (in 1987), and 12 years to reach 6 billion (in 1999). The United Nations projects that the human population will reach 7 billion by 2013.

Thomas Malthus (1766–1834), a British clergyman and economist, was one of the first to recognize that the human population cannot continue to increase indefinitely (see Chapter 17). He pointed out that human population growth is not always desirable (a view contrary to the beliefs of his day and to those of many people even today) and that the human population is capable of increasing faster than the food supply. He maintained that the inevitable consequences of population growth are famine, disease, and war.

The world population increased by over 80 million from 2002 to 2003, reaching 6.3 billion in 2003. This change was not caused by an increase in the birth rate (b). In fact, the world birth rate has actually declined during the past 200 years. The increase in population is due instead to a dramatic decrease in the death rate (d), which has occurred primarily because greater food production, better medical care, and improved sanitation practices have increased the life expectancies of a great majority of the global population. For example, from 1920 to 2000, the death rate in Mexico fell from approximately 40 per 1000 individuals to 4 per 1000, whereas the birth rate dropped from approximately 40 per 1000 individuals to 24 per 1000 (Fig. 51-14).

The human population has reached a turning point. Although our numbers continue to increase, the world per capita growth rate (r) has declined over the past several years, from a peak of 2.2% per year in the mid-1960s to 1.3% per year in 2003. Population experts at the United Nations and the World Bank have projected that the growth rate will continue to slowly decrease until zero population growth is attained. Thus, exponential growth of the human population will end, and the S curve will replace the J curve. Demographers project that **zero population growth,** the point at which the birth rate equals the death rate ($r = 0$), will occur toward the end of the 21st century.

The latest (2003) projections available through the Population Reference Bureau, based on data from the United Nations, forecast that the human population will be 9.2 billion in the year 2050. This is a "medium" projection; such population projections are "what if" exercises. Given certain assumptions about future tendencies in natality, mortality, and dispersal, an area's population can be calculated for a given number of years into the future. Population projections indicate the changes that may occur, but they must be interpreted with care because they vary depending on what assumptions have been made. Small differences in fertility, as well as death rates, produce large differences in population forecasts.

The main unknown factor in any population growth scenario is Earth's carrying capacity. According to Dr. Joel Cohen, who has a joint professorship at Rockefeller University and Columbia University, most published estimates of how many people Earth can support range from 4 billion to 16 billion. These estimates vary so widely because of the assumptions that are made about standard of living, resource consumption, technological innovations, and waste generation. If we want all people to have a high level of material well-being equivalent to the lifestyles common in highly developed countries, then Earth will clearly be able to support far fewer humans than if everyone lives just above the subsistence level. Earth's carrying capacity for the human population, then, is not decided simply by natural environmental constraints. Human choices and values have to be factored into the assessment.

It is also not clear what will happen to the human population if or when the carrying capacity is approached. Optimists suggest that the human population will stabilize because of a decrease in the birth rate. Some experts take a more pessimistic view and predict that the widespread degradation of our environment caused by our ever-expanding numbers will make Earth uninhabitable for humans and other species. These experts contend that a massive wave of human suffering and death will occur. Some experts think the human population has already exceeded the carrying capacity of the environment, a potentially dangerous situation that threatens our long-term survival as a species.

Not all countries have the same growth rate

Although world population figures illustrate overall trends, they do not describe other important aspects of the human population story, such as population differences from country to country. **Human demographics,** the science that deals with human population statistics such as size, density, and distribution, provides interesting information on the populations of various countries. As you probably know, not all parts of the world have the same rates of population increase. Countries can be classified into two groups, highly developed and developing, based on their rates of population growth, degrees of industrialization, and relative prosperity (Table 51-3).

Highly developed countries, such as the United States, Canada, France, Germany, Sweden, Australia, and Japan, have low rates of population growth and are highly industrialized relative to the rest of the world. Highly developed countries have the lowest birth rates in the world. Indeed, some highly developed countries such as Germany have birth rates just below that needed to sustain the population and are thus declining slightly in numbers. Highly developed countries also have low **infant mortality rates** (the number of infant deaths per 1000 live births). The infant mortality rate of the United States was 6.9 in 2003, for example, compared with a world infant mortality rate of 55. Highly developed countries also have longer life expectancies (77 years in the United States versus 67 years worldwide) and higher average GNI PPP per capita ($34,280 in the United States versus $7160 worldwide). GNI PPP per capita is gross national income in purchasing power parity (PPP) divided by midyear population. It indicates the amount of goods and services an average citizen of that particular region or country could buy in the United States.

Developing countries fall into two subcategories: moderately developed and less developed. Mexico, Turkey, Thailand, and most countries of South America are examples of *moderately developed countries.* Their birth rates and infant mortality rates are higher than those of highly developed countries, but they are declining. Moderately developed countries have a medium level of industrialization, and their average GNI PPP per capita is lower than those of highly developed countries.

Bangladesh, Niger, Ethiopia, Laos, and Cambodia are examples of *less developed countries.* These countries have the highest birth rates, the highest infant mortality rates, the lowest life expectancies, and the lowest average GNI PPP per capita in the world.

One way to represent the population growth of a country is to determine the **doubling time,** the amount of time it would take for its population to double in size, assuming its current growth rate did not change. A simplified formula for doubling time (t_d) is $t_d = 70 \div r$. (The actual formula involves calculus and is beyond the scope of this text.)

A look at a country's doubling time identifies it as a highly, moderately, or less developed country: The shorter the doubling time, the less developed the country. At rates of growth in 2000, the doubling time is 26 years for Laos, 29 years for Ethiopia, 46 years for Turkey, 70 years for Thailand, 120 years for the United States, and 204 years for France.

It is also instructive to examine **replacement-level fertility,** the number of children a couple must produce to "replace" themselves. Replacement-level fertility is usually given as 2.1 children in highly developed countries and 2.7 children in developing countries. The number is always greater than 2.0 because some children die before they reach reproductive age. Higher infant mortality rates are the main reason that replacement lev-

TABLE 51-3 Comparison of 2003 Population Data in Developed and Developing Countries

	Developed	Developing	
	(Highly Developed) United States	(Moderately Developed) Brazil	(Less Developed) Ethiopia
Fertility rate	2.0	2.2	5.9
Doubling time rate in 2000	120 years	48 years	29 years
Infant mortality rate	6.9 per 1000	33 per 1000	107 per 1000
Life expectancy at birth	77 years	69 years	42 years
GNI PPP per capita (U.S. $)	$34,280	$7070	$1800
Women using modern contraception	72%	70%	6%

Source: Population Reference Bureau.

els in developing countries are greater than in highly developed countries. The **total fertility rate**—the average number of children born to a woman during her lifetime—is 2.8 worldwide, which is well above replacement levels. However, it has fallen below replacement levels, to 1.5, in the highly developed countries. The total fertility rate in the United States is higher than in other industrialized countries—2.0.

The population in many developing countries is beginning to approach stabilization. The fertility rate must decline for the population to stabilize (Table 51-4; note the general decline in total fertility rate from the 1960s to 2003 in selected developing countries). The total fertility rate in developing countries has decreased from an average of 6.1 children per woman in 1970 to 3.5 in 2003 (excluding China), or 3.1 if China is included.

Although the fertility rates in these countries have declined, it should be remembered that most still exceed replacement-level fertility. Consequently, the populations in these countries are still increasing. Also, even when fertility rates equal replacement-level fertility, population growth will still continue for some time. To understand why this is so, we now examine the age structure of various countries.

TABLE **51-4** Fertility Changes in Selected Developing Countries

Country	Total Fertility Rate 1960–1965	2003
Bangladesh	6.7	3.6
Brazil	6.2	2.2
China	5.9	1.7
Egypt	7.1	3.5
Guatemala	6.9	4.5
India	5.8	3.1
Kenya	8.1	4.4
Mexico	6.8	2.8
Nepal	5.9	4.5
Nigeria	6.9	5.8
Thailand	6.4	1.7

Source: Population Reference Bureau.

The age structure of a country helps predict future population growth

To predict the future growth of a population, it is important to know its **age structure,** which is the number and proportion of people at each age in a population. The number of males and number of females at each age, from birth to death, are represented in an **age structure diagram.**

The overall shape of an age structure diagram indicates whether the population is increasing, stationary, or shrinking. The age structure diagram of a country with a high growth rate (such as Nigeria or Bolivia) is shaped like a pyramid (Fig. 51-15a).

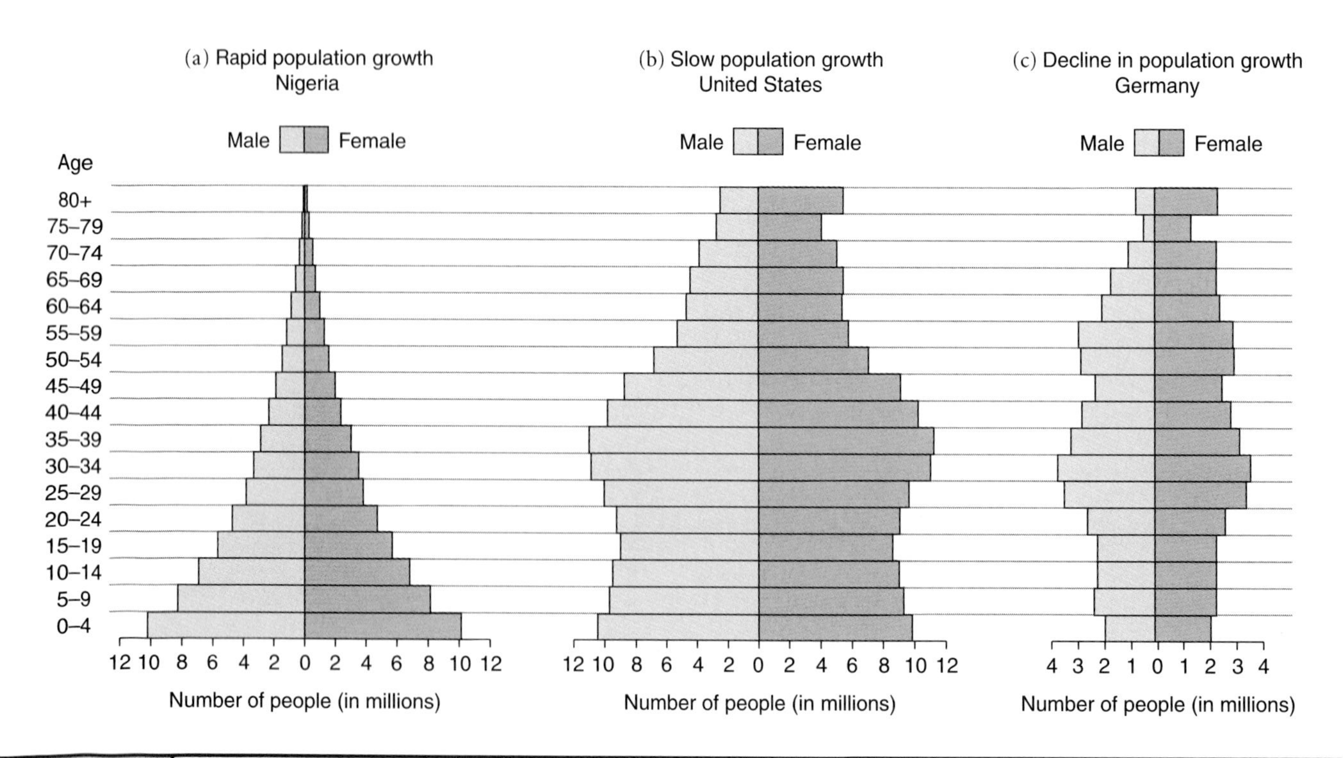

FIGURE **51-15** Age structure diagrams.

These age structure diagrams for **(a)** Nigeria, **(b)** the United States, and **(c)** Germany indicate that less developed countries such as Nigeria have a greater percentage of young people than highly developed countries. As a result, less developed countries are projected to have greater population growth than highly developed countries. *(Reproduced by authors using 1995 data from the Population Reference Bureau)*

Because the largest percentage of the population is in the prereproductive age group (that is, 0 to 14 years of age), the probability of future population growth is great. A strong **population growth momentum** exists because when all these children mature, they will become the parents of the next generation, and this group of parents will be larger than the previous group. Thus, even if the fertility rate of such a country declines to replacement level (that is, couples have smaller families than their parents did), the population will continue to grow for some time. Population growth momentum can have either a positive value (that is, the population will grow) or a negative value (that is, the population will decline). However, it is usually discussed in a positive context, to explain how the future growth of a population is affected by its present age distribution.

In contrast, the more tapered bases of the age structure diagrams of countries with slowly growing, stable, or declining populations indicate a smaller proportion of the population will become the parents of the next generation (Fig. 51-15b and c). The age structure diagram of a stable population, one that is neither growing nor shrinking, demonstrates that the number of people at prereproductive and reproductive ages are approximately the same. Also, a larger percentage of the population is older, that is, postreproductive, than in a rapidly increasing population. Many countries in Europe have stable populations. In a population that is shrinking in size, the prereproductive age group is smaller than either the reproductive or postreproductive group. Germany, Russia, and Bulgaria are examples of countries with slowly shrinking populations.

Worldwide, 30% of the human population is younger than age 15. When these people enter their reproductive years, they have the potential to cause a large increase in the growth rate. Even if the birth rate does not increase, the growth rate will increase simply because more people are reproducing.

Most of the world population increase since 1950 has taken place in developing countries as a result of the younger age structure and the higher-than-replacement-level fertility rates of their populations. In 1950, 66.8% of the world's population was in the developing countries in Africa, Asia (minus Japan), and Latin America. Between 1950 and 2003, the world's population more than doubled in size, but most of that growth occurred in developing countries. As a reflection of this, in 2003 the people in developing countries had increased to about 86% of the world's population. Most of the population increase that will occur during the 21st century will also take place in developing countries, largely the result of their young age structures. These countries, most of which are poor, are least able to support such growth.

Environmental degradation is related to population growth and resource consumption

The relationships among population growth, use of natural resources, and environmental degradation are complex, but we can make two useful generalizations. First, although the amount of resources essential to an individual's survival may be small, a rapidly increasing population (as in developing countries) tends to overwhelm and deplete a country's soils, forests, and other natural resources. Second, in highly developed nations individual resource demands are large, far above requirements for survival. To satisfy their desires rather than their basic needs, people in more affluent nations exhaust resources and degrade the global environment through excessive consumption and "throwaway" lifestyles.

Rapid population growth can cause natural resources to be overexploited. For example, large numbers of poor people must grow crops on land that is inappropriate for farming, such as mountain slopes or some tropical rain forests. Although this practice may provide a short-term solution to the need for food, it does not work in the long run, because when these lands are cleared for farming, their agricultural productivity declines rapidly, and severe environmental deterioration occurs.

The effects of population growth on natural resources are particularly critical in developing countries. The economic growth of developing countries is often tied to the exploitation of natural resources, often for export to highly developed countries. Developing countries are faced with the difficult choice of exploiting natural resources to provide for their expanding populations in the short term (to pay for food or to cover debts) or conserving those resources for future generations. It is instructive to note that the economic growth and development of the United States and of other highly developed nations came about through the exploitation and, in some cases, the destruction of their natural resources.

Although resource issues are clearly related to population size (more people use more resources), an equally, if not more, important factor is a population's resource consumption. People in highly developed countries are extravagant consumers—their use of resources is greatly out of proportion to their numbers. A single child born in a highly developed country such as the United States causes a greater impact on the environment and on resource depletion than a dozen or more children born in a developing country. Many natural resources are needed to provide the automobiles, air conditioners, disposable diapers, cell phones, DVDs, computers, clothes, athletic shoes, furniture, boats, and other "comforts" of life in highly developed nations. Thus the disproportionately large consumption of resources by the United States and other highly developed countries affects natural resources and the environment as much or more than the population explosion in the developing world does.

A country is overpopulated if the level of demand on its resource base results in damage to the environment. If we compare human impact on the environment in developing and highly developed countries, we see that a country can be overpopulated in two ways. **People overpopulation** occurs when the environment is worsening from too many people, even if those people consume few resources per person. People overpopulation is the current problem in many developing nations.

In contrast, **consumption overpopulation** occurs when each individual in a population consumes too large a share of resources. The effect of consumption overpopulation on the environment is the same as that of people overpopulation—pollution and degradation of the environment. Most affluent highly developed nations, including the United States, suffer from consumption overpopulation. Highly developed nations represent only 19.4% of the world's population, yet they consume significantly more than half its resources. According to the Worldwatch Institute, a private research institution in Washington, D.C., highly developed nations account for the lion's share of total resources consumed, as follows: 86% of aluminum used, 76% of timber harvested, 68% of energy produced, 61% of meat eaten, and 42% of fresh water consumed. These nations also generate 75% of the world's pollution and waste.

Review

- What is replacement-level fertility? Why is it greater in developing countries than in highly developed countries?
- How can a single child born in the United States have a greater effect on the environment and natural resources than a dozen children born in Kenya?

BiologyNow™ Assess your understanding of **human populations** by taking the pretest on your BiologyNow CD-ROM.

SUMMARY WITH KEY TERMS

1 Define *population density* and *dispersion,* and describe the main types of population dispersion.

- **Population density** is the number of individuals of a species per unit of area or volume at a given time.
- Population **dispersion** (spacing) may be **clumped dispersion** (clustered in specific parts of the habitat), **uniform dispersion** (evenly spaced), or **random dispersion** (unpredictably spaced).

2 Explain the four factors (natality, mortality, immigration, and emigration) that produce changes in population size, and solve simple problems involving these changes.

- Population size is affected by the average per capita birth rate (b), average per capita death rate (d), and two measures of dispersal: average per capita **immigration** rate (i), and average per capita **emigration** rate (e).
- The **growth rate (*r*)** is the rate of change (increase or decrease) of a population on a per capita basis.
- $r = b - d$ on a global scale (when dispersal is not a factor). Populations increase in size as long as the average per capita birth rate (**natality**) is greater than the average per capita death rate (**mortality**).
- $r = (b - d) + (i - e)$ for a local population (where dispersal is a factor).

3 Define *intrinsic rate of increase* and *carrying capacity,* and explain the differences between J-shaped and S-shaped growth curves.

- **Intrinsic rate of increase (r_{max})** is the maximum rate at which a species or population could increase in number under ideal conditions.
- Although certain populations exhibit an accelerated pattern of growth known as **exponential population growth** for limited periods of time (the J-shaped curve), eventually the growth rate decreases to around zero or becomes negative.
- Population size is modified by limits set by the environment. The **carrying capacity (*K*)** of the environment is the largest population that can be maintained for an indefinite time by a particular environment.
- **Logistic population growth,** when graphed, shows a characteristic S-shaped curve. Seldom do natural populations follow the logistic growth curve very closely.

4 Contrast the influences of density-dependent and density-independent factors on population size, and give examples of each.

- **Density-dependent factors** regulate population growth by affecting a larger proportion of the population as population density rises. Predation, disease, and competition are examples.
- **Density-independent factors** limit population growth but are not influenced by changes in population density. Hurricanes and blizzards are examples.

5 Contrast semelparous and iteroparous reproduction.

- **Semelparous** species expend their energy in a single, immense reproductive effort. **Iteroparous** species exhibit repeated reproductive cycles throughout their lifetimes.

6 Distinguish among species exhibiting an *r* strategy, those with a *K* strategy, and those that do not easily fit either category.

- Although many different combinations of **life history traits** exist, some ecologists recognize two extremes: *r* strategy and *K* strategy.
- An *r* strategy emphasizes a high growth rate. Organisms characterized as ***r*-strategists** often have small body sizes, high reproductive rates, and short **life spans,** and they typically inhabit variable environments.
- A *K* strategy maintains a population near the carrying capacity of the environment. Organisms characterized as ***K*-strategists** often have large body sizes, low reproductive rates, and long life spans, and they typically inhabit stable environments.
- The two strategies oversimplify most life histories. Many species combine *r*-selected and *K*-selected traits, as well as traits that cannot be classified as either *r*-selected or *K*-selected.

7 Describe Type I, Type II, and Type III survivorship curves, and explain how life tables and survivorship curves indicate mortality and survival.

- A **life table** shows the mortality and survival data of a population or **cohort**—a group of individuals of the same age—at different times during their life span.
- **Survivorship** is the probability that a given individual in a population or cohort will survive to a particular age. There are three general **survivorship curves:** Type I survivorship, in which mortality is greatest in old age; Type II survivorship, in which mortality is spread evenly across all age groups; and Type III survivorship, in which mortality is greatest among the young.

8 Define *metapopulation,* and distinguish between source and sink habitats.

- Many species exist as a **metapopulation,** a set of local populations among which individuals that are distributed in distinct habitat patches across a **landscape,** which is a large area of terrain (several to many square kilometers) composed of interacting ecosystems.
- Within a metapopulation, individuals occasionally disperse from one local habitat to another, by emigration and immigration.

SUMMARY WITH KEY TERMS *(continued)*

- **Source habitats** are preferred sites where local reproductive success is greater than local mortality. Surplus individuals disperse from source habitats.
- **Sink habitats** are lower-quality habitats where individuals may suffer death or, if they survive, poor reproductive success.
- If extinction of a local **sink population** occurs, individuals from a **source population** may recolonize the vacant habitat at a later time.

9 Summarize the history of human population growth.

- The world population increased over 80 million from 2002 to 2003, reaching 6.3 billion in 2003.
- Although our numbers continue to increase, the per capita growth rate (r) has declined over the past several years, from a peak in 1965 of about 2% per year to a 2003 growth rate of 1.3% per year.
- Scientists who study **human demographics** (human population statistics), project that the world population will become stationary ($r = 0$, or **zero population growth**) by the end of the 21st century.

10 Explain how highly developed and developing countries differ in population characteristics such as infant mortality rate, total fertility rate, replacement-level fertility, and age structure.

- **Highly developed countries** have the lowest birth rates, lowest **infant mortality rates,** lowest **total fertility rates,** longest life expectancies, and highest GNI PPP per capita (a measure of the amount of goods and services the average citizen could purchase in the United States). **Developing countries** have the highest birth rates, highest infant mortality rates, highest total fertility rates, shortest life expectancies, and lowest GNI PPP per capita.
- The **age structure** of a population greatly influences population dynamics.
- It is possible for a country to have **replacement-level fertility** and still experience population growth if the largest percentage of the population is in the prereproductive years. A young age structure causes a positive **population growth momentum** as the very large prereproductive age group matures and becomes parents.

11 Distinguish between people overpopulation and consumption overpopulation.

- Developing countries tend to have **people overpopulation,** in which population increase degrades the environment even though each individual uses few resources.
- Highly developed countries tend to have **consumption overpopulation,** in which each individual in a slow-growing or stationary population consumes a large share of resources, resulting in environmental degradation.

POST-TEST

1. Population ____________ is the number of individuals of a species per unit of habitat area or volume at a given time. (a) dispersion (b) density (c) survivorship (d) age structure (e) demographics
2. Which of the following patterns of cars parked along a street is an example of uniform dispersion? (a) five cars parked next to one another in the middle, leaving two empty spaces at one end and three empty spaces at the other end (b) five cars parked in the pattern: car, empty space, car, empty space, etc. (c) five cars parked in no discernible pattern, sometimes having empty spaces on each side and sometimes parked next to another car
3. This type of dispersion occurs when individuals are concentrated in certain parts of the habitat. (a) clumped (b) uniform (c) random (d) both a and b (e) both b and c
4. The average per capita birth rate, or ____________, increases population size, whereas the average per capita death rate, or ____________, decreases population size. (a) natality; demography (b) exploitation competition; interference competition (c) mortality; natality (d) total fertility rate; mortality (e) natality; mortality
5. The per capita growth rate of a population where dispersal is not a factor is expressed as (a) $i + e$ (b) $b - d$ (c) dN/dt (d) $rN(K - N)$ (e) $(K - N) \div K$
6. The maximum rate at which a population could increase under ideal conditions is known as its (a) total fertility rate (b) survivorship (c) intrinsic rate of increase (d) doubling time (e) age structure
7. When r is a positive number, the population size is (a) stable (b) increasing (c) decreasing (d) either increasing or decreasing, depending on interference competition (e) either increasing or stable, depending on whether the species is semelparous
8. In a graph of population size versus time, a J-shaped curve is characteristic of (a) exponential population growth (b) logistic population growth (c) zero population growth (d) replacement-level fertility (e) population growth momentum
9. The largest population that can be maintained by a particular environment for an indefinite period is known as a (a) semelparous population (b) population undergoing exponential growth (c) metapopulation (d) population's carrying capacity (e) source population
10. Giant bamboos live many years without reproducing, then send up a huge flowering stalk and die shortly thereafter. Giant bamboo is therefore an example of (a) iteroparity (b) a source population (c) a metapopulation (d) an r strategist (e) semelparity
11. Predation, disease, and competition are examples of ____________ factors. (a) density-dependent (b) density-independent (c) survivorship (d) dispersal (e) semelparous
12. ____________ competition occurs within a population and ____________ competition occurs among populations of different species.(a) interspecific; intraspecific (b) intraspecific; interspecific (c) Type I survivorship; Type II survivorship (d) interference; exploitation (e) exploitation; interference
13. Population experts project that during the 21st century, the human population will (a) increase the most in highly developed countries (b) increase the most in developing countries (c) increase at similar rates in all countries (d) decrease in developing countries but stabilize in highly developed countries (e) increase and then decrease dramatically in all countries

POST-TEST *(continued)*

14. A highly developed country has a (a) long doubling time (b) low infant mortality rate (c) high GNI PPP per capita (d) both a and b are correct (e) all three answers (a, b, and c) are correct
15. The continued growth of a population with a young age structure, even after its fertility rate has declined, is known as (a) population doubling (b) iteroparity (c) population growth momentum (d) *r* selection (e) density dependence
16. The 2003 population of the Netherlands was 16.2 million, and its land area is 15,768 square miles. The 2003 population of the United States was 291.5 million, and its land area is 3,717,796 square miles. Which country has the greater population density?
17. The population of India in 2003 was 1068.6 million, and its growth rate was 1.7% per year. Calculate the 2004 population of India.
18. The world population in 2003 was 6.3 billion, and its annual growth rate was 1.3%. If the birth rate was 22 per 1000 people in the year 2003, what was the death rate, expressed as number per 1000 people?

CRITICAL THINKING

1. How might pigs at a trough be an example of exploitation competition? Of interference competition?
2. Explain why the population size of a species that competes by interference competition is often near the carrying capacity, whereas the population size of a species that competes by exploitation competition is often greater than or below the carrying capacity.
3. A female elephant bears a single offspring every two to four years. Based on this information, which survivorship curve do you think is representative of elephants? Explain your answer.
4. In Bolivia, 40% of the population is younger than age 15, and 4% is older than 65. In Austria, 17% of the population is younger than 15, and 15% is older than 65. Which country will have the highest growth rate over the next two decades? Why?

- Visit our Web site at **http://biology.brookscole.com/solomon7** for links to chapter-related resources on the World Wide Web. Additional online materials relating to this chapter can also be found on our Web site.

BIOLOGY NOW RESOURCES

Active Figures

51-4: Population curves

51-14: Age stricture and the future of human populations

Preparing for an exam? Take a diagnostic test on your BiologyNow CD-ROM.

Post-Test Answers

1. b	2. b	3. a	4. e
5. b	6. c	7. b	8. a
9. d	10. e	11. a	12. b
13. b	14. e	15. c	

16. Netherlands (The density of the Netherlands is 16.2 million ÷ 15,768 mi^2 = 1027 people per mi^2. The density of the United States is 291.5 million ÷ 3,717,796 mi^2 = 78 people per mi^2.)
17. 1086.8 million
    18.9 per 1000

52

Community Ecology

Michael P. Gadomski/Photo Researchers, Inc.

Community in a rotting log. Fallen logs, called "nurse logs," shelter plants and other organisms and enrich the soil as they decay. Photographed in Pennsylvania.

CHAPTER OUTLINE

In Chapter 51 we examined the dynamics of single populations, including the ways individual populations change and what factors affect those changes. In the natural world, however, species do not exist as isolated populations. Rather, most populations are the interacting parts of a complex **community,** which consists of an association of populations of different species that live and interact in the same place at the same time. The definition for *community* is deliberately broad because it refers to ecological categories of various sizes and levels of interaction.

Communities exhibit characteristic properties that populations lack. These properties, known collectively as *community structure* and *community functioning,* include the number and types of species present, the relative abundance of each species, the interactions among different species, community resilience to disturbances, energy and nutrient flow throughout the community, and productivity. **Community ecology** is the description and analysis of patterns and processes within the community. Finding common patterns and processes in a wide variety of communities—for example, a pond community, a pine forest community, and a sagebrush desert community—helps ecologists understand community structure and functioning.

Communities are exceedingly difficult to study, because the large numbers of organisms of many different species interact with one another and are interdependent in a variety of ways. Species compete with one another for food, water, living space, and other resources. (Used in this context, a *resource* is anything from the environment that meets needs of a particular species.) Some organisms kill and eat other organisms. Some species form intimate associations with one another, whereas other species seem only distantly connected. Certain species interact in positive ways, in a process known as *facilitation,* which modifies and enhances the local environment for other species. For example, alpine plants in harsh mountain environments grow faster and larger and reproduce more successfully when other plants are growing nearby. Also, each organism plays one of three main roles in community life: producer, consumer, or decomposer (see Chapter 1). Unraveling the many positive and negative, direct and indirect, interactions of organisms living together as a community is one of the goals of community ecologists.

Communities vary greatly in size, lack precise boundaries, and are rarely completely isolated. They interact with and influence one another in countless ways, even when the interaction is not readily apparent. Furthermore, there are communities nested within larger communities. A forest is a community, but so is a rotting log in that forest (see photograph). Insects, plants, and fungi invade a fallen tree as it undergoes a series of decay steps. First, termites and other wood-boring insects forge tunnels through the bark and wood. Later, other insects, plant roots, and fungi follow and enlarge these openings. Mosses and lichens that establish on the log's surface trap rainwater and extract nutrient minerals, and fungi and bacteria speed decay, providing nutrients for other inhabitants. As decay progresses, small mammals burrow into the wood and eat the fungi, insects, and plants.

Organisms exist in an abiotic (nonliving) environment that is as essential to their lives as their interactions with one another. Minerals, air, water, and sunlight are just as much a part of a honeybee's environment, for example, as the flowers it pollinates and from which it takes nectar. A biological community and its abiotic environment together compose an **ecosystem.** Like communities, ecosystems are broad entities that refer to ecological units of various sizes. Ecosystems consist of all the interactions between the living world and the physical environment.

In this chapter we present some of the questions and challenges that community ecologists face in trying to find common patterns and processes that govern the assembly and persistence of interacting species. Although the living community is emphasized in this chapter, communities and their abiotic environments are inseparably linked. The abiotic components of ecosystems, including energy flow and trophic structure (feeding relationships), nutrient cycling, and climate, are considered in Chapter 53. ■

COMMUNITY STRUCTURE AND FUNCTIONING

Learning Objectives

1. Define *ecological niche,* distinguish between an organism's fundamental niche and its realized niche, and give several examples of limiting resources that might affect an organism's ecological niche.
2. Define *competition,* and distinguish between interspecific and intraspecific competition.
3. Summarize the concepts of the competitive exclusion principle, resource partitioning, and character displacement.
4. Define *predation,* and describe the effects of natural selection on predator–prey relationships.
5. Distinguish among mutualism, commensalism, and parasitism, and give examples of each.
6. Distinguish between keystone species and dominant species.

During the late 1990s, an intricate and fascinating relationship emerged among acorn production, white-footed mice, gypsy moth population growth, and the potential occurrence of Lyme disease in humans (Fig. 52-1). Large-scale experiments con-

FIGURE **52-1** **Connections to the size of an acorn crop.** When there is a bumper crop of acorns, more mice survive and breed in winter, and more deer are attracted to oak forests. Both mice and deer are hosts of ticks that may carry the Lyme disease bacterium to humans. Abundant mice also reduce gypsy moth populations, thereby improving the health of forest trees.

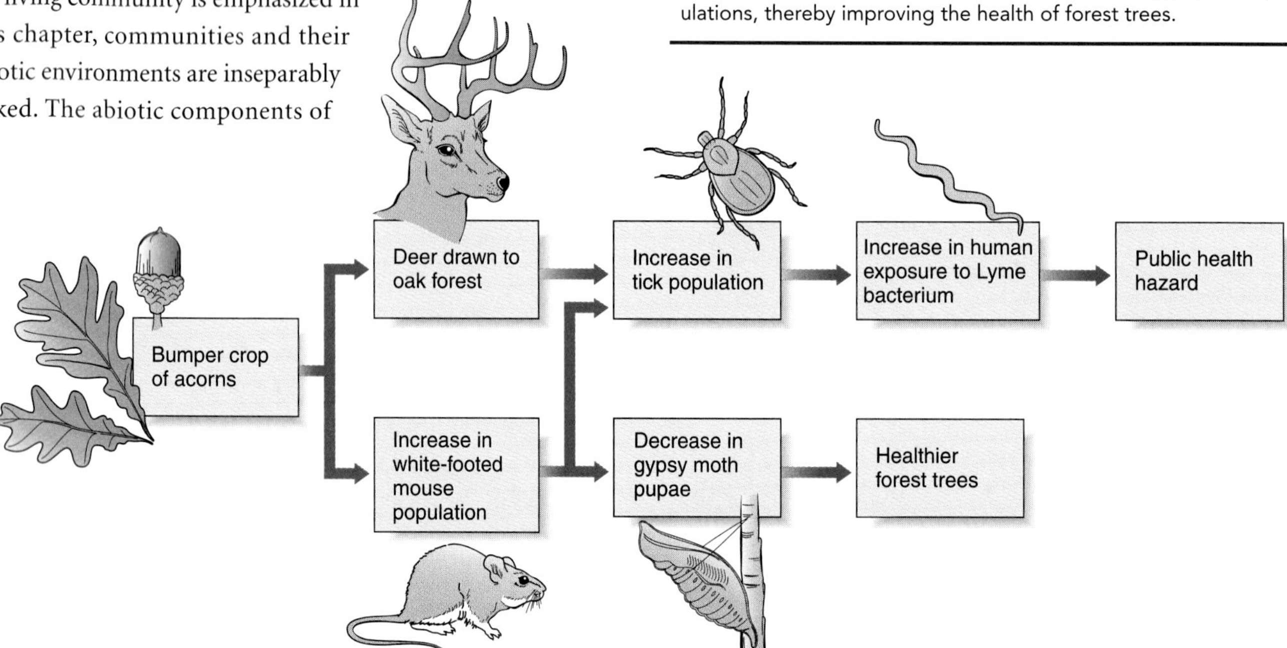

ducted in oak forests of the northeastern United States linked bumper acorn crops, which occur every three to four years, to booming mouse populations (the mice eat the acorns). Because the mice also eat gypsy moth pupae, high acorn conditions also lead to low populations of gypsy moths. This outcome helps the oaks, because gypsy moths cause serious defoliation. However, abundant acorns also attract tick-bearing deer. The ticks' hungry offspring feed on the mice, which often carry the Lyme disease-causing bacterium. The bacterium infects the maturing ticks and, in turn, spreads to humans who are bitten by affected ticks. Thus it may be possible to predict which years pose the greatest potential threat of Lyme disease to humans, based on when oaks are most productive.

An additional complication emerged in the Lyme disease relationships in the early 2000s. Scientists reported that ticks in areas with high biological diversity seldom transmit Lyme disease to humans. In contrast, ticks in areas with less biological diversity, such as where forests are cleared or otherwise degraded by human activities, are more likely to pass Lyme disease to humans. Because the white-footed mouse is one of the last species to disappear in low-diversity areas, it becomes the main food source for ticks, which pick up the disease and transmit it to humans. In high-diversity areas, ticks feed on many species of mammals, birds, and reptiles that may be infected with Lyme disease but do not readily transmit it to ticks. Thus many of the ticks remain disease free. Lyme disease is an excellent example of how human activities—in this case, transforming the land and reducing the number of species living in an area—affect natural interactions to our detriment.

Within a community, no species exists independently of other species. As the preceding example shows, the populations of a community interact with and affect one another in complex ways that are not always obvious. Three main types of interactions occur among species in a community: competition, predation, and symbiosis. Before we address these interactions, however, we need to examine the way of life of a given species in its community.

The niche is a species' ecological role in the community

Every species is thought to have its own ecological role within the structure and function of a community; we call this role its **ecological niche.** Although the concept of ecological niche has been in use in ecology since early in the 20th century, Yale ecologist G.E. Hutchinson in 1957 first described the multidimensional nature of the niche that is accepted today. An ecological niche takes into account all biotic and abiotic aspects of the species' existence, that is, all physical, chemical, and biological factors that the species needs to survive, remain healthy, and reproduce. Among other things, the niche includes the local environment in which a species lives—its **habitat.** A niche also encompasses what a species eats, what eats it, what organisms it competes with, and how it interacts with and is influenced by the abiotic components of its environment, such as light, temperature, and moisture. The niche, then, represents the totality of adaptations by a species to its environment, its use of resources, and the lifestyle to which it is suited. Although a complete description of an organism's ecological niche involves many dimensions and is difficult to define precisely, ecologists usually confine their studies to one or a few niche variables, such as feeding behaviors or ability to tolerate temperature extremes.

The ecological niche of a species is far broader in theory than in actuality. A species is usually capable of using much more of its environment's resources or of living in a wider assortment of habitats than it actually does. The potential ecological niche of a species is its **fundamental niche,** but various factors such as competition with other species may exclude it from part of this fundamental niche. Thus the lifestyle a species actually pursues and the resources it actually uses make up its **realized niche.**

An example may help to make this distinction clear. The green anole, a lizard native to Florida and other southeastern states, perches on trees, shrubs, walls, or fences during the day waiting for insect and spider prey (Fig. 52-2a). In the past these little lizards were widespread in Florida. Several years ago, however, a related species, the brown anole, was introduced from Cuba into southern Florida and quickly became common (Fig. 52-2b). Suddenly the green anoles became rare, apparently driven out of their habitat by competition from the slightly larger brown anoles. Careful investigation disclosed, however, that green anoles were still around. They were now confined largely to the vegetation in wetlands and to the foliated crowns of trees, where they were less easily seen.

The habitat portion of the green anole's fundamental niche includes the trunks and crowns of trees, exterior house walls, and many other locations. Once the brown anoles became established in the green anole habitat, they drove the green anoles from all but wetlands and tree crowns; competition between species shrank their realized niche (Fig. 52-2c, d). Because communities consist of numerous species, many of which compete to some extent, the complex interactions among them produce each species' realized niche.

Limiting resources restrict the ecological niche of a species

A species' structural, physiological, and behavioral adaptations determine its tolerance for environmental extremes. If any feature of an environment lies outside the bounds of its tolerance, then the species cannot live there. Just as you would not expect to find a cactus living in a pond, you would not expect water lilies in a desert.

The environmental factors that actually determine a species' ecological niche are difficult to identify. For this reason, the ecological niche concept is largely abstract, although some of its dimensions can be experimentally determined. Any environmental resource that, because it is scarce or unfavorable, tends to restrict the ecological niche of a species is called a **limiting resource.**

Most of the limiting resources that have been studied are simple variables such as the soil's mineral content, temperature extremes, and precipitation amounts. Such investigations have disclosed that any factor exceeding the tolerance of a species, or present in quantities smaller than the required minimum, limits the presence of that species in a community. By their interaction, such factors help define the ecological niche for a species.

FIGURE **52-2**

Effect of competition on an organism's realized niche.

(a) The green anole (*Anolis carolinensis*) is the only anole species native to North America. Males are about 12.5 cm (5 in) long; females are slightly smaller. **(b)** The 15.2-cm (6-in) long brown anole (*Anolis sagrei*) was introduced to Florida. **(c, d)** Positions of the two species along a single niche dimension (in this case, habitat). Species 1 represents the green anole, and Species 2 represents the brown anole. **(c)** The fundamental niches of the two lizards overlap. **(d)** The brown anole outcompetes the green anole in the area where their niches overlap, restricting the niche of the green anole.

(a)

(b)

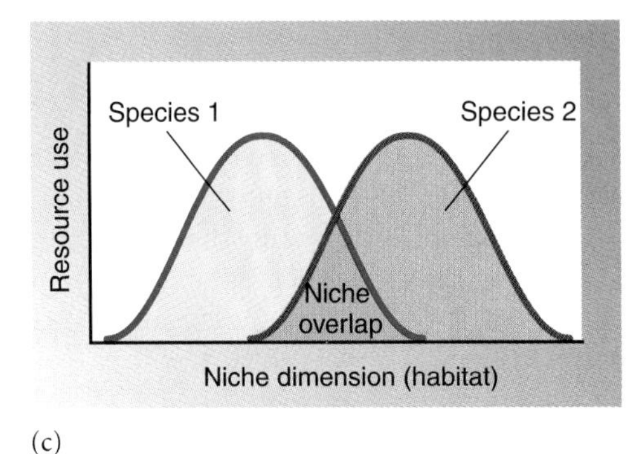

(c)

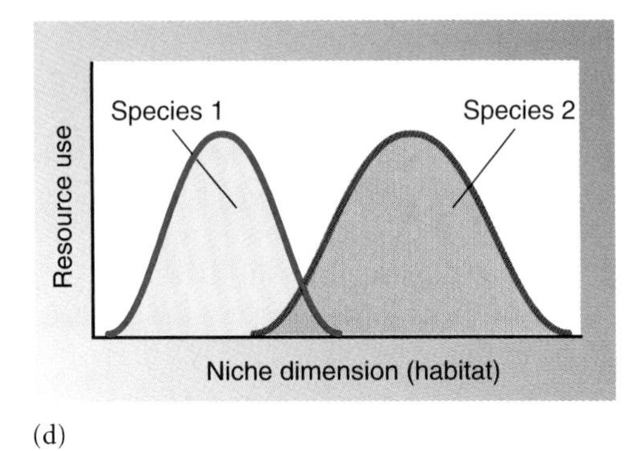

(d)

Limiting resources may affect only part of an organism's life cycle. For example, although adult blue crabs live in fresh or slightly brackish water, they don't become permanently established in such areas, because their larvae (immature forms) require salt water. Similarly, the ring-necked pheasant, a popular game bird native to Eurasia, has been widely introduced in North America but has not become established in the southern United States. The adult birds do well there, but the eggs don't develop properly in the high southern temperatures.

Biotic and abiotic factors influence a species' ecological niche

In the 1960s, the U.S. ecologist Joseph H. Connell investigated biotic and abiotic factors that affect the distribution of two barnacle species in the rocky *intertidal zone* along the coast of Scotland. The intertidal zone is a challenging environment, and organisms in the intertidal zone must tolerate exposure to the drying air during low tides. Barnacles are sessile crustaceans whose bodies are covered by a shell of calcium carbonate (see Figure 29-21a). When the shell is open, feathery appendages extend to filter food from the water.

Along the coast of Scotland, adults of one barnacle species, *Balanus balanoides,* are attached on lower rocks in the intertidal zone than are adults of the other species, *Chthamalus stellatus* (Fig. 52-3). The distributions of the two species do not overlap, although immature larvae of both species live together in the intertidal zone. Connell manipulated the two populations to determine what factors were affecting their distribution. When he removed *Chthamalus* from the upper rocks, *Balanus* barnacles did not expand into the vacant area. Connell's experiments showed that *Chthamalus* is more resistant than *Balanus* to desiccation when the tide retreats. However, when Connell removed *Balanus* from the lower rocks, *Chthamalus* expanded into the lower parts of the intertidal zone. The two species compete for space, and *Balanus,* which is larger and grows faster, outcompetes the smaller *Chthamalus* barnacles.

Competition among barnacle species for space was one of the processes that Connell's research demonstrated. We now examine some of the principles of competition that Connell and other ecologists have revealed in both laboratory and field studies.

Competition is intraspecific or interspecific

Competition occurs when two or more individuals attempt to use the same essential resource, such as food, water, shelter, living space, or sunlight. Because resources are often in limited supply in the environment, their use by one individual decreases the amount available to others (Table 52-1). If a tree in a dense forest grows taller than surrounding trees, for example, it absorbs more of the incoming sunlight. Less sunlight is therefore available for nearby trees that are shaded by the taller tree. Competition occurs among individuals within a population (**intraspecific competition**) or between different species (**interspecific competition**). Intraspecific competition was discussed in Chapter 51.

Ecologists traditionally assumed that competition is the most important determinant of both the number of species found in a community as well as the size of each population. Today ecologists recognize that competition is only one of many interacting biotic and abiotic factors that affect community structure. Furthermore, competition is not always a straightforward, direct

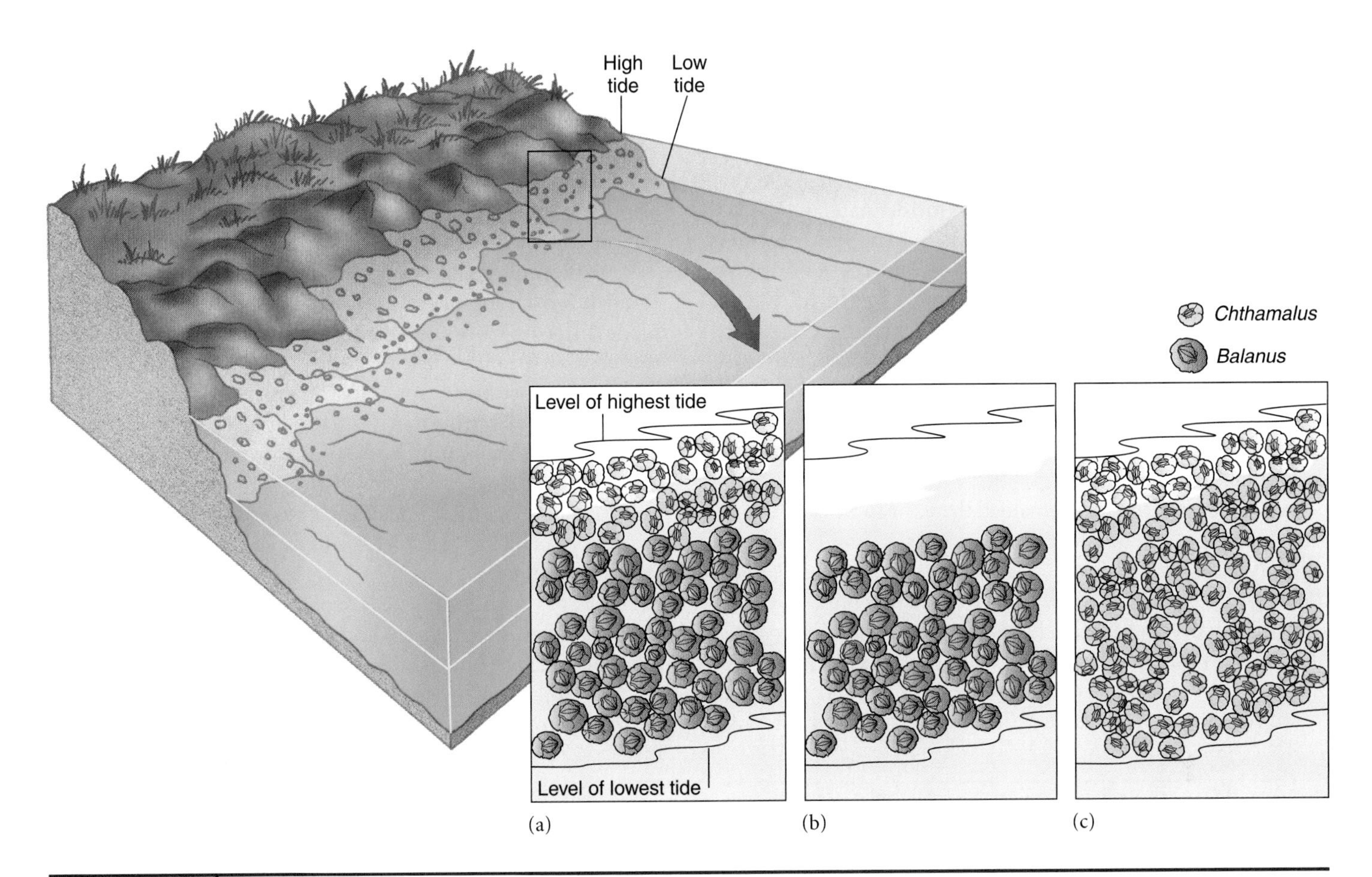

FIGURE **52-3** | **Abiotic and biotic factors that affect barnacle distribution.**
(a) Species of barnacles belonging to two genera, *Chthamalus* and *Balanus,* grow in the intertidal zone of a rocky shore in Scotland. **(b)** When *Chthamalus* individuals were experimentally removed, *Balanus* individuals did not expand into their section of the rock. **(c)** When *Balanus* individuals were experimentally removed, *Chthamalus* individuals spread into the empty area. *(After J.H. Connell, "The Influence of Interspecific Competition and Other Factors on the Distribution of the Barnacle* Chthamalus stellatus," Ecology, *Vol. 42, 1961)*

interaction. A variety of flowering plants, for example, live in a young pine forest and presumably compete with the conifers for such resources as soil moisture and soil nutrient minerals. Their relationship, however, is more complex than simple competition. The flowers produce nectar that is consumed by some insect species that also prey on needle-eating insects, thereby reducing the number of insects feeding on pines. It is therefore difficult to assess the overall effect of flowering plants on pines. If the flowering plants were removed from the community, would the pines grow faster because they were no longer competing for necessary resources? Or would the increased presence of needle-eating insects (caused by fewer predatory insects) inhibit pine growth?

Short-term experiments in which one competing plant species is removed from a forest community often have demonstrated an improved growth for the remaining species. However, very few studies have tested the long-term effects on forest species of the removal of single competing species. These long-term effects may be subtle, indirect, and difficult to ascertain; they may lessen or negate the short-term effects of competition for resources.

TABLE **52-1** **Ecological Interactions among Species**

Interaction	Effect on Species 1	Effect on Species 2
Competition between species 1 and species 2	Harmful	Harmful
Predation of species 2 by species 1	Beneficial	Harmful
Symbiosis		
Mutualism of species 1 and species 2	Beneficial	Beneficial
Commensalism of species 1 with species 2	Beneficial	No effect
Parasitism by species 1 on species 2	Beneficial	Harmful

Competition between species with overlapping niches may lead to competitive exclusion

When two species are similar, as are the green and brown anoles or the two species of barnacles, their fundamental niches may overlap. However, based on experimental and modeling work, many ecologists think that no two species indefinitely occupy the same niche in the same community. According to the **competitive exclusion principle,** it is hypothesized that one species excludes another from its niche as a result of interspecific competition. Although it is possible for different species to compete for some necessary resource without being total competitors, two species with absolutely identical ecological niches cannot coexist. Coexistence occurs, however, if the overlap between the two niches is reduced. In the lizard example, direct competition between the two species was reduced as the brown anole excluded the green anole from most of its former habitat.

The initial evidence that interspecific competition contributes to a species' realized niche came from a series of laboratory experiments by the Russian biologist Georgyi F. Gause in the 1930s (see descriptions of other experiments by Gause in Chapter 51). In one study Gause grew populations of two species of protozoa, *Paramecium aurelia* and the larger *Paramecium caudatum*, in controlled conditions (Fig. 52-4). When grown in separate test tubes, that is, in the absence of the second species, the population of each species quickly increased to a level imposed by the resources and remained there for some time thereafter. When grown together, however, only *P. aurelia* thrived, whereas *P. caudatum* dwindled and eventually died out. Under different sets of culture conditions, *P. caudatum* prevailed over *P. aurelia*. Gause interpreted this to mean that although one set of conditions favored one species, a different set favored the other. Nonetheless, because both species were so similar, in time one or the other would eventually triumph. Similar experiments with competing species of fruit flies, mice, certain beetles, and annual plants have supported Gause's results: One species thrives, and the other eventually dies out.

Competition, then, has adverse effects on species that use a limited resource and may result in competitive exclusion of one or more species. It therefore follows that over time natural selection should favor individuals of each species that avoid or, at least, reduce competition for environmental resources. Reduced competition among coexisting species as a result of each species' niche differing from the others in one or more ways is called **resource partitioning.** Resource partitioning is well documented in animals; documentation includes studies in tropical forests of Central and South America that demonstrate little overlap in the diets of fruit-eating birds, primates, and bats that coexist in the same habitat. Although fruits are the primary food for several hundred bird, primate, and bat species, the wide variety of fruits available have allowed fruit eaters to specialize, reducing competition.

Resource partitioning also may include timing of feeding, location of feeding, nest sites, and other aspects of a species' ecological niche. The late Princeton ecologist Robert MacArthur's study of five North American warbler species is a classic example of resource partitioning (Fig. 52-5). Although initially their niches seemed nearly identical, MacArthur found that individ-

FIGURE **52-4** **Interspecific competition.**

Competition between two species of *Paramecium* was studied by G.F. Gause. The top and middle graphs show how a population of each species of *Paramecium* flourishes when grown alone. The bottom graph shows how they grow together, in competition with each other. *(Adapted from G.F. Gause,* The Struggle for Existence, *Williams & Wilkins, Baltimore, 1934)*

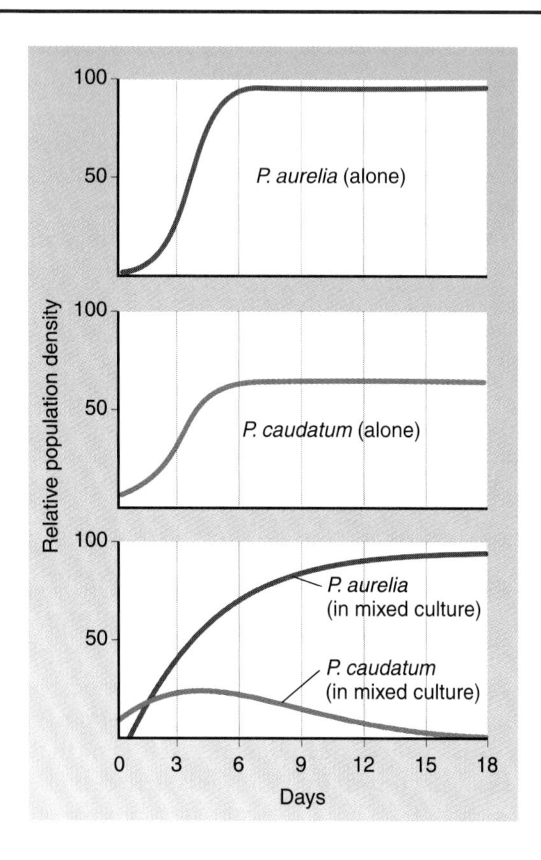

FIGURE **52-5** **Resource partitioning.**

Each warbler species spends at least half its foraging time in its designated area of a spruce tree, thereby reducing the competition among warbler species. *(After R.H. MacArthur, "Population Ecology of Some Warblers of Northeastern Coniferous Forests,"* Ecology, *Vol. 39, 1958)*

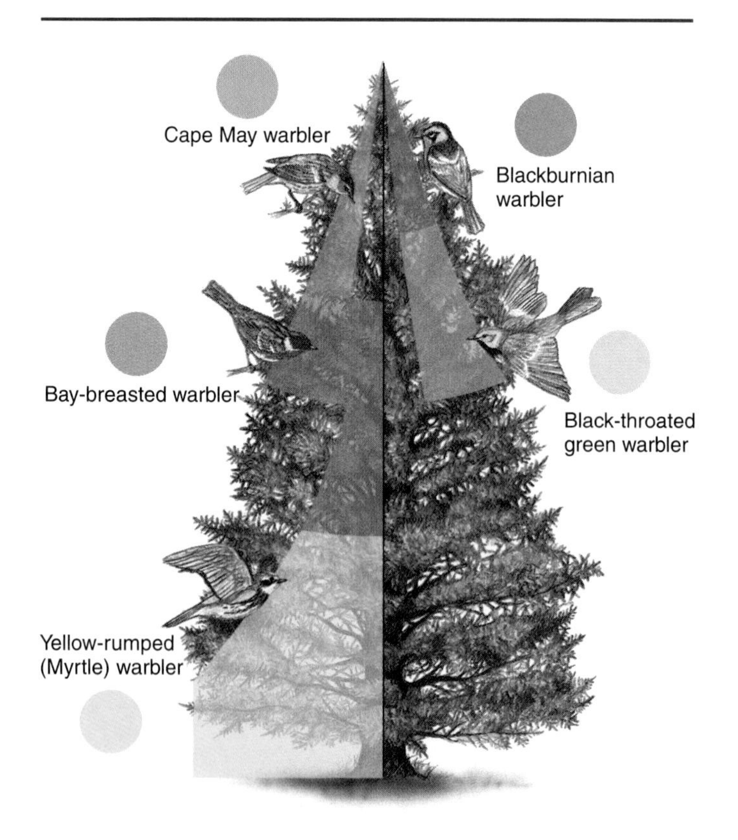

uals of each species spend most of their feeding time in different portions of the spruces and other conifer trees they frequent. They also move in different directions through the canopy, consume different combinations of insects, and nest at slightly different times.

Difference in root depth is an example of resource partitioning in plants. For example, three common annuals found in certain abandoned fields are smartweed, Indian mallow, and bristly foxtail. Smartweed roots extend deep into the soil, Indian mallow roots grow to a medium depth, and bristly foxtail roots are shallow. This difference reduces competition for the same soil resources—water and nutrient minerals—by allowing the plants to exploit different portions of the resource. Soil depth is not the only example of resource partitioning of soil resources in plants. In 2002, biologists reported in the journal *Nature* that the type of nitrogen absorbed from the soil as well as the timing of nitrogen use vary among different plants in an arctic plant community.

Apparent contradictions to the competitive exclusion principle exist. In Florida, native and introduced (nonnative) fish seem to coexist in identical niches. Similarly, botanists have observed closely competitive plant species in the same community. Although such situations seem to contradict the competitive exclusion principle, the realized niches of these organisms may differ in some way that ecologists do not yet understand, as with the warblers before MacArthur studied them.

Character displacement is an adaptive consequence of interspecific competition

Sometimes populations of two similar species occur together in some locations and separately in others. Where their geographic distributions overlap, the two species tend to differ more in their structural, ecological, and behavioral characteristics than they do where each occurs in separate geographic areas. Such divergence in traits in two similar species living in the same geographic area is known as **character displacement.** Biologists think character displacement reduces competition between two species because their differences give them somewhat different ecological niches in the same environment.

There are several well-documented examples of character displacement between two closely related species. For example, the flowers of two *Solanum* species in Mexico are quite similar in areas where either one or the other occurs. However, where their distributions overlap the two species differ significantly in flower size and are pollinated by different kinds of bees. In other words, character displacement reduces interspecific competition, in this case for the same animal pollinator.

The bill sizes of Darwin's finches provide another example of character displacement (Fig. 52-6). On large islands in the Galapagos where the medium ground finch (*Geospiza fortis*) and the small ground finch (*Geospiza fuliginosa*) occur together, their bill depths are distinctive. *G. fuliginosa* has a smaller bill depth that enables it to crack small seeds, whereas *G. fortis* has a larger bill depth that enables it to crack medium-sized seeds. However, *G. fortis* and *G. fuliginosa* also live on separate islands. Where the two species live separately, bill depths are about the

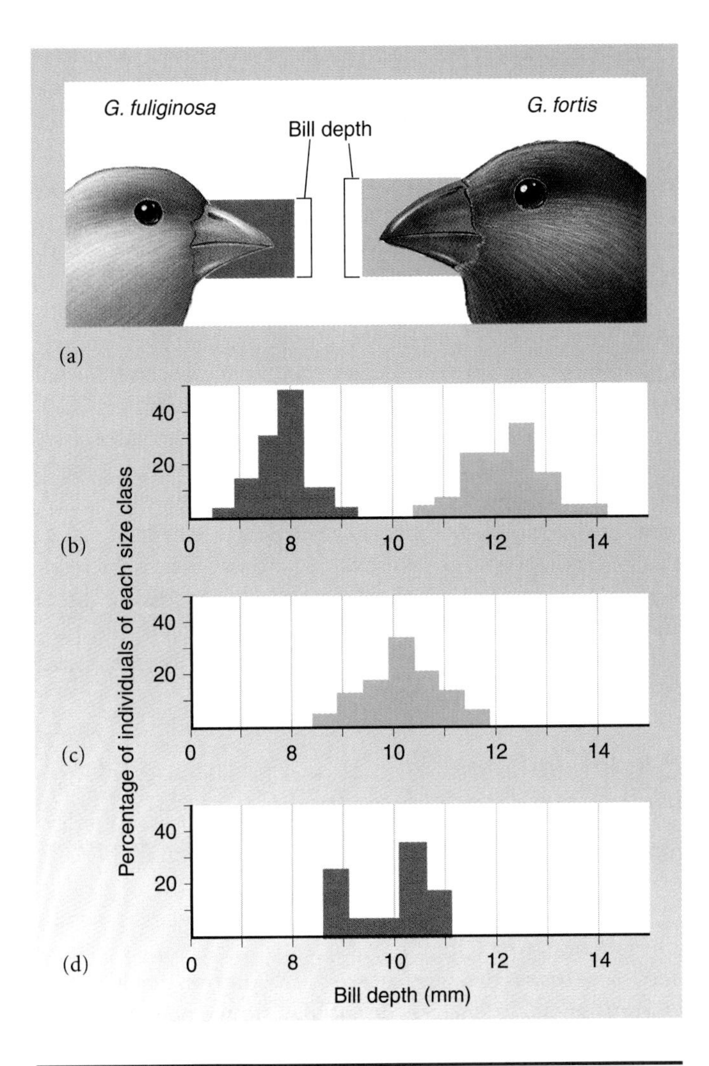

FIGURE **52-6** **Character displacement.**

(a) Bill depth in two closely related species of Galapagos Island finches, *Geospiza fuliginosa* and *Geospiza fortis.* **(b)** Where the two species are found on the same island, *G. fuliginosa (pink)* has a smaller average bill depth than *G. fortis (blue).* Where they occur on separate islands **(c, d),** the average bill depths of each are similar. *(After D. Lack,* Darwin's Finches, *Cambridge University Press, Cambridge, 1947)*

same intermediate size, perhaps because there is no competition from the other species.

Although these examples of the coexistence of similar species are explained in terms of character displacement, the character displacement hypothesis has only been demonstrated in field experiments in a few instances.

Natural selection shapes the body forms and behaviors of both predator and prey

Predation is the consumption of one species, the *prey,* by another, the *predator* (see Table 52-1). Predation includes animals eating other animals, as well as animals eating plants *(herbivory.)* Predation has resulted in an evolutionary "arms race," with the evolution of predator strategies (more efficient ways to catch prey) and prey strategies (better ways to escape the predator). A predator that is more efficient at catching prey exerts a strong selective force on its prey, which over time may evolve some sort of countermeasure that reduces the probability of being captured. The countermeasure acquired by the prey in turn acts as a strong selective force on the predator. This type of interdependent evolution of two interacting species is known as **coevolution** (see Chapter 35).

We now consider several adaptations related to predator–prey interactions. These include predator strategies (pursuit and ambush) and prey strategies (plant defenses and animal defenses). Keep in mind as you read these descriptions that such strategies are not "chosen" by the respective predators or prey. New traits arise randomly in a population as a result of genetic changes (see Chapters 17 and 18). Some new traits are beneficial, some are harmful, and others have no effect. Beneficial strategies, or traits, persist in a population because such characteristics make the individuals that have them well suited to thrive and reproduce. In contrast, characteristics that make the individuals that have them poorly suited to their environment tend to disappear in a population.

Carmela Leszczynski/Animals Animals

FIGURE **52-7** **Ambush.**
A yellow crab spider (*Misumena vatia*) blends into its surroundings, waiting for an unwary insect to visit the flower. An effective predator strategy, ambush relies on surprising the prey.

Pursuit and ambush are two predator strategies

A brown pelican sights its prey—a fish—while in flight. Less than 2 seconds after diving into the water at a speed as great as 72 km/h (45 mph), it has its catch. Orcas (formerly known as killer whales), which hunt in packs, often herd salmon or tuna into a cove so that they are easier to catch. Any trait that increases hunting efficiency, such as the speed of brown pelicans or the intelligence of orcas, favors predators that pursue their prey. Because these carnivores must process information quickly during the pursuit of prey, their brains are generally larger, relative to body size, than those of the prey they pursue.

Ambush is another effective way to catch prey. The yellow crab spider, for example, is the same color as the white or yellow flowers on which it hides (Fig. 52-7). This camouflage keeps unwary insects that visit the flower for nectar from noticing the spider until it is too late. It also fools birds that prey on the crab spider.

Predators that *attract* prey are particularly effective at ambushing. For example, a diverse group of deep-sea fishes called anglerfish possess rodlike luminescent lures close to their mouths to attract prey.

Chemical protection is an effective plant defense against herbivores

Plants cannot escape predators by fleeing, but they have several adaptations that protect them from being eaten. The presence of spines, thorns, tough leathery leaves, or even thick wax on leaves discourages foraging herbivores from grazing. Other plants produce an array of protective chemicals that are unpalatable or even toxic to herbivores. The active ingredients in such plants as marijuana and tobacco affect hormone activity and nerve, muscle, liver, and kidney functions and may discourage foraging by herbivores. Interestingly, many of the chemical defenses in plants are useful to humans. India's neem tree, for example, contains valuable chemicals effective against more than 100 species of herbivorous insects, mites, and nematodes. Nicotine from tobacco, pyrethrum from chrysanthemum, and rotenone from the derris plant are other examples of chemicals extracted and used as insecticides.

Milkweeds are an excellent example of the biochemical coevolution between plants and herbivores (Fig. 52-8). Milkweeds produce alkaloids and cardiac glycosides, chemicals that are poisonous to all animals except a small group of insects. These insects evolved the ability to either tolerate or metabolize the milkweed toxins. They eat milkweeds and avoid competition from other herbivorous insects because few others tolerate milkweed toxins. Predators also learn to avoid these insects, which accumulate the toxins in their tissues and therefore become toxic themselves. The black, white, and yellow coloration of the monarch caterpillar, a milkweed feeder, clearly announces its toxicity to predators that have learned to associate bright colors with illness. Conspicuous colors or patterns, which advertise a species' unpalatability to potential predators, are known as **aposematic coloration** (ap′uh-suh-mat′ik; from the Greek *apo*, away, and *semat*, a mark or sign), or **warning coloration.**

Animals possess various defensive adaptations to avoid predators

Many animals, such as prairie voles and woodchucks, flee from predators by rapidly running to their underground burrows. Others have mechanical defenses, such as the barbed quills of a porcupine and the shell of a pond turtle. To discourage predators, the porcupine fish inflates itself to three times its normal size by pumping water into its stomach (see Fig. 30-13a). Some animals live in groups—for example, a herd of antelope, colony of honeybees, school of anchovies, or flock of pigeons. Because a group has so many eyes, ears, and noses watching, listening, and smelling for predators, this social behavior decreases the likelihood of a predator catching any one of them unaware.

Wild & Natural/Animals Animals/Earth Scenes

FIGURE **52-8** **Plant chemical defenses.**

Toxic chemicals protect the common milkweed (*Asclepias syriaca*). Its leaves are poisonous to most herbivores except monarch caterpillars (*Danaus plexippus*, shown) and a few other insects. Monarch caterpillars have bright aposematic coloration. Photographed in Michigan.

Ronald A. Nussbaum/Museum of Zoology, University of Michigan

FIGURE **52-9** **Cryptic coloration.**

The smooth-skinned leaf-tailed gecko (*Uroplatus malama*), which is about 12 cm (4.7 in) long, hunts for insects by night and sleeps pressed against tree bark by day. It is virtually invisible when it sleeps. Photographed in a rain forest in southern Madagascar.

Chemical defenses are also common among animal prey. The South American poison arrow frog (*Dendrobates*) has poison glands in its skin. Its bright aposematic coloration prompts avoidance by experienced predators (see Fig. 30-16a). Snakes and other animals that have tried once to eat a poisonous frog do not repeat their mistake! Other examples of aposematic coloration occur in the striped skunk, which sprays acrid chemicals from its anal glands, and the bombardier beetle, which spews harsh chemicals at potential predators (see Fig. 6-8).

Some animals have **cryptic coloration,** colors or markings that help them hide from predators by blending into their physical surroundings. Certain caterpillars resemble twigs so closely that you would never guess they were animals unless they moved. Pipefish are slender green fish that are almost perfectly camouflaged in green eelgrass. The smooth-skinned leaf-tailed gecko, discovered in the late 1990s in southern Madagascar, resembles dead leaves not only in color but also in the pattern of leaf veins (Fig. 52-9). Such cryptic coloration has been preserved and accentuated by means of natural selection. Predators are less likely to capture animals with cryptic coloration. Such animals are therefore more likely to live to maturity and produce offspring that also carry the genes for cryptic coloration.

Sometimes a defenseless species (a *mimic*) is protected from predation by its resemblance to a species that is dangerous in some way (a *model*). Such a strategy is known as **Batesian mimicry.** Many examples of this phenomenon exist. For example, a harmless scarlet king snake looks so much like a venomous coral snake that predators may avoid it (Fig. 52-10 on page 1033).

In **Müllerian mimicry,** different species (comodels), all of which are poisonous, harmful, or distasteful, resemble one another. Although their harmfulness protects them as individual species, their similarity in appearance works as an added advantage. Potential predators more easily learn a single common aposematic coloration. Scientists hypothesize that viceroy and monarch butterflies are an example of Müllerian mimicry (see *Focus On: Batesian Butterflies Disproved*).

Symbiosis involves a close association between species

Symbiosis is any intimate relationship or association between members of two or more species. Usually symbiosis involves one species living on or in another species. The partners of a symbiotic relationship, called **symbionts,** may benefit from, be unaffected by, or be harmed by the relationship (see Table 52-1). Most of the thousands, or perhaps even millions, of symbiotic associations are products of coevolution. Symbiosis takes three forms: mutualism, commensalism, and parasitism.

PROCESS OF SCIENCE

The monarch butterfly (*Danaus plexippus*) is an attractive insect found throughout much of North America (see figure, left side). As a caterpillar, it feeds exclusively on milkweed leaves. The milky white liquid produced by the milkweed plant contains poisons that the insect tolerates but that remain in its tissues for life. When a young bird encounters and tries to eat its first monarch butterfly, it sickens and vomits. Thereafter, the bird avoids eating the distinctively marked insect.

Many people confuse the viceroy butterfly (*Limenitis archippus*, see figure, right side) with the monarch. The viceroy, which is found throughout most of North America, is approximately the same size, and the color and markings of its wings are almost identical to those of the monarch. As caterpillars, viceroys eat willow and poplar leaves, which apparently do not contain poisonous substances.

During the past century, it was thought that the viceroy butterfly was a tasty food for birds but that its close resemblance to monarchs gave it some protection against being eaten. In other words, birds that had learned to associate the distinctive markings and coloration of the monarch butterfly with its bad taste tended to avoid viceroys because they were similarly marked. The viceroy butterfly was therefore considered a classic example of Batesian mimicry.

In the journal *Nature* in 1991, ecologists David Ritland and Lincoln Brower of the University of Florida reported the results of an experiment that tested the long-held notion that birds like the taste of viceroys but avoid eating them because of their resemblance to monarchs. They removed the wings of different kinds of butterflies—monarchs, viceroys, and several tasty species—and fed the seemingly identical wingless bodies to red-winged blackbirds. The results were surprising: Monarchs and viceroys were equally distasteful to the birds.

As a result of this work, ecologists are reevaluating the evolutionary significance of different types of mimicry. Rather than being an example of Batesian mimicry, monarchs and viceroys may be an example of Müllerian mimicry, in which two or more different species that are distasteful or poisonous have come to resemble one another during the course of evolution. This likeness provides an adaptive advantage because predators learn quickly to avoid all butterflies with the coloration and markings of monarchs and viceroys. As a result, fewer butterflies of either species die, and more individuals survive to reproduce.

The butterfly study provides us with a useful reminder about the process of science. Expansion of knowledge in science is an ongoing enterprise, and newly acquired evidence helps scientists to reevaluate current models or ideas. Thus scientific knowledge and understanding is not static, but continually changing.

Thomas C. Emmel

Müllerian mimicry. Evidence suggests that monarch *(left)* and viceroy *(right)* butterflies are an example of Müllerian mimicry, in which two or more poisonous, harmful, or distasteful organisms resemble each other.

Benefits are shared in mutualism

Mutualism is a symbiotic relationship in which both partners benefit. Mutualism is either obligate (essential for the survival of both species) or facultative (when either partner can live alone under certain conditions).

The association between **nitrogen-fixing bacteria** of the genus *Rhizobium* and legumes (plants such as peas, beans, and clover) is an example of mutualism (see Fig. 53-8a). Nitrogen-fixing bacteria, which live inside nodules on the roots of legumes, supply the plants with most of the nitrogen they require to manufacture such nitrogen-containing compounds as chlorophylls, proteins, and nucleic acids. The legumes supply sugars and other energy-rich organic molecules to their bacterial symbionts.

Another example of mutualism is the association between reef-building animals and dinoflagellates called **zooxanthellae** (see Chapters 24, 28, and 54). These symbiotic algae live inside cells of the coral polyp (the coral forms a vacuole around the algal cell), where they photosynthesize and provide the animal with carbon and nitrogen compounds as well as oxygen (Fig. 52-11). Zooxanthellae stimulate the growth of corals, causing calcium carbonate skeletons to form around their bodies much faster when the algae are present. The corals, in turn, supply the zooxanthellae with waste products such as ammonia, which the algae use to make nitrogen compounds for both partners.

Mycorrhizae are mutualistic associations between fungi and the roots of plants. The association is common: Biologists think about 80% of all plant species have mycorrhizae. The fungus, which grows around and into the root as well as into the surrounding soil, absorbs essential nutrient minerals, especially phosphorus, from the soil and provides them to the plant. In return, the plant provides the fungus with organic molecules pro-

Suzanne L. and Joseph T. Collins/Photo Researchers, Inc.

(a)

Suzanne L. Collins/Photo Researchers, Inc.

(b)

FIGURE **52-10** | **Batesian mimicry.**

In this example, **(a)** the scarlet king snake (*Lampropeltis triangulum elapsoides*) is the mimic, and **(b)** the eastern coral snake (*Micrurus fulvius fulvius*) is the model. Note that the red and yellow warning colors touch on the coral snake but do not touch on the harmless mimic. Scarlet king snakes are 36 to 51 cm (14 to 20 in) in length, whereas eastern coral snakes are 51 to 76 cm (20 to 30 in).

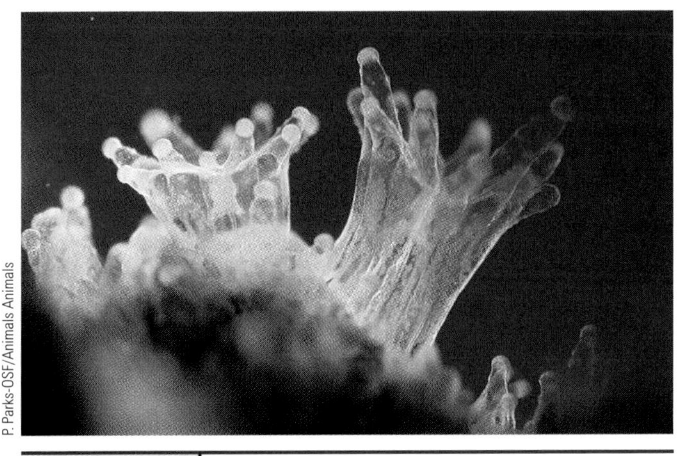
P. Parks-OSF/Animals Animals

FIGURE **52-11** | **Mutualism.**

The greenish-brown specks in these polyps of stony coral (*Pocillopora*) are zooxanthellae, algae that live symbiotically within the coral's translucent cells and supply the coral with carbon and nitrogen compounds. In return, the coral provides the zooxanthellae with nitrogen in the form of ammonia.

duced by photosynthesis. Plants grow more vigorously in the presence of mycorrhizal fungi (see Figs. 25-15 and 34-10), and they better tolerate environmental stressors such as drought and high soil temperatures. Indeed, some plants can't maintain themselves under natural conditions if the fungi with which they normally form mycorrhizae are not present.

Commensalism is taking without harming

Commensalism is a type of symbiosis in which one species benefits and the other one is neither harmed nor helped. One example of commensalism is the relationship between social insects and scavengers, such as mites, beetles, or millipedes, that live in the social insects' nests. Certain types of silverfish, for example, move along in permanent association with marching columns of army ants and share the food caught in the ant raids. The army ants derive no apparent benefit or harm from the silverfish.

Another example of commensalism is the relationship between a host tree and its epiphytes, which are smaller plants, such as orchids, ferns, or Spanish moss, attached to the host's branches (Fig. 52-12). The epiphyte anchors itself to the tree but does not obtain nutrients or water directly from it. Living on the tree enables it to obtain adequate light, water (as rainfall dripping down the branches), and required minerals (washed out of the tree's leaves by rainfall). Thus the epiphyte benefits from the association, whereas the tree is apparently unaffected. (Epiphytes harm their host, however, if they are present in a large enough number to block sunlight from the host's leaves; in this instance, the relationship is no longer commensalism.)

Parasitism is taking at another's expense

Parasitism is a symbiotic relationship in which one member, the *parasite*, benefits, whereas the other, the *host*, is adversely affected. The parasite obtains nourishment from its host. A parasite rarely kills its host directly but may weaken it, rendering it more vulnerable to predators, competitors, or abiotic stressors. When a parasite causes disease and sometimes the death of a host, it is called a **pathogen.**

Some parasites, such as ticks, live outside the host's body; other parasites, such as tapeworms, live within the host. Parasitism is a successful lifestyle; by one estimate, more than two thirds of all species are parasites, and more than 100 species of parasites live in or on the human species alone! Examples of human parasites include *Entamoeba histolytica,* an amoeba that causes amoebic dysentery; *Plasmodium,* an apicomplexan that causes malaria; and a variety of parasitic worms, such as blood flukes, tapeworms, pinworms, and hookworms.

Jeff Lepore/ Photo Researchers, Inc.

FIGURE **52-12** **Commensalism.**

Spanish moss (*Tillandsia usneoides*) is a gray-colored epiphyte that hangs suspended from larger plants in the southeastern United States. Spanish moss is not a moss but a flowering plant in the pineapple family. Although it is often 6 m (20 ft) or longer, it is nonparasitic and does not harm the host tree. Photographed in autumn in North Carolina, hanging from a bald cypress whose foliage has turned color.

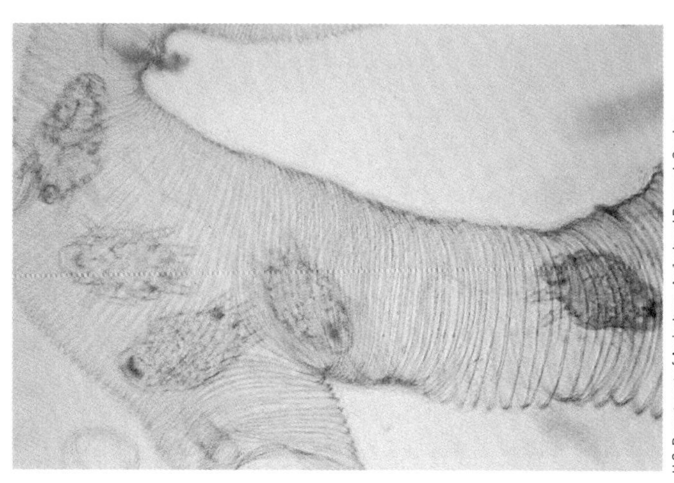

U.S. Department of Agriculture, Agricultural Research Service

FIGURE **52-13** **Parasitism.**

Microscopic tracheal mites (*Acarapis woodi*) live in the tracheal tubes of honeybees, clogging their airways so they cannot breathe efficiently. Tracheal mites also suck the bees' circulatory fluid, weakening and eventually killing them. Entomologists think the larger varroa mites *(not shown)*, which also feed on the circulatory fluid, are more devastating to honeybee populations than tracheal mites. Other species of mites may transmit viruses to their honeybee hosts.

Since the 1980s, wild and domestic honeybees in the United States have been dying off. Although habitat loss and pesticide use have contributed to the problem, tracheal mites (Fig. 52-13) and larger varroa mites have been a major reason for the honeybee decline. The number of commercial colonies has fallen by about 50% during the past several decades. Honeybees pollinate up to $10 billion of apples, almonds, and other crops each year and produce about $250 million of honey, so their decline is a major threat to U.S. agriculture. Entomologists are searching for mite-resistant bees among both North American and European honeybees; the entomologists plan to breed the mite-resistant bees with local honeybees to incorporate genetic resistance into the local bees. Farmers are also experimenting with mite-resistant bee species other than honeybees to pollinate crops.

Keystone species and dominant species affect the character of a community

Certain species, called **keystone species,** are crucial in determining the nature of the entire community, that is, its species composition and ecosystem functioning. Other species of a community depend on or are greatly affected by the keystone species. Keystone species are usually not the most abundant species in the community. Although present in relatively small numbers, the individuals of a keystone species profoundly influence the entire community because they often affect the amount of available food, water, or some other resource. Thus the impact of keystone species is greatly disproportionate to their abundance. Identifying and protecting keystone species are crucial goals of conservation biology (see Chapter 55), because if a keystone species disappears from a community, many other species in that community may become more common, more rare, or even disappear.

One problem with the concept of keystone species is that it is difficult to measure all the direct and indirect impacts of a keystone species on an ecosystem. Consequently, most evidence for the existence of keystone species is based on indirect observations rather than on experimental manipulations. For example, consider the fig tree. Because fig trees produce a continuous crop of fruits, they may be keystone species in tropical rain forests of Central and South America. Fruit-eating monkeys, birds, bats, and other fruit-eating vertebrates of the forest do not normally consume large quantities of figs in their diets. During that time of the year when other fruits are less plentiful, however, fig trees become important in sustaining fruit-eating vertebrates. It is therefore assumed that, should the fig trees disappear, most of the fruit-eating vertebrates would also disappear. In turn, should the fruit eaters disappear, the spatial distribution of other fruit-bearing plants would become more limited because the fruit eaters help disperse their seeds. Thus, protecting fig trees in such tropical rainforest ecosystems probably increases the likelihood that monkeys, birds, bats, and many other tree species will survive. The question is whether this anecdotal evidence of fig trees as keystone species is strong enough for policymakers to grant special protection to fig trees.

Many keystone species are top predators such as the gray wolf. Where wolves were hunted to extinction, the populations of elk, deer, and other larger herbivores increased exponentially. As these herbivores overgrazed the vegetation, many plant species that could not tolerate such grazing pressure disappeared. Smaller animals such as rodents, rabbits, and insects declined in number because the plants they depended on for food were now less abundant. The number of foxes, hawks, owls, and badgers that prey on these small animals decreased, as did the number of ravens, eagles, and other scavengers that eat wolf kill. Thus the disappearance of the wolf resulted in communities with considerably less biological diversity.

The reintroduction of wolves to Yellowstone National Park in 1995 has given ecologists a unique opportunity to study the impacts of a keystone species. The wolf's return has already caused substantial changes for other residents in the park. The top predator's effects have ranged from altering relationships among predator and prey species to transforming vegetation profiles. Coyotes are potential prey for wolves, and wolf packs have decimated some coyote populations. A reduction in coyotes has allowed populations of the coyotes' prey, such as ground squirrel and chipmunks, to increase. Scavengers such as ravens, bald eagles, and grizzly bears have benefited from dining on scraps from wolf kills.

The pruning effect that wolves will have on prey populations such as elk should cause broader, long-term impacts. Yellowstone's booming elk population of 35,000 eventually may decline by as much as 20%, which would relieve heavy grazing pressure on plants and encourage a more lush and varied plant composition. Richer vegetation should support more herbivores such as beaver and snow hare, which in turn will support small predators such as fox, badger, and marten.

Dominant species influence a community as a result of their greater size or abundance

In contrast to keystone species, which have a large impact that is out of proportion to their abundance, **dominant species** greatly affect the community of which they are a part, because they are very common. Trees are the dominant species of forests because they change the local environment. Trees provide shade, which changes both the light and moisture availability on the forest floor. Trees provide numerous habitats and *microhabitats* (such as a hole in a tree trunk) for other species. Forest trees also provide food for many organisms and therefore play a large role in energy flow through the forest ecosystem. Similarly, cordgrass (*Spartina*) is the dominant species in salt marshes, prairie grass in grasslands, and kelp in kelp beds. Animals are also dominant species. For example, corals are dominant species in reefs, and cattle are dominant species in overgrazed rangelands. Typically, a community has one or a few dominant species, and most other species are relatively rare.

Review

- How are acorns, gypsy moths, and Lyme disease related?
- Why is an organism's realized niche usually narrower, or more restricted, than its fundamental niche?
- What is the principle of competitive exclusion?
- How has natural selection affected predator–prey relationships?
- What are the three kinds of symbiosis?

BiologyNow™ Assess your understanding of **community structure and functioning** by taking the pretest on your BiologyNow CD-ROM.

COMMUNITY BIODIVERSITY

Learning Objective

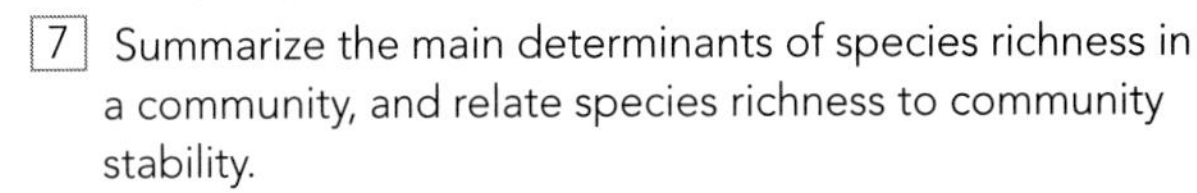

7 Summarize the main determinants of species richness in a community, and relate species richness to community stability.

Species richness and species diversity vary greatly from one community to another and are influenced by many biotic and abiotic factors. **Species richness** is the number of species in a community. Tropical rain forests and coral reefs are examples of communities with extremely high species richness. In contrast, geographically isolated islands and mountaintops exhibit low species richness.

The term **species diversity** is sometimes used interchangeably with species richness, but more precisely, species diversity is a measure of the relative importance of each species within a community, based on its abundance, productivity, or size. Ecologists have developed various mathematical expressions to represent species diversity. These *diversity indices* enable ecologists to compare species diversity in different communities. Conservation biologists use diversity indices as part of a comprehensive approach to saving biodiversity.

Ecologists seek to explain why some communities have more species than others

What determines the number of species in a community? No single conclusive answer exists, but several explanations appear plausible. These include the structural complexity of habitats, geographic isolation, habitat stress, closeness to the margins of adjacent communities, dominance of one species over others, and geological history. Although these and other environmental factors have positive or negative effects on species richness, there are exceptions and variations in every single explanation. Some explanations vary at different spatial scales: An explanation that seems to work at a large geographic scale (such as a continent) may not work at a smaller local scale (such as a meadow).

In many habitats, species richness is related to the structural complexity of habitats. In terrestrial environments, the types of plants growing in an area typically determine structural complexity. A structurally complex community such as a forest offers a greater variety of potential ecological niches than does a sim-

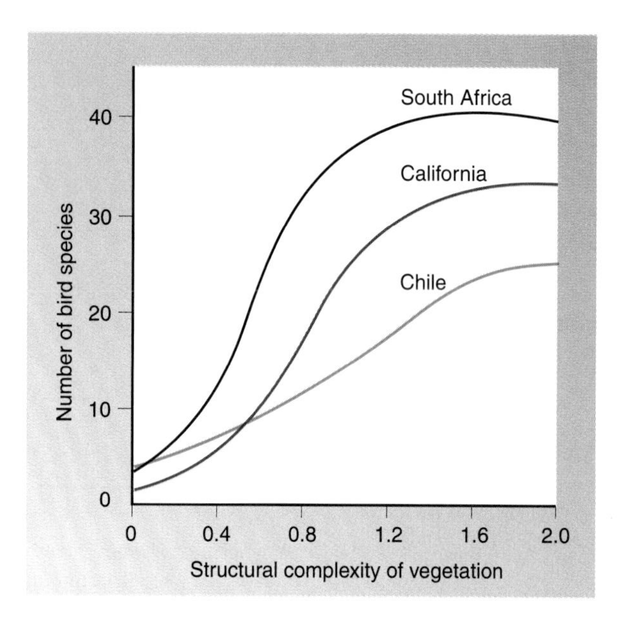

FIGURE **52-14** **Effect of community complexity on species richness.**

Data were compiled in comparable chaparral habitats (shrubby and woody areas) in Chile, South Africa, and California. The structural complexity of vegetation *(X-axis)* is a numerically assigned gradient of habitats, based on height and density of vegetation, from low complexity (very dry scrub) to high complexity (woodland). A community in which the vegetation is structurally complex generally provides birds with more kinds of food and hiding places than a community with a lower structural complexity. Chile has a lower total number of bird species than either California or South Africa because Chile is isolated from the rest of South America by the Andes Mountains to the east and extremely arid deserts to the north. In contrast, the chaparral habitats in California and South Africa are not isolated from the rest of their continents. *(After M.L. Cody, and J.M. Diamond, eds.,* Ecology and Evolution of Communities. *Harvard University, Cambridge, 1975)*

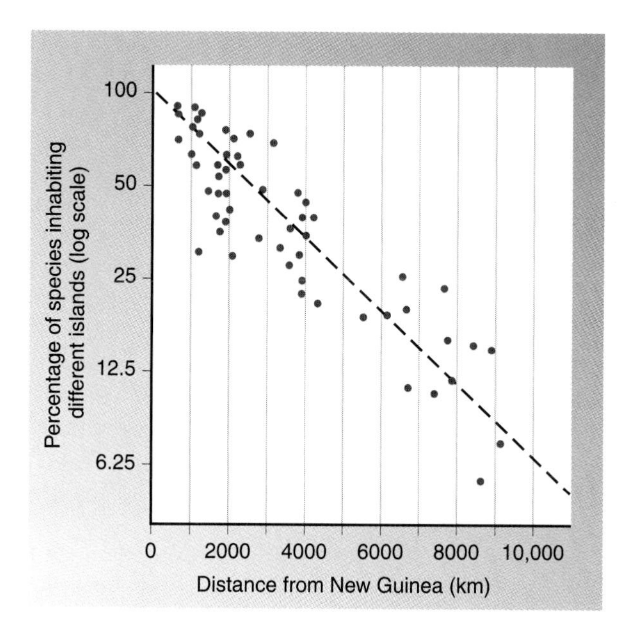

ACTIVE FIGURE **52-15** **The distance effect.**

This graph shows that the percentage of South Pacific bird species found on each island in the South Pacific is related to its distance from New Guinea, which is a source of colonizing species for these islands. (New Guinea has 100 percent of the bird species living in that region.) Note that species richness declines as the distance from New Guinea increases. *(After J.M. Diamond, "Biogeographic Kinetics: Estimation of Relaxation Times for Avifaunas of Southwest Pacific Islands,"* Proceedings of the National Academy of Sciences, *Vol. 69, 1972)*

BiologyNow™ Explore **the effects of area and distance on biodiversity** by clicking on this figure on your BiologyNow CD-ROM.

ple community, such as an arid desert or a grassland (Fig. 52-14). An already complex habitat such as a coral reef may become even more complex if species potentially capable of filling vacant ecological niches evolve or migrate into the community, because these species create "opportunities" for additional species. Thus it appears species richness is self-perpetuating to some degree.

Species richness is inversely related to the geographic isolation of a community. Isolated island communities are generally much less diverse than are communities in similar environments found on continents. This is due partly to the *distance effect,* the difficulty encountered by many species in reaching and successfully colonizing the island (Fig. 52-15). Also, sometimes species become locally extinct as a result of random events. In isolated habitats such as islands or mountaintops, locally extinct species are not readily replaced. Isolated areas are usually small and have fewer potential ecological niches.

Generally, species richness is inversely related to the environmental stress of a habitat. Only those species capable of tolerating extreme conditions live in an environmentally stressed community. Thus the species richness of a highly polluted stream is low compared with that of a nearby pristine stream. Similarly, the species richness of high-latitude (farther from the equator) communities exposed to harsh climates is lower than that of lower-latitude (closer to the equator) communities with milder climates (Fig. 52-16). This observation, known as the *species richness–energy hypothesis,* suggests that different latitudes affect species richness because of variations in solar energy. Greater energy may permit more species to coexist in a given region. Although the equatorial countries of Colombia, Ecuador, and Peru occupy only 2% of Earth's land, they contain a remarkable 45,000 native plant species. The continental United States and Canada, with a significantly larger land area, host a total of 19,000 native plant species. Ecuador alone contains more than 1300 native species of birds—twice as many as the United States and Canada combined.

Species richness is usually greater at the margins of distinct communities than in their centers. This is because an **ecotone,** a transitional zone where two or more communities meet, contains all or most of the ecological niches of the adjacent communities as well as some that are unique to the ecotone (see

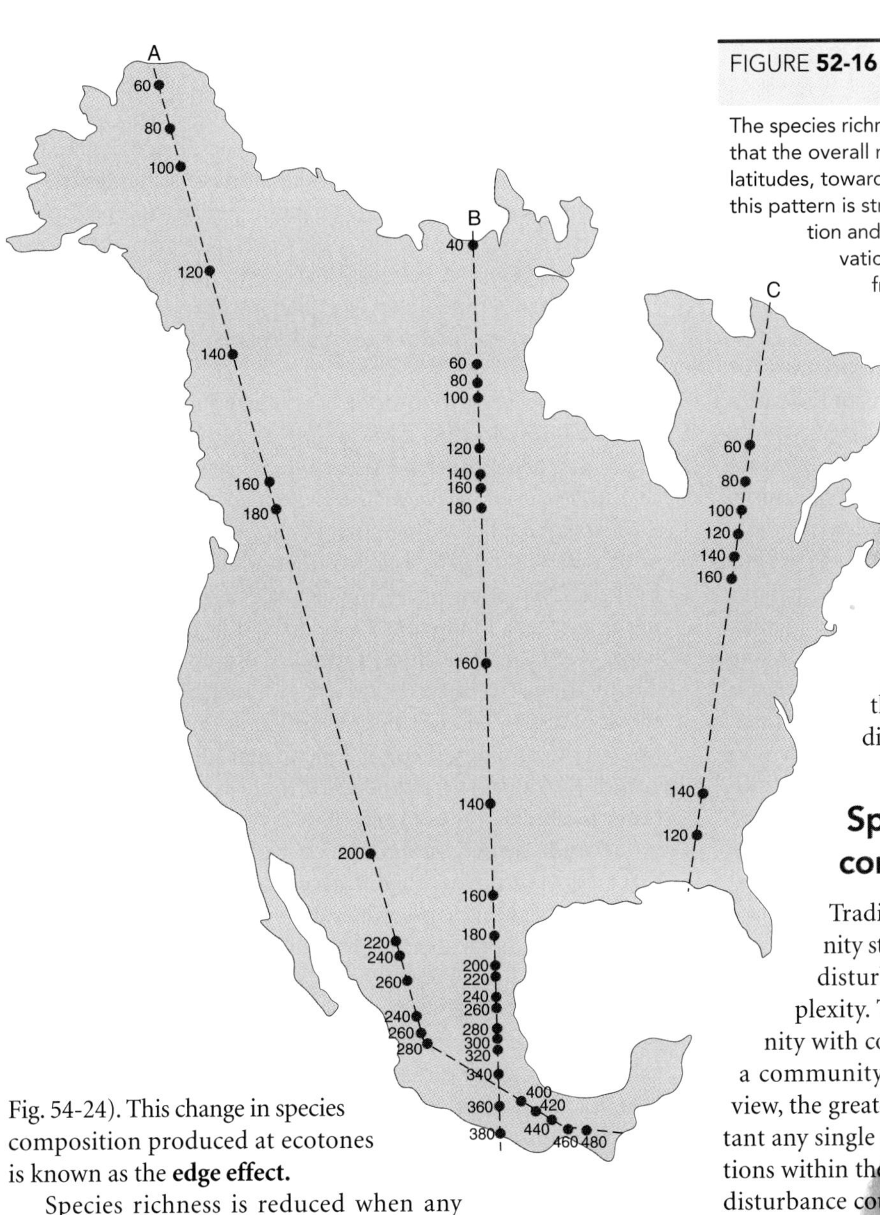

FIGURE **52-16** **Effect of latitude on species richness of breeding birds in North America.**

The species richness for three north–south transects are shown. Note that the overall number of breeding bird species is greater at lower latitudes, toward the equator, than at higher latitudes. However, this pattern is strongly modified by other factors, such as precipitation and surface features (such as mountains). Similar observations have been made in many groups of organisms, from plants to primates, in both terrestrial and marine environments. *(After R.E. Cook, "Variation in Species Density of North American Birds,"* Systematic Zoology, *Vol. 18, 1969)*

Fig. 54-24). This change in species composition produced at ecotones is known as the **edge effect.**

Species richness is reduced when any one species enjoys a position of dominance within a community; a dominant species may appropriate a disproportionate share of available resources, thus crowding out, or outcompeting, other species. Ecologist James H. Brown of the University of New Mexico has addressed species composition and richness in experiments conducted since 1977 in the Chihuahuan desert of southeastern Arizona. In one experiment, the removal of three dominant species, all kangaroo rats, from several plots resulted in an increased diversity of other rodent species. This increase was ascribed both to lowered competition for food and also to an altered habitat, because the abundance of grass species increased dramatically after the removal of the kangaroo rats.

Species richness is greatly affected by geological history. Scientists think tropical rain forests are old, stable communities that have undergone relatively few widespread disturbances through Earth's history. (In ecology, a **disturbance** is any event in time that disrupts community or population structure.) During this time, myriad species evolved in tropical rain forests. In contrast, glaciers have repeatedly altered temperate and arctic regions during Earth's history. An area recently vacated by glaciers will have low species richness because few species have had a chance to enter it and become established. The idea that older, more stable habitats have greater species richness than habitats subjected to frequent, widespread disturbances is known as the *time hypothesis.*

Species richness may promote community stability

Traditionally, most ecologists assumed that community stability—the ability of a community to withstand disturbances—is a consequence of community complexity. That is, ecologists hypothesized that a community with considerable species richness is more stable than a community with less species richness. According to this view, the greater the species richness, the less critically important any single species should be. With many possible interactions within the community, it appears unlikely that any single disturbance could affect enough components of the system to make a significant difference in its functioning.

Supporting evidence for this hypothesis is found in the fact that destructive outbreaks of pests are more common in cultivated fields, which are low-diversity communities, than in natural communities with greater species richness. As another example, the almost complete loss of the American chestnut tree to the chestnut blight fungus had little ecological impact on the moderately diverse Appalachian woodlands of which it was formerly a part.

Ongoing studies by David Tilman of the University of Minnesota and John Downing of the University of Iowa have strengthened the link between species richness and community stability. In their initial study, reported in the journal *Nature* in 1994, they established and monitored 207 plots of Minnesota grasslands for seven years. During the study period, Minnesota's worst drought in 50 years occurred (1987–1988). The ecologists found that those plots with the greatest number of plant species lost less ground cover, as measured by dry weight, and recovered faster than species-poor plots. Later studies by Tilman and his colleagues supported these conclusions and showed a similar effect of species richness on community stability during nondrought years. Similar work by almost three dozen ecologists at eight grassland sites in Europe, published in 1999 in the journal *Science,* also supports the link between species richness and community stability.

Some scientists do not agree with the conclusions of Tilman and other research groups that the more species the better, in terms of community stability. These critics argue it is very difficult to separate species number from other factors that could affect productivity. They suggest it would be better to start with established ecosystems and study what happens to their productivity when plants are removed.

Another observation that adds a layer of complexity to the species richness–community stability debate is that populations of individual species within a diverse community may vary significantly from year to year. It may seem paradoxical that flux within populations of individual species relates to the stability of the entire community. When you consider all the interactions among the organisms in a community, however, it is obvious that some species benefit at the expense of others. If one species declines in a given year, other species that compete with it may flourish. Thus ecosystems with greater species richness seem more likely to contain species that are resistant to any given disturbance.

Review

- How is the species richness of a community related to geographic isolation?
- How is the species richness of a community related to the structural complexity of habitats?
- How is the species richness of a community related to the environmental stress of a habitat?

BiologyNow™ Assess your understanding of **community biodiversity** by taking the pretest on your BiologyNow CD-ROM.

COMMUNITY DEVELOPMENT

Learning Objectives

8 Define *ecological succession,* and distinguish between primary and secondary succession.

9 Describe the intermediate disturbance hypothesis.

10 Discuss the two traditional views of the nature of communities, Clements's organismic model and Gleason's individualistic model.

A community does not spring into existence full blown but develops gradually through a series of stages, each dominated by different organisms. The process of community development over time, which involves species in one stage being replaced by different species, is called **succession.** An area is initially colonized by certain early-successional species that give way over time to others, which in turn give way much later to late-successional species.

Succession is usually described in terms of the changes in the species composition of an area's vegetation, although each successional stage also has its own characteristic kinds of animals and other species. The time involved in ecological succession is on the order of tens, hundreds, or thousands of years, not the millions of years involved in the evolutionary time scale.

Ecologists distinguish between two types of succession, primary and secondary. **Primary succession** is the change in species composition over time in a habitat that was not previously inhabited by organisms. No soil exists when primary succession begins. Bare rock surfaces, such as recently formed volcanic lava and rock scraped clean by glaciers, are examples of sites where primary succession might take place.

The Indonesian island of Krakatoa has provided scientists with a perfect long-term study of primary succession in a tropical rain forest. In 1883 a volcanic eruption destroyed all life on the island. Ecologists have surveyed the ecosystem in the more than 100 years since the devastation to document the return of life forms. As of the 1990s, ecologists had found that the progress of primary succession was extremely slow, in part because of Krakatoa's isolation (recall the distance effect discussed earlier in the chapter). Many species are limited in their ability to disperse over water. Krakatoa's forest, for example, may have only one tenth the tree species richness of undisturbed tropical rain forest of nearby islands. The lack of plant diversity has in turn limited the number of colonizing animal species. In a forested area of Krakatoa where zoologists would expect more than 100 butterfly species, for example, there are only two species.

Secondary succession is the change in species composition that takes place after some disturbance removes the existing vegetation; soil is already present at these sites. Abandoned agricultural fields or open areas produced by forest fires are common examples of sites where secondary succession occurs. During the summer of 1988, wildfires burned approximately one third of Yellowstone National Park. This natural disaster provided a valuable chance for ecologists to study secondary succession in areas that had been forests. After the conflagration, gray ash covered the forest floor, and most of the trees, although standing, were charred and dead. Secondary succession in Yellowstone has occurred rapidly since 1988. Less than one year later, in the spring of 1989, trout lily and other herbs sprouted and covered much of the ground. By 1998, a young forest of knee-high to shoulder-high lodgepole pines dominated the area. Douglas fir seedlings also began appearing in 1998. Ecologists continue to monitor the changes in Yellowstone as secondary succession unfolds.

Disturbance influences succession and species richness

Early studies suggested that succession inevitably progressed to a stable and persistent community, known as a *climax community,* which was determined solely by climate. Periodic disturbances, such as fires or floods, were not thought to exert much influence on climax communities. If the climax community was disturbed in any way, it would return to a self-sustaining, stable equilibrium in time.

This traditional view of stability has fallen out of favor. The apparent end-point stability of species composition in a "climax" forest is probably the result of how long trees live relative to the human life span. It is now recognized that forest communities never reach a state of permanent equilibrium but instead exist in a state of continual disturbance. The species composition and relative abundance of each species varies in a mature community over a range of environmental gradients, although the community retains a relatively uniform appearance overall.

Because all communities are exposed to periodic disturbances, both natural and human-induced, ecologists have long tried to understand the effects of disturbance on species richness. A significant advance was the development of the **intermediate disturbance hypothesis** by Joseph H. Connell. When he examined species richness in tropical rain forests and coral reefs, he proposed that species richness is greatest at moderate levels of disturbance (Fig. 52-17). At a moderate level of disturbance, the community is a mosaic of habitat patches at different stages of succession; a range of sites exists, from those that were recently disturbed to those that have not been disturbed for many years. When disturbances are frequent or intense, only those species best adapted to earlier stages of succession persist, whereas low levels of disturbance allow late-successional species to dominate to such a degree that other species disappear. For example, when periodic wildfires are suppressed in forests, reducing disturbance to a low level, some of the "typical" forest herbs decline in number or even disappear.

One of the difficulties with the intermediate disturbance hypothesis is defining precisely what constitutes an "intermediate" level of disturbance. Despite this problem, the intermediate disturbance hypothesis has important ramifications for conservation biology because it tells us that we can't maintain a particular community simply by creating a reserve around it. Both natural and human-induced disturbances will cause changes in species composition, and humans may have to intentionally intervene to maintain the species richness of the original community. What kind and how much human intervention are necessary to maintain species richness are controversial.

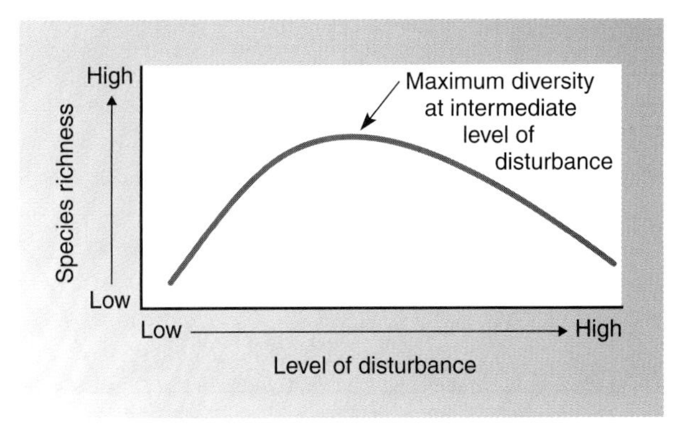

FIGURE **52-17** **Intermediate disturbance hypothesis.**

Species richness is greatest at an intermediate level of disturbance. *(After J.H. Connell, "Diversity in Tropical Rain Forests and Coral Reefs,"* Science, *Vol. 199, 1978)*

Ecologists continue to study community structure

PROCESS OF SCIENCE

One of the major issues in community ecology, from the early 1900s to the present, is the nature of communities. Are communities highly organized systems of predictable species, or are they abstractions produced by the minds of ecologists?

U.S. botanist Frederick E. Clements (1874–1945) was struck by the worldwide uniformity of large tracts of vegetation, for example, tropical rain forests in South America, Africa, and Southeast Asia. He also noted that even though the species composition of a community in a particular habitat may be different from that of a community in a habitat with a similar climate elsewhere in the world, overall the components of the two communities are usually similar. He viewed communities as something like "superorganisms," whose member species cooperated with one another in a manner that resembled the cooperation of the parts of an individual organism's body. Clements's view was that a community went through certain stages of development, like the embryonic stages of an organism, and eventually reached an adult state; the developmental process was succession, and the adult state was the climax community. This cooperative view of the community, called the **organismic model,** stresses the interaction of the members, which tend to cluster in tightly knit groups within discrete community boundaries (Fig. 52-18a).

Opponents of the organismic model, particularly U.S. ecologist Henry A. Gleason (1882–1975), held that biological interactions are less important in the production of communities than are environmental gradients (such as climate and soil) or even chance. Indeed, the concept of a community is questionable. It may be a classification category with no reality, reflecting little more than the tendency of organisms with similar environmental requirements to live in similar places. This school of thought, called the **individualistic model,** emphasizes species individuality, with each species having its own particular abiotic living requirements. It holds that communities are therefore not interdependent associations of organisms. Rather, each species is independently distributed across a continuum of areas that meets its own individual requirements (Fig. 52-18b).

Debates such as this one over the nature of communities are an integral part of the scientific process because they fuel discussion and research that lead to a better understanding of broad scientific principles. Studies testing the organismic and individualistic hypotheses of communities do not support Clements's interactive concept of communities as discrete units. Instead, most studies favor the individualistic model. As shown in Figure 52-18c, tree species in Wisconsin forests are distributed in a gradient from wet to dry environments. Also, studies of plant and animal movements during the past 14,000 years support the individualistic model, because it appears that individual species, not entire communities, became redistributed in response to changes in climate.

Review

- What is succession?
- How does primary succession differ from secondary succession?

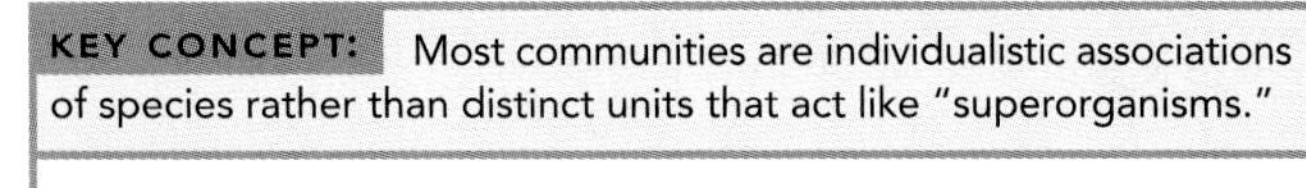

KEY CONCEPT: Most communities are individualistic associations of species rather than distinct units that act like "superorganisms."

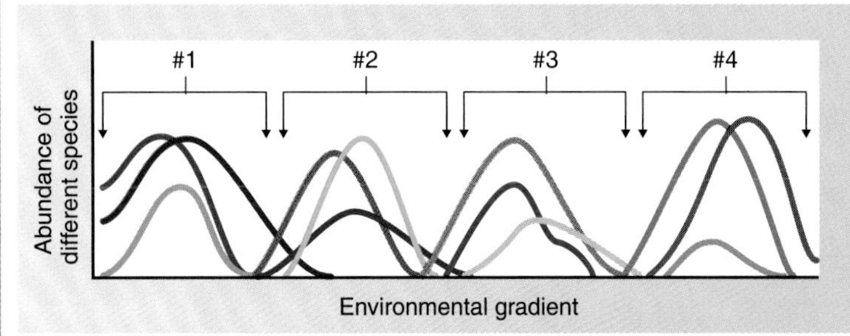

(a) Organismic model

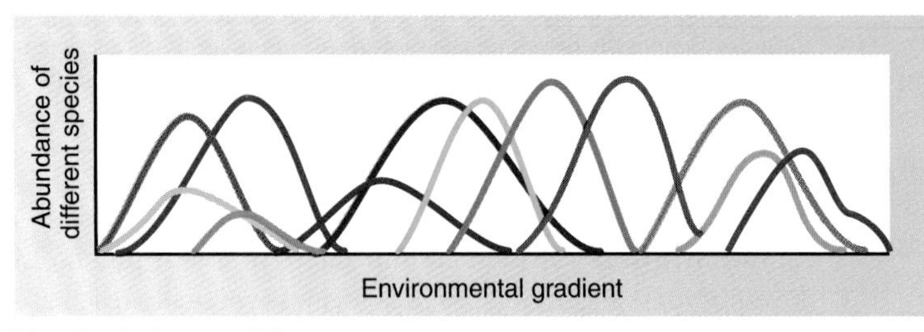

(b) Individualistic model

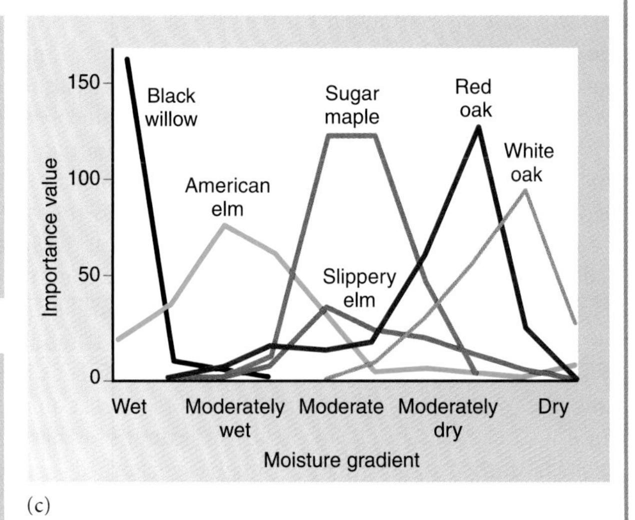

(c)

FIGURE **52-18** | **Nature of communities.**

(a) The hypothesized organismic model, in which communities are organized as distinct units. Each of the four communities shown (numbered brackets) consists of an assemblage of distinct species (each colored curve represents a single species). The arrows indicate ecotones, regions of transition along community boundaries. **(b)** The hypothesized individualistic model, in which there is a more random assemblage of species along a gradient of environmental conditions. **(c)** Tree species in Wisconsin forests, which are distributed along a moisture gradient, more closely resemble the individualistic model. The "importance value" *(Y-axis)* of each species in a given location combines three aspects (population density, frequency, and size). *(c, Adapted from J.T. Curtis,* The Vegetation of Wisconsin, *University of Wisconsin Press, Madison, Wisconsin, 1959)*

- What is Connell's intermediate disturbance hypothesis?
- How do the organismic and individualistic models of the nature of communities differ?

BiologyNow™ Assess your understanding of **community development** by taking the pretest on your BiologyNow CD-ROM.

SUMMARY WITH KEY TERMS

1 Define *ecological niche,* distinguish between an organism's fundamental niche and its realized niche, and give several examples of limiting resources that might affect an organism's ecological niche.

- The distinctive lifestyle and role of an organism in a community is its **ecological niche.** An organism's ecological niche takes into account all abiotic and biotic aspects of the organism's existence. An organism's **habitat** (where it lives) is one of the parameters used to describe the niche.
- Organisms potentially exploit more resources and play a broader role in the life of their community than they actually do. The potential ecological niche for an organism is its **fundamental niche,** whereas the niche it actually occupies is its **realized niche.**
- An organism's **limiting resources,** such as the mineral content of soil, temperature extremes, and amount of precipitation, tend to restrict its realized niche.

2 Define *competition,* and distinguish between interspecific and intraspecific competition.

- **Competition** occurs when two or more individuals attempt to use the same essential resource, such as food, water, shelter, living space, or sunlight.
- Competition occurs among individuals within a population (**intraspecific competition**) or between different species (**interspecific competition**).

3 Summarize the concepts of the competitive exclusion principle, resource partitioning, and character displacement.

- According to the **competitive exclusion principle,** two species can't occupy the same niche in the same community for an indefinite period; one species is excluded by another as a result of competition for a limiting resource.

- Some species reduce competition by **resource partitioning,** in which differences in resource use evolve. Competition among some species is reduced by **character displacement,** in which their structural, ecological, and behavioral characteristics diverge where their ranges overlap.

4 Define *predation,* and describe the effects of natural selection on predator–prey relationships.

- **Predation** is the consumption of one species (the prey) by another (the predator). During **coevolution** between predator and prey, the predator evolves more efficient ways to catch prey, and the prey evolves better ways to escape the predator.
- Two effective predator strategies are pursuit and ambush.
- Plant adaptations that protect them from being eaten include spines; thorns; tough, leathery leaves; and chemicals that are unpalatable or toxic to herbivores.
- Many animals flee from predators, some have mechanical defenses, and some associate in groups. Some animals exhibit **cryptic coloration** that helps them hide from predators by blending into their surroundings.
- Some animals that have chemical defenses exhibit **aposematic coloration,** also called **warning coloration.** In **Batesian mimicry,** a harmless or edible species resembles another species that is dangerous in some way. Predators avoid the mimic as well as the model. In **Müllerian mimicry,** several different species, all of which are poisonous, harmful, or distasteful, resemble one another. Predators learn to avoid their common aposematic coloration.

5 Distinguish among mutualism, commensalism, and parasitism, and give examples of each.

- **Symbiosis** is any intimate or long-term association between two or more species. The three types of symbiosis are mutualism, commensalism, and parasitism.
- In **mutualism,** both partners benefit. Three examples of mutualism are **nitrogen-fixing bacteria** and legumes, **zooxanthellae** and corals, and **mycorrhizae** (fungi and plant roots).
- In **commensalism,** one organism benefits and the other is unaffected. Two examples of commensalism are silverfish and army ants, and epiphytes and larger plants.
- In **parasitism,** one organism (the parasite) benefits while the other (the host) is harmed. One example of parasitism is mites that grow in or on honeybees. Some parasites are **pathogens** that cause disease.

6 Distinguish between keystone species and dominant species.

- **Keystone species** are present in relatively small numbers but are crucial in determining the species composition and ecosystem functioning of the entire community.
- In contrast to keystone species, which have an impact that is out of proportion to their abundance, **dominant species** greatly affect the community of which they are a part because they are very common.

7 Summarize the main determinants of species richness in a community, and relate species richness to community stability.

- Community complexity is expressed in terms of **species richness,** the number of species within a community, and **species diversity,** a measure of the relative importance of each species within a community based on abundance, productivity, or size.
- Species richness is often great when a habitat has structural complexity; when a community is not isolated (the distance effect) or severely stressed; when more energy is available (the species richness-energy hypothesis); in **ecotones** (transition zones between communities); and in communities with long histories without major **disturbances,** events that disrupt community or population structure.
- Several studies suggest that species richness may promote community stability.

8 Define *ecological succession,* and distinguish between primary and secondary succession.

- **Succession** is the orderly replacement of one community by another.
- **Primary succession** occurs in an area that has not previously been inhabited (such as bare rock). **Secondary succession** begins in an area where there was a pre-existing community and well-formed soil (such as abandoned farmland).

9 Describe the intermediate disturbance hypothesis.

- Disturbance affects succession and species richness. According to the **intermediate disturbance hypothesis,** species richness is greatest at moderate levels of disturbance, which create a mosaic of habitat patches at different stages of succession.

10 Discuss the two traditional views of the nature of communities, Clements's organismic model and Gleason's individualistic model.

- The **organismic model** views a community as a "superorganism" that goes through certain stages of development (succession) toward adulthood (climax). In this view, biological interactions are primarily responsible for species composition, and organisms are highly interdependent.
- Most ecologists support the **individualistic model,** which challenges the concept of a highly interdependent community. According to this model, abiotic environmental factors are the primary determinants of species composition in a community, and organisms are largely independent of each other.

POST-TEST

1. An association of populations of different species living together in one area is a(an) (a) organismic model (b) ecological niche (c) ecotone (d) community (e) habitat
2. Community structure and community functioning include all of the following *except* (a) community resilience to disturbances (b) the reproductive success of an individual organism (c) the interactions among different species (d) the number and types of species present and the relative abundance of each species (e) energy flow throughout the community
3. A limiting resource does all of the following *except* (a) tends to restrict the ecological niche of a species (b) is in short supply relative to a species' need for it (c) limits the presence of a species

POST-TEST (continued)

in a given community (d) results in an intermediate disturbance (e) may be limiting for only part of an organism's life cycle

4. Monarch and viceroy butterflies are probably an example of (a) Batesian mimicry (b) character displacement (c) resource partitioning (d) Müllerian mimicry (e) cryptic coloration
5. A symbiotic association in which organisms are beneficial to one another is known as (a) predation (b) interspecific competition (c) intraspecific competition (d) commensalism (e) mutualism
6. A species' ____________ is the totality of its adaptations, its use of resources, and its lifestyle. (a) habitat (b) ecotone (c) ecological niche (d) competitive exclusion (e) coevolution
7. Competition with other species helps to determine an organism's (a) ecotone (b) fundamental niche (c) realized niche (d) limiting resource (e) ecosystem
8. "Complete competitors cannot coexist" is a statement of the principle of (a) primary succession (b) limiting resources (c) Müllerian mimicry (d) competitive exclusion (e) character displacement
9. The ____________ signifies that species richness is greater where two communities meet than at the center of either community. (a) edge effect (b) fundamental niche (c) character displacement (d) realized niche (e) limiting resource
10. Primary succession occurs on (a) bare rock (b) newly cooled lava (c) abandoned farmland (d) both a and b are examples (e) a, b, and c are examples
11. The tendency for two similar species to differ from one another more markedly in areas where they occur together is known as (a) Müllerian mimicry (b) Batesian mimicry (c) resource partitioning (d) competitive exclusion (e) character displacement
12. An unpalatable species demonstrates its threat to potential predators by displaying (a) character displacement (b) limiting resources (c) cryptic coloration (d) aposematic coloration (e) competitive exclusion
13. An ecologist studying several forest-dwelling, insect-eating bird species does not find any evidence of interspecific competition. The most likely explanation is (a) lack of a keystone species (b) low species richness (c) pronounced intraspecific competition (d) coevolution of predator–prey strategies (e) resource partitioning
14. Support for the individualistic model of community structure includes (a) the decline of honeybees due to two species of parasitic mites (b) the identification of fig trees as a keystone species in tropical forests (c) the competitive exclusion of one *Paramecium* species by another (d) the distribution of trees along a moisture gradient in Wisconsin forests (e) the effects of the removal of a dominant rodent species from an Arizona desert
15. Connell's hypothesis to explain the effect of disturbance on species richness is known as the (a) distance effect (b) intermediate disturbance hypothesis (c) species richness–energy hypothesis (d) time hypothesis (e) edge effect

CRITICAL THINKING

1. In what symbiotic relationships are humans involved?
2. Describe the ecological niche of humans. Do you think our realized niche has changed during the past 1000 years? Why or why not?
3. How is the vertical distribution of barnacles in Scotland an example of competitive exclusion?
4. In your opinion, are humans a dominant species or a keystone species? Explain your answer.
5. Examine the top and middle graphs in Figure 52-4. Are these examples of exponential or logistic population growth? Where is K in each graph? (You may need to refer to Chapter 51 to answer these questions.)
6. Many plants that produce nodules for nitrogen-fixing bacteria are common on disturbed sites. Explain how these plants might simultaneously compete with and facilitate other plant species.

- Visit our Web site at **http://biology.brookscole.com/solomon7** for links to chapter-related resources on the World Wide Web. Additional online materials relating to this chapter can also be found on our Web site.

BIOLOGY NOW RESOURCES

Active Figure

52-15 How area and distance affect island biodiversity

Preparing for an exam? Take a diagnostic test on your BiologyNow CD-ROM.

Post-Test Answers

1. d	2. b	3. d	4. d
5. e	6. c	7. c	8. d
9. a	10. d	11. e	12. d
13. e	14. d	15. b	

Ecosystems and the Biosphere

USDA Forest Service

Catchment at Hubbard Brook Experimental Forest. Catchments measure the quantity, timing, and quality of water flowing from a forested watershed.

CHAPTER OUTLINE

Almost completely isolated from everything in the universe but sunlight, our planet Earth has often been compared to a vast spaceship whose life support system consists of the communities of organisms that inhabit it, plus energy from the sun. These organisms produce oxygen, transfer energy, and recycle water and nutrient minerals with great efficiency. Yet none of these ecological processes would be possible without the abiotic (nonliving) environment of spaceship Earth. As the sun warms the planet, it powers the hydrologic cycle (causes precipitation), drives ocean currents and atmospheric circulation patterns, and produces much of the climate to which organisms have adapted. The sun also supplies the energy that almost all organisms use to carry on life processes.

The science of ecology deals with the abiotic environment as well as with living organisms. Individual communities and their abiotic environments are **ecosystems.** An ecosystem encompasses all the interactions among organisms living together in a particular place, and among those organisms and their abiotic environment. Earth, which encompasses the **biosphere** (all of Earth's communities) and its interactions with Earth's water, soil, rock, and atmosphere, is the largest ecosystem.

Ecologists conduct detailed ecosystem studies in laboratory simulations and in the field to measure such processes as energy flow, the cycling of nutrients, and the effects of natural and human-induced disturbances (such as air pollution, tree harvesting, and land use changes). Some ecosystem studies, such as those performed at the Hubbard Brook Experimental Forest (HBEF), a 3100-hectare (7750-acre) reserve in the White Mountain National Forest in New Hampshire, are long term. Beginning in the early 1960s and continuing to the present, HBEF has been the site of numerous studies that address the hydrology (for example, precipitation, surface runoff, and groundwater flow), biology, geology, and chemistry of forests and associated aquatic ecosystems. The National Science Foundation (NSF) has designated HBEF as one of its 24 long-term ecological research sites.

Many of the experiments at HBEF are based on field observations. For example, detailed surveys of the salamander populations in forest communities were originally done in 1970 and were repeated in recent years. Other studies involve manipulative experiments. In 1978, scientists added dilute sulfuric acid to a small stream in HBEF to study the chemical and biological effects of acidification. This experiment was of practical value, because **acid deposition,** a form of air pollution, has acidified numerous lakes and streams in industrialized countries.

Several researchers have studied the effects on HBEF stream ecosystems of **deforestation,** the clearance of large expanses of forest for agriculture or other uses. When a forest is removed, the total amount of water and nutrient minerals that flows into streams increases drastically. The photograph on page 1043 shows a concrete dam called a *catchment,* constructed across a stream to measure the flow of water and chemical components out of the ecosystem. Typically, outflow is measured in two separate ecosystems, one of which serves as a control and one of which is experimentally manipulated. These studies demonstrate that deforestation causes soil erosion and leaching of essential nutrient minerals that result in decreased soil fertility. The summer temperatures in streams running through deforested areas are higher than in shady streams running through uncut forests. Many stream dwellers do not fare well in deforested areas, in part because they are adapted to cold temperatures.

Detailed studies such as those at HBEF provide ecologists with insights into how ecological processes function in individual ecosystems. Ecologists compare these data with similar information from other ecosystem studies to develop generalized insights into how ecosystems are structured and how they function. Long-term ecosystem experiments enable ecologists to evaluate and predict the effects of environmental change, including human-induced change.

Ecosystem experiments also contribute to our practical knowledge about how to maintain water quality, wildlife habitat, and productive forests. **Ecosystem management,** a conservation approach that emphasizes restoring and maintaining the quality of an entire ecosystem rather than the conservation of individual species, makes use of such knowledge.

This chapter develops three key concepts about ecosystems. First, energy flow in ecosystems is linear. Second, matter moves in numerous cycles within an ecosystem, that is, from one organism to another and from organisms to the abiotic environment and back again. Third, the abiotic environment—including solar radiation, the atmosphere, the ocean, climate, and fire—helps shape the biotic component of ecosystems. ■

ENERGY FLOW THROUGH ECOSYSTEMS

Learning Objectives

1. Summarize the concept of energy flow through a food web.
2. Explain typical pyramids of numbers, biomass, and energy.
3. Distinguish between gross primary productivity and net primary productivity.

The passage of energy in a one-way direction through an ecosystem is known as **energy flow.** Energy enters an ecosystem as radiant energy (sunlight), a tiny portion (less than 1%) of which producers trap and use during photosynthesis. The energy, now in chemical form, is stored in the bonds of organic (carbon-containing) molecules such as glucose. When cellular respiration breaks these molecules apart, energy becomes available (in the form of ATP) to do work such as repairing tissues, producing body heat, moving about, or reproducing. As the work is accomplished, energy escapes the organisms and dissipates into the environment as heat. Ultimately, this heat energy radiates into space. Thus, once an organism has used energy, it is unavailable for reuse (Fig. 53-1; see also the discussion of the second law of thermodynamics in Chapter 6).

In an ecosystem, energy flow occurs in **food chains,** in which energy from food passes from one organism to the next in a sequence. Producers form the beginning of the food chain by capturing the sun's energy through photosynthesis. Herbivores (and omnivores) eat plants, obtaining the chemical energy of the producers' molecules as well as building materials from which they construct their own tissues. Herbivores are in turn consumed by carnivores and omnivores, who reap the energy stored in the herbivores' molecules. At every step, decomposers (saprotrophs) break down organic molecules in the remains (carcasses and body wastes) of all members of the food chain. (See *Focus On: Food Chains and Poisons in the Environment* for a discussion of how certain toxins pass through food chains.)

Simple food chains such as just described rarely occur in nature, because few organisms eat just one kind, or are eaten by just one other kind of organism. More typically, the flow of energy and materials through ecosystems takes place in accordance with a range of food choices for each organism. In an ecosystem of average complexity, hundreds of alternative pathways

ACTIVE FIGURE **53-1** ▶ **Energy flow through ecosystems.**
Energy enters ecosystems from an external source (the sun) and exits as heat loss. As stipulated by the second law of thermodynamics, most of the energy acquired by a given trophic level is released into the environment as heat and is therefore unavailable to the next trophic level.

BiologyNow™ Learn more about **how trophic levels interact** by clicking on this figure on your BiologyNow CD-ROM.

Certain toxic substances, including some pesticides, radioactive isotopes, heavy metals such as mercury, and industrial chemicals such as PCBs, enter food chains. The effects of the pesticide DDT on some bird species first drew attention to the problem. Falcons, pelicans, bald eagles, ospreys, and many other birds are very sensitive to traces of DDT in their tissues. A substantial body of scientific evidence indicates that one effect of DDT on these birds is that they lay eggs with extremely thin, fragile shells that usually break during incubation, causing the chicks' deaths. In 1962 U.S. biologist Rachel Carson published *Silent Spring,* which heightened public awareness about the dangers of DDT and other pesticides. After 1972, the year DDT was banned in the United States, the reproductive success of many birds gradually improved.

The impact of DDT on birds is the result of three characteristics of DDT (and other toxins that cause problems in food webs): its persistence, bioaccumulation, and biological magnification. Some toxins are extremely stable and may take many years to be broken down into less toxic forms. The **persistence** of synthetic pesticides and industrial chemicals is a result of their novel chemical structures. Natural decomposers such as bacteria do not have ways to degrade these toxins, which therefore accumulate in the environment and increase in concentration as they pass from one trophic level to another.

When an organism does not metabolize (break down) or excrete a persistent toxin, it simply gets stored, usually in fatty tissues. Over time, the organism may accumulate high concentrations of the toxin. The buildup of such a toxin in an organism's body is known as **bioaccumulation.**

Organisms at higher trophic levels in food webs tend to store greater concentrations of bioaccumulated toxins in their bodies than do those at lower levels. The increase in concentration as the toxin passes through successive levels of the food web is known as **biological magnification.**

As an example of the concentrating characteristic of persistent toxins, consider a food chain studied in a Long Island salt marsh that was sprayed with DDT over a period of years for mosquito control. The concentration of DDT in water during this period was extremely dilute, 0.00005 parts per million (ppm). The algae and other plankton, which took up and accumulated the toxin, contained 0.04 ppm of DDT. Each shrimp grazing on the plankton ingested and concentrated the pesticide in its tissues, to 0.16 ppm. Eels that ate shrimp laced with pesticide ended up with a pesticide level of 0.28 ppm, whereas other predaceous fishes contained 2.07 ppm of DDT. The top carnivores, ring-billed gulls, had a pesticide value of 75.5 ppm from eating contaminated fishes. Although this example involved a bird at the top of the food chain, it is important to recognize that *all* top carnivores, from fishes to humans, are at risk from biological magnification of persistent toxins. Because of this risk, currently approved pesticides have been tested to ensure they do not persist and accumulate in the environment.

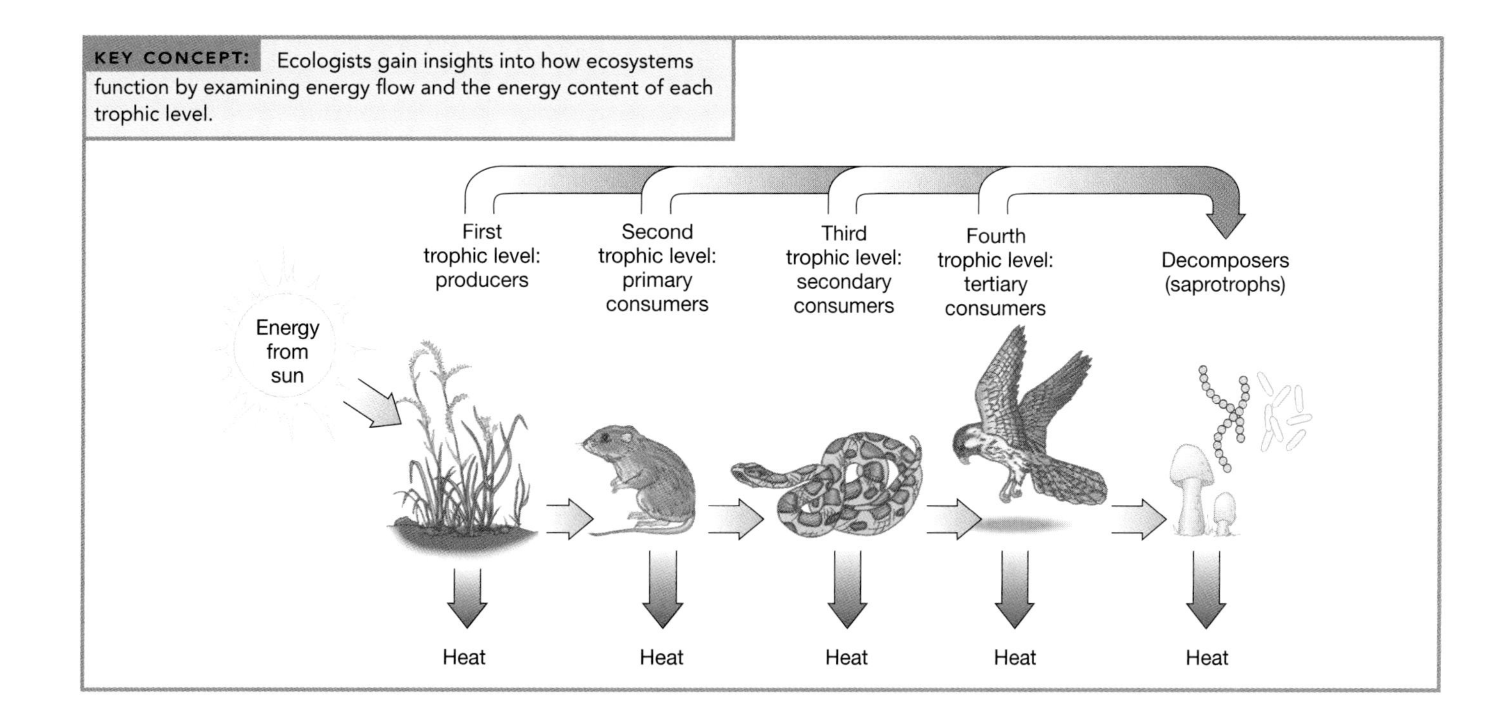

are possible. Thus a **food web,** which is a complex of interconnected food chains in an ecosystem, is a more realistic model of the flow of energy and materials through ecosystems (Fig. 53-2).

Food webs are divided into **trophic levels** (from the Greek *tropho,* which means nourishment). Producers (organisms that photosynthesize) occupy the first trophic level, primary consumers (herbivores) occupy the second, secondary consumers (carnivores and omnivores) the third, and so on (see Fig. 53-1).

Because food webs are descriptions of "who eats whom," they indicate the negative effects predators have on their prey. For example, consider a simple food chain: grass ⟶ field mouse ⟶ owl. The owl, which kills and eats mice, obviously exerts a negative effect on the mouse population; in like manner, field mice, which eat grass seeds, reduce the grass population.

A trophic level in a food web also influences other trophic levels to which it is not directly linked. Producers and top carnivores do not exert direct effects on one another, yet each indirectly affects the other. In our example, the owls help the grasses by keeping the population of seed-eating mice under control. Likewise, the grasses benefit owls by supporting a population of mice on which the owl population feeds. These indirect interactions may be as important in food web dynamics as direct, predator–prey interactions.

The most important thing to remember about energy flow in ecosystems is that it is linear, or one-way. That is, as long as it is not used to do biological work, energy moves along a food web from one trophic level to the next trophic level. Once an organism has used energy, however, it is lost as heat and is unavailable to any other organism in the ecosystem.

FIGURE **53-2** **A food web at the edge of an eastern deciduous forest.**

This diagram is greatly simplified compared with what actually happens in nature. Groups of species are lumped into single categories such as "spiders"; many species are not included; and numerous links in the web are not shown.

Ecological pyramids illustrate how ecosystems work

Ecologists sometimes compare trophic levels by determining the number of organisms, the biomass, or the relative energy found at each level. This information is presented graphically as **ecological pyramids.** The base of each ecological pyramid represents the producers, the next level is the primary consumers (herbivores), the level above that is the secondary consumers (carnivores), and so on. The relative area of each bar of the pyramid is proportional to what is being demonstrated.

A **pyramid of numbers** shows the number of organisms at each trophic level in a given ecosystem, with a larger area illustrating greater numbers for that section of the pyramid. In most pyramids of numbers, fewer organisms occupy each successive trophic level. Thus in African grasslands the number of herbivores, such as zebras and wildebeests, is greater than the number of carnivores, such as lions. Inverted pyramids of numbers, in which higher trophic levels have more organisms than lower trophic levels, are often observed among decomposers, parasites, and herbivorous insects. One tree provides food for thou-

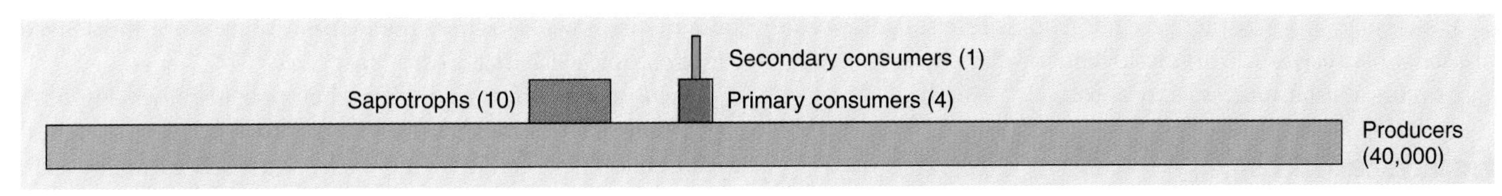

(a) Tropical forest in Panama

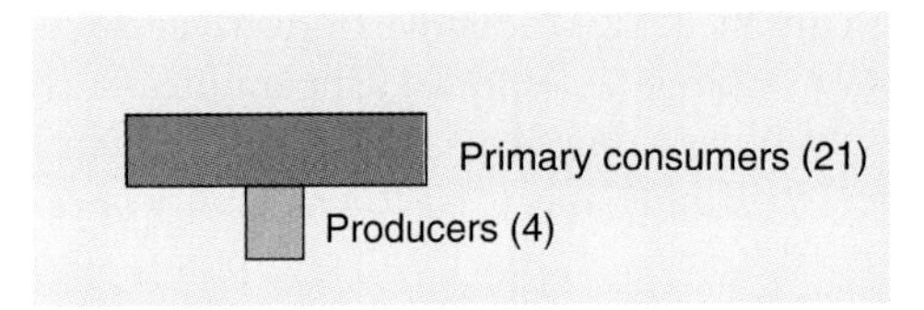

(b) Plankton in English Channel

FIGURE 53-3 **Pyramids of biomass.**

These pyramids are based on the biomass at each trophic level and generally have a pyramid shape with a large base and progressively smaller areas for each succeeding trophic level. Biomass values are in grams of dry weight per square meter. **(a)** A pyramid of biomass for a tropical forest in Panama. **(b)** An inverted biomass pyramid, such as that for plankton in the English Channel, occurs when a highly productive lower trophic level experiences high rates of turnover. Plankton are free-floating, mainly microscopic algae and invertebrate animals that eat algae. *(a, b, Adapted from E.P. Odum,* Fundamentals of Ecology, *3rd ed., W.B. Saunders Company, Philadelphia, 1971, and based on studies by F.B. Golley and G.I. Child [a] and H.W. Harvey [b])*

sands of leaf-eating insects, for example. Pyramids of numbers are of limited usefulness, because they do not indicate the biomass of the organisms at each level and they do not indicate the amount of energy transferred from one level to another.

A **pyramid of biomass** illustrates the total biomass at each successive trophic level. **Biomass** is a quantitative estimate of the total mass, or amount, of living material; it indicates the amount of fixed energy at a particular time. Biomass units of measure vary: Biomass may be represented as total volume, dry weight, or live weight. Typically, these pyramids illustrate a progressive reduction of biomass in succeeding trophic levels (Fig. 53-3a). Assuming an average biomass reduction of about 90% for each trophic level,[1] 10,000 kg of grass should support 1000 kg of grasshoppers, which in turn support 100 kg of frogs. Using this logic, the biomass of frog eaters (such as snakes) could only weigh, at most, about 10 kg. From this brief exercise, you see that although carnivores do not eat producers, a large producer biomass is required to support carnivores via a food web.

Occasionally we find an inverted pyramid of biomass in which the primary consumers outweigh the producers (Fig. 53-3b). In these instances, herbivores such as fish and zooplankton (protozoa, tiny crustaceans, and immature stages of many aquatic animals) consume large numbers of producers, which are usually unicellular algae that are short-lived and reproduce quickly. Thus, although at any point in time relatively few algae are present, the rate of biomass production of the primary consumers is much less than that of the producers.

A **pyramid of energy** indicates the energy content, often expressed as kilocalories per square meter per year, of the biomass of each trophic level. A common method ecologists use to measure energy content is to burn a sample of tissue in a calorimeter; the heat released during combustion is measured to determine the energy content of the organic material in the sample. Energy pyramids, which always have large bases and get progressively smaller through succeeding trophic levels, show that most energy dissipates into the environment when there is a transition from one trophic level to the next. Less energy reaches each successive trophic level from the level beneath it because those organisms at the lower level use some of the energy to perform work, while some of it is lost (Fig. 53-4). (Remember, no biological process is 100% efficient.) The second law of thermo-

[1] The 90% reduction in biomass is an approximation; actual biomass reduction from one trophic level to the next varies widely in nature.

FIGURE 53-4 **Pyramid of energy.**

A pyramid of energy for Silver Springs, Florida, represents energy flow, the functional basis of ecosystem structure. Energy values are in kilocalories per square meter per year. Note the substantial loss of usable energy from one trophic level to the next. (Energy in producers represents gross primary productivity.) The Silver Spring ecosystem is complex, but tape grass (producers), spiral-shelled snails (primary consumers), young river turtles (secondary consumers), gar (tertiary consumers), and bacteria and fungi (saprotrophs) are representative organisms. When they are young, river turtles are carnivores and consume snails, aquatic insects, and worms; as adults, river turtles are herbivores. *(Based on H.T. Odum, "Trophic Structure and Productivity of Silver Springs, Florida,"* Ecological Monographs, *Vol. 27, 1957)*

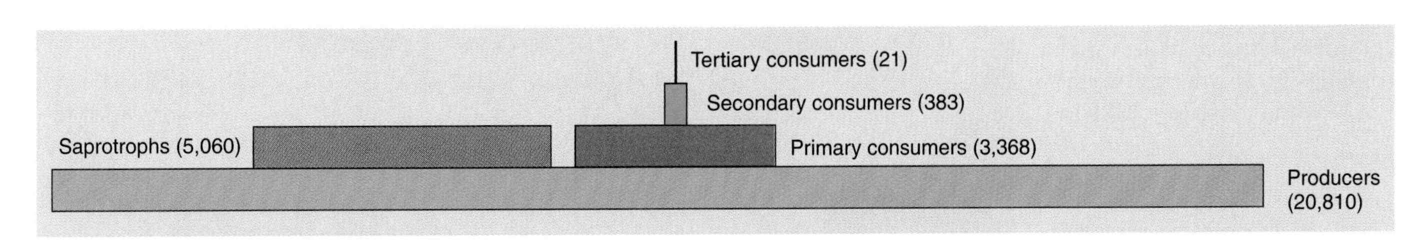

dynamics explains why there are few trophic levels: Food webs are short because of the dramatic reduction in energy content that occurs at each successive trophic level.

Ecosystems vary in productivity

The **gross primary productivity (GPP)**of an ecosystem is the rate at which energy is captured during photosynthesis.[2] Thus, GPP is the total amount of photosynthetic energy captured in a given period. Of course, plants and other producers must respire to provide energy for their life processes, and cellular respiration acts as a drain on photosynthetic output. Energy that remains in plant tissues after cellular respiration has occurred is called **net primary productivity (NPP).** That is, NPP is the amount of biomass (the energy stored in plant tissues) found in excess of that broken down by a plant's cellular respiration for normal daily activities. NPP represents the rate at which this organic matter is actually incorporated into plant tissues to produce growth.

Net primary productivity (plant growth per unit area per unit time)	=	gross primary productivity (total photosynthesis per unit area per unit time)	−	plant respiration (per unit area per unit time)

Only the energy represented by net primary productivity is available for consumers, and of this energy only a portion is actually used by them. Both GPP and NPP are expressed as energy per unit area per unit time (for example, kilocalories of energy fixed by photosynthesis per square meter per year) or as dry weight (that is, grams of carbon incorporated into tissue per square meter per year).

Many factors may interact to determine productivity. Some plants are more efficient than others in fixing carbon. Environmental factors are also important. These include the availability of solar energy, nutrient minerals, and water; other climate factors; the degree of maturity of the community; and the severity of human modification of the environment. Many of these factors are difficult to assess.

Ecologists use different methods to measure primary productivity, depending on whether GPP or NPP is being assessed. Methods also vary from one type of ecosystem to another. On land, ecologists might cut, dry, and weigh plants at the end of a growing season to measure NPP. Dry weights are measured, because the water content of different species varies considerably. This method would not be useful to measure NPP in the ocean, however, where the main producers are short-lived microscopic algae that are heavily grazed.

Ecosystems differ strikingly in their productivities (Fig. 53-5 and Table 53-1). On land, tropical rain forests have the highest productivity, probably owing to the abundant rainfall, warm temperatures, and intense sunlight. As you might expect, tundra with its short, cool growing season and deserts with their lack of precipitation are the least productive terrestrial ecosystems. In ecosystems with comparable annual temperatures (such as temperate deciduous forest, temperate grassland, and temperate desert), water availability affects NPP. Availability of essential nutrient minerals such as nitrogen and phosphorus also affects NPP.

Wetlands (swamps and marshes) connect terrestrial and aquatic environments and are extremely productive. The most productive aquatic ecosystems are algal beds, coral reefs, and estuaries. The lack of available nutrient minerals in the sunlit region of the open ocean makes this area extremely unproductive, equivalent to an aquatic desert. Earth's major aquatic and terrestrial ecosystems are discussed in Chapter 54.

The relationship of productivity to biological diversity is complex

Conventional ecological theory once held that the more productive the ecosystem, the greater the biodiversity it supported. Now ecologists are seeing a recurring pattern worldwide: Ecosystems are more diverse as productivity increases, but at some point diversity actually declines with increasing productivity. For example, the resource-poor depths of the Atlantic Ocean's

FIGURE **53-5** **A measure of Earth's primary productivity.**
The data are from a satellite launched as part of NASA's Mission to Planet Earth in 1997. The satellite measured the amount of plant life on land as well as the concentration of phytoplankton (algae) in the ocean. On land, the most productive areas, such as tropical rain forests, are dark green, whereas the least productive ecosystems (deserts) are orange. In the ocean and other aquatic ecosystems, the most productive regions are red, followed by orange, yellow, green, and blue (the least productive). Data are not available for the gray area.

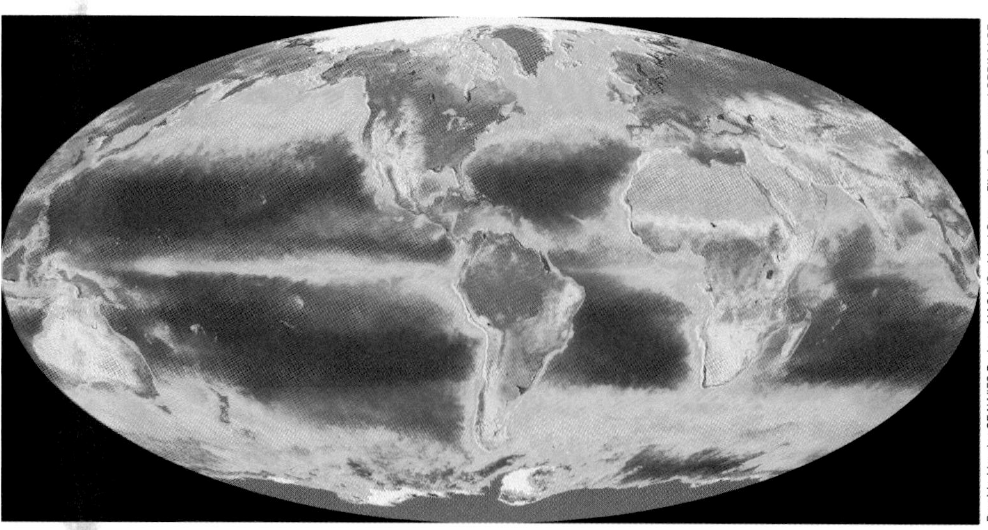

Provided by the SEAWIFS Project, NASA/Goddard Space Flight Center and ORBIMAGE

[2] Gross and net primary productivities are referred to as primary because plants and other producers occupy the first position in food webs.

TABLE 53-1 Net Primary Productivities (NPP) for Selected Ecosystems*

Ecosystem	Average NPP (g dry matter/m^2/year)
Algal beds and reefs	2500
Tropical rain forest	2200
Swamp and marsh	2000
Estuaries	1500
Temperate evergreen forest	1300
Temperate deciduous forest	1200
Savanna	900
Boreal (northern) forest	800
Woodland and shrubland	700
Agricultural land	650
Temperate grassland	600
Upwelling zones in ocean	500
Lake and stream	250
Arctic and alpine tundra	140
Open ocean	125
Desert and semidesert scrub	90
Extreme desert (rock, sand, ice)	3

*Based on R.H. Whittaker, *Communities and Ecosystems,* 2nd ed., Macmillian, New York, 1975.

abyssal plain have a higher species richness than the productive shallow waters near the coasts, and intermediate depths exhibit the greatest diversity. Ecologists have little data to help explain the pattern, which holds true with rodents in Israel, birds in South America, and large mammals in Africa. Mathematical ecosystem models, however, suggest that a *patchy distribution of resources* reduces competition and allows a greater variety of organisms to coexist (see Chapter 51).

The bad news for global biodiversity is that humans are constantly enriching our environment, for example, with nitrogen and phosphorus inputs from fossil fuels, fertilizers, and livestock. This continual fertilization may make the Earth's ecosystems more and more productive, a shift that some ecologists think could cost the world a substantial loss of biodiversity. Other factors that affect species richness were discussed in Chapter 52.

Humans consume an increasingly greater percentage of global primary productivity

Humans consume far more of Earth's resources than do any of the other millions of animal species. Peter Vitousek and coworkers at Stanford University calculated in 1986 how much of the global NPP is appropriated for the human economy. When both direct and indirect human impacts are accounted for, Vitousek estimated that humans use 32% (a conservative estimate) to 40% (a more likely estimate) of land-based annual NPP. Since 1986, scientists have done a lot of research on global ecology, resulting in improved data sets. In 2001 Stuart Rojstaczer and coworkers at Duke University used current satellite-based data to determine a conservative estimate of land-based annual NPP appropriation by humans at 32%.

The take-home message from Vitousek's and Rojstaczer's research is simple. Essentially, human use of global productivity is competing with other species' needs for energy. Our use of so much of the world's productivity may contribute to the loss, through extinction or genetic impoverishment, of many species that have unique roles in maintaining functional ecosystems. Clearly, at these levels of consumption and exploitation of the Earth's resources, human population growth, which was discussed in Chapter 51, seriously threatens the planet's ability to support all its occupants.

Review

- How does energy flow through a food web such as a deciduous forest?
- What are trophic levels, and how are they related to ecological pyramids?
- How do gross primary productivity (GPP) and net primary productivity (NPP) differ?

BiologyNow™ Assess your understanding of **energy flow through ecosystems** by taking the pretest on your BiologyNow CD-ROM.

CYCLES OF MATTER IN ECOSYSTEMS

Learning Objective

4 Describe the main steps in each of these biogeochemical cycles: the carbon, nitrogen, phosphorus, and hydrologic cycles.

Matter moves in numerous cycles from one part of an ecosystem to another—that is, from one organism to another and from living organisms to the abiotic environment and back again. We call these cycles of matter **biogeochemical cycles** because they involve biological, geologic, and chemical interactions. With respect to matter, Earth is essentially a **closed system** (see Chapter 6); for all practical purposes, matter cannot escape from Earth's boundaries. The materials organisms use cannot be "lost," although this matter can end up in locations outside the reach of organisms for a long period. Usually materials are reused and often recycled both within and among ecosystems.

We discuss four different biogeochemical cycles of matter—carbon, nitrogen, phosphorus, and water—as representative of all biogeochemical cycles. These four cycles are particularly important to organisms because they involve materials used to make the chemical components of cells. Carbon, nitrogen, and water have gaseous components and so cycle over large distances of the atmosphere with relative ease. Phosphorus does not have a gaseous phase, and, as a result, only local cycling of phosphorus occurs easily.

Carbon dioxide is the pivotal molecule in the carbon cycle

Carbon must be available to organisms because proteins, nucleic acids, lipids, carbohydrates, and other molecules essential to life contain carbon. Carbon is present in the atmosphere as the gas carbon dioxide (CO_2), which makes up approximately 0.04% of the atmosphere. It is also present in the ocean and fresh water as dissolved carbon dioxide, that is, carbonate (CO_3^{2-}) and bicarbonate (HCO_3^-); other forms of dissolved inorganic carbon; and dissolved organic carbon from decay processes. Carbon is also present in rocks such as limestone ($CaCO_3$). The global movement of carbon between the abiotic environment, including the atmosphere and ocean, and organisms is known as the **carbon cycle** (Fig. 53-6). Refer to the figure as you read the following description.

As shown in ①, during photosynthesis, plants, algae, and cyanobacteria remove carbon dioxide from the air and fix, or incorporate, it into complex organic compounds such as glucose. Plants use much of the glucose to make cellulose, starch, amino acids, nucleic acids, and other compounds. Thus photosynthesis incorporates carbon from the abiotic environment into the biological compounds of producers.

② Many of these compounds are used as fuel for cellular respiration by the producer that made them, by a consumer that eats the producer, or by a decomposer that breaks down the remains of the producer or consumer. The process of cellular respiration returns carbon dioxide to the atmosphere. A similar carbon cycle occurs in aquatic ecosystems between aquatic organisms and dissolved carbon dioxide in the water (not shown in the figure).

Sometimes the carbon in biological molecules is not recycled back to the abiotic environment for some time. A large amount of carbon is stored in the wood of trees, where it may stay for several hundred years or even longer. ③ In addition, millions of years ago vast coal beds formed from the bodies of ancient trees that were buried and subjected to anaerobic conditions before they had fully decayed. ④ Similarly, the oils of unicellular marine organisms probably gave rise to the underground deposits of oil and natural gas that accumulated in the geologic past. Coal, oil, and natural gas, called **fossil fuels** because they formed from the remains of ancient organisms, are vast depositories of carbon compounds, the end products of photosynthesis that occurred millions of years ago. Fossil fuels are nonrenewable resources; that is, Earth has a finite, or limited, supply of them. Although natural forces still form fossil fuels today, they are forming too slowly to replace the fossil fuel reserve we are using.

⑤ The process of burning, or combustion, may return the carbon in coal, oil, natural gas, and wood to the atmosphere. In combustion, organic molecules are rapidly oxidized (combined with oxygen) and converted to carbon dioxide and water with an accompanying release of light and heat.

⑥ An even greater amount of carbon that is stored for millions of years is incorporated into the shells of marine organisms. When these organisms die, their shells sink to the ocean floor, and sediments cover them, forming seabed deposits thousands of meters thick. The deposits are eventually cemented together to form limestone, a sedimentary rock. Earth's crust is dynamically active, and over millions of years, sedimentary rock on the bottom of the sea floor may lift to form land surfaces. The

ACTIVE FIGURE **53-6** **A simplified diagram of the carbon cycle.**

All but a tiny fraction of Earth's estimated 10^{23} g of carbon is buried in sedimentary rocks and fossil fuel deposits. The values shown for some of the active pools in the global carbon budget are expressed as 10^{15} g of carbon. For example, the soil contains an estimated 1500×10^{15} g of carbon. *(Values from W.H. Schlesinger,* Biogeochemistry: An Analysis of Global Change, *2nd ed., Academic Press, San Diego, 1997, and based on several sources)*

BiologyNow™ Explore **the carbon cycle and global warming** by clicking on this figure on your BiologyNow CD-ROM.

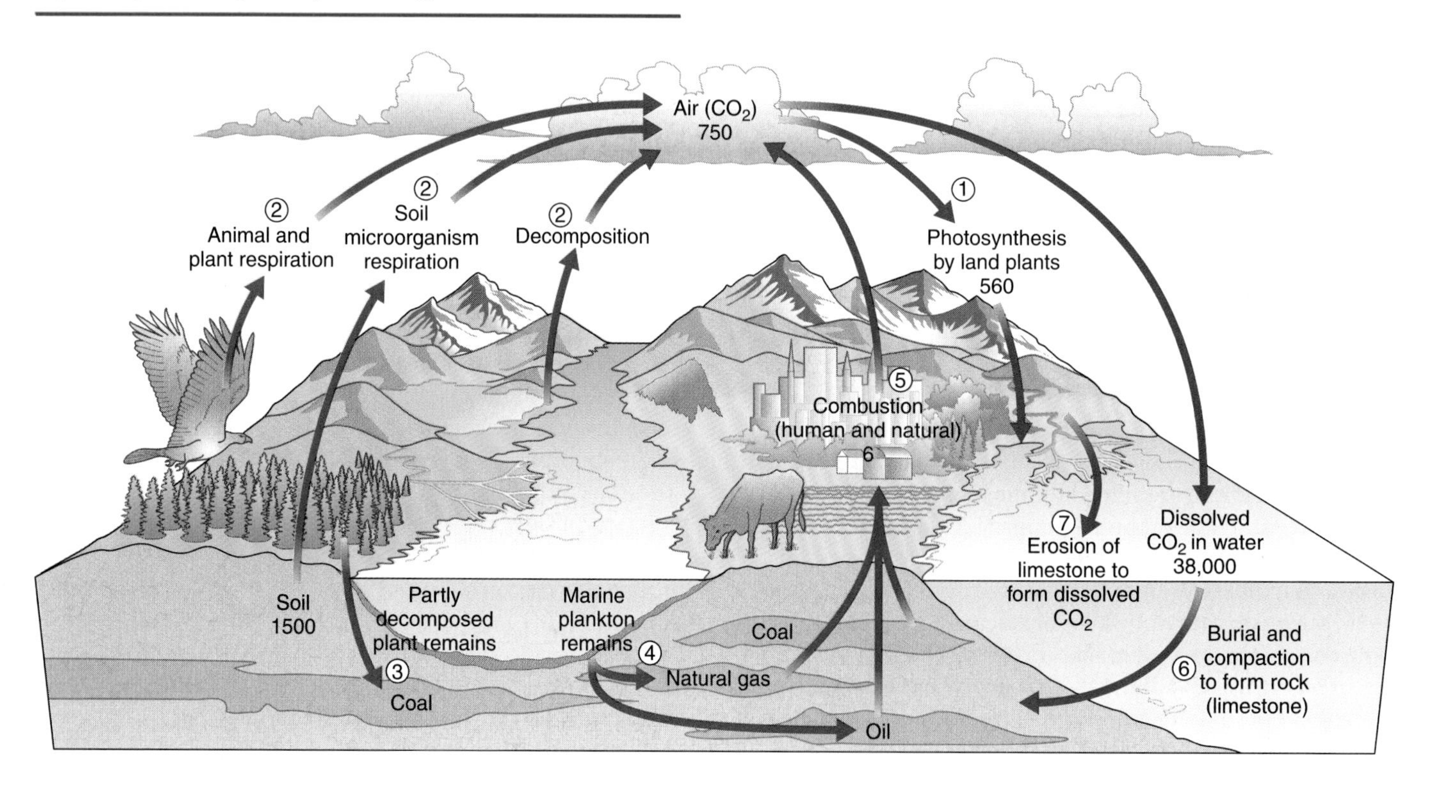

summit of Mount Everest, for example, consists of sedimentary rock. ⑦ When the process of geologic uplift exposes limestone, chemical and physical weathering processes slowly erode it away. This returns carbon to the water and atmosphere, where it is available to participate in the carbon cycle once again.

Thus photosynthesis removes carbon from the abiotic environment and incorporates it into biological molecules. Cellular respiration, combustion, and erosion of limestone return carbon to the water and atmosphere of the abiotic environment.

Human activities have disturbed the global carbon budget

Before the Industrial Revolution around 1850, the global carbon cycle was in a steady state. Enormous amounts of carbon moved to and from the atmosphere, ocean, and terrestrial ecosystems, but these movements within the global carbon cycle just about canceled out one another. Since 1850, our industrial society has required a lot of energy, and we have burned increasing amounts of fossil fuels—coal, oil, and natural gas to obtain this energy. This trend, along with a greater combustion of wood as a fuel and the burning of large sections of tropical forest, has released CO_2 into the atmosphere at a rate greater than the natural carbon cycle can handle.

The level of CO_2 increased dramatically beginning in the last half of the 20th century (see Fig. 55-11), and this rise of CO_2 in the atmosphere may cause human-induced changes in climate called *global warming.* Global warming could result in a rise in sea level, changes in precipitation patterns, death of forests, extinction of organisms, and problems for agriculture. It could force the displacement of thousands or even millions of people, particularly from coastal areas. A more thorough discussion of increasing atmospheric CO_2 and the potential impact of global warming is found in Chapter 55.

Bacteria are essential to the nitrogen cycle

Nitrogen is crucial for all organisms because it is an essential part of proteins, nucleic acids, and chlorophyll. Because Earth's atmosphere is about 78% nitrogen gas (N_2), it would appear there could be no possible shortage of nitrogen for organisms. However, molecular nitrogen is so stable that it does not readily combine with other elements. Therefore, the N_2 molecule must be broken apart before the nitrogen atoms combine with other elements to form proteins, nucleic acids, and chlorophyll. Chemical reactions that break up N_2 and combine nitrogen with such elements as oxygen and hydrogen require a great deal of energy.

The **nitrogen cycle,** in which nitrogen cycles between the abiotic environment and organisms, has five steps: nitrogen fixation, nitrification, assimilation, ammonification, and denitrification (Fig. 53-7). Bacteria are exclusively involved in all these steps except assimilation. Refer to the figure as you read the following paragraphs, which describe the nitrogen cycle that occurs on land; a similar cycle occurs in aquatic ecosystems.

In ①, the first step in the nitrogen cycle, biological **nitrogen fixation** involves conversion of gaseous nitrogen (N_2) to ammonia (NH_3). This process is called *nitrogen fixation* because nitrogen is fixed into a form that organisms use. Combustion, volcanic action, lightning discharges, and industrial

FIGURE **53-7** **A simplified diagram of the nitrogen cycle.** The largest pool of nitrogen, estimated at 3.9×10^{21} g, is in the atmosphere. The values shown for selected nitrogen fluxes in the global nitrogen budget are expressed as 10^{12} g of nitrogen per year and represent terrestrial values. For example, each year humans fix an estimated 100×10^{12} g of nitrogen. *(Values from W.H. Schlesinger,* Biogeochemistry: An Analysis of Global Change, *2nd ed. Academic Press, San Diego, 1997, and based on several sources)*

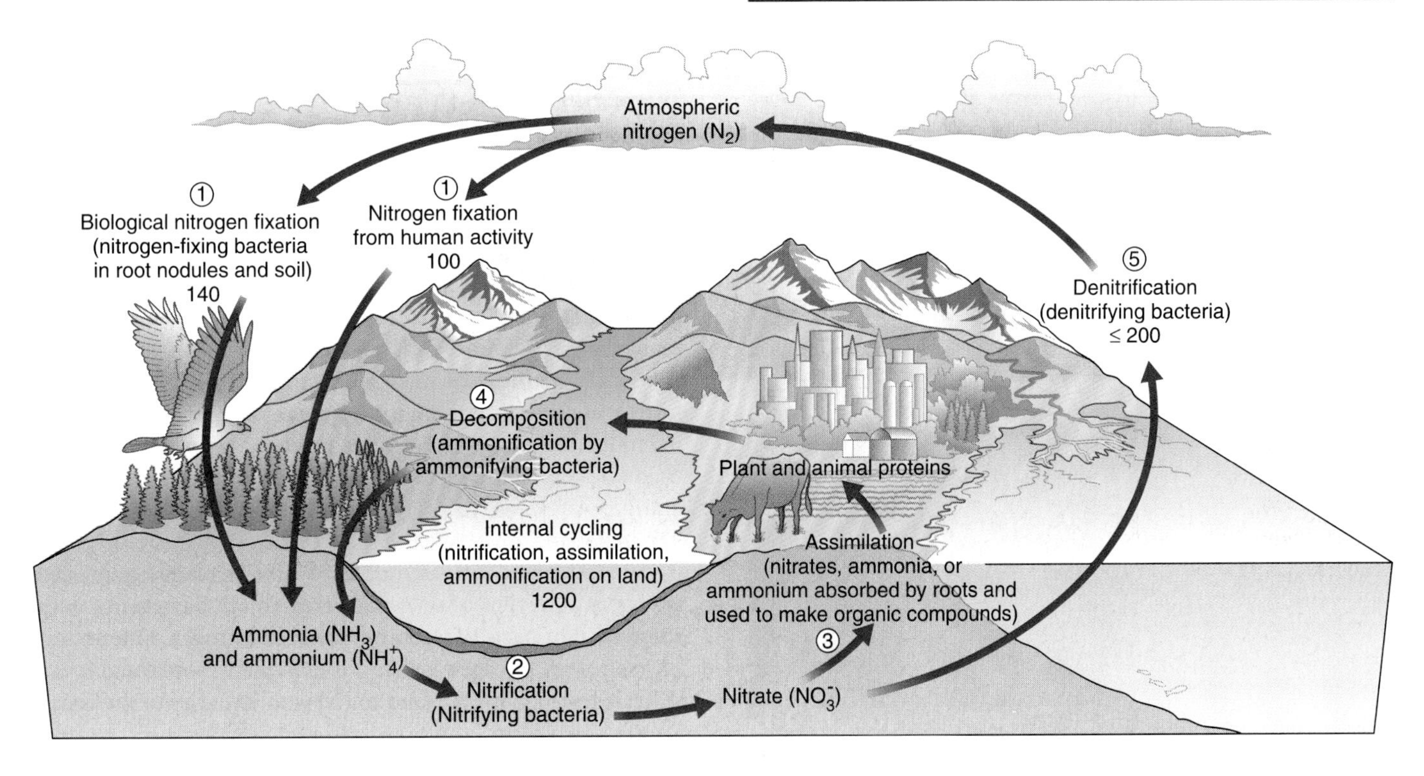

processes also fix nitrogen as nitrate; each of these supplies enough energy to break apart molecular nitrogen. *Nitrogen-fixing bacteria,* including cyanobacteria and certain other free-living and symbiotic bacteria, carry on biological nitrogen fixation in soil and aquatic environments. Nitrogen-fixing bacteria employ an enzyme called **nitrogenase** to break up molecular nitrogen and combine the resulting nitrogen atoms with hydrogen.

Because nitrogenase functions only in the absence of oxygen, the bacteria that fix nitrogen insulate the enzyme from oxygen in some way. Some nitrogen-fixing bacteria live beneath layers of oxygen-excluding slime on the roots of several plant species. Other important nitrogen-fixing bacteria, in the genus *Rhizobium,* live in oxygen-excluding swellings, or **nodules,** on the roots of legumes such as beans and peas and some woody plants (Fig. 53-8). The nodules are reddish-brown because they contain **leghemoglobin,** an oxygen-binding protein similar in structure to hemoglobin. Even though the nodules exclude much of the oxygen, the small amount that is present binds to leghemoglobin. Leghemoglobin supplies oxygen the respiring bacteria need.

In aquatic environments, cyanobacteria perform most nitrogen fixation. Filamentous cyanobacteria have special oxygen-excluding cells called **heterocysts** that function to fix nitrogen (see Fig. 23-16). Some water ferns have cavities in which cyanobacteria live, in a manner comparable to the way *Rhizobium* lives in root nodules of legumes. Other cyanobacteria fix nitrogen in symbiotic association with cycads or some other terrestrial plants, or as the photosynthetic partner of certain lichens.

The reduction of nitrogen gas to ammonia by nitrogenase is a remarkable accomplishment by living organisms, achieved without the tremendous heat, pressure, and energy required during the manufacture of commercial fertilizers. Even so, nitrogen-fixing bacteria must consume the energy in 12 g of glucose (or the equivalent) to fix a single gram of nitrogen biologically.

FIGURE **53-8**

Root nodules and nitrogen fixation.

Root nodules on the roots of a pea plant provide an oxygen-free environment for nitrogen-fixing *Rhizobium* bacteria that live in them.

② The second step of the nitrogen cycle is **nitrification,** the conversion of ammonia (NH_3) or ammonium (NH_4^+, formed when water reacts with ammonia) to nitrate (NO_3^-). Soil bacteria are responsible for the two-step process of nitrification. First, the soil bacteria *Nitrosomonas* and *Nitrococcus* convert ammonia to nitrite (NO_2^-). Then the soil bacterium *Nitrobacter* oxidizes nitrite to nitrate. The process of nitrification furnishes these bacteria, called *nitrifying bacteria,* with energy.

③ In the third step, called **assimilation,** roots absorb ammonia (NH_3), ammonium (NH_4^+), or nitrate (NO_3^-) that nitrogen fixation and nitrification formed and incorporate the nitrogen into proteins, nucleic acids, and chlorophyll. When animals consume plant tissues, they assimilate nitrogen by taking in plant nitrogen compounds and converting them to animal nitrogen compounds.

④ The fourth step is **ammonification,** which is the conversion of organic nitrogen compounds into ammonia (NH_3) and ammonium ions (NH_4^+). Ammonification begins when organisms produce nitrogen-containing wastes such as urea in urine and uric acid in the wastes of birds (see Fig. 46-1). These substances, along with the nitrogen compounds in dead organisms, are decomposed, releasing the nitrogen into the abiotic environment as ammonia (NH_3). The bacteria that perform ammonification in both the soil and aquatic environments are called *ammonifying bacteria.* The ammonia produced by ammonification is available for the processes of nitrification and assimilation. As a matter of fact, most available nitrogen in the soil derives from the recycling of organic nitrogen by ammonification.

⑤ The fifth step of the nitrogen cycle is **denitrification,** which is the reduction of nitrate (NO_3^-) to gaseous nitrogen (N_2). Denitrifying bacteria reverse the action of nitrogen-fixing and nitrifying bacteria by returning nitrogen to the atmosphere as nitrogen gas. Denitrifying bacteria are anaerobic and therefore live and grow best where there is little or no free oxygen. For example, they are found deep in the soil near the water table, an environment that is nearly oxygen free.

Human activities have changed the global nitrogen budget

Human activities have disturbed the balance of the global nitrogen cycle. During the 20th century, humans more than doubled the amount of fixed nitrogen (nitrogen that has been chemically combined with hydrogen, oxygen, or carbon) entering the global nitrogen cycle. The excess nitrogen is seriously altering many terrestrial and aquatic ecosystems.

Large quantities of nitrogen fertilizer, both ammonia and nitrate, are produced from nitrogen gas for agriculture. According to the International Fertilizer Industry Association, farmers used 140 million metric tons of fertilizer worldwide in 2000; this amount was equal to 23.1 kg (51 lb) of fertilizer per person. The increasing use of fertilizer has resulted in higher crop yields.

In 1997 Peter Vitousek and seven other ecologists published an extensive review of scientific research on the negative environmental impacts of human-produced nitrogen. One serious problem with nitrogen is that it is extremely mobile and is easily transferred from the land to rivers to estuaries to the ocean.

Thus the overuse of commercial fertilizer on the land causes water quality problems that may help explain long-term declines in many coastal fisheries. The amount of nitrate or ammonium in most aquatic ecosystems is in limited supply and therefore limits the growth of algae. Rain washes fertilizer into rivers and lakes, where it stimulates the growth of algae, some of which are toxic. As these algae die, their decomposition by bacteria robs the water of dissolved oxygen, which in turn causes other aquatic organisms, including many fishes, to die of suffocation.

Nitrates from fertilizer also leach (dissolve and wash down) through the soil and contaminate groundwater. Many people who live in rural areas drink groundwater, and groundwater contaminated by nitrates is dangerous, particularly for infants and small children.

Another human activity that affects the nitrogen cycle is the combustion of fossil fuels. When fossil fuels are burned, the nitrogen locked in organic compounds in the fuel is chemically altered and transferred to the atmosphere. In addition, the high temperature of combustion converts some atmospheric nitrogen (N_2) to **nitrogen oxides,** trace gases in the atmosphere produced by chemical interactions between nitrogen and oxygen. Automobile exhaust is one of the main sources of nitrogen oxides.

Nitrogen oxides exacerbate several serious environmental problems. Nitrogen oxides are a necessary ingredient in the production of **photochemical smog,** a mixture of several air pollutants that injure plant tissues, irritate eyes, and cause respiratory problems in humans. Nitrogen oxides also react with water in the atmosphere to form nitric acid (HNO_3) and nitrous acid (HNO_2). When these and other acids leave the atmosphere as *acid deposition,* they decrease the pH of surface waters (lakes and streams) and soils. Acid deposition has been linked to declining animal populations in aquatic ecosystems. On land, acid deposition alters soil chemistry so that certain essential nutrient minerals such as calcium and potassium wash out of the soil and are therefore unavailable for plants. Nitrous oxide (N_2O), one of the nitrogen oxides, retains heat in the atmosphere (like CO_2) and so promotes global warming. Nitrous oxide also contributes to the depletion of ozone in the stratosphere. (See Chapter 2 for a discussion of acids and pH, and Chapter 55 for a discussion of global warming and stratospheric ozone depletion.)

The phosphorus cycle lacks a gaseous component

In the **phosphorus cycle,** phosphorus, which does not exist in a gaseous state and therefore does not enter the atmosphere, cycles from the land to sediments in the ocean and back to the land (Fig. 53-9). Refer to the figure as you read the following description. ① As water runs over rocks containing phosphorus, it gradually erodes the surface and carries off inorganic phosphate (PO_4^{3-}).

② The erosion of phosphorus rocks releases phosphate into the soil, where it is taken up by roots in the form of inorganic phosphates. Once in cells, phosphates are incorporated into a variety of biological molecules, including nucleic acids, ATP, and the phospholipids that make up cell membranes. Animals obtain most of their required phosphorus from the food they eat, although in some places drinking water may contain a substantial amount of inorganic phosphate. Phosphate released by decomposers becomes part of the pool of inorganic phosphate in the soil that plants reuse. Thus, like carbon and nitrogen, phosphorus moves through the food web as one organism consumes another.

③ Phosphorus cycles through aquatic ecosystems in much the same way as through terrestrial ecosystems. Dissolved phosphate enters aquatic ecosystems through absorption by algae and aquatic plants, which zooplankton and larger organisms consume. In turn, a variety of fishes and mollusks eat the zooplank-

FIGURE **53-9** **A simplified diagram of the phosphorus cycle.**

Some values of the global phosphorus budget are given, in units of 10^{12} g phosphorus per year. For example, each year an estimated 60×10^{12} g of phosphorus cycles from the soil to terrestrial organisms and back to the soil. *(Values from W.H. Schlesinger,* Biogeochemistry: An Analysis of Global Change, *2nd ed., Academic Press, San Diego, 1997, and based on several sources)*

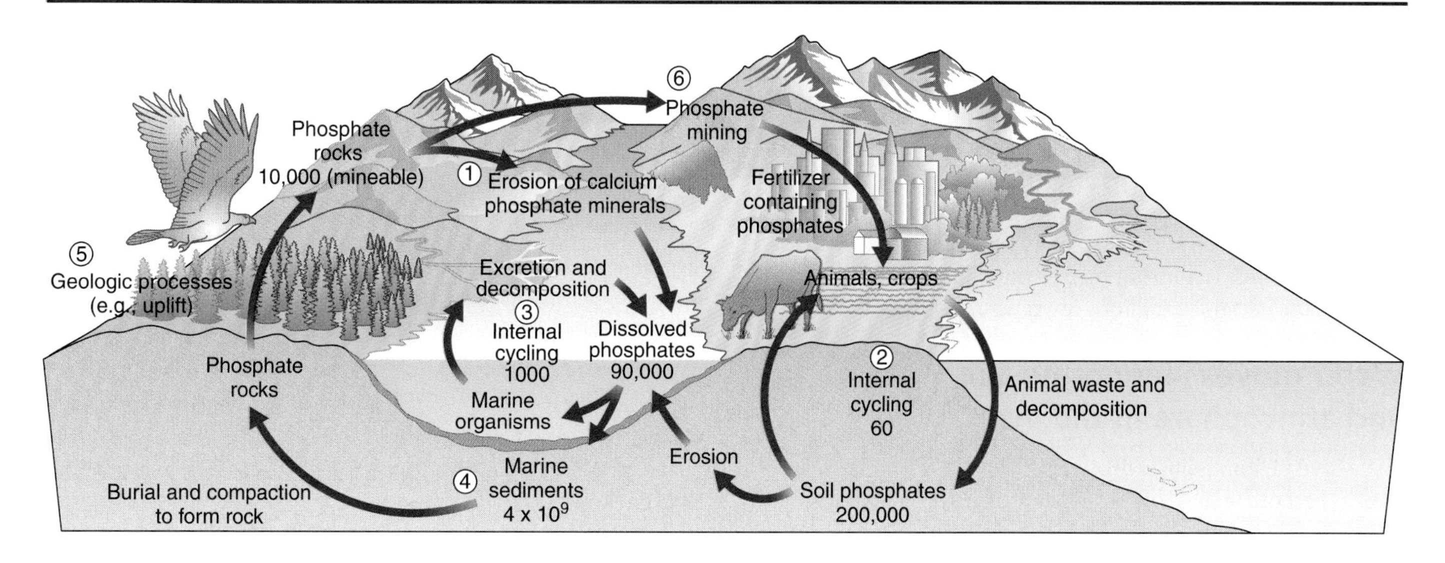

ton. Ultimately, decomposers break down wastes and dead organisms to release inorganic phosphate into the water, making it available for use again by aquatic producers.

Some phosphate in the aquatic food web finds its way back to the land. Sea birds, which eat a tiny fraction of fish and aquatic invertebrates, may defecate on the land where they roost. Guano, the manure of sea birds, contains large amounts of phosphate and nitrate. Once on land, roots may absorb these minerals. The phosphorus contained in guano may enter terrestrial food webs in this way, although the amounts involved are quite small.

④ Phosphate can be lost for varying time periods from biological cycles. Streams and rivers carry some phosphate to the ocean, where it is deposited on the sea floor and remains for millions of years. ⑤ The geologic process of uplift may someday expose these seafloor sediments as new land surfaces, from which phosphate will be once again eroded. ⑥ Phosphate deposits, including guano, are also mined for agricultural use in phosphate fertilizers.

Humans affect the natural cycling of phosphorus

You have seen that once phosphorus moves from terrestrial to aquatic ecosystems, its return to land is very slow. Certain human activities accelerate the long-term loss of phosphorus from terrestrial ecosystems. For example, phosphate is mined and used to make fertilizer, which may be applied to an Iowan farmer's soil. Corn grown there, which contains phosphate absorbed from the soil, may be used to fatten cattle in an Illinois feedlot. Part of the phosphate that roots of the corn plants absorb thus ends up in the feedlot wastes, much of which is used as fertilizer and may eventually wash into the Mississippi River. Beef from the Illinois cattle may be consumed by people living far away, in New York City, for instance, where more of the phosphate ends up in human wastes and is flushed down toilets into the New York City sewer system. Sewage treatment rarely removes phosphates, and so they cause water quality problems in rivers and lakes. To compensate for the steady loss of phosphate from their land, farmers must add phosphate fertilizer to their fields. That fertilizer is extracted from the large deposits of phosphate rock in Florida, Tennessee, and several other states.

In natural terrestrial communities, very little phosphorus is lost from the cycle, but few communities today are in a natural state, that is, unaltered in some way by humans. Land-denuding practices, such as the clearcutting of timber, and erosion of agricultural and residential lands accelerate phosphorus loss from the soil. For practical purposes, phosphorus that washes from the land into the ocean is permanently lost from the terrestrial phosphorus cycle (and from further human use), because it remains in the ocean for millions of years.

Water moves among the ocean, land, and atmosphere in the hydrologic cycle

Life would be impossible without water, which makes up a substantial part of the mass of most organisms. All species, from bacteria to plants and animals, use water as a medium for chemical reactions as well as for the transport of materials within and among cells. (Recall from Chapter 2 that water has many unique properties that help shape the continents, moderate climate, and allow organisms to survive.)

Water continuously circulates from the ocean to the atmosphere to the land and back to the ocean. It provides a renewable supply of purified water for terrestrial organisms. This complex cycle, known as the **hydrologic cycle,** results in a balance between water in the ocean, on the land, and in the atmosphere (Fig. 53-10). Refer to the figure as you read the following description. ① Water moves from the atmosphere to the land and ocean in the form of precipitation (rain, sleet, snow, or hail). ② When water evaporates from the ocean surface and from soil, streams, rivers, and lakes, it eventually condenses and forms clouds in the atmosphere. ③ In addition, **transpiration,** the loss of water vapor from land plants, adds a considerable amount of water vapor to the atmosphere. Roughly 97% of the water a plant absorbs from the soil is transported to the leaves, where it is lost by transpiration.

Water may evaporate from land and re-enter the atmosphere directly. ④ Alternatively, it may flow in rivers and streams to coastal **estuaries,** where fresh water meets the ocean. The movement of surface water from land to ocean is called *runoff,* and the area of land drained by runoff is called a *watershed.* ⑤ Water also percolates (seeps) downward in the soil to become *groundwater,* where it is trapped and held for a time. The underground caverns and porous layers of rock in which groundwater is stored are called *aquifers.* Groundwater may reside in the ground for hundreds to many thousands of years, but eventually it supplies water to the soil, to streams and rivers, to plants, and to the ocean. The human removal of more groundwater than precipitation or melting snow recharges, called *aquifer depletion,* eliminates groundwater as a water resource.

Regardless of its physical form (solid, liquid, or vapor) or location, every molecule of water eventually moves through the hydrologic cycle. Tremendous quantities of water cycle annually between Earth and its atmosphere. The volume of water entering the atmosphere from the ocean each year is estimated at about 425,000 km^3. Approximately 90% of this water re-enters the ocean directly as precipitation over water; the remainder falls on land. As is true of the other cycles, water (in the form of glaciers, polar ice caps, and certain groundwater) can be lost from the cycle for thousands of years.

Review

- Why is the cycling of matter essential to the long-term continuance of life?
- What are the roles of the following processes in the carbon cycle: photosynthesis, cellular respiration, combustion, and erosion?
- What are the five steps in the nitrogen cycle?
- How does phosphorus cycle without a gaseous component?

BiologyNow™ Assess your understanding of **cycles of matter in ecosystems** by taking the pretest on your BiologyNow CD-ROM.

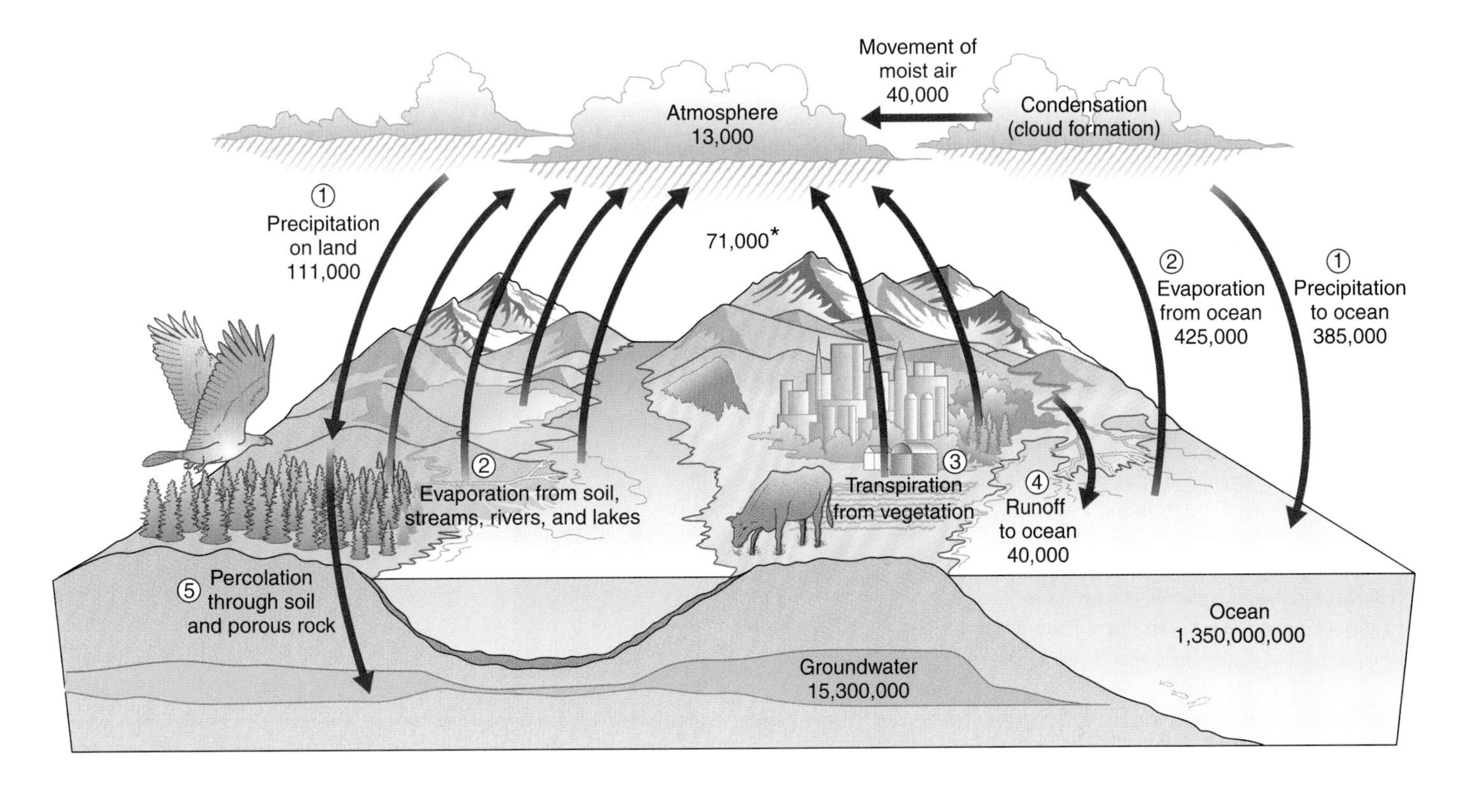

FIGURE **53-10** **A simplified diagram of the hydrologic cycle.**

The global water budget values shown for pools are expressed as cubic kilometers; values for fluxes (movements associated with arrows) are in cubic kilometers per year. The starred value (71,000 km^3/yr) is the sum of both transpiration and evaporation from soil, streams, rivers, and lakes. *(Values from W.H. Schlesinger,* Biogeochemistry: An Analysis of Global Change, *2nd ed., Academic Press, San Diego, 1997, and based on several sources)*

ECOSYSTEM REGULATION FROM THE BOTTOM UP AND THE TOP DOWN

Learning Objective

5 Distinguish between bottom up and top down processes in ecosystem regulation.

One question that ecologists have recently considered is which processes—bottom-up or top-down—are more significant in the regulation of various ecosystems. Both energy flow and cycles of matter are involved in bottom-up and top-down processes. **Bottom-up processes** are based on food webs that, as you know, always have producers at the first (lowest) trophic level. In a sense, the biogeochemical cycles that regenerate nutrients such as nitrates and phosphates for producers to assimilate are located "under" the first trophic level. Thus bottom-up processes regulate ecosystem function by nutrient cycling and other aspects of the abiotic environment. If bottom-up processes dominate an ecosystem, the availability of resources such as nutrient minerals controls the number of producers, which in turn controls the number of herbivores, which controls the number of carnivores. (Recall from Chapter 52 that *limiting resources* tend to restrict the ecological niches of organisms, thereby affecting population size.)

Bottom-up processes apparently predominate in certain aquatic ecosystems in which nitrogen or phosphorus is limiting. An experiment in which phosphorus was added to a phosphorus-deficient river (the Kaparuk River in Alaska) resulted in an increase in algae, followed over time by increased populations of aquatic insects, other invertebrates, and fish.

In contrast, **top-down processes** regulate ecosystem function by trophic interactions, particularly from the highest trophic level—by consumers eating producers. Ecosystem regulation by top-down processes occurs because carnivores eat herbivores, which in turn eat producers, which in turn affects levels of nutrient minerals. If top-down processes dominate an ecosystem, the effects of an increase in the population of top predators cascade down the food web through the herbivores and producers. In fact, top-down processes are also known as a *trophic cascade.* A change in the feeding preferences of killer whales off the coast of Alaska provides an excellent example of a trophic cascade. A few decades ago, killer whales began preying on sea otters, causing a sharp decline in their population. As sea otters have declined, the number of sea urchins, which sea otters eat, has increased. Sea urchins eat kelp, the producers at the base of the food web; the increase in the number of sea urchins has caused a decline in kelp populations.

Top-down regulation appears to predominate in ecosystems with few trophic levels and low species richness. Such ecosystems may have only one or a few species of dominant herbivores, but those species have a strong impact on the producer populations. An excellent example of top-down regulation was dis-

FOCUS ON LIFE WITHOUT THE SUN

In 1977 an oceanographic expedition aboard the research submersible Alvin studied the Galapagos Rift, a deep cleft in the ocean floor off the coast of Ecuador. The expedition revealed, on the floor of the abyss, a series of **hydrothermal vents** where sea water apparently had penetrated and been heated by the hot rocks below. During its time within Earth, the water had also become charged with mineral compounds, including hydrogen sulfide (H_2S), which is toxic to most species.

At the tremendous depths of greater than 2500 m (8200 ft) found in the Galapagos Rift, there is no light for photosynthesis. But hydrothermal vents support a rich variety of life forms (see figure), in contrast with the surrounding

D. Foster, Science VU-WHOI/Visuals Unlimited

A hydrothermal vent community. Chemoautotrophic bacteria living in the tissues of these tube worms (*Riftia pachyptila*) extract energy from hydrogen sulfide to manufacture organic compounds. Because these worms lack digestive systems, they depend on the organic compounds provided by the endosymbiotic bacteria, along with materials filtered from the surrounding water and digested extracellularly. Also visible in the photograph are some filter-feeding mollusks *(yellow)* and a crab *(white)*.

cussed in Chapter 51: Reindeer introduced to the Pribilof Islands of Alaska overgrazed the vegetation until the plants were almost wiped out.

Review

- Biologists think the reintroduction of wolves to Yellowstone National Park, which began in 1995, will ultimately result in a more varied and lush plant composition. Is this an example of bottom-up or top-down processes? Why?

BiologyNow™ Assess your understanding of **ecosystem regulation from the bottom up and the top down,** by taking the pretest on your BiologyNow CD-ROM.

ABIOTIC FACTORS IN ECOSYSTEMS

Learning Objectives

6 Summarize the effects of solar energy on Earth's temperatures.

7 Discuss the roles of solar energy and the Coriolis effect in the production of global air and water flow patterns.

8 Give three causes of regional precipitation differences.

9 Discuss the effects of fire on certain ecosystems.

We have seen how ecosystems depend on the abiotic environment to supply energy and essential materials (in biogeochemical cycles). Abiotic factors such as solar radiation, the atmosphere, the ocean, climate, and fire also affect ecosystems. For a given abiotic factor, each organism living in an ecosystem has an optimal range in which it survives and reproduces. Water and temperature are probably the two abiotic factors that most affect organisms in ecosystems.

The sun warms Earth

The sun makes life on Earth possible. It warms the planet to habitable temperatures. Without the sun's energy, the temperature on planet Earth would approach absolute zero (−273°C), and all water would be frozen, even in the ocean. The sun powers the hydrologic cycle, carbon cycle, and other biogeochemical cycles and is the primary determinant of climate. Photosynthetic organisms capture the sun's energy and use it to make organic compounds that almost all forms of life require. Most of our fuels, such as wood, oil, coal, and natural gas, represent solar energy captured by photosynthetic organisms. Without the sun, almost all life on Earth would cease (see *Focus On: Life without the Sun* for an interesting exception).

The sun's energy, which is the product of a massive nuclear fusion reaction, is emitted into space in the form of electromagnetic radiation, especially ultraviolet, visible, and infrared radiation. About one billionth of the total energy the sun releases strikes the atmosphere, and of this tiny trickle of energy, a minute part operates the biosphere. On average, clouds and surfaces—especially snow, ice, and the ocean—immediately reflect away 30% of the solar radiation that falls on Earth (Fig. 53-11). Earth's surface and atmosphere absorb the remaining 70%, which runs the water cycle, drives winds and ocean currents, powers photosynthesis, and warms the planet. Ultimately, the continual radiation of long-wave infrared (heat) energy returns all this energy to space. If heat gains did not exactly balance losses, Earth would heat up or cool down.

Temperature changes with latitude

The most significant local variations in Earth's temperature are produced because the sun's energy does not uniformly reach all

"desert" of the abyssal floor. Many of the species in these oases of life are not found in other habitats. For example, giant, blood-red tube worms almost 3 m (10 ft) in length cluster in great numbers around the vents. Other animals around the hydrothermal vents include unique species of clams, crabs, barnacles, and mussels. Since 1977 hydrothermal vent communities, some as large as football fields, have been identified at many other oceanic sites, such as mid-ocean ridges.

Scientists initially wondered what energy source sustains these organisms. Most deep-sea communities depend on the organic matter that drifts down from the surface waters; in other words, they rely on energy ultimately derived from photosynthesis. Hydrothermal vent communities, however, are too densely clustered and too productive to be dependent on chance encounters with organic material from surface waters. Instead, chemoautotrophic bacteria occupy the base of the food web in these aquatic oases. These bacteria have enzymes that catalyze the oxidation of hydrogen sulfide to water and sulfur or sulfate. Such chemical reactions are exergonic and provide the energy required to fix CO_2, which is dissolved in the water, into organic compounds. Many of the animals consume the bacteria directly by filter-feeding, but others, such as the giant tube worms, get their energy from bacteria that live in their tissues.

Scientists continue to generate questions about hydrothermal vent communities. How, for example, do the organisms find and colonize vents, which are ephemeral and widely scattered on the ocean floor? How have the inhabitants of these communities adapted to survive the harsh living conditions, including high pressure, high temperatures, and toxic chemicals? As vent community research continues, scientists hope to discover answers to these and other questions.

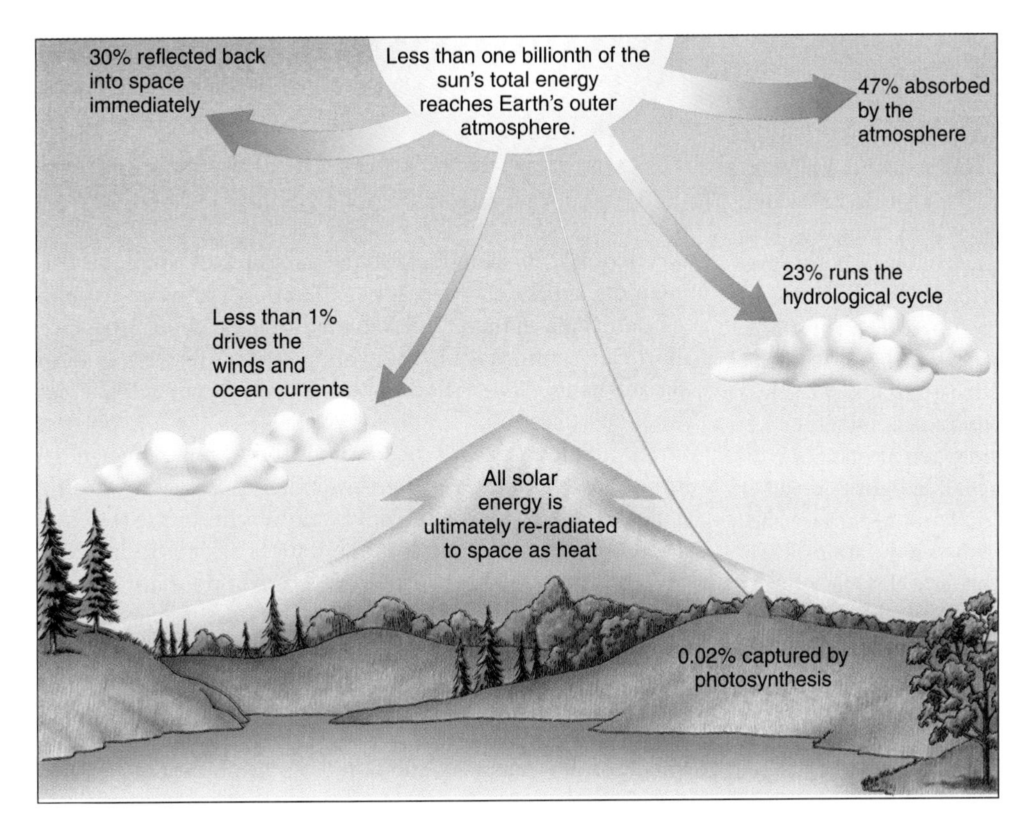
30% reflected back into space immediately
Less than one billionth of the sun's total energy reaches Earth's outer atmosphere.
47% absorbed by the atmosphere
23% runs the hydrological cycle
Less than 1% drives the winds and ocean currents
All solar energy is ultimately re-radiated to space as heat
0.02% captured by photosynthesis

FIGURE **53-11**

The fate of solar radiation that reaches Earth.

Most of the energy released by the sun never reaches Earth. The solar energy that does reach Earth warms the planet's surface, drives the hydrologic and other biogeochemical cycles, produces our climate, and powers almost all life through the process of photosynthesis.

places. Our planet's roughly spherical shape and the tilted angle of its axis produce a lot of variation in the exposure of the surface to sunlight. The sun's rays strike almost vertically near the equator, concentrating the energy and producing warmer temperatures. Near the poles the sun's rays strike more obliquely and, as a result, are spread over a larger surface area. Also, rays of light entering the atmosphere obliquely near the poles must pass through a deeper envelope of air than those entering near the equator. This causes more of the sun's energy to be scattered and reflected back into space, which further lowers temperatures near the poles. Thus the solar energy that reaches polar regions is less concentrated and produces lower temperatures.

Temperature changes with season

Earth's inclination on its axis (23.5 degrees from a line drawn perpendicular to the orbital plane) primarily determines the seasons. During half of the year (March 21 to September 22) the Northern Hemisphere tilts toward the sun, concentrating the sunlight and making the days longer (Fig. 53-12). During the other half of the year (September 22 to March 21) the Northern Hemisphere tilts away from the sun, giving it a lower concentration of sunlight and shorter days. The orientation of the Southern Hemisphere is just the opposite at these times. Summer in the Northern Hemisphere corresponds to winter in the Southern Hemisphere.

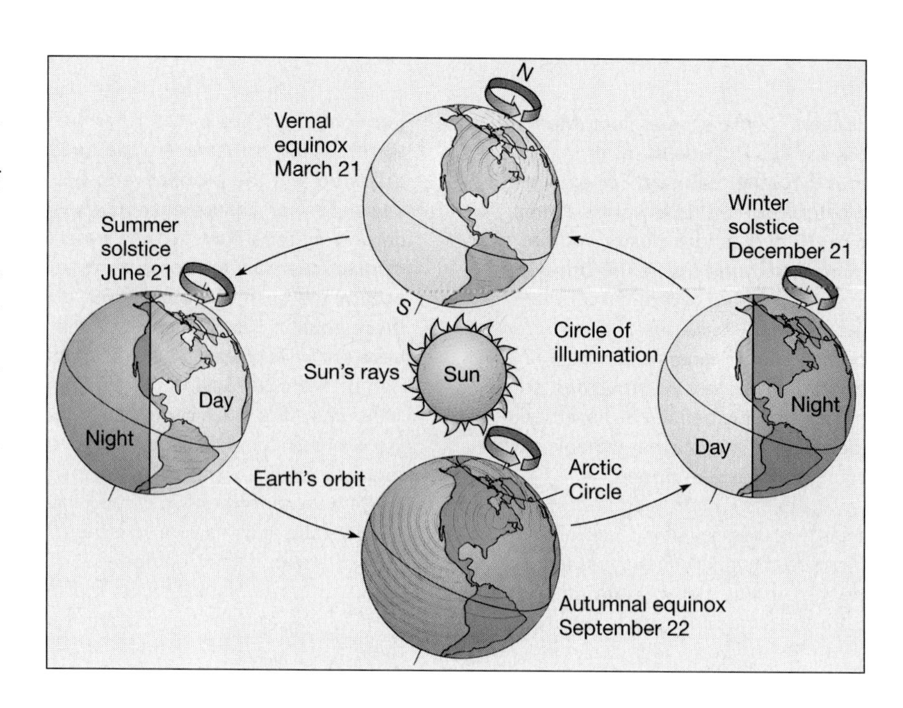

FIGURE **53-12** | **Seasonal changes in temperature.**
Earth's inclination on its axis remains the same as it travels around the sun. Thus the sun's rays hit the Northern Hemisphere obliquely during winter months and more directly during summer months. In the Southern Hemisphere, the sun's rays are oblique during its winter, which corresponds to the Northern Hemisphere's summer. At the equator, the sun's rays are approximately vertical on March 21 and September 22.

The atmosphere contains several gases essential to organisms

The atmosphere is an invisible layer of gases that envelops Earth. Oxygen (21%) and nitrogen (78%) are the predominant gases in the atmosphere, accounting for about 99% of dry air; other gases, including argon, carbon dioxide, neon, and helium, make up the remaining 1%. In addition, water vapor and trace amounts of various air pollutants, such as methane, ozone, dust particles, pollen, microorganisms and chlorofluorocarbons (CFCs) are present. Atmospheric oxygen is essential to plants, animals, and other organisms that respire aerobically, whereas plants and other photosynthetic organisms also require carbon dioxide. The atmosphere becomes less dense as it extends outward into space; most of its mass is found near Earth's surface.

The atmosphere performs several ecologically important functions. It protects the surface from most of the sun's ultraviolet radiation and x-rays and from lethal amounts of cosmic rays from space. Without this atmospheric shielding, life as we know it would cease. Although the atmosphere protects Earth from high-energy radiation, it allows visible light and some infrared radiation to penetrate, and they warm the surface and the lower atmosphere. This interaction between the atmosphere and solar energy is responsible for weather and climate.

Organisms depend on the atmosphere, but they also help maintain and, in certain instances, modify its composition. For example, atmospheric oxygen increased to its present level as a result of billions of years of photosynthesis. Today an approximate balance between oxygen-producing photosynthesis and oxygen-using aerobic respiration helps maintain the level of atmospheric oxygen.

The sun drives global atmospheric circulation

In large measure, differences in temperature that are due to variations in the amount of solar energy at different locations on Earth drive the circulation of the atmosphere. The warm surface near the equator heats the air with which it comes into contact, causing this air to expand and rise. As the warm air rises, it flows away from the equator, cools, and sinks again (Fig. 53-13). Much of it recirculates back to the same areas it left, but the remainder splits and flows in two directions, toward the poles. The air chills enough to sink to the surface at about 30 degrees north and south latitudes. Similar upward movements of warm air and its subsequent flow toward the poles occur at higher latitudes, farther from the equator. At the poles, the cold polar air sinks and flows toward the lower latitudes, generally beneath the sheets of warm air that simultaneously flow toward the poles. The constant motion of air transfers heat from the equator toward the poles, and as the air returns it cools the land over which it passes. This continuous turnover does not equalize temperatures over Earth's surface, but it does moderate them.

The atmosphere exhibits complex horizontal movements

In addition to global circulation patterns, the atmosphere exhibits complex horizontal movements that are commonly referred to as **winds.** The nature of wind, with its turbulent gusts, eddies, and lulls, is complex and difficult to understand or predict. It results in part from differences in atmospheric pressure and from the rotation of Earth.

The gases that constitute the atmosphere have weight and exert a pressure that is, at sea level, about 1013 millibars (14.7 lb/in^2). Air pressure is variable, however, and changes with altitude, temperature, and humidity. Winds tend to blow from areas of high atmospheric pressure to areas of low pressure; the

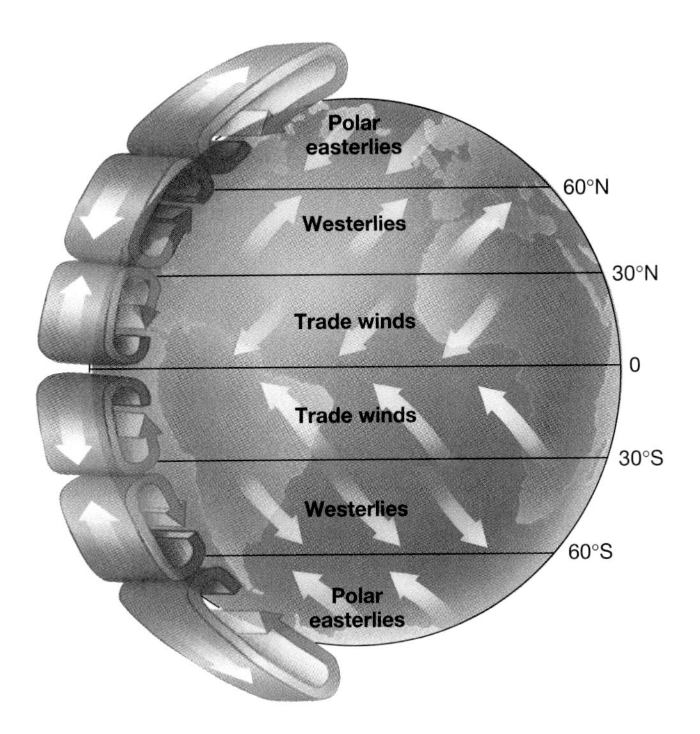

FIGURE **53-13** | **Atmospheric circulation.**

The greatest solar energy input occurs at the equator, heating air most strongly in that area. The air rises and travels poleward *(left)* but is cooled in the process so that much of it descends again around 30 degrees latitude in both hemispheres. At higher latitudes the patterns of air movement are more complex.

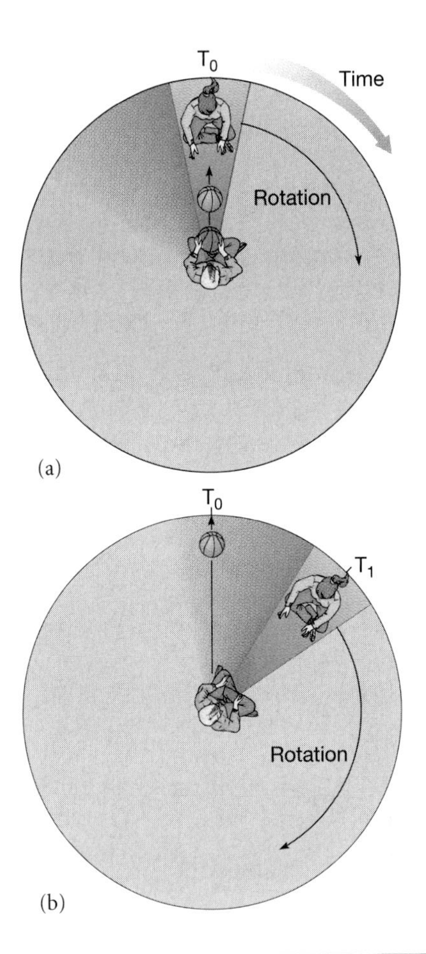

FIGURE **53-14** | **The Coriolis effect as demonstrated by a merry-go-round.**

The center of the merry-go-round *(shown from above)* corresponds to the South Pole, and the outer edge to the equator. **(a)** If, while sitting on the merry-go-round, you throw a ball to a friend (also on the merry-go-round) at time T_0, when the merry-go-round is rotating clockwise, the ball at time T_1 **(b)** appears to curve to the left instead of going straight. Because of Earth's rotation, winds and ocean currents curve to the left (counterclockwise) in the Southern Hemisphere and to the right (clockwise) in the Northern Hemisphere.

greater the difference between the high and low pressure areas, the stronger the wind.

Earth's rotation influences the direction that wind blows. Because Earth rotates from west to east, wind swerves to the right in the Northern Hemisphere and to the left in the Southern Hemisphere. This tendency of moving air to be deflected from its path by Earth's rotation is known as the **Coriolis effect.** The Coriolis effect is visualized by imagining that you and a friend are sitting about 3 m apart on a merry-go-round that is turning clockwise (Fig. 53-14). Suppose you throw a ball directly to your friend. By the time the ball reaches the place where your friend was, he or she is no longer in that spot. The ball will have swerved far to the left of your friend. This is how the Coriolis effect works in the Southern Hemisphere. To visualize how the Coriolis effect works in the Northern Hemisphere, imagine that you and your friend are sitting on the same merry-go-round, only this time it is moving counterclockwise. Now, when you throw the ball, it will appear to swerve far to the right of your friend.

The global ocean covers most of Earth's surface

The global ocean is a huge body of salt water that surrounds the continents and covers almost three fourths of Earth's surface. It is a single, continuous body of water, but geographers divide it into four sections separated by the continents: the Pacific, Atlantic, Indian, and Arctic Oceans. The Pacific Ocean is the largest by far: It covers one third of Earth's surface and contains more than half of Earth's water.

Winds drive surface ocean currents

The persistent prevailing winds blowing over the ocean produce mass movements of surface ocean water known as **ocean currents** (Fig. 53-15). The prevailing winds generate circular ocean currents called *gyres.* For example, in the North Atlantic, the tropical trade winds tend to blow toward the west, whereas the westerlies in the midlatitudes blow toward the east (see Fig. 53-13). This helps establish a clockwise gyre in the North Atlantic. Thus surface ocean currents and winds tend to move in the same direction, although there are many variations on this general rule.

The Coriolis effect is partly responsible for the paths that surface ocean currents travel. Earth's rotation from west to east causes surface ocean currents to swerve to the right in the North-

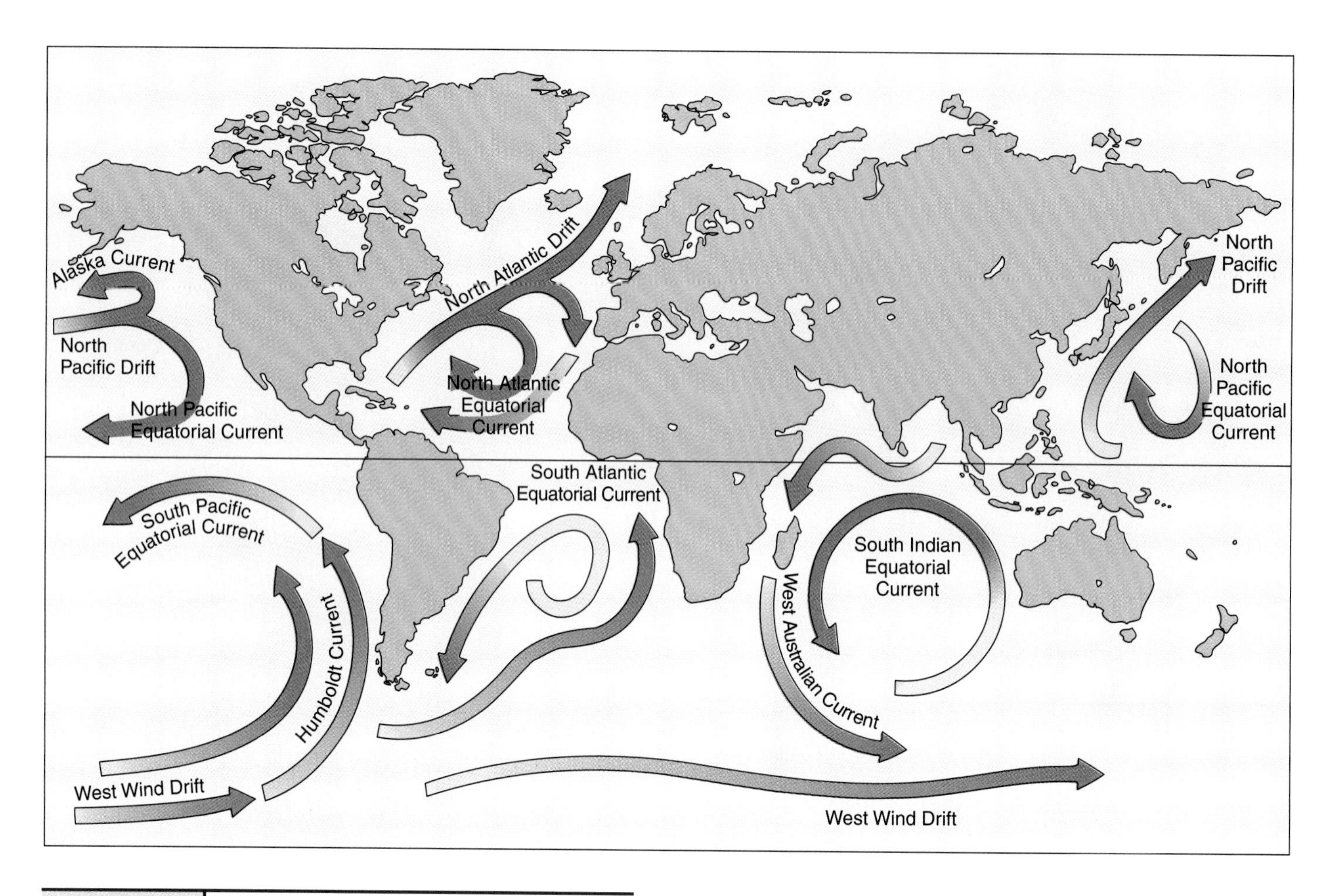

FIGURE **53-15** | **Major surface ocean currents.**

The basic pattern of ocean currents is caused largely by winds. The main ocean current flow—clockwise in the Northern Hemisphere and counterclockwise flow in the Southern Hemisphere—results partly from the Coriolis effect.

ern Hemisphere, producing a clockwise gyre of water currents. In the Southern Hemisphere, ocean currents swerve to the left, producing a counterclockwise gyre.

The ocean interacts with the atmosphere

The ocean and the atmosphere are strongly linked, with wind from the atmosphere affecting the ocean currents and heat from the ocean affecting atmospheric circulation. One of the best examples of the interaction between ocean and atmosphere is the **El Niño–Southern Oscillation (ENSO)** event. ENSO is a periodic warming of surface waters of the tropical East Pacific that alters both oceanic and atmospheric circulation patterns and results in unusual weather in areas far from the tropical Pacific. Normally, westward-blowing trade winds restrict the warmest waters to the western Pacific (near Australia). Every three to seven years, however, the trade winds weaken and the warm water mass expands eastward to South America, raising surface temperatures in the East Pacific 3 or 4 degrees C above average. Ocean currents, which normally flow westward in this area, slow down, stop altogether, or even reverse and go eastward. The phenomenon is called El Niño (Spanish for "the child") because the warming usually reaches the fishing grounds off Peru just before Christmas. Most ENSOs last from one to two years.

An ENSO event changes biological productivity in parts of the ocean. The warmer sea surface temperatures and accompanying changes in ocean circulation patterns off the west coast of South America prevent colder, nutrient-laden deeper waters from **upwelling** (coming to the surface) (Fig. 53-16). The lack of nutrients in the water results in a severe decrease in the populations of anchovies and many other marine fishes. For example, during the 1982 to 1983 ENSO, one of the worst ever recorded, the anchovy population decreased by 99%. Other species, such as shrimp and scallops, thrive during an ENSO event. Along the Pacific coast of North America, ENSO shifts the distribution of tropical fishes northward and even affects the salmon run in Alaska.

Climate profoundly affects organisms

Climate is the average weather conditions, plus extremes (records), that occur in a given place over a period of years. The two most important factors that determine an area's climate are temperature (both average temperature and temperature extremes) and precipitation (both average precipitation and seasonal distribution). Other climate factors include wind, humidity, fog, cloud cover, and lightning-caused wildfires. Unlike weather, which changes rapidly, climate generally changes slowly, over hundreds or thousands of years.

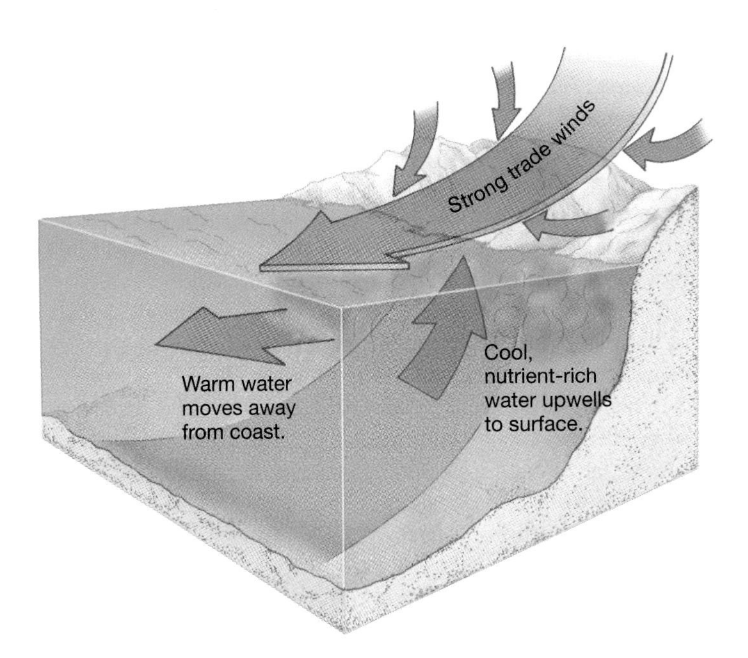

FIGURE **53-16** | **Upwelling.**

Coastal upwelling, where deeper waters come to the surface, occurs in the Pacific Ocean along the South American coast. Upwelling provides nutrients for large numbers of microscopic algae, which in turn support a complex food web. Coastal upwelling weakens considerably during years with El Niño–Southern Oscillation (ENSO) events, temporarily reducing fish populations.

Day-to-day variations, day-to-night variations, and seasonal variations are also important dimensions of climate that affect organisms. Latitude, elevation, topography, vegetation, distance from the ocean or other large bodies of water, and location on a continent or other landmass all influence temperature, precipitation, and other aspects of climate.

Earth has many different climates, and because they are relatively constant for many years, organisms have adapted to them. The wide variety of organisms on Earth evolved in part because of the large number of different climates, ranging from cold, snow-covered, polar climates to hot, tropical climates where it rains almost every day.

Air and water movements and surface features affect precipitation patterns

Precipitation varies from one location to another and has a profound effect on the distribution and kinds of organisms present. One of the driest places on Earth is in the Atacama Desert in Chile, where the average annual rainfall is 0.05 cm (0.02 in). In contrast, Mount Waialeale in Hawaii, Earth's wettest spot, receives an average annual precipitation of 1200 cm (472 in).

Differences in precipitation depend on several factors. The heavy-rainfall areas of the tropics result mainly from the uplifting of moisture-laden air. High surface-water temperatures (recall the enormous amount of solar energy striking the equator) cause the evaporation of vast quantities of water from tropical parts of the ocean. Prevailing winds blow the resulting moist warmer air over landmasses. Land surfaces warmed by the sun heat the air and cause moist air to rise. As it rises, the air cools and its moisture-holding ability decreases. (Cool air holds less water vapor than warm air.) When air reaches its saturation point, it cannot hold any additional water vapor, and clouds form and water is released as precipitation. The air eventually returns to the surface on both sides of the equator near the Tropics of Cancer and Capricorn (latitudes 23.5 degrees north and south, respectively). By then, most of the moisture has precipitated so that dry air returns to the equator. This dry air makes little biological difference over the ocean, but its lack of moisture produces some of the great subtropical deserts, such as the Sahara.

Air is also dried during long journeys over landmasses. Near the windward (side from which the prevailing wind blows) coasts of continents, rainfall may be heavy. However, in the temperate zones—the areas between the tropical and polar zones—continental interiors are usually dry because they are far from the ocean that replenishes water vapor in the air passing over it.

Mountains force air to rise, removing moisture from humid air. As it gains altitude, the air cools, clouds form, and precipitation typically occurs, primarily on the windward slopes of the mountains. As the air mass moves down on the other side of the mountain, it is warmed and clouds evaporate, thereby lessening the chance of precipitation of any remaining moisture. This situation exists on the West Coast of North America, where precipitation falls on the western slopes of mountains that are close to the coast. The dry lands on the sides of the mountains away from the prevailing wind (in this case, east of the mountain range) are called **rain shadows** (Fig. 53-17).

Microclimates are local variations in climate

Differences in elevation, in the steepness and direction of slopes, and therefore in exposure to sunlight and prevailing winds, may produce local variations in climate known as **microclimates,** which are sometimes quite different from their overall surround-

FIGURE **53-17** | **Rain shadow.**

A rain shadow is arid or semiarid land that occurs on the leeward side of a mountain. Prevailing winds blow warm, moist air from the windward side of the mountain. Air cools as it rises, releasing precipitation so that dry air descends on the leeward side. Such a rain shadow occurs east of the Cascade Range in Washington State. The west side of Olympic National Park receives more than 500 cm of precipitation annually, whereas the east side gets 40 to 50 cm.

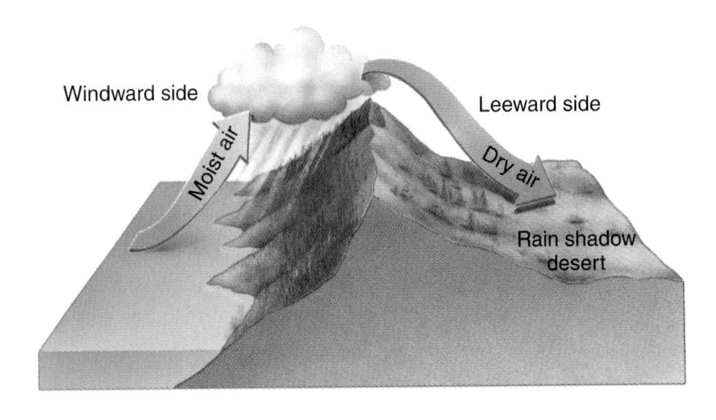

ings. Patches of sun and shade on a forest floor, for example, produce a variety of microclimates for plants, animals, and microorganisms living there. The microclimate of an organism's habitat is of primary importance because that is the climate an organism actually experiences and must cope with. (Keep in mind, however, that the regional climate in which microclimates are located largely affect them.)

Sometimes organisms modify their own microclimate. For instance, trees modify the local climate within a forest so that in summer the temperature is usually lower, and the relative humidity greater, than outside the forest. The temperature and humidity beneath the litter of the forest floor differ still more; in the summer the microclimate of this area is considerably cooler and moister than the surrounding forest. As another example, many desert-dwelling animals burrow to avoid surface climate conditions that would kill them in minutes. The cooler daytime microclimate in their burrows permits them to survive until night, when the surface cools off and they come out to forage or hunt.

Fires are a common disturbance in some ecosystems

Wildfires, which are fires started by lightning, are an important ecological force in many geographic areas. Those areas most prone to wildfires have wet seasons followed by dry seasons. Vegetation that grows and accumulates during the wet season dries out enough during the dry season to burn easily. When lightning hits vegetation or ground litter, it ignites the dry organic material, and a fire spreads through the area.

Fires have several effects on organisms. First, combustion frees the nutrient minerals that were locked in dry organic matter. The ashes remaining after a fire are rich in potassium, phosphorus, calcium, and other nutrient minerals essential for plant growth. With the arrival of precipitation, vegetation flourishes following a fire. Second, fire removes plant cover and exposes the soil. This change stimulates the germination and establishment of seeds requiring bare soil, as well as encourages the growth of shade-intolerant plants. Third, fire causes increased soil erosion because it removes plant cover, leaving the soil more vulnerable to wind and water.

Some plants are adapted to fire. Grasses and some shrubs, for example, have underground stems and buds that are typically unaffected by a fire sweeping over them. After the aerial parts have been killed by fire, the underground parts send up new sprouts. The availability of nutrients in the ash speeds the regrowth of these plants. Fire-adapted trees such as bur oak and ponderosa pine have a thick bark that is resistant to fire. (In contrast, fire-sensitive trees, such as many hardwoods, have a thin bark.) Certain trees such as jack pine and lodgepole pine depend on fire for successful reproduction, because the heat of the fire opens the cones so that the seeds are released.

Fires were a part of the natural environment long before humans appeared, and organisms in many terrestrial ecosystems have adapted to fires. African savanna, California chaparral, North American grasslands, and ponderosa pine forests of the western United States are some fire-adapted ecosystems (see Chapter 54). Fire helps maintain grasses as the dominant vegetation in grasslands by removing fire-sensitive hardwood trees.

The influence of fire on plants became even more pronounced once humans evolved. Because humans deliberately and accidentally set fires, fire became more common. Humans set fires for many reasons: to provide the grasses and shrubs that many game animals require; to clear the land for agriculture and human development, and to reduce enemy cover in times of war.

Humans also try to prevent fires, and sometimes this effort has disastrous consequences. When fire is excluded from a fire-adapted ecosystem, deadwood and other plant litter accumulate. As a result, when a fire does occur it is much more destructive. The sometimes deadly wildfires in Colorado and other western states and provinces are blamed in part on decades of suppressing fires in the region. Prevention of fire also converts grassland to woody vegetation and facilitates the invasion of fire-sensitive trees into fire-adapted forests.

Controlled burning is a tool of ecological management in which the undergrowth and plant litter are deliberately burned under controlled conditions before they have accumulated to dangerous levels (Fig. 53-18). Controlled burns, which were initiated in U.S. national forests beginning in the 1960s, are also used to suppress fire-sensitive trees, thereby maintaining the natural fire-adapted ecosystem. However, little scientific evidence exists on when controlled burning is appropriate or how often it should be carried out. Moreover, there are major practical and political problems in implementing controlled burns.

Review

- What basic forces determine the circulation of the atmosphere?
- What basic forces produce the main ocean currents?
- What are some of the factors that produce regional differences in precipitation?

BiologyNow™ Assess your understanding of **abiotic factors in ecosystems** by taking the pretest on your BiologyNow CD-ROM.

FIGURE **53-18** | **Fire as a tool of ecological management.**
Here, a controlled burn helps maintain a ponderosa pine (*Pinus ponderosa*) stand in Oregon.

Joan Landsberg, U.S. Department of Agriculture Forest Service

SUMMARY WITH KEY TERMS

1 Summarize the concept of energy flow through a food web.

- **Energy flow** through an ecosystem is linear, from the sun to producer to consumer to decomposer. Much of this energy is converted to heat as it moves from one organism to another, so organisms occupying the next **trophic level** cannot use it.
- Trophic relationships may be expressed as **food chains** or, more realistically, as **food webs,** which show the many alternative pathways that energy may take among the producers, consumers, and decomposers of an ecosystem.

2 Explain typical pyramids of numbers, biomass, and energy.

- **Ecological pyramids** typically express the progressive reduction in numbers of organisms, biomass, and energy found in successive trophic levels.
- A **pyramid of numbers** shows the number of organisms at each trophic level in a given ecosystem. A **pyramid of biomass** shows the total biomass at each successive trophic level. A **pyramid of energy** indicates the energy content of the biomass of each trophic level.

3 Distinguish between gross primary productivity and net primary productivity.

- **Gross primary productivity (GPP)** of an ecosystem is the rate at which photosynthesis captures energy. **Net primary productivity (NPP)** is the energy that remains (as biomass) after plants and other producers carry out cellular respiration.

4 Describe the main steps in each of these biogeochemical cycles: the carbon, nitrogen, phosphorus, and hydrologic cycles.

- Carbon dioxide is the important gas of the **carbon cycle.** Carbon enters plants, algae, and cyanobacteria as CO_2, which photosynthesis incorporates into organic molecules. Cellular respiration, combustion, and erosion of limestone return CO_2 to the water and atmosphere, making it available to producers again.
- The **nitrogen cycle** has five steps. **Nitrogen fixation** is the conversion of nitrogen gas to ammonia. **Nitrification** is the conversion of ammonia or ammonium to nitrate. **Assimilation** is the conversion of nitrates, ammonia, or ammonium to proteins, chlorophyll, and other nitrogen-containing compounds by plants; the conversion of plant proteins into animal proteins is also assimilation. **Ammonification** is the conversion of organic nitrogen to ammonia and ammonium ions. **Denitrification** is the conversion of nitrate to nitrogen gas.
- The **phosphorus cycle** has no biologically important gaseous compounds. Phosphorus erodes from rock as inorganic phosphate, which the roots of plants absorb from the soil. Animals obtain the phosphorus they need from their diets. Decomposers release inorganic phosphate into the environment. When phosphorus washes into the ocean and is subsequently deposited in sea beds, it is lost from biological cycles for millions of years.
- The **hydrologic cycle,** which continually renews the supply of water essential to life, involves an exchange of water between the land, ocean, atmosphere, and organisms. Water enters the atmosphere by evaporation and **transpiration** and leaves the atmosphere as precipitation. On land, water filters through the ground or runs off to lakes, rivers, and the ocean. Aquifers are underground caverns and porous layers of rock in which groundwater is stored. Runoff is the movement of surface water from land to ocean.

5 Distinguish between bottom-up and top-down processes in ecosystem regulation.

- If **bottom-up processes** dominate an ecosystem, the availability of resources such as nutrient minerals controls the number of producers (that is, the lowest trophic level), which in turn controls the number of herbivores, which in turn controls the number of carnivores.
- **Top-down processes** regulate ecosystems from the highest trophic level—by consumers eating producers. If top-down processes dominate an ecosystem, an increase in the number of top predators cascades down the food web through the herbivores and producers.

6 Summarize the effects of solar energy on Earth's temperatures.

- Sunlight is the primary (almost sole) source of energy available to the biosphere. Of the solar energy that reaches Earth, 30% is immediately reflected away; the atmosphere and surface absorb the remaining 70%. Ultimately, all absorbed solar energy is reradiated into space as infrared (heat) radiation.
- A combination of Earth's roughly spherical shape and the tilted angle of its axis concentrates solar energy at the equator and dilutes it at the poles; the tropics are hotter and less variable in climate than are temperate and polar areas. The inclination of Earth's axis primarily determines the seasons.

7 Discuss the roles of solar energy and the Coriolis effect in the production of global air and water flow patterns.

- Visible light and some infrared radiation warm the surface and the lower part of the atmosphere.
- Atmospheric heat transferred from the equator to the poles produces both a movement of warm air toward the poles and of cool air toward the equator, thus moderating the extremes of global climate.
- Winds result in part from differences in atmospheric pressure and from the **Coriolis effect,** the tendency of moving air or water, as a result of Earth's rotation, to be deflected to the right in the Northern Hemisphere and to the left in the Southern hemisphere.
- Surface **ocean currents** result in part from prevailing winds and the Coriolis effect.

8 Give three causes of regional precipitation differences.

- Latitude, elevation, topography, vegetation, distance from the ocean or other large bodies of water, and location on a continent or other landmass influence precipitation.
- Precipitation is greatest where warm air passes over the ocean, absorbing moisture, and then cools, such as when mountains force humid air upward. Deserts develop in the **rain shadows** of mountain ranges or in continental interiors.

9 Discuss the effects of fire on certain ecosystems.

- Fire frees the nutrient minerals locked in dry organic matter, removes plant cover and exposes the soil, and increases soil erosion.
- Many ecosystems, such as savanna, chaparral, grasslands, and certain forests, contain fire-adapted organisms.

POST-TEST

1. A community and its abiotic environment best defines a(an) (a) biogeochemical cycle (b) biosphere (c) ecosystem (d) food web (e) trophic level
2. The movement of matter is ___________ in ecosystems, and the movement of energy is ___________. (a) linear; linear (b) linear; cyclic (c) cyclic; cyclic (d) cyclic; linear (e) cyclic; linear or cyclic
3. A complex of interconnected food chains in an ecosystem is called a(an): (a) ecosystem (b) pyramid of numbers (c) pyramid of biomass (d) biosphere (e) food web
4. Which of the following shows the correct flow of energy through ecosystems? (a) sun ⟶ primary consumer ⟶ secondary consumer ⟶ producer (b) sun ⟶ producer ⟶ secondary consumer ⟶ primary consumer (c) sun ⟶ secondary consumer ⟶ primary consumer ⟶ producer (d) sun ⟶ producer ⟶ primary consumer ⟶ secondary consumer (e) sun ⟶ primary consumer ⟶ producer ⟶ secondary consumer
5. The quantitative estimate of the total amount of living material is called (a) biomass (b) energy flow (c) gross primary productivity (d) plant respiration (e) net primary productivity
6. Which of the following equations shows the relationship between gross primary productivity (GPP) and net primary productivity (NPP)? (a) GPP = NPP − photosynthesis (b) NPP = GPP − photosynthesis (c) GPP = NPP − plant respiration (d) NPP = GPP − plant respiration (e) NPP = GPP − animal respiration
7. Which of the following processes increase(s) the amount of atmospheric carbon in the carbon cycle? (a) photosynthesis (b) cellular respiration (c) combustion (d) both a and c (e) both b and c
8. In the carbon cycle, carbon is found in (a) limestone rock (b) oil, coal, and natural gas (c) living organisms (d) the atmosphere (e) all of these contain carbon
9. In the nitrogen cycle, gaseous nitrogen is converted to ammonia during (a) nitrogen fixation (b) nitrification (c) assimilation (d) ammonification (e) denitrification
10. The conversion of ammonia to nitrate, known as ___________, is a two-step process performed by soil bacteria. (a) nitrogen fixation (b) nitrification (c) assimilation (d) ammonification (e) denitrification
11. This biogeochemical cycle does not have a gaseous component but cycles from the land to sediments in the ocean and back to the land. (a) carbon cycle (b) nitrogen cycle (c) phosphorus cycle (d) hydrologic cycle (e) neither a nor c have a gaseous component
12. Which of the following processes is *not* directly involved in the hydrologic cycle? (a) transpiration (b) evaporation (c) precipitation (d) nitrification (e) condensation
13. The ___________, which results from the rotation of Earth, displaces the paths of atmospheric and oceanic currents to the right in the Northern Hemisphere and to the left in the Southern Hemisphere. (a) upwelling (b) prevailing wind (c) ocean current (d) El Niño–Southern Oscillation (e) Coriolis effect
14. The periodic warming of surface waters of the tropical East Pacific that alters both oceanic and atmospheric circulation patterns is known as (a) upwelling (b) prevailing wind (c) ocean current (d) El Niño–Southern Oscillation (e) Coriolis effect
15. A mountain range may produce a downwind arid (a) upwelling (b) rain shadow (c) ocean current (d) microclimate (e) ecological pyramid
16. Which of the following is an example of top-down processes? (a) killer whales in the North Pacific have reduced the number of sea otters, thereby causing an increase in the number of sea urchins and a decline in kelp populations (b) experiments at Hubbard Brook have addressed the hydrology of forests and associated aquatic ecosystems (c) one tree provides food for thousands of insects (d) in African grasslands, the number of herbivores is greater than the number of carnivores (e) the bioaccumulation of DDT in a Long Island salt marsh

CRITICAL THINKING

1. Describe the simplest stable ecosystem you can imagine.
2. How might a food web change if all decomposers were eliminated from it?
3. What would happen to the nitrogen cycle if all bacteria were absent? Explain your answer.
4. What is the energy source that powers the wind?
5. How do ocean currents affect climate on land?
6. Would the microclimate of an ant be the same as that of an elephant living in the same area? Why or why not?

- Visit our Web site at **http://biology.brookscole.com/solomon7** for links to chapter-related resources on the World Wide Web. Additional online materials relating to this chapter can also be found on our Web site.

BIOLOGY NOW RESOURCES

Active Figures

53-1 Interaction of Trophic Levels

53-6 The Carbon Cycle and Global Warming

Preparing for an exam? Take a diagnostic test on your BiologyNow CD-ROM.

Post-Test Answers

1. c	2. d	3. e	4. d
5. a	6. d	7. e	8. e
9. a	10. b	11. c	12. d
13. e	14. d	15. b	16. a

Ecology and the Geography of Life

Black-tailed prairie dog and coyote. Prairie dogs (*Cynomys ludovicianus*) never wander far from their burrows, which they use to escape from predators, such as the coyote (*Canis latrans*).

CHAPTER OUTLINE

- **Biomes**
- **Aquatic ecosystems**
- **Ecotones**
- **Biogeography**

In Chapter 53 you learned that climate, particularly temperature and precipitation, provides Earth with many different environments. *Natural selection* affects an organism's ability to survive and reproduce in a given environment (see Chapter 17). In natural selection, both **abiotic** (nonliving) and **biotic** (living) factors eliminate the least-fit individuals in a given population. Over time, succeeding generations of organisms that live in each biome or major aquatic ecosystem become better adapted to local environmental conditions.

Black-tailed prairie dogs are superbly adapted to their environment. Their teeth and digestive tracts are modified to eat and easily digest the seeds and leaves of grasses that grow in great profusion on the Great Plains. Their enlarged pair of incisors, for example, can nibble grasses and gnaw through seeds. Prairie dogs also eat insects and underground roots and tubers.

Black-tailed prairie dogs, which live in the grasslands of western North America, from southern Canada to northern Mexico, have many behavioral adaptations that protect them from coyotes, eagles, hawks, badgers, ferrets, and other predators. For example, prairie dogs are social animals and live in large colonies of about 500 individuals. The eyes of every individual in the colony watch for potential danger, and when they see it, prairie dogs call out to warn the rest of the colony. In fact, they are called "dogs" not because they are related to dogs (they are not), but because their warning calls are similar to a dog's bark.

Black-tailed prairie dogs are burrowing animals. When danger approaches—a coyote, perhaps—each prairie dog dives into its underground home (see photograph). Their burrows have at least two openings and consist of an elaborate network of long tunnels with several chambers. One room may serve as a nursery for the young, one as a sleeping chamber, one as a toilet chamber, and one, which is close to an entrance, as a listening chamber. High piles of excavated soil surround the burrow entrances and help prevent flooding during rainstorms.

To survive winter, black-tailed prairie dogs hibernate in their burrows. Their metabolism slows, and they subsist on the stored fat in their bodies. They do not hibernate deeply, however, and may leave their burrows to look for food when the weather warms.

Some ranchers view prairie dogs as a nuisance because they dig burrows in fields and pastures. Ranchers fear that livestock will step into the burrow entrances and injure their legs. (Horses and cattle rarely step into the entrances, however.) Ranchers also think prairie dogs compete with livestock for grasses. Despite extensive trapping and poisoning, however, these resourceful animals persist.

Like prairie dogs, each species has evolved structural, behavioral, and physiological adaptations for its own particular environment and lifestyle. As we examine Earth's major terrestrial and aquatic ecosystems, including the species characteristic of each, think about the variety of adaptations that natural selection has produced in organisms in response to their particular environments. ■

KEY CONCEPT: The distribution of the world's biomes is largely the result of climate patterns, which determine an area's water (measured as precipitation) and energy (measured as temperature).

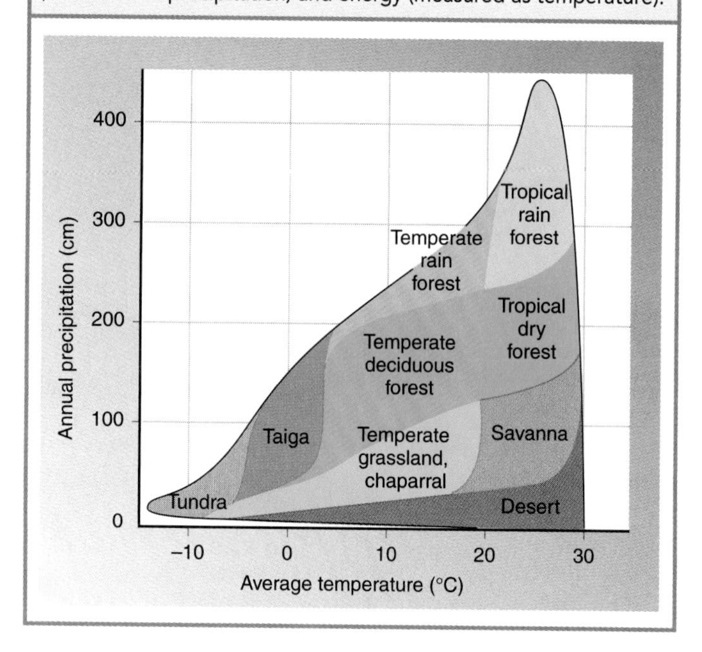

FIGURE **54-1** **Using precipitation and temperature to identify biomes.**

Factors such as soil type, fire, and seasonality of climate affect whether temperate grassland or chaparral develops. *(Adapted from R.H. Whittaker,* Communities and Ecosystems, *2nd ed., Macmillan, New York, 1975)*

BIOMES

Learning Objectives

1. Define *biome,* and briefly describe the nine major terrestrial biomes, giving attention to the climate, soil, and characteristic plants and animals of each.
2. Describe at least one human effect on each of the biomes discussed.

A **biome** is a large, relatively distinct terrestrial region that has similar climate, soil, plants, and animals regardless of where it occurs. Because it covers such a large geographic area, a biome encompasses many interacting landscapes. Recall from Chapter 51 that a **landscape** is a large land area (several to many square kilometers) composed of interacting ecosystems.

Biomes largely correspond to major climate zones, with temperature and precipitation being most important (Fig. 54-1). Near the poles, temperature is generally the overriding climate factor, whereas in tropical and temperate regions, precipitation becomes more significant than temperature (Fig. 54-2 on page 1068). Other abiotic factors to which biomes are sensitive include temperature extremes as well as rapid temperature changes, floods, droughts, strong winds, and fires (see section on fires in Chapter 53).

Nine major biomes are discussed: tundra, taiga, temperate rain forest, temperate deciduous forest, temperate grassland, chaparral, desert, savanna, and tropical rain forest (Fig. 54-3 on page 1069). Although we discuss each biome as a distinct entity, biomes intergrade into one another at their boundaries.

Tundra is the cold, boggy plains of the far north

Tundra (also called **arctic tundra**) occurs in extreme northern latitudes wherever snow melts seasonally (Fig. 54-4 on page 1069; also see Fig. 54-3). The Southern Hemisphere has no equivalent of the arctic tundra because it has no land in the proper latitudes. A similar ecosystem located in the higher elevations of mountains, above the tree line, is called **alpine tundra** to distinguish it from arctic tundra (see *Focus On: The Distribution of Vegetation on Mountains*).

Arctic tundra has long, harsh winters and extremely short summers. Although the growing season, with its warmer temperatures, is as short as 50 days (depending on location), the days are long. Above the Arctic Circle the sun does not set at all for many days in midsummer, although the amount of light at midnight is only one tenth that at noon. There is little precipitation (10 to 25 cm, or 4 to 10 in, per year) over much of the tundra, with most of it falling during summer months.

Tundra soils tend to be geologically young, because most were formed only after the last ice age.[1] These soils are usually nutrient poor and have little organic litter (dead leaves and stems, animal droppings, and remains of organisms) in the uppermost layer of soil. Although the soil surface melts during the summer, tundra has a layer of permanently frozen ground called **permafrost** that varies in depth and thickness. Because permafrost interferes with drainage, the thawed upper zone of soil is usually waterlogged during the summer. Permafrost limits the depth to which roots penetrate, preventing the establishment

[1] Glacier ice, which occupied about 29% of Earth's land during the last ice age, began retreating about 17,000 years ago. Today, glacier ice occupies about 10% of the land surface.

Hiking up a mountain is similar to traveling toward the North Pole with respect to the major life zones encountered *(see figure)*. This elevation–latitude similarity occurs because as one climbs a mountain, the temperature drops just as it does when one travels north; the temperature drops about 6°C (11°F) with each 1000-m increase in elevation. The types of species growing on the mountain change as the temperatures change.

Deciduous trees, which shed their leaves every autumn, may cover the base of a mountain in Colorado, for example. At higher elevations, where the climate is colder and more severe, a coniferous *subalpine forest* resembling northern taiga grows. Spruces and firs are the dominant trees here. Higher still, the forest thins, and the trees become smaller, gnarled, and shrublike. These twisted, shrublike trees, called **krummholz,** are found at their elevational limit (the *treeline*). The exact elevation at which the treeline occurs depends on the latitude and distance from the ocean. In the Rocky Mountains, between 35° and 50° north latitude, the treeline drops 100 m with each 1 degree latitude northward.

Above the tree line, where the climate is quite cold, a kind of tundra occurs, with vegetation composed of grasses, sedges, and small tufted plants, most of which are hardy perennials. Some alpine plants (for example, buttercups) are lowland species that have adapted to the alpine environment, whereas other plants (for example, mountain douglasia) live exclusively in the mountains. This tundra is called *alpine tundra* to distinguish it from arctic tundra. At the top of the mountain, a permanent ice or snow cap might be found, similar to the nearly lifeless polar land areas.

Important environmental differences exist between high elevations and high latitudes that affect the types of organisms found in each place. Alpine tundra typically lacks permafrost and receives more precipitation than does arctic tundra. High elevations of temperate mountains do not have the great extremes of day length that are associated with the changing seasons in high-latitude biomes. The intensity of solar radiation is greater at high elevations than at high latitudes. At high elevations, the sun's rays pass through less atmosphere, which results in greater exposure to ultraviolet radiation (less is filtered out by the atmosphere) than occurs at high latitudes.

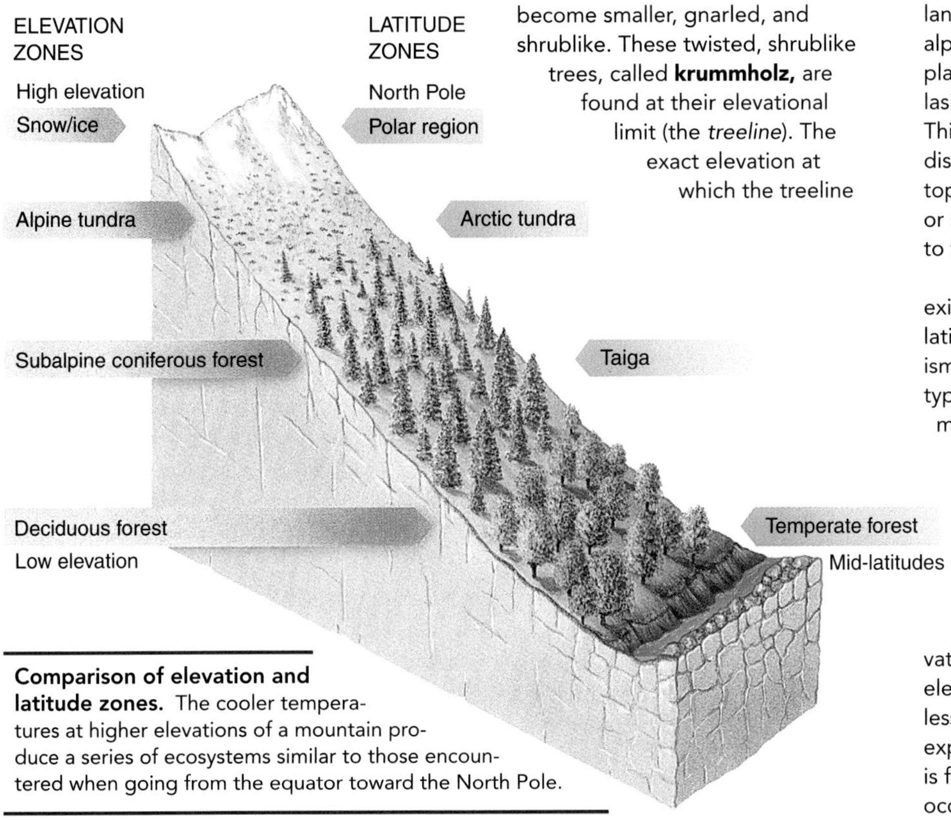

Comparison of elevation and latitude zones. The cooler temperatures at higher elevations of a mountain produce a series of ecosystems similar to those encountered when going from the equator toward the North Pole.

of most woody species, which tend to root deeply. Limited precipitation, combined with low temperatures, flat topography (surface features), and permafrost, produces a landscape of broad, shallow lakes, sluggish streams, and bogs.

Low species richness and low primary productivity characterize tundra. Few plant species occur, but individual species often exist in great numbers. Mosses, lichens (such as reindeer moss), grasses, and grasslike sedges dominate tundra vegetation; most of these short plants are herbaceous perennials that live 20 to 100 years. No readily recognizable trees or shrubs grow except in sheltered locations, although dwarf willows, dwarf birches, and other dwarf trees are common. Tundra plants seldom grow taller than 30 cm (12 in) in open areas.

Year-round animal life of the tundra includes voles, weasels, arctic foxes, gray wolves, snowshoe hares, ptarmigan, snowy owls, musk oxen, and lemmings (see the discussion of lemming population cycles in Chapter 51). In the summer, caribou migrate north to the tundra to graze on sedges, grasses, and dwarf willow. Dozens of bird species also migrate north in summer to nest and feed on abundant insects. Mosquitoes, blackflies, and deerflies survive the winter as eggs or pupae and occur in great numbers during summer weeks. There are no reptiles or amphibians. (Recall from Chapter 30 that amphibians and reptiles are *ectothermic.* Reptiles and amphibians may be excluded from the cold temperatures of the tundra because their body temperature fluctuates with the temperature of the surrounding environment.)

Tundra regenerates quite slowly after it has been disturbed. Even casual use by hikers causes damage. Long-lasting injury, likely to persist for hundreds of years, was done to large por-

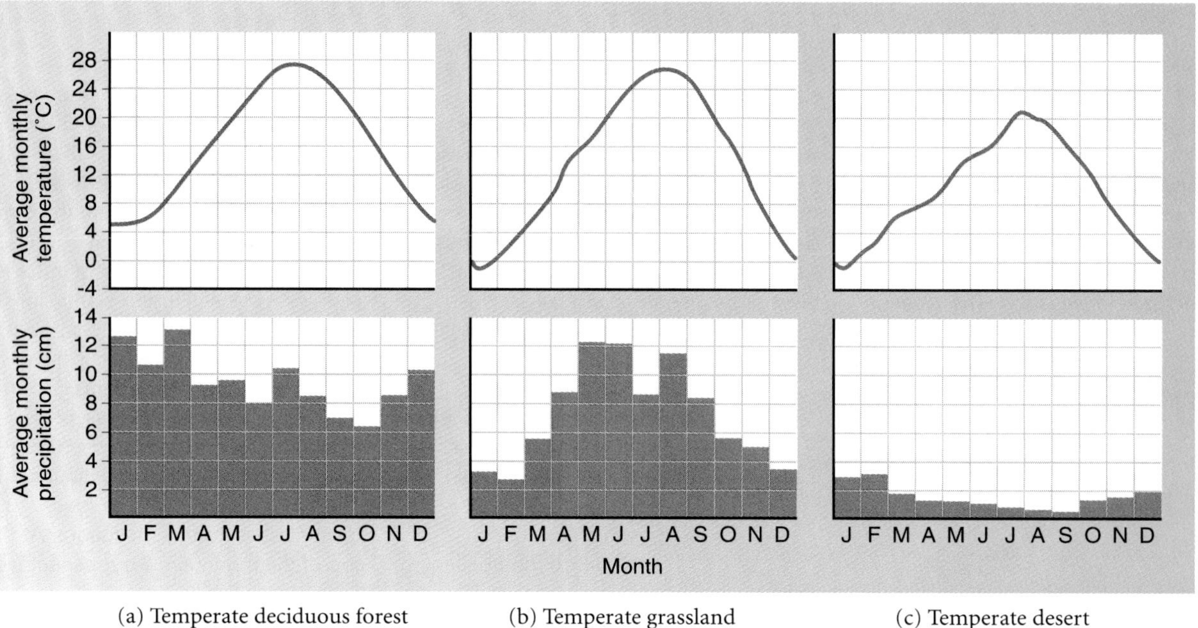

(a) Temperate deciduous forest (Nashville, Tennessee)

(b) Temperate grassland (Lawrence, Kansas)

(c) Temperate desert (Reno, Nevada)

FIGURE **54-2** **Significance of precipitation in temperate biomes.**

Average monthly temperature (top) is approximately the same in each location in North America. However, precipitation (bottom) varies a great deal, resulting in **(a)** deciduous forest where precipitation is plentiful, **(b)** grassland where it is less plentiful and more seasonal, and **(c)** desert where it is quite low.

tions of the arctic tundra as a result of oil exploration and military use.

Taiga is the evergreen forest of the north

Just south of the tundra is the **taiga,** or **boreal forest,** which stretches across both North America and Eurasia. Taiga is the world's largest biome, covering approximately 11% of Earth's land (Fig. 54-5; also see Fig. 54-3). A biome comparable to the taiga is not found in the Southern Hemisphere, because it has no land at the corresponding latitudes. Winters are extremely cold and severe, although not as harsh as in the tundra. The growing season of the boreal forest is somewhat longer than that of the tundra. Taiga receives little precipitation, perhaps 50 cm (20 in) per year, and its soil is typically acidic, mineral poor, and has a deep layer of partly decomposed conifer needles at the surface. (Conifers are cone-bearing evergreens.) Permafrost is patchy and, where found, is often deep underneath the soil. Taiga contains numerous ponds and lakes in water-filled depressions that grinding ice sheets dug during the last ice age.

Black and white spruces, balsam fir, eastern larch, and other conifers dominate the taiga, but **deciduous** trees such as aspen or birch, which shed their leaves in autumn, form striking stands. Conifers have many drought-resistant adaptations, such as needle-like leaves with a minimal surface area to reduce water loss (see Chapter 32). Such an adaptation enables conifers to withstand the "drought" of the northern winter months, when roots don't absorb water because the ground is frozen. Natural selection also favors conifers in the taiga because, being evergreen, they resume photosynthesis as soon as warmer temperatures return.

Animal life of the boreal forest includes some larger species such as caribou (which migrate from the tundra to the taiga for winter), wolves, bears, and moose. However, most animal life is medium-sized to small and includes rodents, rabbits, and furbearing predators such as lynx, sable, and mink. Most species of birds are seasonally abundant but migrate to warmer climates for winter. Insects are numerous, but there are few amphibians and reptiles except in the southern taiga.

Most of the taiga is not suitable for agriculture because of its short growing season and mineral-poor soil. Taiga trees, which are harvested primarily by clearcutting, are currently the primary source of the world's industrial wood and wood fiber. The annual loss of boreal forests is estimated to encompass an area twice as large as the Amazonian rain forests of Brazil. About 1 million hectares (2.5 million acres) of Canadian forests are logged annually, and extensive tracts of Siberian forests in Russia are harvested, although exact estimates are unavailable. Alaska's taiga is also at risk, because the U.S. government may increase logging on public lands in the future.

Temperate rain forest has cool weather, dense fog, and high precipitation

Coniferous **temperate rain forest** grows on the northwest coast of North America. Similar vegetation exists in southeastern Australia and in southwestern South America (see Fig. 54-3). Annual precipitation in this biome is high, from 200 to 380 cm (80 to 150 in); condensation of water from dense coastal fogs augments the annual precipitation. The proximity of temperate rain forest to the coastline moderates the temperature so that seasonal

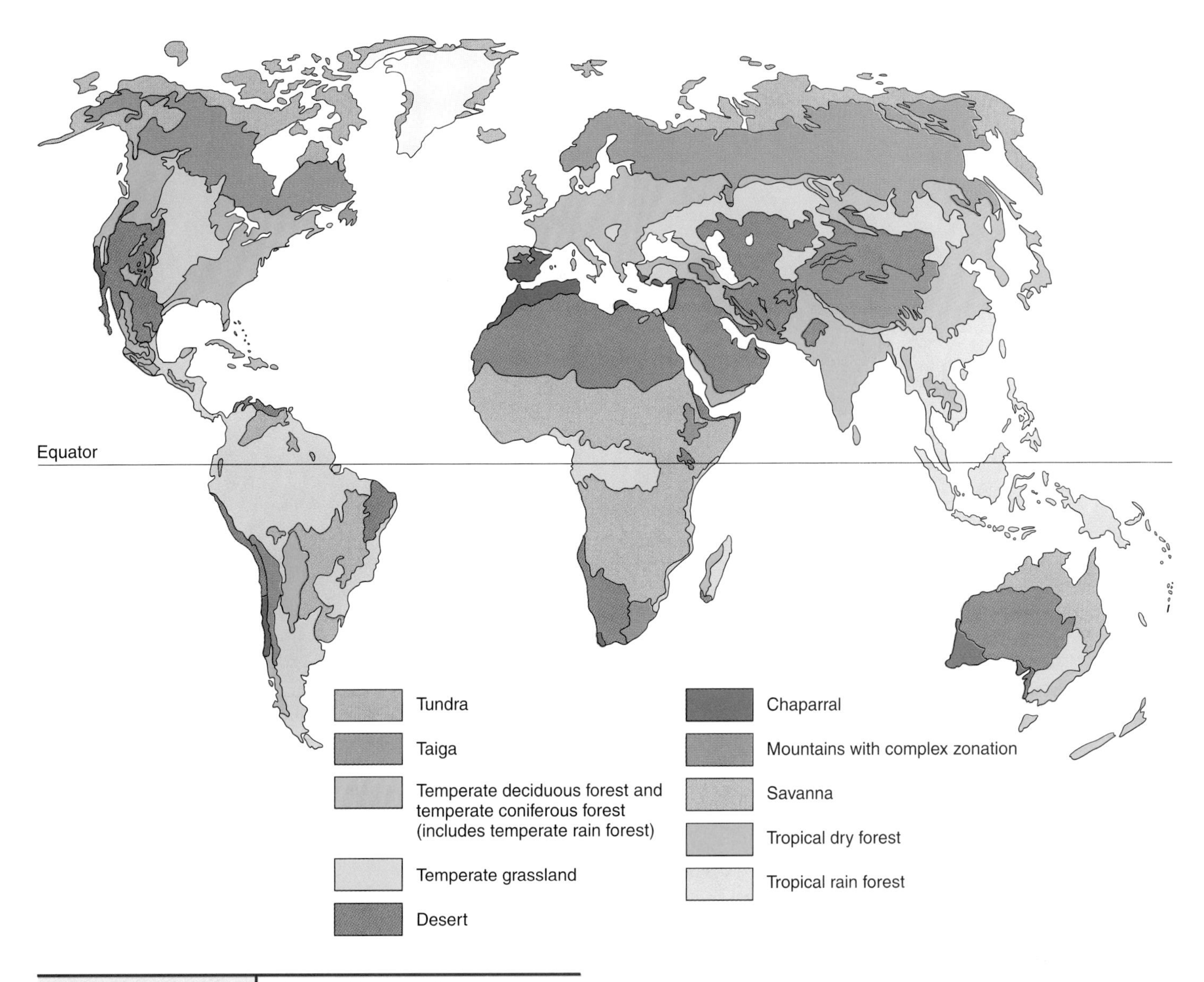

ACTIVE FIGURE **54-3** | **The world's major biomes.**

This highly simplified diagram shows sharp boundaries between biomes. Biomes actually intergrade at their boundaries, sometimes over large areas. Note that forested mountains such as the Appalachian Mountains are keyed according to their predominant vegetation type, whereas mountains with variable vegetation are keyed as "Mountains with complex zonation." *(Based on data from the World Wildlife Fund)*

Biology Now™ Learn more about **biomes** by clicking on this figure on your BiologyNow CD-ROM.

FIGURE **54-4** | **Arctic tundra.**

Because of its short growing season and permafrost, only small, hardy plants grow in the northernmost biome that encircles the Arctic Ocean. Photographed during autumn in Northwest Territories, Canada.

FIGURE **54-5** | **Taiga.**
Taiga is coniferous forest that occurs in cold regions of the Northern Hemisphere adjacent to the tundra. Photographed in Yukon, Canada.

fluctuation is narrow; winters are mild, and summers are cool. Temperate rain forest has a relatively nutrient-poor soil, although its organic content may be high. Cool temperatures slow the activity of bacterial and fungal decomposers. Thus needles and large fallen branches and trunks accumulate on the ground as litter that takes many years to decay and release nutrient minerals to the soil.

The dominant vegetation type in the North American temperate rain forest is large evergreen trees such as western hemlock, Douglas fir, Sitka spruce, and western red cedar. Temperate rain forest is rich in epiphytic vegetation, which are smaller plants that grow nonparasitically on the trunks and branches of large trees (Fig. 54-6). Epiphytes in this biome are mainly mosses, lichens, and ferns, all of which also carpet the ground. Deciduous shrubs such as vine maple grow wherever a break in the overlying canopy occurs. Squirrels, wood rats, mule deer, elk, numerous bird species (such as jays, nuthatches, and chickadees), and several species of reptiles (such as painted turtles and western terrestrial garter snakes) and amphibians (such as Pacific giant salamanders and Pacific treefrogs) are common temperate rainforest animals.

Temperate rain forest is one of the richest wood producers in the world, supplying us with lumber and pulpwood. It is also one of the most complex ecosystems in terms of species richness. Care must be taken to avoid overharvesting original old-growth forest, because such an ecosystem takes hundreds of years to develop. When the logging industry harvests old-growth forest, it typically replants the area with a monoculture (a single species) of trees that it harvests in 40- to 100-year cycles. Thus the old-growth forest ecosystem, once harvested, never has a chance to redevelop. A small fraction of the original, old-growth temperate rain forest in Washington, Oregon, and northern California remains untouched. These stable forest ecosystems provide biological habitat for many species, including 40 endangered and threatened species.

FIGURE **54-6** | **Temperate rain forest.**
Large amounts of precipitation characterize temperate rain forest. Note the epiphytes hanging from the branches of coniferous trees. Photographed in Olympic National Park in Washington State.

Temperate deciduous forest has a canopy of broad-leaf trees

Seasonality (hot summers and cold winters) is characteristic of **temperate deciduous forest,** which occurs in temperate areas where precipitation ranges from about 75 to 126 cm (30 to 50 in) annually (see Fig. 54-3). Typically, the soil of a temperate deciduous forest consists of both a topsoil rich in organic material and a deep, clay-rich lower layer. As organic materials decay, nutrient mineral ions are released. If roots of living trees do not absorb these ions, they leach into the clay, where they may be retained.

Broad-leaf hardwood trees, such as oak, hickory, maple, and beech, that lose their foliage annually dominate temperate deciduous forests of the northeastern and middle eastern United States (Fig. 54-7). In southern areas, the number of broad-leaf evergreen trees, such as magnolia, increases. The trees of the temperate deciduous forest form a dense canopy that overlies saplings and shrubs.

Temperate deciduous forests originally contained a variety of large mammals such as mountain lions, wolves, bison, and other species now regionally extinct, plus deer, bears, and many

Barbara Miller/Biological Photo Service

FIGURE **54-7** **Temperate deciduous forest.**

The broad-leaf trees that dominate the temperate deciduous forest shed their leaves before winter. Photographed during autumn in Pennsylvania.

Harvey Payne

FIGURE **54-8** **Temperate grassland.**

The Nature Conservancy owns this tallgrass prairie preserve in Oklahoma. Like other moist temperate grasslands, it is mostly treeless but contains a profusion of grasses and other herbaceous flowering plants. As bison graze on the plants, they affect community structure and diversity.

small mammals and birds (such as wild turkeys, blue jays, and scarlet tanagers). Both reptiles (such as box turtles and rat snakes) and amphibians (such as spotted salamanders and wood frogs) abounded, together with a denser and more varied insect life than exists today.

In Europe and North America, logging and land clearing for farms, tree plantations, and cities have removed much of the original temperate deciduous forest. Where it has regenerated, temperate deciduous forest is often in a seminatural state, that is, highly modified by humans for recreation, livestock foraging, timber harvest, and other uses. Although these returning forests do not have the biological diversity of virgin stands, many forest organisms have successfully become re-established.

Worldwide, temperate deciduous forest was among the first biomes to be converted to agricultural use. In Europe and Asia, many soils that originally supported temperate deciduous forest have been cultivated by traditional agricultural methods for thousands of years without substantial loss in fertility. During the 20th century, however, intensive agricultural practices were adopted; these, along with overgrazing and deforestation, contributed to the degradation of some agricultural lands.

Temperate grasslands occur in areas of moderate precipitation

Summers are hot, winters are cold, fires help to shape the landscape, and rainfall is often uncertain in **temperate grasslands** (see Fig. 54-3). Annual precipitation averages 25 to 75 cm (10 to 30 in). In grasslands with less precipitation, nutrient minerals tend to accumulate in a marked layer just below the topsoil. These nutrient minerals tend to leach out of the soil in areas with more precipitation. Grassland soil contains considerable organic material because surface parts of many grasses die off each winter and contribute to the organic content of the soil, whereas the roots and rhizomes (underground stems) survive underground. The roots and rhizomes that eventually die also contribute to the soil's organic material. Many grasses are sod formers: Their roots and rhizomes form a thick, continuous underground mat.

Moist temperate grasslands, also known as *tallgrass prairies,* occur in the United States in Iowa, western Minnesota, eastern Nebraska, and parts of other midwestern states and across Canada's prairie provinces. Although few trees grow except near rivers and streams, grasses, some as tall as 2 m (6.5 ft), grow in great profusion in the deep, rich soil (Fig. 54-8). Before most of this area was converted to arable land, it was covered with herds of grazing animals, particularly bison. The principal predators were wolves, although in sparser, drier areas coyotes took their place. Smaller fauna included prairie dogs and their predators (foxes, black-footed ferrets, and birds of prey such as prairie falcons), western meadowlarks, bobolinks, reptiles (such as gopher snakes and short-horned lizards), and great numbers of insects.

Shortgrass prairies, in which the dominant grasses are less than 0.5 m (1.6 ft) tall, are temperate grasslands that receive less precipitation than the moister grasslands just described but more precipitation than deserts. In the United States, shortgrass prairies occur in the eastern half of Montana, the western half of South Dakota, and parts of other midwestern states, as well as western Alberta in Canada. The plants grow in less abundance than in the moister grasslands, and occasionally some bare soil is exposed. Native grasses of shortgrass prairies are drought resistant.

The North American grassland, particularly the tallgrass prairie, was so well suited to agriculture that little of it remains. More than 90% has vanished under the plow, and the remainder is so fragmented that almost nowhere can we see even an approximation of what European settlers saw when they settled in the Midwest. Today, the tallgrass prairie is considered North America's rarest biome. It is not surprising that the American Midwest, the Ukraine, and other moist temperate grasslands became the breadbaskets of the world, because they provide ideal growing conditions for crops such as corn and wheat, which are also grasses.

Chaparral is a thicket of evergreen shrubs and small trees

Some hilly temperate environments have mild winters with abundant rainfall, combined with extremely dry summers. Such Mediterranean climates, as they are called, occur not only in the area around the Mediterranean Sea but also in California, Western Australia, portions of Chile, and South Africa (see Fig. 54-3). In southern California this environment is known as **chaparral.** This vegetation type is also known as maquis in the Mediterranean region, mallee scrub in Australia, matorral in Chile, and Cape scrub in Africa. Chaparral soil is thin and infertile. Frequent fires occur naturally in this environment, particularly in late summer and autumn.

Chaparral vegetation looks strikingly similar in different areas of the world, even though the individual species are quite different. A dense growth of evergreen shrubs, often of drought-resistant pine or scrub oak trees, dominates chaparral (Fig. 54-9). During the rainy winter season the landscape may be lush and green, but during the hot, dry summer the plants lie dormant. Trees and shrubs often have hard, small, leathery leaves that resist water loss. Many plants are also fire adapted and grow best in the months following a fire. Such growth is possible because fire releases nutrient minerals that were tied up in aboveground parts of plants that burned. The fire doesn't kill the underground parts of some plants and the seeds of many others, however. With the new availability of essential nutrient minerals, plants sprout vigorously during winter rains. Mule deer, wood rats, brush rabbits, skinks and other lizards, and many species of birds (such as Anna's hummingbird, scrub jay, and bushtit) are common animals of the chaparral.

FIGURE **54-9** | **Chaparral.**

Chaparral, which consists primarily of drought-resistant evergreen shrubs and small trees, develops where hot, dry summers alternate with mild, rainy winters. Photographed in the Santa Lucia Mountains, California.

Edward Ely/Biological Photo Service

Fires, which occur at irregular intervals in California chaparral vegetation, are often quite costly because they consume expensive homes built on the hilly chaparral landscape. Unfortunately, efforts to control naturally occurring fires sometimes backfire. Denser, thicker vegetation tends to accumulate when periodic fires are prevented; then, when a fire does occur, it is much more severe. Removing the chaparral vegetation, whose roots hold the soil in place, also causes problems—witness the mudslides that sometimes occur during winter rains in these areas.

Deserts are arid ecosystems

Deserts are dry areas found in temperate *(cold deserts)* and subtropical or tropical regions *(warm deserts)* (see Fig. 54-3). North America has four distinct deserts. The Great Basin Desert in Nevada, Utah, and neighboring states is a cold desert dominated by sagebrush. The Mojave Desert in Nevada and California is a warm desert known for its Joshua trees. The Chihuahuan Desert, home of century plants (agaves; see Fig. 51-7), is a warm desert found in Texas and New Mexico as well as Mexico. The warm Sonoran Desert, with its many species of cacti, is found in Arizona, California, and Mexico (Fig. 54-6).

The low water vapor content of the desert atmosphere leads to daily temperature extremes of heat and cold, so that a major change in temperature occurs in each 24-hour period. Deserts vary greatly depending on the amount of precipitation they receive, which is generally less than 25 cm (10 in) per year. A few deserts are so dry that virtually no plant life occurs in them. As a result of sparse vegetation, desert soil is low in organic material but often high in mineral content, particularly the salts NaCl, $CaCO_3$, and $CaSO_4$. In some regions, such as areas of Utah and Nevada, the naturally occurring concentration of certain soil minerals reaches toxic levels for many plants.

Desert vegetation includes both perennials and, after a rain, flowering annuals. Plants in North American deserts include cacti, yuccas, Joshua trees, and sagebrushes. Desert plants tend to have reduced leaves or no leaves, an adaptation that conserves water. In cacti such as the giant saguaro, for example, the stem carries out photosynthesis and also expands accordion-style to store water; the leaves are modified into spines, which discourage herbivores. Other desert plants shed their leaves for most of the year, growing only during the brief moist season. Many desert plants have defensive spines, thorns, or toxins to resist the heavy grazing pressure often experienced in this food- and water-deficient environment.

Desert animals tend to be small. During the heat of the day, they remain under cover or return to shelter periodically, whereas at night they come out to forage or hunt. In addition to desert-adapted insects, there are many specialized desert reptiles (such as desert iguanas, desert tortoises, and rattlesnakes) and a few

Willard Clay/Dembinsky Photo Associates

FIGURE **54-10** **Desert.**

Summer rainfall characterizes the warmer deserts of North America, such as the Sonoran Desert shown here. The Sonoran Desert contains many species of cacti, including the large, treelike saguaro, which grows 15 to 18 m (50 to 60 ft) tall. Photographed in Arizona.

desert-adapted amphibians (such as western spadefoot toads). Mammals include such rodents as the American kangaroo rat, which does not need to drink water but subsists solely on the water content of its food (primarily seeds and insects). American deserts are also home to jackrabbits, and Australian deserts have kangaroos. Carnivores such as the African fennec fox and some birds of prey, especially owls, live on rodents and rabbits. During the driest months of the year, many desert insects, amphibians, reptiles, and mammals tunnel underground, where they remain inactive; this period of dormancy is known as *aestivation.*

Humans have altered North American deserts in several ways. Off-road vehicles damage desert vegetation, which sometimes takes years to recover. Tank crews that trained for World War II in California's Mojave Desert left tracks in 1940 that are still visible today. People who drive across the desert in four-wheel-drive vehicles also inflict environmental damage. When the top layer of desert soil is disturbed, erosion occurs more readily, and less vegetation grows to support native animals. Another problem is that certain cacti and desert tortoises have become rare as a result of poaching. Houses, factories, and farms built in desert areas require vast quantities of water, which must be imported from distant areas. Irrigation of desert soils often causes them to become salty and unfit for crops or native vegetation. Increased groundwater consumption by many desert cities has caused groundwater levels to drop. Aquifer depletion in U.S. deserts is particularly critical in southern Arizona and southwestern New Mexico.

Savanna is a tropical grassland with scattered trees

The **savanna** biome is a tropical grassland with widely scattered clumps of low trees (Fig. 54-11). Savanna is found in areas of relatively low or seasonal rainfall with prolonged dry periods (see Fig. 54-3). Temperatures in savannas vary little throughout the year, and precipitation regulates seasons, not temperature as in temperate grasslands. Annual precipitation is 85 to 150 cm (34 to 60 in). Savanna soil is low in essential nutrient minerals, in part because it is strongly leached. Savanna soil is often rich in aluminum, which resists leaching, and in places the aluminum reaches levels that are toxic to many plants. Although the African savanna is best known, savanna also occurs in South America and northern Australia.

FIGURE **54-11** **Savanna.**

In this photograph of African savanna in Tanzania, the smaller animals in the foreground are Thomson's gazelles (*Gazella thomsonii*), and the larger ones near the trees are wildebeests (*Connochaetes taurinus*).

Carlyn Iverson

Wide expanses of grasses, interrupted by occasional trees, characterize savanna. Trees such as *Acacia* bristle with thorns that provide protection against herbivores. Both trees and grasses have fire-adapted features such as extensive underground root systems that enable them to survive seasonal droughts as well as periodic fires that sweep through the savanna.

The world's greatest assemblage of hoofed mammals occurs in the African savanna. Here live great herds of herbivores, including wildebeests, antelopes, giraffes, zebras, and elephants. Large predators, such as lions and hyenas, kill and scavenge the herds. In areas of seasonally varying rainfall, the herds and their predators may migrate annually.

Savanna is rapidly being converted to rangeland for cattle and other domesticated animals, which are replacing the big herds of wild animals. The problem is particularly acute in Africa, which has the most rapidly growing human population of any continent. In some places severe overgrazing by domestic animals has contributed to the conversion of marginal savanna into desert, a process known as **desertification.** In desertification, the reduced grass cover caused by overgrazing allows wind and water to erode the soil, removing the topsoil. Desertification is a progressive process that reduces the productivity of the land, decreasing its ability to support crops or livestock. (Desertification is not restricted to the savanna biome. Temperate grasslands and tropical dry forests can also be degraded to desert.)

There are two basic types of tropical forests

There are many kinds of tropical forests, but ecologists generally classify them as one of two types: tropical dry forests or tropical rain forests. **Tropical dry forests** occur in regions subjected to a wet season and a dry season (usually two to three months each year). Annual precipitation is 150 to 200 cm (60 to 80 in). During the dry season, many tropical trees shed their leaves and remain dormant, much as temperate trees do during the winter. Sometimes a dense understory of grasses and herbs develops during the dry season when the canopy is open. India, Brazil, Thailand, and Mexico are some of the countries that have tropical dry forests (see Fig. 54-3). Tropical dry forests intergrade with savanna on their dry edges and with tropical rain forests on their wet edges. Logging and overgrazing by domestic animals have fragmented and degraded tropical dry forests.

The annual precipitation of **tropical rain forests** is 200 to 450 cm (80 to 180 in). Much of this precipitation, which occurs almost daily, comes from locally recycled water that enters the atmosphere by transpiration from the forest's own trees. Tropical rain forests are often located in areas with ancient, highly weathered, mineral-poor soil. Little organic matter accumulates in such soils. Because temperatures are high and soil moisture is abundant year round, decay organisms and detritus-feeding ants and termites decompose organic litter quite rapidly. Vast networks of roots and mycorrhizae quickly absorb nutrient minerals from decomposing materials. Thus, nutrient minerals of tropical rain forests are tied up in the vegetation rather than in the soil.

Tropical rain forests are found in Central and South America, Africa, and Southeast Asia (see Fig. 54-3). Tropical rain forest is very productive despite the scarcity of nutrient minerals in the soil. Its plants, stimulated by abundant solar energy and precipitation, capture considerable energy by photosynthesis. Of all the biomes, the tropical rain forest is unrivaled in species richness. Local factors affecting rainforest species composition include varying soil fertility and topography (for example, valleys have more plant diversity than hills).

Most trees of tropical rain forests are evergreen flowering plants (Fig. 54-12a). A fully developed rain forest has three or

Frans Lanting/Minden Pictures

(a)

Mark Moffett/Minden Pictures

(b)

FIGURE **54-12**

Tropical rain forest.

(a) A broad view of tropical rainforest vegetation along a riverbank in Southeast Asia. Except at riverbanks, tropical rain forest has a closed canopy that admits little light to the forest floor. **(b)** Thick, epiphyte-covered lianas grow into the canopy using a tree trunk for support. Photographed in Costa Rica.

more distinct stories of vegetation. The topmost story consists of the crowns of the oldest, tallest trees, some 50 m (164 ft) or more in height; these trees are exposed to direct sunlight and are subject to the warmest temperatures, lowest humidities, and strongest winds. The middle story, which reaches a height of 30 to 40 m (100 to 130 ft), forms a continuous canopy of leaves overhead that lets in little sunlight for the support of the sparse understory. Only 2% to 3% of the light bathing the forest canopy reaches the forest understory, which is exposed to very little wind and is relatively cool and humid. The understory itself consists of both smaller plants specialized for life in shade and seedlings of taller trees. Vegetation of tropical rain forests is not dense at ground level except near stream banks or where a fallen tree has opened the canopy.

Tropical rainforest trees support extensive epiphytic communities of smaller plants such as orchids and bromeliads. Although epiphytes grow in crotches of branches, on bark, or even on the leaves of their hosts, most only use their host trees for physical support, not for nourishment.

Because little light penetrates to the understory, many plants living there are adapted to climb already established host trees. Lianas (woody tropical vines), some as thick as a human thigh, twist up through the branches of tropical rainforest trees (Fig. 54-12b). Once in the canopy, lianas grow from the upper branches of one forest tree to another, connecting the tops of the trees together and providing a walkway for many of the canopy's residents. Lianas and herbaceous vines provide nectar and fruit for many tree-dwelling animals.

Not counting bacteria and other soil-dwelling organisms, about 90% of tropical rainforest organisms live in the middle and upper canopies. Rainforest animals include the most abundant and varied insect, reptile, and amphibian fauna on Earth. Birds, too, are diverse, with some specialized to consume fruits (such as parrots) and others to consume nectar (such as hummingbirds and sunbirds). Most rainforest mammals, such as sloths and monkeys, live only in the trees and never climb down to the ground surface. Some large ground-dwelling mammals, including elephants, are also found in tropical rain forests.

Unless strong conservation measures are initiated soon, human population growth and agricultural and industrial expansion in tropical countries will spell the end of tropical rain forests by the middle of the 22nd century. Biologists know that many rainforest species will become extinct before they are even identified and scientifically described. Tropical rainforest destruction is discussed in detail in Chapter 55.

Review

- What climate and soil factors produce the major biomes?
- What representative organisms are found in each of these forest biomes: (1) taiga, (2) temperate deciduous forest, (3) temperate rain forest, (4) tropical rain forest?
- In which biome do you live? Does it match the description given in this text? If not, explain the discrepancy.
- How does tundra compare to desert? How does temperate grassland compare to savanna?

BiologyNow™ Assess your understanding of **biomes** by taking the pretest on your BiologyNow CD-ROM.

AQUATIC ECOSYSTEMS

Learning Objectives

3 Explain the important environmental factors that affect aquatic ecosystems.

4 Distinguish among plankton, nekton, and benthos.

5 Briefly describe the various freshwater, estuarine, and marine ecosystems, giving attention to the environmental characteristics and representative organisms of each.

6 Describe at least one human effect on each of the aquatic ecosystems described.

Aquatic "biomes" do not exist, in the sense that aquatic ecologists do not distinguish aquatic ecosystems based on the dominant form of vegetation. Aquatic ecosystems are classified primarily on abiotic factors, such as salinity, that help determine an aquatic life zone's boundaries. **Salinity,** the concentration of dissolved salts (such as sodium chloride) in a body of water, affects the kinds of organisms present in aquatic ecosystems, as does the amount of dissolved oxygen. Water greatly interferes with the penetration of light, so floating aquatic organisms that photosynthesize remain near the water's surface, and vegetation attached to the bottom grows only in shallow water. In addition, low levels of essential nutrient minerals often limit the number and distribution of organisms in certain aquatic environments. Other abiotic determinants of species composition in aquatic ecosystems include water depth, temperature, pH, and presence or absence of waves and currents.

Aquatic ecosystems contain three main ecological categories of organisms: free-floating plankton, strongly swimming nekton, and bottom-dwelling benthos. **Plankton** are usually small or microscopic organisms that are relatively feeble swimmers. For the most part, they are carried about at the mercy of currents and waves. They are unable to swim far horizontally, but some species are capable of large vertical migrations and are found at different depths of water at different times of the day or at different seasons. Plankton are generally subdivided into two major categories: phytoplankton and zooplankton. **Phytoplankton** (photosynthetic cyanobacteria and free-floating algae) are producers that form the base of most aquatic food webs. **Zooplankton** are nonphotosynthetic organisms that include protozoa, tiny crustaceans, and the larval stages of many animals. **Nekton** are larger, more strongly swimming organisms such as fishes, turtles, and whales. **Benthos** are bottom-dwelling organisms that fix themselves to one spot (sponges, oysters, and barnacles), burrow into the sand (many worms and echinoderms), or simply walk or swim about on the bottom (crayfish, aquatic insect larvae, and brittle stars).

Freshwater ecosystems are closely linked to land and marine ecosystems

Freshwater ecosystems include streams and rivers (flowing-water ecosystems), ponds and lakes (standing-water ecosystems), and marshes and swamps (freshwater wetlands). Each type of freshwater ecosystem has its own specific abiotic conditions and characteristic organisms. Although freshwater ecosystems occupy a relatively small portion—about 2%—of Earth's surface, they are important in the hydrologic cycle: They assist in recycling precipitation that flows as surface runoff to the ocean (see discussion of hydrologic cycle in Chapter 53). Large bodies of fresh water also moderate daily and seasonal temperature fluctuations on nearby land. Freshwater habitats also provide homes for large numbers of species.

Streams and rivers are flowing-water ecosystems

Many different conditions exist along the length of a stream or river (Fig. 54-13). The nature of a **flowing-water ecosystem** changes greatly from its source (where it begins) to its mouth (where it empties into another body of water). Headwater streams (small streams that are the sources of a river) are usually shallow, clear, cold, swiftly flowing, and highly oxygenated. In contrast, rivers downstream from the headwaters are wider and deeper, cloudy (that is, they contain suspended particulates), not as cold, slower flowing, and less oxygenated. Surrounding forest may shade certain parts of the stream or river, whereas other parts may be exposed to direct sunlight. Along parts of a stream or river, groundwater wells up through sediments on the bottom; this local input of water moderates the water temperature so that summer temperatures are cooler and winter temperatures are warmer than in adjacent parts of the flowing-water ecosystem.

The kinds of organisms in flowing-water ecosystems vary greatly from one stream to another, depending primarily on the strength of the current. In streams with fast currents, inhabitants have adaptations such as suckers to attach themselves to rocks so they are not swept away. The larvae of blackflies, for example, attach themselves with suction disks located on the ends of their abdomens. Some stream inhabitants, such as immature water-penny beetles, have flattened bodies that enable them to slip under or between rocks. The water-penny beetle larva gets its common name from its flattened, nearly circular shape. Alternatively, inhabitants such as the brown trout are streamlined and muscular enough to swim in the current. Organisms in large, slow-moving streams and rivers do not need such adaptations, although they are typically streamlined, as are most aquatic organisms, to lessen resistance during movement through water.

Unlike other freshwater ecosystems, streams and rivers depend on land for much of their energy. In headwater streams, up to 99% of the energy input comes from detritus (dead organic material such as leaves) carried from the land into streams and rivers by wind or surface runoff. Downstream, rivers contain more producers and therefore depend slightly less on detritus as a source of energy than do the headwaters.

Human activities have several adverse impacts on rivers and streams, including water pollution and the effects of dams built to contain the water of rivers or streams. Dams change the nature of flowing-water ecosystems, both upstream and downstream from the dam location. A dam causes water to back up, flooding large areas of land and forming a reservoir, which destroys terrestrial habitat. Below the dam, the once-powerful river is reduced to a relative trickle, altering the flowing-water ecosystem.

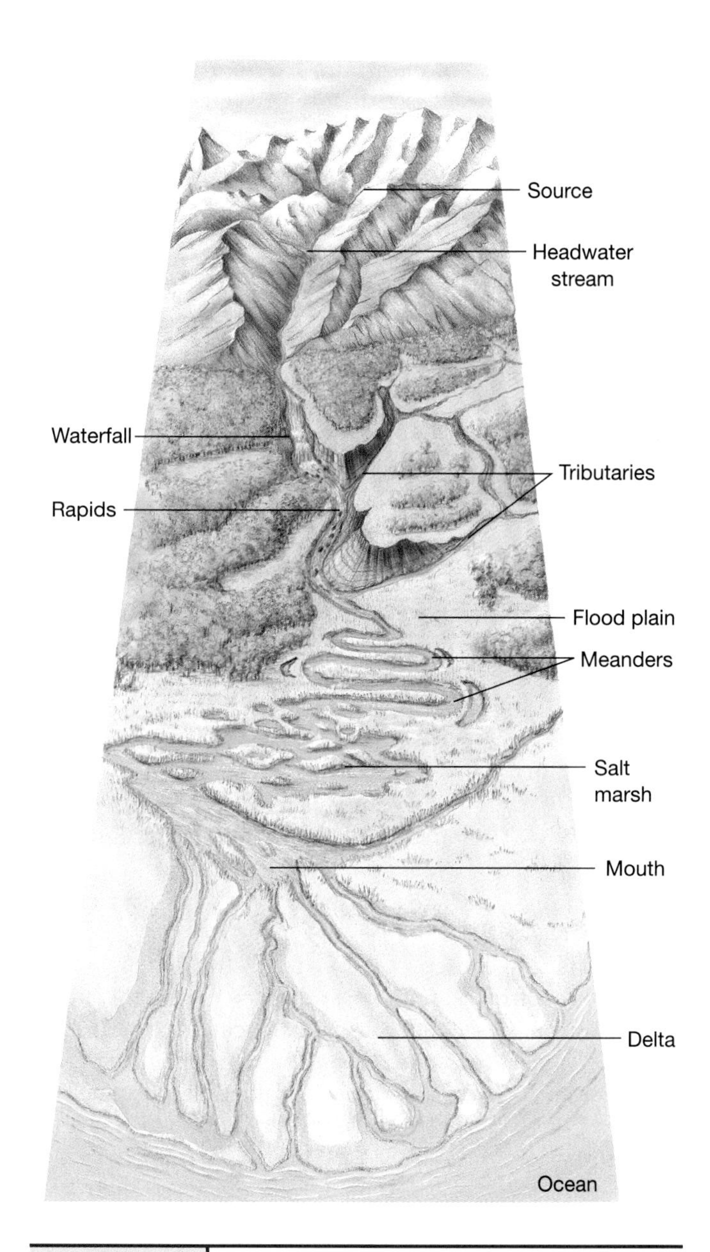

FIGURE **54-13** | **Features of a typical river.**
The river begins at a source, often high in mountains and fed by melting snows or glaciers. Headwater streams flow downstream rapidly, often over rocks (as rapids) or bluffs (as waterfalls). Along the way, tributaries feed into the river, adding to its flow. As the river's course levels out, the river flows more slowly and winds from side to side, forming meanders. The flood plain is the relatively flat area on either side of the river that is subject to flooding. Near the ocean, the river may form a salt marsh where fresh water from the river and salt water from the ocean mix. Sediments deposited by the slow-moving river as it empties into the ocean form the delta, a fertile, low-lying plain at the river's mouth.

Pollution alters the physical environment of a flowing-water ecosystem and changes the biotic component downstream from the pollution source. For example, fertilizer runoff from midwestern fields and manure runoff from livestock operations in such states as Iowa, Wisconsin, and Illinois eventually find their way into the Mississippi River and, from there, into the Gulf of Mexico. These nutrients have created a huge "dead zone" in the Gulf of Mexico some 17,500 km^2 (7000 mi^2) in area. The dead zone extends from the seafloor up into the water column, sometimes to within a few meters of the surface. It generally persists from April or May, as snowmelt and spring rains flow from the Mississippi River into the Gulf, to September. Only anaerobic bacteria exist in the dead zone, because the water does not contain enough dissolved oxygen to support fishes, shrimp, or other aquatic organisms. This low-oxygen condition, known as **hypoxia,** occurs when algae grow rapidly owing to the presence of nutrients in the water. When these algae die, they sink to the bottom, and bacteria decompose them. The bacteria's metabolic activities deplete the water of dissolved oxygen, leaving too little for other aquatic life. Pollution in flowing-water ecosystems has caused hypoxia in more than 60 coastal areas around the world.

Ponds and lakes are standing-water ecosystems

Zonation characterizes **standing-water ecosystems.** A large lake has three basic zones: the littoral, limnetic, and profundal zones (Fig. 54-14). Smaller lakes and ponds typically lack a profundal zone. The **littoral zone** is a shallow water area along the shore of a lake or pond. It includes rooted, emergent vegetation, such as cattails and burreeds, plus several deeper-dwelling aquatic plants and algae. The littoral zone is the most productive zone of the lake. Photosynthesis is greatest in the littoral zone, in part because light is abundant and because the littoral zone receives nutrient inputs from surrounding land that stimulate the growth of plants and algae. Animals of the littoral zone include frogs and their tadpoles, turtles, worms, crayfish and other crustaceans, insect larvae, and many fishes such as perch, carp, and bass. Surface dwellers such as water striders and whirligig beetles are found in the quieter areas.

FIGURE **54-14** | **Zonation in a large lake.**

A lake is a standing-water ecosystem surrounded by land. Here, vegetation around the lake is not drawn to scale.

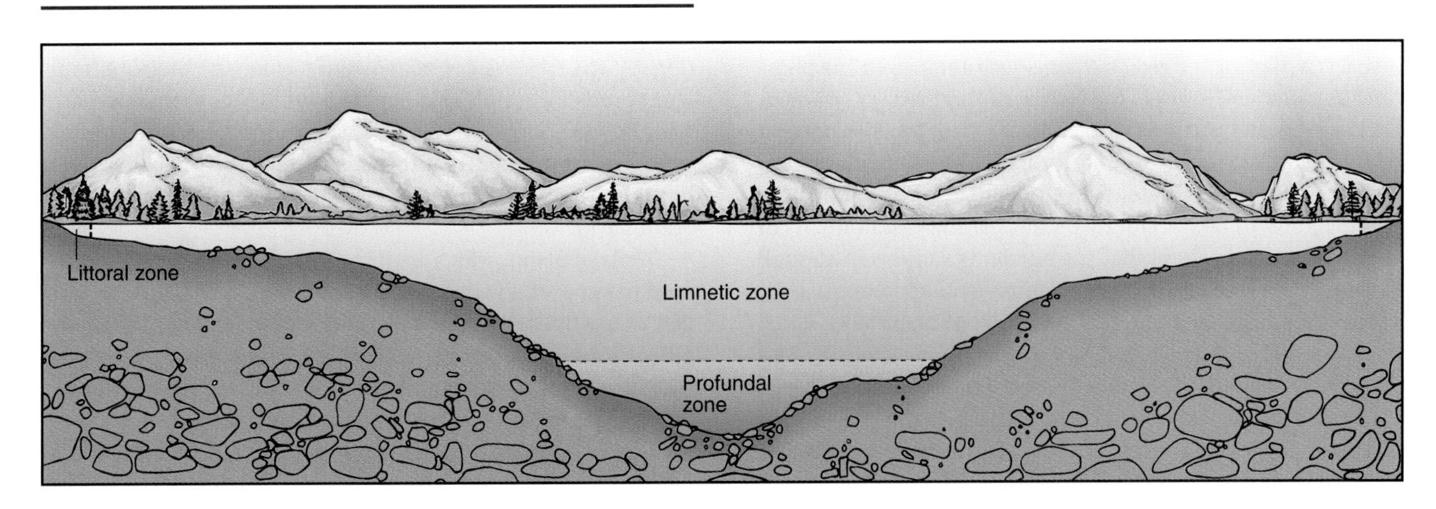

The **limnetic zone** is the open water beyond the littoral zone, that is, away from the shore; it extends down as far as sunlight penetrates to permit photosynthesis. The main organisms of the limnetic zone are microscopic phytoplankton and zooplankton. Larger fishes also spend some of their time in the limnetic zone, although they may visit the littoral zone to feed and reproduce. Owing to its depth, less vegetation grows here than in the littoral zone.

Beneath the limnetic zone of a large lake is the **profundal zone.** Because light does not penetrate effectively to this depth, plants and algae do not live here. Food drifts into the profundal zone from the littoral and limnetic zones. Bacteria decompose dead plants and animals that reach the profundal zone, liberating nutrient minerals. These minerals are not effectively recycled, because no photosynthetic organisms are present to absorb them and incorporate them into the food web. As a result, the profundal zone tends to be both mineral rich and anaerobic (oxygen deficient), with few organisms other than anaerobic bacteria occupying it.

Thermal stratification in temperate lakes The marked layering of large temperate lakes caused by light penetration is accentuated by **thermal stratification,** in which the temperature changes sharply with depth (Fig. 54-15a). Thermal stratification occurs because the summer sunlight penetrates and warms surface waters, making them less dense. (Recall from Chapter 2 that the density of water is greatest at 4°C; both above and below this temperature, water is less dense.) In summer, cool (and therefore more dense) water remains at the lake bottom and is separated from warm (and therefore less dense) water above by an abrupt temperature transition called the **thermocline.** Seasonal distribution of temperature and oxygen (more oxygen dissolves in water at cooler temperatures) affects the distribution of fish in the lake.

In temperate lakes, falling temperatures in autumn cause a mixing of the lake waters called the **fall turnover** (① in Fig. 54-15b). As surface water cools to 4°C, its density increases, and eventually it sinks and displaces the less dense, warmer, mineral-rich water beneath. Warmer water then rises to the surface where it, in turn, cools and sinks. This process of cooling and sinking continues until the lake reaches a uniform temper-

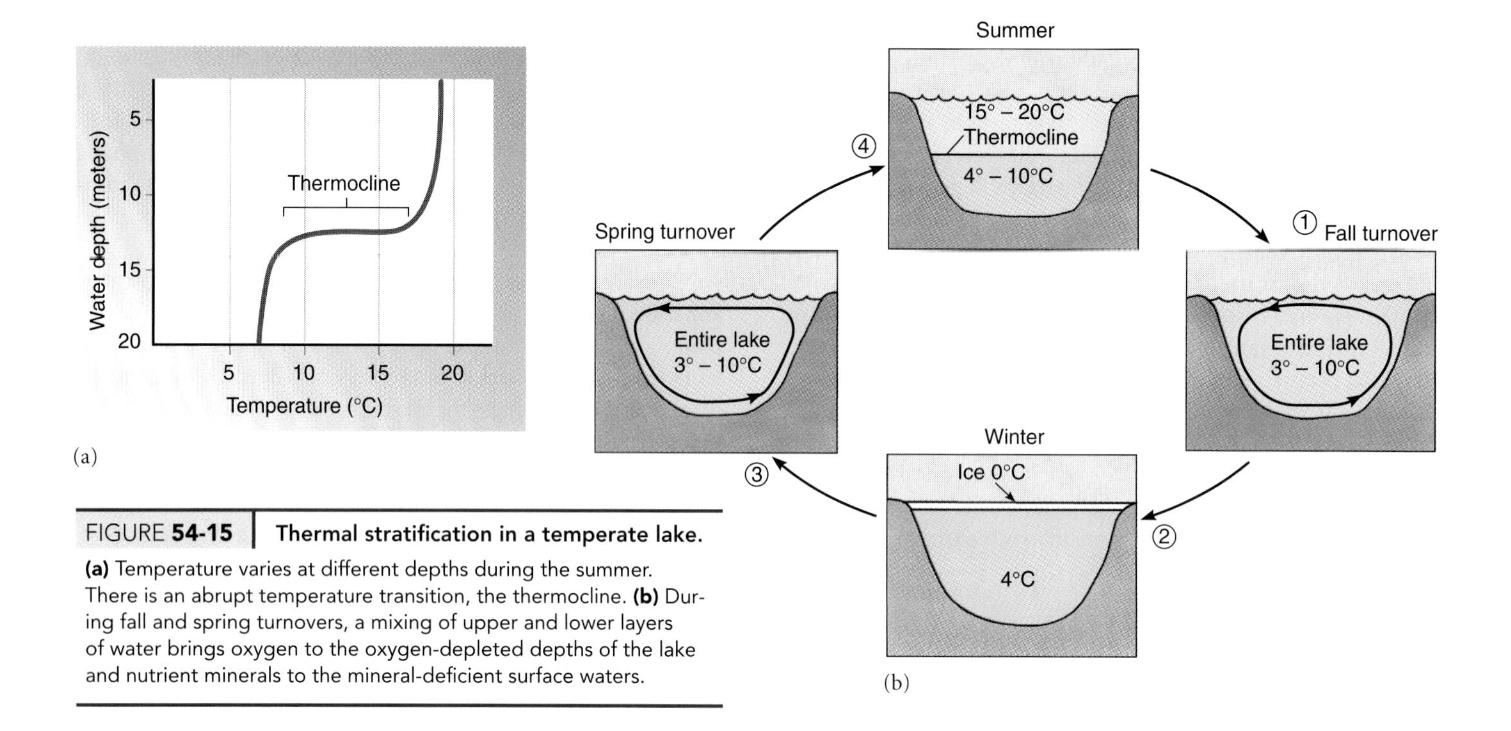

FIGURE **54-15** | **Thermal stratification in a temperate lake.**

(a) Temperature varies at different depths during the summer. There is an abrupt temperature transition, the thermocline. **(b)** During fall and spring turnovers, a mixing of upper and lower layers of water brings oxygen to the oxygen-depleted depths of the lake and nutrient minerals to the mineral-deficient surface waters.

ature throughout. (Because little seasonal temperature variation occurs in the tropics, such turnovers are uncommon there.)

When winter comes, ②, surface water cools to below 4°C, its temperature of greatest density, and, if it is cold enough, ice forms. Ice, which forms at 0°C, is less dense than cold water; thus ice forms on the surface, and the water on the lake bottom is warmer than the ice on the surface.

In the spring, ③, a **spring turnover** occurs as ice melts and surface water reaches 4°C. Surface water again sinks to the bottom, and bottom water returns to the surface. As summer arrives, ④, thermal stratification occurs once again.

The mixing of deeper, nutrient-rich water with surface, nutrient-poor water during the fall and spring turnovers brings essential nutrient minerals to the surface and oxygenated water to the bottom. The sudden presence of large amounts of essential nutrient minerals in surface waters encourages the development of large algal and cyanobacterial populations, which may form temporary **blooms** in the fall and spring.

Increased nutrients and algal growth The presence of high levels of plant and algal nutrients such as nitrogen and phosphorus cause *enrichment,* the fertilization of a body of water. Excess amounts of these nutrients get into waterways from sewage and from fertilizer runoff from lawns and fields. An enriched pond or lake is said to be **eutrophic.** The water in a eutrophic pond or lake is cloudy because of the vast numbers of algae and cyanobacteria that the nutrients support. Although eutrophic lakes contain large populations of aquatic animals, the species composition is different from that of unenriched lakes. For example, an unenriched lake in the northeastern United States may contain pike, sturgeon, and whitefish in the deeper, colder part of the lake where there is a higher concentration of dissolved oxygen. In contrast, the deeper, colder levels of water in eutrophic lakes are depleted of dissolved oxygen because of the greater amount of decomposition on the lake floor. Therefore, fishes such as pike, sturgeon, and whitefish die out, and fishes such as catfish and carp that tolerate lower concentrations of dissolved oxygen replace them.

Eutrophication is reversible and has declined sharply in North America since the 1970s with the passage of legislation to limit the phosphate content of detergents and with the construction of better sewage treatment plants. According to the U.S. Environmental Protection Agency, the leading source of water quality impairment of U.S. surface waters is agriculture. Fertilizer runoff, as well as animal wastes and plant residues in waterways, still cause enrichment problems.

Freshwater wetlands are transitional between aquatic and terrestrial ecosystems

Freshwater wetlands, which shallow water usually covers for at least part of the year, have characteristic soils and water-tolerant vegetation. They include marshes, in which grasslike plants dominate, and swamps, in which woody trees or shrubs dominate (Fig. 54-16). Freshwater wetlands also include hardwood bottomland forests (lowlands along streams and rivers that are periodically flooded), prairie potholes (small, shallow ponds that formed when glacial ice melted at the end of the last ice age), and peat moss bogs (peat-accumulating wetlands where sphagnum moss dominates).

Wetland plants, which are highly productive, provide enough food to support a wide variety of organisms. Wetlands are valued as a wildlife habitat for migratory waterfowl and many other bird species, beaver, otter, muskrat, and game fishes. Wetlands are holding areas for excess water when rivers flood their banks. The floodwater stored in wetlands then drains slowly back into the rivers, providing a steady flow of water throughout the year. Wetlands also serve as groundwater recharging areas. One of their most important roles is to trap and hold

Gregory G. Dimijian/Photo Researchers, Inc.

FIGURE **54-16** | **Freshwater swamp.**

Freshwater swamps are inland areas permanently saturated or covered by water. Trees, such as bald cypress *(shown)*, dominate freshwater swamps. In this wetland, photographed in northeast Texas, a floating carpet of tiny aquatic plants covers the water's surface.

pollutants in the flooded soil, thereby cleansing and purifying the water. Such important environmental functions as these are known as **ecosystem services.**

At one time wetlands were considered wastelands, areas to be filled in or drained so that farms, housing developments, and industrial plants could be built on them. Wetlands are also breeding places for mosquitoes and therefore were viewed as a menace to public health. The crucial ecosystem services that wetlands provide are widely recognized today, and wetlands have some legal protection. Agriculture, pollution, engineering (dams), and urban and suburban development still threaten wetlands, however. In the United States, wetlands have been steadily shrinking by an estimated 23,675 hectares (58,500 acres) per year since 1985. In the contiguous 48 states, only 42 million hectares (104 million acres) remain of the 89.4 million hectares (221 million acres) of wetlands that originally existed during colonial times.

Estuaries occur where fresh water and salt water meet

Where the ocean meets the land, there may be one of several kinds of ecosystems: a rocky shore, a sandy beach, an intertidal mud flat, or a tidal estuary. An **estuary** is a coastal body of water, partly surrounded by land, with access to the open ocean and a large supply of fresh water from rivers. Water levels in an estuary rise and fall with the tides, and salinity fluctuates with tidal cycles, the time of year, and precipitation. Salinity also changes gradually within the estuary, from unsalty fresh water at the river entrance to salty ocean water at the mouth of the estuary. Because estuaries undergo marked daily, seasonal, and annual variations in temperature, salinity, and other physical properties, estuarine organisms have a wide tolerance to such changes.

Estuaries are among the most fertile ecosystems in the world, often having much greater productivities than the adjacent ocean or freshwater river (see Table 53-1). This high productivity is the result of four factors. One, the action of tides promotes a rapid circulation of nutrients and helps remove waste products. Two, nutrient minerals are transported from land into streams and rivers that empty into the estuary. Three, a high level of light penetrates the shallow water. Four, the presence of many plants provides an extensive photosynthetic carpet and also mechanically traps detritus, forming the basis of detritus food webs. Most commercially important fishes and shellfish spend their larval stages in estuaries among the protective tangle of decaying stems.

Temperate estuaries usually contain **salt marshes,** shallow wetlands in which salt-tolerant grasses dominate (Fig. 54-17). Uninformed people have often seen salt marshes as worthless, empty stretches of land. As a result, people have used them as dumps, severely polluting them, or filled them with dredged bottom material to form artificial land for residential and industrial development. A large part of the estuarine environment is lost in this way, along with many of its ecosystem services. These include biological habitats, sediment and pollution trapping, groundwater supply, and storm buffering (salt marshes absorb much of the energy of a storm surge and therefore prevent flood damage elsewhere).

Mangrove forests, the tropical equivalent of salt marshes, cover perhaps 70% of tropical and subtropical coastal mud flats where tidal waves fluctuate (Fig. 54-18). Like salt marshes, mangrove forests provide valuable ecosystem services. Mangrove roots stabilize the sediments, preventing coastal erosion and providing a barrier against the ocean during storms. Their interlacing roots are breeding grounds and nurseries for commercially important fish and shellfish species, such as blue crabs, shrimp, mullet, and spotted sea trout. Mangrove branches are nesting sites for many species of birds, such as pelicans, herons, egrets, and roseate spoonbills. Mangroves are under assault from

FIGURE **54-17** | **Salt marsh.**

In salt marshes, salt water and fresh water mix. Cordgrass (*Spartina alterniflora*) is the dominant vegetation in this salt marsh on the Atlantic coast.

Valerie Giles/ Photo Researchers, Inc.

FIGURE **54-18** | **Mangroves.**
Red mangroves (*Rhizophora mangle*) have stiltlike roots that support the tree. Many animals live in the complex root systems of mangrove forests. Photographed at low tide along the coast of Florida, near Miami.

coastal development, including aquaculture facilities, and unsustainable logging. Some countries, such as the Philippines, Bangladesh, and Guinea-Bissau, have cut down more than two thirds of their mangrove forests.

Marine ecosystems dominate Earth's surface

Although lakes and the ocean are comparable in many ways, they have many differences. Depths of even the deepest lakes do not approach those of the ocean, which has areas that extend more than 6 km (3.6 mi) below the sunlit surface. Tides and currents profoundly influence the ocean. Gravitational pulls of both sun and moon produce two tides a day throughout the ocean, but the height of those tides varies with season, local topography, and phases of the moon (full moons and new moons cause the highest tides).

The immense and extremely complex marine environment is subdivided into several zones: the intertidal zone, the benthic (ocean floor) environment, and the pelagic (ocean water) environment (Fig. 54-19). The pelagic environment is in turn divided into two provinces—the neritic province and the oceanic province.

The intertidal zone is transitional between land and ocean

The **intertidal zone** is the shoreline area between low and high tide. Although high levels of light and nutrients, together with an abundance of oxygen, make the intertidal zone a biologically productive environment, it is also a stressful one. If an intertidal beach is sandy, inhabitants must contend with a constantly shifting environment that threatens to engulf them and gives them scant protection against wave action. Consequently, most sand-dwelling organisms, such as mole crabs, are continual and active burrowers. Because they follow the tides up and down the beach, most do not have any notable adaptations to survive drying or exposure.

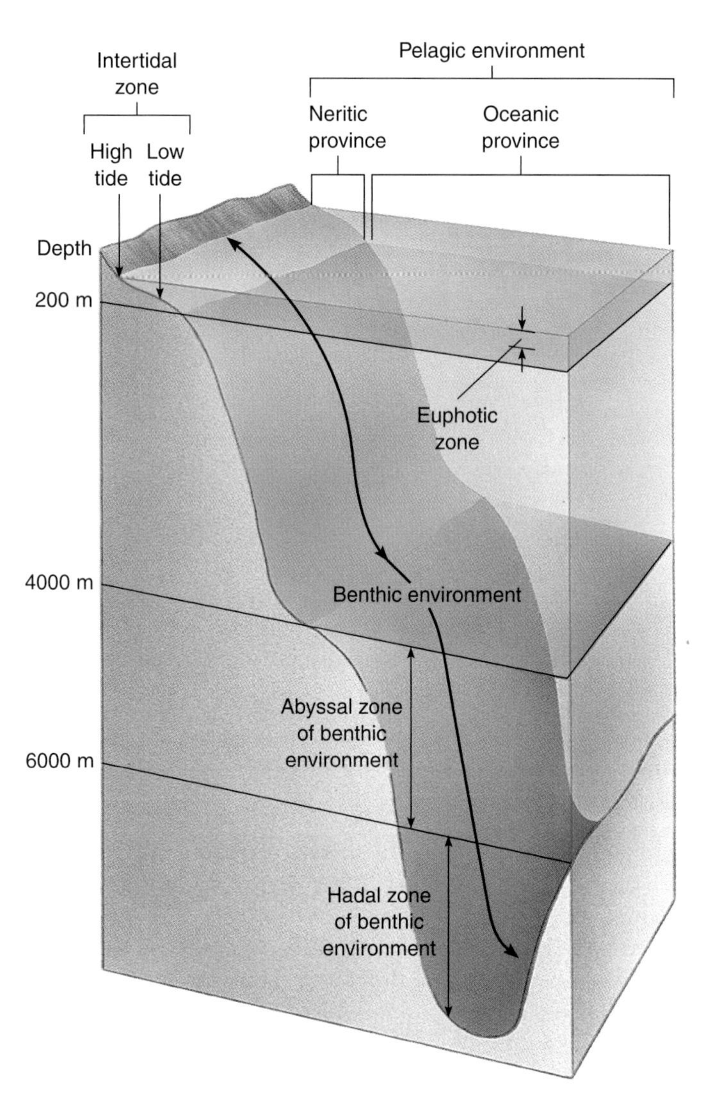

FIGURE **54-19** | **Zonation in the ocean.**
The ocean has three main life zones: the intertidal zone, the benthic environment, and the pelagic environment. The pelagic environment consists of the neritic and oceanic provinces. (The slopes of the ocean floor are not as steep as shown; they are exaggerated to save space.)

A rocky shore provides a fine anchorage for seaweeds and invertebrate animals. However, it is exposed to constant wave action when immersed during high tides and to drying and temperature changes when exposed to the air during low tides. A typical rocky-shore inhabitant has some way of sealing in moisture, perhaps by closing its shell, if it has one, plus a powerful means of anchoring itself to rocks. Mussels, for example, have horny, threadlike anchors, and barnacles have special cement glands. Rocky-shore intertidal algae (seaweeds) usually have thick, gummy polysaccharide coats, which dry out slowly when exposed, and flexible bodies not easily broken by wave action (Fig. 54-20). Some rocky-shore community inhabitants hide in burrows or crevices at low tide.

Seagrass beds, kelp forests, and coral reefs are part of the benthic environment

The **benthic environment** is the ocean floor. It is divided into zones based on distance from land, light availability, and depth. The benthic environment consists of sediments (mostly sand and mud) in which many marine animals, such as worms and clams, burrow. Bacteria are common in marine sediments, and living bacteria are found in ocean sediments more than 500 m (1625 ft) below the ocean floor at several different sites in the Pacific Ocean. The **abyssal zone** is that part of the benthic environment that extends from a depth of 4000 to 6000 m (2.5 to 3.7 mi.). (In Chapter 53, *Focus On: Life Without the Sun* describes some of the unusual organisms in hydrothermal vents in the abyssal zone.) The **hadal zone** is that part of the benthic environment deeper than 6000 m.

Here we describe benthic communities in shallow ocean waters—seagrass beds, kelp forests, and coral reefs. **Sea grasses** are flowering plants that have adapted to complete submersion in ocean water (Fig. 54-21). They are not true grasses. Sea grasses only live in shallow water, to depths of 10 m (33 ft.) where they receive enough light to photosynthesize efficiently. Extensive beds of sea grasses occur in quiet temperate, subtropical, and tropical waters; no sea grasses live in polar waters. Eel grass is the most widely distributed sea grass along the coasts of North America. The most common sea grasses in South Florida and the Caribbean Sea are turtle grass and manatee grass. Sea grasses have a high primary productivity and are therefore ecologically important in shallow marine areas. Their roots and rhizomes stabilize the sediments, reducing surface erosion. Sea grasses provide food and habitat for many marine organisms. In temperate waters, ducks and geese eat sea grasses, whereas in tropical waters, manatees, green turtles, parrot fish, sturgeon fish, and sea urchins eat them. These herbivores consume only about 5% of the sea grasses. The remaining 95% eventually enters the detritus food web when the sea grasses die and bacteria decompose them. In turn, a variety of animals such as mud shrimp, lug worms, and mullet (a type of fish) consume the bacteria.

Kelps, which may reach lengths of 60 m (200 ft.), are the largest brown algae (see Fig. 24-14b). Kelps are common in cooler temperate marine waters of both Northern and Southern Hemispheres. They are especially abundant in relatively shallow waters (depths of about 25 m, or 82 ft) along rocky coastlines. Kelps are photosynthetic and are therefore the primary food producers for the kelp forest ecosystem. Kelp forests also provide habitats for many marine animals. Tube worms, sponges, sea cucumbers, clams, crabs, fishes (such as tuna), and mammals (such as sea otters) find refuge in the algal fronds. Some animals eat the fronds, but kelps are mainly consumed in the detritus food web. Bacteria that decompose the kelp remains provide food for sponges, tunicates, worms, clams, and snails. Kelp beds support a diversity of life that almost rivals that found in coral reefs.

Coral reefs, which are built from accumulated layers of calcium carbonate ($CaCO_3$), are found in warm (usually higher than 21°C), shallow seawater. The living portions of coral reefs grow in shallow waters where light penetrates. Many coral reefs

FIGURE **54-20** | **Seaweeds in a rocky intertidal zone.**

Sea palms (*Postelsia*), which are 50 to 75 cm (20 to 30 in) tall, are common on the rocky Pacific coast from Vancouver Island to California. Their bases are firmly attached to the rocky substrate, enabling them to withstand heavy surf action. Photographed at low tide.

William E. Ferguson

FIGURE **54-21** | **Seagrass bed.**

Turtle grasses (*Thalassia*) have numerous invertebrates and algal epiphytes attached to their leaves. These shallow underwater meadows are ecologically important for shelter and food for many organisms. Photographed off the coast of Mexico.

Ron Phillips

are composed principally of red coralline algae that require light for photosynthesis. Coral animals also require light for the large number of symbiotic dinoflagellates, known as **zooxanthellae,** that live and photosynthesize in their tissues (see Fig. 52-11). Although species of coral without zooxanthellae exist, only those with zooxanthellae build reefs. In addition to obtaining food from the zooxanthellae living inside them, coral animals capture food at night; they use their stinging tentacles to paralyze small animals that drift nearby. Coral reefs grow slowly in warm, shallow water, as coral organisms build on the calcareous remains of countless organisms before them. The waters in which coral reefs are found are often poor in nutrients. Other factors favor high productivity, however, including the presence of symbiotic zooxanthellae, warm temperatures, and plenty of sunlight.

Coral reef ecosystems are the most diverse of all marine environments and contain hundreds or even thousands of species of fishes and invertebrates, such as giant clams, sea urchins, sea stars, sponges, brittle stars, sea fans, and shrimp (Fig. 54-22). The Great Barrier Reef, along the northeastern coast of Australia, occupies only 0.1% of the ocean's surface, but 8% of the world's fish species live there. Many are brightly colored, which advertises the fact that they are poisonous. The complex multitude of relationships and interactions that occur at coral reefs is comparable only to tropical rain forests among terrestrial ecosystems. As in the rain forest, competition is intense, particularly for light and space to grow.

Coral reefs are ecologically important because they both provide habitat for a wide variety of marine organisms and protect coastlines from shoreline erosion. They also provide humans with seafood, pharmaceuticals, and recreational/tourism dollars. Although coral formations are important ecosystems, they are being degraded and destroyed. Of 109 countries with large reef formations, 90 are damaging them. According to the UN Environment Program, 27% of the world's coral reefs are at high risk. Coral reefs of southeastern Asia, which contain the most species of all coral reefs, are the most threatened of any region.

In some areas, silt washing downstream from clearcut inland forests has smothered reefs under a layer of sediment. Some scientists hypothesize that high salinity resulting from the diversion of fresh water to supply the growing human population is killing Florida reefs. Overfishing, pollution from sewage discharge and agricultural runoff, oil spills, boat groundings, fishing with dynamite or cyanide, hurricane damage, disease, coral bleaching, land reclamation, tourism, and the mining of corals for building material are also taking a heavy toll. (Coral bleaching is discussed in Chapter 28.)

Marine biologists are especially concerned about the more than 1 million scuba divers and snorkelers that visit coral reefs each year. Most divers and snorkelers are unaware of how vulnerable coral reefs and their associated organisms are to human interactions. Simply touching or squeezing many reef dwellers kills or injures them. When a diver or snorkeler accidentally kicks or grabs the reef, pieces break off. Divers also stir up the bottom sediments, which suffocate the coral animals.

The neritic province consists of shallow waters close to shore

The **neritic province** is open ocean that overlies the continental shelves, that is, the ocean floor from the shoreline to a depth of 200 m (650 ft). Organisms that live in the neritic province all are floaters or swimmers. The upper reaches of the neritic province make up the **euphotic zone,** which extends from the surface to a depth of approximately 100 m. Enough light penetrates the euphotic zone to support photosynthesis.

Large numbers of phytoplankton, particularly diatoms in cooler waters and dinoflagellates in warmer waters, produce food by photosynthesis and are thus the base of food webs. Zooplankton (including tiny crustaceans, jellyfish, comb jellies, protists such as foraminiferans, and larvae of barnacles, sea urchins, worms, and crabs) feed on phytoplankton. Plankton-eating nekton such as herring, sardines, squid, manta rays, and baleen whales consume zooplankton. These in turn become prey for carnivorous nekton such as sharks, tuna, dolphins, and toothed whales. Nekton are mostly confined to the shallower neritic waters (less than 60 m, or 195 ft, deep) because that is where their food is.

FIGURE **54-22** | **Coral reef organisms.**

A panoramic view of a coral reef in the Indian Ocean off the coast of the Maldives shows the many animals that live on and around coral reefs.

Denise Tackett/Tom Stack & Associates

The oceanic province makes up most of the ocean

The average depth of the world's ocean is 4000 m (2.4 mi). The **oceanic province** is that part of the open ocean that covers the deep ocean basin, that is, the ocean floor at depths more than 200 m. It is the largest marine environment and contains about 75% of the ocean's water. Cold temperatures, high hydrostatic pressure, and an absence of sunlight characterize the oceanic province; these environmental conditions are uniform throughout the year.

Most organisms of the oceanic province depend on **marine snow,** organic debris that drifts down into the **aphotic** ("without light") **region** from the upper, lighted regions. Organisms of this little-known realm are filter feeders, scavengers, or predators. Many are invertebrates, some of which attain great sizes. The giant squid, for example, measures up to 18 m (59 ft) in length, including its tentacles. Fishes of the oceanic province are strikingly adapted to darkness and food scarcity. For example, the gulper eel's huge jaws enable it to swallow large prey (Fig. 54-23). (An organism that encounters food infrequently needs to eat as much as possible when it has the chance.) Many animals of the oceanic province have illuminated organs that enable them to see one another for mating or to capture food. Adapted to drifting or slow swimming, they often have reduced bone and muscle mass.

Human activities are harming the ocean

Because the ocean is so vast, it is hard to visualize that human activities could affect, much less harm, it. They do, however. Development of resorts, cities, industries, and agriculture along coasts alters or destroys many coastal ecosystems, including mangrove forests, salt marshes, seagrass beds, and coral reefs. Coastal and marine ecosystems receive pollution from land, from rivers emptying into the ocean, and from atmospheric contaminants that enter the ocean via precipitation. Disease-causing viruses and bacteria from human sewage contaminate shellfish and other seafood and pose an increasing threat to public health. Millions of tons of trash, including plastic, fishing nets, and packaging materials, end up in coastal and marine ecosystems; some of this trash entangles and kills marine organisms. Less visible contaminants of the ocean include fertilizers, pesticides, heavy metals, and synthetic chemicals from agriculture and industry.

Offshore mining and oil drilling pollute the neritic province with oil and other contaminants. Millions of ships dump oily ballast and other wastes overboard in the neritic and oceanic provinces. Fishing is highly mechanized, and new technologies can remove every single fish in a targeted area of the ocean. Scallop dredges and shrimp trawls are dragged across the benthic environment, destroying entire communities with a single swipe.

Conservation groups and government agencies have made numerous recommendations to protect and manage the ocean's resources. The following are a few proposals:

1. All countries should develop strict policies to protect their coastal and marine ecosystems.
2. Governments should use existing scientific knowledge to develop immediate plans to manage fish stocks; governments should support scientific research to expand our knowledge in this and other critical fields of marine science.
3. Environmental education should be part of the curriculum in every country and should include a strong marine component.
4. Protected coastal and marine areas should be established and carefully monitored around the world.
5. Reduction of marine pollution from land-based sources (sewage discharges, agricultural pollutants, and industrial effluents) should be a top priority of every government.
6. Pollution from ships and offshore installations should be prohibited, and this prohibition should be effectively enforced.

Most countries recognize the importance of the ocean to life on this planet, but few have the resources or programs to

FIGURE **54-23** **Gulper eel from the oceanic province.**

The gulper eel (*Saccopharynx lavenbergi*) uses its "trap-door" jaws to swallow prey as large as itself. The tail, of which only a small portion is shown, makes up most of the length of a gulper eel's body. Gulper eels grow to 1.8 m (6 ft). Shown is a live specimen, photographed in an aquarium aboard ship after being captured at a depth of 1500 m (4921 ft) off Southern California.

Bruce H. Robison, Monterey Bay Aquarium

protect and manage the ocean effectively. The actions just listed will require global, regional, and local cooperation, billions of dollars, and years of effort if they are to be realized.

Review

- What are plankton, nekton, and benthos?
- What environmental factors are most important in determining the adaptations of organisms that live in aquatic environments?
- How do you distinguish between freshwater wetlands and estuaries? Between flowing-water and standing-water ecosystems?
- What are the four main marine environments?
- Which aquatic ecosystem is often compared to tropical rain forests? Why?

BiologyNow™ Assess your understanding of **aquatic ecosystems** by taking the pretest on your BiologyNow CD-ROM.

ECOTONES

Learning Objective

7 Define and describe some of the features of an ecotone.

We have discussed the various terrestrial biomes and aquatic ecosystems as if they were distinct and separate entities, but within landscapes, ecosystems intergrade with one another at their boundaries. You learned in Chapter 52 that the transition zone where two communities or biomes meet and intergrade is called an **ecotone.** Ecotones range in size from quite small, such as the area where an agricultural field meets a woodland or where a stream flows through a forest, to continental in scope. For example, at the border between tundra and taiga an extensive ecotone exists that consists of tundra vegetation interspersed with small, scattered conifers. Such ecotones provide habitat diversity, and a greater variety and density of organisms than either adjacent ecosystem often populate ecotones (Fig. 54-24).

Ecologists who study ecotones look for adaptations that enable organisms to survive there. They also examine the relationship between biological diversity and ecotones, and how ecotones change over time. Long-term studies of ecotones have revealed they are far from static. The ecotone boundary between desert and semiarid grassland in southern New Mexico, for example, has moved during the past 50 years as the desert ecosystem has expanded into the grassland. Many scientists think ecotones will show the first measurable responses to global climate change.

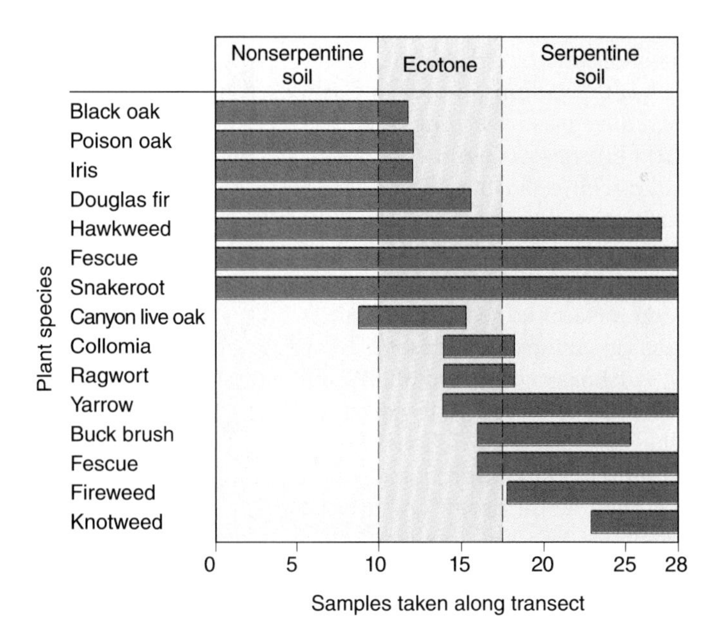

FIGURE **54-24** **Ecotones and species richness.**

Ecotones typically have greater species richness than the communities they connect. Shown are the various plant species found in two communities (yellow and blue) and in the ecotone between them (green). The communities, which are in southwestern Oregon, are defined largely by soil conditions. Nonserpentine soils (samples 1 to 10) are "normal" soils. Serpentine soils (samples 18 to 28) contain high levels of elements such as chromium, nickel, and magnesium that are toxic to many plants. Note that the nonserpentine soil community contains 8 species, the serpentine soil community contains 10 species, and the ecotone contains 13 species. *(Modified from C.D. White, "Vegetation-Soil Chemistry Correlations in Serpentine Ecosystems," dissertation, University of Oregon, Eugene, 1971. Reprinted with permission of Dr. Charles D. White)*

Review

- What is an ecotone?
- Where do ecotones occur?

BiologyNow™ Assess your understanding of **ecotones** by taking the pretest on your BiologyNow CD-ROM.

BIOGEOGRAPHY

Learning Objective

8 Define *biogeography,* and briefly describe Wallace's biogeographic realms.

The study of the geographic distribution of plants and animals is called **biogeography** (see Chapter 17). Biogeographers search for patterns in geographic distribution and try to explain how such patterns arose, including where populations originated, how they spread, and when. Biogeographers recognize that geological and climate changes such as mountain building, continental drift (see Chapter 17), and periods of extensive glaciation influence the distribution of species. Biogeography is linked to evolutionary history and provides insights into how organisms may have interacted in ancient ecosystems. Studying biogeography helps us relate ancient ecosystems to modern ones, because ancient ecosystems form a continuum with modern ecosystems.

One of the basic tenets of biogeography is that each species originated only once. The particular place where this occurred is known as the species' **center of origin.** The center of origin is not a single point but the distribution of the population when the new species originated. From its center of origin, each species spreads until a barrier of some kind halts them; examples of barriers include the ocean, a desert, or a mountain range; unfavorable climate; or the presence of organisms that compete for food or shelter.

Most plant and animal species have characteristic geographic distributions. The **range** of a particular species is that portion of Earth in which it is found. The range of some species may be a relatively small area. For example, wombats, the marsupial equivalent of groundhogs, are found only in drier parts of southeastern Australia and nearby islands. Such localized, native species are said to be **endemic,** that is, they are not found anywhere else in the world. In contrast, some species have a nearly worldwide distribution and occur on more than one continent or throughout much of the ocean. Such species are said to be **cosmopolitan.**

One of the early observations of biogeographers is that the ranges of different species do not include everywhere that they *could* survive. Central Africa has elephants, gorillas, chimpanzees, lions, antelopes, umbrella trees, and guapiruvu trees, whereas areas in South America with a similar climate have none of these. These animals and plants originated in Africa after continental drift had already separated the supercontinent Pangaea into several landmasses. The organisms could not expand their range into South America because the Atlantic Ocean was an impassible barrier. Likewise, the ocean was a barrier to South American monkeys, sloths, tapirs, balsa trees, and snakewood trees, none of which are found in Africa.

Land areas are divided into six biogeographic realms

As the various continents were explored and their organisms studied, biologists observed that the world could be divided into major blocks of vegetation, such as forests, grasslands, and deserts, and that these vegetation types corresponded to specific climates (see Fig. 54-3). The relationship between animal distribution, geography, and climate was not deduced until 1876. At that time, Alfred Wallace, who discovered the same theory of evolution by natural selection as Charles Darwin (see Chapter 17), divided Earth's land areas into six major biogeographic realms: the Palearctic, Nearctic, Neotropical, Ethiopian, Oriental, and Australian (Fig. 54-25). A major barrier separates each of the six biogeographic realms from the others and helps maintain each region's biological distinctiveness. Many biologists quickly embraced Wallace's classification, which is still considered valid today. However, human activities, such as the intentional and unintentional introduction of foreign species, are contributing to a homogenization of the biogeographic realms (see Chapter 55).

Refer to Figure 54-25 as we briefly consider some of the characteristic animals in each realm. The *Nearctic* and *Palearctic realms* are more closely related than the other regions, especially in their northern parts where they share many animals such as wolves, hares, and caribou. This similarity may be due to a land bridge that has periodically connected Siberia and Alaska. This bridge was present late in the Pleistocene epoch, about 10,000 years ago. Animals adapted to cold environments may have dispersed between Asia and North America along this bridge.

The *Neotropical realm* was almost completely isolated from the Nearctic realm and other landmasses for most of the past 70 million years. During this time, many marsupial species evolved. The isthmus of Panama, which formed a dry land connection about 3 million years ago, linked North and South America and provided a route for dispersal. Only three species, the opossum, armadillo, and porcupine, are descendants of animals that survived the northward dispersal from South America, but many species, such as the tapir and llama, are descendants of animals that survived the southward dispersal from North America. Competition from these species caused many of South America's marsupial species to go extinct.

The Sahara Desert separates the *Ethiopian realm,* which contains the most varied vertebrates of all six realms, from other landmasses. Some overlap exists between the Ethiopian and *Oriental realms* because a land bridge with a moist climate linked Africa to Asia during the Miocene and Pliocene epochs. The Oriental realm has the fewest endemic species of all the tropical realms.

The *Australian realm* has not had a land connection with other regions for more than 85 million years. It has no native placental mammals, and marsupials and monotremes, including the duck-billed platypus and the spiny anteater, dominate it. Adaptive radiation of the marsupials during their long period of isolation led to species with ecological niches similar to those of placental mammals of other realms (see Fig. 30-25).

Review

- What is biogeography?
- Which biogeographic realm has been separated from the other biogeographic realms for the longest period? What animals characterize this biogeographic realm?

BiologyNow™ Assess your understanding of **biogeography** by taking the pretest on your BiologyNow CD-ROM.

FIGURE **54-25** **Wallace's biogeographic realms.**

Certain unique species characterize each of the six biogeographic realms.

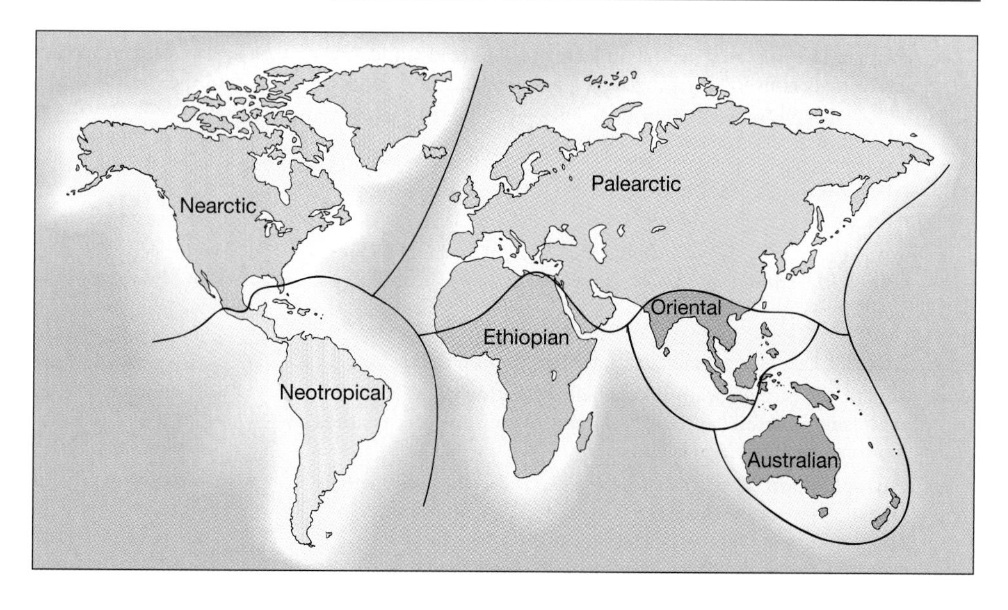

SUMMARY WITH KEY TERMS

1 Define *biome*, and briefly describe the nine major terrestrial biomes, giving attention to the climate, soil, and characteristic plants and animals of each.

- A **biome** is a large, relatively distinct terrestrial region with characteristic climate, soil, plants, and animals. Each biome encompasses a number of interacting landscapes. Temperature and precipitation are important abiotic factors that influence biome distribution.
- A frozen layer of subsoil **(permafrost),** and low-growing vegetation that is adapted to extreme cold and a short growing season characterize **tundra,** the northernmost biome. Coniferous trees adapted to cold winters, a short growing season, and acidic, mineral-poor soil dominate the **taiga,** or **boreal forest.**
- Large conifers dominate **temperate rain forest,** which receives high precipitation. **Temperate deciduous forest** occurs where precipitation is relatively high and soils are rich in organic matter. Broad-leaf trees that lose their leaves seasonally dominate temperate deciduous forest.
- **Temperate grassland** typically has a deep, mineral-rich soil and has moderate but uncertain precipitation. Temperate grassland is well suited to growing grain crops. Thickets of small-leaf evergreen shrubs and trees and a climate of wet, mild winters and dry summers characterize **chaparral.**
- **Desert,** found in both temperate (cold deserts) and subtropical or tropical regions (warm deserts) with low levels of precipitation, contains organisms with specialized water-conserving adaptations.
- Tropical grassland, called **savanna,** has widely scattered trees interspersed with grassy areas. Savanna occurs in tropical areas with low or seasonal rainfall.
- Mineral-poor soil and high rainfall that is evenly distributed throughout the year characterize **tropical rain forest.** Tropical rain forest has high species richness and high productivity.

2 Describe at least one human effect on each of the biomes discussed.

- Oil exploration and military exercises result in long-lasting damage to tundra. Clearcut logging destroys taiga and temperate rain forests. Temperate deciduous forests are removed for logging and to clear land for farms, tree plantations, and land development. Human population growth and its accompanying agricultural and industrial expansion threaten most of the world's tropical rain forests.
- Farmland has replaced most temperate grasslands; savannas are increasingly converted to rangeland for cattle. Development of hilly chaparral results in mudslides and costly fires. Land development in deserts reduces wildlife habitat.

3 Explain the important environmental factors that affect aquatic ecosystems.

- In aquatic ecosystems, important environmental factors include **salinity,** the amount of dissolved oxygen, and the availability of light for photosynthesis.

4 Distinguish among plankton, nekton, and benthos.

- Aquatic life is ecologically divided into **plankton** (free-floating organisms), **nekton** (strongly swimming organisms), and **benthos** (bottom-dwelling organisms).
- **Phytoplankton** are photosynthetic algae and cyanobacteria that form the base of the food web in most aquatic communities. **Zooplankton** are nonphotosynthetic organisms that include protozoa, tiny crustaceans, and the larval stages of many animals.

5 Briefly describe the various freshwater, estuarine, and marine ecosystems, giving attention to the environmental characteristics and representative organisms of each.

- Freshwater ecosystems include flowing-water ecosystems (streams and rivers), standing water ecosystems (ponds and lakes), and freshwater wetlands. In **flowing-water ecosystems** the water flows in a current. Flowing-water ecosystems have few phytoplankton and depend on detritus from the land for much of their energy.
- Large **standing-water ecosystems** (freshwater lakes) are divided into zones on the basis of water depth. The marginal **littoral zone** contains emergent vegetation and algae and is very productive. The **limnetic zone** is open water away from the shore that extends as far down as sunlight penetrates. Organisms in the limnetic zone include phytoplankton, zooplankton, and larger fishes. The deep, dark **profundal zone** holds little life other than bacterial decomposers.
- **Freshwater wetlands,** lands that are transitional between freshwater and terrestrial ecosystems, are usually covered at least part of the year by shallow water and have characteristic soils and vegetation. Freshwater wetlands perform many valuable **ecosystem services.**
- An **estuary** is a coastal body of water, partly surrounded by land, with access to both the ocean and a large supply of fresh water from rivers. Water levels rise and fall with the tides, and salinity fluctuates with tidal cycles, the time of year, and precipitation. Temperate estuaries usually contain **salt marshes,** whereas **mangrove forests** dominate tropical coastlines.
- Four important marine environments are the intertidal zone, the benthic environment, the neritic province, and the oceanic province. The **intertidal zone** is the shoreline area between low and high tides. Organisms of the intertidal zone have adaptations to resist wave action and the extremes of being covered by water (high tide) and exposed to air (low tide).
- The **benthic environment** is the ocean floor. **Sea grasses, kelps,** and **coral reefs** are important benthic communities in shallow ocean waters.
- The **neritic province** is open ocean from the shoreline to a depth of 200 m. Organisms that live in the neritic province are all floaters or swimmers. Phytoplankton are the base of the food web in the **euphotic zone,** where enough light penetrates to support photosynthesis.
- The **oceanic province** is that part of the open ocean that is deeper than 200 m. The uniform environment is one of darkness, cold temperature, and high pressure. Animal inhabitants of the oceanic province are either predators or scavengers that subsist on **marine snow,** detritus that drifts down from other areas of the ocean.

6 Describe at least one human effect on each of the aquatic ecosystems described.

- Water pollution and dams adversely affect flowing-water ecosystems. Increased nutrients, supplied by human activities, stimulate algal growth, resulting in **eutrophic** ponds and lakes. Agriculture, pollution, and land development threaten wetlands and estuaries. Pollution, coastal development, offshore mining and oil drilling, and overfishing threaten marine ecosystems.

7 Define and describe some of the features of an ecotone.

- An **ecotone** is the transition zone where two communities or biomes meet and intergrade. Ecotones provide habitat diversity and are often populated by a greater variety and density of organisms than either adjacent ecosystem.

SUMMARY WITH KEY TERMS *(continued)*

8 Define *biogeography,* and briefly describe Wallace's biogeographic realms.

- **Biogeography** is the study of the geographic distribution of plants and animals, including where populations came from, how they got there, and when. Each species originated only once, at its **center of origin.** From its center of origin, each species spreads until a physical, environmental, or biological barrier halts it. The **range** of a particular species is that portion of the Earth in which it is found.
- Alfred Wallace divided Earth's land areas into six major biogeographic realms: the Palearctic, Nearctic, Neotropical, Ethiopian, Oriental, and Australian. Each realm is biologically distinctive because a mountain range, desert, ocean, or other barrier separates it from the others. Today, human activities are contributing to a homogenization of the biogeographic realms.

POST-TEST

1. The northernmost biome, known as ________________, typically has little precipitation, a short growing season, and permafrost. (a) chaparral (b) taiga (c) tundra (d) northern deciduous forest (e) boreal forest
2. South of tundra is the ________________, which consists of coniferous forests with many lakes. (a) chaparral (b) taiga (c) alpine tundra (d) northern deciduous forest (e) permafrost
3. Forests of the northeastern and middle eastern United States, which have broad-leaf hardwood trees that lose their foliage annually, are called (a) temperate deciduous forests (b) tropical deciduous forests (c) northern coniferous forests (d) temperate rain forests (e) tropical rain forests
4. The deepest, richest soil in the world occurs in (a) temperate rain forest (b) tropical rain forest (c) savanna (d) temperate grassland (e) chaparral
5. This biome, with its thicket of evergreen shrubs and small trees, is found in areas with Mediterranean climates. (a) temperate rain forest (b) tropical rain forest (c) savanna (d) temperate grassland (e) chaparral
6. This biome is a tropical grassland interspersed with widely spaced trees. (a) temperate rain forest (b) tropical rain forest (c) savanna (d) temperate grassland (e) chaparral
7. This biome has the greatest species richness. (a) temperate rain forest (b) tropical rain forest (c) savanna (d) temperate grassland (e) chaparral
8. Organisms in aquatic environments fall into three categories: free-floating ________________, strongly swimming ________________, and bottom-dwelling ________________. (a) nekton; benthos; plankton (b) nekton; plankton; benthos (c) plankton; benthos; nekton (d) plankton; nekton; benthos (e) benthos; nekton; plankton
9. Temperate-zone lakes are thermally stratified, with warm and cold layers separated by a transitional (a) aphotic region (b) thermocline (c) barrier reef (d) ecotone (e) littoral zone
10. Emergent vegetation grows in the ________________ zone of freshwater lakes. (a) littoral (b) limnetic (c) profundal (d) neritic (e) intertidal
11. This coastal body of water has access to the open ocean and a large supply of fresh water from rivers. (a) intertidal zone (b) estuary (c) freshwater wetland (d) neritic province (e) standing-water ecosystem
12. The________________ is open ocean from the shoreline to a depth of 200 m. (a) benthic environment (b) intertidal zone (c) neritic province (d) oceanic province (e) aphotic region
13. Sea grasses (a) occur in shallow water of the ocean's benthic environment (b) may reach lengths of 60 m (c) contain symbiotic algae known as zooxanthellae (d) are common on rocky shores in the intertidal zone (e) are the main producers in freshwater wetlands
14. The transition zone where two ecosystems or biomes meet and intergrade is called a(an): (a) biosphere (b) aphotic region (c) thermocline (d) biogeographic realm (e) ecotone
15. Which biogeographic realm has been separated from the other landmasses for more than 85 million years? (a) Ethiopian (b) Palearctic (c) Nearctic (d) Oriental (e) Australian

CRITICAL THINKING

1. When a black-tailed prairie dog or other small animal dies, other prairie dogs bury it. Develop a hypothesis to explain how this behavior may be adaptive. How would you test this hypothesis?
2. In which biomes would migration be most common? Hibernation? Aestivation? Explain your answers.
3. Develop a hypothesis to explain why animals adapted to the desert are usually small. How would you test your hypothesis?
4. Why do most of the animals of the tropical rain forest live in trees?
5. What would happen to the organisms in a river with a fast current if a dam were built? Would there be any differences in habitat if the dam were upstream or downstream of the organisms in question? Explain your answers.

- Visit our Web site at **http://biology.brookscole.com/solomon7** for links to chapter-related resources on the World Wide Web. Additional online materials relating to this chapter can also be found on our Web site.

BIOLOGY NOW RESOURCES

Active Figure

54-3: Biomes

Preparing for an exam? Take a diagnostic test on your BiologyNow CD-ROM.

Post-Test Answers

1. c	2. b	3. a	4. d
5. e	6. c	7. b	8. d
9. b	10. a	11. b	12. c
13. a	14. e	15. e	

55 Humans in the Environment

Steve Nesbitt, Florida Fish and Wildlife Conservation Commission.

Lucky, the whooping crane (*Grus americana*). Photographed with one of his parents in Florida.

CHAPTER OUTLINE

The whooping crane is a magnificent North American bird named for its loud, penetrating voice. It stands about 1.4 m (4.5 ft) tall, has a wingspan of perhaps 2.3 m (7.5 ft), and lives for 50 or so years. In 1860, scientists estimate that whooping cranes numbered about 1400. By 1941, however, they were highly endangered. Their number had dropped to only 16 in the wild, and their breeding range had shrunk from much of central North America to one remote location in the Northwest Territories of Canada. Whooping crane numbers declined for several reasons, most important of which is the loss of their wetland habitat. Humans fill in and drain wetlands, obtaining land for farms, housing developments, and industrial plants (see Chapter 54). Wetlands also supply the human demand for fresh water.

Thanks to intensive conservation efforts, the original wild flock of whooping cranes still exists and now contains about 320 birds. The wild population spends summer in Canada and migrates to the Texas coast to winter. This population increase is the result of joint efforts by conservation organizations and government agencies in Canada and the United States. As impressive as the increase seems, however, the fate of the whooping crane remains uncertain.

Concerned that bad weather, disease, or some other natural or human-caused disaster could wipe out the single wild population, in the 1990s biologists began reintroducing a second population of whooping cranes to the eastern part of North America. To establish this population, captive-bred whooping cranes are released in central Florida, where a population of related sandhill cranes with similar habitat requirements resides. Biologists use *soft-release techniques,* in which young birds (8 to 10 months old) are initially placed in release pens until they gradually acclimate to the environment. The pens provide some protection against predators such as bobcats; in addition, the pens are constructed so they can be moved about as habitat conditions, such as water levels, change. This population in Florida does not migrate. A breeding pair of whooping cranes in the nonmigratory population made history in 2002 when they successfully reproduced. An eagle killed one of the offspring but the other chick, named Lucky, survived (see photograph). Lucky is the first whooping crane to be produced by captive-reared, released parents. He is also the first whooping crane hatched in the wild in the United States since 1939.

The Whooping Crane Eastern Partnership started a new migrating flock of whooping cranes in 2001. Each year until perhaps 2006, ultralight flyers will guide the flock from their summer breeding ground in central Wisconsin to their wintering site on Florida's gulf coast. After that, the core population should be experienced enough to migrate without being led. The Wisconsin/Florida reintroduction project depends on the support of private landowners along the migratory route, as well as conservationists and biologists.

Whooping cranes are an excellent example of the kinds of injuries humans inflict on biological diversity. They also exemplify human efforts to maintain the world's declining biological diversity. The human species (*Homo sapiens*) has been present on Earth for about 800,000 years (see Chapter 21), which is a brief span of time compared with the age of our planet, some 4.6 billion years. Despite our relatively short tenure on Earth, our biological impact on other species is unparalleled. Our numbers have increased dramatically—the human population reached 6.3 billion in mid-2003—and we have expanded our biological range, moving into almost every habitat on Earth.

Wherever we have gone, we have altered the environment and shaped it to meet our needs. In only a few generations we have transformed the face of Earth, placed a great strain on Earth's resources and resilience, and profoundly affected other species. As a result of these changes, many people are concerned with **environmental sustainability,** the ability to meet humanity's current needs without compromising the ability of future generations to meet their needs.

The impact of humans on the environment merits special study in biology, not merely because we ourselves are humans but also because our impact on the rest of the biosphere is so extensive. Many environmental concerns exist today, too many for us to consider in a single chapter. The rapidly expanding human population underlies and exacerbates all environmental problems. The increasing population is placing an unsustainable stress on the environment, as humans consume ever-increasing quantities of food and water, use more and more energy and raw materials, and produce enormous amounts of waste and pollution. Because the human population crisis was discussed in Chapter 51, we conclude this text by focusing our attention on four other serious environmental issues that affect the biosphere: declining biological diversity, deforestation, global warming, and ozone depletion in the stratosphere. ■

THE BIODIVERSITY CRISIS

Learning Objectives

1. Distinguish among threatened species, endangered species, and extinct species.
2. Discuss at least four causes of declining biological diversity, and identify the most important cause.
3. Define *conservation biology,* and compare in situ and ex situ conservation measures.
4. Describe the benefits and shortcomings of the Endangered Species Act and the Convention on International Trade in Endangered Species of Wild Flora and Fauna.

Extinction, the death of a species, occurs when the last individual member of a species dies (see Chapter 19). Although extinction is a natural biological process, human activities have greatly accelerated it. The burgeoning human population has forced us to spread into almost all areas of Earth. Whenever humans invade an area, the habitats of many plants and animals are disrupted or destroyed, which can contribute to their extinction. For example, the dusky seaside sparrow, a small bird that was found only in the marshes of the St. Johns River in Florida, became extinct in 1987, largely because of human destruction of its habitat (Fig. 55-1).

Biological diversity, also called **biodiversity,** is the variety of living organisms considered at three levels: genetic diversity, species diversity, and ecosystem diversity. *Genetic diversity* is the

FIGURE **55-1** **An extinct species.**
The dusky seaside sparrow (*Ammospiza nigrescens*) became extinct in 1987, largely owing to human destruction of its habitat in Florida. The dusky seaside sparrow was 15 cm (6 in) long.

U.S. Fish and Wildlife Service

Over the last several decades, many of the world's frog populations have dwindled or disappeared. North America, Central and South America, and Australia all have experienced dramatic population declines. At least 14 species of Australian rainforest frogs have become endangered or extinct since 1980. In the United States, as many as 38% of its 242 native amphibian species are declining in numbers. In assessing possible causes, researchers have observed that the declines are not limited to areas with obvious habitat destruction, such as drainage of wetlands where frogs live, or degradation from pollutants. Some remote, pristine locations also show dramatic declines in amphibians: Populations of all seven native species of frogs and toads in Yosemite National Park have declined. No single factor seems responsible for these declines. Potential factors for which there is strong evidence include pollutants, increased UV radiation, infectious diseases, and climate warming.

Agricultural chemicals are implicated in amphibian declines in California's Sierra Nevada Mountains. Frog populations on the eastern slopes are relatively healthy, but about eight species are declining on the western slopes, where prevailing winds carry residues of 15 different pesticides from the Central Valley's agricultural region. Agricultural chemicals are also implicated in amphibian declines on the Eastern Shore of Maryland and in Ontario, Canada.

Another possible culprit is increased UV radiation caused by ozone thinning. Species suffering declines may have a limited ability to repair DNA damage caused by natural UV radiation. Researchers at Oregon State University exposed the eggs of three frog species to natural radiation. Egg survival was high for the Pacific tree frog, which had the greatest repair ability and is not in decline. Egg survival was much less (only 45 to 65%) for the western toad and Cascades frog, both of which have declining populations; egg survival increased dramatically when eggs were shielded from UV radiation.

Infectious diseases may explain some of the declines. In Australia, data suggest that a fungus (a chytrid; see Chapter 25) is responsible for massive declines in more than 12 species. Laboratory studies show that the chytrid kills healthy frogs. This fungus may also contribute to some of the frog declines observed in Central America and the United States.

Golden toads and about 20 other frog species that have disappeared in mountain areas of Costa Rica may have succumbed to climate warming. In recent years, increasing global temperatures have reduced moisture levels in the cloud forests of Costa Rica's central highlands. The organisms in these tropical mountain forests depend on the regular formation of clouds and mist, particularly during the dry winter season. As the equatorial Pacific Ocean surface has warmed in recent years, scientists have correlated less frequent dry-season mist with declines in the abundance of many animal species, including the frogs that disappeared.

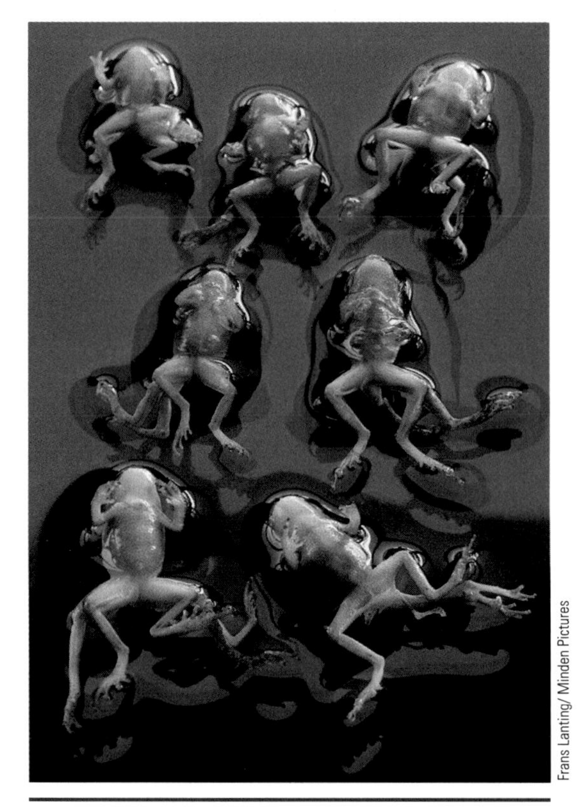
Frans Lanting/ Minden Pictures

Frog deformities. Pollution, ultraviolet radiation, and parasites are implicated in the recent widespread increase in developmental abnormalities in amphibians.

Amphibian deformities

In 1995, while on a field trip to a local pond, schoolchildren in Minnesota discovered that almost half of the leopard frogs they caught were deformed. The children found frogs with extra legs, extra toes, eyes located on the shoulder or back, deformed jaws, missing legs, missing toes, and missing eyes (see photograph). Deformed frogs usually die early, before they reproduce. Predators easily catch frogs with extra or missing legs. Since the children's discovery made headlines, 46 states have reported abnormally large numbers of deformities (as high as 60% in some local populations) in more than 60 amphibian species. Canada has also reported frog deformities, as have other countries.

Biologists have investigated many possible causes that produce amphibian deformities during development. These include exposure to chemicals such as pesticides, to increased ultraviolet (UV) light caused by thinning of the ozone layer, and to parasites. Several pesticides affect normal development in frog embryos. Lab experiments using traces of the pesticide *S*-methoprene and its breakdown products have caused frog deformities similar to those in Vermont ponds close to where *S*-methoprene is used to control mosquitoes and fleas. Ultraviolet radiation causes mutations in DNA. Laboratory tests have also demonstrated that infecting tadpoles with a parasitic flatworm causes them to develop deformities like the ones observed in field sites in Santa Clara County, California. No single factor explains the deformities found in all locations. Furthermore, human activities, such as habitat loss and air and water pollution, may exacerbate the problem. An amphibian stressed by pesticide residues, high UV, or drought, for example, may be more susceptible to a parasite.

genetic variety within a species, both among individuals within a given population and among geographically separate populations. (An individual species may have hundreds of genetically distinct populations.) *Species diversity* includes all the different species of archaebacteria, eubacteria, protists, plants, fungi, and animals on Earth. *Ecosystem diversity* is the variety of interactions within and among the different ecosystems. Biologists must consider all three levels of biological diversity as they address the human impact on biodiversity. For example, the disappearance of populations (that is, a decline in genetic diversity) indicates an increased risk that a species will become extinct (a decline in species diversity).

Biological diversity is currently decreasing at an alarming rate (see *Focus On: Declining Amphibian Populations* on the facing page). Conservation biologists estimate that species are presently becoming extinct at a rate 50 to 500 times the natural rate of background extinctions. According to the UN Global Biodiversity Assessment, more than 31,000 plant and animal species are currently threatened with extinction. Other surveys indicate that this number is seriously underestimated. For example, the World Conservation Union Red List of Threatened Plants lists about 34,000 species of plants currently threatened with extinction. Because Red Lists are not yet available for many tropical countries, it is difficult to know the true scale of the biodiversity crisis.

To dramatize the problem, Harvard University biologist Edward O. Wilson has compiled a list of the "Hundred Heartbeat Club," consisting of critically endangered animal species that have populations of 100 or fewer individuals and are just a few "heartbeats" from extinction. Members of the Hundred Heartbeat Club include the Javan rhinoceros, the Chinese river dolphin, the Hawaiian crow, and the Philippine eagle.

Many biologists fear we have entered the greatest period of mass extinction in Earth's history, but the current situation differs from previous periods of mass extinction in several respects. First, its cause is directly attributable to human activities. Second, it is occurring in a tremendously compressed period (just a few decades, as opposed to hundreds of thousands of years), much faster than rates of speciation (or replacement). Perhaps even more sobering, larger numbers of plant species are becoming extinct today than in previous mass extinctions. Because plants are the base of terrestrial food webs, extinction of animals that depend on plants isn't far behind. It is crucial that we determine how the loss of biodiversity affects the stability and functioning of ecosystems, which make up our life support system.

According to the U.S. Endangered Species Act, a species is designated as **endangered** when its numbers are so severely reduced that it is in imminent danger of extinction throughout all or a significant part of its range. (The area in which a particular species is found is its **range.**) Unless humans intervene, an endangered species will probably become extinct.

When extinction is less imminent but the population of a particular species is quite small, the species is classified as **threatened.** A threatened species is likely to become endangered in the foreseeable future, throughout all or a significant part of its range. Endangered or threatened species represent a decline in biological diversity, because they have severely diminished genetic diversity. Endangered and threatened species are at greater risk of extinction than species with greater genetic variability, because long-term survival and evolution depend on genetic diversity (see the section on genetic drift in Chapter 18).

Human activities contribute to declining biological diversity

Species become endangered and extinct for a variety of reasons, including the destruction or modification of habitats and the production of pollution. Humans also upset the delicate balance of organisms in a given area by introducing new, foreign species or by controlling native pests or predators. Illegal hunting and uncontrolled commercial harvesting are also factors. Table 55-1 summarizes an extensive study of the human threats to biological diversity in the United States. These data illustrate the overall importance of habitat loss and degradation. Some interesting variations among the groups of organisms are also apparent. Pollution, for example, is a very serious threat to fishes, reptiles, and invertebrates but a relatively minor threat to plants.

To save species, we must protect their habitats

Most species facing extinction today are endangered by destruction of natural habitats. Building roads, parking lots, and buildings; clearing forests to grow crops or graze domestic animals; and logging forests for timber all take their toll on natural habitats. Draining marshes converts aquatic habitats to terrestrial ones, whereas building dams and canals floods terrestrial habitats (Fig. 55-2). Habitat destruction threatens the survival of species because most organisms require a particular type of en-

TABLE 55-1 Percentages of Imperiled* U.S. Species That Are Threatened by Various Human Activities

Activity	All Species (1880)†	Plants (1055)	Mammals (85)	Birds (98)	Reptiles (38)	Fishes (213)	Invertebrates (331)
Habitat loss/ degradation	85‡	81	89	90	97	94	87
Exotic species	49	57	27	69	37	53	27
Pollution	24	7	19	22	53	66	45
Overexploitation	17	10	45	33	66	13	23

*This includes species classified as imperiled by The Nature Conservancy and all species listed as endangered or threatened under the Endangered Species Act or formally proposed for listing.

†Numbers in parentheses are the total number of species evaluated. The "All Species" category represents about 75% of imperiled species in the United States.

‡Because many of the species are affected by more than one human activity, the percentages in each column do not add up to 100.

Source: Adapted from Table 2 in D.S. Wilcove, et al., "Quantifying Threats to Imperiled Species in the United States," *BioScience,* Vol. 48, No. 8, Aug. 1998.

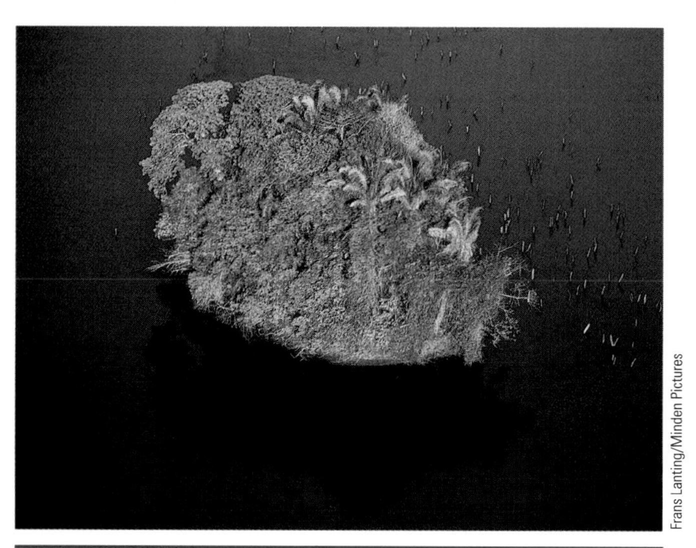

Frans Lanting/Minden Pictures

FIGURE **55-2** | **Habitat destruction.**

This tiny island lies in the Panama Canal. It was once a hilltop in a forest flooded when the Canal was built.

vironment, and habitat destruction reduces their biological range and ability to survive.

Humans often leave small, isolated patches of natural landscape that roads, fields, and buildings completely surround. Like a landmass surrounded by water, an isolated habitat surrounded by an expanse of unsuitable territory is referred to as an *island* (recall the discussion of isolated island communities in Chapter 52). Species from the surrounding "developed" landscape may intrude into the isolated habitat. Species that prefer the isolated habitat may occur in greatly reduced numbers or, if they require a large patch of undisturbed habitat, disappear altogether. As a result, habitat fragments often support only a fraction of the species found in the original, unaltered environment. Extensive evidence documents the effects of **habitat fragmentation,** the breakup of large areas of habitat into small, isolated patches.

Human activities that produce acid precipitation and other forms of pollution indirectly modify habitats left undisturbed and in their natural state. Acid precipitation has contributed to the decline of large stands of forest trees and to the biological death of many freshwater lakes, for example, in the Adirondack Mountains and in Nova Scotia. Other types of pollutants also adversely affect organisms. Such pollutants include industrial and agricultural chemicals, organic pollutants from sewage, acid wastes seeping from mines, and thermal pollution from the heated wastewater of industrial plants.

To save native species, we must control invasions of invasive species

Biotic pollution, the introduction of a foreign species into an area where it is not native, often upsets the balance among the organisms living in that area. If the foreign species causes economic or environmental harm, it is known as an **invasive species.** An invasive species may prey on native species or compete with them for food or habitat. Generally, a foreign competitor or predator harms local organisms more than do native competitors or predators. (Most invasive species lack natural agents, such as parasites, predators, and competitors, that would otherwise control them. Also, without a shared evolutionary history, most native species typically are less equipped to cope with invasive species.) Although foreign species sometimes spread into new areas on their own, humans are usually responsible for such introductions, either knowingly or accidentally.

One of North America's greatest biological threats is the zebra mussel, a native of the Caspian Sea. It was probably introduced by a foreign ship that flushed ballast water into the Great Lakes in 1985 or 1986. Since then, the tiny freshwater mussel, which clusters in extraordinary densities, has massed on hulls of boats, piers, buoys, water intake systems and, most damaging of all, on native clam and mussel shells. The zebra mussel's strong appetite for algae, phytoplankton, and zooplankton is also cutting into the food supply of native fishes, mussels, and clams, threatening their survival. By 1994 the zebra mussel had spread from the Great Lakes into the Mississippi River. It is currently found as far south as New Orleans, as far north as Minnesota and Quebec as far east as the Hudson River in New York. According to the U.S. Coast Guard, the United States spends about $5 billion each year to control the spread of the zebra mussel and to repair damage such as clogged pipes (Fig. 55-3).

In the mid-1980s an aggressive aquarium-bred strain of alga known as *Caulerpa* was accidentally released into the Mediterranean Sea when a seaside aquarium cleaned out its tanks. The alga has densely carpeted the floor of the Mediterranean,

FIGURE **55-3** | **Zebra mussels clog a pipe.**

Zebra mussels (*Dreissena polymorpha*) have caused billions of dollars in damage in addition to displacing native clams and mussels.

Illinois Department of Conservation

crowding out biologically diverse seafloor communities of native sea grasses, sponges, corals, sea fans, anemones, sea stars, and lobsters. *Caulerpa* is also toxic to many Mediterranean species. Other regions far from the Mediterranean are concerned that *Caulerpa* could cause havoc to their marine ecosystems. *Caulerpa* is now established off the coast of Australia and, more recently, appeared along the California coast.

Islands are particularly susceptible to the introduction of invasive species. In Hawaii, the introduction of sheep has imperiled both the mamane tree (because the sheep eat it) and a species of honeycreeper, an endemic bird that relies on the tree for food. Hawaii's plants evolved in the absence of herbivorous mammals and therefore have no defenses against introduced sheep, pigs, goats, and deer (see discussion of plant adaptations against herbivores in Chapter 52.)

Other human activities affect biodiversity directly or indirectly

Sometimes species become endangered or extinct as a result of deliberate efforts to eradicate or control their numbers, often because they prey on game animals or livestock. In the past, ranchers, hunters, and government agents decimated populations of large predators such as the wolf, mountain lion, and grizzly bear. Some animals are killed because their lifestyles cause problems for humans. The Carolina parakeet, a beautiful green, red, and yellow bird endemic to the southeastern United States, was extinct by 1920, exterminated by farmers because it ate fruit from their trees.

Ranchers and farmers poisoned and trapped prairie dogs and pocket gophers so extensively that they disappeared from most of their original range. Because numbers of prairie dogs sharply decreased, the black-footed ferret, a natural predator of these animals, became endangered. By the winter of 1985 to 1986, only 10 such ferrets were known to exist: 4 in Wyoming and 6 in captivity. A successful captive-breeding program enabled biologists to release black-footed ferrets in parts of South Dakota, Wyoming, and Montana, although the Wyoming population died out in an outbreak of plague. Recovery efforts continue in Colorado, Montana, and New Mexico, although ferrets remain vulnerable to human development of the prairie habitat throughout these areas.

Unregulated hunting, or overhunting, has caused the extinction of certain species in the past but is now strictly controlled in most countries. The passenger pigeon was one of the most common birds in North America in the early 1800s, but a century of overhunting resulted in its extinction in the early 1900s.

Illegal commercial hunting (poaching) continues to endanger a number of larger animals such as the tiger, cheetah, and snow leopard, whose beautiful furs are quite valuable. Rhinoceroses are slaughtered for their horns (used for ceremonial dagger handles in the Middle East and for purported medicinal purposes in Asian medicine). Bears are killed for their gallbladders (which Asian doctors use to treat ailments ranging from indigestion to hemorrhoids). Bush meat—meat from rare primates, elephants, anteaters, and such—is sold to urban restaurants. Although laws protect these animals, demand for their products on the black market has promoted illegal hunting, particularly in impoverished countries where a sale of contraband products can support a family for months.

Commercial harvest removes live organisms from the wild. Most organisms that are commercially harvested end up in pet stores. Several million birds are commercially harvested each year for the pet trade, but, many die in transit, and many more die from improper treatment in their owners' homes. At least 40 parrot species are now threatened or endangered, in part because of commercial harvest. Although it is illegal to capture endangered animals from the wild, a thriving black market exists, mainly because collectors in the United States, Europe, and Japan pay extremely large sums for rare tropical birds (Figure 55-4).

Commercial harvest also threatens plants. A number of unique or rare plants have been so extensively collected from the wild that they are now classified as endangered. These include certain carnivorous plants, cacti, and orchids. In contrast, carefully monitored and regulated commercial use of animal and plant resources creates an economic incentive to ensure that these resources do not disappear.

Where is the problem of declining biological diversity greatest?

Declining biological diversity is a concern throughout the United States. However, according to a 2002 study by the conservation

FIGURE **55-4** | **Illegal commercial harvesting.**

These hyacinth macaws (*Anodorhynchus hyacinthus*) were seized in French Guiana in South America as part of the illegal animal trade there. The hyacinth macaw population is much reduced in South America. Hyacinth macaws, which are the largest parrot species in the world, are typically about 100 cm (40 in) long.

Jany Sauvanet/NHPA

organization Nature Serve, declining biological diversity is most serious in Hawaii, with 63% of its native species at risk, and California, with 29% of its species at risk. Other states with a significant number of imperiled species are Nevada, Alabama, Utah, and Florida.

As serious as declining biological diversity is in the United States, it is even more serious abroad. Ecosystem loss and degradation are occurring in many places around the world, but tropical rain forests are being destroyed faster than almost all other ecosystems. The forests are giving way to human settlements, banana plantations, oil and mineral explorations, and other human activities (discussed later in the chapter). Tropical rain forests are home to thousands or even millions of the world's species. Many species in tropical rain forests are endemic; the clearing of tropical rain forests therefore contributes to their extinction.

Conservation biology addresses the issue of declining biological diversity

Conservation biology is the scientific study of how humans impact organisms and of the development of ways to protect biological diversity. Conservation biology is a broad discipline that ranges from safeguarding populations of endangered species, to preserving entire ecosystems and landscapes. It includes two problem-solving approaches that save organisms from extinction: in situ and ex situ conservation.

In situ conservation, which includes the establishment of parks and reserves, concentrates on preserving biological diversity in nature. A high priority of in situ conservation is identifying and protecting sites that harbor a great deal of diversity. With increasing demands on land, however, in situ conservation can't preserve all types of biological diversity. Sometimes only ex situ conservation can save a species.

Ex situ conservation conserves individual species in human-controlled settings. Breeding captive species in zoos and storing seeds of genetically diverse plant crops are examples of ex situ conservation (see *Focus On: Seed Banks* in Chapter 35).

In situ conservation is the best way to preserve biological diversity

Conservation biologists maintain that protecting animal and plant habitats—that is, conserving and managing the ecosystem as a whole—is the single best way to protect biological diversity. Many nations appreciate the need to protect their biological heritage and for this reason have set aside areas for wildlife habitats. Such natural ecosystems offer the best strategy for the long-term protection and preservation of biological diversity. Currently more than 3000 national parks, marine sanctuaries, wildlife refuges, forests, and other areas are protected throughout the world. These encompass some 1 billion hectares, an area almost as large as Canada.

Many protected areas have multiple uses that sometimes conflict with the goal of preserving species. National parks provide recreational needs, for example, whereas national forests are used for logging, grazing, and mineral extraction. The mineral rights to many wildlife refuges are privately owned, and some wildlife refuges have had oil, gas, and other mineral development.

Protected areas are not always effective in preserving biological diversity, particularly in developing countries where biological diversity is greatest, because there is little money or expertise to manage them. Another shortcoming of the world's protected areas is that many are in lightly populated mountain areas, tundra, and the driest deserts, places that often have spectacular scenery but relatively few kinds of species. (Such remote areas are often designated reserves because they are unsuitable for commercial development.) In contrast, ecosystems in which biological diversity is greatest often receive little attention. Protected areas are urgently needed in tropical rain forests, the tropical grasslands and savannas of Brazil and Australia, and dry forests that are widely scattered around the world. Desert organisms are underprotected in northern Africa and Argentina, and the species of many islands and temperate river basins also need protection.

These areas are part of what biologists have identified as the world's 25 *biodiversity hotspots* (Fig. 55-5). The hotspots collectively make up 1.4% of Earth's land but contain as many as 44% of all vascular plant species, 29% of the world's endemic bird species, 27% of endemic mammals species, 38% of endemic reptile species, and 53% of endemic amphibian species. Many hotspots are also densely populated by humans.

Restoring damaged or destroyed habitats is the goal of restoration ecology

Although preserving habitats is an important part of conservation biology, the realities of our world, including the fact that the land-hungry human population continues to increase, dictate a variety of other conservation measures. Sometimes scientists reclaim disturbed lands and convert them into areas with high biological diversity. **Restoration ecology,** in which the principles of ecology are used to return a degraded environment as close as possible to its former state, is an important part of in situ conservation.

Since the 1930s the University of Wisconsin–Madison Arboretum has carried out one of the most famous examples of ecological restoration (Fig. 55-6). Several distinct natural communities have been carefully developed on damaged agricultural land. These communities include a tallgrass prairie, a xeric (dry) prairie, and several types of pine and maple forests native to Wisconsin.

Restoration of disturbed lands not only creates biological habitats but also has additional benefits such as the regeneration of soil that agriculture or mining damaged. The disadvantages of restoration ecology include the time and expense required to restore an area. Nonetheless, restoration ecology is an important aspect of conservation biology, as it is thought that restoration will deter many extinctions.

Ex situ conservation attempts to save species on the brink of extinction

Zoos, aquaria, and botanical gardens often attempt to save certain endangered species from extinction. Eggs are collected from

KEY CONCEPT: Certain areas of the world—biodiversity hotspots—are critically important because they contain a disproportionate number of the world's endemic species.

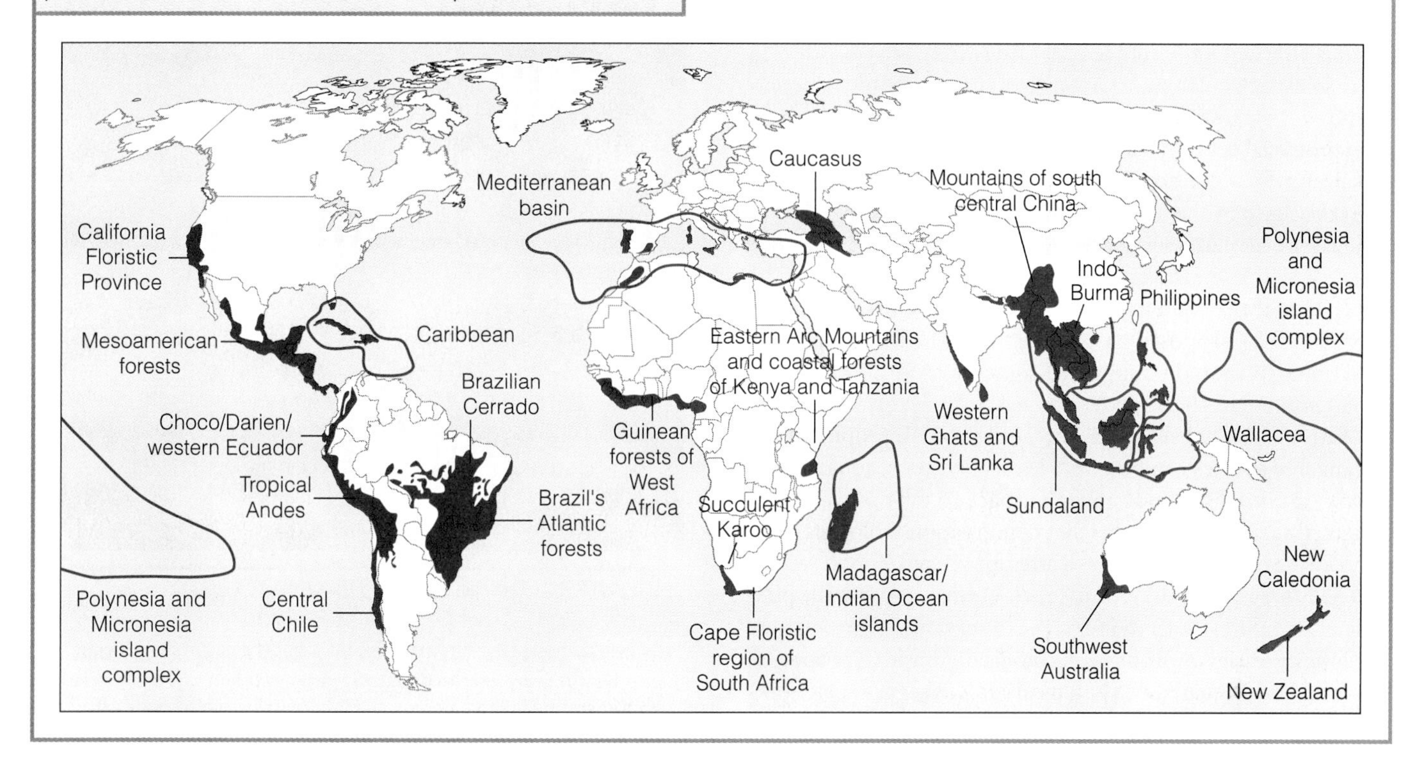

FIGURE **55-5** **Biodiversity hotspots.**

Rich in endemic species, these hotspots are under pressure from the number of humans living in them. *(Data from Conservation International)*

FIGURE **55-6** **Restoring damaged lands.**

(a) The restoration of the prairie by the University of Wisconsin–Madison Arboretum was at an early stage in 1935. The men are digging holes to plant prairie grass sod. **(b)** The prairie as it looks today. This picture was taken at about the same location as the 1935 photograph.

nature, or the remaining few animals are captured and bred in zoos and other research facilities.

Special techniques, such as artificial insemination and host mothering, increase the number of offspring. In **artificial insemination,** sperm is collected from a suitable male of a rare species and used to impregnate a female, perhaps located in another zoo in a different city or even in another country. In **host mothering,** a female of a rare species is treated with fertility drugs, which cause her to produce multiple eggs. Some of these eggs are collected, fertilized with sperm, and surgically implanted into females of a related but less rare species, who later give birth to offspring of the rare species (see *Focus On: Novel Origins* in

Courtesy of the University of Wisconsin–Madison Arboretum

(a)

Courtesy of Virginia Kline

(b)

Chapter 48). Plans are underway to clone endangered species, such as the giant panda, that do not reproduce well in captivity. Another technique involves hormone patches, which are being developed to stimulate reproduction in endangered birds; the patch is attached under the female bird's wing.

A few spectacular successes have occurred in captive breeding programs, in which large enough numbers of a species have been produced to reestablish small populations in the wild. Conservation efforts, for example, let the bald eagle make a remarkable comeback in the lower 48 states, from 417 nesting pairs in 1963 to more than 6000 pairs in 2003. In 1994 the bald eagle was removed from the endangered list and transferred to the less critical threatened list. In 1999, the U.S. Fish and Wildlife Service (FWS) proposed its removal from the threatened list, an action that should be approved sometime soon.

Attempting to save a species on the brink of extinction is usually extremely expensive; therefore, only a small proportion of endangered species can be saved. Moreover, zoos, aquaria, and botanical gardens do not have the space to try to save all endangered species. This means that conservation biologists must set priorities on which species to attempt to save. Zoos have traditionally focused on large, charismatic animals because the public is more interested in them. Such conservation efforts ignore millions of less glamorous but ecologically important species. However, zoos, aquaria, and botanical gardens serve a useful purpose because they educate the public about the value of conservation biology. Clearly, controlling unchecked human development so that species do not become endangered in the first place is a more effective way to protect and maintain natural habitat.

Conservation organizations are essential to conservation biology

Conservation organizations are an important part of the effort to maintain biological diversity. They educate policymakers and the public about the importance of biological diversity. In certain instances, conservation organizations serve as catalysts, galvanizing public support for important biodiversity preservation efforts. They also provide financial support for conservation projects, from basic research to the purchase of land that is a critical habitat for a particular organism or group of organisms.

The World Conservation Union (IUCN)[1] helps countries with conservation biology projects. It and other conservation organizations are currently assessing how effective established wildlife refuges are in maintaining biological diversity. The World Wildlife Fund[2] and Brazil's National Institute for Amazon Research are conducting the Minimum Critical Size of Ecosystems Project, a long-term study of the effects of fragmentation on the Amazonian rain forest in Brazil (Fig. 55-7). In addition, IUCN and the World Wildlife Fund have identified which biomes and ecosystems do not yet have protected areas.

[1] Formerly called the International Union for Conservation of Nature and Natural Resources, the World Conservation Union still uses the abbreviation IUCN.

[2] Outside of the United States, Canada, and Australia, the World Wildlife Fund is known as the World Wide Fund for Nature.

R.O. Bierregaard

FIGURE **55-7** **Minimum Critical Size of Ecosystems Project.**

Shown are 1-hectare and 10-hectare plots of a long-term study underway in Brazil on the effects of fragmentation on Amazonian rain forest. Plots with an area of 100 hectares are also under study, along with identically sized sections of intact forest, which are controls. Preliminary data indicate that the smaller forest fragments don't maintain their ecological integrity. For example, large trees near forest edges often die or are damaged from exposure to wind, desiccation (from lateral exposure of the forest fragment to sunlight), and invasion by parasitic woody vines.

The Endangered Species Act provides some legal protection for species and habitats

In 1973 the **Endangered Species Act (ESA)** was passed in the United States, authorizing the FWS to protect endangered and threatened species in the United States and abroad. Many other countries now have similar legislation. The FWS conducts a detailed study of a species to determine if it should be listed as endangered or threatened. Since the passage of the ESA, more than 1200 species in the United States have been listed as endangered or threatened (Table 55-2). The ESA provides legal protection to listed species so that their danger of extinction is reduced. For example, the act makes it illegal to sell or buy any product made from an endangered or threatened species.

The ESA requires officials of the FWS to select critical habitats and design a recovery plan for each species listed. The recovery plan includes an estimate of the current population size, an analysis of what factors contributed to its endangerment, and a list of activities that may help the population recover. As of early 2001, there were 746 recovery plans. Some recovery plans cover more than one species, whereas a few species have two or more recovery plans for different parts of their ranges. The 1995 reintroduction of gray wolves to Yellowstone National Park is part of the recovery plan for that species, and in 2003 the wolf

TABLE 55-2 Organisms Listed as Endangered or Threatened in the United States, 2003

Type of Organism	Number of Endangered Species	Number of Threatened Species
Mammals	65	9
Birds	78	14
Reptiles	14	22
Amphibians	12	9
Fishes	71	44
Snails	21	11
Clams	62	8
Crustaceans	18	3
Insects	35	9
Spiders	12	0
Flowering plants	571	144
Conifers and cycads	2	1
Ferns and other plants	24	2

Source: U.S. Fish and Wildlife Service.

was removed from the endangered list and placed on the less critical threatened list.

The Endangered Species Act, which was amended in 1982, 1985, and 1988, is considered one of the strongest pieces of environmental legislation in the United States, in part because species are designated as endangered or threatened entirely on biological grounds. Currently, economic considerations can't influence the designation of endangered or threatened species. Biologists generally agree that as a result of passage of the ESA in 1973, fewer species became extinct than would have had the law never been passed.

The ESA is also one of the most controversial pieces of environmental legislation. For example, the ESA does not provide compensation for private property owners who suffer financial losses because they can't develop their land if a threatened or endangered species lives there. The ESA has also interfered with some federally funded development projects.

The ESA was scheduled for congressional reauthorization in 1992 but has been entangled since then in disagreements between conservation advocates and those who support private property rights. Conservation advocates think the ESA does not do enough to save endangered species, whereas those who own land on which rare species live think the law goes too far and infringes on property rights. Some critics—notably business interests and private property owners—view the ESA as an impediment to economic progress. Those who defend the ESA point out that of 34,000 past cases of endangered species versus development, only 21 cases weren't resolved by some sort of compromise. When the black-footed ferret was reintroduced on the Wyoming prairie, for example, it was classified as an "experimental, nonessential species" so that its reintroduction would not block ranching and mining in the area. Thus the ferret release program obtained the support of local landowners, support that was deemed crucial to ferret survival in nature.

This type of compromise is essential to the success of saving endangered species because, according to the U.S. General Accounting Office, more than 90% of endangered species live on at least some privately owned lands. Critics of the ESA think the law should be changed so that private landowners are given economic incentives to protect endangered species living on their lands. For example, tax cuts for property owners who are good land stewards could make the presence of endangered species on their properties an asset instead of a liability.

Defenders of the ESA agree that it is not perfect. Only eight U.S. species have recovered enough to be delisted, that is, no longer classified as endangered or threatened. However, the FWS says that hundreds of listed species are stable or improving; they expect as many as several dozen additional species to be delisted over the next several years. The law is geared more to saving a few popular or unique endangered species rather than the much larger number of less glamorous species that perform valuable **ecosystem services.** Yet it is the less glamorous organisms such as plants, fungi, and insects that play central roles in ecosystems and contribute most to their functioning.

Conservationists would like to see the ESA strengthened in such a way as to manage whole ecosystems and maintain complete biological diversity rather than attempting to save endangered species as isolated entities. This approach offers collective protection to many declining species rather than to single species.

International agreements protect species and habitats

At the international level, 160 countries participate in the **Convention on International Trade in Endangered Species of Wild Flora and Fauna (CITES),** which went into effect in 1975. Originally drawn up to protect endangered animals and plants considered valuable in the highly lucrative international wildlife trade, CITES bans hunting, capturing, and selling of endangered or threatened species and regulates trade of organisms listed as potentially threatened. Unfortunately, enforcement of this treaty varies from country to country, and even where enforcement exists, the penalties are not very severe. As a result, illegal trade in rare, commercially valuable species continues.

The goals of CITES often stir up controversy over such issues as who actually owns the world's wildlife, and whether global conservation concerns take precedence over competing local interests. These conflicts often highlight socioeconomic differences between wealthy consumers of CITES products and poor people who trade the endangered organisms.

The case of the African elephant illustrates these controversies. Listed as an endangered species since 1989 to halt the slaughter of elephants for the ivory trade, the species seems to have recovered in southern Africa (Namibia, Botswana, and Zimbabwe). When elephant populations grow too large for their habitat, they root out and knock over so many smaller trees that the forest habitat can support fewer other species. Organizations such as the Humane Society in the United States are developing a birth control vaccine to reduce elephant births. However, the African people living near the elephants want to cull the herd periodically to sell elephant meat, hides, and ivory for profit. In

1997, CITES transferred elephant populations in Namibia, Botswana, and Zimbabwe to a less restrictive, potentially threatened listing to allow trade of stockpiled ivory to Japan. Some conservationists oppose the resumption of ivory trade, although others think that cooperating with local hunters and traders serves conservation goals better than treating these groups as outlaws.

As of late 2003, 187 nations had ratified the Convention on Biological Diversity treaty, a product of the 1992 Earth Summit that is now considered binding. (The United States signed it in 1993 but has not ratified it.) Under the conditions of the treaty, each signatory nation must inventory its own biodiversity and develop a **national conservation strategy,** a detailed plan for managing and preserving the biological diversity of that specific country.

Review

- Which organism is more likely to become extinct, an endangered species or a threatened species? Explain your answer.
- How does habitat fragmentation contribute to declining biological diversity?
- How do invasive species contribute to the biodiversity crisis?
- Which type of conservation measure, in situ or ex situ, helps the greatest number of species? Why?

BiologyNow™ Assess your understanding of **the biodiversity crisis** by taking the pretest on your BiologyNow CD-ROM.

DEFORESTATION

Learning Objectives

5 Discuss the ecosystem services of forests, and describe the consequences of deforestation.

6 State at least three reasons why forests are disappearing today.

The most serious problem facing the world's forests is **deforestation,** which is defined as the temporary or permanent clearance of forests for agriculture or other uses (Fig. 55-8). The World Commission on Forests and Sustainable Development, formed after the Earth Summit in 1992, released its final report in 1999, in which it concluded that 15 million hectares (37 million acres) of forest are destroyed *each year.* Human destruction of forest includes setting fires to clear land, expanding agriculture, constructing roads in forests, and harvesting trees.

When forests are destroyed, they no longer make valuable contributions to the environment or to the people who depend on them. Deforestation increases soil erosion and thus decreases soil fertility. Soil erosion causes increased sedimentation of waterways, which harms downstream aquatic ecosystems by reducing light penetration, covering aquatic organisms, and filling in waterways. Uncontrolled soil erosion, particularly on steep deforested slopes, causes mud flows that endanger human lives and property and reduces production of hydroelectric power as silt builds up behind dams. In drier areas, deforestation can lead to the formation of deserts.

Deforestation contributes to the loss of biological diversity. In particular, many tropical species have limited ranges within a forest, so they are especially vulnerable to habitat destruction or modification. Migratory species, such as birds and butterflies, also suffer because of tropical deforestation.

Gary Braasch/Tony Stone Images

FIGURE **55-8** | **Deforestation.**
Aerial view of clear-cut areas in the Gifford Pinchot National Forest in southwest Washington state. The lines are roads built at taxpayer expense to haul away logs.

By trapping and absorbing precipitation, forests, particularly on hillsides and mountains, help protect nearby lowlands from floods. When a forest is cut down, the watershed can't absorb and hold water as well, and the total amount of surface runoff flowing into rivers and streams increases. This not only causes soil erosion but puts lowland areas at extreme risk of flooding.

Deforestation may affect regional and global climate changes. Transpiring trees release substantial amounts of moisture into the air. This moisture falls back to the surface in the hydrologic cycle (see Chapter 53). When a large forest is removed, rainfall may decline, and droughts may become common in that region. Studies suggest that the local climate has become drier in parts of Brazil where tracts of the rain forest have been burned. Temperatures may also rise slightly in a deforested area, because there is less evaporative cooling from the trees.

Moreover, deforestation may increase global temperature by releasing carbon originally stored in the trees into the atmosphere as carbon dioxide, which enables the air to retain heat (discussed later in the chapter). Researchers think the carbon in forests is released immediately if the trees are burned or more slowly when unburned parts decay. If trees are harvested and logs are removed, roughly one half of the forest carbon remains as dead materials (branches, twigs, roots, and leaves) that decompose, releasing carbon dioxide. When an old-growth forest is harvested, researchers estimate that it takes about 200 years for the replacement forest to accumulate the amount of carbon that was stored in the original forest.

Where and why are forests disappearing?

During the past 1000 years, deciduous forests in temperate areas were largely cleared for housing and agriculture. Today, however, deforestation in the tropics is proceeding much more rapidly and over a much larger area. Most of the remaining undisturbed tropical rain forests, in the Amazon and Congo River basins of South America and Africa, are being cleared and burned at a rate unprecedented in human history. Tropical rain forests are also being destroyed at an extremely rapid rate in southern Asia, Indonesia, Central America, and the Philippines.

Exact figures on rates of tropical forest destruction are unavailable. The Food and Agriculture Organization (FAO) of the United Nations, which provides world statistics on forest cover, released its most recent assessment of deforestation in 117 tropical countries in 1999. The FAO estimated an average annual forest loss of 0.7% per year, and some areas, such as continental Southeast Asia, experienced a forest loss estimated as high as 1.6% per year. If this rate of deforestation—which represents an annual loss of 12.6 million hectares (31.1 million acres)—continues, tropical forests will be nearly gone in the first half of the 22nd century.

Several studies show a strong statistical correlation between population growth and deforestation. More people need more food, so they clear forests for agricultural expansion. However, tropical deforestation is a complex problem that can't be attributed simply to population pressures. The main causes of deforestation vary from place to place, and a variety of economic, social, and governmental factors interact to cause deforestation. Government policies sometimes provide incentives that favor the removal of forests. For example, in the late 1950s the Brazilian government constructed the Belem-Brasilia Highway, which cut through the Amazon Basin and opened the Amazonian frontier for settlement. Sometimes economic conditions encourage deforestation. The farmer who converts more forest to pasture can maintain a larger herd of cattle, which is a good hedge against inflation.

Keeping in mind that tropical deforestation is a complex problem, three agents are probably the most immediate causes of deforestation in tropical rain forests: subsistence agriculture, commercial logging, and cattle ranching. Other reasons for the destruction of tropical forests include the development of hydroelectric power, which inundates large areas of forest; mining, particularly when ore smelters burn charcoal produced from rainforest trees; and plantation-style agriculture of crops such as citrus fruits and bananas. The main cause of deforestation in tropical dry forests is fuel wood consumption.

Subsistence agriculture, in which a family produces enough food to feed itself, accounts for perhaps 60% of tropical deforestation. In many developing countries where tropical rain forests are located, the majority of people do not own the land that they live and work on. For example, in Brazil only 5% of the farmers own 70% of the land. Most subsistence farmers have no place to go except into the forest, which they clear to grow food. Land reform in Brazil, Madagascar, Mexico, the Philippines, Thailand, and many other countries would make the land owned by a few available to everyone, thereby easing the pressure of subsistence farmers on tropical forests. This scenario is unlikely, however, because wealthy landowners have more economic and political clout than impoverished peasants.

Subsistence farmers often follow loggers' access roads until they find a suitable spot. They first cut down the trees and allow them to dry, then they burn the area (Fig. 55-9) and plant crops immediately after burning. This is known as **slash-and-burn agriculture.** Yields from the first crop are often quite high, because the nutrients that were in the trees are now available in the soil. However, soil productivity declines rapidly, and subsequent crops are poor. In a few years the farmer must move to a new part of the forest and repeat the process. Cattle ranchers often claim the abandoned land for grazing, because land that is not fertile enough to support crops can still support livestock.

Dr. Nigel Smith/Earth Scenes

FIGURE **55-9** | **Clearing the forest.**
Tropical rain forest in Brazil is burned to provide agricultural land. This type of cultivation is known as slash-and burn agriculture.

Slash-and-burn agriculture done on a small scale, with periods of 20 to 100 years between cycles, is sustainable. The forest regrows rapidly after a few years of farming. But when millions of people try to obtain a living in this way, the land is not allowed to lie uncultivated long enough to recover.

About 20% of tropical deforestation is the result of commercial logging, and vast tracts of tropical rain forests, particularly in Southeast Asia, are harvested for export abroad. Most tropical countries allow commercial logging to proceed much faster than is sustainable, because it supplies them with much-needed revenues. In the final analysis, tropical deforestation does not contribute to economic development; rather, it reduces or destroys the value of an important natural resource.

Approximately 12% of tropical rainforest destruction is carried out to provide open rangelands for cattle. Cattle ranching is particularly important in Central America. Much of the beef raised on these ranches, which foreign companies often own, is exported to fast-food restaurant chains in North America and Europe. After the forests are cleared, cattle graze on the land for up to perhaps 20 years, after which time the soil fertility is depleted. When this occurs, shrubby plants, or scrub savanna, take over the range.

Review

- What are three ecosystem services that forests provide?
- What are two reasons for tropical deforestation?

Biology Now™ Assess your understanding of **deforestation** by taking the pretest on your BiologyNow CD-ROM.

GLOBAL WARMING

Learning Objectives

7 Name at least three greenhouse gases, and explain how greenhouse gases contribute to global warming.

8 Describe how global warming may affect sea level, precipitation patterns, organisms (including humans), and food production.

Assessment of Earth's average temperature is based on daily measurements taken at several thousand land-based meteorological stations around the world, as well as on data from weather balloons, orbiting satellites, transoceanic ships, and hundreds of sea-surface buoys with temperature sensors. Earth's average surface temperature increased 0.6°C (1.1 degree F) during the 20th century, and the 1990s was the warmest decade of the century (Fig. 55-10). The early 2000s continued the warming trend.

Other evidence also suggests an increase in global temperature. For example, several studies have documented that spring in the Northern Hemisphere now comes about 6 days earlier than it did in 1959, and autumn is delayed by 5 days. (Spring is determined when buds of specific plants open, and autumn when leaves of specific plants turn color and fall.) Since 1949 the United States has experienced an increased frequency of extreme heat stress events, which are extremely hot, humid days during summer months; medical records show that heat-related deaths among elderly and other vulnerable people increase during these events. The sea level has risen slightly (0.1 to 0.2 m, or 4 to 7.9 in, during the 20th century), mountain glaciers in nonpolar regions have retreated, and extreme weather events such as severe rainstorms have occurred with increasing frequency in certain regions.

FIGURE **55-10** **Mean annual global temperature, 1960 to 2002.**

Data are presented as surface temperatures (°C) for 1960, 1965, and every year thereafter. The measurements, which naturally fluctuate, clearly show the warming trend of the last several decades. *(Surface Air Temperature Analysis, Goddard Institute for Space Studies, NASA)*

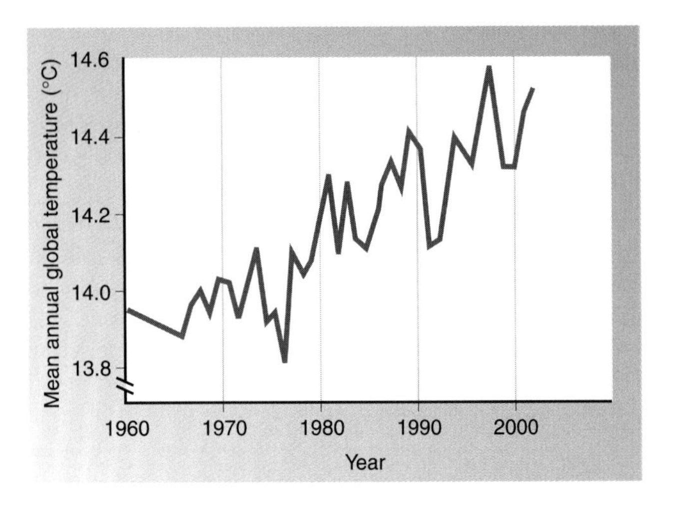

Scientists around the world have studied **global warming** for the past 50 years. As the evidence has accumulated, those most qualified to address the issue have reached a strong consensus that the 21st century will experience significant climate change, and human activities will be responsible for much of this change.

In response to this growing scientific consensus, governments around the world organized the UN Intergovernmental Panel on Climate Change (IPCC). With input from hundreds of climate experts, the IPCC provides the most definitive scientific statement about global warming. The 2001 IPCC Third Assessment Report concluded that human-produced air pollutants continue to change the atmosphere, causing most of the warming observed in the past 50 years. Scientists can identify the human influence on climate change despite questions about *how much* of the recent warming stems from natural variations. The IPCC report projects a 1.4- to 5.8-degree C (2.5- to 10.4-degree F) increase in global temperature by the year 2100, although the warming will probably not occur uniformly from region to region. Thus, Earth may become warmer during the 21st century than it has been for hundreds of thousands of years.

PROCESS OF SCIENCE

Almost all climate experts agree with the IPCC's assessment that the warming trend has already begun and will continue throughout the 21st century. However, scientists are uncertain over how rapidly the warming will proceed, how severe it will be, and where it will be most pronounced. (Recall that uncertainty and debate are part of the scientific process and that scientists can never claim to know a "final answer.") As a result of these uncertainties, many people, including policymakers, are confused about what we should do. Yet the stakes are quite high because human-induced global warming has the potential to disrupt Earth's climate for a very long time.

Greenhouse gases cause global warming

Carbon dioxide (CO_2) and certain other trace gases, including methane (CH_4), surface ozone (O_3),[3] nitrous oxide (N_2O), and chlorofluorocarbons (CFCs), are accumulating in the atmosphere as a result of human activities (Table 55-3). The concentration of atmospheric CO_2 has increased from about 288 parts per million (ppm) approximately 200 years ago (before the Industrial Revolution began) to 373 ppm in 2002 (Fig. 55-11). Burning carbon-containing fossil fuels—coal, oil, and natural gas—accounts for about three fourths of human-made carbon dioxide emissions to the atmosphere. Land conversion, such as when forests are logged or burned, also releases carbon dioxide.

[3] Surface ozone (more precisely, ozone in the troposphere) is a greenhouse gas as well as a component of photochemical smog. Ozone in the upper atmosphere, the stratosphere, provides an important planetary service that is discussed later in this chapter.

TABLE 55-3	Increase in Selected Atmospheric Greenhouse Gases, Preindustrial to Present	
Gas	**Estimated Preindustrial Concentration**	**Present Concentration**
Carbon dioxide	288 ppm*	373 ppm§
Methane	848 ppb†	1780 ppb
Nitrous oxide	285 ppb	315 ppb
CFC-12	0 ppt‡	541 ppt
CFC-11	0 ppt	262 ppt

*ppm = parts per million.

†ppb = parts per billion.

‡ppt = parts per trillion.

§Derived from in situ sampling at Mauna Loa, Hawaii. All other data from Mace Head, Ireland, monitoring site.

Source: Carbon Dioxide Information Analysis Center, Environmental Sciences Division, Oak Ridge National Laboratory.

FIGURE **55-11** **Carbon dioxide in the atmosphere, 1958 to 2002.**

Note the steady increase in the concentration of atmospheric carbon dioxide since 1958, when measurements began at the Mauna Loa Observatory, Hawaii. This location is far from urban areas where factories, power plants, and motor vehicles emit carbon dioxide. The fluctuations correspond to the annual cycle of photosynthesis in the Northern Hemisphere: winter (a high level of carbon dioxide), when plants are not actively growing and absorbing carbon dioxide, and summer (a low level of carbon dioxide), when they are growing and absorbing carbon dioxide. *(C.D. Keeling and T.P. Whorf, Scripps Institution of Oceanography, University of California, La Jolla, California)*

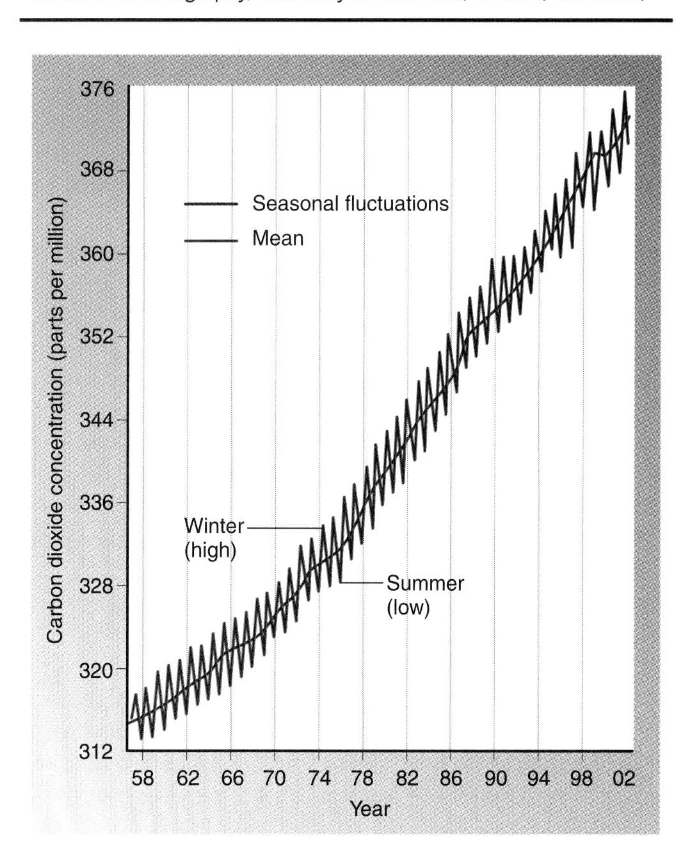

(The burning itself releases carbon dioxide into the atmosphere, and because trees normally remove carbon dioxide from the atmosphere during photosynthesis, tree removal prevents this process.) The levels of the other trace gases associated with global warming are also rising.

Global warming occurs because these gases absorb infrared radiation, that is, heat, in the atmosphere. This absorption slows the natural heat flow into space, warming the lower atmosphere. Some of the heat from the lower atmosphere is transferred to the ocean and raises its temperature as well. This retention of heat in the atmosphere is a natural phenomenon that has made Earth habitable for its millions of species. However, as human activities increase the atmospheric concentration of these gases, the atmosphere and ocean continue to warm, and the overall global temperature rises.

Because carbon dioxide and other gases trap the sun's radiation somewhat like glass does in a greenhouse, the natural trapping of heat in the atmosphere is called the **greenhouse effect,** and the gases that absorb infrared radiation are known as **greenhouse gases.** The additional warming produced when increased levels of gases absorb additional infrared radiation is called the **enhanced greenhouse effect** (Fig. 55-12).

Although current rates of fossil fuel combustion and deforestation are high, causing the carbon dioxide level in the atmosphere to increase markedly, scientists think the warming trend is slower than the increasing level of carbon dioxide might indicate. The reason is that water requires more heat to raise its temper-

ACTIVE FIGURE **55-12** **Enhanced greenhouse effect.**

Visible light passes through the atmosphere to Earth's surface. However, carbon dioxide and other greenhouse gases direct the longer wavelength heat (infrared radiation), which would normally be radiated to space, back to the surface. Therefore, the buildup of greenhouse gases in the atmosphere results in global warming.

BiologyNow™ Watch **the development of the greenhouse effect** by clicking on this figure on your BiologyNow CD-ROM.

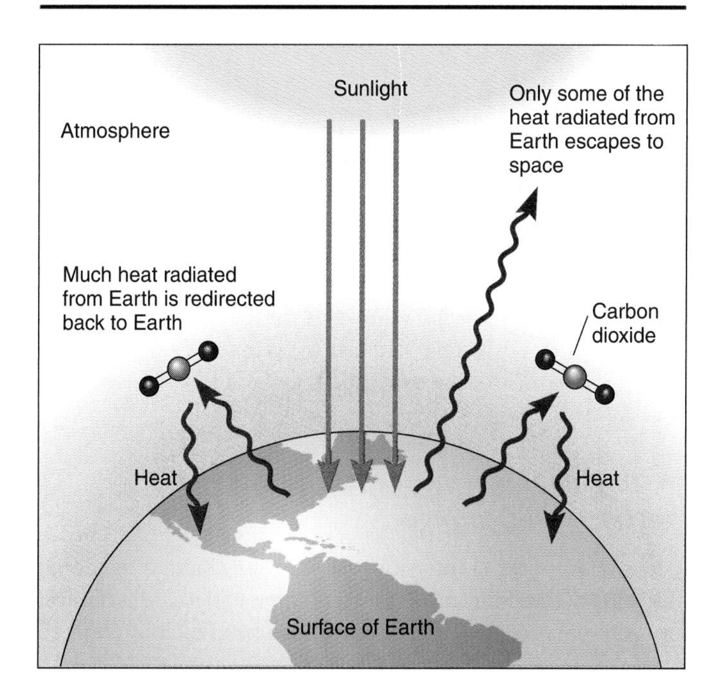

ature than gases in the atmosphere do (recall the high specific heat of water discussed in Chapter 2). As a result, the ocean takes longer to warm than the atmosphere to absorb heat, and most climate scientists think that warming will be more pronounced in the second half of the 21st century than in the first half.

What are the probable effects of global warming?

We now consider some of the probable effects of global warming, including changes in sea level; changes in precipitation patterns; effects on organisms, including humans; and effects on agriculture. These changes will persist for many centuries because many greenhouse gases remain in the atmosphere for hundreds of years. Furthermore, even after greenhouse gas concentrations have stabilized, scientists think Earth's mean surface temperature will continue to rise, because the ocean adjusts to climate change on a delayed time scale.

With global warming, the global average sea level is rising

If Earth's overall temperature increases by just a few degrees, there could be a major thawing of glaciers and the polar ice caps. During the Antarctic summer of 2002, most of the Larsen B ice shelf, an area roughly the size of Rhode Island, broke off the Antarctic Peninsula. (Antarctic ice shelves are thick sheets of floating ice that are fed mainly by glaciers that flow off the land.) This loss of ice coincided with a decades-long trend of atmospheric warming in the Antarctic. A study from the 1970s to the 1990s showed that the area of ice-covered ocean in the Arctic has also retreated. Mountain glaciers around the world are melting at accelerating rates.

In addition to sea-level rise caused by the retreat of glaciers and thawing of polar ice, the sea level will probably rise because of thermal expansion of the warming ocean. Water, like other substances, expands as it warms. As noted earlier, during the 20th century sea level rose between 0.1 and 0.2 m, mostly from thermal expansion.

The IPCC estimates that sea level will rise an additional 0.5 m (20 in) by 2100. Such an increase will flood low-lying coastal areas such as parts of southern Louisiana and South Florida. Coastal areas that are not inundated will more likely suffer erosion and other damage from more frequent and more intense weather events such as hurricanes. These likely effects are certainly a cause for concern, particularly because about two thirds of the world's population lives within 150 km (93 mi) of a coastline. Countries particularly at risk include Bangladesh, Egypt, Vietnam, Mozambique, and many island nations such as the Maldives.

With global warming, precipitation patterns will change

Global computer simulations of weather changes as global warming occurs indicate that precipitation patterns will change, causing some areas such as midlatitude continental interiors to have more frequent droughts. At the same time, more intense precipitation events (heavier snowstorms and rainstorms) may more frequently flood other areas. Changes in precipitation patterns could affect the availability and quality of fresh water in many places. Researchers project that areas that are currently arid or semiarid, such as the Sahel region just south of the Sahara Desert, will have the most troublesome water shortages as the climate changes. Closer to home, water experts predict water shortages in the American West, because warmer winter temperatures will cause more precipitation to fall as rain rather than snow; melting snow currently provides 70% of stream flows in the West during summer months.

The frequency and intensity of storms over warm surface waters may also increase. Scientists at the National Oceanic and Atmospheric Administration developed a computer model that examines how global warming may affect hurricanes. When the model was run with a sea surface temperature 2.2°C warmer than today, more intense hurricanes resulted. (The question of whether hurricanes will occur *more frequently* in a warmer climate remains uncertain.) Changes in storm frequency and intensity are expected because as the atmosphere warms, more water evaporates, which in turn releases more energy into the atmosphere (recall the discussion of water's heat of vaporization in Chapter 2). This energy generates more powerful storms.

With global warming, the ranges of organisms are changing

Biologists are examining some of the effects of global warming on organisms. For example, researchers determined that populations of zooplankton in the California Current have declined 80% since 1951, apparently because the current has warmed slightly. (The California Current flows from Oregon southward along the California coast.) The decline in zooplankton has affected the entire ecosystem's food web, and populations of seabirds and plankton-eating fishes have also declined. As in all studies of effects of warming on organisms in nature, researchers are unsure of the relative contributions of human production of greenhouse gases and of natural fluctuations in the climate; despite this uncertainty, the observed changes are real.

As temperatures have risen in Antarctic waters during the past two decades, a similar decline in shrimp-like krill has reduced Adélie penguin populations. Because there are fewer krill, the birds don't get enough food. Warmer temperatures in Antarctica—during the past 50 years the average annual temperature on the Antarctic Peninsula has increased 2.6 degrees C (5 degrees F)—have also contributed to reproductive failure in Adélie penguins. The birds normally lay their eggs in snow-free rocky outcrops, but the warmer temperatures have caused more snowfall (recall that warmer air holds more moisture), which melts when the birds incubate the eggs. The melted snow forms cold pools of slush that kill the developing chick embryos.

Each species reacts to changes in temperature differently. In response to global warming, some species will undoubtedly become extinct, particularly those with narrow temperature requirements, those confined to small reserves or parks, and those living in fragile ecosystems. Other species may survive in greatly reduced numbers and ranges. Ecosystems considered most vulnerable to species loss in the short term are polar seas, coral reefs and atolls, prairie wetlands, coastal wetlands, tundra, taiga, tropical forests, and mountains, particularly alpine tundra.

Biologists generally agree that global warming will have an especially severe impact on plants, which can't migrate as quickly as animals when environmental conditions change. Although wind and animals disperse seeds, sometimes over long distances, the speed of seed dispersal has definite limitations. During past climate warmings, such as during the glacial retreat that took place some 12,000 years ago, tree species are thought to have dispersed from 4 to 200 km (2.5 to 124 mi) per century. If the Earth warms by the projected 1.4 to 5.8 degrees C during the 21st century, the ideal ranges for some temperate tree species (the environment where they grow the best) may shift northward as much as 480 km (300 mi) (Fig. 55-13). Moreover, soil characteristics, water availability, competition with other plant species, and habitat fragmentation all affect the rate at which plants move into a new area.

In response to global warming, some species may disperse into new environments or adapt to the changing conditions in their present habitats. Global warming may not affect certain species, whereas other species may emerge as winners, with greatly expanded numbers and ranges. Those considered most likely to prosper include weeds, pests, and disease-carrying organisms, all of which are generalists that are already common in many different environments.

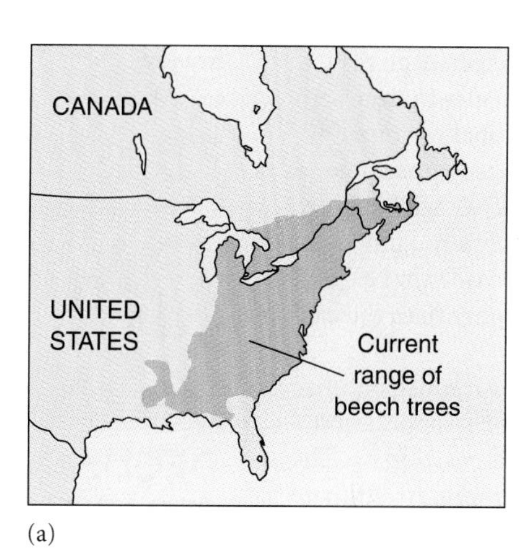

(a)

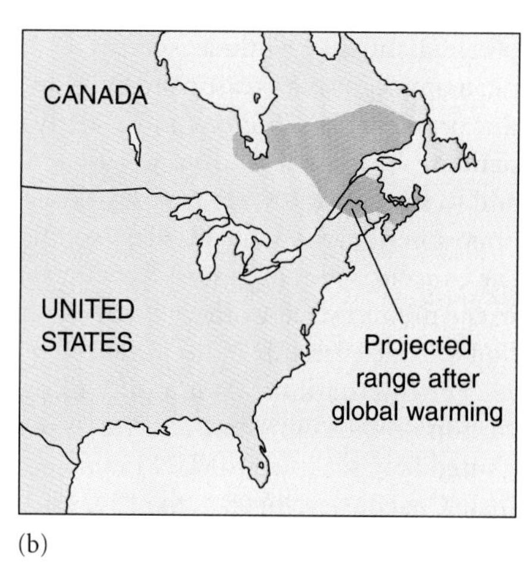

(b)

FIGURE **55-13** **Climate change and beech trees in North America.**

(a) Present geographic range of American beech trees. These large shade trees produce edible beechnuts, which are an important food source for wildlife. **(b)** One projected range of beech trees after global warming occurs. This figure is based on a model of climate change that the U.S. Geophysical Fluid Dynamics Laboratory (GFDL) developed. *(Adapted from M.B. Davis, and C. Zabinski, "Changes in Geographical Range Resulting from Greenhouse Warming: Effects on Biodiversity in Forests,"* Global Warming and Biological Diversity, *edited by R.L. Peters, and T.E. Lovejoy. Yale University Press, New Haven, Connecticut, 1992)*

Global warming will have a more pronounced effect on human health in developing countries

Data linking climate warming and human health problems are accumulating. Scientists hypothesize that the increase in carbon dioxide in the atmosphere and the resultant more frequent and more severe heat waves during summer months will increase the number of heat-related illnesses and deaths.

Climate warming may also affect human health indirectly. Mosquitoes and other disease carriers could expand their range into the newly warm areas and spread malaria, dengue fever, yellow fever, Rift Valley fever, and viral encephalitis. As many as 50 to 80 million additional cases of malaria could occur annually in tropical, subtropical, and temperate areas. According to the World Health Organization, during 1998, the warmest year on record, the incidence of malaria, Rift Valley fever, and cholera surged in developing countries.

Highly developed countries are less vulnerable to such disease outbreaks because of better housing (which keeps mosquitoes outside), medical care, pest control, and public health measures such as water treatment plants. Texas reported a few cases of dengue fever in the late 1990s, for example, whereas nearby Mexico had thousands of cases during that period.

Global warming will probably affect agriculture

Global warming may increase problems for agriculture, which is already challenged to provide enough food for a growing human population without irreparably damaging the environment. Several studies show that the rising sea level will inundate river deltas, which are some of the world's best agricultural lands. The Nile River (Egypt), Mississippi River (United States), and Yangtze River (China) are examples of river deltas that have been studied. Certain agricultural pests and disease-causing organisms will probably proliferate. As mentioned earlier, global warming may also increase the frequency and duration of droughts and, in some areas, crop-damaging floods.

On a regional scale, current global-warming models forecast that agricultural productivity will increase in some areas and decline in others. Models suggest that Canada and Russia will increase their agricultural productivity in a warmer climate, whereas tropical and subtropical regions, where many of the world's poorest people live, will decline in agricultural productivity. Central America and Southeast Asia may experience some of the greatest declines in agricultural productivity.

There are many possible ways to deal with global warming

Despite certain gaps in the knowledge of global warming, the present understanding of the changing global climate and its potential impact on human society and other species gives people many legitimate reasons to develop strategies to deal with this problem. The amount and severity of global warming depend on how much additional greenhouse gas emissions we add to the atmosphere. Many studies assume we will stabilize atmospheric carbon dioxide at 550 ppm, which is roughly twice the concentration of atmospheric carbon dioxide that existed in the preindustrial world and about 50% higher than the carbon dioxide currently in the atmosphere.

The international community recognizes it must stabilize carbon dioxide emissions. At least 174 nations, including the United States, signed the UN Framework Convention on Climate Change developed at the 1992 Earth Summit. Its ultimate goal was to stabilize greenhouse gas concentrations in the atmosphere at levels low enough to prevent dangerous human influences on the climate. However, the details of how to accomplish that goal were left to future conferences.

At the 1996 UN Climate Change Convention held in Geneva, Switzerland, highly developed countries agreed to establish legally binding timetables to cut emissions of greenhouse gases. Representatives from 160 countries met in Kyoto, Japan, in December 1997, to develop the **Kyoto Protocol.** As of late 2003, 119 countries had ratified the treaty. The United States signed it in 1998, but in 2001 the administration of President George W. Bush withdrew the United States from that commitment. The Kyoto Protocol will go into force when enough countries to represent 55% of 1990 global emissions of greenhouse gases have ratified it. As of October 2003, 119 countries, enough to represent 44% of 1990 emissions, had ratified the treaty. It is noteworthy that Australia, Russia, and the United States have not ratified the treaty, whereas Canada, the European Union, and Japan have. (As we go to press, Russia has indicated it may ratify the Kyoto Protocol, which would bring the treaty into force without U.S. participation.)

Review

- What is the enhanced greenhouse effect?
- How do greenhouse gases cause the enhanced greenhouse effect?
- What are some of the significant problems that global warming may cause during the 21st century?

BiologyNow™ Assess your understanding of **global warming** by taking the pretest on your BiologyNow CD-ROM.

DECLINING STRATOSPHERIC OZONE

Learning Objectives

9 Distinguish between surface ozone and stratospheric ozone.

10 Cite the causes and potential effects of ozone destruction in the stratosphere.

Ozone (O_3) is a form of oxygen that is a human-made pollutant in the lower atmosphere but a naturally produced, essential part of the stratosphere (see Fig. 20-6). The **stratosphere,** which encircles the planet some 10 to 45 km (6 to 28 mi) above the surface, contains a layer of ozone that shields the surface from much of the ultraviolet radiation from the sun (Fig. 55-14). If ozone disappeared from the stratosphere, Earth would become unlivable for most forms of life. Ozone in the lower atmosphere is converted back to oxygen in a few days and so does not replenish the ozone depleted in the stratosphere.

A slight thinning in the ozone layer over Antarctica forms naturally for a few months each year. In 1985, however, scientists observed a greater thinning than usual. This increased thinning, which begins each September, is commonly referred

ACTIVE FIGURE **55-14** **Ultraviolet radiation and the ozone layer.**

(a) Stratospheric ozone absorbs 99% of incoming ultraviolet radiation, effectively shielding Earth's surface. **(b)** When stratospheric ozone is reduced, more high-energy ultraviolet radiation penetrates the atmosphere to the surface, where it harms organisms.

BiologyNow™ Learn more about **ultraviolet radiation and the ozone layer** by clicking on this figure on your BiologyNow CD-ROM.

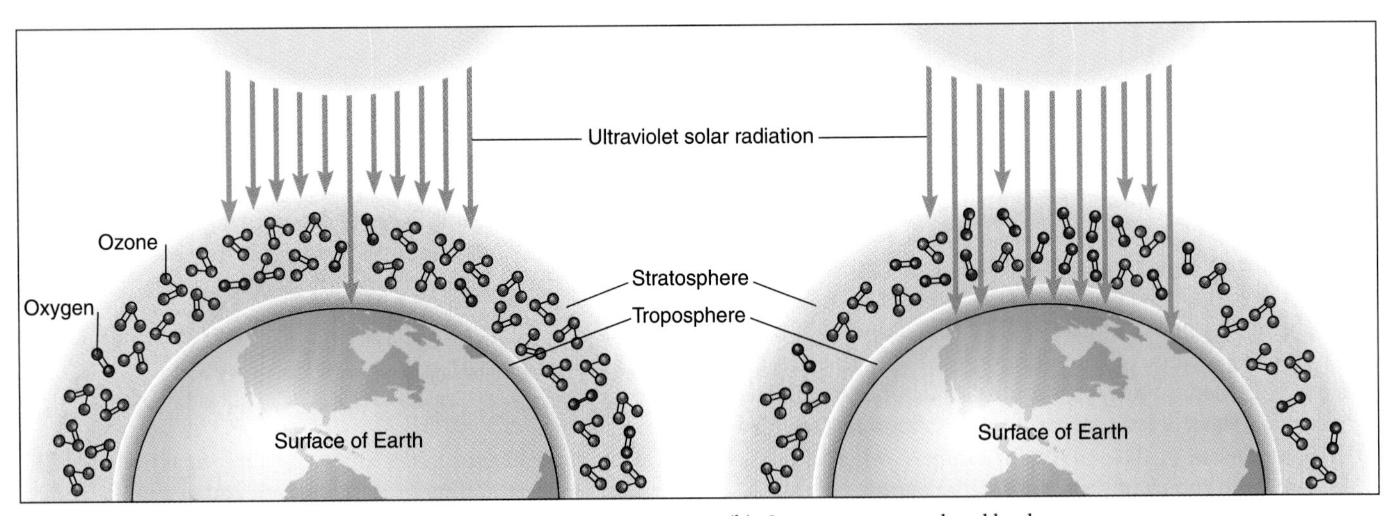

(a) Ozone present at normal levels

(b) Ozone present at reduced levels

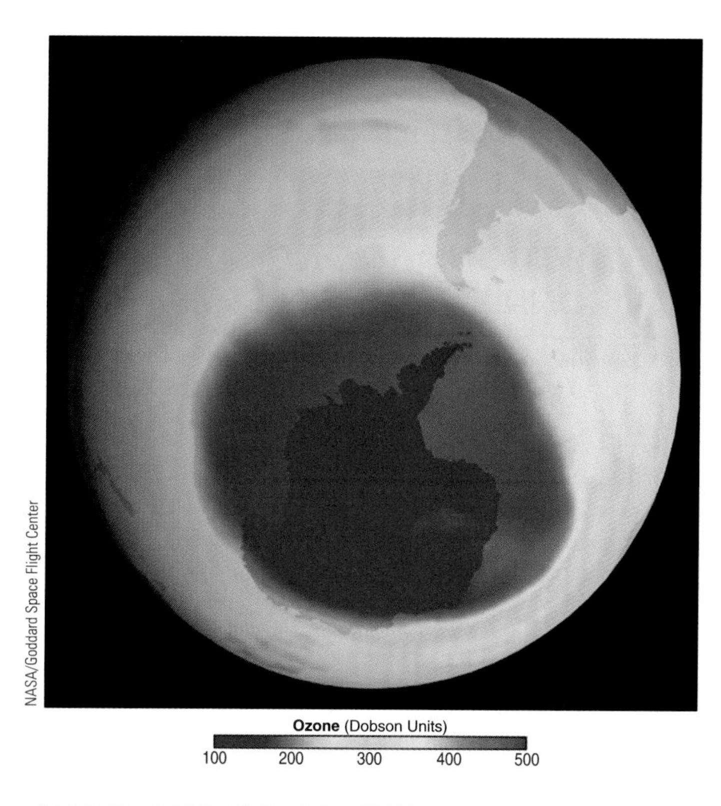

FIGURE **55-15** **Ozone thinning.**

A computer-generated image of part of the Southern Hemisphere, taken on September 11, 2003, reveals the ozone thinning (purple and blue areas). The ozone-thin area is not stationary but is moved about by air currents. (Dobson units measure stratospheric ozone. They are named after British scientist Gordon Dobson, who studied ozone levels over Antarctica during the late 1950s.)

to as the "ozone hole" (Fig. 55-15). There, ozone levels decrease as much as 67% each year. During the 1990s the ozone-thinned area continued to grow, and in 2000 it had reached the record size of 28.3 million km^2 (11.3 million mi^2), larger than the North American continent.

In addition, worldwide levels of stratospheric ozone have been falling for several decades. According to the National Center for Atmospheric Research, since the 1970s ozone levels over Europe and North America have dropped almost 10%.

Certain chemicals destroy stratospheric ozone

Both chlorine- and bromine-containing substances catalyze ozone destruction. The primary chemicals responsible for ozone loss in the stratosphere are a group of chlorine compounds called *chlorofluorocarbons (CFCs).* Chlorofluorocarbons have been used as propellants in aerosol cans, coolants in air conditioners and refrigerators, foam-blowing agents for insulation and packaging, and solvents and cleaners for the electronics industry. Additional compounds that also attack ozone include halons (found in many fire extinguishers); methyl bromide (a pesticide); methyl chloroform (an industrial solvent); and carbon tetrachloride (used in many industrial processes, including the manufacture of pesticides and dyes).

After release into the lower troposphere, CFCs and similar compounds slowly drift up to the stratosphere, where ultraviolet radiation breaks them down, releasing chlorine. Similarly, the breakdown of halons and methyl bromide releases bromine. The thinning in the ozone layer that was discovered over Antarctica occurs annually between September and November (spring in the Southern Hemisphere). At this time, two important conditions occur: Sunlight returns to the polar region, and the *circumpolar vortex,* a mass of cold air that circulates around the southern polar region and isolates it from the warmer air in the rest of the planet, is well developed. The cold air causes polar stratospheric clouds to form; these clouds contain ice crystals to which chlorine and bromine adhere, making them available to destroy ozone. The sunlight promotes the chemical reaction in which chlorine or bromine breaks ozone molecules apart, converting them into oxygen molecules. The chemical reaction in which ozone is destroyed does not alter the chlorine or bromine, and thus a single chlorine or bromine atom breaks down many thousands of ozone molecules. The chlorine and bromine remain in the stratosphere for many years. When the circumpolar vortex breaks up, the ozone-depleted air spreads northward, diluting ozone levels in the stratosphere over South America, New Zealand, and Australia.

Ozone depletion harms organisms

With depletion of the ozone layer, more ultraviolet radiation reaches the surface. Excessive exposure to UV radiation is linked to human health problems, such as cataracts, skin cancer, and a weakened immune system. The lens of the eye contains transparent proteins that are replaced at a very slow rate. Exposure to excessive UV radiation damages these proteins, and over time, the damage accumulates so that the lens becomes cloudy, forming a cataract. Cataracts can be surgically treated, but millions of people in developing nations can't afford the operation and so remain partially or totally blind.

Scientists are concerned that increased levels of UV radiation may disrupt ecosystems. For example, the productivity of Antarctic phytoplankton, the microscopic drifting algae that are the base of the Antarctic food web, has declined from increased exposure to UVB. (UVB is one of three types of UV radiation; see the section on vertebrate skin in Chapter 38.) Research shows that surface UV radiation inhibits photosynthesis in these phytoplankton. Biologists have also documented direct damage to natural populations of Antarctic fish. Increased DNA mutations in icefish eggs and larvae (young fish) were matched to increased levels of UV radiation; researchers are currently studying if these mutations lessen the animals' ability to survive.

There is also concern that high levels of UV radiation may damage crops and forests, but the effects of UVB radiation on plants are very complex and have not been adequately studied. Plants interact with a large number of other species in both natural ecosystems and agricultural ecosystems, and the effects of UV radiation on each of these organisms in turn affect plants indirectly. For example, exposure to higher levels of UVB radiation may increase wheat yields by inhibiting fungi that cause disease in wheat. In contrast, exposure to higher levels of UV radiation may decrease cucumber yields because it increases the incidence of disease.

International cooperation can prevent significant depletion of the ozone layer

In 1987, representatives from many countries met in Montreal to sign the **Montreal Protocol,** an agreement that originally stipulated a 50% reduction of CFC production by 1998. Since then, more than 150 countries have signed this agreement. After scientists reported that decreases in stratospheric ozone occurred over the heavily populated midlatitudes of the Northern Hemisphere in all seasons, the Montreal Protocol was modified to include stricter measures to limit CFC production. Industrial companies that manufacture CFCs quickly developed substitutes.

Production of CFCs, carbon tetrachloride, and methyl chloroform production was completely phased out in the United States and other highly developed countries in 1996, except for a relatively small amount exported to developing countries. Existing stockpiles could be used after the deadline, however. Developing countries will phase out CFC use by 2005. Methyl bromide will be phased out in highly developed countries, which are responsible for 80% of the global use of that chemical, by 2005. Hydrochlorofluorocarbons (HCFCs) will be phased out in 2030.

Satellite measurements taken in 1997 provided the first evidence that the levels of ozone-depleting chemicals were starting to decline in the stratosphere. However, two chemicals (CFC-12 and halon-1211) may have increased and therefore still threaten ozone recovery. Although highly developed countries no longer manufacture CFC-12, it continues to leak into the atmosphere from old refrigerators and vehicle air conditioners discarded in those countries. Also, developing countries such as China, India, Korea, and Mexico continue to produce CFC-12 and halon-1211. An international fund known as the Montreal Multilateral Fund is available to developing countries during their transition from ozone-depleting chemicals to safer alternatives.

CFCs are extremely stable, and those being used today probably will continue to deplete stratospheric ozone for at least 50 years. Assuming that countries continue to adhere to the Montreal Protocol, however, scientists expect human-exacerbated ozone thinning will gradually decline over time.

Review

- What environmental service does the stratospheric ozone layer provide?
- How does ozone depletion occur?
- What are some of the consequences of the thinning ozone layer?

BiologyNow™ Assess your understanding of **declining stratospheric ozone** by taking the pretest on your BiologyNow CD-ROM.

CONNECTIONS AMONG ENVIRONMENTAL PROBLEMS

Learning Objective

11 Describe how an environmental problem such as the biodiversity crisis is related to human population growth.

We have pointed out several connections among the four environmental problems discussed in this chapter. For example, deforestation, global warming, and ozone depletion will likely cause future reductions in biological diversity. Similarly, any environmental problem you think of, even if not discussed in this chapter, is related to other environmental concerns. Most of all, however, environmental problems are connected to overpopulation. Stated simply, human population growth is outstripping Earth's resources and straining our ecological support systems.

We have seen that the rate of human population growth is greatest in developing countries but that highly developed nations are overpopulated as well, in the sense that they have a high per capita consumption of resources (see Chapter 51). **Consumption overpopulation,** the disproportionately large consumption of resources by people in highly developed countries, affects the environment as much as does the population explosion in the developing world.

Humans share much in common with the fate of other organisms on the planet. We are not immune to the environmental damage we have produced. We differ from other organisms, however, in our capacity to reflect on the consequences of our actions and to alter our behavior accordingly. Humans, both individually and collectively, can bring about change. The widespread substitution of constructive change for destructive change is the key to ensuring the biosphere's survival.

Review

- How is human population growth related to destruction of natural habitats?
- How are CFCs related to both stratospheric ozone depletion and global warming?

BiologyNow™ Assess your understanding of **connections among environmental problems** by taking the pretest on your BiologyNow CD-ROM.

SUMMARY WITH KEY TERMS

1 Distinguish among threatened species, endangered species, and extinct species.

- **Biological diversity** is the variety of organisms considered at three levels: genetic diversity, species diversity, and ecosystem diversity. Biological diversity is decreasing worldwide.
- A species becomes **extinct** when its last individual member dies. A species whose severely reduced numbers throughout all or a significant part of its **range** put it in imminent danger of extinction is classified as an **endangered species.** When extinction is less imminent but the population is quite small, a species is classified as a **threatened species.**

2 Discuss at least four causes of declining biological diversity, and identify the most important cause.

- Human activities that reduce biological diversity include habitat loss and **habitat fragmentation,** pollution, introduction of **invasive species,** pest and predator control, illegal commercial hunting, and commercial harvest. Of these, habitat loss and fragmentation are the most significant.
- In the United States, declining biological diversity is most serious in Hawaii and California. Worldwide, tropical rain forests are being destroyed faster than almost all other ecosystems.

3 Define *conservation biology,* and compare in situ and ex situ conservation measures.

- **Conservation biology** is the study of how humans affect organisms and of the development of ways to protect biological diversity. Conservation biology is a broad discipline that ranges from safeguarding populations of endangered species to preserving entire ecosystems and landscapes.
- Efforts to preserve biological diversity in the wild are known as **in situ conservation.** In situ conservation is urgently needed in the world's 25 biodiversity hotspots.
- **Ex situ conservation** involves conserving individual species in human-controlled settings.

4 Describe the benefits and shortcomings of the Endangered Species Act and the Convention on International Trade in Endangered Species of Wild Flora and Fauna.

- The **Endangered Species Act (ESA)** authorizes the U.S. Fish and Wildlife Service to protect from extinction endangered and threatened species, both in the United States and abroad. The ESA is controversial; it does not compensate private property owners for financial losses because they can't develop their land if a threatened or endangered species lives there. Conservationists would like to strengthen the ESA so as to manage whole ecosystems and maintain complete biological diversity rather than attempting to save endangered species as isolated entities.
- At the international level, the **Convention on International Trade in Endangered Species of Wild Flora and Fauna (CITES)** protects endangered animals and plants considered valuable in the highly lucrative international wildlife trade. Enforcement of this treaty varies from country to country; where enforcement exists, penalties are not very severe, and so illegal trade in rare, commercially valuable species continues.

5 Discuss the ecosystem services of forests, and describe the consequences of deforestation.

- Forests provide many **ecosystem services,** including watershed protection, soil erosion prevention, climate moderation, protection from flooding, and wildlife habitat.
- The greatest problem facing forests today is **deforestation,** the temporary or permanent clearance of forests for agriculture or other uses. Deforestation increases soil erosion and thus decreases soil fertility. Deforestation contributes to loss of biological diversity. When a forest is cut down, the watershed can't absorb and hold water as well, and the total amount of surface runoff flowing into rivers and streams increases. Deforestation may affect regional and global climate changes and may contribute to an increase in global temperature.

6 State at least three reasons why forests are disappearing today.

- Forests are destroyed to provide subsistence farmers with agricultural land; to produce timber; to provide open rangeland for cattle; and to supply fuel wood. **Subsistence agriculture,** in which a family produces enough food to feed itself, accounts for perhaps 60% of tropical deforestation.

7 Name at least three greenhouse gases, and explain how greenhouse gases contribute to global warming.

- Build up of carbon dioxide, methane, surface ozone, nitrous oxide, and chlorofluorocarbons causes the **greenhouse effect,** in which the atmosphere retains heat and warms Earth's surface.
- Increased levels of CO_2 and other **greenhouse gases** in the atmosphere are causing concerns about an **enhanced greenhouse effect,** which is additional warming produced by increased levels of gases that absorb infrared radiation.

8 Describe how global warming may affect sea level, precipitation patterns, organisms (including humans), and food production.

- During the 21st century, **global warming** may cause a rise in sea level. Precipitation patterns may change, resulting in more frequent droughts in some areas and more frequent flooding in other areas.
- Global warming is causing some species to shift their ranges; biologists think global warming will cause some species to go extinct, some to be unaffected, and others to expand their numbers and ranges. Data linking climate warming and human health problems (particularly in developing countries) are accumulating. Problems for agriculture include increased flooding, increased droughts, and declining agricultural productivity in tropical and subtropical areas.

9 Distinguish between surface ozone and stratospheric ozone.

- **Ozone (O_3)** is a form of oxygen that is a human-made pollutant in the lower atmosphere but a naturally produced, essential part of the stratosphere. The **stratosphere,** which encircles Earth some 10 to 45 km above the surface, contains a layer of ozone that shields the surface from much of the damaging ultraviolet radiation from the sun.

10 Cite the causes and potential effects of ozone destruction in the stratosphere.

- The total amount of ozone in the stratosphere is declining (**ozone depletion**), and large areas of ozone thinning develop over Antarctica and the Arctic each year. Chlorofluorocarbons and similar chlorine- and bromine-containing compounds attack the ozone layer.
- Excessive exposure to UV radiation is linked to a number of human health problems, including cataracts, skin cancer, and a weakened immune system. Scientists are concerned that increased levels of UV radiation may disrupt ecosystems, such as the Antarctic food web. There is also concern that high levels of UV radiation may damage crops and forests.

11 Describe how an environmental problem such as the biodiversity crisis is related to human population growth.

- Overpopulation increases destruction of natural habitats, as people convert natural environments to agricultural land and log forests for needed lumber. **Consumption overpopulation**—which is the disproportionately large consumption of resources by people in highly developed countries—harms the environment as much as the population explosion in the developing world.

POST-TEST

1. Which of the following statements about extinction is *not* correct? (a) extinction is the permanent loss of a species (b) extinction is a natural biological process (c) once a species is extinct, it never reappears (d) human activities have little impact on extinctions (e) thousands of plant and animal species are currently threatened with extinction
2. An endangered species (a) is severely reduced in number (b) is in imminent danger of becoming extinct throughout all or a significant part of its range (c) usually does not have reduced genetic variability (d) is not in danger of extinction in the foreseeable future (e) both a and b are correct
3. The most important reason for declining biological diversity is (a) air pollution (b) introduction of foreign (invasive) species (c) habitat destruction and fragmentation (d) illegal commercial hunting (e) commercial harvesting
4. In situ conservation (a) includes breeding captive species in zoos (b) includes seed storage of genetically diverse crops (c) concentrates on preserving biological diversity in the wild (d) focuses exclusively on large, charismatic animals (e) both a and b are correct
5. Restoration ecology (a) is the study of how humans impact organisms (b) returns a degraded environment as close as possible to its former state (c) is an example of ex situ conservation (d) has been used to successfully reverse the decline in amphibian populations (e) is an important provision of the Endangered Species Act
6. Which of the following is linked to the decline of amphibian populations? (a) agricultural chemicals (b) increased UV radiation (c) infectious diseases (d) global climate warming (e) all of these
7. When forests are destroyed (a) soil fertility increases (b) soil erosion decreases (c) some forest organisms decline in number (d) the danger of flooding in nearby lowlands declines (e) carbon dioxide is removed from the atmosphere and stored in forest plants
8. About 60% of tropical deforestation is the result of (a) commercial logging (b) cattle ranching (c) hydroelectric dams (d) mining (e) subsistence agriculture
9. If deforestation of tropical forests continues at present estimated rates, the forests will (a) disappear in a few hundred years (b) disappear in the next 20 years (c) disappear by the first half of the 22nd century (d) regrow and replenish themselves (e) persist into the foreseeable future
10. Which of the following gases contributes to both global warming and thinning of the ozone layer? (a) CO_2 (b) CH_4 (c) surface O_3 (d) CFCs (e) N_2O
11. Global warming occurs because (a) carbon dioxide and other greenhouse gases react chemically to produce excess heat (b) Earth has too many greenhouses and other glassed buildings (c) volcanic eruptions produce large quantities of sulfur and other greenhouse gases (d) carbon dioxide and other greenhouse gases trap infrared radiation in the atmosphere (e) carbon dioxide and other greenhouse gases allow excess heat to pass out of the atmosphere
12. What gas is a human-made pollutant in the lower (surface) atmosphere but a natural and beneficial gas in the stratosphere? (a) CO_2 (b) CH_4 (c) O_3 (d) CFCs (e) N_2O
13. Where is stratospheric ozone depletion most pronounced? (a) over Antarctica (b) over the equator (c) over South America (d) over North America and Europe (e) over Alaska and Siberia
14. The Montreal Protocol is an agreement to (a) phase out greenhouse gas emissions (b) curtail CFC production (c) design a recovery plan for each endangered species (d) curtail deforestation (e) prevent the selling of products made from endangered or threatened species

CRITICAL THINKING

1. If half of the world's biological diversity were to disappear, how would it affect your life?
2. Because new species will eventually evolve to replace those that humans are driving to extinction, why is declining biological diversity such a threat to us?
3. Why might captive breeding programs that reintroduce species into natural environments fail?
4. If you were given the task of developing a policy for the United States to deal with global climate change during the next 50 years, what would you propose? Explain your answer.
5. A more descriptive name for *Homo sapiens* is "*Homo dangerous.*" Explain this specific epithet, given what you have learned in this chapter.

- Visit our Web site at **http://biology.brookscole.com/solomon7** for links to chapter-related resources on the World Wide Web. Additional online materials relating to this chapter can also be found on our Web site.

BIOLOGY NOW RESOURCES

Active Figures

55-12: The greenhouse effect

55-14: Ultraviolet radiation and the ozone layer

Preparing for an exam? Take a diagnostic test on your BiologyNow CD-ROM.

Post-Test Answers

1. d	2. e	3. c	4. c
5. b	6. e	7. c	8. e
9. c	10. d	11. d	12. c
13. a	14. b		

Appendix A

Periodic Table of the Elements

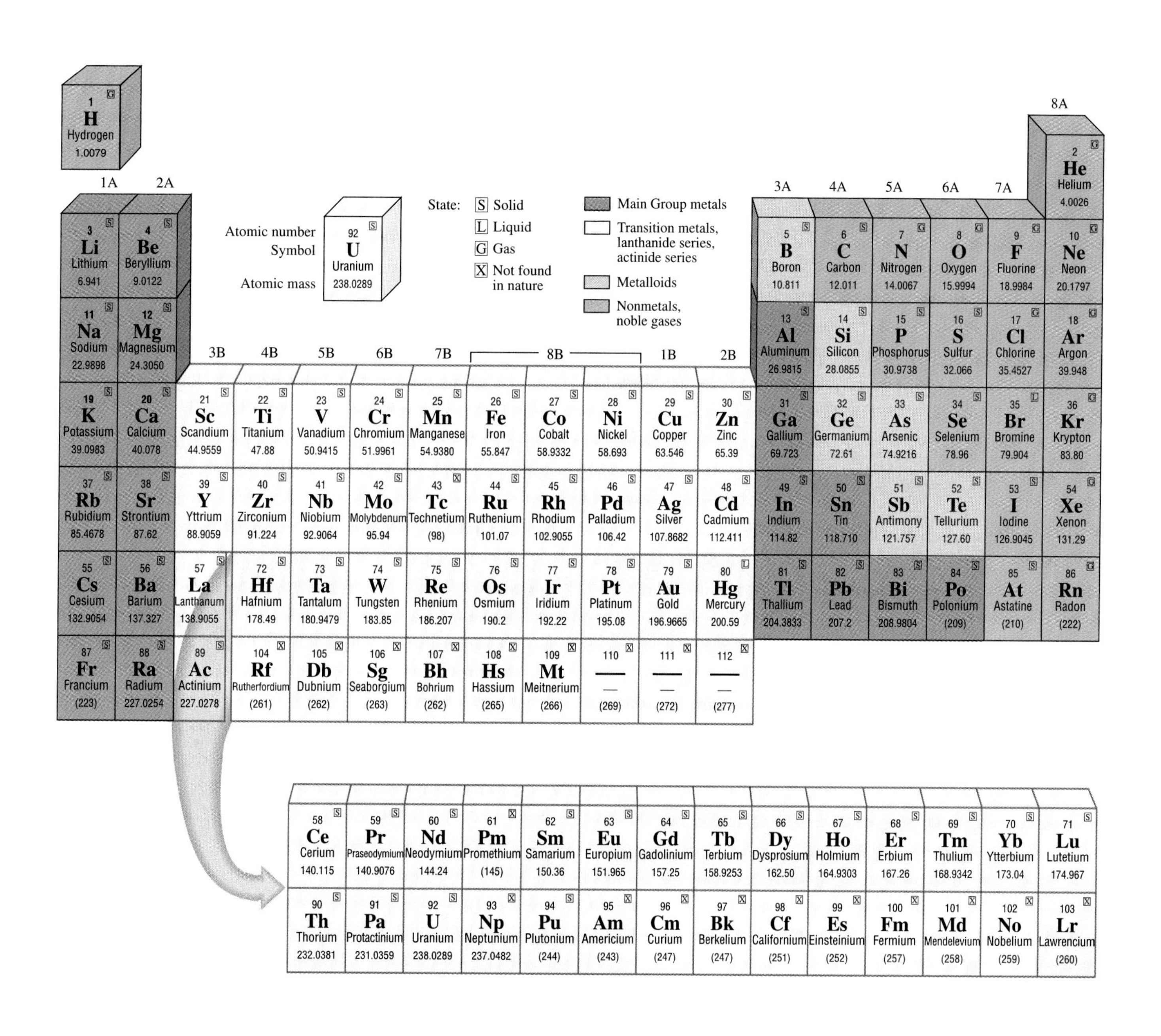

Appendix B
Classification of Organisms

The system of cataloging organisms used in this book is described in Chapter 1 and in Part 5. In this seventh edition of *Biology,* we use the three-domain and six-kingdom classification. The three domains are **Eubacteria** (or simply **Bacteria**), **Archaea,** and **Eukarya** (eukaryotes). The six kingdoms are Eubacteria (which corresponds to Domain Eubacteria), Archaebacteria (which corresponds to Domain Archaea), Protista, Plantae, Fungi, and Animalia. In this classification overview, we have omitted many groups, especially extinct ones. We have omitted the viruses from this survey, because they do not fit into any of the three domains or six kingdoms.

Domain Eubacteria and domain Archaea are made up of prokaryotic organisms. They are distinguished from eukaryotic organisms by their smaller ribosomes and absence of membranous organelles, including a discrete nucleus surrounded by a nuclear envelope. Prokaryotes reproduce mainly asexually by binary fission. When present, flagella are simple and solid; they do not have the 9+2 microfilament structure typical of eukaryotes.

KINGDOM EUBACTERIA

Very large, diverse group of prokaryotic organisms. Typically unicellular, but some form colonies or filaments. Mainly heterotrophic, but some groups are photosynthetic or chemosynthetic. Bacteria are nonmotile or move by beating flagella. Typically have peptidoglycan in their cell walls. More than 10,000 species.

Bacterial nomenclature and taxonomic practices are controversial and changing. In the system used in this edition, eubacteria are classified in three main groups: (1) bacteria with a Gram-negative type of cell wall, (2) bacteria with a Gram-positive type of cell wall, and (3) bacteria with no cell walls (mycoplasmas). See Chapter 23, Table 23-3.

Gram-negative bacteria. Thin cell wall containing peptidoglycan. Include *proteobacteria, chlamydias, spirochetes, cyanobacteria.*

Gram-positive bacteria. Thick cell wall of peptidoglycan. Nonphotosynthetic. Many produce spores. Include *actinomycetes, lactic acid bacteria, mycobacteria, streptococci, staphylococci, clostridia.*

Bacteria that lack cell walls. Extremely small bacteria bounded by a plasma membrane. *Mycoplasmas.*

KINGDOM ARCHAEBACTERIA

Prokaryotes with cell walls lacking peptidoglycan and unique cell membrane structure. Also distinguished by their ribosomal RNA, lipid structure, and specific enzymes. Archaebacteria are found in extreme environments—hot springs, sea vents, dry and salty seashores, boiling mud, and near ash-ejecting volcanoes. Three main types are *methanogens, extreme halophiles,* and *extreme thermophiles.*

Methanogens. Anaerobes that produce methane gas from simple carbon compounds.

Extreme Halophiles. Inhabit saturated salt solutions.

Extreme Thermophiles. Grow at 70°C or higher, some thrive above the boiling point.

KINGDOM PROTISTA

Primarily unicellular or simple multicellular eukaryotic organisms that do not form tissues and that exhibit relatively little division of labor. Most modes of nutrition occur in this kingdom. Life cycles may include both sexually and asexually reproducing phases and may be extremely complex, especially in parasitic forms. Locomotion is by cilia, flagella, amoeboid movement, or by other means. Flagella and cilia have 9 + 2 structure.

Excavates

Anaerobic zooflagellates. Have atypical mitochondria or lack them. Most are endosymbionts. Include diplomonads and trichonymphs.

Phylum Retortamonada *Diplomonads.* Excavates with one or two nuclei, no mitochondria, no Golgi complex, and up to eight flagella.

Phylum Axostylata *Trichonymphs.* Excavates with hundreds of flagella; live in the guts of termites and wood-eating cockroaches.

Discicristates

Zooflagellates with disc-shaped cristae in their mitochondria. Include euglenoids and trypanosomes.

Phylum Euglenozoa *Euglenoids and trypanosomes* Unicellular flagellates; some free-living and some pathogenic. Many are heterotrophic, but some (about one third) are photosynthetic.

Alveolates

Unicellular protists with alveoli, flattened vesicles located inside the plasma membrane. Include ciliates, dinoflagellates, and apicomplexans.

Phylum Alveolata, subphylum Ciliophora. *Ciliates.* Unicellular protests that move by means of cilia. Reproduction is asexual by binary fission or sexual by conjugation. About 7200 species.*

* We provide the number of species for a given phylum when a reliable estimate is known.

Phylum Alveolata, subphylum Dinoflagellata. *Dinoflagellates.* Unicellular (some colonial), photosynthetic, biflagellate. Cell walls, composed of overlapping cell plates, contain cellulose. Contain chlorophylls *a* and *c* and carotenoids, including fucoxanthin. About 2100 species.

Phylum Alveolata, subphylum Apicomplexa. *Apicomplexans.* Parasitic unicellular protists that lack specific structures for locomotion. At some stage in life cycle, they develop spores (small infective agents). Some pathogenic. About 3900 species.

Heterokonts

Diverse group; all have motile cells with two flagella, one with tiny, hairlike projections off the shaft. Include water molds, diatoms, golden algae, and brown algae.

Phylum Oomycota. *Water molds.* Consist of branched, coenocytic mycelia. Cellulose and/or chitin in cell walls. Produce biflagellate asexual spores. Sexual stage involves production of oospores. Some parasitic. About 580 species.

Phylum Bacillariophyta. *Diatoms.* Unicellular (some colonial), photosynthetic. Most nonmotile, but some move by gliding. Cell walls composed of silica rather than cellulose. Contain chlorophylls *a* and *c* and carotenoids, including fucoxanthin. About 5600 species.

Phylum Chrysophyta. *Golden algae.* Unicellular (some colonial), photosynthetic, biflagellate (some lack flagella). Cells covered by tiny scales of either silica or calcium carbonate. Contain chlorophylls *a* and *c* and carotenoids, including fucoxanthin. About 500 species.

Phylum Phaeophyta. *Brown algae.* Multicellular, often quite large (kelps). Photosynthetic; contain chlorophylls *a* and *c* and carotenoids, including fucoxanthin. Biflagellate reproductive cells. About 1500 species.

Cercozoa

Amoeboid cells, often with hard outer shells through which cytoplasmic projections extend. Include foraminiferans and actinopods.

Phylum Foraminifera. *Foraminiferans.* Unicellular protists that produce calcareous tests (shells) with pores through which cytoplasmic projections extend, forming a sticky net to entangle prey.

Phylum Actinopoda. *Actinopods.* Unicellular protists that produce axopods (long, filamentous cytoplasmic projections) that protrude through pores in their siliceous shells.

"Plants"

Photosynthetic organisms with chloroplasts bounded by outer and inner membranes. This monophyletic group includes organisms currently placed in two kingdoms, kingdom Plantae (land plants) and kingdom Protista (red algae and green algae).

Phylum Rhodophyta. *Red algae.* Most multicellular (some unicellular), mainly marine. Some (coralline algae) have bodies impregnated with calcium carbonate. No motile cells. Photosynthetic; contain chlorophyll *a*, carotenoids, phycocyanin, and phycoerythrin. About 4000 species.

Phylum Chlorophyta. *Green algae.* Unicellular, colonial, siphonous, and multicellular forms. Some motile and flagellated. Photosynthetic; contain chlorophylls *a* and *b* and carotenoids. About 7000 species.

Amoebozoa

Amoeboid protists that lack tests and move by means of lobose pseudopodia. Include amoebas, plasmodial slime molds, and cellular slime molds.

Phylum Rhizopoda. *Amoebas.* Free-living and parasitic unicellular protists whose movement and capture of food are associated with pseudopodia.

Phylum Myxomycota. *Plasmodial slime molds.* Spend part of life cycle as a thin, streaming, multinucleate plasmodium that creeps along on decaying leaves or wood. Flagellated or amoeboid reproductive cells; form spores in sporangia. About 500 species.

Phylum Dictyostelida *Cellular slime molds.* Vegetative (nonreproductive) form unicellular; move by pseudopods. Amoeba-like cells aggregate to form a multicellular pseudoplasmodium that eventually develops into a fruiting body that bears spores. About 70 species.

Opisthokonts

Monophyletic group that includes members of three kingdoms: kingdom Fungi, kingdom Animalia, and kingdom Protista (choanoflagellates).

Phylum Choanoflagellata. *Choanoflagellates.* Zooflagellates with a single flagellum surrounded at the base by a collar of microvilli. May be related to the common ancestor of animals.

KINGDOM PLANTAE

Multicellular eukaryotic organisms with differentiated tissues and organs. Cell walls contain cellulose. Cells frequently contain large vacuoles; photosynthetic pigments in plastids. Photosynthetic pigments are chlorophylls *a* and *b*, and carotenoids. Nonmotile. Reproduce both asexually and sexually, with alternation of gametophyte (n) and sporophyte ($2n$) generations.

Phylum Bryophyta. *Mosses.* Nonvascular plants that lack xylem and phloem. Marked alternation of generations with dominant gametophyte generation. Motile sperm. The gametophytes generally form a dense green mat consisting of individual plants. About 9000 species.

Phylum Hepatophyta. *Liverworts.* Nonvascular plants that lack xylem and phloem. Marked alternation of generations with dominant gametophyte generation. Motile sperm. The gametophytes of certain species have a flat, liver-like thallus; other species are more mosslike in appearance. About 6000 species.

Phylum Anthocerotophyta. *Hornworts.* Nonvascular plants that lack xylem and phloem. Marked alternation of generations with dominant gametophyte generation. Motile sperm. The gametophyte is a small, flat, green thallus with scalloped edges. Spores are produced on an erect, hornlike stalk. About 100 species.

Phylum Polypodiophyta. *Ferns.* Vascular plants with a dominant sporophyte generation. Generally homosporous. Gametophyte is free-living and photosynthetic. Reproduce by spores. Motile sperm. About 11,000 species.

Phylum Psilotophyta. *Whisk ferns.* Vascular plants with a dominant sporophyte generation. Homosporous. Stem is distinctive because it branches dichotomously; plant lacks true roots and leaves. The gametophyte is subterranean and nonphotosynthetic and forms a mycorrhizal relationship with a fungus. Motile sperm. About 12 species.

Phylum Equisetophyta. *Horsetails.* Vascular plants with hollow, jointed stems and reduced, scalelike leaves. Although modern representatives are small, some extinct species were treelike. Homosporous. Gametophyte a tiny photosynthetic plant. Motile sperm. About 15 species.

Phylum Lycophyta. *Club mosses.* Sporophyte plants are vascular with branching rhizomes and upright stems that bear microphylls. Although modern representatives are small, some extinct species were treelike. Some homosporous, others heterosporous. Motile sperm. About 1000 species.

Phylum Pinophyta. *Conifers.* Heterosporous vascular plants with woody tissues (trees and shrubs) and needle-shaped or scalelike leaves. Most are evergreen. Seeds are usually borne naked on the surface of cone scales. Nutritive tissue in the seed is haploid female gametophyte tissue. Nonmotile sperm. About 550 species.

Phylum Cycadophyta. *Cycads.* Heterosporous, vascular, dioecious plants that are small and shrubby or larger and palmlike. Produce naked seeds in conspicuous cones. Flagellate sperm. About 140 species.

Phylum Ginkgophyta. *Ginkgo.* Broad-leaved deciduous trees that bear naked seeds directly on branches. Dioecious. Contain vascular tissues. Flagellate sperm. The ginkgo tree is the only living representative. One species.

Phylum Gnetophyta. *Gnetophytes.* Woody shrubs, vines, or small trees that bear naked seeds in cones. Contain vascular tissues. Possess many features similar to flowering plants. About 70 species.

Phylum Magnoliophyta. *Flowering plants* or *angiosperms.* Largest, most successful group of plants. Heterosporous; dominant sporophytes with extremely reduced gametophytes. Contain vascular tissues. Bear flowers, fruits, and seeds (enclosed in a fruit; seeds contain endosperm as nutritive tissue). Double fertilization. More than 235,000 species.

KINGDOM FUNGI

All eukaryotic, mainly multicellular organisms with cell walls containing chitin. Body form often a mycelium. Cells usually haploid or dikaryotic, with brief diploid period following fertilization. Heterotrophs that secrete digestive enzymes onto food source and then absorb the predigested food. Most decomposers, but some parasites. Reproduce by means of spores, which may be produced sexually or asexually. No flagellate stages except in chytrids.

Phylum Chytridiomycota. *Chytridiomycetes* or *chytrids.* Parasites and decomposers found mainly in fresh water. Motile cells (gametes and zoospores) contain a single, posterior flagellum. Reproduce both sexually and asexually. About 750 species.

Phylum Zygomycota. *Zygomycetes (molds).* Produce sexual resting spores called *zygospores,* and nonmotile asexual spores in a sporangium. Hyphae are coenocytic. Many are heterothallic (two mating types). About 800 species.

Phylum Ascomycota. *Ascomycetes* or *sac fungi (yeasts, molds, morels, truffles).* Sexual reproduction involves formation of ascospores in little sacs called *asci.* Asexual reproduction involves production of spores called *conidia,* which pinch off from conidiophores. Hyphae usually have perforated septa. About 30,000 species.

Phylum Basidiomycota. *Basidiomycetes* or *club fungi (mushrooms, bracket fungi, puffballs).* Sexual reproduction involves formation of basidiospores on a basidium. Asexual reproduction uncommon. Heterothallic. Hyphae usually have perforated septa. About 25,000 species.

Phylum Deuteromycota. *Deuteromycetes* or *imperfect fungi (molds).* Sexual stage has not been observed. Most reproduce only by conidia. About 25,000 species.

KINGDOM ANIMALIA

Eukaryotic, multicellular, heterotrophs with differentiated cells. In most animals, cells are organized to form tissues, tissues form organs, and tissues and organs form specialized organ systems that carry on specific functions. Most have a well-developed nervous system and respond adaptively to changes in their environment. Most are capable of locomotion during some time in their life cycle. Most diploid and reproduce sexually; a flagellate haploid sperm unites with a large, nonmotile haploid egg forming a diploid zygote that undergoes cleavage.

Phylum Porifera. *Sponges.* Mainly marine; solitary or colonial. Body bears many pores through which water circulates. Food is filtered from the water by collar cells (choanocytes). Asexual reproduction by budding; external sexual reproduction in which sperm are released and swim to internal egg. Larva is motile. About 10,000 species.

Phylum Cnidaria. *Hydras, jellyfish, sea anemones, corals.* Marine, with a few freshwater species; solitary or colonial; polyp and medusa forms. Radial symmetry. Tentacles surrounding mouth. Stinging cells (cnidocytes) contain stinging structures called *nematocysts.* Planula larva. About 10,000 species.

Phylum Ctenophora. *Comb jellies.* Marine; free-swimming. Biradial symmetry. Two tentacles and eight longitudinal rows of cilia resembling combs; animal moves by means of these bands of cilia. About 100 species.

Protostomes

Coelomates (have true coelom, a body cavity completely lined with mesoderm). Spiral, determinate cleavage, and the mouth typically develops from the blastopore. Two branches of protostomes are Lophotrochozoa and Ecdysozoa.

Lophotrochozoa

Phylum Platyhelminthes. *Flatworms.* Acoelomate (no body cavity); region between body wall and internal organs filled with tissue. Planarians are free-living; flukes and tapeworms are parasitic. Body dorsoventrally flattened; cephalization; three tissue layers. Simple nervous system with ganglia in head region. Excretory organs are protonephridia with flame cells. About 20,000 species.

Phylum Nemertea. *Proboscis worms* (also called *ribbon worms*). Long, dorsoventrally flattened body with complex proboscis used for defense and for capturing prey. Functionally acoelomate, but have a small true coelom in the proboscis. Definite organ systems. Complete digestive tract. Circulatory system with blood. About 1000 species.

Phylum Mollusca. *Snails, clams, squids, octopods.* Unsegmented, soft-bodied animals usually covered by a dorsal shell. Have a ventral, muscular foot. Most organs located above foot in visceral mass. A shell-secreting mantle covers the visceral mass and forms a mantle cavity, which contains gills. Trochophore and/or veliger larva. About 50,000 species.

Phylum Annelida. *Segmented worms: Polychaetes, earthworms, leeches.* Both body wall and internal organs are segmented. Body segments separated by septa. Some have nonjointed appendages. Setae used in locomotion. Closed circulatory system; metanephridia; specialized regions of digestive tract. Trochophore larva. About 15,000 species.

Phylum Brachiopoda. *Lamp shells.* One of the lophophorate phyla. Marine; body enclosed between two shells. About 350 species.

Phylum Phoronida. One of the lophophorate phyla. Tube-dwelling marine worms. About 12 species.

Phylum Bryozoa. One of the lophophorate phyla. Mainly marine; sessile colonies produced by asexual budding. About 5000 species.

Phylum Rotifera. *Wheel animals.* Aquatic, microscopic, wormlike animals. Anterior end has ciliated crown that looks like a wheel when the cilia beat. Posterior end tapers to a foot. Characterized by pseudocoelom (body cavity not completely lined with mesoderm). Constant number of cells. About 1800 species.

Ecdysozoa

Animals in this group characterized by ecdysis (molting)

Phylum Nematoda. *Roundworms.* (*Ascaris, hookworms, pinworms*). Slender, elongated, cylindrical worms; covered with cuticle. Characterized by pseudocoelom. Free-living and parasitic forms. About 15,000 species.

Phylum Arthropoda. *Arachnids (spiders, mites, ticks), crustaceans (lobsters, crabs, shrimp), insects, centipedes, millipedes.* Segmented animals with paired, jointed appendages and a hard exoskeleton made of chitin. Open circulatory system with dorsal heart. Hemocoel occupies most of body cavity, and coelom is reduced. More than 1 million species.

Deuterostomes

Coelomates with radial, indeterminate cleavage. Blastopore develops into anus, and mouth forms from a second opening.

Phylum Echinodermata. *Sea stars, sea urchins, sand dollars, sea cucumbers.* Marine animals. Pentaradial symmetry as adults; bilateral symmetry as larvae. Endoskeleton of small, calcareous plates. Water vascular system; tube feet for locomotion. About 7000 species.

Phylum Hemichordata. *Acorn worms.* Marine animals with a ring of cilia around mouth. Anterior muscular proboscis is connected by a collar region to a long, wormlike body. The larval form resembles an echinoderm larva. About 85 species.

Phylum Chordata. *Subphylum Urochordata (tunicates), subphylum Cephalochordata (lancelets), subphylum Vertebrata (fishes, amphibians, reptiles, birds, mammals).* Notochord; pharyngeal gill slits; dorsal, tubular nerve cord, and postanal tail present at some time in life cycle. About 48,000 species.

Appendix C
Understanding Biological Terms

Your task of mastering new terms will be greatly simplified if you learn to dissect each new word. Many terms can be divided into a prefix, the part of the word that precedes the main root, the word root itself, and often a suffix, a word ending that may add to or modify the meaning of the root. As you progress in your study of biology, you will learn to recognize the more common prefixes, word roots, and suffixes. Such recognition will help you analyze new terms so that you can more readily determine their meaning and will also help you remember them.

Prefixes

a-, ab- from, away, apart (abduct, move away from the midline of the body)
a-, an-, un- less, lack, not (asymmetrical, not symmetrical)
ad- (also **af-, ag-, an-, ap-**) to, toward (adduct, move toward the midline of the body)
allo- different (allometric growth, different rates of growth for different parts of the body during development)
ambi- both sides (ambidextrous, able to use either hand)
andro- a man (androecium, the male portion of a flower)
anis- unequal (anisogamy, sexual reproduction in which the gametes are of unequal sizes)
ante- forward, before (anteflexion, bending forward)
anti- against (antibody, proteins that have the capacity to react against foreign substances in the body)
auto- self (autotroph, organism that manufactures its own food)
bi- two (biennial, a plant that takes two years to complete its life cycle)
bio- life (biology, the study of life)
circum-, circ- around (circumcision, a cutting around)
co-, con- with, together (congenital, existing with or before birth)
contra- against (contraception, against conception)
cyt- cell (cytology, the study of cells)
di- two (disaccharide, a compound made of two sugar molecules chemically combined)
dis- apart (dissect, cut apart)
ecto- outside (ectoplasm, outer layer of cytoplasm)
end-, endo- within, inner (endoplasmic reticulum, a network of membranes found within the cytoplasm)
epi- on, upon (epidermis, upon the dermis)
ex-, e-, ef- out from, out of (extension, a straightening out)
extra- outside, beyond (extraembryonic membrane, a membrane that encircles and protects the embryo)
gravi- heavy (gravitropism, growth of a plant in response to gravity)
hemi- half (cerebral hemisphere, lateral half of the cerebrum)
hetero- other, different (heterozygous, having unlike members of a gene pair)
homeo- unchanging, steady (homeostasis, reaching a steady state)
homo-, hom- same (homologous, corresponding in structure; homozygous, having identical members of a gene pair)
hyper- excessive, above normal (hypersecretion, excessive secretion)
hypo- under, below, deficient (hypotonic, a solution whose osmotic pressure is less than that of a solution with which it is compared)
in-, im- not (incomplete flower, a flower that does not have one or more of the four main parts)
inter- between, among (interstitial, situated between parts)
intra- within (intracellular, within the cell)
iso- equal, like (isotonic, equal osmotic concentration)
macro- large (macronucleus, a large, polyploid nucleus found in ciliates)
mal- bad, abnormal (malnutrition, poor nutrition)
mega- large, great (megakaryocyte, giant cell of bone marrow)
meso- middle (mesoderm, middle tissue layer of the animal embryo)
meta- after, beyond (metaphase, the stage of mitosis after prophase)
micro- small (microscope, instrument for viewing small objects)
mono- one (monocot, a group of flowering plants with one cotyledon, or seed leaf, in the seed)
oligo- small, few, scant (oligotrophic lake, a lake deficient in nutrients and organisms)
oo- egg (oocyte, cell that gives rise to an egg cell)
paedo- a child (paedomorphosis, the preservation of a juvenile characteristic in an adult)
para- near, beside, beyond (paracentral, near the center)
peri- around (pericardial membrane, membrane that surrounds the heart)
photo- light (phototropism, growth of a plant in response to the direction of light)
poly- many, much, multiple, complex (polysaccharide, a carbohydrate composed of many simple sugars)
post- after, behind (postnatal, after birth)
pre- before (prenatal, before birth)
pseudo- false (pseudopod, a temporary protrusion of a cell, i.e., "false foot")
retro- backward (retroperitoneal, located behind the peritoneum)
semi- half (semilunar, half-moon)
sub- under (subcutaneous tissue, tissue immediately under the skin)

super, supra- above (suprarenal, above the kidney)
sym- with, together (sympatric speciation, evolution of a new species within the same geographical region as the parent species)
syn- with, together (syndrome, a group of symptoms that occur together and characterize a disease)
trans- across, beyond (transport, carry across)

Suffixes

-able, -ible able (viable, able to live)
-ad used in anatomy to form adverbs of direction (cephalad, toward the head)
-asis, -asia, -esis condition or state of (euthanasia, state of "good death")
-cide kill, destroy (biocide, substance that kills living things)
-emia condition of blood (anemia, a blood condition in which there is a lack of red blood cells)
-gen something produced or generated or something that produces or generates (pathogen, an organism that produces disease)
-gram record, write (electrocardiogram, a record of the electrical activity of the heart)
-graph record, write (electrocardiograph, an instrument for recording the electrical activity of the heart)
-ic adjective-forming suffix that means *of* or *pertaining to* (ophthalmic, of or pertaining to the eye)
-itis inflammation of (appendicitis, inflammation of the appendix)
-logy study or science of (cytology, study of cells)
-oid like, in the form of (thyroid, in the form of a shield, referring to the shape of the thyroid gland)
-oma tumor (carcinoma, a malignant tumor)
-osis indicates disease (psychosis, a mental disease)
-pathy disease (dermopathy, disease of the skin)
-phyll leaf (mesophyll, the middle tissue of the leaf)
-scope instrument for viewing or observing (microscope, instrument for viewing small objects)

Some Common Word Roots

abscis cut off (abscission, the falling off of leaves or other plant parts)
angi, angio vessel (angiosperm, a plant that produces seeds enclosed within a fruit or "vessel")
apic tip, apex (apical meristem, area of cell division located at the tips of plant stems and roots)
arthr joint (arthropods, invertebrate animals with jointed legs and segmented bodies)
aux grow, enlarge (auxin, a plant hormone involved in growth and development)
blast a formative cell, germ layer (osteoblast, cell that gives rise to bone cells)
brachi arm (brachial artery, blood vessel that supplies the arm)
bry grow, swell (embryo, an organism in the early stages of development)
cardi heart (cardiac, pertaining to the heart)
carot carrot (carotene, a yellow, orange, or red pigment in plants)
cephal head (cephalad, toward the head)
cerebr brain (cerebral, pertaining to the brain)
cervic, cervix neck (cervical, pertaining to the neck)
chlor green (chlorophyll, a green pigment found in plants)
chondr cartilage (chondrocyte, a cartilage cell)
chrom color (chromosome, deeply staining body in nucleus)
cili small hair (cilium, a short, fine cytoplasmic hair projecting from the surface of a cell)
coleo a sheath (coleoptile, a protective sheath that encircles the stem in grass seedlings)
conjug joined together (conjugation, a sexual phenomenon in certain protists)
cran skull (cranial, pertaining to the skull)
decid falling off (deciduous, a plant that sheds its leaves at the end of the growing season)
dehis split (dehiscent fruit, a fruit that splits open at maturity)
derm skin (dermatology, study of the skin)
ecol dwelling, house (ecology, the study of organisms in relation to their environment, i.e., "their house")
enter intestine (enterobacteria, a group of bacteria that include species that inhabit the intestines of humans and other animals)
evol to unroll (evolution, descent with modification, or gradual directional change)
fil a thread (filament, the thin stalk of the stamen in flowers)
gamet a wife or husband (gametangium, the part of a plant, protist, or fungus that produces reproductive cells)
gastr stomach (gastrointestinal tract, the digestive tract)
glyc, glyco sweet, sugar (glycogen, storage form of glucose)
gon seed (gonad, an organ that produces gametes)
gutt a drop (guttation, loss of water as liquid "drops" from plants)
gymn naked (gymnosperm, a plant that produces seeds that are not enclosed with a fruit, i.e., "naked")
hem blood (hemoglobin, the pigment of red blood cells)
hepat liver (hepatic, of or pertaining to the liver)
hist tissue (histology, study of tissues)
hydr water (hydrolysis, a breakdown reaction involving water)
leuk white (leukocyte, white blood cell)
menin membrane (meninges, the three membranes that envelop the brain and spinal cord)
morph form (morphogenesis, development of body form)
my, myo muscle (myocardium, muscle layer of the heart)
myc a fungus (mycelium, the vegetative body of a fungus)
nephr kidney (nephron, microscopic unit of the kidney)
neur, nerv nerve (neuromuscular, involving both the nerves and muscles)
occiput back part of the head (occipital, back region of the head)
ost bone (osteology, study of bones)
path disease (pathologist, one who studies disease processes)
ped, pod foot (bipedal, walking on two feet)
pell skin (pellicle, a flexible covering over the body of certain protists)
phag eat (phagocytosis, process by which certain cells ingest particles and foreign matter)
phil love (hydrophilic, a substance that attracts, i.e., "loves," water)
phloe bark of a tree (phloem, food-conducting tissue in plants that corresponds to bark in woody plants)
phyt plant (xerophyte, a plant adapted to xeric, or dry, conditions)
plankt wandering (plankton, microscopic aquatic protists that float or drift passively)
rhiz root (rhizome, a horizontal, underground stem that superficially resembles a root)
scler hard (sclerenchyma, cells that provide strength and support in the plant body)

sipho a tube (siphonous, a type of tubular body form found in certain algae)

som body (chromosome, deeply staining body in the nucleus)

sor heap (sorus, a cluster or "heap" of sporangia in a fern)

spor seed (spore, a reproductive cell that gives rise to individual offspring in plants, protists, and fungi)

stom a mouth (stoma, a small pore, i.e., "mouth," in the epidermis of plants)

thigm a touch (thigmotropism, plant growth in response to touch)

thromb clot (thrombus, a clot within a blood vessel)

tropi turn (thigmotropism, growth of a plant in response to contact with a solid object, as when a tendril "turns" or wraps around a wire fence)

visc pertaining to an internal organ or body cavity (viscera, internal organs)

xanth yellow (xanthophyll, a yellowish pigment found in plants)

xyl wood (xylem, water-conducting tissue in plant, the "wood" of woody plants)

zoo an animal (zoology, the science of animals)

Appendix D
Abbreviations

The biological sciences use a great many abbreviations and with good reason. Many technical terms in biology and biological chemistry are both long and difficult to pronounce. Yet it can be difficult for beginners, when confronted with something like NADPH or EPSP, to understand the reference. Here, for your ready reference, are some of the common abbreviations used in biology.

A adenine
ABA abscisic acid
ACTH adrenocorticotropic hormone
AD Alzheimer's disease
ADA adenosine deaminase
ADH antidiuretic hormone
ADP adenosine diphosphate
AIDS acquired immunodeficiency syndrome
AMP adenosine monophosphate
amu atomic mass unit (dalton)
APC anaphase-promoting complex *or* antigen-presenting cell
ATP adenosine triphosphate
AV node or valve atrioventricular node or valve (of heart)
B lymphocyte or B cell lymphocyte responsible for antibody-mediated immunity
BAC bacterial artificial chromosome
BH brain hormone (of insects)
BMR basal metabolic rate
bya billion years ago
C cytosine
C_3 three-carbon pathway for carbon fixation (Calvin cycle)
C_4 four-carbon pathway for carbon fixation (Hatch-Slack pathway)
CAM crassulacean acid metabolism
cAMP cyclic adenosine monophosphate
CAP catabolite gene activator protein
CD4 T cell T helper cell (T_H); has a surface marker designated CD4
CD8 T cell T cell with a surface marker designated CD8; includes T cytotoxic cells
Cdk cyclin-dependent protein kinase
cDNA complementary deoxyribonucleic acid
CFCs chlorofluorocarbons
CFTR cystic fibrosis transmembrane conductance regulator
CITES the Convention on International Trade in Endangered Species of Wild Flora and Fauna
CNS central nervous system
CO cardiac output
CoA coenzyme A
COPD chronic obstructive pulmonary disease
CP creatine phosphate
CPR cardiopulmonary resuscitation
CR conditioned response
CS conditioned stimulus
CSF cerebrospinal fluid
CVS cardiovascular system
DAG diacylglycerol
DNA deoxyribonucleic acid
DOC dissolved organic carbon
E_A activation energy (of an enzyme)
ECG electrocardiogram
ECM extracellular matrix
EEG electroencephalogram
EKG electrocardiogram
EM electron microscope or micrograph
ENSO El Niño–Southern Oscillation
EPSP excitatory postsynaptic potential (of a neuron)
ER endoplasmic reticulum
ES cells embryonic stem cells
F_1 first filial generation
F_2 second filial generation
Fab portion the part of an antibody that binds to an antigen
Factor VIII blood-clotting factor (absent in hemophiliacs)
FAD/$FADH_2$ flavin adenine dinucleotide (oxidized and reduced forms, respectively)
FAP fixed action pattern
Fc portion the part of an antibody that interacts with cells of the immune system
FISH fluorescent in-situ hybridization
FSH follicle-stimulating hormone
G guanine
G_1 phase first gap phase (of the cell cycle)
G_2 phase second gap phase (of the cell cycle)
G3P glyceraldehyde-3-phosphate
G protein cell-signaling molecule that requires GTP
GA_3 gibberellin
GABA gamma-aminobutyric acid
GH growth hormone (somatotropin)
GnRH gonadotropin-releasing hormone
GTP guanosine triphosphate
HB hemoglobin
HBEF Hubbard Brook Experimental Forest
HBO_2 oxyhemoglobin
HCFCs hydrochlorofluorocarbons
hCG human chorionic gonadotropin
HD Huntington disease
HDL high-density lipoprotein

HFCs hydrofluorocarbons
hGH human growth hormone
HIV human immunodeficiency virus
HLA human leukocyte antigen
IAA indole acetic acid (natural auxin)
Ig immunoglobulin, as in IgA, IgG, etc.
IGF insulin-like growth factor
IP_3 inositol trisphosphate
IPCC United Nations Intergovernmental Panel on Climate Change
IPSP inhibitory postsynaptic potential (of a neuron)
IUCN World Conservation Union
IUD intrauterine device
JH juvenile hormone (of insects)
kb kilobase
LDH lactate dehydrogenase enzyme
LDL low-density lipoprotein
LH luteinizing hormone
LM light microscope or micrograph
LSD lysergic acid diethylamide
LTP long-term potentiation
MAO monoamine oxidase
MAPs microtubule-associated proteins
MHC major histocompatibility complex
MI myocardial infarction
MPF mitosis-promoting factor
MRI magnetic resonance imaging
mRNA messenger RNA
MSAFP maternal serum α-fetoprotein
mtDNA mitochondrial DNA
MTOC microtubule organizing center
mya million years ago
9 + 2 structure cilium or flagellum (of a eukaryote)
9 × 3 structure centriole or basal body (of a eukaryote)
n*, 2*n the chromosome number of a gamete and of a zygote, respectively
NAD^+/NADH nicotinamide adenine dinucleotide (oxidized and reduced forms, respectively)
$NADP^+$/NADPH nicotinamide adenine dinucleotide phosphate (oxidized and reduced forms, respectively)
NAG *N*-acetyl glucosamine
NK cell natural killer cell
NMDA *N*-methyl-D aspartate (an artificial ligand)
NSF National Science Foundation
P generation parental generation
P53 a tumor suppressor gene
P680 reaction center of photosystem II
P700 reaction center of photosystem I
PABA para-aminobenzoic acid
PAMPs pathogen-associated molecular patterns
PCR polymerase chain reaction
PEP phosphoenolpyruvate
Pfr phytochrome (form that absorbs far red light)
PGA phosphoglycerate
PID pelvic inflammatory disease
PIF3 phytochrome-interacting factor-3
PKU phenylketonuria
PNS peripheral nervous system
pre-mRNA precursor messenger RNA (in eukaryotes)
Pr phytochrome (form that absorbs red light)
PTH parathyroid hormone
RAS reticular activating system
RBC red blood cell (erythrocyte)
REM sleep rapid eye movement sleep
RFLP restriction fragment length polymorphism
RNA ribonucleic acid
rRNA ribosomal RNA
rubisco ribulose bisphosphate carboxylase/oxygenase
RuBP ribulose bisphosphate
S phase DNA synthetic phase (of the cell cycle)
SA node sinoatrial node (of heart)
SCID severe combined immunodeficiency
SEM scanning electron microscope or micrograph
snRNP small nuclear ribonucleoprotein complex
SSB protein single-strand DNA-binding protein
ssp subspecies
STD sexually transmitted disease
STR short tandem repeat
T thymine
T lymphocyte or T cell lymphocyte responsible for cell-mediated immunity
T_C lymphocyte T cytotoxic cell
T_H lymphocyte T helper cell (CD4 T cell)
TATA box base sequence in eukaryotic promoter
TCA cycle tricarboxylic acid cycle (synonym for citric acid cycle)
TCR T-cell receptor
TEM transmission electron microscope or micrograph
Tm tubular transport maximum
TNF tumor necrosis factor
tRNA transfer RNA
U uracil
UPE upstream promoter element
UR unconditioned response
US unconditioned stimulus
UV light ultraviolet light
WBC white blood cell (leukocyte)

Glossary

abiotic factors Elements of the nonliving, physical environment that affect a particular organism. Compare with *biotic factors.*

abscisic acid (ab-sis′ik) A plant hormone involved in dormancy and responses to stress.

abscission (ab-sizh′en) The normal (usually seasonal) fall of leaves or other plant parts, such as fruits or flowers.

abscission layer The area at the base of the petiole where the leaf will break away from the stem. Also known as *abscission zone.*

absorption (ab-sorp′shun) (1) The movement of nutrients and other substances through the wall of the digestive tract and into the blood or lymph. (2) The process by which chlorophyll takes up light for photosynthesis.

absorption spectrum A graph of the amount of light at specific wavelengths that is absorbed as light passes through a substance. Each type of molecule has a characteristic absorption spectrum. Compare with *action spectrum.*

accessory fruit A fruit consisting primarily of tissue other than ovary tissue, e.g., apple, pear. Compare with *aggregate, simple,* and *multiple fruits.*

acclimitization Adjustment to seasonal changes.

acetyl coenzyme A (acetyl CoA) (as′uh-teel) A key intermediate compound in metabolism; consists of a two-carbon acetyl group covalently bonded to coenzyme A.

acetyl group A two-carbon group derived from acetic acid (acetate).

acetylcholine (ah′see-til-koh′leen) A common neurotransmitter released by cholinergic neurons, including motor neurons.

achene (a-keen′) A simple, dry fruit with one seed in which the fruit wall is separate from the seed coat, e.g., sunflower fruit.

acid A substance that is a hydrogen ion (proton) donor; acids unite with bases to form salts. Compare with *base.*

acidic solution A solution in which the concentration of hydrogen ions $[H^+]$ exceeds the concentration of hydroxide ions $[OH^-]$. An acidic solution has a pH less than 7. Compare with *basic solution* and *neutral solution.*

acid precipitation Precipitation that is acidic as a result of both sulfur and nitrogen oxides forming acids when they react with water in the atmosphere.

acoelomate (a-seel′oh-mate) An animal lacking a body cavity (coelom). Compare with *coelomate* and *pseudocoelomate.*

acquired immune responses See *specific immune responses.*

acquired immunodeficiency syndrome (AIDS) A serious, potentially fatal disease caused by the human immunodeficiency virus (HIV).

acromegaly (ak″roh-meg′ah-lee) A condition characterized by overgrowth of the extremities of the skeleton, fingers, toes, jaws, and nose. It may be produced by excessive secretion of growth hormone by the anterior pituitary gland.

acrosome reaction (ak′roh-sohm) A series of events in which the acrosome, a caplike structure covering the head of a sperm cell, releases proteolytic (protein-digesting) enzymes and undergoes other changes that permit the sperm to penetrate the outer coverings of the egg.

actin (ak′tin) The protein of which microfilaments consist. Actin, together with the protein myosin, is responsible for muscle contraction.

actin filaments Thin filaments consisting mainly of the protein actin; actin and myosin filaments make up the myofibrils of muscle fibers.

actinopods (ak-tin′o-podz) Protozoa characterized by axopods that protrude through pores in their shells. See *radiolarians.*

action potential An electric signal resulting from depolarization of the plasma membrane in a neuron or muscle cell. Compare with *resting potential.*

action spectrum A graph of the effectiveness of light at specific wavelengths in promoting a light-requiring reaction. Compare with *absorption spectrum.*

activation energy (E_A) The kinetic energy required to initiate a chemical reaction.

activator protein A positive regulatory protein that stimulates transcription when bound to DNA. Compare with *repressor protein.*

active immunity Immunity that develops as a result of exposure to antigens; it can occur naturally after recovery from a disease or can be artificially induced by immunization with a vaccine. Compare with *passive immunity.*

active site A specific region of an enzyme (generally near the surface) that accepts one or more substrates and catalyzes a chemical reaction. Compare with *allosteric site.*

active transport Transport of a substance across a membrane that does not rely on the potential energy of a concentration gradient for the substance being transported and therefore requires an additional energy source (often ATP); includes carrier-mediated active transport, endocytosis, and exocytosis. Compare with *diffusion* and *facilitated diffusion.*

adaptation (1) An evolutionary modification that improves an organism′s chances of survival and reproductive success. (2) A decline in the response of a receptor subjected to repeated or prolonged stimulation.

adaptive immune responses See *specific immune responses.*

adaptive radiation The evolution of a large number of related species from an unspecialized ancestral organism.

adaptive zone A new ecological opportunity that was not exploited by an ancestral organism; used by evolutionary biologists to explain the ecological paths along which different taxa evolve.

addiction Physical dependence on a drug, generally based on physiological changes that take place in response to the drug; when the drug is withheld, the addict may suffer characteristic withdrawal symptoms.

adenine (ad′eh-neen) A nitrogenous purine base that is a component of nucleic acids and ATP.
adenosine triphosphate (ATP) (a-den′oh-seen) An organic compound containing adenine, ribose, and three phosphate groups; of prime importance for energy transfers in cells.
adhering junction A type of anchoring junction between cells; connects epithelial cells.
adhesion The property of sticking to some other substance. Compare with *cohesion.*
adipose tissue (ad′i-pohs) Tissue in which fat is stored.
adrenal cortex (ah-dree′nul kor′teks) The outer region of each adrenal gland; secretes steroid hormones, including mineralocorticoids and glucocorticoids.
adrenal glands (ah-dree′nul) Paired endocrine glands, one located just superior to each kidney; secrete hormones that help regulate metabolism and help the body cope with stress.
adrenal medulla (ah-dree′nul meh-dull′uh) The inner region of each adrenal gland; secretes epinephrine and norepinephrine.
adrenergic neuron (ad-ren-er′jik) A neuron that releases norepinephrine or epinephrine as a neurotransmitter. Compare with *cholinergic neuron.*
adventitious (ad′ven-tish′us) Of plant organs, such as roots or buds, that arise in an unusual position on a plant.
aerobe Organism that grows or metabolizes only in the presence of molecular oxygen. Compare with *anaerobe.*
aerobic (air-oh′bik) Growing or metabolizing only in the presence of molecular oxygen. Compare with *anaerobic.*
aerobic respiration See *respiration.*
afferent (af′fer-ent) Leading toward some point of reference. Compare with *efferent.*
afferent neurons Neurons that transmit action potentials from sensory receptors to the brain or spinal cord. Compare with *efferent neurons.*
age structure The number and proportion of people at each age in a population. Age structure diagrams represent the number of males and females at each age, from birth to death, in the population.
aggregate fruit A fruit that develops from a single flower with many separate carpels, e.g., raspberry. Compare with *simple, accessory,* and *multiple fruits.*
aggregated distribution See *clumped dispersion.*
aging Progressive changes in development in an adult organism.
agnathans (ag-na′thanz) Jawless fishes; historical class of vertebrates, including lampreys, hagfishes, and many extinct forms.
albinism (al′bih-niz-em) A hereditary inability to form melanin pigment, resulting in light coloration.
AIDS See *acquired immunodeficiency syndrome.*
albumin (al-bew′min) A class of protein found in most animal tissues; a fraction of plasma proteins.
aldehyde An organic molecule containing a carbonyl group bonded to at least one hydrogen atom. Compare with *ketone.*
aldosterone (al-dos′tur-ohn) A steroid hormone produced by the vertebrate adrenal cortex; stimulates sodium reabsorption. See *mineralocorticoids.*
algae (al′gee) (sing., *alga*) An informal group of unicellular, or simple multicellular, photosynthetic protists that are important producers in aquatic ecosystems.
allantois (a-lan′toe-iss) An extraembryonic membrane of reptiles, birds, and mammals that stores the embryo's nitrogenous wastes; most of the allantois is detached at hatching or birth.
allele frequency The proportion of a specific allele in the population.
alleles (al-leels′) Genes governing variation of the same character that occupy corresponding positions (loci) on homologous chromosomes; alternative forms of a gene.
allelopathy (uh-leel′uh-path″ee) An adaptation in which toxic substances secreted by roots or shed leaves inhibit the establishment of competing plants nearby.
allergen A substance that stimulates an allergic reaction.
allergy A hypersensitivity to some substance in the environment, manifested as hay fever, skin rash, asthma, food allergies, etc.
all-or-none law The principle that neurons transmit an impulse in a similar way no matter how weak or strong the stimulus; the neuron either transmits an action potential (all) or does not (none).
allopatric speciation (al-oh-pa′trik) Speciation that occurs when one population becomes geographically separated from the rest of the species and subsequently evolves. Compare with *sympatric speciation.*
allopolyploid (al″oh-pol′ee-ploid) A polyploid whose chromosomes are derived from two species. Compare with *autopolyploid.*
allosteric regulators Substances that affect protein function by binding to allosteric sites.
allosteric site (al-oh-steer′ik) A site on an enzyme other than the active site, to which a specific substance binds, thereby changing the shape and activity of the enzyme. Compare with *active site.*
alpha (α) helix A regular, coiled type of secondary structure of a polypeptide chain, maintained by hydrogen bonds. Compare with *beta (β)-pleated sheet.*
alpine tundra An ecosystem located in the higher elevations of mountains, above the tree line and below the snow line. Compare with *tundra.*
alternation of generations A type of life cycle characteristic of plants and a few algae and fungi in which they spend part of their life in a multicellular n gametophyte stage and part in a multicellular $2n$ sporophyte stage.
altruistic behavior Behavior in which one individual helps another, seemingly at its own risk or expense.
alveolus (al-vee′o-lus) (pl., *alveoli*) (1) An air sac of the lung through which gas exchange with the blood takes place. (2) Saclike unit of some glands, e.g., mammary glands. (3) One of several flattened vesicles located just inside the plasma membrane in certain protists.
Alzheimer's disease (AD) A progressive, degenerative brain disorder characterized by amyloid plaques and neurofibrillary tangles.
amino acid (uh-mee′no) An organic compound containing an amino group ($—NH_2$) and a carboxyl group (—COOH); may be joined by peptide bonds to form a polypeptide chain.
amino group A weakly basic functional group; abbreviated $—NH_2$.
aminoacyl-tRNA (uh-mee″no-ace′seel) Molecule consisting of an amino acid covalently linked to a transfer RNA.
aminoacyl-tRNA synthetase One of a family of enzymes, each responsible for covalently linking an amino acid to its specific transfer RNA.
ammonification (uh-moe″nuh-fah-kay′shun) The conversion of nitrogen-containing organic compounds to ammonia (NH_3) by certain soil bacteria (ammonifying bacteria); part of the nitrogen cycle.
amniocentesis (am″nee-oh-sen-tee′sis) Sampling of the amniotic fluid surrounding a fetus to obtain information about its development and genetic makeup. Compare with *chorionic villus sampling.*
amnion (am′nee-on) In terrestrial vertebrates, an extraembryonic membrane that forms a fluid-filled sac for the protection of the developing embryo.
amniotes Terrestrial vertebrates: reptiles, birds, and mammals; animals whose embryos are enclosed by an amnion.
amoeba (a-mee′ba) (pl., *amoebas*) A unicellular protozoon that moves by means of pseudopodia.
amphibians Members of vertebrate class that includes salamanders, frogs, and caecilians.

amphipathic molecule (am″fih-pa′thik) A molecule containing both hydrophobic and hydrophilic regions.

ampulla Any small, saclike extension, e.g., the expanded structure at the end of each semicircular canal of the ear.

amylase (am′-uh-laze) Starch-digesting enzyme, e.g., human salivary amylase or pancreatic amylase.

amyloplasts See *leukoplasts.*

anabolic steroids Synthetic androgens that increase muscle mass, physical strength, endurance, and aggressiveness, but cause serious side effects; these drugs are often abused.

anabolism (an-ab′oh-lizm) The aspect of metabolism in which simpler substances are combined to form more complex substances, resulting in the storage of energy, the production of new cell materials, and growth. Compare with *catabolism.*

anaerobe Organism that grows or metabolizes only in the absence of molecular oxygen. See *facultative anaerobe* and *obligate anaerobe.* Compare with *aerobe.*

anaerobic (an″air-oh′bik) Growing or metabolizing only in the absence of molecular oxygen. Compare with *aerobic.*

anaerobic respiration See *respiration.*

anaphase (an′uh-faze) The stage of mitosis, and of meiosis I and II, in which the chromosomes move to opposite poles of the cell; anaphase occurs after metaphase and before telophase.

anaphylaxis (an″uh-fih-lak′sis) An acute allergic reaction following sensitization to a foreign substance or other substance.

ancestral characters See *shared ancestral characters.*

androgen (an′dro-jen) Any substance that has masculinizing properties, such as a sex hormone. See *testosterone.*

androgen-binding protein (ABP) A protein produced by Sertoli cells in the testes; binds and concentrates testosterone.

anemia (uh-nee′mee-uh) A deficiency of hemoglobin or red blood cells.

aneuploidy (an′you-ploy-dee) Any chromosomal aberration in which there are either extra or missing copies of certain chromosomes.

angiosperms (an′jee-oh-spermz″) The traditional name for flowering plants, a very large (about 235,000 species), diverse phylum of plants that form flowers for sexual reproduction and produce seeds enclosed in fruits; include monocots and dicots.

angiotensin I (an-jee-o-ten′sin) A polypeptide produced by the action of renin on a plasma protein (angiotensinogen).

angiotensin II A peptide hormone formed by the action of angiotensin-converting enzyme on angiotensin I; stimulates aldosterone secretion by the adrenal cortex.

animal pole The nonyolky, metabolically active pole of a vertebrate or echinoderm egg. Compare with *vegetal pole.*

anion (an′eye-on) A particle with one or more units of negative charge, such as a chloride ion (Cl^-) or hydroxide ion (OH^-). Compare with *cation.*

anisogamy (an″eye-sog′uh-me) Sexual reproduction involving motile gametes of similar form but dissimilar size. Compare with *isogamy* and *oogamy.*

annelid (an′eh-lid) A member of phylum Annelida; segmented worm such as earthworm.

annual plant A plant that completes its entire life cycle in one year or less. Compare with *perennial* and *biennial.*

antenna complex The arrangement of chlorophyll, accessory pigments, and pigment-binding proteins into light-gathering units in the thylakoid membranes of photoautotrophic eukaryotes. See *reaction center* and *photosystem.*

antennae (sing., *antenna*) Sensory structures characteristic of some arthropod groups.

anterior Toward the head end of a bilaterally symmetrical animal. Compare with *posterior.*

anther (an′thur) The part of the stamen in flowers that produces microspores and, ultimately, pollen grains.

antheridium (an″thur-id′ee-im) (pl., *antheridia*) In plants, the multicellular male gametangium (sex organ) that produces sperm cells. Compare with *archegonium.*

anthropoid (an′thra-poid) A member of a suborder of primates that includes monkeys, apes, and humans.

antibody (an′tee-bod″ee) A specific protein (immunoglobulin) that recognizes and binds to specific antigens; produced by plasma cells.

antibody-mediated immunity A type of specific immune response in which B cells differentiate into plasma cells and produce antibodies that bind with foreign antigens, leading to the destruction of pathogens. Compare with *cell-mediated immunity.*

anticodon (an″tee-koh′don) A sequence of three nucleotides in transfer RNA that is complementary to, and combines with, the three nucleotide codon on messenger RNA, thus helping to specify the addition of a particular amino acid to the end of a growing polypeptide.

antidiuretic hormone (ADH) (an″ty-dy-uh-ret′ik) A hormone secreted by the posterior lobe of the pituitary that controls the rate of water reabsorption by the kidney.

antigen (an′tih-jen) Any molecule, usually a protein or large carbohydrate, that is specifically recognized as foreign by cells of the immune system.

antigen–antibody complex The combination of antigen and antibody molecules.

antigen-presenting cell (APC) A cell that displays foreign antigens as well as its own surface proteins. Dendritic cells, macrophages, and B cells are APCs.

antimicrobial peptides Soluble molecules that destroy pathogens.

antion-cogene A gene (also known as a *tumor suppressor gene*) whose normal role is to block cell division in response to certain growth inhibiting factors; when mutated, may contribute to the formation of a cancer cell. Compare with *oncogene.*

antioxidants Certain enzymes (e.g., catalase and peroxidase), vitamins, and other substances that destroy free radicals and other reactive molecules. Compare with *oxidants.*

anus (ay′nus) The distal end and outlet of the digestive tract.

aorta (ay-or′tah) The largest and main systemic artery of the vertebrate body; arises from the left ventricle and branches to distribute blood to all parts of the body except the lungs.

aphotic region (ay-fote′ik) The lower layer of the ocean (deeper than 100 m or so) where light does not penetrate.

apical dominance (ape′ih-kl) The inhibition of lateral buds by a shoot tip.

apical meristem (mehr′ih-stem) An area of dividing tissue, located at the tip of a shoot or root, that gives rise to primary tissues; apical meristems cause an increase in the length of the plant body. Compare with *lateral meristems.*

apicomplexans A group of parasitic protozoa that lack structures for locomotion and that produce sporozoites as infective agents; malaria is caused by an apicomplexan. Also called *sporozoa.*

apoenzyme (ap″oh-en′zime) Protein portion of an enzyme; requires the presence of a specific coenzyme to become a complete functional enzyme.

apomixis (ap″uh-mix′us) A type of reproduction in which fruits and seeds are formed asexually.

apoplast A continuum consisting of the interconnected, porous plant cell walls, along which water moves freely. Compare with *symplast.*

apoptosis (ap-uh-toe′sis) Programmed cell death; apoptosis is a normal part of an organism's development and maintenance. Compare with *necrosis.*

aposematic coloration The conspicuous coloring of a poisonous or distasteful organism that enables potential predators to easily see and recognize it. Also called *warning coloration.* Compare with *cryptic coloration.*

arachnids (ah-rack′nids) Eight-legged arthropods such as spiders, scorpions, ticks, and mites.

arachnoid The middle of the three meningeal layers that cover and protect the brain and spinal cord; see *pia mater* and *dura mater.*

archaebacteria (ar″kuh-bak-teer′ee-uh) Prokaryotic organisms with a number of features, such as the absence of peptidoglycan in their cell walls, that set them apart from the rest of the bacteria. Compare with *eubacteria.*

archegonium (ar′ke-go′nee-um) (pl., *archegonia*) In plants, the multicellular female gametangium (sex organ) that contains an egg. Compare with *antheridium.*

archenteron (ark-en′ter-on) The central cavity of the gastrula stage of embryonic development that is lined with endoderm; primitive digestive system.

arctic tundra See *tundra.*

arterial pulse See *pulse, arterial.*

arteriole (ar-teer′ee-ole) A very small artery. Vasoconstriction and vasodilation of arterioles help regulate blood pressure.

artery A thick-walled blood vessel that carries blood away from a heart chamber and toward the body organs. Compare with *vein.*

arthropod (ar′throh-pod) An invertebrate that belongs to phylum Arthropoda; characterized by a hard exoskeleton, a segmented body, and paired, jointed appendages.

artificial insemination The impregnation of a female by artificially introducing sperm from a male.

artificial selection The selection by humans of traits that are desirable in plants or animals, and breeding only those individuals that have the desired traits.

ascocarp (ass′koh-karp) The fruiting body of an ascomycete.

ascomycete (ass″koh-my′seat) Member of a phylum of fungi characterized by the production of nonmotile asexual conidia and sexual ascospores.

ascospore (ass′koh-spor) One of a set of sexual spores, usually eight, contained in a special spore case (an ascus) of an ascomycete.

ascus (ass′kus) A saclike spore case in ascomycetes that contains sexual spores called *ascospores.*

asexual reproduction Reproduction in which there is no fusion of gametes and in which the genetic makeup of parent and of offspring is usually identical. Compare with *sexual reproduction.*

assimilation (of nitrogen) The conversion of inorganic nitrogen (nitrate, NO_3^-, or ammonia, NH_3) to the organic molecules of living things; part of the nitrogen cycle.

association areas Areas of the brain that link sensory and motor areas; responsible for thought, learning, memory, language abilities, judgment, and personality.

association neuron See *interneuron.*

assortative mating Sexual reproduction in which individuals pair nonrandomly, i.e., select mates on the basis of phenotype.

asters Clusters of microtubules radiating out from the poles in dividing cells that have centrioles.

astrocyte A type of glial cell; some are phagocytic; others regulate the composition of the extracellular fluid in the central nervous system.

atherosclerosis (ath″ur-oh-skle-row′sis) A progressive disease in which lipid deposits accumulate in the inner lining of arteries, leading eventually to impaired circulation and heart disease.

atom The smallest quantity of an element that retains the chemical properties of that element.

atomic mass The total number of protons and neutrons in an atom; expressed in atomic mass units or daltons.

atomic mass unit (amu) The approximate mass of a proton or neutron; also called a *dalton.*

atomic number The number of protons in the atomic nucleus of an atom, which uniquely identifies the element to which the atom corresponds.

ATP See *adenosine triphosphate.*

ATP synthase Large enzyme complex that catalyzes the formation of ATP from ADP and inorganic phosphate by chemiosmosis; located in the inner mitochondrial membrane, the thylakoid membrane of chloroplasts, and the plasma membrane of bacteria.

atrial natriuretic peptide (ANP) A hormone released by the atrium of the heart; helps regulate sodium excretion and lowers blood pressure.

atrioventricular (AV) node (ay″tree-oh-ven-trik′you-lur) Mass of specialized cardiac tissue that receives an impulse from the sinoatrial node (pacemaker) and conducts it to the ventricles.

atrioventricular (AV) valve (of the heart) A valve between each atrium and its ventricle that prevents backflow of blood. The right AV valve is the tricuspid valve, the left AV valve is the mitral valve.

atrium (of the heart) (ay′tree-um) A heart chamber that receives blood from the veins.

australopithecines Early hominids that lived between about 4.4 and 1.25 million years ago, based on fossil evidence. Includes several species in two genera, *Ardipithecus* and *Australopithecus.*

autocrine regulation A type of regulation in which a signaling molecule (e.g., a hormone) is secreted into interstitial fluid and then acts on the cells that produce it. Compare with *paracrine regulation.*

autoimmune disease (aw″toh-ih-mune′) A disease in which the body produces antibodies against its own cells or tissues. Also called *autoimmunity.*

autonomic nervous system (aw-tuh-nom′ik) The portion of the peripheral nervous system that controls the visceral functions of the body, e.g., regulates smooth muscle, cardiac muscle, and glands. Its divisions are the sympathetic and parasympathetic nervous systems. Compare with *somatic nervous system.*

autopolyploid A polyploid whose chromosomes are derived from a single species. Compare with *allopolyploid.*

autoradiography Method for detecting of radioactive decay; radiation causes the appearance of dark silver grains in special x-ray film.

autosome (aw′toh-sohm) A chromosome other than the sex (X and Y) chromosomes.

autotroph (aw′toh-trof) An organism that synthesizes complex organic compounds from simple inorganic raw materials; also called producer or primary producer. Compare with *heterotroph.* See *chemoautotroph* and *photoautotroph.*

auxin (awk′sin) A plant hormone involved in various aspects of growth and development, such as stem elongation, apical dominance, and root formation on cuttings, e.g., indole acetic acid (IAA).

avirulence Properties that render an infectious agent nonlethal, that is, unable to cause disease in its host. Compare with *virulence.*

Avogadro's number The number of units (6.02×10^{23}) present in one mole of any substance.

axillary bud A bud in the axil of a leaf. Compare with *terminal bud.*

axon (aks′on) The long extension of the neuron that transmits nerve impulses away from the cell body. Compare with *dendrite.*

axopods (aks′o-podz) Long, filamentous, cytoplasmic projections characteristic of actinopods.

B cell (B lymphocyte) The type of white blood cell responsible for antibody-mediated immunity. When stimulated, B cells differentiate to become plasma cells that produce antibodies. Compare with *T cell.*

bacillus (bah-sill′us) (pl., *bacilli*) A rod-shaped bacterium. Compare with *coccus, spirillum, vibrio,* and *spirochete.*

background extinction The continuous, low-level extinction of species that has occurred throughout much of the history of life. Compare with *mass extinction.*

bacteria (bak-teer′ee-uh) A general term for two groups of unicellular, prokaryotic microorganisms, the archaebacteria and eubacteria. Most bacteria are decomposers, but some are parasites and others are autotrophs.

bacterial artificial chromosome (BAC) Genetically engineered segments of chromosomal DNA with the ability to carry large segments of foreign DNA

bacteriophage (bak-teer′ee-oh-fayj) A virus that infects a bacterium (literally, "bacteria eater"). Also called *phage.*

balanced polymorphism (pol″ee-mor′fizm) The presence in a population of two or more genetic variants that are maintained in a stable frequency over several generations.

bark The outermost covering over woody stems and roots; consists of all plant tissues located outside the vascular cambium.

baroreceptors (bare″oh-ree-sep′torz) Receptors within certain blood vessels that are stimulated by changes in blood pressure.

Barr body A condensed and inactivated X chromosome appearing as a distinctive dense spot in the interphase nucleus of certain cells of female mammals.

basal body (bay′sl) Structure involved in the organization and anchorage of a cilium or flagellum. Structurally similar to a centriole; each is in the form of a cylinder composed of nine triplets of microtubules (9×3 structure).

basal metabolic rate (BMR) The amount of energy expended by the body at resting conditions, when no food is being digested and no voluntary muscular work is being performed.

base (1) A substance that is a hydrogen ion (proton) acceptor; bases unite with acids to form salts. Compare with *acid.* (2) A nitrogenous base in a nucleotide or nucleic acid. See *purines* and *pyrimidines.*

basement membrane The thin, noncell layer of an epithelial membrane that attaches to the underlying tissues; consists of tiny fibers and polysaccharides produced by the epithelial cells.

base-substitution mutation A change in one base pair in DNA. See *missense mutation* and *nonsense mutation.*

basic solution A solution in which the concentration of hydroxide ions $[OH^-]$ exceeds the concentration of hydrogen ions $[H^+]$. A basic solution has pH greater than 7. Compare with *acidic solution* and *neutral solution.*

basidiocarp (ba-sid′ee-o-karp) The fruiting body of a basidiomycete, e.g., a mushroom.

basidiomycete (ba-sid″ee-o-my′seat) Member of a phylum of fungi characterized by the production of sexual basidiospores.

basidiospore (ba-sid′ee-o-spor) One of a set of sexual spores, usually four, borne on a basidium of a basidiomycete.

basidium (ba-sid′ee-um) The clublike spore-producing organ of basidiomycetes that bears sexual spores called *basidiospores.*

basilar membrane The multicellular tissue in the inner ear that separates the cochlear duct from the tympanic canal; the sensory cells of the organ of Corti rest on this membrane.

Batesian mimicry (bate′see-un mim′ih-kree) The resemblance of a harmless or palatable species to one that is dangerous, unpalatable, or poisonous. Compare with *Müllerian mimicry.*

behavioral ecology The scientific study of behavior in natural environments from the evolutionary perspective.

behavioral isolation A prezygotic reproductive isolating mechanism in which reproduction between similar species is prevented because each group exhibits its own characteristic courtship behavior; also called *sexual isolation.*

bellwether species An organism that provides an early warning of environmental damage. Examples include lichens, which are very sensitive to air pollution, and amphibians, which are sensitive to a wide variety of environmental stressors.

benthos (ben′thos) Bottom-dwelling sea organisms that fix themselves to one spot, burrow into the sediment, or simply walk about on the ocean floor.

berry A simple, fleshy fruit in which the fruit wall is soft throughout, e.g., tomato, banana, grape.

beta (β) oxidation Process by which fatty acids are converted to acetyl CoA before entry into the citric acid cycle.

beta (β)-pleated sheet A regular, folded, sheetlike type of protein secondary structure, resulting from hydrogen bonding between two different polypeptide chains or two regions of the same polypeptide chain. Compare with *alpha (α) helix.*

biennial plant (by-en ee-ul) A plant that takes two years to complete its life cycle. Compare with *annual* and *perennial.*

bilateral symmetry A body shape with right and left halves that are approximately mirror images of one another. Compare with *radial symmetry.*

bile The fluid secreted by the liver; emulsifies fats.

binary fission (by′nare-ee fish′un) Equal division of a cell or organism into two; a type of asexual reproduction.

binomial system of nomenclature (by-nome′ee-ul) System of naming a species by the combination of the genus name and a specific epithet.

bioaccumulation The buildup of a persistent toxic substance, such as certain pesticides, in an organism's body.

biodiversity See *biological diversity.*

biogenic amines A class of neurotransmitters that includes norepinephrine, serotonin, and dopamine.

biogeochemical cycle (bye″o-jee″o-kem′ee-kl) Process by which matter cycles from the living world to the nonliving, physical environment and back again, e.g., the carbon cycle, the nitrogen cycle, and the phosphorus cycle.

biogeography The study of the past and present geographic distributions of organisms.

bioinformatics The storage, retrieval, and comparison of DNA sequences within a given species and among different species.

biological clocks Mechanisms by which activities of organisms are adapted to regularly recurring changes in the environment. See *circadian rhythm.*

biological diversity The variety of living organisms considered at three levels: genetic diversity, species diversity, and ecosystem diversity. Also called *biodiversity.*

biological magnification The increased concentration of toxic chemicals, such as PCBs, heavy metals, and certain pesticides, in the tissues of organisms at higher trophic levels in food webs.

biological species concept See *species.*

biomass (bye′o-mas) A quantitative estimate of the total mass, or amount, of living material in a particular ecosystem.

biome (by′ohm) A large, relatively distinct terrestrial region characterized by a similar climate, soil, plants, and animals, regardless of where it occurs on Earth.

bioremediation A method to clean up a hazardous waste site that uses microorganisms to break down toxic pollutants, or plants to selectively accumulate toxins.

biosphere All of Earth's living organisms, collectively.

biotic factors Elements of the living world that affect a particular organism, that is, its relationships with other organisms. Compare with *abiotic factors.*
biotic potential See *intrinsic rate of increase.*
bipedal Walking on two feet.
bipolar cell A type of neuron in the retina of the eye; receives input from the photoreceptors (rods and cones) and synapses on ganglion cells.
biramous appendages Appendages with two jointed branches at their ends; characteristic of crustaceans.
bivalent (by-vale′ent or biv′ah-lent) See *tetrad.*
blade (1) The thin, expanded part of a leaf. (2) The flat, leaflike structure of certain multicellular algae.
blastocoel (blas′toh-seel) The fluid-filled cavity of a blastula.
blastocyst The mammalian blastula. See *blastula.*
blastodisc A small disc of cytoplasm at the animal pole of a reptile or bird egg; cleavage is restricted to the blastodisc (meroblastic cleavage).
blastopore (blas′toh-pore) The primitive opening into the body cavity of an early embryo that may become the mouth (in protostomes) or anus (in deuterostomes) of the adult organism.
blastula (blas′tew-lah) In animal development, a hollow ball of cells produced by cleavage of a fertilized ovum. In mammalian development, known as a blastocyst.
blood A fluid, circulating connective tissue that transports nutrients and other materials through the bodies of many types of animals.
blood pressure The force exerted by blood against the inner walls of the blood vessels.
bloom The sporadic occurrence of huge numbers of algae in freshwater and marine ecosystems.
body mass index (BMI) An index of weight in relation to height; calculated by dividing the square of the weight (square kilograms) by height (meters).
Bohr effect Increased oxyhemoglobin dissociation due to lowered pH; occurs as carbon dioxide concentration increases.
bolting The production of a tall flower stalk by a plant that grows vegetatively as a rosette (growth habit with a short stem and a circular cluster of leaves).
bond energy The energy required to break a particular chemical bond.
bone tissue Principal vertebrate skeletal tissue; a type of connective tissue.
boreal forest (bor′ee-uhl) See *taiga.*
bottleneck A sudden decrease in a population size caused by adverse environmental factors; may result in genetic drift; also called *genetic bottleneck* or *population bottleneck.*
bottom-up processes Control of ecosystem function by nutrient cycles and other parts of the abiotic environment. Compare with *top-down processes.*
Bowman's capsule A double-walled sac of cells that surrounds the glomerulus of each nephron.
brachiopods (bray′kee-oh-pods) The phylum of solitary marine invertebrates having a pair of shells, and internally, a pair of coiled arms with ciliated tentacles; one of the lophophorate phyla.
brain A concentration of nervous tissue that controls neural function; in vertebrates, the anterior, enlarged portion of the central nervous system.
brain stem The part of the vertebrate brain that includes the medulla, pons, and midbrain.
branchial Pertaining to the gills or gill region.
branching evolution See *cladogenesis.*
bronchiole (bronk′ee-ole) Air duct in the lung that branches from a bronchus; divides to form air sacs (alveoli).
bronchus (bronk′us) (pl., *bronchi*) One branch of the trachea and its immediate branches within the lung.
brown alga One of a phylum of predominantly marine algae that are multicellular and contain the pigments chlorophyll *a* and *c*, and carotenoids, including fucoxanthin.
bryophytes (bry′oh-fites) Nonvascular plants including mosses, liverworts, and hornworts.
bryozoans Animals belonging to phylum Bryozoa, one of the three lophophorate phyla; form sessile colonies by asexual budding.
bud An undeveloped shoot that develops into flowers, stems, or leaves. Buds are enclosed in bud scales.
bud scale A modified leaf that covers and protects a dormant bud.
bud scale scar Scar on a twig left when a bud scale abscises from the terminal bud.
budding Asexual reproduction in which a small part of the parent's body separates from the rest and develops into a new individual; characteristic of yeasts and certain other organisms.
buffer A substance in a solution that tends to lessen the change in hydrogen ion concentration (pH) that otherwise would be produced by adding an acid or base.
bulb A globose, fleshy, underground bud that consists of a short stem with fleshy leaves, e.g., onion.
bundle scar Marks on a leaf scar left when vascular bundles of the petiole break during leaf abscission.
bundle sheath cells Tightly packed cells that form a sheath around the veins of a leaf.
bundle sheath extension Support cells that extend from the bundle sheath of a leaf vein toward the upper and/or lower epidermis.
buttress root A bracelike root at the base of certain trees that provides upright support.
C_3 plant Plant that carries out carbon fixation solely by the Calvin cycle. Compare with *C_4 plant* and *CAM plant.*
C_4 plant Plant that fixes carbon initially by the Hatch-Slack pathway, in which the reaction of CO_2 with phosphoenolpyruvate is catalyzed by PEP carboxylase in leaf mesophyll cells; the products are transferred to the bundle sheath cells, where the Calvin cycle takes place. Compare with *C_3 plant* and *CAM plant.*
calcitonin (kal-sih-toh′nin) A hormone secreted by the thyroid gland that rapidly lowers the calcium content in the blood.
callus (kal′us) Undifferentiated tissue formed on an explant (excised tissue or organ) in plant tissue culture.
calmodulin A calcium-binding protein; when bound, it alters the activity of certain enzymes or transport proteins.
calorie The amount of heat energy required to raise the temperature of 1g of water 1°C; equivalent to 4.184 joules. Compare with *kilocalorie.*
Calvin cycle Cyclic series of reactions in the chloroplast stroma in photosynthesis; fixes carbon dioxide and produces carbohydrate. See *C_3 plant.*
calyx (kay′liks) The collective term for the sepals of a flower.
cambium See *lateral meristems.*
Cambrian explosion A span of 40 million years, from about 565 to 525 million years ago (mya), during which many new animal groups appeared in the fossil record.
CAM plant Plant that carries out crassulacean acid metabolism; carbon is initially fixed into organic acids at night in the reaction of CO_2 and phosphoenolpyruvate, catalyzed by PEP carboxylase; during the day the acids break down to yield CO_2, which enters the Calvin cycle. Compare with *C_3 plant* and *C_4 plant.*
cAMP See *cyclic AMP.*
cancer cells See *malignant.*
CAP See *catabolite gene activator protein.*

capillaries (kap′i-lare-eez) Microscopic blood vessels in the tissues that permit exchange of materials between cells and blood.
capillary action The ability of water to move in small-diameter tubes as a consequence of its cohesive and adhesive properties.
capping See *mRNA cap.*
capsid Protein coat surrounding the nucleic acid of a virus.
capsule (1) The portion of the moss sporophyte that contains spores. (2) A simple, dry, dehiscent fruit that opens along many sutures or pores to release seeds. (3) A gelatinous coat that surrounds some bacteria.
carbohydrate Compound containing carbon, hydrogen, and oxygen, in the approximate ratio of C:2H:O, e.g., sugars, starch, and cellulose.
carbon cycle The worldwide circulation of carbon from the abiotic environment into living things and back into the abiotic environment.
carbon fixation reactions Reduction reactions of photosynthesis in which carbon from carbon dioxide becomes incorporated into organic molecules, leading to the production of carbohydrate. See *Calvin cycle.*
carbonyl group A polar functional group consisting of a carbon attached to an oxygen by a double bond; found in aldehydes and ketones.
carboxyl group A weakly acidic functional group; abbreviated —COOH.
carcinogen (kar-sin′oh-jen) An agent that causes cancer or accelerates its development.
cardiac cycle One complete heartbeat.
cardiac muscle Involuntary, striated type of muscle found in the vertebrate heart. Compare with *smooth muscle* and *skeletal muscle.*
cardiac output The volume of blood pumped by the left ventricle into the aorta in one minute.
cardiovascular disease Disease of the heart or blood vessels; the leading cause of death in most industrial societies.
carnivore (kar′ni-vor) An animal that feeds on other animals; flesh-eater; also called *secondary* or *tertiary consumer.* Secondary consumers eat primary consumers (herbivores), whereas tertiary consumers eat secondary consumers.
carotenoids (ka-rot′n-oidz) A group of yellow to orange plant pigments synthesized from isoprene subunits; include carotenes and xanthophylls.
carpel (kar′pul) The female reproductive unit of a flower; carpels bear ovules. Compare with *pistil.*
carrier-mediated active transport Transport across a membrane of a substance from a region of low concentration to a region of high concentration; requires both a transport protein with a binding site for the specific substance and an energy source (often ATP).
carrier-mediated transport Any form of transport across a membrane that uses a membrane-bound transport protein with a binding site for a specific substance; includes both facilitated diffusion and carrier-mediated active transport.
carrying capacity The largest population that a particular habitat can support and sustain for an indefinite period, assuming there are no changes in the environment.
cartilage A flexible skeletal tissue of vertebrates; a type of connective tissue.
caryopsis See *grain.*
Casparian strip (kas-pare′ee-un) A band of waterproof material around the radial and transverse walls of endodermal root cells.
caspase Any of a group of proteolytic enzymes that are active in the early stages of apoptosis.
catabolism The aspect of metabolism in which complex substances are broken down to form simpler substances; catabolic reactions are particularly important in releasing chemical energy stored by the cell. Compare with *anabolism.*
catabolite gene activator protein (CAP) A positively acting regulator that becomes active when bound to cAMP; active CAP stimulates transcription of the *lac* operon and other operons that code for enzymes used in catabolic pathways. Also known as *cyclic AMP receptor protein (CRP).*
catalyst (kat′ah-list) A substance that increases the speed at which a chemical reaction occurs without being used up in the reaction. Enzymes are biological catalysts.
catecholamine (cat″eh-kole′-ah-meen) A class of compounds including dopamine, epinephrine, and norepinephrine; these compounds serve as neurotransmitters and hormones.
cation A particle with one or more units of positive charge, such as a hydrogen ion (H^+) or calcium ion (Ca^{2+}). Compare with *anion.*
cDNA library A collection of recombinant plasmids that contain complementary DNA (cDNA) copies of mRNA templates. The cDNA, which lacks introns, is synthesized by reverse transcriptase. Compare with *genomic DNA library.*
cell The basic structural and functional unit of life, which consists of living material bounded by a membrane.
cell cycle Cyclic series of events in the life of a dividing eukaryotic cell; consists of mitosis, cytokinesis, and the stages of interphase.
cell determination See *determination.*
cell differentiation See *differentiation.*
cell fractionation The technique used to separate the components of cells by subjecting them to centrifugal force. See *differential centrifugation* and *density gradient centrifugation.*
cell-mediated immunity A type of specific immune response carried out by T cells. Compare with *antibody-mediated immunity.*
cell plate The structure that forms during cytokinesis in plants, separating the two daughter cells produced by mitosis.
cell signaling Mechanisms of communication between cells. Cells signal one another with secreted signaling molecules, or a signaling molecule on one cell combines with a receptor on another cell. See *signal transduction.*
cell theory The theory that the cell is the basic unit of life, of which all living things are composed, and that all cells are derived from pre-existing cells.
cell wall The structure outside the plasma membrane of certain cells; may contain cellulose (plant cells), chitin (most fungal cells), peptidoglycan and/or lipopolysaccharide (most bacterial cells), or other material.
cellular respiration See *respiration.*
cellular slime mold A phylum of fungus-like protists whose feeding stage consists of unicellular, amoeboid organisms that aggregate to form a pseudoplasmodium during reproduction.
cellulose (sel′yoo-lohs) A structural polysaccharide consisting of beta glucose subunits; the main constituent of plant primary cell walls.
Cenozoic era A geological era that began about 65 million years ago and extends to the present time.
center of origin The geographic area where a given species originated.
central nervous system (CNS) In vertebrates, the brain and spinal cord. Compare with *peripheral nervous system (PNS).*
centrifuge A device used to separate cells or their components by subjecting them to centrifugal force.
centriole (sen′tree-ohl) One of a pair of small, cylindrical organelles lying at right angles to each other near the nucleus in the cytoplasm of animal cells and certain protist and plant cells; each centriole is

in the form of a cylinder composed of nine triplets of microtubules (9×3 structure).

centromere (sen′tro-meer) A specialized constricted region of a chromatid; contains the kinetochore. In cells at prophase and metaphase, sister chromatids are joined in the vicinity of their centromeres.

cephalization The evolution of a head; the concentration of nervous tissue and sense organs at the front end of the animal.

cephalochordates Members of the chordate subphylum that includes the lancelets.

cerebellum (ser-eh-bel′um) A convoluted subdivision of the vertebrate brain concerned with coordination of muscular movements, muscle tone, and balance.

cerebral cortex (ser-ee′brul kor′tex) The outer layer of the cerebrum composed of gray matter and consisting mainly of nerve cell bodies.

cerebrospinal fluid (CSF) The fluid that bathes the central nervous system of vertebrates.

cerebrum (ser-ee′brum) A large, convoluted subdivision of the vertebrate brain; in humans, it functions as the center for learning, voluntary movement, and interpretation of sensation.

chaos The tendency of a simple system to exhibit complex, erratic dynamics; used by some ecologists to model the state of flux displayed by some populations.

chaparral (shap″uh-ral′) A biome with a Mediterranean climate (mild, moist winters and hot, dry summers). Chaparral vegetation is characterized by drought-resistant, small-leaved evergreen shrubs and small trees.

chaperones See *molecular chaperones.*

character displacement The tendency for two similar species to diverge (become more different) in areas where their ranges overlap; reduces interspecific competition.

Chargaff's rules A relationship in DNA molecules based on nucleotide composition data; the number of adenines equals the number of thymines, and the number of guanines equals the number of cytosines.

chelicerae (keh-lis′er-ee) The first pair of appendages in certain arthropods; clawlike appendages located immediately anterior to the mouth and used to manipulate food into the mouth.

chemical bond A force of attraction between atoms in a compound. See *covalent bond, hydrogen bond,* and *ionic bond.*

chemical compound Two or more elements combined in a fixed ratio.

chemical evolution The origin of life from nonliving matter.

chemical formula A representation of the composition of a compound; the elements are indicated by chemical symbols with subscripts to indicate their ratios. See *molecular formula, structural formula,* and *simplest formula.*

chemical symbol The abbreviation for an element; usually the first letter (or first and second letters) of the English or Latin name.

chemiosmosis Process by which phosphorylation of ADP to form ATP is coupled to the transfer of electrons down an electron transport chain; the electron transport chain powers proton pumps that produce a proton gradient across the membrane; ATP is formed as protons diffuse through transmembrane channels in ATP synthase.

chemoautotroph (kee″moh-aw′toh-trof) Organism that obtains energy from inorganic compounds and synthesizes organic compounds from inorganic raw materials; includes some bacteria. Compare with *photoautotroph, photoheterotroph,* and *chemoheterotroph.*

chemoheterotroph (kee″moh-het′ur-oh-trof) Organism that uses organic compounds as a source of energy and carbon; includes animals, fungi, and many bacteria. Compare with *photoautotroph, photoheterotroph,* and *chemoautotroph.*

chemoreceptor (kee″moh-ree-sep′tor) A sensory receptor that responds to chemical stimuli.

chemotroph (kee′moh-trof) Organism that uses organic compounds or inorganic substances, such as iron, nitrate, ammonia, or sulfur, as sources of energy. Compare with *phototroph.* See *chemoautotroph* and *chemoheterotroph.*

chiasma (ky-az′muh) (pl., *chiasmata*) An X-shaped site in a tetrad (bivalent) usually marking the location where homologous (nonsister) chromatids previously crossed over.

chimera (ky meer″ uh) An organism consisting of two or more kinds of genetically dissimilar cells.

chitin (ky′tin) A nitrogen-containing structural polysaccharide that forms the exoskeleton of insects and the cell walls of many fungi.

chlorophyll (klor′oh-fil) A group of light-trapping green pigments found in most photosynthetic organisms.

chlorophyll-binding proteins About 15 different proteins associated with chlorophyll molecules in the thylakoid membrane.

chloroplasts (klor′oh-plastz) Membranous organelles that are the sites of photosynthesis in eukaryotes; occur in some plant and algal cells.

cholinergic neuron (kohl″in-air′jik) A neuron that releases acetylcholine as a neurotransmitter. Compare with *adrenergic neuron.*

chondrichthyes (kon-drik′-thees) The class of cartilaginous fishes that includes the sharks, rays, and skates.

chondrocytes Cartilage cells.

chordates (kor′dates) Deuterostome animals that, at some time in their lives, have a cartilaginous, dorsal skeletal structure called a *notochord;* a dorsal, tubular, nerve cord; pharyngeal gill grooves; and a postanal tail.

chorion (kor′ee-on) An extraembryonic membrane in reptiles, birds, and mammals that forms an outer cover around the embryo, and in mammals contributes to the formation of the placenta.

chorionic villus sampling (CVS) (kor″ee-on′ik) Study of extraembryonic cells that are genetically identical to the cells of an embryo, making it possible to assess its genetic makeup. Compare with *amniocentesis.*

choroid layer A layer of cells filled with black pigment that absorbs light and prevents reflected light from blurring the image that falls on the retina; the layer of the eyeball outside the retina.

chromatid (kroh′mah-tid) One of the two identical halves of a duplicated chromosome; the two chromatids that make up a chromosome are referred to as *sister chromatids.*

chromatin (kro′mah-tin) The complex of DNA and protein that makes up eukaryotic chromosomes.

chromoplasts Pigment-containing plastids; usually found in flowers and fruits.

chromosomes Structures in the cell nucleus that consist of chromatin and contain the genes. The chromosomes become visible under the microscope as distinct structures during cell division.

chylomicrons (kie-low-my′kronz) Protein-covered fat droplets produced in the intestinal cells; they enter the lymphatic system and are transported to the blood.

chytrid See *chytridiomycete.*

chytridiomycete (ki-trid″ee-o-my′seat) A member of a phylum of fungi characterized by the production of flagellate cells at some stage in their life history. Also called *chytrid.*

ciliate (sil′e-ate) A unicellular protozoon covered by many short cilia.

cilium (sil′ee-um) (pl., *cilia*) One of many short, hairlike structures that project from the surface of some eukaryotic cells and are used for locomotion or movement of materials across the cell surface.

circadian rhythm (sir-kay′dee-un) An internal rhythm that approximates the 24-hour day. See *biological clocks.*

circulatory system The body system that functions in internal transport and protects the body from disease.

cisternae (sing., *cisterna*) Stacks of flattened membranous sacs that make up the Golgi complex.
citrate (citric acid) A six-carbon organic acid.
citric acid cycle Series of chemical reactions in aerobic respiration in which acetyl coenzyme A is completely degraded to carbon dioxide and water with the release of metabolic energy that is used to produce ATP; also known as the *Krebs cycle* and the *tricarboxylic acid (TCA) cycle.*
clade A taxon containing a common ancestor and all the taxa descended from it; a monophyletic group.
cladistics An approach to classification based on recency of common ancestry rather than degree of structural similarity. Also called *phylogenetic systematics.* Compare with *phenetics* and *evolutionary systematics.*
cladogram A branching diagram that illustrates taxonomic relationships based on the principles of cladistics.
class A taxonomic category made up of related orders.
classical conditioning A type of learning in which an association is formed between some normal response to a stimulus and a new stimulus, after which the new stimulus elicits the response.
cleavage Series of mitotic cell divisions, without growth, that converts the zygote to a multicellular blastula.
cleavage furrow A constricted region of the cytoplasm that forms and progressively deepens during cytokinesis of animal cells, thereby separating the two daughter cells.
cline Gradual change in phenotype and genotype frequencies among contiguous populations that is the result of an environmental gradient.
clitoris (klit′o-ris) A small, erectile structure at the anterior part of the vulva in female mammals; homologous to the male penis.
cloaca (klow-a′ka) An exit chamber in some animals that receives digestive wastes and urine; may also serve as an exit for gametes.
clonal selection Lymphocyte activation in which a specific antigen causes activation, cell division, and differentiation only in cells that express receptors with which the antigen binds.
clone (1) A population of cells descended by mitotic division from a single ancestral cell. (2) A population of genetically identical organisms asexually propagated from a single individual. Also see *DNA cloning.*
cloning The process of forming a clone.
closed circulatory system A type of circulatory system in which the blood flows through a continuous circuit of blood vessels; characteristic of annelids, cephalopods, and vertebrates. Compare with *open circulatory system.*
closed system An entity that doesn't exchange energy with its surroundings. Compare with *open system.*
club mosses A phylum of seedless vascular plants with a life cycle similar to ferns.
clumped dispersion The spatial distribution pattern of a population in which individuals are more concentrated in specific parts of the habitat. Also called *aggregated distribution* and *patchiness.* Compare with *random dispersion* and *uniform dispersion.*
cnidarians (ni-dah′ree-anz) Phylum of animals that have stinging cells called *cnidocytes,* two tissue layers, and radial symmetry; include hydras and jellyfish.
cnidocytes Stinging cells characteristic of cnidarians.
coated pit A depression in the plasma membrane, the cytosolic side of which is coated with the protein clathrin; important in receptor-mediated endocytosis.
cochlea (koke′lee-ah) The structure of the inner ear of mammals that contains the auditory receptors (organ of Corti).
coccus (kok′us) (pl., *cocci*) A bacterium with a spherical shape. Compare with *bacillus, spirillum, vibrio,* and *spirochete.*
codominance (koh″dom′in-ants) Condition in which two alleles of a locus are expressed in a heterozygote.
codon (koh′don) A triplet of mRNA nucleotides. The 64 possible codons collectively constitute a universal genetic code in which each codon specifies an amino acid in a polypeptide, or a signal to either start or terminate polypeptide synthesis.
coelacanths A genus of lobe-finned fish that have survived to the present day.
coelom (see′lum) The main body cavity of most animals; a true coelom is lined with mesoderm. Compare with *pseudocoelom.*
coelomate (seel′oh-mate) Animal that has a true coelom. Compare with *acoelomate* and *pseudocoelomate.*
coenocyte (see′no-site) An organism consisting of a multinucleate cell, i.e., the nuclei are not separated from one another by septa.
coenzyme (koh-en′zime) An organic cofactor for an enzyme; generally participates in the reaction by transferring some component, such as electrons or part of a substrate molecule.
coenzyme A (CoA) Organic cofactor responsible for transferring groups derived from organic acids.
coevolution The reciprocal adaptation of two or more species that occurs as a result of their close interactions over a long period.
cofactor A nonprotein substance needed by an enzyme for normal activity; some cofactors are inorganic (usually metal ions); others are organic (coenzymes).
cohesion The property of sticking together. Compare with *adhesion.*
cohort A group of individuals of the same age.
colchicine A drug that blocks the division of eukaryotic cells by binding to tubulin subunits, which make up the microtubules of which the major component of the mitotic spindle consists.
coleoptile (kol-ee-op′tile) A protective sheath that encloses the young stem in certain monocots.
collagens (kol′ah-gen) Proteins found in the collagen fibers of connective tissues.
collecting duct A tube in the kidney that receives filtrate from several nephrons and conducts it to the renal pelvis.
collenchyma (kol-en′kih-mah) Living cells with moderately but unevenly thickened primary cell walls; collenchyma cells help support the herbaceous plant body.
commensalism (kuh-men′sul-iz-m) A type of symbiosis in which one organism benefits and the other one is neither harmed nor helped. Compare with *mutualism* and *parasitism.*
commercial harvest The collection of commercially important organisms from the wild. Examples include the commercial harvest of parrots (for the pet trade) and cacti (for houseplants).
community An association of populations of different species living together in a defined habitat with some degree of interdependence. Compare with *ecosystem.*
community ecology The description and analysis of patterns and processes within the community.
compact bone Dense, hard bone tissue found mainly near the surfaces of a bone.
companion cell A cell in the phloem of flowering plants that governs loading and unloading sugar into the sieve tube element for translocation.
competition The interaction among two or more individuals that attempt to use the same essential resource, such as food, water, sunlight, or living space. See *interspecific* and *intraspecific competition.* See *interference* and *exploitation competition.*
competitive exclusion principle The concept that no two species with identical living requirements can occupy the same ecological niche indefinitely.

competitive inhibitor A substance that binds to the active site of an enzyme, thus lowering the rate of the reaction catalyzed by the enzyme. Compare with *noncompetitive inhibitor.*

complement A group of proteins in blood and other body fluids that are activated by an antigen-antibody complex, and then destroy pathogens.

complementary DNA (cDNA) DNA synthesized by reverse transcriptase, using RNA as a template.

complete flower A flower that has all four parts: sepals, petals, stamens, and carpels. Compare with *incomplete flower.*

compound eye An eye, such as that of an insect, consisting of many light-sensitive units called *ommatidia.*

concentration gradient A difference in the concentration of a substance from one point to another, as for example, across a cell membrane.

condensation synthesis A reaction in which two monomers are combined covalently through the removal of the equivalent of a water molecule. Compare with *hydrolysis.*

cone (1) In botany, a reproductive structure in many gymnosperms that produces either microspores or megaspores. (2) In zoology, one of the conical photoreceptive cells of the retina that is particularly sensitive to bright light and, by distinguishing light of various wavelengths, mediates color vision. Compare with *rod.*

conidiophore (kah-nid′e-o-for″) A specialized hypha that bears conidia.

conidium (kah-nid′e-um) (pl., *conidia*) An asexual spore that is usually formed at the tip of a specialized hypha called a *conidiophore.*

conifer (kon′ih-fur) Any of a large phylum of gymnosperms that are woody trees and shrubs with needle-like, mostly evergreen, leaves and with seeds in cones.

conjugation (kon″jew-gay′shun) (1) A sexual process in certain protists that involves exchange or fusion of a cell with another cell. (2) A mechanism for DNA exchange in bacteria that involves cell-to-cell contact.

connective tissue Animal tissue consisting mostly of intercellular substance (fibers scattered through a matrix) in which the cells are embedded, e.g., bone.

conservation biology A multidisciplinary science that focuses on the study of how humans impact organisms and on the development of ways to protect biological diversity.

constitutive gene A gene that is constantly transcribed.

consumer See *heterotroph.*

consumption overpopulation A situation in which each individual in a human population consumes too large a share of resources; results in pollution, environmental degradation, and resource depletion. Compare with *people overpopulation.*

contest competition See *interference competition.*

continental drift The theory that continents were once joined together and later split and drifted apart.

contraception Any method used to intentionally prevent pregnancy.

contractile root (kun-trak′til) A specialized type of root that contracts and pulls a bulb or corm deeper into the soil.

contractile vacuole A membrane-bounded organelle found in certain freshwater protists, such as *Paramecium;* appears to have an osmoregulatory function.

control group In a scientific experiment, a group in which the experimental variable is kept constant. The control provides a standard of comparison used to verify the results of the experiment.

controlled mating A mating in which the genotypes of the parents are known.

convergent circuit (kun-vur′jent) A neural pathway in which a postsynaptic neuron is controlled by signals coming from two or more presynaptic neurons. Compare with *divergent circuit.*

convergent evolution (kun-vur′jent) The independent evolution of structural or functional similarity in two or more distantly related species, usually as a result of adaptations to similar environments.

corepressor Substance that binds to a repressor protein, converting it to its active form, which is capable of preventing transcription.

Coriolis effect (kor″e-o′lis) The tendency of moving air or water to be deflected from its path to the right in the Northern Hemisphere and to the left in the Southern Hemisphere. Caused by the direction of Earth's rotation.

cork cambium (kam′bee-um) A lateral meristem that produces cork cells and cork parenchyma; cork cambium and the tissues it produces make up the outer bark of a woody plant. Compare with *vascular cambium.*

cork cell A cell in the bark that is produced outwardly by the cork cambium; cork cells are dead at maturity and function for protection and reduction of water loss.

cork parenchyma (par-en′kih-mah) One or more layers of parenchyma cells produced inwardly by the cork cambium.

corm A short, thickened underground stem specialized for food storage and asexual reproduction, e.g., crocus, gladiolus.

cornea (kor′nee-ah) The transparent covering of an eye.

corolla (kor-ohl′ah) Collectively, the petals of a flower.

corpus callosum (kah-loh′sum) In mammals, a large bundle of nerve fibers interconnecting the two cerebral hemispheres.

corpus luteum (loo′tee″um) The temporary endocrine tissue in the ovary that develops from the ruptured follicle after ovulation; secretes progesterone and estrogen.

cortex (kor′tex) (1) The outer part of an organ, such as the cortex of the kidney. Compare with *medulla.* (2) The tissue between the epidermis and vascular tissue in the stems and roots of many herbaceous plants.

cortical reaction Process occurring after fertilization that prevents additional sperm from entering the egg; also known as the "slow block to polyspermy."

cosmid cloning vector A cloning vector with features of both bacteriophages and plasmids, and with the ability to carry large segments of foreign DNA.

cosmopolitan species Species that have a nearly worldwide distribution and occur on more than one continent or throughout much of the ocean. Compare with *endemic species.*

cotransport The active transport of a substance from a region of low concentration to a region of high concentration by coupling its transport to the transport of a substance down its concentration gradient.

cotyledon (kot″uh-lee′dun) The seed leaf of a plant embryo, which may contain food stored for germination.

cotylosaurs The first reptiles; also known as *stem reptiles.*

countercurrent exchange system A biological mechanism that enables maximum exchange between two fluids. The two fluids must be flowing in opposite directions and have a concentration gradient between them.

coupled reactions A set of reactions in which an exergonic reaction provides the free energy required to drive an endergonic reaction; energy coupling generally occurs through a common intermediate.

covalent bond The chemical bond involving shared pairs of electrons; may be single, double, or triple (with one, two, or three shared pairs of electrons, respectively). Compare with *ionic bond* and *hydrogen bond.*

covalent compound A compound in which atoms are held together by covalent bonds; covalent compounds consist of molecules. Compare with *ionic compound.*

cranial nerves The 10 to 12 pairs of nerves in vertebrates that emerge directly from the brain.

cranium The bony framework that protects the brain in vertebrates.
crassulacean acid metabolism See *CAM plant.*
creatine phosphate An energy-storing compound in muscle cells.
cretinism (kree′tin-izm) A chronic condition caused by lack of thyroid secretion during fetal development and early childhood; results in retarded physical and mental development if untreated.
cri-du-chat A human genetic disease caused by losing part of the short arm of chromosome 5 and characterized by mental retardation, a cry that sounds like a kitten mewing, and by death in infancy or childhood.
cristae (kris′tee) (sing., *crista*) Shelflike or finger-like inward projections of the inner membrane of a mitochondrion.
Cro-Magnons Prehistoric humans *(Homo sapiens)* with modern features (tall, erect, lacking a heavy brow) who lived in Europe some 30,000 years ago.
cross bridges The connections between myosin and actin filaments in muscle fibers; formed by the binding of myosin heads to active sites on actin filaments.
crossing-over A process in which genetic material (DNA) is exchanged between paired, homologous chromosomes.
CRP See *catabolite gene activator protein.*
cryptic coloration Colors or markings that help some organisms hide from predators by blending into their physical surroundings. Compare with *aposematic coloration.*
cryptochrome A proteinaceous pigment that strongly absorbs blue light: implicated in resetting the biological clock in plants, fruit flies, and mice.
ctenophores (ten′oh-forz) Phylum of marine animals (comb jellies) whose bodies consist of two layers of cells enclosing a gelatinous mass. The outer surface is covered with comblike rows of cilia, by which the animal moves.
cuticle (kew′tih-kl) (1) A noncell, waxy covering over the epidermis of the aerial parts of plants that reduces water loss. (2) The outer covering of some animals, such as roundworms.
cyanobacteria (sy-an″oh-bak-teer′ee-uh) Prokaryotic photosynthetic microorganisms that possess chlorophyll and produce oxygen during photosynthesis. Formerly known as blue-green algae.
cycad (sih′kad) Any of a phylum of gymnosperms that live mainly in tropical and semitropical regions and have stout stems (to 20 m in height) and fernlike leaves.
cyclic AMP (cAMP) A form of adenosine monophosphate in which the phosphate is part of a ring-shaped structure; acts as a regulatory molecule and second messenger in organisms ranging from bacteria to humans.
cyclic AMP receptor protein See *catabolite gene activator protein.*
cyclic electron transport In photosynthesis, the cyclic flow of electrons through Photosystem I; ATP is formed by chemiosmosis, but no photolysis of water occurs, and O_2 and NADPH are not produced. Compare with *noncyclic electron transport.*
cyclin-dependent kinases (Cdks) Protein kinases involved in controlling the cell cycle.
cyclins Regulatory proteins whose levels oscillate during the cell cycle; activate cyclin-dependent kinases.
cystic fibrosis A genetic disease with an autosomal recessive inheritance pattern; characterized by secretion of abnormally thick mucus, particularly in the respiratory and digestive systems.
cytochromes (sy′toh-kromz) Iron-containing heme proteins of an electron transport system.
cytokines Signaling proteins that regulate interactions between cells in the immune system. Important groups include interferons, interleukins, tumor necrosis factors, and chemokines.
cytokinesis (sy″toh-kih-nee′sis) Stage of cell division in which the cytoplasm divides to form two daughter cells.
cytokinin (sy″toh-ky′nin) A plant hormone involved in various aspects of plant growth and development, such as cell division and delay of senescence.
cytoplasm The plasma membrane and cell contents with the exception of the nucleus.
cytosine A nitrogenous pyrimidine base that is a component of nucleic acids.
cytoskeleton The dynamic internal network of protein fibers that includes microfilaments, intermediate filaments, and microtubules.
cytosol The fluid component of the cytoplasm in which the organelles are suspended.
cytotoxic T cell See *T cytotoxic cell.*
dalton See *atomic mass unit (amu).*
day-neutral plant A plant whose flowering is not controlled by variations in day length that occur with changing seasons. Compare with *long-day, short-day,* and *intermediate-day plants.*
deamination (dee-am-ih-nay′shun) The removal of an amino group ($—NH_2$) from an amino acid or other organic compound.
decarboxylation A reaction in which a molecule of CO_2 is removed from a carboxyl group of an organic acid.
deciduous A term describing a plant that sheds leaves or other structures at regular intervals; e.g., during autumn. Compare with *evergreen.*
decomposers Microbial heterotrophs that break down dead organic material and use the decomposition products as a source of energy. Also called *saprotrophs* or *saprobes.*
deductive reasoning The reasoning that operates from generalities to specifics and can make relationships among data more apparent. Compare with *inductive reasoning.* See *hypothetico-deductive approach.*
deforestation The temporary or permanent removal of forest for agriculture or other uses.
dehydrogenation (dee-hy″dro-jen-ay′shun) A form of oxidation in which hydrogen atoms are removed from a molecule.
deletion (1) A chromosome abnormality in which part of a chromosome is missing, e.g., cri-du-chat. (2) The loss of one or more base pairs from DNA, which can result in a frameshift mutation.
demographics The science that deals with human population statistics, such as size, density, and distribution.
denature (dee-nay′ture) To alter the physical properties and three-dimensional structure of a protein, nucleic acid, or other macromolecule by treating it with excess heat, strong acids, or strong bases.
dendrite (den′drite) A branch of a neuron that receives and conducts nerve impulses toward the cell body. Compare with *axon.*
dendritic cells A set of immune cells present in many tissues that capture antigens and present them to T cells.
dendrochronology (den″dro-kruh-naal′uh-gee) A method of dating using the annual rings of trees.
denitrification (dee-nie″tra-fuh-kay′shun) The conversion of nitrate (NO_3^-) to nitrogen gas (N_2) by certain bacteria (denitrifying bacteria) in the soil; part of the nitrogen cycle.
dense connective tissue A type of tissue that may be irregular, as in the dermis of the skin, or regular, as in tendons.
density-dependent factor An environmental factor whose effects on a population change as population density changes; tends to retard population growth as population density increases and enhance population growth as population density decreases. Compare with *density-independent factor.*
density gradient centrifugation Procedure in which cell components are placed in a layer on top of a density gradient, usually a sucrose solution and water. Cell structures migrate during centrifugation, forming a band at the position in the gradient where their own density equals that of the sucrose solution.

density-independent factor An environmental factor that affects the size of a population but is not influenced by changes in population density. Compare with *density-dependent factor.*

deoxyribonucleic acid (DNA) Double-stranded nucleic acid; contains genetic information coded in specific sequences of its constituent nucleotides.

deoxyribose Pentose sugar lacking a hydroxyl (—OH) group on carbon-2′; a constituent of DNA.

depolarization (dee-pol″ar-ih-zay′shun) A decrease in the charge difference across a plasma membrane; may result in an action potential in a neuron or muscle cell.

derived characters See *shared derived characters.*

dermal tissue system The tissue that forms the outer covering over a plant; the epidermis or periderm.

dermis (dur′mis) The layer of dense connective tissue beneath the epidermis in the skin of vertebrates.

desert A temperate or tropical biome in which lack of precipitation limits plant growth.

desertification The degradation of once-fertile land into nonproductive desert; caused partly by soil erosion, deforestation, and overgrazing by domestic animals.

desmosomes (dez′moh-somz) Button-like plaques, present on two opposing cell surfaces, that hold the cells together by means of protein filaments that span the intercellular space.

determinate growth Growth of limited duration, as for example, in flowers and leaves. Compare with *indeterminate growth.*

determination The developmental process by which one or more cells become progressively committed to a particular fate. Determination is a series of molecular events usually leading to differentiation. Also called *cell determination.*

detritivore (duh-try′tuh-vore) An organism, such as an earthworm or crab, that consumes fragments of freshly dead or decomposing organisms; also called *detritus feeder.*

detritus (duh-try′tus) Organic debris from decomposing organisms.

detritus feeder See *detritivore.*

deuteromycetes (doo″ter-o-my′seats) An artificial grouping of fungi characterized by the absence of sexual reproduction but usually having other traits similar to ascomycetes; also called *imperfect fungi.*

deuterostome (doo′ter-oh-stome) Major division of the animal kingdom in which the anus develops from the blastopore; includes the echinoderms and chordates. Compare with *protostome.*

development All the progressive changes that take place throughout the life of an organism.

diabetes mellitus (mel′i-tus) The most common endocrine disorder. In Type I diabetes, there is a marked decrease in the number of beta cells in the pancreas, resulting in insulin deficiency. In the more common Type II diabetes, insulin receptors on target cells do not bind with insulin (insulin resistance).

diacylglycerol (DAG) (di″as-il-glis′er-ol) A lipid consisting of glycerol combined chemically with two fatty acids; also called *diglyceride.* Compare with *monoacylglycerol* and *triacylglycerol.*

dialysis The diffusion of certain solutes across a selectively permeable membrane.

diaphragm In mammals, the muscular floor of the chest cavity; contracts during inhalation, expanding the chest cavity.

diastole (di-ass′toh-lee) Phase of the cardiac cycle in which the heart is relaxed. Compare with *systole.*

diatom (die′eh-tom″) A usually unicellular alga that is covered by an ornate, siliceous shell consisting of two overlapping halves; an important component of plankton in both marine and fresh waters.

dichotomous branching (di-kaut′uh-mus) In botany, a type of branching in which one part always divides into two more or less equal parts.

dicot (dy′kot) One of the two classes of flowering plants; dicot seeds contain two cotyledons, or seed leaves. Compare with *monocot.*

diencephalon See *forebrain.*

differential centrifugation Separation of cell particles according to their mass, size, or density. In differential centrifugation, the supernatant is spun at successively higher revolutions per minute.

differential gene expression The expression of different subsets of genes at different times and in different cells during development.

differentiated cell A specialized cell; carries out unique activities, expresses a specific set of proteins, and usually has a recognizable appearance.

differentiation (dif″ah-ren-she-ay′shun) Development toward a more mature state; a process changing a young, relatively unspecialized cell to a more specialized cell. Also called *cell differentiation.*

diffusion The net movement of particles (atoms, molecules, or ions) from a region of higher concentration to a region of lower concentration (i.e., down a concentration gradient), resulting from random motion. Compare with *facilitated diffusion* and *active transport.*

digestion The breakdown of food to smaller molecules.

diglyceride See *diacylglycerol.*

dihybrid cross (dy-hy′brid) A genetic cross that takes into account the behavior of alleles of two loci. Compare with *monohybrid cross.*

dikaryotic (dy-kare-ee-ot′ik) Condition of having two nuclei per cell (i.e., $n + n$), characteristic of certain fungal hyphae. Compare with *monokaryotic.*

dimer An association of two monomers (e.g., a disaccharide or a dipeptide).

dinoflagellate (dy″noh-flaj′eh-late) A unicellular, biflagellate, typically marine alga that is an important component of plankton; usually photosynthetic.

dioecious (dy-ee′shus) Having male and female reproductive structures on separate plants; compare with *monoecious.*

dipeptide See *peptide.*

diploid (dip′loyd) The condition of having two sets of chromosomes per nucleus. Compare with *haploid* and *polyploid.*

diplomonads Small, mostly parasitic zooflagellates with one or two nuclei, no mitochondria, and one to four flagella

direct fitness An individual's reproductive success, measured by the number of viable offspring it produces. Compare with *inclusive fitness.*

directed evolution See *in vitro evolution.*

directional selection The gradual replacement of one phenotype with another because of environmental change that favors phenotypes at one of the extremes of the normal distribution. Compare with *stabilizing selection* and *disruptive selection.*

disaccharide (dy-sak′ah-ride) A sugar produced by covalently linking two monosaccharides (e.g., maltose or sucrose).

disomy The normal condition in which both members of a chromosome pair are present in a diploid cell or organism. Compare with *monosomy* and *trisomy.*

dispersal The movement of individuals among populations. See *immigration and emigration.*

dispersion The pattern of distribution in space of the individuals of a population relative to their neighbors; may be clumped, random, or uniform.

disruptive selection A special type of directional selection in which changes in the environment favor two or more variant phenotypes at the expense of the mean. Compare with *stabilizing selection* and *directional selection.*

distal Remote; farther from the point of reference. Compare with *proximal.*

distal convoluted tubule The part of the renal tubule that extends from the loop of Henle to the collecting duct. Compare with *proximal convoluted tubule.*

disturbance In ecology, any event that disrupts community or population structure.

divergent circuit A neural pathway in which a presynaptic neuron stimulates many postsynaptic neurons. Compare with *convergent circuit.*

diving reflex A group of physiological mechanisms, such as decrease in metabolic rate, that are activated when a mammal dives to its limit.

division A taxonomic category below that of kingdom, comparable to a phylum; often used in classifying plants, fungi, and certain protists.

dizygotic twins Twins that arise from the separate fertilization of two eggs; commonly known as *fraternal twins.* Compare with *monozygotic twins.*

DNA See *deoxyribonucleic acid.*

DNA cloning The process of selectively amplifying DNA sequences so their structure and function can be studied.

DNA-dependent RNA polymerase See *RNA polymerase.*

DNA fingerprinting See *DNA typing.*

DNA ligase Enzyme that catalyzes the joining of the 5′ and 3′ ends of two DNA fragments; essential in DNA replication and used in recombinant DNA technology.

DNA methylation A process in which gene inactivation is perpetuated by enzymes that add methyl groups to DNA.

DNA microarray A diagnostic test involving thousands of DNA molecules placed on a glass slide or chip.

DNA polymerases Family of enzymes that catalyze the synthesis of DNA from a DNA template, by adding nucleotides to a growing 3′ end.

DNA profiling See *DNA typing.*

DNA provirus Double-stranded DNA molecule that is an intermediate in the life cycle of an RNA tumor virus (retrovirus).

DNA replication The process by which DNA is duplicated; ordinarily a semiconservative process in which a double helix gives rise to two double helices, each with an "old" strand and a newly synthesized strand.

DNA sequencing Procedure by which the sequence of nucleotides in DNA is determined.

domain (1) A structural and functional region of a protein. (2) A taxonomic category that includes one or more kingdoms.

DNA typing The analysis of DNA extracted from an individual, which is unique to that individual; also called DNA fingerprinting or DNA profiling.

dominance hierarchy A linear "pecking order" into which animals in a population may organize according to status; regulates aggressive behavior within the population.

dominant allele (al-leel′) An allele that is always expressed when it is present, regardless of whether it is homozygous or heterozygous. Compare with *recessive allele.*

dopamine A neurotransmitter of the biogenic amine group.

dormancy A temporary period of arrested growth in plants or plant parts such as spores, seeds, bulbs, and buds.

dorsal (dor′sl) Toward the uppermost surface or back of an animal. Compare with *ventral.*

dosage compensation Genetic mechanism by which the expression of X-linked genes in mammals is made equivalent in XX females and XY males by rendering all but one X chromosome inactive.

double fertilization A process in the flowering plant life cycle in which there are two fertilizations; one fertilization results in formation of a zygote, whereas the second results in formation of endosperm.

doubling time The amount of time it takes for a population to double in size, assuming that its current rate of increase does not change.

Down syndrome An inherited condition in which individuals have abnormalities of the face, eyelids, tongue, and other parts of the body, and are physically and mentally retarded; usually results from trisomy of chromosome 21.

drupe (droop) A simple, fleshy fruit in which the inner wall of the fruit is hard and stony, e.g., peach, cherry.

duodenum (doo″o-dee′num) The portion of the small intestine into which the contents of the stomach first enter.

duplication An abnormality in which a set of chromosomes contains more than one copy of a particular chromosomal segment; the translocation form of Down syndrome is an example.

dura mater The tough, outer meningeal layer that covers and protects the brain and spinal cord. Also see *arachnoid* and *pia mater.*

dynamic equilibrium The condition of a chemical reaction when the rate of change in one direction is exactly the same as the rate of change in the opposite direction, i.e., the concentrations of the reactants and products are not changing, and the difference in free energy between reactants and products is zero.

ecdysone (ek′dih-sone) See *molting hormone.*

Ecdysozoa A branch of the protostomes that includes animals that molt, such as the rotifers, nematodes, and arthropods.

echinoderms (eh-kine′oh-derms) Phylum of spiny-skinned marine deuterostome invertebrates characterized by a water vascular system and tube feet; include sea stars, sea urchins, and sea cucumbers.

echolocation Determination of the position of objects by detecting echos of high-pitched sounds emitted by an animal; a type of sensory system used by bats and dolphins.

ecological niche See *niche.*

ecological pyramid A graphical representation of the relative energy value at each trophic level. See *pyramid of biomass* and *pyramid of energy.*

ecological succession See *succession.*

ecology (ee-kol′uh-jee) A discipline of biology that studies the interrelations among living things and their environments.

ecosystem (ee′koh-sis-tem) The interacting system that encompasses a community and its nonliving, physical environment. Compare with *community.*

ecosystem management A conservation focus that emphasizes restoring and maintaining ecosystem quality rather than the conservation of individual species.

ecosystem services Important environmental services, such as clean air to breathe, clean water to drink, and fertile soil in which to grow crops, that ecosystems provide.

ecotone The transition zone where two communities meet and intergrade.

ectoderm (ek′toh-derm) The outer germ layer of the early embryo; gives rise to the skin and nervous system. Compare with *mesoderm* and *endoderm.*

ectotherm An animal whose temperature fluctuates with that of the environment; may use behavioral adaptations to regulate temperature; sometimes referred to as *cold-blooded.* Compare with *endotherm.*

edge effect The ecological phenomenon in which ecotones between adjacent communities often contain a greater number of species or greater population densities of certain species than either adjacent community.

effector A muscle or gland that contracts or secretes in direct response to nerve impulses.

efferent (ef′fur-ent) Leading away from some point of reference. Compare with *afferent.*

efferent neurons Neurons that transmit action potentials from the brain or spinal cord to muscles or glands. Compare with *afferent neurons.*

ejaculation (ee-jak″yoo-lay′shun) A sudden expulsion, as in the ejection of semen from the penis.

electrolyte A substance that dissociates into ions when dissolved in water; the resulting solution can conduct an electrical current.

electron A particle with one unit of negative charge and negligible mass, located outside the atomic nucleus. Compare with *neutron* and *proton.*

electron configuration The arrangement of electrons around the atom. In a Bohr model, the electron configuration is depicted as a series of concentric circles.

electron microscope A microscope capable of producing high-resolution, highly magnified images through the use of an electron beam (rather than light). Transmission electron microscopes (TEMs) produce images of thin sections; scanning electron microscopes (SEMs) produce images of surfaces.

electron shell Group of orbitals of electrons with similar energies.

electron transport system A series of chemical reactions during which hydrogens or their electrons are passed along an electron transport chain from one acceptor molecule to another, with the release of energy.

electronegativity A measure of an atom's attraction for electrons.

electrophoresis, gel See *gel electrophoresis.*

electroreceptor A receptor that responds to electrical stimuli.

element A substance that can't be changed to a simpler substance by a normal chemical reaction.

elimination Ejection of undigested food from the body. Compare with *excretion.*

El Niño–Southern Oscillation (ENSO) (el nee′nyo) A recurring climatic phenomenon that involves a surge of warm water in the Pacific Ocean and unusual weather patterns elsewhere in the world.

elongation (in protein synthesis) Cyclic process by which amino acids are added one by one to a growing polypeptide chain. See *initiation* and *termination.*

embryo (em′bree-oh) (1) A young organism before it emerges from the egg, seed, or body of its mother. (2) Developing human until the end of the second month, after which it is referred to as a fetus. (3) In plants, the young sporophyte produced following fertilization and subsequent development of the zygote.

embryo sac The female gametophyte generation in flowering plants.

embryo transfer See *host mothering.*

emigration A type of migration in which individuals leave a population and thus decrease its size. Compare with *immigration.*

enantiomers (en-an′tee-oh-merz) Two isomeric chemical compounds that are mirror images.

endangered species A species whose numbers are so severely reduced that it is in imminent danger of extinction throughout all or part of its range. Compare with *threatened species.*

Endangered Species Act A U.S. law that authorizes the U.S. Fish and Wildlife Service to protect from extinction all endangered and threatened species in the United States and abroad.

endemic species Localized, native species that are not found anywhere else in the world. Compare with *cosmopolitan species.*

endergonic reaction (end′er-gon″ik) A nonspontaneous reaction; a reaction requiring a net input of free energy. Compare with *exergonic reaction.*

endocrine gland (en′doh-crin) A gland that secretes hormones directly into the blood or tissue fluid instead of into ducts. Compare with *exocrine gland.*

endocrine system The body system that helps regulate metabolic activities; consists of ductless glands and tissues that secrete hormones.

endocytosis (en″doh-sy-toh′sis) The active transport of substances into the cell by the formation of invaginated regions of the plasma membrane that pinch off and become cytoplasmic vesicles. Compare with *exocytosis.*

endoderm (en′doh-derm) The inner germ layer of the early embryo; becomes the lining of the digestive tract and the structures that develop from the digestive tract—liver, lungs, and pancreas. Compare with *ectoderm* and *mesoderm.*

endodermis (en″doh-der′mis) The innermost layer of the plant root cortex. Endodermal cells have a waterproof Casparian strip around their radial and transverse walls that ensures that water and minerals enter the xylem only by passing through the endoderm cells.

endolymph (en′doh-limf) The fluid of the membranous labyrinth and cochlear duct of the ear.

endomembrane system See *internal membrane system.*

endometrium (en″doh-mee′tree-um) The uterine lining.

endoplasmic reticulum (ER) (en′doh-plaz″mik reh-tik′yoo-lum) An interconnected network of internal membranes in eukaryotic cells enclosing a compartment, the ER lumen. Rough ER has ribosomes attached to the cytosolic surface; smooth ER, a site of lipid biosynthesis, lacks ribosomes.

endorphins (en-dor′finz) Neuropeptides released by certain brain neurons; block pain signals.

endoskeleton (en″doh-skel′eh-ton) Bony and/or cartilaginous structures within the body that provide support. Compare with *exoskeleton.*

endosperm (en′doh-sperm) The $3n$ nutritive tissue that is formed at some point in the development of all angiosperm seeds.

endospore A resting cell formed by certain bacteria; highly resistant to heat, radiation, and disinfectants.

endosymbiont (en″doe-sim′bee-ont) An organism that lives inside the body of another kind of organism. Endosymbionts may benefit their host (mutualism) or harm their host (parasitism).

endosymbiont theory The theory that certain organelles such as mitochondria and chloroplasts originated as symbiotic prokaryotes that lived inside other, free-living, prokaryotic cells.

endothelium (en-doh-theel′ee-um) The tissue that lines the cavities of the heart, blood vessels, and lymph vessels.

endotherm (en′doh-therm) An animal that uses metabolic energy to maintain a constant body temperature despite variations in environmental temperature; e.g., birds and mammals. Compare with *ectotherm.*

endotoxin A poisonous substance in the cell walls of Gram-negative bacteria. Compare with *exotoxin.*

end product inhibition See *feedback inhibition.*

energy The capacity to do work; expressed in kilojoules or kilocalories.

energy of activation See *activation energy.*

enhanced greenhouse effect See *greenhouse effect.*

enhancers Regulatory DNA sequences that can be located long distances away from the actual coding regions of a gene.

enkephalins (en-kef′ah-linz) Neuropeptides released by certain brain neurons that block pain signals.

enterocoely (en′ter-oh-seely) The process by which the coelom forms as a cavity within mesoderm produced by outpocketings of the primitive gut (archenteron); characteristic of many deuterostomes. Compare with *schizocoely.*

enthalpy The total potential energy of a system; sometimes referred to as the "heat content of the system."

entropy (en′trop-ee) Disorderliness; a quantitative measure of the amount of the random, disordered energy that is unavailable to do work.

environmental resistance Unfavorable environmental conditions, such as crowding, that prevent organisms from reproducing indefinitely at their intrinsic rate of increase.

environmental sustainability The ability to meet humanity's current needs without compromising the ability of future generations to meet their needs.

enzyme (en′zime) An organic catalyst (usually a protein) that accelerates a specific chemical reaction by lowering the activation energy required for that reaction.

enzyme-substrate complex The temporary association between enzyme and substrate that forms during the course of a catalyzed reaction; also called *ES complex.*

eosinophil (ee-oh-sin′oh-fil) A type of white blood cell whose cytoplasmic granules absorb acidic stains; functions in parasitic infestations and allergic reactions.

epidermis (ep-ih-dur′mis) (1) An outer layer of cells that covers the body of plants and functions primarily for protection. (2) The outer layer of vertebrate skin.

epididymis (ep-ih-did′ih-mis) (pl., *epididymides*) A coiled tube that receives sperm from the testis and conveys it to the vas deferens.

epiglottis A thin, flexible structure that guards the entrance to the larynx, preventing food from entering the airway during swallowing.

epinephrine (ep-ih-nef′rin) Hormone produced by the adrenal medulla; stimulates the sympathetic nervous system.

epistasis (ep′ih-sta-sis) Condition in which certain alleles of one locus alter the expression of alleles of a different locus.

epithelial tissue (ep-ih-theel′ee-al) The type of animal tissue that covers body surfaces, lines body cavities, and forms glands; also called *epithelium.*

epoch The smallest unit of geological time; a subdivision of a period.

equilibrium See *dynamic equilibrium, genetic equilibrium,* and *punctuated equilibrium.*

era One of the main divisions of geological time; eras are subdivided into periods.

erythroblastosis fetalis (eh-rith′row-blas-toe″sis fi-tal′is) Serious condition in which Rh^+ red blood cells (which bear antigen D) of a fetus are destroyed by maternal anti-D antibodies.

erythrocyte (eh-rith′row-site) A vertebrate red blood cell; contains hemoglobin, which transports oxygen.

erythropoietin (eh-rith″row-poy′ih-tin) A peptide hormone secreted mainly by kidney cells; stimulates red blood cell production.

ES complex See *enzyme-substrate complex.*

essential nutrient A nutrient that must be provided in the diet because the body can't make it or can't make it in sufficient quantities to meet nutritional needs, e.g., essential amino acids and essential fatty acids.

ester linkage Covalent linkage formed by the reaction of a carboxyl group and a hydroxyl group, with the removal of the equivalent of a water molecule; the linkage includes an oxygen atom bonded to a carbonyl group.

estivation A state of torpor caused by lack of food or water during periods of high temperature. Compare with *hibernation.*

estrogens (es′troh-jens) Female sex hormones produced by the ovary; promote the development and maintenance of female reproductive structures and of secondary sex characteristics.

estuary (es′choo-wear-ee) A coastal body of water that connects to an ocean, in which fresh water from the land mixes with salt water.

ethology (ee-thol′oh-jee) The study of animal behavior under natural conditions from the point of view of adaptation.

ethyl alcohol A two-carbon alcohol.

ethylene (eth′ih-leen) A gaseous plant hormone involved in various aspects of plant growth and development, such as leaf abscission and fruit ripening.

eubacteria (yoo″bak-teer′ee-ah) Prokaryotes other than the archaebacteria. Compare with *archaebacteria.*

euchromatin (yoo-croh′mah-tin) A loosely coiled chromatin that is generally capable of transcription. Compare with *heterochromatin.*

euglenoids (yoo-glee′noids) A group of mostly freshwater, flagellate, unicellular algae that move by means of an anterior flagellum and are usually photosynthetic.

eukaryote (yoo″kar′ee-ote) An organism whose cells have nuclei and other membrane-bounded organelles. Includes protists, fungi, plants, and animals. Compare with *prokaryote.*

euphotic zone The upper reaches of the ocean, in which enough light penetrates to support photosynthesis.

eustachian tube (yoo-stay′shee-un) The auditory tube passing between the middle ear cavity and the pharynx in vertebrates; permits the equalization of pressure on the tympanic membrane.

eutrophic lake A lake enriched with nutrients such as nitrate and phosphate and consequently overgrown with plants or algae.

evergreen A plant that sheds leaves over a long period, so that some leaves are always present. Compare with *deciduous.*

evolution Any cumulative genetic changes in a population from generation to generation. Evolution leads to differences in populations and explains the origin of all the organisms that exist today or have ever existed.

evolutionary species concept An alternative to the biological species concept in which, for a population to be declared a separate species, it must have undergone evolution long enough for statistically significant differences to emerge. Compare *species.*

evolutionary systematics An approach to classification that considers both evolutionary relationships and the extent of divergence that has occurred since a group branched from an ancestral group. Compare with *cladistics* and *phenetics.*

excitatory postsynaptic potential (EPSP) A change in membrane potential that brings a neuron closer to the firing level. Compare with *inhibitory postsynaptic potential (IPSP).*

excretion (ek-skree′shun) The discharge from the body of a waste product of metabolism (not to be confused with the elimination of undigested food materials). Compare with *elimination.*

excretory system The body system in animals that functions in osmoregulation and in the discharge of metabolic wastes.

exergonic reaction (ex′er-gon″ik) A reaction characterized by a release of free energy. Also called *spontaneous reaction.* Compare with *endergonic reaction.*

exocrine gland (ex′oh-crin) A gland that excretes its products through a duct that opens onto a free surface such as the skin (e.g., sweat glands). Compare with *endocrine gland.*

exocytosis (ex″oh-sy-toh′sis) The active transport of materials out of the cell by fusion of cytoplasmic vesicles with the plasma membrane. Compare with *endocytosis.*

exon (1) A protein-coding region of a eukaryotic gene. (2) The mRNA transcribed from such a region. Compare with *intron.*

exoskeleton (ex″oh-skel′eh-ton) An external skeleton, such as the shell of mollusks or outer covering of arthropods; provides protection and sites of attachment for muscles. Compare with *endoskeleton.*

exotoxin A poisonous substance released by certain bacteria. Compare with *endotoxin.*

explicit memory Factual knowledge of people, places, or objects; requires conscious recall of the information.

exploitation competition An intraspecific competition in which all the individuals in a population "share" the limited resource equally, so that at high population densities, none of them obtains an adequate amount. Also called *scramble competition.* Compare with *interference competition.*
exponential population growth The accelerating population growth rate that occurs when optimal conditions allow a constant per capita growth rate. Compare with *logistic population growth.*
ex situ conservation Conservation efforts that involve conserving individual species in human-controlled settings, such as zoos. Compare with *in situ conservation.*
exteroceptor (ex′tur-oh-sep″tor) One of the sense organs that receives sensory stimuli from the outside world, such as the eyes or touch receptors. Compare with *interoceptor.*
extinction The elimination of a species; occurs when the last individual member of a species dies.
extracellular matrix (ECM) A network of proteins and carbohydrates that surrounds many animal cells.
extraembryonic membranes Multicellular membranous structures that develop from the germ layers of a terrestrial vertebrate embryo but are not part of the embryo itself. See *chorion, amnion, allantois,* and *yolk sac.*
F_1 generation (first filial generation) The first generation of hybrid offspring resulting from a cross between parents from two different true-breeding lines.
F_2 generation (second filial generation) The offspring of the F_1 generation.
facilitated diffusion The passive transport of ions or molecules by a specific carrier protein in a membrane. As in simple diffusion, net transport is down a concentration gradient, and no additional energy has to be supplied. Compare with *diffusion* and *active transport.*
facilitation A process in which a neuron is brought closer to its threshold level by stimulation from various presynaptic neurons.
facultative anaerobe An organism capable of carrying out aerobic respiration but able to switch to fermentation when oxygen is unavailable; e.g., yeast. Compare with *obligate anaerobe.*
FAD/$FADH_2$ Oxidized and reduced forms, respectively, of flavin adenine dinucleotide, a coenzyme that transfers electrons (as hydrogen) in metabolism, including cellular respiration.
fallopian tube See *oviduct.*
family A taxonomic category made up of related genera.
fatty acid A lipid that is an organic acid containing a long hydrocarbon chain, with no double bonds (saturated fatty acid), one double bond (monounsaturated fatty acid), or two or more double bonds (polyunsaturated fatty acid); components of triacylglycerols, and phospholipids, as well as monoacylglycerols and diacylglycerols.
fecundity The potential capacity of an individual to produce offspring.
feedback inhibition A type of enzyme regulation in which the accumulation of the product of a reaction inhibits an earlier reaction in the sequence; also known as *end product inhibition.*
fermentation An anaerobic process by which ATP is produced by a series of redox reactions in which organic compounds serve both as electron donors and terminal electron acceptors.
fern One of a phylum of seedless vascular plants that reproduce by spores produced in sporangia; ferns have an alternation of generations between the dominant sporophyte and the gametophyte (prothallus).
fertilization The fusion of two n gametes; results in the formation of a $2n$ zygote. Compare with *double fertilization.*
fetus The unborn human offspring from the third month of pregnancy to birth.
fiber (1) In plants, a type of sclerenchyma cell; fibers are long, tapered cells with thick walls. Compare with *sclereid.* (2) In animals, an elongated cell such as a muscle or nerve cell. (3) In animals, the microscopic, threadlike protein and carbohydrate complexes scattered through the matrix of connective tissues.
fibrin An insoluble protein formed from the plasma protein fibrinogen during blood clotting.
fibroblasts Connective tissue cells that produce the fibers and the protein and carbohydrate complexes of the matrix of connective tissues.
fibronectins Glycoproteins of the extracellular matrix that bind to integrins (receptor proteins in the plasma membrane).
fibrous root system A root system consisting of several adventitious roots of approximately equal size that arise from the base of the stem. Compare with *taproot system.*
Fick's law of diffusion A physical law governing rates of gas exchange in animal respiratory systems; states that the rate of diffusion of a substance across a membrane is directly proportional to the surface area and to the difference in pressure between the two sides.
filament In flowering plants, the thin stalk of a stamen; the filament bears an anther at its tip.
first law of thermodynamics The law of conservation of energy, which states that the total energy of any closed system (any object plus its surroundings, i.e., the universe) remains constant. Compare with *second law of thermodynamics.*
fitness See *direct fitness.*
fixed action pattern (FAP) An innate behavior triggered by a sign stimulus.
flagellum (flah-jel′um) (pl., *flagella*) A long, whiplike structure extending from certain cells and used in locomotion. (1) Eukaryote flagella consist of two central single microtubules surrounded by nine double microtubules (9 + 2 structure), all covered by a plasma membrane. (2) Prokaryote flagella are filaments rotated by special structures located in the plasma membrane and cell wall.
flame cells Collecting cells that have cilia; part of the osmoregulatory system of flatworms.
flavin adenine dinucleotide See *FAD/$FADH_2$.*
flowering plants See *angiosperms.*
flowing-water ecosystem A river or stream ecosystem.
fluid-mosaic model The currently accepted model of the plasma membrane and other cell membranes, in which protein molecules "float" in a fluid phospholipid bilayer.
fluorescence The emission of light of a longer wavelength (lower energy) than the light originally absorbed.
fluorescent in situ hybridization (FISH) A technique to detect specific DNA segments by hybridization directly to chromosomes; visualized microscopically by using a fluorescent dye.
follicle (fol′i-kl) (1) A simple, dry, dehiscent fruit that splits open at maturity along one suture to liberate the seeds. (2) A small sac of cells in the mammalian ovary that contains a maturing egg. (3) The pocket in the skin from which a hair grows.
follicle-stimulating hormone (FSH) A gonadotropic hormone secreted by the anterior lobe of the pituitary gland; stimulates follicle development in the ovaries of females and sperm production in the testes of males.
food chain The series of organisms through which energy flows in an ecosystem. Each organism in the series eats or decomposes the preceding organism in the chain. See *food web.*
food web A complex interconnection of all the food chains in an ecosystem.

foramen magnum The opening in the vertebrate skull through which the spinal cord passes.

foraminiferan (for″am-in-if′er-an) A marine protozoon that produces a shell, or test, that encloses an amoeboid body.

forebrain In the early embryo, one of the three divisions of the developing vertebrate brain; subdivides to form the telencephalon, which gives rise to the cerebrum, and the diencephalon, which gives rise to the thalamus and hypothalamus. Compare with *midbrain* and *hindbrain.*

forest decline A gradual deterioration (and often death) of many trees in a forest; can be caused by a combination of factors, such as acid precipitation, toxic heavy metals, and surface-level ozone.

fossil Parts or traces of an ancient organism usually preserved in rock.

fossil fuel Combustible deposits in Earth's crust that are composed of the remnants of prehistoric organisms that existed millions of years ago, e.g., oil, natural gas, and coal.

founder cell A cell from which a particular cell lineage is derived.

founder effect Genetic drift that results from a small population colonizing a new area.

fovea (foe′vee-ah) The area of sharpest vision in the retina; cone cells are concentrated here.

fragile site A weak point at a specific location on a chromosome where part of a chromatid appears attached to the rest of the chromosome by a thin thread of DNA.

fragile X syndrome A human genetic disorder caused by a fragile site that occurs near the tip on the X chromosome; effects range from mild learning disabilities to severe mental retardation and hyperactivity.

frameshift mutation A mutation that results when one or two nucleotide pairs are inserted into or deleted from the DNA. The change causes the mRNA transcribed from the mutated DNA to have an altered reading frame such that all codons downstream from the mutation are changed.

free energy The maximum amount of energy available to do work under the conditions of a biochemical reaction.

free radicals Toxic, highly reactive compounds that have unpaired electrons that bond with other compounds in the cell, interfering with normal function.

frequency-dependent selection Selection in which the relative fitness of different genotypes is related to how frequently they occur in the population.

freshwater wetlands Land that is transitional between freshwater and terrestrial ecosystems and is covered with water for at least part of the year; e.g., marshes and swamps.

frontal lobes In mammals, the anterior part of the cerebrum.

fruit In flowering plants, a mature, ripened ovary. Fruits contain seeds and usually provide seed protection and dispersal.

fruiting body A multicellular structure that contains the sexual spores of certain fungi; refers to the ascocarp of an ascomycete and the basidiocarp of a basidiomycete.

fucoxanthin (few″koh-zan′thin) The brown carotenoid pigment found in brown algae, golden algae, diatoms, and dinoflagellates.

functional genomics The study of the roles of genes in cells.

functional group A group of atoms that confers distinctive properties on an organic molecule (or region of a molecule) to which it is attached, e.g., hydroxyl, carbonyl, carboxyl, amino, phosphate, and sulfhydryl groups.

fundamental niche The potential ecological niche that an organism could occupy if there were no competition from other species. Compare with *realized niche.*

fungus (pl., *fungi*) A heterotrophic eukaryote with chitinous cell walls and a body usually in the form of a mycelium of branched, threadlike hyphae. Most fungi are decomposers; some are parasitic.

G protein One of a group of proteins that bind GTP and are involved in the transfer of signals across the plasma membrane.

G_1 phase The first gap phase within the interphase stage of the cell cycle; G_1 occurs before DNA synthesis (S phase) begins. Compare with *S* and *G_2 phases.*

G_2 phase Second gap phase within the interphase stage of the cell cycle; G_2 occurs after DNA synthesis (S phase) and before mitosis. Compare with *S* and *G_1 phases.*

gallbladder A small sac that stores bile.

gametangium (gam″uh-tan′gee-um) Special multicellular or unicellular structure of plants, protists, and fungi in which gametes are formed.

gamete (gam′eet) A sex cell; in plants and animals, an egg or sperm. In sexual reproduction, the union of gametes results in the formation of a zygote. The chromosome number of a gamete is designated *n.* Species that are not polyploid have haploid gametes and diploid zygotes.

gametic isolation (gam-ee′tik) A prezygotic reproductive isolating mechanism in which sexual reproduction between two closely related species can't occur because of chemical differences in the gametes.

gametogenesis The process of gamete formation. See *spermatogenesis* and *oogenesis.*

gametophyte generation (gam-ee′toh-fite) The *n,* gamete-producing stage in the life cycle of a plant. Compare with *sporophyte generation.*

gamma-aminobutyric acid (GABA) A neurotransmitter that has an inhibitory effect.

ganglion (gang′glee-on) (pl., *ganglia*) A mass of neuron cell bodies.

ganglion cell A type of neuron in the retina of the eye; receives input from bipolar cells.

gap junction Structure consisting of specialized regions of the plasma membrane of two adjacent cells; contains numerous pores that allow the passage of certain small molecules and ions between them.

gastrin (gas′trin) A hormone released by the stomach mucosa; stimulates the gastric glands to secrete pepsinogen.

gastrovascular cavity A central digestive cavity with a single opening that functions as both mouth and anus; characteristic of cnidarians and flatworms.

gastrula (gas′troo-lah) A three-layered embryo formed by the process of gastrulation.

gastrulation (gas-troo-lay′shun) Process in embryonic development during which the three germ layers (ectoderm, mesoderm, and endoderm) form.

gel electrophoresis Procedure by which proteins or nucleic acids are separated on the basis of size and charge as they migrate through a gel in an electrical field.

gene A segment of DNA that serves as a unit of hereditary information; includes a transcribable DNA sequence (plus associated sequences regulating its transcription) that yields a protein or RNA product with a specific function.

gene amplification The developmental process in which certain cells produce multiple copies of a gene by selective replication, thus allowing for increased synthesis of the gene product. Compare with *nuclear equivalence* and *genomic rearrangement.*

gene flow The movement of alleles between local populations due to the migration of individuals; can have significant evolutionary consequences.

gene locus See *locus.*

gene pool All the alleles of all the genes present in a freely interbreeding population.
gene therapy Any of a variety of methods designed to correct a disease or alleviate its symptoms through the introduction of genes into the affected person's cells.
genetic bottleneck See *bottleneck.*
genetic code See *codon.*
genetic counseling Medical and genetic information provided to couples who are concerned about the risk of abnormality in their children.
genetic drift A random change in allele frequency in a small breeding population.
genetic engineering Manipulation of genes, often through recombinant DNA technology.
genetic equilibrium The condition of a population that is not undergoing evolutionary change, i.e., in which allele and genotype frequencies do not change from one generation to the next. See *Hardy-Weinberg principle.*
genetic polymorphism (pol″ee-mor′fizm) The presence in a population of two or more alleles for a given gene locus.
genetic probe A single-stranded nucleic acid (either DNA or RNA) used to identify a complementary sequence by hydrogen-bonding to it.
genetic recombination See *recombination, genetic.*
genetic screening A systematic search through a population for individuals with a genotype or karyotype that might cause a serious genetic disease in them or their offspring.
genome (jee′nome) Originally, all the genetic material in a cell or individual organism. The term is used more than one way, depending on context: e.g., an organism's haploid genome is all the DNA contained in one haploid set of its chromosomes, and its mitochondrial genome is all the DNA in a mitochondrion. See *human genome.*
genomic DNA library A collection of recombinant plasmids in which all the DNA in the genome is represented. Compare with *cDNA library.*
genomic imprinting See *imprinting,* first definition.
genomic rearrangement A physical change in the structure of one or more genes that occurs during the development of an organism and leads to an alteration in gene expression; compare with *nuclear equivalence* and *gene amplification.*
genotype (jeen′oh-type) The genetic makeup of an individual. Compare with *phenotype.*
genotype frequency The proportion of a particular genotype in the population.
genus (jee′nus) A taxonomic category made up of related species.
germination Resumption of growth of an embryo or spore; occurs when a seed or spore sprouts.
germ layers In animals, three embryonic tissue layers: endoderm, mesoderm, and ectoderm.
germ line cell In animals, a cell that is part of the line of cells that will ultimately undergo meiosis to form gametes. Compare with *somatic cell.*
germplasm Any plant or animal material that may be used in breeding; includes seeds, plants, and plant tissues of traditional crop varieties and the sperm and eggs of traditional livestock breeds.
gibberellin (jib″ur-el′lin) A plant hormone involved in many aspects of plant growth and development, such as stem elongation, flowering, and seed germination.
gills (1) The respiratory organs characteristic of many aquatic animals, usually thin-walled projections from the body surface or from some part of the digestive tract; (2) The spore-bearing, platelike structures under the caps of mushrooms.
ginkgo (ging′ko) A member of an ancient gymnosperm group that consists of a single living representative (*Ginkgo biloba*), a hardy, deciduous tree with broad, fan-shaped leaves and naked, fleshy seeds (on female trees).
gland See *endocrine gland* and *exocrine gland.*
glial cells (glee′ul) In nervous tissue, cells that support and nourish neurons.
globulin (glob′yoo-lin) One of a class of proteins in blood plasma, some of which (gamma globulins) function as antibodies.
glomerulus (glom-air′yoo-lus) The cluster of capillaries at the proximal end of a nephron; the glomerulus is surrounded by Bowman's capsule.
glucagon (gloo′kah-gahn) A hormone secreted by the pancreas that stimulates glycogen breakdown, thereby increasing the concentration of glucose in the blood. Compare with *insulin.*
glucose A hexose aldehyde sugar that is central to many metabolic processes.
glutamate An amino acid that functions as the major excitatory neurotransmitter in the vertebrate brain.
glyceraldehyde-3-phosphate (G3P) Phosphorylated three-carbon compound that is an important intermediate in glycolysis and in the Calvin cycle.
glycerol A three-carbon alcohol with a hydroxyl group on each carbon; a component of triacylglycerols and phospholipids, as well as monoacylglycerols and diacylglycerols.
glycocalyx (gly″koh-kay′lix) A coating on the outside of an animal cell, formed by the polysaccharide portions of glycoproteins and glycolipids associated with the plasma membrane.
glycogen (gly′koh-jen) The principal storage polysaccharide in animal cells; formed from glucose and stored primarily in the liver and, to a lesser extent, in muscle cells.
glycolipid A lipid with covalently attached carbohydrates.
glycolysis (gly-kol′ih-sis) The first stage of cellular respiration, literally the "splitting of sugar." The metabolic conversion of glucose into pyruvate, accompanied by the production of ATP.
glycoprotein (gly′koh-pro-teen) A protein with covalently attached carbohydrates.
glycosidic linkage Covalent linkage joining two sugars; includes an oxygen atom bonded to a carbon of each sugar.
glyoxysomes (gly-ox′ih-somz) Membrane-bounded structures in cells of certain plant seeds; contain a large array of enzymes that convert stored fat to sugar.
gnetophyte (nee′toe-fite) One of a small phylum of unusual gymnosperms that have some features similar to flowering plants.
goblet cells Unicellular glands that secrete mucus.
goiter (goy′ter) An enlargement of the thyroid gland.
golden alga A member of a phylum of algae, most of which are biflagellate, unicellular, and contain pigments, including chlorophyll *a* and *c* and carotenoids, including fucoxanthin.
Golgi complex (goal′jee) Organelle composed of stacks of flattened, membranous sacs. Mainly responsible for modifying, packaging, and sorting proteins that will be secreted or targeted to other organelles of the internal membrane system or to the plasma membrane; also called *Golgi body* or *Golgi apparatus.*
gonad (goh′nad) A gamete-producing gland; an ovary or a testis.
gonadotropin-releasing hormone (GnRH) A hormone secreted by the hypothalamus that stimulates the anterior pituitary to secrete the gonadotropic hormones: follicle-stimulating hormone (FSH) and luteinizing hormone (LH).
gonadotropic hormones (go-nad-oh-troh′pic) Hormones produced by the anterior pituitary gland that stimulate the testes and ovaries; include follicle-stimulating hormone (FSH) and luteinizing hormone (LH).

graded potential A local change in electrical potential that varies in magnitude depending on the strength of the applied stimulus.

gradualism The idea that evolution occurs by a slow, steady accumulation of genetic changes over time. Compare with *punctuated equilibrium.*

graft rejection An immune response directed against a transplanted tissue or organ.

grain A simple, dry, one-seeded fruit in which the fruit wall is fused to the seed coat, e.g., corn and wheat kernels. Also called *caryopsis.*

granulosa cells In mammals, cells that surround the developing oocyte and are part of the follicle; produce estrogens and inhibin.

granum (pl., *grana*) A stack of thylakoids within a chloroplast.

gravitropism (grav″ih-troh′pizm) Growth of a plant in response to gravity.

gray crescent The grayish area of cytoplasm that marks the region where gastrulation begins in an amphibian embryo.

gray matter Nervous tissue in the brain and spinal cord that contains cell bodies, dendrites, and unmyelinated axons. Compare with *white matter.*

green alga A member of a diverse phylum of algae that contain the same pigments as plants (chlorophylls *a* and *b* and carotenoids).

greenhouse effect The natural global warming of Earth's atmosphere caused by the presence of carbon dioxide and other gases that trap the sun's radiation. The additional warming produced when increased levels of greenhouse gases absorb infrared radiation is known as the *enhanced greenhouse effect.*

greenhouse gases Trace gases in the atmosphere that allow the sun's energy to penetrate to Earth's surface but do not allow as much of it to escape as heat.

gross primary productivity The rate at which energy accumulates (is assimilated) in an ecosystem during photosynthesis. Compare with *net primary productivity.*

ground state The lowest energy state of an atom.

ground tissue system All tissues in the plant body other than the dermal tissue system and vascular tissue system; consists of parenchyma, collenchyma, and sclerenchyma.

growth factors A group of more than 50 extracellular peptides that signal certain cells to grow and divide.

growth hormone (GH) A hormone secreted by the anterior lobe of the pituitary gland; stimulates growth of body tissues; also called *somatotropin.*

growth rate The rate of change of a population's size on a per capita basis.

guanine (gwan′een) A nitrogenous purine base that is a component of nucleic acids and GTP.

guanosine triphosphate (GTP) An energy transfer molecule similar to ATP that releases free energy with the hydrolysis of its terminal phosphate group.

guard cell One of a pair of epidermal cells that adjust their shape to form a stomatal pore for gas exchange.

guttation (gut-tay′shun) The appearance of water droplets on leaves, forced out through leaf pores by root pressure.

gymnosperm (jim′noh-sperm) Any of a group of seed plants in which the seeds are not enclosed in an ovary; gymnosperms frequently bear their seeds in cones. Includes four phyla: conifers, cycads, ginkgoes, and gnetophytes.

habitat The natural environment or place where an organism, population, or species lives.

habitat fragmentation The division of habitats that formerly occupied large, unbroken areas into smaller pieces by roads, fields, cities, and other human land-transforming activities.

habitat isolation A prezygotic reproductive isolating mechanism in which reproduction between similar species is prevented because they live and breed in different habitats

habituation (hab-it″yoo-ay′shun) A type of learning in which an animal becomes accustomed to a repeated, irrelevant stimulus and no longer responds to it.

hair cell A vertebrate mechanoreceptor found in the lateral line of fishes, the vestibular apparatus, semicircular canals, and cochlea.

half-life The period of time required for a radioisotope to change into a different material.

haploid (hap′loyd) The condition of having one set of chromosomes per nucleus. Compare with *diploid* and *polyploid.*

"hard-wiring" Refers to how neurons signal one another, how they connect, and how they carry out basic functions such as regulating heart rate, blood pressure, and sleep–wake cycles.

Hardy-Weinberg principle The mathematical prediction that allele frequencies do not change from generation to generation in a large population in the absence of microevolutionary processes (mutation, genetic drift, gene flow, natural selection).

Hatch-Slack pathway See *C_4 plant.*

haustorium (hah-stor′ee-um) (pl., *haustoria*) In parasitic fungi, a specialized hypha that penetrates a host cell and obtains nourishment from the cytoplasm.

Haversian canals (ha-vur′zee-un) Channels extending through the matrix of bone; contain blood vessels and nerves.

heat The total amount of kinetic energy in a sample of a substance.

heat energy The thermal energy that flows from an object with a higher temperature to an object with a lower temperature.

heat of vaporization The amount of heat energy that must be supplied to change one gram of a substance from the liquid phase to the vapor phase.

helicases Enzymes that unwind the two strands of a DNA double helix.

heliotropism The ability of leaves or flowers of certain plants to follow the sun by aligning themselves either perpendicular or parallel to the sun's rays; also called *solar tracking.*

helper T cell See *T helper cell.*

hemichordates A phylum of sedentary, wormlike deuterostomes.

hemizygous (hem″ih-zy′gus) Possessing only one allele for a particular locus; a human male is hemizygous for all X-linked genes. Compare with *homozygous* and *heterozygous.*

hemocoel Blood cavity characteristic of animals with an open circulatory system.

hemocyanin A hemolymph pigment that transports oxygen in some mollusks and arthropods.

hemoglobin (hee′moh-gloh″bin) The red, iron-containing protein pigment in blood that transports oxygen and carbon dioxide and aids in regulation of pH.

hemolymph (hee′moh-limf) The fluid that bathes the tissues in animals with an open circulatory system, e.g., arthropods and most mollusks.

hemophilia (hee″-moh-feel′ee-ah) A hereditary disease in which blood does not clot properly; the form known as *hemophilia A* has an X-linked, recessive inheritance pattern.

Hensen's node See *primitive streak.*

hepatic (heh-pat′ik) Pertaining to the liver.

hepatic portal system The portion of the circulatory system that carries blood from the intestine through the liver.

herbivore (erb′uh-vore) An animal that feeds on plants or algae. Also called *primary consumer.*

hermaphrodite (her-maf′roh-dite) An organism that has both male and female sex organs.

heterochromatin (het″ur-oh-kroh′mah-tin) Highly coiled and compacted chromatin in an inactive state. Compare with *euchromatin.*

heterocyst (het′ur-oh-sist″) An oxygen-excluding cell of cyanobacteria that is the site of nitrogen fixation.

heterogametic A term describing an individual that produces two classes of gametes with respect to their sex chromosome constitutions. Human males (XY) are heterogametic, producing X and Y sperm. Compare with *homogametic.*

heterospory (het″ur-os′pur-ee) Production of two types of *n* spores, microspores (male) and megaspores (female). Compare with *homospory.*

heterothallic (het″ur-oh-thal′ik) Pertaining to certain algae and fungi that have two mating types; only by combining a plus strain and a minus strain can sexual reproduction occur. Compare with *homothallic.*

heterotroph (het′ur-oh-trof) An organism that cannot synthesize its own food from inorganic raw materials and therefore must obtain energy and body-building materials from other organisms. Also called *consumer.* Compare with *autotroph.* See *chemoheterotroph* and *photoheterotroph.*

heterozygote advantage A phenomenon in which the heterozygous condition confers some special advantage on an individual that either homozygous condition does not (i.e., *Aa* has a higher degree of fitness than does *AA* or *aa*).

heterozygous (het-ur′oh-zye′gus) Having a pair of unlike alleles for a particular locus. Compare with *homozygous.*

hexose A monosaccharide containing six carbon atoms.

hibernation Long-term torpor in response to winter cold and scarcity of food. Compare with *estivation.*

high-density lipoprotein (HDL) See *lipoprotein.*

histamine (his′tah-meen) Substance released from mast cells that is involved in allergic and inflammatory reactions.

hindbrain In the early embryo, one of the three divisions of the developing vertebrate brain; subdivides to form the metencephalon, which gives rise to the cerebellum and pons, and the myelencephalon, which gives rise to the medulla. Compare with *forebrain* and *midbrain.*

histones (his′tones) Small, positively charged (basic) proteins in the cell nucleus that bind to the negatively charged DNA. See *nucleosomes.*

holdfast The basal structure for attachment to solid surfaces found in multicellular algae.

holoblastic cleavage A cleavage pattern in which the entire embryo cleaves; characteristic of eggs with little or moderate yolk (isolecithal or moderately telolecithal), e.g., the eggs of echinoderms, amphioxus, and mammals. Compare with *meroblastic cleavage.*

homeobox A DNA sequence of approximately 180 base pairs found in many homeotic genes and some other genes that are important in development; genes containing homeobox sequences code for certain transcription factors.

homeodomain A functional region of certain transcription factors; consists of approximately 60 amino acids specified by a homeobox DNA sequence and includes a recognition alpha helix, which binds to specific DNA sequences and affects their transcription.

home range A geographic area that an animal seldom or never leaves.

homeostasis (home″ee-oh-stay′sis) The balanced internal environment of the body; the automatic tendency of an organism to maintain such a steady state.

homeotic gene (home″ee-ah′tik) A gene that controls the formation of specific structures during development. Such genes were originally identified through insect mutants in which one body part is substituted for another.

hominid (hah′min-id) Any of a group of extinct and living humans.

hominoid (hah′min-oid) The apes and hominids.

homogametic Term describing an individual that produces gametes with identical sex chromosome constitutions. Human females (XX) are homogametic, producing all X eggs. Compare with *heterogametic.*

homologous chromosomes (hom-ol′ah-gus) Chromosomes that are similar in morphology and genetic constitution. In humans there are 23 pairs of homologous chromosomes; one member of each pair is inherited from the mother, and the other from the father.

homologous features See *homology.*

homology Similarity in different species that results from their derivation from a common ancestor. The features that exhibit such similarity are called *homologous features.* Compare with *homoplasy.*

homoplastic features See *homoplasy.*

homoplasy Similarity in the characters in different species that is due to convergent evolution, not common descent. Characters that exhibit such similarity are called *homoplastic features.* Compare with *homology.*

homospory (hoh″mos′pur-ee) Production of one type of *n* spore that gives rise to a bisexual gametophyte. Compare with *heterospory.*

homothallic (hoh″moh-thal′ik) Pertaining to certain algae and fungi that are self-fertile. Compare with *heterothallic.*

homozygous (hoh″moh-zy′gous) Having a pair of identical alleles for a particular locus. Compare with *heterozygous.*

hormone An organic chemical messenger in multicellular organisms that is produced in one part of the body and often transported to another part where it signals cells to alter some aspect of metabolism.

hornwort A phylum of spore-producing, nonvascular, thallose plants with a life cycle similar to mosses.

horsetail A phylum of seedless vascular plants with a life cycle similar to ferns.

host mothering The introduction of an embryo from one species into the uterus of another species, where it implants and develops; the host mother subsequently gives birth and may raise the offspring as her own.

***Hox* genes** Clusters of homeobox-containing genes that specify the anterior–posterior axis of various animals during development.

human chorionic gonadotropin (hCG) A hormone secreted by cells surrounding the early embryo; signals the mother's corpus luteum to continue to function.

human genetics The science of inherited variation in humans.

human genome The totality of genetic information in human cells; includes the DNA content of both the nucleus and mitochondria. See *genome.*

human immunodeficiency virus (HIV) The retrovirus that causes AIDS (acquired immunodeficiency syndrome).

human leukocyte antigen (HLA) See *major histocompatibility complex.*

humus (hew′mus) Organic matter in various stages of decomposition in the soil; gives soil a dark brown or black color.

Huntington disease A genetic disease that has an autosomal dominant inheritance pattern and causes mental and physical deterioration.

hybrid The offspring of two genetically dissimilar parents.

hybrid breakdown A postzygotic reproductive isolating mechanism in which, although an interspecific hybrid is fertile and produces a second (F_2) generation, the F_2 has defects that prevent it from successfully reproducing.

hybrid inviability A postzygotic reproductive isolating mechanism in which the embryonic development of an interspecific hybrid is aborted.

hybridization (1) Interbreeding between members of two different taxa; (2) Interbreeding between genetically dissimilar parents; (3) In molecular biology, complementary base pairing between nucleic acid (DNA or RNA) strands from different sources.

hybrid sterility A postzygotic reproductive isolating mechanism in which an interspecific hybrid can't reproduce successfully.

hybrid vigor The genetic superiority of an F_1 hybrid over either parent, caused by the presence of heterozygosity for a number of different loci.

hybrid zone An area of overlap between two closely related populations, subspecies, or species, in which interbreeding occurs.

hydration Process of association of a substance with the partial positive and/or negative charges of water molecules.

hydrocarbon An organic compound composed solely of hydrogen and carbon atoms.

hydrogen bond A weak attractive force existing between a hydrogen atom with a partial positive charge and an electronegative atom (usually oxygen or nitrogen) with a partial negative charge. Compare with *covalent bond* and *ionic bond.*

hydrologic cycle The water cycle, which includes evaporation, precipitation, and flow to the ocean; supplies terrestrial organisms with a continual supply of fresh water.

hydrolysis Reaction in which a covalent bond between two subunits is broken through the addition of the equivalent of a water molecule; a hydrogen atom is added to one subunit and a hydroxyl group to the other. Compare with *condensation synthesis.*

hydrophilic Interacting readily with water; having a greater affinity for water molecules than they have for each other. Compare with *hydrophobic.*

hydrophobic Not readily interacting with water; having less affinity for water molecules than they have for each other. Compare with *hydrophilic.*

hydroponics (hy″dra-paun′iks) Growing plants in an aerated solution of dissolved inorganic minerals; i.e., without soil.

hydrostatic skeleton A type of skeleton found in some invertebrates in which contracting muscles push against a tube of fluid.

hydroxide ion An anion (negatively charged particle) consisting of oxygen and hydrogen; usually written OH^-.

hydroxyl group (hy-drok′sil) Polar functional group; abbreviated —OH.

hyperpolarize To change the membrane potential so that the inside of the cell becomes more negative than its resting potential.

hypertonic A term referring to a solution having an osmotic pressure (or solute concentration) greater than that of the solution with which it is compared. Compare with *hypotonic* and *isotonic.*

hypha (hy′fah) (pl., *hyphae*) One of the threadlike filaments composing the mycelium of a water mold or fungus.

hypocotyl (hy′poh-kah″tl) The part of the axis of a plant embryo or seedling below the point of attachment of the cotyledons.

hypothalamus (hy-poh-thal′uh-mus) Part of the mammalian brain that regulates the pituitary gland, the autonomic system, emotional responses, body temperature, water balance, and appetite; located below the thalamus.

hypothesis A testable statement about the nature of an observation or relationship. Compare with *theory* and *principle.*

hypothetico-deductive approach Emphasizes the use of deductive reasoning to test hypotheses. Compare with *hypothetico-inductive approach.* See *deductive reasoning.*

hypothetico-inductive approach Emphasizes the use of inductive reasoning to discover new general principles. Compare with *hypothetico-deductive approach.* See *inductive reasoning.*

hypotonic A term referring to a solution having an osmotic pressure (or solute concentration) less than that of the solution with which it is compared. Compare with *hypertonic* and *isotonic.*

hypotrichs A group of dorsoventrally flattened ciliates that exhibit an unusual creeping-darting locomotion.

illuviation The deposition of material leached from the upper layers of soil into the lower layers.

imaginal discs Paired structures in an insect larva that develop into specific adult structures during complete metamorphosis.

imago (ih-may′go) The adult form of an insect.

imbibition (im″bi-bish′en) The absorption of water by a seed prior to germination.

immigration A type of migration in which individuals enter a population and thus increase its size. Compare with *emigration.*

immune response Process of recognizing foreign macromolecules and mounting a response aimed at eliminating them. See *specific* and *nonspecific immune responses; primary* and *secondary immune responses.*

immunoglobulin (im-yoon″oh-glob′yoo-lin) See *antibody.*

imperfect flower A flower that lacks either stamens or carpels. Compare with *perfect flower.*

imperfect fungi See *deuteromycetes.*

implantation The embedding of a developing embryo in the inner lining (endometrium) of the uterus.

implicit memory The unconscious memory for perceptual and motor skills, e.g., riding a bicycle.

imprinting (1) The expression of a gene based on its parental origin; also called *genomic imprinting.* (2) A type of learning by which a young bird or mammal forms a strong social attachment to an individual (usually a parent) or object within a few hours after hatching or birth.

inborn error of metabolism A metabolic disorder caused by the mutation of a gene that codes for an enzyme needed for a biochemical pathway.

inbreeding The mating of genetically similar individuals. Homozygosity increases with each successive generation of inbreeding. Compare with *outbreeding.*

inbreeding depression The phenomenon in which inbred offspring of genetically similar individuals have lower fitness (e.g., decline in fertility and high juvenile mortality) than do noninbred individuals.

inclusive fitness The total of an individual's direct and indirect fitness; includes the genes contributed directly to offspring and those contributed indirectly by kin selection. Compare with *direct fitness.* See *kin selection.*

incomplete dominance A condition in which neither member of a pair of contrasting alleles is completely expressed when the other is present.

incomplete flower A flower that lacks one or more of the four parts: sepals, petals, stamens, and/or carpels. Compare with *complete flower.*

independent assortment, principle of The genetic principle, first noted by Gregor Mendel, that states that the alleles of unlinked loci are randomly distributed to gametes.

indeterminate growth Unrestricted growth, as for example, in stems and roots. Compare with *determinate growth.*

index fossils Fossils restricted to a narrow period of geological time and found in the same sedimentary layers in different geographic areas.

indoleacetic acid See *auxin.*

induced fit Conformational change in the active site of an enzyme that occurs when it binds to its substrate.

inducer A molecule that binds to a repressor protein, converting it to its inactive form, which is unable to prevent transcription.

inducible operon An operon that is normally inactive because a repressor molecule is attached to its operator; transcription is activated when an inducer binds to the repressor, making it incapable of binding to the operator, e.g., the *lac* operon of *Escherichia coli.* Compare with *repressible operon.*

induction The process by which the differentiation of a cell or group of cells is influenced by interactions with neighboring cells.

inductive reasoning The reasoning that uses specific examples to draw a general conclusion or discover a general principle. Compare with *deductive reasoning.* See *hypothetico-inductive approach.*

infant mortality rate The number of infant deaths per 1000 live births. (A child is an infant during its first two years of life.)

inflammatory response The response of body tissues to injury or infection, characterized clinically by heat, swelling, redness, and pain, and physiologically by increased dilation of blood vessels and increased phagocytosis.

inflorescence A cluster of flowers on a common floral stalk.

ingestion The process of taking food (or other material) into the body.

inhibin A hormone that inhibits FSH secretion; produced by Sertoli cells in the testes and by granulosa cells in the ovaries.

inhibitory postsynaptic potential (IPSP) A change in membrane potential that takes a neuron farther from the firing level. Compare with *excitatory postsynaptic potential (EPSP).*

initiation (of protein synthesis) The first steps of protein synthesis, in which the large and small ribosomal subunits and other components of the translation machinery bind to the 5′ end of mRNA. See *elongation* and *termination.*

initiation codon See *start codon.*

innate behavior Behavior that is inherited and typical of the species; also called *instinct.*

innate immune responses See *nonspecific immune responses.*

inner cell mass The cluster of cells in the early mammalian embryo that gives rise to the embryo proper.

inorganic compound A simple substance that does not contain a carbon backbone. Compare with *organic compound.*

inositol trisphosphate (IP_3) A second messenger that increases intracellular calcium concentration and activates enzymes.

insight learning A complex learning process in which an animal adapts past experience to solve a new problem that may involve different stimuli.

in situ conservation Conservation efforts that concentrate on preserving biological diversity in the wild. Compare with *ex situ conservation.*

instinct See *innate behavior.*

instrumental conditioning See *operant conditioning.*

insulin (in′suh-lin) A hormone secreted by the pancreas that lowers blood-glucose concentration. Compare with *glucagon.*

insulin-like growth factors (IGF) Somatomedins; proteins that mediate responses to growth hormone.

insulin resistance See *diabetes mellitus.*

insulin shock A condition in which the blood glucose concentration is so low that the individual may appear intoxicated, or may become unconscious and even die; caused by the injection of too much insulin or by certain metabolic malfunctions.

integrins Receptor proteins that bind to specific proteins in the extracellular matrix and to membrane proteins on adjacent cells; transmit signals into the cell from the extracellular matrix.

integral membrane protein A protein that is tightly associated with the lipid bilayer of a biological membrane; a transmembrane integral protein spans the bilayer. Compare with *peripheral membrane protein.*

integration The process of summing (adding and subtracting) incoming neural signals.

integumentary system (in-teg″yoo-men′tar-ee) The body's covering, including the skin and its nails, glands, hair, and other associated structures.

integuments The outer cell layers that surround the megasporangium of an ovule; develop into the seed coat.

intercellular substance In connective tissues, the combination of matrix and fibers in which the cells are embedded.

interference competition Intraspecific competition in which certain dominant individuals obtain an adequate supply of the limited resource at the expense of other individuals in the population. Also called *contest competition.* Compare with *exploitation competition.*

interferons (in″tur-feer′onz) Cytokines produced by animal cells when challenged by a virus; prevent viral reproduction and enable cells to resist a variety of viruses.

interkinesis The stage between meiosis I and meiosis II. Interkinesis is usually brief; the chromosomes may decondense, reverting at least partially to an interphase-like state, but DNA synthesis and chromosome duplication do not occur.

interleukins A diverse group of cytokines produced mainly by macrophages and lymphocytes.

intermediate-day plant A plant that flowers when it is exposed to days and nights of intermediate length but does not flower when the daylength is too long or too short. Compare with *long-day, short-day,* and *day-neutral plants.*

intermediate disturbance hypothesis In community ecology, the idea that species richness is greatest at moderate levels of disturbance, which create a mosaic of habitat patches at different stages of succession.

intermediate filaments Cytoplasmic fibers that are part of the cytoskeletal network and are intermediate in size between microtubules and microfilaments.

internal membrane system The group of membranous structures in eukaryotic cells that interact through direct connections by vesicles; includes the endoplasmic reticulum, outer membrane of the nuclear envelope, Golgi complex, lysosomes, and the plasma membrane; also called *endomembrane system.*

interneuron (in″tur-noor′on) A nerve cell that carries impulses from one nerve cell to another and is not directly associated with either an effector or a sensory receptor. Also known as an *association neuron.*

internode The region on a stem between two successive nodes. Compare with *node.*

interoceptor (in′tur-oh-sep″tor) A sense organ within a body organ that transmits information regarding chemical composition, pH, osmotic pressure, or temperature. Compare with *exteroceptor.*

interphase The stage of the cell cycle between successive mitotic divisions; its subdivisions are the G_1 (first gap), S (DNA synthesis), and G_2 (second gap) phases.

interspecific competition The interaction between members of different species that vie for the same resource in an ecosystem (e.g., food or living space). Compare with *intraspecific competition.*

interstitial cells (of testis) The cells between the seminiferous tubules that secrete testosterone.

interstitial fluid The fluid that bathes the tissues of the body; also called *tissue fluid.*

intertidal zone The marine shoreline area between the high-tide mark and the low-tide mark.

intraspecific competition The interaction between members of the same species that vie for the same resource in an ecosystem (e.g., food or living space). Compare with *interspecific competition.*

intrinsic rate of increase The theoretical maximum rate of increase in population size occurring under optimal environmental conditions. Also called *biotic potential.*

intron A non-protein-coding region of a eukaryotic gene and also of the pre-mRNA transcribed from such a region. Introns do not appear in mRNA. Compare with *exon.*

invasive species A foreign species that, when introduced into an area where it is not native, upsets the balance among the organisms living there and causes economic or environmental harm.

invertebrate An animal without a backbone (vertebral column); invertebrates account for about 95% of animal species.

in vitro Occurring outside a living organism (literally "in glass"). Compare with *in vivo*.

in vitro evolution Test tube experiments that demonstrate that RNA molecules in the RNA world could have catalyzed the many different chemical reactions needed for life. Also called *directed evolution*.

in vitro fertilization The fertilization of eggs in a test tube.

in vivo Occurring in a living organism. Compare with *in vitro*.

ion An atom or group of atoms bearing one or more units of electrical charge, either positive (cation) or negative (anion).

ion channels Channels for the passage of ions through a membrane; formed by specific membrane proteins.

ionic bond The chemical attraction between a cation and an anion. Compare with *covalent bond* and *hydrogen bond*.

ionic compound A substance consisting of cations and anions, which are attracted by their opposite charges; ionic compounds do not consist of molecules. Compare with *covalent compound*.

ionization The dissociation of a substance to yield ions, e.g., the ionization of water yields H^+ and OH^-.

iris The pigmented portion of the vertebrate eye.

iron-sulfur world hypothesis The hypothesis that simple organic molecules that are the precursors of life originated at hydrothermal vents in the deep-ocean floor. Compare with *prebiotic broth hypothesis*.

irreversible inhibitor A substance that permanently inactivates an enzyme. Compare with *reversible inhibitor*.

islets of Langerhans (eye′lets of Lahng′er-hanz) The endocrine portion of the pancreas that secretes glucagon and insulin, hormones that regulate the concentration of glucose in the blood.

isogamy (eye-sog′uh-me) Sexual reproduction involving motile gametes of similar form and size. Compare with *anisogamy* and *oogamy*.

isolecithal egg An egg containing a relatively small amount of uniformly distributed yolk. Compare with *telolecithal egg*.

isomer (eye′soh-mer) One of two or more chemical compounds having the same chemical formula, but different structural formulas, e.g., structural and geometrical isomers and enantiomers.

isoprene units Five-carbon hydrocarbon monomers that make up certain lipids such as carotenoids and steroids.

isotonic (eye″soh-ton′ik) A term applied to solutions that have identical concentrations of solute molecules and hence the same osmotic pressure. Compare with *hypertonic* and *hypotonic*.

isotope (eye′suh-tope) An alternate form of an element with a different number of neutrons, but the same number of protons and electrons. See *radioisotopes*.

iteroparity The condition of having repeated reproductive cycles throughout a lifetime. Compare with *semelparity*.

jelly coat One of the acellular coverings of the eggs of certain animals, such as echinoderms.

joint The junction between two or more bones of the skeleton.

joule A unit of energy, equivalent to 0.239 calorie.

juvenile hormone (JH) An arthropod hormone that preserves juvenile structure during a molt. Without it, metamorphosis toward the adult form takes place.

juxtaglomerular apparatus (juks″tah-glo-mer′yoo-lar) A structure in the kidney that secretes renin in response to a decrease in blood pressure.

***K* selection** A reproductive strategy recognized by some ecologists in which a species typically has a large body size, slow development, long life span, and does not devote a large proportion of its metabolic energy to the production of offspring. Compare with *r selection*.

karyotype (kare′ee-oh-type) The chromosomal composition of an individual.

keratin (kare′ah-tin) A horny, water-insoluble protein found in the epidermis of vertebrates and in nails, feathers, hair, and horns.

ketone An organic molecule containing a carbonyl group bonded to two carbon atoms. Compare with *aldehyde*.

keystone species A species whose presence in an ecosystem largely determines the species composition and functioning of that ecosystem.

kidney The paired vertebrate organ important in excretion of metabolic wastes and in osmoregulation.

killer T cell See *T cytotoxic cell*.

kilobase (kb) 1000 bases or base pairs of a nucleic acid.

kilocalorie The amount of heat required to raise the temperature of 1 kg of water 1°C; also called *Calorie*, which is equivalent to 1000 calories.

kilojoule 1000 joules. See *joule*.

kinases Enzymes that catalyze the transfer of phosphate groups from ATP to acceptor molecules. See *protein kinases*.

kinetic energy Energy of motion. Compare with *potential energy*.

kinetochore (kin-eh′toh-kore) The portion of the chromosome centromere to which the mitotic spindle fibers attach.

kingdom A broad taxonomic category made up of related phyla; many biologists currently recognize six kingdoms of living organisms.

kin selection A type of natural selection that favors altruistic behavior toward relatives (kin), thereby ensuring that, although the chances of an individual's survival are lessened, some of its genes will survive through successful reproduction of close relatives; increases inclusive fitness.

Klinefelter syndrome Inherited condition in which the affected individual is a sterile male with an XXY karyotype.

Koch's postulates A set of guidelines used to demonstrate that a specific pathogen causes specific disease symptoms.

Krebs cycle See *citric acid cycle*.

krummholz The gnarled, shrublike growth habit found in trees at high elevations, near their upper limit of distribution.

labyrinth The system of interconnecting canals of the inner ear of vertebrates.

labyrinthodonts The first successful group of tetrapods.

lactate (lactic acid) A three-carbon organic acid.

lactation (lak-tay′shun) The production or release of milk from the breast.

lacteal (lak′tee-al) One of the many lymphatic vessels in the intestinal villi that absorb fat.

lagging strand A strand of DNA that is synthesized as a series of short segments, called *Okazaki fragments*, which are then covalently joined by DNA ligase. Compare with *leading strand*.

lamins Polypeptides attached to the inner surface of the nuclear envelope that provide a type of skeletal framework.

landscape A large land area (several to many square kilometers) composed of interacting ecosystems.

landscape ecology The subdiscipline in ecology that studies the connections in a heterogeneous landscape.

large intestine The portion of the digestive tract of humans (and other vertebrates) consisting of the cecum, colon, rectum, and anus.

larva (pl., *larvae*) An immature form in the life history of some animals; may be unlike the parent.

larynx (lare′inks) The organ at the upper end of the trachea that contains the vocal cords.

lateral meristems Areas of localized cell division on the side of a plant that give rise to secondary tissues. Lateral meristems, including the vascular cambium and the cork cambium, cause an increase in the girth of the plant body. Compare with *apical meristem*.

leaching The process by which dissolved materials are washed away or carried with water down through the various layers of the soil.

leader sequence Noncoding sequence of nucleotides in mRNA that is transcribed from the region that precedes (is upstream to) the coding region.

leading strand Strand of DNA that is synthesized continuously. Compare with *lagging strand.*

learning A change in the behavior of an animal that results from experience.

legume (leg′yoom) (1) A simple, dry fruit that splits open at maturity along two sutures to release seeds. (2) Any member of the pea family, e.g., pea, bean, peanut, alfalfa.

lek A small territory in which males compete for females.

lens The oval, transparent structure located behind the iris of the vertebrate eye; bends incoming light rays and brings them to a focus on the retina.

lenticels (len′tih-sels) Porous swellings of cork cells in the stems of woody plants; facilitate the exchange of gases.

leptin A hormone produced by adipose tissue that signals brain centers about the status of energy stores.

leukocytes (loo′koh-sites) White blood cells; colorless amoeboid cells that defend the body against disease-causing organisms.

leukoplasts Colorless plastids; include amyloplasts, which are used for starch storage in cells of roots and tubers.

lichen (ly′ken) A compound organism consisting of a symbiotic fungus and an alga or cyanobacterium.

life history traits Significant features of a species' life cycle, particularly traits that influence survival and reproduction.

life span The maximum duration of life for an individual of a species.

life table A table showing mortality and survival data by age of a population or cohort.

ligament (lig′uh-ment) A connective tissue cable or strap that connects bones to each other or holds other organs in place.

ligand A molecule that binds to a specific site in a receptor or other protein.

light-dependent reactions Reactions of photosynthesis in which light energy absorbed by chlorophyll is used to synthesize ATP and usually NADPH. Includes *cyclic electron transport* and *noncyclic electron transport.*

lignin (lig′nin) A substance found in many plant cell walls that confers rigidity and strength, particularly in woody tissues.

limbic system In vertebrates, an action system of the brain. In humans, plays a role in emotional responses, motivation, autonomic function, and sexual response.

limiting resource An environmental resource that, because it is scarce or unfavorable, tends to restrict the ecological niche of an organism.

limnetic zone (lim-net′ik) The open water away from the shore of a lake or pond extending down as far as sunlight penetrates. Compare with *littoral zone* and *profundal zone.*

linkage The tendency for a group of genes located on the same chromosome to be inherited together in successive generations.

lipase (lip′ase) A fat-digesting enzyme.

lipid Any of a group of organic compounds that are insoluble in water but soluble in nonpolar solvents; lipids serve as energy storage and are important components of cell membranes.

lipoprotein (lip-oh-proh′ teen) A large molecular complex consisting of lipids and protein; transports lipids in the blood. High-density lipoproteins (HDLs) transport cholesterol to the liver; low-density lipoproteins (LDLs) deliver cholesterol to many cells of the body.

littoral zone (lit′or-ul) The region of shallow water along the shore of a lake or pond. Compare with *limnetic zone* and *profundal zone.*

liver A large, complex organ that secretes bile, helps maintain homeostasis by removing or adding nutrients to the blood, and performs many other metabolic functions.

liverworts A phylum of spore-producing, nonvascular, thallose or leafy plants with a life cycle similar to mosses.

local hormones See *local regulators.*

local regulators Prostaglandins (a group of local hormones), growth factors, cytokines, and other soluble molecules that act on nearby cells by paracrine regulation or act on the cells that produce them (autocrine regulation).

locus The place on the chromosome at which the gene for a given trait occurs, i.e., a segment of the chromosomal DNA containing information that controls some feature of the organism; also called *gene locus.*

logistic population growth Population growth that initially occurs at a constant rate of increase over time (i.e., exponential) but then levels out as the carrying capacity of the environment is approached. Compare with *exponential population growth.*

long-day plant A plant that flowers in response to shortening nights; also called *short-night plant.* Compare with *short-day, intermediate-day,* and *day-neutral plants.*

long-night plant See *short-day plant.*

long-term potentiation (LTP) Long-lasting increase in the strength of synaptic connections that occurs in response to a series of high-frequency electrical stimuli. Compare with *long-term synaptic depresssion (LTD).*

long-term synaptic depression (LTD) Long-lasting decrease in the strength of synaptic connections that occurs in response to low-frequency stimulation of neurons. Compare with *long-term potentiation (LTP).*

loop of Henle (Hen′lee) The U-shaped loop of a mammalian kidney tubule, which extends down into the renal medulla.

loose connective tissue A type of connective tissue that is widely distributed in the body; consists of fibers strewn through a semi-fluid matrix.

lophoporate phyla Three related invertebrate protostome phyla, characterized by a ciliated ring of tentacles that surrounds the mouth.

Lophotrochozoa A branch of the protostomes that includes the flatworms, nemerteans (proboscis worms), mollusks, annelids, and the lophophorate phyla.

low-density lipoprotein (LDL) See *lipoprotein.*

lumen (loo′men) (1) The space enclosed by a membrane, such as the lumen of the endoplasmic reticulum or the thylakoid lumen. (2) The cavity or channel within a tube or tubular organ, such as a blood vessel or the digestive tract. (3) The space left within a plant cell after the cell's living material dies, as in tracheids.

lung An internal respiratory organ that functions in gas exchange; enables an animal to breathe air.

luteinizing hormone (LH) (loot′eh-ny-zing) Gonadotropic hormone secreted by the anterior pituitary; stimulates ovulation and maintains the corpus luteum in the ovaries of females; stimulates testosterone production in the testes of males.

lymph (limf) The colorless fluid within the lymphatic vessels that is derived from blood plasma; contains white blood cells; ultimately lymph is returned to the blood.

lymph node A mass of lymph tissue surrounded by a connective tissue capsule; manufactures lymphocytes and filters lymph.

lymphatic system A subsystem of the cardiovascular system; returns excess interstitial fluid (lymph) to the circulation; defends the body against disease organisms.

lymphocyte (lim′foh-site) White blood cell with nongranular cytoplasm that governs immune responses. See *B cell* and *T cell.*

lysis (ly′sis) The process of disintegration of a cell or some other structure.

lysogenic conversion The change in properties of bacteria that results from the presence of a prophage.

lysosomes (ly′soh-somes) Intracellular organelles present in many animal cells; contain a variety of hydrolytic enzymes.

lysozyme An enzyme found in many tissues and in tears and other body fluids; attacks the cell wall of many Gram-positive bacteria.

macroevolution Large-scale evolutionary events over long time spans. Macroevolution results in phenotypic changes in populations that are significant enough to warrant their placement in taxonomic groups at the species level and higher. Compare with *microevolution.*

macromolecule A very large organic molecule, such as a protein or nucleic acid.

macronucleus A large nucleus found, along with one or several micronuclei, in ciliates. The macronucleus regulates metabolism and growth. Compare with *micronucleus.*

macronutrient An essential element that is required in fairly large amounts for normal growth. Compare with *micronutrient.*

macrophage (mak′roh-faje) A large phagocytic cell capable of ingesting and digesting bacteria and cell debris. Macrophages are also antigen-presenting cells.

major histocompatibility complex (MHC) A group of membrane proteins, present on the surface of most cells, that are slightly different in each individual. In humans, the MHC is called the *HLA (human leukocyte antigen) group.*

malignant cells Cancer cells; tumor cells that are able to invade tissue and metastasize.

malnutrition Poor nutritional status; results from dietary intake that is either below or above required needs.

Malpighian tubules (mal-pig′ee-an) The excretory organs of many arthropods.

mammals The class of vertebrates characterized by hair, mammary glands, a diaphragm, and differentiation of teeth.

mandible (man′dih-bl) (1) The lower jaw of vertebrates. (2) Jaw-like, external mouthparts of insects.

mangrove forest A tidal wetland dominated by mangrove trees, in which the salinity fluctuates between that of sea water and fresh water.

mantle In the mollusk, a fold of tissue that covers the visceral mass and that usually produces a shell.

marine snow The organic debris (plankton, dead organisms, fecal material, etc.) that "rains" into the dark area of the oceanic province from the lighted region above; the primary food of most organisms that live in the ocean's depths.

marsupials (mar-soo′pee-uls) A subclass of mammals, characterized by the presence of an abdominal pouch in which the young, which are born in a very undeveloped condition, are carried for some time after birth.

mass extinction The extinction of numerous species during a relatively short period of geological time. Compare with *background extinction.*

mast cell A type of cell found in connective tissue; contains histamine and is important in an inflammatory response and in allergic reactions.

maternal effect genes Genes of the mother that are transcribed during oogenesis and subsequently affect the development of the embryo. Compare with *zygotic genes.*

matrix (may′triks) (1) In cell biology, the interior of the compartment enclosed by the inner mitochondrial membrane. (2) In zoology, nonliving material secreted by and surrounding connective tissue cells; contains a network of microscopic fibers.

matter Anything that has mass and takes up space.

maxillae Appendages used for manipulating food; characteristic of crustaceans.

mechanical isolation A prezygotic reproductive isolating mechanism in which fusion of the gametes of two species is prevented by morphological or anatomical differences.

mechanoreceptor (meh-kan′oh-ree-sep″tor) A sensory cell or organ that perceives mechanical stimuli, e.g., touch, pressure, gravity, stretching, or movement.

medulla (meh-dul′uh) (1) The inner part of an organ, such as the medulla of the kidney. Compare with *cortex.* (2) The most posterior part of the vertebrate brain, lying next to the spinal cord.

medusa A jellyfish-like animal; a free-swimming, umbrella-shaped stage in the life cycle of certain cnidarians. Compare with *polyp.*

megaphyll (meg′uh-fil) Type of leaf found in horsetails, ferns, gymnosperms, and angiosperms; contains multiple vascular strands (i.e., complex venation). Compare with *microphyll.*

megaspore (meg′uh-spor) The *n* spore in heterosporous plants that gives rise to a female gametophyte. Compare with *microspore.*

meiosis (my-oh′sis) Process in which a 2*n* cell undergoes two successive nuclear divisions (meiosis I and meiosis II), potentially producing four *n* nuclei; leads to the formation of gametes in animals and spores in plants.

melanin A dark pigment present in many animals; contributes to the color of the skin.

melanin concentrating hormone (MCH) A neuropeptide signaling molecule that helps regulate energy homeostasis.

melanocortins A group of peptides that appear to decrease appetite in response to increased fat stores.

melatonin (mel-ah-toh′ nin) A hormone secreted by the pineal gland that plays a role in setting circadian rhythms.

memory cell B or T cell (lymphocyte) that permits rapid mobilization of immune response on second or subsequent exposure to a particular antigen. Memory B cells continue to produce antibodies after the immune system overcomes an infection.

meninges (meh-nin′jeez) (sing., *meninx*) The three membranes that protect the brain and spinal cord: the dura mater, arachnoid, and pia mater.

menopause The period (usually occurring between 45 and 55 years of age) in women when the recurring menstrual cycle ceases.

menstrual cycle (men′stroo-ul) In the human female, the monthly sequence of events that prepares the body for pregnancy.

menstruation (men-stroo-ay′shun) The monthly discharge of blood and degenerated uterine lining in the human female; marks the beginning of each menstrual cycle.

meristem (mer′ih-stem) A localized area of mitotic cell division in the plant body. See *apical meristem* and *lateral meristems.*

meroblastic cleavage Cleavage pattern observed in the telolecithal eggs of reptiles and birds, in which cleavage is restricted to a small disc of cytoplasm at the animal pole. Compare with *holoblastic cleavage.*

mesencephalon See *midbrain.*

mesenchyme (mes′en-kime) A loose, often jelly-like connective tissue containing undifferentiated cells; found in the embryos of vertebrates and the adults of some invertebrates.

mesoderm (mez′oh-derm) The middle germ layer of the early embryo; gives rise to connective tissue, muscle, bone, blood vessels, kidneys, and many other structures. Compare with *ectoderm* and *endoderm.*

mesophyll (mez′oh-fil) Photosynthetic tissue in the interior of a leaf; sometimes differentiated into palisade mesophyll and spongy mesophyll.

Mesozoic era That part of geological time extending from roughly 248 to 65 million years ago.

messenger RNA (mRNA) RNA that specifies the amino acid sequence of a protein; transcribed from DNA.

metabolic pathway A series of chemical reactions in which the product of one reaction becomes the substrate of the next reaction.

metabolic rate Energy use by an organism per unit time. See *basal metabolic rate.*

metabolism The sum of all the chemical processes that occur within a cell or organism: the transformations by which energy and matter are made available for use by the organism. See *anabolism* and *catabolism.*

metamorphosis (met″ah-mor′fuh-sis) Transition from one developmental stage to another, such as from a larva to an adult.

metanephridia (sing., *metanephridium*) The excretory organs of annelids and mollusks; each consists of a tubule open at both ends; at one end a ciliated funnel opens into the coelom, and the other end opens to the outside of the body.

metaphase (met′ah-faze) The stage of mitosis, and of meiosis I and II, in which the chromosomes line up on the equatorial plane of the cell. Occurs after prophase and before anaphase.

metapopulation A population that is divided into several local populations among which individuals occasionally disperse.

metastasis (met-tas′tuh-sis) The spreading of cancer cells from one organ or part of the body to another.

metencephalon See *hindbrain.*

methyl group A nonpolar functional group; abbreviated $—CH_3$.

microclimate Local variations in climate produced by differences in elevation, in the steepness and direction of slopes, and in exposure to prevailing winds.

microevolution Small-scale evolutionary change caused by changes in allele or genotype frequencies that occur within a population over a few generations. Compare with *macroevolution.*

microfilaments Thin fibers consisting of actin protein subunits; form part of the cytoskeleton.

microfossils Ancient traces (fossils) of microscopic life.

microglia Phagocytic glial cells found in the CNS.

micronucleus One or more smaller nuclei found, along with the macronucleus, in ciliates. The micronucleus is involved in sexual reproduction. Compare with *macronucleus.*

micronutrient An essential element that is required in trace amounts for normal growth. Compare with *macronutrient.*

microphyll (mi′kro-fil) Type of leaf found in club mosses; contains one vascular strand (i.e., simple venation). Compare with *megaphyll.*

microsphere A protobiont produced by adding water to abiotically formed polypeptides.

microspore (mi′kro-spor) The *n* spore in heterosporous plants that gives rise to a male gametophyte. Compare with *megaspore.*

microtubules (my-kroh-too′bewls) Hollow cylindrical fibers consisting of tubulin protein subunits; major components of the cytoskeleton and found in mitotic spindles, cilia, flagella, centrioles, and basal bodies.

microtubule-associated proteins (MAPs) Include structural proteins that help regulate microtubule assembly and cross-link microtubules to other cytoskeletal polymers; and motors, such as kinesin and dynein, that use ATP to produce movement.

microtubule-organizing center (MTOC) The region of the cell from which microtubules are anchored and possibly assembled. The MTOCs of many organisms (including animals, but not flowering plants or most gymnosperms) contain a pair of centrioles.

microvilli (sing., *microvillus*) Minute projections of the plasma membrane that increase the surface area of the cell; found mainly in cells concerned with absorption or secretion, such as those lining the intestine or the kidney tubules.

midbrain In vertebrate embryos, one of the three divisions of the developing brain. Also called *mesencephalon.* Compare with *forebrain* and *hindbrain.*

middle lamella The layer composed of pectin polysaccharides that serves to cement together the primary cell walls of adjacent plant cells.

midvein The main, or central, vein of a leaf.

migration (1) The periodic or seasonal movement of an organism (individual or population) from one place to another, usually over a long distance. See *dispersal.* (2) In evolutionary biology, a movement of individuals that results in a transfer of alleles from one population to another. See *gene flow.*

mineralocorticoids (min″ur-al-oh-kor′tih-koidz) Hormones produced by the adrenal cortex that regulate mineral metabolism and, indirectly, fluid balance. The principal mineralocorticoid is aldosterone.

minerals Inorganic nutrients ingested as salts dissolved in food and water.

missense mutation A type of base-substitution mutation that causes one amino acid to be substituted for another in the resulting protein product. Compare with *nonsense mutation.*

mitochondria (my″toh-kon′dree-ah) (sing., *mitochondrion*) Intracellular organelles that are the sites of oxidative phosphorylation in eukaryotes; include an outer membrane and an inner membrane.

mitochondrial DNA (mtDNA) DNA present in mitochondria that is transmitted maternally, from mothers to their offspring. Mitochondrial DNA mutates more rapidly than nuclear DNA.

mitosis (my-toh′sis) The division of the cell nucleus resulting in two daughter nuclei, each with the same number of chromosomes as the parent nucleus; mitosis consists of four phases: prophase, metaphase, anaphase, and telophase. Cytokinesis usually overlaps the telophase stage.

mitotic spindle Structure consisting mainly of microtubules that provides the framework for chromosome movement during cell division.

mitral valve See *atrioventricular valve.*

mobile genetic element See *transposon.*

modern synthesis A comprehensive, unified explanation of evolution based on combining previous theories, especially of Mendelian genetics, with Darwin's theory of evolution by natural selection; also called the *synthetic theory of evolution.*

mole The atomic mass of an element or the molecular mass of a compound, expressed in grams; one mole of any substance has 6.02×10^{23} units (Avogadro's number).

molecular anthropology The branch of science that compares genetic material from individuals of regional human populations to help unravel the origin and migrations of modern humans.

molecular chaperones Proteins that help other proteins fold properly. Although not dictating the folding pattern, chaperones make the process more efficient.

molecular clock analysis A comparison of the DNA nucleotide sequences of related organisms to estimate when they diverged from one another during the course of evolution.

molecular formula The type of chemical formula that gives the actual numbers of each type of atom in a molecule. Compare with *simplest formula* and *structural formula.*

molecular mass The sum of the atomic masses of the atoms that make up a single molecule of a compound; expressed in atomic mass units (amu) or daltons.

molecule The smallest particle of a covalently bonded element or compound; two or more atoms joined by covalent bonds.

mollusks A phylum of coelomate protostome animals characterized by a soft body, visceral mass, mantle, and foot.

molting The shedding and replacement of an outer covering such as an exoskeleton.

molting hormone A steroid hormone that stimulates growth and molting in insects. Also called *ecdysone.*

monoacylglycerol (mon″o-as″-il-glis′er-ol) Lipid consisting of glycerol combined chemically with a single fatty acid. Also called *monoglyceride.* Compare with *diacylglycerol* and *triacylglycerol.*

monocot (mon′oh-kot) One of the two classes of flowering plants; monocot seeds contain a single cotyledon, or seed leaf. Compare with *dicot.*

monoclonal antibodies Identical antibody molecules produced by cells cloned from a single cell.

monocyte (mon′oh-site) A type of white blood cell; a large, phagocytic, nongranular leukocyte that enters the tissues and differentiates into a macrophage.

monoecious (mon-ee′shus) Having male and female reproductive parts in separate flowers or cones on the same plant; compare with *dioecious.*

monogamy A mating system in which a male animal mates with a single female during a breeding season.

monoglyceride See *monoacylglycerol.*

monohybrid cross A genetic cross that takes into account the behavior of alleles of a single locus. Compare with *dihybrid cross.*

monokaryotic (mon″o-kare-ee-ot′ik) The condition of having a single *n* nucleus per cell, characteristic of certain fungal hyphae. Compare with *dikaryotic.*

monomer (mon′oh-mer) A molecule that can link with other similar molecules; two monomers join to form a dimer, whereas many form a polymer. Monomers are small (e.g., sugars or amino acids) or large (e.g., tubulin or actin proteins).

monophyletic group (mon″oh-fye-let′ik) A group made up of organisms that evolved from a common ancestor. Compare with *polyphyletic group* and *paraphyletic group.*

monosaccharide (mon-oh-sak′ah-ride) A sugar that can't be degraded by hydrolysis to a simpler sugar (*e.g., glucose or fructose*).

monosomy The condition in which only one member of a chromosome pair is present and the other is missing. Compare with *trisomy* and *disomy.*

monotremes (mon′oh-treems) Egg-laying mammals such as the duck-billed platypus of Australia.

monounsaturated fatty acid See *fatty acid.*

monozygotic twins Genetically identical twins that arise from the division of a single fertilized egg; commonly known as *identical twins.* Compare with *dizygotic twins.*

morphogen Any chemical agent thought to govern the processes of cell differentiation and pattern formation that lead to morphogenesis.

morphogenesis (mor-foh-jen′eh-sis) The development of the form and structures of an organism and its parts; proceeds by a series of steps known as *pattern formation.*

mortality The rate at which individuals die; the average per capita death rate.

morula (mor′yoo-lah) An early embryo consisting of a solid ball of cells.

mosaic development A rigid developmental pattern in which the fates of cells become restricted early in development. Compare with *regulative development.*

mosses A phylum of spore-producing nonvascular plants with an alternation of generations in which the dominant *n* gametophyte alternates with a 2*n* sporophyte that remains attached to the gametophyte.

motor neuron An efferent neuron that transmits impulses away from the central nervous system to skeletal muscle.

motor unit All the skeletal muscle fibers that are stimulated by a single motor neuron.

mRNA cap An unusual nucleotide, 7-methylguanylate, that is added to the 5′ end of a eukaryotic messenger RNA. Capping enables eukaryotic ribosomes to bind to mRNA.

mucosa (mew-koh′suh) See *mucous membrane.*

mucous membrane A type of epithelial membrane that lines a body cavity that opens to the outside of the body, e.g., the digestive and respiratory tracts; also called *mucosa.*

mucus (mew′cus) A sticky secretion composed of covalently linked protein and carbohydrate; serves to lubricate body parts and trap particles of dirt and other contaminants. (The adjectival form is spelled *mucous.*)

Müllerian mimicry (mul-ler′ee-un mim′ih-kree) The resemblance of dangerous, unpalatable, or poisonous species to one another so that they are more easily recognized by potential predators. Compare with *Batesian mimicry.*

multiple alleles (al-leels′) Three or more alleles of a single locus (in a population), such as the alleles governing the ABO series of blood types.

multiple fruit A fruit that develops from many ovaries of many separate flowers, e.g., pineapple. Compare with *simple, aggregate,* and *accessory fruits.*

muscle (1) A tissue specialized for contraction. (2) An organ that produces movement by contraction.

mutagen (mew′tah-jen) Any agent capable of entering the cell and producing mutations.

mutation Any change in DNA; may include a change in the nucleotide base pairs of a gene, a rearrangement of genes within the chromosomes so that their interactions produce different effects, or a change in the chromosomes themselves.

mutualism (1) In ecology, a symbiotic relationship in which both partners benefit from the association. Compare with *parasitism* and *commensalism.* (2) In animal behavior, cooperative behavior in which each animal in the group benefits.

mycelium (my-seel′ee-um) (pl., *mycelia*) The vegetative body of most fungi and certain protists (water molds); consists of a branched network of hyphae.

mycorrhizae (my″kor-rye′zee) Mutualistic associations of fungi and plant roots that aid in the plant's absorption of essential minerals from the soil.

mycotoxins Poisonous chemical compounds produced by fungi, e.g., aflatoxins that harm the liver and are known carcinogens.

myelencephalon See *hindbrain.*

myelin sheath (my′eh-lin) The white, fatty material that forms a sheath around the axons of certain nerve cells, which are then called *myelinated fibers.*

myocardial infarction (MI) Heart attack; serious consequence occurring when the heart muscle receives insufficient oxygen.

myofibrils (my-oh-fy′brilz) Tiny threadlike structures in the cytoplasm of striated and cardiac muscle that are responsible for contractions of the cell; contain myofilaments.

myofilament (my-oh-fil′uh-ment) One of the filaments making up a myofibril; the structural unit of muscle proteins in a muscle cell. See *myosin filaments* and *actin filaments.*

myoglobin (my′oh-glob″bin) A hemoglobin-like, oxygen-transferring protein found in muscle.

myosin (my′oh-sin) A protein that, together with actin, is responsible for muscle contraction.

myosin filaments Thick filaments consisting mainly of the protein myosin; actin and myosin filaments make up the myofibrils of muscle fibers.

n The chromosome number of a gamete. The chromosome number of a zygote is $2n$. If an organism is not polyploid, the n gametes are haploid and the $2n$ zygotes are diploid.

NAD^+/NADH Oxidized and reduced forms, respectively, of nicotinamide adenine dinucleotide, a coenzyme that transfers electrons (as hydrogen), particularly in catabolic pathways, including cellular respiration.

$NADP^+$/NADPH Oxidized and reduced forms, respectively, of nicotinamide adenine dinucleotide phosphate, a coenzyme that acts as an electron (hydrogen) transfer agent, particularly in anabolic pathways, including photosynthesis.

nanoplankton Extremely minute (< 20 μm in length) algae that are major producers in the ocean because of their great abundance; part of phytoplankton.

nastic movement A temporary, reversible movement of a plant organ in response to external stimuli; movement is caused by changes in the turgor of certain cells.

natality The rate at which individuals produce offspring; the average per capita birth rate.

natural selection The mechanism of evolution proposed by Charles Darwin; the tendency of organisms that have favorable adaptations to their environment to survive and become the parents of the next generation. Evolution occurs when natural selection results in changes in allele frequencies in a population.

natural killer cell (NK cell) A large, granular lymphocyte that functions in both nonspecific and specific immune responses; releases cytokines and proteolytic enzymes that target tumor cells and cells infected with viruses and other pathogens.

necrosis Uncontrolled cell death that causes inflammation and damages other cells. Compare with *apoptosis.*

nectary (nek′ter-ee) A gland or other structure that secretes nectar.

negative feedback mechanism A homeostatic mechanism in which a change in some condition triggers a response that counteracts, or reverses, the changed condition, restoring homeostasis, e.g., how mammals maintain body temperature. Compare with *positive feedback mechanism.*

nekton (nek′ton) Free-swimming aquatic organisms such as fish and turtles. Compare with *plankton.*

nematocyst (nem-at′oh-sist) A stinging structure found within cnidocytes (stinging cells) in cnidarians; used for anchorage, defense, and capturing prey.

nematodes The phylum of animals commonly known as *roundworms.*

nemerteans The phylum of animals commonly known as *ribbon worms;* each has a proboscis (tubular feeding organ) for capturing prey.

neonate Newborn individual.

neoplasm See *tumor.*

nephridial organ (neh-frid′ee-al) The excretory organ of many invertebrates; consists of simple or branching tubes that usually open to the outside of the body through pores; also called *nephridium.*

nephron (nef′ron) The functional, microscopic unit of the vertebrate kidney.

neritic province (ner-ih′tik) Ocean water that extends from the shoreline to where the bottom reaches a depth of 200 m. Compare with *oceanic province.*

nerve A bundle of axons (or dendrites) wrapped in connective tissue that conveys impulses between the central nervous system and some other part of the body.

nerve net A system of interconnecting nerve cells found in cnidarians and echinoderms.

nervous tissue A type of animal tissue specialized for transmitting electrical and chemical signals.

nest parasitism A behavior practiced by brown-headed cowbirds and certain other bird species in which females lay their eggs in the nests of other bird species and leave all parenting jobs to the hosts.

net primary productivity The energy that remains in an ecosystem (as biomass) after cellular respiration has occurred; net primary productivity equals gross primary productivity minus respiration. Compare with *gross primary productivity.*

neural crest (noor′ul) A group of cells along the neural tube that migrate and form various parts of the embryo, including parts of the peripheral nervous system.

neural plasticity The ability of the nervous system to change in response to experience.

neural plate See *neural tube.*

neural transmission See *transmission, neural.*

neural tube The hollow, longitudinal structure in the early vertebrate embryo that gives rise to the brain and spinal cord. The neural tube forms from the neural plate, a flattened, thickened region of the ectoderm that rolls up and sinks below the surface.

neuroendocrine cells Neurons that produce neurohormones.

neurohormones Hormones produced by neuroendocrine cells; transported down axons and released into interstitial fluid; typically diffuse into capillaries and are transported by the blood; common in invertebrates; in vertebrates, neurohormones are produced by the hypothalamus.

neuron (noor′on) A nerve cell; a conducting cell of the nervous system that typically consists of a cell body, dendrites, and an axon.

neuropeptide The group of peptides produced in neural tissue that function as signaling molecules; many are neurotransmitters.

neuropeptide Y A signaling molecule produced by the hypothalamus that increases appetite and slows metabolism; helps restore energy homeostasis when leptin levels and food intake are low.

neurotransmitter A chemical signal used by neurons to transmit impulses across a synapse.

neutral solution A solution of pH 7; there are equal concentrations of hydrogen ions $[H^+]$ and hydroxide ions $[OH^-]$. Compare with *acidic solution* and *basic solution.*

neutral variation Variation that does not appear to confer any selective advantage or disadvantage to the organism.

neutron (noo′tron) An electrically neutral particle with a mass of 1 atomic mass unit (amu) found in the atomic nucleus. Compare with *proton* and *electron.*

neutrophil (new′truh-fil) A type of granular leukocyte important in immune responses; a type of phagocyte that engulfs and destroys bacteria and foreign matter.

niche (nich) The totality of an organism's adaptations, its use of resources, and the lifestyle to which it is fitted in its community; how an organism uses materials in its environment as well as how it interacts with other organisms; also called *ecological niche.* See *fundamental niche* and *realized niche.*

nicotinamide adenine dinucleotide See NAD^+/NADH.

nicotinamide adenine dinucleotide phosphate See $NADP^+$/NADPH.

nitric oxide (NO) A gaseous signaling molecule; a neurotransmitter.

nitrification (nie″tra-fuh-kay′shun) The conversion of ammonia (NH_3) to nitrate (NO_3^-) by certain bacteria (nitrifying bacteria) in the soil; part of the nitrogen cycle.

nitrogen cycle The worldwide circulation of nitrogen from the abiotic environment into living things and back into the abiotic environment.

nitrogen fixation The conversion of atmospheric nitrogen (N_2) to ammonia (NH_3) by certain bacteria; part of the nitrogen cycle.

nitrogenase (nie-traa'jen-ase) The enzyme responsible for nitrogen fixation under anaerobic conditions.

nociceptors (no'sih-sep-torz) Pain receptors; free endings of certain sensory neurons whose stimulation is perceived as pain.

node The area on a stem where each leaf is attached. Compare with *internode.*

nodules Swellings on the roots of plants, such as legumes, in which symbiotic nitrogen-fixing bacteria *(Rhizobium)* live.

noncompetitive inhibitor A substance that lowers the rate at which an enzyme catalyzes a reaction but does not bind to the active site. Compare with *competitive inhibitor.*

noncyclic electron transport In photosynthesis, the linear flow of electrons, produced by photolysis of water, through Photosystems I and II; results in the formation of ATP (by chemiosmosis), NADPH, and O_2. Compare with *cyclic electron transport.*

nondisjunction Abnormal separation of sister chromatids or of homologous chromosomes caused by their failure to disjoin (move apart) properly during mitosis or meiosis.

nonpolar covalent bond Chemical bond formed by the equal sharing of electrons between atoms of approximately equal electronegativity. Compare with *polar covalent bond.*

nonpolar molecule Molecule that does not have a positively charged end and a negatively charged end; nonpolar molecules are generally insoluble in water. Compare with *polar molecule.*

nonsense mutation A base substitution mutation that results in an amino acid–specifying codon being changed to a termination (stop) codon; when the abnormal mRNA is translated, the resulting protein is usually truncated and nonfunctional. Compare with *missense mutation.*

nonspecific immune responses Mechanisms such as physical barriers (e.g., the skin) and phagocytosis that provide immediate and general protection against pathogens. Also called *innate immunity.* Compare with *specific immune responses.*

norepinephrine (nor-ep-ih-nef'rin) A neurotransmitter that is also a hormone secreted by the adrenal medulla.

Northern blot A technique in which RNA fragments, previously separated by gel electrophoresis, are transferred to a nitrocellulose membrane and detected by autoradiography or chemical luminescence. Compare with *Southern blot* and *Western blotting.*

notochord (no'toe-kord) The flexible, longitudinal rod in the anteroposterior axis that serves as an internal skeleton in the embryos of all chordates and in the adults of some.

nuclear area Region of a bacterial cell that contains DNA but is not enclosed by a membrane.

nuclear envelope The double membrane system that encloses the cell nucleus of eukaryotes.

nuclear equivalence The concept that the nuclei of all differentiated cells of an adult organism are genetically identical to each other and to the nucleus of the zygote from which they were derived. Compare with *genomic rearrangement* and *gene amplification.*

nuclear pores Structures in the nuclear envelope that allow passage of certain materials between the cell nucleus and the cytoplasm.

nucleolus (new-klee'oh-lus) (pl., *nucleoli*) Specialized structure in the cell nucleus formed from regions of several chromosomes; site of assembly of the ribosomal subunits.

nucleoplasm The contents of the cell nucleus.

nucleoside triphosphate Molecule consisting of a nitrogenous base, a pentose sugar, and three phosphate groups, e.g., adenosine triphosphate (ATP).

nucleosomes (new'klee-oh-somz) Repeating units of chromatin structure, each consisting of a length of DNA wound around a complex of eight histone molecules. Adjacent nucleosomes are connected by a DNA linker region associated with another histone protein.

nucleotide (noo'klee-oh-tide) A molecule consisting of one or more phosphate groups, a five-carbon sugar (ribose or deoxyribose), and a nitrogenous base (purine or pyrimidine).

nucleus (new'klee-us) (pl., *nuclei*) (1) The central region of an atom, containing the protons and neutrons. (2) A cell organelle in eukaryotes that contains the DNA and serves as the control center of the cell. (3) A mass of nerve cell bodies in the central nervous system.

nut A simple, dry fruit that contains a single seed and is surrounded by a hard fruit wall.

nutrients The chemical substances in food that are used as components for synthesizing needed materials and/or as energy sources.

nutrition The process of taking in and using food (nutrients).

obesity Excess accumulation of body fat; a person is considered obese if the body mass index (BMI) is 30 or more.

obligate anaerobe An organism that grows only in the absence of oxygen. Compare with *facultative anaerobe.*

occipital lobes Posterior areas of the mammalian cerebrum; interpret visual stimuli from the retina of the eye.

oceanic province That part of the open ocean that overlies an ocean bottom deeper than 200 m. Compare with *neritic province.*

Okazaki fragment One of many short segments of DNA, each 100 to 1000 nucleotides long, that must be joined by DNA ligase to form the lagging strand in DNA replication.

olfactory epithelium Tissue containing odor-sensing neurons.

oligodendrocyte A type of glial cell that forms myelin sheaths around neurons in the CNS.

ommatidium (om''ah-tid'ee-um) (pl., *ommatidia*) One of the light-detecting units of a compound eye, consisting of a lens and a crystalline cone that focus light onto photoreceptors called *retinular cells.*

omnivore (om'nih-vore) An animal that eats a variety of plant and animal materials.

oncogene (on'koh-jeen) An abnormally functioning gene implicated in causing cancer. Compare with *proto-oncogene* and *antioncogene.*

oocytes (oh'oh-sites) Meiotic cells that give rise to egg cells (ova).

oogamy (oh-og'uh-me) The fertilization of a large, nonmotile female gamete by a small, motile male gamete. Compare with *isogamy* and *anisogamy.*

oogenesis (oh''oh-jen'eh-sis) Production of female gametes (eggs) by meiosis. Compare with *spermatogenesis.*

oospore A thick-walled, resistant spore formed from a zygote during sexual reproduction in water molds.

open circulatory system A type of circulatory system in which the blood bathes the tissues directly; characteristic of arthropods and many mollusks. Compare with *closed circulatory system.*

open system An entity that exchanges energy with its surroundings. Compare with *closed system.*

operant conditioning A type of learning in which an animal is rewarded or punished for performing a behavior it discovers by chance; also called *instrumental conditioning.*

operator site One of the control regions of an operon; the DNA segment to which a repressor binds, thereby inhibiting the transcription of the adjacent structural genes of the operon.

operculum In bony fishes, a protective flap of the body wall that covers the gills.

operon (op'er-on) In prokaryotes, a group of structural genes that are coordinately controlled and transcribed as a single message, plus their adjacent regulatory elements.

optimal foraging The process of obtaining food in a manner that maximizes benefits and/or minimizes costs.

orbital Region in which electrons occur in an atom or molecule.

order A taxonomic category made up of related families.

organ A specialized structure, such as the heart or liver, made up of tissues and adapted to perform a specific function or group of functions.

organ of Corti The structure within the inner ear of vertebrates that contains receptor cells that sense sound vibrations.

organ system An organized group of tissues and organs that work together to perform a specialized set of functions, e.g., the digestive system or circulatory system.

organelle One of the specialized structures within the cell, such as the mitochondria, Golgi complex, ribosomes, or contractile vacuole; many organelles are membrane-bounded.

organic compound A compound consisting of a backbone made up of carbon atoms. Compare with *inorganic compound.*

organism Any living system consisting of one or more cells.

organismic respiration See *respiration.*

organogenesis The process of organ formation.

orgasm (or′gazm) The climax of sexual excitement.

origin of replication A specific site on the DNA where replication begins.

osmoconformer An animal in which the salt concentration of body fluids varies with changes in surrounding seawater. Compare with *osmoregulator.*

osmoregulation (oz″moh-reg-yoo-lay′shun) The active regulation of the osmotic pressure of body fluids so that they do not become excessively dilute or excessively concentrated.

osmoregulator An animal that maintains an optimal salt concentration in its body fluids despite changes in salinity of its surroundings. Compare with *osmoconformer.*

osmosis (oz-moh′sis) The net movement of water (the principal solvent in biological systems) by diffusion through a selectively permeable membrane from a region of higher concentration of water (a hypotonic solution) to a region of lower concentration of water (a hypertonic solution).

osmotic pressure The pressure that must be exerted on the hypertonic side of a selectively permeable membrane to prevent diffusion of water (by osmosis) from the side containing pure water.

osteichthyes (os″tee-ick′thees) Historically, the vertebrate class of bony fishes. Biologists now divide bony fishes into three classes: Actinopterygii, the ray-finned fishes; Actinistia, the lobe-finned fishes, and Dipnoi, the lung fishes.

osteoblast (os′tee-oh-blast) A type of bone cell that secretes the protein matrix of bone. Also see *osteocyte.*

osteoclast (os′tee-oh-clast) Large, multinucleate cell that helps sculpt and remodel bones by dissolving and removing part of the bony substance.

osteocyte (os′tee-oh-site) A mature bone cell; an osteoblast that has become embedded within the bone matrix and occupies a lacuna.

osteon (os′tee-on) The spindle-shaped unit of bone composed of concentric layers of osteocytes organized around a central Haversian canal containing blood vessels and nerves.

otoliths (oh′toe-liths) Small calcium-carbonate crystals in the saccule and utricle of the inner ear; sense gravity and are important in static equilibrium.

outbreeding The mating of individuals of unrelated strains. Compare with *inbreeding.*

ovary (oh′var-ee) (1) In animals, one of the paired female gonads responsible for producing eggs and sex hormones. (2) In flowering plants, the base of the carpel that contains ovules; ovaries develop into fruits after fertilization.

oviduct (oh′vih-dukt) The tube that carries ova from the ovary to the uterus, cloaca, or body exterior. Also called *fallopian tube* or *uterine tube.*

oviparous (oh-vip′ur-us) Bearing young in the egg stage of development; egg-laying. Compare with *viviparous* and *ovoviviparous.*

ovoviviparous (oh′voh-vih-vip″ur-us) A type of development in which the young hatch from eggs incubated inside the mother's body. Compare with *viviparous* and *oviparous.*

ovulation (ov-u-lay′shun) The release of an ovum from the ovary.

ovule (ov′yool) The structure (i.e., megasporangium) in the plant ovary that develops into the seed following fertilization.

ovum (pl., *ova*) Female gamete of an animal.

oxaloacetate Four-carbon compound; important intermediate in the citric acid cycle and in the C_4 and CAM pathways of carbon fixation in photosynthesis.

oxidants Highly reactive molecules such as free radicals, peroxides, and superoxides that are produced during normal cell processes that require oxygen; can damage DNA and other molecules by snatching electrons. Compare with *antioxidants.*

oxidation The loss of one or more electrons (or hydrogen atoms) by an atom, ion, or molecule. Compare with *reduction.*

oxidative phosphorylation (fos″for-ih-lay′shun) The production of ATP using energy derived from the transfer of electrons in the electron transport system of mitochondria; occurs by chemiosmosis.

oxygen-carrying capacity The maximum amount of oxygen transported by hemoglobin.

oxygen debt The oxygen necessary to metabolize the lactic acid produced during strenuous exercise.

oxygen-hemoglobin dissociation curve A curve depicting the percentage saturation of hemoglobin with oxygen, as a function of certain variables such as oxygen concentration, carbon dioxide concentration, or pH.

oxyhemoglobin Hemoglobin that has combined with oxygen.

oxytocin (ok″see-tow′sin) Hormone secreted by the hypothalamus and released by the posterior lobe of the pituitary gland; stimulates contraction of the pregnant uterus and the ducts of mammary glands.

ozone A blue gas, O_3, with a distinctive odor that is a human-made pollutant near Earth's surface (in the troposphere) but a natural and essential component of the stratosphere.

P generation (parental generation) Members of two different true-breeding lines that are crossed to produce the F_1 generation.

P680 Chlorophyll *a* molecules that serve as the reaction center of Photosystem II, transferring photoexcited electrons to a primary acceptor; named by their absorption peak at 680 nm.

P700 Chlorophyll *a* molecules that serve as the reaction center of Photosystem I, transferring photoexcited electrons to a primary acceptor; named by their absorption peak at 700 nm.

pacemaker (of the heart) See *sinoatrial (SA) node.*

Pacinian corpuscle (pah-sin′-ee-an kor′pus-el) A receptor located in the dermis of the skin that responds to pressure.

paedomorphosis Retention of juvenile or larval features in a sexually mature animal.

pair bond A stable relationship between animals of opposite sex that ensures cooperative behavior in mating and rearing the young.

paleoanthropology (pay″lee-o-an-thro-pol′uh-gee) The study of human evolution.

Paleozoic era That part of geological time extending from roughly 570 to 248 million years ago.

palindromic Reading the same forward and backward; DNA sequences are palindromic when the base sequence of one strand reads the same as its complement when both are read in the 5′ to 3′ direction.

palisade mesophyll (mez′oh-fil) The vertically stacked, columnar mesophyll cells near the upper epidermis in certain leaves. Compare with *spongy mesophyll.*

pancreas (pan′kree-us) Large gland located in the vertebrate abdominal cavity. The pancreas produces pancreatic juice containing digestive enzymes; also serves as an endocrine gland, secreting the hormones insulin and glucagon.

panspermia The idea that life did not originate on Earth, but began elsewhere in the galaxy and drifted through space to Earth.

parabronchi (sing., *parabronchus*) Thin-walled ducts in the lungs of birds; gases are exchanged across their walls.

paracrine regulation A type of regulation in which a signal molecule (e.g., certain hormones) diffuses through interstitial fluid and acts on nearby target cells. Compare with *autocrine regulation.*

paraphyletic group A group of organisms made up of a common ancestor and some, but not all, of its descendants. Compare with *monophyletic group* and *polyphyletic group.*

parapodia (par″uh-poh′dee-ah) (sing., *parapodium*) Paired, thickly bristled paddle-like appendages extending laterally from each segment of polychaete worms.

parasite A heterotrophic organism that obtains nourishment from the living tissue of another organism (the host).

parasitism (par′uh-si-tiz″m) A symbiotic relationship in which one member (the parasite) benefits and the other (the host) is adversely affected. Compare with *commensalism* and *mutualism.*

parasympathetic nervous system A division of the autonomic nervous system concerned with the control of the internal organs; functions to conserve or restore energy. Compare with *sympathetic nervous system.*

parathyroid glands Small, pea-sized glands closely adjacent to the thyroid gland; their secretion regulates calcium and phosphate metabolism.

parathyroid hormone (PTH) A hormone secreted by the parathyroid glands; regulates calcium and phosphate metabolism.

parenchyma (par-en′kih-mah) Highly variable living plant cells that have thin primary walls; function in photosynthesis, the storage of nutrients, and/or secretion.

parsimony The principle based on the experience that the simplest explanation is most probably the correct one.

parthenogenesis (par″theh-noh-jen′eh-sis) The development of an unfertilized egg into an adult organism; common among honeybees, wasps, and certain other arthropods.

parturition (par″to-rish′un) The birth process.

partial pressure (of a gas) The pressure exerted by a gas in a mixture, which is the same pressure it would exert if alone. For example, the partial pressure of atmospheric oxygen (P_{O_2}) is 160 mm Hg at sea level.

passive immunity Temporary immunity that depends on the presence of immunoglobulins produced by another organism. Compare with *active immunity.*

passive ion channel A channel in the plasma membrane that permits the passage of specific ions such as Na^+, K^+, or Cl^-.

patch clamp technique A method that allows researchers to study the ion channels of a tiny patch of membrane by tightly sealing a micropipette to the patch and measuring the flow of ions through the channels.

patchiness See *clumped dispersion.*

pathogen (path′oh-gen) An organism, usually a microorganism, capable of producing disease.

pathogen-associated molecular patterns (PAMPs) Molecules on bacteria and other pathogens that combine with Toll-like receptors on macrophages, stimulating them to produce cytokines.

pattern formation See *morphogenesis.*

pedigree A chart constructed to show an inheritance pattern within a family through multiple generations.

peduncle The stalk of a flower or inflorescence.

pellicle A flexible outer covering of protein; characteristic of certain protists, e.g., ciliates and euglenoids.

penis The male sexual organ of copulation in reptiles, mammals, and a few birds.

pentose A sugar molecule containing five carbons.

people overpopulation A situation in which there are too many people in a given geographic area; results in pollution, environmental degradation, and resource depletion. Compare with *consumption overpopulation.*

pepsin (pep′sin) An enzyme produced in the stomach that initiates digestion of protein.

pepsinogen The precursor of pepsin; secreted by chief cells in the gastric glands of the stomach

peptide (pep′tide) A compound consisting of a chain of amino acid groups linked by peptide bonds. A dipeptide consists of two amino acids, a polypeptide of many.

peptide bond A distinctive covalent carbon-to-nitrogen bond that links amino acids in peptides and proteins.

peptidoglycan (pep″tid-oh-gly′kan) A modified protein or peptide having an attached carbohydrate; component of the eubacterial cell wall.

peptidyl transferase The ribosomal enzyme that catalyzes the formation of a peptide bond.

perennial plant (purr-en′ee-ul) A woody or herbaceous plant that grows year after year, i.e., lives more than two years. Compare with *annual* and *biennial.*

perfect flower A flower that has both stamens and carpels. Compare with *imperfect* flower.

pericentriolar material Fibrils surrounding the centrioles in the microtubule organizing centers in cells of animals and other organisms having centrioles.

pericycle (pehr′eh-sy″kl) A layer of meristematic cells typically found between the endodermis and phloem in roots.

periderm (pehr′ih-durm) The outer bark of woody stems and roots; composed of cork cells, cork cambium, and cork parenchyma, along with traces of primary tissues.

period An interval of geological time that is a subdivision of an era. Each period is divided into epochs.

peripheral membrane protein A protein associated with one of the surfaces of a biological membrane. Compare with *integral membrane protein.*

peripheral nervous system (PNS) In vertebrates, the nerves and receptors that lie outside the central nervous system. Compare with *central nervous system (CNS).*

peristalsis (pehr″ih-stal′sis) Rhythmic waves of muscular contraction and relaxation in the walls of hollow tubular organs, such as the ureter or parts of the digestive tract, that serve to move the contents through the tube.

permafrost Permanently frozen subsoil characteristic of frigid areas such as the tundra.

peroxisomes (pehr-ox′ih-somz) In eukaryotic cells, membrane-bounded organelles containing enzymes that produce or degrade hydrogen peroxide.

persistence A characteristic of certain chemicals that are extremely stable and may take many years to be broken down into simpler forms by natural processes.

petal One of the parts of the flower attached inside the whorl of sepals; petals are usually colored.

petiole (pet′ee-ohl) The part of a leaf that attaches to a stem.

pH The negative logarithm of the hydrogen ion concentration of a solution (expressed as moles per liter). Neutral pH is 7, values less than 7 are acidic, and those greater than 7 are basic.

phage See *bacteriophage.*

phagocytosis (fag″oh-sy-toh′sis) Literally, "cell eating"; a type of endocytosis by which certain cells engulf food particles, microorganisms, foreign matter, or other cells.

pharmacogenetics A new field of gene-based medicine in which drugs are personalized to match a patient's genetic makeup.

pharynx (fair′inks) Part of the digestive tract. In complex vertebrates it is bounded anteriorly by the mouth and nasal cavities and posteriorly by the esophagus and larynx; the throat region in humans.

phenetics (feh-neh′tiks) An approach to classification based on measurable similarities in phenotypic characters, without consideration of homology or other evolutionary relationships. Compare with *cladistics* and *evolutionary systematics.*

phenotype (fee′noh-type) The physical or chemical expression of an organism's genes. Compare with *genotype.*

phenotype frequency The proportion of a particular phenotype in the population.

phenylketonuria (PKU) (fee″nl-kee″toh-noor′ee-ah) An inherited disease in which there is a deficiency of the enzyme that normally converts phenylalanine to tyrosine; results in mental retardation if untreated.

pheromone (fer′oh-mone) A substance secreted by an organism to the external environment that influences the development or behavior of other members of the same species.

phloem (flo′em) The vascular tissue that conducts dissolved sugar and other organic compounds in plants.

phosphate group A weakly acidic functional group that can release one or two hydrogen ions.

phosphodiester linkage Covalent linkage between two nucleotides in a strand of DNA or RNA; includes a phosphate group bonded to the sugars of two adjacent nucleotides.

phosphoenolpyruvate (PEP) Three-carbon phosphorylated compound that is an important intermediate in glycolysis and is a reactant in the initial carbon fixation step in C_4 and CAM photosynthesis.

phosphoglycerate (PGA) Phosphorylated three-carbon compound that is an important metabolic intermediate.

phospholipids (fos″foh-lip′idz) Lipids in which two fatty acids and a phosphorus-containing group are attached to glycerol; major components of cell membranes.

phosphorus cycle The worldwide circulation of phosphorus from the abiotic environment into living things and back into the abiotic environment.

phosphorylation (fos″for-ih-lay′shun) The introduction of a phosphate group into an organic molecule. See *kinases.*

photoautotroph An organism that obtains energy from light and synthesizes organic compounds from inorganic raw materials; includes plants, algae, and some bacteria. Compare with *photoheterotroph, chemoautotroph,* and *chemoheterotroph.*

photoheterotroph An organism that can carry out photosynthesis to obtain energy but cannot fix carbon dioxide and therefore requires organic compounds as a carbon source; includes some bacteria. Compare with *photoautotroph, chemoautotroph,* and *chemoheterotroph.*

photolysis (foh-tol′uh-sis) The photochemical splitting of water in the light-dependent reactions of photosynthesis, catalyzed by a specific enzyme.

photon (foh′ton) A particle of electromagnetic radiation; one quantum of radiant energy.

photoperiodism (foh″teh-peer′ee-o-dizm) The physiological response (such as flowering) of plants to variations in the length of daylight and darkness.

photophosphorylation (foh″toh-fos-for-ih-lay′shun) The production of ATP in photosynthesis.

photoreceptor (foh″toh-ree-sep′tor) (1) A sense organ specialized to detect light. (2) A pigment that absorbs light before triggering a physiological response.

photorespiration (foh″toh-res-pur-ay′shun) The process that reduces the efficiency of photosynthesis in C_3 plants during hot spells in summer; consumes oxygen and produces carbon dioxide through the degradation of Calvin cycle intermediates.

photosynthesis The biological process that captures light energy and transforms it into the chemical energy of organic molecules (e.g., carbohydrates), which are manufactured from carbon dioxide and water.

photosystem One of two photosynthetic units responsible for capturing light energy and transferring excited electrons; photosystem I strongly absorbs light of about 700 nm, whereas photosystem II strongly absorbs light of about 680 nm. See *antenna complex* and *reaction center.*

phototroph (foh′toh-trof) Organism that uses light as a source of energy. Compare with *chemotroph.* See *photoautotroph* and *photoheterotroph.*

phototropism (foh″toh-troh′pizm) The growth of a plant in response to the direction of light.

phycocyanin (fy″koh-sy-ah′nin) A blue pigment found in cyanobacteria and red algae.

phycoerythrin (fy″koh-ee-rih′thrin) A red pigment found in cyanobacteria and red algae.

phylogenetic systematics See *cladistics.*

phylogenetic tree A branching diagram that shows lines of descent among a group of related species.

phylogeny (fy-loj′en-ee) The complete evolutionary history of a group of organisms.

phylum (fy′lum) A taxonomic grouping of related, similar classes; a category beneath the kingdom and above the class.

phytochemicals Compounds found in plants that play important roles in preventing certain diseases; some function as antioxidants.

phytochrome (fy′toh-krome) A blue-green, proteinaceous pigment involved in a wide variety of physiological responses to light; occurs in two interchangeable forms depending on the ratio of red to far-red light.

phytoplankton (fy″toh-plank′tun) Microscopic floating algae and cyanobacteria that are the base of most aquatic food webs. Compare with *zooplankton.* See *plankton* and *nanoplankton.*

pia mater (pee′a may′ter) The inner membrane covering the brain and spinal cord; the innermost of the meninges; also see *dura mater* and *arachnoid.*

pigment A substance that selectively absorbs light of specific wavelengths.

pili (pie′lie) (sing., *pilus*) Hairlike structures on the surface of many bacteria. Function in conjugation or attachment.

pineal gland (pie-nee′al) Endocrine gland located in the brain.

pinocytosis (pin″oh-sy-toh′sis) Cell drinking; a type of endocytosis by which cells engulf and absorb droplets of liquids.

pioneer The first organism to colonize an area and begin the first stage of succession.

pistil The female reproductive organ of a flower; consists of either a single carpel or two or more fused carpels. See *carpel.*

pith The innermost tissue in the stems and roots of many herbaceous plants; primarily a storage tissue.

pituitary gland (pi-too′ih-tehr″ee) An endocrine gland located below the hypothalamus; secretes several hormones that influence a wide range of physiological processes.

placenta (plah-sen′tah) The partly fetal and partly maternal organ whereby materials are exchanged between fetus and mother in the uterus of placental mammals.

placoderms (plak′oh-durms) A group of extinct jawed fishes.

plankton Free-floating, mainly microscopic aquatic organisms found in the upper layers of the water; consisting of phytoplankton and zooplankton. Compare with *nekton.*

planula larva (plan′yoo-lah) A ciliated larval form found in cnidarians.

plasma The fluid portion of blood in which red blood cells, white blood cells, and platelets are suspended.

plasma membrane The selectively permeable surface membrane that encloses the cell contents and through which all materials entering or leaving the cell must pass.

plasma cell Cell that secretes antibodies; a differentiated B lymphocyte (B cell).

plasma proteins Proteins such as albumins, globulins, and fibrinogen that circulate in the blood plasma.

plasmid (plaz′mid) Small, circular, double-stranded DNA molecule that carries genes separate from the main DNA of a cell.

plasmodesmata (sing., *plasmodesma*) Cytoplasmic channels connecting adjacent plant cells and allowing for the movement of molecules and ions between cells.

plasmodial slime mold (plaz-moh′dee-uhl) A fungus-like protist whose feeding stage consists of a plasmodium.

plasmodium (plaz-moh′dee-um) A multinucleate mass of living matter that moves and feeds in an amoeboid fashion.

plasmolysis (plaz-mol′ih-sis) The shrinkage of cytoplasm and the pulling away of the plasma membrane from the cell wall when a plant cell (or other walled cell) loses water, usually in a hypertonic environment.

plastids (plas′tidz) A family of membrane-bounded organelles occurring in photosynthetic eukaryotic cells; include chloroplasts, chromoplasts, and amyloplasts and other leukoplasts.

platelets (playt′lets) Cell fragments in vertebrate blood that function in clotting; also called *thrombocytes.*

platyhelminthes The phylum of acoelomate animals commonly known as *flatworms.*

pleiotropy The ability of a single gene to have multiple effects..

plesiomorphic characters See *shared ancestral characters.*

pleural membrane (ploor′ul) The membrane that lines the thoracic cavity and envelops each lung.

ploidy The number of chromosome sets in a nucleus or cell. See *haploid, diploid,* and *polyploid.*

plumule (ploom′yool) The embryonic shoot apex, or terminal bud, located above the point of attachment of the cotyledon(s).

pluripotent (ploor-i-poh′tent) A term describing a stem cell that can divide to give rise to many types of cells in an organism. Compare with *totipotent.*

pneumatophore (noo-mat′uh-for″) Roots that extend up out of the water in swampy areas and are thought to provide aeration between the atmosphere and submerged roots.

polar body A small *n* cell produced during oogenesis in female animals that does not develop into a functional ovum.

polar covalent bond Chemical bond formed by the sharing of electrons between atoms that differ in electronegativity; the end of the bond near the more electronegative atom has a partial negative charge, the other end has a partial positive charge. Compare with *nonpolar covalent bond.*

polar molecule Molecule that has one end with a partial positive charge and the other with a partial negative charge; polar molecules are generally soluble in water. Compare with *nonpolar molecule.*

polar nucleus In flowering plants, one of two *n* cells in the embryo sac that fuse with a sperm during double fertilization to form the 3*n* endosperm.

pollen grain The immature male gametophyte of seed plants (gymnosperms and angiosperms) that produces sperm capable of fertilization.

pollen tube In gymnosperms and flowering plants, a tube or extension that forms after germination of the pollen grain and through which male gametes (sperm cells) pass into the ovule.

pollination (pol″uh-nay′shen) In seed plants, the transfer of pollen from the male to the female part of the plant.

polyadenylation (pol″ee-a-den-uh-lay′shun) That part of eukaryotic mRNA processing in which multiple adenine-containing nucleotides (a poly-A tail) are added to the 3′ end of the molecule.

polyandry A mating system in which a female mates with several males during a breeding season. Compare with *polygyny.*

poly-A tail See *polyadenylation.*

polygyny A mating system in which a male animal mates with many females during a breeding season. Compare with *polyandry.*

polygenic inheritance (pol″ee-jen′ik) Inheritance in which several independently assorting or loosely linked nonallelic genes modify the intensity of a trait or contribute to the phenotype in additive fashion.

polymer (pol′ih-mer) A molecule built up from repeating subunits of the same general type (monomers); examples include proteins, nucleic acids, or polysaccharides.

polymerase chain reaction (PCR) A method by which a targeted DNA fragment is amplified in vitro to produce millions of copies.

polymorphism (pol″ee-mor′fizm) (1) The existence of two or more phenotypically different individuals within a population. (2) the presence of detectable variation in the genomes of different individuals in a population.

polyp (pol′ip) A hydra-like animal; the sessile stage of the life cycle of certain cnidarians. Compare with *medusa.*

polypeptide See *peptide.*

polyphyletic group (pol″ee-fye-let′ik) A group made up of organisms that evolved from two or more different ancestors. Compare with *monophyletic group* and *paraphyletic group.*

polyploid (pol′ee-ployd) The condition of having more than two sets of chromosomes per nucleus. Compare with *diploid* and *haploid.*

polyribosome A complex consisting of a number of ribosomes attached to an mRNA during translation; also known as a *polysome.*

polysaccharide (pol-ee-sak′ah-ride) A carbohydrate consisting of many monosaccharide subunits (e.g., starch, glycogen, and cellulose).

polysome See *polyribosome.*

polyspermy The fertilization of an egg by more than one sperm.

polytene A term describing a giant chromosome consisting of many (usually > 1000) parallel DNA double helices. Polytene chromosomes are typically found in cells of the salivary glands and some other tissues of certain insects, such as the fruit fly, *Drosophila.*

polyunsaturated fatty acid See *fatty acid.*

pons (ponz) The white bulge that is the part of the brain stem between the medulla and the midbrain; connects various parts of the brain.

population A group of organisms of the same species that live in a defined geographic area at the same time.

population bottleneck See *bottleneck.*

population crash An abrupt decline in the size of a population.

population density The number of individuals of a species per unit of area or volume at a given time.

population dynamics The study of changes in populations, such as how and why population numbers change over time.

population ecology That branch of biology that deals with the numbers of a particular species that are found in an area and how and why those numbers change (or remain fixed) over time.

population genetics The study of genetic variability in a population and of the forces that act on it.

population growth momentum The continued growth of a population after fertility rates have declined, as a result of a population's young age structure.

poriferans Sponges; members of phylum Porifera.

positive feedback mechanism A homeostatic mechanism in which a change in some condition triggers a response that intensifies the changing condition. Compare with *negative feedback mechanism.*

posterior Toward the tail end of a bilaterally symmetrical animal. Compare with *anterior.*

postsynaptic neuron A neuron that transmits an impulse away from a synapse. Compare with *presynaptic neuron.*

postzygotic barrier One of several reproductive isolating mechanisms that prevent gene flow between species after fertilization has taken place; e.g., hybrid inviability, hybrid sterility, and hybrid breakdown. Compare with *prezygotic barrier.*

potential energy Stored energy; energy that can do work as a consequence of its position or state. Compare with *kinetic energy.*

potentiation A form of synaptic enhancement (increase in neurotransmitter release) that can last for several minutes; occurs when a presynaptic neuron continues to transmit action potentials at a high rate for a minute or longer.

preadaptation A novel evolutionary change in a pre-existing biological structure that enables it to have a different function; feathers, which evolved from reptilian scales, represent a preadaptation for flight

prebiotic broth hypothesis The hypothesis that simple organic molecules that are the precursors of life originated and accumulated at Earth's surface, in shallow seas or on rock or clay surfaces. Compare with *iron-sulfur world hypothesis.*

Precambrian time Geological time from the traces of life (3.8 bya) to the beginning of the Paleozoic era, (543 mya).

predation Relationship in which one organism (the predator) kills and devours another organism (the prey).

pre-mRNA RNA precursor to mRNA in eukaryotes; contains both introns and exons.

pressure-flow hypothesis The mechanism by which dissolved sugar is thought to be transported in phloem; caused by a pressure gradient between the source (where sugar is loaded into the phloem) and the sink (where sugar is removed from phloem).

prenatal Pertaining to the time before birth.

presynaptic neuron A neuron that transmits an impulse to a synapse. Compare with *postsynaptic neuron.*

prezygotic barrier One of several reproductive isolating mechanisms that interfere with fertilization between male and female gametes of different species; e.g., temporal isolation, habitat isolation, behavioral isolation, mechanical isolation, and gametic isolation. Compare with *postzygotic barrier.*

primary consumer See *herbivore.*

primary growth An increase in the length of a plant that occurs at the tips of the shoots and roots due to the activity of apical meristems. Compare with *secondary growth.*

primary mycelium A mycelium in which the cells are monokaryotic and haploid; a mycelium that grows from either an ascospore or a basidiospore. Compare with *secondary mycelium.*

primary producer See *autotroph.*

primary immune response The response of the immune system to first exposure to an antigen. Compare with *secondary immune response.*

primary structure (of a protein) The complete sequence of amino acids in a polypeptide chain, beginning at the amino end and ending at the carboxyl end. Compare with *secondary, tertiary,* and *quaternary protein structure.*

primary succession An ecological succession that occurs on land that has not previously been inhabited by plants; no soil is present initially. See *succession.* Compare with *secondary succession.*

primates Mammals that share such traits as flexible hands and feet with five digits; a strong social organization; and front-facing eyes. Includes lemurs, tarsiers, monkeys, apes, and humans.

primer See *RNA primer.*

primitive groove See *primitive streak.*

primitive streak Dynamic, constantly changing structure that forms at the midline of the blastodisc in birds, mammals, and some other vertebrates, and is active in gastrulation. The anterior end of the primitive streak is Hensen's node.

primosome A complex of proteins responsible for synthesizing the RNA primers required in DNA synthesis.

principle A scientific theory that has withstood repeated testing and has the highest level of scientific confidence. Compare with *hypothesis* and *theory.*

prion (pri′-on) An infectious agent that consists only of protein.

producer See *autotroph.*

product Substance formed by a chemical reaction. Compare with *reactant.*

product rule The rule for combining the probabilities of independent events by multiplying their individual probabilities. Compare with *sum rule.*

profundal zone (pro-fun′dl) The deepest zone of a large lake, located below the level of penetration by sunlight. Compare with *littoral zone* and *limnetic zone.*

progesterone (pro-jes′ter-own) A steroid hormone secreted by the ovary (mainly by the corpus luteum) and placenta; stimulates the uterus (to prepare the endometrium for implanation) and breasts (for milk secretion).

progymnosperm (pro-jim′noh-sperm) An extinct group of plants that may have been the ancestors of gymnosperms.

prokaryote (pro-kar′ee-ote) A cell that lacks a nucleus and other membrane-enclosed organelles; includes the bacteria and archae (kingdoms Eubacteria and Archaebacteria). Compare with *eukaryote.*

promoter The nucleotide sequence in DNA to which RNA polymerase attaches to begin transcription.

prophage (pro′faj) Bacteriophage nucleic acid that is inserted into the bacterial DNA.

prophase The first stage of mitosis and of meiosis I and meiosis II. During prophase the chromosomes become visible as distinct structures, the nuclear envelope breaks down, and a spindle forms. Meiotic prophase I is complex and includes synapsis of homologous chromosomes and crossing-over.

proplastids Organelles that are plastid precursors; may mature into various specialized plastids, including chloroplasts, chromoplasts, or leukoplasts.

proprioceptors (pro″pree-oh-sep′torz) Receptors in muscles, tendons, and joints that respond to changes in movement, tension, and position; enable an animal to perceive the position of its body.

prop root An adventitious root that arises from the stem and provides additional support for a plant such as corn.

prostaglandins (pros″tah-glan′dinz) A group of local regulators derived from fatty acids; synthesized by most cells of the body and produce a wide variety of effects; sometimes called local *hormones.*

prostate gland A gland in male animals that produces an alkaline secretion that is part of the semen.

protein A large, complex organic compound composed of covalently linked amino acid subunits; contains carbon, hydrogen, oxygen, nitrogen, and sulfur.

protein kinase One of a group of enzymes that activate or inactivate other proteins by phosphorylating (adding phosphate groups to) them.

Proterozoic era The period of Earth's history that began approximately 2.5 billion years ago and ended 543 million years ago; marked by the accumulation of oxygen and the appearance of the first multicellular eukaryotic life forms.

prothallus (pro-thal′us) (pl., *prothalli*) The free-living, *n* gametophyte in ferns and other seedless vascular plants.

protist (pro′tist) One of a vast kingdom of eukaryotic organisms, primarily unicellular or simple multicellular; mostly aquatic.

protobionts (pro″toh-by′ontz) Assemblages of organic polymers that spontaneously form under certain conditions. Protobionts may have been involved in chemical evolution.

proton A particle present in the nuclei of all atoms that has one unit of positive charge and a mass of one atomic mass unit (amu). Compare with *electron* and *neutron.*

protonema (pro″toh-nee′mah) (pl., *protonemata*) In mosses, a filament of *n* cells that grows from a spore and develops into leafy moss gametophytes.

protonephridia (pro″toh-nef-rid′ee-ah) (sing., *protonephridium*) The flame-cell excretory organs of flatworms and some other simple invertebrates.

proto-oncogene A gene that normally promotes cell division in response to the presence of certain growth factors; when mutated it may become an oncogene, possibly leading to the formation of a cancer cell. Compare with *oncogene.*

protostome (pro′toh-stome) A major division of the animal kingdom in which the blastopore develops into the mouth, and the anus forms secondarily; includes the annelids, arthropods, and mollusks. Compare with *deuterostome.*

protozoa (proh″toh-zoh′a) (sing., *protozoon*) An informal group of unicellular, animal-like protists, including amoebas, foraminiferans, actinopods, ciliates, flagellates, and apicomplexans. (The adjectival form is *protozoan.*)

provirus (pro-vy′rus) A part of a virus, consisting of nucleic acid only, that was inserted into a host genome. See *DNA provirus.*

proximal Closer to the point of reference. Compare with *distal.*

proximate causes (of behavior) The immediate causes of behavior, such as genetic, developmental, and physiological processes that permit the animal to carry out a specific behavior. Compare with *ultimate causes of behavior.*

proximal convoluted tubule The part of the renal tubule that extends from Bowman's capsule to the loop of Henle. Compare with *distal convoluted tubule.*

pseudocoelom (sue″doh-see′lom) A body cavity between the mesoderm and endoderm; derived from the blastocoel. Compare with *coelom.*

pseudocoelomate (sue″doh-seel′oh-mate) An animal having a pseudocoelom. Compare with *coelomate* and *acoelomate.*

pseudoplasmodium (sue″doe-plaz-moh′dee-um) In cellular slime molds, an aggregation of amoeboid cells that forms a spore-producing fruiting body during reproduction.

pseudopodium (sue″doe-poe′dee-um) (pl., *pseudopodia*) A temporary extension of an amoeboid cell that is used for feeding and locomotion.

puff In a polytene chromosome, a decondensed region that is a site of intense RNA synthesis.

pulmonary circulation The part of the circulatory system that delivers blood to and from the lungs for oxygenation. Compare with *systemic circulation.*

pulse, arterial The alternate expansion and recoil of an artery.

pulvinus (pul-vy′nus) A special structure, often located at the base of the petiole, that functions in leaf movement by changes in turgor.

punctuated equilibrium The idea that evolution proceeds with periods of little or no genetic change, followed by very active phases, so that major adaptations or clusters of adaptations appear suddenly in the fossil record. Compare with *gradualism.*

Punnett square The grid structure, first developed by Reginald Punnett, that allows direct calculation of the probabilities of occurrence of all possible offspring of a genetic cross.

pupa (pew′pah) (pl., *pupae*) A stage in the development of an insect, between the larva and the imago (adult); a form that neither moves nor feeds, and may be in a cocoon.

purines (pure′eenz) Nitrogenous bases with carbon and nitrogen atoms in two attached rings, e.g., adenine and guanine; components of nucleic acids, ATP, GTP, NAD^+, and certain other biologically active substances. Compare with *pyrimidines.*

pyramid of biomass An ecological pyramid that illustrates the total biomass, as, for example, the total dry weight, of all organisms at each trophic level in an ecosystem.

pyramid of energy An ecological pyramid that shows the energy flow through each trophic level of an ecosystem.

pyrimidines (pyr-im′ih-deenz) Nitrogenous bases, each composed of a single ring of carbon and nitrogen atoms, e.g., thymine, cytosine, and uracil; components of nucleic acids. Compare with *purines.*

pyruvate (pyruvic acid) A three-carbon compound; the end product of glycolysis.

quadrupedal (kwad′roo-ped″ul) Walking on all fours.

quantitative trait A trait that shows continuous variation in a population (e.g., human height) and typically has a polygenic inheritance pattern.

quaternary structure (of a protein) The overall conformation of a protein produced by the interaction of two or more polypeptide chains. Compare with *primary, secondary,* and *tertiary protein structure.*

***r* selection** A reproductive strategy recognized by some ecologists, in which a species typically has a small body size, rapid development, short life span, and devotes a large proportion of its metabolic energy to the production of offspring. Compare with *K selection.*

radial cleavage The pattern of blastomere production in which the cells are located directly above or below one another; characteristic of early deuterostome embryos. Compare with *spiral cleavage.*

radial symmetry A body plan in which any section through the mouth and down the length of the body divides the body into similar halves. Jellyfish and other cnidarians have radial symmetry. Compare with *bilateral symmetry.*

radicle (rad′ih-kl) The embryonic root of a seed plant.

radioactive decay The process in which a radioactive element emits radiation and, as a result, its nucleus changes into the nucleus of a different element.

radioisotopes Unstable isotopes that spontaneously emit radiation; also called *radioactive isotopes.*

radiolarians Those actinopods that secrete elaborate shells of silica (glass).

radula (rad′yoo-lah) A rasplike structure in the digestive tract of chitons, snails, squids, and certain other mollusks.

rain shadow An area that has very little precipitation, found on the downwind side of a mountain range. Deserts often occur in rain shadows.

random dispersion The spatial distribution pattern of a population in which the presence of one individual has no effect on the distribution of other individuals. Compare with *clumped dispersion* and *uniform dispersion.*

range The area where a particular species occurs.

ray A chain of parenchyma cells (one to many cells thick) that functions for lateral transport in stems and roots of woody plants.

ray-finned fishes A class (Actinopterygii) of modern bony fishes; contains about 95% of living fish species.

reabsorption The selective removal of certain substances from the glomerular filtrate by the renal tubules and collecting ducts of the kidney, and their return into the blood.

reactant Substance that participates in a chemical reaction. Compare with *product.*

reaction center The portion of a photosystem that includes chlorophyll *a* molecules capable of transferring electrons to a primary electron acceptor, which is the first of several electron acceptors in a series. See *antenna complex* and *photosystem.*

realized niche The lifestyle that an organism actually pursues, including the resources that it actually uses. An organism's realized niche is narrower than its fundamental niche because of interspecific competition. Compare with *fundamental niche.*

receptacle The end of a flower stalk where the flower parts (sepals, petals, stamens, and carpels) are attached.

reception Process of detecting a stimulus.

receptor down-regulation The process by which some hormone receptors decrease in number, thereby suppressing the sensitivity of target cells to the hormone. Compare with *receptor up-regulation.*

receptor-mediated endocytosis A type of endocytosis in which extracellular molecules become bound to specific receptors on the cell surface and then enter the cytoplasm enclosed in vesicles.

receptor up-regulation The process by which some hormone receptors increase in number, thereby increasing the sensitivity of the target cells to the hormone. Compare with *receptor down-regulation.*

recessive allele (al-leel′) An allele that is not expressed in the heterozygous state. Compare with *dominant allele.*

recombinant DNA Any DNA molecule made by combining genes from different organisms.

recombination, genetic The appearance of new gene combinations. Recombination in eukaryotes generally results from meiotic events, either crossing-over or shuffling of chromosomes.

red alga A member of a diverse phylum of algae that contain the pigments chlorophyll *a,* carotenoids, phycocyanin, and phycoerythrin.

red blood cell (RBC) See *erythrocyte.*

redox reaction (ree′dox) The chemical reaction in which one or more electrons are transferred from one substance (the substance that becomes oxidized) to another (the substance that becomes reduced). See *oxidation* and *reduction.*

red tide A red or brown coloration of ocean water caused by a population explosion, or bloom, of dinoflagellates.

reduction The gain of one or more electrons (or hydrogen atoms) by an atom, ion, or molecule. Compare with *oxidation.*

reflex action An automatic, involuntary response to a given stimulus that generally functions to restore homeostasis.

refractory period The brief period that elapses after the response of a neuron or muscle fiber, during which it can't respond to another stimulus.

regulative development The very plastic developmental pattern in which each individual cell of an early embryo retains totipotency. Compare with *mosaic development.*

regulatory gene Gene that turns the transcription of other genes on or off.

regulon A group of operons that are coordinately controlled.

renal (ree′nl) Pertaining to the kidney.

renal pelvis The funnel-shaped chamber of the kidney that receives urine from the collecting ducts; urine then moves into the ureters.

renin (reh′nin) An enzyme released by the kidney in response to a decrease in blood pressure; activates a pathway leading to production of angiotensin II, a hormone that increases aldosterone release; aldosterone increases blood pressure.

replacement-level fertility The number of children a couple must produce to "replace" themselves. The average number is greater than two, because some children die before reaching reproductive age.

replication fork Y-shaped structure produced during the semiconservative replication of DNA.

replication See *DNA replication.*

repolarization The process of returning membrane potential to its resting level.

repressible operon An operon that can be normally active is be controlled by a repressor protein, which becomes active when it binds to a corepressor; the active repressor binds to the operator, making the operon transcriptionally inactive. Compare with *inducible operon.*

repressor protein A negative regulatory protein that inhibits transcription when bound to DNA; some repressors require a corepressor to be active; some other repressors become inactive when bound to an inducer molecule. Compare with *activator protein.*

reproduction The process by which new individuals are produced. See *asexual reproduction* and *sexual reproduction.*

reproductive isolating mechanisms The reproductive barriers that prevent a species from interbreeding with another species; as a result, each species' gene pool is isolated from other species. See *prezygotic barrier* and *postzygotic barrier.*

reptiles A class of vertebrates characterized by dry skin with horny scales and adaptations for terrestrial reproduction; include turtles, snakes, and alligators; reptiles are not a monophyletic group.

residual capacity The volume of air that remains in the lungs at the end of a normal exhalation.

resin A viscous organic material that certain plants produce and secrete into specialized ducts; may play a role in deterring disease organisms or plant-eating insects.

resolution See *resolving power.*

resolving power The ability of a microscope to show fine detail, defined as the minimum distance between two points at which they are seen as separate images; also called *resolution.*

resource partitioning The reduction of competition for environmental resources such as food that occurs among coexisting species as a result of each species' niche differing from the others in one or more ways.

respiration (1) Cellular respiration is the process by which cells generate ATP through a series of redox reactions. In aerobic respiration the terminal electron acceptor is molecular oxygen; in anaerobic respiration the terminal acceptor is an inorganic molecule other than oxygen. (2) Organismic respiration is the process of gas exchange between a complex animal and its environment, generally through a specialized respiratory surface, such as a lung or gill.

respiratory centers Centers in the medulla and pons that regulate breathing.

resting potential The membrane potential (difference in electrical charge between the two sides of the plasma membrane) of a neuron in which no action potential is occurring. The typical resting potential is about -70 millivolts. Compare with *action potential.*

restoration ecology The scientific field that uses the principles of ecology to help return a degraded environment as closely as possible to its former undisturbed state.

restriction enzyme One of a class of enzymes that cleave DNA at specific base sequences; produced by bacteria to degrade foreign DNA; used in recombinant DNA technology.

restriction map A physical map of DNA in which sites cut by specific restriction enzymes serve as landmarks.

reticular activating system (RAS) (reh-tik′yoo-lur) A diffuse network of neurons in the brain stem; responsible for maintaining consciousness.

retina (ret′ih-nah) The innermost of the three layers (retina, choroid layer, and sclera) of the eyeball, which is continuous with the optic nerve and contains the light-sensitive rod and cone cells.

retrovirus (ret′roh-vy″rus) An RNA virus that uses reverse transcriptase to produce a DNA intermediate, known as a *DNA provirus,* in the host cell. See *DNA provirus.*

reverse transcriptase An enzyme produced by retroviruses that catalyzes the production of DNA using RNA as a template.

reversible inhibitor A substance that forms weak bonds with an enzyme, temporarily interfering with its function; a reversible inhibitor is either competitive or noncompetitive. Compare with *irreversible inhibitor.*

Rh factors Red blood cell antigens, known as *D antigens,* first identified in *Rhesus* monkeys. People who have these antigens are Rh^+; people lacking them are Rh^-. See *erythroblastosis fetalis.*

rhizome (ry′zome) A horizontal underground stem that bears leaves and buds and often serves as a storage organ and a means of asexual reproduction, e.g., iris.

rhodopsin (rho-dop′sin) Visual purple; a light-sensitive pigment found in the rod cells of the vertebrate eye; a similar molecule is employed by certain bacteria in the capture of light energy to make ATP.

ribonucleic acid (RNA) A family of single-stranded nucleic acids that function mainly in protein synthesis.

ribosomal RNA (rRNA) See *ribosomes.*

ribosomes (ry′boh-sohms) Organelles that are part of the protein synthesis machinery of both prokaryotic and eukaryotic cells; consist of a larger and smaller subunit, each composed of ribosomal RNA (rRNA) and ribosomal proteins.

ribozyme (ry′boh-zime) A molecule of RNA that has catalytic properties.

ribulose bisphosphate (RuBP) A five-carbon phosphorylated compound with a high energy potential that reacts with carbon dioxide in the initial step of the Calvin cycle.

ribulose bisphosphate carboxylase See *rubisco.*

RNA polymerase An enzyme that catalyzes the synthesis of RNA from a DNA template. Also called *DNA-dependent RNA polymerase.*

RNA primer The sequence of about five RNA nucleotides that are synthesized during DNA replication to provide a 3′ end to which DNA polymerase adds nucleotides. The RNA primer is later degraded and replaced with DNA.

RNA world A model that proposes that, during the evolution of cells, RNA was the first informational molecule to evolve, followed at a later time by proteins and DNA.

rod One of the rod-shaped, light-sensitive cells of the retina that are particularly sensitive to dim light and mediate black and white vision. Compare with *cone.*

root cap A covering of cells over the root tip that protects the delicate meristematic tissue directly behind it.

root graft The process of roots from two different plants growing together and becoming permanently attached to one another.

root hair An extension, or outgrowth, of a root epidermal cell. Root hairs increase the absorptive capacity of roots.

root pressure The pressure in xylem sap that occurs as a result of the active absorption of mineral ions followed by the osmotic uptake of water into roots from the soil.

root system The underground portion of a plant that anchors it in the soil and absorbs water and dissolved minerals.

rough ER See *endoplasmic reticulum.*

rubisco The common name of ribulose bisphosphate carboxylase, the enzyme that catalyzes the fixation of carbon dioxide in the Calvin cycle.

rugae (roo′jee) Folds, such as those in the lining of the stomach.

runner See *stolon.*

S phase Stage in interphase of the cell cycle during which DNA and other chromosomal constituents are synthesized. Compare with G_1 and G_2 *phases.*

saccule The structure within the vestibule of the inner vertebrate ear that along with the utricle houses the receptors of static equilibrium.

salinity The concentration of dissolved salts (e.g., sodium chloride) in a body of water.

salivary glands Accessory digestive glands found in vertebrates and some invertebrates; in humans there are three pairs.

salt An ionic compound consisting of an anion other than a hydroxide ion and a cation other than a hydrogen ion. A salt is formed by the reaction between an acid and a base.

salt marsh A wetland dominated by grasses in which the salinity fluctuates between that of sea water and fresh water; salt marshes are usually located in estuaries.

saltatory conduction The transmission of a neural impulse along a myelinated neuron; ion activity at one node depolarizes the next node along the axon.

saprobe See *decomposer.*

saprotroph (sap′roh-trof) See *decomposer.*

sarcolemma (sar″koh-lem′mah) The muscle cell plasma membrane.

sarcomere (sar′koh-meer) A segment of a striated muscle cell located between adjacent Z-lines that serves as a unit of contraction.

sarcoplasmic reticulum The system of vesicles in a muscle cell that surrounds the myofibrils and releases calcium in muscle contraction; a modified endoplasmic reticulum.

saturated fatty acid See *fatty acid.*

savanna (suh-van′uh) A tropical grassland containing scattered trees; found in areas of low rainfall or seasonal rainfall with prolonged dry periods.

scaffolding proteins Nonhistone proteins that help maintain the structure of a chromosome.

schizocoely (skiz′oh-seely) The process of coelom formation in which the mesoderm splits into two layers, forming a cavity between them; characteristic of protostomes. Compare with *enterocoely.*

Schwann cells Supporting cells found in nervous tissue outside the central nervous system; produce the myelin sheath around peripheral neurons.

sclera (skler′ah) The outer coat of the eyeball; a tough, opaque sheet of connective tissue that protects the inner structures and helps maintain the rigidity of the eyeball.

sclereid (skler′id) In plants, a sclerenchyma cell that is variable in shape but typically not long and tapered. Compare with *fiber.*

sclerenchyma (skler-en′kim-uh) Cells that provide strength and support in the plant body, are often dead at maturity, and have extremely thick walls; includes fibers and sclereids.

scramble competition See *exploitation competition.*

scrotum (skroh′tum) The external sac of skin found in most male mammals that contains the testes and their accessory organs.

secondary consumer See *carnivore.*

secondary growth An increase in the girth of a plant due to the activity of the vascular cambium and cork cambium; secondary growth results in the production of secondary tissues, i.e., wood and bark. Compare with *primary growth.*

secondary immune response The rapid production of antibodies induced by a second exposure to an antigen several days, weeks, or even months after the initial exposure. Compare with *primary response.*

secondary mycelium A dikaryotic mycelium formed by the fusion of two primary hyphae. Compare with *primary mycelium.*

secondary structure (of a protein) A regular geometric shape produced by hydrogen bonding between the atoms of the uniform polypeptide backbone; includes the alpha helix and the beta-pleated sheet. Compare with *primary, tertiary,* and *quaternary protein structure.*

secondary succession An ecological succession that takes place after some disturbance destroys the existing vegetation; soil is already present. See *succession.* Compare with *primary succession.*

second law of thermodynamics The physical law stating that the total amount of entropy in the universe continually increases. Compare with *first law of thermodynamics.*

second messenger A substance, e.g. cyclic AMP or calcium ions, that relays a message from a hormone bound to a cell-surface receptor; leads to some change in the cell.

secretory vesicles Small cytoplasmic vesicles that move substances from an internal membrane system to the plasma membrane.

seed A plant reproductive body consisting of a young, multicellular plant and nutritive tissue (food reserves), enclosed by a seed coat.

seed coat The outer protective covering of a seed.

seed fern An extinct group of seed-bearing woody plants with fernlike leaves; seed ferns probably descended from progymnosperms and gave rise to cycads and possibly ginkgoes.

segmentation genes In *Drosophila,* genes transcribed in the embryo that are responsible for generating a repeating pattern of body segments within the embryo and adult fly.

segregation, principle of The genetic principle, first noted by Gregor Mendel, that states that two alleles of a locus become separated into different gametes.

selectively permeable membrane A membrane that allows some substances to cross it more easily than others. Biological membranes are generally permeable to water, but restrict the passage of many solutes.

self-incompatibility A genetic condition in which the pollen cannot fertilize the same flower or flowers on the same plant.

semelparity The condition of having a single reproductive effort in a lifetime. Compare with *iteroparity.*

semen The fluid consisting of sperm suspended in various glandular secretions that is ejaculated from the penis during orgasm.

semicircular canals The passages in the vertebrate inner ear containing structures that control the sense of equilibrium (balance).

semiconservative replication See *DNA replication.*

semilunar valves Valves between the ventricles of the heart and the arteries that carry blood away from the heart; aortic and pulmonary valves.

seminal vesicles (1) In mammals, glandular sacs that secrete a component of semen (seminal fluid). (2) In some invertebrates, structures that store sperm.

seminiferous tubules (sem-ih-nif′er-ous) Coiled tubules in the testes in which spermatogenesis takes place in male vertebrates.

senescence (se-nes′cents) The aging process.

sensory neuron A neuron that transmits an impulse from a receptor to the central nervous system.

sensory receptor A cell (or part of a cell) specialized to detect specific energy stimuli in the environment.

sepal (see′pul) One of the outermost parts of a flower, usually leaflike in appearance, that protect the flower as a bud.

septum (pl., *septa*) A cross wall or partition, e.g., the walls that divide a hypha into cells.

sequencing See *DNA sequencing.*

serotonin A neurotransmitter of the biogenic amine group.

Sertoli cells (sur-tole′ee) Supporting cells of the tubules of the testis.

sessile (ses′sile) Permanently attached to one location, e.g., coral animals.

setae (sing., *seta*) Bristle-like structures that aid in annelid locomotion.

set point A normal condition maintained by homeostatic mechanisms.

sex-influenced trait A genetic trait that is expressed differently in males and females.

sex-linked gene A gene carried on a sex chromosome. In mammals almost all sex-linked genes are borne on the X chromosome, i.e., are X-linked.

sexual dimorphism Marked phenotypic differences between the two sexes of the same species.

sexual isolation See *behavioral isolation.*

sexual reproduction A type of reproduction in which two gametes (usually, but not necessarily, contributed by two different parents) fuse to form a zygote. Compare with *asexual reproduction.*

sexual selection A type of natural selection that occurs when individuals of a species vary in their ability to compete for mates; individuals with reproductive advantages are selected over others of the same sex.

shade avoidance The tendency of plants that are adapted to high light intensities to grow taller when they are closely surrounded by other plants.

shared ancestral characters Traits that were present in an ancestral species that have remained essentially unchanged; suggest a distant common ancestor. Also called *plesiomorphic characters.* Compare with *shared derived characters.*

shared derived characters Homologous traits found in two or more taxa that are present in their most recent common ancestor but not in earlier common ancestors. Also called synapomorphic characters. Compare with *shared ancestral characters.*

shoot system The above-ground portion of a plant, such as the stem and leaves.

short-day plant A plant that flowers in response to lengthening nights; also called *long-night plant.* Compare with *long-day, intermediate-day,* and *day-neutral plants.*

short-night plant See *long-day plant.*

short tandem repeats (STRs) Molecular markers that are short sequences of repetitive DNA; because STRs vary in length from one individual to another, they are useful in identifying individuals with a high degree of certainty.

sickle cell anemia An inherited form of anemia in which there is abnormality in the hemoglobin beta chains; the inheritance pattern is autosomal recessive.

sieve tube elements Cells that conduct dissolved sugar in the phloem of flowering plants.

signaling molecule See *cell signaling.*

signal transduction A process in which a cell converts and amplifies an extracellular signal into an intracellular signal that affects some function in the cell. Also see *cell signaling.*

sign stimulus Any stimulus that elicits a fixed action pattern in an animal.

simple fruit A fruit that develops from a single ovary. Compare with *aggregate, accessory,* and *multiple fruits.*

simplest formula A type of chemical formula that gives the smallest whole-number ratio of the component atoms. Compare with *molecular formula* and *structural formula.*

single-strand DNA-binding proteins (SSBs) Proteins involved in DNA replication that bind to single DNA strands and prevent the double helix from reforming until the strands are copied.

sink habitat A lower-quality habitat in which local reproductive success is less than local mortality. Compare with *source habitat.*

sinoatrial (SA) node The mass of specialized cardiac muscle in which the impulse triggering the heartbeat originates; the pacemaker of the heart.

skeletal muscle The voluntary striated muscle of vertebrates, so called because it usually is directly or indirectly attached to some part of the skeleton. Compare with *cardiac muscle* and *smooth muscle.*

slash-and-burn agriculture A type of agriculture in which tropical rain forest is cut down, allowed to dry, and burned. The crops that are planted immediately afterward thrive because the ashes provide nutrients; in a few years, however, the soil is depleted and the land must be abandoned.

slow block to polyspermy See *cortical reaction.*

small intestine Portion of the vertebrate digestive tract that extends from the stomach to the large intestine.

small nuclear ribonucleoprotein complexes (snRNP) Aggregations of RNA and protein responsible for binding to pre-mRNA in eukaryotes catalyzes the excision of introns and the splicing of exons.

smooth ER See *endoplasmic reticulum.*

smooth muscle Involuntary muscle tissue that lacks transverse striations; found mainly in sheets surrounding hollow organs, such as the intestine. Compare with *cardiac muscle* and *skeletal muscle.*

social behavior Interaction of two or more animals, usually of the same species.

sociobiology The branch of biology that focuses on the evolution of social behavior through natural selection.

sodium-potassium pump Active transport system that transports sodium ions out of, and potassium ions into, cells.

soil erosion The wearing away or removal of soil from the land; although soil erosion occurs naturally from precipitation and runoff, human activities (such as clearing the land) accelerate it.

solar tracking See *heliotropism.*

solute A dissolved substance. Compare with *solvent.*

solvent Substance capable of dissolving other substances. Compare with *solute.*

somatic cell In animals, a cell of the body not involved in formation of gametes. Compare with *germ line cell.*

somatic nervous system That part of the vertebrate peripheral nervous system that keeps the body in adjustment with the external environment; includes sensory receptors on the body surface and within the muscles, and the nerves that link them with the central nervous system. Compare with *autonomic nervous system.*

somatomedins See *insulin-like growth factors.*

somatotropin See *growth hormone.*

sonogram See *ultrasound imaging.*

soredium (sor-id′e-um) (pl., *soredia*) In lichens, a type of asexual reproductive structure that consists of a cluster of algal cells surrounded by fungal hyphae.

sorus (soh′rus) (pl., *sori*) In ferns, a cluster of spore-producing sporangia.

source habitat A good habitat in which local reproductive success is greater than local mortality. Surplus individuals in a source habitat may disperse to other habitats. Compare with *sink habitat.*

Southern blot A technique in which DNA fragments, previously separated by gel electrophoresis, are transferred to a nitrocellulose or nylon membrane and detected by autoradiography or chemical luminescence. Compare with *Northern blot* and *Western blotting.*

speciation Evolution of a new species.

species According to the biological species concept, one or more populations whose members are capable of interbreeding in nature to produce fertile offspring and do not interbreed with members of other species. Compare with *evolutionary species concept.*

species diversity A measure of the relative importance of each species within a comunity; represents a combination of species richness and species evenness.

species richness The number of species in a community.

specific immune responses Defense mechanisms that target specific macromolecules associated with a pathogen. Includes cell-mediated immunity and antibody-mediated immunity. Also known as *acquired* or *adaptive immune responses.*

specific epithet The second part of the name of a species; designates a specific species belonging to that genus.

specific heat The amount of heat energy that must be supplied to raise the temperature of 1 g of a substance 1°C.

sperm The motile, *n* male reproductive cell of animals and some plants and protists; also called a *spermatozoan.*

spermatid (spur′ma-tid) An immature sperm cell.

spermatocyte (spur-mah′toh-site) A meiotic cell that gives rise to spermatids and ultimately to mature sperm cells.

spermatogenesis (spur″mah-toh-jen′eh-sis) The production of male gametes (sperm) by meiosis and subsequent cell differentiation. Compare with *oogenesis.*

spermatozoan (spur-mah-toh-zoh′un) See *sperm.*

sphincter (sfink′tur) A group of circularly arranged muscle fibers, the contractions of which close an opening, e.g., the pyloric sphincter at the exit of the stomach.

spinal cord In vertebrates, the dorsal, tubular nerve cord.

spinal nerves In vertebrates, the nerves that emerge from the spinal cord.

spindle See *mitotic spindle.*

spine A leaf that is modified for protection, such as a cactus spine.

spiracle (speer′ih-kl) An opening for gas exchange, such as the opening of a trachea on the body surface of an insect.

spiral cleavage A distinctive spiral pattern of blastomere production in an early protostome embryo. Compare with *radial cleavage.*

spirillum (pl., *spirilla*) A long, rigid, helical bacterium. Compare with *spirochete, vibrio, bacillus,* and *coccus.*

spirochete A long, flexible, helical bacterium. Compare with *spirillum, vibrio, bacillus,* and *coccus.*

spleen An abdominal organ located just below the diaphragm that removes worn-out blood cells and bacteria from the blood and plays a role in immunity.

spongy mesophyll (mez′oh-fil) The loosely arranged mesophyll cells near the lower epidermis in certain leaves. Compare with *palisade mesophyll.*

spontaneous reaction See *exergonic reaction.*

sporangium (spor-an′jee-um) (pl., *sporangia*) A spore case, found in plants, certain protists, and fungi.

spore A reproductive cell that gives rise to individual offspring in plants, fungi, and certain algae and protozoa.

sporophyll (spor′oh-fil) A leaflike structure that bears spores.

sporophyte generation (spor′oh-fite) The $2n$, spore-producing stage in the life cycle of a plant. Compare with *gametophyte generation.*

sporozoa See *apicomplexans.*

sporozoite The infective sporelike state in apicomplexans.

stabilizing selection Natural selection that acts against extreme phenotypes and favors intermediate variants; associated with a population well adapted to its environment. Compare with *directional selection* and *disruptive selection.*

stamen (stay′men) The male part of a flower; consists of a filament and anther.

standing-water ecosystem A lake or pond ecosystem.

starch A polysaccharide composed of alpha glucose subunits; made by plants for energy storage.

start codon The codon AUG, which signals the beginning of translation of messenger RNA. Compare with *stop codon.*

stasis Long periods in the fossil record in which there is little or no evolutionary change.

statocyst (stat′oh-sist) An invertebrate sense organ containing one or more granules (statoliths); senses gravity and motion.

statoliths (stat′uh-liths) Granules of loose sand or calcium carbonate found in statocysts.

stele The cylinder in the center of roots and stems that contains the vascular tissue.

stem cell A relatively undifferentiated cell capable of repeated cell division. At each division at least one of the daughter cells usually remains a stem cell, whereas the other may differentiate as a specific cell type.

sterilization A procedure that renders an individual incapable of producing offspring; the most common surgical procedures are vasectomy in the male and tubal ligation in the female.

stereocilia Hairlike projections of hair cells; microvilli that contain actin filaments.

steroids (steer′oids) Complex molecules containing carbon atoms arranged in four attached rings, three of which contain six carbon atoms each and the fourth of which contains five; e.g., cholesterol and certain hormones, including the male and female sex hormones of vertebrates.

stigma The portion of the carpel where pollen grains land during pollination (and before fertilization).

stipe A short stalk or stemlike structure that is a part of the body of certain multicellular algae.

stipule (stip′yule) One of a pair of scalelike or leaflike structures found at the base of certain leaves.

stolon (stow′lon) An above-ground, horizontal stem with long internodes; stolons often form buds that develop into separate plants, e.g., strawberry; also called *runner.*

stomach Muscular region of the vertebrate digestive tract, extending from the esophagus to the small intestine.

stomata (sing., *stoma*) Small pores located in the epidermis of plants that provide for gas exchange for photosynthesis; each stoma is flanked by two guard cells, which are responsible for its opening and closing.

stop codon Any of the three codons in mRNA that do not code for an amino acid (UAA, UAG, or UGA) but signals the termination of translation. Compare with *start codon.*

stratosphere The layer of the atmosphere between the troposphere and the mesosphere. It contains a thin ozone layer that protects life by filtering out much of the sun's ultraviolet radiation.

stratum basale (strat′um bah-say′lee) The deepest sublayer of the human epidermis, consisting of cells that continuously divide. Compare with *stratum corneum.*

stratum corneum The most superficial sublayer of the human epidermis. Compare with *stratum basale.*

strobilus (stroh′bil-us) (pl., *strobili*) In certain plants, a conelike structure that bears spore-producing sporangia.

stroke volume The volume of blood pumped by one ventricle during one contraction.

stroma A fluid space of the chloroplast, enclosed by the chloroplast inner membrane and surrounding the thylakoids; site of the reactions of the Calvin cycle.

stromatolite (stroh-mat′oh-lite) A column-like rock that consists of many minute layers of prokaryotic cells, usually cyanobacteria.

structural formula A type of chemical formula that shows the spatial arrangement of the atoms in a molecule. Compare with *simplest formula* and *molecular formula.*

structural isomer One of two or more chemical compounds having the same chemical formula, but differing in the covalent arrangement of their atoms, e.g., glucose and fructose.

style The neck connecting the stigma to the ovary of a carpel.

subsidiary cell In plants, a structurally distinct epidermal cell associated with a guard cell.

substance P A peptide neurotransmitter released by certain sensory neurons in pain pathways; signals the brain regarding painful stimuli; also stimulates other structures including smooth muscle in the digestive tract.

substrate A substance on which an enzyme acts; a reactant in an enzymatically catalyzed reaction.

succession The sequence of changes in the species composition of a community over time. See *primary succession* and *secondary succession.*

sucker A shoot that develops adventitiously from a root; a type of asexual reproduction.

sulcus (sul′kus) (pl., *sulci*) A groove, trench, or depression, especially one occurring on the surface of the brain, separating the convolutions.

sulfhydryl group Functional group abbreviated —SH; found in organic compounds called thiols.

sum rule The rule for combining the probabilities of mutually exclusive events by adding their individual probabilities. Compare with *product rule.*

summation The process of adding together excitatory postsynaptic potentials (EPSPs).

suppressor T cell T lymphocyte that suppresses the immune response.

supraorbital ridge (soop″rah-or′bit-ul) The prominent bony ridge above the eye socket; ape skulls have prominent supraorbital ridges.

surface tension The attraction that the molecules at the surface of a liquid may have for each other.

survivorship The probability that a given individual in a population or cohort will survive to a particular age; usually presented as a survivorship curve.

survivorship curve A graph of the number of surviving individuals of a cohort, from birth to the maximum age attained by any individual.

suspensor (suh-spen′sur) In plant embryo development, a multicellular structure that anchors the embryo and aids in nutrient absorption from the endosperm.

sustainability See *environmental sustainability.*

swim bladder The hydrostatic organ in bony fishes that permits the fish to hover at a given depth.

symbiosis (sim-bee-oh′sis) An intimate relationship between two or more organisms of different species. See *commensalism, mutualism,* and *parasitism.*

sympathetic nervous system A division of the autonomic nervous system; its general effect is to mobilize energy, especially during stress situations; prepares the body for fight-or-flight response. Compare with *parasympathetic nervous system.*

sympatric speciation (sim-pa′trik) The evolution of a new species within the same geographical region as the parental species. Compare with *allopatric speciation.*

symplast A continuum consisting of the cytoplasm of many plant cells, connected from one cell to the next by plasmodesmata. Compare with *apoplast.*

synapomorphic characters See *shared derived characters.*

synapse (sin′aps) The junction between two neurons or between a neuron and an effector (muscle or gland).

synapsis (sin-ap′sis) The process of physical association of homologous chromosomes during prophase I of meiosis.

synaptic enhancement An increase in neurotransmitter release thought to occur as a result of calcium ion accumulation inside the presynaptic neuron.

synaptic plasticity The ability of synapses to change in response to certain types of stimuli. Synaptic changes occur during learning and memory storage.

synaptonemal complex The structure, visible with the electron microscope, produced when homologous chromosomes undergo synapsis.

syngamy (sin′gah-mee) The union of the gametes in sexual reproduction.

synthetic theory of evolution See *modern synthesis.*

systematics The scientific study of the diversity of organisms and their evolutionary relationships. Taxonomy is an aspect of systematics. See *taxonomy.*

systemic anaphylaxis A rapid, widespread allergic reaction that can lead to death.

systemic circulation The part of the circulatory system that delivers blood to and from the tissues and organs of the body. Compare with *pulmonary circulation.*

systole (sis′tuh-lee) The phase of the cardiac cycle when the heart is contracting. Compare with *diastole.*

T cell (T lymphocyte) The type of white blood cell responsible for a wide variety of immune functions, particularly cell-mediated immunity. T cells are processed in the thymus. Compare with *B cell.*

T cytotoxic cell (T_C) T lymphocyte that destroys cancer cells and other pathogenic cells on contact. Also known as *CD8 T cell* and *killer T cell.*

T helper cell (T_H) T lymphocyte that activates B cells (B lymphocytes) and stimulates T cytotoxic cell production. Also known as *CD4 T cell.*

T tubules Transverse tubules; system of inward extensions of the muscle fiber plasma membrane.

taiga (tie′gah) The northern coniferous forest biome found primarily in Canada, northern Europe, and Siberia; also called *boreal forest.*

taproot system A root system consisting of a prominant main root with smaller lateral roots branching off it; a taproot develops directly from the embryonic radicle. Compare with *fibrous root system.*

target cell or tissue A cell or tissue with receptors that bind a hormone.

TATA box A component of a eukaryotic promoter region; consists of a sequence of bases located about 30 base pairs upstream from the transcription initiation site.

taxon A formal taxonomic group at any level, e.g., phylum or genus.

taxonomy (tax-on′ah-mee) The science of naming, describing, and classifying organisms; see *systematics.*

Tay-Sachs disease A serious genetic disease in which abnormal lipid metabolism in the brain causes mental deterioration in affected infants and young children; inheritance pattern is autosomal recessive.

tectorial membrane (tek-tor′ee-ul) The roof membrane of the organ of Corti in the cochlea of the ear.

telencephalon See *forebrain.*

telolecithal egg An egg with a large amount of yolk, concentrated at the vegetal pole. Compare with *isolecithal egg.*

telophase (teel′oh-faze or tel′oh-faze) The last stage of mitosis and of meiosis I and II when, having reached the poles, chromosomes become decondensed, and a nuclear envelope forms around each group.

temperate deciduous forest A forest biome that occurs in temperate areas where annual precipitation ranges from about 75 cm to 125 cm.

temperate grassland A grassland characterized by hot summers, cold winters, and less rainfall than is found in a temperate deciduous forest biome.

temperate rain forest A coniferous biome characterized by cool weather, dense fog, and high precipitation, e.g., the north Pacific coast of North America.

temperate virus A virus that integrates into the host DNA as a prophage.

temperature The average kinetic energy of the particles in a sample of a substance.

temporal isolation A prezygotic reproductive isolating mechanism in which genetic exchange is prevented between similar species because they reproduce at different times of the day, season, or year.

tendon A connective tissue structure that joins a muscle to another muscle, or a muscle to a bone. Tendons transmit the force generated by a muscle.

tendril A leaf or stem that is modified for holding or attaching onto objects.

tension–cohesion model The mechanism by which water and dissolved inorganic minerals are thought to be transported in xylem; water is pulled upward under tension because of transpiration while maintaining an unbroken column in xylem because of cohesion; also called *transpiration–cohesion model.*

teratogen Any agent capable of interfering with normal morphogenesis in an embryo, thereby causing malformations; examples include radiation, certain chemicals, and certain infectious agents.

terminal bud A bud at the tip of a stem. Compare with *axillary bud.*

termination (of protein synthesis) The final stage of protein synthesis, which occurs when a termination (stop) codon is reached, causing the completed polypeptide chain to be released from the ribosome. See *initiation* and *elongation.*

termination codon See *stop codon.*

territoriality Behavior pattern in which one organism (usually a male) stakes out a territory of its own and defends it against intrusion by other members of the same species and sex.

tertiary consumer See *carnivore.*

tertiary structure (of a protein) (tur′she-air″ee) The overall three-dimensional shape of a polypeptide that is determined by interactions involving the amino acid side chains. Compare with *primary, secondary,* and *quaternary protein structure.*

test A shell.

test cross The genetic cross in which either an F_1 individual, or an individual of unknown genotype, is mated to a homozygous recessive individual.

testis (tes′tis) (pl., *testes*) The male gonad that produces sperm and the male hormone testosterone; in humans and certain other mammals; the testes are located in the scrotum.

testosterone (tes-tos′ter-own) The principal male sex hormone (androgen); a steroid hormone produced by the interstitial cells of the testes; stimulates spermatogenesis and is responsible for primary and secondary sex characteristics in the male.

tetrad The chromosome complex formed by the synapsis of a pair of homologous chromosomes (i.e., four chromatids) during meiotic prophase I; also known as a *bivalent.*

tetrapods (tet′rah-podz) Four-limbed vertebrates: the amphibians, reptiles, birds, and mammals.

thalamus (thal′uh-mus) The part of the vertebrate brain that serves as a main relay center, transmitting information between the spinal cord and the cerebrum.

thallus (thal′us) (pl., *thalli*) The simple body of an alga, fungus, or nonvascular plant that lacks root, stems, or leaves, e.g., a liverwort thallus or a lichen thallus.

theca cells The layer of connective tissue cells that surrounds the granulosa cells in an ovarian follicle; stimulated by luteinizing hormone (LH) to produce androgens, which are converted to estrogen in the granulosa cells.

theory A widely accepted explanation supported by a large body of observations and experiments. A good theory relates facts that appear unrelated; it predicts new facts and suggests new relationships. Compare with *hypothesis* and *principle.*

therapsids (ther-ap′sids) A group of mammal-like reptiles of the Permian period; gave rise to the mammals.

thermal stratification The marked layering (separation into warm and cold layers) of temperate lakes during the summer. See *thermocline.*

thermocline (thur′moh-kline) A marked and abrupt temperature transition in temperate lakes between warm surface water and cold deeper water. See *thermal stratification.*

thermodynamics Principles governing energy transfer (often expressed in terms of heat transfer). See *first law of thermodynamics* and *second law of thermodynamics.*

thermoreceptor A sensory receptor that responds to heat.

thigmomorphogenesis (thig″moh-mor-foh-jen′uh-sis) An alteration of plant growth in response to mechanical stimuli such as wind, rain, hail, and contact with passing animals.

thigmotropism (thig′moh-troh′pizm) Plant growth in response to contact with a solid object, such as the twining of plant tendrils.

threatened species A species in which the population is small enough for it to be at risk of becoming extinct throughout all or part of its range, but not so small that it is in imminent danger of extinction. Compare with *endangered species.*

threshold level The potential that a neuron or other excitable cell must reach for an action potential to be initiated.

thrombocytes See *platelets.*

thylakoid lumen See *thylakoids.*

thylakoids (thy′lah-koidz) An interconnected system of flattened, saclike membranous structures inside the chloroplast.

thymine (thy′meen) A nitrogenous pyrimidine base found in DNA.

thymus gland (thy′mus) An endocrine gland that functions as part of the lymphatic system; process T cells; important in cell-mediated immunity.

thyroid gland An endocrine gland that lies anterior to the trachea and releases hormones that regulate the rate of metabolism.

thyroid hormones Hormones, including thyroxin, secreted by the thyroid gland; stimulate rate of metabolism.

tidal volume The volume of air moved into and out of the lungs with each normal resting breath.

tight junctions Specialized structures that form between some animal cells, producing a tight seal that prevents materials from passing through the spaces between the cells.

tissue A group of closely associated, similar cells that work together to carry out specific functions.

tissue culture The growth of tissue or cells in a synthetic growth medium under sterile conditions.

tissue engineering A developing technology that is striving to grow human tissues and organs (for transplantation) in cell cultures.

tissue fluid See *interstitial fluid.*

tolerance A decreased response to a drug over time.

tonoplast The membrane surrounding a vacuole.

top-down processes Control of ecosystem function by trophic interactions, particularly from the highest trophic level. Compare *bottom-up processes.*

topoisomerases (toe-poe-eye-sahm′er-ases) Enzymes that relieve twists and kinks in a DNA molecule by breaking and rejoining the strands.

torpor An energy-conserving state of low metabolic rate and inactivity. See *estivation* and *hibernation.*

torsion The twisting of the visceral mass characteristic of gastropod mollusks.

total fertility rate The average number of children born to a woman during her lifetime.

totipotent (toh-ti-poh′tent) A term describing a cell or nucleus that contains the complete set of genetic instructions required to direct the normal development of an entire organism. Compare with *pluripotent.*

trace element An element required by an organism in very small amounts.

trachea (tray′kee-uh)(pl., *tracheae*) (1) Principal thoracic air duct of terrestrial vertebrates; windpipe. (2) One of the microscopic air ducts (or tracheal tubes) branching throughout the body of most terrestrial arthopods and some terrestrial mollusks.

tracheal tubes See *trachea.*

tracheid (tray′kee-id) A type of water-conducting and supporting cell in the xylem of vascular plants.

tract A bundle of nerve fibers within the central nervous system.

transcription The synthesis of RNA from a DNA template.

transcription factors DNA-binding proteins that regulate transcription in eukaryotes; include positively acting activators and negatively acting repressors.

transduction (1) The transfer of a genetic fragment from one cell to another, e.g., from one bacterium to another, by a virus. (2) In the nervous system, the conversion of energy of a stimulus to electrical signals.

transfer RNA (tRNA) RNA molecules that bind to specific amino acids and serve as adapter molecules in protein synthesis. The tRNA anticodons bind to complementary mRNA codons.

transformation (1) The incorporation of genetic material into a cell, thereby changing its phenotype. (2) The conversion of a normal cell to a malignant cell.

transgenic organism A plant or animal that has foreign DNA incorporated into its genome.

translation The conversion of information provided by mRNA into a specific sequence of amino acids in a polypeptide chain; process also requires transfer RNA and ribosomes.

translocation (1) The movement of organic materials (dissolved food) in the phloem of a plant. (2) Chromosome abnormality in which part of one chromosome has become attached to another. (3) Part of the elongation cycle of protein synthesis in which a transfer RNA attached to the growing polypeptide chain is transferred from the A site to the P site.

transmembrane protein An integral membrane protein that spans the lipid bilayer.

transmission, neural The conduction of a neural impulse along a neuron or from one neuron to another.

transpiration The loss of water vapor from the aerial surfaces of a plant (i.e., leaves and stems).

transpiration–cohesion model See *tension–cohesion model.*

transport vesicles Small cytoplasmic vesicles that move substances from one membrane system to another.

transposable element See *transposon.*

transposon (tranz-poze′on) A DNA segment that is capable of moving from one chromosome to another or to different sites within the same chromosome; also called a *transposable element* or *mobile genetic element.*

transverse tubules See *T tubules.*

triacylglycerol (try-ace″il-glis′er-ol) The main storage lipid of organisms, consisting of a glycerol combined chemically with three fatty acids; also called *triglyceride.* Compare with *monoacylglycerol* and *diacylglycerol.*

tricarboxylic acid (TCA) cycle See *citric acid cycle.*

trichocyst (trik′oh-sist) A cell organelle found in certain ciliates that discharges a threadlike structure that aids in trapping and holding prey.

trichome (try′kohm) A hair or other appendage growing out from the epidermis of a plant.

tricuspid valve See *atrioventricular valve.*

triglyceride See *triacylglycerol.*

triose A sugar molecule containing three carbons.

triplet A sequence of three nucleotides that serves as the basic unit of genetic information.

triplet code The sequences of three nucleotides that compose the codons, the units of genetic information in mRNA that specify the order of amino acids in a polypeptide chain.

trisomy (try′sohm-ee) The condition in which each chromosome has two copies, except for one, which is present in triplicate. Compare with *monosomy* and *disomy.*

trochophore larva (troh′koh-for) A larval form found in mollusks and many polychaetes.

trophic level (troh′fik) Each sequential step of matter and energy in a food web, from producers to primary, secondary, or tertiary consumers; each organism is assigned to a trophic level based on its primary source of nourishment.

trophoblast (troh′foh-blast) The outer cell layer of a late blastocyst, which in placental mammals gives rise to the chorion and to the fetal contribution to the placenta.

tropic hormone (trow′pic) A hormone, produced by one endocrine gland, that targets another endocrine gland.

tropical dry forest A tropical forest where enough precipitation falls to support trees but not enough to support the lush vegetation of a tropical rain forest; often occurs in areas with pronounced rainy and dry seasons.

tropical rain forest A lush, species-rich forest biome that occurs in tropical areas where the climate is very moist throughout the year. Tropical rain forests are also characterized by old, infertile soils.

tropism (troh′pizm) In plants, a directional growth response that is elicited by an environmental stimulus.

tropomyosin (troh-poh-my′oh-sin) A regulatory muscle protein involved in contraction.

true-breeding strain A genetic strain of an organism in which all individuals are homozygous at the loci under consideration.

tube feet Structures characteristic of echinoderms; function in locomotion and feeding.

tuber A thickened end of a rhizome that is fleshy and enlarged for food storage, e.g., white potato.

tubular transport maximum (Tm) The maximum rate at which a substance is reabsorbed from the renal tubules of the kidney.

tumor A mass of tissue that grows in an uncontrolled manner; a neoplasm.

tumor necrosis factors (TNFs) Cytokines that kill tumor cells and stimulate immune cells to initiate an inflammatory response.

tumor suppressor gene See *antioncogene.*

tundra (tun′dra) A treeless biome between the taiga in the south and the polar ice cap in the north that consists of boggy plains covered by lichens and small plants. Also called *arctic tundra.* Compare with *alpine tundra.*

tunicates Chordates belonging to subphylum Urochordata; sea squirts.

turgor pressure (tur′gor) Hydrostatic pressure that develops within a walled cell, such as a plant cell, when the osmotic pressure of the cell's contents is greater than the osmotic pressure of the surrounding fluid.

Turner syndrome An inherited condition in which only one sex chromosome (an X chromosome) is present in cells; karyotype is designated XO; affected individuals are sterile females.

tyrosine kinase An enzyme that phosphorylates the tyrosine part of proteins.

tyrosine kinase receptor A plasma membrane receptor that phosphorylates the tyrosine part of proteins; when a ligand binds to the receptor, the conformation of the receptor changes and it may phosphorylate itself as well as other molecules; important in immune function and serves as a receptor for insulin.

ultimate causes (of behavior) Evolutionary explanations for why a certain behavior occurs. Compare with *proximate causes of behavior.*

ultrasound imaging A technique in which high-frequency sound waves (ultrasound) are used to provide an image (sonogram) of an internal structure.

ultrastructure The fine detail of a cell, generally only observable by use of an electron microscope.

umbilical cord In placental mammals, the organ that connects the embryo to the placenta.

uniform dispersion The spatial distribution pattern of a population in which individuals are regularly spaced. Compare with *random dispersion* and *clumped dispersion.*

unsaturated fatty acid See *fatty acid.*

upstream promoter elements (UPEs) Components of a eukaryotic promoter, found upstream of the RNA polymerase-binding site; the strength of a promoter is affected by the number and type of UPEs present.

upwelling An upward movement of water that brings nutrients from the ocean depths to the surface. Where upwelling occurs, the ocean is very productive.

uracil (yur′ah-sil) A nitrogenous pyrimidine base found in RNA.

urea (yur-ee′ah) The principal nitrogenous excretory product of mammals; one of the water-soluble end products of protein metabolism.

ureter (yur′ih-tur) One of the paired tubular structures that conducts urine from the kidney to the bladder.

urethra (yoo-ree′thruh) The tube that conducts urine from the bladder to the outside of the body.

uric acid (yoor′ik) The principal nitrogenous excretory product of insects, birds, and reptiles; a relatively insoluble end product of protein metabolism; also occurs in mammals as an end product of purine metabolism.

urinary bladder An organ that receives urine from the ureters and temporarily stores it.

urinary system The body system in vertebrates that consists of kidneys, urinary bladder, and associated ducts.

urochordates A subphylum of chordates; includes the tunicates.

uterine tube (yoo′tur-in) See *oviduct.*

uterus (yoo′tur-us) The hollow, muscular organ of the female reproductive tract in which the fetus undergoes development.

utricle The structure within the vestibule of the vertebrate inner ear that, along with the saccule, houses the receptors of static equilibrium.

vaccine (vak-seen′) A commercially produced, weakened or killed antigen associated with a particular disease that stimulates the body to make antibodies.

vacuole (vak′yoo-ole) A fluid-filled, membrane-enclosed sac found within the cytoplasm; may function in storage, digestion, or water elimination.

vagina The elastic, muscular tube, extending from the cervix to its orifice, that receives the penis during sexual intercourse and serves as the birth canal.

valence electrons The electrons in the outer electron shell, known as the *valence shell,* of an atom; in the formation of a chemical bond an atom can accept electrons into its valence shell, or donate or share valence electrons.

van der Waals interactions Weak attractive forces between atoms; caused by interactions among fluctuating charges.
vas deferens (vas def′ur-enz) (pl., *vasa deferentia*) One of the paired sperm ducts that connects the epididymis of the testis to the ejaculatory duct.
vascular cambium A lateral meristem that produces secondary xylem (wood) and secondary phloem (inner bark). Compare with *cork cambium.*
vascular tissue system The tissues specialized for translocation of materials throughout the plant body, i.e., the xylem and phloem.
vasoconstriction Narrowing of the diameter of blood vessels.
vasodilation Expansion of the diameter of blood vessels.
vector (1) Any carrier or means of transfer. (2) Agent, e.g., a plasmid or virus, that transfers genetic information. (3) Agent that transfers a parasite from one host to another.
vegetal pole The yolky pole of a vertebrate or echinoderm egg. Compare with *animal pole.*
vein (1) A blood vessel that carries blood from the tissues toward a chamber of the heart (compare with *artery*). (2) A strand of vascular tissue that is part of the network of conducting tissue in a leaf.
veliger larva The larval stage of many marine gastropods (snails) and bivalves (e.g., clams); often is a second larval stage that develops after the trochophore larva.
ventilation The process of actively moving air or water over a respiratory surface.
ventral Toward the lowermost surface or belly of an animal. Compare with *dorsal.*
ventricle (1) A cavity in an organ. (2) One of the several cavities of the brain. (3) One of the chambers of the heart that receives blood from an atrium.
vernalization (vur″nul-uh-zay′shun) The induction of flowering by a low temperature treatment.
vertebrates A subphylum of chordates that possess a bony vertebral column; includes fishes, amphibians, reptiles, birds, and mammals.
vesicle (ves′ih-kl) Any small sac, especially a small, spherical, membrane-enclosed compartment, within the cytoplasm.
vessel element A type of water-conducting cell in the xylem of vascular plants.
vestibular apparatus Collectively, the saccule, utricle, and semicircular canals of the inner ear.
vestigial (ves-tij′ee-ul) Rudimentary; an evolutionary remnant of a formerly functional structure.
vibrio A spirillum (spiral-shaped bacterium) that is shaped like a comma. Compare with *spirillum, spirochete, bacillus,* and *coccus.*
villus (pl., *villi*) A multicellular, minute, elongated projection, from the surface of an epithelial membrane, e.g., villi of the mucosa of the small intestine.
viroid (vy′roid) A tiny, naked, infectious particle consisting only of nucleic acid.
virulence Properties that render an infectious agent pathogenic (and often lethal) to its host. Compare with *avirulence.*
virus A tiny pathogen consisting of a core of nucleic acid usually encased in protein and capable of infecting living cells; a virus is characterized by total dependence on a living host.
viscera (vis′ur-uh) The internal body organs, especially those located in the abdominal or thoracic cavities.
visceral mass The concentration of body organs (viscera) located above the foot in mollusks.
vital capacity The maximum volume of air a person exhales after filling the lungs to the maximum extent.
vitamin A complex organic molecule required in very small amounts for normal metabolic functioning.
vitelline envelope An acellular covering of the eggs of certain animals (e.g., echinoderms), located just outside the plasma membrane.
viviparous (vih-vip′er-us) Bearing living young that develop within the body of the mother. Compare with *oviparous* and *ovoviviparous.*
voltage-activated ion channels Ion channels in the plasma membrane of neurons that are regulated by changes in voltage. Also called *voltage-gated channels.*
vomeronasal organ In mammals, an organ in the epithelium of the nose, made up of specialized chemoreceptor cells that detect pheromones.
vulva The external genital structures of the female.
warning coloration See *aposematic coloration.*
water mold A fungus-like protist with a body consisting of a coenocytic mycelium that reproduces asexually by forming motile zoospores and sexually by forming oospores.
water potential Free energy of water; the water potential of pure water is zero and that of solutions is a negative value. Differences in water potential are used to predict the direction of water movement (always from a region of less negative water potential to a region of more negative water potential).
water vascular system Unique hydraulic system of echinoderms; functions in locomotion and feeding.
wavelength The distance from one wave peak to the next; the energy of electromagnetic radiation is inversely proportional to its wavelength.
weathering processes Chemical or physical processes that help form soil from rock; during weathering processes, the rock is gradually broken into smaller and smaller pieces.
Western blotting A technique in which proteins, previously separated by gel electrophoresis, are transferred to paper. A specific labeled antibody is generally used to mark the location of a particular protein. Compare with *Southern blot* and *Northern blot.*
whisk ferns One of a phylum of seedless vascular plants with a life cycle similar to ferns.
white matter Nervous tissue in the brain and spinal cord that contains myelinated axons. Compare with *gray matter.*
wild type The phenotypically normal (naturally occurring) form of a gene or organism.
wobble The ability of some tRNA anticodons to associate with more than one mRNA codon; in these cases the 5′ base of the anticodon is capable of forming hydrogen bonds with more than one kind of base in the 3′ position of the codon.
work Any change in the state or motion of matter.
X-linked gene A gene carried on an X chromosome.
x-ray diffraction A technique for determining the spatial arrangement of the components of a crystal.
xylem (zy′lem) The vascular tissue that conducts water and dissolved minerals in plants.
XYY karyotype Chromosome constitution that causes affected individuals (who are fertile males) to be unusually tall, with severe acne.
yeast A unicellular fungus (ascomycete) that reproduces asexually by budding or fission and sexually by ascospores.
yolk sac One of the extraembryonic membranes; a pouchlike outgrowth of the digestive tract of embryos of certain vertebrates (e.g., birds) that grows around the yolk and digests it. Embryonic blood cells are formed in the mammalian yolk sac, which lacks yolk.
zero population growth Point at which the birth rate equals the death rate. A population with zero population growth does not change in size.
zona pellucida (pel-loo′sih-duh) The thick, transparent covering that surrounds the plasma membrane of a mammalian ovum.

zooflagellate A unicellular, nonphotosynthetic protozoon that has one or more long, whiplike flagella.

zooplankton (zoh″oh-plank′tun) The nonphotosynthetic organisms present in plankton, e.g., protozoa, tiny crustaceans, and the larval stages of many animals. See *plankton.* Compare with *phytoplankton.*

zoospore (zoh′oh-spore) A flagellated motile spore produced asexually by certain algae, chytrids, and water molds and other protists.

zooxanthellae (zoh″oh-zan-thel′ee) (sing., *zooxanthella*) Endosymbiotic, photosynthetic dinoflagellates found in certain marine invertebrates; their mutualistic relationship with corals enhances the corals' reef-building ability.

zygomycetes (zy″gah-my′seats) Fungi characterized by the production of nonmotile asexual spores and sexual zygospores.

zygosporangium (zy″gah-spor-an′gee-um) A thick-walled sporangium containing a zygospore.

zygospore (zy′gah-spor) A sexual spore produced by a zygomycete.

zygote The $2n$ cell that results from the union of n gametes in sexual reproduction. Species that are not polyploid have haploid gametes and diploid zygotes.

zygotic genes Genes that are transcribed after fertilization, either in the zygote or in the embryo. Compare with *maternal effect genes.*

CREDITS

This page constitutes an extension of the copyright page. We have made every effort to trace the ownership of all copyrighted material and to secure permission from copyright holders. In the event of any question arising as to the use of any material, we will be pleased to make the necessary corrections in future printings. Thanks are due to the following authors, publishers, and agents for permission to use the material indicated.

Photo Credits

Chapter 1 **1:** © Jim Olive/Peter Arnold, Inc.; **3** (top left): © Mike Abbey/visuals Unlimited; (middle left): © McMurray Photography; **4** (top right): © A.B. Dowsett/Science Photo Library/ Photo researchers, Inc.; (bottom left): © David Dennis/Tom Stack & Associates; (bottom right): © David Dennis/tom Stack & Associates; **5** (top left): © Cabisco/Visuals Unlimited; (bottom left): © L.E. Gilbert/Biological Photo Service; (top right): © McMurray Photography; **8:** © Jon Wilson/Science Photo Library/Photo researchers, Inc.; **9:** © T. Whittaker/Dembinsky Photo Associates; **11** (top right): © R. Robinson/Visuals Unlimited; (middle left): © CNRI/Science Photo Library/Photo Researchers, Inc.; (center): © David M. Phillips/Visuals Unlimited; (middle right):© Ulf Sjostedt/FPG International; (bottom left): © John Arnaldi; (bottom right): © McMurray Photography; **12** (top left): © J. Serrao/ Photo Researchers, Inc,; (bottom,. all): © Jack Jeffrey; **14:** © Mark Moffett/Minden Pictures; **17:** © Tom McHugh/Photo Researchers, Inc.

Chapter 2 **22:** © Frans Lanting/Minden Pictures; **25:** © Peter J. Bryant/Biological Photo Service; **30:** © D.W. Fawcett; **33** (top left): Courtesy of Diane R. Nelson; (top right): Robert O. Schuster, courtesy of Diane R. Nelson; **34:** Dennis Drenner; **35** (top left): Woodbridge Wilson, National Park Service; (middle left): Gary R. Bonner; (bottom left): © Barbara O'Donnell/Biological Photo Service

Chapter 3 **41:** © Momatiuk Eastcott/The Image Works; **43:** Dennis Drenner; **49:** © Ed Reschke; **50** (top left): © Omikron/ Photo Researchers, Inc.; (bottom right): © Dwight R. Kuhn; **51:** From Garrett, R.H, and C.M. Grisham, Biochemistry, 2nd edition, Saunders College Publishing, Philadelphia, 1999, p. 240

Chapter 4 **66:** Courtesy of Dr. John M. Murray, Dept. of Cell & Developmental Biology, University of Pennsylvania; **69** (top right): From Hookeís Micrographica , 1665; (all others): Jim Solliday/Biological Photo Service; 71(all): Courtesy of T.K. Maugel/ University of Maryland; **73:** © M. Wurtz/Photo Researchers, Inc.; **74** (top left): D.W. Fawcett; (top right): D.W. Fawcett & R. Bolender; (bottom left): E.H. Newcomb & W.P. Wergin/Biological Photo Service; (bottom center): R. Bolender & D.W. Fawcett; (bottom right): © D.W. Fawcett/Visuals Unlimited; **75** (top left): D.W. Fawcett; (top right): D.W. Fawcett & R. Bolender; (bottom left): R. Bolender & D.W. Fawcett/Visuals Unlimited; (bottom center): B.F. King/Biological Photo service; (bottom right): D.W. Fawcett; **76** (top left): Courtesy of Kenneth Miller, Brown University; (top right): Dr. Susumu Ito, Harvard Mecial School; **77** (left): D.W. Fawcett; (right): © R. Kess, G. Shih/Visuals Unlimited; **78:** © L. Sims/Visuals Unlimited; **80:** © R. Bolender & D.W. Fawcett/Visuals Unlimited; **82:** D.W. Fawcett & R. Bolender; **83:** © Don Fawcett/Photo Researchers, Inc.; **84** (top left): © E.H. Newcomb & S.E. Frederick/Biological Photo Service; (bottom right): © M.I. Walker/Photo Researchers, Inc.; **85:** D.W. Fawcett; **86:** E.H. Newcomb & W.P. Wergin/Biological Photo Service; **87:** Nancy Kedersha; **88:** © B.F. King/Biological Photo Service; **89** (both): W.L. Dentler/Biological Photo Service; **90** (top left): Nancy Kedersha/ ImmunoGen, Inc.; (top right): K.G. Murti/Visuals Unlimited; **91:** Biophoto Associates

Chapter 5 **95:** Nancy Kedersha; **97:** © Omikron/Photo Researchers, Inc.; **100:** D.W. Fawcett; **105** (all): Courtesy of Dr. R.F. Baker, University of California Medical School; **106** (all): Dennis Drenner; **110:** A. Ichikawa, from D.W. Fawcett; **111:** D.W. Fawcett; **112:** From Perry, M.M., and A.B. Gilbert, Journal of Cell Science 39:257-272, 1979; **115:** D.W. Fawcett; **116** (top center): G.E. Palade; (bottom left): D.W. Fawcett; (bottom right): E. Anderson, Journal of Morphology 156:339-366, 1978; **117:** © E.H. Newcomb/Biological Photo Service

Chapter 6 **121:** © Barbara Gerlach/Visuals Unlimited; **129:** Thomas Eisner and Daniel Aneshansley/Cornell University; **130:** © AP/Wide World Photos; **131** (both): Courtesy of Thomas A. Steitz; **132:** From Smith, R.B., and L.J. Siegel, Windows into the Earth: The Geologic Story of Yellowstone and Grand Teton National Parks. Oxford University Press, Oxford, 2000

Chapter 7 **137:** © Renee Lynn/Photo Researchers, Inc.; **147:** Leonard Lee Rue III/Animals Animals Earth Scenes; **149:** © R. Bhatnagar/Visuals Unlimited

Chapter 9 **156:** © Skip Moody/Dembinsky Photo Associates; **159:** © M. Eichelberger/Visuals Unlimited; **161:** © T.E. Adams/ Visuals Unlimited; **162:** © Bernd Wittich/Visuals Unlimited; **171:** © Robert W. Domm/Visuals Unlimited

Chapter 9 **174:** Alexey Khodjakov, Wadsworth Center, Albany, NY; **175:** Courtesy of Oncor, Inc.; **176** (top right): D.E. Olins and A.L. Olins; (bottom left): Courtesy of U. Laemmli, from Cell Vol. 12, p. 817,1988, copyright by Cell Press; **177** (top left): © K.G. Murti/Visuals Unlimited; **179** (left, all but last): © Michael Abbey/Photo Researchers, Inc.; (right, all): © Ed Reschke; (bottom right): © Carolina Biological Supply/Phototake; **180** (top left): E.J. Du Praw; (bottom right): © CNRI/Phototake; **182** (top left): T.E. Schroeder/Biological Photo Service; (middle right): E.H. Newcomb & B.A. Palevitz/Biological Photo Service; **187** (top, all): © Clare Hasenkampf/Biological Photo Service; (bottom right): D. Von Wettstein, Proceedings of the National Academy of Sciences, Vol.68,1971,pp 851-855; **188:** Courtesy of J. Kezer

Chapter 10 **193:** Corbis/Bettmann; **209:** © Eunice Percy/ Animals Animals/Earth Scenes; **213:** © Renee Stockdale/Animals Animals/Earth Scenes

Chapter 11 **218:** © Dr. Gopal Murti/Visuals Unlimited; **222:** Dr. S. Dover, Division of Biomolecular Sciences, Kings College, London; **230:** H.J. Kriegstein and D.S. Hogness, 1974, Proc. Nat. Acad. Sci. USA, 71:135-139

Chapter 12 **234:** © Professor Oscar Miller/Science Photo Library/Photo Researchers, Inc. **246:** Courtesy of Dr. Barbara Hamakalo, University of California, Irvine

Chapter 13 **255:** Courtesy of David J. Goldhamer, University of Pennsylvania School of Medicine; **262:** Courtesy of S.C. Schultz, G.C. Shiels, and T.A. Steitz, Yale University

Chapter 14 **272:** © Keith V. Wood/Visuals Unlimited; **275:** © Dr. Stanley Cohen/Science Photo Library/Photo Researchers, Inc.; **279:** © Jean-Claude Revy/ISM/Phototake; **281:** Microchemical Core Facility, San Diego State University; **282:** © Rosenfeld Images Ltd/Science Photo Library/Photo Researchers, Inc.; **283:** © David Parker/Science Photo Library/Photo Researchers, Inc.; **284:** R.L. Brinster, University of Pennsylvania Medical School; **285:** Pentti Vänskä, Helsingin Sanomat International Edition; **286** (bottom left): U.S. Department of Agriculture; (bottom right): Courtesy of Peter Beyer

Chapter 15 **290:** Courtesy of Bob Boston, Washington University School of Medicine; **292** (top left): SIU/Peter Arnold, Inc.; (top right): Oklahoma State University; **299** (top left): © Richard Hutchings/Photo Researchers, Inc.; (right): © CNRI/ Science Photo Library/Photo Researchers, Inc.; **301:** © Science/ Visuals Unlimited; **303** (top right): Abraham Menashe

Chapter 16 **312:** © Stephen W. Paddock; **321:** Courtesy of Steve Paddock, Jam Langeland, Sean Carroll, Howard Hughes Medical Institute, University of Wisconsin; **322:** Dr. Thomas Kaufman; **324:** Courtesy of Dr. John Sultson, Medical Research council; **328** (all): Courtesy of Jose Luis Riechmann and Elliot Meyerowitz

Chapter 17 **333:** The Granger Collection, New York; **336** (left): Eric Sander; (top right): John Arnaldi; **337:** BIOS/Peter Arnold, Inc.; **338:** George Gilchrist; **339:** Tom Till; **340** (top left): Carolina Biological Supply Company/Phototake; (middle left): © Kenneth Murray/Photo Researchers, Inc.; (bottom left): © Alfred Paieka/Science Photo Library/Photo Researchers, Inc.; (top right): © A.J. Copley/Visuals Unlimited; (bottom right): © Scott Berner/Visuals Unlimited; **345** (top left): © Kjell Sandved/Visuals Unlimited; (top right): © Gunter Ziesler/Peter Arnold, Inc.; (middle left): © Mandal Ranjit/Photo Researchers, Inc.; (bottom left): © E.R. Degginger/Animals Animals; (bottom right): © J.D. Cunningham/Visuals Unlimited

Chapter 18 **353:** © G.I. Bernard/Animals Animals; **359:** www.comma.fi

Chapter 19 **367:** © Gary Retherford/Photo Researchers, Inc.; **369** (top left): © L. & D. Klein/Photo Researchers, Inc.; (top right): © Rod Planck/Photo Researchers, Inc.; (bottom left): © R. Brown/ Vireo; **370:** © John Eastcott/Yva Momatiuk/Animals Animals; **371:** Steinhart Aquarium, Tom McHugh/Photo Researchers, Inc.; **372** (top left): © Victoria McCormick/Animals Animals; (bottom left): © Tom and Pat Leeson; (bottom right): © Kent and Donna Dannen; **375** (all): Courtesy of Ole Seehausen/Leiden University, The Netherlands, and Hull University, United Kingdom; **378:** Richard Herrman; **379:** © Jane Burton/Bruce Coleman, Inc.; **380** (all): © Jack Jeffrey Photography

Chapter 20 **385:** William E. Ferguson; **386:** Courtesy of Readerís Digest Books. Drawing by H.K. Wimmer; **388:** Steven Brooke and Richard LeDuc; **390** (top left): © Fred Bavendam/ Peter Arnold, Inc.; (bottom left): © Stanley M. Awramik/Biological Photo Service; **395** (all): Chip Clark; **396:** No. GEO85638c, Field Museum of Natural History, Chicago; **399:** Paul Soreno, Reprinted with permission of Discover; **400:** © Portia Sloan

Chapter 1 **404:** © John Reader/Photo Researchers, Inc.; **407** (top left): © Frans Lanting/Minden Pictures; (bottom left): C.C. Lockwood/DRK Photo; (bottom right): © S. Meyers/Okapia/ Photo Researchers, Inc.; **409** (top left): © Joe McDonald/Visuals Unlimited; (top right): © BIOS/Peter Arnold, Inc.; (bottom left): © Nancy Adams/Tom Stack & Associates; (bottom right): © K. & K. Ammann/Bruce Coleman, Inc.; **413:** Ken Mowbray; **414:** © Omikron/Photo Researchers, Inc.

Chapter 22 **419:** © Larry Tackett (tackettproductions.com); **426:** © Jean Philippe Varin/Jacana/Photo Researchers, Inc.

Chapter 23 **435:** © Dr. Karl Lounatmaa/Science Photo Library/Photo Researchers, Inc.; **437** (above): © Omikron/Photo Researchers, Inc.; (center): © Biozentrum/Science Photo Library/ Photo Researchers, Inc.; **437** (bottom): © Lee D. Simon/Science Source/Photo Researchers, Inc.; **438:** © Oliver Meckes/MPL-Tubingen/Photo Researchers, Inc.; **440:** Centers for Disease Control and Prevention U.S. Department of Health and Human Services; **442** (bottom left): Kenneth M. Corbett; (bottom right): © Jack M. Bostrack/Visuals Unlimited; **445** (top left): © David M. Phillips/Visuals Unlimited; (top center) © David M. Phillips/ Visuals Unlimited; (top right): © David M. Phillips/Visuals Unlimited; **447:** © Oliver Meckes/Gelderblom/Photo Researchers, Inc.; **448:** © Manfred Kage/Peter Arnold; **449:** © Alfred Pasieka/ Peter Arnold, Inc.; **451:** © Helen E. Carr/Biological Photo Service; **452** (top left): © David Scharf/Peter Arnold, Inc.; (2nd from top) From Grilicone, P.L., and Pangborn, J., Journal of Bacteriology 124:1558, 1975; (3rd from top): © David M. Phillips/Visuals Unlimited; (4th from top): © Charles W. Stratton/Visuals Unlimited; **453** (top): © David M. Phillips/Visuals Unlimited; (2nd from top): © David M. Phillips/Visuals Unlimited; (bottom): © Dr. Karl Lounatmaa/Science Photo Library/Photo Researchers, Inc.; **454:** Dennis Drenner; **455:** © George J. Wilder/Visuals Unlimited

Chapter 24 **458:** © Gopal Murti/Science Photo Library/ Photo Researchers, Inc.; **463** (left): © E. White/Visuals Unlimited; (right): © M.A. Abbey/Visuals Unlimited; **464** (above): © Biophoto Associates; (right): © Biophoto Associates/Photo Researchers, Inc.; **465** (left): © Robert Brons/Biological Photo Service; (right): © Cabisco/Visuals Unlimited; **467** (left): Courtesy of T.K. Maugel, University of Maryland; (right): From Freshwater Algae: Their Microscopic world Explored. Bristol, United Kingdom, Biopress, 1995. Photo by Hilda Canter-Lund; **469:** © James W. Richardson/visuals Unlimited; **470:** © Philip Harrington/The Stock Market; **471** (left): © Philip Sze/Visuals Unlimited; (right): Dr. Elizabeth Venrick/Scripps Institution of Oceanography; **472** (top right): © Gregory Ochocki/Photo Researchers, Inc.; (top left): © J.R. Waaland/Biological Photo Service; (bottom left): Biophoto Associates; (bottom right): © Lynn McLaren/Photo Researchers, Inc.; **473** (above): © Robert Brons/Biological Photo Service; (bottom left): © Philip Sze/Visuals Unlimited; (bottom right): © D. Gotshall/Visuals Unlimited; **474** (top left): Biophoto Associates; (top right): © J. Robert Waaland/Biological Photo Service; (bottom left): J.M. Kingsbury; (bottom right): James W. Perry; 476(above, all): © M.I. Walker/Science Source/Photo Researchers, Inc.; (below): © Michael Abbey/Photo Researchers, Inc.; **477** (top left): R. Calentine/CBR Images; (top right): R. Calentine/ CBR Images; (bottom, all)): © Cabisco/Visuals Unlimited

Chapter 25 **481:** © Jeff Lepore/Photo Researchers, Inc.; **482** (left): Dennis Drenner; (right): G.T. Cole/University of Texas/ Biological Photo Service; **484:** John Taylor, University of California-Berkeley; **485:** © Cabisco/Visuals Unlimited; **487:** © Biophoto Associates/Photo Researchers, Inc.; **489** (top left): © Ed Reschke; (top right): © Robert & Linda Mitchell; **490** (top): Dennis Drenner; (middle left): © Richard D. Poe/Visuals Unlimited; (middle right): © Jeffrey Lepore/Photo Researchers, Inc.; (bottom): © Biophoto Associates/Photo Researchers, Inc.; **492:** Fred M. Rhoades; **493** (below): Courtesy of Randy Molina, U.S. Forest Service; **494** (bottom left): © Richard Shiell/Dembinsky Photo Associates; (bottom right): © John D. cunningham/Visuals Unlimited; **495:** James W. Richardson/CBR Images; **496** (top left): © Kathy Merrifield/Photo Researchers, Inc; (top right) © Runk/ Schoenberger, from Grant Heilman

Chapter 26 **499:** © Sidney Karp/Photo/Nats, Inc.; **503** (bottom left): © Rod Planck/Dembinsky Photo Associates; **503** (bottom center): © Rod Planck/Dembinsky Photo Associates; (bottom right): Robert A. Ross; **505** (above): David Cavagnaro; **506:** Courtesy of David J. Cove, University of Leeds; **508** (bottom left): © Ed Reschke; (bottom center): © W. Omerod/Visuals Unlimited; (bottom right): Carlyn Iverson; **509** (both): Marion Lobstein; **511** (top left): John Arnaldi; (top right): © J. Robert Waaland/Biological Photo Service; (below): © Ed Reschke; **512:** Courtesy of Hans Steur, The Netherlands

Chapter 27 **517:** © Holt Studios International (Silvestre Silva)/Photo Researchers, Inc.; **519:** © Geoffrey Bryant/The National Audubon society Collection/Photo Researchers, Inc.; **521:** © Walt Anderson/Visuals Unlimited; **522** (above): © Walter H. Hodge/Peter Arnold, Inc.; (below): Marion Lobstein; **523** (top left): © John D. Cunningham/Visuals Unlimited; (top right): David Cavagnaro; (below): © Robert and Linda Mitchell; **524** (bottom left): © John Gerlach/Tom Stack & Associates; (bottom right): © Richard H. Gross; **526** (top left): James Mauseth, University of Texas; (top right): Marion Lobstein; **530:** David Dilcher and Ge Sun; **531** (above): Courtesy of David Dilcher; (below): David Littschwager and Susan Middleton with Environmental Defense

Chapter 28 **534:** © Charles V. Angelo/Photo Researchers, Inc.; **542:** Charles Seaborn/Odyssey Productions, Chicago; **544** (bottom left): Robert Brons/Biological Photo Service; (bottom center): © Brian Parker/Tom Stack & Associates; (bottom right): © Mike Bacon/Tom Stack & Associates; **545** (top left): © Robert & Linda Mitchell; (center): Biophoto Associates/Photo Researchers, Inc.; (bottom left): © David M. Phillips/Visuals Unlimited; **547** (top right): Photo by Ove Hoegh-Guldberg, Centre for Marine Studies, University of Queensland; **547** © David Wrobel/Biological Photo Service

Chapter 29 **550:** © Marty Snyderman/visuals Unlimited; **552:** © T.E. Adams/Peter Arnold, Inc.; **553** (top, both) U.S. Centers for Disease Control & Prevention; **553** (bottom right): © Cath Ellis/Science Photo Library/Photo Researchers, Inc.; **555** (both): © Kjell Sandved; **556** (bottom right): © Jeffrey Rotman/Peter Arnold, Inc.; (top right): © Kjell Sandved; (2nd from top): © E.R. Degginger/Dembinsky Photo Associates; (top center): © Tom McHugh/Photo Researchers, Inc.; (top) © Kjell B. Sandved; **557:** © Bruce Watkins/Animals Animals/EarthScenes; **561** (bottom left): © T.E. Adams/Visuals Unlimited; (bottom right): St. Bartholomew's Hospital/Science Photo Library/Photo Researchers, Inc.; **562** (top left): © Andrew J. Martinez/Photo Researchers, Inc.; (top right): © M.I. Walker/Photo Researchers, Inc.; (below): © John Walsh/Science Photo Library/Photo Researchers, Inc.; **563** (below): © T.E. Adams/Photo Researchers, Inc.; **566** (top): © Dwight Kuhn; (bottom): © John R. MacGregor/Peter Arnold, Inc.; **568** (top left): ©Milton H. Tierney, Jr./Visuals Unlimited; (below): © Steve Maslowski/Visuals Unlimited; (top right): © K.H. Kjeldsen/Science Photo Library/Photo Researchers, Inc.; **569** (bottom left): © Hal Harrison/Grant Heilman Photography; (bottom right); © Fred Bavendam/Peter Arnold, Inc.; **570:** Photo by Tim Dyes, Grand Cayman

Chapter 30 **575:** © Ed Robinson/Tom Stack & Associates; **576** (top): D.J. Wrobel, Monterey Bay Aquarium; (middle left): © Marc Chamberlain/Getty/Stone; (bottom left): D.J. Wrobel, Monterey Bay Aquarium; (bottom right): Peter Scoones/Seaphot Ltd.; (middle right): © Robert Dunne/Photo Researchers, Inc.; **577:** Charles Seaborn/Odyssey Productions, Chicago; **581** (top left): Robert Shupak; (top right): Richard A. Cloney/University of Washington; **585** (top left): © Tom Stack/Tom Stack & Associates; (right): Courtesy of Dr. Kiyoko Uehara; **586** (top): © Jeffrey Rotman/Peter Arnold, Inc.; (bottom): © Kelvin Aitken/Peter Arnold, Inc.; **588** (middle left): © Fred Bavendam/Peter Arnold, Inc.; (top left): © David Hall/Photo Researchers, Inc.; (top right): © Jeffrey Rotman/Peter Arnold, Inc.; (bottom right): Mark Erdmann; **589** (bottom left): Photograph by Logan, courtesy of Department of Library Services, American Museum of Natural History; (middle right): © Gerald and Buff Corsi/Tom Stack & Associates; (bottom right): © Roy Morsch/Corbis/Stock Market; **592** (top left): Carlyn Iverson; (bottom left): © BIOS/Peter Arnold, Inc.; (bottom right): © Frans Lanting/Minden Pictures; **593** (top left): Chinese Academy of Sciences; (top right): From a painting by Rudolph Freund, courtesy of Carnegie Museum of Natural History; **594** (top left): © Alan G. Nelson/Dembinsky Photo Associates; (top center): © Gary Meszaros/Dembinsky Photo Associates; (top right): Smithsonian, Vol. 31, Feb. 2001/Tom Blagden; **595:** Painting by John C. Germann, Department of LIbrary Services, American Museum of Natural History; **596** (top): © Tom McHugh/Photo Researchers, Inc.; (middle left): © John Cancalosi/Peter Arnold, Inc.; (bottom left): Robert Anderson, reprinted with permission of Hubbard Scientific company; **597** (bottom left): McMurray Photography; (top right): © James D. Watt/Animals Animals/Earth Scenes; (bottom right): © Michio Hoshino/Minden Pictures

Chapter 31 **601:** © Runk/Schoenberger from Grant Heilman; **602** (top right): Carlyn Iverson; (bottom right): © Joyce Photographers/Photo researchers, Inc.; **606** (top): © Phil Gates/Biological Photo Service; (2nd fromtop): Dennis Drenner; (3rd from top): Dennis Drenner; (bottom): © Ed Reschke; **607** (both): James Mauseth, University of Texas; **609** (top): © John D. Cunningham/Visuals Unlimited; (2nd & 3rd from top): James Mauseth, University of Texas; (bottom): © J. Robert Waaland/Biological Photo Service; **611** (top): James Mauseth, University of Texas; (2nd from top): Dwight R. Kuhn; (3rd from top): Biophoto Associates; (bottom): Dennis Drenner; **613:** Dennis Drenner; **614:** James Mauseth, University of Texas

Chapter 32 **617:** © David Sieren/Visuals Unlimited; **620:** © Alan L. Detrick/Photo Researchers, Inc.; **621** (top): © Dwight R. Kuhn; (bottom): © Phil Gates/Biological Photo Service; **622** (top): © Ed Reschke; (bottom): © Dwight R. Kuhn; **623:** © John D. Cunningham/Visuals Unlimited; **627** (top, both): Carlyn Iverson; (bottom): © Ed Reschke/Peter Arnold, Inc.; **628:** James Mauseth, University of Texas; **629** (top left): © Barbara Gerlach/Dembinsky Photo Associates; (top center): © Stephen P. Parker/Photo Researchers, Inc.; (top right): Dennis Drenner; (bottom left): Dennis Drenner; (bottom right): James Mauseth, University of Texas; **630:** © Bill Lea/Dembinsky Photo Associates

Chapter 33 **633:** © Michael Fairchild/Peter Arnold, Inc.; **635** (both): © Ed Reschke; **636** F33.03a (both): © Ed Reschke; **638:** © Carolina Biological Supply Company/Phototake; **639:** Dennis Drenner; **640** (top): James Mauseth, University of Texas; (bottom): Carlyn Iverson; **641:** Dennis Drenner; **647** (top left): © Dwight R. Kuhn; (top right): M.H. Zimmerman, Science, Vol. 133, pp 73–79 (Fig. 41), 13 Jan. 1961. © 2002 by the American Association for the Advancement of Science; **649** (both): © Ed Reschke

Chapter 34 **650:** © R. Calentine/Visuals Unlimited; **652** (top left): © Runk/Rannels/Grant Heilman Photography; **652:** Dennis Drenner; **653** (both): © Ed Reschke; **655** (top left): James Mauseth, University of Texas; (right): Dennis Drenner; **657** (top left): Dr. Linda R. Berg; (top right): John Arnaldi; (middle left): © Robert & Linda Mitchell; (middle right): John Arnaldi; (bottom left): Courtesy of Judith Jernstedt; **659** (top left): © Robert Knauft/Biology Media/Photo Researchers, Inc.; (top right): © Cabisco/Visuals Unlimited; **664** (top left): © Hank Morgan/Photo Researchers, Inc.; (bottom): USDA/Natural Resources Conservation Service

Chapter 35 **668:** © Skip Moody/Dembinsky Photo associates; **670:** Dr. Elliot Meyerowitz, California Institute of Technology; **672** (both): Thomas Eisner, Cornell University; **673** (top left): © Dan Dempster/Dembinsky Photo Associates; (top right): © Merlin Tuttle/Bat conservation International/Photo Researchers, Inc.; **674:** © Dr. Jeremy Burgess/Science Photo Library/Photo Researchers, Inc.; **676:** Courtesy of Nordiska Genbanken, Alnarp, Svenge; **677:** James Mauseth, University of Texas; **679:** Dennis Drenner; **681** (top left): © John Serrao/Visuals Unlimited; (top right): © DPA/Dembinsky Photo Associates; (bottom left): Marion Lobstein; **683:** © Jerome Wexler/Photo Researchers, Inc.

Chapter 36 **687:** © Dwight Kuhn; **688:** Marion Lobstein; **692:** Cary Wolinsky; **693** (both): Dennis Drenner; **694** (both): Dennis Drenner; **695** (top left): © Runk/Schoenberger/Grant Heilman; (bottom right): Dennis Drenner; **700:** © Joe Eakes, Color Advantage/Visuals Unlimtied; **701** (top left): Courtesy of B.O. Phinney, University of California, Los Angeles; (top right): Robert E. Lyons; **702:** Courtesy of Dr. Richard M. Amasino, University of Wisconsin; **703** (bottom left): Illustration Services; (bottom right): Courtesy of M.G. Neuffer

Chapter 37 **709:** © Fritz Polking/Dembinsky Photo Associates; **712** (all): © Ed Reschke; **713** (all): © Ed Reschke; **714** (all): © Ed Reschke; **715** (top left): Dennis Drenner; (2nd from top): © Ed Reschke; (3rd from top): Dennis Drenner; (bottom left): © Ed Reschke; **718:** © Moredum Animal Health LTD/Photo Researchers, Inc.; **719:** © Ed Reschke; **723** (bottom left): © Breck P. Kent/Animals Animals/Earth Scenes; (bottom right): © Frans Lanting/Minden Pictures

Chapter 38 **728:** © Y. Momatiuk/Photo Researchers, Inc.; **731:** © Judy Davidson/Science Photo Library/Photo Researchers, Inc.; **732:** © Patrica Jordan/Peter Arnold, Inc.; **736** (bottom left): D.W. Fawcett; (bottom right): © Ed Reschke; **737:** © Don Fawcett/Science Source/Photo Researchers, Inc.

Chapter 39 **744:** © Nancy Kedersha/UCLA/Science Photo Library/Photo Researchers, Inc.; **747:** © E.R. Lewis/Biological Photo Service; **749:** © Jeffrey L. Rotman/Peter Arnold, Inc.; **755:** © J.F. Gennaro/Photo Researchers, Inc.

Chapter 40 **762:** © Volker Steger/Peter Arnold, Inc.; **769:** © M.I. Walker/Photo Researchers, Inc.; **771** (top right): © Fred Hossler/Visuals Unlimited; (bottom right): © Science Pictures Limited/Science Photo Library/Photo Researchers, Inc.

Chapter 41 **786:** © Merlin D. Tuttle/Bat conservation International/Photo Researchers, Inc.; **787:** © Carmela Leszczynski/Animals Animals; **790:** © Ed Reschke; **797:** © Ed Reschke; **799:** © Terry Ashley/Tom Stack & Associates; **800** (top left): © David Scharf/Peter Arnold, Inc.; (middle left): Courtesy of J. Gould, Princeton University; **802:** © Lennart Nilsson, from The Incredible Machine, p. 279

Chapter 42 **807:** © Ted Horowitz/Corbis/Stock Market; **813:** © Lennart Nilssen/Boehringer Ingelheim International, GmbH; **814:** © Lennart Nilssen/Boehringer Ingelheim International, CmbH; **818** (top left): © Ed Reschke; (bottom left): © Don Fawcett/Visuals Unlimited; **822** (bottom left): © Cabisco/Visuals Unlimited; (bottom right): © Sloop-Ober/Visuals Unlimted

Chapter 43 **831:** © NIBSC/Science Photo Library/Photo Researchers, Inc.; **835** (all): © Lennart Nilssen, Boehringer Ingelheim International GmbH; **848** (all): © Lennart Nilssen, Boehringer Ingelheim Internationl GmbH

Chapter 44 **857:** © Skip Moody/Dembinsky Photo Associates; **859:** Courtesy of Dr. James L. Nation and Stain Technology, Vol. 58, 1983; **860:** © Bernard Photo Production/Animals Animals; **864** (left): © R.G. Kessel/visuals Unlimited; (right): Courtesy of Drs. Peter Gehr, marianne Bachofen, and Ewald R. Wiebel; **870** (top left): © Bruce Watkins/Animals Animals/Earth Scenes; (bottom right): © David Nunuk/Science Photo LibraryPhoto Researchers, Inc.; **872** (both): © Martin Rotker/Taurus Photos

Chapter 45 **875:** © McMurray Photography; **877** (top left): © Darwin Dale/Photo Researchers, Inc.; (top right): © Tom McHugh/Photo Researchers, Inc.; (bottom left): © Carmela Leszczynski/Animals Animals/Earth Scenes; (bottom right): © Frans Lanting/Minden Pictures; **882** (top left): © G. Shih-R. Kessel/Visuals Unlimited; (bottom right): Courtesy of J.D. Hoskings, W.G. Henk, and Y.Z. Abdelbaki, from the American Journal of Veterinary Research, Vol. 43, No. 10; **891:** P. Pittet/United Nations Food and Agricultural Organization; **892:** Courtesy of John Sholtis/Rockefeller University, New York City

Chapter 46 **896:** McMurray Photography; **903:** © CNRI/Science Photo Library/Photo Researchers, Inc.

Chapter 47 **913:** © Shark Song/M. Kasmers/Dembinsky Photo Associates; **926:** © John Paul Kay/Peter Arnold, Inc.; **929:** © Ed Reschke

Chapter 48 **936:** © Bruce Watkins/Animals Animals ; **937:** © Richard Campbell/Biological Photo Service; **938** (top left): © Carmela Leszczynski/Animals Animals; (top right): © Fritz Polking/Dembinsky Photo Associates; **940:** © Custom Medical Stock Photo; **941:** © Tony Brain/Science Photo Library/Photo Researchers, Inc.; **945:** Biophoto Associates; **948:** SIU/Visuals Unlimited; **953:** © David Scharf/Peter Arnold, Inc.; **955:** Courtesy of Dan Atchinson; **956:** © McMurray Photography

Chapter 49 **962:** Lennart Nilsson, from A Child is Born, Dell Publishing Co, 1989; **964** (bottom left): Frank J. Longo; (bottom right): From E. Anderson, Reprinted from The Journal of Cell Biology, Vol. 37, 1968, by copyright permission of the Rockefeller University Press; **966:** Carolina Biological Supply Company/Phototake; **973** (all): Lennart Nilsson, from Being Born, 1992, pp 14,15,17, The Putnam Publishing Group; **974:** Courtesy of Carnegie Institute of Washington; **975:** © Petit Format/Nestle/Photo Researchers, Inc.; **978:** Courtesy of GE Medical Systems

Chapter 50 **981:** © E.S. Ross; **985** (bottom left): McMurray Photography; (top right): © J.H. Dick/VIREO; **986:** © H. Cruickshank/VIREO; **987:** © BIOS/Peter Arnold, Inc.; **989:** Kenneth Lohmann; **991** (top left): McMurray Photography; (top right): © R. Lindholm/Visuals Unlimited; (middle right): April Ottey, Chimpanzee and Human Communication Institute, Central Washington University; **993** (top left): © Gerald Lacz/Peter Arnold, Inc.; (bottom left): McMurray Photography; **994:** © Treat Davidson/Photo Researchers, Inc.; **997** (top left): © Sid Bart/Photo Researchers, Inc.; (middle left): © Arthur Morris/Visuals Unlimited; **998** (bottom left) © Dominique Braud/Dembinsky Photo Associates; (middle right): © M. Reardon/Photo Researchers, Inc.; (bottom right): McMurray Photography; (top left): © David M. Dennis/Animals Animals/Earth Scenes; **1000:** © Wu Wal Ping/Bruce Coleman, Inc.

Chapter 51 **1003:** © Jim Brandenberg/Minden Pictures; **1005** (middle left): © Andrew G. Wood/Photo Researchers, Inc.; (bottom left): © Robert Hernandez/Photo Researchers, Inc.; **1009:** © Tom McHugh/Photo Researchers, Inc.; **1011:** Courtesy of Thomas W. Schoener; **1012:** © R. Gustafson/Visuals Unlimited; **1015:** © Adam Jones/Photo Researchers, Inc.

Chapter 52 **1023:** © Michael P. Gadomski/Photo Researchers, Inc.; **1026** (top left): © Ed Kanze/Dembinsky Photo Associates; **1026:** © Robert Clay/Visuals Unlimited; **1030:** © Carmela Leszczynski/Animals Animals; **1031** (top left): © Wild & Natural/Animals Animals/Earth Scenes; (top right): Ronald A. Nussbaum/Museum of Zoology, University of Michigan; **1032:** Thomas C. Emmel; **1033** (top left): © Suzanne L. and Joseph T. Collins/Photo Researchers, Inc.; (top right): © Suzanne L. Collins/Photo Researchers, Inc.; (middle left): © P. Parks-OSF/Animals Animals; **1034** (top left): © Jeff Lepore/Photo Researchers, Inc.; (top right): U.S. Department of Agriculture, Agricultural Research Service

Chapter 3 **1043:** USDA Forest Service; **1048:** Provided by the SEAWIFS Project, NASA/Goddard Space Flight Center and ORBIMAGE; **1052:** © Hugh Spencer/Photo Researchers, Inc.; **1056:** © D. Foster/Science VU-WHOI/Visuals Unlimited; **1062:** Joan Landsberg, U.S. Department of Agriculture, Forest Service

Chapter 54 **1065:** © Jim Brandenburg/Minden Pictures; **1069:** © Eastcott/Momatiuk/Earth Scenes; **1070** (top left): © Beth Davidow/Visuals Unlimited; (top right): © Terry Donnelly/Dembinsky Photo Associates; **1071** (top left): © Barbara Miller/Biological Photo Service; (top right): Harvey Payne; **1072:** © Edward Ely/Biological Photo Service; **1073** (top left): © Willard Clay/Dembinsky Photo Associates; (bottom right): Carlyn Iverson; **1074** (bottom left): © Frans Lanting/Minden Pictures; (bottom right): © Mark Moffett/Minden Pictures; **1079** (top left): © Gregory C. Dimijian/Photo Researchers, Inc.; (bottom right): © Valerie Giles/Photo Researchers, Inc.; **1080:** © Patti Murray/Earth Scenes; **1081** (bottom left): William E. Ferguson; (bottom right): Photo by Ron Phillips; **1082:** © Denise Tackett/Tom Stack & Associates; **1083:** Bruce H. Robison, Monterey Bay Aquarium

Chapter 55 **1088:** Photo by Steve Nesbitt, Florida Fish & Wildlife Conservation Commission; **1089:** U.S. Fish and Wildlife Service; **1090:** © Frans Lanting/Minden Pictures; **1092** (top left): © Frans Lanting/Minden Pictures; (bottom right): Illinois Department of Conservation; **1093:** Jany Sauvanet/NHPA; **1095** (bottom left): Courtesy of the University of Wisconsin-Madison Arboretum; (bottom right): Courtesy of Virginia Kline; **1096:** R.O. Bierregaard; **1098:** © Gary Braasch/Getty/Stone; **1099:** © Dr. Nidel Smith/Earth Scenes; **1105:** NASA/Goddard Space Flight Center

Figure Credits

Figure 14-11 From *An Introduction to Genetic Analysis, 7th Edition* by Griffiths, et al., p. 389. Copyright © 2000 by W. H. Freeman and Company. Reprinted with permission.

Figure 16-13b From *Developmental Biology* by V. Walbot & N. Holder, Fig 22-6a. Copyright © 1987 McGraw-Hill Companies, Inc. Reprinted with permission.

Figure 18-2 Adapted with permission from "An Experimental Study of Inbreeding Depression in a Natural Habitat," by J. A. Jimenez, et al., in *Science*, Vol. 266 (October 14, 1994). Copyright © 1994 American Association for the Advancement of Science.

Figure 18-6b Adapted with permission from "Frequency-Dependent Natural Selection in the Handedness of Scale-Eating Cinchild Fish," by M. Hori in *Science*, Vol. 260 (April 9, 1993). Copyright © 1993 American Association for the Advancement of Science.

Figure 22-5 Reprinted with permission from C. Vila, et al., *Science*, Vol. 276 (June 13, 1997). Copyright © 1997 American Association for the Advancement of Science.

Figure 24-7d By H. Machemer in K. G. Grell, *Protozoology*, p. 304. Copyright © 1973 Springer-Verlag.

Figure 36-2 From C. Starr and R. Taggart, *Biology, 9th Edition*, p. 547, art by Raychel Ciemma. Copyright © 2001 Brooks/Cole, a division of Thomson Learning.

Figure 51-5b From D. A. Spiller and T. W. Schoener, "Lizards Reduce Spider Species Richness by Excluding Rare Species," *Ecology*, Vol. 79, No. 2. Copyright © 1998 Ecological Society of America. Reprinted with permission.

Figure 51-8 From W. J. Leverich and D. A. Levin, "Age-Specific Surviorship and Reproduction in Phlox drummondii," *American Naturalist 113*:6 (1979), p. 1148. Copyright © 1979 University of Chicago Press. Reprinted with permission.

Figure 54-24 Reprinted with permission from Dr. Charles D. White

Index

Note: A *t* following a page number indicates tabular material, an *f* following a page number indicates a figure, and a *b* following a page number indicates a boxed feature. Page numbers beginning with "A" are in the appendix.

Scientific Measurement

Some Common Prefixes

		Examples
kilo	1000	a kilogram is 1000 grams
centi	0.01	a centimeter is 0.01 meter
milli	0.001	a milliliter is 0.001 liter
micro (μ)	one-millionth	a micrometer is 0.000001 (one-millionth) of a meter
nano (n)	one-billionth	a nanogram is 10^{-9} (one-billionth) of a gram
pico (p)	one-trillionth	a picogram is 10^{-12} (one-trillionth) of a gram

The relationship between mass and volume of water (at 20°C)

$1 \text{ g} = 1 \text{ cm}^3 = 1 \text{ mL}$

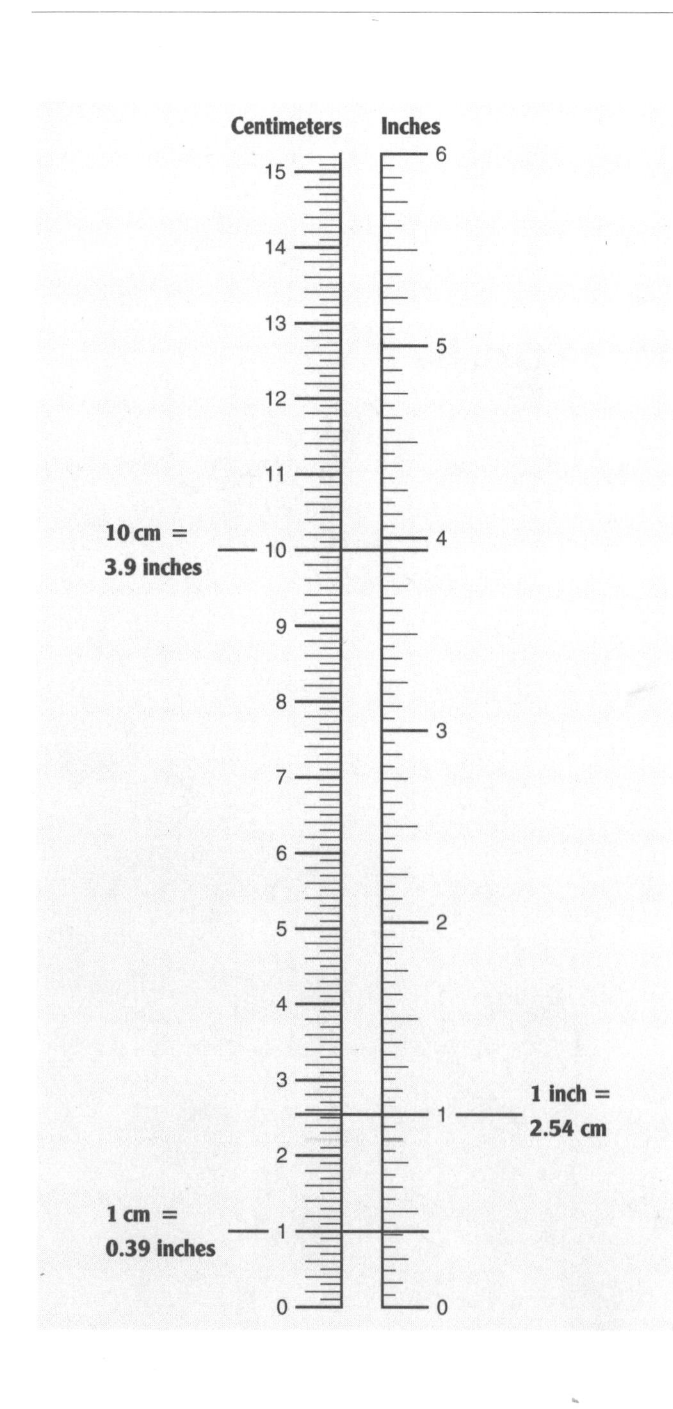

Some Common Units of Length

Unit	Abbreviation	Equivalent
meter	m	approximately 39 in
centimeter	cm	10^{-2} m
millimeter	mm	10^{-3} m
micrometer	μm	10^{-6} m
nanometer	nm	10^{-9} m

Length Conversions

1 in = 2.5 cm	1 mm = 0.039 in
1 ft = 30 cm	1 cm = 0.39 in
1 yd = 0.9 m	1 m = 39 in
1 mi = 1.6 km	1 m = 1.094 yd
	1 km = 0.6 mi

To convert	*Multiply by*	*To obtain*
inches	2.54	centimeters
feet	30	centimeters
centimeters	0.39	inches
millimeters	0.039	inches

Standard Metric Units

		Abbreviation
Standard unit of mass	gram	g
Standard unit of length	meter	m
Standard unit of volume	liter	L